Building Construction Costs with RSMeans Data

Stephen C. Plotner, Senior Editor

2017
75th annual edition

Engineering Director
Bob Mewis, CCP

Contributing Editors
Christopher Babbitt
Adrian C. Charest, PE
Cheryl Elsmore
John Gomes (13,41)
Derrick Hale (2, 31, 32, 33, 34, 35, 44, 46)
Wafaa Hamitou (11, 12)
Joseph Kelble (14, 23)
Charles Kibbee (1, 4)
Robert J. Kuchta (8)
Gerard Lafond
Michael Landry

Thomas Lane (6, 7)
Genevieve Medeiros
Elisa Mello
Ken Monty
Chris Morris (26, 27, 28, 48)
Dale Morris (21, 22)
Marilyn Phelan, AIA (9, 10)
Stephen C. Plotner (3, 5)
Stephen Rosenberg
Matthew Sorrentino
Kevin Souza
Tim Tonello
David Yazbek

Product Manager
Andrea Sillah

Production Manager
Debbie Panarelli

Production
Jonathan Forgit
Mary Lou Geary
Sharon Larsen
Sheryl Rose

Technical Support
Judy Abbruzzese
Gary L. Hoitt

Cover Design
Blaire Gaddis

Numbers in italics are the divisional responsibilities for each editor. Please contact the designated editor directly with any questions.

Gordian RSMeans Data
Construction Publishers & Consultants
1099 Hingham Street, Suite 201
Rockland, MA 02370
United States of America
1-800-448-8182
www.RSMeans.com

Printed in the United States of America
ISSN 0068-3531
ISBN 978-1-943215-48-5

0017 | $247.99 per copy (in United States)
Price is subject to change without prior notice.

Related Data and Services

This data set is aimed primarily at estimating commercial and industrial projects or large multi-family housing projects costing $3,500,000 and up. For civil engineering structures such as bridges, dams, highways, or the like, please refer to *Heavy Construction Costs with RSMeans Data*. For procurement and construction cost data, such as Job Order Contracting and Change Order Management, please refer to Gordian's other offerings.

Our engineers, who are experts in RSMeans data, suggest the following products and services as companion information resources to *Building Construction Costs with RSMeans Data*:

Construction Cost Data
Assemblies Costs with RSMeans Data 2017
Square Foot Costs with RSMeans Data 2017

Reference Books
Estimating Building Costs
RSMeans Estimating Handbook
Green Building: Project Planning & Estimating
How to Estimate with RSMeans Data
Plan Reading & Material Takeoff
Project Scheduling & Management for Construction

Seminars and In-House Training
Unit Price Estimating
Training for our online estimating solution
Practical Project Management for Construction Professionals
Scheduling with MSProject for Construction Professionals
Mechanical & Electrical Estimating

Online Store
Visit www.RSMeans.com for the most reliable and current resources available on the market. Learn more about our more than 20 data sets available in Online, Book, eBook, and CD formats. Our library of reference books is also available, along with professional development seminars aimed at improving cost estimating, project management, administration, and facilities management skills.

RSMeans Electronic Data
Receive the most up-to-date cost information with our online estimating solution. This web-based service is quick, intuitive, and easy to use, giving you instant access to our comprehensive database. Learn more at: **www.RSMeans.com/Online**.

Custom Solutions
Building owners, facility managers, building product manufacturers, attorneys, and even insurance firms across the public and private sectors have engaged in custom solutions to solve their estimating needs including:

Custom Cost Engineering Solutions
Knowing the lifetime maintenance cost of your building, how much it will cost to build a specific building type in different locations across the country and globe, and if the estimate from which you are basing a major decision is accurate are all imperative for building owners and their facility managers.

Market & Custom Analytics
Taking a construction product to market requires accurate market research in order to make critical business development decisions. Once the product goes to market, one way to differentiate your product from the competition is to make any cost saving claims you have to offer.

Third-Party Legal Resources
Natural disasters can lead to reconstruction and repairs just as new developments and expansions can lead to issues of eminent domain. In these cases where one party must pay, it is plausible for legal disputes over costs and estimates to arise.

Construction Costs for Software Applications
More than 25 unit price and assemblies cost databases are available through a number of leading estimating and facilities management software partners.

RSMeans data is also available to federal, state, and local government agencies as multi-year, multi-seat licenses.

For information on our current partners, call 1-800-448-8182.

Table of Contents

Foreword   iv

How the Cost Data Is Built: An Overview   v

Estimating with RSMeans Unit Price Cost Data   vii

How to Use the Cost Data: The Details   ix

Unit Price Section   1

How RSMeans Unit Price Data Works   4

Reference Section
    Construction Equipment Rental Costs   711
    Crew Listings   723
    Historical Cost Indexes   760
    City Cost Indexes   761
    Location Factors   804
    Reference Tables   810
    Change Orders   873
    Square Foot Costs   876
    Abbreviations   877

Index   881

Other Data and Services   925

Labor Trade Rates including
Overhead & Profit   Inside Back Cover

Foreword

Who We Are

Gordian was founded in the spirit of innovation and a strong commitment to help our clients reach and exceed their construction goals. In 1982, Gordian's Chairman and Founder, Harry H. Mellon, invented the family of contracting systems known as Job Order Contracting while serving as Chief Engineer at the Supreme Headquarters Allied Powers Europe in Mons, Belgium. Mr. Mellon realized facility and infrastructure owners could greatly benefit from the time and cost saving advantages of this cutting-edge construction procurement solution, and, in 1990, he started Gordian to provide these products and services for that purpose.

Since 1942, RSMeans data has been available for construction cost estimating information and consulting throughout North America, and is the industry standard for estimating. In 2014, Gordian combined two industry-leading construction cost data companies, enhancing the fortitude of its data with the addition of RSMeans. Today, Gordian draws on its highly specialized engineers, software and unique proprietary data sets to solve the construction information, planning and management needs of people in building construction, building products manufacturing, education, healthcare, retail, insurance, legal and government.

Our Offerings

Gordian is the leading provider of construction cost data, software and services to organizations pursuing efficiency and effectiveness throughout all phases of the construction lifecycle. From Planning, Design, and Procurement through Construction and, finally, Operations, Gordian's portfolio of solutions delivers accurate and up-to-date data to meet the needs of contractors, architects, engineers, facility owners, managers and others in the construction industry. Our products and services include Sightlines planning solutions, estimating solutions with RSMeans data and Job Order Contracting offerings from basic online services to enterprise solutions.

Our Commitment

Today at Gordian, we do more than talk about the quality of our data and the usefulness of its application. We stand behind all of our RSMeans data—from historical cost indexes to construction materials and techniques—to craft current costs and predict future trends.

If you have any questions about our products or services, please call us toll-free at 800-448-8182 or visit our website at: www.TheGordianGroup.com.

How the Cost Data Is Built: An Overview

A Powerful Construction Tool

You now have one of the most powerful construction tools available today. A successful project is built on the foundation of an accurate and dependable estimate. This tool will enable you to construct such an estimate.

For the casual user the information is designed to be:

- quickly and easily understood so you can get right to your estimate.
- filled with valuable information so you can understand the necessary factors that go into a cost estimate.

For the professional user, the information is designed to be:

- a handy reference that can be quickly referred to for key costs.
- a comprehensive, fully reliable source of current construction costs and productivity rates so you'll be prepared to estimate any project.
- a source for preliminary project costs, product selections, and alternate materials and methods.

To meet all of these requirements, we have organized the information into the following clearly defined sections.

Estimating with RSMeans Unit Price Cost Data*

Please refer to these steps for guidance on completing an estimate using RSMeans unit price cost data.

How to Use the Cost Data: The Details

This section contains an in-depth explanation of how the information is arranged and how you can use it to determine a reliable construction cost estimate. It includes how we develop our cost figures and how to prepare your estimate.

Unit Prices*

All cost data has been divided into 50 divisions according to the MasterFormat system of classification and numbering.

Assemblies*

The cost data in this section has been organized in an "Assemblies" format. These assemblies are the functional elements of a building and are arranged according to the 7 elements of the UNIFORMAT II classification system. For a complete explanation of a typical "Assembly", see "How RSMeans Assemblies Data Works."

Residential Models*

Model buildings for four classes of construction—economy, average, custom, and luxury—are developed and shown with complete costs per square foot.

Commercial/Industrial/ Institutional Models*

This section contains complete costs for 77 typical model buildings expressed as costs per square foot.

Green Commercial/Industrial/ Institutional Models*

This section contains complete costs for 25 green model buildings expressed as costs per square foot.

References*

This section includes information on Equipment Rental Costs, Crew Listings, Historical Cost Indexes, City Cost Indexes, Location Factors, Reference Tables, and Change Orders, as well as a listing of abbreviations.

- **Equipment Rental Costs:** Included are the average costs to rent and operate hundreds of pieces of construction equipment.
- **Crew Listings:** This section lists all the crews referenced in the cost data. A crew is composed of more than one trade classification and/or the addition of power equipment to any trade classification. Power equipment is included in the cost of the crew. Costs are shown both with bare labor rates and with the installing contractor's overhead and profit added. For each, the total crew cost per eight-hour day and the composite cost per labor-hour are listed.
- **Historical Cost Indexes:** These indexes provide you with data to adjust construction costs over time.
- **City Cost Indexes:** All costs in this data set are U.S. national averages. Costs vary by region. You can adjust for this by CSI Division to over 900 locations throughout the U.S. and Canada by using this data.
- **Location Factors:** You can adjust total project costs to over 900 locations throughout the U.S. and Canada by using the weighted number, which applies across all divisions.
- **Reference Tables:** At the beginning of selected major classifications in the Unit Prices are reference numbers indicators. These numbers refer you to related information in the Reference Section. In this section, you'll find reference tables, explanations, and estimating information that support how we develop the unit price data, technical data, and estimating procedures.
- **Change Orders:** This section includes information on the factors that influence the pricing of change orders.
- **Abbreviations:** A listing of abbreviations used throughout this information, along with the terms they represent, is included.

Index (printed versions only)

A comprehensive listing of all terms and subjects will help you quickly find what you need when you are not sure where it occurs in MasterFormat.

Conclusion

This information is designed to be as comprehensive and easy to use as possible.

The Construction Specifications Institute (CSI) and Construction Specifications Canada (CSC) have produced the 2014 edition of MasterFormat®, a system of titles and numbers used extensively to organize construction information.

All unit prices in the RSMeans cost data are now arranged in the 50-division MasterFormat® 2014 system.

---

* Not all information is available in all data sets

Note: The material prices in RSMeans cost data are "contractor's prices." They are the prices that contractors can expect to pay at the lumberyards, suppliers'/distributors' warehouses, etc. Small orders of specialty items would be higher than the costs shown, while very large orders, such as truckload lots, would be less. The variation would depend on the size, timing, and negotiating power of the contractor. The labor costs are primarily for new construction or major renovation rather than repairs or minor alterations. With reasonable exercise of judgment, the figures can be used for any building work.

Estimating with RSMeans Unit Price Cost Data

Following these steps will allow you to complete an accurate estimate using RSMeans Unit Price cost data.

1. Scope Out the Project

- Think through the project and identify the CSI Divisions needed in your estimate.
- Identify the individual work tasks that will need to be covered in your estimate.
- The Unit Price data has been divided into 50 Divisions according to the CSI MasterFormat® 2014.
- In printed versions, the Unit Price Section Table of Contents on page 1 may also be helpful when scoping out your project.
- Experienced estimators find it helpful to begin with Division 2 and continue through completion. Division 1 can be estimated after the full project scope is known.

2. Quantify

- Determine the number of units required for each work task that you identified.
- Experienced estimators include an allowance for waste in their quantities. (Waste is not included in our Unit Price line items unless otherwise stated.)

3. Price the Quantities

- Use the search tools available to locate individual Unit Price line items for your estimate.
- Reference Numbers indicated within a Unit Price section refer to additional information that you may find useful.
- The crew indicates who is performing the work for that task. Crew codes are expanded in the Crew Listings in the Reference Section to include all trades and equipment that comprise the crew.
- The Daily Output is the amount of work the crew is expected to complete in one day.
- The Labor-Hours value is the amount of time it will take for the crew to install one unit of work.
- The abbreviated Unit designation indicates the unit of measure upon which the crew, productivity, and prices are based.
- Bare Costs are shown for materials, labor, and equipment needed to complete the Unit Price line item. Bare costs do not include waste, project overhead, payroll insurance, payroll taxes, main office overhead, or profit.
- The Total Incl O&P cost is the billing rate or invoice amount of the installing contractor or subcontractor who performs the work for the Unit Price line item.

4. Multiply

- Multiply the total number of units needed for your project by the Total Incl O&P cost for each Unit Price line item.
- Be careful that your take off unit of measure matches the unit of measure in the Unit column.
- The price you calculate is an estimate for a completed item of work.
- Keep scoping individual tasks, determining the number of units required for those tasks, matching each task with individual Unit Price line items, and multiplying quantities by Total Incl O&P costs.
- An estimate completed in this manner is priced as if a subcontractor, or set of subcontractors, is performing the work. The estimate does not yet include Project Overhead or Estimate Summary components such as general contractor markups on subcontracted work, general contractor office overhead and profit, contingency, and location factors.

5. Project Overhead

- Include project overhead items from Division 1–General Requirements.
- These items are needed to make the job run. They are typically, but not always, provided by the general contractor. Items include, but are not limited to, field personnel, insurance, performance bond, permits, testing, temporary utilities, field office and storage facilities, temporary scaffolding and platforms, equipment mobilization and demobilization, temporary roads and sidewalks, winter protection, temporary barricades and fencing, temporary security, temporary signs, field engineering and layout, final cleaning, and commissioning.
- Each item should be quantified and matched to individual Unit Price line items in Division 1, then priced and added to your estimate.

- An alternate method of estimating project overhead costs is to apply a percentage of the total project cost, usually 5% to 15% with an average of 10% (see General Conditions).
- Include other project related expenses in your estimate such as:
  - Rented equipment not itemized in the Crew Listings
  - Rubbish handling throughout the project (see section 02 41 19.19)

6. Estimate Summary

- Include sales tax as required by laws of your state or county.
- Include the general contractor's markup on self-performed work, usually 5% to 15% with an average of 10%.
- Include the general contractor's markup on subcontracted work, usually 5% to 15% with an average of 10%.
- Include the general contractor's main office overhead and profit:
  - RSMeans data provides general guidelines on the general contractor's main office overhead (see section 01 31 13.60 and Reference Number R013113-50).
  - Markups will depend on the size of the general contractor's operations, projected annual revenue, the level of risk, and the level of competition in the local area and for this project in particular.
- Include a contingency, usually 3% to 5%, if appropriate.
- Adjust your estimate to the project's location by using the City Cost Indexes or the Location Factors in the Reference Section:
  - Look at the rules in "How to Use the City Cost Indexes" to see how to apply the Indexes for your location.
  - When the proper Index or Factor has been identified for the project's location, convert it to a multiplier by dividing it by 100, then multiply that multiplier by your estimated total cost. The original estimated total cost will now be adjusted up or down from the national average to a total that is appropriate for your location.

Editors' Note:
We urge you to spend time reading and understanding the supporting material. An accurate estimate requires experience, knowledge, and careful calculation. The more you know about how we at RSMeans developed the data, the more accurate your estimate will be. In addition, it is important to take into consideration the reference material such as Equipment Listings, Crew Listings, City Cost Indexes, Location Factors, and Reference Tables.

How to Use the Cost Data: The Details

What's Behind the Numbers? The Development of Cost Data

Gordian's RSMeans staff continually monitors developments in the construction industry in order to ensure reliable, thorough, and up-to-date cost information. While overall construction costs may vary relative to general economic conditions, price fluctuations within the industry are dependent upon many factors. Individual price variations may, in fact, be opposite to overall economic trends. Therefore, costs are constantly tracked, and complete updates are performed yearly. Also, new items are frequently added in response to changes in materials and methods.

Costs—$ (U.S.)

All costs represent U.S. national averages and are given in U.S. dollars. The City Cost Index (CCI) with RSMeans data can be used to adjust costs to a particular location. The CCI for Canada can be used to adjust U.S. national averages to local costs in Canadian dollars. No exchange rate conversion is necessary because it has already been factored in.

G The processes or products identified by the green symbol in our publications have been determined to be environmentally responsible and/or resource-efficient solely by Gordian's RSMeans engineering staff. The inclusion of the green symbol does not represent compliance with any specific industry association or standard.

Material Costs

Gordian's RSMeans staff contacts manufacturers, dealers, distributors, and contractors all across the U.S. and Canada to determine national average material costs. If you have access to current material costs for your specific location, you may wish to make adjustments to reflect differences from the national average. Included within material costs are fasteners for a normal installation. Gordian's RSMeans engineers use manufacturers' recommendations, written specifications, and/or standard construction practices for the sizing and spacing of fasteners. Adjustments to material costs may be required for your specific application or location. The manufacturer's warranty is assumed. Extended warranties are not included in the material costs. **Material costs do not include sales tax.**

Labor Costs

Labor costs are based upon a mathematical average of trade-specific wages in 30 major U.S. cities. The type of wage (union, open shop, or residential) is identified on the inside back cover of printed publications or is selected by the estimator when using the electronic products. Markups for the wages can also be found on the inside back cover of printed publications and/or under the labor references found in the electronic products.

- If wage rates in your area vary from those used, or if rate increases are expected within a given year, labor costs should be adjusted accordingly.

Labor costs reflect productivity based on actual working conditions. In addition to actual installation, these figures include time spent during a normal weekday on tasks, such as material receiving and handling, mobilization at site, site movement, breaks, and cleanup.

Productivity data is developed over an extended period so as not to be influenced by abnormal variations and reflects a typical average.

Equipment Costs

Equipment costs include not only rental, but also operating costs for equipment under normal use. The operating costs include parts and labor for routine servicing, such as repair and replacement of pumps, filters, and worn lines. Normal operating expendables, such as fuel, lubricants, tires, and electricity (where applicable), are also included. Extraordinary operating expendables with highly variable wear patterns, such as diamond bits and blades, are excluded. These costs are included under materials. Equipment rental rates are obtained from industry sources throughout North America—contractors, suppliers, dealers, manufacturers, and distributors.

Rental rates can also be treated as reimbursement costs for contractor-owned equipment. Owned equipment costs include depreciation, loan payments, interest, taxes, insurance, storage, and major repairs.

Equipment costs do not include operators' wages.

Equipment Cost/Day—The cost of equipment required for each crew is included in the Crew Listings in the Reference Section (small tools that are considered essential everyday tools are not listed out separately). The Crew Listings itemize specialized tools and heavy equipment along with labor trades. The daily cost of itemized equipment included in a crew is based on dividing the weekly bare rental rate by 5 (number of working days per week), then adding the hourly operating cost times 8 (the number of hours per day). This Equipment Cost/Day is shown in the last column of the Equipment Rental Costs in the Reference Section.

Mobilization, Demobilization—The cost to move construction equipment from an equipment yard or rental company to the job site and back again is not included in equipment costs. Mobilization (to the site) and demobilization (from the site) costs can be found in the Unit Price Section. If a piece of equipment is already at the job site, it is not appropriate to utilize mobilization or demobilization costs again in an estimate.

Overhead and Profit

Total Cost including O&P for the installing contractor is shown in the last column of the Unit Price and/or Assemblies. This figure is the sum of the bare material cost plus 10% for profit, the bare labor cost plus total overhead and profit, and the bare equipment cost plus 10% for profit. Details for the calculation of overhead and profit on labor are shown on the inside back cover of the printed product and in the Reference Section of the electronic product.

General Conditions

Cost data in this data set is presented in two ways: Bare Costs and Total Cost including O&P (Overhead and Profit). General Conditions, or General Requirements, of the contract should also be added to the Total Cost including O&P when applicable. Costs for General Conditions are listed in Division 1 of the Unit Price Section and in the Reference Section.

General Conditions for the installing contractor may range from 0% to 10% of the Total Cost including O&P. For the general or prime contractor, costs for General Conditions may range from 5% to 15% of the Total Cost including O&P, with a figure of 10% as the most typical allowance. If applicable, the Assemblies and Models sections use costs that include the installing contractor's overhead and profit (O&P).

Factors Affecting Costs

Costs can vary depending upon a number of variables. Here's a listing of some factors that affect costs and points to consider.

Quality—The prices for materials and the workmanship upon which productivity is based represent sound construction work. They are also in line with industry standard and manufacturer specifications and are frequently used by federal, state, and local governments.

Overtime—We have made no allowance for overtime. If you anticipate premium time or work beyond normal working hours, be sure to make an appropriate adjustment to your labor costs.

Productivity—The productivity, daily output, and labor-hour figures for each line item are based on working an eight-hour day in daylight hours in moderate temperatures, and up to a 14 foot working height unless otherwise indicated. For work that extends beyond normal work hours or is performed under adverse conditions, productivity may decrease.

Size of Project—The size, scope of work, and type of construction project will have a significant impact on cost. Economies of scale can reduce costs for large projects. Unit costs can often run higher for small projects.

Location—Material prices are for metropolitan areas. However, in dense urban areas, traffic and site storage limitations may increase costs. Beyond a 20-mile radius of metropolitan areas, extra trucking or transportation charges may also increase the material costs slightly. On the other hand, lower wage rates may be in effect. Be sure to consider both of these factors when preparing an estimate, particularly if the job site is located in a central city or remote rural location. In addition, highly specialized subcontract items may require travel and per-diem expenses for mechanics.

Other Factors—

- season of year
- contractor management
- weather conditions
- local union restrictions
- building code requirements
- availability of:
  - adequate energy
  - skilled labor
  - building materials
- owner's special requirements/restrictions
- safety requirements
- environmental considerations
- access

Unpredictable Factors—General business conditions influence "in-place" costs of all items. Substitute materials and construction methods may have to be employed. These may affect the installed cost and/or life cycle costs. Such factors may be difficult to evaluate and cannot necessarily be predicted on the basis of the job's location in a particular section of the country. Thus, where these factors apply, you may find significant but unavoidable cost variations for which you will have to apply a measure of judgment to your estimate.

Rounding of Costs

In printed publications only, all unit prices in excess of $5.00 have been rounded to make them easier to use and still maintain adequate precision of the results.

How Subcontracted Items Affect Costs

A considerable portion of all large construction jobs is usually subcontracted. In fact, the percentage done by subcontractors is constantly increasing and may run over 90%. Since the workers employed by these companies do nothing else but install their particular product, they soon become experts in that line. The result is, installation by these firms is accomplished so efficiently that the total in-place cost, even adding the general contractor's overhead and profit, is no more, and often less, than if the principal contractor had handled the installation. Companies that deal with construction specialties are anxious to have their product perform well and, consequently, the installation will be the best possible.

Contingencies

The allowance for contingencies generally provides for unforeseen construction difficulties. On alterations or repair jobs, 20% is not too much. If drawings are final and only field contingencies are being considered, 2% or 3% is probably sufficient, and often nothing needs to be added. Contractually, changes in plans will be covered by extras. The contractor should consider inflationary price trends and possible material shortages during the course of the job. These escalation factors are dependent upon both economic conditions and the anticipated time between the estimate and actual construction. If drawings are not complete or approved, or a budget cost is wanted, it is wise to add 5% to 10%. Contingencies, then, are a matter of judgment.

Important Estimating Considerations

The productivity, or daily output, of each craftsman or crew assumes a well-managed job where tradesmen with the proper tools and equipment, along with the appropriate construction materials, are present. Included are daily set-up and cleanup time, break time, and plan layout time. Unless otherwise indicated, time for material movement on site (for items

that can be transported by hand) of up to 200' into the building and to the first or second floor is also included. If material has to be transported by other means, over greater distances, or to higher floors, an additional allowance should be considered by the estimator.

While horizontal movement is typically a sole function of distances, vertical transport introduces other variables that can significantly impact productivity. In an occupied building, the use of elevators (assuming access, size, and required protective measures are acceptable) must be understood at the time of the estimate. For new construction, hoist wait and cycle times can easily be 15 minutes and may result in scheduled access extending beyond the normal work day. Finally, all vertical transport will impose strict weight limits likely to preclude the use of any motorized material handling.

The productivity, or daily output, also assumes installation that meets manufacturer/designer/standard specifications. A time allowance for quality control checks, minor adjustments, and any task required to ensure the proper function or operation is also included. For items that require connections to services, time is included for positioning, leveling, securing the unit, and for making all the necessary connections (and start up where applicable), ensuring a complete installation. Estimating of the services themselves (electrical, plumbing, water, steam, hydraulics, dust collection, etc.) is separate.

In some cases, the estimator must consider the use of a crane and an appropriate crew for the installation of large or heavy items. For those situations where a crane is not included in the assigned crew and as part of the line item cost, equipment rental costs, mobilization and demobilization costs, and operator and support personnel costs must be considered.

Labor-Hours

The labor-hours expressed in this publication are derived by dividing the total daily labor-hours for the crew by the daily output. Based on average installation time and the assumptions listed above, the labor-hours include: direct labor, indirect labor, and nonproductive time. A typical day for a craftsman might include, but is not limited to:

- Direct Work
  - ☐ Measuring and layout
  - ☐ Preparing materials
  - ☐ Actual installation
  - ☐ Quality assurance/quality control
- Indirect Work
  - ☐ Reading plans or specifications
  - ☐ Preparing space
  - ☐ Receiving materials
  - ☐ Material movement
  - ☐ Giving or receiving instruction
  - ☐ Miscellaneous
- Non-Work
  - ☐ Chatting
  - ☐ Personal issues
  - ☐ Breaks
  - ☐ Interruptions (i.e., sickness, weather, material or equipment shortages, etc.)

If any of the items for a typical day do not apply to the particular work or project situation, the estimator should make any necessary adjustments.

Final Checklist

Estimating can be a straightforward process provided you remember the basics. Here's a checklist of some of the steps you should remember to complete before finalizing your estimate.

Did you remember to:

- factor in the City Cost Index for your locale?
- take into consideration which items have been marked up and by how much?
- mark up the entire estimate sufficiently for your purposes?
- read the background information on techniques and technical matters that could impact your project time span and cost?
- include all components of your project in the final estimate?
- double check your figures for accuracy?
- call Gordian's RSMeans staff if you have any questions about your estimate or the data you've used? Remember, Gordian stands behind all of our products, including our extensive RSMeans data solutions. If you have any questions about your estimate, about the costs you've used from our data, or even about the technical aspects of the job that may affect your estimate, feel free to call the Gordian RSMeans editors at 1-800-448-8182.

Free Quarterly Updates

Stay up-to-date throughout the year with free RSMeans cost data updates four times a year. Sign up online to make sure you have access to the newest data. Every quarter we provide the city cost adjustment factors for hundreds of cities and updated national average pricing for key materials.

Register at: **http://info.TheGordianGroup.com/2017-quarterly-updates.html**.

Unit Price Section

Table of Contents

Sect. No.   **Page**

General Requirements
01 11	Summary of Work	10
01 21	Allowances	11
01 31	Project Management and Coordination	12
01 32	Construction Progress Documentation	14
01 41	Regulatory Requirements	14
01 45	Quality Control	15
01 51	Temporary Utilities	17
01 52	Construction Facilities	17
01 54	Construction Aids	18
01 55	Vehicular Access and Parking	22
01 56	Temporary Barriers and Enclosures	23
01 58	Project Identification	25
01 66	Product Storage and Handling Requirements	25
01 71	Examination and Preparation	25
01 74	Cleaning and Waste Management	25
01 76	Protecting Installed Construction	25
01 91	Commissioning	26

Existing Conditions
02 21	Surveys	28
02 32	Geotechnical Investigations	28
02 41	Demolition	29
02 42	Removal and Salvage of Construction Materials	35
02 43	Structure Moving	38
02 56	Site Containment	38
02 58	Snow Control	39
02 65	Underground Storage Tank Rmvl.	39
02 81	Transportation and Disposal of Hazardous Materials	40
02 82	Asbestos Remediation	40
02 83	Lead Remediation	44
02 85	Mold Remediation	46
02 91	Chemical Sampling, Testing and Analysis	47

Concrete
03 01	Maintenance of Concrete	50
03 05	Common Work Results for Concrete	50
03 11	Concrete Forming	52
03 15	Concrete Accessories	60
03 21	Reinforcement Bars	68
03 22	Fabric and Grid Reinforcing	73
03 23	Stressed Tendon Reinforcing	73
03 24	Fibrous Reinforcing	74
03 30	Cast-In-Place Concrete	75
03 31	Structural Concrete	77
03 35	Concrete Finishing	80
03 37	Specialty Placed Concrete	83
03 39	Concrete Curing	83
03 41	Precast Structural Concrete	84
03 45	Precast Architectural Concrete	86
03 47	Site-Cast Concrete	86
03 48	Precast Concrete Specialties	87
03 51	Cast Roof Decks	87
03 52	Lightweight Concrete Roof Insulation	87
03 53	Concrete Topping	88
03 54	Cast Underlayment	88
03 62	Non-Shrink Grouting	88
03 63	Epoxy Grouting	89
03 81	Concrete Cutting	89
03 82	Concrete Boring	89

Masonry
04 01	Maintenance of Masonry	94
04 05	Common Work Results for Masonry	95
04 21	Clay Unit Masonry	99
04 22	Concrete Unit Masonry	104
04 23	Glass Unit Masonry	110
04 24	Adobe Unit Masonry	111
04 25	Unit Masonry Panels	111
04 27	Multiple-Wythe Unit Masonry	111
04 41	Dry-Placed Stone	112
04 43	Stone Masonry	112
04 51	Flue Liner Masonry	116
04 54	Refractory Brick Masonry	116
04 57	Masonry Fireplaces	117
04 71	Manufactured Brick Masonry	117
04 72	Cast Stone Masonry	117
04 73	Manufactured Stone Masonry	118

Metals
05 01	Maintenance of Metals	120
05 05	Common Work Results for Metals	120
05 12	Structural Steel Framing	127
05 14	Structural Aluminum Framing	135
05 15	Wire Rope Assemblies	136
05 21	Steel Joist Framing	138
05 31	Steel Decking	141
05 35	Raceway Decking Assemblies	143
05 41	Structural Metal Stud Framing	143
05 42	Cold-Formed Metal Joist Framing	147
05 44	Cold-Formed Metal Trusses	153
05 51	Metal Stairs	154
05 52	Metal Railings	156
05 53	Metal Gratings	157
05 54	Metal Floor Plates	159
05 55	Metal Stair Treads and Nosings	160
05 56	Metal Castings	161
05 58	Formed Metal Fabrications	161
05 71	Decorative Metal Stairs	163
05 73	Decorative Metal Railings	164
05 75	Decorative Formed Metal	164

Wood, Plastics & Composites
06 05	Common Work Results for Wood, Plastics, and Composites	166
06 11	Wood Framing	180
06 12	Structural Panels	188
06 13	Heavy Timber Construction	189
06 15	Wood Decking	190
06 16	Sheathing	190
06 17	Shop-Fabricated Structural Wood	193
06 18	Glued-Laminated Construction	193
06 22	Millwork	196
06 25	Prefinished Paneling	208
06 26	Board Paneling	209
06 43	Wood Stairs and Railings	209
06 44	Ornamental Woodwork	212
06 48	Wood Frames	213
06 49	Wood Screens and Exterior Wood Shutters	214
06 51	Structural Plastic Shapes and Plates	214
06 52	Plastic Structural Assemblies	216
06 63	Plastic Railings	217
06 65	Plastic Trim	217
06 80	Composite Fabrications	218
06 81	Composite Railings	219

Thermal & Moisture Protection
07 01	Operation and Maint. of Thermal and Moisture Protection	222
07 05	Common Work Results for Thermal and Moisture Protection	222
07 11	Dampproofing	224
07 12	Built-up Bituminous Waterproofing	224
07 13	Sheet Waterproofing	225
07 16	Cementitious and Reactive Waterproofing	225
07 17	Bentonite Waterproofing	225
07 19	Water Repellents	226
07 21	Thermal Insulation	226
07 22	Roof and Deck Insulation	230
07 24	Exterior Insulation and Finish Systems	231
07 25	Weather Barriers	232
07 26	Vapor Retarders	232
07 27	Air Barriers	233
07 31	Shingles and Shakes	233
07 32	Roof Tiles	235
07 33	Natural Roof Coverings	236
07 41	Roof Panels	237
07 42	Wall Panels	238
07 44	Faced Panels	239
07 46	Siding	240
07 51	Built-Up Bituminous Roofing	243
07 52	Modified Bituminous Membrane Roofing	244
07 53	Elastomeric Membrane Roofing	245
07 54	Thermoplastic Membrane Roofing	246
07 55	Protected Membrane Roofing	247
07 56	Fluid-Applied Roofing	247
07 57	Coated Foamed Roofing	247
07 58	Roll Roofing	248
07 61	Sheet Metal Roofing	248
07 65	Flexible Flashing	249
07 71	Roof Specialties	251
07 72	Roof Accessories	255
07 76	Roof Pavers	257
07 81	Applied Fireproofing	257
07 84	Firestopping	258
07 91	Preformed Joint Seals	260
07 92	Joint Sealants	261
07 95	Expansion Control	262

Openings
08 05	Common Work Results for Openings	264
08 11	Metal Doors and Frames	265
08 12	Metal Frames	266
08 13	Metal Doors	267
08 14	Wood Doors	270
08 16	Composite Doors	276
08 17	Integrated Door Opening Assemblies	277
08 31	Access Doors and Panels	278
08 32	Sliding Glass Doors	279
08 33	Coiling Doors and Grilles	280
08 34	Special Function Doors	281
08 36	Panel Doors	283
08 38	Traffic Doors	285
08 41	Entrances and Storefronts	285
08 42	Entrances	287
08 43	Storefronts	287
08 44	Curtain Wall and Glazed Assemblies	288
08 45	Translucent Wall and Roof Assemblies	288
08 51	Metal Windows	289
08 52	Wood Windows	290
08 53	Plastic Windows	295
08 54	Composite Windows	298
08 56	Special Function Windows	298
08 62	Unit Skylights	298
08 63	Metal-Framed Skylights	298
08 71	Door Hardware	299
08 74	Access Control Hardware	309
08 75	Window Hardware	310
08 79	Hardware Accessories	310
08 81	Glass Glazing	310
08 83	Mirrors	313
08 84	Plastic Glazing	313
08 87	Glazing Surface Films	314
08 88	Special Function Glazing	315
08 91	Louvers	315
08 95	Vents	316

Finishes
09 01	Maintenance of Finishes	318
09 05	Common Work Results for Finishes	318
09 21	Plaster and Gypsum Board Assemblies	320

Table of Contents (cont.)

Sect. No.		Page
09 22	Supports for Plaster and Gypsum Board	322
09 23	Gypsum Plastering	326
09 24	Cement Plastering	327
09 25	Other Plastering	327
09 26	Veneer Plastering	327
09 28	Backing Boards and Underlayments	328
09 29	Gypsum Board	328
09 30	Tiling	332
09 31	Thin-Set Tiling	333
09 32	Mortar-Bed Tiling	334
09 34	Waterproofing-Membrane Tiling	336
09 51	Acoustical Ceilings	336
09 53	Acoustical Ceiling Suspension Assemblies	337
09 54	Specialty Ceilings	338
09 57	Special Function Ceilings	339
09 61	Flooring Treatment	339
09 62	Specialty Flooring	339
09 63	Masonry Flooring	340
09 64	Wood Flooring	341
09 65	Resilient Flooring	343
09 66	Terrazzo Flooring	345
09 67	Fluid-Applied Flooring	347
09 68	Carpeting	348
09 69	Access Flooring	349
09 72	Wall Coverings	350
09 74	Flexible Wood Sheets	351
09 77	Special Wall Surfacing	351
09 81	Acoustic Insulation	353
09 84	Acoustic Room Components	353
09 91	Painting	354
09 93	Staining and Transparent Finishing	367
09 96	High-Performance Coatings	367
09 97	Special Coatings	369

Specialties
10 05	Common Work Results for Specialties	372
10 11	Visual Display Units	372
10 13	Directories	375
10 14	Signage	376
10 17	Telephone Specialties	378
10 21	Compartments and Cubicles	378
10 22	Partitions	382
10 26	Wall and Door Protection	385
10 28	Toilet, Bath, and Laundry Accessories	386
10 31	Manufactured Fireplaces	389
10 32	Fireplace Specialties	390
10 35	Stoves	391
10 43	Emergency Aid Specialties	391
10 44	Fire Protection Specialties	391
10 51	Lockers	392
10 55	Postal Specialties	393
10 56	Storage Assemblies	394
10 57	Wardrobe and Closet Specialties	395
10 71	Exterior Protection	395
10 73	Protective Covers	396
10 74	Manufactured Exterior Specialties	397
10 75	Flagpoles	398
10 81	Pest Control Devices	399
10 86	Security Mirrors and Domes	399
10 88	Scales	399

Equipment
11 05	Common Work Results for Equipment	402
11 11	Vehicle Service Equipment	404
11 12	Parking Control Equipment	404
11 13	Loading Dock Equipment	405
11 14	Pedestrian Control Equipment	406
11 21	Retail and Service Equipment	407
11 22	Banking Equipment	408
11 30	Residential Equipment	410
11 32	Unit Kitchens	412
11 41	Foodservice Storage Equipment	412

Sect. No.		Page
11 42	Food Preparation Equipment	414
11 43	Food Delivery Carts and Conveyors	415
11 44	Food Cooking Equipment	415
11 46	Food Dispensing Equipment	416
11 48	Foodservice Cleaning and Disposal Equipment	417
11 52	Audio-Visual Equipment	417
11 53	Laboratory Equipment	419
11 57	Vocational Shop Equipment	420
11 61	Broadcast, Theater, and Stage Equipment	420
11 62	Musical Equipment	421
11 66	Athletic Equipment	421
11 67	Recreational Equipment	423
11 68	Play Field Equipment and Structures	423
11 71	Medical Sterilizing Equipment	425
11 72	Examination and Treatment Equipment	425
11 73	Patient Care Equipment	426
11 74	Dental Equipment	426
11 76	Operating Room Equipment	427
11 77	Radiology Equipment	427
11 78	Mortuary Equipment	428
11 81	Facility Maintenance Equipment	428
11 82	Facility Solid Waste Handling Equipment	428
11 91	Religious Equipment	430
11 92	Agricultural Equipment	431
11 97	Security Equipment	431
11 98	Detention Equipment	431

Furnishings
12 05	Common Work Results for Furnishings	434
12 21	Window Blinds	434
12 22	Curtains and Drapes	435
12 23	Interior Shutters	436
12 24	Window Shades	436
12 32	Manufactured Wood Casework	437
12 35	Specialty Casework	439
12 36	Countertops	440
12 46	Furnishing Accessories	443
12 48	Rugs and Mats	443
12 51	Office Furniture	444
12 52	Seating	445
12 54	Hospitality Furniture	445
12 55	Detention Furniture	446
12 56	Institutional Furniture	446
12 61	Fixed Audience Seating	448
12 63	Stadium and Arena Seating	448
12 67	Pews and Benches	448
12 92	Interior Planters and Artificial Plants	448
12 93	Interior Public Space Furnishings	449

Special Construction
13 05	Common Work Results for Special Construction	452
13 11	Swimming Pools	455
13 12	Fountains	457
13 17	Tubs and Pools	457
13 18	Ice Rinks	457
13 21	Controlled Environment Rooms	458
13 24	Special Activity Rooms	460
13 28	Athletic and Recreational Special Construction	460
13 31	Fabric Structures	461
13 34	Fabricated Engineered Structures	462
13 36	Towers	469
13 42	Building Modules	469
13 47	Facility Protection	469
13 48	Sound, Vibration, and Seismic Control	470
13 49	Radiation Protection	470
13 53	Meteorological Instrumentation	472

Sect. No.		Page
Conveying Equipment		
14 11	Manual Dumbwaiters	474
14 12	Electric Dumbwaiters	474
14 21	Electric Traction Elevators	474
14 24	Hydraulic Elevators	475
14 27	Custom Elevator Cabs and Doors	476
14 28	Elevator Equipment and Controls	477
14 31	Escalators	478
14 32	Moving Walks	479
14 42	Wheelchair Lifts	479
14 45	Vehicle Lifts	479
14 91	Facility Chutes	480
14 92	Pneumatic Tube Systems	480

Fire Suppression
21 05	Common Work Results for Fire Suppression	484
21 11	Facility Fire-Suppression Water-Service Piping	484
21 12	Fire-Suppression Standpipes	485
21 13	Fire-Suppression Sprinkler Systems	486
21 21	Carbon-Dioxide Fire-Extinguishing Systems	487
21 22	Clean-Agent Fire-Extinguishing Systems	487
21 31	Centrifugal Fire Pumps	488

Plumbing
22 01	Operation and Maintenance of Plumbing	490
22 05	Common Work Results for Plumbing	491
22 07	Plumbing Insulation	497
22 11	Facility Water Distribution	498
22 13	Facility Sanitary Sewerage	506
22 14	Facility Storm Drainage	509
22 31	Domestic Water Softeners	510
22 33	Electric Domestic Water Heaters	511
22 34	Fuel-Fired Domestic Water Heaters	511
22 35	Domestic Water Heat Exchangers	512
22 41	Residential Plumbing Fixtures	513
22 42	Commercial Plumbing Fixtures	516
22 45	Emergency Plumbing Fixtures	520
22 47	Drinking Fountains and Water Coolers	520
22 51	Swimming Pool Plumbing Systems	522
22 52	Fountain Plumbing Systems	522
22 66	Chemical-Waste Systems for Lab. and Healthcare Facilities	522

Heating Ventilation Air Conditioning
23 05	Common Work Results for HVAC	526
23 07	HVAC Insulation	528
23 09	Instrumentation and Control for HVAC	529
23 13	Facility Fuel-Storage Tanks	530
23 21	Hydronic Piping and Pumps	532
23 22	Steam and Condensate Piping and Pumps	535
23 31	HVAC Ducts and Casings	536
23 33	Air Duct Accessories	537
23 34	HVAC Fans	539
23 37	Air Outlets and Inlets	542
23 38	Ventilation Hoods	545
23 41	Particulate Air Filtration	545
23 42	Gas-Phase Air Filtration	546
23 43	Electronic Air Cleaners	546
23 51	Breechings, Chimneys, and Stacks	546
23 52	Heating Boilers	547
23 54	Furnaces	549
23 55	Fuel-Fired Heaters	549
23 56	Solar Energy Heating Equipment	551
23 57	Heat Exchangers for HVAC	552
23 62	Packaged Compressor and Condenser Units	552
23 63	Refrigerant Condensers	552
23 64	Packaged Water Chillers	553
23 65	Cooling Towers	554

Table of Contents (cont.)

Sect. No.		Page
23 73	Indoor Central-Station Air-Handling Units	555
23 74	Packaged Outdoor HVAC Equipment	556
23 81	Decentralized Unitary HVAC Equipment	556
23 82	Convection Heating and Cooling Units	558
23 83	Radiant Heating Units	560
23 84	Humidity Control Equipment	562

Electrical

26 01	Operation and Maintenance of Electrical Systems	564
26 05	Common Work Results for Electrical	565
26 09	Instrumentation and Control for Electrical Systems	582
26 12	Medium-Voltage Transformers	583
26 22	Low-Voltage Transformers	583
26 24	Switchboards and Panelboards	584
26 25	Enclosed Bus Assemblies	587
26 27	Low-Voltage Distribution Equipment	588
26 28	Low-Voltage Circuit Protective Devices	589
26 29	Low-Voltage Controllers	590
26 32	Packaged Generator Assemblies	590
26 33	Battery Equipment	591
26 35	Power Filters and Conditioners	591
26 51	Interior Lighting	592
26 52	Emergency Lighting	594
26 53	Exit Signs	594
26 54	Classified Location Lighting	595
26 55	Special Purpose Lighting	595
26 56	Exterior Lighting	595
26 61	Lighting Systems and Accessories	597
26 71	Electrical Machines	600

Communications

27 13	Communications Backbone Cabling	602
27 41	Audio-Video Systems	602
27 51	Distributed Audio-Video Communications Systems	603
27 52	Healthcare Communications and Monitoring Systems	604
27 53	Distributed Systems	604

Electronic Safety & Security

28 13	Access Control	606
28 16	Intrusion Detection	606
28 23	Video Surveillance	606
28 31	Fire Detection and Alarm	607
28 33	Gas Detection and Alarm	608
28 39	Mass Notification Systems	609

Sect. No.		Page

Earthwork

31 05	Common Work Results for Earthwork	612
31 06	Schedules for Earthwork	612
31 11	Clearing and Grubbing	613
31 13	Selective Tree and Shrub Removal and Trimming	614
31 14	Earth Stripping and Stockpiling	615
31 22	Grading	615
31 23	Excavation and Fill	616
31 25	Erosion and Sedimentation Controls	637
31 31	Soil Treatment	637
31 32	Soil Stabilization	638
31 33	Rock Stabilization	638
31 36	Gabions	638
31 37	Riprap	638
31 41	Shoring	639
31 43	Concrete Raising	640
31 45	Vibroflotation and Densification	640
31 46	Needle Beams	640
31 48	Underpinning	641
31 52	Cofferdams	641
31 56	Slurry Walls	642
31 62	Driven Piles	642
31 63	Bored Piles	644

Exterior Improvements

32 01	Operation and Maintenance of Exterior Improvements	650
32 06	Schedules for Exterior Improvements	650
32 11	Base Courses	651
32 12	Flexible Paving	652
32 13	Rigid Paving	653
32 14	Unit Paving	654
32 16	Curbs, Gutters, Sidewalks, and Driveways	655
32 17	Paving Specialties	656
32 18	Athletic and Recreational Surfacing	658
32 31	Fences and Gates	659
32 32	Retaining Walls	662
32 33	Site Furnishings	664
32 34	Fabricated Bridges	665
32 35	Screening Devices	666
32 84	Planting Irrigation	666
32 91	Planting Preparation	667
32 92	Turf and Grasses	668
32 93	Plants	669
32 94	Planting Accessories	671
32 96	Transplanting	671

Sect. No.		Page

Utilities

33 01	Operation and Maintenance of Utilities	674
33 05	Common Work Results for Utilities	674
33 11	Water Utility Distribution Piping	675
33 12	Water Utility Distribution Equipment	677
33 16	Water Utility Storage Tanks	678
33 21	Water Supply Wells	679
33 31	Sanitary Utility Sewerage Piping	679
33 36	Utility Septic Tanks	680
33 41	Storm Utility Drainage Piping	681
33 42	Culverts	683
33 44	Storm Utility Water Drains	683
33 46	Subdrainage	684
33 49	Storm Drainage Structures	685
33 51	Natural-Gas Distribution	685
33 52	Liquid Fuel Distribution	686
33 71	Electrical Utility Transmission and Distribution	687
33 81	Communications Structures	688

Transportation

34 01	Operation and Maintenance of Transportation	690
34 11	Rail Tracks	690
34 41	Roadway Signaling and Control Equipment	691
34 71	Roadway Construction	691
34 72	Railway Construction	693

Waterway & Marine

| 35 20 | Waterway and Marine Construction and Equipment | 696 |
| 35 51 | Floating Construction | 696 |

Material Processing & Handling Equipment

| 41 21 | Conveyors | 698 |
| 41 22 | Cranes and Hoists | 698 |

Pollution Control Equipment

| 44 11 | Particulate Control Equipment | 702 |

Water and Wastewater Equipment

| 46 07 | Packaged Water and Wastewater Treatment Equipment | 706 |

Electrical Power Generation

| 48 15 | Wind Energy Electrical Power Generation Equipment | 708 |

3

How RSMeans Unit Price Data Works

All RSMeans unit price data is organized in the same way.

03 30 Cast-In-Place Concrete

03 30 53 – Miscellaneous Cast-In-Place Concrete

03 30 53.40 Concrete In Place		Crew	Daily Output	Labor-Hours	Unit	Material	2017 Bare Costs Labor	Equipment	Total	Total Incl O&P
0010	**CONCRETE IN PLACE** R033105-10									
0020	Including forms (4 uses), Grade 60 rebar, concrete (Portland cement R033105-20									
0050	Type I), placement and finishing unless otherwise indicated R033105-50									
0300	Beams (3500 psi), 5 kip per L.F., 10' span R033105-65	C-14A	15.62	12.804	C.Y.	330	635	53.50	1,018.50	1,400
0350	25' span R033105-70	"	18.55	10.782		345	535	45	925	1,250
0500	Chimney foundations (5000 psi), over 5 C.Y. R033105-85	C-14C	32.22	3.476		167	164	.87	331.87	435
0510	(3500 psi), under 5 C.Y.	"	23.71	4.724		193	222	1.18	416.18	555
0700	Columns, square (4000 psi), 12" x 12", up to 1% reinforcing by area	C-14A	11.96	16.722		365	825	69.50	1,259.50	1,750
3540	Equipment pad (3000 psi), 3' x 3' x 6" thick	C-14H	45	1.067	Ea.	49	51.50	.63	101.13	133
3550	4' x 4' x 6" thick		30	1.600		73.50	77.50	.94	151.94	200
3560	5' x 5' x 8" thick		18	2.667		131	129	1.57	261.57	345
3570	6' x 6' x 8" thick		14	3.429		177	166	2.02	345.02	450
3580	8' x 8' x 10" thick		8	6		380	290	3.54	673.54	870
3590	10' x 10' x 12" thick		5	9.600		655	465	5.65	1,125.65	1,425

It is important to understand the structure of RSMeans Unit Price data, so that you can find information easily and use it correctly.

---

1 Line Numbers

Line Numbers consist of 12 characters, which identify a unique location in the database for each task. The first 6 or 8 digits conform to the Construction Specifications Institute MasterFormat® 2014. The remainder of the digits are a further breakdown in order to arrange items in understandable groups of similar tasks. Line numbers are consistent across all of our publications, so a line number in any of our products will always refer to the same item of work.

2 Descriptions

Descriptions are shown in a hierarchical structure to make them readable. In order to read a complete description, read up through the indents to the top of the section. Include everything that is above and to the left that is not contradicted by information below. For instance, the complete description for line 03 30 53.40 3550 is "Concrete in place, including forms (4 uses), Grade 60 rebar, concrete (Portland cement Type 1), placement and finishing unless otherwise indicated; Equipment pad (3000 psi), 4' × 4' × 6" thick."

3 RSMeans Data

When using **RSMeans data**, it is important to read through an entire section to ensure that you use the data that most closely matches your work. Note that sometimes there is additional information shown in the section that may improve your price. There are frequently lines that further describe, add to, or adjust data for specific situations.

4 Reference Information

Gordian's RSMeans engineers have created **reference** information to assist you in your estimate. **If** there is information that applies to a section, it will be indicated at the start of the section. The Reference Section is located in the back of the data set.

5 Crews

Crews include labor and/or equipment necessary to accomplish each task. In this case, Crew C-14H is used. Gordian's RSMeans staff selects a crew to represent the workers and equipment that are

typically used for that task. In this case, Crew C-14H consists of one carpenter foreman (outside), two carpenters, one rodman, one laborer, one cement finisher, and one gas engine vibrator. Details of all crews can be found in the reference section.

Crews - Standard

Crew No.	Bare Costs		Incl. Subs O&P		Cost Per Labor-Hour	
Crew C-14H	Hr.	Daily	Hr.	Daily	Bare Costs	Incl. O&P
1 Carpenter Foreman (outside)	$51.25	$410.00	$78.50	$628.00	$48.33	$73.78
2 Carpenters	49.25	788.00	75.45	1207.20		
1 Rodman (reinf.)	54.30	434.40	83.70	669.60		
1 Laborer	39.15	313.20	60.00	480.00		
1 Cement Finisher	46.80	374.40	69.60	556.80		
1 Gas Engine Vibrator		28.25		31.07	.59	.65
48 L.H., Daily Totals		$2348.25		$3572.68	$48.92	$74.43

6 Daily Output

The **Daily Output** is the amount of work that the crew can do in a normal 8-hour workday, including mobilization, layout, movement of materials, and cleanup. In this case, crew C-14H can install thirty 4' × 4' × 6" thick concrete pads in a day. Daily output is variable and based on many factors, including the size of the job, location, and environmental conditions. RSMeans data represents work done in daylight (or adequate lighting) and temperate conditions.

7 Labor-Hours

The figure in the **Labor-Hours** column is the amount of labor required to perform one unit of work–in this case the amount of labor required to construct one 4' × 4' equipment pad. This figure is calculated by dividing the number of hours of labor in the crew by the daily output (48 labor hours divided by 30 pads = 1.6 hours of labor per pad). Multiply 1.600 times 60 to see the value in minutes: 60 × 1.6 = 96

minutes. Note: the labor-hour figure is not dependent on the crew size. A change in crew size will result in a corresponding change in daily output, but the labor-hours per unit of work will not change.

8 Unit of Measure

All RSMeans unit cost data includes the typical **Unit of Measure** used for estimating that item. For concrete-in-place the typical unit is cubic yards (C.Y.) or each (Ea.). For installing broadloom carpet it is square yard, and for gypsum board it is square foot. The estimator needs to take special care that the unit in the data matches the unit in the take-off. Unit conversions may be found in the Reference Section.

9 Bare Costs

Bare Costs are the costs of materials, labor, and equipment that the installing contractor pays. They represent the cost, in U.S. dollars, for one unit of work. They do not include any markups for profit or labor burden.

10 Bare Total

The **Total column** represents the total bare cost for the installing contractor, in U.S. dollars. In this case, the sum of $73.50 for material + $77.50 for labor + $.94 for equipment is $151.94.

11 Total Incl O&P

The **Total Incl O&P column** is the total cost, including overhead and profit, that the installing contractor will charge the customer. This represents the cost of materials plus 10% profit, the cost of labor plus labor burden and 10% profit, and the cost of equipment plus 10% profit. It does not include the general contractor's overhead and profit. Note: See the inside back cover of the printed product or the reference section of the electronic product for details of how the labor burden is calculated.

National Average

*The RSMeans data in our print publications represents a "national average" cost. This data should be modified to the project location using the **City Cost Indexes** or **Location Factors** tables found in the Reference Section. Use the location factors to adjust estimate totals if the project covers multiple trades. Use the city cost indexes (CCI) for single trade*

projects or projects where a more detailed analysis is required. All figures in the two tables are derived from the same research. The last row of data in the CCI, the weighted average, is the same as the numbers reported for each location in the location factor table.

How RSMeans Unit Price Data Works (Continued)

①

Line Number	Description	Qty	Unit	Material	Labor	Equipment	SubContract	Estimate Total
	Project Name: Pre-Engineered Steel Building			**Architect: As Shown**				
	Location: Anywhere, USA						**01/01/17**	**STD**
03 30 53.40 3940	Strip footing, 12" x 24", reinforced	34	C.Y.	$5,304.00	$3,740.00	$19.72	$0.00	
03 30 53.40 3950	Strip footing, 12" x 36", reinforced	15	C.Y.	$2,250.00	$1,320.00	$7.05	$0.00	
03 11 13.65 3000	Concrete slab edge forms	500	L.F.	$120.00	$1,245.00	$0.00	$0.00	
03 22 11.10 0200	Welded wire fabric reinforcing	150	C.S.F.	$2,820.00	$4,200.00	$0.00	$0.00	
03 31 13.35 0300	Ready mix concrete, 4000 psi for slab on grade	278	C.Y.	$34,750.00	$0.00	$0.00	$0.00	
03 31 13.70 4300	Place, strike off & consolidate concrete slab	278	C.Y.	$0.00	$4,948.40	$141.78	$0.00	
03 35 13.30 0250	Machine float & trowel concrete slab	15,000	S.F.	$0.00	$9,300.00	$450.00	$0.00	
03 15 16.20 0140	Cut control joints in concrete slab	950	L.F.	$47.50	$399.00	$85.50	$0.00	
03 39 23.13 0300	Sprayed concrete curing membrane	150	C.S.F.	$1,905.00	$990.00	$0.00	$0.00	
Division 03	**Subtotal**			**$47,196.50**	**$26,142.40**	**$704.05**	**$0.00**	**$74,042.95**
08 36 13.10 2650	Manual 10' x 10' steel sectional overhead door	8	Ea.	$10,000.00	$3,520.00	$0.00	$0.00	
08 36 13.10 2860	Insulation and steel back panel for OH door	800	S.F.	$3,800.00	$0.00	$0.00	$0.00	
Division 08	**Subtotal**			**$13,800.00**	**$3,520.00**	**$0.00**	**$0.00**	**$17,320.00**
13 34 19.50 1100	Pre-Engineered Steel Building, 100' x 150' x 24'	15,000	SF Flr.	$0.00	$0.00	$0.00	$367,500.00	
13 34 19.50 6050	Framing for PESB door opening, 3' x 7'	4	Opng.	$0.00	$0.00	$0.00	$2,240.00	
13 34 19.50 6100	Framing for PESB door opening, 10' x 10'	8	Opng.	$0.00	$0.00	$0.00	$9,200.00	
13 34 19.50 6200	Framing for PESB window opening, 4' x 3'	6	Opng.	$0.00	$0.00	$0.00	$3,300.00	
13 34 19.50 5750	PESB door, 3' x 7', single leaf	4	Opng.	$2,460.00	$696.00	$0.00	$0.00	
13 34 19.50 7750	PESB sliding window, 4' x 3' with screen	6	Opng.	$2,490.00	$597.00	$66.00	$0.00	
13 34 19.50 6550	PESB gutter, eave type, 26 ga., painted	300	L.F.	$2,190.00	$816.00	$0.00	$0.00	
13 34 19.50 8650	PESB roof vent, 12" wide x 10' long	15	Ea.	$547.50	$3,255.00	$0.00	$0.00	
13 34 19.50 6900	PESB insulation, vinyl faced, 4" thick	27,400	S.F.	$12,878.00	$9,316.00	$0.00	$0.00	
Division 13	**Subtotal**			**$20,565.50**	**$14,680.00**	**$66.00**	**$382,240.00**	**$417,551.50**
	Subtotal			$81,562.00	$44,342.40	$770.05	$382,240.00	$508,914.45
Division 01 **②**	General Requirements @ 7%			5,709.34	3,103.97	53.90	26,756.80	
	Estimate Subtotal			$87,271.34	$47,446.37	$823.95	$408,996.80	$508,914.45
③	Sales Tax @ 5%			4,363.57		41.20	10,224.92	
	Subtotal A			91,634.91	47,446.37	865.15	419,221.72	
④	GC O & P			9,163.49	25,573.59	86.52	41,922.17	
	Subtotal B			100,798.40	73,019.96	951.67	461,143.89	$635,913.92
⑤	Contingency @ 5%							31,795.70
	Subtotal C							$667,709.61
⑥	Bond @ $12/1000 +10% O&P							8,813.77
	Subtotal D							$676,523.38
⑦	Location Adjustment Factor				102.30			15,560.04
	Grand Total							**$692,083.42**

This estimate is based on an interactive spreadsheet. A copy of this spreadsheet is located at **www.RSMeans.com/2017samples.** You are free to download it and adjust it to your methodology.

Sample Estimate

This sample demonstrates the elements of an estimate, including a tally of the RSMeans data lines and a summary of the markups on a contractor's work to arrive at a total cost to the owner. The Location Factor with RSMeans data is added at the bottom of the estimate to adjust the cost of the work to a specific location.

① Work Performed

The body of the estimate shows the RSMeans data selected, including the line number, a brief description of each item, its take-off unit and quantity, and the bare costs of materials, labor, and equipment. This estimate also includes a column titled "SubContract." This data is taken from the column "Total Incl O&P" and represents the total that a subcontractor would charge a general contractor for the work, including the sub's markup for overhead and profit.

② Division 1, General Requirements

This is the first division numerically but the last division estimated. Division 1 includes project-wide needs provided by the general contractor. These requirements vary by project but may include temporary facilities and utilities, security, testing, project cleanup, etc. For small projects a percentage can be used, typically between 5% and 15% of project cost. For large projects the costs may be itemized and priced individually.

③ Sales Tax

If the work is subject to state or local sales taxes, the amount must be added to the estimate. Sales tax may be added to material costs, equipment costs, and subcontracted work. In this case, sales tax was added in all three categories. It was assumed that approximately half the subcontracted work would be material cost, so the tax was applied to 50% of the subcontract total.

④ GC O&P

This entry represents the general contractor's markup on material, labor, equipment, and subcontractor costs. Our standard markup on materials, equipment, and subcontracted work is 10%. In this estimate, the markup on the labor performed by the GC's workers uses "Skilled Workers Average" shown in Column F on the table "Installing Contractor's Overhead & Profit," which can be found on the inside-back cover of the printed product or in the Reference Section of the electronic product.

⑤ Contingency

A factor for contingency may be added to any estimate to represent the cost of unknowns that may occur between the time that the estimate is performed and the time the project is constructed. The amount of the allowance will depend on the stage of design at which the estimate is done and the contractor's assessment of the risk involved. Refer to section 01 21 16.50 for contingency allowances.

⑥ Bonds

Bond costs should be added to the estimate. The figures here represent a typical performance bond, ensuring the owner that if the general contractor does not complete the obligations in the construction contract the bonding company will pay the cost for completion of the work.

⑦ Location Adjustment

RSMeans published data is based on national average costs. If necessary, adjust the total cost of the project using a location factor from the "Location Factor" table or the "City Cost Index" table. Use location factors if the work is general, covering multiple trades. If the work is by a single trade (e.g., masonry) use the more specific data found in the "City Cost Indexes."

Estimating Tips
01 20 00 Price and Payment Procedures

- Allowances that should be added to estimates to cover contingencies and job conditions that are not included in the national average material and labor costs are shown in Section 01 21.
- When estimating historic preservation projects (depending on the condition of the existing structure and the owner's requirements), a 15%–20% contingency or allowance is recommended, regardless of the stage of the drawings.

01 30 00 Administrative Requirements

- Before determining a final cost estimate, it is a good practice to review all the items listed in Subdivisions 01 31 and 01 32 to make final adjustments for items that may need customizing to specific job conditions.
- Requirements for initial and periodic submittals can represent a significant cost to the General Requirements of a job. Thoroughly check the submittal specifications when estimating a project to determine any costs that should be included.

01 40 00 Quality Requirements

- All projects will require some degree of quality control. This cost is not included in the unit cost of construction listed in each division. Depending upon the terms of the contract, the various costs of inspection and testing can be the responsibility of either the owner or the contractor. Be sure to include the required costs in your estimate.

01 50 00 Temporary Facilities and Controls

- Barricades, access roads, safety nets, scaffolding, security, and many more requirements for the execution of a safe project are elements of direct cost. These costs can easily be overlooked when preparing an estimate. When looking through the major classifications of this subdivision, determine which items apply to each division in your estimate.
- Construction equipment rental costs can be found in the Reference Section in Section 01 54 33. Operators' wages are not included in equipment rental costs.
- Equipment mobilization and demobilization costs are not included in equipment rental costs and must be considered separately.

- The cost of small tools provided by the installing contractor for his workers is covered in the "Overhead" column on the "Installing Contractor's Overhead and Profit" table that lists labor trades, base rates, and markups and, therefore, is included in the "Total Incl. O&P" cost of any unit price line item.

01 70 00 Execution and Closeout Requirements

- When preparing an estimate, thoroughly read the specifications to determine the requirements for Contract Closeout. Final cleaning, record documentation, operation and maintenance data, warranties and bonds, and spare parts and maintenance materials can all be elements of cost for the completion of a contract. Do not overlook these in your estimate.

Reference Numbers

Reference numbers are shown at the beginning of some major classifications. These numbers refer to related items in the Reference Section. The reference information may be an estimating procedure, an alternate pricing method, or technical information.

Note: Not all subdivisions listed here necessarily appear. ∎

Did you know?

Our online estimating solution gives you the same access to RSMeans' data with 24/7 access:

- Quickly locate costs in the searchable database.
- Build cost lists, estimates, and reports in minutes.
- Adjust costs to any location in the U.S. and Canada with the click of a button.

Start your free trial today at
www.RSMeansOnline.com

01 11 31 – Professional Consultants

01 11 31.10 Architectural Fees		Crew	Daily Output	Labor-Hours	Unit	Material	2017 Bare Costs Labor	Equipment	Total	Total Incl O&P
0010	**ARCHITECTURAL FEES** R011110-10									
0020	For new construction									
0060	Minimum				Project				4.90%	4.90%
0090	Maximum								16%	16%
0100	For alteration work, to $500,000, add to new construction fee								50%	50%
0150	Over $500,000, add to new construction fee								25%	25%
2000	For "Greening" of building [G]				▼				3%	3%

01 11 31.20 Construction Management Fees										
0010	**CONSTRUCTION MANAGEMENT FEES**									
0020	$1,000,000 job, minimum				Project				4.50%	4.50%
0050	Maximum								7.50%	7.50%
0060	For work to $100,000								10%	10%
0070	To $250,000								9%	9%
0090	To $1,000,000								6%	6%
0100	To $5,000,000								5%	5%
0110	To $10,000,000								4%	4%
0300	$50,000,000 job, minimum								2.50%	2.50%
0350	Maximum				▼				4%	4%

01 11 31.30 Engineering Fees										
0010	**ENGINEERING FEES** R011110-30									
0020	Educational planning consultant, minimum				Project				.50%	.50%
0100	Maximum				"				2.50%	2.50%
0200	Electrical, minimum				Contrct				4.10%	4.10%
0300	Maximum								10.10%	10.10%
0400	Elevator & conveying systems, minimum								2.50%	2.50%
0500	Maximum								5%	5%
0600	Food service & kitchen equipment, minimum								8%	8%
0700	Maximum								12%	12%
0800	Landscaping & site development, minimum								2.50%	2.50%
0900	Maximum								6%	6%
1000	Mechanical (plumbing & HVAC), minimum								4.10%	4.10%
1100	Maximum				▼				10.10%	10.10%
1200	Structural, minimum				Project				1%	1%
1300	Maximum				"				2.50%	2.50%

01 11 31.50 Models										
0010	**MODELS**									
0500	2 story building, scaled 100' x 200', simple materials and details				Ea.	4,600			4,600	5,050
0510	Elaborate materials and details				"	30,000			30,000	33,000

01 11 31.75 Renderings										
0010	**RENDERINGS** Color, matted, 20" x 30", eye level,									
0020	1 building, minimum				Ea.	2,200			2,200	2,425
0050	Average					3,100			3,100	3,400
0100	Maximum					5,000			5,000	5,500
1000	5 buildings, minimum					4,100			4,100	4,500
1100	Maximum					8,100			8,100	8,900
2000	Aerial perspective, color, 1 building, minimum					3,000			3,000	3,300
2100	Maximum					8,200			8,200	9,025
3000	5 buildings, minimum					6,000			6,000	6,600
3100	Maximum				▼	12,100			12,100	13,300

01 21 Allowances

01 21 16 – Contingency Allowances

01 21 16.50 Contingencies	Crew	Daily Output	Labor-Hours	Unit	Material	2017 Bare Costs Labor	Equipment	Total	Total Incl O&P
0010 **CONTINGENCIES**, Add to estimate									
0020 Conceptual stage				Project				20%	20%
0050 Schematic stage								15%	15%
0100 Preliminary working drawing stage (Design Dev.)								10%	10%
0150 Final working drawing stage								3%	3%

01 21 53 – Factors Allowance

01 21 53.60 Security Factors

	Crew	Daily Output	Labor-Hours	Unit	Material	Labor	Equipment	Total	Total Incl O&P
0010 **SECURITY FACTORS**									
0100 Additional costs due to security requirements									
0110 Daily search of personnel, supplies, equipment and vehicles									
0120 Physical search, inventory and doc of assets, at entry				Costs		30%			
0130 At entry and exit						50%			
0140 Physical search, at entry						6.25%			
0150 At entry and exit						12.50%			
0160 Electronic scan search, at entry						2%			
0170 At entry and exit						4%			
0180 Visual inspection only, at entry						.25%			
0190 At entry and exit						.50%			
0200 ID card or display sticker only, at entry						.12%			
0210 At entry and exit						.25%			
0220 Day 1 as described below, then visual only for up to 5 day job duration									
0230 Physical search, inventory and doc of assets, at entry				Costs		5%			
0240 At entry and exit						10%			
0250 Physical search, at entry						1.25%			
0260 At entry and exit						2.50%			
0270 Electronic scan search, at entry						.42%			
0280 At entry and exit						.83%			
0290 Day 1 as described below, then visual only for 6-10 day job duration									
0300 Physical search, inventory and doc of assets, at entry				Costs		2.50%			
0310 At entry and exit						5%			
0320 Physical search, at entry						.63%			
0330 At entry and exit						1.25%			
0340 Electronic scan search, at entry						.21%			
0350 At entry and exit						.42%			
0360 Day 1 as described below, then visual only for 11-20 day job duration									
0370 Physical search, inventory and doc of assets, at entry				Costs		1.25%			
0380 At entry and exit						2.50%			
0390 Physical search, at entry						.31%			
0400 At entry and exit						.63%			
0410 Electronic scan search, at entry						.10%			
0420 At entry and exit						.21%			
0430 Beyond 20 days, costs are negligible									
0440 Escort required to be with tradesperson during work effort				Costs		6.25%			

01 21 55 – Job Conditions Allowance

01 21 55.50 Job Conditions

	Crew	Daily Output	Labor-Hours	Unit	Material	Labor	Equipment	Total	Total Incl O&P
0010 **JOB CONDITIONS** Modifications to applicable									
0020 cost summaries									
0100 Economic conditions, favorable, deduct				Project				2%	2%
0200 Unfavorable, add								5%	5%
0300 Hoisting conditions, favorable, deduct								2%	2%
0400 Unfavorable, add								5%	5%
0500 General Contractor management, experienced, deduct								2%	2%

For customer support on your Building Construction Costs with RSMeans Data, call 800.448.8182.

11

01 21 Allowances

01 21 55 – Job Conditions Allowance

01 21 55.50 Job Conditions	Crew	Daily Output	Labor-Hours	Unit	Material	2017 Bare Costs Labor	Equipment	Total	Total Incl O&P
0600	Inexperienced, add				Project				10%	10%
0700	Labor availability, surplus, deduct								1%	1%
0800	Shortage, add								10%	10%
0900	Material storage area, available, deduct								1%	1%
1000	Not available, add								2%	2%
1100	Subcontractor availability, surplus, deduct								5%	5%
1200	Shortage, add								12%	12%
1300	Work space, available, deduct								2%	2%
1400	Not available, add								5%	5%

01 21 57 – Overtime Allowance

01 21 57.50 Overtime

0010	**OVERTIME** for early completion of projects or where	R012909-90								
0020	labor shortages exist, add to usual labor, up to				Costs		100%			

01 21 63 – Taxes

01 21 63.10 Taxes

0010	**TAXES**	R012909-80								
0020	Sales tax, State, average				%	5.08%				
0050	Maximum	R012909-85					7.50%				
0200	Social Security, on first $118,500 of wages						7.65%			
0300	Unemployment, combined Federal and State, minimum						.60%			
0350	Average						9.60%			
0400	Maximum						12%			

01 31 Project Management and Coordination

01 31 13 – Project Coordination

01 31 13.20 Field Personnel

		Crew	Daily Output	Labor-Hours	Unit	Material	Labor	Equipment	Total	Total Incl O&P
0010	**FIELD PERSONNEL**									
0020	Clerk, average				Week	475			475	730
0100	Field engineer, minimum					1,125			1,125	1,725
0120	Average					1,475			1,475	2,275
0140	Maximum					1,675			1,675	2,575
0160	General purpose laborer, average					1,575			1,575	2,400
0180	Project manager, minimum					2,100			2,100	3,225
0200	Average					2,400			2,400	3,700
0220	Maximum					2,750			2,750	4,225
0240	Superintendent, minimum					2,050			2,050	3,150
0260	Average					2,250			2,250	3,450
0280	Maximum					2,550			2,550	3,925
0290	Timekeeper, average					1,300			1,300	2,000

01 31 13.30 Insurance

0010	**INSURANCE**	R013113-40								
0020	Builders risk, standard, minimum				Job				.24%	.24%
0050	Maximum	R013113-50							.64%	.64%
0200	All-risk type, minimum								.25%	.25%
0250	Maximum	R013113-60							.62%	.62%
0400	Contractor's equipment floater, minimum				Value				.50%	.50%
0450	Maximum				"				1.50%	1.50%
0600	Public liability, average				Job				2.02%	2.02%
0800	Workers' compensation & employer's liability, average									
0850	by trade, carpentry, general				Payroll		13.93%			

01 31 Project Management and Coordination

01 31 13 – Project Coordination

01 31 13.30 Insurance

		Crew	Daily Output	Labor-Hours	Unit	Material	2017 Bare Costs Labor	2017 Bare Costs Equipment	Total	Total Incl O&P
0900	Clerical				Payroll		.46%			
0950	Concrete						12.44%			
1000	Electrical						5.52%			
1050	Excavation						9.03%			
1100	Glazing						12.91%			
1150	Insulation						11.13%			
1200	Lathing						8.67%			
1250	Masonry						13.73%			
1300	Painting & decorating						11.68%			
1350	Pile driving						14.53%			
1400	Plastering						10.73%			
1450	Plumbing						6.94%			
1500	Roofing						30.73%			
1550	Sheet metal work (HVAC)						8.82%			
1600	Steel erection, structural						27.45%			
1650	Tile work, interior ceramic						8.85%			
1700	Waterproofing, brush or hand caulking						7.01%			
1800	Wrecking						20.85%			
2000	Range of 35 trades in 50 states, excl. wrecking & clerical, min.						1.41%			
2100	Average						12.25%			
2200	Maximum						108.79%			

01 31 13.40 Main Office Expense

		Crew	Daily Output	Labor-Hours	Unit	Material	Labor	Equipment	Total	Total Incl O&P
0010	**MAIN OFFICE EXPENSE** Average for General Contractors R013113-50									
0020	As a percentage of their annual volume									
0125	Annual volume under 1 million dollars				% Vol.				17.50%	
0145	Up to 2.5 million dollars								8%	
0150	Up to 4.0 million dollars								6.80%	
0200	Up to 7.0 million dollars								5.60%	
0250	Up to 10 million dollars								5.10%	
0300	Over 10 million dollars								3.90%	

01 31 13.50 General Contractor's Mark-Up

		Crew	Daily Output	Labor-Hours	Unit	Material	Labor	Equipment	Total	Total Incl O&P
0010	**GENERAL CONTRACTOR'S MARK-UP** on Change Orders									
0200	Extra work, by subcontractors, add				%				10%	10%
0250	By General Contractor, add								15%	15%
0400	Omitted work, by subcontractors, deduct all but								5%	5%
0450	By General Contractor, deduct all but								7.50%	7.50%
0600	Overtime work, by subcontractors, add								15%	15%
0650	By General Contractor, add								10%	10%

01 31 13.60 General Contractor's Main Office Overhead

		Crew	Daily Output	Labor-Hours	Unit	Material	Labor	Equipment	Total	Total Incl O&P
0010	**GENERAL CONTRACTOR'S MAIN OFFICE OVERHEAD** R013113-50									
0020	As percent of direct costs, minimum				%				5%	
0050	Average								13%	
0100	Maximum								30%	

01 31 13.80 Overhead and Profit

		Crew	Daily Output	Labor-Hours	Unit	Material	Labor	Equipment	Total	Total Incl O&P
0010	**OVERHEAD & PROFIT** Allowance to add to items in this									
0020	book that do not include Subs O&P, average				%				25%	
0100	Allowance to add to items in this book that									
0110	do include Subs O&P, minimum				%				5%	5%
0150	Average								10%	10%
0200	Maximum								15%	15%
0300	Typical, by size of project, under $100,000								30%	
0350	$500,000 project								25%	

01 31 Project Management and Coordination

01 31 13 – Project Coordination

01 31 13.80 Overhead and Profit

		Crew	Daily Output	Labor-Hours	Unit	Material	2017 Bare Costs Labor	Equipment	Total	Total Incl O&P
0400	$2,000,000 project				%					20%
0450	Over $10,000,000 project									15%

01 31 13.90 Performance Bond

		Crew	Daily Output	Labor-Hours	Unit	Material	2017 Bare Costs Labor	Equipment	Total	Total Incl O&P
0010	**PERFORMANCE BOND**	R013113-80								
0020	For buildings, minimum				Job				.60%	.60%
0100	Maximum				"				2.50%	2.50%

01 32 Construction Progress Documentation

01 32 13 – Scheduling of Work

01 32 13.50 Scheduling of Work

		Crew	Daily Output	Labor-Hours	Unit	Material	2017 Bare Costs Labor	Equipment	Total	Total Incl O&P
0010	**SCHEDULING**									
0020	Critical path, as % of architectural fee, minimum				%				.50%	.50%
0100	Maximum				"				1%	1%
0300	Computer-update, micro, no plots, minimum				Ea.				455	500
0400	Including plots, maximum				"				1,450	1,600
0600	Rule of thumb, CPM scheduling, small job ($10 Million)				Job				.05%	.05%
0650	Large job ($50 Million +)								.03%	.03%
0700	Including cost control, small job								.08%	.08%
0750	Large job								.04%	.04%

01 32 33 – Photographic Documentation

01 32 33.50 Photographs

		Crew	Daily Output	Labor-Hours	Unit	Material	2017 Bare Costs Labor	Equipment	Total	Total Incl O&P
0010	**PHOTOGRAPHS**									
0020	8" x 10", 4 shots, 2 prints ea., std. mounting				Set	535			535	590
0100	Hinged linen mounts					540			540	595
0200	8" x 10", 4 shots, 2 prints each, in color					495			495	545
0300	For I.D. slugs, add to all above					5			5	5.50
0500	Aerial photos, initial fly-over, 5 shots, digital images					410			410	450
0550	10 shots, digital images, 1 print					450			450	495
0600	For each additional print from fly-over					205			205	226
0700	For full color prints, add					40%				
0750	Add for traffic control area				Ea.	330			330	365
0900	For over 30 miles from airport, add per				Mile	6.65			6.65	7.35
1500	Time lapse equipment, camera and projector, buy				Ea.	2,700			2,700	2,975
1550	Rent per month				"	1,275			1,275	1,400
1700	Cameraman and processing, black & white				Day	1,250			1,250	1,375
1720	Color				"	1,450			1,450	1,600

01 41 Regulatory Requirements

01 41 26 – Permit Requirements

01 41 26.50 Permits

		Crew	Daily Output	Labor-Hours	Unit	Material	2017 Bare Costs Labor	Equipment	Total	Total Incl O&P
0010	**PERMITS**									
0020	Rule of thumb, most cities, minimum				Job				.50%	.50%
0100	Maximum				"				2%	2%

01 45 Quality Control

01 45 23 – Testing and Inspecting Services

01 45 23.50 Testing	Crew	Daily Output	Labor-Hours	Unit	Material	2017 Bare Costs Labor	Equipment	Total	Total Incl O&P
0010 **TESTING** and Inspecting Services									
0015 For concrete building costing $1,000,000, minimum				Project				4,725	5,200
0020 Maximum								38,000	41,800
0050 Steel building, minimum								4,725	5,200
0070 Maximum								14,800	16,300
0100 For building costing, $10,000,000, minimum								30,100	33,100
0150 Maximum				▼				48,200	53,000
0200 Asphalt testing, compressive strength Marshall stability, set of 3				Ea.				145	165
0220 Density, set of 3								86	95
0250 Extraction, individual tests on sample								136	150
0300 Penetration								41	45
0350 Mix design, 5 specimens								182	200
0360 Additional specimen								36	40
0400 Specific gravity								41	45
0420 Swell test								64	70
0450 Water effect and cohesion, set of 6								182	200
0470 Water effect and plastic flow								64	70
0600 Concrete testing, aggregates, abrasion, ASTM C 131								136	150
0650 Absorption, ASTM C 127								42	46
0800 Petrographic analysis, ASTM C 295								775	850
0900 Specific gravity, ASTM C 127								50	55
1000 Sieve analysis, washed, ASTM C 136								59	65
1050 Unwashed								59	65
1200 Sulfate soundness								114	125
1300 Weight per cubic foot								36	40
1500 Cement, physical tests, ASTM C 150								320	350
1600 Chemical tests, ASTM C 150								245	270
1800 Compressive test, cylinder, delivered to lab, ASTM C 39								12	13
1900 Picked up by lab, minimum								14	15
1950 Average								18	20
2000 Maximum								27	30
2200 Compressive strength, cores (not incl. drilling), ASTM C 42				▼				36	40
2250 Core drilling, 4" diameter (plus technician)				Inch				23	25
2260 Technician for core drilling				Hr.				45	50
2300 Patching core holes				Ea.				22	24
2400 Drying shrinkage at 28 days								236	260
2500 Flexural test beams, ASTM C 78								59	65
2600 Mix design, one batch mix								259	285
2650 Added trial batches								120	132
2800 Modulus of elasticity, ASTM C 469								164	180
2900 Tensile test, cylinders, ASTM C 496								45	50
3000 Water-Cement ratio curve, 3 batches								141	155
3100 4 batches								186	205
3300 Masonry testing, absorption, per 5 brick, ASTM C 67								45	50
3350 Chemical resistance, per 2 brick								50	55
3400 Compressive strength, per 5 brick, ASTM C 67								68	75
3420 Efflorescence, per 5 brick, ASTM C 67								68	75
3440 Imperviousness, per 5 brick								87	96
3470 Modulus of rupture, per 5 brick								86	95
3500 Moisture, block only								32	35
3550 Mortar, compressive strength, set of 3								23	25
4100 Reinforcing steel, bend test								55	61
4200 Tensile test, up to #8 bar				▼				36	40

01 45 23 – Testing and Inspecting Services

01 45 23.50 Testing		Crew	Daily Output	Labor-Hours	Unit	Material	2017 Bare Costs Labor	Equipment	Total	Total Incl O&P
4220	#9 to #11 bar				Ea.				41	45
4240	#14 bar and larger								64	70
4400	Soil testing, Atterberg limits, liquid and plastic limits								59	65
4510	Hydrometer analysis								109	120
4530	Specific gravity, ASTM D 354								44	48
4600	Sieve analysis, washed, ASTM D 422								55	60
4700	Unwashed, ASTM D 422								59	65
4710	Consolidation test (ASTM D 2435), minimum								250	275
4715	Maximum								430	475
4720	Density and classification of undisturbed sample								73	80
4735	Soil density, nuclear method, ASTM D 2922								35	38.50
4740	Sand cone method, ASTM D 1556								27	30
4750	Moisture content, ASTM D 2216								9	10
4780	Permeability test, double ring infiltrometer								500	550
4800	Permeability, var. or constant head, undist., ASTM D 2434								227	250
4850	Recompacted								250	275
4900	Proctor compaction, 4" standard mold, ASTM D 698								123	135
4950	6" modified mold								68	75
5100	Shear tests, triaxial, minimum								410	450
5150	Maximum								545	600
5300	Direct shear, minimum, ASTM D 3080								320	350
5350	Maximum								410	450
5550	Technician for inspection, per day, earthwork								320	350
5650	Bolting								400	440
5750	Roofing								480	530
5790	Welding				▼				480	530
5820	Non-destructive metal testing, dye penetrant				Day				310	340
5840	Magnetic particle								310	340
5860	Radiography								450	495
5880	Ultrasonic				▼				310	340
6000	Welding certification, minimum				Ea.				91	100
6100	Maximum				"				250	275
7000	Underground storage tank									
7500	Volumetric tightness test, <=12,000 gal.				Ea.				435	480
7510	<=30,000 gal.				"				615	675
7600	Vadose zone (soil gas) sampling, 10-40 samples, min.				Day				1,375	1,500
7610	Maximum				"				2,275	2,500
7700	Ground water monitoring incl. drilling 3 wells, min.				Total				4,550	5,000
7710	Maximum				"				6,375	7,000
8000	X-ray concrete slabs				Ea.				182	200
9000	Thermographic testing, for bldg envelope heat loss, average 2,000 S.F. [G]				"				500	500

01 51 Temporary Utilities

01 51 13 – Temporary Electricity

01 51 13.80 Temporary Utilities

01 51 13.80 Temporary Utilities	Crew	Daily Output	Labor-Hours	Unit	Material	2017 Bare Costs Labor	Equipment	Total	Total Incl O&P
0010 **TEMPORARY UTILITIES**									
0350 Lighting, lamps, wiring, outlets, 40,000 S.F. building, 8 strings	1 Elec	34	.235	CSF Flr	5	13.30		18.30	25.50
0360 16 strings	"	17	.471		10	26.50		36.50	51
0400 Power for temp lighting only, 6.6 KWH, per month								.92	1.01
0430 11.8 KWH, per month								1.65	1.82
0450 23.6 KWH, per month								3.30	3.63
0600 Power for job duration incl. elevator, etc., minimum								47	51.50
0650 Maximum								110	121
1000 Toilet, portable, see Equip. Rental 01 54 33 in Reference Section									

01 52 Construction Facilities

01 52 13 – Field Offices and Sheds

01 52 13.20 Office and Storage Space

01 52 13.20 Office and Storage Space	Crew	Daily Output	Labor-Hours	Unit	Material	2017 Bare Costs Labor	Equipment	Total	Total Incl O&P
0010 **OFFICE AND STORAGE SPACE**									
0020 Office trailer, furnished, no hookups, 20' x 8', buy	2 Skwk	1	16	Ea.	10,800	825		11,625	13,100
0250 Rent per month					195			195	214
0300 32' x 8', buy	2 Skwk	.70	22.857		16,000	1,175		17,175	19,400
0350 Rent per month					250			250	275
0400 50' x 10', buy	2 Skwk	.60	26.667		24,700	1,375		26,075	29,300
0450 Rent per month					355			355	390
0500 50' x 12', buy	2 Skwk	.50	32		30,200	1,650		31,850	35,700
0550 Rent per month					440			440	485
0700 For air conditioning, rent per month, add					49			49	54
0800 For delivery, add per mile				Mile	11			11	12.10
0890 Delivery each way				Ea.	1,650			1,650	1,800
0900 Bunk house trailer, 8' x 40' duplex dorm with kitchen, no hookups, buy	2 Carp	1	16		86,000	790		86,790	95,500
0910 9 man with kitchen and bath, no hookups, buy		1	16		90,500	790		91,290	100,500
0920 18 man sleeper with bath, no hookups, buy		1	16		95,000	790		95,790	105,500
1000 Portable buildings, prefab, on skids, economy, 8' x 8'		265	.060	S.F.	24	2.97		26.97	31
1100 Deluxe, 8' x 12'		150	.107	"	19.60	5.25		24.85	29.50
1200 Storage boxes, 20' x 8', buy	2 Skwk	1.80	8.889	Ea.	3,025	455		3,480	4,025
1250 Rent per month					81			81	89
1300 40' x 8', buy	2 Skwk	1.40	11.429		3,900	590		4,490	5,175
1350 Rent per month					110			110	121
5000 Air supported structures, see Section 13 31 13.13									

01 52 13.40 Field Office Expense

01 52 13.40 Field Office Expense	Crew	Daily Output	Labor-Hours	Unit	Material	2017 Bare Costs Labor	Equipment	Total	Total Incl O&P
0010 **FIELD OFFICE EXPENSE**									
0100 Office equipment rental average				Month	205			205	226
0120 Office supplies, average				"	82			82	90
0125 Office trailer rental, see Section 01 52 13.20									
0140 Telephone bill; avg. bill/month incl. long dist.				Month	86			86	94.50
0160 Lights & HVAC				"	161			161	177

For customer support on your Building Construction Costs with RSMeans Data, call 800.448.8182.

17

01 54 Construction Aids

01 54 09 – Protection Equipment

01 54 09.50 Personnel Protective Equipment	Crew	Daily Output	Labor-Hours	Unit	Material	2017 Bare Costs Labor	Equipment	Total	Total Incl O&P
0010 **PERSONNEL PROTECTIVE EQUIPMENT**									
0015 Hazardous waste protection									
0020 Respirator mask only, full face, silicone				Ea.	345			345	380
0030 Half face, silicone					65			65	71.50
0040 Respirator cartridges, 2 req'd/mask, dust or asbestos					4.31			4.31	4.74
0050 Chemical vapor					3.03			3.03	3.34
0060 Combination vapor and dust					8.90			8.90	9.80
0100 Emergency escape breathing apparatus, 5 minutes					715			715	790
0110 10 minutes					860			860	945
0150 Self contained breathing apparatus with full face piece, 30 minutes					2,325			2,325	2,550
0160 60 minutes					3,725			3,725	4,100
0200 Encapsulating suits, limited use, level A					2,000			2,000	2,200
0210 Level B				▼	460			460	505
0300 Over boots, latex				Pr.	7.65			7.65	8.40
0310 PVC					33			33	36
0320 Neoprene					29			29	32
0400 Gloves, nitrile/PVC					77.50			77.50	85
0410 Neoprene coated				▼	44			44	48.50

01 54 09.60 Safety Nets

	Crew	Daily Output	Labor-Hours	Unit	Material	Labor	Equipment	Total	Total Incl O&P
0010 **SAFETY NETS**									
0020 No supports, stock sizes, nylon, 3-1/2" mesh				S.F.	2.91			2.91	3.20
0100 Polypropylene, 6" mesh					1.48			1.48	1.63
0200 Small mesh debris nets, 1/4" mesh, stock sizes					.51			.51	.56
0220 Combined 3-1/2" mesh and 1/4" mesh, stock sizes					4.66			4.66	5.15
0300 Rental, 4" mesh, stock sizes, 3 months					.66			.66	.73
0320 6 month rental					.94			.94	1.03
0340 12 months				▼	1.28			1.28	1.41

01 54 16 – Temporary Hoists

01 54 16.50 Weekly Forklift Crew

	Crew	Daily Output	Labor-Hours	Unit	Material	Labor	Equipment	Total	Total Incl O&P
0010 **WEEKLY FORKLIFT CREW**									
0100 All-terrain forklift, 45' lift, 35' reach, 9000 lb. capacity	A-3P	.20	40	Week		2,050	2,600	4,650	5,950

01 54 19 – Temporary Cranes

01 54 19.50 Daily Crane Crews

	Crew	Daily Output	Labor-Hours	Unit	Material	Labor	Equipment	Total	Total Incl O&P
0010 **DAILY CRANE CREWS** for small jobs, portal to portal									
0100 12-ton truck-mounted hydraulic crane	A-3H	1	8	Day		445	770	1,215	1,525
0200 25-ton	A-3I	1	8			445	905	1,350	1,675
0300 40-ton	A-3J	1	8			445	1,350	1,795	2,175
0400 55-ton	A-3K	1	16			830	1,600	2,430	3,025
0500 80-ton	A-3L	1	16	▼		830	2,400	3,230	3,900
0900 If crane is needed on a Saturday, Sunday or Holiday									
0910 At time-and-a-half, add				Day		50%			
0920 At double time, add				"		100%			

01 54 19.60 Monthly Tower Crane Crew

	Crew	Daily Output	Labor-Hours	Unit	Material	Labor	Equipment	Total	Total Incl O&P
0010 **MONTHLY TOWER CRANE CREW**, excludes concrete footing									
0100 Static tower crane, 130' high, 106' jib, 6200 lb. capacity	A-3N	.05	176	Month		9,800	25,700	35,500	43,000

01 54 23 – Temporary Scaffolding and Platforms

01 54 23.60 Pump Staging

	Crew	Daily Output	Labor-Hours	Unit	Material	Labor	Equipment	Total	Total Incl O&P
0010 **PUMP STAGING**, Aluminum R015423-20									
0200 24' long pole section, buy				Ea.	420			420	460
0300 18' long pole section, buy				↓	325			325	355

01 54 Construction Aids

01 54 23 – Temporary Scaffolding and Platforms

01 54 23.60 Pump Staging		Crew	Daily Output	Labor-Hours	Unit	Material	2017 Bare Costs Labor	Equipment	Total	Total Incl O&P
0400	12' long pole section, buy				Ea.	219			219	241
0500	6' long pole section, buy					115			115	127
0600	6' long splice joint section, buy					85.50			85.50	94
0700	Pump jack, buy					174			174	192
0900	Foldable brace, buy					68.50			68.50	75.50
1000	Workbench/back safety rail support, buy					92			92	101
1100	Scaffolding planks/workbench, 14" wide x 24' long, buy					765			765	840
1200	Plank end safety rail, buy					350			350	385
1250	Safety net, 22' long, buy				▼	420			420	460
1300	System in place, 50' working height, per use based on 50 uses	2 Carp	84.80	.189	C.S.F.	6.95	9.30		16.25	22
1400	100 uses		84.80	.189		3.49	9.30		12.79	18.10
1500	150 uses	▼	84.80	.189	▼	2.34	9.30		11.64	16.80

01 54 23.70 Scaffolding

01 54 23.70 Scaffolding		Crew	Daily Output	Labor-Hours	Unit	Material	2017 Bare Costs Labor	Equipment	Total	Total Incl O&P
0010	**SCAFFOLDING** R015423-10									
0015	Steel tube, regular, no plank, labor only to erect & dismantle									
0090	Building exterior, wall face, 1 to 5 stories, 6'-4" x 5' frames	3 Carp	8	3	C.S.F.		148		148	226
0200	6 to 12 stories	4 Carp	8	4			197		197	300
0301	13 to 20 stories	5 Clab	8	5			196		196	300
0460	Building interior, wall face area, up to 16' high	3 Carp	12	2			98.50		98.50	151
0560	16' to 40' high		10	2.400	▼		118		118	181
0800	Building interior floor area, up to 30' high		150	.160	C.C.F.		7.90		7.90	12.05
0900	Over 30' high	4 Carp	160	.200	"		9.85		9.85	15.10
0906	Complete system for face of walls, no plank, material only rent/mo				C.S.F.	35			35	38.50
0908	Interior spaces, no plank, material only rent/mo				C.C.F.	4.28			4.28	4.71
0910	Steel tubular, heavy duty shoring, buy									
0920	Frames 5' high 2' wide				Ea.	88			88	97
0925	5' high 4' wide					101			101	111
0930	6' high 2' wide					102			102	112
0935	6' high 4' wide				▼	117			117	129
0940	Accessories									
0945	Cross braces				Ea.	17			17	18.70
0950	U-head, 8" x 8"					21			21	23
0955	J-head, 4" x 8"					15.20			15.20	16.70
0960	Base plate, 8" x 8"					17			17	18.70
0965	Leveling jack				▼	36			36	39.50
1000	Steel tubular, regular, buy									
1100	Frames 3' high 5' wide				Ea.	89			89	97.50
1150	5' high 5' wide					94.50			94.50	104
1200	6'-4" high 5' wide					140			140	154
1350	7'-6" high 6' wide					170			170	187
1500	Accessories, cross braces					16.50			16.50	18.15
1550	Guardrail post					17.60			17.60	19.35
1600	Guardrail 7' section					8			8	8.80
1650	Screw jacks & plates					21.50			21.50	24
1700	Sidearm brackets					28			28	30.50
1750	8" casters					31			31	34
1800	Plank 2" x 10" x 16'-0"					59.50			59.50	65.50
1900	Stairway section					275			275	305
1910	Stairway starter bar					32			32	35
1920	Stairway inside handrail					53			53	58
1930	Stairway outside handrail					86			86	94.50
1940	Walk-thru frame guardrail				▼	41.50			41.50	46

01 54 23.70 Scaffolding	Crew	Daily Output	Labor-Hours	Unit	Material	2017 Bare Costs Labor	Equipment	Total	Total Incl O&P
2000 Steel tubular, regular, rent/mo.									
2100 Frames 3' high 5' wide				Ea.	5			5	5.50
2150 5' high 5' wide					5			5	5.50
2200 6'-4" high 5' wide					5.40			5.40	5.95
2250 7'-6" high 6' wide					10			10	11
2500 Accessories, cross braces					1			1	1.10
2550 Guardrail post					1			1	1.10
2600 Guardrail 7' section					1			1	1.10
2650 Screw jacks & plates					2			2	2.20
2700 Sidearm brackets					2			2	2.20
2750 8" casters					8			8	8.80
2800 Outrigger for rolling tower					3			3	3.30
2850 Plank 2" x 10" x 16'-0"					10			10	11
2900 Stairway section					33			33	36.50
2940 Walk-thru frame guardrail					2.50			2.50	2.75
3000 Steel tubular, heavy duty shoring, rent/mo.									
3250 5' high 2' & 4' wide				Ea.	8.50			8.50	9.35
3300 6' high 2' & 4' wide					8.50			8.50	9.35
3500 Accessories, cross braces					1			1	1.10
3600 U-head, 8" x 8"					2.50			2.50	2.75
3650 J-head, 4" x 8"					2.50			2.50	2.75
3700 Base plate, 8" x 8"					1			1	1.10
3750 Leveling jack					2.50			2.50	2.75
5700 Planks, 2" x 10" x 16'-0", labor only to erect & remove to 50' H	3 Carp	72	.333			16.40		16.40	25
5800 Over 50' high	4 Carp	80	.400			19.70		19.70	30
6000 Heavy duty shoring for elevated slab forms to 8'-2" high, floor area									
6100 Labor only to erect & dismantle	4 Carp	16	2	C.S.F.		98.50		98.50	151
6110 Materials only, rent/mo.				"	44.50			44.50	49
6500 To 14'-8" high									
6600 Labor only to erect & dismantle	4 Carp	10	3.200	C.S.F.		158		158	241
6610 Materials only, rent/mo				"	65			65	71.50

01 54 23.75 Scaffolding Specialties

	Crew	Daily Output	Labor-Hours	Unit	Material	2017 Bare Costs Labor	Equipment	Total	Total Incl O&P
0010 **SCAFFOLDING SPECIALTIES**									
1200 Sidewalk bridge, heavy duty steel posts & beams, including									
1210 parapet protection & waterproofing (material cost is rent/month)									
1220 8' to 10' wide, 2 posts	3 Carp	15	1.600	L.F.	44	79		123	170
1230 3 posts	"	10	2.400	"	68.50	118		186.50	256
1500 Sidewalk bridge using tubular steel scaffold frames including									
1510 planking (material cost is rent/month)	3 Carp	45	.533	L.F.	8.55	26.50		35.05	49.50
1600 For 2 uses per month, deduct from all above					50%				
1700 For 1 use every 2 months, add to all above					100%				
1900 Catwalks, 20" wide, no guardrails, 7' span, buy				Ea.	175			175	193
2000 10' span, buy					251			251	276
3720 Putlog, standard, 8' span, with hangers, buy					61			61	67
3730 Rent per month					20			20	22
3750 12' span, buy					108			108	119
3755 Rent per month					25			25	27.50
3760 Trussed type, 16' span, buy					249			249	274
3770 Rent per month					30			30	33
3790 22' span, buy					255			255	281
3795 Rent per month					40			40	44
3800 Rolling ladders with handrails, 30" wide, buy, 2 step					280			280	305

01 54 23 – Temporary Scaffolding and Platforms

01 54 23.75 Scaffolding Specialties	Crew	Daily Output	Labor-Hours	Unit	Material	2017 Bare Costs Labor	Equipment	Total	Total Incl O&P
4000	7 step				Ea.	790			790	870
4050	10 step					1,125			1,125	1,225
4100	Rolling towers, buy, 5' wide, 7' long, 10' high					1,225			1,225	1,350
4200	For additional 5' high sections, to buy				↓	222			222	244
4300	Complete incl. wheels, railings, outriggers,									
4350	21' high, to buy				Ea.	2,075			2,075	2,275
4400	Rent/month = 5% of purchase cost				"	200			200	220
5000	Motorized work platform, mast climber									
5050	Base unit, 50' W, less than 100' tall, rent/mo				Ea.	3,025			3,025	3,325
5100	Less than 200' tall, rent/mo					3,650			3,650	4,025
5150	Less than 300' tall, rent/mo					4,525			4,525	4,975
5200	Less than 400' tall, rent/mo				↓	5,350			5,350	5,900
5250	Set up and demob, per unit, less than 100' tall	B-68F	16.60	1.446	C.S.F.		75.50	18.40	93.90	137
5300	Less than 200' tall		25	.960			50	12.25	62.25	90.50
5350	Less than 300' tall		30	.800			41.50	10.20	51.70	75
5400	Less than 400' tall	↓	33	.727	↓		38	9.25	47.25	68.50
5500	Mobilization (price includes freight in and out) per unit				Ea.				1,100	1,225

01 54 23.80 Staging Aids

		Crew	Daily Output	Labor-Hours	Unit	Material	Labor	Equipment	Total	Total Incl O&P
0010	**STAGING AIDS** and fall protection equipment									
0100	Sidewall staging bracket, tubular, buy				Ea.	52.50			52.50	57.50
0110	Cost each per day, based on 250 days use				Day	.21			.21	.23
0200	Guard post, buy				Ea.	54.50			54.50	60
0210	Cost each per day, based on 250 days use				Day	.22			.22	.24
0300	End guard chains, buy per pair				Pair	40			40	44
0310	Cost per set per day, based on 250 days use				Day	.21			.21	.23
1000	Roof shingling bracket, steel, buy				Ea.	9.80			9.80	10.80
1010	Cost each per day, based on 250 days use				Day	.04			.04	.04
1100	Wood bracket, buy				Ea.	23			23	25.50
1110	Cost each per day, based on 250 days use				Day	.09			.09	.10
2000	Ladder jack, aluminum, buy per pair				Pair	126			126	139
2010	Cost per pair per day, based on 250 days use				Day	.50			.50	.56
2100	Steel siderail jack, buy per pair				Pair	97.50			97.50	107
2110	Cost per pair per day, based on 250 days use				Day	.39			.39	.43
3000	Laminated wood plank, 2" x 10" x 16', buy				Ea.	59.50			59.50	65.50
3010	Cost each per day, based on 250 days use				Day	.24			.24	.26
3100	Aluminum scaffolding plank, 20" wide x 24' long, buy				Ea.	790			790	870
3110	Cost each per day, based on 250 days use				Day	3.16			3.16	3.48
4000	Nylon full body harness, lanyard and rope grab				Ea.	182			182	200
4010	Cost each per day, based on 250 days use				Day	.73			.73	.80
4100	Rope for safety line, 5/8" x 100' nylon, buy				Ea.	58			58	64
4110	Cost each per day, based on 250 days use				Day	.23			.23	.26
4200	Permanent U-Bolt roof anchor, buy				Ea.	35.50			35.50	39.50
4300	Temporary (one use) roof ridge anchor, buy				"	32			32	35.50
5000	Installation (setup and removal) of staging aids									
5010	Sidewall staging bracket	2 Carp	64	.250	Ea.		12.30		12.30	18.85
5020	Guard post with 2 wood rails	"	64	.250			12.30		12.30	18.85
5030	End guard chains, set	1 Carp	64	.125			6.15		6.15	9.45
5100	Roof shingling bracket		96	.083			4.10		4.10	6.30
5200	Ladder jack	↓	64	.125			6.15		6.15	9.45
5300	Wood plank, 2" x 10" x 16'	2 Carp	80	.200			9.85		9.85	15.10
5310	Aluminum scaffold plank, 20" x 24'	"	40	.400			19.70		19.70	30
5410	Safety rope	1 Carp	40	.200	↓		9.85		9.85	15.10

For customer support on your Building Construction Costs with RSMeans Data, call 800.448.8182.

21

01 54 Construction Aids

01 54 23 – Temporary Scaffolding and Platforms

01 54 23.80 Staging Aids	Crew	Daily Output	Labor-Hours	Unit	Material	2017 Bare Costs Labor	Equipment	Total	Total Incl O&P
5420	Permanent U-Bolt roof anchor (install only)	2 Carp	40	.400	Ea.		19.70		19.70	30
5430	Temporary roof ridge anchor (install only)	1 Carp	64	.125	↓		6.15		6.15	9.45

01 54 26 – Temporary Swing Staging

01 54 26.50 Swing Staging

		Crew	Daily Output	Labor-Hours	Unit	Material	Labor	Equipment	Total	Total Incl O&P
0010	**SWING STAGING**, 500 lb. cap., 2' wide to 24' long, hand operated									
0020	steel cable type, with 60' cables, buy				Ea.	5,275			5,275	5,825
0030	Rent per month				"	530			530	580
0600	Lightweight (not for masons) 24' long for 150' height,									
0610	manual type, buy				Ea.	11,000			11,000	12,100
0620	Rent per month					1,100			1,100	1,200
0700	Powered, electric or air, to 150' high, buy					27,700			27,700	30,500
0710	Rent per month					1,950			1,950	2,125
0780	To 300' high, buy					28,100			28,100	30,900
0800	Rent per month					1,975			1,975	2,175
1000	Bosun's chair or work basket 3' x 3.5', to 300' high, electric, buy					11,200			11,200	12,400
1010	Rent per month				↓	785			785	865
2200	Move swing staging (setup and remove)	E-4	2	16	Move		875	65	940	1,525

01 54 36 – Equipment Mobilization

01 54 36.50 Mobilization

		Crew	Daily Output	Labor-Hours	Unit	Material	Labor	Equipment	Total	Total Incl O&P
0010	**MOBILIZATION** (Use line item again for demobilization) R015436-50									
0015	Up to 25 mi. haul dist. (50 mi. RT for mob/demob crew)									
1200	Small equipment, placed in rear of, or towed by pickup truck	A-3A	4	2	Ea.		102	35	137	193
1300	Equipment hauled on 3-ton capacity towed trailer	A-3Q	2.67	3			153	61.50	214.50	299
1400	20-ton capacity	B-34U	2	8			385	232	617	840
1500	40-ton capacity	B-34N	2	8			395	365	760	1,000
1600	50-ton capacity	B-34V	1	24			1,225	1,000	2,225	2,975
1700	Crane, truck-mounted, up to 75 ton (driver only)	1 Eqhv	4	2			111		111	169
1800	Over 75 ton (with chase vehicle)	A-3E	2.50	6.400			325	56	381	550
2400	Crane, large lattice boom, requiring assembly	B-34W	.50	144	↓		6,950	7,325	14,275	18,600
2500	For each additional 5 miles haul distance, add						10%	10%		
3000	For large pieces of equipment, allow for assembly/knockdown									
3001	For mob/demob of vibrofloatation equip, see Section 31 45 13.10									
3100	For mob/demob of micro-tunneling equip, see Section 33 05 23.19									
3200	For mob/demob of pile driving equip, see Section 31 62 19.10									
3300	For mob/demob of caisson drilling equip, see Section 31 63 26.13									

01 55 Vehicular Access and Parking

01 55 23 – Temporary Roads

01 55 23.50 Roads and Sidewalks

		Crew	Daily Output	Labor-Hours	Unit	Material	Labor	Equipment	Total	Total Incl O&P
0010	**ROADS AND SIDEWALKS** Temporary									
0050	Roads, gravel fill, no surfacing, 4" gravel depth	B-14	715	.067	S.Y.	3.44	2.78	.51	6.73	8.60
0100	8" gravel depth	"	615	.078	"	6.90	3.24	.60	10.74	13.15
1000	Ramp, 3/4" plywood on 2" x 6" joists, 16" O.C.	2 Carp	300	.053	S.F.	1.56	2.63		4.19	5.75
1100	On 2" x 10" joists, 16" O.C.	"	275	.058	"	2.29	2.87		5.16	6.90

01 56 Temporary Barriers and Enclosures

01 56 13 – Temporary Air Barriers

01 56 13.60 Tarpaulins

		Crew	Daily Output	Labor-Hours	Unit	Material	2017 Bare Costs Labor	2017 Bare Costs Equipment	Total	Total Incl O&P
0010	**TARPAULINS**									
0020	Cotton duck, 10 oz. to 13.13 oz. per S.Y., 6' x 8'				S.F.	.88			.88	.97
0050	30' x 30'					.59			.59	.65
0100	Polyvinyl coated nylon, 14 oz. to 18 oz., minimum					1.42			1.42	1.56
0150	Maximum					1.43			1.43	1.57
0200	Reinforced polyethylene 3 mils thick, white					.04			.04	.04
0300	4 mils thick, white, clear or black					.12			.12	.13
0400	5.5 mils thick, clear					.19			.19	.21
0500	White, fire retardant					.56			.56	.62
0600	12 mils, oil resistant, fire retardant					.47			.47	.52
0700	8.5 mils, black					.22			.22	.24
0710	Woven polyethylene, 6 mils thick					.19			.19	.21
0730	Polyester reinforced w/integral fastening system, 11 mils thick					.22			.22	.24
0740	Mylar polyester, non-reinforced, 7 mils thick					1.18			1.18	1.30

01 56 13.90 Winter Protection

		Crew	Daily Output	Labor-Hours	Unit	Material	2017 Bare Costs Labor	2017 Bare Costs Equipment	Total	Total Incl O&P
0010	**WINTER PROTECTION**									
0100	Framing to close openings	2 Clab	500	.032	S.F.	.44	1.25		1.69	2.40
0200	Tarpaulins hung over scaffolding, 8 uses, not incl. scaffolding		1500	.011		.25	.42		.67	.92
0250	Tarpaulin polyester reinf. w/integral fastening system, 11 mils thick		1600	.010		.22	.39		.61	.84
0300	Prefab fiberglass panels, steel frame, 8 uses		1200	.013		2.45	.52		2.97	3.50

01 56 16 – Temporary Dust Barriers

01 56 16.10 Dust Barriers, Temporary

		Crew	Daily Output	Labor-Hours	Unit	Material	2017 Bare Costs Labor	2017 Bare Costs Equipment	Total	Total Incl O&P
0010	**DUST BARRIERS, TEMPORARY**									
0020	Spring loaded telescoping pole & head, to 12', erect and dismantle	1 Clab	240	.033	Ea.		1.30		1.30	2
0025	Cost per day (based upon 250 days)				Day	.26			.26	.29
0030	To 21', erect and dismantle	1 Clab	240	.033	Ea.		1.30		1.30	2
0035	Cost per day (based upon 250 days)				Day	.58			.58	.63
0040	Accessories, caution tape reel, erect and dismantle	1 Clab	480	.017	Ea.		.65		.65	1
0045	Cost per day (based upon 250 days)				Day	.37			.37	.40
0060	Foam rail and connector, erect and dismantle	1 Clab	240	.033	Ea.		1.30		1.30	2
0065	Cost per day (based upon 250 days)				Day	.12			.12	.13
0070	Caution tape	1 Clab	384	.021	C.L.F.	3.13	.82		3.95	4.69
0080	Zipper, standard duty		60	.133	Ea.	8	5.20		13.20	16.80
0090	Heavy duty		48	.167	"	9.75	6.55		16.30	21
0100	Polyethylene sheet, 4 mil		37	.216	Sq.	2.58	8.45		11.03	15.80
0110	6 mil		37	.216	"	3.73	8.45		12.18	17.05
1000	Dust partition, 6 mil polyethylene, 1" x 3" frame	2 Carp	2000	.008	S.F.	.31	.39		.70	.94
1080	2" x 4" frame	"	2000	.008	"	.34	.39		.73	.97
1085	Negative air machine, 1800 CFM				Ea.	805			805	885
1090	Adhesive strip application, 2" width	1 Clab	192	.042	C.L.F.	7.25	1.63		8.88	10.45

01 56 23 – Temporary Barricades

01 56 23.10 Barricades

		Crew	Daily Output	Labor-Hours	Unit	Material	2017 Bare Costs Labor	2017 Bare Costs Equipment	Total	Total Incl O&P
0010	**BARRICADES**									
0020	5' high, 3 rail @ 2" x 8", fixed	2 Carp	20	.800	L.F.	5.90	39.50		45.40	67
0150	Movable	"	30	.533	"	4.88	26.50		31.38	45.50
0300	Stock units, 6' high, 8' wide, plain, buy				Ea.	390			390	430
0350	With reflective tape, buy				"	405			405	445
0400	Break-a-way 3" PVC pipe barricade									
0410	with 3 ea. 1' x 4' reflectorized panels, buy				Ea.	106			106	117
0500	Barricades, plastic, 8" x 24" wide, foldable					66			66	72.50
0800	Traffic cones, PVC, 18" high					13.65			13.65	15

01 56 Temporary Barriers and Enclosures

01 56 23 – Temporary Barricades

01 56 23.10 Barricades		Crew	Daily Output	Labor-Hours	Unit	Material	2017 Bare Costs Labor	2017 Bare Costs Equipment	Total	Total Incl O&P
0850	28" high				Ea.	21			21	23
1000	Guardrail, wooden, 3' high, 1" x 6", on 2" x 4" posts	2 Carp	200	.080	L.F.	1.28	3.94		5.22	7.45
1100	2" x 6", on 4" x 4" posts	"	165	.097		2.48	4.78		7.26	10.05
1200	Portable metal with base pads, buy					16.10			16.10	17.70
1250	Typical installation, assume 10 reuses	2 Carp	600	.027		2.60	1.31		3.91	4.87
1300	Barricade tape, polyethylene, 7 mil, 3" wide x 500' long roll				Ea.	25			25	27.50
3000	Detour signs, set up and remove									
3010	Reflective aluminum, MUTCD, 24" x 24", post mounted	1 Clab	20	.400	Ea.	2.48	15.65		18.13	26.50
4000	Roof edge portable barrier stands and warning flags, 50 uses	1 Rohe	9100	.001	L.F.	.07	.03		.10	.13
4010	100 uses	"	9100	.001	"	.04	.03		.07	.09
5000	Barricades, see Section 01 54 33.40									

01 56 26 – Temporary Fencing

01 56 26.50 Temporary Fencing		Crew	Daily Output	Labor-Hours	Unit	Material	2017 Bare Costs Labor	2017 Bare Costs Equipment	Total	Total Incl O&P
0010	**TEMPORARY FENCING**									
0020	Chain link, 11 ga., 4' high	2 Clab	400	.040	L.F.	1.64	1.57		3.21	4.20
0100	6' high		300	.053		3.80	2.09		5.89	7.40
0200	Rented chain link, 6' high, to 1000' (up to 12 mo.)		400	.040		2.95	1.57		4.52	5.65
0250	Over 1000' (up to 12 mo.)		300	.053		3	2.09		5.09	6.50
0350	Plywood, painted, 2" x 4" frame, 4' high	A-4	135	.178		6.25	8.30		14.55	19.55
0400	4" x 4" frame, 8' high	"	110	.218		12	10.20		22.20	29
0500	Wire mesh on 4" x 4" posts, 4' high	2 Carp	100	.160		10.20	7.90		18.10	23.50
0550	8' high	"	80	.200		15.40	9.85		25.25	32

01 56 29 – Temporary Protective Walkways

01 56 29.50 Protection		Crew	Daily Output	Labor-Hours	Unit	Material	2017 Bare Costs Labor	2017 Bare Costs Equipment	Total	Total Incl O&P
0010	**PROTECTION**									
0020	Stair tread, 2" x 12" planks, 1 use	1 Carp	75	.107	Tread	5.20	5.25		10.45	13.75
0100	Exterior plywood, 1/2" thick, 1 use		65	.123		1.88	6.05		7.93	11.35
0200	3/4" thick, 1 use		60	.133		2.81	6.55		9.36	13.15
2200	Sidewalks, 2" x 12" planks, 2 uses		350	.023	S.F.	.87	1.13		2	2.67
2300	Exterior plywood, 2 uses, 1/2" thick		750	.011		.31	.53		.84	1.15
2400	5/8" thick		650	.012		.39	.61		1	1.36
2500	3/4" thick		600	.013		.47	.66		1.13	1.53

01 56 32 – Temporary Security

01 56 32.50 Watchman		Crew	Daily Output	Labor-Hours	Unit	Material	2017 Bare Costs Labor	2017 Bare Costs Equipment	Total	Total Incl O&P
0010	**WATCHMAN**									
0020	Service, monthly basis, uniformed person, minimum				Hr.				25	27.50
0100	Maximum								45.50	50
0200	Person and command dog, minimum								31	34
0300	Maximum								54.50	60
0500	Sentry dog, leased, with job patrol (yard dog), 1 dog				Week				290	320
0600	2 dogs				"				390	430
0800	Purchase, trained sentry dog, minimum				Ea.				1,375	1,500
0900	Maximum				"				2,725	3,000

01 58 Project Identification

01 58 13 – Temporary Project Signage

01 58 13.50 Signs	Crew	Daily Output	Labor-Hours	Unit	Material	2017 Bare Costs Labor	Equipment	Total	Total Incl O&P
0010 **SIGNS**									
0020 High intensity reflectorized, no posts, buy				Ea.	26			26	28.50

01 66 Product Storage and Handling Requirements

01 66 19 – Material Handling

01 66 19.10 Material Handling

	Crew	Daily Output	Labor-Hours	Unit	Material	2017 Bare Costs Labor	Equipment	Total	Total Incl O&P
0010 **MATERIAL HANDLING**									
0020 Above 2nd story, via stairs, per CY of material per floor	2 Clab	145	.110	C.Y.		4.32		4.32	6.60
0030 Via elevator, per CY of material		240	.067			2.61		2.61	4
0050 Distances greater than 200', per CY of material per each addl 200'		300	.053			2.09		2.09	3.20

01 71 Examination and Preparation

01 71 23 – Field Engineering

01 71 23.13 Construction Layout

	Crew	Daily Output	Labor-Hours	Unit	Material	2017 Bare Costs Labor	Equipment	Total	Total Incl O&P
0010 **CONSTRUCTION LAYOUT**									
1100 Crew for layout of building, trenching or pipe laying, 2 person crew	A-6	1	16	Day		795	41.50	836.50	1,275
1200 3 person crew	A-7	1	24			1,300	41.50	1,341.50	2,025
1400 Crew for roadway layout, 4 person crew	A-8	1	32			1,675	41.50	1,716.50	2,600

01 71 23.19 Surveyor Stakes

	Crew	Daily Output	Labor-Hours	Unit	Material	2017 Bare Costs Labor	Equipment	Total	Total Incl O&P
0010 **SURVEYOR STAKES**									
0020 Hardwood, 1" x 1" x 48" long				C	70			70	77
0100 2" x 2" x 18" long					78			78	86
0150 2" x 2" x 24" long					150			150	165

01 74 Cleaning and Waste Management

01 74 13 – Progress Cleaning

01 74 13.20 Cleaning Up

	Crew	Daily Output	Labor-Hours	Unit	Material	2017 Bare Costs Labor	Equipment	Total	Total Incl O&P
0010 **CLEANING UP**									
0020 After job completion, allow, minimum				Job				.30%	.30%
0040 Maximum				"				1%	1%
0050 Cleanup of floor area, continuous, per day, during const.	A-5	24	.750	M.S.F.	2.27	30	2.27	34.54	50.50
0100 Final by GC at end of job	"	11.50	1.565	"	2.40	62	4.73	69.13	103
0200 Rubbish removal, see Section 02 41 19.19									

01 76 Protecting Installed Construction

01 76 13 – Temporary Protection of Installed Construction

01 76 13.20 Temporary Protection

	Crew	Daily Output	Labor-Hours	Unit	Material	2017 Bare Costs Labor	Equipment	Total	Total Incl O&P
0010 **TEMPORARY PROTECTION**									
0020 Flooring, 1/8" tempered hardboard, taped seams	2 Carp	1500	.011	S.F.	.45	.53		.98	1.31
0030 Peel away carpet protection	1 Clab	3200	.003	"	.13	.10		.23	.29

01 91 Commissioning

01 91 13 – General Commissioning Requirements

01 91 13.50 Building Commissioning	Crew	Daily Output	Labor-Hours	Unit	Material	2017 Bare Costs Labor	Equipment	Total	Total Incl O&P
0010 **BUILDING COMMISSIONING**									
0100 Systems operation and verification during turnover				%				.25%	.25%
0150 Including all systems subcontractors								.50%	.50%
0200 Systems design assistance, operation, verification and training								.50%	.50%
0250 Including all systems subcontractors				▼				1%	1%

For customer support on your Building Construction Costs with RSMeans Data, call 800.448.8182.

Estimating Tips

02 30 00 Subsurface Investigation

In preparing estimates on structures involving earthwork or foundations, all information concerning soil characteristics should be obtained. Look particularly for hazardous waste, evidence of prior dumping of debris, and previous stream beds.

02 40 00 Demolition and Structure Moving

The costs shown for selective demolition do not include rubbish handling or disposal. These items should be estimated separately using RSMeans data or other sources.

- Historic preservation often requires that the contractor remove materials from the existing structure, rehab them, and replace them. The estimator must be aware of any related measures and precautions that must be taken when doing selective demolition and cutting and patching. Requirements may include special handling and storage, as well as security.

- In addition to Subdivision 02 41 00, you can find selective demolition items in each division. Example: Roofing demolition is in Division 7.
- Absent of any other specific reference, an approximate demolish-in-place cost can be obtained by halving the new-install labor cost. To remove for reuse, allow the entire new-install labor figure.

02 40 00 Building Deconstruction

This section provides costs for the careful dismantling and recycling of most low-rise building materials.

02 50 00 Containment of Hazardous Waste

This section addresses on-site hazardous waste disposal costs.

02 80 00 Hazardous Material Disposal/Remediation

This subdivision includes information on hazardous waste handling, asbestos remediation, lead remediation, and mold remediation. See reference numbers R028213-20 and R028319-60 for further guidance in using these unit price lines.

02 90 00 Monitoring Chemical Sampling, Testing Analysis

This section provides costs for on-site sampling and testing hazardous waste.

Reference Numbers

Reference numbers are shown at the beginning of some major classifications. These numbers refer to related items in the Reference Section. The reference information may be an estimating procedure, an alternate pricing method, or technical information.

Note: Not all subdivisions listed here necessarily appear. ■

02 21 Surveys

02 21 13 – Site Surveys

02 21 13.09 Topographical Surveys

02 21 13.09 Topographical Surveys	Crew	Daily Output	Labor-Hours	Unit	Material	2017 Bare Costs Labor	2017 Bare Costs Equipment	Total	Total Incl O&P
0010 **TOPOGRAPHICAL SURVEYS**									
0020 Topographical surveying, conventional, minimum	A-7	3.30	7.273	Acre	19.35	395	12.60	426.95	635
0100 Maximum	A-8	.60	53.333	"	58	2,800	69.50	2,927.50	4,425

02 21 13.13 Boundary and Survey Markers

02 21 13.13 Boundary and Survey Markers	Crew	Daily Output	Labor-Hours	Unit	Material	Labor	Equipment	Total	Total Incl O&P
0010 **BOUNDARY AND SURVEY MARKERS**									
0300 Lot location and lines, large quantities, minimum	A-7	2	12	Acre	34	650	21	705	1,050
0320 Average	"	1.25	19.200		53	1,025	33	1,111	1,675
0400 Small quantities, maximum	A-8	1	32	▼	72.50	1,675	41.50	1,789	2,675
0600 Monuments, 3' long	A-7	10	2.400	Ea.	40	130	4.15	174.15	247
0800 Property lines, perimeter, cleared land	"	1000	.024	L.F.	.06	1.30	.04	1.40	2.10
0900 Wooded land	A-8	875	.037	"	.08	1.92	.05	2.05	3.07

02 21 13.16 Aerial Surveys

02 21 13.16 Aerial Surveys	Crew	Daily Output	Labor-Hours	Unit	Material	Labor	Equipment	Total	Total Incl O&P
0010 **AERIAL SURVEYS**									
1500 Aerial surveying, including ground control, minimum fee, 10 acres				Total				4,700	4,700
1510 100 acres								9,400	9,400
1550 From existing photography, deduct								1,625	1,625
1600 2' contours, 10 acres				Acre				470	470
1850 100 acres								94	94
2000 1000 acres								90	90
2050 10,000 acres				▼				85	85

02 32 Geotechnical Investigations

02 32 13 – Subsurface Drilling and Sampling

02 32 13.10 Boring and Exploratory Drilling

02 32 13.10 Boring and Exploratory Drilling	Crew	Daily Output	Labor-Hours	Unit	Material	Labor	Equipment	Total	Total Incl O&P
0010 **BORING AND EXPLORATORY DRILLING**									
0020 Borings, initial field stake out & determination of elevations	A-6	1	16	Day		795	41.50	836.50	1,275
0100 Drawings showing boring details				Total		335		335	425
0200 Report and recommendations from P.E.						775		775	970
0300 Mobilization and demobilization	B-55	4	6	▼		245	263	508	665
0350 For over 100 miles, per added mile		450	.053	Mile		2.18	2.34	4.52	5.90
0600 Auger holes in earth, no samples, 2-1/2" diameter		78.60	.305	L.F.		12.45	13.40	25.85	33.50
0650 4" diameter		67.50	.356			14.50	15.60	30.10	39
0800 Cased borings in earth, with samples, 2-1/2" diameter		55.50	.432		11	17.65	18.95	47.60	60
0850 4" diameter	▼	32.60	.736		17.80	30	32.50	80.30	101
1000 Drilling in rock, "BX" core, no sampling	B-56	34.90	.458			20.50	43.50	64	79
1050 With casing & sampling		31.70	.505		11	23	47.50	81.50	99
1200 "NX" core, no sampling		25.92	.617	▼		28	58.50	86.50	107
1250 With casing and sampling	▼	25	.640	▼	13.35	29	60.50	102.85	125
1400 Borings, earth, drill rig and crew with truck mounted auger	B-55	1	24	Day		980	1,050	2,030	2,650
1450 Rock using crawler type drill	B-56	1	16	"		720	1,525	2,245	2,775
1500 For inner city borings add, minimum								10%	10%
1510 Maximum								20%	20%

02 32 19 – Exploratory Excavations

02 32 19.10 Test Pits

02 32 19.10 Test Pits	Crew	Daily Output	Labor-Hours	Unit	Material	Labor	Equipment	Total	Total Incl O&P
0010 **TEST PITS**									
0020 Hand digging, light soil	1 Clab	4.50	1.778	C.Y.		69.50		69.50	107
0100 Heavy soil	"	2.50	3.200			125		125	192
0120 Loader-backhoe, light soil	B-11M	28	.571			26.50	14.10	40.60	56
0130 Heavy soil	"	20	.800	▼		37	19.70	56.70	78
1000 Subsurface exploration, mobilization				Mile				6.75	8.40

02 32 Geotechnical Investigations

02 32 19 – Exploratory Excavations

02 32 19.10 Test Pits	Crew	Daily Output	Labor-Hours	Unit	Material	2017 Bare Costs Labor	Equipment	Total	Total Incl O&P
1010	Difficult access for rig, add				Hr.				260	320
1020	Auger borings, drill rig, incl. samples				L.F.				26.50	33
1030	Hand auger								31.50	40
1050	Drill and sample every 5', split spoon				▼				31.50	40
1060	Extra samples				Ea.				36	45.50

02 41 Demolition

02 41 13 – Selective Site Demolition

02 41 13.15 Hydrodemolition

		Crew	Daily Output	Labor-Hours	Unit	Material	Labor	Equipment	Total	Total Incl O&P
0010	**HYDRODEMOLITION** R024119-10									
0015	Hydrodemolition, concrete pavement									
0120	20,000 PSI, Crew to include loader/vacuum truck as required									
0130	2" depth	B-5	1000	.056	S.F.		2.44	1.49	3.93	5.35
0410	4" depth		800	.070			3.05	1.87	4.92	6.70
0420	6" depth	▼	600	.093	▼		4.06	2.49	6.55	8.95

02 41 13.17 Demolish, Remove Pavement and Curb

		Crew	Daily Output	Labor-Hours	Unit	Material	Labor	Equipment	Total	Total Incl O&P
0010	**DEMOLISH, REMOVE PAVEMENT AND CURB** R024119-10									
5010	Pavement removal, bituminous roads, up to 3" thick	B-38	690	.058	S.Y.		2.60	1.83	4.43	5.95
5050	4" to 6" thick		420	.095			4.27	3.01	7.28	9.80
5100	Bituminous driveways		640	.063			2.80	1.98	4.78	6.45
5200	Concrete to 6" thick, hydraulic hammer, mesh reinforced		255	.157			7.05	4.96	12.01	16.15
5300	Rod reinforced		200	.200	▼		8.95	6.30	15.25	20.50
5400	Concrete, 7" to 24" thick, plain		33	1.212	C.Y.		54.50	38.50	93	125
5500	Reinforced	▼	24	1.667	"		74.50	52.50	127	172
5600	With hand held air equipment, bituminous, to 6" thick	B-39	1900	.025	S.F.		1.05	.11	1.16	1.72
5700	Concrete to 6" thick, no reinforcing		1600	.030			1.24	.13	1.37	2.05
5800	Mesh reinforced		1400	.034			1.42	.15	1.57	2.34
5900	Rod reinforced	▼	765	.063	▼		2.60	.28	2.88	4.29
6000	Curbs, concrete, plain	B-6	360	.067	L.F.		2.87	1.02	3.89	5.50
6100	Reinforced		275	.087			3.76	1.33	5.09	7.20
6200	Granite		360	.067			2.87	1.02	3.89	5.50
6300	Bituminous	▼	528	.045	▼		1.96	.69	2.65	3.75

02 41 13.23 Utility Line Removal

		Crew	Daily Output	Labor-Hours	Unit	Material	Labor	Equipment	Total	Total Incl O&P
0010	**UTILITY LINE REMOVAL**									
0015	No hauling, abandon catch basin or manhole	B-6	7	3.429	Ea.		148	52.50	200.50	283
0020	Remove existing catch basin or manhole, masonry		4	6			259	91.50	350.50	495
0030	Catch basin or manhole frames and covers, stored		13	1.846			79.50	28	107.50	152
0040	Remove and reset	▼	7	3.429			148	52.50	200.50	283
0900	Hydrants, fire, remove only	B-21A	5	8			395	87	482	695
0950	Remove and reset	"	2	20	▼		990	218	1,208	1,750
2900	Pipe removal, sewer/water, no excavation, 12" diameter	B-6	175	.137	L.F.		5.90	2.09	7.99	11.30
2930	15"-18" diameter	B-12Z	150	.160			7.15	10.25	17.40	22
2960	21"-24" diameter		120	.200			8.95	12.80	21.75	27.50
3000	27"-36" diameter	▼	90	.267			11.90	17.05	28.95	37
3200	Steel, welded connections, 4" diameter	B-6	160	.150			6.45	2.29	8.74	12.35
3300	10" diameter	"	80	.300	▼		12.95	4.58	17.53	25

02 41 13.30 Minor Site Demolition

		Crew	Daily Output	Labor-Hours	Unit	Material	Labor	Equipment	Total	Total Incl O&P
0010	**MINOR SITE DEMOLITION** R024119-10									
0100	Roadside delineators, remove only	B-80	175	.183	Ea.		8	3.85	11.85	16.45
0110	Remove and reset	"	100	.320	"		14.05	6.75	20.80	29

For customer support on your Building Construction Costs with RSMeans Data, call 800.448.8182.

29

02 41 Demolition

02 41 13 – Selective Site Demolition

02 41 13.30 Minor Site Demolition

		Crew	Daily Output	Labor-Hours	Unit	Material	2017 Bare Costs		Total	Total Incl O&P
							Labor	Equipment		
0800	Guiderail, corrugated steel, remove only	B-80A	100	.240	L.F.		9.40	2.70	12.10	17.35
0850	Remove and reset	"	40	.600	"		23.50	6.75	30.25	43.50
0860	Guide posts, remove only	B-80B	120	.267	Ea.		11.25	2.06	13.31	19.40
0870	Remove and reset	B-55	50	.480	"		19.60	21	40.60	53
1000	Masonry walls, block, solid	B-5	1800	.031	C.F.		1.35	.83	2.18	2.98
1200	Brick, solid		900	.062			2.71	1.66	4.37	5.95
1400	Stone, with mortar		900	.062			2.71	1.66	4.37	5.95
1500	Dry set		1500	.037			1.63	1	2.63	3.58
1600	Median barrier, precast concrete, remove and store	B-3	430	.112	L.F.		4.91	5.95	10.86	14
1610	Remove and reset	"	390	.123	"		5.40	6.60	12	15.50
4000	Sidewalk removal, bituminous, 2" thick	B-6	350	.069	S.Y.		2.96	1.05	4.01	5.65
4010	2-1/2" thick		325	.074			3.18	1.13	4.31	6.10
4050	Brick, set in mortar		185	.130			5.60	1.98	7.58	10.75
4100	Concrete, plain, 4"		160	.150			6.45	2.29	8.74	12.35
4110	Plain, 5"		140	.171			7.40	2.62	10.02	14.15
4120	Plain, 6"		120	.200			8.60	3.05	11.65	16.50
4200	Mesh reinforced, concrete, 4"		150	.160			6.90	2.44	9.34	13.20
4210	5" thick		131	.183			7.90	2.80	10.70	15.10
4220	6" thick		112	.214			9.25	3.27	12.52	17.70
4300	Slab on grade removal, plain	B-5	45	1.244	C.Y.		54	33	87	119
4310	Mesh reinforced		33	1.697			74	45.50	119.50	163
4320	Rod reinforced		25	2.240			97.50	60	157.50	215
4400	For congested sites or small quantities, add up to								200%	200%
4450	For disposal on site, add	B-11A	232	.069			3.20	5.95	9.15	11.40
4500	To 5 miles, add	B-34D	76	.105			4.79	9.25	14.04	17.40

02 41 13.33 Railtrack Removal

		Crew	Daily Output	Labor-Hours	Unit	Material	Labor	Equipment	Total	Total Incl O&P
0010	**RAILTRACK REMOVAL**									
3500	Railroad track removal, ties and track	B-13	330	.170	L.F.		7.30	2.04	9.34	13.40
3600	Ballast	B-14	500	.096	C.Y.		3.98	.73	4.71	6.90
3700	Remove and re-install, ties & track using new bolts & spikes		50	.960	L.F.		40	7.30	47.30	69
3800	Turnouts using new bolts and spikes		1	48	Ea.		2,000	365	2,365	3,450

02 41 13.60 Selective Demolition Fencing

		Crew	Daily Output	Labor-Hours	Unit	Material	Labor	Equipment	Total	Total Incl O&P
0010	**SELECTIVE DEMOLITION FENCING** R024119-10									
1600	Fencing, barbed wire, 3 strand	2 Clab	430	.037	L.F.		1.46		1.46	2.23
1650	5 strand	"	280	.057			2.24		2.24	3.43
1700	Chain link, posts & fabric, 8' to 10' high, remove only	B-6	445	.054			2.32	.82	3.14	4.44

02 41 16 – Structure Demolition

02 41 16.13 Building Demolition

		Crew	Daily Output	Labor-Hours	Unit	Material	Labor	Equipment	Total	Total Incl O&P
0010	**BUILDING DEMOLITION** Large urban projects, incl. 20 mi. haul R024119-10									
0011	No foundation or dump fees, C.F. is vol. of building standing									
0020	Steel	B-8	21500	.003	C.F.		.14	.15	.29	.38
0050	Concrete		15300	.004			.19	.21	.40	.52
0080	Masonry		20100	.003			.15	.16	.31	.40
0100	Mixture of types		20100	.003			.15	.16	.31	.40
0500	Small bldgs, or single bldgs, no salvage included, steel	B-3	14800	.003			.14	.17	.31	.41
0600	Concrete		11300	.004			.19	.23	.42	.53
0650	Masonry		14800	.003			.14	.17	.31	.41
0700	Wood		14800	.003			.14	.17	.31	.41
0750	For buildings with no interior walls, deduct								30%	30%
1000	Demoliton single family house, one story, wood 1600 S.F.	B-3	1	48	Ea.		2,125	2,575	4,700	6,025
1020	3200 S.F.		.50	96			4,225	5,125	9,350	12,100
1200	Demoliton two family house, two story, wood 2400 S.F.		.67	71.964			3,175	3,850	7,025	9,050

02 41 Demolition

02 41 16 – Structure Demolition

02 41 16.13 Building Demolition

		Crew	Daily Output	Labor-Hours	Unit	Material	2017 Bare Costs Labor	2017 Bare Costs Equipment	Total	Total Incl O&P
1220	4200 S.F.	B-3	.38	128	Ea.		5,625	6,850	12,475	16,100
1300	Demoliton three family house, three story, wood 3200 S.F.		.50	96			4,225	5,125	9,350	12,100
1320	5400 S.F.		.30	160			7,050	8,550	15,600	20,100
5000	For buildings with no interior walls, deduct								30%	30%

02 41 16.15 Explosive/Implosive Demolition

		Crew	Daily Output	Labor-Hours	Unit	Material	2017 Bare Costs Labor	2017 Bare Costs Equipment	Total	Total Incl O&P
0010	**EXPLOSIVE/IMPLOSIVE DEMOLITION** R024119-10									
0011	Large projects,									
0020	No disposal fee based on building volume, steel building	B-5B	16900	.003	C.F.		.14	.16	.30	.39
0100	Concrete building		16900	.003			.14	.16	.30	.39
0200	Masonry building		16900	.003			.14	.16	.30	.39
0400	Disposal of material, minimum	B-3	445	.108	C.Y.		4.75	5.75	10.50	13.55
0500	Maximum	"	365	.132	"		5.80	7.05	12.85	16.55

02 41 16.17 Building Demolition Footings and Foundations

		Crew	Daily Output	Labor-Hours	Unit	Material	2017 Bare Costs Labor	2017 Bare Costs Equipment	Total	Total Incl O&P
0010	**BUILDING DEMOLITION FOOTINGS AND FOUNDATIONS** R024119-10									
0200	Floors, concrete slab on grade,									
0240	4" thick, plain concrete	B-13L	5000	.003	S.F.		.18	.40	.58	.71
0280	Reinforced, wire mesh		4000	.004			.22	.50	.72	.89
0300	Rods		4500	.004			.20	.45	.65	.79
0400	6" thick, plain concrete		4000	.004			.22	.50	.72	.89
0420	Reinforced, wire mesh		3200	.005			.28	.63	.91	1.11
0440	Rods		3600	.004			.25	.56	.81	.99
1000	Footings, concrete, 1' thick, 2' wide		300	.053	L.F.		2.97	6.70	9.67	11.90
1080	1'-6" thick, 2' wide		250	.064			3.56	8.05	11.61	14.25
1120	3' wide		200	.080			4.46	10.10	14.56	17.85
1140	2' thick, 3' wide		175	.091			5.10	11.50	16.60	20.50
1200	Average reinforcing, add								10%	10%
1220	Heavy reinforcing, add								20%	20%
2000	Walls, block, 4" thick	B-13L	8000	.002	S.F.		.11	.25	.36	.45
2040	6" thick		6000	.003			.15	.34	.49	.59
2080	8" thick		4000	.004			.22	.50	.72	.89
2100	12" thick		3000	.005			.30	.67	.97	1.19
2200	For horizontal reinforcing, add								10%	10%
2220	For vertical reinforcing, add								20%	20%
2400	Concrete, plain concrete, 6" thick	B-13L	4000	.004			.22	.50	.72	.89
2420	8" thick		3500	.005			.25	.58	.83	1.02
2440	10" thick		3000	.005			.30	.67	.97	1.19
2500	12" thick		2500	.006			.36	.81	1.17	1.43
2600	For average reinforcing, add								10%	10%
2620	For heavy reinforcing, add								20%	20%
4000	For congested sites or small quantities, add up to								200%	200%
4200	Add for disposal, on site	B-11A	232	.069	C.Y.		3.20	5.95	9.15	11.40
4250	To five miles	B-30	220	.109	"		5.25	10.25	15.50	19.20

02 41 19 – Selective Demolition

02 41 19.13 Selective Building Demolition

		Crew	Daily Output	Labor-Hours	Unit	Material	2017 Bare Costs Labor	2017 Bare Costs Equipment	Total	Total Incl O&P
0010	**SELECTIVE BUILDING DEMOLITION**									
0020	Costs related to selective demolition of specific building components									
0025	are included under Common Work Results (XX 05)									
0030	in the component's appropriate division.									

02 41 19.16 Selective Demolition, Cutout

		Crew	Daily Output	Labor-Hours	Unit	Material	2017 Bare Costs Labor	2017 Bare Costs Equipment	Total	Total Incl O&P
0010	**SELECTIVE DEMOLITION, CUTOUT** R024119-10									
0020	Concrete, elev. slab, light reinforcement, under 6 C.F.	B-9	65	.615	C.F.		24.50	3.28	27.78	41

02 41 Demolition

02 41 19 – Selective Demolition

02 41 19.16 Selective Demolition, Cutout

		Crew	Daily Output	Labor-Hours	Unit	Material	2017 Bare Costs Labor	Equipment	Total	Total Incl O&P
0050	Light reinforcing, over 6 C.F.	B-9	75	.533	C.F.		21	2.84	23.84	35.50
0200	Slab on grade to 6" thick, not reinforced, under 8 S.F.		85	.471	S.F.		18.60	2.51	21.11	31.50
0250	8 – 16 S.F.	↓	175	.229	"		9.05	1.22	10.27	15.20
0255	For over 16 S.F. see Line 02 41 16.17 0400									
0600	Walls, not reinforced, under 6 C.F.	B-9	60	.667	C.F.		26.50	3.55	30.05	44.50
0650	6 – 12 C.F.	"	80	.500	"		19.80	2.67	22.47	33.50
0655	For over 12 C.F. see Line 02 41 16.17 2500									
1000	Concrete, elevated slab, bar reinforced, under 6 C.F.	B-9	45	.889	C.F.		35	4.74	39.74	59
1050	Bar reinforced, over 6 C.F.		50	.800	"		31.50	4.26	35.76	53
1200	Slab on grade to 6" thick, bar reinforced, under 8 S.F.		75	.533	S.F.		21	2.84	23.84	35.50
1250	8 – 16 S.F.	↓	150	.267	"		10.55	1.42	11.97	17.70
1255	For over 16 S.F. see Line 02 41 16.17 0440									
1400	Walls, bar reinforced, under 6 C.F.	B-9	50	.800	C.F.		31.50	4.26	35.76	53
1450	6 – 12 C.F.	"	70	.571	"		22.50	3.05	25.55	38
1455	For over 12 C.F. see Lines 02 41 16.17 2500 and 2600									
2000	Brick, to 4 S.F. opening, not including toothing									
2040	4" thick	B-9	30	1.333	Ea.		52.50	7.10	59.60	89
2060	8" thick		18	2.222			88	11.85	99.85	148
2080	12" thick		10	4			158	21.50	179.50	266
2400	Concrete block, to 4 S.F. opening, 2" thick		35	1.143			45	6.10	51.10	76
2420	4" thick		30	1.333			52.50	7.10	59.60	89
2440	8" thick		27	1.481			58.50	7.90	66.40	98.50
2460	12" thick		24	1.667			66	8.90	74.90	111
2600	Gypsum block, to 4 S.F. opening, 2" thick		80	.500			19.80	2.67	22.47	33.50
2620	4" thick		70	.571			22.50	3.05	25.55	38
2640	8" thick		55	.727			29	3.88	32.88	48.50
2800	Terra cotta, to 4 S.F. opening, 4" thick		70	.571			22.50	3.05	25.55	38
2840	8" thick		65	.615			24.50	3.28	27.78	41
2880	12" thick.	↓	50	.800	↓		31.50	4.26	35.76	53
3000	Toothing masonry cutouts, brick, soft old mortar	1 Brhe	40	.200	V.L.F.		7.75		7.75	11.95
3100	Hard mortar		30	.267			10.35		10.35	15.90
3200	Block, soft old mortar		70	.114			4.43		4.43	6.80
3400	Hard mortar	↓	50	.160	↓		6.20		6.20	9.55
6000	Walls, interior, not including re-framing,									
6010	openings to 5 S.F.									
6100	Drywall to 5/8" thick	1 Clab	24	.333	Ea.		13.05		13.05	20
6200	Paneling to 3/4" thick		20	.400			15.65		15.65	24
6300	Plaster, on gypsum lath		20	.400			15.65		15.65	24
6340	On wire lath	↓	14	.571	↓		22.50		22.50	34.50
7000	Wood frame, not including re-framing, openings to 5 S.F.									
7200	Floors, sheathing and flooring to 2" thick	1 Clab	5	1.600	Ea.		62.50		62.50	96
7310	Roofs, sheathing to 1" thick, not including roofing		6	1.333			52		52	80
7410	Walls, sheathing to 1" thick, not including siding	↓	7	1.143	↓		44.50		44.50	68.50

02 41 19.18 Selective Demolition, Disposal Only

		Crew	Daily Output	Labor-Hours	Unit	Material	2017 Bare Costs Labor	Equipment	Total	Total Incl O&P
0010	**SELECTIVE DEMOLITION, DISPOSAL ONLY** R024119-10									
0015	Urban bldg w/salvage value allowed									
0020	Including loading and 5 mile haul to dump									
0200	Steel frame	B-3	430	.112	C.Y.		4.91	5.95	10.86	14
0300	Concrete frame		365	.132			5.80	7.05	12.85	16.55
0400	Masonry construction		445	.108			4.75	5.75	10.50	13.55
0500	Wood frame	↓	247	.194	↓		8.55	10.40	18.95	24.50

02 41 Demolition

02 41 19 – Selective Demolition

02 41 19.19 Selective Demolition		Crew	Daily Output	Labor-Hours	Unit	Material	2017 Bare Costs Labor	Equipment	Total	Total Incl O&P
0010	**SELECTIVE DEMOLITION**, Rubbish Handling R024119-10									
0020	The following are to be added to the demolition prices									
0050	The following are components for a complete chute system									
0100	Top chute circular steel, 4' long, 18" diameter R024119-30	B-1C	15	1.600	Ea.	272	63.50	30.50	366	430
0102	23" diameter		15	1.600		295	63.50	30.50	389	455
0104	27" diameter		15	1.600		320	63.50	30.50	414	480
0106	30" diameter		15	1.600		340	63.50	30.50	434	505
0108	33" diameter		15	1.600		365	63.50	30.50	459	530
0110	36" diameter		15	1.600		385	63.50	30.50	479	555
0112	Regular chute, 18" diameter		15	1.600		204	63.50	30.50	298	355
0114	23" diameter		15	1.600		227	63.50	30.50	321	380
0116	27" diameter		15	1.600		249	63.50	30.50	343	405
0118	30" diameter		15	1.600		261	63.50	30.50	355	420
0120	33" diameter		15	1.600		295	63.50	30.50	389	455
0122	36" diameter		15	1.600		320	63.50	30.50	414	480
0124	Control door chute, 18" diameter		15	1.600		385	63.50	30.50	479	555
0126	23" diameter		15	1.600		410	63.50	30.50	504	580
0128	27" diameter		15	1.600		430	63.50	30.50	524	605
0130	30" diameter		15	1.600		455	63.50	30.50	549	630
0132	33" diameter		15	1.600		475	63.50	30.50	569	655
0134	36" diameter		15	1.600		500	63.50	30.50	594	680
0136	Chute liners, 14 ga., 18-30" diameter		15	1.600		209	63.50	30.50	303	360
0138	33-36" diameter		15	1.600		261	63.50	30.50	355	420
0140	17% thinner chute, 30" diameter		15	1.600		220	63.50	30.50	314	375
0142	33% thinner chute, 30" diameter		15	1.600		166	63.50	30.50	260	315
0144	Top chute cover	1 Clab	24	.333		143	13.05		156.05	177
0146	Door chute cover	"	24	.333		143	13.05		156.05	177
0148	Top chute trough	2 Clab	12	1.333		475	52		527	605
0150	Bolt down frame & counter weights, 250 lb.	B-1	4	6		4,175	239		4,414	4,975
0152	500 lb.		4	6		6,175	239		6,414	7,175
0154	750 lb.		4	6		8,925	239		9,164	10,200
0156	1000 lb.		2.67	8.989		9,800	360		10,160	11,400
0158	1500 lb.		2.67	8.989		12,600	360		12,960	14,400
0160	Chute warning light system, 5 stories	B-1C	4	6		8,900	239	114	9,253	10,300
0162	10 stories	"	2	12		14,200	480	228	14,908	16,600
0164	Dust control device for dumpsters	1 Clab	8	1		133	39		172	206
0166	Install or replace breakaway cord		8	1		25	39		64	87.50
0168	Install or replace warning sign		16	.500		10	19.60		29.60	41
0600	Dumpster, weekly rental, 1 dump/week, 6 C.Y. capacity (2 Tons)				Week	415			415	455
0700	10 C.Y. capacity (3 Tons)					480			480	530
0725	20 C.Y. capacity (5 Tons) R024119-20					565			565	625
0800	30 C.Y. capacity (7 Tons)					730			730	800
0840	40 C.Y. capacity (10 Tons)					775			775	850
2000	Load, haul, dump and return, 0 – 50' haul, hand carried	2 Clab	24	.667	C.Y.		26		26	40
2005	Wheeled		37	.432			16.95		16.95	26
2040	0 – 100' haul, hand carried		16.50	.970			38		38	58
2045	Wheeled		25	.640			25		25	38.50
2050	Forklift	A-3R	25	.320			16.30	6.85	23.15	32
2080	Haul and return, add per each extra 100' haul, hand carried	2 Clab	35.50	.451			17.65		17.65	27
2085	Wheeled		54	.296			11.60		11.60	17.80
2120	For travel in elevators, up to 10 floors, add		140	.114			4.47		4.47	6.85
2130	0 – 50' haul, incl. up to 5 riser stairs, hand carried		23	.696			27		27	41.50

02 41 Demolition

02 41 19 – Selective Demolition

02 41 19.19 Selective Demolition

		Crew	Daily Output	Labor-Hours	Unit	Material	2017 Bare Costs Labor	Equipment	Total	Total Incl O&P
2135	Wheeled	2 Clab	35	.457	C.Y.		17.90		17.90	27.50
2140	6 – 10 riser stairs, hand carried		22	.727			28.50		28.50	43.50
2145	Wheeled		34	.471			18.40		18.40	28
2150	11 – 20 riser stairs, hand carried		20	.800			31.50		31.50	48
2155	Wheeled		31	.516			20		20	31
2160	21 – 40 riser stairs, hand carried		16	1			39		39	60
2165	Wheeled		24	.667			26		26	40
2170	0 – 100' haul, incl. 5 riser stairs, hand carried		15	1.067			42		42	64
2175	Wheeled		23	.696			27		27	41.50
2180	6 – 10 riser stairs, hand carried		14	1.143			44.50		44.50	68.50
2185	Wheeled		21	.762			30		30	45.50
2190	11 – 20 riser stairs, hand carried		12	1.333			52		52	80
2195	Wheeled		18	.889			35		35	53.50
2200	21 – 40 riser stairs, hand carried		8	2			78.50		78.50	120
2205	Wheeled		12	1.333			52		52	80
2210	Haul and return, add per each extra 100' haul, hand carried		35.50	.451			17.65		17.65	27
2215	Wheeled		54	.296	↓		11.60		11.60	17.80
2220	For each additional flight of stairs, up to 5 risers, add		550	.029	Flight		1.14		1.14	1.75
2225	6 – 10 risers, add		275	.058			2.28		2.28	3.49
2230	11 – 20 risers, add		138	.116			4.54		4.54	6.95
2235	21 – 40 risers, add	↓	69	.232	↓		9.10		9.10	13.90
3000	Loading & trucking, including 2 mile haul, chute loaded	B-16	45	.711	C.Y.	·	29.50	14.30	43.80	60.50
3040	Hand loading truck, 50' haul	"	48	.667			27.50	13.40	40.90	57
3080	Machine loading truck	B-17	120	.267			11.65	6.30	17.95	24.50
5000	Haul, per mile, up to 8 C.Y. truck	B-34B	1165	.007			.31	.55	.86	1.08
5100	Over 8 C.Y. truck	"	1550	.005	↓		.24	.41	.65	.81

02 41 19.20 Selective Demolition, Dump Charges

		Crew	Daily Output	Labor-Hours	Unit	Material	2017 Bare Costs Labor	Equipment	Total	Total Incl O&P
0010	**SELECTIVE DEMOLITION, DUMP CHARGES** R024119-10									
0020	Dump charges, typical urban city, tipping fees only									
0100	Building construction materials				Ton	74			74	81
0200	Trees, brush, lumber					63			63	69.50
0300	Rubbish only					63			63	69.50
0500	Reclamation station, usual charge				↓	74			74	81

02 41 19.21 Selective Demolition, Gutting

		Crew	Daily Output	Labor-Hours	Unit	Material	2017 Bare Costs Labor	Equipment	Total	Total Incl O&P
0010	**SELECTIVE DEMOLITION, GUTTING** R024119-10									
0020	Building interior, including disposal, dumpster fees not included									
0500	Residential building									
0560	Minimum	B-16	400	.080	SF Flr.		3.30	1.61	4.91	6.80
0580	Maximum	"	360	.089	"		3.67	1.78	5.45	7.55
0900	Commercial building									
1000	Minimum	B-16	350	.091	SF Flr.		3.77	1.84	5.61	7.75
1020	Maximum	"	250	.128	"		5.30	2.57	7.87	10.90

02 41 19.25 Selective Demolition, Saw Cutting

		Crew	Daily Output	Labor-Hours	Unit	Material	2017 Bare Costs Labor	Equipment	Total	Total Incl O&P
0010	**SELECTIVE DEMOLITION, SAW CUTTING** R024119-10									
0015	Asphalt, up to 3" deep	B-89	1050	.015	L.F.	.11	.73	.42	1.26	1.68
0020	Each additional inch of depth	"	1800	.009		.04	.42	.25	.71	.95
1200	Masonry walls, hydraulic saw, brick, per inch of depth	B-89B	300	.053		.04	2.54	2.53	5.11	6.65
1220	Block walls, solid, per inch of depth	"	250	.064		.04	3.05	3.03	6.12	8
2000	Brick or masonry w/hand held saw, per inch of depth	A-1	125	.064		.03	2.51	.59	3.13	4.53
5000	Wood sheathing to 1" thick, on walls	1 Carp	200	.040			1.97		1.97	3.02
5020	On roof	"	250	.032	↓		1.58		1.58	2.41

02 41 Demolition

02 41 19 – Selective Demolition

02 41 19.27 Selective Demolition, Torch Cutting		Crew	Daily Output	Labor-Hours	Unit	Material	2017 Bare Costs Labor	Equipment	Total	Total Incl O&P
0010	**SELECTIVE DEMOLITION, TORCH CUTTING**	R024119-10								
0020	Steel, 1" thick plate	E-25	333	.024	L.F.	.84	1.35	.04	2.23	3.21
0040	1" diameter bar	"	600	.013	Ea.	.14	.75	.02	.91	1.42
1000	Oxygen lance cutting, reinforced concrete walls									
1040	12" to 16" thick walls	1 Clab	10	.800	L.F.		31.50		31.50	48
1080	24" thick walls	"	6	1.333	"		52		52	80

02 42 Removal and Salvage of Construction Materials

02 42 10 – Building Deconstruction

02 42 10.10 Estimated Salvage Value or Savings

			Crew	Daily Output	Labor-Hours	Unit	Material	2017 Bare Costs Labor	Equipment	Total	Total Incl O&P
0010	**ESTIMATED SALVAGE VALUE OR SAVINGS**										
0015	Excludes material handling, packaging, container costs and										
0020	transportation for salvage or disposal										
0050	All Items in Section 02 42 10.10 are credit deducts and not costs										
0100	Copper Wire Salvage Value	G				Lb.				1.60	1.60
0110	Disposal Savings	G								.04	.04
0200	Copper Pipe Salvage Value	G								2.50	2.50
0210	Disposal Savings	G								.05	.05
0300	Steel Pipe Salvage Value	G								.06	.06
0310	Disposal Savings	G								.03	.03
0400	Cast Iron Pipe Salvage Value	G								.03	.03
0410	Disposal Savings	G								.01	.01
0500	Steel Doors or Windows Salvage Value	G								.06	.06
0510	Aluminum	G								.55	.55
0520	Disposal Savings	G								.03	.03
0600	Aluminum Siding Salvage Value	G								.49	.49
0630	Disposal Savings	G								.03	.03
0640	Wood Siding (no lead or asbestos)	G				C.Y.				12	12
0800	Clean Concrete Disposal Savings	G				Ton				62	62
0850	Asphalt Shingles Disposal Savings	G				"				60	60
1000	Wood wall framing clean salvage value	G				M.B.F.				55	55
1010	Painted	G								44	44
1020	Floor framing	G								55	55
1030	Painted	G								44	44
1050	Roof framing	G								55	55
1060	Painted	G								44	44
1100	Wood beams salvage value	G								55	55
1200	Wood framing and beams disposal savings	G				Ton				66	66
1220	Wood sheathing and sub-base flooring	G								72.50	72.50
1230	Wood wall paneling (1/4 inch thick)	G								66	66
1300	Wood panel 3/4-1 inch thick low salvage value	G				S.F.				.55	.55
1350	High salvage value	G				"				2.20	2.20
1400	Disposal savings	G				Ton				66	66
1500	Flooring tongue and groove 25/32 inch thick low salvage value	G				S.F.				.55	.55
1530	High salvage value	G				"				1.10	1.10
1560	Disposal savings	G				Ton				66	66
1600	Drywall or sheet rock salvage value	G								22	22
1650	Disposal savings	G								66	66

For customer support on your Building Construction Costs with RSMeans Data, call 800.448.8182.

35

02 42 10 – Building Deconstruction

02 42 10.20 Deconstruction of Building Components		Crew	Daily Output	Labor-Hours	Unit	Material	2017 Bare Costs Labor	Equipment	Total	Total Incl O&P
0010	**DECONSTRUCTION OF BUILDING COMPONENTS**									
0012	Buildings one or two stories only									
0015	Excludes material handling, packaging, container costs and									
0020	transportation for salvage or disposal									
0050	Deconstruction of plumbing fixtures									
0100	Wall hung or countertop lavatory	G	2 Clab	16	1	Ea.		39		39	60
0110	Single or double compartment kitchen sink	G		14	1.143			44.50		44.50	68.50
0120	Wall hung urinal	G		14	1.143			44.50		44.50	68.50
0130	Floor mounted	G		8	2			78.50		78.50	120
0140	Floor mounted water closet	G		16	1			39		39	60
0150	Wall hung	G		14	1.143			44.50		44.50	68.50
0160	Water fountain, free standing	G		16	1			39		39	60
0170	Wall hung or deck mounted	G		12	1.333			52		52	80
0180	Bathtub, steel or fiberglass	G		10	1.600			62.50		62.50	96
0190	Cast iron	G		8	2			78.50		78.50	120
0200	Shower, single	G		6	2.667			104		104	160
0210	Group	G		7	2.286			89.50		89.50	137
0300	Deconstruction of electrical fixtures									
0310	Surface mounted incandescent fixtures	G	2 Clab	48	.333	Ea.		13.05		13.05	20
0320	Fluorescent, 2 lamp	G		32	.500			19.60		19.60	30
0330	4 lamp	G		24	.667			26		26	40
0340	Strip fluorescent, 1 lamp	G		40	.400			15.65		15.65	24
0350	2 lamp	G		32	.500			19.60		19.60	30
0400	Recessed drop-in fluorescent fixture, 2 lamp	G		27	.593			23		23	35.50
0410	4 lamp	G		18	.889			35		35	53.50
0500	Deconstruction of appliances									
0510	Cooking stoves	G	2 Clab	26	.615	Ea.		24		24	37
0520	Dishwashers	G	"	26	.615	"		24		24	37
0600	Deconstruction of millwork and trim									
0610	Cabinets, wood	G	2 Carp	40	.400	L.F.		19.70		19.70	30
0620	Countertops	G		100	.160	"		7.90		7.90	12.05
0630	Wall paneling, 1 inch thick	G		500	.032	S.F.		1.58		1.58	2.41
0640	Ceiling trim	G		500	.032	L.F.		1.58		1.58	2.41
0650	Wainscoting	G		500	.032	S.F.		1.58		1.58	2.41
0660	Base, 3/4" to 1" thick	G		600	.027	L.F.		1.31		1.31	2.01
0700	Deconstruction of doors and windows									
0710	Doors, wrap, interior, wood, single, no closers	G	2 Carp	21	.762	Ea.	4.75	37.50		42.25	63
0720	Double	G		13	1.231		9.50	60.50		70	103
0730	Solid core, single, exterior or interior	G		10	1.600		4.75	79		83.75	126
0740	Double	G		8	2		9.50	98.50		108	161
0810	Windows, wrap, wood, single									
0812	with no casement or cladding	G	2 Carp	21	.762	Ea.	4.75	37.50		42.25	63
0820	with casement and/or cladding	G	"	18	.889	"	4.75	44		48.75	72.50
0900	Deconstruction of interior finishes									
0910	Drywall for recycling	G	2 Clab	1775	.009	S.F.		.35		.35	.54
0920	Plaster wall, first floor	G		1775	.009			.35		.35	.54
0930	Second floor	G		1330	.012			.47		.47	.72
1000	Deconstruction of roofing and accessories									
1010	Built-up roofs	G	2 Clab	570	.028	S.F.		1.10		1.10	1.68
1020	Gutters, fascia and rakes	G	"	1140	.014	L.F.		.55		.55	.84
2000	Deconstruction of wood components									
2010	Roof sheeting	G	2 Clab	570	.028	S.F.		1.10		1.10	1.68

02 42 Removal and Salvage of Construction Materials

02 42 10 – Building Deconstruction

02 42 10.20 Deconstruction of Building Components

		Crew	Daily Output	Labor-Hours	Unit	Material	2017 Bare Costs Labor	Equipment	Total	Total Incl O&P
2020	Main roof framing	G	2 Clab	760	.021	L.F.		.82		.82	1.26
2030	Porch roof framing	G	↓	445	.036			1.41		1.41	2.16
2040	Beams 4" x 8"	G	B-1	375	.064			2.55		2.55	3.91
2050	4" x 10"	G		300	.080			3.19		3.19	4.88
2055	4" x 12"	G		250	.096			3.82		3.82	5.85
2060	6" x 8"	G		250	.096			3.82		3.82	5.85
2065	6" x 10"	G		200	.120			4.78		4.78	7.30
2070	6" x 12"	G		170	.141			5.60		5.60	8.60
2075	8" x 12"	G		126	.190			7.60		7.60	11.60
2080	10" x 12"	G	↓	100	.240			9.55		9.55	14.65
2100	Ceiling joists	G	2 Clab	800	.020			.78		.78	1.20
2150	Wall framing, interior	G		1230	.013	↓		.51		.51	.78
2160	Sub-floor	G		2000	.008	S.F.		.31		.31	.48
2170	Floor joists	G		2000	.008	L.F.		.31		.31	.48
2200	Wood siding (no lead or asbestos)	G		1300	.012	S.F.		.48		.48	.74
2300	Wall framing, exterior	G		1600	.010	L.F.		.39		.39	.60
2400	Stair risers	G		53	.302	Ea.		11.80		11.80	18.10
2500	Posts	G	↓	800	.020	L.F.		.78		.78	1.20
3000	Deconstruction of exterior brick walls										
3010	Exterior brick walls, first floor	G	2 Clab	200	.080	S.F.		3.13		3.13	4.80
3020	Second floor	G		64	.250	"		9.80		9.80	15
3030	Brick chimney	G	↓	100	.160	C.F.		6.25		6.25	9.60
4000	Deconstruction of concrete										
4010	Slab on grade, 4" thick, plain concrete	G	B-9	500	.080	S.F.		3.16	.43	3.59	5.30
4020	Wire mesh reinforced	G		470	.085			3.37	.45	3.82	5.65
4030	Rod reinforced	G		400	.100			3.96	.53	4.49	6.65
4110	Foundation wall, 6" thick, plain concrete	G		160	.250			9.90	1.33	11.23	16.60
4120	8" thick	G		140	.286			11.30	1.52	12.82	18.95
4130	10" thick	G	↓	120	.333	↓		13.20	1.78	14.98	22
9000	Deconstruction process, support equipment as needed										
9010	Daily use, portal to portal, 12-ton truck-mounted hydraulic crane crew	G	A-3H	1	8	Day		445	770	1,215	1,525
9020	Daily use, skid steer and operator	G	A-3C	1	8			410	310	720	955
9030	Daily use, backhoe 48 H.P., operator and labor	G	"	1	8	↓		410	310	720	955

02 42 10.30 Deconstruction Material Handling

		Crew	Daily Output	Labor-Hours	Unit	Material	2017 Bare Costs Labor	Equipment	Total	Total Incl O&P
0010	**DECONSTRUCTION MATERIAL HANDLING**										
0012	Buildings one or two stories only										
0100	Clean and stack brick on pallet	G	2 Clab	1200	.013	Ea.		.52		.52	.80
0200	Haul 50' and load rough lumber up to 2" x 8" size	G		2000	.008	"		.31		.31	.48
0210	Lumber larger than 2" x 8"	G		3200	.005	B.F.		.20		.20	.30
0300	Finish wood for recycling stack and wrap per pallet	G		8	2	Ea.	38	78.50		116.50	162
0350	Light fixtures			6	2.667		68.50	104		172.50	235
0375	Windows			6	2.667		64.50	104		168.50	231
0400	Miscellaneous materials		↓	8	2	↓	19	78.50		97.50	141
1000	See Section 02 41 19.19 for bulk material handling										

For customer support on your Building Construction Costs with RSMeans Data, call 800.448.8182.

37

02 43 Structure Moving

02 43 13 – Structure Relocation

02 43 13.13 Building Relocation		Crew	Daily Output	Labor-Hours	Unit	Material	2017 Bare Costs Labor	Equipment	Total	Total Incl O&P
0010	**BUILDING RELOCATION**									
0011	One day move, up to 24' wide									
0020	Reset on existing foundation				Total				11,500	11,500
0040	Wood or steel frame bldg., based on ground floor area G	B-4	185	.259	S.F.		10.50	2.77	13.27	19.15
0060	Masonry bldg., based on ground floor area G	"	137	.350			14.20	3.73	17.93	25.50
0200	For 24' to 42' wide, add				↓				15%	15%

02 56 Site Containment

02 56 13 – Waste Containment

02 56 13.10 Containment of Hazardous Waste

		Crew	Daily Output	Labor-Hours	Unit	Material	2017 Bare Costs Labor	Equipment	Total	Total Incl O&P
0010	**CONTAINMENT OF HAZARDOUS WASTE**									
0020	OSHA Hazard level C									
0030	OSHA Hazard level D decrease labor and equipment, deduct						45%	45%		
0035	OSHA Hazard level B increase labor and equipment, add						22%	22%		
0040	OSHA Hazard level A increase labor and equipment, add						71%	71%		
0100	Excavation of contaminated soil & waste									
0105	Includes one respirator filter and two disposable suits per work day									
0110	3/4 C.Y. excavator to 10 feet deep	B-12F	51	.314	B.C.Y.	1.61	14.90	13.20	29.71	39
0120	Labor crew to 6' deep	B-2	19	2.105		10.80	83.50		94.30	140
0130	6' - 12' deep	"	12	3.333	↓	17.10	132		149.10	221
0200	Move contaminated soil/waste up to 150' on-site with 2.5 C.Y. loader	B-10T	300	.040	L.C.Y.	.27	1.95	1.85	4.07	5.30
0210	300'	"	186	.065	"	.44	3.15	2.99	6.58	8.55
0300	Secure burial cell construction									
0310	Various liner and cover materials									
0400	Very low density polyethylene (VLDPE)									
0410	50 mil top cover	B-47H	4000	.008	S.F.	.53	.42	.07	1.02	1.29
0420	80 mil liner	"	4000	.008	"	.68	.42	.07	1.17	1.46
0500	Chlorosulfonated polyethylene									
0510	36 mil hypalon top cover	B-47H	4000	.008	S.F.	2	.42	.07	2.49	2.91
0520	45 mil hypalon liner	"	4000	.008	"	2.36	.42	.07	2.85	3.31
0600	Polyvinyl chloride (PVC)									
0610	60 mil top cover	B-47H	4000	.008	S.F.	1.10	.42	.07	1.59	1.92
0620	80 mil liner	"	4000	.008	"	1.25	.42	.07	1.74	2.09
0700	Rough textured H.D. polyethylene (HDPE)									
0710	40 mil top cover	B-47H	4000	.008	S.F.	.53	.42	.07	1.02	1.29
0720	60 mil top cover		4000	.008		.60	.42	.07	1.09	1.37
0722	60 mil liner		4000	.008		.56	.42	.07	1.05	1.33
0730	80 mil liner	↓	3800	.008	↓	.69	.44	.07	1.20	1.51
1000	3/4" crushed stone, 6" deep ballast around liner	B-6	30	.800	L.C.Y.	26	34.50	12.20	72.70	94.50
1100	Hazardous waste, ballast cover with common borrow material	B-63	56	.714		12.80	29.50	3.06	45.36	63
1110	Mixture of common borrow & topsoil		56	.714		19	29.50	3.06	51.56	70
1120	Bank sand		56	.714		18.60	29.50	3.06	51.16	69.50
1130	Medium priced clay		44	.909		24	38	3.89	65.89	88.50
1140	Mixture of common borrow & medium priced clay	↓	56	.714	↓	18.50	29.50	3.06	51.06	69.50

02 58 Snow Control

02 58 13 – Snow Fencing

02 58 13.10 Snow Fencing System

	Crew	Daily Output	Labor-Hours	Unit	Material	2017 Bare Costs Labor	2017 Bare Costs Equipment	Total	Total Incl O&P
0010 **SNOW FENCING SYSTEM**									
7001 Snow fence on steel posts 10' O.C., 4' high	B-1	500	.048	L.F.	.46	1.91		2.37	3.43

02 65 Underground Storage Tank Removal

02 65 10 – Underground Tank and Contaminated Soil Removal

02 65 10.30 Removal of Underground Storage Tanks

		Crew	Daily Output	Labor-Hours	Unit	Material	2017 Bare Costs Labor	2017 Bare Costs Equipment	Total	Total Incl O&P
0010 **REMOVAL OF UNDERGROUND STORAGE TANKS** R026510-20										
0011 Petroleum storage tanks, non-leaking										
0100 Excavate & load onto trailer										
0110 3000 gal. to 5000 gal. tank	G	B-14	4	12	Ea.		500	91.50	591.50	860
0120 6000 gal. to 8000 gal. tank	G	B-3A	3	13.333			560	320	880	1,200
0130 9000 gal. to 12000 gal. tank	G	"	2	20			840	485	1,325	1,800
0190 Known leaking tank, add					%				100%	100%
0200 Remove sludge, water and remaining product from tank bottom										
0201 of tank with vacuum truck										
0300 3000 gal. to 5000 gal. tank	G	A-13	5	1.600	Ea.		81.50	148	229.50	285
0310 6000 gal. to 8000 gal. tank	G		4	2			102	184	286	355
0320 9000 gal. to 12000 gal. tank	G		3	2.667			136	246	382	475
0390 Dispose of sludge off-site, average					Gal.				6.25	6.80
0400 Insert inert solid CO₂ "dry ice" into tank										
0401 For cleaning/transporting tanks (1.5 lb./100 gal. cap)	G	1 Clab	500	.016	Lb.	1.33	.63		1.96	2.42
0403 Insert solid carbon dioxide, 1.5 lb./100 gal.	G	"	400	.020	"	1.33	.78		2.11	2.66
0503 Disconnect and remove piping	G	1 Plum	160	.050	L.F.		3.09		3.09	4.66
0603 Transfer liquids, 10% of volume	G	"	1600	.005	Gal.		.31		.31	.47
0703 Cut accessway into underground storage tank	G	1 Clab	5.33	1.501	Ea.		59		59	90
0813 Remove sludge, wash and wipe tank, 500 gal.	G	1 Plum	8	1			62		62	93.50
0823 3,000 gal.	G		6.67	1.199			74		74	112
0833 5,000 gal.	G		6.15	1.301			80.50		80.50	121
0843 8,000 gal.	G		5.33	1.501			93		93	140
0853 10,000 gal.	G		4.57	1.751			108		108	163
0863 12,000 gal.	G		4.21	1.900			117		117	177
1020 Haul tank to certified salvage dump, 100 miles round trip										
1023 3000 gal. to 5000 gal. tank					Ea.				760	830
1026 6000 gal. to 8000 gal. tank									880	960
1029 9,000 gal. to 12,000 gal. tank									1,050	1,150
1100 Disposal of contaminated soil to landfill										
1110 Minimum					C.Y.				145	160
1111 Maximum					"				400	440
1120 Disposal of contaminated soil to										
1121 bituminous concrete batch plant										
1130 Minimum					C.Y.				80	88
1131 Maximum					"				115	125
1203 Excavate, pull, & load tank, backfill hole, 8,000 gal. +	G	B-12C	.50	32	Ea.		1,525	2,250	3,775	4,775
1213 Haul tank to certified dump, 100 miles rt, 8,000 gal. +	G	B-34K	1	8			365	920	1,285	1,575
1223 Excavate, pull, & load tank, backfill hole, 500 gal.	G	B-11C	1	16			740	365	1,105	1,525
1233 Excavate, pull, & load tank, backfill hole, 3,000 – 5,000 gal.	G	B-11M	.50	32			1,475	790	2,265	3,125
1243 Haul tank to certified dump, 100 miles rt, 500 gal.	G	B-34L	1	8			410	217	627	855
1253 Haul tank to certified dump, 100 miles rt, 3,000 – 5,000 gal.	G	B-34M	1	8			410	270	680	910
2010 Decontamination of soil on site incl poly tarp on top/bottom										
2011 Soil containment berm and chemical treatment										
2020 Minimum	G	B-11C	100	.160	C.Y.	7.80	7.40	3.66	18.86	24

02 65 Underground Storage Tank Removal

02 65 10 – Underground Tank and Contaminated Soil Removal

02 65 10.30 Removal of Underground Storage Tanks		Crew	Daily Output	Labor-Hours	Unit	Material	2017 Bare Costs Labor	Equipment	Total	Total Incl O&P
2021	Maximum	G B-11C	100	.160	C.Y.	10.10	7.40	3.66	21.16	26.50
2050	Disposal of decontaminated soil, minimum								135	150
2055	Maximum				↓				400	440

02 81 Transportation and Disposal of Hazardous Materials

02 81 20 – Hazardous Waste Handling

02 81 20.10 Hazardous Waste Cleanup/Pickup/Disposal

		Crew	Daily Output	Labor-Hours	Unit	Material	2017 Bare Costs Labor	Equipment	Total	Total Incl O&P
0010	**HAZARDOUS WASTE CLEANUP/PICKUP/DISPOSAL**									
0100	For contractor rental equipment, i.e., dozer,									
0110	Front end loader, dump truck, etc., see 01 54 33 Reference Section									
1000	Solid pickup									
1100	55 gal. drums				Ea.				240	265
1120	Bulk material, minimum				Ton				190	210
1130	Maximum				"				595	655
1200	Transportation to disposal site									
1220	Truckload = 80 drums or 25 C.Y. or 18 tons									
1260	Minimum				Mile				3.95	4.45
1270	Maximum				"				7.25	7.35
3000	Liquid pickup, vacuum truck, stainless steel tank									
3100	Minimum charge, 4 hours									
3110	1 compartment, 2200 gallon				Hr.				140	155
3120	2 compartment, 5000 gallon				"				200	225
3400	Transportation in 6900 gallon bulk truck				Mile				7.95	8.75
3410	In teflon lined truck				"				10.20	11.25
5000	Heavy sludge or dry vacuumable material				Hr.				140	160
6000	Dumpsite disposal charge, minimum				Ton				140	155
6020	Maximum				"				415	455

02 82 Asbestos Remediation

02 82 13 – Asbestos Abatement

02 82 13.39 Asbestos Remediation Plans and Methods

		Crew	Daily Output	Labor-Hours	Unit	Material	2017 Bare Costs Labor	Equipment	Total	Total Incl O&P
0010	**ASBESTOS REMEDIATION PLANS AND METHODS**									
0100	Building Survey-Commercial Building				Ea.				2,200	2,400
0200	Asbestos Abatement Remediation Plan				"				1,350	1,475

02 82 13.41 Asbestos Abatement Equipment

		Crew	Daily Output	Labor-Hours	Unit	Material	2017 Bare Costs Labor	Equipment	Total	Total Incl O&P
0010	**ASBESTOS ABATEMENT EQUIPMENT** R028213-20									
0011	Equipment and supplies, buy									
0200	Air filtration device, 2000 CFM				Ea.	870			870	955
0250	Large volume air sampling pump, minimum					335			335	370
0260	Maximum					345			345	380
0300	Airless sprayer unit, 2 gun					2,175			2,175	2,375
0350	Light stand, 500 watt				↓	38			38	42
0400	Personal respirators									
0410	Negative pressure, 1/2 face, dual operation, min.				Ea.	26			26	28.50
0420	Maximum					29			29	32
0450	P.A.P.R., full face, minimum					127			127	140
0460	Maximum					169			169	185
0470	Supplied air, full face, incl. air line, minimum					173			173	191
0480	Maximum				↓	415			415	455

02 82 Asbestos Remediation

02 82 13 – Asbestos Abatement

02 82 13.41 Asbestos Abatement Equipment

		Crew	Daily Output	Labor-Hours	Unit	Material	2017 Bare Costs Labor	Equipment	Total	Total Incl O&P
0500	Personnel sampling pump				Ea.	218			218	240
1500	Power panel, 20 unit, incl. GFI					500			500	550
1600	Shower unit, including pump and filters					1,100			1,100	1,200
1700	Supplied air system (type C)					3,475			3,475	3,825
1750	Vacuum cleaner, HEPA, 16 gal., stainless steel, wet/dry					440			440	485
1760	55 gallon					1,400			1,400	1,525
1800	Vacuum loader, 9 – 18 ton/hr.					97,000			97,000	107,000
1900	Water atomizer unit, including 55 gal. drum					289			289	320
2000	Worker protection, whole body, foot, head cover & gloves, plastic					9			9	9.90
2500	Respirator, single use					25.50			25.50	28.50
2550	Cartridge for respirator					4.98			4.98	5.50
2570	Glove bag, 7 mil, 50" x 64"					9.45			9.45	10.40
2580	10 mil, 44" x 60"					5.80			5.80	6.40
2590	6 mil, 44" x 60"					5.80			5.80	6.40
3000	HEPA vacuum for work area, minimum					320			320	350
3050	Maximum					840			840	925
6000	Disposable polyethylene bags, 6 mil, 3 C.F.					.85			.85	.94
6300	Disposable fiber drums, 3 C.F.					18.75			18.75	20.50
6400	Pressure sensitive caution labels, 3" x 5"					3.58			3.58	3.94
6450	11" x 17"					7.55			7.55	8.30
6500	Negative air machine, 1800 CFM				▼	805			805	885

02 82 13.42 Preparation of Asbestos Containment Area

		Crew	Daily Output	Labor-Hours	Unit	Material	2017 Bare Costs Labor	Equipment	Total	Total Incl O&P
0010	**PREPARATION OF ASBESTOS CONTAINMENT AREA**									
0100	Pre-cleaning, HEPA vacuum and wet wipe, flat surfaces	A-9	12000	.005	S.F.	.02	.29		.31	.48
0200	Protect carpeted area, 2 layers 6 mil poly on 3/4" plywood	"	1000	.064		2.02	3.53		5.55	7.70
0300	Separation barrier, 2" x 4" @ 16", 1/2" plywood ea. side, 8' high	2 Carp	400	.040		2.59	1.97		4.56	5.85
0310	12' high		320	.050		2.54	2.46		5	6.55
0320	16' high		200	.080		2.52	3.94		6.46	8.80
0400	Personnel decontam. chamber, 2" x 4" @ 16", 3/4" ply ea. side		280	.057		3.16	2.81		5.97	7.80
0450	Waste decontam. chamber, 2" x 4" studs @ 16", 3/4" ply ea. side	▼	360	.044	▼	3.16	2.19		5.35	6.85
0500	Cover surfaces with polyethylene sheeting									
0501	Including glue and tape									
0550	Floors, each layer, 6 mil	A-9	8000	.008	S.F.	.04	.44		.48	.73
0551	4 mil		9000	.007		.03	.39		.42	.64
0560	Walls, each layer, 6 mil		6000	.011		.04	.59		.63	.95
0561	4 mil	▼	7000	.009	▼	.03	.50		.53	.81
0570	For heights above 14', add						20%			
0575	For heights above 20', add						30%			
0580	For fire retardant poly, add					100%				
0590	For large open areas, deduct					10%	20%			
0600	Seal floor penetrations with foam firestop to 36 sq. in.	2 Carp	200	.080	Ea.	11.25	3.94		15.19	18.45
0610	36 sq. in. to 72 sq. in.		125	.128		22.50	6.30		28.80	34.50
0615	72 sq. in. to 144 sq. in.		80	.200		45	9.85		54.85	64.50
0620	Wall penetrations, to 36 sq. in.		180	.089		11.25	4.38		15.63	19.10
0630	36 sq. in. to 72 sq. in.		100	.160		22.50	7.90		30.40	37
0640	72 sq. in. to 144 sq. in.	▼	60	.267	▼	45	13.15		58.15	69.50
0800	Caulk seams with latex	1 Carp	230	.035	L.F.	.17	1.71		1.88	2.81
0900	Set up neg. air machine, 1-2k CFM/25 M.C.F. volume	1 Asbe	4.30	1.860	Ea.		102		102	159
0950	Set up and remove portable shower unit	2 Asbe	4	4	"		220		220	340

For customer support on your Building Construction Costs with RSMeans Data, call 800.448.8182.

41

02 82 13.43 Bulk Asbestos Removal		Crew	Daily Output	Labor-Hours	Unit	Material	2017 Bare Costs Labor	Equipment	Total	Total Incl O&P
0010	**BULK ASBESTOS REMOVAL**									
0020	Includes disposable tools and 2 suits and 1 respirator filter/day/worker									
0100	Beams, W 10 x 19	A-9	235	.272	L.F.	.78	15		15.78	24.50
0110	W 12 x 22		210	.305		.88	16.80		17.68	27
0120	W 14 x 26		180	.356		1.02	19.60		20.62	31.50
0130	W 16 x 31		160	.400		1.15	22		23.15	36
0140	W 18 x 40		140	.457		1.31	25		26.31	40.50
0150	W 24 x 55		110	.582		1.67	32		33.67	52
0160	W 30 x 108		85	.753		2.16	41.50		43.66	67
0170	W 36 x 150		72	.889		2.55	49		51.55	79
0200	Boiler insulation		480	.133	S.F.	.45	7.35		7.80	11.90
0210	With metal lath, add				%				50%	50%
0300	Boiler breeching or flue insulation	A-9	520	.123	S.F.	.35	6.80		7.15	10.95
0310	For active boiler, add				%				100%	100%
0400	Duct or AHU insulation	A-10B	440	.073	S.F.	.21	4.01		4.22	6.50
0500	Duct vibration isolation joints, up to 24 sq. in. duct	A-9	56	1.143	Ea.	3.28	63		66.28	102
0520	25 sq. in. to 48 sq. in. duct		48	1.333		3.83	73.50		77.33	118
0530	49 sq. in. to 76 sq. in. duct		40	1.600		4.60	88		92.60	142
0600	Pipe insulation, air cell type, up to 4" diameter pipe		900	.071	L.F.	.20	3.92		4.12	6.30
0610	4" to 8" diameter pipe		800	.080		.23	4.41		4.64	7.10
0620	10" to 12" diameter pipe		700	.091		.26	5.05		5.31	8.15
0630	14" to 16" diameter pipe		550	.116		.33	6.40		6.73	10.30
0650	Over 16" diameter pipe		650	.098	S.F.	.28	5.45		5.73	8.75
0700	With glove bag up to 3" diameter pipe		200	.320	L.F.	9.40	17.65		27.05	38
1000	Pipe fitting insulation up to 4" diameter pipe		320	.200	Ea.	.57	11		11.57	17.80
1100	6" to 8" diameter pipe		304	.211		.60	11.60		12.20	18.70
1110	10" to 12" diameter pipe		192	.333		.96	18.35		19.31	29.50
1120	14" to 16" diameter pipe		128	.500		1.44	27.50		28.94	44.50
1130	Over 16" diameter pipe		176	.364	S.F.	1.04	20		21.04	32
1200	With glove bag, up to 8" diameter pipe		75	.853	L.F.	6.45	47		53.45	80
2000	Scrape foam fireproofing from flat surface		2400	.027	S.F.	.08	1.47		1.55	2.36
2100	Irregular surfaces		1200	.053		.15	2.94		3.09	4.74
3000	Remove cementitious material from flat surface		1800	.036		.10	1.96		2.06	3.16
3100	Irregular surface		1000	.064		.13	3.53		3.66	5.65
4000	Scrape acoustical coating/fireproofing, from ceiling		3200	.020		.06	1.10		1.16	1.77
5000	Remove VAT and mastic from floor by hand		2400	.027		.08	1.47		1.55	2.36
5100	By machine	A-11	4800	.013		.04	.73	.01	.78	1.19
5150	For 2 layers, add				%				50%	50%
6000	Remove contaminated soil from crawl space by hand	A-9	400	.160	C.F.	.46	8.80		9.26	14.20
6100	With large production vacuum loader	A-12	700	.091	"	.26	5.05	1.05	6.36	9.30
7000	Radiator backing, not including radiator removal	A-9	1200	.053	S.F.	.15	2.94		3.09	4.74
8000	Cement-asbestos transite board and cement wall board	2 Asbe	1000	.016		.16	.88		1.04	1.54
8100	Transite shingle siding	A-10B	750	.043		.22	2.35		2.57	3.90
8200	Shingle roofing	"	2000	.016		.08	.88		.96	1.46
8250	Built-up, no gravel, non-friable	B-2	1400	.029		.08	1.13		1.21	1.82
8260	Bituminous flashing	1 Rofc	300	.027		.08	1.15		1.23	2.05
8300	Asbestos millboard, flat board and VAT contaminated plywood	2 Asbe	1000	.016		.08	.88		.96	1.46
9000	For type B (supplied air) respirator equipment, add				%				10%	10%

02 82 13.44 Demolition In Asbestos Contaminated Area

		Crew	Daily Output	Labor-Hours	Unit	Material	Labor	Equipment	Total	Total Incl O&P
0010	**DEMOLITION IN ASBESTOS CONTAMINATED AREA**									
0200	Ceiling, including suspension system, plaster and lath	A-9	2100	.030	S.F.	.09	1.68		1.77	2.71
0210	Finished plaster, leaving wire lath		585	.109		.31	6.05		6.36	9.70

02 82 Asbestos Remediation

02 82 13 – Asbestos Abatement

02 82 13.44 Demolition In Asbestos Contaminated Area

		Crew	Daily Output	Labor-Hours	Unit	Material	2017 Bare Costs Labor	2017 Bare Costs Equipment	Total	Total Incl O&P
0220	Suspended acoustical tile	A-9	3500	.018	S.F.	.05	1.01		1.06	1.63
0230	Concealed tile grid system		3000	.021		.06	1.18		1.24	1.90
0240	Metal pan grid system		1500	.043		.12	2.35		2.47	3.78
0250	Gypsum board		2500	.026		.07	1.41		1.48	2.27
0260	Lighting fixtures up to 2' x 4'	↓	72	.889	Ea.	2.55	49		51.55	79
0400	Partitions, non load bearing									
0410	Plaster, lath, and studs	A-9	690	.093	S.F.	.91	5.10		6.01	8.95
0450	Gypsum board and studs	"	1390	.046	"	.13	2.54		2.67	4.09
9000	For type B (supplied air) respirator equipment, add				%				10%	10%

02 82 13.45 OSHA Testing

		Crew	Daily Output	Labor-Hours	Unit	Material	2017 Bare Costs Labor	2017 Bare Costs Equipment	Total	Total Incl O&P
0010	**OSHA TESTING**									
0100	Certified technician, minimum				Day				200	220
0110	Maximum								300	330
0121	Industrial hygienist, minimum	1 Asbe	1.75	4.571			252		252	390
0130	Maximum				↓				400	440
0200	Asbestos sampling and PCM analysis, NIOSH 7400, minimum	1 Asbe	8	1	Ea.	16	55		71	103
0210	Maximum		4	2		27.50	110		137.50	202
1000	Cleaned area samples		8	1		201	55		256	305
1100	PCM air sample analysis, NIOSH 7400, minimum		8	1		15.05	55		70.05	102
1110	Maximum	↓	4	2		2.22	110		112.22	173
1200	TEM air sample analysis, NIOSH 7402, minimum								80	106
1210	Maximum				↓				360	450

02 82 13.46 Decontamination of Asbestos Containment Area

		Crew	Daily Output	Labor-Hours	Unit	Material	2017 Bare Costs Labor	2017 Bare Costs Equipment	Total	Total Incl O&P
0010	**DECONTAMINATION OF ASBESTOS CONTAINMENT AREA**									
0100	Spray exposed substrate with surfactant (bridging)									
0200	Flat surfaces	A-9	6000	.011	S.F.	.40	.59		.99	1.35
0250	Irregular surfaces		4000	.016	"	.40	.88		1.28	1.81
0300	Pipes, beams, and columns		2000	.032	L.F.	.60	1.76		2.36	3.40
1000	Spray encapsulate polyethylene sheeting		8000	.008	S.F.	.40	.44		.84	1.13
1100	Roll down polyethylene sheeting		8000	.008	"		.44		.44	.69
1500	Bag polyethylene sheeting		400	.160	Ea.	1	8.80		9.80	14.80
2000	Fine clean exposed substrate, with nylon brush		2400	.027	S.F.		1.47		1.47	2.28
2500	Wet wipe substrate		4800	.013			.73		.73	1.14
2600	Vacuum surfaces, fine brush	↓	6400	.010	↓		.55		.55	.86
3000	Structural demolition									
3100	Wood stud walls	A-9	2800	.023	S.F.		1.26		1.26	1.96
3500	Window manifolds, not incl. window replacement		4200	.015			.84		.84	1.31
3600	Plywood carpet protection	↓	2000	.032	↓		1.76		1.76	2.74
4000	Remove custom decontamination facility	A-10A	8	3	Ea.	15	166		181	274
4100	Remove portable decontamination facility	3 Asbe	12	2	"	14.05	110		124.05	186
5000	HEPA vacuum, shampoo carpeting	A-9	4800	.013	S.F.	.10	.73		.83	1.25
9000	Final cleaning of protected surfaces	A-10A	8000	.003	"		.17		.17	.26

02 82 13.47 Asbestos Waste Pkg., Handling, and Disp.

		Crew	Daily Output	Labor-Hours	Unit	Material	2017 Bare Costs Labor	2017 Bare Costs Equipment	Total	Total Incl O&P
0010	**ASBESTOS WASTE PACKAGING, HANDLING, AND DISPOSAL**									
0100	Collect and bag bulk material, 3 C.F. bags, by hand	A-9	400	.160	Ea.	.85	8.80		9.65	14.65
0200	Large production vacuum loader	A-12	880	.073		1	4.01	.84	5.85	8.25
1000	Double bag and decontaminate	A-9	960	.067		.85	3.67		4.52	6.65
2000	Containerize bagged material in drums, per 3 C.F. drum	"	800	.080		18.75	4.41		23.16	27.50
3000	Cart bags 50' to dumpster	2 Asbe	400	.040	↓		2.20		2.20	3.42
5000	Disposal charges, not including haul, minimum				C.Y.				61	67
5020	Maximum				"				355	395
9000	For type B (supplied air) respirator equipment, add				%				10%	10%

02 82 Asbestos Remediation

02 82 13 – Asbestos Abatement

02 82 13.48 Asbestos Encapsulation With Sealants		Crew	Daily Output	Labor-Hours	Unit	Material	2017 Bare Costs Labor	Equipment	Total	Total Incl O&P
0010	**ASBESTOS ENCAPSULATION WITH SEALANTS**									
0100	Ceilings and walls, minimum	A-9	21000	.003	S.F.	.40	.17		.57	.70
0110	Maximum		10600	.006		.50	.33		.83	1.07
0200	Columns and beams, minimum		13300	.005		.40	.27		.67	.85
0210	Maximum		5325	.012		.60	.66		1.26	1.69
0300	Pipes to 12" diameter including minor repairs, minimum		800	.080	L.F.	.50	4.41		4.91	7.40
0310	Maximum		400	.160	"	1.20	8.80		10	15

02 83 Lead Remediation

02 83 19 – Lead-Based Paint Remediation

02 83 19.21 Lead Paint Remediation Plans and Methods

		Crew	Daily Output	Labor-Hours	Unit	Material	Labor	Equipment	Total	Total Incl O&P
0010	**LEAD PAINT REMEDIATION PLANS AND METHODS**									
0100	Building Survey-Commercial Building				Ea.				2,050	2,250
0200	Lead Abatement Remediation Plan								1,225	1,350
0300	Lead Paint Testing, AAS Analysis								51	56
0400	Lead Paint Testing, X-Ray Fluorescence								51	56

02 83 19.23 Encapsulation of Lead-Based Paint

		Crew	Daily Output	Labor-Hours	Unit	Material	Labor	Equipment	Total	Total Incl O&P
0010	**ENCAPSULATION OF LEAD-BASED PAINT**									
0020	Interior, brushwork, trim, under 6"	1 Pord	240	.033	L.F.	2.32	1.39		3.71	4.65
0030	6" to 12" wide		180	.044		3.03	1.85		4.88	6.15
0040	Balustrades		300	.027		2.02	1.11		3.13	3.90
0050	Pipe to 4" diameter		500	.016		1.21	.67		1.88	2.34
0060	To 8" diameter		375	.021		1.52	.89		2.41	3.01
0070	To 12" diameter		250	.032		2.22	1.33		3.55	4.45
0080	To 16" diameter		170	.047		3.33	1.96		5.29	6.60
0090	Cabinets, ornate design		200	.040	S.F.	3.03	1.67		4.70	5.85
0100	Simple design		250	.032	"	2.32	1.33		3.65	4.56
0110	Doors, 3' x 7', both sides, incl. frame & trim									
0120	Flush	1 Pord	6	1.333	Ea.	28.50	55.50		84	115
0130	French, 10 – 15 lite		3	2.667		6.05	111		117.05	175
0140	Panel		4	2		35.50	83.50		119	165
0150	Louvered		2.75	2.909		32.50	121		153.50	219
0160	Windows, per interior side, per 15 S.F.									
0170	1 to 6 lite	1 Pord	14	.571	Ea.	20	24		44	58
0180	7 to 10 lite		7.50	1.067		22	44.50		66.50	91.50
0190	12 lite		5.75	1.391		30.50	58		88.50	121
0200	Radiators		8	1		70.50	41.50		112	141
0210	Grilles, vents		275	.029	S.F.	2.02	1.21		3.23	4.05
0220	Walls, roller, drywall or plaster		1000	.008		.61	.33		.94	1.17
0230	With spunbonded reinforcing fabric		720	.011		.71	.46		1.17	1.48
0240	Wood		800	.010		.71	.42		1.13	1.41
0250	Ceilings, roller, drywall or plaster		900	.009		.71	.37		1.08	1.34
0260	Wood		700	.011		.81	.48		1.29	1.61
0270	Exterior, brushwork, gutters and downspouts		300	.027	L.F.	1.92	1.11		3.03	3.79
0280	Columns		400	.020	S.F.	1.41	.83		2.24	2.81
0290	Spray, siding		600	.013	"	1.01	.56		1.57	1.95
0300	Miscellaneous									
0310	Electrical conduit, brushwork, to 2" diameter	1 Pord	500	.016	L.F.	1.21	.67		1.88	2.34
0320	Brick, block or concrete, spray		500	.016	S.F.	1.21	.67		1.88	2.34
0330	Steel, flat surfaces and tanks to 12"		500	.016		1.21	.67		1.88	2.34

02 83 Lead Remediation

02 83 19 – Lead-Based Paint Remediation

02 83 19.23 Encapsulation of Lead-Based Paint	Crew	Daily Output	Labor-Hours	Unit	Material	2017 Bare Costs Labor	Equipment	Total	Total Incl O&P
0340	Beams, brushwork	1 Pord	400	.020	S.F.	1.41	.83		2.24	2.81
0350	Trusses	▼	400	.020	▼	1.41	.83		2.24	2.81

02 83 19.26 Removal of Lead-Based Paint

	02 83 19.26 Removal of Lead-Based Paint	Crew	Daily Output	Labor-Hours	Unit	Material	Labor	Equipment	Total	Total Incl O&P
0010	**REMOVAL OF LEAD-BASED PAINT** R028319-60									
0011	By chemicals, per application									
0050	Baseboard, to 6" wide	1 Pord	64	.125	L.F.	.80	5.20		6	8.75
0070	To 12" wide		32	.250	"	1.47	10.45		11.92	17.35
0200	Balustrades, one side		28	.286	S.F.	1.50	11.90		13.40	19.65
1400	Cabinets, simple design		32	.250		1.37	10.45		11.82	17.25
1420	Ornate design		25	.320		1.60	13.35		14.95	22
1600	Cornice, simple design		60	.133		1.60	5.55		7.15	10.15
1620	Ornate design		20	.400		5.55	16.70		22.25	31
2800	Doors, one side, flush		84	.095		1.73	3.97		5.70	7.90
2820	Two panel		80	.100		1.37	4.17		5.54	7.80
2840	Four panel		45	.178	▼	1.43	7.40		8.83	12.75
2880	For trim, one side, add		64	.125	L.F.	.77	5.20		5.97	8.70
3000	Fence, picket, one side		30	.267	S.F.	1.37	11.10		12.47	18.30
3200	Grilles, one side, simple design		30	.267		1.37	11.10		12.47	18.30
3220	Ornate design		25	.320	▼	1.47	13.35		14.82	21.50
3240	Handrails		90	.089	L.F.	1.37	3.71		5.08	7.10
4400	Pipes, to 4" diameter		90	.089		1.73	3.71		5.44	7.50
4420	To 8" diameter		50	.160		3.47	6.65		10.12	13.85
4440	To 12" diameter		36	.222		5.20	9.25		14.45	19.70
4460	To 16" diameter		20	.400	▼	6.95	16.70		23.65	32.50
4500	For hangers, add		40	.200	Ea.	2.53	8.35		10.88	15.40
4800	Siding		90	.089	S.F.	1.31	3.71		5.02	7.05
5000	Trusses, open		55	.145	SF Face	2.02	6.05		8.07	11.35
6200	Windows, one side only, double hung, 1/1 light, 24" x 48" high		4	2	Ea.	24	83.50		107.50	153
6220	30" x 60" high		3	2.667		32.50	111		143.50	204
6240	36" x 72" high		2.50	3.200		38.50	133		171.50	243
6280	40" x 80" high		2	4		47.50	167		214.50	305
6400	Colonial window, 6/6 light, 24" x 48" high		2	4		47.50	167		214.50	305
6420	30" x 60" high		1.50	5.333		62.50	222		284.50	405
6440	36" x 72" high		1	8		94	335		429	610
6480	40" x 80" high		1	8		94	335		429	610
6600	8/8 light, 24" x 48" high		2	4		47.50	167		214.50	305
6620	40" x 80" high		1	8		94	335		429	610
6800	12/12 light, 24" x 48" high		1	8		94	335		429	610
6820	40" x 80" high	▼	.75	10.667	▼	126	445		571	810
6840	Window frame & trim items, included in pricing above									
7000	Hand scraping and HEPA vacuum, less than 4 S.F.	1 Pord	8	1	Ea.	.85	41.50		42.35	64
8000	Collect and bag bulk material, 3 C.F. bags, by hand	"	30	.267	"	.85	11.10		11.95	17.75

For customer support on your Building Construction Costs with RSMeans Data, call 800.448.8182.

45

02 85 Mold Remediation

02 85 16 – Mold Remediation Preparation and Containment

02 85 16.40 Mold Remediation Plans and Methods

		Crew	Daily Output	Labor-Hours	Unit	Material	2017 Bare Costs Labor	Equipment	Total	Total Incl O&P
0010	**MOLD REMEDIATION PLANS AND METHODS**									
0020	Initial inspection, areas to 2500 S.F.				Total				268	295
0030	Areas to 5000 S.F.								440	485
0032	Areas to 10000 S.F.								440	485
0040	Testing, air sample each								126	139
0050	Swab sample								115	127
0060	Tape sample								125	140
0070	Post remediation air test								126	139
0080	Mold abatement plan, area to 2500 S.F.								1,225	1,325
0090	Areas to 5000 S.F.								1,625	1,775
0095	Areas to 10000 S.F.								2,550	2,800
0100	Packup & removal of contents, average 3 bedroom home, excl storage								8,150	8,975
0110	Average 5 bedroom home, excl storage				▼				15,300	16,800
0600	For demolition in mold contaminated areas, see Section 02 85 33.50									
0610	For personal protection equipment, see Section 02 82 13.41									

02 85 16.50 Preparation of Mold Containment Area

		Crew	Daily Output	Labor-Hours	Unit	Material	2017 Bare Costs Labor	Equipment	Total	Total Incl O&P
0010	**PREPARATION OF MOLD CONTAINMENT AREA**									
0100	Pre-cleaning, HEPA vacuum and wet wipe, flat surfaces	A-9	12000	.005	S.F.	.02	.29		.31	.48
0300	Separation barrier, 2" x 4" @ 16", 1/2" plywood ea. side, 8' high	2 Carp	400	.040		3.43	1.97		5.40	6.80
0310	12' high		320	.050		3.43	2.46		5.89	7.55
0320	16' high		200	.080		2.32	3.94		6.26	8.60
0400	Personnel decontam. chamber, 2" x 4" @ 16", 3/4" ply ea. side		280	.057		4.44	2.81		7.25	9.20
0450	Waste decontam. chamber, 2" x 4" studs @ 16", 3/4" ply each side	▼	360	.044	▼	4.44	2.19		6.63	8.25
0500	Cover surfaces with polyethylene sheeting									
0501	Including glue and tape									
0550	Floors, each layer, 6 mil	A-9	8000	.008	S.F.	.04	.44		.48	.73
0551	4 mil		9000	.007		.03	.39		.42	.64
0560	Walls, each layer, 6 mil		6000	.011		.04	.59		.63	.95
0561	4 mil	▼	7000	.009	▼	.03	.50		.53	.81
0570	For heights above 14', add						20%			
0575	For heights above 20', add						30%			
0580	For fire retardant poly, add					100%				
0590	For large open areas, deduct					10%	20%			
0600	Seal floor penetrations with foam firestop to 36 sq. in.	2 Carp	200	.080	Ea.	11.25	3.94		15.19	18.45
0610	36 sq. in. to 72 sq. in.		125	.128		22.50	6.30		28.80	34.50
0615	72 sq. in. to 144 sq. in.		80	.200		45	9.85		54.85	64.50
0620	Wall penetrations, to 36 sq. in.		180	.089		11.25	4.38		15.63	19.10
0630	36 sq. in. to 72 sq. in.		100	.160		22.50	7.90		30.40	37
0640	72 sq. in. to 144 sq. in.	▼	60	.267	▼	45	13.15		58.15	69.50
0800	Caulk seams with latex caulk	1 Carp	230	.035	L.F.	.17	1.71		1.88	2.81
0900	Set up neg. air machine, 1-2k CFM/25 M.C.F. volume	1 Asbe	4.30	1.860	Ea.		102		102	159

02 85 33 – Removal and Disposal of Materials with Mold

02 85 33.50 Demolition in Mold Contaminated Area

		Crew	Daily Output	Labor-Hours	Unit	Material	2017 Bare Costs Labor	Equipment	Total	Total Incl O&P
0010	**DEMOLITION IN MOLD CONTAMINATED AREA**									
0200	Ceiling, including suspension system, plaster and lath	A-9	2100	.030	S.F.	.09	1.68		1.77	2.71
0210	Finished plaster, leaving wire lath		585	.109		.31	6.05		6.36	9.70
0220	Suspended acoustical tile		3500	.018		.05	1.01		1.06	1.63
0230	Concealed tile grid system		3000	.021		.06	1.18		1.24	1.90
0240	Metal pan grid system		1500	.043		.12	2.35		2.47	3.78
0250	Gypsum board		2500	.026		.07	1.41		1.48	2.27
0255	Plywood		2500	.026	▼	.07	1.41		1.48	2.27
0260	Lighting fixtures up to 2' x 4'	▼	72	.889	Ea.	2.55	49		51.55	79

02 85 Mold Remediation

02 85 33 – Removal and Disposal of Materials with Mold

02 85 33.50 Demolition in Mold Contaminated Area	Crew	Daily Output	Labor-Hours	Unit	Material	2017 Bare Costs Labor	Equipment	Total	Total Incl O&P
0400	Partitions, non load bearing								
0410	Plaster, lath, and studs	A-9	690	.093	S.F.	.91	5.10		6.01	8.95
0450	Gypsum board and studs		1390	.046		.13	2.54		2.67	4.09
0465	Carpet & pad		1390	.046		.13	2.54		2.67	4.09
0600	Pipe insulation, air cell type, up to 4" diameter pipe		900	.071	L.F.	.20	3.92		4.12	6.30
0610	4" to 8" diameter pipe		800	.080		.23	4.41		4.64	7.10
0620	10" to 12" diameter pipe		700	.091		.26	5.05		5.31	8.15
0630	14" to 16" diameter pipe		550	.116		.33	6.40		6.73	10.30
0650	Over 16" diameter pipe		650	.098	S.F.	.28	5.45		5.73	8.75
9000	For type B (supplied air) respirator equipment, add				%				10%	10%

02 91 Chemical Sampling, Testing and Analysis

02 91 10 – Monitoring, Sampling, Testing and Analysis

02 91 10.10 Monitoring, Chemical Sampling, Testing and Analysis

		Crew	Daily Output	Labor-Hours	Unit	Material	2017 Bare Costs Labor	Equipment	Total	Total Incl O&P
0010	**MONITORING, CHEMICAL SAMPLING, TESTING AND ANALYSIS**									
0015	Field sampling of waste									
0100	Field samples, sample collection, sludge	1 Skwk	32	.250	Ea.		12.85		12.85	19.80
0110	Contaminated soils	"	32	.250	"		12.85		12.85	19.80
0200	Vials and bottles									
0210	32 oz. clear wide mouth jar (case of 12)				Ea.	44			44	48.50
0220	32 oz. Boston round bottle (case of 12)					38.50			38.50	42
0230	32 oz. HDPE bottle (case of 12)					37			37	40.50
0300	Laboratory analytical services									
0310	Laboratory testing 13 metals				Ea.	212			212	233
0312	13 metals + mercury					212			212	233
0314	8 metals					162			162	178
0316	Mercury only					55.50			55.50	61
0318	Single metal (only Cs, Li, Sr, Ta)					45.50			45.50	50
0320	Single metal (excludes Hg, Cs, Li, Sr, Ta)					45.50			45.50	50
0400	Hydrocarbons standard					227			227	250
0410	Hydrocarbons fingerprint					253			253	278
0500	Radioactivity gross alpha					155			155	171
0510	Gross alpha & beta					155			155	171
0520	Radium 226					101			101	111
0530	Radium 228					131			131	144
0540	Radon					149			149	164
0550	Uranium					131			131	144
0600	Volatile organics without GC/MS					158			158	174
0610	Volatile organics including GC/MS					305			305	335
0630	Synthetic organic compounds					1,025			1,025	1,150
0640	Herbicides					217			217	239
0650	Pesticides					153			153	168
0660	PCB's					141			141	156

For customer support on your Building Construction Costs with RSMeans Data, call 800.448.8182.

47

Division Notes

	CREW	DAILY OUTPUT	LABOR-HOURS	UNIT	BARE COSTS				TOTAL INCL O&P
					MAT.	LABOR	EQUIP.	TOTAL	

Estimating Tips
General

- Carefully check all the plans and specifications. Concrete often appears on drawings other than structural drawings, including mechanical and electrical drawings for equipment pads. The cost of cutting and patching is often difficult to estimate. See Subdivision 03 81 for Concrete Cutting, Subdivision 02 41 19.16 for Cutout Demolition, Subdivision 03 05 05.10 for Concrete Demolition, and Subdivision 02 41 19.19 for Rubbish Handling (handling, loading, and hauling of debris).

- Always obtain concrete prices from suppliers near the job site. A volume discount can often be negotiated, depending upon competition in the area. Remember to add for waste, particularly for slabs and footings on grade.

03 10 00 Concrete Forming and Accessories

- A primary cost for concrete construction is forming. Most jobs today are constructed with prefabricated forms. The selection of the forms best suited for the job and the total square feet of forms required for efficient concrete forming and placing are key elements in estimating concrete construction. Enough forms must be available for erection to make efficient use of the concrete placing equipment and crew.

- Concrete accessories for forming and placing depend upon the systems used. Study the plans and specifications to ensure that all special accessory requirements have been included in the cost estimate, such as anchor bolts, inserts, and hangers.

- Included within costs for forms-in-place are all necessary bracing and shoring.

03 20 00 Concrete Reinforcing

- Ascertain that the reinforcing steel supplier has included all accessories, cutting, bending, and an allowance for lapping, splicing, and waste. A good rule of thumb is 10% for lapping, splicing, and waste. Also, 10% waste should be allowed for welded wire fabric.

- The unit price items in the subdivisions for Reinforcing In Place, Glass Fiber Reinforcing, and Welded Wire Fabric include the labor to install accessories such as beam and slab bolsters, high chairs, and bar ties and tie wire. The material cost for these accessories is not included; they may be obtained from the Accessories Subdivisions.

03 30 00 Cast-In-Place Concrete

- When estimating structural concrete, pay particular attention to requirements for concrete additives, curing methods, and surface treatments. Special consideration for climate, hot or cold, must be included in your estimate. Be sure to include requirements for concrete placing equipment and concrete finishing.

- For accurate concrete estimating, the estimator must consider each of the following major components individually: forms, reinforcing steel, ready-mix concrete, placement of the concrete, and finishing of the top surface. For faster estimating, Subdivision 03 30 53.40 for Concrete-In-Place can be used; here, various items of concrete work are presented that include the costs of all five major components (unless specifically stated otherwise).

03 40 00 Precast Concrete
03 50 00 Cast Decks and Underlayment

- The cost of hauling precast concrete structural members is often an important factor. For this reason, it is important to get a quote from the nearest supplier. It may become economically feasible to set up precasting beds on the site if the hauling costs are prohibitive.

Reference Numbers

Reference numbers are shown at the beginning of some major classifications. These numbers refer to related items in the Reference Section. The reference information may be an estimating procedure, an alternate pricing method, or technical information.

Note: Not all subdivisions listed here necessarily appear. ■

Did you know?

Our online estimating solution gives you the same access to RSMeans' data with 24/7 access:

- Quickly locate costs in the searchable database.

- Build cost lists, estimates, and reports in minutes.

- Adjust costs to any location in the U.S. and Canada with the click of a button.

Start your free trial today at
www.RSMeansOnline.com

03 01 Maintenance of Concrete

03 01 30 – Maintenance of Cast-In-Place Concrete

03 01 30.62 Concrete Patching

03 01 30.62 Concrete Patching	Crew	Daily Output	Labor-Hours	Unit	Material	2017 Bare Costs Labor	Equipment	Total	Total Incl O&P
0010 **CONCRETE PATCHING**									
0100 Floors, 1/4" thick, small areas, regular grout	1 Cefi	170	.047	S.F.	1.56	2.20		3.76	5
0150 Epoxy grout	"	100	.080	"	8.30	3.74		12.04	14.65
2000 Walls, including chipping, cleaning and epoxy grout									
2100 1/4" deep	1 Cefi	65	.123	S.F.	7.25	5.75		13	16.55
2150 1/2" deep		50	.160		14.50	7.50		22	27
2200 3/4" deep	↓	40	.200	↓	22	9.35		31.35	38

03 05 Common Work Results for Concrete

03 05 05 – Selective Demolition for Concrete

03 05 05.10 Selective Demolition, Concrete

	Crew	Daily Output	Labor-Hours	Unit	Material	2017 Bare Costs Labor	Equipment	Total	Total Incl O&P
0010 **SELECTIVE DEMOLITION, CONCRETE** R024119-10									
0012 Excludes saw cutting, torch cutting, loading or hauling									
0050 Break into small pieces, reinf. less than 1% of cross-sectional area	B-9	24	1.667	C.Y.		66	8.90	74.90	111
0060 Reinforcing 1% to 2% of cross-sectional area		16	2.500			99	13.35	112.35	167
0070 Reinforcing more than 2% of cross-sectional area	↓	8	5	↓		198	26.50	224.50	335
0150 Remove whole pieces, up to 2 tons per piece	E-18	36	1.111	Ea.		60.50	30	90.50	132
0160 2 – 5 tons per piece		30	1.333			72.50	36	108.50	159
0170 5 – 10 tons per piece		24	1.667			91	45	136	199
0180 10 – 15 tons per piece	↓	18	2.222			121	59.50	180.50	264
0250 Precast unit embedded in masonry, up to 1 C.F.	D-1	16	1			43.50		43.50	66.50
0260 1 – 2 C.F.		12	1.333			57.50		57.50	88.50
0270 2 – 5 C.F.		10	1.600			69.50		69.50	106
0280 5 – 10 C.F.	↓	8	2	↓		86.50		86.50	133
0990 For hydrodemolition see Section 02 41 13.15									

03 05 13 – Basic Concrete Materials

03 05 13.20 Concrete Admixtures and Surface Treatments

	Crew	Daily Output	Labor-Hours	Unit	Material	2017 Bare Costs Labor	Equipment	Total	Total Incl O&P
0010 **CONCRETE ADMIXTURES AND SURFACE TREATMENTS**									
0040 Abrasives, aluminum oxide, over 20 tons				Lb.	1.92			1.92	2.11
0050 1 to 20 tons					2.05			2.05	2.26
0070 Under 1 ton					2.14			2.14	2.35
0100 Silicon carbide, black, over 20 tons					2.93			2.93	3.22
0110 1 to 20 tons					3.10			3.10	3.41
0120 Under 1 ton				↓	3.23			3.23	3.55
0200 Air entraining agent, .7 to 1.5 oz. per bag, 55 gallon drum				Gal.	14.30			14.30	15.70
0220 5 gallon pail					19.65			19.65	21.50
0300 Bonding agent, acrylic latex, 250 S.F. per gallon, 5 gallon pail					23.50			23.50	26
0320 Epoxy resin, 80 S.F. per gallon, 4 gallon case				↓	63.50			63.50	70
0400 Calcium chloride, 50 lb. bags, TL lots				Ton	825			825	905
0420 Less than truckload lots				Bag	17.75			17.75	19.55
0500 Carbon black, liquid, 2 to 8 lb. per bag of cement				Lb.	9.30			9.30	10.20
0600 Colored pigments, integral, 2 to 10 lb. per bag of cement, subtle colors					2.50			2.50	2.75
0610 Standard colors					3.36			3.36	3.70
0620 Premium colors				↓	5.55			5.55	6.10
0920 Dustproofing compound, 250 S.F./gal., 5 gallon pail				Gal.	7.10			7.10	7.80
1010 Epoxy based, 125 S.F./gal., 5 gallon pail				"	57			57	63
1100 Hardeners, metallic, 55 lb. bags, natural (grey)				Lb.	.83			.83	.92
1200 Colors					1.45			1.45	1.60
1300 Non-metallic, 55 lb. bags, natural grey					.43			.43	.47
1320 Colors				↓	.93			.93	1.03

03 05 13 – Basic Concrete Materials

03 05 13.20 Concrete Admixtures and Surface Treatments		Crew	Daily Output	Labor-Hours	Unit	Material	2017 Bare Costs Labor	Equipment	Total	Total Incl O&P
1550	Release agent, for tilt slabs, 5 gallon pail				Gal.	17.90			17.90	19.65
1570	For forms, 5 gallon pail					11.65			11.65	12.80
1590	Concrete release agent for forms, 100% biodegradable, zero VOC, 5 gal pail [G]					21			21	23
1595	55 gallon drum [G]					17.75			17.75	19.50
1600	Sealer, hardener and dustproofer, epoxy-based, 125 S.F./gal., 5 gallon unit					57			57	63
1620	3 gallon unit					63			63	69
1630	Sealer, solvent-based, 250 S.F./gal., 55 gallon drum					21			21	23
1640	5 gallon pail					31.50			31.50	35
1650	Sealer, water based, 350 S.F., 55 gallon drum					20.50			20.50	22.50
1660	5 gallon pail					23.50			23.50	26
1900	Set retarder, 100 S.F./gal., 1 gallon pail				▼	6.30			6.30	6.95
2000	Waterproofing, integral 1 lb. per bag of cement				Lb.	2.16			2.16	2.38
2100	Powdered metallic, 40 lbs. per 100 S.F., standard colors					2.57			2.57	2.83
2120	Premium colors				▼	3.60			3.60	3.96
3000	For integral colored pigments, 2500 psi (5 bag mix)									
3100	Standard colors, 1.8 lbs. per bag, add				C.Y.	22.50			22.50	25
3200	9.4 lbs. per bag, add					118			118	129
3400	Premium colors, 1.8 lbs. per bag, add					30			30	33.50
3500	7.5 lbs. per bag, add					126			126	139
3700	Ultra premium colors, 1.8 lbs. per bag, add					50			50	55
3800	7.5 lbs. per bag, add				▼	207			207	228
6000	Concrete ready mix additives, recycled coal fly ash, mixed at plant [G]				Ton	58			58	63.50
6010	Recycled blast furnace slag, mixed at plant [G]				"	90.50			90.50	99.50

03 05 13.25 Aggregate

		Crew	Daily Output	Labor-Hours	Unit	Material	Labor	Equipment	Total	Total Incl O&P
0010	**AGGREGATE** R033105-20									
0100	Lightweight vermiculite or perlite, 4 C.F. bag, C.L. lots [G]				Bag	24.50			24.50	27
0150	L.C.L. lots [G]				"	27			27	30
0250	Sand & stone, loaded at pit, crushed bank gravel				Ton	21.50			21.50	23.50
0350	Sand, washed, for concrete R033105-50					21			21	23.50
0400	For plaster or brick					21			21	23.50
0450	Stone, 3/4" to 1-1/2"					19.65			19.65	21.50
0470	Round, river stone					36			36	39.50
0500	3/8" roofing stone & 1/2" pea stone					27.50			27.50	30.50
0550	For trucking 10-mile round trip, add to the above	B-34B	117	.068			3.11	5.50	8.61	10.75
0600	For trucking 30-mile round trip, add to the above	"	72	.111	▼		5.05	8.95	14	17.45
0850	Sand & stone, loaded at pit, crushed bank gravel				C.Y.	30			30	33
0950	Sand, washed, for concrete					29.50			29.50	32.50
1000	For plaster or brick					29.50			29.50	32.50
1050	Stone, 3/4" to 1-1/2"					37.50			37.50	41.50
1055	Round, river stone					38.50			38.50	42.50
1100	3/8" roofing stone & 1/2" pea stone					28.50			28.50	31
1150	For trucking 10-mile round trip, add to the above	B-34B	78	.103			4.67	8.25	12.92	16.10
1200	For trucking 30-mile round trip, add to the above	"	48	.167	▼		7.60	13.40	21	26
1310	Onyx chips, 50 lb. bags				Cwt.	32			32	35
1330	Quartz chips, 50 lb. bags					32			32	35
1410	White marble, 3/8" to 1/2", 50 lb. bags				▼	9.50			9.50	10.45
1430	3/4", bulk				Ton	146			146	160

03 05 13.30 Cement

		Crew	Daily Output	Labor-Hours	Unit	Material	Labor	Equipment	Total	Total Incl O&P
0010	**CEMENT** R033105-20									
0240	Portland, Type I/II, T.L. lots, 94 lb. bags				Bag	11.30			11.30	12.45
0250	L.T.L./L.C.L. lots R033105-30				"	12.55			12.55	13.85
0300	Trucked in bulk, per Cwt.				Cwt.	7.35			7.35	8.10

03 05 Common Work Results for Concrete

03 05 13 – Basic Concrete Materials

03 05 13.30 Cement		Crew	Daily Output	Labor-Hours	Unit	Material	2017 Bare Costs Labor	2017 Bare Costs Equipment	Total	Total Incl O&P
0400	Type III, high early strength, T.L. lots, 94 lb. bags R033105-40				Bag	12.30			12.30	13.55
0420	L.T.L. or L.C.L. lots					13.70			13.70	15.05
0500	White, type III, high early strength, T.L. or C.L. lots, bags R033105-50					24.50			24.50	27
0520	L.T.L. or L.C.L. lots					27.50			27.50	30
0600	White, type I, T.L. or C.L. lots, bags					27			27	29.50
0620	L.T.L. or L.C.L. lots					28			28	31

03 05 13.80 Waterproofing and Dampproofing

0010	**WATERPROOFING AND DAMPPROOFING**									
0050	Integral waterproofing, add to cost of regular concrete				C.Y.	12.95			12.95	14.25

03 05 13.85 Winter Protection

0010	**WINTER PROTECTION**									
0012	For heated ready mix, add				C.Y.	4.45			4.45	4.90
0100	Temporary heat to protect concrete, 24 hours	2 Clab	50	.320	M.S.F.	200	12.55		212.55	239
0200	Temporary shelter for slab on grade, wood frame/polyethylene sheeting									
0201	Build or remove, light framing for short spans	2 Carp	10	1.600	M.S.F.	300	79		379	450
0210	Large framing for long spans	"	3	5.333	"	395	263		658	830
0500	Electrically heated pads, 110 volts, 15 watts per S.F., buy				S.F.	11.15			11.15	12.30
0600	20 watts per S.F., buy					14.85			14.85	16.35
0710	Electrically heated pads, 15 watts/S.F., 20 uses					.56			.56	.61

03 11 Concrete Forming

03 11 13 – Structural Cast-In-Place Concrete Forming

03 11 13.20 Forms In Place, Beams and Girders

03 11 13.20 Forms In Place, Beams and Girders		Crew	Daily Output	Labor-Hours	Unit	Material	2017 Bare Costs Labor	2017 Bare Costs Equipment	Total	Total Incl O&P
0010	**FORMS IN PLACE, BEAMS AND GIRDERS** R031113-40									
0500	Exterior spandrel, job-built plywood, 12" wide, 1 use	C-2	225	.213	SFCA	3.18	10.20		13.38	19.15
0550	2 use R031113-60		275	.175		1.68	8.35		10.03	14.65
0600	3 use		295	.163		1.27	7.80		9.07	13.35
0650	4 use		310	.155		1.03	7.40		8.43	12.50
1000	18" wide, 1 use		250	.192		2.86	9.20		12.06	17.25
1050	2 use		275	.175		1.58	8.35		9.93	14.55
1100	3 use		305	.157		1.15	7.55		8.70	12.80
1150	4 use		315	.152		.94	7.30		8.24	12.25
1500	24" wide, 1 use		265	.181		2.63	8.70		11.33	16.20
1550	2 use		290	.166		1.49	7.95		9.44	13.80
1600	3 use		315	.152		1.05	7.30		8.35	12.35
1650	4 use		325	.148		.86	7.05		7.91	11.80
2000	Interior beam, job-built plywood, 12" wide, 1 use		300	.160		3.50	7.65		11.15	15.60
2050	2 use		340	.141		1.70	6.75		8.45	12.20
2100	3 use		364	.132		1.40	6.30		7.70	11.25
2150	4 use		377	.127		1.14	6.10		7.24	10.60
2500	24" wide, 1 use		320	.150		2.69	7.20		9.89	13.95
2550	2 use		365	.132		1.52	6.30		7.82	11.30
2600	3 use		385	.125		1.07	5.95		7.02	10.30
2650	4 use		395	.122		.87	5.80		6.67	9.85
3000	Encasing steel beam, hung, job-built plywood, 1 use		325	.148		3.33	7.05		10.38	14.50
3050	2 use		390	.123		1.83	5.90		7.73	11.05
3100	3 use		415	.116		1.33	5.55		6.88	9.95
3150	4 use		430	.112		1.08	5.35		6.43	9.40
3500	Bottoms only, to 30" wide, job-built plywood, 1 use		230	.209		4.28	10		14.28	20
3550	2 use		265	.181		2.40	8.70		11.10	15.95

03 11 Concrete Forming

03 11 13 – Structural Cast-In-Place Concrete Forming

03 11 13.20 Forms In Place, Beams and Girders

		Crew	Daily Output	Labor-Hours	Unit	Material	2017 Bare Costs Labor	Equipment	Total	Total Incl O&P
3600	3 use	C-2	280	.171	SFCA	1.71	8.20		9.91	14.50
3650	4 use		290	.166		1.39	7.95		9.34	13.70
4000	Sides only, vertical, 36" high, job-built plywood, 1 use		335	.143		5.40	6.85		12.25	16.45
4050	2 use		405	.119		2.98	5.70		8.68	11.95
4100	3 use		430	.112		2.16	5.35		7.51	10.60
4150	4 use		445	.108		1.76	5.15		6.91	9.85
4500	Sloped sides, 36" high, 1 use		305	.157		5.25	7.55		12.80	17.30
4550	2 use		370	.130		2.92	6.20		9.12	12.70
4600	3 use		405	.119		2.09	5.70		7.79	11
4650	4 use		425	.113		1.70	5.40		7.10	10.15
5000	Upstanding beams, 36" high, 1 use		225	.213		6.60	10.20		16.80	23
5050	2 use		255	.188		3.68	9		12.68	17.85
5100	3 use		275	.175		2.67	8.35		11.02	15.75
5150	4 use		280	.171		2.16	8.20		10.36	15

03 11 13.25 Forms In Place, Columns

				Crew	Daily Output	Labor-Hours	Unit	Material	2017 Bare Costs Labor	Equipment	Total	Total Incl O&P
0010	**FORMS IN PLACE, COLUMNS**	R031113-40										
0500	Round fiberglass, 4 use per mo., rent, 12" diameter			C-1	160	.200	L.F.	9.85	9.35		19.20	25
0550	16" diameter	R031113-60			150	.213		11.75	9.95		21.70	28
0600	18" diameter				140	.229		13.15	10.70		23.85	31
0650	24" diameter				135	.237		16.35	11.10		27.45	35
0700	28" diameter				130	.246		18.25	11.50		29.75	37.50
0800	30" diameter				125	.256		19.05	11.95		31	39.50
0850	36" diameter				120	.267		25.50	12.45		37.95	47
1500	Round fiber tube, recycled paper, 1 use, 8" diameter		G		155	.206		1.91	9.65		11.56	16.90
1550	10" diameter		G		155	.206		2.56	9.65		12.21	17.60
1600	12" diameter		G		150	.213		3.05	9.95		13	18.60
1650	14" diameter		G		145	.221		4.26	10.30		14.56	20.50
1700	16" diameter		G		140	.229		5.25	10.70		15.95	22
1720	18" diameter		G		140	.229		6.15	10.70		16.85	23
1750	20" diameter		G		135	.237		7.70	11.10		18.80	25.50
1800	24" diameter		G		130	.246		9.60	11.50		21.10	28
1850	30" diameter		G		125	.256		14.55	11.95		26.50	34.50
1900	36" diameter		G		115	.278		17.85	13		30.85	39.50
1950	42" diameter		G		100	.320		43	14.95		57.95	70.50
2000	48" diameter		G		85	.376		50.50	17.60		68.10	83
2200	For seamless type, add							15%				
3000	Round, steel, 4 use per mo., rent, regular duty, 14" diameter		G	C-1	145	.221	L.F.	18.45	10.30		28.75	36.50
3050	16" diameter		G		125	.256		18.80	11.95		30.75	39
3100	Heavy duty, 20" diameter		G		105	.305		20.50	14.25		34.75	44.50
3150	24" diameter		G		85	.376		22.50	17.60		40.10	51.50
3200	30" diameter		G		70	.457		25.50	21.50		47	60.50
3250	36" diameter		G		60	.533		27.50	25		52.50	68
3300	48" diameter		G		50	.640		41	30		71	91
3350	60" diameter		G		45	.711		50.50	33		83.50	107
4500	For second and succeeding months, deduct							50%				
5000	Job-built plywood, 8" x 8" columns, 1 use			C-1	165	.194	SFCA	2.44	9.05		11.49	16.60
5050	2 use				195	.164		1.40	7.65		9.05	13.30
5100	3 use				210	.152		.98	7.10		8.08	11.95
5150	4 use				215	.149		.80	6.95		7.75	11.55
5500	12" x 12" columns, 1 use				180	.178		2.37	8.30		10.67	15.35
5550	2 use				210	.152		1.30	7.10		8.40	12.35
5600	3 use				220	.145		.95	6.80		7.75	11.45

For customer support on your Building Construction Costs with RSMeans Data, call 800.448.8182.

53

03 11 Concrete Forming

03 11 13 – Structural Cast-In-Place Concrete Forming

03 11 13.25 Forms In Place, Columns

		Crew	Daily Output	Labor-Hours	Unit	Material	2017 Bare Costs Labor	Equipment	Total	Total Incl O&P
5650	4 use	C-1	225	.142	SFCA	.77	6.65		7.42	11.05
6000	16" x 16" columns, 1 use		185	.173		2.39	8.10		10.49	15
6050	2 use		215	.149		1.27	6.95		8.22	12.05
6100	3 use		230	.139		.96	6.50		7.46	11
6150	4 use		235	.136		.78	6.35		7.13	10.60
6500	24" x 24" columns, 1 use		190	.168		2.71	7.85		10.56	15.05
6550	2 use		216	.148		1.49	6.90		8.39	12.25
6600	3 use		230	.139		1.08	6.50		7.58	11.15
6650	4 use		238	.134		.88	6.30		7.18	10.60
7000	36" x 36" columns, 1 use		200	.160		2.41	7.50		9.91	14.10
7050	2 use		230	.139		1.36	6.50		7.86	11.45
7100	3 use		245	.131		.96	6.10		7.06	10.40
7150	4 use	▼	250	.128	▼	.78	6		6.78	10
7400	Steel framed plywood, based on 50 uses of purchased									
7420	forms, and 4 uses of bracing lumber									
7500	8" x 8" column	C-1	340	.094	SFCA	2.17	4.40		6.57	9.15
7550	10" x 10"		350	.091		1.89	4.27		6.16	8.65
7600	12" x 12"		370	.086		1.61	4.04		5.65	7.95
7650	16" x 16"		400	.080		1.25	3.74		4.99	7.10
7700	20" x 20"		420	.076		1.11	3.56		4.67	6.65
7750	24" x 24"		440	.073		.80	3.40		4.20	6.10
7755	30" x 30"		440	.073		1.02	3.40		4.42	6.30
7760	36" x 36"	▼	460	.070	▼	.89	3.25		4.14	5.95

03 11 13.30 Forms In Place, Culvert

		Crew	Daily Output	Labor-Hours	Unit	Material	2017 Bare Costs Labor	Equipment	Total	Total Incl O&P
0010	**FORMS IN PLACE, CULVERT** R031113-40									
0015	5' to 8' square or rectangular, 1 use	C-1	170	.188	SFCA	4.22	8.80		13.02	18.15
0050	2 use R031113-60		180	.178		2.53	8.30		10.83	15.55
0100	3 use		190	.168		1.97	7.85		9.82	14.20
0150	4 use	▼	200	.160	▼	1.69	7.50		9.19	13.30

03 11 13.35 Forms In Place, Elevated Slabs

		Crew	Daily Output	Labor-Hours	Unit	Material	2017 Bare Costs Labor	Equipment	Total	Total Incl O&P
0010	**FORMS IN PLACE, ELEVATED SLABS** R031113-40									
1000	Flat plate, job-built plywood, to 15' high, 1 use R031113-60	C-2	470	.102	S.F.	3.65	4.89		8.54	11.50
1050	2 use		520	.092		2.01	4.42		6.43	8.95
1100	3 use		545	.088		1.46	4.22		5.68	8.05
1150	4 use		560	.086		1.19	4.11		5.30	7.60
1500	15' to 20' high ceilings, 4 use		495	.097		1.34	4.64		5.98	8.55
1600	21' to 35' high ceilings, 4 use		450	.107		1.66	5.10		6.76	9.65
2000	Flat slab, drop panels, job-built plywood, to 15' high, 1 use		449	.107		4.22	5.10		9.32	12.50
2050	2 use		509	.094		2.32	4.52		6.84	9.45
2100	3 use		532	.090		1.69	4.32		6.01	8.45
2150	4 use		544	.088		1.37	4.23		5.60	8
2250	15' to 20' high ceilings, 4 use		480	.100		2.48	4.79		7.27	10.10
2350	20' to 35' high ceilings, 4 use		435	.110		2.80	5.30		8.10	11.20
3000	Floor slab hung from steel beams, 1 use		485	.099		2.93	4.74		7.67	10.45
3050	2 use		535	.090		2.26	4.30		6.56	9.10
3100	3 use		550	.087		2.03	4.18		6.21	8.65
3150	4 use		565	.085		1.92	4.07		5.99	8.35
3500	Floor slab, with 1-way joist pans, 1 use		415	.116		6.35	5.55		11.90	15.50
3550	2 use		445	.108		4.45	5.15		9.60	12.80
3600	3 use		475	.101		3.81	4.84		8.65	11.60
3650	4 use		500	.096		3.49	4.60		8.09	10.90
4500	With 2-way waffle domes, 1 use	▼	405	.119	▼	6.50	5.70		12.20	15.85

03 11 13 – Structural Cast-In-Place Concrete Forming

03 11 13.35 Forms In Place, Elevated Slabs

		Crew	Daily Output	Labor-Hours	Unit	Material	2017 Bare Costs Labor	Equipment	Total	Total Incl O&P
4520	2 use	C-2	450	.107	S.F.	4.60	5.10		9.70	12.90
4530	3 use		460	.104		3.97	5		8.97	12
4550	4 use		470	.102		3.65	4.89		8.54	11.50
5000	Box out for slab openings, over 16" deep, 1 use		190	.253	SFCA	4.55	12.10		16.65	23.50
5050	2 use		240	.200	"	2.51	9.60		12.11	17.45
5500	Shallow slab box outs, to 10 S.F.		42	1.143	Ea.	11.50	54.50		66	96.50
5550	Over 10 S.F. (use perimeter)		600	.080	L.F.	1.54	3.83		5.37	7.55
6000	Bulkhead forms for slab, with keyway, 1 use, 2 piece		500	.096		2.13	4.60		6.73	9.40
6100	3 piece (see also edge forms)		460	.104		2.32	5		7.32	10.20
6200	Slab bulkhead form, 4-1/2" high, exp metal, w/keyway & stakes G	C-1	1200	.027		.93	1.25		2.18	2.93
6210	5-1/2" high G		1100	.029		1.17	1.36		2.53	3.37
6215	7-1/2" high G		960	.033		1.31	1.56		2.87	3.83
6220	9-1/2" high G		840	.038		1.43	1.78		3.21	4.30
6500	Curb forms, wood, 6" to 12" high, on elevated slabs, 1 use		180	.178	SFCA	1.67	8.30		9.97	14.60
6550	2 use		205	.156		.92	7.30		8.22	12.20
6600	3 use		220	.145		.67	6.80		7.47	11.15
6650	4 use		225	.142		.54	6.65		7.19	10.80
7000	Edge forms to 6" high, on elevated slab, 4 use		500	.064	L.F.	.20	2.99		3.19	4.80
7070	7" to 12" high, 1 use		162	.198	SFCA	1.27	9.25		10.52	15.55
7080	2 use		198	.162		.70	7.55		8.25	12.30
7090	3 use		222	.144		.51	6.75		7.26	10.85
7101	4 use		350	.091		.20	4.27		4.47	6.75
7500	Depressed area forms to 12" high, 4 use		300	.107	L.F.	1.02	4.98		6	8.75
7550	12" to 24" high, 4 use		175	.183		1.38	8.55		9.93	14.60
8000	Perimeter deck and rail for elevated slabs, straight		90	.356		12.85	16.60		29.45	39.50
8050	Curved		65	.492		17.65	23		40.65	54.50
8500	Void forms, round plastic, 8" high x 3" diameter G		450	.071	Ea.	2.02	3.32		5.34	7.30
8550	4" diameter G		425	.075		2.29	3.52		5.81	7.90
8600	6" diameter G		400	.080		4.09	3.74		7.83	10.25
8650	8" diameter G		375	.085		6.75	3.99		10.74	13.55

03 11 13.40 Forms In Place, Equipment Foundations

		Crew	Daily Output	Labor-Hours	Unit	Material	2017 Bare Costs Labor	Equipment	Total	Total Incl O&P
0010	**FORMS IN PLACE, EQUIPMENT FOUNDATIONS** R031113-40									
0020	1 use	C-2	160	.300	SFCA	3.40	14.35		17.75	25.50
0050	2 use R031113-60		190	.253		1.87	12.10		13.97	20.50
0100	3 use		200	.240		1.36	11.50		12.86	19.10
0150	4 use		205	.234		1.11	11.20		12.31	18.40

03 11 13.45 Forms In Place, Footings

		Crew	Daily Output	Labor-Hours	Unit	Material	2017 Bare Costs Labor	Equipment	Total	Total Incl O&P
0010	**FORMS IN PLACE, FOOTINGS** R031113-40									
0020	Continuous wall, plywood, 1 use	C-1	375	.085	SFCA	6.85	3.99		10.84	13.60
0050	2 use R031113-60		440	.073		3.76	3.40		7.16	9.35
0100	3 use		470	.068		2.74	3.18		5.92	7.90
0150	4 use		485	.066		2.23	3.08		5.31	7.15
0500	Dowel supports for footings or beams, 1 use		500	.064	L.F.	.90	2.99		3.89	5.55
1000	Integral starter wall, to 4" high, 1 use		400	.080		.93	3.74		4.67	6.75
1500	Keyway, 4 use, tapered wood, 2" x 4"	1 Carp	530	.015		.22	.74		.96	1.38
1550	2" x 6"		500	.016		.31	.79		1.10	1.55
2000	Tapered plastic		530	.015		1.30	.74		2.04	2.57
2250	For keyway hung from supports, add		150	.053		.90	2.63		3.53	5
3000	Pile cap, square or rectangular, job-built plywood, 1 use	C-1	290	.110	SFCA	2.73	5.15		7.88	10.90
3050	2 use		346	.092		1.50	4.32		5.82	8.25
3100	3 use		371	.086		1.09	4.03		5.12	7.35
3150	4 use		383	.084		.89	3.90		4.79	7

For customer support on your Building Construction Costs with RSMeans Data, call 800.448.8182.

55

03 11 Concrete Forming

03 11 13 – Structural Cast-In-Place Concrete Forming

03 11 13.45 Forms In Place, Footings

		Crew	Daily Output	Labor-Hours	Unit	Material	2017 Bare Costs Labor	Equipment	Total	Total Incl O&P
4000	Triangular or hexagonal, 1 use	C-1	225	.142	SFCA	3.19	6.65		9.84	13.70
4050	2 use		280	.114		1.75	5.35		7.10	10.15
4100	3 use		305	.105		1.27	4.90		6.17	8.90
4150	4 use		315	.102		1.04	4.75		5.79	8.40
5000	Spread footings, job-built lumber, 1 use		305	.105		2.20	4.90		7.10	9.90
5050	2 use		371	.086		1.22	4.03		5.25	7.50
5100	3 use		401	.080		.88	3.73		4.61	6.65
5150	4 use		414	.077		.72	3.61		4.33	6.35
6000	Supports for dowels, plinths or templates, 2' x 2' footing		25	1.280	Ea.	5.50	60		65.50	97.50
6050	4' x 4' footing		22	1.455		11	68		79	116
6100	8' x 8' footing		20	1.600		22	75	·	97	139
6150	12' x 12' footing		17	1.882		36	88		124	175
7000	Plinths, job-built plywood, 1 use		250	.128	SFCA	3.73	6		9.73	13.25
7100	4 use		270	.119	"	1.22	5.55		6.77	9.85

03 11 13.47 Forms In Place, Gas Station Forms

			Crew	Daily Output	Labor-Hours	Unit	Material	2017 Bare Costs Labor	Equipment	Total	Total Incl O&P
0010	**FORMS IN PLACE, GAS STATION FORMS**										
0050	Curb fascia, with template, 12 ga. steel, left in place, 9" high	G	1 Carp	50	.160	L.F.	13.75	7.90		21.65	27
1000	Sign or light bases, 18" diameter, 9" high	G		9	.889	Ea.	86.50	44		130.50	162
1050	30" diameter, 13" high	G		8	1		138	49.50		187.50	227
2000	Island forms, 10' long, 9" high, 3'-6" wide	G	C-1	10	3.200		385	150		535	650
2050	4' wide	G		9	3.556		395	166		561	690
2500	20' long, 9" high, 4' wide	G		6	5.333		635	249		884	1,075
2550	5' wide	G		5	6.400		665	299		964	1,200

03 11 13.50 Forms In Place, Grade Beam

			Crew	Daily Output	Labor-Hours	Unit	Material	2017 Bare Costs Labor	Equipment	Total	Total Incl O&P
0010	**FORMS IN PLACE, GRADE BEAM**	R031113-40									
0020	Job-built plywood, 1 use		C-2	530	.091	SFCA	3.10	4.34		7.44	10.05
0050	2 use	R031113-60		580	.083		1.71	3.96		5.67	7.95
0100	3 use			600	.080		1.24	3.83		5.07	7.20
0150	4 use			605	.079		1.01	3.80		4.81	6.90

03 11 13.55 Forms In Place, Mat Foundation

			Crew	Daily Output	Labor-Hours	Unit	Material	2017 Bare Costs Labor	Equipment	Total	Total Incl O&P
0010	**FORMS IN PLACE, MAT FOUNDATION**	R031113-40									
0020	Job-built plywood, 1 use		C-2	290	.166	SFCA	2.90	7.95		10.85	15.35
0050	2 use	R031113-60		310	.155		1.21	7.40		8.61	12.70
0100	3 use			330	.145		.78	6.95		7.73	11.50
0120	4 use			350	.137		.72	6.55		7.27	10.85

03 11 13.65 Forms In Place, Slab On Grade

			Crew	Daily Output	Labor-Hours	Unit	Material	2017 Bare Costs Labor	Equipment	Total	Total Incl O&P
0010	**FORMS IN PLACE, SLAB ON GRADE**	R031113-40									
1000	Bulkhead forms w/keyway, wood, 6" high, 1 use		C-1	510	.063	L.F.	1.01	2.93		3.94	5.60
1050	2 uses	R031113-60		400	.080		.55	3.74		4.29	6.35
1100	4 uses			350	.091		.33	4.27		4.60	6.90
1400	Bulkhead form for slab, 4-1/2" high, exp metal, incl keyway & stakes	G		1200	.027		.93	1.25		2.18	2.93
1410	5-1/2" high	G		1100	.029		1.17	1.36		2.53	3.37
1420	7-1/2" high	G		960	.033		1.31	1.56		2.87	3.83
1430	9-1/2" high	G		840	.038		1.43	1.78		3.21	4.30
2000	Curb forms, wood, 6" to 12" high, on grade, 1 use			215	.149	SFCA	2.62	6.95		9.57	13.55
2050	2 use			250	.128		1.45	6		7.45	10.75
2100	3 use			265	.121		1.05	5.65		6.70	9.80
2150	4 use			275	.116		.85	5.45		6.30	9.30
3000	Edge forms, wood, 4 use, on grade, to 6" high			600	.053	L.F.	.34	2.49		2.83	4.19
3050	7" to 12" high			435	.074	SFCA	.82	3.44		4.26	6.15
3060	Over 12"			350	.091	"	.91	4.27		5.18	7.55
3500	For depressed slabs, 4 use, to 12" high			300	.107	L.F.	.74	4.98		5.72	8.45

03 11 Concrete Forming

03 11 13 – Structural Cast-In-Place Concrete Forming

03 11 13.65 Forms In Place, Slab On Grade

		Crew	Daily Output	Labor-Hours	Unit	Material	2017 Bare Costs Labor	Equipment	Total	Total Incl O&P
3550	To 24" high	C-1	175	.183	L.F.	.98	8.55		9.53	14.20
4000	For slab blockouts, to 12" high, 1 use		200	.160		.80	7.50		8.30	12.35
4050	To 24" high, 1 use		120	.267		1.02	12.45		13.47	20
4100	Plastic (extruded), to 6" high, multiple use, on grade	▼	800	.040	▼	6.65	1.87		8.52	10.20
5000	Screed, 24 ga. metal key joint, see Section 03 15 16.30									
5020	Wood, incl. wood stakes, 1" x 3"	C-1	900	.036	L.F.	.83	1.66		2.49	3.46
5050	2" x 4"		900	.036	"	.82	1.66		2.48	3.45
6000	Trench forms in floor, wood, 1 use		160	.200	SFCA	1.84	9.35		11.19	16.30
6050	2 use		175	.183		1.01	8.55		9.56	14.20
6100	3 use		180	.178		.74	8.30		9.04	13.55
6150	4 use		185	.173	▼	.60	8.10		8.70	13.05
8760	Void form, corrugated fiberboard, 4" x 12", 4' long	[G]	3000	.011	S.F.	3.14	.50		3.64	4.21
8770	6" x 12", 4' long	▼	3000	.011		3.75	.50		4.25	4.89
8780	1/4" thick hardboard protective cover for void form	2 Carp	1500	.011	▼	.63	.53		1.16	1.50

03 11 13.85 Forms In Place, Walls

		Crew	Daily Output	Labor-Hours	Unit	Material	2017 Bare Costs Labor	Equipment	Total	Total Incl O&P
0010	**FORMS IN PLACE, WALLS** R031113-10									
0100	Box out for wall openings, to 16" thick, to 10 S.F.	C-2	24	2	Ea.	26.50	96		122.50	176
0150	Over 10 S.F. (use perimeter) R031113-40	"	280	.171	L.F.	2.27	8.20		10.47	15.10
0250	Brick shelf, 4" w, add to wall forms, use wall area above shelf									
0260	1 use R031113-60	C-2	240	.200	SFCA	2.44	9.60		12.04	17.40
0300	2 use		275	.175		1.34	8.35		9.69	14.25
0350	4 use		300	.160	▼	.97	7.65		8.62	12.80
0500	Bulkhead, wood with keyway, 1 use, 2 piece	▼	265	.181	L.F.	2.13	8.70		10.83	15.65
0600	Bulkhead forms with keyway, 1 piece expanded metal, 8" wall [G]	C-1	1000	.032		1.31	1.50		2.81	3.73
0610	10" wall [G]		800	.040		1.43	1.87		3.30	4.43
0620	12" wall [G]	▼	525	.061	▼	1.72	2.85		4.57	6.25
0700	Buttress, to 8' high, 1 use	C-2	350	.137	SFCA	4.45	6.55		11	14.95
0750	2 use		430	.112		2.45	5.35		7.80	10.90
0800	3 use		460	.104		1.79	5		6.79	9.60
0850	4 use		480	.100	▼	1.47	4.79		6.26	8.95
1000	Corbel or haunch, to 12" wide, add to wall forms, 1 use		150	.320	L.F.	2.20	15.35		17.55	26
1050	2 use		170	.282		1.21	13.50		14.71	22
1100	3 use		175	.274		.88	13.15		14.03	21
1150	4 use		180	.267		.72	12.75		13.47	20.50
2000	Wall, job-built plywood, to 8' high, 1 use		370	.130	SFCA	2.56	6.20		8.76	12.30
2050	2 use		435	.110		1.64	5.30		6.94	9.90
2100	3 use		495	.097		1.19	4.64		5.83	8.40
2150	4 use		505	.095		.97	4.55		5.52	8
2400	Over 8' to 16' high, 1 use		280	.171		2.85	8.20		11.05	15.75
2450	2 use		345	.139		1.33	6.65		7.98	11.65
2500	3 use		375	.128		.95	6.15		7.10	10.45
2550	4 use		395	.122		.78	5.80		6.58	9.75
2700	Over 16' high, 1 use		235	.204		2.69	9.80		12.49	17.95
2750	2 use		290	.166		1.48	7.95		9.43	13.80
2800	3 use		315	.152		1.07	7.30		8.37	12.40
2850	4 use		330	.145		.87	6.95		7.82	11.60
4000	Radial, smooth curved, job-built plywood, 1 use		245	.196		2.57	9.40		11.97	17.25
4050	2 use		300	.160		1.41	7.65		9.06	13.30
4100	3 use		325	.148		1.03	7.05		8.08	12
4150	4 use		335	.143		.84	6.85		7.69	11.40
4200	Below grade, job-built plywood, 1 use		225	.213		2.69	10.20		12.89	18.60
4210	2 use	▼	225	.213	▼	1.48	10.20		11.68	17.30

For customer support on your Building Construction Costs with RSMeans Data, call 800.448.8182.

57

03 11 13 – Structural Cast-In-Place Concrete Forming

03 11 13.85 Forms In Place, Walls

		Crew	Daily Output	Labor-Hours	Unit	Material	2017 Bare Costs Labor	Equipment	Total	Total Incl O&P
4220	3 use	C-2	225	.213	SFCA	1.23	10.20		11.43	17
4230	4 use		225	.213		.88	10.20		11.08	16.60
4300	Curved, 2' chords, job-built plywood, to 8' high, 1 use		290	.166		2.14	7.95		10.09	14.50
4350	2 use		355	.135		1.18	6.50		7.68	11.20
4400	3 use		385	.125		.86	5.95		6.81	10.10
4450	4 use		400	.120		.70	5.75		6.45	9.55
4500	Over 8' to 16' high, 1 use		290	.166		.94	7.95		8.89	13.20
4525	2 use		355	.135		.52	6.50		7.02	10.45
4550	3 use		385	.125		.37	5.95		6.32	9.55
4575	4 use		400	.120		.31	5.75		6.06	9.15
4600	Retaining wall, battered, job-built plyw'd, to 8' high, 1 use		300	.160		2.02	7.65		9.67	13.95
4650	2 use		355	.135		1.11	6.50		7.61	11.10
4700	3 use		375	.128		.81	6.15		6.96	10.30
4750	4 use		390	.123		.66	5.90		6.56	9.75
4900	Over 8' to 16' high, 1 use		240	.200		2.20	9.60		11.80	17.10
4950	2 use		295	.163		1.21	7.80		9.01	13.30
5000	3 use		305	.157		.88	7.55		8.43	12.50
5050	4 use		320	.150		.72	7.20		7.92	11.80
5500	For gang wall forming, 192 S.F. sections, deduct					10%	10%			
5550	384 S.F. sections, deduct					20%	20%			
7500	Lintel or sill forms, 1 use	1 Carp	30	.267		3.03	13.15		16.18	23.50
7520	2 use		34	.235		1.67	11.60		13.27	19.60
7540	3 use		36	.222		1.21	10.95		12.16	18.10
7560	4 use		37	.216	SFCA	.99	10.65		11.64	17.40
7800	Modular prefabricated plywood, based on 20 uses of purchased									
7820	forms, and 4 uses of bracing lumber									
7860	To 8' high	C-2	800	.060	SFCA	.94	2.87		3.81	5.45
8060	Over 8' to 16' high		600	.080		1	3.83		4.83	6.95
8600	Pilasters, 1 use		270	.178		3	8.50		11.50	16.35
8620	2 use		330	.145		1.65	6.95		8.60	12.45
8640	3 use		370	.130		1.20	6.20		7.40	10.80
8660	4 use		385	.125		.97	5.95		6.92	10.20
9010	Steel framed plywood, based on 50 uses of purchased									
9020	forms, and 4 uses of bracing lumber									
9060	To 8' high	C-2	600	.080	SFCA	.72	3.83		4.55	6.65
9260	Over 8' to 16' high		450	.107		.72	5.10		5.82	8.65
9460	Over 16' to 20' high		400	.120		.72	5.75		6.47	9.60
9475	For elevated walls, add						10%			
9480	For battered walls, 1 side battered, add					10%	10%			
9485	For battered walls, 2 sides battered, add					15%	15%			

03 11 16 – Architectural Cast-in-Place Concrete Forming

03 11 16.13 Concrete Form Liners

		Crew	Daily Output	Labor-Hours	Unit	Material	2017 Bare Costs Labor	Equipment	Total	Total Incl O&P
0010	**CONCRETE FORM LINERS**									
5750	Liners for forms (add to wall forms), ABS plastic									
5800	Aged wood, 4" wide, 1 use	1 Carp	256	.031	SFCA	3.47	1.54		5.01	6.20
5820	2 use		256	.031		1.91	1.54		3.45	4.46
5830	3 use		256	.031		1.39	1.54		2.93	3.89
5840	4 use		256	.031		1.13	1.54		2.67	3.60
5900	Fractured rope rib, 1 use		192	.042		5.05	2.05		7.10	8.70
5925	2 use		192	.042		2.77	2.05		4.82	6.20
5950	3 use		192	.042		2.01	2.05		4.06	5.35
6000	4 use		192	.042		1.63	2.05		3.68	4.94

03 11 Concrete Forming

03 11 16 – Architectural Cast-in-Place Concrete Forming

03 11 16.13 Concrete Form Liners

		Crew	Daily Output	Labor-Hours	Unit	Material	2017 Bare Costs Labor	2017 Bare Costs Equipment	Total	Total Incl O&P
6100	Ribbed, 3/4" deep x 1-1/2" O.C., 1 use	1 Carp	224	.036	SFCA	5.05	1.76		6.81	8.25
6125	2 use		224	.036		2.77	1.76		4.53	5.75
6150	3 use		224	.036		2.01	1.76		3.77	4.90
6200	4 use		224	.036		1.63	1.76		3.39	4.49
6300	Rustic brick pattern, 1 use		224	.036		3.47	1.76		5.23	6.50
6325	2 use		224	.036		1.91	1.76		3.67	4.79
6350	3 use		224	.036		1.39	1.76		3.15	4.22
6400	4 use		224	.036		1.13	1.76		2.89	3.93
6500	3/8" striated, random, 1 use		224	.036		3.47	1.76		5.23	6.50
6525	2 use		224	.036		1.91	1.76		3.67	4.79
6550	3 use		224	.036		1.39	1.76		3.15	4.22
6600	4 use		224	.036		1.13	1.76		2.89	3.93
6850	Random vertical rustication, 1 use		384	.021		6.60	1.03		7.63	8.80
6900	2 use		384	.021		3.62	1.03		4.65	5.55
6925	3 use		384	.021		2.64	1.03		3.67	4.47
6950	4 use		384	.021		2.14	1.03		3.17	3.93
7050	Wood, beveled edge, 3/4" deep, 1 use		384	.021	L.F.	.16	1.03		1.19	1.75
7100	1" deep, 1 use		384	.021	"	.29	1.03		1.32	1.89
7200	4" wide aged cedar, 1 use		256	.031	SFCA	3.47	1.54		5.01	6.20
7300	4" variable depth rough cedar		224	.036	"	5.05	1.76		6.81	8.25

03 11 19 – Insulating Concrete Forming

03 11 19.10 Insulating Forms, Left In Place

			Crew	Daily Output	Labor-Hours	Unit	Material	2017 Bare Costs Labor	2017 Bare Costs Equipment	Total	Total Incl O&P
0010	**INSULATING FORMS, LEFT IN PLACE**										
0020	S.F. is for exterior face, but includes forms for both faces (total R22)										
2000	4" wall, straight block, 16" x 48" (5.33 S.F.)	G	2 Carp	90	.178	Ea.	20.50	8.75		29.25	36
2010	90 corner block, exterior 16" x 38" x 22" (6.67 S.F.)	G		75	.213		24.50	10.50		35	43
2020	45 corner block, exterior 16" x 34" x 18" (5.78 S.F.)	G		75	.213		24	10.50		34.50	42.50
2100	6" wall, straight block, 16" x 48" (5.33 S.F.)	G		90	.178		21	8.75		29.75	36.50
2110	90 corner block, exterior 16" x 32" x 24" (6.22 S.F.)	G		75	.213		27	10.50		37.50	45.50
2120	45 corner block, exterior 16" x 26" x 18" (4.89 S.F.)	G		75	.213		23.50	10.50		34	42
2130	Brick ledge block, 16" x 48" (5.33 S.F.)	G		80	.200		26	9.85		35.85	43.50
2140	Taper top block, 16" x 48" (5.33 S.F.)	G		80	.200		24.50	9.85		34.35	42
2200	8" wall, straight block, 16" x 48" (5.33 S.F.)	G		90	.178		22	8.75		30.75	38
2210	90 corner block, exterior 16" x 34" x 26" (6.67 S.F.)	G		75	.213		29.50	10.50		40	48.50
2220	45 corner block, exterior 16" x 28" x 20" (5.33 S.F.)	G		75	.213		24.50	10.50		35	43
2230	Brick ledge block, 16" x 48" (5.33 S.F.)	G		80	.200		27.50	9.85		37.35	45
2240	Taper top block, 16" x 48" (5.33 S.F.)	G		80	.200		25.50	9.85		35.35	43

03 11 19.60 Roof Deck Form Boards

			Crew	Daily Output	Labor-Hours	Unit	Material	2017 Bare Costs Labor	2017 Bare Costs Equipment	Total	Total Incl O&P
0010	**ROOF DECK FORM BOARDS**	R051223-50									
0050	Includes bulb tee sub-purlins @ 32-5/8" O.C.										
0070	Non-asbestos fiber cement, 5/16" thick		C-13	2950	.008	S.F.	2.92	.43	.04	3.39	3.96
0100	Fiberglass, 1" thick			2700	.009		3.64	.47	.05	4.16	4.82
0500	Wood fiber, 1" thick	G		2700	.009		2.28	.47	.05	2.80	3.32

03 11 23 – Permanent Stair Forming

03 11 23.75 Forms In Place, Stairs

			Crew	Daily Output	Labor-Hours	Unit	Material	2017 Bare Costs Labor	2017 Bare Costs Equipment	Total	Total Incl O&P
0010	**FORMS IN PLACE, STAIRS**	R031113-40									
0015	(Slant length x width), 1 use		C-2	165	.291	S.F.	5.90	13.95		19.85	28
0050	2 use	R031113-60		170	.282		3.34	13.50		16.84	24
0100	3 use			180	.267		2.50	12.75		15.25	22.50
0150	4 use			190	.253		2.08	12.10		14.18	21
1000	Alternate pricing method (1.0 L.F./S.F.), 1 use			100	.480	LF Rsr	5.90	23		28.90	41.50

For customer support on your Building Construction Costs with RSMeans Data, call 800.448.8182.

59

03 11 Concrete Forming

03 11 23 – Permanent Stair Forming

03 11 23.75 Forms In Place, Stairs		Crew	Daily Output	Labor-Hours	Unit	Material	2017 Bare Costs Labor	Equipment	Total	Total Incl O&P
1050	2 use	C-2	105	.457	LF Rsr	3.34	22		25.34	37
1100	3 use		110	.436		2.50	21		23.50	35
1150	4 use		115	.417		2.08	20		22.08	33
2000	Stairs, cast on sloping ground (length x width), 1 use		220	.218	S.F.	2.28	10.45		12.73	18.50
2025	2 use		232	.207		1.26	9.90		11.16	16.60
2050	3 use		244	.197		.91	9.40		10.31	15.45
2100	4 use		256	.188		.74	9		9.74	14.55

03 15 Concrete Accessories

03 15 05 – Concrete Forming Accessories

03 15 05.12 Chamfer Strips

		Crew	Daily Output	Labor-Hours	Unit	Material	2017 Bare Costs Labor	Equipment	Total	Total Incl O&P
0010	**CHAMFER STRIPS**									
2000	Polyvinyl chloride, 1/2" wide with leg	1 Carp	535	.015	L.F.	.68	.74		1.42	1.88
2200	3/4" wide with leg		525	.015		.74	.75		1.49	1.96
2400	1" radius with leg		515	.016		.78	.76		1.54	2.03
2800	2" radius with leg		500	.016		1.50	.79		2.29	2.86
5000	Wood, 1/2" wide		535	.015		.13	.74		.87	1.27
5200	3/4" wide		525	.015		.16	.75		.91	1.33
5400	1" wide		515	.016		.29	.76		1.05	1.49

03 15 05.15 Column Form Accessories

			Crew	Daily Output	Labor-Hours	Unit	Material	2017 Bare Costs Labor	Equipment	Total	Total Incl O&P
0010	**COLUMN FORM ACCESSORIES**										
1000	Column clamps, adjustable to 24" x 24", buy	G				Set	160			160	176
1100	Rent per month	G					11.15			11.15	12.25
1300	For sizes to 30" x 30", buy	G					187			187	206
1400	Rent per month	G					13.05			13.05	14.35
1600	For sizes to 36" x 36", buy	G					233			233	256
1700	Rent per month	G					16.10			16.10	17.70
2000	Bar type with wedges, 36" x 36", buy	G					145			145	160
2100	Rent per month	G					10.15			10.15	11.15
2300	48" x 48", buy	G					200			200	220
2400	Rent per month	G					14			14	15.40
3000	Scissor type with wedges, 36" x 36", buy	G					125			125	138
3100	Rent per month	G					12.50			12.50	13.75
3300	60" x 60", buy	G					178			178	195
3400	Rent per month	G					17.75			17.75	19.55
4000	Friction collars 2'-6" diam., buy	G					2,700			2,700	2,975
4100	Rent per month	G					189			189	208
4300	4'-0" diam., buy	G					3,300			3,300	3,625
4400	Rent per month	G					231			231	254

03 15 05.30 Hangers

			Crew	Daily Output	Labor-Hours	Unit	Material	2017 Bare Costs Labor	Equipment	Total	Total Incl O&P
0010	**HANGERS**										
0020	Slab and beam form										
0500	Banding iron										
0550	3/4" x 22 ga., 14 L.F. per lb. or 1/2" x 14 ga., 7 L.F. per lb.	G				Lb.	1.38			1.38	1.52
1000	Fascia ties, coil type, to 24" long	G				C	465			465	515
1500	Frame ties to 8-1/8"	G					545			545	600
1550	8-1/8" to 10-1/8"	G					575			575	635
5000	Snap tie hanger, to 30" overall length, 3000#	G					465			465	515
5050	To 36" overall length	G					525			525	580
5100	To 48" overall length	G					645			645	710

03 15 Concrete Accessories

03 15 05 – Concrete Forming Accessories

03 15 05.30 Hangers

		Crew	Daily Output	Labor-Hours	Unit	Material	2017 Bare Costs Labor	Equipment	Total	Total Incl O&P
5500	Steel beam hanger									
5600	Flange to 8-1/8"	G				C	545			545	600
5650	8-1/8" to 10-1/8"	G				"	575			575	635
5900	Coil threaded rods, continuous, 1/2" diameter	G				L.F.	1.41			1.41	1.55
6000	Tie hangers to 30" overall length, 4000#	G				C	530			530	580
6100	To 36" overall length	G					590			590	645
6500	Tie back hanger, up to 12-1/8" flange	G				↓	1,475			1,475	1,625
8500	Wire, black annealed, 15 gage	G				Cwt.	154			154	169
8600	16 ga	G				"	170			170	187

03 15 05.70 Shores

		Crew	Daily Output	Labor-Hours	Unit	Material	2017 Bare Costs Labor	Equipment	Total	Total Incl O&P
0010	**SHORES**									
0020	Erect and strip, by hand, horizontal members									
0500	Aluminum joists and stringers	G	2 Carp	60	.267	Ea.		13.15		13.15	20
0600	Steel, adjustable beams	G		45	.356			17.50		17.50	27
0700	Wood joists			50	.320			15.75		15.75	24
0800	Wood stringers			30	.533			26.50		26.50	40
1000	Vertical members to 10' high	G		55	.291			14.35		14.35	22
1050	To 13' high	G		50	.320			15.75		15.75	24
1100	To 16' high	G		45	.356	↓		17.50		17.50	27
1500	Reshoring	G		1400	.011	S.F.	.61	.56		1.17	1.53
1600	Flying truss system	G	C-17D	9600	.009	SFCA		.46	.08	.54	.79
1760	Horizontal, aluminum joists, 6-1/4" high x 5' to 21' span, buy	G				L.F.	16.25			16.25	17.85
1770	Beams, 7-1/4" high x 4' to 30' span	G				"	19			19	21
1810	Horizontal, steel beam, W8x10, 7' span, buy	G				Ea.	61			61	67.50
1830	10' span	G					71			71	78
1920	15' span	G					122			122	135
1940	20' span	G					172			172	189
1970	Steel stringer, W8x10, 4' to 16' span, buy	G				L.F.	7.10			7.10	7.85
3000	Rent for job duration, aluminum joist @ 2' O.C., per mo.	G				SF Flr.	.41			.41	.45
3050	Steel W8x10	G					.18			.18	.20
3060	Steel adjustable	G				↓	.18			.18	.20
3500	#1 post shore, steel, 5'-7" to 9'-6" high, 10,000# cap., buy	G				Ea.	155			155	171
3550	#2 post shore, 7'-3" to 12'-10" high, 7800# capacity	G					180			180	198
3600	#3 post shore, 8'-10" to 16'-1" high, 3800# capacity	G				↓	196			196	216
5010	Frame shoring systems, steel, 12,000#/leg, buy										
5040	Frame, 2' wide x 6' high	G				Ea.	102			102	112
5250	X-brace	G					17			17	18.70
5550	Base plate	G					17			17	18.70
5600	Screw jack	G					36			36	39.50
5650	U-head, 8" x 8"	G				↓	21			21	23

03 15 05.75 Sleeves and Chases

		Crew	Daily Output	Labor-Hours	Unit	Material	2017 Bare Costs Labor	Equipment	Total	Total Incl O&P
0010	**SLEEVES AND CHASES**									
0100	Plastic, 1 use, 12" long, 2" diameter	1 Carp	100	.080	Ea.	1.62	3.94		5.56	7.85
0150	4" diameter		90	.089		4.54	4.38		8.92	11.70
0200	6" diameter		75	.107		7.80	5.25		13.05	16.60
0250	12" diameter	↓	60	.133	↓	23.50	6.55		30.05	36

03 15 05.80 Snap Ties

		Crew	Daily Output	Labor-Hours	Unit	Material	2017 Bare Costs Labor	Equipment	Total	Total Incl O&P
0010	**SNAP TIES**, 8-1/4" L&W (Lumber and wedge)									
0100	2250 lb., w/flat washer, 8" wall	G				C	93			93	102
0150	10" wall	G					135			135	149
0200	12" wall	G					140			140	154
0250	16" wall	G				↓	155			155	171

03 15 05.80 Snap Ties

		Crew	Daily Output	Labor-Hours	Unit	Material	2017 Bare Costs Labor	Equipment	Total	Total Incl O&P
0300	18" wall	G			C	161			161	177
0500	With plastic cone, 8" wall	G				82			82	90
0550	10" wall	G				85			85	93.50
0600	12" wall	G				92			92	101
0650	16" wall	G				101			101	111
0700	18" wall	G				104			104	114
1000	3350 lb., w/flat washer, 8" wall	G				168			168	185
1100	10" wall	G				184			184	202
1150	12" wall	G				188			188	207
1200	16" wall	G				216			216	238
1250	18" wall	G				225			225	248
1500	With plastic cone, 8" wall	G				136			136	150
1550	10" wall	G				149			149	164
1600	12" wall	G				153			153	168
1650	16" wall	G				175			175	193
1700	18" wall	G				181			181	199

03 15 05.85 Stair Tread Inserts

		Crew	Daily Output	Labor-Hours	Unit	Material	2017 Bare Costs Labor	Equipment	Total	Total Incl O&P
0010	**STAIR TREAD INSERTS**									
0105	Cast nosing insert, abrasive surface, pre-drilled, includes screws									
0110	Aluminum, 3" wide x 3' long	1 Cefi	32	.250	Ea.	58	11.70		69.70	81
0120	4' long		31	.258		74	12.10		86.10	99
0130	5' long		30	.267		91	12.50		103.50	119
0135	Extruded nosing insert, black abrasive strips, continuous anchor									
0140	Aluminum, 3" wide x 3' long	1 Cefi	64	.125	Ea.	41	5.85		46.85	53.50
0150	4' long		60	.133		58	6.25		64.25	73
0160	5' long		56	.143		65	6.70		71.70	81.50
0165	Extruded nosing insert, black abrasive strips, pre-drilled, incl. screws									
0170	Aluminum, 3" wide x 3' long	1 Cefi	32	.250	Ea.	51.50	11.70		63.20	74
0180	4' long		31	.258		67	12.10		79.10	91.50
0190	5' long		30	.267		78.50	12.50		91	105

03 15 05.95 Wall and Foundation Form Accessories

		Crew	Daily Output	Labor-Hours	Unit	Material	2017 Bare Costs Labor	Equipment	Total	Total Incl O&P
0010	**WALL AND FOUNDATION FORM ACCESSORIES**									
2000	Footings, turnbuckle form aligner	G			Ea.	16.65			16.65	18.30
2050	Spreaders for footer, adjustable	G			"	25.50			25.50	28
3000	Form oil, up to 1200 S.F. per gallon coverage				Gal.	13.55			13.55	14.90
3050	Up to 800 S.F. per gallon				"	20.50			20.50	22.50
3500	Form patches, 1-3/4" diameter				C	28			28	31
3550	2-3/4" diameter				"	48			48	53
4000	Nail stakes, 3/4" diameter, 18" long	G			Ea.	3.40			3.40	3.74
4050	24" long	G				4.39			4.39	4.83
4200	30" long	G				5.65			5.65	6.20
4250	36" long	G				6.80			6.80	7.45

03 15 13 – Waterstops

03 15 13.50 Waterstops

		Crew	Daily Output	Labor-Hours	Unit	Material	2017 Bare Costs Labor	Equipment	Total	Total Incl O&P
0010	**WATERSTOPS**, PVC and Rubber									
0020	PVC, ribbed 3/16" thick, 4" wide	1 Carp	155	.052	L.F.	1.36	2.54		3.90	5.40
0050	6" wide		145	.055		2.33	2.72		5.05	6.70
0500	With center bulb, 6" wide, 3/16" thick		135	.059		2.34	2.92		5.26	7.05
0550	3/8" thick		130	.062		4.27	3.03		7.30	9.35
0600	9" wide x 3/8" thick		125	.064		6.95	3.15		10.10	12.50
0800	Dumbbell type, 6" wide, 3/16" thick		150	.053		1.98	2.63		4.61	6.20
0850	3/8" thick		145	.055		3.75	2.72		6.47	8.30

03 15 13 – Waterstops

03 15 13.50 Waterstops	Crew	Daily Output	Labor-Hours	Unit	Material	2017 Bare Costs Labor	Equipment	Total	Total Incl O&P
1000	9" wide, 3/8" thick, plain	1 Carp	130	.062	L.F.	5.80	3.03		8.83	11.05
1050	Center bulb		130	.062		8.25	3.03		11.28	13.75
1250	Ribbed type, split, 3/16" thick, 6" wide		145	.055		1.94	2.72		4.66	6.30
1300	3/8" thick		130	.062		4.49	3.03		7.52	9.60
2000	Rubber, flat dumbbell, 3/8" thick, 6" wide		145	.055		11.55	2.72		14.27	16.85
2050	9" wide		135	.059		15.70	2.92		18.62	21.50
2500	Flat dumbbell split, 3/8" thick, 6" wide		145	.055		1.94	2.72		4.66	6.30
2550	9" wide		135	.059		4.49	2.92		7.41	9.40
3000	Center bulb, 1/4" thick, 6" wide		145	.055		10.25	2.72		12.97	15.40
3050	9" wide		135	.059		21.50	2.92		24.42	28
3500	Center bulb split, 3/8" thick, 6" wide		145	.055		9.85	2.72		12.57	15
3550	9" wide		135	.059		17.05	2.92		19.97	23
5000	Waterstop fittings, rubber, flat									
5010	Dumbbell or center bulb, 3/8" thick,									
5200	Field union, 6" wide	1 Carp	50	.160	Ea.	29.50	7.90		37.40	44.50
5250	9" wide		50	.160		32.50	7.90		40.40	48
5500	Flat cross, 6" wide		30	.267		47	13.15		60.15	72
5550	9" wide		30	.267		64.50	13.15		77.65	91
6000	Flat tee, 6" wide		30	.267		45	13.15		58.15	69.50
6050	9" wide		30	.267		59	13.15		72.15	85
6500	Flat ell, 6" wide		40	.200		44	9.85		53.85	63.50
6550	9" wide		40	.200		54.50	9.85		64.35	75
7000	Vertical tee, 6" wide		25	.320		36.50	15.75		52.25	64
7050	9" wide		25	.320		44	15.75		59.75	72.50
7500	Vertical ell, 6" wide		35	.229		31.50	11.25		42.75	52.50
7550	9" wide		35	.229		38	11.25		49.25	59.50

03 15 16 – Concrete Construction Joints

03 15 16.20 Control Joints, Saw Cut

		Crew	Daily Output	Labor-Hours	Unit	Material	Labor	Equipment	Total	Total Incl O&P
0010	**CONTROL JOINTS, SAW CUT**									
0100	Sawcut control joints in green concrete									
0120	1" depth	C-27	2000	.008	L.F.	.04	.37	.08	.49	.68
0140	1-1/2" depth		1800	.009		.05	.42	.09	.56	.77
0160	2" depth		1600	.010		.07	.47	.10	.64	.89
0180	Sawcut joint reservoir in cured concrete									
0182	3/8" wide x 3/4" deep, with single saw blade	C-27	1000	.016	L.F.	.05	.75	.15	.95	1.34
0184	1/2" wide x 1" deep, with double saw blades		900	.018		.10	.83	.17	1.10	1.54
0186	3/4" wide x 1-1/2" deep, with double saw blades		800	.020		.21	.94	.19	1.34	1.83
0190	Water blast joint to wash away laitance, 2 passes	C-29	2500	.003			.13	.03	.16	.22
0200	Air blast joint to blow out debris and air dry, 2 passes	C-28	2000	.004			.19	.01	.20	.29
0300	For backer rod, see Section 07 91 23.10									
0340	For joint sealant, see Section 03 15 16.30 or 07 92 13.20									
0900	For replacement of joint sealant, see Section 07 01 90.81									

03 15 16.30 Expansion Joints

			Crew	Daily Output	Labor-Hours	Unit	Material	Labor	Equipment	Total	Total Incl O&P
0010	**EXPANSION JOINTS**										
0020	Keyed, cold, 24 ga., incl. stakes, 3-1/2" high	G	1 Carp	200	.040	L.F.	.87	1.97		2.84	3.98
0050	4-1/2" high	G		200	.040		.93	1.97		2.90	4.04
0100	5-1/2" high	G		195	.041		1.17	2.02		3.19	4.39
0150	7-1/2" high	G		190	.042		1.31	2.07		3.38	4.62
0160	9-1/2" high	G		185	.043		1.43	2.13		3.56	4.83
0300	Poured asphalt, plain, 1/2" x 1"		1 Clab	450	.018		.85	.70		1.55	2.01
0350	1" x 2"			400	.020		3.38	.78		4.16	4.92
0500	Neoprene, liquid, cold applied, 1/2" x 1"			450	.018		2.28	.70		2.98	3.58

03 15 Concrete Accessories

03 15 16 – Concrete Construction Joints

03 15 16.30 Expansion Joints

	03 15 16.30 Expansion Joints		Crew	Daily Output	Labor-Hours	Unit	Material	2017 Bare Costs Labor	Equipment	Total	Total Incl O&P
0550	1" x 2"		1 Clab	400	.020	L.F.	9.10	.78		9.88	11.25
0700	Polyurethane, poured, 2 part, 1/2" x 1"			400	.020		1.38	.78		2.16	2.72
0750	1" x 2"			350	.023		5.55	.90		6.45	7.45
0900	Rubberized asphalt, hot or cold applied, 1/2" x 1"			450	.018		.44	.70		1.14	1.55
0950	1" x 2"			400	.020		1.74	.78		2.52	3.11
1100	Hot applied, fuel resistant, 1/2" x 1"			450	.018		.66	.70		1.36	1.80
1150	1" x 2"		▼	400	.020		2.61	.78		3.39	4.07
2000	Premolded, bituminous fiber, 1/2" x 6"		1 Carp	375	.021		.41	1.05		1.46	2.06
2050	1" x 12"			300	.027		1.97	1.31		3.28	4.18
2140	Concrete expansion joint, recycled paper and fiber, 1/2" x 6"	G		390	.021		.43	1.01		1.44	2.03
2150	1/2" x 12"	G		360	.022		.87	1.09		1.96	2.64
2250	Cork with resin binder, 1/2" x 6"			375	.021		1.18	1.05		2.23	2.91
2300	1" x 12"			300	.027		3.30	1.31		4.61	5.65
2500	Neoprene sponge, closed cell, 1/2" x 6"			375	.021		2.41	1.05		3.46	4.26
2550	1" x 12"			300	.027		8.45	1.31		9.76	11.25
2750	Polyethylene foam, 1/2" x 6"			375	.021		.64	1.05		1.69	2.31
2800	1" x 12"			300	.027		2.19	1.31		3.50	4.42
3000	Polyethylene backer rod, 3/8" diameter			460	.017		.03	.86		.89	1.35
3050	3/4" diameter			460	.017		.07	.86		.93	1.38
3100	1" diameter			460	.017		.12	.86		.98	1.44
3500	Polyurethane foam, with polybutylene, 1/2" x 1/2"			475	.017		1.16	.83		1.99	2.55
3550	1" x 1"			450	.018		2.94	.88		3.82	4.57
3750	Polyurethane foam, regular, closed cell, 1/2" x 6"			375	.021		.86	1.05		1.91	2.56
3800	1" x 12"			300	.027		3.06	1.31		4.37	5.40
4000	Polyvinyl chloride foam, closed cell, 1/2" x 6"			375	.021		2.33	1.05		3.38	4.17
4050	1" x 12"			300	.027		8	1.31		9.31	10.80
4250	Rubber, gray sponge, 1/2" x 6"			375	.021		1.96	1.05		3.01	3.77
4300	1" x 12"			300	.027		7.05	1.31		8.36	9.75
4400	Redwood heartwood, 1" x 4"			400	.020		1.16	.99		2.15	2.79
4450	1" x 6"		▼	375	.021	▼	1.75	1.05		2.80	3.54
5000	For installation in walls, add							75%			
5250	For installation in boxouts, add							25%			

03 15 19 – Cast-In Concrete Anchors

03 15 19.05 Anchor Bolt Accessories

	03 15 19.05 Anchor Bolt Accessories		Crew	Daily Output	Labor-Hours	Unit	Material	Labor	Equipment	Total	Total Incl O&P
0010	**ANCHOR BOLT ACCESSORIES**										
0015	For anchor bolts set in fresh concrete, see Section 03 15 19.10										
8150	Anchor bolt sleeve, plastic, 1" diam. bolts		1 Carp	60	.133	Ea.	12.95	6.55		19.50	24.50
8500	1-1/2" diameter			28	.286		19.20	14.05		33.25	42.50
8600	2" diameter			24	.333		19.45	16.40		35.85	46.50
8650	3" diameter		▼	20	.400	▼	35.50	19.70		55.20	69

03 15 19.10 Anchor Bolts

	03 15 19.10 Anchor Bolts		Crew	Daily Output	Labor-Hours	Unit	Material	Labor	Equipment	Total	Total Incl O&P
0010	**ANCHOR BOLTS**										
0015	Made from recycled materials										
0025	Single bolts installed in fresh concrete, no templates										
0030	Hooked w/nut and washer, 1/2" diameter, 8" long	G	1 Carp	132	.061	Ea.	1.38	2.99		4.37	6.10
0040	12" long	G		131	.061		1.53	3.01		4.54	6.30
0050	5/8" diameter, 8" long	G		129	.062		3.35	3.05		6.40	8.35
0060	12" long	G		127	.063		4.12	3.10		7.22	9.30
0070	3/4" diameter, 8" long	G		127	.063		4.12	3.10		7.22	9.30
0080	12" long	G	▼	125	.064		5.15	3.15		8.30	10.50
0090	2-bolt pattern, including job-built 2-hole template, per set										
0100	J-type, incl. hex nut & washer, 1/2" diameter x 6" long	G	1 Carp	21	.381	Set	5.60	18.75		24.35	34.50

For customer support on your Building Construction Costs with RSMeans Data, call 800.448.8182.

03 15 19.10 Anchor Bolts			Crew	Daily Output	Labor-Hours	Unit	Material	2017 Bare Costs Labor	Equipment	Total	Total Incl O&P
0110	12" long	G	1 Carp	21	.381	Set	6.20	18.75		24.95	35.50
0120	18" long	G		21	.381		7.10	18.75		25.85	36.50
0130	3/4" diameter x 8" long	G		20	.400		11.40	19.70		31.10	42.50
0140	12" long	G		20	.400		13.45	19.70		33.15	45
0150	18" long	G		20	.400		16.55	19.70		36.25	48
0160	1" diameter x 12" long	G		19	.421		22	20.50		42.50	56
0170	18" long	G		19	.421		25.50	20.50		46	60.50
0180	24" long	G		19	.421		30.50	20.50		51	65.50
0190	36" long	G		18	.444		41	22		63	78.50
0200	1-1/2" diameter x 18" long	G		17	.471		42	23		65	81.50
0210	24" long	G		16	.500		49	24.50		73.50	91.50
0300	L-type, incl. hex nut & washer, 3/4" diameter x 12" long	G		20	.400		13.60	19.70		33.30	45
0310	18" long	G		20	.400		16.55	19.70		36.25	48
0320	24" long	G		20	.400		19.50	19.70		39.20	51.50
0330	30" long	G		20	.400		24	19.70		43.70	56.50
0340	36" long	G		20	.400		27	19.70		46.70	59.50
0350	1" diameter x 12" long	G		19	.421		21	20.50		41.50	55
0360	18" long	G		19	.421		25.50	20.50		46	60
0370	24" long	G		19	.421		30.50	20.50		51	65.50
0380	30" long	G		19	.421		35.50	20.50		56	71
0390	36" long	G		18	.444		40.50	22		62.50	78
0400	42" long	G		18	.444		48.50	22		70.50	86.50
0410	48" long	G		18	.444		54	22		76	92.50
0420	1-1/4" diameter x 18" long	G		18	.444		30	22		52	66.50
0430	24" long	G		18	.444		35	22		57	72
0440	30" long	G		17	.471		40	23		63	79.50
0450	36" long	G		17	.471		45	23		68	85
0460	42" long	G	2 Carp	32	.500		50.50	24.50		75	93
0470	48" long	G		32	.500		57	24.50		81.50	101
0480	54" long	G		31	.516		67	25.50		92.50	113
0490	60" long	G		31	.516		73	25.50		98.50	120
0500	1-1/2" diameter x 18" long	G		33	.485		46	24		70	87
0510	24" long	G		32	.500		53	24.50		77.50	96
0520	30" long	G		31	.516		59.50	25.50		85	105
0530	36" long	G		30	.533		68	26.50		94.50	115
0540	42" long	G		30	.533		77	26.50		103.50	125
0550	48" long	G		29	.552		86.50	27		113.50	137
0560	54" long	G		28	.571		105	28		133	158
0570	60" long	G		28	.571		114	28		142	169
0580	1-3/4" diameter x 18" long	G		31	.516		75	25.50		100.50	122
0590	24" long	G		30	.533		87.50	26.50		114	136
0600	30" long	G		29	.552		101	27		128	153
0610	36" long	G		28	.571		115	28		143	169
0620	42" long	G		27	.593		129	29		158	187
0630	48" long	G		26	.615		141	30.50		171.50	202
0640	54" long	G		26	.615		174	30.50		204.50	239
0650	60" long	G		25	.640		188	31.50		219.50	256
0660	2" diameter x 24" long	G		27	.593		112	29		141	168
0670	30" long	G		27	.593		125	29		154	183
0680	36" long	G		26	.615		137	30.50		167.50	198
0690	42" long	G		25	.640		153	31.50		184.50	217
0700	48" long	G		24	.667		175	33		208	244
0710	54" long	G		23	.696		208	34.50		242.50	282

03 15 19.10 Anchor Bolts		Crew	Daily Output	Labor-Hours	Unit	Material	2017 Bare Costs Labor	Equipment	Total	Total Incl O&P
0720	60" long	G 2 Carp	23	.696	Set	224	34.50		258.50	299
0730	66" long	G	22	.727		239	36		275	320
0740	72" long	G ▼	21	.762	▼	261	37.50		298.50	345
1000	4-bolt pattern, including job-built 4-hole template, per set									
1100	J-type, incl. hex nut & washer, 1/2" diameter x 6" long	G 1 Carp	19	.421	Set	8.05	20.50		28.55	41
1110	12" long	G	19	.421		9.25	20.50		29.75	42
1120	18" long	G	18	.444		11.10	22		33.10	45.50
1130	3/4" diameter x 8" long	G	17	.471		19.60	23		42.60	57
1140	12" long	G	17	.471		23.50	23		46.50	61.50
1150	18" long	G	17	.471		30	23		53	68.50
1160	1" diameter x 12" long	G	16	.500		40.50	24.50		65	82
1170	18" long	G	15	.533		48.50	26.50		75	93
1180	24" long	G	15	.533		58	26.50		84.50	104
1190	36" long	G	15	.533		78.50	26.50		105	127
1200	1-1/2" diameter x 18" long	G	13	.615		80.50	30.50		111	135
1210	24" long	G	12	.667		95	33		128	156
1300	L-type, incl. hex nut & washer, 3/4" diameter x 12" long	G	17	.471		24	23		47	62
1310	18" long	G	17	.471		30	23		53	68.50
1320	24" long	G	17	.471		36	23		59	75
1330	30" long	G	16	.500		44.50	24.50		69	86.50
1340	36" long	G	16	.500		50.50	24.50		75	93
1350	1" diameter x 12" long	G	16	.500		39	24.50		63.50	80.50
1360	18" long	G	15	.533		47.50	26.50		74	92.50
1370	24" long	G	15	.533		58	26.50		84.50	104
1380	30" long	G	15	.533		68	26.50		94.50	115
1390	36" long	G	15	.533		77.50	26.50		104	125
1400	42" long	G	14	.571		93.50	28		121.50	146
1410	48" long	G	14	.571		104	28		132	158
1420	1-1/4" diameter x 18" long	G	14	.571		56.50	28		84.50	106
1430	24" long	G	14	.571		67	28		95	117
1440	30" long	G	13	.615		77	30.50		107.50	131
1450	36" long	G ▼	13	.615		87	30.50		117.50	142
1460	42" long	G 2 Carp	25	.640		98	31.50		129.50	157
1470	48" long	G	24	.667		111	33		144	173
1480	54" long	G	23	.696		130	34.50		164.50	197
1490	60" long	G	23	.696		143	34.50		177.50	210
1500	1-1/2" diameter x 18" long	G	25	.640		89	31.50		120.50	147
1510	24" long	G	24	.667		103	33		136	165
1520	30" long	G	23	.696		116	34.50		150.50	181
1530	36" long	G	22	.727		133	36		169	201
1540	42" long	G	22	.727		151	36		187	221
1550	48" long	G	21	.762		169	37.50		206.50	244
1560	54" long	G	20	.800		206	39.50		245.50	287
1570	60" long	G	20	.800		225	39.50		264.50	310
1580	1-3/4" diameter x 18" long	G	22	.727		147	36		183	216
1590	24" long	G	21	.762		171	37.50		208.50	247
1600	30" long	G	21	.762		199	37.50		236.50	277
1610	36" long	G	20	.800		227	39.50		266.50	310
1620	42" long	G	19	.842		254	41.50		295.50	345
1630	48" long	G	18	.889		279	44		323	370
1640	54" long	G	18	.889		345	44		389	445
1650	60" long	G	17	.941		375	46.50		421.50	480
1660	2" diameter x 24" long	G ▼	19	.842	▼	220	41.50		261.50	305

03 15 19 – Cast-In Concrete Anchors

03 15 19.10 Anchor Bolts		Crew	Daily Output	Labor-Hours	Unit	Material	2017 Bare Costs Labor	Equipment	Total	Total Incl O&P
1670	30" long	G	2 Carp	18	.889	Set	248	44		292	340
1680	36" long	G		18	.889		272	44		316	365
1690	42" long	G		17	.941		305	46.50		351.50	405
1700	48" long	G		16	1		350	49.50		399.50	455
1710	54" long	G		15	1.067		415	52.50		467.50	535
1720	60" long	G		15	1.067		445	52.50		497.50	570
1730	66" long	G		14	1.143		475	56.50		531.50	605
1740	72" long	G		14	1.143		520	56.50		576.50	655
1990	For galvanized, add					Ea.	75%				

03 15 19.20 Dovetail Anchor System		Crew	Daily Output	Labor-Hours	Unit	Material	2017 Bare Costs Labor	Equipment	Total	Total Incl O&P
0010	**DOVETAIL ANCHOR SYSTEM**										
0500	Dovetail anchor slot, galvanized, foam-filled, 26 ga.	G	1 Carp	425	.019	L.F.	1.17	.93		2.10	2.71
0600	24 ga.	G		400	.020		1.59	.99		2.58	3.26
0625	22 ga.	G		400	.020		1.86	.99		2.85	3.56
0900	Stainless steel, foam-filled, 26 ga.	G		375	.021		1.87	1.05		2.92	3.67
1200	Dovetail brick anchor, corrugated, galvanized, 3-1/2" long, 16 ga.	G	1 Bric	10.50	.762	C	40.50	36.50		77	101
1300	12 ga.	G		10.50	.762		60	36.50		96.50	122
1500	Seismic, galvanized, 3-1/2" long, 16 ga.	G		10.50	.762		67.50	36.50		104	130
1600	12 ga.	G		10.50	.762		79.50	36.50		116	144
2000	Dovetail cavity wall, corrugated, galvanized, 5-1/2" long, 16 ga.	G		10.50	.762		41	36.50		77.50	101
2100	12 ga.	G		10.50	.762		59.50	36.50		96	122
3000	Dovetail furring anchors, corrugated, galvanized, 1-1/2" long, 16 ga.	G		10.50	.762		24	36.50		60.50	82.50
3100	12 ga.	G		10.50	.762		31	36.50		67.50	90
6000	Dovetail stone panel anchors, galvanized, 1/8" x 1" wide, 3-1/2" long	G		10.50	.762		99.50	36.50		136	165
6100	1/4" x 1" wide	G		10.50	.762		164	36.50		200.50	236

03 15 19.30 Inserts		Crew	Daily Output	Labor-Hours	Unit	Material	2017 Bare Costs Labor	Equipment	Total	Total Incl O&P
0010	**INSERTS**										
1000	Inserts, slotted nut type for 3/4" bolts, 4" long	G	1 Carp	84	.095	Ea.	20	4.69		24.69	29.50
2100	6" long	G		84	.095		23	4.69		27.69	32
2150	8" long	G		84	.095		30.50	4.69		35.19	40.50
2200	Slotted, strap type, 4" long	G		84	.095		22	4.69		26.69	31
2250	6" long	G		84	.095		24.50	4.69		29.19	34
2300	8" long	G		84	.095		32.50	4.69		37.19	43
2350	Strap for slotted insert, 4" long	G		84	.095		11.90	4.69		16.59	20.50
4100	6" long	G		84	.095		12.95	4.69		17.64	21.50
4150	8" long	G		84	.095		16.15	4.69		20.84	25
4200	10" long	G		84	.095		19.60	4.69		24.29	28.50
7000	Loop ferrule type										
7100	1/4" diameter bolt	G	1 Carp	84	.095	Ea.	2.32	4.69		7.01	9.75
7350	7/8" diameter bolt	G	"	84	.095	"	10.80	4.69		15.49	19.05
9000	Wedge type										
9100	For 3/4" diameter bolt	G	1 Carp	60	.133	Ea.	9.70	6.55		16.25	20.50
9800	Cut washers, black										
9900	3/4" bolt	G				Ea.	1.21			1.21	1.33
9950	For galvanized inserts, add						30%				

03 15 19.45 Machinery Anchors		Crew	Daily Output	Labor-Hours	Unit	Material	2017 Bare Costs Labor	Equipment	Total	Total Incl O&P
0010	**MACHINERY ANCHORS**, heavy duty, incl. sleeve, floating base nut,										
0020	lower stud & coupling nut, fiber plug, connecting stud, washer & nut.										
0030	For flush mounted embedment in poured concrete heavy equip. pads.										
0200	Stud & bolt, 1/2" diameter	G	E-16	40	.400	Ea.	52.50	22	3.24	77.74	98.50
0300	5/8" diameter	G		35	.457		61	25.50	3.71	90.21	113
0500	3/4" diameter	G		30	.533		72.50	29.50	4.33	106.33	134

03 15 Concrete Accessories

03 15 19 – Cast-In Concrete Anchors

03 15 19.45 Machinery Anchors		Crew	Daily Output	Labor-Hours	Unit	Material	2017 Bare Costs Labor	Equipment	Total	Total Incl O&P
0600	7/8" diameter	G	E-16	25	.640	Ea.	82.50	35.50	5.20	123.20	155
0800	1" diameter	G		20	.800		88.50	44	6.50	139	178
0900	1-1/4" diameter	G		15	1.067		118	59	8.65	185.65	238

03 21 Reinforcement Bars

03 21 05 – Reinforcing Steel Accessories

03 21 05.10 Rebar Accessories

			Crew	Daily Output	Labor-Hours	Unit	Material	2017 Bare Costs Labor	Equipment	Total	Total Incl O&P
0010	**REBAR ACCESSORIES**										
0030	Steel & plastic made from recycled materials										
0100	Beam bolsters (BB), lower, 1-1/2" high, plain steel	G				C.L.F.	35			35	38.50
0102	Galvanized	G					42			42	46
0104	Stainless tipped legs	G					460			460	505
0106	Plastic tipped legs	G					51			51	56
0108	Epoxy dipped	G					88			88	97
0110	2" high, plain	G					44			44	48.50
0120	Galvanized	G					53			53	58
0140	Stainless tipped legs	G					470			470	515
0160	Plastic tipped legs	G					60			60	66
0162	Epoxy dipped	G					100			100	110
0200	Upper (BBU), 1-1/2" high, plain steel	G					85			85	93.50
0210	3" high	G					96			96	106
0500	Slab bolsters, continuous (SB), 1" high, plain steel	G					30			30	33
0502	Galvanized	G					36			36	39.50
0504	Stainless tipped legs	G					455			455	500
0506	Plastic tipped legs	G					44			44	48.50
0510	2" high, plain steel	G					38			38	42
0515	Galvanized	G					45.50			45.50	50
0520	Stainless tipped legs	G					465			465	510
0525	Plastic tipped legs	G					51			51	56
0530	For bolsters with wire runners (SBR), add	G					41			41	45
0540	For bolsters with plates (SBP), add	G					97			97	107
0700	Bag ties, 16 ga., plain, 4" long	G				C	4			4	4.40
0710	5" long	G					5			5	5.50
0720	6" long	G					4			4	4.40
0730	7" long	G					5			5	5.50
1200	High chairs, individual (HC), 3" high, plain steel	G					62			62	68
1202	Galvanized	G					74.50			74.50	82
1204	Stainless tipped legs	G					485			485	535
1206	Plastic tipped legs	G					67			67	73.50
1210	5" high, plain	G					90			90	99
1212	Galvanized	G					108			108	119
1214	Stainless tipped legs	G					515			515	565
1216	Plastic tipped legs	G					99			99	109
1220	8" high, plain	G					132			132	145
1222	Galvanized	G					158			158	174
1224	Stainless tipped legs	G					555			555	615
1226	Plastic tipped legs	G					146			146	161
1230	12" high, plain	G					315			315	345
1232	Galvanized	G					380			380	415
1234	Stainless tipped legs	G					740			740	815
1236	Plastic tipped legs	G					345			345	380

03 21 Reinforcement Bars

03 21 05 – Reinforcing Steel Accessories

03 21 05.10 Rebar Accessories		Crew	Daily Output	Labor-Hours	Unit	Material	2017 Bare Costs Labor	Equipment	Total	Total Incl O&P
1400	Individual high chairs, with plate (HCP), 5" high	G				C	188			188	207
1410	8" high	G					260			260	286
1500	Bar chair (BC), 1-1/2" high, plain steel	G					40			40	44
1520	Galvanized	G					45			45	49.50
1530	Stainless tipped legs	G					465			465	510
1540	Plastic tipped legs	G					43			43	47.50
1700	Continuous high chairs (CHC), legs 8" O.C., 4" high, plain steel	G				C.L.F.	54			54	59.50
1705	Galvanized	G					65			65	71.50
1710	Stainless tipped legs	G					480			480	525
1715	Plastic tipped legs	G					72			72	79
1718	Epoxy dipped	G					93			93	102
1720	6" high, plain	G					74			74	81.50
1725	Galvanized	G					89			89	97.50
1730	Stainless tipped legs	G					500			500	550
1735	Plastic tipped legs	G					99			99	109
1738	Epoxy dipped	G					127			127	140
1740	8" high, plain	G					105			105	116
1745	Galvanized	G					126			126	139
1750	Stainless tipped legs	G					530			530	585
1755	Plastic tipped legs	G					125			125	137
1758	Epoxy dipped	G					161			161	177
1900	For continuous bottom wire runners, add	G					29			29	32
1940	For continuous bottom plate, add	G					199			199	219
2200	Screed chair base, 1/2" coil thread diam., 2-1/2" high, plain steel	G				C	335			335	370
2210	Galvanized	G					405			405	445
2220	5-1/2" high, plain	G					405			405	445
2250	Galvanized	G					485			485	535
2300	3/4" coil thread diam., 2-1/2" high, plain steel	G					420			420	465
2310	Galvanized	G					505			505	555
2320	5-1/2" high, plain steel	G					520			520	570
2350	Galvanized	G					625			625	685
2400	Screed holder, 1/2" coil thread diam. for pipe screed, plain steel, 6" long	G					405			405	445
2420	12" long	G					625			625	685
2500	3/4" coil thread diam. for pipe screed, plain steel, 6" long	G					585			585	640
2520	12" long	G					935			935	1,025
2700	Screw anchor for bolts, plain steel, 3/4" diameter x 4" long	G					565			565	620
2720	1" diameter x 6" long	G					930			930	1,025
2740	1-1/2" diameter x 8" long	G					1,175			1,175	1,300
2800	Screw anchor eye bolts, 3/4" x 3" long	G					2,950			2,950	3,250
2820	1" x 3-1/2" long	G					3,950			3,950	4,350
2840	1-1/2" x 6" long	G					12,100			12,100	13,300
2900	Screw anchor bolts, 3/4" x 9" long	G					1,700			1,700	1,850
2920	1" x 12" long	G					3,250			3,250	3,575
3001	Slab lifting inserts, single pickup, galv, 3/4" diam., 5" high	G					1,825			1,825	2,000
3010	6" high	G					1,850			1,850	2,025
3030	7" high	G					1,875			1,875	2,075
3100	1" diameter, 5-1/2" high	G					1,925			1,925	2,125
3120	7" high	G					1,975			1,975	2,175
3200	Double pickup lifting inserts, 1" diameter, 5-1/2" high	G					3,700			3,700	4,075
3220	7" high	G					4,100			4,100	4,525
3330	1-1/2" diameter, 8" high	G					5,175			5,175	5,675
3800	Subgrade chairs, #4 bar head, 3-1/2" high	G					40			40	44
3850	12" high	G					47			47	51.50

03 21 05 – Reinforcing Steel Accessories

03 21 05.10 Rebar Accessories		Crew	Daily Output	Labor-Hours	Unit	Material	2017 Bare Costs Labor	Equipment	Total	Total Incl O&P
3900	#6 bar head, 3-1/2" high	G			C	40			40	44
3950	12" high	G				47			47	51.50
4200	Subgrade stakes, no nail holes, 3/4" diameter, 12" long	G				305			305	335
4250	24" long	G				370			370	410
4300	7/8" diameter, 12" long	G				390			390	430
4350	24" long	G			↓	660			660	725
4500	Tie wire, 16 ga. annealed steel	G			Cwt.	170			170	187

03 21 05.75 Splicing Reinforcing Bars

			Crew	Daily Output	Labor-Hours	Unit	Material	Labor	Equipment	Total	Total Incl O&P
0010	**SPLICING REINFORCING BARS**	R032110-70									
0020	Including holding bars in place while splicing										
0100	Standard, self-aligning type, taper threaded, #4 bars	G	C-25	190	.168	Ea.	6.25	7.30		13.55	18.55
0105	#5 bars	G		170	.188		7.60	8.15		15.75	21.50
0110	#6 bars	G		150	.213		8.80	9.20		18	24.50
0120	#7 bars	G		130	.246		10.25	10.65		20.90	28.50
0300	#8 bars	G	↓	115	.278		17.45	12		29.45	38.50
0305	#9 bars	G	C-5	105	.533		19.05	28.50	6.40	53.95	72
0310	#10 bars	G		95	.589		21	32	7.10	60.10	80
0320	#11 bars	G		85	.659		22.50	35.50	7.95	65.95	88
0330	#14 bars	G		65	.862		33.50	46.50	10.35	90.35	119
0340	#18 bars	G		45	1.244		51.50	67	14.95	133.45	176
0500	Transition self-aligning, taper threaded, #18-14	G		45	1.244		54	67	14.95	135.95	179
0510	#18-11	G		45	1.244		55	67	14.95	136.95	180
0520	#14-11	G		65	.862		36.50	46.50	10.35	93.35	122
0540	#11-10	G		85	.659		25	35.50	7.95	68.45	90.50
0550	#10-9	G	↓	95	.589		24	32	7.10	63.10	82.50
0560	#9-8	G	C-25	105	.305		21.50	13.15		34.65	45
0580	#8-7	G		115	.278		20	12		32	41.50
0590	#7-6	G		130	.246		13.95	10.65		24.60	32.50
0600	Position coupler for curved bars, taper threaded, #4 bars	G		160	.200		27.50	8.65		36.15	44
0610	#5 bars	G		145	.221		28.50	9.55		38.05	47
0620	#6 bars	G		130	.246		34.50	10.65		45.15	55
0630	#7 bars	G		110	.291		36.50	12.55		49.05	60.50
0640	#8 bars	G	↓	100	.320		38	13.80		51.80	63.50
0650	#9 bars	G	C-5	90	.622		41.50	33.50	7.50	82.50	105
0660	#10 bars	G		80	.700		44.50	37.50	8.40	90.40	116
0670	#11 bars	G		70	.800		46.50	43	9.60	99.10	128
0680	#14 bars	G		55	1.018		58	55	12.25	125.25	162
0690	#18 bars	G		40	1.400		83.50	75.50	16.85	175.85	227
0700	Transition position coupler for curved bars, taper threaded, #18-14	G		40	1.400		92	75.50	16.85	184.35	236
0710	#18-11	G		40	1.400		93.50	75.50	16.85	185.85	238
0720	#14-11	G		55	1.018		65	55	12.25	132.25	169
0730	#11-10	G		70	.800		52.50	43	9.60	105.10	134
0740	#10-9	G	↓	80	.700		50.50	37.50	8.40	96.40	123
0750	#9-8	G	C-25	90	.356		47	15.35		62.35	76
0760	#8-7	G		100	.320		43	13.80		56.80	69.50
0770	#7-6	G		110	.291		41	12.55		53.55	65
0800	Sleeve type w/grout filler, for precast concrete, #6 bars	G		72	.444		22.50	19.20		41.70	55.50
0802	#7 bars	G		64	.500		27	21.50		48.50	64.50
0805	#8 bars	G		56	.571		32	24.50		56.50	75
0807	#9 bars	G		48	.667		37	29		66	87
0810	#10 bars	G	C-5	40	1.400		45.50	75.50	16.85	137.85	185
0900	#11 bars	G		32	1.750		50	94.50	21	165.50	223

03 21 Reinforcement Bars

03 21 05 – Reinforcing Steel Accessories

03 21 05.75 Splicing Reinforcing Bars

			Crew	Daily Output	Labor-Hours	Unit	Material	2017 Bare Costs Labor	Equipment	Total	Total Incl O&P
0920	#14 bars	G	C-5	24	2.333	Ea.	78	126	28	232	310
1000	Sleeve type w/ferrous filler, for critical structures, #6 bars	G	C-25	72	.444		67.50	19.20		86.70	105
1210	#7 bars	G		64	.500		68.50	21.50		90	110
1220	#8 bars	G		56	.571		72	24.50		96.50	119
1230	#9 bars	G	C-5	48	1.167		74	63	14.05	151.05	193
1240	#10 bars	G		40	1.400		79	75.50	16.85	171.35	222
1250	#11 bars	G		32	1.750		95.50	94.50	21	211	273
1260	#14 bars	G		24	2.333		120	126	28	274	355
1270	#18 bars	G		16	3.500		174	189	42	405	525
2000	Weldable half coupler, taper threaded, #4 bars	G	E-16	120	.133		8.90	7.35	1.08	17.33	23.50
2100	#5 bars	G		112	.143		10.45	7.90	1.16	19.51	26
2200	#6 bars	G		104	.154		16.65	8.50	1.25	26.40	34
2300	#7 bars	G		96	.167		19.35	9.20	1.35	29.90	38.50
2400	#8 bars	G		88	.182		20	10.05	1.47	31.52	40.50
2500	#9 bars	G		80	.200		22	11.05	1.62	34.67	44.50
2600	#10 bars	G		72	.222		22.50	12.30	1.80	36.60	47.50
2700	#11 bars	G		64	.250		24.50	13.85	2.03	40.38	51.50
2800	#14 bars	G		56	.286		28	15.80	2.32	46.12	60
2900	#18 bars	G		48	.333		45.50	18.45	2.70	66.65	83.50

03 21 11 – Plain Steel Reinforcement Bars

03 21 11.60 Reinforcing In Place

			Crew	Daily Output	Labor-Hours	Unit	Material	2017 Bare Costs Labor	Equipment	Total	Total Incl O&P
0010	**REINFORCING IN PLACE**, 50-60 ton lots, A615 Grade 60 R032110-10										
0020	Includes labor, but not material cost, to install accessories										
0030	Made from recycled materials										
0100	Beams & Girders, #3 to #7	G	4 Rodm	1.60	20	Ton	940	1,075		2,015	2,700
0150	#8 to #18 R032110-20	G		2.70	11.852		940	645		1,585	2,025
0200	Columns, #3 to #7	G		1.50	21.333		940	1,150		2,090	2,800
0250	#8 to #18	G		2.30	13.913		940	755		1,695	2,200
0300	Spirals, hot rolled, 8" to 15" diameter	G		2.20	14.545		1,550	790		2,340	2,925
0320	15" to 24" diameter R032110-40	G		2.20	14.545		1,475	790		2,265	2,850
0330	24" to 36" diameter	G		2.30	13.913		1,400	755		2,155	2,725
0340	36" to 48" diameter R032110-50	G		2.40	13.333		1,325	725		2,050	2,600
0360	48" to 64" diameter	G		2.50	12.800		1,475	695		2,170	2,700
0380	64" to 84" diameter R032110-70	G		2.60	12.308		1,550	670		2,220	2,725
0390	84" to 96" diameter	G		2.70	11.852		1,600	645		2,245	2,775
0400	Elevated slabs, #4 to #7 R032110-80	G		2.90	11.034		940	600		1,540	1,950
0500	Footings, #4 to #7	G		2.10	15.238		940	825		1,765	2,300
0550	#8 to #18	G		3.60	8.889		940	485		1,425	1,775
0600	Slab on grade, #3 to #7	G		2.30	13.913		940	755		1,695	2,200
0700	Walls, #3 to #7	G		3	10.667		940	580		1,520	1,925
0750	#8 to #18	G		4	8		940	435		1,375	1,700
0900	For other than 50 – 60 ton lots										
1000	Under 10 ton job, #3 to #7, add						25%	10%			
1010	#8 to #18, add						20%	10%			
1050	10 – 50 ton job, #3 to #7, add						10%				
1060	#8 to #18, add						5%				
1100	60 – 100 ton job, #3 to #7, deduct						5%				
1110	#8 to #18, deduct						10%				
1150	Over 100 ton job, #3 to #7, deduct						10%				
1160	#8 to #18, deduct						15%				
1200	Reinforcing in place, A615 Grade 75, add	G				Ton	90			90	99
1220	Grade 90, add						120			120	132

03 21 Reinforcement Bars

03 21 11 – Plain Steel Reinforcement Bars

03 21 11.60 Reinforcing In Place		Crew	Daily Output	Labor-Hours	Unit	Material	2017 Bare Costs Labor	Equipment	Total	Total Incl O&P
2000	Unloading & sorting, add to above	C-5	100	.560	Ton		30	6.75	36.75	54
2200	Crane cost for handling, 90 picks/day, up to 1.5 Tons/bundle, add to above		135	.415			22.50	4.99	27.49	40
2210	1.0 Ton/bundle		92	.609			33	7.30	40.30	58.50
2220	0.5 Ton/bundle		35	1.600			86	19.25	105.25	153
2400	Dowels, 2 feet long, deformed, #3 [G]	2 Rodm	520	.031	Ea.	.39	1.67		2.06	3.01
2410	#4 [G]		480	.033		.69	1.81		2.50	3.55
2420	#5 [G]		435	.037		1.08	2		3.08	4.27
2430	#6 [G]		360	.044		1.55	2.41		3.96	5.45
2450	Longer and heavier dowels, add [G]		725	.022	Lb.	.52	1.20		1.72	2.42
2500	Smooth dowels, 12" long, 1/4" or 3/8" diameter [G]		140	.114	Ea.	.72	6.20		6.92	10.35
2520	5/8" diameter [G]		125	.128		1.25	6.95		8.20	12.10
2530	3/4" diameter [G]		110	.145		1.55	7.90		9.45	13.85
2600	Dowel sleeves for CIP concrete, 2-part system									
2610	Sleeve base, plastic, for 5/8" smooth dowel sleeve, fasten to edge form	1 Rodm	200	.040	Ea.	.53	2.17		2.70	3.93
2615	Sleeve, plastic, 12" long, for 5/8" smooth dowel, snap onto base		400	.020		1.33	1.09		2.42	3.13
2620	Sleeve base, for 3/4" smooth dowel sleeve		175	.046		.53	2.48		3.01	4.41
2625	Sleeve, 12" long, for 3/4" smooth dowel		350	.023		1.21	1.24		2.45	3.24
2630	Sleeve base, for 1" smooth dowel sleeve		150	.053		.68	2.90		3.58	5.20
2635	Sleeve, 12" long, for 1" smooth dowel		300	.027		1.42	1.45		2.87	3.79
2700	Dowel caps, visual warning only, plastic, #3 to #8	2 Rodm	800	.020		.29	1.09		1.38	1.99
2720	#8 to #18		750	.021		.74	1.16		1.90	2.61
2750	Impalement protective, plastic, #4 to #9		800	.020		1.62	1.09		2.71	3.45

03 21 13 – Galvanized Reinforcement Steel Bars

03 21 13.10 Galvanized Reinforcing

0010	**GALVANIZED REINFORCING**									
0150	Add to plain steel rebar pricing for galvanized rebar				Ton	450			450	495

03 21 16 – Epoxy-Coated Reinforcement Steel Bars

03 21 16.10 Epoxy-Coated Reinforcing

0010	**EPOXY-COATED REINFORCING**									
0100	Add to plain steel rebar pricing for epoxy-coated rebar				Ton	770			770	845

03 21 21 – Composite Reinforcement Bars

03 21 21.11 Glass Fiber-Reinforced Polymer Reinf. Bars

0010	**GLASS FIBER-REINFORCED POLYMER REINFORCEMENT BARS**									
0020	Includes labor, but not material cost, to install accessories									
0050	#2 bar, .043 lb./L.F.	4 Rodm	9500	.003	L.F.	.40	.18		.58	.72
0100	#3 bar, .092 lb./L.F.		9300	.003		.49	.19		.68	.83
0150	#4 bar, .160 lb./L.F.		9100	.004		.71	.19		.90	1.07
0200	#5 bar, .258 lb./L.F.		8700	.004		1.09	.20		1.29	1.51
0250	#6 bar, .372 lb./L.F.		8300	.004		1.41	.21		1.62	1.87
0300	#7 bar, .497 lb./L.F.		7900	.004		1.84	.22		2.06	2.36
0350	#8 bar, .620 lb./L.F.		7400	.004		2.38	.23		2.61	2.98
0400	#9 bar, .800 lb./L.F.		6800	.005		3.10	.26		3.36	3.80
0450	#10 bar, 1.08 lb./L.F.		5800	.006		3.75	.30		4.05	4.59
0500	For bends, add per bend				Ea.	1.48			1.48	1.63

03 22 Fabric and Grid Reinforcing

03 22 11 – Plain Welded Wire Fabric Reinforcing

03 22 11.10 Plain Welded Wire Fabric

		Crew	Daily Output	Labor-Hours	Unit	Material	2017 Bare Costs Labor	2017 Bare Costs Equipment	Total	Total Incl O&P
0010	**PLAIN WELDED WIRE FABRIC** ASTM A185 R032205-30									
0020	Includes labor, but not material cost, to install accessories									
0030	Made from recycled materials									
0050	Sheets									
0100	6 x 6 - W1.4 x W1.4 (10 x 10) 21 lb. per C.S.F.	G	2 Rodm	35	.457	C.S.F.	14.45	25		39.45	54.50
0200	6 x 6 - W2.1 x W2.1 (8 x 8) 30 lb. per C.S.F.	G		31	.516		18.80	28		46.80	63.50
0300	6 x 6 - W2.9 x W2.9 (6 x 6) 42 lb. per C.S.F.	G		29	.552		24.50	30		54.50	73
0400	6 x 6 - W4 x W4 (4 x 4) 58 lb. per C.S.F.	G		27	.593		34	32		66	86.50
0500	4 x 4 - W1.4 x W1.4 (10 x 10) 31 lb. per C.S.F.	G		31	.516		21.50	28		49.50	67
0600	4 x 4 - W2.1 x W2.1 (8 x 8) 44 lb. per C.S.F.	G		29	.552		27	30		57	75.50
0650	4 x 4 - W2.9 x W2.9 (6 x 6) 61 lb. per C.S.F.	G		27	.593		43	32		75	97
0700	4 x 4 - W4 x W4 (4 x 4) 85 lb. per C.S.F.	G	▼	25	.640	▼	54	35		89	113
0750	Rolls									
0800	2 x 2 - #14 galv., 21 lb./C.S.F., beam & column wrap	G	2 Rodm	6.50	2.462	C.S.F.	44.50	134		178.50	255
0900	2 x 2 - #12 galv. for gunite reinforcing	G	"	6.50	2.462	"	67	134		201	280

03 22 13 – Galvanized Welded Wire Fabric Reinforcing

03 22 13.10 Galvanized Welded Wire Fabric

		Crew	Daily Output	Labor-Hours	Unit	Material	2017 Bare Costs Labor	2017 Bare Costs Equipment	Total	Total Incl O&P
0010	**GALVANIZED WELDED WIRE FABRIC**									
0100	Add to plain welded wire pricing for galvanized welded wire					Lb.	.23			.23	.25

03 22 16 – Epoxy-Coated Welded Wire Fabric Reinforcing

03 22 16.10 Epoxy-Coated Welded Wire Fabric

		Crew	Daily Output	Labor-Hours	Unit	Material	2017 Bare Costs Labor	2017 Bare Costs Equipment	Total	Total Incl O&P
0010	**EPOXY-COATED WELDED WIRE FABRIC**									
0100	Add to plain welded wire pricing for epoxy-coated welded wire					Lb.	.39			.39	.42

03 23 Stressed Tendon Reinforcing

03 23 05 – Prestressing Tendons

03 23 05.50 Prestressing Steel

		Crew	Daily Output	Labor-Hours	Unit	Material	2017 Bare Costs Labor	2017 Bare Costs Equipment	Total	Total Incl O&P
0010	**PRESTRESSING STEEL** R034136-90									
0100	Grouted strand, in beams, post-tensioned in field, 50' span, 100 kip	G	C-3	1200	.053	Lb.	2.82	2.69	.09	5.60	7.30
0150	300 kip	G		2700	.024		1.15	1.19	.04	2.38	3.15
0300	100' span, 100 kip	G		1700	.038		2.82	1.90	.07	4.79	6.10
0350	300 kip	G		3200	.020		2.41	1.01	.04	3.46	4.25
0500	200' span, 100 kip	G		2700	.024		2.82	1.19	.04	4.05	4.98
0550	300 kip	G		3500	.018		2.41	.92	.03	3.36	4.10
0800	Grouted bars, in beams, 50' span, 42 kip	G		2600	.025		1.09	1.24	.04	2.37	3.15
0850	143 kip	G		3200	.020		1.04	1.01	.04	2.09	2.74
1000	75' span, 42 kip	G		3200	.020		1.11	1.01	.04	2.16	2.81
1050	143 kip	G	▼	4200	.015		.92	.77	.03	1.72	2.22
1200	Ungrouted strand, in beams, 50' span, 100 kip	G	C-4	1275	.025		.59	1.38	.02	1.99	2.80
1250	300 kip	G		1475	.022		.59	1.19	.02	1.80	2.50
1400	100' span, 100 kip	G		1500	.021		.59	1.17	.02	1.78	2.47
1450	300 kip	G		1650	.019		.59	1.06	.02	1.67	2.31
1600	200' span, 100 kip	G		1500	.021		.59	1.17	.02	1.78	2.47
1650	300 kip	G		1700	.019		.59	1.03	.02	1.64	2.26
1800	Ungrouted bars, in beams, 50' span, 42 kip	G		1400	.023		.47	1.25	.02	1.74	2.47
1850	143 kip	G		1700	.019		.47	1.03	.02	1.52	2.13
2000	75' span, 42 kip	G		1800	.018		.47	.97	.02	1.46	2.04
2050	143 kip	G		2200	.015		.47	.80	.01	1.28	1.77
2220	Ungrouted single strand, 100' elevated slab, 25 kip	G		1200	.027		.59	1.46	.03	2.08	2.93
2250	35 kip	G	▼	1475	.022	▼	.59	1.19	.02	1.80	2.50

03 23 05 – Prestressing Tendons

03 23 05.50 Prestressing Steel	Crew	Daily Output	Labor-Hours	Unit	Material	2017 Bare Costs Labor	Equipment	Total	Total Incl O&P
3000	Slabs on grade, 0.5-inch diam. non-bonded strands, HDPE sheathed,								
3050	attached dead-end anchors, loose stressing-end anchors								
3100	25' x 30' slab, strands @ 36" O.C., placing	2 Rodm	2940	.005	S.F.	.63	.30		.93	1.15
3105	Stressing	C-4A	3750	.004			.23	.01	.24	.37
3110	42" O.C., placing	2 Rodm	3200	.005		.56	.27		.83	1.03
3115	Stressing	C-4A	4040	.004			.22	.01	.23	.34
3120	48" O.C., placing	2 Rodm	3510	.005		.49	.25		.74	.91
3125	Stressing	C-4A	4390	.004			.20	.01	.21	.31
3150	25' x 40' slab, strands @ 36" O.C., placing	2 Rodm	3370	.005		.61	.26		.87	1.07
3155	Stressing	C-4A	4360	.004			.20	.01	.21	.32
3160	42" O.C., placing	2 Rodm	3760	.004		.52	.23		.75	.94
3165	Stressing	C-4A	4820	.003			.18	.01	.19	.29
3170	48" O.C., placing	2 Rodm	4090	.004		.47	.21		.68	.84
3175	Stressing	C-4A	5190	.003			.17	.01	.18	.27
3200	30' x 30' slab, strands @ 36" O.C., placing	2 Rodm	3260	.005		.60	.27		.87	1.07
3205	Stressing	C-4A	4190	.004			.21	.01	.22	.33
3210	42" O.C., placing	2 Rodm	3530	.005		.54	.25		.79	.98
3215	Stressing	C-4A	4500	.004			.19	.01	.20	.31
3220	48" O.C., placing	2 Rodm	3840	.004		.48	.23		.71	.88
3225	Stressing	C-4A	4850	.003			.18	.01	.19	.29
3230	30' x 40' slab, strands @ 36" O.C., placing	2 Rodm	3780	.004		.58	.23		.81	.99
3235	Stressing	C-4A	4920	.003			.18	.01	.19	.28
3240	42" O.C., placing	2 Rodm	4190	.004		.51	.21		.72	.88
3245	Stressing	C-4A	5410	.003			.16	.01	.17	.26
3250	48" O.C., placing	2 Rodm	4520	.004		.46	.19		.65	.81
3255	Stressing	C-4A	5790	.003			.15	.01	.16	.24
3260	30' x 50' slab, strands @ 36" O.C., placing	2 Rodm	4300	.004		.55	.20		.75	.92
3265	Stressing	C-4A	5650	.003			.15	.01	.16	.25
3270	42" O.C., placing	2 Rodm	4720	.003		.49	.18		.67	.82
3275	Stressing	C-4A	6150	.003			.14	.01	.15	.23
3280	48" O.C., placing	2 Rodm	5240	.003		.43	.17		.60	.73
3285	Stressing	C-4A	6760	.002			.13	.01	.14	.21

03 24 Fibrous Reinforcing

03 24 05 – Reinforcing Fibers

03 24 05.30 Synthetic Fibers

					Unit	Material	Labor	Equipment	Total	Total Incl O&P
0010	**SYNTHETIC FIBERS**									
0100	Synthetic fibers, add to concrete				Lb.	4.40			4.40	4.84
0110	1-1/2 lb. per C.Y.				C.Y.	6.80			6.80	7.50

03 24 05.70 Steel Fibers

					Unit	Material	Labor	Equipment	Total	Total Incl O&P
0010	**STEEL FIBERS**									
0140	ASTM A850, Type V, continuously deformed, 1-1/2" long x 0.045" diam.									
0150	Add to price of ready mix concrete	G			Lb.	1.13			1.13	1.24
0205	Alternate pricing, dosing at 5 lb. per C.Y., add to price of RMC	G			C.Y.	5.65			5.65	6.20
0210	10 lb. per C.Y.	G				11.30			11.30	12.45
0215	15 lb. per C.Y.	G				16.95			16.95	18.65
0220	20 lb. per C.Y.	G				22.50			22.50	25
0225	25 lb. per C.Y.	G				28.50			28.50	31
0230	30 lb. per C.Y.	G				34			34	37.50
0235	35 lb. per C.Y.	G				39.50			39.50	43.50
0240	40 lb. per C.Y.	G				45			45	49.50

03 24 Fibrous Reinforcing

03 24 05 – Reinforcing Fibers

03 24 05.70 Steel Fibers

		Crew	Daily Output	Labor-Hours	Unit	Material	2017 Bare Costs Labor	Equipment	Total	Total Incl O&P
0250	50 lb. per C.Y.	G			C.Y.	56.50			56.50	62
0275	75 lb. per C.Y.	G				85			85	93
0300	100 lb. per C.Y.	G		↓		113			113	124

03 30 Cast-In-Place Concrete

03 30 53 – Miscellaneous Cast-In-Place Concrete

03 30 53.40 Concrete In Place

			Crew	Daily Output	Labor-Hours	Unit	Material	2017 Bare Costs Labor	Equipment	Total	Total Incl O&P
0010	**CONCRETE IN PLACE**										
0020	Including forms (4 uses), Grade 60 rebar, concrete (Portland cement	R033105-10									
0050	Type I), placement and finishing unless otherwise indicated	R033105-20									
0300	Beams (3500 psi), 5 kip per L.F., 10' span	R033105-50	C-14A	15.62	12.804	C.Y.	330	635	53.50	1,018.50	1,400
0350	25' span	R033105-65	"	18.55	10.782		345	535	45	925	1,250
0500	Chimney foundations (5000 psi), over 5 C.Y.	R033105-70	C-14C	32.22	3.476		167	164	.87	331.87	435
0510	(3500 psi), under 5 C.Y.	R033105-85	"	23.71	4.724		193	222	1.18	416.18	555
0700	Columns, square (4000 psi), 12" x 12", up to 1% reinforcing by area		C-14A	11.96	16.722		365	825	69.50	1,259.50	1,750
0720	Up to 2% reinforcing by area			10.13	19.743		560	975	82.50	1,617.50	2,200
0740	Up to 3% reinforcing by area			9.03	22.148		825	1,100	92.50	2,017.50	2,675
0800	16" x 16", up to 1% reinforcing by area			16.22	12.330		293	610	51.50	954.50	1,300
0820	Up to 2% reinforcing by area			12.57	15.911		480	785	66.50	1,331.50	1,800
0840	Up to 3% reinforcing by area			10.25	19.512		730	965	81.50	1,776.50	2,375
0900	24" x 24", up to 1% reinforcing by area			23.66	8.453		252	420	35.50	707.50	955
0920	Up to 2% reinforcing by area			17.71	11.293		430	560	47	1,037	1,375
0940	Up to 3% reinforcing by area			14.15	14.134		670	700	59	1,429	1,875
1000	36" x 36", up to 1% reinforcing by area			33.69	5.936		226	293	25	544	725
1020	Up to 2% reinforcing by area			23.32	8.576		380	425	36	841	1,100
1040	Up to 3% reinforcing by area			17.82	11.223		625	555	47	1,227	1,600
1100	Columns, round (4000 psi), tied, 12" diameter, up to 1% reinforcing by area			20.97	9.537		340	470	40	850	1,150
1120	Up to 2% reinforcing by area			15.27	13.098		535	645	54.50	1,234.50	1,625
1140	Up to 3% reinforcing by area			12.11	16.515		790	815	69	1,674	2,200
1200	16" diameter, up to 1% reinforcing by area			31.49	6.351		310	315	26.50	651.50	855
1220	Up to 2% reinforcing by area			19.12	10.460		510	515	43.50	1,068.50	1,400
1240	Up to 3% reinforcing by area			13.77	14.524		745	720	60.50	1,525.50	1,975
1300	20" diameter, up to 1% reinforcing by area			41.04	4.873		305	241	20.50	566.50	730
1320	Up to 2% reinforcing by area			24.05	8.316		485	410	34.50	929.50	1,200
1340	Up to 3% reinforcing by area			17.01	11.758		735	580	49	1,364	1,750
1400	24" diameter, up to 1% reinforcing by area			51.85	3.857		285	191	16.10	492.10	625
1420	Up to 2% reinforcing by area			27.06	7.391		480	365	31	876	1,125
1440	Up to 3% reinforcing by area			18.29	10.935		720	540	45.50	1,305.50	1,675
1500	36" diameter, up to 1% reinforcing by area			75.04	2.665		285	132	11.10	428.10	530
1520	Up to 2% reinforcing by area			37.49	5.335		460	264	22.50	746.50	935
1540	Up to 3% reinforcing by area		↓	22.84	8.757		700	435	36.50	1,171.50	1,475
1900	Elevated slab (4000 psi), flat slab with drops, 125 psf Sup. Load, 20' span		C-14B	38.45	5.410		275	267	21.50	563.50	740
1950	30' span			50.99	4.079		290	201	16.35	507.35	650
2100	Flat plate, 125 psf Sup. Load, 15' span			30.24	6.878		254	340	27.50	621.50	830
2150	25' span			49.60	4.194		263	207	16.80	486.80	625
2300	Waffle const., 30" domes, 125 psf Sup. Load, 20' span			37.07	5.611		281	277	22.50	580.50	760
2350	30' span			44.07	4.720		262	233	18.95	513.95	665
2500	One way joists, 30" pans, 125 psf Sup. Load, 15' span			27.38	7.597		340	375	30.50	745.50	985
2550	25' span			31.15	6.677		320	330	27	677	885
2700	One way beam & slab, 125 psf Sup. Load, 15' span			20.59	10.102		273	500	40.50	813.50	1,100
2750	25' span		↓	28.36	7.334		259	360	29.50	648.50	875

03 30 53 – Miscellaneous Cast-In-Place Concrete

03 30 53.40 Concrete In Place	Crew	Daily Output	Labor-Hours	Unit	Material	2017 Bare Costs Labor	2017 Bare Costs Equipment	Total	Total Incl O&P
2900	Two way beam & slab, 125 psf Sup. Load, 15' span	C-14B	24.04	8.652	C.Y.	263	425	34.50	722.50	980
2950	25' span	↓	35.87	5.799	↓	230	286	23.50	539.50	715
3100	Elevated slabs, flat plate, including finish, not									
3110	including forms or reinforcing									
3150	Regular concrete (4000 psi), 4" slab	C-8	2613	.021	S.F.	1.67	.94	.31	2.92	3.59
3200	6" slab		2585	.022		2.43	.95	.31	3.69	4.45
3250	2-1/2" thick floor fill		2685	.021		1.09	.91	.30	2.30	2.91
3300	Lightweight, 110# per C.F., 2-1/2" thick floor fill		2585	.022		1.48	.95	.31	2.74	3.40
3400	Cellular concrete, 1-5/8" fill, under 5000 S.F.		2000	.028		1	1.22	.40	2.62	3.40
3450	Over 10,000 S.F.		2200	.025		.96	1.11	.37	2.44	3.14
3500	Add per floor for 3 to 6 stories high		31800	.002			.08	.03	.11	.15
3520	For 7 to 20 stories high	↓	21200	.003	↓		.12	.04	.16	.21
3540	Equipment pad (3000 psi), 3' x 3' x 6" thick	C-14H	45	1.067	Ea.	49	51.50	.63	101.13	133
3550	4' x 4' x 6" thick		30	1.600		73.50	77.50	.94	151.94	200
3560	5' x 5' x 8" thick		18	2.667		131	129	1.57	261.57	345
3570	6' x 6' x 8" thick		14	3.429		177	166	2.02	345.02	450
3580	8' x 8' x 10" thick		8	6		380	290	3.54	673.54	870
3590	10' x 10' x 12" thick	↓	5	9.600	↓	655	465	5.65	1,125.65	1,425
3800	Footings (3000 psi), spread under 1 C.Y.	C-14C	28	4	C.Y.	181	188	1	370	490
3825	1 C.Y. to 5 C.Y.		43	2.605		218	123	.65	341.65	430
3850	Over 5 C.Y.	↓	75	1.493		201	70.50	.37	271.87	330
3900	Footings, strip (3000 psi), 18" x 9", unreinforced	C-14L	40	2.400		143	110	.70	253.70	325
3920	18" x 9", reinforced	C-14C	35	3.200		165	151	.80	316.80	415
3925	20" x 10", unreinforced	C-14L	45	2.133		139	98	.62	237.62	305
3930	20" x 10", reinforced	C-14C	40	2.800		157	132	.70	289.70	375
3935	24" x 12", unreinforced	C-14L	55	1.745		137	80	.51	217.51	274
3940	24" x 12", reinforced	C-14C	48	2.333		156	110	.58	266.58	340
3945	36" x 12", unreinforced	C-14L	70	1.371		133	63	.40	196.40	242
3950	36" x 12", reinforced	C-14C	60	1.867		150	88	.47	238.47	300
4000	Foundation mat (3000 psi), under 10 C.Y.		38.67	2.896		218	136	.72	354.72	450
4050	Over 20 C.Y.	↓	56.40	1.986		193	93.50	.50	287	355
4200	Wall, free-standing (3000 psi), 8" thick, 8' high	C-14D	45.83	4.364		178	214	18.20	410.20	540
4250	14' high		27.26	7.337		207	360	30.50	597.50	810
4260	12" thick, 8' high		64.32	3.109		161	152	12.95	325.95	425
4270	14' high		40.01	4.999		171	245	21	437	585
4300	15" thick, 8' high		80.02	2.499		156	122	10.40	288.40	370
4350	12' high		51.26	3.902		156	191	16.25	363.25	485
4500	18' high	↓	48.85	4.094	↓	174	201	17.05	392.05	515
4520	Handicap access ramp (4000 psi), railing both sides, 3' wide	C-14H	14.58	3.292	L.F.	330	159	1.94	490.94	610
4525	5' wide		12.22	3.928		340	190	2.32	532.32	670
4530	With 6" curb and rails both sides, 3' wide		8.55	5.614		340	271	3.31	614.31	795
4535	5' wide	↓	7.31	6.566	↓	345	315	3.87	663.87	870
4650	Slab on grade (3500 psi), not including finish, 4" thick	C-14E	60.75	1.449	C.Y.	142	70	.46	212.46	264
4700	6" thick	"	92	.957	"	137	46	.31	183.31	222
4701	Thickened slab edge (3500 psi), for slab on grade poured									
4702	monolithically with slab; depth is in addition to slab thickness;									
4703	formed vertical outside edge, earthen bottom and inside slope									
4705	8" deep x 8" wide bottom, unreinforced	C-14L	2190	.044	L.F.	3.85	2.01	.01	5.87	7.30
4710	8" x 8", reinforced	C-14C	1670	.067		6.05	3.16	.02	9.23	11.50
4715	12" deep x 12" wide bottom, unreinforced	C-14L	1800	.053		7.90	2.45	.02	10.37	12.45
4720	12" x 12", reinforced	C-14C	1310	.086		11.95	4.02	.02	15.99	19.30
4725	16" deep x 16" wide bottom, unreinforced	C-14L	1440	.067		13.45	3.06	.02	16.53	19.50
4730	16" x 16", reinforced	C-14C	1120	.100	↓	18.25	4.71	.03	22.99	27

03 30 Cast-In-Place Concrete

03 30 53 – Miscellaneous Cast-In-Place Concrete

03 30 53.40 Concrete In Place

		Crew	Daily Output	Labor-Hours	Unit	Material	2017 Bare Costs Labor	2017 Bare Costs Equipment	Total	Total Incl O&P
4735	20" deep x 20" wide bottom, unreinforced	C-14L	1150	.083	L.F.	20.50	3.83	.02	24.35	28.50
4740	20" x 20", reinforced	C-14C	920	.122		26.50	5.75	.03	32.28	38
4745	24" deep x 24" wide bottom, unreinforced	C-14L	930	.103		29	4.73	.03	33.76	39.50
4750	24" x 24", reinforced	C-14C	740	.151	▼	37	7.10	.04	44.14	51.50
4751	Slab on grade (3500 psi), incl. troweled finish, not incl. forms									
4760	or reinforcing, over 10,000 S.F., 4" thick	C-14F	3425	.021	S.F.	1.58	.93	.01	2.52	3.14
4820	6" thick		3350	.021		2.30	.96	.01	3.27	3.97
4840	8" thick		3184	.023		3.15	1.01	.01	4.17	4.99
4900	12" thick		2734	.026		4.73	1.17	.01	5.91	6.95
4950	15" thick	▼	2505	.029	▼	5.95	1.28	.01	7.24	8.50
5000	Slab on grade (3000 psi), incl. broom finish, not incl. forms									
5001	or reinforcing, 4" thick	C-14G	2873	.019	S.F.	1.53	.85	.01	2.39	2.98
5010	6" thick		2590	.022		2.40	.95	.01	3.36	4.08
5020	8" thick	▼	2320	.024	▼	3.13	1.06	.01	4.20	5.05
5200	Lift slab in place above the foundation, incl. forms, reinforcing,									
5210	concrete (4000 psi) and columns, over 20,000 S.F. per floor	C-14B	2113	.098	S.F.	7.25	4.85	.39	12.49	15.85
5250	10,000 S.F. to 20,000 S.F. per floor		1650	.126		7.90	6.20	.51	14.61	18.75
5300	Under 10,000 S.F. per floor	▼	1500	.139	▼	8.55	6.85	.56	15.96	20.50
5500	Lightweight, ready mix, including screed finish only,									
5510	not including forms or reinforcing									
5550	1:4 (2500 psi) for structural roof decks	C-14B	260	.800	C.Y.	166	39.50	3.21	208.71	247
5600	1:6 (3000 psi) for ground slab with radiant heat	C-14F	92	.783		168	35	.31	203.31	237
5650	1:3:2 (2000 psi) with sand aggregate, roof deck	C-14B	260	.800		164	39.50	3.21	206.71	245
5700	Ground slab (2000 psi)	C-14F	107	.673		164	30	.26	194.26	226
5900	Pile caps (3000 psi), incl. forms and reinf., sq. or rect., under 10 C.Y.	C-14C	54.14	2.069		187	97.50	.52	285.02	355
5950	Over 10 C.Y.		75	1.493		175	70.50	.37	245.87	300
6000	Triangular or hexagonal, under 10 C.Y.		53	2.113		140	99.50	.53	240.03	305
6050	Over 10 C.Y.	▼	85	1.318		155	62	.33	217.33	266
6200	Retaining walls (3000 psi), gravity, 4' high see Section 32 32	C-14D	66.20	3.021		157	148	12.60	317.60	415
6250	10' high		125	1.600		150	78.50	6.65	235.15	292
6300	Cantilever, level backfill loading, 8' high		70	2.857		168	140	11.90	319.90	410
6350	16' high	▼	91	2.198	▼	162	108	9.15	279.15	355
6800	Stairs (3500 psi), not including safety treads, free standing, 3'-6" wide	C-14H	83	.578	LF Nose	6	28	.34	34.34	49.50
6850	Cast on ground		125	.384	"	5.05	18.55	.23	23.83	34.50
7000	Stair landings, free standing		200	.240	S.F.	4.82	11.60	.14	16.56	23
7050	Cast on ground	▼	475	.101	"	3.85	4.88	.06	8.79	11.75

03 31 Structural Concrete

03 31 13 – Heavyweight Structural Concrete

03 31 13.25 Concrete, Hand Mix

		Crew	Daily Output	Labor-Hours	Unit	Material	2017 Bare Costs Labor	2017 Bare Costs Equipment	Total	Total Incl O&P
0010	**CONCRETE, HAND MIX** for small quantities or remote areas									
0050	Includes bulk local aggregate, bulk sand, bagged Portland									
0060	cement (Type I) and water, using gas powered cement mixer									
0125	2500 psi	C-30	135	.059	C.F.	3.69	2.32	1.22	7.23	8.95
0130	3000 psi		135	.059		3.96	2.32	1.22	7.50	9.25
0135	3500 psi		135	.059		4.11	2.32	1.22	7.65	9.45
0140	4000 psi		135	.059		4.30	2.32	1.22	7.84	9.65
0145	4500 psi		135	.059		4.50	2.32	1.22	8.04	9.85
0150	5000 psi		135	.059	▼	4.80	2.32	1.22	8.34	10.20
0300	Using pre-bagged dry mix and wheelbarrow (80-lb. bag = 0.6 C.F.)									
0340	4000 psi	1 Clab	48	.167	C.F.	6.50	6.55		13.05	17.15

For customer support on your Building Construction Costs with RSMeans Data, call 800.448.8182.

77

03 31 13 – Heavyweight Structural Concrete

03 31 13.30 Concrete, Volumetric Site-Mixed

	Crew	Daily Output	Labor-Hours	Unit	Material	2017 Bare Costs Labor	2017 Bare Costs Equipment	Total	Total Incl O&P
0010 **CONCRETE, VOLUMETRIC SITE-MIXED**									
0015 Mixed on-site in volumetric truck									
0020 Includes local aggregate, sand, Portland cement (Type I) and water									
0025 Excludes all additives and treatments									
0100 3000 psi, 1 C.Y. mixed and discharged				C.Y.	194			194	214
0110 2 C.Y.					148			148	163
0120 3 C.Y.					132			132	145
0130 4 C.Y.					119			119	131
0140 5 C.Y.					111			111	122
0200 For truck holding/waiting time past first 2 on-site hours, add				Hr.	88			88	97
0210 For trip charge beyond first 20 miles, each way, add				Mile	3.50			3.50	3.85
0220 For each additional increase of 500 psi, add				Ea.	4.17			4.17	4.59

03 31 13.35 Heavyweight Concrete, Ready Mix

	Crew	Daily Output	Labor-Hours	Unit	Material	2017 Bare Costs Labor	2017 Bare Costs Equipment	Total	Total Incl O&P
0010 **HEAVYWEIGHT CONCRETE, READY MIX**, delivered R033105-10									
0012 Includes local aggregate, sand, Portland cement (Type I) and water									
0015 Excludes all additives and treatments R033105-20									
0020 2000 psi				C.Y.	112			112	124
0100 2500 psi R033105-30					116			116	127
0150 3000 psi					118			118	130
0200 3500 psi R033105-40					121			121	133
0300 4000 psi					125			125	137
0350 4500 psi R033105-50					128			128	141
0400 5000 psi					131			131	144
0411 6000 psi					135			135	148
0412 8000 psi					142			142	156
0413 10,000 psi					149			149	164
0414 12,000 psi					156			156	172
1000 For high early strength (Portland cement Type III), add					10%				
1010 For structural lightweight with regular sand, add					25%				
1300 For winter concrete (hot water), add					4.45			4.45	4.90
1410 For mid-range water reducer, add					3.33			3.33	3.66
1420 For high-range water reducer/superplasticizer, add					6			6	6.60
1430 For retarder, add					3.23			3.23	3.55
1440 For non-Chloride accelerator, add					5.70			5.70	6.25
1450 For Chloride accelerator, per 1%, add					3			3	3.30
1460 For fiber reinforcing, synthetic (1 lb./C.Y.), add					6.70			6.70	7.35
1500 For Saturday delivery, add					8.50			8.50	9.35
1510 For truck holding/waiting time past 1st hour per load, add				Hr.	96.50			96.50	106
1520 For short load (less than 4 C.Y.), add per load				Ea.	74.50			74.50	82
2000 For all lightweight aggregate, add				C.Y.	45%				

03 31 13.70 Placing Concrete

	Crew	Daily Output	Labor-Hours	Unit	Material	2017 Bare Costs Labor	2017 Bare Costs Equipment	Total	Total Incl O&P
0010 **PLACING CONCRETE** R033105-70									
0020 Includes labor and equipment to place, level (strike off) and consolidate									
0050 Beams, elevated, small beams, pumped	C-20	60	1.067	C.Y.		45	14.40	59.40	84.50
0100 With crane and bucket	C-7	45	1.600			68.50	25.50	94	133
0200 Large beams, pumped	C-20	90	.711			30	9.60	39.60	56
0250 With crane and bucket	C-7	65	1.108			47.50	17.80	65.30	91.50
0400 Columns, square or round, 12" thick, pumped	C-20	60	1.067			45	14.40	59.40	84.50
0450 With crane and bucket	C-7	40	1.800			77	29	106	149
0600 18" thick, pumped	C-20	90	.711			30	9.60	39.60	56
0650 With crane and bucket	C-7	55	1.309			56	21	77	109
0800 24" thick, pumped	C-20	92	.696			29.50	9.40	38.90	55

03 31 Structural Concrete

03 31 13 – Heavyweight Structural Concrete

03 31 13.70 Placing Concrete

		Crew	Daily Output	Labor-Hours	Unit	Material	2017 Bare Costs Labor	2017 Bare Costs Equipment	Total	Total Incl O&P
0850	With crane and bucket	C-7	70	1.029	C.Y.		44	16.50	60.50	85
1000	36" thick, pumped	C-20	140	.457			19.25	6.15	25.40	36.50
1050	With crane and bucket	C-7	100	.720			31	11.55	42.55	59.50
1400	Elevated slabs, less than 6" thick, pumped	C-20	140	.457			19.25	6.15	25.40	36.50
1450	With crane and bucket	C-7	95	.758			32.50	12.15	44.65	63
1500	6" to 10" thick, pumped	C-20	160	.400			16.85	5.40	22.25	31.50
1550	With crane and bucket	C-7	110	.655			28	10.50	38.50	54
1600	Slabs over 10" thick, pumped	C-20	180	.356			15	4.79	19.79	28.50
1650	With crane and bucket	C-7	130	.554			23.50	8.90	32.40	46
1900	Footings, continuous, shallow, direct chute	C-6	120	.400			16.30	.47	16.77	25.50
1950	Pumped	C-20	150	.427			18	5.75	23.75	34
2000	With crane and bucket	C-7	90	.800			34.50	12.85	47.35	66
2100	Footings, continuous, deep, direct chute	C-6	140	.343			14	.40	14.40	22
2150	Pumped	C-20	160	.400			16.85	5.40	22.25	31.50
2200	With crane and bucket	C-7	110	.655			28	10.50	38.50	54
2400	Footings, spread, under 1 C.Y., direct chute	C-6	55	.873			35.50	1.03	36.53	55
2450	Pumped	C-20	65	.985			41.50	13.25	54.75	77.50
2500	With crane and bucket	C-7	45	1.600			68.50	25.50	94	133
2600	Over 5 C.Y., direct chute	C-6	120	.400			16.30	.47	16.77	25.50
2650	Pumped	C-20	150	.427			18	5.75	23.75	34
2700	With crane and bucket	C-7	100	.720			31	11.55	42.55	59.50
2900	Foundation mats, over 20 C.Y., direct chute	C-6	350	.137			5.60	.16	5.76	8.70
2950	Pumped	C-20	400	.160			6.75	2.16	8.91	12.60
3000	With crane and bucket	C-7	300	.240			10.25	3.85	14.10	19.90
3200	Grade beams, direct chute	C-6	150	.320			13.05	.38	13.43	20.50
3250	Pumped	C-20	180	.356			15	4.79	19.79	28.50
3300	With crane and bucket	C-7	120	.600			25.50	9.65	35.15	49.50
3500	High rise, for more than 5 stories, pumped, add per story	C-20	2100	.030			1.29	.41	1.70	2.41
3510	With crane and bucket, add per story	C-7	2100	.034			1.47	.55	2.02	2.84
3700	Pile caps, under 5 C.Y., direct chute	C-6	90	.533			21.50	.63	22.13	33.50
3750	Pumped	C-20	110	.582			24.50	7.85	32.35	46
3800	With crane and bucket	C-7	80	.900			38.50	14.45	52.95	74.50
3850	Pile cap, 5 C.Y. to 10 C.Y., direct chute	C-6	175	.274			11.20	.32	11.52	17.40
3900	Pumped	C-20	200	.320			13.50	4.31	17.81	25.50
3950	With crane and bucket	C-7	150	.480			20.50	7.70	28.20	40
4000	Over 10 C.Y., direct chute	C-6	215	.223			9.10	.26	9.36	14.15
4050	Pumped	C-20	240	.267			11.25	3.59	14.84	21
4100	With crane and bucket	C-7	185	.389			16.65	6.25	22.90	32.50
4300	Slab on grade, up to 6" thick, direct chute	C-6	110	.436			17.80	.51	18.31	27.50
4350	Pumped	C-20	130	.492			21	6.65	27.65	39
4400	With crane and bucket	C-7	110	.655			28	10.50	38.50	54
4600	Over 6" thick, direct chute	C-6	165	.291			11.85	.34	12.19	18.45
4650	Pumped	C-20	185	.346			14.60	4.66	19.26	27
4700	With crane and bucket	C-7	145	.497			21.50	7.95	29.45	41.50
4900	Walls, 8" thick, direct chute	C-6	90	.533			21.50	.63	22.13	33.50
4950	Pumped	C-20	100	.640			27	8.65	35.65	50.50
5000	With crane and bucket	C-7	80	.900			38.50	14.45	52.95	74.50
5050	12" thick, direct chute	C-6	100	.480			19.55	.57	20.12	30.50
5100	Pumped	C-20	110	.582			24.50	7.85	32.35	46
5200	With crane and bucket	C-7	90	.800			34.50	12.85	47.35	66
5300	15" thick, direct chute	C-6	105	.457			18.65	.54	19.19	29
5350	Pumped	C-20	120	.533			22.50	7.20	29.70	42.50
5400	With crane and bucket	C-7	95	.758			32.50	12.15	44.65	63

03 31 Structural Concrete

03 31 13 – Heavyweight Structural Concrete

03 31 13.70 Placing Concrete

03 31 13.70 Placing Concrete		Crew	Daily Output	Labor-Hours	Unit	Material	2017 Bare Costs Labor	Equipment	Total	Total Incl O&P
5600	Wheeled concrete dumping, add to placing costs above									
5610	Walking cart, 50' haul, add	C-18	32	.281	C.Y.		11.05	1.81	12.86	18.95
5620	150' haul, add		24	.375			14.75	2.42	17.17	25
5700	250' haul, add		18	.500			19.70	3.22	22.92	33.50
5800	Riding cart, 50' haul, add	C-19	80	.113			4.43	1.21	5.64	8.15
5810	150' haul, add		60	.150			5.90	1.61	7.51	10.80
5900	250' haul, add		45	.200			7.85	2.15	10	14.40
6000	Concrete in-fill for pan-type metal stairs and landings. Manual placement									
6010	includes up to 50' horizontal haul from point of concrete discharge.									
6100	Stair pan treads, 2" deep									
6110	Flights in 1st floor level up/down from discharge point	C-8A	3200	.015	S.F.		.63		.63	.96
6120	2nd floor level		2500	.019			.81		.81	1.22
6130	3rd floor level		2000	.024			1.01		1.01	1.53
6140	4th floor level		1800	.027			1.12		1.12	1.70
6200	Intermediate stair landings, pan-type 4" deep									
6210	Flights in 1st floor level up/down from discharge point	C-8A	2000	.024	S.F.		1.01		1.01	1.53
6220	2nd floor level		1500	.032			1.35		1.35	2.04
6230	3rd floor level		1200	.040			1.68		1.68	2.55
6240	4th floor level		1000	.048			2.02		2.02	3.06

03 35 Concrete Finishing

03 35 13 – High-Tolerance Concrete Floor Finishing

03 35 13.30 Finishing Floors, High Tolerance

03 35 13.30 Finishing Floors, High Tolerance		Crew	Daily Output	Labor-Hours	Unit	Material	2017 Bare Costs Labor	Equipment	Total	Total Incl O&P
0010	**FINISHING FLOORS, HIGH TOLERANCE**									
0012	Finishing of fresh concrete flatwork requires that concrete									
0013	first be placed, struck off & consolidated									
0015	Basic finishing for various unspecified flatwork									
0100	Bull float only	C-10	4000	.006	S.F.		.27		.27	.40
0125	Bull float & manual float		2000	.012			.53		.53	.80
0150	Bull float, manual float, & broom finish, w/edging & joints		1850	.013			.57		.57	.86
0200	Bull float, manual float & manual steel trowel		1265	.019			.84		.84	1.26
0210	For specified Random Access Floors in ACI Classes 1, 2, 3 and 4 to achieve									
0215	Composite Overall Floor Flatness and Levelness values up to FF35/FL25									
0250	Bull float, machine float & machine trowel (walk-behind)	C-10C	1715	.014	S.F.		.62	.03	.65	.96
0300	Power screed, bull float, machine float & trowel (walk-behind)	C-10D	2400	.010			.44	.05	.49	.71
0350	Power screed, bull float, machine float & trowel (ride-on)	C-10E	4000	.006			.27	.06	.33	.47
0352	For specified Random Access Floors in ACI Classes 5, 6, 7 and 8 to achieve									
0354	Composite Overall Floor Flatness and Levelness values up to FF50/FL50									
0356	Add for two-dimensional restraightening after power float	C-10	6000	.004	S.F.		.18		.18	.27
0358	For specified Random or Defined Access Floors in ACI Class 9 to achieve									
0360	Composite Overall Floor Flatness and Levelness values up to FF100/FL100									
0362	Add for two-dimensional restraightening after bull float & power float	C-10	3000	.008	S.F.		.35		.35	.53
0364	For specified Superflat Defined Access Floors in ACI Class 9 to achieve									
0366	Minimum Floor Flatness and Levelness values of FF100/FL100									
0368	Add for 2-dim'l restraightening after bull float, power float, power trowel	C-10	2000	.012	S.F.		.53		.53	.80

03 35 16 – Heavy-Duty Concrete Floor Finishing

03 35 16.30 Finishing Floors, Heavy-Duty

03 35 16.30 Finishing Floors, Heavy-Duty		Crew	Daily Output	Labor-Hours	Unit	Material	2017 Bare Costs Labor	Equipment	Total	Total Incl O&P
0010	**FINISHING FLOORS, HEAVY-DUTY**									
1800	Floor abrasives, dry shake on fresh concrete, .25 psf, aluminum oxide	1 Cefi	850	.009	S.F.	.54	.44		.98	1.24
1850	Silicon carbide		850	.009		.81	.44		1.25	1.54

03 35 Concrete Finishing

03 35 16 – Heavy-Duty Concrete Floor Finishing

03 35 16.30 Finishing Floors, Heavy-Duty

		Crew	Daily Output	Labor-Hours	Unit	Material	2017 Bare Costs Labor	Equipment	Total	Total Incl O&P
2000	Floor hardeners, dry shake, metallic, light service, .50 psf	1 Cefi	850	.009	S.F.	.52	.44		.96	1.22
2050	Medium service, .75 psf		750	.011		.78	.50		1.28	1.60
2100	Heavy service, 1.0 psf		650	.012		1.04	.58		1.62	2
2150	Extra heavy, 1.5 psf		575	.014		1.56	.65		2.21	2.69
2300	Non-metallic, light service, .50 psf		850	.009		.16	.44		.60	.83
2350	Medium service, .75 psf		750	.011		.24	.50		.74	1.01
2400	Heavy service, 1.00 psf		650	.012		.33	.58		.91	1.22
2450	Extra heavy, 1.50 psf		575	.014		.49	.65		1.14	1.51
2800	Trap rock wearing surface, dry shake, for monolithic floors									
2810	2.0 psf	C-10B	1250	.032	S.F.	.02	1.35	.20	1.57	2.29
3800	Dustproofing, liquid, for cured concrete, solvent-based, 1 coat	1 Cefi	1900	.004		.16	.20		.36	.47
3850	2 coats		1300	.006		.59	.29		.88	1.08
4000	Epoxy-based, 1 coat		1500	.005		.15	.25		.40	.54
4050	2 coats		1500	.005		.30	.25		.55	.70

03 35 19 – Colored Concrete Finishing

03 35 19.30 Finishing Floors, Colored

		Crew	Daily Output	Labor-Hours	Unit	Material	2017 Bare Costs Labor	Equipment	Total	Total Incl O&P
0010	**FINISHING FLOORS, COLORED**									
3000	Floor coloring, dry shake on fresh concrete (0.6 psf)	1 Cefi	1300	.006	S.F.	.41	.29		.70	.88
3050	(1.0 psf)	"	625	.013	"	.68	.60		1.28	1.64
3100	Colored dry shake powder only				Lb.	.68			.68	.75
3600	1/2" topping using 0.6 psf dry shake powdered color	C-10B	590	.068	S.F.	5.20	2.86	.43	8.49	10.55
3650	1.0 psf dry shake powdered color	"	590	.068	"	5.50	2.86	.43	8.79	10.85

03 35 23 – Exposed Aggregate Concrete Finishing

03 35 23.30 Finishing Floors, Exposed Aggregate

		Crew	Daily Output	Labor-Hours	Unit	Material	2017 Bare Costs Labor	Equipment	Total	Total Incl O&P
0010	**FINISHING FLOORS, EXPOSED AGGREGATE**									
1600	Exposed local aggregate finish, seeded on fresh concrete, 3 lb. per S.F.	1 Cefi	625	.013	S.F.	.20	.60		.80	1.11
1650	4 lb. per S.F.	"	465	.017	"	.33	.81		1.14	1.57

03 35 29 – Tooled Concrete Finishing

03 35 29.30 Finishing Floors, Tooled

		Crew	Daily Output	Labor-Hours	Unit	Material	2017 Bare Costs Labor	Equipment	Total	Total Incl O&P
0010	**FINISHING FLOORS, TOOLED**									
4400	Stair finish, fresh concrete, float finish	1 Cefi	275	.029	S.F.		1.36		1.36	2.02
4500	Steel trowel finish		200	.040			1.87		1.87	2.78
4600	Silicon carbide finish, dry shake on fresh concrete, .25 psf		150	.053		.54	2.50		3.04	4.30

03 35 29.60 Finishing Walls

		Crew	Daily Output	Labor-Hours	Unit	Material	2017 Bare Costs Labor	Equipment	Total	Total Incl O&P
0010	**FINISHING WALLS**									
0020	Break ties and patch voids	1 Cefi	540	.015	S.F.	.04	.69		.73	1.08
0050	Burlap rub with grout		450	.018		.04	.83		.87	1.29
0100	Carborundum rub, dry		270	.030			1.39		1.39	2.06
0150	Wet rub		175	.046			2.14		2.14	3.18
0300	Bush hammer, green concrete	B-39	1000	.048			1.99	.21	2.20	3.27
0350	Cured concrete	"	650	.074			3.06	.33	3.39	5.05
0500	Acid etch	1 Cefi	575	.014		.14	.65		.79	1.12
0600	Float finish, 1/16" thick	"	300	.027		.39	1.25		1.64	2.29
0700	Sandblast, light penetration	E-11	1100	.029		.52	1.28	.20	2	2.84
0750	Heavy penetration	"	375	.085		1.04	3.76	.58	5.38	7.85
0850	Grind form fins flush	1 Clab	700	.011	L.F.		.45		.45	.69

03 35 Concrete Finishing

03 35 33 – Stamped Concrete Finishing

03 35 33.50 Slab Texture Stamping

	03 35 33.50 Slab Texture Stamping	Crew	Daily Output	Labor-Hours	Unit	Material	2017 Bare Costs Labor	Equipment	Total	Total Incl O&P
0010	**SLAB TEXTURE STAMPING**									
0050	Stamping requires that concrete first be placed, struck off, consolidated,									
0060	bull floated and free of bleed water. Decorative stamping tasks include:									
0100	Step 1 - first application of dry shake colored hardener	1 Cefi	6400	.001	S.F.	.43	.06		.49	.56
0110	Step 2 - bull float		6400	.001			.06		.06	.09
0130	Step 3 - second application of dry shake colored hardener		6400	.001		.21	.06		.27	.32
0140	Step 4 - bull float, manual float & steel trowel	3 Cefi	1280	.019			.88		.88	1.31
0150	Step 5 - application of dry shake colored release agent	1 Cefi	6400	.001		.10	.06		.16	.20
0160	Step 6 - place, tamp & remove mats	3 Cefi	2400	.010		.88	.47		1.35	1.67
0170	Step 7 - touch up edges, mat joints & simulated grout lines	1 Cefi	1280	.006			.29		.29	.44
0300	Alternate stamping estimating method includes all tasks above	4 Cefi	800	.040		1.62	1.87		3.49	4.56
0400	Step 8 - pressure wash @ 3000 psi after 24 hours	1 Cefi	1600	.005			.23		.23	.35
0500	Step 9 - roll 2 coats cure/seal compound when dry	"	800	.010		.63	.47		1.10	1.39

03 35 43 – Polished Concrete Finishing

03 35 43.10 Polished Concrete Floors

	03 35 43.10 Polished Concrete Floors	Crew	Daily Output	Labor-Hours	Unit	Material	2017 Bare Costs Labor	Equipment	Total	Total Incl O&P
0010	**POLISHED CONCRETE FLOORS** R033543-10									
0015	Processing of cured concrete to include grinding, honing,									
0020	and polishing of interior floors with 22" segmented diamond									
0025	planetary floor grinder (2 passes in different directions per grit)									
0100	Removal of pre-existing coatings, dry, with carbide discs using									
0105	dry vacuum pick-up system, final hand sweeping									
0110	Glue, adhesive or tar	J-4	1.60	15	M.S.F.	21	665	138	824	1,175
0120	Paint, epoxy, 1 coat		3.60	6.667		21	295	61.50	377.50	535
0130	2 coats		1.80	13.333		21	590	123	734	1,050
0200	Grinding and edging, wet, including wet vac pick-up and auto									
0205	scrubbing between grit changes									
0210	40-grit diamond/metal matrix	J-4A	1.60	20	M.S.F.	33.50	860	286	1,179.50	1,650
0220	80-grit diamond/metal matrix		2	16		33.50	690	229	952.50	1,325
0230	120-grit diamond/metal matrix		2.40	13.333		33.50	575	191	799.50	1,100
0240	200-grit diamond/metal matrix		2.80	11.429		33.50	490	164	687.50	955
0300	Spray on dye or stain (1 coat)	1 Cefi	16	.500		224	23.50		247.50	281
0400	Spray on densifier/hardener (2 coats)	"	8	1		365	47		412	470
0410	Auto scrubbing after 2nd coat, when dry	J-4B	16	.500			19.60	14.75	34.35	46.50
0500	Honing and edging, wet, including wet vac pick-up and auto									
0505	scrubbing between grit changes									
0510	100-grit diamond/resin matrix	J-4A	2.80	11.429	M.S.F.	33.50	490	164	687.50	955
0520	200-grit diamond/resin matrix	"	2.80	11.429	"	33.50	490	164	687.50	955
0530	Dry, including dry vacuum pick-up system, final hand sweeping									
0540	400-grit diamond/resin matrix	J-4A	2.80	11.429	M.S.F.	33.50	490	164	687.50	955
0600	Polishing and edging, dry, including dry vac pick-up and hand									
0605	sweeping between grit changes									
0610	800-grit diamond/resin matrix	J-4A	2.80	11.429	M.S.F.	33.50	490	164	687.50	955
0620	1500-grit diamond/resin matrix		2.80	11.429		33.50	490	164	687.50	955
0630	3000-grit diamond/resin matrix		2.80	11.429		33.50	490	164	687.50	955
0700	Auto scrubbing after final polishing step	J-4B	16	.500			19.60	14.75	34.35	46.50

03 37 Specialty Placed Concrete

03 37 13 - Shotcrete

03 37 13.30 Gunite (Dry-Mix)	Crew	Daily Output	Labor-Hours	Unit	Material	2017 Bare Costs Labor	Equipment	Total	Total Incl O&P
0010 **GUNITE (DRY-MIX)**									
0020 Typical in place, 1" layers, no mesh included	C-16	2000	.028	S.F.	.39	1.22	.19	1.80	2.49
0100 Mesh for gunite 2 x 2, #12	2 Rodm	800	.020		.67	1.09		1.76	2.41
0150 #4 reinforcing bars @ 6" each way	"	500	.032		1.57	1.74		3.31	4.41
0300 Typical in place, including mesh, 2" thick, flat surfaces	C-16	1000	.056		1.46	2.45	.38	4.29	5.75
0350 Curved surfaces		500	.112		1.46	4.89	.76	7.11	9.85
0500 4" thick, flat surfaces		750	.075		2.24	3.26	.51	6.01	7.95
0550 Curved surfaces		350	.160		2.24	7	1.08	10.32	14.25
0900 Prepare old walls, no scaffolding, good condition	C-10	1000	.024			1.06		1.06	1.59
0950 Poor condition	"	275	.087			3.86		3.86	5.80
1100 For high finish requirement or close tolerance, add						50%			
1150 Very high						110%			

03 37 13.60 Shotcrete (Wet-Mix)	Crew	Daily Output	Labor-Hours	Unit	Material	2017 Bare Costs Labor	Equipment	Total	Total Incl O&P
0010 **SHOTCRETE (WET-MIX)**									
0020 Wet mix, placed @ up to 12 C.Y. per hour, 3000 psi	C-8C	80	.600	C.Y.	127	26	5.50	158.50	186
0100 Up to 35 C.Y. per hour	C-8E	240	.200	"	115	8.55	2.10	125.65	141
1010 Fiber reinforced, 1" thick	C-8C	1740	.028	S.F.	.93	1.19	.25	2.37	3.11
1020 2" thick		900	.053		1.86	2.30	.49	4.65	6.10
1030 3" thick		825	.058		2.79	2.51	.53	5.83	7.50
1040 4" thick		750	.064		3.72	2.76	.59	7.07	8.95

03 39 Concrete Curing

03 39 13 - Water Concrete Curing

03 39 13.50 Water Curing	Crew	Daily Output	Labor-Hours	Unit	Material	2017 Bare Costs Labor	Equipment	Total	Total Incl O&P
0010 **WATER CURING**									
0015 With burlap, 4 uses assumed, 7.5 oz.	2 Clab	55	.291	C.S.F.	14.75	11.40		26.15	33.50
0100 10 oz.	"	55	.291	"	26.50	11.40		37.90	46.50
0400 Curing blankets, 1" to 2" thick, buy				S.F.	.26			.26	.29

03 39 23 - Membrane Concrete Curing

03 39 23.13 Chemical Compound Membrane Concrete Curing	Crew	Daily Output	Labor-Hours	Unit	Material	2017 Bare Costs Labor	Equipment	Total	Total Incl O&P
0010 **CHEMICAL COMPOUND MEMBRANE CONCRETE CURING**									
0300 Sprayed membrane curing compound	2 Clab	95	.168	C.S.F.	12.70	6.60		19.30	24
0700 Curing compound, solvent based, 400 S.F./gal., 55 gallon lots				Gal.	21			21	23
0720 5 gallon lots					31.50			31.50	35
0800 Curing compound, water based, 250 S.F./gal., 55 gallon lots					20.50			20.50	22.50
0820 5 gallon lots					23.50			23.50	26

03 39 23.23 Sheet Membrane Concrete Curing	Crew	Daily Output	Labor-Hours	Unit	Material	2017 Bare Costs Labor	Equipment	Total	Total Incl O&P
0010 **SHEET MEMBRANE CONCRETE CURING**									
0200 Curing blanket, burlap/poly, 2-ply	2 Clab	70	.229	C.S.F.	26.50	8.95		35.45	43

03 41 Precast Structural Concrete

03 41 13 – Precast Concrete Hollow Core Planks

03 41 13.50 Precast Slab Planks

		Crew	Daily Output	Labor-Hours	Unit	Material	2017 Bare Costs Labor	Equipment	Total	Total Incl O&P
0010	**PRECAST SLAB PLANKS** R034105-30									
0020	Prestressed roof/floor members, grouted, solid, 4" thick	C-11	2400	.030	S.F.	7.30	1.62	.83	9.75	11.60
0050	6" thick		2800	.026		8.20	1.39	.71	10.30	12.05
0100	Hollow, 8" thick		3200	.023		9.15	1.21	.63	10.99	12.70
0150	10" thick		3600	.020		9.50	1.08	.56	11.14	12.80
0200	12" thick		4000	.018		9.85	.97	.50	11.32	12.95

03 41 16 – Precast Concrete Slabs

03 41 16.20 Precast Concrete Channel Slabs

		Crew	Daily Output	Labor-Hours	Unit	Material	2017 Bare Costs Labor	Equipment	Total	Total Incl O&P
0010	**PRECAST CONCRETE CHANNEL SLABS**									
0335	Lightweight concrete channel slab, long runs, 2-3/4" thick	C-12	1575	.030	S.F.	10.40	1.49	.36	12.25	14.15
0375	3-3/4" thick		1550	.031		10.70	1.52	.37	12.59	14.50
0475	4-3/4" thick		1525	.031		11.95	1.54	.38	13.87	15.85
1275	Short pieces, 2-3/4" thick		785	.061		15.60	3	.73	19.33	22.50
1375	3-3/4" thick		770	.062		16.05	3.05	.74	19.84	23
1475	4-3/4" thick		762	.063		17.90	3.09	.75	21.74	25

03 41 16.50 Precast Lightweight Concrete Plank

		Crew	Daily Output	Labor-Hours	Unit	Material	2017 Bare Costs Labor	Equipment	Total	Total Incl O&P
0010	**PRECAST LIGHTWEIGHT CONCRETE PLANK**									
0015	Lightweight plank, nailable, T&G, 2" thick	C-12	1800	.027	S.F.	9.25	1.31	.32	10.88	12.55
0150	For premium ceiling finish, add				"	10%				
0200	For sloping roofs, slope over 4 in 12, add						25%			
0250	Slope over 6 in 12, add						150%			

03 41 23 – Precast Concrete Stairs

03 41 23.50 Precast Stairs

		Crew	Daily Output	Labor-Hours	Unit	Material	2017 Bare Costs Labor	Equipment	Total	Total Incl O&P
0010	**PRECAST STAIRS**									
0020	Precast concrete treads on steel stringers, 3' wide	C-12	75	.640	Riser	151	31.50	7.65	190.15	222
0300	Front entrance, 5' wide with 48" platform, 2 risers		16	3	Flight	575	147	36	758	895
0350	5 risers		12	4		910	196	47.50	1,153.50	1,350
0500	6' wide, 2 risers		15	3.200		635	157	38	830	980
0550	5 risers		11	4.364		1,000	214	52	1,266	1,475
0700	7' wide, 2 risers		14	3.429		810	168	41	1,019	1,200
0750	5 risers		10	4.800		1,350	235	57.50	1,642.50	1,900
1200	Basement entrance stairwell, 6 steps, incl. steel bulkhead door	B-51	22	2.182		1,700	88	9.90	1,797.90	2,025
1250	14 steps	"	11	4.364		2,850	176	19.75	3,045.75	3,425

03 41 33 – Precast Structural Pretensioned Concrete

03 41 33.10 Precast Beams

		Crew	Daily Output	Labor-Hours	Unit	Material	2017 Bare Costs Labor	Equipment	Total	Total Incl O&P
0010	**PRECAST BEAMS** R034105-30									
0011	L-shaped, 20' span, 12" x 20"	C-11	32	2.250	Ea.	3,750	121	62.50	3,933.50	4,400
0060	18" x 36"		24	3		5,125	162	83.50	5,370.50	6,000
0100	24" x 44"		22	3.273		6,150	177	91	6,418	7,175
0150	30' span, 12" x 36"		24	3		6,950	162	83.50	7,195.50	8,000
0200	18" x 44"		20	3.600		8,450	194	100	8,744	9,725
0250	24" x 52"		16	4.500		10,100	243	125	10,468	11,700
0400	40' span, 12" x 52"		20	3.600		10,800	194	100	11,094	12,300
0450	18" x 52"		16	4.500		12,000	243	125	12,368	13,700
0500	24" x 52"		12	6		13,500	325	167	13,992	15,600
1200	Rectangular, 20' span, 12" x 20"		32	2.250		3,650	121	62.50	3,833.50	4,275
1250	18" x 36"		24	3		4,500	162	83.50	4,745.50	5,300
1300	24" x 44"		22	3.273		5,375	177	91	5,643	6,325
1400	30' span, 12" x 36"		24	3		6,075	162	83.50	6,320.50	7,025
1450	18" x 44"		20	3.600		7,325	194	100	7,619	8,475
1500	24" x 52"		16	4.500		8,825	243	125	9,193	10,200

03 41 Precast Structural Concrete

03 41 33 – Precast Structural Pretensioned Concrete

03 41 33.10 Precast Beams

03 41 33.10 Precast Beams		Crew	Daily Output	Labor-Hours	Unit	Material	2017 Bare Costs Labor	Equipment	Total	Total Incl O&P
1600	40' span, 12" x 52"	C-11	20	3.600	Ea.	9,025	194	100	9,319	10,400
1650	18" x 52"		16	4.500		10,300	243	125	10,668	11,800
1700	24" x 52"		12	6		11,800	325	167	12,292	13,600
2000	"T" shaped, 20' span, 12" x 20"		32	2.250		4,275	121	62.50	4,458.50	4,975
2050	18" x 36"		24	3		5,750	162	83.50	5,995.50	6,675
2100	24" x 44"		22	3.273		6,900	177	91	7,168	8,000
2200	30' span, 12" x 36"		24	3		7,900	162	83.50	8,145.50	9,050
2250	18" x 44"		20	3.600		9,575	194	100	9,869	10,900
2300	24" x 52"		16	4.500		11,500	243	125	11,868	13,100
2500	40' span, 12" x 52"		20	3.600		13,000	194	100	13,294	14,700
2550	18" x 52"		16	4.500		13,800	243	125	14,168	15,700
2600	24" x 52"	▼	12	6	▼	15,300	325	167	15,792	17,500

03 41 33.15 Precast Columns

03 41 33.15 Precast Columns		Crew	Daily Output	Labor-Hours	Unit	Material	Labor	Equipment	Total	Total Incl O&P
0010	**PRECAST COLUMNS** R034105-30									
0020	Rectangular to 12' high, 16" x 16"	C-11	120	.600	L.F.	214	32.50	16.65	263.15	305
0050	24" x 24"		96	.750		289	40.50	21	350.50	405
0300	24' high, 28" x 28"		192	.375		335	20	10.40	365.40	415
0350	36" x 36"	▼	144	.500	▼	450	27	13.90	490.90	555

03 41 33.25 Precast Joists

03 41 33.25 Precast Joists		Crew	Daily Output	Labor-Hours	Unit	Material	Labor	Equipment	Total	Total Incl O&P
0010	**PRECAST JOISTS** R034105-30									
0015	40 psf L.L., 6" deep for 12' spans	C-12	600	.080	L.F.	30	3.92	.95	34.87	40
0050	8" deep for 16' spans		575	.083		50	4.09	1	55.09	62.50
0100	10" deep for 20' spans		550	.087		87.50	4.27	1.04	92.81	104
0150	12" deep for 24' spans	▼	525	.091	▼	120	4.48	1.09	125.57	140

03 41 33.60 Precast Tees

03 41 33.60 Precast Tees		Crew	Daily Output	Labor-Hours	Unit	Material	Labor	Equipment	Total	Total Incl O&P
0010	**PRECAST TEES** R034105-30									
0020	Quad tee, short spans, roof	C-11	7200	.010	S.F.	9.55	.54	.28	10.37	11.70
0050	Floor		7200	.010		9.55	.54	.28	10.37	11.70
0200	Double tee, floor members, 60' span		8400	.009		10.75	.46	.24	11.45	12.80
0250	80' span		8000	.009		14.85	.49	.25	15.59	17.40
0300	Roof members, 30' span		4800	.015		10.30	.81	.42	11.53	13.10
0350	50' span		6400	.011		10.50	.61	.31	11.42	12.90
0400	Wall members, up to 55' high		3600	.020		14.25	1.08	.56	15.89	18
0500	Single tee roof members, 40' span		3200	.023		14.30	1.21	.63	16.14	18.40
0550	80' span		5120	.014		14.75	.76	.39	15.90	17.90
0600	100' span		6000	.012		22	.65	.33	22.98	25.50
0650	120' span	▼	6000	.012	▼	23.50	.65	.33	24.48	27.50
1000	Double tees, floor members									
1100	Lightweight, 20" x 8' wide, 45' span	C-11	20	3.600	Ea.	3,775	194	100	4,069	4,575
1150	24" x 8' wide, 50' span		18	4		4,200	216	111	4,527	5,100
1200	32" x 10' wide, 60' span		16	4.500		6,300	243	125	6,668	7,450
1250	Standard weight, 12" x 8' wide, 20' span		22	3.273		1,525	177	91	1,793	2,075
1300	16" x 8' wide, 25' span		20	3.600		1,900	194	100	2,194	2,525
1350	18" x 8' wide, 30' span		20	3.600		2,275	194	100	2,569	2,950
1400	20" x 8' wide, 45' span		18	4		3,425	216	111	3,752	4,250
1450	24" x 8' wide, 50' span		16	4.500		3,800	243	125	4,168	4,725
1500	32" x 10' wide, 60' span	▼	14	5.143	▼	5,725	278	143	6,146	6,900
2000	Roof members									
2050	Lightweight, 20" x 8' wide, 40' span	C-11	20	3.600	Ea.	3,350	194	100	3,644	4,125
2100	24" x 8' wide, 50' span		18	4		4,200	216	111	4,527	5,100
2150	32" x 10' wide, 60' span		16	4.500		6,300	243	125	6,668	7,450
2200	Standard weight, 12" x 8' wide, 30' span		22	3.273		2,275	177	91	2,543	2,925

For customer support on your Building Construction Costs with RSMeans Data, call 800.448.8182.

85

03 41 Precast Structural Concrete

03 41 33 – Precast Structural Pretensioned Concrete

03 41 33.60 Precast Tees		Crew	Daily Output	Labor-Hours	Unit	Material	2017 Bare Costs		Total	Total Incl O&P
							Labor	Equipment		
2250	16" x 8' wide, 30' span	C-11	20	3.600	Ea.	2,400	194	100	2,694	3,075
2300	18" x 8' wide, 30' span		20	3.600		2,525	194	100	2,819	3,200
2350	20" x 8' wide, 40' span		18	4		3,050	216	111	3,377	3,825
2400	24" x 8' wide, 50' span		16	4.500		3,800	243	125	4,168	4,725
2450	32" x 10' wide, 60' span		14	5.143		5,725	278	143	6,146	6,900

03 45 Precast Architectural Concrete

03 45 13 – Faced Architectural Precast Concrete

03 45 13.50 Precast Wall Panels

03 45 13.50 Precast Wall Panels		Crew	Daily Output	Labor-Hours	Unit	Material	Labor	Equipment	Total	Total Incl O&P
0010	**PRECAST WALL PANELS** R034513-10									
0050	Uninsulated, smooth gray									
0150	Low rise, 4' x 8' x 4" thick	C-11	320	.225	S.F.	25	12.15	6.25	43.40	54.50
0210	8' x 8', 4" thick		576	.125		25	6.75	3.47	35.22	42.50
0250	8' x 16' x 4" thick		1024	.070		24.50	3.80	1.95	30.25	35.50
0600	High rise, 4' x 8' x 4" thick		288	.250		25	13.50	6.95	45.45	57
0650	8' x 8' x 4" thick		512	.141		25	7.60	3.91	36.51	44
0700	8' x 16' x 4" thick		768	.094		24.50	5.05	2.60	32.15	38
0750	10' x 20', 6" thick		1400	.051		42	2.78	1.43	46.21	52
0800	Insulated panel, 2" polystyrene, add					1.14			1.14	1.25
0850	2" urethane, add					.76			.76	.84
1200	Finishes, white, add					3.13			3.13	3.45
1250	Exposed aggregate, add					.70			.70	.77
1300	Granite faced, domestic, add					30.50			30.50	33.50
1350	Brick faced, modular, red, add					4.53			4.53	4.99
2200	Fiberglass reinforced cement with urethane core									
2210	R20, 8' x 8', 5" plain finish	E-2	750	.075	S.F.	26	4.02	2.23	32.25	37.50
2220	Exposed aggregate or brick finish	"	600	.093	"	38.50	5.05	2.79	46.34	53.50

03 47 Site-Cast Concrete

03 47 13 – Tilt-Up Concrete

03 47 13.50 Tilt-Up Wall Panels

03 47 13.50 Tilt-Up Wall Panels		Crew	Daily Output	Labor-Hours	Unit	Material	Labor	Equipment	Total	Total Incl O&P
0010	**TILT-UP WALL PANELS** R034713-20									
0015	Wall panel construction, walls only, 5-1/2" thick	C-14	1600	.090	S.F.	5.90	4.34	1.01	11.25	14.25
0100	7-1/2" thick		1550	.093		7.35	4.48	1.04	12.87	16.10
0500	Walls and columns, 5-1/2" thick walls, 12" x 12" columns		1565	.092		8.75	4.44	1.03	14.22	17.55
0550	7-1/2" thick wall, 12" x 12" columns		1370	.105		10.65	5.05	1.18	16.88	21
0800	Columns only, site precast, 12" x 12"		200	.720	L.F.	20.50	35	8.10	63.60	85
0850	16" x 16"		105	1.371	"	30	66	15.40	111.40	151

03 48 Precast Concrete Specialties

03 48 43 – Precast Concrete Trim

03 48 43.40 Precast Lintels

		Crew	Daily Output	Labor-Hours	Unit	Material	2017 Bare Costs Labor	Equipment	Total	Total Incl O&P
0010	**PRECAST LINTELS**, smooth gray, prestressed, stock units only									
0800	4" wide, 8" high, x 4' long	D-10	28	1.143	Ea.	30	55	15.55	100.55	134
0850	8' long		24	1.333		75	64	18.15	157.15	200
1000	6" wide, 8" high, x 4' long		26	1.231		44	59	16.75	119.75	157
1050	10' long		22	1.455		110	70	19.80	199.80	250
1200	8" wide, 8" high, x 4' long		24	1.333		51.50	64	18.15	133.65	175
1250	12' long		20	1.600		168	77	22	267	325
1275	For custom sizes, types, colors, or finishes of precast lintels, add					150%				

03 48 43.90 Precast Window Sills

		Crew	Daily Output	Labor-Hours	Unit	Material	2017 Bare Costs Labor	Equipment	Total	Total Incl O&P
0010	**PRECAST WINDOW SILLS**									
0600	Precast concrete, 4" tapers to 3", 9" wide	D-1	70	.229	L.F.	17.40	9.90		27.30	34.50
0650	11" wide	"	60	.267	"	28.50	11.55		40.05	49

03 51 Cast Roof Decks

03 51 13 – Cementitious Wood Fiber Decks

03 51 13.50 Cementitious/Wood Fiber Planks

			Crew	Daily Output	Labor-Hours	Unit	Material	2017 Bare Costs Labor	Equipment	Total	Total Incl O&P
0010	**CEMENTITIOUS/WOOD FIBER PLANKS**	R051223-50									
0050	Plank, beveled edge, 1" thick		2 Carp	1000	.016	S.F.	3.12	.79		3.91	4.64
0100	1-1/2" thick			975	.016		4.05	.81		4.86	5.70
0150	T & G, 2" thick			950	.017		3.43	.83		4.26	5.05
0200	2-1/2" thick			925	.017		3.74	.85		4.59	5.40
0250	3" thick			900	.018		4.24	.88		5.12	6
1000	Bulb tee, sub-purlin and grout, 6' span, add		E-1	5000	.005		2.14	.26	.03	2.43	2.80
1100	8' span		"	4200	.006		2.14	.31	.03	2.48	2.88

03 51 16 – Gypsum Concrete Roof Decks

03 51 16.50 Gypsum Roof Deck

		Crew	Daily Output	Labor-Hours	Unit	Material	2017 Bare Costs Labor	Equipment	Total	Total Incl O&P
0010	**GYPSUM ROOF DECK**									
1000	Poured gypsum, 2" thick	C-8	6000	.009	S.F.	1.52	.41	.13	2.06	2.44
1100	3" thick	"	4800	.012	"	2.27	.51	.17	2.95	3.45

03 52 Lightweight Concrete Roof Insulation

03 52 16 – Lightweight Insulating Concrete

03 52 16.13 Lightweight Cellular Insulating Concrete

			Crew	Daily Output	Labor-Hours	Unit	Material	2017 Bare Costs Labor	Equipment	Total	Total Incl O&P
0010	**LIGHTWEIGHT CELLULAR INSULATING CONCRETE**	R035216-10									
0020	Portland cement and foaming agent	G	C-8	50	1.120	C.Y.	123	49	16.15	188.15	227

03 52 16.16 Lightweight Aggregate Insulating Concrete

			Crew	Daily Output	Labor-Hours	Unit	Material	2017 Bare Costs Labor	Equipment	Total	Total Incl O&P
0010	**LIGHTWEIGHT AGGREGATE INSULATING CONCRETE**	R035216-10									
0100	Poured vermiculite or perlite, field mix,										
0110	1:6 field mix	G	C-8	50	1.120	C.Y.	267	49	16.15	332.15	385
0200	Ready mix, 1:6 mix, roof fill, 2" thick	G		10000	.006	S.F.	1.48	.24	.08	1.80	2.09
0250	3" thick	G		7700	.007		2.23	.32	.10	2.65	3.05
0400	Expanded volcanic glass rock, 1" thick	G	2 Carp	1500	.011		.52	.53		1.05	1.38
0450	3" thick	G	"	1200	.013		1.56	.66		2.22	2.73

03 53 Concrete Topping

03 53 16 – Iron-Aggregate Concrete Topping

03 53 16.50 Floor Topping

03 53 16.50 Floor Topping	Crew	Daily Output	Labor-Hours	Unit	Material	2017 Bare Costs Labor	Equipment	Total	Total Incl O&P
0010 **FLOOR TOPPING**									
0400 Integral topping/finish, on fresh concrete, using 1:1:2 mix, 3/16" thick	C-10B	1000	.040	S.F.	.12	1.69	.26	2.07	2.96
0450 1/2" thick		950	.042		.31	1.78	.27	2.36	3.33
0500 3/4" thick		850	.047		.47	1.99	.30	2.76	3.84
0600 1" thick		750	.053		.62	2.25	.34	3.21	4.46
0800 Granolithic topping, on fresh or cured concrete, 1:1:1-1/2 mix, 1/2" thick		590	.068		.35	2.86	.43	3.64	5.20
0820 3/4" thick		580	.069		.52	2.91	.44	3.87	5.45
0850 1" thick		575	.070		.69	2.94	.44	4.07	5.70
0950 2" thick		500	.080		1.39	3.38	.51	5.28	7.20
1200 Heavy duty, 1:1:2, 3/4" thick, preshrunk, gray, 20 M.S.F.		320	.125		.86	5.30	.80	6.96	9.80
1300 100 M.S.F.	▼	380	.105	▼	.47	4.44	.67	5.58	7.95

03 54 Cast Underlayment

03 54 13 – Gypsum Cement Underlayment

03 54 13.50 Poured Gypsum Underlayment

03 54 13.50 Poured Gypsum Underlayment	Crew	Daily Output	Labor-Hours	Unit	Material	2017 Bare Costs Labor	Equipment	Total	Total Incl O&P
0010 **POURED GYPSUM UNDERLAYMENT**									
0400 Underlayment, gypsum based, self-leveling 2500 psi, pumped, 1/2" thick	C-8	24000	.002	S.F.	.38	.10	.03	.51	.61
0500 3/4" thick		20000	.003		.57	.12	.04	.73	.86
0600 1" thick	▼	16000	.004		.76	.15	.05	.96	1.12
1400 Hand placed, 1/2" thick	C-18	450	.020		.38	.79	.13	1.30	1.77
1500 3/4" thick	"	300	.030	▼	.57	1.18	.19	1.94	2.65

03 54 16 – Hydraulic Cement Underlayment

03 54 16.50 Cement Underlayment

03 54 16.50 Cement Underlayment	Crew	Daily Output	Labor-Hours	Unit	Material	2017 Bare Costs Labor	Equipment	Total	Total Incl O&P
0010 **CEMENT UNDERLAYMENT**									
2510 Underlayment, P.C. based, self-leveling, 4100 psi, pumped, 1/4" thick	C-8	20000	.003	S.F.	1.63	.12	.04	1.79	2.03
2520 1/2" thick		19000	.003		3.27	.13	.04	3.44	3.84
2530 3/4" thick		18000	.003		4.90	.14	.04	5.08	5.65
2540 1" thick		17000	.003		6.55	.14	.05	6.74	7.45
2550 1-1/2" thick	▼	15000	.004		9.80	.16	.05	10.01	11.10
2560 Hand placed, 1/2" thick	C-18	450	.020		3.27	.79	.13	4.19	4.94
2610 Topping, P.C. based, self-leveling, 6100 psi, pumped, 1/4" thick	C-8	20000	.003		2.40	.12	.04	2.56	2.87
2620 1/2" thick		19000	.003		4.81	.13	.04	4.98	5.55
2630 3/4" thick		18000	.003		7.20	.14	.04	7.38	8.20
2660 1" thick		17000	.003		9.60	.14	.05	9.79	10.80
2670 1-1/2" thick	▼	15000	.004		14.40	.16	.05	14.61	16.15
2680 Hand placed, 1/2" thick	C-18	450	.020	▼	4.81	.79	.13	5.73	6.65

03 62 Non-Shrink Grouting

03 62 13 – Non-Metallic Non-Shrink Grouting

03 62 13.50 Grout, Non-Metallic Non-Shrink

03 62 13.50 Grout, Non-Metallic Non-Shrink	Crew	Daily Output	Labor-Hours	Unit	Material	2017 Bare Costs Labor	Equipment	Total	Total Incl O&P
0010 **GROUT, NON-METALLIC NON-SHRINK**									
0300 Non-shrink, non-metallic, 1" deep	1 Cefi	35	.229	S.F.	6.70	10.70		17.40	23.50
0350 2" deep	"	25	.320	"	13.40	15		28.40	37.50

03 62 16 – Metallic Non-Shrink Grouting

03 62 16.50 Grout, Metallic Non-Shrink

03 62 16.50 Grout, Metallic Non-Shrink	Crew	Daily Output	Labor-Hours	Unit	Material	2017 Bare Costs Labor	Equipment	Total	Total Incl O&P
0010 **GROUT, METALLIC NON-SHRINK**									
0020 Column & machine bases, non-shrink, metallic, 1" deep	1 Cefi	35	.229	S.F.	11.25	10.70		21.95	28.50
0050 2" deep	"	25	.320	"	22.50	15		37.50	47

03 63 Epoxy Grouting

03 63 05 – Grouting of Dowels and Fasteners

03 63 05.10 Epoxy Only

		Crew	Daily Output	Labor-Hours	Unit	Material	2017 Bare Costs Labor	Equipment	Total	Total Incl O&P
0010	**EPOXY ONLY**									
1500	Chemical anchoring, epoxy cartridge, excludes layout, drilling, fastener									
1530	For fastener 3/4" diam. x 6" embedment	2 Skwk	72	.222	Ea.	5.15	11.45		16.60	23.50
1535	1" diam. x 8" embedment		66	.242		7.75	12.45		20.20	27.50
1540	1-1/4" diam. x 10" embedment		60	.267		15.45	13.70		29.15	38
1545	1-3/4" diam. x 12" embedment		54	.296		26	15.25		41.25	52
1550	14" embedment		48	.333		31	17.15		48.15	60.50
1555	2" diam. x 12" embedment		42	.381		41	19.60		60.60	75.50
1560	18" embedment	▼	32	.500	▼	51.50	25.50		77	96

03 81 Concrete Cutting

03 81 13 – Flat Concrete Sawing

03 81 13.50 Concrete Floor/Slab Cutting

		Crew	Daily Output	Labor-Hours	Unit	Material	2017 Bare Costs Labor	Equipment	Total	Total Incl O&P
0010	**CONCRETE FLOOR/SLAB CUTTING**									
0050	Includes blade cost, layout and set-up time									
0300	Saw cut concrete slabs, plain, up to 3" deep	B-89	1060	.015	L.F.	.13	.72	.42	1.27	1.68
0320	Each additional inch of depth		3180	.005		.04	.24	.14	.42	.56
0400	Mesh reinforced, up to 3" deep		980	.016		.14	.78	.45	1.37	1.83
0420	Each additional inch of depth		2940	.005		.05	.26	.15	.46	.61
0500	Rod reinforced, up to 3" deep		800	.020		.18	.95	.55	1.68	2.24
0520	Each additional inch of depth	▼	2400	.007	▼	.06	.32	.18	.56	.74

03 81 13.75 Concrete Saw Blades

		Crew	Daily Output	Labor-Hours	Unit	Material	2017 Bare Costs Labor	Equipment	Total	Total Incl O&P
0010	**CONCRETE SAW BLADES**									
3000	Blades for saw cutting, included in cutting line items									
3020	Diamond, 12" diameter				Ea.	230			230	253
3040	18" diameter					425			425	465
3080	24" diameter					705			705	775
3120	30" diameter					995			995	1,100
3160	36" diameter					1,325			1,325	1,450
3200	42" diameter				▼	2,350			2,350	2,600

03 81 16 – Track Mounted Concrete Wall Sawing

03 81 16.50 Concrete Wall Cutting

		Crew	Daily Output	Labor-Hours	Unit	Material	2017 Bare Costs Labor	Equipment	Total	Total Incl O&P
0010	**CONCRETE WALL CUTTING**									
0750	Includes blade cost, layout and set-up time									
0800	Concrete walls, hydraulic saw, plain, per inch of depth	B-89B	250	.064	L.F.	.04	3.05	3.03	6.12	8
0820	Rod reinforcing, per inch of depth	"	150	.107	"	.06	5.10	5.05	10.21	13.25

03 82 Concrete Boring

03 82 13 – Concrete Core Drilling

03 82 13.10 Core Drilling

		Crew	Daily Output	Labor-Hours	Unit	Material	2017 Bare Costs Labor	Equipment	Total	Total Incl O&P
0010	**CORE DRILLING**									
0015	Includes bit cost, layout and set-up time									
0020	Reinforced concrete slab, up to 6" thick									
0100	1" diameter core	B-89A	17	.941	Ea.	.18	42.50	6.75	49.43	73
0150	For each additional inch of slab thickness in same hole, add		1440	.011		.03	.50	.08	.61	.89
0200	2" diameter core		16.50	.970		.29	44	6.95	51.24	75.50
0250	For each additional inch of slab thickness in same hole, add		1080	.015		.05	.67	.11	.83	1.20
0300	3" diameter core	▼	16	1		.39	45.50	7.15	53.04	78

03 82 13.10 Core Drilling	Crew	Daily Output	Labor-Hours	Unit	Material	2017 Bare Costs Labor	Equipment	Total	Total Incl O&P
0350	For each additional inch of slab thickness in same hole, add	B-89A	720	.022	Ea.	.06	1.01	.16	1.23	1.80
0500	4" diameter core		15	1.067		.50	48.50	7.65	56.65	83
0550	For each additional inch of slab thickness in same hole, add		480	.033		.08	1.51	.24	1.83	2.67
0700	6" diameter core		14	1.143		.77	52	8.20	60.97	89.50
0750	For each additional inch of slab thickness in same hole, add		360	.044		.13	2.01	.32	2.46	3.58
0900	8" diameter core		13	1.231		1.05	56	8.80	65.85	96.50
0950	For each additional inch of slab thickness in same hole, add		288	.056		.17	2.52	.40	3.09	4.50
1100	10" diameter core		12	1.333		1.47	60.50	9.55	71.52	105
1150	For each additional inch of slab thickness in same hole, add		240	.067		.24	3.02	.48	3.74	5.45
1300	12" diameter core		11	1.455		1.78	66	10.40	78.18	114
1350	For each additional inch of slab thickness in same hole, add		206	.078		.30	3.52	.56	4.38	6.35
1500	14" diameter core		10	1.600		2.07	72.50	11.45	86.02	126
1550	For each additional inch of slab thickness in same hole, add		180	.089		.35	4.03	.64	5.02	7.30
1700	18" diameter core		9	1.778		2.80	80.50	12.75	96.05	141
1750	For each additional inch of slab thickness in same hole, add		144	.111		.47	5.05	.80	6.32	9.15
1754	24" diameter core		8	2		3.97	90.50	14.30	108.77	159
1756	For each additional inch of slab thickness in same hole, add	▼	120	.133		.66	6.05	.95	7.66	11.10
1760	For horizontal holes, add to above				▼		20%	20%		
1770	Prestressed hollow core plank, 8" thick									
1780	1" diameter core	B-89A	17.50	.914	Ea.	.24	41.50	6.55	48.29	71
1790	For each additional inch of plank thickness in same hole, add		3840	.004		.03	.19	.03	.25	.35
1794	2" diameter core		17.25	.928		.38	42	6.65	49.03	72
1796	For each additional inch of plank thickness in same hole, add		2880	.006		.05	.25	.04	.34	.48
1800	3" diameter core		17	.941		.52	42.50	6.75	49.77	73.50
1810	For each additional inch of plank thickness in same hole, add		1920	.008		.06	.38	.06	.50	.72
1820	4" diameter core		16.50	.970		.67	44	6.95	51.62	76
1830	For each additional inch of plank thickness in same hole, add		1280	.013		.08	.57	.09	.74	1.06
1840	6" diameter core		15.50	1.032		1.03	47	7.40	55.43	81.50
1850	For each additional inch of plank thickness in same hole, add		960	.017		.13	.76	.12	1.01	1.43
1860	8" diameter core		15	1.067		1.40	48.50	7.65	57.55	84
1870	For each additional inch of plank thickness in same hole, add		768	.021		.17	.94	.15	1.26	1.80
1880	10" diameter core		14	1.143		1.96	52	8.20	62.16	90.50
1890	For each additional inch of plank thickness in same hole, add		640	.025		.24	1.13	.18	1.55	2.21
1900	12" diameter core		13.50	1.185		2.37	53.50	8.50	64.37	94.50
1910	For each additional inch of plank thickness in same hole, add	▼	548	.029	▼	.30	1.32	.21	1.83	2.59
3000	Bits for core drilling, included in drilling line items									
3010	Diamond, premium, 1" diameter				Ea.	73			73	80
3020	2" diameter					115			115	127
3030	3" diameter					156			156	171
3040	4" diameter					201			201	221
3060	6" diameter					310			310	340
3080	8" diameter					420			420	460
3110	10" diameter					585			585	645
3120	12" diameter					710			710	780
3140	14" diameter					830			830	910
3180	18" diameter					1,125			1,125	1,225
3240	24" diameter				▼	1,575			1,575	1,750

03 82 Concrete Boring

03 82 16 – Concrete Drilling

03 82 16.10 Concrete Impact Drilling	Crew	Daily Output	Labor-Hours	Unit	Material	2017 Bare Costs Labor	Equipment	Total	Total Incl O&P
0010 **CONCRETE IMPACT DRILLING**									
0020 Includes bit cost, layout and set-up time, no anchors									
0050 Up to 4" deep in concrete/brick floors/walls									
0100 Holes, 1/4" diameter	1 Carp	75	.107	Ea.	.07	5.25		5.32	8.15
0150 For each additional inch of depth in same hole, add		430	.019		.02	.92		.94	1.42
0200 3/8" diameter		63	.127		.06	6.25		6.31	9.65
0250 For each additional inch of depth in same hole, add		340	.024		.01	1.16		1.17	1.80
0300 1/2" diameter		50	.160		.06	7.90		7.96	12.10
0350 For each additional inch of depth in same hole, add		250	.032		.02	1.58		1.60	2.43
0400 5/8" diameter		48	.167		.10	8.20		8.30	12.70
0450 For each additional inch of depth in same hole, add		240	.033		.03	1.64		1.67	2.54
0500 3/4" diameter		45	.178		.13	8.75		8.88	13.55
0550 For each additional inch of depth in same hole, add		220	.036		.03	1.79		1.82	2.78
0600 7/8" diameter		43	.186		.18	9.15		9.33	14.25
0650 For each additional inch of depth in same hole, add		210	.038		.04	1.88		1.92	2.92
0700 1" diameter		40	.200		.16	9.85		10.01	15.25
0750 For each additional inch of depth in same hole, add		190	.042		.04	2.07		2.11	3.22
0800 1-1/4" diameter		38	.211		.29	10.35		10.64	16.20
0850 For each additional inch of depth in same hole, add		180	.044		.07	2.19		2.26	3.43
0900 1-1/2" diameter		35	.229		.44	11.25		11.69	17.75
0950 For each additional inch of depth in same hole, add	▼	165	.048	▼	.11	2.39		2.50	3.78
1000 For ceiling installations, add						40%			

For customer support on your Building Construction Costs with RSMeans Data, call 800.448.8182.

91

Division Notes

	CREW	DAILY OUTPUT	LABOR-HOURS	UNIT	BARE COSTS				TOTAL INCL O&P
					MAT.	LABOR	EQUIP.	TOTAL	

Estimating Tips
04 05 00 Common Work Results for Masonry

- The terms mortar and grout are often used interchangeably—and incorrectly. Mortar is used to bed masonry units, seal the entry of air and moisture, provide architectural appearance, and allow for size variations in the units. Grout is used primarily in reinforced masonry construction and to bond the masonry to the reinforcing steel. Common mortar types are M (2500 psi), S (1800 psi), N (750 psi), and O (350 psi), and they conform to ASTM C270. Grout is either fine or coarse and conforms to ASTM C476, and in-place strengths generally exceed 2500 psi. Mortar and grout are different components of masonry construction and are placed by entirely different methods. An estimator should be aware of their unique uses and costs.

- Mortar is included in all assembled masonry line items. The mortar cost, part of the assembled masonry material cost, includes all ingredients, all labor, and all equipment required. Please see reference number R040513-10.

- Waste, specifically the loss/droppings of mortar and the breakage of brick and block, is included in all unit cost lines that include mortar and masonry units in this division. A factor of 25% is added for mortar and 3% for brick and concrete masonry units.

- Scaffolding or staging is not included in any of the Division 4 costs. Refer to Subdivision 01 54 23 for scaffolding and staging costs.

04 20 00 Unit Masonry

- The most common types of unit masonry are brick and concrete masonry. The major classifications of brick are building brick (ASTM C62), facing brick (ASTM C216), glazed brick, fire brick, and pavers. Many varieties of texture and appearance can exist within these classifications, and the estimator would be wise to check local custom and availability within the project area. For repair and remodeling jobs, matching the existing brick may be the most important criteria.

- Brick and concrete block are priced by the piece and then converted into a price per square foot of wall. Openings less than two square feet are generally ignored by the estimator because any savings in units used are offset by the cutting and trimming required.

- It is often difficult and expensive to find and purchase small lots of historic brick. Costs can vary widely. Many design issues affect costs, selection of mortar mix, and repairs or replacement of masonry materials. Cleaning techniques must be reflected in the estimate.

- All masonry walls, whether interior or exterior, require bracing. The cost of bracing walls during construction should be included by the estimator, and this bracing must remain in place until permanent bracing is complete. Permanent bracing of masonry walls is accomplished by masonry itself, in the form of pilasters or abutting wall corners, or by anchoring the walls to the structural frame. Accessories in the form of anchors, anchor slots, and ties are used, but their supply and installation can be by different trades. For instance, anchor slots on spandrel beams and columns are supplied and welded in place by the steel fabricator, but the ties from the slots into the masonry are installed by the bricklayer. Regardless of the installation method, the estimator must be certain that these accessories are accounted for in pricing.

Reference Numbers
Reference numbers are shown at the beginning of some major classifications. These numbers refer to related items in the Reference Section. The reference information may be an estimating procedure, an alternate pricing method, or technical information.

Note: Not all subdivisions listed here necessarily appear. ■

Did you know?
Our online estimating solution gives you the same access to RSMeans' data with 24/7 access:
- Quickly locate costs in the searchable database.
- Build cost lists, estimates, and reports in minutes.
- Adjust costs to any location in the U.S. and Canada with the click of a button.

Start your free trial today at
www.RSMeansOnline.com

04 01 20 – Maintenance of Unit Masonry

04 01 20.20 Pointing Masonry	Crew	Daily Output	Labor-Hours	Unit	Material	2017 Bare Costs Labor	Equipment	Total	Total Incl O&P
0010 **POINTING MASONRY**									
0300 Cut and repoint brick, hard mortar, running bond	1 Bric	80	.100	S.F.	.56	4.78		5.34	7.95
0320 Common bond		77	.104		.56	4.96		5.52	8.25
0360 Flemish bond		70	.114		.59	5.45		6.04	9.05
0400 English bond		65	.123		.59	5.90		6.49	9.70
0600 Soft old mortar, running bond		100	.080		.56	3.82		4.38	6.50
0620 Common bond		96	.083		.56	3.98		4.54	6.70
0640 Flemish bond		90	.089		.59	4.24		4.83	7.20
0680 English bond		82	.098		.59	4.66		5.25	7.80
0700 Stonework, hard mortar		140	.057	L.F.	.75	2.73		3.48	5
0720 Soft old mortar		160	.050	"	.75	2.39		3.14	4.49
1000 Repoint, mask and grout method, running bond		95	.084	S.F.	.75	4.02		4.77	7
1020 Common bond		90	.089		.75	4.24		4.99	7.35
1040 Flemish bond		86	.093		.79	4.44		5.23	7.70
1060 English bond		77	.104		.79	4.96		5.75	8.50
2000 Scrub coat, sand grout on walls, thin mix, brushed		120	.067		3.27	3.18		6.45	8.50
2020 Troweled		98	.082		4.56	3.90		8.46	11

04 01 20.30 Pointing CMU

	Crew	Daily Output	Labor-Hours	Unit	Material	Labor	Equipment	Total	Total Incl O&P
0010 **POINTING CMU**									
0300 Cut and repoint block, hard mortar, running bond	1 Bric	190	.042	S.F.	.23	2.01		2.24	3.34
0310 Stacked bond		200	.040		.23	1.91		2.14	3.19
0600 Soft old mortar, running bond		230	.035		.23	1.66		1.89	2.80
0610 Stacked bond		245	.033		.23	1.56		1.79	2.65

04 01 20.40 Sawing Masonry

	Crew	Daily Output	Labor-Hours	Unit	Material	Labor	Equipment	Total	Total Incl O&P
0010 **SAWING MASONRY**									
0050 Brick or block by hand, per inch depth	A-1	125	.064	L.F.	.03	2.51	.59	3.13	4.53

04 01 20.41 Unit Masonry Stabilization

	Crew	Daily Output	Labor-Hours	Unit	Material	Labor	Equipment	Total	Total Incl O&P
0010 **UNIT MASONRY STABILIZATION**									
0100 Structural repointing method									
0110 Cut/grind mortar joint	1 Bric	240	.033	L.F.		1.59		1.59	2.45
0120 Clean and mask joint		2500	.003		.11	.15		.26	.36
0130 Epoxy paste and 1/4" FRP rod		240	.033		1.77	1.59		3.36	4.39
0132 3/8" FRP rod		160	.050		2.50	2.39		4.89	6.40
0140 Remove masking		14400	.001			.03		.03	.04
0300 Structural fabric method									
0310 Primer	1 Bric	600	.013	S.F.	.93	.64		1.57	2
0320 Apply filling/leveling paste		720	.011		.73	.53		1.26	1.62
0330 Epoxy, glass fiber fabric		720	.011		8.35	.53		8.88	10
0340 Carbon fiber fabric		720	.011		20	.53		20.53	23

04 01 30 – Unit Masonry Cleaning

04 01 30.20 Cleaning Masonry

	Crew	Daily Output	Labor-Hours	Unit	Material	Labor	Equipment	Total	Total Incl O&P
0010 **CLEANING MASONRY**									
0200 By chemical, brush and rinse, new work, light construction dust	D-1	1000	.016	S.F.	.06	.69		.75	1.13
0220 Medium construction dust		800	.020		.09	.87		.96	1.43
0240 Heavy construction dust, drips or stains		600	.027		.13	1.15		1.28	1.91
0260 Low pressure wash and rinse, light restoration, light soil		800	.020		.14	.87		1.01	1.48
0270 Average soil, biological staining		400	.040		.21	1.73		1.94	2.89
0280 Heavy soil, biological and mineral staining, paint		330	.048		.28	2.10		2.38	3.54
0300 High pressure wash and rinse, heavy restoration, light soil		600	.027		.11	1.15		1.26	1.89
0310 Average soil, biological staining		400	.040		.17	1.73		1.90	2.84
0320 Heavy soil, biological and mineral staining, paint		250	.064		.22	2.77		2.99	4.50

04 01 Maintenance of Masonry

04 01 30 – Unit Masonry Cleaning

04 01 30.20 Cleaning Masonry

		Crew	Daily Output	Labor-Hours	Unit	Material	2017 Bare Costs Labor	Equipment	Total	Total Incl O&P
0400	High pressure wash, water only, light soil	C-29	500	.016	S.F.		.63	.13	.76	1.11
0420	Average soil, biological staining		375	.021			.84	.18	1.02	1.47
0440	Heavy soil, biological and mineral staining, paint		250	.032			1.25	.26	1.51	2.21
0800	High pressure water and chemical, light soil		450	.018		.17	.70	.15	1.02	1.41
0820	Average soil, biological staining		300	.027		.25	1.04	.22	1.51	2.12
0840	Heavy soil, biological and mineral staining, paint		200	.040		.34	1.57	.33	2.24	3.13
1200	Sandblast, wet system, light soil	J-6	1750	.018		.35	.79	.12	1.26	1.72
1220	Average soil, biological staining		1100	.029		.52	1.26	.20	1.98	2.70
1240	Heavy soil, biological and mineral staining, paint		700	.046		.70	1.98	.31	2.99	4.12
1400	Dry system, light soil		2500	.013		.35	.56	.09	1	1.32
1420	Average soil, biological staining		1750	.018		.52	.79	.12	1.43	1.91
1440	Heavy soil, biological and mineral staining, paint		1000	.032		.70	1.39	.22	2.31	3.11
1800	For walnut shells, add					.73			.73	.80
1820	For corn chips, add					.74			.74	.81
2000	Steam cleaning, light soil	A-1H	750	.011			.42	.10	.52	.75
2020	Average soil, biological staining		625	.013			.50	.12	.62	.90
2040	Heavy soil, biological and mineral staining		375	.021			.84	.20	1.04	1.50
4000	Add for masking doors and windows	1 Clab	800	.010		.07	.39		.46	.68
4200	Add for pedestrian protection				Job				10%	10%

04 01 30.60 Brick Washing

			Crew	Daily Output	Labor-Hours	Unit	Material	2017 Bare Costs Labor	Equipment	Total	Total Incl O&P
0010	**BRICK WASHING**	R040130-10									
0012	Acid cleanser, smooth brick surface		1 Bric	560	.014	S.F.	.05	.68		.73	1.10
0050	Rough brick			400	.020		.06	.96		1.02	1.54
0060	Stone, acid wash			600	.013		.08	.64		.72	1.07
1000	Muriatic acid, price per gallon in 5 gallon lots					Gal.	9.70			9.70	10.70

04 05 Common Work Results for Masonry

04 05 05 – Selective Demolition for Masonry

04 05 05.10 Selective Demolition

			Crew	Daily Output	Labor-Hours	Unit	Material	2017 Bare Costs Labor	Equipment	Total	Total Incl O&P
0010	**SELECTIVE DEMOLITION**	R024119-10									
0200	Bond beams, 8" block with #4 bar		2 Clab	32	.500	L.F.		19.60		19.60	30
0300	Concrete block walls, unreinforced, 2" thick			1200	.013	S.F.		.52		.52	.80
0310	4" thick			1150	.014			.54		.54	.83
0320	6" thick			1100	.015			.57		.57	.87
0330	8" thick			1050	.015			.60		.60	.91
0340	10" thick			1000	.016			.63		.63	.96
0360	12" thick			950	.017			.66		.66	1.01
0380	Reinforced alternate courses, 2" thick			1130	.014			.55		.55	.85
0390	4" thick			1080	.015			.58		.58	.89
0400	6" thick			1035	.015			.61		.61	.93
0410	8" thick			990	.016			.63		.63	.97
0420	10" thick			940	.017			.67		.67	1.02
0430	12" thick			890	.018			.70		.70	1.08
0440	Reinforced alternate courses & vertically 48" OC, 4" thick			900	.018			.70		.70	1.07
0450	6" thick			850	.019			.74		.74	1.13
0460	8" thick			800	.020			.78		.78	1.20
0480	10" thick			750	.021			.84		.84	1.28
0490	12" thick			700	.023			.90		.90	1.37
1000	Chimney, 16" x 16", soft old mortar		1 Clab	55	.145	C.F.		5.70		5.70	8.75
1020	Hard mortar			40	.200			7.85		7.85	12
1030	16" x 20", soft old mortar			55	.145			5.70		5.70	8.75

For customer support on your Building Construction Costs with RSMeans Data, call 800.448.8182.

95

04 05 Common Work Results for Masonry

04 05 05 – Selective Demolition for Masonry

04 05 05.10 Selective Demolition	Crew	Daily Output	Labor-Hours	Unit	Material	2017 Bare Costs Labor	Equipment	Total	Total Incl O&P
1040	Hard mortar	1 Clab	40	.200	C.F.		7.85		7.85	12
1050	16" x 24", soft old mortar		55	.145			5.70		5.70	8.75
1060	Hard mortar		40	.200			7.85		7.85	12
1080	20" x 20", soft old mortar		55	.145			5.70		5.70	8.75
1100	Hard mortar		40	.200			7.85		7.85	12
1110	20" x 24", soft old mortar		55	.145			5.70		5.70	8.75
1120	Hard mortar		40	.200			7.85		7.85	12
1140	20" x 32", soft old mortar		55	.145			5.70		5.70	8.75
1160	Hard mortar		40	.200			7.85		7.85	12
1200	48" x 48", soft old mortar		55	.145			5.70		5.70	8.75
1220	Hard mortar	▼	40	.200	▼		7.85		7.85	12
1250	Metal, high temp steel jacket, 24" diameter	E-2	130	.431	V.L.F.		23	12.85	35.85	51.50
1260	60" diameter	"	60	.933			50.50	28	78.50	112
1280	Flue lining, up to 12" x 12"	1 Clab	200	.040			1.57		1.57	2.40
1282	Up to 24" x 24"		150	.053			2.09		2.09	3.20
2000	Columns, 8" x 8", soft old mortar		48	.167			6.55		6.55	10
2020	Hard mortar		40	.200			7.85		7.85	12
2060	16" x 16", soft old mortar		16	.500			19.60		19.60	30
2100	Hard mortar		14	.571			22.50		22.50	34.50
2140	24" x 24", soft old mortar		8	1			39		39	60
2160	Hard mortar		6	1.333			52		52	80
2200	36" x 36", soft old mortar		4	2			78.50		78.50	120
2220	Hard mortar		3	2.667	▼		104		104	160
2230	Alternate pricing method, soft old mortar		30	.267	C.F.		10.45		10.45	16
2240	Hard mortar	▼	23	.348	"		13.60		13.60	21
3000	Copings, precast or masonry, to 8" wide									
3020	Soft old mortar	1 Clab	180	.044	L.F.		1.74		1.74	2.67
3040	Hard mortar	"	160	.050	"		1.96		1.96	3
3100	To 12" wide									
3120	Soft old mortar	1 Clab	160	.050	L.F.		1.96		1.96	3
3140	Hard mortar	"	140	.057	"		2.24		2.24	3.43
4000	Fireplace, brick, 30" x 24" opening									
4020	Soft old mortar	1 Clab	2	4	Ea.		157		157	240
4040	Hard mortar		1.25	6.400			251		251	385
4100	Stone, soft old mortar		1.50	5.333			209		209	320
4120	Hard mortar		1	8	▼		315		315	480
5000	Veneers, brick, soft old mortar		140	.057	S.F.		2.24		2.24	3.43
5020	Hard mortar		125	.064			2.51		2.51	3.84
5050	Glass block, up to 4" thick		500	.016			.63		.63	.96
5100	Granite and marble, 2" thick		180	.044			1.74		1.74	2.67
5120	4" thick		170	.047			1.84		1.84	2.82
5140	Stone, 4" thick		180	.044			1.74		1.74	2.67
5160	8" thick		175	.046	▼		1.79		1.79	2.74
5400	Alternate pricing method, stone, 4" thick		60	.133	C.F.		5.20		5.20	8
5420	8" thick	▼	85	.094	"		3.68		3.68	5.65

04 05 13 – Masonry Mortaring

04 05 13.10 Cement

		Crew	Daily Output	Labor-Hours	Unit	Material	2017 Bare Costs Labor	Equipment	Total	Total Incl O&P
0010	**CEMENT**									
0100	Masonry, 70 lb. bag, T.L. lots				Bag	13.55			13.55	14.90
0150	L.T.L. lots					14.35			14.35	15.80
0200	White, 70 lb. bag, T.L. lots					17.70			17.70	19.50
0250	L.T.L. lots				▼	17.75			17.75	19.55

04 05 13 – Masonry Mortaring

04 05 13.20 Lime	Crew	Daily Output	Labor-Hours	Unit	Material	2017 Bare Costs Labor	Equipment	Total	Total Incl O&P
0010 **LIME**									
0020 Masons, hydrated, 50 lb. bag, T.L. lots				Bag	10.25			10.25	11.30
0050 L.T.L. lots					11.30			11.30	12.40
0200 Finish, double hydrated, 50 lb. bag, T.L. lots					9.40			9.40	10.35
0250 L.T.L. lots				↓	10.35			10.35	11.35

04 05 13.23 Surface Bonding Masonry Mortaring

	Crew	Daily Output	Labor-Hours	Unit	Material	Labor	Equipment	Total	Total Incl O&P
0010 **SURFACE BONDING MASONRY MORTARING**									
0020 Gray or white colors, not incl. block work	1 Bric	540	.015	S.F.	.15	.71		.86	1.25

04 05 13.30 Mortar

	Crew	Daily Output	Labor-Hours	Unit	Material	Labor	Equipment	Total	Total Incl O&P
0010 **MORTAR** R040513-10									
0020 With masonry cement									
0100 Type M, 1:1:6 mix	1 Brhe	143	.056	C.F.	5.95	2.17		8.12	9.85
0200 Type N, 1:3 mix		143	.056		6.05	2.17		8.22	10
0300 Type O, 1:3 mix		143	.056		3.80	2.17		5.97	7.50
0400 Type PM, 1:1:6 mix, 2500 psi		143	.056		6	2.17		8.17	9.95
0500 Type S, 1/2:1:4 mix	↓	143	.056	↓	5.55	2.17		7.72	9.45
2000 With Portland cement and lime									
2100 Type M, 1:1/4:3 mix	1 Brhe	143	.056	C.F.	9.35	2.17		11.52	13.65
2200 Type N, 1:1:6 mix, 750 psi		143	.056		7.20	2.17		9.37	11.30
2300 Type O, 1:2:9 mix (Pointing Mortar)		143	.056		8.10	2.17		10.27	12.25
2400 Type PL, 1:1/2:4 mix, 2500 psi		143	.056		6	2.17		8.17	9.95
2600 Type S, 1:1/2:4 mix, 1800 psi	↓	143	.056		8.50	2.17		10.67	12.70
2650 Pre-mixed, type S or N					5.55			5.55	6.15
2700 Mortar for glass block	1 Brhe	143	.056	↓	11.90	2.17		14.07	16.40
2900 Mortar for fire brick, dry mix, 10 lb. pail				Ea.	26.50			26.50	29

04 05 13.91 Masonry Restoration Mortaring

	Crew	Daily Output	Labor-Hours	Unit	Material	Labor	Equipment	Total	Total Incl O&P
0010 **MASONRY RESTORATION MORTARING**									
0020 Masonry restoration mix				Lb.	.26			.26	.29
0050 White				"	.25			.25	.28

04 05 13.93 Mortar Pigments

	Crew	Daily Output	Labor-Hours	Unit	Material	Labor	Equipment	Total	Total Incl O&P
0010 **MORTAR PIGMENTS**, 50 lb. bags (2 bags per M bricks) R040513-10									
0020 Color admixture, range 2 to 10 lb. per bag of cement, light colors				Lb.	5.55			5.55	6.10
0050 Medium colors					7.40			7.40	8.15
0100 Dark colors				↓	15.50			15.50	17.05

04 05 13.95 Sand

	Crew	Daily Output	Labor-Hours	Unit	Material	Labor	Equipment	Total	Total Incl O&P
0010 **SAND**, screened and washed at pit									
0020 For mortar, per ton				Ton	21			21	23.50
0050 With 10 mile haul					38			38	41.50
0100 With 30 mile haul				↓	72			72	79.50
0200 Screened and washed, at the pit				C.Y.	29.50			29.50	32.50
0250 With 10 mile haul					52.50			52.50	58
0300 With 30 mile haul				↓	100			100	110

04 05 13.98 Mortar Admixtures

	Crew	Daily Output	Labor-Hours	Unit	Material	Labor	Equipment	Total	Total Incl O&P
0010 **MORTAR ADMIXTURES**									
0020 Waterproofing admixture, per quart (1 qt. to 2 bags of masonry cement)				Qt.	3.22			3.22	3.54

04 05 16 – Masonry Grouting

04 05 16.30 Grouting

	Crew	Daily Output	Labor-Hours	Unit	Material	Labor	Equipment	Total	Total Incl O&P
0010 **GROUTING**									
0011 Bond beams & lintels, 8" deep, 6" thick, 0.15 C.F. per L.F.	D-4	1480	.022	L.F.	.70	.95	.09	1.74	2.33
0020 8" thick, 0.2 C.F. per L.F.		1400	.023	↓	1.13	1.01	.09	2.23	2.88

04 05 16 – Masonry Grouting

04 05 16.30 Grouting

04 05 16.30 Grouting		Crew	Daily Output	Labor-Hours	Unit	Material	2017 Bare Costs Labor	Equipment	Total	Total Incl O&P
0050	10" thick, 0.25 C.F. per L.F.	D-4	1200	.027	L.F.	1.17	1.18	.11	2.46	3.21
0060	12" thick, 0.3 C.F. per L.F.		1040	.031		1.40	1.36	.13	2.89	3.76
0200	Concrete block cores, solid, 4" thk., by hand, 0.067 C.F.per S.F. of wall	D-8	1100	.036	S.F.	.31	1.61		1.92	2.81
0210	6" thick, pumped, 0.175 C.F. per S.F.	D-4	720	.044		.82	1.96	.18	2.96	4.10
0250	8" thick, pumped, 0.258 C.F. per S.F.		680	.047		1.21	2.07	.20	3.48	4.73
0300	10" thick, pumped, 0.340 C.F. per S.F.		660	.048		1.59	2.14	.20	3.93	5.25
0350	12" thick, pumped, 0.422 C.F. per S.F.		640	.050		1.97	2.20	.21	4.38	5.75
0500	Cavity walls, 2" space, pumped, 0.167 C.F. per S.F. of wall		1700	.019		.78	.83	.08	1.69	2.22
0550	3" space, 0.250 C.F. per S.F.		1200	.027		1.17	1.18	.11	2.46	3.21
0600	4" space, 0.333 C.F. per S.F.		1150	.028		1.56	1.23	.12	2.91	3.72
0700	6" space, 0.500 C.F. per S.F.		800	.040		2.34	1.76	.17	4.27	5.45
0800	Door frames, 3' x 7' opening, 2.5 C.F. per opening		60	.533	Opng.	11.70	23.50	2.21	37.41	51.50
0850	6' x 7' opening, 3.5 C.F. per opening		45	.711	"	16.35	31.50	2.95	50.80	69.50
2000	Grout, C476, for bond beams, lintels and CMU cores		350	.091	C.F.	4.68	4.03	.38	9.09	11.70

04 05 19 – Masonry Anchorage and Reinforcing

04 05 19.05 Anchor Bolts

04 05 19.05 Anchor Bolts		Crew	Daily Output	Labor-Hours	Unit	Material	2017 Bare Costs Labor	Equipment	Total	Total Incl O&P
0010	**ANCHOR BOLTS**									
0015	Installed in fresh grout in CMU bond beams or filled cores, no templates									
0020	Hooked, with nut and washer, 1/2" diam., 8" long	1 Bric	132	.061	Ea.	1.38	2.89		4.27	5.95
0030	12" long		131	.061		1.53	2.92		4.45	6.15
0040	5/8" diameter, 8" long		129	.062		3.35	2.96		6.31	8.25
0050	12" long		127	.063		4.12	3.01		7.13	9.15
0060	3/4" diameter, 8" long		127	.063		4.12	3.01		7.13	9.15
0070	12" long		125	.064		5.15	3.06		8.21	10.35

04 05 19.16 Masonry Anchors

04 05 19.16 Masonry Anchors		Crew	Daily Output	Labor-Hours	Unit	Material	2017 Bare Costs Labor	Equipment	Total	Total Incl O&P
0010	**MASONRY ANCHORS**									
0020	For brick veneer, galv., corrugated, 7/8" x 7", 22 Ga.	1 Bric	10.50	.762	C	15.35	36.50		51.85	73
0100	24 Ga.		10.50	.762		9.90	36.50		46.40	67
0150	16 Ga.		10.50	.762		29	36.50		65.50	88
0200	Buck anchors, galv., corrugated, 16 ga., 2" bend, 8" x 2"		10.50	.762		59	36.50		95.50	121
0250	8" x 3"		10.50	.762		61.50	36.50		98	124
0660	Cavity wall, Z-type, galvanized, 6" long, 1/8" diam.		10.50	.762		25.50	36.50		62	84
0670	3/16" diameter		10.50	.762		31	36.50		67.50	90
0680	1/4" diameter		10.50	.762		39	36.50		75.50	99
0850	8" long, 3/16" diameter		10.50	.762		26	36.50		62.50	84.50
0855	1/4" diameter		10.50	.762		47	36.50		83.50	108
1000	Rectangular type, galvanized, 1/4" diameter, 2" x 6"		10.50	.762		73	36.50		109.50	137
1050	4" x 6"		10.50	.762		88.50	36.50		125	153
1100	3/16" diameter, 2" x 6"		10.50	.762		45.50	36.50		82	106
1150	4" x 6"		10.50	.762		52	36.50		88.50	113
1500	Rigid partition anchors, plain, 8" long, 1" x 1/8"		10.50	.762		230	36.50		266.50	310
1550	1" x 1/4"		10.50	.762		271	36.50		307.50	355
1580	1-1/2" x 1/8"		10.50	.762		253	36.50		289.50	335
1600	1-1/2" x 1/4"		10.50	.762		315	36.50		351.50	405
1650	2" x 1/8"		10.50	.762		299	36.50		335.50	385
1700	2" x 1/4"		10.50	.762		395	36.50		431.50	490

04 05 19.26 Masonry Reinforcing Bars

04 05 19.26 Masonry Reinforcing Bars		Crew	Daily Output	Labor-Hours	Unit	Material	2017 Bare Costs Labor	Equipment	Total	Total Incl O&P
0010	**MASONRY REINFORCING BARS** R040519-50									
0015	Steel bars A615, placed horiz., #3 & #4 bars	1 Bric	450	.018	Lb.	.47	.85		1.32	1.83
0020	#5 & #6 bars		800	.010		.47	.48		.95	1.25
0050	Placed vertical, #3 & #4 bars		350	.023		.47	1.09		1.56	2.20
0060	#5 & #6 bars		650	.012		.47	.59		1.06	1.42

04 05 Common Work Results for Masonry

04 05 19 - Masonry Anchorage and Reinforcing

04 05 19.26 Masonry Reinforcing Bars

	04 05 19.26 Masonry Reinforcing Bars	Crew	Daily Output	Labor-Hours	Unit	Material	2017 Bare Costs Labor	Equipment	Total	Total Incl O&P
0200	Joint reinforcing, regular truss, to 6" wide, mill std galvanized	1 Bric	30	.267	C.L.F.	22.50	12.75		35.25	44.50
0250	12" wide		20	.400		26.50	19.10		45.60	58.50
0400	Cavity truss with drip section, to 6" wide		30	.267		22	12.75		34.75	43.50
0450	12" wide		20	.400		25.50	19.10		44.60	57.50

04 05 23 - Masonry Accessories

04 05 23.13 Masonry Control and Expansion Joints

	04 05 23.13 Masonry Control and Expansion Joints	Crew	Daily Output	Labor-Hours	Unit	Material	2017 Bare Costs Labor	Equipment	Total	Total Incl O&P
0010	**MASONRY CONTROL AND EXPANSION JOINTS**									
0020	Rubber, for double wythe 8" minimum wall (Brick/CMU)	1 Bric	400	.020	L.F.	2.03	.96		2.99	3.70
0025	"T" shaped		320	.025		1.27	1.19		2.46	3.24
0030	Cross-shaped for CMU units		280	.029		1.51	1.36		2.87	3.76
0050	PVC, for double wythe 8" minimum wall (Brick/CMU)		400	.020		1.53	.96		2.49	3.15
0120	"T" shaped		320	.025		.79	1.19		1.98	2.71
0160	Cross-shaped for CMU units		280	.029		.98	1.36		2.34	3.18

04 05 23.19 Masonry Cavity Drainage, Weepholes, and Vents

	04 05 23.19 Masonry Cavity Drainage, Weepholes, and Vents	Crew	Daily Output	Labor-Hours	Unit	Material	2017 Bare Costs Labor	Equipment	Total	Total Incl O&P
0010	**MASONRY CAVITY DRAINAGE, WEEPHOLES, AND VENTS**									
0020	Extruded aluminum, 4" deep, 2-3/8" x 8-1/8"	1 Bric	30	.267	Ea.	36.50	12.75		49.25	59.50
0050	5" x 8-1/8"		25	.320		47.50	15.30		62.80	75.50
0100	2-1/4" x 25"		25	.320		82	15.30		97.30	114
0150	5" x 16-1/2"		22	.364		60	17.35		77.35	92.50
0175	5" x 24"		22	.364		85.50	17.35		102.85	121
0200	6" x 16-1/2"		22	.364		91.50	17.35		108.85	128
0250	7-3/4" x 16-1/2"		20	.400		83	19.10		102.10	121
0400	For baked enamel finish, add					35%				
0500	For cast aluminum, painted, add					60%				
1000	Stainless steel ventilators, 6" x 6"	1 Bric	25	.320		205	15.30		220.30	249
1050	8" x 8"		24	.333		228	15.90		243.90	276
1100	12" x 12"		23	.348		263	16.60		279.60	315
1150	12" x 6"		24	.333		248	15.90		263.90	297
1200	Foundation block vent, galv., 1-1/4" thk, 8" high, 16" long, no damper		30	.267		16.20	12.75		28.95	37.50
1250	For damper, add					3.28			3.28	3.61

04 05 23.95 Wall Plugs

	04 05 23.95 Wall Plugs	Crew	Daily Output	Labor-Hours	Unit	Material	2017 Bare Costs Labor	Equipment	Total	Total Incl O&P
0010	**WALL PLUGS** (for nailing to brickwork)									
0020	25 ga., galvanized, plain	1 Bric	10.50	.762	C	26.50	36.50		63	85
0050	Wood filled	"	10.50	.762	"	66.50	36.50		103	129

04 21 Clay Unit Masonry

04 21 13 - Brick Masonry

04 21 13.13 Brick Veneer Masonry

	04 21 13.13 Brick Veneer Masonry	Crew	Daily Output	Labor-Hours	Unit	Material	2017 Bare Costs Labor	Equipment	Total	Total Incl O&P
0010	**BRICK VENEER MASONRY**, T.L. lots, excl. scaff., grout & reinforcing R042110-20									
0015	Material costs incl. 3% brick and 25% mortar waste									
0020	Standard, select common, 4" x 2-2/3" x 8" (6.75/S.F.)	D-8	1.50	26.667	M	640	1,175		1,815	2,500
0050	Red, 4" x 2-2/3" x 8", running bond		1.50	26.667		615	1,175		1,790	2,475
0100	Full header every 6th course (7.88/S.F.) R042110-50		1.45	27.586		615	1,225		1,840	2,550
0150	English, full header every 2nd course (10.13/S.F.)		1.40	28.571		610	1,250		1,860	2,625
0200	Flemish, alternate header every course (9.00/S.F.)		1.40	28.571		610	1,250		1,860	2,625
0250	Flemish, alt. header every 6th course (7.13/S.F.)		1.45	27.586		615	1,225		1,840	2,550
0300	Full headers throughout (13.50/S.F.)		1.40	28.571		610	1,250		1,860	2,625
0350	Rowlock course (13.50/S.F.)		1.35	29.630		610	1,300		1,910	2,700
0400	Rowlock stretcher (4.50/S.F.)		1.40	28.571		620	1,250		1,870	2,625
0450	Soldier course (6.75/S.F.)		1.40	28.571		615	1,250		1,865	2,625

For customer support on your Building Construction Costs with RSMeans Data, call 800.448.8182.

99

04 21 Clay Unit Masonry

04 21 13 - Brick Masonry

04 21 13.13 Brick Veneer Masonry	Crew	Daily Output	Labor-Hours	Unit	Material	2017 Bare Costs Labor	Equipment	Total	Total Incl O&P
0500	Sailor course (4.50/S.F.)	D-8	1.30	30.769	M	620	1,350		1,970	2,775
0601	Buff or gray face, running bond, (6.75/S.F.)		1.50	26.667		615	1,175		1,790	2,475
0700	Glazed face, 4" x 2-2/3" x 8", running bond		1.40	28.571		2,025	1,250		3,275	4,175
0750	Full header every 6th course (7.88/S.F.)		1.35	29.630		1,925	1,300		3,225	4,125
1000	Jumbo, 6" x 4" x 12" (3.00/S.F.)		1.30	30.769		1,900	1,350		3,250	4,175
1051	Norman, 4" x 2-2/3" x 12" (4.50/S.F.)		1.45	27.586		1,200	1,225		2,425	3,200
1100	Norwegian, 4" x 3-1/5" x 12" (3.75/S.F.)		1.40	28.571		1,550	1,250		2,800	3,650
1150	Economy, 4" x 4" x 8" (4.50 per S.F.)		1.40	28.571		1,000	1,250		2,250	3,050
1201	Engineer, 4" x 3-1/5" x 8" (5.63/S.F.)		1.45	27.586		710	1,225		1,935	2,650
1251	Roman, 4" x 2" x 12" (6.00/S.F.)		1.50	26.667		1,225	1,175		2,400	3,150
1300	S.C.R., 6" x 2-2/3" x 12" (4.50/S.F.)		1.40	28.571		1,400	1,250		2,650	3,500
1350	Utility, 4" x 4" x 12" (3.00/S.F.)	▼	1.08	37.037		1,750	1,625		3,375	4,450
1360	For less than truck load lots, add					15%				
1400	For battered walls, add						30%			
1450	For corbels, add						75%			
1500	For curved walls, add						30%			
1550	For pits and trenches, deduct				▼		20%			
1999	Alternate method of figuring by square foot									
2000	Standard, sel. common, 4" x 2-2/3" x 8" (6.75/S.F.)	D-8	230	.174	S.F.	4.34	7.70		12.04	16.55
2020	Red, 4" x 2-2/3" x 8", running bond		220	.182		4.15	8.05		12.20	16.90
2050	Full header every 6th course (7.88/S.F.)		185	.216		4.83	9.55		14.38	20
2100	English, full header every 2nd course (10.13/S.F.)		140	.286		6.20	12.60		18.80	26
2150	Flemish, alternate header every course (9.00/S.F.)		150	.267		5.50	11.80		17.30	24
2200	Flemish, alt. header every 6th course (7.13/S.F.)		205	.195		4.38	8.60		12.98	18.05
2250	Full headers throughout (13.50/S.F.)		105	.381		8.25	16.85		25.10	35
2300	Rowlock course (13.50/S.F.)		100	.400		8.25	17.65		25.90	36
2350	Rowlock stretcher (4.50/S.F.)		310	.129		2.79	5.70		8.49	11.80
2400	Soldier course (6.75/S.F.)		200	.200		4.15	8.85		13	18.15
2450	Sailor course (4.50/S.F.)		290	.138		2.79	6.10		8.89	12.40
2600	Buff or gray face, running bond (6.75/S.F.)		220	.182		4.38	8.05		12.43	17.15
2700	Glazed face brick, running bond		210	.190		12.95	8.40		21.35	27
2750	Full header every 6th course (7.88/S.F.)		170	.235		15.15	10.40		25.55	32.50
3000	Jumbo, 6" x 4" x 12" running bond (3.00/S.F.)		435	.092		5.15	4.06		9.21	11.95
3050	Norman, 4" x 2-2/3" x 12" running bond (4.5/S.F.)		320	.125		6.40	5.50		11.90	15.55
3100	Norwegian, 4" x 3-1/5" x 12" (3.75/S.F.)		375	.107		5.70	4.71		10.41	13.50
3150	Economy, 4" x 4" x 8" (4.50/S.F.)		310	.129		4.49	5.70		10.19	13.70
3200	Engineer, 4" x 3-1/5" x 8" (5.63/S.F.)		260	.154		3.98	6.80		10.78	14.85
3250	Roman, 4" x 2" x 12" (6.00/S.F.)		250	.160		7.20	7.05		14.25	18.75
3300	S.C.R., 6" x 2-2/3" x 12" (4.50/S.F.)		310	.129		6.20	5.70		11.90	15.60
3350	Utility, 4" x 4" x 12" (3.00/S.F.)	▼	360	.111		5.10	4.91		10.01	13.15
3360	For less than truck load lots, add					.10%				
3370	For battered walls, add						30%			
3380	For corbels, add						75%			
3400	For cavity wall construction, add						15%			
3450	For stacked bond, add						10%			
3500	For interior veneer construction, add						15%			
3510	For pits and trenches, deduct						20%			
3550	For curved walls, add				▼		30%			

For customer support on your Building Construction Costs with RSMeans Data, call 800.448.8182.

04 21 Clay Unit Masonry

04 21 13 – Brick Masonry

04 21 13.14 Thin Brick Veneer

	Crew	Daily Output	Labor-Hours	Unit	Material	2017 Bare Costs Labor	Equipment	Total	Total Incl O&P
0010 **THIN BRICK VENEER**									
0015 Material costs incl. 3% brick and 25% mortar waste									
0020 On & incl. metal panel support sys, modular, 2-2/3" x 5/8" x 8", red	D-7	92	.174	S.F.	9.35	7.15		16.50	21
0100 Closure, 4" x 5/8" x 8"		110	.145		9.10	6		15.10	18.85
0110 Norman, 2-2/3" x 5/8" x 12"		110	.145		9.15	6		15.15	18.90
0120 Utility, 4" x 5/8" x 12"		125	.128		8.85	5.25		14.10	17.55
0130 Emperor, 4" x 3/4" x 16"		175	.091		10	3.76		13.76	16.55
0140 Super emperor, 8" x 3/4" x 16"		195	.082		10.30	3.37		13.67	16.35
0150 For L shaped corners with 4" return, add				L.F.	9.25			9.25	10.20
0200 On masonry/plaster back-up, modular, 2-2/3" x 5/8" x 8", red	D-7	137	.117	S.F.	4.36	4.80		9.16	11.90
0210 Closure, 4" x 5/8" x 8"		165	.097		4.12	3.98		8.10	10.45
0220 Norman, 2-2/3" x 5/8" x 12"		165	.097		4.16	3.98		8.14	10.50
0230 Utility, 4" x 5/8" x 12"		185	.086		3.88	3.55		7.43	9.50
0240 Emperor, 4" x 3/4" x 16"		260	.062		5	2.53		7.53	9.25
0250 Super emperor, 8" x 3/4" x 16"		285	.056		5.35	2.31		7.66	9.25
0260 For L shaped corners with 4" return, add				L.F.	9.25			9.25	10.20
0270 For embedment into pre-cast concrete panels, add				S.F.	14.40			14.40	15.85

04 21 13.15 Chimney

	Crew	Daily Output	Labor-Hours	Unit	Material	2017 Bare Costs Labor	Equipment	Total	Total Incl O&P
0010 **CHIMNEY**, excludes foundation, scaffolding, grout and reinforcing									
0100 Brick, 16" x 16", 8" flue	D-1	18.20	.879	V.L.F.	25	38		63	86
0150 16" x 20" with one 8" x 12" flue		16	1		39	43.50		82.50	110
0200 16" x 24" with two 8" x 8" flues		14	1.143		57	49.50		106.50	139
0250 20" x 20" with one 12" x 12" flue		13.70	1.168		46.50	50.50		97	129
0300 20" x 24" with two 8" x 12" flues		12	1.333		65	57.50		122.50	160
0350 20" x 32" with two 12" x 12" flues		10	1.600		82.50	69.50		152	197

04 21 13.18 Columns

	Crew	Daily Output	Labor-Hours	Unit	Material	2017 Bare Costs Labor	Equipment	Total	Total Incl O&P
0010 **COLUMNS**, solid, excludes scaffolding, grout and reinforcing R042110-10									
0050 Brick, 8" x 8", 9 brick per V.L.F.	D-1	56	.286	V.L.F.	5.30	12.35		17.65	25
0100 12" x 8", 13.5 brick per V.L.F.		37	.432		7.90	18.70		26.60	37.50
0200 12" x 12", 20 brick per V.L.F.		25	.640		11.75	27.50		39.25	55.50
0300 16" x 12", 27 brick per V.L.F.		19	.842		15.85	36.50		52.35	73.50
0400 16" x 16", 36 brick per V.L.F.		14	1.143		21	49.50		70.50	99.50
0500 20" x 16", 45 brick per V.L.F.		11	1.455		26.50	63		89.50	126
0600 20" x 20", 56 brick per V.L.F.		9	1.778		33	77		110	154
0700 24" x 20", 68 brick per V.L.F.		7	2.286		40	99		139	196
0800 24" x 24", 81 brick per V.L.F.		6	2.667		47.50	115		162.50	230
1000 36" x 36", 182 brick per V.L.F.		3	5.333		107	231		338	475

04 21 13.30 Oversized Brick

	Crew	Daily Output	Labor-Hours	Unit	Material	2017 Bare Costs Labor	Equipment	Total	Total Incl O&P
0010 **OVERSIZED BRICK**, excludes scaffolding, grout and reinforcing									
0100 Veneer, 4" x 2.25" x 16"	D-8	387	.103	S.F.	5.30	4.57		9.87	12.85
0102 8" x 2.25" x 16", multicell		265	.151		16.65	6.65		23.30	28.50
0105 4" x 2.75" x 16"		412	.097		5.50	4.29		9.79	12.65
0107 8" x 2.75" x 16", multicell		295	.136		16.70	6		22.70	27.50
0110 4" x 4" x 16"		460	.087		3.61	3.84		7.45	9.85
0120 4" x 8" x 16"		533	.075		4.25	3.32		7.57	9.75
0122 4" x 8" x 16" multicell		327	.122		15.65	5.40		21.05	25.50
0125 Loadbearing, 6" x 4" x 16", grouted and reinforced		387	.103		10.75	4.57		15.32	18.80
0130 8" x 4" x 16", grouted and reinforced		327	.122		11.80	5.40		17.20	21.50
0132 10" x 4" x 16", grouted and reinforced		327	.122		24	5.40		29.40	34.50
0135 6" x 8" x 16", grouted and reinforced		440	.091		13.95	4.02		17.97	21.50
0140 8" x 8" x 16", grouted and reinforced		400	.100		14.90	4.42		19.32	23

For customer support on your Building Construction Costs with RSMeans Data, call 800.448.8182.

101

04 21 13 – Brick Masonry

04 21 13.30 Oversized Brick

		Crew	Daily Output	Labor-Hours	Unit	Material	2017 Bare Costs Labor	2017 Bare Costs Equipment	Total	Total Incl O&P
0145	Curtainwall/reinforced veneer, 6" x 4" x 16"	D-8	387	.103	S.F.	15.35	4.57		19.92	24
0150	8" x 4" x 16"		327	.122		18.70	5.40		24.10	29
0152	10" x 4" x 16"		327	.122		26.50	5.40		31.90	37.50
0155	6" x 8" x 16"		440	.091		19.05	4.02		23.07	27
0160	8" x 8" x 16"	↓	400	.100		26.50	4.42		30.92	36.50
0200	For 1 to 3 slots in face, add					15%				
0210	For 4 to 7 slots in face, add					25%				
0220	For bond beams, add					20%				
0230	For bullnose shapes, add					20%				
0240	For open end knockout, add					10%				
0250	For white or gray color group, add					10%				
0260	For 135 degree corner, add					250%				

04 21 13.35 Common Building Brick

		Crew	Daily Output	Labor-Hours	Unit	Material	2017 Bare Costs Labor	2017 Bare Costs Equipment	Total	Total Incl O&P
0010	**COMMON BUILDING BRICK**, C62, TL lots, material only R042110-20									
0020	Standard				M	500			500	550
0050	Select				"	520			520	570

04 21 13.40 Structural Brick

		Crew	Daily Output	Labor-Hours	Unit	Material	2017 Bare Costs Labor	2017 Bare Costs Equipment	Total	Total Incl O&P
0010	**STRUCTURAL BRICK** C652, Grade SW, incl. mortar, scaffolding not incl.									
0100	Standard unit, 4-5/8" x 2-3/4" x 9-5/8"	D-8	245	.163	S.F.	4.23	7.20		11.43	15.75
0120	Bond beam		225	.178		4.23	7.85		12.08	16.75
0140	V cut bond beam		225	.178		4.23	7.85		12.08	16.75
0160	Stretcher quoin, 5-5/8" x 2-3/4" x 9-5/8"		245	.163		7.85	7.20		15.05	19.70
0180	Corner quoin		245	.163		7.85	7.20		15.05	19.70
0200	Corner, 45 deg, 4-5/8" x 2-3/4" x 10-7/16"	↓	235	.170	↓	7.80	7.50		15.30	20

04 21 13.45 Face Brick

		Crew	Daily Output	Labor-Hours	Unit	Material	2017 Bare Costs Labor	2017 Bare Costs Equipment	Total	Total Incl O&P
0010	**FACE BRICK** Material Only, C216, TL lots R042110-20									
0300	Standard modular, 4" x 2-2/3" x 8"				M	490			490	540
0450	Economy, 4" x 4" x 8"					860			860	945
0510	Economy, 4" x 4" x 12"					1,375			1,375	1,500
0550	Jumbo, 6" x 4" x 12"					1,550			1,550	1,700
0610	Jumbo, 8" x 4" x 12"					1,550			1,550	1,700
0650	Norwegian, 4" x 3-1/5" x 12"					1,350			1,350	1,500
0710	Norwegian, 6" x 3-1/5" x 12"					1,675			1,675	1,850
0850	Standard glazed, plain colors, 4" x 2-2/3" x 8"					1,750			1,750	1,925
1000	Deep trim shades, 4" x 2-2/3" x 8"					2,250			2,250	2,475
1080	Jumbo utility, 4" x 4" x 12"					1,525			1,525	1,675
1120	4" x 8" x 8"					1,950			1,950	2,150
1140	4" x 8" x 16"					5,775			5,775	6,350
1260	Engineer, 4" x 3-1/5" x 8"					580			580	635
1350	King, 4" x 2-3/4" x 10"					535			535	590
1770	Standard modular, double glazed, 4" x 2-2/3" x 8"					2,650			2,650	2,925
1850	Jumbo, colored glazed ceramic, 6" x 4" x 12"					2,800			2,800	3,075
2050	Jumbo utility, glazed, 4" x 4" x 12"					5,250			5,250	5,775
2100	4" x 8" x 8"					6,200			6,200	6,800
2150	4" x 16" x 8"					7,250			7,250	7,975
2170	For less than truck load lots, add					15			15	16.50
2180	For buff or gray brick, add				↓	16			16	17.60

04 21 Clay Unit Masonry

04 21 26 – Glazed Structural Clay Tile Masonry

04 21 26.10 Structural Facing Tile

	04 21 26.10 Structural Facing Tile	Crew	Daily Output	Labor-Hours	Unit	Material	2017 Bare Costs Labor	Equipment	Total	Total Incl O&P
0010	**STRUCTURAL FACING TILE**, std. colors, excl. scaffolding, grout, reinforcing									
0020	6T series, 5-1/3" x 12", 2.3 pieces per S.F., glazed 1 side, 2" thick	D-8	225	.178	S.F.	8.65	7.85		16.50	21.50
0100	4" thick		220	.182		11.75	8.05		19.80	25.50
0150	Glazed 2 sides		195	.205		15.60	9.05		24.65	31
0250	6" thick		210	.190		17.60	8.40		26	32.50
0300	Glazed 2 sides		185	.216		21	9.55		30.55	37.50
0400	8" thick		180	.222		23	9.80		32.80	40.50
0500	Special shapes, group 1		400	.100	Ea.	7.50	4.42		11.92	15.05
0550	Group 2		375	.107		12.15	4.71		16.86	20.50
0600	Group 3		350	.114		15.60	5.05		20.65	25
0650	Group 4		325	.123		32.50	5.45		37.95	44
0700	Group 5		300	.133		38.50	5.90		44.40	51.50
0750	Group 6		275	.145		52.50	6.40		58.90	67.50
1000	Fire rated, 4" thick, 1 hr. rating		210	.190	S.F.	16.75	8.40		25.15	31.50
1300	Acoustic, 4" thick		210	.190	"	35.50	8.40		43.90	52.50
2000	8W series, 8" x 16", 1.125 pieces per S.F.									
2050	2" thick, glazed 1 side	D-8	360	.111	S.F.	10.10	4.91		15.01	18.70
2100	4" thick, glazed 1 side		345	.116		13.85	5.10		18.95	23
2150	Glazed 2 sides		325	.123		16.50	5.45		21.95	26.50
2200	6" thick, glazed 1 side		330	.121		23.50	5.35		28.85	34.50
2250	8" thick, glazed 1 side		310	.129		24.50	5.70		30.20	36
2500	Special shapes, group 1		300	.133	Ea.	14.45	5.90		20.35	25
2550	Group 2		280	.143		19.30	6.30		25.60	30.50
2600	Group 3		260	.154		20.50	6.80		27.30	33
2650	Group 4		250	.160		43	7.05		50.05	58
2700	Group 5		240	.167		38.50	7.35		45.85	53.50
2750	Group 6		230	.174		83	7.70		90.70	103
3000	4" thick, glazed 1 side		345	.116	S.F.	13.35	5.10		18.45	22.50
3100	Acoustic, 4" thick		345	.116	"	19.10	5.10		24.20	29
3120	4W series, 8" x 8", 2.25 pieces per S.F.									
3125	2" thick, glazed 1 side	D-8	360	.111	S.F.	9.90	4.91		14.81	18.45
3130	4" thick, glazed 1 side		345	.116		11.65	5.10		16.75	20.50
3135	Glazed 2 sides		325	.123		16.05	5.45		21.50	26
3140	6" thick, glazed 1 side		330	.121		16.60	5.35		21.95	26.50
3150	8" thick, glazed 1 side		310	.129		24	5.70		29.70	35.50
3155	Special shapes, group 1		300	.133	Ea.	7.75	5.90		13.65	17.60
3160	Group 2		280	.143	"	8.95	6.30		15.25	19.55
3200	For designer colors, add					25%				
3300	For epoxy mortar joints, add				S.F.	1.79			1.79	1.97

04 21 29 – Terra Cotta Masonry

04 21 29.10 Terra Cotta Masonry Components

	04 21 29.10 Terra Cotta Masonry Components	Crew	Daily Output	Labor-Hours	Unit	Material	Labor	Equipment	Total	Total Incl O&P
0010	**TERRA COTTA MASONRY COMPONENTS**									
0020	Coping, split type, not glazed, 9" wide	D-1	90	.178	L.F.	15	7.70		22.70	28.50
0100	13" wide		80	.200		21	8.65		29.65	36.50
0200	Coping, split type, glazed, 9" wide		90	.178		12.95	7.70		20.65	26
0250	13" wide		80	.200		17.05	8.65		25.70	32
0500	Partition or back-up blocks, scored, in C.L. lots									
0700	Non-load bearing 12" x 12", 3" thick, special order	D-8	550	.073	S.F.	17.75	3.21		20.96	24.50
0750	4" thick, standard		500	.080		5.90	3.53		9.43	11.95
0800	6" thick		450	.089		7.90	3.93		11.83	14.70
0850	8" thick		400	.100		9.95	4.42		14.37	17.75
1000	Load bearing, 12" x 12", 4" thick, in walls		500	.080		6.30	3.53		9.83	12.35

04 21 Clay Unit Masonry

04 21 29 – Terra Cotta Masonry

04 21 29.10 Terra Cotta Masonry Components	Crew	Daily Output	Labor-Hours	Unit	Material	2017 Bare Costs Labor	Equipment	Total	Total Incl O&P
1050	In floors	D-8	750	.053	S.F.	6.30	2.36		8.66	10.50
1200	6" thick, in walls		450	.089		8.40	3.93		12.33	15.30
1250	In floors		675	.059		8.40	2.62		11.02	13.30
1400	8" thick, in walls		400	.100		10.45	4.42		14.87	18.30
1450	In floors		575	.070		10.45	3.07		13.52	16.25
1600	10" thick, in walls, special order		350	.114		25.50	5.05		30.55	36
1650	In floors, special order		500	.080		25.50	3.53		29.03	33.50
1800	12" thick, in walls, special order		300	.133		25.50	5.90		31.40	37
1850	In floors, special order		450	.089		25.50	3.93		29.43	34
2000	For reinforcing with steel rods, add to above					15%	5%			
2100	For smooth tile instead of scored, add					4.05			4.05	4.46
2200	For L.C.L. quantities, add					10%	10%			

04 21 29.20 Terra Cotta Tile

		Crew	Daily Output	Labor-Hours	Unit	Material	Labor	Equipment	Total	Total Incl O&P
0010	**TERRA COTTA TILE**, on walls, dry set, 1/2" thick									
0100	Square, hexagonal or lattice shapes, unglazed	1 Tilf	135	.059	S.F.	4.74	2.72		7.46	9.25
0300	Glazed, plain colors		130	.062		7.25	2.82		10.07	12.15
0400	Intense colors		125	.064		8.50	2.93		11.43	13.70

04 22 Concrete Unit Masonry

04 22 10 – Concrete Masonry Units

04 22 10.11 Autoclave Aerated Concrete Block

			Crew	Daily Output	Labor-Hours	Unit	Material	Labor	Equipment	Total	Total Incl O&P
0010	**AUTOCLAVE AERATED CONCRETE BLOCK**, excl. scaffolding, grout & reinforcing										
0050	Solid, 4" x 8" x 24", incl. mortar	G	D-8	600	.067	S.F.	1.62	2.94		4.56	6.30
0060	6" x 8" x 24"	G		600	.067		2.44	2.94		5.38	7.20
0070	8" x 8" x 24"	G		575	.070		3.26	3.07		6.33	8.30
0080	10" x 8" x 24"	G		575	.070		3.98	3.07		7.05	9.10
0090	12" x 8" x 24"	G		550	.073		4.88	3.21		8.09	10.30

04 22 10.12 Chimney Block

		Crew	Daily Output	Labor-Hours	Unit	Material	Labor	Equipment	Total	Total Incl O&P
0010	**CHIMNEY BLOCK**, excludes scaffolding, grout and reinforcing									
0220	1 piece, with 8" x 8" flue, 16" x 16"	D-1	28	.571	V.L.F.	18.85	24.50		43.35	59
0230	2 piece, 16" x 16"		26	.615		23.50	26.50		50	66.50
0240	2 piece, with 8" x 12" flue, 16" x 20"		24	.667		32.50	29		61.50	80

04 22 10.14 Concrete Block, Back-Up

		Crew	Daily Output	Labor-Hours	Unit	Material	Labor	Equipment	Total	Total Incl O&P
0010	**CONCRETE BLOCK, BACK-UP**, C90, 2000 psi R042210-20									
0020	Normal weight, 8" x 16" units, tooled joint 1 side									
0050	Not-reinforced, 2000 psi, 2" thick	D-8	475	.084	S.F.	1.56	3.72		5.28	7.40
0200	4" thick		460	.087		1.88	3.84		5.72	7.95
0300	6" thick		440	.091		2.45	4.02		6.47	8.90
0350	8" thick		400	.100		2.62	4.42		7.04	9.70
0400	10" thick		330	.121		3.10	5.35		8.45	11.65
0450	12" thick	D-9	310	.155		4.27	6.70		10.97	15
1000	Reinforced, alternate courses, 4" thick	D-8	450	.089		2.05	3.93		5.98	8.30
1100	6" thick		430	.093		2.62	4.11		6.73	9.20
1150	8" thick		395	.101		2.82	4.47		7.29	10
1200	10" thick		320	.125		3.26	5.50		8.76	12.10
1250	12" thick	D-9	300	.160		4.44	6.90		11.34	15.55

04 22 Concrete Unit Masonry

04 22 10 – Concrete Masonry Units

04 22 10.16 Concrete Block, Bond Beam

		Crew	Daily Output	Labor-Hours	Unit	Material	2017 Bare Costs Labor	2017 Bare Costs Equipment	Total	Total Incl O&P
0010	**CONCRETE BLOCK, BOND BEAM**, C90, 2000 psi									
0020	Not including grout or reinforcing									
0125	Regular block, 6" thick	D-8	584	.068	L.F.	2.69	3.03		5.72	7.60
0130	8" high, 8" thick	"	565	.071		2.80	3.13		5.93	7.90
0150	12" thick	D-9	510	.094		3.97	4.07		8.04	10.60
0525	Lightweight, 6" thick	D-8	592	.068		2.92	2.98		5.90	7.80
0530	8" high, 8" thick	"	575	.070		3.51	3.07		6.58	8.60
0550	12" thick	D-9	520	.092		4.74	4		8.74	11.35
2000	Including grout and 2 #5 bars									
2100	Regular block, 8" high, 8" thick	D-8	300	.133	L.F.	4.90	5.90		10.80	14.45
2150	12" thick	D-9	250	.192		6.70	8.30		15	20
2500	Lightweight, 8" high, 8" thick	D-8	305	.131		5.60	5.80		11.40	15.10
2550	12" thick	D-9	255	.188		7.45	8.15		15.60	21

04 22 10.18 Concrete Block, Column

		Crew	Daily Output	Labor-Hours	Unit	Material	2017 Bare Costs Labor	2017 Bare Costs Equipment	Total	Total Incl O&P
0010	**CONCRETE BLOCK, COLUMN** or pilaster									
0050	Including vertical reinforcing (4-#4 bars) and grout									
0160	1 piece unit, 16" x 16"	D-1	26	.615	V.L.F.	16.30	26.50		42.80	59
0170	2 piece units, 16" x 20"		24	.667		21	29		50	67.50
0180	20" x 20"		22	.727		27	31.50		58.50	78
0190	22" x 24"		18	.889		40.50	38.50		79	104
0200	20" x 32"		14	1.143		47	49.50		96.50	128

04 22 10.19 Concrete Block, Insulation Inserts

		Crew	Daily Output	Labor-Hours	Unit	Material	2017 Bare Costs Labor	2017 Bare Costs Equipment	Total	Total Incl O&P
0010	**CONCRETE BLOCK, INSULATION INSERTS**									
0100	Styrofoam, plant installed, add to block prices									
0200	8" x 16" units, 6" thick				S.F.	1.22			1.22	1.34
0250	8" thick					1.37			1.37	1.51
0300	10" thick					1.42			1.42	1.56
0350	12" thick					1.57			1.57	1.73
0500	8" x 8" units, 8" thick					1.22			1.22	1.34
0550	12" thick					1.42			1.42	1.56

04 22 10.23 Concrete Block, Decorative

		Crew	Daily Output	Labor-Hours	Unit	Material	2017 Bare Costs Labor	2017 Bare Costs Equipment	Total	Total Incl O&P
0010	**CONCRETE BLOCK, DECORATIVE**, C90, 2000 psi									
0020	Embossed, simulated brick face									
0100	8" x 16" units, 4" thick	D-8	400	.100	S.F.	2.81	4.42		7.23	9.90
0200	8" thick		340	.118		3.06	5.20		8.26	11.35
0250	12" thick		300	.133		5.15	5.90		11.05	14.75
0400	Embossed both sides									
0500	8" thick	D-8	300	.133	S.F.	4.13	5.90		10.03	13.60
0550	12" thick	"	275	.145	"	5.40	6.40		11.80	15.85
1000	Fluted high strength									
1100	8" x 16" x 4" thick, flutes 1 side,	D-8	345	.116	S.F.	3.98	5.10		9.08	12.30
1150	Flutes 2 sides		335	.119		4.57	5.25		9.82	13.15
1200	8" thick		300	.133		5.35	5.90		11.25	14.95
1250	For special colors, add					.64			.64	.71
1400	Deep grooved, smooth face									
1450	8" x 16" x 4" thick	D-8	345	.116	S.F.	2.56	5.10		7.66	10.70
1500	8" thick	"	300	.133		4.05	5.90		9.95	13.50
2000	Formblock, incl. inserts & reinforcing									
2100	8" x 16" x 8" thick	D-8	345	.116	S.F.	3.57	5.10		8.67	11.85
2150	12" thick	"	310	.129	"	4.73	5.70		10.43	13.95
2500	Ground face									

04 22 10 – Concrete Masonry Units

04 22 10.23 Concrete Block, Decorative	Crew	Daily Output	Labor-Hours	Unit	Material	2017 Bare Costs Labor	Equipment	Total	Total Incl O&P
2600	8" x 16" x 4" thick	D-8	345	.116	S.F.	4.18	5.10		9.28	12.50
2650	6" thick		325	.123		4.61	5.45		10.06	13.45
2700	8" thick		300	.133		5.05	5.90		10.95	14.65
2750	12" thick	D-9	265	.181		6.55	7.85		14.40	19.25
2900	For special colors, add, minimum					15%				
2950	For special colors, add, maximum					45%				
4000	Slump block									
4100	4" face height x 16" x 4" thick	D-1	165	.097	S.F.	4.21	4.20		8.41	11.10
4150	6" thick		160	.100		6.25	4.33		10.58	13.50
4200	8" thick		155	.103		6.05	4.47		10.52	13.50
4250	10" thick		140	.114		11.55	4.95		16.50	20.50
4300	12" thick		130	.123		13.10	5.35		18.45	22.50
4400	6" face height x 16" x 6" thick		155	.103		5.80	4.47		10.27	13.20
4450	8" thick		150	.107		8.75	4.62		13.37	16.75
4500	10" thick		130	.123		13.40	5.35		18.75	23
4550	12" thick		120	.133		14.05	5.75		19.80	24.50
5000	Split rib profile units, 1" deep ribs, 8 ribs									
5100	8" x 16" x 4" thick	D-8	345	.116	S.F.	3.96	5.10		9.06	12.25
5150	6" thick		325	.123		4.49	5.45		9.94	13.30
5200	8" thick		300	.133		5.05	5.90		10.95	14.65
5250	12" thick	D-9	275	.175		5.90	7.55		13.45	18.10
5400	For special deeper colors, 4" thick, add					1.32			1.32	1.45
5450	12" thick, add					1.38			1.38	1.52
5600	For white, 4" thick, add					1.32			1.32	1.45
5650	6" thick, add					1.34			1.34	1.48
5700	8" thick, add					1.39			1.39	1.53
5750	12" thick, add					1.44			1.44	1.58
6000	Split face									
6100	8" x 16" x 4" thick	D-8	350	.114	S.F.	3.68	5.05		8.73	11.80
6150	6" thick		325	.123		4.18	5.45		9.63	12.95
6200	8" thick		300	.133		4.71	5.90		10.61	14.25
6250	12" thick	D-9	270	.178		5.70	7.70		13.40	18.10
6300	For scored, add					.39			.39	.43
6400	For special deeper colors, 4" thick, add					.64			.64	.71
6450	6" thick, add					.77			.77	.85
6500	8" thick, add					.79			.79	.87
6550	12" thick, add					.81			.81	.89
6650	For white, 4" thick, add					1.31			1.31	1.44
6700	6" thick, add					1.32			1.32	1.45
6750	8" thick, add					1.33			1.33	1.47
6800	12" thick, add					1.38			1.38	1.52
7000	Scored ground face, 2 to 5 scores									
7100	8" x 16" x 4" thick	D-8	340	.118	S.F.	8.25	5.20		13.45	17.10
7150	6" thick		310	.129		9.15	5.70		14.85	18.85
7200	8" thick		290	.138		10.35	6.10		16.45	21
7250	12" thick	D-9	265	.181		13.85	7.85		21.70	27.50
8000	Hexagonal face profile units, 8" x 16" units									
8100	4" thick, hollow	D-8	340	.118	S.F.	3.82	5.20		9.02	12.20
8200	Solid		340	.118		4.90	5.20		10.10	13.40
8300	6" thick, hollow		310	.129		3.96	5.70		9.66	13.10
8350	8" thick, hollow		290	.138		4.53	6.10		10.63	14.35
8500	For stacked bond, add						26%			
8550	For high rise construction, add per story	D-8	67.80	.590	M.S.F.		26		26	40

For customer support on your Building Construction Costs with RSMeans Data, call 800.448.8182.

04 22 Concrete Unit Masonry

04 22 10 – Concrete Masonry Units

04 22 10.23 Concrete Block, Decorative

		Crew	Daily Output	Labor-Hours	Unit	Material	2017 Bare Costs Labor	Equipment	Total	Total Incl O&P
8600	For scored block, add					10%				
8650	For honed or ground face, per face, add				Ea.	1.18			1.18	1.29
8700	For honed or ground end, per end, add				"	1.18			1.18	1.29
8750	For bullnose block, add					10%				
8800	For special color, add					13%				

04 22 10.24 Concrete Block, Exterior

		Crew	Daily Output	Labor-Hours	Unit	Material	2017 Bare Costs Labor	Equipment	Total	Total Incl O&P
0010	**CONCRETE BLOCK, EXTERIOR**, C90, 2000 psi									
0020	Reinforced alt courses, tooled joints 2 sides									
0100	Normal weight, 8" x 16" x 6" thick	D-8	395	.101	S.F.	2.40	4.47		6.87	9.55
0200	8" thick		360	.111		3.71	4.91		8.62	11.65
0250	10" thick	↓	290	.138		4.30	6.10		10.40	14.10
0300	12" thick	D-9	250	.192		5	8.30		13.30	18.30
0500	Lightweight, 8" x 16" x 6" thick	D-8	450	.089		3.32	3.93		7.25	9.70
0600	8" thick		430	.093		3.75	4.11		7.86	10.40
0650	10" thick	↓	395	.101		4.14	4.47		8.61	11.45
0700	12" thick	D-9	350	.137	↓	4.27	5.95		10.22	13.85

04 22 10.26 Concrete Block Foundation Wall

		Crew	Daily Output	Labor-Hours	Unit	Material	2017 Bare Costs Labor	Equipment	Total	Total Incl O&P
0010	**CONCRETE BLOCK FOUNDATION WALL**, C90/C145									
0050	Normal-weight, cut joints, horiz joint reinf, no vert reinf.									
0200	Hollow, 8" x 16" x 6" thick	D-8	455	.088	S.F.	3.07	3.88		6.95	9.35
0250	8" thick		425	.094		3.29	4.16		7.45	10
0300	10" thick	↓	350	.114		3.77	5.05		8.82	11.90
0350	12" thick	D-9	300	.160		4.97	6.90		11.87	16.10
0500	Solid, 8" x 16" block, 6" thick	D-8	440	.091		3.13	4.02		7.15	9.65
0550	8" thick	"	415	.096		4.37	4.26		8.63	11.35
0600	12" thick	D-9	350	.137	↓	6.25	5.95		12.20	16.05

04 22 10.28 Concrete Block, High Strength

		Crew	Daily Output	Labor-Hours	Unit	Material	2017 Bare Costs Labor	Equipment	Total	Total Incl O&P
0010	**CONCRETE BLOCK, HIGH STRENGTH**									
0050	Hollow, reinforced alternate courses, 8" x 16" units									
0200	3500 psi, 4" thick	D-8	440	.091	S.F.	2.38	4.02		6.40	8.80
0250	6" thick		395	.101		2.31	4.47		6.78	9.45
0300	8" thick	↓	360	.111		3.62	4.91		8.53	11.55
0350	12" thick	D-9	250	.192		4.90	8.30		13.20	18.20
0500	5000 psi, 4" thick	D-8	440	.091		2.09	4.02		6.11	8.50
0550	6" thick		395	.101		2.98	4.47		7.45	10.20
0600	8" thick	↓	360	.111		4.24	4.91		9.15	12.20
0650	12" thick	D-9	300	.160	↓	4.91	6.90		11.81	16.05
1000	For 75% solid block, add					30%				
1050	For 100% solid block, add					50%				

04 22 10.30 Concrete Block, Interlocking

		Crew	Daily Output	Labor-Hours	Unit	Material	2017 Bare Costs Labor	Equipment	Total	Total Incl O&P
0010	**CONCRETE BLOCK, INTERLOCKING**									
0100	Not including grout or reinforcing									
0200	8" x 16" units, 2,000 psi, 8" thick	D-1	245	.065	S.F.	3.10	2.83		5.93	7.75
0300	12" thick		220	.073		4.59	3.15		7.74	9.90
0350	16" thick	↓	185	.086		6.90	3.74		10.64	13.35
0400	Including grout & reinforcing, 8" thick	D-4	245	.131		8.15	5.75	.54	14.44	18.40
0450	12" thick		220	.145		9.85	6.40	.60	16.85	21.50
0500	16" thick	↓	185	.173	↓	12.35	7.65	.72	20.72	26

For customer support on your Building Construction Costs with RSMeans Data, call 800.448.8182.

107

04 22 10.32 Concrete Block, Lintels

	Crew	Daily Output	Labor-Hours	Unit	Material	2017 Bare Costs Labor	Equipment	Total	Total Incl O&P
0010 **CONCRETE BLOCK, LINTELS**, C90, normal weight									
0100 Including grout and horizontal reinforcing									
0200 8" x 8" x 8", 1 #4 bar	D-4	300	.107	L.F.	3.88	4.70	.44	9.02	11.95
0250 2 #4 bars		295	.108		4.09	4.78	.45	9.32	12.30
0400 8" x 16" x 8", 1 #4 bar		275	.116		3.95	5.15	.48	9.58	12.75
0450 2 #4 bars		270	.119		4.17	5.25	.49	9.91	13.10
1000 12" x 8" x 8", 1 #4 bar		275	.116		5.30	5.15	.48	10.93	14.25
1100 2 #4 bars		270	.119		5.55	5.25	.49	11.29	14.65
1150 2 #5 bars		270	.119		5.75	5.25	.49	11.49	14.90
1200 2 #6 bars		265	.121		6.05	5.30	.50	11.85	15.35
1500 12" x 16" x 8", 1 #4 bar		250	.128		6	5.65	.53	12.18	15.80
1600 2 #3 bars		245	.131		6	5.75	.54	12.29	16
1650 2 #4 bars		245	.131		6.20	5.75	.54	12.49	16.20
1700 2 #5 bars		240	.133		6.40	5.90	.55	12.85	16.65

04 22 10.33 Lintel Block

	Crew	Daily Output	Labor-Hours	Unit	Material	2017 Bare Costs Labor	Equipment	Total	Total Incl O&P
0010 **LINTEL BLOCK**									
3481 Lintel block 6" x 8" x 8"	D-1	300	.053	Ea.	1.32	2.31		3.63	5
3501 6" x 16" x 8"		275	.058		2.10	2.52		4.62	6.20
3521 8" x 8" x 8"		275	.058		1.20	2.52		3.72	5.20
3561 8" x 16" x 8"		250	.064		2.10	2.77		4.87	6.55

04 22 10.34 Concrete Block, Partitions

	Crew	Daily Output	Labor-Hours	Unit	Material	2017 Bare Costs Labor	Equipment	Total	Total Incl O&P
0010 **CONCRETE BLOCK, PARTITIONS**, excludes scaffolding R042210-20									
0100 Acoustical slotted block									
0200 4" thick, type A-1	D-8	315	.127	S.F.	4.67	5.60		10.27	13.80
0210 8" thick		275	.145		6.05	6.40		12.45	16.60
0250 8" thick, type Q		275	.145		10.30	6.40		16.70	21.50
0260 4" thick, type RSC		315	.127		7.25	5.60		12.85	16.60
0270 6" thick		295	.136		7.25	6		13.25	17.15
0280 8" thick		275	.145		7.25	6.40		13.65	17.85
0290 12" thick		250	.160		7.25	7.05		14.30	18.80
0300 8" thick, type RSR		275	.145		7.25	6.40		13.65	17.85
0400 8" thick, type RSC/RF		275	.145		7.80	6.40		14.20	18.50
0410 10" thick		260	.154		8.30	6.80		15.10	19.60
0420 12" thick		250	.160		9	7.05		16.05	21
0430 12" thick, type RSC/RF-4		250	.160		11.40	7.05		18.45	23.50
0500 NRC .60 type R, 8" thick		265	.151		8.90	6.65		15.55	20
0600 NRC .65 type RR, 8" thick		265	.151		8.45	6.65		15.10	19.55
0700 NRC .65 type 4R-RF, 8" thick		265	.151		9.10	6.65		15.75	20.50
0710 NRC .70 type R, 12" thick		245	.163		9.60	7.20		16.80	21.50
1000 Lightweight block, tooled joints, 2 sides, hollow									
1100 Not reinforced, 8" x 16" x 4" thick	D-8	440	.091	S.F.	1.90	4.02		5.92	8.30
1150 6" thick		410	.098		2.69	4.31		7	9.60
1200 8" thick		385	.104		3.28	4.59		7.87	10.65
1250 10" thick		370	.108		3.98	4.78		8.76	11.75
1300 12" thick	D-9	350	.137		4.19	5.95		10.14	13.75
2000 Not reinforced, 8" x 24" x 4" thick, hollow		460	.104		1.39	4.52		5.91	8.45
2100 6" thick		440	.109		1.93	4.72		6.65	9.40
2150 8" thick		415	.116		2.40	5		7.40	10.35
2200 10" thick		385	.125		2.91	5.40		8.31	11.50
2250 12" thick		365	.132		3.09	5.70		8.79	12.15
2800 Solid, not reinforced, 8" x 16" x 2" thick	D-8	440	.091		1.72	4.02		5.74	8.10
2900 4" thick		420	.095		1.70	4.21		5.91	8.30

04 22 10 – Concrete Masonry Units

04 22 10.34 Concrete Block, Partitions

		Crew	Daily Output	Labor-Hours	Unit	Material	2017 Bare Costs Labor	Equipment	Total	Total Incl O&P
2950	6" thick	D-8	390	.103	S.F.	3.18	4.53		7.71	10.45
3000	8" thick		365	.110		3.59	4.84		8.43	11.40
3050	10" thick	▼	350	.114		3.97	5.05		9.02	12.10
3100	12" thick	D-9	330	.145	▼	4.10	6.30		10.40	14.20
4000	Regular block, tooled joints, 2 sides, hollow									
4100	Not reinforced, 8" x 16" x 4" thick	D-8	430	.093	S.F.	1.79	4.11		5.90	8.25
4150	6" thick		400	.100		2.35	4.42		6.77	9.40
4200	8" thick		375	.107		2.52	4.71		7.23	10
4250	10" thick	▼	360	.111		3	4.91		7.91	10.85
4300	12" thick	D-9	340	.141		4.18	6.10		10.28	14
4500	Reinforced alternate courses, 8" x 16" x 4" thick	D-8	425	.094		1.95	4.16		6.11	8.55
4550	6" thick		395	.101		2.52	4.47		6.99	9.65
4600	8" thick		370	.108		2.72	4.78		7.50	10.35
4650	10" thick	▼	355	.113		4.20	4.98		9.18	12.25
4700	12" thick	D-9	335	.143		4.35	6.20		10.55	14.35
4900	Solid, not reinforced, 2" thick	D-8	435	.092		1.50	4.06		5.56	7.90
5000	3" thick		430	.093		1.40	4.11		5.51	7.85
5050	4" thick		415	.096		2.05	4.26		6.31	8.80
5100	6" thick		385	.104		2.41	4.59		7	9.70
5150	8" thick	▼	360	.111		3.61	4.91		8.52	11.50
5200	12" thick	D-9	325	.148		5.50	6.40		11.90	15.85
5500	Solid, reinforced alternate courses, 4" thick	D-8	420	.095		2.21	4.21		6.42	8.90
5550	6" thick		380	.105		2.55	4.65		7.20	9.95
5600	8" thick	▼	355	.113		3.76	4.98		8.74	11.80
5650	12" thick	D-9	320	.150	▼	4.44	6.50		10.94	14.90

04 22 10.38 Concrete Brick

		Crew	Daily Output	Labor-Hours	Unit	Material	2017 Bare Costs Labor	Equipment	Total	Total Incl O&P
0010	**CONCRETE BRICK**, C55, grade N, type 1									
0100	Regular, 4" x 2-1/4" x 8"	D-8	660	.061	Ea.	.58	2.68		3.26	4.76
0125	Rusticated, 4" x 2-1/4" x 8"		660	.061		.65	2.68		3.33	4.83
0150	Frog, 4" x 2-1/4" x 8"		660	.061		.63	2.68		3.31	4.81
0200	Double, 4" x 4-7/8" x 8"	▼	535	.075	▼	1.02	3.30		4.32	6.20

04 22 10.42 Concrete Block, Screen Block

		Crew	Daily Output	Labor-Hours	Unit	Material	2017 Bare Costs Labor	Equipment	Total	Total Incl O&P
0010	**CONCRETE BLOCK, SCREEN BLOCK**									
0200	8" x 16", 4" thick	D-8	330	.121	S.F.	2.45	5.35		7.80	10.95
0300	8" thick		270	.148		3.88	6.55		10.43	14.30
0350	12" x 12", 4" thick		290	.138		4.13	6.10		10.23	13.90
0500	8" thick	▼	250	.160	▼	7.10	7.05		14.15	18.65

04 22 10.44 Glazed Concrete Block

		Crew	Daily Output	Labor-Hours	Unit	Material	2017 Bare Costs Labor	Equipment	Total	Total Incl O&P
0010	**GLAZED CONCRETE BLOCK** C744									
0100	Single face, 8" x 16" units, 2" thick	D-8	360	.111	S.F.	10.25	4.91		15.16	18.80
0200	4" thick		345	.116		10.50	5.10		15.60	19.45
0250	6" thick		330	.121		11.35	5.35		16.70	21
0300	8" thick		310	.129		12.25	5.70		17.95	22.50
0350	10" thick	▼	295	.136		13.90	6		19.90	24.50
0400	12" thick	D-9	280	.171		14.80	7.40		22.20	27.50
0700	Double face, 8" x 16" units, 4" thick	D-8	340	.118		14.55	5.20		19.75	24
0750	6" thick		320	.125		18.05	5.50		23.55	28.50
0800	8" thick		300	.133	▼	18.90	5.90		24.80	30
1000	Jambs, bullnose or square, single face, 8" x 16", 2" thick		315	.127	Ea.	19.50	5.60		25.10	30
1050	4" thick		285	.140	"	20	6.20		26.20	31.50
1200	Caps, bullnose or square, 8" x 16", 2" thick		420	.095	L.F.	18.10	4.21		22.31	26.50
1250	4" thick	▼	380	.105	"	19.40	4.65		24.05	28.50

04 22 Concrete Unit Masonry

04 22 10 – Concrete Masonry Units

04 22 10.44 Glazed Concrete Block		Crew	Daily Output	Labor-Hours	Unit	Material	2017 Bare Costs Labor	Equipment	Total	Total Incl O&P
1256	Corner, bullnose or square, 2" thick	D-8	280	.143	Ea.	21	6.30		27.30	33
1258	4" thick		270	.148		22.50	6.55		29.05	35
1260	6" thick		260	.154		28.50	6.80		35.30	42
1270	8" thick		250	.160		35	7.05		42.05	49.50
1280	10" thick		240	.167		35.50	7.35		42.85	50.50
1290	12" thick		230	.174		37	7.70		44.70	53
1500	Cove base, 8" x 16", 2" thick		315	.127	L.F.	9	5.60		14.60	18.55
1550	4" thick		285	.140		9.10	6.20		15.30	19.55
1600	6" thick		265	.151		9.75	6.65		16.40	21
1650	8" thick		245	.163		10.25	7.20		17.45	22.50

04 23 Glass Unit Masonry

04 23 13 – Vertical Glass Unit Masonry

04 23 13.10 Glass Block		Crew	Daily Output	Labor-Hours	Unit	Material	2017 Bare Costs Labor	Equipment	Total	Total Incl O&P
0010	**GLASS BLOCK**									
0100	Plain, 4" thick, under 1,000 S.F., 6" x 6"	D-8	115	.348	S.F.	28.50	15.35		43.85	55
0150	8" x 8"		160	.250		16.05	11.05		27.10	34.50
0160	end block		160	.250		60.50	11.05		71.55	83.50
0170	90 deg corner		160	.250		56	11.05		67.05	78.50
0180	45 deg corner		160	.250		46	11.05		57.05	67.50
0200	12" x 12"		175	.229		24	10.10		34.10	42
0210	4" x 8"		160	.250		28.50	11.05		39.55	48.50
0220	6" x 8"		160	.250		21	11.05		32.05	40
0300	1,000 to 5,000 S.F., 6" x 6"		135	.296		28	13.10		41.10	50.50
0350	8" x 8"		190	.211		15.70	9.30		25	31.50
0400	12" x 12"		215	.186		23.50	8.20		31.70	38.50
0410	4" x 8"		215	.186		28	8.20		36.20	43
0420	6" x 8"		215	.186		20.50	8.20		28.70	35
0500	Over 5,000 S.F., 6" x 6"		145	.276		27	12.20		39.20	49
0550	8" x 8"		215	.186		15.25	8.20		23.45	29.50
0600	12" x 12"		240	.167		22.50	7.35		29.85	36.50
0610	4" x 8"		240	.167		27	7.35		34.35	41.50
0620	6" x 8"		240	.167		19.80	7.35		27.15	33.50
0700	For solar reflective blocks, add					100%				
1000	Thinline, plain, 3-1/8" thick, under 1,000 S.F., 6" x 6"	D-8	115	.348	S.F.	23	15.35		38.35	48.50
1050	8" x 8"		160	.250		12.75	11.05		23.80	31
1200	Over 5,000 S.F., 6" x 6"		145	.276		22	12.20		34.20	43
1250	8" x 8"		215	.186		12.20	8.20		20.40	26
1400	For cleaning block after installation (both sides), add		1000	.040		.16	1.77		1.93	2.90
4000	Accessories									
4100	Anchors, 20 ga. galv., 1-3/4" wide x 24" long				Ea.	5.15			5.15	5.70
4200	Emulsion asphalt				Gal.	11.40			11.40	12.50
4300	Expansion joint, fiberglass				L.F.	.64			.64	.70
4400	Steel mesh, double galvanized				"	1.01			1.01	1.11

04 24 Adobe Unit Masonry

04 24 16 – Manufactured Adobe Unit Masonry

04 24 16.06 Adobe Brick

04 24 16.06 Adobe Brick		Crew	Daily Output	Labor-Hours	Unit	Material	2017 Bare Costs Labor	Equipment	Total	Total Incl O&P
0010	**ADOBE BRICK**, Semi-stabilized, with cement mortar									
0060	Brick, 10" x 4" x 14", 2.6/S.F. ☐G	D-8	560	.071	S.F.	4.80	3.16		7.96	10.15
0080	12" x 4" x 16", 2.3/S.F. ☐G		580	.069		7.05	3.05		10.10	12.45
0100	10" x 4" x 16", 2.3/S.F. ☐G		590	.068		6.55	2.99		9.54	11.85
0120	8" x 4" x 16", 2.3/S.F. ☐G		560	.071		5	3.16		8.16	10.35
0140	4" x 4" x 16", 2.3/S.F. ☐G		540	.074		4.89	3.27		8.16	10.45
0160	6" x 4" x 16", 2.3/S.F. ☐G		540	.074		4.54	3.27		7.81	10.05
0180	4" x 4" x 12", 3.0/S.F. ☐G		520	.077		5.25	3.40		8.65	11.05
0200	8" x 4" x 12", 3.0/S.F. ☐G	▼	520	.077	▼	4.35	3.40		7.75	10.05

04 25 Unit Masonry Panels

04 25 20 – Pre-Fabricated Masonry Panels

04 25 20.10 Brick and Epoxy Mortar Panels

04 25 20.10 Brick and Epoxy Mortar Panels		Crew	Daily Output	Labor-Hours	Unit	Material	2017 Bare Costs Labor	Equipment	Total	Total Incl O&P
0010	**BRICK AND EPOXY MORTAR PANELS**									
0020	Prefabricated brick & epoxy mortar, 4" thick, minimum	C-11	775	.093	S.F.	8	5	2.58	15.58	19.85
0100	Maximum	"	500	.144		10	7.75	4	21.75	28
0200	For 2" concrete back-up, add					50%				
0300	For 1" urethane & 3" concrete back-up, add				▼	70%				

04 27 Multiple-Wythe Unit Masonry

04 27 10 – Multiple-Wythe Masonry

04 27 10.20 Cavity Walls

04 27 10.20 Cavity Walls		Crew	Daily Output	Labor-Hours	Unit	Material	2017 Bare Costs Labor	Equipment	Total	Total Incl O&P
0010	**CAVITY WALLS**, brick and CMU, includes joint reinforcing and ties									
0200	4" face brick, 4" block	D-8	165	.242	S.F.	6.15	10.70		16.85	23
0400	6" block		145	.276		6.55	12.20		18.75	26
0600	8" block	▼	125	.320	▼	6.60	14.15		20.75	29

04 27 10.30 Brick Walls

04 27 10.30 Brick Walls		Crew	Daily Output	Labor-Hours	Unit	Material	2017 Bare Costs Labor	Equipment	Total	Total Incl O&P
0010	**BRICK WALLS**, including mortar, excludes scaffolding R042110-20									
0020	Estimating by number of brick									
0140	Face brick, 4" thick wall, 6.75 brick/S.F.	D-8	1.45	27.586	M	605	1,225		1,830	2,550
0150	Common brick, 4" thick wall, 6.75 brick/S.F.		1.60	25		615	1,100		1,715	2,375
0204	8" thick, 13.50 brick/S.F.		1.80	22.222		635	980		1,615	2,200
0250	12" thick, 20.25 brick/S.F.		1.90	21.053		640	930		1,570	2,125
0304	16" thick, 27.00 brick/S.F.		2	20		645	885		1,530	2,050
0500	Reinforced, face brick, 4" thick wall, 6.75 brick/S.F.		1.40	28.571		625	1,250		1,875	2,650
0520	Common brick, 4" thick wall, 6.75 brick/S.F.		1.55	25.806		635	1,150		1,785	2,450
0550	8" thick, 13.50 brick/S.F.		1.75	22.857		660	1,000		1,660	2,275
0600	12" thick, 20.25 brick/S.F.		1.85	21.622		660	955		1,615	2,200
0650	16" thick, 27.00 brick/S.F.	▼	1.95	20.513	▼	670	905		1,575	2,150
0790	Alternate method of figuring by square foot									
0800	Face brick, 4" thick wall, 6.75 brick/S.F.	D-8	215	.186	S.F.	4.07	8.20		12.27	17.15
0850	Common brick, 4" thick wall, 6.75 brick/S.F.		240	.167		4.14	7.35		11.49	15.85
0900	8" thick, 13.50 brick/S.F.		135	.296		8.55	13.10		21.65	29.50
1000	12" thick, 20.25 brick/S.F.		95	.421		12.90	18.60		31.50	42.50
1050	16" thick, 27.00 brick/S.F.		75	.533		17.45	23.50		40.95	55
1200	Reinforced, face brick, 4" thick wall, 6.75 brick/S.F.		210	.190		4.23	8.40		12.63	17.60
1220	Common brick, 4" thick wall, 6.75 brick/S.F.		235	.170		4.30	7.50		11.80	16.25
1250	8" thick, 13.50 brick/S.F.		130	.308		8.90	13.60		22.50	31
1300	12" thick, 20.25 brick/S.F.	▼	90	.444		13.40	19.65		33.05	44.50

04 27 Multiple-Wythe Unit Masonry

04 27 10 – Multiple-Wythe Masonry

04 27 10.30 Brick Walls		Crew	Daily Output	Labor-Hours	Unit	Material	2017 Bare Costs Labor	Equipment	Total	Total Incl O&P
1350	16" thick, 27.00 brick/S.F.	D-8	70	.571	S.F.	18.10	25		43.10	59

04 27 10.40 Steps										
0010	**STEPS**									
0012	Entry steps, select common brick	D-1	.30	53.333	M	520	2,300		2,820	4,125

04 41 Dry-Placed Stone

04 41 10 – Dry Placed Stone

04 41 10.10 Rough Stone Wall

			Crew	Daily Output	Labor-Hours	Unit	Material	2017 Bare Costs Labor	Equipment	Total	Total Incl O&P
0011	**ROUGH STONE WALL**, Dry										
0012	Dry laid (no mortar), under 18" thick	G	D-1	60	.267	C.F.	13	11.55		24.55	32
0100	Random fieldstone, under 18" thick	G	D-12	60	.533		13	23.50		36.50	50.50
0150	Over 18" thick	G	"	63	.508	↓	15.60	22		37.60	51
0500	Field stone veneer	G	D-8	120	.333	S.F.	12.70	14.70		27.40	36.50
0510	Valley stone veneer	G	↓	120	.333		12.70	14.70		27.40	36.50
0520	River stone veneer	G	↓	120	.333	↓	12.70	14.70		27.40	36.50
0600	Rubble stone walls, in mortar bed, up to 18" thick	G	D-11	75	.320	C.F.	15.70	14.55		30.25	40

04 43 Stone Masonry

04 43 10 – Masonry with Natural and Processed Stone

04 43 10.05 Ashlar Veneer

		Crew	Daily Output	Labor-Hours	Unit	Material	2017 Bare Costs Labor	Equipment	Total	Total Incl O&P
0011	**ASHLAR VENEER** +/- 4" thk, random or random rectangular									
0150	Sawn face, split joints, low priced stone	D-8	140	.286	S.F.	12.05	12.60		24.65	32.50
0200	Medium priced stone		130	.308		13.70	13.60		27.30	36
0300	High priced stone		120	.333		18.25	14.70		32.95	42.50
0600	Seam face, split joints, medium price stone		125	.320		19.35	14.15		33.50	43
0700	High price stone		120	.333		19.10	14.70		33.80	43.50
1000	Split or rock face, split joints, medium price stone		125	.320		11.55	14.15		25.70	34
1100	High price stone	↓	120	.333	↓	17.55	14.70		32.25	42

04 43 10.10 Bluestone

		Crew	Daily Output	Labor-Hours	Unit	Material	2017 Bare Costs Labor	Equipment	Total	Total Incl O&P
0010	**BLUESTONE**, cut to size									
0500	Sills, natural cleft, 10" wide to 6' long, 1-1/2" thick	D-11	70	.343	L.F.	13.75	15.60		29.35	39
0550	2" thick	"	63	.381		15.05	17.30		32.35	43
1000	Stair treads, natural cleft, 12" wide, 6' long, 1-1/2" thick	D-10	115	.278		12.85	13.35	3.79	29.99	39
1050	2" thick		105	.305		13.90	14.65	4.15	32.70	42.50
1100	Smooth finish, 1-1/2" thick		115	.278		12.85	13.35	3.79	29.99	39
1150	2" thick		105	.305		13.90	14.65	4.15	32.70	42.50
1300	Thermal finish, 1-1/2" thick		115	.278		12.85	13.35	3.79	29.99	39
1350	2" thick	↓	105	.305	↓	13.90	14.65	4.15	32.70	42.50

04 43 10.45 Granite

		Crew	Daily Output	Labor-Hours	Unit	Material	2017 Bare Costs Labor	Equipment	Total	Total Incl O&P
0010	**GRANITE**, cut to size									
0050	Veneer, polished face, 3/4" to 1-1/2" thick									
0150	Low price, gray, light gray, etc.	D-10	130	.246	S.F.	27.50	11.80	3.35	42.65	52.50
0180	Medium price, pink, brown, etc.		130	.246		30.50	11.80	3.35	45.65	55.50
0220	High price, red, black, etc.	↓	130	.246	↓	43	11.80	3.35	58.15	69
0300	1-1/2" to 2-1/2" thick, veneer									
0350	Low price, gray, light gray, etc.	D-10	130	.246	S.F.	29	11.80	3.35	44.15	54
0500	Medium price, pink, brown, etc.		130	.246		34	11.80	3.35	49.15	59.50
0550	High price, red, black, etc.	↓	130	.246	↓	53	11.80	3.35	68.15	80.50

04 43 10 – Masonry with Natural and Processed Stone

04 43 10.45 Granite

		Crew	Daily Output	Labor-Hours	Unit	Material	2017 Bare Costs Labor	Equipment	Total	Total Incl O&P
0700	2-1/2" to 4" thick, veneer									
0750	Low price, gray, light gray, etc.	D-10	110	.291	S.F.	39	13.95	3.96	56.91	69
0850	Medium price, pink, brown, etc.		110	.291		45	13.95	3.96	62.91	75.50
0950	High price, red, black, etc.	▼	110	.291		64	13.95	3.96	81.91	96.50
1000	For bush hammered finish, deduct					5%				
1050	Coarse rubbed finish, deduct					10%				
1100	Honed finish, deduct					5%				
1150	Thermal finish, deduct				▼	18%				
2450	For radius under 5', add				L.F.	100%				
2500	Steps, copings, etc., finished on more than one surface									
2550	Low price, gray, light gray, etc.	D-10	50	.640	C.F.	94	30.50	8.70	133.20	160
2575	Medium price, pink, brown, etc.		50	.640		122	30.50	8.70	161.20	191
2600	High price, red, black, etc.	▼	50	.640	▼	150	30.50	8.70	189.20	222
2800	Pavers, 4" x 4" x 4" blocks, split face and joints									
2850	Low price, gray, light gray, etc.	D-11	80	.300	S.F.	13.30	13.65		26.95	35.50
2875	Medium price, pinks, browns, etc.		80	.300		21.50	13.65		35.15	44.50
2900	High price, red, black, etc.	▼	80	.300		29.50	13.65		43.15	53.50
4000	Soffits, 2" thick, low price, gray, light gray	D-13	35	1.371		39	64	12.45	115.45	154
4050	Medium price, pink, brown, etc.		35	1.371		65.50	64	12.45	141.95	184
4100	High price, red, black, etc.		35	1.371		92.50	64	12.45	168.95	214
4200	Low price, gray, light gray, etc.		35	1.371		64	64	12.45	140.45	182
4250	Medium price, pink, brown, etc.		35	1.371		92	64	12.45	168.45	213
4300	High price, red, black, etc.	▼	35	1.371	▼	120	64	12.45	196.45	244
5000	Reclaimed or antique									
5010	Treads, up to 12" wide	D-10	100	.320	L.F.	42.50	15.35	4.36	62.21	75.50
5020	Up to 18" wide		100	.320		38.50	15.35	4.36	58.21	71
5030	Capstone, size varies		50	.640	▼	30.50	30.50	8.70	69.70	90
5040	Posts	▼	30	1.067	V.L.F.	30.50	51	14.50	96	128

04 43 10.50 Lightweight Natural Stone

			Crew	Daily Output	Labor-Hours	Unit	Material	2017 Bare Costs Labor	Equipment	Total	Total Incl O&P
0011	**LIGHTWEIGHT NATURAL STONE** Lava type										
0100	Veneer, rubble face, sawed back, irregular shapes	G	D-10	130	.246	S.F.	8.95	11.80	3.35	24.10	31.50
0200	Sawed face and back, irregular shapes	G		130	.246	"	8.95	11.80	3.35	24.10	31.50
1000	Reclaimed or antique, barn or foundation stone		▼	1	32	Ton	395	1,525	435	2,355	3,275

04 43 10.55 Limestone

		Crew	Daily Output	Labor-Hours	Unit	Material	2017 Bare Costs Labor	Equipment	Total	Total Incl O&P
0010	**LIMESTONE**, cut to size									
0020	Veneer facing panels									
0500	Texture finish, light stick, 4-1/2" thick, 5' x 12'	D-4	300	.107	S.F.	20.50	4.70	.44	25.64	30
0750	5" thick, 5' x 14' panels	D-10	275	.116		21.50	5.60	1.58	28.68	34.50
1000	Sugarcube finish, 2" thick, 3' x 5' panels		275	.116		29	5.60	1.58	36.18	42.50
1050	3" thick, 4' x 9' panels		275	.116		25.50	5.60	1.58	32.68	39
1200	4" thick, 5' x 11' panels		275	.116		31.50	5.60	1.58	38.68	45
1400	Sugarcube, textured finish, 4-1/2" thick, 5' x 12'		275	.116		32.50	5.60	1.58	39.68	46.50
1450	5" thick, 5' x 14' panels		275	.116	▼	34	5.60	1.58	41.18	47.50
2000	Coping, sugarcube finish, top & 2 sides		30	1.067	C.F.	68.50	51	14.50	134	170
2100	Sills, lintels, jambs, trim, stops, sugarcube finish, simple		20	1.600		68.50	77	22	167.50	218
2150	Detailed		20	1.600	▼	68.50	77	22	167.50	218
2300	Steps, extra hard, 14" wide, 6" rise	▼	50	.640	L.F.	25.50	30.50	8.70	64.70	84.50
3000	Quoins, plain finish, 6" x 12" x 12"	D-12	25	1.280	Ea.	41	56		97	132
3050	6" x 16" x 24"	"	25	1.280	"	55	56		111	147

For customer support on your Building Construction Costs with RSMeans Data, call 800.448.8182.

113

04 43 10.60 Marble

		Crew	Daily Output	Labor-Hours	Unit	Material	2017 Bare Costs Labor	2017 Bare Costs Equipment	Total	Total Incl O&P
0011	**MARBLE**, ashlar, split face, +/- 4" thick, random									
0040	Lengths 1' to 4' & heights 2" to 7-1/2", average	D-8	175	.229	S.F.	18.30	10.10		28.40	35.50
0100	Base, polished, 3/4" or 7/8" thick, polished, 6" high	D-10	65	.492	L.F.	12	23.50	6.70	42.20	56.50
0300	Carvings or bas-relief, from templates, simple design		80	.400	S.F.	150	19.20	5.45	174.65	201
0350	Intricate design	↓	80	.400	"	350	19.20	5.45	374.65	415
0600	Columns, cornices, mouldings, etc.									
0650	Hand or special machine cut, simple design	D-10	35	.914	C.F.	55	44	12.45	111.45	141
0700	Intricate design	"	35	.914	"	288	44	12.45	344.45	395
1000	Facing, polished finish, cut to size, 3/4" to 7/8" thick									
1050	Carrara or equal	D-10	130	.246	S.F.	22.50	11.80	3.35	37.65	47
1100	Arabescato or equal		130	.246		39.50	11.80	3.35	54.65	65.50
1300	1-1/4" thick, Botticino Classico or equal		125	.256		24	12.30	3.48	39.78	49
1350	Statuarietto or equal		125	.256		41.50	12.30	3.48	57.28	68
1500	2" thick, Crema Marfil or equal		120	.267		48.50	12.80	3.63	64.93	77
1550	Cafe Pinta or equal	↓	120	.267	↓	70	12.80	3.63	86.43	101
1700	Rubbed finish, cut to size, 4" thick									
1740	Average	D-10	100	.320	S.F.	41.50	15.35	4.36	61.21	74
1780	Maximum	"	100	.320	"	71.50	15.35	4.36	91.21	107
2200	Window sills, 6" x 3/4" thick	D-1	85	.188	L.F.	11.10	8.15		19.25	25
2500	Flooring, polished tiles, 12" x 12" x 3/8" thick									
2510	Thin set, Giallo Solare or equal	D-11	90	.267	S.F.	17.50	12.10		29.60	38
2600	Sky Blue or equal		90	.267		16	12.10		28.10	36.50
2700	Mortar bed, Giallo Solare or equal		65	.369		17.50	16.75		34.25	45.50
2740	Sky Blue or equal	↓	65	.369		16	16.75		32.75	43.50
2780	Travertine, 3/8" thick, Sierra or equal	D-10	130	.246		9.40	11.80	3.35	24.55	32
2790	Silver or equal	"	130	.246		26	11.80	3.35	41.15	50.50
2800	Patio tile, non-slip, 1/2" thick, flame finish	D-11	75	.320	↓	11.15	14.55		25.70	35
2900	Shower or toilet partitions, 7/8" thick partitions									
3050	3/4" or 1-1/4" thick stiles, polished 2 sides, average	D-11	75	.320	S.F.	46.50	14.55		61.05	73.50
3201	Soffits, add to above prices				"	20%	100%			
3210	Stairs, risers, 7/8" thick x 6" high	D-10	115	.278	L.F.	15.60	13.35	3.79	32.74	42
3360	Treads, 12" wide x 1-1/4" thick	"	115	.278	"	44	13.35	3.79	61.14	73
3500	Thresholds, 3' long, 7/8" thick, 4" to 5" wide, plain	D-12	24	1.333	Ea.	35.50	58.50		94	129
3550	Beveled		24	1.333	"	71.50	58.50		130	169
3700	Window stools, polished, 7/8" thick, 5" wide	↓	85	.376	L.F.	21.50	16.50		38	49.50

04 43 10.75 Sandstone or Brownstone

		Crew	Daily Output	Labor-Hours	Unit	Material	2017 Bare Costs Labor	2017 Bare Costs Equipment	Total	Total Incl O&P
0011	**SANDSTONE OR BROWNSTONE**									
0100	Sawed face veneer, 2-1/2" thick, to 2' x 4' panels	D-10	130	.246	S.F.	20	11.80	3.35	35.15	44
0150	4" thick, to 3'-6" x 8' panels		100	.320		20	15.35	4.36	39.71	50.50
0300	Split face, random sizes	↓	100	.320	↓	13.40	15.35	4.36	33.11	43
0350	Cut stone trim (limestone)									
0360	Ribbon stone, 4" thick, 5' pieces	D-8	120	.333	Ea.	160	14.70		174.70	199
0370	Cove stone, 4" thick, 5' pieces		105	.381		161	16.85		177.85	203
0380	Cornice stone, 10" to 12" wide		90	.444		199	19.65		218.65	249
0390	Band stone, 4" thick, 5' pieces		145	.276		109	12.20		121.20	139
0410	Window and door trim, 3" to 4" wide		160	.250		93.50	11.05		104.55	120
0420	Key stone, 18" long	↓	60	.667	↓	93.50	29.50		123	149

04 43 10.80 Slate		Crew	Daily Output	Labor-Hours	Unit	Material	2017 Bare Costs Labor	2017 Bare Costs Equipment	Total	Total Incl O&P
0010	**SLATE**									
0040	Pennsylvania - blue gray to black									
0050	Vermont - unfading green, mottled green & purple, gray & purple									
0100	Virginia - blue black									
0200	Exterior paving, natural cleft, 1" thick									
0250	6" x 6" Pennsylvania	D-12	100	.320	S.F.	7.15	14		21.15	29.50
0300	Vermont		100	.320		10.70	14		24.70	33.50
0350	Virginia		100	.320		14.95	14		28.95	38
0500	24" x 24", Pennsylvania		120	.267		13.80	11.65		25.45	33
0550	Vermont		120	.267		26.50	11.65		38.15	47
0600	Virginia		120	.267		21.50	11.65		33.15	42
0700	18" x 30" Pennsylvania		120	.267		15.65	11.65		27.30	35
0750	Vermont		120	.267		26.50	11.65		38.15	47
0800	Virginia		120	.267		19.30	11.65		30.95	39
1000	Interior flooring, natural cleft, 1/2" thick									
1100	6" x 6" Pennsylvania	D-12	100	.320	S.F.	4.24	14		18.24	26
1150	Vermont		100	.320		9.40	14		23.40	32
1200	Virginia		100	.320		11.80	14		25.80	34.50
1300	24" x 24" Pennsylvania		120	.267		8.20	11.65		19.85	27
1350	Vermont		120	.267		21.50	11.65		33.15	41.50
1400	Virginia		120	.267		15.60	11.65		27.25	35
1500	18" x 24" Pennsylvania		120	.267		8.20	11.65		19.85	27
1550	Vermont		120	.267		17.10	11.65		28.75	37
1600	Virginia		120	.267		15.85	11.65		27.50	35.50
2000	Facing panels, 1-1/4" thick, to 4' x 4' panels									
2100	Natural cleft finish, Pennsylvania	D-10	180	.178	S.F.	36	8.55	2.42	46.97	55
2110	Vermont		180	.178		29	8.55	2.42	39.97	47
2120	Virginia		180	.178		35	8.55	2.42	45.97	54.50
2150	Sand rubbed finish, surface, add					10.80			10.80	11.85
2200	Honed finish, add					7.80			7.80	8.55
2500	Ribbon, natural cleft finish, 1" thick, to 9 S.F.	D-10	80	.400		13.90	19.20	5.45	38.55	51
2550	Sand rubbed finish		80	.400		18.80	19.20	5.45	43.45	56
2600	Honed finish		80	.400		17.50	19.20	5.45	42.15	55
2700	1-1/2" thick		78	.410		18.05	19.70	5.60	43.35	56
2750	Sand rubbed finish		78	.410		24	19.70	5.60	49.30	62
2800	Honed finish		78	.410		22.50	19.70	5.60	47.80	61
2850	2" thick		76	.421		21.50	20	5.75	47.25	61.50
2900	Sand rubbed finish		76	.421		30	20	5.75	55.75	70.50
2950	Honed finish		76	.421		27.50	20	5.75	53.25	68
3100	Stair landings, 1" thick, black, clear	D-1	65	.246		21	10.65		31.65	40
3200	Ribbon	"	65	.246		23.50	10.65		34.15	42
3500	Stair treads, sand finish, 1" thick x 12" wide									
3550	Under 3 L.F.	D-10	85	.376	L.F.	23.50	18.05	5.10	46.65	58.50
3600	3 L.F. to 6 L.F.	"	120	.267	"	25	12.80	3.63	41.43	51
3700	Ribbon, sand finish, 1" thick x 12" wide									
3750	To 6 L.F.	D-10	120	.267	L.F.	21	12.80	3.63	37.43	47
4000	Stools or sills, sand finish, 1" thick, 6" wide	D-12	160	.200		12.20	8.75		20.95	27
4100	Honed finish		160	.200		11.65	8.75		20.40	26.50
4200	10" wide		90	.356		18.80	15.55		34.35	44.50
4250	Honed finish		90	.356		17.50	15.55		33.05	43.50
4400	2" thick, 6" wide		140	.229		19.60	10		29.60	37
4450	Honed finish		140	.229		18.65	10		28.65	36

04 43 Stone Masonry

04 43 10 – Masonry with Natural and Processed Stone

04 43 10.80 Slate

		Crew	Daily Output	Labor-Hours	Unit	Material	2017 Bare Costs Labor	Equipment	Total	Total Incl O&P
4600	10" wide	D-12	90	.356	L.F.	30.50	15.55		46.05	58
4650	Honed finish	↓	90	.356		29	15.55		44.55	56
4800	For lengths over 3', add				↓	25%				

04 43 10.85 Window Sill

		Crew	Daily Output	Labor-Hours	Unit	Material	2017 Bare Costs Labor	Equipment	Total	Total Incl O&P
0010	**WINDOW SILL**									
0020	Bluestone, thermal top, 10" wide, 1-1/2" thick	D-1	85	.188	S.F.	9.60	8.15		17.75	23
0050	2" thick		75	.213	"	9.60	9.25		18.85	25
0100	Cut stone, 5" x 8" plain		48	.333	L.F.	12.55	14.45		27	36
0200	Face brick on edge, brick, 8" wide		80	.200		5.45	8.65		14.10	19.30
0400	Marble, 9" wide, 1" thick		85	.188		8.95	8.15		17.10	22.50
0900	Slate, colored, unfading, honed, 12" wide, 1" thick		85	.188		8.95	8.15		17.10	22.50
0950	2" thick	↓	70	.229	↓	8.95	9.90		18.85	25

04 51 Flue Liner Masonry

04 51 10 – Clay Flue Lining

04 51 10.10 Flue Lining

		Crew	Daily Output	Labor-Hours	Unit	Material	2017 Bare Costs Labor	Equipment	Total	Total Incl O&P
0010	**FLUE LINING**, including mortar									
0020	Clay, 8" x 8"	D-1	125	.128	V.L.F.	5.85	5.55		11.40	14.90
0100	8" x 12"		103	.155		8.45	6.70		15.15	19.65
0200	12" x 12"		93	.172		11	7.45		18.45	23.50
0300	12" x 18"		84	.190		22.50	8.25		30.75	37.50
0400	18" x 18"		75	.213		29	9.25		38.25	46
0500	20" x 20"		66	.242		43.50	10.50		54	63.50
0600	24" x 24"		56	.286		55.50	12.35		67.85	80
1000	Round, 18" diameter		66	.242		39.50	10.50		50	59.50
1100	24" diameter	↓	47	.340	↓	76	14.75		90.75	106

04 54 Refractory Brick Masonry

04 54 10 – Refractory Brick Work

04 54 10.10 Fire Brick

		Crew	Daily Output	Labor-Hours	Unit	Material	2017 Bare Costs Labor	Equipment	Total	Total Incl O&P
0010	**FIRE BRICK**									
0012	Low duty, 2000°F, 9" x 2-1/2" x 4-1/2"	D-1	.60	26.667	M	1,700	1,150		2,850	3,650
0050	High duty, 3000°F	"	.60	26.667	"	2,700	1,150		3,850	4,750

04 54 10.20 Fire Clay

		Crew	Daily Output	Labor-Hours	Unit	Material	2017 Bare Costs Labor	Equipment	Total	Total Incl O&P
0010	**FIRE CLAY**									
0020	Gray, high duty, 100 lb. bag				Bag	36			36	39.50
0050	100 lb. drum, premixed (400 brick per drum)				Drum	43			43	47.50

04 57 Masonry Fireplaces

04 57 10 – Brick or Stone Fireplaces

04 57 10.10 Fireplace		Crew	Daily Output	Labor-Hours	Unit	Material	2017 Bare Costs Labor	Equipment	Total	Total Incl O&P
0010	**FIREPLACE**									
0100	Brick fireplace, not incl. foundations or chimneys									
0110	30" x 29" opening, incl. chamber, plain brickwork	D-1	.40	40	Ea.	570	1,725		2,295	3,275
0200	Fireplace box only (110 brick)	"	2	8	"	162	345		507	710
0300	For elaborate brickwork and details, add					35%	35%			
0400	For hearth, brick & stone, add	D-1	2	8	Ea.	213	345		558	765
0410	For steel, damper, cleanouts, add		4	4		17.55	173		190.55	285
0600	Plain brickwork, incl. metal circulator		.50	32		900	1,375		2,275	3,125
0800	Face brick only, standard size, 8" x 2-2/3" x 4"		.30	53.333	M	600	2,300		2,900	4,200

04 71 Manufactured Brick Masonry

04 71 10 – Simulated or Manufactured Brick

04 71 10.10 Simulated Brick

		Crew	Daily Output	Labor-Hours	Unit	Material	2017 Bare Costs Labor	Equipment	Total	Total Incl O&P
0010	**SIMULATED BRICK**									
0020	Aluminum, baked on colors	1 Carp	200	.040	S.F.	4.50	1.97		6.47	7.95
0050	Fiberglass panels		200	.040		8.45	1.97		10.42	12.25
0100	Urethane pieces cemented in mastic		150	.053		8.60	2.63		11.23	13.45
0150	Vinyl siding panels		200	.040		10.05	1.97		12.02	14.05
0160	Cement base, brick, incl. mastic	D-1	100	.160		9.30	6.90		16.20	21
0170	Corner		50	.320	V.L.F.	23.50	13.85		37.35	47.50
0180	Stone face, incl. mastic		100	.160	S.F.	10.40	6.90		17.30	22
0190	Corner		50	.320	V.L.F.	29.50	13.85		43.35	54

04 72 Cast Stone Masonry

04 72 10 – Cast Stone Masonry Features

04 72 10.10 Coping

		Crew	Daily Output	Labor-Hours	Unit	Material	2017 Bare Costs Labor	Equipment	Total	Total Incl O&P
0010	**COPING**, stock units									
0050	Precast concrete, 10" wide, 4" tapers to 3-1/2", 8" wall	D-1	75	.213	L.F.	20.50	9.25		29.75	37
0100	12" wide, 3-1/2" tapers to 3", 10" wall		70	.229		22.50	9.90		32.40	39.50
0110	14" wide, 4" tapers to 3-1/2", 12" wall		65	.246		25.50	10.65		36.15	44.50
0150	16" wide, 4" tapers to 3-1/2", 14" wall		60	.267		27.50	11.55		39.05	48.50
0250	Precast concrete corners		40	.400	Ea.	36	17.30		53.30	66
0300	Limestone for 12" wall, 4" thick		90	.178	L.F.	14.80	7.70		22.50	28
0350	6" thick		80	.200		22	8.65		30.65	38
0500	Marble, to 4" thick, no wash, 9" wide		90	.178		12.90	7.70		20.60	26
0550	12" wide		80	.200		18	8.65		26.65	33
0700	Terra cotta, 9" wide		90	.178		6.45	7.70		14.15	18.95
0750	12" wide		80	.200		8.80	8.65		17.45	23
0800	Aluminum, for 12" wall		80	.200		9	8.65		17.65	23

04 72 20 – Cultured Stone Veneer

04 72 20.10 Cultured Stone Veneer Components

		Crew	Daily Output	Labor-Hours	Unit	Material	2017 Bare Costs Labor	Equipment	Total	Total Incl O&P
0010	**CULTURED STONE VENEER COMPONENTS**									
0110	On wood frame and sheathing substrate, random sized cobbles, corner stones	D-8	70	.571	V.L.F.	10.40	25		35.40	50.50
0120	Field stones		140	.286	S.F.	7.55	12.60		20.15	27.50
0130	Random sized flats, corner stones		70	.571	V.L.F.	10.70	25		35.70	51
0140	Field stones		140	.286	S.F.	8.90	12.60		21.50	29
0150	Horizontal lined ledgestones, corner stones		75	.533	V.L.F.	10.40	23.50		33.90	47.50
0160	Field stones		150	.267	S.F.	7.55	11.80		19.35	26.50
0170	Random shaped flats, corner stones		65	.615	V.L.F.	10.40	27		37.40	53.50

04 72 Cast Stone Masonry

04 72 20 – Cultured Stone Veneer

04 72 20.10 Cultured Stone Veneer Components	Crew	Daily Output	Labor-Hours	Unit	Material	2017 Bare Costs Labor	2017 Bare Costs Equipment	Total	Total Incl O&P
0180	Field stones	D-8	150	.267	S.F.	7.55	11.80		19.35	26.50
0190	Random shaped/textured face, corner stones		65	.615	V.L.F.	10.40	27		37.40	53.50
0200	Field stones		130	.308	S.F.	7.55	13.60		21.15	29.50
0210	Random shaped river rock, corner stones		65	.615	V.L.F.	10.40	27		37.40	53.50
0220	Field stones		130	.308	S.F.	7.55	13.60		21.15	29.50
0240	On concrete or CMU substrate, random sized cobbles, corner stones		70	.571	V.L.F.	9.75	25		34.75	49.50
0250	Field stones		140	.286	S.F.	7.20	12.60		19.80	27.50
0260	Random sized flats, corner stones		70	.571	V.L.F.	10.05	25		35.05	50
0270	Field stones		140	.286	S.F.	8.55	12.60		21.15	29
0280	Horizontal lined ledgestones, corner stones		75	.533	V.L.F.	9.75	23.50		33.25	46.50
0290	Field stones		150	.267	S.F.	7.20	11.80		19	26
0300	Random shaped flats, corner stones		70	.571	V.L.F.	9.75	25		34.75	49.50
0310	Field stones		140	.286	S.F.	7.20	12.60		19.80	27.50
0320	Random shaped/textured face, corner stones		65	.615	V.L.F.	9.75	27		36.75	52.50
0330	Field stones		130	.308	S.F.	7.20	13.60		20.80	29
0340	Random shaped river rock, corner stones		65	.615	V.L.F.	9.75	27		36.75	52.50
0350	Field stones	▼	130	.308	S.F.	7.20	13.60		20.80	29
0360	Cultured stone veneer, #15 felt weather resistant barrier	1 Clab	3700	.002	Sq.	5.30	.08		5.38	6
0370	Expanded metal lath, diamond, 2.5 lb./S.Y., galvanized	1 Lath	85	.094	S.Y.	3.02	4.51		7.53	9.95
0390	Water table or window sill, 18" long	1 Bric	80	.100	Ea.	10	4.78		14.78	18.35

04 73 Manufactured Stone Masonry

04 73 20 – Simulated or Manufactured Stone

04 73 20.10 Simulated Stone

		Crew	Daily Output	Labor-Hours	Unit	Material	2017 Bare Costs Labor	2017 Bare Costs Equipment	Total	Total Incl O&P
0010	**SIMULATED STONE**									
0100	Insulated fiberglass panels, 5/8" ply backer	L-4	200	.120	S.F.	11	5.60		16.60	21

Estimating Tips

05 05 00 Common Work Results for Metals

- Nuts, bolts, washers, connection angles, and plates can add a significant amount to both the tonnage of a structural steel job and the estimated cost. As a rule of thumb, add 10% to the total weight to account for these accessories.

- Type 2 steel construction, commonly referred to as "simple construction," consists generally of field-bolted connections with lateral bracing supplied by other elements of the building, such as masonry walls or x-bracing. The estimator should be aware, however, that shop connections may be accomplished by welding or bolting. The method may be particular to the fabrication shop and may have an impact on the estimated cost.

05 10 00 Structural Steel

- Steel items can be obtained from two sources: a fabrication shop or a metals service center. Fabrication shops can fabricate items under more controlled conditions than crews in the field can. They are also more efficient and can produce items more economically. Metal service centers serve as a source of long mill shapes to both fabrication shops and contractors.

- Most line items in this structural steel subdivision, and most items in 05 50 00 Metal Fabrications, are indicated as being shop fabricated. The bare material cost for these shop fabricated items is the "Invoice Cost" from the shop and includes the mill base price of steel plus mill extras, transportation to the shop, shop drawings and detailing where warranted, shop fabrication and handling, sandblasting and a shop coat of primer paint, all necessary structural bolts, and delivery to the job site. The bare labor cost and bare equipment cost for these shop fabricated items are for field installation or erection.

- Line items in Subdivision 05 12 23.40 Lightweight Framing, and other items scattered in Division 5, are indicated as being field fabricated. The bare material cost for these field fabricated items is the "Invoice Cost" from the metals service center and includes the mill base price of steel plus mill extras, transportation to the metals service center, material handling, and delivery of long lengths of mill shapes to the job site. Material costs for structural bolts and welding rods should be added to the estimate. The bare labor cost and bare equipment cost for these items are for both field fabrication and field installation or erection, and include time for cutting, welding, and drilling in the fabricated metal items. Drilling into concrete and fasteners to fasten field fabricated items to other work are not included and should be added to the estimate.

05 20 00 Steel Joist Framing

- In any given project the total weight of open web steel joists is determined by the loads to be supported and the design. However, economies can be realized in minimizing the amount of labor used to place the joists. This is done by maximizing the joist spacing, and therefore minimizing the number of joists required to be installed on the job. Certain spacings and locations may be required by the design, but in other cases maximizing the spacing and keeping it as uniform as possible will keep the costs down.

05 30 00 Steel Decking

- The takeoff and estimating of a metal deck involves more than the area of the floor or roof and the type of deck specified or shown on the drawings. Many different sizes and types of openings may exist. Small openings for individual pipes or conduits may be drilled after the floor/roof is installed, but larger openings may require special deck lengths as well as reinforcing or structural support. The estimator should determine who will be supplying this reinforcing. Additionally, some deck terminations are part of the deck package, such as screed angles and pour stops, and others will be part of the steel contract, such as angles attached to structural members and cast-in-place angles and plates. The estimator must ensure that all pieces are accounted for in the complete estimate.

05 50 00 Metal Fabrications

- The most economical steel stairs are those that use common materials, standard details, and most importantly, a uniform and relatively simple method of field assembly. Commonly available A36/A992 channels and plates are very good choices for the main stringers of the stairs, as are angles and tees for the carrier members. Risers and treads are usually made by specialty shops, and it is most economical to use a typical detail in as many places as possible. The stairs should be pre-assembled and shipped directly to the site. The field connections should be simple and straightforward enough to be accomplished efficiently, and with minimum equipment and labor.

Reference Numbers

Reference numbers are shown at the beginning of some major classifications. These numbers refer to related items in the Reference Section. The reference information may be an estimating procedure, an alternate pricing method, or technical information.

Note: Not all subdivisions listed here necessarily appear. ■

Did you know?

Our online estimating solution gives you the same access to RSMeans' data with 24/7 access:

- Quickly locate costs in the searchable database.
- Build cost lists, estimates, and reports in minutes.
- Adjust costs to any location in the U.S. and Canada with the click of a button.

Start your free trial today at
www.RSMeansOnline.com

05 01 Maintenance of Metals

05 01 10 – Maintenance of Structural Metal Framing

05 01 10.51 Cleaning of Structural Metal Framing	Crew	Daily Output	Labor-Hours	Unit	Material	2017 Bare Costs Labor	Equipment	Total	Total Incl O&P
0010 **CLEANING OF STRUCTURAL METAL FRAMING**									
6125 Steel surface treatments, PDCA guidelines									
6170 Wire brush, hand (SSPC-SP2)	1 Psst	400	.020	S.F.	.03	.86		.89	1.48
6180 Power tool (SSPC-SP3)	"	700	.011		.09	.49		.58	.93
6215 Pressure washing, up to 5000 psi, 5000 – 15,000 S.F./day	1 Pord	10000	.001			.03		.03	.05
6220 Steam cleaning, 600 psi @ 300 degree F, 1250 – 2500 S.F./day		2000	.004			.17		.17	.25
6225 Water blasting, up to 25,000 psi, 1750 – 3500 S.F./day		2500	.003			.13		.13	.20
6230 Brush-off blast (SSPC-SP7)	E-11	1750	.018		.17	.81	.12	1.10	1.62
6235 Com'l blast (SSPC-SP6), loose scale, fine pwder rust, 2.0#/S.F. sand		1200	.027		.35	1.17	.18	1.70	2.46
6240 Tight mill scale, little/no rust, 3.0#/S.F. sand		1000	.032		.52	1.41	.22	2.15	3.07
6245 Exist coat blistered/pitted, 4.0#/S.F. sand		875	.037		.70	1.61	.25	2.56	3.62
6250 Exist coat badly pitted/nodules, 6.7#/S.F. sand		825	.039		1.17	1.71	.26	3.14	4.31
6255 Near white blast (SSPC-SP10), loose scale, fine rust, 5.6#/S.F. sand		450	.071		.97	3.13	.48	4.58	6.60
6260 Tight mill scale, little/no rust, 6.9#/S.F. sand		325	.098		1.20	4.34	.67	6.21	9
6265 Exist coat blistered/pitted, 9.0#/S.F. sand		225	.142		1.57	6.25	.96	8.78	12.85
6270 Exist coat badly pitted/nodules, 11.3#/S.F. sand		150	.213		1.97	9.40	1.45	12.82	18.80

05 05 Common Work Results for Metals

05 05 05 – Selective Demolition for Metals

05 05 05.10 Selective Demolition, Metals

	Crew	Daily Output	Labor-Hours	Unit	Material	2017 Bare Costs Labor	Equipment	Total	Total Incl O&P
0010 **SELECTIVE DEMOLITION, METALS** R024119-10									
0015 Excludes shores, bracing, cutting, loading, hauling, dumping									
0020 Remove nuts only up to 3/4" diameter	1 Sswk	480	.017	Ea.		.91		.91	1.51
0030 7/8" to 1-1/4" diameter		240	.033			1.81		1.81	3.02
0040 1-3/8" to 2" diameter		160	.050			2.72		2.72	4.53
0060 Unbolt and remove structural bolts up to 3/4" diameter		240	.033			1.81		1.81	3.02
0070 7/8" to 2" diameter		160	.050			2.72		2.72	4.53
0140 Light weight framing members, remove whole or cut up, up to 20 lb.		240	.033			1.81		1.81	3.02
0150 21 – 40 lb.	2 Sswk	210	.076			4.14		4.14	6.90
0160 41 – 80 lb.	3 Sswk	180	.133			7.25		7.25	12.05
0170 81 – 120 lb.	4 Sswk	150	.213			11.60		11.60	19.30
0230 Structural members, remove whole or cut up, up to 500 lb.	E-19	48	.500			27	22.50	49.50	68
0240 1/4 – 2 tons	E-18	36	1.111			60.50	30	90.50	132
0250 2 – 5 tons	E-24	30	1.067			57.50	22.50	80	119
0260 5 – 10 tons	E-20	24	2.667			144	54.50	198.50	294
0270 10 – 15 tons	E-2	18	3.111			168	93	261	375
0340 Fabricated item, remove whole or cut up, up to 20 lb.	1 Sswk	96	.083			4.52		4.52	7.55
0350 21 – 40 lb.	2 Sswk	84	.190			10.35		10.35	17.25
0360 41 – 80 lb.	3 Sswk	72	.333			18.10		18.10	30
0370 81 – 120 lb.	4 Sswk	60	.533			29		29	48.50
0380 121 – 500 lb.	E-19	48	.500			27	22.50	49.50	68
0390 501 – 1000 lb.	"	36	.667			36	30	66	91
0500 Steel roof decking, uncovered, bare	B-2	5000	.008	S.F.		.32		.32	.48

05 05 13 – Shop-Applied Coatings for Metal

05 05 13.50 Paints and Protective Coatings

	Crew	Daily Output	Labor-Hours	Unit	Material	2017 Bare Costs Labor	Equipment	Total	Total Incl O&P
0010 **PAINTS AND PROTECTIVE COATINGS**									
5900 Galvanizing structural steel in shop, under 1 ton R050516-30				Ton	540			540	595
5950 1 ton to 20 tons					495			495	545
6000 Over 20 tons					450			450	495

05 05 Common Work Results for Metals

05 05 19 – Post-Installed Concrete Anchors

05 05 19.10 Chemical Anchors

		Crew	Daily Output	Labor-Hours	Unit	Material	2017 Bare Costs Labor	2017 Bare Costs Equipment	Total	Total Incl O&P
0010	**CHEMICAL ANCHORS**									
0020	Includes layout & drilling									
1430	Chemical anchor, w/rod & epoxy cartridge, 3/4" diam. x 9-1/2" long	B-89A	27	.593	Ea.	9.80	27	4.24	41.04	56.50
1435	1" diameter x 11-3/4" long		24	.667		18.25	30	4.77	53.02	72
1440	1-1/4" diameter x 14" long		21	.762		37	34.50	5.45	76.95	100
1445	1-3/4" diameter x 15" long		20	.800		65	36	5.75	106.75	133
1450	18" long		17	.941		78	42.50	6.75	127.25	158
1455	2" diameter x 18" long		16	1		103	45.50	7.15	155.65	190
1460	24" long	▼	15	1.067	▼	134	48.50	7.65	190.15	229

05 05 19.20 Expansion Anchors

			Crew	Daily Output	Labor-Hours	Unit	Material	2017 Bare Costs Labor	2017 Bare Costs Equipment	Total	Total Incl O&P
0010	**EXPANSION ANCHORS**										
0100	Anchors for concrete, brick or stone, no layout and drilling										
0200	Expansion shields, zinc, 1/4" diameter, 1-5/16" long, single	G	1 Carp	90	.089	Ea.	.45	4.38		4.83	7.20
0300	1-3/8" long, double	G		85	.094		.55	4.64		5.19	7.70
0400	3/8" diameter, 1-1/2" long, single	G		85	.094		.68	4.64		5.32	7.85
0500	2" long, double	G		80	.100		1.22	4.93		6.15	8.90
0600	1/2" diameter, 2-1/16" long, single	G		80	.100		1.25	4.93		6.18	8.95
0700	2-1/2" long, double	G		75	.107		2.35	5.25		7.60	10.65
0800	5/8" diameter, 2-5/8" long, single	G		75	.107		2.16	5.25		7.41	10.45
0900	2-3/4" long, double	G		70	.114		3.31	5.65		8.96	12.25
1000	3/4" diameter, 2-3/4" long, single	G		70	.114		3.27	5.65		8.92	12.20
1100	3-15/16" long, double	G	▼	65	.123	▼	6.15	6.05		12.20	16.10
2100	Hollow wall anchors for gypsum wall board, plaster or tile										
2300	1/8" diameter, short	G	1 Carp	160	.050	Ea.	.27	2.46		2.73	4.07
2400	Long	G		150	.053		.30	2.63		2.93	4.35
2500	3/16" diameter, short	G		150	.053		.47	2.63		3.10	4.54
2600	Long	G		140	.057		.64	2.81		3.45	5
2700	1/4" diameter, short	G		140	.057		.64	2.81		3.45	5
2800	Long	G		130	.062		.70	3.03		3.73	5.40
3000	Toggle bolts, bright steel, 1/8" diameter, 2" long	G		85	.094		.22	4.64		4.86	7.35
3100	4" long	G		80	.100		.27	4.93		5.20	7.85
3200	3/16" diameter, 3" long	G		80	.100		.29	4.93		5.22	7.85
3300	6" long	G		75	.107		.38	5.25		5.63	8.45
3400	1/4" diameter, 3" long	G		75	.107		.36	5.25		5.61	8.45
3500	6" long	G		70	.114		.54	5.65		6.19	9.20
3600	3/8" diameter, 3" long	G		70	.114		.96	5.65		6.61	9.65
3700	6" long	G		60	.133		1.38	6.55		7.93	11.55
3800	1/2" diameter, 4" long	G		60	.133		1.94	6.55		8.49	12.20
3900	6" long	G	▼	50	.160	▼	2.48	7.90		10.38	14.80
4000	Nailing anchors										
4100	Nylon nailing anchor, 1/4" diameter, 1" long		1 Carp	3.20	2.500	C	22	123		145	213
4200	1-1/2" long			2.80	2.857		26	141		167	245
4300	2" long			2.40	3.333		30.50	164		194.50	286
4400	Metal nailing anchor, 1/4" diameter, 1" long	G		3.20	2.500		20.50	123		143.50	212
4500	1-1/2" long	G		2.80	2.857		26	141		167	245
4600	2" long	G	▼	2.40	3.333	▼	31.50	164		195.50	287
5000	Screw anchors for concrete, masonry,										
5100	stone & tile, no layout or drilling included										
5700	Lag screw shields, 1/4" diameter, short	G	1 Carp	90	.089	Ea.	.37	4.38		4.75	7.10
5800	Long	G		85	.094		.43	4.64		5.07	7.55
5900	3/8" diameter, short	G		85	.094		.72	4.64		5.36	7.90
6000	Long	G	▼	80	.100		.95	4.93		5.88	8.60

05 05 19 – Post-Installed Concrete Anchors

05 05 19.20 Expansion Anchors

			Crew	Daily Output	Labor-Hours	Unit	Material	2017 Bare Costs Labor	Equipment	Total	Total Incl O&P
6100	1/2" diameter, short	G	1 Carp	80	.100	Ea.	1.06	4.93		5.99	8.70
6200	Long	G		75	.107		1.35	5.25		6.60	9.55
6300	5/8" diameter, short	G		70	.114		1.64	5.65		7.29	10.40
6400	Long	G		65	.123		2.05	6.05		8.10	11.55
6600	Lead, #6 & #8, 3/4" long	G		260	.031		.20	1.52		1.72	2.54
6700	#10 - #14, 1-1/2" long	G		200	.040		.41	1.97		2.38	3.47
6800	#16 & #18, 1-1/2" long	G		160	.050		.43	2.46		2.89	4.24
6900	Plastic, #6 & #8, 3/4" long			260	.031		.04	1.52		1.56	2.36
7000	#8 & #10, 7/8" long			240	.033		.04	1.64		1.68	2.55
7100	#10 & #12, 1" long			220	.036		.05	1.79		1.84	2.80
7200	#14 & #16, 1-1/2" long		↓	160	.050	↓	.08	2.46		2.54	3.86
8000	Wedge anchors, not including layout or drilling										
8050	Carbon steel, 1/4" diameter, 1-3/4" long	G	1 Carp	150	.053	Ea.	.45	2.63		3.08	4.51
8100	3-1/4" long	G		140	.057		.59	2.81		3.40	4.96
8150	3/8" diameter, 2-1/4" long	G		145	.055		.53	2.72		3.25	4.74
8200	5" long	G		140	.057		.93	2.81		3.74	5.35
8250	1/2" diameter, 2-3/4" long	G		140	.057		1.05	2.81		3.86	5.45
8300	7" long	G		125	.064		1.80	3.15		4.95	6.80
8350	5/8" diameter, 3-1/2" long	G		130	.062		2.03	3.03		5.06	6.85
8400	8-1/2" long	G		115	.070		4.32	3.43		7.75	10
8450	3/4" diameter, 4-1/4" long	G		115	.070		3.09	3.43		6.52	8.65
8500	10" long	G		95	.084		7.05	4.15		11.20	14.10
8550	1" diameter, 6" long	G		100	.080		9.70	3.94		13.64	16.75
8575	9" long	G		85	.094		12.60	4.64		17.24	21
8600	12" long	G		75	.107		13.65	5.25		18.90	23
8650	1-1/4" diameter, 9" long	G		70	.114		25.50	5.65		31.15	37
8700	12" long	G	↓	60	.133	↓	33	6.55		39.55	46
8750	For type 303 stainless steel, add						350%				
8800	For type 316 stainless steel, add						450%				
8950	Self-drilling concrete screw, hex washer head, 3/16" diam. x 1-3/4" long	G	1 Carp	300	.027	Ea.	.21	1.31		1.52	2.24
8960	2-1/4" long	G		250	.032		.25	1.58		1.83	2.69
8970	Phillips flat head, 3/16" diam. x 1-3/4" long	G		300	.027		.21	1.31		1.52	2.24
8980	2-1/4" long	G	↓	250	.032		.24	1.58		1.82	2.67

05 05 21 – Fastening Methods for Metal

05 05 21.10 Cutting Steel

			Crew	Daily Output	Labor-Hours	Unit	Material	2017 Bare Costs Labor	Equipment	Total	Total Incl O&P
0010	**CUTTING STEEL**										
0020	Hand burning, incl. preparation, torch cutting & grinding, no staging										
0050	Steel to 1/4" thick		E-25	400	.020	L.F.	.19	1.13	.03	1.35	2.12
0100	1/2" thick			320	.025		.35	1.41	.04	1.80	2.78
0150	3/4" thick			260	.031		.58	1.73	.05	2.36	3.58
0200	1" thick		↓	200	.040	↓	.84	2.25	.06	3.15	4.74

05 05 21.15 Drilling Steel

			Crew	Daily Output	Labor-Hours	Unit	Material	2017 Bare Costs Labor	Equipment	Total	Total Incl O&P
0010	**DRILLING STEEL**										
1910	Drilling & layout for steel, up to 1/4" deep, no anchor										
1920	Holes, 1/4" diameter		1 Sswk	112	.071	Ea.	.10	3.88		3.98	6.55
1925	For each additional 1/4" depth, add			336	.024		.10	1.29		1.39	2.26
1930	3/8" diameter			104	.077		.09	4.18		4.27	7.05
1935	For each additional 1/4" depth, add			312	.026		.09	1.39		1.48	2.42
1940	1/2" diameter			96	.083		.11	4.52		4.63	7.70
1945	For each additional 1/4" depth, add			288	.028		.11	1.51		1.62	2.64
1950	5/8" diameter			88	.091		.16	4.94		5.10	8.45
1955	For each additional 1/4" depth, add			264	.030		.16	1.65		1.81	2.92

05 05 Common Work Results for Metals

05 05 21 – Fastening Methods for Metal

05 05 21.15 Drilling Steel

		Crew	Daily Output	Labor-Hours	Unit	Material	2017 Bare Costs Labor	Equipment	Total	Total Incl O&P
1960	3/4" diameter	1 Sswk	80	.100	Ea.	.20	5.45		5.65	9.25
1965	For each additional 1/4" depth, add		240	.033		.20	1.81		2.01	3.24
1970	7/8" diameter		72	.111		.26	6.05		6.31	10.35
1975	For each additional 1/4" depth, add		216	.037		.26	2.01		2.27	3.64
1980	1" diameter		64	.125		.23	6.80		7.03	11.55
1985	For each additional 1/4" depth, add		192	.042		.23	2.26		2.49	4.02
1990	For drilling up, add						40%			

05 05 21.90 Welding Steel

		Crew	Daily Output	Labor-Hours	Unit	Material	2017 Bare Costs Labor	Equipment	Total	Total Incl O&P
0010	**WELDING STEEL**, Structural R050521-20									
0020	Field welding, 1/8" E6011, cost per welder, no operating engineer	E-14	8	1	Hr.	4.71	56.50	16.25	77.46	117
0200	With 1/2 operating engineer	E-13	8	1.500		4.71	82	16.25	102.96	155
0300	With 1 operating engineer	E-12	8	2		4.71	107	16.20	127.91	194
0500	With no operating engineer, 2# weld rod per ton	E-14	8	1	Ton	4.71	56.50	16.25	77.46	117
0600	8# E6011 per ton	"	2	4		18.85	225	65	308.85	465
0800	With one operating engineer per welder, 2# E6011 per ton	E-12	8	2		4.71	107	16.20	127.91	194
0900	8# E6011 per ton	"	2	8		18.85	430	65	513.85	775
1200	Continuous fillet, down welding									
1300	Single pass, 1/8" thick, 0.1#/L.F.	E-14	150	.053	L.F.	.24	3	.87	4.11	6.20
1400	3/16" thick, 0.2#/L.F.		75	.107		.47	6	1.73	8.20	12.40
1500	1/4" thick, 0.3#/L.F.		50	.160		.71	9	2.60	12.31	18.65
1610	5/16" thick, 0.4#/L.F.		38	.211		.94	11.85	3.42	16.21	24.50
1800	3 passes, 3/8" thick, 0.5#/L.F.		30	.267		1.18	15	4.33	20.51	31
2010	4 passes, 1/2" thick, 0.7#/L.F.		22	.364		1.65	20.50	5.90	28.05	42.50
2200	5 to 6 passes, 3/4" thick, 1.3#/L.F.		12	.667		3.06	37.50	10.80	51.36	78
2400	8 to 11 passes, 1" thick, 2.4#/L.F.		6	1.333		5.65	75	21.50	102.15	155
2600	For vertical joint welding, add						20%			
2700	Overhead joint welding, add						300%			
2900	For semi-automatic welding, obstructed joints, deduct						5%			
3000	Exposed joints, deduct						15%			
4000	Cleaning and welding plates, bars, or rods									
4010	to existing beams, columns, or trusses	E-14	12	.667	L.F.	1.18	37.50	10.80	49.48	75.50

05 05 23 – Metal Fastenings

05 05 23.10 Bolts and Hex Nuts

			Crew	Daily Output	Labor-Hours	Unit	Material	2017 Bare Costs Labor	Equipment	Total	Total Incl O&P
0010	**BOLTS & HEX NUTS**, Steel, A307										
0100	1/4" diameter, 1/2" long	G	1 Sswk	140	.057	Ea.	.06	3.10		3.16	5.20
0200	1" long	G		140	.057		.07	3.10		3.17	5.25
0300	2" long	G		130	.062		.10	3.34		3.44	5.65
0400	3" long	G		130	.062		.15	3.34		3.49	5.70
0500	4" long	G		120	.067		.17	3.62		3.79	6.25
0600	3/8" diameter, 1" long	G		130	.062		.14	3.34		3.48	5.70
0700	2" long	G		130	.062		.18	3.34		3.52	5.75
0800	3" long	G		120	.067		.24	3.62		3.86	6.30
0900	4" long	G		120	.067		.30	3.62		3.92	6.40
1000	5" long	G		115	.070		.38	3.78		4.16	6.70
1100	1/2" diameter, 1-1/2" long	G		120	.067		.40	3.62		4.02	6.50
1200	2" long	G		120	.067		.46	3.62		4.08	6.55
1300	4" long	G		115	.070		.75	3.78		4.53	7.15
1400	6" long	G		110	.073		1.05	3.95		5	7.75
1500	8" long	G		105	.076		1.38	4.14		5.52	8.40
1600	5/8" diameter, 1-1/2" long	G		120	.067		.85	3.62		4.47	7
1700	2" long	G		120	.067		.94	3.62		4.56	7.10
1800	4" long	G		115	.070		1.34	3.78		5.12	7.80

For customer support on your Building Construction Costs with RSMeans Data, call 800.448.8182.

123

05 05 Common Work Results for Metals

05 05 23 – Metal Fastenings

05 05 23.10 Bolts and Hex Nuts

			Crew	Daily Output	Labor-Hours	Unit	Material	2017 Bare Costs Labor	Equipment	Total	Total Incl O&P
1900	6" long	G	1 Sswk	110	.073	Ea.	1.72	3.95		5.67	8.50
2000	8" long	G		105	.076		2.55	4.14		6.69	9.70
2100	10" long	G		100	.080		3.20	4.34		7.54	10.75
2200	3/4" diameter, 2" long	G		120	.067		1.15	3.62		4.77	7.30
2300	4" long	G		110	.073		1.65	3.95		5.60	8.40
2400	6" long	G		105	.076		2.12	4.14		6.26	9.25
2500	8" long	G		95	.084		3.20	4.57		7.77	11.10
2600	10" long	G		85	.094		4.20	5.10		9.30	13.10
2700	12" long	G		80	.100		4.92	5.45		10.37	14.45
2800	1" diameter, 3" long	G		105	.076		2.67	4.14		6.81	9.85
2900	6" long	G		90	.089		3.92	4.83		8.75	12.35
3000	12" long	G		75	.107		7.05	5.80		12.85	17.45
3100	For galvanized, add						75%				
3200	For stainless, add						350%				

05 05 23.25 High Strength Bolts

			Crew	Daily Output	Labor-Hours	Unit	Material	2017 Bare Costs Labor	Equipment	Total	Total Incl O&P
0010	**HIGH STRENGTH BOLTS** R050523-10										
0020	A325 Type 1, structural steel, bolt-nut-washer set										
0100	1/2" diameter x 1-1/2" long	G	1 Sswk	130	.062	Ea.	1.20	3.34		4.54	6.85
0120	2" long	G		125	.064		1.30	3.48		4.78	7.25
0150	3" long	G		120	.067		1.83	3.62		5.45	8.05
0170	5/8" diameter x 1-1/2" long	G		125	.064		1.96	3.48		5.44	7.95
0180	2" long	G		120	.067		2.11	3.62		5.73	8.35
0190	3" long	G		115	.070		2.62	3.78		6.40	9.20
0200	3/4" diameter x 2" long	G		120	.067		3.12	3.62		6.74	9.50
0220	3" long	G		115	.070		3.77	3.78		7.55	10.45
0250	4" long	G		110	.073		4.64	3.95		8.59	11.70
0300	6" long	G		105	.076		6.05	4.14		10.19	13.55
0350	8" long	G		95	.084		12.10	4.57		16.67	21
0360	7/8" diameter x 2" long	G		115	.070		4.06	3.78		7.84	10.75
0365	3" long	G		110	.073		4.81	3.95		8.76	11.90
0370	4" long	G		105	.076		5.80	4.14		9.94	13.30
0380	6" long	G		100	.080		7.40	4.34		11.74	15.40
0390	8" long	G		90	.089		11.80	4.83		16.63	21
0400	1" diameter x 2" long	G		105	.076		5.40	4.14		9.54	12.85
0420	3" long	G		100	.080		6.10	4.34		10.44	13.95
0450	4" long	G		95	.084		6.95	4.57		11.52	15.25
0500	6" long	G		90	.089		9.25	4.83		14.08	18.25
0550	8" long	G		85	.094		16.20	5.10		21.30	26.50
0600	1-1/4" diameter x 3" long	G		85	.094		11.05	5.10		16.15	20.50
0650	4" long	G		80	.100		12	5.45		17.45	22.50
0700	6" long	G		75	.107		15.60	5.80		21.40	27
0750	8" long	G		70	.114		19.90	6.20		26.10	32.50
1020	A490, bolt-nut-washer set										
1170	5/8" diameter x 1-1/2" long	G	1 Sswk	125	.064	Ea.	4.49	3.48		7.97	10.75
1180	2" long	G		120	.067		5.35	3.62		8.97	11.95
1190	3" long	G		115	.070		6.55	3.78		10.33	13.50
1200	3/4" diameter x 2" long	G		120	.067		4.09	3.62		7.71	10.55
1220	3" long	G		115	.070		4.84	3.78		8.62	11.60
1250	4" long	G		110	.073		5.65	3.95		9.60	12.80
1300	6" long	G		105	.076		8.30	4.14		12.44	16
1350	8" long	G		95	.084		14.10	4.57		18.67	23
1360	7/8" diameter x 2" long	G		115	.070		6.15	3.78		9.93	13.05

124

05 05 Common Work Results for Metals

05 05 23 – Metal Fastenings

05 05 23.25 High Strength Bolts

			Crew	Daily Output	Labor-Hours	Unit	Material	2017 Bare Costs Labor	Equipment	Total	Total Incl O&P
1365	3" long	G	1 Sswk	110	.073	Ea.	7.25	3.95		11.20	14.55
1370	4" long	G		105	.076		9	4.14		13.14	16.80
1380	6" long	G		100	.080		12.65	4.34		16.99	21
1390	8" long	G		90	.089		18.45	4.83		23.28	28.50
1400	1" diameter x 2" long	G		105	.076		8.15	4.14		12.29	15.85
1420	3" long	G		100	.080		9.85	4.34		14.19	18.10
1450	4" long	G		95	.084		11.35	4.57		15.92	20
1500	6" long	G		90	.089		15.25	4.83		20.08	25
1550	8" long	G		85	.094		24	5.10		29.10	35
1600	1-1/4" diameter x 3" long	G		85	.094		37.50	5.10		42.60	49.50
1650	4" long	G		80	.100		43.50	5.45		48.95	56.50
1700	6" long	G		75	.107		60.50	5.80		66.30	76
1750	8" long	G		70	.114		79	6.20		85.20	97.50

05 05 23.30 Lag Screws

			Crew	Daily Output	Labor-Hours	Unit	Material	2017 Bare Costs Labor	Equipment	Total	Total Incl O&P
0010	**LAG SCREWS**										
0020	Steel, 1/4" diameter, 2" long	G	1 Carp	200	.040	Ea.	.11	1.97		2.08	3.14
0100	3/8" diameter, 3" long	G		150	.053		.34	2.63		2.97	4.39
0200	1/2" diameter, 3" long	G		130	.062		.70	3.03		3.73	5.40
0300	5/8" diameter, 3" long	G		120	.067		1.20	3.28		4.48	6.35

05 05 23.35 Machine Screws

			Crew	Daily Output	Labor-Hours	Unit	Material	2017 Bare Costs Labor	Equipment	Total	Total Incl O&P
0010	**MACHINE SCREWS**										
0020	Steel, round head, #8 x 1" long	G	1 Carp	4.80	1.667	C	5.05	82		87.05	132
0110	#8 x 2" long	G		2.40	3.333		7.55	164		171.55	260
0200	#10 x 1" long	G		4	2		5.95	98.50		104.45	158
0300	#10 x 2" long	G		2	4		12.05	197		209.05	315

05 05 23.50 Powder Actuated Tools and Fasteners

			Crew	Daily Output	Labor-Hours	Unit	Material	2017 Bare Costs Labor	Equipment	Total	Total Incl O&P
0010	**POWDER ACTUATED TOOLS & FASTENERS**										
0020	Stud driver, .22 caliber, single shot					Ea.	151			151	166
0100	.27 caliber, semi automatic, strip					"	460			460	505
0300	Powder load, single shot, .22 cal, power level 2, brown					C	5.60			5.60	6.20
0400	Strip, .27 cal, power level 4, red						8.10			8.10	8.90
0600	Drive pin, .300 x 3/4" long	G	1 Carp	4.80	1.667		4.36	82		86.36	131
0700	.300 x 3" long with washer	G	"	4	2		13.30	98.50		111.80	166

05 05 23.55 Rivets

			Crew	Daily Output	Labor-Hours	Unit	Material	2017 Bare Costs Labor	Equipment	Total	Total Incl O&P
0010	**RIVETS**										
0100	Aluminum rivet & mandrel, 1/2" grip length x 1/8" diameter	G	1 Carp	4.80	1.667	C	8.80	82		90.80	136
0200	3/16" diameter	G		4	2		12.95	98.50		111.45	165
0300	Aluminum rivet, steel mandrel, 1/8" diameter	G		4.80	1.667		10.35	82		92.35	137
0400	3/16" diameter	G		4	2		16.60	98.50		115.10	169
0500	Copper rivet, steel mandrel, 1/8" diameter	G		4.80	1.667		9.95	82		91.95	137
0800	Stainless rivet & mandrel, 1/8" diameter	G		4.80	1.667		28	82		110	157
0900	3/16" diameter	G		4	2		44	98.50		142.50	200
1000	Stainless rivet, steel mandrel, 1/8" diameter	G		4.80	1.667		16	82		98	144
1100	3/16" diameter	G		4	2		26	98.50		124.50	180
1200	Steel rivet and mandrel, 1/8" diameter	G		4.80	1.667		8.55	82		90.55	135
1300	3/16" diameter	G		4	2		13.70	98.50		112.20	166
1400	Hand riveting tool, standard					Ea.	76.50			76.50	84
1500	Deluxe						395			395	435
1600	Power riveting tool, standard						505			505	555
1700	Deluxe						1,650			1,650	1,825

For customer support on your Building Construction Costs with RSMeans Data, call 800.448.8182.

125

05 05 Common Work Results for Metals

05 05 23 – Metal Fastenings

05 05 23.70 Structural Blind Bolts

		Crew	Daily Output	Labor-Hours	Unit	Material	2017 Bare Costs Labor	2017 Bare Costs Equipment	Total	Total Incl O&P
0010	**STRUCTURAL BLIND BOLTS**									
0100	1/4" diameter x 1/4" grip	G 1 Sswk	240	.033	Ea.	1.73	1.81		3.54	4.92
0150	1/2" grip	G	216	.037		1.86	2.01		3.87	5.40
0200	3/8" diameter x 1/2" grip	G	232	.034		2.44	1.87		4.31	5.80
0250	3/4" grip	G	208	.038		2.56	2.09		4.65	6.30
0300	1/2" diameter x 1/2" grip	G	224	.036		5.85	1.94		7.79	9.65
0350	3/4" grip	G	200	.040		8.20	2.17		10.37	12.60
0400	5/8" diameter x 3/4" grip	G	216	.037		10.30	2.01		12.31	14.70
0450	1" grip	G	192	.042		11.85	2.26		14.11	16.80

05 05 23.80 Vibration and Bearing Pads

		Crew	Daily Output	Labor-Hours	Unit	Material	2017 Bare Costs Labor	2017 Bare Costs Equipment	Total	Total Incl O&P
0010	**VIBRATION & BEARING PADS**									
0300	Laminated synthetic rubber impregnated cotton duck, 1/2" thick	2 Sswk	24	.667	S.F.	72.50	36		108.50	141
0400	1" thick		20	.800		142	43.50		185.50	229
0600	Neoprene bearing pads, 1/2" thick		24	.667		27.50	36		63.50	91
0700	1" thick		20	.800		55	43.50		98.50	133
0900	Fabric reinforced neoprene, 5000 psi, 1/2" thick		24	.667		12	36		48	73.50
1000	1" thick		20	.800		24	43.50		67.50	99
1200	Felt surfaced vinyl pads, cork and sisal, 5/8" thick		24	.667		30.50	36		66.50	94
1300	1" thick		20	.800		55	43.50		98.50	133
1500	Teflon bonded to 10 ga. carbon steel, 1/32" layer		24	.667		54	36		90	120
1600	3/32" layer		24	.667		81	36		117	150
1800	Bonded to 10 ga. stainless steel, 1/32" layer		24	.667		95.50	36		131.50	166
1900	3/32" layer		24	.667		126	36		162	200
2100	Circular machine leveling pad & stud				Kip	6.75			6.75	7.45

05 05 23.85 Weld Shear Connectors

		Crew	Daily Output	Labor-Hours	Unit	Material	2017 Bare Costs Labor	2017 Bare Costs Equipment	Total	Total Incl O&P
0010	**WELD SHEAR CONNECTORS**									
0020	3/4" diameter, 3-3/16" long	G E-10	960	.017	Ea.	.53	.92	.42	1.87	2.59
0030	3-3/8" long	G	950	.017		.56	.93	.42	1.91	2.62
0200	3-7/8" long	G	945	.017		.60	.94	.42	1.96	2.69
0300	4-3/16" long	G	935	.017		.63	.95	.43	2.01	2.74
0500	4-7/8" long	G	930	.017		.70	.95	.43	2.08	2.83
0600	5-3/16" long	G	920	.017		.73	.96	.43	2.12	2.88
0800	5-3/8" long	G	910	.018		.74	.97	.44	2.15	2.91
0900	6-3/16" long	G	905	.018		.81	.98	.44	2.23	3.01
1000	7-3/16" long	G	895	.018		1	.99	.45	2.44	3.24
1100	8-3/16" long	G	890	.018		1.10	.99	.45	2.54	3.36
1500	7/8" diameter, 3-11/16" long	G	920	.017		.86	.96	.43	2.25	3.03
1600	4-3/16" long	G	910	.018		.93	.97	.44	2.34	3.12
1700	5-3/16" long	G	905	.018		1.05	.98	.44	2.47	3.28
1800	6-3/16" long	G	895	.018		1.17	.99	.45	2.61	3.43
1900	7-3/16" long	G	890	.018		1.30	.99	.45	2.74	3.58
2000	8-3/16" long	G	880	.018		1.42	1.01	.45	2.88	3.75

05 05 23.87 Weld Studs

		Crew	Daily Output	Labor-Hours	Unit	Material	2017 Bare Costs Labor	2017 Bare Costs Equipment	Total	Total Incl O&P
0010	**WELD STUDS**									
0020	1/4" diameter, 2-11/16" long	G E-10	1120	.014	Ea.	.35	.79	.36	1.50	2.10
0100	4-1/8" long	G	1080	.015		.33	.82	.37	1.52	2.14
0200	3/8" diameter, 4-1/8" long	G	1080	.015		.38	.82	.37	1.57	2.20
0300	6-1/8" long	G	1040	.015		.49	.85	.38	1.72	2.38
0400	1/2" diameter, 2-1/8" long	G	1040	.015		.35	.85	.38	1.58	2.23
0500	3-1/8" long	G	1025	.016		.43	.86	.39	1.68	2.34
0600	4-1/8" long	G	1010	.016		.50	.88	.40	1.78	2.45

05 05 Common Work Results for Metals

05 05 23 – Metal Fastenings

05 05 23.87 Weld Studs

		Crew	Daily Output	Labor-Hours	Unit	Material	2017 Bare Costs Labor	Equipment	Total	Total Incl O&P
0700	5-5/16" long [G]	E-10	990	.016	Ea.	.62	.89	.40	1.91	2.61
0800	6-1/8" long [G]		975	.016		.67	.91	.41	1.99	2.70
0900	8-1/8" long [G]		960	.017		.95	.92	.42	2.29	3.04
1000	5/8" diameter, 2-11/16" long [G]		1000	.016		.61	.88	.40	1.89	2.59
1010	4-3/16" long [G]		990	.016		.76	.89	.40	2.05	2.77
1100	6-9/16" long [G]		975	.016		.99	.91	.41	2.31	3.05
1200	8-3/16" long [G]		960	.017		1.33	.92	.42	2.67	3.46

05 05 23.90 Welding Rod

		Crew	Daily Output	Labor-Hours	Unit	Material	2017 Bare Costs Labor	Equipment	Total	Total Incl O&P
0010	**WELDING ROD**									
0020	Steel, type 6011, 1/8" diam., less than 500#				Lb.	2.35			2.35	2.59
0100	500# to 2,000#					2.12			2.12	2.33
0200	2,000# to 5,000#					1.99			1.99	2.19
0300	5/32" diameter, less than 500#					2.42			2.42	2.66
0310	500# to 2,000#					2.18			2.18	2.40
0320	2,000# to 5,000#					2.05			2.05	2.25
0400	3/16" diam., less than 500#					2.46			2.46	2.71
0500	500# to 2,000#					2.22			2.22	2.44
0600	2,000# to 5,000#					2.09			2.09	2.30
0620	Steel, type 6010, 1/8" diam., less than 500#					2.41			2.41	2.65
0630	500# to 2,000#					2.17			2.17	2.39
0640	2,000# to 5,000#					2.04			2.04	2.24
0650	Steel, type 7018 Low Hydrogen, 1/8" diam., less than 500#					2.44			2.44	2.69
0660	500# to 2,000#					2.20			2.20	2.42
0670	2,000# to 5,000#					2.07			2.07	2.27
0700	Steel, type 7024 Jet Weld, 1/8" diam., less than 500#					2.50			2.50	2.75
0710	500# to 2,000#					2.25			2.25	2.48
0720	2,000# to 5,000#					2.12			2.12	2.33
1550	Aluminum, type 4043 TIG, 1/8" diam., less than 10#					5.10			5.10	5.65
1560	10# to 60#					4.61			4.61	5.05
1570	Over 60#					4.33			4.33	4.77
1600	Aluminum, type 5356 TIG, 1/8" diam., less than 10#					5.45			5.45	6
1610	10# to 60#					4.90			4.90	5.40
1620	Over 60#					4.61			4.61	5.05
1900	Cast iron, type 8 Nickel, 1/8" diam., less than 500#					22			22	24
1910	500# to 1,000#					19.75			19.75	21.50
1920	Over 1,000#					18.55			18.55	20.50
2000	Stainless steel, type 316/316L, 1/8" diam., less than 500#					7.05			7.05	7.75
2100	500# to 1000#					6.35			6.35	6.95
2220	Over 1000#					5.95			5.95	6.55

05 12 Structural Steel Framing

05 12 23 – Structural Steel for Buildings

05 12 23.05 Canopy Framing

		Crew	Daily Output	Labor-Hours	Unit	Material	2017 Bare Costs Labor	Equipment	Total	Total Incl O&P
0010	**CANOPY FRAMING**									
0020	6" and 8" members, shop fabricated [G]	E-4	3000	.011	Lb.	1.57	.58	.04	2.19	2.75

05 12 23.10 Ceiling Supports

		Crew	Daily Output	Labor-Hours	Unit	Material	2017 Bare Costs Labor	Equipment	Total	Total Incl O&P
0010	**CEILING SUPPORTS**									
1000	Entrance door/folding partition supports, shop fabricated [G]	E-4	60	.533	L.F.	26	29	2.17	57.17	80
1100	Linear accelerator door supports [G]		14	2.286		119	125	9.30	253.30	350
1200	Lintels or shelf angles, hung, exterior hot dipped galv. [G]		267	.120		17.90	6.55	.49	24.94	31

For customer support on your Building Construction Costs with RSMeans Data, call 800.448.8182.

127

05 12 23.10 Ceiling Supports		Crew	Daily Output	Labor-Hours	Unit	Material	2017 Bare Costs Labor	Equipment	Total	Total Incl O&P
1250	Two coats primer paint instead of galv.	G	E-4	267	.120	L.F.	15.50	6.55	.49	22.54	28.50
1400	Monitor support, ceiling hung, expansion bolted	G		4	8	Ea.	415	440	32.50	887.50	1,225
1450	Hung from pre-set inserts	G		6	5.333		445	292	21.50	758.50	1,000
1600	Motor supports for overhead doors	G		4	8		211	440	32.50	683.50	1,000
1700	Partition support for heavy folding partitions, without pocket	G		24	1.333	L.F.	59.50	73	5.40	137.90	193
1750	Supports at pocket only	G		12	2.667		119	146	10.85	275.85	385
2000	Rolling grilles & fire door supports	G		34	.941		51	51.50	3.82	106.32	146
2100	Spider-leg light supports, expansion bolted to ceiling slab	G		8	4	Ea.	170	219	16.25	405.25	570
2150	Hung from pre-set inserts	G		12	2.667		183	146	10.85	339.85	460
2400	Toilet partition support	G		36	.889	L.F.	59.50	48.50	3.61	111.61	150
2500	X-ray travel gantry support	G		12	2.667	"	204	146	10.85	360.85	480

05 12 23.15 Columns, Lightweight

		Crew	Daily Output	Labor-Hours	Unit	Material	Labor	Equipment	Total	Total Incl O&P
0010	**COLUMNS, LIGHTWEIGHT**									
1000	Lightweight units (lally), 3-1/2" diameter	E-2	780	.072	L.F.	5.50	3.87	2.14	11.51	14.70
1050	4" diameter	"	900	.062	"	6.75	3.35	1.86	11.96	14.95
5800	Adjustable jack post, 8' maximum height, 2-3/4" diameter	G			Ea.	50			50	55.50
5850	4" diameter	G			"	80.50			80.50	88.50

05 12 23.17 Columns, Structural

			Crew	Daily Output	Labor-Hours	Unit	Material	Labor	Equipment	Total	Total Incl O&P
0010	**COLUMNS, STRUCTURAL**	R051223-10									
0015	Made from recycled materials										
0020	Shop fab'd for 100-ton, 1-2 story project, bolted connections										
0800	Steel, concrete filled, extra strong pipe, 3-1/2" diameter		E-2	660	.085	L.F.	43.50	4.57	2.53	50.60	57.50
0830	4" diameter			780	.072		48	3.87	2.14	54.01	61.50
0890	5" diameter			1020	.055		57.50	2.96	1.64	62.10	69.50
0930	6" diameter			1200	.047		76	2.52	1.39	79.91	89
0940	8" diameter			1100	.051		76	2.74	1.52	80.26	89.50
1100	For galvanizing, add					Lb.	.25			.25	.27
1300	For web ties, angles, etc., add per added lb.		1 Sswk	945	.008		1.31	.46		1.77	2.21
1500	Steel pipe, extra strong, no concrete, 3" to 5" diameter	G	E-2	16000	.004		1.31	.19	.10	1.60	1.86
1600	6" to 12" diameter	G		14000	.004		1.31	.22	.12	1.65	1.92
1700	Steel pipe, extra strong, no concrete, 3" diameter x 12'-0"	G		60	.933	Ea.	161	50.50	28	239.50	289
1750	4" diameter x 12'-0"	G		58	.966		236	52	29	317	375
1800	6" diameter x 12'-0"	G		54	1.037		450	56	31	537	620
1850	8" diameter x 14'-0"	G		50	1.120		795	60.50	33.50	889	1,000
1900	10" diameter x 16'-0"	G		48	1.167		1,150	63	35	1,248	1,425
1950	12" diameter x 18'-0"	G		45	1.244		1,550	67	37	1,654	1,850
3300	Structural tubing, square, A500GrB, 4" to 6" square, light section	G		11270	.005	Lb.	1.31	.27	.15	1.73	2.04
3600	Heavy section	G		32000	.002	"	1.31	.09	.05	1.45	1.65
4000	Concrete filled, add					L.F.	4.45			4.45	4.89
4500	Structural tubing, square, 4" x 4" x 1/4" x 12'-0"	G	E-2	58	.966	Ea.	216	52	29	297	355
4550	6" x 6" x 1/4" x 12'-0"	G		54	1.037		355	56	31	442	515
4600	8" x 8" x 3/8" x 14'-0"	G		50	1.120		765	60.50	33.50	859	980
4650	10" x 10" x 1/2" x 16'-0"	G		48	1.167		1,425	63	35	1,523	1,725
5100	Structural tubing, rect., 5" to 6" wide, light section	G		8000	.007	Lb.	1.31	.38	.21	1.90	2.28
5200	Heavy section	G		12000	.005		1.31	.25	.14	1.70	2
5300	7" to 10" wide, light section	G		15000	.004		1.31	.20	.11	1.62	1.89
5400	Heavy section	G		18000	.003		1.31	.17	.09	1.57	1.81
5500	Structural tubing, rect., 5" x 3" x 1/4" x 12'-0"	G		58	.966	Ea.	210	52	29	291	345
5550	6" x 4" x 5/16" x 12'-0"	G		54	1.037		330	56	31	417	485
5600	8" x 4" x 3/8" x 12'-0"	G		54	1.037		480	56	31	567	650
5650	10" x 6" x 3/8" x 14'-0"	G		50	1.120		765	60.50	33.50	859	980
5700	12" x 8" x 1/2" x 16'-0"	G		48	1.167		1,425	63	35	1,523	1,700

05 12 Structural Steel Framing

05 12 23 – Structural Steel for Buildings

05 12 23.17 Columns, Structural		Crew	Daily Output	Labor-Hours	Unit	Material	2017 Bare Costs Labor	Equipment	Total	Total Incl O&P
6800	W Shape, A992 steel, 2 tier, W8 x 24	G	E-2	1080	.052	L.F.	34.50	2.79	1.55	38.84	44
6850	W8 x 31	G		1080	.052		44.50	2.79	1.55	48.84	55
6900	W8 x 48	G		1032	.054		69	2.92	1.62	73.54	82.50
6950	W8 x 67	G		984	.057		96.50	3.07	1.70	101.27	113
7000	W10 x 45	G		1032	.054		65	2.92	1.62	69.54	78
7050	W10 x 68	G		984	.057		98	3.07	1.70	102.77	115
7100	W10 x 112	G		960	.058		161	3.14	1.74	165.88	185
7150	W12 x 50	G		1032	.054		72	2.92	1.62	76.54	86
7200	W12 x 87	G		984	.057		125	3.07	1.70	129.77	145
7250	W12 x 120	G		960	.058		173	3.14	1.74	177.88	197
7300	W12 x 190	G		912	.061		274	3.31	1.83	279.14	305
7350	W14 x 74	G		984	.057		107	3.07	1.70	111.77	124
7400	W14 x 120	G		960	.058		173	3.14	1.74	177.88	197
7450	W14 x 176	G	▼	912	.061	▼	254	3.31	1.83	259.14	286
8090	For projects 75 to 99 tons, add					%	10%				
8092	50 to 74 tons, add						20%				
8094	25 to 49 tons, add						30%	10%			
8096	10 to 24 tons, add						50%	25%			
8098	2 to 9 tons, add						75%	50%			
8099	Less than 2 tons, add					▼	100%	100%			

05 12 23.18 Corner Guards

		Crew	Daily Output	Labor-Hours	Unit	Material	Labor	Equipment	Total	Total Incl O&P
0010	**CORNER GUARDS**									
0020	Steel angle w/anchors, 1" x 1" x 1/4", 1.5#/L.F.	2 Carp	160	.100	L.F.	7.85	4.93		12.78	16.20
0100	2" x 2" x 1/4" angles, 3.2#/L.F.		150	.107		10.70	5.25		15.95	19.80
0200	3" x 3" x 5/16" angles, 6.1#/L.F.		140	.114		15.50	5.65		21.15	25.50
0300	4" x 4" x 5/16" angles, 8.2#/L.F.	▼	120	.133		18.80	6.55		25.35	30.50
0350	For angles drilled and anchored to masonry, add					15%	120%			
0370	Drilled and anchored to concrete, add					20%	170%			
0400	For galvanized angles, add					35%				
0450	For stainless steel angles, add				▼	100%				

05 12 23.20 Curb Edging

		Crew	Daily Output	Labor-Hours	Unit	Material	Labor	Equipment	Total	Total Incl O&P
0010	**CURB EDGING**									
0020	Steel angle w/anchors, shop fabricated, on forms, 1" x 1", 0.8#/L.F.	G E-4	350	.091	L.F.	1.68	5	.37	7.05	10.60
0100	2" x 2" angles, 3.92#/L.F.	G	330	.097		6.60	5.30	.39	12.29	16.55
0200	3" x 3" angles, 6.1#/L.F.	G	300	.107		10.55	5.85	.43	16.83	22
0300	4" x 4" angles, 8.2#/L.F.	G	275	.116		13.85	6.40	.47	20.72	26.50
1000	6" x 4" angles, 12.3#/L.F.	G	250	.128		20.50	7	.52	28.02	35
1050	Steel channels with anchors, on forms, 3" channel, 5#/L.F.	G	290	.110		8.30	6.05	.45	14.80	19.70
1100	4" channel, 5.4#/L.F.	G	270	.119		8.90	6.50	.48	15.88	21
1200	6" channel, 8.2#/L.F.	G	255	.125		13.85	6.90	.51	21.26	27
1300	8" channel, 11.5#/L.F.	G	225	.142		19.05	7.80	.58	27.43	34.50
1400	10" channel, 15.3#/L.F.	G	180	.178		25	9.75	.72	35.47	44.50
1500	12" channel, 20.7#/L.F.	G	▼	140	.229		33.50	12.55	.93	46.98	59
2000	For curved edging, add				▼	35%	10%			

05 12 23.40 Lightweight Framing

		Crew	Daily Output	Labor-Hours	Unit	Material	Labor	Equipment	Total	Total Incl O&P
0010	**LIGHTWEIGHT FRAMING** R051223-35									
0015	Made from recycled materials									
0200	For load-bearing steel studs see Section 05 41 13.30									
0400	Angle framing, field fabricated, 4" and larger R051223-45	G E-3	440	.055	Lb.	.76	3	.30	4.06	6.15
0450	Less than 4" angles	G	265	.091	"	.79	4.98	.49	6.26	9.70
0460	1/2" x 1/2" x 1/8"	G	200	.120	L.F.	.16	6.60	.65	7.41	11.90
0462	3/4" x 3/4" x 1/8"	G	▼	160	.150	▼	.44	8.25	.81	9.50	15.10

05 12 23.40 Lightweight Framing		Crew	Daily Output	Labor-Hours	Unit	Material	2017 Bare Costs Labor	Equipment	Total	Total Incl O&P
0464	1" x 1" x 1/8"	G	E-3	135	.178	L.F.	.63	9.75	.96	11.34	18.05
0466	1-1/4" x 1-1/4" x 3/16"	G		115	.209		1.16	11.45	1.13	13.74	21.50
0468	1-1/2" x 1-1/2" x 3/16"	G		100	.240		1.42	13.20	1.30	15.92	25
0470	2" x 2" x 1/4"	G		90	.267		2.51	14.65	1.44	18.60	29
0472	2-1/2" x 2-1/2" x 1/4"	G		72	.333		3.22	18.30	1.80	23.32	36
0474	3" x 2" x 3/8"	G		65	.369		4.64	20.50	2	27.14	41.50
0476	3" x 3" x 3/8"	G		57	.421	↓	5.65	23	2.28	30.93	47.50
0600	Channel framing, field fabricated, 8" and larger	G		500	.048	Lb.	.79	2.64	.26	3.69	5.55
0650	Less than 8" channels	G		335	.072	"	.79	3.94	.39	5.12	7.85
0660	C2 x 1.78	G		115	.209	L.F.	1.40	11.45	1.13	13.98	22
0662	C3 x 4.1	G		80	.300		3.22	16.50	1.62	21.34	33
0664	C4 x 5.4	G		66	.364		4.25	20	1.97	26.22	40.50
0666	C5 x 6.7	G		57	.421		5.25	23	2.28	30.53	47
0668	C6 x 8.2	G		55	.436		6.25	24	2.36	32.61	49.50
0670	C7 x 9.8	G		40	.600		7.70	33	3.25	43.95	67
0672	C8 x 11.5	G		36	.667		9.05	36.50	3.61	49.16	75
0710	Structural bar tee, field fabricated, 3/4" x 3/4" x 1/8"	G		160	.150		.44	8.25	.81	9.50	15.10
0712	1" x 1" x 1/8"	G		135	.178		.63	9.75	.96	11.34	18.05
0714	1-1/2" x 1-1/2" x 1/4"	G		114	.211		1.84	11.55	1.14	14.53	22.50
0716	2" x 2" x 1/4"	G		89	.270		2.51	14.80	1.46	18.77	29
0718	2-1/2" x 2-1/2" x 3/8"	G		72	.333		4.64	18.30	1.80	24.74	37.50
0720	3" x 3" x 3/8"	G		57	.421		5.65	23	2.28	30.93	47.50
0730	Structural zee, field fabricated, 1-1/4" x 1-3/4" x 1-3/4"	G		114	.211		.60	11.55	1.14	13.29	21
0732	2-11/16" x 3" x 2-11/16"	G		114	.211		1.40	11.55	1.14	14.09	22
0734	3-1/16" x 4" x 3-1/16"	G		133	.180		2.12	9.90	.98	13	19.95
0736	3-1/4" x 5" x 3-1/4"	G		133	.180		2.89	9.90	.98	13.77	21
0738	3-1/2" x 6" x 3-1/2"	G		160	.150		4.35	8.25	.81	13.41	19.40
0740	Junior beam, field fabricated, 3"	G		80	.300		4.48	16.50	1.62	22.60	34
0742	4"	G		72	.333		6.05	18.30	1.80	26.15	39
0744	5"	G		67	.358		7.85	19.70	1.94	29.49	44
0746	6"	G		62	.387		9.85	21.50	2.09	33.44	48.50
0748	7"	G		57	.421		12.05	23	2.28	37.33	54.50
0750	8"	G		53	.453	↓	14.45	25	2.45	41.90	60
1000	Continuous slotted channel framing system, shop fab, simple framing	G	2 Sswk	2400	.007	Lb.	4.06	.36		4.42	5.05
1200	Complex framing	G	"	1600	.010		4.59	.54		5.13	5.95
1300	Cross bracing, rods, shop fabricated, 3/4" diameter	G	E-3	700	.034		1.57	1.88	.19	3.64	5.05
1310	7/8" diameter	G		850	.028		1.57	1.55	.15	3.27	4.49
1320	1" diameter	G		1000	.024		1.57	1.32	.13	3.02	4.07
1330	Angle, 5" x 5" x 3/8"	G		2800	.009		1.57	.47	.05	2.09	2.57
1350	Hanging lintels, shop fabricated	G	↓	850	.028		1.57	1.55	.15	3.27	4.49
1380	Roof frames, shop fabricated, 3'-0" square, 5' span	G	E-2	4200	.013		1.57	.72	.40	2.69	3.34
1400	Tie rod, not upset, 1-1/2" to 4" diameter, with turnbuckle	G	2 Sswk	800	.020		1.70	1.09		2.79	3.68
1420	No turnbuckle	G		700	.023		1.64	1.24		2.88	3.87
1500	Upset, 1-3/4" to 4" diameter, with turnbuckle	G		800	.020		1.70	1.09		2.79	3.68
1520	No turnbuckle	G	↓	700	.023	↓	1.64	1.24		2.88	3.87

05 12 23.45 Lintels

			Crew	Daily Output	Labor-Hours	Unit	Material	Labor	Equipment	Total	Total Incl O&P
0010	**LINTELS**										
0015	Made from recycled materials										
0020	Plain steel angles, shop fabricated, under 500 lb.	G	1 Bric	550	.015	Lb.	1.01	.69		1.70	2.18
0100	500 to 1000 lb.	G		640	.013		.98	.60		1.58	2
0200	1,000 to 2,000 lb.	G		640	.013		.96	.60		1.56	1.97
0300	2,000 to 4,000 lb.	G	↓	640	.013	↓	.93	.60		1.53	1.94

05 12 23.45 Lintels

		Crew	Daily Output	Labor-Hours	Unit	Material	2017 Bare Costs Labor	Equipment	Total	Total Incl O&P
0500	For built-up angles and plates, add to above	G			Lb.	1.31			1.31	1.44
0700	For engineering, add to above					.13			.13	.14
0900	For galvanizing, add to above, under 500 lb.					.30			.30	.33
0950	500 to 2,000 lb.					.27			.27	.30
1000	Over 2,000 lb.					.25			.25	.27
2000	Steel angles, 3-1/2" x 3", 1/4" thick, 2'-6" long	G	1 Bric	47	.170	Ea.	14.15	8.15		22.30	28
2100	4'-6" long	G		26	.308		25.50	14.70		40.20	50.50
2600	4" x 3-1/2", 1/4" thick, 5'-0" long	G		21	.381		32.50	18.20		50.70	64
2700	9'-0" long	G		12	.667		58.50	32		90.50	114

05 12 23.60 Pipe Support Framing

		Crew	Daily Output	Labor-Hours	Unit	Material	2017 Bare Costs Labor	Equipment	Total	Total Incl O&P
0010	**PIPE SUPPORT FRAMING**									
0020	Under 10#/L.F., shop fabricated	G	E-4	3900	.008	Lb.	1.76	.45	.03	2.24	2.72
0200	10.1 to 15#/L.F.	G		4300	.007		1.73	.41	.03	2.17	2.61
0400	15.1 to 20#/L.F.	G		4800	.007		1.70	.37	.03	2.10	2.51
0600	Over 20#/L.F.	G		5400	.006		1.68	.33	.02	2.03	2.42

05 12 23.65 Plates

		Crew	Daily Output	Labor-Hours	Unit	Material	2017 Bare Costs Labor	Equipment	Total	Total Incl O&P
0010	**PLATES** R051223-80									
0015	Made from recycled materials									
0020	For connections & stiffener plates, shop fabricated									
0050	1/8" thick (5.1 lb./S.F.)	G				S.F.	6.70			6.70	7.35
0100	1/4" thick (10.2 lb./S.F.)	G					13.35			13.35	14.70
0300	3/8" thick (15.3 lb./S.F.)	G					20			20	22
0400	1/2" thick (20.4 lb./S.F.)	G					26.50			26.50	29.50
0450	3/4" thick (30.6 lb./S.F.)	G					40			40	44
0500	1" thick (40.8 lb./S.F.)	G					53.50			53.50	59
2000	Steel plate, warehouse prices, no shop fabrication									
2100	1/4" thick (10.2 lb./S.F.)	G				S.F.	6.05			6.05	6.65

05 12 23.70 Stressed Skin Steel Roof and Ceiling System

		Crew	Daily Output	Labor-Hours	Unit	Material	2017 Bare Costs Labor	Equipment	Total	Total Incl O&P
0010	**STRESSED SKIN STEEL ROOF & CEILING SYSTEM**									
0020	Double panel flat roof, spans to 100'	G	E-2	1150	.049	S.F.	10.50	2.62	1.45	14.57	17.40
0100	Double panel convex roof, spans to 200'	G		960	.058		17.05	3.14	1.74	21.93	26
0200	Double panel arched roof, spans to 300'	G		760	.074		26	3.97	2.20	32.17	38

05 12 23.75 Structural Steel Members

		Crew	Daily Output	Labor-Hours	Unit	Material	2017 Bare Costs Labor	Equipment	Total	Total Incl O&P
0010	**STRUCTURAL STEEL MEMBERS** R051223-10									
0015	Made from recycled materials									
0020	Shop fab'd for 100-ton, 1-2 story project, bolted connections									
0100	Beam or girder, W 6 x 9	G	E-2	600	.093	L.F.	13	5.05	2.79	20.84	25.50
0120	x 15	G		600	.093		21.50	5.05	2.79	29.34	35
0140	x 20	G		600	.093		29	5.05	2.79	36.84	42.50
0300	W 8 x 10	G		600	.093		14.40	5.05	2.79	22.24	27
0320	x 15	G		600	.093		21.50	5.05	2.79	29.34	35
0350	x 21	G		600	.093		30.50	5.05	2.79	38.34	44.50
0360	x 24	G		550	.102		34.50	5.50	3.04	43.04	50
0370	x 28	G		550	.102		40.50	5.50	3.04	49.04	56.50
0500	x 31	G		550	.102		44.50	5.50	3.04	53.04	61
0520	x 35	G		550	.102		50.50	5.50	3.04	59.04	67.50
0540	x 48	G		550	.102		69	5.50	3.04	77.54	88
0600	W 10 x 12	G		600	.093		17.30	5.05	2.79	25.14	30.50
0620	x 15	G		600	.093		21.50	5.05	2.79	29.34	35
0700	x 22	G		600	.093		31.50	5.05	2.79	39.34	46
0720	x 26	G		600	.093		37.50	5.05	2.79	45.34	52
0740	x 33	G		550	.102		47.50	5.50	3.04	56.04	64.50

05 12 23.75 Structural Steel Members		Crew	Daily Output	Labor-Hours	Unit	Material	2017 Bare Costs Labor	Equipment	Total	Total Incl O&P
0900	x 49	G	E-2	550	.102	L.F.	70.50	5.50	3.04	79.04	89.50
1100	W 12 x 16	G		880	.064		23	3.43	1.90	28.33	33
1300	x 22	G		880	.064		31.50	3.43	1.90	36.83	42.50
1500	x 26	G		880	.064		37.50	3.43	1.90	42.83	48.50
1520	x 35	G		810	.069		50.50	3.73	2.06	56.29	64
1560	x 50	G		750	.075		72	4.02	2.23	78.25	88.50
1580	x 58	G		750	.075		83.50	4.02	2.23	89.75	101
1700	x 72	G		640	.088		104	4.72	2.61	111.33	125
1740	x 87	G		640	.088		125	4.72	2.61	132.33	149
1900	W 14 x 26	G		990	.057		37.50	3.05	1.69	42.24	48
2100	x 30	G		900	.062		43.50	3.35	1.86	48.71	55
2300	x 34	G		810	.069		49	3.73	2.06	54.79	62.50
2320	x 43	G		810	.069		62	3.73	2.06	67.79	76.50
2340	x 53	G		800	.070		76.50	3.77	2.09	82.36	92.50
2360	x 74	G		760	.074		107	3.97	2.20	113.17	126
2380	x 90	G		740	.076		130	4.08	2.26	136.34	152
2500	x 120	G		720	.078		173	4.19	2.32	179.51	199
2700	W 16 x 26	G		1000	.056		37.50	3.02	1.67	42.19	47.50
2900	x 31	G		900	.062		44.50	3.35	1.86	49.71	56.50
3100	x 40	G		800	.070		57.50	3.77	2.09	63.36	72
3120	x 50	G		800	.070		72	3.77	2.09	77.86	88
3140	x 67	G		760	.074		96.50	3.97	2.20	102.67	115
3300	W 18 x 35	G	E-5	960	.083		50.50	4.52	1.88	56.90	65
3500	x 40	G		960	.083		57.50	4.52	1.88	63.90	73
3520	x 46	G		960	.083		66.50	4.52	1.88	72.90	82.50
3700	x 50	G		912	.088		72	4.76	1.97	78.73	89.50
3900	x 55	G		912	.088		79.50	4.76	1.97	86.23	97
3920	x 65	G		900	.089		93.50	4.82	2	100.32	113
3940	x 76	G		900	.089		110	4.82	2	116.82	131
3960	x 86	G		900	.089		124	4.82	2	130.82	146
3980	x 106	G		900	.089		153	4.82	2	159.82	178
4100	W 21 x 44	G		1064	.075		63.50	4.08	1.69	69.27	78.50
4300	x 50	G		1064	.075		72	4.08	1.69	77.77	88
4500	x 62	G		1036	.077		89.50	4.19	1.74	95.43	107
4700	x 68	G		1036	.077		98	4.19	1.74	103.93	117
4720	x 83	G		1000	.080		120	4.34	1.80	126.14	141
4740	x 93	G		1000	.080		134	4.34	1.80	140.14	156
4760	x 101	G		1000	.080		146	4.34	1.80	152.14	169
4780	x 122	G		1000	.080		176	4.34	1.80	182.14	202
4900	W 24 x 55	G		1110	.072		79.50	3.91	1.62	85.03	95
5100	x 62	G		1110	.072		89.50	3.91	1.62	95.03	107
5300	x 68	G		1110	.072		98	3.91	1.62	103.53	116
5500	x 76	G		1110	.072		110	3.91	1.62	115.53	129
5700	x 84	G		1080	.074		121	4.02	1.67	126.69	141
5720	x 94	G		1080	.074		136	4.02	1.67	141.69	157
5740	x 104	G		1050	.076		150	4.13	1.72	155.85	174
5760	x 117	G		1050	.076		169	4.13	1.72	174.85	195
5780	x 146	G		1050	.076		210	4.13	1.72	215.85	241
5800	W 27 x 84	G		1190	.067		121	3.64	1.51	126.15	141
5900	x 94	G		1190	.067		136	3.64	1.51	141.15	157
5920	x 114	G		1150	.070		164	3.77	1.57	169.34	189
5940	x 146	G		1150	.070		210	3.77	1.57	215.34	240
5960	x 161	G		1150	.070		232	3.77	1.57	237.34	263

05 12 23 – Structural Steel for Buildings

05 12 23.75 Structural Steel Members		Crew	Daily Output	Labor-Hours	Unit	Material	2017 Bare Costs Labor	Equipment	Total	Total Incl O&P
6100	W 30 x 99	G E-5	1200	.067	L.F.	143	3.61	1.50	148.11	165
6300	x 108	G	1200	.067		156	3.61	1.50	161.11	179
6500	x 116	G	1160	.069		167	3.74	1.55	172.29	192
6520	x 132	G	1160	.069		190	3.74	1.55	195.29	217
6540	x 148	G	1160	.069		213	3.74	1.55	218.29	243
6560	x 173	G	1120	.071		249	3.87	1.61	254.48	282
6580	x 191	G	1120	.071		275	3.87	1.61	280.48	315
6700	W 33 x 118	G	1176	.068		170	3.69	1.53	175.22	195
6900	x 130	G	1134	.071		187	3.82	1.59	192.41	214
7100	x 141	G	1134	.071		203	3.82	1.59	208.41	232
7120	x 169	G	1100	.073		244	3.94	1.64	249.58	276
7140	x 201	G	1100	.073		290	3.94	1.64	295.58	330
7300	W 36 x 135	G	1170	.068		195	3.71	1.54	200.25	222
7500	x 150	G	1170	.068		216	3.71	1.54	221.25	246
7600	x 170	G	1150	.070		245	3.77	1.57	250.34	278
7700	x 194	G	1125	.071		280	3.85	1.60	285.45	320
7900	x 231	G	1125	.071		335	3.85	1.60	340.45	375
7920	x 262	G	1035	.077		380	4.19	1.74	385.93	425
8100	x 302	G	1035	.077		435	4.19	1.74	440.93	490
8490	For projects 75 to 99 tons, add					10%				
8492	50 to 74 tons, add					20%				
8494	25 to 49 tons, add					30%	10%			
8496	10 to 24 tons, add					50%	25%			
8498	2 to 9 tons, add					75%	50%			
8499	Less than 2 tons, add					100%	100%			

05 12 23.77 Structural Steel Projects										
0010	**STRUCTURAL STEEL PROJECTS** R050516-30									
0015	Made from recycled materials									
0020	Shop fab'd for 100-ton, 1-2 story project, bolted connections									
0200	Apartments, nursing homes, etc., 1 to 2 stories R050523-10	G E-5	10.30	7.767	Ton	2,625	420	175	3,220	3,750
0300	3 to 6 stories	G "	10.10	7.921		2,675	430	178	3,283	3,850
0400	7 to 15 stories R051223-10	G E-6	14.20	9.014		2,725	490	140	3,355	3,950
0500	Over 15 stories	G "	13.90	9.209		2,825	500	143	3,468	4,100
0700	Offices, hospitals, etc., steel bearing, 1 to 2 stories R051223-20	G E-5	10.30	7.767		2,625	420	175	3,220	3,750
0800	3 to 6 stories	G E-6	14.40	8.889		2,675	480	138	3,293	3,900
0900	7 to 15 stories R051223-25	G	14.20	9.014		2,725	490	140	3,355	3,950
1000	Over 15 stories	G	13.90	9.209		2,825	500	143	3,468	4,100
1100	For multi-story masonry wall bearing construction, add R051223-30						30%			
1300	Industrial bldgs., 1 story, beams & girders, steel bearing	G E-5	12.90	6.202		2,625	335	140	3,100	3,575
1400	Masonry bearing	G "	10	8		2,625	435	180	3,240	3,775
1500	Industrial bldgs., 1 story, under 10 tons,									
1510	steel from warehouse, trucked	G E-2	7.50	7.467	Ton	3,150	400	223	3,773	4,350
1600	1 story with roof trusses, steel bearing	G E-5	10.60	7.547		3,100	410	170	3,680	4,250
1700	Masonry bearing	G "	8.30	9.639		3,100	525	217	3,842	4,500
1900	Monumental structures, banks, stores, etc., simple connections	G E-6	13	9.846		2,625	535	152	3,312	3,925
2000	Moment/composite connections	G "	9	14.222		4,350	770	220	5,340	6,300
2200	Churches, simple connections	G E-5	11.60	6.897		2,450	375	155	2,980	3,450
2300	Moment/composite connections	G "	5.20	15.385		3,250	835	345	4,430	5,325
2800	Power stations, fossil fuels, simple connections	G E-6	11	11.636		2,625	630	180	3,435	4,100
2900	Moment/composite connections	G	5.70	22.456		3,925	1,225	350	5,500	6,700
2950	Nuclear fuels, non-safety steel, simple connections	G	7	18.286		2,625	990	283	3,898	4,800
3000	Moment/composite connections	G	5.50	23.273		3,925	1,250	360	5,535	6,800

05 12 Structural Steel Framing

05 12 23 – Structural Steel for Buildings

05 12 23.77 Structural Steel Projects

		Crew	Daily Output	Labor-Hours	Unit	Material	2017 Bare Costs Labor	Equipment	Total	Total Incl O&P
3040	Safety steel, simple connections	G	E-6	2.50	51.200	Ton	3,825	2,775	795	7,395	9,625
3070	Moment/composite connections	G	↓	1.50	85.333		5,025	4,625	1,325	10,975	14,600
3100	Roof trusses, simple connections	G	E-5	13	6.154		3,675	335	139	4,149	4,725
3200	Moment/composite connections	G		8.30	9.639		4,450	525	217	5,192	6,000
3210	Schools, simple connections	G		14.50	5.517		2,625	299	124	3,048	3,500
3220	Moment/composite connections	G	↓	8.30	9.639		3,825	525	217	4,567	5,300
3400	Welded construction, simple commercial bldgs., 1 to 2 stories	G	E-7	7.60	10.526		2,675	570	254	3,499	4,175
3500	7 to 15 stories	G	E-9	8.30	15.422		3,100	835	280	4,215	5,075
3700	Welded rigid frame, 1 story, simple connections	G	E-7	15.80	5.063		2,725	274	122	3,121	3,575
3800	Moment/composite connections	G	"	5.50	14.545	↓	3,550	790	350	4,690	5,575
3810	Fabrication shop costs (incl in project bare material cost, above)										
3820	Mini mill base price, Grade A992	G				Ton	710			710	780
3830	Mill extras plus delivery to warehouse						275			275	305
3835	Delivery from warehouse to fabrication shop						105			105	116
3840	Shop extra for shop drawings and detailing						315			315	345
3850	Shop fabricating and handling						940			940	1,025
3860	Shop sandblasting and primer coat of paint						145			145	160
3870	Shop delivery to the job site					↓	135			135	149
3880	Total material cost, shop fabricated, primed, delivered						2,625			2,625	2,875
3900	High strength steel mill spec extras:										
3950	A529, A572 (50 ksi) and A36: same as A992 steel (no extra)										
4000	Add to A992 price for A572 (60, 65 ksi)	G				Ton	80			80	88
4100	A242 and A588 Weathering	G				"	80			80	88
4200	Mill size extras for W-Shapes: 0 to 30 plf: no extra charge										
4210	Member sizes 31 to 65 plf, add	G				Ton	5			5	5.50
4220	Member sizes 66 to 100 plf, add	G					60			60	66
4230	Member sizes 101 to 387 plf, add	G				↓	145			145	160
4300	Column base plates, light, up to 150 lb.	G	2 Sswk	2000	.008	Lb.	1.44	.43		1.87	2.31
4400	Heavy, over 150 lb.	G	E-2	7500	.007	"	1.51	.40	.22	2.13	2.56
4600	Castellated beams, light sections, to 50#/L.F., simple connections	G		10.70	5.234	Ton	2,750	282	156	3,188	3,650
4700	Moment/composite connections	G		7	8		3,025	430	239	3,694	4,300
4900	Heavy sections, over 50 plf, simple connections	G		11.70	4.786		2,875	258	143	3,276	3,750
5000	Moment/composite connections	G	↓	7.80	7.179		3,150	385	214	3,749	4,325
5390	For projects 75 to 99 tons, add						10%				
5392	50 to 74 tons, add						20%				
5394	25 to 49 tons, add						30%	10%			
5396	10 to 24 tons, add						50%	25%			
5398	2 to 9 tons, add						75%	50%			
5399	Less than 2 tons, add						100%	100%			

05 12 23.78 Structural Steel Secondary Members

		Crew	Daily Output	Labor-Hours	Unit	Material	2017 Bare Costs Labor	Equipment	Total	Total Incl O&P
0010	**STRUCTURAL STEEL SECONDARY MEMBERS**										
0015	Made from recycled materials										
0020	Shop fabricated for 20-ton girt/purlin framing package, materials only										
0100	Girts/purlins, C/Z-shapes, includes clips and bolts										
0110	6" x 2-1/2" x 2-1/2", 16 ga., 3.0 lb./L.F.					L.F.	3.54			3.54	3.89
0115	14 ga., 3.5 lb./L.F.						4.13			4.13	4.54
0120	8" x 2-3/4" x 2-3/4", 16 ga., 3.4 lb./L.F.						4.01			4.01	4.41
0125	14 ga., 4.1 lb./L.F.						4.84			4.84	5.30
0130	12 ga., 5.6 lb./L.F.						6.60			6.60	7.25
0135	10" x 3-1/2" x 3-1/2", 14 ga., 4.7 lb./L.F.						5.55			5.55	6.10
0140	12 ga., 6.7 lb./L.F.						7.90			7.90	8.70
0145	12" x 3-1/2" x 3-1/2", 14 ga., 5.3 lb./L.F.						6.25			6.25	6.90

05 12 Structural Steel Framing

05 12 23 – Structural Steel for Buildings

05 12 23.78 Structural Steel Secondary Members

05 12 23.78 Structural Steel Secondary Members	Crew	Daily Output	Labor-Hours	Unit	Material	2017 Bare Costs Labor	Equipment	Total	Total Incl O&P
0150 12 ga., 7.4 lb./L.F.				L.F.	8.75			8.75	9.60
0200 Eave struts, C-shape, includes clips and bolts									
0210 6" x 4" x 3", 16 ga., 3.1 lb./L.F.				L.F.	3.66			3.66	4.02
0215 14 ga., 3.9 lb./L.F.					4.60			4.60	5.05
0220 8" x 4" x 3", 16 ga., 3.5 lb./L.F.					4.13			4.13	4.54
0225 14 ga., 4.4 lb./L.F.					5.20			5.20	5.70
0230 12 ga., 6.2 lb./L.F.					7.30			7.30	8.05
0235 10" x 5" x 3", 14 ga., 5.2 lb./L.F.					6.15			6.15	6.75
0240 12 ga., 7.3 lb./L.F.					8.60			8.60	9.45
0245 12" x 5" x 4", 14 ga., 6.0 lb./L.F.					7.10			7.10	7.80
0250 12 ga., 8.4 lb./L.F.					9.90			9.90	10.90
0300 Rake/base angle, excludes concrete drilling and expansion anchors									
0310 2" x 2", 14 ga., 1.0 lb./L.F.	2 Sswk	640	.025	L.F.	1.18	1.36		2.54	3.56
0315 3" x 2", 14 ga., 1.3 lb./L.F.		535	.030		1.53	1.62		3.15	4.40
0320 3" x 3", 14 ga., 1.6 lb./L.F.		500	.032		1.89	1.74		3.63	4.98
0325 4" x 3", 14 ga., 1.8 lb./L.F.		480	.033		2.12	1.81		3.93	5.35
0600 Installation of secondary members, erection only									
0610 Girts, purlins, eave struts, 16 ga., 6" deep	E-18	100	.400	Ea.		22	10.75	32.75	47.50
0615 8" deep		80	.500			27.50	13.45	40.95	59.50
0620 14 ga., 6" deep		80	.500			27.50	13.45	40.95	59.50
0625 8" deep		65	.615			33.50	16.55	50.05	73
0630 10" deep		55	.727			39.50	19.55	59.05	86.50
0635 12" deep		50	.800			43.50	21.50	65	95
0640 12 ga., 8" deep		50	.800			43.50	21.50	65	95
0645 10" deep		45	.889			48.50	24	72.50	106
0650 12" deep		40	1			54.50	27	81.50	119
0900 For less than 20-ton job lots									
0905 For 15 to 19 tons, add				%	10%				
0910 For 10 to 14 tons, add					25%				
0915 For 5 to 9 tons, add					50%	50%	50%		
0920 For 1 to 4 tons, add					75%	75%	75%		
0925 For less than 1 ton, add					100%	100%	100%		

05 12 23.80 Subpurlins

05 12 23.80 Subpurlins	Crew	Daily Output	Labor-Hours	Unit	Material	2017 Bare Costs Labor	Equipment	Total	Total Incl O&P
0010 **SUBPURLINS** R051223-50									
0015 Made from recycled materials									
0020 Bulb tees, shop fabricated, painted, 32-5/8" O.C., 40 psf L.L.									
0200 Type 218, max 10'-2" span, 3.19 plf, 2-1/8" high x 2-1/8" wide G	E-1	3100	.008	S.F.	1.72	.42	.04	2.18	2.61
1420 For 24-5/8" spacing, add					33%	33%			
1430 For 48-5/8" spacing, deduct					33%	33%			

05 14 Structural Aluminum Framing

05 14 23 – Non-Exposed Structural Aluminum Framing

05 14 23.05 Aluminum Shapes

05 14 23.05 Aluminum Shapes		Crew	Daily Output	Labor-Hours	Unit	Material	2017 Bare Costs Labor	Equipment	Total	Total Incl O&P
0010 **ALUMINUM SHAPES**										
0015 Made from recycled materials										
0020 Structural shapes, 1" to 10" members, under 1 ton	G	E-2	4000	.014	Lb.	2.92	.75	.42	4.09	4.90
0050 1 to 5 tons	G		4300	.013		2.68	.70	.39	3.77	4.52
0100 Over 5 tons	G		4600	.012		2.55	.66	.36	3.57	4.28
0300 Extrusions, over 5 tons, stock shapes	G		1330	.042		3.35	2.27	1.26	6.88	8.75
0400 Custom shapes	G		1330	.042		4.38	2.27	1.26	7.91	9.90

05 15 16.05 Accessories for Steel Wire Rope		Crew	Daily Output	Labor-Hours	Unit	Material	2017 Bare Costs Labor	Equipment	Total	Total Incl O&P
0010	**ACCESSORIES FOR STEEL WIRE ROPE**									
0015	Made from recycled materials									
1500	Thimbles, heavy duty, 1/4"	G	E-17	160	.100	Ea.	.37	5.55		5.92	9.60
1510	1/2"	G		160	.100		1.63	5.55		7.18	11
1520	3/4"	G		105	.152		3.72	8.45		12.17	18.15
1530	1"	G		52	.308		7.45	17		24.45	36.50
1540	1-1/4"	G		38	.421		11.45	23.50		34.95	51.50
1550	1-1/2"	G		13	1.231		32	68		100	149
1560	1-3/4"	G		8	2		66.50	111		177.50	257
1570	2"	G		6	2.667		96.50	147		243.50	350
1580	2-1/4"	G		4	4		131	221		352	515
1600	Clips, 1/4" diameter	G		160	.100		1.98	5.55		7.53	11.40
1610	3/8" diameter	G		160	.100		2.17	5.55		7.72	11.60
1620	1/2" diameter	G		160	.100		3.49	5.55		9.04	13.05
1630	3/4" diameter	G		102	.157		5.65	8.65		14.30	20.50
1640	1" diameter	G		64	.250		9.45	13.85		23.30	33.50
1650	1-1/4" diameter	G		35	.457		15.45	25.50		40.95	59
1670	1-1/2" diameter	G		26	.615		21	34		55	79.50
1680	1-3/4" diameter	G		16	1		48.50	55.50		104	146
1690	2" diameter	G		12	1.333		54	73.50		127.50	183
1700	2-1/4" diameter	G		10	1.600		79.50	88.50		168	235
1800	Sockets, open swage, 1/4" diameter	G		160	.100		26	5.55		31.55	37.50
1810	1/2" diameter	G		77	.208		37.50	11.50		49	60
1820	3/4" diameter	G		19	.842		58	46.50		104.50	142
1830	1" diameter	G		9	1.778		104	98.50		202.50	278
1840	1-1/4" diameter	G		5	3.200		144	177		321	455
1850	1-1/2" diameter	G		3	5.333		315	295		610	840
1860	1-3/4" diameter	G		3	5.333		560	295		855	1,100
1870	2" diameter	G		1.50	10.667		850	590		1,440	1,925
1900	Closed swage, 1/4" diameter	G		160	.100		15.35	5.55		20.90	26
1910	1/2" diameter	G		104	.154		26.50	8.50		35	43
1920	3/4" diameter	G		32	.500		39.50	27.50		67	89.50
1930	1" diameter	G		15	1.067		69.50	59		128.50	175
1940	1-1/4" diameter	G		7	2.286		105	126		231	325
1950	1-1/2" diameter	G		4	4		190	221		411	580
1960	1-3/4" diameter	G		3	5.333		280	295		575	800
1970	2" diameter	G		2	8		545	440		985	1,325
2000	Open spelter, galv., 1/4" diameter	G		160	.100		35	5.55		40.55	47.50
2010	1/2" diameter	G		70	.229		36.50	12.65		49.15	61
2020	3/4" diameter	G		26	.615		55	34		89	117
2030	1" diameter	G		10	1.600		152	88.50		240.50	315
2040	1-1/4" diameter	G		5	3.200		218	177		395	535
2050	1-1/2" diameter	G		4	4		460	221		681	880
2060	1-3/4" diameter	G		2	8		805	440		1,245	1,625
2070	2" diameter	G		1.20	13.333		930	735		1,665	2,250
2080	2-1/2" diameter	G		1	16		1,700	885		2,585	3,350
2100	Closed spelter, galv., 1/4" diameter	G		160	.100		28.50	5.55		34.05	40.50
2110	1/2" diameter	G		88	.182		30.50	10.05		40.55	51
2120	3/4" diameter	G		30	.533		46.50	29.50		76	100
2130	1" diameter	G		13	1.231		99	68		167	222
2140	1-1/4" diameter	G		7	2.286		158	126		284	385
2150	1-1/2" diameter	G		6	2.667		340	147		487	620
2160	1-3/4" diameter	G		2.80	5.714		455	315		770	1,025

05 15 16 – Steel Wire Rope Assemblies

05 15 16.05 Accessories for Steel Wire Rope

		Crew	Daily Output	Labor-Hours	Unit	Material	2017 Bare Costs Labor	Equipment	Total	Total Incl O&P
2170	2" diameter	G	E-17	2	8	Ea.	560	440		1,000	1,350
2200	Jaw & jaw turnbuckles, 1/4" x 4"	G		160	.100		12	5.55		17.55	22.50
2250	1/2" x 6"	G		96	.167		15.15	9.20		24.35	32
2260	1/2" x 9"	G		77	.208		20	11.50		31.50	41
2270	1/2" x 12"	G		66	.242		22.50	13.40		35.90	47.50
2300	3/4" x 6"	G		38	.421		29.50	23.50		53	71.50
2310	3/4" x 9"	G		30	.533		33	29.50		62.50	85
2320	3/4" x 12"	G		28	.571		42.50	31.50		74	99
2330	3/4" x 18"	G		23	.696		50.50	38.50		89	120
2350	1" x 6"	G		17	.941		57.50	52		109.50	150
2360	1" x 12"	G		13	1.231		63	68		131	183
2370	1" x 18"	G		10	1.600		94.50	88.50		183	251
2380	1" x 24"	G		9	1.778		104	98.50		202.50	279
2400	1-1/4" x 12"	G		7.	2.286		106	126		232	330
2410	1-1/4" x 18"	G		6.50	2.462		131	136		267	370
2420	1-1/4" x 24"	G		5.60	2.857		177	158		335	455
2450	1-1/2" x 12"	G		5.20	3.077		232	170		402	540
2460	1-1/2" x 18"	G		4	4		248	221		469	645
2470	1-1/2" x 24"	G		3.20	5		335	277		612	825
2500	1-3/4" x 18"	G		3.20	5		505	277		782	1,025
2510	1-3/4" x 24"	G		2.80	5.714		575	315		890	1,150
2550	2" x 24"	G		1.60	10		775	555		1,330	1,775

05 15 16.50 Steel Wire Rope

		Crew	Daily Output	Labor-Hours	Unit	Material	2017 Bare Costs Labor	Equipment	Total	Total Incl O&P
0010	**STEEL WIRE ROPE**									
0015	Made from recycled materials									
0020	6 x 19, bright, fiber core, 5000' rolls, 1/2" diameter	G				L.F.	.87			.87	.95
0050	Steel core	G					1.14			1.14	1.25
0100	Fiber core, 1" diameter	G					2.92			2.92	3.21
0150	Steel core	G					3.33			3.33	3.66
0300	6 x 19, galvanized, fiber core, 1/2" diameter	G					1.28			1.28	1.40
0350	Steel core	G					1.46			1.46	1.61
0400	Fiber core, 1" diameter	G					3.74			3.74	4.11
0450	Steel core	G					3.92			3.92	4.31
0500	6 x 7, bright, IPS, fiber core, <500 L.F. w/acc., 1/4" diameter	G	E-17	6400	.003		1.10	.14		1.24	1.44
0510	1/2" diameter	G		2100	.008		2.69	.42		3.11	3.66
0520	3/4" diameter	G		960	.017		4.87	.92		5.79	6.90
0550	6 x 19, bright, IPS, IWRC, <500 L.F. w/acc., 1/4" diameter	G		5760	.003		.96	.15		1.11	1.31
0560	1/2" diameter	G		1730	.009		1.55	.51		2.06	2.56
0570	3/4" diameter	G		770	.021		2.69	1.15		3.84	4.88
0580	1" diameter	G		420	.038		4.56	2.11		6.67	8.50
0590	1-1/4" diameter	G		290	.055		7.55	3.05		10.60	13.45
0600	1-1/2" diameter	G		192	.083		9.30	4.61		13.91	17.95
0610	1-3/4" diameter	G	E-18	240	.167		14.80	9.10	4.48	28.38	36
0620	2" diameter	G		160	.250		19	13.65	6.70	39.35	51
0630	2-1/4" diameter	G		160	.250		25.50	13.65	6.70	45.85	58
0650	6 x 37, bright, IPS, IWRC, <500 L.F. w/acc., 1/4" diameter	G	E-17	6400	.003		1.10	.14		1.24	1.44
0660	1/2" diameter	G		1730	.009		1.87	.51		2.38	2.90
0670	3/4" diameter	G		770	.021		3.02	1.15		4.17	5.25
0680	1" diameter	G		430	.037		4.79	2.06		6.85	8.70
0690	1-1/4" diameter	G		290	.055		7.25	3.05		10.30	13.05
0700	1-1/2" diameter	G		190	.084		10.35	4.66		15.01	19.15
0710	1-3/4" diameter	G	E-18	260	.154		16.45	8.40	4.14	28.99	36.50

05 15 Wire Rope Assemblies

05 15 16 – Steel Wire Rope Assemblies

05 15 16.50 Steel Wire Rope

			Crew	Daily Output	Labor-Hours	Unit	Material	2017 Bare Costs Labor	Equipment	Total	Total Incl O&P
0720	2" diameter	G	E-18	200	.200	L.F.	21.50	10.90	5.40	37.80	47.50
0730	2-1/4" diameter	G	↓	160	.250		28	13.65	6.70	48.35	61
0800	6 x 19 & 6 x 37, swaged, 1/2" diameter	G	E-17	1220	.013		2.52	.73		3.25	3.99
0810	9/16" diameter	G		1120	.014		2.93	.79		3.72	4.55
0820	5/8" diameter	G		930	.017		3.48	.95		4.43	5.40
0830	3/4" diameter	G		640	.025		4.43	1.38		5.81	7.20
0840	7/8" diameter	G		480	.033		5.60	1.84		7.44	9.20
0850	1" diameter	G		350	.046		6.80	2.53		9.33	11.70
0860	1-1/8" diameter	G		288	.056		8.40	3.07		11.47	14.35
0870	1-1/4" diameter	G		230	.070		10.15	3.85		14	17.60
0880	1-3/8" diameter	G	↓	192	.083		11.75	4.61		16.36	20.50
0890	1-1/2" diameter	G	E-18	300	.133	↓	14.25	7.25	3.58	25.08	31.50

05 15 16.60 Galvanized Steel Wire Rope and Accessories

			Crew	Daily Output	Labor-Hours	Unit	Material	Labor	Equipment	Total	Total Incl O&P
0010	**GALVANIZED STEEL WIRE ROPE & ACCESSORIES**										
0015	Made from recycled materials										
3000	Aircraft cable, galvanized, 7 x 7 x 1/8"	G	E-17	5000	.003	L.F.	.19	.18		.37	.51
3100	Clamps, 1/8"	G	"	125	.128	Ea.	1.41	7.10		8.51	13.35

05 15 16.70 Temporary Cable Safety Railing

			Crew	Daily Output	Labor-Hours	Unit	Material	Labor	Equipment	Total	Total Incl O&P
0010	**TEMPORARY CABLE SAFETY RAILING**, Each 100' strand incl.										
0020	2 eyebolts, 1 turnbuckle, 100' cable, 2 thimbles, 6 clips										
0025	Made from recycled materials										
0100	One strand using 1/4" cable & accessories	G	2 Sswk	4	4	C.L.F.	152	217		369	530
0200	1/2" cable & accessories	G	"	2	8	"	325	435		760	1,075

05 21 Steel Joist Framing

05 21 13 – Deep Longspan Steel Joist Framing

05 21 13.50 Deep Longspan Joists

			Crew	Daily Output	Labor-Hours	Unit	Material	Labor	Equipment	Total	Total Incl O&P
0010	**DEEP LONGSPAN JOISTS**										
3010	DLH series, 40-ton job lots, bolted cross bridging, shop primer										
3015	Made from recycled materials										
3040	Spans to 144' (shipped in 2 pieces)	G	E-7	13	6.154	Ton	2,150	335	148	2,633	3,075
3200	52DLH11, 26 lb./L.F.	G		2000	.040	L.F.	27	2.17	.97	30.14	34
3220	52DLH16, 45 lb./L.F.	G		2000	.040		48.50	2.17	.97	51.64	57.50
3240	56DLH11, 26 lb./L.F.	G		2000	.040		28	2.17	.97	31.14	35.50
3260	56DLH16, 46 lb./L.F.	G		2000	.040		49.50	2.17	.97	52.64	59
3280	60DLH12, 29 lb./L.F.	G		2000	.040		31	2.17	.97	34.14	39
3300	60DLH17, 52 lb./L.F.	G		2000	.040		56	2.17	.97	59.14	66
3320	64DLH12, 31 lb./L.F.	G		2200	.036		33.50	1.97	.88	36.35	40.50
3340	64DLH17, 52 lb./L.F.	G		2200	.036		56	1.97	.88	58.85	65.50
3360	68DLH13, 37 lb./L.F.	G		2200	.036		40	1.97	.88	42.85	48
3380	68DLH18, 61 lb./L.F.	G		2200	.036		65.50	1.97	.88	68.35	76
3400	72DLH14, 41 lb./L.F.	G		2200	.036		44	1.97	.88	46.85	52.50
3420	72DLH19, 70 lb./L.F.	G	↓	2200	.036	↓	75.50	1.97	.88	78.35	87
3500	For less than 40-ton job lots										
3502	For 30 to 39 tons, add					%	10%				
3504	20 to 29 tons, add						20%				
3506	10 to 19 tons, add						30%				
3507	5 to 9 tons, add						50%	25%			
3508	1 to 4 tons, add						75%	50%			
3509	Less than 1 ton, add						100%	100%			

05 21 13 – Deep Longspan Steel Joist Framing

05 21 13.50 Deep Longspan Joists		Crew	Daily Output	Labor-Hours	Unit	Material	2017 Bare Costs Labor	Equipment	Total	Total Incl O&P
4010	SLH series, 40-ton job lots, bolted cross bridging, shop primer									
4040	Spans to 200' (shipped in 3 pieces)	G E-7	13	6.154	Ton	2,225	335	148	2,708	3,150
4200	80SLH15, 40 lb./L.F.	G	1500	.053	L.F.	44.50	2.89	1.29	48.68	55
4220	80SLH20, 75 lb./L.F.	G	1500	.053		83.50	2.89	1.29	87.68	98
4240	88SLH16, 46 lb./L.F.	G	1500	.053		51	2.89	1.29	55.18	62.50
4260	88SLH21, 89 lb./L.F.	G	1500	.053		99	2.89	1.29	103.18	115
4280	96SLH17, 52 lb./L.F.	G	1500	.053		58	2.89	1.29	62.18	69.50
4300	96SLH22, 102 lb./L.F.	G	1500	.053		113	2.89	1.29	117.18	131
4320	104SLH18, 59 lb./L.F.	G	1800	.044		65.50	2.41	1.07	68.98	77
4340	104SLH23, 109 lb./L.F.	G	1800	.044		121	2.41	1.07	124.48	138
4360	112SLH19, 67 lb./L.F.	G	1800	.044		74.50	2.41	1.07	77.98	87
4380	112SLH24, 131 lb./L.F.	G	1800	.044		146	2.41	1.07	149.48	165
4400	120SLH20, 77 lb./L.F.	G	1800	.044		85.50	2.41	1.07	88.98	99
4420	120SLH25, 152 lb./L.F.	G	1800	.044		169	2.41	1.07	172.48	191
6100	For less than 40-ton job lots									
6102	For 30 to 39 tons, add				%	10%				
6104	20 to 29 tons, add					20%				
6106	10 to 19 tons, add					30%				
6107	5 to 9 tons, add					50%	25%			
6108	1 to 4 tons, add					75%	50%			
6109	Less than 1 ton, add					100%	100%			

05 21 16 – Longspan Steel Joist Framing

05 21 16.50 Longspan Joists		Crew	Daily Output	Labor-Hours	Unit	Material	2017 Bare Costs Labor	Equipment	Total	Total Incl O&P
0010	**LONGSPAN JOISTS**									
2000	LH series, 40-ton job lots, bolted cross bridging, shop primer									
2015	Made from recycled materials									
2040	Longspan joists, LH series, up to 96'	G E-7	13	6.154	Ton	2,050	335	148	2,533	2,950
2200	18LH04, 12 lb./L.F.	G	1400	.057	L.F.	12.30	3.10	1.38	16.78	20
2220	18LH08, 19 lb./L.F.	G	1400	.057		19.50	3.10	1.38	23.98	28
2240	20LH04, 12 lb./L.F.	G	1400	.057		12.30	3.10	1.38	16.78	20
2260	20LH08, 19 lb./L.F.	G	1400	.057		19.50	3.10	1.38	23.98	28
2280	24LH05, 13 lb./L.F.	G	1400	.057		13.35	3.10	1.38	17.83	21
2300	24LH10, 23 lb./L.F.	G	1400	.057		23.50	3.10	1.38	27.98	32.50
2320	28LH06, 16 lb./L.F.	G	1800	.044		16.40	2.41	1.07	19.88	23
2340	28LH11, 25 lb./L.F.	G	1800	.044		25.50	2.41	1.07	28.98	33
2360	32LH08, 17 lb./L.F.	G	1800	.044		17.45	2.41	1.07	20.93	24.50
2380	32LH13, 30 lb./L.F.	G	1800	.044		31	2.41	1.07	34.48	39
2400	36LH09, 21 lb./L.F.	G	1800	.044		21.50	2.41	1.07	24.98	28.50
2420	36LH14, 36 lb./L.F.	G	1800	.044		37	2.41	1.07	40.48	45.50
2440	40LH10, 21 lb./L.F.	G	2200	.036		21.50	1.97	.88	24.35	27.50
2460	40LH15, 36 lb./L.F.	G	2200	.036		37	1.97	.88	39.85	44.50
2480	44LH11, 22 lb./L.F.	G	2200	.036		22.50	1.97	.88	25.35	29
2500	44LH16, 42 lb./L.F.	G	2200	.036		43	1.97	.88	45.85	51.50
2520	48LH11, 22 lb./L.F.	G	2200	.036		22.50	1.97	.88	25.35	29
2540	48LH16, 42 lb./L.F.	G	2200	.036		43	1.97	.88	45.85	51.50
2600	For less than 40-ton job lots									
2602	For 30 to 39 tons, add				%	10%				
2604	20 to 29 tons, add					20%				
2606	10 to 19 tons, add					30%				
2607	5 to 9 tons, add					50%	25%			
2608	1 to 4 tons, add					75%	50%			
2609	Less than 1 ton, add					100%	100%			

05 21 Steel Joist Framing

05 21 16 – Longspan Steel Joist Framing

05 21 16.50 Longspan Joists		Crew	Daily Output	Labor-Hours	Unit	Material	2017 Bare Costs Labor	Equipment	Total	Total Incl O&P
6000	For welded cross bridging, add						30%			

05 21 19 – Open Web Steel Joist Framing

05 21 19.10 Open Web Joists

			Crew	Daily Output	Labor-Hours	Unit	Material	2017 Bare Costs Labor	Equipment	Total	Total Incl O&P
0010	**OPEN WEB JOISTS**										
0015	Made from recycled materials										
0050	K series, 40-ton lots, horiz. bridging, spans to 30', shop primer	G	E-7	12	6.667	Ton	1,750	360	161	2,271	2,700
0130	8K1, 5.1 lb./L.F.	G		1200	.067	L.F.	4.46	3.61	1.61	9.68	12.60
0140	10K1, 5.0 lb./L.F.	G		1200	.067		4.38	3.61	1.61	9.60	12.50
0160	12K3, 5.7 lb./L.F.	G		1500	.053		4.99	2.89	1.29	9.17	11.65
0180	14K3, 6.0 lb./L.F.	G		1500	.053		5.25	2.89	1.29	9.43	11.95
0200	16K3, 6.3 lb./L.F.	G		1800	.044		5.50	2.41	1.07	8.98	11.15
0220	16K6, 8.1 lb./L.F.	G		1800	.044		7.10	2.41	1.07	10.58	12.90
0240	18K5, 7.7 lb./L.F.	G		2000	.040		6.75	2.17	.97	9.89	12
0260	18K9, 10.2 lb./L.F.	G		2000	.040		8.95	2.17	.97	12.09	14.40
0440	K series, 30' to 50' spans	G		17	4.706	Ton	1,800	255	114	2,169	2,525
0500	20K5, 8.2 lb./L.F.	G		2000	.040	L.F.	7.40	2.17	.97	10.54	12.70
0520	20K9, 10.8 lb./L.F.	G		2000	.040		9.70	2.17	.97	12.84	15.30
0540	22K5, 8.8 lb./L.F.	G		2000	.040		7.90	2.17	.97	11.04	13.30
0560	22K9, 11.3 lb./L.F.	G		2000	.040		10.15	2.17	.97	13.29	15.80
0580	24K6, 9.7 lb./L.F.	G		2200	.036		8.75	1.97	.88	11.60	13.80
0600	24K10, 13.1 lb./L.F.	G		2200	.036		11.80	1.97	.88	14.65	17.15
0620	26K6, 10.6 lb./L.F.	G		2200	.036		9.55	1.97	.88	12.40	14.70
0640	26K10, 13.8 lb./L.F.	G		2200	.036		12.40	1.97	.88	15.25	17.85
0660	28K8, 12.7 lb./L.F.	G		2400	.033		11.45	1.81	.80	14.06	16.40
0680	28K12, 17.1 lb./L.F.	G		2400	.033		15.40	1.81	.80	18.01	21
0700	30K8, 13.2 lb./L.F.	G		2400	.033		11.90	1.81	.80	14.51	16.90
0720	30K12, 17.6 lb./L.F.	G		2400	.033		15.85	1.81	.80	18.46	21
0800	For less than 40-ton job lots										
0802	For 30 to 39 tons, add					%	10%				
0804	20 to 29 tons, add						20%				
0806	10 to 19 tons, add						30%				
0807	5 to 9 tons, add						50%	25%			
0808	1 to 4 tons, add						75%	50%			
0809	Less than 1 ton, add						100%	100%			
1010	CS series, 40-ton job lots, horizontal bridging, shop primer										
1040	Spans to 30'	G	E-7	12	6.667	Ton	1,900	360	161	2,421	2,875
1100	10CS2, 7.5 lb./L.F.	G		1200	.067	L.F.	7.15	3.61	1.61	12.37	15.50
1120	12CS2, 8.0 lb./L.F.	G		1500	.053		7.60	2.89	1.29	11.78	14.50
1140	14CS2, 8.0 lb./L.F.	G		1500	.053		7.60	2.89	1.29	11.78	14.50
1160	16CS2, 8.5 lb./L.F.	G		1800	.044		8.10	2.41	1.07	11.58	14
1180	16CS4, 14.5 lb./L.F.	G		1800	.044		13.80	2.41	1.07	17.28	20.50
1200	18CS2, 9.0 lb./L.F.	G		2000	.040		8.55	2.17	.97	11.69	14
1220	18CS4, 15.0 lb./L.F.	G		2000	.040		14.25	2.17	.97	17.39	20.50
1240	20CS2, 9.5 lb./L.F.	G		2000	.040		9.05	2.17	.97	12.19	14.55
1260	20CS4, 16.5 lb./L.F.	G		2000	.040		15.70	2.17	.97	18.84	22
1280	22CS2, 10.0 lb./L.F.	G		2000	.040		9.50	2.17	.97	12.64	15.05
1300	22CS4, 16.5 lb./L.F.	G		2000	.040		15.70	2.17	.97	18.84	22
1320	24CS2, 10.0 lb./L.F.	G		2200	.036		9.50	1.97	.88	12.35	14.65
1340	24CS4, 16.5 lb./L.F.	G		2200	.036		15.70	1.97	.88	18.55	21.50
1360	26CS2, 10.0 lb./L.F.	G		2200	.036		9.50	1.97	.88	12.35	14.65
1380	26CS4, 16.5 lb./L.F.	G		2200	.036		15.70	1.97	.88	18.55	21.50
1400	28CS2, 10.5 lb./L.F.	G		2400	.033		10	1.81	.80	12.61	14.80

05 21 Steel Joist Framing

05 21 19 – Open Web Steel Joist Framing

05 21 19.10 Open Web Joists

		Crew	Daily Output	Labor-Hours	Unit	Material	2017 Bare Costs Labor	Equipment	Total	Total Incl O&P
1420	28CS4, 16.5 lb./L.F. **G**	E-7	2400	.033	L.F.	15.70	1.81	.80	18.31	21
1440	30CS2, 11.0 lb./L.F. **G**		2400	.033		10.45	1.81	.80	13.06	15.35
1460	30CS4, 16.5 lb./L.F. **G**	↓	2400	.033	↓	15.70	1.81	.80	18.31	21
1500	For less than 40-ton job lots									
1502	For 30 to 39 tons, add				%	10%				
1504	20 to 29 tons, add					20%				
1506	10 to 19 tons, add					30%				
1507	5 to 9 tons, add					50%	25%			
1508	1 to 4 tons, add					75%	50%			
1509	Less than 1 ton, add				↓	100%	100%			
6200	For shop prime paint other than mfrs. standard, add					20%				
6300	For bottom chord extensions, add per chord **G**				Ea.	40			40	44
6400	Individual steel bearing plate, 6" x 6" x 1/4" with J-hook **G**	1 Bric	160	.050	"	7.85	2.39		10.24	12.30

05 21 23 – Steel Joist Girder Framing

05 21 23.50 Joist Girders

		Crew	Daily Output	Labor-Hours	Unit	Material	2017 Bare Costs Labor	Equipment	Total	Total Incl O&P
0010	**JOIST GIRDERS**									
0015	Made from recycled materials									
7020	Joist girders, 40-ton job lots, shop primer **G**	E-5	13	6.154	Ton	1,850	335	139	2,324	2,725
7100	For less than 40-ton job lots									
7102	For 30 to 39 tons, add				Ton	10%				
7104	20 to 29 tons, add					20%				
7106	10 to 19 tons, add					30%				
7107	5 to 9 tons, add					50%	25%			
7108	1 to 4 tons, add					75%	50%			
7109	Less than 1 ton, add					100%	100%			
8000	Trusses, 40-ton job lots, shop fabricated WT chords, shop primer **G**	E-5	11	7.273	↓	6,050	395	164	6,609	7,475
8100	For less than 40-ton job lots									
8102	For 30 to 39 tons, add				Ton	10%				
8104	20 to 29 tons, add					20%				
8106	10 to 19 tons, add					30%				
8107	5 to 9 tons, add					50%	25%			
8108	1 to 4 tons, add					75%	50%			
8109	Less than 1 ton, add				↓	100%	100%			

05 31 Steel Decking

05 31 13 – Steel Floor Decking

05 31 13.50 Floor Decking

		Crew	Daily Output	Labor-Hours	Unit	Material	2017 Bare Costs Labor	Equipment	Total	Total Incl O&P
0010	**FLOOR DECKING** R053100-10									
0015	Made from recycled materials									
5100	Non-cellular composite decking, galvanized, 1-1/2" deep, 16 ga. **G**	E-4	3500	.009	S.F.	3.89	.50	.04	4.43	5.15
5120	18 ga. **G**		3650	.009		3.15	.48	.04	3.67	4.30
5140	20 ga. **G**		3800	.008		2.50	.46	.03	2.99	3.56
5200	2" deep, 22 ga. **G**		3860	.008		2.18	.45	.03	2.66	3.20
5300	20 ga. **G**		3600	.009		2.40	.49	.04	2.93	3.49
5400	18 ga. **G**		3380	.009		3.08	.52	.04	3.64	4.30
5500	16 ga. **G**		3200	.010		3.85	.55	.04	4.44	5.20
5700	3" deep, 22 ga. **G**		3200	.010		2.38	.55	.04	2.97	3.56
5800	20 ga. **G**		3000	.011		2.64	.58	.04	3.26	3.92
5900	18 ga. **G**		2850	.011		3.28	.62	.05	3.95	4.69
6000	16 ga. **G**	↓	2700	.012	↓	4.37	.65	.05	5.07	5.95

For customer support on your Building Construction Costs with RSMeans Data, call 800.448.8182.

141

05 31 23 – Steel Roof Decking

05 31 23.50 Roof Decking		Crew	Daily Output	Labor-Hours	Unit	Material	2017 Bare Costs Labor	Equipment	Total	Total Incl O&P
0010	**ROOF DECKING**									
0015	Made from recycled materials									
2100	Open type, 1-1/2" deep, Type B, wide rib, galv., 22 ga., under 50 sq.	G E-4	4500	.007	S.F.	2.32	.39	.03	2.74	3.23
2200	50-500 squares	G	4900	.007		1.80	.36	.03	2.19	2.61
2400	Over 500 squares	G	5100	.006		1.67	.34	.03	2.04	2.44
2600	20 ga., under 50 squares	G	3865	.008		2.70	.45	.03	3.18	3.77
2650	50-500 squares	G	4170	.008		2.16	.42	.03	2.61	3.11
2700	Over 500 squares	G	4300	.007		1.95	.41	.03	2.39	2.85
2900	18 ga., under 50 squares	G	3800	.008		3.49	.46	.03	3.98	4.65
2950	50-500 squares	G	4100	.008		2.79	.43	.03	3.25	3.81
3000	Over 500 squares	G	4300	.007		2.51	.41	.03	2.95	3.47
3050	16 ga., under 50 squares	G	3700	.009		4.71	.47	.04	5.22	6.05
3060	50-500 squares	G	4000	.008		3.77	.44	.03	4.24	4.92
3100	Over 500 squares	G	4200	.008		3.39	.42	.03	3.84	4.46
3200	3" deep, Type N, 22 ga., under 50 squares	G	3600	.009		3.40	.49	.04	3.93	4.59
3250	50-500 squares	G	3800	.008		2.72	.46	.03	3.21	3.80
3260	over 500 squares	G	4000	.008		2.45	.44	.03	2.92	3.46
3300	20 ga., under 50 squares	G	3400	.009		3.67	.52	.04	4.23	4.94
3350	50-500 squares	G	3600	.009		2.94	.49	.04	3.47	4.08
3360	over 500 squares	G	3800	.008		2.64	.46	.03	3.13	3.72
3400	18 ga., under 50 squares	G	3200	.010		4.75	.55	.04	5.34	6.20
3450	50-500 squares	G	3400	.009		3.80	.52	.04	4.36	5.10
3460	over 500 squares	G	3600	.009		3.42	.49	.04	3.95	4.61
3500	16 ga., under 50 squares	G	3000	.011		6.25	.58	.04	6.87	7.90
3550	50-500 squares	G	3200	.010		5	.55	.04	5.59	6.45
3560	over 500 squares	G	3400	.009		4.51	.52	.04	5.07	5.85
3700	4-1/2" deep, Type J, 20 ga., over 50 squares	G	2700	.012		4.11	.65	.05	4.81	5.65
3800	18 ga.	G	2460	.013		5.40	.71	.05	6.16	7.20
3900	16 ga.	G	2350	.014		7.05	.75	.06	7.86	9.05
4100	6" deep, Type H, 18 ga., over 50 squares	G	2000	.016		6.50	.88	.07	7.45	8.70
4200	16 ga.	G	1930	.017		8.15	.91	.07	9.13	10.55
4300	14 ga.	G	1860	.017		10.45	.94	.07	11.46	13.15
4500	7-1/2" deep, Type H, 18 ga., over 50 squares	G	1690	.019		7.70	1.04	.08	8.82	10.30
4600	16 ga.	G	1590	.020		9.65	1.10	.08	10.83	12.55
4700	14 ga.	G	1490	.021		12	1.18	.09	13.27	15.25
4800	For painted instead of galvanized, deduct					5%				
5000	For acoustical perforated with fiberglass insulation, add				S.F.	25%				
5100	For type F intermediate rib instead of type B wide rib, add	G				25%				
5150	For type A narrow rib instead of type B wide rib, add	G				25%				

05 31 33 – Steel Form Decking

05 31 33.50 Form Decking		Crew	Daily Output	Labor-Hours	Unit	Material	2017 Bare Costs Labor	Equipment	Total	Total Incl O&P
0010	**FORM DECKING**									
0015	Made from recycled materials									
6100	Slab form, steel, 28 ga., 9/16" deep, Type UFS, uncoated	G E-4	4000	.008	S.F.	1.74	.44	.03	2.21	2.68
6200	Galvanized	G	4000	.008		1.54	.44	.03	2.01	2.46
6220	24 ga., 1" deep, Type UF1X, uncoated	G	3900	.008		1.67	.45	.03	2.15	2.63
6240	Galvanized	G	3900	.008		1.97	.45	.03	2.45	2.96
6300	24 ga., 1-5/16" deep, Type UFX, uncoated	G	3800	.008		1.78	.46	.03	2.27	2.76
6400	Galvanized	G	3800	.008		2.09	.46	.03	2.58	3.11
6500	22 ga., 1-5/16" deep, uncoated	G	3700	.009		2.25	.47	.04	2.76	3.31
6600	Galvanized	G	3700	.009		2.30	.47	.04	2.81	3.36
6700	22 ga., 2" deep, uncoated	G	3600	.009		2.93	.49	.04	3.46	4.07

05 31 Steel Decking

05 31 33 – Steel Form Decking

05 31 33.50 Form Decking		Crew	Daily Output	Labor-Hours	Unit	Material	2017 Bare Costs Labor	Equipment	Total	Total Incl O&P
6800	Galvanized	G E-4	3600	.009	S.F.	2.87	.49	.04	3.40	4.01
7000	Sheet metal edge closure form, 12" wide with 2 bends, galvanized									
7100	18 ga.	G E-14	360	.022	L.F.	4.74	1.25	.36	6.35	7.70
7200	16 ga.	G "	360	.022	"	6.45	1.25	.36	8.06	9.55

05 35 Raceway Decking Assemblies

05 35 13 – Steel Cellular Decking

05 35 13.50 Cellular Decking

		Crew	Daily Output	Labor-Hours	Unit	Material	2017 Bare Costs Labor	Equipment	Total	Total Incl O&P
0010	**CELLULAR DECKING**									
0015	Made from recycled materials									
0200	Cellular units, galv., 1-1/2" deep, Type BC, 20-20 ga., over 15 squares	G E-4	1460	.022	S.F.	9.15	1.20	.09	10.44	12.15
0250	18-20 ga.	G	1420	.023		10.35	1.24	.09	11.68	13.55
0300	18-18 ga.	G	1390	.023		10.65	1.26	.09	12	13.90
0320	16-18 ga.	G	1360	.024		12.70	1.29	.10	14.09	16.20
0340	16-16 ga.	G	1330	.024		14.10	1.32	.10	15.52	17.80
0400	3" deep, Type NC, galvanized, 20-20 ga.	G	1375	.023		10.05	1.28	.09	11.42	13.30
0500	18-20 ga.	G	1350	.024		12.15	1.30	.10	13.55	15.60
0600	18-18 ga.	G	1290	.025		12.10	1.36	.10	13.56	15.70
0700	16-18 ga.	G	1230	.026		13.65	1.43	.11	15.19	17.50
0800	16-16 ga.	G	1150	.028		14.85	1.53	.11	16.49	19
1000	4-1/2" deep, Type JC, galvanized, 18-20 ga.	G	1100	.029		14	1.59	.12	15.71	18.20
1100	18-18 ga.	G	1040	.031		13.90	1.69	.12	15.71	18.25
1200	16-18 ga.	G	980	.033		15.65	1.79	.13	17.57	20.50
1300	16-16 ga.	G	935	.034		17.10	1.88	.14	19.12	22
1500	For acoustical deck, add					15%				
1700	For cells used for ventilation, add					15%				
1900	For multi-story or congested site, add						50%			
8000	Metal deck and trench, 2" thick, 20 ga., combination									
8010	60% cellular, 40% non-cellular, inserts and trench	G R-4	1100	.036	S.F.	18.25	2.01	.12	20.38	23.50

05 41 Structural Metal Stud Framing

05 41 13 – Load-Bearing Metal Stud Framing

05 41 13.05 Bracing

		Crew	Daily Output	Labor-Hours	Unit	Material	2017 Bare Costs Labor	Equipment	Total	Total Incl O&P
0010	**BRACING**, shear wall X-bracing, per 10' x 10' bay, one face									
0015	Made of recycled materials									
0120	Metal strap, 20 ga. x 4" wide	G 2 Carp	18	.889	Ea.	18.90	44		62.90	88
0130	6" wide	G	18	.889		31.50	44		75.50	102
0160	18 ga. x 4" wide	G	16	1		33	49.50		82.50	112
0170	6" wide	G	16	1		48.50	49.50		98	129
0410	Continuous strap bracing, per horizontal row on both faces									
0420	Metal strap, 20 ga. x 2" wide, studs 12" O.C.	G 1 Carp	7	1.143	C.L.F.	57	56.50		113.50	149
0430	16" O.C.	G	8	1		57	49.50		106.50	139
0440	24" O.C.	G	10	.800		57	39.50		96.50	124
0450	18 ga. x 2" wide, studs 12" O.C.	G	6	1.333		79	65.50		144.50	188
0460	16" O.C.	G	7	1.143		79	56.50		135.50	173
0470	24" O.C.	G	8	1		79	49.50		128.50	163

For customer support on your Building Construction Costs with RSMeans Data, call 800.448.8182.

143

05 41 Structural Metal Stud Framing

05 41 13 – Load-Bearing Metal Stud Framing

05 41 13.10 Bridging

		Crew	Daily Output	Labor-Hours	Unit	Material	2017 Bare Costs Labor	Equipment	Total	Total Incl O&P
0010	**BRIDGING**, solid between studs w/1-1/4" leg track, per stud bay									
0015	Made from recycled materials									
0200	Studs 12" O.C., 18 ga. x 2-1/2" wide	G 1 Carp	125	.064	Ea.	.95	3.15		4.10	5.90
0210	3-5/8" wide	G	120	.067		1.16	3.28		4.44	6.30
0220	4" wide	G	120	.067		1.22	3.28		4.50	6.40
0230	6" wide	G	115	.070		1.60	3.43		5.03	7
0240	8" wide	G	110	.073		1.97	3.58		5.55	7.65
0300	16 ga. x 2-1/2" wide	G	115	.070		1.21	3.43		4.64	6.60
0310	3-5/8" wide	G	110	.073		1.47	3.58		5.05	7.10
0320	4" wide	G	110	.073		1.57	3.58		5.15	7.25
0330	6" wide	G	105	.076		2.01	3.75		5.76	7.95
0340	8" wide	G	100	.080		2.51	3.94		6.45	8.80
1200	Studs 16" O.C., 18 ga. x 2-1/2" wide	G	125	.064		1.22	3.15		4.37	6.15
1210	3-5/8" wide	G	120	.067		1.49	3.28		4.77	6.70
1220	4" wide	G	120	.067		1.57	3.28		4.85	6.75
1230	6" wide	G	115	.070		2.05	3.43		5.48	7.50
1240	8" wide	G	110	.073		2.52	3.58		6.10	8.30
1300	16 ga. x 2-1/2" wide	G	115	.070		1.55	3.43		4.98	6.95
1310	3-5/8" wide	G	110	.073		1.88	3.58		5.46	7.55
1320	4" wide	G	110	.073		2.01	3.58		5.59	7.70
1330	6" wide	G	105	.076		2.57	3.75		6.32	8.60
1340	8" wide	G	100	.080		3.22	3.94		7.16	9.60
2200	Studs 24" O.C., 18 ga. x 2-1/2" wide	G	125	.064		1.77	3.15		4.92	6.75
2210	3-5/8" wide	G	120	.067		2.15	3.28		5.43	7.40
2220	4" wide	G	120	.067		2.27	3.28		5.55	7.55
2230	6" wide	G	115	.070		2.96	3.43		6.39	8.50
2240	8" wide	G	110	.073		3.65	3.58		7.23	9.50
2300	16 ga. x 2-1/2" wide	G	115	.070		2.24	3.43		5.67	7.70
2310	3-5/8" wide	G	110	.073		2.72	3.58		6.30	8.50
2320	4" wide	G	110	.073		2.91	3.58		6.49	8.70
2330	6" wide	G	105	.076		3.72	3.75		7.47	9.85
2340	8" wide	G	100	.080		4.65	3.94		8.59	11.15
3000	Continuous bridging, per row									
3100	16 ga. x 1-1/2" channel thru studs 12" O.C.	G 1 Carp	6	1.333	C.L.F.	51.50	65.50		117	158
3110	16" O.C.	G	7	1.143		51.50	56.50		108	143
3120	24" O.C.	G	8.80	.909		51.50	45		96.50	126
4100	2" x 2" angle x 18 ga., studs 12" O.C.	G	7	1.143		80.50	56.50		137	175
4110	16" O.C.	G	9	.889		80.50	44		124.50	156
4120	24" O.C.	G	12	.667		80.50	33		113.50	139
4200	16 ga., studs 12" O.C.	G	5	1.600		101	79		180	232
4210	16" O.C.	G	7	1.143		101	56.50		157.50	197
4220	24" O.C.	G	10	.800		101	39.50		140.50	172

05 41 13.25 Framing, Boxed Headers/Beams

		Crew	Daily Output	Labor-Hours	Unit	Material	2017 Bare Costs Labor	Equipment	Total	Total Incl O&P
0010	**FRAMING, BOXED HEADERS/BEAMS**									
0015	Made from recycled materials									
0200	Double, 18 ga. x 6" deep	G 2 Carp	220	.073	L.F.	5.50	3.58		9.08	11.55
0210	8" deep	G	210	.076		6.05	3.75		9.80	12.40
0220	10" deep	G	200	.080		7.40	3.94		11.34	14.20
0230	12" deep	G	190	.084		8.05	4.15		12.20	15.20
0300	16 ga. x 8" deep	G	180	.089		7	4.38		11.38	14.40
0310	10" deep	G	170	.094		8.45	4.64		13.09	16.35
0320	12" deep	G	160	.100		9.15	4.93		14.08	17.65

05 41 Structural Metal Stud Framing

05 41 13 – Load-Bearing Metal Stud Framing

05 41 13.25 Framing, Boxed Headers/Beams

		Crew	Daily Output	Labor-Hours	Unit	Material	2017 Bare Costs Labor	Equipment	Total	Total Incl O&P
0400	14 ga. x 10" deep [G]	2 Carp	140	.114	L.F.	9.75	5.65		15.40	19.30
0410	12" deep [G]		130	.123		10.65	6.05		16.70	21
1210	Triple, 18 ga. x 8" deep [G]		170	.094		8.75	4.64		13.39	16.75
1220	10" deep [G]		165	.097		10.60	4.78		15.38	18.95
1230	12" deep [G]		160	.100		11.60	4.93		16.53	20.50
1300	16 ga. x 8" deep [G]		145	.110		10.15	5.45		15.60	19.50
1310	10" deep [G]		140	.114		12.15	5.65		17.80	22
1320	12" deep [G]		135	.119		13.25	5.85		19.10	23.50
1400	14 ga. x 10" deep [G]		115	.139		13.25	6.85		20.10	25
1410	12" deep [G]		110	.145		14.65	7.15		21.80	27

05 41 13.30 Framing, Stud Walls

		Crew	Daily Output	Labor-Hours	Unit	Material	2017 Bare Costs Labor	Equipment	Total	Total Incl O&P
0010	**FRAMING, STUD WALLS** w/top & bottom track, no openings,									
0020	Headers, beams, bridging or bracing									
0025	Made from recycled materials									
4100	8' high walls, 18 ga. x 2-1/2" wide, studs 12" O.C. [G]	2 Carp	54	.296	L.F.	9.15	14.60		23.75	32.50
4110	16" O.C. [G]		77	.208		7.35	10.25		17.60	24
4120	24" O.C. [G]		107	.150		5.50	7.35		12.85	17.35
4130	3-5/8" wide, studs 12" O.C. [G]		53	.302		10.85	14.85		25.70	35
4140	16" O.C. [G]		76	.211		8.70	10.35		19.05	25.50
4150	24" O.C. [G]		105	.152		6.55	7.50		14.05	18.70
4160	4" wide, studs 12" O.C. [G]		52	.308		11.30	15.15		26.45	35.50
4170	16" O.C. [G]		74	.216		9.05	10.65		19.70	26.50
4180	24" O.C. [G]		103	.155		6.80	7.65		14.45	19.20
4190	6" wide, studs 12" O.C. [G]		51	.314		14.40	15.45		29.85	39.50
4200	16" O.C. [G]		73	.219		11.55	10.80		22.35	29.50
4210	24" O.C. [G]		101	.158		8.70	7.80		16.50	21.50
4220	8" wide, studs 12" O.C. [G]		50	.320		17.45	15.75		33.20	43
4230	16" O.C. [G]		72	.222		14	10.95		24.95	32
4240	24" O.C. [G]		100	.160		10.60	7.90		18.50	23.50
4300	16 ga. x 2-1/2" wide, studs 12" O.C. [G]		47	.340		10.85	16.75		27.60	37.50
4310	16" O.C. [G]		68	.235		8.60	11.60		20.20	27
4320	24" O.C. [G]		94	.170		6.35	8.40		14.75	19.80
4330	3-5/8" wide, studs 12" O.C. [G]		46	.348		12.95	17.15		30.10	40
4340	16" O.C. [G]		66	.242		10.25	11.95		22.20	29.50
4350	24" O.C. [G]		92	.174		7.55	8.55		16.10	21.50
4360	4" wide, studs 12" O.C. [G]		45	.356		13.55	17.50		31.05	42
4370	16" O.C. [G]		65	.246		10.75	12.10		22.85	30.50
4380	24" O.C. [G]		90	.178		7.95	8.75		16.70	22
4390	6" wide, studs 12" O.C. [G]		44	.364		17.05	17.90		34.95	46.50
4400	16" O.C. [G]		64	.250		13.55	12.30		25.85	34
4410	24" O.C. [G]		88	.182		10.05	8.95		19	25
4420	8" wide, studs 12" O.C. [G]		43	.372		21	18.35		39.35	51
4430	16" O.C. [G]		63	.254		16.60	12.50		29.10	37.50
4440	24" O.C. [G]		86	.186		12.30	9.15		21.45	27.50
5100	10' high walls, 18 ga. x 2-1/2" wide, studs 12" O.C. [G]		54	.296		11	14.60		25.60	34.50
5110	16" O.C. [G]		77	.208		8.70	10.25		18.95	25.50
5120	24" O.C. [G]		107	.150		6.40	7.35		13.75	18.35
5130	3-5/8" wide, studs 12" O.C. [G]		53	.302		13	14.85		27.85	37.50
5140	16" O.C. [G]		76	.211		10.30	10.35		20.65	27.50
5150	24" O.C. [G]		105	.152		7.60	7.50		15.10	19.85
5160	4" wide, studs 12" O.C. [G]		52	.308		13.55	15.15		28.70	38
5170	16" O.C. [G]		74	.216		10.75	10.65		21.40	28

05 41 Structural Metal Stud Framing

05 41 13 – Load-Bearing Metal Stud Framing

05 41 13.30 Framing, Stud Walls			Crew	Daily Output	Labor-Hours	Unit	Material	2017 Bare Costs Labor	Equipment	Total	Total Incl O&P
5180	24" O.C.	G	2 Carp	103	.155	L.F.	7.95	7.65		15.60	20.50
5190	6" wide, studs 12" O.C.	G		51	.314		17.25	15.45		32.70	42.50
5200	16" O.C.	G		73	.219		13.70	10.80		24.50	31.50
5210	24" O.C.	G		101	.158		10.15	7.80		17.95	23
5220	8" wide, studs 12" O.C.	G		50	.320		21	15.75		36.75	47
5230	16" O.C.	G		72	.222		16.55	10.95		27.50	35
5240	24" O.C.	G		100	.160		12.30	7.90		20.20	25.50
5300	16 ga. x 2-1/2" wide, studs 12" O.C.	G		47	.340		13.10	16.75		29.85	40
5310	16" O.C.	G		68	.235		10.30	11.60		21.90	29
5320	24" O.C.	G		94	.170		7.45	8.40		15.85	21
5330	3-5/8" wide, studs 12" O.C.	G		46	.348		15.60	17.15		32.75	43
5340	16" O.C.	G		66	.242		12.25	11.95		24.20	32
5350	24" O.C.	G		92	.174		8.90	8.55		17.45	23
5360	4" wide, studs 12" O.C.	G		45	.356		16.35	17.50		33.85	45
5370	16" O.C.	G		65	.246		12.85	12.10		24.95	32.50
5380	24" O.C.	G		90	.178		9.35	8.75		18.10	23.50
5390	6" wide, studs 12" O.C.	G		44	.364		20.50	17.90		38.40	50
5400	16" O.C.	G		64	.250		16.15	12.30		28.45	36.50
5410	24" O.C.	G		88	.182		11.80	8.95		20.75	26.50
5420	8" wide, studs 12" O.C.	G		43	.372		25	18.35		43.35	55.50
5430	16" O.C.	G		63	.254		19.80	12.50		32.30	41
5440	24" O.C.	G		86	.186		14.45	9.15		23.60	30
6190	12' high walls, 18 ga. x 6" wide, studs 12" O.C.	G		41	.390		20	19.20		39.20	51.50
6200	16" O.C.	G		58	.276		15.80	13.60		29.40	38.50
6210	24" O.C.	G		81	.198		11.55	9.75		21.30	27.50
6220	8" wide, studs 12" O.C.	G		40	.400		24.50	19.70		44.20	56.50
6230	16" O.C.	G		57	.281		19.15	13.80		32.95	42
6240	24" O.C.	G		80	.200		14	9.85		23.85	30.50
6390	16 ga. x 6" wide, studs 12" O.C.	G		35	.457		24	22.50		46.50	61
6400	16" O.C.	G		51	.314		18.80	15.45		34.25	44
6410	24" O.C.	G		70	.229		13.55	11.25		24.80	32
6420	8" wide, studs 12" O.C.	G		34	.471		29.50	23		52.50	68
6430	16" O.C.	G		50	.320		23	15.75		38.75	49.50
6440	24" O.C.	G		69	.232		16.60	11.40		28	36
6530	14 ga. x 3-5/8" wide, studs 12" O.C.	G		34	.471		22.50	23		45.50	60.50
6540	16" O.C.	G		48	.333		17.75	16.40		34.15	44.50
6550	24" O.C.	G		65	.246		12.75	12.10		24.85	32.50
6560	4" wide, studs 12" O.C.	G		33	.485		24	24		48	63
6570	16" O.C.	G		47	.340		18.75	16.75		35.50	46
6580	24" O.C.	G		64	.250		13.50	12.30		25.80	33.50
6730	12 ga. x 3-5/8" wide, studs 12" O.C.	G		31	.516		31.50	25.50		57	73.50
6740	16" O.C.	G		43	.372		24.50	18.35		42.85	54.50
6750	24" O.C.	G		59	.271		17.15	13.35		30.50	39.50
6760	4" wide, studs 12" O.C.	G		30	.533		33.50	26.50		60	77
6770	16" O.C.	G		42	.381		26	18.75		44.75	57
6780	24" O.C.	G		58	.276		18.30	13.60		31.90	41
7390	16' high walls, 16 ga. x 6" wide, studs 12" O.C.	G		33	.485		31	24		55	70.50
7400	16" O.C.	G		48	.333		24	16.40		40.40	51.50
7410	24" O.C.	G		67	.239		17.05	11.75		28.80	37
7420	8" wide, studs 12" O.C.	G		32	.500		38	24.50		62.50	79.50
7430	16" O.C.	G		47	.340		29.50	16.75		46.25	58
7440	24" O.C.	G		66	.242		21	11.95		32.95	41.50
7560	14 ga. x 4" wide, studs 12" O.C.	G		31	.516		31	25.50		56.50	73

05 41 Structural Metal Stud Framing

05 41 13 – Load-Bearing Metal Stud Framing

05 41 13.30 Framing, Stud Walls

			Crew	Daily Output	Labor-Hours	Unit	Material	2017 Bare Costs Labor	Equipment	Total	Total Incl O&P
7570	16" O.C.	G	2 Carp	45	.356	L.F.	24	17.50		41.50	53.50
7580	24" O.C.	G		61	.262		17	12.90		29.90	38.50
7590	6" wide, studs 12" O.C.	G		30	.533		39	26.50		65.50	83
7600	16" O.C.	G		44	.364		30	17.90		47.90	60.50
7610	24" O.C.	G		60	.267		21.50	13.15		34.65	43.50
7760	12 ga. x 4" wide, studs 12" O.C.	G		29	.552		44	27		71	89.50
7770	16" O.C.	G		40	.400		33.50	19.70		53.20	67
7780	24" O.C.	G		55	.291		23.50	14.35		37.85	47.50
7790	6" wide, studs 12" O.C.	G		28	.571		55.50	28		83.50	104
7800	16" O.C.	G		39	.410		42.50	20		62.50	77.50
7810	24" O.C.	G		54	.296		29.50	14.60		44.10	55
8590	20' high walls, 14 ga. x 6" wide, studs 12" O.C.	G		29	.552		48	27		75	94
8600	16" O.C.	G		42	.381		37	18.75		55.75	69
8610	24" O.C.	G		57	.281		26	13.80		39.80	49.50
8620	8" wide, studs 12" O.C.	G		28	.571		52	28		80	100
8630	16" O.C.	G		41	.390		40	19.20		59.20	73.50
8640	24" O.C.	G		56	.286		28.50	14.05		42.55	52.50
8790	12 ga. x 6" wide, studs 12" O.C.	G		27	.593		68	29		97	120
8800	16" O.C.	G		37	.432		52	21.50		73.50	90
8810	24" O.C.	G		51	.314		36	15.45		51.45	63
8820	8" wide, studs 12" O.C.	G		26	.615		82.50	30.50		113	138
8830	16" O.C.	G		36	.444		63	22		85	103
8840	24" O.C.	G		50	.320		43.50	15.75		59.25	72

05 42 Cold-Formed Metal Joist Framing

05 42 13 – Cold-Formed Metal Floor Joist Framing

05 42 13.05 Bracing

			Crew	Daily Output	Labor-Hours	Unit	Material	2017 Bare Costs Labor	Equipment	Total	Total Incl O&P
0010	**BRACING**, continuous, per row, top & bottom										
0015	Made from recycled materials										
0120	Flat strap, 20 ga. x 2" wide, joists at 12" O.C.	G	1 Carp	4.67	1.713	C.L.F.	60	84.50		144.50	195
0130	16" O.C.	G		5.33	1.501		57.50	74		131.50	177
0140	24" O.C.	G		6.66	1.201		55.50	59		114.50	152
0150	18 ga. x 2" wide, joists at 12" O.C.	G		4	2		78.50	98.50		177	238
0160	16" O.C.	G		4.67	1.713		77	84.50		161.50	214
0170	24" O.C.	G		5.33	1.501		75.50	74		149.50	196

05 42 13.10 Bridging

			Crew	Daily Output	Labor-Hours	Unit	Material	2017 Bare Costs Labor	Equipment	Total	Total Incl O&P
0010	**BRIDGING**, solid between joists w/1-1/4" leg track, per joist bay										
0015	Made from recycled materials										
0230	Joists 12" O.C., 18 ga. track x 6" wide	G	1 Carp	80	.100	Ea.	1.60	4.93		6.53	9.30
0240	8" wide	G		75	.107		1.97	5.25		7.22	10.20
0250	10" wide	G		70	.114		2.46	5.65		8.11	11.30
0260	12" wide	G		65	.123		2.78	6.05		8.83	12.35
0330	16 ga. track x 6" wide	G		70	.114		2.01	5.65		7.66	10.80
0340	8" wide	G		65	.123		2.51	6.05		8.56	12.05
0350	10" wide	G		60	.133		3.10	6.55		9.65	13.45
0360	12" wide	G		55	.145		3.58	7.15		10.73	14.90
0440	14 ga. track x 8" wide	G		60	.133		3.13	6.55		9.68	13.50
0450	10" wide	G		55	.145		3.89	7.15		11.04	15.25
0460	12" wide	G		50	.160		4.49	7.90		12.39	17
0550	12 ga. track x 10" wide	G		45	.178		5.70	8.75		14.45	19.65
0560	12" wide	G		40	.200		5.80	9.85		15.65	21.50

For customer support on your Building Construction Costs with RSMeans Data, call 800.448.8182.

147

05 42 Cold-Formed Metal Joist Framing

05 42 13 – Cold-Formed Metal Floor Joist Framing

05 42 13.10 Bridging

		Crew	Daily Output	Labor-Hours	Unit	Material	2017 Bare Costs Labor	Equipment	Total	Total Incl O&P
1230	16" O.C., 18 ga. track x 6" wide	G	1 Carp	80	.100	Ea.	2.05	4.93		6.98	9.80
1240	8" wide	G		75	.107		2.52	5.25		7.77	10.85
1250	10" wide	G		70	.114		3.15	5.65		8.80	12.05
1260	12" wide	G		65	.123		3.56	6.05		9.61	13.20
1330	16 ga. track x 6" wide	G		70	.114		2.57	5.65		8.22	11.45
1340	8" wide	G		65	.123		3.22	6.05		9.27	12.85
1350	10" wide	G		60	.133		3.98	6.55		10.53	14.40
1360	12" wide	G		55	.145		4.59	7.15		11.74	16
1440	14 ga. track x 8" wide	G		60	.133		4.01	6.55		10.56	14.45
1450	10" wide	G		55	.145		4.98	7.15		12.13	16.45
1460	12" wide	G		50	.160		5.75	7.90		13.65	18.40
1550	12 ga. track x 10" wide	G		45	.178		7.30	8.75		16.05	21.50
1560	12" wide	G		40	.200		7.45	9.85		17.30	23.50
2230	24" O.C., 18 ga. track x 6" wide	G		80	.100		2.96	4.93		7.89	10.80
2240	8" wide	G		75	.107		3.65	5.25		8.90	12.05
2250	10" wide	G		70	.114		4.56	5.65		10.21	13.60
2260	12" wide	G		65	.123		5.15	6.05		11.20	14.95
2330	16 ga. track x 6" wide	G		70	.114		3.72	5.65		9.37	12.70
2340	8" wide	G		65	.123		4.65	6.05		10.70	14.40
2350	10" wide	G		60	.133		5.75	6.55		12.30	16.40
2360	12" wide	G		55	.145		6.65	7.15		13.80	18.25
2440	14 ga. track x 8" wide	G		60	.133		5.80	6.55		12.35	16.45
2450	10" wide	G		55	.145		7.20	7.15		14.35	18.90
2460	12" wide	G		50	.160		8.35	7.90		16.25	21
2550	12 ga. track x 10" wide	G		45	.178		10.55	8.75		19.30	25
2560	12" wide	G		40	.200		10.80	9.85		20.65	27

05 42 13.25 Framing, Band Joist

		Crew	Daily Output	Labor-Hours	Unit	Material	2017 Bare Costs Labor	Equipment	Total	Total Incl O&P
0010	**FRAMING, BAND JOIST** (track) fastened to bearing wall										
0015	Made from recycled materials										
0220	18 ga. track x 6" deep	G	2 Carp	1000	.016	L.F.	1.30	.79		2.09	2.64
0230	8" deep	G		920	.017		1.61	.86		2.47	3.08
0240	10" deep	G		860	.019		2.01	.92		2.93	3.61
0320	16 ga. track x 6" deep	G		900	.018		1.64	.88		2.52	3.14
0330	8" deep	G		840	.019		2.05	.94		2.99	3.69
0340	10" deep	G		780	.021		2.53	1.01		3.54	4.33
0350	12" deep	G		740	.022		2.92	1.06		3.98	4.84
0430	14 ga. track x 8" deep	G		750	.021		2.55	1.05		3.60	4.42
0440	10" deep	G		720	.022		3.17	1.09		4.26	5.15
0450	12" deep	G		700	.023		3.66	1.13		4.79	5.75
0540	12 ga. track x 10" deep	G		670	.024		4.64	1.18		5.82	6.90
0550	12" deep	G		650	.025		4.75	1.21		5.96	7.05

05 42 13.30 Framing, Boxed Headers/Beams

		Crew	Daily Output	Labor-Hours	Unit	Material	2017 Bare Costs Labor	Equipment	Total	Total Incl O&P
0010	**FRAMING, BOXED HEADERS/BEAMS**										
0015	Made from recycled materials										
0200	Double, 18 ga. x 6" deep	G	2 Carp	220	.073	L.F.	5.50	3.58		9.08	11.55
0210	8" deep	G		210	.076		6.05	3.75		9.80	12.40
0220	10" deep	G		200	.080		7.40	3.94		11.34	14.20
0230	12" deep	G		190	.084		8.05	4.15		12.20	15.20
0300	16 ga. x 8" deep	G		180	.089		7	4.38		11.38	14.40
0310	10" deep	G		170	.094		8.45	4.64		13.09	16.35
0320	12" deep	G		160	.100		9.15	4.93		14.08	17.65
0400	14 ga. x 10" deep	G		140	.114		9.75	5.65		15.40	19.30

05 42 Cold-Formed Metal Joist Framing

05 42 13 – Cold-Formed Metal Floor Joist Framing

05 42 13.30 Framing, Boxed Headers/Beams

			Crew	Daily Output	Labor-Hours	Unit	Material	2017 Bare Costs Labor	Equipment	Total	Total Incl O&P
0410	12" deep	G	2 Carp	130	.123	L.F.	10.65	6.05		16.70	21
0500	12 ga. x 10" deep	G		110	.145		12.85	7.15		20	25
0510	12" deep	G		100	.160		14.15	7.90		22.05	27.50
1210	Triple, 18 ga. x 8" deep	G		170	.094		8.75	4.64		13.39	16.75
1220	10" deep	G		165	.097		10.60	4.78		15.38	18.95
1230	12" deep	G		160	.100		11.60	4.93		16.53	20.50
1300	16 ga. x 8" deep	G		145	.110		10.15	5.45		15.60	19.50
1310	10" deep	G		140	.114		12.15	5.65		17.80	22
1320	12" deep	G		135	.119		13.25	5.85		19.10	23.50
1400	14 ga. x 10" deep	G		115	.139		14.10	6.85		20.95	26
1410	12" deep	G		110	.145		15.50	7.15		22.65	28
1500	12 ga. x 10" deep	G		90	.178		18.75	8.75		27.50	34
1510	12" deep	G	↓	85	.188	↓	21	9.25		30.25	37

05 42 13.40 Framing, Joists

			Crew	Daily Output	Labor-Hours	Unit	Material	2017 Bare Costs Labor	Equipment	Total	Total Incl O&P
0010	**FRAMING, JOISTS**, no band joists (track), web stiffeners, headers,										
0020	Beams, bridging or bracing										
0025	Made from recycled materials										
0030	Joists (2" flange) and fasteners, materials only										
0220	18 ga. x 6" deep	G				L.F.	1.69			1.69	1.86
0230	8" deep	G					1.98			1.98	2.18
0240	10" deep	G					2.34			2.34	2.58
0320	16 ga. x 6" deep	G					2.07			2.07	2.28
0330	8" deep	G					2.47			2.47	2.71
0340	10" deep	G					2.89			2.89	3.18
0350	12" deep	G					3.28			3.28	3.60
0430	14 ga. x 8" deep	G					3.10			3.10	3.41
0440	10" deep	G					3.57			3.57	3.93
0450	12" deep	G					4.06			4.06	4.47
0540	12 ga. x 10" deep	G					5.20			5.20	5.70
0550	12" deep	G				↓	5.90			5.90	6.50
1010	Installation of joists to band joists, beams & headers, labor only										
1220	18 ga. x 6" deep		2 Carp	110	.145	Ea.		7.15		7.15	10.95
1230	8" deep			90	.178			8.75		8.75	13.40
1240	10" deep			80	.200			9.85		9.85	15.10
1320	16 ga. x 6" deep			95	.168			8.30		8.30	12.70
1330	8" deep			70	.229			11.25		11.25	17.25
1340	10" deep			60	.267			13.15		13.15	20
1350	12" deep			55	.291			14.35		14.35	22
1430	14 ga. x 8" deep			65	.246			12.10		12.10	18.55
1440	10" deep			45	.356			17.50		17.50	27
1450	12" deep			35	.457			22.50		22.50	34.50
1540	12 ga. x 10" deep			40	.400			19.70		19.70	30
1550	12" deep		↓	30	.533	↓		26.50		26.50	40

05 42 13.45 Framing, Web Stiffeners

			Crew	Daily Output	Labor-Hours	Unit	Material	2017 Bare Costs Labor	Equipment	Total	Total Incl O&P
0010	**FRAMING, WEB STIFFENERS** at joist bearing, fabricated from										
0020	Stud piece (1-5/8" flange) to stiffen joist (2" flange)										
0025	Made from recycled materials										
2120	For 6" deep joist, with 18 ga. x 2-1/2" stud	G	1 Carp	120	.067	Ea.	.92	3.28		4.20	6.05
2130	3-5/8" stud	G		110	.073		1.08	3.58		4.66	6.70
2140	4" stud	G		105	.076		1.12	3.75		4.87	7
2150	6" stud	G		100	.080		1.42	3.94		5.36	7.60
2160	8" stud	G	↓	95	.084	↓	1.71	4.15		5.86	8.25

05 42 13.45 Framing, Web Stiffeners		Crew	Daily Output	Labor-Hours	Unit	Material	2017 Bare Costs Labor	Equipment	Total	Total Incl O&P
2220	8" deep joist, with 2-1/2" stud	G	1 Carp	120	.067	Ea.	1.23	3.28		4.51	6.40
2230	3-5/8" stud	G		110	.073		1.45	3.58		5.03	7.10
2240	4" stud	G		105	.076		1.50	3.75		5.25	7.40
2250	6" stud	G		100	.080		1.90	3.94		5.84	8.15
2260	8" stud	G		95	.084		2.29	4.15		6.44	8.85
2320	10" deep joist, with 2-1/2" stud	G		110	.073		1.53	3.58		5.11	7.20
2330	3-5/8" stud	G		100	.080		1.79	3.94		5.73	8
2340	4" stud	G		95	.084		1.86	4.15		6.01	8.40
2350	6" stud	G		90	.089		2.36	4.38		6.74	9.30
2360	8" stud	G		85	.094		2.84	4.64		7.48	10.20
2420	12" deep joist, with 2-1/2" stud	G		110	.073		1.84	3.58		5.42	7.50
2430	3-5/8" stud	G		100	.080		2.16	3.94		6.10	8.45
2440	4" stud	G		95	.084		2.24	4.15		6.39	8.80
2450	6" stud	G		90	.089		2.84	4.38		7.22	9.80
2460	8" stud	G		85	.094		3.42	4.64		8.06	10.85
3130	For 6" deep joist, with 16 ga. x 3-5/8" stud	G		100	.080		1.34	3.94		5.28	7.50
3140	4" stud	G		95	.084		1.40	4.15		5.55	7.90
3150	6" stud	G		90	.089		1.75	4.38		6.13	8.65
3160	8" stud	G		85	.094		2.14	4.64		6.78	9.45
3230	8" deep joist, with 3-5/8" stud	G		100	.080		1.80	3.94		5.74	8.05
3240	4" stud	G		95	.084		1.88	4.15		6.03	8.40
3250	6" stud	G		90	.089		2.35	4.38		6.73	9.30
3260	8" stud	G		85	.094		2.87	4.64		7.51	10.25
3330	10" deep joist, with 3-5/8" stud	G		85	.094		2.22	4.64		6.86	9.55
3340	4" stud	G		80	.100		2.32	4.93		7.25	10.10
3350	6" stud	G		75	.107		2.91	5.25		8.16	11.25
3360	8" stud	G		70	.114		3.55	5.65		9.20	12.50
3430	12" deep joist, with 3-5/8" stud	G		85	.094		2.68	4.64		7.32	10.05
3440	4" stud	G		80	.100		2.80	4.93		7.73	10.65
3450	6" stud	G		75	.107		3.50	5.25		8.75	11.90
3460	8" stud	G		70	.114		4.28	5.65		9.93	13.30
4230	For 8" deep joist, with 14 ga. x 3-5/8" stud	G		90	.089		2.22	4.38		6.60	9.15
4240	4" stud	G		85	.094		2.35	4.64		6.99	9.70
4250	6" stud	G		80	.100		2.95	4.93		7.88	10.80
4260	8" stud	G		75	.107		3.15	5.25		8.40	11.50
4330	10" deep joist, with 3-5/8" stud	G		75	.107		2.76	5.25		8.01	11.10
4340	4" stud	G		70	.114		2.91	5.65		8.56	11.80
4350	6" stud	G		65	.123		3.65	6.05		9.70	13.30
4360	8" stud	G		60	.133		3.90	6.55		10.45	14.35
4430	12" deep joist, with 3-5/8" stud	G		75	.107		3.32	5.25		8.57	11.70
4440	4" stud	G		70	.114		3.50	5.65		9.15	12.45
4450	6" stud	G		65	.123		4.40	6.05		10.45	14.15
4460	8" stud	G		60	.133		4.70	6.55		11.25	15.20
5330	For 10" deep joist, with 12 ga. x 3-5/8" stud	G		65	.123		3.97	6.05		10.02	13.65
5340	4" stud	G		60	.133		4.23	6.55		10.78	14.70
5350	6" stud	G		55	.145		5.35	7.15		12.50	16.85
5360	8" stud	G		50	.160		6.45	7.90		14.35	19.15
5430	12" deep joist, with 3-5/8" stud	G		65	.123		4.78	6.05		10.83	14.55
5440	4" stud	G		60	.133		5.10	6.55		11.65	15.65
5450	6" stud	G		55	.145		6.45	7.15		13.60	18.05
5460	8" stud	G		50	.160		7.80	7.90		15.70	20.50

05 42 Cold-Formed Metal Joist Framing

05 42 23 – Cold-Formed Metal Roof Joist Framing

05 42 23.05 Framing, Bracing

05 42 23.05 Framing, Bracing		Crew	Daily Output	Labor-Hours	Unit	Material	2017 Bare Costs Labor	Equipment	Total	Total Incl O&P
0010	**FRAMING, BRACING**									
0015	Made from recycled materials									
0020	Continuous bracing, per row									
0100	16 ga. x 1-1/2" channel thru rafters/trusses @ 16" O.C.	G	1 Carp	4.50	1.778	C.L.F.	51.50	87.50	139	191
0120	24" O.C.	G		6	1.333		51.50	65.50	117	158
0300	2" x 2" angle x 18 ga., rafters/trusses @ 16" O.C.	G		6	1.333		80.50	65.50	146	190
0320	24" O.C.	G		8	1		80.50	49.50	130	164
0400	16 ga., rafters/trusses @ 16" O.C.	G		4.50	1.778		101	87.50	188.50	245
0420	24" O.C.	G	▼	6.50	1.231	▼	101	60.50	161.50	204

05 42 23.10 Framing, Bridging

05 42 23.10 Framing, Bridging		Crew	Daily Output	Labor-Hours	Unit	Material	2017 Bare Costs Labor	Equipment	Total	Total Incl O&P
0010	**FRAMING, BRIDGING**									
0015	Made from recycled materials									
0020	Solid, between rafters w/1-1/4" leg track, per rafter bay									
1200	Rafters 16" O.C., 18 ga. x 4" deep	G	1 Carp	60	.133	Ea.	1.57	6.55	8.12	11.75
1210	6" deep	G		57	.140		2.05	6.90	8.95	12.85
1220	8" deep	G		55	.145		2.52	7.15	9.67	13.75
1230	10" deep	G		52	.154		3.15	7.60	10.75	15.05
1240	12" deep	G		50	.160		3.56	7.90	11.46	15.95
2200	24" O.C., 18 ga. x 4" deep	G		60	.133		2.27	6.55	8.82	12.55
2210	6" deep	G		57	.140		2.96	6.90	9.86	13.85
2220	8" deep	G		55	.145		3.65	7.15	10.80	14.95
2230	10" deep	G		52	.154		4.56	7.60	12.16	16.60
2240	12" deep	G	▼	50	.160	▼	5.15	7.90	13.05	17.70

05 42 23.50 Framing, Parapets

05 42 23.50 Framing, Parapets		Crew	Daily Output	Labor-Hours	Unit	Material	2017 Bare Costs Labor	Equipment	Total	Total Incl O&P
0010	**FRAMING, PARAPETS**									
0015	Made from recycled materials									
0100	3' high installed on 1st story, 18 ga. x 4" wide studs, 12" O.C.	G	2 Carp	100	.160	L.F.	5.70	7.90	13.60	18.30
0110	16" O.C.	G		150	.107		4.85	5.25	10.10	13.40
0120	24" O.C.	G		200	.080		4.01	3.94	7.95	10.45
0200	6" wide studs, 12" O.C.	G		100	.160		7.30	7.90	15.20	20
0210	16" O.C.	G		150	.107		6.25	5.25	11.50	14.90
0220	24" O.C.	G		200	.080		5.15	3.94	9.09	11.75
1100	Installed on 2nd story, 18 ga. x 4" wide studs, 12" O.C.	G		95	.168		5.70	8.30	14	18.95
1110	16" O.C.	G		145	.110		4.85	5.45	10.30	13.70
1120	24" O.C.	G		190	.084		4.01	4.15	8.16	10.75
1200	6" wide studs, 12" O.C.	G		95	.168		7.30	8.30	15.60	21
1210	16" O.C.	G		145	.110		6.25	5.45	11.70	15.20
1220	24" O.C.	G		190	.084		5.15	4.15	9.30	12.05
2100	Installed on gable, 18 ga. x 4" wide studs, 12" O.C.	G		85	.188		5.70	9.25	14.95	20.50
2110	16" O.C.	G		130	.123		4.85	6.05	10.90	14.65
2120	24" O.C.	G		170	.094		4.01	4.64	8.65	11.50
2200	6" wide studs, 12" O.C.	G		85	.188		7.30	9.25	16.55	22.50
2210	16" O.C.	G		130	.123		6.25	6.05	12.30	16.15
2220	24" O.C.	G	▼	170	.094	▼	5.15	4.64	9.79	12.80

05 42 23.60 Framing, Roof Rafters

05 42 23.60 Framing, Roof Rafters		Crew	Daily Output	Labor-Hours	Unit	Material	2017 Bare Costs Labor	Equipment	Total	Total Incl O&P
0010	**FRAMING, ROOF RAFTERS**									
0015	Made from recycled materials									
0100	Boxed ridge beam, double, 18 ga. x 6" deep	G	2 Carp	160	.100	L.F.	5.50	4.93	10.43	13.60
0110	8" deep	G		150	.107		6.05	5.25	11.30	14.70
0120	10" deep	G		140	.114		7.40	5.65	13.05	16.75
0130	12" deep	G	▼	130	.123	▼	8.05	6.05	14.10	18.15

05 42 Cold-Formed Metal Joist Framing

05 42 23 – Cold-Formed Metal Roof Joist Framing

05 42 23.60 Framing, Roof Rafters

			Crew	Daily Output	Labor-Hours	Unit	Material	2017 Bare Costs Labor	2017 Bare Costs Equipment	Total	Total Incl O&P
0200	16 ga. x 6" deep	G	2 Carp	150	.107	L.F.	6.20	5.25		11.45	14.90
0210	8" deep	G		140	.114		7	5.65		12.65	16.30
0220	10" deep	G		130	.123		8.45	6.05		14.50	18.55
0230	12" deep	G		120	.133		9.15	6.55		15.70	20
1100	Rafters, 2" flange, material only, 18 ga. x 6" deep	G					1.69			1.69	1.86
1110	8" deep	G					1.98			1.98	2.18
1120	10" deep	G					2.34			2.34	2.58
1130	12" deep	G					2.70			2.70	2.97
1200	16 ga. x 6" deep	G					2.07			2.07	2.28
1210	8" deep	G					2.47			2.47	2.71
1220	10" deep	G					2.89			2.89	3.18
1230	12" deep	G					3.28			3.28	3.60
2100	Installation only, ordinary rafter to 4:12 pitch, 18 ga. x 6" deep		2 Carp	35	.457	Ea.		22.50		22.50	34.50
2110	8" deep			30	.533			26.50		26.50	40
2120	10" deep			25	.640			31.50		31.50	48.50
2130	12" deep			20	.800			39.50		39.50	60.50
2200	16 ga. x 6" deep			30	.533			26.50		26.50	40
2210	8" deep			25	.640			31.50		31.50	48.50
2220	10" deep			20	.800			39.50		39.50	60.50
2230	12" deep			15	1.067			52.50		52.50	80.50
8100	Add to labor, ordinary rafters on steep roofs							25%			
8110	Dormers & complex roofs							50%			
8200	Hip & valley rafters to 4:12 pitch							25%			
8210	Steep roofs							50%			
8220	Dormers & complex roofs							75%			
8300	Hip & valley jack rafters to 4:12 pitch							50%			
8310	Steep roofs							75%			
8320	Dormers & complex roofs							100%			

05 42 23.70 Framing, Soffits and Canopies

			Crew	Daily Output	Labor-Hours	Unit	Material	2017 Bare Costs Labor	2017 Bare Costs Equipment	Total	Total Incl O&P
0010	**FRAMING, SOFFITS & CANOPIES**										
0015	Made from recycled materials										
0130	Continuous ledger track @ wall, studs @ 16" O.C., 18 ga. x 4" wide	G	2 Carp	535	.030	L.F.	1.05	1.47		2.52	3.41
0140	6" wide	G		500	.032		1.36	1.58		2.94	3.91
0150	8" wide	G		465	.034		1.68	1.69		3.37	4.45
0160	10" wide	G		430	.037		2.10	1.83		3.93	5.10
0230	Studs @ 24" O.C., 18 ga. x 4" wide	G		800	.020		1	.99		1.99	2.61
0240	6" wide	G		750	.021		1.30	1.05		2.35	3.04
0250	8" wide	G		700	.023		1.61	1.13		2.74	3.49
0260	10" wide	G		650	.025		2.01	1.21		3.22	4.07
1000	Horizontal soffit and canopy members, material only										
1030	1-5/8" flange studs, 18 ga. x 4" deep	G				L.F.	1.34			1.34	1.48
1040	6" deep	G					1.70			1.70	1.87
1050	8" deep	G					2.05			2.05	2.26
1140	2" flange joists, 18 ga. x 6" deep	G					1.93			1.93	2.13
1150	8" deep	G					2.27			2.27	2.49
1160	10" deep	G					2.68			2.68	2.94
4030	Installation only, 18 ga., 1-5/8" flange x 4" deep		2 Carp	130	.123	Ea.		6.05		6.05	9.30
4040	6" deep			110	.145			7.15		7.15	10.95
4050	8" deep			90	.178			8.75		8.75	13.40
4140	2" flange, 18 ga. x 6" deep			110	.145			7.15		7.15	10.95
4150	8" deep			90	.178			8.75		8.75	13.40
4160	10" deep			80	.200			9.85		9.85	15.10

For customer support on your Building Construction Costs with RSMeans Data, call 800.448.8182.

05 42 Cold-Formed Metal Joist Framing

05 42 23 – Cold-Formed Metal Roof Joist Framing

05 42 23.70 Framing, Soffits and Canopies		Crew	Daily Output	Labor-Hours	Unit	Material	2017 Bare Costs Labor	Equipment	Total	Total Incl O&P
6010	Clips to attach fascia to rafter tails, 2" x 2" x 18 ga. angle	G 1 Carp	120	.067	Ea.	.95	3.28		4.23	6.10
6020	16 ga. angle	G "	100	.080	↓	1.20	3.94		5.14	7.35

05 44 Cold-Formed Metal Trusses

05 44 13 – Cold-Formed Metal Roof Trusses

05 44 13.60 Framing, Roof Trusses

		Crew	Daily Output	Labor-Hours	Unit	Material	2017 Bare Costs Labor	Equipment	Total	Total Incl O&P
0010	**FRAMING, ROOF TRUSSES**									
0015	Made from recycled materials									
0020	Fabrication of trusses on ground, Fink (W) or King Post, to 4:12 pitch									
0120	18 ga. x 4" chords, 16' span	G 2 Carp	12	1.333	Ea.	62.50	65.50		128	170
0130	20' span	G	11	1.455		78.50	71.50		150	196
0140	24' span	G	11	1.455		94	71.50		165.50	213
0150	28' span	G	10	1.600		110	79		189	242
0160	32' span	G	10	1.600		125	79		204	259
0250	6" chords, 28' span	G	9	1.778		139	87.50		226.50	287
0260	32' span	G	9	1.778		159	87.50		246.50	310
0270	36' span	G	8	2		179	98.50		277.50	350
0280	40' span	G	8	2		199	98.50		297.50	370
1120	5:12 to 8:12 pitch, 18 ga. x 4" chords, 16' span	G	10	1.600		71.50	79		150.50	200
1130	20' span	G	9	1.778		89.50	87.50		177	233
1140	24' span	G	9	1.778		108	87.50		195.50	252
1150	28' span	G	8	2		125	98.50		223.50	289
1160	32' span	G	8	2		143	98.50		241.50	310
1250	6" chords, 28' span	G	7	2.286		159	113		272	345
1260	32' span	G	7	2.286		182	113		295	370
1270	36' span	G	6	2.667		204	131		335	425
1280	40' span	G	6	2.667		227	131		358	450
2120	9:12 to 12:12 pitch, 18 ga. x 4" chords, 16' span	G	8	2		89.50	98.50		188	250
2130	20' span	G	7	2.286		112	113		225	295
2140	24' span	G	7	2.286		134	113		247	320
2150	28' span	G	6	2.667		157	131		288	375
2160	32' span	G	6	2.667		179	131		310	400
2250	6" chords, 28' span	G	5	3.200		199	158		357	460
2260	32' span	G	5	3.200		227	158		385	490
2270	36' span	G	4	4		256	197		453	580
2280	40' span	G	4	4		284	197		481	610
5120	Erection only of roof trusses, to 4:12 pitch, 16' span	F-6	48	.833			39	11.95	50.95	72
5130	20' span		46	.870			40.50	12.45	52.95	75.50
5140	24' span		44	.909			42.50	13	55.50	79
5150	28' span		42	.952			44.50	13.65	58.15	82.50
5160	32' span		40	1			46.50	14.30	60.80	87
5170	36' span		38	1.053			49	15.05	64.05	91.50
5180	40' span		36	1.111			51.50	15.90	67.40	96.50
5220	5:12 to 8:12 pitch, 16' span		42	.952			44.50	13.65	58.15	82.50
5230	20' span		40	1			46.50	14.30	60.80	87
5240	24' span		38	1.053			49	15.05	64.05	91.50
5250	28' span		36	1.111			51.50	15.90	67.40	96.50
5260	32' span		34	1.176			54.50	16.85	71.35	102
5270	36' span		32	1.250			58	17.90	75.90	109
5280	40' span		30	1.333			62	19.10	81.10	116
5320	9:12 to 12:12 pitch, 16' span	↓	36	1.111	↓		51.50	15.90	67.40	96.50

For customer support on your Building Construction Costs with RSMeans Data, call 800.448.8182.

153

05 44 Cold-Formed Metal Trusses

05 44 13 – Cold-Formed Metal Roof Trusses

05 44 13.60 Framing, Roof Trusses

		Crew	Daily Output	Labor-Hours	Unit	Material	2017 Bare Costs Labor	Equipment	Total	Total Incl O&P
5330	20' span	F-6	34	1.176	Ea.	54.50	16.85		71.35	102
5340	24' span		32	1.250		58	17.90		75.90	109
5350	28' span		30	1.333		62	19.10		81.10	116
5360	32' span		28	1.429		66.50	20.50		87	124
5370	36' span		26	1.538		71.50	22		93.50	133
5380	40' span		24	1.667		77.50	24		101.50	145

05 51 Metal Stairs

05 51 13 – Metal Pan Stairs

05 51 13.50 Pan Stairs

			Crew	Daily Output	Labor-Hours	Unit	Material	2017 Bare Costs Labor	Equipment	Total	Total Incl O&P
0010	**PAN STAIRS**, shop fabricated, steel stringers										
0015	Made from recycled materials										
0200	Metal pan tread for concrete in-fill, picket rail, 3'-6" wide	G	E-4	35	.914	Riser	600	50	3.71	653.71	750
0300	4'-0" wide	G		30	1.067		670	58.50	4.33	732.83	835
0350	Wall rail, both sides, 3'-6" wide	G		53	.604		395	33	2.45	430.45	495
1500	Landing, steel pan, conventional	G		160	.200	S.F.	79.50	10.95	.81	91.26	107
1600	Pre-erected	G		255	.125	"	142	6.90	.51	149.41	168
1700	Pre-erected, steel pan tread, 3'-6" wide, 2 line pipe rail	G	E-2	87	.644	Riser	660	34.50	19.20	713.70	805

05 51 16 – Metal Floor Plate Stairs

05 51 16.50 Floor Plate Stairs

			Crew	Daily Output	Labor-Hours	Unit	Material	2017 Bare Costs Labor	Equipment	Total	Total Incl O&P
0010	**FLOOR PLATE STAIRS**, shop fabricated, steel stringers										
0015	Made from recycled materials										
0400	Cast iron tread and pipe rail, 3'-6" wide	G	E-4	35	.914	Riser	640	50	3.71	693.71	795
0500	Checkered plate tread, industrial, 3'-6" wide	G		28	1.143		395	62.50	4.64	462.14	545
0550	Circular, for tanks, 3'-0" wide	G		33	.970		445	53	3.94	501.94	585
0600	For isolated stairs, add							100%			
0800	Custom steel stairs, 3'-6" wide, economy	G	E-4	35	.914		600	50	3.71	653.71	750
0810	Medium priced	G		30	1.067		795	58.50	4.33	857.83	975
0900	Deluxe	G		20	1.600		995	87.50	6.50	1,089	1,250
1100	For 4' wide stairs, add							5%	5%		
1300	For 5' wide stairs, add							10%	10%		

05 51 19 – Metal Grating Stairs

05 51 19.50 Grating Stairs

			Crew	Daily Output	Labor-Hours	Unit	Material	2017 Bare Costs Labor	Equipment	Total	Total Incl O&P
0010	**GRATING STAIRS**, shop fabricated, steel stringers, safety nosing on treads										
0015	Made from recycled materials										
0020	Grating tread and pipe railing, 3'-6" wide	G	E-4	35	.914	Riser	395	50	3.71	448.71	525
0100	4'-0" wide	G	"	30	1.067	"	515	58.50	4.33	577.83	665

05 51 23 – Metal Fire Escapes

05 51 23.25 Fire Escapes

			Crew	Daily Output	Labor-Hours	Unit	Material	2017 Bare Costs Labor	Equipment	Total	Total Incl O&P
0010	**FIRE ESCAPES**, shop fabricated										
0200	2' wide balcony, 1" x 1/4" bars 1-1/2" O.C., with railing	G	2 Sswk	10	1.600	L.F.	58	87		145	209
0400	1st story cantilevered stair, standard, with railing	G		.50	32	Ea.	2,400	1,750		4,150	5,550
0500	Cable counterweighted, with railing	G		.40	40	"	2,225	2,175		4,400	6,075
0700	36" x 40" platform & fixed stair, with railing	G		.40	40	Flight	1,075	2,175		3,250	4,800
0900	For 3'-6" wide escapes, add to above							100%	150%		

05 51 Metal Stairs

05 51 23 – Metal Fire Escapes

05 51 23.50 Fire Escape Stairs

		Crew	Daily Output	Labor-Hours	Unit	Material	2017 Bare Costs Labor	2017 Bare Costs Equipment	Total	Total Incl O&P
0010	**FIRE ESCAPE STAIRS**, portable									
0100	Portable ladder				Ea.	112			112	123

05 51 33 – Metal Ladders

05 51 33.13 Vertical Metal Ladders

			Crew	Daily Output	Labor-Hours	Unit	Material	2017 Bare Costs Labor	2017 Bare Costs Equipment	Total	Total Incl O&P
0010	**VERTICAL METAL LADDERS**, shop fabricated										
0015	Made from recycled materials										
0020	Steel, 20" wide, bolted to concrete, with cage	G	E-4	50	.640	V.L.F.	67	35	2.60	104.60	135
0100	Without cage	G		85	.376		36	20.50	1.53	58.03	75.50
0300	Aluminum, bolted to concrete, with cage	G		50	.640		128	35	2.60	165.60	201
0400	Without cage	G		85	.376		51	20.50	1.53	73.03	92

05 51 33.16 Inclined Metal Ladders

			Crew	Daily Output	Labor-Hours	Unit	Material	2017 Bare Costs Labor	2017 Bare Costs Equipment	Total	Total Incl O&P
0010	**INCLINED METAL LADDERS**, shop fabricated										
0015	Made from recycled materials										
3900	Industrial ships ladder, steel, 24" W, grating treads, 2 line pipe rail	G	E-4	30	1.067	Riser	214	58.50	4.33	276.83	335
4000	Aluminum	G	"	30	1.067	"	299	58.50	4.33	361.83	430

05 51 33.23 Alternating Tread Ladders

		Crew	Daily Output	Labor-Hours	Unit	Material	2017 Bare Costs Labor	2017 Bare Costs Equipment	Total	Total Incl O&P
0010	**ALTERNATING TREAD LADDERS**, shop fabricated									
0015	Made from recycled materials									
0800	Alternating tread ladders, 68-degree angle of incline									
0810	8 foot vertical rise, steel, 149 lb., standard paint color	B-68G	3	5.333	Ea.	2,300	290	102	2,692	3,125
0820	Non-standard paint color		3	5.333		2,650	290	102	3,042	3,525
0830	Galvanized		3	5.333		2,675	290	102	3,067	3,550
0840	Stainless		3	5.333		3,875	290	102	4,267	4,875
0850	Aluminum, 87 lb.		3	5.333		2,825	290	102	3,217	3,725
1010	10 foot vertical rise, steel, 181 lb., standard paint color		2.75	5.818		2,775	315	111	3,201	3,700
1020	Non-standard paint color		2.75	5.818		3,175	315	111	3,601	4,150
1030	Galvanized		2.75	5.818		3,225	315	111	3,651	4,200
1040	Stainless		2.75	5.818		4,675	315	111	5,101	5,800
1050	Aluminum, 103 lb.		2.75	5.818		3,425	315	111	3,851	4,400
1210	12 foot vertical rise, steel, 245 lb., standard paint color		2.50	6.400		3,250	350	122	3,722	4,300
1220	Non-standard paint color		2.50	6.400		3,700	350	122	4,172	4,800
1230	Galvanized		2.50	6.400		3,800	350	122	4,272	4,900
1240	Stainless		2.50	6.400		5,450	350	122	5,922	6,725
1250	Aluminum, 103 lb.		2.50	6.400		4,000	350	122	4,472	5,125
1410	14 foot vertical rise, steel, 281 lb., standard paint color		2.25	7.111		3,725	385	136	4,246	4,900
1420	Non-standard paint color		2.25	7.111		4,225	385	136	4,746	5,450
1430	Galvanized		2.25	7.111		4,350	385	136	4,871	5,600
1440	Stainless		2.25	7.111		6,250	385	136	6,771	7,675
1450	Aluminum, 136 lb.		2.25	7.111		4,575	385	136	5,096	5,825
1610	16 foot vertical rise, steel, 317 lb., standard paint color		2	8		4,225	435	153	4,813	5,550
1620	Non-standard paint color		2	8		4,725	435	153	5,313	6,100
1630	Galvanized		2	8		4,925	435	153	5,513	6,300
1640	Stainless		2	8		7,050	435	153	7,638	8,650
1650	Aluminum, 153 lb.		2	8		5,150	435	153	5,738	6,575

For customer support on your Building Construction Costs with RSMeans Data, call 800.448.8182.

155

05 52 Metal Railings

05 52 13 – Pipe and Tube Railings

05 52 13.50 Railings, Pipe		Crew	Daily Output	Labor-Hours	Unit	Material	2017 Bare Costs Labor	Equipment	Total	Total Incl O&P
0010	**RAILINGS, PIPE**, shop fab'd, 3'-6" high, posts @ 5' O.C.									
0015	Made from recycled materials									
0020	Aluminum, 2 rail, satin finish, 1-1/4" diameter	G	E-4	160	.200	L.F.	52	10.95	.81	63.76	76
0030	Clear anodized	G		160	.200		64	10.95	.81	75.76	89.50
0040	Dark anodized	G		160	.200		71	10.95	.81	82.76	97
0080	1-1/2" diameter, satin finish	G		160	.200		61	10.95	.81	72.76	86
0090	Clear anodized	G		160	.200		68.50	10.95	.81	80.26	94.50
0100	Dark anodized	G		160	.200		75.50	10.95	.81	87.26	102
0140	Aluminum, 3 rail, 1-1/4" diam., satin finish	G		137	.234		67.50	12.80	.95	81.25	97
0150	Clear anodized	G		137	.234		84	12.80	.95	97.75	115
0160	Dark anodized	G		137	.234		92.50	12.80	.95	106.25	125
0200	1-1/2" diameter, satin finish	G		137	.234		80	12.80	.95	93.75	111
0210	Clear anodized	G		137	.234		91.50	12.80	.95	105.25	124
0220	Dark anodized	G		137	.234		100	12.80	.95	113.75	133
0500	Steel, 2 rail, on stairs, primed, 1-1/4" diameter	G		160	.200		26	10.95	.81	37.76	47.50
0520	1-1/2" diameter	G		160	.200		28	10.95	.81	39.76	50
0540	Galvanized, 1-1/4" diameter	G		160	.200		35	10.95	.81	46.76	57.50
0560	1-1/2" diameter	G		160	.200		39.50	10.95	.81	51.26	62.50
0580	Steel, 3 rail, primed, 1-1/4" diameter	G		137	.234		38.50	12.80	.95	52.25	65
0600	1-1/2" diameter	G		137	.234		40.50	12.80	.95	54.25	67
0620	Galvanized, 1-1/4" diameter	G		137	.234		54	12.80	.95	67.75	82
0640	1-1/2" diameter	G		137	.234		62.50	12.80	.95	76.25	91.50
0700	Stainless steel, 2 rail, 1-1/4" diam. #4 finish	G		137	.234		114	12.80	.95	127.75	148
0720	High polish	G		137	.234		184	12.80	.95	197.75	225
0740	Mirror polish	G		137	.234		230	12.80	.95	243.75	276
0760	Stainless steel, 3 rail, 1-1/2" diam., #4 finish	G		120	.267		171	14.60	1.08	186.68	214
0770	High polish	G		120	.267		284	14.60	1.08	299.68	335
0780	Mirror finish	G		120	.267		345	14.60	1.08	360.68	405
0900	Wall rail, alum. pipe, 1-1/4" diam., satin finish	G		213	.150		25	8.25	.61	33.86	42
0905	Clear anodized	G		213	.150		31	8.25	.61	39.86	48.50
0910	Dark anodized	G		213	.150		37	8.25	.61	45.86	55
0915	1-1/2" diameter, satin finish	G		213	.150		28	8.25	.61	36.86	45.50
0920	Clear anodized	G		213	.150		35	8.25	.61	43.86	53
0925	Dark anodized	G		213	.150		43.50	8.25	.61	52.36	62.50
0930	Steel pipe, 1-1/4" diameter, primed	G		213	.150		15.50	8.25	.61	24.36	31.50
0935	Galvanized	G		213	.150		22.50	8.25	.61	31.36	39.50
0940	1-1/2" diameter	G		176	.182		16	9.95	.74	26.69	35
0945	Galvanized	G		213	.150		22.50	8.25	.61	31.36	39.50
0955	Stainless steel pipe, 1-1/2" diam., #4 finish	G		107	.299		91	16.40	1.21	108.61	129
0960	High polish	G		107	.299		185	16.40	1.21	202.61	233
0965	Mirror polish	G		107	.299		219	16.40	1.21	236.61	270
2000	2-line pipe rail (1-1/2" T&B) with 1/2" pickets @ 4-1/2" O.C.,										
2005	attached handrail on brackets										
2010	42" high aluminum, satin finish, straight & level	G	E-4	120	.267	L.F.	266	14.60	1.08	281.68	320
2050	42" high steel, primed, straight & level	G	"	120	.267		131	14.60	1.08	146.68	170
4000	For curved and level rails, add						10%	10%			
4100	For sloped rails for stairs, add						30%	30%			

05 52 16 – Industrial Railings

05 52 16.50 Railings, Industrial		Crew	Daily Output	Labor-Hours	Unit	Material	2017 Bare Costs Labor	Equipment	Total	Total Incl O&P
0010	**RAILINGS, INDUSTRIAL**, shop fab'd, 3'-6" high, posts @ 5' O.C.									
0020	2 rail, 3'-6" high, 1-1/2" pipe	G	E-4	255	.125	L.F.	39.50	6.90	.51	46.91	55.50
0100	2" angle rail	G	"	255	.125		35.50	6.90	.51	42.91	51

05 52 Metal Railings

05 52 16 – Industrial Railings

05 52 16.50 Railings, Industrial		Crew	Daily Output	Labor-Hours	Unit	Material	2017 Bare Costs Labor	Equipment	Total	Total Incl O&P
0200	For 4" high kick plate, 10 ga., add	G				L.F.	5.60			5.60	6.15
0300	1/4" thick, add	G					7.35			7.35	8.05
0500	For curved level rails, add						10%	10%			
0550	For sloped rails for stairs, add					↓	30%	30%			

05 53 Metal Gratings

05 53 13 – Bar Gratings

05 53 13.10 Floor Grating, Aluminum

			Crew	Daily Output	Labor-Hours	Unit	Material	2017 Bare Costs Labor	Equipment	Total	Total Incl O&P
0010	**FLOOR GRATING, ALUMINUM**, field fabricated from panels										
0015	Made from recycled materials										
0110	Bearing bars @ 1-3/16" O.C., cross bars @ 4" O.C.,										
0111	Up to 300 S.F., 1" x 1/8" bar	G	E-4	900	.036	S.F.	17.05	1.95	.14	19.14	22
0112	Over 300 S.F.	G		850	.038		15.50	2.06	.15	17.71	20.50
0113	1-1/4" x 1/8" bar, up to 300 S.F.	G		800	.040		17.90	2.19	.16	20.25	23.50
0114	Over 300 S.F.	G		1000	.032		16.25	1.75	.13	18.13	21
0122	1-1/4" x 3/16" bar, up to 300 S.F.	G		750	.043		27.50	2.34	.17	30.01	34.50
0124	Over 300 S.F.	G		1000	.032		25	1.75	.13	26.88	30.50
0132	1-1/2" x 3/16" bar, up to 300 S.F.	G		700	.046		31	2.50	.19	33.69	39
0134	Over 300 S.F.	G		1000	.032		28.50	1.75	.13	30.38	34
0136	1-3/4" x 3/16" bar, up to 300 S.F.	G		500	.064		34	3.51	.26	37.77	43.50
0138	Over 300 S.F.	G		1000	.032		31	1.75	.13	32.88	37
0146	2-1/4" x 3/16" bar, up to 300 S.F.	G		600	.053		43	2.92	.22	46.14	52.50
0148	Over 300 S.F.	G		1000	.032		39	1.75	.13	40.88	46
0162	Cross bars @ 2" O.C., 1" x 1/8", up to 300 S.F.	G		600	.053		30	2.92	.22	33.14	38
0164	Over 300 S.F.	G		1000	.032		27.50	1.75	.13	29.38	33
0172	1-1/4" x 3/16" bar, up to 300 S.F.	G		600	.053		49	2.92	.22	52.14	59
0174	Over 300 S.F.	G		1000	.032		44.50	1.75	.13	46.38	52
0182	1-1/2" x 3/16" bar, up to 300 S.F.	G		600	.053		56.50	2.92	.22	59.64	67
0184	Over 300 S.F.	G		1000	.032		51	1.75	.13	52.88	59.50
0186	1-3/4" x 3/16" bar, up to 300 S.F.	G		600	.053		61	2.92	.22	64.14	72
0188	Over 300 S.F.	G	↓	1000	.032	↓	55.50	1.75	.13	57.38	64
0200	For straight cuts, add					L.F.	4.30			4.30	4.73
0300	For curved cuts, add						5.30			5.30	5.85
0400	For straight banding, add	G					5.50			5.50	6.05
0500	For curved banding, add	G					6.60			6.60	7.25
0600	For aluminum checkered plate nosings, add	G					7.15			7.15	7.85
0700	For straight toe plate, add	G					10.80			10.80	11.90
0800	For curved toe plate, add	G					12.60			12.60	13.85
1000	For cast aluminum abrasive nosings, add	G				↓	10.50			10.50	11.55
1200	Expanded aluminum, .65# per S.F.	G	E-4	1050	.030	S.F.	19.90	1.67	.12	21.69	25
1400	Extruded I bars are 10% less than 3/16" bars										
1600	Heavy duty, all extruded plank, 3/4" deep, 1.8# per S.F.	G	E-4	1100	.029	S.F.	26	1.59	.12	27.71	31.50
1700	1-1/4" deep, 2.9# per S.F.	G		1000	.032		30	1.75	.13	31.88	36
1800	1-3/4" deep, 4.2# per S.F.	G		925	.035		41.50	1.90	.14	43.54	49.50
1900	2-1/4" deep, 5.0# per S.F.	G	↓	875	.037	↓	60.50	2	.15	62.65	70
2100	For safety serrated surface, add						15%				

05 53 13.70 Floor Grating, Steel

			Crew	Daily Output	Labor-Hours	Unit	Material	2017 Bare Costs Labor	Equipment	Total	Total Incl O&P
0010	**FLOOR GRATING, STEEL**, field fabricated from panels										
0015	Made from recycled materials										
0300	Platforms, to 12' high, rectangular	G	E-4	3150	.010	Lb.	3.21	.56	.04	3.81	4.51
0400	Circular	G	"	2300	.014	"	4.01	.76	.06	4.83	5.75

05 53 13.70 Floor Grating, Steel

		Crew	Daily Output	Labor-Hours	Unit	Material	2017 Bare Costs Labor	Equipment	Total	Total Incl O&P
0410	Painted bearing bars @ 1-3/16"									
0412	Cross bars @ 4" O.C., 3/4" x 1/8" bar, up to 300 S.F.	E-2	500	.112	S.F.	7.80	6.05	3.34	17.19	22
0414	Over 300 S.F.		750	.075		7.10	4.02	2.23	13.35	16.80
0422	1-1/4" x 3/16", up to 300 S.F.		400	.140		12.15	7.55	4.18	23.88	30
0424	Over 300 S.F.		600	.093		11.05	5.05	2.79	18.89	23.50
0432	1-1/2" x 3/16", up to 300 S.F.		400	.140		14.10	7.55	4.18	25.83	32.50
0434	Over 300 S.F.		600	.093		12.85	5.05	2.79	20.69	25.50
0436	1-3/4" x 3/16", up to 300 S.F.		400	.140		17.70	7.55	4.18	29.43	36.50
0438	Over 300 S.F.		600	.093		16.10	5.05	2.79	23.94	29
0452	2-1/4" x 3/16", up to 300 S.F.		300	.187		21	10.05	5.55	36.60	45.50
0454	Over 300 S.F.		450	.124		19.15	6.70	3.71	29.56	36
0462	Cross bars @ 2" O.C., 3/4" x 1/8", up to 300 S.F.		500	.112		15.15	6.05	3.34	24.54	30
0464	Over 300 S.F.		750	.075		12.60	4.02	2.23	18.85	23
0472	1-1/4" x 3/16", up to 300 S.F.		400	.140		19.80	7.55	4.18	31.53	39
0474	Over 300 S.F.		600	.093		16.50	5.05	2.79	24.34	29.50
0482	1-1/2" x 3/16", up to 300 S.F.		400	.140		22.50	7.55	4.18	34.23	41.50
0484	Over 300 S.F.		600	.093		18.60	5.05	2.79	26.44	31.50
0486	1-3/4" x 3/16", up to 300 S.F.		400	.140		33.50	7.55	4.18	45.23	54
0488	Over 300 S.F.		600	.093		28	5.05	2.79	35.84	42
0502	2-1/4" x 3/16", up to 300 S.F.		300	.187		33	10.05	5.55	48.60	58.50
0504	Over 300 S.F.		450	.124		27.50	6.70	3.71	37.91	45
0690	For galvanized grating, add					25%				
0800	For straight cuts, add				L.F.	6.15			6.15	6.75
0900	For curved cuts, add					7.80			7.80	8.60
1000	For straight banding, add					6.60			6.60	7.25
1100	For curved banding, add					8.70			8.70	9.55
1200	For checkered plate nosings, add					7.65			7.65	8.40
1300	For straight toe or kick plate, add					13.70			13.70	15.05
1400	For curved toe or kick plate, add					15.50			15.50	17.05
1500	For abrasive nosings, add					10.15			10.15	11.15
1510	For stair treads, see Section 05 55 13.50									
1600	For safety serrated surface, bearing bars @ 1-3/16" O.C., add					15%				
1700	Bearing bars @ 15/16" O.C., add					25%				
2000	Stainless steel gratings, close spaced, 1" x 1/8" bars, up to 300 S.F.	E-4	450	.071	S.F.	75.50	3.90	.29	79.69	90.50
2100	Standard spacing, 3/4" x 1/8" bars		500	.064		84	3.51	.26	87.77	98
2200	1-1/4" x 3/16" bars		400	.080		82	4.38	.32	86.70	98
2400	Expanded steel grating, at ground, 3.0# per S.F.		900	.036		8.05	1.95	.14	10.14	12.30
2500	3.14# per S.F.		900	.036		6.70	1.95	.14	8.79	10.75
2600	4.0# per S.F.		850	.038		7.90	2.06	.15	10.11	12.30
2650	4.27# per S.F.		850	.038		9	2.06	.15	11.21	13.50
2700	5.0# per S.F.		800	.040		13.80	2.19	.16	16.15	19
2800	6.25# per S.F.		750	.043		17.80	2.34	.17	20.31	23.50
2900	7.0# per S.F.		700	.046		19.65	2.50	.19	22.34	26
3100	For flattened expanded steel grating, add					8%				
3300	For elevated installation above 15', add						15%			

05 53 16.50 Grating Planks

		Crew	Daily Output	Labor-Hours	Unit	Material	2017 Bare Costs Labor	Equipment	Total	Total Incl O&P
0010	**GRATING PLANKS**, field fabricated from planks									
0020	Aluminum, 9-1/2" wide, 14 ga., 2" rib	E-4	950	.034	L.F.	27.50	1.85	.14	29.49	33.50
0200	Galvanized steel, 9-1/2" wide, 14 ga., 2-1/2" rib		950	.034		16.80	1.85	.14	18.79	21.50
0300	4" rib		950	.034		18.50	1.85	.14	20.49	23.50
0500	12 ga., 2-1/2" rib		950	.034		16.75	1.85	.14	18.74	21.50

05 53 Metal Gratings

05 53 16 – Plank Gratings

05 53 16.50 Grating Planks		Crew	Daily Output	Labor-Hours	Unit	Material	2017 Bare Costs Labor	Equipment	Total	Total Incl O&P
0600	3" rib	G E-4	950	.034	L.F.	22.50	1.85	.14	24.49	27.50
0800	Stainless steel, type 304, 16 ga., 2" rib	G ↓	950	.034	↓	38.50	1.85	.14	40.49	45
0900	Type 316	G ↓	950	.034	↓	57	1.85	.14	58.99	66

05 53 19 – Floor Grating Frame

05 53 19.30 Grating Frame		Crew	Daily Output	Labor-Hours	Unit	Material	2017 Bare Costs Labor	Equipment	Total	Total Incl O&P
0010	**GRATING FRAME**, field fabricated									
0020	Aluminum, for gratings 1" to 1-1/2" deep	G 1 Sswk	70	.114	L.F.	3.41	6.20		9.61	14.10
0100	For each corner, add	G			Ea.	5.10			5.10	5.65

05 54 Metal Floor Plates

05 54 13 – Floor Plates

05 54 13.20 Checkered Plates

		Crew	Daily Output	Labor-Hours	Unit	Material	2017 Bare Costs Labor	Equipment	Total	Total Incl O&P
0010	**CHECKERED PLATES**, steel, field fabricated									
0015	Made from recycled materials									
0020	1/4" & 3/8", 2000 to 5000 S.F., bolted	G E-4	2900	.011	Lb.	.71	.60	.04	1.35	1.84
0100	Welded	G ↓	4400	.007	"	.67	.40	.03	1.10	1.43
0300	Pit or trench cover and frame, 1/4" plate, 2' to 3' wide	G ↓	100	.320	S.F.	8.80	17.55	1.30	27.65	40
0400	For galvanizing, add	G			Lb.	.28			.28	.31
0500	Platforms, 1/4" plate, no handrails included, rectangular	G E-4	4200	.008	↓	3.21	.42	.03	3.66	4.26
0600	Circular	G "	2500	.013	↓	4.01	.70	.05	4.76	5.65

05 54 13.70 Trench Covers

		Crew	Daily Output	Labor-Hours	Unit	Material	2017 Bare Costs Labor	Equipment	Total	Total Incl O&P
0010	**TRENCH COVERS**, field fabricated									
0020	Cast iron grating with bar stops and angle frame, to 18" wide	G 1 Sswk	20	.400	L.F.	213	21.50		234.50	270
0100	Frame only (both sides of trench), 1" grating	G ↓	45	.178	↓	1.19	9.65		10.84	17.40
0150	2" grating	G ↓	35	.229	↓	2.76	12.40		15.16	23.50
0200	Aluminum, stock units, including frames and									
0210	3/8" plain cover plate, 4" opening	G E-4	205	.156	L.F.	19.65	8.55	.63	28.83	36.50
0300	6" opening	G ↓	185	.173		24.50	9.50	.70	34.70	43.50
0400	10" opening	G ↓	170	.188		34	10.30	.76	45.06	55.50
0500	16" opening	G ↓	155	.206		48.50	11.30	.84	60.64	73
0700	Add per inch for additional widths to 24"	G				2.04			2.04	2.24
0900	For custom fabrication, add					50%				
1100	For 1/4" plain cover plate, deduct					12%				
1500	For cover recessed for tile, 1/4" thick, deduct					12%				
1600	3/8" thick, add					5%				
1800	For checkered plate cover, 1/4" thick, deduct					12%				
1900	3/8" thick, add					2%				
2100	For slotted or round holes in cover, 1/4" thick, add					3%				
2200	3/8" thick, add					4%				
2300	For abrasive cover, add					12%				

05 55 13 – Metal Stair Treads

05 55 13.50 Stair Treads		Crew	Daily Output	Labor-Hours	Unit	Material	2017 Bare Costs Labor	Equipment	Total	Total Incl O&P
0010	**STAIR TREADS**, stringers and bolts not included									
3000	Diamond plate treads, steel, 1/8" thick									
3005	Open riser, black enamel									
3010	9" deep x 36" long	G	2 Sswk	48	.333	Ea.	91.50	18.10		109.60	130
3020	42" long	G		48	.333		96.50	18.10		114.60	136
3030	48" long	G		48	.333		100	18.10		118.10	140
3040	11" deep x 36" long	G		44	.364		96.50	19.75		116.25	139
3050	42" long	G		44	.364		102	19.75		121.75	145
3060	48" long	G		44	.364		108	19.75		127.75	152
3110	Galvanized, 9" deep x 36" long	G		48	.333		144	18.10		162.10	189
3120	42" long	G		48	.333		154	18.10		172.10	199
3130	48" long	G		48	.333		163	18.10		181.10	209
3140	11" deep x 36" long	G		44	.364		149	19.75		168.75	197
3150	42" long	G		44	.364		163	19.75		182.75	212
3160	48" long	G	▼	44	.364	▼	169	19.75		188.75	219
3200	Closed riser, black enamel									
3210	12" deep x 36" long	G	2 Sswk	40	.400	Ea.	110	21.50		131.50	157
3220	42" long	G		40	.400		119	21.50		140.50	167
3230	48" long	G		40	.400		126	21.50		147.50	174
3240	Galvanized, 12" deep x 36" long	G		40	.400		173	21.50		194.50	227
3250	42" long	G		40	.400		189	21.50		210.50	244
3260	48" long	G	▼	40	.400	▼	198	21.50		219.50	254
4000	Bar grating treads									
4005	Steel, 1-1/4" x 3/16" bars, anti-skid nosing, black enamel									
4010	8-5/8" deep x 30" long	G	2 Sswk	48	.333	Ea.	54	18.10		72.10	89.50
4020	36" long	G		48	.333		63.50	18.10		81.60	99.50
4030	48" long	G		48	.333		97.50	18.10		115.60	137
4040	10-15/16" deep x 36" long	G		44	.364		70	19.75		89.75	110
4050	48" long	G		44	.364		100	19.75		119.75	143
4060	Galvanized, 8-5/8" deep x 30" long	G		48	.333		62	18.10		80.10	98.50
4070	36" long	G		48	.333		73.50	18.10		91.60	111
4080	48" long	G		48	.333		108	18.10		126.10	149
4090	10-15/16" deep x 36" long	G		44	.364		86.50	19.75		106.25	128
4100	48" long	G	▼	44	.364	▼	112	19.75		131.75	156
4200	Aluminum, 1-1/4" x 3/16" bars, serrated, with nosing									
4210	7-5/8" deep x 18" long	G	2 Sswk	52	.308	Ea.	50	16.70		66.70	83
4220	24" long	G		52	.308		59	16.70		75.70	93
4230	30" long	G		52	.308		68.50	16.70		85.20	103
4240	36" long	G		52	.308		177	16.70		193.70	223
4250	8-13/16" deep x 18" long	G		48	.333		67	18.10		85.10	104
4260	24" long	G		48	.333		99.50	18.10		117.60	139
4270	30" long	G		48	.333		115	18.10		133.10	156
4280	36" long	G		48	.333		194	18.10		212.10	244
4290	10" deep x 18" long	G		44	.364		130	19.75		149.75	176
4300	30" long	G		44	.364		173	19.75		192.75	223
4310	36" long	G	▼	44	.364	▼	210	19.75		229.75	264
5000	Channel grating treads									
5005	Steel, 14 ga., 2-1/2" thick, galvanized									
5010	9" deep x 36" long	G	2 Sswk	48	.333	Ea.	103	18.10		121.10	143
5020	48" long	G	"	48	.333	"	139	18.10		157.10	182

05 55 Metal Stair Treads and Nosings

05 55 19 – Metal Stair Tread Covers

05 55 19.50 Stair Tread Covers for Renovation		Crew	Daily Output	Labor-Hours	Unit	Material	2017 Bare Costs Labor	Equipment	Total	Total Incl O&P
0010	**STAIR TREAD COVERS FOR RENOVATION**									
0205	Extruded tread cover with nosing, pre-drilled, includes screws									
0210	Aluminum with black abrasive strips, 9" wide x 3' long	1 Carp	24	.333	Ea.	111	16.40		127.40	147
0220	4' long		22	.364		146	17.90		163.90	189
0230	5' long		20	.400		182	19.70		201.70	231
0240	11" wide x 3' long		24	.333		144	16.40		160.40	183
0250	4' long		22	.364		185	17.90		202.90	231
0260	5' long	↓	20	.400	↓	238	19.70		257.70	292
0305	Black abrasive strips with yellow front strips									
0310	Aluminum, 9" wide x 3' long	1 Carp	24	.333	Ea.	120	16.40		136.40	157
0320	4' long		22	.364		160	17.90		177.90	204
0330	5' long		20	.400		203	19.70		222.70	253
0340	11" wide x 3' long		24	.333		154	16.40		170.40	195
0350	4' long		22	.364		199	17.90		216.90	247
0360	5' long	↓	20	.400	↓	256	19.70		275.70	310
0405	Black abrasive strips with photoluminescent front strips									
0410	Aluminum, 9" wide x 3' long	1 Carp	24	.333	Ea.	146	16.40		162.40	185
0420	4' long		22	.364		181	17.90		198.90	227
0430	5' long		20	.400		226	19.70		245.70	278
0440	11" wide x 3' long		24	.333		161	16.40		177.40	202
0450	4' long		22	.364		214	17.90		231.90	264
0460	5' long	↓	20	.400	↓	268	19.70		287.70	325

05 56 Metal Castings

05 56 13 – Metal Construction Castings

05 56 13.50 Construction Castings			Crew	Daily Output	Labor-Hours	Unit	Material	2017 Bare Costs Labor	Equipment	Total	Total Incl O&P
0010	**CONSTRUCTION CASTINGS**										
0020	Manhole covers and frames, see Section 33 44 13.13										
0100	Column bases, cast iron, 16" x 16", approx. 65 lb.	G	E-4	46	.696	Ea.	142	38	2.82	182.82	223
0200	32" x 32", approx. 256 lb.	G	"	23	1.391		520	76	5.65	601.65	710
0400	Cast aluminum for wood columns, 8" x 8"	G	1 Carp	32	.250		43.50	12.30		55.80	67
0500	12" x 12"	G	"	32	.250	↓	70	12.30		82.30	96.50
0600	Miscellaneous C.I. castings, light sections, less than 150 lb.	G	E-4	3200	.010	Lb.	8.95	.55	.04	9.54	10.80
1100	Heavy sections, more than 150 lb.	G		4200	.008		5.15	.42	.03	5.60	6.45
1300	Special low volume items	G	↓	3200	.010		11.25	.55	.04	11.84	13.30
1500	For ductile iron, add	G				↓	100%				

05 58 Formed Metal Fabrications

05 58 13 – Column Covers

05 58 13.05 Column Covers			Crew	Daily Output	Labor-Hours	Unit	Material	2017 Bare Costs Labor	Equipment	Total	Total Incl O&P
0010	**COLUMN COVERS**										
0015	Made from recycled materials										
0020	Excludes structural steel, light ga. metal framing, misc. metals, sealants										
0100	Round covers, 2 halves with 2 vertical joints for backer rod and sealant										
0110	Up to 12' high, no horizontal joints										
0120	12" diameter, 0.125" aluminum, anodized/painted finish	G	2 Sswk	32	.500	V.L.F.	31	27		58	80
0130	Type 304 stainless steel, 16 gauge, #4 brushed finish	G		32	.500		44.50	27		71.50	94.50
0140	Type 316 stainless steel, 16 gauge, #4 brushed finish	G		32	.500		55	27		82	106
0150	18" diameter, aluminum	G	↓	32	.500	↓	47	27		74	97

05 58 13 – Column Covers

05 58 13.05 Column Covers		Crew	Daily Output	Labor-Hours	Unit	Material	2017 Bare Costs Labor	Equipment	Total	Total Incl O&P
0160	Type 304 stainless steel	G 2 Sswk	32	.500	V.L.F.	67	27		94	120
0170	Type 316 stainless steel	G	32	.500		82.50	27		109.50	136
0180	24" diameter, aluminum	G	32	.500		62.50	27		89.50	114
0190	Type 304 stainless steel	G	32	.500		89.50	27		116.50	144
0200	Type 316 stainless steel	G	32	.500		110	27		137	167
0210	30" diameter, aluminum	G	30	.533		78	29		107	134
0220	Type 304 stainless steel	G	30	.533		112	29		141	172
0230	Type 316 stainless steel	G	30	.533		137	29		166	200
0240	36" diameter, aluminum	G	30	.533		93.50	29		122.50	152
0250	Type 304 stainless steel	G	30	.533		134	29		163	197
0260	Type 316 stainless steel	G	30	.533		165	29		194	230
0400	Up to 24' high, 2 stacked sections with 1 horizontal joint									
0410	18" diameter, aluminum	G 2 Sswk	28	.571	V.L.F.	49	31		80	106
0450	Type 304 stainless steel	G	28	.571		70.50	31		101.50	129
0460	Type 316 stainless steel	G	28	.571		86.50	31		117.50	147
0470	24" diameter, aluminum	G	28	.571		65.50	31		96.50	124
0480	Type 304 stainless steel	G	28	.571		94	31		125	155
0490	Type 316 stainless steel	G	28	.571		115	31		146	179
0500	30" diameter, aluminum	G	24	.667		82	36		118	151
0510	Type 304 stainless steel	G	24	.667		117	36		153	190
0520	Type 316 stainless steel	G	24	.667		144	36		180	219
0530	36" diameter, aluminum	G	24	.667		98	36		134	169
0540	Type 304 stainless steel	G	24	.667		141	36		177	216
0550	Type 316 stainless steel	G	24	.667		173	36		209	251

05 58 21 – Formed Chain

05 58 21.05 Alloy Steel Chain

05 58 21.05 Alloy Steel Chain		Crew	Daily Output	Labor-Hours	Unit	Material	2017 Bare Costs Labor	Equipment	Total	Total Incl O&P
0010	**ALLOY STEEL CHAIN**, Grade 80, for lifting									
0015	Self-colored, cut lengths, 1/4"	G E-17	4	4	C.L.F.	895	221		1,116	1,350
0020	3/8"	G	2	8		1,100	440		1,540	1,950
0030	1/2"	G	1.20	13.333		1,725	735		2,460	3,125
0040	5/8"	G	.72	22.222		2,850	1,225		4,075	5,175
0050	3/4"	G E-18	.48	83.333		3,650	4,550	2,250	10,450	14,000
0060	7/8"	G	.40	100		6,650	5,450	2,700	14,800	19,200
0070	1"	G	.35	114		8,500	6,225	3,075	17,800	22,900
0080	1-1/4"	G	.24	166		13,100	9,100	4,475	26,675	34,300
0110	Hook, Grade 80, Clevis slip, 1/4"	G			Ea.	30.50			30.50	33.50
0120	3/8"	G				38			38	42
0130	1/2"	G				61			61	67.50
0140	5/8"	G				98			98	108
0150	3/4"	G				132			132	146
0160	Hook, Grade 80, eye/sling w/hammerlock coupling, 15 Ton	G				390			390	425
0170	22 Ton	G				950			950	1,050
0180	37 Ton	G				3,000			3,000	3,300

05 58 23 – Formed Metal Guards

05 58 23.90 Window Guards

05 58 23.90 Window Guards		Crew	Daily Output	Labor-Hours	Unit	Material	2017 Bare Costs Labor	Equipment	Total	Total Incl O&P
0010	**WINDOW GUARDS**, shop fabricated									
0015	Expanded metal, steel angle frame, permanent	G E-4	350	.091	S.F.	22.50	5	.37	27.87	34
0025	Steel bars, 1/2" x 1/2", spaced 5" O.C.	G "	290	.110	"	15.75	6.05	.45	22.25	28
0030	Hinge mounted, add	G			Opng.	45.50			45.50	50
0040	Removable type, add	G			"	29			29	31.50
0050	For galvanized guards, add				S.F.	35%				
0070	For pivoted or projected type, add					105%	40%			

05 58 Formed Metal Fabrications

05 58 23 – Formed Metal Guards

05 58 23.90 Window Guards		Crew	Daily Output	Labor- Hours	Unit	Material	2017 Bare Costs Labor	Equipment	Total	Total Incl O&P
0100	Mild steel, stock units, economy	G	E-4	405	.079	S.F.	6.20	4.33	.32	10.85	14.35
0200	Deluxe	G		405	.079		12.60	4.33	.32	17.25	21.50
0400	Woven wire, stock units, 3/8" channel frame, 3' x 5' opening	G		40	.800	Opng.	167	44	3.25	214.25	261
0500	4' x 6' opening	G		38	.842		267	46	3.42	316.42	375
0800	Basket guards for above, add	G					231			231	254
1000	Swinging guards for above, add	G					79			79	86.50

05 58 25 – Formed Lamp Posts

05 58 25.40 Lamp Posts

0010	**LAMP POSTS**										
0020	Aluminum, 7' high, stock units, post only	G	1 Carp	16	.500	Ea.	87	24.50		111.50	133
0100	Mild steel, plain	G	"	16	.500	"	74.50	24.50		99	120

05 71 Decorative Metal Stairs

05 71 13 – Fabricated Metal Spiral Stairs

05 71 13.50 Spiral Stairs

0010	**SPIRAL STAIRS**										
1805	Shop fabricated, custom ordered										
1810	Aluminum, 5'-0" diameter, plain units	G	E-4	45	.711	Riser	530	39	2.89	571.89	655
1820	Fancy units	G		45	.711		925	39	2.89	966.89	1,100
1900	Cast iron, 4'-0" diameter, plain units	G		45	.711		665	39	2.89	706.89	800
1920	Fancy units	G		25	1.280		1,150	70	5.20	1,225.20	1,400
2000	Steel, industrial checkered plate, 4' diameter	G		45	.711		440	39	2.89	481.89	555
2200	6' diameter	G		40	.800		620	44	3.25	667.25	755
3100	Spiral stair kits, 12 stacking risers to fit exact floor height										
3110	Steel, flat metal treads, primed, 3'-6" diameter	G	2 Carp	1.60	10	Flight	1,300	495		1,795	2,175
3120	4'-0" diameter	G		1.45	11.034		1,475	545		2,020	2,425
3130	4'-6" diameter	G		1.35	11.852		1,625	585		2,210	2,700
3140	5'-0" diameter	G		1.25	12.800		1,775	630		2,405	2,950
3210	Galvanized, 3'-6" diameter	G		1.60	10		1,800	495		2,295	2,750
3220	4'-0" diameter	G		1.45	11.034		2,150	545		2,695	3,175
3230	4'-6" diameter	G		1.35	11.852		2,350	585		2,935	3,500
3240	5'-0" diameter	G		1.25	12.800		2,575	630		3,205	3,825
3310	Checkered plate tread, primed, 3'-6" diameter	G		1.45	11.034		1,550	545		2,095	2,525
3320	4'-0" diameter	G		1.35	11.852		1,725	585		2,310	2,800
3330	4'-6" diameter	G		1.25	12.800		1,900	630		2,530	3,050
3340	5'-0" diameter	G		1.15	13.913		2,075	685		2,760	3,325
3410	Galvanized, 3'-6" diameter	G		1.45	11.034		2,150	545		2,695	3,200
3420	4'-0" diameter	G		1.35	11.852		2,500	585		3,085	3,625
3430	4'-6" diameter	G		1.25	12.800		2,700	630		3,330	3,950
3440	5'-0" diameter	G		1.15	13.913		2,900	685		3,585	4,250
3510	Red oak covers on flat metal treads, 3'-6" diameter			1.35	11.852		2,350	585		2,935	3,500
3520	4'-0" diameter			1.25	12.800		2,800	630		3,430	4,050
3530	4'-6" diameter			1.15	13.913		3,025	685		3,710	4,375
3540	5'-0" diameter			1.05	15.238		3,250	750		4,000	4,725

05 73 Decorative Metal Railings

05 73 16 – Wire Rope Decorative Metal Railings

05 73 16.10 Cable Railings		Crew	Daily Output	Labor-Hours	Unit	Material	2017 Bare Costs Labor	Equipment	Total	Total Incl O&P
0010	**CABLE RAILINGS**, with 316 stainless steel 1 x 19 cable, 3/16" diameter									
0015	Made from recycled materials									
0100	1-3/4" diameter stainless steel posts x 42" high, cables 4" OC	G 2 Sswk	25	.640	L.F.	43	35		78	106

05 73 23 – Ornamental Railings

05 73 23.50 Railings, Ornamental

		Crew	Daily Output	Labor-Hours	Unit	Material	2017 Bare Costs Labor	Equipment	Total	Total Incl O&P
0010	**RAILINGS, ORNAMENTAL**, 3'-6" high, posts @ 6' O.C.									
0020	Bronze or stainless, hand forged, plain	G 2 Sswk	24	.667	L.F.	260	36		296	345
0100	Fancy	G	18	.889		520	48.50		568.50	650
0200	Aluminum, panelized, plain	G	24	.667		12.20	36		48.20	74
0300	Fancy	G	18	.889		25.50	48.50		74	109
0400	Wrought iron, hand forged, plain	G	24	.667		94	36		130	165
0500	Fancy	G	18	.889		231	48.50		279.50	335
0550	Steel, panelized, plain	G	24	.667		18.65	36		54.65	81
0560	Fancy	G	18	.889		28	48.50		76.50	112
0600	Composite metal/wood/glass, plain		18	.889		146	48.50		194.50	241
0700	Fancy		12	1.333		291	72.50		363.50	440

05 75 Decorative Formed Metal

05 75 13 – Columns

05 75 13.10 Aluminum Columns

		Crew	Daily Output	Labor-Hours	Unit	Material	2017 Bare Costs Labor	Equipment	Total	Total Incl O&P
0010	**ALUMINUM COLUMNS**									
0015	Made from recycled materials									
0020	Aluminum, extruded, stock units, no cap or base, 6" diameter	G E-4	240	.133	L.F.	12.50	7.30	.54	20.34	26.50
0100	8" diameter	G	170	.188		15.95	10.30	.76	27.01	35.50
0200	10" diameter	G	150	.213		20.50	11.70	.87	33.07	43
0300	12" diameter	G	140	.229		34	12.55	.93	47.48	59
0400	15" diameter	G	120	.267		44	14.60	1.08	59.68	74
0410	Caps and bases, plain, 6" diameter	G			Set	24.50			24.50	27
0420	8" diameter	G				30			30	33
0430	10" diameter	G				45			45	49.50
0440	12" diameter	G				64.50			64.50	71
0450	15" diameter	G				95.50			95.50	105
0460	Caps, ornamental, plain	G				272			272	300
0470	Fancy	G				1,425			1,425	1,550
0500	For square columns, add to column prices above				L.F.	50%				
0700	Residential, flat, 8' high, plain	G E-4	20	1.600	Ea.	97.50	87.50	6.50	191.50	260
0720	Fancy	G	20	1.600		190	87.50	6.50	284	360
0740	Corner type, plain	G	20	1.600		168	87.50	6.50	262	340
0760	Fancy	G	20	1.600		330	87.50	6.50	424	520

05 75 13.20 Columns, Ornamental

			Crew	Daily Output	Labor-Hours	Unit	Material	2017 Bare Costs Labor	Equipment	Total	Total Incl O&P
0010	**COLUMNS, ORNAMENTAL**, shop fabricated	R051223-10									
6400	Mild steel, flat, 9" wide, stock units, painted, plain		G E-4	160	.200	V.L.F.	9.15	10.95	.81	20.91	29
6450	Fancy		G	160	.200		17.75	10.95	.81	29.51	38.50
6500	Corner columns, painted, plain		G	160	.200		15.75	10.95	.81	27.51	36.50
6550	Fancy		G	160	.200		31	10.95	.81	42.76	53.50

Estimating Tips
06 05 00 Common Work Results for Wood, Plastics, and Composites

- Common to any wood-framed structure are the accessory connector items such as screws, nails, adhesives, hangers, connector plates, straps, angles, and hold-downs. For typical wood-framed buildings, such as residential projects, the aggregate total for these items can be significant, especially in areas where seismic loading is a concern. For floor and wall framing, the material cost is based on 10 to 25 lbs. of accessory connectors per MBF. Hold-downs, hangers, and other connectors should be taken off by the piece.

  Included with material costs are fasteners for a normal installation. Gordian's RSMeans engineers use manufacturers' recommendations, written specifications, and/or standard construction practice for the sizing and spacing of fasteners. Prices for various fasteners are shown for informational purposes only. Adjustments should be made if unusual fastening conditions exist.

06 10 00 Carpentry

- Lumber is a traded commodity and therefore sensitive to supply and demand in the marketplace. Even with

"budgetary" estimating of wood-framed projects, it is advisable to call local suppliers for the latest market pricing.

- The common quantity unit for wood-framed projects is "thousand board feet" (MBF). A board foot is a volume of wood—1" x 1' x 1' or 144 cubic inches. Board-foot quantities are generally calculated using nominal material dimensions— dressed sizes are ignored. Board foot per lineal foot of any stick of lumber can be calculated by dividing the nominal cross-sectional area by 12. As an example, 2,000 lineal feet of 2 x 12 equates to 4 MBF by dividing the nominal area, 2 x 12, by 12, which equals 2, and multiplying by 2,000 to give 4,000 board feet. This simple rule applies to all nominal dimensioned lumber.

- Waste is an issue of concern at the quantity takeoff for any area of construction. Framing lumber is sold in even foot lengths, i.e., 8', 10', 12', 14', 16' and depending on spans, wall heights, and the grade of lumber, waste is inevitable. A rule of thumb for lumber waste is 5%–10% depending on material quality and the complexity of the framing.

- Wood in various forms and shapes is used in many projects, even where the main structural framing is steel, concrete, or masonry. Plywood as a back-up partition material and 2x

boards used as blocking and cant strips around roof edges are two common examples. The estimator should ensure that the costs of all wood materials are included in the final estimate.

06 20 00 Finish Carpentry

- It is necessary to consider the grade of workmanship when estimating labor costs for erecting millwork and an interior finish. In practice, there are three grades: premium, custom, and economy. The RSMeans daily output for base and case moldings is in the range of 200 to 250 L.F. per carpenter per day. This is appropriate for most average custom-grade projects. For premium projects, an adjustment to productivity of 25%–50% should be made, depending on the complexity of the job.

Reference Numbers

Reference numbers are shown at the beginning of some major classifications. These numbers refer to related items in the Reference Section. The reference information may be an estimating procedure, an alternate pricing method, or technical information.

Note: Not all subdivisions listed here necessarily appear. ■

06 05 05.10 Selective Demolition Wood Framing	Crew	Daily Output	Labor-Hours	Unit	Material	2017 Bare Costs Labor	Equipment	Total	Total Incl O&P
0010 **SELECTIVE DEMOLITION WOOD FRAMING** R024119-10									
0100 Timber connector, nailed, small	1 Clab	96	.083	Ea.		3.26		3.26	5
0110 Medium		60	.133			5.20		5.20	8
0120 Large		48	.167			6.55		6.55	10
0130 Bolted, small		48	.167			6.55		6.55	10
0140 Medium		32	.250			9.80		9.80	15
0150 Large		24	.333			13.05		13.05	20
2958 Beams, 2" x 6"	2 Clab	1100	.015	L.F.		.57		.57	.87
2960 2" x 8"		825	.019			.76		.76	1.16
2965 2" x 10"		665	.024			.94		.94	1.44
2970 2" x 12"		550	.029			1.14		1.14	1.75
2972 2" x 14"		470	.034			1.33		1.33	2.04
2975 4" x 8"	B-1	413	.058			2.31		2.31	3.55
2980 4" x 10"		330	.073			2.90		2.90	4.44
2985 4" x 12"		275	.087			3.48		3.48	5.35
3000 6" x 8"		275	.087			3.48		3.48	5.35
3040 6" x 10"		220	.109			4.34		4.34	6.65
3080 6" x 12"		185	.130			5.15		5.15	7.90
3120 8" x 12"		140	.171			6.85		6.85	10.45
3160 10" x 12"		110	.218			8.70		8.70	13.30
3162 Alternate pricing method		1.10	21.818	M.B.F.		870		870	1,325
3170 Blocking, in 16" OC wall framing, 2" x 4"	1 Clab	600	.013	L.F.		.52		.52	.80
3172 2" x 6"		400	.020			.78		.78	1.20
3174 In 24" OC wall framing, 2" x 4"		600	.013			.52		.52	.80
3176 2" x 6"		400	.020			.78		.78	1.20
3178 Alt method, wood blocking removal from wood framing		.40	20	M.B.F.		785		785	1,200
3179 Wood blocking removal from steel framing		.36	22.222	"		870		870	1,325
3180 Bracing, let in, 1" x 3", studs 16" OC		1050	.008	L.F.		.30		.30	.46
3181 Studs 24" OC		1080	.007			.29		.29	.44
3182 1" x 4", studs 16" OC		1050	.008			.30		.30	.46
3183 Studs 24" OC		1080	.007			.29		.29	.44
3184 1" x 6", studs 16" OC		1050	.008			.30		.30	.46
3185 Studs 24" OC		1080	.007			.29		.29	.44
3186 2" x 3", studs 16" OC		800	.010			.39		.39	.60
3187 Studs 24" OC		830	.010			.38		.38	.58
3188 2" x 4", studs 16" OC		800	.010			.39		.39	.60
3189 Studs 24" OC		830	.010			.38		.38	.58
3190 2" x 6", studs 16" OC		800	.010			.39		.39	.60
3191 Studs 24" OC		830	.010			.38		.38	.58
3192 2" x 8", studs 16" OC		800	.010			.39		.39	.60
3193 Studs 24" OC		830	.010			.38		.38	.58
3194 "T" shaped metal bracing, studs at 16" OC		1060	.008			.30		.30	.45
3195 Studs at 24" OC		1200	.007			.26		.26	.40
3196 Metal straps, studs at 16" OC		1200	.007			.26		.26	.40
3197 Studs at 24" OC		1240	.006			.25		.25	.39
3200 Columns, round, 8' to 14' tall		40	.200	Ea.		7.85		7.85	12
3202 Dimensional lumber sizes	2 Clab	1.10	14.545	M.B.F.		570		570	875
3250 Blocking, between joists	1 Clab	320	.025	Ea.		.98		.98	1.50
3252 Bridging, metal strap, between joists		320	.025	Pr.		.98		.98	1.50
3254 Wood, between joists		320	.025	"		.98		.98	1.50
3260 Door buck, studs, header & access., 8' high 2" x 4" wall, 3' wide		32	.250	Ea.		9.80		9.80	15
3261 4' wide		32	.250			9.80		9.80	15
3262 5' wide		32	.250			9.80		9.80	15

06 05 05 – Selective Demolition for Wood, Plastics, and Composites

06 05 05.10 Selective Demolition Wood Framing	Crew	Daily Output	Labor-Hours	Unit	Material	2017 Bare Costs Labor	Equipment	Total	Total Incl O&P
3263	6' wide	1 Clab	32	.250	Ea.		9.80		9.80	15
3264	8' wide		30	.267			10.45		10.45	16
3265	10' wide		30	.267			10.45		10.45	16
3266	12' wide		30	.267			10.45		10.45	16
3267	2" x 6" wall, 3' wide		32	.250			9.80		9.80	15
3268	4' wide		32	.250			9.80		9.80	15
3269	5' wide		32	.250			9.80		9.80	15
3270	6' wide		32	.250			9.80		9.80	15
3271	8' wide		30	.267			10.45		10.45	16
3272	10' wide		30	.267			10.45		10.45	16
3273	12' wide		30	.267			10.45		10.45	16
3274	Window buck, studs, header & access., 8' high 2" x 4" wall, 2' wide		24	.333			13.05		13.05	20
3275	3' wide		24	.333			13.05		13.05	20
3276	4' wide		24	.333			13.05		13.05	20
3277	5' wide		24	.333			13.05		13.05	20
3278	6' wide		24	.333			13.05		13.05	20
3279	7' wide		24	.333			13.05		13.05	20
3280	8' wide		22	.364			14.25		14.25	22
3281	10' wide		22	.364			14.25		14.25	22
3282	12' wide		22	.364			14.25		14.25	22
3283	2" x 6" wall, 2' wide		24	.333			13.05		13.05	20
3284	3' wide		24	.333			13.05		13.05	20
3285	4' wide		24	.333			13.05		13.05	20
3286	5' wide		24	.333			13.05		13.05	20
3287	6' wide		24	.333			13.05		13.05	20
3288	7' wide		24	.333			13.05		13.05	20
3289	8' wide		22	.364			14.25		14.25	22
3290	10' wide		22	.364			14.25		14.25	22
3291	12' wide		22	.364			14.25		14.25	22
3360	Deck or porch decking		825	.010	L.F.		.38		.38	.58
3400	Fascia boards, 1" x 6"		500	.016			.63		.63	.96
3440	1" x 8"		450	.018			.70		.70	1.07
3480	1" x 10"		400	.020			.78		.78	1.20
3490	2" x 6"		450	.018			.70		.70	1.07
3500	2" x 8"		400	.020			.78		.78	1.20
3510	2" x 10"		350	.023			.90		.90	1.37
3610	Furring, on wood walls or ceiling		4000	.002	S.F.		.08		.08	.12
3620	On masonry or concrete walls or ceiling		1200	.007	"		.26		.26	.40
3800	Headers over openings, 2 @ 2" x 6"		110	.073	L.F.		2.85		2.85	4.36
3840	2 @ 2" x 8"		100	.080			3.13		3.13	4.80
3880	2 @ 2" x 10"		90	.089			3.48		3.48	5.35
3885	Alternate pricing method		.26	30.651	M.B.F.		1,200		1,200	1,850
3920	Joists, 1" x 4"		1250	.006	L.F.		.25		.25	.38
3930	1" x 6"		1135	.007			.28		.28	.42
3940	1" x 8"		1000	.008			.31		.31	.48
3950	1" x 10"		895	.009			.35		.35	.54
3960	1" x 12"		765	.010			.41		.41	.63
4200	2" x 4"	2 Clab	1000	.016			.63		.63	.96
4230	2" x 6"		970	.016			.65		.65	.99
4240	2" x 8"		940	.017			.67		.67	1.02
4250	2" x 10"		910	.018			.69		.69	1.05
4280	2" x 12"		880	.018			.71		.71	1.09
4281	2" x 14"		850	.019			.74		.74	1.13

06 05 05.10 Selective Demolition Wood Framing		Crew	Daily Output	Labor-Hours	Unit	Material	2017 Bare Costs Labor	Equipment	Total	Total Incl O&P
4282	Composite joists, 9-1/2"	2 Clab	960	.017	L.F.		.65		.65	1
4283	11-7/8"		930	.017			.67		.67	1.03
4284	14"		897	.018			.70		.70	1.07
4285	16"		865	.019			.72		.72	1.11
4290	Wood joists, alternate pricing method		1.50	10.667	M.B.F.		420		420	640
4500	Open web joist, 12" deep		500	.032	L.F.		1.25		1.25	1.92
4505	14" deep		475	.034			1.32		1.32	2.02
4510	16" deep		450	.036			1.39		1.39	2.13
4520	18" deep		425	.038			1.47		1.47	2.26
4530	24" deep		400	.040			1.57		1.57	2.40
4550	Ledger strips, 1" x 2"	1 Clab	1200	.007			.26		.26	.40
4560	1" x 3"		1200	.007			.26		.26	.40
4570	1" x 4"		1200	.007			.26		.26	.40
4580	2" x 2"		1100	.007			.28		.28	.44
4590	2" x 4"		1000	.008			.31		.31	.48
4600	2" x 6"		1000	.008			.31		.31	.48
4601	2" x 8" or 2" x 10"		800	.010			.39		.39	.60
4602	4" x 6"		600	.013			.52		.52	.80
4604	4" x 8"		450	.018			.70		.70	1.07
5400	Posts, 4" x 4"	2 Clab	800	.020			.78		.78	1.20
5405	4" x 6"		550	.029			1.14		1.14	1.75
5410	4" x 8"		440	.036			1.42		1.42	2.18
5425	4" x 10"		390	.041			1.61		1.61	2.46
5430	4" x 12"		350	.046			1.79		1.79	2.74
5440	6" x 6"		400	.040			1.57		1.57	2.40
5445	6" x 8"		350	.046			1.79		1.79	2.74
5450	6" x 10"		320	.050			1.96		1.96	3
5455	6" x 12"		290	.055			2.16		2.16	3.31
5480	8" x 8"		300	.053			2.09		2.09	3.20
5500	10" x 10"		240	.067			2.61		2.61	4
5660	Tongue and groove floor planks		2	8	M.B.F.		315		315	480
5682	Rafters, ordinary, 16" OC, 2" x 4"		880	.018	S.F.		.71		.71	1.09
5683	2" x 6"		840	.019			.75		.75	1.14
5684	2" x 8"		820	.020			.76		.76	1.17
5685	2" x 10"		820	.020			.76		.76	1.17
5686	2" x 12"		810	.020			.77		.77	1.19
5687	24" OC, 2" x 4"		1170	.014			.54		.54	.82
5688	2" x 6"		1117	.014			.56		.56	.86
5689	2" x 8"		1091	.015			.57		.57	.88
5690	2" x 10"		1091	.015			.57		.57	.88
5691	2" x 12"		1077	.015			.58		.58	.89
5795	Rafters, ordinary, 2" x 4" (alternate method)		862	.019	L.F.		.73		.73	1.11
5800	2" x 6" (alternate method)		850	.019			.74		.74	1.13
5840	2" x 8" (alternate method)		837	.019			.75		.75	1.15
5855	2" x 10" (alternate method)		825	.019			.76		.76	1.16
5865	2" x 12" (alternate method)		812	.020			.77		.77	1.18
5870	Sill plate, 2" x 4"	1 Clab	1170	.007			.27		.27	.41
5871	2" x 6"		780	.010			.40		.40	.62
5872	2" x 8"		586	.014			.53		.53	.82
5873	Alternate pricing method		.78	10.256	M.B.F.		400		400	615
5885	Ridge board, 1" x 4"	2 Clab	900	.018	L.F.		.70		.70	1.07
5886	1" x 6"		875	.018			.72		.72	1.10
5887	1" x 8"		850	.019			.74		.74	1.13

06 05 05 – Selective Demolition for Wood, Plastics, and Composites

06 05 05.10 Selective Demolition Wood Framing		Crew	Daily Output	Labor-Hours	Unit	Material	2017 Bare Costs Labor	Equipment	Total	Total Incl O&P
5888	1" x 10"	2 Clab	825	.019	L.F.		.76		.76	1.16
5889	1" x 12"		800	.020			.78		.78	1.20
5890	2" x 4"		900	.018			.70		.70	1.07
5892	2" x 6"		875	.018			.72		.72	1.10
5894	2" x 8"		850	.019			.74		.74	1.13
5896	2" x 10"		825	.019			.76		.76	1.16
5898	2" x 12"		800	.020			.78		.78	1.20
6050	Rafter tie, 1" x 4"		1250	.013			.50		.50	.77
6052	1" x 6"		1135	.014			.55		.55	.85
6054	2" x 4"		1000	.016			.63		.63	.96
6056	2" x 6"	▼	970	.016			.65		.65	.99
6070	Sleepers, on concrete, 1" x 2"	1 Clab	4700	.002			.07		.07	.10
6075	1" x 3"		4000	.002			.08		.08	.12
6080	2" x 4"		3000	.003			.10		.10	.16
6085	2" x 6"	▼	2600	.003	▼		.12		.12	.18
6086	Sheathing from roof, 5/16"	2 Clab	1600	.010	S.F.		.39		.39	.60
6088	3/8"		1525	.010			.41		.41	.63
6090	1/2"		1400	.011			.45		.45	.69
6092	5/8"		1300	.012			.48		.48	.74
6094	3/4"		1200	.013			.52		.52	.80
6096	Board sheathing from roof		1400	.011			.45		.45	.69
6100	Sheathing, from walls, 1/4"		1200	.013			.52		.52	.80
6110	5/16"		1175	.014			.53		.53	.82
6120	3/8"		1150	.014			.54		.54	.83
6130	1/2"		1125	.014			.56		.56	.85
6140	5/8"		1100	.015			.57		.57	.87
6150	3/4"		1075	.015			.58		.58	.89
6152	Board sheathing from walls		1500	.011			.42		.42	.64
6158	Subfloor/roof deck, with boards		2200	.007			.28		.28	.44
6159	Subfloor/roof deck, with tongue & groove boards		2000	.008			.31		.31	.48
6160	Plywood, 1/2" thick		768	.021			.82		.82	1.25
6162	5/8" thick		760	.021			.82		.82	1.26
6164	3/4" thick		750	.021			.84		.84	1.28
6165	1-1/8" thick	▼	720	.022			.87		.87	1.33
6166	Underlayment, particle board, 3/8" thick	1 Clab	780	.010			.40		.40	.62
6168	1/2" thick		768	.010			.41		.41	.63
6170	5/8" thick		760	.011			.41		.41	.63
6172	3/4" thick	▼	750	.011	▼		.42		.42	.64
6200	Stairs and stringers, straight run	2 Clab	40	.400	Riser		15.65		15.65	24
6240	With platforms, winders or curves	"	26	.615	"		24		24	37
6300	Components, tread	1 Clab	110	.073	Ea.		2.85		2.85	4.36
6320	Riser		80	.100	"		3.92		3.92	6
6390	Stringer, 2" x 10"		260	.031	L.F.		1.20		1.20	1.85
6400	2" x 12"		260	.031			1.20		1.20	1.85
6410	3" x 10"		250	.032			1.25		1.25	1.92
6420	3" x 12"	▼	250	.032			1.25		1.25	1.92
6590	Wood studs, 2" x 3"	2 Clab	3076	.005			.20		.20	.31
6600	2" x 4"		2000	.008			.31		.31	.48
6640	2" x 6"	▼	1600	.010	▼		.39		.39	.60
6720	Wall framing, including studs, plates and blocking, 2" x 4"	1 Clab	600	.013	S.F.		.52		.52	.80
6740	2" x 6"		480	.017	"		.65		.65	1
6750	Headers, 2" x 4"		1125	.007	L.F.		.28		.28	.43
6755	2" x 6"	▼	1125	.007			.28		.28	.43

For customer support on your Building Construction Costs with RSMeans Data, call 800.448.8182.

169

06 05 Common Work Results for Wood, Plastics, and Composites

06 05 05 – Selective Demolition for Wood, Plastics, and Composites

06 05 05.10 Selective Demolition Wood Framing

		Crew	Daily Output	Labor-Hours	Unit	Material	2017 Bare Costs Labor	Equipment	Total	Total Incl O&P
6760	2" x 8"	1 Clab	1050	.008	L.F.		.30		.30	.46
6765	2" x 10"		1050	.008			.30		.30	.46
6770	2" x 12"		1000	.008			.31		.31	.48
6780	4" x 10"		525	.015			.60		.60	.91
6785	4" x 12"		500	.016			.63		.63	.96
6790	6" x 8"		560	.014			.56		.56	.86
6795	6" x 10"		525	.015			.60		.60	.91
6797	6" x 12"		500	.016			.63		.63	.96
7000	Trusses									
7050	12' span	2 Clab	74	.216	Ea.		8.45		8.45	12.95
7150	24' span	F-3	66	.606			30.50	8.70	39.20	56.50
7200	26' span		64	.625			31.50	8.95	40.45	58.50
7250	28' span		62	.645			32.50	9.25	41.75	60
7300	30' span		58	.690			35	9.90	44.90	64.50
7350	32' span		56	.714			36	10.25	46.25	66.50
7400	34' span		54	.741			37.50	10.60	48.10	68.50
7450	36' span		52	.769			39	11	50	71.50
8000	Soffit, T & G wood	1 Clab	520	.015	S.F.		.60		.60	.92
8010	Hardboard, vinyl or aluminum	"	640	.013			.49		.49	.75
8030	Plywood	2 Carp	315	.051			2.50		2.50	3.83
9500	See Section 02 41 19.19 for rubbish handling									

06 05 05.20 Selective Demolition Millwork and Trim

		Crew	Daily Output	Labor-Hours	Unit	Material	2017 Bare Costs Labor	Equipment	Total	Total Incl O&P
0010	**SELECTIVE DEMOLITION MILLWORK AND TRIM** R024119-10									
1000	Cabinets, wood, base cabinets, per L.F.	2 Clab	80	.200	L.F.		7.85		7.85	12
1020	Wall cabinets, per L.F.	"	80	.200	"		7.85		7.85	12
1060	Remove and reset, base cabinets	2 Carp	18	.889	Ea.		44		44	67
1070	Wall cabinets		20	.800			39.50		39.50	60.50
1072	Oven cabinet, 7' high		11	1.455			71.50		71.50	110
1074	Cabinet door, up to 2' high	1 Clab	66	.121			4.75		4.75	7.25
1076	2' - 4' high	"	46	.174			6.80		6.80	10.45
1100	Steel, painted, base cabinets	2 Clab	60	.267	L.F.		10.45		10.45	16
1120	Wall cabinets		60	.267	"		10.45		10.45	16
1200	Casework, large area		320	.050	S.F.		1.96		1.96	3
1220	Selective		200	.080	"		3.13		3.13	4.80
1500	Counter top, straight runs		200	.080	L.F.		3.13		3.13	4.80
1510	L, U or C shapes		120	.133			5.20		5.20	8
1550	Remove and reset, straight runs	2 Carp	50	.320			15.75		15.75	24
1560	L, U or C shape	"	40	.400			19.70		19.70	30
2000	Paneling, 4' x 8' sheets	2 Clab	2000	.008	S.F.		.31		.31	.48
2100	Boards, 1" x 4"		700	.023			.90		.90	1.37
2120	1" x 6"		750	.021			.84		.84	1.28
2140	1" x 8"		800	.020			.78		.78	1.20
3000	Trim, baseboard, to 6" wide		1200	.013	L.F.		.52		.52	.80
3040	Greater than 6" and up to 12" wide		1000	.016			.63		.63	.96
3080	Remove and reset, minimum	2 Carp	400	.040			1.97		1.97	3.02
3090	Maximum	"	300	.053			2.63		2.63	4.02
3100	Ceiling trim	2 Clab	1000	.016			.63		.63	.96
3120	Chair rail		1200	.013			.52		.52	.80
3140	Railings with balusters		240	.067			2.61		2.61	4
3160	Wainscoting		700	.023	S.F.		.90		.90	1.37

06 05 23 – Wood, Plastic, and Composite Fastenings

06 05 23.10 Nails

		Crew	Daily Output	Labor-Hours	Unit	Material	2017 Bare Costs Labor	Equipment	Total	Total Incl O&P
0010	**NAILS**, material only, based upon 50# box purchase									
0020	Copper nails, plain				Lb.	11			11	12.10
0400	Stainless steel, plain					8.45			8.45	9.25
0500	Box, 3d to 20d, bright					1.49			1.49	1.64
0520	Galvanized					2.08			2.08	2.29
0600	Common, 3d to 60d, plain					1.25			1.25	1.38
0700	Galvanized					2.08			2.08	2.29
0800	Aluminum					9.90			9.90	10.85
1000	Annular or spiral thread, 4d to 60d, plain					2.79			2.79	3.07
1200	Galvanized					3.13			3.13	3.44
1400	Drywall nails, plain					1.48			1.48	1.63
1600	Galvanized					1.91			1.91	2.10
1800	Finish nails, 4d to 10d, plain					1.29			1.29	1.42
2000	Galvanized					2.07			2.07	2.28
2100	Aluminum					6.80			6.80	7.50
2300	Flooring nails, hardened steel, 2d to 10d, plain					3.53			3.53	3.88
2400	Galvanized					4.32			4.32	4.75
2500	Gypsum lath nails, 1-1/8", 13 ga. flathead, blued					2.53			2.53	2.78
2600	Masonry nails, hardened steel, 3/4" to 3" long, plain					2.22			2.22	2.44
2700	Galvanized					3.60			3.60	3.96
2900	Roofing nails, threaded, galvanized					2.32			2.32	2.55
3100	Aluminum					6.15			6.15	6.75
3300	Compressed lead head, threaded, galvanized					2.90			2.90	3.19
3600	Siding nails, plain shank, galvanized					2.20			2.20	2.42
3800	Aluminum					5.85			5.85	6.45
5000	Add to prices above for cement coating					.15			.15	.17
5200	Zinc or tin plating					.20			.20	.22
5500	Vinyl coated sinkers, 8d to 16d					2.56			2.56	2.82

06 05 23.40 Sheet Metal Screws

		Crew	Daily Output	Labor-Hours	Unit	Material	2017 Bare Costs Labor	Equipment	Total	Total Incl O&P
0010	**SHEET METAL SCREWS**									
0020	Steel, standard, #8 x 3/4", plain				C	3.13			3.13	3.44
0100	Galvanized					2.46			2.46	2.71
0300	#10 x 1", plain					4.98			4.98	5.50
0400	Galvanized					3.54			3.54	3.89
0600	With washers, #14 x 1", plain					9.15			9.15	10.10
0700	Galvanized					11.70			11.70	12.85
0900	#14 x 2", plain					18.90			18.90	21
1000	Galvanized					18.90			18.90	21
1500	Self-drilling, with washers, (pinch point) #8 x 3/4", plain					7.80			7.80	8.55
1600	Galvanized					7.80			7.80	8.55
1800	#10 x 3/4", plain					7.50			7.50	8.25
1900	Galvanized					7.50			7.50	8.25
3000	Stainless steel w/aluminum or neoprene washers, #14 x 1", plain					9.20			9.20	10.10
3100	#14 x 2", plain					11.50			11.50	12.65

06 05 23.50 Wood Screws

		Crew	Daily Output	Labor-Hours	Unit	Material	2017 Bare Costs Labor	Equipment	Total	Total Incl O&P
0010	**WOOD SCREWS**									
0020	#8, 1" long, steel				C	4.28			4.28	4.71
0100	Brass					10.75			10.75	11.80
0200	#8, 2" long, steel					5.65			5.65	6.20
0300	Brass					18.85			18.85	20.50
0400	#10, 1" long, steel					4.72			4.72	5.20
0500	Brass					14.50			14.50	15.95

06 05 23 – Wood, Plastic, and Composite Fastenings

06 05 23.50 Wood Screws		Crew	Daily Output	Labor-Hours	Unit	Material	2017 Bare Costs Labor	Equipment	Total	Total Incl O&P
0600	#10, 2" long, steel				C	6.85			6.85	7.50
0700	Brass					23.50			23.50	25.50
0800	#10, 3" long, steel					10.05			10.05	11.05
1000	#12, 2" long, steel					8.80			8.80	9.70
1100	Brass					31			31	34
1500	#12, 3" long, steel					11.30			11.30	12.40
2000	#12, 4" long, steel				▼	19.70			19.70	21.50

06 05 23.60 Timber Connectors										
0010	**TIMBER CONNECTORS**									
0020	Add up cost of each part for total cost of connection									
0100	Connector plates, steel, with bolts, straight	2 Carp	75	.213	Ea.	27.50	10.50		38	46.50
0110	Tee, 7 ga.		50	.320		35	15.75		50.75	62.50
0120	T- Strap, 14 ga., 12" x 8" x 2"		50	.320		35	15.75		50.75	62.50
0150	Anchor plates, 7 ga., 9" x 7"	▼	75	.213		27.50	10.50		38	46.50
0200	Bolts, machine, sq. hd. with nut & washer, 1/2" diameter, 4" long	1 Carp	140	.057		.75	2.81		3.56	5.15
0300	7-1/2" long		130	.062		1.37	3.03		4.40	6.15
0500	3/4" diameter, 7-1/2" long		130	.062		3.20	3.03		6.23	8.15
0610	Machine bolts, w/nut, washer, 3/4" diam., 15" L, HD's & beam hangers		95	.084	▼	5.95	4.15		10.10	12.90
0800	Drilling bolt holes in timber, 1/2" diameter		450	.018	Inch		.88		.88	1.34
0900	1" diameter		350	.023	"		1.13		1.13	1.72
1100	Framing anchor, angle, 3" x 3" x 1-1/2", 12 ga		175	.046	Ea.	2.56	2.25		4.81	6.25
1150	Framing anchors, 18 ga., 4-1/2" x 2-3/4"		175	.046		2.56	2.25		4.81	6.25
1160	Framing anchors, 18 ga., 4-1/2" x 3"		175	.046		2.56	2.25		4.81	6.25
1170	Clip anchors plates, 18 ga., 12" x 1-1/8"		175	.046		2.56	2.25		4.81	6.25
1250	Holdowns, 3 ga. base, 10 ga. body		8	1		26	49.50		75.50	105
1260	Holdowns, 7 ga. 11-1/16" x 3-1/4"		8	1		26	49.50		75.50	105
1270	Holdowns, 7 ga. 14-3/8" x 3-1/8"		8	1		26	49.50		75.50	105
1275	Holdowns, 12 ga. 8" x 2-1/2"		8	1		26	49.50		75.50	105
1300	Joist and beam hangers, 18 ga. galv., for 2" x 4" joist		175	.046		.80	2.25		3.05	4.33
1400	2" x 6" to 2" x 10" joist		165	.048		1.36	2.39		3.75	5.15
1600	16 ga. galv., 3" x 6" to 3" x 10" joist		160	.050		3.02	2.46		5.48	7.10
1700	3" x 10" to 3" x 14" joist		160	.050		4.88	2.46		7.34	9.10
1800	4" x 6" to 4" x 10" joist		155	.052		3.08	2.54		5.62	7.30
1900	4" x 10" to 4" x 14" joist		155	.052		5.30	2.54		7.84	9.75
2000	Two-2" x 6" to two-2" x 10" joists		150	.053		4	2.63		6.63	8.40
2100	Two-2" x 10" to two-2" x 14" joists		150	.053		4.48	2.63		7.11	8.95
2300	3/16" thick, 6" x 8" joist		145	.055		65	2.72		67.72	75.50
2400	6" x 10" joist		140	.057		67.50	2.81		70.31	79
2500	6" x 12" joist		135	.059		64.50	2.92		67.42	75.50
2700	1/4" thick, 6" x 14" joist	▼	130	.062		67	3.03		70.03	78
2900	Plywood clips, extruded aluminum H clip, for 3/4" panels					.22			.22	.24
3000	Galvanized 18 ga. back-up clip					.18			.18	.20
3200	Post framing, 16 ga. galv. for 4" x 4" base, 2 piece	1 Carp	130	.062		14.50	3.03		17.53	20.50
3300	Cap		130	.062		22.50	3.03		25.53	29.50
3500	Rafter anchors, 18 ga. galv., 1-1/2" wide, 5-1/4" long		145	.055		.48	2.72		3.20	4.69
3600	10-3/4" long		145	.055		1.43	2.72		4.15	5.75
3800	Shear plates, 2-5/8" diameter		120	.067		2.44	3.28		5.72	7.75
3900	4" diameter		115	.070		5.70	3.43		9.13	11.55
4000	Sill anchors, embedded in concrete or block, 25-1/2" long		115	.070		12.70	3.43		16.13	19.25
4100	Spike grids, 3" x 6"		120	.067		.96	3.28		4.24	6.10
4400	Split rings, 2-1/2" diameter		120	.067		2.02	3.28		5.30	7.25
4500	4" diameter	▼	110	.073	▼	2.98	3.58		6.56	8.80

06 05 23 – Wood, Plastic, and Composite Fastenings

06 05 23.60 Timber Connectors	Crew	Daily Output	Labor-Hours	Unit	Material	2017 Bare Costs Labor	Equipment	Total	Total Incl O&P
4550	Tie plate, 20 ga., 7" x 3-1/8"	1 Carp	110	.073	Ea.	2.98	3.58		6.56	8.80
4560	5" x 4-1/8"		110	.073		2.98	3.58		6.56	8.80
4575	Twist straps, 18 ga., 12" x 1-1/4"		110	.073		2.98	3.58		6.56	8.80
4580	16" x 1-1/4"		110	.073		2.98	3.58		6.56	8.80
4600	Strap ties, 20 ga., 2-1/16" wide, 12-13/16" long		180	.044		.87	2.19		3.06	4.31
4700	16 ga., 1-3/8" wide, 12" long		180	.044		.87	2.19		3.06	4.31
4800	1-1/4" wide, 21-5/8" long		160	.050		2.88	2.46		5.34	6.95
5000	Toothed rings, 2-5/8" or 4" diameter		90	.089	↓	2.05	4.38		6.43	8.95
5200	Truss plates, nailed, 20 ga., up to 32' span	↓	17	.471	Truss	12.85	23		35.85	49.50
5400	Washers, 2" x 2" x 1/8"				Ea.	.38			.38	.42
5500	3" x 3" x 3/16"				"	1.14			1.14	1.25
6000	Angles and gussets, painted									
6012	7 ga., 3-1/4" x 3-1/4" x 2-1/2" long	1 Carp	1.90	4.211	C	1,075	207		1,282	1,500
6014	3-1/4" x 3-1/4" x 5" long		1.90	4.211		2,100	207		2,307	2,625
6016	3-1/4" x 3-1/4" x 7-1/2" long		1.85	4.324		3,975	213		4,188	4,675
6018	5-3/4" x 5-3/4" x 2-1/2" long		1.85	4.324		2,550	213		2,763	3,150
6020	5-3/4" x 5-3/4" x 5" long		1.85	4.324		4,075	213		4,288	4,800
6022	5-3/4" x 5-3/4" x 7-1/2" long		1.80	4.444		6,000	219		6,219	6,950
6024	3 ga., 4-1/4" x 4-1/4" x 3" long		1.85	4.324		2,725	213		2,938	3,300
6026	4-1/4" x 4-1/4" x 6" long		1.85	4.324		5,850	213		6,063	6,750
6028	4-1/4" x 4-1/4" x 9" long		1.80	4.444		6,575	219		6,794	7,550
6030	7-1/4" x 7-1/4" x 3" long		1.80	4.444		4,700	219		4,919	5,500
6032	7-1/4" x 7-1/4" x 6" long		1.80	4.444		6,350	219		6,569	7,300
6034	7-1/4" x 7-1/4" x 9" long	↓	1.75	4.571	↓	14,200	225		14,425	15,900
6036	Gussets									
6038	7 ga., 8-1/8" x 8-1/8" x 2-3/4" long	1 Carp	1.80	4.444	C	4,850	219		5,069	5,650
6040	3 ga., 9-3/4" x 9-3/4" x 3-1/4" long	"	1.80	4.444	"	7,275	219		7,494	8,325
6101	Beam hangers, polymer painted									
6102	Bolted, 3 ga. (W x H x L)									
6104	3-1/4" x 9" x 12" top flange	1 Carp	1	8	C	19,100	395		19,495	21,600
6106	5-1/4" x 9" x 12" top flange		1	8		20,000	395		20,395	22,600
6108	5-1/4" x 11" x 11-3/4" top flange		1	8		20,800	395		21,195	23,400
6110	6-7/8" x 9" x 12" top flange		1	8		20,600	395		20,995	23,300
6112	6-7/8" x 11" x 13-1/2" top flange		1	8		24,300	395		24,695	27,400
6114	8-7/8" x 11" x 15-1/2" top flange	↓	1	8	↓	23,600	395		23,995	26,500
6116	Nailed, 3 ga. (W x H x L)									
6118	3-1/4" x 10-1/2" x 10" top flange	1 Carp	1.80	4.444	C	20,500	219		20,719	22,900
6120	3-1/4" x 10-1/2" x 12" top flange		1.80	4.444		21,400	219		21,619	23,900
6122	5-1/4" x 9-1/2" x 10" top flange		1.80	4.444		22,300	219		22,519	24,900
6124	5-1/4" x 9-1/2" x 12" top flange		1.80	4.444		22,900	219		23,119	25,500
6128	6-7/8" x 8-1/2" x 12" top flange	↓	1.80	4.444	↓	20,600	219		20,819	23,000
6134	Saddle hangers, glu-lam (W x H x L)									
6136	3-1/4" x 10-1/2" x 5-1/4" x 6" saddle	1 Carp	.50	16	C	15,600	790		16,390	18,400
6138	3-1/4" x 10-1/2" x 6-7/8" x 6" saddle		.50	16		16,600	790		17,390	19,400
6140	3-1/4" x 10-1/2" x 8-7/8" x 6" saddle		.50	16		17,400	790		18,190	20,300
6142	3-1/4" x 19-1/2" x 5-1/4" x 10-1/8" saddle		.40	20		17,400	985		18,385	20,700
6144	3-1/4" x 19-1/2" x 6-7/8" x 10-1/8" saddle		.40	20		16,800	985		17,785	20,000
6146	3-1/4" x 19-1/2" x 8-7/8" x 10-1/8" saddle		.40	20		17,400	985		18,385	20,600
6148	5-1/4" x 9-1/2" x 5-1/4" x 12" saddle		.50	16		18,400	790		19,190	21,400
6150	5-1/4" x 9-1/2" x 6-7/8" x 9" saddle		.50	16		20,400	790		21,190	23,600
6152	5-1/4" x 10-1/2" x spec x 12" saddle		.50	16		22,000	790		22,790	25,400
6154	5-1/4" x 18" x 5-1/4" x 12-1/8" saddle		.40	20		18,700	985		19,685	22,100
6156	5-1/4" x 18" x 6-7/8" x 12-1/8" saddle	↓	.40	20		20,600	985		21,585	24,200

06 05 23.60 Timber Connectors		Crew	Daily Output	Labor-Hours	Unit	Material	2017 Bare Costs Labor	2017 Bare Costs Equipment	Total	Total Incl O&P
6158	5-1/4" x 18" x spec x 12-1/8" saddle	1 Carp	.40	20	C	22,400	985		23,385	26,200
6160	6-7/8" x 8-1/2" x 6-7/8" x 12" saddle		.50	16		22,000	790		22,790	25,400
6162	6-7/8" x 8-1/2" x 8-7/8" x 12" saddle		.50	16		22,800	790		23,590	26,300
6164	6-7/8" x 10-1/2" x spec x 12" saddle		.50	16		21,900	790		22,690	25,300
6166	6-7/8" x 18" x 6-7/8" x 13-3/4" saddle		.40	20		21,800	985		22,785	25,500
6168	6-7/8" x 18" x 8-7/8" x 13-3/4" saddle		.40	20		23,500	985		24,485	27,300
6170	6-7/8" x 18" x spec x 13-3/4" saddle		.40	20		24,400	985		25,385	28,300
6172	8-7/8" x 18" x spec x 15-3/4" saddle		.40	20		38,300	985		39,285	43,700
6201	Beam and purlin hangers, galvanized, 12 ga.									
6202	Purlin or joist size, 3" x 8"	1 Carp	1.70	4.706	C	2,025	232		2,257	2,575
6204	3" x 10"		1.70	4.706		2,250	232		2,482	2,825
6206	3" x 12"		1.65	4.848		2,525	239		2,764	3,150
6208	3" x 14"		1.65	4.848		2,675	239		2,914	3,325
6210	3" x 16"		1.65	4.848		2,850	239		3,089	3,500
6212	4" x 8"		1.65	4.848		2,025	239		2,264	2,600
6214	4" x 10"		1.65	4.848		2,250	239		2,489	2,850
6216	4" x 12"		1.60	5		2,675	246		2,921	3,300
6218	4" x 14"		1.60	5		2,800	246		3,046	3,450
6220	4" x 16"		1.60	5		2,975	246		3,221	3,650
6222	6" x 8"		1.60	5		2,700	246		2,946	3,325
6224	6" x 10"		1.55	5.161		2,725	254		2,979	3,400
6226	6" x 12"		1.55	5.161		4,675	254		4,929	5,525
6228	6" x 14"		1.50	5.333		4,950	263		5,213	5,850
6230	6" x 16"		1.50	5.333		5,275	263		5,538	6,200
6250	Beam seats									
6252	Beam size, 5-1/4" wide									
6254	5" x 7" x 1/4"	1 Carp	1.80	4.444	C	7,950	219		8,169	9,075
6256	6" x 7" x 3/8"		1.80	4.444		8,725	219		8,944	9,925
6258	7" x 7" x 3/8"		1.80	4.444		9,400	219		9,619	10,600
6260	8" x 7" x 3/8"		1.80	4.444		11,100	219		11,319	12,500
6262	Beam size, 6-7/8" wide									
6264	5" x 9" x 1/4"	1 Carp	1.80	4.444	C	9,425	219		9,644	10,700
6266	6" x 9" x 3/8"		1.80	4.444		12,400	219		12,619	13,900
6268	7" x 9" x 3/8"		1.80	4.444		12,200	219		12,419	13,700
6270	8" x 9" x 3/8"		1.80	4.444		14,500	219		14,719	16,200
6272	Special beams, over 6-7/8" wide									
6274	5" x 10" x 3/8"	1 Carp	1.80	4.444	C	12,500	219		12,719	14,100
6276	6" x 10" x 3/8"		1.80	4.444		14,700	219		14,919	16,500
6278	7" x 10" x 3/8"		1.80	4.444		15,500	219		15,719	17,300
6280	8" x 10" x 3/8"		1.75	4.571		15,900	225		16,125	17,800
6282	5-1/4" x 12" x 5/16"		1.75	4.571		12,300	225		12,525	13,800
6284	6-1/2" x 12" x 3/8"		1.75	4.571		20,400	225		20,625	22,700
6286	5-1/4" x 16" x 5/16"		1.70	4.706		18,100	232		18,332	20,300
6288	6-1/2" x 16" x 3/8"		1.70	4.706		24,200	232		24,432	27,100
6290	5-1/4" x 20" x 5/16"		1.70	4.706		22,000	232		22,232	24,600
6292	6-1/2" x 20" x 3/8"		1.65	4.848		28,600	239		28,839	31,900
6300	Column bases									
6302	4 x 4, 16 ga.	1 Carp	1.80	4.444	C	775	219		994	1,200
6306	7 ga.		1.80	4.444		2,900	219		3,119	3,500
6308	4 x 6, 16 ga.		1.80	4.444		1,750	219		1,969	2,250
6312	7 ga.		1.80	4.444		2,850	219		3,069	3,450
6314	6 x 6, 16 ga.		1.75	4.571		2,300	225		2,525	2,875
6318	7 ga.		1.75	4.571		3,650	225		3,875	4,350

06 05 23.60 Timber Connectors		Crew	Daily Output	Labor-Hours	Unit	Material	2017 Bare Costs Labor	Equipment	Total	Total Incl O&P
6320	6 x 8, 7 ga.	1 Carp	1.70	4.706	C	3,150	232		3,382	3,825
6322	6 x 10, 7 ga.		1.70	4.706		3,400	232		3,632	4,075
6324	6 x 12, 7 ga.		1.70	4.706		3,675	232		3,907	4,400
6326	8 x 8, 7 ga.		1.65	4.848		6,200	239		6,439	7,200
6330	8 x 10, 7 ga.		1.65	4.848		7,225	239		7,464	8,325
6332	8 x 12, 7 ga.		1.60	5		7,850	246		8,096	9,000
6334	10 x 10, 3 ga.		1.60	5		8,000	246		8,246	9,200
6336	10 x 12, 3 ga.		1.60	5		9,200	246		9,446	10,500
6338	12 x 12, 3 ga.		1.55	5.161		10,000	254		10,254	11,400
6350	Column caps, painted, 3 ga.									
6352	3-1/4" x 3-5/8"	1 Carp	1.80	4.444	C	10,400	219		10,619	11,700
6354	3-1/4" x 5-1/2"		1.80	4.444		10,400	219		10,619	11,700
6356	3-5/8" x 3-5/8"		1.80	4.444		8,475	219		8,694	9,650
6358	3-5/8" x 5-1/2"		1.80	4.444		8,475	219		8,694	9,650
6360	5-1/4" x 5-1/2"		1.75	4.571		10,200	225		10,425	11,500
6362	5-1/4" x 7-1/2"		1.75	4.571		10,900	225		11,125	12,300
6364	5-1/2" x 3-5/8"		1.75	4.571		10,200	225		10,425	11,500
6366	5-1/2" x 5-1/2"		1.75	4.571		10,200	225		10,425	11,500
6368	5-1/2" x 7-1/2"		1.70	4.706		11,700	232		11,932	13,300
6370	6-7/8" x 5-1/2"		1.70	4.706		12,200	232		12,432	13,900
6372	6-7/8" x 6-7/8"		1.70	4.706		12,200	232		12,432	13,900
6374	6-7/8" x 7-1/2"		1.70	4.706		12,200	232		12,432	13,900
6376	7-1/2" x 5-1/2"		1.65	4.848		12,800	239		13,039	14,500
6378	7-1/2" x 7-1/2"		1.65	4.848		12,800	239		13,039	14,500
6380	8-7/8" x 5-1/2"		1.60	5		13,500	246		13,746	15,300
6382	8-7/8" x 7-1/2"		1.60	5		13,500	246		13,746	15,300
6384	9-1/2" x 5-1/2"		1.60	5		18,300	246		18,546	20,500
6400	Floor tie anchors, polymer paint									
6402	10 ga., 3" x 37-1/2"	1 Carp	1.80	4.444	C	4,875	219		5,094	5,675
6404	3-1/2" x 45-1/2"		1.75	4.571		5,100	225		5,325	5,950
6406	3 ga., 3-1/2" x 56"		1.70	4.706		8,975	232		9,207	10,200
6410	Girder hangers									
6412	6" wall thickness, 4" x 6"	1 Carp	1.80	4.444	C	2,725	219		2,944	3,325
6414	4" x 8"		1.80	4.444		3,050	219		3,269	3,675
6416	8" wall thickness, 4" x 6"		1.80	4.444		3,050	219		3,269	3,675
6418	4" x 8"		1.80	4.444		3,050	219		3,269	3,675
6420	Hinge connections, polymer painted									
6422	3/4" thick top plate									
6424	5-1/4" x 12" w/5" x 5" top	1 Carp	1	8	C	35,800	395		36,195	40,000
6426	5-1/4" x 15" w/6" x 6" top		.80	10		38,100	495		38,595	42,700
6428	5-1/4" x 18" w/7" x 7" top		.70	11.429		40,100	565		40,665	45,000
6430	5-1/4" x 26" w/9" x 9" top		.60	13.333		42,600	655		43,255	47,900
6432	1" thick top plate									
6434	6-7/8" x 14" w/5" x 5" top	1 Carp	.80	10	C	43,700	495		44,195	48,800
6436	6-7/8" x 17" w/6" x 6" top		.80	10		48,600	495		49,095	54,500
6438	6-7/8" x 21" w/7" x 7" top		.70	11.429		53,000	565		53,565	59,500
6440	6-7/8" x 31" w/9" x 9" top		.60	13.333		58,000	655		58,655	64,500
6442	1-1/4" thick top plate									
6444	8-7/8" x 16" w/5" x 5" top	1 Carp	.60	13.333	C	54,500	655		55,155	61,000
6446	8-7/8" x 21" w/6" x 6" top		.50	16		60,000	790		60,790	67,000
6448	8-7/8" x 26" w/7" x 7" top		.40	20		68,000	985		68,985	76,000
6450	8-7/8" x 39" w/9" x 9" top		.30	26.667		84,500	1,325		85,825	95,000
6460	Holddowns									

For customer support on your Building Construction Costs with RSMeans Data, call 800.448.8182.

175

06 05 23.60 Timber Connectors	Crew	Daily Output	Labor-Hours	Unit	Material	2017 Bare Costs Labor	Equipment	Total	Total Incl O&P
6462	Embedded along edge									
6464	26" long, 12 ga.	1 Carp	.90	8.889	C	1,175	440		1,615	1,975
6466	35" long, 12 ga.		.85	9.412		1,625	465		2,090	2,500
6468	35" long, 10 ga.		.85	9.412		1,700	465		2,165	2,575
6470	Embedded away from edge									
6472	Medium duty, 12 ga.									
6474	18-1/2" long	1 Carp	.95	8.421	C	825	415		1,240	1,550
6476	23-3/4" long		.90	8.889		960	440		1,400	1,725
6478	28" long		.85	9.412		980	465		1,445	1,775
6480	35" long		.85	9.412		1,350	465		1,815	2,175
6482	Heavy duty, 10 ga.									
6484	28" long	1 Carp	.85	9.412	C	1,750	465		2,215	2,625
6486	35" long	"	.85	9.412	"	1,900	465		2,365	2,800
6490	Surface mounted (W x H)									
6492	2-1/2" x 5-3/4", 7 ga.	1 Carp	1	8	C	2,025	395		2,420	2,850
6494	2-1/2" x 8", 12 ga.		1	8		1,225	395		1,620	1,950
6496	2-7/8" x 6-3/8", 7 ga.		1	8		4,775	395		5,170	5,850
6498	2-7/8" x 12-1/2", 3 ga.		1	8		4,500	395		4,895	5,550
6500	3-3/16" x 9-3/8", 10 ga.		1	8		3,000	395		3,395	3,900
6502	3-1/2" x 11-5/8", 3 ga.		1	8		5,550	395		5,945	6,700
6504	3-1/2" x 14-3/4", 3 ga.		1	8		7,075	395		7,470	8,400
6506	3-1/2" x 16-1/2", 3 ga.		1	8		8,450	395		8,845	9,900
6508	3-1/2" x 20-1/2", 3 ga.		.90	8.889		9,200	440		9,640	10,800
6510	3-1/2" x 24-1/2", 3 ga.		.90	8.889		11,000	440		11,440	12,800
6512	4-1/4" x 20-3/4", 3 ga.		.90	8.889		7,625	440		8,065	9,075
6520	Joist hangers									
6522	Sloped, field adjustable, 18 ga.									
6524	2" x 6"	1 Carp	1.65	4.848	C	545	239		784	965
6526	2" x 8"		1.65	4.848		985	239		1,224	1,450
6528	2" x 10" and up		1.65	4.848		1,425	239		1,664	1,925
6530	3" x 10" and up		1.60	5		1,175	246		1,421	1,675
6532	4" x 10" and up		1.55	5.161		1,350	254		1,604	1,875
6536	Skewed 45°, 16 ga.									
6538	2" x 4"	1 Carp	1.75	4.571	C	1,075	225		1,300	1,525
6540	2" x 6" or 2" x 8"		1.65	4.848		980	239		1,219	1,450
6542	2" x 10" or 2" x 12"		1.65	4.848		1,025	239		1,264	1,500
6544	2" x 14" or 2" x 16"		1.60	5		1,825	246		2,071	2,375
6546	(2) 2" x 6" or (2) 2" x 8"		1.60	5		1,650	246		1,896	2,175
6548	(2) 2" x 10" or (2) 2" x 12"		1.55	5.161		1,750	254		2,004	2,325
6550	(2) 2" x 14" or (2) 2" x 16"		1.50	5.333		2,775	263		3,038	3,450
6552	4" x 6" or 4" x 8"		1.60	5		1,375	246		1,621	1,875
6554	4" x 10" or 4" x 12"		1.55	5.161		1,600	254		1,854	2,175
6556	4" x 14" or 4" x 16"		1.55	5.161		2,475	254		2,729	3,125
6560	Skewed 45°, 14 ga.									
6562	(2) 2" x 6" or (2) 2" x 8"	1 Carp	1.60	5	C	1,900	246		2,146	2,450
6564	(2) 2" x 10" or (2) 2" x 12"		1.55	5.161		2,625	254		2,879	3,275
6566	(2) 2" x 14" or (2) 2" x 16"		1.50	5.333		3,850	263		4,113	4,625
6568	4" x 6" or 4" x 8"		1.60	5		2,075	246		2,321	2,675
6570	4" x 10" or 4" x 12"		1.55	5.161		2,375	254		2,629	3,000
6572	4" x 14" or 4" x 16"		1.55	5.161		3,175	254		3,429	3,900
6590	Joist hangers, heavy duty 12 ga., galvanized									
6592	2" x 4"	1 Carp	1.75	4.571	C	1,275	225		1,500	1,750
6594	2" x 6"		1.65	4.848		1,400	239		1,639	1,900

For customer support on your Building Construction Costs with RSMeans Data, call 800.448.8182.

06 05 23.60 Timber Connectors	Crew	Daily Output	Labor-Hours	Unit	Material	2017 Bare Costs Labor	Equipment	Total	Total Incl O&P
6595	2" x 6", 16 ga.	1 Carp	1.65	4.848	C	1,325	239		1,564	1,825
6596	2" x 8"		1.65	4.848		2,100	239		2,339	2,700
6597	2" x 8", 16 ga.		1.65	4.848		2,000	239		2,239	2,575
6598	2" x 10"		1.65	4.848		2,175	239		2,414	2,750
6600	2" x 12"		1.65	4.848		2,650	239		2,889	3,300
6602	2" x 14"		1.65	4.848		2,775	239		3,014	3,425
6604	2" x 16"		1.65	4.848		2,925	239		3,164	3,600
6606	3" x 4"		1.65	4.848		1,775	239		2,014	2,325
6608	3" x 6"		1.65	4.848		2,375	239		2,614	2,975
6610	3" x 8"		1.65	4.848		2,400	239		2,639	3,025
6612	3" x 10"		1.60	5		2,775	246		3,021	3,425
6614	3" x 12"		1.60	5		3,325	246		3,571	4,050
6616	3" x 14"		1.60	5		3,900	246		4,146	4,675
6618	3" x 16"		1.60	5		4,300	246		4,546	5,125
6620	(2) 2" x 4"		1.75	4.571		2,075	225		2,300	2,625
6622	(2) 2" x 6"		1.60	5		2,500	246		2,746	3,100
6624	(2) 2" x 8"		1.60	5		2,550	246		2,796	3,175
6626	(2) 2" x 10"		1.55	5.161		2,750	254		3,004	3,425
6628	(2) 2" x 12"		1.55	5.161		3,525	254		3,779	4,275
6630	(2) 2" x 14"		1.50	5.333		3,550	263		3,813	4,325
6632	(2) 2" x 16"		1.50	5.333		3,600	263		3,863	4,350
6634	4" x 4"		1.65	4.848		1,500	239		1,739	2,025
6636	4" x 6"		1.60	5		1,650	246		1,896	2,200
6638	4" x 8"		1.60	5		1,900	246		2,146	2,475
6640	4" x 10"		1.55	5.161		2,350	254		2,604	2,975
6642	4" x 12"		1.55	5.161		2,500	254		2,754	3,125
6644	4" x 14"		1.55	5.161		2,975	254		3,229	3,675
6646	4" x 16"		1.55	5.161		3,275	254		3,529	4,000
6648	(3) 2" x 10"		1.50	5.333		3,600	263		3,863	4,375
6650	(3) 2" x 12"		1.50	5.333		4,025	263		4,288	4,825
6652	(3) 2" x 14"		1.45	5.517		4,325	272		4,597	5,175
6654	(3) 2" x 16"		1.45	5.517		4,400	272		4,672	5,250
6656	6" x 6"		1.60	5		1,950	246		2,196	2,500
6658	6" x 8"		1.60	5		2,000	246		2,246	2,575
6660	6" x 10"		1.55	5.161		2,400	254		2,654	3,025
6662	6" x 12"		1.55	5.161		2,725	254		2,979	3,400
6664	6" x 14"		1.50	5.333		3,425	263		3,688	4,150
6666	6" x 16"	↓	1.50	5.333	↓	4,025	263		4,288	4,825
6690	Knee braces, galvanized, 12 ga.									
6692	Beam depth, 10" x 15" x 5' long	1 Carp	1.80	4.444	C	5,225	219		5,444	6,075
6694	15" x 22-1/2" x 7' long		1.70	4.706		6,000	232		6,232	6,950
6696	22-1/2" x 28-1/2" x 8' long		1.60	5		6,450	246		6,696	7,450
6698	28-1/2" x 36" x 10' long		1.55	5.161		6,725	254		6,979	7,800
6700	36" x 42" x 12' long	↓	1.50	5.333	↓	7,425	263		7,688	8,550
6710	Mudsill anchors									
6714	2" x 4" or 3" x 4"	1 Carp	115	.070	C	1,425	3.43		1,428.43	1,550
6716	2" x 6" or 3" x 6"		115	.070		1,425	3.43		1,428.43	1,550
6718	Block wall, 13-1/4" long		115	.070		85	3.43		88.43	99
6720	21-1/4" long	↓	115	.070	↓	126	3.43		129.43	144
6730	Post bases, 12 ga. galvanized									
6732	Adjustable, 3-9/16" x 3-9/16"	1 Carp	1.30	6.154	C	1,075	305		1,380	1,650
6734	3-9/16" x 5-1/2"		1.30	6.154		2,025	305		2,330	2,700
6736	4" x 4"	↓	1.30	6.154		930	305		1,235	1,500

06 05 23.60 Timber Connectors		Crew	Daily Output	Labor-Hours	Unit	Material	2017 Bare Costs Labor	2017 Bare Costs Equipment	Total	Total Incl O&P
6738	4" x 6"	1 Carp	1.30	6.154	C	2,700	305		3,005	3,450
6740	5-1/2" x 5-1/2"		1.30	6.154		3,325	305		3,630	4,125
6742	6" x 6"		1.30	6.154		3,325	305		3,630	4,125
6744	Elevated, 3-9/16" x 3-1/4"		1.30	6.154		1,150	305		1,455	1,750
6746	5-1/2" x 3-5/16"		1.30	6.154		1,650	305		1,955	2,275
6748	5-1/2" x 5"		1.30	6.154		2,450	305		2,755	3,175
6750	Regular, 3-9/16" x 3-3/8"		1.30	6.154		880	305		1,185	1,425
6752	4" x 3-3/8"		1.30	6.154		1,250	305		1,555	1,850
6754	18 ga., 5-1/4" x 3-1/8"		1.30	6.154		1,300	305		1,605	1,900
6755	5-1/2" x 3-3/8"		1.30	6.154		1,300	305		1,605	1,900
6756	5-1/2" x 5-3/8"		1.30	6.154		1,875	305		2,180	2,525
6758	6" x 3-3/8"		1.30	6.154		2,250	305		2,555	2,950
6760	6" x 5-3/8"		1.30	6.154		2,500	305		2,805	3,225
6762	Post combination cap/bases									
6764	3-9/16" x 3-9/16"	1 Carp	1.20	6.667	C	460	330		790	1,000
6766	3-9/16" x 5-1/2"		1.20	6.667		1,050	330		1,380	1,650
6768	4" x 4"		1.20	6.667		2,075	330		2,405	2,800
6770	5-1/2" x 5-1/2"		1.20	6.667		1,150	330		1,480	1,775
6772	6" x 6"		1.20	6.667		4,150	330		4,480	5,075
6774	7-1/2" x 7-1/2"		1.20	6.667		4,550	330		4,880	5,525
6776	8" x 8"		1.20	6.667		4,725	330		5,055	5,700
6790	Post-beam connection caps									
6792	Beam size 3-9/16"									
6794	12 ga. post, 4" x 4"	1 Carp	1	8	C	2,900	395		3,295	3,775
6796	4" x 6"		1	8		3,850	395		4,245	4,850
6798	4" x 8"		1	8		5,700	395		6,095	6,850
6800	16 ga. post, 4" x 4"		1	8		1,200	395		1,595	1,900
6802	4" x 6"		1	8		2,000	395		2,395	2,775
6804	4" x 8"		1	8		3,350	395		3,745	4,275
6805	18 ga. post, 2-7/8" x 3"		1	8		3,350	395		3,745	4,275
6806	Beam size 5-1/2"									
6808	12 ga. post, 6" x 4"	1 Carp	1	8	C	3,350	395		3,745	4,275
6810	6" x 6"		1	8		5,450	395		5,845	6,600
6812	6" x 8"		1	8		3,750	395		4,145	4,725
6816	16 ga. post, 6" x 4"		1	8		1,875	395		2,270	2,675
6818	6" x 6"		1	8		2,000	395		2,395	2,775
6820	Beam size 7-1/2"									
6822	12 ga. post, 8" x 4"	1 Carp	1	8	C	4,775	395		5,170	5,850
6824	8" x 6"		1	8		5,025	395		5,420	6,125
6826	8" x 8"		1	8		7,550	395		7,945	8,900
6840	Purlin anchors, embedded									
6842	Heavy duty, 10 ga.									
6844	Straight, 28" long	1 Carp	1.60	5	C	1,425	246		1,671	1,950
6846	35" long		1.50	5.333		1,775	263		2,038	2,350
6848	Twisted, 28" long		1.60	5		1,425	246		1,671	1,950
6850	35" long		1.50	5.333		1,775	263		2,038	2,350
6852	Regular duty, 12 ga.									
6854	Straight, 18-1/2" long	1 Carp	1.80	4.444	C	820	219		1,039	1,225
6856	23-3/4" long		1.70	4.706		1,025	232		1,257	1,475
6858	29" long		1.60	5		1,050	246		1,296	1,525
6860	35" long		1.50	5.333		1,450	263		1,713	2,000
6862	Twisted, 18" long		1.80	4.444		820	219		1,039	1,225
6866	28" long		1.60	5		980	246		1,226	1,450

06 05 23.60 Timber Connectors		Crew	Daily Output	Labor-Hours	Unit	Material	2017 Bare Costs Labor	Equipment	Total	Total Incl O&P
6868	35" long	1 Carp	1.50	5.333	C	1,450	263		1,713	2,000
6870	Straight, plastic coated									
6872	23-1/2" long	1 Carp	1.60	5	C	2,025	246		2,271	2,600
6874	26-7/8" long		1.60	5		2,375	246		2,621	2,975
6876	32-1/2" long		1.50	5.333		2,525	263		2,788	3,175
6878	35-7/8" long		1.50	5.333		2,625	263		2,888	3,300
6890	Purlin hangers, painted									
6892	12 ga., 2" x 6"	1 Carp	1.80	4.444	C	1,850	219		2,069	2,375
6894	2" x 8"		1.80	4.444		2,025	219		2,244	2,550
6896	2" x 10"		1.80	4.444		2,175	219		2,394	2,725
6898	2" x 12"		1.75	4.571		2,325	225		2,550	2,925
6900	2" x 14"		1.75	4.571		2,500	225		2,725	3,100
6902	2" x 16"		1.75	4.571		2,650	225		2,875	3,275
6904	3" x 6"		1.70	4.706		1,875	232		2,107	2,400
6906	3" x 8"		1.70	4.706		2,025	232		2,257	2,575
6908	3" x 10"		1.70	4.706		2,175	232		2,407	2,750
6910	3" x 12"		1.65	4.848		2,500	239		2,739	3,125
6912	3" x 14"		1.65	4.848		2,675	239		2,914	3,300
6914	3" x 16"		1.65	4.848		2,825	239		3,064	3,475
6916	4" x 6"		1.65	4.848		1,875	239		2,114	2,425
6918	4" x 8"		1.65	4.848		2,025	239		2,264	2,600
6920	4" x 10"		1.65	4.848		2,200	239		2,439	2,775
6922	4" x 12"		1.60	5		2,600	246		2,846	3,225
6924	4" x 14"		1.60	5		2,750	246		2,996	3,400
6926	4" x 16"		1.60	5		2,925	246		3,171	3,575
6928	6" x 6"		1.60	5		2,475	246		2,721	3,100
6930	6" x 8"		1.60	5		2,625	246		2,871	3,250
6932	6" x 10"		1.55	5.161		2,675	254		2,929	3,350
6934	double 2" x 6"		1.70	4.706		2,025	232		2,257	2,575
6936	double 2" x 8"		1.70	4.706		2,175	232		2,407	2,750
6938	double 2" x 10"		1.70	4.706		2,350	232		2,582	2,925
6940	double 2" x 12"		1.65	4.848		2,500	239		2,739	3,125
6942	double 2" x 14"		1.65	4.848		2,675	239		2,914	3,300
6944	double 2" x 16"		1.65	4.848		2,825	239		3,064	3,475
6960	11 ga., 4" x 6"		1.65	4.848		3,700	239		3,939	4,450
6962	4" x 8"		1.65	4.848		3,975	239		4,214	4,750
6964	4" x 10"		1.65	4.848		4,250	239		4,489	5,050
6966	6" x 6"		1.60	5		3,750	246		3,996	4,500
6968	6" x 8"		1.60	5		4,025	246		4,271	4,800
6970	6" x 10"		1.55	5.161		4,300	254		4,554	5,125
6972	6" x 12"		1.55	5.161		4,575	254		4,829	5,425
6974	6" x 14"		1.55	5.161		4,850	254		5,104	5,725
6976	6" x 16"		1.50	5.333		5,125	263		5,388	6,050
6978	7 ga., 8" x 6"		1.60	5		4,075	246		4,321	4,850
6980	8" x 8"		1.60	5		4,350	246		4,596	5,150
6982	8" x 10"		1.55	5.161		4,625	254		4,879	5,475
6984	8" x 12"		1.55	5.161		4,900	254		5,154	5,800
6986	8" x 14"		1.50	5.333		5,175	263		5,438	6,100
6988	8" x 16"		1.50	5.333		5,450	263		5,713	6,400
7000	Strap connectors, galvanized									
7002	12 ga., 2-1/16" x 36"	1 Carp	1.55	5.161	C	1,175	254		1,429	1,700
7004	2-1/16" x 47"		1.50	5.333		1,625	263		1,888	2,200
7005	10 ga., 2-1/16" x 72"		1.50	5.333		1,725	263		1,988	2,300

06 05 23 – Wood, Plastic, and Composite Fastenings

06 05 23.60 Timber Connectors		Crew	Daily Output	Labor-Hours	Unit	Material	2017 Bare Costs Labor	Equipment	Total	Total Incl O&P
7006	7 ga., 2-1/16" x 34"	1 Carp	1.55	5.161	C	2,900	254		3,154	3,600
7008	2-1/16" x 45"		1.50	5.333		3,825	263		4,088	4,600
7010	3 ga., 3" x 32"		1.55	5.161		4,925	254		5,179	5,825
7012	3" x 41"		1.55	5.161		5,125	254		5,379	6,025
7014	3" x 50"		1.50	5.333		7,825	263		8,088	9,000
7016	3" x 59"		1.50	5.333		9,475	263		9,738	10,800
7018	3-1/2" x 68"	↓	1.45	5.517	↓	9,650	272		9,922	11,000
7030	Tension ties									
7032	19-1/8" long, 16 ga., 3/4" anchor bolt	1 Carp	1.80	4.444	C	1,350	219		1,569	1,800
7034	20" long, 12 ga., 1/2" anchor bolt		1.80	4.444		1,750	219		1,969	2,250
7036	20" long, 12 ga., 3/4" anchor bolt		1.80	4.444		1,750	219		1,969	2,250
7038	27-3/4" long, 12 ga., 3/4" anchor bolt	↓	1.75	4.571	↓	3,100	225		3,325	3,775
7050	Truss connectors, galvanized									
7052	Adjustable hanger									
7054	18 ga., 2" x 6"	1 Carp	1.65	4.848	C	530	239		769	945
7056	4" x 6"		1.65	4.848		695	239		934	1,125
7058	16 ga., 4" x 10"		1.60	5		1,025	246		1,271	1,500
7060	(2) 2" x 10"	↓	1.60	5	↓	1,025	246		1,271	1,500
7062	Connectors to plate									
7064	16 ga., 2" x 4" plate	1 Carp	1.80	4.444	C	520	219		739	905
7066	2" x 6" plate	"	1.80	4.444	"	670	219		889	1,075
7068	Hip jack connector									
7070	14 ga.	1 Carp	1.50	5.333	C	3,225	263		3,488	3,950

06 05 23.80 Metal Bracing

0010	METAL BRACING									
0302	Let-in, "T" shaped, 22 ga. galv. steel, studs at 16" O.C.	1 Carp	580	.014	L.F.	.83	.68		1.51	1.95
0402	Studs at 24" O.C.		600	.013		.83	.66		1.49	1.92
0502	Steel straps, 16 ga. galv. steel, studs at 16" O.C.		600	.013		1.07	.66		1.73	2.19
0602	Studs at 24" O.C.	↓	620	.013	↓	1.07	.64		1.71	2.15

06 11 Wood Framing

06 11 10 – Framing with Dimensional, Engineered or Composite Lumber

06 11 10.01 Forest Stewardship Council Certification

0010	FOREST STEWARDSHIP COUNCIL CERTIFICATION									
0020	For Forest Stewardship Council (FSC) cert dimension lumber, add	G				65%				

06 11 10.02 Blocking

0010	BLOCKING									
2600	Miscellaneous, to wood construction									
2620	2" x 4"	1 Carp	.17	47.059	M.B.F.	615	2,325		2,940	4,225
2625	Pneumatic nailed		.21	38.095		620	1,875		2,495	3,550
2660	2" x 8"		.27	29.630		655	1,450		2,105	2,950
2665	Pneumatic nailed	↓	.33	24.242	↓	660	1,200		1,860	2,550
2720	To steel construction									
2740	2" x 4"	1 Carp	.14	57.143	M.B.F.	615	2,825		3,440	4,975
2780	2" x 8"	"	.21	38.095	"	655	1,875		2,530	3,600

06 11 10.04 Wood Bracing

0010	WOOD BRACING									
0012	Let-in, with 1" x 6" boards, studs @ 16" O.C.	1 Carp	150	.053	L.F.	.76	2.63		3.39	4.85
0202	Studs @ 24" O.C.	"	230	.035	"	.76	1.71		2.47	3.45

For customer support on your Building Construction Costs with RSMeans Data, call 800.448.8182.

06 11 Wood Framing

06 11 10 – Framing with Dimensional, Engineered or Composite Lumber

06 11 10.06 Bridging

		Crew	Daily Output	Labor-Hours	Unit	Material	2017 Bare Costs Labor	2017 Bare Costs Equipment	Total	Total Incl O&P
0010	**BRIDGING**									
0012	Wood, for joists 16" O.C., 1" x 3"	1 Carp	130	.062	Pr.	.68	3.03		3.71	5.40
0017	Pneumatic nailed		170	.047		.73	2.32		3.05	4.36
0102	2" x 3" bridging		130	.062		.68	3.03		3.71	5.40
0107	Pneumatic nailed		170	.047		.70	2.32		3.02	4.32
0302	Steel, galvanized, 18 ga., for 2" x 10" joists at 12" O.C.		130	.062		1.98	3.03		5.01	6.80
0352	16" O.C.		135	.059		1.86	2.92		4.78	6.50
0402	24" O.C.		140	.057		1.94	2.81		4.75	6.45
0602	For 2" x 14" joists at 16" O.C.		130	.062		1.40	3.03		4.43	6.20
0902	Compression type, 16" O.C., 2" x 8" joists		200	.040		.94	1.97		2.91	4.06
1002	2" x 12" joists		200	.040		.94	1.97		2.91	4.06

06 11 10.10 Beam and Girder Framing

		Crew	Daily Output	Labor-Hours	Unit	Material	2017 Bare Costs Labor	2017 Bare Costs Equipment	Total	Total Incl O&P
0010	**BEAM AND GIRDER FRAMING** R061110-30									
3500	Single, 2" x 6"	2 Carp	.70	22.857	M.B.F.	630	1,125		1,755	2,425
3505	Pneumatic nailed		.81	19.704		635	970		1,605	2,175
3520	2" x 8"		.86	18.605		655	915		1,570	2,125
3525	Pneumatic nailed		1	16.048		660	790		1,450	1,925
3540	2" x 10"		1	16		825	790		1,615	2,100
3545	Pneumatic nailed		1.16	13.793		830	680		1,510	1,975
3560	2" x 12"		1.10	14.545		875	715		1,590	2,075
3565	Pneumatic nailed		1.28	12.539		880	620		1,500	1,925
3580	2" x 14"		1.17	13.675		950	675		1,625	2,075
3585	Pneumatic nailed		1.36	11.791		955	580		1,535	1,950
3600	3" x 8"		1.10	14.545		1,400	715		2,115	2,650
3620	3" x 10"		1.25	12.800		1,400	630		2,030	2,525
3640	3" x 12"		1.35	11.852		1,400	585		1,985	2,450
3660	3" x 14"		1.40	11.429		1,400	565		1,965	2,400
3680	4" x 8"	F-3	2.66	15.038		1,275	760	215	2,250	2,775
3700	4" x 10"		3.16	12.658		1,450	640	181	2,271	2,775
3720	4" x 12"		3.60	11.111		1,350	560	159	2,069	2,500
3740	4" x 14"		3.96	10.101		1,350	510	145	2,005	2,425
4000	Double, 2" x 6"	2 Carp	1.25	12.800		630	630		1,260	1,650
4005	Pneumatic nailed		1.45	11.034		635	545		1,180	1,525
4020	2" x 8"		1.60	10		655	495		1,150	1,475
4025	Pneumatic nailed		1.86	8.621		660	425		1,085	1,375
4040	2" x 10"		1.92	8.333		825	410		1,235	1,525
4045	Pneumatic nailed		2.23	7.185		830	355		1,185	1,450
4060	2" x 12"		2.20	7.273		875	360		1,235	1,525
4065	Pneumatic nailed		2.55	6.275		880	310		1,190	1,450
4080	2" x 14"		2.45	6.531		950	320		1,270	1,550
4085	Pneumatic nailed		2.84	5.634		955	277		1,232	1,475
5000	Triple, 2" x 6"		1.65	9.697		630	480		1,110	1,425
5005	Pneumatic nailed		1.91	8.377		635	415		1,050	1,325
5020	2" x 8"		2.10	7.619		655	375		1,030	1,300
5025	Pneumatic nailed		2.44	6.568		660	325		985	1,225
5040	2" x 10"		2.50	6.400		825	315		1,140	1,400
5045	Pneumatic nailed		2.90	5.517		830	272		1,102	1,325
5060	2" x 12"		2.85	5.614		875	276		1,151	1,400
5065	Pneumatic nailed		3.31	4.840		880	238		1,118	1,325
5080	2" x 14"		3.15	5.079		950	250		1,200	1,425
5085	Pneumatic nailed		3.35	4.770		955	235		1,190	1,400

06 11 Wood Framing

06 11 10 – Framing with Dimensional, Engineered or Composite Lumber

06 11 10.12 Ceiling Framing

		Crew	Daily Output	Labor-Hours	Unit	Material	2017 Bare Costs Labor	Equipment	Total	Total Incl O&P
0010	**CEILING FRAMING**									
6400	Suspended, 2" x 3"	2 Carp	.50	32	M.B.F.	790	1,575		2,365	3,300
6450	2" x 4"		.59	27.119		615	1,325		1,940	2,725
6500	2" x 6"		.80	20		630	985		1,615	2,200
6550	2" x 8"	↓	.86	18.605	↓	655	915		1,570	2,125

06 11 10.14 Posts and Columns

		Crew	Daily Output	Labor-Hours	Unit	Material	2017 Bare Costs Labor	Equipment	Total	Total Incl O&P
0010	**POSTS AND COLUMNS**									
0400	4" x 4"	2 Carp	.52	30.769	M.B.F.	1,350	1,525		2,875	3,800
0420	4" x 6"		.55	29.091		1,500	1,425		2,925	3,850
0440	4" x 8"		.59	27.119		1,275	1,325		2,600	3,450
0460	6" x 6"		.65	24.615		1,750	1,200		2,950	3,775
0480	6" x 8"		.70	22.857		1,925	1,125		3,050	3,850
0500	6" x 10"	↓	.75	21.333	↓	1,350	1,050		2,400	3,100

06 11 10.18 Joist Framing

		Crew	Daily Output	Labor-Hours	Unit	Material	2017 Bare Costs Labor	Equipment	Total	Total Incl O&P
0010	**JOIST FRAMING** R061110-30									
2650	Joists, 2" x 4"	2 Carp	.83	19.277	M.B.F.	615	950		1,565	2,125
2655	Pneumatic nailed		.96	16.667		620	820		1,440	1,925
2680	2" x 6"		1.25	12.800		630	630		1,260	1,650
2685	Pneumatic nailed		1.44	11.111		635	545		1,180	1,525
2700	2" x 8"		1.46	10.959		655	540		1,195	1,550
2705	Pneumatic nailed		1.68	9.524		660	470		1,130	1,450
2720	2" x 10"		1.49	10.738		825	530		1,355	1,725
2725	Pneumatic nailed		1.71	9.357		830	460		1,290	1,625
2740	2" x 12"		1.75	9.143		875	450		1,325	1,650
2745	Pneumatic nailed		2.01	7.960		880	390		1,270	1,575
2760	2" x 14"		1.79	8.939		950	440		1,390	1,725
2765	Pneumatic nailed		2.06	7.767		955	385		1,340	1,625
2780	3" x 6"		1.39	11.511		1,300	565		1,865	2,300
2790	3" x 8"		1.90	8.421		1,400	415		1,815	2,175
2800	3" x 10"		1.95	8.205		1,400	405		1,805	2,175
2820	3" x 12"		1.80	8.889		1,400	440		1,840	2,225
2840	4" x 6"		1.60	10		1,500	495		1,995	2,400
2860	4" x 10"		2	8		1,450	395		1,845	2,200
2880	4" x 12"		1.80	8.889	↓	1,350	440		1,790	2,150
3000	Composite wood joist 9-1/2" deep		.90	17.778	M.L.F.	1,675	875		2,550	3,200
3010	11-1/2" deep		.88	18.182		1,900	895		2,795	3,475
3020	14" deep		.82	19.512		2,275	960		3,235	3,975
3030	16" deep		.78	20.513		3,650	1,000		4,650	5,550
4000	Open web joist 12" deep		.88	18.182	↓	3,600	895		4,495	5,325
4002	Per linear foot		880	.018	L.F.	3.60	.90		4.50	5.35
4004	Treated, per linear foot		880	.018	"	4.46	.90		5.36	6.30
4010	14" deep		.82	19.512	M.L.F.	3,650	960		4,610	5,475
4012	Per linear foot		820	.020	L.F.	3.64	.96		4.60	5.45
4014	Treated, per linear foot		820	.020	"	4.65	.96		5.61	6.55
4020	16" deep		.78	20.513	M.L.F.	3,900	1,000		4,900	5,850
4022	Per linear foot		780	.021	L.F.	3.90	1.01		4.91	5.85
4024	Treated, per linear foot		780	.021	"	5.05	1.01		6.06	7.10
4030	18" deep		.74	21.622	M.L.F.	3,800	1,075		4,875	5,800
4032	Per linear foot		740	.022	L.F.	3.80	1.06		4.86	5.80
4034	Treated, per linear foot		740	.022	"	5.10	1.06		6.16	7.25
6000	Composite rim joist, 1-1/4" x 9-1/2"		.90	17.778	M.L.F.	1,900	875		2,775	3,425
6010	1-1/4" x 11-1/2"	↓	.88	18.182	↓	2,300	895		3,195	3,925

06 11 Wood Framing

06 11 10 – Framing with Dimensional, Engineered or Composite Lumber

06 11 10.18 Joist Framing	Crew	Daily Output	Labor-Hours	Unit	Material	2017 Bare Costs Labor	Equipment	Total	Total Incl O&P
6020	1-1/4" x 14-1/2"	2 Carp	.82	19.512	M.L.F.	2,700	960		3,660	4,450
6030	1-1/4" x 16-1/2"	↓	.78	20.513	↓	2,925	1,000		3,925	4,775

06 11 10.24 Miscellaneous Framing

		Crew	Daily Output	Labor-Hours	Unit	Material	Labor	Equipment	Total	Total Incl O&P
0010	**MISCELLANEOUS FRAMING**									
8500	Firestops, 2" x 4"	2 Carp	.51	31.373	M.B.F.	615	1,550		2,165	3,050
8505	Pneumatic nailed		.62	25.806		620	1,275		1,895	2,625
8520	2" x 6"		.60	26.667		630	1,325		1,955	2,700
8525	Pneumatic nailed		.73	21.858		635	1,075		1,710	2,350
8540	2" x 8"		.60	26.667		655	1,325		1,980	2,725
8560	2" x 12"		.70	22.857		875	1,125		2,000	2,700
8600	Nailers, treated, wood construction, 2" x 4"		.53	30.189		790	1,475		2,265	3,150
8605	Pneumatic nailed		.64	25.157		795	1,250		2,045	2,775
8620	2" x 6"		.75	21.333		815	1,050		1,865	2,500
8625	Pneumatic nailed		.90	17.778		820	875		1,695	2,250
8640	2" x 8"		.93	17.204		795	845		1,640	2,175
8645	Pneumatic nailed		1.12	14.337		800	705		1,505	1,950
8660	Steel construction, 2" x 4"		.50	32		795	1,575		2,370	3,300
8680	2" x 6"		.70	22.857		815	1,125		1,940	2,625
8700	2" x 8"		.87	18.391		795	905		1,700	2,275
8760	Rough bucks, treated, for doors or windows, 2" x 6"		.40	40		815	1,975		2,790	3,925
8765	Pneumatic nailed		.48	33.333		820	1,650		2,470	3,425
8780	2" x 8"		.51	31.373		795	1,550		2,345	3,250
8785	Pneumatic nailed		.61	26.144		800	1,300		2,100	2,850
8800	Stair stringers, 2" x 10"		.22	72.727		825	3,575		4,400	6,375
8820	2" x 12"		.26	61.538		875	3,025		3,900	5,625
8840	3" x 10"		.31	51.613		1,400	2,550		3,950	5,450
8860	3" x 12"		.38	42.105	↓	1,400	2,075		3,475	4,725
8870	Laminated structural lumber, 1-1/4" x 11-1/2"		130	.123	L.F.	2.31	6.05		8.36	11.85
8880	1-1/4" x 14-1/2"	↓	130	.123	"	2.67	6.05		8.72	12.25

06 11 10.26 Partitions

		Crew	Daily Output	Labor-Hours	Unit	Material	Labor	Equipment	Total	Total Incl O&P
0010	**PARTITIONS**									
0020	Single bottom and double top plate, no waste, std. & better lumber									
0180	2" x 4" studs, 8' high, studs 12" O.C.	2 Carp	80	.200	L.F.	4.49	9.85		14.34	20
0185	12" O.C., pneumatic nailed		96	.167		4.53	8.20		12.73	17.60
0200	16" O.C.		100	.160		3.68	7.90		11.58	16.10
0205	16" O.C., pneumatic nailed		120	.133		3.71	6.55		10.26	14.15
0300	24" O.C.		125	.128		2.86	6.30		9.16	12.80
0305	24" O.C., pneumatic nailed		150	.107		2.88	5.25		8.13	11.20
0380	10' high, studs 12" O.C.		80	.200		5.30	9.85		15.15	21
0385	12" O.C., pneumatic nailed		96	.167		5.35	8.20		13.55	18.50
0400	16" O.C.		100	.160		4.29	7.90		12.19	16.75
0405	16" O.C., pneumatic nailed		120	.133		4.33	6.55		10.88	14.80
0500	24" O.C.		125	.128		3.27	6.30		9.57	13.25
0505	24" O.C., pneumatic nailed		150	.107		3.30	5.25		8.55	11.70
0580	12' high, studs 12" O.C.		65	.246		6.15	12.10		18.25	25.50
0585	12" O.C., pneumatic nailed		78	.205		6.20	10.10		16.30	22.50
0600	16" O.C.		80	.200		4.90	9.85		14.75	20.50
0605	16" O.C., pneumatic nailed		96	.167		4.94	8.20		13.14	18.05
0700	24" O.C.		100	.160		3.68	7.90		11.58	16.10
0705	24" O.C., pneumatic nailed		120	.133		3.71	6.55		10.26	14.15
0780	2" x 6" studs, 8' high, studs 12" O.C.		70	.229		6.90	11.25		18.15	25
0785	12" O.C., pneumatic nailed	↓	84	.190	↓	6.95	9.40		16.35	22

06 11 10.26 Partitions

		Crew	Daily Output	Labor-Hours	Unit	Material	2017 Bare Costs Labor	2017 Bare Costs Equipment	Total	Total Incl O&P
0800	16" O.C.	2 Carp	90	.178	L.F.	5.65	8.75		14.40	19.60
0805	16" O.C., pneumatic nailed		108	.148		5.70	7.30		13	17.45
0900	24" O.C.		115	.139		4.40	6.85		11.25	15.35
0905	24" O.C., pneumatic nailed		138	.116		4.43	5.70		10.13	13.60
0980	10' high, studs 12" O.C.		70	.229		8.15	11.25		19.40	26.50
0985	12" O.C., pneumatic nailed		84	.190		8.25	9.40		17.65	23.50
1000	16" O.C.		90	.178		6.60	8.75		15.35	20.50
1005	16" O.C., pneumatic nailed		108	.148		6.65	7.30		13.95	18.50
1100	24" O.C.		115	.139		5	6.85		11.85	16.05
1105	24" O.C., pneumatic nailed		138	.116		5.05	5.70		10.75	14.30
1180	12' high, studs 12" O.C.		55	.291		9.40	14.35		23.75	32.50
1185	12" O.C., pneumatic nailed		66	.242		9.50	11.95		21.45	29
1200	16" O.C.		70	.229		7.55	11.25		18.80	25.50
1205	16" O.C., pneumatic nailed		84	.190		7.60	9.40		17	22.50
1300	24" O.C.		90	.178		5.65	8.75		14.40	19.60
1305	24" O.C., pneumatic nailed		108	.148		5.70	7.30		13	17.45
1400	For horizontal blocking, 2" x 4", add		600	.027		.41	1.31		1.72	2.46
1500	2" x 6", add		600	.027		.63	1.31		1.94	2.70
1600	For openings, add		250	.064			3.15		3.15	4.83
1700	Headers for above openings, material only, add				M.B.F.	660			660	725

06 11 10.28 Porch or Deck Framing

		Crew	Daily Output	Labor-Hours	Unit	Material	2017 Bare Costs Labor	2017 Bare Costs Equipment	Total	Total Incl O&P
0010	**PORCH OR DECK FRAMING**									
0100	Treated lumber, posts or columns, 4" x 4"	2 Carp	390	.041	L.F.	1.37	2.02		3.39	4.61
0110	4" x 6"		275	.058		2.06	2.87		4.93	6.65
0120	4" x 8"		220	.073		3.84	3.58		7.42	9.75
0130	Girder, single, 4" x 4"		675	.024		1.37	1.17		2.54	3.30
0140	4" x 6"		600	.027		2.06	1.31		3.37	4.27
0150	4" x 8"		525	.030		3.84	1.50		5.34	6.55
0160	Double, 2" x 4"		625	.026		1.08	1.26		2.34	3.12
0170	2" x 6"		600	.027		1.67	1.31		2.98	3.84
0180	2" x 8"		575	.028		2.17	1.37		3.54	4.49
0190	2" x 10"		550	.029		2.86	1.43		4.29	5.35
0200	2" x 12"		525	.030		4.06	1.50		5.56	6.75
0210	Triple, 2" x 4"		575	.028		1.62	1.37		2.99	3.89
0220	2" x 6"		550	.029		2.50	1.43		3.93	4.94
0230	2" x 8"		525	.030		3.26	1.50		4.76	5.90
0240	2" x 10"		500	.032		4.29	1.58		5.87	7.15
0250	2" x 12"		475	.034		6.10	1.66		7.76	9.25
0260	Ledger, bolted 4' O.C., 2" x 4"		400	.040		.70	1.97		2.67	3.79
0270	2" x 6"		395	.041		.98	2		2.98	4.14
0280	2" x 8"		390	.041		1.22	2.02		3.24	4.44
0290	2" x 10"		385	.042		1.55	2.05		3.60	4.85
0300	2" x 12"		380	.042		2.14	2.07		4.21	5.55
0310	Joists, 2" x 4"		1250	.013		.54	.63		1.17	1.57
0320	2" x 6"		1250	.013		.83	.63		1.46	1.89
0330	2" x 8"		1100	.015		1.09	.72		1.81	2.29
0340	2" x 10"		900	.018		1.43	.88		2.31	2.91
0350	2" x 12"		875	.018		1.71	.90		2.61	3.27
0360	Railings and trim, 1" x 4"	1 Carp	300	.027		.50	1.31		1.81	2.56
0370	2" x 2"		300	.027		.36	1.31		1.67	2.40
0380	2" x 4"		300	.027		.53	1.31		1.84	2.60
0390	2" x 6"		300	.027		.82	1.31		2.13	2.91

06 11 10 – Framing with Dimensional, Engineered or Composite Lumber

06 11 10.28 Porch or Deck Framing

		Crew	Daily Output	Labor-Hours	Unit	Material	2017 Bare Costs Labor	Equipment	Total	Total Incl O&P
0400	Decking, 1" x 4"	1 Carp	275	.029	S.F.	2.37	1.43		3.80	4.79
0410	2" x 4"		300	.027		1.79	1.31		3.10	3.98
0420	2" x 6"		320	.025		1.77	1.23		3	3.83
0430	5/4" x 6"	▼	320	.025	▼	2.06	1.23		3.29	4.15
0440	Balusters, square, 2" x 2"	2 Carp	660	.024	L.F.	.36	1.19		1.55	2.23
0450	Turned, 2" x 2"		420	.038		.48	1.88		2.36	3.39
0460	Stair stringer, 2" x 10"		130	.123		1.43	6.05		7.48	10.85
0470	2" x 12"		130	.123		1.71	6.05		7.76	11.20
0480	Stair treads, 1" x 4"		140	.114		2.37	5.65		8.02	11.20
0490	2" x 4"		140	.114		.54	5.65		6.19	9.20
0500	2" x 6"		160	.100		.81	4.93		5.74	8.45
0510	5/4" x 6"		160	.100	▼	.96	4.93		5.89	8.60
0520	Turned handrail post, 4" x 4"		64	.250	Ea.	32.50	12.30		44.80	55
0530	Lattice panel, 4' x 8', 1/2"		1600	.010	S.F.	.82	.49		1.31	1.65
0535	3/4"		1600	.010	"	1.22	.49		1.71	2.09
0540	Cedar, posts or columns, 4" x 4"		390	.041	L.F.	4.02	2.02		6.04	7.50
0550	4" x 6"		275	.058		7.30	2.87		10.17	12.40
0560	4" x 8"		220	.073		10.45	3.58		14.03	16.95
0800	Decking, 1" x 4"		550	.029		2.02	1.43		3.45	4.42
0810	2" x 4"		600	.027		3.99	1.31		5.30	6.40
0820	2" x 6"		640	.025		7.45	1.23		8.68	10.10
0830	5/4" x 6"		640	.025		4.80	1.23		6.03	7.20
0840	Railings and trim, 1" x 4"		600	.027		2.02	1.31		3.33	4.24
0860	2" x 4"		600	.027		3.99	1.31		5.30	6.40
0870	2" x 6"		600	.027		7.45	1.31		8.76	10.20
0920	Stair treads, 1" x 4"		140	.114		2.02	5.65		7.67	10.85
0930	2" x 4"		140	.114		3.99	5.65		9.64	13
0940	2" x 6"		160	.100		7.45	4.93		12.38	15.75
0950	5/4" x 6"		160	.100		4.80	4.93		9.73	12.85
0980	Redwood, posts or columns, 4" x 4"		390	.041		6.45	2.02		8.47	10.15
0990	4" x 6"		275	.058		12.60	2.87		15.47	18.25
1000	4" x 8"	▼	220	.073	▼	23.50	3.58		27.08	31.50
1240	Decking, 1" x 4"	1 Carp	275	.029	S.F.	3.97	1.43		5.40	6.55
1260	2" x 6"		340	.024		7.50	1.16		8.66	10.05
1270	5/4" x 6"	▼	320	.025	▼	4.77	1.23		6	7.15
1280	Railings and trim, 1" x 4"	2 Carp	600	.027	L.F.	1.17	1.31		2.48	3.30
1310	2" x 6"		600	.027		7.50	1.31		8.81	10.25
1420	Alternative decking, wood/plastic composite, 5/4" x 6" [G]		640	.025		3.42	1.23		4.65	5.65
1440	1" x 4" square edge fir		550	.029		2.38	1.43		3.81	4.81
1450	1" x 4" tongue and groove fir		450	.036		1.52	1.75		3.27	4.35
1460	1" x 4" mahogany		550	.029		2.16	1.43		3.59	4.57
1462	5/4" x 6" PVC	▼	550	.029	▼	3.91	1.43		5.34	6.50
1465	Framing, porch or deck, alt deck fastening, screws, add	1 Carp	240	.033	S.F.		1.64		1.64	2.51
1470	Accessories, joist hangers, 2" x 4"		160	.050	Ea.	.80	2.46		3.26	4.65
1480	2" x 6" through 2" x 12"	▼	150	.053		1.36	2.63		3.99	5.50
1530	Post footing, incl excav, backfill, tube form & concrete, 4' deep, 8" dia	F-7	12	2.667		13.55	118		131.55	196
1540	10" diameter		11	2.909		19.70	129		148.70	219
1550	12" diameter	▼	10	3.200	▼	26.50	141		167.50	246

06 11 10.30 Roof Framing

		Crew	Daily Output	Labor-Hours	Unit	Material	2017 Bare Costs Labor	Equipment	Total	Total Incl O&P
0010	**ROOF FRAMING**									
5250	Composite rafter, 9-1/2" deep	2 Carp	575	.028	L.F.	1.69	1.37		3.06	3.96
5260	11-1/2" deep		575	.028	"	1.91	1.37		3.28	4.20

06 11 10 – Framing with Dimensional, Engineered or Composite Lumber

06 11 10.30 Roof Framing

		Crew	Daily Output	Labor-Hours	Unit	Material	2017 Bare Costs Labor	2017 Bare Costs Equipment	Total	Total Incl O&P
6070	Fascia boards, 2" x 8"	2 Carp	.30	53.333	M.B.F.	655	2,625		3,280	4,750
6080	2" x 10"		.30	53.333		825	2,625		3,450	4,925
7000	Rafters, to 4 in 12 pitch, 2" x 6"		1	16		630	790		1,420	1,900
7060	2" x 8"		1.26	12.698		655	625		1,280	1,675
7300	Hip and valley rafters, 2" x 6"		.76	21.053		630	1,025		1,655	2,300
7360	2" x 8"		.96	16.667		655	820		1,475	1,975
7540	Hip and valley jacks, 2" x 6"		.60	26.667		630	1,325		1,955	2,700
7600	2" x 8"		.65	24.615		655	1,200		1,855	2,575
7780	For slopes steeper than 4 in 12, add						30%			
7790	For dormers or complex roofs, add						50%			
7800	Rafter tie, 1" x 4", #3	2 Carp	.27	59.259	M.B.F.	1,475	2,925		4,400	6,100
7820	Ridge board, #2 or better, 1" x 6"		.30	53.333		1,525	2,625		4,150	5,700
7840	1" x 8"		.37	43.243		1,900	2,125		4,025	5,350
7860	1" x 10"		.42	38.095		1,975	1,875		3,850	5,050
7880	2" x 6"		.50	32		630	1,575		2,205	3,125
7900	2" x 8"		.60	26.667		655	1,325		1,980	2,725
7920	2" x 10"		.66	24.242		825	1,200		2,025	2,725
7940	Roof cants, split, 4" x 4"		.86	18.605		1,350	915		2,265	2,875
7960	6" x 6"		1.80	8.889		1,750	440		2,190	2,600
7980	Roof curbs, untreated, 2" x 6"		.52	30.769		630	1,525		2,155	3,025
8000	2" x 12"		.80	20		875	985		1,860	2,475

06 11 10.32 Sill and Ledger Framing

		Crew	Daily Output	Labor-Hours	Unit	Material	2017 Bare Costs Labor	2017 Bare Costs Equipment	Total	Total Incl O&P
0010	**SILL AND LEDGER FRAMING**									
0020	Extruded polystyrene sill sealer, 5-1/2" wide	1 Carp	1600	.005	L.F.	.14	.25		.39	.53
4482	Ledgers, nailed, 2" x 4"	2 Carp	.50	32	M.B.F.	615	1,575		2,190	3,100
4484	2" x 6"		.60	26.667		630	1,325		1,955	2,700
4486	Bolted, not including bolts, 3" x 8"		.65	24.615		1,400	1,200		2,600	3,375
4488	3" x 12"		.70	22.857		1,400	1,125		2,525	3,275
4490	Mud sills, redwood, construction grade, 2" x 4"		.59	27.119		3,400	1,325		4,725	5,775
4492	2" x 6"		.78	20.513		3,400	1,000		4,400	5,300
4500	Sills, 2" x 4"		.40	40		605	1,975		2,580	3,700
4520	2" x 6"		.55	29.091		620	1,425		2,045	2,875
4540	2" x 8"		.67	23.881		645	1,175		1,820	2,500
4600	Treated, 2" x 4"		.36	44.444		780	2,200		2,980	4,200
4620	2" x 6"		.50	32		805	1,575		2,380	3,300
4640	2" x 8"		.60	26.667		785	1,325		2,110	2,850
4700	4" x 4"		.60	26.667		1,000	1,325		2,325	3,100
4720	4" x 6"		.70	22.857		1,000	1,125		2,125	2,825
4740	4" x 8"		.80	20		1,400	985		2,385	3,050
4760	4" x 10"		.87	18.391		1,450	905		2,355	2,975

06 11 10.34 Sleepers

		Crew	Daily Output	Labor-Hours	Unit	Material	2017 Bare Costs Labor	2017 Bare Costs Equipment	Total	Total Incl O&P
0010	**SLEEPERS**									
0300	On concrete, treated, 1" x 2"	2 Carp	.39	41.026	M.B.F.	1,725	2,025		3,750	5,000
0320	1" x 3"		.50	32		1,900	1,575		3,475	4,500
0340	2" x 4"		.99	16.162		925	795		1,720	2,250
0360	2" x 6"		1.30	12.308		945	605		1,550	1,975

06 11 10.36 Soffit and Canopy Framing

		Crew	Daily Output	Labor-Hours	Unit	Material	2017 Bare Costs Labor	2017 Bare Costs Equipment	Total	Total Incl O&P
0010	**SOFFIT AND CANOPY FRAMING**									
1300	Canopy or soffit framing, 1" x 4"	2 Carp	.30	53.333	M.B.F.	1,475	2,625		4,100	5,650
1340	1" x 8"		.50	32		1,900	1,575		3,475	4,500
1360	2" x 4"		.41	39.024		615	1,925		2,540	3,625
1400	2" x 8"		.67	23.881		655	1,175		1,830	2,525

For customer support on your Building Construction Costs with RSMeans Data, call 800.448.8182.

06 11 Wood Framing

06 11 10 – Framing with Dimensional, Engineered or Composite Lumber

06 11 10.36 Soffit and Canopy Framing	Crew	Daily Output	Labor-Hours	Unit	Material	2017 Bare Costs Labor	Equipment	Total	Total Incl O&P
1420	3" x 4"	2 Carp	.50	32	M.B.F.	1,175	1,575		2,750	3,725
1460	3" x 8"	↓	.60	26.667	↓	1,400	1,325		2,725	3,550

06 11 10.38 Treated Lumber Framing Material

		Crew	Daily Output	Labor-Hours	Unit	Material	2017 Bare Costs Labor	Equipment	Total	Total Incl O&P
0010	**TREATED LUMBER FRAMING MATERIAL**									
0100	2" x 4"				M.B.F.	780			780	860
0110	2" x 6"					805			805	885
0120	2" x 8"					785			785	860
0130	2" x 10"					825			825	910
0140	2" x 12"					985			985	1,075
0200	4" x 4"					1,000			1,000	1,100
0210	4" x 6"					1,000			1,000	1,100
0220	4" x 8"				↓	1,400			1,400	1,550

06 11 10.40 Wall Framing

		Crew	Daily Output	Labor-Hours	Unit	Material	2017 Bare Costs Labor	Equipment	Total	Total Incl O&P
0010	**WALL FRAMING** R061110-30									
5860	Headers over openings, 2" x 6"	2 Carp	.36	44.444	M.B.F.	630	2,200		2,830	4,050
5865	2" x 6", pneumatic nailed		.43	37.209		635	1,825		2,460	3,500
5880	2" x 8"		.45	35.556		655	1,750		2,405	3,400
5885	2" x 8", pneumatic nailed		.54	29.630		660	1,450		2,110	2,950
5900	2" x 10"		.53	30.189		825	1,475		2,300	3,175
5905	2" x 10", pneumatic nailed		.67	23.881		830	1,175		2,005	2,725
5920	2" x 12"		.60	26.667		875	1,325		2,200	2,975
5925	2" x 12", pneumatic nailed		.72	22.222		880	1,100		1,980	2,650
5940	4" x 12"		.76	21.053		1,350	1,025		2,375	3,075
5945	4" x 12", pneumatic nailed		.92	17.391		1,350	855		2,205	2,775
5960	6" x 12"		.84	19.048		1,400	940		2,340	2,975
5965	6" x 12", pneumatic nailed		1.01	15.873		1,425	780		2,205	2,750
6000	Plates, untreated, 2" x 3"		.43	37.209		790	1,825		2,615	3,675
6005	2" x 3", pneumatic nailed		.52	30.769		795	1,525		2,320	3,200
6020	2" x 4"		.53	30.189		615	1,475		2,090	2,950
6025	2" x 4", pneumatic nailed		.67	23.881		620	1,175		1,795	2,475
6040	2" x 6"		.75	21.333		630	1,050		1,680	2,300
6045	2" x 6", pneumatic nailed		.90	17.778		635	875		1,510	2,050
6120	Studs, 8' high wall, 2" x 3"		.60	26.667		790	1,325		2,115	2,875
6125	2" x 3", pneumatic nailed		.72	22.222		795	1,100		1,895	2,550
6140	2" x 4"		.92	17.391		615	855		1,470	1,975
6145	2" x 4", pneumatic nailed		1.10	14.493		620	715		1,335	1,775
6160	2" x 6"		1	16		630	790		1,420	1,900
6165	2" x 6", pneumatic nailed		1.20	13.333		635	655		1,290	1,700
6180	3" x 4"		.80	20		1,175	985		2,160	2,800
6185	3" x 4", pneumatic nailed	↓	.96	16.667	↓	1,175	820		1,995	2,550
8200	For 12' high walls, deduct						5%			
8220	For stub wall, 6' high, add						20%			
8240	3' high, add						40%			
8250	For second story & above, add						5%			
8300	For dormer & gable, add						15%			

06 11 10.42 Furring

		Crew	Daily Output	Labor-Hours	Unit	Material	2017 Bare Costs Labor	Equipment	Total	Total Incl O&P
0010	**FURRING**									
0012	Wood strips, 1" x 2", on walls, on wood	1 Carp	550	.015	L.F.	.26	.72		.98	1.38
0015	On wood, pneumatic nailed		710	.011		.26	.56		.82	1.13
0300	On masonry		495	.016		.28	.80		1.08	1.53
0400	On concrete		260	.031		.28	1.52		1.80	2.63
0600	1" x 3", on walls, on wood	↓	550	.015	↓	.42	.72		1.14	1.56

06 11 10 – Framing with Dimensional, Engineered or Composite Lumber

06 11 10.42 Furring

		Crew	Daily Output	Labor-Hours	Unit	Material	2017 Bare Costs Labor	Equipment	Total	Total Incl O&P
0605	On wood, pneumatic nailed	1 Carp	710	.011	L.F.	.42	.56		.98	1.31
0700	On masonry		495	.016		.45	.80		1.25	1.71
0800	On concrete		260	.031		.45	1.52		1.97	2.81
0850	On ceilings, on wood		350	.023		.42	1.13		1.55	2.18
0855	On wood, pneumatic nailed		450	.018		.42	.88		1.30	1.80
0900	On masonry		320	.025		.45	1.23		1.68	2.38
0950	On concrete		210	.038		.45	1.88		2.33	3.36

06 11 10.44 Grounds

		Crew	Daily Output	Labor-Hours	Unit	Material	2017 Bare Costs Labor	Equipment	Total	Total Incl O&P
0010	**GROUNDS**									
0020	For casework, 1" x 2" wood strips, on wood	1 Carp	330	.024	L.F.	.26	1.19		1.45	2.11
0100	On masonry		285	.028		.28	1.38		1.66	2.43
0200	On concrete		250	.032		.28	1.58		1.86	2.72
0400	For plaster, 3/4" deep, on wood		450	.018		.26	.88		1.14	1.62
0500	On masonry		225	.036		.28	1.75		2.03	2.99
0600	On concrete		175	.046		.28	2.25		2.53	3.76
0700	On metal lath		200	.040		.28	1.97		2.25	3.33

06 12 10 – Structural Insulated Panels

06 12 10.10 OSB Faced Panels

			Crew	Daily Output	Labor-Hours	Unit	Material	2017 Bare Costs Labor	Equipment	Total	Total Incl O&P
0010	**OSB FACED PANELS**										
0100	Structural insul. panels, 7/16" OSB both faces, EPS insul, 3-5/8" T	G	F-3	2075	.019	S.F.	3.65	.97	.28	4.90	5.80
0110	5-5/8" thick	G		1725	.023		4.10	1.17	.33	5.60	6.65
0120	7-3/8" thick	G		1425	.028		4.45	1.42	.40	6.27	7.50
0130	9-3/8" thick	G		1125	.036		4.75	1.80	.51	7.06	8.55
0140	7/16" OSB one face, EPS insul, 3-5/8" thick	G		2175	.018		3.75	.93	.26	4.94	5.85
0150	5-5/8" thick	G		1825	.022		4.35	1.11	.31	5.77	6.85
0160	7-3/8" thick	G		1525	.026		4.85	1.33	.38	6.56	7.80
0170	9-3/8" thick	G		1225	.033		5.35	1.65	.47	7.47	8.95
0190	7/16" OSB - 1/2" GWB faces, EPS insul, 3-5/8" T	G		2075	.019		3.45	.97	.28	4.70	5.60
0200	5-5/8" thick	G		1725	.023		4.10	1.17	.33	5.60	6.65
0210	7-3/8" thick	G		1425	.028		4.65	1.42	.40	6.47	7.70
0220	9-3/8" thick	G		1125	.036		5.25	1.80	.51	7.56	9.10
0240	7/16" OSB - 1/2" MRGWB faces, EPS insul, 3-5/8" T	G		2075	.019		3.55	.97	.28	4.80	5.70
0250	5-5/8" thick	G		1725	.023		4.25	1.17	.33	5.75	6.85
0260	7-3/8" thick	G		1425	.028		4.65	1.42	.40	6.47	7.70
0270	9-3/8" thick	G		1125	.036		5.35	1.80	.51	7.66	9.20
0300	For 1/2" GWB added to OSB skin, add	G					1.40			1.40	1.54
0310	For 1/2" MRGWB added to OSB skin, add	G					1.40			1.40	1.54
0320	For one T1-11 skin, add to OSB-OSB	G					1.95			1.95	2.15
0330	For one 19/32" CDX skin, add to OSB-OSB	G					1.50			1.50	1.65
0500	Structural insulated panel, 7/16" OSB both sides, straw core										
0510	4-3/8" T, walls (w/sill, splines, plates)	G	F-6	2400	.017	S.F.	7.55	.78	.24	8.57	9.75
0520	Floors (w/splines)	G		2400	.017		7.55	.78	.24	8.57	9.75
0530	Roof (w/splines)	G		2400	.017		7.55	.78	.24	8.57	9.75
0550	7-7/8" T, walls (w/sill, splines, plates)	G		2400	.017		11.40	.78	.24	12.42	14
0560	Floors (w/splines)	G		2400	.017		11.40	.78	.24	12.42	14
0570	Roof (w/splines)	G		2400	.017		11.40	.78	.24	12.42	14

188

For customer support on your Building Construction Costs with RSMeans Data, call 800.448.8182.

06 12 Structural Panels

06 12 19 - Composite Shearwall Panels

06 12 19.10 Steel and Wood Composite Shearwall Panels	Crew	Daily Output	Labor-Hours	Unit	Material	2017 Bare Costs Labor	Equipment	Total	Total Incl O&P
0010 **STEEL & WOOD COMPOSITE SHEARWALL PANELS**									
0020 Anchor bolts, 36" long (must be placed in wet concrete)	1 Carp	150	.053	Ea.	34.50	2.63		37.13	42
0030 On concrete, 2 x 4 & 2 x 6 walls, 7' - 10' high, 360 lb. shear, 12" wide	2 Carp	8	2		460	98.50		558.50	655
0040 715 lb. shear, 15" wide		8	2		510	98.50		608.50	710
0050 1860 lb. shear, 18" wide		8	2		525	98.50		623.50	730
0060 2780 lb. shear, 21" wide		8	2		560	98.50		658.50	765
0070 3790 lb. shear, 24" wide		8	2		640	98.50		738.50	850
0080 2 x 6 walls, 11' - 13' high, 1180 lb. shear, 18" wide		6	2.667		645	131		776	910
0090 1555 lb. shear, 21" wide		6	2.667		720	131		851	990
0100 2280 lb. shear, 24" wide		6	2.667		805	131		936	1,075
0110 For installing above on wood floor frame, add									
0120 Coupler nuts, threaded rods, bolts, shear transfer plate kit	1 Carp	16	.500	Ea.	65	24.50		89.50	109
0130 Framing anchors, angle (2 required)	"	96	.083	"	2.56	4.10		6.66	9.10
0140 For blocking see Section 06 11 10.02									
0150 For installing above, first floor to second floor, wood floor frame, add									
0160 Add stack option to first floor wall panel				Ea.	71.50			71.50	78.50
0170 Threaded rods, bolts, shear transfer plate kit	1 Carp	16	.500		75	24.50		99.50	120
0180 Framing anchors, angle (2 required)	"	96	.083		2.56	4.10		6.66	9.10
0190 For blocking see section 06 11 10.02									
0200 For installing stacked panels, balloon framing									
0210 Add stack option to first floor wall panel				Ea.	71.50			71.50	78.50
0220 Threaded rods, bolts kit	1 Carp	16	.500	"	45	24.50		69.50	87

06 13 Heavy Timber Construction

06 13 23 - Heavy Timber Framing

06 13 23.10 Heavy Framing

		Crew	Daily Output	Labor-Hours	Unit	Material	2017 Bare Costs Labor	Equipment	Total	Total Incl O&P
0010	**HEAVY FRAMING**									
0020	Beams, single 6" x 10"	2 Carp	1.10	14.545	M.B.F.	1,550	715		2,265	2,825
0100	Single 8" x 16"		1.20	13.333		1,925	655		2,580	3,125
0200	Built from 2" lumber, multiple 2" x 14"		.90	17.778		940	875		1,815	2,375
0210	Built from 3" lumber, multiple 3" x 6"		.70	22.857		1,300	1,125		2,425	3,150
0220	Multiple 3" x 8"		.80	20		1,400	985		2,385	3,025
0230	Multiple 3" x 10"		.90	17.778		1,400	875		2,275	2,900
0240	Multiple 3" x 12"		1	16		1,400	790		2,190	2,750
0250	Built from 4" lumber, multiple 4" x 6"		.80	20		1,475	985		2,460	3,125
0260	Multiple 4" x 8"		.90	17.778		1,275	875		2,150	2,750
0270	Multiple 4" x 10"		1	16		1,450	790		2,240	2,775
0280	Multiple 4" x 12"		1.10	14.545		1,325	715		2,040	2,550
0290	Columns, structural grade, 1500f, 4" x 4"		.60	26.667		1,275	1,325		2,600	3,400
0300	6" x 6"		.65	24.615		1,450	1,200		2,650	3,450
0400	8" x 8"		.70	22.857		1,200	1,125		2,325	3,050
0500	10" x 10"		.75	21.333		1,625	1,050		2,675	3,375
0600	12" x 12"		.80	20		1,550	985		2,535	3,200
0800	Floor planks, 2" thick, T & G, 2" x 6"		1.05	15.238		1,600	750		2,350	2,925
0900	2" x 10"		1.10	14.545		1,625	715		2,340	2,900
1100	3" thick, 3" x 6"		1.05	15.238		1,625	750		2,375	2,925
1200	3" x 10"		1.10	14.545		1,650	715		2,365	2,925
1400	Girders, structural grade, 12" x 12"		.80	20		1,550	985		2,535	3,200
1500	10" x 16"		1	16		2,550	790		3,340	4,000
2300	Roof purlins, 4" thick, structural grade		1.05	15.238		1,275	750		2,025	2,550

For customer support on your Building Construction Costs with RSMeans Data, call 800.448.8182.

189

06 15 Wood Decking

06 15 16 – Wood Roof Decking

06 15 16.10 Solid Wood Roof Decking

		Crew	Daily Output	Labor-Hours	Unit	Material	2017 Bare Costs Labor	Equipment	Total	Total Incl O&P
0010	**SOLID WOOD ROOF DECKING**									
0350	Cedar planks, 2" thick	2 Carp	350	.046	S.F.	6.50	2.25		8.75	10.55
0400	3" thick		320	.050		9.70	2.46		12.16	14.45
0500	4" thick		250	.064		12.95	3.15		16.10	19.10
0550	6" thick		200	.080		19.45	3.94		23.39	27.50
0650	Douglas fir, 2" thick		350	.046		2.55	2.25		4.80	6.25
0700	3" thick		320	.050		3.83	2.46		6.29	8
0800	4" thick		250	.064		5.10	3.15		8.25	10.45
0850	6" thick		200	.080		7.65	3.94		11.59	14.45
0950	Hemlock, 2" thick		350	.046		2.60	2.25		4.85	6.30
1000	3" thick		320	.050		3.89	2.46		6.35	8.05
1100	4" thick		250	.064		5.20	3.15		8.35	10.55
1150	6" thick		200	.080		7.80	3.94		11.74	14.60
1250	Western white spruce, 2" thick		350	.046		1.66	2.25		3.91	5.25
1300	3" thick		320	.050		2.49	2.46		4.95	6.50
1400	4" thick		250	.064		3.32	3.15		6.47	8.50
1450	6" thick		200	.080		4.97	3.94		8.91	11.50

06 15 23 – Laminated Wood Decking

06 15 23.10 Laminated Roof Deck

		Crew	Daily Output	Labor-Hours	Unit	Material	2017 Bare Costs Labor	Equipment	Total	Total Incl O&P
0010	**LAMINATED ROOF DECK**									
0020	Pine or hemlock, 3" thick	2 Carp	425	.038	S.F.	6.25	1.85		8.10	9.75
0100	4" thick		325	.049		8.15	2.42		10.57	12.65
0300	Cedar, 3" thick		425	.038		7.10	1.85		8.95	10.65
0400	4" thick		325	.049		9.50	2.42		11.92	14.15
0600	Fir, 3" thick		425	.038		5.55	1.85		7.40	8.95
0700	4" thick		325	.049		7.55	2.42		9.97	12

06 16 Sheathing

06 16 13 – Insulating Sheathing

06 16 13.10 Insulating Sheathing

			Crew	Daily Output	Labor-Hours	Unit	Material	2017 Bare Costs Labor	Equipment	Total	Total Incl O&P
0010	**INSULATING SHEATHING**										
0020	Expanded polystyrene, 1#/C.F. density, 3/4" thick, R2.89	G	2 Carp	1400	.011	S.F.	.35	.56		.91	1.25
0030	1" thick, R3.85	G		1300	.012		.42	.61		1.03	1.39
0040	2" thick, R7.69	G		1200	.013		.68	.66		1.34	1.76
0050	Extruded polystyrene, 15 PSI compressive strength, 1" thick, R5	G		1300	.012		.77	.61		1.38	1.77
0060	2" thick, R10	G		1200	.013		.95	.66		1.61	2.06
0070	Polyisocyanurate, 2#/C.F. density, 3/4" thick	G		1400	.011		.58	.56		1.14	1.50
0080	1" thick	G		1300	.012		.59	.61		1.20	1.58
0090	1-1/2" thick	G		1250	.013		.75	.63		1.38	1.79
0100	2" thick	G		1200	.013		.91	.66		1.57	2.01

06 16 23 – Subflooring

06 16 23.10 Subfloor

			Crew	Daily Output	Labor-Hours	Unit	Material	2017 Bare Costs Labor	Equipment	Total	Total Incl O&P
0010	**SUBFLOOR**	R061636-20									
0011	Plywood, CDX, 1/2" thick		2 Carp	1500	.011	SF Flr.	.63	.53		1.16	1.50
0015	Pneumatic nailed			1860	.009		.63	.42		1.05	1.34
0100	5/8" thick			1350	.012		.78	.58		1.36	1.75
0105	Pneumatic nailed			1674	.010		.78	.47		1.25	1.58
0200	3/4" thick			1250	.013		.94	.63		1.57	2
0205	Pneumatic nailed			1550	.010		.94	.51		1.45	1.81
0300	1-1/8" thick, 2-4-1 including underlayment			1050	.015		2.10	.75		2.85	3.46

06 16 Sheathing

06 16 23 – Subflooring

06 16 23.10 Subfloor

		Crew	Daily Output	Labor-Hours	Unit	Material	2017 Bare Costs Labor	Equipment	Total	Total Incl O&P
0440	With boards, 1" x 6", S4S, laid regular	2 Carp	900	.018	SF Flr.	1.64	.88		2.52	3.15
0450	1" x 8", laid regular		1000	.016		2.01	.79		2.80	3.42
0460	Laid diagonal		850	.019		2.01	.93		2.94	3.63
0500	1" x 10", laid regular		1100	.015		2.05	.72		2.77	3.36
0600	Laid diagonal	↓	900	.018	↓	2.05	.88		2.93	3.60
8990	Subfloor adhesive, 3/8" bead	1 Carp	2300	.003	L.F.	.11	.17		.28	.38

06 16 26 – Underlayment

06 16 26.10 Wood Product Underlayment

		Crew	Daily Output	Labor-Hours	Unit	Material	2017 Bare Costs Labor	Equipment	Total	Total Incl O&P
0010	**WOOD PRODUCT UNDERLAYMENT** R061636-20									
0015	Plywood, underlayment grade, 1/4" thick	2 Carp	1500	.011	S.F.	.95	.53		1.48	1.85
0018	Pneumatic nailed		1860	.009		.95	.42		1.37	1.69
0030	3/8" thick		1500	.011		1.05	.53		1.58	1.97
0070	Pneumatic nailed		1860	.009		1.05	.42		1.47	1.81
0100	1/2" thick		1450	.011		1.25	.54		1.79	2.21
0105	Pneumatic nailed		1798	.009		1.25	.44		1.69	2.05
0200	5/8" thick		1400	.011		1.38	.56		1.94	2.38
0205	Pneumatic nailed		1736	.009		1.38	.45		1.83	2.22
0300	3/4" thick		1300	.012		1.49	.61		2.10	2.57
0305	Pneumatic nailed		1612	.010		1.49	.49		1.98	2.39
0500	Particle board, 3/8" thick G		1500	.011		.40	.53		.93	1.25
0505	Pneumatic nailed G		1860	.009		.40	.42		.82	1.09
0600	1/2" thick G		1450	.011		.45	.54		.99	1.33
0605	Pneumatic nailed G		1798	.009		.45	.44		.89	1.17
0800	5/8" thick G		1400	.011		.54	.56		1.10	1.45
0805	Pneumatic nailed G		1736	.009		.54	.45		.99	1.29
0900	3/4" thick G		1300	.012		.70	.61		1.31	1.70
0905	Pneumatic nailed G		1612	.010		.70	.49		1.19	1.52
1100	Hardboard, underlayment grade, 4' x 4', .215" thick G	↓	1500	.011	↓	.63	.53		1.16	1.50

06 16 33 – Wood Board Sheathing

06 16 33.10 Board Sheathing

		Crew	Daily Output	Labor-Hours	Unit	Material	2017 Bare Costs Labor	Equipment	Total	Total Incl O&P
0009	**BOARD SHEATHING**									
0010	Roof, 1" x 6" boards, laid horizontal	2 Carp	725	.022	S.F.	1.64	1.09		2.73	3.48
0020	On steep roof		520	.031		1.64	1.52		3.16	4.13
0040	On dormers, hips, & valleys		480	.033		1.64	1.64		3.28	4.32
0050	Laid diagonal		650	.025		1.64	1.21		2.85	3.67
0070	1" x 8" boards, laid horizontal		875	.018		2.01	.90		2.91	3.59
0080	On steep roof		635	.025		2.01	1.24		3.25	4.11
0090	On dormers, hips, & valleys		580	.028		2.05	1.36		3.41	4.34
0100	Laid diagonal	↓	725	.022		2.01	1.09		3.10	3.88
0110	Skip sheathing, 1" x 4", 7" OC	1 Carp	1200	.007		.61	.33		.94	1.18
0120	1" x 6", 9" OC		1450	.006		.75	.27		1.02	1.25
0180	Tongue and groove sheathing/decking, 1" x 6"		1000	.008		1.80	.39		2.19	2.58
0190	2" x 6"	↓	1000	.008		3.86	.39		4.25	4.85
0200	Walls, 1" x 6" boards, laid regular	2 Carp	650	.025		1.64	1.21		2.85	3.67
0210	Laid diagonal		585	.027		1.64	1.35		2.99	3.87
0220	1" x 8" boards, laid regular		765	.021		2.01	1.03		3.04	3.79
0230	Laid diagonal	↓	650	.025	↓	2.01	1.21		3.22	4.07

For customer support on your Building Construction Costs with RSMeans Data, call 800.448.8182.

191

06 16 36 – Wood Panel Product Sheathing

06 16 36.10 Sheathing		Crew	Daily Output	Labor-Hours	Unit	Material	2017 Bare Costs Labor	Equipment	Total	Total Incl O&P
0010	**SHEATHING** R061636-20									
0012	Plywood on roofs, CDX									
0030	5/16" thick	2 Carp	1600	.010	S.F.	.59	.49		1.08	1.40
0035	Pneumatic nailed R061110-30		1952	.008		.59	.40		.99	1.27
0050	3/8" thick		1525	.010		.61	.52		1.13	1.46
0055	Pneumatic nailed		1860	.009		.61	.42		1.03	1.32
0100	1/2" thick		1400	.011		.63	.56		1.19	1.55
0105	Pneumatic nailed		1708	.009		.63	.46		1.09	1.40
0200	5/8" thick		1300	.012		.78	.61		1.39	1.79
0205	Pneumatic nailed		1586	.010		.78	.50		1.28	1.62
0300	3/4" thick		1200	.013		.94	.66		1.60	2.04
0305	Pneumatic nailed		1464	.011		.94	.54		1.48	1.85
0500	Plywood on walls, with exterior CDX, 3/8" thick		1200	.013		.61	.66		1.27	1.68
0505	Pneumatic nailed		1488	.011		.61	.53		1.14	1.48
0600	1/2" thick		1125	.014		.63	.70		1.33	1.76
0605	Pneumatic nailed		1395	.011		.63	.56		1.19	1.56
0700	5/8" thick		1050	.015		.78	.75		1.53	2.01
0705	Pneumatic nailed		1302	.012		.78	.61		1.39	1.79
0800	3/4" thick		975	.016		.94	.81		1.75	2.27
0805	Pneumatic nailed		1209	.013		.94	.65		1.59	2.03
1000	For shear wall construction, add						20%			
1200	For structural 1 exterior plywood, add				S.F.	10%				
3000	Wood fiber, regular, no vapor barrier, 1/2" thick	2 Carp	1200	.013		.64	.66		1.30	1.71
3100	5/8" thick		1200	.013		.70	.66		1.36	1.78
3300	No vapor barrier, in colors, 1/2" thick		1200	.013		.81	.66		1.47	1.90
3400	5/8" thick		1200	.013		.85	.66		1.51	1.95
3600	With vapor barrier one side, white, 1/2" thick		1200	.013		.60	.66		1.26	1.67
3700	Vapor barrier 2 sides, 1/2" thick		1200	.013		.84	.66		1.50	1.93
3800	Asphalt impregnated, 25/32" thick		1200	.013		.33	.66		.99	1.37
3850	Intermediate, 1/2" thick		1200	.013		.28	.66		.94	1.32
4500	Oriented strand board, on roof, 7/16" thick Ⓖ		1460	.011		.38	.54		.92	1.25
4505	Pneumatic nailed Ⓖ		1780	.009		.38	.44		.82	1.10
4550	1/2" thick Ⓖ		1400	.011		.38	.56		.94	1.28
4555	Pneumatic nailed Ⓖ		1736	.009		.38	.45		.83	1.12
4600	5/8" thick Ⓖ		1300	.012		.58	.61		1.19	1.57
4605	Pneumatic nailed Ⓖ		1586	.010		.58	.50		1.08	1.40
4610	On walls, 7/16" thick Ⓖ		1200	.013		.38	.66		1.04	1.43
4615	Pneumatic nailed Ⓖ		1488	.011		.38	.53		.91	1.23
4620	1/2" thick Ⓖ		1195	.013		.38	.66		1.04	1.43
4625	Pneumatic nailed Ⓖ		1325	.012		.38	.59		.97	1.33
4630	5/8" thick Ⓖ		1050	.015		.58	.75		1.33	1.79
4635	Pneumatic nailed Ⓖ		1302	.012		.58	.61		1.19	1.57
4700	Oriented strand board, factory laminated W.R. barrier, on roof, 1/2" thick Ⓖ		1400	.011		.71	.56		1.27	1.64
4705	Pneumatic nailed Ⓖ		1736	.009		.71	.45		1.16	1.48
4720	5/8" thick Ⓖ		1300	.012		.86	.61		1.47	1.88
4725	Pneumatic nailed Ⓖ		1586	.010		.86	.50		1.36	1.71
4730	5/8" thick, T&G Ⓖ		1150	.014		.86	.69		1.55	2
4735	Pneumatic nailed, T&G Ⓖ		1400	.011		.86	.56		1.42	1.81
4740	On walls, 7/16" thick Ⓖ		1200	.013		.61	.66		1.27	1.68
4745	Pneumatic nailed Ⓖ		1488	.011		.61	.53		1.14	1.48
4750	1/2" thick Ⓖ		1195	.013		.71	.66		1.37	1.79
4755	Pneumatic nailed Ⓖ		1325	.012		.71	.59		1.30	1.69

06 16 Sheathing

06 16 36 – Wood Panel Product Sheathing

06 16 36.10 Sheathing		Crew	Daily Output	Labor-Hours	Unit	Material	2017 Bare Costs Labor	Equipment	Total	Total Incl O&P
4800	Joint sealant tape, 3-1/2"	2 Carp	7600	.002	L.F.	.28	.10		.38	.47
4810	Joint sealant tape, 6"	↓	7600	.002	"	.41	.10		.51	.61

06 16 43 – Gypsum Sheathing

06 16 43.10 Gypsum Sheathing

		Crew	Daily Output	Labor-Hours	Unit	Material	Labor	Equipment	Total	Total Incl O&P
0010	**GYPSUM SHEATHING**									
0020	Gypsum, weatherproof, 1/2" thick	2 Carp	1125	.014	S.F.	.47	.70		1.17	1.59
0040	With embedded glass mats	"	1100	.015	"	.70	.72		1.42	1.87

06 17 Shop-Fabricated Structural Wood

06 17 33 – Wood I-Joists

06 17 33.10 Wood and Composite I-Joists

		Crew	Daily Output	Labor-Hours	Unit	Material	Labor	Equipment	Total	Total Incl O&P
0010	**WOOD AND COMPOSITE I-JOISTS**									
0100	Plywood webs, incl. bridging & blocking, panels 24" O.C.									
1200	15' to 24' span, 50 psf live load	F-5	2400	.013	SF Flr.	1.90	.66		2.56	3.11
1300	55 psf live load		2250	.014		2.15	.71		2.86	3.45
1400	24' to 30' span, 45 psf live load		2600	.012		2.56	.61		3.17	3.76
1500	55 psf live load	↓	2400	.013	↓	4.12	.66		4.78	5.55

06 17 53 – Shop-Fabricated Wood Trusses

06 17 53.10 Roof Trusses

		Crew	Daily Output	Labor-Hours	Unit	Material	Labor	Equipment	Total	Total Incl O&P
0010	**ROOF TRUSSES**									
0100	Fink (W) or King post type, 2'-0" O.C.									
0200	Metal plate connected, 4 in 12 slope									
0210	24' to 29' span	F-3	3000	.013	SF Flr.	1.57	.67	.19	2.43	2.97
0300	30' to 43' span		3000	.013		2.18	.67	.19	3.04	3.64
0400	44' to 60' span	↓	3000	.013		2.28	.67	.19	3.14	3.75
0700	Glued and nailed, add				↓	50%				

06 18 Glued-Laminated Construction

06 18 13 – Glued-Laminated Beams

06 18 13.20 Laminated Framing

		Crew	Daily Output	Labor-Hours	Unit	Material	Labor	Equipment	Total	Total Incl O&P
0010	**LAMINATED FRAMING**									
0020	30 lb., short term live load, 15 lb. dead load									
0200	Straight roof beams, 20' clear span, beams 8' O.C.	F-3	2560	.016	SF Flr.	2.14	.79	.22	3.15	3.81
0300	Beams 16' O.C.		3200	.013		1.56	.63	.18	2.37	2.89
0500	40' clear span, beams 8' O.C.		3200	.013		4.06	.63	.18	4.87	5.65
0600	Beams 16' O.C.	↓	3840	.010		3.35	.53	.15	4.03	4.65
0800	60' clear span, beams 8' O.C.	F-4	2880	.017		6.95	.84	.37	8.16	9.35
0900	Beams 16' O.C.	"	3840	.013		5.20	.63	.28	6.11	6.95
1100	Tudor arches, 30' to 40' clear span, frames 8' O.C.	F-3	1680	.024		9.05	1.20	.34	10.59	12.15
1200	Frames 16' O.C.	"	2240	.018		7.10	.90	.26	8.26	9.45
1400	50' to 60' clear span, frames 8' O.C.	F-4	2200	.022		9.75	1.09	.49	11.33	12.95
1500	Frames 16' O.C.		2640	.018		8.30	.91	.41	9.62	11
1700	Radial arches, 60' clear span, frames 8' O.C.		1920	.025		9.15	1.25	.56	10.96	12.60
1800	Frames 16' O.C.		2880	.017		7.25	.84	.37	8.46	9.65
2000	100' clear span, frames 8' O.C.		1600	.030		9.40	1.50	.67	11.57	13.40
2100	Frames 16' O.C.		2400	.020		8.30	1	.45	9.75	11.15
2300	120' clear span, frames 8' O.C.		1440	.033		12.50	1.67	.75	14.92	17.10
2400	Frames 16' O.C.	↓	1920	.025	↓	11.45	1.25	.56	13.26	15.10

For customer support on your Building Construction Costs with RSMeans Data, call 800.448.8182.

193

06 18 13.20 Laminated Framing		Crew	Daily Output	Labor-Hours	Unit	Material	2017 Bare Costs Labor	Equipment	Total	Total Incl O&P
2600	Bowstring trusses, 20' O.C., 40' clear span	F-3	2400	.017	SF Flr.	5.65	.84	.24	6.73	7.80
2700	60' clear span	F-4	3600	.013		5.10	.67	.30	6.07	6.95
2800	100' clear span		4000	.012		7.20	.60	.27	8.07	9.15
2900	120' clear span		3600	.013		7.65	.67	.30	8.62	9.80
3000	For less than 1000 B.F., add					20%				
3050	For over 5000 B.F., deduct					10%				
3100	For premium appearance, add to S.F. prices					5%				
3300	For industrial type, deduct					15%				
3500	For stain and varnish, add					5%				
3900	For 3/4" laminations, add to straight					25%				
4100	Add to curved					15%				
4300	Alternate pricing method: (use nominal footage of									
4310	components). Straight beams, camber less than 6"	F-3	3.50	11.429	M.B.F.	3,000	580	164	3,744	4,350
4400	Columns, including hardware		2	20		3,225	1,000	286	4,511	5,425
4600	Curved members, radius over 32'		2.50	16		3,300	810	229	4,339	5,100
4700	Radius 10' to 32'		3	13.333		3,275	675	191	4,141	4,825
4900	For complicated shapes, add maximum					100%				
5100	For pressure treating, add to straight					35%				
5200	Add to curved					45%				
6000	Laminated veneer members, southern pine or western species									
6050	1-3/4" wide x 5-1/2" deep	2 Carp	480	.033	L.F.	3.24	1.64		4.88	6.05
6100	9-1/2" deep		480	.033		4.59	1.64		6.23	7.55
6150	14" deep		450	.036		7.10	1.75		8.85	10.50
6200	18" deep		450	.036		9.90	1.75		11.65	13.60
6300	Parallel strand members, southern pine or western species									
6350	1-3/4" wide x 9-1/4" deep	2 Carp	480	.033	L.F.	4.81	1.64		6.45	7.80
6400	11-1/4" deep		450	.036		5.20	1.75		6.95	8.40
6450	14" deep		400	.040		7.80	1.97		9.77	11.60
6500	3-1/2" wide x 9-1/4" deep		480	.033		16.10	1.64		17.74	20
6550	11-1/4" deep		450	.036		19.90	1.75		21.65	24.50
6600	14" deep		400	.040		22.50	1.97		24.47	27.50
6650	7" wide x 9-1/4" deep		450	.036		33.50	1.75		35.25	39.50
6700	11-1/4" deep		420	.038		42	1.88		43.88	49.50
6750	14" deep		400	.040		50	1.97		51.97	58
8000	Straight beams									
8102	20' span									
8104	3-1/8" x 9"	F-3	30	1.333	Ea.	141	67.50	19.10	227.60	279
8106	X 10-1/2"		30	1.333		164	67.50	19.10	250.60	305
8108	X 12"		30	1.333		188	67.50	19.10	274.60	330
8110	X 13-1/2"		30	1.333		211	67.50	19.10	297.60	355
8112	X 15"		29	1.379		235	69.50	19.75	324.25	385
8114	5-1/8" x 10-1/2"		30	1.333		270	67.50	19.10	356.60	420
8116	X 12"		30	1.333		310	67.50	19.10	396.60	465
8118	X 13-1/2"		30	1.333		345	67.50	19.10	431.60	505
8120	X 15"		29	1.379		385	69.50	19.75	474.25	555
8122	X 16-1/2"		29	1.379		425	69.50	19.75	514.25	595
8124	X 18"		29	1.379		460	69.50	19.75	549.25	640
8126	X 19-1/2"		29	1.379		500	69.50	19.75	589.25	680
8128	X 21"		28	1.429		540	72	20.50	632.50	730
8130	X 22-1/2"		28	1.429		580	72	20.50	672.50	770
8132	X 24"		28	1.429		615	72	20.50	707.50	815
8134	6-3/4" x 12"		29	1.379		405	69.50	19.75	494.25	575
8136	X 13-1/2"		29	1.379		455	69.50	19.75	544.25	630

06 18 13.20 Laminated Framing		Crew	Daily Output	Labor-Hours	Unit	Material	2017 Bare Costs Labor	Equipment	Total	Total Incl O&P
8138	X 15"	F-3	29	1.379	Ea.	510	69.50	19.75	599.25	690
8140	X 16-1/2"		28	1.429		560	72	20.50	652.50	750
8142	X 18"		28	1.429		610	72	20.50	702.50	805
8144	X 19-1/2"		28	1.429		660	72	20.50	752.50	860
8146	X 21"		27	1.481		710	75	21	806	920
8148	X 22-1/2"		27	1.481		760	75	21	856	975
8150	X 24"		27	1.481		810	75	21	906	1,025
8152	X 25-1/2"		27	1.481		865	75	21	961	1,100
8154	X 27"		26	1.538		915	78	22	1,015	1,150
8156	X 28-1/2"		26	1.538		965	78	22	1,065	1,200
8158	X 30"	▼	26	1.538	▼	1,025	78	22	1,125	1,275
8200	30' span									
8250	3-1/8" x 9"	F-3	30	1.333	Ea.	211	67.50	19.10	297.60	355
8252	X 10-1/2"		30	1.333		247	67.50	19.10	333.60	395
8254	X 12"		30	1.333		282	67.50	19.10	368.60	435
8256	X 13-1/2"		30	1.333		315	67.50	19.10	401.60	475
8258	X 15"		29	1.379		350	69.50	19.75	439.25	520
8260	5-1/8" x 10-1/2"		30	1.333		405	67.50	19.10	491.60	570
8262	X 12"		30	1.333		460	67.50	19.10	546.60	635
8264	X 13-1/2"		30	1.333		520	67.50	19.10	606.60	695
8266	X 15"		29	1.379		580	69.50	19.75	669.25	765
8268	X 16-1/2"		29	1.379		635	69.50	19.75	724.25	830
8270	X 18"		29	1.379		695	69.50	19.75	784.25	895
8272	X 19-1/2"		29	1.379		750	69.50	19.75	839.25	955
8274	X 21"		28	1.429		810	72	20.50	902.50	1,025
8276	X 22-1/2"		28	1.429		865	72	20.50	957.50	1,100
8278	X 24"		28	1.429		925	72	20.50	1,017.50	1,150
8280	6-3/4" x 12"		29	1.379		610	69.50	19.75	699.25	800
8282	X 13-1/2"		29	1.379		685	69.50	19.75	774.25	885
8284	X 15"		29	1.379		760	69.50	19.75	849.25	965
8286	X 16-1/2"		28	1.429		835	72	20.50	927.50	1,050
8288	X 18"		28	1.429		915	72	20.50	1,007.50	1,125
8290	X 19-1/2"		28	1.429		990	72	20.50	1,082.50	1,225
8292	X 21"		27	1.481		1,075	75	21	1,171	1,325
8294	X 22-1/2"		27	1.481		1,150	75	21	1,246	1,400
8296	X 24"		27	1.481		1,225	75	21	1,321	1,500
8298	X 25-1/2"		27	1.481		1,300	75	21	1,396	1,575
8300	X 27"		26	1.538		1,375	78	22	1,475	1,650
8302	X 28-1/2"		26	1.538		1,450	78	22	1,550	1,750
8304	X 30"	▼	26	1.538	▼	1,525	78	22	1,625	1,825
8400	40' span									
8402	3-1/8" x 9"	F-3	30	1.333	Ea.	282	67.50	19.10	368.60	435
8404	X 10-1/2"		30	1.333		330	67.50	19.10	416.60	485
8406	X 12"		30	1.333		375	67.50	19.10	461.60	540
8408	X 13-1/2"		30	1.333		425	67.50	19.10	511.60	590
8410	X 15"		29	1.379		470	69.50	19.75	559.25	645
8412	5-1/8" x 10-1/2"		30	1.333		540	67.50	19.10	626.60	720
8414	X 12"		30	1.333		615	67.50	19.10	701.60	805
8416	X 13-1/2"		30	1.333		695	67.50	19.10	781.60	890
8418	X 15"		29	1.379		770	69.50	19.75	859.25	980
8420	X 16-1/2"		29	1.379		850	69.50	19.75	939.25	1,075
8422	X 18"		29	1.379		925	69.50	19.75	1,014.25	1,150
8424	X 19-1/2"		29	1.379		1,000	69.50	19.75	1,089.25	1,225

For customer support on your Building Construction Costs with RSMeans Data, call 800.448.8182.

195

06 18 Glued-Laminated Construction

06 18 13 – Glued-Laminated Beams

06 18 13.20 Laminated Framing

		Crew	Daily Output	Labor-Hours	Unit	Material	2017 Bare Costs Labor	Equipment	Total	Total Incl O&P
8426	X 21"	F-3	28	1.429	Ea.	1,075	72	20.50	1,167.50	1,300
8428	X 22-1/2"		28	1.429		1,150	72	20.50	1,242.50	1,400
8430	X 24"		28	1.429		1,225	72	20.50	1,317.50	1,475
8432	6-3/4" x 12"		29	1.379		810	69.50	19.75	899.25	1,025
8434	X 13-1/2"		29	1.379		915	69.50	19.75	1,004.25	1,125
8436	X 15"		29	1.379		1,025	69.50	19.75	1,114.25	1,250
8438	X 16-1/2"		28	1.429		1,125	72	20.50	1,217.50	1,350
8440	X 18"		28	1.429		1,225	72	20.50	1,317.50	1,475
8442	X 19-1/2"		28	1.429		1,325	72	20.50	1,417.50	1,575
8444	X 21"		27	1.481		1,425	75	21	1,521	1,725
8446	X 22-1/2"		27	1.481		1,525	75	21	1,621	1,825
8448	X 24"		27	1.481		1,625	75	21	1,721	1,925
8450	X 25-1/2"		27	1.481		1,725	75	21	1,821	2,050
8452	X 27"		26	1.538		1,825	78	22	1,925	2,150
8454	X 28-1/2"		26	1.538		1,925	78	22	2,025	2,275
8456	X 30"		26	1.538		2,025	78	22	2,125	2,375

06 22 Millwork

06 22 13 – Standard Pattern Wood Trim

06 22 13.15 Moldings, Base

		Crew	Daily Output	Labor-Hours	Unit	Material	2017 Bare Costs Labor	Equipment	Total	Total Incl O&P
0010	**MOLDINGS, BASE**									
5100	Classic profile, 5/8" x 5-1/2", finger jointed and primed	1 Carp	250	.032	L.F.	1.49	1.58		3.07	4.05
5105	Poplar		240	.033		1.57	1.64		3.21	4.24
5110	Red oak		220	.036		2.41	1.79		4.20	5.40
5115	Maple		220	.036		3.64	1.79		5.43	6.75
5120	Cherry		220	.036		4.32	1.79		6.11	7.50
5125	3/4" x 7-1/2", finger jointed and primed		250	.032		1.89	1.58		3.47	4.49
5130	Poplar		240	.033		2.51	1.64		4.15	5.25
5135	Red oak		220	.036		3.50	1.79		5.29	6.60
5140	Maple		220	.036		4.84	1.79		6.63	8.05
5145	Cherry		220	.036		5.75	1.79		7.54	9.10
5150	Modern profile, 5/8" x 3-1/2", finger jointed and primed		250	.032		.91	1.58		2.49	3.41
5155	Poplar		240	.033		1	1.64		2.64	3.61
5160	Red oak		220	.036		1.65	1.79		3.44	4.55
5165	Maple		220	.036		2.55	1.79		4.34	5.55
5170	Cherry		220	.036		2.56	1.79		4.35	5.55
5175	Ogee profile, 7/16" x 3", finger jointed and primed		250	.032		.63	1.58		2.21	3.10
5180	Poplar		240	.033		.66	1.64		2.30	3.23
5185	Red oak		220	.036		.93	1.79		2.72	3.76
5200	9/16" x 3-1/2", finger jointed and primed		250	.032		.66	1.58		2.24	3.13
5205	Pine		240	.033		1.13	1.64		2.77	3.75
5210	Red oak		220	.036		2.80	1.79		4.59	5.80
5215	9/16" x 4-1/2", red oak		220	.036		4.19	1.79		5.98	7.35
5220	5/8" x 3-1/2", finger jointed and primed		250	.032		.91	1.58		2.49	3.41
5225	Poplar		240	.033		1	1.64		2.64	3.61
5230	Red oak		220	.036		1.65	1.79		3.44	4.55
5235	Maple		220	.036		2.55	1.79		4.34	5.55
5240	Cherry		220	.036		2.75	1.79		4.54	5.75
5245	5/8" x 4", finger jointed and primed		250	.032		1.17	1.58		2.75	3.70
5250	Poplar		240	.033		1.29	1.64		2.93	3.93
5255	Red oak		220	.036		1.85	1.79		3.64	4.77

06 22 13 – Standard Pattern Wood Trim

06 22 13.15 Moldings, Base

		Crew	Daily Output	Labor-Hours	Unit	Material	2017 Bare Costs Labor	Equipment	Total	Total Incl O&P
5260	Maple	1 Carp	220	.036	L.F.	2.85	1.79		4.64	5.85
5265	Cherry		220	.036		2.99	1.79		4.78	6.05
5270	Rectangular profile, oak, 3/8" x 1-1/4"		260	.031		1.29	1.52		2.81	3.74
5275	1/2" x 2-1/2"		255	.031		2.35	1.55		3.90	4.95
5280	1/2" x 3-1/2"		250	.032		2.78	1.58		4.36	5.45
5285	1" x 6"		240	.033		4.21	1.64		5.85	7.15
5290	1" x 8"		240	.033		5.80	1.64		7.44	8.90
5295	Pine, 3/8" x 1-3/4"		260	.031		.47	1.52		1.99	2.84
5300	7/16" x 2-1/2"		255	.031		.83	1.55		2.38	3.28
5305	1" x 6"		240	.033		.83	1.64		2.47	3.42
5310	1" x 8"		240	.033		1.06	1.64		2.70	3.67
5315	Shoe, 1/2" x 3/4", primed		260	.031		.49	1.52		2.01	2.86
5320	Pine		240	.033		.33	1.64		1.97	2.87
5325	Poplar		240	.033		.39	1.64		2.03	2.94
5330	Red oak		220	.036		.53	1.79		2.32	3.32
5335	Maple		220	.036		.70	1.79		2.49	3.51
5340	Cherry		220	.036		.80	1.79		2.59	3.62
5345	11/16" x 1-1/2", pine		240	.033		.70	1.64		2.34	3.28
5350	Caps, 11/16" x 1-3/8", pine		240	.033		.55	1.64		2.19	3.11
5355	3/4" x 1-3/4", finger jointed and primed		260	.031		.71	1.52		2.23	3.10
5360	Poplar		240	.033		1.01	1.64		2.65	3.62
5365	Red oak		220	.036		1.21	1.79		3	4.07
5370	Maple		220	.036		1.58	1.79		3.37	4.48
5375	Cherry		220	.036		3.29	1.79		5.08	6.35
5380	Combination base & shoe, 9/16" x 3-1/2" & 1/2" x 3/4", pine		125	.064		1.46	3.15		4.61	6.45
5385	Three piece oak, 6" high		80	.100		5.95	4.93		10.88	14.10
5390	Including 3/4" x 1" base shoe		70	.114		6.45	5.65		12.10	15.70
5395	Flooring cant strip, 3/4" x 3/4", pre-finished pine		260	.031		.51	1.52		2.03	2.88
5400	For pre-finished, stain and clear coat, add					.55			.55	.61
5405	Clear coat only, add					.45			.45	.50

06 22 13.30 Moldings, Casings

		Crew	Daily Output	Labor-Hours	Unit	Material	2017 Bare Costs Labor	Equipment	Total	Total Incl O&P
0010	**MOLDINGS, CASINGS**									
0085	Apron, 9/16" x 2-1/2", pine	1 Carp	250	.032	L.F.	1.63	1.58		3.21	4.20
0090	5/8" x 2-1/2", pine		250	.032		1.63	1.58		3.21	4.20
0110	5/8" x 3-1/2", pine		220	.036		1.86	1.79		3.65	4.78
0300	Band, 11/16" x 1-1/8", pine		270	.030		.75	1.46		2.21	3.06
0310	11/16" x 1-1/2", finger jointed and primed		270	.030		.76	1.46		2.22	3.07
0320	Pine		270	.030		.96	1.46		2.42	3.29
0330	11/16" x 1-3/4", finger jointed and primed		270	.030		.95	1.46		2.41	3.28
0350	Pine		270	.030		1.16	1.46		2.62	3.51
0355	Beaded, 3/4" x 3-1/2", finger jointed and primed		220	.036		1.11	1.79		2.90	3.96
0360	Poplar		220	.036		1.11	1.79		2.90	3.96
0365	Red oak		220	.036		1.65	1.79		3.44	4.55
0370	Maple		220	.036		2.55	1.79		4.34	5.55
0375	Cherry		220	.036		3.08	1.79		4.87	6.15
0380	3/4" x 4", finger jointed and primed		220	.036		1.20	1.79		2.99	4.06
0385	Poplar		220	.036		1.45	1.79		3.24	4.33
0390	Red oak		220	.036		2.02	1.79		3.81	4.96
0395	Maple		220	.036		2.65	1.79		4.44	5.65
0400	Cherry		220	.036		3.33	1.79		5.12	6.40
0405	3/4" x 5-1/2", finger jointed and primed		200	.040		1.27	1.97		3.24	4.42
0410	Poplar		200	.040		1.93	1.97		3.90	5.15

06 22 Millwork

06 22 13 – Standard Pattern Wood Trim

06 22 13.30 Moldings, Casings

		Crew	Daily Output	Labor-Hours	Unit	Material	2017 Bare Costs Labor	Equipment	Total	Total Incl O&P
0415	Red oak	1 Carp	200	.040	L.F.	2.84	1.97		4.81	6.15
0420	Maple		200	.040		3.73	1.97		5.70	7.10
0425	Cherry		200	.040		4.30	1.97		6.27	7.75
0430	Classic profile, 3/4" x 2-3/4", finger jointed and primed		250	.032		.82	1.58		2.40	3.31
0435	Poplar		250	.032		1	1.58		2.58	3.51
0440	Red oak		250	.032		1.46	1.58		3.04	4.01
0445	Maple		250	.032		2.09	1.58		3.67	4.71
0450	Cherry		250	.032		2.37	1.58		3.95	5
0455	Fluted, 3/4" x 3-1/2", poplar		220	.036		1.11	1.79		2.90	3.96
0460	Red oak		220	.036		1.65	1.79		3.44	4.55
0465	Maple		220	.036		2.55	1.79		4.34	5.55
0470	Cherry		220	.036		3.08	1.79		4.87	6.15
0475	3/4" x 4", poplar		220	.036		1.44	1.79		3.23	4.32
0480	Red oak		220	.036		2.03	1.79		3.82	4.97
0485	Maple		220	.036		2.65	1.79		4.44	5.65
0490	Cherry		220	.036		3.33	1.79		5.12	6.40
0495	3/4" x 5-1/2", poplar		200	.040		1.93	1.97		3.90	5.15
0500	Red oak		200	.040		2.84	1.97		4.81	6.15
0505	Maple		200	.040		3.73	1.97		5.70	7.10
0510	Cherry		200	.040		4.30	1.97		6.27	7.75
0515	3/4" x 7-1/2", poplar		190	.042		1.31	2.07		3.38	4.62
0520	Red oak		190	.042		3.93	2.07		6	7.50
0525	Maple		190	.042		5.65	2.07		7.72	9.40
0530	Cherry		190	.042		6.75	2.07		8.82	10.65
0535	3/4" x 9-1/2", poplar		180	.044		4.12	2.19		6.31	7.90
0540	Red oak		180	.044		6.55	2.19		8.74	10.55
0545	Maple		180	.044		9	2.19		11.19	13.25
0550	Cherry		180	.044		9.80	2.19		11.99	14.15
0555	Modern profile, 9/16" x 2-1/4", poplar		250	.032		.66	1.58		2.24	3.13
0560	Red oak		250	.032		.78	1.58		2.36	3.27
0565	11/16" x 2-1/2", finger jointed & primed		250	.032		.84	1.58		2.42	3.33
0570	Pine		250	.032		1.28	1.58		2.86	3.82
0575	3/4" x 2-1/2", poplar		250	.032		.88	1.58		2.46	3.38
0580	Red oak		250	.032		1.19	1.58		2.77	3.72
0585	Maple		250	.032		1.80	1.58		3.38	4.39
0590	Cherry		250	.032		2.27	1.58		3.85	4.91
0595	Mullion, 5/16" x 2", pine		270	.030		.86	1.46		2.32	3.18
0600	9/16" x 2-1/2", finger jointed and primed		250	.032		.98	1.58		2.56	3.49
0605	Pine		250	.032		1.28	1.58		2.86	3.82
0610	Red oak		250	.032		3.28	1.58		4.86	6
0615	1-1/16" x 3-3/4", red oak		220	.036		6.95	1.79		8.74	10.35
0620	Ogee, 7/16" x 2-1/2", poplar		250	.032		.57	1.58		2.15	3.04
0625	Red oak		250	.032		.78	1.58		2.36	3.27
0630	9/16" x 2-1/4", finger jointed and primed		250	.032		.53	1.58		2.11	2.99
0635	Poplar		250	.032		.57	1.58		2.15	3.04
0640	Red oak		250	.032		.78	1.58		2.36	3.27
0645	11/16" x 2-1/2", finger jointed and primed		250	.032		.72	1.58		2.30	3.20
0700	Pine		250	.032		1.41	1.58		2.99	3.96
0701	Red oak		250	.032		2.70	1.58		4.28	5.40
0730	11/16" x 3-1/2", finger jointed and primed		220	.036		1.20	1.79		2.99	4.06
0750	Pine		220	.036		1.72	1.79		3.51	4.63
0755	3/4" x 2-1/2", finger jointed and primed		250	.032		.85	1.58		2.43	3.34
0760	Poplar		250	.032		.88	1.58		2.46	3.38

198

For customer support on your Building Construction Costs with RSMeans Data, call 800.448.8182.

06 22 Millwork

06 22 13 – Standard Pattern Wood Trim

06 22 13.30 Moldings, Casings		Crew	Daily Output	Labor-Hours	Unit	Material	2017 Bare Costs Labor	Equipment	Total	Total Incl O&P
0765	Red oak	1 Carp	250	.032	L.F.	1.19	1.58		2.77	3.72
0770	Maple		250	.032		1.80	1.58		3.38	4.39
0775	Cherry		250	.032		2.27	1.58		3.85	4.91
0780	3/4" x 3-1/2", finger jointed and primed		220	.036		.94	1.79		2.73	3.77
0785	Poplar		220	.036		1.11	1.79		2.90	3.96
0790	Red oak		220	.036		1.65	1.79		3.44	4.55
0795	Maple		220	.036		2.55	1.79		4.34	5.55
0800	Cherry		220	.036		3.08	1.79		4.87	6.15
4700	Square profile, 1" x 1", teak		215	.037		2.39	1.83		4.22	5.45
4800	Rectangular profile, 1" x 3", teak		200	.040		6.40	1.97		8.37	10.05

06 22 13.35 Moldings, Ceilings

06 22 13.35 Moldings, Ceilings		Crew	Daily Output	Labor-Hours	Unit	Material	2017 Bare Costs Labor	Equipment	Total	Total Incl O&P
0010	**MOLDINGS, CEILINGS**									
0600	Bed, 9/16" x 1-3/4", pine	1 Carp	270	.030	L.F.	1.06	1.46		2.52	3.40
0650	9/16" x 2", pine		270	.030		1.14	1.46		2.60	3.49
0710	9/16" x 1-3/4", oak		270	.030		2.35	1.46		3.81	4.82
1200	Cornice, 9/16" x 1-3/4", pine		270	.030		.96	1.46		2.42	3.29
1300	9/16" x 2-1/4", pine		265	.030		1.27	1.49		2.76	3.67
1350	Cove, 1/2" x 2-1/4", poplar		265	.030		1.02	1.49		2.51	3.40
1360	Red oak		265	.030		1.36	1.49		2.85	3.77
1370	Hard maple		265	.030		1.87	1.49		3.36	4.33
1380	Cherry		265	.030		2.28	1.49		3.77	4.78
2400	9/16" x 1-3/4", pine		270	.030		.97	1.46		2.43	3.31
2500	11/16" x 2-3/4", pine		265	.030		1.80	1.49		3.29	4.26
2510	Crown, 5/8" x 5/8", poplar		300	.027		.44	1.31		1.75	2.49
2520	Red oak		300	.027		.52	1.31		1.83	2.58
2530	Hard maple		300	.027		.70	1.31		2.01	2.78
2540	Cherry		300	.027		.74	1.31		2.05	2.82
2600	9/16" x 3-5/8", pine		250	.032		2.02	1.58		3.60	4.63
2700	11/16" x 4-1/4", pine		250	.032		2.85	1.58		4.43	5.55
2705	Oak		250	.032		6.30	1.58		7.88	9.35
2710	3/4" x 1-3/4", poplar		270	.030		.71	1.46		2.17	3.02
2720	Red oak		270	.030		1.05	1.46		2.51	3.39
2730	Hard maple		270	.030		1.39	1.46		2.85	3.77
2740	Cherry		270	.030		1.71	1.46		3.17	4.12
2750	3/4" x 2", poplar		270	.030		.94	1.46		2.40	3.27
2760	Red oak		270	.030		1.26	1.46		2.72	3.62
2770	Hard maple		270	.030		1.68	1.46		3.14	4.09
2780	Cherry		270	.030		1.88	1.46		3.34	4.31
2790	3/4" x 2-3/4", poplar		265	.030		1.02	1.49		2.51	3.40
2800	Red oak		265	.030		1.57	1.49		3.06	4
2810	Hard maple		265	.030		2.17	1.49		3.66	4.66
2820	Cherry		265	.030		2.40	1.49		3.89	4.92
2830	3/4" x 3-1/2", poplar		250	.032		1.30	1.58		2.88	3.84
2840	Red oak		250	.032		1.96	1.58		3.54	4.56
2850	Hard maple		250	.032		2.56	1.58		4.14	5.20
2860	Cherry		250	.032		3	1.58		4.58	5.70
2870	FJP poplar		250	.032		.92	1.58		2.50	3.42
2880	3/4" x 5", poplar		245	.033		1.93	1.61		3.54	4.58
2890	Red oak		245	.033		2.85	1.61		4.46	5.60
2900	Hard maple		245	.033		3.74	1.61		5.35	6.55
2910	Cherry		245	.033		4.32	1.61		5.93	7.20
2920	FJP poplar		245	.033		1.32	1.61		2.93	3.91

06 22 13 – Standard Pattern Wood Trim

06 22 13.35 Moldings, Ceilings

		Crew	Daily Output	Labor-Hours	Unit	Material	2017 Bare Costs Labor	2017 Bare Costs Equipment	Total	Total Incl O&P
2930	3/4" x 6-1/4", poplar	1 Carp	240	.033	L.F.	2.30	1.64		3.94	5.05
2940	Red oak		240	.033		3.47	1.64		5.11	6.30
2950	Hard maple		240	.033		4.53	1.64		6.17	7.50
2960	Cherry		240	.033		5.35	1.64		6.99	8.40
2970	7/8" x 8-3/4", poplar		220	.036		3.57	1.79		5.36	6.65
2980	Red oak		220	.036		4.76	1.79		6.55	8
2990	Hard maple		220	.036		6.80	1.79		8.59	10.20
3000	Cherry		220	.036		7.65	1.79		9.44	11.15
3010	1" x 7-1/4", poplar		220	.036		3.78	1.79		5.57	6.90
3020	Red oak		220	.036		5.90	1.79		7.69	9.20
3030	Hard maple		220	.036		8.15	1.79		9.94	11.70
3040	Cherry		220	.036		9.50	1.79		11.29	13.20
3050	1-1/16" x 4-1/4", poplar		250	.032		2.29	1.58		3.87	4.93
3060	Red oak		250	.032		2.69	1.58		4.27	5.35
3070	Hard maple		250	.032		3.56	1.58		5.14	6.30
3080	Cherry		250	.032		5.10	1.58		6.68	8
3090	Dentil crown, 3/4" x 5", poplar		250	.032		1.93	1.58		3.51	4.53
3100	Red oak		250	.032		2.85	1.58		4.43	5.55
3110	Hard maple		250	.032		3.74	1.58		5.32	6.50
3120	Cherry		250	.032		4.32	1.58		5.90	7.15
3130	Dentil piece for above, 1/2" x 1/2", poplar		300	.027		2.73	1.31		4.04	5
3140	Red oak		300	.027		3.14	1.31		4.45	5.45
3150	Hard maple		300	.027		3.57	1.31		4.88	5.95
3160	Cherry		300	.027		3.89	1.31		5.20	6.30

06 22 13.40 Moldings, Exterior

		Crew	Daily Output	Labor-Hours	Unit	Material	2017 Bare Costs Labor	2017 Bare Costs Equipment	Total	Total Incl O&P
0010	**MOLDINGS, EXTERIOR**									
0100	Band board, cedar, rough sawn, 1" x 2"	1 Carp	300	.027	L.F.	.59	1.31		1.90	2.66
0110	1" x 3"		300	.027		.88	1.31		2.19	2.98
0120	1" x 4"		250	.032		1.18	1.58		2.76	3.70
0130	1" x 6"		250	.032		1.76	1.58		3.34	4.35
0140	1" x 8"		225	.036		2.35	1.75		4.10	5.25
0150	1" x 10"		225	.036		2.92	1.75		4.67	5.90
0160	1" x 12"		200	.040		3.51	1.97		5.48	6.90
0240	STK, 1" x 2"		300	.027		.54	1.31		1.85	2.60
0250	1" x 3"		300	.027		.59	1.31		1.90	2.66
0260	1" x 4"		250	.032		.66	1.58		2.24	3.14
0270	1" x 6"		250	.032		1.12	1.58		2.70	3.65
0280	1" x 8"		225	.036		1.71	1.75		3.46	4.56
0290	1" x 10"		225	.036		2.29	1.75		4.04	5.20
0300	1" x 12"		200	.040		3.69	1.97		5.66	7.10
0310	Pine, #2, 1" x 2"		300	.027		.27	1.31		1.58	2.31
0320	1" x 3"		300	.027		.43	1.31		1.74	2.48
0330	1" x 4"		250	.032		.52	1.58		2.10	2.99
0340	1" x 6"		250	.032		.79	1.58		2.37	3.28
0350	1" x 8"		225	.036		1.29	1.75		3.04	4.10
0360	1" x 10"		225	.036		1.68	1.75		3.43	4.53
0370	1" x 12"		200	.040		2.06	1.97		4.03	5.30
0380	D & better, 1" x 2"		300	.027		.44	1.31		1.75	2.49
0390	1" x 3"		300	.027		.67	1.31		1.98	2.74
0400	1" x 4"		250	.032		.89	1.58		2.47	3.39
0410	1" x 6"		250	.032		1.16	1.58		2.74	3.69
0420	1" x 8"		225	.036		1.62	1.75		3.37	4.47

06 22 Millwork

06 22 13 – Standard Pattern Wood Trim

06 22 13.40 Moldings, Exterior		Crew	Daily Output	Labor-Hours	Unit	Material	2017 Bare Costs Labor	Equipment	Total	Total Incl O&P
0430	1" x 10"	1 Carp	225	.036	L.F.	2.04	1.75		3.79	4.92
0440	1" x 12"		200	.040		2.64	1.97		4.61	5.90
0450	Redwood, clear all heart, 1" x 2"		300	.027		.64	1.31		1.95	2.71
0460	1" x 3"		300	.027		.95	1.31		2.26	3.05
0470	1" x 4"		250	.032		1.19	1.58		2.77	3.72
0480	1" x 6"		252	.032		1.78	1.56		3.34	4.36
0490	1" x 8"		225	.036		2.36	1.75		4.11	5.30
0500	1" x 10"		225	.036		3.97	1.75		5.72	7.05
0510	1" x 12"		200	.040		4.76	1.97		6.73	8.25
0530	Corner board, cedar, rough sawn, 1" x 2"		225	.036		.59	1.75		2.34	3.33
0540	1" x 3"		225	.036		.88	1.75		2.63	3.65
0550	1" x 4"		200	.040		1.18	1.97		3.15	4.31
0560	1" x 6"		200	.040		1.76	1.97		3.73	4.96
0570	1" x 8"		200	.040		2.35	1.97		4.32	5.60
0580	1" x 10"		175	.046		2.92	2.25		5.17	6.65
0590	1" x 12"		175	.046		3.51	2.25		5.76	7.30
0670	STK, 1" x 2"		225	.036		.54	1.75		2.29	3.27
0680	1" x 3"		225	.036		.59	1.75		2.34	3.33
0690	1" x 4"		200	.040		.65	1.97		2.62	3.73
0700	1" x 6"		200	.040		1.12	1.97		3.09	4.26
0710	1" x 8"		200	.040		1.71	1.97		3.68	4.90
0720	1" x 10"		175	.046		2.29	2.25		4.54	5.95
0730	1" x 12"		175	.046		3.69	2.25		5.94	7.50
0740	Pine, #2, 1" x 2"		225	.036		.27	1.75		2.02	2.98
0750	1" x 3"		225	.036		.43	1.75		2.18	3.15
0760	1" x 4"		200	.040		.52	1.97		2.49	3.60
0770	1" x 6"		200	.040		.79	1.97		2.76	3.89
0780	1" x 8"		200	.040		1.29	1.97		3.26	4.44
0790	1" x 10"		175	.046		1.68	2.25		3.93	5.30
0800	1" x 12"		175	.046		2.06	2.25		4.31	5.70
0810	D & better, 1" x 2"		225	.036		.44	1.75		2.19	3.16
0820	1" x 3"		225	.036		.67	1.75		2.42	3.41
0830	1" x 4"		200	.040		.89	1.97		2.86	4
0840	1" x 6"		200	.040		1.16	1.97		3.13	4.30
0850	1" x 8"		200	.040		1.62	1.97		3.59	4.81
0860	1" x 10"		175	.046		2.04	2.25		4.29	5.70
0870	1" x 12"		175	.046		2.64	2.25		4.89	6.35
0880	Redwood, clear all heart, 1" x 2"		225	.036		.64	1.75		2.39	3.38
0890	1" x 3"		225	.036		.95	1.75		2.70	3.72
0900	1" x 4"		200	.040		1.19	1.97		3.16	4.33
0910	1" x 6"		200	.040		1.78	1.97		3.75	4.98
0920	1" x 8"		200	.040		2.36	1.97		4.33	5.60
0930	1" x 10"		175	.046		3.97	2.25		6.22	7.80
0940	1" x 12"		175	.046		4.76	2.25		7.01	8.70
0950	Cornice board, cedar, rough sawn, 1" x 2"		330	.024		.59	1.19		1.78	2.48
0960	1" x 3"		290	.028		.88	1.36		2.24	3.05
0970	1" x 4"		250	.032		1.18	1.58		2.76	3.70
0980	1" x 6"		250	.032		1.76	1.58		3.34	4.35
0990	1" x 8"		200	.040		2.35	1.97		4.32	5.60
1000	1" x 10"		180	.044		2.92	2.19		5.11	6.55
1010	1" x 12"		180	.044		3.51	2.19		5.70	7.20
1020	STK, 1" x 2"		330	.024		.54	1.19		1.73	2.42
1030	1" x 3"		290	.028		.59	1.36		1.95	2.73

06 22 13 – Standard Pattern Wood Trim

06 22 13.40 Moldings, Exterior		Crew	Daily Output	Labor-Hours	Unit	Material	2017 Bare Costs Labor	Equipment	Total	Total Incl O&P
1040	1" x 4"	1 Carp	250	.032	L.F.	.66	1.58		2.24	3.14
1050	1" x 6"		250	.032		1.12	1.58		2.70	3.65
1060	1" x 8"		200	.040		1.71	1.97		3.68	4.90
1070	1" x 10"		180	.044		2.29	2.19		4.48	5.85
1080	1" x 12"		180	.044		3.69	2.19		5.88	7.40
1500	Pine, #2, 1" x 2"		330	.024		.27	1.19		1.46	2.13
1510	1" x 3"		290	.028		.29	1.36		1.65	2.40
1600	1" x 4"		250	.032		.52	1.58		2.10	2.99
1700	1" x 6"		250	.032		.79	1.58		2.37	3.28
1800	1" x 8"		200	.040		1.29	1.97		3.26	4.44
1900	1" x 10"		180	.044		1.68	2.19		3.87	5.20
2000	1" x 12"		180	.044		2.06	2.19		4.25	5.60
2020	D & better, 1" x 2"		330	.024		.44	1.19		1.63	2.31
2030	1" x 3"		290	.028		.67	1.36		2.03	2.81
2040	1" x 4"		250	.032		.89	1.58		2.47	3.39
2050	1" x 6"		250	.032		1.16	1.58		2.74	3.69
2060	1" x 8"		200	.040		1.62	1.97		3.59	4.81
2070	1" x 10"		180	.044		2.04	2.19		4.23	5.60
2080	1" x 12"		180	.044		2.64	2.19		4.83	6.25
2090	Redwood, clear all heart, 1" x 2"		330	.024		.64	1.19		1.83	2.53
2100	1" x 3"		290	.028		.95	1.36		2.31	3.12
2110	1" x 4"		250	.032		1.19	1.58		2.77	3.72
2120	1" x 6"		250	.032		1.78	1.58		3.36	4.37
2130	1" x 8"		200	.040		2.36	1.97		4.33	5.60
2140	1" x 10"		180	.044		3.97	2.19		6.16	7.70
2150	1" x 12"		180	.044		4.76	2.19		6.95	8.60
2160	3 piece, 1" x 2", 1" x 4", 1" x 6", rough sawn cedar		80	.100		3.54	4.93		8.47	11.45
2180	STK cedar		80	.100		2.32	4.93		7.25	10.10
2200	#2 pine		80	.100		1.58	4.93		6.51	9.30
2210	D & better pine		80	.100		2.49	4.93		7.42	10.30
2220	Clear all heart redwood		80	.100		3.61	4.93		8.54	11.50
2230	1" x 8", 1" x 10", 1" x 12", rough sawn cedar		65	.123		8.75	6.05		14.80	18.90
2240	STK cedar		65	.123		7.65	6.05		13.70	17.75
2300	#2 pine		65	.123		4.99	6.05		11.04	14.80
2320	D & better pine		65	.123		6.25	6.05		12.30	16.20
2330	Clear all heart redwood		65	.123		11.05	6.05		17.10	21.50
2340	Door/window casing, cedar, rough sawn, 1" x 2"		275	.029		.59	1.43		2.02	2.84
2350	1" x 3"		275	.029		.88	1.43		2.31	3.16
2360	1" x 4"		250	.032		1.18	1.58		2.76	3.70
2370	1" x 6"		250	.032		1.76	1.58		3.34	4.35
2380	1" x 8"		230	.035		2.35	1.71		4.06	5.20
2390	1" x 10"		230	.035		2.92	1.71		4.63	5.85
2395	1" x 12"		210	.038		3.51	1.88		5.39	6.75
2410	STK, 1" x 2"		275	.029		.54	1.43		1.97	2.78
2420	1" x 3"		275	.029		.59	1.43		2.02	2.84
2430	1" x 4"		250	.032		.66	1.58		2.24	3.14
2440	1" x 6"		250	.032		1.12	1.58		2.70	3.65
2450	1" x 8"		230	.035		1.71	1.71		3.42	4.50
2460	1" x 10"		230	.035		2.29	1.71		4	5.15
2470	1" x 12"		210	.038		3.69	1.88		5.57	6.95
2550	Pine, #2, 1" x 2"		275	.029		.27	1.43		1.70	2.49
2560	1" x 3"		275	.029		.43	1.43		1.86	2.66
2570	1" x 4"		250	.032		.52	1.58		2.10	2.99

For customer support on your Building Construction Costs with RSMeans Data, call 800.448.8182.

06 22 13 – Standard Pattern Wood Trim

06 22 13.40 Moldings, Exterior		Crew	Daily Output	Labor-Hours	Unit	Material	2017 Bare Costs Labor	Equipment	Total	Total Incl O&P
2580	1" x 6"	1 Carp	250	.032	L.F.	.79	1.58		2.37	3.28
2590	1" x 8"		230	.035		1.29	1.71		3	4.04
2600	1" x 10"		230	.035		1.68	1.71		3.39	4.47
2610	1" x 12"		210	.038		2.06	1.88		3.94	5.15
2620	Pine, D & better, 1" x 2"		275	.029		.44	1.43		1.87	2.67
2630	1" x 3"		275	.029		.67	1.43		2.10	2.92
2640	1" x 4"		250	.032		.89	1.58		2.47	3.39
2650	1" x 6"		250	.032		1.16	1.58		2.74	3.69
2660	1" x 8"		230	.035		1.62	1.71		3.33	4.41
2670	1" x 10"		230	.035		2.04	1.71		3.75	4.86
2680	1" x 12"		210	.038		2.64	1.88		4.52	5.75
2690	Redwood, clear all heart, 1" x 2"		275	.029		.64	1.43		2.07	2.89
2695	1" x 3"		275	.029		.95	1.43		2.38	3.23
2710	1" x 4"		250	.032		1.19	1.58		2.77	3.72
2715	1" x 6"		250	.032		1.78	1.58		3.36	4.37
2730	1" x 8"		230	.035		2.36	1.71		4.07	5.20
2740	1" x 10"		230	.035		3.97	1.71		5.68	7
2750	1" x 12"		210	.038		4.76	1.88		6.64	8.10
3500	Bellyband, pine, 11/16" x 4-1/4"		250	.032		3.23	1.58		4.81	5.95
3610	Brickmold, pine, 1-1/4" x 2"		200	.040		2.44	1.97		4.41	5.70
3620	FJP, 1-1/4" x 2"		200	.040		1.13	1.97		3.10	4.26
5100	Fascia, cedar, rough sawn, 1" x 2"		275	.029		.59	1.43		2.02	2.84
5110	1" x 3"		275	.029		.88	1.43		2.31	3.16
5120	1" x 4"		250	.032		1.18	1.58		2.76	3.70
5200	1" x 6"		250	.032		1.76	1.58		3.34	4.35
5300	1" x 8"		230	.035		2.35	1.71		4.06	5.20
5310	1" x 10"		230	.035		2.92	1.71		4.63	5.85
5320	1" x 12"		210	.038		3.51	1.88		5.39	6.75
5400	2" x 4"		220	.036		1.16	1.79		2.95	4.01
5500	2" x 6"		220	.036		1.74	1.79		3.53	4.65
5600	2" x 8"		200	.040		2.31	1.97		4.28	5.55
5700	2" x 10"		180	.044		2.88	2.19		5.07	6.50
5800	2" x 12"		170	.047		6.85	2.32		9.17	11.10
6120	STK, 1" x 2"		275	.029		.54	1.43		1.97	2.78
6130	1" x 3"		275	.029		.59	1.43		2.02	2.84
6140	1" x 4"		250	.032		.66	1.58		2.24	3.14
6150	1" x 6"		250	.032		1.12	1.58		2.70	3.65
6160	1" x 8"		230	.035		1.71	1.71		3.42	4.50
6170	1" x 10"		230	.035		2.29	1.71		4	5.15
6180	1" x 12"		210	.038		3.69	1.88		5.57	6.95
6185	2" x 2"		260	.031		.66	1.52		2.18	3.04
6190	Pine, #2, 1" x 2"		275	.029		.27	1.43		1.70	2.49
6200	1" x 3"		275	.029		.43	1.43		1.86	2.66
6210	1" x 4"		250	.032		.52	1.58		2.10	2.99
6220	1" x 6"		250	.032		.79	1.58		2.37	3.28
6230	1" x 8"		230	.035		1.29	1.71		3	4.04
6240	1" x 10"		230	.035		1.68	1.71		3.39	4.47
6250	1" x 12"		210	.038		2.06	1.88		3.94	5.15
6260	D & better, 1" x 2"		275	.029		.44	1.43		1.87	2.67
6270	1" x 3"		275	.029		.67	1.43		2.10	2.92
6280	1" x 4"		250	.032		.89	1.58		2.47	3.39
6290	1" x 6"		250	.032		1.16	1.58		2.74	3.69
6300	1" x 8"		230	.035		1.62	1.71		3.33	4.41

06 22 13.40 Moldings, Exterior		Crew	Daily Output	Labor-Hours	Unit	Material	2017 Bare Costs Labor	Equipment	Total	Total Incl O&P
6310	1" x 10"	1 Carp	230	.035	L.F.	2.04	1.71		3.75	4.86
6312	1" x 12"		210	.038		2.64	1.88		4.52	5.75
6330	Southern yellow, 1-1/4" x 5"		240	.033		2.60	1.64		4.24	5.35
6340	1-1/4" x 6"		240	.033		2.53	1.64		4.17	5.30
6350	1-1/4" x 8"		215	.037		3.13	1.83		4.96	6.25
6360	1-1/4" x 12"		190	.042		4.59	2.07		6.66	8.25
6370	Redwood, clear all heart, 1" x 2"		275	.029		.64	1.43		2.07	2.89
6380	1" x 3"		275	.029		1.19	1.43		2.62	3.50
6390	1" x 4"		250	.032		1.19	1.58		2.77	3.72
6400	1" x 6"		250	.032		1.78	1.58		3.36	4.37
6410	1" x 8"		230	.035		2.36	1.71		4.07	5.20
6420	1" x 10"		230	.035		3.97	1.71		5.68	7
6430	1" x 12"		210	.038		4.76	1.88		6.64	8.10
6440	1-1/4" x 5"		240	.033		1.86	1.64		3.50	4.56
6450	1-1/4" x 6"		240	.033		2.23	1.64		3.87	4.96
6460	1-1/4" x 8"		215	.037		3.54	1.83		5.37	6.70
6470	1-1/4" x 12"		190	.042		7.15	2.07		9.22	11.05
6580	Frieze, cedar, rough sawn, 1" x 2"		275	.029		.59	1.43		2.02	2.84
6590	1" x 3"		275	.029		.88	1.43		2.31	3.16
6600	1" x 4"		250	.032		1.18	1.58		2.76	3.70
6610	1" x 6"		250	.032		1.76	1.58		3.34	4.35
6620	1" x 8"		250	.032		2.35	1.58		3.93	5
6630	1" x 10"		225	.036		2.92	1.75		4.67	5.90
6640	1" x 12"		200	.040		3.48	1.97		5.45	6.85
6650	STK, 1" x 2"		275	.029		.54	1.43		1.97	2.78
6660	1" x 3"		275	.029		.59	1.43		2.02	2.84
6670	1" x 4"		250	.032		.66	1.58		2.24	3.14
6680	1" x 6"		250	.032		1.12	1.58		2.70	3.65
6690	1" x 8"		250	.032		1.71	1.58		3.29	4.29
6700	1" x 10"		225	.036		2.29	1.75		4.04	5.20
6710	1" x 12"		200	.040		3.69	1.97		5.66	7.10
6790	Pine, #2, 1" x 2"		275	.029		.27	1.43		1.70	2.49
6800	1" x 3"		275	.029		.43	1.43		1.86	2.66
6810	1" x 4"		250	.032		.52	1.58		2.10	2.99
6820	1" x 6"		250	.032		.79	1.58		2.37	3.28
6830	1" x 8"		250	.032		1.29	1.58		2.87	3.83
6840	1" x 10"		225	.036		1.68	1.75		3.43	4.53
6850	1" x 12"		200	.040		2.06	1.97		4.03	5.30
6860	D & better, 1" x 2"		275	.029		.44	1.43		1.87	2.67
6870	1" x 3"		275	.029		.67	1.43		2.10	2.92
6880	1" x 4"		250	.032		.89	1.58		2.47	3.39
6890	1" x 6"		250	.032		1.16	1.58		2.74	3.69
6900	1" x 8"		250	.032		1.62	1.58		3.20	4.20
6910	1" x 10"		225	.036		2.04	1.75		3.79	4.92
6920	1" x 12"		200	.040		2.64	1.97		4.61	5.90
6930	Redwood, clear all heart, 1" x 2"		275	.029		.64	1.43		2.07	2.89
6940	1" x 3"		275	.029		.95	1.43		2.38	3.23
6950	1" x 4"		250	.032		1.19	1.58		2.77	3.72
6960	1" x 6"		250	.032		1.78	1.58		3.36	4.37
6970	1" x 8"		250	.032		2.36	1.58		3.94	5
6980	1" x 10"		225	.036		3.97	1.75		5.72	7.05
6990	1" x 12"		200	.040		4.76	1.97		6.73	8.25
7000	Grounds, 1" x 1", cedar, rough sawn		300	.027		.30	1.31		1.61	2.34

06 22 13 – Standard Pattern Wood Trim

06 22 13.40 Moldings, Exterior		Crew	Daily Output	Labor-Hours	Unit	Material	2017 Bare Costs Labor	Equipment	Total	Total Incl O&P
7010	STK	1 Carp	300	.027	L.F.	.33	1.31		1.64	2.37
7020	Pine, #2		300	.027		.17	1.31		1.48	2.20
7030	D & better		300	.027		.27	1.31		1.58	2.31
7050	Redwood		300	.027		.39	1.31		1.70	2.44
7060	Rake/verge board, cedar, rough sawn, 1" x 2"		225	.036		.59	1.75		2.34	3.33
7070	1" x 3"		225	.036		.88	1.75		2.63	3.65
7080	1" x 4"		200	.040		1.18	1.97		3.15	4.31
7090	1" x 6"		200	.040		1.76	1.97		3.73	4.96
7100	1" x 8"		190	.042		2.35	2.07		4.42	5.75
7110	1" x 10"		190	.042		2.92	2.07		4.99	6.40
7120	1" x 12"		180	.044		3.51	2.19		5.70	7.20
7130	STK, 1" x 2"		225	.036		.54	1.75		2.29	3.27
7140	1" x 3"		225	.036		.59	1.75		2.34	3.33
7150	1" x 4"		200	.040		.66	1.97		2.63	3.75
7160	1" x 6"		200	.040		1.12	1.97		3.09	4.26
7170	1" x 8"		190	.042		1.71	2.07		3.78	5.05
7180	1" x 10"		190	.042		2.29	2.07		4.36	5.70
7190	1" x 12"		180	.044		3.69	2.19		5.88	7.40
7200	Pine, #2, 1" x 2"		225	.036		.27	1.75		2.02	2.98
7210	1" x 3"		225	.036		.43	1.75		2.18	3.15
7220	1" x 4"		200	.040		.52	1.97		2.49	3.60
7230	1" x 6"		200	.040		.79	1.97		2.76	3.89
7240	1" x 8"		190	.042		1.29	2.07		3.36	4.60
7250	1" x 10"		190	.042		1.68	2.07		3.75	5.05
7260	1" x 12"		180	.044		2.06	2.19		4.25	5.60
7340	D & better, 1" x 2"		225	.036		.44	1.75		2.19	3.16
7350	1" x 3"		225	.036		.67	1.75		2.42	3.41
7360	1" x 4"		200	.040		.89	1.97		2.86	4
7370	1" x 6"		200	.040		1.16	1.97		3.13	4.30
7380	1" x 8"		190	.042		1.62	2.07		3.69	4.97
7390	1" x 10"		190	.042		2.04	2.07		4.11	5.40
7400	1" x 12"		180	.044		2.64	2.19		4.83	6.25
7410	Redwood, clear all heart, 1" x 2"		225	.036		.64	1.75		2.39	3.38
7420	1" x 3"		225	.036		.95	1.75		2.70	3.72
7430	1" x 4"		200	.040		1.19	1.97		3.16	4.33
7440	1" x 6"		200	.040		1.78	1.97		3.75	4.98
7450	1" x 8"		190	.042		2.36	2.07		4.43	5.80
7460	1" x 10"		190	.042		3.97	2.07		6.04	7.55
7470	1" x 12"		180	.044		4.76	2.19		6.95	8.60
7480	2" x 4"		200	.040		2.29	1.97		4.26	5.55
7490	2" x 6"		182	.044		3.43	2.17		5.60	7.10
7500	2" x 8"		165	.048		4.56	2.39		6.95	8.65
7630	Soffit, cedar, rough sawn, 1" x 2"	2 Carp	440	.036		.59	1.79		2.38	3.39
7640	1" x 3"		440	.036		.88	1.79		2.67	3.71
7650	1" x 4"		420	.038		1.18	1.88		3.06	4.16
7660	1" x 6"		420	.038		1.76	1.88		3.64	4.81
7670	1" x 8"		420	.038		2.35	1.88		4.23	5.45
7680	1" x 10"		400	.040		2.92	1.97		4.89	6.25
7690	1" x 12"		400	.040		3.51	1.97		5.48	6.90
7700	STK, 1" x 2"		440	.036		.54	1.79		2.33	3.33
7710	1" x 3"		440	.036		.59	1.79		2.38	3.39
7720	1" x 4"		420	.038		.66	1.88		2.54	3.60
7730	1" x 6"		420	.038		1.12	1.88		3	4.11

06 22 Millwork

06 22 13 – Standard Pattern Wood Trim

06 22 13.40 Moldings, Exterior

		Crew	Daily Output	Labor-Hours	Unit	Material	2017 Bare Costs Labor	Equipment	Total	Total Incl O&P
7740	1" x 8"	2 Carp	420	.038	L.F.	1.71	1.88		3.59	4.75
7750	1" x 10"		400	.040		2.29	1.97		4.26	5.55
7760	1" x 12"		400	.040		3.69	1.97		5.66	7.10
7770	Pine, #2, 1" x 2"		440	.036		.27	1.79		2.06	3.04
7780	1" x 3"		440	.036		.43	1.79		2.22	3.21
7790	1" x 4"		420	.038		.52	1.88		2.40	3.45
7800	1" x 6"		420	.038		.79	1.88		2.67	3.74
7810	1" x 8"		420	.038		1.29	1.88		3.17	4.29
7820	1" x 10"		400	.040		1.68	1.97		3.65	4.87
7830	1" x 12"		400	.040		2.06	1.97		4.03	5.30
7840	D & better, 1" x 2"		440	.036		.44	1.79		2.23	3.22
7850	1" x 3"		440	.036		.67	1.79		2.46	3.47
7860	1" x 4"		420	.038		.89	1.88		2.77	3.85
7870	1" x 6"		420	.038		1.16	1.88		3.04	4.15
7880	1" x 8"		420	.038		1.62	1.88		3.50	4.66
7890	1" x 10"		400	.040		2.04	1.97		4.01	5.25
7900	1" x 12"		400	.040		2.64	1.97		4.61	5.90
7910	Redwood, clear all heart, 1" x 2"		440	.036		.64	1.79		2.43	3.44
7920	1" x 3"		440	.036		.95	1.79		2.74	3.78
7930	1" x 4"		420	.038		1.19	1.88		3.07	4.18
7940	1" x 6"		420	.038		1.78	1.88		3.66	4.83
7950	1" x 8"		420	.038		2.36	1.88		4.24	5.45
7960	1" x 10"		400	.040		3.97	1.97		5.94	7.40
7970	1" x 12"		400	.040		4.76	1.97		6.73	8.25
8050	Trim, crown molding, pine, 11/16" x 4-1/4"	1 Carp	250	.032		4.54	1.58		6.12	7.40
8060	Back band, 11/16" x 1-1/16"		250	.032		1	1.58		2.58	3.51
8070	Insect screen frame stock, 1-1/16" x 1-3/4"		395	.020		2.40	1		3.40	4.17
8080	Dentils, 2-1/2" x 2-1/2" x 4", 6" O.C.		30	.267		1.31	13.15		14.46	21.50
8100	Fluted, 5-1/2"		165	.048		5.05	2.39		7.44	9.20
8110	Stucco bead, 1-3/8" x 1-5/8"		250	.032		2.51	1.58		4.09	5.15

06 22 13.45 Moldings, Trim

		Crew	Daily Output	Labor-Hours	Unit	Material	2017 Bare Costs Labor	Equipment	Total	Total Incl O&P
0010	**MOLDINGS, TRIM**									
0200	Astragal, stock pine, 11/16" x 1-3/4"	1 Carp	255	.031	L.F.	1.42	1.55		2.97	3.93
0250	1-5/16" x 2-3/16"		240	.033		2.10	1.64		3.74	4.82
0800	Chair rail, stock pine, 5/8" x 2-1/2"		270	.030		1.47	1.46		2.93	3.86
0900	5/8" x 3-1/2"		240	.033		2.30	1.64		3.94	5.05
1000	Closet pole, stock pine, 1-1/8" diameter		200	.040		1.11	1.97		3.08	4.24
1100	Fir, 1-5/8" diameter		200	.040		2.20	1.97		4.17	5.45
3300	Half round, stock pine, 1/4" x 1/2"		270	.030		.24	1.46		1.70	2.50
3350	1/2" x 1"		255	.031		.73	1.55		2.28	3.17
3400	Handrail, fir, single piece, stock, hardware not included									
3450	1-1/2" x 1-3/4"	1 Carp	80	.100	L.F.	2.56	4.93		7.49	10.35
3470	Pine, 1-1/2" x 1-3/4"		80	.100		2.40	4.93		7.33	10.20
3500	1-1/2" x 2-1/2"		76	.105		2.46	5.20		7.66	10.65
3600	Lattice, stock pine, 1/4" x 1-1/8"		270	.030		.35	1.46		1.81	2.63
3700	1/4" x 1-3/4"		250	.032		.69	1.58		2.27	3.17
3800	Miscellaneous, custom, pine, 1" x 1"		270	.030		.40	1.46		1.86	2.68
3850	1" x 2"		265	.030		.80	1.49		2.29	3.16
3900	1" x 3"		240	.033		1.20	1.64		2.84	3.83
4100	Birch or oak, nominal 1" x 1"		240	.033		.39	1.64		2.03	2.94
4200	Nominal 1" x 3"		215	.037		1.18	1.83		3.01	4.10
4400	Walnut, nominal 1" x 1"		215	.037		.62	1.83		2.45	3.50

For customer support on your Building Construction Costs with RSMeans Data, call 800.448.8182.

06 22 13 – Standard Pattern Wood Trim

06 22 13.45 Moldings, Trim

		Crew	Daily Output	Labor-Hours	Unit	Material	2017 Bare Costs Labor	2017 Bare Costs Equipment	Total	Total Incl O&P
4500	Nominal 1" x 3"	1 Carp	200	.040	L.F.	1.87	1.97		3.84	5.10
4700	Teak, nominal 1" x 1"		215	.037		2.69	1.83		4.52	5.75
4800	Nominal 1" x 3"		200	.040		8.05	1.97		10.02	11.85
4900	Quarter round, stock pine, 1/4" x 1/4"		275	.029		.24	1.43		1.67	2.45
4950	3/4" x 3/4"		255	.031		.50	1.55		2.05	2.92
5600	Wainscot moldings, 1-1/8" x 9/16", 2' high, minimum		76	.105	S.F.	12	5.20		17.20	21
5700	Maximum		65	.123	"	16.40	6.05		22.45	27.50

06 22 13.50 Moldings, Window and Door

		Crew	Daily Output	Labor-Hours	Unit	Material	2017 Bare Costs Labor	2017 Bare Costs Equipment	Total	Total Incl O&P
0010	**MOLDINGS, WINDOW AND DOOR**									
2800	Door moldings, stock, decorative, 1-1/8" wide, plain	1 Carp	17	.471	Set	47.50	23		70.50	88
2900	Detailed		17	.471	"	91.50	23		114.50	137
2960	Clear pine door jamb, no stops, 11/16" x 4-9/16"		240	.033	L.F.	5.30	1.64		6.94	8.35
3150	Door trim set, 1 head and 2 sides, pine, 2-1/2" wide		12	.667	Opng.	24	33		57	76.50
3170	3-1/2" wide		11	.727	"	29	36		65	87
3250	Glass beads, stock pine, 3/8" x 1/2"		275	.029	L.F.	.33	1.43		1.76	2.55
3270	3/8" x 7/8"		270	.030		.42	1.46		1.88	2.70
4850	Parting bead, stock pine, 3/8" x 3/4"		275	.029		.44	1.43		1.87	2.67
4870	1/2" x 3/4"		255	.031		.42	1.55		1.97	2.83
5000	Stool caps, stock pine, 11/16" x 3-1/2"		200	.040		2.07	1.97		4.04	5.30
5100	1-1/16" x 3-1/4"		150	.053		3.24	2.63		5.87	7.60
5300	Threshold, oak, 3' long, inside, 5/8" x 3-5/8"		32	.250	Ea.	6.50	12.30		18.80	26
5400	Outside, 1-1/2" x 7-5/8"		16	.500	"	45	24.50		69.50	87
5900	Window trim sets, including casings, header, stops,									
5910	stool and apron, 2-1/2" wide, FJP	1 Carp	13	.615	Opng.	31.50	30.50		62	81
5950	Pine		10	.800		37	39.50		76.50	102
6000	Oak		6	1.333		65.50	65.50		131	173

06 22 13.60 Moldings, Soffits

		Crew	Daily Output	Labor-Hours	Unit	Material	2017 Bare Costs Labor	2017 Bare Costs Equipment	Total	Total Incl O&P
0010	**MOLDINGS, SOFFITS**									
0200	Soffits, pine, 1" x 4"	2 Carp	420	.038	L.F.	.49	1.88		2.37	3.41
0210	1" x 6"		420	.038		.75	1.88		2.63	3.70
0220	1" x 8"		420	.038		1.25	1.88		3.13	4.25
0230	1" x 10"		400	.040		1.63	1.97		3.60	4.81
0240	1" x 12"		400	.040		2.01	1.97		3.98	5.25
0250	STK cedar, 1" x 4"		420	.038		.63	1.88		2.51	3.56
0260	1" x 6"		420	.038		1.09	1.88		2.97	4.07
0270	1" x 8"		420	.038		1.68	1.88		3.56	4.72
0280	1" x 10"		400	.040		2.24	1.97		4.21	5.50
0290	1" x 12"		400	.040		3.64	1.97		5.61	7
1000	Exterior AC plywood, 1/4" thick		400	.040	S.F.	1.01	1.97		2.98	4.13
1050	3/8" thick		400	.040		1.05	1.97		3.02	4.18
1100	1/2" thick		400	.040		1.25	1.97		3.22	4.40
1150	Polyvinyl chloride, white, solid	1 Carp	230	.035		2.15	1.71		3.86	4.99
1160	Perforated	"	230	.035		2.15	1.71		3.86	4.99
1170	Accessories, "J" channel 5/8"	2 Carp	700	.023	L.F.	.49	1.13		1.62	2.26

06 25 13 – Prefinished Hardboard Paneling

06 25 13.10 Paneling, Hardboard		Crew	Daily Output	Labor-Hours	Unit	Material	2017 Bare Costs Labor	Equipment	Total	Total Incl O&P
0010	**PANELING, HARDBOARD**									
0050	Not incl. furring or trim, hardboard, tempered, 1/8" thick [G]	2 Carp	500	.032	S.F.	.45	1.58		2.03	2.91
0100	1/4" thick [G]		500	.032		.64	1.58		2.22	3.11
0300	Tempered pegboard, 1/8" thick [G]		500	.032		.44	1.58		2.02	2.89
0400	1/4" thick [G]		500	.032		.70	1.58		2.28	3.18
0600	Untempered hardboard, natural finish, 1/8" thick [G]		500	.032		.44	1.58		2.02	2.89
0700	1/4" thick [G]		500	.032		.53	1.58		2.11	2.99
0900	Untempered pegboard, 1/8" thick [G]		500	.032		.48	1.58		2.06	2.94
1000	1/4" thick [G]		500	.032		.48	1.58		2.06	2.94
1200	Plastic faced hardboard, 1/8" thick [G]		500	.032		.62	1.58		2.20	3.09
1300	1/4" thick [G]		500	.032		.84	1.58		2.42	3.33
1500	Plastic faced pegboard, 1/8" thick [G]		500	.032		.62	1.58		2.20	3.09
1600	1/4" thick [G]		500	.032		.80	1.58		2.38	3.29
1800	Wood grained, plain or grooved, 1/8" thick [G]		500	.032		.65	1.58		2.23	3.13
1900	1/4" thick [G]		425	.038	↓	1.42	1.85		3.27	4.40
2100	Moldings, wood grained MDF		500	.032	L.F.	.41	1.58		1.99	2.86
2200	Pine	↓	425	.038	"	1.40	1.85		3.25	4.38

06 25 16 – Prefinished Plywood Paneling

06 25 16.10 Paneling, Plywood		Crew	Daily Output	Labor-Hours	Unit	Material	2017 Bare Costs Labor	Equipment	Total	Total Incl O&P
0010	**PANELING, PLYWOOD** R061636-20									
2400	Plywood, prefinished, 1/4" thick, 4' x 8' sheets									
2410	with vertical grooves. Birch faced, economy	2 Carp	500	.032	S.F.	1.47	1.58		3.05	4.03
2420	Average		420	.038		1.20	1.88		3.08	4.19
2430	Custom		350	.046		1.27	2.25		3.52	4.85
2600	Mahogany, African		400	.040		2.45	1.97		4.42	5.70
2700	Philippine (Lauan)		500	.032		.65	1.58		2.23	3.13
2900	Oak		500	.032		1.34	1.58		2.92	3.88
3000	Cherry		400	.040		2	1.97		3.97	5.20
3200	Rosewood		320	.050		3.10	2.46		5.56	7.20
3400	Teak		400	.040		3.12	1.97		5.09	6.45
3600	Chestnut		375	.043		5.30	2.10		7.40	9
3800	Pecan		400	.040		2.50	1.97		4.47	5.75
3900	Walnut, average		500	.032		2.47	1.58		4.05	5.15
3950	Custom		400	.040		5.05	1.97		7.02	8.55
4000	Plywood, prefinished, 3/4" thick, stock grades, economy		320	.050		1.44	2.46		3.90	5.35
4100	Average		224	.071		4.93	3.52		8.45	10.80
4300	Architectural grade, custom		224	.071		5.40	3.52		8.92	11.30
4400	Luxury		160	.100		5.30	4.93		10.23	13.35
4600	Plywood, "A" face, birch, VC, 1/2" thick, natural		450	.036		1.85	1.75		3.60	4.72
4700	Select		450	.036		1.95	1.75		3.70	4.83
4900	Veneer core, 3/4" thick, natural		320	.050		2.25	2.46		4.71	6.25
5000	Select		320	.050		2.53	2.46		4.99	6.55
5200	Lumber core, 3/4" thick, natural		320	.050		3.14	2.46		5.60	7.20
5500	Plywood, knotty pine, 1/4" thick, A2 grade		450	.036		1.85	1.75		3.60	4.72
5600	A3 grade		450	.036		2.09	1.75		3.84	4.98
5800	3/4" thick, veneer core, A2 grade		320	.050		2.31	2.46		4.77	6.30
5900	A3 grade		320	.050		2.39	2.46		4.85	6.40
6100	Aromatic cedar, 1/4" thick, plywood		400	.040		2.25	1.97		4.22	5.50
6200	1/4" thick, particle board	↓	400	.040	↓	1.15	1.97		3.12	4.29

06 25 Prefinished Paneling

06 25 26 – Panel System

06 25 26.10 Panel Systems

		Crew	Daily Output	Labor-Hours	Unit	Material	2017 Bare Costs Labor	Equipment	Total	Total Incl O&P
0010	**PANEL SYSTEMS**									
0100	Raised panel, eng. wood core w/wood veneer, std., paint grade	2 Carp	300	.053	S.F.	12.15	2.63		14.78	17.35
0110	Oak veneer		300	.053		24	2.63		26.63	30.50
0120	Maple veneer		300	.053		30	2.63		32.63	37
0130	Cherry veneer		300	.053		37.50	2.63		40.13	45.50
0300	Class I fire rated, paint grade		300	.053		13.50	2.63		16.13	18.85
0310	Oak veneer		300	.053		30	2.63		32.63	37
0320	Maple veneer		300	.053		41	2.63		43.63	49
0330	Cherry veneer		300	.053		49	2.63		51.63	58
0510	Beadboard, 5/8" MDF, standard, primed		300	.053		9.35	2.63		11.98	14.30
0520	Oak veneer, unfinished		300	.053		14.70	2.63		17.33	20
0530	Maple veneer, unfinished		300	.053		15.80	2.63		18.43	21.50
0610	Rustic paneling, 5/8" MDF, standard, maple veneer, unfinished	↓	300	.053	↓	18.50	2.63		21.13	24.50

06 26 Board Paneling

06 26 13 – Profile Board Paneling

06 26 13.10 Paneling, Boards

		Crew	Daily Output	Labor-Hours	Unit	Material	2017 Bare Costs Labor	Equipment	Total	Total Incl O&P
0010	**PANELING, BOARDS**									
6400	Wood board paneling, 3/4" thick, knotty pine	2 Carp	300	.053	S.F.	2.05	2.63		4.68	6.30
6500	Rough sawn cedar		300	.053		3.30	2.63		5.93	7.65
6700	Redwood, clear, 1" x 4" boards		300	.053		4.96	2.63		7.59	9.45
6900	Aromatic cedar, closet lining, boards	↓	275	.058	↓	2.42	2.87		5.29	7.05

06 43 Wood Stairs and Railings

06 43 13 – Wood Stairs

06 43 13.20 Prefabricated Wood Stairs

		Crew	Daily Output	Labor-Hours	Unit	Material	2017 Bare Costs Labor	Equipment	Total	Total Incl O&P
0010	**PREFABRICATED WOOD STAIRS**									
0100	Box stairs, prefabricated, 3'-0" wide									
0110	Oak treads, up to 14 risers	2 Carp	39	.410	Riser	92.50	20		112.50	133
0600	With pine treads for carpet, up to 14 risers	"	39	.410	"	59.50	20		79.50	96.50
1100	For 4' wide stairs, add				Flight	25%				
1550	Stairs, prefabricated stair handrail with balusters	1 Carp	30	.267	L.F.	82	13.15		95.15	111
1700	Basement stairs, prefabricated, pine treads									
1710	Pine risers, 3' wide, up to 14 risers	2 Carp	52	.308	Riser	59.50	15.15		74.65	88.50
4000	Residential, wood, oak treads, prefabricated		1.50	10.667	Flight	1,200	525		1,725	2,125
4200	Built in place	↓	.44	36.364	"	2,250	1,800		4,050	5,225
4400	Spiral, oak, 4'-6" diameter, unfinished, prefabricated,									
4500	incl. railing, 9' high	2 Carp	1.50	10.667	Flight	3,300	525		3,825	4,450

06 43 13.40 Wood Stair Parts

		Crew	Daily Output	Labor-Hours	Unit	Material	2017 Bare Costs Labor	Equipment	Total	Total Incl O&P
0010	**WOOD STAIR PARTS**									
0020	Pin top balusters, 1-1/4", oak, 34"	1 Carp	96	.083	Ea.	5.20	4.10		9.30	12
0030	38"		96	.083		5.85	4.10		9.95	12.75
0040	42"		96	.083		6.35	4.10		10.45	13.30
0050	Poplar, 34"		96	.083		3.37	4.10		7.47	10
0060	38"		96	.083		3.99	4.10		8.09	10.70
0070	42"		96	.083		8.80	4.10		12.90	15.95
0080	Maple, 34"		96	.083		4.90	4.10		9	11.70
0090	38"	↓	96	.083	↓	5.60	4.10		9.70	12.50

For customer support on your Building Construction Costs with RSMeans Data, call 800.448.8182.

209

06 43 13.40 Wood Stair Parts		Crew	Daily Output	Labor- Hours	Unit	Material	2017 Bare Costs Labor	Equipment	Total	Total Incl O&P
0100	42"	1 Carp	96	.083	Ea.	6.50	4.10		10.60	13.45
0130	Primed, 34"		96	.083		3.70	4.10		7.80	10.35
0140	38"		96	.083		4.25	4.10		8.35	11
0150	42"		96	.083		4.89	4.10		8.99	11.70
0180	Box top balusters, 1-1/4", oak, 34"		60	.133		8.95	6.55		15.50	19.90
0190	38"		60	.133		9.95	6.55		16.50	21
0200	42"		60	.133		10.95	6.55		17.50	22
0210	Poplar, 34"		60	.133		6.25	6.55		12.80	16.95
0220	38"		60	.133		6.95	6.55		13.50	17.70
0230	42"		60	.133		7.50	6.55		14.05	18.30
0240	Maple, 34"		60	.133		8.25	6.55		14.80	19.15
0250	38"		60	.133		9	6.55		15.55	19.95
0260	42"		60	.133		10	6.55		16.55	21
0290	Primed, 34"		60	.133		7	6.55		13.55	17.75
0300	38"		60	.133		8	6.55		14.55	18.85
0310	42"		60	.133	▼	8.90	6.55		15.45	19.80
0340	Square balusters, cut from lineal stock, pine, 1-1/16" x 1-1/16"		180	.044	L.F.	1.50	2.19		3.69	5
0350	1-5/16" x 1-5/16"		180	.044		2.10	2.19		4.29	5.65
0360	1-5/8" x 1-5/8"		180	.044	▼	3.30	2.19		5.49	7
0370	Turned newel, oak, 3-1/2" square, 48" high		8	1	Ea.	95	49.50		144.50	181
0380	62" high		8	1		101	49.50		150.50	187
0390	Poplar, 3-1/2" square, 48" high		8	1		52	49.50		101.50	133
0400	62" high		8	1		66	49.50		115.50	148
0410	Maple, 3-1/2" square, 48" high		8	1		65	49.50		114.50	147
0420	62" high		8	1		86	49.50		135.50	170
0430	Square newel, oak, 3-1/2" square, 48" high		8	1		59	49.50		108.50	141
0440	58" high		8	1		68.50	49.50		118	151
0450	Poplar, 3-1/2" square, 48" high		8	1		35.50	49.50		85	115
0460	58" high		8	1		43.50	49.50		93	124
0470	Maple, 3" square, 48" high		8	1		54	49.50		103.50	135
0480	58" high		8	1	▼	68	49.50		117.50	151
0490	Railings, oak, economy		96	.083	L.F.	9.45	4.10		13.55	16.70
0500	Average		96	.083		13.50	4.10		17.60	21
0510	Custom		96	.083		17.95	4.10		22.05	26
0520	Maple, economy		96	.083		11.90	4.10		16	19.35
0530	Average		96	.083		13.50	4.10		17.60	21
0540	Custom		96	.083		18.55	4.10		22.65	27
0550	Oak, for bending rail, economy		48	.167		26	8.20		34.20	41
0560	Average		48	.167		28.50	8.20		36.70	44
0570	Custom		48	.167		32	8.20		40.20	47.50
0580	Maple, for bending rail, economy		48	.167		29.50	8.20		37.70	45
0590	Average		48	.167		30.50	8.20		38.70	46
0600	Custom		48	.167	▼	33.50	8.20		41.70	49.50
0610	Risers, oak, 3/4" x 8", 36" long		80	.100	Ea.	13	4.93		17.93	22
0620	42" long		70	.114		15.15	5.65		20.80	25.50
0630	48" long		63	.127		17.30	6.25		23.55	28.50
0640	54" long		56	.143		19.50	7.05		26.55	32.50
0650	60" long		50	.160		21.50	7.90		29.40	36
0660	72" long		42	.190		26	9.40		35.40	43
0670	Poplar, 3/4" x 8", 36" long		80	.100		12.50	4.93		17.43	21.50
0680	42" long		71	.113		14.55	5.55		20.10	24.50
0690	48" long		63	.127		16.65	6.25		22.90	28
0700	54" long	▼	56	.143	▼	18.70	7.05		25.75	31.50

For customer support on your Building Construction Costs with RSMeans Data, call 800.448.8182.

06 43 Wood Stairs and Railings

06 43 13 – Wood Stairs

06 43 13.40 Wood Stair Parts		Crew	Daily Output	Labor-Hours	Unit	Material	2017 Bare Costs Labor	Equipment	Total	Total Incl O&P
0710	60" long	1 Carp	50	.160	Ea.	21	7.90		28.90	35
0720	72" long		42	.190		25	9.40		34.40	42
0730	Pine, 1" x 8", 36" long		80	.100		3.76	4.93		8.69	11.70
0740	42" long		70	.114		4.39	5.65		10.04	13.45
0750	48" long		63	.127		5	6.25		11.25	15.10
0760	54" long		56	.143		5.65	7.05		12.70	17
0770	60" long		50	.160		6.25	7.90		14.15	18.95
0780	72" long		42	.190		7.50	9.40		16.90	22.50
0790	Treads, oak, no returns, 1-1/32" x 11-1/2" x 36" long		32	.250		29.50	12.30		41.80	51.50
0800	42" long		32	.250		34.50	12.30		46.80	57
0810	48" long		32	.250		39.50	12.30		51.80	62
0820	54" long		32	.250		44	12.30		56.30	67.50
0830	60" long		32	.250		49	12.30		61.30	73
0840	72" long		32	.250		59	12.30		71.30	84
0850	Mitred return one end, 1-1/32" x 11-1/2" x 36" long		24	.333		39	16.40		55.40	68
0860	42" long		24	.333		45.50	16.40		61.90	75
0870	48" long		24	.333		52	16.40		68.40	82
0880	54" long		24	.333		58.50	16.40		74.90	89.50
0890	60" long		24	.333		65	16.40		81.40	96.50
0900	72" long		24	.333		78	16.40		94.40	111
0910	Mitred return two ends, 1-1/32" x 11-1/2" x 36" long		12	.667		49	33		82	105
0920	42" long		12	.667		57	33		90	114
0930	48" long		12	.667		65.50	33		98.50	123
0940	54" long		12	.667		73.50	33		106.50	132
0950	60" long		12	.667		81.50	33		114.50	141
0960	72" long		12	.667		98	33		131	159
0970	Starting step, oak, 48", bullnose		8	1		172	49.50		221.50	265
0980	Double end bullnose		8	1	▼	254	49.50		303.50	355
1030	Skirt board, pine, 1" x 10"		55	.145	L.F.	1.63	7.15		8.78	12.75
1040	1" x 12"		52	.154	"	2.01	7.60		9.61	13.80
1050	Oak landing tread, 1-1/16" thick		54	.148	S.F.	9	7.30		16.30	21
1060	Oak cove molding		96	.083	L.F.	1	4.10		5.10	7.40
1070	Oak stringer molding		96	.083	"	4	4.10		8.10	10.70
1090	Rail bolt, 5/16" x 3-1/2"		48	.167	Ea.	2.75	8.20		10.95	15.65
1100	5/16" x 4-1/2"		48	.167		2.75	8.20		10.95	15.65
1120	Newel post anchor		16	.500		13	24.50		37.50	52
1130	Tapered plug, 1/2"		240	.033		1	1.64		2.64	3.61
1140	1"	▼	240	.033	▼	.99	1.64		2.63	3.60

06 43 16 – Wood Railings

06 43 16.10 Wood Handrails and Railings

		Crew	Daily Output	Labor-Hours	Unit	Material	2017 Bare Costs Labor	Equipment	Total	Total Incl O&P
0010	**WOOD HANDRAILS AND RAILINGS**									
0020	Custom design, architectural grade, hardwood, plain	1 Carp	38	.211	L.F.	12.65	10.35		23	30
0100	Shaped		30	.267		63	13.15		76.15	89.50
0300	Stock interior railing with spindles 4" O.C., 4' long		40	.200		41	9.85		50.85	60
0400	8' long	▼	48	.167	▼	41	8.20		49.20	57.50

For customer support on your Building Construction Costs with RSMeans Data, call 800.448.8182.

211

06 44 19 – Wood Grilles

06 44 19.10 Grilles		Crew	Daily Output	Labor-Hours	Unit	Material	2017 Bare Costs Labor	Equipment	Total	Total Incl O&P
0010	**GRILLES** and panels, hardwood, sanded									
0020	2' x 4' to 4' x 8', custom designs, unfinished, economy	1 Carp	38	.211	S.F.	62	10.35		72.35	84
0050	Average		30	.267		70	13.15		83.15	97
0100	Custom		19	.421		72	20.50		92.50	111

06 44 33 – Wood Mantels

06 44 33.10 Fireplace Mantels

		Crew	Daily Output	Labor-Hours	Unit	Material	Labor	Equipment	Total	Total Incl O&P
0010	**FIREPLACE MANTELS**									
0015	6" molding, 6' x 3'-6" opening, plain, paint grade	1 Carp	5	1.600	Opng.	450	79		529	615
0100	Ornate, oak		5	1.600		600	79		679	780
0300	Prefabricated pine, colonial type, stock, deluxe		2	4		1,500	197		1,697	1,950
0400	Economy		3	2.667		705	131		836	975

06 44 33.20 Fireplace Mantel Beam

		Crew	Daily Output	Labor-Hours	Unit	Material	Labor	Equipment	Total	Total Incl O&P
0010	**FIREPLACE MANTEL BEAM**									
0020	Rough texture wood, 4" x 8"	1 Carp	36	.222	L.F.	8.75	10.95		19.70	26.50
0100	4" x 10"		35	.229	"	10.30	11.25		21.55	28.50
0300	Laminated hardwood, 2-1/4" x 10-1/2" wide, 6' long		5	1.600	Ea.	125	79		204	258
0400	8' long		5	1.600	"	145	79		224	280
0600	Brackets for above, rough sawn		12	.667	Pr.	12.05	33		45.05	64
0700	Laminated		12	.667	"	17.20	33		50.20	69.50

06 44 39 – Wood Posts and Columns

06 44 39.10 Decorative Beams

		Crew	Daily Output	Labor-Hours	Unit	Material	Labor	Equipment	Total	Total Incl O&P
0010	**DECORATIVE BEAMS**									
0020	Rough sawn cedar, non-load bearing, 4" x 4"	2 Carp	180	.089	L.F.	1.35	4.38		5.73	8.20
0100	4" x 6"		170	.094		1.92	4.64		6.56	9.20
0200	4" x 8"		160	.100		2.52	4.93		7.45	10.30
0300	4" x 10"		150	.107		3.99	5.25		9.24	12.45
0400	4" x 12"		140	.114		5.05	5.65		10.70	14.15
0500	8" x 8"		130	.123		5.20	6.05		11.25	15

06 44 39.20 Columns

		Crew	Daily Output	Labor-Hours	Unit	Material	Labor	Equipment	Total	Total Incl O&P
0010	**COLUMNS**									
0050	Aluminum, round colonial, 6" diameter	2 Carp	80	.200	V.L.F.	20	9.85		29.85	37
0100	8" diameter		62.25	.257		23	12.65		35.65	44.50
0200	10" diameter		55	.291		23	14.35		37.35	47.50
0250	Fir, stock units, hollow round, 6" diameter		80	.200		30	9.85		39.85	48
0300	8" diameter		80	.200		39	9.85		48.85	58
0350	10" diameter		70	.229		48	11.25		59.25	70.50
0360	12" diameter		65	.246		55.50	12.10		67.60	79.50
0400	Solid turned, to 8' high, 3-1/2" diameter		80	.200		9.90	9.85		19.75	26
0500	4-1/2" diameter		75	.213		12.10	10.50		22.60	29.50
0600	5-1/2" diameter		70	.229		16.35	11.25		27.60	35.50
0800	Square columns, built-up, 5" x 5"		65	.246		14.50	12.10		26.60	34.50
0900	Solid, 3-1/2" x 3-1/2"		130	.123		9.90	6.05		15.95	20
1600	Hemlock, tapered, T & G, 12" diam., 10' high		100	.160		40	7.90		47.90	56
1700	16' high		65	.246		74.50	12.10		86.60	100
1900	14" diameter, 10' high		100	.160		116	7.90		123.90	139
2000	18' high		65	.246		105	12.10		117.10	134
2200	18" diameter, 12' high		65	.246		168	12.10		180.10	204
2300	20' high		50	.320		120	15.75		135.75	156
2500	20" diameter, 14' high		40	.400		184	19.70		203.70	232
2600	20' high		35	.457		173	22.50		195.50	226
2800	For flat pilasters, deduct					33%				

06 44 Ornamental Woodwork

06 44 39 – Wood Posts and Columns

06 44 39.20 Columns	Crew	Daily Output	Labor-Hours	Unit	Material	2017 Bare Costs Labor	Equipment	Total	Total Incl O&P
3000	For splitting into halves, add				Ea.	110			110	121
4000	Rough sawn cedar posts, 4" x 4"	2 Carp	250	.064	V.L.F.	3.80	3.15		6.95	9
4100	4" x 6"		235	.068		6.80	3.35		10.15	12.65
4200	6" x 6"		220	.073		9.90	3.58		13.48	16.40
4300	8" x 8"		200	.080		20.50	3.94		24.44	28.50

06 48 Wood Frames

06 48 13 – Exterior Wood Door Frames

06 48 13.10 Exterior Wood Door Frames and Accessories

		Crew	Daily Output	Labor-Hours	Unit	Material	2017 Bare Costs Labor	Equipment	Total	Total Incl O&P
0010	**EXTERIOR WOOD DOOR FRAMES AND ACCESSORIES**									
0400	Exterior frame, incl. ext. trim, pine, 5/4 x 4-9/16" deep	2 Carp	375	.043	L.F.	6.75	2.10		8.85	10.65
0420	5-3/16" deep		375	.043		8.30	2.10		10.40	12.30
0440	6-9/16" deep		375	.043		9.75	2.10		11.85	13.95
0600	Oak, 5/4 x 4-9/16" deep		350	.046		21	2.25		23.25	26.50
0620	5-3/16" deep		350	.046		23	2.25		25.25	29
0640	6-9/16" deep		350	.046		21.50	2.25		23.75	27
1000	Sills, 8/4 x 8" deep, oak, no horns		100	.160		6.65	7.90		14.55	19.35
1020	2" horns		100	.160		20.50	7.90		28.40	34.50
1040	3" horns		100	.160		20.50	7.90		28.40	34.50
1100	8/4 x 10" deep, oak, no horns		90	.178		6.40	8.75		15.15	20.50
1120	2" horns		90	.178		26.50	8.75		35.25	42.50
1140	3" horns		90	.178		26.50	8.75		35.25	42.50
2000	Wood frame & trim, ext, colonial, 3' opng, fluted pilasters, flat head		22	.727	Ea.	515	36		551	620
2010	Dentil head		21	.762		590	37.50		627.50	705
2020	Ram's head		20	.800		695	39.50		734.50	825
2100	5'-4" opening, in-swing, fluted pilasters, flat head		17	.941		470	46.50		516.50	585
2120	Ram's head		15	1.067		1,400	52.50		1,452.50	1,625
2140	Out swing, fluted pilasters, flat head		17	.941		495	46.50		541.50	615
2160	Ram's head		15	1.067		1,575	52.50		1,627.50	1,800
2400	6'-0" opening, in-swing, fluted pilasters, flat head		16	1		495	49.50		544.50	620
2420	Ram's head		10	1.600		1,475	79		1,554	1,750
2460	Out-swing, fluted pilasters, flat head		16	1		500	49.50		549.50	625
2480	Ram's head		10	1.600		1,475	79		1,554	1,750
2600	For two sidelights, flat head, add		30	.533	Opng.	230	26.50		256.50	293
2620	Ram's head, add		20	.800	"	850	39.50		889.50	995
2700	Custom birch frame, 3'-0" opening		16	1	Ea.	248	49.50		297.50	350
2750	6'-0" opening		16	1		370	49.50		419.50	480
2900	Exterior, modern, plain trim, 3' opng., in-swing, FJP		26	.615		47.50	30.50		78	99
2920	Fir		24	.667		56	33		89	112
2940	Oak		22	.727		65	36		101	127

06 48 16 – Interior Wood Door Frames

06 48 16.10 Interior Wood Door Jamb and Frames

		Crew	Daily Output	Labor-Hours	Unit	Material	2017 Bare Costs Labor	Equipment	Total	Total Incl O&P
0010	**INTERIOR WOOD DOOR JAMB AND FRAMES**									
3000	Interior frame, pine, 11/16" x 3-5/8" deep	2 Carp	375	.043	L.F.	4.50	2.10		6.60	8.15
3020	4-9/16" deep		375	.043		5.05	2.10		7.15	8.75
3200	Oak, 11/16" x 3-5/8" deep		350	.046		10.25	2.25		12.50	14.75
3220	4-9/16" deep		350	.046		10.30	2.25		12.55	14.75
3240	5-3/16" deep		350	.046		14.95	2.25		17.20	19.90
3400	Walnut, 11/16" x 3-5/8" deep		350	.046		9.40	2.25		11.65	13.80
3420	4-9/16" deep		350	.046		9.90	2.25		12.15	14.35

For customer support on your Building Construction Costs with RSMeans Data, call 800.448.8182.

213

06 48 Wood Frames

06 48 16 – Interior Wood Door Frames

06 48 16.10 Interior Wood Door Jamb and Frames		Crew	Daily Output	Labor-Hours	Unit	Material	2017 Bare Costs Labor	Equipment	Total	Total Incl O&P
3440	5-3/16" deep	2 Carp	350	.046	L.F.	10	2.25		12.25	14.45
3600	Pocket door frame		16	1	Ea.	82.50	49.50		132	166
3800	Threshold, oak, 5/8" x 3-5/8" deep		200	.080	L.F.	3.57	3.94		7.51	10
3820	4-5/8" deep		190	.084		4.03	4.15		8.18	10.80
3840	5-5/8" deep	▼	180	.089	▼	6.70	4.38		11.08	14.10

06 49 Wood Screens and Exterior Wood Shutters

06 49 19 – Exterior Wood Shutters

06 49 19.10 Shutters, Exterior

		Crew	Daily Output	Labor-Hours	Unit	Material	2017 Bare Costs Labor	Equipment	Total	Total Incl O&P
0010	**SHUTTERS, EXTERIOR**									
0012	Aluminum, louvered, 1'-4" wide, 3'-0" long	1 Carp	10	.800	Pr.	201	39.50		240.50	283
0400	6'-8" long		9	.889		355	44		399	455
1000	Pine, louvered, primed, each 1'-2" wide, 3'-3" long		10	.800		215	39.50		254.50	297
1100	4'-7" long		10	.800		276	39.50		315.50	365
1500	Each 1'-6" wide, 3'-3" long		10	.800		255	39.50		294.50	340
1600	4'-7" long		10	.800		315	39.50		354.50	410
1620	Cedar, louvered, 1'-2" wide, 5'-7" long		10	.800		325	39.50		364.50	420
1630	Each 1'-4" wide, 2'-2" long		10	.800		176	39.50		215.50	255
1670	4'-3" long		10	.800		294	39.50		333.50	385
1690	5'-11" long		10	.800		350	39.50		389.50	445
1700	Door blinds, 6'-9" long, each 1'-3" wide		9	.889		385	44		429	490
1710	1'-6" wide		9	.889		440	44		484	550
1720	Cedar, solid raised panel, each 1'-4" wide, 3'-3" long		10	.800		330	39.50		369.50	425
1740	4'-3" long		10	.800		390	39.50		429.50	485
1770	5'-11" long		10	.800		530	39.50		569.50	640
1800	Door blinds, 6'-9" long, each 1'-3" wide		9	.889		535	44		579	655
1900	1'-6" wide		9	.889		630	44		674	760
2500	Polystyrene, solid raised panel, each 1'-4" wide, 3'-3" long		10	.800		76	39.50		115.50	145
2700	4'-7" long		10	.800		105	39.50		144.50	177
4500	Polystyrene, louvered, each 1'-2" wide, 3'-3" long		10	.800		36	39.50		75.50	100
4750	5'-3" long		10	.800		59	39.50		98.50	126
6000	Vinyl, louvered, each 1'-2" x 4'-7" long		10	.800		64	39.50		103.50	131
6200	Each 1'-4" x 6'-8" long	▼	9	.889	▼	83.50	44		127.50	159

06 51 Structural Plastic Shapes and Plates

06 51 13 – Plastic Lumber

06 51 13.10 Recycled Plastic Lumber

			Crew	Daily Output	Labor-Hours	Unit	Material	2017 Bare Costs Labor	Equipment	Total	Total Incl O&P
0010	**RECYCLED PLASTIC LUMBER**										
4000	Sheeting, recycled plastic, black or white, 4' x 8' x 1/8"	G	2 Carp	1100	.015	S.F.	1.10	.72		1.82	2.31
4010	4' x 8' x 3/16"	G		1100	.015		1.70	.72		2.42	2.97
4020	4' x 8' x 1/4"	G		950	.017		1.96	.83		2.79	3.43
4030	4' x 8' x 3/8"	G		950	.017		3.40	.83		4.23	5
4040	4' x 8' x 1/2"	G		900	.018		4.45	.88		5.33	6.25
4050	4' x 8' x 5/8"	G		900	.018		6.70	.88		7.58	8.75
4060	4' x 8' x 3/4"	G	▼	850	.019	▼	8.40	.93		9.33	10.65
4070	Add for colors	G				Ea.	5%				
8500	100% recycled plastic, var colors, NLB, 2" x 2"	G				L.F.	1.91			1.91	2.10
8510	2" x 4"	G					3.95			3.95	4.35
8520	2" x 6"	G				▼	6.20			6.20	6.80

06 51 Structural Plastic Shapes and Plates

06 51 13 – Plastic Lumber

06 51 13.10 Recycled Plastic Lumber		Crew	Daily Output	Labor-Hours	Unit	Material	2017 Bare Costs Labor	Equipment	Total	Total Incl O&P
8530	2" x 8"	G			L.F.	8.50			8.50	9.35
8540	2" x 10"	G				12.40			12.40	13.60
8550	5/4" x 4"	G				4			4	4.40
8560	5/4" x 6"	G				5.70			5.70	6.25
8570	1" x 6"	G				3.11			3.11	3.42
8580	1/2" x 8"	G				3.83			3.83	4.21
8590	2" x 10" T & G	G				12.20			12.20	13.40
8600	3" x 10" T & G	G				18			18	19.80
8610	Add for premium colors	G				20%				

06 51 13.12 Structural Plastic Lumber

		Crew	Daily Output	Labor-Hours	Unit	Material	2017 Bare Costs Labor	Equipment	Total	Total Incl O&P
0010	**STRUCTURAL PLASTIC LUMBER**									
1320	Plastic lumber, posts or columns, 4" x 4"	2 Carp	390	.041	L.F.	10.30	2.02		12.32	14.45
1325	4" x 6"		275	.058		13.30	2.87		16.17	19
1330	4" x 8"		220	.073		19.55	3.58		23.13	27
1340	Girder, single, 4" x 4"		675	.024		10.30	1.17		11.47	13.15
1345	4" x 6"		600	.027		13.30	1.31		14.61	16.60
1350	4" x 8"		525	.030		19.55	1.50		21.05	24
1352	Double, 2" x 4"		625	.026		9.50	1.26		10.76	12.40
1354	2" x 6"		600	.027		12.85	1.31		14.16	16.15
1356	2" x 8"		575	.028		16.80	1.37		18.17	20.50
1358	2" x 10"		550	.029		20.50	1.43		21.93	24.50
1360	2" x 12"		525	.030		25	1.50		26.50	30
1362	Triple, 2" x 4"		575	.028		14.25	1.37		15.62	17.80
1364	2" x 6"		550	.029		19.30	1.43		20.73	23
1366	2" x 8"		525	.030		25	1.50		26.50	30.50
1368	2" x 10"		500	.032		31	1.58		32.58	36.50
1370	2" x 12"		475	.034		37.50	1.66		39.16	43.50
1372	Ledger, bolted 4' O.C., 2" x 4"		400	.040		4.91	1.97		6.88	8.40
1374	2" x 6"		550	.029		6.50	1.43		7.93	9.35
1376	2" x 8"		390	.041		8.55	2.02		10.57	12.50
1378	2" x 10"		385	.042		10.40	2.05		12.45	14.55
1380	2" x 12"		380	.042		12.60	2.07		14.67	17.05
1382	Joists, 2" x 4"		1250	.013		4.75	.63		5.38	6.20
1384	2" x 6"		1250	.013		6.45	.63		7.08	8
1386	2" x 8"		1100	.015		8.40	.72		9.12	10.35
1388	2" x 10"		500	.032		10.35	1.58		11.93	13.80
1390	2" x 12"		875	.018		12.45	.90		13.35	15.10
1392	Railings and trim, 5/4" x 4"	1 Carp	300	.027		4.15	1.31		5.46	6.60
1394	2" x 2"		300	.027		2.47	1.31		3.78	4.73
1396	2" x 4"		300	.027		4.73	1.31		6.04	7.20
1398	2" x 6"		300	.027		6.40	1.31		7.71	9.05

06 52 Plastic Structural Assemblies

06 52 10 – Fiberglass Structural Assemblies

06 52 10.10 Castings, Fiberglass

		Crew	Daily Output	Labor-Hours	Unit	Material	2017 Bare Costs Labor	Equipment	Total	Total Incl O&P
0010	**CASTINGS, FIBERGLASS**									
0100	Angle, 1" x 1" x 1/8" thick	2 Sswk	240	.067	L.F.	2.07	3.62		5.69	8.35
0120	3" x 3" x 1/4" thick		200	.080		7.05	4.34		11.39	15.05
0140	4" x 4" x 1/4" thick		200	.080		9	4.34		13.34	17.15
0160	4" x 4" x 3/8" thick		200	.080		13.15	4.34		17.49	22
0180	6" x 6" x 1/2" thick		160	.100		32	5.45		37.45	44
1000	Flat sheet, 1/8" thick		140	.114	S.F.	6.85	6.20		13.05	17.90
1020	1/4" thick		120	.133		11.95	7.25		19.20	25
1040	3/8" thick		100	.160		19.05	8.70		27.75	35.50
1060	1/2" thick		80	.200		22.50	10.85		33.35	42.50
2000	Handrail, 42" high, 2" diam. rails pickets 5' O.C.		32	.500	L.F.	54	27		81	105
3000	Round bar, 1/4" diam.		240	.067		.51	3.62		4.13	6.60
3020	1/2" diam.		200	.080		1.45	4.34		5.79	8.85
3040	3/4" diam.		200	.080		3.58	4.34		7.92	11.20
3060	1" diam.		160	.100		6.30	5.45		11.75	16
3080	1-1/4" diam.		160	.100		10.95	5.45		16.40	21
3100	1-1/2" diam.		140	.114		14.15	6.20		20.35	26
3500	Round tube, 1" diam. x 1/8" thick		240	.067		2.85	3.62		6.47	9.20
3520	2" diam. x 1/4" thick		200	.080		9.30	4.34		13.64	17.45
3540	3" diam. x 1/4" thick		160	.100		11.95	5.45		17.40	22
4000	Square bar, 1/2" square		240	.067		4.61	3.62		8.23	11.10
4020	1" square		200	.080		6.50	4.34		10.84	14.40
4040	1-1/2" square		160	.100		12.50	5.45		17.95	23
4500	Square tube, 1" x 1" x 1/8" thick		240	.067		3.14	3.62		6.76	9.50
4520	2" x 2" x 1/8" thick		200	.080		6.30	4.34		10.64	14.20
4540	3" x 3" x 1/4" thick		160	.100		12.75	5.45		18.20	23
5000	Threaded rod, 3/8" diam.		320	.050		4.69	2.72		7.41	9.70
5020	1/2" diam.		320	.050		5.80	2.72		8.52	10.90
5040	5/8" diam.		280	.057		6.05	3.10		9.15	11.80
5060	3/4" diam.		280	.057		6.90	3.10		10	12.75
6000	Wide flange beam, 4" x 4" x 1/4" thick		120	.133		13.25	7.25		20.50	26.50
6020	6" x 6" x 1/4" thick		100	.160		22.50	8.70		31.20	39
6040	8" x 8" x 3/8" thick		80	.200		37.50	10.85		48.35	59.50

06 52 10.20 Fiberglass Stair Treads

		Crew	Daily Output	Labor-Hours	Unit	Material	2017 Bare Costs Labor	Equipment	Total	Total Incl O&P
0010	**FIBERGLASS STAIR TREADS**									
0100	24" wide	2 Sswk	52	.308	Ea.	32.50	16.70		49.20	64
0140	30" wide		52	.308		40.50	16.70		57.20	72.50
0180	36" wide		52	.308		49	16.70		65.70	82
0220	42" wide		52	.308		57	16.70		73.70	91

06 52 10.30 Fiberglass Grating

		Crew	Daily Output	Labor-Hours	Unit	Material	2017 Bare Costs Labor	Equipment	Total	Total Incl O&P
0010	**FIBERGLASS GRATING**									
0100	Molded, green (for mod. corrosive environment)									
0140	1" x 4" mesh, 1" thick	2 Sswk	400	.040	S.F.	8.50	2.17		10.67	12.95
0180	1-1/2" square mesh, 1" thick		400	.040		17.65	2.17		19.82	23
0220	1-1/4" thick		400	.040		10.20	2.17		12.37	14.80
0260	1-1/2" thick		400	.040		24	2.17		26.17	29.50
0300	2" square mesh, 2" thick		320	.050		28.50	2.72		31.22	36
1000	Orange (for highly corrosive environment)									
1040	1" x 4" mesh, 1" thick	2 Sswk	400	.040	S.F.	17.70	2.17		19.87	23
1080	1-1/2" square mesh, 1" thick		400	.040		19.60	2.17		21.77	25
1120	1-1/4" thick		400	.040		21.50	2.17		23.67	27
1160	1-1/2" thick		400	.040		21.50	2.17		23.67	27.50

For customer support on your Building Construction Costs with RSMeans Data, call 800.448.8182.

06 52 Plastic Structural Assemblies

06 52 10 - Fiberglass Structural Assemblies

06 52 10.30 Fiberglass Grating	Crew	Daily Output	Labor-Hours	Unit	Material	2017 Bare Costs Labor	Equipment	Total	Total Incl O&P
1200	2" square mesh, 2" thick	2 Sswk	320	.050	S.F.	21.50	2.72		24.22	28.50
3000	Pultruded, green (for mod. corrosive environment)									
3040	1" O.C. bar spacing, 1" thick	2 Sswk	400	.040	S.F.	15.75	2.17		17.92	21
3080	1-1/2" thick		320	.050		15.50	2.72		18.22	21.50
3120	1-1/2" O.C. bar spacing, 1" thick		400	.040		13.90	2.17		16.07	18.90
3160	1-1/2" thick		400	.040		18.65	2.17		20.82	24
4000	Grating support legs, fixed height, no base				Ea.	49.50			49.50	54
4040	With base					43.50			43.50	48
4080	Adjustable to 60"					66			66	72.50

06 52 10.40 Fiberglass Floor Grating

		Crew	Daily Output	Labor-Hours	Unit	Material	2017 Bare Costs Labor	Equipment	Total	Total Incl O&P
0010	**FIBERGLASS FLOOR GRATING**									
0100	Reinforced polyester, fire retardant, 1" x 4" grid, 1" thick	E-4	510	.063	S.F.	14.45	3.44	.25	18.14	22
0200	1-1/2" x 6" mesh, 1-1/2" thick		500	.064		16.80	3.51	.26	20.57	24.50
0300	With grit surface, 1-1/2" x 6" grid, 1-1/2" thick		500	.064		17.20	3.51	.26	20.97	25

06 63 Plastic Railings

06 63 10 - Plastic (PVC) Railings

06 63 10.10 Plastic Railings

		Crew	Daily Output	Labor-Hours	Unit	Material	2017 Bare Costs Labor	Equipment	Total	Total Incl O&P
0010	**PLASTIC RAILINGS**									
0100	Horizontal PVC handrail with balusters, 3-1/2" wide, 36" high	1 Carp	96	.083	L.F.	26.50	4.10		30.60	35.50
0150	42" high		96	.083		30	4.10		34.10	39.50
0200	Angled PVC handrail with balusters, 3-1/2" wide, 36" high		72	.111		31	5.45		36.45	42.50
0250	42" high		72	.111		33.50	5.45		38.95	45.50
0300	Post sleeve for 4 x 4 post		96	.083		13.55	4.10		17.65	21
0400	Post cap for 4 x 4 post, flat profile		48	.167	Ea.	13.90	8.20		22.10	28
0450	Newel post style profile		48	.167		25	8.20		33.20	40
0500	Raised corbeled profile		48	.167		37	8.20		45.20	53
0550	Post base trim for 4 x 4 post		96	.083		19.20	4.10		23.30	27.50

06 65 Plastic Trim

06 65 10 - PVC Trim

06 65 10.10 PVC Trim, Exterior

		Crew	Daily Output	Labor-Hours	Unit	Material	2017 Bare Costs Labor	Equipment	Total	Total Incl O&P
0010	**PVC TRIM, EXTERIOR**									
0100	Cornerboards, 5/4" x 6" x 6"	1 Carp	240	.033	L.F.	7.20	1.64		8.84	10.45
0110	Door/window casing, 1" x 4"		200	.040		1.39	1.97		3.36	4.55
0120	1" x 6"		200	.040		2.06	1.97		4.03	5.30
0130	1" x 8"		195	.041		2.72	2.02		4.74	6.10
0140	1" x 10"		195	.041		3.44	2.02		5.46	6.90
0150	1" x 12"		190	.042		4.22	2.07		6.29	7.80
0160	5/4" x 4"		195	.041		2.05	2.02		4.07	5.35
0170	5/4" x 6"		195	.041		3.02	2.02		5.04	6.40
0180	5/4" x 8"		190	.042		3.99	2.07		6.06	7.55
0190	5/4" x 10"		190	.042		5.10	2.07		7.17	8.80
0200	5/4" x 12"		185	.043		5.35	2.13		7.48	9.15
0210	Fascia, 1" x 4"		250	.032		1.39	1.58		2.97	3.94
0220	1" x 6"		250	.032		2.06	1.58		3.64	4.68
0230	1" x 8"		225	.036		2.72	1.75		4.47	5.65
0240	1" x 10"		225	.036		3.44	1.75		5.19	6.45
0250	1" x 12"		200	.040		4.22	1.97		6.19	7.65

For customer support on your Building Construction Costs with RSMeans Data, call 800.448.8182.

217

06 65 Plastic Trim

06 65 10 – PVC Trim

	06 65 10.10 PVC Trim, Exterior	Crew	Daily Output	Labor-Hours	Unit	Material	2017 Bare Costs Labor	Equipment	Total	Total Incl O&P
0260	5/4" x 4"	1 Carp	240	.033	L.F.	2.05	1.64		3.69	4.77
0270	5/4" x 6"		240	.033		3.02	1.64		4.66	5.85
0280	5/4" x 8"		215	.037		3.99	1.83		5.82	7.20
0290	5/4" x 10"		215	.037		5.10	1.83		6.93	8.40
0300	5/4" x 12"		190	.042		5.35	2.07		7.42	9.10
0310	Frieze, 1" x 4"		250	.032		1.39	1.58		2.97	3.94
0320	1" x 6"		250	.032		2.06	1.58		3.64	4.68
0330	1" x 8"		225	.036		2.72	1.75		4.47	5.65
0340	1" x 10"		225	.036		3.44	1.75		5.19	6.45
0350	1" x 12"		200	.040		4.22	1.97		6.19	7.65
0360	5/4" x 4"		240	.033		2.05	1.64		3.69	4.77
0370	5/4" x 6"		240	.033		3.02	1.64		4.66	5.85
0380	5/4" x 8"		215	.037		3.99	1.83		5.82	7.20
0390	5/4" x 10"		215	.037		5.10	1.83		6.93	8.40
0400	5/4" x 12"		190	.042		5.35	2.07		7.42	9.10
0410	Rake, 1" x 4"		200	.040		1.39	1.97		3.36	4.55
0420	1" x 6"		200	.040		2.06	1.97		4.03	5.30
0430	1" x 8"		190	.042		2.72	2.07		4.79	6.15
0440	1" x 10"		190	.042		3.44	2.07		5.51	6.95
0450	1" x 12"		180	.044		4.22	2.19		6.41	8
0460	5/4" x 4"		195	.041		2.05	2.02		4.07	5.35
0470	5/4" x 6"		195	.041		3.02	2.02		5.04	6.40
0480	5/4" x 8"		185	.043		3.99	2.13		6.12	7.65
0490	5/4" x 10"		185	.043		5.10	2.13		7.23	8.85
0500	5/4" x 12"		175	.046		5.35	2.25		7.60	9.35
0510	Rake trim, 1" x 4"		225	.036		1.39	1.75		3.14	4.21
0520	1" x 6"		225	.036		2.06	1.75		3.81	4.95
0560	5/4" x 4"		220	.036		2.05	1.79		3.84	5
0570	5/4" x 6"	↓	220	.036		3.02	1.79		4.81	6.05
0610	Soffit, 1" x 4"	2 Carp	420	.038		1.39	1.88		3.27	4.40
0620	1" x 6"		420	.038		2.06	1.88		3.94	5.15
0630	1" x 8"		420	.038		2.72	1.88		4.60	5.85
0640	1" x 10"		400	.040		3.44	1.97		5.41	6.80
0650	1" x 12"		400	.040		4.22	1.97		6.19	7.65
0660	5/4" x 4"		410	.039		2.05	1.92		3.97	5.20
0670	5/4" x 6"		410	.039		3.02	1.92		4.94	6.25
0680	5/4" x 8"		410	.039		3.99	1.92		5.91	7.35
0690	5/4" x 10"		390	.041		5.10	2.02		7.12	8.70
0700	5/4" x 12"	↓	390	.041	↓	5.35	2.02		7.37	9

06 80 Composite Fabrications

06 80 10 – Composite Decking

06 80 10.10 Woodgrained Composite Decking

		Crew	Daily Output	Labor-Hours	Unit	Material	2017 Bare Costs Labor	Equipment	Total	Total Incl O&P
0010	**WOODGRAINED COMPOSITE DECKING**									
0100	Woodgrained composite decking, 1" x 6"	2 Carp	640	.025	L.F.	3.72	1.23		4.95	6
0110	Grooved edge		660	.024		3.87	1.19		5.06	6.10
0120	2" x 6"		640	.025		3.81	1.23		5.04	6.10
0130	Encased, 1" x 6"		640	.025		3.81	1.23		5.04	6.10
0140	Grooved edge		660	.024		3.96	1.19		5.15	6.20
0150	2" x 6"	↓	640	.025	↓	5.35	1.23		6.58	7.75

06 81 Composite Railings

06 81 10 – Encased Railings

06 81 10.10 Encased Composite Railings	Crew	Daily Output	Labor-Hours	Unit	Material	2017 Bare Costs Labor	Equipment	Total	Total Incl O&P
0010 **ENCASED COMPOSITE RAILINGS**									
0100 Encased composite railing, 6' long, 36" high, incl. balusters	1 Carp	16	.500	Ea.	169	24.50		193.50	224
0110 42" high, incl. balusters		16	.500		190	24.50		214.50	247
0120 8' long, 36" high, incl. balusters		12	.667		169	33		202	237
0130 42" high, incl. balusters		12	.667		189	33		222	259
0140 Accessories, post sleeve, 4" x 4", 39" long		32	.250		25	12.30		37.30	46.50
0150 96" long		24	.333		95	16.40		111.40	129
0160 6" x 6", 39" long		32	.250		52.50	12.30		64.80	76.50
0170 96" long		24	.333		149	16.40		165.40	189
0180 Accessories, post skirt, 4" x 4"		96	.083		4.34	4.10		8.44	11.05
0190 6" x 6"		96	.083		5.30	4.10		9.40	12.10
0200 Post cap, 4" x 4", flat		48	.167		6.65	8.20		14.85	19.90
0210 Pyramid		48	.167		6.65	8.20		14.85	19.90
0220 Post cap, 6" x 6", flat		48	.167		10.80	8.20		19	24.50
0230 Pyramid		48	.167		10.95	8.20		19.15	24.50

For customer support on your Building Construction Costs with RSMeans Data, call 800.448.8182.

219

Division Notes

	CREW	DAILY OUTPUT	LABOR-HOURS	UNIT	BARE COSTS				TOTAL INCL O&P
					MAT.	LABOR	EQUIP.	TOTAL	

Estimating Tips
07 10 00 Dampproofing and Waterproofing

- Be sure of the job specifications before pricing this subdivision. The difference in cost between waterproofing and dampproofing can be great. Waterproofing will hold back standing water. Dampproofing prevents the transmission of water vapor. Also included in this section are vapor retarding membranes.

07 20 00 Thermal Protection

- Insulation and fireproofing products are measured by area, thickness, volume or R-value. Specifications may give only what the specific R-value should be in a certain situation. The estimator may need to choose the type of insulation to meet that R-value.

07 30 00 Steep Slope Roofing
07 40 00 Roofing and Siding Panels

- Many roofing and siding products are bought and sold by the square. One square is equal to an area that measures 100 square feet.

  This simple change in unit of measure could create a large error if the estimator is not observant. Accessories necessary for a complete installation must be figured into any calculations for both material and labor.

07 50 00 Membrane Roofing
07 60 00 Flashing and Sheet Metal
07 70 00 Roofing and Wall Specialties and Accessories

- The items in these subdivisions compose a roofing system. No one component completes the installation, and all must be estimated. Built-up or single-ply membrane roofing systems are made up of many products and installation trades. Wood blocking at roof perimeters or penetrations, parapet coverings, reglets, roof drains, gutters, downspouts, sheet metal flashing, skylights, smoke vents, and roof hatches all need to be considered along with the roofing material. Several different installation trades will need to work together on the roofing system. Inherent difficulties in the scheduling and coordination of various trades must be accounted for when estimating labor costs.

07 90 00 Joint Protection

- To complete the weather-tight shell, the sealants and caulkings must be estimated. Where different materials meet—at expansion joints, at flashing penetrations, and at hundreds of other locations throughout a construction project—caulking and sealants provide another line of defense against water penetration. Often, an entire system is based on the proper location and placement of caulking or sealants. The detailed drawings that are included as part of a set of architectural plans show typical locations for these materials. When caulking or sealants are shown at typical locations, this means the estimator must include them for all the locations where this detail is applicable. Be careful to keep different types of sealants separate, and remember to consider backer rods and primers if necessary.

Reference Numbers

Reference numbers are shown at the beginning of some major classifications. These numbers refer to related items in the Reference Section. The reference information may be an estimating procedure, an alternate pricing method, or technical information.

Note: Not all subdivisions listed here necessarily appear. ∎

Did you know?

Our online estimating solution gives you the same access to RSMeans' data with 24/7 access:

- Quickly locate costs in the searchable database.
- Build cost lists, estimates, and reports in minutes.
- Adjust costs to any location in the U.S. and Canada with the click of a button.

Start your free trial today at
www.RSMeansOnline.com

07 01 Operation and Maint. of Thermal and Moisture Protection

07 01 50 – Maintenance of Membrane Roofing

07 01 50.10 Roof Coatings

07 01 50.10 Roof Coatings	Crew	Daily Output	Labor-Hours	Unit	Material	2017 Bare Costs Labor	Equipment	Total	Total Incl O&P
0010 **ROOF COATINGS**									
0012 Asphalt, brush grade, material only				Gal.	9			9	9.90
0200 Asphalt base, fibered aluminum coating [G]					8.65			8.65	9.50
0300 Asphalt primer, 5 gallon					7.50			7.50	8.25
0600 Coal tar pitch, 200 lb. barrels				Ton	1,250			1,250	1,375
0700 Tar roof cement, 5 gal. lots				Gal.	14.25			14.25	15.65
0800 Glass fibered roof & patching cement, 5 gallon				"	8.65			8.65	9.50
0900 Reinforcing glass membrane, 450 S.F./roll				Ea.	59.50			59.50	65.50
1000 Neoprene roof coating, 5 gal., 2 gal./sq.				Gal.	33.50			33.50	36.50
1100 Roof patch & flashing cement, 5 gallon					8.75			8.75	9.60
1200 Roof resaturant, glass fibered, 3 gal./sq.					8.85			8.85	9.75

07 01 90 – Maintenance of Joint Protection

07 01 90.81 Joint Sealant Replacement

07 01 90.81	Crew	Daily Output	Labor-Hours	Unit	Material	2017 Bare Costs Labor	Equipment	Total	Total Incl O&P
0010 **JOINT SEALANT REPLACEMENT**									
0050 Control joints in concrete floors/slabs									
0100 Option 1 for joints with hard dry sealant									
0110 Step 1: Sawcut to remove 95% of old sealant									
0112 1/4" wide x 1/2" deep, with single saw blade	C-27	4800	.003	L.F.	.02	.16	.03	.21	.29
0114 3/8" wide x 3/4" deep, with single saw blade		4000	.004		.03	.19	.04	.26	.35
0116 1/2" wide x 1" deep, with double saw blades		3600	.004		.06	.21	.04	.31	.42
0118 3/4" wide x 1-1/2" deep, with double saw blades		3200	.005		.12	.23	.05	.40	.53
0120 Step 2: Water blast joint faces and edges	C-29	2500	.003			.13	.03	.16	.22
0130 Step 3: Air blast joint faces and edges	C-28	2000	.004			.19	.01	.20	.29
0140 Step 4: Sand blast joint faces and edges	E-11	2000	.016			.70	.11	.81	1.25
0150 Step 5: Air blast joint faces and edges	C-28	2000	.004			.19	.01	.20	.29
0200 Option 2 for joints with soft pliable sealant									
0210 Step 1: Plow joint with rectangular blade	B-62	2600	.009	L.F.		.40	.07	.47	.68
0220 Step 2: Sawcut to re-face joint faces									
0222 1/4" wide x 1/2" deep, with single saw blade	C-27	2400	.007	L.F.	.02	.31	.06	.39	.55
0224 3/8" wide x 3/4" deep, with single saw blade		2000	.008		.04	.37	.08	.49	.68
0226 1/2" wide x 1" deep, with double saw blades		1800	.009		.08	.42	.09	.59	.80
0228 3/4" wide x 1-1/2" deep, with double saw blades		1600	.010		.16	.47	.10	.73	.98
0230 Step 3: Water blast joint faces and edges	C-29	2500	.003			.13	.03	.16	.22
0240 Step 4: Air blast joint faces and edges	C-28	2000	.004			.19	.01	.20	.29
0250 Step 5: Sand blast joint faces and edges	E-11	2000	.016			.70	.11	.81	1.25
0260 Step 6: Air blast joint faces and edges	C-28	2000	.004			.19	.01	.20	.29
0290 For saw cutting new control joints, see Section 03 15 16.20									
8910 For backer rod, see Section 07 91 23.10									
8920 For joint sealant, see Section 03 15 16.30 or 07 92 13.20									

07 05 Common Work Results for Thermal and Moisture Protection

07 05 05 – Selective Demolition for Thermal and Moisture Protection

07 05 05.10 Selective Demo., Thermal and Moist. Protection

07 05 05.10		Crew	Daily Output	Labor-Hours	Unit	Material	2017 Bare Costs Labor	Equipment	Total	Total Incl O&P
0010 **SELECTIVE DEMO., THERMAL AND MOISTURE PROTECTION**										
0020 Caulking/sealant, to 1" x 1" joint	R024119-10	1 Clab	600	.013	L.F.		.52		.52	.80
0120 Downspouts, including hangers			350	.023	"		.90		.90	1.37
0220 Flashing, sheet metal			290	.028	S.F.		1.08		1.08	1.66
0420 Gutters, aluminum or wood, edge hung			240	.033	L.F.		1.30		1.30	2
0520 Built-in			100	.080	"		3.13		3.13	4.80
0620 Insulation, air/vapor barrier			3500	.002	S.F.		.09		.09	.14

07 05 05.10 Selective Demo., Thermal and Moist. Protection	Crew	Daily Output	Labor-Hours	Unit	Material	2017 Bare Costs Labor	2017 Bare Costs Equipment	Total	Total Incl O&P
0670	Batts or blankets	1 Clab	1400	.006	C.F.		.22		.22	.34
0720	Foamed or sprayed in place	2 Clab	1000	.016	B.F.		.63		.63	.96
0770	Loose fitting	1 Clab	3000	.003	C.F.		.10		.10	.16
0870	Rigid board		3450	.002	B.F.		.09		.09	.14
1120	Roll roofing, cold adhesive		12	.667	Sq.		26		26	40
1170	Roof accessories, adjustable metal chimney flashing		9	.889	Ea.		35		35	53.50
1325	Plumbing vent flashing		32	.250	"		9.80		9.80	15
1375	Ridge vent strip, aluminum		310	.026	L.F.		1.01		1.01	1.55
1620	Skylight to 10 S.F.		8	1	Ea.		39		39	60
2120	Roof edge, aluminum soffit and fascia		570	.014	L.F.		.55		.55	.84
2170	Concrete coping, up to 12" wide	2 Clab	160	.100			3.92		3.92	6
2220	Drip edge	1 Clab	1000	.008			.31		.31	.48
2270	Gravel stop		950	.008			.33		.33	.51
2370	Sheet metal coping, up to 12" wide		240	.033			1.30		1.30	2
2470	Roof insulation board, over 2" thick	B-2	7800	.005	B.F.		.20		.20	.31
2520	Up to 2" thick	"	3900	.010	S.F.		.41		.41	.62
2620	Roof ventilation, louvered gable vent	1 Clab	16	.500	Ea.		19.60		19.60	30
2670	Remove, roof hatch	G-3	15	2.133			104		104	159
2675	Rafter vents	1 Clab	960	.008			.33		.33	.50
2720	Soffit vent and/or fascia vent		575	.014	L.F.		.54		.54	.83
2775	Soffit vent strip, aluminum, 3" to 4" wide		160	.050			1.96		1.96	3
2820	Roofing accessories, shingle moulding, to 1" x 4"		1600	.005			.20		.20	.30
2870	Cant strip	B-2	2000	.020			.79		.79	1.21
2920	Concrete block walkway	1 Clab	230	.035			1.36		1.36	2.09
3070	Roofing, felt paper, #15		70	.114	Sq.		4.47		4.47	6.85
3125	#30 felt		30	.267	"		10.45		10.45	16
3170	Asphalt shingles, 1 layer	B-2	3500	.011	S.F.		.45		.45	.69
3180	2 layers		1750	.023	"		.90		.90	1.39
3370	Modified bitumen		26	1.538	Sq.		61		61	93.50
3420	Built-up, no gravel, 3 ply		25	1.600			63.50		63.50	97
3470	4 ply		21	1.905			75.50		75.50	115
3620	5 ply		1600	.025	S.F.		.99		.99	1.52
3720	5 ply, with gravel		890	.045			1.78		1.78	2.72
3725	Loose gravel removal		5000	.008			.32		.32	.48
3730	Embedded gravel removal		2000	.020			.79		.79	1.21
3870	Fiberglass sheet		1200	.033			1.32		1.32	2.02
4120	Slate shingles		1900	.021			.83		.83	1.28
4170	Ridge shingles, clay or slate		2000	.020	L.F.		.79		.79	1.21
4320	Single ply membrane, attached at seams		52	.769	Sq.		30.50		30.50	46.50
4370	Ballasted		75	.533			21		21	32.50
4420	Fully adhered		39	1.026			40.50		40.50	62
4550	Roof hatch, 2'-6" x 3'-0"	1 Clab	10	.800	Ea.		31.50		31.50	48
4670	Wood shingles	B-2	2200	.018	S.F.		.72		.72	1.10
4820	Sheet metal roofing	"	2150	.019			.74		.74	1.13
4970	Siding, horizontal wood clapboards	1 Clab	380	.021			.82		.82	1.26
5025	Exterior insulation finish system	"	120	.067			2.61		2.61	4
5070	Tempered hardboard, remove and reset	1 Carp	380	.021			1.04		1.04	1.59
5120	Tempered hardboard sheet siding	"	375	.021			1.05		1.05	1.61
5170	Metal, corner strips	1 Clab	850	.009	L.F.		.37		.37	.56
5225	Horizontal strips		444	.018	S.F.		.71		.71	1.08
5320	Vertical strips		400	.020			.78		.78	1.20
5520	Wood shingles		350	.023			.90		.90	1.37
5620	Stucco siding		360	.022			.87		.87	1.33

07 05 Common Work Results for Thermal and Moisture Protection

07 05 05 – Selective Demolition for Thermal and Moisture Protection

07 05 05.10 Selective Demo., Thermal and Moist. Protection		Crew	Daily Output	Labor-Hours	Unit	Material	2017 Bare Costs Labor	Equipment	Total	Total Incl O&P
5670	Textured plywood	1 Clab	725	.011	S.F.		.43		.43	.66
5720	Vinyl siding		510	.016	↓		.61		.61	.94
5770	Corner strips		900	.009	L.F.		.35		.35	.53
5870	Wood, boards, vertical		400	.020	S.F.		.78		.78	1.20
5880	Steel siding, corrugated/ribbed	▼	402.50	.020	"		.78		.78	1.19
5920	Waterproofing, protection/drain board	2 Clab	3900	.004	B.F.		.16		.16	.25
5970	Over 1/2" thick		1750	.009	S.F.		.36		.36	.55
6020	To 1/2" thick	▼	2000	.008	"		.31		.31	.48

07 11 Dampproofing

07 11 13 – Bituminous Dampproofing

07 11 13.10 Bituminous Asphalt Coating

		Crew	Daily Output	Labor-Hours	Unit	Material	2017 Bare Costs Labor	Equipment	Total	Total Incl O&P
0010	**BITUMINOUS ASPHALT COATING**									
0030	Brushed on, below grade, 1 coat	1 Rofc	665	.012	S.F.	.22	.52		.74	1.13
0100	2 coat		500	.016		.45	.69		1.14	1.66
0300	Sprayed on, below grade, 1 coat		830	.010		.22	.42		.64	.96
0400	2 coat	▼	500	.016	↓	.44	.69		1.13	1.65
0500	Asphalt coating, with fibers				Gal.	8.65			8.65	9.50
0600	Troweled on, asphalt with fibers, 1/16" thick	1 Rofc	500	.016	S.F.	.38	.69		1.07	1.58
0700	1/8" thick		400	.020		.66	.86		1.52	2.20
1000	1/2" thick		350	.023	▼	2.16	.99		3.15	4.06

07 11 16 – Cementitious Dampproofing

07 11 16.20 Cementitious Parging

		Crew	Daily Output	Labor-Hours	Unit	Material	2017 Bare Costs Labor	Equipment	Total	Total Incl O&P
0010	**CEMENTITIOUS PARGING**									
0020	Portland cement, 2 coats, 1/2" thick	D-1	250	.064	S.F.	.37	2.77		3.14	4.67
0100	Waterproofed Portland cement, 1/2" thick, 2 coats	"	250	.064	"	4.04	2.77		6.81	8.70

07 12 Built-up Bituminous Waterproofing

07 12 13 – Built-Up Asphalt Waterproofing

07 12 13.20 Membrane Waterproofing

		Crew	Daily Output	Labor-Hours	Unit	Material	2017 Bare Costs Labor	Equipment	Total	Total Incl O&P
0010	**MEMBRANE WATERPROOFING**									
0012	On slabs, 1 ply, felt, mopped	G-1	3000	.019	S.F.	.42	.75	.18	1.35	1.94
0100	On slabs, 1 ply, glass fiber fabric, mopped		2100	.027		.48	1.07	.25	1.80	2.64
0300	On slabs, 2 ply, felt, mopped		2500	.022		.85	.90	.21	1.96	2.69
0400	On slabs, 2 ply, glass fiber fabric, mopped		1650	.034		1.05	1.37	.32	2.74	3.83
0600	On slabs, 3 ply, felt, mopped		2100	.027		1.27	1.07	.25	2.59	3.51
0700	On slabs, 3 ply, glass fiber fabric, mopped	▼	1550	.036		1.44	1.46	.34	3.24	4.43
0710	Asphaltic hardboard protection board, 1/8" thick	2 Rofc	500	.032		.67	1.38		2.05	3.09
1000	EPS membrane protection board, 1/4"		3500	.005		.34	.20		.54	.71
1050	3/8" thick		3500	.005		.37	.20		.57	.74
1060	1/2" thick		3500	.005	▼	.40	.20		.60	.78
1070	Fiberglass fabric, black, 20/10 mesh	▼	116	.138	Sq.	15.85	5.95		21.80	27.50

07 13 Sheet Waterproofing

07 13 53 – Elastomeric Sheet Waterproofing

07 13 53.10 Elastomeric Sheet Waterproofing and Access.	Crew	Daily Output	Labor-Hours	Unit	Material	2017 Bare Costs Labor	Equipment	Total	Total Incl O&P
0010 **ELASTOMERIC SHEET WATERPROOFING AND ACCESS.**									
0090 EPDM, plain, 45 mils thick	2 Rofc	580	.028	S.F.	1.45	1.19		2.64	3.62
0100 60 mils thick		570	.028		1.50	1.21		2.71	3.71
0300 Nylon reinforced sheets, 45 mils thick		580	.028		1.56	1.19		2.75	3.74
0400 60 mils thick	↓	570	.028	↓	1.67	1.21		2.88	3.90
0600 Vulcanizing splicing tape for above, 2" wide				C.L.F.	63			63	69
0700 4" wide				"	121			121	134
0900 Adhesive, bonding, 60 S.F. per gal.				Gal.	26.50			26.50	29
1000 Splicing, 75 S.F. per gal.				"	38			38	42
1200 Neoprene sheets, plain, 45 mils thick	2 Rofc	580	.028	S.F.	1.88	1.19		3.07	4.09
1300 60 mils thick		570	.028		2.12	1.21		3.33	4.39
1500 Nylon reinforced, 45 mils thick		580	.028		2.15	1.19		3.34	4.39
1600 60 mils thick		570	.028		2.78	1.21		3.99	5.10
1800 120 mils thick	↓	500	.032	↓	5.65	1.38		7.03	8.55
1900 Adhesive, splicing, 150 S.F. per gal. per coat				Gal.	38			38	42
2100 Fiberglass reinforced, fluid applied, 1/8" thick	2 Rofc	500	.032	S.F.	1.76	1.38		3.14	4.29
2200 Polyethylene and rubberized asphalt sheets, 60 mils thick		550	.029		.86	1.26		2.12	3.08
2400 Polyvinyl chloride sheets, plain, 10 mils thick		580	.028		.15	1.19		1.34	2.19
2500 20 mils thick		570	.028		.20	1.21		1.41	2.28
2700 30 mils thick	↓	560	.029	↓	.25	1.23		1.48	2.38
3000 Adhesives, trowel grade, 40-100 S.F. per gal.				Gal.	24			24	26.50
3100 Brush grade, 100-250 S.F. per gal.				"	21.50			21.50	24
3300 Bitumen modified polyurethane, fluid applied, 55 mils thick	2 Rofc	665	.024	S.F.	.99	1.04		2.03	2.85

07 16 Cementitious and Reactive Waterproofing

07 16 16 – Crystalline Waterproofing

07 16 16.20 Cementitious Waterproofing

	Crew	Daily Output	Labor-Hours	Unit	Material	2017 Bare Costs Labor	Equipment	Total	Total Incl O&P
0010 **CEMENTITIOUS WATERPROOFING**									
0020 1/8" application, sprayed on	G-2A	1000	.024	S.F.	.77	.92	.67	2.36	3.09
0050 4 coat cementitious metallic slurry	1 Cefi	1.20	6.667	C.S.F.	35	310		345	505

07 17 Bentonite Waterproofing

07 17 13 – Bentonite Panel Waterproofing

07 17 13.10 Bentonite

	Crew	Daily Output	Labor-Hours	Unit	Material	2017 Bare Costs Labor	Equipment	Total	Total Incl O&P
0010 **BENTONITE**									
0020 Panels, 4' x 4', 3/16" thick	1 Rofc	625	.013	S.F.	1.62	.55		2.17	2.72
0100 Rolls, 3/8" thick, with geotextile fabric both sides	"	550	.015	"	1.58	.63		2.21	2.81
0300 Granular bentonite, 50 lb. bags (.625 C.F.)				Bag	17.80			17.80	19.60
0400 3/8" thick, troweled on	1 Rofc	475	.017	S.F.	.89	.73		1.62	2.22
0500 Drain board, expanded polystyrene, 1-1/2" thick	1 Rohe	1600	.005		.39	.16		.55	.70
0510 2" thick		1600	.005		.52	.16		.68	.84
0520 3" thick		1600	.005		.78	.16		.94	1.13
0530 4" thick		1600	.005		1.04	.16		1.20	1.41
0600 With filter fabric, 1-1/2" thick		1600	.005		.45	.16		.61	.77
0625 2" thick		1600	.005		.58	.16		.74	.91
0650 3" thick		1600	.005		.84	.16		1	1.20
0675 4" thick	↓	1600	.005	↓	1.10	.16		1.26	1.48

For customer support on your Building Construction Costs with RSMeans Data, call 800.448.8182.

225

07 19 Water Repellents

07 19 19 – Silicone Water Repellents

07 19 19.10 Silicone Based Water Repellents		Crew	Daily Output	Labor-Hours	Unit	Material	2017 Bare Costs Labor	Equipment	Total	Total Incl O&P
0010	**SILICONE BASED WATER REPELLENTS**									
0020	Water base liquid, roller applied	2 Rofc	7000	.002	S.F.	.49	.10		.59	.71
0200	Silicone or stearate, sprayed on CMU, 1 coat	1 Rofc	4000	.002		.37	.09		.46	.56
0300	2 coats	"	3000	.003	▼	.74	.12		.86	1.02

07 21 Thermal Insulation

07 21 13 – Board Insulation

07 21 13.10 Rigid Insulation

			Crew	Daily Output	Labor-Hours	Unit	Material	2017 Bare Costs Labor	Equipment	Total	Total Incl O&P
0010	**RIGID INSULATION**, for walls										
0040	Fiberglass, 1.5#/C.F., unfaced, 1" thick, R4.1	G	1 Carp	1000	.008	S.F.	.30	.39		.69	.93
0060	1-1/2" thick, R6.2	G		1000	.008		.39	.39		.78	1.03
0080	2" thick, R8.3	G		1000	.008		.49	.39		.88	1.14
0120	3" thick, R12.4	G		800	.010		.57	.49		1.06	1.38
0370	3#/C.F., unfaced, 1" thick, R4.3	G		1000	.008		.52	.39		.91	1.17
0390	1-1/2" thick, R6.5	G		1000	.008		.78	.39		1.17	1.46
0400	2" thick, R8.7	G		890	.009		1.05	.44		1.49	1.84
0420	2-1/2" thick, R10.9	G		800	.010		1.10	.49		1.59	1.96
0440	3" thick, R13	G		800	.010		1.59	.49		2.08	2.50
0520	Foil faced, 1" thick, R4.3	G		1000	.008		.90	.39		1.29	1.59
0540	1-1/2" thick, R6.5	G		1000	.008		1.35	.39		1.74	2.09
0560	2" thick, R8.7	G		890	.009		1.70	.44		2.14	2.55
0580	2-1/2" thick, R10.9	G		800	.010		2	.49		2.49	2.95
0600	3" thick, R13	G	▼	800	.010	▼	2.25	.49		2.74	3.23
1600	Isocyanurate, 4' x 8' sheet, foil faced, both sides										
1610	1/2" thick	G	1 Carp	800	.010	S.F.	.31	.49		.80	1.09
1620	5/8" thick	G		800	.010		.33	.49		.82	1.11
1630	3/4" thick	G		800	.010		.39	.49		.88	1.18
1640	1" thick	G		800	.010		.52	.49		1.01	1.32
1650	1-1/2" thick	G		730	.011		.60	.54		1.14	1.49
1660	2" thick	G		730	.011		.76	.54		1.30	1.67
1670	3" thick	G		730	.011		1.90	.54		2.44	2.92
1680	4" thick	G		730	.011		2.15	.54		2.69	3.20
1700	Perlite, 1" thick, R2.77	G		800	.010		.45	.49		.94	1.25
1750	2" thick, R5.55	G		730	.011		.78	.54		1.32	1.69
1900	Extruded polystyrene, 25 PSI compressive strength, 1" thick, R5	G		800	.010		.57	.49		1.06	1.38
1940	2" thick, R10	G		730	.011		1.14	.54		1.68	2.08
1960	3" thick, R15	G		730	.011		1.59	.54		2.13	2.58
2100	Expanded polystyrene, 1" thick, R3.85	G		800	.010		.26	.49		.75	1.04
2120	2" thick, R7.69	G		730	.011		.52	.54		1.06	1.40
2140	3" thick, R11.49	G	▼	730	.011	▼	.78	.54		1.32	1.69

07 21 13.13 Foam Board Insulation

			Crew	Daily Output	Labor-Hours	Unit	Material	2017 Bare Costs Labor	Equipment	Total	Total Incl O&P
0010	**FOAM BOARD INSULATION**										
0600	Polystyrene, expanded, 1" thick, R4	G	1 Carp	680	.012	S.F.	.26	.58		.84	1.18
0700	2" thick, R8	G	"	675	.012	"	.52	.58		1.10	1.46

07 21 16 – Blanket Insulation

07 21 16.10 Blanket Insulation for Floors/Ceilings

			Crew	Daily Output	Labor-Hours	Unit	Material	2017 Bare Costs Labor	Equipment	Total	Total Incl O&P
0010	**BLANKET INSULATION FOR FLOORS/CEILINGS**										
0020	Including spring type wire fasteners										
2000	Fiberglass, blankets or batts, paper or foil backing										
2100	3-1/2" thick, R13	G	1 Carp	700	.011	S.F.	.40	.56		.96	1.30

07 21 Thermal Insulation

07 21 16 – Blanket Insulation

07 21 16.10 Blanket Insulation for Floors/Ceilings		Crew	Daily Output	Labor-Hours	Unit	Material	2017 Bare Costs Labor	Equipment	Total	Total Incl O&P
2150	6-1/4" thick, R19	G 1 Carp	600	.013	S.F.	.52	.66		1.18	1.58
2210	9-1/2" thick, R30	G	500	.016		.75	.79		1.54	2.04
2220	12" thick, R38	G	475	.017		1.08	.83		1.91	2.46
3000	Unfaced, 3-1/2" thick, R13	G	600	.013		.34	.66		1	1.38
3010	6-1/4" thick, R19	G	500	.016		.40	.79		1.19	1.65
3020	9-1/2" thick, R30	G	450	.018		.62	.88		1.50	2.02
3030	12" thick, R38	G	425	.019		.77	.93		1.70	2.27

07 21 16.20 Blanket Insulation for Walls

			Daily Output	Labor-Hours	Unit	Material	Labor	Equipment	Total	Total Incl O&P
0010	**BLANKET INSULATION FOR WALLS**									
0020	Kraft faced fiberglass, 3-1/2" thick, R11, 15" wide	G 1 Carp	1350	.006	S.F.	.32	.29		.61	.80
0030	23" wide	G	1600	.005		.32	.25		.57	.73
0060	R13, 11" wide	G	1150	.007		.33	.34		.67	.89
0080	15" wide	G	1350	.006		.33	.29		.62	.81
0100	23" wide	G	1600	.005		.33	.25		.58	.74
0110	R15, 11" wide	G	1150	.007		.51	.34		.85	1.09
0120	15" wide	G	1350	.006		.51	.29		.80	1.01
0130	23" wide	G	1600	.005		.51	.25		.76	.94
0140	6" thick, R19, 11" wide	G	1150	.007		.45	.34		.79	1.03
0160	15" wide	G	1350	.006		.45	.29		.74	.95
0180	23" wide	G	1600	.005		.45	.25		.70	.88
0182	R21, 11" wide	G	1150	.007		.68	.34		1.02	1.28
0184	15" wide	G	1350	.006		.68	.29		.97	1.20
0186	23" wide	G	1600	.005		.68	.25		.93	1.13
0188	9" thick, R30, 11" wide	G	985	.008		.75	.40		1.15	1.44
0200	15" wide	G	1150	.007		.75	.34		1.09	1.36
0220	23" wide	G	1350	.006		.75	.29		1.04	1.28
0230	12" thick, R38, 11" wide	G	985	.008		1.08	.40		1.48	1.80
0240	15" wide	G	1150	.007		1.08	.34		1.42	1.72
0260	23" wide	G	1350	.006		1.08	.29		1.37	1.64
0410	Foil faced fiberglass, 3-1/2" thick, R13, 11" wide	G	1150	.007		.48	.34		.82	1.06
0420	15" wide	G	1350	.006		.48	.29		.77	.98
0440	23" wide	G	1600	.005		.48	.25		.73	.91
0442	R15, 11" wide	G	1150	.007		.50	.34		.84	1.08
0444	15" wide	G	1350	.006		.50	.29		.79	1
0446	23" wide	G	1600	.005		.50	.25		.75	.93
0448	6" thick, R19, 11" wide	G	1150	.007		.64	.34		.98	1.23
0460	15" wide	G	1350	.006		.64	.29		.93	1.15
0480	23" wide	G	1600	.005		.64	.25		.89	1.08
0482	R21, 11" wide	G	1150	.007		.66	.34		1	1.26
0484	15" wide	G	1350	.006		.66	.29		.95	1.18
0486	23" wide	G	1600	.005		.66	.25		.91	1.11
0488	9" thick, R30, 11" wide	G	985	.008		.97	.40		1.37	1.68
0500	15" wide	G	1150	.007		.97	.34		1.31	1.60
0550	23" wide	G	1350	.006		.97	.29		1.26	1.52
0560	12" thick, R38, 11" wide	G	985	.008		1.19	.40		1.59	1.92
0570	15" wide	G	1150	.007		1.19	.34		1.53	1.84
0580	23" wide	G	1350	.006		1.19	.29		1.48	1.76
0620	Unfaced fiberglass, 3-1/2" thick, R13, 11" wide	G	1150	.007		.34	.34		.68	.90
0820	15" wide	G	1350	.006		.34	.29		.63	.82
0830	23" wide	G	1600	.005		.34	.25		.59	.75
0832	R15, 11" wide	G	1150	.007		.47	.34		.81	1.05
0834	15" wide	G	1350	.006		.47	.29		.76	.97

For customer support on your Building Construction Costs with RSMeans Data, call 800.448.8182.

227

07 21 Thermal Insulation

07 21 16 – Blanket Insulation

07 21 16.20 Blanket Insulation for Walls		Crew	Daily Output	Labor-Hours	Unit	Material	2017 Bare Costs Labor	Equipment	Total	Total Incl O&P
0836	23" wide	G	1 Carp	1600	.005	S.F.	.47	.25		.72	.90
0838	6" thick, R19, 11" wide	G		1150	.007		.40	.34		.74	.97
0860	15" wide	G		1150	.007		.40	.34		.74	.97
0880	23" wide	G		1350	.006		.40	.29		.69	.89
0882	R21, 11" wide	G		1150	.007		.51	.34		.85	1.09
0886	15" wide	G		1350	.006		.51	.29		.80	1.01
0888	23" wide	G		1600	.005		.51	.25		.76	.94
0890	9" thick, R30, 11" wide	G		985	.008		.62	.40		1.02	1.29
0900	15" wide	G		1150	.007		.62	.34		.96	1.21
0920	23" wide	G		1350	.006		.62	.29		.91	1.13
0930	12" thick, R38, 11" wide	G		985	.008		.77	.40		1.17	1.46
0940	15" wide	G		1000	.008		.77	.39		1.16	1.45
0960	23" wide	G	▼	1150	.007	▼	.77	.34		1.11	1.38
1300	Wall or ceiling insulation, mineral wool batts										
1320	3-1/2" thick, R15	G	1 Carp	1600	.005	S.F.	.70	.25		.95	1.15
1340	5-1/2" thick, R23	G		1600	.005		1.10	.25		1.35	1.59
1380	7-1/4" thick, R30	G		1350	.006		1.45	.29		1.74	2.05
1700	Non-rigid insul, recycled blue cotton fiber, unfaced batts, R13, 16" wide	G		1600	.005		.99	.25		1.24	1.47
1710	R19, 16" wide	G		1600	.005	▼	1.35	.25		1.60	1.87
1850	Friction fit wire insulation supports, 16" O.C.	▼		960	.008	Ea.	.07	.41		.48	.71

07 21 19 – Foamed In Place Insulation

07 21 19.10 Masonry Foamed In Place Insulation

			Crew	Daily Output	Labor-Hours	Unit	Material	Labor	Equipment	Total	Total Incl O&P
0010	**MASONRY FOAMED IN PLACE INSULATION**										
0100	Amino-plast foam, injected into block core, 6" block	G	G-2A	6000	.004	Ea.	.16	.15	.11	.42	.55
0110	8" block	G		5000	.005		.20	.18	.13	.51	.67
0120	10" block	G		4000	.006		.25	.23	.17	.65	.83
0130	12" block	G		3000	.008	▼	.33	.31	.22	.86	1.11
0140	Injected into cavity wall	G	▼	13000	.002	B.F.	.06	.07	.05	.18	.24
0150	Preparation, drill holes into mortar joint every 4 VLF, 5/8" diameter		1 Clab	960	.008	Ea.		.33		.33	.50
0160	7/8" diameter			680	.012			.46		.46	.71
0170	Patch drilled holes, 5/8" diameter			1800	.004		.04	.17		.21	.31
0180	7/8" diameter		▼	1200	.007	▼	.05	.26		.31	.46

07 21 23 – Loose-Fill Insulation

07 21 23.10 Poured Loose-Fill Insulation

			Crew	Daily Output	Labor-Hours	Unit	Material	Labor	Equipment	Total	Total Incl O&P
0010	**POURED LOOSE-FILL INSULATION**										
0020	Cellulose fiber, R3.8 per inch	G	1 Carp	200	.040	C.F.	.69	1.97		2.66	3.78
0021	4" thick	G		1000	.008	S.F.	.17	.39		.56	.78
0022	6" thick	G		800	.010	"	.28	.49		.77	1.06
0080	Fiberglass wool, R4 per inch	G		200	.040	C.F.	.55	1.97		2.52	3.62
0081	4" thick	G		600	.013	S.F.	.19	.66		.85	1.22
0082	6" thick	G		400	.020	"	.26	.99		1.25	1.80
0100	Mineral wool, R3 per inch	G		200	.040	C.F.	.45	1.97		2.42	3.52
0101	4" thick	G		600	.013	S.F.	.15	.66		.81	1.17
0102	6" thick	G		400	.020	"	.23	.99		1.22	1.76
0300	Polystyrene, R4 per inch	G		200	.040	C.F.	1.60	1.97		3.57	4.78
0301	4" thick	G		600	.013	S.F.	.53	.66		1.19	1.59
0302	6" thick	G		400	.020	"	.80	.99		1.79	2.39
0400	Perlite, R2.78 per inch	G		200	.040	C.F.	5.30	1.97		7.27	8.80
0401	4" thick	G		1000	.008	S.F.	1.76	.39		2.15	2.54
0402	6" thick	G		800	.010	"	2.65	.49		3.14	3.66

07 21 Thermal Insulation

07 21 23 – Loose-Fill Insulation

07 21 23.20 Masonry Loose-Fill Insulation		Crew	Daily Output	Labor-Hours	Unit	Material	2017 Bare Costs Labor	Equipment	Total	Total Incl O&P
0010	**MASONRY LOOSE-FILL INSULATION**, vermiculite or perlite	G								
0100	In cores of concrete block, 4" thick wall, .115 C.F./S.F.	G	D-1	4800	.003	S.F.	.61	.14		.75	.89
0200	6" thick wall, .175 C.F./S.F.	G		3000	.005		.93	.23		1.16	1.37
0300	8" thick wall, .258 C.F./S.F.	G		2400	.007		1.36	.29		1.65	1.94
0400	10" thick wall, .340 C.F./S.F.	G		1850	.009		1.80	.37		2.17	2.56
0500	12" thick wall, .422 C.F./S.F.	G		1200	.013	▼	2.23	.58		2.81	3.35
0600	Poured cavity wall, vermiculite or perlite, water repellent	G	▼	250	.064	C.F.	5.30	2.77		8.07	10.05
0700	Foamed in place, urethane in 2-5/8" cavity	G	G-2A	1035	.023	S.F.	1.36	.88	.65	2.89	3.65
0800	For each 1" added thickness, add	G	"	2372	.010	"	.52	.39	.28	1.19	1.51

07 21 26 – Blown Insulation

07 21 26.10 Blown Insulation

		Crew	Daily Output	Labor-Hours	Unit	Material	Labor	Equipment	Total	Total Incl O&P
0010	**BLOWN INSULATION** Ceilings, with open access										
0020	Cellulose, 3-1/2" thick, R13	G	G-4	5000	.005	S.F.	.24	.19	.07	.50	.63
0030	5-3/16" thick, R19	G		3800	.006		.35	.25	.10	.70	.88
0050	6-1/2" thick, R22	G		3000	.008		.45	.32	.12	.89	1.12
0100	8-11/16" thick, R30	G		2600	.009		.61	.37	.14	1.12	1.38
0120	10-7/8" thick, R38	G		1800	.013		.78	.53	.20	1.51	1.88
1000	Fiberglass, 5.5" thick, R11	G		3800	.006		.19	.25	.10	.54	.69
1050	6" thick, R12	G		3000	.008		.26	.32	.12	.70	.91
1100	8.8" thick, R19	G		2200	.011		.33	.43	.16	.92	1.21
1200	10" thick, R22	G		1800	.013		.38	.53	.20	1.11	1.45
1300	11.5" thick, R26	G		1500	.016		.46	.64	.24	1.34	1.74
1350	13" thick, R30	G		1400	.017		.53	.68	.26	1.47	1.91
1450	16" thick, R38	G		1145	.021		.68	.83	.32	1.83	2.37
1500	20" thick, R49	G	▼	920	.026	▼	.89	1.04	.39	2.32	3

07 21 27 – Reflective Insulation

07 21 27.10 Reflective Insulation Options

		Crew	Daily Output	Labor-Hours	Unit	Material	Labor	Equipment	Total	Total Incl O&P
0010	**REFLECTIVE INSULATION OPTIONS**										
0020	Aluminum foil on reinforced scrim	G	1 Carp	19	.421	C.S.F.	15	20.50		35.50	48.50
0100	Reinforced with woven polyolefin	G		19	.421		22	20.50		42.50	56.50
0500	With single bubble air space, R8.8	G		15	.533		26	26.50		52.50	69
0600	With double bubble air space, R9.8	G	▼	15	.533	▼	31.50	26.50		58	74.50

07 21 29 – Sprayed Insulation

07 21 29.10 Sprayed-On Insulation

		Crew	Daily Output	Labor-Hours	Unit	Material	Labor	Equipment	Total	Total Incl O&P
0010	**SPRAYED-ON INSULATION**										
0020	Fibrous/cementitious, finished wall, 1" thick, R3.7	G	G-2	2050	.012	S.F.	.34	.48	.06	.88	1.17
0100	Attic, 5.2" thick, R19	G		1550	.015	"	.42	.64	.09	1.15	1.52
0200	Fiberglass, R4 per inch, vertical	G		1600	.015	B.F.	.19	.62	.08	.89	1.24
0210	Horizontal	G	▼	1200	.020	"	.19	.83	.11	1.13	1.58
0300	Closed cell, spray polyurethane foam, 2 pounds per cubic foot density										
0310	1" thick	G	G-2A	6000	.004	S.F.	.52	.15	.11	.78	.94
0320	2" thick	G		3000	.008		1.04	.31	.22	1.57	1.89
0330	3" thick	G		2000	.012		1.55	.46	.34	2.35	2.83
0335	3-1/2" thick	G		1715	.014		1.81	.53	.39	2.73	3.30
0340	4" thick	G		1500	.016		2.07	.61	.45	3.13	3.77
0350	5" thick	G		1200	.020		2.59	.76	.56	3.91	4.72
0355	5-1/2" thick	G		1090	.022		2.85	.84	.62	4.31	5.20
0360	6" thick	G	▼	1000	.024	▼	3.11	.92	.67	4.70	5.65

07 22 16.10 Roof Deck Insulation		Crew	Daily Output	Labor-Hours	Unit	Material	2017 Bare Costs Labor	Equipment	Total	Total Incl O&P
0010	**ROOF DECK INSULATION**, fastening excluded									
0016	Asphaltic cover board, fiberglass lined, 1/8" thick	1 Rofc	1400	.006	S.F.	.47	.25		.72	.94
0018	1/4" thick		1400	.006		.94	.25		1.19	1.45
0020	Fiberboard low density, 1/2" thick R1.39	G	1300	.006		.33	.27		.60	.81
0030	1" thick R2.78	G	1040	.008		.56	.33		.89	1.18
0080	1-1/2" thick R4.17	G	1040	.008		.85	.33		1.18	1.50
0100	2" thick R5.56	G	1040	.008		1.10	.33		1.43	1.77
0110	Fiberboard high density, 1/2" thick R1.3	G	1300	.006		.30	.27		.57	.78
0120	1" thick R2.5	G	1040	.008		.56	.33		.89	1.18
0130	1-1/2" thick R3.8	G	1040	.008		.86	.33		1.19	1.51
0200	Fiberglass, 3/4" thick R2.78	G	1300	.006		.61	.27		.88	1.12
0400	15/16" thick R3.70	G	1300	.006		.81	.27		1.08	1.34
0460	1-1/16" thick R4.17	G	1300	.006		1.04	.27		1.31	1.59
0600	1-5/16" thick R5.26	G	1300	.006		1.40	.27		1.67	1.99
0650	2-1/16" thick R8.33	G	1040	.008		1.50	.33		1.83	2.21
0700	2-7/16" thick R10	G	1040	.008		1.70	.33		2.03	2.43
0800	Gypsum cover board, fiberglass mat facer, 1/4" thick		1400	.006		.47	.25		.72	.94
0810	1/2" thick		1300	.006		.57	.27		.84	1.08
0820	5/8" thick		1200	.007		.61	.29		.90	1.16
0830	Primed fiberglass mat facer, 1/4" thick		1400	.006		.48	.25		.73	.95
0840	1/2" thick		1300	.006		.57	.27		.84	1.08
0850	5/8" thick		1200	.007		.60	.29		.89	1.15
1650	Perlite, 1/2" thick R1.32	G	1365	.006		.26	.25		.51	.72
1655	3/4" thick R2.08	G	1040	.008		.37	.33		.70	.97
1660	1" thick R2.78	G	1040	.008		.53	.33		.86	1.14
1670	1-1/2" thick R4.17	G	1040	.008		.78	.33		1.11	1.42
1680	2" thick R5.56	G	910	.009		1.07	.38		1.45	1.82
1685	2-1/2" thick R6.67	G	910	.009		1.45	.38		1.83	2.24
1690	Tapered for drainage	G	1040	.008	B.F.	1.05	.33		1.38	1.72
1700	Polyisocyanurate, 2#/C.F. density, 3/4" thick	G	1950	.004	S.F.	.42	.18		.60	.76
1705	1" thick	G	1820	.004		.43	.19		.62	.79
1715	1-1/2" thick	G	1625	.005		.59	.21		.80	1.01
1725	2" thick	G	1430	.006		.75	.24		.99	1.24
1735	2-1/2" thick	G	1365	.006		1.01	.25		1.26	1.54
1745	3" thick	G	1300	.006		1.12	.27		1.39	1.68
1755	3-1/2" thick	G	1300	.006		1.74	.27		2.01	2.36
1765	Tapered for drainage	G	1820	.004	B.F.	.59	.19		.78	.97
1900	Extruded polystyrene									
1910	15 PSI compressive strength, 1" thick, R5	G	1 Rofc	1950	.004	S.F.	.61	.18	.79	.97
1920	2" thick, R10	G	1625	.005		.79	.21		1	1.23
1930	3" thick, R15	G	1300	.006		1.59	.27		1.86	2.19
1932	4" thick, R20	G	1300	.006		2.14	.27		2.41	2.80
1934	Tapered for drainage	G	1950	.004	B.F.	.63	.18		.81	.99
1940	25 PSI compressive strength, 1" thick, R5	G	1950	.004	S.F.	.69	.18		.87	1.06
1942	2" thick, R10	G	1625	.005		1.31	.21		1.52	1.80
1944	3" thick, R15	G	1300	.006		2	.27		2.27	2.65
1946	4" thick, R20	G	1300	.006		2.76	.27		3.03	3.49
1948	Tapered for drainage	G	1950	.004	B.F.	.62	.18		.80	.98
1950	40 PSI compressive strength, 1" thick, R5	G	1950	.004	S.F.	.54	.18		.72	.89
1952	2" thick, R10	G	1625	.005		1.03	.21		1.24	1.49
1954	3" thick, R15	G	1300	.006		1.49	.27		1.76	2.08
1956	4" thick, R20	G	1300	.006		1.94	.27		2.21	2.59
1958	Tapered for drainage	G	1820	.004	B.F.	.77	.19		.96	1.17

07 22 Roof and Deck Insulation

07 22 16 - Roof Board Insulation

07 22 16.10 Roof Deck Insulation		Crew	Daily Output	Labor-Hours	Unit	Material	2017 Bare Costs Labor	Equipment	Total	Total Incl O&P
1960	60 PSI compressive strength, 1" thick, R5	G	1 Rofc	1885	.004	S.F.	.75	.18		.93	1.14
1962	2" thick, R10	G		1560	.005		1.43	.22		1.65	1.95
1964	3" thick, R15	G		1270	.006		2.33	.27		2.60	3.02
1966	4" thick, R20	G		1235	.006		2.89	.28		3.17	3.66
1968	Tapered for drainage	G		1820	.004	B.F.	.97	.19		1.16	1.39
2010	Expanded polystyrene, 1#/C.F. density, 3/4" thick, R2.89	G		1950	.004	S.F.	.20	.18		.38	.51
2020	1" thick, R3.85	G		1950	.004		.26	.18		.44	.59
2100	2" thick, R7.69	G		1625	.005		.52	.21		.73	.93
2110	3" thick, R11.49	G		1625	.005		.78	.21		.99	1.22
2120	4" thick, R15.38	G		1625	.005		1.04	.21		1.25	1.50
2130	5" thick, R19.23	G		1495	.005		1.30	.23		1.53	1.82
2140	6" thick, R23.26	G		1495	.005		1.56	.23		1.79	2.11
2150	Tapered for drainage	G		1950	.004	B.F.	.54	.18		.72	.89
2400	Composites with 2" EPS										
2410	1" fiberboard	G	1 Rofc	1325	.006	S.F.	1.40	.26		1.66	1.98
2420	7/16" oriented strand board	G		1040	.008		1.13	.33		1.46	1.80
2430	1/2" plywood	G		1040	.008		1.40	.33		1.73	2.10
2440	1" perlite	G		1040	.008		1.14	.33		1.47	1.81
2450	Composites with 1-1/2" polyisocyanurate										
2460	1" fiberboard	G	1 Rofc	1040	.008	S.F.	1.09	.33		1.42	1.76
2470	1" perlite	G		1105	.007		.97	.31		1.28	1.60
2480	7/16" oriented strand board	G		1040	.008		.85	.33		1.18	1.50
3000	Fastening alternatives, coated screws, 2" long			3744	.002	Ea.	.06	.09		.15	.23
3010	4" long			3120	.003		.11	.11		.22	.31
3020	6" long			2675	.003		.19	.13		.32	.43
3030	8" long			2340	.003		.28	.15		.43	.56
3040	10" long			1872	.004		.47	.18		.65	.83
3050	Pre-drill and drive wedge spike, 2-1/2"			1248	.006		.40	.28		.68	.91
3060	3-1/2"			1101	.007		.57	.31		.88	1.16
3070	4-1/2"			936	.009		.64	.37		1.01	1.33
3075	3" galvanized deck plates			7488	.001		.08	.05		.13	.17
3080	Spot mop asphalt		G-1	295	.190	Sq.	5.65	7.65	1.79	15.09	21
3090	Full mop asphalt		"	192	.292		11.35	11.75	2.75	25.85	35.50
3110	Low-rise polyurethane adhesive, from 5 gallon kit, 12" OC beads		1 Rofc	45	.178		33	7.65		40.65	49.50
3120	6" OC beads			32	.250		66	10.80		76.80	91
3130	4" OC beads			30	.267		99	11.50		110.50	129

07 24 Exterior Insulation and Finish Systems

07 24 13 - Polymer-Based Exterior Insulation and Finish System

07 24 13.10 Exterior Insulation and Finish Systems

		Crew	Daily Output	Labor-Hours	Unit	Material	2017 Bare Costs Labor	Equipment	Total	Total Incl O&P
0010	**EXTERIOR INSULATION AND FINISH SYSTEMS**										
0095	Field applied, 1" EPS insulation	G	J-1	390	.103	S.F.	1.95	4.43	.35	6.73	9.20
0100	With 1/2" cement board sheathing	G		268	.149		2.73	6.45	.50	9.68	13.20
0105	2" EPS insulation	G		390	.103		2.21	4.43	.35	6.99	9.45
0110	With 1/2" cement board sheathing	G		268	.149		2.99	6.45	.50	9.94	13.50
0115	3" EPS insulation	G		390	.103		2.47	4.43	.35	7.25	9.75
0120	With 1/2" cement board sheathing	G		268	.149		3.25	6.45	.50	10.20	13.80
0125	4" EPS insulation	G		390	.103		2.73	4.43	.35	7.51	10.05
0130	With 1/2" cement board sheathing	G		268	.149		4.29	6.45	.50	11.24	14.90
0140	Premium finish add			1265	.032		.35	1.36	.11	1.82	2.56
0150	Heavy duty reinforcement add			914	.044		.85	1.89	.15	2.89	3.93

For customer support on your Building Construction Costs with RSMeans Data, call 800.448.8182.

231

07 24 Exterior Insulation and Finish Systems

07 24 13 – Polymer-Based Exterior Insulation and Finish System

07 24 13.10 Exterior Insulation and Finish Systems	Crew	Daily Output	Labor-Hours	Unit	Material	2017 Bare Costs Labor	Equipment	Total	Total Incl O&P
0160	2.5#/S.Y. metal lath substrate add	1 Lath	75	.107	S.Y.	3.02	5.10		8.12	10.85
0170	3.4#/S.Y. metal lath substrate add	"	75	.107	"	4.40	5.10		9.50	12.40
0180	Color or texture change	J-1	1265	.032	S.F.	.82	1.36	.11	2.29	3.07
0190	With substrate leveling base coat	1 Plas	530	.015		.83	.69		1.52	1.95
0210	With substrate sealing base coat	1 Pord	1224	.007	↓	.12	.27		.39	.54
0370	V groove shape in panel face				L.F.	.66			.66	.73
0380	U groove shape in panel face				"	.85			.85	.94
0440	For higher than one story, add						25%			

07 25 Weather Barriers

07 25 10 – Weather Barriers or Wraps

07 25 10.10 Weather Barriers

		Crew	Daily Output	Labor-Hours	Unit	Material	2017 Bare Costs Labor	Equipment	Total	Total Incl O&P
0010	**WEATHER BARRIERS**									
0400	Asphalt felt paper, 15#	1 Carp	37	.216	Sq.	5.30	10.65		15.95	22
0401	Per square foot	"	3700	.002	S.F.	.05	.11		.16	.22
0450	Housewrap, exterior, spun bonded polypropylene									
0470	Small roll	1 Carp	3800	.002	S.F.	.15	.10		.25	.33
0480	Large roll	"	4000	.002	"	.14	.10		.24	.30
2100	Asphalt felt roof deck vapor barrier, class 1 metal decks	1 Rofc	37	.216	Sq.	22	9.35		31.35	40
2200	For all other decks	"	37	.216		16.65	9.35		26	34
2800	Asphalt felt, 50% recycled content, 15 lb., 4 sq. per roll	1 Carp	36	.222		5.50	10.95		16.45	23
2810	30 lb., 2 sq. per roll	"	36	.222	↓	11	10.95		21.95	29
3000	Building wrap, spun bonded polyethylene	2 Carp	8000	.002	S.F.	.16	.10		.26	.33

07 26 Vapor Retarders

07 26 10 – Above-Grade Vapor Retarders

07 26 10.10 Vapor Retarders

			Crew	Daily Output	Labor-Hours	Unit	Material	2017 Bare Costs Labor	Equipment	Total	Total Incl O&P
0010	**VAPOR RETARDERS**										
0020	Aluminum and kraft laminated, foil 1 side	G	1 Carp	37	.216	Sq.	11.95	10.65		22.60	29.50
0100	Foil 2 sides	G		37	.216		14	10.65		24.65	31.50
0600	Polyethylene vapor barrier, standard, 2 mil	G		37	.216		1.56	10.65		12.21	18
0700	4 mil	G		37	.216		2.58	10.65		13.23	19.15
0900	6 mil	G		37	.216		3.73	10.65		14.38	20.50
1200	10 mil	G		37	.216		7.90	10.65		18.55	25
1300	Clear reinforced, fire retardant, 8 mil	G		37	.216		10.75	10.65		21.40	28
1350	Cross laminated type, 3 mil	G		37	.216		7.75	10.65		18.40	25
1400	4 mil	G		37	.216		8.25	10.65		18.90	25.50
1800	Reinf. waterproof, 2 mil polyethylene backing, 1 side			37	.216		6.25	10.65		16.90	23
1900	2 sides			37	.216		8.10	10.65		18.75	25
2400	Waterproofed kraft with sisal or fiberglass fibers		↓	37	.216	↓	13.20	10.65		23.85	31

07 27 Air Barriers

07 27 13 – Modified Bituminous Sheet Air Barriers

07 27 13.10 Modified Bituminous Sheet Air Barrier

07 27 13.10 Modified Bituminous Sheet Air Barrier	Crew	Daily Output	Labor-Hours	Unit	Material	2017 Bare Costs Labor	Equipment	Total	Total Incl O&P
0010 **MODIFIED BITUMINOUS SHEET AIR BARRIER**									
0100 SBS modified sheet laminated to polyethylene sheet, 40 mils, 4" wide	1 Carp	1200	.007	L.F.	.37	.33		.70	.91
0120 6" wide		1100	.007		.51	.36		.87	1.11
0140 9" wide		1000	.008		.71	.39		1.10	1.38
0160 12" wide	↓	900	.009	↓	.91	.44		1.35	1.67
0180 18" wide	2 Carp	1700	.009	S.F.	.85	.46		1.31	1.64
0200 36" wide	"	1800	.009		.83	.44		1.27	1.58
0220 Adhesive for above	1 Carp	1400	.006	↓	.35	.28		.63	.81

07 27 26 – Fluid-Applied Membrane Air Barriers

07 27 26.10 Fluid Applied Membrane Air Barrier

	Crew	Daily Output	Labor-Hours	Unit	Material	2017 Bare Costs Labor	Equipment	Total	Total Incl O&P
0010 **FLUID APPLIED MEMBRANE AIR BARRIER**									
0100 Spray applied vapor barrier, 25 S.F./gallon	1 Pord	1375	.006	S.F.	.01	.24		.25	.39

07 31 Shingles and Shakes

07 31 13 – Asphalt Shingles

07 31 13.10 Asphalt Roof Shingles

	Crew	Daily Output	Labor-Hours	Unit	Material	2017 Bare Costs Labor	Equipment	Total	Total Incl O&P
0010 **ASPHALT ROOF SHINGLES**									
0100 Standard strip shingles									
0150 Inorganic, class A, 25 year	1 Rofc	5.50	1.455	Sq.	81.50	63		144.50	197
0155 Pneumatic nailed		7	1.143		81.50	49.50		131	174
0200 30 year		5	1.600		95	69		164	222
0205 Pneumatic nailed	↓	6.25	1.280	↓	95	55		150	199
0250 Standard laminated multi-layered shingles									
0300 Class A, 240-260 lb./square	1 Rofc	4.50	1.778	Sq.	109	76.50		185.50	250
0305 Pneumatic nailed		5.63	1.422		109	61.50		170.50	224
0350 Class A, 250-270 lb./square		4	2		109	86.50		195.50	267
0355 Pneumatic nailed	↓	5	1.600	↓	109	69		178	237
0400 Premium, laminated multi-layered shingles									
0450 Class A, 260-300 lb./square	1 Rofc	3.50	2.286	Sq.	158	98.50		256.50	340
0455 Pneumatic nailed		4.37	1.831		158	79		237	310
0500 Class A, 300-385 lb./square		3	2.667		242	115		357	460
0505 Pneumatic nailed		3.75	2.133		242	92		334	420
0800 #15 felt underlayment		64	.125		5.30	5.40		10.70	15
0825 #30 felt underlayment		58	.138		10.35	5.95		16.30	21.50
0850 Self adhering polyethylene and rubberized asphalt underlayment		22	.364	↓	77	15.70		92.70	111
0900 Ridge shingles		330	.024	L.F.	2.25	1.05		3.30	4.26
0905 Pneumatic nailed	↓	412.50	.019	"	2.25	.84		3.09	3.90
1000 For steep roofs (7 to 12 pitch or greater), add						50%			

07 31 16 – Metal Shingles

07 31 16.10 Aluminum Shingles

	Crew	Daily Output	Labor-Hours	Unit	Material	2017 Bare Costs Labor	Equipment	Total	Total Incl O&P
0010 **ALUMINUM SHINGLES**									
0020 Mill finish, .019" thick	1 Carp	5	1.600	Sq.	220	79		299	365
0100 .020" thick	"	5	1.600		249	79		328	395
0300 For colors, add				↓	21			21	23
0600 Ridge cap, .024" thick	1 Carp	170	.047	L.F.	3.73	2.32		6.05	7.65
0700 End wall flashing, .024" thick		170	.047		2.18	2.32		4.50	5.95
0900 Valley section, .024" thick		170	.047		3.70	2.32		6.02	7.60
1000 Starter strip, .024" thick		400	.020		1.76	.99		2.75	3.45
1200 Side wall flashing, .024" thick		170	.047		2.13	2.32		4.45	5.90
1500 Gable flashing, .024" thick	↓	400	.020	↓	1.73	.99		2.72	3.41

07 31 Shingles and Shakes

07 31 16 – Metal Shingles

07 31 16.20 Steel Shingles

		Crew	Daily Output	Labor-Hours	Unit	Material	2017 Bare Costs Labor	Equipment	Total	Total Incl O&P
0010	**STEEL SHINGLES**									
0012	Galvanized, 26 ga.	1 Rots	2.20	3.636	Sq.	345	158		503	650
0200	24 ga.	"	2.20	3.636		340	158		498	645
0300	For colored galvanized shingles, add					57			57	62.50

07 31 26 – Slate Shingles

07 31 26.10 Slate Roof Shingles

			Crew	Daily Output	Labor-Hours	Unit	Material	2017 Bare Costs Labor	Equipment	Total	Total Incl O&P
0010	**SLATE ROOF SHINGLES** R073126-20										
0100	Buckingham Virginia black, 3/16" - 1/4" thick	G	1 Rots	1.75	4.571	Sq.	550	198		748	940
0200	1/4" thick	G		1.75	4.571		550	198		748	940
0900	Pennsylvania black, Bangor, #1 clear	G		1.75	4.571		495	198		693	880
1200	Vermont, unfading, green, mottled green	G		1.75	4.571		495	198		693	880
1300	Semi-weathering green & gray	G		1.75	4.571		360	198		558	730
1400	Purple	G		1.75	4.571		435	198		633	815
1500	Black or gray	G		1.75	4.571		475	198		673	860
1600	Red	G		1.75	4.571		1,175	198		1,373	1,600
1700	Variegated purple			1.75	4.571		425	198		623	805
2700	Ridge shingles, slate			200	.040	L.F.	10	1.73		11.73	13.95

07 31 29 – Wood Shingles and Shakes

07 31 29.13 Wood Shingles

		Crew	Daily Output	Labor-Hours	Unit	Material	2017 Bare Costs Labor	Equipment	Total	Total Incl O&P
0010	**WOOD SHINGLES**									
0012	16" No. 1 red cedar shingles, 5" exposure, on roof	1 Carp	2.50	3.200	Sq.	282	158		440	550
0015	Pneumatic nailed		3.25	2.462		282	121		403	495
0200	7-1/2" exposure, on walls		2.05	3.902		188	192		380	500
0205	Pneumatic nailed		2.67	2.996		188	148		336	430
0300	18" No. 1 red cedar perfections, 5-1/2" exposure, on roof		2.75	2.909		250	143		393	495
0305	Pneumatic nailed		3.57	2.241		250	110		360	445
0500	7-1/2" exposure, on walls		2.25	3.556		184	175		359	470
0505	Pneumatic nailed		2.92	2.740		184	135		319	410
0600	Resquared and rebutted, 5-1/2" exposure, on roof		3	2.667		288	131		419	515
0605	Pneumatic nailed		3.90	2.051		288	101		389	470
0900	7-1/2" exposure, on walls		2.45	3.265		211	161		372	480
0905	Pneumatic nailed		3.18	2.516		211	124		335	425
1000	Add to above for fire retardant shingles					56			56	61.50
1060	Preformed ridge shingles	1 Carp	400	.020	L.F.	5	.99		5.99	7
2000	White cedar shingles, 16" long, extras, 5" exposure, on roof		2.40	3.333	Sq.	195	164		359	465
2005	Pneumatic nailed		3.12	2.564		195	126		321	410
2050	5" exposure on walls		2	4		195	197		392	515
2055	Pneumatic nailed		2.60	3.077		195	152		347	445
2100	7-1/2" exposure, on walls		2	4		139	197		336	455
2105	Pneumatic nailed		2.60	3.077		139	152		291	385
2150	"B" grade, 5" exposure on walls		2	4		165	197		362	480
2155	Pneumatic nailed		2.60	3.077		165	152		317	415
2300	For 15# organic felt underlayment on roof, 1 layer, add		64	.125		5.30	6.15		11.45	15.30
2400	2 layers, add		32	.250		10.60	12.30		22.90	30.50
2600	For steep roofs (7/12 pitch or greater), add to above						50%			
2700	Panelized systems, No.1 cedar shingles on 5/16" CDX plywood									
2800	On walls, 8' strips, 7" or 14" exposure	2 Carp	700	.023	S.F.	6	1.13		7.13	8.30
3500	On roofs, 8' strips, 7" or 14" exposure	1 Carp	3	2.667	Sq.	620	131		751	880
3505	Pneumatic nailed	"	4	2	"	620	98.50		718.50	830

07 31 Shingles and Shakes

07 31 29 – Wood Shingles and Shakes

07 31 29.16 Wood Shakes

	07 31 29.16 Wood Shakes	Crew	Daily Output	Labor-Hours	Unit	Material	2017 Bare Costs Labor	Equipment	Total	Total Incl O&P
0010	**WOOD SHAKES**									
1100	Hand-split red cedar shakes, 1/2" thick x 24" long, 10" exp. on roof	1 Carp	2.50	3.200	Sq.	287	158		445	555
1105	Pneumatic nailed		3.25	2.462		287	121		408	500
1110	3/4" thick x 24" long, 10" exp. on roof		2.25	3.556		287	175		462	585
1115	Pneumatic nailed		2.92	2.740		287	135		422	520
1200	1/2" thick, 18" long, 8-1/2" exp. on roof		2	4		179	197		376	495
1205	Pneumatic nailed		2.60	3.077		179	152		331	430
1210	3/4" thick x 18" long, 8-1/2" exp. on roof		1.80	4.444		179	219		398	530
1215	Pneumatic nailed		2.34	3.419		179	168		347	455
1255	10" exposure on walls		2	4		173	197		370	490
1260	10" exposure on walls, pneumatic nailed		2.60	3.077		173	152		325	420
1700	Add to above for fire retardant shakes, 24" long					56			56	61.50
1800	18" long					56			56	61.50
1810	Ridge shakes	1 Carp	350	.023	L.F.	5.75	1.13		6.88	8.05

07 32 Roof Tiles

07 32 13 – Clay Roof Tiles

07 32 13.10 Clay Tiles

	07 32 13.10 Clay Tiles	Crew	Daily Output	Labor-Hours	Unit	Material	Labor	Equipment	Total	Total Incl O&P
0010	**CLAY TILES**, including accessories									
0300	Flat shingle, interlocking, 15", 166 pcs/sq, fireflashed blend	3 Rots	6	4	Sq.	470	173		643	810
0500	Terra cotta red		6	4		520	173		693	865
0600	Roman pan and top, 18", 102 pcs/sq, fireflashed blend		5.50	4.364		505	189		694	875
0640	Terra cotta red	1 Rots	2.40	3.333		570	145		715	870
1100	Barrel mission tile, 18", 166 pcs/sq, fireflashed blend	3 Rots	5.50	4.364		415	189		604	775
1140	Terra cotta red		5.50	4.364		420	189		609	780
1700	Scalloped edge flat shingle, 14", 145 pcs/sq, fireflashed blend		6	4		1,150	173		1,323	1,575
1800	Terra cotta red		6	4		1,050	173		1,223	1,450
3010	#15 felt underlayment	1 Rofc	64	.125		5.30	5.40		10.70	15
3020	#30 felt underlayment		58	.138		10.35	5.95		16.30	21.50
3040	Polyethylene and rubberized asph. underlayment		22	.364		77	15.70		92.70	111

07 32 16 – Concrete Roof Tiles

07 32 16.10 Concrete Tiles

	07 32 16.10 Concrete Tiles	Crew	Daily Output	Labor-Hours	Unit	Material	Labor	Equipment	Total	Total Incl O&P
0010	**CONCRETE TILES**									
0020	Corrugated, 13" x 16-1/2", 90 per sq., 950 lb. per sq.									
0050	Earthtone colors, nailed to wood deck	1 Rots	1.35	5.926	Sq.	105	257		362	550
0150	Blues		1.35	5.926		105	257		362	550
0200	Greens		1.35	5.926		106	257		363	550
0250	Premium colors		1.35	5.926		106	257		363	550
0500	Shakes, 13" x 16-1/2", 90 per sq., 950 lb. per sq.									
0600	All colors, nailed to wood deck	1 Rots	1.50	5.333	Sq.	126	231		357	535
1500	Accessory pieces, ridge & hip, 10" x 16-1/2", 8 lb. each	"	120	.067	Ea.	3.80	2.89		6.69	9.10
1700	Rake, 6-1/2" x 16-3/4", 9 lb. each					3.80			3.80	4.18
1800	Mansard hip, 10" x 16-1/2", 9.2 lb. each					3.80			3.80	4.18
1900	Hip starter, 10" x 16-1/2", 10.5 lb. each					10.50			10.50	11.55
2000	3 or 4 way apex, 10" each side, 11.5 lb. each					12			12	13.20

For customer support on your Building Construction Costs with RSMeans Data, call 800.448.8182.

235

07 32 Roof Tiles

07 32 19 – Metal Roof Tiles

07 32 19.10 Metal Roof Tiles		Crew	Daily Output	Labor-Hours	Unit	Material	2017 Bare Costs Labor	Equipment	Total	Total Incl O&P
0010	**METAL ROOF TILES**									
0020	Accessories included, .032" thick aluminum, mission tile	1 Carp	2.50	3.200	Sq.	830	158		988	1,150
0200	Spanish tiles	"	3	2.667	"	550	131		681	805

07 33 Natural Roof Coverings

07 33 63 – Vegetated Roofing

07 33 63.10 Green Roof Systems

07 33 63.10 Green Roof Systems			Crew	Daily Output	Labor-Hours	Unit	Material	2017 Bare Costs Labor	Equipment	Total	Total Incl O&P
0010	**GREEN ROOF SYSTEMS**										
0020	Soil mixture for green roof 30% sand, 55% gravel, 15% soil										
0100	Hoist and spread soil mixture 4 inch depth up to five stories tall roof	G	B-13B	4000	.014	S.F.	.24	.60	.27	1.11	1.48
0150	6 inch depth	G		2667	.021		.35	.90	.40	1.65	2.21
0200	8 inch depth	G		2000	.028		.47	1.21	.54	2.22	2.95
0250	10 inch depth	G		1600	.035		.59	1.51	.67	2.77	3.69
0300	12 inch depth	G		1335	.042		.71	1.81	.81	3.33	4.43
0310	Alt. man-made soil mix, hoist & spread, 4" deep up to 5 stories tall roof	G		4000	.014		1.92	.60	.27	2.79	3.33
0350	Mobilization 55 ton crane to site	G	1 Eqhv	3.60	2.222	Ea.		124		124	187
0355	Hoisting cost to five stories per day (Avg. 28 picks per day)	G	B-13B	1	56	Day		2,400	1,075	3,475	4,850
0360	Mobilization or demobilization, 100 ton crane to site driver & escort	G	A-3E	2.50	6.400	Ea.		325	56	381	550
0365	Hoisting cost six to ten stories per day (Avg. 21 picks per day)	G	B-13C	1	56	Day		2,400	1,800	4,200	5,650
0370	Hoist and spread soil mixture 4 inch depth six to ten stories tall roof	G		4000	.014	S.F.	.24	.60	.45	1.29	1.68
0375	6 inch depth	G		2667	.021		.35	.90	.68	1.93	2.51
0380	8 inch depth	G		2000	.028		.47	1.21	.90	2.58	3.35
0385	10 inch depth	G		1600	.035		.59	1.51	1.13	3.23	4.19
0390	12 inch depth	G		1335	.042		.71	1.81	1.35	3.87	5.05
0400	Green roof edging treated lumber 4" x 4", no hoisting included	G	2 Carp	400	.040	L.F.	1.37	1.97		3.34	4.53
0410	4" x 6"	G		400	.040		2.06	1.97		4.03	5.30
0420	4" x 8"	G		360	.044		3.84	2.19		6.03	7.60
0430	4" x 6" double stacked	G		300	.053		4.11	2.63		6.74	8.55
0500	Green roof edging redwood lumber 4" x 4", no hoisting included	G		400	.040		6.45	1.97		8.42	10.05
0510	4" x 6"	G		400	.040		12.60	1.97		14.57	16.85
0520	4" x 8"	G		360	.044		23.50	2.19		25.69	29.50
0530	4" x 6" double stacked	G		300	.053		25	2.63		27.63	31.50
0550	Components, not including membrane or insulation:										
0560	Fluid applied rubber membrane, reinforced, 215 mil thick	G	G-5	350	.114	S.F.	.30	4.47	.52	5.29	8.50
0570	Root barrier	G	2 Rofc	775	.021		.70	.89		1.59	2.28
0580	Moisture retention barrier and reservoir	G	"	900	.018		2.66	.77		3.43	4.23
0600	Planting sedum, light soil, potted, 2-1/4" diameter, two per S.F.	G	1 Clab	420	.019		5.70	.75		6.45	7.40
0610	one per S.F.	G	"	840	.010		2.85	.37		3.22	3.71
0630	Planting sedum mat per S.F. including shipping (4000 S.F. min)	G	4 Clab	4000	.008		6.85	.31		7.16	8
0640	Installation sedum mat system (no soil required) per S.F. (4000 S.F. min)	G	"	4000	.008		9.50	.31		9.81	10.95
0645	Note: pricing of sedum mats shipped in full truck loads (4000-5000 S.F.)										

07 41 Roof Panels

07 41 13 – Metal Roof Panels

07 41 13.10 Aluminum Roof Panels	Crew	Daily Output	Labor-Hours	Unit	Material	2017 Bare Costs Labor	Equipment	Total	Total Incl O&P
0010 **ALUMINUM ROOF PANELS**									
0020 Corrugated or ribbed, .0155" thick, natural	G-3	1200	.027	S.F.	1	1.30		2.30	3.09
0300 Painted		1200	.027		1.45	1.30		2.75	3.59
0400 Corrugated, .018" thick, on steel frame, natural finish		1200	.027		1.25	1.30		2.55	3.37
0600 Painted		1200	.027		1.55	1.30		2.85	3.70
0700 Corrugated, on steel frame, natural, .024" thick		1200	.027		1.80	1.30		3.10	3.97
0800 Painted		1200	.027		2.20	1.30		3.50	4.41
0900 .032" thick, natural		1200	.027		2.59	1.30		3.89	4.84
1200 Painted		1200	.027		3.20	1.30		4.50	5.50
1300 V-Beam, on steel frame construction, .032" thick, natural		1200	.027		2.62	1.30		3.92	4.87
1500 Painted		1200	.027		3.33	1.30		4.63	5.65
1600 .040" thick, natural		1200	.027		3.25	1.30		4.55	5.55
1800 Painted		1200	.027		3.93	1.30		5.23	6.30
1900 .050" thick, natural		1200	.027		3.85	1.30		5.15	6.25
2100 Painted		1200	.027		4.63	1.30		5.93	7.10
2200 For roofing on wood frame, deduct		4600	.007		.08	.34		.42	.61
2400 Ridge cap, .032" thick, natural		800	.040	L.F.	3.10	1.94		5.04	6.40

07 41 13.20 Steel Roofing Panels

07 41 13.20 Steel Roofing Panels	Crew	Daily Output	Labor-Hours	Unit	Material	2017 Bare Costs Labor	Equipment	Total	Total Incl O&P
0010 **STEEL ROOFING PANELS**									
0012 Corrugated or ribbed, on steel framing, 30 ga. galv	G-3	1100	.029	S.F.	1.65	1.41		3.06	3.99
0100 28 ga.		1050	.030		1.70	1.48		3.18	4.14
0300 26 ga.		1000	.032		2	1.56		3.56	4.58
0400 24 ga.		950	.034		2.90	1.64		4.54	5.70
0600 Colored, 28 ga.		1050	.030		1.75	1.48		3.23	4.20
0700 26 ga.		1000	.032		2.05	1.56		3.61	4.64
0710 Flat profile, 1-3/4" standing seams, 10" wide, standard finish, 26 ga.		1000	.032		3.80	1.56		5.36	6.55
0715 24 ga.		950	.034		4.45	1.64		6.09	7.40
0720 22 ga.		900	.036		5.50	1.73		7.23	8.70
0725 Zinc aluminum alloy finish, 26 ga.		1000	.032		3.10	1.56		4.66	5.80
0730 24 ga.		950	.034		3.60	1.64		5.24	6.45
0735 22 ga.		900	.036		4.10	1.73		5.83	7.15
0740 12" wide, standard finish, 26 ga.		1000	.032		3.85	1.56		5.41	6.60
0745 24 ga.		950	.034		4.98	1.64		6.62	8
0750 Zinc aluminum alloy finish, 26 ga.		1000	.032		4.30	1.56		5.86	7.10
0755 24 ga.		950	.034		3.60	1.64		5.24	6.45
0840 Flat profile, 1" x 3/8" batten, 12" wide, standard finish, 26 ga.		1000	.032		3.35	1.56		4.91	6.05
0845 24 ga.		950	.034		3.95	1.64		5.59	6.85
0850 22 ga.		900	.036		4.75	1.73		6.48	7.90
0855 Zinc aluminum alloy finish, 26 ga.		1000	.032		3.25	1.56		4.81	5.95
0860 24 ga.		950	.034		3.70	1.64		5.34	6.60
0865 22 ga.		900	.036		4.20	1.73		5.93	7.25
0870 16-1/2" wide, standard finish, 24 ga.		950	.034		3.90	1.64		5.54	6.80
0875 22 ga.		900	.036		4.40	1.73		6.13	7.50
0880 Zinc aluminum alloy finish, 24 ga.		950	.034		3.40	1.64		5.04	6.25
0885 22 ga.		900	.036		3.85	1.73		5.58	6.90
0890 Flat profile, 2" x 2" batten, 12" wide, standard finish, 26 ga.		1000	.032		3.90	1.56		5.46	6.65
0895 24 ga.		950	.034		4.70	1.64		6.34	7.65
0900 22 ga.		900	.036		5.70	1.73		7.43	8.90
0905 Zinc aluminum alloy finish, 26 ga.		1000	.032		3.70	1.56		5.26	6.45
0910 24 ga.		950	.034		4.15	1.64		5.79	7.10
0915 22 ga.		900	.036		4.80	1.73		6.53	7.95
0920 16-1/2" wide, standard finish, 24 ga.		950	.034		4.30	1.64		5.94	7.25

07 41 Roof Panels

07 41 13 – Metal Roof Panels

07 41 13.20 Steel Roofing Panels

	07 41 13.20 Steel Roofing Panels	Crew	Daily Output	Labor-Hours	Unit	Material	2017 Bare Costs Labor	Equipment	Total	Total Incl O&P
0925	22 ga.	G-3	900	.036	S.F.	4.95	1.73		6.68	8.10
0930	Zinc aluminum alloy finish, 24 ga.		950	.034		3.85	1.64		5.49	6.75
0935	22 ga.		900	.036		4.40	1.73		6.13	7.50
1200	Ridge, galvanized, 10" wide Ⓖ		800	.040	L.F.	3.20	1.94		5.14	6.50
1203	14" wide Ⓖ	2 Shee	316	.051		3.84	2.94		6.78	8.70
1205	18" wide Ⓖ	"	308	.052		4.48	3.02		7.50	9.55
1210	20" wide Ⓖ	G-3	750	.043		4.20	2.07		6.27	7.80

07 41 33 – Plastic Roof Panels

07 41 33.10 Fiberglass Panels

		Crew	Daily Output	Labor-Hours	Unit	Material	2017 Bare Costs Labor	Equipment	Total	Total Incl O&P
0010	**FIBERGLASS PANELS**									
0012	Corrugated panels, roofing, 8 oz. per S.F.	G-3	1000	.032	S.F.	2.24	1.56		3.80	4.84
0100	12 oz. per S.F.		1000	.032		4.32	1.56		5.88	7.15
0300	Corrugated siding, 6 oz. per S.F.		880	.036		1.93	1.77		3.70	4.83
0400	8 oz. per S.F.		880	.036		2.24	1.77		4.01	5.15
0500	Fire retardant		880	.036		3.80	1.77		5.57	6.90
0600	12 oz. siding, textured		880	.036		3.80	1.77		5.57	6.90
0700	Fire retardant		880	.036		4.45	1.77		6.22	7.60
0900	Flat panels, 6 oz. per S.F., clear or colors		880	.036		2.45	1.77		4.22	5.40
1100	Fire retardant, class A		880	.036		3.50	1.77		5.27	6.55
1300	8 oz. per S.F., clear or colors		880	.036		2.55	1.77		4.32	5.50
1700	Sandwich panels, fiberglass, 1-9/16" thick, panels to 20 S.F.		180	.178		35	8.65		43.65	52
1900	As above, but 2-3/4" thick, panels to 100 S.F.		265	.121		25	5.85		30.85	36.50

07 42 Wall Panels

07 42 13 – Metal Wall Panels

07 42 13.10 Mansard Panels

		Crew	Daily Output	Labor-Hours	Unit	Material	2017 Bare Costs Labor	Equipment	Total	Total Incl O&P
0010	**MANSARD PANELS**									
0600	Aluminum, stock units, straight surfaces	1 Shee	115	.070	S.F.	4.30	4.04		8.34	10.95
0700	Concave or convex surfaces, add		75	.107	"	2.20	6.20		8.40	11.90
0800	For framing, to 5' high, add		115	.070	L.F.	3.75	4.04		7.79	10.35
0900	Soffits, to 1' wide		125	.064	S.F.	2.50	3.72		6.22	8.45

07 42 13.20 Aluminum Siding Panels

		Crew	Daily Output	Labor-Hours	Unit	Material	2017 Bare Costs Labor	Equipment	Total	Total Incl O&P
0010	**ALUMINUM SIDING PANELS**									
0012	Corrugated, on steel framing, .019" thick, natural finish	G-3	775	.041	S.F.	1.65	2.01		3.66	4.89
0100	Painted		775	.041		1.80	2.01		3.81	5.05
0400	Farm type, .021" thick on steel frame, natural		775	.041		1.70	2.01		3.71	4.94
0600	Painted		775	.041		1.80	2.01		3.81	5.05
0700	Industrial type, corrugated, on steel, .024" thick, mill		775	.041		2.35	2.01		4.36	5.65
0900	Painted		775	.041		2.50	2.01		4.51	5.80
1000	.032" thick, mill		775	.041		2.50	2.01		4.51	5.80
1200	Painted		775	.041		3	2.01		5.01	6.35
1300	V-Beam, on steel frame, .032" thick, mill		775	.041		2.83	2.01		4.84	6.20
1500	Painted		775	.041		3.15	2.01		5.16	6.55
1600	.040" thick, mill		775	.041		3.40	2.01		5.41	6.80
1800	Painted		775	.041		3.95	2.01		5.96	7.40
1900	.050" thick, mill		775	.041		4	2.01		6.01	7.45
2100	Painted		775	.041		4.73	2.01		6.74	8.25
2200	Ribbed, 3" profile, on steel frame, .032" thick, natural		775	.041		2.60	2.01		4.61	5.95
2400	Painted		775	.041		3.05	2.01		5.06	6.45
2500	.040" thick, natural		775	.041		3	2.01		5.01	6.35

07 42 Wall Panels

07 42 13 – Metal Wall Panels

07 42 13.20 Aluminum Siding Panels

		Crew	Daily Output	Labor-Hours	Unit	Material	2017 Bare Costs Labor	Equipment	Total	Total Incl O&P
2700	Painted	G-3	775	.041	S.F.	3.47	2.01		5.48	6.90
2750	.050" thick, natural		775	.041		3.42	2.01		5.43	6.85
2760	Painted		775	.041		3.94	2.01		5.95	7.40
3300	For siding on wood frame, deduct from above	↓	2800	.011	↓	.09	.56		.65	.95
3400	Screw fasteners, aluminum, self tapping, neoprene washer, 1"				M	210			210	231
3600	Stitch screws, self tapping, with neoprene washer, 5/8"				"	158			158	174
3630	Flashing, sidewall, .032" thick	G-3	800	.040	L.F.	3	1.94		4.94	6.30
3650	End wall, .040" thick		800	.040		3.50	1.94		5.44	6.85
3670	Closure strips, corrugated, .032" thick		800	.040		.95	1.94		2.89	4.03
3680	Ribbed, 4" or 8", .032" thick		800	.040		.94	1.94		2.88	4.01
3690	V-beam, .040" thick	↓	800	.040	↓	1.25	1.94		3.19	4.36
3800	Horizontal, colored clapboard, 8" wide, plain	2 Carp	515	.031	S.F.	2.50	1.53		4.03	5.10
3810	Insulated		515	.031		2.85	1.53		4.38	5.50
4000	Vertical board & batten, colored, non-insulated	↓	515	.031		2.10	1.53		3.63	4.65
4200	For simulated wood design, add				↓	.15			.15	.17
4300	Corners for above, outside	2 Carp	515	.031	V.L.F.	3.60	1.53		5.13	6.30
4500	Inside corners	"	515	.031	"	1.75	1.53		3.28	4.27

07 42 13.30 Steel Siding

		Crew	Daily Output	Labor-Hours	Unit	Material	2017 Bare Costs Labor	Equipment	Total	Total Incl O&P
0010	**STEEL SIDING**									
0020	Beveled, vinyl coated, 8" wide	1 Carp	265	.030	S.F.	1.85	1.49		3.34	4.32
0050	10" wide	"	275	.029		1.95	1.43		3.38	4.34
0080	Galv, corrugated or ribbed, on steel frame, 30 ga.	G-3	800	.040		1.25	1.94		3.19	4.36
0100	28 ga.		795	.040		1.35	1.96		3.31	4.49
0300	26 ga.		790	.041		1.83	1.97		3.80	5.05
0400	24 ga.		785	.041		2	1.98		3.98	5.25
0600	22 ga.		770	.042		2.25	2.02		4.27	5.55
0700	Colored, corrugated/ribbed, on steel frame, 10 yr. finish, 28 ga.		800	.040		2.05	1.94		3.99	5.25
0900	26 ga.		795	.040		2.17	1.96		4.13	5.40
1000	24 ga.		790	.041		2.43	1.97		4.40	5.70
1020	20 ga.		785	.041		3.08	1.98		5.06	6.40
1200	Factory sandwich panel, 26 ga., 1" insulation, galvanized		380	.084		5.40	4.09		9.49	12.15
1300	Colored 1 side		380	.084		6.95	4.09		11.04	13.90
1500	Galvanized 2 sides		380	.084		8.25	4.09		12.34	15.35
1600	Colored 2 sides		380	.084		8.50	4.09		12.59	15.60
1800	Acrylic paint face, regular paint liner	↓	380	.084		6.25	4.09		10.34	13.15
1900	For 2" thick polystyrene, add					1			1	1.10
2000	22 ga., galv, 2" insulation, baked enamel exterior	G-3	360	.089		12.25	4.32		16.57	20
2100	Polyvinylidene exterior finish	"	360	.089	↓	13	4.32		17.32	21

07 44 Faced Panels

07 44 73 – Metal Faced Panels

07 44 73.10 Metal Faced Panels and Accessories

		Crew	Daily Output	Labor-Hours	Unit	Material	2017 Bare Costs Labor	Equipment	Total	Total Incl O&P
0010	**METAL FACED PANELS AND ACCESSORIES**									
0400	Textured aluminum, 4' x 8' x 5/16" plywood backing, single face	2 Shee	375	.043	S.F.	4.13	2.48		6.61	8.35
0600	Double face		375	.043		5.40	2.48		7.88	9.75
0700	4' x 10' x 5/16" plywood backing, single face		375	.043		4.40	2.48		6.88	8.65
0900	Double face		375	.043		5.90	2.48		8.38	10.30
1000	4' x 12' x 5/16" plywood backing, single face		375	.043		4.40	2.48		6.88	8.65
1300	Smooth aluminum, 1/4" plywood panel, fluoropolymer finish, double face		375	.043		6.15	2.48		8.63	10.55
1350	Clear anodized finish, double face		375	.043		10.20	2.48		12.68	15
1400	Double face textured aluminum, structural panel, 1" EPS insulation	↓	375	.043	↓	5.80	2.48		8.28	10.20

07 44 Faced Panels

07 44 73 – Metal Faced Panels

07 44 73.10 Metal Faced Panels and Accessories	Crew	Daily Output	Labor-Hours	Unit	Material	2017 Bare Costs Labor	Equipment	Total	Total Incl O&P
1500	Accessories, outside corner	1 Shee	175	.046	L.F.	1.92	2.65		4.57	6.15
1600	Inside corner		175	.046		1.38	2.65		4.03	5.60
1800	Batten mounting clip		200	.040		.50	2.32		2.82	4.10
1900	Low profile batten		480	.017		.62	.97		1.59	2.16
2100	High profile batten		480	.017		1.43	.97		2.40	3.05
2200	Water table		200	.040		2.16	2.32		4.48	5.95
2400	Horizontal joint connector		200	.040		1.70	2.32		4.02	5.40
2500	Corner cap		200	.040		1.88	2.32		4.20	5.60
2700	H - moulding		480	.017		1.27	.97		2.24	2.88

07 46 Siding

07 46 23 – Wood Siding

07 46 23.10 Wood Board Siding

		Crew	Daily Output	Labor-Hours	Unit	Material	2017 Bare Costs Labor	Equipment	Total	Total Incl O&P
0010	**WOOD BOARD SIDING**									
3200	Wood, cedar bevel, A grade, 1/2" x 6"	1 Carp	295	.027	S.F.	4.38	1.34		5.72	6.85
3300	1/2" x 8"		330	.024		6.70	1.19		7.89	9.20
3500	3/4" x 10", clear grade		375	.021		6.65	1.05		7.70	8.90
3600	"B" grade		375	.021		3.79	1.05		4.84	5.80
3800	Cedar, rough sawn, 1" x 4", A grade, natural		220	.036		6.55	1.79		8.34	10
3900	Stained		220	.036		6.70	1.79		8.49	10.10
4100	1" x 12", board & batten, #3 & Btr., natural		420	.019		4.48	.94		5.42	6.35
4200	Stained		420	.019		4.81	.94		5.75	6.75
4400	1" x 8" channel siding, #3 & Btr., natural		330	.024		4.57	1.19		5.76	6.85
4500	Stained		330	.024		4.92	1.19		6.11	7.25
4700	Redwood, clear, beveled, vertical grain, 1/2" x 4"		220	.036		4.20	1.79		5.99	7.35
4750	1/2" x 6"		295	.027		4.80	1.34		6.14	7.35
4800	1/2" x 8"		330	.024		5.20	1.19		6.39	7.55
5000	3/4" x 10"		375	.021		4.95	1.05		6	7.05
5200	Channel siding, 1" x 10", B grade		375	.021		4.60	1.05		5.65	6.65
5250	Redwood, T&G boards, B grade, 1" x 4"		220	.036		3.30	1.79		5.09	6.35
5270	1" x 8"		330	.024		5.25	1.19		6.44	7.65
5400	White pine, rough sawn, 1" x 8", natural		330	.024		2.19	1.19		3.38	4.24
5500	Stained		330	.024		2.19	1.19		3.38	4.24

07 46 29 – Plywood Siding

07 46 29.10 Plywood Siding Options

		Crew	Daily Output	Labor-Hours	Unit	Material	2017 Bare Costs Labor	Equipment	Total	Total Incl O&P
0010	**PLYWOOD SIDING OPTIONS**									
0900	Plywood, medium density overlaid, 3/8" thick	2 Carp	750	.021	S.F.	1.24	1.05		2.29	2.97
1000	1/2" thick		700	.023		1.50	1.13		2.63	3.37
1100	3/4" thick		650	.025		1.84	1.21		3.05	3.88
1600	Texture 1-11, cedar, 5/8" thick, natural		675	.024		2.67	1.17		3.84	4.73
1700	Factory stained		675	.024		2.80	1.17		3.97	4.87
1900	Texture 1-11, fir, 5/8" thick, natural		675	.024		1.45	1.17		2.62	3.39
2000	Factory stained		675	.024		1.80	1.17		2.97	3.77
2050	Texture 1-11, S.Y.P., 5/8" thick, natural		675	.024		1.37	1.17		2.54	3.30
2100	Factory stained		675	.024		1.44	1.17		2.61	3.37
2200	Rough sawn cedar, 3/8" thick, natural		675	.024		1.25	1.17		2.42	3.17
2300	Factory stained		675	.024		1.50	1.17		2.67	3.44
2500	Rough sawn fir, 3/8" thick, natural		675	.024		.90	1.17		2.07	2.78
2600	Factory stained		675	.024		1.04	1.17		2.21	2.93
2800	Redwood, textured siding, 5/8" thick		675	.024		1.92	1.17		3.09	3.90

07 46 Siding

07 46 29 – Plywood Siding

07 46 29.10 Plywood Siding Options	Crew	Daily Output	Labor-Hours	Unit	Material	2017 Bare Costs Labor	Equipment	Total	Total Incl O&P
3000 Polyvinyl chloride coated, 3/8" thick	2 Carp	750	.021	S.F.	1.11	1.05		2.16	2.83

07 46 33 – Plastic Siding

07 46 33.10 Vinyl Siding

	Crew	Daily Output	Labor-Hours	Unit	Material	2017 Bare Costs Labor	Equipment	Total	Total Incl O&P
0010 **VINYL SIDING**									
3995 Clapboard profile, woodgrain texture, .048 thick, double 4	2 Carp	495	.032	S.F.	1.06	1.59		2.65	3.61
4000 Double 5		550	.029		1.06	1.43		2.49	3.36
4005 Single 8		495	.032		1.37	1.59		2.96	3.95
4010 Single 10		550	.029		1.65	1.43		3.08	4
4015 .044 thick, double 4		495	.032		1.04	1.59		2.63	3.59
4020 Double 5		550	.029		1.06	1.43		2.49	3.36
4025 .042 thick, double 4		495	.032		1.06	1.59		2.65	3.61
4030 Double 5		550	.029		1.06	1.43		2.49	3.36
4035 Cross sawn texture, .040 thick, double 4		495	.032		.72	1.59		2.31	3.24
4040 Double 5		550	.029		.66	1.43		2.09	2.92
4045 Smooth texture, .042 thick, double 4		495	.032		.79	1.59		2.38	3.31
4050 Double 5		550	.029		.79	1.43		2.22	3.06
4055 Single 8		495	.032		.79	1.59		2.38	3.31
4060 Cedar texture, .044 thick, double 4		495	.032		1.13	1.59		2.72	3.69
4065 Double 6		600	.027		1.13	1.31		2.44	3.26
4070 Dutch lap profile, woodgrain texture, .048 thick, double 5		550	.029		1.08	1.43		2.51	3.38
4075 .044 thick, double 4.5		525	.030		1.08	1.50		2.58	3.49
4080 .042 thick, double 4.5		525	.030		.91	1.50		2.41	3.31
4085 .040 thick, double 4.5		525	.030		.72	1.50		2.22	3.10
4100 Shake profile, 10" wide		400	.040		3.66	1.97		5.63	7.05
4105 Vertical pattern, .046 thick, double 5		550	.029		1.57	1.43		3	3.92
4110 .044 thick, triple 3		550	.029		1.76	1.43		3.19	4.13
4115 .040 thick, triple 4		550	.029		1.66	1.43		3.09	4.02
4120 .040 thick, triple 2.66		550	.029		1.76	1.43		3.19	4.13
4125 Insulation, fan folded extruded polystyrene, 1/4"		2000	.008		.29	.39		.68	.92
4130 3/8"		2000	.008	▼	.32	.39		.71	.95
4135 Accessories, J channel, 5/8" pocket		700	.023	L.F.	.50	1.13		1.63	2.27
4140 3/4" pocket		695	.023		.55	1.13		1.68	2.34
4145 1-1/4" pocket		680	.024		.85	1.16		2.01	2.71
4150 Flexible, 3/4" pocket		600	.027		2.37	1.31		3.68	4.61
4155 Under sill finish trim		500	.032		.55	1.58		2.13	3.01
4160 Vinyl starter strip		700	.023		.66	1.13		1.79	2.44
4165 Aluminum starter strip		700	.023		.29	1.13		1.42	2.04
4170 Window casing, 2-1/2" wide, 3/4" pocket		510	.031		1.69	1.55		3.24	4.23
4175 Outside corner, woodgrain finish, 4" face, 3/4" pocket		700	.023		2.11	1.13		3.24	4.05
4180 5/8" pocket		700	.023		2.11	1.13		3.24	4.05
4185 Smooth finish, 4" face, 3/4" pocket		700	.023		2.11	1.13		3.24	4.05
4190 7/8" pocket		690	.023		2.01	1.14		3.15	3.97
4195 1-1/4" pocket		700	.023		1.41	1.13		2.54	3.28
4200 Soffit and fascia, 1' overhang, solid		120	.133		4.67	6.55		11.22	15.20
4205 Vented		120	.133		4.67	6.55		11.22	15.20
4207 18" overhang, solid		110	.145		5.45	7.15		12.60	16.95
4208 Vented		110	.145		5.45	7.15		12.60	16.95
4210 2' overhang, solid		100	.160		6.25	7.90		14.15	18.90
4215 Vented		100	.160		6.25	7.90		14.15	18.90
4217 3' overhang, solid		100	.160		7.80	7.90		15.70	20.50
4218 Vented	▼	100	.160	▼	7.80	7.90		15.70	20.50
4220 Colors for siding and soffits, add				S.F.	.15			.15	.17

07 46 Siding

07 46 33 – Plastic Siding

07 46 33.10 Vinyl Siding

		Crew	Daily Output	Labor-Hours	Unit	Material	2017 Bare Costs Labor	Equipment	Total	Total Incl O&P
4225	Colors for accessories and trim, add				L.F.	.31			.31	.34

07 46 33.20 Polypropylene Siding

		Crew	Daily Output	Labor-Hours	Unit	Material	2017 Bare Costs Labor	Equipment	Total	Total Incl O&P
0010	**POLYPROPYLENE SIDING**									
4090	Shingle profile, random grooves, double 7	2 Carp	400	.040	S.F.	3.21	1.97		5.18	6.55
4092	Cornerpost for above	1 Carp	365	.022	L.F.	13	1.08		14.08	15.95
4095	Triple 5	2 Carp	400	.040	S.F.	3.21	1.97		5.18	6.55
4097	Cornerpost for above	1 Carp	365	.022	L.F.	12.20	1.08		13.28	15.05
5000	Staggered butt, double 7"	2 Carp	400	.040	S.F.	3.66	1.97		5.63	7.05
5002	Cornerpost for above	1 Carp	365	.022	L.F.	13	1.08		14.08	15.95
5010	Half round, double 6-1/4"	2 Carp	360	.044	S.F.	3.66	2.19		5.85	7.40
5020	Shake profile, staggered butt, double 9"	"	510	.031	"	3.66	1.55		5.21	6.40
5022	Cornerpost for above	1 Carp	365	.022	L.F.	9.70	1.08		10.78	12.30
5030	Straight butt, double 7"	2 Carp	400	.040	S.F.	3.66	1.97		5.63	7.05
5032	Cornerpost for above	1 Carp	365	.022	L.F.	12.85	1.08		13.93	15.80
6000	Accessories, J channel, 5/8" pocket	2 Carp	700	.023		.50	1.13		1.63	2.27
6010	3/4" pocket		695	.023		.55	1.13		1.68	2.34
6020	1-1/4" pocket		680	.024		.85	1.16		2.01	2.71
6030	Aluminum starter strip		700	.023		.29	1.13		1.42	2.04

07 46 46 – Fiber Cement Siding

07 46 46.10 Fiber Cement Siding

		Crew	Daily Output	Labor-Hours	Unit	Material	2017 Bare Costs Labor	Equipment	Total	Total Incl O&P
0010	**FIBER CEMENT SIDING**									
0020	Lap siding, 5/16" thick, 6" wide, 4-3/4" exposure, smooth texture	2 Carp	415	.039	S.F.	1.29	1.90		3.19	4.32
0025	Woodgrain texture		415	.039		1.29	1.90		3.19	4.32
0030	7-1/2" wide, 6-1/4" exposure, smooth texture		425	.038		1.61	1.85		3.46	4.61
0035	Woodgrain texture		425	.038		1.61	1.85		3.46	4.61
0040	8" wide, 6-3/4" exposure, smooth texture		425	.038		1.21	1.85		3.06	4.17
0045	Rough sawn texture		425	.038		1.21	1.85		3.06	4.17
0050	9-1/2" wide, 8-1/4" exposure, smooth texture		440	.036		1.28	1.79		3.07	4.15
0055	Woodgrain texture		440	.036		1.28	1.79		3.07	4.15
0060	12" wide, 10-3/8" exposure, smooth texture		455	.035		2.06	1.73		3.79	4.92
0065	Woodgrain texture		455	.035		2.06	1.73		3.79	4.92
0070	Panel siding, 5/16" thick, smooth texture		750	.021		1.31	1.05		2.36	3.05
0075	Stucco texture		750	.021		1.31	1.05		2.36	3.05
0080	Grooved woodgrain texture		750	.021		1.31	1.05		2.36	3.05
0085	V - grooved woodgrain texture		750	.021		1.31	1.05		2.36	3.05
0088	Shingle siding, 48" x 15-1/4" panels, 7" exposure		700	.023		4.15	1.13		5.28	6.30
0090	Wood starter strip		400	.040	L.F.	.44	1.97		2.41	3.50

07 46 73 – Soffit

07 46 73.10 Soffit Options

		Crew	Daily Output	Labor-Hours	Unit	Material	2017 Bare Costs Labor	Equipment	Total	Total Incl O&P
0010	**SOFFIT OPTIONS**									
0012	Aluminum, residential, .020" thick	1 Carp	210	.038	S.F.	2.05	1.88		3.93	5.15
0100	Baked enamel on steel, 16 or 18 ga.		105	.076		6	3.75		9.75	12.35
0300	Polyvinyl chloride, white, solid		230	.035		2.15	1.71		3.86	4.99
0400	Perforated		230	.035		2.15	1.71		3.86	4.99
0500	For colors, add					.15			.15	.17

07 51 Built-Up Bituminous Roofing

07 51 13 – Built-Up Asphalt Roofing

07 51 13.10 Built-Up Roofing Components

		Crew	Daily Output	Labor-Hours	Unit	Material	2017 Bare Costs Labor	2017 Bare Costs Equipment	Total	Total Incl O&P
0010	**BUILT-UP ROOFING COMPONENTS**									
0012	Asphalt saturated felt, #30, 2 square per roll	1 Rofc	58	.138	Sq.	10.35	5.95		16.30	21.50
0200	#15, 4 sq. per roll, plain or perforated, not mopped		58	.138		5.30	5.95		11.25	15.95
0300	Roll roofing, smooth, #65		15	.533		10.15	23		33.15	50
0500	#90		12	.667		44.50	29		73.50	98
0520	Mineralized		12	.667		35.50	29		64.50	88
0540	D.C. (double coverage), 19" selvage edge		10	.800		54	34.50		88.50	118
0580	Adhesive (lap cement)				Gal.	8.45			8.45	9.30
0800	Steep, flat or dead level asphalt, 10 ton lots, packaged				Ton	945			945	1,050

07 51 13.13 Cold-Applied Built-Up Asphalt Roofing

		Crew	Daily Output	Labor-Hours	Unit	Material	2017 Bare Costs Labor	2017 Bare Costs Equipment	Total	Total Incl O&P
0010	**COLD-APPLIED BUILT-UP ASPHALT ROOFING**									
0020	3 ply system, installation only (components listed below)	G-5	50	.800	Sq.		31.50	3.67	35.17	57
0100	Spunbond poly. fabric, 1.35 oz./S.Y., 36"W, 10.8 sq./roll				Ea.	140			140	154
0500	Base & finish coat, 3 gal./sq., 5 gal./can				Gal.	8			8	8.80
0600	Coating, ceramic granules, 1/2 sq./bag				Ea.	23			23	25.50
0700	Aluminum, 2 gal./sq.				Gal.	12.90			12.90	14.15
0800	Emulsion, fibered or non-fibered, 4 gal./sq.				"	6.50			6.50	7.15

07 51 13.20 Built-Up Roofing Systems

		Crew	Daily Output	Labor-Hours	Unit	Material	2017 Bare Costs Labor	2017 Bare Costs Equipment	Total	Total Incl O&P
0010	**BUILT-UP ROOFING SYSTEMS** R075113-20									
0120	Asphalt flood coat with gravel/slag surfacing, not including									
0140	Insulation, flashing or wood nailers									
0200	Asphalt base sheet, 3 plies #15 asphalt felt, mopped	G-1	22	2.545	Sq.	101	103	24	228	310
0350	On nailable decks		21	2.667		104	107	25	236	325
0500	4 plies #15 asphalt felt, mopped		20	2.800		139	113	26.50	278.50	375
0550	On nailable decks		19	2.947		123	119	28	270	370
0700	Coated glass base sheet, 2 plies glass (type IV), mopped		22	2.545		112	103	24	239	325
0850	3 plies glass, mopped		20	2.800		136	113	26.50	275.50	370
0950	On nailable decks		19	2.947		128	119	28	275	375
1100	4 plies glass fiber felt (type IV), mopped		20	2.800		168	113	26.50	307.50	405
1150	On nailable decks		19	2.947		152	119	28	299	400
1200	Coated & saturated base sheet, 3 plies #15 asph. felt, mopped		20	2.800		113	113	26.50	252.50	345
1250	On nailable decks		19	2.947		105	119	28	252	350
1300	4 plies #15 asphalt felt, mopped		22	2.545		132	103	24	259	345
2000	Asphalt flood coat, smooth surface									
2200	Asphalt base sheet & 3 plies #15 asphalt felt, mopped	G-1	24	2.333	Sq.	108	94	22	224	300
2400	On nailable decks		23	2.435		99.50	98	23	220.50	305
2600	4 plies #15 asphalt felt, mopped		24	2.333		126	94	22	242	325
2700	On nailable decks		23	2.435		118	98	23	239	325
2900	Coated glass fiber base sheet, mopped, and 2 plies of									
2910	glass fiber felt (type IV)	G-1	25	2.240	Sq.	107	90	21	218	295
3100	On nailable decks		24	2.333		101	94	22	217	295
3200	3 plies, mopped		23	2.435		131	98	23	252	335
3300	On nailable decks		22	2.545		123	103	24	250	335
3800	4 plies glass fiber felt (type IV), mopped		23	2.435		155	98	23	276	365
3900	On nailable decks		22	2.545		147	103	24	274	365
4000	Coated & saturated base sheet, 3 plies #15 asph. felt, mopped		24	2.333		109	94	22	225	305
4200	On nailable decks		23	2.435		101	98	23	222	305
4300	4 plies #15 organic felt, mopped		22	2.545		127	103	24	254	340
4500	Coal tar pitch with gravel/slag surfacing									
4600	4 plies #15 tarred felt, mopped	G-1	21	2.667	Sq.	201	107	25	333	430
4800	3 plies glass fiber felt (type IV), mopped	"	19	2.947	"	166	119	28	313	415
5000	Coated glass fiber base sheet, and 2 plies of									

07 51 Built-Up Bituminous Roofing

07 51 13 – Built-Up Asphalt Roofing

07 51 13.20 Built-Up Roofing Systems

07 51 13.20 Built-Up Roofing Systems	Crew	Daily Output	Labor-Hours	Unit	Material	2017 Bare Costs Labor	Equipment	Total	Total Incl O&P
5010	glass fiber felt (type IV), mopped	G-1	19	2.947	Sq.	170	119	28	317	420
5300	On nailable decks		18	3.111		148	125	29.50	302.50	410
5600	4 plies glass fiber felt (type IV), mopped		21	2.667		230	107	25	362	465
5800	On nailable decks		20	2.800		208	113	26.50	347.50	450

07 51 13.30 Cants

		Crew	Daily Output	Labor-Hours	Unit	Material	Labor	Equipment	Total	Total Incl O&P
0010	**CANTS**									
0012	Lumber, treated, 4" x 4" cut diagonally	1 Rofc	325	.025	L.F.	1.83	1.06		2.89	3.82
0300	Mineral or fiber, trapezoidal, 1" x 4" x 48"		325	.025		.30	1.06		1.36	2.14
0400	1-1/2" x 5-5/8" x 48"		325	.025		.48	1.06		1.54	2.34

07 51 13.40 Felts

		Crew	Daily Output	Labor-Hours	Unit	Material	Labor	Equipment	Total	Total Incl O&P
0010	**FELTS**									
0012	Glass fibered roofing felt, #15, not mopped	1 Rofc	58	.138	Sq.	10.85	5.95		16.80	22
0300	Base sheet, #80, channel vented		58	.138		42.50	5.95		48.45	57
0400	#70, coated		58	.138		17.80	5.95		23.75	29.50
0500	Cap, #87, mineral surfaced		58	.138		81	5.95		86.95	99
0600	Flashing membrane, #65		16	.500		10.15	21.50		31.65	47.50
0800	Coal tar fibered, #15, no mopping		58	.138		18	5.95		23.95	30
0900	Asphalt felt, #15, 4 sq. per roll, no mopping		58	.138		5.30	5.95		11.25	15.95
1100	#30, 2 sq. per roll		58	.138		10.35	5.95		16.30	21.50
1200	Double coated, #33		58	.138		11.10	5.95		17.05	22.50
1400	#40, base sheet		58	.138		11.10	5.95		17.05	22.50
1450	Coated and saturated		58	.138		12.25	5.95		18.20	23.50
1500	Tarred felt, organic, #15, 4 sq. rolls		58	.138		13.40	5.95		19.35	25
1550	#30, 2 sq. roll		58	.138		26.50	5.95		32.45	39
1700	Add for mopping above felts, per ply, asphalt, 24 lb. per sq.	G-1	192	.292		11.35	11.75	2.75	25.85	35.50
1800	Coal tar mopping, 30 lb. per sq.		186	.301		18.90	12.15	2.84	33.89	44.50
1900	Flood coat, with asphalt, 60 lb. per sq.		60	.933		28.50	37.50	8.80	74.80	105
2000	With coal tar, 75 lb. per sq.		56	1		47.50	40.50	9.45	97.45	131

07 51 13.50 Walkways for Built-Up Roofs

		Crew	Daily Output	Labor-Hours	Unit	Material	Labor	Equipment	Total	Total Incl O&P
0010	**WALKWAYS FOR BUILT-UP ROOFS**									
0020	Asphalt impregnated, 3' x 6' x 1/2" thick	1 Rofc	400	.020	S.F.	1.79	.86		2.65	3.44
0100	3' x 3' x 3/4" thick	"	400	.020		5.15	.86		6.01	7.15
0300	Concrete patio blocks, 2" thick, natural	1 Clab	115	.070		3.37	2.72		6.09	7.90
0400	Colors	"	115	.070		3.73	2.72		6.45	8.25

07 52 Modified Bituminous Membrane Roofing

07 52 13 – Atactic-Polypropylene-Modified Bituminous Membrane Roofing

07 52 13.10 APP Modified Bituminous Membrane

		Crew	Daily Output	Labor-Hours	Unit	Material	Labor	Equipment	Total	Total Incl O&P
0010	**APP MODIFIED BITUMINOUS MEMBRANE** R075213-30									
0020	Base sheet, #15 glass fiber felt, nailed to deck	1 Rofc	58	.138	Sq.	12.15	5.95		18.10	23.50
0030	Spot mopped to deck	G-1	295	.190		16.50	7.65	1.79	25.94	33
0040	Fully mopped to deck	"	192	.292		22	11.75	2.75	36.50	47.50
0050	#15 organic felt, nailed to deck	1 Rofc	58	.138		6.60	5.95		12.55	17.35
0060	Spot mopped to deck	G-1	295	.190		10.95	7.65	1.79	20.39	27
0070	Fully mopped to deck	"	192	.292		16.65	11.75	2.75	31.15	41.50
2100	APP mod., smooth surf. cap sheet, poly. reinf., torched, 160 mils	G-5	2100	.019	S.F.	.74	.75	.09	1.58	2.18
2150	170 mils		2100	.019		.75	.75	.09	1.59	2.20
2200	Granule surface cap sheet, poly. reinf., torched, 180 mils		2000	.020		.94	.78	.09	1.81	2.46
2250	Smooth surface flashing, torched, 160 mils		1260	.032		.74	1.24	.15	2.13	3.08
2300	170 mils		1260	.032		.75	1.24	.15	2.14	3.10

07 52 Modified Bituminous Membrane Roofing

07 52 13 – Atactic-Polypropylene-Modified Bituminous Membrane Roofing

07 52 13.10 APP Modified Bituminous Membrane	Crew	Daily Output	Labor-Hours	Unit	Material	2017 Bare Costs Labor	2017 Bare Costs Equipment	Total	Total Incl O&P
2350 Granule surface flashing, torched, 180 mils	G-5	1260	.032	S.F.	.94	1.24	.15	2.33	3.30
2400 Fibrated aluminum coating	1 Rofc	3800	.002	↓	.09	.09		.18	.25
2450 Seam heat welding	"	205	.039	L.F.	.09	1.68		1.77	2.96

07 52 16 – Styrene-Butadiene-Styrene Modified Bituminous Membrane Roofing

07 52 16.10 SBS Modified Bituminous Membrane

		Crew	Daily Output	Labor-Hours	Unit	Material	2017 Bare Costs Labor	2017 Bare Costs Equipment	Total	Total Incl O&P
0010	**SBS MODIFIED BITUMINOUS MEMBRANE**									
0080	Mod bit rfng, SBS mod, gran surf cap sheet, poly reinf									
0650	120 to 149 mils thick	G-1	2000	.028	S.F.	1.27	1.13	.26	2.66	3.61
0750	150 to 160 mils	"	2000	.028		1.72	1.13	.26	3.11	4.10
1150	For reflective granules, add					.71	1.02	.28	2.01	2.82
1600	Smooth surface cap sheet, mopped, 145 mils	G-1	2100	.027		.80	1.07	.25	2.12	2.99
1620	Lightweight base sheet, fiberglass reinforced, 35 to 47 mil		2100	.027		.28	1.07	.25	1.60	2.42
1625	Heavyweight base/ply sheet, reinforced, 87 to 120 mil thick	↓	2100	.027		.89	1.07	.25	2.21	3.09
1650	Granulated walkpad, 180 to 220 mils	1 Rofc	400	.020		1.79	.86		2.65	3.44
1700	Smooth surface flashing, 145 mils	G-1	1260	.044		.80	1.79	.42	3.01	4.38
1800	150 mils		1260	.044		.50	1.79	.42	2.71	4.05
1900	Granular surface flashing, 150 mils		1260	.044		.70	1.79	.42	2.91	4.27
2000	160 mils	↓	1260	.044		.73	1.79	.42	2.94	4.30
2010	Elastomeric asphalt primer	1 Rofc	2600	.003		.16	.13		.29	.41
2015	Roofing asphalt, 30 lb. per square	G-1	19000	.003		.14	.12	.03	.29	.39
2020	Cold process adhesive, 20 to 30 mils thick	1 Rofc	750	.011		.24	.46		.70	1.04
2025	Self adhering vapor retarder, 30 to 45 mils thick	G-5	2150	.019	↓	1.02	.73	.09	1.84	2.45
2050	Seam heat welding	1 Rofc	205	.039	L.F.	.09	1.68		1.77	2.96

07 53 Elastomeric Membrane Roofing

07 53 16 – Chlorosulfonate-Polyethylene Roofing

07 53 16.10 Chlorosulfonated Polyethylene Roofing

		Crew	Daily Output	Labor-Hours	Unit	Material	2017 Bare Costs Labor	2017 Bare Costs Equipment	Total	Total Incl O&P
0010	**CHLOROSULFONATED POLYETHYLENE ROOFING**									
0800	Chlorosulfonated polyethylene (CSPE)									
0900	45 mils, heat welded seams, plate attachment	G-5	35	1.143	Sq.	256	44.50	5.25	305.75	365
1100	Heat welded seams, plate attachment and ballasted		26	1.538		266	60	7.05	333.05	400
1200	60 mils, heat welded seams, plate attachment		35	1.143		330	44.50	5.25	379.75	445
1300	Heat welded seams, plate attachment and ballasted	↓	26	1.538	↓	340	60	7.05	407.05	485

07 53 23 – Ethylene-Propylene-Diene-Monomer Roofing

07 53 23.20 Ethylene-Propylene-Diene-Monomer Roofing

		Crew	Daily Output	Labor-Hours	Unit	Material	2017 Bare Costs Labor	2017 Bare Costs Equipment	Total	Total Incl O&P
0010	**ETHYLENE-PROPYLENE-DIENE-MONOMER ROOFING (EPDM)**									
3500	Ethylene-propylene-diene-monomer (EPDM), 45 mils, 0.28 psf									
3600	Loose-laid & ballasted with stone (10 psf)	G-5	51	.784	Sq.	84.50	30.50	3.60	118.60	149
3700	Mechanically attached		35	1.143		79	44.50	5.25	128.75	169
3800	Fully adhered with adhesive	↓	26	1.538	↓	112	60	7.05	179.05	233
4500	60 mils, 0.40 psf									
4600	Loose-laid & ballasted with stone (10 psf)	G-5	51	.784	Sq.	95.50	30.50	3.60	129.60	161
4700	Mechanically attached		35	1.143		89	44.50	5.25	138.75	180
4800	Fully adhered with adhesive	↓	26	1.538		122	60	7.05	189.05	244
4810	45 mil, 0.28 psf, membrane only					50.50			50.50	55.50
4820	60 mil, 0.40 psf, membrane only				↓	59.50			59.50	65.50
4850	Seam tape for membrane, 3" x 100' roll				Ea.	43.50			43.50	48
4900	Batten strips, 10' sections					3.45			3.45	3.80
4910	Cover tape for batten strips, 6" x 100' roll				↓	184			184	203
4930	Plate anchors				M	80			80	88

07 53 Elastomeric Membrane Roofing

07 53 23 – Ethylene-Propylene-Diene-Monomer Roofing

07 53 23.20 Ethylene-Propylene-Diene-Monomer Roofing	Crew	Daily Output	Labor-Hours	Unit	Material	2017 Bare Costs Labor	Equipment	Total	Total Incl O&P
4970	Adhesive for fully adhered systems, 60 S.F./gal.				Gal.	20			20	22

07 53 29 – Polyisobutylene Roofing

07 53 29.10 Polyisobutylene Roofing

		Crew	Daily Output	Labor-Hours	Unit	Material	Labor	Equipment	Total	Total Incl O&P
0010	POLYISOBUTYLENE ROOFING									
7500	Polyisobutylene (PIB), 100 mils, 0.57 psf									
7600	Loose-laid & ballasted with stone/gravel (10 psf)	G-5	51	.784	Sq.	205	30.50	3.60	239.10	282
7700	Partially adhered with adhesive		35	1.143		245	44.50	5.25	294.75	350
7800	Hot asphalt attachment		35	1.143		235	44.50	5.25	284.75	340
7900	Fully adhered with contact cement		26	1.538		255	60	7.05	322.05	390

07 54 Thermoplastic Membrane Roofing

07 54 19 – Polyvinyl-Chloride Roofing

07 54 19.10 Polyvinyl-Chloride Roofing (PVC)

		Crew	Daily Output	Labor-Hours	Unit	Material	Labor	Equipment	Total	Total Incl O&P
0010	POLYVINYL-CHLORIDE ROOFING (PVC)									
8200	Heat welded seams									
8700	Reinforced, 48 mils, 0.33 psf									
8750	Loose-laid & ballasted with stone/gravel (12 psf)	G-5	51	.784	Sq.	119	30.50	3.60	153.10	187
8800	Mechanically attached		35	1.143		113	44.50	5.25	162.75	206
8850	Fully adhered with adhesive		26	1.538		156	60	7.05	223.05	282
8860	Reinforced, 60 mils, 0.40 psf									
8870	Loose-laid & ballasted with stone/gravel (12 psf)	G-5	51	.784	Sq.	120	30.50	3.60	154.10	188
8880	Mechanically attached		35	1.143		114	44.50	5.25	163.75	208
8890	Fully adhered with adhesive		26	1.538		157	60	7.05	224.05	283

07 54 23 – Thermoplastic-Polyolefin Roofing

07 54 23.10 Thermoplastic Polyolefin Roofing (T.P.O.)

		Crew	Daily Output	Labor-Hours	Unit	Material	Labor	Equipment	Total	Total Incl O&P
0010	THERMOPLASTIC POLYOLEFIN ROOFING (T.P.O.)									
0100	45 mil, loose laid & ballasted with stone (1/2 ton/sq.)	G-5	51	.784	Sq.	87.50	30.50	3.60	121.60	152
0120	Fully adhered		25	1.600		78.50	62.50	7.35	148.35	200
0140	Mechanically attached		34	1.176		77.50	46	5.40	128.90	170
0160	Self adhered		35	1.143		85	44.50	5.25	134.75	175
0180	60 mil membrane, heat welded seams, ballasted		50	.800		101	31.50	3.67	136.17	168
0200	Fully adhered		25	1.600		91.50	62.50	7.35	161.35	215
0220	Mechanically attached		34	1.176		95	46	5.40	146.40	188
0240	Self adhered		35	1.143		104	44.50	5.25	153.75	197

07 54 30 – Ketone Ethylene Ester Roofing

07 54 30.10 Ketone Ethylene Ester Roofing

		Crew	Daily Output	Labor-Hours	Unit	Material	Labor	Equipment	Total	Total Incl O&P
0010	KETONE ETHYLENE ESTER ROOFING									
0100	Ketone ethylene ester roofing, 50 mil, fully adhered	G-5	26	1.538	Sq.	193	60	7.05	260.05	320
0120	Mechanically attached		35	1.143		125	44.50	5.25	174.75	219
0140	Ballasted with stone		51	.784		131	30.50	3.60	165.10	201
0160	50 mil, fleece backed, adhered w/hot asphalt	G-1	26	2.154		154	87	20.50	261.50	340
0180	Accessories, pipe boot	1 Rofc	32	.250	Ea.	24.50	10.80		35.30	45.50
0200	Pre-formed corners		32	.250	"	8.50	10.80		19.30	27.50
0220	Ketone clad metal, including up to 4 bends		330	.024	S.F.	3.93	1.05		4.98	6.10
0240	Walkway pad	2 Rofc	800	.020	"	4.12	.86		4.98	6
0260	Stripping material	1 Rofc	310	.026	L.F.	.95	1.11		2.06	2.94

07 55 Protected Membrane Roofing

07 55 10 – Protected Membrane Roofing Components

07 55 10.10 Protected Membrane Roofing Components	Crew	Daily Output	Labor-Hours	Unit	Material	2017 Bare Costs Labor	Equipment	Total	Total Incl O&P
0010 **PROTECTED MEMBRANE ROOFING COMPONENTS**									
0100 Choose roofing membrane from 07 50									
0120 Then choose roof deck insulation from 07 22									
0130 Filter fabric	2 Rofc	10000	.002	S.F.	.09	.07		.16	.22
0140 Ballast, 3/8" - 1/2" in place	G-1	36	1.556	Ton	18.95	62.50	14.70	96.15	143
0150 3/4" - 1-1/2" in place	"	36	1.556	"	18.95	62.50	14.70	96.15	143
0200 2" concrete blocks, natural	1 Clab	115	.070	S.F.	3.37	2.72		6.09	7.90
0210 Colors	"	115	.070	"	3.73	2.72		6.45	8.25

07 56 Fluid-Applied Roofing

07 56 10 – Fluid-Applied Roofing Elastomers

07 56 10.10 Elastomeric Roofing

	Crew	Daily Output	Labor-Hours	Unit	Material	2017 Bare Costs Labor	Equipment	Total	Total Incl O&P
0010 **ELASTOMERIC ROOFING**									
0020 Acrylic, 44% solids, 2 coats, on corrugated metal	2 Rofc	2400	.007	S.F.	.56	.29		.85	1.11
0025 On smooth metal		3000	.005		.45	.23		.68	.89
0030 On foam or modified bitumen		1500	.011		.90	.46		1.36	1.77
0035 On concrete		1500	.011		.90	.46		1.36	1.77
0040 On tar and gravel		1500	.011		.90	.46		1.36	1.77
0045 36% solids, 2 coats, on corrugated metal		2400	.007		.59	.29		.88	1.14
0050 On smooth metal		3000	.005		.47	.23		.70	.91
0055 On foam or modified bitumen		1500	.011		.95	.46		1.41	1.82
0060 On concrete		1500	.011		.95	.46		1.41	1.82
0065 On tar and gravel		1500	.011		.95	.46		1.41	1.82
0070 Primer if required, 2 coats on corrugated metal		2400	.007		.50	.29		.79	1.04
0075 On smooth metal		3000	.005		.50	.23		.73	.94
0080 On foam or modified bitumen		1500	.011		.74	.46		1.20	1.60
0085 On concrete		1500	.011		.56	.46		1.02	1.39
0090 On tar & gravel/rolled roof		1500	.011		.74	.46		1.20	1.60
0110 Acrylic rubber, fluid applied, 20 mils thick	G-5	2000	.020		2.20	.78	.09	3.07	3.85
0120 50 mils, reinforced		1200	.033		3.25	1.30	.15	4.70	5.95
0130 For walking surface, add		900	.044		1.20	1.74	.20	3.14	4.50
0300 Neoprene, fluid applied, 20 mil thick, not reinforced	G-1	1135	.049		1.35	1.99	.47	3.81	5.40
0600 Non-woven polyester, reinforced		960	.058		1.48	2.35	.55	4.38	6.25
0700 5 coat neoprene deck, 60 mil thick, under 10,000 S.F.		325	.172		4.50	6.95	1.63	13.08	18.55
0900 Over 10,000 S.F.		625	.090		4.50	3.61	.85	8.96	12.05

07 57 Coated Foamed Roofing

07 57 13 – Sprayed Polyurethane Foam Roofing

07 57 13.10 Sprayed Polyurethane Foam Roofing (S.P.F.)

	Crew	Daily Output	Labor-Hours	Unit	Material	2017 Bare Costs Labor	Equipment	Total	Total Incl O&P
0010 **SPRAYED POLYURETHANE FOAM ROOFING (S.P.F.)**									
0100 Primer for metal substrate (when required)	G-2A	3000	.008	S.F.	.49	.31	.22	1.02	1.29
0200 Primer for non-metal substrate (when required)		3000	.008		.19	.31	.22	.72	.96
0300 Closed cell spray, polyurethane foam, 3 lb. per C.F. density, 1", R6.7		15000	.002		.65	.06	.04	.75	.87
0400 2", R13.4		13125	.002		1.30	.07	.05	1.42	1.60
0500 3", R18.6		11485	.002		1.96	.08	.06	2.10	2.34
0550 4", R24.8		10080	.002		2.61	.09	.07	2.77	3.09
0700 Spray-on silicone coating		2500	.010		1.24	.37	.27	1.88	2.26
0800 Warranty 5-20 year manufacturer's								.15	.15
0900 Warranty 20 year, no dollar limit								.20	.20

07 58 Roll Roofing

07 58 10 – Asphalt Roll Roofing

07 58 10.10 Roll Roofing	Crew	Daily Output	Labor-Hours	Unit	Material	2017 Bare Costs Labor	Equipment	Total	Total Incl O&P
0010 **ROLL ROOFING**									
0100 Asphalt, mineral surface									
0200 1 ply #15 organic felt, 1 ply mineral surfaced									
0300 Selvage roofing, lap 19", nailed & mopped	G-1	27	2.074	Sq.	76.50	83.50	19.60	179.60	248
0400 3 plies glass fiber felt (type IV), 1 ply mineral surfaced									
0500 Selvage roofing, lapped 19", mopped	G-1	25	2.240	Sq.	132	90	21	243	320
0600 Coated glass fiber base sheet, 2 plies of glass fiber									
0700 Felt (type IV), 1 ply mineral surfaced selvage									
0800 Roofing, lapped 19", mopped	G-1	25	2.240	Sq.	139	90	21	250	330
0900 On nailable decks	"	24	2.333	"	127	94	22	243	325
1000 3 plies glass fiber felt (type III), 1 ply mineral surfaced									
1100 Selvage roofing, lapped 19", mopped	G-1	25	2.240	Sq.	132	90	21	243	320

07 61 Sheet Metal Roofing

07 61 13 – Standing Seam Sheet Metal Roofing

07 61 13.10 Standing Seam Sheet Metal Roofing, Field Fab.

	Crew	Daily Output	Labor-Hours	Unit	Material	2017 Bare Costs Labor	Equipment	Total	Total Incl O&P
0010 **STANDING SEAM SHEET METAL ROOFING, FIELD FABRICATED**									
0400 Copper standing seam roofing, over 10 squares, 16 oz., 125 lb. per sq.	1 Shee	1.30	6.154	Sq.	990	355		1,345	1,650
0600 18 oz., 140 lb. per sq.		1.20	6.667		1,100	385		1,485	1,825
0700 20 oz., 150 lb. per sq.		1.10	7.273		1,325	420		1,745	2,100
1200 For abnormal conditions or small areas, add					25%	100%			
1300 For lead-coated copper, add					25%				

07 61 16 – Batten Seam Sheet Metal Roofing

07 61 16.10 Batten Seam Sheet Metal Roofing, Field Fabricated

	Crew	Daily Output	Labor-Hours	Unit	Material	2017 Bare Costs Labor	Equipment	Total	Total Incl O&P
0010 **BATTEN SEAM SHEET METAL ROOFING, FIELD FABRICATED**									
0012 Copper batten seam roofing, over 10 sq., 16 oz., 130 lb. per sq.	1 Shee	1.10	7.273	Sq.	1,250	420		1,670	2,025
0020 Lead batten seam roofing, 5 lb. per S.F.		1.20	6.667		1,400	385		1,785	2,125
0100 Zinc/copper alloy batten seam roofing, .020" thick		1.20	6.667		1,325	385		1,710	2,050
0200 Copper roofing, batten seam, over 10 sq., 18 oz., 145 lb. per sq.		1	8		1,400	465		1,865	2,250
0300 20 oz., 160 lb. per sq.		1	8		1,650	465		2,115	2,525
0500 Stainless steel batten seam roofing, type 304, 28 ga.		1.20	6.667		630	385		1,015	1,275
0600 26 ga.		1.15	6.957		675	405		1,080	1,350
0800 Zinc, copper alloy roofing, batten seam, .027" thick		1.15	6.957		1,675	405		2,080	2,450
0900 .032" thick		1.10	7.273		1,900	420		2,320	2,750
1000 .040" thick		1.05	7.619		2,375	440		2,815	3,300

07 61 19 – Flat Seam Sheet Metal Roofing

07 61 19.10 Flat Seam Sheet Metal Roofing, Field Fabricated

	Crew	Daily Output	Labor-Hours	Unit	Material	2017 Bare Costs Labor	Equipment	Total	Total Incl O&P
0010 **FLAT SEAM SHEET METAL ROOFING, FIELD FABRICATED**									
0900 Copper flat seam roofing, over 10 squares, 16 oz., 115 lb./sq.	1 Shee	1.20	6.667	Sq.	920	385		1,305	1,600
0950 18 oz., 130 lb./sq.		1.15	6.957		1,025	405		1,430	1,750
1000 20 oz., 145 lb./sq.		1.10	7.273		1,225	420		1,645	2,000
1008 Zinc flat seam roofing, .020" thick		1.20	6.667		1,125	385		1,510	1,850
1010 .027" thick		1.15	6.957		1,425	405		1,830	2,200
1020 .032" thick		1.12	7.143		1,625	415		2,040	2,400
1030 .040" thick		1.05	7.619		2,025	440		2,465	2,900
1100 Lead flat seam roofing, 5 lb. per S.F.		1.30	6.154		1,200	355		1,555	1,850

07 65 Flexible Flashing

07 65 10 – Sheet Metal Flashing

07 65 10.10 Sheet Metal Flashing and Counter Flashing

	07 65 10.10 Sheet Metal Flashing and Counter Flashing	Crew	Daily Output	Labor-Hours	Unit	Material	2017 Bare Costs Labor	Equipment	Total	Total Incl O&P
0010	**SHEET METAL FLASHING AND COUNTER FLASHING**									
0011	Including up to 4 bends									
0020	Aluminum, mill finish, .013" thick	1 Rofc	145	.055	S.F.	.83	2.38		3.21	4.96
0030	.016" thick		145	.055		.98	2.38		3.36	5.15
0060	.019" thick		145	.055		1.27	2.38		3.65	5.45
0100	.032" thick		145	.055		1.38	2.38		3.76	5.55
0200	.040" thick		145	.055		2.35	2.38		4.73	6.65
0300	.050" thick		145	.055		2.75	2.38		5.13	7.10
0325	Mill finish 5" x 7" step flashing, .016" thick		1920	.004	Ea.	.15	.18		.33	.48
0350	Mill finish 12" x 12" step flashing, .016" thick		1600	.005	"	.55	.22		.77	.98
0400	Painted finish, add				S.F.	.33			.33	.36
1000	Mastic-coated 2 sides, .005" thick	1 Rofc	330	.024		1.86	1.05		2.91	3.83
1100	.016" thick		330	.024		2.05	1.05		3.10	4.04
1600	Copper, 16 oz. sheets, under 1000 lb.		115	.070		7.90	3		10.90	13.80
1700	Over 4000 lb.		155	.052		7.90	2.23		10.13	12.50
1900	20 oz. sheets, under 1000 lb.		110	.073		10.50	3.14		13.64	16.90
2000	Over 4000 lb.		145	.055		10	2.38		12.38	15
2200	24 oz. sheets, under 1000 lb.		105	.076		14.75	3.29		18.04	22
2300	Over 4000 lb.		135	.059		14	2.56		16.56	19.75
2500	32 oz. sheets, under 1000 lb.		100	.080		19	3.45		22.45	27
2600	Over 4000 lb.		130	.062		18.05	2.66		20.71	24.50
2700	W shape for valleys, 16 oz., 24" wide		100	.080	L.F.	16.25	3.45		19.70	24
5800	Lead, 2.5 lb. per S.F., up to 12" wide		135	.059	S.F.	5.50	2.56		8.06	10.40
5900	Over 12" wide		135	.059		4	2.56		6.56	8.75
8900	Stainless steel sheets, 32 ga.		155	.052		3.35	2.23		5.58	7.50
9000	28 ga.		155	.052		4.25	2.23		6.48	8.45
9100	26 ga.		155	.052		4.50	2.23		6.73	8.75
9200	24 ga.		155	.052		5	2.23		7.23	9.30
9290	For mechanically keyed flashing, add					40%				
9320	Steel sheets, galvanized, 20 ga.	1 Rofc	130	.062	S.F.	1.27	2.66		3.93	5.90
9322	22 ga.		135	.059		1.23	2.56		3.79	5.70
9324	24 ga.		140	.057		.93	2.47		3.40	5.20
9326	26 ga.		148	.054		.81	2.33		3.14	4.85
9328	28 ga.		155	.052		.70	2.23		2.93	4.56
9340	30 ga.		160	.050		.59	2.16		2.75	4.32
9400	Terne coated stainless steel, .015" thick, 28 ga		155	.052		8.10	2.23		10.33	12.70
9500	.018" thick, 26 ga		155	.052		9.05	2.23		11.28	13.75
9600	Zinc and copper alloy (brass), .020" thick		155	.052		10.25	2.23		12.48	15.10
9700	.027" thick		155	.052		12.25	2.23		14.48	17.30
9800	.032" thick		155	.052		15.50	2.23		17.73	21
9900	.040" thick		155	.052		20.50	2.23		22.73	26.50

07 65 12 – Fabric and Mastic Flashings

07 65 12.10 Fabric and Mastic Flashing and Counter Flashing

	07 65 12.10 Fabric and Mastic Flashing and Counter Flashing	Crew	Daily Output	Labor-Hours	Unit	Material	2017 Bare Costs Labor	Equipment	Total	Total Incl O&P
0010	**FABRIC AND MASTIC FLASHING AND COUNTER FLASHING**									
1300	Asphalt flashing cement, 5 gallon				Gal.	9.75			9.75	10.75
4900	Fabric, asphalt-saturated cotton, specification grade	1 Rofc	35	.229	S.Y.	3.05	9.85		12.90	20
5000	Utility grade		35	.229		1.45	9.85		11.30	18.35
5300	Close-mesh fabric, saturated, 17 oz. per S.Y.		35	.229		2.20	9.85		12.05	19.15
5500	Fiberglass, resin-coated		35	.229		1	9.85		10.85	17.85
8500	Shower pan, bituminous membrane, 7 oz.		155	.052	S.F.	1.75	2.23		3.98	5.70

For customer support on your Building Construction Costs with RSMeans Data, call 800.448.8182.

249

07 65 Flexible Flashing

07 65 13 – Laminated Sheet Flashing

07 65 13.10 Laminated Sheet Flashing

	07 65 13.10 Laminated Sheet Flashing	Crew	Daily Output	Labor-Hours	Unit	Material	2017 Bare Costs Labor	Equipment	Total	Total Incl O&P
0010	**LAMINATED SHEET FLASHING**, Including up to 4 bends									
0500	Aluminum, fabric-backed 2 sides, mill finish, .004" thick	1 Rofc	330	.024	S.F.	1.60	1.05		2.65	3.54
0700	.005" thick		330	.024		1.85	1.05		2.90	3.82
0750	Mastic-backed, self adhesive		460	.017		3.60	.75		4.35	5.25
0800	Mastic-coated 2 sides, .004" thick		330	.024		1.60	1.05		2.65	3.54
2800	Copper, paperbacked 1 side, 2 oz.		330	.024		1.84	1.05		2.89	3.80
2900	3 oz.		330	.024		2.20	1.05		3.25	4.20
3100	Paperbacked 2 sides, 2 oz.		330	.024		2.07	1.05		3.12	4.06
3150	3 oz.		330	.024		2.30	1.05		3.35	4.31
3200	5 oz.		330	.024		3.38	1.05		4.43	5.50
3250	7 oz.		330	.024		6.70	1.05		7.75	9.15
3400	Mastic-backed 2 sides, copper, 2 oz.		330	.024		2	1.05		3.05	3.98
3500	3 oz.		330	.024		2.50	1.05		3.55	4.53
3700	5 oz.		330	.024		3.80	1.05		4.85	5.95
3800	Fabric-backed 2 sides, copper, 2 oz.		330	.024		2.10	1.05		3.15	4.09
4000	3 oz.		330	.024		2.80	1.05		3.85	4.86
4100	5 oz.		330	.024		3.90	1.05		4.95	6.05
4300	Copper-clad stainless steel, .015" thick, under 500 lb.		115	.070		6.50	3		9.50	12.25
4400	Over 2000 lb.		155	.052		6.75	2.23		8.98	11.25
4600	.018" thick, under 500 lb.		100	.080		7.95	3.45		11.40	14.60
4700	Over 2000 lb.		145	.055		7.75	2.38		10.13	12.60
8550	Shower pan, 3 ply copper and fabric, 3 oz.		155	.052		4	2.23		6.23	8.20
8600	7 oz.		155	.052		4.20	2.23		6.43	8.40
9300	Stainless steel, paperbacked 2 sides, .005" thick		330	.024		3.92	1.05		4.97	6.10

07 65 19 – Plastic Sheet Flashing

07 65 19.10 Plastic Sheet Flashing and Counter Flashing

	07 65 19.10 Plastic Sheet Flashing and Counter Flashing	Crew	Daily Output	Labor-Hours	Unit	Material	2017 Bare Costs Labor	Equipment	Total	Total Incl O&P
0010	**PLASTIC SHEET FLASHING AND COUNTER FLASHING**									
7300	Polyvinyl chloride, black, 10 mil	1 Rofc	285	.028	S.F.	.21	1.21		1.42	2.29
7400	20 mil		285	.028		.26	1.21		1.47	2.35
7600	30 mil		285	.028		.34	1.21		1.55	2.43
7700	60 mil		285	.028		.82	1.21		2.03	2.96
7900	Black or white for exposed roofs, 60 mil		285	.028		1.80	1.21		3.01	4.04
8060	PVC tape, 5" x 45 mils, for joint covers, 100 L.F./roll				Ea.	175			175	193
8850	Polyvinyl chloride, 30 mil	1 Rofc	160	.050	S.F.	1.50	2.16		3.66	5.30

07 65 23 – Rubber Sheet Flashing

07 65 23.10 Rubber Sheet Flashing and Counter Flashing

	07 65 23.10 Rubber Sheet Flashing and Counter Flashing	Crew	Daily Output	Labor-Hours	Unit	Material	2017 Bare Costs Labor	Equipment	Total	Total Incl O&P
0010	**RUBBER SHEET FLASHING AND COUNTER FLASHING**									
4810	EPDM 90 mils, 1" diameter pipe flashing	1 Rofc	32	.250	Ea.	19.50	10.80		30.30	40
4820	2" diameter		30	.267		19.45	11.50		30.95	41
4830	3" diameter		28	.286		21	12.35		33.35	44
4840	4" diameter		24	.333		24	14.40		38.40	51
4850	6" diameter		22	.364		24	15.70		39.70	53
8100	Rubber, butyl, 1/32" thick		285	.028	S.F.	2	1.21		3.21	4.26
8200	1/16" thick		285	.028		2.75	1.21		3.96	5.10
8300	Neoprene, cured, 1/16" thick		285	.028		2.60	1.21		3.81	4.92
8400	1/8" thick		285	.028		6.10	1.21		7.31	8.75

07 65 Flexible Flashing

07 65 26 – Self-Adhering Sheet Flashing

07 65 26.10 Self-Adhering Sheet or Roll Flashing	Crew	Daily Output	Labor-Hours	Unit	Material	2017 Bare Costs Labor	Equipment	Total	Total Incl O&P
0010 **SELF-ADHERING SHEET OR ROLL FLASHING**									
0020 Self-adhered flashing, 25 mil cross laminated HDPE, 4" wide	1 Rofc	960	.008	L.F.	.20	.36		.56	.83
0040 6" wide		896	.009		.30	.39		.69	.98
0060 9" wide		832	.010		.44	.42		.86	1.20
0080 12" wide		768	.010		.59	.45		1.04	1.41

07 71 Roof Specialties

07 71 19 – Manufactured Gravel Stops and Fasciae

07 71 19.10 Gravel Stop

		Crew	Daily Output	Labor-Hours	Unit	Material	2017 Bare Costs Labor	Equipment	Total	Total Incl O&P
0010	**GRAVEL STOP**									
0020	Aluminum, .050" thick, 4" face height, mill finish	1 Shee	145	.055	L.F.	6.55	3.20		9.75	12.10
0080	Duranodic finish		145	.055		7.40	3.20		10.60	13.05
0100	Painted		145	.055		7.50	3.20		10.70	13.15
0300	6" face height		135	.059		6.70	3.44		10.14	12.65
0350	Duranodic finish		135	.059		7.95	3.44		11.39	14
0400	Painted		135	.059		8.75	3.44		12.19	14.90
0600	8" face height		125	.064		7.65	3.72		11.37	14.10
0650	Duranodic finish		125	.064		9.15	3.72		12.87	15.75
0700	Painted		125	.064		9.75	3.72		13.47	16.45
0900	12" face height, .080" thick, 2 piece		100	.080		10.80	4.64		15.44	18.95
0950	Duranodic finish		100	.080		11	4.64		15.64	19.20
1000	Painted		100	.080		13.50	4.64		18.14	22
1200	Copper, 16 oz., 3" face height		145	.055		24.50	3.20		27.70	31.50
1300	6" face height		135	.059		33.50	3.44		36.94	42.50
1350	Galv steel, 24 ga., 4" leg, plain, with continuous cleat, 4" face		145	.055		6.35	3.20		9.55	11.90
1360	6" face height		145	.055		6.55	3.20		9.75	12.10
1500	Polyvinyl chloride, 6" face height		135	.059		5.75	3.44		9.19	11.60
1600	9" face height		125	.064		6.75	3.72		10.47	13.15
1800	Stainless steel, 24 ga., 6" face height		135	.059		15.50	3.44		18.94	22.50
1900	12" face height		100	.080		23	4.64		27.64	32
2100	20 ga., 6" face height		135	.059		19.50	3.44		22.94	27
2200	12" face height		100	.080		27.50	4.64		32.14	37

07 71 19.30 Fascia

		Crew	Daily Output	Labor-Hours	Unit	Material	2017 Bare Costs Labor	Equipment	Total	Total Incl O&P
0010	**FASCIA**									
0100	Aluminum, reverse board and batten, .032" thick, colored, no furring incl	1 Shee	145	.055	S.F.	6.85	3.20		10.05	12.45
0300	Steel, galv and enameled, stock, no furring, long panels		145	.055		5.30	3.20		8.50	10.75
0600	Short panels		115	.070		5.95	4.04		9.99	12.75

07 71 23 – Manufactured Gutters and Downspouts

07 71 23.10 Downspouts

		Crew	Daily Output	Labor-Hours	Unit	Material	2017 Bare Costs Labor	Equipment	Total	Total Incl O&P
0010	**DOWNSPOUTS**									
0020	Aluminum, embossed, .020" thick, 2" x 3"	1 Shee	190	.042	L.F.	1.08	2.44		3.52	4.93
0100	Enameled		190	.042		1.42	2.44		3.86	5.30
0300	.024" thick, 2" x 3"		180	.044		2.10	2.58		4.68	6.25
0400	3" x 4"		140	.057		2.42	3.32		5.74	7.75
0600	Round, corrugated aluminum, 3" diameter, .020" thick		190	.042		2.03	2.44		4.47	5.95
0700	4" diameter, .025" thick		140	.057		3	3.32		6.32	8.40
0900	Wire strainer, round, 2" diameter		155	.052	Ea.	1.90	3		4.90	6.70
1000	4" diameter		155	.052		2.50	3		5.50	7.35
1200	Rectangular, perforated, 2" x 3"		145	.055		2.30	3.20		5.50	7.45
1300	3" x 4"		145	.055		3.30	3.20		6.50	8.55

07 71 23 – Manufactured Gutters and Downspouts

07 71 23.10 Downspouts		Crew	Daily Output	Labor-Hours	Unit	Material	2017 Bare Costs Labor	Equipment	Total	Total Incl O&P
1500	Copper, round, 16 oz., stock, 2" diameter	1 Shee	190	.042	L.F.	7.70	2.44		10.14	12.20
1600	3" diameter		190	.042		8.15	2.44		10.59	12.70
1800	4" diameter		145	.055		9.85	3.20		13.05	15.75
1900	5" diameter		130	.062		15.85	3.57		19.42	23
2100	Rectangular, corrugated copper, stock, 2" x 3"		190	.042		8.80	2.44		11.24	13.45
2200	3" x 4"		145	.055		10.50	3.20		13.70	16.45
2400	Rectangular, plain copper, stock, 2" x 3"		190	.042		11.25	2.44		13.69	16.15
2500	3" x 4"		145	.055		14	3.20		17.20	20.50
2700	Wire strainers, rectangular, 2" x 3"		145	.055	Ea.	17.20	3.20		20.40	24
2800	3" x 4"		145	.055		17.95	3.20		21.15	24.50
3000	Round, 2" diameter		145	.055		6.50	3.20		9.70	12.05
3100	3" diameter		145	.055		7.25	3.20		10.45	12.90
3300	4" diameter		145	.055		12.25	3.20		15.45	18.40
3400	5" diameter		115	.070		22.50	4.04		26.54	31
3600	Lead-coated copper, round, stock, 2" diameter		190	.042	L.F.	22.50	2.44		24.94	28
3700	3" diameter		190	.042		23	2.44		25.44	28.50
3900	4" diameter		145	.055		23.50	3.20		26.70	31
4000	5" diameter, corrugated		130	.062		23	3.57		26.57	31
4200	6" diameter, corrugated		105	.076		30.50	4.42		34.92	40.50
4300	Rectangular, corrugated, stock, 2" x 3"		190	.042		15.75	2.44		18.19	21
4500	Plain, stock, 2" x 3"		190	.042		24	2.44		26.44	30
4600	3" x 4"		145	.055		33.50	3.20		36.70	41.50
4800	Steel, galvanized, round, corrugated, 2" or 3" diameter, 28 ga.		190	.042		2.15	2.44		4.59	6.10
4900	4" diameter, 28 ga.		145	.055		2.59	3.20		5.79	7.75
5100	5" diameter, 26 ga.		130	.062		3.78	3.57		7.35	9.60
5400	6" diameter, 28 ga.		105	.076		4.25	4.42		8.67	11.45
5500	26 ga.		105	.076		3.60	4.42		8.02	10.70
5700	Rectangular, corrugated, 28 ga., 2" x 3"		190	.042		2	2.44		4.44	5.95
5800	3" x 4"		145	.055		1.85	3.20		5.05	6.95
6000	Rectangular, plain, 28 ga., galvanized, 2" x 3"		190	.042		3.75	2.44		6.19	7.85
6100	3" x 4"		145	.055		4.12	3.20		7.32	9.45
6300	Epoxy painted, 24 ga., corrugated, 2" x 3"		190	.042		2.37	2.44		4.81	6.35
6400	3" x 4"		145	.055		2.88	3.20		6.08	8.05
6600	Wire strainers, rectangular, 2" x 3"		145	.055	Ea.	17.10	3.20		20.30	23.50
6700	3" x 4"		145	.055		17.85	3.20		21.05	24.50
6900	Round strainers, 2" or 3" diameter		145	.055		3.95	3.20		7.15	9.25
7000	4" diameter		145	.055		5.95	3.20		9.15	11.45
7800	Stainless steel tubing, schedule 5, 2" x 3" or 3" diameter		190	.042	L.F.	33.50	2.44		35.94	40.50
7900	3" x 4" or 4" diameter		145	.055		45	3.20		48.20	54.50
8100	4" x 5" or 5" diameter		135	.059		101	3.44		104.44	116
8200	Vinyl, rectangular, 2" x 3"		210	.038		2.15	2.21		4.36	5.75
8300	Round, 2-1/2"		220	.036		1.35	2.11		3.46	4.72

07 71 23.20 Downspout Elbows

		Crew	Daily Output	Labor-Hours	Unit	Material	Labor	Equipment	Total	Total Incl O&P
0010	**DOWNSPOUT ELBOWS**									
0020	Aluminum, embossed, 2" x 3", .020" thick	1 Shee	100	.080	Ea.	.96	4.64		5.60	8.15
0100	Enameled		100	.080		1.75	4.64		6.39	9.05
0200	Embossed, 3" x 4", .025" thick		100	.080		4.55	4.64		9.19	12.10
0300	Enameled		100	.080		3.85	4.64		8.49	11.35
0400	Embossed, corrugated, 3" diameter, .020" thick		100	.080		2.75	4.64		7.39	10.15
0500	4" diameter, .025" thick		100	.080		5.90	4.64		10.54	13.60
0600	Copper, 16 oz., 2" diameter		100	.080		9.50	4.64		14.14	17.55
0700	3" diameter		100	.080		9.15	4.64		13.79	17.15

07 71 Roof Specialties

07 71 23 – Manufactured Gutters and Downspouts

07 71 23.20 Downspout Elbows

		Crew	Daily Output	Labor-Hours	Unit	Material	2017 Bare Costs Labor	2017 Bare Costs Equipment	Total	Total Incl O&P
0800	4" diameter	1 Shee	100	.080	Ea.	14	4.64		18.64	22.50
1000	Rectangular, 2" x 3" corrugated		100	.080		9.65	4.64		14.29	17.70
1100	3" x 4" corrugated		100	.080		13.35	4.64		17.99	22
1300	Vinyl, 2-1/2" diameter, 45 or 75 degree bend		100	.080		3.99	4.64		8.63	11.50
1400	Tee Y junction	↓	75	.107	↓	13	6.20		19.20	24

07 71 23.30 Gutters

		Crew	Daily Output	Labor-Hours	Unit	Material	2017 Bare Costs Labor	2017 Bare Costs Equipment	Total	Total Incl O&P
0010	**GUTTERS**									
0012	Aluminum, stock units, 5" K type, .027" thick, plain	1 Shee	125	.064	L.F.	2.86	3.72		6.58	8.85
0100	Enameled		125	.064		2.86	3.72		6.58	8.85
0300	5" K type, .032" thick, plain		125	.064		3.53	3.72		7.25	9.60
0400	Enameled		125	.064		3.64	3.72		7.36	9.70
0700	Copper, half round, 16 oz., stock units, 4" wide		125	.064		9.60	3.72		13.32	16.25
0900	5" wide		125	.064		7.40	3.72		11.12	13.85
1000	6" wide		118	.068		12.10	3.94		16.04	19.30
1200	K type, 16 oz., stock, 5" wide		125	.064		8.10	3.72		11.82	14.60
1300	6" wide		125	.064		8.60	3.72		12.32	15.15
1500	Lead coated copper, 16 oz., half round, stock, 4" wide		125	.064		15	3.72		18.72	22
1600	6" wide		118	.068		22.50	3.94		26.44	30.50
1800	K type, stock, 5" wide		125	.064		18.50	3.72		22.22	26
1900	6" wide		125	.064		19.90	3.72		23.62	27.50
2100	Copper clad stainless steel, K type, 5" wide		125	.064		7.75	3.72		11.47	14.25
2200	6" wide		125	.064		9.80	3.72		13.52	16.50
2400	Steel, galv, half round or box, 28 ga., 5" wide, plain		125	.064		2.15	3.72		5.87	8.05
2500	Enameled		125	.064		2.23	3.72		5.95	8.15
2700	26 ga., stock, 5" wide		125	.064		2.45	3.72		6.17	8.40
2800	6" wide	↓	125	.064		2.83	3.72		6.55	8.80
3000	Vinyl, O.G., 4" wide	1 Carp	115	.070		1.33	3.43		4.76	6.70
3100	5" wide		115	.070		1.62	3.43		5.05	7.05
3200	4" half round, stock units	↓	115	.070	↓	1.38	3.43		4.81	6.75
3250	Joint connectors				Ea.	3			3	3.30
3300	Wood, clear treated cedar, fir or hemlock, 3" x 4"	1 Carp	100	.080	L.F.	10	3.94		13.94	17.05
3400	4" x 5"	"	100	.080	"	16.75	3.94		20.69	24.50
5000	Accessories, end cap, K type, aluminum 5"	1 Shee	625	.013	Ea.	.71	.74		1.45	1.92
5010	6"		625	.013		1.54	.74		2.28	2.83
5020	Copper, 5"		625	.013		3.25	.74		3.99	4.72
5030	6"		625	.013		3.75	.74		4.49	5.25
5040	Lead coated copper, 5"		625	.013		12.50	.74		13.24	14.90
5050	6"		625	.013		13.50	.74		14.24	16
5060	Copper clad stainless steel, 5"		625	.013		3	.74		3.74	4.44
5070	6"		625	.013		3.60	.74		4.34	5.10
5080	Galvanized steel, 5"		625	.013		1.35	.74		2.09	2.63
5090	6"	↓	625	.013		2.20	.74		2.94	3.56
5100	Vinyl, 4"	1 Carp	625	.013		13.25	.63		13.88	15.55
5110	5"	"	625	.013		13.25	.63		13.88	15.55
5120	Half round, copper, 4"	1 Shee	625	.013		4.55	.74		5.29	6.15
5130	5"		625	.013		4.55	.74		5.29	6.15
5140	6"		625	.013		7.75	.74		8.49	9.70
5150	Lead coated copper, 5"		625	.013		14.20	.74		14.94	16.75
5160	6"		625	.013		21.50	.74		22.24	24.50
5170	Copper clad stainless steel, 5"		625	.013		4.50	.74		5.24	6.10
5180	6"		625	.013		4.50	.74		5.24	6.10
5190	Galvanized steel, 5"	↓	625	.013	↓	2.30	.74		3.04	3.67

07 71 23 – Manufactured Gutters and Downspouts

07 71 23.30 Gutters

07 71 23.30 Gutters		Crew	Daily Output	Labor-Hours	Unit	Material	2017 Bare Costs Labor	Equipment	Total	Total Incl O&P
5200	6"	1 Shee	625	.013	Ea.	2.85	.74		3.59	4.28
5210	Outlet, aluminum, 2" x 3"		420	.019		.65	1.11		1.76	2.41
5220	3" x 4"		420	.019		1.05	1.11		2.16	2.85
5230	2-3/8" round		420	.019		.55	1.11		1.66	2.30
5240	Copper, 2" x 3"		420	.019		6.55	1.11		7.66	8.90
5250	3" x 4"		420	.019		7.90	1.11		9.01	10.40
5260	2-3/8" round		420	.019		4.75	1.11		5.86	6.95
5270	Lead coated copper, 2" x 3"		420	.019		25.50	1.11		26.61	29.50
5280	3" x 4"		420	.019		29	1.11		30.11	33.50
5290	2-3/8" round		420	.019		25.50	1.11		26.61	29.50
5300	Copper clad stainless steel, 2" x 3"		420	.019		6.50	1.11		7.61	8.85
5310	3" x 4"		420	.019		8	1.11		9.11	10.50
5320	2-3/8" round		420	.019		4.75	1.11		5.86	6.95
5330	Galvanized steel, 2" x 3"		420	.019		3.25	1.11		4.36	5.25
5340	3" x 4"		420	.019		5	1.11		6.11	7.20
5350	2-3/8" round		420	.019		4	1.11		5.11	6.10
5360	K type mitres, aluminum		65	.123		3.10	7.15		10.25	14.35
5370	Copper		65	.123		17.50	7.15		24.65	30
5380	Lead coated copper		65	.123		54.50	7.15		61.65	71
5390	Copper clad stainless steel		65	.123		27.50	7.15		34.65	41.50
5400	Galvanized steel		65	.123		23	7.15		30.15	36
5420	Half round mitres, copper		65	.123		64	7.15		71.15	81.50
5430	Lead coated copper		65	.123		90	7.15		97.15	110
5440	Copper clad stainless steel		65	.123		56.50	7.15		63.65	73
5450	Galvanized steel		65	.123		25.50	7.15		32.65	39
5460	Vinyl mitres and outlets		65	.123		10	7.15		17.15	22
5470	Sealant		940	.009	L.F.	.01	.49		.50	.77
5480	Soldering		96	.083	"	.25	4.84		5.09	7.70

07 71 23.35 Gutter Guard

		Crew	Daily Output	Labor-Hours	Unit	Material	2017 Bare Costs Labor	Equipment	Total	Total Incl O&P
0010	**GUTTER GUARD**									
0020	6" wide strip, aluminum mesh	1 Carp	500	.016	L.F.	2.55	.79		3.34	4.02
0100	Vinyl mesh	"	500	.016	"	2.75	.79		3.54	4.24

07 71 26 – Reglets

07 71 26.10 Reglets and Accessories

		Crew	Daily Output	Labor-Hours	Unit	Material	2017 Bare Costs Labor	Equipment	Total	Total Incl O&P
0010	**REGLETS AND ACCESSORIES**									
0020	Reglet, aluminum, .025" thick, in parapet	1 Carp	225	.036	L.F.	2.05	1.75		3.80	4.94
0300	16 oz. copper		225	.036		6.50	1.75		8.25	9.85
0400	Galvanized steel, 24 ga.		225	.036		1.38	1.75		3.13	4.20
0600	Stainless steel, .020" thick		225	.036		3.85	1.75		5.60	6.90
0900	Counter flashing for above, 12" wide, .032" aluminum	1 Shee	150	.053		2.20	3.10		5.30	7.15
1200	16 oz. copper		150	.053		6.25	3.10		9.35	11.65
1300	Galvanized steel, 26 ga.		150	.053		1.30	3.10		4.40	6.15
1500	Stainless steel, .020" thick		150	.053		6.30	3.10		9.40	11.70

07 71 29 – Manufactured Roof Expansion Joints

07 71 29.10 Expansion Joints

		Crew	Daily Output	Labor-Hours	Unit	Material	2017 Bare Costs Labor	Equipment	Total	Total Incl O&P
0010	**EXPANSION JOINTS**									
0300	Butyl or neoprene center with foam insulation, metal flanges									
0400	Aluminum, .032" thick for openings to 2-1/2"	1 Rofc	165	.048	L.F.	11.75	2.09		13.84	16.50
0600	For joint openings to 3-1/2"		165	.048		11.75	2.09		13.84	16.50
0610	For joint openings to 5"		165	.048		13.75	2.09		15.84	18.70
0620	For joint openings to 8"		165	.048		16.50	2.09		18.59	21.50

07 71 Roof Specialties

07 71 29 – Manufactured Roof Expansion Joints

07 71 29.10 Expansion Joints

		Crew	Daily Output	Labor-Hours	Unit	Material	2017 Bare Costs Labor	Equipment	Total	Total Incl O&P
0700	Copper, 16 oz. for openings to 2-1/2"	1 Rofc	165	.048	L.F.	18.50	2.09		20.59	24
0900	For joint openings to 3-1/2"		165	.048		19.50	2.09		21.59	25
0910	For joint openings to 5"		165	.048		21.50	2.09		23.59	27
0920	For joint openings to 8"		165	.048		24.50	2.09		26.59	30.50
1000	Galvanized steel, 26 ga. for openings to 2-1/2"		165	.048		10	2.09		12.09	14.55
1200	For joint openings to 3-1/2"		165	.048		10	2.09		12.09	14.55
1210	For joint openings to 5"		165	.048		11.50	2.09		13.59	16.20
1220	For joint openings to 8"		165	.048		15	2.09		17.09	20
1300	Lead-coated copper, 16 oz. for openings to 2-1/2"		165	.048		34	2.09		36.09	41
1500	For joint openings to 3-1/2"		165	.048		34	2.09		36.09	41
1600	Stainless steel, .018", for openings to 2-1/2"		165	.048		13.50	2.09		15.59	18.40
1800	For joint openings to 3-1/2"		165	.048		13.50	2.09		15.59	18.40
1810	For joint openings to 5"		165	.048		14.75	2.09		16.84	19.80
1820	For joint openings to 8"		165	.048		21.50	2.09		23.59	27
1900	Neoprene, double-seal type with thick center, 4-1/2" wide		125	.064		15	2.76		17.76	21
1950	Polyethylene bellows, with galv steel flat flanges		100	.080		6.50	3.45		9.95	13
1960	With galvanized angle flanges		100	.080		7	3.45		10.45	13.55
2000	Roof joint with extruded aluminum cover, 2"	1 Shee	115	.070		28.50	4.04		32.54	37.50
2100	Roof joint, plastic curbs, foam center, standard	1 Rofc	100	.080		13.50	3.45		16.95	20.50
2200	Large	"	100	.080		17.50	3.45		20.95	25
2500	Roof to wall joint with extruded aluminum cover	1 Shee	115	.070		28.50	4.04		32.54	37.50
2700	Wall joint, closed cell foam on PVC cover, 9" wide	1 Rofc	125	.064		5.50	2.76		8.26	10.75
2800	12" wide	"	115	.070		6.55	3		9.55	12.30

07 71 43 – Drip Edge

07 71 43.10 Drip Edge, Rake Edge, Ice Belts

		Crew	Daily Output	Labor-Hours	Unit	Material	2017 Bare Costs Labor	Equipment	Total	Total Incl O&P
0010	**DRIP EDGE, RAKE EDGE, ICE BELTS**									
0020	Aluminum, .016" thick, 5" wide, mill finish	1 Carp	400	.020	L.F.	.57	.99		1.56	2.14
0100	White finish		400	.020		.63	.99		1.62	2.20
0200	8" wide, mill finish		400	.020		1.50	.99		2.49	3.16
0300	Ice belt, 28" wide, mill finish		100	.080		7.75	3.94		11.69	14.60
0310	Vented, mill finish		400	.020		1.99	.99		2.98	3.70
0320	Painted finish		400	.020		2.24	.99		3.23	3.97
0400	Galvanized, 5" wide		400	.020		.60	.99		1.59	2.17
0500	8" wide, mill finish		400	.020		.82	.99		1.81	2.41
0510	Rake edge, aluminum, 1-1/2" x 1-1/2"		400	.020		.33	.99		1.32	1.87
0520	3-1/2" x 1-1/2"		400	.020		.45	.99		1.44	2.01

07 72 Roof Accessories

07 72 23 – Relief Vents

07 72 23.10 Roof Vents

		Crew	Daily Output	Labor-Hours	Unit	Material	2017 Bare Costs Labor	Equipment	Total	Total Incl O&P
0010	**ROOF VENTS**									
0020	Mushroom shape, for built-up roofs, aluminum	1 Rofc	30	.267	Ea.	65	11.50		76.50	91
0100	PVC, 6" high	"	30	.267	"	21.50	11.50		33	43

07 72 26 – Ridge Vents

07 72 26.10 Ridge Vents and Accessories

		Crew	Daily Output	Labor-Hours	Unit	Material	2017 Bare Costs Labor	Equipment	Total	Total Incl O&P
0010	**RIDGE VENTS AND ACCESSORIES**									
0100	Aluminum strips, mill finish	1 Rofc	160	.050	L.F.	2.88	2.16		5.04	6.85
0150	Painted finish		160	.050	"	4.13	2.16		6.29	8.20
0200	Connectors		48	.167	Ea.	4.80	7.20		12	17.55
0300	End caps		48	.167	"	2.08	7.20		9.28	14.55

For customer support on your Building Construction Costs with RSMeans Data, call 800.448.8182.

255

07 72 Roof Accessories

07 72 26 – Ridge Vents

07 72 26.10 Ridge Vents and Accessories

		Crew	Daily Output	Labor-Hours	Unit	Material	2017 Bare Costs Labor	Equipment	Total	Total Incl O&P
0400	Galvanized strips	1 Rofc	160	.050	L.F.	3.63	2.16		5.79	7.65
0430	Molded polyethylene, shingles not included		160	.050	"	2.75	2.16		4.91	6.70
0440	End plugs		48	.167	Ea.	2.08	7.20		9.28	14.55
0450	Flexible roll, shingles not included		160	.050	L.F.	2.35	2.16		4.51	6.25
2300	Ridge vent strip, mill finish	1 Shee	155	.052	"	3.95	3		6.95	8.95

07 72 33 – Roof Hatches

07 72 33.10 Roof Hatch Options

		Crew	Daily Output	Labor-Hours	Unit	Material	2017 Bare Costs Labor	Equipment	Total	Total Incl O&P
0010	**ROOF HATCH OPTIONS**									
0500	2'-6" x 3', aluminum curb and cover	G-3	10	3.200	Ea.	950	156		1,106	1,300
0520	Galvanized steel curb and aluminum cover		10	3.200		800	156		956	1,125
0540	Galvanized steel curb and cover		10	3.200		695	156		851	1,000
0600	2'-6" x 4'-6", aluminum curb and cover		9	3.556		1,150	173		1,323	1,550
0800	Galvanized steel curb and aluminum cover		9	3.556		900	173		1,073	1,250
0900	Galvanized steel curb and cover		9	3.556		940	173		1,113	1,300
1100	4' x 4' aluminum curb and cover		8	4		1,700	194		1,894	2,175
1120	Galvanized steel curb and aluminum cover		8	4		1,625	194		1,819	2,100
1140	Galvanized steel curb and cover		8	4		1,200	194		1,394	1,625
1200	2'-6" x 8'-0", aluminum curb and cover		6.60	4.848		1,950	236		2,186	2,500
1400	Galvanized steel curb and aluminum cover		6.60	4.848		1,800	236		2,036	2,325
1500	Galvanized steel curb and cover		6.60	4.848		1,450	236		1,686	1,950
1800	For plexiglass panels, 2'-6" x 3'-0", add to above					450			450	495

07 72 36 – Smoke Vents

07 72 36.10 Smoke Hatches

		Crew	Daily Output	Labor-Hours	Unit	Material	2017 Bare Costs Labor	Equipment	Total	Total Incl O&P
0010	**SMOKE HATCHES**									
0200	For 3'-0" long, add to roof hatches from Section 07 72 33.10				Ea.	25%	5%			
0250	For 4'-0" long, add to roof hatches from Section 07 72 33.10					20%	5%			
0300	For 8'-0" long, add to roof hatches from Section 07 72 33.10					10%	5%			

07 72 36.20 Smoke Vent Options

		Crew	Daily Output	Labor-Hours	Unit	Material	2017 Bare Costs Labor	Equipment	Total	Total Incl O&P
0010	**SMOKE VENT OPTIONS**									
0100	4' x 4' aluminum cover and frame	G-3	13	2.462	Ea.	2,050	120		2,170	2,425
0200	Galvanized steel cover and frame		13	2.462		1,550	120		1,670	1,875
0300	4' x 8' aluminum cover and frame		8	4		2,650	194		2,844	3,225
0400	Galvanized steel cover and frame		8	4		2,200	194		2,394	2,725

07 72 53 – Snow Guards

07 72 53.10 Snow Guard Options

		Crew	Daily Output	Labor-Hours	Unit	Material	2017 Bare Costs Labor	Equipment	Total	Total Incl O&P
0010	**SNOW GUARD OPTIONS**									
0100	Slate & asphalt shingle roofs, fastened with nails	1 Rofc	160	.050	Ea.	12.55	2.16		14.71	17.45
0200	Standing seam metal roofs, fastened with set screws		48	.167		16.95	7.20		24.15	31
0300	Surface mount for metal roofs, fastened with solder		48	.167		8.45	7.20		15.65	21.50
0400	Double rail pipe type, including pipe		130	.062	L.F.	34	2.66		36.66	42

07 72 73 – Pitch Pockets

07 72 73.10 Pitch Pockets, Variable Sizes

		Crew	Daily Output	Labor-Hours	Unit	Material	2017 Bare Costs Labor	Equipment	Total	Total Incl O&P
0010	**PITCH POCKETS, VARIABLE SIZES**									
0100	Adjustable, 4" to 7", welded corners, 4" deep	1 Rofc	48	.167	Ea.	16.45	7.20		23.65	30.50
0200	Side extenders, 6"	"	240	.033	"	2.75	1.44		4.19	5.45

07 72 80 – Vents

07 72 80.30 Vent Options

		Crew	Daily Output	Labor-Hours	Unit	Material	2017 Bare Costs Labor	Equipment	Total	Total Incl O&P
0010	**VENT OPTIONS**									
0020	Plastic, for insulated decks, 1 per M.S.F.	1 Rofc	40	.200	Ea.	21	8.65		29.65	37.50
0100	Heavy duty		20	.400		38	17.25		55.25	71.50

07 72 Roof Accessories

07 72 80 - Vents

	07 72 80.30 Vent Options	Crew	Daily Output	Labor-Hours	Unit	Material	2017 Bare Costs Labor	Equipment	Total	Total Incl O&P
0300	Aluminum	1 Rofc	30	.267	Ea.	21	11.50		32.50	42.50
0800	Polystyrene baffles, 12" wide for 16" O.C. rafter spacing	1 Carp	90	.089		.44	4.38		4.82	7.20
0900	For 24" O.C. rafter spacing	"	110	.073	↓	.78	3.58		4.36	6.35

07 76 Roof Pavers

07 76 16 - Roof Decking Pavers

07 76 16.10 Roof Pavers and Supports

		Crew	Daily Output	Labor-Hours	Unit	Material	2017 Bare Costs Labor	Equipment	Total	Total Incl O&P
0010	**ROOF PAVERS AND SUPPORTS**									
1000	Roof decking pavers, concrete blocks, 2" thick, natural	1 Clab	115	.070	S.F.	3.37	2.72		6.09	7.90
1100	Colors		115	.070	"	3.73	2.72		6.45	8.25
1200	Support pedestal, bottom cap		960	.008	Ea.	3	.33		3.33	3.80
1300	Top cap		960	.008		4.80	.33		5.13	5.80
1400	Leveling shims, 1/16"		1920	.004		1.20	.16		1.36	1.57
1500	1/8"		1920	.004		1.20	.16		1.36	1.57
1600	Buffer pad		960	.008	↓	2.50	.33		2.83	3.25
1700	PVC legs (4" SDR 35)		2880	.003	Inch	.13	.11		.24	.31
2000	Alternate pricing method, system in place	↓	101	.079	S.F.	7	3.10		10.10	12.45

07 81 Applied Fireproofing

07 81 16 - Cementitious Fireproofing

07 81 16.10 Sprayed Cementitious Fireproofing

		Crew	Daily Output	Labor-Hours	Unit	Material	2017 Bare Costs Labor	Equipment	Total	Total Incl O&P
0010	**SPRAYED CEMENTITIOUS FIREPROOFING**									
0050	Not including canvas protection, normal density									
0100	Per 1" thick, on flat plate steel	G-2	3000	.008	S.F.	.56	.33	.04	.93	1.16
0200	Flat decking		2400	.010		.56	.41	.06	1.03	1.29
0400	Beams		1500	.016		.56	.66	.09	1.31	1.71
0500	Corrugated or fluted decks		1250	.019		.84	.79	.11	1.74	2.24
0700	Columns, 1-1/8" thick		1100	.022		.63	.90	.12	1.65	2.18
0800	2-3/16" thick		700	.034		1.35	1.42	.19	2.96	3.84
0900	For canvas protection, add	↓	5000	.005	↓	.10	.20	.03	.33	.44
1000	Not including canvas protection, high density									
1100	Per 1" thick, on flat plate steel	G-2	3000	.008	S.F.	2.16	.33	.04	2.53	2.92
1110	On flat decking		2400	.010		2.16	.41	.06	2.63	3.05
1120	On beams		1500	.016		2.16	.66	.09	2.91	3.47
1130	Corrugated or fluted decks		1250	.019		2.16	.79	.11	3.06	3.69
1140	Columns, 1-1/8" thick		1100	.022		2.43	.90	.12	3.45	4.16
1150	2-3/16" thick		1100	.022		4.85	.90	.12	5.87	6.85
1170	For canvas protection, add	↓	5000	.005	↓	.10	.20	.03	.33	.44
1200	Not including canvas protection, retrofitting									
1210	Per 1" thick, on flat plate steel	G-2	1500	.016	S.F.	.48	.66	.09	1.23	1.63
1220	On flat decking		1200	.020		.48	.83	.11	1.42	1.90
1230	On beams		750	.032		.48	1.32	.18	1.98	2.72
1240	Corrugated or fluted decks		625	.038		.72	1.59	.21	2.52	3.42
1250	Columns, 1-1/8" thick		550	.044		.54	1.81	.24	2.59	3.59
1260	2-3/16" thick		500	.048		1.08	1.99	.27	3.34	4.47
1400	Accessories, preliminary spattered texture coat	↓	4500	.005		.04	.22	.03	.29	.40
1410	Bonding agent	1 Plas	1000	.008	↓	.10	.37		.47	.66
1500	Intumescent epoxy fireproofing on wire mesh, 3/16" thick									
1550	1 hour rating, exterior use	G-2	136	.176	S.F.	7.50	7.30	.98	15.78	20.50

07 81 Applied Fireproofing

07 81 16 – Cementitious Fireproofing

07 81 16.10 Sprayed Cementitious Fireproofing	Crew	Daily Output	Labor-Hours	Unit	Material	2017 Bare Costs Labor	Equipment	Total	Total Incl O&P
1551	2 hour rating, exterior use	G-2	136	.176	S.F.	7.50	7.30	.98	15.78	20.50
1600	Magnesium oxychloride, 35# to 40# density, 1/4" thick		3000	.008		1.60	.33	.04	1.97	2.31
1650	1/2" thick		2000	.012		3.20	.50	.07	3.77	4.34
1700	60# to 70# density, 1/4" thick		3000	.008		2.10	.33	.04	2.47	2.86
1750	1/2" thick		2000	.012		4.25	.50	.07	4.82	5.50
2000	Vermiculite cement, troweled or sprayed, 1/4" thick		3000	.008		1.45	.33	.04	1.82	2.15
2050	1/2" thick		2000	.012		2.85	.50	.07	3.42	3.96

07 84 Firestopping

07 84 13 – Penetration Firestopping

07 84 13.10 Firestopping

		Crew	Daily Output	Labor-Hours	Unit	Material	2017 Bare Costs Labor	Equipment	Total	Total Incl O&P
0010	**FIRESTOPPING** R078413-30									
0100	Metallic piping, non insulated									
0110	Through walls, 2" diameter	1 Carp	16	.500	Ea.	3.60	24.50		28.10	41.50
0120	4" diameter		14	.571		6.30	28		34.30	50
0130	6" diameter		12	.667		9.25	33		42.25	60.50
0140	12" diameter		10	.800		18.45	39.50		57.95	81
0150	Through floors, 2" diameter		32	.250		1.80	12.30		14.10	21
0160	4" diameter		28	.286		3.15	14.05		17.20	25
0170	6" diameter		24	.333		4.60	16.40		21	30
0180	12" diameter		20	.400		9.25	19.70		28.95	40
0190	Metallic piping, insulated									
0200	Through walls, 2" diameter	1 Carp	16	.500	Ea.	6.30	24.50		30.80	44.50
0210	4" diameter		14	.571		9.25	28		37.25	53
0220	6" diameter		12	.667		12.40	33		45.40	64
0230	12" diameter		10	.800		21	39.50		60.50	84
0240	Through floors, 2" diameter		32	.250		3.15	12.30		15.45	22.50
0250	4" diameter		28	.286		4.60	14.05		18.65	26.50
0260	6" diameter		24	.333		6.20	16.40		22.60	32
0270	12" diameter		20	.400		10.60	19.70		30.30	41.50
0280	Non metallic piping, non insulated									
0290	Through walls, 2" diameter	1 Carp	12	.667	Ea.	45.50	33		78.50	101
0300	4" diameter		10	.800		87	39.50		126.50	156
0310	6" diameter		8	1		170	49.50		219.50	263
0330	Through floors, 2" diameter		16	.500		30	24.50		54.50	70.50
0340	4" diameter		6	1.333		55.50	65.50		121	162
0350	6" diameter		6	1.333		103	65.50		168.50	214
0370	Ductwork, insulated & non insulated, round									
0380	Through walls, 6" diameter	1 Carp	12	.667	Ea.	12.40	33		45.40	64
0390	12" diameter		10	.800		21	39.50		60.50	84
0400	18" diameter		8	1		25.50	49.50		75	104
0410	Through floors, 6" diameter		16	.500		6.20	24.50		30.70	44.50
0420	12" diameter		14	.571		10.60	28		38.60	54.50
0430	18" diameter		12	.667		12.85	33		45.85	64.50
0440	Ductwork, insulated & non insulated, rectangular									
0450	With stiffener/closure angle, through walls, 6" x 12"	1 Carp	8	1	Ea.	16.20	49.50		65.70	93.50
0460	12" x 24"		6	1.333		32.50	65.50		98	137
0470	24" x 48"		4	2		65	98.50		163.50	223
0480	With stiffener/closure angle, through floors, 6" x 12"		10	.800		8.10	39.50		47.60	69.50
0490	12" x 24"		8	1		16.20	49.50		65.70	93.50
0500	24" x 48"		6	1.333		32.50	65.50		98	137

07 84 13 – Penetration Firestopping

07 84 13.10 Firestopping	Crew	Daily Output	Labor-Hours	Unit	Material	2017 Bare Costs Labor	Equipment	Total	Total Incl O&P
0510	Multi trade openings								
0520	Through walls, 6" x 12"	1 Carp	2	4	Ea.	45	197		242	350
0530	12" x 24"	"	1	8		147	395		542	765
0540	24" x 48"	2 Carp	1	16		525	790		1,315	1,775
0550	48" x 96"	"	.75	21.333		1,975	1,050		3,025	3,775
0560	Through floors, 6" x 12"	1 Carp	2	4		49.50	197		246.50	355
0570	12" x 24"	"	1	8		68	395		463	680
0580	24" x 48"	2 Carp	.75	21.333		138	1,050		1,188	1,750
0590	48" x 96"	"	.50	32	↓	390	1,575		1,965	2,850
0600	Structural penetrations, through walls								
0610	Steel beams, W8 x 10	1 Carp	8	1	Ea.	68.50	49.50		118	151
0620	W12 x 14		6	1.333		99	65.50		164.50	210
0630	W21 x 44		5	1.600		138	79		217	273
0640	W36 x 135		3	2.667		219	131		350	440
0650	Bar joists, 18" deep		6	1.333		40.50	65.50		106	146
0660	24" deep		6	1.333		54	65.50		119.50	160
0670	36" deep		5	1.600		81	79		160	210
0680	48" deep	↓	4	2	↓	108	98.50		206.50	269
0690	Construction joints, floor slab at exterior wall								
0700	Precast, brick, block or drywall exterior								
0710	2" wide joint	1 Carp	125	.064	L.F.	9.50	3.15		12.65	15.30
0720	4" wide joint	"	75	.107	"	14.85	5.25		20.10	24.50
0730	Metal panel, glass or curtain wall exterior								
0740	2" wide joint	1 Carp	40	.200	L.F.	9.50	9.85		19.35	25.50
0750	4" wide joint	"	25	.320	"	14.85	15.75		30.60	40.50
0760	Floor slab to drywall partition								
0770	Flat joint	1 Carp	100	.080	L.F.	13.20	3.94		17.14	20.50
0780	Fluted joint		50	.160		15.05	7.90		22.95	28.50
0790	Etched fluted joint	↓	75	.107	↓	18.50	5.25		23.75	28.50
0800	Floor slab to concrete/masonry partition								
0810	Flat joint	1 Carp	75	.107	L.F.	13.20	5.25		18.45	22.50
0820	Fluted joint	"	50	.160	"	15.05	7.90		22.95	28.50
0830	Concrete/CMU wall joints								
0840	1" wide	1 Carp	100	.080	L.F.	18.85	3.94		22.79	26.50
0850	2" wide		75	.107		23.50	5.25		28.75	33.50
0860	4" wide	↓	50	.160	↓	33.50	7.90		41.40	48.50
0870	Concrete/CMU floor joints								
0880	1" wide	1 Carp	200	.040	L.F.	18.85	1.97		20.82	23.50
0890	2" wide		150	.053		23.50	2.63		26.13	29.50
0900	4" wide	↓	100	.080	↓	33.50	3.94		37.44	42.50

For customer support on your Building Construction Costs with RSMeans Data, call 800.448.8182.

259

07 91 13 – Compression Seals

07 91 13.10 Compression Seals

07 91 13.10 Compression Seals	Crew	Daily Output	Labor-Hours	Unit	Material	2017 Bare Costs Labor	Equipment	Total	Total Incl O&P
0010 **COMPRESSION SEALS**									
4900 O-ring type cord, 1/4"	1 Bric	472	.017	L.F.	.43	.81		1.24	1.72
4910 1/2"		440	.018		1.21	.87		2.08	2.67
4920 3/4"		424	.019		2.43	.90		3.33	4.06
4930 1"		408	.020		4.10	.94		5.04	5.95
4940 1-1/4"		384	.021		7.50	.99		8.49	9.80
4950 1-1/2"		368	.022		9.25	1.04		10.29	11.75
4960 1-3/4"		352	.023		18.75	1.09		19.84	22
4970 2"		344	.023		22.50	1.11		23.61	26.50

07 91 16 – Joint Gaskets

07 91 16.10 Joint Gaskets

	Crew	Daily Output	Labor-Hours	Unit	Material	2017 Bare Costs Labor	Equipment	Total	Total Incl O&P
0010 **JOINT GASKETS**									
4400 Joint gaskets, neoprene, closed cell w/adh, 1/8" x 3/8"	1 Bric	240	.033	L.F.	.32	1.59		1.91	2.80
4500 1/4" x 3/4"		215	.037		.62	1.78		2.40	3.41
4700 1/2" x 1"		200	.040		1.45	1.91		3.36	4.54
4800 3/4" x 1-1/2"		165	.048		1.60	2.31		3.91	5.30

07 91 23 – Backer Rods

07 91 23.10 Backer Rods

	Crew	Daily Output	Labor-Hours	Unit	Material	2017 Bare Costs Labor	Equipment	Total	Total Incl O&P
0010 **BACKER RODS**									
0030 Backer rod, polyethylene, 1/4" diameter	1 Bric	4.60	1.739	C.L.F.	2.37	83		85.37	131
0050 1/2" diameter		4.60	1.739		4.28	83		87.28	133
0070 3/4" diameter		4.60	1.739		6.75	83		89.75	135
0090 1" diameter		4.60	1.739		12.10	83		95.10	141

07 91 26 – Joint Fillers

07 91 26.10 Joint Fillers

	Crew	Daily Output	Labor-Hours	Unit	Material	2017 Bare Costs Labor	Equipment	Total	Total Incl O&P
0010 **JOINT FILLERS**									
4360 Butyl rubber filler, 1/4" x 1/4"	1 Bric	290	.028	L.F.	.23	1.32		1.55	2.28
4365 1/2" x 1/2"		250	.032		.90	1.53		2.43	3.34
4370 1/2" x 3/4"		210	.038		1.35	1.82		3.17	4.29
4375 3/4" x 3/4"		230	.035		2.03	1.66		3.69	4.78
4380 1" x 1"		180	.044		2.71	2.12		4.83	6.25
4390 For coloring, add					12%				
4980 Polyethylene joint backing, 1/4" x 2"	1 Bric	2.08	3.846	C.L.F.	13.20	184		197.20	298
4990 1/4" x 6"		1.28	6.250	"	29	298		327	490
5600 Silicone, room temp vulcanizing foam seal, 1/4" x 1/2"		1312	.006	L.F.	.46	.29		.75	.96
5610 1/2" x 1/2"		656	.012		.93	.58		1.51	1.92
5620 1/2" x 3/4"		442	.018		1.39	.86		2.25	2.86
5630 3/4" x 3/4"		328	.024		2.08	1.16		3.24	4.08
5640 1/8" x 1"		1312	.006		.46	.29		.75	.96
5650 1/8" x 3"		442	.018		1.39	.86		2.25	2.86
5670 1/4" x 3"		295	.027		2.78	1.30		4.08	5.05
5680 1/4" x 6"		148	.054		5.55	2.58		8.13	10.05
5690 1/2" x 6"		82	.098		11.10	4.66		15.76	19.35
5700 1/2" x 9"		52.50	.152		16.65	7.30		23.95	29.50
5710 1/2" x 12"		33	.242		22	11.60		33.60	42.50

07 92 Joint Sealants

07 92 13 – Elastomeric Joint Sealants

07 92 13.20 Caulking and Sealant Options

		Crew	Daily Output	Labor-Hours	Unit	Material	2017 Bare Costs Labor	Equipment	Total	Total Incl O&P
0010	**CAULKING AND SEALANT OPTIONS**									
0050	Latex acrylic based, bulk				Gal.	27			27	29.50
0055	Bulk in place 1/4" x 1/4" bead	1 Bric	300	.027	L.F.	.08	1.27		1.35	2.05
0060	1/4" x 3/8"		294	.027		.14	1.30		1.44	2.15
0065	1/4" x 1/2"		288	.028		.19	1.33		1.52	2.25
0075	3/8" x 3/8"		284	.028		.21	1.35		1.56	2.30
0080	3/8" x 1/2"		280	.029		.28	1.36		1.64	2.41
0085	3/8" x 5/8"		276	.029		.35	1.38		1.73	2.52
0095	3/8" x 3/4"		272	.029		.42	1.40		1.82	2.62
0100	1/2" x 1/2"		275	.029		.38	1.39		1.77	2.55
0105	1/2" x 5/8"		269	.030		.47	1.42		1.89	2.70
0110	1/2" x 3/4"		263	.030		.56	1.45		2.01	2.85
0115	1/2" x 7/8"		256	.031		.66	1.49		2.15	3.02
0120	1/2" x 1"		250	.032		.75	1.53		2.28	3.18
0125	3/4" x 3/4"		244	.033		.85	1.57		2.42	3.34
0130	3/4" x 1"		225	.036		1.13	1.70		2.83	3.85
0135	1" x 1"	↓	200	.040	↓	1.50	1.91		3.41	4.59
0190	Cartridges				Gal.	33			33	36
0200	11 fl. oz. cartridge				Ea.	2.82			2.82	3.10
0500	1/4" x 1/2"	1 Bric	288	.028	L.F.	.23	1.33		1.56	2.29
0600	1/2" x 1/2"		275	.029		.46	1.39		1.85	2.65
0800	3/4" x 3/4"		244	.033		1.04	1.57		2.61	3.55
0900	3/4" x 1"		225	.036		1.38	1.70		3.08	4.13
1000	1" x 1"	↓	200	.040	↓	1.73	1.91		3.64	4.84
1400	Butyl based, bulk				Gal.	37			37	40.50
1500	Cartridges				"	40.50			40.50	44.50
1700	1/4" x 1/2", 154 L.F./gal.	1 Bric	288	.028	L.F.	.24	1.33		1.57	2.30
1800	1/2" x 1/2", 77 L.F./gal.	"	275	.029	"	.48	1.39		1.87	2.67
2300	Polysulfide compounds, 1 component, bulk				Gal.	84			84	92.50
2600	1 or 2 component, in place, 1/4" x 1/4", 308 L.F./gal.	1 Bric	300	.027	L.F.	.27	1.27		1.54	2.26
2700	1/2" x 1/4", 154 L.F./gal.		288	.028		.55	1.33		1.88	2.64
2900	3/4" x 3/8", 68 L.F./gal.		272	.029		1.24	1.40		2.64	3.52
3000	1" x 1/2", 38 L.F./gal.	↓	250	.032	↓	2.21	1.53		3.74	4.78
3200	Polyurethane, 1 or 2 component				Gal.	53.50			53.50	58.50
3500	Bulk, in place, 1/4" x 1/4"	1 Bric	300	.027	L.F.	.17	1.27		1.44	2.15
3655	1/2" x 1/4"		288	.028		.35	1.33		1.68	2.42
3800	3/4" x 3/8"		272	.029		.78	1.40		2.18	3.02
3900	1" x 1/2"	↓	250	.032	↓	1.38	1.53		2.91	3.87
4100	Silicone rubber, bulk				Gal.	54			54	59.50
4200	Cartridges				"	53			53	58.50

07 92 16 – Rigid Joint Sealants

07 92 16.10 Rigid Joint Sealants

		Crew	Daily Output	Labor-Hours	Unit	Material	2017 Bare Costs Labor	Equipment	Total	Total Incl O&P
0010	**RIGID JOINT SEALANTS**									
5800	Tapes, sealant, PVC foam adhesive, 1/16" x 1/4"				C.L.F.	10			10	11
5900	1/16" x 1/2"					10			10	11
5950	1/16" x 1"					16			16	17.60
6000	1/8" x 1/2"				↓	9.50			9.50	10.45

For customer support on your Building Construction Costs with RSMeans Data, call 800.448.8182.

261

07 92 Joint Sealants

07 92 19 – Acoustical Joint Sealants

07 92 19.10 Acoustical Sealant	Crew	Daily Output	Labor-Hours	Unit	Material	2017 Bare Costs Labor	Equipment	Total	Total Incl O&P
0010 **ACOUSTICAL SEALANT**									
0020 Acoustical sealant, elastomeric, cartridges				Ea.	8.50			8.50	9.35
0025 In place, 1/4" x 1/4"	1 Bric	300	.027	L.F.	.35	1.27		1.62	2.34
0030 1/4" x 1/2"		288	.028		.69	1.33		2.02	2.80
0035 1/2" x 1/2"		275	.029		1.39	1.39		2.78	3.67
0040 1/2" x 3/4"		263	.030		2.08	1.45		3.53	4.52
0045 3/4" x 3/4"		244	.033		3.13	1.57		4.70	5.85
0050 1" x 1"		200	.040		5.55	1.91		7.46	9.05

07 95 Expansion Control

07 95 13 – Expansion Joint Cover Assemblies

07 95 13.50 Expansion Joint Assemblies

07 95 13.50 Expansion Joint Assemblies	Crew	Daily Output	Labor-Hours	Unit	Material	2017 Bare Costs Labor	Equipment	Total	Total Incl O&P
0010 **EXPANSION JOINT ASSEMBLIES**									
0200 Floor cover assemblies, 1" space, aluminum	1 Sswk	38	.211	L.F.	19	11.45		30.45	40
0300 Bronze		38	.211		60.50	11.45		71.95	85.50
0500 2" space, aluminum		38	.211		18.50	11.45		29.95	39.50
0600 Bronze		38	.211		58.50	11.45		69.95	83.50
0800 Wall and ceiling assemblies, 1" space, aluminum		38	.211		16.75	11.45		28.20	37.50
0900 Bronze		38	.211		51.50	11.45		62.95	76
1100 2" space, aluminum		38	.211		17.10	11.45		28.55	38
1200 Bronze		38	.211		52.50	11.45		63.95	76.50
1400 Floor to wall assemblies, 1" space, aluminum		38	.211		19.50	11.45		30.95	40.50
1500 Bronze or stainless		38	.211		62	11.45		73.45	87
1700 Gym floor angle covers, aluminum, 3" x 3" angle		46	.174		19.50	9.45		26.95	37.50
1800 3" x 4" angle		46	.174		23	9.45		32.45	41.50
2000 Roof closures, aluminum, flat roof, low profile, 1" space		57	.140		24.50	7.60		32.10	39.50
2100 High profile		57	.140		26	7.60		33.60	41
2300 Roof to wall, low profile, 1" space		57	.140		23.50	7.60		31.10	38
2400 High profile		57	.140		26	7.60		33.60	41

Estimating Tips
08 10 00 Doors and Frames
All exterior doors should be addressed for their energy conservation (insulation and seals).

- Most metal doors and frames look alike, but there may be significant differences among them. When estimating these items, be sure to choose the line item that most closely compares to the specification or door schedule requirements regarding:
  - ☐ type of metal
  - ☐ metal gauge
  - ☐ door core material
  - ☐ fire rating
  - ☐ finish
- Wood and plastic doors vary considerably in price. The primary determinant is the veneer material. Lauan, birch, and oak are the most common veneers. Other variables include the following:
  - ☐ hollow or solid core
  - ☐ fire rating
  - ☐ flush or raised panel
  - ☐ finish
- Door pricing includes bore for cylindrical locksets and mortise for hinges.

08 30 00 Specialty Doors and Frames
- There are many varieties of special doors, and they are usually priced per each. Add frames, hardware, or operators required for a complete installation.

08 40 00 Entrances, Storefronts, and Curtain Walls
- Glazed curtain walls consist of the metal tube framing and the glazing material. The cost data in this subdivision is presented for the metal tube framing alone or the composite wall. If your estimate requires a detailed takeoff of the framing, be sure to add the glazing cost and any tints.

08 50 00 Windows
- Steel windows are unglazed and aluminum can be glazed or unglazed. Some metal windows are priced without glass. Refer to 08 80 00 Glazing for glass pricing. The grade C indicates commercial grade windows, usually ASTM C-35.
- All wood windows and vinyl are priced preglazed. The glazing is insulating glass. Add the cost of screens and grills if required and not already included.

08 70 00 Hardware
- Hardware costs add considerably to the cost of a door. The most efficient method to determine the hardware requirements for a project is to review the door and hardware schedule together. One type of door may have different hardware, depending on the door usage.
- Door hinges are priced by the pair, with most doors requiring 1-1/2 pairs per door. The hinge prices do not include installation labor, because it is included in door installation.

Hinges are classified according to the frequency of use, base material, and finish.

08 80 00 Glazing
- Different openings require different types of glass. The most common types are:
  - ☐ float
  - ☐ tempered
  - ☐ insulating
  - ☐ impact-resistant
  - ☐ ballistic-resistant
- Most exterior windows are glazed with insulating glass. Entrance doors and window walls, where the glass is less than 18" from the floor, are generally glazed with tempered glass. Interior windows and some residential windows are glazed with float glass.
- Coastal communities require the use of impact-resistant glass, dependent on wind speed.
- The insulation or 'u' value is a strong consideration, along with solar heat gain, to determine total energy efficiency.

Reference Numbers
Reference numbers are shown at the beginning of some major classifications. These numbers refer to related items in the Reference Section. The reference information may be an estimating procedure, an alternate pricing method, or technical information.

Note: Not all subdivisions listed here necessarily appear. ■

Did you know?

Our online estimating solution gives you the same access to RSMeans' data with 24/7 access:

- Quickly locate costs in the searchable database.
- Build cost lists, estimates, and reports in minutes.
- Adjust costs to any location in the U.S. and Canada with the click of a button.

Start your free trial today at
www.RSMeansOnline.com

08 05 05.10 Selective Demolition Doors

		Crew	Daily Output	Labor-Hours	Unit	Material	2017 Bare Costs Labor	Equipment	Total	Total Incl O&P
0010	**SELECTIVE DEMOLITION DOORS** R024119-10									
0200	Doors, exterior, 1-3/4" thick, single, 3' x 7' high	1 Clab	16	.500	Ea.		19.60		19.60	30
0220	Double, 6' x 7' high		12	.667			26		26	40
0500	Interior, 1-3/8" thick, single, 3' x 7' high		20	.400			15.65		15.65	24
0520	Double, 6' x 7' high		16	.500			19.60		19.60	30
0700	Bi-folding, 3' x 6'-8" high		20	.400			15.65		15.65	24
0720	6' x 6'-8" high		18	.444			17.40		17.40	26.50
0900	Bi-passing, 3' x 6'-8" high		16	.500			19.60		19.60	30
0940	6' x 6'-8" high		14	.571			22.50		22.50	34.50
0960	Interior metal door 1-3/4" thick, 3'-0" x 6'-8"		18	.444			17.40		17.40	26.50
0980	Interior metal door 1-3/4" thick, 3'-0" x 7'-0"		18	.444			17.40		17.40	26.50
1000	Interior wood door 1-3/4" thick, 3'-0" x 6'-8"		20	.400			15.65		15.65	24
1020	Interior wood door 1-3/4" thick, 3'-0" x 7'-0"	▼	20	.400			15.65		15.65	24
1100	Door demo, floor door	2 Sswk	5	3.200			174		174	290
1500	Remove and reset, hollow core	1 Carp	8	1			49.50		49.50	75.50
1520	Solid		6	1.333			65.50		65.50	101
2000	Frames, including trim, metal	▼	8	1			49.50		49.50	75.50
2200	Wood	2 Carp	32	.500			24.50		24.50	37.50
3000	Special doors, counter doors		6	2.667			131		131	201
3100	Double acting		10	1.600			79		79	121
3200	Floor door (trap type), or access type		8	2			98.50		98.50	151
3300	Glass, sliding, including frames		12	1.333			65.50		65.50	101
3400	Overhead, commercial, 12' x 12' high		4	4			197		197	300
3440	up to 20' x 16' high		3	5.333			263		263	400
3445	up to 35' x 30' high		1	16			790		790	1,200
3500	Residential, 9' x 7' high		8	2			98.50		98.50	151
3540	16' x 7' high		7	2.286			113		113	172
3620	Large		2.50	6.400			315		315	485
3700	Roll-up grille		5	3.200			158		158	241
3800	Revolving door, no wire or connections		2	8			395		395	605
3900	Storefront swing door	▼	3	5.333	SF Surf		263		263	400
6600	Demo flexible transparent strip entrance	3 Shee	115	.209	SF Surf		12.10		12.10	18.55
7100	Remove double swing pneumatic doors, openers and sensors	2 Skwk	.50	32	Opng.		1,650		1,650	2,525
7110	Remove automatic operators, industrial, sliding doors, to 12' wide	"	.40	40	"		2,050		2,050	3,175
7550	Hangar door demo	2 Sswk	330	.048	S.F.		2.63		2.63	4.39
7570	Remove shock absorbing door	"	1.90	8.421	Opng.		455		455	760

08 05 05.20 Selective Demolition of Windows

		Crew	Daily Output	Labor-Hours	Unit	Material	2017 Bare Costs Labor	Equipment	Total	Total Incl O&P
0010	**SELECTIVE DEMOLITION OF WINDOWS** R024119-10									
0200	Aluminum, including trim, to 12 S.F.	1 Clab	16	.500	Ea.		19.60		19.60	30
0240	To 25 S.F.		11	.727			28.50		28.50	43.50
0280	To 50 S.F.		5	1.600			62.50		62.50	96
0320	Storm windows/screens, to 12 S.F.		27	.296			11.60		11.60	17.80
0360	To 25 S.F.		21	.381			14.90		14.90	23
0400	To 50 S.F.		16	.500	▼		19.60		19.60	30
0600	Glass, up to 10 S.F. per window		200	.040	S.F.		1.57		1.57	2.40
0620	Over 10 S.F. per window		150	.053	"		2.09		2.09	3.20
1000	Steel, including trim, to 12 S.F.		13	.615	Ea.		24		24	37
1020	To 25 S.F.		9	.889			35		35	53.50
1040	To 50 S.F.		4	2			78.50		78.50	120
2000	Wood, including trim, to 12 S.F.		22	.364			14.25		14.25	22
2020	To 25 S.F.		18	.444			17.40		17.40	26.50
2060	To 50 S.F.	▼	13	.615	▼		24		24	37

264

For customer support on your Building Construction Costs with RSMeans Data, call 800.448.8182.

08 05 Common Work Results for Openings

08 05 05 – Selective Demolition for Openings

08 05 05.20 Selective Demolition of Windows		Crew	Daily Output	Labor-Hours	Unit	Material	2017 Bare Costs Labor	Equipment	Total	Total Incl O&P
2065	To 180 S.F.	1 Clab	8	1	Ea.		39		39	60
4300	Remove bay/bow window	2 Carp	6	2.667	↓		131		131	201
4410	Remove skylight, plstc domes, flush/curb mtd	"	395	.041	S.F.		2		2	3.06
4420	Remove skylight, plstc/glass up to 2' x 3'	1 Carp	15	.533	Ea.		26.50		26.50	40
4440	Remove skylight, plstc/glass up to 4' x 6'	2 Carp	10	1.600			79		79	121
4480	Remove roof window up to 3' x 4'	1 Carp	8	1			49.50		49.50	75.50
4500	Remove roof window up to 4' x 6'	2 Carp	6	2.667			131		131	201
5020	Remove and reset window, up to a 2' x 2' window	1 Carp	6	1.333			65.50		65.50	101
5040	Up to a 3' x 3' window	↓	4	2			98.50		98.50	151
5080	Up to a 4' x 5' window	↓	2	4	↓		197		197	300

08 11 Metal Doors and Frames

08 11 16 – Aluminum Doors and Frames

08 11 16.10 Entrance Doors

		Crew	Daily Output	Labor-Hours	Unit	Material	2017 Bare Costs Labor	Equipment	Total	Total Incl O&P
0010	**ENTRANCE DOORS** and frame, aluminum, narrow stile									
0011	Including standard hardware, clear finish, no glass									
0012	Top and bottom offset pivots, 1/4" beveled glass stops, threshold									
0013	Dead bolt lock with inside thumb screw, standard push pull									
0020	3'-0" x 7'-0" opening	2 Sswk	2	8	Ea.	950	435		1,385	1,775
0025	Anodizing aluminum entr. door & frame, add					107			107	117
0030	3'-6" x 7'-0" opening	2 Sswk	2	8		920	435		1,355	1,725
0100	3'-0" x 10'-0" opening, 3' high transom		1.80	8.889		1,325	485		1,810	2,275
0200	3'-6" x 10'-0" opening, 3' high transom		1.80	8.889		1,400	485		1,885	2,325
0280	5'-0" x 7'-0" opening		2	8		1,550	435		1,985	2,425
0300	6'-0" x 7'-0" opening		1.30	12.308	↓	1,300	670		1,970	2,550
0301	6'-0" x 7'-0" opening		1.30	12.308	Pr.	1,300	670		1,970	2,550
0400	6'-0" x 10'-0" opening, 3' high transom		1.10	14.545	"	1,725	790		2,515	3,225
0520	3'-0" x 7'-0" opening, wide stile		2	8	Ea.	1,025	435		1,460	1,850
0540	3'-6" x 7'-0" opening		2	8		1,275	435		1,710	2,150
0560	5'-0" x 7'-0" opening		2	8	↓	1,650	435		2,085	2,525
0580	6'-0" x 7'-0" opening		1.30	12.308	Pr.	1,725	670		2,395	3,025
0600	7'-0" x 7'-0" opening	↓	1	16	"	1,875	870		2,745	3,500
1200	For non-standard size, add				Leaf	80%				
1250	For installation of non-standard size, add						20%			
1300	Light bronze finish, add				Leaf	36%				
1400	Dark bronze finish, add					25%				
1500	For black finish, add					40%				
1600	Concealed panic device, add				↓	980			980	1,075
1700	Electric striker release, add				Opng.	294			294	325
1800	Floor check, add				Leaf	680			680	745
1900	Concealed closer, add				"	555			555	610

08 11 63 – Metal Screen and Storm Doors and Frames

08 11 63.23 Aluminum Screen and Storm Doors and Frames

		Crew	Daily Output	Labor-Hours	Unit	Material	2017 Bare Costs Labor	Equipment	Total	Total Incl O&P
0010	**ALUMINUM SCREEN AND STORM DOORS AND FRAMES**									
0020	Combination storm and screen									
0420	Clear anodic coating, 2'-8" wide	2 Carp	14	1.143	Ea.	208	56.50		264.50	315
0440	3'-0" wide	"	14	1.143	"	180	56.50		236.50	284
0500	For 7'-0" door height, add					8%				
1020	Mill finish, 2'-8" wide	2 Carp	14	1.143	Ea.	245	56.50		301.50	355
1040	3'-0" wide	"	14	1.143	↓	265	56.50		321.50	380

For customer support on your Building Construction Costs with RSMeans Data, call 800.448.8182.

265

08 11 Metal Doors and Frames

08 11 63 – Metal Screen and Storm Doors and Frames

08 11 63.23 Aluminum Screen and Storm Doors and Frames	Crew	Daily Output	Labor-Hours	Unit	Material	2017 Bare Costs Labor	Equipment	Total	Total Incl O&P
1100	For 7'-0" door, add				Ea.	8%				
1520	White painted, 2'-8" wide	2 Carp	14	1.143		288	56.50		344.50	400
1540	3'-0" wide	"	14	1.143		300	56.50		356.50	415
1600	For 7'-0" door, add					8%				
2000	Wood door & screen, see Section 08 14 33.20									

08 12 Metal Frames

08 12 13 – Hollow Metal Frames

08 12 13.13 Standard Hollow Metal Frames

			Crew	Daily Output	Labor-Hours	Unit	Material	2017 Bare Costs Labor	Equipment	Total	Total Incl O&P
0010	**STANDARD HOLLOW METAL FRAMES**										
0020	16 ga., up to 5-3/4" jamb depth										
0025	3'-0" x 6'-8" single	G	2 Carp	16	1	Ea.	149	49.50		198.50	240
0028	3'-6" wide, single	G		16	1		163	49.50		212.50	255
0030	4'-0" wide, single	G		16	1		162	49.50		211.50	254
0040	6'-0" wide, double	G		14	1.143		207	56.50		263.50	315
0045	8'-0" wide, double	G		14	1.143		217	56.50		273.50	325
0100	3'-0" x 7'-0" single	G		16	1		157	49.50		206.50	249
0110	3'-6" wide, single	G		16	1		168	49.50		217.50	261
0112	4'-0" wide, single	G		16	1		168	49.50		217.50	261
0140	6'-0" wide, double	G		14	1.143		198	56.50		254.50	305
0145	8'-0" wide, double	G		14	1.143		234	56.50		290.50	345
1000	16 ga., up to 4-7/8" deep, 3'-0" x 7'-0" single	G		16	1		180	49.50		229.50	273
1140	6'-0" wide, double	G		14	1.143		203	56.50		259.50	310
1200	16 ga., 8-3/4" deep, 3'-0" x 7'-0" single	G		16	1		202	49.50		251.50	298
1240	6'-0" wide, double	G		14	1.143		237	56.50		293.50	345
2800	14 ga., up to 3-7/8" deep, 3'-0" x 7'-0" single	G		16	1		185	49.50		234.50	280
2840	6'-0" wide, double	G		14	1.143		219	56.50		275.50	325
3000	14 ga., up to 5-3/4" deep, 3'-0" x 6'-8" single	G		16	1		155	49.50		204.50	247
3002	3'-6" wide, single	G		16	1		208	49.50		257.50	305
3005	4'-0" wide, single	G		16	1		214	49.50		263.50	310
3600	up to 5-3/4" jamb depth, 4'-0" x 7'-0" single	G		15	1.067		185	52.50		237.50	285
3620	6'-0" wide, double	G		12	1.333		237	65.50		302.50	360
3640	8'-0" wide, double	G		12	1.333		247	65.50		312.50	375
3700	8'-0" high, 4'-0" wide, single	G		15	1.067		237	52.50		289.50	340
3740	8'-0" wide, double	G		12	1.333		293	65.50		358.50	420
4000	6-3/4" deep, 4'-0" x 7'-0" single	G		15	1.067		225	52.50		277.50	330
4020	6'-0" wide, double	G		12	1.333		277	65.50		342.50	405
4040	8'-0" wide, double	G		12	1.333		292	65.50		357.50	420
4100	8'-0" high, 4'-0" wide, single	G		15	1.067		280	52.50		332.50	385
4140	8'-0" wide, double	G		12	1.333		325	65.50		390.50	460
4400	8-3/4" deep, 4'-0" x 7'-0", single	G		15	1.067		254	52.50		306.50	360
4440	8'-0" wide, double	G		12	1.333		315	65.50		380.50	445
4500	4'-0" x 8'-0", single	G		15	1.067		289	52.50		341.50	400
4540	8'-0" wide, double	G		12	1.333		360	65.50		425.50	495
4900	For welded frames, add						65			65	71.50
5400	14 ga., "B" label, up to 5-3/4" deep, 4'-0" x 7'-0" single	G	2 Carp	15	1.067		213	52.50		265.50	315
5440	8'-0" wide, double	G		12	1.333		279	65.50		344.50	405
5800	6-3/4" deep, 7'-0" high, 4'-0" wide, single	G		15	1.067		221	52.50		273.50	325
5840	8'-0" wide, double	G		12	1.333		390	65.50		455.50	530
6200	8-3/4" deep, 4'-0" x 7'-0" single	G		15	1.067		299	52.50		351.50	410
6240	8'-0" wide, double	G		12	1.333		380	65.50		445.50	520

08 12 Metal Frames

08 12 13 - Hollow Metal Frames

08 12 13.13 Standard Hollow Metal Frames

		Crew	Daily Output	Labor-Hours	Unit	Material	2017 Bare Costs Labor	Equipment	Total	Total Incl O&P
6300	For "A" label use same price as "B" label									
6400	For baked enamel finish, add					30%	15%			
6500	For galvanizing, add					20%				
6600	For hospital stop, add				Ea.	300			300	330
6620	For hospital stop, stainless steel, add				"	395			395	435
7900	Transom lite frames, fixed, add	2 Carp	155	.103	S.F.	55	5.10		60.10	68.50
8000	Movable, add	"	130	.123	"	70	6.05		76.05	86.50

08 12 13.25 Channel Metal Frames

			Crew	Daily Output	Labor-Hours	Unit	Material	Labor	Equipment	Total	Total Incl O&P
0010	**CHANNEL METAL FRAMES**										
0020	Steel channels with anchors and bar stops										
0100	6" channel @ 8.2#/L.F., 3' x 7' door, weighs 150#	G	E-4	13	2.462	Ea.	236	135	10	381	495
0200	8" channel @ 11.5#/L.F., 6' x 8' door, weighs 275#	G		9	3.556		435	195	14.45	644.45	815
0300	8' x 12' door, weighs 400#	G		6.50	4.923		630	270	20	920	1,150
0400	10" channel @ 15.3#/L.F., 10' x 10' door, weighs 500#	G		6	5.333		785	292	21.50	1,098.50	1,375
0500	12' x 12' door, weighs 600#	G		5.50	5.818		945	320	23.50	1,288.50	1,600
0600	12" channel @ 20.7#/L.F., 12' x 12' door, weighs 825#	G		4.50	7.111		1,300	390	29	1,719	2,100
0700	12' x 16' door, weighs 1000#	G		4	8		1,575	440	32.50	2,047.50	2,500
0800	For frames without bar stops, light sections, deduct						15%				
0900	Heavy sections, deduct						10%				

08 13 Metal Doors

08 13 13 - Hollow Metal Doors

08 13 13.13 Standard Hollow Metal Doors

			Crew	Daily Output	Labor-Hours	Unit	Material	Labor	Equipment	Total	Total Incl O&P
0010	**STANDARD HOLLOW METAL DOORS**	R081313-20									
0015	Flush, full panel, hollow core										
0017	When noted doors are prepared but do not include glass or louvers										
0020	1-3/8" thick, 20 ga., 2'-0" x 6'-8"	G	2 Carp	20	.800	Ea.	340	39.50		379.50	435
0040	2'-8" x 6'-8"	G		18	.889		350	44		394	450
0060	3'-0" x 6'-8"	G		17	.941		355	46.50		401.50	460
0100	3'-0" x 7'-0"	G		17	.941		365	46.50		411.50	475
0120	For vision lite, add						98			98	108
0140	For narrow lite, add						105			105	116
0320	Half glass, 20 ga., 2'-0" x 6'-8"	G	2 Carp	20	.800		505	39.50		544.50	615
0340	2'-8" x 6'-8"	G		18	.889		515	44		559	635
0360	3'-0" x 6'-8"	G		17	.941		510	46.50		556.50	630
0400	3'-0" x 7'-0"	G		17	.941		605	46.50		651.50	735
0410	1-3/8" thick, 18 ga., 2'-0" x 6'-8"	G		20	.800		405	39.50		444.50	505
0420	3'-0" x 6'-8"	G		17	.941		410	46.50		456.50	520
0425	3'-0" x 7'-0"	G		17	.941		430	46.50		476.50	545
0450	For vision lite, add						98			98	108
0452	For narrow lite, add						105			105	116
0460	Half glass, 18 ga., 2'-0" x 6'-8"	G	2 Carp	20	.800		560	39.50		599.50	680
0465	2'-8" x 6'-8"	G		18	.889		580	44		624	705
0470	3'-0" x 6'-8"	G		17	.941		570	46.50		616.50	695
0475	3'-0" x 7'-0"	G		17	.941		585	46.50		631.50	715
0500	Hollow core, 1-3/4" thick, full panel, 20 ga., 2'-8" x 6'-8"	G		18	.889		425	44		469	530
0520	3'-0" x 6'-8"	G		17	.941		425	46.50		471.50	540
0640	3'-0" x 7'-0"	G		17	.941		460	46.50		506.50	575
0680	4'-0" x 7'-0"	G		15	1.067		635	52.50		687.50	775
0700	4'-0" x 8'-0"	G		13	1.231		750	60.50		810.50	915
1000	18 ga., 2'-8" x 6'-8"	G		17	.941		505	46.50		551.50	625

08 13 13 – Hollow Metal Doors

08 13 13.13 Standard Hollow Metal Doors		Crew	Daily Output	Labor-Hours	Unit	Material	2017 Bare Costs Labor	Equipment	Total	Total Incl O&P
1020	3'-0" x 6'-8"	G	2 Carp	16	1	Ea.	480	49.50		529.50	605
1120	3'-0" x 7'-0"	G		17	.941		520	46.50		566.50	645
1180	4'-0" x 7'-0"	G		14	1.143		635	56.50		691.50	780
1200	4'-0" x 8'-0"	G		17	.941		740	46.50		786.50	885
1212	For vision lite, add						98			98	108
1214	For narrow lite, add						105			105	116
1230	Half glass, 20 ga., 2'-8" x 6'-8"	G	2 Carp	20	.800		580	39.50		619.50	700
1240	3'-0" x 6'-8"	G		18	.889		580	44		624	700
1260	3'-0" x 7'-0"	G		18	.889		595	44		639	720
1280	Embossed panel, 1-3/4" thick, poly core, 20 ga., 3'-0" x 7'-0"			18	.889		420	44		464	525
1290	Half glass, 1-3/4" thick, poly core, 20 ga., 3'-0" x 7'-0"	G		18	.889		615	44		659	740
1320	18 ga., 2'-8" x 6'-8"	G		18	.889		640	44		684	770
1340	3'-0" x 6'-8"	G		17	.941		635	46.50		681.50	765
1360	3'-0" x 7'-0"	G		17	.941		645	46.50		691.50	780
1380	4'-0" x 7'-0"	G		15	1.067		795	52.50		847.50	955
1400	4'-0" x 8'-0"	G		14	1.143		895	56.50		951.50	1,075
1500	Flush full panel, 16 ga., steel hollow core										
1520	2'-0" x 6'-8"	G	2 Carp	20	.800	Ea.	570	39.50		609.50	685
1530	2'-8" x 6'-8"	G		20	.800		570	39.50		609.50	685
1540	3'-0" x 6'-8"	G		20	.800		555	39.50		594.50	670
1560	2'-8" x 7'-0"	G		18	.889		580	44		624	705
1570	3'-0" x 7'-0"	G		18	.889		560	44		604	685
1580	3'-6" x 7'-0"	G		18	.889		665	44		709	800
1590	4'-0" x 7'-0"	G		18	.889		730	44		774	870
1600	2'-8" x 8'-0"	G		18	.889		715	44		759	850
1620	3'-0" x 8'-0"	G		18	.889		730	44		774	865
1630	3'-6" x 8'-0"	G		18	.889		810	44		854	955
1640	4'-0" x 8'-0"	G		18	.889		850	44		894	1,000
1720	Insulated, 1-3/4" thick, full panel, 18 ga., 3'-0" x 6'-8"	G		15	1.067		485	52.50		537.50	615
1740	2'-8" x 7'-0"	G		16	1		505	49.50		554.50	630
1760	3'-0" x 7'-0"	G		15	1.067		505	52.50		557.50	635
1800	4'-0" x 8'-0"	G		13	1.231		780	60.50		840.50	955
1820	Half glass, 18 ga., 3'-0" x 6'-8"	G		16	1		635	49.50		684.50	770
1840	2'-8" x 7'-0"	G		17	.941		665	46.50		711.50	800
1860	3'-0" x 7'-0"	G		16	1		690	49.50		739.50	835
1900	4'-0" x 8'-0"	G		14	1.143		685	56.50		741.50	835
2000	For vision lite, add						98			98	108
2010	For narrow lite, add						105			105	116
8100	For bottom louver, add						290			290	320
8110	For baked enamel finish, add						30%	15%			
8120	For galvanizing, add						20%				

08 13 13.15 Metal Fire Doors

0010	**METAL FIRE DOORS**	R081313-20									
0015	Steel, flush, "B" label, 90 minute										
0020	Full panel, 20 ga., 2'-0" x 6'-8"		2 Carp	20	.800	Ea.	425	39.50		464.50	525
0040	2'-8" x 6'-8"			18	.889		440	44		484	550
0060	3'-0" x 6'-8"			17	.941		440	46.50		486.50	555
0080	3'-0" x 7'-0"			17	.941		455	46.50		501.50	570
0140	18 ga., 3'-0" x 6'-8"			16	1		505	49.50		554.50	630
0160	2'-8" x 7'-0"			17	.941		535	46.50		581.50	655
0180	3'-0" x 7'-0"			16	1		515	49.50		564.50	645
0200	4'-0" x 7'-0"			15	1.067		650	52.50		702.50	795

08 13 Metal Doors

08 13 13 – Hollow Metal Doors

08 13 13.15 Metal Fire Doors

		Crew	Daily Output	Labor-Hours	Unit	Material	2017 Bare Costs Labor	Equipment	Total	Total Incl O&P
0220	For "A" label, 3 hour, 18 ga., use same price as "B" label									
0240	For vision lite, add				Ea.	163			163	179
0300	Full panel, 16 ga., 2'-0" x 6'-8"	2 Carp	20	.800		555	39.50		594.50	670
0310	2'-8" x 6'-8"		18	.889		550	44		594	670
0320	3'-0" x 6'-8"		17	.941		545	46.50		591.50	670
0350	2'-8" x 7'-0"		17	.941		570	46.50		616.50	695
0360	3'-0" x 7'-0"		16	1		555	49.50		604.50	685
0370	4'-0" x 7'-0"		15	1.067		715	52.50		767.50	865
0520	Flush, "B" label, 90 min., egress core, 20 ga., 2'-0" x 6'-8"		18	.889		665	44		709	800
0540	2'-8" x 6'-8"		17	.941		670	46.50		716.50	805
0560	3'-0" x 6'-8"		16	1		675	49.50		724.50	815
0580	3'-0" x 7'-0"		16	1		700	49.50		749.50	845
0640	Flush, "A" label, 3 hour, egress core, 18 ga., 3'-0" x 6'-8"		15	1.067		730	52.50		782.50	880
0660	2'-8" x 7'-0"		16	1		765	49.50		814.50	915
0680	3'-0" x 7'-0"		15	1.067		750	52.50		802.50	900
0700	4'-0" x 7'-0"	↓	14	1.143	↓	910	56.50		966.50	1,075

08 13 13.20 Residential Steel Doors

			Crew	Daily Output	Labor-Hours	Unit	Material	2017 Bare Costs Labor	Equipment	Total	Total Incl O&P
0010	**RESIDENTIAL STEEL DOORS**										
0020	Prehung, insulated, exterior										
0030	Embossed, full panel, 2'-8" x 6'-8"	G	2 Carp	17	.941	Ea.	320	46.50		366.50	425
0040	3'-0" x 6'-8"	G		15	1.067		280	52.50		332.50	390
0060	3'-0" x 7'-0"	G		15	1.067		350	52.50		402.50	465
0070	5'-4" x 6'-8", double	G		8	2		650	98.50		748.50	865
0220	Half glass, 2'-8" x 6'-8"	G		17	.941		335	46.50		381.50	435
0240	3'-0" x 6'-8"	G		16	1		335	49.50		384.50	440
0260	3'-0" x 7'-0"	G		16	1		395	49.50		444.50	510
0270	5'-4" x 6'-8", double	G		8	2		710	98.50		808.50	935
1320	Flush face, full panel, 2'-8" x 6'-8"	G		16	1		287	49.50		336.50	390
1340	3'-0" x 6'-8"	G		15	1.067		287	52.50		339.50	395
1360	3'-0" x 7'-0"	G		15	1.067		310	52.50		362.50	420
1380	5'-4" x 6'-8", double	G		8	2		575	98.50		673.50	780
1420	Half glass, 2'-8" x 6'-8"	G		17	.941		350	46.50		396.50	455
1440	3'-0" x 6'-8"	G		16	1		350	49.50		399.50	460
1460	3'-0" x 7'-0"	G		16	1		435	49.50		484.50	550
1480	5'-4" x 6'-8", double	G	↓	8	2		660	98.50		758.50	875
1500	Sidelight, full lite, 1'-0" x 6'-8" with grille	G					270			270	297
1510	1'-0" x 6'-8", low e	G					290			290	320
1520	1'-0" x 6'-8", half lite	G					300			300	330
1530	1'-0" x 6'-8", half lite, low e	G					292			292	320
2300	Interior, residential, closet, bi-fold, 2'-0" x 6'-8"	G	2 Carp	16	1		175	49.50		224.50	269
2330	3'-0" wide	G		16	1		205	49.50		254.50	300
2360	4'-0" wide	G		15	1.067		280	52.50		332.50	390
2400	5'-0" wide	G		14	1.143		325	56.50		381.50	445
2420	6'-0" wide	G	↓	13	1.231	↓	300	60.50		360.50	425

08 13 13.25 Doors Hollow Metal

			Crew	Daily Output	Labor-Hours	Unit	Material	2017 Bare Costs Labor	Equipment	Total	Total Incl O&P
0010	**DOORS HOLLOW METAL**										
0500	Exterior, commercial, flush, 20 ga., 1-3/4" x 7'-0" x 2'-6" wide	G	2 Carp	15	1.067	Ea.	395	52.50		447.50	515
0530	2'-8" wide	G		15	1.067		445	52.50		497.50	570
0560	3'-0" wide	G		14	1.143		445	56.50		501.50	575
1000	18 ga., 1-3/4" x 7'-0" x 2'-6" wide	G		15	1.067		505	52.50		557.50	635
1030	2'-8" wide	G		15	1.067		510	52.50		562.50	640
1060	3'-0" wide	G	↓	14	1.143	↓	495	56.50		551.50	630

08 13 Metal Doors

08 13 13 – Hollow Metal Doors

08 13 13.25 Doors Hollow Metal		Crew	Daily Output	Labor-Hours	Unit	Material	2017 Bare Costs Labor	Equipment	Total	Total Incl O&P
1500	16 ga., 1-3/4" x 7'-0" x 2'-6" wide	G	2 Carp	15	1.067	Ea.	575	52.50		627.50	710
1530	2'-8" wide	G		15	1.067		580	52.50		632.50	715
1560	3'-0" wide	G		14	1.143		575	56.50		631.50	715
1590	3'-6" wide	G		14	1.143		665	56.50		721.50	815
2900	Fire door, "A" label, 18 gauge, 1-3/4" x 2'-6" x 7'-0"	G		15	1.067		660	52.50		712.50	805
2930	2'-8" wide	G		15	1.067		680	52.50		732.50	830
2960	3'-0" wide	G		14	1.143		655	56.50		711.50	805
2990	3'-6" wide	G		14	1.143		745	56.50		801.50	905
3100	"B" label, 2'-6" wide	G		15	1.067		600	52.50		652.50	735
3130	2'-8" wide	G		15	1.067		600	52.50		652.50	740
3160	3'-0" wide	G		14	1.143		595	56.50		651.50	735

08 13 16 – Aluminum Doors

08 13 16.10 Commercial Aluminum Doors

		Crew	Daily Output	Labor-Hours	Unit	Material	2017 Bare Costs Labor	Equipment	Total	Total Incl O&P
0010	**COMMERCIAL ALUMINUM DOORS**, flush, no glazing									
5000	Flush panel doors, pair of 2'-6" x 7'-0"	2 Sswk	2	8	Pr.	1,525	435		1,960	2,400
5050	3'-0" x 7'-0", single		2.50	6.400	Ea.	860	350		1,210	1,525
5100	Pair of 3'-0" x 7'-0"		2	8	Pr.	1,625	435		2,060	2,525
5150	3'-6" x 7'-0", single		2.50	6.400	Ea.	1,025	350		1,375	1,700

08 14 Wood Doors

08 14 13 – Carved Wood Doors

08 14 13.10 Types of Wood Doors, Carved

		Crew	Daily Output	Labor-Hours	Unit	Material	2017 Bare Costs Labor	Equipment	Total	Total Incl O&P
0010	**TYPES OF WOOD DOORS, CARVED**									
3000	Solid wood, 1-3/4" thick stile and rail									
3020	Mahogany, 3'-0" x 7'-0", six panel	2 Carp	14	1.143	Ea.	1,300	56.50		1,356.50	1,500
3030	With two lites		10	1.600		1,975	79		2,054	2,300
3040	3'-6" x 8'-0", six panel		10	1.600		1,550	79		1,629	1,850
3050	With two lites		8	2		2,700	98.50		2,798.50	3,125
3100	Pine, 3'-0" x 7'-0", six panel		14	1.143		560	56.50		616.50	700
3110	With two lites		10	1.600		825	79		904	1,025
3120	3'-6" x 8'-0", six panel		10	1.600		1,000	79		1,079	1,225
3130	With two lites		8	2		1,900	98.50		1,998.50	2,250
3200	Red oak, 3'-0" x 7'-0", six panel		14	1.143		1,850	56.50		1,906.50	2,100
3210	With two lites		10	1.600		2,550	79		2,629	2,925
3220	3'-6" x 8'-0", six panel		10	1.600		2,575	79		2,654	2,950
3230	With two lites		8	2		3,525	98.50		3,623.50	4,050
4000	Hand carved door, mahogany									
4020	3'-0" x 7'-0", simple design	2 Carp	14	1.143	Ea.	1,750	56.50		1,806.50	2,000
4030	Intricate design		11	1.455		3,700	71.50		3,771.50	4,175
4040	3'-6" x 8'-0", simple design		10	1.600		3,000	79		3,079	3,425
4050	Intricate design		8	2		3,700	98.50		3,798.50	4,225
4400	For custom finish, add					475			475	525
4600	Side light, mahogany, 7'-0" x 1'-6" wide, 4 lites	2 Carp	18	.889		1,100	44		1,144	1,275
4610	6 lites		14	1.143		2,550	56.50		2,606.50	2,875
4620	8'-0" x 1'-6" wide, 4 lites		14	1.143		1,800	56.50		1,856.50	2,050
4630	6 lites		10	1.600		2,100	79		2,179	2,425
4640	Side light, oak, 7'-0" x 1'-6" wide, 4 lites		18	.889		1,200	44		1,244	1,400
4650	6 lites		14	1.143		2,100	56.50		2,156.50	2,375
4660	8'-0" x 1'-6" wide, 4 lites		14	1.143		1,100	56.50		1,156.50	1,275
4670	6 lites		10	1.600		2,100	79		2,179	2,425

For customer support on your Building Construction Costs with RSMeans Data, call 800.448.8182.

08 14 16 – Flush Wood Doors

08 14 16.09 Smooth Wood Doors	Crew	Daily Output	Labor-Hours	Unit	Material	2017 Bare Costs Labor	Equipment	Total	Total Incl O&P
0010 **SMOOTH WOOD DOORS**									
0015 Flush, interior, hollow core									
0025 Lauan face, 1-3/8", 3'-0" x 6'-8"	2 Carp	17	.941	Ea.	55	46.50		101.50	132
0030 4'-0" x 6'-8"		16	1		126	49.50		175.50	215
0080 1-3/4", 2'-0" x 6'-8"		17	.941		54	46.50		100.50	131
0108 3'-0" x 7'-0"		16	1		145	49.50		194.50	236
0112 Pair of 3'-0" x 7'-0"		9	1.778	Pr.	128	87.50		215.50	275
0140 Birch face, 1-3/8", 2'-6" x 6'-8"		17	.941	Ea.	90	46.50		136.50	170
0180 3'-0" x 6'-8"		17	.941		97.50	46.50		144	178
0200 4'-0" x 6'-8"		16	1		158	49.50		207.50	250
0202 1-3/4", 2'-0" x 6'-8"		17	.941		60.50	46.50		107	138
0210 3'-0" x 7'-0"		16	1		139	49.50		188.50	229
0214 Pair of 3'-0" x 7'-0"		9	1.778	Pr.	266	87.50		353.50	425
0220 Oak face, 1-3/8", 2'-0" x 6'-8"		17	.941	Ea.	104	46.50		150.50	185
0280 3'-0" x 6'-8"		17	.941		115	46.50		161.50	198
0300 4'-0" x 6'-8"		16	1		139	49.50		188.50	229
0305 1-3/4", 2'-6" x 6'-8"		17	.941		110	46.50		156.50	192
0310 3'-0" x 7'-0"		16	1		240	49.50		289.50	340
0320 Walnut face, 1-3/8", 2'-0" x 6'-8"		17	.941		182	46.50		228.50	272
0340 2'-6" x 6'-8"		17	.941		189	46.50		235.50	279
0380 3'-0" x 6'-8"		17	.941		197	46.50		243.50	287
0400 4'-0" x 6'-8"		16	1		216	49.50		265.50	315
0430 For 7'-0" high, add					26			26	28.50
0440 For 8'-0" high, add					38			38	42
0480 For prefinishing, clear, add					47			47	51.50
0500 For prefinishing, stain, add					59			59	65
1320 M.D. overlay on hardboard, 1-3/8", 2'-0" x 6'-8"	2 Carp	17	.941		120	46.50		166.50	203
1340 2'-6" x 6'-8"		17	.941		121	46.50		167.50	204
1380 3'-0" x 6'-8"		17	.941		130	46.50		176.50	214
1400 4'-0" x 6'-8"		16	1		185	49.50		234.50	280
1420 For 7'-0" high, add					17			17	18.70
1440 For 8'-0" high, add					34			34	37.50
1720 H.P. plastic laminate, 1-3/8", 2'-0" x 6'-8"	2 Carp	16	1		275	49.50		324.50	380
1740 2'-6" x 6'-8"		16	1		270	49.50		319.50	375
1780 3'-0" x 6'-8"		15	1.067		299	52.50		351.50	410
1800 4'-0" x 6'-8"		14	1.143		395	56.50		451.50	520
1820 For 7'-0" high, add					17			17	18.70
1840 For 8'-0" high, add					34			34	37.50
2020 Particle core, lauan face, 1-3/8", 2'-6" x 6'-8"	2 Carp	15	1.067		97	52.50		149.50	188
2040 3'-0" x 6'-8"		14	1.143		99	56.50		155.50	195
2080 3'-0" x 7'-0"		13	1.231		103	60.50		163.50	206
2085 4'-0" x 7'-0"		12	1.333		129	65.50		194.50	243
2110 1-3/4", 3'-0" x 7'-0"		13	1.231		145	60.50		205.50	253
2120 Birch face, 1-3/8", 2'-6" x 6'-8"		15	1.067		108	52.50		160.50	200
2140 3'-0" x 6'-8"		14	1.143		119	56.50		175.50	217
2180 3'-0" x 7'-0"		13	1.231		129	60.50		189.50	235
2200 4'-0" x 7'-0"		12	1.333		146	65.50		211.50	262
2205 1-3/4", 3'-0" x 7'-0"		13	1.231		129	60.50		189.50	235
2220 Oak face, 1-3/8", 2'-6" x 6'-8"		15	1.067		122	52.50		174.50	215
2240 3'-0" x 6'-8"		14	1.143		134	56.50		190.50	233
2280 3'-0" x 7'-0"		13	1.231		140	60.50		200.50	247
2300 4'-0" x 7'-0"		12	1.333		164	65.50		229.50	281

08 14 16 – Flush Wood Doors

08 14 16.09 Smooth Wood Doors

		Crew	Daily Output	Labor-Hours	Unit	Material	2017 Bare Costs Labor	Equipment	Total	Total Incl O&P
2305	1-3/4", 3'-0" x 7'-0"	2 Carp	13	1.231	Ea.	210	60.50		270.50	325
2320	Walnut face, 1-3/8", 2'-0" x 6'-8"		15	1.067		129	52.50		181.50	223
2340	2'-6" x 6'-8"		14	1.143		146	56.50		202.50	247
2380	3'-0" x 6'-8"		13	1.231		164	60.50		224.50	273
2400	4'-0" x 6'-8"		12	1.333		216	65.50		281.50	340
2440	For 8'-0" high, add					43			43	47.50
2460	For 8'-0" high walnut, add					38			38	42
2720	For prefinishing, clear, add					38			38	42
2740	For prefinishing, stain, add					56			56	61.50
3320	M.D. overlay on hardboard, 1-3/8", 2'-6" x 6'-8"	2 Carp	14	1.143		106	56.50		162.50	203
3340	3'-0" x 6'-8"		13	1.231		115	60.50		175.50	220
3380	3'-0" x 7'-0"		12	1.333		117	65.50		182.50	230
3400	4'-0" x 7'-0"		10	1.600		155	79		234	292
3440	For 8'-0" height, add					37			37	40.50
3460	For solid wood core, add					42			42	46
3720	H.P. plastic laminate, 1-3/8", 2'-6" x 6'-8"	2 Carp	13	1.231		160	60.50		220.50	269
3740	3'-0" x 6'-8"		12	1.333		185	65.50		250.50	305
3780	3'-0" x 7'-0"		11	1.455		190	71.50		261.50	320
3800	4'-0" x 7'-0"		8	2		225	98.50		323.50	400
3840	For 8'-0" height, add					37			37	40.50
3860	For solid wood core, add					42			42	46
4000	Exterior, flush, solid core, birch, 1-3/4" x 2'-6" x 7'-0"	2 Carp	15	1.067		185	52.50		237.50	285
4020	2'-8" wide		15	1.067		163	52.50		215.50	260
4040	3'-0" wide		14	1.143		214	56.50		270.50	320
4100	Oak faced 1-3/4" x 2'-6" x 7'-0"		15	1.067		225	52.50		277.50	330
4120	2'-8" wide		15	1.067		245	52.50		297.50	350
4140	3'-0" wide		14	1.143		240	56.50		296.50	350
4200	Walnut faced, 1-3/4" x 2'-6" x 7'-0"		15	1.067		330	52.50		382.50	445
4220	2'-8" wide		15	1.067		335	52.50		387.50	450
4240	3'-0" wide		14	1.143		325	56.50		381.50	445
4300	For 6'-8" high door, deduct from 7'-0" door					18			18	19.80
5000	Wood doors, for vision lite, add					98			98	108
5010	Wood doors, for narrow lite, add					105			105	116
5015	Wood doors, for bottom (or top) louver, add					290			290	320

08 14 16.20 Wood Fire Doors

		Crew	Daily Output	Labor-Hours	Unit	Material	2017 Bare Costs Labor	Equipment	Total	Total Incl O&P
0010	**WOOD FIRE DOORS**									
0020	Particle core, 7 face plys, "B" label,									
0040	1 hour, birch face, 1-3/4" x 2'-6" x 6'-8"	2 Carp	14	1.143	Ea.	420	56.50		476.50	545
0080	3'-0" x 6'-8"		13	1.231		410	60.50		470.50	545
0090	3'-0" x 7'-0"		12	1.333		440	65.50		505.50	585
0100	4'-0" x 7'-0"		12	1.333		550	65.50		615.50	705
0140	Oak face, 2'-6" x 6'-8"		14	1.143		465	56.50		521.50	595
0180	3'-0" x 6'-8"		13	1.231		475	60.50		535.50	620
0190	3'-0" x 7'-0"		12	1.333		490	65.50		555.50	640
0200	4'-0" x 7'-0"		12	1.333		610	65.50		675.50	770
0240	Walnut face, 2'-6" x 6'-8"		14	1.143		490	56.50		546.50	625
0280	3'-0" x 6'-8"		13	1.231		505	60.50		565.50	650
0290	3'-0" x 7'-0"		12	1.333		540	65.50		605.50	695
0300	4'-0" x 7'-0"		12	1.333		690	65.50		755.50	860
0440	M.D. overlay on hardboard, 2'-6" x 6'-8"		15	1.067		330	52.50		382.50	445
0480	3'-0" x 6'-8"		14	1.143		390	56.50		446.50	515
0490	3'-0" x 7'-0"		13	1.231		415	60.50		475.50	550

08 14 16 – Flush Wood Doors

08 14 16.20 Wood Fire Doors	Crew	Daily Output	Labor-Hours	Unit	Material	2017 Bare Costs Labor	Equipment	Total	Total Incl O&P
0500	4'-0" x 7'-0"	2 Carp	12	1.333	Ea.	440	65.50		505.50	585
0740	90 minutes, birch face, 1-3/4" x 2'-6" x 6'-8"		14	1.143		330	56.50		386.50	445
0780	3'-0" x 6'-8"		13	1.231		320	60.50		380.50	445
0790	3'-0" x 7'-0"		12	1.333		380	65.50		445.50	515
0800	4'-0" x 7'-0"		12	1.333		525	65.50		590.50	680
0840	Oak face, 2'-6" x 6'-8"		14	1.143		460	56.50		516.50	590
0880	3'-0" x 6'-8"		13	1.231		470	60.50		530.50	610
0890	3'-0" x 7'-0"		12	1.333		470	65.50		535.50	615
0900	4'-0" x 7'-0"		12	1.333		625	65.50		690.50	790
0940	Walnut face, 2'-6" x 6'-8"		14	1.143		420	56.50		476.50	545
0980	3'-0" x 6'-8"		13	1.231		435	60.50		495.50	575
0990	3'-0" x 7'-0"		12	1.333		495	65.50		560.50	645
1000	4'-0" x 7'-0"		12	1.333		650	65.50		715.50	815
1140	M.D. overlay on hardboard, 2'-6" x 6'-8"		15	1.067		390	52.50		442.50	510
1180	3'-0" x 6'-8"		14	1.143		400	56.50		456.50	525
1190	3'-0" x 7'-0"		13	1.231		415	60.50		475.50	550
1200	4'-0" x 7'-0"		12	1.333		470	65.50		535.50	615
1240	For 8'-0" height, add					81			81	89
1260	For 8'-0" height walnut, add					93			93	102
2200	Custom architectural "B" label, flush, 1-3/4" thick, birch,									
2210	Solid core									
2220	2'-6" x 7'-0"	2 Carp	15	1.067	Ea.	330	52.50		382.50	445
2260	3'-0" x 7'-0"		14	1.143		335	56.50		391.50	455
2300	4'-0" x 7'-0"		13	1.231		435	60.50		495.50	575
2420	4'-0" x 8'-0"		11	1.455		445	71.50		516.50	595
2480	For oak veneer, add					50%				
2500	For walnut veneer, add					75%				

08 14 33 – Stile and Rail Wood Doors

08 14 33.10 Wood Doors Paneled	Crew	Daily Output	Labor-Hours	Unit	Material	2017 Bare Costs Labor	Equipment	Total	Total Incl O&P
0010	**WOOD DOORS PANELED**									
0020	Interior, six panel, hollow core, 1-3/8" thick									
0040	Molded hardboard, 2'-0" x 6'-8"	2 Carp	17	.941	Ea.	62	46.50		108.50	139
0060	2'-6" x 6'-8"		17	.941		64	46.50		110.50	142
0070	2'-8" x 6'-8"		17	.941		67	46.50		113.50	145
0080	3'-0" x 6'-8"		17	.941		72	46.50		118.50	150
0140	Embossed print, molded hardboard, 2'-0" x 6'-8"		17	.941		64	46.50		110.50	142
0160	2'-6" x 6'-8"		17	.941		64	46.50		110.50	142
0180	3'-0" x 6'-8"		17	.941		72	46.50		118.50	150
0540	Six panel, solid, 1-3/8" thick, pine, 2'-0" x 6'-8"		15	1.067		155	52.50		207.50	252
0560	2'-6" x 6'-8"		14	1.143		170	56.50		226.50	273
0580	3'-0" x 6'-8"		13	1.231		145	60.50		205.50	253
1020	Two panel, bored rail, solid, 1-3/8" thick, pine, 1'-6" x 6'-8"		16	1		270	49.50		319.50	375
1040	2'-0" x 6'-8"		15	1.067		355	52.50		407.50	470
1060	2'-6" x 6'-8"		14	1.143		400	56.50		456.50	525
1340	Two panel, solid, 1-3/8" thick, fir, 2'-0" x 6'-8"		15	1.067		160	52.50		212.50	257
1360	2'-6" x 6'-8"		14	1.143		210	56.50		266.50	315
1380	3'-0" x 6'-8"		13	1.231		415	60.50		475.50	550
1740	Five panel, solid, 1-3/8" thick, fir, 2'-0" x 6'-8"		15	1.067		280	52.50		332.50	390
1760	2'-6" x 6'-8"		14	1.143		420	56.50		476.50	545
1780	3'-0" x 6'-8"		13	1.231		420	60.50		480.50	555

08 14 33.20 Wood Doors Residential	Crew	Daily Output	Labor-Hours	Unit	Material	2017 Bare Costs Labor	Equipment	Total	Total Incl O&P
0010 **WOOD DOORS RESIDENTIAL**									
0200 Exterior, combination storm & screen, pine									
0260 2'-8" wide	2 Carp	10	1.600	Ea.	315	79		394	465
0280 3'-0" wide		9	1.778		320	87.50		407.50	485
0300 7'-1" x 3'-0" wide		9	1.778		365	87.50		452.50	540
0400 Full lite, 6'-9" x 2'-6" wide		11	1.455		325	71.50		396.50	465
0420 2'-8" wide		10	1.600		325	79		404	480
0440 3'-0" wide		9	1.778		335	87.50		422.50	500
0500 7'-1" x 3'-0" wide		9	1.778		365	87.50		452.50	535
0700 Dutch door, pine, 1-3/4" x 2'-8" x 6'-8", 6 panel		12	1.333		745	65.50		810.50	920
0720 Half glass		10	1.600		930	79		1,009	1,150
0800 3'-0" wide, 6 panel		12	1.333		745	65.50		810.50	920
0820 Half glass		10	1.600		880	79		959	1,075
1000 Entrance door, colonial, 1-3/4" x 6'-8" x 2'-8" wide		16	1		560	49.50		609.50	690
1020 6 panel pine, 3'-0" wide		15	1.067		435	52.50		487.50	560
1100 8 panel pine, 2'-8" wide		16	1		610	49.50		659.50	745
1120 3'-0" wide	▼	15	1.067		600	52.50		652.50	740
1200 For tempered safety glass lites (min of 2), add					80			80	88
1300 Flush, birch, solid core, 1-3/4" x 6'-8" x 2'-8" wide	2 Carp	16	1		111	49.50		160.50	198
1320 3'-0" wide		15	1.067		119	52.50		171.50	211
1350 7'-0" x 2'-8" wide		16	1		133	49.50		182.50	223
1360 3'-0" wide	▼	15	1.067		155	52.50		207.50	251
1380 For tempered safety glass lites, add				▼	108			108	118
1930 For dutch door with shelf, add					140%				
2700 Interior, closet, bi-fold, w/hardware, no frame or trim incl.									
2720 Flush, birch, 2'-6" x 6'-8"	2 Carp	13	1.231	Ea.	71	60.50		131.50	171
2740 3'-0" wide		13	1.231		73	60.50		133.50	173
2760 4'-0" wide		12	1.333		113	65.50		178.50	225
2780 5'-0" wide		11	1.455		113	71.50		184.50	234
2800 6'-0" wide		10	1.600		132	79		211	266
3000 Raised panel pine, 6'-6" or 6'-8" x 2'-6" wide		13	1.231		214	60.50		274.50	330
3020 3'-0" wide		13	1.231		281	60.50		341.50	405
3040 4'-0" wide		12	1.333		310	65.50		375.50	440
3060 5'-0" wide		11	1.455		395	71.50		466.50	545
3080 6'-0" wide		10	1.600		435	79		514	595
3200 Louvered, pine, 6'-6" or 6'-8" x 2'-6" wide		13	1.231		152	60.50		212.50	260
3220 3'-0" wide		13	1.231		228	60.50		288.50	345
3240 4'-0" wide		12	1.333		244	65.50		309.50	370
3260 5'-0" wide		11	1.455		272	71.50		343.50	410
3280 6'-0" wide	▼	10	1.600	▼	300	79		379	450
4400 Bi-passing closet, incl. hardware and frame, no trim incl.									
4420 Flush, lauan, 6'-8" x 4'-0" wide	2 Carp	12	1.333	Opng.	180	65.50		245.50	299
4440 5'-0" wide		11	1.455		193	71.50		264.50	325
4460 6'-0" wide		10	1.600		217	79		296	360
4600 Flush, birch, 6'-8" x 4'-0" wide		12	1.333		248	65.50		313.50	375
4620 5'-0" wide		11	1.455		216	71.50		287.50	345
4640 6'-0" wide		10	1.600		293	79		372	440
4800 Louvered, pine, 6'-8" x 4'-0" wide		12	1.333		490	65.50		555.50	635
4820 5'-0" wide		11	1.455		535	71.50		606.50	700
4840 6'-0" wide		10	1.600		630	79		709	815
5000 Paneled, pine, 6'-8" x 4'-0" wide		12	1.333		475	65.50		540.50	620
5020 5'-0" wide	▼	11	1.455	▼	550	71.50		621.50	715

08 14 Wood Doors

08 14 33 – Stile and Rail Wood Doors

08 14 33.20 Wood Doors Residential

		Crew	Daily Output	Labor-Hours	Unit	Material	2017 Bare Costs Labor	Equipment	Total	Total Incl O&P
5040	6'-0" wide	2 Carp	10	1.600	Opng.	735	79		814	930
5042	8'-0" wide	↓	12	1.333	↓	950	65.50		1,015.50	1,150
6100	Folding accordion, closet, including track and frame									
6120	Vinyl, 2 layer, stock	2 Carp	10	1.600	Ea.	67	79		146	195
6140	Woven mahogany and vinyl, stock		10	1.600		60	79		139	187
6160	Wood slats with vinyl overlay, stock		10	1.600		151	79		230	287
6180	Economy vinyl, stock		10	1.600		43.50	79		122.50	169
6200	Rigid PVC	↓	10	1.600	↓	59.50	79		138.50	187
7310	Passage doors, flush, no frame included									
7320	Hardboard, hollow core, 1-3/8" x 6'-8" x 1'-6" wide	2 Carp	18	.889	Ea.	42	44		86	113
7330	2'-0" wide		18	.889		44.50	44		88.50	116
7340	2'-6" wide		18	.889		51	44		95	123
7350	2'-8" wide		18	.889		52	44		96	124
7360	3'-0" wide		17	.941		54	46.50		100.50	131
7420	Lauan, hollow core, 1-3/8" x 6'-8" x 1'-6" wide		18	.889		37	44		81	108
7440	2'-0" wide		18	.889		36	44		80	107
7450	2'-4" wide		18	.889		39.50	44		83.50	111
7460	2'-6" wide		18	.889		40	44		84	111
7480	2'-8" wide		18	.889		41.50	44		85.50	113
7500	3'-0" wide		17	.941		43.50	46.50		90	119
7700	Birch, hollow core, 1-3/8" x 6'-8" x 1'-6" wide		18	.889		43.50	44		87.50	115
7720	2'-0" wide		18	.889		38.50	44		82.50	110
7740	2'-6" wide		18	.889		51.50	44		95.50	124
7760	2'-8" wide		18	.889		51.50	44		95.50	124
7780	3'-0" wide		17	.941		52.50	46.50		99	129
8000	Pine louvered, 1-3/8" x 6'-8" x 1'-6" wide		19	.842		112	41.50		153.50	187
8020	2'-0" wide		18	.889		131	44		175	211
8040	2'-6" wide		18	.889		149	44		193	230
8060	2'-8" wide		18	.889		156	44		200	239
8080	3'-0" wide		17	.941		168	46.50		214.50	256
8300	Pine paneled, 1-3/8" x 6'-8" x 1'-6" wide		19	.842		121	41.50		162.50	197
8320	2'-0" wide		18	.889		163	44		207	246
8330	2'-4" wide		18	.889		178	44		222	263
8340	2'-6" wide		18	.889		184	44		228	269
8360	2'-8" wide		18	.889		184	44		228	269
8380	3'-0" wide	↓	17	.941	↓	196	46.50		242.50	286

08 14 35 – Torrified Doors

08 14 35.10 Torrified Exterior Doors

		Crew	Daily Output	Labor-Hours	Unit	Material	2017 Bare Costs Labor	Equipment	Total	Total Incl O&P
0010	**TORRIFIED EXTERIOR DOORS**									
0020	Wood doors made from torrified wood, exterior									
0030	All doors require a finish be applied, all glass is insulated									
0040	All doors require pilot holes for all fasteners									
0100	6 panel, paint grade poplar, 1-3/4" x 3'-0" x 6'-8"	2 Carp	12	1.333	Ea.	1,075	65.50		1,140.50	1,275
0120	Half glass 3'-0" x 6'-8"	"	12	1.333		1,175	65.50		1,240.50	1,400
0200	Side lite, full glass, 1-3/4" x 1'-2" x 6'-8"					905			905	995
0220	Side lite, half glass, 1-3/4" x 1'-2" x 6'-8"					905			905	995
0300	Raised face, 2 panel, paint grade poplar, 1-3/4" x 3'-0" x 7'-0"	2 Carp	12	1.333		1,275	65.50		1,340.50	1,500
0320	Side lite, raised face, half glass, 1-3/4" x 1'-2" x 7'-0"					1,050			1,050	1,175
0500	6 panel, Fir, 1-3/4" x 3'-0" x 6'-8"	2 Carp	12	1.333		1,550	65.50		1,615.50	1,800
0520	Half glass 3'-0" x 6'-8"	"	12	1.333		1,650	65.50		1,715.50	1,925
0600	Side lite, full glass, 1-3/4" x 1'-2" x 6'-8"					1,150			1,150	1,275
0620	Side lite, half glass, 1-3/4" x 1'-2" x 6'-8"		↓			1,450			1,450	1,600

08 14 Wood Doors

08 14 35 – Torrified Doors

08 14 35.10 Torrified Exterior Doors		Crew	Daily Output	Labor-Hours	Unit	Material	2017 Bare Costs Labor	Equipment	Total	Total Incl O&P
0700	6 panel, Mahogany, 1-3/4" x 3'-0" x 6'-8"	2 Carp	12	1.333	Ea.	1,625	65.50		1,690.50	1,875
0800	Side lite, full glass, 1-3/4" x 1'-2" x 6'-8"					1,225			1,225	1,350
0820	Side lite, half glass, 1-3/4" x 1'-2" x 6'-8"		↓			1,200			1,200	1,325

08 14 40 – Interior Cafe Doors

08 14 40.10 Cafe Style Doors

		Crew	Daily Output	Labor-Hours	Unit	Material	2017 Bare Costs Labor	Equipment	Total	Total Incl O&P
0010	**CAFE STYLE DOORS**									
6520	Interior cafe doors, 2'-6" opening, stock, panel pine	2 Carp	16	1	Ea.	203	49.50		252.50	299
6540	3'-0" opening	"	16	1	"	251	49.50		300.50	350
6550	Louvered pine									
6560	2'-6" opening	2 Carp	16	1	Ea.	169	49.50		218.50	262
8000	3'-0" opening		16	1		195	49.50		244.50	291
8010	2'-6" opening, hardwood		16	1		272	49.50		321.50	375
8020	3'-0" opening	↓	16	1	↓	298	49.50		347.50	400

08 16 Composite Doors

08 16 13 – Fiberglass Doors

08 16 13.10 Entrance Doors, Fibrous Glass

			Crew	Daily Output	Labor-Hours	Unit	Material	2017 Bare Costs Labor	Equipment	Total	Total Incl O&P
0010	**ENTRANCE DOORS, FIBROUS GLASS**										
0020	Exterior, fiberglass, door, 2'-8" wide x 6'-8" high	G	2 Carp	15	1.067	Ea.	281	52.50		333.50	390
0040	3'-0" wide x 6'-8" high	G		15	1.067		274	52.50		326.50	380
0060	3'-0" wide x 7'-0" high	G		15	1.067		490	52.50		542.50	615
0080	3'-0" wide x 6'-8" high, with two lites	G		15	1.067		340	52.50		392.50	455
0100	3'-0" wide x 8'-0" high, with two lites	G		15	1.067		525	52.50		577.50	660
0110	Half glass, 3'-0" wide x 6'-8" high	G		15	1.067		450	52.50		502.50	575
0120	3'-0" wide x 6'-8" high, low e	G		15	1.067		490	52.50		542.50	615
0130	3'-0" wide x 8'-0" high	G		15	1.067		585	52.50		637.50	725
0140	3'-0" wide x 8'-0" high, low e	G	↓	15	1.067		675	52.50		727.50	825
0150	Side lights, 1'-0" wide x 6'-8" high	G					281			281	310
0160	1'-0" wide x 6'-8" high, low e	G					289			289	320
0180	1'-0" wide x 6'-8" high, full glass	G					340			340	370
0190	1'-0" wide x 6'-8" high, low e	G				↓	360			360	395

08 16 14 – French Doors

08 16 14.10 Exterior Doors With Glass Lites

		Crew	Daily Output	Labor-Hours	Unit	Material	2017 Bare Costs Labor	Equipment	Total	Total Incl O&P
0010	**EXTERIOR DOORS WITH GLASS LITES**									
0020	French, Fir, 1-3/4", 3'-0" wide x 6'-8" high	2 Carp	12	1.333	Ea.	620	65.50		685.50	780
0025	Double		12	1.333		1,250	65.50		1,315.50	1,450
0030	Maple, 1-3/4", 3'-0" wide x 6'-8" high		12	1.333		695	65.50		760.50	865
0035	Double		12	1.333		1,400	65.50		1,465.50	1,625
0040	Cherry, 1-3/4", 3'-0" wide x 6'-8" high		12	1.333		810	65.50		875.50	990
0045	Double		12	1.333		1,625	65.50		1,690.50	1,875
0100	Mahogany, 1-3/4", 3'-0" wide x 8'-0" high		10	1.600		825	79		904	1,025
0105	Double		10	1.600		1,650	79		1,729	1,950
0110	Fir, 1-3/4", 3'-0" wide x 8'-0" high		10	1.600		1,225	79		1,304	1,475
0115	Double		10	1.600		2,475	79		2,554	2,850
0120	Oak, 1-3/4", 3'-0" wide x 8'-0" high		10	1.600		1,875	79		1,954	2,200
0125	Double	↓	10	1.600	↓	3,750	79		3,829	4,250

08 17 Integrated Door Opening Assemblies

08 17 13 – Integrated Metal Door Opening Assemblies

08 17 13.20 Stainless Steel Doors and Frames		Crew	Daily Output	Labor-Hours	Unit	Material	2017 Bare Costs Labor	Equipment	Total	Total Incl O&P
0010	**STAINLESS STEEL DOORS AND FRAMES**									
0500	Stainless steel, prehung door, foam core, 14 ga, 3'-0" x 7'-0"	G	2 Carp	5	3.200	Ea.	3,400	158		3,558	4,000
0600	Stainless steel, prehung double door, foam core, 14 ga, 3'-0" x 7'-0"	G	"	4	4	"	6,700	197		6,897	7,675

08 17 23 – Integrated Wood Door Opening Assemblies

08 17 23.10 Pre-Hung Doors

		Crew	Daily Output	Labor-Hours	Unit	Material	Labor	Equipment	Total	Total Incl O&P
0010	**PRE-HUNG DOORS**									
0300	Exterior, wood, comb. storm & screen, 6'-9" x 2'-6" wide	2 Carp	15	1.067	Ea.	299	52.50		351.50	410
0320	2'-8" wide		15	1.067		300	52.50		352.50	410
0340	3'-0" wide		15	1.067		310	52.50		362.50	420
0360	For 7'-0" high door, add					32			32	35
1600	Entrance door, flush, birch, solid core									
1620	4-5/8" solid jamb, 1-3/4" x 6'-8" x 2'-8" wide	2 Carp	16	1	Ea.	305	49.50		354.50	410
1640	3'-0" wide		16	1		395	49.50		444.50	510
1642	5-5/8" jamb		16	1		335	49.50		384.50	440
1680	For 7'-0" high door, add					25			25	27.50
2000	Entrance door, colonial, 6 panel pine									
2020	4-5/8" solid jamb, 1-3/4" x 6'-8" x 2'-8" wide	2 Carp	16	1	Ea.	645	49.50		694.50	785
2040	3'-0" wide	"	16	1		675	49.50		724.50	820
2060	For 7'-0" high door, add					55			55	60
2200	For 5-5/8" solid jamb, add					45			45	49.50
4000	Interior, passage door, 4-5/8" solid jamb									
4400	Lauan, flush, solid core, 1-3/8" x 6'-8" x 2'-6" wide	2 Carp	17	.941	Ea.	188	46.50		234.50	277
4420	2'-8" wide		17	.941		188	46.50		234.50	277
4440	3'-0" wide		16	1		204	49.50		253.50	300
4600	Hollow core, 1-3/8" x 6'-8" x 2'-6" wide		17	.941		129	46.50		175.50	213
4620	2'-8" wide		17	.941		131	46.50		177.50	215
4640	3'-0" wide		16	1		143	49.50		192.50	233
4700	For 7'-0" high door, add					38			38	42
5000	Birch, flush, solid core, 1-3/8" x 6'-8" x 2'-6" wide	2 Carp	17	.941		279	46.50		325.50	375
5020	2'-8" wide		17	.941		203	46.50		249.50	294
5040	3'-0" wide		16	1		305	49.50		354.50	410
5200	Hollow core, 1-3/8" x 6'-8" x 2'-6" wide		17	.941		230	46.50		276.50	325
5220	2'-8" wide		17	.941		249	46.50		295.50	345
5240	3'-0" wide		16	1		248	49.50		297.50	350
5280	For 7'-0" high door, add					32			32	35
5500	Hardboard paneled, 1-3/8" x 6'-8" x 2'-6" wide	2 Carp	17	.941		151	46.50		197.50	237
5520	2'-8" wide		17	.941		158	46.50		204.50	244
5540	3'-0" wide		16	1		156	49.50		205.50	248
6000	Pine paneled, 1-3/8" x 6'-8" x 2'-6" wide		17	.941		267	46.50		313.50	365
6020	2'-8" wide		17	.941		284	46.50		330.50	380
6040	3'-0" wide		16	1		292	49.50		341.50	395

For customer support on your Building Construction Costs with RSMeans Data, call 800.448.8182.

277

08 31 13.10 Types of Framed Access Doors	Crew	Daily Output	Labor-Hours	Unit	Material	2017 Bare Costs Labor	Equipment	Total	Total Incl O&P
0010 **TYPES OF FRAMED ACCESS DOORS**									
1000 Fire rated door with lock									
1100 Metal, 12" x 12"	1 Carp	10	.800	Ea.	158	39.50		197.50	234
1150 18" x 18"		9	.889		237	44		281	330
1200 24" x 24"		9	.889		335	44		379	435
1250 24" x 36"		8	1		340	49.50		389.50	450
1300 24" x 48"		8	1		465	49.50		514.50	585
1350 36" x 36"		7.50	1.067		535	52.50		587.50	670
1400 48" x 48"		7.50	1.067		650	52.50		702.50	795
1600 Stainless steel, 12" x 12"		10	.800		355	39.50		394.50	450
1650 18" x 18"		9	.889		475	44		519	585
1700 24" x 24"		9	.889		570	44		614	695
1750 24" x 36"	↓	8	1	↓	725	49.50		774.50	875
2000 Flush door for finishing									
2100 Metal 8" x 8"	1 Carp	10	.800	Ea.	37.50	39.50		77	102
2150 12" x 12"	"	10	.800	"	45.50	39.50		85	111
3000 Recessed door for acoustic tile									
3100 Metal, 12" x 12"	1 Carp	4.50	1.778	Ea.	87	87.50		174.50	230
3150 12" x 24"		4.50	1.778		104	87.50		191.50	248
3200 24" x 24"		4	2		116	98.50		214.50	278
3250 24" x 36"	↓	4	2	↓	173	98.50		271.50	340
4000 Recessed door for drywall									
4100 Metal 12" x 12"	1 Carp	6	1.333	Ea.	87	65.50		152.50	197
4150 12" x 24"		5.50	1.455		116	71.50		187.50	237
4200 24" x 36"	↓	5	1.600	↓	173	79		252	310
6000 Standard door									
6100 Metal, 8" x 8"	1 Carp	10	.800	Ea.	45	39.50		84.50	110
6150 12" x 12"		10	.800		54.50	39.50		94	121
6200 18" x 18"		9	.889		76.50	44		120.50	151
6250 24" x 24"		9	.889		92	44		136	168
6300 24" x 36"		8	1		136	49.50		185.50	225
6350 36" x 36"		8	1		155	49.50		204.50	246
6500 Stainless steel, 8" x 8"		10	.800		82.50	39.50		122	152
6550 12" x 12"		10	.800		102	39.50		141.50	173
6600 18" x 18"		9	.889		220	44		264	310
6650 24" x 24"	↓	9	.889	↓	290	44		334	385

08 31 13.20 Bulkhead/Cellar Doors

	Crew	Daily Output	Labor-Hours	Unit	Material	2017 Bare Costs Labor	Equipment	Total	Total Incl O&P
0010 **BULKHEAD/CELLAR DOORS**									
0020 Steel, not incl. sides, 44" x 62"	1 Carp	5.50	1.455	Ea.	610	71.50		681.50	780
0100 52" x 73"		5.10	1.569		885	77.50		962.50	1,100
0500 With sides and foundation plates, 57" x 45" x 24"		4.70	1.702		920	84		1,004	1,125
0600 42" x 49" x 51"	↓	4.30	1.860	↓	935	91.50		1,026.50	1,175

08 31 13.30 Commercial Floor Doors

	Crew	Daily Output	Labor-Hours	Unit	Material	2017 Bare Costs Labor	Equipment	Total	Total Incl O&P
0010 **COMMERCIAL FLOOR DOORS**									
0020 Aluminum tile, steel frame, one leaf, 2' x 2' opng.	2 Sswk	3.50	4.571	Opng.	880	248		1,128	1,375
0050 3'-6" x 3'-6" opening		3.50	4.571		1,600	248		1,848	2,175
0500 Double leaf, 4' x 4' opening		3	5.333		1,750	290		2,040	2,400
0550 5' x 5' opening	↓	3	5.333	↓	3,100	290		3,390	3,875

08 31 13.35 Industrial Floor Doors

	Crew	Daily Output	Labor-Hours	Unit	Material	2017 Bare Costs Labor	Equipment	Total	Total Incl O&P
0010 **INDUSTRIAL FLOOR DOORS**									
0020 Steel 300 psf L.L., single leaf, 2' x 2', 175#	2 Sswk	6	2.667	Opng.	800	145		945	1,125
0050 3' x 3' opening, 300#	↓	5.50	2.909	↓	1,075	158		1,233	1,450

08 31 Access Doors and Panels

08 31 13 – Access Doors and Frames

08 31 13.35 Industrial Floor Doors

		Crew	Daily Output	Labor-Hours	Unit	Material	2017 Bare Costs Labor	2017 Bare Costs Equipment	Total	Total Incl O&P
0300	Double leaf, 4' x 4' opening, 455#	2 Sswk	5	3.200	Opng.	2,275	174		2,449	2,800
0350	5' x 5' opening, 645#		4.50	3.556		3,300	193		3,493	3,950
1000	Aluminum, 300 psf L.L., single leaf, 2' x 2', 60#		6	2.667		850	145		995	1,175
1050	3' x 3' opening, 100#		5.50	2.909		1,300	158		1,458	1,700
1500	Double leaf, 4' x 4' opening, 160#		5	3.200		2,100	174		2,274	2,600
1550	5' x 5' opening, 235#		4.50	3.556		2,800	193		2,993	3,400
2000	Aluminum, 150 psf L.L., single leaf, 2' x 2', 60#		6	2.667		760	145		905	1,075
2050	3' x 3' opening, 95#		5.50	2.909		1,200	158		1,358	1,600
2500	Double leaf, 4' x 4' opening, 150#		5	3.200		1,450	174		1,624	1,900
2550	5' x 5' opening, 230#	↓	4.50	3.556	↓	1,950	193		2,143	2,450

08 31 13.40 Kennel Doors

		Crew	Daily Output	Labor-Hours	Unit	Material	2017 Bare Costs Labor	2017 Bare Costs Equipment	Total	Total Incl O&P
0010	**KENNEL DOORS**									
0020	2 way, swinging type, 13" x 19" opening	2 Carp	11	1.455	Opng.	90	71.50		161.50	209
0100	17" x 29" opening		11	1.455		99	71.50		170.50	219
0200	9" x 9" opening, electronic with accessories	↓	11	1.455	↓	120	71.50		191.50	242

08 32 Sliding Glass Doors

08 32 13 – Sliding Aluminum-Framed Glass Doors

08 32 13.10 Sliding Aluminum Doors

		Crew	Daily Output	Labor-Hours	Unit	Material	2017 Bare Costs Labor	2017 Bare Costs Equipment	Total	Total Incl O&P
0010	**SLIDING ALUMINUM DOORS**									
0350	Aluminum, 5/8" tempered insulated glass, 6' wide									
0400	Premium	2 Carp	4	4	Ea.	1,550	197		1,747	2,000
0450	Economy		4	4		820	197		1,017	1,200
0500	8' wide, premium		3	5.333		1,675	263		1,938	2,225
0550	Economy		3	5.333		1,450	263		1,713	1,975
0600	12' wide, premium		2.50	6.400		3,000	315		3,315	3,775
0650	Economy		2.50	6.400		1,575	315		1,890	2,200
4000	Aluminum, baked on enamel, temp glass, 6'-8" x 10'-0" wide		4	4		1,100	197		1,297	1,500
4020	Insulating glass, 6'-8" x 6'-0" wide		4	4		955	197		1,152	1,350
4040	8'-0" wide		3	5.333		1,125	263		1,388	1,625
4060	10'-0" wide		2	8		1,375	395		1,770	2,100
4080	Anodized, temp glass, 6'-8" x 6'-0" wide		4	4		455	197		652	800
4100	8'-0" wide		3	5.333		575	263		838	1,025
4120	10'-0" wide	↓	2	8	↓	645	395		1,040	1,325
5000	Aluminum sliding glass door system									
5010	Sliding door 4' wide opening single side	2 Carp	2	8	Ea.	6,000	395		6,395	7,200
5015	8' wide opening single side		2	8		9,000	395		9,395	10,500
5020	Telescoping glass door system, 4' wide opening biparting		2	8		5,100	395		5,495	6,200
5025	8' wide opening biparting		2	8		6,000	395		6,395	7,200
5030	Folding glass door, 4' wide opening biparting		2	8		8,000	395		8,395	9,400
5035	8' wide opening biparting		2	8		10,000	395		10,395	11,600
5040	ICU-CCU sliding telescoping glass door, 4' x 7', single side opening		2	8		3,050	395		3,445	3,950
5045	8' x 7', single side opening	↓	2	8	↓	4,625	395		5,020	5,700

08 32 19 – Sliding Wood-Framed Glass Doors

08 32 19.15 Sliding Glass Vinyl-Clad Wood Doors

			Crew	Daily Output	Labor-Hours	Unit	Material	2017 Bare Costs Labor	2017 Bare Costs Equipment	Total	Total Incl O&P
0010	**SLIDING GLASS VINYL-CLAD WOOD DOORS**										
0020	Glass, sliding vinyl clad, insul. glass, 6'-0" x 6'-8"	G	2 Carp	4	4	Opng.	1,525	197		1,722	1,975
0025	6'-0" x 6'-10" high	G		4	4		1,700	197		1,897	2,150
0030	6'-0" x 8'-0" high	G		4	4		2,050	197		2,247	2,550
0050	5'-0" x 6'-8" high	G	↓	4	4	↓	1,575	197		1,772	2,050

08 32 Sliding Glass Doors

08 32 19 – Sliding Wood-Framed Glass Doors

08 32 19.15 Sliding Glass Vinyl-Clad Wood Doors		Crew	Daily Output	Labor-Hours	Unit	Material	2017 Bare Costs Labor	2017 Bare Costs Equipment	Total	Total Incl O&P
0100	8'-0" x 6'-10" high	G	2 Carp	4	4	Opng.	2,025	197		2,222	2,525
0500	4 leaf, 9'-0" x 6'-10" high	G		3	5.333		3,350	263		3,613	4,100
0600	12'-0" x 6'-10" high	G		3	5.333		4,025	263		4,288	4,825

08 33 Coiling Doors and Grilles

08 33 13 – Coiling Counter Doors

08 33 13.10 Counter Doors, Coiling Type

		Crew	Daily Output	Labor-Hours	Unit	Material	2017 Bare Costs Labor	2017 Bare Costs Equipment	Total	Total Incl O&P
0010	**COUNTER DOORS, COILING TYPE**									
0020	Manual, incl. frame and hardware, galv. stl., 4' roll-up, 6' long	2 Carp	2	8	Opng.	1,250	395		1,645	1,975
0300	Galvanized steel, UL label		1.80	8.889		1,300	440		1,740	2,100
0600	Stainless steel, 4' high roll-up, 6' long		2	8		2,150	395		2,545	2,950
0700	10' long		1.80	8.889		2,600	440		3,040	3,525
2000	Aluminum, 4' high, 4' long		2.20	7.273		1,600	360		1,960	2,300
2020	6' long		2	8		1,900	395		2,295	2,675
2040	8' long		1.90	8.421		2,175	415		2,590	3,000
2060	10' long		1.80	8.889		2,200	440		2,640	3,100
2080	14' long		1.40	11.429		2,800	565		3,365	3,950
2100	6' high, 4' long		2	8		1,925	395		2,320	2,725
2120	6' long		1.60	10		1,750	495		2,245	2,675
2140	10' long		1.40	11.429		2,300	565		2,865	3,375

08 33 16 – Coiling Counter Grilles

08 33 16.10 Coiling Grilles

		Crew	Daily Output	Labor-Hours	Unit	Material	2017 Bare Costs Labor	2017 Bare Costs Equipment	Total	Total Incl O&P
0010	**COILING GRILLES**									
0015	Aluminum, manual, incl. frame, mill finish									
0020	Top coiling, 4' high, 4' long	2 Sswk	3.20	5	Opng.	1,650	272		1,922	2,250
0030	6' long		3.20	5		1,700	272		1,972	2,325
0040	8' long		2.40	6.667		2,125	360		2,485	2,925
0050	12' long		2.40	6.667		2,575	360		2,935	3,425
0060	16' long		1.60	10		2,750	545		3,295	3,925
0070	6' high, 4' long		3.20	5		1,925	272		2,197	2,550
0080	6' long		3.20	5		1,875	272		2,147	2,500
0090	8' long		2.40	6.667		1,875	360		2,235	2,650
0100	12' long		1.60	10		2,450	545		2,995	3,600
0110	16' long		1.20	13.333		2,925	725		3,650	4,425
0200	Side coiling, 8' high, 12' long		.60	26.667		2,375	1,450		3,825	5,050
0220	18' long		.50	32		4,600	1,750		6,350	7,975
0240	24' long		.40	40		3,850	2,175		6,025	7,850
0260	12' high, 12' long		.50	32		3,475	1,750		5,225	6,700
0280	18' long		.40	40		5,425	2,175		7,600	9,575
0300	24' long		.28	57.143		6,850	3,100		9,950	12,700

08 33 23 – Overhead Coiling Doors

08 33 23.10 Coiling Service Doors

		Crew	Daily Output	Labor-Hours	Unit	Material	2017 Bare Costs Labor	2017 Bare Costs Equipment	Total	Total Incl O&P
0010	**COILING SERVICE DOORS** Steel, manual, 20 ga., incl. hardware									
0050	8' x 8' high	2 Sswk	1.60	10	Ea.	1,225	545		1,770	2,225
0100	10' x 10' high		1.40	11.429		2,000	620		2,620	3,225
0200	20' x 10' high		1	16		3,200	870		4,070	4,975
0300	12' x 12' high		1.20	13.333		2,000	725		2,725	3,400
0400	20' x 12' high		.90	17.778		2,250	965		3,215	4,075
0500	14' x 14' high		.80	20		3,225	1,075		4,300	5,350
0600	20' x 16' high		.60	26.667		3,925	1,450		5,375	6,725

For customer support on your Building Construction Costs with RSMeans Data, call 800.448.8182.

08 33 Coiling Doors and Grilles

08 33 23 – Overhead Coiling Doors

	08 33 23.10 Coiling Service Doors	Crew	Daily Output	Labor-Hours	Unit	Material	2017 Bare Costs Labor	Equipment	Total	Total Incl O&P
0700	10' x 20' high	2 Sswk	.50	32	Ea.	2,775	1,750		4,525	5,950
1000	12' x 12', crank operated, crank on door side		.80	20		1,875	1,075		2,950	3,875
1100	Crank thru wall		.70	22.857		2,175	1,250		3,425	4,450
1300	For vision panel, add					380			380	420
1600	3' x 7' pass door within rolling steel door, new construction					2,025			2,025	2,225
1700	Existing construction	2 Sswk	2	8		2,150	435		2,585	3,100
2000	Class A fire doors, manual, 20 ga., 8' x 8' high		1.40	11.429		1,675	620		2,295	2,875
2100	10' x 10' high		1.10	14.545		2,300	790		3,090	3,850
2200	20' x 10' high		.80	20		4,700	1,075		5,775	6,975
2300	12' x 12' high		1	16		3,625	870		4,495	5,425
2400	20' x 12' high		.80	20		5,100	1,075		6,175	7,400
2500	14' x 14' high		.60	26.667		3,975	1,450		5,425	6,800
2600	20' x 16' high		.50	32		6,200	1,750		7,950	9,725
2700	10' x 20' high		.40	40		4,900	2,175		7,075	9,025
3000	For 18 ga. doors, add				S.F.	1.80			1.80	1.98
3300	For enamel finish, add				"	2.20			2.20	2.42
3600	For safety edge bottom bar, pneumatic, add				L.F.	24			24	26
3700	Electric, add					45			45	49.50
4000	For weatherstripping, extruded rubber, jambs, add					15.70			15.70	17.25
4100	Hood, add					8.80			8.80	9.70
4200	Sill, add					6.05			6.05	6.65
4500	Motor operators, to 14' x 14' opening	2 Sswk	5	3.200	Ea.	1,200	174		1,374	1,625
4600	Over 14' x 14', jack shaft type	"	5	3.200		1,125	174		1,299	1,525
4700	For fire door, additional fusible link, add					31			31	34

08 34 Special Function Doors

08 34 13 – Cold Storage Doors

	08 34 13.10 Doors for Cold Area Storage	Crew	Daily Output	Labor-Hours	Unit	Material	2017 Bare Costs Labor	Equipment	Total	Total Incl O&P
0010	**DOORS FOR COLD AREA STORAGE**									
0020	Single, 20 ga. galvanized steel									
0300	Horizontal sliding, 5' x 7', manual operation, 3.5" thick	2 Carp	2	8	Ea.	3,325	395		3,720	4,250
0400	4" thick		2	8		3,300	395		3,695	4,250
0500	6" thick		2	8		3,500	395		3,895	4,450
0800	5' x 7', power operation, 2" thick		1.90	8.421		5,700	415		6,115	6,900
0900	4" thick		1.90	8.421		5,800	415		6,215	7,000
1000	6" thick		1.90	8.421		6,600	415		7,015	7,875
1300	9' x 10', manual operation, 2" insulation		1.70	9.412		4,550	465		5,015	5,700
1400	4" insulation		1.70	9.412		4,775	465		5,240	5,950
1500	6" insulation		1.70	9.412		5,700	465		6,165	6,950
1800	Power operation, 2" insulation		1.60	10		7,825	495		8,320	9,375
1900	4" insulation		1.60	10		8,000	495		8,495	9,550
2000	6" insulation		1.70	9.412		9,000	465		9,465	10,600
2300	For stainless steel face, add					25%				
3000	Hinged, lightweight, 3' x 7'-0", 2" thick	2 Carp	2	8	Ea.	1,450	395		1,845	2,175
3050	4" thick		1.90	8.421		1,800	415		2,215	2,600
3300	Polymer doors, 3' x 7'-0"		1.90	8.421		1,375	415		1,790	2,125
3350	6" thick		1.40	11.429		2,375	565		2,940	3,475
3600	Stainless steel, 3' x 7'-0", 4" thick		1.90	8.421		1,775	415		2,190	2,575
3650	6" thick		1.40	11.429		3,000	565		3,565	4,150
3900	Painted, 3' x 7'-0", 4" thick		1.90	8.421		1,300	415		1,715	2,050
3950	6" thick		1.40	11.429		2,375	565		2,940	3,475

08 34 Special Function Doors

08 34 13 – Cold Storage Doors

08 34 13.10 Doors for Cold Area Storage

		Crew	Daily Output	Labor-Hours	Unit	Material	2017 Bare Costs Labor	Equipment	Total	Total Incl O&P
5000	Bi-parting, electric operated									
5010	6' x 8' opening, galv. faces, 4" thick for cooler	2 Carp	.80	20	Opng.	7,350	985		8,335	9,600
5050	For freezer, 4" thick		.80	20		8,100	985		9,085	10,400
5300	For door buck framing and door protection, add		2.50	6.400		655	315		970	1,200
6000	Galvanized batten door, galvanized hinges, 4' x 7'		2	8		1,875	395		2,270	2,650
6050	6' x 8'		1.80	8.889		2,525	440		2,965	3,450
6500	Fire door, 3 hr., 6' x 8', single slide		.80	20		8,400	985		9,385	10,800
6550	Double, bi-parting		.70	22.857		12,400	1,125		13,525	15,300

08 34 36 – Darkroom Doors

08 34 36.10 Various Types of Darkroom Doors

		Crew	Daily Output	Labor-Hours	Unit	Material	2017 Bare Costs Labor	Equipment	Total	Total Incl O&P
0010	**VARIOUS TYPES OF DARKROOM DOORS**									
0015	Revolving, standard, 2 way, 36" diameter	2 Carp	3.10	5.161	Opng.	2,900	254		3,154	3,600
0020	41" diameter		3.10	5.161		3,025	254		3,279	3,725
0050	3 way, 51" diameter		1.40	11.429		3,850	565		4,415	5,075
1000	4 way, 49" diameter		1.40	11.429		3,975	565		4,540	5,225
2000	Hinged safety, 2 way, 41" diameter		2.30	6.957		3,750	345		4,095	4,650
2500	3 way, 51" diameter		1.40	11.429		4,075	565		4,640	5,350
3000	Pop out safety, 2 way, 41" diameter		3.10	5.161		4,050	254		4,304	4,875
4000	3 way, 51" diameter		1.40	11.429		4,925	565		5,490	6,250
5000	Wheelchair-type, pop out, 51" diameter		1.40	11.429		5,525	565		6,090	6,925
5020	72" diameter		.90	17.778		9,025	875		9,900	11,300
9300	For complete darkrooms, see Section 13 21 53.50									

08 34 53 – Security Doors and Frames

08 34 53.20 Steel Door

		Crew	Daily Output	Labor-Hours	Unit	Material	2017 Bare Costs Labor	Equipment	Total	Total Incl O&P
0010	**STEEL DOOR** flush with ballistic core and welded frame both 14 ga.									
0120	UL 752 Level 3, 1-3/4", 3'-0" x 7'-0"	2 Carp	1.50	10.667	Opng.	2,525	525		3,050	3,575
0125	1-3/4", 3'-6" x 7'-0"		1.50	10.667		2,750	525		3,275	3,825
0130	1-3/4", 4'-0" x 7'-0"		1.20	13.333		2,950	655		3,605	4,250
0150	UL 752 Level 8, 1-3/4", 3'-0" x 7'-0"		1.50	10.667		12,800	525		13,325	14,900
0155	1-3/4", 3'-6" x 7'-0"		1.50	10.667		13,900	525		14,425	16,100
0160	1-3/4", 4'-0" x 7'-0"		1.20	13.333		14,100	655		14,755	16,500
1000	Safe room sliding door and hardware, 1-3/4", 3'-0" x 7'-0" UL 752 Level 3		.50	32		23,500	1,575		25,075	28,300
1050	Safe room swinging door and hardware, 1-3/4", 3'-0" x 7'-0" UL 752 Level 3		.50	32		28,100	1,575		29,675	33,300

08 34 53.30 Wood Ballistic Doors

		Crew	Daily Output	Labor-Hours	Unit	Material	2017 Bare Costs Labor	Equipment	Total	Total Incl O&P
0010	**WOOD BALLISTIC DOORS** with frames and hardware									
0050	Wood, 1-3/4", 3'-0" x 7'-0" UL 752 Level 3	2 Carp	1.50	10.667	Opng.	2,500	525		3,025	3,550

08 34 56 – Security Gates

08 34 56.10 Gates

		Crew	Daily Output	Labor-Hours	Unit	Material	2017 Bare Costs Labor	Equipment	Total	Total Incl O&P
0010	**GATES**									
0015	Driveway gates include mounting hardware									
0500	Wood, security gate, driveway, dual, 10' wide	H-4	.80	25	Opng.	3,900	1,150		5,050	6,050
0505	12' wide		.80	25		4,300	1,150		5,450	6,475
0510	15' wide		.80	25		5,100	1,150		6,250	7,350
0600	Steel, security gate, driveway, single, 10' wide		.80	25		2,000	1,150		3,150	3,950
0605	12' wide		.80	25		2,100	1,150		3,250	4,050
0620	Steel, security gate, driveway, dual, 12' wide		.80	25		1,950	1,150		3,100	3,900
0625	14' wide		.80	25		2,100	1,150		3,250	4,050
0630	16' wide		.80	25		2,400	1,150		3,550	4,400
0700	Aluminum, security gate, driveway, dual, 10' wide		.80	25		3,000	1,150		4,150	5,050
0705	12' wide		.80	25		3,600	1,150		4,750	5,700
0710	16' wide		.80	25		4,800	1,150		5,950	7,025

08 34 Special Function Doors

08 34 56 - Security Gates

08 34 56.10 Gates

		Crew	Daily Output	Labor-Hours	Unit	Material	2017 Bare Costs Labor	Equipment	Total	Total Incl O&P
1000	Security gate, driveway, opener 12 VDC				Ea.	800			800	880
1010	Wireless					1,850			1,850	2,025
1020	Security gate, driveway, opener 24 VDC					1,350			1,350	1,475
1030	Wireless					1,500			1,500	1,650
1040	Security gate, driveway, opener 12 VDC, solar panel 10 watt	1 Elec	2	4		380	226		606	760
1050	20 watt	"	2	4		530	226		756	925

08 34 59 - Vault Doors and Day Gates

08 34 59.10 Secure Storage Doors

		Crew	Daily Output	Labor-Hours	Unit	Material	2017 Bare Costs Labor	Equipment	Total	Total Incl O&P
0010	**SECURE STORAGE DOORS**									
0020	Door and frame, 32" x 78", clear opening									
0100	1 hour test, 32" door, weighs 750 lb.	2 Sswk	1.50	10.667	Opng.	6,625	580		7,205	8,275
0200	2 hour test, 32" door, weighs 950 lb.		1.30	12.308		8,200	670		8,870	10,100
0250	40" door, weighs 1130 lb.		1	16		9,250	870		10,120	11,700
0300	4 hour test, 32" door, weighs 1025 lb.		1.20	13.333		8,650	725		9,375	10,700
0350	40" door, weighs 1140 lb.		.90	17.778		9,975	965		10,940	12,600
0600	For time lock, two movement, add	1 Elec	2	4	Ea.	1,800	226		2,026	2,325
0800	Day gate, painted, steel, 32" wide	2 Sswk	1.50	10.667		2,000	580		2,580	3,175
0850	40" wide		1.40	11.429		2,100	620		2,720	3,325
0900	Aluminum, 32" wide		1.50	10.667		3,200	580		3,780	4,500
0950	40" wide		1.40	11.429		3,400	620		4,020	4,775
2050	Security vault door, class I, 3' wide, 3-1/2" thick	E-24	.19	166	Opng.	15,700	9,025	3,500	28,225	35,900
2100	Class II, 3' wide, 7" thick		.19	166		18,600	9,025	3,500	31,125	39,100
2150	Class III, 9R, 3' wide, 10" thick		.13	250		23,500	13,500	5,275	42,275	53,500
2160	Class V, type 1, 40" door		2.48	12.903	Ea.	7,125	700	272	8,097	9,250
2170	Class V, type 2, 40" door		2.48	12.903		6,850	700	272	7,822	8,950
2180	Day gate for class V vault	2 Sswk	2	8		2,025	435		2,460	2,950

08 34 73 - Sound Control Door Assemblies

08 34 73.10 Acoustical Doors

		Crew	Daily Output	Labor-Hours	Unit	Material	2017 Bare Costs Labor	Equipment	Total	Total Incl O&P
0010	**ACOUSTICAL DOORS**									
0020	Including framed seals, 3' x 7', wood, 40 STC rating	2 Carp	1.50	10.667	Ea.	3,000	525		3,525	4,100
0100	Steel, 41 STC rating		1.50	10.667		3,400	525		3,925	4,550
0200	45 STC rating		1.50	10.667		3,800	525		4,325	4,975
0300	48 STC rating		1.50	10.667		4,400	525		4,925	5,650
0400	52 STC rating		1.50	10.667		5,000	525		5,525	6,300

08 36 Panel Doors

08 36 13 - Sectional Doors

08 36 13.10 Overhead Commercial Doors

		Crew	Daily Output	Labor-Hours	Unit	Material	2017 Bare Costs Labor	Equipment	Total	Total Incl O&P
0010	**OVERHEAD COMMERCIAL DOORS**									
1000	Stock, sectional, heavy duty, wood, 1-3/4" thick, 8' x 8' high	2 Carp	2	8	Ea.	1,150	395		1,545	1,875
1100	10' x 10' high		1.80	8.889		1,625	440		2,065	2,450
1200	12' x 12' high		1.50	10.667		2,225	525		2,750	3,250
1300	Chain hoist, 14' x 14' high		1.30	12.308		3,400	605		4,005	4,650
1400	12' x 16' high		1	16		3,350	790		4,140	4,900
1500	20' x 8' high		1.30	12.270		2,750	605		3,355	3,950
1600	20' x 16' high		.65	24.615		5,700	1,200		6,900	8,125
1800	Center mullion openings, 8' high		4	4		1,225	197		1,422	1,650
1900	20' high		2	8		2,050	395		2,445	2,850
2100	For medium duty custom door, deduct					5%	5%			
2150	For medium duty stock doors, deduct					10%	5%			

08 36 Panel Doors

08 36 13 – Sectional Doors

08 36 13.10 Overhead Commercial Doors

		Crew	Daily Output	Labor-Hours	Unit	Material	2017 Bare Costs Labor	Equipment	Total	Total Incl O&P
2300	Fiberglass and aluminum, heavy duty, sectional, 12' x 12' high	2 Carp	1.50	10.667	Ea.	3,000	525		3,525	4,125
2450	Chain hoist, 20' x 20' high		.50	32		7,225	1,575		8,800	10,400
2600	Steel, 24 ga. sectional, manual, 8' x 8' high		2	8		910	395		1,305	1,600
2650	10' x 10' high		1.80	8.889		1,250	440		1,690	2,050
2700	12' x 12' high		1.50	10.667		1,475	525		2,000	2,425
2800	Chain hoist, 20' x 14' high		.70	22.857		3,375	1,125		4,500	5,425
2850	For 1-1/4" rigid insulation and 26 ga. galv.									
2860	back panel, add				S.F.	4.75			4.75	5.25
2900	For electric trolley operator, 1/3 H.P., to 12' x 12', add	1 Carp	2	4	Ea.	1,075	197		1,272	1,475
2950	Over 12' x 12', 1/2 H.P., add		1	8		1,150	395		1,545	1,875
2980	Overhead, for row of clear lites, add		1	8		150	395		545	770

08 36 13.20 Residential Garage Doors

		Crew	Daily Output	Labor-Hours	Unit	Material	2017 Bare Costs Labor	Equipment	Total	Total Incl O&P
0010	**RESIDENTIAL GARAGE DOORS**									
0050	Hinged, wood, custom, double door, 9' x 7'	2 Carp	4	4	Ea.	820	197		1,017	1,200
0070	16' x 7'		3	5.333		1,250	263		1,513	1,775
0200	Overhead, sectional, incl. hardware, fiberglass, 9' x 7', standard		5	3.200		975	158		1,133	1,325
0220	Deluxe		5	3.200		1,150	158		1,308	1,525
0300	16' x 7', standard		6	2.667		1,575	131		1,706	1,925
0320	Deluxe		6	2.667		2,150	131		2,281	2,550
0500	Hardboard, 9' x 7', standard		8	2		685	98.50		783.50	905
0520	Deluxe		8	2		815	98.50		913.50	1,050
0600	16' x 7', standard		6	2.667		1,225	131		1,356	1,550
0620	Deluxe		6	2.667		1,425	131		1,556	1,775
0700	Metal, 9' x 7', standard		8	2		835	98.50		933.50	1,075
0720	Deluxe		6	2.667		960	131		1,091	1,250
0800	16' x 7', standard		6	2.667		1,025	131		1,156	1,325
0820	Deluxe		5	3.200		1,400	158		1,558	1,800
0900	Wood, 9' x 7', standard		8	2		970	98.50		1,068.50	1,225
0920	Deluxe		8	2		2,150	98.50		2,248.50	2,500
1000	16' x 7', standard		6	2.667		1,625	131		1,756	2,000
1020	Deluxe		6	2.667		3,025	131		3,156	3,525
1800	Door hardware, sectional	1 Carp	4	2		350	98.50		448.50	535
1810	Door tracks only		4	2		164	98.50		262.50	330
1820	One side only		7	1.143		120	56.50		176.50	218
4000	For electric operator, economy, add		8	1		425	49.50		474.50	545
4100	Deluxe, including remote control		8	1		610	49.50		659.50	750
4500	For transmitter/receiver control, add to operator				Total	110			110	121
4600	Transmitters, additional				"	60			60	66

08 36 19 – Multi-Leaf Vertical Lift Doors

08 36 19.10 Sectional Vertical Lift Doors

		Crew	Daily Output	Labor-Hours	Unit	Material	2017 Bare Costs Labor	Equipment	Total	Total Incl O&P
0010	**SECTIONAL VERTICAL LIFT DOORS**									
0020	Motorized, 14 ga. steel, incl. frame and control panel									
0050	16' x 16' high	L-10	.50	48	Ea.	21,500	2,650	1,150	25,300	29,300
0100	10' x 20' high		1.30	18.462		35,000	1,025	440	36,465	40,600
0120	15' x 20' high		1.30	18.462		42,800	1,025	440	44,265	49,200
0140	20' x 20' high		1	24		50,000	1,325	575	51,900	58,000
0160	25' x 20' high		1	24		56,500	1,325	575	58,400	65,000
0170	32' x 24' high		.75	32		48,700	1,775	765	51,240	57,000
0180	20' x 25' high		1	24		58,000	1,325	575	59,900	66,500
0200	25' x 25' high		.70	34.286		70,500	1,900	820	73,220	81,500
0220	25' x 30' high		.70	34.286		75,500	1,900	820	78,220	87,000
0240	30' x 30' high		.70	34.286		86,000	1,900	820	88,720	99,000

08 36 Panel Doors

08 36 19 – Multi-Leaf Vertical Lift Doors

08 36 19.10 Sectional Vertical Lift Doors	Crew	Daily Output	Labor-Hours	Unit	Material	2017 Bare Costs Labor	Equipment	Total	Total Incl O&P
0260 35' x 30' high	L-10	.70	34.286	Ea.	92,500	1,900	820	95,220	106,000

08 38 Traffic Doors

08 38 13 – Flexible Strip Doors

08 38 13.10 Flexible Transparent Strip Doors

	Crew	Daily Output	Labor-Hours	Unit	Material	2017 Bare Costs Labor	Equipment	Total	Total Incl O&P
0010 **FLEXIBLE TRANSPARENT STRIP DOORS**									
0100 12" strip width, 2/3 overlap	3 Shee	135	.178	SF Surf	7.65	10.30		17.95	24
0200 Full overlap		115	.209		9.60	12.10		21.70	29
0220 8" strip width, 1/2 overlap		140	.171		6.25	9.95		16.20	22
0240 Full overlap		120	.200		7.85	11.60		19.45	26.50
0300 Add for suspension system, header mount				L.F.	9.20			9.20	10.10
0400 Wall mount				"	9.50			9.50	10.45

08 38 19 – Rigid Traffic Doors

08 38 19.20 Double Acting Swing Doors

	Crew	Daily Output	Labor-Hours	Unit	Material	2017 Bare Costs Labor	Equipment	Total	Total Incl O&P
0010 **DOUBLE ACTING SWING DOORS**									
0020 Including frame, closer, hardware and vision panel									
1000 Polymer, 7'-0" high, 4'-0" wide	2 Carp	4.20	3.810	Pr.	2,100	188		2,288	2,575
1025 6'-0" wide		4	4		2,200	197		2,397	2,725
1050 6'-8" wide		4	4		2,400	197		2,597	2,950
2000 3/4" thick, stainless steel									
2010 Stainless steel, 7' high opening, 4' wide	2 Carp	4	4	Pr.	2,600	197		2,797	3,150
2050 7' wide	"	3.80	4.211	"	2,800	207		3,007	3,400

08 38 19.30 Shock Absorbing Doors

	Crew	Daily Output	Labor-Hours	Unit	Material	2017 Bare Costs Labor	Equipment	Total	Total Incl O&P
0010 **SHOCK ABSORBING DOORS**									
0020 Rigid, no frame, 1-1/2" thick, 5' x 7'	2 Sswk	1.90	8.421	Opng.	1,525	455		1,980	2,425
0100 8' x 8'		1.80	8.889		2,000	485		2,485	3,000
0500 Flexible, no frame, insulated, .16" thick, economy, 5' x 7'		2	8		1,750	435		2,185	2,650
0600 Deluxe		1.90	8.421		2,625	455		3,080	3,625
1000 8' x 8' opening, economy		2	8		2,750	435		3,185	3,750
1100 Deluxe		1.90	8.421		3,500	455		3,955	4,600

08 41 Entrances and Storefronts

08 41 13 – Aluminum-Framed Entrances and Storefronts

08 41 13.20 Tube Framing

	Crew	Daily Output	Labor-Hours	Unit	Material	2017 Bare Costs Labor	Equipment	Total	Total Incl O&P
0010 **TUBE FRAMING**, For window walls and storefronts, aluminum stock									
0050 Plain tube frame, mill finish, 1-3/4" x 1-3/4"	2 Glaz	103	.155	L.F.	10.30	7.45		17.75	22.50
0150 1-3/4" x 4"		98	.163		13.80	7.80		21.60	27
0200 1-3/4" x 4-1/2"		95	.168		16.60	8.05		24.65	30.50
0250 2" x 6"		89	.180		24.50	8.60		33.10	40
0350 4" x 4"		87	.184		28	8.80		36.80	44
0400 4-1/2" x 4-1/2"		85	.188		29.50	9		38.50	46
0450 Glass bead		240	.067		3.21	3.19		6.40	8.40
1000 Flush tube frame, mill finish, 1/4" glass, 1-3/4" x 4", open header		80	.200		13.80	9.55		23.35	29.50
1050 Open sill		82	.195		11.25	9.35		20.60	26.50
1100 Closed back header		83	.193		19.75	9.20		28.95	35.50
1150 Closed back sill		85	.188		18.90	9		27.90	34.50
1160 Tube fmg., spandrel cover both sides, alum 1" wide	1 Sswk	85	.094	S.F.	101	5.10		106.10	120
1170 Tube fmg., spandrel cover both sides, alum 2" wide	"	85	.094	"	40	5.10		45.10	52.50

08 41 13 – Aluminum-Framed Entrances and Storefronts

08 41 13.20 Tube Framing

08 41 13.20 Tube Framing	Crew	Daily Output	Labor-Hours	Unit	Material	2017 Bare Costs Labor	Equipment	Total	Total Incl O&P
1200 Vertical mullion, one piece	2 Glaz	75	.213	L.F.	20.50	10.20		30.70	38
1250 Two piece		73	.219		21.50	10.50		32	39.50
1300 90° or 180° vertical corner post		75	.213		33	10.20		43.20	52
1400 1-3/4" x 4-1/2", open header		80	.200		16.45	9.55		26	32.50
1450 Open sill		82	.195		13.90	9.35		23.25	29.50
1500 Closed back header		83	.193		19.90	9.20		29.10	36
1550 Closed back sill		85	.188		19.50	9		28.50	35
1600 Vertical mullion, one piece		75	.213		22	10.20		32.20	39.50
1650 Two piece		73	.219		23	10.50		33.50	41.50
1700 90° or 180° vertical corner post		75	.213		23.50	10.20		33.70	41.50
2000 Flush tube frame, mil fin.,ins. glass w/thml brk, 2" x 4-1/2", open header		75	.213		16.70	10.20		26.90	34
2050 Open sill		77	.208		14.05	9.95		24	30.50
2100 Closed back header		78	.205		15.70	9.80		25.50	32
2150 Closed back sill		80	.200		15.30	9.55		24.85	31.50
2200 Vertical mullion, one piece		70	.229		17.70	10.95		28.65	36
2250 Two piece		68	.235		19.15	11.25		30.40	38
2300 90° or 180° vertical corner post		70	.229		18.30	10.95		29.25	36.50
5000 Flush tube frame, mill fin., thermal brk., 2-1/4" x 4-1/2", open header		74	.216		17.55	10.35		27.90	35
5050 Open sill		75	.213		15.50	10.20		25.70	32.50
5100 Vertical mullion, one piece		69	.232		18.40	11.10		29.50	37
5150 Two piece		67	.239		22	11.40		33.40	41.50
5200 90° or 180° vertical corner post		69	.232		19.65	11.10		30.75	38.50
6980 Door stop (snap in)	↓	380	.042	↓	3.52	2.01		5.53	6.95
7000 For joints, 90°, clip type, add				Ea.	26.50			26.50	29
7050 Screw spline joint, add					24.50			24.50	27
7100 For joint other than 90°, add				↓	51.50			51.50	57
8000 For bronze anodized aluminum, add					15%				
8020 For black finish, add					30%				
8050 For stainless steel materials, add					350%				
8100 For monumental grade, add					53%				
8150 For steel stiffener, add	2 Glaz	200	.080	L.F.	11.80	3.82		15.62	18.80
8200 For 2 to 5 stories, add per story				Story		8%			

08 41 19 – Stainless-Steel-Framed Entrances and Storefronts

08 41 19.10 Stainless-Steel and Glass Entrance Unit

	Crew	Daily Output	Labor-Hours	Unit	Material	2017 Bare Costs Labor	Equipment	Total	Total Incl O&P
0010 **STAINLESS-STEEL AND GLASS ENTRANCE UNIT**, narrow stiles									
0020 3' x 7' opening, including hardware, minimum	2 Sswk	1.60	10	Opng.	7,200	545		7,745	8,825
0050 Average		1.40	11.429		7,700	620		8,320	9,500
0100 Maximum	↓	1.20	13.333		8,200	725		8,925	10,200
1000 For solid bronze entrance units, statuary finish, add					64%				
1100 Without statuary finish, add				↓	45%				
2000 Balanced doors, 3' x 7', economy	2 Sswk	.90	17.778	Ea.	9,800	965		10,765	12,400
2100 Premium	"	.70	22.857	"	16,000	1,250		17,250	19,700

08 41 26 – All-Glass Entrances and Storefronts

08 41 26.10 Window Walls Aluminum, Stock

	Crew	Daily Output	Labor-Hours	Unit	Material	2017 Bare Costs Labor	Equipment	Total	Total Incl O&P
0010 **WINDOW WALLS ALUMINUM, STOCK**, including glazing									
0020 Minimum	H-2	160	.150	S.F.	48	6.75		54.75	63.50
0050 Average		140	.171		66	7.70		73.70	84.50
0100 Maximum	↓	110	.218	↓	177	9.80		186.80	210
0500 For translucent sandwich wall systems, see Section 07 41 33.10									
0850 Cost of the above walls depends on material,									
0860 finish, repetition, and size of units.									
0870 The larger the opening, the lower the S.F. cost									

286

For customer support on your Building Construction Costs with RSMeans Data, call 800.448.8182.

08 42 Entrances

08 42 26 – All-Glass Entrances

08 42 26.10 Swinging Glass Doors	Crew	Daily Output	Labor-Hours	Unit	Material	2017 Bare Costs Labor	Equipment	Total	Total Incl O&P
0010 **SWINGING GLASS DOORS**									
0020 Including hardware, 1/2" thick, tempered, 3' x 7' opening	2 Glaz	2	8	Opng.	2,300	380		2,680	3,100
0100 6' x 7' opening	"	1.40	11.429	"	4,500	545		5,045	5,775

08 42 33 – Revolving Door Entrances

08 42 33.10 Circular Rotating Entrance Doors	Crew	Daily Output	Labor-Hours	Unit	Material	2017 Bare Costs Labor	Equipment	Total	Total Incl O&P
0010 **CIRCULAR ROTATING ENTRANCE DOORS**, Aluminum									
0020 6'-10" to 7' high, stock units, minimum	4 Sswk	.75	42.667	Opng.	30,000	2,325		32,325	36,900
0050 Average		.60	53.333		35,000	2,900		37,900	43,300
0100 Maximum		.45	71.111		41,700	3,850		45,550	52,000
1000 Stainless steel		.30	105		47,400	5,750		53,150	61,500
1100 Solid bronze		.15	213		48,400	11,600		60,000	72,500
1500 For automatic controls, add	2 Elec	2	8		15,200	455		15,655	17,400

08 42 36 – Balanced Door Entrances

08 42 36.10 Balanced Entrance Doors	Crew	Daily Output	Labor-Hours	Unit	Material	2017 Bare Costs Labor	Equipment	Total	Total Incl O&P
0010 **BALANCED ENTRANCE DOORS**									
0020 Hardware & frame, alum. & glass, 3' x 7', econ.	2 Sswk	.90	17.778	Ea.	6,850	965		7,815	9,125
0150 Premium	"	.70	22.857	"	8,200	1,250		9,450	11,100

08 43 Storefronts

08 43 13 – Aluminum-Framed Storefronts

08 43 13.10 Aluminum-Framed Entrance Doors and Frames	Crew	Daily Output	Labor-Hours	Unit	Material	2017 Bare Costs Labor	Equipment	Total	Total Incl O&P
0010 **ALUMINUM-FRAMED ENTRANCE DOORS AND FRAMES**									
0015 Standard hardware and glass stops but no glass									
0020 Entrance door, 3' x 7' opening, clear anodized finish	2 Sswk	7	2.286	Opng.	585	124		709	850
0040 Bronze finish		7	2.286		630	124		754	900
0060 Black finish		7	2.286		655	124		779	925
0200 3'-6" x 7'-0", mill finish		7	2.286		660	124		784	930
0220 Bronze finish		7	2.286		775	124		899	1,050
0240 Black finish		7	2.286		820	124		944	1,100
0500 6' x 7' opening, clear finish		6	2.667		890	145		1,035	1,225
0520 Bronze finish		6	2.667		965	145		1,110	1,300
0540 Black finish		6	2.667		1,050	145		1,195	1,400
0600 Door frame for above doors 3'-0" x 7'-0", mill finish		6	2.667		425	145		570	710
0620 Bronze finish		6	2.667		485	145		630	770
0640 Black finish		6	2.667		530	145		675	825
0700 3'-6" x 7'-0", mill finish		6	2.667		350	145		495	625
0720 Bronze finish		6	2.667		350	145		495	625
0740 Black finish		6	2.667		350	145		495	625
0800 6'-0" x 7'-0", mill finish		6	2.667		350	145		495	625
0820 Bronze finish		6	2.667		355	145		500	630
0840 Black finish		6	2.667		385	145		530	660
1000 With 3' high transom above, 3' x 7' opening, clear finish		5.50	2.909		525	158		683	840
1050 Bronze finish		5.50	2.909		535	158		693	855
1100 Black finish		5.50	2.909		565	158		723	890
1300 3'-6" x 7'-0" opening, clear finish		5.50	2.909		375	158		533	675
1320 Bronze finish		5.50	2.909		390	158		548	695
1340 Black finish		5.50	2.909		400	158		558	705
1500 6' x 7' opening, clear finish		5.50	2.909		630	158		788	955
1550 Bronze finish		5.50	2.909		650	158		808	980
1600 Black finish		5.50	2.909		715	158		873	1,050

08 43 Storefronts

08 43 13 – Aluminum-Framed Storefronts

08 43 13.20 Storefront Systems	Crew	Daily Output	Labor-Hours	Unit	Material	2017 Bare Costs Labor	Equipment	Total	Total Incl O&P
0010 **STOREFRONT SYSTEMS**, aluminum frame clear 3/8" plate glass									
0020　incl. 3' x 7' door with hardware (400 sq. ft. max. wall)									
0500　　Wall height to 12' high, commercial grade	2 Glaz	150	.107	S.F.	23.50	5.10		28.60	33.50
0600　　　Institutional grade		130	.123		28.50	5.90		34.40	40.50
0700　　　Monumental grade		115	.139		41.50	6.65		48.15	55.50
1000　　6' x 7' door with hardware, commercial grade		135	.119		31.50	5.65		37.15	43
1100　　　Institutional grade		115	.139		29.50	6.65		36.15	42.50
1200　　　Monumental grade		100	.160		57	7.65		64.65	74.50
1500　　For bronze anodized finish, add					15%				
1600　　For black anodized finish, add					36%				
1700　　For stainless steel framing, add to monumental					78%				

08 43 29 – Sliding Storefronts

08 43 29.10 Sliding Panels

	Crew	Daily Output	Labor-Hours	Unit	Material	2017 Bare Costs Labor	Equipment	Total	Total Incl O&P
0010 **SLIDING PANELS**									
0020　Mall fronts, aluminum & glass, 15' x 9' high	2 Glaz	1.30	12.308	Opng.	3,700	590		4,290	4,975
0100　　24' x 9' high		.70	22.857		5,300	1,100		6,400	7,500
0200　　48' x 9' high, with fixed panels		.90	17.778		9,600	850		10,450	11,900
0500　　For bronze finish, add					17%				

08 44 Curtain Wall and Glazed Assemblies

08 44 13 – Glazed Aluminum Curtain Walls

08 44 13.10 Glazed Curtain Walls

	Crew	Daily Output	Labor-Hours	Unit	Material	2017 Bare Costs Labor	Equipment	Total	Total Incl O&P
0010 **GLAZED CURTAIN WALLS**, aluminum, stock, including glazing									
0020　Minimum	H-1	205	.156	S.F.	38.50	7.95		46.45	55
0050　Average, single glazed		195	.164		55	8.40		63.40	74
0150　Average, double glazed		180	.178		71	9.10		80.10	92.50
0200　Maximum		160	.200		186	10.20		196.20	220

08 45 Translucent Wall and Roof Assemblies

08 45 10 – Translucent Roof Assemblies

08 45 10.10 Skyroofs

	Crew	Daily Output	Labor-Hours	Unit	Material	2017 Bare Costs Labor	Equipment	Total	Total Incl O&P
0010 **SKYROOFS**									
1200　Skylights, circular, clear, double glazed acrylic									
1230　　30" diameter	2 Carp	3	5.333	Ea.	3,000	263		3,263	3,700
1250　　60" diameter		3	5.333		4,000	263		4,263	4,800
1290　　96" diameter		2	8		5,000	395		5,395	6,100
1300　Skylight Barrel Vault, clear, double glazed, acrylic									
1330　　3'-0" x 12'-0"	G-3	3	10.667	Ea.	5,000	520		5,520	6,300
1350　　4'-0" x 12'-0"		3	10.667		5,500	520		6,020	6,850
1390　　5'-0" x 12'-0"		2	16		6,000	780		6,780	7,800
1400　Skylight Pyramid, aluminum frame, clear low-E laminated glass									
1410　The glass is installed in the frame except where noted									
1430　　Square, 3' x 3'	G-3	3	10.667	Ea.	6,000	520		6,520	7,400
1440　　4' x 4'		3	10.667		7,000	520		7,520	8,500
1450　　5' x 5', glass must be field installed		3	10.667		8,000	520		8,520	9,600
1460　　6' x 6', glass must be field installed		2	16		10,000	780		10,780	12,200
1550　Install pre-cut laminated glass in aluminum frame on a flat roof	2 Glaz	55	.291	SF Surf		13.90		13.90	21
1560　Install pre-cut laminated glass in aluminum frame on a sloped roof	"	40	.400	"		19.10		19.10	29

08 51 Metal Windows

08 51 13 – Aluminum Windows

08 51 13.10 Aluminum Sash

		Crew	Daily Output	Labor-Hours	Unit	Material	2017 Bare Costs Labor	Equipment	Total	Total Incl O&P
0010	**ALUMINUM SASH**									
0020	Stock, grade C, glaze & trim not incl., casement	2 Sswk	200	.080	S.F.	39.50	4.34		43.84	50.50
0050	Double hung		200	.080		39.50	4.34		43.84	51
0100	Fixed casement		200	.080		17.40	4.34		21.74	26.50
0150	Picture window		200	.080		18.65	4.34		22.99	28
0200	Projected window		200	.080		35.50	4.34		39.84	47
0250	Single hung		200	.080		16.80	4.34		21.14	26
0300	Sliding		200	.080		22	4.34		26.34	31.50
1000	Mullions for above, tubular		240	.067	L.F.	6.20	3.62		9.82	12.85
2000	Custom aluminum sash, grade HC, glazing not included		140	.114	S.F.	40	6.20		46.20	54.50

08 51 13.20 Aluminum Windows

		Crew	Daily Output	Labor-Hours	Unit	Material	2017 Bare Costs Labor	Equipment	Total	Total Incl O&P
0010	**ALUMINUM WINDOWS**, incl. frame and glazing, commercial grade									
1000	Stock units, casement, 3'-1" x 3'-2" opening	2 Sswk	10	1.600	Ea.	380	87		467	565
1050	Add for storms					122			122	134
1600	Projected, with screen, 3'-1" x 3'-2" opening	2 Sswk	10	1.600		360	87		447	540
1700	Add for storms					119			119	131
2000	4'-5" x 5'-3" opening	2 Sswk	8	2		410	109		519	630
2100	Add for storms					128			128	141
2500	Enamel finish windows, 3'-1" x 3'-2"	2 Sswk	10	1.600		365	87		452	550
2600	4'-5" x 5'-3"		8	2		415	109		524	635
3000	Single hung, 2' x 3' opening, enameled, standard glazed		10	1.600		212	87		299	380
3100	Insulating glass		10	1.600		257	87		344	430
3300	2'-8" x 6'-8" opening, standard glazed		8	2		370	109		479	585
3400	Insulating glass		8	2		480	109		589	710
3700	3'-4" x 5'-0" opening, standard glazed		9	1.778		305	96.50		401.50	495
3800	Insulating glass		9	1.778		340	96.50		436.50	535
3890	Awning type, 3' x 3' opening, standard glass		14	1.143		430	62		492	580
3900	Insulating glass		14	1.143		460	62		522	610
3910	3' x 4' opening, standard glass		10	1.600		500	87		587	695
3920	Insulating glass		10	1.600		575	87		662	780
3930	3' x 5'-4" opening, standard glass		10	1.600		600	87		687	805
3940	Insulating glass		10	1.600		710	87		797	925
3950	4' x 5'-4" opening, standard glass		9	1.778		660	96.50		756.50	885
3960	Insulating glass		9	1.778		785	96.50		881.50	1,025
4000	Sliding aluminum, 3' x 2' opening, standard glazed		10	1.600		220	87		307	385
4100	Insulating glass		10	1.600		236	87		323	405
4300	5' x 3' opening, standard glazed		9	1.778		335	96.50		431.50	530
4400	Insulating glass		9	1.778		390	96.50		486.50	590
4600	8' x 4' opening, standard glazed		6	2.667		355	145		500	635
4700	Insulating glass		6	2.667		575	145		720	875
5000	9' x 5' opening, standard glazed		4	4		540	217		757	955
5100	Insulating glass		4	4		850	217		1,067	1,300
5500	Sliding, with thermal barrier and screen, 6' x 4', 2 track		8	2		725	109		834	975
5700	4 track		8	2		910	109		1,019	1,175
6000	For above units with bronze finish, add					15%				
6200	For installation in concrete openings, add					8%				

08 51 13.30 Impact Resistant Aluminum Windows

		Crew	Daily Output	Labor-Hours	Unit	Material	2017 Bare Costs Labor	Equipment	Total	Total Incl O&P
0010	**IMPACT RESISTANT ALUMINUM WINDOWS**, incl. frame and glazing									
0100	Single hung, impact resistant, 2'-8" x 5'-0"	2 Carp	9	1.778	Ea.	1,250	87.50		1,337.50	1,500
0120	3'-0" x 5'-0"		9	1.778		1,300	87.50		1,387.50	1,575
0130	4'-0" x 5'-0"		9	1.778		1,300	87.50		1,387.50	1,575
0250	Horizontal slider, impact resistant, 5'-5" x 5'-2"		9	1.778		1,550	87.50		1,637.50	1,850

08 51 Metal Windows

08 51 23 – Steel Windows

08 51 23.10 Steel Sash

		Crew	Daily Output	Labor-Hours	Unit	Material	2017 Bare Costs Labor	Equipment	Total	Total Incl O&P
0010	**STEEL SASH** Custom units, glazing and trim not included R085123-10									
0100	Casement, 100% vented	2 Sswk	200	.080	S.F.	67.50	4.34		71.84	81.50
0200	50% vented		200	.080		55	4.34		59.34	68
0300	Fixed		200	.080		29.50	4.34		33.84	40
1000	Projected, commercial, 40% vented		200	.080		52.50	4.34		56.84	65
1100	Intermediate, 50% vented		200	.080		60	4.34		64.34	73.50
1500	Industrial, horizontally pivoted		200	.080		55.50	4.34		59.84	68.50
1600	Fixed		200	.080		32	4.34		36.34	43
2000	Industrial security sash, 50% vented		200	.080		60	4.34		64.34	73.50
2100	Fixed		200	.080		48.50	4.34		52.84	61
2500	Picture window		200	.080		31	4.34		35.34	41.50
3000	Double hung		200	.080		60.50	4.34		64.84	74
5000	Mullions for above, open interior face		240	.067	L.F.	10.65	3.62		14.27	17.75
5100	With interior cover		240	.067	"	17.60	3.62		21.22	25.50

08 51 23.20 Steel Windows

		Crew	Daily Output	Labor-Hours	Unit	Material	2017 Bare Costs Labor	Equipment	Total	Total Incl O&P
0010	**STEEL WINDOWS** Stock, including frame, trim and insul. glass									
0020	See Section 13 34 19.50									
1000	Custom units, double hung, 2'-8" x 4'-6" opening R085123-10	2 Sswk	12	1.333	Ea.	705	72.50		777.50	900
1100	2'-4" x 3'-9" opening		12	1.333		585	72.50		657.50	760
1500	Commercial projected, 3'-9" x 5'-5" opening		10	1.600		1,225	87		1,312	1,500
1600	6'-9" x 4'-1" opening		7	2.286		1,625	124		1,749	1,975
2000	Intermediate projected, 2'-9" x 4'-1" opening		12	1.333		685	72.50		757.50	875
2100	4'-1" x 5'-5" opening		10	1.600		1,400	87		1,487	1,700

08 51 23.40 Basement Utility Windows

		Crew	Daily Output	Labor-Hours	Unit	Material	2017 Bare Costs Labor	Equipment	Total	Total Incl O&P
0010	**BASEMENT UTILITY WINDOWS**									
0015	1'-3" x 2'-8"	1 Carp	16	.500	Ea.	139	24.50		163.50	191
1100	1'-7" x 2'-8"	"	16	.500	"	143	24.50		167.50	195

08 51 66 – Metal Window Screens

08 51 66.10 Screens

		Crew	Daily Output	Labor-Hours	Unit	Material	2017 Bare Costs Labor	Equipment	Total	Total Incl O&P
0010	**SCREENS**									
0020	For metal sash, aluminum or bronze mesh, flat screen	2 Sswk	1200	.013	S.F.	4.40	.72		5.12	6.05
0500	Wicket screen, inside window		1000	.016		6.70	.87		7.57	8.80
0800	Security screen, aluminum frame with stainless steel cloth		1200	.013		24	.72		24.72	27.50
0900	Steel grate, painted, on steel frame		1600	.010		13.20	.54		13.74	15.40
1000	Screens for solar louvers		160	.100		24.50	5.45		29.95	36
4000	See Section 05 58 23.90 for window guards									

08 52 Wood Windows

08 52 10 – Plain Wood Windows

08 52 10.20 Awning Window

		Crew	Daily Output	Labor-Hours	Unit	Material	2017 Bare Costs Labor	Equipment	Total	Total Incl O&P
0010	**AWNING WINDOW**, Including frame, screens and grilles									
0100	34" x 22", insulated glass	1 Carp	10	.800	Ea.	276	39.50		315.50	365
0200	Low E glass		10	.800		298	39.50		337.50	390
0300	40" x 28", insulated glass		9	.889		315	44		359	410
0400	Low E glass		9	.889		345	44		389	445
0500	48" x 36", insulated glass		8	1		470	49.50		519.50	590
0600	Low E glass		8	1		495	49.50		544.50	620

08 52 10.40 Casement Window

		Crew	Daily Output	Labor-Hours	Unit	Material	2017 Bare Costs Labor	2017 Bare Costs Equipment	Total	Total Incl O&P
0010	**CASEMENT WINDOW**, including frame, screen and grilles									
0100	2'-0" x 3'-0" H, dbl. insulated glass [G]	1 Carp	10	.800	Ea.	274	39.50		313.50	360
0150	Low E glass [G]		10	.800		266	39.50		305.50	355
0200	2'-0" x 4'-6" high, double insulated glass [G]		9	.889		375	44		419	480
0250	Low E glass [G]		9	.889		375	44		419	480
0260	Casement 4'-2" x 4'-2" double insulated glass [G]		11	.727		890	36		926	1,025
0270	4'-0" x 4'-0" Low E glass [G]		11	.727		550	36		586	655
0290	6'-4" x 5'-7" Low E glass [G]		9	.889		1,125	44		1,169	1,325
0300	2'-4" x 6'-0" high, double insulated glass [G]		8	1		445	49.50		494.50	565
0350	Low E glass [G]		8	1		440	49.50		489.50	560
0522	Vinyl clad, premium, double insulated glass, 2'-0" x 3'-0" [G]		10	.800		276	39.50		315.50	365
0524	2'-0" x 4'-0" [G]		9	.889		320	44		364	415
0525	2'-0" x 5'-0" [G]		8	1		380	49.50		429.50	495
0528	2'-0" x 6'-0" [G]		8	1		390	49.50		439.50	505
0600	3'-0" x 5'-0" [G]		8	1		675	49.50		724.50	820
0700	4'-0" x 3'-0" [G]		8	1		740	49.50		789.50	890
0710	4'-0" x 4'-0" [G]		8	1		635	49.50		684.50	775
0720	4'-8" x 4'-0" [G]		8	1		700	49.50		749.50	845
0730	4'-8" x 5'-0" [G]		6	1.333		800	65.50		865.50	980
0740	4'-8" x 6'-0" [G]		6	1.333		900	65.50		965.50	1,100
0750	6'-0" x 4'-0" [G]		6	1.333		820	65.50		885.50	1,000
0800	6'-0" x 5'-0" [G]		6	1.333		900	65.50		965.50	1,100
0900	5'-6" x 5'-6" [G]	2 Carp	15	1.067		1,450	52.50		1,502.50	1,650
2000	Bay, casement units, 8' x 5', w/screens, dbl. insul. glass		2.50	6.400	Opng.	1,625	315		1,940	2,275
2100	Low E glass		2.50	6.400	"	1,725	315		2,040	2,350
8190	For installation, add per leaf				Ea.		15%			
8200	For multiple leaf units, deduct for stationary sash									
8220	2' high				Ea.	23.50			23.50	25.50
8240	4'-6" high					26.50			26.50	29
8260	6' high					35			35	38.50

08 52 10.50 Double Hung

		Crew	Daily Output	Labor-Hours	Unit	Material	2017 Bare Costs Labor	2017 Bare Costs Equipment	Total	Total Incl O&P
0010	**DOUBLE HUNG**, Including frame, screens and grilles R085216-10									
0100	2'-0" x 3'-0" high, low E insul. glass [G]	1 Carp	10	.800	Ea.	193	39.50		232.50	273
0200	3'-0" x 4'-0" high, double insulated glass [G]		9	.889		283	44		327	375
0300	4'-0" x 4'-6" high, low E insulated glass [G]		8	1		330	49.50		379.50	440

08 52 10.55 Picture Window

		Crew	Daily Output	Labor-Hours	Unit	Material	2017 Bare Costs Labor	2017 Bare Costs Equipment	Total	Total Incl O&P
0010	**PICTURE WINDOW**, Including frame and grilles									
0100	3'-6" x 4'-0" high, dbl. insulated glass	2 Carp	12	1.333	Ea.	435	65.50		500.50	580
0150	Low E glass		12	1.333		445	65.50		510.50	590
0200	4'-0" x 4'-6" high, double insulated glass		11	1.455		550	71.50		621.50	715
0250	Low E glass		11	1.455		530	71.50		601.50	695
0300	5'-0" x 4'-0" high, double insulated glass		11	1.455		580	71.50		651.50	750
0350	Low E glass		11	1.455		605	71.50		676.50	775
0400	6'-0" x 4'-6" high, double insulated glass		10	1.600		640	79		719	820
0450	Low E glass		10	1.600		640	79		719	825

08 52 10.65 Wood Sash

		Crew	Daily Output	Labor-Hours	Unit	Material	2017 Bare Costs Labor	2017 Bare Costs Equipment	Total	Total Incl O&P
0010	**WOOD SASH**, Including glazing but not trim									
0050	Custom, 5'-0" x 4'-0", 1" dbl. glazed, 3/16" thick lites	2 Carp	3.20	5	Ea.	233	246		479	630
0100	1/4" thick lites		5	3.200		263	158		421	530
0200	1" thick, triple glazed		5	3.200		430	158		588	715
0300	7'-0" x 4'-6" high, 1" double glazed, 3/16" thick lites		4.30	3.721		430	183		613	750

08 52 10 – Plain Wood Windows

08 52 10.65 Wood Sash		Crew	Daily Output	Labor-Hours	Unit	Material	2017 Bare Costs Labor	Equipment	Total	Total Incl O&P
0400	1/4" thick lites	2 Carp	4.30	3.721	Ea.	490	183		673	815
0500	1" thick, triple glazed		4.30	3.721		560	183		743	895
0600	8'-6" x 5'-0" high, 1" double glazed, 3/16" thick lites		3.50	4.571		580	225		805	985
0700	1/4" thick lites		3.50	4.571		640	225		865	1,050
0800	1" thick, triple glazed		3.50	4.571		675	225		900	1,100
0900	Window frames only, based on perimeter length				L.F.	4.13			4.13	4.54
1200	Window sill, stock, per lineal foot					8.65			8.65	9.50
1250	Casing, stock					3.38			3.38	3.72

08 52 10.70 Sliding Windows			Crew	Daily Output	Labor-Hours	Unit	Material	Labor	Equipment	Total	Total Incl O&P
0010	**SLIDING WINDOWS**										
0100	3'-0" x 3'-0" high, double insulated	G	1 Carp	10	.800	Ea.	293	39.50		332.50	380
0120	Low E glass	G		10	.800		320	39.50		359.50	410
0200	4'-0" x 3'-6" high, double insulated	G		9	.889		380	44		424	485
0220	Low E glass	G		9	.889		385	44		429	485
0300	6'-0" x 5'-0" high, double insulated	G		8	1		505	49.50		554.50	630
0320	Low E glass	G		8	1		550	49.50		599.50	680

08 52 13 – Metal-Clad Wood Windows

08 52 13.10 Awning Windows, Metal-Clad

			Crew	Daily Output	Labor-Hours	Unit	Material	Labor	Equipment	Total	Total Incl O&P
0010	**AWNING WINDOWS, METAL-CLAD**										
2000	Metal-clad, awning deluxe, double insulated glass, 34" x 22"		1 Carp	9	.889	Ea.	255	44		299	350
2050	36" x 25"			9	.889		276	44		320	370
2100	40" x 22"			9	.889		295	44		339	390
2150	40" x 30"			9	.889		340	44		384	440
2200	48" x 28"			8	1		350	49.50		399.50	460
2250	60" x 36"			8	1		375	49.50		424.50	485

08 52 13.20 Casement Windows, Metal-Clad

			Crew	Daily Output	Labor-Hours	Unit	Material	Labor	Equipment	Total	Total Incl O&P
0010	**CASEMENT WINDOWS, METAL-CLAD**										
0100	Metal-clad, deluxe, dbl. insul. glass, 2'-0" x 3'-0" high	G	1 Carp	10	.800	Ea.	285	39.50		324.50	375
0120	2'-0" x 4'-0" high	G		9	.889		320	44		364	415
0130	2'-0" x 5'-0" high	G		8	1		330	49.50		379.50	440
0140	2'-0" x 6'-0" high	G		8	1		370	49.50		419.50	480
0300	Metal-clad, casement, bldrs mdl, 6'-0" x 4'-0", dbl. insltd gls, 3 panels		2 Carp	10	1.600		1,200	79		1,279	1,450
0310	9'-0" x 4'-0", 4 panels			8	2		1,550	98.50		1,648.50	1,850
0320	10'-0" x 5'-0", 5 panels			7	2.286		2,100	113		2,213	2,500
0330	12'-0" x 6'-0", 6 panels			6	2.667		2,700	131		2,831	3,150

08 52 13.30 Double-Hung Windows, Metal-Clad

			Crew	Daily Output	Labor-Hours	Unit	Material	Labor	Equipment	Total	Total Incl O&P
0010	**DOUBLE-HUNG WINDOWS, METAL-CLAD**										
0100	Metal-clad, deluxe, dbl. insul. glass, 2'-6" x 3'-0" high	G	1 Carp	10	.800	Ea.	280	39.50		319.50	370
0120	3'-0" x 3'-6" high	G		10	.800		320	39.50		359.50	410
0140	3'-0" x 4'-0" high	G		9	.889		335	44		379	435
0160	3'-0" x 4'-6" high	G		9	.889		350	44		394	450
0180	3'-0" x 5'-0" high	G		8	1		380	49.50		429.50	490
0200	3'-6" x 6'-0" high	G		8	1		455	49.50		504.50	580

08 52 13.35 Picture and Sliding Windows Metal-Clad

			Crew	Daily Output	Labor-Hours	Unit	Material	Labor	Equipment	Total	Total Incl O&P
0010	**PICTURE AND SLIDING WINDOWS METAL-CLAD**										
2000	Metal-clad, dlx picture, dbl. insul. glass, 4'-0" x 4'-0" high		2 Carp	12	1.333	Ea.	380	65.50		445.50	520
2100	4'-0" x 6'-0" high			11	1.455		560	71.50		631.50	725
2200	5'-0" x 6'-0" high			10	1.600		620	79		699	800
2300	6'-0" x 6'-0" high			10	1.600		710	79		789	900
2400	Metal-clad, dlx sliding, double insulated glass, 3'-0" x 3'-0" high	G	1 Carp	10	.800		330	39.50		369.50	420
2420	4'-0" x 3'-6" high	G		9	.889		400	44		444	505

08 52 Wood Windows

08 52 13 – Metal-Clad Wood Windows

08 52 13.35 Picture and Sliding Windows Metal-Clad	Crew	Daily Output	Labor-Hours	Unit	Material	2017 Bare Costs Labor	Equipment	Total	Total Incl O&P
2440	5'-0" x 4'-0" high	G	1 Carp	9	.889	Ea.	480	44		524	590
2460	6'-0" x 5'-0" high	G	▼	8	1	▼	745	49.50		794.50	895

08 52 13.40 Bow and Bay Windows, Metal-Clad

		Crew	Daily Output	Labor-Hours	Unit	Material	Labor	Equipment	Total	Total Incl O&P
0010	**BOW AND BAY WINDOWS, METAL-CLAD**									
0100	Metal-clad, deluxe, dbl. insul. glass, 8'-0" x 5'-0" high, 4 panels	2 Carp	10	1.600	Ea.	1,700	79		1,779	2,000
0120	10'-0" x 5'-0" high, 5 panels		8	2		1,825	98.50		1,923.50	2,150
0140	10'-0" x 6'-0" high, 5 panels		7	2.286		2,150	113		2,263	2,550
0160	12'-0" x 6'-0" high, 6 panels		6	2.667		2,925	131		3,056	3,425
0400	Double hung, bldrs. model, bay, 8' x 4' high, dbl. insulated glass		10	1.600		1,350	79		1,429	1,600
0440	Low E glass		10	1.600		1,450	79		1,529	1,725
0460	9'-0" x 5'-0" high, double insulated glass		6	2.667		1,450	131		1,581	1,800
0480	Low E glass		6	2.667		1,525	131		1,656	1,875
0500	Metal-clad, deluxe, dbl. insul. glass, 7'-0" x 4'-0" high		10	1.600		1,300	79		1,379	1,550
0520	8'-0" x 4'-0" high		8	2		1,350	98.50		1,448.50	1,625
0540	8'-0" x 5'-0" high		7	2.286		1,375	113		1,488	1,700
0560	9'-0" x 5'-0" high	▼	6	2.667	▼	1,475	131		1,606	1,825

08 52 16 – Plastic-Clad Wood Windows

08 52 16.10 Bow Window

		Crew	Daily Output	Labor-Hours	Unit	Material	Labor	Equipment	Total	Total Incl O&P
0010	**BOW WINDOW** Including frames, screens, and grilles									
0020	End panels operable									
1000	Bow type, casement, wood, bldrs. mdl., 8' x 5' dbl. insltd glass, 4 panel	2 Carp	10	1.600	Ea.	1,575	79		1,654	1,850
1050	Low E glass		10	1.600		1,325	79		1,404	1,575
1100	10'-0" x 5'-0", double insulated glass, 6 panels		6	2.667		1,350	131		1,481	1,700
1200	Low E glass, 6 panels		6	2.667		1,450	131		1,581	1,800
1300	Vinyl clad, bldrs. model, double insulated glass, 6'-0" x 4'-0", 3 panel		10	1.600		1,050	79		1,129	1,275
1340	9'-0" x 4'-0", 4 panel		8	2		1,425	98.50		1,523.50	1,725
1380	10'-0" x 6'-0", 5 panels		7	2.286		2,350	113		2,463	2,750
1420	12'-0" x 6'-0", 6 panels		6	2.667		3,075	131		3,206	3,575
2000	Bay window, 8' x 5', dbl. insul glass		10	1.600		1,925	79		2,004	2,225
2050	Low E glass		10	1.600		2,325	79		2,404	2,675
2100	12'-0" x 6'-0", double insulated glass, 6 panels		6	2.667		2,400	131		2,531	2,825
2200	Low E glass		6	2.667		3,250	131		3,381	3,775
2280	6'-0" x 4'-0"		11	1.455		1,250	71.50		1,321.50	1,475
2300	Vinyl clad, premium, double insulated glass, 8'-0" x 5'-0"		10	1.600		1,800	79		1,879	2,100
2340	10'-0" x 5'-0"		8	2		2,400	98.50		2,498.50	2,800
2380	10'-0" x 6'-0"		7	2.286		2,800	113		2,913	3,250
2420	12'-0" x 6'-0"		6	2.667		3,350	131		3,481	3,900
3300	Vinyl clad, premium, double insulated glass, 7'-0" x 4'-6"		10	1.600		1,400	79		1,479	1,650
3340	8'-0" x 4'-6"		8	2		1,425	98.50		1,523.50	1,700
3380	8'-0" x 5'-0"		7	2.286		1,500	113		1,613	1,825
3420	9'-0" x 5'-0"	▼	6	2.667	▼	1,525	131		1,656	1,875

08 52 16.15 Awning Window Vinyl-Clad

		Crew	Daily Output	Labor-Hours	Unit	Material	Labor	Equipment	Total	Total Incl O&P
0010	**AWNING WINDOW VINYL-CLAD** Including frames, screens, and grilles									
0240	Vinyl-clad, 34" x 22"	1 Carp	10	.800	Ea.	267	39.50		306.50	355
0280	36" x 28"		9	.889		305	44		349	405
0300	36" x 36"		9	.889		350	44		394	450
0340	40" x 22"		10	.800		295	39.50		334.50	385
0360	48" x 28"		8	1		370	49.50		419.50	485
0380	60" x 36"	▼	8	1	▼	520	49.50		569.50	645

08 52 16 – Plastic-Clad Wood Windows

08 52 16.30 Palladian Windows

		Crew	Daily Output	Labor-Hours	Unit	Material	2017 Bare Costs Labor	2017 Bare Costs Equipment	Total	Total Incl O&P
0010	**PALLADIAN WINDOWS**									
0020	Vinyl-clad, double insulated glass, including frame and grilles									
0040	3'-2" x 2'-6" high	2 Carp	11	1.455	Ea.	1,275	71.50		1,346.50	1,500
0060	3'-2" x 4'-10"		11	1.455		1,750	71.50		1,821.50	2,025
0080	3'-2" x 6'-4"		10	1.600		1,750	79		1,829	2,050
0100	4'-0" x 4'-0"		10	1.600		1,575	79		1,654	1,850
0120	4'-0" x 5'-4"	3 Carp	10	2.400		1,875	118		1,993	2,250
0140	4'-0" x 6'-0"		9	2.667		1,875	131		2,006	2,275
0160	4'-0" x 7'-4"		9	2.667		2,125	131		2,256	2,550
0180	5'-5" x 4'-10"		9	2.667		2,300	131		2,431	2,725
0200	5'-5" x 6'-10"		9	2.667		2,650	131		2,781	3,100
0220	5'-5" x 7'-9"		9	2.667		2,800	131		2,931	3,300
0240	6'-0" x 7'-11"		8	3		3,575	148		3,723	4,175
0260	8'-0" x 6'-0"		8	3		3,200	148		3,348	3,725

08 52 16.35 Double-Hung Window

			Crew	Daily Output	Labor-Hours	Unit	Material	2017 Bare Costs Labor	2017 Bare Costs Equipment	Total	Total Incl O&P
0010	**DOUBLE-HUNG WINDOW** Including frames, screens, and grilles										
0300	Vinyl clad, premium, double insulated glass, 2'-6" x 3'-0"	G	1 Carp	10	.800	Ea.	340	39.50		379.50	430
0305	2'-6" x 4'-0"	G		10	.800		375	39.50		414.50	475
0400	3'-0" x 3'-6"	G		10	.800		340	39.50		379.50	430
0500	3'-0" x 4'-0"	G		9	.889		395	44		439	500
0600	3'-0" x 4'-6"	G		9	.889		420	44		464	525
0700	3'-0" x 5'-0"	G		8	1		450	49.50		499.50	570
0790	3'-4" x 5'-0"	G		8	1		450	49.50		499.50	570
0800	3'-6" x 6'-0"	G		8	1		505	49.50		554.50	630
0820	4'-0" x 5'-0"	G		7	1.143		565	56.50		621.50	705
0830	4'-0" x 6'-0"	G		7	1.143		715	56.50		771.50	870

08 52 16.40 Transom Windows

		Crew	Daily Output	Labor-Hours	Unit	Material	2017 Bare Costs Labor	2017 Bare Costs Equipment	Total	Total Incl O&P
0010	**TRANSOM WINDOWS**									
0050	Vinyl clad, premium, double insulated glass, 32" x 8"	1 Carp	16	.500	Ea.	191	24.50		215.50	248
0100	36" x 8"		16	.500		206	24.50		230.50	265
0110	36" x 12"		16	.500		221	24.50		245.50	281
0200	44" x 48"		12	.667		590	33		623	700
1000	Vinyl clad, premium, dbl. insul. glass, 4'-0" x 4'-0"	2 Carp	12	1.333		530	65.50		595.50	685
1100	4'-0" x 6'-0"		11	1.455		975	71.50		1,046.50	1,175
1200	5'-0" x 6'-0"		10	1.600		1,075	79		1,154	1,325
1300	6'-0" x 6'-0"		10	1.600		1,150	79		1,229	1,375

08 52 16.70 Vinyl Clad, Premium, Dbl. Insulated Glass

			Crew	Daily Output	Labor-Hours	Unit	Material	2017 Bare Costs Labor	2017 Bare Costs Equipment	Total	Total Incl O&P
0010	**VINYL CLAD, PREMIUM, DBL. INSULATED GLASS**										
1000	Sliding, 3'-0" x 3'-0"	G	1 Carp	10	.800	Ea.	625	39.50		664.50	745
1050	4'-0" x 3'-6"	G		9	.889		690	44		734	825
1100	5'-0" x 4'-0"	G		9	.889		920	44		964	1,075
1150	6'-0" x 5'-0"	G		8	1		1,175	49.50		1,224.50	1,350

08 52 50 – Window Accessories

08 52 50.10 Window Grille or Muntin

		Crew	Daily Output	Labor-Hours	Unit	Material	2017 Bare Costs Labor	2017 Bare Costs Equipment	Total	Total Incl O&P
0010	**WINDOW GRILLE OR MUNTIN**, snap in type									
0020	Standard pattern interior grilles									
2000	Wood, awning window, glass size, 28" x 16" high	1 Carp	30	.267	Ea.	30.50	13.15		43.65	53.50
2060	44" x 24" high		32	.250		44	12.30		56.30	67.50
2100	Casement, glass size, 20" x 36" high		30	.267		33.50	13.15		46.65	57
2180	20" x 56" high		32	.250		46.50	12.30		58.80	70
2200	Double hung, glass size, 16" x 24" high		24	.333	Set	56	16.40		72.40	86.50

08 52 Wood Windows

08 52 50 – Window Accessories

08 52 50.10 Window Grille or Muntin		Crew	Daily Output	Labor-Hours	Unit	Material	2017 Bare Costs Labor	Equipment	Total	Total Incl O&P
2280	32" x 32" high	1 Carp	34	.235	Set	140	11.60		151.60	172
2500	Picture, glass size, 48" x 48" high		30	.267	Ea.	128	13.15		141.15	161
2580	60" x 68" high		28	.286	"	196	14.05		210.05	238
2600	Sliding, glass size, 14" x 36" high		24	.333	Set	37.50	16.40		53.90	66.50
2680	36" x 36" high		22	.364	"	45	17.90		62.90	77

08 52 66 – Wood Window Screens

08 52 66.10 Wood Screens

		Crew	Daily Output	Labor-Hours	Unit	Material	Labor	Equipment	Total	Total Incl O&P
0010	**WOOD SCREENS**									
0020	Over 3 S.F., 3/4" frames	2 Carp	375	.043	S.F.	4.75	2.10		6.85	8.45
0100	1-1/8" frames	"	375	.043	"	8.45	2.10		10.55	12.50

08 52 69 – Wood Storm Windows

08 52 69.10 Storm Windows

			Crew	Daily Output	Labor-Hours	Unit	Material	Labor	Equipment	Total	Total Incl O&P
0010	**STORM WINDOWS**, aluminum residential										
0300	Basement, mill finish, incl. fiberglass screen										
0320	1'-10" x 1'-0" high	G	2 Carp	30	.533	Ea.	36	26.50		62.50	79.50
0340	2'-9" x 1'-6" high	G		30	.533		39	26.50		65.50	83
0360	3'-4" x 2'-0" high	G		30	.533		46	26.50		72.50	90.50
1600	Double-hung, combination, storm & screen										
2000	Clear anodic coating, 2'-0" x 3'-5" high	G	2 Carp	30	.533	Ea.	97	26.50		123.50	147
2020	2'-6" x 5'-0" high	G		28	.571		124	28		152	179
2040	4'-0" x 6'-0" high	G		25	.640		135	31.50		166.50	198
2400	White painted, 2'-0" x 3'-5" high	G		30	.533		95	26.50		121.50	145
2420	2'-6" x 5'-0" high	G		28	.571		100	28		128	153
2440	4'-0" x 6'-0" high	G		25	.640		115	31.50		146.50	176
2600	Mill finish, 2'-0" x 3'-5" high	G		30	.533		90	26.50		116.50	139
2620	2'-6" x 5'-0" high	G		28	.571		95	28		123	148
2640	4'-0" x 6'-8" high	G		25	.640		115	31.50		146.50	176

08 53 Plastic Windows

08 53 13 – Vinyl Windows

08 53 13.20 Vinyl Single Hung Windows

			Crew	Daily Output	Labor-Hours	Unit	Material	Labor	Equipment	Total	Total Incl O&P
0010	**VINYL SINGLE HUNG WINDOWS**, insulated glass										
0020	Grids, low e, J fin, extension jambs										
0130	25" x 41"	G	2 Carp	20	.800	Ea.	210	39.50		249.50	292
0140	25" x 49"	G		18	.889		220	44		264	310
0150	25" x 57"	G		17	.941		230	46.50		276.50	325
0160	25" x 65"	G		16	1		250	49.50		299.50	350
0170	29" x 41"	G		18	.889		210	44		254	298
0180	29" x 53"	G		18	.889		230	44		274	320
0190	29" x 57"	G		17	.941		240	46.50		286.50	335
0200	29" x 65"	G		16	1		250	49.50		299.50	350
0210	33" x 41"	G		20	.800		225	39.50		264.50	310
0220	33" x 53"	G		18	.889		245	44		289	335
0230	33" x 57"	G		17	.941		250	46.50		296.50	345
0240	33" x 65"	G		16	1		260	49.50		309.50	360
0250	37" x 41"	G		20	.800		250	39.50		289.50	335
0260	37" x 53"	G		18	.889		275	44		319	370
0270	37" x 57"	G		17	.941		285	46.50		331.50	385
0280	37" x 65"	G		16	1		300	49.50		349.50	405

08 53 13.30 Vinyl Double Hung Windows		Crew	Daily Output	Labor-Hours	Unit	Material	2017 Bare Costs Labor	Equipment	Total	Total Incl O&P
0010	**VINYL DOUBLE HUNG WINDOWS**, insulated glass									
0100	Grids, low E, J fin, ext. jambs, 21" x 53"	G	2 Carp	18	.889	Ea.	220	44		264	310
0102	21" x 37"	G		18	.889		230	44		274	320
0104	21" x 41"	G		18	.889		240	44		284	330
0106	21" x 49"	G		18	.889		250	44		294	340
0110	21" x 57"	G		17	.941		270	46.50		316.50	370
0120	21" x 65"	G		16	1		285	49.50		334.50	390
0128	25" x 37"	G		20	.800		240	39.50		279.50	325
0130	25" x 41"	G		20	.800		250	39.50		289.50	335
0140	25" x 49"	G		18	.889		260	44		304	355
0145	25" x 53"	G		18	.889		275	44		319	370
0150	25" x 57"	G		17	.941		280	46.50		326.50	380
0160	25" x 65"	G		16	1		295	49.50		344.50	400
0162	25" x 69"	G		16	1		300	49.50		349.50	405
0164	25" x 77"	G		16	1		325	49.50		374.50	435
0168	29" x 37"	G		18	.889		250	44		294	340
0170	29" x 41"	G		18	.889		260	44		304	355
0172	29" x 49"	G		18	.889		270	44		314	365
0180	29" x 53"	G		18	.889		280	44		324	375
0190	29" x 57"	G		17	.941		290	46.50		336.50	390
0200	29" x 65"	G		16	1		310	49.50		359.50	415
0202	29" x 69"	G		16	1		315	49.50		364.50	420
0205	29" x 77"	G		16	1		340	49.50		389.50	450
0208	33" x 37"	G		20	.800		265	39.50		304.50	355
0210	33" x 41"	G		20	.800		270	39.50		309.50	360
0215	33" x 49"	G		20	.800		285	39.50		324.50	375
0220	33" x 53"	G		18	.889		290	44		334	385
0230	33" x 57"	G		17	.941		300	46.50		346.50	400
0240	33" x 65"	G		16	1		320	49.50		369.50	425
0242	33" x 69"	G		16	1		330	49.50		379.50	440
0246	33" x 77"	G		16	1		350	49.50		399.50	460
0250	37" x 41"	G		20	.800		295	39.50		334.50	385
0255	37" x 49"	G		20	.800		297	39.50		336.50	385
0260	37" x 53"	G		18	.889		320	44		364	415
0270	37" x 57"	G		17	.941		340	46.50		386.50	445
0280	37" x 65"	G		16	1		360	49.50		409.50	470
0282	37" x 69"	G		16	1		365	49.50		414.50	475
0286	37" x 77"	G		16	1		380	49.50		429.50	495
0300	Solid vinyl, average quality, double insulated glass, 2'-0" x 3'-0"	G	1 Carp	10	.800		291	39.50		330.50	380
0310	3'-0" x 4'-0"	G		9	.889		214	44		258	300
0330	Premium, double insulated glass, 2'-6" x 3'-0"	G		10	.800		276	39.50		315.50	365
0340	3'-0" x 3'-6"	G		9	.889		310	44		354	405
0350	3'-0" x 4"-0"	G		9	.889		330	44		374	430
0360	3'-0" x 4'-6"	G		9	.889		335	44		379	435
0370	3'-0" x 5'-0"	G		8	1		360	49.50		409.50	470
0380	3'-6" x 6'-0"	G		8	1		395	49.50		444.50	510

08 53 13.40 Vinyl Casement Windows

08 53 13.40 Vinyl Casement Windows		Crew	Daily Output	Labor-Hours	Unit	Material	Labor	Equipment	Total	Total Incl O&P
0010	**VINYL CASEMENT WINDOWS**, insulated glass									
0015	Grids, low E, J fin, extension jambs, screens									
0100	One lite, 21" x 41"	G	2 Carp	20	.800	Ea.	300	39.50		339.50	390
0110	21" x 47"	G		20	.800		330	39.50		369.50	425
0120	21" x 53"	G		20	.800		350	39.50		389.50	445

08 53 Plastic Windows

08 53 13 – Vinyl Windows

08 53 13.40 Vinyl Casement Windows		Crew	Daily Output	Labor-Hours	Unit	Material	2017 Bare Costs Labor	Equipment	Total	Total Incl O&P
0128	24" x 35" G	2 Carp	19	.842	Ea.	285	41.50		326.50	380
0130	24" x 41" G		19	.842		320	41.50		361.50	415
0140	24" x 47" G		19	.842		340	41.50		381.50	440
0150	24" x 53" G		19	.842		365	41.50		406.50	465
0158	28" x 35" G		19	.842		310	41.50		351.50	405
0160	28" x 41" G		19	.842		340	41.50		381.50	440
0170	28" x 47" G		19	.842		360	41.50		401.50	460
0180	28" x 53" G		19	.842		390	41.50		431.50	495
0184	28" x 59" G		19	.842		400	41.50		441.50	505
0188	Two lites, 33" x 35" G		18	.889		475	44		519	590
0190	33" x 41" G		18	.889		510	44		554	625
0200	33" x 47" G		18	.889		540	44		584	660
0210	33" x 53" G		18	.889		580	44		624	705
0212	33" x 59" G		18	.889		610	44		654	735
0215	33" x 72" G		18	.889		640	44		684	770
0220	41" x 41" G		18	.889		550	44		594	670
0230	41" x 47" G		18	.889		585	44		629	710
0240	41" x 53" G		17	.941		625	46.50		671.50	760
0242	41" x 59" G		17	.941		660	46.50		706.50	795
0246	41" x 72" G		17	.941		690	46.50		736.50	830
0250	47" x 41" G		17	.941		560	46.50		606.50	685
0260	47" x 47" G		17	.941		600	46.50		646.50	730
0270	47" x 53" G		17	.941		640	46.50		686.50	775
0272	47" x 59" G		17	.941		695	46.50		741.50	835
0280	56" x 41" G		15	1.067		600	52.50		652.50	740
0290	56" x 47" G		15	1.067		640	52.50		692.50	785
0300	56" x 53" G		15	1.067		695	52.50		747.50	845
0302	56" x 59" G		15	1.067		720	52.50		772.50	870
0310	56" x 72" G		15	1.067		775	52.50		827.50	935
0340	Solid vinyl, premium, double insulated glass, 2'-0" x 3'-0" high G	1 Carp	10	.800		270	39.50		309.50	360
0360	2'-0" x 4'-0" high G		9	.889		299	44		343	395
0380	2'-0" x 5'-0" high G		8	1		335	49.50		384.50	445

08 53 13.50 Vinyl Picture Windows

08 53 13.50 Vinyl Picture Windows		Crew	Daily Output	Labor-Hours	Unit	Material	2017 Bare Costs Labor	Equipment	Total	Total Incl O&P
0010	**VINYL PICTURE WINDOWS**, insulated glass									
0120	Grids, low e, J fin, ext. jambs, 47" x 35"	2 Carp	12	1.333	Ea.	305	65.50		370.50	435
0130	47" x 41"		12	1.333		400	65.50		465.50	540
0140	47" x 47"		12	1.333		350	65.50		415.50	485
0150	47" x 53"		11	1.455		375	71.50		446.50	525
0160	71" x 35"		11	1.455		400	71.50		471.50	550
0170	71" x 41"		11	1.455		420	71.50		491.50	570
0180	71" x 47"		11	1.455		450	71.50		521.50	605

08 54 Composite Windows

08 54 13 – Fiberglass Windows

08 54 13.10 Fiberglass Single Hung Windows		Crew	Daily Output	Labor-Hours	Unit	Material	2017 Bare Costs Labor	Equipment	Total	Total Incl O&P
0010	**FIBERGLASS SINGLE HUNG WINDOWS**									
0100	Grids, low E, 18" x 24"	G	2 Carp	18	.889	Ea.	335	44		379	435
0110	18" x 40"	G		17	.941		340	46.50		386.50	445
0130	24" x 40"	G		20	.800		360	39.50		399.50	455
0230	36" x 36"	G		17	.941		370	46.50		416.50	475
0250	36" x 48"	G		20	.800		405	39.50		444.50	505
0260	36" x 60"	G		18	.889		445	44		489	555
0280	36" x 72"	G		16	1		470	49.50		519.50	590
0290	48" x 40"	G		16	1		470	49.50		519.50	590

08 56 Special Function Windows

08 56 63 – Detention Windows

08 56 63.13 Visitor Cubicle Windows

0010	**VISITOR CUBICLE WINDOWS**	Crew	Daily Output	Labor-Hours	Unit	Material	Labor	Equipment	Total	Total Incl O&P
4000	Visitor cubicle, vision panel, no intercom	E-4	2	16	Ea.	3,350	875	65	4,290	5,225

08 62 Unit Skylights

08 62 13 – Domed Unit Skylights

08 62 13.20 Skylights

			Crew	Daily Output	Labor-Hours	Unit	Material	Labor	Equipment	Total	Total Incl O&P
0010	**SKYLIGHTS,** flush or curb mounted										
2120	Ventilating insulated plexiglass dome with										
2130	curb mounting, 36" x 36"	G	G-3	12	2.667	Ea.	480	130		610	730
2150	52" x 52"	G		12	2.667		670	130		800	935
2160	28" x 52"	G		10	3.200		490	156		646	780
2170	36" x 52"	G		10	3.200		545	156		701	835
2180	For electric opening system, add	G					315			315	345
2300	Insulated safety glass with aluminum frame	G	G-3	160	.200	S.F.	94	9.70		103.70	118

08 63 Metal-Framed Skylights

08 63 13 – Domed Metal-Framed Skylights

08 63 13.20 Skylight Rigid Metal-Framed

		Crew	Daily Output	Labor-Hours	Unit	Material	Labor	Equipment	Total	Total Incl O&P
0010	**SKYLIGHT RIGID METAL-FRAMED** skylight framing is aluminum									
0050	Fixed acrylic double domes, curb mount, 25-1/2" x 25-1/2"	G-3	10	3.200	Ea.	210	156		366	470
0060	25-1/2" x 33-1/2"		10	3.200		242	156		398	505
0070	25-1/2" x 49-1/2"		6	5.333		279	259		538	700
0080	33-1/2" x 33-1/2"		8	4		305	194		499	635
0090	37-1/2" x 25-1/2"		8	4		250	194		444	575
0100	37-1/2" x 37-1/2"		8	4		245	194		439	570
0110	37-1/2" x 49-1/2"		6	5.333		420	259		679	855
0120	49-1/2" x 33-1/2"		6	5.333		375	259		634	805
0130	49-1/2" x 49-1/2"		6	5.333		470	259		729	910
1000	Fixed tempered glass, curb mount, 17-1/2" x 33-1/2"		10	3.200		170	156		326	425
1020	17-1/2" x 49-1/2"		6	5.333		190	259		449	605
1030	25-1/2" x 25-1/2"		10	3.200		170	156		326	425
1040	25-1/2" x 33-1/2"		10	3.200		200	156		356	460
1050	25-1/2" x 37-1/2"		8	4		210	194		404	530
1060	25-1/2" x 49-1/2"		6	5.333		222	259		481	640

08 63 Metal-Framed Skylights

08 63 13 – Domed Metal-Framed Skylights

08 63 13.20 Skylight Rigid Metal-Framed		Crew	Daily Output	Labor-Hours	Unit	Material	2017 Bare Costs Labor	Equipment	Total	Total Incl O&P
1070	25-1/2" x 73-1/2"	G-3	6	5.333	Ea.	375	259		634	805
1080	33-1/2" x 33-1/2"		8	4		235	194		429	555
2000	Manual vent tempered glass & screen, curb, 25-1/2" x 25-1/2"		10	3.200		410	156		566	690
2020	25-1/2" x 37-1/2"		10	3.200		465	156		621	750
2030	25-1/2" x 49-1/2"		8	4		505	194		699	855
2040	33-1/2" x 33-1/2"		8	4		535	194		729	890
2050	33-1/2" x 49-1/2"		6	5.333		675	259		934	1,150
2060	37-1/2" x 37-1/2"		6	5.333		615	259		874	1,075
2070	49-1/2" x 49-1/2"		6	5.333		790	259		1,049	1,275
3000	Electric vent tempered glass, curb mount, 25-1/2" x 25-1/2"		10	3.200		960	156		1,116	1,300
3020	25-1/2" x 37-1/2"		10	3.200		1,050	156		1,206	1,400
3030	25-1/2" x 49-1/2"		8	4		1,125	194		1,319	1,525
3040	33-1/2" x 33-1/2"		8	4		1,125	194		1,319	1,550
3050	33-1/2" x 49-1/2"		6	5.333		1,225	259		1,484	1,725
3060	37-1/2" x 37-1/2"		6	5.333		1,200	259		1,459	1,700
3070	49-1/2" x 49-1/2"		6	5.333		1,325	259		1,584	1,850

08 71 Door Hardware

08 71 13 – Automatic Door Operators

08 71 13.10 Automatic Openers Commercial

		Crew	Daily Output	Labor-Hours	Unit	Material	Labor	Equipment	Total	Total Incl O&P
0010	**AUTOMATIC OPENERS COMMERCIAL**									
0020	Pneumatic, incl opener, motion sens, control box, tubing, compressor									
0050	For single swing door, per opening	2 Skwk	.80	20	Ea.	4,550	1,025		5,575	6,575
0100	Pair, per opening		.50	32	Opng.	7,375	1,650		9,025	10,700
1000	For single sliding door, per opening		.60	26.667		4,975	1,375		6,350	7,575
1300	Bi-parting pair		.50	32		7,475	1,650		9,125	10,800
1420	Electronic door opener incl motion sens, 12 V control box, motor									
1450	For single swing door, per opening	2 Skwk	.80	20	Opng.	3,675	1,025		4,700	5,625
1500	Pair, per opening		.50	32		6,800	1,650		8,450	10,000
1600	For single sliding door, per opening		.60	26.667		4,500	1,375		5,875	7,050
1700	Bi-parting pair		.50	32		5,475	1,650		7,125	8,550
1750	Handicap actuator buttons, 2, including 12 V DC wiring, add	1 Carp	1.50	5.333	Pr.	470	263		733	915

08 71 13.20 Automatic Openers Industrial

		Crew	Daily Output	Labor-Hours	Unit	Material	Labor	Equipment	Total	Total Incl O&P
0010	**AUTOMATIC OPENERS INDUSTRIAL**									
0015	Sliding doors up to 6' wide	2 Skwk	.60	26.667	Opng.	5,900	1,375		7,275	8,600
0200	To 12' wide	"	.40	40	"	7,075	2,050		9,125	11,000
0400	Over 12' wide, add per L.F. of excess				L.F.	800			800	880
1000	Swing doors, to 5' wide	2 Skwk	.80	20	Ea.	3,500	1,025		4,525	5,425
1860	Add for controls, wall pushbutton, 3 button		4	4		240	206		446	580
1870	Control pull cord		4.30	3.721		195	191		386	510
1880	For addl elec eye for sliding door operation, one side, add					9%				

08 71 20 – Hardware

08 71 20.10 Bolts, Flush

		Crew	Daily Output	Labor-Hours	Unit	Material	Labor	Equipment	Total	Total Incl O&P
0010	**BOLTS, FLUSH**									
0020	Standard, concealed	1 Carp	7	1.143	Ea.	25.50	56.50		82	114
0800	Automatic fire exit	"	5	1.600		287	79		366	435
1600	Electrified dead bolt	1 Elec	3	2.667		160	151		311	400
3000	Barrel, brass, 2" long	1 Carp	40	.200		8	9.85		17.85	24
3020	4" long		40	.200		13.20	9.85		23.05	29.50
3060	6" long		40	.200		26	9.85		35.85	43.50

08 71 20.15 Hardware		Crew	Daily Output	Labor-Hours	Unit	Material	2017 Bare Costs Labor	Equipment	Total	Total Incl O&P
0010	**HARDWARE**									
0020	Average percentage for hardware, total job cost									
0500	Total hardware for building, average distribution				Job	85%	15%			
1000	Door hardware, apartment, interior	1 Carp	4	2	Door	465	98.50		563.50	665
1300	Average, door hardware, motel/hotel interior, with access card		4	2		615	98.50		713.50	830
1500	Hospital bedroom, average quality		4	2		675	98.50		773.50	895
2000	High quality		3	2.667		765	131		896	1,050
2100	Pocket door		6	1.333	Ea.	110	65.50		175.50	222
2250	School, single exterior, incl. lever, incl. panic device		3	2.667	Door	1,325	131		1,456	1,675
2500	Single interior, regular use, lever included		3	2.667		655	131		786	925
2550	Avg., door hdwe., school, classroom, ANSI F84, lever handle		3	2.667		875	131		1,006	1,175
2600	Avg., door hdwe. set, school, classroom, ANSI F88, incl. lever		3	2.667		935	131		1,066	1,225
2850	Stairway, single interior		3	2.667		615	131		746	880
3100	Double exterior, with panic device		2	4	Pr.	2,475	197		2,672	3,025
6020	Add for fire alarm door holder, electro-magnetic	1 Elec	4	2	Ea.	104	113		217	284

08 71 20.20 Door Protectors

08 71 20.20 Door Protectors		Crew	Daily Output	Labor-Hours	Unit	Material	2017 Bare Costs Labor	Equipment	Total	Total Incl O&P
0010	**DOOR PROTECTORS**									
0020	1-3/4" x 3/4" U channel	2 Carp	80	.200	L.F.	28	9.85		37.85	46
0021	1-3/4" x 1-1/4" U channel		80	.200	"	30	9.85		39.85	48
1000	Tear drop, spring-stl, 8" high x 19" long		15	1.067	Ea.	130	52.50		182.50	224
1010	8" high x 32" long		15	1.067		240	52.50		292.50	345
1100	Tear drop, stainless stl., 8" high x 19" long		15	1.067		390	52.50		442.50	510
1200	8" high x 32" long		15	1.067		460	52.50		512.50	585

08 71 20.30 Door Closers

08 71 20.30 Door Closers		Crew	Daily Output	Labor-Hours	Unit	Material	2017 Bare Costs Labor	Equipment	Total	Total Incl O&P
0010	**DOOR CLOSERS** adjustable backcheck, multiple mounting									
0015	and rack and pinion									
0020	Standard Regular Arm	1 Carp	6	1.333	Ea.	201	65.50		266.50	320
0040	Hold open arm		6	1.333		200	65.50		265.50	320
0100	Fusible link		6.50	1.231		170	60.50		230.50	280
0210	Light duty, regular arm		6	1.333		114	65.50		179.50	227
0220	Parallel arm		6	1.333		138	65.50		203.50	253
0230	Hold open arm		6	1.333		122	65.50		187.50	236
0240	Fusible link arm		6	1.333		155	65.50		220.50	272
0250	Medium duty, regular arm		6	1.333		123	65.50		188.50	237
0500	Surface mount regular arm		6.50	1.231		160	60.50		220.50	269
0550	Fusible link		6.50	1.231		151	60.50		211.50	259
1520	Overhead concealed, all sizes, regular arm		5.50	1.455		205	71.50		276.50	335
1525	Concealed arm		5	1.600		320	79		399	470
1530	Concealed in door, all sizes, regular arm		5.50	1.455		330	71.50		401.50	475
1535	Concealed arm		5	1.600		260	79		339	405
1560	Floor concealed, all sizes, single acting		2.20	3.636		510	179		689	835
1565	Double acting		2.20	3.636		500	179		679	825
1610	Hold open arm		6	1.333		440	65.50		505.50	580
1620	Double acting, standard arm		6	1.333		790	65.50		855.50	970
1630	Hold open arm		6	1.333		795	65.50		860.50	975
1640	Floor, center hung, single acting, bottom arm		6	1.333		430	65.50		495.50	570
1650	Double acting		6	1.333		485	65.50		550.50	635
1660	Offset hung, single acting, bottom arm		6	1.333		575	65.50		640.50	730
2000	Backcheck and adjustable power, hinge face mount									
5000	For cast aluminum cylinder, deduct				Ea.	35			35	38.50
5010	For delayed action surface mounted, add	1 Carp	6	1.333		37	65.50		102.50	142
5040	For delayed action, add					50			50	55

08 71 Door Hardware

08 71 20 – Hardware

08 71 20.30 Door Closers

		Crew	Daily Output	Labor-Hours	Unit	Material	2017 Bare Costs Labor	Equipment	Total	Total Incl O&P
5080	For fusible link arm, add				Ea.	40			40	44
5120	For shock absorbing arm, add					50			50	55
5160	For spring power adjustment, add					40			40	44
6000	Closer-holder, hinge face mount, all sizes, exposed arm	1 Carp	6.50	1.231		210	60.50		270.50	325
6500	Electro magnetic closer/holder									
6510	Single point, no detector	1 Carp	4	2	Ea.	530	98.50		628.50	730
6515	Including detector		4	2		695	98.50		793.50	915
6520	Multi-point, no detector		4	2		920	98.50		1,018.50	1,150
6524	Including detector		4	2		1,375	98.50		1,473.50	1,675
6550	Electric automatic operators									
6555	Operator	1 Carp	4	2	Ea.	2,050	98.50		2,148.50	2,400
6570	Wall plate actuator		4	2		226	98.50		324.50	400
7000	Electronic closer-holder, hinge facemount, concealed arm		5	1.600		420	79		499	580
7400	With built-in detector		5	1.600		605	79		684	785
8000	Surface mounted, standard duty, parallel arm, primed, traditional		6	1.333		208	65.50		273.50	330
8030	Light duty		6	1.333		153	65.50		218.50	269
8042	Extra duty parallel arm		6	1.333		123	65.50		188.50	237
8044	Hold open arm		6	1.333		163	65.50		228.50	281
8046	Positive stop arm		6	1.333		222	65.50		287.50	345
8050	Heavy duty		6	1.333		245	65.50		310.50	370
8052	Heavy duty, regular arm		6	1.333		224	65.50		289.50	345
8054	Top jamb mount		6	1.333		224	65.50		289.50	345
8056	Extra duty parallel arm		6	1.333		225	65.50		290.50	350
8058	Hold open arm		6	1.333		274	65.50		339.50	400
8060	Positive stop arm		6	1.333		249	65.50		314.50	375
8062	Fusible link arm		6	1.333		270	65.50		335.50	400
8080	Universal heavy duty, regular arm		6	1.333		235	65.50		300.50	360
8084	Parallel arm		6	1.333		235	65.50		300.50	360
8088	Extra duty, parallel arm		6	1.333		270	65.50		335.50	400
8090	Hold open arm		6	1.333		278	65.50		343.50	405
8094	Positive stop arm		6	1.333		287	65.50		352.50	415
8100	Standard duty, parallel arm, modern		6	1.333		257	65.50		322.50	385
8150	Heavy duty		6	1.333		290	65.50		355.50	420

08 71 20.36 Panic Devices

		Crew	Daily Output	Labor-Hours	Unit	Material	2017 Bare Costs Labor	Equipment	Total	Total Incl O&P
0010	**PANIC DEVICES** R087110-10									
0015	Touch bars various styles									
0040	Single door exit only, rim device, wide stile									
0050	Economy US28	1 Carp	5	1.600	Ea.	239	79		318	385
0060	Standard duty US28		5	1.600		675	79		754	860
0065	US26D		5	1.600		710	79		789	905
0070	US10		5	1.600		735	79		814	925
0075	US3		5	1.600		810	79		889	1,000
0080	Night Latch, Economy US28		5	1.600		325	79		404	480
0085	Standard duty, Night Latch US28		5	1.600		975	79		1,054	1,200
0090	US26D		5	1.600		1,075	79		1,154	1,325
0095	US10		5	1.600		1,100	79		1,179	1,325
0100	US3		5	1.600		1,200	79		1,279	1,450
0500	Single door exit only, rim device, narrow stile									
0540	Economy US28	1 Carp	5	1.600	Ea.	465	79		544	630
0550	Standard duty US28		5	1.600		845	79		924	1,050
0565	US26D		5	1.600		910	79		989	1,125
0570	US10		5	1.600		755	79		834	950

For customer support on your Building Construction Costs with RSMeans Data, call 800.448.8182.

301

08 71 Door Hardware

08 71 20 – Hardware

08 71 20.36 Panic Devices

		Crew	Daily Output	Labor-Hours	Unit	Material	2017 Bare Costs Labor	Equipment	Total	Total Incl O&P
0575	US3	1 Carp	5	1.600	Ea.	950	79		1,029	1,175
0580	Economy with night latch US28		5	1.600		330	79		409	480
0585	Standard duty US28		5	1.600		885	79		964	1,100
0590	US26D		5	1.600		1,050	79		1,129	1,300
0595	US10		5	1.600		890	79		969	1,100
0600	US32D		5	1.600		1,200	79		1,279	1,450
1040	Single door exit only, surface vertical rod									
1050	Surface Vertical Rod, Economy US28	1 Carp	4	2	Ea.	525	98.50		623.50	730
1060	Surface Vertical Rod, Standard duty US28		4	2		975	98.50		1,073.50	1,225
1065	US26D		4	2		1,125	98.50		1,223.50	1,400
1070	US10		4	2		1,050	98.50		1,148.50	1,300
1075	US3		4	2		1,200	98.50		1,298.50	1,475
1080	US32D		4	2		1,225	98.50		1,323.50	1,475
1500	Single door exit, rim, narrow stile, economy, surface vertical rod									
1540	Surface Vertical Rod, Narrow, Economy US28	1 Carp	4	2	Ea.	420	98.50		518.50	615
2040	Single door exit only, concealed vertical rod									
2042	These devices will require separate and extra door preparation									
2060	Concealed Rod, standard duty US28	1 Carp	4	2	Ea.	1,025	98.50		1,123.50	1,275
2065	US26D		4	2		1,150	98.50		1,248.50	1,400
2070	US10		4	2		1,325	98.50		1,423.50	1,600
2560	Concealed Rod, narrow, standard duty US28		4	2		990	98.50		1,088.50	1,250
2565	US26D		4	2		1,075	98.50		1,173.50	1,350
2567	US26		4	2		1,225	98.50		1,323.50	1,500
2570	US10		4	2		1,025	98.50		1,123.50	1,275
2575	US3		4	2		1,225	98.50		1,323.50	1,500
3040	Single door exit only, mortise device, wide stile									
3042	These devices must be combined with a mortise lock									
3060	Mortise, standard duty US28	1 Carp	4	2	Ea.	925	98.50		1,023.50	1,175
3065	US26D		4	2		470	98.50		568.50	665
3067	US26		4	2		1,150	98.50		1,248.50	1,425
3070	US10		4	2		1,025	98.50		1,123.50	1,275
3075	US3		4	2		1,050	98.50		1,148.50	1,300
3080	US32D		4	2		970	98.50		1,068.50	1,225
3500	Single door exit only, mortise device, narrow stile									
3510	These devices must be combined with a mortise lock									
3540	Mortise, economy US28	1 Carp	4	2	Ea.	395	98.50		493.50	585

08 71 20.40 Lockset

		Crew	Daily Output	Labor-Hours	Unit	Material	2017 Bare Costs Labor	Equipment	Total	Total Incl O&P
0010	**LOCKSET**, Standard duty									
0020	Non-keyed, passage, w/sect. trim	1 Carp	12	.667	Ea.	72.50	33		105.50	130
0100	Privacy		12	.667		80	33		113	139
0400	Keyed, single cylinder function		10	.800		113	39.50		152.50	185
0420	Hotel (see also Section 08 71 20.15)		8	1		200	49.50		249.50	296
0500	Lever handled, keyed, single cylinder function		10	.800		126	39.50		165.50	199
1000	Heavy duty with sectional trim, non-keyed, passages		12	.667		125	33		158	189
1100	Privacy		12	.667		155	33		188	222
1400	Keyed, single cylinder function		10	.800		185	39.50		224.50	265
1420	Hotel		8	1		500	49.50		549.50	625
1600	Communicating		10	.800		300	39.50		339.50	390
1690	For re-core cylinder, add					50			50	55
3800	Cipher lockset w/key pad (security item)	1 Carp	13	.615		975	30.50		1,005.50	1,125
3900	Cipher lockset with dial for swinging doors (security item)		13	.615		1,800	30.50		1,830.50	2,050
3920	with dial for swinging doors & drill resistant plate (security item)		12	.667		2,200	33		2,233	2,475

08 71 20 – Hardware

08 71 20.40 Lockset	Crew	Daily Output	Labor-Hours	Unit	Material	2017 Bare Costs Labor	Equipment	Total	Total Incl O&P
3950	Cipher lockset with dial for safe/vault door (security item)	1 Carp	12	.667	Ea.	1,600	33		1,633	1,800
3980	Keyless, pushbutton type									
4000	Residential/light commercial, deadbolt, standard	1 Carp	9	.889	Ea.	140	44		184	221
4010	Heavy duty		9	.889		230	44		274	320
4020	Industrial, heavy duty, with deadbolt		9	.889		415	44		459	520
4030	Key override		9	.889		420	44		464	525
4040	Lever activated handle		9	.889		430	44		474	540
4050	Key override		9	.889		445	44		489	555
4060	Double sided pushbutton type		8	1		780	49.50		829.50	935
4070	Key override		8	1		805	49.50		854.50	960

08 71 20.41 Dead Locks

		Crew	Daily Output	Labor-Hours	Unit	Material	Labor	Equipment	Total	Total Incl O&P
0010	**DEAD LOCKS**									
0011	Mortise heavy duty outside key (security item)	1 Carp	9	.889	Ea.	175	44		219	260
0020	Double cylinder		9	.889		175	44		219	260
0100	Medium duty, outside key		10	.800		110	39.50		149.50	182
0110	Double cylinder		10	.800		120	39.50		159.50	193
1000	Tubular, standard duty, outside key		10	.800		50	39.50		89.50	116
1010	Double cylinder		10	.800		65	39.50		104.50	132
1200	Night latch, outside key		10	.800		50	39.50		89.50	116

08 71 20.42 Mortise Locksets

		Crew	Daily Output	Labor-Hours	Unit	Material	Labor	Equipment	Total	Total Incl O&P
0010	**MORTISE LOCKSETS**, Comm., wrought knobs & full escutcheon trim									
0015	Assumes mortise is cut									
0020	Non-keyed, passage, Grade 3	1 Carp	9	.889	Ea.	155	44		199	238
0030	Grade 1		8	1		405	49.50		454.50	520
0040	Privacy set, Grade 3		9	.889		169	44		213	253
0050	Grade 1		8	1		455	49.50		504.50	575
0100	Keyed, office/entrance/apartment, Grade 2		8	1		197	49.50		246.50	293
0110	Grade 1		7	1.143		515	56.50		571.50	655
0120	Single cylinder, typical, Grade 3		8	1		189	49.50		238.50	284
0130	Grade 1		7	1.143		500	56.50		556.50	635
0200	Hotel, room, Grade 3		7	1.143		190	56.50		246.50	295
0210	Grade 1 (see also Section 08 71 20.15)		6	1.333		510	65.50		575.50	660
0300	Double cylinder, Grade 3		8	1		225	49.50		274.50	325
0310	Grade 1		7	1.143		515	56.50		571.50	655
1000	Wrought knobs and sectional trim, non-keyed, passage, Grade 3		10	.800		130	39.50		169.50	204
1010	Grade 1		9	.889		405	44		449	510
1040	Privacy, Grade 3		10	.800		150	39.50		189.50	226
1050	Grade 1		9	.889		455	44		499	565
1100	Keyed, entrance, office/apartment, Grade 3		9	.889		220	44		264	310
1103	Install lockset		6.92	1.156		220	57		277	330
1110	Grade 1		8	1		520	49.50		569.50	645
1120	Single cylinder, Grade 3		9	.889		225	44		269	315
1130	Grade 1		8	1		500	49.50		549.50	625
2000	Cast knobs and full escutcheon trim									
2010	Non-keyed, passage, Grade 3	1 Carp	9	.889	Ea.	275	44		319	370
2020	Grade 1		8	1		380	49.50		429.50	495
2040	Privacy, Grade 3		9	.889		320	44		364	415
2050	Grade 1		8	1		440	49.50		489.50	560
2120	Keyed, single cylinder, Grade 3		8	1		330	49.50		379.50	440
2123	Mortise lock		6.15	1.301		330	64		394	465
2130	Grade 1		7	1.143		525	56.50		581.50	665
3000	Cast knob and sectional trim, non-keyed, passage, Grade 3		10	.800		210	39.50		249.50	292

08 71 Door Hardware

08 71 20 – Hardware

08 71 20.42 Mortise Locksets

		Crew	Daily Output	Labor-Hours	Unit	Material	2017 Bare Costs Labor	Equipment	Total	Total Incl O&P
3010	Grade 1	1 Carp	10	.800	Ea.	385	39.50		424.50	485
3040	Privacy, Grade 3		10	.800		225	39.50		264.50	310
3050	Grade 1		10	.800		440	39.50		479.50	545
3100	Keyed, office/entrance/apartment, Grade 3		9	.889		255	44		299	350
3110	Grade 1		9	.889		580	44		624	705
3120	Single cylinder, Grade 3		9	.889		260	44		304	355
3130	Grade 1		9	.889		525	44		569	645
3190	For re-core cylinder, add					80			80	88

08 71 20.45 Peepholes

		Crew	Daily Output	Labor-Hours	Unit	Material	2017 Bare Costs Labor	Equipment	Total	Total Incl O&P
0010	**PEEPHOLES**									
2010	Peephole	1 Carp	32	.250	Ea.	15.80	12.30		28.10	36.50
2020	Peephole, wide view	"	32	.250	"	16.50	12.30		28.80	37

08 71 20.50 Door Stops

		Crew	Daily Output	Labor-Hours	Unit	Material	2017 Bare Costs Labor	Equipment	Total	Total Incl O&P
0010	**DOOR STOPS**									
0020	Holder & bumper, floor or wall	1 Carp	32	.250	Ea.	35.50	12.30		47.80	58
1300	Wall bumper, 4" diameter, with rubber pad, aluminum		32	.250		12.30	12.30		24.60	32.50
1600	Door bumper, floor type, aluminum		32	.250		8.70	12.30		21	28.50
1900	Plunger type, door mounted		32	.250		30	12.30		42.30	52

08 71 20.55 Push-Pull Plates

		Crew	Daily Output	Labor-Hours	Unit	Material	2017 Bare Costs Labor	Equipment	Total	Total Incl O&P
0010	**PUSH-PULL PLATES**									
0090	Push plate, 0.050 thick, 3" x 12", aluminum	1 Carp	12	.667	Ea.	6.25	33		39.25	57.50
0100	4" x 16"		12	.667		12.90	33		45.90	64.50
0110	6" x 16"		12	.667		8.50	33		41.50	60
0120	8" x 16"		12	.667		9.90	33		42.90	61.50
0200	Push plate, 0.050 thick, 3" x 12", brass		12	.667		14	33		47	66
0210	4" x 16"		12	.667		17.50	33		50.50	70
0220	6" x 16"		12	.667		27	33		60	80
0230	8" x 16"		12	.667		35	33		68	89
0250	Push plate, 0.050 thick, 3" x 12", satin brass		12	.667		14.25	33		47.25	66
0260	4" x 16"		12	.667		17.70	33		50.70	70
0270	6" x 16"		12	.667		27	33		60	80
0280	8" x 16"		12	.667		35	33		68	89
0490	Push plate, 0.050 thick, 3" x 12", bronze		12	.667		17	33		50	69
0500	4" x 16"		12	.667		25.50	33		58.50	78.50
0510	6" x 16"		12	.667		32	33		65	85.50
0520	8" x 16"		12	.667		38.50	33		71.50	93
0740	Push plate, 0.050 thick, 3" x 12", stainless steel		12	.667		12	33		45	63.50
0760	6" x 16"		12	.667		18.65	33		51.65	71
0780	8" x 16"		12	.667		23.50	33		56.50	76.50
0790	Push plate, 0.050 thick, 3" x 12", satin stainless steel		12	.667		7	33		40	58
0820	6" x 16"		12	.667		11.70	33		44.70	63.50
0830	8" x 16"		12	.667		16	33		49	68
0980	Pull plate, 0.050 thick, 3" x 12", aluminum		12	.667		26.50	33		59.50	79.50
1050	Pull plate, 0.050 thick, 3" x 12", brass		12	.667		41	33		74	95.50
1080	Pull plate, 0.050 thick, 3" x 12", bronze		12	.667		51.50	33		84.50	107
1180	Pull plate, 0.050 thick, 3" x 12", stainless steel		12	.667		47.50	33		80.50	103
1250	Pull plate, 0.050 thick, 3" x 12", chrome		12	.667		46.50	33		79.50	102
1500	Pull handle and push bar, aluminum		11	.727		122	36		158	189
2000	Bronze		10	.800		164	39.50		203.50	241

08 71 20.60 Entrance Locks

		Crew	Daily Output	Labor-Hours	Unit	Material	2017 Bare Costs Labor	Equipment	Total	Total Incl O&P
0010	**ENTRANCE LOCKS**									
0015	Cylinder, grip handle deadlocking latch	1 Carp	9	.889	Ea.	175	44		219	260

304

For customer support on your Building Construction Costs with RSMeans Data, call 800.448.8182.

08 71 Door Hardware

08 71 20 – Hardware

08 71 20.60 Entrance Locks		Crew	Daily Output	Labor-Hours	Unit	Material	2017 Bare Costs Labor	Equipment	Total	Total Incl O&P
0020	Deadbolt	1 Carp	8	1	Ea.	190	49.50		239.50	285
0100	Push and pull plate, dead bolt	↓	8	1		225	49.50		274.50	325
0900	For handicapped lever, add				↓	150			150	165

08 71 20.65 Thresholds		Crew	Daily Output	Labor-Hours	Unit	Material	Labor	Equipment	Total	Total Incl O&P
0010	**THRESHOLDS**									
0011	Threshold 3' long saddles aluminum	1 Carp	48	.167	L.F.	9.70	8.20		17.90	23.50
0100	Aluminum, 8" wide, 1/2" thick		12	.667	Ea.	49	33		82	104
0500	Bronze		60	.133	L.F.	44	6.55		50.55	58.50
0600	Bronze, panic threshold, 5" wide, 1/2" thick		12	.667	Ea.	158	33		191	224
0700	Rubber, 1/2" thick, 5-1/2" wide		20	.400		40	19.70		59.70	74
0800	2-3/4" wide	↓	20	.400	↓	45.50	19.70		65.20	80
1950	ADA Compliant Thresholds									
2300	Threshold, aluminum 4" wide x 36" long	1 Carp	12	.667	Ea.	34.50	33		67.50	88
2310	4" wide x 48" long		12	.667		42.50	33		75.50	97.50
2360	6" wide x 36" long		12	.667		55	33		88	111
2370	6" wide x 48" long	↓	12	.667	↓	70.50	33		103.50	129

08 71 20.70 Floor Checks		Crew	Daily Output	Labor-Hours	Unit	Material	Labor	Equipment	Total	Total Incl O&P
0010	**FLOOR CHECKS**									
0020	For over 3' wide doors single acting	1 Carp	2.50	3.200	Ea.	745	158		903	1,050
0500	Double acting	"	2.50	3.200	"	860	158		1,018	1,175

08 71 20.75 Door Hardware Accessories		Crew	Daily Output	Labor-Hours	Unit	Material	Labor	Equipment	Total	Total Incl O&P
0010	**DOOR HARDWARE ACCESSORIES**									
0050	Door closing coordinator, 36" (for paired openings up to 56")	1 Carp	8	1	Ea.	105	49.50		154.50	192
0060	48" (for paired openings up to 84")		8	1		113	49.50		162.50	200
0070	56" (for paired openings up to 96")	↓	8	1	↓	124	49.50		173.50	212

08 71 20.80 Hasps		Crew	Daily Output	Labor-Hours	Unit	Material	Labor	Equipment	Total	Total Incl O&P
0010	**HASPS**, steel assembly									
0015	3"	1 Carp	26	.308	Ea.	5.25	15.15		20.40	29
0020	4-1/2"		13	.615		6.65	30.50		37.15	54
0040	6"	↓	12.50	.640	↓	9.70	31.50		41.20	59

08 71 20.90 Hinges		Crew	Daily Output	Labor-Hours	Unit	Material	Labor	Equipment	Total	Total Incl O&P
0010	**HINGES**									
0012	Full mortise, avg. freq., steel base, USP, 4-1/2" x 4-1/2"				Pr.	39.50			39.50	43.50
0100	5" x 5", USP					59			59	65
0200	6" x 6", USP					124			124	137
0400	Brass base, 4-1/2" x 4-1/2", US10					62.50			62.50	69
0500	5" x 5", US10					89			89	98
0600	6" x 6", US10					164			164	180
0800	Stainless steel base, 4-1/2" x 4-1/2", US32				↓	77.50			77.50	85
0900	For non removable pin, add (security item)				Ea.	5.55			5.55	6.10
0910	For floating pin, driven tips, add					3.40			3.40	3.74
0930	For hospital type tip on pin, add					14.20			14.20	15.60
0940	For steeple type tip on pin, add				↓	19.30			19.30	21
0950	Full mortise, high frequency, steel base, 3-1/2" x 3-1/2", US26D				Pr.	30.50			30.50	33.50
1000	4-1/2" x 4-1/2", USP					66.50			66.50	73
1100	5" x 5", USP					52.50			52.50	57.50
1200	6" x 6", USP					139			139	153
1400	Brass base, 3-1/2" x 3-1/2", US4					53.50			53.50	59
1430	4-1/2" x 4-1/2", US10					76			76	83.50
1500	5" x 5", US10					127			127	139
1600	6" x 6", US10					171			171	188

08 71 20.90 Hinges

		Crew	Daily Output	Labor-Hours	Unit	Material	2017 Bare Costs Labor	Equipment	Total	Total Incl O&P
1800	Stainless steel base, 4-1/2" x 4-1/2", US32				Pr.	104			104	115
1810	5" x 4-1/2", US32				↓	140			140	154
1930	For hospital type tip on pin, add				Ea.	14.60			14.60	16.05
1950	Full mortise, low frequency, steel base, 3-1/2" x 3-1/2", US26D				Pr.	27.50			27.50	30
2000	4-1/2" x 4-1/2", USP					23.50			23.50	25.50
2100	5" x 5", USP					49			49	54
2200	6" x 6", USP					94			94	103
2300	4-1/2" x 4-1/2", US3					17.50			17.50	19.25
2310	5" x 5", US3					42.50			42.50	46.50
2400	Brass bass, 4-1/2" x 4-1/2", US10					55			55	60.50
2500	5" x 5", US10					79			79	86.50
2800	Stainless steel base, 4-1/2" x 4-1/2", US32				↓	76.50			76.50	84.50

08 71 20.91 Special Hinges

		Crew	Daily Output	Labor-Hours	Unit	Material	2017 Bare Costs Labor	Equipment	Total	Total Incl O&P
0010	**SPECIAL HINGES**									
0015	Paumelle, high frequency									
0020	Steel base, 6" x 4-1/2", US10				Pr.	145			145	160
0100	Brass base, 5" x 4-1/2", US10				Ea.	248			248	273
0200	Paumelle, average frequency, steel base, 4-1/2" x 3-1/2", US10				Pr.	98.50			98.50	108
0400	Olive knuckle, low frequency, brass base, 6" x 4-1/2", US10				Ea.	144			144	159
1000	Electric hinge with concealed conductor, average frequency									
1010	Steel base, 4-1/2" x 4-1/2", US26D				Pr.	315			315	345
1100	Bronze base, 4-1/2" x 4-1/2", US26D				"	335			335	365
1200	Electric hinge with concealed conductor, high frequency									
1210	Steel base, 4-1/2" x 4-1/2", US26D				Pr.	265			265	292
1600	Double weight, 800 lb., steel base, removable pin, 5" x 6", USP					440			440	485
1700	Steel base-welded pin, 5" x 6", USP					170			170	187
1800	Triple weight, 2000 lb., steel base, welded pin, 5" x 6", USP					570			570	625
2000	Pivot reinf., high frequency, steel base, 7-3/4" door plate, USP					151			151	167
2200	Bronze base, 7-3/4" door plate, US10				↓	238			238	262
3000	Swing clear, full mortise, full or half surface, high frequency,									
3010	Steel base, 5" high, USP				Pr.	144			144	159
3200	Swing clear, full mortise, average frequency									
3210	Steel base, 4-1/2" high, USP				Pr.	130			130	143
4000	Wide throw, average frequency, steel base, 4-1/2" x 6", USP				↓	95.50			95.50	105
4200	High frequency, steel base, 4-1/2" x 6", USP				↓	113			113	125
4600	Spring hinge, single acting, 6" flange, steel				Ea.	53			53	58
4700	Brass					97.50			97.50	107
4900	Double acting, 6" flange, steel					84			84	92.50
4950	Brass				↓	137			137	151
8000	Continuous hinges									
8010	Steel, piano, 2" x 72"	1 Carp	20	.400	Ea.	23	19.70		42.70	55.50
8020	Brass, piano, 1-1/16" x 30"		30	.267		9	13.15		22.15	30
8030	Acrylic, piano, 1-3/4" x 12"		40	.200		16	9.85		25.85	32.50
9000	Continuous hinge, steel, full mortise, heavy duty, 96"	↓	2	4	↓	630	197		827	995

08 71 20.95 Kick Plates

		Crew	Daily Output	Labor-Hours	Unit	Material	2017 Bare Costs Labor	Equipment	Total	Total Incl O&P
0010	**KICK PLATES**									
0020	Stainless steel, .050", 16 ga., 8" x 28", US32	1 Carp	15	.533	Ea.	40	26.50		66.50	84
0030	8" x 30"		15	.533		43	26.50		69.50	87.50
0040	8" x 34"		15	.533		48.50	26.50		75	93
0050	10" x 28"		15	.533		80	26.50		106.50	128
0060	10" x 30"		15	.533		86	26.50		112.50	135
0070	10" x 34"		15	.533	↓	96.50	26.50		123	146

08 71 Door Hardware

08 71 20 – Hardware

08 71 20.95 Kick Plates		Crew	Daily Output	Labor-Hours	Unit	Material	2017 Bare Costs Labor	Equipment	Total	Total Incl O&P
0080	Mop/Kick, 4" x 28"	1 Carp	15	.533	Ea.	35.50	26.50		62	79.50
0090	4" x 30"		15	.533		38	26.50		64.50	81.50
0100	4" x 34"		15	.533		43	26.50		69.50	87.50
0110	6" x 28"		15	.533		45	26.50		71.50	89.50
0120	6" x 30"		15	.533		49.50	26.50		76	94.50
0130	6" x 34"		15	.533		55.50	26.50		82	101
0500	Bronze, .050", 8" x 28"		15	.533		68.50	26.50		95	115
0510	8" x 30"		15	.533		68.50	26.50		95	115
0520	8" x 34"		15	.533		76.50	26.50		103	125
0530	10" x 28"		15	.533		79	26.50		105.50	127
0540	10" x 30"		15	.533		80	26.50		106.50	128
0550	10" x 34"		15	.533		95.50	26.50		122	145
0560	Mop/Kick, 4" x 28"		15	.533		34.50	26.50		61	78
0570	4" x 30"		15	.533		38	26.50		64.50	81.50
0580	4" x 34"		15	.533		39	26.50		65.50	82.50
0590	6" x 28"		15	.533		48.50	26.50		75	93
0600	6" x 30"		15	.533		54.50	26.50		81	100
0610	6" x 34"		15	.533		59	26.50		85.50	105
1000	Acrylic, .125", 8" x 26"		15	.533		28	26.50		54.50	71
1010	8" x 36"		15	.533		38	26.50		64.50	82
1020	8" x 42"		15	.533		45	26.50		71.50	89.50
1030	10" x 26"		15	.533		35	26.50		61.50	78.50
1040	10" x 36"		15	.533		48	26.50		74.50	93
1050	10" x 42"		15	.533		68.50	26.50		95	116
1060	Mop/Kick, 4" x 26"		15	.533		17	26.50		43.50	58.50
1070	4" x 36"		15	.533		24	26.50		50.50	66.50
1080	4" x 42"		15	.533		27	26.50		53.50	69.50
1090	6" x 26"		15	.533		23	26.50		49.50	65.50
1100	6" x 36"		15	.533		34	26.50		60.50	77.50
1110	6" x 42"		15	.533		39	26.50		65.50	83
1220	Brass, .050", 8" x 26"		15	.533		56	26.50		82.50	102
1230	8" x 36"		15	.533		75	26.50		101.50	123
1240	8" x 42"		15	.533		86	26.50		112.50	135
1250	10" x 26"		15	.533		71	26.50		97.50	118
1260	10" x 36"		15	.533		91	26.50		117.50	140
1270	10" x 42"		15	.533		105	26.50		131.50	156
1320	Mop/Kick, 4" x 26"		15	.533		28	26.50		54.50	71
1330	4" x 36"		15	.533		39	26.50		65.50	83
1340	4" x 42"		15	.533		44	26.50		70.50	88.50
1350	6" x 26"		15	.533		38	26.50		64.50	82
1360	6" x 36"		15	.533		48	26.50		74.50	93
1370	6" x 42"		15	.533		55	26.50		81.50	101
1800	Aluminum, .050", 8" x 26"		15	.533		35	26.50		61.50	78.50
1810	8" x 36"		15	.533		40	26.50		66.50	84
1820	8" x 42"		15	.533		47	26.50		73.50	91.50
1830	10" x 26"		15	.533		36	26.50		62.50	79.50
1840	10" x 36"		15	.533		50	26.50		76.50	95
1850	10" x 42"		15	.533		59	26.50		85.50	105
1860	Mop/Kick, 4" x 26"		15	.533		15	26.50		41.50	56.50
1870	4" x 36"		15	.533		20	26.50		46.50	62
1880	4" x 42"		15	.533		24	26.50		50.50	66.50
1890	6" x 26"		15	.533		22	26.50		48.50	64
1900	6" x 36"		15	.533		30	26.50		56.50	73

08 71 20 – Hardware

08 71 20.95 Kick Plates	Crew	Daily Output	Labor-Hours	Unit	Material	2017 Bare Costs Labor	Equipment	Total	Total Incl O&P
1910	6" x 42"	1 Carp	15	.533	Ea.	35	26.50		61.50	78.50

08 71 21 – Astragals

08 71 21.10 Exterior Mouldings, Astragals

		Crew	Daily Output	Labor-Hours	Unit	Material	2017 Bare Costs Labor	Equipment	Total	Total Incl O&P
0010	**EXTERIOR MOULDINGS, ASTRAGALS**									
0400	One piece, overlapping cadmium plated steel, flat, 3/16" x 2"	1 Carp	90	.089	L.F.	4	4.38		8.38	11.10
0600	Prime coated steel, flat, 1/8" x 3"		90	.089		5.75	4.38		10.13	13.05
0800	Stainless steel, flat, 3/32" x 1-5/8"		90	.089		16	4.38		20.38	24.50
1000	Aluminum, flat, 1/8" x 2"		90	.089		4.10	4.38		8.48	11.20
1200	Nail on, "T" extrusion		120	.067		1.90	3.28		5.18	7.15
1300	Vinyl bulb insert		105	.076		2.50	3.75		6.25	8.50
1600	Screw on, "T" extrusion		90	.089		3.75	4.38		8.13	10.85
1700	Vinyl insert		75	.107		4.50	5.25		9.75	13
2000	"L" extrusion, neoprene bulbs		75	.107		4.10	5.25		9.35	12.55
2100	Neoprene sponge insert		75	.107		6.90	5.25		12.15	15.65
2200	Magnetic		75	.107		10.60	5.25		15.85	19.70
2400	Spring hinged security seal, with cam		75	.107		6.80	5.25		12.05	15.55
2600	Spring loaded locking bolt, vinyl insert		45	.178		9.20	8.75		17.95	23.50
2800	Neoprene sponge strip, "Z" shaped, aluminum		60	.133		8.30	6.55		14.85	19.20
2900	Solid neoprene strip, nail on aluminum strip		90	.089		4.05	4.38		8.43	11.15
3000	One piece stile protection									
3020	Neoprene fabric loop, nail on aluminum strips	1 Carp	60	.133	L.F.	1.10	6.55		7.65	11.25
3110	Flush mounted aluminum extrusion, 1/2" x 1-1/4"		60	.133		6.90	6.55		13.45	17.65
3140	3/4" x 1-3/8"		60	.133		4.10	6.55		10.65	14.55
3160	1-1/8" x 1-3/4"		60	.133		4.70	6.55		11.25	15.20
3300	Mortise, 9/16" x 3/4"		60	.133		4.10	6.55		10.65	14.55
3320	13/16" x 1-3/8"		60	.133		4.30	6.55		10.85	14.80
3600	Spring bronze strip, nail on type		105	.076		1.85	3.75		5.60	7.80
3620	Screw on, with retainer		75	.107		2.70	5.25		7.95	11
3800	Flexible stainless steel housing, pile insert, 1/2" door		105	.076		7.25	3.75		11	13.75
3820	3/4" door		105	.076		8.10	3.75		11.85	14.65
4000	Extruded aluminum retainer, flush mount, pile insert		105	.076		2.25	3.75		6	8.25
4080	Mortise, felt insert		90	.089		4.55	4.38		8.93	11.70
4160	Mortise with spring, pile insert		90	.089		3.40	4.38		7.78	10.45
4400	Rigid vinyl retainer, mortise, pile insert		105	.076		2.70	3.75		6.45	8.70
4600	Wool pile filler strip, aluminum backing		105	.076		2.70	3.75		6.45	8.70
5000	Two piece overlapping astragal, extruded aluminum retainer									
5010	Pile insert	1 Carp	60	.133	L.F.	3.35	6.55		9.90	13.75
5020	Vinyl bulb insert		60	.133		1.85	6.55		8.40	12.10
5040	Vinyl flap insert		60	.133		3.65	6.55		10.20	14.05
5060	Solid neoprene flap insert		60	.133		6.55	6.55		13.10	17.25
5080	Hypalon rubber flap insert		60	.133		6.65	6.55		13.20	17.35
5090	Snap on cover, pile insert		60	.133		9.45	6.55		16	20.50
5400	Magnetic aluminum, surface mounted		60	.133		23	6.55		29.55	35
5500	Interlocking aluminum, 5/8" x 1" neoprene bulb insert		45	.178		5.70	8.75		14.45	19.65
5600	Adjustable aluminum, 9/16" x 21/32", pile insert		45	.178		17.55	8.75		26.30	32.50
5800	Magnetic, adjustable, 9/16" x 21/32"		45	.178		23	8.75		31.75	38.50
6000	Two piece stile protection									
6010	Cloth backed rubber loop, 1" gap, nail on aluminum strips	1 Carp	45	.178	L.F.	4.35	8.75		13.10	18.20
6040	Screw on aluminum strips		45	.178		6.55	8.75		15.30	20.50
6100	1-1/2" gap, screw on aluminum extrusion		45	.178		5.85	8.75		14.60	19.85
6240	Vinyl fabric loop, slotted aluminum extrusion, 1" gap		45	.178		2.20	8.75		10.95	15.80
6300	1-1/4" gap		45	.178		6.20	8.75		14.95	20

08 71 Door Hardware

08 71 25 – Weatherstripping

08 71 25.10 Mechanical Seals, Weatherstripping

		Crew	Daily Output	Labor-Hours	Unit	Material	2017 Bare Costs Labor	2017 Bare Costs Equipment	Total	Total Incl O&P
0010	**MECHANICAL SEALS, WEATHERSTRIPPING**									
1000	Doors, wood frame, interlocking, for 3' x 7' door, zinc	1 Carp	3	2.667	Opng.	44	131		175	250
1100	Bronze		3	2.667		56	131		187	263
1300	6' x 7' opening, zinc		2	4		54	197		251	360
1400	Bronze	↓	2	4	↓	65	197		262	370
1700	Wood frame, spring type, bronze									
1800	3' x 7' door	1 Carp	7.60	1.053	Opng.	23.50	52		75.50	106
1900	6' x 7' door	"	7	1.143	"	31	56.50		87.50	120
2200	Metal frame, spring type, bronze									
2300	3' x 7' door	1 Carp	3	2.667	Opng.	47.50	131		178.50	253
2400	6' x 7' door	"	2.50	3.200	"	52	158		210	298
2500	For stainless steel, spring type, add					133%				
2700	Metal frame, extruded sections, 3' x 7' door, aluminum	1 Carp	3	2.667	Opng.	28	131		159	232
2800	Bronze		3	2.667		82	131		213	291
3100	6' x 7' door, aluminum		1.50	5.333		35	263		298	440
3200	Bronze	↓	1.50	5.333	↓	137	263		400	550
3500	Threshold weatherstripping									
3650	Door sweep, flush mounted, aluminum	1 Carp	25	.320	Ea.	19	15.75		34.75	45
3700	Vinyl		25	.320		18	15.75		33.75	44
5000	Garage door bottom weatherstrip, 12' aluminum, clear		14	.571		25	28		53	70.50
5010	Bronze		14	.571		90	28		118	142
5050	Bottom protection, rubber		14	.571		38	28		66	84.50
5100	Threshold	↓	14	.571	↓	74.50	28		102.50	125

08 74 Access Control Hardware

08 74 13 – Card Key Access Control Hardware

08 74 13.50 Card Key Access

		Crew	Daily Output	Labor-Hours	Unit	Material	2017 Bare Costs Labor	2017 Bare Costs Equipment	Total	Total Incl O&P
0010	**CARD KEY ACCESS**									
0020	Computerized system, processor, proximity reader and cards									
0030	Does not include door hardware, lockset or wiring									
0040	Card key system for 1 door				Ea.	1,350			1,350	1,475
0060	Card key system for 2 doors					2,175			2,175	2,400
0080	Card key system for 4 doors					2,700			2,700	2,975
0100	Processor for card key access system					940			940	1,025
0160	Magnetic lock for electric access, 600 pound holding force					205			205	226
0170	Magnetic lock for electric access, 1200 pound holding force					205			205	226
0200	Proximity card reader				↓	140			140	154

08 74 19 – Biometric Identity Access Control Hardware

08 74 19.50 Biometric Identity Access

		Crew	Daily Output	Labor-Hours	Unit	Material	2017 Bare Costs Labor	2017 Bare Costs Equipment	Total	Total Incl O&P
0010	**BIOMETRIC IDENTITY ACCESS**									
0220	Hand geometry scanner, mem of 512 users, excl striker/power	1 Elec	3	2.667	Ea.	2,200	151		2,351	2,650
0230	Memory upgrade for, adds 9,700 user profiles		8	1		320	56.50		376.50	435
0240	Adds 32,500 user profiles		8	1		650	56.50		706.50	800
0250	Prison type, memory of 256 users, excl striker/power		3	2.667		2,650	151		2,801	3,150
0260	Memory upgrade for, adds 3,300 user profiles		8	1		300	56.50		356.50	415
0270	Adds 9,700 user profiles		8	1		480	56.50		536.50	615
0280	Adds 27,900 user profiles		8	1		650	56.50		706.50	800
0290	All weather, mem of 512 users, excl striker/power		3	2.667		4,000	151		4,151	4,625
0300	Facial & fingerprint scanner, combination unit, excl striker/power		3	2.667		4,300	151		4,451	4,950
0310	Access for, for initial setup, excl striker/power	↓	3	2.667	↓	1,200	151		1,351	1,550

08 75 Window Hardware

08 75 30 – Weatherstripping

08 75 30.10 Mechanical Weather Seals	Crew	Daily Output	Labor-Hours	Unit	Material	2017 Bare Costs Labor	Equipment	Total	Total Incl O&P
0010 **MECHANICAL WEATHER SEALS**, Window, double hung, 3' x 5'									
0020 Zinc	1 Carp	7.20	1.111	Opng.	21	54.50		75.50	107
0100 Bronze		7.20	1.111		40	54.50		94.50	128
0500 As above but heavy duty, zinc		4.60	1.739		20	85.50		105.50	153
0600 Bronze	↓	4.60	1.739	↓	70	85.50		155.50	208

08 79 Hardware Accessories

08 79 13 – Key Storage Equipment

08 79 13.10 Key Cabinets

	Crew	Daily Output	Labor-Hours	Unit	Material	2017 Bare Costs Labor	Equipment	Total	Total Incl O&P
0010 **KEY CABINETS**									
0020 Wall mounted, 60 key capacity	1 Carp	20	.400	Ea.	93.50	19.70		113.20	133
0400 Wall mounted, 30 key capacity	1 Clab	50	.160		73.50	6.25		79.75	90.50
0500 Wall mounted, 80 key capacity	"	40	.200	↓	118	7.85		125.85	141

08 79 20 – Door Accessories

08 79 20.10 Door Hardware Accessories

	Crew	Daily Output	Labor-Hours	Unit	Material	2017 Bare Costs Labor	Equipment	Total	Total Incl O&P
0010 **DOOR HARDWARE ACCESSORIES**									
0140 Door bolt, surface, 4"	1 Carp	32	.250	Ea.	15	12.30		27.30	35.50
0160 Door latch	"	12	.667	"	8.55	33		41.55	60
0200 Sliding closet door									
0220 Track and hanger, single	1 Carp	10	.800	Ea.	65	39.50		104.50	132
0240 Double		8	1		80	49.50		129.50	164
0260 Door guide, single		48	.167		30	8.20		38.20	45.50
0280 Double		48	.167		40	8.20		48.20	56.50
0600 Deadbolt and lock cover plate, brass or stainless steel		30	.267		28	13.15		41.15	51
0620 Hole cover plate, brass or chrome		35	.229		8	11.25		19.25	26
2240 Mortise lockset, passage, lever handle		9	.889		160	44		204	243
4000 Security chain, standard	↓	18	.444	↓	11.50	22		33.50	46

08 81 Glass Glazing

08 81 10 – Float Glass

08 81 10.10 Various Types and Thickness of Float Glass

	Crew	Daily Output	Labor-Hours	Unit	Material	2017 Bare Costs Labor	Equipment	Total	Total Incl O&P
0010 **VARIOUS TYPES AND THICKNESS OF FLOAT GLASS** R088110-10									
0020 3/16" plain	2 Glaz	130	.123	S.F.	6.90	5.90		12.80	16.55
0200 Tempered, clear		130	.123		7.10	5.90		13	16.75
0300 Tinted		130	.123		6.90	5.90		12.80	16.55
0600 1/4" thick, clear, plain		120	.133		10.05	6.35		16.40	21
0700 Tinted		120	.133		9.15	6.35		15.50	19.75
0800 Tempered, clear		120	.133		6.90	6.35		13.25	17.30
0900 Tinted		120	.133		11.70	6.35		18.05	22.50
1600 3/8" thick, clear, plain		75	.213		10.80	10.20		21	27.50
1700 Tinted		75	.213		16	10.20		26.20	33
1800 Tempered, clear		75	.213		17.10	10.20		27.30	34.50
1900 Tinted		75	.213		19.40	10.20		29.60	37
2200 1/2" thick, clear, plain		55	.291		18.45	13.90		32.35	41.50
2300 Tinted		55	.291		28	13.90		41.90	52
2400 Tempered, clear		55	.291		25.50	13.90		39.40	49
2500 Tinted		55	.291		27	13.90		40.90	50.50
2800 5/8" thick, clear, plain		45	.356		28	17		45	57
2900 Tempered, clear	↓	45	.356		32	17		49	61

08 81 Glass Glazing

08 81 10 – Float Glass

08 81 10.10 Various Types and Thickness of Float Glass

		Crew	Daily Output	Labor-Hours	Unit	Material	2017 Bare Costs Labor	Equipment	Total	Total Incl O&P
3200	3/4" thick, clear, plain	2 Glaz	35	.457	S.F.	36	22		58	73
3300	Tempered, clear		35	.457		42	22		64	79.50
3600	1" thick, clear, plain	↓	30	.533		60	25.50		85.50	105
8900	For low emissivity coating for 3/16" & 1/4" only, add to above				↓	18%				

08 81 13 – Decorative Glass Glazing

08 81 13.10 Beveled Glass

0010	**BEVELED GLASS**, with design patterns									
0020	Simple pattern	2 Glaz	150	.107	S.F.	61.50	5.10		66.60	75.50
0050	Intricate pattern	"	125	.128	"	134	6.10		140.10	156

08 81 13.30 Sandblasted Glass

0010	**SANDBLASTED GLASS**, float glass									
0020	1/8" thick	2 Glaz	160	.100	S.F.	11.25	4.78		16.03	19.70
0100	3/16" thick		130	.123		12.40	5.90		18.30	22.50
0500	1/4" thick		120	.133		12.90	6.35		19.25	24
0600	3/8" thick	↓	75	.213	↓	13.80	10.20		24	30.50

08 81 17 – Fire Glass

08 81 17.10 Fire Resistant Glass

0010	**FIRE RESISTANT GLASS**									
0020	Fire Glass Minimum	2 Glaz	40	.400	S.F.	38	19.10		57.10	71
0030	Mid Range		40	.400		80	19.10		99.10	117
0050	High End	↓	40	.400	↓	355	19.10		374.10	420

08 81 20 – Vision Panels

08 81 20.10 Full Vision

0010	**FULL VISION**, window system with 3/4" glass mullions									
0020	Up to 10' high	H-2	130	.185	S.F.	65.50	8.30		73.80	84.50
0100	10' to 20' high, minimum		110	.218		69.50	9.80		79.30	91.50
0150	Average		100	.240		75	10.80		85.80	99
0200	Maximum	↓	80	.300	↓	84	13.50		97.50	113

08 81 25 – Glazing Variables

08 81 25.10 Applications of Glazing

0010	**APPLICATIONS OF GLAZING**	R088110-10								
0600	For glass replacement, add				S.F.		100%			
0700	For gasket settings, add				L.F.	5.85			5.85	6.45
0900	For sloped glazing, add				S.F.		26%			
2000	Fabrication, polished edges, 1/4" thick				Inch	.56			.56	.62
2100	1/2" thick					1.32			1.32	1.45
2500	Mitered edges, 1/4" thick					1.32			1.32	1.45
2600	1/2" thick				↓	2.19			2.19	2.41

08 81 30 – Insulating Glass

08 81 30.10 Reduce Heat Transfer Glass

0010	**REDUCE HEAT TRANSFER GLASS**		R088110-10								
0015	2 lites 1/8" float, 1/2" thk under 15 S.F.										
0020	Clear	G	2 Glaz	95	.168	S.F.	10.50	8.05		18.55	24
0100	Tinted	G		95	.168		14.65	8.05		22.70	28.50
0200	2 lites 3/16" float, for 5/8" thk unit, 15 to 30 S.F., clear	G		90	.178		13.95	8.50		22.45	28.50
0300	Tinted	G		90	.178		14	8.50		22.50	28.50
0400	1" thk, dbl. glazed, 1/4" float, 30 to 70 S.F., clear	G		75	.213		17.60	10.20		27.80	35
0500	Tinted	G		75	.213		24	10.20		34.20	42
0600	1" thk, dbl. glazed, 1/4" float, 1/4" wire			75	.213		24.50	10.20		34.70	42.50
0700	1/4" float, 1/4" tempered			75	.213		32	10.20		42.20	51

For customer support on your Building Construction Costs with RSMeans Data, call 800.448.8182.

311

08 81 Glass Glazing

08 81 30 – Insulating Glass

08 81 30.10 Reduce Heat Transfer Glass

08 81 30.10 Reduce Heat Transfer Glass		Crew	Daily Output	Labor-Hours	Unit	Material	2017 Bare Costs Labor	Equipment	Total	Total Incl O&P
0800	1/4" wire, 1/4" tempered	2 Glaz	75	.213	S.F.	31	10.20		41.20	50
2000	Both lites, light & heat reflective [G]		85	.188		32	9		41	48.50
2500	Heat reflective, film inside, 1" thick unit, clear [G]		85	.188		28	9		37	44.50
2600	Tinted [G]		85	.188		29	9		38	45.50
3000	Film on weatherside, clear, 1/2" thick unit [G]		95	.168		20	8.05		28.05	34.50
3100	5/8" thick unit [G]		90	.178		20	8.50		28.50	35
3200	1" thick unit [G]		85	.188		27.50	9		36.50	44

08 81 35 – Translucent Glass

08 81 35.10 Obscure Glass

08 81 35.10 Obscure Glass		Crew	Daily Output	Labor-Hours	Unit	Material	2017 Bare Costs Labor	Equipment	Total	Total Incl O&P
0010	**OBSCURE GLASS**									
0020	1/8" thick, textured	2 Glaz	140	.114	S.F.	11.80	5.45		17.25	21.50
0100	Color		125	.128		14	6.10		20.10	24.50
0300	7/32" thick, textured		120	.133		12.95	6.35		19.30	24
0400	Color		105	.152		16.25	7.30		23.55	29

08 81 35.20 Patterned Glass

08 81 35.20 Patterned Glass		Crew	Daily Output	Labor-Hours	Unit	Material	2017 Bare Costs Labor	Equipment	Total	Total Incl O&P
0010	**PATTERNED GLASS**, colored									
0020	1/8" thick	2 Glaz	140	.114	S.F.	9.60	5.45		15.05	18.85
0300	7/32" thick	"	120	.133	"	12.05	6.35		18.40	23

08 81 45 – Sheet Glass

08 81 45.10 Window Glass, Sheet

08 81 45.10 Window Glass, Sheet		Crew	Daily Output	Labor-Hours	Unit	Material	2017 Bare Costs Labor	Equipment	Total	Total Incl O&P
0010	**WINDOW GLASS, SHEET** gray									
0020	1/8" thick	2 Glaz	160	.100	S.F.	6.15	4.78		10.93	14.05
0200	1/4" thick	"	130	.123	"	7.35	5.90		13.25	17.05

08 81 50 – Spandrel Glass

08 81 50.10 Glass for Non Vision Areas

08 81 50.10 Glass for Non Vision Areas		Crew	Daily Output	Labor-Hours	Unit	Material	2017 Bare Costs Labor	Equipment	Total	Total Incl O&P
0010	**GLASS FOR NON VISION AREAS**, 1/4" thick standard colors									
0020	Up to 1,000 S.F.	2 Glaz	110	.145	S.F.	17.85	6.95		24.80	30.50
0200	1,000 to 2,000 S.F.	"	120	.133	"	16.10	6.35		22.45	27.50
0300	For custom colors, add				Total	10%				
0500	For 3/8" thick, add				S.F.	12.15			12.15	13.35
1000	For double coated, 1/4" thick, add					4.34			4.34	4.77
1200	For insulation on panels, add					7.10			7.10	7.80
2000	Panels, insulated, with aluminum backed fiberglass, 1" thick	2 Glaz	120	.133		17.20	6.35		23.55	28.50
2100	2" thick	"	120	.133		21	6.35		27.35	32.50

08 81 55 – Window Glass

08 81 55.10 Sheet Glass

08 81 55.10 Sheet Glass		Crew	Daily Output	Labor-Hours	Unit	Material	2017 Bare Costs Labor	Equipment	Total	Total Incl O&P
0010	**SHEET GLASS** (window), clear float, stops, putty bed									
0015	1/8" thick, clear float	2 Glaz	480	.033	S.F.	3.78	1.59		5.37	6.60
0500	3/16" thick, clear		480	.033		6.10	1.59		7.69	9.10
0600	Tinted		480	.033		7.60	1.59		9.19	10.75
0700	Tempered		480	.033		9.40	1.59		10.99	12.75

08 81 65 – Wire Glass

08 81 65.10 Glass Reinforced With Wire

08 81 65.10 Glass Reinforced With Wire		Crew	Daily Output	Labor-Hours	Unit	Material	2017 Bare Costs Labor	Equipment	Total	Total Incl O&P
0010	**GLASS REINFORCED WITH WIRE**									
0012	1/4" thick rough obscure	2 Glaz	135	.119	S.F.	24	5.65		29.65	35
1000	Polished wire, 1/4" thick, diamond, clear		135	.119		28	5.65		33.65	39.50
1500	Pinstripe, obscure		135	.119		47	5.65		52.65	60.50

08 83 Mirrors

08 83 13 – Mirrored Glass Glazing

08 83 13.10 Mirrors

		Crew	Daily Output	Labor-Hours	Unit	Material	2017 Bare Costs Labor	Equipment	Total	Total Incl O&P
0010	**MIRRORS**, No frames, wall type, 1/4" plate glass, polished edge									
0100	Up to 5 S.F.	2 Glaz	125	.128	S.F.	9.70	6.10		15.80	19.95
0200	Over 5 S.F.		160	.100		9.45	4.78		14.23	17.70
0500	Door type, 1/4" plate glass, up to 12 S.F.		160	.100		9	4.78		13.78	17.20
1000	Float glass, up to 10 S.F., 1/8" thick		160	.100		6.05	4.78		10.83	13.95
1100	3/16" thick		150	.107		7.45	5.10		12.55	15.95
1500	12" x 12" wall tiles, square edge, clear		195	.082		2.51	3.92		6.43	8.70
1600	Veined		195	.082		6.35	3.92		10.27	12.90
2000	1/4" thick, stock sizes, one way transparent		125	.128		20	6.10		26.10	31.50
2010	Bathroom, unframed, laminated		160	.100		14.25	4.78		19.03	23

08 83 13.15 Reflective Glass

		Crew	Daily Output	Labor-Hours	Unit	Material	2017 Bare Costs Labor	Equipment	Total	Total Incl O&P
0010	**REFLECTIVE GLASS**									
0100	1/4" float with fused metallic oxide fixed [G]	2 Glaz	115	.139	S.F.	17.30	6.65		23.95	29
0500	1/4" float glass with reflective applied coating [G]	"	115	.139	"	14	6.65		20.65	25.50

08 84 Plastic Glazing

08 84 10 – Plexiglass Glazing

08 84 10.10 Plexiglass Acrylic

		Crew	Daily Output	Labor-Hours	Unit	Material	2017 Bare Costs Labor	Equipment	Total	Total Incl O&P
0010	**PLEXIGLASS ACRYLIC**, clear, masked,									
0020	1/8" thick, cut sheets	2 Glaz	170	.094	S.F.	12.20	4.50		16.70	20.50
0200	Full sheets		195	.082		5.15	3.92		9.07	11.60
0500	1/4" thick, cut sheets		165	.097		14.30	4.64		18.94	23
0600	Full sheets		185	.086		9.20	4.13		13.33	16.40
0900	3/8" thick, cut sheets		155	.103		20	4.93		24.93	29.50
1000	Full sheets		180	.089		15.30	4.25		19.55	23.50
1300	1/2" thick, cut sheets		135	.119		28.50	5.65		34.15	40
1400	Full sheets		150	.107		20	5.10		25.10	30
1700	3/4" thick, cut sheets		115	.139		70.50	6.65		77.15	87.50
1800	Full sheets		130	.123		40.50	5.90		46.40	53.50
2100	1" thick, cut sheets		105	.152		79	7.30		86.30	98
2200	Full sheets		125	.128		49	6.10		55.10	63.50
3000	Colored, 1/8" thick, cut sheets		170	.094		18.35	4.50		22.85	27
3200	Full sheets		195	.082		11.20	3.92		15.12	18.25
3500	1/4" thick, cut sheets		165	.097		20.50	4.64		25.14	29.50
3600	Full sheets		185	.086		14.30	4.13		18.43	22
4000	Mirrors, untinted, cut sheets, 1/8" thick		185	.086		12.25	4.13		16.38	19.80
4200	1/4" thick		180	.089		16.30	4.25		20.55	24.50

08 84 20 – Polycarbonate

08 84 20.10 Thermoplastic

		Crew	Daily Output	Labor-Hours	Unit	Material	2017 Bare Costs Labor	Equipment	Total	Total Incl O&P
0010	**THERMOPLASTIC**, clear, masked, cut sheets									
0020	1/8" thick	2 Glaz	170	.094	S.F.	14.30	4.50		18.80	22.50
0500	3/16" thick		165	.097		17.35	4.64		21.99	26
1000	1/4" thick		155	.103		17.35	4.93		22.28	26.50
1500	3/8" thick		150	.107		27	5.10		32.10	37.50

For customer support on your Building Construction Costs with RSMeans Data, call 800.448.8182.

313

08 87 Glazing Surface Films

08 87 13 – Solar Control Films

08 87 13.10 Solar Films On Glass

		Crew	Daily Output	Labor-Hours	Unit	Material	2017 Bare Costs Labor	Equipment	Total	Total Incl O&P
0010	**SOLAR FILMS ON GLASS** (glass not included)									
1000	Bronze, 20% VLT	H-2	180	.133	S.F.	.80	6		6.80	10.05
1020	50% VLT		180	.133		1.35	6		7.35	10.65
1050	Neutral, 20% VLT		180	.133		1.35	6		7.35	10.65
1100	Silver, 15% VLT		180	.133		.90	6		6.90	10.15
1120	35% VLT		180	.133		3.25	6		9.25	12.75
1150	68% VLT		180	.133		3.80	6		9.80	13.35
3000	One way mirror with night vision 5% in and 15% out VLT		180	.133		1.65	6		7.65	10.95

08 87 16 – Glass Safety Films

08 87 16.10 Safety Films On Glass

		Crew	Daily Output	Labor-Hours	Unit	Material	2017 Bare Costs Labor	Equipment	Total	Total Incl O&P
0010	**SAFETY FILMS ON GLASS** (glass not included)									
0015	Safety film helps hold glass together when broken									
0050	Safety film, clear, 2 mil	H-2	180	.133	S.F.	1.31	6		7.31	10.60
0100	Safety film, clear, 4 mil		180	.133		1.81	6		7.81	11.15
0110	Safety film, tinted, 4 mil		180	.133		1.40	6		7.40	10.70
0150	Safety film, black tint, 8 mil		180	.133		2.13	6		8.13	11.50

08 87 26 – Bird Control Film

08 87 26.10 Bird Control Film

		Crew	Daily Output	Labor-Hours	Unit	Material	2017 Bare Costs Labor	Equipment	Total	Total Incl O&P
0010	**BIRD CONTROL FILM**									
0050	Patterned, adhered to glass	2 Glaz	180	.089	S.F.	20	4.25		24.25	28.50
0200	Decals small, adhered to glass	1 Glaz	50	.160	Ea.	2.50	7.65		10.15	14.40
0250	Decals large, adhered to glass		50	.160	"	9	7.65		16.65	21.50
0300	Decals set of 4, adhered to glass		25	.320	Set	8	15.30		23.30	32.50
0350	Decals set of 6, adhered to glass		20	.400	"	20	19.10		39.10	51
0400	Bird control film tape		10	.800	Roll	22	38		60	82

08 87 33 – Electrically Tinted Window Film

08 87 33.20 Window Film

		Crew	Daily Output	Labor-Hours	Unit	Material	2017 Bare Costs Labor	Equipment	Total	Total Incl O&P
0010	**WINDOW FILM** adhered on glass (glass not included)									
0015	Film is pre-wired and can be trimmed									
0100	Window film 4 S.F. adhered on glass	1 Glaz	200	.040	S.F.	242	1.91		243.91	269
0120	6 S.F.		180	.044		360	2.12		362.12	405
0160	9 S.F.		170	.047		540	2.25		542.25	600
0180	10 S.F.		150	.053		605	2.55		607.55	670
0200	12 S.F.		144	.056		720	2.66		722.66	795
0220	14 S.F.		140	.057		840	2.73		842.73	930
0240	16 S.F.		128	.063		960	2.99		962.99	1,050
0260	18 S.F.		126	.063		1,075	3.03		1,078.03	1,200
0280	20 S.F.		120	.067		1,200	3.19		1,203.19	1,325

08 87 53 – Security Films On Glass

08 87 53.10 Security Film Adhered On Glass

		Crew	Daily Output	Labor-Hours	Unit	Material	2017 Bare Costs Labor	Equipment	Total	Total Incl O&P
0010	**SECURITY FILM ADHERED ON GLASS** (glass not included)									
0020	Security film, clear, 7 mil	1 Glaz	200	.040	S.F.	1.18	1.91		3.09	4.21
0030	8 mil		200	.040		1.95	1.91		3.86	5.05
0040	9 mil		200	.040		2.10	1.91		4.01	5.20
0050	10 mil		200	.040		2.25	1.91		4.16	5.40
0060	12 mil		200	.040		2.10	1.91		4.01	5.20
0075	15 mil		200	.040		2.45	1.91		4.36	5.60
0100	Security film, sealed with structural adhesive, 7 mil		180	.044		5.65	2.12		7.77	9.45
0110	8 mil		180	.044		6.55	2.12		8.67	10.45
0140	14 mil		180	.044		7.15	2.12		9.27	11.10

08 88 Special Function Glazing

08 88 40 – Acoustical Glass Units

08 88 40.10 Sound Reduction Units

		Crew	Daily Output	Labor-Hours	Unit	Material	2017 Bare Costs Labor	Equipment	Total	Total Incl O&P
0010	**SOUND REDUCTION UNITS**, 1 lite at 3/8", 1 lite at 3/16"									
0020	For 1" thick	2 Glaz	100	.160	S.F.	35	7.65		42.65	50
0100	For 4" thick	"	80	.200	"	59.50	9.55		69.05	80

08 88 56 – Ballistics-Resistant Glazing

08 88 56.10 Laminated Glass

		Crew	Daily Output	Labor-Hours	Unit	Material	2017 Bare Costs Labor	Equipment	Total	Total Incl O&P
0010	**LAMINATED GLASS**									
0020	Clear float .03" vinyl 1/4" thick	2 Glaz	90	.178	S.F.	12.75	8.50		21.25	27
0100	3/8" thick		78	.205		27.50	9.80		37.30	45.50
0200	.06" vinyl, 1/2" thick		65	.246		28	11.75		39.75	49
1000	5/8" thick		90	.178		30.50	8.50		39	46.50
2000	Bullet-resisting, 1-3/16" thick, to 15 S.F.		16	1		114	48		162	199
2100	Over 15 S.F.		16	1		128	48		176	213
2500	2-1/4" thick, to 15 S.F.		12	1.333		185	63.50		248.50	300
2600	Over 15 S.F.		12	1.333		175	63.50		238.50	290
2700	Level 2 (.357 magnum), NIJ and UL		12	1.333		85.50	63.50		149	191
2750	Level 3A (.44 magnum) NIJ, UL 3		12	1.333		90.50	63.50		154	197
2800	Level 4 (AK-47) NIJ, UL 7 & 8		12	1.333		125	63.50		188.50	235
2850	Level 5 (M-16) UL		12	1.333		129	63.50		192.50	239
2900	Level 3 (7.62 Armor Piercing) NIJ, UL 4 & 5		12	1.333		151	63.50		214.50	263

08 91 Louvers

08 91 19 – Fixed Louvers

08 91 19.10 Aluminum Louvers

		Crew	Daily Output	Labor-Hours	Unit	Material	2017 Bare Costs Labor	Equipment	Total	Total Incl O&P
0010	**ALUMINUM LOUVERS**									
0020	Aluminum with screen, residential, 8" x 8"	1 Carp	38	.211	Ea.	20	10.35		30.35	38
0100	12" x 12"		38	.211		16	10.35		26.35	33.50
0200	12" x 18"		35	.229		20	11.25		31.25	39.50
0250	14" x 24"		30	.267		29	13.15		42.15	52
0300	18" x 24"		27	.296		32	14.60		46.60	57.50
0500	24" x 30"		24	.333		61.50	16.40		77.90	92.50
0700	Triangle, adjustable, small		20	.400		58.50	19.70		78.20	94.50
0800	Large		15	.533		80.50	26.50		107	129
1200	Extruded aluminum, see Section 23 37 15.40									
2100	Midget, aluminum, 3/4" deep, 1" diameter	1 Carp	85	.094	Ea.	.77	4.64		5.41	7.95
2150	3" diameter		60	.133		2.73	6.55		9.28	13.05
2200	4" diameter		50	.160		5.10	7.90		13	17.65
2250	6" diameter		30	.267		4.10	13.15		17.25	24.50

08 91 26 – Door Louvers

08 91 26.10 Steel Louvers, 18 Gauge, Fixed Blade

		Crew	Daily Output	Labor-Hours	Unit	Material	2017 Bare Costs Labor	Equipment	Total	Total Incl O&P
0010	**STEEL LOUVERS, 18 GAUGE, FIXED BLADE**									
0050	12" x 12", with powder coat	1 Carp	20	.400	Ea.	80	19.70		99.70	118
0055	18" x 12"		20	.400		86	19.70		105.70	125
0060	18" x 18"		20	.400		105	19.70		124.70	146
0065	24" x 12"		20	.400		112	19.70		131.70	153
0070	24" x 18"		20	.400		124	19.70		143.70	166
0075	24" x 24"		20	.400		149	19.70		168.70	194
0100	12" x 12", galvanized		20	.400		74	19.70		93.70	112
0105	18" x 12"		20	.400		87.50	19.70		107.20	127
0115	24" x 12"		20	.400		104	19.70		123.70	144
0125	24" x 24"		20	.400		144	19.70		163.70	188

08 91 Louvers

08 91 26 – Door Louvers

08 91 26.10 Steel Louvers, 18 Gauge, Fixed Blade

	08 91 26.10 Steel Louvers, 18 Gauge, Fixed Blade	Crew	Daily Output	Labor-Hours	Unit	Material	2017 Bare Costs Labor	Equipment	Total	Total Incl O&P
0300	6" x 12", painted	1 Carp	20	.400	Ea.	52	19.70		71.70	87
0320	10" x 16"		20	.400		82	19.70		101.70	120
0340	12" x 22"		20	.400		92	19.70		111.70	131
0360	14" x 14"		20	.400		103	19.70		122.70	143
0400	18" x 18"		20	.400		132	19.70		151.70	175
0420	20" x 26"		20	.400		142	19.70		161.70	186
0440	22" x 22"		20	.400		135	19.70		154.70	179
0460	26" x 26"		20	.400		173	19.70		192.70	220
3000	Fire rated									
3010	12" x 12"	1 Carp	10	.800	Ea.	204	39.50		243.50	286
3050	18" x 18"		10	.800		210	39.50		249.50	292
3100	24" x 12"		10	.800		244	39.50		283.50	330
3120	24" x 18"		10	.800		300	39.50		339.50	390
3130	24" x 24"		10	.800		293	39.50		332.50	385

08 91 26.20 Aluminum Louvers

	08 91 26.20 Aluminum Louvers	Crew	Daily Output	Labor-Hours	Unit	Material	Labor	Equipment	Total	Total Incl O&P
0010	ALUMINUM LOUVERS, 18 ga, fixed blade, clear anodized									
0050	6" x 12"	1 Carp	20	.400	Ea.	65	19.70		84.70	102
0060	10" x 10"		20	.400		75	19.70		94.70	113
0065	10" x 14"		20	.400		90	19.70		109.70	129
0080	12" x 22"		20	.400		117	19.70		136.70	159
0090	14" x 14"		20	.400		116	19.70		135.70	158
0100	18" x 10"		20	.400		88	19.70		107.70	127
0110	18" x 18"		20	.400		135	19.70		154.70	179
0120	22" x 22"		20	.400		188	19.70		207.70	237
0130	26" x 26"		20	.400		238	19.70		257.70	292

08 95 Vents

08 95 13 – Soffit Vents

08 95 13.10 Wall Louvers

	08 95 13.10 Wall Louvers	Crew	Daily Output	Labor-Hours	Unit	Material	Labor	Equipment	Total	Total Incl O&P
0010	WALL LOUVERS									
2340	Baked enamel finish	1 Carp	200	.040	L.F.	7.05	1.97		9.02	10.80
2400	Under eaves vent, aluminum, mill finish, 16" x 4"		48	.167	Ea.	1.86	8.20		10.06	14.65
2500	16" x 8"		48	.167	"	2.23	8.20		10.43	15.05

08 95 16 – Wall Vents

08 95 16.10 Louvers

	08 95 16.10 Louvers	Crew	Daily Output	Labor-Hours	Unit	Material	Labor	Equipment	Total	Total Incl O&P
0010	LOUVERS									
0020	Redwood, 2'-0" diameter, full circle	1 Carp	16	.500	Ea.	200	24.50		224.50	258
0100	Half circle		16	.500		183	24.50		207.50	239
0200	Octagonal		16	.500		150	24.50		174.50	203
0300	Triangular, 5/12 pitch, 5'-0" at base		16	.500		210	24.50		234.50	269
7000	Vinyl gable vent, 8" x 8"		38	.211		15	10.35		25.35	32.50
7020	12" x 12"		38	.211		28	10.35		38.35	47
7080	12" x 18"		35	.229		37	11.25		48.25	58
7200	18" x 24"		30	.267		45	13.15		58.15	69.50

Estimating Tips
General
- Room Finish Schedule: A complete set of plans should contain a room finish schedule. If one is not available, it would be well worth the time and effort to obtain one.

09 20 00 Plaster and Gypsum Board
- Lath is estimated by the square yard plus a 5% allowance for waste. Furring, channels, and accessories are measured by the linear foot. An extra foot should be allowed for each accessory miter or stop.
- Plaster is also estimated by the square yard. Deductions for openings vary by preference, from zero deduction to 50% of all openings over 2 feet in width. The estimator should allow one extra square foot for each linear foot of horizontal interior or exterior angle located below the ceiling level. Also, double the areas of small radius work.
- Drywall accessories, studs, track, and acoustical caulking are all measured by the linear foot. Drywall taping is figured by the square foot. Gypsum wallboard is estimated by the square foot. No material deductions should be made for door or window openings under 32 S.F.

09 60 00 Flooring
- Tile and terrazzo areas are taken off on a square foot basis. Trim and base materials are measured by the linear foot. Accent tiles are listed per each. Two basic methods of installation are used. Mud set is approximately 30% more expensive than thin set. The cost of grout is included with tile unit price lines unless otherwise noted. In terrazzo work, be sure to include the linear footage of embedded decorative strips, grounds, machine rubbing, and power cleanup.
- Wood flooring is available in strip, parquet, or block configuration. The latter two types are set in adhesives with quantities estimated by the square foot. The laying pattern will influence labor costs and material waste. In addition to the material and labor for laying wood floors, the estimator must make allowances for sanding and finishing these areas, unless the flooring is prefinished.
- Sheet flooring is measured by the square yard. Roll widths vary, so consideration should be given to use the most economical width, as waste must be figured into the total quantity. Consider also the installation methods available—direct glue down or stretched. Direct glue-down installation is assumed with sheet carpet unit price lines unless otherwise noted.

09 70 00 Wall Finishes
- Wall coverings are estimated by the square foot. The area to be covered is measured—length by height of the wall above the baseboards—to calculate the square footage of each wall. This figure is divided by the number of square feet in the single roll which is being used. Deduct, in full, the areas of openings such as doors and windows. Where a pattern match is required allow 25%–30% waste.

09 80 00 Acoustic Treatment
- Acoustical systems fall into several categories. The takeoff of these materials should be by the square foot of area with a 5% allowance for waste. Do not forget about scaffolding, if applicable, when estimating these systems.

09 90 00 Painting and Coating
- A major portion of the work in painting involves surface preparation. Be sure to include cleaning, sanding, filling, and masking costs in the estimate.
- Protection of adjacent surfaces is not included in painting costs. When considering the method of paint application, an important factor is the amount of protection and masking required. These must be estimated separately and may be the determining factor in choosing the method of application.

Reference Numbers
Reference numbers are shown at the beginning of some major classifications. These numbers refer to related items in the Reference Section. The reference information may be an estimating procedure, an alternate pricing method, or technical information.

Note: Not all subdivisions listed here necessarily appear. ■

Did you know?
Our online estimating solution gives you the same access to RSMeans' data with 24/7 access:
- Quickly locate costs in the searchable database.
- Build cost lists, estimates, and reports in minutes.
- Adjust costs to any location in the U.S. and Canada with the click of a button.

Start your free trial today at
www.RSMeansOnline.com

09 01 Maintenance of Finishes

09 01 60 – Maintenance of Flooring

09 01 60.10 Carpet Maintenance

		Crew	Daily Output	Labor-Hours	Unit	Material	2017 Bare Costs Labor	Equipment	Total	Total Incl O&P
0010	**CARPET MAINTENANCE**									
0020	Steam clean, per cleaning, routine maintenance	1 Clab	3000	.003	S.F.	.07	.10		.17	.24
0500	Stain removal	"	2000	.004	"	.10	.16		.26	.35

09 01 70 – Maintenance of Wall Finishes

09 01 70.10 Gypsum Wallboard Repairs

		Crew	Daily Output	Labor-Hours	Unit	Material	2017 Bare Costs Labor	Equipment	Total	Total Incl O&P
0010	**GYPSUM WALLBOARD REPAIRS**									
0100	Fill and sand, pin/nail holes	1 Carp	960	.008	Ea.		.41		.41	.63
0110	Screw head pops		480	.017			.82		.82	1.26
0120	Dents, up to 2" square		48	.167		.01	8.20		8.21	12.60
0130	2" to 4" square		24	.333		.03	16.40		16.43	25
0140	Cut square, patch, sand and finish, holes, up to 2" square		12	.667		.04	33		33.04	50.50
0150	2" to 4" square		11	.727		.09	36		36.09	55
0160	4" to 8" square		10	.800		.25	39.50		39.75	61
0170	8" to 12" square		8	1		.49	49.50		49.99	76
0180	12" to 32" square		6	1.333		1.61	65.50		67.11	103
0210	16" by 48"		5	1.600		2.77	79		81.77	124
0220	32" by 48"		4	2		4.54	98.50		103.04	156
0230	48" square		3.50	2.286		6.40	113		119.40	179
0240	60" square		3.20	2.500		10	123		133	200
0500	Skim coat surface with joint compound		1600	.005	S.F.	.03	.25		.28	.41
0510	Prepare, retape and refinish joints		60	.133	L.F.	.63	6.55		7.18	10.75

09 05 Common Work Results for Finishes

09 05 05 – Selective Demolition for Finishes

09 05 05.10 Selective Demolition, Ceilings

		Crew	Daily Output	Labor-Hours	Unit	Material	2017 Bare Costs Labor	Equipment	Total	Total Incl O&P
0010	**SELECTIVE DEMOLITION, CEILINGS** R024119-10									
0200	Ceiling, gypsum wall board, furred and nailed or screwed	2 Clab	800	.020	S.F.		.78		.78	1.20
0220	On metal frame		760	.021			.82		.82	1.26
0240	On suspension system, including system		720	.022			.87		.87	1.33
1000	Plaster, lime and horse hair, on wood lath, incl. lath		700	.023			.90		.90	1.37
1020	On metal lath		570	.028			1.10		1.10	1.68
1100	Gypsum, on gypsum lath		720	.022			.87		.87	1.33
1120	On metal lath		500	.032			1.25		1.25	1.92
1200	Suspended ceiling, mineral fiber, 2' x 2' or 2' x 4'		1500	.011			.42		.42	.64
1250	On suspension system, incl. system		1200	.013			.52		.52	.80
1500	Tile, wood fiber, 12" x 12", glued		900	.018			.70		.70	1.07
1540	Stapled		1500	.011			.42		.42	.64
1580	On suspension system, incl. system		760	.021			.82		.82	1.26
2000	Wood, tongue and groove, 1" x 4"		1000	.016			.63		.63	.96
2040	1" x 8"		1100	.015			.57		.57	.87
2400	Plywood or wood fiberboard, 4' x 8' sheets		1200	.013			.52		.52	.80

09 05 05.20 Selective Demolition, Flooring

		Crew	Daily Output	Labor-Hours	Unit	Material	2017 Bare Costs Labor	Equipment	Total	Total Incl O&P
0010	**SELECTIVE DEMOLITION, FLOORING** R024119-10									
0200	Brick with mortar	2 Clab	475	.034	S.F.		1.32		1.32	2.02
0400	Carpet, bonded, including surface scraping		2000	.008			.31		.31	.48
0440	Scrim applied		8000	.002			.08		.08	.12
0480	Tackless		9000	.002			.07		.07	.11
0550	Carpet tile, releasable adhesive		5000	.003			.13		.13	.19
0560	Permanent adhesive		1850	.009			.34		.34	.52
0600	Composition, acrylic or epoxy		400	.040			1.57		1.57	2.40

09 05 05 – Selective Demolition for Finishes

09 05 05.20 Selective Demolition, Flooring		Crew	Daily Output	Labor-Hours	Unit	Material	2017 Bare Costs Labor	Equipment	Total	Total Incl O&P
0700	Concrete, scarify skin	A-1A	225	.036	S.F.		1.83	.98	2.81	3.90
0800	Resilient, sheet goods	2 Clab	1400	.011			.45		.45	.69
0820	For gym floors	"	900	.018			.70		.70	1.07
0850	Vinyl or rubber cove base	1 Clab	1000	.008	L.F.		.31		.31	.48
0860	Vinyl or rubber cove base, molded corner	"	1000	.008	Ea.		.31		.31	.48
0870	For glued and caulked installation, add to labor						50%			
0900	Vinyl composition tile, 12" x 12"	2 Clab	1000	.016	S.F.		.63		.63	.96
2000	Tile, ceramic, thin set		675	.024			.93		.93	1.42
2020	Mud set		625	.026			1		1	1.54
2200	Marble, slate, thin set		675	.024			.93		.93	1.42
2220	Mud set		625	.026			1		1	1.54
2600	Terrazzo, thin set		450	.036			1.39		1.39	2.13
2620	Mud set		425	.038			1.47		1.47	2.26
2640	Terrazzo, cast in place		300	.053			2.09		2.09	3.20
3000	Wood, block, on end	1 Carp	400	.020			.99		.99	1.51
3200	Parquet		450	.018			.88		.88	1.34
3400	Strip flooring, interior, 2-1/4" x 25/32" thick		325	.025			1.21		1.21	1.86
3500	Exterior, porch flooring, 1" x 4"		220	.036			1.79		1.79	2.74
3800	Subfloor, tongue and groove, 1" x 6"		325	.025			1.21		1.21	1.86
3820	1" x 8"		430	.019			.92		.92	1.40
3840	1" x 10"		520	.015			.76		.76	1.16
4000	Plywood, nailed		600	.013			.66		.66	1.01
4100	Glued and nailed		400	.020			.99		.99	1.51
4200	Hardboard, 1/4" thick		760	.011			.52		.52	.79
8000	Remove flooring, bead blast, simple floor plan	A-1A	1000	.008		.05	.41	.22	.68	.93
8100	Complex floor plan		400	.020		.05	1.03	.55	1.63	2.25
8150	Mastic only		1500	.005		.05	.27	.15	.47	.64

09 05 05.30 Selective Demolition, Walls and Partitions

09 05 05.30		Crew	Daily Output	Labor-Hours	Unit	Material	2017 Bare Costs Labor	Equipment	Total	Total Incl O&P
0010	**SELECTIVE DEMOLITION, WALLS AND PARTITIONS** R024119-10									
0020	Walls, concrete, reinforced	B-39	120	.400	C.F.		16.60	1.78	18.38	27.50
0025	Plain	"	160	.300			12.45	1.33	13.78	20.50
0100	Brick, 4" to 12" thick	B-9	220	.182			7.20	.97	8.17	12.05
0200	Concrete block, 4" thick		1150	.035	S.F.		1.38	.19	1.57	2.31
0280	8" thick		1050	.038			1.51	.20	1.71	2.53
0300	Exterior stucco 1" thick over mesh		3200	.013			.49	.07	.56	.83
1000	Gypsum wallboard, nailed or screwed	1 Clab	1000	.008			.31		.31	.48
1010	2 layers		400	.020			.78		.78	1.20
1020	Glued and nailed		900	.009			.35		.35	.53
1500	Fiberboard, nailed		900	.009			.35		.35	.53
1520	Glued and nailed		800	.010			.39		.39	.60
1568	Plenum barrier, sheet lead		300	.027			1.04		1.04	1.60
2000	Movable walls, metal, 5' high		300	.027			1.04		1.04	1.60
2020	8' high		400	.020			.78		.78	1.20
2200	Metal or wood studs, finish 2 sides, fiberboard	B-1	520	.046			1.84		1.84	2.82
2250	Lath and plaster		260	.092			3.68		3.68	5.65
2300	Gypsum wallboard		520	.046			1.84		1.84	2.82
2350	Plywood		450	.053			2.12		2.12	3.25
2800	Paneling, 4' x 8' sheets	1 Clab	475	.017			.66		.66	1.01
3000	Plaster, lime and horsehair, on wood lath		400	.020			.78		.78	1.20
3020	On metal lath		335	.024			.93		.93	1.43
3400	Gypsum or perlite, on gypsum lath		410	.020			.76		.76	1.17
3420	On metal lath		300	.027			1.04		1.04	1.60

09 05 Common Work Results for Finishes

09 05 05 – Selective Demolition for Finishes

09 05 05.30 Selective Demolition, Walls and Partitions

		Crew	Daily Output	Labor-Hours	Unit	Material	2017 Bare Costs Labor	Equipment	Total	Total Incl O&P
3450	Plaster, interior gypsum, acoustic, or cement	1 Clab	60	.133	S.Y.		5.20		5.20	8
3500	Stucco, on masonry		145	.055			2.16		2.16	3.31
3510	Commercial 3-coat		80	.100			3.92		3.92	6
3520	Interior stucco	↓	25	.320	↓		12.55		12.55	19.20
3600	Plywood, one side	B-1	1500	.016	S.F.		.64		.64	.98
3750	Terra cotta block and plaster, to 6" thick	"	175	.137			5.45		5.45	8.35
3760	Tile, ceramic, on walls, thin set	1 Clab	300	.027			1.04		1.04	1.60
3765	Mud set		250	.032	↓		1.25		1.25	1.92
3800	Toilet partitions, slate or marble		5	1.600	Ea.		62.50		62.50	96
3820	Metal or plastic	↓	8	1	"		39		39	60
5000	Wallcovering, vinyl	1 Pape	700	.011	S.F.		.48		.48	.72
5010	With release agent		1500	.005			.22		.22	.34
5025	Wallpaper, 2 layers or less, by hand		250	.032			1.34		1.34	2.02
5035	3 layers or more		165	.048			2.03		2.03	3.06
5040	Designer	↓	480	.017	↓		.70		.70	1.05

09 05 71 – Acoustic Underlayment

09 05 71.10 Acoustical Underlayment

		Crew	Daily Output	Labor-Hours	Unit	Material	2017 Bare Costs Labor	Equipment	Total	Total Incl O&P
0010	**ACOUSTICAL UNDERLAYMENT**									
4000	Nylon matting 0.4" thick, with carbon black spinerette									
4010	plus polyester fabric, on floor	D-7	1600	.010	S.F.	1.41	.41		1.82	2.16
4200	Fiberglass reinf. backer board underlayment, 7/16" thick, on floor	"	1500	.011	"	2.78	.44		3.22	3.71

09 21 Plaster and Gypsum Board Assemblies

09 21 13 – Plaster Assemblies

09 21 13.10 Plaster Partition Wall

		Crew	Daily Output	Labor-Hours	Unit	Material	2017 Bare Costs Labor	Equipment	Total	Total Incl O&P
0010	**PLASTER PARTITION WALL**									
0400	Stud walls, 3.4 lb. metal lath, 3 coat gypsum plaster, 2 sides									
0600	2" x 4" wood studs, 16" O.C.	J-2	315	.152	S.F.	2.48	6.70	.43	9.61	13.20
0700	2-1/2" metal studs, 25 ga., 12" O.C.		325	.148		2.52	6.50	.42	9.44	12.95
0800	3-5/8" metal studs, 25 ga., 16" O.C.	↓	320	.150	↓	3.22	6.60	.42	10.24	13.85
0900	Gypsum lath, 2 coat vermiculite plaster, 2 sides									
1000	2" x 4" wood studs, 16" O.C.	J-2	355	.135	S.F.	3.85	5.95	.38	10.18	13.55
1200	2-1/2" metal studs, 25 ga., 12" O.C.		365	.132		3.42	5.80	.37	9.59	12.80
1300	3-5/8" metal studs, 25 ga., 16" O.C.		360	.133	↓	3.51	5.85	.37	9.73	13

09 21 16 – Gypsum Board Assemblies

09 21 16.23 Gypsum Board Shaft Wall Assemblies

		Crew	Daily Output	Labor-Hours	Unit	Material	2017 Bare Costs Labor	Equipment	Total	Total Incl O&P
0010	**GYPSUM BOARD SHAFT WALL ASSEMBLIES**									
0020	Cavity type on 25 ga. J-track & C-H studs, 24" O.C.									
0030	1" thick coreboard wall liner on shaft side									
0040	2-hour assembly with double layer									
0060	5/8" f.r. gyp bd on rm side, 2-1/2" J-track & C-H studs	2 Carp	220	.073	S.F.	2.98	3.58		6.56	8.80
0065	4" J-track & C-H studs		220	.073		3.14	3.58		6.72	8.95
0070	6" J-track & C-H studs	↓	220	.073	↓	3.36	3.58		6.94	9.20
0100	3-hour assembly with triple layer									
0300	5/8" f.r. gyp bd on rm side, 2-1/2" J-track & C-H studs	2 Carp	180	.089	S.F.	2.63	4.38		7.01	9.60
0305	4" J-track & C-H studs		180	.089		2.79	4.38		7.17	9.75
0310	6" J-track & C-H studs	↓	180	.089	↓	3.01	4.38		7.39	10
0400	4-hour assembly, 1" coreboard, 5/8" fire rated gypsum board									
0600	and 3/4" galv. metal furring channels, 24" O.C., with									
0700	Dbl. layer 5/8" f.r. gyp bd on rm side, 2-1/2" trk. & C-H studs	2 Carp	110	.145	S.F.	3.36	7.15		10.51	14.65

09 21 Plaster and Gypsum Board Assemblies

09 21 16 – Gypsum Board Assemblies

09 21 16.23 Gypsum Board Shaft Wall Assemblies	Crew	Daily Output	Labor-Hours	Unit	Material	2017 Bare Costs Labor	Equipment	Total	Total Incl O&P
0705	4" J-track & C-H studs	2 Carp	110	.145	S.F.	3.14	7.15		10.29	14.40
0710	6" J-track & C-H studs	↓	110	.145		3.36	7.15		10.51	14.65
0900	For taping & finishing, add per side	1 Carp	1050	.008	↓	.05	.38		.43	.62
1000	For insulation, see Section 07 21									
5200	For work over 8' high, add	2 Carp	3060	.005	S.F.		.26		.26	.39
5300	For distribution cost over 3 stories high, add per story	"	6100	.003	"		.13		.13	.20

09 21 16.33 Partition Wall

		Crew	Daily Output	Labor-Hours	Unit	Material	2017 Bare Costs Labor	Equipment	Total	Total Incl O&P
0010	**PARTITION WALL** Stud wall, 8' to 12' high									
0050	1/2", interior, gypsum board, std, tape & finish 2 sides									
0500	Installed on and incl., 2" x 4" wood studs, 16" O.C.	2 Carp	310	.052	S.F.	1.23	2.54		3.77	5.25
1000	Metal studs, NLB, 25 ga., 16" O.C., 3-5/8" wide		350	.046		1.15	2.25		3.40	4.72
1200	6" wide		330	.048		1.28	2.39		3.67	5.05
1400	Water resistant, on 2" x 4" wood studs, 16" O.C.		310	.052		1.39	2.54		3.93	5.40
1600	Metal studs, NLB, 25 ga., 16" O.C., 3-5/8" wide		350	.046		1.31	2.25		3.56	4.89
1800	6" wide		330	.048		1.44	2.39		3.83	5.25
2000	Fire res., 2 layers, 1-1/2 hr., on 2" x 4" wood studs, 16" O.C.		210	.076		2.03	3.75		5.78	8
2200	Metal studs, NLB, 25 ga., 16" O.C., 3-5/8" wide		250	.064		1.95	3.15		5.10	7
2400	6" wide		230	.070		2.08	3.43		5.51	7.55
2600	Fire & water res., 2 layers, 1-1/2 hr., 2" x 4" studs, 16" O.C.		210	.076		2.03	3.75		5.78	8
2800	Metal studs, NLB, 25 ga., 16" O.C., 3-5/8" wide		250	.064		1.95	3.15		5.10	7
3000	6" wide	↓	230	.070	↓	2.08	3.43		5.51	7.55
3200	5/8", interior, gypsum board, standard, tape & finish 2 sides									
3400	Installed on and including 2" x 4" wood studs, 16" O.C.	2 Carp	300	.053	S.F.	1.25	2.63		3.88	5.40
3600	24" O.C.		330	.048		1.15	2.39		3.54	4.92
3800	Metal studs, NLB, 25 ga., 16" O.C., 3-5/8" wide		340	.047		1.17	2.32		3.49	4.84
4000	6" wide		320	.050		1.30	2.46		3.76	5.20
4200	24" O.C., 3-5/8" wide		360	.044		1.07	2.19		3.26	4.53
4400	6" wide		340	.047		1.16	2.32		3.48	4.83
4800	Water resistant, on 2" x 4" wood studs, 16" O.C.		300	.053		1.43	2.63		4.06	5.60
5000	24" O.C.		330	.048		1.33	2.39		3.72	5.10
5200	Metal studs, NLB, 25 ga. 16" O.C., 3-5/8" wide		340	.047		1.35	2.32		3.67	5.05
5400	6" wide		320	.050		1.48	2.46		3.94	5.40
5600	24" O.C., 3-5/8" wide		360	.044		1.25	2.19		3.44	4.73
5800	6" wide		340	.047		1.34	2.32		3.66	5.05
6000	Fire resistant, 2 layers, 2 hr., on 2" x 4" wood studs, 16" O.C.		205	.078		1.89	3.84		5.73	7.95
6200	24" O.C.		235	.068		1.89	3.35		5.24	7.20
6400	Metal studs, NLB, 25 ga., 16" O.C., 3-5/8" wide		245	.065		1.95	3.22		5.17	7.05
6600	6" wide		225	.071		2.04	3.50		5.54	7.60
6800	24" O.C., 3-5/8" wide		265	.060		1.81	2.97		4.78	6.55
7000	6" wide		245	.065		1.90	3.22		5.12	7
7200	Fire & water resistant, 2 layers, 2 hr., 2" x 4" studs, 16" O.C.		205	.078		1.99	3.84		5.83	8.10
7400	24" O.C.		235	.068		1.89	3.35		5.24	7.20
7600	Metal studs, NLB, 25 ga., 16" O.C., 3-5/8" wide		245	.065		1.91	3.22		5.13	7.05
7800	6" wide		225	.071		2.04	3.50		5.54	7.60
8000	24" O.C., 3-5/8" wide		265	.060		1.81	2.97		4.78	6.55
8200	6" wide	↓	245	.065	↓	1.90	3.22		5.12	7
8600	1/2" blueboard, mesh tape both sides									
8620	Installed on and including 2" x 4" wood studs, 16" O.C.	2 Carp	300	.053	S.F.	1.25	2.63		3.88	5.40
8640	Metal studs, NLB, 25 ga., 16" O.C., 3-5/8" wide		340	.047		1.17	2.32		3.49	4.84
8660	6" wide	↓	320	.050	↓	1.30	2.46		3.76	5.20
8800	Hospital security partition, 5/8" fiber reinf. high abuse gyp. bd.									
8810	Mtl. studs, NLB, 20 ga., 16" O.C., 3-5/8" wide, w/sec. mesh, gyp. bd.	2 Carp	208	.077	S.F.	4.30	3.79		8.09	10.55

09 21 Plaster and Gypsum Board Assemblies

09 21 16 – Gypsum Board Assemblies

09 21 16.33 Partition Wall

		Crew	Daily Output	Labor-Hours	Unit	Material	2017 Bare Costs Labor	Equipment	Total	Total Incl O&P
9000	Exterior, 1/2" gypsum sheathing, 1/2" gypsum finished, interior,									
9100	including foil faced insulation, metal studs, 20 ga.									
9200	16" O.C., 3-5/8" wide	2 Carp	290	.055	S.F.	1.81	2.72		4.53	6.15
9400	6" wide		270	.059		2.01	2.92		4.93	6.70
9600	Partitions, for work over 8' high, add		1530	.010			.52		.52	.79

09 22 Supports for Plaster and Gypsum Board

09 22 03 – Fastening Methods for Finishes

09 22 03.20 Drilling Plaster/Drywall

		Crew	Daily Output	Labor-Hours	Unit	Material	2017 Bare Costs Labor	Equipment	Total	Total Incl O&P
0010	**DRILLING PLASTER/DRYWALL**									
1100	Drilling & layout for drywall/plaster walls, up to 1" deep, no anchor									
1200	Holes, 1/4" diameter	1 Carp	150	.053	Ea.	.01	2.63		2.64	4.03
1300	3/8" diameter		140	.057		.01	2.81		2.82	4.32
1400	1/2" diameter		130	.062		.01	3.03		3.04	4.65
1500	3/4" diameter		120	.067		.02	3.28		3.30	5.05
1600	1" diameter		110	.073		.02	3.58		3.60	5.50
1700	1-1/4" diameter		100	.080		.04	3.94		3.98	6.10
1800	1-1/2" diameter		90	.089		.05	4.38		4.43	6.75
1900	For ceiling installations, add						40%			

09 22 13 – Metal Furring

09 22 13.13 Metal Channel Furring

		Crew	Daily Output	Labor-Hours	Unit	Material	2017 Bare Costs Labor	Equipment	Total	Total Incl O&P
0010	**METAL CHANNEL FURRING**									
0030	Beams and columns, 7/8" channels, galvanized, 12" O.C.	1 Lath	155	.052	S.F.	.43	2.47		2.90	4.14
0050	16" O.C.		170	.047		.35	2.25		2.60	3.73
0070	24" O.C.		185	.043		.24	2.07		2.31	3.33
0100	Ceilings, on steel, 7/8" channels, galvanized, 12" O.C.		210	.038		.39	1.83		2.22	3.13
0300	16" O.C.		290	.028		.35	1.32		1.67	2.35
0400	24" O.C.		420	.019		.24	.91		1.15	1.61
0600	1-5/8" channels, galvanized, 12" O.C.		190	.042		.53	2.02		2.55	3.57
0700	16" O.C.		260	.031		.47	1.47		1.94	2.70
0900	24" O.C.		390	.021		.32	.98		1.30	1.80
0930	7/8" channels with sound isolation clips, 12" O.C.		120	.067		1.89	3.19		5.08	6.80
0940	16" O.C.		100	.080		1.42	3.83		5.25	7.20
0950	24" O.C.		165	.048		.95	2.32		3.27	4.48
0960	1-5/8" channels, galvanized, 12" O.C.		110	.073		2.02	3.48		5.50	7.40
0970	16" O.C.		100	.080		1.52	3.83		5.35	7.30
0980	24" O.C.		155	.052		1.01	2.47		3.48	4.77
1000	Walls, 7/8" channels, galvanized, 12" O.C.		235	.034		.39	1.63		2.02	2.84
1200	16" O.C.		265	.030		.35	1.45		1.80	2.53
1300	24" O.C.		350	.023		.24	1.10		1.34	1.88
1500	1-5/8" channels, galvanized, 12" O.C.		210	.038		.53	1.83		2.36	3.28
1600	16" O.C.		240	.033		.47	1.60		2.07	2.88
1800	24" O.C.		305	.026		.32	1.26		1.58	2.21
1920	7/8" channels with sound isolation clips, 12" O.C.		125	.064		1.89	3.07		4.96	6.60
1940	16" O.C.		100	.080		1.42	3.83		5.25	7.20
1950	24" O.C.		150	.053		.95	2.55		3.50	4.82
1960	1-5/8" channels, galvanized, 12" O.C.		115	.070		2.02	3.33		5.35	7.15
1970	16" O.C.		95	.084		1.52	4.03		5.55	7.60
1980	24" O.C.		140	.057		1.01	2.74		3.75	5.15

09 22 16 – Non-Structural Metal Framing

09 22 16.13 Non-Structural Metal Stud Framing	Crew	Daily Output	Labor-Hours	Unit	Material	2017 Bare Costs Labor	Equipment	Total	Total Incl O&P
0010 **NON-STRUCTURAL METAL STUD FRAMING**									
1600 Non-load bearing, galv., 8' high, 25 ga. 1-5/8" wide, 16" O.C.	1 Carp	619	.013	S.F.	.27	.64		.91	1.27
1610 24" O.C.		950	.008		.20	.41		.61	.86
1620 2-1/2" wide, 16" O.C.		613	.013		.36	.64		1	1.37
1630 24" O.C.		938	.009		.27	.42		.69	.93
1640 3-5/8" wide, 16" O.C.		600	.013		.40	.66		1.06	1.45
1650 24" O.C.		925	.009		.30	.43		.73	.98
1660 4" wide, 16" O.C.		594	.013		.45	.66		1.11	1.51
1670 24" O.C.		925	.009		.33	.43		.76	1.02
1680 6" wide, 16" O.C.		588	.014		.53	.67		1.20	1.61
1690 24" O.C.		906	.009		.40	.43		.83	1.11
1700 20 ga. studs, 1-5/8" wide, 16" O.C.		494	.016		.34	.80		1.14	1.60
1710 24" O.C.		763	.010		.26	.52		.78	1.07
1720 2-1/2" wide, 16" O.C.		488	.016		.44	.81		1.25	1.72
1730 24" O.C.		750	.011		.33	.53		.86	1.17
1740 3-5/8" wide, 16" O.C.		481	.017		.50	.82		1.32	1.80
1750 24" O.C.		738	.011		.37	.53		.90	1.23
1760 4" wide, 16" O.C.		475	.017		.60	.83		1.43	1.93
1770 24" O.C.		738	.011		.45	.53		.98	1.32
1780 6" wide, 16" O.C.		469	.017		.71	.84		1.55	2.08
1790 24" O.C.		725	.011		.54	.54		1.08	1.42
2000 Non-load bearing, galv., 10' high, 25 ga. 1-5/8" wide, 16" O.C.		495	.016		.25	.80		1.05	1.50
2100 24" O.C.		760	.011		.19	.52		.71	1
2200 2-1/2" wide, 16" O.C.		490	.016		.34	.80		1.14	1.60
2250 24" O.C.		750	.011		.25	.53		.78	1.08
2300 3-5/8" wide, 16" O.C.		480	.017		.38	.82		1.20	1.68
2350 24" O.C.		740	.011		.28	.53		.81	1.13
2400 4" wide, 16" O.C.		475	.017		.42	.83		1.25	1.74
2450 24" O.C.		740	.011		.31	.53		.84	1.16
2500 6" wide, 16" O.C.		470	.017		.50	.84		1.34	1.83
2550 24" O.C.		725	.011		.37	.54		.91	1.24
2600 20 ga. studs, 1-5/8" wide, 16" O.C.		395	.020		.32	1		1.32	1.89
2650 24" O.C.		610	.013		.24	.65		.89	1.25
2700 2-1/2" wide, 16" O.C.		390	.021		.41	1.01		1.42	2
2750 24" O.C.		600	.013		.30	.66		.96	1.34
2800 3-5/8" wide, 16" O.C.		385	.021		.47	1.02		1.49	2.09
2850 24" O.C.		590	.014		.35	.67		1.02	1.40
2900 4" wide, 16" O.C.		380	.021		.57	1.04		1.61	2.22
2950 24" O.C.		590	.014		.42	.67		1.09	1.48
3000 6" wide, 16" O.C.		375	.021		.68	1.05		1.73	2.35
3050 24" O.C.		580	.014		.50	.68		1.18	1.59
3060 Non-load bearing, galv., 12' high, 25 ga. 1-5/8" wide, 16" O.C.		413	.019		.24	.95		1.19	1.73
3070 24" O.C.		633	.013		.18	.62		.80	1.15
3080 2-1/2" wide, 16" O.C.		408	.020		.32	.97		1.29	1.84
3090 24" O.C.		625	.013		.24	.63		.87	1.23
3100 3-5/8" wide, 16" O.C.		400	.020		.36	.99		1.35	1.91
3110 24" O.C.		617	.013		.26	.64		.90	1.27
3120 4" wide, 16" O.C.		396	.020		.40	.99		1.39	1.96
3130 24" O.C.		617	.013		.30	.64		.94	1.30
3140 6" wide, 16" O.C.		392	.020		.48	1.01		1.49	2.07
3150 24" O.C.		604	.013		.35	.65		1	1.39
3160 20 ga. studs, 1-5/8" wide, 16" O.C.		329	.024		.31	1.20		1.51	2.17

For customer support on your Building Construction Costs with RSMeans Data, call 800.448.8182.

323

09 22 Supports for Plaster and Gypsum Board

09 22 16 – Non-Structural Metal Framing

09 22 16.13 Non-Structural Metal Stud Framing

		Crew	Daily Output	Labor-Hours	Unit	Material	2017 Bare Costs Labor	Equipment	Total	Total Incl O&P
3170	24" O.C.	1 Carp	508	.016	S.F.	.23	.78		1.01	1.44
3180	2-1/2" wide, 16" O.C.		325	.025		.40	1.21		1.61	2.29
3190	24" O.C.		500	.016		.29	.79		1.08	1.53
3200	3-5/8" wide, 16" O.C.		321	.025		.45	1.23		1.68	2.38
3210	24" O.C.		492	.016		.33	.80		1.13	1.59
3220	4" wide, 16" O.C.		317	.025		.55	1.24		1.79	2.50
3230	24" O.C.		492	.016		.40	.80		1.20	1.67
3240	6" wide, 16" O.C.		313	.026		.65	1.26		1.91	2.64
3250	24" O.C.		483	.017		.47	.82		1.29	1.77
5000	For load bearing studs, see Section 05 41 13.30									

09 22 26 – Suspension Systems

09 22 26.13 Ceiling Suspension Systems

		Crew	Daily Output	Labor-Hours	Unit	Material	2017 Bare Costs Labor	Equipment	Total	Total Incl O&P
0010	**CEILING SUSPENSION SYSTEMS** for gypsum board or plaster									
8000	Suspended ceilings, including carriers									
8200	1-1/2" carriers, 24" O.C. with:									
8300	7/8" channels, 16" O.C.	1 Lath	275	.029	S.F.	.57	1.39		1.96	2.69
8320	24" O.C.		310	.026		.46	1.24		1.70	2.33
8400	1-5/8" channels, 16" O.C.		205	.039		.69	1.87		2.56	3.53
8420	24" O.C.		250	.032		.53	1.53		2.06	2.86
8600	2" carriers, 24" O.C. with:									
8700	7/8" channels, 16" O.C.	1 Lath	250	.032	S.F.	.62	1.53		2.15	2.96
8720	24" O.C.		285	.028		.51	1.34		1.85	2.55
8800	1-5/8" channels, 16" O.C.		190	.042		.74	2.02		2.76	3.81
8820	24" O.C.		225	.036		.59	1.70		2.29	3.16

09 22 36 – Lath

09 22 36.13 Gypsum Lath

		Crew	Daily Output	Labor-Hours	Unit	Material	2017 Bare Costs Labor	Equipment	Total	Total Incl O&P
0010	**GYPSUM LATH** R092000-50									
0020	Plain or perforated, nailed, 3/8" thick	1 Lath	85	.094	S.Y.	3.15	4.51		7.66	10.10
0100	1/2" thick		80	.100		2.52	4.79		7.31	9.85
0300	Clipped to steel studs, 3/8" thick		75	.107		3.15	5.10		8.25	11
0400	1/2" thick		70	.114		2.52	5.45		7.97	10.85
1500	For ceiling installations, add		216	.037			1.77		1.77	2.63
1600	For columns and beams, add		170	.047			2.25		2.25	3.34

09 22 36.23 Metal Lath

		Crew	Daily Output	Labor-Hours	Unit	Material	2017 Bare Costs Labor	Equipment	Total	Total Incl O&P
0010	**METAL LATH** R092000-50									
0020	Diamond, expanded, 2.5 lb. per S.Y., painted				S.Y.	4.09			4.09	4.50
0100	Galvanized					3.02			3.02	3.32
0300	3.4 lb. per S.Y., painted					4.29			4.29	4.72
0400	Galvanized					4.40			4.40	4.84
0600	For 15# asphalt sheathing paper, add					.48			.48	.52
0900	Flat rib, 1/8" high, 2.75 lb., painted					3.62			3.62	3.98
1000	Foil backed					3.83			3.83	4.21
1200	3.4 lb. per S.Y., painted					4.54			4.54	4.99
1300	Galvanized					4.56			4.56	5
1500	For 15# asphalt sheathing paper, add					.48			.48	.52
1800	High rib, 3/8" high, 3.4 lb. per S.Y., painted					4.52			4.52	4.97
1900	Galvanized					4.13			4.13	4.54
2400	3/4" high, painted, .60 lb. per S.F.				S.F.	.71			.71	.78
2500	.75 lb. per S.F.				"	1.54			1.54	1.69
2800	Stucco mesh, painted, 3.6 lb.				S.Y.	4.14			4.14	4.55
3000	K-lath, perforated, absorbent paper, regular					4.42			4.42	4.86

For customer support on your Building Construction Costs with RSMeans Data, call 800.448.8182.

09 22 36.23 Metal Lath

	Crew	Daily Output	Labor-Hours	Unit	Material	2017 Bare Costs Labor	Equipment	Total	Total Incl O&P
3100	Heavy duty				S.Y.	5.20			5.20	5.75
3300	Waterproof, heavy duty, grade B backing					5.10			5.10	5.60
3400	Fire resistant backing					5.65			5.65	6.20
3600	2.5 lb. diamond painted, on wood framing, on walls	1 Lath	85	.094		4.09	4.51		8.60	11.15
3700	On ceilings		75	.107		4.09	5.10		9.19	12.05
3900	3.4 lb. diamond painted, on wood framing, on walls		80	.100		4.54	4.79		9.33	12.10
4000	On ceilings		70	.114		4.54	5.45		9.99	13.10
4200	3.4 lb. diamond painted, wired to steel framing		75	.107		4.54	5.10		9.64	12.55
4300	On ceilings		60	.133		4.54	6.40		10.94	14.45
4500	Columns and beams, wired to steel		40	.200		4.54	9.60		14.14	19.20
4600	Cornices, wired to steel		35	.229		4.54	10.95		15.49	21
4800	Screwed to steel studs, 2.5 lb.		80	.100		4.09	4.79		8.88	11.60
4900	3.4 lb.		75	.107		4.29	5.10		9.39	12.25
5100	Rib lath, painted, wired to steel, on walls, 2.5 lb.		75	.107		3.62	5.10		8.72	11.55
5200	3.4 lb.		70	.114		4.52	5.45		9.97	13.05
5400	4.0 lb.	▼	65	.123		5.75	5.90		11.65	15.05
5500	For self-furring lath, add					.11			.11	.12
5700	Suspended ceiling system, incl. 3.4 lb. diamond lath, painted	1 Lath	15	.533		4.49	25.50		29.99	43
5800	Galvanized	"	15	.533	▼	4.54	25.50		30.04	43
6000	Hollow metal stud partitions, 3.4 lb. painted lath both sides									
6010	Non-load bearing, 25 ga., w/rib lath, 2-1/2" studs, 12" O.C.	1 Lath	20.30	.394	S.Y.	12.90	18.90		31.80	42
6300	16" O.C.		21.10	.379		12.05	18.15		30.20	40.50
6350	24" O.C.		22.70	.352		11.25	16.90		28.15	37.50
6400	3-5/8" studs, 16" O.C.		19.50	.410		12.45	19.65		32.10	42.50
6600	24" O.C.		20.40	.392		11.55	18.80		30.35	40.50
6700	4" studs, 16" O.C.		20.40	.392		12.85	18.80		31.65	42
6900	24" O.C.		21.60	.370		11.85	17.75		29.60	39.50
7000	6" studs, 16" O.C.		19.50	.410		13.55	19.65		33.20	44
7100	24" O.C.		21.10	.379		12.35	18.15		30.50	40.50
7200	L.B. partitions, 16 ga., w/rib lath, 2-1/2" studs, 16" O.C.		20	.400		13.55	19.15		32.70	43.50
7300	3-5/8" studs, 16 ga.		19.70	.406		15.30	19.45		34.75	46
7500	4" studs, 16 ga.		19.50	.410		15.85	19.65		35.50	46.50
7600	6" studs, 16 ga.	▼	18.70	.428	▼	18.85	20.50		39.35	51

09 22 36.43 Security Mesh

	Crew	Daily Output	Labor-Hours	Unit	Material	2017 Bare Costs Labor	Equipment	Total	Total Incl O&P
0010	**SECURITY MESH**, expanded metal, flat, screwed to framing									
0100	On walls, 3/4", 1.76 lb./S.F.	2 Carp	1500	.011	S.F.	2.01	.53		2.54	3.02
0110	1-1/2", 1.14 lb./S.F.		1600	.010		1.54	.49		2.03	2.44
0200	On ceilings, 3/4", 1.76 lb./S.F.		1350	.012		2.01	.58		2.59	3.10
0210	1-1/2", 1.14 lb./S.F.	▼	1450	.011	▼	1.54	.54		2.08	2.52

09 22 36.83 Accessories, Plaster

	Crew	Daily Output	Labor-Hours	Unit	Material	2017 Bare Costs Labor	Equipment	Total	Total Incl O&P
0010	**ACCESSORIES, PLASTER**									
0020	Casing bead, expanded flange, galvanized	1 Lath	2.70	2.963	C.L.F.	53.50	142		195.50	269
0200	Foundation weep screed, galvanized	"	2.70	2.963		52	142		194	267
0900	Channels, cold rolled, 16 ga., 3/4" deep, galvanized					39.50			39.50	43.50
1200	1-1/2" deep, 16 ga., galvanized					52.50			52.50	58
1620	Corner bead, expanded bullnose, 3/4" radius, #10, galvanized	1 Lath	2.60	3.077		26.50	147		173.50	247
1650	#1, galvanized		2.55	3.137		48.50	150		198.50	275
1670	Expanded wing, 2-3/4" wide, #1, galvanized		2.65	3.019		40	145		185	258
1700	Inside corner (corner rite), 3" x 3", painted		2.60	3.077		20.50	147		167.50	241
1750	Strip-ex, 4" wide, painted		2.55	3.137		23	150		173	247
1800	Expansion joint, 3/4" grounds, limited expansion, galv., 1 piece		2.70	2.963		73	142		215	290
2100	Extreme expansion, galvanized, 2 piece	▼	2.60	3.077	▼	151	147		298	385

09 23 Gypsum Plastering

09 23 13 - Acoustical Gypsum Plastering

09 23 13.10 Perlite or Vermiculite Plaster

		Crew	Daily Output	Labor-Hours	Unit	Material	2017 Bare Costs Labor	Equipment	Total	Total Incl O&P
0010	**PERLITE OR VERMICULITE PLASTER** R092000-50									
0020	In 100 lb. bags, under 200 bags				Bag	17.65			17.65	19.40
0100	Over 200 bags				"	16.85			16.85	18.50
0300	2 coats, no lath included, on walls	J-1	92	.435	S.Y.	5.85	18.75	1.47	26.07	36
0400	On ceilings	"	79	.506		5.85	22	1.71	29.56	41.50
0600	On and incl. 3/8" gypsum lath, on metal studs	J-2	84	.571		10	25	1.61	36.61	50.50
0700	On ceilings	"	70	.686		10	30	1.93	41.93	58
0900	3 coats, no lath included, on walls	J-1	74	.541		6.40	23.50	1.82	31.72	44
1000	On ceilings	"	63	.635		6.40	27.50	2.14	36.04	50.50
1200	On and incl. painted metal lath, on metal studs	J-2	72	.667		10.90	29.50	1.87	42.27	58
1300	On ceilings		61	.787		10.90	34.50	2.21	47.61	66.50
1500	On and incl. suspended metal lath ceiling		37	1.297		10.90	57	3.65	71.55	101
1700	For irregular or curved surfaces, add to above						30%			
1800	For columns and beams, add to above						50%			
1900	For soffits, add to ceiling prices						40%			

09 23 20 - Gypsum Plaster

09 23 20.10 Gypsum Plaster On Walls and Ceilings

		Crew	Daily Output	Labor-Hours	Unit	Material	2017 Bare Costs Labor	Equipment	Total	Total Incl O&P
0010	**GYPSUM PLASTER ON WALLS AND CEILINGS** R092000-50									
0020	80# bag, less than 1 ton				Bag	15.80			15.80	17.40
0100	Over 1 ton				"	13.80			13.80	15.20
0300	2 coats, no lath included, on walls	J-1	105	.381	S.Y.	3.71	16.45	1.28	21.44	30
0400	On ceilings	"	92	.435		3.71	18.75	1.47	23.93	33.50
0600	On and incl. 3/8" gypsum lath on steel, on walls	J-2	97	.495		6.85	22	1.39	30.24	41.50
0700	On ceilings	"	83	.578		6.85	25.50	1.63	33.98	47.50
0900	3 coats, no lath included, on walls	J-1	87	.460		5.35	19.85	1.55	26.75	37.50
1000	On ceilings	"	78	.513		5.35	22	1.73	29.08	41
1200	On and including painted metal lath, on wood studs	J-2	86	.558		10.65	24.50	1.57	36.72	50
1300	On ceilings	"	76.50	.627		10.65	27.50	1.76	39.91	55
1600	For irregular or curved surfaces, add						30%			
1800	For columns & beams, add						50%			

09 23 20.20 Gauging Plaster

		Crew	Daily Output	Labor-Hours	Unit	Material	2017 Bare Costs Labor	Equipment	Total	Total Incl O&P
0010	**GAUGING PLASTER** R092000-50									
0020	100 lb. bags, less than 1 ton				Bag	19.75			19.75	21.50
0100	Over 1 ton				"	18.70			18.70	20.50

09 23 20.30 Keenes Cement

		Crew	Daily Output	Labor-Hours	Unit	Material	2017 Bare Costs Labor	Equipment	Total	Total Incl O&P
0010	**KEENES CEMENT** R092000-50									
0020	In 100 lb. bags, less than 1 ton				Bag	22.50			22.50	24.50
0100	Over 1 ton				"	20.50			20.50	22.50
0300	Finish only, add to plaster prices, standard	J-1	215	.186	S.Y.	1.98	8.05	.63	10.66	14.90
0400	High quality	"	144	.278	"	2	12	.94	14.94	21

09 24 Cement Plastering

09 24 23 – Cement Stucco

09 24 23.40 Stucco

		Crew	Daily Output	Labor-Hours	Unit	Material	2017 Bare Costs Labor	Equipment	Total	Total Incl O&P
0010	**STUCCO** R092000-50									
0015	3 coats 7/8" thick, float finish, with mesh, on wood frame	J-2	63	.762	S.Y.	8	33.50	2.14	43.64	61
0100	On masonry construction, no mesh incl.	J-1	67	.597		3.29	26	2.01	31.30	44.50
0300	For trowel finish, add	1 Plas	170	.047			2.16		2.16	3.24
0600	For coloring, add	J-1	685	.058		.41	2.52	.20	3.13	4.45
0700	For special texture, add	"	200	.200		1.44	8.65	.67	10.76	15.25
0900	For soffits, add	J-2	155	.310		2.22	13.60	.87	16.69	24
1000	Stucco, with bonding agent, 3 coats, on walls, no mesh incl.	J-1	200	.200		4.20	8.65	.67	13.52	18.30
1200	Ceilings		180	.222		3.72	9.60	.75	14.07	19.30
1300	Beams		80	.500		3.72	21.50	1.69	26.91	38.50
1500	Columns		100	.400		3.72	17.25	1.35	22.32	31.50
1600	Mesh, galvanized, nailed to wood, 1.8 lb.	1 Lath	60	.133		7.15	6.40		13.55	17.35
1800	3.6 lb.		55	.145		4.14	6.95		11.09	14.85
1900	Wired to steel, galvanized, 1.8 lb.		53	.151		7.15	7.25		14.40	18.60
2100	3.6 lb.		50	.160		4.14	7.65		11.79	15.90

09 25 Other Plastering

09 25 23 – Lime Based Plastering

09 25 23.10 Venetian Plaster

		Crew	Daily Output	Labor-Hours	Unit	Material	2017 Bare Costs Labor	Equipment	Total	Total Incl O&P
0010	**VENETIAN PLASTER**									
0100	Walls, 1 coat primer, roller applied	1 Plas	950	.008	S.F.	.16	.39		.55	.76
0200	Plaster, 3 coats, incl. sanding	2 Plas	700	.023		.46	1.05		1.51	2.07
0210	For pigment, light colors add per S.F. plaster					.02			.02	.02
0220	For pigment, dark colors add per S.F. plaster					.05			.05	.06
0300	For sealer/wax coat incl. burnishing, add	1 Plas	300	.027		.41	1.22		1.63	2.29

09 26 Veneer Plastering

09 26 13 – Gypsum Veneer Plastering

09 26 13.20 Blueboard

		Crew	Daily Output	Labor-Hours	Unit	Material	2017 Bare Costs Labor	Equipment	Total	Total Incl O&P
0010	**BLUEBOARD** For use with thin coat									
0100	plaster application see Section 09 26 13.80									
1000	3/8" thick, on walls or ceilings, standard, no finish included	2 Carp	1900	.008	S.F.	.34	.41		.75	1.01
1100	With thin coat plaster finish		875	.018		.45	.90		1.35	1.88
1400	On beams, columns, or soffits, standard, no finish included		675	.024		.39	1.17		1.56	2.22
1450	With thin coat plaster finish		475	.034		.50	1.66		2.16	3.09
3000	1/2" thick, on walls or ceilings, standard, no finish included		1900	.008		.35	.41		.76	1.03
3100	With thin coat plaster finish		875	.018		.46	.90		1.36	1.89
3300	Fire resistant, no finish included		1900	.008		.35	.41		.76	1.03
3400	With thin coat plaster finish		875	.018		.46	.90		1.36	1.89
3450	On beams, columns, or soffits, standard, no finish included		675	.024		.40	1.17		1.57	2.23
3500	With thin coat plaster finish		475	.034		.51	1.66		2.17	3.11
3700	Fire resistant, no finish included		675	.024		.40	1.17		1.57	2.23
3800	With thin coat plaster finish		475	.034		.51	1.66		2.17	3.11
5000	5/8" thick, on walls or ceilings, fire resistant, no finish included		1900	.008		.33	.41		.74	1
5100	With thin coat plaster finish		875	.018		.44	.90		1.34	1.87
5500	On beams, columns, or soffits, no finish included		675	.024		.38	1.17		1.55	2.21
5600	With thin coat plaster finish		475	.034		.49	1.66		2.15	3.08
6000	For high ceilings, over 8' high, add		3060	.005			.26		.26	.39
6500	For over 3 stories high, add per story		6100	.003			.13		.13	.20

09 26 Veneer Plastering

09 26 13 – Gypsum Veneer Plastering

09 26 13.80 Thin Coat Plaster	Crew	Daily Output	Labor-Hours	Unit	Material	2017 Bare Costs Labor	Equipment	Total	Total Incl O&P
0010 **THIN COAT PLASTER**									
0012 1 coat veneer, not incl. lath	J-1	3600	.011	S.F.	.11	.48	.04	.63	.88
1000 In 50 lb. bags				Bag	15.05			15.05	16.55

09 28 Backing Boards and Underlayments

09 28 13 – Cementitious Backing Boards

09 28 13.10 Cementitious Backerboard

	Crew	Daily Output	Labor-Hours	Unit	Material	2017 Bare Costs Labor	Equipment	Total	Total Incl O&P
0010 **CEMENTITIOUS BACKERBOARD**									
0070 Cementitious backerboard, on floor, 3' x 4' x 1/2" sheets	2 Carp	525	.030	S.F.	.86	1.50		2.36	3.25
0080 3' x 5' x 1/2" sheets		525	.030		.81	1.50		2.31	3.19
0090 3' x 6' x 1/2" sheets		525	.030		.78	1.50		2.28	3.16
0100 3' x 4' x 5/8" sheets		525	.030		.99	1.50		2.49	3.39
0110 3' x 5' x 5/8" sheets		525	.030		.98	1.50		2.48	3.38
0120 3' x 6' x 5/8" sheets		525	.030		.96	1.50		2.46	3.35
0150 On wall, 3' x 4' x 1/2" sheets		350	.046		.86	2.25		3.11	4.40
0160 3' x 5' x 1/2" sheets		350	.046		.81	2.25		3.06	4.34
0170 3' x 6' x 1/2" sheets		350	.046		.78	2.25		3.03	4.31
0180 3' x 4' x 5/8" sheets		350	.046		.99	2.25		3.24	4.54
0190 3' x 5' x 5/8" sheets		350	.046		.98	2.25		3.23	4.53
0200 3' x 6' x 5/8" sheets		350	.046		.96	2.25		3.21	4.50
0250 On counter, 3' x 4' x 1/2" sheets		180	.089		.86	4.38		5.24	7.65
0260 3' x 5' x 1/2" sheets		180	.089		.81	4.38		5.19	7.60
0270 3' x 6' x 1/2" sheets		180	.089		.78	4.38		5.16	7.55
0300 3' x 4' x 5/8" sheets		180	.089		.99	4.38		5.37	7.80
0310 3' x 5' x 5/8" sheets		180	.089		.98	4.38		5.36	7.80
0320 3' x 6' x 5/8" sheets		180	.089		.96	4.38		5.34	7.75

09 29 Gypsum Board

09 29 10 – Gypsum Board Panels

09 29 10.30 Gypsum Board

	Crew	Daily Output	Labor-Hours	Unit	Material	2017 Bare Costs Labor	Equipment	Total	Total Incl O&P
0010 **GYPSUM BOARD** on walls & ceilings R092910-10									
0100 Nailed or screwed to studs unless otherwise noted									
0110 1/4" thick, on walls or ceilings, standard, no finish included	2 Carp	1330	.012	S.F.	.37	.59		.96	1.32
0115 1/4" thick, on walls or ceilings, flexible, no finish included		1050	.015		.53	.75		1.28	1.73
0117 1/4" thick, on columns or soffits, flexible, no finish included		1050	.015		.53	.75		1.28	1.73
0130 1/4" thick, standard, no finish included, less than 800 S.F.		510	.031		.37	1.55		1.92	2.78
0150 3/8" thick, on walls, standard, no finish included		2000	.008		.36	.39		.75	1
0200 On ceilings, standard, no finish included		1800	.009		.36	.44		.80	1.07
0250 On beams, columns, or soffits, no finish included		675	.024		.36	1.17		1.53	2.19
0300 1/2" thick, on walls, standard, no finish included		2000	.008		.34	.39		.73	.97
0350 Taped and finished (level 4 finish)		965	.017		.39	.82		1.21	1.68
0390 With compound skim coat (level 5 finish)		775	.021		.44	1.02		1.46	2.04
0400 Fire resistant, no finish included		2000	.008		.37	.39		.76	1.01
0450 Taped and finished (level 4 finish)		965	.017		.42	.82		1.24	1.71
0490 With compound skim coat (level 5 finish)		775	.021		.47	1.02		1.49	2.08
0500 Water resistant, no finish included		2000	.008		.42	.39		.81	1.06
0550 Taped and finished (level 4 finish)		965	.017		.47	.82		1.29	1.76
0590 With compound skim coat (level 5 finish)		775	.021		.52	1.02		1.54	2.13
0600 Prefinished, vinyl, clipped to studs		900	.018		.51	.88		1.39	1.90

09 29 10.30 Gypsum Board	Crew	Daily Output	Labor-Hours	Unit	Material	2017 Bare Costs Labor	Equipment	Total	Total Incl O&P
0700 Mold resistant, no finish included	2 Carp	2000	.008	S.F.	.44	.39		.83	1.08
0710 Taped and finished (level 4 finish)		965	.017		.49	.82		1.31	1.79
0720 With compound skim coat (level 5 finish)		775	.021		.54	1.02		1.56	2.15
1000 On ceilings, standard, no finish included		1800	.009		.34	.44		.78	1.04
1050 Taped and finished (level 4 finish)		765	.021		.39	1.03		1.42	2.01
1090 With compound skim coat (level 5 finish)		610	.026		.44	1.29		1.73	2.46
1100 Fire resistant, no finish included		1800	.009		.37	.44		.81	1.08
1150 Taped and finished (level 4 finish)		765	.021		.42	1.03		1.45	2.04
1195 With compound skim coat (level 5 finish)		610	.026		.47	1.29		1.76	2.50
1200 Water resistant, no finish included		1800	.009		.42	.44		.86	1.13
1250 Taped and finished (level 4 finish)		765	.021		.47	1.03		1.50	2.09
1290 With compound skim coat (level 5 finish)		610	.026		.52	1.29		1.81	2.55
1310 Mold resistant, no finish included		1800	.009		.44	.44		.88	1.15
1320 Taped and finished (level 4 finish)		765	.021		.49	1.03		1.52	2.12
1330 With compound skim coat (level 5 finish)		610	.026		.54	1.29		1.83	2.57
1350 Sag resistant, no finish included		1600	.010		.36	.49		.85	1.15
1360 Taped and finished (level 4 finish)		765	.021		.41	1.03		1.44	2.03
1370 With compound skim coat (level 5 finish)		610	.026		.46	1.29		1.75	2.48
1500 On beams, columns, or soffits, standard, no finish included		675	.024		.39	1.17		1.56	2.22
1550 Taped and finished (level 4 finish)		540	.030		.39	1.46		1.85	2.67
1590 With compound skim coat (level 5 finish)		475	.034		.44	1.66		2.10	3.02
1600 Fire resistant, no finish included		675	.024		.37	1.17		1.54	2.20
1650 Taped and finished (level 4 finish)		540	.030		.42	1.46		1.88	2.70
1690 With compound skim coat (level 5 finish)		475	.034		.47	1.66		2.13	3.06
1700 Water resistant, no finish included		675	.024		.48	1.17		1.65	2.32
1750 Taped and finished (level 4 finish)		540	.030		.47	1.46		1.93	2.75
1790 With compound skim coat (level 5 finish)		475	.034		.52	1.66		2.18	3.11
1800 Mold resistant, no finish included		675	.024		.51	1.17		1.68	2.35
1810 Taped and finished (level 4 finish)		540	.030		.49	1.46		1.95	2.78
1820 With compound skim coat (level 5 finish)		475	.034		.54	1.66		2.20	3.13
1850 Sag resistant, no finish included		675	.024		.41	1.17		1.58	2.25
1860 Taped and finished (level 4 finish)		540	.030		.41	1.46		1.87	2.69
1870 With compound skim coat (level 5 finish)		475	.034		.46	1.66		2.12	3.04
2000 5/8" thick, on walls, standard, no finish included		2000	.008		.35	.39		.74	.99
2050 Taped and finished (level 4 finish)		965	.017		.40	.82		1.22	1.69
2090 With compound skim coat (level 5 finish)		775	.021		.45	1.02		1.47	2.05
2100 Fire resistant, no finish included		2000	.008		.36	.39		.75	1
2150 Taped and finished (level 4 finish)		965	.017		.41	.82		1.23	1.70
2195 With compound skim coat (level 5 finish)		775	.021		.46	1.02		1.48	2.06
2200 Water resistant, no finish included		2000	.008		.44	.39		.83	1.08
2250 Taped and finished (level 4 finish)		965	.017		.49	.82		1.31	1.79
2290 With compound skim coat (level 5 finish)		775	.021		.54	1.02		1.56	2.15
2300 Prefinished, vinyl, clipped to studs		900	.018		.79	.88		1.67	2.21
2510 Mold resistant, no finish included		2000	.008		.49	.39		.88	1.14
2520 Taped and finished (level 4 finish)		965	.017		.54	.82		1.36	1.84
2530 With compound skim coat (level 5 finish)		775	.021		.59	1.02		1.61	2.21
3000 On ceilings, standard, no finish included		1800	.009		.35	.44		.79	1.06
3050 Taped and finished (level 4 finish)		765	.021		.40	1.03		1.43	2.02
3090 With compound skim coat (level 5 finish)		615	.026		.45	1.28		1.73	2.45
3100 Fire resistant, no finish included		1800	.009		.36	.44		.80	1.07
3150 Taped and finished (level 4 finish)		765	.021		.41	1.03		1.44	2.03
3190 With compound skim coat (level 5 finish)		615	.026		.46	1.28		1.74	2.46
3200 Water resistant, no finish included		1800	.009		.44	.44		.88	1.15

09 29 10.30 Gypsum Board	Crew	Daily Output	Labor-Hours	Unit	Material	2017 Bare Costs Labor	Equipment	Total	Total Incl O&P
3250 Taped and finished (level 4 finish)	2 Carp	765	.021	S.F.	.49	1.03		1.52	2.12
3290 With compound skim coat (level 5 finish)		615	.026		.54	1.28		1.82	2.55
3300 Mold resistant, no finish included		1800	.009		.49	.44		.93	1.21
3310 Taped and finished (level 4 finish)		765	.021		.54	1.03		1.57	2.17
3320 With compound skim coat (level 5 finish)		615	.026		.59	1.28		1.87	2.61
3500 On beams, columns, or soffits, no finish included		675	.024		.40	1.17		1.57	2.23
3550 Taped and finished (level 4 finish)		475	.034		.46	1.66		2.12	3.04
3590 With compound skim coat (level 5 finish)		380	.042		.52	2.07		2.59	3.75
3600 Fire resistant, no finish included		675	.024		.41	1.17		1.58	2.25
3650 Taped and finished (level 4 finish)		475	.034		.47	1.66		2.13	3.05
3690 With compound skim coat (level 5 finish)		380	.042		.46	2.07		2.53	3.68
3700 Water resistant, no finish included		675	.024		.51	1.17		1.68	2.35
3750 Taped and finished (level 4 finish)		475	.034		.54	1.66		2.20	3.13
3790 With compound skim coat (level 5 finish)		380	.042		.56	2.07		2.63	3.80
3800 Mold resistant, no finish included		675	.024		.56	1.17		1.73	2.41
3810 Taped and finished (level 4 finish)		475	.034		.59	1.66		2.25	3.19
3820 With compound skim coat (level 5 finish)		380	.042		.62	2.07		2.69	3.86
4000 Fireproofing, beams or columns, 2 layers, 1/2" thick, incl finish		330	.048		.83	2.39		3.22	4.58
4010 Mold resistant		330	.048		.97	2.39		3.36	4.73
4050 5/8" thick		300	.053		.81	2.63		3.44	4.91
4060 Mold resistant		300	.053		1.07	2.63		3.70	5.20
4100 3 layers, 1/2" thick		225	.071		1.25	3.50		4.75	6.75
4110 Mold resistant		225	.071		1.46	3.50		4.96	6.95
4150 5/8" thick		210	.076		1.22	3.75		4.97	7.10
4160 Mold resistant		210	.076		1.61	3.75		5.36	7.50
5050 For 1" thick coreboard on columns		480	.033		.78	1.64		2.42	3.37
5100 For foil-backed board, add					.17			.17	.19
5200 For work over 8' high, add	2 Carp	3060	.005			.26		.26	.39
5270 For textured spray, add	2 Lath	1600	.010		.04	.48		.52	.75
5300 For distribution cost over 3 stories high, add per story	2 Carp	6100	.003			.13		.13	.20
5350 For finishing inner corners, add		950	.017	L.F.	.10	.83		.93	1.38
5355 For finishing outer corners, add		1250	.013		.24	.63		.87	1.23
5500 For acoustical sealant, add per bead	1 Carp	500	.016		.04	.79		.83	1.26
5550 Sealant, 1 quart tube				Ea.	7.05			7.05	7.80
6000 Gypsum sound dampening panels									
6010 1/2" thick on walls, multi-layer, light weight, no finish included	2 Carp	1500	.011	S.F.	2.08	.53		2.61	3.10
6015 Taped and finished (level 4 finish)		725	.022		2.13	1.09		3.22	4.01
6020 With compound skim coat (level 5 finish)		580	.028		2.18	1.36		3.54	4.48
6025 5/8" thick on walls, for wood studs, no finish included		1500	.011		2.21	.53		2.74	3.24
6030 Taped and finished (level 4 finish)		725	.022		2.26	1.09		3.35	4.15
6035 With compound skim coat (level 5 finish)		580	.028		2.31	1.36		3.67	4.62
6040 For metal stud, no finish included		1500	.011		2.20	.53		2.73	3.23
6045 Taped and finished (level 4 finish)		725	.022		2.25	1.09		3.34	4.14
6050 With compound skim coat (level 5 finish)		580	.028		2.30	1.36		3.66	4.61
6055 Abuse resist, no finish included		1500	.011		3.95	.53		4.48	5.15
6060 Taped and finished (level 4 finish)		725	.022		4	1.09		5.09	6.05
6065 With compound skim coat (level 5 finish)		580	.028		4.05	1.36		5.41	6.55
6070 Shear rated, no finish included		1500	.011		4.95	.53		5.48	6.25
6075 Taped and finished (level 4 finish)		725	.022		5	1.09		6.09	7.15
6080 With compound skim coat (level 5 finish)		580	.028		5.05	1.36		6.41	7.65
6085 For SCIF applications, no finish included		1500	.011		4.96	.53		5.49	6.25
6090 Taped and finished (level 4 finish)		725	.022		5	1.09		6.09	7.15
6095 With compound skim coat (level 5 finish)		580	.028		5.05	1.36		6.41	7.65

09 29 Gypsum Board

09 29 10 – Gypsum Board Panels

	09 29 10.30 Gypsum Board	Crew	Daily Output	Labor-Hours	Unit	Material	2017 Bare Costs Labor	Equipment	Total	Total Incl O&P
6100	1-3/8" thick on walls, THX Certified, no finish included	2 Carp	1500	.011	S.F.	8.75	.53		9.28	10.40
6105	Taped and finished (level 4 finish)		725	.022		8.80	1.09		9.89	11.30
6110	With compound skim coat (level 5 finish)		580	.028		8.85	1.36		10.21	11.80
6115	5/8" thick on walls, score & snap installation, no finish included		2000	.008		1.96	.39		2.35	2.76
6120	Taped and finished (level 4 finish)		965	.017		2.01	.82		2.83	3.46
6125	With compound skim coat (level 5 finish)		775	.021		2.06	1.02		3.08	3.82
7020	5/8" thick on ceilings, for wood joists, no finish included		1200	.013		2.21	.66		2.87	3.44
7025	Taped and finished (level 4 finish)		510	.031		2.26	1.55		3.81	4.85
7030	With compound skim coat (level 5 finish)		410	.039		2.31	1.92		4.23	5.50
7035	For metal joists, no finish included		1200	.013		2.20	.66		2.86	3.43
7040	Taped and finished (level 4 finish)		510	.031		2.25	1.55		3.80	4.84
7045	With compound skim coat (level 5 finish)		410	.039		2.30	1.92		4.22	5.45
7050	Abuse resist, no finish included		1200	.013		3.95	.66		4.61	5.35
7055	Taped and finished (level 4 finish)		510	.031		4	1.55		5.55	6.75
7060	With compound skim coat (level 5 finish)		410	.039		4.05	1.92		5.97	7.40
7065	Shear rated, no finish included		1200	.013		4.95	.66		5.61	6.45
7070	Taped and finished (level 4 finish)		510	.031		5	1.55		6.55	7.85
7075	With compound skim coat (level 5 finish)		410	.039		5.05	1.92		6.97	8.50
7080	For SCIF applications, no finish included		1200	.013		4.96	.66		5.62	6.45
7085	Taped and finished (level 4 finish)		510	.031		5	1.55		6.55	7.85
7090	With compound skim coat (level 5 finish)		410	.039		5.05	1.92		6.97	8.50
8010	5/8" thick on ceilings, score & snap installation, no finish included		1600	.010		1.96	.49		2.45	2.91
8015	Taped and finished (level 4 finish)		680	.024		2.01	1.16		3.17	3.99
8020	With compound skim coat (level 5 finish)	▼	545	.029	▼	2.06	1.45		3.51	4.48

09 29 10.50 High Abuse Gypsum Board

		Crew	Daily Output	Labor-Hours	Unit	Material	2017 Bare Costs Labor	Equipment	Total	Total Incl O&P
0010	**HIGH ABUSE GYPSUM BOARD**, fiber reinforced, nailed or									
0100	screwed to studs unless otherwise noted									
0110	1/2" thick, on walls, no finish included	2 Carp	1800	.009	S.F.	.77	.44		1.21	1.52
0120	Taped and finished (level 4 finish)		870	.018		.82	.91		1.73	2.29
0130	With compound skim coat (level 5 finish)		700	.023		.87	1.13		2	2.68
0150	On ceilings, no finish included		1620	.010		.77	.49		1.26	1.60
0160	Taped and finished (level 4 finish)		690	.023		.82	1.14		1.96	2.65
0170	With compound skim coat (level 5 finish)		550	.029		.87	1.43		2.30	3.15
0210	5/8" thick, on walls, no finish included		1800	.009		.82	.44		1.26	1.57
0220	Taped and finished (level 4 finish)		870	.018		.87	.91		1.78	2.34
0230	With compound skim coat (level 5 finish)		700	.023		.92	1.13		2.05	2.73
0250	On ceilings, no finish included		1620	.010		.82	.49		1.31	1.65
0260	Taped and finished (level 4 finish)		690	.023		.87	1.14		2.01	2.70
0270	With compound skim coat (level 5 finish)		550	.029		.92	1.43		2.35	3.20
0272	5/8" thick, on roof sheathing for 1-hour rating		1620	.010		.82	.49		1.31	1.65
0310	5/8" thick, on walls, very high impact, no finish included		1800	.009		1	.44		1.44	1.77
0320	Taped and finished (level 4 finish)		870	.018		1.05	.91		1.96	2.54
0330	With compound skim coat (level 5 finish)		700	.023		1.10	1.13		2.23	2.93
0350	On ceilings, no finish included		1620	.010		1	.49		1.49	1.85
0360	Taped and finished (level 4 finish)		690	.023		1.05	1.14		2.19	2.90
0370	With compound skim coat (level 5 finish)	▼	550	.029	▼	1.10	1.43		2.53	3.40
0400	High abuse, gypsum core, paper face									
0410	1/2" thick, on walls, no finish included	2 Carp	1800	.009	S.F.	.63	.44		1.07	1.36
0420	Taped and finished (level 4 finish)		870	.018		.68	.91		1.59	2.13
0430	With compound skim coat (level 5 finish)		700	.023		.73	1.13		1.86	2.52
0450	On ceilings, no finish included		1620	.010		.63	.49		1.12	1.44
0460	Taped and finished (level 4 finish)		690	.023		.68	1.14		1.82	2.49

09 29 Gypsum Board

09 29 10 – Gypsum Board Panels

09 29 10.50 High Abuse Gypsum Board

		Crew	Daily Output	Labor-Hours	Unit	Material	2017 Bare Costs Labor	Equipment	Total	Total Incl O&P
0470	With compound skim coat (level 5 finish)	2 Carp	550	.029	S.F.	.73	1.43		2.16	2.99
0510	5/8" thick, on walls, no finish included		1800	.009		.67	.44		1.11	1.41
0520	Taped and finished (level 4 finish)		870	.018		.72	.91		1.63	2.18
0530	With compound skim coat (level 5 finish)		700	.023		.77	1.13		1.90	2.57
0550	On ceilings, no finish included		1620	.010		.67	.49		1.16	1.49
0560	Taped and finished (level 4 finish)		690	.023		.72	1.14		1.86	2.54
0570	With compound skim coat (level 5 finish)		550	.029		.77	1.43		2.20	3.04
1000	For high ceilings, over 8' high, add		2750	.006			.29		.29	.44
1010	For distribution cost over 3 stories high, add per story		5500	.003			.14		.14	.22

09 29 15 – Gypsum Board Accessories

09 29 15.10 Accessories, Gypsum Board

		Crew	Daily Output	Labor-Hours	Unit	Material	2017 Bare Costs Labor	Equipment	Total	Total Incl O&P
0010	**ACCESSORIES, GYPSUM BOARD**									
0020	Casing bead, galvanized steel	1 Carp	2.90	2.759	C.L.F.	24	136		160	235
0100	Vinyl		3	2.667		22.50	131		153.50	226
0300	Corner bead, galvanized steel, 1" x 1"		4	2		15.90	98.50		114.40	169
0400	1-1/4" x 1-1/4"		3.50	2.286		17	113		130	191
0600	Vinyl		4	2		20.50	98.50		119	174
0900	Furring channel, galv. steel, 7/8" deep, standard		2.60	3.077		33.50	152		185.50	269
1000	Resilient		2.55	3.137		25	155		180	265
1100	J trim, galvanized steel, 1/2" wide		3	2.667		22.50	131		153.50	226
1120	5/8" wide		2.95	2.712		32	134		166	241
1140	L trim, galvanized		3	2.667		19.40	131		150.40	223
1150	U trim, galvanized		2.95	2.712		23.50	134		157.50	231
1160	Screws #6 x 1" A				M	10.25			10.25	11.30
1170	#6 x 1-5/8" A				"	15.10			15.10	16.60
1200	For stud partitions, see Sections 05 41 13.30 and 09 22 16.13									
1500	Z stud, galvanized steel, 1-1/2" wide	1 Carp	2.60	3.077	C.L.F.	41	152		193	277
1600	2" wide	"	2.55	3.137	"	64	155		219	310

09 30 Tiling

09 30 13 – Ceramic Tiling

09 30 13.20 Ceramic Tile Repairs

		Crew	Daily Output	Labor-Hours	Unit	Material	2017 Bare Costs Labor	Equipment	Total	Total Incl O&P
0010	**CERAMIC TILE REPAIRS**									
1000	Grout removal, carbide tipped, rotary grinder	1 Clab	240	.033	L.F.		1.30		1.30	2
1100	Regrout tile 4-1/2 x 4-1/2, or larger, wall	1 Tilf	100	.080	S.F.	.14	3.67		3.81	5.60
1150	Floor		125	.064		.15	2.93		3.08	4.53
1200	Seal tile and grout		360	.022			1.02		1.02	1.51

09 30 13.45 Ceramic Tile Accessories

		Crew	Daily Output	Labor-Hours	Unit	Material	2017 Bare Costs Labor	Equipment	Total	Total Incl O&P
0010	**CERAMIC TILE ACCESSORIES**									
0100	Spacers, 1/8"				C	1.98			1.98	2.18
1310	Sealer for natural stone tile, installed	1 Tilf	650	.012	S.F.	.05	.56		.61	.90

09 30 29 – Metal Tiling

09 30 29.10 Metal Tile

		Crew	Daily Output	Labor-Hours	Unit	Material	2017 Bare Costs Labor	Equipment	Total	Total Incl O&P
0010	**METAL TILE** 4' x 4' sheet, 24 ga., tile pattern, nailed									
0200	Stainless steel	2 Carp	512	.031	S.F.	28.50	1.54		30.04	34
0400	Aluminized steel	"	512	.031	"	15.40	1.54		16.94	19.25

09 30 Tiling

09 30 95 – Tile & Stone Setting Materials and Specialties

09 30 95.10 Moisture Resistant, Anti-Fracture Membrane	Crew	Daily Output	Labor-Hours	Unit	Material	2017 Bare Costs Labor	Equipment	Total	Total Incl O&P
0010 **MOISTURE RESISTANT, ANTI-FRACTURE MEMBRANE**									
0200 Elastomeric membrane, 1/16" thick	D-7	275	.058	S.F.	1.05	2.39		3.44	4.71

09 31 Thin-Set Tiling

09 31 13 – Thin-Set Ceramic Tiling

09 31 13.10 Thin-Set Ceramic Tile

		Crew	Daily Output	Labor-Hours	Unit	Material	Labor	Equipment	Total	Total Incl O&P
0010	**THIN-SET CERAMIC TILE**									
0020	Backsplash, average grade tiles	1 Tilf	50	.160	S.F.	2.80	7.35		10.15	14
0022	Custom grade tiles		50	.160		5.60	7.35		12.95	17.05
0024	Luxury grade tiles		50	.160		11.20	7.35		18.55	23
0026	Economy grade tiles		50	.160		2.56	7.35		9.91	13.70
0100	Base, using 1' x 4" high piece with 1" x 1" tiles	D-7	128	.125	L.F.	5	5.15		10.15	13.10
0300	For 6" high base, 1" x 1" tile face, add					1.23			1.23	1.35
0400	For 2" x 2" tile face, add to above					.72			.72	.79
0700	Cove base, 4-1/4" x 4-1/4"	D-7	128	.125		4.08	5.15		9.23	12.10
1000	6" x 4-1/4" high		137	.117		4.30	4.80		9.10	11.85
1300	Sanitary cove base, 6" x 4-1/4" high		124	.129		4.63	5.30		9.93	12.95
1600	6" x 6" high		117	.137		5.35	5.60		10.95	14.25
1800	Bathroom accessories, average (soap dish, tooth brush holder)		82	.195	Ea.	9.75	8		17.75	22.50
1900	Bathtub, 5', rec. 4-1/4" x 4-1/4" tile wainscot, adhesive set 6' high		2.90	5.517		175	227		402	525
2100	7' high wainscot		2.50	6.400		204	263		467	615
2200	8' high wainscot		2.20	7.273		233	299		532	700
2500	Bullnose trim, 4-1/4" x 4-1/4"		128	.125	L.F.	4.11	5.15		9.26	12.10
2800	2" x 6"		124	.129	"	4.27	5.30		9.57	12.55
3300	Ceramic tile, porcelain type, 1 color, color group 2, 1" x 1"		183	.087	S.F.	6.40	3.59		9.99	12.40
3310	2" x 2" or 2" x 1"		190	.084		6.65	3.46		10.11	12.45
3350	For random blend, 2 colors, add					1			1	1.10
3360	4 colors, add					1.50			1.50	1.65
3370	For color group 3, add					.65			.65	.72
3380	For abrasive non-slip tile, add					.44			.44	.48
4300	Specialty tile, 4-1/4" x 4-1/4" x 1/2", decorator finish	D-7	183	.087		12.65	3.59		16.24	19.30
4500	Add for epoxy grout, 1/16" joint, 1" x 1" tile		800	.020		.68	.82		1.50	1.97
4600	2" x 2" tile		820	.020		.62	.80		1.42	1.87
4610	Add for epoxy grout, 1/8" joint, 8" x 8" x 3/8" tile, add		900	.018		1.41	.73		2.14	2.63
4800	Pregrouted sheets, walls, 4-1/4" x 4-1/4", 6" x 4-1/4"									
4810	and 8-1/2" x 4-1/4", 4 S.F. sheets, silicone grout	D-7	240	.067	S.F.	5.60	2.74		8.34	10.25
5100	Floors, unglazed, 2 S.F. sheets,									
5110	urethane adhesive	D-7	180	.089	S.F.	2.03	3.65		5.68	7.65
5400	Walls, interior, 4-1/4" x 4-1/4" tile		190	.084		2.61	3.46		6.07	8
5500	6" x 4-1/4" tile		190	.084		3.15	3.46		6.61	8.60
5700	8-1/2" x 4-1/4" tile		190	.084		5.55	3.46		9.01	11.25
5800	6" x 6" tile		175	.091		3.59	3.76		7.35	9.50
5810	8" x 8" tile		170	.094		4.83	3.87		8.70	11.05
5820	12" x 12" tile		160	.100		4.63	4.11		8.74	11.20
5830	16" x 16" tile		150	.107		5.15	4.38		9.53	12.15
6000	Decorated wall tile, 4-1/4" x 4-1/4", color group 1		270	.059		3.67	2.43		6.10	7.65
6100	Color group 4		180	.089		52.50	3.65		56.15	63
9300	Ceramic tiles, recycled glass, standard colors, 2" x 2" thru 6" x 6" G		190	.084		22	3.46		25.46	29.50
9310	6" x 6" G		175	.091		22	3.76		25.76	30
9320	8" x 8" G		170	.094		23.50	3.87		27.37	32

For customer support on your Building Construction Costs with RSMeans Data, call 800.448.8182.

333

09 31 13 – Thin-Set Ceramic Tiling

09 31 13.10 Thin-Set Ceramic Tile

			Crew	Daily Output	Labor-Hours	Unit	Material	2017 Bare Costs Labor	Equipment	Total	Total Incl O&P
9330	12" x 12"	G	D-7	160	.100	S.F.	23.50	4.11		27.61	32
9340	Earthtones, 2" x 2" to 4" x 8"	G		190	.084		26	3.46		29.46	33.50
9350	6" x 6"	G		175	.091		26	3.76		29.76	34
9360	8" x 8"	G		170	.094		27	3.87		30.87	35.50
9370	12" x 12"	G		160	.100		27	4.11		31.11	35.50
9380	Deep colors, 2" x 2" to 4" x 8"	G		190	.084		30.50	3.46		33.96	39
9390	6" x 6"	G		175	.091		30.50	3.76		34.26	39
9400	8" x 8"	G		170	.094		32	3.87		35.87	41.50
9410	12" x 12"	G		160	.100		32	4.11		36.11	41.50

09 31 33 – Thin-Set Stone Tiling

09 31 33.10 Tiling, Thin-Set Stone

		Crew	Daily Output	Labor-Hours	Unit	Material	Labor	Equipment	Total	Total Incl O&P
0010	**TILING, THIN-SET STONE**									
3000	Floors, natural clay, random or uniform, color group 1	D-7	183	.087	S.F.	4.30	3.59		7.89	10.10
3100	Color group 2		183	.087		5.90	3.59		9.49	11.85
3255	Floors, glazed, 6" x 6", color group 1		300	.053		5.25	2.19		7.44	9.05
3260	8" x 8" tile		300	.053		5.25	2.19		7.44	9
3270	12" x 12" tile		290	.055		5.85	2.27		8.12	9.75
3280	16" x 16" tile		280	.057		7.50	2.35		9.85	11.75
3281	18" x 18" tile		270	.059		6.85	2.43		9.28	11.15
3282	20" x 20" tile		260	.062		9.95	2.53		12.48	14.70
3283	24" x 24" tile		250	.064		11.30	2.63		13.93	16.35
3285	Border, 6" x 12" tile		200	.080		11.55	3.29		14.84	17.60
3290	3" x 12" tile		200	.080		10.60	3.29		13.89	16.55

09 32 13 – Mortar-Bed Ceramic Tiling

09 32 13.10 Ceramic Tile

		Crew	Daily Output	Labor-Hours	Unit	Material	Labor	Equipment	Total	Total Incl O&P
0010	**CERAMIC TILE**									
0050	Base, using 1' x 4" high pc. with 1" x 1" tiles	D-7	82	.195	L.F.	5.25	8		13.25	17.70
0600	Cove base, 4-1/4" x 4-1/4" high		91	.176		4.21	7.20		11.41	15.35
0900	6" x 4-1/4" high		100	.160		4.43	6.55		10.98	14.65
1200	Sanitary cove base, 6" x 4-1/4" high		93	.172		4.76	7.05		11.81	15.75
1500	6" x 6" high		84	.190		5.50	7.80		13.30	17.65
2400	Bullnose trim, 4-1/4" x 4-1/4"		82	.195		4.20	8		12.20	16.50
2700	2" x 6" bullnose trim		84	.190		4.34	7.80		12.14	16.35
6210	Wall tile, 4-1/4" x 4-1/4", better grade	1 Tilf	50	.160	S.F.	9.50	7.35		16.85	21.50
6240	2" x 2"		50	.160		6.50	7.35		13.85	18.05
6250	6" x 6"		55	.145		10.15	6.65		16.80	21
6260	8" x 8"		60	.133		9.40	6.10		15.50	19.35
6300	Exterior walls, frostproof, 4-1/4" x 4-1/4"	D-7	102	.157		7.25	6.45		13.70	17.50
6400	1-3/8" x 1-3/8"		93	.172		6.60	7.05		13.65	17.80
6600	Crystalline glazed, 4-1/4" x 4-1/4", plain		100	.160		4.66	6.55		11.21	14.90
6700	4-1/4" x 4-1/4", scored tile		100	.160		6.50	6.55		13.05	16.90
6900	6" x 6" plain		93	.172		7	7.05		14.05	18.20
7000	For epoxy grout, 1/16" joints, 4-1/4" tile, add		800	.020		.41	.82		1.23	1.67
7200	For tile set in dry mortar, add		1735	.009			.38		.38	.56
7300	For tile set in Portland cement mortar, add		290	.055		.18	2.27		2.45	3.55

334

For customer support on your Building Construction Costs with RSMeans Data, call 800.448.8182.

09 32 Mortar-Bed Tiling

09 32 16 – Mortar-Bed Quarry Tiling

09 32 16.10 Quarry Tile

		Crew	Daily Output	Labor-Hours	Unit	Material	2017 Bare Costs Labor	Equipment	Total	Total Incl O&P
0010	**QUARRY TILE**									
0100	Base, cove or sanitary, to 5" high, 1/2" thick	D-7	110	.145	L.F.	6.45	6		12.45	15.95
0300	Bullnose trim, red, 6" x 6" x 1/2" thick		120	.133		5.40	5.50		10.90	14.10
0400	4" x 4" x 1/2" thick		110	.145		5	6		11	14.35
0600	4" x 8" x 1/2" thick, using 8" as edge		130	.123		5.40	5.05		10.45	13.40
0700	Floors, 1,000 S.F. lots, red, 4" x 4" x 1/2" thick		120	.133	S.F.	8.70	5.50		14.20	17.70
0900	6" x 6" x 1/2" thick		140	.114		8.45	4.70		13.15	16.25
1000	4" x 8" x 1/2" thick		130	.123		9.95	5.05		15	18.45
1300	For waxed coating, add					.76			.76	.84
1500	For non-standard colors, add					.46			.46	.51
1600	For abrasive surface, add					.52			.52	.57
1800	Brown tile, imported, 6" x 6" x 3/4"	D-7	120	.133		7.75	5.50		13.25	16.65
1900	8" x 8" x 1"		110	.145		9.50	6		15.50	19.30
2100	For thin set mortar application, deduct		700	.023			.94		.94	1.39
2200	For epoxy grout & mortar, 6" x 6" x 1/2", add		350	.046		2.32	1.88		4.20	5.35
2700	Stair tread, 6" x 6" x 3/4", plain		50	.320		8.40	13.15		21.55	29
2800	Abrasive		47	.340		7.85	14		21.85	29.50
3000	Wainscot, 6" x 6" x 1/2", thin set, red		105	.152		5.95	6.25		12.20	15.85
3100	Non-standard colors		105	.152		5.90	6.25		12.15	15.80
3300	Window sill, 6" wide, 3/4" thick		90	.178	L.F.	8.30	7.30		15.60	20
3400	Corners		80	.200	Ea.	6.55	8.20		14.75	19.40

09 32 23 – Mortar-Bed Glass Mosaic Tiling

09 32 23.10 Glass Mosaics

		Crew	Daily Output	Labor-Hours	Unit	Material	2017 Bare Costs Labor	Equipment	Total	Total Incl O&P
0010	**GLASS MOSAICS** 3/4" tile on 12" sheets, standard grout									
0300	Color group 1 & 2	D-7	73	.219	S.F.	20	9		29	35.50
0350	Color group 3		73	.219		22	9		31	37.50
0400	Color group 4		73	.219		27.50	9		36.50	43.50
0450	Color group 5		73	.219		30.50	9		39.50	47
0500	Color group 6		73	.219		40.50	9		49.50	58
0600	Color group 7		73	.219		41	9		50	58.50
0700	Color group 8, golds, silvers & specialties		64	.250		41.50	10.25		51.75	61.50
1020	1" tile on 12" sheets, opalescent finish		73	.219		18.10	9		27.10	33.50
1040	1" x 2" tile on 12" sheet, blend		73	.219		20.50	9		29.50	36
1060	2" tile on 12" sheet, blend		73	.219		17.30	9		26.30	32.50
1080	5/8" x random tile, linear, on 12" sheet, blend		73	.219		25.50	9		34.50	41.50
1600	Dots on 12" sheet		73	.219		25.50	9		34.50	42
1700	For glass mosaic tiles set in dry mortar, add		290	.055		.45	2.27		2.72	3.86
1720	For glass mosaic tiles set in Portland cement mortar, add		290	.055		.01	2.27		2.28	3.37
1730	For polyblend sanded tile grout		96.15	.166	Lb.	2.19	6.85		9.04	12.55

For customer support on your Building Construction Costs with RSMeans Data, call 800.448.8182.

335

09 34 Waterproofing-Membrane Tiling

09 34 13 – Waterproofing-Membrane Ceramic Tiling

09 34 13.10 Ceramic Tile Waterproofing Membrane	Crew	Daily Output	Labor-Hours	Unit	Material	2017 Bare Costs Labor	Equipment	Total	Total Incl O&P
0010 **CERAMIC TILE WATERPROOFING MEMBRANE**									
0020 On floors, including thinset									
0030 Fleece laminated polyethylene grid, 1/8" thick	D-7	250	.064	S.F.	2.28	2.63		4.91	6.40
0040 5/16" thick	"	250	.064	"	2.60	2.63		5.23	6.75
0050 On walls, including thinset									
0060 Fleece laminated polyethylene sheet, 8 mil thick	D-7	480	.033	S.F.	2.28	1.37		3.65	4.53
0070 Accessories, including thinset									
0080 Joint and corner sheet, 4 mils thick, 5" wide	1 Tilf	240	.033	L.F.	1.35	1.53		2.88	3.75
0090 7-1/4" wide		180	.044		1.71	2.04		3.75	4.90
0100 10" wide		120	.067	↓	2.08	3.06		5.14	6.80
0110 Pre-formed corners, inside		32	.250	Ea.	7.50	11.45		18.95	25.50
0120 Outside		32	.250		7.65	11.45		19.10	25.50
0130 2" flanged floor drain with 6" stainless steel grate		16	.500	↓	370	23		393	445
0140 EPS, sloped shower floor		480	.017	S.F.	4.97	.76		5.73	6.60
0150 Curb	↓	32	.250	L.F.	14.05	11.45		25.50	32.50

09 51 Acoustical Ceilings

09 51 13 – Acoustical Panel Ceilings

09 51 13.10 Ceiling, Acoustical Panel

	Crew	Daily Output	Labor-Hours	Unit	Material	2017 Bare Costs Labor	Equipment	Total	Total Incl O&P
0010 **CEILING, ACOUSTICAL PANEL**									
0100 Fiberglass boards, film faced, 2' x 2' or 2' x 4', 5/8" thick	1 Carp	625	.013	S.F.	1.29	.63		1.92	2.39
0120 3/4" thick		600	.013		2.66	.66		3.32	3.94
0130 3" thick, thermal, R11	↓	450	.018	↓	3.50	.88		4.38	5.20

09 51 14 – Acoustical Fabric-Faced Panel Ceilings

09 51 14.10 Ceiling, Acoustical Fabric-Faced Panel

	Crew	Daily Output	Labor-Hours	Unit	Material	2017 Bare Costs Labor	Equipment	Total	Total Incl O&P
0010 **CEILING, ACOUSTICAL FABRIC-FACED PANEL**									
0100 Glass cloth faced fiberglass, 3/4" thick	1 Carp	500	.016	S.F.	2.81	.79		3.60	4.30
0120 1" thick		485	.016		3.41	.81		4.22	4.99
0130 1-1/2" thick, nubby face	↓	475	.017	↓	2.67	.83		3.50	4.21

09 51 23 – Acoustical Tile Ceilings

09 51 23.10 Suspended Acoustic Ceiling Tiles

	Crew	Daily Output	Labor-Hours	Unit	Material	2017 Bare Costs Labor	Equipment	Total	Total Incl O&P
0010 **SUSPENDED ACOUSTIC CEILING TILES**, not including									
0100 suspension system									
1110 Mineral fiber tile, lay-in, 2' x 2' or 2' x 4', 5/8" thick, fine texture	1 Carp	625	.013	S.F.	.96	.63		1.59	2.03
1115 Rough textured		625	.013		.89	.63		1.52	1.95
1125 3/4" thick, fine textured		600	.013		2.07	.66		2.73	3.29
1130 Rough textured		600	.013		1.65	.66		2.31	2.83
1135 Fissured		600	.013		2.04	.66		2.70	3.25
1150 Tegular, 5/8" thick, fine textured		470	.017		1.08	.84		1.92	2.47
1155 Rough textured		470	.017		1.32	.84		2.16	2.73
1165 3/4" thick, fine textured		450	.018		2.25	.88		3.13	3.82
1170 Rough textured		450	.018		1.45	.88		2.33	2.94
1175 Fissured	↓	450	.018		2.21	.88		3.09	3.77
1185 For plastic film face, add					.83			.83	.91
1190 For fire rating, add					.48			.48	.53
3720 Mineral fiber, 24" x 24" or 48", reveal edge, painted, 5/8" thick	1 Carp	600	.013		1.29	.66		1.95	2.43
3740 3/4" thick		575	.014		1.54	.69		2.23	2.74
5020 66 – 78% recycled content, 3/4" thick G		600	.013		2.13	.66		2.79	3.35
5040 Mylar, 42% recycled content, 3/4" thick G		600	.013		5	.66		5.66	6.50
6000 Remove & install new ceiling tiles, min fiber, 2' x 2' or 2' x 4', 5/8"thk.	↓	335	.024	↓	.96	1.18		2.14	2.86

09 51 Acoustical Ceilings

09 51 23 – Acoustical Tile Ceilings

09 51 23.30 Suspended Ceilings, Complete

		Crew	Daily Output	Labor-Hours	Unit	Material	2017 Bare Costs Labor	Equipment	Total	Total Incl O&P
0010	**SUSPENDED CEILINGS, COMPLETE**, including standard									
0100	suspension system but not incl. 1-1/2" carrier channels									
0600	Fiberglass ceiling board, 2' x 4' x 5/8", plain faced	1 Carp	500	.016	S.F.	2.08	.79		2.87	3.50
0700	Offices, 2' x 4' x 3/4"		380	.021		3.45	1.04		4.49	5.40
0800	Mineral fiber, on 15/16" T bar susp. 2' x 2' x 3/4" lay-in board		345	.023		3.09	1.14		4.23	5.15
0810	2' x 4' x 5/8" tile		380	.021		2.46	1.04		3.50	4.29
0820	Tegular, 2' x 2' x 5/8" tile on 9/16" grid		250	.032		2.53	1.58		4.11	5.20
0830	2' x 4' x 3/4" tile		275	.029		2.75	1.43		4.18	5.20
0900	Luminous panels, prismatic, acrylic		255	.031		3.70	1.55		5.25	6.45
1200	Metal pan with acoustic pad, steel		75	.107		5.35	5.25		10.60	13.90
1300	Painted aluminum		75	.107		3.37	5.25		8.62	11.75
1500	Aluminum, degreased finish		75	.107		5.35	5.25		10.60	13.95
1600	Stainless steel		75	.107		10.15	5.25		15.40	19.20
1800	Tile, Z bar suspension, 5/8" mineral fiber tile		150	.053		2.35	2.63		4.98	6.60
1900	3/4" mineral fiber tile	↓	150	.053	↓	2.41	2.63		5.04	6.65
2400	For strip lighting, see Section 26 51 13.50									
2500	For rooms under 500 S.F., add				S.F.		25%			

09 51 33 – Acoustical Metal Pan Ceilings

09 51 33.10 Ceiling, Acoustical Metal Pan

		Crew	Daily Output	Labor-Hours	Unit	Material	2017 Bare Costs Labor	Equipment	Total	Total Incl O&P
0010	**CEILING, ACOUSTICAL METAL PAN**									
0100	Metal panel, lay-in, 2' x 2', sq. edge	1 Carp	500	.016	S.F.	10.05	.79		10.84	12.25
0110	Tegular edge		500	.016		13.85	.79		14.64	16.45
0120	2' x 4', sq. edge		500	.016		13.55	.79		14.34	16.10
0130	Tegular edge		500	.016		13.85	.79		14.64	16.45
0140	Perforated alum. clip-in, 2' x 2'		500	.016		13.70	.79		14.49	16.25
0150	2' x 4'	↓	500	.016	↓	11.15	.79		11.94	13.45

09 51 53 – Direct-Applied Acoustical Ceilings

09 51 53.10 Ceiling Tile

		Crew	Daily Output	Labor-Hours	Unit	Material	2017 Bare Costs Labor	Equipment	Total	Total Incl O&P
0010	**CEILING TILE**, stapled or cemented									
0100	12" x 12" or 12" x 24", not including furring									
0600	Mineral fiber, vinyl coated, 5/8" thick	1 Carp	300	.027	S.F.	2.34	1.31		3.65	4.58
0700	3/4" thick		300	.027		3.03	1.31		4.34	5.35
0900	Fire rated, 3/4" thick, plain faced		300	.027		1.36	1.31		2.67	3.51
1000	Plastic coated face		300	.027		2.07	1.31		3.38	4.29
1200	Aluminum faced, 5/8" thick, plain		300	.027		1.83	1.31		3.14	4.02
3700	Wall application of above, add	↓	1000	.008			.39		.39	.60
3900	For ceiling primer, add					.13			.13	.14
4000	For ceiling cement, add				↓	.40			.40	.44

09 53 Acoustical Ceiling Suspension Assemblies

09 53 23 – Metal Acoustical Ceiling Suspension Assemblies

09 53 23.30 Ceiling Suspension Systems

			Crew	Daily Output	Labor-Hours	Unit	Material	2017 Bare Costs Labor	Equipment	Total	Total Incl O&P
0010	**CEILING SUSPENSION SYSTEMS** for boards and tile										
0050	Class A suspension system, 15/16" T bar, 2' x 4' grid		1 Carp	800	.010	S.F.	.79	.49		1.28	1.62
0300	2' x 2' grid			650	.012		1.02	.61		1.63	2.05
0310	25% recycled steel, 2' x 4' grid	G		800	.010		.83	.49		1.32	1.66
0320	2' x 2' grid	G	↓	650	.012		1.04	.61		1.65	2.07
0350	For 9/16" grid, add						.16			.16	.18
0360	For fire rated grid, add					↓	.09			.09	.10

For customer support on your Building Construction Costs with RSMeans Data, call 800.448.8182.

337

09 53 Acoustical Ceiling Suspension Assemblies

09 53 23 – Metal Acoustical Ceiling Suspension Assemblies

09 53 23.30 Ceiling Suspension Systems	Crew	Daily Output	Labor-Hours	Unit	Material	2017 Bare Costs Labor	Equipment	Total	Total Incl O&P
0370	For colored grid, add				S.F.	.21			.21	.23
0400	Concealed Z bar suspension system, 12" module	1 Carp	520	.015		.89	.76		1.65	2.14
0600	1-1/2" carrier channels, 4' O.C., add	"	470	.017		.12	.84		.96	1.41
0700	Carrier channels for ceilings with									
0900	recessed lighting fixtures, add	1 Carp	460	.017	S.F.	.22	.86		1.08	1.55
1040	Hanging wire, 12 ga., 4' long		65	.123	C.S.F.	.37	6.05		6.42	9.70
1080	8' long		65	.123	"	.74	6.05		6.79	10.10
3000	Seismic ceiling bracing, IBC Site Class D, Occupancy Category II									
3050	For ceilings less than 2500 S.F.									
3060	Seismic clips at attached walls	1 Carp	180	.044	Ea.	1.09	2.19		3.28	4.55
3100	For ceilings greater than 2500 S.F., add									
3120	Seismic clips, joints at cross tees	1 Carp	120	.067	Ea.	3.17	3.28		6.45	8.55
3140	At cross tees and mains, mains field cut	"	60	.133	"	3.17	6.55		9.72	13.55
3200	Compression posts, telescopic, attached to structure above									
3210	To 30" high	1 Carp	26	.308	Ea.	38.50	15.15		53.65	65.50
3220	30" to 48" high		25.50	.314		42	15.45		57.45	70
3230	48" to 84" high		25	.320		51	15.75		66.75	80
3240	84" to 102" high		24.50	.327		58	16.10		74.10	88.50
3250	102" to 120" high		24	.333		83	16.40		99.40	117
3260	120" to 144" high		24	.333		92	16.40		108.40	126
3300	Stabilizer bars									
3310	12" long	1 Carp	240	.033	Ea.	1.07	1.64		2.71	3.69
3320	24" long		235	.034		1.02	1.68		2.70	3.69
3330	36" long		230	.035		.98	1.71		2.69	3.70
3340	48" long		220	.036		.81	1.79		2.60	3.63
3400	Wire support for light fixtures, per L.F. height to structure above									
3410	Less than 10 lb.	1 Carp	400	.020	L.F.	.29	.99		1.28	1.83
3420	10 lb. to 56 lb.	"	240	.033	"	.58	1.64		2.22	3.15

09 54 Specialty Ceilings

09 54 16 – Luminous Ceilings

09 54 16.10 Ceiling, Luminous

		Crew	Daily Output	Labor-Hours	Unit	Material	2017 Bare Costs Labor	Equipment	Total	Total Incl O&P
0010	**CEILING, LUMINOUS**									
0020	Translucent lay-in panels, 2' x 2'	1 Carp	500	.016	S.F.	23	.79		23.79	26
0030	2' x 6'	"	500	.016	"	17.80	.79		18.59	21

09 54 23 – Linear Metal Ceilings

09 54 23.10 Metal Ceilings

		Crew	Daily Output	Labor-Hours	Unit	Material	2017 Bare Costs Labor	Equipment	Total	Total Incl O&P
0010	**METAL CEILINGS**									
0015	Solid alum. planks, 3-1/4" x 12', open reveal	1 Carp	500	.016	S.F.	2.35	.79		3.14	3.80
0020	Closed reveal		500	.016		3	.79		3.79	4.51
0030	7-1/4" x 12', open reveal		500	.016		4	.79		4.79	5.60
0040	Closed reveal		500	.016		5.10	.79		5.89	6.80
0050	Metal, open cell, 2' x 2', 6" cell		500	.016		8.70	.79		9.49	10.75
0060	8" cell		500	.016		9.60	.79		10.39	11.75
0070	2' x 4', 6" cell		500	.016		5.70	.79		6.49	7.45
0080	8" cell		500	.016		5.70	.79		6.49	7.45

09 54 26 – Suspended Wood Ceilings

09 54 26.10 Wood Ceilings

		Crew	Daily Output	Labor-Hours	Unit	Material	2017 Bare Costs Labor	Equipment	Total	Total Incl O&P
0010	**WOOD CEILINGS**									
1000	4" - 6" wood slats on heavy duty 15/16" T-bar grid	2 Carp	250	.064	S.F.	25	3.15		28.15	32.50

For customer support on your Building Construction Costs with RSMeans Data, call 800.448.8182.

09 54 Specialty Ceilings

09 54 33 - Decorative Panel Ceilings

09 54 33.20 Metal Panel Ceilings	Crew	Daily Output	Labor-Hours	Unit	Material	2017 Bare Costs Labor	Equipment	Total	Total Incl O&P
0010 **METAL PANEL CEILINGS**									
0020 Lay-in or screwed to furring, not including grid									
0100 Tin ceilings, 2' x 2' or 2' x 4', bare steel finish	2 Carp	300	.053	S.F.	2.45	2.63		5.08	6.70
0120 Painted white finish		300	.053	"	3.63	2.63		6.26	8
0140 Copper, chrome or brass finish		300	.053	L.F.	6.40	2.63		9.03	11.05
0200 Cornice molding, 2-1/2" to 3-1/2" wide, 4' long, bare steel finish		200	.080	S.F.	2.17	3.94		6.11	8.45
0220 Painted white finish		200	.080		2.84	3.94		6.78	9.15
0240 Copper, chrome or brass finish		200	.080		4.01	3.94		7.95	10.45
0320 5" to 6-1/2" wide, 4' long, bare steel finish		150	.107		3.36	5.25		8.61	11.75
0340 Painted white finish		150	.107		4.09	5.25		9.34	12.55
0360 Copper, chrome or brass finish		150	.107		6.50	5.25		11.75	15.20
0420 Flat molding, 3-1/2" to 5" wide, 4' long, bare steel finish		250	.064		3.95	3.15		7.10	9.20
0440 Painted white finish		250	.064		4.06	3.15		7.21	9.30
0460 Copper, chrome or brass finish	↓	250	.064	↓	7.10	3.15		10.25	12.65

09 57 Special Function Ceilings

09 57 53 - Security Ceiling Assemblies

09 57 53.10 Ceiling Assem., Security, Radio Freq. Shielding	Crew	Daily Output	Labor-Hours	Unit	Material	2017 Bare Costs Labor	Equipment	Total	Total Incl O&P
0010 **CEILING ASSEMBLY, SECURITY, RADIO FREQUENCY SHIELDING**									
0020 Prefabricated, galvanized steel	2 Carp	375	.043	SF Surf	5.30	2.10		7.40	9
0050 5 oz., copper ceiling panel	↓	155	.103		4.48	5.10		9.58	12.75
0110 12 oz., copper ceiling panel	↓	140	.114	↓	9.05	5.65		14.70	18.55
0250 Ceiling hangers	E-1	45	.533	Ea.	32	28.50	2.89	63.39	84.50
0300 Shielding transition, 11 ga. preformed angles	"	1365	.018	L.F.	32	.95	.10	33.05	36.50

09 61 Flooring Treatment

09 61 19 - Concrete Floor Staining

09 61 19.40 Floors, Interior	Crew	Daily Output	Labor-Hours	Unit	Material	2017 Bare Costs Labor	Equipment	Total	Total Incl O&P
0010 **FLOORS, INTERIOR**									
0300 Acid stain and sealer									
0310 Stain, one coat	1 Pord	650	.012	S.F.	.12	.51		.63	.91
0320 Two coats		570	.014		.25	.59		.84	1.15
0330 Acrylic sealer, one coat		2600	.003		.22	.13		.35	.43
0340 Two coats	↓	1400	.006	↓	.44	.24		.68	.85

09 62 Specialty Flooring

09 62 19 - Laminate Flooring

09 62 19.10 Floating Floor	Crew	Daily Output	Labor-Hours	Unit	Material	2017 Bare Costs Labor	Equipment	Total	Total Incl O&P
0010 **FLOATING FLOOR**									
8300 Floating floor, laminate, wood pattern strip, complete	1 Clab	133	.060	S.F.	4.30	2.35		6.65	8.35
8310 Components, T & G wood composite strips					3.81			3.81	4.19
8320 Film					.17			.17	.18
8330 Foam					.26			.26	.29
8340 Adhesive					.48			.48	.53
8350 Installation kit				↓	.19			.19	.21
8360 Trim, 2" wide x 3' long				L.F.	4.30			4.30	4.73
8370 Reducer moulding				"	5.70			5.70	6.25

09 62 Specialty Flooring

09 62 23 – Bamboo Flooring

09 62 23.10 Flooring, Bamboo		Crew	Daily Output	Labor-Hours	Unit	Material	2017 Bare Costs Labor	Equipment	Total	Total Incl O&P
0010	**FLOORING, BAMBOO**									
8600	Flooring, wood, bamboo strips, unfinished, 5/8" x 4" x 3'	G	1 Carp	255	.031	S.F.	4.69	1.55		6.24	7.50
8610	5/8" x 4" x 4'	G		275	.029		4.87	1.43		6.30	7.55
8620	5/8" x 4" x 6'	G		295	.027		5.35	1.34		6.69	7.90
8630	Finished, 5/8" x 4" x 3'	G		255	.031		5.15	1.55		6.70	8.05
8640	5/8" x 4" x 4'	G		275	.029		5.40	1.43		6.83	8.15
8650	5/8" x 4" x 6'	G		295	.027		4.94	1.34		6.28	7.50
8660	Stair treads, unfinished, 1-1/16" x 11-1/2" x 4'	G		18	.444	Ea.	45	22		67	83
8670	Finished, 1-1/16" x 11-1/2" x 4'	G		18	.444		82	22		104	124
8680	Stair risers, unfinished, 5/8" x 7-1/2" x 4'	G		18	.444		16.60	22		38.60	52
8690	Finished, 5/8" x 7-1/2" x 4'	G		18	.444		31.50	22		53.50	68
8700	Stair nosing, unfinished, 6' long	G		16	.500		37	24.50		61.50	78
8710	Finished, 6' long	G		16	.500		42	24.50		66.50	83.50

09 62 29 – Cork Flooring

09 62 29.10 Cork Tile Flooring		Crew	Daily Output	Labor-Hours	Unit	Material	2017 Bare Costs Labor	Equipment	Total	Total Incl O&P
0010	**CORK TILE FLOORING**									
2200	Cork tile, standard finish, 1/8" thick	G	1 Tilf	315	.025	S.F.	6.80	1.16		7.96	9.20
2250	3/16" thick	G		315	.025		5.80	1.16		6.96	8.10
2300	5/16" thick	G		315	.025		8.35	1.16		9.51	10.95
2350	1/2" thick	G		315	.025		10.95	1.16		12.11	13.75
2500	Urethane finish, 1/8" thick	G		315	.025		8.10	1.16		9.26	10.65
2550	3/16" thick	G		315	.025		8.25	1.16		9.41	10.80
2600	5/16" thick	G		315	.025		8.85	1.16		10.01	11.50
2650	1/2" thick	G		315	.025		12.15	1.16		13.31	15.10

09 63 Masonry Flooring

09 63 13 – Brick Flooring

09 63 13.10 Miscellaneous Brick Flooring		Crew	Daily Output	Labor-Hours	Unit	Material	2017 Bare Costs Labor	Equipment	Total	Total Incl O&P
0010	**MISCELLANEOUS BRICK FLOORING**									
0020	Acid-proof shales, red, 8" x 3-3/4" x 1-1/4" thick		D-7	.43	37.209	M	710	1,525		2,235	3,050
0050	2-1/4" thick		D-1	.40	40		985	1,725		2,710	3,725
0200	Acid-proof clay brick, 8" x 3-3/4" x 2-1/4" thick	G		.40	40		950	1,725		2,675	3,700
0250	9" x 4-1/2" x 3"	G		95	.168	S.F.	4.28	7.30		11.58	15.90
0260	Cast ceramic, pressed, 4" x 8" x 1/2", unglazed		D-7	100	.160		6.45	6.55		13	16.85
0270	Glazed			100	.160		8.60	6.55		15.15	19.25
0280	Hand molded flooring, 4" x 8" x 3/4", unglazed			95	.168		8.55	6.90		15.45	19.65
0290	Glazed			95	.168		10.75	6.90		17.65	22
0300	8" hexagonal, 3/4" thick, unglazed			85	.188		9.35	7.75		17.10	22
0310	Glazed			85	.188		16.90	7.75		24.65	30
0400	Heavy duty industrial, cement mortar bed, 2" thick, not incl. brick		D-1	80	.200		1.13	8.65		9.78	14.55
0450	Acid-proof joints, 1/4" wide		"	65	.246		1.49	10.65		12.14	18.05
0500	Pavers, 8" x 4", 1" to 1-1/4" thick, red		D-7	95	.168		3.76	6.90		10.66	14.40
0510	Ironspot		"	95	.168		5.30	6.90		12.20	16.10
0540	1-3/8" to 1-3/4" thick, red		D-1	95	.168		3.63	7.30		10.93	15.20
0560	Ironspot			95	.168		5.25	7.30		12.55	17
0580	2-1/4" thick, red			90	.178		3.69	7.70		11.39	15.90
0590	Ironspot			90	.178		5.75	7.70		13.45	18.15
0700	Paver, adobe brick, 6" x 12", 1/2" joint	G		42	.381		1.37	16.50		17.87	27
0710	Mexican red, 12" x 12"	G	1 Tilf	48	.167		1.72	7.65		9.37	13.25
0720	Saltillo, 12" x 12"	G	"	48	.167		1.49	7.65		9.14	13

09 63 Masonry Flooring

09 63 13 – Brick Flooring

09 63 13.10 Miscellaneous Brick Flooring

		Crew	Daily Output	Labor-Hours	Unit	Material	2017 Bare Costs Labor	Equipment	Total	Total Incl O&P
0800	For sidewalks and patios with pavers, see Section 32 14 16.10									
0870	For epoxy joints, add	D-1	600	.027	S.F.	2.87	1.15		4.02	4.93
0880	For Furan underlayment, add	"	600	.027		2.38	1.15		3.53	4.39
0890	For waxed surface, steam cleaned, add	A-1H	1000	.008	↓	.20	.31	.08	.59	.78

09 63 40 – Stone Flooring

09 63 40.10 Marble

		Crew	Daily Output	Labor-Hours	Unit	Material	2017 Bare Costs Labor	Equipment	Total	Total Incl O&P
0010	**MARBLE**									
0020	Thin gauge tile, 12" x 6", 3/8", white Carara	D-7	60	.267	S.F.	15.70	10.95		26.65	33.50
0100	Travertine		60	.267		10.45	10.95		21.40	28
0200	12" x 12" x 3/8", thin set, floors		60	.267		10.45	10.95		21.40	28
0300	On walls		52	.308	↓	9.85	12.65		22.50	29.50
1000	Marble threshold, 4" wide x 36" long x 5/8" thick, white	↓	60	.267	Ea.	10.45	10.95		21.40	28

09 63 40.20 Slate Tile

		Crew	Daily Output	Labor-Hours	Unit	Material	2017 Bare Costs Labor	Equipment	Total	Total Incl O&P
0010	**SLATE TILE**									
0020	Vermont, 6" x 6" x 1/4" thick, thin set	D-7	180	.089	S.F.	7.60	3.65		11.25	13.80
0200	See also Section 32 14 40.10									

09 64 Wood Flooring

09 64 16 – Wood Block Flooring

09 64 16.10 End Grain Block Flooring

		Crew	Daily Output	Labor-Hours	Unit	Material	2017 Bare Costs Labor	Equipment	Total	Total Incl O&P
0010	**END GRAIN BLOCK FLOORING**									
0020	End grain flooring, coated, 2" thick	1 Carp	295	.027	S.F.	3.64	1.34		4.98	6.05
0400	Natural finish, 1" thick, fir		125	.064		3.76	3.15		6.91	8.95
0600	1-1/2" thick, pine		125	.064		3.69	3.15		6.84	8.90
0700	2" thick, pine	↓	125	.064	↓	4.58	3.15		7.73	9.90

09 64 19 – Wood Composition Flooring

09 64 19.10 Wood Composition

		Crew	Daily Output	Labor-Hours	Unit	Material	2017 Bare Costs Labor	Equipment	Total	Total Incl O&P
0010	**WOOD COMPOSITION** Gym floors									
0100	2-1/4" x 6-7/8" x 3/8", on 2" grout setting bed	D-7	150	.107	S.F.	6.25	4.38		10.63	13.40
0200	Thin set, on concrete	"	250	.064		5.70	2.63		8.33	10.20
0300	Sanding and finishing, add	1 Carp	200	.040	↓	.86	1.97		2.83	3.97

09 64 23 – Wood Parquet Flooring

09 64 23.10 Wood Parquet

		Crew	Daily Output	Labor-Hours	Unit	Material	2017 Bare Costs Labor	Equipment	Total	Total Incl O&P
0010	**WOOD PARQUET** flooring									
5200	Parquetry, 5/16" thk, no finish, oak, plain pattern	1 Carp	160	.050	S.F.	5.45	2.46		7.91	9.70
5300	Intricate pattern		100	.080		9.80	3.94		13.74	16.80
5500	Teak, plain pattern		160	.050		6.30	2.46		8.76	10.70
5600	Intricate pattern		100	.080		10.70	3.94		14.64	17.80
5650	13/16" thick, select grade oak, plain pattern		160	.050		9.75	2.46		12.21	14.50
5700	Intricate pattern		100	.080		17.25	3.94		21.19	25
5800	Custom parquetry, including finish, plain pattern		100	.080		17.05	3.94		20.99	25
5900	Intricate pattern		50	.160		25	7.90		32.90	39.50
6700	Parquetry, prefinished white oak, 5/16" thick, plain pattern		160	.050		7.95	2.46		10.41	12.45
6800	Intricate pattern		100	.080		8.55	3.94		12.49	15.45
7000	Walnut or teak, parquetry, plain pattern		160	.050		8.40	2.46		10.86	13
7100	Intricate pattern	↓	100	.080	↓	13.50	3.94		17.44	21
7200	Acrylic wood parquet blocks, 12" x 12" x 5/16",									
7210	Irradiated, set in epoxy	1 Carp	160	.050	S.F.	10.40	2.46		12.86	15.20

09 64 Wood Flooring

09 64 29 – Wood Strip and Plank Flooring

09 64 29.10 Wood

09 64 29.10 Wood	Crew	Daily Output	Labor-Hours	Unit	Material	2017 Bare Costs Labor	Equipment	Total	Total Incl O&P
0010 **WOOD**									
0020 Fir, vertical grain, 1" x 4", not incl. finish, grade B & better	1 Carp	255	.031	S.F.	2.85	1.55		4.40	5.50
0100 Grade C & better		255	.031		2.68	1.55		4.23	5.30
4000 Maple, strip, 25/32" x 2-1/4", not incl. finish, select		170	.047		4.90	2.32		7.22	8.95
4100 #2 & better		170	.047		4.79	2.32		7.11	8.80
4300 33/32" x 3-1/4", not incl. finish, #1 grade		170	.047		4.65	2.32		6.97	8.65
4400 #2 & better	↓	170	.047	↓	4.14	2.32		6.46	8.10
4600 Oak, white or red, 25/32" x 2-1/4", not incl. finish									
4700 #1 common	1 Carp	170	.047	S.F.	3.39	2.32		5.71	7.30
4900 Select quartered, 2-1/4" wide		170	.047		4.29	2.32		6.61	8.25
5000 Clear		170	.047		4.19	2.32		6.51	8.15
6100 Prefinished, white oak, prime grade, 2-1/4" wide		170	.047		5.20	2.32		7.52	9.25
6200 3-1/4" wide		185	.043		5.65	2.13		7.78	9.45
6400 Ranch plank		145	.055		7.20	2.72		9.92	12.10
6500 Hardwood blocks, 9" x 9", 25/32" thick		160	.050		6.15	2.46		8.61	10.50
7400 Yellow pine, 3/4" x 3-1/8", T & G, C & better, not incl. finish	↓	200	.040		1.65	1.97		3.62	4.84
7500 Refinish wood floor, sand, 2 coats poly, wax, soft wood	1 Clab	400	.020		.92	.78		1.70	2.21
7600 Hardwood		130	.062		1.37	2.41		3.78	5.20
7800 Sanding and finishing, 2 coats polyurethane	↓	295	.027	↓	.92	1.06		1.98	2.64
7900 Subfloor and underlayment, see Section 06 16									
8015 Transition molding, 2-1/4" wide, 5' long	1 Carp	19.20	.417	Ea.	10.85	20.50		31.35	43.50

09 64 66 – Wood Athletic Flooring

09 64 66.10 Gymnasium Flooring

09 64 66.10 Gymnasium Flooring	Crew	Daily Output	Labor-Hours	Unit	Material	2017 Bare Costs Labor	Equipment	Total	Total Incl O&P
0010 **GYMNASIUM FLOORING**									
0600 Gym floor, in mastic, over 2 ply felt, #2 & better									
0700 25/32" thick maple	1 Carp	100	.080	S.F.	4.12	3.94		8.06	10.60
0900 33/32" thick maple		98	.082		5.50	4.02		9.52	12.20
1000 For 1/2" corkboard underlayment, add	↓	750	.011		.99	.53		1.52	1.90
1300 For #1 grade maple, add				↓	.52			.52	.57
1600 Maple flooring, over sleepers, #2 & better									
1700 25/32" thick	1 Carp	85	.094	S.F.	4.79	4.64		9.43	12.35
1900 33/32" thick	"	83	.096		5.60	4.75		10.35	13.40
2000 For #1 grade, add					.56			.56	.62
2200 For 3/4" subfloor, add	1 Carp	350	.023		1.24	1.13		2.37	3.08
2300 With two 1/2" subfloors, 25/32" thick	"	69	.116	↓	6	5.70		11.70	15.35
2500 Maple, incl. finish, #2 & btr., 25/32" thick, on rubber									
2600 Sleepers, with two 1/2" subfloors	1 Carp	76	.105	S.F.	6.45	5.20		11.65	15
2800 With steel spline, double connection to channels	"	73	.110		6.85	5.40		12.25	15.80
2900 For 33/32" maple, add					.73			.73	.80
3100 For #1 grade maple, add					.56			.56	.62
3500 For termite proofing all of the above, add					.30			.30	.33
3700 Portable hardwood, prefinished panels	1 Carp	83	.096		8.60	4.75		13.35	16.70
3720 Insulated with polystyrene, 1" thick, add		165	.048		.73	2.39		3.12	4.46
3750 Running tracks, Sitka spruce surface, 25/32" x 2-1/4"		62	.129		15.70	6.35		22.05	27
3770 3/4" plywood surface, finished	↓	100	.080	↓	3.72	3.94		7.66	10.15

09 65 Resilient Flooring

09 65 10 – Resilient Tile Underlayment

09 65 10.10 Latex Underlayment

		Crew	Daily Output	Labor-Hours	Unit	Material	2017 Bare Costs Labor	Equipment	Total	Total Incl O&P
0010	**LATEX UNDERLAYMENT**									
3600	Latex underlayment, 1/8" thk., cementitious for resilient flooring	1 Tilf	160	.050	S.F.	1.25	2.29		3.54	4.78
4000	Liquid, fortified				Gal.	31			31	34

09 65 13 – Resilient Base and Accessories

09 65 13.13 Resilient Base

		Crew	Daily Output	Labor-Hours	Unit	Material	2017 Bare Costs Labor	Equipment	Total	Total Incl O&P
0010	**RESILIENT BASE**									
0690	1/8" vinyl base, 2-1/2" H, straight or cove, standard colors	1 Tilf	315	.025	L.F.	.63	1.16		1.79	2.42
0700	4" high		315	.025		1.17	1.16		2.33	3.02
0710	6" high		315	.025	↓	1.30	1.16		2.46	3.16
0720	Corners, 2-1/2" high		315	.025	Ea.	2.05	1.16		3.21	3.99
0730	4" high		315	.025		2.23	1.16		3.39	4.18
0740	6" high		315	.025	↓	2.53	1.16		3.69	4.51
0800	1/8" rubber base, 2-1/2" H, straight or cove, standard colors		315	.025	L.F.	1.01	1.16		2.17	2.84
1100	4" high		315	.025		1.14	1.16		2.30	2.98
1110	6" high		315	.025	↓	1.87	1.16		3.03	3.79
1150	Corners, 2-1/2" high		315	.025	Ea.	2.19	1.16		3.35	4.14
1153	4" high		315	.025		2.27	1.16		3.43	4.23
1155	6" high	↓	315	.025	↓	2.77	1.16		3.93	4.78
1450	For premium color/finish add					50%				
1500	Millwork profile	1 Tilf	315	.025	L.F.	5.90	1.16		7.06	8.20

09 65 13.23 Resilient Stair Treads and Risers

		Crew	Daily Output	Labor-Hours	Unit	Material	2017 Bare Costs Labor	Equipment	Total	Total Incl O&P
0010	**RESILIENT STAIR TREADS AND RISERS**									
0300	Rubber, molded tread, 12" wide, 5/16" thick, black	1 Tilf	115	.070	L.F.	15	3.19		18.19	21
0400	Colors		115	.070		15.35	3.19		18.54	21.50
0600	1/4" thick, black		115	.070		13.35	3.19		16.54	19.45
0700	Colors		115	.070		14.95	3.19		18.14	21
0900	Grip strip safety tread, colors, 5/16" thick		115	.070		20	3.19		23.19	26.50
1000	3/16" thick		120	.067	↓	14.75	3.06		17.81	21
1200	Landings, smooth sheet rubber, 1/8" thick		120	.067	S.F.	8.15	3.06		11.21	13.55
1300	3/16" thick		120	.067	"	9	3.06		12.06	14.45
1500	Nosings, 3" wide, 3/16" thick, black		140	.057	L.F.	4.71	2.62		7.33	9.10
1600	Colors		140	.057		5.30	2.62		7.92	9.70
1800	Risers, 7" high, 1/8" thick, flat		250	.032		8.15	1.47		9.62	11.15
1900	Coved		250	.032		9.35	1.47		10.82	12.50
2100	Vinyl, molded tread, 12" wide, colors, 1/8" thick		115	.070		5.85	3.19		9.04	11.15
2200	1/4" thick		115	.070	↓	7.35	3.19		10.54	12.80
2300	Landing material, 1/8" thick		200	.040	S.F.	6.20	1.83		8.03	9.50
2400	Riser, 7" high, 1/8" thick, coved		175	.046	L.F.	2.93	2.10		5.03	6.35
2500	Tread and riser combined, 1/8" thick	↓	80	.100	"	10.05	4.59		14.64	17.85

09 65 16 – Resilient Sheet Flooring

09 65 16.10 Rubber and Vinyl Sheet Flooring

		Crew	Daily Output	Labor-Hours	Unit	Material	2017 Bare Costs Labor	Equipment	Total	Total Incl O&P
0010	**RUBBER AND VINYL SHEET FLOORING**									
5500	Linoleum, sheet goods [G]	1 Tilf	360	.022	S.F.	3.59	1.02		4.61	5.45
5900	Rubber, sheet goods, 36" wide, 1/8" thick		120	.067		7.10	3.06		10.16	12.35
5950	3/16" thick		100	.080		10.45	3.67		14.12	16.95
6000	1/4" thick		90	.089		12.30	4.08		16.38	19.60
8000	Vinyl sheet goods, backed, .065" thick, plain pattern/colors		250	.032		3.69	1.47		5.16	6.25
8050	Intricate pattern/colors		200	.040		4.34	1.83		6.17	7.50
8100	.080" thick, plain pattern/colors		230	.035		4.12	1.59		5.71	6.90
8150	Intricate pattern/colors		200	.040		6.45	1.83		8.28	9.80
8200	.125" thick, plain pattern/colors	↓	230	.035		3.87	1.59		5.46	6.65

For customer support on your Building Construction Costs with RSMeans Data, call 800.448.8182.

343

09 65 Resilient Flooring

09 65 16 – Resilient Sheet Flooring

09 65 16.10 Rubber and Vinyl Sheet Flooring

	09 65 16.10 Rubber and Vinyl Sheet Flooring	Crew	Daily Output	Labor-Hours	Unit	Material	2017 Bare Costs Labor	Equipment	Total	Total Incl O&P
8250	Intricate pattern/colors	1 Tilf	200	.040	S.F.	7.50	1.83		9.33	10.95
8400	For welding seams, add		100	.080	L.F.	.23	3.67		3.90	5.70
8450	For integral cove base, add	↓	175	.046	"	.83	2.10		2.93	4.02
8700	Adhesive cement, 1 gallon per 200 to 300 S.F.				Gal.	30.50			30.50	33.50
8800	Asphalt primer, 1 gallon per 300 S.F.					14.60			14.60	16.05
8900	Emulsion, 1 gallon per 140 S.F.				↓	19.60			19.60	21.50

09 65 19 – Resilient Tile Flooring

09 65 19.19 Vinyl Composition Tile Flooring

	09 65 19.19 Vinyl Composition Tile Flooring	Crew	Daily Output	Labor-Hours	Unit	Material	Labor	Equipment	Total	Total Incl O&P
0010	**VINYL COMPOSITION TILE FLOORING**									
7000	Vinyl composition tile, 12" x 12", 1/16" thick	1 Tilf	500	.016	S.F.	1.22	.73		1.95	2.43
7050	Embossed		500	.016		2.25	.73		2.98	3.57
7100	Marbleized		500	.016		2.25	.73		2.98	3.57
7150	Solid		500	.016		2.91	.73		3.64	4.29
7200	3/32" thick, embossed		500	.016		1.47	.73		2.20	2.71
7250	Marbleized		500	.016		2.59	.73		3.32	3.94
7300	Solid		500	.016		2.41	.73		3.14	3.74
7350	1/8" thick, marbleized		500	.016		2.44	.73		3.17	3.77
7400	Solid		500	.016		1.55	.73		2.28	2.80
7450	Conductive	↓	500	.016	↓	6.05	.73		6.78	7.75

09 65 19.23 Vinyl Tile Flooring

	09 65 19.23 Vinyl Tile Flooring	Crew	Daily Output	Labor-Hours	Unit	Material	Labor	Equipment	Total	Total Incl O&P
0010	**VINYL TILE FLOORING**									
7500	Vinyl tile, 12" x 12", 3/32" thick, standard colors/patterns	1 Tilf	500	.016	S.F.	3.71	.73		4.44	5.15
7550	1/8" thick, standard colors/patterns		500	.016		5.20	.73		5.93	6.80
7600	1/8" thick, premium colors/patterns		500	.016		7	.73		7.73	8.80
7650	Solid colors		500	.016		3.27	.73		4	4.69
7700	Marbleized or Travertine pattern		500	.016		6.05	.73		6.78	7.75
7750	Florentine pattern		500	.016		6.45	.73		7.18	8.20
7800	Premium colors/patterns	↓	500	.016	↓	6.25	.73		6.98	7.95

09 65 19.33 Rubber Tile Flooring

	09 65 19.33 Rubber Tile Flooring	Crew	Daily Output	Labor-Hours	Unit	Material	Labor	Equipment	Total	Total Incl O&P
0010	**RUBBER TILE FLOORING**									
6050	Rubber tile, marbleized colors, 12" x 12", 1/8" thick	1 Tilf	400	.020	S.F.	5.70	.92		6.62	7.60
6100	3/16" thick		400	.020		9.05	.92		9.97	11.30
6300	Special tile, plain colors, 1/8" thick		400	.020		7.95	.92		8.87	10.10
6350	3/16" thick		400	.020		9.05	.92		9.97	11.30
6410	Raised, radial or square, .5 mm black		400	.020		7.65	.92		8.57	9.75
6430	.5 mm colored		400	.020		8.25	.92		9.17	10.45
6450	For golf course, skating rink, etc., 1/4" thick	↓	275	.029	↓	10.15	1.33		11.48	13.20

09 65 33 – Conductive Resilient Flooring

09 65 33.10 Conductive Rubber and Vinyl Flooring

	09 65 33.10 Conductive Rubber and Vinyl Flooring	Crew	Daily Output	Labor-Hours	Unit	Material	Labor	Equipment	Total	Total Incl O&P
0010	**CONDUCTIVE RUBBER AND VINYL FLOORING**									
1700	Conductive flooring, rubber tile, 1/8" thick	1 Tilf	315	.025	S.F.	6.95	1.16		8.11	9.40
1800	Homogeneous vinyl tile, 1/8" thick	"	315	.025	"	6.70	1.16		7.86	9.15

09 65 66 – Resilient Athletic Flooring

09 65 66.10 Resilient Athletic Flooring

	09 65 66.10 Resilient Athletic Flooring	Crew	Daily Output	Labor-Hours	Unit	Material	Labor	Equipment	Total	Total Incl O&P
0010	**RESILIENT ATHLETIC FLOORING**									
1000	Recycled rubber rolled goods, for weight rooms, 3/8" thk.	1 Tilf	315	.025	S.F.	2.74	1.16		3.90	4.74
2000	Vinyl sheet flooring, 1/4" thk.	"	315	.025	"	4.05	1.16		5.21	6.20

09 66 Terrazzo Flooring

09 66 13 – Portland Cement Terrazzo Flooring

09 66 13.10 Portland Cement Terrazzo		Crew	Daily Output	Labor-Hours	Unit	Material	2017 Bare Costs Labor	Equipment	Total	Total Incl O&P
0010	**PORTLAND CEMENT TERRAZZO**, cast-in-place	R096613-10								
0020	Cove base, 6" high, 16 ga. zinc	1 Mstz	20	.400	L.F.	3.30	18.40		21.70	31
0100	Curb, 6" high and 6" wide		6	1.333		5.65	61.50		67.15	97
0300	Divider strip for floors, 14 ga., 1-1/4" deep, zinc		375	.021		1.47	.98		2.45	3.07
0400	Brass		375	.021		2.49	.98		3.47	4.19
0600	Heavy top strip 1/4" thick, 1-1/4" deep, zinc		300	.027		2.16	1.23		3.39	4.20
0900	Galv. bottoms, brass		300	.027		2.68	1.23		3.91	4.77
1200	For thin set floors, 16 ga., 1/2" x 1/2", zinc		350	.023		1.29	1.05		2.34	2.98
1300	Brass	↓	350	.023	↓	2.75	1.05		3.80	4.59
1500	Floor, bonded to concrete, 1-3/4" thick, gray cement	J-3	75	.213	S.F.	3.38	9.05	4.02	16.45	21.50
1600	White cement, mud set		75	.213		3.76	9.05	4.02	16.83	22
1800	Not bonded, 3" total thickness, gray cement		70	.229		4.17	9.70	4.31	18.18	23.50
1900	White cement, mud set	↓	70	.229	↓	4.86	9.70	4.31	18.87	24.50
2100	For Venetian terrazzo, 1" topping, add					50%	50%			
2200	For heavy duty abrasive terrazzo, add					50%	50%			
2700	Monolithic terrazzo, 1/2" thick									
2710	10' panels	J-3	125	.128	S.F.	3.18	5.40	2.41	10.99	14.20
3000	Stairs, cast-in-place, pan filled treads		30	.533	L.F.	3.61	22.50	10.05	36.16	48.50
3100	Treads and risers	↓	14	1.143	"	6.25	48.50	21.50	76.25	102
3300	For stair landings, add to floor prices						50%			
3400	Stair stringers and fascia	J-3	30	.533	S.F.	5.30	22.50	10.05	37.85	50.50
3600	For abrasive metal nosings on stairs, add		150	.107	L.F.	9.50	4.52	2.01	16.03	19.35
3700	For abrasive surface finish, add		600	.027	S.F.	1.58	1.13	.50	3.21	3.97
3900	For raised abrasive strips, add		150	.107	L.F.	1.37	4.52	2.01	7.90	10.40
4000	Wainscot, bonded, 1-1/2" thick		30	.533	S.F.	4	22.50	10.05	36.55	49
4200	1/4" thick	↓	40	.400	"	5.60	16.95	7.55	30.10	39.50
4300	Stone chips, onyx gemstone, per 50 lb. bag				Bag	16.80			16.80	18.50

09 66 16 – Terrazzo Floor Tile

09 66 16.10 Tile or Terrazzo Base

		Crew	Daily Output	Labor-Hours	Unit	Material	Labor	Equipment	Total	Total Incl O&P
0010	**TILE OR TERRAZZO BASE**									
0020	Scratch coat only	1 Mstz	150	.053	S.F.	.44	2.45		2.89	4.12
0500	Scratch and brown coat only	"	75	.107	"	.84	4.90		5.74	8.15

09 66 16.13 Portland Cement Terrazzo Floor Tile

		Crew	Daily Output	Labor-Hours	Unit	Material	Labor	Equipment	Total	Total Incl O&P
0010	**PORTLAND CEMENT TERRAZZO FLOOR TILE**									
1200	Floor tiles, non-slip, 1" thick, 12" x 12"	D-1	60	.267	S.F.	20.50	11.55		32.05	40.50
1300	1-1/4" thick, 12" x 12"		60	.267		21	11.55		32.55	41
1500	16" x 16"		50	.320		23	13.85		36.85	46.50
1600	1-1/2" thick, 16" x 16"	↓	45	.356		21	15.40		36.40	46.50
1800	For Venetian terrazzo, add					6.30			6.30	6.90
1900	For white cement, add				↓	.59			.59	.65

09 66 16.16 Plastic Matrix Terrazzo Floor Tile

		Crew	Daily Output	Labor-Hours	Unit	Material	Labor	Equipment	Total	Total Incl O&P
0010	**PLASTIC MATRIX TERRAZZO FLOOR TILE**									
0100	12" x 12", 3/16" thick, floor tiles w/marble chips	1 Tilf	500	.016	S.F.	7.30	.73		8.03	9.15
0200	12" x 12", 3/16" thick, floor tiles w/glass chips		500	.016		7.90	.73		8.63	9.75
0300	12" x 12", 3/16" thick, floor tiles w/recycled content	↓	500	.016	↓	6.20	.73		6.93	7.95

09 66 16.30 Terrazzo, Precast

		Crew	Daily Output	Labor-Hours	Unit	Material	Labor	Equipment	Total	Total Incl O&P
0010	**TERRAZZO, PRECAST**									
0020	Base, 6" high, straight	1 Mstz	70	.114	L.F.	12.35	5.25		17.60	21.50
0100	Cove		60	.133		13.70	6.15		19.85	24
0300	8" high, straight		60	.133		12.25	6.15		18.40	22.50
0400	Cove	↓	50	.160	↓	18.05	7.35		25.40	31

09 66 Terrazzo Flooring

09 66 16 – Terrazzo Floor Tile

09 66 16.30 Terrazzo, Precast	Crew	Daily Output	Labor-Hours	Unit	Material	2017 Bare Costs Labor	Equipment	Total	Total Incl O&P
0600	For white cement, add				L.F.	.46			.46	.51
0700	For 16 ga. zinc toe strip, add					1.81			1.81	1.99
0900	Curbs, 4" x 4" high	1 Mstz	40	.200		33.50	9.20		42.70	50
1000	8" x 8" high	"	30	.267		38.50	12.25		50.75	60.50
2400	Stair treads, 1-1/2" thick, non-slip, three line pattern	2 Mstz	70	.229		41.50	10.50		52	61.50
2500	Nosing and two lines		70	.229		41.50	10.50		52	61.50
2700	2" thick treads, straight		60	.267		48	12.25		60.25	71
2800	Curved		50	.320		62.50	14.70		77.20	91
3000	Stair risers, 1" thick, to 6" high, straight sections		60	.267		11.45	12.25		23.70	31
3100	Cove		50	.320		15.90	14.70		30.60	39.50
3300	Curved, 1" thick, to 6" high, vertical		48	.333		22	15.30		37.30	46.50
3400	Cove		38	.421		41	19.35		60.35	73.50
3600	Stair tread and riser, single piece, straight, smooth surface		60	.267		54	12.25		66.25	77
3700	Non skid surface		40	.400		69.50	18.40		87.90	104
3900	Curved tread and riser, smooth surface		40	.400		75	18.40		93.40	111
4000	Non skid surface		32	.500		95	23		118	138
4200	Stair stringers, notched, 1" thick		25	.640		30.50	29.50		60	77.50
4300	2" thick		22	.727	↓	36.50	33.50		70	89.50
4500	Stair landings, structural, non-slip, 1-1/2" thick		85	.188	S.F.	33.50	8.65		42.15	50
4600	3" thick	↓	75	.213		47.50	9.80		57.30	67
4800	Wainscot, 12" x 12" x 1" tiles	1 Mstz	12	.667		7.30	30.50		37.80	53.50
4900	16" x 16" x 1-1/2" tiles	"	8	1	↓	14.50	46		60.50	84

09 66 23 – Resinous Matrix Terrazzo Flooring

09 66 23.13 Polyacrylate Mod. Cementitious Terrazzo Flr.

		Crew	Daily Output	Labor-Hours	Unit	Material	Labor	Equipment	Total	Total Incl O&P
0010	**POLYACRYLATE MODIFIED CEMENTITIOUS TERRAZZO FLOORING**									
3150	Polyacrylate, 1/4" thick, granite chips	C-6	735	.065	S.F.	3.55	2.66	.08	6.29	8.05
3170	Recycled porcelain		480	.100		4.60	4.08	.12	8.80	11.40
3200	3/8" thick, granite chips		620	.077		4.66	3.16	.09	7.91	10.05
3220	Recycled porcelain	↓	480	.100	↓	6.55	4.08	.12	10.75	13.55

09 66 23.16 Epoxy-Resin Terrazzo Flooring

		Crew	Daily Output	Labor-Hours	Unit	Material	Labor	Equipment	Total	Total Incl O&P
0010	**EPOXY-RESIN TERRAZZO FLOORING**									
1800	Epoxy terrazzo, 1/4" thick, chemical resistant, granite chips	J-3	200	.080	S.F.	6.05	3.39	1.51	10.95	13.35
1900	Recycled porcelain		150	.107		9.20	4.52	2.01	15.73	19.05
2500	Epoxy terrazzo, 1/4" thick, granite chips		200	.080		5.25	3.39	1.51	10.15	12.45
2550	Average		175	.091		5.60	3.87	1.72	11.19	13.85
2600	Recycled aggregate		150	.107		5.90	4.52	2.01	12.43	15.40
2650	Epoxy terrazzo, 3/8" thick, marble chips		200	.080		5.90	3.39	1.51	10.80	13.20
2675	Glass or mother of pearl	↓	200	.080	↓	6.90	3.39	1.51	11.80	14.30

09 66 33 – Conductive Terrazzo Flooring

09 66 33.10 Conductive Terrazzo

		Crew	Daily Output	Labor-Hours	Unit	Material	Labor	Equipment	Total	Total Incl O&P
0010	**CONDUCTIVE TERRAZZO**									
2400	Bonded conductive floor for hospitals	J-3	90	.178	S.F.	4.89	7.55	3.35	15.79	20.50

09 66 33.13 Conductive Epoxy-Resin Terrazzo

		Crew	Daily Output	Labor-Hours	Unit	Material	Labor	Equipment	Total	Total Incl O&P
0010	**CONDUCTIVE EPOXY-RESIN TERRAZZO**									
2100	Epoxy terrazzo, 1/4" thick, conductive, granite chips	J-3	100	.160	S.F.	8.15	6.80	3.01	17.96	22.50
2200	Recycled porcelain	"	90	.178	"	10.60	7.55	3.35	21.50	26.50

09 66 33.19 Conductive Plastic-Matrix Terrazzo Flooring

		Crew	Daily Output	Labor-Hours	Unit	Material	Labor	Equipment	Total	Total Incl O&P
0010	**CONDUCTIVE PLASTIC-MATRIX TERRAZZO FLOORING**									
3300	Conductive, 1/4" thick, granite chips	C-6	450	.107	S.F.	7.70	4.35	.13	12.18	15.25
3330	Recycled porcelain		305	.157		10.20	6.40	.19	16.79	21
3350	3/8" thick, granite chips	↓	365	.132		10.10	5.35	.16	15.61	19.40

09 66 Terrazzo Flooring

09 66 33 – Conductive Terrazzo Flooring

09 66 33.19 Conductive Plastic-Matrix Terrazzo Flooring	Crew	Daily Output	Labor-Hours	Unit	Material	2017 Bare Costs Labor	Equipment	Total	Total Incl O&P
3370 Recycled porcelain	C-6	255	.188	S.F.	13	7.65	.22	20.87	26
3450 Granite, conductive, 1/4" thick, 20% chip		695	.069		9.40	2.81	.08	12.29	14.75
3470 50% chip		420	.114		12.10	4.66	.13	16.89	20.50
3500 3/8" thick, 20% chip		695	.069		13.70	2.81	.08	16.59	19.50
3520 50% chip		380	.126		16.60	5.15	.15	21.90	26.50

09 67 Fluid-Applied Flooring

09 67 13 – Elastomeric Liquid Flooring

09 67 13.13 Elastomeric Liquid Flooring

		Crew	Daily Output	Labor-Hours	Unit	Material	2017 Bare Costs Labor	Equipment	Total	Total Incl O&P
0010	ELASTOMERIC LIQUID FLOORING									
0020	Cementitious acrylic, 1/4" thick	C-6	520	.092	S.F.	1.66	3.76	.11	5.53	7.70
0100	3/8" thick	"	450	.107		2.10	4.35	.13	6.58	9.10
0200	Methyl methachrylate, 1/4" thick	C-8A	3000	.016		6.50	.67		7.17	8.15
0210	1/8" thick	"	3000	.016		5.50	.67		6.17	7.05
0300	Cupric oxychloride, on bond coat, simple configs and patterns	C-6	480	.100		3.58	4.08	.12	7.78	10.25
0400	Complex configurations and patterns		420	.114		6	4.66	.13	10.79	13.85
2400	Mastic, hot laid, 2 coat, 1-1/2" thick, std., simple configs and patterns		690	.070		4.15	2.84	.08	7.07	9
2500	Maximum		520	.092		5.35	3.76	.11	9.22	11.70
2700	Acid-proof, minimum		605	.079		5.35	3.23	.09	8.67	10.90
2800	Maximum		350	.137		7.40	5.60	.16	13.16	16.80
3000	Neoprene, troweled on, 1/4" thick, minimum		545	.088		4.09	3.59	.10	7.78	10.05
3100	Maximum		430	.112		5.55	4.55	.13	10.23	13.25
4300	Polyurethane, with suspended vinyl chips, clear		1065	.045		7.60	1.84	.05	9.49	11.20
4500	Pigmented		860	.056		11.05	2.27	.07	13.39	15.70

09 67 23 – Resinous Flooring

09 67 23.23 Resinous Flooring

		Crew	Daily Output	Labor-Hours	Unit	Material	2017 Bare Costs Labor	Equipment	Total	Total Incl O&P
0010	RESINOUS FLOORING									
1200	Heavy duty epoxy topping, 1/4" thick,									
1300	500 to 1,000 S.F.	C-6	420	.114	S.F.	5.65	4.66	.13	10.44	13.50
1500	1,000 to 2,000 S.F.		450	.107		5.15	4.35	.13	9.63	12.45
1600	Over 10,000 S.F.		480	.100		4.75	4.08	.12	8.95	11.60

09 67 26 – Quartz Flooring

09 67 26.26 Quartz Flooring

		Crew	Daily Output	Labor-Hours	Unit	Material	2017 Bare Costs Labor	Equipment	Total	Total Incl O&P
0010	QUARTZ FLOORING									
0600	Epoxy, with colored quartz chips, broadcast, 3/8" thick	C-6	675	.071	S.F.	2.81	2.90	.08	5.79	7.60
0700	1/2" thick		490	.098		4.05	3.99	.12	8.16	10.70
0900	Troweled, minimum		560	.086		3.57	3.49	.10	7.16	9.35
1000	Maximum		480	.100		5.40	4.08	.12	9.60	12.25
3600	Polyester, with colored quartz chips, 1/16" thick, minimum		1065	.045		3.21	1.84	.05	5.10	6.40
3700	Maximum		560	.086		4.24	3.49	.10	7.83	10.05
3900	1/8" thick, minimum		810	.059		3.74	2.42	.07	6.23	7.85
4000	Maximum		675	.071		4.81	2.90	.08	7.79	9.80
4200	Polyester, heavy duty, compared to epoxy, add		2590	.019		1.51	.76	.02	2.29	2.83

09 67 66 – Fluid-Applied Athletic Flooring

09 67 66.10 Polyurethane

		Crew	Daily Output	Labor-Hours	Unit	Material	2017 Bare Costs Labor	Equipment	Total	Total Incl O&P
0010	POLYURETHANE									
4400	Thermoset, prefabricated in place, indoor									
4500	3/8" thick for basketball, gyms, etc.	1 Tilf	100	.080	S.F.	5.50	3.67		9.17	11.50
4600	1/2" thick for professional sports		95	.084		7.35	3.86		11.21	13.85

For customer support on your Building Construction Costs with RSMeans Data, call 800.448.8182.

347

09 67 Fluid-Applied Flooring

09 67 66 – Fluid-Applied Athletic Flooring

09 67 66.10 Polyurethane		Crew	Daily Output	Labor-Hours	Unit	Material	2017 Bare Costs Labor	2017 Bare Costs Equipment	Total	Total Incl O&P
4700	Outdoor, 1/4" thick, smooth, for tennis	1 Tilf	100	.080	S.F.	5.75	3.67		9.42	11.75
5000	Poured in place, indoor, with finish, 1/4" thick		80	.100		4.07	4.59		8.66	11.30
5050	3/8" thick		65	.123		4.94	5.65		10.59	13.85
5100	1/2" thick		50	.160		6.75	7.35		14.10	18.35

09 68 Carpeting

09 68 05 – Carpet Accessories

09 68 05.11 Flooring Transition Strip

		Crew	Daily Output	Labor-Hours	Unit	Material	2017 Bare Costs Labor	2017 Bare Costs Equipment	Total	Total Incl O&P
0010	**FLOORING TRANSITION STRIP**									
0107	Clamp down brass divider, 12' strip, vinyl to carpet	1 Tilf	31.25	.256	Ea.	14.65	11.75		26.40	33.50
0117	Vinyl to hard surface	"	31.25	.256	"	14.65	11.75		26.40	33.50

09 68 10 – Carpet Pad

09 68 10.10 Commercial Grade Carpet Pad

		Crew	Daily Output	Labor-Hours	Unit	Material	2017 Bare Costs Labor	2017 Bare Costs Equipment	Total	Total Incl O&P
0010	**COMMERCIAL GRADE CARPET PAD**									
9000	Sponge rubber pad, 20 oz./sq. yd.	1 Tilf	150	.053	S.Y.	4.65	2.45		7.10	8.75
9100	40 to 62 oz./sq. yd.		150	.053		8.75	2.45		11.20	13.30
9200	Felt pad, 20 oz./sq. yd.		150	.053		5.90	2.45		8.35	10.15
9300	32 to 56 oz./sq. yd.		150	.053		10.60	2.45		13.05	15.30
9400	Bonded urethane pad, 2.7 density		150	.053		5.95	2.45		8.40	10.20
9500	13.0 density		150	.053		8	2.45		10.45	12.45
9600	Prime urethane pad, 2.7 density		150	.053		3.51	2.45		5.96	7.50
9700	13.0 density		150	.053		6.60	2.45		9.05	10.90

09 68 13 – Tile Carpeting

09 68 13.10 Carpet Tile

		Crew	Daily Output	Labor-Hours	Unit	Material	2017 Bare Costs Labor	2017 Bare Costs Equipment	Total	Total Incl O&P
0010	**CARPET TILE**									
0100	Tufted nylon, 18" x 18", hard back, 20 oz.	1 Tilf	80	.100	S.Y.	23.50	4.59		28.09	32.50
0110	26 oz.		80	.100		26	4.59		30.59	35.50
0200	Cushion back, 20 oz.		80	.100		23.50	4.59		28.09	33
0210	26 oz.		80	.100		36.50	4.59		41.09	47
1100	Tufted, 24" x 24", hard back, 24 oz. nylon		80	.100		31	4.59		35.59	41
1180	35 oz.		80	.100		36	4.59		40.59	47
5060	42 oz.		80	.100		47	4.59		51.59	59

09 68 16 – Sheet Carpeting

09 68 16.10 Sheet Carpet

		Crew	Daily Output	Labor-Hours	Unit	Material	2017 Bare Costs Labor	2017 Bare Costs Equipment	Total	Total Incl O&P
0010	**SHEET CARPET**									
0700	Nylon, level loop, 26 oz., light to medium traffic	1 Tilf	75	.107	S.Y.	21.50	4.89		26.39	31.50
0720	28 oz., light to medium traffic		75	.107		29.50	4.89		34.39	40
0900	32 oz., medium traffic		75	.107		39.50	4.89		44.39	51
1100	40 oz., medium to heavy traffic		75	.107		45	4.89		49.89	57
2920	Nylon plush, 30 oz., medium traffic		75	.107		31	4.89		35.89	41.50
3000	36 oz., medium traffic		75	.107		35	4.89		39.89	46
3100	42 oz., medium to heavy traffic		70	.114		45	5.25		50.25	57.50
3200	46 oz., medium to heavy traffic		70	.114		52.50	5.25		57.75	66
3300	54 oz., heavy traffic		70	.114		58	5.25		63.25	72
3340	60 oz., heavy traffic		70	.114		61	5.25		66.25	75.50
3665	Olefin, 24 oz., light to medium traffic		75	.107		19	4.89		23.89	28.50
3670	26 oz., medium traffic		75	.107		13.10	4.89		17.99	21.50
3680	28 oz., medium to heavy traffic		75	.107		23.50	4.89		28.39	33.50
3700	32 oz., medium to heavy traffic		75	.107		27	4.89		31.89	37

09 68 Carpeting

09 68 16 – Sheet Carpeting

09 68 16.10 Sheet Carpet		Crew	Daily Output	Labor-Hours	Unit	Material	2017 Bare Costs Labor	Equipment	Total	Total Incl O&P
3730	42 oz., heavy traffic	1 Tilf	70	.114	S.Y.	28.50	5.25		33.75	39
4110	Wool, level loop, 40 oz., medium traffic		70	.114		106	5.25		111.25	125
4500	50 oz., medium to heavy traffic		70	.114		100	5.25		105.25	118
4700	Patterned, 32 oz., medium to heavy traffic		70	.114		91	5.25		96.25	108
4900	48 oz., heavy traffic		70	.114		100	5.25		105.25	118
5000	For less than full roll (approx. 1500 S.F.), add					25%				
5100	For small rooms, less than 12' wide, add						25%			
5200	For large open areas (no cuts), deduct						25%			
5600	For bound carpet baseboard, add	1 Tilf	300	.027	L.F.	3	1.22		4.22	5.10
5610	For stairs, not incl. price of carpet, add	"	30	.267	Riser		12.25		12.25	18.15
5620	For borders and patterns, add to labor						18%			
8950	For tackless, stretched installation, add padding from 09 68 10.10 to above									
9850	For brand-named specific fiber, add				S.Y.	25%				

09 68 20 – Athletic Carpet

09 68 20.10 Indoor Athletic Carpet

		Crew	Daily Output	Labor-Hours	Unit	Material	2017 Bare Costs Labor	Equipment	Total	Total Incl O&P
0010	**INDOOR ATHLETIC CARPET**									
3700	Polyethylene, in rolls, no base incl., landscape surfaces	1 Tilf	275	.029	S.F.	4.23	1.33		5.56	6.65
3800	Nylon action surface, 1/8" thick		275	.029		3.84	1.33		5.17	6.20
3900	1/4" thick		275	.029		5.55	1.33		6.88	8.10
4000	3/8" thick		275	.029		6.95	1.33		8.28	9.65
4100	Golf tee surface with foam back		235	.034		6.90	1.56		8.46	9.90
4200	Practice putting, knitted nylon surface		235	.034		5.85	1.56		7.41	8.70
4300	Synthetic turf, 3/8" thick		90	.089		4.54	4.08		8.62	11.05
4350	Interlocking 2' x 2' squares, 3/8" thick		210	.038		4.69	1.75		6.44	7.75
4400	1/2" thick		190	.042		5.05	1.93		6.98	8.40
5500	Polyvinyl chloride, sheet goods for gyms, 1/4" thick		80	.100		6.50	4.59		11.09	13.95
5600	3/8" thick		60	.133		9.55	6.10		15.65	19.55

09 69 Access Flooring

09 69 13 – Rigid-Grid Access Flooring

09 69 13.10 Access Floors

		Crew	Daily Output	Labor-Hours	Unit	Material	2017 Bare Costs Labor	Equipment	Total	Total Incl O&P
0010	**ACCESS FLOORS**									
0015	Access floor pkg. including panel, pedestal, & stringers 1500 lbs load									
0100	Package pricing, conc. fill panels, no fin., 6" ht.	4 Carp	750	.043	S.F.	16.05	2.10		18.15	21
0105	12" ht.		750	.043		16.65	2.10		18.75	21.50
0110	18" ht.		750	.043		16.45	2.10		18.55	21.50
0115	24" ht.		750	.043		17.30	2.10		19.40	22.50
0120	Package pricing, steel panels, no fin., 6" ht.		750	.043		12.70	2.10		14.80	17.15
0125	12" ht.		750	.043		15.55	2.10		17.65	20.50
0130	18" ht.		750	.043		15.65	2.10		17.75	20.50
0135	24" ht.		750	.043		15.75	2.10		17.85	20.50
0140	Package pricing, wood core panels, no fin., 6" ht.		750	.043		11.65	2.10		13.75	16
0145	12" ht.		750	.043		11.80	2.10		13.90	16.20
0150	18" ht.		750	.043		11.95	2.10		14.05	16.30
0155	24" ht.		750	.043		12.05	2.10		14.15	16.45
0160	Pkg. pricing, conc. fill pnls, no stringers, no fin., 6" ht, 1250 lbs load		700	.046		10.45	2.25		12.70	14.95
0165	12" ht.		700	.046		10.25	2.25		12.50	14.70
0170	18" ht.		700	.046		10.80	2.25		13.05	15.30
0175	24" ht.		700	.046		11.35	2.25		13.60	15.90
0250	Panels, 2' x 2' conc. fill, no fin.	2 Carp	500	.032		20.50	1.58		22.08	25.50

09 69 Access Flooring

09 69 13 – Rigid-Grid Access Flooring

09 69 13.10 Access Floors

		Crew	Daily Output	Labor-Hours	Unit	Material	2017 Bare Costs Labor	Equipment	Total	Total Incl O&P
0255	With 1/8" high pressure laminate	2 Carp	500	.032	S.F.	62	1.58		63.58	70.50
0260	Metal panels with 1/8" high pressure laminate		500	.032		73	1.58		74.58	83
0265	Wood core panels with 1/8" high pressure laminate		500	.032		48	1.58		49.58	55.50
0400	Aluminum panels, no fin.		500	.032		33.50	1.58		35.08	39
0600	For carpet covering, add					9.05			9.05	10
0700	For vinyl floor covering, add					9.35			9.35	10.30
0900	For high pressure laminate covering, add					7.70			7.70	8.45
0910	For snap on stringer system, add	2 Carp	1000	.016		1.59	.79		2.38	2.96
1000	Machine cutouts after initial installation	1 Carp	50	.160	Ea.	20	7.90		27.90	34
1050	Pedestals, 6" to 12"	2 Carp	85	.188		8.65	9.25		17.90	23.50
1100	Air conditioning grilles, 4" x 12"	1 Carp	17	.471		72.50	23		95.50	116
1150	4" x 18"	"	14	.571		99.50	28		127.50	153
1200	Approach ramps, steel	2 Carp	60	.267	S.F.	26	13.15		39.15	49
1300	Aluminum	"	40	.400	"	34	19.70		53.70	67.50
1500	Handrail, 2 rail, aluminum	1 Carp	15	.533	L.F.	114	26.50		140.50	165

09 72 Wall Coverings

09 72 13 – Cork Wall Coverings

09 72 13.10 Covering, Cork Wall

		Crew	Daily Output	Labor-Hours	Unit	Material	2017 Bare Costs Labor	Equipment	Total	Total Incl O&P
0010	**COVERING, CORK WALL**									
0600	Cork tiles, light or dark, 12" x 12" x 3/16"	1 Pape	240	.033	S.F.	4.33	1.39		5.72	6.85
0700	5/16" thick		235	.034		3.16	1.42		4.58	5.65
0900	1/4" basketweave		240	.033		3.50	1.39		4.89	5.95
1000	1/2" natural, non-directional pattern		240	.033		6.75	1.39		8.14	9.55
1100	3/4" natural, non-directional pattern		240	.033		11.95	1.39		13.34	15.25
1200	Granular surface, 12" x 36", 1/2" thick		385	.021		1.34	.87		2.21	2.78
1300	1" thick		370	.022		1.71	.90		2.61	3.25
1500	Polyurethane coated, 12" x 12" x 3/16" thick		240	.033		4.16	1.39		5.55	6.70
1600	5/16" thick		235	.034		6.60	1.42		8.02	9.40
1800	Cork wallpaper, paperbacked, natural		480	.017		1.90	.70		2.60	3.14
1900	Colors		480	.017		2.93	.70		3.63	4.27

09 72 16 – Vinyl-Coated Fabric Wall Coverings

09 72 16.13 Flexible Vinyl Wall Coverings

		Crew	Daily Output	Labor-Hours	Unit	Material	2017 Bare Costs Labor	Equipment	Total	Total Incl O&P
0010	**FLEXIBLE VINYL WALL COVERINGS**									
3000	Vinyl wall covering, fabric-backed, lightweight, type 1 (12-15 oz./S.Y.)	1 Pape	640	.013	S.F.	1	.52		1.52	1.89
3300	Medium weight, type 2 (20-24 oz./S.Y.)		480	.017		.91	.70		1.61	2.05
3400	Heavy weight, type 3 (28 oz./S.Y.)		435	.018		1.40	.77		2.17	2.70
3600	Adhesive, 5 gal. lots (18 S.Y./gal.)				Gal.	12.10			12.10	13.30

09 72 16.16 Rigid-Sheet Vinyl Wall Coverings

		Crew	Daily Output	Labor-Hours	Unit	Material	2017 Bare Costs Labor	Equipment	Total	Total Incl O&P
0010	**RIGID-SHEET VINYL WALL COVERINGS**									
0100	Acrylic, modified, semi-rigid PVC, .028" thick	2 Carp	330	.048	S.F.	1.28	2.39		3.67	5.05
0110	.040" thick	"	320	.050	"	1.91	2.46		4.37	5.85

09 72 19 – Textile Wall Coverings

09 72 19.10 Textile Wall Covering

		Crew	Daily Output	Labor-Hours	Unit	Material	2017 Bare Costs Labor	Equipment	Total	Total Incl O&P
0010	**TEXTILE WALL COVERING**, including sizing; add 10-30% waste @ takeoff									
0020	Silk	1 Pape	640	.013	S.F.	4.23	.52		4.75	5.45
0030	Cotton		640	.013		6.75	.52		7.27	8.25
0040	Linen		640	.013		1.89	.52		2.41	2.87
0050	Blend		640	.013		3.03	.52		3.55	4.12
0060	Linen wall covering, paper backed									

09 72 Wall Coverings

09 72 19 - Textile Wall Coverings

09 72 19.10 Textile Wall Covering

		Crew	Daily Output	Labor-Hours	Unit	Material	2017 Bare Costs Labor	Equipment	Total	Total Incl O&P
0070	Flame treatment				S.F.	1.09			1.09	1.20
0080	Stain resistance treatment					1.82			1.82	2
0090	Grass cloth, natural fabric	G	1 Pape	400	.020		1.92	.84		2.76	3.37
0100	Grass cloths with lining paper	G		400	.020		.90	.84		1.74	2.25
0110	Premium texture/color	G		350	.023		3.15	.96		4.11	4.91

09 72 20 - Natural Fiber Wall Covering

09 72 20.10 Natural Fiber Wall Covering

		Crew	Daily Output	Labor-Hours	Unit	Material	2017 Bare Costs Labor	Equipment	Total	Total Incl O&P
0010	**NATURAL FIBER WALL COVERING**, including sizing; add 10-30% waste @ takeoff									
0015	Bamboo	1 Pape	640	.013	S.F.	2.31	.52		2.83	3.33
0030	Burlap		640	.013		2.07	.52		2.59	3.07
0045	Jute		640	.013		1.32	.52		1.84	2.24
0060	Sisal		640	.013		1.49	.52		2.01	2.43

09 72 23 - Wallpapering

09 72 23.10 Wallpaper

		Crew	Daily Output	Labor-Hours	Unit	Material	2017 Bare Costs Labor	Equipment	Total	Total Incl O&P
0010	**WALLPAPER** including sizing; add 10-30% waste @ takeoff R097223-10									
0050	Aluminum foil	1 Pape	275	.029	S.F.	1.06	1.22		2.28	3.01
0100	Copper sheets, .025" thick, vinyl backing		240	.033		5.65	1.39		7.04	8.35
0300	Phenolic backing		240	.033		7.35	1.39		8.74	10.20
2400	Gypsum-based, fabric-backed, fire resistant									
2500	for masonry walls, 21 oz./S.Y.	1 Pape	800	.010	S.F.	.77	.42		1.19	1.48
2700	Small quantities		640	.013		.70	.52		1.22	1.56
3700	Wallpaper, average workmanship, solid pattern, low cost paper		640	.013		.59	.52		1.11	1.44
3900	Basic patterns (matching required), avg. cost paper		535	.015		1.17	.63		1.80	2.23
4000	Paper at $85 per double roll, quality workmanship		435	.018		1.82	.77		2.59	3.16

09 74 Flexible Wood Sheets

09 74 16 - Flexible Wood Veneers

09 74 16.10 Veneer, Flexible Wood

		Crew	Daily Output	Labor-Hours	Unit	Material	2017 Bare Costs Labor	Equipment	Total	Total Incl O&P
0010	**VENEER, FLEXIBLE WOOD**									
0100	Flexible wood veneer, 1/32" thick, plain woods	1 Pape	100	.080	S.F.	2.49	3.35		5.84	7.80
0110	Exotic woods	"	95	.084	"	3.76	3.52		7.28	9.45

09 77 Special Wall Surfacing

09 77 30 - Fiberglass Reinforced Panels

09 77 30.10 Fiberglass Reinforced Plastic Panels

		Crew	Daily Output	Labor-Hours	Unit	Material	2017 Bare Costs Labor	Equipment	Total	Total Incl O&P
0010	**FIBERGLASS REINFORCED PLASTIC PANELS**, .090" thick									
0020	On walls, adhesive mounted, embossed surface	2 Carp	640	.025	S.F.	1.17	1.23		2.40	3.18
0030	Smooth surface		640	.025		1.44	1.23		2.67	3.47
0040	Fire rated, embossed surface		640	.025		2.09	1.23		3.32	4.19
0050	Nylon rivet mounted, on drywall, embossed surface		480	.033		1.17	1.64		2.81	3.80
0060	Smooth surface		480	.033		1.44	1.64		3.08	4.09
0070	Fire rated, embossed surface		480	.033		2.09	1.64		3.73	4.81
0080	On masonry, embossed surface		320	.050		1.17	2.46		3.63	5.05
0090	Smooth surface		320	.050		1.44	2.46		3.90	5.35
0100	Fire rated, embossed surface		320	.050		2.09	2.46		4.55	6.05
0110	Nylon rivet and adhesive mounted, on drywall, embossed surface		240	.067		1.32	3.28		4.60	6.50
0120	Smooth surface		240	.067		1.32	3.28		4.60	6.50
0130	Fire rated, embossed surface		240	.067		2.30	3.28		5.58	7.60

09 77 Special Wall Surfacing

09 77 30 – Fiberglass Reinforced Panels

09 77 30.10 Fiberglass Reinforced Plastic Panels	Crew	Daily Output	Labor-Hours	Unit	Material	2017 Bare Costs Labor	Equipment	Total	Total Incl O&P
0140	On masonry, embossed surface	2 Carp	190	.084	S.F.	1.32	4.15		5.47	7.80
0150	Smooth surface		190	.084		1.32	4.15		5.47	7.80
0160	Fire rated, embossed surface	↓	190	.084	↓	2.30	4.15		6.45	8.90
0170	For moldings, add	1 Carp	250	.032	L.F.	.27	1.58		1.85	2.71
0180	On ceilings, for lay in grid system, embossed surface		400	.020	S.F.	1.17	.99		2.16	2.80
0190	Smooth surface		400	.020		1.44	.99		2.43	3.09
0200	Fire rated, embossed surface	↓	400	.020	↓	2.09	.99		3.08	3.81

09 77 43 – Panel Systems

09 77 43.20 Slatwall Panels and Accessories

		Crew	Daily Output	Labor-Hours	Unit	Material	Labor	Equipment	Total	Total Incl O&P
0010	**SLATWALL PANELS AND ACCESSORIES**									
0100	Slatwall panel, 4' x 8' x 3/4" T, MDF, paint grade	1 Carp	500	.016	S.F.	1.49	.79		2.28	2.85
0110	Melamine finish		500	.016		2.13	.79		2.92	3.55
0120	High pressure plastic laminate finish		500	.016		3.78	.79		4.57	5.35
0125	Wood veneer	↓	500	.016		3.80	.79		4.59	5.40
0130	Aluminum channel inserts, add				↓	2.64			2.64	2.90
0200	Accessories, corner forms, 8' L				L.F.	6.25			6.25	6.90
0210	T-connector, 8' L					8.70			8.70	9.60
0220	J-mold, 8' L					1.25			1.25	1.38
0230	Edge cap, 8' L					1.56			1.56	1.72
0240	Finish end cap, 8' L				↓	3.37			3.37	3.71
0300	Display hook, metal, 4" L				Ea.	.43			.43	.47
0310	6" L					.48			.48	.53
0320	8" L					.52			.52	.57
0330	10" L					.58			.58	.64
0340	12" L					.65			.65	.72
0350	Acrylic, 4" L					1.07			1.07	1.18
0360	6" L					1.08			1.08	1.19
0400	Waterfall hanger, metal, 12" - 16"					3.85			3.85	4.24
0410	Acrylic					11.50			11.50	12.65
0500	Shelf bracket, metal, 8"					1.87			1.87	2.06
0510	10"					2.06			2.06	2.27
0520	12"					2.30			2.30	2.53
0530	14"					2.50			2.50	2.75
0540	16"					2.93			2.93	3.22
0550	Acrylic, 8"					3.82			3.82	4.20
0560	10"					4.11			4.11	4.52
0570	12"					4.61			4.61	5.05
0580	14"					5.55			5.55	6.10
0600	Shelf, acrylic, 12" x 16" x 1/4"					22			22	24.50
0610	12" x 24" x 1/4"				↓	25			25	27.50

09 81 Acoustic Insulation

09 81 13 – Acoustic Board Insulation

09 81 13.10 Acoustic Board Insulation	Crew	Daily Output	Labor-Hours	Unit	Material	2017 Bare Costs Labor	Equipment	Total	Total Incl O&P
0010 **ACOUSTIC BOARD INSULATION**									
0020 Cellulose fiber board, 1/2" thk.	1 Carp	800	.010	S.F.	.88	.49		1.37	1.72

09 81 16 – Acoustic Blanket Insulation

09 81 16.10 Sound Attenuation Blanket

	Crew	Daily Output	Labor-Hours	Unit	Material	2017 Bare Costs Labor	Equipment	Total	Total Incl O&P
0010 **SOUND ATTENUATION BLANKET**									
0020 Blanket, 1" thick	1 Carp	925	.009	S.F.	.26	.43		.69	.94
0500 1-1/2" thick		920	.009		.27	.43		.70	.96
1000 2" thick		915	.009		.39	.43		.82	1.09
1500 3" thick		910	.009		.56	.43		.99	1.28
2000 Wall hung, STC 18 – 21, 1" thick, 4' x 20'	2 Carp	22	.727	Ea.	400	36		436	495
2010 10' x 20'	"	19	.842		995	41.50		1,036.50	1,175
2020 Wall hung, STC 27 – 28, 3" thick, 4' x 20'	3 Carp	12	2		575	98.50		673.50	780
2030 10' x 20'	"	9	2.667		1,425	131		1,556	1,775
3000 Thermal or acoustical batt above ceiling, 2" thick	1 Carp	900	.009	S.F.	.51	.44		.95	1.23
3100 3" thick		900	.009		.77	.44		1.21	1.52
3200 4" thick		900	.009		.96	.44		1.40	1.73
3400 Urethane plastic foam, open cell, on wall, 2" thick	2 Carp	2050	.008		3.17	.38		3.55	4.08
3500 3" thick		1550	.010		4.21	.51		4.72	5.40
3600 4" thick		1050	.015		5.90	.75		6.65	7.65
3700 On ceiling, 2" thick		1700	.009		3.16	.46		3.62	4.19
3800 3" thick		1300	.012		4.21	.61		4.82	5.55
3900 4" thick		900	.018		5.90	.88		6.78	7.85

09 84 Acoustic Room Components

09 84 13 – Fixed Sound-Absorptive Panels

09 84 13.10 Fixed Panels

	Crew	Daily Output	Labor-Hours	Unit	Material	2017 Bare Costs Labor	Equipment	Total	Total Incl O&P
0010 **FIXED PANELS** Perforated steel facing, painted with									
0100 Fiberglass or mineral filler, no backs, 2-1/4" thick, modular									
0200 space units, ceiling or wall hung, white or colored	1 Carp	100	.080	S.F.	9.45	3.94		13.39	16.45
0300 Fiberboard sound deadening panels, 1/2" thick	"	600	.013	"	.34	.66		1	1.38
0500 Fiberglass panels, 4' x 8' x 1" thick, with									
0600 glass cloth face for walls, cemented	1 Carp	155	.052	S.F.	8.75	2.54		11.29	13.55
0700 1-1/2" thick, dacron covered, inner aluminum frame,									
0710 wall mounted	1 Carp	300	.027	S.F.	9.10	1.31		10.41	12
0900 Mineral fiberboard panels, fabric covered, 30" x 108",									
1000 3/4" thick, concealed spline, wall mounted	1 Carp	150	.053	S.F.	6.55	2.63		9.18	11.20

09 84 36 – Sound-Absorbing Ceiling Units

09 84 36.10 Barriers

	Crew	Daily Output	Labor-Hours	Unit	Material	2017 Bare Costs Labor	Equipment	Total	Total Incl O&P
0010 **BARRIERS** Plenum									
0600 Aluminum foil, fiberglass reinf., parallel with joists	1 Carp	275	.029	S.F.	1.13	1.43		2.56	3.43
0700 Perpendicular to joists		180	.044		1.13	2.19		3.32	4.59
0900 Aluminum mesh, kraft paperbacked		275	.029		.81	1.43		2.24	3.08
0970 Fiberglass batts, kraft faced, 3-1/2" thick		1400	.006		.38	.28		.66	.85
0980 6" thick		1300	.006		.67	.30		.97	1.20
1000 Sheet lead, 1 lb., 1/64" thick, perpendicular to joists		150	.053		6.55	2.63		9.18	11.20
1100 Vinyl foam reinforced, 1/8" thick, 1.0 lb. per S.F.		150	.053		4.58	2.63		7.21	9.05

For customer support on your Building Construction Costs with RSMeans Data, call 800.448.8182.

353

09 91 03.20 Sanding

09 91 03.20 Sanding		Crew	Daily Output	Labor-Hours	Unit	Material	2017 Bare Costs Labor	2017 Bare Costs Equipment	Total	Total Incl O&P
0010	**SANDING** and puttying interior trim, compared to	R099100-10								
0100	Painting 1 coat, on quality work				L.F.		100%			
0300	Medium work						50%			
0400	Industrial grade						25%			
0500	Surface protection, placement and removal									
0510	Surface protection, placement and removal, basic drop cloths	1 Pord	6400	.001	S.F.		.05		.05	.08
0520	Masking with paper		800	.010		.07	.42		.49	.71
0530	Volume cover up (using plastic sheathing or building paper)		16000	.001			.02		.02	.03

09 91 03.30 Exterior Surface Preparation

09 91 03.30 Exterior Surface Preparation		Crew	Daily Output	Labor-Hours	Unit	Material	2017 Bare Costs Labor	2017 Bare Costs Equipment	Total	Total Incl O&P
0010	**EXTERIOR SURFACE PREPARATION**	R099100-10								
0015	Doors, per side, not incl. frames or trim									
0020	Scrape & sand									
0030	Wood, flush	1 Pord	616	.013	S.F.		.54		.54	.82
0040	Wood, detail		496	.016			.67		.67	1.02
0050	Wood, louvered		280	.029			1.19		1.19	1.80
0060	Wood, overhead		616	.013			.54		.54	.82
0070	Wire brush									
0080	Metal, flush	1 Pord	640	.013	S.F.		.52		.52	.79
0090	Metal, detail		520	.015			.64		.64	.97
0100	Metal, louvered		360	.022			.93		.93	1.40
0110	Metal or fibr., overhead		640	.013			.52		.52	.79
0120	Metal, roll up		560	.014			.60		.60	.90
0130	Metal, bulkhead		640	.013			.52		.52	.79
0140	Power wash, based on 2500 lb. operating pressure									
0150	Metal, flush	A-1H	2240	.004	S.F.		.14	.03	.17	.25
0160	Metal, detail		2120	.004			.15	.04	.19	.27
0170	Metal, louvered		2000	.004			.16	.04	.20	.28
0180	Metal or fibr., overhead		2400	.003			.13	.03	.16	.23
0190	Metal, roll up		2400	.003			.13	.03	.16	.23
0200	Metal, bulkhead		2200	.004			.14	.03	.17	.26
0400	Windows, per side, not incl. trim									
0410	Scrape & sand									
0420	Wood, 1-2 lite	1 Pord	320	.025	S.F.		1.04		1.04	1.57
0430	Wood, 3-6 lite		280	.029			1.19		1.19	1.80
0440	Wood, 7-10 lite		240	.033			1.39		1.39	2.10
0450	Wood, 12 lite		200	.040			1.67		1.67	2.52
0460	Wood, Bay/Bow		320	.025			1.04		1.04	1.57
0470	Wire brush									
0480	Metal, 1-2 lite	1 Pord	480	.017	S.F.		.70		.70	1.05
0490	Metal, 3-6 lite		400	.020			.83		.83	1.26
0500	Metal, Bay/Bow		480	.017			.70		.70	1.05
0510	Power wash, based on 2500 lb. operating pressure									
0520	1-2 lite	A-1H	4400	.002	S.F.		.07	.02	.09	.13
0530	3-6 lite		4320	.002			.07	.02	.09	.13
0540	7-10 lite		4240	.002			.07	.02	.09	.13
0550	12 lite		4160	.002			.08	.02	.10	.14
0560	Bay/Bow		4400	.002			.07	.02	.09	.13
0600	Siding, scrape and sand, light=10-30%, med.=30-70%									
0610	Heavy=70-100% of surface to sand									
0650	Texture 1-11, light	1 Pord	480	.017	S.F.		.70		.70	1.05
0660	Med.		440	.018			.76		.76	1.14
0670	Heavy		360	.022			.93		.93	1.40

09 91 Painting

09 91 03 – Paint Restoration

09 91 03.30 Exterior Surface Preparation

		Crew	Daily Output	Labor-Hours	Unit	Material	2017 Bare Costs Labor	2017 Bare Costs Equipment	Total	Total Incl O&P
0680	Wood shingles, shakes, light	1 Pord	440	.018	S.F.		.76		.76	1.14
0690	Med.		360	.022			.93		.93	1.40
0700	Heavy		280	.029			1.19		1.19	1.80
0710	Clapboard, light		520	.015			.64		.64	.97
0720	Med.		480	.017			.70		.70	-1.05
0730	Heavy	▼	400	.020	▼		.83		.83	1.26
0740	Wire brush									
0750	Aluminum, light	1 Pord	600	.013	S.F.		.56		.56	.84
0760	Med.		520	.015			.64		.64	.97
0770	Heavy	▼	440	.018	▼		.76		.76	1.14
0780	Pressure wash, based on 2500 lb. operating pressure									
0790	Stucco	A-1H	3080	.003	S.F.		.10	.02	.12	.19
0800	Aluminum or vinyl		3200	.003			.10	.02	.12	.18
0810	Siding, masonry, brick & block	▼	2400	.003	▼		.13	.03	.16	.23
1300	Miscellaneous, wire brush									
1310	Metal, pedestrian gate	1 Pord	100	.080	S.F.		3.34		3.34	5.05
1320	Aluminum chain link, both sides		250	.032			1.33		1.33	2.01
1400	Existing galvanized surface, clean and prime, prep for painting	▼	380	.021	▼	.13	.88		1.01	1.47
8000	For chemical washing, see Section 04 01 30									
8010	For steam cleaning, see Section 04 01 30.20									
8020	For sand blasting, see Sections 03 35 29.60 and 05 01 10.51									

09 91 03.40 Interior Surface Preparation

		Crew	Daily Output	Labor-Hours	Unit	Material	2017 Bare Costs Labor	2017 Bare Costs Equipment	Total	Total Incl O&P
0010	**INTERIOR SURFACE PREPARATION** R099100-10									
0020	Doors, per side, not incl. frames or trim									
0030	Scrape & sand									
0040	Wood, flush	1 Pord	616	.013	S.F.		.54		.54	.82
0050	Wood, detail		496	.016			.67		.67	1.02
0060	Wood, louvered	▼	280	.029	▼		1.19		1.19	1.80
0070	Wire brush									
0080	Metal, flush	1 Pord	640	.013	S.F.		.52		.52	.79
0090	Metal, detail		520	.015			.64		.64	.97
0100	Metal, louvered	▼	360	.022			.93		.93	1.40
0110	Hand wash									
0120	Wood, flush	1 Pord	2160	.004	S.F.		.15		.15	.23
0130	Wood, detail		2000	.004			.17		.17	.25
0140	Wood, louvered		1360	.006			.25		.25	.37
0150	Metal, flush		2160	.004			.15		.15	.23
0160	Metal, detail		2000	.004			.17		.17	.25
0170	Metal, louvered	▼	1360	.006	▼		.25		.25	.37
0400	Windows, per side, not incl. trim									
0410	Scrape & sand									
0420	Wood, 1-2 lite	1 Pord	360	.022	S.F.		.93		.93	1.40
0430	Wood, 3-6 lite		320	.025			1.04		1.04	1.57
0440	Wood, 7-10 lite		280	.029			1.19		1.19	1.80
0450	Wood, 12 lite		240	.033			1.39		1.39	2.10
0460	Wood, Bay/Bow	▼	360	.022			.93		.93	1.40
0470	Wire brush									
0480	Metal, 1-2 lite	1 Pord	520	.015	S.F.		.64		.64	.97
0490	Metal, 3-6 lite		440	.018			.76		.76	1.14
0500	Metal, Bay/Bow		520	.015	▼		.64		.64	.97
0600	Walls, sanding, light=10-30%, medium=30-70%,									
0610	heavy=70-100% of surface to sand									

09 91 03 – Paint Restoration

09 91 03.40 Interior Surface Preparation

		Crew	Daily Output	Labor-Hours	Unit	Material	2017 Bare Costs Labor	Equipment	Total	Total Incl O&P
0650	Walls, sand									
0660	Gypsum board or plaster, light	1 Pord	3077	.003	S.F.		.11		.11	.16
0670	Gypsum board or plaster, medium		2160	.004			.15		.15	.23
0680	Gypsum board or plaster, heavy		923	.009			.36		.36	.55
0690	Wood, T&G, light		2400	.003			.14		.14	.21
0700	Wood, T&G, med.		1600	.005			.21		.21	.31
0710	Wood, T&G, heavy	↓	800	.010	↓		.42		.42	.63
0720	Walls, wash									
0730	Gypsum board or plaster	1 Pord	3200	.003	S.F.		.10		.10	.16
0740	Wood, T&G		3200	.003			.10		.10	.16
0750	Masonry, brick & block, smooth		2800	.003			.12		.12	.18
0760	Masonry, brick & block, coarse	↓	2000	.004	↓		.17		.17	.25
8000	For chemical washing, see Section 04 01 30									
8010	For steam cleaning, see Section 04 01 30.20									
8020	For sand blasting, see Sections 03 35 29.60 and 05 01 10.51									

09 91 13 – Exterior Painting

09 91 13.30 Fences

		Crew	Daily Output	Labor-Hours	Unit	Material	2017 Bare Costs Labor	Equipment	Total	Total Incl O&P
0010	**FENCES** R099100-20									
0100	Chain link or wire metal, one side, water base									
0110	Roll & brush, first coat	1 Pord	960	.008	S.F.	.07	.35		.42	.60
0120	Second coat		1280	.006		.07	.26		.33	.46
0130	Spray, first coat		2275	.004		.07	.15		.22	.30
0140	Second coat	↓	2600	.003	↓	.07	.13		.20	.27
0150	Picket, water base									
0160	Roll & brush, first coat	1 Pord	865	.009	S.F.	.07	.39		.46	.66
0170	Second coat		1050	.008		.07	.32		.39	.56
0180	Spray, first coat		2275	.004		.07	.15		.22	.30
0190	Second coat	↓	2600	.003		.07	.13		.20	.27
0200	Stockade, water base									
0210	Roll & brush, first coat	1 Pord	1040	.008	S.F.	.07	.32		.39	.56
0220	Second coat		1200	.007		.07	.28		.35	.50
0230	Spray, first coat		2275	.004		.07	.15		.22	.30
0240	Second coat	↓	2600	.003	↓	.07	.13		.20	.27

09 91 13.42 Miscellaneous, Exterior

		Crew	Daily Output	Labor-Hours	Unit	Material	2017 Bare Costs Labor	Equipment	Total	Total Incl O&P
0010	**MISCELLANEOUS, EXTERIOR** R099100-20									
0015	For painting metals, see Section 09 97 13.23									
0100	Railing, ext., decorative wood, incl. cap & baluster									
0110	Newels & spindles @ 12" O.C.									
0120	Brushwork, stain, sand, seal & varnish									
0130	First coat	1 Pord	90	.089	L.F.	.90	3.71		4.61	6.60
0140	Second coat	"	120	.067	"	.90	2.78		3.68	5.20
0150	Rough sawn wood, 42" high, 2" x 2" verticals, 6" O.C.									
0160	Brushwork, stain, each coat	1 Pord	90	.089	L.F.	.29	3.71		4	5.90
0170	Wrought iron, 1" rail, 1/2" sq. verticals									
0180	Brushwork, zinc chromate, 60" high, bars 6" O.C.									
0190	Primer	1 Pord	130	.062	L.F.	.88	2.57		3.45	4.84
0200	Finish coat		130	.062		1.13	2.57		3.70	5.10
0210	Additional coat	↓	190	.042	↓	1.32	1.76		3.08	4.10
0220	Shutters or blinds, single panel, 2' x 4', paint all sides									
0230	Brushwork, primer	1 Pord	20	.400	Ea.	.68	16.70		17.38	26
0240	Finish coat, exterior latex		20	.400		.56	16.70		17.26	25.50
0250	Primer & 1 coat, exterior latex	↓	13	.615		1.09	25.50		26.59	39.50

09 91 13 – Exterior Painting

09 91 13.42 Miscellaneous, Exterior

		Crew	Daily Output	Labor-Hours	Unit	Material	2017 Bare Costs Labor	Equipment	Total	Total Incl O&P
0260	Spray, primer	1 Pord	35	.229	Ea.	.99	9.55		10.54	15.50
0270	Finish coat, exterior latex		35	.229		1.20	9.55		10.75	15.70
0280	Primer & 1 coat, exterior latex	↓	20	.400	↓	1.06	16.70		17.76	26
0290	For louvered shutters, add				S.F.	10%				
0300	Stair stringers, exterior, metal									
0310	Roll & brush, zinc chromate, to 14", each coat	1 Pord	320	.025	L.F.	.38	1.04		1.42	1.98
0320	Rough sawn wood, 4" x 12"									
0330	Roll & brush, exterior latex, each coat	1 Pord	215	.037	L.F.	.08	1.55		1.63	2.43
0340	Trellis/lattice, 2" x 2" @ 3" O.C. with 2" x 8" supports									
0350	Spray, latex, per side, each coat	1 Pord	475	.017	S.F.	.08	.70		.78	1.15
0450	Decking, ext., sealer, alkyd, brushwork, sealer coat		1140	.007		.10	.29		.39	.55
0460	1st coat		1140	.007		.11	.29		.40	.56
0470	2nd coat		1300	.006		.08	.26		.34	.48
0500	Paint, alkyd, brushwork, primer coat		1140	.007		.11	.29		.40	.56
0510	1st coat		1140	.007		.14	.29		.43	.59
0520	2nd coat		1300	.006		.10	.26		.36	.50
0600	Sand paint, alkyd, brushwork, 1 coat	↓	150	.053	↓	.14	2.22		2.36	3.51

09 91 13.60 Siding Exterior

		Crew	Daily Output	Labor-Hours	Unit	Material	2017 Bare Costs Labor	Equipment	Total	Total Incl O&P
0010	**SIDING EXTERIOR**, Alkyd (oil base) R099100-10									
0450	Steel siding, oil base, paint 1 coat, brushwork	2 Pord	2015	.008	S.F.	.11	.33		.44	.62
0500	Spray R099100-20		4550	.004		.16	.15		.31	.40
0800	Paint 2 coats, brushwork		1300	.012		.22	.51		.73	1.01
1000	Spray		2750	.006		.18	.24		.42	.57
1200	Stucco, rough, oil base, paint 2 coats, brushwork		1300	.012		.22	.51		.73	1.01
1400	Roller		1625	.010		.23	.41		.64	.87
1600	Spray		2925	.005		.24	.23		.47	.61
1800	Texture 1-11 or clapboard, oil base, primer coat, brushwork		1300	.012		.15	.51		.66	.93
2000	Spray		4550	.004		.15	.15		.30	.38
2400	Paint 2 coats, brushwork		810	.020		.32	.82		1.14	1.59
2600	Spray		2600	.006		.35	.26		.61	.78
3400	Stain 2 coats, brushwork		950	.017		.19	.70		.89	1.27
4000	Spray		3050	.005		.21	.22		.43	.56
4200	Wood shingles, oil base primer coat, brushwork		1300	.012		.14	.51		.65	.92
4400	Spray		3900	.004		.13	.17		.30	.40
5000	Paint 2 coats, brushwork		810	.020		.27	.82		1.09	1.53
5200	Spray		2275	.007		.25	.29		.54	.72
6500	Stain 2 coats, brushwork		950	.017		.19	.70		.89	1.27
7000	Spray	↓	2660	.006		.26	.25		.51	.67
8000	For latex paint, deduct					10%				
8100	For work over 12' H, from pipe scaffolding, add						15%			
8200	For work over 12' H, from extension ladder, add						25%			
8300	For work over 12' H, from swing staging, add				↓		35%			

09 91 13.62 Siding, Misc.

		Crew	Daily Output	Labor-Hours	Unit	Material	2017 Bare Costs Labor	Equipment	Total	Total Incl O&P
0010	**SIDING, MISC.**, latex paint R099100-10									
0100	Aluminum siding									
0110	Brushwork, primer R099100-20	2 Pord	2275	.007	S.F.	.06	.29		.35	.51
0120	Finish coat, exterior latex		2275	.007		.06	.29		.35	.50
0130	Primer & 1 coat exterior latex		1300	.012		.13	.51		.64	.91
0140	Primer & 2 coats exterior latex	↓	975	.016	↓	.19	.68		.87	1.24
0150	Mineral fiber shingles									
0160	Brushwork, primer	2 Pord	1495	.011	S.F.	.15	.45		.60	.83
0170	Finish coat, industrial enamel	↓	1495	.011	↓	.18	.45		.63	.87

09 91 13 – Exterior Painting

09 91 13.62 Siding, Misc.

		Crew	Daily Output	Labor-Hours	Unit	Material	2017 Bare Costs Labor	Equipment	Total	Total Incl O&P
0180	Primer & 1 coat enamel	2 Pord	810	.020	S.F.	.33	.82		1.15	1.60
0190	Primer & 2 coats enamel		540	.030		.50	1.24		1.74	2.42
0200	Roll, primer		1625	.010		.16	.41		.57	.80
0210	Finish coat, industrial enamel		1625	.010		.19	.41		.60	.83
0220	Primer & 1 coat enamel		975	.016		.36	.68		1.04	1.42
0230	Primer & 2 coats enamel		650	.025		.55	1.03		1.58	2.15
0240	Spray, primer		3900	.004		.13	.17		.30	.40
0250	Finish coat, industrial enamel		3900	.004		.16	.17		.33	.44
0260	Primer & 1 coat enamel		2275	.007		.29	.29		.58	.76
0270	Primer & 2 coats enamel		1625	.010		.45	.41		.86	1.11
0280	Waterproof sealer, first coat		4485	.004		.10	.15		.25	.33
0290	Second coat	↓	5235	.003	↓	.09	.13		.22	.29
0300	Rough wood incl. shingles, shakes or rough sawn siding									
0310	Brushwork, primer	2 Pord	1280	.013	S.F.	.14	.52		.66	.94
0320	Finish coat, exterior latex		1280	.013		.10	.52		.62	.90
0330	Primer & 1 coat exterior latex		960	.017		.24	.70		.94	1.31
0340	Primer & 2 coats exterior latex		700	.023		.34	.95		1.29	1.82
0350	Roll, primer		2925	.005		.18	.23		.41	.54
0360	Finish coat, exterior latex		2925	.005		.12	.23		.35	.48
0370	Primer & 1 coat exterior latex		1790	.009		.31	.37		.68	.90
0380	Primer & 2 coats exterior latex		1300	.012		.43	.51		.94	1.24
0390	Spray, primer		3900	.004		.15	.17		.32	.43
0400	Finish coat, exterior latex		3900	.004		.10	.17		.27	.37
0410	Primer & 1 coat exterior latex		2600	.006		.25	.26		.51	.67
0420	Primer & 2 coats exterior latex		2080	.008		.35	.32		.67	.86
0430	Waterproof sealer, first coat		4485	.004		.18	.15		.33	.41
0440	Second coat	↓	4485	.004	↓	.10	.15		.25	.33
0450	Smooth wood incl. butt, T&G, beveled, drop or B&B siding									
0460	Brushwork, primer	2 Pord	2325	.007	S.F.	.10	.29		.39	.54
0470	Finish coat, exterior latex		1280	.013		.10	.52		.62	.90
0480	Primer & 1 coat exterior latex		800	.020		.20	.83		1.03	1.48
0490	Primer & 2 coats exterior latex		630	.025		.30	1.06		1.36	1.94
0500	Roll, primer		2275	.007		.11	.29		.40	.56
0510	Finish coat, exterior latex		2275	.007		.11	.29		.40	.56
0520	Primer & 1 coat exterior latex		1300	.012		.22	.51		.73	1.01
0530	Primer & 2 coats exterior latex		975	.016		.33	.68		1.01	1.40
0540	Spray, primer		4550	.004		.09	.15		.24	.31
0550	Finish coat, exterior latex		4550	.004		.10	.15		.25	.33
0560	Primer & 1 coat exterior latex		2600	.006		.18	.26		.44	.59
0570	Primer & 2 coats exterior latex		1950	.008		.28	.34		.62	.82
0580	Waterproof sealer, first coat		5230	.003		.10	.13		.23	.30
0590	Second coat	↓	5980	.003	↓	.10	.11		.21	.28
0600	For oil base paint, add					10%				

09 91 13.70 Doors and Windows, Exterior

		Crew	Daily Output	Labor-Hours	Unit	Material	2017 Bare Costs Labor	Equipment	Total	Total Incl O&P
0010	**DOORS AND WINDOWS, EXTERIOR** R099100-10									
0100	Door frames & trim, only									
0110	Brushwork, primer R099100-20	1 Pord	512	.016	L.F.	.06	.65		.71	1.05
0120	Finish coat, exterior latex		512	.016		.07	.65		.72	1.06
0130	Primer & 1 coat, exterior latex		300	.027		.13	1.11		1.24	1.83
0140	Primer & 2 coats, exterior latex	↓	265	.030	↓	.20	1.26		1.46	2.12
0150	Doors, flush, both sides, incl. frame & trim									
0160	Roll & brush, primer	1 Pord	10	.800	Ea.	4.70	33.50		38.20	55.50

09 91 Painting

09 91 13 – Exterior Painting

09 91 13.70 Doors and Windows, Exterior		Crew	Daily Output	Labor-Hours	Unit	Material	2017 Bare Costs Labor	Equipment	Total	Total Incl O&P
0170	Finish coat, exterior latex	1 Pord	10	.800	Ea.	5.35	33.50		38.85	56.50
0180	Primer & 1 coat, exterior latex		7	1.143		10.05	47.50		57.55	83
0190	Primer & 2 coats, exterior latex		5	1.600		15.40	66.50		81.90	118
0200	Brushwork, stain, sealer & 2 coats polyurethane		4	2		32	83.50		115.50	162
0210	Doors, French, both sides, 10-15 lite, incl. frame & trim									
0220	Brushwork, primer	1 Pord	6	1.333	Ea.	2.35	55.50		57.85	86.50
0230	Finish coat, exterior latex		6	1.333		2.68	55.50		58.18	87
0240	Primer & 1 coat, exterior latex		3	2.667		5.05	111		116.05	174
0250	Primer & 2 coats, exterior latex		2	4		7.55	167		174.55	260
0260	Brushwork, stain, sealer & 2 coats polyurethane		2.50	3.200		11.40	133		144.40	214
0270	Doors, louvered, both sides, incl. frame & trim									
0280	Brushwork, primer	1 Pord	7	1.143	Ea.	4.70	47.50		52.20	77
0290	Finish coat, exterior latex		7	1.143		5.35	47.50		52.85	78
0300	Primer & 1 coat, exterior latex		4	2		10.05	83.50		93.55	137
0310	Primer & 2 coats, exterior latex		3	2.667		15.10	111		126.10	185
0320	Brushwork, stain, sealer & 2 coats polyurethane		4.50	1.778		32	74		106	148
0330	Doors, panel, both sides, incl. frame & trim									
0340	Roll & brush, primer	1 Pord	6	1.333	Ea.	4.70	55.50		60.20	89
0350	Finish coat, exterior latex		6	1.333		5.35	55.50		60.85	90
0360	Primer & 1 coat, exterior latex		3	2.667		10.05	111		121.05	179
0370	Primer & 2 coats, exterior latex		2.50	3.200		15.10	133		148.10	218
0380	Brushwork, stain, sealer & 2 coats polyurethane		3	2.667		32	111		143	204
0400	Windows, per ext. side, based on 15 S.F.									
0410	1 to 6 lite									
0420	Brushwork, primer	1 Pord	13	.615	Ea.	.93	25.50		26.43	39.50
0430	Finish coat, exterior latex		13	.615		1.06	25.50		26.56	39.50
0440	Primer & 1 coat, exterior latex		8	1		1.98	41.50		43.48	65
0450	Primer & 2 coats, exterior latex		6	1.333		2.98	55.50		58.48	87.50
0460	Stain, sealer & 1 coat varnish		7	1.143		4.50	47.50		52	77
0470	7 to 10 lite									
0480	Brushwork, primer	1 Pord	11	.727	Ea.	.93	30.50		31.43	47
0490	Finish coat, exterior latex		11	.727		1.06	30.50		31.56	47
0500	Primer & 1 coat, exterior latex		7	1.143		1.98	47.50		49.48	74
0510	Primer & 2 coats, exterior latex		5	1.600		2.98	66.50		69.48	104
0520	Stain, sealer & 1 coat varnish		6	1.333		4.50	55.50		60	89
0530	12 lite									
0540	Brushwork, primer	1 Pord	10	.800	Ea.	.93	33.50		34.43	51.50
0550	Finish coat, exterior latex		10	.800		1.06	33.50		34.56	51.50
0560	Primer & 1 coat, exterior latex		6	1.333		1.98	55.50		57.48	86
0570	Primer & 2 coats, exterior latex		5	1.600		2.98	66.50		69.48	104
0580	Stain, sealer & 1 coat varnish		6	1.333		4.41	55.50		59.91	89
0590	For oil base paint, add					10%				

09 91 13.80 Trim, Exterior

		Crew	Daily Output	Labor-Hours	Unit	Material	2017 Bare Costs Labor	Equipment	Total	Total Incl O&P
0010	**TRIM, EXTERIOR** R099100-10									
0100	Door frames & trim (see Doors, interior or exterior)									
0110	Fascia, latex paint, one coat coverage R099100-20									
0120	1" x 4", brushwork	1 Pord	640	.013	L.F.	.02	.52		.54	.81
0130	Roll		1280	.006		.02	.26		.28	.42
0140	Spray		2080	.004		.02	.16		.18	.26
0150	1" x 6" to 1" x 10", brushwork		640	.013		.07	.52		.59	.87
0160	Roll		1230	.007		.08	.27		.35	.50
0170	Spray		2100	.004		.06	.16		.22	.31

09 91 13 – Exterior Painting

09 91 13.80 Trim, Exterior

		Crew	Daily Output	Labor-Hours	Unit	Material	2017 Bare Costs Labor	Equipment	Total	Total Incl O&P
0180	1" x 12", brushwork	1 Pord	640	.013	L.F.	.07	.52		.59	.87
0190	Roll		1050	.008		.08	.32		.40	.57
0200	Spray		2200	.004		.06	.15		.21	.30
0210	Gutters & downspouts, metal, zinc chromate paint									
0220	Brushwork, gutters, 5", first coat	1 Pord	640	.013	L.F.	.40	.52		.92	1.23
0230	Second coat		960	.008		.38	.35		.73	.93
0240	Third coat		1280	.006		.30	.26		.56	.73
0250	Downspouts, 4", first coat		640	.013		.40	.52		.92	1.23
0260	Second coat		960	.008		.38	.35		.73	.93
0270	Third coat		1280	.006		.30	.26		.56	.73
0280	Gutters & downspouts, wood									
0290	Brushwork, gutters, 5", primer	1 Pord	640	.013	L.F.	.06	.52		.58	.86
0300	Finish coat, exterior latex		640	.013		.06	.52		.58	.86
0310	Primer & 1 coat exterior latex		400	.020		.13	.83		.96	1.41
0320	Primer & 2 coats exterior latex		325	.025		.20	1.03		1.23	1.77
0330	Downspouts, 4", primer		640	.013		.06	.52		.58	.86
0340	Finish coat, exterior latex		640	.013		.06	.52		.58	.86
0350	Primer & 1 coat exterior latex		400	.020		.13	.83		.96	1.41
0360	Primer & 2 coats exterior latex		325	.025		.10	1.03		1.13	1.66
0370	Molding, exterior, up to 14" wide									
0380	Brushwork, primer	1 Pord	640	.013	L.F.	.07	.52		.59	.87
0390	Finish coat, exterior latex		640	.013		.08	.52		.60	.87
0400	Primer & 1 coat exterior latex		400	.020		.16	.83		.99	1.43
0410	Primer & 2 coats exterior latex		315	.025		.16	1.06		1.22	1.77
0420	Stain & fill		1050	.008		.11	.32		.43	.61
0430	Shellac		1850	.004		.14	.18		.32	.43
0440	Varnish		1275	.006		.10	.26		.36	.50

09 91 13.90 Walls, Masonry (CMU), Exterior

		Crew	Daily Output	Labor-Hours	Unit	Material	2017 Bare Costs Labor	Equipment	Total	Total Incl O&P
0010	**WALLS, MASONRY (CMU), EXTERIOR** R099100-10									
0360	Concrete masonry units (CMU), smooth surface									
0370	Brushwork, latex, first coat	1 Pord	640	.013	S.F.	.07	.52		.59	.87
0380	Second coat		960	.008		.06	.35		.41	.58
0390	Waterproof sealer, first coat		736	.011		.27	.45		.72	.97
0400	Second coat		1104	.007		.27	.30		.57	.75
0410	Roll, latex, paint, first coat		1465	.005		.09	.23		.32	.43
0420	Second coat		1790	.004		.06	.19		.25	.35
0430	Waterproof sealer, first coat		1680	.005		.27	.20		.47	.59
0440	Second coat		2060	.004		.27	.16		.43	.53
0450	Spray, latex, paint, first coat		1950	.004		.07	.17		.24	.33
0460	Second coat		2600	.003		.05	.13		.18	.25
0470	Waterproof sealer, first coat		2245	.004		.27	.15		.42	.51
0480	Second coat		2990	.003		.27	.11		.38	.46
0490	Concrete masonry unit (CMU), porous									
0500	Brushwork, latex, first coat	1 Pord	640	.013	S.F.	.14	.52		.66	.95
0510	Second coat		960	.008		.07	.35		.42	.60
0520	Waterproof sealer, first coat		736	.011		.27	.45		.72	.97
0530	Second coat		1104	.007		.27	.30		.57	.75
0540	Roll latex, first coat		1465	.005		.11	.23		.34	.46
0550	Second coat		1790	.004		.07	.19		.26	.36
0560	Waterproof sealer, first coat		1680	.005		.27	.20		.47	.59
0570	Second coat		2060	.004		.27	.16		.43	.53
0580	Spray latex, first coat		1950	.004		.08	.17		.25	.35

09 91 Painting

09 91 13 – Exterior Painting

09 91 13.90 Walls, Masonry (CMU), Exterior

		Crew	Daily Output	Labor-Hours	Unit	Material	2017 Bare Costs Labor	Equipment	Total	Total Incl O&P
0590	Second coat	1 Pord	2600	.003	S.F.	.05	.13		.18	.25
0600	Waterproof sealer, first coat		2245	.004		.27	.15		.42	.51
0610	Second coat	▼	2990	.003	▼	.27	.11		.38	.46

09 91 23 – Interior Painting

09 91 23.20 Cabinets and Casework

		Crew	Daily Output	Labor-Hours	Unit	Material	2017 Bare Costs Labor	Equipment	Total	Total Incl O&P
0010	**CABINETS AND CASEWORK** R099100-10									
1000	Primer coat, oil base, brushwork	1 Pord	650	.012	S.F.	.06	.51		.57	.84
2000	Paint, oil base, brushwork, 1 coat R099100-20		650	.012		.11	.51		.62	.89
3000	Stain, brushwork, wipe off		650	.012		.10	.51		.61	.87
4000	Shellac, 1 coat, brushwork		650	.012		.12	.51		.63	.90
4500	Varnish, 3 coats, brushwork, sand after 1st coat	▼	325	.025		.25	1.03		1.28	1.82
5000	For latex paint, deduct				▼	10%				
6300	Strip, prep and refinish wood furniture									
6310	Remove paint using chemicals, wood furniture	1 Pord	28	.286	S.F.	1.60	11.90		13.50	19.75
6320	Prep for painting, sanding		75	.107		.24	4.45		4.69	6.95
6350	Stain and wipe, brushwork		600	.013		.10	.56		.66	.94
6355	Spray applied		900	.009		.08	.37		.45	.65
6360	Sealer or varnish, brushwork		1080	.007		.08	.31		.39	.56
6365	Spray applied		2100	.004		.08	.16		.24	.33
6370	Paint, primer, brushwork		720	.011		.06	.46		.52	.77
6375	Spray applied		2100	.004		.06	.16		.22	.31
6380	Finish coat, brushwork		810	.010		.11	.41		.52	.74
6385	Spray applied	▼	2100	.004	▼	.11	.16		.27	.36

09 91 23.33 Doors and Windows, Interior Alkyd (Oil Base)

		Crew	Daily Output	Labor-Hours	Unit	Material	2017 Bare Costs Labor	Equipment	Total	Total Incl O&P
0010	**DOORS AND WINDOWS, INTERIOR ALKYD (OIL BASE)** R099100-10									
0500	Flush door & frame, 3' x 7', oil, primer, brushwork	1 Pord	10	.800	Ea.	3.74	33.50		37.24	54.50
1000	Paint, 1 coat R099100-20		10	.800		3.91	33.50		37.41	55
1400	Stain, brushwork, wipe off		18	.444		2	18.55		20.55	30
1600	Shellac, 1 coat, brushwork		25	.320		2.52	13.35		15.87	23
1800	Varnish, 3 coats, brushwork, sand after 1st coat		9	.889		5.15	37		42.15	61.50
2000	Panel door & frame, 3' x 7', oil, primer, brushwork		6	1.333		2.45	55.50		57.95	86.50
2200	Paint, 1 coat		6	1.333		3.91	55.50		59.41	88.50
2600	Stain, brushwork, panel door, 3' x 7', not incl. frame		16	.500		2	21		23	33.50
2800	Shellac, 1 coat, brushwork		22	.364		2.52	15.15		17.67	26
3000	Varnish, 3 coats, brushwork, sand after 1st coat	▼	7.50	1.067	▼	5.15	44.50		49.65	72.50
4400	Windows, including frame and trim, per side									
4600	Colonial type, 6/6 lites, 2' x 3', oil, primer, brushwork	1 Pord	14	.571	Ea.	.39	24		24.39	36.50
5800	Paint, 1 coat		14	.571		.62	24		24.62	36.50
6200	3' x 5' opening, 6/6 lites, primer coat, brushwork		12	.667		.97	28		28.97	43
6400	Paint, 1 coat		12	.667		1.55	28		29.55	43.50
6800	4' x 8' opening, 6/6 lites, primer coat, brushwork		8	1		2.06	41.50		43.56	65.50
7000	Paint, 1 coat		8	1		3.30	41.50		44.80	66.50
8000	Single lite type, 2' x 3', oil base, primer coat, brushwork		33	.242		.39	10.10		10.49	15.70
8200	Paint, 1 coat		33	.242		.62	10.10		10.72	15.95
8600	3' x 5' opening, primer coat, brushwork		20	.400		.97	16.70		17.67	26
8800	Paint, 1 coat		20	.400		1.55	16.70		18.25	26.50
9200	4' x 8' opening, primer coat, brushwork		14	.571		2.06	24		26.06	38.50
9400	Paint, 1 coat	▼	14	.571	▼	3.30	24		27.30	39.50

For customer support on your Building Construction Costs with RSMeans Data, call 800.448.8182.

361

09 91 23.35 Doors and Windows, Interior Latex		Crew	Daily Output	Labor-Hours	Unit	Material	2017 Bare Costs Labor	Equipment	Total	Total Incl O&P
0010	**DOORS & WINDOWS, INTERIOR LATEX**	R099100-10								
0100	Doors, flush, both sides, incl. frame & trim									
0110	Roll & brush, primer	R099100-20	1 Pord	10	.800	Ea.	4.25	33.50		37.75	55
0120	Finish coat, latex			10	.800		5.45	33.50		38.95	56.50
0130	Primer & 1 coat latex			7	1.143		9.70	47.50		57.20	82.50
0140	Primer & 2 coats latex			5	1.600		14.80	66.50		81.30	117
0160	Spray, both sides, primer			20	.400		4.47	16.70		21.17	30
0170	Finish coat, latex			20	.400		5.70	16.70		22.40	31.50
0180	Primer & 1 coat latex			11	.727		10.25	30.50		40.75	57.50
0190	Primer & 2 coats latex			8	1		15.70	41.50		57.20	80.50
0200	Doors, French, both sides, 10-15 lite, incl. frame & trim									
0210	Roll & brush, primer		1 Pord	6	1.333	Ea.	2.12	55.50		57.62	86.50
0220	Finish coat, latex			6	1.333		2.72	55.50		58.22	87
0230	Primer & 1 coat latex			3	2.667		4.84	111		115.84	173
0240	Primer & 2 coats latex			2	4		7.40	167		174.40	260
0260	Doors, louvered, both sides, incl. frame & trim									
0270	Roll & brush, primer		1 Pord	7	1.143	Ea.	4.25	47.50		51.75	76.50
0280	Finish coat, latex			7	1.143		5.45	47.50		52.95	78
0290	Primer & 1 coat, latex			4	2		9.45	83.50		92.95	136
0300	Primer & 2 coats, latex			3	2.667		15.15	111		126.15	185
0320	Spray, both sides, primer			20	.400		4.47	16.70		21.17	30
0330	Finish coat, latex			20	.400		5.70	16.70		22.40	31.50
0340	Primer & 1 coat, latex			11	.727		10.25	30.50		40.75	57.50
0350	Primer & 2 coats, latex			8	1		16.05	41.50		57.55	80.50
0360	Doors, panel, both sides, incl. frame & trim									
0370	Roll & brush, primer		1 Pord	6	1.333	Ea.	4.47	55.50		59.97	89
0380	Finish coat, latex			6	1.333		5.45	55.50		60.95	90
0390	Primer & 1 coat, latex			3	2.667		9.70	111		120.70	179
0400	Primer & 2 coats, latex			2.50	3.200		15.15	133		148.15	218
0420	Spray, both sides, primer			10	.800		4.47	33.50		37.97	55.50
0430	Finish coat, latex			10	.800		5.70	33.50		39.20	57
0440	Primer & 1 coat, latex			5	1.600		10.25	66.50		76.75	112
0450	Primer & 2 coats, latex			4	2		16.05	83.50		99.55	144
0460	Windows, per interior side, based on 15 S.F.									
0470	1 to 6 lite									
0480	Brushwork, primer		1 Pord	13	.615	Ea.	.84	25.50		26.34	39.50
0490	Finish coat, enamel			13	.615		1.07	25.50		26.57	39.50
0500	Primer & 1 coat enamel			8	1		1.91	41.50		43.41	65
0510	Primer & 2 coats enamel			6	1.333		2.99	55.50		58.49	87.50
0530	7 to 10 lite									
0540	Brushwork, primer		1 Pord	11	.727	Ea.	.84	30.50		31.34	47
0550	Finish coat, enamel			11	.727		1.07	30.50		31.57	47
0560	Primer & 1 coat enamel			7	1.143		1.91	47.50		49.41	74
0570	Primer & 2 coats enamel			5	1.600		2.99	66.50		69.49	104
0590	12 lite									
0600	Brushwork, primer		1 Pord	10	.800	Ea.	.84	33.50		34.34	51.50
0610	Finish coat, enamel			10	.800		1.07	33.50		34.57	51.50
0620	Primer & 1 coat enamel			6	1.333		1.91	55.50		57.41	86
0630	Primer & 2 coats enamel			5	1.600		2.99	66.50		69.49	104
0650	For oil base paint, add						10%				

09 91 23.39 Doors and Windows, Interior Latex, Zero Voc

		Crew	Daily Output	Labor-Hours	Unit	Material	2017 Bare Costs Labor	Equipment	Total	Total Incl O&P
0010	**DOORS & WINDOWS, INTERIOR LATEX, ZERO VOC**									
0100	Doors flush, both sides, incl. frame & trim									
0110	Roll & brush, primer	G	1 Pord	10	.800	Ea.	5.05	33.50		38.55	56
0120	Finish coat, latex	G		10	.800		5.75	33.50		39.25	57
0130	Primer & 1 coat latex	G		7	1.143		10.80	47.50		58.30	84
0140	Primer & 2 coats latex	G		5	1.600		16.20	66.50		82.70	119
0160	Spray, both sides, primer	G		20	.400		5.30	16.70		22	31
0170	Finish coat, latex	G		20	.400		6.05	16.70		22.75	31.50
0180	Primer & 1 coat latex	G		11	.727		11.40	30.50		41.90	58.50
0190	Primer & 2 coats latex	G	↓	8	1	↓	17.15	41.50		58.65	82
0200	Doors, French, both sides, 10-15 lite, incl. frame & trim									
0210	Roll & brush, primer	G	1 Pord	6	1.333	Ea.	2.52	55.50		58.02	87
0220	Finish coat, latex	G		6	1.333		2.88	55.50		58.38	87
0230	Primer & 1 coat latex	G		3	2.667		5.40	111		116.40	174
0240	Primer & 2 coats latex	G	↓	2	4	↓	8.10	167		175.10	261
0360	Doors, panel, both sides, incl. frame & trim									
0370	Roll & brush, primer	G	1 Pord	6	1.333	Ea.	5.30	55.50		60.80	90
0380	Finish coat, latex	G		6	1.333		5.75	55.50		61.25	90.50
0390	Primer & 1 coat, latex	G		3	2.667		10.80	111		121.80	180
0400	Primer & 2 coats, latex	G		2.50	3.200		16.55	133		149.55	219
0420	Spray, both sides, primer	G		10	.800		5.30	33.50		38.80	56.50
0430	Finish coat, latex	G		10	.800		6.05	33.50		39.55	57
0440	Primer & 1 coat, latex	G		5	1.600		11.40	66.50		77.90	114
0450	Primer & 2 coats, latex	G	↓	4	2	↓	17.55	83.50		101.05	145
0460	Windows, per interior side, based on 15 S.F.									
0470	1 to 6 lite									
0480	Brushwork, primer	G	1 Pord	13	.615	Ea.	.99	25.50		26.49	39.50
0490	Finish coat, enamel	G		13	.615		1.14	25.50		26.64	40
0500	Primer & 1 coat enamel	G		8	1		2.13	41.50		43.63	65.50
0510	Primer & 2 coats enamel	G	↓	6	1.333	↓	3.27	55.50		58.77	87.50

09 91 23.40 Floors, Interior

		Crew	Daily Output	Labor-Hours	Unit	Material	2017 Bare Costs Labor	Equipment	Total	Total Incl O&P
0010	**FLOORS, INTERIOR** R099100-10									
0100	Concrete paint, latex									
0110	Brushwork									
0120	1st coat	1 Pord	975	.008	S.F.	.14	.34		.48	.67
0130	2nd coat		1150	.007		.09	.29		.38	.54
0140	3rd coat	↓	1300	.006	↓	.07	.26		.33	.47
0150	Roll									
0160	1st coat	1 Pord	2600	.003	S.F.	.19	.13		.32	.40
0170	2nd coat		3250	.002		.11	.10		.21	.27
0180	3rd coat	↓	3900	.002	↓	.08	.09		.17	.22
0190	Spray									
0200	1st coat	1 Pord	2600	.003	S.F.	.16	.13		.29	.37
0210	2nd coat		3250	.002		.09	.10		.19	.25
0220	3rd coat	↓	3900	.002	↓	.07	.09		.16	.21

09 91 23.44 Anti-Slip Floor Treatments

		Crew	Daily Output	Labor-Hours	Unit	Material	2017 Bare Costs Labor	Equipment	Total	Total Incl O&P
0010	**ANTI-SLIP FLOOR TREATMENTS**									
1000	Walking surface treatment, ADA compliant, mop on and rinse									
1100	For tile, terrazzo, stone or smooth concrete	1 Pord	4000	.002	S.F.	.33	.08		.41	.50
1110	For marble		4000	.002		.23	.08		.31	.38
1120	For wood		4000	.002		.21	.08		.29	.36
1130	For baths and showers	↓	500	.016	↓	.38	.67		1.05	1.43

For customer support on your Building Construction Costs with RSMeans Data, call 800.448.8182.

363

09 91 23.44 Anti-Slip Floor Treatments

		Crew	Daily Output	Labor-Hours	Unit	Material	2017 Bare Costs Labor	Equipment	Total	Total Incl O&P
2000	Granular additive for paint or sealer, add to paint cost				S.F.	.02			.02	.02

09 91 23.52 Miscellaneous, Interior

		Crew	Daily Output	Labor-Hours	Unit	Material	2017 Bare Costs Labor	Equipment	Total	Total Incl O&P
0010	**MISCELLANEOUS, INTERIOR** R099100-10									
2400	Floors, conc./wood, oil base, primer/sealer coat, brushwork	2 Pord	1950	.008	S.F.	.09	.34		.43	.62
2450	Roller		5200	.003		.10	.13		.23	.30
2600	Spray		6000	.003		.10	.11		.21	.27
2650	Paint 1 coat, brushwork		1950	.008		.10	.34		.44	.63
2800	Roller		5200	.003		.11	.13		.24	.31
2850	Spray		6000	.003		.11	.11		.22	.29
3000	Stain, wood floor, brushwork, 1 coat		4550	.004		.10	.15		.25	.32
3200	Roller		5200	.003		.10	.13		.23	.30
3250	Spray		6000	.003		.10	.11		.21	.28
3400	Varnish, wood floor, brushwork		4550	.004		.08	.15		.23	.31
3450	Roller		5200	.003		.09	.13		.22	.29
3600	Spray	▼	6000	.003	▼	.09	.11		.20	.27
3650	For anti skid, see Section 09 91 23.44									
3800	Grilles, per side, oil base, primer coat, brushwork	1 Pord	520	.015	S.F.	.13	.64		.77	1.11
3850	Spray		1140	.007		.13	.29		.42	.59
3920	Paint 2 coats, brushwork		325	.025		.40	1.03		1.43	1.99
3940	Spray	▼	650	.012	▼	.46	.51		.97	1.27
4600	Miscellaneous surfaces, metallic paint, spray applied									
4610	Water based, non-tintable, warm silver	1 Pord	1140	.007	S.F.	.78	.29		1.07	1.29
4620	Rusted iron		1140	.007		1	.29		1.29	1.54
4630	Low VOC, tintable	▼	1140	.007	▼	.35	.29		.64	.82
5000	Pipe, 1" - 4" diameter, primer or sealer coat, oil base, brushwork	2 Pord	1250	.013	L.F.	.10	.53		.63	.92
5100	Spray		2165	.007		.09	.31		.40	.57
5350	Paint 2 coats, brushwork		775	.021		.20	.86		1.06	1.52
5400	Spray		1240	.013		.22	.54		.76	1.05
6300	13" - 16" diameter, primer or sealer coat, brushwork		310	.052		.38	2.15		2.53	3.67
6350	Spray		540	.030		.43	1.24		1.67	2.34
6500	Paint 2 coats, brushwork		195	.082		.79	3.42		4.21	6
6550	Spray	▼	310	.052	▼	.87	2.15		3.02	4.21
7000	Trim, wood, incl. puttying, under 6" wide									
7200	Primer coat, oil base, brushwork	1 Pord	650	.012	L.F.	.03	.51		.54	.81
7250	Paint, 1 coat, brushwork		650	.012		.05	.51		.56	.83
7450	3 coats		325	.025		.15	1.03		1.18	1.71
7500	Over 6" wide, primer coat, brushwork		650	.012		.06	.51		.57	.84
7550	Paint, 1 coat, brushwork		650	.012		.10	.51		.61	.88
7650	3 coats		325	.025	▼	.30	1.03		1.33	1.88
8000	Cornice, simple design, primer coat, oil base, brushwork		650	.012	S.F.	.06	.51		.57	.84
8250	Paint, 1 coat		650	.012		.10	.51		.61	.88
8350	Ornate design, primer coat		350	.023		.06	.95		1.01	1.51
8400	Paint, 1 coat		350	.023		.10	.95		1.05	1.55
8600	Balustrades, primer coat, oil base, brushwork		520	.015		.06	.64		.70	1.04
8650	Paint, 1 coat		520	.015		.10	.64		.74	1.08
8900	Trusses and wood frames, primer coat, oil base, brushwork		800	.010		.06	.42		.48	.70
8950	Spray		1200	.007		.07	.28		.35	.49
9220	Paint 2 coats, brushwork		500	.016		.20	.67		.87	1.23
9240	Spray		600	.013		.22	.56		.78	1.08
9260	Stain, brushwork, wipe off		600	.013		.10	.56		.66	.94
9280	Varnish, 3 coats, brushwork	▼	275	.029		.25	1.21		1.46	2.10
9350	For latex paint, deduct				▼	10%				

364

For customer support on your Building Construction Costs with RSMeans Data, call 800.448.8182.

09 91 23 – Interior Painting

09 91 23.62 Electrostatic Painting

		Crew	Daily Output	Labor-Hours	Unit	Material	2017 Bare Costs Labor	2017 Bare Costs Equipment	Total	Total Incl O&P
0010	**ELECTROSTATIC PAINTING**									
0100	In shop									
0200	Flat surfaces (lockers, casework, elevator doors, etc.)									
0300	One coat	1 Pord	200	.040	S.F.	.61	1.67		2.28	3.19
0400	Two coats	"	120	.067	"	.89	2.78		3.67	5.20
0500	Irregular surfaces (furniture, door frames, etc.)									
0600	One coat	1 Pord	150	.053	S.F.	.61	2.22		2.83	4.03
0700	Two coats	"	100	.080	"	.89	3.34		4.23	6.05
0800	On site									
0900	Flat surfaces (lockers, casework, elevator doors, etc.)									
1000	One coat	1 Pord	150	.053	S.F.	.61	2.22		2.83	4.03
1100	Two coats	"	100	.080	"	.89	3.34		4.23	6.05
1200	Irregular surfaces (furniture, door frames, etc.)									
1300	One coat	1 Pord	115	.070	S.F.	.61	2.90		3.51	5.05
1400	Two coats	↓	70	.114		.89	4.77		5.66	8.20
2000	Anti-microbial coating, hospital application	↓	150	.053	↓	.05	2.22		2.27	3.42

09 91 23.72 Walls and Ceilings, Interior

		Crew	Daily Output	Labor-Hours	Unit	Material	2017 Bare Costs Labor	2017 Bare Costs Equipment	Total	Total Incl O&P
0010	**WALLS AND CEILINGS, INTERIOR** R099100-10									
0100	Concrete, drywall or plaster, latex, primer or sealer coat R099100-20									
0200	Smooth finish, brushwork	1 Pord	1150	.007	S.F.	.06	.29		.35	.51
0240	Roller		1350	.006		.06	.25		.31	.44
0280	Spray		2750	.003		.06	.12		.18	.24
0300	Sand finish, brushwork		975	.008		.06	.34		.40	.59
0340	Roller		1150	.007		.06	.29		.35	.51
0380	Spray		2275	.004		.06	.15		.21	.28
0800	Paint 2 coats, smooth finish, brushwork		680	.012		.14	.49		.63	.89
0840	Roller		800	.010		.14	.42		.56	.78
0880	Spray		1625	.005		.13	.21		.34	.45
0900	Sand finish, brushwork		605	.013		.14	.55		.69	.98
0940	Roller		1020	.008		.14	.33		.47	.64
0980	Spray		1700	.005		.13	.20		.33	.44
1200	Paint 3 coats, smooth finish, brushwork		510	.016		.21	.65		.86	1.22
1240	Roller		650	.012		.21	.51		.72	1
1280	Spray		850	.009		.19	.39		.58	.80
1600	Glaze coating, 2 coats, spray, clear		1200	.007		.49	.28		.77	.96
1640	Multicolor	↓	1200	.007	↓	1.05	.28		1.33	1.57
1660	Painting walls, complete, including surface prep, primer &									
1670	2 coats finish, on drywall or plaster, with roller	1 Pord	325	.025	S.F.	.21	1.03		1.24	1.78
1700	For oil base paint, add				↓	10%				
1800	For ceiling installations, add				↓		25%			
2000	Masonry or concrete block, primer/sealer, latex paint									
2100	Primer, smooth finish, brushwork	1 Pord	1000	.008	S.F.	.15	.33		.48	.66
2110	Roller		1150	.007		.11	.29		.40	.56
2180	Spray		2400	.003		.10	.14		.24	.32
2200	Sand finish, brushwork		850	.009		.11	.39		.50	.71
2210	Roller		975	.008		.11	.34		.45	.64
2280	Spray		2050	.004		.10	.16		.26	.36
2400	Finish coat, smooth finish, brush		1100	.007		.08	.30		.38	.55
2410	Roller		1300	.006		.08	.26		.34	.48
2480	Spray		2400	.003		.07	.14		.21	.29
2500	Sand finish, brushwork		950	.008		.08	.35		.43	.62
2510	Roller	↓	1090	.007	↓	.08	.31		.39	.55

09 91 23 – Interior Painting

09 91 23.72 Walls and Ceilings, Interior

		Crew	Daily Output	Labor-Hours	Unit	Material	2017 Bare Costs Labor	Equipment	Total	Total Incl O&P
2580	Spray	1 Pord	2040	.004	S.F.	.07	.16		.23	.33
2800	Primer plus one finish coat, smooth brush		525	.015		.30	.64		.94	1.29
2810	Roller		615	.013		.19	.54		.73	1.03
2880	Spray		1200	.007		.17	.28		.45	.61
2900	Sand finish, brushwork		450	.018		.19	.74		.93	1.33
2910	Roller		515	.016		.19	.65		.84	1.19
2980	Spray		1025	.008		.17	.33		.50	.68
3200	Primer plus 2 finish coats, smooth, brush		355	.023		.27	.94		1.21	1.72
3210	Roller		415	.019		.27	.80		1.07	1.51
3280	Spray		800	.010		.24	.42		.66	.89
3300	Sand finish, brushwork		305	.026		.27	1.09		1.36	1.95
3310	Roller		350	.023		.27	.95		1.22	1.74
3380	Spray		675	.012		.24	.49		.73	1.01
3600	Glaze coating, 3 coats, spray, clear		900	.009		.70	.37		1.07	1.33
3620	Multicolor		900	.009		1.23	.37		1.60	1.91
4000	Block filler, 1 coat, brushwork		425	.019		.12	.78		.90	1.32
4100	Silicone, water repellent, 2 coats, spray		2000	.004		.37	.17		.54	.66
4120	For oil base paint, add					10%				
8200	For work 8' - 15' H, add						10%			
8300	For work over 15' H, add						20%			
8400	For light textured surfaces, add						10%			
8410	Heavy textured, add						25%			

09 91 23.74 Walls and Ceilings, Interior, Zero VOC Latex

			Crew	Daily Output	Labor-Hours	Unit	Material	2017 Bare Costs Labor	Equipment	Total	Total Incl O&P
0010	**WALLS AND CEILINGS, INTERIOR, ZERO VOC LATEX**										
0100	Concrete, dry wall or plaster, latex, primer or sealer coat										
0200	Smooth finish, brushwork	G	1 Pord	1150	.007	S.F.	.07	.29		.36	.51
0240	Roller	G		1350	.006		.06	.25		.31	.44
0280	Spray	G		2750	.003		.05	.12		.17	.23
0300	Sand finish, brushwork	G		975	.008		.06	.34		.40	.59
0340	Roller	G		1150	.007		.07	.29		.36	.52
0380	Spray	G		2275	.004		.05	.15		.20	.28
0800	Paint 2 coats, smooth finish, brushwork	G		680	.012		.15	.49		.64	.90
0840	Roller	G		800	.010		.15	.42		.57	.80
0880	Spray	G		1625	.005		.13	.21		.34	.45
0900	Sand finish, brushwork	G		605	.013		.14	.55		.69	.99
0940	Roller	G		1020	.008		.15	.33		.48	.66
0980	Spray	G		1700	.005		.13	.20		.33	.44
1200	Paint 3 coats, smooth finish, brushwork	G		510	.016		.21	.65		.86	1.23
1240	Roller	G		650	.012		.23	.51		.74	1.02
1280	Spray	G		850	.009		.20	.39		.59	.80
1800	For ceiling installations, add	G						25%			
8200	For work 8' - 15' H, add							10%			
8300	For work over 15' H, add							20%			

09 91 23.75 Dry Fall Painting

			Crew	Daily Output	Labor-Hours	Unit	Material	2017 Bare Costs Labor	Equipment	Total	Total Incl O&P
0010	**DRY FALL PAINTING**	R099100-10									
0100	Sprayed on walls, gypsum board or plaster										
0220	One coat	R099100-20	1 Pord	2600	.003	S.F.	.08	.13		.21	.28
0250	Two coats			1560	.005		.16	.21		.37	.49
0280	Concrete or textured plaster, one coat			1560	.005		.08	.21		.29	.41
0310	Two coats			1300	.006		.16	.26		.42	.56
0340	Concrete block, one coat			1560	.005		.08	.21		.29	.41
0370	Two coats			1300	.006		.16	.26		.42	.56

09 91 Painting

09 91 23 – Interior Painting

09 91 23.75 Dry Fall Painting	Crew	Daily Output	Labor-Hours	Unit	Material	2017 Bare Costs Labor	Equipment	Total	Total Incl O&P
0400 Wood, one coat	1 Pord	877	.009	S.F.	.08	.38		.46	.66
0430 Two coats	↓	650	.012	↓	.16	.51		.67	.94
0440 On ceilings, gypsum board or plaster									
0470 One coat	1 Pord	1560	.005	S.F.	.08	.21		.29	.41
0500 Two coats		1300	.006		.16	.26		.42	.56
0530 Concrete or textured plaster, one coat		1560	.005		.08	.21		.29	.41
0560 Two coats		1300	.006		.16	.26		.42	.56
0570 Structural steel, bar joists or metal deck, one coat		1560	.005		.08	.21		.29	.41
0580 Two coats	↓	1040	.008	↓	.16	.32		.48	.65

09 93 Staining and Transparent Finishing

09 93 23 – Interior Staining and Finishing

09 93 23.10 Varnish

	Crew	Daily Output	Labor-Hours	Unit	Material	2017 Bare Costs Labor	Equipment	Total	Total Incl O&P
0010 **VARNISH**									
0012 1 coat + sealer, on wood trim, brush, no sanding included	1 Pord	400	.020	S.F.	.07	.83		.90	1.34
0020 1 coat + sealer, on wood trim, brush, no sanding included, no VOC		400	.020		.22	.83		1.05	1.50
0100 Hardwood floors, 2 coats, no sanding included, roller	↓	1890	.004	↓	.15	.18		.33	.44

09 96 High-Performance Coatings

09 96 23 – Graffiti-Resistant Coatings

09 96 23.10 Graffiti Resistant Treatments

	Crew	Daily Output	Labor-Hours	Unit	Material	2017 Bare Costs Labor	Equipment	Total	Total Incl O&P
0010 **GRAFFITI RESISTANT TREATMENTS**, sprayed on walls									
0100 Non-sacrificial, permanent non-stick coating, clear, on metals	1 Pord	2000	.004	S.F.	2.06	.17		2.23	2.52
0200 Concrete		2000	.004		2.35	.17		2.52	2.83
0300 Concrete block		2000	.004		3.03	.17		3.20	3.59
0400 Brick		2000	.004		3.44	.17		3.61	4.03
0500 Stone		2000	.004		3.44	.17		3.61	4.03
0600 Unpainted wood		2000	.004		3.97	.17		4.14	4.62
2000 Semi-permanent cross linking polymer primer, on metals		2000	.004		.65	.17		.82	.96
2100 Concrete		2000	.004		.78	.17		.95	1.11
2200 Concrete block		2000	.004		.97	.17		1.14	1.32
2300 Brick		2000	.004		.78	.17		.95	1.11
2400 Stone		2000	.004		.78	.17		.95	1.11
2500 Unpainted wood		2000	.004		1.08	.17		1.25	1.44
3000 Top coat, on metals		2000	.004		.55	.17		.72	.85
3100 Concrete		2000	.004		.62	.17		.79	.94
3200 Concrete block		2000	.004		.87	.17		1.04	1.21
3300 Brick		2000	.004		.73	.17		.90	1.05
3400 Stone		2000	.004		.73	.17		.90	1.05
3500 Unpainted wood		2000	.004		.87	.17		1.04	1.21
5000 Sacrificial, water based, on metal		2000	.004		.32	.17		.49	.61
5100 Concrete		2000	.004		.32	.17		.49	.61
5200 Concrete block		2000	.004		.32	.17		.49	.61
5300 Brick		2000	.004		.32	.17		.49	.61
5400 Stone		2000	.004		.32	.17		.49	.61
5500 Unpainted wood	↓	2000	.004	↓	.32	.17		.49	.61
8000 Cleaner for use after treatment									
8100 Towels or wipes, per package of 30				Ea.	.63			.63	.70
8200 Aerosol spray, 24 oz. can				"	18			18	19.80

For customer support on your Building Construction Costs with RSMeans Data, call 800.448.8182.

367

09 96 High-Performance Coatings

09 96 23 – Graffiti-Resistant Coatings

09 96 23.10 Graffiti Resistant Treatments	Crew	Daily Output	Labor-Hours	Unit	Material	2017 Bare Costs Labor	Equipment	Total	Total Incl O&P
8500	Graffiti removal with chemicals								
8510	Brush on, spray rinse off, on brick, masonry or stone	A-1H	200	.040	S.F.	.32	1.57	.38	2.27	3.16
8520	On smooth concrete	"	400	.020		.26	.78	.19	1.23	1.69
8530	Wipe on, wipe off, on plastic or painted metal	1 Clab	1500	.005		.63	.21		.84	1.02
8540	Brush on, wipe off, on painted wood	"	500	.016		.35	.63		.98	1.34

09 96 46 – Intumescent Painting

09 96 46.10 Coatings, Intumescent

		Crew	Daily Output	Labor-Hours	Unit	Material	Labor	Equipment	Total	Total Incl O&P
0010	**COATINGS, INTUMESCENT**, spray applied									
0100	On exterior structural steel, 0.25" d.f.t.	1 Pord	475	.017	S.F.	.47	.70		1.17	1.57
0150	0.51" d.f.t.		350	.023		.47	.95		1.42	1.95
0200	0.98" d.f.t.		280	.029		.47	1.19		1.66	2.31
0300	On interior structural steel, 0.108" d.f.t.		300	.027		.45	1.11		1.56	2.18
0350	0.310" d.f.t.		150	.053		.45	2.22		2.67	3.86
0400	0.670" d.f.t.		100	.080		.45	3.34		3.79	5.55

09 96 53 – Elastomeric Coatings

09 96 53.10 Coatings, Elastomeric

		Crew	Daily Output	Labor-Hours	Unit	Material	Labor	Equipment	Total	Total Incl O&P
0010	**COATINGS, ELASTOMERIC**									
0020	High build, water proof, one coat system									
0100	Concrete, brush	1 Pord	650	.012	S.F.	.27	.51		.78	1.07
0110	Roll		1650	.005		.27	.20		.47	.61
0120	Spray		2600	.003		.27	.13		.40	.49
0200	Concrete block, brush		600	.013		.32	.56		.88	1.20
0210	Roll		1400	.006		.32	.24		.56	.72
0220	Spray		1900	.004		.32	.18		.50	.63
0300	Stucco, brush		400	.020		.44	.83		1.27	1.75
0310	Roll		1000	.008		.44	.33		.77	.99
0320	Spray		1500	.005		.44	.22		.66	.83

09 96 56 – Epoxy Coatings

09 96 56.20 Wall Coatings

		Crew	Daily Output	Labor-Hours	Unit	Material	Labor	Equipment	Total	Total Incl O&P
0010	**WALL COATINGS**									
0100	Acrylic glazed coatings, matte	1 Pord	525	.015	S.F.	.32	.64		.96	1.31
0200	Gloss		305	.026		.66	1.09		1.75	2.38
0300	Epoxy coatings, solvent based		525	.015		.41	.64		1.05	1.41
0400	Water based		170	.047		.29	1.96		2.25	3.28
0600	Exposed aggregate, troweled on, 1/16" to 1/4", solvent based		235	.034		.63	1.42		2.05	2.83
0700	Water based (epoxy or polyacrylate)		130	.062		1.36	2.57		3.93	5.35
0900	1/2" to 5/8" aggregate, solvent based		130	.062		1.22	2.57		3.79	5.20
1000	Water based		80	.100		2.12	4.17		6.29	8.65
1500	Exposed aggregate, sprayed on, 1/8" aggregate, solvent based		295	.027		.58	1.13		1.71	2.35
1600	Water based		145	.055		1.07	2.30		3.37	4.65
1800	High build epoxy, 50 mil, solvent based		390	.021		.69	.86		1.55	2.05
1900	Water based		95	.084		1.17	3.51		4.68	6.60
2100	Laminated epoxy with fiberglass, solvent based		295	.027		.74	1.13		1.87	2.52
2200	Water based		145	.055		1.35	2.30		3.65	4.96
2400	Sprayed perlite or vermiculite, 1/16" thick, solvent based		2935	.003		.28	.11		.39	.48
2500	Water based		640	.013		.75	.52		1.27	1.62
2700	Vinyl plastic wall coating, solvent based		735	.011		.34	.45		.79	1.05
2800	Water based		240	.033		.84	1.39		2.23	3.02
3000	Urethane on smooth surface, 2 coats, solvent based		1135	.007		.32	.29		.61	.79
3100	Water based		665	.012		.61	.50		1.11	1.43
3600	Ceramic-like glazed coating, cementitious, solvent based		440	.018		.49	.76		1.25	1.68

09 96 High-Performance Coatings

09 96 56 – Epoxy Coatings

09 96 56.20 Wall Coatings		Crew	Daily Output	Labor-Hours	Unit	Material	2017 Bare Costs Labor	Equipment	Total	Total Incl O&P
3700	Water based	1 Pord	345	.023	S.F.	.83	.97		1.80	2.37
3900	Resin base, solvent based		640	.013		.40	.52		.92	1.23
4000	Water based		330	.024		.59	1.01		1.60	2.18

09 97 Special Coatings

09 97 13 – Steel Coatings

09 97 13.23 Exterior Steel Coatings

		Crew	Daily Output	Labor-Hours	Unit	Material	2017 Bare Costs Labor	Equipment	Total	Total Incl O&P
0010	**EXTERIOR STEEL COATINGS** R050516-30									
6100	Cold galvanizing, brush in field	1 Psst	1100	.007	S.F.	.22	.31		.53	.77
6510	Paints & protective coatings, sprayed in field									
6520	Alkyds, primer	2 Psst	3600	.004	S.F.	.09	.19		.28	.42
6540	Gloss topcoats		3200	.005		.08	.22		.30	.45
6560	Silicone alkyd		3200	.005		.17	.22		.39	.55
6610	Epoxy, primer		3000	.005		.28	.23		.51	.70
6630	Intermediate or topcoat		2800	.006		.27	.25		.52	.71
6650	Enamel coat		2800	.006		.34	.25		.59	.79
6700	Epoxy ester, primer		2800	.006		.48	.25		.73	.94
6720	Topcoats		2800	.006		.23	.25		.48	.66
6810	Latex primer		3600	.004		.06	.19		.25	.39
6830	Topcoats		3200	.005		.07	.22		.29	.44
6910	Universal primers, one part, phenolic, modified alkyd		2000	.008		.39	.34		.73	1.01
6940	Two part, epoxy spray		2000	.008		.40	.34		.74	1.02
7000	Zinc rich primers, self cure, spray, inorganic		1800	.009		.88	.38		1.26	1.62
7010	Epoxy, spray, organic		1800	.009		.27	.38		.65	.95
7020	Above one story, spray painting simple structures, add						25%			
7030	Intricate structures, add						50%			

For customer support on your Building Construction Costs with RSMeans Data, call 800.448.8182.

369

Division Notes

	CREW	DAILY OUTPUT	LABOR-HOURS	UNIT	BARE COSTS				TOTAL INCL O&P
					MAT.	LABOR	EQUIP.	TOTAL	

Estimating Tips
General

- The items in this division are usually priced per square foot or each.

- Many items in Division 10 require some type of support system or special anchors that are not usually furnished with the item. The required anchors must be added to the estimate in the appropriate division.

- Some items in Division 10, such as lockers, may require assembly before installation. Verify the amount of assembly required. Assembly can often exceed installation time.

10 20 00 Interior Specialties

- Support angles and blocking are not included in the installation of toilet compartments, shower/dressing compartments, or cubicles. Appropriate line items from Division 5 or 6 may need to be added to support the installations.

- Toilet partitions are priced by the stall. A stall consists of a side wall, pilaster, and door with hardware. Toilet tissue holders and grab bars are extra.

- The required acoustical rating of a folding partition can have a significant impact on costs. Verify the sound transmission coefficient rating of the panel priced against the specification requirements.

- Grab bar installation does not include supplemental blocking or backing to support the required load. When grab bars are installed at an existing facility, provisions must be made to attach the grab bars to a solid structure.

Reference Numbers

Reference numbers are shown at the beginning of some major classifications. These numbers refer to related items in the Reference Section. The reference information may be an estimating procedure, an alternate pricing method, or technical information.

Note: Not all subdivisions listed here necessarily appear. ■

Did you know?

Our online estimating solution gives you the same access to RSMeans' data with 24/7 access:

- Quickly locate costs in the searchable database.

- Build cost lists, estimates, and reports in minutes.

- Adjust costs to any location in the U.S. and Canada with the click of a button.

Start your free trial today at
www.RSMeansOnline.com

10 05 Common Work Results for Specialties

10 05 05 – Selective Demolition for Specialties

10 05 05.10 Selective Demolition, Specialties	Crew	Daily Output	Labor-Hours	Unit	Material	2017 Bare Costs Labor	Equipment	Total	Total Incl O&P
0010 **SELECTIVE DEMOLITION, SPECIALTIES**									
1100 Boards and panels, wall mounted	2 Clab	15	1.067	Ea.		42		42	64
1105 Demolition, mirror, wall mounted	1 Clab	30	.267			10.45		10.45	16
1200 Cases, for directory and/or bulletin boards, including doors	2 Clab	24	.667			26		26	40
1850 Shower partitions, cabinet or stall, including base and door	"	8	2			78.50		78.50	120
1855 Shower receptor, terrazzo or concrete	1 Clab	14	.571	↓		22.50		22.50	34.50
1900 Curtain track or rod, hospital type, ceiling mounted or suspended	"	220	.036	L.F.		1.42		1.42	2.18
1910 Toilet cubicles, remove	2 Clab	8	2	Ea.		78.50		78.50	120
1930 Urinal screen, remove	1 Clab	12	.667	"		26		26	40
2650 Wall guard, misc. wall or corner protection	"	320	.025	L.F.		.98		.98	1.50
2750 Access floor, metal panel system, including pedestals, covering	2 Clab	850	.019	S.F.		.74		.74	1.13
3050 Fireplace, prefab, freestanding or wall hung, including hood and screen	1 Clab	2	4	Ea.		157		157	240
3054 Chimney top, simulated brick, 4' high	"	15	.533			21		21	32
3200 Stove, woodburning, cast iron	2 Clab	2	8			315		315	480
3440 Weathervane, residential	1 Clab	12	.667			26		26	40
3500 Flagpole, groundset, to 70' high, excluding base/foundation	K-1	1	16			745	270	1,015	1,425
3555 To 30' high	"	2.50	6.400			299	108	407	575
4300 Letter, signs or plaques, exterior on wall	1 Clab	20	.400			15.65		15.65	24
4310 Signs, street, reflective aluminum, including post and bracket		60	.133			5.20		5.20	8
4320 Door signs interior on door 6" x 6", selective demolition	↓	20	.400			15.65		15.65	24
4550 Turnstiles, manual or electric	2 Clab	2	8	↓		315		315	480
5050 Lockers	1 Clab	15	.533	Opng.		21		21	32
5250 Cabinets, recessed	Q-12	12	1.333	Ea.		71		71	107
5260 Mail boxes, horiz., key lock, front loading, remove	1 Carp	34	.235	"		11.60		11.60	17.75
5350 Awning, fabric, including frame	2 Clab	100	.160	S.F.		6.25		6.25	9.60
6050 Partition, woven wire		1400	.011			.45		.45	.69
6100 Folding gate, security, door or window		500	.032			1.25		1.25	1.92
6580 Acoustic air wall		650	.025	↓		.96		.96	1.48
7550 Telephone enclosure, exterior, post mounted		3	5.333	Ea.		209		209	320
8850 Scale, platform, excludes foundation or pit	↓	.25	64	"		2,500		2,500	3,850

10 11 Visual Display Units

10 11 13 – Chalkboards

10 11 13.13 Fixed Chalkboards

	Crew	Daily Output	Labor-Hours	Unit	Material	2017 Bare Costs Labor	Equipment	Total	Total Incl O&P
0010 **FIXED CHALKBOARDS** Porcelain enamel steel									
3900 Wall hung									
4000 Aluminum frame and chalktrough									
4200 3' x 4'	2 Carp	16	1	Ea.	252	49.50		301.50	355
4300 3' x 5'		15	1.067		325	52.50		377.50	435
4500 4' x 8'		14	1.143		440	56.50		496.50	570
4600 4' x 12'	↓	13	1.231	↓	575	60.50		635.50	725
4700 Wood frame and chalktrough									
4800 3' x 4'	2 Carp	16	1	Ea.	199	49.50		248.50	295
5000 3' x 5'		15	1.067		243	52.50		295.50	350
5100 4' x 5'		14	1.143		255	56.50		311.50	365
5300 4' x 8'	↓	13	1.231		320	60.50		380.50	445
5400 Liquid chalk, white porcelain enamel, wall hung									
5420 Deluxe units, aluminum trim and chalktrough									
5450 4' x 4'	2 Carp	16	1	Ea.	277	49.50		326.50	380
5500 4' x 8'		14	1.143		410	56.50		466.50	535
5550 4' x 12'	↓	12	1.333		560	65.50		625.50	720

10 11 13 – Chalkboards

10 11 13.13 Fixed Chalkboards

		Crew	Daily Output	Labor-Hours	Unit	Material	2017 Bare Costs Labor	Equipment	Total	Total Incl O&P
5700	Wood trim and chalktrough									
5900	4' x 4'	2 Carp	16	1	Ea.	715	49.50		764.50	860
6000	4' x 6'		15	1.067		810	52.50		862.50	970
6200	4' x 8'	↓	14	1.143		970	56.50		1,026.50	1,150
6300	Liquid chalk, felt tip markers					2.28			2.28	2.51
6500	Erasers					2			2	2.20
6600	Board cleaner, 8 oz. bottle				↓	6.50			6.50	7.15

10 11 13.23 Modular-Support-Mounted Chalkboards

		Crew	Daily Output	Labor-Hours	Unit	Material	2017 Bare Costs Labor	Equipment	Total	Total Incl O&P
0010	**MODULAR-SUPPORT-MOUNTED CHALKBOARDS**									
0400	Sliding chalkboards									
0450	Vertical, one sliding board with back panel, wall mounted									
0500	8' x 4'	2 Carp	8	2	Ea.	2,225	98.50		2,323.50	2,600
0520	8' x 8'		7.50	2.133		3,225	105		3,330	3,700
0540	8' x 12'	↓	7	2.286	↓	4,175	113		4,288	4,775
0600	Two sliding boards, with back panel									
0620	8' x 4'	2 Carp	8	2	Ea.	3,425	98.50		3,523.50	3,925
0640	8' x 8'		7.50	2.133		5,025	105		5,130	5,700
0660	8' x 12'	↓	7	2.286	↓	8,275	113		8,388	9,275
0700	Horizontal, two track									
0800	4' x 8', 2 sliding panels	2 Carp	8	2	Ea.	1,950	98.50		2,048.50	2,300
0820	4' x 12', 2 sliding panels		7.50	2.133		2,550	105		2,655	2,950
0840	4' x 16', 4 sliding panels	↓	7	2.286	↓	3,425	113		3,538	3,950
0900	Four track, four sliding panels									
0920	4' x 8'	2 Carp	8	2	Ea.	3,150	98.50		3,248.50	3,600
0940	4' x 12'		7.50	2.133		4,100	105		4,205	4,675
0960	4' x 16'	↓	7	2.286		5,350	113		5,463	6,050
1200	Vertical, motor operated									
1400	One sliding panel with back panel									
1450	10' x 4'	2 Carp	4	4	Ea.	5,350	197		5,547	6,175
1500	10' x 10'		3.75	4.267		6,450	210		6,660	7,400
1550	10' x 16'	↓	3.50	4.571	↓	7,575	225		7,800	8,700
1700	Two sliding panels with back panel									
1750	10' x 4'	2 Carp	4	4	Ea.	9,475	197		9,672	10,700
1800	10' x 10'		3.75	4.267		10,600	210		10,810	12,000
1850	10' x 16'	↓	3.50	4.571	↓	12,600	225		12,825	14,200
2000	Three sliding panels with back panel									
2100	10' x 4'	2 Carp	4	4	Ea.	13,200	197		13,397	14,800
2150	10' x 10'		3.75	4.267		14,700	210		14,910	16,400
2200	10' x 16'	↓	3.50	4.571	↓	17,500	225		17,725	19,500
2400	For projection screen, glass beaded, add				S.F.	4.77			4.77	5.25
2500	For remote control, 1 panel control, add				Ea.	360			360	395
2600	2 panel control, add				"	620			620	680
2800	For units without back panels, deduct				S.F.	5.05			5.05	5.55
2850	For liquid chalk porcelain panels, add				"	5.15			5.15	5.70
3000	Swing leaf, any comb. of chalkboard & cork, aluminum frame									
3100	Floor style, 6 panels									
3150	30" x 40" panels				Ea.	1,550			1,550	1,700
3200	48" x 40" panels				"	2,600			2,600	2,875
3300	Wall mounted, 6 panels									
3400	30" x 40" panels	2 Carp	16	1	Ea.	1,475	49.50		1,524.50	1,700
3450	48" x 40" panels	"	16	1	"	1,800	49.50		1,849.50	2,075
3600	Extra panels for swing leaf units									

10 11 13 – Chalkboards

10 11 13.23 Modular-Support-Mounted Chalkboards

		Crew	Daily Output	Labor-Hours	Unit	Material	2017 Bare Costs Labor	Equipment	Total	Total Incl O&P
3700	30" x 40" panels				Ea.	294			294	325
3750	48" x 40" panels				"	365			365	400

10 11 13.43 Portable Chalkboards

		Crew	Daily Output	Labor-Hours	Unit	Material	2017 Bare Costs Labor	Equipment	Total	Total Incl O&P
0010	**PORTABLE CHALKBOARDS**									
0100	Freestanding, reversible									
0120	Economy, wood frame, 4' x 6'									
0140	Chalkboard both sides				Ea.	650			650	715
0160	Chalkboard one side, cork other side				"	600			600	660
0200	Standard, lightweight satin finished aluminum, 4' x 6'									
0220	Chalkboard both sides				Ea.	670			670	735
0240	Chalkboard one side, cork other side				"	605			605	665
0300	Deluxe, heavy duty extruded aluminum, 4' x 6'									
0320	Chalkboard both sides				Ea.	1,075			1,075	1,175
0340	Chalkboard one side, cork other side				"	1,100			1,100	1,225

10 11 16 – Markerboards

10 11 16.53 Electronic Markerboards

		Crew	Daily Output	Labor-Hours	Unit	Material	2017 Bare Costs Labor	Equipment	Total	Total Incl O&P
0010	**ELECTRONIC MARKERBOARDS**									
0100	Wall hung or free standing, 3' x 4' to 4' x 6'	2 Carp	8	2	S.F.	85	98.50		183.50	245
0150	5' x 6' to 4' x 8'		8	2	"	62	98.50		160.50	219
0500	Interactive projection module for existing whiteboards		8	2	Ea.	1,300	98.50		1,398.50	1,575

10 11 23 – Tackboards

10 11 23.10 Fixed Tackboards

		Crew	Daily Output	Labor-Hours	Unit	Material	2017 Bare Costs Labor	Equipment	Total	Total Incl O&P
0010	**FIXED TACKBOARDS**									
0020	Cork sheets, unbacked, no frame, 1/4" thick	2 Carp	290	.055	S.F.	1.96	2.72		4.68	6.30
0100	1/2" thick		290	.055		4.34	2.72		7.06	8.95
0300	Fabric-face, no frame, on 7/32" cork underlay		290	.055		7.10	2.72		9.82	11.95
0400	On 1/4" cork on 1/4" hardboard		290	.055		8.95	2.72		11.67	14
0600	With edges wrapped		290	.055		10.85	2.72		13.57	16.10
0700	On 7/16" fire retardant core		290	.055		6.70	2.72		9.42	11.50
0900	With edges wrapped		290	.055		8.70	2.72		11.42	13.70
1000	Designer fabric only, cut to size					2.70			2.70	2.97
1200	1/4" vinyl cork, on 1/4" hardboard, no frame	2 Carp	290	.055		8.70	2.72		11.42	13.70
1300	On 1/4" coreboard		290	.055		5.45	2.72		8.17	10.10
2000	For map and display rail, economy, add		385	.042	L.F.	3.06	2.05		5.11	6.50
2100	Deluxe, add		350	.046	"	4.70	2.25		6.95	8.60
2120	Prefabricated, 1/4" cork, 3' x 5' with aluminum frame		16	1	Ea.	135	49.50		184.50	224
2140	Wood frame		16	1		159	49.50		208.50	251
2160	4' x 4' with aluminum frame		16	1		152	49.50		201.50	243
2180	Wood frame		16	1		196	49.50		245.50	292
2200	4' x 8' with aluminum frame		14	1.143		279	56.50		335.50	390
2210	Wood frame		14	1.143		290	56.50		346.50	405
2220	4' x 12' with aluminum frame		12	1.333		415	65.50		480.50	555
2230	Bulletin board case, single glass door, with lock									
2240	36" x 24", economy	2 Carp	12	1.333	Ea.	320	65.50		385.50	450
2250	Deluxe		12	1.333		370	65.50		435.50	510
2260	42" x 30", economy		12	1.333		380	65.50		445.50	520
2270	Deluxe		12	1.333		470	65.50		535.50	615
2300	Glass enclosed cabinets, alum., cork panel, hinged doors									
2400	3' x 3', 1 door	2 Carp	12	1.333	Ea.	605	65.50		670.50	765
2500	4' x 4', 2 door		11	1.455		1,000	71.50		1,071.50	1,200
2600	4' x 7', 3 door		10	1.600		1,775	79		1,854	2,075

10 11 Visual Display Units

10 11 23 – Tackboards

10 11 23.10 Fixed Tackboards

		Crew	Daily Output	Labor-Hours	Unit	Material	2017 Bare Costs Labor	Equipment	Total	Total Incl O&P
2800	4' x 10', 4 door	2 Carp	8	2	Ea.	2,350	98.50		2,448.50	2,750
2900	For lights, add per door opening	1 Elec	13	.615		165	35		200	234
3100	Horizontal sliding units, 4 doors, 4' x 8', 8' x 4'	2 Carp	9	1.778		1,950	87.50		2,037.50	2,275
3200	4' x 12'		7	2.286		2,550	113		2,663	2,975
3400	8 doors, 4' x 16'		5	3.200		3,450	158		3,608	4,025
3500	4' x 24'	▼	4	4	▼	4,650	197		4,847	5,400

10 11 23.20 Control Boards

		Crew	Daily Output	Labor-Hours	Unit	Material	2017 Bare Costs Labor	Equipment	Total	Total Incl O&P
0010	**CONTROL BOARDS**									
0020	Magnetic, porcelain finish, 18" x 24", framed	2 Carp	8	2	Ea.	199	98.50		297.50	370
0100	24" x 36"		7.50	2.133		268	105		373	455
0200	36" x 48"		7	2.286		350	113		463	555
0300	48" x 72"		6	2.667		670	131		801	935
0400	48" x 96"	▼	5	3.200	▼	1,075	158		1,233	1,425
1000	Hospital patient display board, 4-color custom design									
1010	Porcelain steel dry erase board, 36" x 24"	2 Carp	7.50	2.133	Ea.	238	105		343	420

10 13 Directories

10 13 10 – Building Directories

10 13 10.10 Directory Boards

		Crew	Daily Output	Labor-Hours	Unit	Material	2017 Bare Costs Labor	Equipment	Total	Total Incl O&P
0010	**DIRECTORY BOARDS**									
0050	Plastic, glass covered, 30" x 20"	2 Carp	3	5.333	Ea.	198	263		461	615
0100	36" x 48"		2	8		845	395		1,240	1,525
0300	Grooved cork, 30" x 20"		3	5.333		405	263		668	845
0400	36" x 48"		2	8		555	395		950	1,225
0600	Black felt, 30" x 20"		3	5.333		239	263		502	665
0700	36" x 48"		2	8		465	395		860	1,125
0900	Outdoor, weatherproof, black plastic, 36" x 24"		2	8		760	395		1,155	1,450
1000	36" x 36"		1.50	10.667		880	525		1,405	1,775
1800	Indoor, economy, open face, 18" x 24"		7	2.286		166	113		279	355
1900	24" x 36"		7	2.286		154	113		267	340
2000	36" x 24"		6	2.667		154	131		285	370
2100	36" x 48"		6	2.667		251	131		382	475
2400	Building directory, alum., black felt panels, 1 door, 24" x 18"		4	4		315	197		512	645
2500	36" x 24"		3.50	4.571		385	225		610	770
2600	48" x 32"		3	5.333		610	263		873	1,075
2700	2 door, 36" x 48"		2.50	6.400		655	315		970	1,200
2800	36" x 60"		2	8		860	395		1,255	1,550
2900	48" x 60"	▼	1	16		970	790		1,760	2,275
3100	For bronze enamel finish, add					15%				
3200	For bronze anodized finish, add					25%				
3400	For illuminated directory, single door unit, add				▼	138			138	151
3500	For 6" header panel, 6 letters per foot, add				L.F.	21.50			21.50	23.50
6050	Building directory, electronic display, alum. frame, wall mounted	2 Carp	32	.500	S.F.	2,500	24.50		2,524.50	2,825
6100	Free standing	"	60	.267	"	3,800	13.15		3,813.15	4,225

10 14 19 – Dimensional Letter Signage

10 14 19.10 Exterior Signs

		Crew	Daily Output	Labor-Hours	Unit	Material	2017 Bare Costs Labor	Equipment	Total	Total Incl O&P
0010	**EXTERIOR SIGNS**									
0020	Letters, 2" high, 3/8" deep, cast bronze	1 Carp	24	.333	Ea.	26	16.40		42.40	54
0140	1/2" deep, cast aluminum		18	.444		26	22		48	62.50
0160	Cast bronze		32	.250		34	12.30		46.30	56
0300	6" high, 5/8" deep, cast aluminum		24	.333		30	16.40		46.40	58
0400	Cast bronze		24	.333		64	16.40		80.40	95
0600	8" high, 3/4" deep, cast aluminum		14	.571		33.50	28		61.50	79.50
0700	Cast bronze		20	.400		88	19.70		107.70	127
0900	10" high, 1" deep, cast aluminum		18	.444		50	22		72	88.50
1000	Cast bronze		18	.444		110	22		132	155
1200	12" high, 1-1/4" deep, cast aluminum		12	.667		54.50	33		87.50	111
1500	Cast bronze		18	.444		143	22		165	192
1600	14" high, 2-5/16" deep, cast aluminum		12	.667		97	33		130	158
1800	Fabricated stainless steel, 6" high, 2" deep		20	.400		42.50	19.70		62.20	77
1900	12" high, 3" deep		18	.444		72.50	22		94.50	114
2100	18" high, 3" deep		12	.667		111	33		144	173
2200	24" high, 4" deep		10	.800		224	39.50		263.50	310
2700	Acrylic, on high density foam, 12" high, 2" deep		20	.400		19.60	19.70		39.30	51.50
2800	18" high, 2" deep		18	.444		38	22		60	75
3900	Plaques, custom, 20" x 30", for up to 450 letters, cast aluminum	2 Carp	4	4		1,850	197		2,047	2,325
4000	Cast bronze		4	4		1,875	197		2,072	2,350
4200	30" x 36", up to 900 letters, cast aluminum		3	5.333		2,725	263		2,988	3,400
4300	Cast bronze		3	5.333		4,175	263		4,438	5,000
4500	36" x 48", for up to 1300 letters, cast bronze		2	8		4,700	395		5,095	5,775
4800	Signs, reflective alum. directional signs, dbl. face, 2-way, w/bracket		30	.533		132	26.50		158.50	185
4900	4-way		30	.533		249	26.50		275.50	315
5100	Exit signs, 24 ga. alum., 14" x 12" surface mounted	1 Carp	30	.267		50	13.15		63.15	75
5200	10" x 7"		20	.400		30.50	19.70		50.20	63.50
5400	Bracket mounted, double face, 12" x 10"		30	.267		51.50	13.15		64.65	76.50
5500	Sticky back, stock decals, 14" x 10"	1 Clab	50	.160		27.50	6.25		33.75	39.50
6400	Replacement sign faces, 6" or 8"	"	50	.160		66	6.25		72.25	82
8000	Internally illuminated, custom									
8100	On pedestal, 84" x 30" x 12"	L-7	.50	56	Ea.	12,400	2,650		15,050	17,800

10 14 23 – Panel Signage

10 14 23.13 Engraved Panel Signage

		Crew	Daily Output	Labor-Hours	Unit	Material	2017 Bare Costs Labor	Equipment	Total	Total Incl O&P
0010	**ENGRAVED PANEL SIGNAGE**, interior									
1010	Flexible door sign, adhesive back, w/Braille, 5/8" letters, 4" x 4"	1 Clab	32	.250	Ea.	34	9.80		43.80	52
1050	6" x 6"		32	.250		49.50	9.80		59.30	69.50
1100	8" x 2"		32	.250		35	9.80		44.80	53.50
1150	8" x 4"		32	.250		44	9.80		53.80	63.50
1200	8" x 8"		32	.250		60	9.80		69.80	81
1250	12" x 2"		32	.250		35	9.80		44.80	53.50
1300	12" x 6"		32	.250		39	9.80		48.80	58
1350	12" x 12"		32	.250		155	9.80		164.80	185
1500	Graphic symbols, 2" x 2"		32	.250		12	9.80		21.80	28
1550	6" x 6"		32	.250		31	9.80		40.80	49
1600	8" x 8"		32	.250		39	9.80		48.80	57.50
2010	Corridor, stock acrylic, 2-sided, with mounting bracket, 2" x 8"	1 Carp	24	.333		26	16.40		42.40	53.50
2020	2" x 10"		24	.333		35	16.40		51.40	63.50
2050	3" x 8"		24	.333		28.50	16.40		44.90	56.50
2060	3" x 10"		24	.333		36.50	16.40		52.90	65
2070	3" x 12"		24	.333		33	16.40		49.40	61

10 14 Signage

10 14 23 – Panel Signage

10 14 23.13 Engraved Panel Signage

		Crew	Daily Output	Labor-Hours	Unit	Material	2017 Bare Costs Labor	Equipment	Total	Total Incl O&P
2100	4" x 8"	1 Carp	24	.333	Ea.	22.50	16.40		38.90	49.50
2110	4" x 10"		24	.333		40	16.40		56.40	69
2120	4" x 12"		24	.333		50	16.40		66.40	80
7000	Wayfinding signage, custom									
7010	Plastic, flexible, 1/8" thick, incl. mounting									
7020	6" x 6"	1 Carp	32	.250	Ea.	16.90	12.30		29.20	37.50
7030	6" x 12"		32	.250		35	12.30		47.30	57.50
7040	6" x 24"		28	.286		70	14.05		84.05	98.50
7050	8" x 8"		32	.250		31	12.30		43.30	53
7060	8" x 16"		28	.286		62	14.05		76.05	89.50
7070	8" x 24"		26	.308		93	15.15		108.15	125
7080	12" x 12"		28	.286		70	14.05		84.05	98.50
7090	12" x 24"		24	.333		140	16.40		156.40	179
7100	12" x 36"		20	.400		209	19.70		228.70	260
7210	Weather resistant, engraved and color filled									
7220	8" x 8"	1 Carp	32	.250	Ea.	38.50	12.30		50.80	61.50
7230	8" x 16"		28	.286		61.50	14.05		75.55	89
7240	8" x 24"		26	.308		93.50	15.15		108.65	126
7250	12" x 12"		28	.286		65.50	14.05		79.55	93.50
7260	12" x 24"		24	.333		119	16.40		135.40	155
7270	12" x 36"		20	.400		192	19.70		211.70	242
7280	16" x 32"	2 Carp	60	.267		213	13.15		226.15	255
7290	16" x 48"	"	60	.267		355	13.15		368.15	410
7320	For engraved letters, 1/2" high, add					.56			.56	.62
7330	1" high, add					.70			.70	.77
7340	2" high, add					1.27			1.27	1.40
7350	3" high, add					1.88			1.88	2.07
7360	For engraved graphic symbols, 3" high, add					14.30			14.30	15.75
7370	7" high, add					16.65			16.65	18.30
7380	10" high, add					23			23	25.50
9990	Replace interior labels	1 Carp	12	.667		26	33		59	79.50
9991	Replace interior plaques	"	12	.667		84.50	33		117.50	144

10 14 26 – Post and Panel/Pylon Signage

10 14 26.10 Post and Panel Signage

		Crew	Daily Output	Labor-Hours	Unit	Material	2017 Bare Costs Labor	Equipment	Total	Total Incl O&P
0010	**POST AND PANEL SIGNAGE**, Alum., incl. (2) posts,									
0011	2-sided panel (blank), concrete footings per post									
0025	24"wx18"h	B-1	25	.960	Ea.	915	38		953	1,050
0050	36"wx24"h		25	.960		1,050	38		1,088	1,200
0100	57"wx36"h		25	.960		2,025	38		2,063	2,300
0150	81"wx46"h		25	.960		2,225	38		2,263	2,475

10 14 53 – Traffic Signage

10 14 53.20 Traffic Signs

		Crew	Daily Output	Labor-Hours	Unit	Material	2017 Bare Costs Labor	Equipment	Total	Total Incl O&P
0010	**TRAFFIC SIGNS**									
0012	Stock, 24" x 24", no posts, .080" alum. reflectorized	B-80	70	.457	Ea.	88	20	9.60	117.60	138
0100	High intensity		70	.457		101	20	9.60	130.60	152
0300	30" x 30", reflectorized		70	.457		126	20	9.60	155.60	179
0400	High intensity		70	.457		136	20	9.60	165.60	190
0600	Guide and directional signs, 12" x 18", reflectorized		70	.457		40.50	20	9.60	70.10	86
0700	High intensity		70	.457		55	20	9.60	84.60	102
0900	18" x 24", stock signs, reflectorized		70	.457		48	20	9.60	77.60	94
1000	High intensity		70	.457		53.50	20	9.60	83.10	99.50
1200	24" x 24", stock signs, reflectorized		70	.457		58.50	20	9.60	88.10	105

For customer support on your Building Construction Costs with RSMeans Data, call 800.448.8182.

377

10 14 Signage

10 14 53 – Traffic Signage

10 14 53.20 Traffic Signs

	10 14 53.20 Traffic Signs	Crew	Daily Output	Labor-Hours	Unit	Material	2017 Bare Costs Labor	Equipment	Total	Total Incl O&P
1300	High intensity	B-80	70	.457	Ea.	64	20	9.60	93.60	111
1500	Add to above for steel posts, galvanized, 10'-0" upright, bolted		200	.160		32.50	7	3.37	42.87	50.50
1600	12'-0" upright, bolted		140	.229	▼	38.50	10.05	4.81	53.36	63
1800	Highway road signs, aluminum, over 20 S.F., reflectorized		350	.091	S.F.	35	4.01	1.92	40.93	46.50
2000	High intensity		350	.091		35	4.01	1.92	40.93	46.50
2200	Highway, suspended over road, 80 S.F. min., reflectorized		165	.194		35	8.50	4.08	47.58	56
2300	High intensity		165	.194	▼	33.50	8.50	4.08	46.08	54.50
9000	Replace directional sign	▼	6	5.333	Ea.	136	234	112	482	630

10 17 Telephone Specialties

10 17 16 – Telephone Enclosures

10 17 16.10 Commercial Telephone Enclosures

	10 17 16.10 Commercial Telephone Enclosures	Crew	Daily Output	Labor-Hours	Unit	Material	2017 Bare Costs Labor	Equipment	Total	Total Incl O&P
0010	**COMMERCIAL TELEPHONE ENCLOSURES**									
0300	Shelf type, wall hung, recessed	2 Carp	5	3.200	Ea.	805	158		963	1,125
0400	Surface mount	"	5	3.200	"	1,675	158		1,833	2,075

10 21 Compartments and Cubicles

10 21 13 – Toilet Compartments

10 21 13.13 Metal Toilet Compartments

	10 21 13.13 Metal Toilet Compartments	Crew	Daily Output	Labor-Hours	Unit	Material	2017 Bare Costs Labor	Equipment	Total	Total Incl O&P
0010	**METAL TOILET COMPARTMENTS**									
0110	Cubicles, ceiling hung									
0200	Powder coated steel	2 Carp	4	4	Ea.	555	197		752	910
0500	Stainless steel	"	4	4		1,150	197		1,347	1,550
0600	For handicap units, incl. 52" grab bars, add				▼	450			450	495
0900	Floor and ceiling anchored									
1000	Powder coated steel	2 Carp	5	3.200	Ea.	570	158		728	870
1300	Stainless steel	"	5	3.200		1,275	158		1,433	1,650
1400	For handicap units, incl. 52" grab bars, add				▼	370			370	405
1610	Floor anchored									
1700	Powder coated steel	2 Carp	7	2.286	Ea.	620	113		733	850
2000	Stainless steel	"	7	2.286		1,475	113		1,588	1,800
2100	For handicap units, incl. 52" grab bars, add				▼	335			335	370
2200	For juvenile units, deduct					42			42	46.50
2450	Floor anchored, headrail braced									
2500	Powder coated steel	2 Carp	6	2.667	Ea.	380	131		511	620
2804	Stainless steel	"	6	2.667		1,025	131		1,156	1,325
2900	For handicap units, incl. 52" grab bars, add					370			370	410
3000	Wall hung partitions, powder coated steel	2 Carp	7	2.286		650	113		763	885
3300	Stainless steel	"	7	2.286		1,675	113		1,788	2,025
3400	For handicap units, incl. 52" grab bars, add				▼	370			370	410
4000	Screens, entrance, floor mounted, 58" high, 48" wide									
4200	Powder coated steel	2 Carp	15	1.067	Ea.	242	52.50		294.50	345
4500	Stainless steel	"	15	1.067	"	930	52.50		982.50	1,100
4650	Urinal screen, 18" wide									
4704	Powder coated steel	2 Carp	6.15	2.602	Ea.	209	128		337	425
5004	Stainless steel	"	6.15	2.602	"	535	128		663	785
5100	Floor mounted, headrail braced									
5300	Powder coated steel	2 Carp	8	2	Ea.	230	98.50		328.50	405
5600	Stainless steel	"	8	2	"	590	98.50		688.50	795

10 21 13 – Toilet Compartments

10 21 13.13 Metal Toilet Compartments

	10 21 13.13 Metal Toilet Compartments	Crew	Daily Output	Labor-Hours	Unit	Material	2017 Bare Costs Labor	Equipment	Total	Total Incl O&P
5750	Pilaster, flush									
5800	Powder coated steel	2 Carp	10	1.600	Ea.	267	79		346	415
6100	Stainless steel		10	1.600		610	79		689	790
6300	Post braced, powder coated steel		10	1.600		170	79		249	310
6600	Stainless steel		10	1.600		450	79		529	615
6700	Wall hung, bracket supported									
6800	Powder coated steel	2 Carp	10	1.600	Ea.	170	79		249	310
7100	Stainless steel		10	1.600		293	79		372	440
7400	Flange supported, powder coated steel		10	1.600		114	79		193	246
7700	Stainless steel		10	1.600		315	79		394	470
7800	Wedge type, powder coated steel		10	1.600		134	79		213	269
8100	Stainless steel		10	1.600		590	79		669	770

10 21 13.14 Metal Toilet Compartment Components

	10 21 13.14 Metal Toilet Compartment Components	Crew	Daily Output	Labor-Hours	Unit	Material	2017 Bare Costs Labor	Equipment	Total	Total Incl O&P
0010	**METAL TOILET COMPARTMENT COMPONENTS**									
0100	Pilasters									
0110	Overhead braced, powder coated steel, 7" wide x 82" high	2 Carp	22.20	.721	Ea.	78.50	35.50		114	141
0120	Stainless steel		22.20	.721		136	35.50		171.50	205
0130	Floor braced, powder coated steel, 7" wide x 70" high		23.30	.687		130	34		164	195
0140	Stainless steel		23.30	.687		261	34		295	340
0150	Ceiling hung, powder coated steel, 7" wide x 83" high		13.30	1.203		140	59.50		199.50	245
0160	Stainless steel		13.30	1.203		281	59.50		340.50	400
0170	Wall hung, powder coated steel, 3" wide x 58" high		18.90	.847		142	41.50		183.50	220
0180	Stainless steel		18.90	.847		219	41.50		260.50	305
0200	Panels									
0210	Powder coated steel, 31" wide x 58" high	2 Carp	18.90	.847	Ea.	142	41.50		183.50	220
0220	Stainless steel		18.90	.847		375	41.50		416.50	480
0230	Powder coated steel, 53" wide x 58" high		18.90	.847		176	41.50		217.50	257
0240	Stainless steel		18.90	.847		475	41.50		516.50	590
0250	Powder coated steel, 63" wide x 58" high		18.90	.847		213	41.50		254.50	298
0260	Stainless steel		18.90	.847		520	41.50		561.50	635
0300	Doors									
0310	Powder coated steel, 24" wide x 58" high	2 Carp	14.10	1.135	Ea.	147	56		203	247
0320	Stainless steel		14.10	1.135		310	56		366	425
0330	Powder coated steel, 26" wide x 58" high		14.10	1.135		149	56		205	250
0340	Stainless steel		14.10	1.135		315	56		371	435
0350	Powder coated steel, 28" wide x 58" high		14.10	1.135		170	56		226	273
0360	Stainless steel		14.10	1.135		330	56		386	445
0370	Powder coated steel, 36" wide x 58" high		14.10	1.135		183	56		239	287
0380	Stainless steel		14.10	1.135		395	56		451	520
0400	Headrails									
0410	For powder coated steel, 62" long	2 Carp	65	.246	Ea.	22.50	12.10		34.60	43
0420	Stainless steel		65	.246		22.50	12.10		34.60	43
0430	For powder coated steel, 84" long		50	.320		32	15.75		47.75	59.50
0440	Stainless steel		50	.320		32	15.75		47.75	59
0450	For powder coated steel, 120" long		30	.533		43	26.50		69.50	87.50
0460	Stainless steel		30	.533		43	26.50		69.50	87.50

10 21 13.16 Plastic-Laminate-Clad Toilet Compartments

	10 21 13.16 Plastic-Laminate-Clad Toilet Compartments	Crew	Daily Output	Labor-Hours	Unit	Material	2017 Bare Costs Labor	Equipment	Total	Total Incl O&P
0010	**PLASTIC-LAMINATE-CLAD TOILET COMPARTMENTS**									
0110	Cubicles, ceiling hung									
0300	Plastic laminate on particle board	2 Carp	4	4	Ea.	540	197		737	895
0600	For handicap units, incl. 52" grab bars, add				"	450			450	495
0900	Floor and ceiling anchored									

10 21 Compartments and Cubicles

10 21 13 – Toilet Compartments

10 21 13.16 Plastic-Laminate-Clad Toilet Compartments

		Crew	Daily Output	Labor-Hours	Unit	Material	2017 Bare Costs Labor	Equipment	Total	Total Incl O&P
1100	Plastic laminate on particle board	2 Carp	5	3.200	Ea.	865	158		1,023	1,200
1400	For handicap units, incl. 52" grab bars, add				"	370			370	405
1610	Floor mounted									
1800	Plastic laminate on particle board	2 Carp	7	2.286	Ea.	545	113		658	770
2450	Floor mounted, headrail braced									
2600	Plastic laminate on particle board	2 Carp	6	2.667	Ea.	810	131		941	1,100
3400	For handicap units, incl. 52" grab bars, add					370			370	410
4300	Entrance screen, floor mtd., plas. lam., 58" high, 48" wide	2 Carp	15	1.067		645	52.50		697.50	790
4800	Urinal screen, 18" wide, ceiling braced, plastic laminate		8	2		204	98.50		302.50	375
5400	Floor mounted, headrail braced		8	2		204	98.50		302.50	375
5900	Pilaster, flush, plastic laminate		10	1.600		540	79		619	710
6400	Post braced, plastic laminate		10	1.600		320	79		399	475
6700	Wall hung, bracket supported									
6900	Plastic laminate on particle board	2 Carp	10	1.600	Ea.	97.50	79		176.50	228
7450	Flange supported									
7500	Plastic laminate on particle board	2 Carp	10	1.600	Ea.	245	79		324	390

10 21 13.17 Plastic-Lam. Clad Toilet Compart. Components

		Crew	Daily Output	Labor-Hours	Unit	Material	2017 Bare Costs Labor	Equipment	Total	Total Incl O&P
0010	**PLASTIC-LAMINATE CLAD TOILET COMPARTMENT COMPONENTS**									
0100	Pilasters									
0110	Overhead braced, 7" wide x 82" high	2 Carp	22.20	.721	Ea.	103	35.50		138.50	168
0130	Floor anchored, 7" wide x 70" high		23.30	.687		107	34		141	170
0150	Ceiling hung, 7" wide x 83" high		13.30	1.203		110	59.50		169.50	211
0180	Wall hung, 3" wide x 58" high		18.90	.847		96	41.50		137.50	169
0200	Panels									
0210	31" wide x 58" high	2 Carp	18.90	.847	Ea.	153	41.50		194.50	232
0230	51" wide x 58" high		18.90	.847		204	41.50		245.50	289
0250	63" wide x 58" high		18.90	.847		242	41.50		283.50	330
0300	Doors									
0310	24" wide x 58" high	2 Carp	14.10	1.135	Ea.	150	56		206	251
0330	26" wide x 58" high		14.10	1.135		154	56		210	256
0350	28" wide x 58" high		14.10	1.135		159	56		215	261
0370	36" wide x 58" high		14.10	1.135		203	56		259	310
0400	Headrails									
0410	62" long	2 Carp	65	.246	Ea.	22.50	12.10		34.60	43.50
0430	84" long		60	.267		31	13.15		44.15	54.50
0450	120" long		30	.533		42.50	26.50		69	86.50

10 21 13.19 Plastic Toilet Compartments

		Crew	Daily Output	Labor-Hours	Unit	Material	2017 Bare Costs Labor	Equipment	Total	Total Incl O&P
0010	**PLASTIC TOILET COMPARTMENTS**									
0110	Cubicles, ceiling hung									
0250	Phenolic	2 Carp	4	4	Ea.	910	197		1,107	1,300
0600	For handicap units, incl. 52" grab bars, add				"	450			450	495
0900	Floor and ceiling anchored									
1050	Phenolic	2 Carp	5	3.200	Ea.	845	158		1,003	1,175
1400	For handicap units, incl. 52" grab bars, add				"	370			370	405
1610	Floor mounted									
1750	Phenolic	2 Carp	7	2.286	Ea.	750	113		863	995
2100	For handicap units, incl. 52" grab bars, add					335			335	370
2200	For juvenile units, deduct					42			42	46.50
2450	Floor mounted, headrail braced									
2550	Phenolic	2 Carp	6	2.667	Ea.	750	131		881	1,025

10 21 Compartments and Cubicles

10 21 13 – Toilet Compartments

10 21 13.20 Plastic Toilet Compartment Components

		Crew	Daily Output	Labor-Hours	Unit	Material	2017 Bare Costs Labor	Equipment	Total	Total Incl O&P
0010	**PLASTIC TOILET COMPARTMENT COMPONENTS**									
0100	Pilasters									
0110	Overhead braced, polymer plastic, 7" wide x 82" high	2 Carp	22.20	.721	Ea.	132	35.50		167.50	200
0120	Phenolic		22.20	.721		150	35.50		185.50	220
0130	Floor braced, polymer plastic, 7" wide x 70" high		23.30	.687		161	34		195	229
0140	Phenolic		23.30	.687		148	34		182	214
0150	Ceiling hung, polymer plastic, 7" wide x 83" high		13.30	1.203		178	59.50		237.50	287
0160	Phenolic		13.30	1.203		173	59.50		232.50	281
0180	Wall hung, phenolic, 3" wide x 58" high	▼	18.90	.847	▼	98.50	41.50		140	173
0200	Panels									
0203	Polymer plastic, 18" wide x 55" high	2 Carp	18.90	.847	Ea.	274	41.50		315.50	365
0206	Phenolic, 18" wide x 58" high		18.90	.847		244	41.50		285.50	335
0210	Polymer plastic, 31" wide x 55" high		18.90	.847		325	41.50		366.50	425
0220	Phenolic, 31" wide x 58" high		18.90	.847		278	41.50		319.50	370
0223	Polymer plastic, 48" wide x 55" high		18.90	.847		520	41.50		561.50	640
0226	Phenolic, 48" wide x 58" high		18.90	.847		465	41.50		506.50	575
0230	Polymer plastic, 51" wide x 55" high		18.90	.847		440	41.50		481.50	550
0240	Phenolic, 51" wide x 58" high		18.90	.847		400	41.50		441.50	505
0250	Polymer plastic, 63" wide x 55" high		18.90	.847		570	41.50		611.50	695
0260	Phenolic, 63" wide x 58" high	▼	18.90	.847	▼	450	41.50		491.50	560
0300	Doors									
0310	Polymer plastic, 24" wide x 55" high	2 Carp	14.10	1.135	Ea.	221	56		277	330
0320	Phenolic, 24" wide x 58" high		14.10	1.135		300	56		356	415
0330	Polymer plastic, 26" wide x 55" high		14.10	1.135		236	56		292	345
0340	Phenolic, 26" wide x 58" high		14.10	1.135		325	56		381	445
0350	Polymer plastic, 28" wide x 55" high		14.10	1.135		249	56		305	360
0360	Phenolic, 28" wide x 58" high		14.10	1.135		340	56		396	455
0370	Polymer plastic, 36" wide x 55" high		14.10	1.135		305	56		361	420
0380	Phenolic, 36" wide x 58" high	▼	14.10	1.135	▼	435	56		491	565
0400	Headrails									
0410	For polymer plastic, 62" long	2 Carp	65	.246	Ea.	23	12.10		35.10	44
0420	Phenolic		65	.246		22.50	12.10		34.60	43.50
0430	For polymer plastic, 84" long		50	.320		33	15.75		48.75	60.50
0440	Phenolic		50	.320		34.50	15.75		50.25	62
0450	For polymer plastic, 120" long		30	.533		44	26.50		70.50	88.50
0460	Phenolic	▼	30	.533	▼	43.50	26.50		70	88

10 21 13.40 Stone Toilet Compartments

		Crew	Daily Output	Labor-Hours	Unit	Material	2017 Bare Costs Labor	Equipment	Total	Total Incl O&P
0010	**STONE TOILET COMPARTMENTS**									
0100	Cubicles, ceiling hung, marble	2 Marb	2	8	Ea.	1,800	375		2,175	2,550
0600	For handicap units, incl. 52" grab bars, add	♿				450			450	495
0800	Floor & ceiling anchored, marble	2 Marb	2.50	6.400		1,975	300		2,275	2,625
1400	For handicap units, incl. 52" grab bars, add	♿				370			370	405
1600	Floor mounted, marble	2 Marb	3	5.333		1,225	250		1,475	1,725
2400	Floor mounted, headrail braced, marble	"	3	5.333		1,150	250		1,400	1,650
2900	For handicap units, incl. 52" grab bars, add	♿				370			370	410
4100	Entrance screen, floor mounted marble, 58" high, 48" wide	2 Marb	9	1.778		795	83.50		878.50	1,000
4600	Urinal screen, 18" wide, ceiling braced, marble	D-1	6	2.667	▼	755	115		870	1,000
5100	Floor mounted, headrail braced									
5200	Marble	D-1	6	2.667	Ea.	645	115		760	885
5700	Pilaster, flush, marble		9	1.778		840	77		917	1,050
6200	Post braced, marble	▼	9	1.778	▼	825	77		902	1,025

For customer support on your Building Construction Costs with RSMeans Data, call 800.448.8182.

381

10 21 Compartments and Cubicles

10 21 23 – Cubicle Curtains and Track

10 21 23.16 Cubicle Track and Hardware	Crew	Daily Output	Labor-Hours	Unit	Material	2017 Bare Costs Labor	Equipment	Total	Total Incl O&P
0010 **CUBICLE TRACK AND HARDWARE**									
0020 Curtain track, box channel, ceiling mounted	1 Carp	135	.059	L.F.	6.60	2.92		9.52	11.70
0100 Suspended	"	100	.080	"	8.85	3.94		12.79	15.80
0300 Curtains, nylon mesh tops, fire resistant, 11 oz. per lineal yard									
0310 Polyester oxford cloth, 9' ceiling height	1 Carp	425	.019	L.F.	16.85	.93		17.78	19.95
0500 8' ceiling height		425	.019		6.65	.93		7.58	8.70
0550 Polyester, antimicrobial, 9' ceiling height		425	.019		19.65	.93		20.58	23
0560 8' ceiling height		425	.019		17.50	.93		18.43	20.50
0700 Designer oxford cloth		425	.019		6.65	.93		7.58	8.70
0800 I.V. track systems									
0820 I.V. track, oval	1 Carp	135	.059	L.F.	7.90	2.92		10.82	13.15
0830 I.V. trolley		32	.250	Ea.	41	12.30		53.30	64
0840 I.V. pendant (tree, 5 hook)		32	.250	"	179	12.30		191.30	216

10 22 Partitions

10 22 13 – Wire Mesh Partitions

10 22 13.10 Partitions, Woven Wire

	Crew	Daily Output	Labor-Hours	Unit	Material	2017 Bare Costs Labor	Equipment	Total	Total Incl O&P
0010 **PARTITIONS, WOVEN WIRE** for tool or stockroom enclosures									
0100 Channel frame, 1-1/2" diamond mesh, 10 ga. wire, painted									
0300 Wall panels, 4'-0" wide, 7' high	2 Carp	25	.640	Ea.	144	31.50		175.50	207
0400 8' high		23	.696		165	34.50		199.50	235
0600 10' high		18	.889		195	44		239	282
0700 For 5' wide panels, add					5%				
0900 Ceiling panels, 10' long, 2' wide	2 Carp	25	.640		125	31.50		156.50	187
1000 4' wide		15	1.067		189	52.50		241.50	289
1200 Panel with service window & shelf, 5' wide, 7' high		20	.800		365	39.50		404.50	465
1300 8' high		15	1.067		460	52.50		512.50	585
1500 Sliding doors, full height, 3' wide, 7' high		6	2.667		470	131		601	715
1600 10' high		5	3.200		505	158		663	795
1800 6' wide sliding door, 7' full height		5	3.200		645	158		803	950
1900 10' high		4	4		800	197		997	1,175
2100 Swinging doors, 3' wide, 7' high, no transom		6	2.667		297	131		428	525
2200 7' high, 3' transom		5	3.200		360	158		518	635

10 22 16 – Folding Gates

10 22 16.10 Security Gates

	Crew	Daily Output	Labor-Hours	Unit	Material	2017 Bare Costs Labor	Equipment	Total	Total Incl O&P
0010 **SECURITY GATES**									
0015 For roll up type, see Section 08 33 13.10									
0300 Scissors type folding gate, ptd. steel, single, 6-1/2' high, 5-1/2' wide	2 Sswk	4	4	Opng.	231	217		448	615
0350 6-1/2' wide		4	4		248	217		465	630
0400 7-1/2' wide		4	4		244	217		461	630
0600 Double gate, 8' high, 8' wide		2.50	6.400		395	350		745	1,025
0650 10' wide		2.50	6.400		435	350		785	1,050
0700 12' wide		2	8		625	435		1,060	1,400
0750 14' wide		2	8		635	435		1,070	1,425
0900 Door gate, folding steel, 4' wide, 61" high		4	4		141	217		358	515
1000 71" high		4	4		169	217		386	545
1200 81" high		4	4		195	217		412	575
1300 Window gates, 2' to 4' wide, 31" high		4	4		83	217		300	450
1500 55" high		3.75	4.267		122	232		354	520
1600 79" high		3.50	4.571		144	248		392	575

10 22 Partitions

10 22 19 – Demountable Partitions

10 22 19.43 Demountable Composite Partitions	Crew	Daily Output	Labor-Hours	Unit	Material	2017 Bare Costs Labor	Equipment	Total	Total Incl O&P
0010 **DEMOUNTABLE COMPOSITE PARTITIONS**, add for doors									
0100 Do not deduct door openings from total L.F.									
0900 Demountable gypsum system on 2" to 2-1/2"									
1000 Steel studs, 9' high, 3" to 3-3/4" thick									
1200 Vinyl clad gypsum	2 Carp	48	.333	L.F.	60	16.40		76.40	91
1300 Fabric clad gypsum		44	.364		150	17.90		167.90	193
1500 Steel clad gypsum		40	.400		167	19.70		186.70	214
1600 1.75 system, aluminum framing, vinyl clad hardboard,									
1800 Paper honeycomb core panel, 1-3/4" to 2-1/2" thick									
1900 9' high	2 Carp	48	.333	L.F.	101	16.40		117.40	136
2100 7' high		60	.267		90.50	13.15		103.65	120
2200 5' high		80	.200		76.50	9.85		86.35	99
2250 Unitized gypsum system									
2300 Unitized panel, 9' high, 2" to 2-1/2" thick									
2350 Vinyl clad gypsum	2 Carp	48	.333	L.F.	130	16.40		146.40	168
2400 Fabric clad gypsum	"	44	.364	"	214	17.90		231.90	263
2500 Unitized mineral fiber system									
2510 Unitized panel, 9' high, 2-1/4" thick, aluminum frame									
2550 Vinyl clad mineral fiber	2 Carp	48	.333	L.F.	129	16.40		145.40	167
2600 Fabric clad mineral fiber	"	44	.364	"	193	17.90		210.90	240
2800 Movable steel walls, modular system									
2900 Unitized panels, 9' high, 48" wide									
3100 Baked enamel, pre-finished	2 Carp	60	.267	L.F.	146	13.15		159.15	181
3200 Fabric clad steel		56	.286	"	212	14.05		226.05	255
5310 Trackless wall, cork finish, semi-acoustic, 1-5/8" thick, unsealed		325	.049	S.F.	38.50	2.42		40.92	46
5320 Sealed		190	.084		42.50	4.15		46.65	53.50
5330 Acoustic, 2" thick, unsealed		305	.052		36.50	2.58		39.08	44
5340 Sealed		225	.071		56	3.50		59.50	67
5500 For acoustical partitions, add, unsealed					2.36			2.36	2.60
5550 Sealed					11			11	12.10
5700 For doors, see Section 08 16									
5800 For door hardware, see Section 08 71									
6100 In-plant modular office system, w/prehung hollow core door									
6200 3" thick polystyrene core panels									
6250 12' x 12', 2 wall	2 Clab	3.80	4.211	Ea.	4,200	165		4,365	4,875
6300 4 wall		1.90	8.421		6,400	330		6,730	7,550
6350 16' x 16', 2 wall		3.60	4.444		6,375	174		6,549	7,275
6400 4 wall		1.80	8.889		8,650	350		9,000	10,100

10 22 23 – Portable Partitions, Screens, and Panels

10 22 23.13 Wall Screens

	Crew	Daily Output	Labor-Hours	Unit	Material	Labor	Equipment	Total	Total Incl O&P
0010 **WALL SCREENS**, divider panels, free standing, fiber core									
0020 Fabric face straight									
0100 3'-0" long, 4'-0" high	2 Carp	100	.160	L.F.	124	7.90		131.90	148
0200 5'-0" high		90	.178		108	8.75		116.75	132
0500 6'-0" high		75	.213		112	10.50		122.50	139
0900 5'-0" long, 4'-0" high		175	.091		76	4.50		80.50	90.50
1000 5'-0" high		150	.107		80.50	5.25		85.75	96.50
1500 6"-0" high		125	.128		101	6.30		107.30	121
1600 6'-0" long, 5'-0" high		162	.099		101	4.86		105.86	118
3200 Economical panels, fabric face, 4'-0" long, 5'-0" high		132	.121		51.50	5.95		57.45	65.50
3250 6'-0" high		112	.143		57.50	7.05		64.55	74
3300 5'-0" long, 5'-0" high		150	.107		55.50	5.25		60.75	69

For customer support on your Building Construction Costs with RSMeans Data, call 800.448.8182.

383

10 22 Partitions

10 22 23 – Portable Partitions, Screens, and Panels

10 22 23.13 Wall Screens		Crew	Daily Output	Labor-Hours	Unit	Material	2017 Bare Costs Labor	Equipment	Total	Total Incl O&P
3350	6'-0" high	2 Carp	125	.128	L.F.	53	6.30		59.30	68
3450	Acoustical panels, 60 to 90 NRC, 3'-0" long, 5'-0" high		90	.178		78.50	8.75		87.25	100
3550	6'-0" high		75	.213		89.50	10.50		100	115
3600	5'-0" long, 5'-0" high		150	.107		62	5.25		67.25	76.50
3650	6'-0" high		125	.128		69.50	6.30		75.80	86
3700	6'-0" long, 5'-0" high		162	.099		54	4.86		58.86	67
3750	6'-0" high		138	.116		82	5.70		87.70	99
3800	Economy acoustical panels, 40 NRC, 4'-0" long, 5'-0" high		132	.121		51.50	5.95		57.45	65.50
3850	6'-0" high		112	.143		57.50	7.05		64.55	74
3900	5'-0" long, 6'-0" high		125	.128		53	6.30		59.30	68
3950	6'-0" long, 5'-0" high		162	.099		48.50	4.86		53.36	61
4000	Metal chalkboard, 6'-6" high, chalkboard, 1 side		125	.128		123	6.30		129.30	145
4100	Metal chalkboard, 2 sides		120	.133		140	6.55		146.55	164
4300	Tackboard, both sides		123	.130		111	6.40		117.40	132

10 22 33 – Accordion Folding Partitions

10 22 33.10 Partitions, Accordion Folding

		Crew	Daily Output	Labor-Hours	Unit	Material	2017 Bare Costs Labor	Equipment	Total	Total Incl O&P
0010	**PARTITIONS, ACCORDION FOLDING**									
0100	Vinyl covered, over 150 S.F., frame not included									
0300	Residential, 1.25 lb. per S.F., 8' maximum height	2 Carp	300	.053	S.F.	26	2.63		28.63	32.50
0400	Commercial, 1.75 lb. per S.F., 8' maximum height		225	.071		29.50	3.50		33	38
0600	2 lb. per S.F., 17' maximum height		150	.107		30.50	5.25		35.75	41.50
0700	Industrial, 4 lb. per S.F., 20' maximum height		75	.213		45.50	10.50		56	66
0900	Acoustical, 3 lb. per S.F., 17' maximum height		100	.160		32.50	7.90		40.40	47.50
1200	5 lb. per S.F., 20' maximum height		95	.168		45	8.30		53.30	62
1300	5.5 lb. per S.F., 17' maximum height		90	.178		53	8.75		61.75	71.50
1400	Fire rated, 4.5 psf, 20' maximum height		160	.100		53	4.93		57.93	65.50
1500	Vinyl clad wood or steel, electric operation, 5.0 psf		160	.100		64.50	4.93		69.43	78.50
1900	Wood, non-acoustic, birch or mahogany, to 10' high		300	.053		34.50	2.63		37.13	42

10 22 39 – Folding Panel Partitions

10 22 39.10 Partitions, Folding Panel

		Crew	Daily Output	Labor-Hours	Unit	Material	2017 Bare Costs Labor	Equipment	Total	Total Incl O&P
0010	**PARTITIONS, FOLDING PANEL**, acoustic, wood									
0100	Vinyl faced, to 18' high, 6 psf, economy trim	2 Carp	60	.267	S.F.	58.50	13.15		71.65	84.50
0150	Standard trim		45	.356		70	17.50		87.50	104
0200	Premium trim		30	.533		90	26.50		116.50	140
0400	Plastic laminate or hardwood finish, standard trim		60	.267		60.50	13.15		73.65	86.50
0500	Premium trim		30	.533		64.50	26.50		91	111
0600	Wood, low acoustical type, 4.5 psf, to 14' high		50	.320		44	15.75		59.75	72
1100	Steel, acoustical, 9 to 12 lb. per S.F., vinyl faced, standard trim		60	.267		62.50	13.15		75.65	88.50
1200	Premium trim		30	.533		76	26.50		102.50	124
1700	Aluminum framed, acoustical, to 12' high, 5.5 psf, standard trim		60	.267		42	13.15		55.15	66
1800	Premium trim		30	.533		50.50	26.50		77	96
2000	6.5 lb. per S.F., standard trim		60	.267		44	13.15		57.15	68.50
2100	Premium trim		30	.533		54.50	26.50		81	100

10 22 43 – Sliding Partitions

10 22 43.10 Partitions, Sliding

		Crew	Daily Output	Labor-Hours	Unit	Material	2017 Bare Costs Labor	Equipment	Total	Total Incl O&P
0010	**PARTITIONS, SLIDING**									
0020	Acoustic air wall, 1-5/8" thick, standard trim	2 Carp	375	.043	S.F.	34	2.10		36.10	40.50
0100	Premium trim		365	.044		58.50	2.16		60.66	67.50
0300	2-1/4" thick, standard trim		360	.044		38	2.19		40.19	45.50
0400	Premium trim		330	.048		67	2.39		69.39	77
0600	For track type, add to above				L.F.	124			124	137

10 22 Partitions

10 22 43 – Sliding Partitions

10 22 43.10 Partitions, Sliding	Crew	Daily Output	Labor-Hours	Unit	Material	2017 Bare Costs Labor	Equipment	Total	Total Incl O&P
0700	Overhead track type, acoustical, 3" thick, 11 psf, standard trim	2 Carp	350	.046	S.F.	86	2.25		88.25	98
0800	Premium trim	"	300	.053	"	103	2.63		105.63	118

10 26 Wall and Door Protection

10 26 13 – Corner Guards

10 26 13.20 Corner Protection

		Crew	Daily Output	Labor-Hours	Unit	Material	2017 Bare Costs Labor	Equipment	Total	Total Incl O&P
0010	**CORNER PROTECTION**									
0100	Stainless steel, 16 ga., adhesive mount, 3-1/2" leg	1 Carp	80	.100	L.F.	23.50	4.93		28.43	33
0200	12 ga. stainless, adhesive mount	"	80	.100		21.50	4.93		26.43	31.50
0300	For screw mount, add						10%			
0500	Vinyl acrylic, adhesive mount, 3" leg	1 Carp	128	.063		9.65	3.08		12.73	15.30
0550	1-1/2" leg		160	.050		5.05	2.46		7.51	9.35
0600	Screw mounted, 3" leg		80	.100		10.35	4.93		15.28	18.90
0650	1-1/2" leg		100	.080		4.52	3.94		8.46	11
0700	Clear plastic, screw mounted, 2-1/2"		60	.133		4.68	6.55		11.23	15.20
1000	Vinyl cover, alum. retainer, surface mount, 3" x 3"		48	.167		10.90	8.20		19.10	24.50
1050	2" x 2"		48	.167		9.75	8.20		17.95	23.50
1100	Flush mounted, 3" x 3"		32	.250		20.50	12.30		32.80	41.50
1150	2" x 2"		32	.250		16.70	12.30		29	37.50

10 26 16 – Bumper Guards

10 26 16.10 Guard, Bumper

		Crew	Daily Output	Labor-Hours	Unit	Material	2017 Bare Costs Labor	Equipment	Total	Total Incl O&P
0010	**GUARD, BUMPER**									
1200	Bed bumper, vinyl acrylic, alum. retainer, 21" long	1 Carp	10	.800	Ea.	42	39.50		81.50	107
1300	53" long with aligner		9	.889	"	105	44		149	183
1400	Bumper, vinyl cover, alum. retain., cush. mnt., 1-1/2" x 2-3/4"		80	.100	L.F.	14.20	4.93		19.13	23
1500	2" x 4-1/4"		80	.100		20.50	4.93		25.43	30.50
1600	Surface mounted, 1-3/4" x 3-5/8"		80	.100		11.95	4.93		16.88	20.50

10 26 16.16 Protective Corridor Handrails

		Crew	Daily Output	Labor-Hours	Unit	Material	2017 Bare Costs Labor	Equipment	Total	Total Incl O&P
0010	**PROTECTIVE CORRIDOR HANDRAILS**									
3000	Handrail/bumper, vinyl cover, alum. retainer									
3010	Bracket mounted, flat rail, 5-1/2"	1 Carp	80	.100	L.F.	18.30	4.93		23.23	27.50
3100	6-1/2"		80	.100		23	4.93		27.93	32.50
3200	Bronze bracket, 1-3/4" diam. rail		80	.100		16.50	4.93		21.43	25.50
4000	Handrail, with antimicrobial copper alloy, #6 finish, 1-1/2" OD		80	.100		7.80	4.93		12.73	16.15

10 26 23 – Protective Wall Covering

10 26 23.10 Wall Covering, Protective

		Crew	Daily Output	Labor-Hours	Unit	Material	2017 Bare Costs Labor	Equipment	Total	Total Incl O&P
0010	**WALL COVERING, PROTECTIVE**									
0400	Rub rail, vinyl, adhesive mounted	1 Carp	185	.043	L.F.	8.95	2.13		11.08	13.10
0500	Neoprene, aluminum backing, 1-1/2" x 2"		110	.073		9.10	3.58		12.68	15.50
1000	Trolley rail, PVC, clipped to wall, 5" high		185	.043		8.95	2.13		11.08	13.05
1050	8" high		180	.044		15.10	2.19		17.29	20
1700	Bumper rail, stainless steel, flat bar on brackets, 4" x 1/4"	2 Skwk	120	.133		36	6.85		42.85	50
1775	Wall end guard, stainless steel, 16 ga, 36" tall, screwed to studs	1 Skwk	30	.267	Ea.	24.50	13.70		38.20	48
2000	Crash rail, vinyl cover, alum. retainer, 1" x 4"	1 Carp	110	.073	L.F.	11.05	3.58		14.63	17.65
2100	1" x 8"		90	.089		18.70	4.38		23.08	27
2150	Vinyl inserts, aluminum plate, 1" x 2-1/2"		110	.073		14.90	3.58		18.48	22
2200	1" x 5"		90	.089		24	4.38		28.38	33

10 26 Wall and Door Protection

10 26 23 – Protective Wall Covering

10 26 23.13 Impact Resistant Wall Protection

		Crew	Daily Output	Labor-Hours	Unit	Material	2017 Bare Costs Labor	Equipment	Total	Total Incl O&P
0010	**IMPACT RESISTANT WALL PROTECTION**									
0100	Vinyl wall protection, complete instl. incl. panels, trim, adhesive									
0110	.040" thk., std. colors	2 Carp	320	.050	S.F.	14.40	2.46		16.86	19.60
0120	.040" thk., element patterns		300	.053		16.20	2.63		18.83	22
0130	.060" thk., std. colors		300	.053		16.20	2.63		18.83	22
0140	.060" thk., element patterns	↓	280	.057	↓	17.80	2.81		20.61	24
1750	Wallguard stainless steel baseboard, 12" tall, adhesive applied	2 Skwk	260	.062	L.F.	33	3.17		36.17	41.50
1775	Wall protection, stainless steel, 16 ga, 48" x 36" tall, screwed to studs	"	500	.032	S.F.	8.50	1.65		10.15	11.90

10 26 33 – Door and Frame Protection

10 26 33.10 Protection, Door and Frame

		Crew	Daily Output	Labor-Hours	Unit	Material	2017 Bare Costs Labor	Equipment	Total	Total Incl O&P
0010	**PROTECTION, DOOR AND FRAME**									
0100	Door frame guard, vinyl, 3"x3"x4'	1 Carp	30	.267	Ea.	82	13.15		95.15	110
0110	Door frame guard, vinyl, 3"x3"x8'		26	.308		127	15.15		142.15	163
0120	Door frame guard, stainless steel, 3"x3"x4'	↓	30	.267		415	13.15		428.15	475
0500	Steel door track/wheel guards, 4'-0" high	E-4	22	1.455	↓	109	79.50	5.90	194.40	260

10 28 Toilet, Bath, and Laundry Accessories

10 28 13 – Toilet Accessories

10 28 13.13 Commercial Toilet Accessories

		Crew	Daily Output	Labor-Hours	Unit	Material	2017 Bare Costs Labor	Equipment	Total	Total Incl O&P
0010	**COMMERCIAL TOILET ACCESSORIES**									
0200	Curtain rod, stainless steel, 5' long, 1" diameter	1 Carp	13	.615	Ea.	28	30.50		58.50	77
0300	1-1/4" diameter		13	.615		32.50	30.50		63	82
0350	Chrome, 1" diameter		13	.615	↓	28	30.50		58.50	77.50
0360	For vinyl curtain, add		1950	.004	S.F.	1.03	.20		1.23	1.44
0400	Diaper changing station, horizontal, wall mounted, plastic		10	.800	Ea.	243	39.50		282.50	330
0420	Vertical		10	.800		216	39.50		255.50	298
0430	Oval shaped		10	.800		216	39.50		255.50	298
0440	Recessed, with stainless steel flange	↓	6	1.333	↓	510	65.50		575.50	660
0500	Dispenser units, combined soap & towel dispensers,									
0510	Mirror and shelf, flush mounted	1 Carp	10	.800	Ea.	310	39.50		349.50	405
0600	Towel dispenser and waste receptacle,									
0610	18 gallon capacity	1 Carp	10	.800	Ea.	320	39.50		359.50	415
0800	Grab bar, straight, 1-1/4" diameter, stainless steel, 18" long		24	.333		30	16.40		46.40	58
0900	24" long		23	.348		29.50	17.15		46.65	58.50
1000	30" long		22	.364		32.50	17.90		50.40	63.50
1100	36" long		20	.400		35	19.70		54.70	68.50
1105	42" long		20	.400		36.50	19.70		56.20	70
1120	Corner, 36" long		20	.400		93.50	19.70		113.20	133
1200	1-1/2" diameter, 24" long		23	.348		31.50	17.15		48.65	60.50
1300	36" long		20	.400		34	19.70		53.70	67.50
1310	42" long		18	.444		37.50	22		59.50	75
1500	Tub bar, 1-1/4" diameter, 24" x 36"		14	.571		103	28		131	156
1600	Plus vertical arm		12	.667		93.50	33		126.50	154
1900	End tub bar, 1" diameter, 90° angle, 16" x 32"		12	.667		104	33		137	165
2010	Tub/shower/toilet, 2-wall, 36" x 24"		12	.667		97.50	33		130.50	158
2110	Antimicrobial copper alloy finish, straight, 18" long		24	.333		71.50	16.40		87.90	104
2120	24" long		23	.348		78.50	17.15		95.65	112
2130	36" long		20	.400		93.50	19.70		113.20	133
2140	48" long		19	.421		107	20.50		127.50	150
2300	Hand dryer, surface mounted, electric, 115 volt, 20 amp		4	2		490	98.50		588.50	690

10 28 Toilet, Bath, and Laundry Accessories

10 28 13 – Toilet Accessories

10 28 13.13 Commercial Toilet Accessories	Crew	Daily Output	Labor-Hours	Unit	Material	2017 Bare Costs Labor	Equipment	Total	Total Incl O&P
2400	230 volt, 10 amp	1 Carp	4	2	Ea.	655	98.50		753.50	870
2450	Hand dryer, touch free, 1400 watt, 81,000 rpm		4	2		1,050	98.50		1,148.50	1,300
2600	Hat and coat strip, stainless steel, 4 hook, 36" long		24	.333		68	16.40		84.40	100
2700	6 hook, 60" long		20	.400		124	19.70		143.70	167
3000	Mirror, with stainless steel 3/4" square frame, 18" x 24"		20	.400		47	19.70		66.70	82
3100	36" x 24"		15	.533		111	26.50		137.50	162
3200	48" x 24"		10	.800		168	39.50		207.50	246
3300	72" x 24"		6	1.333		283	65.50		348.50	410
3500	With 5" stainless steel shelf, 18" x 24"		20	.400		193	19.70		212.70	242
3600	36" x 24"		15	.533		244	26.50		270.50	310
3700	48" x 24"		10	.800		257	39.50		296.50	345
3800	72" x 24"		6	1.333		259	65.50		324.50	385
4100	Mop holder strip, stainless steel, 5 holders, 48" long		20	.400		80	19.70		99.70	118
4200	Napkin/tampon dispenser, recessed		15	.533		685	26.50		711.50	790
4220	Semi-recessed		6.50	1.231		315	60.50		375.50	440
4250	Napkin receptacle, recessed		6.50	1.231		180	60.50		240.50	291
4300	Robe hook, single, regular		96	.083		21	4.10		25.10	30
4400	Heavy duty, concealed mounting		56	.143		23.50	7.05		30.55	37
4600	Soap dispenser, chrome, surface mounted, liquid		20	.400		50	19.70		69.70	85
5000	Recessed stainless steel, liquid		10	.800		169	39.50		208.50	246
5600	Shelf, stainless steel, 5" wide, 18 ga., 24" long		24	.333		86	16.40		102.40	120
5700	48" long		16	.500		157	24.50		181.50	210
5800	8" wide shelf, 18 ga., 24" long		22	.364		75.50	17.90		93.40	111
5900	48" long		14	.571		141	28		169	199
6000	Toilet seat cover dispenser, stainless steel, recessed		20	.400		183	19.70		202.70	232
6050	Surface mounted		15	.533		36.50	26.50		63	80
6100	Toilet tissue dispenser, surface mounted, SS, single roll		30	.267		19.85	13.15		33	42
6200	Double roll		24	.333		25	16.40		41.40	52.50
6240	Plastic, twin/jumbo dbl. roll		24	.333		28.50	16.40		44.90	56
6400	Towel bar, stainless steel, 18" long		23	.348		44.50	17.15		61.65	75
6500	30" long	↓	21	.381	↓	61	18.75		79.75	96
6610	Antimicrobial copper alloy finish, 3/4" round, straight, w/o mounting				L.F.	11.40			11.40	12.55
6620	Antimicrobial copper alloy finish, 1" round, straight, w/o mounting				"	13.70			13.70	15.05
6630	24" long, including mounting	1 Carp	23	.348	Ea.	97	17.15		114.15	133
6700	Towel dispenser, stainless steel, surface mounted		16	.500		45	24.50		69.50	87
6800	Flush mounted, recessed		10	.800		96.50	39.50		136	167
6900	Plastic, touchless, battery operated		16	.500		94	24.50		118.50	141
7000	Towel holder, hotel type, 2 guest size		20	.400		62	19.70		81.70	98
7200	Towel shelf, stainless steel, 24" long, 8" wide		20	.400		66.50	19.70		86.20	103
7400	Tumbler holder, for tumbler only		30	.267		19.10	13.15		32.25	41
7410	Tumbler holder, recessed		20	.400		8	19.70		27.70	39
7500	Soap, tumbler & toothbrush		30	.267		19.60	13.15		32.75	41.50
7510	Tumbler & toothbrush holder		20	.400		13.65	19.70		33.35	45
7700	Wall urn ash receiver, surface mount, 11" long		12	.667		95	33		128	155
7800	7-1/2" long		18	.444		97	22		119	141
8000	Waste receptacles, stainless steel, with top, 13 gallon		10	.800		335	39.50		374.50	430
8100	36 gallon		8	1		415	49.50		464.50	535
9996	Bathroom access., grab bar, straight, 1-1/2" dia, SS, 42" long install only	↓	18	.444	↓		22		22	33.50

For customer support on your Building Construction Costs with RSMeans Data, call 800.448.8182.

387

10 28 16 – Bath Accessories

10 28 16.20 Medicine Cabinets

		Crew	Daily Output	Labor-Hours	Unit	Material	2017 Bare Costs Labor	Equipment	Total	Total Incl O&P
0010	**MEDICINE CABINETS**									
0020	With mirror, sst frame, 16" x 22", unlighted	1 Carp	14	.571	Ea.	107	28		135	161
0100	Wood frame		14	.571		140	28		168	197
0300	Sliding mirror doors, 20" x 16" x 4-3/4", unlighted		7	1.143		128	56.50		184.50	226
0400	24" x 19" x 8-1/2", lighted		5	1.600		189	79		268	330
0600	Triple door, 30" x 32", unlighted, plywood body		7	1.143		350	56.50		406.50	470
0700	Steel body		7	1.143		375	56.50		431.50	495
0900	Oak door, wood body, beveled mirror, single door		7	1.143		196	56.50		252.50	300
1000	Double door		6	1.333		415	65.50		480.50	555
1200	Hotel cabinets, stainless, with lower shelf, unlighted		10	.800		200	39.50		239.50	280
1300	Lighted		5	1.600		310	79		389	465

10 28 19 – Tub and Shower Enclosures

10 28 19.10 Partitions, Shower

		Crew	Daily Output	Labor-Hours	Unit	Material	2017 Bare Costs Labor	Equipment	Total	Total Incl O&P
0010	**PARTITIONS, SHOWER** floor mounted, no plumbing									
0400	Cabinet, one piece, fiberglass, 32" x 32"	2 Carp	5	3.200	Ea.	540	158		698	835
0420	36" x 36"		5	3.200		655	158		813	960
0440	36" x 48"		5	3.200		1,375	158		1,533	1,775
0460	Acrylic, 32" x 32"		5	3.200		350	158		508	625
0480	36" x 36"		5	3.200		1,050	158		1,208	1,400
0500	36" x 48"		5	3.200		1,475	158		1,633	1,875
0520	Shower door for above, clear plastic, 24" wide	1 Carp	8	1		192	49.50		241.50	288
0540	28" wide		8	1		213	49.50		262.50	310
0560	Tempered glass, 24" wide		8	1		205	49.50		254.50	300
0580	28" wide		8	1		235	49.50		284.50	335
2400	Glass stalls, with doors, no receptors, chrome on brass	2 Shee	3	5.333		1,700	310		2,010	2,350
2700	Anodized aluminum	"	4	4		1,175	232		1,407	1,650
2900	Marble shower stall, stock design, with shower door	2 Marb	1.20	13.333		2,400	625		3,025	3,600
3000	With curtain		1.30	12.308		2,125	575		2,700	3,200
3200	Receptors, precast terrazzo, 32" x 32"		14	1.143		350	53.50		403.50	470
3300	48" x 34"		9.50	1.684		465	79		544	630
3500	Plastic, simulated terrazzo receptor, 32" x 32"		14	1.143		160	53.50		213.50	259
3600	32" x 48"		12	1.333		230	62.50		292.50	350
3800	Precast concrete, colors, 32" x 32"		14	1.143		218	53.50		271.50	320
3900	48" x 48"		8	2		279	93.50		372.50	450
4100	Shower doors, economy plastic, 24" wide	1 Shee	9	.889		144	51.50		195.50	237
4200	Tempered glass door, economy		8	1		267	58		325	385
4400	Folding, tempered glass, aluminum frame		6	1.333		420	77.50		497.50	580
4500	Sliding, tempered glass, 48" opening		6	1.333		575	77.50		652.50	750
4700	Deluxe, tempered glass, chrome on brass frame, 42" to 44"		8	1		415	58		473	545
4800	39" to 48" wide		1	8		700	465		1,165	1,475
4850	On anodized aluminum frame, obscure glass		2	4		575	232		807	985
4900	Clear glass		1	8		700	465		1,165	1,475
5100	Shower enclosure, tempered glass, anodized alum. frame									
5120	2 panel & door, corner unit, 32" x 32"	1 Shee	2	4	Ea.	1,100	232		1,332	1,575
5140	Neo-angle corner unit, 16" x 24" x 16"	"	2	4		1,125	232		1,357	1,575
5200	Shower surround, 3 wall, polypropylene, 32" x 32"	1 Carp	4	2		520	98.50		618.50	720
5220	PVC, 32" x 32"		4	2		405	98.50		503.50	595
5240	Fiberglass		4	2		420	98.50		518.50	610
5250	2 wall, polypropylene, 32" x 32"		4	2		325	98.50		423.50	510
5270	PVC		4	2		395	98.50		493.50	585
5290	Fiberglass		4	2		400	98.50		498.50	590
5300	Tub doors, tempered glass & frame, obscure glass	1 Shee	8	1		234	58		292	345

10 28 Toilet, Bath, and Laundry Accessories

10 28 19 – Tub and Shower Enclosures

10 28 19.10 Partitions, Shower

10 28 19.10 Partitions, Shower	Crew	Daily Output	Labor-Hours	Unit	Material	2017 Bare Costs Labor	Equipment	Total	Total Incl O&P
5400	Clear glass	1 Shee	6	1.333	Ea.	545	77.50		622.50	720
5600	Chrome plated, brass frame, obscure glass		8	1		310	58		368	430
5700	Clear glass		6	1.333		755	77.50		832.50	955
5900	Tub/shower enclosure, temp. glass, alum. frame, obscure glass		2	4		420	232		652	815
6200	Clear glass		1.50	5.333		875	310		1,185	1,425
6500	On chrome-plated brass frame, obscure glass		2	4		580	232		812	990
6600	Clear glass		1.50	5.333		1,225	310		1,535	1,825
6800	Tub surround, 3 wall, polypropylene	1 Carp	4	2		264	98.50		362.50	440
6900	PVC		4	2		390	98.50		488.50	580
7000	Fiberglass, obscure glass		4	2		410	98.50		508.50	600
7100	Clear glass		3	2.667		695	131		826	965

10 28 23 – Laundry Accessories

10 28 23.13 Built-In Ironing Boards

		Crew	Daily Output	Labor-Hours	Unit	Material	2017 Bare Costs Labor	Equipment	Total	Total Incl O&P
0010	**BUILT-IN IRONING BOARDS**									
0020	Including cabinet, board & light, 42"	1 Carp	2	4	Ea.	455	197		652	800
0100	46"	"	1.50	5.333	"	520	263		783	970

10 31 Manufactured Fireplaces

10 31 13 – Manufactured Fireplace Chimneys

10 31 13.10 Fireplace Chimneys

		Crew	Daily Output	Labor-Hours	Unit	Material	2017 Bare Costs Labor	Equipment	Total	Total Incl O&P
0010	**FIREPLACE CHIMNEYS**									
0500	Chimney dbl. wall, all stainless, over 8'-6", 7" diam., add to fireplace	1 Carp	33	.242	V.L.F.	87.50	11.95		99.45	114
0600	10" diameter, add to fireplace		32	.250		112	12.30		124.30	143
0700	12" diameter, add to fireplace		31	.258		172	12.70		184.70	209
0800	14" diameter, add to fireplace		30	.267		227	13.15		240.15	270
1000	Simulated brick chimney top, 4' high, 16" x 16"		10	.800	Ea.	440	39.50		479.50	545
1100	24" x 24"		7	1.143	"	560	56.50		616.50	700

10 31 13.20 Chimney Accessories

		Crew	Daily Output	Labor-Hours	Unit	Material	2017 Bare Costs Labor	Equipment	Total	Total Incl O&P
0010	**CHIMNEY ACCESSORIES**									
0020	Chimney screens, galv., 13" x 13" flue	1 Bric	8	1	Ea.	58.50	48		106.50	138
0050	24" x 24" flue		5	1.600		124	76.50		200.50	255
0200	Stainless steel, 13" x 13" flue		8	1		97.50	48		145.50	181
0250	20" x 20" flue		5	1.600		152	76.50		228.50	285
2400	Squirrel and bird screens, galvanized, 8" x 8" flue		16	.500		60	24		84	103
2450	13" x 13" flue		12	.667		62.50	32		94.50	118

10 31 16 – Manufactured Fireplace Forms

10 31 16.10 Fireplace Forms

		Crew	Daily Output	Labor-Hours	Unit	Material	2017 Bare Costs Labor	Equipment	Total	Total Incl O&P
0010	**FIREPLACE FORMS**									
1800	Fireplace forms, no accessories, 32" opening	1 Bric	3	2.667	Ea.	720	127		847	985
1900	36" opening		2.50	3.200		915	153		1,068	1,225
2000	40" opening		2	4		1,225	191		1,416	1,625
2100	78" opening		1.50	5.333		1,775	255		2,030	2,350

10 31 23 – Prefabricated Fireplaces

10 31 23.10 Fireplace, Prefabricated

		Crew	Daily Output	Labor-Hours	Unit	Material	2017 Bare Costs Labor	Equipment	Total	Total Incl O&P
0010	**FIREPLACE, PREFABRICATED**, free standing or wall hung									
0100	With hood & screen, painted	1 Carp	1.30	6.154	Ea.	1,525	305		1,830	2,150
0150	Average		1	8		1,750	395		2,145	2,525
0200	Stainless steel		.90	8.889		3,200	440		3,640	4,175
1500	Simulated logs, gas fired, 40,000 BTU, 2' long, manual safety pilot		7	1.143	Set	475	56.50		531.50	605

10 31 Manufactured Fireplaces

10 31 23 – Prefabricated Fireplaces

10 31 23.10 Fireplace, Prefabricated	Crew	Daily Output	Labor-Hours	Unit	Material	2017 Bare Costs Labor	Equipment	Total	Total Incl O&P
1600	Adjustable flame remote pilot	1 Carp	6	1.333	Set	1,250	65.50		1,315.50	1,475
1700	Electric, 1,500 BTU, 1'-6" long, incandescent flame		7	1.143		218	56.50		274.50	325
1800	1,500 BTU, LED flame		6	1.333	↓	335	65.50		400.50	465
2000	Fireplace, built-in, 36" hearth, radiant		1.30	6.154	Ea.	700	305		1,005	1,225
2100	Recirculating, small fan		1	8		905	395		1,300	1,600
2150	Large fan		.90	8.889		2,000	440		2,440	2,875
2200	42" hearth, radiant		1.20	6.667		950	330		1,280	1,550
2300	Recirculating, small fan		.90	8.889		1,225	440		1,665	2,025
2350	Large fan		.80	10		1,400	495		1,895	2,300
2400	48" hearth, radiant		1.10	7.273		2,275	360		2,635	3,050
2500	Recirculating, small fan		.80	10		2,600	495		3,095	3,625
2550	Large fan		.70	11.429		2,700	565		3,265	3,825
3000	See through, including doors		.80	10		2,275	495		2,770	3,250
3200	Corner (2 wall)	↓	1	8	↓	3,700	395		4,095	4,675

10 32 Fireplace Specialties

10 32 13 – Fireplace Dampers

10 32 13.10 Dampers

		Crew	Daily Output	Labor-Hours	Unit	Material	2017 Bare Costs Labor	Equipment	Total	Total Incl O&P
0010	**DAMPERS**									
0800	Damper, rotary control, steel, 30" opening	1 Bric	6	1.333	Ea.	119	63.50		182.50	229
0850	Cast iron, 30" opening		6	1.333		125	63.50		188.50	235
0880	36" opening		6	1.333		127	63.50		190.50	238
0900	48" opening		6	1.333		167	63.50		230.50	282
0920	60" opening		6	1.333		355	63.50		418.50	490
0950	72" opening		5	1.600		425	76.50		501.50	585
1000	84" opening, special order		5	1.600		910	76.50		986.50	1,125
1050	96" opening, special order		4	2		925	95.50		1,020.50	1,175
1200	Steel plate, poker control, 60" opening		8	1		320	48		368	430
1250	84" opening, special order		5	1.600		585	76.50		661.50	765
1400	"Universal" type, chain operated, 32" x 20" opening		8	1		250	48		298	350
1450	48" x 24" opening	↓	5	1.600	↓	375	76.50		451.50	530

10 32 23 – Fireplace Doors

10 32 23.10 Doors

		Crew	Daily Output	Labor-Hours	Unit	Material	2017 Bare Costs Labor	Equipment	Total	Total Incl O&P
0010	**DOORS**									
0400	Cleanout doors and frames, cast iron, 8" x 8"	1 Bric	12	.667	Ea.	47.50	32		79.50	101
0450	12" x 12"		10	.800		89	38		127	157
0500	18" x 24"		8	1		150	48		198	239
0550	Cast iron frame, steel door, 24" x 30"		5	1.600		315	76.50		391.50	465
1600	Dutch Oven door and frame, cast iron, 12" x 15" opening		13	.615		131	29.50		160.50	190
1650	Copper plated, 12" x 15" opening	↓	13	.615	↓	257	29.50		286.50	330

10 35 Stoves

10 35 13 – Heating Stoves

10 35 13.10 Woodburning Stoves		Crew	Daily Output	Labor-Hours	Unit	Material	2017 Bare Costs Labor	Equipment	Total	Total Incl O&P
0010	**WOODBURNING STOVES**									
0015	Cast iron, less than 1500 S.F.	2 Carp	1.30	12.308	Ea.	1,300	605		1,905	2,375
0020	1500 to 2000 S.F.		1	16		2,050	790		2,840	3,450
0030	greater than 2,000 S.F.	↓	.80	20		2,800	985		3,785	4,575
0050	For gas log lighter, add				↓	46			46	50.50

10 43 Emergency Aid Specialties

10 43 13 – Defibrillator Cabinets

10 43 13.05 Defibrillator Cabinets		Crew	Daily Output	Labor-Hours	Unit	Material	2017 Bare Costs Labor	Equipment	Total	Total Incl O&P
0010	**DEFIBRILLATOR CABINETS**, not equipped, stainless steel									
0050	Defibrillator cabinet, stainless steel with strobe & alarm 12" x 27"	1 Carp	10	.800	Ea.	430	39.50		469.50	530
0100	Automatic External Defibrillator	"	30	.267	"	1,375	13.15		1,388.15	1,550

10 44 Fire Protection Specialties

10 44 13 – Fire Protection Cabinets

10 44 13.53 Fire Equipment Cabinets		Crew	Daily Output	Labor-Hours	Unit	Material	2017 Bare Costs Labor	Equipment	Total	Total Incl O&P
0010	**FIRE EQUIPMENT CABINETS**, not equipped, 20 ga. steel box									
0040	Recessed, D.S. glass in door, box size given									
1000	Portable extinguisher, single, 8" x 12" x 27", alum. door & frame	Q-12	8	2	Ea.	163	106		269	340
1100	Steel door and frame		8	2		124	106		230	298
2700	Fire blanket & extinguisher cab, inc blanket, rec stl., 14" x 40" x 8"		7	2.286		197	121		318	400
2800	Fire blanket cab, inc blanket, surf mtd, stl, 15"x10"x5", w/pwdr coat fin	↓	8	2	↓	99	106		205	270
3000	Hose rack assy., 1-1/2" valve & 100' hose, 24" x 40" x 5-1/2"									
3100	Aluminum door and frame	Q-12	6	2.667	Ea.	410	142		552	665
3200	Steel door and frame		6	2.667		271	142		413	510
3300	Stainless steel door and frame	↓	6	2.667	↓	480	142		622	745
4000	Hose rack assy., 2-1/2" x 1-1/2" valve, 100' hose, 24" x 40" x 8"									
4100	Aluminum door and frame	Q-12	6	2.667	Ea.	410	142		552	670
4200	Steel door and frame		6	2.667		279	142		421	520
4300	Stainless steel door and frame	↓	6	2.667	↓	550	142		692	815
5000	Hose rack assy., 2-1/2" x 1-1/2" valve, 100' hose									
5010	and extinguisher, 30" x 40" x 8"									
5100	Aluminum door and frame	Q-12	5	3.200	Ea.	525	170		695	830
5200	Steel door and frame		5	3.200		285	170		455	570
5300	Stainless steel door and frame	↓	5	3.200	↓	565	170		735	875
8000	Valve cabinet for 2-1/2" FD angle valve, 18" x 18" x 8"									
8100	Aluminum door and frame	Q-12	12	1.333	Ea.	181	71		252	305
8200	Steel door and frame		12	1.333		149	71		220	271
8300	Stainless steel door and frame	↓	12	1.333	↓	243	71		314	375

10 44 16 – Fire Extinguishers

10 44 16.13 Portable Fire Extinguishers		Crew	Daily Output	Labor-Hours	Unit	Material	2017 Bare Costs Labor	Equipment	Total	Total Incl O&P
0010	**PORTABLE FIRE EXTINGUISHERS**									
0140	CO_2, with hose and "H" horn, 10 lb.				Ea.	296			296	325
0160	15 lb.					380			380	420
0180	20 lb.				↓	420			420	465
1000	Dry chemical, pressurized									
1040	Standard type, portable, painted, 2-1/2 lb.				Ea.	40.50			40.50	44.50
1060	5 lb.					55			55	60.50
1080	10 lb.				↓	83.50			83.50	92

10 44 Fire Protection Specialties

10 44 16 – Fire Extinguishers

10 44 16.13 Portable Fire Extinguishers

		Crew	Daily Output	Labor-Hours	Unit	Material	2017 Bare Costs Labor	2017 Bare Costs Equipment	Total	Total Incl O&P
1100	20 lb.				Ea.	136			136	149
1120	30 lb.					405			405	445
1300	Standard type, wheeled, 150 lb.					1,200			1,200	1,325
2000	ABC all purpose type, portable, 2-1/2 lb.					22.50			22.50	24.50
2060	5 lb.					26.50			26.50	29
2080	9-1/2 lb.					49			49	54
2100	20 lb.					79.50			79.50	87.50
3500	Halotron 1, 2-1/2 lb.					127			127	140
3600	5 lb.					210			210	231
3700	11 lb.					410			410	450
5000	Pressurized water, 2-1/2 gallon, stainless steel					102			102	112
5060	With anti-freeze					113			113	124
9400	Installation of extinguishers, 12 or more, on nailable surface	1 Carp	30	.267			13.15		13.15	20
9420	On masonry or concrete	"	15	.533	▼		26.50		26.50	40

10 44 16.16 Wheeled Fire Extinguisher Units

		Crew	Daily Output	Labor-Hours	Unit	Material	2017 Bare Costs Labor	2017 Bare Costs Equipment	Total	Total Incl O&P
0010	**WHEELED FIRE EXTINGUISHER UNITS**									
0350	CO₂, portable, with swivel horn									
0360	Wheeled type, cart mounted, 50 lb.				Ea.	1,175			1,175	1,300
0400	100 lb.				"	4,400			4,400	4,825
2200	ABC all purpose type									
2300	Wheeled, 45 lb.				Ea.	735			735	810
2360	150 lb.				"	1,850			1,850	2,025

10 51 Lockers

10 51 13 – Metal Lockers

10 51 13.10 Lockers

		Crew	Daily Output	Labor-Hours	Unit	Material	2017 Bare Costs Labor	2017 Bare Costs Equipment	Total	Total Incl O&P
0011	**LOCKERS** steel, baked enamel, pre-assembled									
0110	Single tier box locker, 12" x 15" x 72"	1 Shee	20	.400	Ea.	234	23		257	293
0120	18" x 15" x 72"		20	.400		248	23		271	310
0130	12" x 18" x 72"		20	.400		240	23		263	300
0140	18" x 18" x 72"		20	.400		294	23		317	360
0410	Double tier, 12" x 15" x 36"		30	.267		233	15.50		248.50	280
0420	18" x 15" x 36"		30	.267		235	15.50		250.50	283
0430	12" x 18" x 36"		30	.267		271	15.50		286.50	320
0440	18" x 18" x 36"		30	.267		230	15.50		245.50	277
0500	Two person, 18" x 15" x 72"		20	.400		305	23		328	370
0510	18" x 18" x 72"		20	.400		330	23		353	400
0520	Duplex, 15" x 15" x 72"		20	.400		335	23		358	405
0530	15" x 21" x 72"		20	.400	▼	370	23		393	445
0600	5 tier box lockers, unassembled		30	.267	Opng.	48.50	15.50		64	76.50
0700	Set up		24	.333		52	19.35		71.35	86.50
0900	6 tier box lockers, unassembled		36	.222		38.50	12.90		51.40	62.50
1000	Set up		30	.267	▼	45.50	15.50		61	73.50
1100	Wire meshed wardrobe, floor mtd., open front varsity type	▼	7.50	1.067	Ea.	280	62		342	405
2400	16-person locker unit with clothing rack									
2500	72" wide x 15" deep x 72" high	1 Shee	15	.533	Ea.	495	31		526	590
2550	18" deep	"	15	.533	"	655	31		686	770
3000	Wall mounted lockers, 4 person, with coat bar									
3100	48" wide x 18" deep x 12" high	1 Shee	20	.400	Ea.	340	23		363	410
3250	Rack w/24 wire mesh baskets		1.50	5.333	Set	410	310		720	925
3260	30 baskets	▼	1.25	6.400	▼	385	370		755	995

10 51 Lockers

10 51 13 - Metal Lockers

10 51 13.10 Lockers		Crew	Daily Output	Labor-Hours	Unit	Material	2017 Bare Costs Labor	Equipment	Total	Total Incl O&P
3270	36 baskets	1 Shee	.95	8.421	Set	470	490		960	1,275
3280	42 baskets	↓	.80	10	↓	455	580		1,035	1,400
3300	For built-in lock with 2 keys, add				Ea.	13.85			13.85	15.25
3600	For hanger rods, add					2.30			2.30	2.53
3650	For number plate kit, 100 plates #1 - #100, add	1 Shee	4	2		82	116		198	268
3700	For locker base, closed front panel		90	.089		7.50	5.15		12.65	16.15
3710	End panel, bolted		36	.222		9.55	12.90		22.45	30.50
3800	For sloping top, 12" wide		24	.333		32.50	19.35		51.85	65.50
3810	15" wide		24	.333		37.50	19.35		56.85	71
3820	18" wide		24	.333		35.50	19.35		54.85	68.50
3850	Sloping top end panel, 12" deep		72	.111		14.75	6.45		21.20	26
3860	15" deep		72	.111		15	6.45		21.45	26.50
3870	18" deep		72	.111		18.15	6.45		24.60	30
3900	For finish end panels, steel, 60" high, 15" deep		12	.667		36	38.50		74.50	98.50
3910	72" high, 12" deep		12	.667		29	38.50		67.50	91
3920	18" deep	↓	12	.667	↓	43.50	38.50		82	107
5000	For "ready to assemble" lockers,									
5010	Add to labor						75%			
5020	Deduct from material					20%				
6000	Heavy duty for detention facility, tamper proof, 14 ga. welded steel, solid									
6100	24" W x 24" D x 74" H, single tier	1 Shee	18	.444	Ea.	535	26		561	630
6110	Double tier		18	.444		535	26		561	625
6120	Triple tier	↓	18	.444	↓	555	26		581	650

10 51 26 - Plastic Lockers

10 51 26.13 Recycled Plastic Lockers

			Crew	Daily Output	Labor-Hours	Unit	Material	Labor	Equipment	Total	Total Incl O&P
0011	**RECYCLED PLASTIC LOCKERS**, 30% recycled										
0110	Single tier box locker, 12" x 12" x 72"	G	1 Shee	8	1	Ea.	455	58		513	590
0120	12" x 15" x 72"	G		8	1		475	58		533	610
0130	12" x 18" x 72"	G		8	1		480	58		538	620
0410	Double tier, 12" x 12" x 72"	G		21	.381		490	22		512	575
0420	12" x 15" x 72"	G		21	.381		510	22		532	595
0430	12" x 18" x 72"	G	↓	21	.381	↓	505	22		527	590

10 51 53 - Locker Room Benches

10 51 53.10 Benches

		Crew	Daily Output	Labor-Hours	Unit	Material	Labor	Equipment	Total	Total Incl O&P
0010	**BENCHES**									
2100	Locker bench, laminated maple, top only	1 Shee	100	.080	L.F.	26.50	4.64		31.14	36
2200	Pedestals, steel pipe		25	.320	Ea.	44	18.60		62.60	77
2250	Plastic, 9.5" top with PVC pedestals	↓	80	.100	L.F.	61.50	5.80		67.30	76.50

10 55 Postal Specialties

10 55 23 - Mail Boxes

10 55 23.10 Commercial Mail Boxes

		Crew	Daily Output	Labor-Hours	Unit	Material	Labor	Equipment	Total	Total Incl O&P
0010	**COMMERCIAL MAIL BOXES**									
0020	Horiz., key lock, 5"H x 6"W x 15"D, alum., rear load	1 Carp	34	.235	Ea.	42.50	11.60		54.10	65
0100	Front loading		34	.235		42.50	11.60		54.10	65
0200	Double, 5"H x 12"W x 15"D, rear loading		26	.308		62.50	15.15		77.65	92
0300	Front loading		26	.308		75	15.15		90.15	106
0500	Quadruple, 10"H x 12"W x 15"D, rear loading		20	.400		102	19.70		121.70	142
0600	Front loading		20	.400		92.50	19.70		112.20	132
0800	Vertical, front load, 15"H x 5"W x 6"D, alum., per compartment	↓	34	.235	↓	45	11.60		56.60	67.50

10 55 Postal Specialties

10 55 23 – Mail Boxes

10 55 23.10 Commercial Mail Boxes	Crew	Daily Output	Labor-Hours	Unit	Material	2017 Bare Costs Labor	Equipment	Total	Total Incl O&P
0900	Bronze, duranodic finish	1 Carp	34	.235	Ea.	48	11.60		59.60	71
1000	Steel, enameled		34	.235		45	11.60		56.60	67.50
1700	Alphabetical directories, 120 names		10	.800		123	39.50		162.50	196
1800	Letter collection box	↓	6	1.333		740	65.50		805.50	910
1830	Lobby collection boxes, aluminum	2 Shee	5	3.200		1,775	186		1,961	2,225
1840	Bronze or stainless	"	4.50	3.556		1,950	206		2,156	2,475
1900	Letter slot, residential	1 Carp	20	.400		80	19.70		99.70	118
2000	Post office type		8	1		120	49.50		169.50	208
2250	Key keeper, single key, aluminum		26	.308		45	15.15		60.15	72.50
2300	Steel, enameled	↓	26	.308	↓	80	15.15		95.15	111

10 56 Storage Assemblies

10 56 13 – Metal Storage Shelving

10 56 13.10 Shelving

		Crew	Daily Output	Labor-Hours	Unit	Material	2017 Bare Costs Labor	Equipment	Total	Total Incl O&P
0010	**SHELVING**									
0020	Metal, industrial, cross-braced, 3' wide, 12" deep	1 Sswk	175	.046	SF Shlf	7.20	2.48		9.68	12.05
0100	24" deep		330	.024		5.20	1.32		6.52	7.90
0300	4' wide, 12" deep		185	.043		6.70	2.35		9.05	11.25
0400	24" deep		380	.021		4.76	1.14		5.90	7.15
1200	Enclosed sides, cross-braced back, 3' wide, 12" deep		175	.046		12.35	2.48		14.83	17.75
1300	24" deep		290	.028		8.35	1.50		9.85	11.70
1500	Fully enclosed, sides and back, 3' wide, 12" deep		150	.053		16.40	2.90		19.30	23
1600	24" deep		255	.031		10.90	1.70		12.60	14.85
1800	4' wide, 12" deep		150	.053		10.35	2.90		13.25	16.25
1900	24" deep		290	.028		8.85	1.50		10.35	12.20
2200	Wide span, 1600 lb. capacity per shelf, 6' wide, 24" deep		380	.021		6.95	1.14		8.09	9.50
2400	36" deep		440	.018		6.10	.99		7.09	8.35
2600	8' wide, 24" deep		440	.018		6.85	.99		7.84	9.20
2800	36" deep	↓	520	.015		4.51	.84		5.35	6.35
4000	Pallet racks, steel frame 5,000 lb. capacity, 8' long, 36" deep	2 Sswk	450	.036		9.35	1.93		11.28	13.50
4200	42" deep		500	.032		8.25	1.74		9.99	11.95
4400	48" deep	↓	520	.031	↓	7.75	1.67		9.42	11.30

10 56 13.20 Parts Bins

		Crew	Daily Output	Labor-Hours	Unit	Material	2017 Bare Costs Labor	Equipment	Total	Total Incl O&P
0010	**PARTS BINS** metal, gray baked enamel finish									
0100	6'-3" high, 3' wide									
0300	12 bins, 18" wide x 12" high, 12" deep	2 Clab	10	1.600	Ea.	335	62.50		397.50	465
0400	24" deep		10	1.600		420	62.50		482.50	555
0600	72 bins, 6" wide x 6" high, 12" deep		8	2		550	78.50		628.50	725
0700	18" deep	↓	8	2	↓	860	78.50		938.50	1,075
1000	7'-3" high, 3' wide									
1200	14 bins, 18" wide x 12" high, 12" deep	2 Clab	10	1.600	Ea.	355	62.50		417.50	485
1300	24" deep		10	1.600		445	62.50		507.50	585
1500	84 bins, 6" wide x 6" high, 12" deep		8	2		965	78.50		1,043.50	1,175
1600	24" deep	↓	8	2	↓	1,175	78.50		1,253.50	1,400

10 57 Wardrobe and Closet Specialties

10 57 13 – Hat and Coat Racks

10 57 13.10 Coat Racks and Wardrobes

		Crew	Daily Output	Labor-Hours	Unit	Material	2017 Bare Costs Labor	Equipment	Total	Total Incl O&P
0010	**COAT RACKS AND WARDROBES**									
0020	Hat & coat rack, floor model, 6 hangers									
0050	Standing, beech wood, 21" x 21" x 72", chrome				Ea.	247			247	272
0100	18 ga. tubular steel, 21" x 21" x 69", wood walnut				"	350			350	385
0500	16 ga. steel frame, 22 ga. steel shelves									
0650	Single pedestal, 30" x 18" x 63"				Ea.	305			305	335
0800	Single face rack, 29" x 18-1/2" x 62"					315			315	345
0900	51" x 18-1/2" x 70"					465			465	510
0910	Double face rack, 39" x 26" x 70"					405			405	445
0920	63" x 26" x 70"					505			505	555
0940	For 2" ball casters, add				Set	88			88	97
1400	Utility hook strips, 3/8" x 2-1/2" x 18", 6 hooks	1 Carp	48	.167	Ea.	56.50	8.20		64.70	74.50
1500	34" long, 12 hooks	"	48	.167	"	70	8.20		78.20	89.50
1650	Wall mounted racks, 16 ga. steel frame, 22 ga. steel shelves									
1850	12" x 15" x 26", 6 hangers	1 Carp	32	.250	Ea.	140	12.30		152.30	173
2000	12" x 15" x 50", 12 hangers	"	32	.250	"	162	12.30		174.30	197
2150	Wardrobe cabinet, steel, baked enamel finish									
2300	36" x 21" x 78", incl. top shelf & hanger rod				Ea.	330			330	365
2400	Wardrobe, 24" x 24" x 76", KD, w/door, hospital, baked enamel steel	1 Carp	2	4		655	197		852	1,025
2500	Hardwood	"	2	4		1,075	197		1,272	1,500

10 57 23 – Closet and Utility Shelving

10 57 23.19 Wood Closet and Utility Shelving

		Crew	Daily Output	Labor-Hours	Unit	Material	2017 Bare Costs Labor	Equipment	Total	Total Incl O&P
0010	**WOOD CLOSET AND UTILITY SHELVING**									
0020	Pine, clear grade, no edge band, 1" x 8"	1 Carp	115	.070	L.F.	3.21	3.43		6.64	8.80
0100	1" x 10"		110	.073		3.99	3.58		7.57	9.90
0200	1" x 12"		105	.076		4.81	3.75		8.56	11.05
0600	Plywood, 3/4" thick with lumber edge, 12" wide		75	.107		1.89	5.25		7.14	10.15
0700	24" wide		70	.114		3.38	5.65		9.03	12.30
0900	Bookcase, clear grade pine, shelves 12" O.C., 8" deep, per S.F. shelf		70	.114	S.F.	10.40	5.65		16.05	20
1000	12" deep shelves		65	.123	"	15.65	6.05		21.70	26.50
1200	Adjustable closet rod and shelf, 12" wide, 3' long		20	.400	Ea.	11.60	19.70		31.30	43
1300	8' long		15	.533	"	19.95	26.50		46.45	62
1500	Prefinished shelves with supports, stock, 8" wide		75	.107	L.F.	4.98	5.25		10.23	13.55
1600	10" wide		70	.114	"	5.70	5.65		11.35	14.85

10 71 Exterior Protection

10 71 13 – Exterior Sun Control Devices

10 71 13.19 Rolling Exterior Shutters

		Crew	Daily Output	Labor-Hours	Unit	Material	2017 Bare Costs Labor	Equipment	Total	Total Incl O&P
0010	**ROLLING EXTERIOR SHUTTERS**									
0020	Roll-up, manual operation, aluminum, 3' x 4', incl. frame	2 Carp	8	2	Ea.	610	98.50		708.50	820
0030	6' x 7'	"	8	2	"	1,225	98.50		1,323.50	1,500

For customer support on your Building Construction Costs with RSMeans Data, call 800.448.8182.

395

10 73 13 – Awnings

10 73 13.10 Awnings, Fabric	Crew	Daily Output	Labor-Hours	Unit	Material	2017 Bare Costs Labor	Equipment	Total	Total Incl O&P
0010 **AWNINGS, FABRIC**									
0020 Including acrylic canvas and frame, standard design									
0100 Door and window, slope, 3' high, 4' wide	1 Carp	4.50	1.778	Ea.	755	87.50		842.50	965
0110 6' wide		3.50	2.286		970	113		1,083	1,250
0120 8' wide		3	2.667		1,200	131		1,331	1,500
0200 Quarter round convex, 4' wide		3	2.667		1,175	131		1,306	1,500
0210 6' wide		2.25	3.556		1,525	175		1,700	1,950
0220 8' wide		1.80	4.444		1,875	219		2,094	2,375
0300 Dome, 4' wide		7.50	1.067		455	52.50		507.50	580
0310 6' wide		3.50	2.286		1,025	113		1,138	1,300
0320 8' wide		2	4		1,800	197		1,997	2,300
0350 Elongated dome, 4' wide		1.33	6.015		1,700	296		1,996	2,325
0360 6' wide		1.11	7.207		2,025	355		2,380	2,775
0370 8' wide		1	8		2,375	395		2,770	3,225
1000 Entry or walkway, peak, 12' long, 4' wide	2 Carp	.90	17.778		5,475	875		6,350	7,375
1010 6' wide		.60	26.667		8,450	1,325		9,775	11,300
1020 8' wide		.40	40		11,700	1,975		13,675	15,800
1100 Radius with dome end, 4' wide		1.10	14.545		4,150	715		4,865	5,675
1110 6' wide		.70	22.857		6,675	1,125		7,800	9,075
1120 8' wide		.50	32		9,500	1,575		11,075	12,800
2000 Retractable lateral arm awning, manual									
2010 To 12' wide, 8'-6" projection	2 Carp	1.70	9.412	Ea.	1,250	465		1,715	2,050
2020 To 14' wide, 8'-6" projection		1.10	14.545		1,450	715		2,165	2,700
2030 To 19' wide, 8'-6" projection		.85	18.824		1,950	925		2,875	3,575
2040 To 24' wide, 8'-6" projection		.67	23.881		2,475	1,175		3,650	4,525
2050 Motor for above, add	1 Carp	2.67	3		1,050	148		1,198	1,400
3000 Patio/deck canopy with frame									
3010 12' wide, 12' projection	2 Carp	2	8	Ea.	1,750	395		2,145	2,525
3020 16' wide, 14' projection	"	1.20	13.333		2,725	655		3,380	4,000
9000 For fire retardant canvas, add					7%				
9010 For lettering or graphics, add					35%				
9020 For painted or coated acrylic canvas, deduct					8%				
9030 For translucent or opaque vinyl canvas, add					10%				
9040 For 6 or more units, deduct					20%	15%			

10 73 16 – Canopies

10 73 16.20 Metal Canopies	Crew	Daily Output	Labor-Hours	Unit	Material	2017 Bare Costs Labor	Equipment	Total	Total Incl O&P
0010 **METAL CANOPIES**									
0020 Wall hung, .032", aluminum, prefinished, 8' x 10'	K-2	1.30	18.462	Ea.	2,375	950	208	3,533	4,400
0300 8' x 20'		1.10	21.818		4,050	1,125	246	5,421	6,575
0500 10' x 10'		1.30	18.462		2,875	950	208	4,033	4,925
0700 10' x 20'		1.10	21.818		4,600	1,125	246	5,971	7,150
1000 12' x 20'		1	24		5,500	1,250	270	7,020	8,350
1360 12' x 30'		.80	30		8,250	1,550	340	10,140	12,000
1700 12' x 40'		.60	40		11,000	2,075	450	13,525	15,900
1900 For free standing units, add					20%	10%			
2300 Aluminum entrance canopies, flat soffit, .032"									
2500 3'-6" x 4'-0", clear anodized	2 Carp	4	4	Ea.	965	197		1,162	1,350
2700 Bronze anodized		4	4		1,700	197		1,897	2,175
3000 Polyurethane painted		4	4		1,375	197		1,572	1,825
3300 4'-6" x 10'-0", clear anodized		2	8		2,650	395		3,045	3,525
3500 Bronze anodized		2	8		3,400	395		3,795	4,325
3700 Polyurethane painted		2	8		2,850	395		3,245	3,725

10 73 Protective Covers

10 73 16 – Canopies

10 73 16.20 Metal Canopies	Crew	Daily Output	Labor-Hours	Unit	Material	2017 Bare Costs Labor	Equipment	Total	Total Incl O&P
4000	Wall downspout, 10 L.F., clear anodized	1 Carp	7	1.143	Ea.	165	56.50		221.50	268
4300	Bronze anodized		7	1.143		288	56.50		344.50	400
4500	Polyurethane painted	▼	7	1.143	▼	248	56.50		304.50	360
7000	Carport, baked vinyl finish, .032", 20' x 10', no foundations, flat panel	K-2	4	6	Car	4,000	310	67.50	4,377.50	4,975
7250	Insulated flat panel		2	12	"	6,300	620	135	7,055	8,075
7500	Walkway cover, to 12' wide, stl., vinyl finish, .032", no fndtns., flat		250	.096	S.F.	22.50	4.95	1.08	28.53	33.50
7750	Arched	▼	200	.120	"	55	6.20	1.35	62.55	72

10 74 Manufactured Exterior Specialties

10 74 23 – Cupolas

10 74 23.10 Wood Cupolas

		Crew	Daily Output	Labor-Hours	Unit	Material	2017 Bare Costs Labor	Equipment	Total	Total Incl O&P
0010	**WOOD CUPOLAS**									
0020	Stock units, pine, painted, 18" sq., 28" high, alum. roof	1 Carp	4.10	1.951	Ea.	207	96		303	375
0100	Copper roof		3.80	2.105		289	104		393	480
0300	23" square, 33" high, aluminum roof		3.70	2.162		440	106		546	650
0400	Copper roof		3.30	2.424		530	119		649	765
0600	30" square, 37" high, aluminum roof		3.70	2.162		600	106		706	825
0700	Copper roof		3.30	2.424		715	119		834	975
0900	Hexagonal, 31" wide, 46" high, copper roof		4	2		930	98.50		1,028.50	1,175
1000	36" wide, 50" high, copper roof	▼	3.50	2.286		1,550	113		1,663	1,900
1200	For deluxe stock units, add to above					25%				
1400	For custom built units, add to above			▼		50%	50%			

10 74 29 – Steeples

10 74 29.10 Prefabricated Steeples

		Crew	Daily Output	Labor-Hours	Unit	Material	2017 Bare Costs Labor	Equipment	Total	Total Incl O&P
0010	**PREFABRICATED STEEPLES**									
4000	Steeples, translucent fiberglass, 30" square, 15' high	F-3	2	20	Ea.	9,400	1,000	286	10,686	12,200
4150	25' high		1.80	22.222		10,600	1,125	320	12,045	13,800
4350	Opaque fiberglass, 24" square, 14' high		2	20		7,650	1,000	286	8,936	10,300
4500	28' high	▼	1.80	22.222		6,400	1,125	320	7,845	9,100
4600	Aluminum, baked finish, 16" square, 14' high					6,200			6,200	6,825
4620	20' high, 3'-6" base					10,800			10,800	11,900
4640	35' high, 8' base					38,800			38,800	42,600
4660	60' high, 14' base					77,500			77,500	85,000
4680	152' high, custom					635,000			635,000	698,500
4700	Porcelain enamel steeples, custom, 40' high	F-3	.50	80		14,300	4,050	1,150	19,500	23,100
4800	60' high	"	.30	133	▼	20,200	6,750	1,900	28,850	34,600

10 74 46 – Window Wells

10 74 46.10 Area Window Wells

		Crew	Daily Output	Labor-Hours	Unit	Material	2017 Bare Costs Labor	Equipment	Total	Total Incl O&P
0010	**AREA WINDOW WELLS**, Galvanized steel									
0020	20 ga., 3'-2" wide, 1' deep	1 Sswk	29	.276	Ea.	16.95	15		31.95	43.50
0100	2' deep		23	.348		31	18.90		49.90	66
0300	16 ga., 3'-2" wide, 1' deep		29	.276		23	15		38	50.50
0400	3' deep		23	.348		47	18.90		65.90	83
0600	Welded grating for above, 15 lb., painted		45	.178		91.50	9.65		101.15	117
0700	Galvanized		45	.178		124	9.65		133.65	153
0900	Translucent plastic cap for above	▼	60	.133	▼	20.50	7.25		27.75	34.50

10 75 Flagpoles

10 75 16 – Ground-Set Flagpoles

10 75 16.10 Flagpoles		Crew	Daily Output	Labor-Hours	Unit	Material	2017 Bare Costs Labor	Equipment	Total	Total Incl O&P
0010	**FLAGPOLES**, ground set									
0050	Not including base or foundation									
0100	Aluminum, tapered, ground set 20' high	K-1	2	8	Ea.	1,075	375	135	1,585	1,900
0200	25' high		1.70	9.412		1,175	440	159	1,774	2,125
0300	30' high		1.50	10.667		1,400	500	180	2,080	2,500
0400	35' high		1.40	11.429		1,850	535	193	2,578	3,050
0500	40' high		1.20	13.333		2,975	625	225	3,825	4,475
0600	50' high		1	16		3,450	745	270	4,465	5,200
0700	60' high		.90	17.778		4,900	830	300	6,030	6,975
0800	70' high		.80	20		8,850	935	340	10,125	11,500
1100	Counterbalanced, internal halyard, 20' high		1.80	8.889		2,725	415	150	3,290	3,775
1200	30' high		1.50	10.667		2,950	500	180	3,630	4,175
1300	40' high		1.30	12.308		6,925	575	208	7,708	8,725
1400	50' high		1	16		9,625	745	270	10,640	12,000
2820	Aluminum, electronically operated, 30' high		1.40	11.429		4,400	535	193	5,128	5,850
2840	35' high		1.30	12.308		5,200	575	208	5,983	6,825
2860	39' high		1.10	14.545		6,325	680	246	7,251	8,275
2880	45' high		1	16		6,700	745	270	7,715	8,775
2900	50' high		.90	17.778		8,600	830	300	9,730	11,100
3000	Fiberglass, tapered, ground set, 23' high		2	8		640	375	135	1,150	1,425
3100	29'-7" high		1.50	10.667		1,325	500	180	2,005	2,425
3200	36'-1" high		1.40	11.429		1,625	535	193	2,353	2,800
3300	39'-5" high		1.20	13.333		1,950	625	225	2,800	3,350
3400	49'-2" high		1	16		4,150	745	270	5,165	6,000
3500	59' high		.90	17.778		4,750	830	300	5,880	6,825
4300	Steel, direct imbedded installation									
4400	Internal halyard, 20' high	K-1	2.50	6.400	Ea.	1,375	299	108	1,782	2,075
4500	25' high		2.50	6.400		2,050	299	108	2,457	2,825
4600	30' high		2.30	6.957		2,450	325	118	2,893	3,300
4700	40' high		2.10	7.619		3,725	355	129	4,209	4,775
4800	50' high		1.90	8.421		4,325	395	142	4,862	5,500
5000	60' high		1.80	8.889		7,200	415	150	7,765	8,725
5100	70' high		1.60	10		7,850	465	169	8,484	9,525
5200	80' high		1.40	11.429		10,100	535	193	10,828	12,100
5300	90' high		1.20	13.333		16,300	625	225	17,150	19,100
5500	100' high		1	16		18,500	745	270	19,515	21,800
6400	Wood poles, tapered, clear vertical grain fir with tilting									
6410	base, not incl. foundation, 4" butt, 25' high	K-1	1.90	8.421	Ea.	1,625	395	142	2,162	2,550
6800	6" butt, 30' high	"	1.30	12.308	"	2,825	575	208	3,608	4,200
7300	Foundations for flagpoles, including									
7400	excavation and concrete, to 35' high poles	C-1	10	3.200	Ea.	705	150		855	1,000
7600	40' to 50' high		3.50	9.143		1,300	425		1,725	2,075
7700	Over 60' high		2	16		1,625	750		2,375	2,950

10 75 23 – Wall-Mounted Flagpoles

10 75 23.10 Flagpoles

		Crew	Daily Output	Labor-Hours	Unit	Material	2017 Bare Costs Labor	Equipment	Total	Total Incl O&P
0010	**FLAGPOLES**, structure mounted									
0100	Fiberglass, vertical wall set, 19'-8" long	K-1	1.50	10.667	Ea.	1,250	500	180	1,930	2,325
0200	23' long		1.40	11.429		1,450	535	193	2,178	2,625
0300	26'-3" long		1.30	12.308		1,950	575	208	2,733	3,250
0800	19'-8" long outrigger		1.30	12.308		1,375	575	208	2,158	2,625
1300	Aluminum, vertical wall set, tapered, with base, 20' high		1.20	13.333		1,125	625	225	1,975	2,450
1400	29'-6" high		1	16		2,850	745	270	3,865	4,550

10 75 Flagpoles

10 75 23 - Wall-Mounted Flagpoles

10 75 23.10 Flagpoles	Crew	Daily Output	Labor-Hours	Unit	Material	2017 Bare Costs Labor	Equipment	Total	Total Incl O&P
2400	Outrigger poles with base, 12' long	K-1	1.30	12.308	Ea.	1,250	575	208	2,033	2,475
2500	14' long	↓	1	16	↓	1,350	745	270	2,365	2,900

10 81 Pest Control Devices

10 81 13 - Bird Control Devices

10 81 13.10 Bird Control Netting

		Crew	Daily Output	Labor-Hours	Unit	Material	Labor	Equipment	Total	Total Incl O&P
0010	**BIRD CONTROL NETTING**									
0020	1/8" square mesh	4 Clab	4000	.008	S.F.	.56	.31		.87	1.10
0100	1/4" square mesh		4000	.008		.61	.31		.92	1.15
0120	1/2" square mesh		4000	.008		.14	.31		.45	.63
0140	5/8" x 3/4" mesh		4000	.008		.08	.31		.39	.57
0160	1-1/4" x 1-1/2" mesh		4000	.008		.16	.31		.47	.66
0200	4" square mesh	↓	4000	.008	↓	.10	.31		.41	.59
1000	Poly clips				Ea.	.23			.23	.25

10 86 Security Mirrors and Domes

10 86 10 - Security Mirrors

10 86 10.10 Exterior Traffic Control Mirrors

		Crew	Daily Output	Labor-Hours	Unit	Material	Labor	Equipment	Total	Total Incl O&P
0010	**EXTERIOR TRAFFIC CONTROL MIRRORS**									
0100	Convex, stainless steel, 20 ga., 26" diameter	1 Carp	12	.667	Ea.	202	33		235	273

10 86 20 - Security Domes

10 86 20.10 Domes

		Crew	Daily Output	Labor-Hours	Unit	Material	Labor	Equipment	Total	Total Incl O&P
0010	**DOMES** for security cameras (CCTV)									
0100	Ceiling mounted, 10" diameter	1 Carp	30	.267	Ea.	11.95	13.15		25.10	33
0110	12" diameter	"	30	.267	"	8.90	13.15		22.05	30

10 88 Scales

10 88 05 - Commercial Scales

10 88 05.10 Scales

		Crew	Daily Output	Labor-Hours	Unit	Material	Labor	Equipment	Total	Total Incl O&P
0010	**SCALES**									
0700	Truck scales, incl. steel weigh bridge,									
0800	not including foundation, pits									
1550	Digital, electronic, 100 ton capacity, steel deck 12' x 10' platform	3 Carp	.20	120	Ea.	14,900	5,900		20,800	25,500
1600	40' x 10' platform		.14	171		29,700	8,450		38,150	45,600
1640	60' x 10' platform		.13	184		39,400	9,100		48,500	57,000
1680	70' x 10' platform	↓	.12	200		41,800	9,850		51,650	61,000
2000	For standard automatic printing device, add					1,400			1,400	1,525
2100	For remote reading electronic system, add					2,800			2,800	3,075
2300	Concrete foundation pits for above, 8' x 6', 5 C.Y. required	C-1	.50	64		1,125	3,000		4,125	5,825
2400	14' x 6' platform, 10 C.Y. required		.35	91.429		1,675	4,275		5,950	8,375
2600	50' x 10' platform, 30 C.Y. required		.25	128		2,250	5,975		8,225	11,700
2700	70' x 10' platform, 40 C.Y. required	↓	.15	213		4,900	9,975		14,875	20,700
2750	Crane scales, dial, 1 ton capacity					1,100			1,100	1,200
2780	5 ton capacity					1,400			1,400	1,525
2800	Digital, 1 ton capacity					1,900			1,900	2,075
2850	10 ton capacity					4,800			4,800	5,300
2900	Low profile electronic warehouse scale,									

10 88 Scales

10 88 05 – Commercial Scales

10 88 05.10 Scales		Crew	Daily Output	Labor-Hours	Unit	Material	2017 Bare Costs Labor	Equipment	Total	Total Incl O&P
3000	not incl. printer, 4' x 4' platform, 10,000 lb. capacity	2 Carp	.30	53.333	Ea.	1,425	2,625		4,050	5,575
3300	5' x 7' platform, 10,000 lb. capacity		.25	64		5,100	3,150		8,250	10,500
3400	20,000 lb. capacity		.20	80		6,275	3,950		10,225	12,900
3500	For printers, incl. time, date & numbering, add					925			925	1,025
3800	Portable, beam type, capacity 1000#, platform 18" x 24"					800			800	880
3900	Dial type, capacity 2000#, platform 24" x 24"					1,500			1,500	1,650
4000	Digital type, capacity 1000#, platform 24" x 30"					2,400			2,400	2,650
4100	Portable contractor truck scales, 50 ton cap., 40' x 10' platform					34,300			34,300	37,700
4200	60' x 10' platform					33,500			33,500	36,900
4400	Heavy-Duty Steel Deck Truck Scales 20' x 10'	3 Carp	.20	120		17,300	5,900		23,200	28,100
4500	Heavy-Duty Steel Deck Truck Scales 80' x 10'		.10	240		49,900	11,800		61,700	73,000
4600	Heavy-Duty Steel Deck Truck Scales 160' x 10'		.04	600		100,000	29,600		129,600	155,000
4700	Heavy-Duty Steel Deck Truck Scales 20' x 12'		.18	137		20,600	6,750		27,350	32,900
4800	Heavy-Duty Steel Deck Truck Scales 80' x 12'		.09	266		59,000	13,100		72,100	85,000
4900	Heavy-Duty Steel Deck Truck Scales 160' x 12'		.04	600		118,000	29,600		147,600	175,000
5000	Heavy-Duty Steel Deck Truck Scales 20' x 14'		.16	150		23,900	7,400		31,300	37,600
5100	Heavy-Duty Steel Deck Truck Scales 80' x 14'		.06	400		68,000	19,700		87,700	104,500
5200	Heavy-Duty Steel Deck Truck Scales 160' x 14'		.03	800		135,500	39,400		174,900	210,000

Estimating Tips
General

- The items in this division are usually priced per square foot or each. Many of these items are purchased by the owner for installation by the contractor. Check the specifications for responsibilities and include time for receiving, storage, installation, and mechanical and electrical hookups in the appropriate divisions.

- Many items in Division 11 require some type of support system that is not usually furnished with the item. Examples of these systems include blocking for the attachment of casework and support angles for ceiling-hung projection screens. The required blocking or supports must be added to the estimate in the appropriate division.

- Some items in Division 11 may require assembly or electrical hookups. Verify the amount of assembly required or the need for a hard electrical connection and add the appropriate costs.

Reference Numbers

Reference numbers are shown at the beginning of some major classifications. These numbers refer to related items in the Reference Section. The reference information may be an estimating procedure, an alternate pricing method, or technical information.

Note: Not all subdivisions listed here necessarily appear. ■

11 05 05.10 Selective Demolition	Crew	Daily Output	Labor-Hours	Unit	Material	2017 Bare Costs Labor	Equipment	Total	Total Incl O&P
0010 **SELECTIVE DEMOLITION**									
0130 Central vacuum, motor unit, residential or commercial	1 Clab	2	4	Ea.		157		157	240
0210 Vault door and frame	2 Skwk	2	8			410		410	635
0215 Day gate, for vault	"	3	5.333			274		274	420
0380 Bank equipment, teller window, bullet resistant	1 Clab	1.20	6.667			261		261	400
0381 Counter	2 Clab	1.50	10.667	Station		420		420	640
0382 Drive-up window, including drawer and glass		1.50	10.667	"		420		420	640
0383 Thru-wall boxes and chests, selective demolition		2.50	6.400	Ea.		251		251	385
0384 Bullet resistant partitions		20	.800	L.F.		31.50		31.50	48
0385 Pneumatic tube system, 2 lane drive-up	L-3	.45	35.556	Ea.		1,900		1,900	2,875
0386 Safety deposit box	1 Clab	50	.160	Opng.		6.25		6.25	9.60
0387 Surveillance system, video, complete	2 Elec	2	8	Ea.		455		455	680
0410 Church equipment, misc movable fixtures	2 Clab	1	16			625		625	960
0412 Steeple, to 28' high	F-3	3	13.333			675	191	866	1,225
0414 40' to 60' high	"	.80	50			2,525	715	3,240	4,650
0510 Library equipment, bookshelves, wood, to 90" high	1 Clab	20	.400	L.F.		15.65		15.65	24
0515 Carrels, hardwood, 36" x 24"	"	9	.889	Ea.		35		35	53.50
0630 Stage equipment, light control panel	1 Elec	1	8	"		455		455	680
0632 Border lights		40	.200	L.F.		11.30		11.30	16.95
0634 Spotlights		8	1	Ea.		56.50		56.50	85
0636 Telescoping platforms and risers	2 Clab	175	.091	SF Stg.		3.58		3.58	5.50
1020 Barber equipment, hydraulic chair	1 Clab	40	.200	Ea.		7.85		7.85	12
1030 Checkout counter, supermarket or warehouse conveyor	2 Clab	18	.889			35		35	53.50
1040 Food cases, refrigerated or frozen	Q-5	6	2.667			149		149	225
1190 Laundry equipment, commercial	L-6	3	4			240		240	360
1360 Movie equipment, lamphouse, to 4000 watt, incl rectifier	1 Elec	4	2			113		113	170
1365 Sound system, incl amplifier	"	1.25	6.400			360		360	545
1410 Air compressor, to 5 H.P.	2 Clab	2.50	6.400			251		251	385
1412 Lubrication equipment, automotive, 3 reel type, incl pump, excl piping	L-4	1	24	Set		1,125		1,125	1,725
1414 Booth, spray paint, complete, to 26' long	"	.80	30	"		1,400		1,400	2,150
1560 Parking equipment, cashier booth	B-22	2	15	Ea.		695	104	799	1,175
1600 Loading dock equipment, dock bumpers, rubber	1 Clab	50	.160	"		6.25		6.25	9.60
1610 Door seal for door perimeter	"	50	.160	L.F.		6.25		6.25	9.60
1611 Loading dock equipment, dock seal for perimeter, selective demolition	2 Clab	13	1.231	"		48		48	74
1620 Platform lifter, fixed, 6' x 8', 5000 lb. capacity	E-16	1.50	10.667	Ea.		590	86.50	676.50	1,075
1630 Dock leveller	"	2	8			440	65	505	805
1640 Lights, single or double arm	1 Elec	8	1			56.50		56.50	85
1650 Shelter, fabric, truck or train	1 Clab	1.50	5.333			209		209	320
1790 Waste handling equipment, commercial compactor	L-4	2	12			560		560	865
1792 Commercial or municipal incinerator, gas	"	2	12			560		560	865
1795 Crematory, excluding building	Q-3	.25	128			7,525		7,525	11,400
1910 Detection equipment, cell bar front	E-4	4	8			440	32.50	472.50	765
1912 Cell door and frame		8	4			219	16.25	235.25	385
1914 Prefab cell, 4' to 5' wide, 7' to 8' high, 7' deep		8	4			219	16.25	235.25	385
1916 Cot, bolted, single		40	.800			44	3.25	47.25	76.50
1918 Visitor cubicle		4	8			440	32.50	472.50	765
2850 Hydraulic gates, canal, flap, knife, slide or sluice, to 18" diameter	L-5A	8	4			221	75.50	296.50	445
2852 19" to 36" diameter		6	5.333			294	101	395	590
2854 37" to 48" diameter		2	16			880	305	1,185	1,750
2856 49" to 60" diameter		1	32			1,775	605	2,380	3,550
2858 Over 60" diameter		.30	106			5,875	2,025	7,900	11,800
3100 Sewage pumping system, prefabricated, to 1000 GPM	C-17D	.20	420			21,900	4,050	25,950	38,100
3110 Sewage treatment, holding tank for recirc chemical water closet	1 Plum	8	1			62		62	93.50

11 05 Common Work Results for Equipment

11 05 05 – Selective Demolition for Equipment

11 05 05.10 Selective Demolition

		Crew	Daily Output	Labor-Hours	Unit	Material	2017 Bare Costs Labor	2017 Bare Costs Equipment	Total	Total Incl O&P
3900	Wastewater treatment system, to 1500 gallons	B-21	2	14	Ea.		640	69	709	1,050
4050	Food storage equipment, walk-in refrigerator/freezer	2 Clab	64	.250	S.F.		9.80		9.80	15
4052	Shelving, stainless steel, 4 tier or dunnage rack	1 Clab	12	.667	Ea.		26		26	40
4100	Food preparation equipment, small countertop		18	.444			17.40		17.40	26.50
4150	Food delivery carts, heated cabinets		18	.444			17.40		17.40	26.50
4200	Cooking equipment, commercial range	Q-1	12	1.333			74		74	112
4250	Hood and ventilation equipment, kitchen exhaust hood, excl fire prot	1 Clab	3	2.667			104		104	160
4255	Fire protection system	Q-1	3	5.333			297		297	450
4300	Food dispensing equipment, countertop items	1 Clab	15	.533			21		21	32
4310	Serving counter	"	65	.123	L.F.		4.82		4.82	7.40
4350	Ice machine, ice cube maker, flakers and storage bins, to 2000 lb./day	Q-1	1.60	10	Ea.		555		555	840
4400	Cleaning and disposal, commercial dishwasher, to 50 racks per hour	L-6	1	12			720		720	1,075
4405	To 275 racks per hour	L-4	1	24			1,125		1,125	1,725
4410	Dishwasher hood	2 Clab	5	3.200			125		125	192
4420	Garbage disposal, commercial, to 5 H.P.	L-1	8	2			118		118	178
4540	Water heater, residential, to 80 gal/day	"	5	3.200			189		189	285
4542	Water softener, automatic	2 Plum	10	1.600			99		99	149
4544	Disappearing stairway, to 15' floor height	2 Clab	6	2.667			104		104	160
4710	Darkroom equipment, light	L-7	10	2.800			133		133	203
4712	Heavy	"	1.50	18.667			885		885	1,350
4720	Doors	2 Clab	3.50	4.571	Opng.		179		179	274
4830	Bowling alley, complete, incl pinsetter, scorer, counters, misc supplies	4 Clab	.40	80	Lane		3,125		3,125	4,800
4840	Health club equipment, circuit training apparatus	2 Clab	2	8	Set		315		315	480
4842	Squat racks	"	10	1.600	Ea.		62.50		62.50	96
4860	School equipment, basketball backstop	L-2	2	8			345		345	530
4862	Table and benches, folding, in wall, 14' long	L-4	4	6			280		280	430
4864	Bleachers, telescoping, to 30 tier	F-5	120	.267	Seat		13.25		13.25	20.50
4865	to 15 tier	"	160	.200	"		9.95		9.95	15.25
4866	Boxing ring, elevated	L-4	.20	120	Ea.		5,600		5,600	8,650
4867	Boxing ring, floor level	"	2	12			560		560	865
4868	Exercise equipment	1 Clab	6	1.333			52		52	80
4870	Gym divider	L-4	1000	.024	S.F.		1.12		1.12	1.73
4875	Scoreboard	R-3	2	10	Ea.		565	69	634	925
4880	Shooting range, incl bullet traps, targets, excl structure	L-9	1	36	Point		1,600		1,600	2,500
5200	Vocational shop equipment	2 Clab	8	2	Ea.		78.50		78.50	120
6200	Fume hood, incl countertop, excl HVAC	"	6	2.667	L.F.		104		104	160
7100	Medical sterilizing, distiller, water, steam heated, 50 gal. capacity	1 Plum	2.80	2.857	Ea.		177		177	266
7200	Medical equipment, surgery table, minor	1 Clab	1	8			315		315	480
7210	Surgical lights, doctor's office, single or double arm	2 Elec	3	5.333			300		300	450
7300	Physical therapy, table	2 Clab	4	4			157		157	240
7310	Whirlpool bath, fixed, incl mixing valves	1 Plum	4	2			124		124	187
7400	Dental equipment, chair, electric or hydraulic	1 Clab	.75	10.667			420		420	640
7410	Central suction system	1 Plum	2	4			247		247	375
7420	Drill console with accessories	1 Clab	3.20	2.500			98		98	150
7430	X-ray unit	"	4	2			78.50		78.50	120
7440	X-ray developer	1 Plum	10	.800			49.50		49.50	74.50

11 05 10 – Equipment Installation

11 05 10.10 Industrial Equipment Installation

		Crew	Daily Output	Labor-Hours	Unit	Material	2017 Bare Costs Labor	2017 Bare Costs Equipment	Total	Total Incl O&P
0010	**INDUSTRIAL EQUIPMENT INSTALLATION**									
0020	Industrial equipment, minimum	E-2	12	4.667	Ton		251	139	390	565
0200	Maximum	"	2	28	"		1,500	835	2,335	3,375

For customer support on your Building Construction Costs with RSMeans Data, call 800.448.8182.

403

11 11 Vehicle Service Equipment

11 11 13 – Compressed-Air Vehicle Service Equipment

11 11 13.10 Compressed Air Equipment

		Crew	Daily Output	Labor-Hours	Unit	Material	2017 Bare Costs Labor	2017 Bare Costs Equipment	Total	Total Incl O&P
0010	**COMPRESSED AIR EQUIPMENT**									
0030	Compressors, electric, 1-1/2 H.P., standard controls	L-4	1.50	16	Ea.	510	745		1,255	1,700
0550	Dual controls		1.50	16		830	745		1,575	2,075
0600	5 H.P., 115/230 volt, standard controls		1	24		1,725	1,125		2,850	3,625
0650	Dual controls	↓	1	24	↓	3,025	1,125		4,150	5,050

11 11 19 – Vehicle Lubrication Equipment

11 11 19.10 Lubrication Equipment

		Crew	Daily Output	Labor-Hours	Unit	Material	2017 Bare Costs Labor	2017 Bare Costs Equipment	Total	Total Incl O&P
0010	**LUBRICATION EQUIPMENT**									
3000	Lube equipment, 3 reel type, with pumps, not including piping	L-4	.50	48	Set	8,875	2,250		11,125	13,200
3100	Hose reel, including hose, oil/lube, 1000 PSI	2 Sswk	2	8	Ea.	425	435		860	1,200
3200	Grease, 5000 PSI		2	8		500	435		935	1,275
3300	Air, 50 feet, 160 PSI		2	8		730	435		1,165	1,525
3350	25 feet, 160 PSI	↓	2	8	↓	435	435		870	1,200

11 11 33 – Vehicle Spray Painting Equipment

11 11 33.10 Spray Painting Equipment

		Crew	Daily Output	Labor-Hours	Unit	Material	2017 Bare Costs Labor	2017 Bare Costs Equipment	Total	Total Incl O&P
0010	**SPRAY PAINTING EQUIPMENT**									
4000	Spray painting booth, 26' long, complete	L-4	.40	60	Ea.	10,400	2,800		13,200	15,800

11 12 Parking Control Equipment

11 12 13 – Parking Key and Card Control Units

11 12 13.10 Parking Control Units

		Crew	Daily Output	Labor-Hours	Unit	Material	2017 Bare Costs Labor	2017 Bare Costs Equipment	Total	Total Incl O&P
0010	**PARKING CONTROL UNITS**									
5100	Card reader	1 Elec	2	4	Ea.	1,350	226		1,576	1,825
5120	Proximity with customer display	2 Elec	1	16		5,925	905		6,830	7,875
6000	Parking control software, basic functionality	1 Elec	.50	16		23,700	905		24,605	27,500
6020	Multi-function	"	.20	40	↓	97,000	2,275		99,275	110,000

11 12 16 – Parking Ticket Dispensers

11 12 16.10 Ticket Dispensers

		Crew	Daily Output	Labor-Hours	Unit	Material	2017 Bare Costs Labor	2017 Bare Costs Equipment	Total	Total Incl O&P
0010	**TICKET DISPENSERS**									
5900	Ticket spitter with time/date stamp, standard	2 Elec	2	8	Ea.	6,550	455		7,005	7,875
5920	Mag stripe encoding	"	2	8	"	18,800	455		19,255	21,400

11 12 26 – Parking Fee Collection Equipment

11 12 26.13 Parking Fee Coin Collection Equipment

		Crew	Daily Output	Labor-Hours	Unit	Material	2017 Bare Costs Labor	2017 Bare Costs Equipment	Total	Total Incl O&P
0010	**PARKING FEE COIN COLLECTION EQUIPMENT**									
5200	Cashier booth, average	B-22	1	30	Ea.	10,500	1,400	207	12,107	14,000
5300	Collector station, pay on foot	2 Elec	.20	80		111,000	4,525		115,525	129,000
5320	Credit card only	"	.50	32	↓	20,500	1,800		22,300	25,200

11 12 26.23 Fee Equipment

		Crew	Daily Output	Labor-Hours	Unit	Material	2017 Bare Costs Labor	2017 Bare Costs Equipment	Total	Total Incl O&P
0010	**FEE EQUIPMENT**									
5600	Fee computer	1 Elec	1.50	5.333	Ea.	12,000	300		12,300	13,700

11 12 33 – Parking Gates

11 12 33.13 Lift Arm Parking Gates

		Crew	Daily Output	Labor-Hours	Unit	Material	2017 Bare Costs Labor	2017 Bare Costs Equipment	Total	Total Incl O&P
0010	**LIFT ARM PARKING GATES**									
5000	Barrier gate with programmable controller	2 Elec	3	5.333	Ea.	3,275	300		3,575	4,050
5020	Industrial		3	5.333		5,150	300		5,450	6,100
5050	Non-programmable, with reader and 12' arm		3	5.333		1,950	300		2,250	2,575
5500	Exit verifier	↓	1	16		19,200	905		20,105	22,500
5700	Full sign, 4" letters	1 Elec	2	4	↓	1,225	226		1,451	1,700

11 12 Parking Control Equipment

11 12 33 – Parking Gates

11 12 33.13 Lift Arm Parking Gates		Crew	Daily Output	Labor-Hours	Unit	Material	2017 Bare Costs Labor	2017 Bare Costs Equipment	Total	Total Incl O&P
5800	Inductive loop	2 Elec	4	4	Ea.	207	226		433	570
5950	Vehicle detector, microprocessor based	1 Elec	3	2.667		435	151		586	705
7100	Traffic spike unit, flush mount, spring loaded, 72" L	B-89	4	4		1,375	190	110	1,675	1,900
7200	Surface mount, 72" L	2 Skwk	10	1.600		2,550	82.50		2,632.50	2,925

11 13 Loading Dock Equipment

11 13 13 – Loading Dock Bumpers

11 13 13.10 Dock Bumpers

		Crew	Daily Output	Labor-Hours	Unit	Material	2017 Bare Costs Labor	2017 Bare Costs Equipment	Total	Total Incl O&P
0010	**DOCK BUMPERS** Bolts not included									
0020	2" x 6" to 4" x 8", average	1 Carp	.30	26.667	M.B.F.	1,350	1,325		2,675	3,475
0050	Bumpers, lam. rubber blocks 4-1/2" thick, 10" high, 14" long		26	.308	Ea.	61.50	15.15		76.65	90.50
0200	24" long		22	.364		95	17.90		112.90	133
0300	36" long		17	.471		130	23		153	179
0500	12" high, 14" long		25	.320		74.50	15.75		90.25	106
0550	24" long		20	.400		117	19.70		136.70	159
0600	36" long		15	.533		150	26.50		176.50	205
0800	Laminated rubber blocks 6" thick, 10" high, 14" long		22	.364		70	17.90		87.90	105
0850	24" long		18	.444		125	22		147	171
0900	36" long		13	.615		132	30.50		162.50	192
0910	20" high, 11" long		13	.615		125	30.50		155.50	184
0920	Extruded rubber bumpers, T section, 22" x 22" x 3" thick		41	.195		61	9.60		70.60	82
0940	Molded rubber bumpers, 24" x 12" x 3" thick		20	.400		58	19.70		77.70	94
1000	Welded installation of above bumpers	E-14	8	1		3.78	56.50	16.25	76.53	116
1100	For drilled anchors, add per anchor	1 Carp	36	.222		6.80	10.95		17.75	24
1300	Steel bumpers, see Section 10 26 13.10									

11 13 16 – Loading Dock Seals and Shelters

11 13 16.10 Dock Seals and Shelters

		Crew	Daily Output	Labor-Hours	Unit	Material	2017 Bare Costs Labor	2017 Bare Costs Equipment	Total	Total Incl O&P
0010	**DOCK SEALS AND SHELTERS**									
3600	Door seal for door perimeter, 12" x 12", vinyl covered	1 Carp	26	.308	L.F.	33.50	15.15		48.65	60
3700	Loading dock, seal for perimeter, 9' x 8', with 12" vinyl	2 Carp	6	2.667	Ea.	1,375	131		1,506	1,725
3900	Folding gates, see Section 10 22 16.10									
6200	Shelters, fabric, for truck or train, scissor arms, minimum	1 Carp	1	8	Ea.	1,875	395		2,270	2,675
6300	Maximum	"	.50	16	"	2,500	790		3,290	3,950

11 13 19 – Stationary Loading Dock Equipment

11 13 19.10 Dock Equipment

		Crew	Daily Output	Labor-Hours	Unit	Material	2017 Bare Costs Labor	2017 Bare Costs Equipment	Total	Total Incl O&P
0010	**DOCK EQUIPMENT**									
2200	Dock boards, heavy duty, 60" x 60", aluminum, 5,000 lb. capacity				Ea.	1,425			1,425	1,575
2700	9,000 lb. capacity					1,300			1,300	1,425
3200	15,000 lb. capacity					1,500			1,500	1,650
4200	Platform lifter, 6' x 6', portable, 3,000 lb. capacity					9,250			9,250	10,200
4250	4,000 lb. capacity					11,400			11,400	12,500
4400	Fixed, 6' x 8', 5,000 lb. capacity	E-16	.70	22.857		9,425	1,275	185	10,885	12,700
4500	Levelers, hinged for trucks, 10 ton capacity, 6' x 8'		1.08	14.815		4,375	820	120	5,315	6,300
4650	7' x 8'		1.08	14.815		5,775	820	120	6,715	7,850
4670	Air bag power operated, 10 ton cap., 6' x 8'		1.08	14.815		4,650	820	120	5,590	6,625
4680	7' x 8'		1.08	14.815		5,350	820	120	6,290	7,400
4700	Hydraulic, 10 ton capacity, 6' x 8'		1.08	14.815		5,350	820	120	6,290	7,375
4800	7' x 8'		1.08	14.815		5,750	820	120	6,690	7,825
5800	Loading dock safety restraints, manual style		1.08	14.815		3,250	820	120	4,190	5,075
5900	Automatic style		1.08	14.815		5,450	820	120	6,390	7,500

11 13 Loading Dock Equipment

11 13 19 – Stationary Loading Dock Equipment

11 13 19.10 Dock Equipment	Crew	Daily Output	Labor-Hours	Unit	Material	2017 Bare Costs Labor	Equipment	Total	Total Incl O&P
6000	Dock leveler, 15 ton capacity									
6100	I beam construction, mechanical, 6' x 8'	E-16	.50	32	Ea.	4,925	1,775	260	6,960	8,650
6150	Hydraulic		.50	32		5,800	1,775	260	7,835	9,600
6200	Formed beam deck construction, mechanical, 6' x 8'		.50	32		3,675	1,775	260	5,710	7,275
6250	Hydraulic		.50	32		5,025	1,775	260	7,060	8,750
6300	Edge of dock leveler, mechanical, 15 ton capacity		2	8		1,100	440	65	1,605	2,000
6301	Edge of dock leveler, mechanical, 10 ton capacity	↓	2	8	↓	1,100	440	65	1,605	2,000
7000	22.5 ton capacity									
7100	Vertical storing dock leveler, hydraulic, 6' x 6'	E-16	.40	40	Ea.	7,875	2,200	325	10,400	12,700

11 13 26 – Loading Dock Lights

11 13 26.10 Dock Lights

		Crew	Daily Output	Labor-Hours	Unit	Material	Labor	Equipment	Total	Total Incl O&P
0010	**DOCK LIGHTS**									
5000	Lights for loading docks, single arm, 24" long	1 Elec	3.80	2.105	Ea.	142	119		261	335
5700	Double arm, 60" long	"	3.80	2.105	"	194	119		313	390

11 14 Pedestrian Control Equipment

11 14 13 – Pedestrian Gates

11 14 13.13 Portable Posts and Railings

		Crew	Daily Output	Labor-Hours	Unit	Material	Labor	Equipment	Total	Total Incl O&P
0010	**PORTABLE POSTS AND RAILINGS**									
0020	Portable for pedestrian traffic control, standard				Ea.	144			144	159
0300	Deluxe posts				"	236			236	260
0600	Ropes for above posts, plastic covered, 1-1/2" diameter				L.F.	19			19	21
0700	Chain core				"	14			14	15.40
1500	Portable security or safety barrier, black with 7' yellow strap				Ea.	255			255	280
1510	12' yellow strap					280			280	310
1550	Sign holder, standard design				↓	85			85	93.50

11 14 13.19 Turnstiles

		Crew	Daily Output	Labor-Hours	Unit	Material	Labor	Equipment	Total	Total Incl O&P
0010	**TURNSTILES**									
0020	One way, 4 arm, 46" diameter, economy, manual	2 Carp	5	3.200	Ea.	1,875	158		2,033	2,325
0100	Electric		1.20	13.333		2,200	655		2,855	3,425
0300	High security, galv., 5'-5" diameter, 7' high, manual		1	16		7,375	790		8,165	9,300
0350	Electric		.60	26.667		8,375	1,325		9,700	11,200
0420	Three arm, 24" opening, light duty, manual		2	8		3,250	395		3,645	4,175
0450	Heavy duty		1.50	10.667		4,875	525		5,400	6,150
0460	Manual, with registering & controls, light duty		2	8		4,350	395		4,745	5,375
0470	Heavy duty		1.50	10.667		4,600	525		5,125	5,850
0480	Electric, heavy duty	↓	1.10	14.545	↓	4,900	715		5,615	6,500
0500	For coin or token operating, add					705			705	775
1200	One way gate with horizontal bars, 5'-5" diameter									
1300	7' high, recreation or transit type	2 Carp	.80	20	Ea.	5,425	985		6,410	7,475
1500	For electronic counter, add				"	260			260	286

11 21 Retail and Service Equipment

11 21 13 – Cash Registers and Checking Equipment

11 21 13.10 Checkout Counter

	11 21 13.10 Checkout Counter	Crew	Daily Output	Labor-Hours	Unit	Material	2017 Bare Costs Labor	Equipment	Total	Total Incl O&P
0010	**CHECKOUT COUNTER**									
0020	Supermarket conveyor, single belt	2 Clab	10	1.600	Ea.	3,325	62.50		3,387.50	3,775
0100	Double belt, power take-away		9	1.778		4,800	69.50		4,869.50	5,375
0400	Double belt, power take-away, incl. side scanning		7	2.286		5,625	89.50		5,714.50	6,325
0800	Warehouse or bulk type	↓	6	2.667	↓	6,675	104		6,779	7,475
1000	Scanning system, 2 lanes, w/registers, scan gun & memory				System	17,500			17,500	19,200
1100	10 lanes, single processor, full scan, with scales				"	166,000			166,000	182,500
2000	Register, restaurant, minimum				Ea.	775			775	855
2100	Maximum					3,275			3,275	3,600
2150	Store, minimum					775			775	855
2200	Maximum				↓	3,275			3,275	3,600

11 21 33 – Checkroom Equipment

11 21 33.10 Clothes Check Equipment

	11 21 33.10 Clothes Check Equipment	Crew	Daily Output	Labor-Hours	Unit	Material	2017 Bare Costs Labor	Equipment	Total	Total Incl O&P
0010	**CLOTHES CHECK EQUIPMENT**									
0030	Clothes check rack, free standing, st. stl., 2-tier, 90 bag capacity				Ea.	1,550			1,550	1,700
0050	Wall mounted, 45 bag capacity	L-2	8	2		845	86.50		931.50	1,075
0100	Garment checking bag, green mesh fabric, 21" H x 17" W with 4.5" hook				↓	25.50			25.50	28

11 21 53 – Barber and Beauty Shop Equipment

11 21 53.10 Barber Equipment

	11 21 53.10 Barber Equipment	Crew	Daily Output	Labor-Hours	Unit	Material	2017 Bare Costs Labor	Equipment	Total	Total Incl O&P
0010	**BARBER EQUIPMENT**									
0020	Chair, hydraulic, movable, minimum	1 Carp	24	.333	Ea.	625	16.40		641.40	715
0050	Maximum	"	16	.500		3,775	24.50		3,799.50	4,200
0200	Wall hung styling station with mirrors, minimum	L-2	8	2		505	86.50		591.50	690
0300	Maximum	"	4	4		2,525	173		2,698	3,050
0500	Sink, hair washing basin, rough plumbing not incl.	1 Plum	8	1		495	62		557	640
1000	Sterilizer, liquid solution for tools					161			161	177
1100	Total equipment, rule of thumb, per chair, minimum	L-8	1	20		2,075	1,025		3,100	3,875
1150	Maximum	"	1	20	↓	5,600	1,025		6,625	7,725

11 21 73 – Commercial Laundry and Dry Cleaning Equipment

11 21 73.13 Dry Cleaning Equipment

	11 21 73.13 Dry Cleaning Equipment	Crew	Daily Output	Labor-Hours	Unit	Material	2017 Bare Costs Labor	Equipment	Total	Total Incl O&P
0010	**DRY CLEANING EQUIPMENT**									
2000	Dry cleaners, electric, 20 lb. capacity, not incl. rough-in	L-1	.20	80	Ea.	34,100	4,725		38,825	44,600
2050	25 lb. capacity		.17	94.118		50,500	5,575		56,075	64,000
2100	30 lb. capacity		.15	106		53,000	6,325		59,325	68,000
2150	60 lb. capacity	↓	.09	177	↓	79,000	10,500		89,500	103,000

11 21 73.16 Drying and Conditioning Equipment

	11 21 73.16 Drying and Conditioning Equipment	Crew	Daily Output	Labor-Hours	Unit	Material	2017 Bare Costs Labor	Equipment	Total	Total Incl O&P
0010	**DRYING AND CONDITIONING EQUIPMENT**									
0100	Dryers, not including rough-in									
1500	Industrial, 30 lb. capacity	1 Plum	2	4	Ea.	3,300	247		3,547	4,025
1600	50 lb. capacity	"	1.70	4.706		3,750	291		4,041	4,575
4700	Lint collector, ductwork not included, 8,000 to 10,000 CFM	Q-10	.30	80	↓	9,575	4,325		13,900	17,100

11 21 73.19 Finishing Equipment

	11 21 73.19 Finishing Equipment	Crew	Daily Output	Labor-Hours	Unit	Material	2017 Bare Costs Labor	Equipment	Total	Total Incl O&P
0010	**FINISHING EQUIPMENT**									
3500	Folders, blankets & sheets, minimum	1 Elec	.17	47.059	Ea.	34,400	2,675		37,075	41,900
3700	King size with automatic stacker		.10	80		62,500	4,525		67,025	75,500
3800	For conveyor delivery, add	↓	.45	17.778		15,500	1,000		16,500	18,600
4900	Spreader feeders, 240V, 2 station	L-6	.70	17.143		58,500	1,025		59,525	66,000
4920	4 station	"	.35	34.286	↓	71,000	2,050		73,050	81,000

11 21 Retail and Service Equipment

11 21 73 - Commercial Laundry and Dry Cleaning Equipment

11 21 73.23 Commercial Ironing Equipment

	11 21 73.23 Commercial Ironing Equipment	Crew	Daily Output	Labor-Hours	Unit	Material	2017 Bare Costs Labor	2017 Bare Costs Equipment	Total	Total Incl O&P
0010	**COMMERCIAL IRONING EQUIPMENT**									
4500	Ironers, institutional, 110", single roll	1 Elec	.20	40	Ea.	33,400	2,275		35,675	40,100
4800	Pressers, low capacity air operated	L-6	1.75	6.857		10,100	410		10,510	11,700
4820	Hand operated		1.75	6.857		9,375	410		9,785	10,900
4840	Ironer 48", 240V		3.50	3.429		117,000	206		117,206	129,000
6600	Hand operated presser		.70	17.143		6,300	1,025		7,325	8,475
6620	Mushroom press 115V		.70	17.143		7,925	1,025		8,950	10,300

11 21 73.26 Commercial Washers and Extractors

	11 21 73.26 Commercial Washers and Extractors	Crew	Daily Output	Labor-Hours	Unit	Material	Labor	Equipment	Total	Total Incl O&P
0010	**COMMERCIAL WASHERS AND EXTRACTORS**, not including rough-in									
6000	Combination washer/extractor, 20 lb. capacity	L-6	1.50	8	Ea.	6,225	480		6,705	7,575
6100	30 lb. capacity		.80	15		9,650	900		10,550	12,000
6200	50 lb. capacity		.68	17.647		11,300	1,050		12,350	14,000
6300	75 lb. capacity		.30	40		21,300	2,400		23,700	27,000
6350	125 lb. capacity		.16	75		28,200	4,500		32,700	37,800
6380	Washer extractor/dryer, 110 lb., 240V		1	12		9,125	720		9,845	11,100
6400	Washer extractor, 135 lb., 240V		1	12		27,000	720		27,720	30,800
6450	Pass through		1	12		71,500	720		72,220	80,000
6500	200 lb. washer extractor		1	12		73,500	720		74,220	82,000
6550	Pass through		1	12		77,500	720		78,220	86,500
6600	Extractor, low capacity		1.75	6.857		7,575	410		7,985	8,950

11 21 73.33 Coin-Operated Laundry Equipment

	11 21 73.33 Coin-Operated Laundry Equipment	Crew	Daily Output	Labor-Hours	Unit	Material	Labor	Equipment	Total	Total Incl O&P
0010	**COIN-OPERATED LAUNDRY EQUIPMENT**									
0990	Dryer, gas fired									
1000	Commercial, 30 lb. capacity, coin operated, single	1 Plum	3	2.667	Ea.	3,425	165		3,590	4,025
1100	Double stacked	"	2	4		8,075	247		8,322	9,250
4860	Coin dry cleaner 20 lb.	L-6	1.75	6.857		29,400	410		29,810	32,900
5290	Clothes washer									
5300	Commercial, coin operated, average	1 Plum	3	2.667	Ea.	1,275	165		1,440	1,650

11 21 83 - Photo Processing Equipment

11 21 83.13 Darkroom Equipment

	11 21 83.13 Darkroom Equipment	Crew	Daily Output	Labor-Hours	Unit	Material	Labor	Equipment	Total	Total Incl O&P
0010	**DARKROOM EQUIPMENT**									
0020	Developing sink, 5" deep, 24" x 48"	Q-1	2	8	Ea.	585	445		1,030	1,325
0050	48" x 52"		1.70	9.412		1,400	525		1,925	2,350
0200	10" deep, 24" x 48"		1.70	9.412		1,700	525		2,225	2,675
0250	24" x 108"		1.50	10.667		3,575	595		4,170	4,825
0500	Dryers, dehumidified filtered air, 36" x 25" x 68" high	L-7	6	4.667		2,350	221		2,571	2,925
3000	Viewing lights, 20" x 24"		6	4.667		289	221		510	660
3100	20" x 24" with color correction		6	4.667		420	221		641	800

11 22 Banking Equipment

11 22 13 - Vault Equipment

11 22 13.16 Safes

	11 22 13.16 Safes	Crew	Daily Output	Labor-Hours	Unit	Material	Labor	Equipment	Total	Total Incl O&P
0010	**SAFES**									
0200	Office, 1 hr. rating, 30" x 18" x 18"				Ea.	2,275			2,275	2,500
0250	40" x 18" x 18"					4,950			4,950	5,450
0300	60" x 36" x 18", double door					9,325			9,325	10,300
0600	Data, 1 hr. rating, 27" x 19" x 16"					5,425			5,425	5,975
0700	63" x 34" x 16"					16,700			16,700	18,400
0750	Diskette, 1 hr., 14" x 12" x 11", inside					4,425			4,425	4,850

For customer support on your Building Construction Costs with RSMeans Data, call 800.448.8182.

11 22 Banking Equipment

11 22 13 – Vault Equipment

11 22 13.16 Safes

		Crew	Daily Output	Labor-Hours	Unit	Material	2017 Bare Costs Labor	Equipment	Total	Total Incl O&P
0800	Money, "B" label, 9" x 14" x 14"				Ea.	570			570	625
0900	Tool resistive, 24" x 24" x 20"					4,175			4,175	4,600
1050	Tool and torch resistive, 24" x 24" x 20"					8,450			8,450	9,300
1150	Jewelers, 23" x 20" x 18"					9,525			9,525	10,500
1200	63" x 25" x 18"					14,400			14,400	15,800
1300	For handling into building, add, minimum	A-2	8.50	2.824			115	25.50	140.50	204
1400	Maximum	"	.78	30.769			1,250	279	1,529	2,225

11 22 16 – Teller and Service Equipment

11 22 16.13 Teller Equipment Systems

		Crew	Daily Output	Labor-Hours	Unit	Material	2017 Bare Costs Labor	Equipment	Total	Total Incl O&P
0010	**TELLER EQUIPMENT SYSTEMS**									
0020	Alarm system, police	2 Elec	1.60	10	Ea.	4,975	565		5,540	6,325
0100	With vault alarm	"	.40	40		19,700	2,275		21,975	25,100
0400	Bullet resistant teller window, 44" x 60"	1 Glaz	.60	13.333		4,475	635		5,110	5,900
0500	48" x 60"	"	.60	13.333		6,175	635		6,810	7,775
3000	Counters for banks, frontal only	2 Carp	1	16	Station	1,950	790		2,740	3,350
3100	Complete with steel undercounter	"	.50	32	"	3,625	1,575		5,200	6,400
4600	Door and frame, bullet-resistant, with vision panel, minimum	2 Sswk	1.10	14.545	Ea.	5,575	790		6,365	7,450
4700	Maximum		1.10	14.545		7,550	790		8,340	9,625
4800	Drive-up window, drawer & mike, not incl. glass, minimum		1	16		7,425	870		8,295	9,625
4900	Maximum		.50	32		9,575	1,750		11,325	13,400
5000	Night depository, with chest, minimum		1	16		7,875	870		8,745	10,100
5100	Maximum		.50	32		11,300	1,750		13,050	15,300
5200	Package receiver, painted		3.20	5		1,375	272		1,647	1,975
5300	Stainless steel		3.20	5		2,350	272		2,622	3,025
5400	Partitions, bullet-resistant, 1-3/16" glass, 8' high	2 Carp	10	1.600	L.F.	201	79		280	340
5450	Acrylic	"	10	1.600	"	380	79		459	540
5500	Pneumatic tube systems, 2 lane drive-up, complete	L-3	.25	64	Total	26,100	3,400		29,500	34,000
5550	With T.V. viewer	"	.20	80	"	50,500	4,275		54,775	62,000
5570	Safety deposit boxes, minimum	1 Sswk	44	.182	Opng.	58	9.85		67.85	80.50
5580	Maximum, 10" x 15" opening		19	.421		123	23		146	173
5590	Teller locker, average		15	.533		1,600	29		1,629	1,800
5600	Pass thru, bullet-res. window, painted steel, 24" x 36"	2 Sswk	1.60	10	Ea.	2,950	545		3,495	4,150
5700	48" x 48"		1.20	13.333		2,700	725		3,425	4,175
5800	72" x 40"		.80	20		4,475	1,075		5,550	6,725
5900	For stainless steel frames, add					20%				
6100	Surveillance system, video camera, complete	2 Elec	1	16	Ea.	9,775	905		10,680	12,200
6110	For each additional camera, add				"	1,000			1,000	1,125
6120	CCTV system, see Section 28 23 13.10									
6200	Twenty-four hour teller, single unit,									
6300	automated deposit, cash and memo	L-3	.25	64	Ea.	45,200	3,400		48,600	55,000
7000	Vault front, see Section 08 34 59.10									

For customer support on your Building Construction Costs with RSMeans Data, call 800.448.8182.

409

11 30 Residential Equipment

11 30 13 – Residential Appliances

11 30 13.15 Cooking Equipment

		Crew	Daily Output	Labor-Hours	Unit	Material	2017 Bare Costs Labor	Equipment	Total	Total Incl O&P
0010	**COOKING EQUIPMENT**									
0020	Cooking range, 30" free standing, 1 oven, minimum	2 Clab	10	1.600	Ea.	485	62.50		547.50	625
0050	Maximum		4	4		1,975	157		2,132	2,425
0150	2 oven, minimum		10	1.600		1,350	62.50		1,412.50	1,575
0200	Maximum	↓	10	1.600		2,675	62.50		2,737.50	3,050
0350	Built-in, 30" wide, 1 oven, minimum	1 Elec	6	1.333		900	75.50		975.50	1,100
0400	Maximum	2 Carp	2	8		1,650	395		2,045	2,425
0500	2 oven, conventional, minimum		4	4		1,350	197		1,547	1,775
0550	1 conventional, 1 microwave, maximum	↓	2	8		2,650	395		3,045	3,500
0700	Free standing, 1 oven, 21" wide range, minimum	2 Clab	10	1.600		515	62.50		577.50	665
0750	21" wide, maximum	"	4	4		610	157		767	910
0900	Countertop cooktops, 4 burner, standard, minimum	1 Elec	6	1.333		340	75.50		415.50	490
0950	Maximum		3	2.667		1,775	151		1,926	2,175
1050	As above, but with grill and griddle attachment, minimum		6	1.333		1,550	75.50		1,625.50	1,825
1100	Maximum		3	2.667		3,825	151		3,976	4,425
1200	Induction cooktop, 30" wide		3	2.667		1,325	151		1,476	1,700
1250	Microwave oven, minimum		4	2		119	113		232	300
1300	Maximum	↓	2	4	↓	475	226		701	865

11 30 13.16 Refrigeration Equipment

		Crew	Daily Output	Labor-Hours	Unit	Material	2017 Bare Costs Labor	Equipment	Total	Total Incl O&P
0010	**REFRIGERATION EQUIPMENT**									
2000	Deep freeze, 15 to 23 C.F., minimum	2 Clab	10	1.600	Ea.	595	62.50		657.50	750
2050	Maximum		5	3.200		785	125		910	1,050
2200	30 C.F., minimum		8	2		690	78.50		768.50	880
2250	Maximum	↓	3	5.333		850	209		1,059	1,250
5200	Icemaker, automatic, 20 lb. per day	1 Plum	7	1.143		1,100	70.50		1,170.50	1,300
5350	51 lb. per day	"	2	4		1,500	247		1,747	2,025
5500	Refrigerator, no frost, 10 C.F. to 12 C.F., minimum	2 Clab	10	1.600		515	62.50		577.50	660
5600	Maximum		6	2.667		535	104		639	750
5750	14 C.F. to 16 C.F., minimum		9	1.778		590	69.50		659.50	755
5800	Maximum		5	3.200		715	125		840	975
5950	18 C.F. to 20 C.F., minimum		8	2		750	78.50		828.50	940
6000	Maximum		4	4		1,750	157		1,907	2,175
6150	21 C.F. to 29 C.F., minimum		7	2.286		1,175	89.50		1,264.50	1,400
6200	Maximum	↓	3	5.333		3,500	209		3,709	4,175
6790	Energy-star qualified, 18 C.F., minimum G	2 Carp	4	4		675	197		872	1,050
6795	Maximum G		2	8		1,600	395		1,995	2,350
6797	21.7 C.F., minimum G		4	4		1,250	197		1,447	1,675
6799	Maximum G		4	4		2,200	197		2,397	2,725

11 30 13.17 Kitchen Cleaning Equipment

		Crew	Daily Output	Labor-Hours	Unit	Material	2017 Bare Costs Labor	Equipment	Total	Total Incl O&P
0010	**KITCHEN CLEANING EQUIPMENT**									
2750	Dishwasher, built-in, 2 cycles, minimum	L-1	4	4	Ea.	315	237		552	700
2800	Maximum		2	8		550	475		1,025	1,325
2950	4 or more cycles, minimum		4	4		415	237		652	815
2960	Average		4	4		555	237		792	965
3000	Maximum		2	8		1,250	475		1,725	2,075
3100	Energy-star qualified, minimum G		4	4		480	237		717	885
3110	Maximum G	↓	2	8		1,650	475		2,125	2,525

11 30 13.18 Waste Disposal Equipment

		Crew	Daily Output	Labor-Hours	Unit	Material	2017 Bare Costs Labor	Equipment	Total	Total Incl O&P
0010	**WASTE DISPOSAL EQUIPMENT**									
1750	Compactor, residential size, 4 to 1 compaction, minimum	1 Carp	5	1.600	Ea.	690	79		769	880
1800	Maximum	"	3	2.667		1,150	131		1,281	1,475
3300	Garbage disposal, sink type, minimum	L-1	10	1.600	↓	97.50	94.50		192	249

11 30 Residential Equipment

11 30 13 – Residential Appliances

11 30 13.18 Waste Disposal Equipment

	Crew	Daily Output	Labor-Hours	Unit	Material	2017 Bare Costs Labor	2017 Bare Costs Equipment	Total	Total Incl O&P
3350	Maximum	L-1	10	1.600	Ea.	246	94.50		340.50	415

11 30 13.19 Kitchen Ventilation Equipment

		Crew	Daily Output	Labor-Hours	Unit	Material	Labor	Equipment	Total	Total Incl O&P
0010	**KITCHEN VENTILATION EQUIPMENT**									
4150	Hood for range, 2 speed, vented, 30" wide, minimum	L-3	5	3.200	Ea.	85.50	171		256.50	355
4200	Maximum		3	5.333		925	284		1,209	1,450
4300	42" wide, minimum		5	3.200		153	171		324	430
4330	Custom		5	3.200		1,600	171		1,771	2,025
4350	Maximum		3	5.333		1,950	284		2,234	2,575
4500	For ventless hood, 2 speed, add					22			22	24
4650	For vented 1 speed, deduct from maximum					66			66	72.50

11 30 13.24 Washers

		Crew	Daily Output	Labor-Hours	Unit	Material	Labor	Equipment	Total	Total Incl O&P
0010	**WASHERS**									
5000	Residential, 4 cycle, average	1 Plum	3	2.667	Ea.	915	165		1,080	1,250
6650	Washing machine, automatic, minimum		3	2.667		565	165		730	870
6700	Maximum		1	8		1,525	495		2,020	2,425
6750	Energy star qualified, front loading, minimum	G	3	2.667		670	165		835	990
6760	Maximum	G	1	8		1,700	495		2,195	2,625
6764	Top loading, minimum	G	3	2.667		575	165		740	885
6766	Maximum	G	3	2.667		1,450	165		1,615	1,825

11 30 13.25 Dryers

		Crew	Daily Output	Labor-Hours	Unit	Material	Labor	Equipment	Total	Total Incl O&P
0010	**DRYERS**									
0500	Gas fired residential, 16 lb. capacity, average	1 Plum	3	2.667	Ea.	700	165		865	1,025
6770	Electric, front loading, energy-star qualified, minimum	G	L-2	3	5.333		645	230		875	1,075
6780	Maximum	G	"	2	8		1,525	345		1,870	2,200
7450	Vent kits for dryers	1 Carp	10	.800		40	39.50		79.50	105

11 30 15 – Miscellaneous Residential Appliances

11 30 15.13 Sump Pumps

		Crew	Daily Output	Labor-Hours	Unit	Material	Labor	Equipment	Total	Total Incl O&P
0010	**SUMP PUMPS**									
6400	Cellar drainer, pedestal, 1/3 H.P., molded PVC base	1 Plum	3	2.667	Ea.	141	165		306	405
6450	Solid brass	"	2	4	"	300	247		547	705
6460	Sump pump, see also Section 22 14 29.16									

11 30 15.23 Water Heaters

		Crew	Daily Output	Labor-Hours	Unit	Material	Labor	Equipment	Total	Total Incl O&P
0010	**WATER HEATERS**									
6900	Electric, glass lined, 30 gallon, minimum	L-1	5	3.200	Ea.	970	189		1,159	1,350
6950	Maximum		3	5.333		1,350	315		1,665	1,950
7100	80 gallon, minimum		2	8		1,800	475		2,275	2,675
7150	Maximum		1	16		2,500	945		3,445	4,175
7180	Gas, glass lined, 30 gallon, minimum	2 Plum	5	3.200		1,725	198		1,923	2,200
7220	Maximum		3	5.333		2,400	330		2,730	3,150
7260	50 gallon, minimum		2.50	6.400		1,875	395		2,270	2,650
7300	Maximum		1.50	10.667		2,600	660		3,260	3,875
7310	Water heater, see also Section 22 33 30.13									

11 30 15.43 Air Quality

		Crew	Daily Output	Labor-Hours	Unit	Material	Labor	Equipment	Total	Total Incl O&P
0010	**AIR QUALITY**									
2450	Dehumidifier, portable, automatic, 15 pint	1 Elec	4	2	Ea.	191	113		304	380
2550	40 pint		3.75	2.133		218	121		339	420
3550	Heater, electric, built-in, 1250 watt, ceiling type, minimum		4	2		123	113		236	305
3600	Maximum		3	2.667		194	151		345	440
3700	Wall type, minimum		4	2		181	113		294	370
3750	Maximum		3	2.667		186	151		337	430
3900	1500 watt wall type, with blower		4	2		181	113		294	370

11 30 Residential Equipment

11 30 15 – Miscellaneous Residential Appliances

11 30 15.43 Air Quality	Crew	Daily Output	Labor-Hours	Unit	Material	2017 Bare Costs Labor	Equipment	Total	Total Incl O&P
3950	3000 watt	1 Elec	3	2.667	Ea.	445	151		596	715
4850	Humidifier, portable, 8 gallons per day					157			157	173
5000	15 gallons per day					224			224	246

11 30 33 – Retractable Stairs

11 30 33.10 Disappearing Stairway

		Crew	Daily Output	Labor-Hours	Unit	Material	2017 Bare Costs Labor	Equipment	Total	Total Incl O&P
0010	**DISAPPEARING STAIRWAY** No trim included									
0100	Custom grade, pine, 8'-6" ceiling, minimum	1 Carp	4	2	Ea.	222	98.50		320.50	395
0150	Average		3.50	2.286		245	113		358	440
0200	Maximum		3	2.667		294	131		425	525
0500	Heavy duty, pivoted, from 7'-7" to 12'-10" floor to floor		3	2.667		1,025	131		1,156	1,325
0600	16'-0" ceiling		2	4		1,600	197		1,797	2,050
0800	Economy folding, pine, 8'-6" ceiling		4	2		179	98.50		277.50	350
0900	9'-6" ceiling		4	2		207	98.50		305.50	380
1100	Automatic electric, aluminum, floor to floor height, 8' to 9'	2 Carp	1	16		9,450	790		10,240	11,600
1400	11' to 12'		.90	17.778		10,000	875		10,875	12,400
1700	14' to 15'		.70	22.857		10,700	1,125		11,825	13,500

11 32 Unit Kitchens

11 32 13 – Metal Unit Kitchens

11 32 13.10 Commercial Unit Kitchens

		Crew	Daily Output	Labor-Hours	Unit	Material	2017 Bare Costs Labor	Equipment	Total	Total Incl O&P
0010	**COMMERCIAL UNIT KITCHENS**									
1500	Combination range, refrigerator and sink, 30" wide, minimum	L-1	2	8	Ea.	1,100	475		1,575	1,925
1550	Maximum		1	16		1,350	945		2,295	2,900
1570	60" wide, average		1.40	11.429		1,475	675		2,150	2,650
1590	72" wide, average		1.20	13.333		1,850	790		2,640	3,200
1600	Office model, 48" wide		2	8		1,975	475		2,450	2,875
1620	Refrigerator and sink only		2.40	6.667		2,650	395		3,045	3,525
1640	Combination range, refrigerator, sink, microwave									
1660	Oven and ice maker	L-1	.80	20	Ea.	4,775	1,175		5,950	7,025

11 41 Foodservice Storage Equipment

11 41 13 – Refrigerated Food Storage Cases

11 41 13.10 Refrigerated Food Cases

		Crew	Daily Output	Labor-Hours	Unit	Material	2017 Bare Costs Labor	Equipment	Total	Total Incl O&P
0010	**REFRIGERATED FOOD CASES**									
0030	Dairy, multi-deck, 12' long	Q-5	3	5.333	Ea.	10,900	299		11,199	12,400
0100	For rear sliding doors, add					1,950			1,950	2,150
0200	Delicatessen case, service deli, 12' long, single deck	Q-5	3.90	4.103		8,075	230		8,305	9,225
0300	Multi-deck, 18 S.F. shelf display		3	5.333		7,950	299		8,249	9,200
0400	Freezer, self-contained, chest-type, 30 C.F.		3.90	4.103		9,475	230		9,705	10,700
0500	Glass door, upright, 78 C.F.		3.30	4.848		9,075	272		9,347	10,400
0600	Frozen food, chest type, 12' long		3.30	4.848		7,900	272		8,172	9,075
0700	Glass door, reach-in, 5 door		3	5.333		9,900	299		10,199	11,400
0800	Island case, 12' long, single deck		3.30	4.848		7,725	272		7,997	8,900
0900	Multi-deck		3	5.333		9,000	299		9,299	10,400
1000	Meat case, 12' long, single deck		3.30	4.848		7,725	272		7,997	8,900
1050	Multi-deck		3.10	5.161		9,975	289		10,264	11,400
1100	Produce, 12' long, single deck		3.30	4.848		6,625	272		6,897	7,675
1200	Multi-deck		3.10	5.161		8,375	289		8,664	9,650

11 41 Foodservice Storage Equipment

11 41 13 – Refrigerated Food Storage Cases

11 41 13.20 Refrigerated Food Storage Equipment	Crew	Daily Output	Labor-Hours	Unit	Material	2017 Bare Costs Labor	Equipment	Total	Total Incl O&P
0010 **REFRIGERATED FOOD STORAGE EQUIPMENT**									
2350 Cooler, reach-in, beverage, 6' long	Q-1	6	2.667	Ea.	3,225	148		3,373	3,775
4300 Freezers, reach-in, 44 C.F.		4	4		5,675	223		5,898	6,550
4500 68 C.F.		3	5.333		4,875	297		5,172	5,800
4600 Freezer, pre-fab, 8' x 8' w/refrigeration	2 Carp	.45	35.556		10,600	1,750		12,350	14,300
4620 8' x 12'		.35	45.714		12,900	2,250		15,150	17,600
4640 8' x 16'		.25	64		15,800	3,150		18,950	22,100
4660 8' x 20'		.17	94.118		23,200	4,625		27,825	32,700
4680 Reach-in, 1 compartment	Q-1	4	4		2,650	223		2,873	3,225
4685 Energy star rated [G]	R-18	7.80	3.333		2,900	166		3,066	3,450
4700 2 compartment	Q-1	3	5.333		4,050	297		4,347	4,900
4705 Energy star rated [G]	R-18	6.20	4.194		3,375	208		3,583	4,025
4710 3 compartment	Q-1	3	5.333		5,300	297		5,597	6,275
4715 Energy star rated [G]	R-18	5.60	4.643		4,775	231		5,006	5,600
8320 Refrigerator, reach-in, 1 compartment		7.80	3.333		2,750	166		2,916	3,275
8325 Energy star rated [G]		7.80	3.333		2,900	166		3,066	3,450
8330 2 compartment		6.20	4.194		4,025	208		4,233	4,750
8335 Energy star rated [G]		6.20	4.194		3,375	208		3,583	4,025
8340 3 compartment		5.60	4.643		5,275	231		5,506	6,150
8345 Energy star rated [G]		5.60	4.643		4,775	231		5,006	5,600
8350 Pre-fab, with refrigeration, 8' x 8'	2 Carp	.45	35.556		6,950	1,750		8,700	10,300
8360 8' x 12'		.35	45.714		8,400	2,250		10,650	12,700
8370 8' x 16'		.25	64		13,000	3,150		16,150	19,100
8380 8' x 20'		.17	94.118		16,400	4,625		21,025	25,200
8390 Pass-thru/roll-in, 1 compartment	R-18	7.80	3.333		4,675	166		4,841	5,400
8400 2 compartment		6.24	4.167		6,625	207		6,832	7,600
8410 3 compartment		5.60	4.643		9,025	231		9,256	10,300
8420 Walk-in, alum, door & floor only, no refrig, 6' x 6' x 7'-6"	2 Carp	1.40	11.429		8,475	565		9,040	10,200
8430 10' x 6' x 7'-6"		.55	29.091		12,200	1,425		13,625	15,600
8440 12' x 14' x 7'-6"		.25	64		16,400	3,150		19,550	22,900
8450 12' x 20' x 7'-6"		.17	94.118		18,900	4,625		23,525	27,900
8460 Refrigerated cabinets, mobile					4,075			4,075	4,475
8470 Refrigerator/freezer, reach-in, 1 compartment	R-18	5.60	4.643		5,500	231		5,731	6,400
8480 2 compartment	"	4.80	5.417		7,550	269		7,819	8,700

11 41 13.30 Wine Cellar

	Crew	Daily Output	Labor-Hours	Unit	Material	Labor	Equipment	Total	Total Incl O&P
0010 **WINE CELLAR**, refrigerated, Redwood interior, carpeted, walk-in type									
0020 6'-8" high, including racks									
0200 80" W x 48" D for 900 bottles	2 Carp	1.50	10.667	Ea.	4,250	525		4,775	5,475
0250 80" W x 72" D for 1300 bottles		1.33	12.030		5,175	590		5,765	6,600
0300 80" W x 94" D for 1900 bottles		1.17	13.675		6,275	675		6,950	7,925
0400 80" W x 124" D for 2500 bottles		1	16		7,375	790		8,165	9,300
0600 Portable cabinets, red oak, reach-in temp. & humidity controlled									
0650 26-5/8"W x 26-1/2"D x 68"H for 235 bottles				Ea.	4,400			4,400	4,825
0660 32"W x 21-1/2"D x 73-1/2"H for 144 bottles					3,550			3,550	3,925
0670 32"W x 29-1/2"D x 73-1/2"H for 288 bottles					4,575			4,575	5,025
0680 39-1/2"W x 29-1/2"D x 86-1/2"H for 440 bottles					4,575			4,575	5,050
0690 52-1/2"W x 29-1/2"D x 73-1/2"H for 468 bottles					5,050			5,050	5,550
0700 52-1/2"W x 29-1/2"D x 86-1/2"H for 572 bottles					5,175			5,175	5,675
0730 Portable, red oak, can be built-in with glass door									
0750 23-7/8"W x 24"D x 34-1/2"H for 50 bottles				Ea.	1,050			1,050	1,150

For customer support on your Building Construction Costs with RSMeans Data, call 800.448.8182.

413

11 41 Foodservice Storage Equipment

11 41 33 – Foodservice Shelving

11 41 33.20 Metal Food Storage Shelving		Crew	Daily Output	Labor-Hours	Unit	Material	2017 Bare Costs Labor	Equipment	Total	Total Incl O&P
0010	**METAL FOOD STORAGE SHELVING**									
8600	Stainless steel shelving, louvered 4-tier, 20" x 3'	1 Clab	6	1.333	Ea.	1,475	52		1,527	1,700
8605	20" x 4'		6	1.333		1,675	52		1,727	1,900
8610	20" x 6'		6	1.333		1,775	52		1,827	2,025
8615	24" x 3'		6	1.333		2,075	52		2,127	2,350
8620	24" x 4'		6	1.333		2,475	52		2,527	2,800
8625	24" x 6'		6	1.333		3,450	52		3,502	3,850
8630	Flat 4-tier, 20" x 3'		6	1.333		1,175	52		1,227	1,375
8635	20" x 4'		6	1.333		1,425	52		1,477	1,625
8640	20" x 5'		6	1.333		1,600	52		1,652	1,850
8645	24" x 3'		6	1.333		1,175	52		1,227	1,375
8650	24" x 4'		6	1.333		2,325	52		2,377	2,625
8655	24" x 6'		6	1.333		2,475	52		2,527	2,775
8700	Galvanized shelving, louvered 4-tier, 20" x 3'		6	1.333		800	52		852	960
8705	20" x 4'		6	1.333		905	52		957	1,075
8710	20" x 6'		6	1.333		960	52		1,012	1,125
8715	24" x 3'		6	1.333		730	52		782	885
8720	24" x 4'		6	1.333		965	52		1,017	1,125
8725	24" x 6'		6	1.333		1,375	52		1,427	1,600
8730	Flat 4-tier, 20" x 3'		6	1.333		735	52		787	890
8735	20" x 4'		6	1.333		650	52		702	795
8740	20" x 6'		6	1.333		955	52		1,007	1,125
8/45	24" x 3'		6	1.333		720	52		772	875
8750	24" x 4'		6	1.333		750	52		802	905
8755	24" x 6'		6	1.333		860	52		912	1,025
8760	Stainless steel dunnage rack, 24" x 3'		8	1		380	39		419	475
8765	24" x 4'		8	1		430	39		469	535
8770	Galvanized dunnage rack, 24" x 3'		8	1		174	39		213	251
8775	24" x 4'		8	1		191	39		230	270

11 42 Food Preparation Equipment

11 42 10 – Commercial Food Preparation Equipment

11 42 10.10 Choppers, Mixers and Misc. Equipment		Crew	Daily Output	Labor-Hours	Unit	Material	2017 Bare Costs Labor	Equipment	Total	Total Incl O&P
0010	**CHOPPERS, MIXERS AND MISC. EQUIPMENT**									
1700	Choppers, 5 pounds	R-18	7	3.714	Ea.	2,350	185		2,535	2,850
1720	16 pounds		5	5.200		2,500	258		2,758	3,150
1740	35 to 40 pounds		4	6.500		3,600	325		3,925	4,450
1840	Coffee brewer, 5 burners	1 Plum	3	2.667		1,375	165		1,540	1,750
1850	Coffee urn, twin 6 gallon urns		2	4		2,400	247		2,647	3,000
1860	Single, 3 gallon		3	2.667		1,800	165		1,965	2,225
3000	Fast food equipment, total package, minimum	6 Skwk	.08	600		214,000	30,900		244,900	282,500
3100	Maximum	"	.07	685		291,500	35,300		326,800	375,000
3800	Food mixers, bench type, 20 quarts	L-7	7	4		3,050	190		3,240	3,650
3850	40 quarts		5.40	5.185		7,075	246		7,321	8,175
3900	60 quarts		5	5.600		12,100	266		12,366	13,700
4040	80 quarts		3.90	7.179		12,800	340		13,140	14,600
4100	Floor type, 20 quarts		15	1.867		3,375	88.50		3,463.50	3,825
4120	60 quarts		14	2		10,500	95		10,595	11,700
4140	80 quarts		12	2.333		16,800	111		16,911	18,700
4160	140 quarts		8.60	3.256		26,800	154		26,954	29,700

414

11 42 Food Preparation Equipment

11 42 10 – Commercial Food Preparation Equipment

11 42 10.10 Choppers, Mixers and Misc. Equipment		Crew	Daily Output	Labor-Hours	Unit	Material	2017 Bare Costs Labor	Equipment	Total	Total Incl O&P
6700	Peelers, small	R-18	8	3.250	Ea.	2,100	161		2,261	2,550
6720	Large	"	6	4.333		4,900	215		5,115	5,700
6800	Pulper/extractor, close coupled, 5 HP	1 Plum	1.90	4.211		3,825	260		4,085	4,625
8580	Slicer with table	R-18	9	2.889	↓	3,925	144		4,069	4,550

11 43 Food Delivery Carts and Conveyors

11 43 13 – Food Delivery Carts

11 43 13.10 Mobile Carts, Racks and Trays

		Crew	Daily Output	Labor-Hours	Unit	Material	2017 Bare Costs Labor	Equipment	Total	Total Incl O&P
0010	**MOBILE CARTS, RACKS AND TRAYS**									
1650	Cabinet, heated, 1 compartment, reach-in	R-18	5.60	4.643	Ea.	3,600	231		3,831	4,300
1655	Pass-thru roll-in		5.60	4.643		3,725	231		3,956	4,450
1660	2 compartment, reach-in	↓	4.80	5.417		9,500	269		9,769	10,800
1670	Mobile					3,625			3,625	3,975
2000	Hospital food cart, hot and cold service, 20 tray capacity					16,100			16,100	17,700
6850	Mobile rack w/pan slide					1,475			1,475	1,625
9180	Tray and silver dispenser, mobile	1 Clab	16	.500	↓	955	19.60		974.60	1,075

11 44 Food Cooking Equipment

11 44 13 – Commercial Ranges

11 44 13.10 Cooking Equipment

		Crew	Daily Output	Labor-Hours	Unit	Material	2017 Bare Costs Labor	Equipment	Total	Total Incl O&P
0010	**COOKING EQUIPMENT**									
0020	Bake oven, gas, one section	Q-1	8	2	Ea.	5,675	111		5,786	6,400
0300	Two sections		7	2.286		9,325	127		9,452	10,500
0600	Three sections	↓	6	2.667		12,200	148		12,348	13,600
0900	Electric convection, single deck	L-7	4	7		5,950	330		6,280	7,050
1300	Broiler, without oven, standard	Q-1	8	2		3,900	111		4,011	4,475
1550	Infrared	L-7	4	7		7,500	330		7,830	8,750
4750	Fryer, with twin baskets, modular model	Q-1	7	2.286		1,350	127		1,477	1,675
5000	Floor model, on 6" legs	"	5	3.200		2,625	178		2,803	3,175
5100	Extra single basket, large					105			105	115
5170	Energy star rated, 50 lb. capacity G	R-18	4	6.500		4,775	325		5,100	5,750
5175	85 lb. capacity G	"	4	6.500		10,000	325		10,325	11,500
5300	Griddle, SS, 24" plate, w/4" legs, elec, 208 V, 3 phase, 3' long	Q-1	7	2.286		1,550	127		1,677	1,900
5550	4' long	"	6	2.667		2,225	148		2,373	2,675
6200	Iced tea brewer	1 Plum	3.44	2.326		780	144		924	1,075
6350	Kettle, w/steam jacket, tilting, w/positive lock, SS, 20 gallons	L-7	7	4		9,150	190		9,340	10,400
6600	60 gallons	"	6	4.667		12,300	221		12,521	13,800
6900	Range, restaurant type, 6 burners and 1 standard oven, 36" wide	Q-1	7	2.286		3,025	127		3,152	3,525
6950	Convection		7	2.286		4,575	127		4,702	5,250
7150	2 standard ovens, 24" griddle, 60" wide		6	2.667		5,225	148		5,373	5,975
7200	1 standard, 1 convection oven		6	2.667		9,500	148		9,648	10,600
7450	Heavy duty, single 34" standard oven, open top		5	3.200		5,600	178		5,778	6,425
7500	Convection oven		5	3.200		6,050	178		6,228	6,950
7700	Griddle top		6	2.667		3,225	148		3,373	3,775
7750	Convection oven	↓	6	2.667		6,850	148		6,998	7,750
7760	Induction cooker, electric	L-7	7	4		1,825	190		2,015	2,325
8850	Steamer, electric 27 KW		7	4		11,500	190		11,690	13,000
9100	Electric, 10 KW or gas 100,000 BTU	↓	5	5.600		6,625	266		6,891	7,700
9150	Toaster, conveyor type, 16-22 slices per minute					1,225			1,225	1,350

For customer support on your Building Construction Costs with RSMeans Data, call 800.448.8182.

415

11 44 Food Cooking Equipment

11 44 13 – Commercial Ranges

11 44 13.10 Cooking Equipment	Crew	Daily Output	Labor-Hours	Unit	Material	2017 Bare Costs Labor	Equipment	Total	Total Incl O&P
9160	Pop-up, 2 slot				Ea.	650			650	715
9200	For deluxe models of above equipment, add					75%				
9400	Rule of thumb: Equipment cost based									
9410	on kitchen work area									
9420	Office buildings, minimum	L-7	77	.364	S.F.	95	17.25		112.25	132
9450	Maximum		58	.483		161	23		184	212
9550	Public eating facilities, minimum		77	.364		125	17.25		142.25	164
9600	Maximum		46	.609		203	29		232	267
9750	Hospitals, minimum		58	.483		128	23		151	176
9800	Maximum	↓	39	.718	↓	236	34		270	310

11 46 Food Dispensing Equipment

11 46 16 – Service Line Equipment

11 46 16.10 Commercial Food Dispensing Equipment

		Crew	Daily Output	Labor-Hours	Unit	Material	2017 Bare Costs Labor	Equipment	Total	Total Incl O&P
0010	**COMMERCIAL FOOD DISPENSING EQUIPMENT**									
1050	Butter pat dispenser	1 Clab	13	.615	Ea.	915	24		939	1,025
1100	Bread dispenser, counter top		13	.615		935	24		959	1,050
1900	Cup and glass dispenser, drop in		4	2		1,100	78.50		1,178.50	1,325
1920	Disposable cup, drop in		16	.500		560	19.60		579.60	645
2650	Dish dispenser, drop in, 12"		11	.727		2,150	28.50		2,178.50	2,400
2660	Mobile	↓	10	.800		3,200	31.50		3,231.50	3,575
3300	Food warmer, counter, 1.2 KW					720			720	790
3550	1.6 KW					2,175			2,175	2,400
3600	Well, hot food, built-in, rectangular, 12" x 20"	R-30	10	2.600		880	120		1,000	1,150
3610	Circular, 7 qt.		10	2.600		420	120		540	640
3620	Refrigerated, 2 compartments		10	2.600		3,200	120		3,320	3,700
3630	3 compartments		9	2.889		3,950	133		4,083	4,550
3640	4 compartments		8	3.250		4,575	150		4,725	5,250
4720	Frost cold plate	↓	9	2.889		19,900	133		20,033	22,100
5700	Hot chocolate dispenser	1 Plum	4	2		1,200	124		1,324	1,500
5750	Ice dispenser 567 pound	Q-1	6	2.667		5,450	148		5,598	6,225
6250	Jet spray dispenser	R-18	4.50	5.778		2,825	287		3,112	3,525
6300	Juice dispenser, concentrate	"	4.50	5.778		1,900	287		2,187	2,500
6690	Milk dispenser, bulk, 2 flavor	R-30	8	3.250		1,825	150		1,975	2,250
6695	3 flavor	"	8	3.250	↓	2,450	150		2,600	2,925
8800	Serving counter, straight	1 Carp	40	.200	L.F.	975	9.85		984.85	1,100
8820	Curved section	"	30	.267	"	1,250	13.15		1,263.15	1,400
8825	Solid surface, see Section 12 36 61.16									
8860	Sneeze guard with lights, 60" L	1 Clab	16	.500	Ea.	350	19.60		369.60	415
8900	Sneeze guard, stainless steel and glass, single sided									
8910	Portable, 48" W				Ea.	325			325	355
8920	Portable, 72" W					350			350	385
8930	Adjustable, 36" W	1 Carp	24	.333		250	16.40		266.40	300
8940	Adjustable, 48" W	"	20	.400		315	19.70		334.70	380
9100	Soft serve ice cream machine, medium	R-18	11	2.364		9,400	117		9,517	10,600
9110	Large	"	9	2.889	↓	21,600	144		21,744	24,000

11 46 Food Dispensing Equipment

11 46 83 – Ice Machines

11 46 83.10 Commercial Ice Equipment

	11 46 83.10 Commercial Ice Equipment	Crew	Daily Output	Labor-Hours	Unit	Material	2017 Bare Costs Labor	Equipment	Total	Total Incl O&P
0010	**COMMERCIAL ICE EQUIPMENT**									
5800	Ice cube maker, 50 pounds per day	Q-1	6	2.667	Ea.	1,750	148		1,898	2,150
5810	65 pounds per day, energy star rated		6	2.667		1,675	148		1,823	2,075
5900	250 pounds per day		1.20	13.333		2,600	740		3,340	3,975
5950	300 pounds per day, remote condensing		1.20	13.333		2,425	740		3,165	3,775
6050	500 pounds per day		4	4		2,775	223		2,998	3,375
6060	With bin		1.20	13.333		3,825	740		4,565	5,325
6070	Modular, with bin and condenser		1.20	13.333		4,075	740		4,815	5,600
6090	1000 pounds per day, with bin		1	16		5,050	890		5,940	6,925
6100	Ice flakers, 300 pounds per day		1.60	10		3,325	555		3,880	4,500
6120	600 pounds per day		.95	16.842		4,875	935		5,810	6,775
6130	1000 pounds per day		.75	21.333		4,750	1,175		5,925	7,025
6140	2000 pounds per day	↓	.65	24.615		22,600	1,375		23,975	27,000
6160	Ice storage bin, 500 pound capacity	Q-5	1	16		1,100	895		1,995	2,550
6180	1000 pound	"	.56	28.571	↓	2,850	1,600		4,450	5,575

11 48 Foodservice Cleaning and Disposal Equipment

11 48 13 – Commercial Dishwashers

11 48 13.10 Dishwashers

	11 48 13.10 Dishwashers		Crew	Daily Output	Labor-Hours	Unit	Material	2017 Bare Costs Labor	Equipment	Total	Total Incl O&P
0010	**DISHWASHERS**										
2700	Dishwasher, commercial, rack type										
2720	10 to 12 racks per hour		Q-1	3.20	5	Ea.	3,425	278		3,703	4,200
2730	Energy star rated, 35 to 40 racks/hour	G		1.30	12.308		5,000	685		5,685	6,525
2740	50 to 60 racks/hour	G	↓	1.30	12.308		11,100	685		11,785	13,200
2800	Automatic, 190 to 230 racks per hour		L-6	.35	34.286		14,600	2,050		16,650	19,200
2820	235 to 275 racks per hour			.25	48		33,700	2,875		36,575	41,400
2840	8,750 to 12,500 dishes per hour		↓	.10	120	↓	59,000	7,200		66,200	76,000
2950	Dishwasher hood, canopy type		L-3A	10	1.200	L.F.	1,050	64		1,114	1,250
2960	Pant leg type		"	2.50	4.800	Ea.	9,150	257		9,407	10,500
5200	Garbage disposal 1.5 HP, 100 GPH		L-1	4.80	3.333		1,825	197		2,022	2,300
5210	3 HP, 120 GPH			4.60	3.478		2,450	206		2,656	2,975
5220	5 HP, 250 GPH		↓	4.50	3.556	↓	3,675	210		3,885	4,375
6750	Pot sink, 3 compartment		1 Plum	7.25	1.103	L.F.	970	68		1,038	1,175
6760	Pot washer, low temp wash/rinse			1.60	5	Ea.	5,450	310		5,760	6,475
6770	High pressure wash, high temperature rinse		↓	1.20	6.667		38,400	410		38,810	42,800
9170	Trash compactor, small, up to 125 lb. compacted weight		L-4	4	6		24,600	280		24,880	27,400
9175	Large, up to 175 lb. compacted weight		"	3	8	↓	29,200	375		29,575	32,700

11 52 Audio-Visual Equipment

11 52 13 – Projection Screens

11 52 13.10 Projection Screens, Wall or Ceiling Hung

	11 52 13.10 Projection Screens, Wall or Ceiling Hung	Crew	Daily Output	Labor-Hours	Unit	Material	2017 Bare Costs Labor	Equipment	Total	Total Incl O&P
0010	**PROJECTION SCREENS, WALL OR CEILING HUNG**, matte white									
0100	Manually operated, economy	2 Carp	500	.032	S.F.	6.45	1.58		8.03	9.45
0300	Intermediate		450	.036		7.85	1.75		9.60	11.35
0400	Deluxe		400	.040	↓	9.75	1.97		11.72	13.75
0600	Electric operated, matte white, 25 S.F., economy		5	3.200	Ea.	1,125	158		1,283	1,500
0700	Deluxe		4	4		1,325	197		1,522	1,775
0900	50 S.F., economy		3	5.333		890	263		1,153	1,375
1000	Deluxe	↓	2	8		2,650	395		3,045	3,525

For customer support on your Building Construction Costs with RSMeans Data, call 800.448.8182.

417

11 52 Audio-Visual Equipment

11 52 13 – Projection Screens

11 52 13.10 Projection Screens, Wall or Ceiling Hung	Crew	Daily Output	Labor-Hours	Unit	Material	2017 Bare Costs Labor	Equipment	Total	Total Incl O&P
1200	Heavy duty, electric operated, 200 S.F.	2 Carp	1.50	10.667	Ea.	4,150	525		4,675	5,375
1300	400 S.F.		1	16		4,950	790		5,740	6,650
1500	Rigid acrylic in wall, for rear projection, 1/4" thick	2 Glaz	30	.533	S.F.	70	25.50		95.50	116
1600	1/2" thick (maximum size 10' x 20')	"	25	.640	"	82	30.50		112.50	137

11 52 16 – Projectors

11 52 16.10 Movie Equipment

		Crew	Daily Output	Labor-Hours	Unit	Material	2017 Bare Costs Labor	Equipment	Total	Total Incl O&P
0010	**MOVIE EQUIPMENT**									
0020	Changeover, minimum				Ea.	505			505	555
0100	Maximum					960			960	1,050
0400	Film transport, incl. platters and autowind, minimum					5,350			5,350	5,875
0500	Maximum					15,200			15,200	16,700
0800	Lamphouses, incl. rectifiers, xenon, 1,000 watt	1 Elec	2	4		7,075	226		7,301	8,125
0900	1,600 watt		2	4		7,550	226		7,776	8,650
1000	2,000 watt		1.50	5.333		8,200	300		8,500	9,475
1100	4,000 watt		1.50	5.333		11,300	300		11,600	12,900
1400	Lenses, anamorphic, minimum					1,350			1,350	1,500
1500	Maximum					3,050			3,050	3,350
1800	Flat 35 mm, minimum					1,175			1,175	1,300
1900	Maximum					1,825			1,825	2,025
2200	Pedestals, for projectors					1,625			1,625	1,800
2300	Console type					11,700			11,700	12,900
2600	Projector mechanisms, incl. soundhead, 35 mm, minimum					12,100			12,100	13,300
2700	Maximum					16,600			16,600	18,300
3000	Projection screens, rigid, in wall, acrylic, 1/4" thick	2 Glaz	195	.082	S.F.	45.50	3.92		49.42	56.50
3100	1/2" thick	"	130	.123	"	53	5.90		58.90	67.50
3300	Electric operated, heavy duty, 400 S.F.	2 Carp	1	16	Ea.	3,225	790		4,015	4,750
3320	Theater projection screens, matte white, including frames	"	200	.080	S.F.	7.25	3.94		11.19	14.05
3400	Also see Section 11 52 13.10									
3700	Sound systems, incl. amplifier, mono, minimum	1 Elec	.90	8.889	Ea.	3,650	505		4,155	4,775
3800	Dolby/Super Sound, maximum		.40	20		19,800	1,125		20,925	23,500
4100	Dual system, 2 channel, front surround, minimum		.70	11.429		5,075	645		5,720	6,550
4200	Dolby/Super Sound, 4 channel, maximum		.40	20		18,400	1,125		19,525	22,000
4500	Sound heads, 35 mm					5,875			5,875	6,450
4900	Splicer, tape system, minimum					815			815	900
5000	Tape type, maximum					1,450			1,450	1,600
5300	Speakers, recessed behind screen, minimum	1 Elec	2	4		1,150	226		1,376	1,625
5400	Maximum	"	1	8		3,375	455		3,830	4,400
5700	Seating, painted steel, upholstered, minimum	2 Carp	35	.457		149	22.50		171.50	199
5800	Maximum	"	28	.571		490	28		518	585
6100	Rewind tables, minimum					2,925			2,925	3,200
6200	Maximum					5,200			5,200	5,725
7000	For automation, varying sophistication, minimum	1 Elec	1	8	System	2,600	455		3,055	3,550
7100	Maximum	2 Elec	.30	53.333	"	6,100	3,025		9,125	11,200

11 52 16.20 Movie Equipment- Digital

		Crew	Daily Output	Labor-Hours	Unit	Material	2017 Bare Costs Labor	Equipment	Total	Total Incl O&P
0010	**MOVIE EQUIPMENT- DIGITAL**									
1000	Digital 2K projection system, 98" DMD	1 Elec	2	4	Ea.	46,800	226		47,026	52,000
1100	OEM lens		2	4		5,875	226		6,101	6,825
2000	Pedestal with power distribution		2	4		2,250	226		2,476	2,825
3000	Software		2	4		1,900	226		2,126	2,425

11 53 Laboratory Equipment

11 53 03 – Laboratory Test Equipment

11 53 03.13 Test Equipment

11 53 03.13 Test Equipment	Crew	Daily Output	Labor-Hours	Unit	Material	2017 Bare Costs Labor	Equipment	Total	Total Incl O&P
0010 **TEST EQUIPMENT**									
1700 Thermometer, electric, portable				Ea.	345			345	380
1800 Titration unit, four 2000 ml reservoirs				"	5,975			5,975	6,575

11 53 13 – Laboratory Fume Hoods

11 53 13.13 Recirculating Laboratory Fume Hoods

	Crew	Daily Output	Labor-Hours	Unit	Material	Labor	Equipment	Total	Total Incl O&P
0010 **RECIRCULATING LABORATORY FUME HOODS**									
0600 Fume hood, with countertop & base, not including HVAC									
0610 Simple, minimum	2 Carp	5.40	2.963	L.F.	550	146		696	830
0620 Complex, including fixtures		2.40	6.667		1,050	330		1,380	1,675
0630 Special, maximum		1.70	9.412		995	465		1,460	1,800
0670 Service fixtures, average				Ea.	297			297	325
0680 For sink assembly with hot and cold water, add	1 Plum	1.40	5.714		740	355		1,095	1,350
0750 Glove box, fiberglass, bacteriological					16,900			16,900	18,600
0760 Controlled atmosphere					19,400			19,400	21,300
0770 Radioisotope					16,900			16,900	18,600
0780 Carcinogenic					16,900			16,900	18,600

11 53 16 – Laboratory Incubators

11 53 16.13 Incubators

	Crew	Daily Output	Labor-Hours	Unit	Material	Labor	Equipment	Total	Total Incl O&P
0010 **INCUBATORS**									
1000 Incubators, minimum				Ea.	3,200			3,200	3,525
1010 Maximum				"	12,700			12,700	13,900

11 53 19 – Laboratory Sterilizers

11 53 19.13 Sterilizers

	Crew	Daily Output	Labor-Hours	Unit	Material	Labor	Equipment	Total	Total Incl O&P
0010 **STERILIZERS**									
0700 Glassware washer, undercounter, minimum	L-1	1.80	8.889	Ea.	6,525	525		7,050	7,975
0710 Maximum	"	1	16		14,200	945		15,145	17,100
1850 Utensil washer-sanitizer	1 Plum	2	4		8,875	247		9,122	10,100

11 53 23 – Laboratory Refrigerators

11 53 23.13 Refrigerators

	Crew	Daily Output	Labor-Hours	Unit	Material	Labor	Equipment	Total	Total Incl O&P
0010 **REFRIGERATORS**									
1200 Blood bank, 28.6 C.F. emergency signal				Ea.	10,800			10,800	11,900
1210 Reach-in, 16.9 C.F.				"	8,325			8,325	9,175

11 53 33 – Emergency Safety Appliances

11 53 33.13 Emergency Equipment

	Crew	Daily Output	Labor-Hours	Unit	Material	Labor	Equipment	Total	Total Incl O&P
0010 **EMERGENCY EQUIPMENT**									
1400 Safety equipment, eye wash, hand held				Ea.	405			405	445
1450 Deluge shower				"	810			810	895

11 53 43 – Service Fittings and Accessories

11 53 43.13 Fittings

	Crew	Daily Output	Labor-Hours	Unit	Material	Labor	Equipment	Total	Total Incl O&P
0010 **FITTINGS**									
1600 Sink, one piece plastic, flask wash, hose, free standing	1 Plum	1.60	5	Ea.	1,975	310		2,285	2,650
1610 Epoxy resin sink, 25" x 16" x 10"	"	2	4	"	223	247		470	620
1950 Utility table, acid resistant top with drawers	2 Carp	30	.533	L.F.	167	26.50		193.50	223
8000 Alternate pricing method: as percent of lab furniture									
8050 Installation, not incl. plumbing & duct work				% Furn.				22%	22%
8100 Plumbing, final connections, simple system								10%	10%
8110 Moderately complex system								15%	15%
8120 Complex system								20%	20%
8150 Electrical, simple system								10%	10%

11 53 Laboratory Equipment

11 53 43 – Service Fittings and Accessories

11 53 43.13 Fittings		Crew	Daily Output	Labor-Hours	Unit	Material	2017 Bare Costs Labor	2017 Bare Costs Equipment	Total	Total Incl O&P
8160	Moderately complex system				% Furn.				20%	20%
8170	Complex system								35%	35%

11 53 53 – Biological Safety Cabinets

11 53 53.10 Pharmacy Cabinets

		Crew	Daily Output	Labor-Hours	Unit	Material	Labor	Equipment	Total	Total Incl O&P
0010	**PHARMACY CABINETS**, vertical flow									
0100	Class II, type B2, 6' L	2 Carp	1.50	10.667	Ea.	14,500	525		15,025	16,700

11 57 Vocational Shop Equipment

11 57 10 – Shop Equipment

11 57 10.10 Vocational School Shop Equipment

		Crew	Daily Output	Labor-Hours	Unit	Material	Labor	Equipment	Total	Total Incl O&P
0010	**VOCATIONAL SCHOOL SHOP EQUIPMENT**									
0020	Benches, work, wood, average	2 Carp	5	3.200	Ea.	685	158		843	995
0100	Metal, average		5	3.200		580	158		738	875
0400	Combination belt & disc sander, 6"		4	4		1,725	197		1,922	2,175
0700	Drill press, floor mounted, 12", 1/2 H.P.		4	4		450	197		647	795
0800	Dust collector, not incl. ductwork, 6" diameter	1 Shee	1.10	7.273		5,100	420		5,520	6,250
0810	Dust collector bag, 20" diameter	"	5	1.600		445	93		538	630
1000	Grinders, double wheel, 1/2 H.P.	2 Carp	5	3.200		214	158		372	475
1300	Jointer, 4", 3/4 H.P.		4	4		1,350	197		1,547	1,775
1600	Kilns, 16 C.F., to 2000°		4	4		1,600	197		1,797	2,050
1900	Lathe, woodworking, 10", 1/2 H.P.		4	4		580	197		777	940
2200	Planer, 13" x 6"		4	4		1,175	197		1,372	1,600
2500	Potter's wheel, motorized		4	4		1,200	197		1,397	1,625
2800	Saws, band, 14", 3/4 H.P.		4	4		955	197		1,152	1,350
3100	Metal cutting band saw, 14"		4	4		2,375	197		2,572	2,925
3400	Radial arm saw, 10", 2 H.P.		4	4		1,275	197		1,472	1,700
3700	Scroll saw, 24"		4	4		685	197		882	1,050
4000	Table saw, 10", 3 H.P.		4	4		2,775	197		2,972	3,350
4300	Welder AC arc, 30 amp capacity		4	4		3,175	197		3,372	3,800

11 61 Broadcast, Theater, and Stage Equipment

11 61 23 – Folding and Portable Stages

11 61 23.10 Portable Stages

		Crew	Daily Output	Labor-Hours	Unit	Material	Labor	Equipment	Total	Total Incl O&P
0010	**PORTABLE STAGES**									
1500	Flooring, portable oak parquet, 3' x 3' sections				S.F.	13.80			13.80	15.20
1600	Cart to carry 225 S.F. of flooring				Ea.	425			425	470
5000	Stages, portable with steps, folding legs, stock, 8" high				SF Stg.	35.50			35.50	39
5100	16" high					53			53	58.50
5200	32" high					54.50			54.50	60
5300	40" high					61			61	67.50
6000	Telescoping platforms, extruded alum., straight, minimum	4 Carp	157	.204		37.50	10.05		47.55	56.50
6100	Maximum		77	.416		53	20.50		73.50	89.50
6500	Pie-shaped, minimum		150	.213		81	10.50		91.50	105
6600	Maximum		70	.457		90.50	22.50		113	134
6800	For 3/4" plywood covered deck, deduct					5.20			5.20	5.70
7000	Band risers, steel frame, plywood deck, minimum	4 Carp	275	.116		31.50	5.75		37.25	43.50
7100	Maximum	"	138	.232		77	11.40		88.40	103
7500	Chairs for above, self-storing, minimum	2 Carp	43	.372	Ea.	116	18.35		134.35	156
7600	Maximum	"	40	.400	"	206	19.70		225.70	257

11 61 Broadcast, Theater, and Stage Equipment

11 61 33 – Rigging Systems and Controls

11 61 33.10 Controls

11 61 33.10 Controls	Crew	Daily Output	Labor-Hours	Unit	Material	2017 Bare Costs Labor	Equipment	Total	Total Incl O&P
0010 **CONTROLS**									
0050 Control boards with dimmers and breakers, minimum	1 Elec	1	8	Ea.	13,300	455		13,755	15,300
0100 Average		.50	16		42,500	905		43,405	48,100
0150 Maximum		.20	40		135,500	2,275		137,775	152,500
8000 Rule of thumb: total stage equipment, minimum	4 Carp	100	.320	SF Stg.	100	15.75		115.75	134
8100 Maximum	"	25	1.280	"	565	63		628	715

11 61 43 – Stage Curtains

11 61 43.10 Curtains

	Crew	Daily Output	Labor-Hours	Unit	Material	2017 Bare Costs Labor	Equipment	Total	Total Incl O&P
0010 **CURTAINS**									
0500 Curtain track, straight, light duty	2 Carp	20	.800	L.F.	29.50	39.50		69	93
0600 Heavy duty		18	.889		64.50	44		108.50	138
0700 Curved sections		12	1.333		190	65.50		255.50	310
1000 Curtains, velour, medium weight		600	.027	S.F.	8.40	1.31		9.71	11.25
1150 Silica based yarn, inherently fire retardant		50	.320	"	15.70	15.75		31.45	41.50

11 62 Musical Equipment

11 62 16 – Carillons

11 62 16.10 Bell Tower Equipment

	Crew	Daily Output	Labor-Hours	Unit	Material	2017 Bare Costs Labor	Equipment	Total	Total Incl O&P
0010 **BELL TOWER EQUIPMENT**									
0300 Carillon, 4 octave (48 bells), with keyboard				System	996,000			996,000	1,095,500
0320 2 octave (24 bells)					468,500			468,500	515,500
0340 3 to 4 bell peal, minimum					117,000			117,000	129,000
0360 Maximum					703,000			703,000	773,000
0380 Cast bronze bell, average				Ea.	105,500			105,500	116,000
0400 Electronic, digital, minimum					17,600			17,600	19,300
0410 With keyboard, maximum					88,000			88,000	96,500

11 66 Athletic Equipment

11 66 13 – Exercise Equipment

11 66 13.10 Physical Training Equipment

	Crew	Daily Output	Labor-Hours	Unit	Material	2017 Bare Costs Labor	Equipment	Total	Total Incl O&P
0010 **PHYSICAL TRAINING EQUIPMENT**									
0020 Abdominal rack, 2 board capacity				Ea.	510			510	565
0050 Abdominal board, upholstered					720			720	790
0200 Bicycle trainer, minimum					500			500	550
0300 Deluxe, electric					4,700			4,700	5,175
0400 Barbell set, chrome plated steel, 25 lb.					289			289	320
0420 100 lb.					555			555	610
0450 200 lb.					760			760	835
0500 Weight plates, cast iron, per lb.				Lb.	3.80			3.80	4.18
0520 Storage rack, 10 station				Ea.	1,000			1,000	1,100
0600 Circuit training apparatus, 12 machines minimum	2 Clab	1.25	12.800	Set	31,800	500		32,300	35,800
0700 Average		1	16		39,700	625		40,325	44,700
0800 Maximum		.75	21.333		47,600	835		48,435	54,000
0820 Dumbbell set, cast iron, with rack and 5 pair					440			440	485
0900 Squat racks	2 Clab	5	3.200	Ea.	900	125		1,025	1,175
1200 Multi-station gym machine, 5 station					5,000			5,000	5,500
1250 9 station					11,100			11,100	12,200
1280 Rowing machine, hydraulic					1,750			1,750	1,925

11 66 Athletic Equipment

11 66 13 – Exercise Equipment

11 66 13.10 Physical Training Equipment

		Crew	Daily Output	Labor-Hours	Unit	Material	2017 Bare Costs Labor	Equipment	Total	Total Incl O&P
1300	Treadmill, manual				Ea.	1,050			1,050	1,150
1320	Motorized					3,525			3,525	3,875
1340	Electronic					4,000			4,000	4,400
1360	Cardio-testing					5,000			5,000	5,500
1400	Treatment/massage tables, minimum					570			570	625
1420	Deluxe, with accessories					780			780	860
4150	Exercise equipment, bicycle trainer					900			900	990
4180	Chinning bar, adjustable, wall mounted	1 Carp	5	1.600		217	79		296	360
4200	Exercise ladder, 16' x 1'-7", suspended	L-2	3	5.333		1,400	230		1,630	1,875
4210	High bar, floor plate attached	1 Carp	4	2		2,125	98.50		2,223.50	2,475
4240	Parallel bars, adjustable		4	2		1,750	98.50		1,848.50	2,075
4270	Uneven parallel bars, adjustable	↓	4	2	↓	3,575	98.50		3,673.50	4,075
4280	Wall mounted, adjustable	L-2	1.50	10.667	Set	955	460		1,415	1,750
4300	Rope, ceiling mounted, 18' long	1 Carp	3.66	2.186	Ea.	196	108		304	380
4330	Side horse, vaulting		5	1.600		1,425	79		1,504	1,700
4360	Treadmill, motorized, deluxe, training type	↓	5	1.600		3,925	79		4,004	4,425
4390	Weight lifting multi-station, minimum	2 Clab	1	16		315	625		940	1,300
4450	Maximum	"	.50	32	↓	14,800	1,250		16,050	18,200

11 66 23 – Gymnasium Equipment

11 66 23.13 Basketball Equipment

		Crew	Daily Output	Labor-Hours	Unit	Material	2017 Bare Costs Labor	Equipment	Total	Total Incl O&P
0010	**BASKETBALL EQUIPMENT**									
1000	Backstops, wall mtd., 6' extended, fixed, minimum	L-2	1	16	Ea.	1,500	690		2,190	2,725
1100	Maximum		1	16		1,800	690		2,490	3,050
1200	Swing up, minimum		1	16		1,700	690		2,390	2,950
1250	Maximum		1	16		2,950	690		3,640	4,325
1300	Portable, manual, heavy duty, spring operated		1.90	8.421		12,800	365		13,165	14,700
1400	Ceiling suspended, stationary, minimum		.78	20.513		4,200	885		5,085	6,000
1450	Fold up, with accessories, maximum	↓	.40	40		6,275	1,725		8,000	9,550
1600	For electrically operated, add	1 Elec	1	8	↓	2,400	455		2,855	3,300
5800	Wall pads, 1-1/2" thick, standard (not fire rated)	2 Carp	640	.025	S.F.	5.35	1.23		6.58	7.80

11 66 23.19 Boxing Ring

		Crew	Daily Output	Labor-Hours	Unit	Material	2017 Bare Costs Labor	Equipment	Total	Total Incl O&P
0010	**BOXING RING**									
4100	Elevated, 22' x 22'	L-4	.10	240	Ea.	5,975	11,200		17,175	23,900
4110	For cellular plastic foam padding, add		.10	240		1,225	11,200		12,425	18,700
4120	Floor level, including posts and ropes only, 20' x 20'		.80	30		4,475	1,400		5,875	7,050
4130	Canvas, 30' x 30'	↓	5	4.800	↓	1,450	224		1,674	1,950

11 66 23.47 Gym Mats

		Crew	Daily Output	Labor-Hours	Unit	Material	2017 Bare Costs Labor	Equipment	Total	Total Incl O&P
0010	**GYM MATS**									
5500	2" thick, naugahyde covered				S.F.	4.99			4.99	5.50
5600	Vinyl/nylon covered					8.30			8.30	9.10
6000	Wrestling mats, 1" thick, heavy duty					5.05			5.05	5.60

11 66 43 – Interior Scoreboards

11 66 43.10 Scoreboards

		Crew	Daily Output	Labor-Hours	Unit	Material	2017 Bare Costs Labor	Equipment	Total	Total Incl O&P
0010	**SCOREBOARDS**									
7000	Baseball, minimum	R-3	1.30	15.385	Ea.	3,600	870	106	4,576	5,375
7200	Maximum		.05	400		20,300	22,600	2,750	45,650	59,500
7300	Football, minimum		.86	23.256		5,175	1,325	160	6,660	7,850
7400	Maximum		.20	100		18,300	5,650	690	24,640	29,400
7500	Basketball (one side), minimum		2.07	9.662		2,500	545	66.50	3,111.50	3,650
7600	Maximum		.30	66.667		7,475	3,775	460	11,710	14,400
7700	Hockey-basketball (four sides), minimum	↓	.25	80		8,925	4,525	550	14,000	17,200

11 66 Athletic Equipment

11 66 43 – Interior Scoreboards

11 66 43.10 Scoreboards		Crew	Daily Output	Labor-Hours	Unit	Material	2017 Bare Costs Labor	Equipment	Total	Total Incl O&P
7800	Maximum	R-3	.15	133	Ea.	15,800	7,550	920	24,270	29,600

11 66 53 – Gymnasium Dividers

11 66 53.10 Divider Curtains

0010	**DIVIDER CURTAINS**									
4500	Gym divider curtain, mesh top, vinyl bottom, manual	L-4	500	.048	S.F.	10	2.24		12.24	14.45
4700	Electric roll up	L-7	400	.070	"	13.25	3.32		16.57	19.65

11 67 Recreational Equipment

11 67 13 – Bowling Alley Equipment

11 67 13.10 Bowling Alleys

0010	**BOWLING ALLEYS** Including alley, pinsetter, scorer,									
0020	Counters and misc. supplies, minimum	4 Carp	.20	160	Lane	40,200	7,875		48,075	56,500
0150	Average		.19	168		48,300	8,300		56,600	65,500
0300	Maximum		.18	177		55,500	8,750		64,250	74,500
0400	Combo table ball rack, add					1,275			1,275	1,400
0600	For automatic scorer, add, minimum					5,825			5,825	6,400
0700	Maximum					10,000			10,000	11,000

11 67 23 – Shooting Range Equipment

11 67 23.10 Shooting Range

0010	**SHOOTING RANGE** Incl. bullet traps, target provisions, controls,									
0100	Separators, ceiling system, etc. Not incl. structural shell									
0200	Commercial	L-9	.64	56.250	Point	30,600	2,500		33,100	37,600
0300	Law enforcement		.28	128		41,600	5,725		47,325	54,500
0400	National Guard armories		.71	50.704		22,400	2,250		24,650	28,200
0500	Reserve training centers		.71	50.704		17,200	2,250		19,450	22,500
0600	Schools and colleges		.32	112		39,400	5,025		44,425	51,000
0700	Major academies		.19	189		57,000	8,450		65,450	75,500
0800	For acoustical treatment, add					10%	10%			
0900	For lighting, add					28%	25%			
1000	For plumbing, add					5%	5%			
1100	For ventilating system, add, minimum					40%	40%			
1200	Add, average					25%	25%			
1300	Add, maximum					35%	35%			

11 68 Play Field Equipment and Structures

11 68 13 – Playground Equipment

11 68 13.10 Free-Standing Playground Equipment

0010	**FREE-STANDING PLAYGROUND EQUIPMENT** See also individual items									
0200	Bike rack, 10' long, permanent	G	B-1	12	2	Ea.	475	79.50		554.50	640
0392	Upper body warm-up station			2.60	9.231		2,750	370		3,120	3,600
0394	Bench stepper station			2.60	9.231		1,525	370		1,895	2,250
0396	Standing push up station			2.60	9.231		1,025	370		1,395	1,700
0398	Upper body stretch station			2.60	9.231		2,175	370		2,545	2,950
0400	Horizontal monkey ladder, 14' long, 6' high			4	6		1,300	239		1,539	1,800
0590	Parallel bars, 10' long			4	6		505	239		744	920
0600	Posts, tether ball set, 2-3/8" O.D.			12	2		350	79.50		429.50	505
0800	Poles, multiple purpose, 10'-6" long			12	2	Pr.	203	79.50		282.50	345
1000	Ground socket for movable posts, 2-3/8" post			10	2.400		106	95.50		201.50	262

11 68 Play Field Equipment and Structures

11 68 13 – Playground Equipment

11 68 13.10 Free-Standing Playground Equipment

		Crew	Daily Output	Labor-Hours	Unit	Material	2017 Bare Costs Labor	Equipment	Total	Total Incl O&P
1100	3-1/2" post	B-1	10	2.400	Pr.	191	95.50		286.50	355
1300	See-saw, spring, steel, 2 units		6	4	Ea.	710	159		869	1,025
1400	4 units		4	6		1,275	239		1,514	1,775
1500	6 units		3	8		1,750	320		2,070	2,425
1700	Shelter, fiberglass golf tee, 3 person		4.60	5.217		4,200	208		4,408	4,950
1900	Slides, stainless steel bed, 12' long, 6' high		3	8		4,075	320		4,395	4,975
2000	20' long, 10' high		2	12		6,375	480		6,855	7,725
2200	Swings, plain seats, 8' high, 4 seats		2	12		1,375	480		1,855	2,250
2300	8 seats		1.30	18.462		2,525	735		3,260	3,900
2500	12' high, 4 seats		2	12		2,125	480		2,605	3,075
2600	8 seats		1.30	18.462		3,525	735		4,260	5,000
2800	Whirlers, 8' diameter		3	8		3,050	320		3,370	3,850
2900	10' diameter		3	8		6,450	320		6,770	7,600

11 68 13.20 Modular Playground

		Crew	Daily Output	Labor-Hours	Unit	Material	2017 Bare Costs Labor	Equipment	Total	Total Incl O&P
0010	**MODULAR PLAYGROUND** Basic components									
0100	Deck, square, steel, 48" x 48"	B-1	1	24	Ea.	635	955		1,590	2,175
0110	Recycled polyurethane		1	24		600	955		1,555	2,125
0120	Triangular, steel, 48" side		1	24		720	955		1,675	2,275
0130	Post, steel, 5" square		18	1.333	L.F.	45.50	53		98.50	132
0140	Aluminum, 2-3/8" square		20	1.200		49.50	48		97.50	128
0150	5" square		18	1.333		43.50	53		96.50	130
0160	Roof, square poly, 54" side		18	1.333	Ea.	1,575	53		1,628	1,825
0170	Wheelchair transfer module, for 3' high deck		3	8	"	3,100	320		3,420	3,900
0180	Guardrail, pipe, 36" high		60	.400	L.F.	237	15.95		252.95	286
0190	Steps, deck-to-deck, three 8" steps		8	3	Ea.	1,175	119		1,294	1,475
0200	Activity panel, crawl through panel		2	12		520	480		1,000	1,300
0210	Alphabet/spelling panel		2	12		595	480		1,075	1,375
0360	With guardrails		3	8		1,800	320		2,120	2,500
0370	Crawl tunnel, straight, 56" long		4	6		1,175	239		1,414	1,675
0380	90°, 4' long		4	6		1,500	239		1,739	2,025
1200	Slide, tunnel, for 56" high deck		8	3		1,975	119		2,094	2,350
1210	Straight, poly		8	3		415	119		534	645
1220	Stainless steel, 54" high deck		6	4		680	159		839	995
1230	Curved, poly, 40" high deck		6	4		780	159		939	1,100
1240	Spiral slide, 56" - 72" high		5	4.800		4,900	191		5,091	5,700
1300	Ladder, vertical, for 24" - 72" high deck		5	4.800		515	191		706	860
1310	Horizontal, 8' long		5	4.800		855	191		1,046	1,225
1320	Corkscrew climber, 6' high		3	8		1,075	320		1,395	1,675
1330	Fire pole for 72" high deck		6	4		265	159		424	535
1340	Bridge, ring climber, 8' long		4	6		2,250	239		2,489	2,850
1350	Suspension		4	6	L.F.	420	239		659	825

11 68 16 – Play Structures

11 68 16.10 Handball/Squash Court

		Crew	Daily Output	Labor-Hours	Unit	Material	2017 Bare Costs Labor	Equipment	Total	Total Incl O&P
0010	**HANDBALL/SQUASH COURT**, outdoor									
0900	Handball or squash court, outdoor, wood	2 Carp	.50	32	Ea.	4,375	1,575		5,950	7,225
1000	Masonry handball/squash court	D-1	.30	53.333	"	25,600	2,300		27,900	31,800

11 68 16.30 Platform/Paddle Tennis Court

		Crew	Daily Output	Labor-Hours	Unit	Material	2017 Bare Costs Labor	Equipment	Total	Total Incl O&P
0010	**PLATFORM/PADDLE TENNIS COURT** Complete with lighting, etc.									
0100	Aluminum slat deck with aluminum frame	B-1	.08	300	Court	64,000	11,900		75,900	88,500
0500	Aluminum slat deck with wood frame	C-1	.12	266		73,000	12,500		85,500	99,000
0800	Aluminum deck heater, add	B-1	1.18	20.339		2,700	810		3,510	4,225
0900	Douglas fir planking with wood frame 2" x 6" x 30'	C-1	.12	266		68,500	12,500		81,000	94,500

11 68 Play Field Equipment and Structures

11 68 16 – Play Structures

11 68 16.30 Platform/Paddle Tennis Court	Crew	Daily Output	Labor-Hours	Unit	Material	2017 Bare Costs Labor	Equipment	Total	Total Incl O&P
1000	Plywood deck with steel frame	C-1	.12	266	Court	68,500	12,500		81,000	94,500
1100	Steel slat deck with wood frame	↓	.12	266	↓	43,300	12,500		55,800	66,500

11 68 33 – Athletic Field Equipment

11 68 33.13 Football Field Equipment

		Crew	Daily Output	Labor-Hours	Unit	Material	Labor	Equipment	Total	Total Incl O&P
0010	**FOOTBALL FIELD EQUIPMENT**									
0020	Goal posts, steel, football, double post	B-1	1.50	16	Pr.	4,025	635		4,660	5,425
0100	Deluxe, single post		1.50	16		3,025	635		3,660	4,300
0300	Football, convertible to soccer		1.50	16		3,625	635		4,260	4,975
0500	Soccer, regulation	↓	2	12	↓	1,775	480		2,255	2,675

11 71 Medical Sterilizing Equipment

11 71 10 – Medical Sterilizers & Distillers

11 71 10.10 Sterilizers and Distillers

		Crew	Daily Output	Labor-Hours	Unit	Material	Labor	Equipment	Total	Total Incl O&P
0010	**STERILIZERS AND DISTILLERS**									
0700	Distiller, water, steam heated, 50 gal. capacity	1 Plum	1.40	5.714	Ea.	20,200	355		20,555	22,700
3010	Portable, top loading, 105 – 135 degree C, 3 to 30 psi, 50 L chamber					9,600			9,600	10,600
3020	Stainless steel basket, 10.7" diam. x 11.8" H					213			213	235
3025	Stainless steel pail, 10.7" diam. x 10.7" H					299			299	330
3050	85 L chamber					16,600			16,600	18,200
3060	Stainless steel basket, 15.3" diam. x 11.5" H					380			380	420
3065	Stainless steel pail, 15.3" diam. x 11" H					610			610	670
5600	Sterilizers, floor loading, 26" x 62" x 42", single door, steam					129,500			129,500	142,500
5650	Double door, steam					218,500			218,500	240,500
5800	General purpose, 20" x 20" x 38", single door					12,500			12,500	13,800
6000	Portable, counter top, steam, minimum					2,800			2,800	3,075
6020	Maximum					4,850			4,850	5,350
6050	Portable, counter top, gas, 17" x 15" x 32-1/2"					41,800			41,800	45,900
6150	Manual washer/sterilizer, 16" x 16" x 26"	1 Plum	2	4	↓	57,000	247		57,247	63,500
6200	Steam generators, electric 10 kW to 180 kW, freestanding									
6250	Minimum	1 Elec	3	2.667	Ea.	10,500	151		10,651	11,700
6300	Maximum	"	.70	11.429		23,100	645		23,745	26,400
8200	Bed pan washer-sanitizer	1 Plum	2	4	↓	7,875	247		8,122	9,050

11 72 Examination and Treatment Equipment

11 72 13 – Examination Equipment

11 72 13.13 Examination Equipment

		Crew	Daily Output	Labor-Hours	Unit	Material	Labor	Equipment	Total	Total Incl O&P
0010	**EXAMINATION EQUIPMENT**									
0300	Blood pressure unit, mercurial, wall				Ea.	161			161	177
0400	Diagnostic set, wall					520			520	575
4400	Scale, physician's, with height rod				↓	355			355	390

11 72 53 – Treatment Equipment

11 72 53.13 Medical Treatment Equipment

		Crew	Daily Output	Labor-Hours	Unit	Material	Labor	Equipment	Total	Total Incl O&P
0010	**MEDICAL TREATMENT EQUIPMENT**									
6300	Exam light, portable, 14" flexible arm				Ea.	192			192	212
6500	Surgery table, minor minimum	1 Sswk	.70	11.429		12,400	620		13,020	14,600
6520	Maximum	"	.50	16		18,400	870		19,270	21,700
6700	Surgical lights, doctor's office, single arm	2 Elec	2	8	↓	2,700	455		3,155	3,625
6750	Dual arm	"	1	16	↓	5,000	905		5,905	6,825

11 73 Patient Care Equipment

11 73 10 – Patient Treatment Equipment

11 73 10.10 Treatment Equipment

		Crew	Daily Output	Labor-Hours	Unit	Material	2017 Bare Costs Labor	Equipment	Total	Total Incl O&P
0010	**TREATMENT EQUIPMENT**									
0750	Exam room furnishings, average per room				Ea.	4,800			4,800	5,275
1800	Heat therapy unit, humidified, 26" x 78" x 28"				"	3,275			3,275	3,600
2100	Hubbard tank with accessories, stainless steel,									
2110	125 GPM at 45 psi water pressure				Ea.	12,400			12,400	13,700
2150	For electric overhead hoist, add					2,700			2,700	2,975
2900	K-Module for heat therapy, 20 oz. capacity, 75°F to 110°F					460			460	505
3600	Paraffin bath, 126°F, auto controlled					212			212	233
3900	Parallel bars for walking training, 12'-0"					1,550			1,550	1,725
4600	Station, dietary, medium, with ice					17,600			17,600	19,300
4700	Medicine					7,925			7,925	8,725
7000	Tables, physical therapy, walk off, electric	2 Carp	3	5.333		2,475	263		2,738	3,125
7150	Standard, vinyl top with base cabinets, minimum		3	5.333		1,075	263		1,338	1,600
7200	Maximum	↓	2	8		5,400	395		5,795	6,525
7250	Table, hospital, adjustable height					1,350			1,350	1,475
8400	Whirlpool bath, mobile, sst, 18" x 24" x 60"					6,025			6,025	6,625
8450	Fixed, incl. mixing valves	1 Plum	2	4	▼	4,775	247		5,022	5,650

11 73 10.20 Bariatric Equipment

		Crew	Daily Output	Labor-Hours	Unit	Material	2017 Bare Costs Labor	Equipment	Total	Total Incl O&P
0010	**BARIATRIC EQUIPMENT**									
5000	Patient lift, electric operated, arm style									
5110	400 lb. capacity				Ea.	1,275			1,275	1,400
5120	450 lb. capacity					2,300			2,300	2,550
5130	600 lb. capacity					3,475			3,475	3,825
5140	700 lb. capacity					4,775			4,775	5,250
5150	1,000 lb. capacity					7,100			7,100	7,800
5200	Overhead, 4-post, 1,000 lb. capacity					10,300			10,300	11,300
5300	Overhead, track type, 450 lb. capacity, not including track					3,375			3,375	3,725
5500	For fabric sling, add					305			305	335
5550	For digital scale, add				▼	815			815	900

11 74 Dental Equipment

11 74 10 – Dental Office Equipment

11 74 10.10 Diagnostic and Treatment Equipment

		Crew	Daily Output	Labor-Hours	Unit	Material	2017 Bare Costs Labor	Equipment	Total	Total Incl O&P
0010	**DIAGNOSTIC AND TREATMENT EQUIPMENT**									
0020	Central suction system, minimum	1 Plum	1.20	6.667	Ea.	1,725	410		2,135	2,525
0100	Maximum	"	.90	8.889		4,950	550		5,500	6,275
0300	Air compressor, minimum	1 Skwk	.80	10		3,325	515		3,840	4,450
0400	Maximum		.50	16		8,500	825		9,325	10,600
0600	Chair, electric or hydraulic, minimum		.50	16		2,450	825		3,275	3,975
0700	Maximum	↓	.25	32		8,075	1,650		9,725	11,400
0800	Doctor's/assistant's stool, minimum					258			258	283
0850	Maximum					750			750	825
1000	Drill console with accessories, minimum	1 Skwk	1.60	5		2,200	257		2,457	2,825
1100	Maximum		1.60	5		5,200	257		5,457	6,125
2000	Light, ceiling mounted, minimum		8	1		1,175	51.50		1,226.50	1,350
2100	Maximum	▼	8	1		2,100	51.50		2,151.50	2,375
2200	Unit light, minimum	2 Skwk	5.33	3.002		715	154		869	1,025
2210	Maximum		5.33	3.002		1,550	154		1,704	1,950
2220	Track light, minimum		3.20	5		1,650	257		1,907	2,200
2230	Maximum	▼	3.20	5		2,850	257		3,107	3,550
2300	Sterilizers, steam portable, minimum				▼	1,300			1,300	1,425

11 74 Dental Equipment

11 74 10 – Dental Office Equipment

11 74 10.10 Diagnostic and Treatment Equipment	Crew	Daily Output	Labor-Hours	Unit	Material	2017 Bare Costs Labor	Equipment	Total	Total Incl O&P
2350	Maximum				Ea.	9,575			9,575	10,500
2600	Steam, institutional					2,900			2,900	3,175
2650	Dry heat, electric, portable, 3 trays					1,325			1,325	1,450
2700	Ultra-sonic cleaner, portable, minimum					460			460	505
2750	Maximum (institutional)					1,425			1,425	1,575
3000	X-ray unit, wall, minimum	1 Skwk	4	2		2,450	103		2,553	2,850
3010	Maximum		4	2		4,700	103		4,803	5,325
3100	Panoramic unit		.60	13.333		17,100	685		17,785	19,900
3105	Deluxe, minimum	2 Skwk	1.60	10		16,600	515		17,115	19,100
3110	Maximum	"	1.60	10		44,700	515		45,215	49,900
3500	Developers, X-ray, average	1 Plum	5.33	1.501		5,525	93		5,618	6,225
3600	Maximum	"	5.33	1.501		9,025	93		9,118	10,100

11 76 Operating Room Equipment

11 76 10 – Operating Room Equipment

11 76 10.10 Surgical Equipment

0010	SURGICAL EQUIPMENT									
5000	Scrub, surgical, stainless steel, single station, minimum	1 Plum	3	2.667	Ea.	4,450	165		4,615	5,150
5100	Maximum					6,575			6,575	7,250
6550	Major surgery table, minimum	1 Sswk	.50	16		25,300	870		26,170	29,300
6570	Maximum		.50	16		29,200	870		30,070	33,600
6600	Hydraulic, hand-held control, general surgery		.60	13.333		31,500	725		32,225	35,800
6650	Stationary, universal		.50	16		41,000	870		41,870	46,600
6800	Surgical lights, major operating room, dual head, minimum	2 Elec	1	16		4,200	905		5,105	5,975
6850	Maximum		1	16		31,100	905		32,005	35,600
6900	Ceiling mount articulation, single arm		1	16		3,875	905		4,780	5,600

11 77 Radiology Equipment

11 77 10 – Radiology Equipment

11 77 10.10 X-Ray Equipment

0010	X-RAY EQUIPMENT									
8700	X-ray, mobile, minimum				Ea.	17,500			17,500	19,300
8750	Maximum					79,500			79,500	87,000
8900	Stationary, minimum					30,800			30,800	33,900
8950	Maximum					229,000			229,000	252,000
9150	Developing processors, minimum					4,575			4,575	5,025
9200	Maximum					12,300			12,300	13,500

For customer support on your Building Construction Costs with RSMeans Data, call 800.448.8182.

427

11 78 Mortuary Equipment

11 78 13 – Mortuary Refrigerators

11 78 13.10 Mortuary and Autopsy Equipment

11 78 13.10 Mortuary and Autopsy Equipment	Crew	Daily Output	Labor-Hours	Unit	Material	2017 Bare Costs Labor	Equipment	Total	Total Incl O&P
0010 **MORTUARY AND AUTOPSY EQUIPMENT**									
0015 Autopsy table, standard	1 Plum	1	8	Ea.	10,200	495		10,695	12,000
0020 Deluxe	"	.60	13.333		15,700	825		16,525	18,600
3200 Mortuary refrigerator, end operated, 2 capacity					9,850			9,850	10,800
3300 6 capacity					15,800			15,800	17,400

11 78 16 – Crematorium Equipment

11 78 16.10 Crematory

	Crew	Daily Output	Labor-Hours	Unit	Material	2017 Bare Costs Labor	Equipment	Total	Total Incl O&P
0010 **CREMATORY**									
1500 Crematory, not including building, 1 place	Q-3	.20	160	Ea.	77,500	9,425		86,925	99,500
1750 2 place	"	.10	320	"	111,000	18,800		129,800	150,500

11 81 Facility Maintenance Equipment

11 81 19 – Vacuum Cleaning Systems

11 81 19.10 Vacuum Cleaning

	Crew	Daily Output	Labor-Hours	Unit	Material	2017 Bare Costs Labor	Equipment	Total	Total Incl O&P
0010 **VACUUM CLEANING**									
0020 Central, 3 inlet, residential	1 Skwk	.90	8.889	Total	1,175	455		1,630	1,975
0200 Commercial		.70	11.429		1,375	590		1,965	2,400
0400 5 inlet system, residential		.50	16		1,675	825		2,500	3,125
0600 7 inlet system, commercial		.40	20		1,900	1,025		2,925	3,675
0800 9 inlet system, residential		.30	26.667		4,050	1,375		5,425	6,575
4010 Rule of thumb: First 1200 S.F., installed								1,425	1,575
4020 For each additional S.F., add				S.F.				.26	.26

11 82 Facility Solid Waste Handling Equipment

11 82 19 – Packaged Incinerators

11 82 19.10 Packaged Gas Fired Incinerators

	Crew	Daily Output	Labor-Hours	Unit	Material	2017 Bare Costs Labor	Equipment	Total	Total Incl O&P
0010 **PACKAGED GAS FIRED INCINERATORS**									
4400 Incinerator, gas, not incl. chimney, elec. or pipe, 50#/hr., minimum	Q-3	.80	40	Ea.	41,200	2,350		43,550	48,900
4420 Maximum		.70	45.714		43,400	2,700		46,100	52,000
4440 200 lb. per hr., minimum (batch type)		.60	53.333		72,500	3,150		75,650	84,500
4460 Maximum (with feeder)		.50	64		80,000	3,775		83,775	93,500
4480 400 lb. per hr., minimum (batch type)		.30	106		79,000	6,275		85,275	96,500
4500 Maximum (with feeder)		.25	128		94,500	7,525		102,025	115,500
4520 800 lb. per hr., with feeder, minimum		.20	160		121,000	9,425		130,425	147,000
4540 Maximum		.17	188		203,000	11,100		214,100	239,500
4560 1,200 lb. per hr., with feeder, minimum		.15	213		148,000	12,600		160,600	182,000
4580 Maximum		.11	290		185,500	17,100		202,600	230,000
4600 2,000 lb. per hr., with feeder, minimum		.10	320		400,500	18,800		419,300	469,000
4620 Maximum		.05	640		647,000	37,700		684,700	768,500
4700 For heat recovery system, add, minimum		.25	128		85,500	7,525		93,025	105,500
4710 Add, maximum		.11	290		267,000	17,100		284,100	319,500
4720 For automatic ash conveyer, add		.50	64		35,600	3,775		39,375	44,800
4750 Large municipal incinerators, incl. stack, minimum		.25	128	Ton/day	21,600	7,525		29,125	35,100
4850 Maximum		.10	320	"	57,500	18,800		76,300	91,500

11 82 Facility Solid Waste Handling Equipment

11 82 26 – Facility Waste Compactors

11 82 26.10 Compactors

	11 82 26.10 Compactors	Crew	Daily Output	Labor-Hours	Unit	Material	2017 Bare Costs Labor	Equipment	Total	Total Incl O&P
0010	**COMPACTORS**									
0020	Compactors, 115 volt, 250#/hr., chute fed	L-4	1	24	Ea.	12,600	1,125		13,725	15,500
0100	Hand fed		2.40	10		16,200	465		16,665	18,500
0300	Multi-bag, 230 volt, 600#/hr., chute fed		1	24		16,100	1,125		17,225	19,500
0400	Hand fed		1	24		16,200	1,125		17,325	19,600
0500	Containerized, hand fed, 2 to 6 C.Y. containers, 250#/hr.		1	24		16,200	1,125		17,325	19,500
0550	For chute fed, add per floor		1	24		1,500	1,125		2,625	3,375
1000	Heavy duty industrial compactor, 0.5 C.Y. capacity		1	24		10,800	1,125		11,925	13,600
1050	1.0 C.Y. capacity		1	24		16,500	1,125		17,625	19,900
1100	3.0 C.Y. capacity		.50	48		27,700	2,250		29,950	33,900
1150	5.0 C.Y. capacity		.50	48		34,900	2,250		37,150	41,900
1200	Combination shredder/compactor (5,000 lb./hr.)		.50	48		67,500	2,250		69,750	78,000
1400	For handling hazardous waste materials, 55 gallon drum packer, std.					21,100			21,100	23,200
1410	55 gallon drum packer w/HEPA filter					26,300			26,300	29,000
1420	55 gallon drum packer w/charcoal & HEPA filter					35,100			35,100	38,600
1430	All of the above made explosion proof, add					1,550			1,550	1,700
5500	Shredder, municipal use, 35 ton per hour					321,000			321,000	353,500
5600	60 ton per hour					684,500			684,500	753,000
5750	Shredder & baler, 50 ton per day					641,500			641,500	706,000
5800	Shredder, industrial, minimum					25,400			25,400	28,000
5850	Maximum					135,500			135,500	149,000
5900	Baler, industrial, minimum					10,100			10,100	11,200
5950	Maximum					591,500			591,500	651,000
6000	Transfer station compactor, with power unit									
6050	and pedestal, not including pit, 50 ton per hour				Ea.	203,000			203,000	223,000

11 82 39 – Medical Waste Disposal Systems

11 82 39.10 Off-Site Disposal

		Crew	Daily Output	Labor-Hours	Unit	Material	2017 Bare Costs Labor	Equipment	Total	Total Incl O&P
0010	**OFF-SITE DISPOSAL**									
0100	Medical waste disposal, Red Bag system, pick up & treat, 200 lb. per week				Week	192			192	211
0110	Per month				Month	735			735	810
0150	Red bags, 7-10 gal., 1.2 mil, pkg of 500				Ea.	71			71	78
0200	15 gal., package of 250					67			67	73.50
0250	33 gal., package of 250					69.50			69.50	76.50
0300	45 gal., package of 100					58.50			58.50	64.50

11 82 39.20 Disposal Carts

		Crew	Daily Output	Labor-Hours	Unit	Material	2017 Bare Costs Labor	Equipment	Total	Total Incl O&P
0010	**DISPOSAL CARTS**									
2010	Medical waste disposal cart, HDPE, w/lid, 28 gal. capacity				Ea.	280			280	310
2020	96 gal. capacity					325			325	360
2030	150 gal. capacity, low profile					505			505	555
2040	200 gal. capacity					975			975	1,075

11 82 39.30 Medical Waste Sanitizers

		Crew	Daily Output	Labor-Hours	Unit	Material	2017 Bare Costs Labor	Equipment	Total	Total Incl O&P
0010	**MEDICAL WASTE SANITIZERS**									
2010	Small, hand loaded, 1.5 C.Y., 225 lb. capacity				Ea.	83,500			83,500	92,000
2020	Medium, cart loaded, 6.25 C.Y., 938 lb. capacity					111,500			111,500	123,000
2030	Large, cart loaded, 15 C.Y., 2250 lb. capacity					133,000			133,000	146,000
3010	Cart, aluminum, 75 lb. capacity					2,200			2,200	2,425
3020	95 lb. capacity					2,400			2,400	2,625
4010	Stainless steel, 173 lb. capacity					3,125			3,125	3,450
4020	232 lb. capacity					3,500			3,500	3,850
4030	Cart lift, hydraulic scissor type					6,225			6,225	6,825
4040	Portable aluminum ramp					1,750			1,750	1,925

11 82 Facility Solid Waste Handling Equipment

11 82 39 – Medical Waste Disposal Systems

11 82 39.30 Medical Waste Sanitizers		Crew	Daily Output	Labor-Hours	Unit	Material	2017 Bare Costs Labor	Equipment	Total	Total Incl O&P
4050	Fold-down steel tracks				Ea.	1,425			1,425	1,575
4060	Pull-out drawer, small					6,700			6,700	7,375
4070	Medium					9,675			9,675	10,600
4080	Large					13,600			13,600	15,000
5000	Medical waste treatment, sanitize, on-site									
5010	Less than 15,000 lb. per month				Lb.	.20			.20	.22
5020	Over 15,000 lb. per month				"	.16			.16	.18

11 91 Religious Equipment

11 91 13 – Baptisteries

11 91 13.10 Baptistry

		Crew	Daily Output	Labor-Hours	Unit	Material	2017 Bare Costs Labor	Equipment	Total	Total Incl O&P
0010	**BAPTISTRY**									
0150	Fiberglass, 3'-6" deep, x 13'-7" long,									
0160	steps at both ends, incl. plumbing, minimum	L-8	1	20	Ea.	5,600	1,025		6,625	7,750
0200	Maximum	"	.70	28.571		9,425	1,475		10,900	12,700
0250	Add for filter, heater and lights					1,850			1,850	2,050

11 91 23 – Sanctuary Equipment

11 91 23.10 Sanctuary Furnishings

		Crew	Daily Output	Labor-Hours	Unit	Material	2017 Bare Costs Labor	Equipment	Total	Total Incl O&P
0010	**SANCTUARY FURNISHINGS**									
0020	Altar, wood, custom design, plain	1 Carp	1.40	5.714	Ea.	2,600	281		2,881	3,300
0050	Deluxe	"	.20	40		12,500	1,975		14,475	16,800
0070	Granite or marble, average	2 Marb	.50	32		13,500	1,500		15,000	17,200
0090	Deluxe	"	.20	80		36,900	3,750		40,650	46,300
0100	Arks, prefabricated, plain	2 Carp	.80	20		9,375	985		10,360	11,800
0130	Deluxe, maximum	"	.20	80		133,000	3,950		136,950	152,500
0500	Reconciliation room, wood, prefabricated, single, plain	1 Carp	.60	13.333		3,250	655		3,905	4,575
0550	Deluxe		.40	20		8,700	985		9,685	11,100
0650	Double, plain		.40	20		6,375	985		7,360	8,500
0700	Deluxe		.20	40		19,000	1,975		20,975	23,900
1000	Lecterns, wood, plain		5	1.600		745	79		824	940
1100	Deluxe		2	4		5,925	197		6,122	6,825
2000	Pulpits, hardwood, prefabricated, plain		2	4		1,575	197		1,772	2,025
2100	Deluxe		1.60	5		10,300	246		10,546	11,800
2500	Railing, hardwood, average		25	.320	L.F.	213	15.75		228.75	258
3000	Seating, individual, oak, contour, laminated		21	.381	Person	182	18.75		200.75	229
3100	Cushion seat		21	.381		165	18.75		183.75	211
3200	Fully upholstered		21	.381		180	18.75		198.75	227
3300	Combination, self-rising		21	.381		320	18.75		338.75	380
3500	For cherry, add					30%				
5000	Wall cross, aluminum, extruded, 2" x 2" section	1 Carp	34	.235	L.F.	244	11.60		255.60	287
5150	4" x 4" section		29	.276		350	13.60		363.60	405
5300	Bronze, extruded, 1" x 2" section		31	.258		480	12.70		492.70	550
5350	2-1/2" x 2-1/2" section		34	.235		730	11.60		741.60	820
5450	Solid bar stock, 1/2" x 3" section		29	.276		960	13.60		973.60	1,075
5600	Fiberglass, stock		34	.235		168	11.60		179.60	202
5700	Stainless steel, 4" deep, channel section		29	.276		775	13.60		788.60	870
5800	4" deep box section		29	.276		1,075	13.60		1,088.60	1,200

11 92 Agricultural Equipment

11 92 16 – Stock Feeders

11 92 16.16 Barns	Crew	Daily Output	Labor-Hours	Unit	Material	2017 Bare Costs Labor	Equipment	Total	Total Incl O&P
0010 **BARNS**									
0015 Swine barn, farrowing pens and equipment	B-1	1000	.024	S.F.	14	.96		14.96	16.85
0020 Gestation pens and equipment		800	.030		12.30	1.19		13.49	15.40
0030 Nursery pens and equipment		1150	.021		10.75	.83		11.58	13.10
0040 Finishing pens and equipment		1400	.017		8.20	.68		8.88	10.05
0120 Poultry barn, cages and equipment		700	.034		18	1.37		19.37	22
0220 Animal barn stall, feed and water equipment		1000	.024		16	.96		16.96	19.05
0230 Manure floor scraper system		1500	.016		3	.64		3.64	4.28
0240 Below slab manure gutter and shuttle stroker		400	.060		10.25	2.39		12.64	14.95
0250 Exhaust system		1400	.017		1.24	.68		1.92	2.41
0260 Milking barn, milking equipment		220	.109		90	4.34		94.34	106
0270 Milk storage equipment		450	.053		27	2.12		29.12	33
0280 Sheep barn, equipment, sheep shear street gates		1000	.024		1,050	.96		1,050.96	1,150
0290 Gate in frame		1000	.024		106	.96		106.96	118
0295 Maternity fence with drinking trough	↓	1000	.024	↓	62	.96		62.96	69.50
0300 Tobacco barn, fruit & tobacco auto. dryer machine /heat pump dryer	Q-20	2.90	6.897	Ea.	6,350	365		6,715	7,525
0310 Tobacco curing generator		4.90	4.082		4,450	217		4,667	5,225
0320 Automatic system controller		6.90	2.899		1,625	154		1,779	2,000
0330 Humidity & temp. transmitters for humidity measurement		6.90	2.899		900	154		1,054	1,225
0340 Handling system for leaf tobacco	↓	5.90	3.390	↓	560	180		740	890

11 97 Security Equipment

11 97 30 – Security Drawers

11 97 30.10 Pass Through Drawer

	Crew	Daily Output	Labor-Hours	Unit	Material	2017 Bare Costs Labor	Equipment	Total	Total Incl O&P
0010 **PASS THROUGH DRAWER**									
0100 Pass-thru drawer for personal items, 18" x 15" x 24"	1 Skwk	2	4	Ea.	2,925	206		3,131	3,525
0110 Including speakers	"	1.50	5.333	"	3,300	274		3,574	4,050

11 98 Detention Equipment

11 98 30 – Detention Cell Equipment

11 98 30.10 Cell Equipment

	Crew	Daily Output	Labor-Hours	Unit	Material	2017 Bare Costs Labor	Equipment	Total	Total Incl O&P
0010 **CELL EQUIPMENT**									
3000 Toilet apparatus including wash basin, average	L-8	1.50	13.333	Ea.	3,450	690		4,140	4,825

For customer support on your Building Construction Costs with RSMeans Data, call 800.448.8182.

431

Division Notes

	CREW	DAILY OUTPUT	LABOR-HOURS	UNIT	BARE COSTS				TOTAL INCL O&P
					MAT.	LABOR	EQUIP.	TOTAL	

Estimating Tips
General

- The items in this division are usually priced per square foot or each. Most of these items are purchased by the owner and installed by the contractor. Do not assume the items in Division 12 will be purchased and installed by the contractor. Check the specifications for responsibilities and include receiving, storage, installation, and mechanical and electrical hookups in the appropriate divisions.

- Some items in this division require some type of support system that is not usually furnished with the item. Examples of these systems include blocking for the attachment of casework and heavy drapery rods. The required blocking must be added to the estimate in the appropriate division.

Reference Numbers

Reference numbers are shown at the beginning of some major classifications. These numbers refer to related items in the Reference Section. The reference information may be an estimating procedure, an alternate pricing method, or technical information.

Note: Not all subdivisions listed here necessarily appear. ■

12 05 Common Work Results for Furnishings

12 05 05 – Selective Demolition for Furnishings

12 05 05.10 Selective Demolition, Interiors	Crew	Daily Output	Labor-Hours	Unit	Material	2017 Bare Costs Labor	Equipment	Total	Total Incl O&P
0010 **SELECTIVE DEMOLITION, INTERIORS**									
1100 Casework, wood base cabinets	2 Clab	24	.667	L.F.		26		26	40
1200 Countertop		96	.167			6.55		6.55	10
3100 Casework, metal base cabinets		20	.800			31.50		31.50	48
3110 Cabinet base trim		400	.040			1.57		1.57	2.40
3120 Countertop, stainless steel or acid proof		80	.200			7.85		7.85	12
3122 custom	1 Carp	48	.167	S.F.		8.20		8.20	12.60
3125 Ceramic tile	1 Clab	120	.067	"		2.61		2.61	4
3127 Trim	1 Tilf	72	.111	L.F.		5.10		5.10	7.55
3130 Wall cabinets, wood, 84" high	2 Clab	30	.533			21		21	32
3500 Laboratory casework, tall storage cabinets, 84" high		40	.400			15.65		15.65	24
3510 Wall cabinets, metal		40	.400			15.65		15.65	24
4830 Floor mats, recessed or link mats	1 Clab	300	.027	S.F.		1.04		1.04	1.60
4832 Skate lock tile		200	.040			1.57		1.57	2.40
4834 Duckboard		300	.027			1.04		1.04	1.60
4920 Blinds, interior, horizontal or vertical		150	.053	L.F.		2.09		2.09	3.20
4922 Wood folding panels		35	.229	Pr.		8.95		8.95	13.70
4924 Shades, interior		700	.011	S.F.		.45		.45	.69
4930 Drapery hardware, traverse rods		35	.229	Ea.		8.95		8.95	13.70
4950 Blast curtains, including hardware		25	.320			12.55		12.55	19.20
5200 Fixed seating, per seat	2 Carp	44	.364			17.90		17.90	27.50
6400 Booth, restaurant	2 Clab	80	.200	L.F.		7.85		7.85	12
7400 Office systems, furniture, cubicle	"	3000	.005	S.F.		.21		.21	.32

12 21 Window Blinds

12 21 13 – Horizontal Louver Blinds

12 21 13.13 Metal Horizontal Louver Blinds

	Crew	Daily Output	Labor-Hours	Unit	Material	2017 Bare Costs Labor	Equipment	Total	Total Incl O&P
0010 **METAL HORIZONTAL LOUVER BLINDS**									
0020 Horizontal, 1" aluminum slats, solid color, stock	1 Carp	590	.014	S.F.	4.61	.67		5.28	6.05
0250 2" aluminum slats, solid color, stock	"	590	.014	"	5.90	.67		6.57	7.50

12 21 13.33 Vinyl Horizontal Louver Blinds

	Crew	Daily Output	Labor-Hours	Unit	Material	2017 Bare Costs Labor	Equipment	Total	Total Incl O&P
0010 **VINYL HORIZONTAL LOUVER BLINDS**									
0100 2" composite, 48" wide, 48" high	1 Carp	30	.267	Ea.	88	13.15		101.15	117
0120 72" high		29	.276		119	13.60		132.60	152
0140 96" high		28	.286		162	14.05		176.05	200
0200 60" wide, 60" high		27	.296		134	14.60		148.60	170
0220 72" high		25	.320		154	15.75		169.75	193
0240 96" high		24	.333		230	16.40		246.40	278
0300 72" wide, 72" high		25	.320		188	15.75		203.75	231
0320 96" high		23	.348		260	17.15		277.15	310
0400 96" wide, 96" high		20	.400		340	19.70		359.70	405
1000 2" faux wood, 48" wide, 48" high		30	.267		66.50	13.15		79.65	93
1020 72" high		29	.276		86.50	13.60		100.10	116
1040 96" high		28	.286		103	14.05		117.05	135
1300 72" wide, 72" high		25	.320		132	15.75		147.75	169
1320 96" high		23	.348		198	17.15		215.15	244
1400 96" wide, 96" high		20	.400		228	19.70		247.70	280

12 21 Window Blinds

12 21 16 – Vertical Louver Blinds

12 21 16.13 Metal Vertical Louver Blinds	Crew	Daily Output	Labor-Hours	Unit	Material	2017 Bare Costs Labor	Equipment	Total	Total Incl O&P
0010 **METAL VERTICAL LOUVER BLINDS**									
1500 Vertical, 3" PVC strips, minimum	1 Carp	460	.017	S.F.	7.40	.86		8.26	9.40
1600 Maximum		400	.020		14	.99		14.99	16.90
1800 4" aluminum slats, minimum		460	.017		9.40	.86		10.26	11.65
1900 Maximum		400	.020		16	.99		16.99	19.10

12 22 Curtains and Drapes

12 22 16 – Drapery Track and Accessories

12 22 16.10 Drapery Hardware

	Crew	Daily Output	Labor-Hours	Unit	Material	2017 Bare Costs Labor	Equipment	Total	Total Incl O&P
0010 **DRAPERY HARDWARE**									
0030 Standard traverse, per foot, minimum	1 Carp	59	.136	L.F.	6.35	6.70		13.05	17.25
0100 Maximum		51	.157	"	12.35	7.75		20.10	25.50
4000 Traverse rods, adjustable, 28" to 48"		22	.364	Ea.	32	17.90		49.90	62.50
4020 48" to 84"		20	.400		40.50	19.70		60.20	74.50
4040 66" to 120"		18	.444		49.50	22		71.50	87.50
4060 84" to 156"		16	.500		56	24.50		80.50	99
4080 100" to 180"		14	.571		65.50	28		93.50	115
4090 156" to 228"		13	.615		78.50	30.50		109	133
4100 228" to 312"		13	.615		90.50	30.50		121	146
4200 Double rods, adjustable, 30" to 48"		9	.889		54.50	44		98.50	127
4220 48" to 86"		9	.889		75.50	44		119.50	150
4240 86" to 150"		8	1		83.50	49.50		133	168
4260 100" to 180"		7	1.143		88	56.50		144.50	183
4300 Curtain rod & brackets, adjustable, 30" to 48"		9	.889		31.50	44		75.50	102
4320 48" to 86"		9	.889		45	44		89	117
4340 86" to 150"		8	1		55.50	49.50		105	137
4360 100" to 180"		7	1.143		69	56.50		125.50	162
4600 Valance, pinch pleated fabric, 12" deep, up to 54" long, minimum					41.50			41.50	45.50
4610 Maximum					103			103	113
4620 Up to 77" long, minimum					73			73	80.50
4630 Maximum					166			166	183
5000 Stationary rods, first 2'					8.45			8.45	9.30
5020 Each additional foot, add				L.F.	4.06			4.06	4.47

12 22 16.20 Blast Curtains

	Crew	Daily Output	Labor-Hours	Unit	Material	2017 Bare Costs Labor	Equipment	Total	Total Incl O&P
0010 **BLAST CURTAINS** per L.F. horizontal opening width, off-white or gray fabric									
0100 Blast curtains, drapery system, complete, including hardware, minimum	1 Carp	10.25	.780	L.F.	246	38.50		284.50	330
0120 Average		10.25	.780		261	38.50		299.50	345
0140 Maximum		10.25	.780		275	38.50		313.50	365

For customer support on your Building Construction Costs with RSMeans Data, call 800.448.8182.

435

12 23 Interior Shutters

12 23 10 – Wood Interior Shutters

12 23 10.10 Wood Interior Shutters

		Crew	Daily Output	Labor-Hours	Unit	Material	2017 Bare Costs Labor	Equipment	Total	Total Incl O&P
0010	**WOOD INTERIOR SHUTTERS**, louvered									
0200	Two panel, 27" wide, 36" high	1 Carp	5	1.600	Set	161	79		240	299
0300	33" wide, 36" high		5	1.600		208	79		287	350
0500	47" wide, 36" high		5	1.600		279	79		358	425
1000	Four panel, 27" wide, 36" high		5	1.600		160	79		239	297
1100	33" wide, 36" high		5	1.600		205	79		284	345
1300	47" wide, 36" high		5	1.600		274	79		353	420
1400	Plantation shutters, 16" x 48"		5	1.600	Ea.	183	79		262	320
1450	16" x 96"		4	2		300	98.50		398.50	480
1460	36" x 96"		3	2.667		660	131		791	925

12 23 10.13 Wood Panels

		Crew	Daily Output	Labor-Hours	Unit	Material	2017 Bare Costs Labor	Equipment	Total	Total Incl O&P
0010	**WOOD PANELS**									
3000	Wood folding panels with movable louvers, 7" x 20" each	1 Carp	17	.471	Pr.	87	23		110	131
3300	8" x 28" each		17	.471		87	23		110	131
3450	9" x 36" each		17	.471		100	23		123	146
3600	10" x 40" each		17	.471		109	23		132	156
4000	Fixed louver type, stock units, 8" x 20" each		17	.471		94	23		117	139
4150	10" x 28" each		17	.471		79.50	23		102.50	123
4300	12" x 36" each		17	.471		94	23		117	139
4450	18" x 40" each		17	.471		134	23		157	184
5000	Insert panel type, stock, 7" x 20" each		17	.471		22	23		45	59.50
5150	8" x 28" each		17	.471		40	23		63	79.50
5300	9" x 36" each		17	.471		51	23		74	91.50
5450	10" x 40" each		17	.471		54.50	23		77.50	95.50
5600	Raised panel type, stock, 10" x 24" each		17	.471		110	23		133	157
5650	12" x 26" each		17	.471		110	23		133	157
5700	14" x 30" each		17	.471		122	23		145	170
5750	16" x 36" each		17	.471		134	23		157	184
6000	For custom built pine, add					22%				
6500	For custom built hardwood blinds, add					42%				

12 24 Window Shades

12 24 13 – Roller Window Shades

12 24 13.10 Shades

		Crew	Daily Output	Labor-Hours	Unit	Material	2017 Bare Costs Labor	Equipment	Total	Total Incl O&P
0010	**SHADES**									
0020	Basswood, roll-up, stain finish, 3/8" slats	1 Carp	300	.027	S.F.	17.85	1.31		19.16	21.50
0200	7/8" slats		300	.027		17.70	1.31		19.01	21.50
0300	Vertical side slide, stain finish, 3/8" slats		300	.027		20	1.31		21.31	24
0400	7/8" slats		300	.027		22	1.31		23.31	26
0500	For fire retardant finishes, add					16%				
0600	For "B" rated finishes, add					20%				
0900	Mylar, single layer, non-heat reflective	1 Carp	685	.012		3.25	.58		3.83	4.46
0910	Mylar, single layer, heat reflective		685	.012		2.52	.58		3.10	3.65
1000	Double layered, heat reflective		685	.012		5.95	.58		6.53	7.45
1100	Triple layered, heat reflective		685	.012		7.25	.58		7.83	8.85
1200	For metal roller instead of wood, add per				Shade	5.10			5.10	5.60
1300	Vinyl coated cotton, standard	1 Carp	685	.012	S.F.	4.13	.58		4.71	5.40
1400	Lightproof decorator shades		685	.012		4.68	.58		5.26	6.05
1500	Vinyl, lightweight, 4 ga.		685	.012		.65	.58		1.23	1.60
1600	Heavyweight, 6 ga.		685	.012		1.99	.58		2.57	3.07
1700	Vinyl laminated fiberglass, 6 ga., translucent		685	.012		3.62	.58		4.20	4.86

For customer support on your Building Construction Costs with RSMeans Data, call 800.448.8182.

12 24 Window Shades

12 24 13 – Roller Window Shades

12 24 13.10 Shades

	12 24 13.10 Shades	Crew	Daily Output	Labor-Hours	Unit	Material	2017 Bare Costs Labor	Equipment	Total	Total Incl O&P
1800	Lightproof	1 Carp	685	.012	S.F.	4.44	.58		5.02	5.75
2000	Polyester, room darkening, with continuous cord, GEI									
2010	36" x 72"	1 Carp	38	.211	Ea.	133	10.35		143.35	162
2020	48" x 72"		28	.286		158	14.05		172.05	196
2030	60" x 72"		23	.348		187	17.15		204.15	232
2040	72" x 72"		19	.421		220	20.50		240.50	274
6011	Solar screening, fiberglass G		85	.094	S.F.	6.70	4.64		11.34	14.50

12 32 Manufactured Wood Casework

12 32 23 – Hardwood Casework

12 32 23.10 Manufactured Wood Casework, Stock Units

	12 32 23.10 Manufactured Wood Casework, Stock Units	Crew	Daily Output	Labor-Hours	Unit	Material	2017 Bare Costs Labor	Equipment	Total	Total Incl O&P
0010	**MANUFACTURED WOOD CASEWORK, STOCK UNITS**									
0300	Built-in drawer units, pine, 18" deep, 32" high, unfinished									
0400	Minimum	2 Carp	53	.302	L.F.	133	14.85		147.85	170
0500	Maximum	"	40	.400	"	157	19.70		176.70	202
0700	Kitchen base cabinets, hardwood, not incl. counter tops,									
0710	24" deep, 35" high, prefinished									
0800	One top drawer, one door below, 12" wide	2 Carp	24.80	.645	Ea.	292	32		324	370
0840	18" wide		23.30	.687		330	34		364	415
0880	24" wide		22.30	.717		400	35.50		435.50	495
1000	Four drawers, 12" wide		24.80	.645		305	32		337	390
1040	18" wide		23.30	.687		345	34		379	425
1060	24" wide		22.30	.717		380	35.50		415.50	470
1200	Two top drawers, two doors below, 27" wide		22	.727		435	36		471	530
1260	36" wide		20.30	.788		510	39		549	620
1300	48" wide		18.90	.847		575	41.50		616.50	700
1500	Range or sink base, two doors below, 30" wide		21.40	.748		390	37		427	485
1540	36" wide		20.30	.788		440	39		479	545
1580	48" wide		18.90	.847		485	41.50		526.50	600
1800	For sink front units, deduct					174			174	191
2000	Corner base cabinets, 36" wide, standard	2 Carp	18	.889		725	44		769	865
2100	Lazy Susan with revolving door	"	16.50	.970		925	48		973	1,100
4000	Kitchen wall cabinets, hardwood, 12" deep with two doors									
4050	12" high, 30" wide	2 Carp	24.80	.645	Ea.	264	32		296	340
4100	36" wide		24	.667		315	33		348	395
4400	15" high, 30" wide		24	.667		269	33		302	345
4440	36" wide		22.70	.705		325	34.50		359.50	410
4700	24" high, 30" wide		23.30	.687		360	34		394	445
4720	36" wide		22.70	.705		395	34.50		429.50	490
5000	30" high, one door, 12" wide		22	.727		253	36		289	335
5040	18" wide		20.90	.766		292	37.50		329.50	380
5060	24" wide		20.30	.788		340	39		379	435
5300	Two doors, 27" wide		19.80	.808		375	40		415	475
5340	36" wide		18.80	.851		455	42		497	565
5380	48" wide		18.40	.870		555	43		598	675
6000	Corner wall, 30" high, 24" wide		18	.889		390	44		434	495
6050	30" wide		17.20	.930		410	46		456	525
6100	36" wide		16.50	.970		470	48		518	590
6500	Revolving Lazy Susan		15.20	1.053		530	52		582	660
7000	Broom cabinet, 84" high, 24" deep, 18" wide		10	1.600		725	79		804	920
7500	Oven cabinets, 84" high, 24" deep, 27" wide		8	2		1,100	98.50		1,198.50	1,350

12 32 Manufactured Wood Casework

12 32 23 – Hardwood Casework

12 32 23.10 Manufactured Wood Casework, Stock Units

		Crew	Daily Output	Labor-Hours	Unit	Material	2017 Bare Costs Labor	Equipment	Total	Total Incl O&P
7750	Valance board trim	2 Carp	396	.040	L.F.	17.55	1.99		19.54	22.50
7780	Toe kick trim	1 Carp	256	.031	"	3.35	1.54		4.89	6.05
7790	Base cabinet corner filler		16	.500	Ea.	47	24.50		71.50	89
7800	Cabinet filler, 3" x 24"		20	.400		18.50	19.70		38.20	50.50
7810	3" x 30"		20	.400		23	19.70		42.70	55.50
7820	3" x 42"		18	.444		32.50	22		54.50	69
7830	3" x 80"		16	.500		61.50	24.50		86	106
7850	Cabinet panel		50	.160	S.F.	10.05	7.90		17.95	23
9000	For deluxe models of all cabinets, add					40%				
9500	For custom built in place, add					25%	10%			
9558	Rule of thumb, kitchen cabinets not including									
9560	appliances & counter top, minimum	2 Carp	30	.533	L.F.	196	26.50		222.50	256
9600	Maximum	"	25	.640	"	425	31.50		456.50	520
9610	For metal cabinets, see Section 12 35 70.13									

12 32 23.30 Manufactured Wood Casework Vanities

		Crew	Daily Output	Labor-Hours	Unit	Material	2017 Bare Costs Labor	Equipment	Total	Total Incl O&P
0010	**MANUFACTURED WOOD CASEWORK VANITIES**									
8000	Vanity bases, 2 doors, 30" high, 21" deep, 24" wide	2 Carp	20	.800	Ea.	340	39.50		379.50	435
8050	30" wide		16	1		405	49.50		454.50	520
8100	36" wide		13.33	1.200		395	59		454	525
8150	48" wide		11.43	1.400		515	69		584	675
9000	For deluxe models of all vanities, add to above					40%				
9500	For custom built in place, add to above					25%	10%			

12 32 23.35 Manufactured Wood Casework Hardware

		Crew	Daily Output	Labor-Hours	Unit	Material	2017 Bare Costs Labor	Equipment	Total	Total Incl O&P
0010	**MANUFACTURED WOOD CASEWORK HARDWARE**									
1000	Catches, minimum	1 Carp	235	.034	Ea.	1.30	1.68		2.98	4
1040	Maximum	"	80	.100	"	7.75	4.93		12.68	16.10
2000	Door/drawer pulls, handles									
2200	Handles and pulls, projecting, metal, minimum	1 Carp	48	.167	Ea.	5.20	8.20		13.40	18.30
2240	Maximum		36	.222		10.65	10.95		21.60	28.50
2300	Wood, minimum		48	.167		5.30	8.20		13.50	18.45
2340	Maximum		36	.222		9.80	10.95		20.75	27.50
2400	Drawer pulls, antimicrobial copper alloy finish		50	.160		19.55	7.90		27.45	33.50
2600	Flush, metal, minimum		48	.167		5.30	8.20		13.50	18.45
2640	Maximum		36	.222		9.90	10.95		20.85	27.50
2900	Drawer knobs, antimicrobial copper alloy finish		50	.160		12.95	7.90		20.85	26.50
3000	Drawer tracks/glides, minimum		48	.167	Pr.	8.95	8.20		17.15	22.50
3040	Maximum		24	.333		26	16.40		42.40	53.50
4000	Cabinet hinges, minimum		160	.050		3.10	2.46		5.56	7.20
4040	Maximum		68	.118		11.80	5.80		17.60	22
7000	Appliance pulls, antimicrobial copper alloy finish		50	.160	L.F.	70	7.90		77.90	89

12 35 Specialty Casework

12 35 50 – Educational/Library Casework

12 35 50.13 Educational Casework

		Crew	Daily Output	Labor-Hours	Unit	Material	2017 Bare Costs Labor	Equipment	Total	Total Incl O&P
0010	**EDUCATIONAL CASEWORK**									
5000	School, 24" deep, metal, 84" high units	2 Carp	15	1.067	L.F.	465	52.50		517.50	590
5150	Counter height units		20	.800		330	39.50		369.50	425
5450	Wood, custom fabricated, 32" high counter		20	.800		240	39.50		279.50	325
5600	Add for counter top		56	.286		25.50	14.05		39.55	49.50
5800	84" high wall units	↓	15	1.067	↓	505	52.50		557.50	635
6000	Laminated plastic finish is same price as wood									

12 35 53 – Laboratory Casework

12 35 53.13 Metal Laboratory Casework

		Crew	Daily Output	Labor-Hours	Unit	Material	2017 Bare Costs Labor	Equipment	Total	Total Incl O&P
0010	**METAL LABORATORY CASEWORK**									
0020	Cabinets, base, door units, metal	2 Carp	18	.889	L.F.	240	44		284	330
0300	Drawer units		18	.889		535	44		579	650
0700	Tall storage cabinets, open, 7' high		20	.800		500	39.50		539.50	610
0900	With glazed doors		20	.800		825	39.50		864.50	970
1300	Wall cabinets, metal, 12-1/2" deep, open		20	.800		199	39.50		238.50	280
1500	With doors	↓	20	.800	↓	390	39.50		429.50	490
6300	Rule of thumb: lab furniture including installation & connection									
6320	High school				S.F.				35	39
6340	College								52	57
6360	Clinical, health care								45	49.50
6380	Industrial				↓				72.50	79.50

12 35 59 – Display Casework

12 35 59.10 Display Cases

		Crew	Daily Output	Labor-Hours	Unit	Material	2017 Bare Costs Labor	Equipment	Total	Total Incl O&P
0010	**DISPLAY CASES** Free standing, all glass									
0020	Aluminum frame, 42" high x 36" wide x 12" deep	2 Carp	8	2	Ea.	1,400	98.50		1,498.50	1,700
0100	70" high x 48" wide x 18" deep	"	6	2.667		3,950	131		4,081	4,525
0500	For wood bases, add					9%				
0600	For hardwood frames, deduct					8%				
0700	For bronze, baked enamel finish, add				↓	10%				
2000	Wall mounted, glass front, aluminum frame									
2010	Non-illuminated, one section 3' x 4' x 1'-4"	2 Carp	5	3.200	Ea.	2,325	158		2,483	2,800
2100	5' x 4' x 1'-4"		5	3.200		2,475	158		2,633	2,950
2200	6' x 4' x 1'-4"		4	4		3,225	197		3,422	3,825
2500	Two sections, 8' x 4' x 1'-4"		2	8		2,550	395		2,945	3,400
2600	10' x 4' x 1'-4"		2	8		3,500	395		3,895	4,475
3000	Three sections, 16' x 4' x 1'-4"	↓	1.50	10.667	↓	4,050	525		4,575	5,275
3500	For fluorescent lights, add				Section	415			415	455
4000	Table exhibit cases, 2' wide, 3' high, 4' long, flat top	2 Carp	5	3.200	Ea.	1,175	158		1,333	1,525
4100	3' wide, 3' high, 4' long, sloping top	"	3	5.333	"	1,225	263		1,488	1,750

12 35 70 – Healthcare Casework

12 35 70.13 Hospital Casework

		Crew	Daily Output	Labor-Hours	Unit	Material	2017 Bare Costs Labor	Equipment	Total	Total Incl O&P
0010	**HOSPITAL CASEWORK**									
0500	Base cabinets, laminated plastic	2 Carp	10	1.600	L.F.	297	79		376	445
1000	Stainless steel	"	10	1.600		545	79		624	720
1200	For all drawers, add					31			31	34
1300	Cabinet base trim, 4" high, enameled steel	2 Carp	200	.080		49.50	3.94		53.44	60.50
1400	Stainless steel		200	.080		99.50	3.94		103.44	115
1450	Countertop, laminated plastic, no backsplash		40	.400		49	19.70		68.70	83.50
1650	With backsplash	↓	40	.400	↓	60.50	19.70		80.20	97
1800	For sink cutout, add		12.20	1.311	Ea.		64.50		64.50	99
1900	Stainless steel counter top	↓	40	.400	L.F.	158	19.70		177.70	204

12 35 Specialty Casework

12 35 70 – Healthcare Casework

12 35 70.13 Hospital Casework

12 35 70.13 Hospital Casework	Crew	Daily Output	Labor-Hours	Unit	Material	2017 Bare Costs Labor	Equipment	Total	Total Incl O&P
2000 For drop-in stainless 43" x 21" sink, add				Ea.	1,100			1,100	1,200
2050 Laminate with antimicrobial finish #4	2 Carp	40	.400	L.F.	33.50	19.70		53.20	67
2500 Wall cabinets, laminated plastic		15	1.067		222	52.50		274.50	325
2600 Enameled steel		15	1.067		273	52.50		325.50	380
2700 Stainless steel		15	1.067		545	52.50		597.50	675
3000 Hospital cabinets, stainless steel with glass door(s), lockable									
3010 One door, 24" W x 18" D x 60" H	2 Clab	18	.889	Ea.	2,875	35		2,910	3,200
3020 Two doors, 36" W x 18" D x 60" H		15	1.067		3,225	42		3,267	3,625
3030 36" W x 24" D x 67" H		15	1.067		3,975	42		4,017	4,450
3040 48" W x 24" D x 66" H		12	1.333		4,600	52		4,652	5,125
3050 48" W x 24" D x 72" H		12	1.333		5,650	52		5,702	6,275
3060 60" W x 24" D x 72" H		9	1.778		6,350	69.50		6,419.50	7,100

12 35 70.16 Nurse Station Casework

12 35 70.16 Nurse Station Casework	Crew	Daily Output	Labor-Hours	Unit	Material	2017 Bare Costs Labor	Equipment	Total	Total Incl O&P
0010 **NURSE STATION CASEWORK**									
2100 Door type, laminated plastic	2 Carp	10	1.600	L.F.	345	79		424	500
2200 Enameled steel		10	1.600		330	79		409	480
2300 Stainless steel		10	1.600		655	79		734	845
2400 For drawer type, add					279			279	305

12 35 80 – Commercial Kitchen Casework

12 35 80.13 Metal Kitchen Casework

12 35 80.13 Metal Kitchen Casework	Crew	Daily Output	Labor-Hours	Unit	Material	2017 Bare Costs Labor	Equipment	Total	Total Incl O&P
0010 **METAL KITCHEN CASEWORK**									
3500 Base cabinets, metal, minimum	2 Carp	30	.533	L.F.	80	26.50		106.50	128
3600 Maximum		25	.640		203	31.50		234.50	272
3700 Wall cabinets, metal, minimum		30	.533		80	26.50		106.50	128
3800 Maximum		25	.640		184	31.50		215.50	251

12 36 Countertops

12 36 16 – Metal Countertops

12 36 16.10 Stainless Steel Countertops

	Crew	Daily Output	Labor-Hours	Unit	Material	2017 Bare Costs Labor	Equipment	Total	Total Incl O&P
0010 **STAINLESS STEEL COUNTERTOPS**									
3200 Stainless steel, custom	1 Carp	24	.333	S.F.	159	16.40		175.40	200
3210 Stainless steel	"	12	.667	L.F.	320	33		353	400

12 36 19 – Wood Countertops

12 36 19.10 Maple Countertops

	Crew	Daily Output	Labor-Hours	Unit	Material	2017 Bare Costs Labor	Equipment	Total	Total Incl O&P
0010 **MAPLE COUNTERTOPS**									
2900 Solid, laminated, 1-1/2" thick, no splash	1 Carp	28	.286	L.F.	78.50	14.05		92.55	108
3000 With square splash		28	.286	"	93.50	14.05		107.55	125
3400 Recessed cutting block with trim, 16" x 20" x 1"		8	1	Ea.	95.50	49.50		145	181

12 36 23 – Plastic Countertops

12 36 23.13 Plastic-Laminate-Clad Countertops

	Crew	Daily Output	Labor-Hours	Unit	Material	2017 Bare Costs Labor	Equipment	Total	Total Incl O&P
0010 **PLASTIC-LAMINATE-CLAD COUNTERTOPS**									
0020 Stock, 24" wide w/backsplash, minimum	1 Carp	30	.267	L.F.	16.35	13.15		29.50	38
0100 Maximum		25	.320		37.50	15.75		53.25	65
0300 Custom plastic, 7/8" thick, aluminum molding, no splash		30	.267		33	13.15		46.15	56.50
0400 Cove splash		30	.267		31	13.15		44.15	54
0600 1-1/4" thick, no splash		28	.286		39	14.05		53.05	64
0700 Square splash		28	.286		43	14.05		57.05	69
0900 Square edge, plastic face, 7/8" thick, no splash		30	.267		33.50	13.15		46.65	57
1000 With splash		30	.267		40	13.15		53.15	64

12 36 Countertops

12 36 23 – Plastic Countertops

12 36 23.13 Plastic-Laminate-Clad Countertops	Crew	Daily Output	Labor-Hours	Unit	Material	2017 Bare Costs Labor	Equipment	Total	Total Incl O&P
1200	For stainless channel edge, 7/8" thick, add				L.F.	3.25			3.25	3.58
1300	1-1/4" thick, add					3.87			3.87	4.26
1500	For solid color suede finish, add				↓	4.14			4.14	4.55
1700	For end splash, add				Ea.	18.65			18.65	20.50
1900	For cut outs, standard, add, minimum	1 Carp	32	.250		12.40	12.30		24.70	32.50
2000	Maximum		8	1	↓	6.20	49.50		55.70	82.50
2100	Postformed, including backsplash and front edge		30	.267	L.F.	10.35	13.15		23.50	31.50
2110	Mitred, add		12	.667	Ea.		33		33	50.50
2200	Built-in place, 25" wide, plastic laminate	↓	25	.320	L.F.	43	15.75		58.75	71.50

12 36 33 – Tile Countertops

12 36 33.10 Ceramic Tile Countertops

		Crew	Daily Output	Labor-Hours	Unit	Material	Labor	Equipment	Total	Total Incl O&P
0010	**CERAMIC TILE COUNTERTOPS**									
2300	Ceramic tile mosaic	1 Carp	25	.320	L.F.	35	15.75		50.75	62.50

12 36 40 – Stone Countertops

12 36 40.10 Natural Stone Countertops

		Crew	Daily Output	Labor-Hours	Unit	Material	Labor	Equipment	Total	Total Incl O&P
0010	**NATURAL STONE COUNTERTOPS**									
2500	Marble, stock, with splash, 1/2" thick, minimum	1 Bric	17	.471	L.F.	43.50	22.50		66	82.50
2700	3/4" thick, maximum		13	.615		109	29.50		138.50	165
2800	Granite, average, 1-1/4" thick, 24" wide, no splash	↓	13.01	.615	↓	147	29.50		176.50	206

12 36 53 – Laboratory Countertops

12 36 53.10 Laboratory Countertops and Sinks

		Crew	Daily Output	Labor-Hours	Unit	Material	Labor	Equipment	Total	Total Incl O&P
0010	**LABORATORY COUNTERTOPS AND SINKS**									
0020	Countertops, epoxy resin, not incl. base cabinets, acid-proof, minimum	2 Carp	82	.195	S.F.	44.50	9.60		54.10	63.50
0030	Maximum		70	.229		47.50	11.25		58.75	70
0040	Stainless steel	↓	82	.195	↓	158	9.60		167.60	188

12 36 61 – Simulated Stone Countertops

12 36 61.16 Solid Surface Countertops

		Crew	Daily Output	Labor-Hours	Unit	Material	Labor	Equipment	Total	Total Incl O&P
0010	**SOLID SURFACE COUNTERTOPS**, Acrylic polymer									
0020	Pricing for orders of 100 L.F. or greater									
0100	25" wide, solid colors	2 Carp	28	.571	L.F.	59	28		87	108
0200	Patterned colors		28	.571		75	28		103	126
0300	Premium patterned colors		28	.571		95	28		123	147
0400	With silicone attached 4" backsplash, solid colors		27	.593		65.50	29		94.50	117
0500	Patterned colors		27	.593		83	29		112	136
0600	Premium patterned colors		27	.593		103	29		132	159
0700	With hard seam attached 4" backsplash, solid colors		23	.696		65.50	34.50		100	125
0800	Patterned colors		23	.696		83	34.50		117.50	144
0900	Premium patterned colors	↓	23	.696	↓	103	34.50		137.50	167
1000	Pricing for order of 51 – 99 L.F.									
1100	25" wide, solid colors	2 Carp	24	.667	L.F.	67.50	33		100.50	125
1200	Patterned colors		24	.667		86	33		119	145
1300	Premium patterned colors		24	.667		109	33		142	171
1400	With silicone attached 4" backsplash, solid colors		23	.696		75.50	34.50		110	136
1500	Patterned colors		23	.696		95.50	34.50		130	158
1600	Premium patterned colors		23	.696		119	34.50		153.50	184
1700	With hard seam attached 4" backsplash, solid colors		20	.800		75.50	39.50		115	144
1800	Patterned colors		20	.800		95.50	39.50		135	166
1900	Premium patterned colors	↓	20	.800	↓	119	39.50		158.50	192
2000	Pricing for order of 1 – 50 L.F.									
2100	25" wide, solid colors	2 Carp	20	.800	L.F.	79.50	39.50		119	148

12 36 61 – Simulated Stone Countertops

12 36 61.16 Solid Surface Countertops

		Crew	Daily Output	Labor-Hours	Unit	Material	2017 Bare Costs Labor	2017 Bare Costs Equipment	Total	Total Incl O&P
2200	Patterned colors	2 Carp	20	.800	L.F.	101	39.50		140.50	172
2300	Premium patterned colors		20	.800		128	39.50		167.50	202
2400	With silicone attached 4" backsplash, solid colors		19	.842		88.50	41.50		130	161
2500	Patterned colors		19	.842		112	41.50		153.50	187
2600	Premium patterned colors		19	.842		140	41.50		181.50	218
2700	With hard seam attached 4" backsplash, solid colors		15	1.067		88.50	52.50		141	178
2800	Patterned colors		15	1.067		112	52.50		164.50	204
2900	Premium patterned colors		15	1.067		140	52.50		192.50	235
3000	Sinks, pricing for order of 100 or greater units									
3100	Single bowl, hard seamed, solid colors, 13" x 17"	1 Carp	3	2.667	Ea.	405	131		536	645
3200	10" x 15"		7	1.143		187	56.50		243.50	291
3300	Cutouts for sinks		8	1			49.50		49.50	75.50
3400	Sinks, pricing for order of 51 – 99 units									
3500	Single bowl, hard seamed, solid colors, 13" x 17"	1 Carp	2.55	3.137	Ea.	465	155		620	745
3600	10" x 15"		6	1.333		215	65.50		280.50	335
3700	Cutouts for sinks		7	1.143			56.50		56.50	86
3800	Sinks, pricing for order of 1 – 50 units									
3900	Single bowl, hard seamed, solid colors, 13" x 17"	1 Carp	2	4	Ea.	545	197		742	900
4000	10" x 15"		4.55	1.758		252	86.50		338.50	410
4100	Cutouts for sinks		5.25	1.524			75		75	115
4200	Cooktop cutouts, pricing for 100 or greater units		4	2		29	98.50		127.50	183
4300	51 – 99 units		3.40	2.353		33	116		149	215
4400	1 – 50 units		3	2.667		39	131		170	244

12 36 61.17 Solid Surface Vanity Tops

		Crew	Daily Output	Labor-Hours	Unit	Material	2017 Bare Costs Labor	2017 Bare Costs Equipment	Total	Total Incl O&P
0010	**SOLID SURFACE VANITY TOPS**									
0015	Solid surface, center bowl, 17" x 19"	1 Carp	12	.667	Ea.	201	33		234	273
0020	19" x 25"		12	.667		206	33		239	277
0030	19" x 31"		12	.667		241	33		274	315
0040	19" x 37"		12	.667		281	33		314	360
0050	22" x 25"		10	.800		370	39.50		409.50	465
0060	22" x 31"		10	.800		430	39.50		469.50	530
0070	22" x 37"		10	.800		500	39.50		539.50	610
0080	22" x 43"		10	.800		575	39.50		614.50	695
0090	22" x 49"		10	.800		625	39.50		664.50	745
0110	22" x 55"		8	1		705	49.50		754.50	850
0120	22" x 61"		8	1		800	49.50		849.50	955
0220	Double bowl, 22" x 61"		8	1		900	49.50		949.50	1,075
0230	Double bowl, 22" x 73"		8	1		980	49.50		1,029.50	1,150
0240	For aggregate colors, add					35%				
0250	For faucets and fittings, see Section 22 41 39.10									

12 36 61.19 Quartz Agglomerate Countertops

		Crew	Daily Output	Labor-Hours	Unit	Material	2017 Bare Costs Labor	2017 Bare Costs Equipment	Total	Total Incl O&P
0010	**QUARTZ AGGLOMERATE COUNTERTOPS**									
0100	25" wide, 4" backsplash, color group A, minimum	2 Carp	15	1.067	L.F.	65	52.50		117.50	152
0110	Maximum		15	1.067		91	52.50		143.50	181
0120	Color group B, minimum		15	1.067		67	52.50		119.50	154
0130	Maximum		15	1.067		95	52.50		147.50	186
0140	Color group C, minimum		15	1.067		78	52.50		130.50	167
0150	Maximum		15	1.067		107	52.50		159.50	199
0160	Color group D, minimum		15	1.067		85	52.50		137.50	174
0170	Maximum		15	1.067		115	52.50		167.50	208

12 46 Furnishing Accessories

12 46 13 – Ash Receptacles

12 46 13.10 Ash/Trash Receivers

		Crew	Daily Output	Labor-Hours	Unit	Material	2017 Bare Costs Labor	Equipment	Total	Total Incl O&P
0010	**ASH/TRASH RECEIVERS**									
1000	Ash urn, cylindrical metal									
1020	8" diameter, 20" high	1 Clab	60	.133	Ea.	145	5.20		150.20	167
1060	10" diameter, 26" high	"	60	.133	"	116	5.20		121.20	136
2000	Combination ash/trash urn, metal									
2020	8" diameter, 20" high	1 Clab	60	.133	Ea.	190	5.20		195.20	217
2050	10" diameter, 26" high	"	60	.133	"	152	5.20		157.20	175

12 46 19 – Clocks

12 46 19.50 Wall Clocks

		Crew	Daily Output	Labor-Hours	Unit	Material	2017 Bare Costs Labor	Equipment	Total	Total Incl O&P
0010	**WALL CLOCKS**									
0080	12" diameter, single face	1 Elec	8	1	Ea.	125	56.50		181.50	222
0100	Double face	"	6.20	1.290	"	310	73		383	450

12 46 33 – Waste Receptacles

12 46 33.13 Trash Receptacles

		Crew	Daily Output	Labor-Hours	Unit	Material	2017 Bare Costs Labor	Equipment	Total	Total Incl O&P
0010	**TRASH RECEPTACLES**									
4000	Trash receptacle, metal									
4020	8" diameter, 15" high	1 Clab	60	.133	Ea.	94	5.20		99.20	111
4040	10" diameter, 18" high		60	.133		176	5.20		181.20	201
5040	16" x 8" x 14" high	↓	60	.133	↓	34.50	5.20		39.70	46
5500	Plastic, with lid									
5520	35 gallon	1 Clab	60	.133	Ea.	160	5.20		165.20	184
5540	45 gallon		60	.133		258	5.20		263.20	291
5550	Plastic recycling barrel, w/lid & wheels, 32 gal.	**G**	60	.133		96	5.20		101.20	114
5560	65 gal.	**G**	60	.133		615	5.20		620.20	690
5570	95 gal.	**G**	60	.133	↓	1,150	5.20		1,155.20	1,275

12 48 Rugs and Mats

12 48 13 – Entrance Floor Mats and Frames

12 48 13.13 Entrance Floor Mats

		Crew	Daily Output	Labor-Hours	Unit	Material	2017 Bare Costs Labor	Equipment	Total	Total Incl O&P
0010	**ENTRANCE FLOOR MATS**									
0020	Recessed, black rubber, 3/8" thick, solid	1 Clab	155	.052	S.F.	24.50	2.02		26.52	29.50
0050	Perforated		155	.052		17.10	2.02		19.12	22
0100	1/2" thick, solid		155	.052		20.50	2.02		22.52	25.50
0150	Perforated		155	.052		22.50	2.02		24.52	27.50
0200	In colors, 3/8" thick, solid		155	.052		19.60	2.02		21.62	24.50
0250	Perforated		155	.052		17.70	2.02		19.72	22.50
0300	1/2" thick, solid		155	.052		28.50	2.02		30.52	34.50
0350	Perforated	↓	155	.052	↓	29	2.02		31.02	35
1225	Recessed, alum. rail, hinged mat, 7/16" thk									
1250	Carpet insert	1 Clab	360	.022	S.F.	49.50	.87		50.37	56
1275	Vinyl insert		360	.022		49.50	.87		50.37	56
1300	Abrasive insert	↓	360	.022	↓	49.50	.87		50.37	56
1325	Recessed, vinyl rail, hinged mat, 7/16" thk									
1350	Carpet insert	1 Clab	360	.022	S.F.	55	.87		55.87	62
1375	Vinyl insert		360	.022		55	.87		55.87	62
1400	Abrasive insert		360	.022		55	.87		55.87	62
2000	Recycled rubber tire tile, 12" x 12" x 3/8" thick	**G**	125	.064		10.20	2.51		12.71	15.10
2510	Natural cocoa fiber, 1/2" thick	**G**	125	.064		8.30	2.51		10.81	13
2520	3/4" thick	**G**	125	.064		8	2.51		10.51	12.65
2530	1" thick	**G**	125	.064	↓	11.70	2.51		14.21	16.70

12 48 Rugs and Mats

12 48 13 - Entrance Floor Mats and Frames

12 48 13.13 Entrance Floor Mats	Crew	Daily Output	Labor-Hours	Unit	Material	2017 Bare Costs Labor	Equipment	Total	Total Incl O&P
3000	Hospital tacky mats, package of 30 with frame				Ea.	63.50			63.50	70
3010	4 packages of 30				"	92			92	101

12 51 Office Furniture

12 51 16 - Case Goods

12 51 16.13 Metal Case Goods

		Crew	Daily Output	Labor-Hours	Unit	Material	Labor	Equipment	Total	Total Incl O&P
0010	**METAL CASE GOODS**									
0020	Desks, 29" high, double pedestal, 30" x 60", metal, minimum				Ea.	515			515	570
0030	Maximum					895			895	985
0600	Desks, single pedestal, 30" x 60", metal, minimum					490			490	535
0620	Maximum					920			920	1,025
0720	Desks, secretarial, 30" x 60", metal, minimum					445			445	490
0730	Maximum					855			855	940
0740	Return, 20" x 42", minimum					415			415	460
0750	Maximum					655			655	720
0940	59" x 12" x 23" high, steel, minimum					345			345	380
0960	Maximum					440			440	485

12 51 16.16 Wood Case Goods

		Crew	Daily Output	Labor-Hours	Unit	Material	Labor	Equipment	Total	Total Incl O&P
0010	**WOOD CASE GOODS**									
0150	Desk, 29" high, double pedestal, 30" x 60"									
0160	Wood, minimum				Ea.	740			740	815
0180	Maximum				"	1,300			1,300	1,425
0630	Single pedestal, 30" x 60"									
0640	Wood, minimum				Ea.	630			630	695
0650	Maximum					905			905	995
0670	Executive return, 24" x 42", with box, file, wood, minimum					430			430	475
0680	Maximum					850			850	935
0790	Desk, 29" high, secretarial, 30" x 60"									
0800	Wood, minimum				Ea.	450			450	495
0810	Maximum					1,350			1,350	1,475
0820	Return, 20" x 42", minimum					365			365	405
0830	Maximum					705			705	775
0900	Desktop organizer, 72" x 14" x 36" high, wood, minimum					174			174	192
0920	Maximum					430			430	470
1110	Furniture, credenza, 29" high, 18" to 22" x 60" to 72"									
1120	Wood, minimum				Ea.	735			735	805
1140	Maximum				"	2,800			2,800	3,075

12 51 23 - Office Tables

12 51 23.33 Conference Tables

		Crew	Daily Output	Labor-Hours	Unit	Material	Labor	Equipment	Total	Total Incl O&P
0010	**CONFERENCE TABLES**									
6050	Boat, 96" x 42", minimum				Ea.	875			875	965
6150	Maximum					3,975			3,975	4,375
6720	Rectangle, 96" x 42", minimum					1,375			1,375	1,525
6740	Maximum					3,175			3,175	3,500

12 52 Seating

12 52 23 – Office Seating

12 52 23.13 Office Chairs	Crew	Daily Output	Labor-Hours	Unit	Material	2017 Bare Costs Labor	Equipment	Total	Total Incl O&P
0010 **OFFICE CHAIRS**									
2000 Standard office chair, executive, minimum				Ea.	281			281	310
2150 Maximum					2,075			2,075	2,275
2200 Management, minimum					250			250	275
2250 Maximum					1,850			1,850	2,050
2280 Task, minimum					187			187	206
2290 Maximum					630			630	690
2300 Arm kit, minimum					72.50			72.50	80
2320 Maximum					119			119	131

12 54 Hospitality Furniture

12 54 13 – Hotel and Motel Furniture

12 54 13.10 Hotel Furniture

	Crew	Daily Output	Labor-Hours	Unit	Material	2017 Bare Costs Labor	Equipment	Total	Total Incl O&P
0010 **HOTEL FURNITURE**									
0020 Standard quality set, minimum				Room	2,575			2,575	2,825
0200 Maximum				"	8,600			8,600	9,475

12 54 16 – Restaurant Furniture

12 54 16.10 Tables, Folding

	Crew	Daily Output	Labor-Hours	Unit	Material	2017 Bare Costs Labor	Equipment	Total	Total Incl O&P
0010 **TABLES, FOLDING** Laminated plastic tops									
1000 Tubular steel legs with glides									
1020 18" x 60", minimum				Ea.	276			276	305
1040 Maximum					1,275			1,275	1,400
1840 36" x 96", minimum					330			330	365
1860 Maximum					2,275			2,275	2,500
2000 Round, wood stained, plywood top, 60" diameter, minimum					223			223	245
2020 Maximum					298			298	330

12 54 16.20 Furniture, Restaurant

	Crew	Daily Output	Labor-Hours	Unit	Material	2017 Bare Costs Labor	Equipment	Total	Total Incl O&P
0010 **FURNITURE, RESTAURANT**									
0020 Bars, built-in, front bar	1 Carp	5	1.600	L.F.	300	79		379	450
0200 Back bar	"	5	1.600	"	218	79		297	360
0300 Booth seating, see Section 12 54 16.70									
2000 Chair, bentwood side chair, metal, minimum				Ea.	106			106	116
2020 Maximum					124			124	136
2600 Upholstered seat & back, arms, minimum					172			172	189
2620 Maximum					490			490	535

12 54 16.70 Booths

	Crew	Daily Output	Labor-Hours	Unit	Material	2017 Bare Costs Labor	Equipment	Total	Total Incl O&P
0010 **BOOTHS**									
1000 Banquet, upholstered seat and back, custom									
1500 Straight, minimum	2 Carp	40	.400	L.F.	208	19.70		227.70	259
1520 Maximum		36	.444		405	22		427	480
1600 "L" or "U" shape, minimum		35	.457		212	22.50		234.50	268
1620 Maximum		30	.533		375	26.50		401.50	455
1800 Upholstered outside finished backs for									
1810 single booths and custom banquets									
1820 Minimum	2 Carp	44	.364	L.F.	24.50	17.90		42.40	54.50
1840 Maximum	"	40	.400	"	72	19.70		91.70	110
3000 Fixed seating, one piece plastic chair and									
3010 plastic laminate table top									
3100 Two seat, 24" x 24" table, minimum	F-7	30	1.067	Ea.	835	47		882	995
3120 Maximum		26	1.231		1,200	54.50		1,254.50	1,375

For customer support on your Building Construction Costs with RSMeans Data, call 800.448.8182.

445

12 54 Hospitality Furniture

12 54 16 – Restaurant Furniture

12 54 16.70 Booths

		Crew	Daily Output	Labor-Hours	Unit	Material	2017 Bare Costs Labor	Equipment	Total	Total Incl O&P
3200	Four seat, 24" x 48" table, minimum	F-7	28	1.143	Ea.	830	50.50		880.50	990
3220	Maximum	↓	24	1.333	↓	1,425	59		1,484	1,650
5000	Mount in floor, wood fiber core with									
5010	plastic laminate face, single booth									
5050	24" wide	F-7	30	1.067	Ea.	320	47		367	425
5100	48" wide	"	28	1.143	"	410	50.50		460.50	530

12 55 Detention Furniture

12 55 13 – Detention Bunks

12 55 13.13 Cots

		Crew	Daily Output	Labor-Hours	Unit	Material	2017 Bare Costs Labor	Equipment	Total	Total Incl O&P
0010	**COTS**									
2500	Bolted, single, painted steel	E-4	20	1.600	Ea.	340	87.50	6.50	434	530
2700	Stainless steel	"	20	1.600	"	985	87.50	6.50	1,079	1,225

12 56 Institutional Furniture

12 56 33 – Classroom Furniture

12 56 33.10 Furniture, School

		Crew	Daily Output	Labor-Hours	Unit	Material	2017 Bare Costs Labor	Equipment	Total	Total Incl O&P
0010	**FURNITURE, SCHOOL**									
0500	Classroom, movable chair & desk type, minimum				Set				73.50	81
0600	Maximum				"				155	171
1000	Chair, molded plastic									
1100	Integral tablet arm, minimum				Ea.	101			101	112
1150	Maximum					186			186	205
2000	Desk, single pedestal, top book compartment, minimum					102			102	112
2020	Maximum					186			186	205
2200	Flip top, minimum					246			246	270
2220	Maximum				↓	300			300	330

12 56 43 – Dormitory Furniture

12 56 43.10 Dormitory Furnishings

		Crew	Daily Output	Labor-Hours	Unit	Material	2017 Bare Costs Labor	Equipment	Total	Total Incl O&P
0010	**DORMITORY FURNISHINGS**									
0300	Bunkable bed, twin, minimum				Ea.	385			385	420
0320	Maximum					600			600	660
1000	Chest, four drawer, minimum					345			345	380
1020	Maximum				↓	685			685	750
1050	Built-in, minimum	2 Carp	13	1.231	L.F.	133	60.50		193.50	240
1150	Maximum		10	1.600		247	79		326	390
1200	Desk top, built-in, laminated plastic, 24" deep, minimum		50	.320		48.50	15.75		64.25	77.50
1300	Maximum		40	.400		145	19.70		164.70	190
1450	30" deep, minimum		50	.320		62	15.75		77.75	92
1550	Maximum		40	.400		271	19.70		290.70	330
1750	Dressing unit, built-in, minimum		12	1.333	↓	200	65.50		265.50	320
1850	Maximum	↓	8	2	↓	600	98.50		698.50	810
8000	Rule of thumb: total cost for furniture, minimum				Student				2,525	2,800
8050	Maximum				"				4,850	5,350

12 56 Institutional Furniture

12 56 51 – Library Furniture

12 56 51.10 Library Furnishings	Crew	Daily Output	Labor-Hours	Unit	Material	2017 Bare Costs Labor	Equipment	Total	Total Incl O&P
0010 **LIBRARY FURNISHINGS**									
0100 Attendant desk, 36" x 62" x 29" high	1 Carp	16	.500	Ea.	1,900	24.50		1,924.50	2,150
0200 Book display, "A" frame display, both sides, 42" x 42" x 60" high		16	.500		1,300	24.50		1,324.50	1,475
0220 Table with bulletin board, 42" x 24" x 49" high		16	.500		710	24.50		734.50	820
0800 Card catalogue, 30 tray unit		16	.500		3,075	24.50		3,099.50	3,425
0840 60 tray unit		16	.500		6,175	24.50		6,199.50	6,850
0880 72 tray unit	2 Carp	16	1		8,200	49.50		8,249.50	9,100
1000 Carrels, single face, initial unit	1 Carp	16	.500		880	24.50		904.50	1,000
1500 Double face, initial unit	2 Carp	16	1		1,475	49.50		1,524.50	1,675
1710 Carrels, hardwood, 36" x 24", minimum	1 Carp	5	1.600		600	79		679	780
1720 Maximum	"	4	2		1,650	98.50		1,748.50	1,975
2700 Card catalog file, 60 trays, complete					8,175			8,175	8,975
2720 Alternate method: each tray					136			136	150
3800 Charging desk, built-in, with counter, plastic laminated top	1 Carp	7	1.143	L.F.	400	56.50		456.50	525
4000 Dictionary stand, stationary		16	.500	Ea.	760	24.50		784.50	875
4020 Revolving		16	.500		216	24.50		240.50	276
4200 Exhibit case, table style, 60" x 28" x 36"		11	.727		2,700	36		2,736	3,000
6010 Bookshelf, metal, 90" high, 10" shelf, double face		11.50	.696	L.F.	268	34.50		302.50	350
6020 Single face		12	.667	"	114	33		147	176
6050 For 8" shelving, subtract from above					10%				
6060 For 12" shelving, add to above					10%				
6070 For 42" high with countertop, subtract from above					20%				
6100 Mobile compacted shelving, hand crank, 9'-0" high									
6110 Double face, including track, 3' section				Ea.	2,100			2,100	2,325
6150 For electrical operation, add					25%				
6200 Magazine shelving, 82" high, 12" deep, single face	1 Carp	11.50	.696	L.F.	169	34.50		203.50	239
6210 Double face	"	11.50	.696	"	225	34.50		259.50	300
7200 Reading table, laminated top, 60" x 36"				Ea.	590			590	650

12 56 70 – Healthcare Furniture

12 56 70.10 Furniture, Hospital

	Crew	Daily Output	Labor-Hours	Unit	Material	2017 Bare Costs Labor	Equipment	Total	Total Incl O&P
0010 **FURNITURE, HOSPITAL**									
0020 Beds, manual, minimum				Ea.	710			710	780
0100 Maximum					2,350			2,350	2,575
0600 All electric hospital beds, minimum					1,900			1,900	2,075
0700 Maximum					3,800			3,800	4,175
0900 Manual, nursing home beds, minimum					725			725	795
1000 Maximum					1,800			1,800	1,975
1020 Overbed table, laminated top, minimum					360			360	400
1040 Maximum					730			730	805
1100 Patient wall systems, not incl. plumbing, minimum				Room	1,425			1,425	1,575
1200 Maximum				"	1,925			1,925	2,125
2000 Geriatric chairs, minimum				Ea.	450			450	495
2020 Maximum				"	750			750	825

For customer support on your Building Construction Costs with RSMeans Data, call 800.448.8182.

447

12 61 Fixed Audience Seating

12 61 13 – Upholstered Audience Seating

12 61 13.13 Auditorium Chairs

12 61 13.13 Auditorium Chairs	Crew	Daily Output	Labor-Hours	Unit	Material	2017 Bare Costs Labor	Equipment	Total	Total Incl O&P
0010 **AUDITORIUM CHAIRS**									
2000 All veneer construction	2 Carp	22	.727	Ea.	248	36		284	330
2200 Veneer back, padded seat		22	.727		258	36		294	340
2350 Fully upholstered, spring seat		22	.727		254	36		290	335
2450 For tablet arms, add					71.50			71.50	79
2500 For fire retardancy, CATB-133, add					33			33	36

12 61 13.23 Lecture Hall Seating

	Crew	Daily Output	Labor-Hours	Unit	Material	2017 Bare Costs Labor	Equipment	Total	Total Incl O&P
0010 **LECTURE HALL SEATING**									
1000 Pedestal type, minimum	2 Carp	22	.727	Ea.	300	36		336	385
1200 Maximum	"	14.50	1.103	"	515	54.50		569.50	650

12 63 Stadium and Arena Seating

12 63 13 – Stadium and Arena Bench Seating

12 63 13.13 Bleachers

	Crew	Daily Output	Labor-Hours	Unit	Material	2017 Bare Costs Labor	Equipment	Total	Total Incl O&P
0010 **BLEACHERS**									
3000 Telescoping, manual to 15 tier, minimum	F-5	65	.492	Seat	109	24.50		133.50	158
3100 Maximum		60	.533		157	26.50		183.50	214
3300 16 to 20 tier, minimum		60	.533		262	26.50		288.50	330
3400 Maximum		55	.582		315	29		344	390
3600 21 to 30 tier, minimum		50	.640		252	32		284	325
3700 Maximum		40	.800		345	40		385	440
3900 For integral power operation, add, minimum	2 Elec	300	.053		51	3.02		54.02	60.50
4000 Maximum	"	250	.064		83	3.62		86.62	97
5000 Benches, folding, in wall, 14' table, 2 benches	L-4	2	12	Set	925	560		1,485	1,900

12 67 Pews and Benches

12 67 13 – Pews

12 67 13.13 Sanctuary Pews

	Crew	Daily Output	Labor-Hours	Unit	Material	2017 Bare Costs Labor	Equipment	Total	Total Incl O&P
0010 **SANCTUARY PEWS**									
1500 Bench type, hardwood, minimum	1 Carp	20	.400	L.F.	101	19.70		120.70	141
1550 Maximum	"	15	.533		176	26.50		202.50	234
1570 For kneeler, add					22.50			22.50	25

12 92 Interior Planters and Artificial Plants

12 92 33 – Interior Planters

12 92 33.10 Planters

	Crew	Daily Output	Labor-Hours	Unit	Material	2017 Bare Costs Labor	Equipment	Total	Total Incl O&P
0010 **PLANTERS**									
1000 Fiberglass, hanging, 12" diameter, 7" high				Ea.	112			112	124
1500 Rectangular, 48" long, 16" high, 15" wide					610			610	670
1650 60" long, 30" high, 28" wide					1,050			1,050	1,150
2000 Round, 12" diameter, 13" high					162			162	178
2050 25" high					214			214	236
5000 Square, 10" side, 20" high					199			199	219
5100 14" side, 15" high					215			215	236
6000 Metal bowl, 32" diameter, 8" high, minimum					575			575	630
6050 Maximum					800			800	880
8750 Wood, fiberglass liner, square									

For customer support on your Building Construction Costs with RSMeans Data, call 800.448.8182.

12 92 Interior Planters and Artificial Plants

12 92 33 – Interior Planters

12 92 33.10 Planters	Crew	Daily Output	Labor-Hours	Unit	Material	2017 Bare Costs Labor	Equipment	Total	Total Incl O&P
8780	14" square, 15" high, minimum				Ea.	430			430	475
8800	Maximum					530			530	580
9400	Plastic cylinder, molded, 10" diameter, 10" high					47			47	51.50
9500	11" diameter, 11" high					89.50			89.50	98

12 93 Interior Public Space Furnishings

12 93 13 – Bicycle Racks

12 93 13.10 Bicycle Racks

		Crew	Daily Output	Labor-Hours	Unit	Material	Labor	Equipment	Total	Total Incl O&P
0010	**BICYCLE RACKS**									
0020	Single side, grid, 1-5/8" OD stl. pipe, w/1/2" bars, galv, 5 bike cap	2 Clab	10	1.600	Ea.	273	62.50		335.50	395
0025	Powder coat finish		10	1.600		269	62.50		331.50	390
0030	Single side, grid, 1-5/8" OD stl. pipe, w/1/2" bars, galv, 9 bike cap		8	2		370	78.50		448.50	525
0035	Powder coat finish		8	2		370	78.50		448.50	530
0040	Single side, grid, 1-5/8" OD stl. pipe, w/1/2" bars, galv, 18 bike cap		4	4		445	157		602	730
0045	Powder coat finish		4	4		445	157		602	730
0050	S curve, 1-7/8" OD stl pipe, 11 ga, galv, 5 bike cap		10	1.600		122	62.50		184.50	230
0055	Powder coat finish		10	1.600		131	62.50		193.50	240
0060	S curve, 1-7/8" OD stl pipe, 11 ga, galv, 7 bike cap		9	1.778		174	69.50		243.50	298
0065	Powder coat finish		9	1.778		176	69.50		245.50	300
0070	S curve, 1-7/8" OD stl pipe, 11 ga, galv, 9 bike cap		8	2		270	78.50		348.50	415
0075	Powder coat finish		8	2		258	78.50		336.50	405
0080	S curve, 1-7/8" OD stl pipe, 11 ga, galv, 11 bike cap		6	2.667		365	104		469	560
0085	Powder coat finish		6	2.667		355	104		459	555

12 93 23 – Trash and Litter Receptacles

12 93 23.10 Trash Receptacles

			Crew	Daily Output	Labor-Hours	Unit	Material	Labor	Equipment	Total	Total Incl O&P
0010	**TRASH RECEPTACLES**										
0020	Fiberglass, 2' square, 18" high		2 Clab	30	.533	Ea.	550	21		571	635
0100	2' square, 2'-6" high			30	.533		745	21		766	850
0300	Circular, 2' diameter, 18" high			30	.533		465	21		486	540
0400	2' diameter, 2'-6" high			30	.533		555	21		576	640
0500	Recycled plastic, var colors, round, 32 gal., 28" x 38" H	G		5	3.200		660	125		785	915
0510	32 gal., 31" x 32" H	G		5	3.200		730	125		855	990
9110	Plastic, with dome lid, 32 gal. capacity			35	.457		62.50	17.90		80.40	96
9120	Recycled plastic slats, plastic dome lid, 32 gal. capacity			35	.457		310	17.90		327.90	370

12 93 23.20 Trash Closure

		Crew	Daily Output	Labor-Hours	Unit	Material	Labor	Equipment	Total	Total Incl O&P
0010	**TRASH CLOSURE**									
0020	Steel with pullover cover, 2'-3" wide, 4'-7" high, 6'-2" long	2 Clab	5	3.200	Ea.	2,100	125		2,225	2,525
0100	10'-1" long		4	4		2,600	157		2,757	3,125
0300	Wood, 10' wide, 6' high, 10' long		1.20	13.333		2,000	520		2,520	3,000

Division Notes

		CREW	DAILY OUTPUT	LABOR-HOURS	UNIT	BARE COSTS				TOTAL INCL O&P
						MAT.	LABOR	EQUIP.	TOTAL	

Estimating Tips
General

- The items and systems in this division are usually estimated, purchased, supplied, and installed as a unit by one or more subcontractors. The estimator must ensure that all parties are operating from the same set of specifications and assumptions, and that all necessary items are estimated and will be provided. Many times the complex items and systems are covered, but the more common ones, such as excavation or a crane, are overlooked for the very reason that everyone assumes nobody could miss them. The estimator should be the central focus and be able to ensure that all systems are complete.

- Another area where problems can develop in this division is at the interface between systems. The estimator must ensure, for instance, that anchor bolts, nuts, and washers are estimated and included for the air-supported structures and pre-engineered buildings to be bolted to their foundations. Utility supply is a common area where essential items or pieces of equipment can be missed or overlooked, because each subcontractor may feel it is another's responsibility. The estimator should also be aware of certain items which may be supplied as part of a package but installed by others, and ensure that the installing contractor's estimate includes the cost of installation. Conversely, the estimator must also ensure that items are not costed by two different subcontractors, resulting in an inflated overall estimate.

13 30 00 Special Structures

- The foundations and floor slab, as well as rough mechanical and electrical, should be estimated, as this work is required for the assembly and erection of the structure. Generally, as noted in the data set, the pre-engineered building comes as a shell. Pricing is based on the size and structural design parameters stated in the reference section. Additional features, such as windows and doors with their related structural framing, must also be included by the estimator. Here again, the estimator must have a clear understanding of the scope of each portion of the work and all the necessary interfaces.

Reference Numbers

Reference numbers are shown at the beginning of some major classifications. These numbers refer to related items in the Reference Section. The reference information may be an estimating procedure, an alternate pricing method, or technical information.

Note: Not all subdivisions listed here necessarily appear. ■

Did you know?

Our online estimating solution gives you the same access to RSMeans' data with 24/7 access:

- Quickly locate costs in the searchable database.

- Build cost lists, estimates, and reports in minutes.

- Adjust costs to any location in the U.S. and Canada with the click of a button.

Start your free trial today at
www.RSMeansOnline.com

13 05 05 – Selective Demolition for Special Construction

13 05 05.10 Selective Demolition, Air Supported Structures	Crew	Daily Output	Labor-Hours	Unit	Material	2017 Bare Costs Labor	Equipment	Total	Total Incl O&P
0010 **SELECTIVE DEMOLITION, AIR SUPPORTED STRUCTURES**									
0020 Tank covers, scrim, dbl. layer, vinyl poly w/hdwe., blower & controls									
0050 Round and rectangular R024119-10	B-2	9000	.004	S.F.		.18		.18	.27
0100 Warehouse structures									
0120 Poly/vinyl fabric, 28 oz., incl. tension cables & inflation system	4 Clab	9000	.004	SF Flr.		.14		.14	.21
0150 Reinforced vinyl, 12 oz., 3000 S.F.	"	5000	.006			.25		.25	.38
0200 12,000 to 24,000 S.F.	8 Clab	20000	.003			.13		.13	.19
0250 Tedlar vinyl fabric, 28 oz. w/liner, to 3000 S.F.	4 Clab	5000	.006			.25		.25	.38
0300 12,000 to 24,000 S.F.	8 Clab	20000	.003			.13		.13	.19
0350 Greenhouse/shelter, woven polyethylene with liner									
0400 3000 S.F.	4 Clab	5000	.006	SF Flr.		.25		.25	.38
0450 12,000 to 24,000 S.F.	8 Clab	20000	.003			.13		.13	.19
0500 Tennis/gymnasium, poly/vinyl fabric, 28 oz., incl. thermal liner	4 Clab	9000	.004			.14		.14	.21
0600 Stadium/convention center, teflon coated fiberglass, incl. thermal liner	9 Clab	40000	.002			.07		.07	.11
0700 Doors, air lock, 15' long, 10' x 10'	2 Carp	1.50	10.667	Ea.		525		525	805
0720 15' x 15'		.80	20			985		985	1,500
0750 Revolving personnel door, 6' diam. x 6'-6" high		1.50	10.667			525		525	805

13 05 05.20 Selective Demolition, Garden Houses

	Crew	Daily Output	Labor-Hours	Unit	Material	Labor	Equipment	Total	Total Incl O&P
0010 **SELECTIVE DEMOLITION, GARDEN HOUSES** R024119-10									
0020 Prefab, wood, excl foundation, average	2 Clab	400	.040	SF Flr.		1.57		1.57	2.40

13 05 05.25 Selective Demolition, Geodesic Domes

	Crew	Daily Output	Labor-Hours	Unit	Material	Labor	Equipment	Total	Total Incl O&P
0010 **SELECTIVE DEMOLITION, GEODESIC DOMES**									
0050 Shell only, interlocking plywood panels, 30' diameter	F-5	3.20	10	Ea.		500		500	760
0060 34' diameter		2.30	13.913			690		690	1,050
0070 39' diameter		2	16			795		795	1,225
0080 45' diameter	F-3	2.20	18.182			920	260	1,180	1,675
0090 55' diameter		2	20			1,000	286	1,286	1,875
0100 60' diameter		2	20			1,000	286	1,286	1,875
0110 65' diameter		1.60	25			1,275	360	1,635	2,325

13 05 05.30 Selective Demolition, Greenhouses

	Crew	Daily Output	Labor-Hours	Unit	Material	Labor	Equipment	Total	Total Incl O&P
0010 **SELECTIVE DEMOLITION, GREENHOUSES** R024119-10									
0020 Resi-type, free standing, excl. foundations, 9' long x 8' wide	2 Clab	160	.100	SF Flr.		3.92		3.92	6
0030 9' long x 11' wide		170	.094			3.68		3.68	5.65
0040 9' long x 14' wide		220	.073			2.85		2.85	4.36
0050 9' long x 17' wide		320	.050			1.96		1.96	3
0060 Lean-to type, 4' wide		64	.250			9.80		9.80	15
0070 7' wide		120	.133			5.20		5.20	8
0080 Geodesic hemisphere, 1/8" plexiglass glazing, 8' diam.		4	4	Ea.		157		157	240
0090 24' diam.		.80	20			785		785	1,200
0100 48' diam.		.40	40			1,575		1,575	2,400

13 05 05.35 Selective Demolition, Hangars

	Crew	Daily Output	Labor-Hours	Unit	Material	Labor	Equipment	Total	Total Incl O&P
0010 **SELECTIVE DEMOLITION, HANGARS**									
0020 T type hangars, prefab, steel, galv roof & walls, incl doors, excl fndtn	E-2	2550	.022	SF Flr.		1.18	.66	1.84	2.64
0030 Circular type, prefab, steel frame, plastic skin, incl foundation, 80' diam	"	.50	112	Total		6,025	3,350	9,375	13,500

13 05 05.45 Selective Demolition, Lightning Protection

	Crew	Daily Output	Labor-Hours	Unit	Material	Labor	Equipment	Total	Total Incl O&P
0010 **SELECTIVE DEMOLITION, LIGHTNING PROTECTION**									
0020 Air terminal & base, copper, 3/8" diam. x 10", to 75' h	1 Clab	16	.500	Ea.		19.60		19.60	30
0030 1/2" diam. x 12", over 75' h		16	.500			19.60		19.60	30
0050 Aluminum, 1/2" diam. x 12", to 75' h		16	.500			19.60		19.60	30
0060 5/8" diam. x 12", over 75' h		16	.500			19.60		19.60	30
0070 Cable, copper, 220 lb. per thousand feet, to 75' high		640	.013	L.F.		.49		.49	.75

13 05 05 – Selective Demolition for Special Construction

13 05 05.45 Selective Demolition, Lightning Protection	Crew	Daily Output	Labor-Hours	Unit	Material	2017 Bare Costs Labor	Equipment	Total	Total Incl O&P
0080	375 lb. per thousand feet, over 75' high	1 Clab	460	.017	L.F.		.68		.68	1.04
0090	Aluminum, 101 lb. per thousand feet, to 75' high		560	.014			.56		.56	.86
0100	199 lb. per thousand feet, over 75' high		480	.017			.65		.65	1
0110	Arrester, 175 V AC, to ground		16	.500	Ea.		19.60		19.60	30
0120	650 V AC, to ground		13	.615	"		24		24	37

13 05 05.50 Selective Demolition, Pre-Engineered Steel Buildings

		Crew	Daily Output	Labor-Hours	Unit	Material	Labor	Equipment	Total	Total Incl O&P
0010	**SELECTIVE DEMOLITION, PRE-ENGINEERED STEEL BUILDINGS**									
0500	Pre-engd. steel bldgs., rigid frame, clear span & multi post, excl. salvage									
0550	3,500 to 7,500 S.F.	L-10	1000	.024	SF Flr.		1.33	.57	1.90	2.78
0600	7,501 to 12,500 S.F.		1500	.016			.89	.38	1.27	1.85
0650	12,501 S.F. or greater		1650	.015			.81	.35	1.16	1.68
0700	Pre-engd. steel building components									
0710	Entrance canopy, including frame 4' x 4'	E-24	8	4	Ea.		216	84	300	450
0720	4' x 8'	"	7	4.571			247	96.50	343.50	510
0730	HM doors, self framing, single leaf	2 Skwk	8	2			103		103	158
0740	Double leaf		5	3.200			165		165	253
0760	Gutter, eave type		600	.027	L.F.		1.37		1.37	2.11
0770	Sash, single slide, double slide or fixed		24	.667	Ea.		34.50		34.50	53
0780	Skylight, fiberglass, to 30 S.F.		16	1			51.50		51.50	79
0785	Roof vents, circular, 12" to 24" diameter		12	1.333			68.50		68.50	106
0790	Continuous, 10' long		8	2			103		103	158
0900	Shelters, aluminum frame									
0910	Acrylic glazing, 3' x 9' x 8' high	2 Skwk	2	8	Ea.		410		410	635
0920	9' x 12' x 8' high	"	1.50	10.667	"		550		550	845

13 05 05.60 Selective Demolition, Silos

		Crew	Daily Output	Labor-Hours	Unit	Material	Labor	Equipment	Total	Total Incl O&P
0010	**SELECTIVE DEMOLITION, SILOS**									
0020	Conc stave, indstrl, conical/sloping bott, excl fndtn, 12' diam., 35' h	E-24	.18	177	Ea.		9,625	3,750	13,375	19,800
0030	16' diam., 45' h		.12	266			14,400	5,625	20,025	29,700
0040	25' diam., 75' h		.08	400			21,600	8,425	30,025	44,600
0050	Steel, factory fabricated, 30,000 gal. cap, painted or epoxy lined	L-5	2	28			1,525	335	1,860	2,900

13 05 05.65 Selective Demolition, Sound Control

		Crew	Daily Output	Labor-Hours	Unit	Material	Labor	Equipment	Total	Total Incl O&P
0010	**SELECTIVE DEMOLITION, SOUND CONTROL** R024119-10									
0120	Acoustical enclosure, 4" thick walls & ceiling panels, 8 lb./S.F.	3 Carp	144	.167	SF Surf		8.20		8.20	12.60
0130	10.5 lb./S.F.		128	.188			9.25		9.25	14.15
0140	Reverb chamber, parallel walls, 4" thick		120	.200			9.85		9.85	15.10
0150	Skewed walls, parallel roof, 4" thick		110	.218			10.75		10.75	16.45
0160	Skewed walls/roof, 4" layer/air space		96	.250			12.30		12.30	18.85
0170	Sound-absorbing panels, painted metal, 2'-6" x 8', under 1,000 S.F.		430	.056			2.75		2.75	4.21
0180	Over 1,000 S.F.		480	.050			2.46		2.46	3.77
0190	Flexible transparent curtain, clear	3 Shee	430	.056			3.24		3.24	4.96
0192	50% clear, 50% foam		430	.056			3.24		3.24	4.96
0194	25% clear, 75% foam		430	.056			3.24		3.24	4.96
0196	100% foam		430	.056			3.24		3.24	4.96
0200	Audio-masking sys., incl. speakers, amplfr., signal gnrtr.									
0205	Ceiling mounted, 5,000 S.F.	2 Elec	4800	.003	S.F.		.19		.19	.28
0210	10,000 S.F.		5600	.003			.16		.16	.24
0220	Plenum mounted, 5,000 S.F.		7600	.002			.12		.12	.18
0230	10,000 S.F.		8800	.002			.10		.10	.15

13 05 05.70 Selective Demolition, Special Purpose Rooms

		Crew	Daily Output	Labor-Hours	Unit	Material	Labor	Equipment	Total	Total Incl O&P
0010	**SELECTIVE DEMOLITION, SPECIAL PURPOSE ROOMS** R024119-10									
0100	Audiometric rooms, under 500 S.F. surface	4 Carp	200	.160	SF Surf		7.90		7.90	12.05
0110	Over 500 S.F. surface	"	240	.133	"		6.55		6.55	10.05

13 05 05.70 Selective Demolition, Special Purpose Rooms

		Crew	Daily Output	Labor-Hours	Unit	Material	2017 Bare Costs Labor	Equipment	Total	Total Incl O&P
0200	Clean rooms, 12' x 12' soft wall, class 100	1 Carp	.30	26.667	Ea.		1,325		1,325	2,000
0210	Class 1000		.30	26.667			1,325		1,325	2,000
0220	Class 10,000		.35	22.857			1,125		1,125	1,725
0230	Class 100,000		.35	22.857			1,125		1,125	1,725
0300	Darkrooms, shell complete, 8' high	2 Carp	220	.073	SF Flr.		3.58		3.58	5.50
0310	12' high		110	.145	"		7.15		7.15	10.95
0350	Darkrooms doors, mini-cylindrical, revolving		4	4	Ea.		197		197	300
0400	Music room, practice modular		140	.114	SF Surf		5.65		5.65	8.60
0500	Refrigeration structures and finishes									
0510	Wall finish, 2 coat Portland cement plaster, 1/2" thick	1 Clab	200	.040	S.F.		1.57		1.57	2.40
0520	Fiberglass panels, 1/8" thick		400	.020			.78		.78	1.20
0530	Ceiling finish, polystyrene plastic, 1" to 2" thick		500	.016			.63		.63	.96
0540	4" thick		450	.018			.70		.70	1.07
0550	Refrigerator, prefab aluminum walk-in, 7'-6" high, 6' x 6' OD	2 Carp	100	.160	SF Flr.		7.90		7.90	12.05
0560	10' x 10' OD		160	.100			4.93		4.93	7.55
0570	Over 150 S.F.		200	.080			3.94		3.94	6.05
0600	Sauna, prefabricated, including heater & controls, 7' high, to 30 S.F.		120	.133			6.55		6.55	10.05
0610	To 40 S.F.		140	.114			5.65		5.65	8.60
0620	To 60 S.F.		175	.091			4.50		4.50	6.90
0630	To 100 S.F.		220	.073			3.58		3.58	5.50
0640	To 130 S.F.		250	.064			3.15		3.15	4.83
0650	Steam bath, heater, timer, head, single, to 140 C.F.	1 Plum	2.20	3.636	Ea.		225		225	340
0660	To 300 C.F.		2.20	3.636			225		225	340
0670	Steam bath, comm. size, w/blow-down assembly, to 800 C.F.		1.80	4.444			275		275	415
0680	To 2500 C.F.		1.60	5			310		310	465
0690	Steam bath, comm. size, multiple, for motels, apts, 500 C.F., 2 baths		2	4			247		247	375
0700	1,000 C.F., 4 baths		1.40	5.714			355		355	535

13 05 05.75 Selective Demolition, Storage Tanks

		Crew	Daily Output	Labor-Hours	Unit	Material	2017 Bare Costs Labor	Equipment	Total	Total Incl O&P
0010	**SELECTIVE DEMOLITION, STORAGE TANKS**									
0500	Steel tank, single wall, above ground, not incl. fdn., pumps or piping									
0510	Single wall, 275 gallon	Q-1	3	5.333	Ea.		297		297	450
0520	550 thru 2,000 gallon	B-34P	2	12			640	320	960	1,325
0530	5,000 thru 10,000 gallon	B-34Q	2	12			650	595	1,245	1,625
0540	15,000 thru 30,000 gallon	B-34S	2	16			905	1,725	2,630	3,275
0600	Steel tank, double wall, above ground not incl. fdn., pumps & piping									
0620	500 thru 2,000 gallon	B-34P	2	12	Ea.		640	320	960	1,325

Note: Row 0510 references R024119-10

13 05 05.85 Selective Demolition, Swimming Pool Equip

		Crew	Daily Output	Labor-Hours	Unit	Material	2017 Bare Costs Labor	Equipment	Total	Total Incl O&P
0010	**SELECTIVE DEMOLITION, SWIMMING POOL EQUIP**									
0020	Diving stand, stainless steel, 3 meter	2 Clab	3	5.333	Ea.		209		209	320
0030	1 meter		5	3.200			125		125	192
0040	Diving board, 16' long, aluminum		5.40	2.963			116		116	178
0050	Fiberglass		5.40	2.963			116		116	178
0070	Ladders, heavy duty, stainless steel, 2 tread		14	1.143			44.50		44.50	68.50
0080	4 tread		12	1.333			52		52	80
0090	Lifeguard chair, stainless steel, fixed		5	3.200			125		125	192
0100	Slide, tubular, fiberglass, aluminum handrails & ladder, 5', straight		4	4			157		157	240
0110	8', curved		6	2.667			104		104	160
0120	10', curved		3	5.333			209		209	320
0130	12' straight, with platform		2.50	6.400			251		251	385
0140	Removable access ramp, stainless steel		4	4			157		157	240
0150	Removable stairs, stainless steel, collapsible		4	4			157		157	240

454

For customer support on your Building Construction Costs with RSMeans Data, call 800.448.8182.

13 05 Common Work Results for Special Construction

13 05 05 – Selective Demolition for Special Construction

13 05 05.90 Selective Demolition, Tension Structures	Crew	Daily Output	Labor-Hours	Unit	Material	2017 Bare Costs Labor	Equipment	Total	Total Incl O&P
0010 **SELECTIVE DEMOLITION, TENSION STRUCTURES**									
0020 Steel/alum. frame, fabric shell, 60' clear span, 6,000 S.F.	B-41	2000	.022	SF Flr.		.89	.16	1.05	1.55
0030 12,000 S.F.		2200	.020			.81	.15	.96	1.40
0040 80' clear span, 20,800 S.F.	↓	2440	.018			.73	.13	.86	1.27
0050 100' clear span, 10,000 S.F.	L-5	4350	.013			.71	.15	.86	1.33
0060 26,000 S.F.		4600	.012			.67	.15	.82	1.26
0070 36,000 S.F.	↓	5000	.011	↓		.61	.13	.74	1.16

13 05 05.95 Selective Demo, X-Ray/Radio Freq Protection

	Crew	Daily Output	Labor-Hours	Unit	Material	2017 Bare Costs Labor	Equipment	Total	Total Incl O&P
0010 **SELECTIVE DEMO, X-RAY/RADIO FREQ PROTECTION**									
0020 Shielding lead, lined door frame, excl. hdwe., 1/16" thick	1 Clab	4.80	1.667	Ea.		65.50		65.50	100
0030 Lead sheets, 1/16" thick R024119-10	2 Clab	270	.059	S.F.		2.32		2.32	3.56
0040 1/8" thick		240	.067			2.61		2.61	4
0050 Lead shielding, 1/4" thick		270	.059			2.32		2.32	3.56
0060 1/2" thick	↓	240	.067	↓		2.61		2.61	4
0070 Lead glass, 1/4" thick, 2.0 mm LE, 12" x 16"	2 Glaz	16	1	Ea.		48		48	73
0080 24" x 36"		8	2			95.50		95.50	146
0090 36" x 60"		4	4			191		191	291
0100 Lead glass window frame, with 1/16" lead & voice passage, 36" x 60"		4	4			191		191	291
0110 Lead glass window frame, 24" x 36"	↓	8	2	↓		95.50		95.50	146
0120 Lead gypsum board, 5/8" thick with 1/16" lead	2 Clab	320	.050	S.F.		1.96		1.96	3
0130 1/8" lead		280	.057			2.24		2.24	3.43
0140 1/32" lead		400	.040	↓		1.57		1.57	2.40
0150 Butt joints, 1/8" lead or thicker, 2" x 7' long batten strip		480	.033	Ea.		1.30		1.30	2
0160 X-ray protection, average radiography room, up to 300 S.F., 1/16" lead, min		.50	32	Total		1,250		1,250	1,925
0170 Maximum		.30	53.333			2,100		2,100	3,200
0180 Deep therapy X-ray room, 250 kV cap, up to 300 S.F., 1/4" lead, min		.20	80			3,125		3,125	4,800
0190 Maximum		.12	133	↓		5,225		5,225	8,000
0880 Radio frequency shielding, prefab or screen-type copper or steel, minimum		360	.044	SF Surf		1.74		1.74	2.67
0890 Average		310	.052			2.02		2.02	3.10
0895 Maximum	↓	290	.055	↓		2.16		2.16	3.31

13 11 Swimming Pools

13 11 13 – Below-Grade Swimming Pools

13 11 13.50 Swimming Pools	Crew	Daily Output	Labor-Hours	Unit	Material	2017 Bare Costs Labor	Equipment	Total	Total Incl O&P
0010 **SWIMMING POOLS** Residential in-ground, vinyl lined									
0020 Concrete sides, w/equip, sand bottom	B-52	300	.187	SF Surf	26	8.50	2.14	36.64	44.50
0100 Metal or polystyrene sides R131113-20	B-14	410	.117		22	4.85	.89	27.74	32.50
0200 Add for vermiculite bottom				↓	1.67			1.67	1.84
0500 Gunite bottom and sides, white plaster finish									
0600 12' x 30' pool	B-52	145	.386	SF Surf	49	17.60	4.42	71.02	85.50
0720 16' x 32' pool		155	.361		44	16.45	4.13	64.58	78
0750 20' x 40' pool	↓	250	.224	↓	39.50	10.20	2.56	52.26	61.50
0810 Concrete bottom and sides, tile finish									
0820 12' x 30' pool	B-52	80	.700	SF Surf	49.50	32	8	89.50	111
0830 16' x 32' pool		95	.589		40.50	27	6.75	74.25	93.50
0840 20' x 40' pool	↓	130	.431	↓	32.50	19.60	4.93	57.03	71
1100 Motel, gunite with plaster finish, incl. medium									
1150 capacity filtration & chlorination	B-52	115	.487	SF Surf	60	22	5.55	87.55	106
1200 Municipal, gunite with plaster finish, incl. high									
1250 capacity filtration & chlorination	B-52	100	.560	SF Surf	77.50	25.50	6.40	109.40	132

For customer support on your Building Construction Costs with RSMeans Data, call 800.448.8182.

455

13 11 Swimming Pools

13 11 13 – Below-Grade Swimming Pools

13 11 13.50 Swimming Pools	Crew	Daily Output	Labor-Hours	Unit	Material	2017 Bare Costs Labor	Equipment	Total	Total Incl O&P
1350	Add for formed gutters				L.F.	114			114	126
1360	Add for stainless steel gutters				"	340			340	370
1600	For water heating system, see Section 23 52 28.10									
1700	Filtration and deck equipment only, as % of total				Total				20%	20%
1800	Deck equipment, rule of thumb, 20' x 40' pool				SF Pool				1.18	1.30
1900	5000 S.F. pool				"				1.73	1.90
3000	Painting pools, preparation + 3 coats, 20' x 40' pool, epoxy	2 Pord	.33	48.485	Total	1,825	2,025		3,850	5,050
3100	Rubber base paint, 18 gallons	"	.33	48.485		1,325	2,025		3,350	4,500
3500	42' x 82' pool, 75 gallons, epoxy paint	3 Pord	.14	171		7,700	7,150		14,850	19,300
3600	Rubber base paint	"	.14	171		5,500	7,150		12,650	16,800

13 11 23 – On-Grade Swimming Pools

13 11 23.50 Swimming Pools

		Crew	Daily Output	Labor-Hours	Unit	Material	2017 Bare Costs Labor	Equipment	Total	Total Incl O&P
0010	**SWIMMING POOLS** Residential above ground, steel construction									
0100	Round, 15' dia.	B-80A	3	8	Ea.	840	315	90	1,245	1,500
0120	18' dia.		2.50	9.600		910	375	108	1,393	1,700
0140	21' dia.		2	12		1,050	470	135	1,655	2,025
0160	24' dia.		1.80	13.333		1,150	520	150	1,820	2,225
0180	27' dia.		1.50	16		1,350	625	180	2,155	2,625
0200	30' dia.		1	24		1,500	940	270	2,710	3,400
0220	Oval, 12' x 24'		2.30	10.435		1,575	410	118	2,103	2,475
0240	15' x 30'		1.80	13.333		1,750	520	150	2,420	2,900
0260	18' x 33'		1	24		1,975	940	270	3,185	3,900

13 11 46 – Swimming Pool Accessories

13 11 46.50 Swimming Pool Equipment

		Crew	Daily Output	Labor-Hours	Unit	Material	2017 Bare Costs Labor	Equipment	Total	Total Incl O&P
0010	**SWIMMING POOL EQUIPMENT**									
0020	Diving stand, stainless steel, 3 meter	2 Carp	.40	40	Ea.	16,100	1,975		18,075	20,700
0300	1 meter		2.70	5.926		9,775	292		10,067	11,200
0600	Diving boards, 16' long, aluminum		2.70	5.926		4,350	292		4,642	5,250
0700	Fiberglass		2.70	5.926		3,325	292		3,617	4,125
0800	14' long, aluminum		2.70	5.926		3,850	292		4,142	4,700
0850	Fiberglass		2.70	5.926		3,325	292		3,617	4,100
1100	Bulkhead, movable, PVC, 8'-2" wide	2 Clab	8	2		2,475	78.50		2,553.50	2,850
1120	7'-9" wide		8	2		2,300	78.50		2,378.50	2,650
1140	7'-3" wide		8	2		2,050	78.50		2,128.50	2,400
1160	6'-9" wide		8	2		2,050	78.50		2,128.50	2,400
1200	Ladders, heavy duty, stainless steel, 2 tread	2 Carp	7	2.286		870	113		983	1,125
1500	4 tread		6	2.667		1,150	131		1,281	1,475
1800	Lifeguard chair, stainless steel, fixed		2.70	5.926		3,650	292		3,942	4,450
1900	Portable					2,825			2,825	3,100
2100	Lights, underwater, 12 volt, with transformer, 300 watt	1 Elec	1	8		350	455		805	1,075
2200	110 volt, 500 watt, standard		1	8		293	455		748	1,000
2400	Low water cutoff type		1	8		293	455		748	1,000
2800	Heaters, see Section 23 52 28.10									
3000	Pool covers, reinforced vinyl	3 Clab	1800	.013	S.F.	1.20	.52		1.72	2.12
3050	Automatic, electric								8.75	9.65
3100	Vinyl, for winter, 400 S.F. max pool surface	3 Clab	3200	.008		.26	.29		.55	.74
3200	With water tubes, 400 S.F. max pool surface	"	3000	.008		.33	.31		.64	.84
3250	Sealed air bubble polyethylene solar blanket, 16 mils					.33			.33	.36
3300	Slides, tubular, fiberglass, aluminum handrails & ladder, 5'-0", straight	2 Carp	1.60	10	Ea.	3,650	495		4,145	4,775
3320	8'-0", curved		3	5.333		7,225	263		7,488	8,350
3400	10'-0", curved		1	16		26,700	790		27,490	30,500
3420	12'-0", straight with platform		1.20	13.333		14,900	655		15,555	17,400

13 11 Swimming Pools

13 11 46 – Swimming Pool Accessories

13 11 46.50 Swimming Pool Equipment	Crew	Daily Output	Labor-Hours	Unit	Material	2017 Bare Costs Labor	Equipment	Total	Total Incl O&P
4500 Hydraulic lift, movable pool bottom, single ram									
4520 Under 1,000 S.F. area	L-9	72	.500	S.F.	160	22.50		182.50	211
4600 Four ram lift, over 1,000 S.F.	"	109	.330	"	130	14.70		144.70	166
5000 Removable access ramp, stainless steel	2 Clab	2	8	Ea.	5,925	315		6,240	7,000

13 12 Fountains

13 12 13 – Exterior Fountains

13 12 13.10 Outdoor Fountains

		Crew	Daily Output	Labor-Hours	Unit	Material	Labor	Equipment	Total	Total Incl O&P
0010	OUTDOOR FOUNTAINS									
0100	Outdoor fountain, 48" high with bowl and figures	2 Clab	2	8	Ea.	570	315		885	1,100
0200	Commercial, concrete or cast stone, 40-60" H, simple		2	8		775	315		1,090	1,325
0220	Average		2	8		1,350	315		1,665	1,950
0240	Ornate		2	8		2,625	315		2,940	3,350
0260	Metal, 72" high		2	8		1,225	315		1,540	1,825
0280	90" high		2	8		1,875	315		2,190	2,525
0300	120" high		2	8		4,950	315		5,265	5,925
0320	Resin or fiberglass, 40-60" H, wall type		2	8		480	315		795	1,000
0340	Waterfall type		2	8		1,200	315		1,515	1,800

13 12 23 – Interior Fountains

13 12 23.10 Indoor Fountains

		Crew	Daily Output	Labor-Hours	Unit	Material	Labor	Equipment	Total	Total Incl O&P
0010	INDOOR FOUNTAINS									
0100	Commercial, floor type, resin or fiberglass, lighted, cascade type	2 Clab	2	8	Ea.	296	315		611	805
0120	Tiered type		2	8		284	315		599	790
0140	Waterfall type		2	8		300	315		615	810

13 17 Tubs and Pools

13 17 33 – Whirlpool Tubs

13 17 33.10 Whirlpool Bath

		Crew	Daily Output	Labor-Hours	Unit	Material	Labor	Equipment	Total	Total Incl O&P
0010	WHIRLPOOL BATH									
6000	Whirlpool, bath with vented overflow, molded fiberglass									
6100	66" x 36" x 24"	Q-1	1	16	Ea.	3,525	890		4,415	5,225

13 18 Ice Rinks

13 18 13 – Ice Rink Floor Systems

13 18 13.50 Ice Skating

		Crew	Daily Output	Labor-Hours	Unit	Material	Labor	Equipment	Total	Total Incl O&P
0010	ICE SKATING Equipment incl. refrigeration, plumbing & cooling									
0020	coils & concrete slab, 85' x 200' rink									
0300	55° system, 5 mos., 100 ton				Total	604,000			604,000	664,000
0700	90° system, 12 mos., 135 ton				"	682,500			682,500	751,000
1200	Subsoil heating system (recycled from compressor), 85' x 200'	Q-7	.27	118	Ea.	42,000	7,025		49,025	57,000
1300	Subsoil insulation, 2 lb. polystyrene with vapor barrier, 85' x 200'	2 Carp	.14	114	"	31,500	5,625		37,125	43,300

13 18 16 – Ice Rink Dasher Boards

13 18 16.50 Ice Rink Dasher Boards

		Crew	Daily Output	Labor-Hours	Unit	Material	Labor	Equipment	Total	Total Incl O&P
0010	ICE RINK DASHER BOARDS									
1000	Dasher boards, 1/2" H.D. polyethylene faced steel frame, 3' acrylic									
1020	screen at sides, 5' acrylic ends, 85' x 200'	F-5	.06	533	Ea.	142,000	26,500		168,500	196,500

For customer support on your Building Construction Costs with RSMeans Data, call 800.448.8182.

457

13 18 Ice Rinks

13 18 16 – Ice Rink Dasher Boards

13 18 16.50 Ice Rink Dasher Boards	Crew	Daily Output	Labor-Hours	Unit	Material	2017 Bare Costs Labor	Equipment	Total	Total Incl O&P
1100	Fiberglass & aluminum construction, same sides and ends	F-5	.06	533	Ea.	163,000	26,500		189,500	219,500

13 21 Controlled Environment Rooms

13 21 13 – Clean Rooms

13 21 13.50 Clean Room Components

		Crew	Daily Output	Labor-Hours	Unit	Material	2017 Bare Costs Labor	Equipment	Total	Total Incl O&P
0010	**CLEAN ROOM COMPONENTS**									
1100	Clean room, soft wall, 12' x 12', Class 100	1 Carp	.18	44.444	Ea.	18,000	2,200		20,200	23,200
1110	Class 1,000		.18	44.444		16,000	2,200		18,200	21,100
1120	Class 10,000		.21	38.095		13,200	1,875		15,075	17,400
1130	Class 100,000		.21	38.095		12,100	1,875		13,975	16,200
2800	Ceiling grid support, slotted channel struts 4'-0" O.C., ea. way				S.F.				5.90	6.50
3000	Ceiling panel, vinyl coated foil on mineral substrate									
3020	Sealed, non-perforated				S.F.				1.27	1.40
4000	Ceiling panel seal, silicone sealant, 150 L.F./gal.	1 Carp	150	.053	L.F.	.35	2.63		2.98	4.41
4100	Two sided adhesive tape	"	240	.033	"	.12	1.64		1.76	2.64
4200	Clips, one per panel				Ea.	1.02			1.02	1.12
6000	HEPA filter, 2' x 4', 99.97% eff., 3" dp beveled frame (silicone seal)					405			405	445
6040	6" deep skirted frame (channel seal)					465			465	510
6100	99.99% efficient, 3" deep beveled frame (silicone seal)					415			415	455
6140	6" deep skirted frame (channel seal)					470			470	520
6200	99.999% efficient, 3" deep beveled frame (silicone seal)					470			470	515
6240	6" deep skirted frame (channel seal)					505			505	555
7000	Wall panel systems, including channel strut framing									
7020	Polyester coated aluminum, particle board				S.F.				18.20	18.20
7100	Porcelain coated aluminum, particle board								32	32
7400	Wall panel support, slotted channel struts, to 12' high								16.35	16.35

13 21 26 – Cold Storage Rooms

13 21 26.50 Refrigeration

		Crew	Daily Output	Labor-Hours	Unit	Material	2017 Bare Costs Labor	Equipment	Total	Total Incl O&P
0010	**REFRIGERATION**									
0020	Curbs, 12" high, 4" thick, concrete	2 Carp	58	.276	L.F.	5.30	13.60		18.90	27
1000	Doors, see Section 08 34 13.10									
2400	Finishes, 2 coat Portland cement plaster, 1/2" thick	1 Plas	48	.167	S.F.	1.44	7.65		9.09	13.10
2500	For galvanized reinforcing mesh, add	1 Lath	335	.024		.98	1.14		2.12	2.77
2700	3/16" thick latex cement	1 Plas	88	.091		2.47	4.17		6.64	8.95
2900	For glass cloth reinforced ceilings, add	"	450	.018		.62	.82		1.44	1.90
3100	Fiberglass panels, 1/8" thick	1 Carp	149.45	.054		3.30	2.64		5.94	7.65
3200	Polystyrene, plastic finish ceiling, 1" thick		274	.029		3.04	1.44		4.48	5.55
3400	2" thick		274	.029		3.45	1.44		4.89	6
3500	4" thick		219	.037		3.81	1.80		5.61	6.95
3800	Floors, concrete, 4" thick	1 Cefi	93	.086		1.53	4.03		5.56	7.70
3900	6" thick	"	85	.094		2.40	4.40		6.80	9.20
4000	Insulation, 1" to 6" thick, cork				B.F.	1.39			1.39	1.53
4100	Urethane					.52			.52	.57
4300	Polystyrene, regular					.57			.57	.63
4400	Bead board					.26			.26	.29
4600	Installation of above, add per layer	2 Carp	657.60	.024	S.F.	.46	1.20		1.66	2.35
4700	Wall and ceiling juncture		298.90	.054	L.F.	2.27	2.64		4.91	6.55
4900	Partitions, galvanized sandwich panels, 4" thick, stock		219.20	.073	S.F.	9.50	3.59		13.09	15.95
5000	Aluminum or fiberglass		219.20	.073	"	10.40	3.59		13.99	16.95
5200	Prefab walk-in, 7'-6" high, aluminum, incl. refrigeration, door & floor									

13 21 Controlled Environment Rooms

13 21 26 – Cold Storage Rooms

13 21 26.50 Refrigeration

		Crew	Daily Output	Labor-Hours	Unit	Material	2017 Bare Costs Labor	Equipment	Total	Total Incl O&P
5210	not incl. partitions, 6' x 6'	2 Carp	54.80	.292	SF Flr.	169	14.40		183.40	208
5500	10' x 10'		82.20	.195		136	9.60		145.60	165
5700	12' x 14'		109.60	.146		122	7.20		129.20	145
5800	12' x 20'		109.60	.146		107	7.20		114.20	128
6100	For 8'-6" high, add					5%				
6300	Rule of thumb for complete units, w/o doors & refrigeration, cooler	2 Carp	146	.110		154	5.40		159.40	177
6400	Freezer		109.60	.146		182	7.20		189.20	211
6600	Shelving, plated or galvanized, steel wire type		360	.044	SF Hor.	13.45	2.19		15.64	18.15
6700	Slat shelf type		375	.043		16.60	2.10		18.70	21.50
6900	For stainless steel shelving, add					300%				
7000	Vapor barrier, on wood walls	2 Carp	1644	.010	S.F.	.21	.48		.69	.96
7200	On masonry walls	"	1315	.012	"	.52	.60		1.12	1.49
7500	For air curtain doors, see Section 23 34 33.10									

13 21 48 – Sound-Conditioned Rooms

13 21 48.10 Anechoic Chambers

		Crew	Daily Output	Labor-Hours	Unit	Material	2017 Bare Costs Labor	Equipment	Total	Total Incl O&P
0010	**ANECHOIC CHAMBERS** Standard units, 7' ceiling heights									
0100	Area for pricing is net inside dimensions									
0300	200 cycles per second cutoff, 25 S.F. floor area				SF Flr.	1,650			1,650	1,825
0400	50 S.F.								1,050	1,150
0600	75 S.F.								1,000	1,100
0700	100 S.F.					1,250			1,250	1,375
0900	For 150 cycles per second cutoff, add to 100 S.F. room								30%	30%
1000	For 100 cycles per second cutoff, add to 100 S.F. room								45%	45%

13 21 48.15 Audiometric Rooms

		Crew	Daily Output	Labor-Hours	Unit	Material	2017 Bare Costs Labor	Equipment	Total	Total Incl O&P
0010	**AUDIOMETRIC ROOMS**									
0020	Under 500 S.F. surface	4 Carp	98	.327	SF Surf	53.50	16.10		69.60	83.50
0100	Over 500 S.F. surface	"	120	.267	"	52.50	13.15		65.65	77.50

13 21 53 – Darkrooms

13 21 53.50 Darkrooms

		Crew	Daily Output	Labor-Hours	Unit	Material	2017 Bare Costs Labor	Equipment	Total	Total Incl O&P
0010	**DARKROOMS**									
0020	Shell, complete except for door, 64 S.F., 8' high	2 Carp	128	.125	SF Flr.	53.50	6.15		59.65	68.50
0100	12' high		64	.250		70	12.30		82.30	96
0500	120 S.F. floor, 8' high		120	.133		39	6.55		45.55	53
0600	12' high		60	.267		53	13.15		66.15	78.50
0800	240 S.F. floor, 8' high		120	.133		28	6.55		34.55	41
0900	12' high		60	.267		39	13.15		52.15	63
1200	Mini-cylindrical, revolving, unlined, 4' diameter		3.50	4.571	Ea.	2,725	225		2,950	3,350
1400	5'-6" diameter		2.50	6.400		5,675	315		5,990	6,700
1600	Add for lead lining, inner cylinder, 1/32" thick					1,650			1,650	1,825
1700	1/16" thick					4,450			4,450	4,875
1800	Add for lead lining, inner and outer cylinder, 1/32" thick					3,050			3,050	3,375
1900	1/16" thick					6,825			6,825	7,500
2000	For darkroom door, see Section 08 34 36.10									

13 21 56 – Music Rooms

13 21 56.50 Music Rooms

		Crew	Daily Output	Labor-Hours	Unit	Material	2017 Bare Costs Labor	Equipment	Total	Total Incl O&P
0010	**MUSIC ROOMS**									
0020	Practice room, modular, perforated steel, under 500 S.F.	2 Carp	70	.229	SF Surf	34.50	11.25		45.75	55.50
0100	Over 500 S.F.	"	80	.200	"	29	9.85		38.85	47

For customer support on your Building Construction Costs with RSMeans Data, call 800.448.8182.

459

13 24 Special Activity Rooms

13 24 16 – Saunas

13 24 16.50 Saunas and Heaters		Crew	Daily Output	Labor-Hours	Unit	Material	2017 Bare Costs Labor	Equipment	Total	Total Incl O&P
0010	**SAUNAS AND HEATERS**									
0020	Prefabricated, incl. heater & controls, 7' high, 6' x 4', C/C	L-7	2.20	12.727	Ea.	5,350	605		5,955	6,825
0050	6' x 4', C/P		2	14		4,825	665		5,490	6,325
0400	6' x 5', C/C		2	14		6,025	665		6,690	7,650
0450	6' x 5', C/P		2	14		5,350	665		6,015	6,900
0600	6' x 6', C/C		1.80	15.556		6,325	735		7,060	8,075
0650	6' x 6', C/P		1.80	15.556		5,600	735		6,335	7,300
0800	6' x 9', C/C		1.60	17.500		8,200	830		9,030	10,300
0850	6' x 9', C/P		1.60	17.500		7,150	830		7,980	9,150
1000	8' x 12', C/C		1.10	25.455		10,900	1,200		12,100	13,900
1050	8' x 12', C/P		1.10	25.455		9,675	1,200		10,875	12,600
1200	8' x 8', C/C		1.40	20		8,600	950		9,550	10,900
1250	8' x 8', C/P		1.40	20		7,650	950		8,600	9,875
1400	8' x 10', C/C		1.20	23.333		9,400	1,100		10,500	12,000
1450	8' x 10', C/P		1.20	23.333		8,300	1,100		9,400	10,800
1600	10' x 12', C/C		1	28		12,600	1,325		13,925	15,900
1650	10' x 12', C/P	▼	1	28		11,400	1,325		12,725	14,500
1700	Door only, cedar, 2'x6', with 1' x 4' tempered insulated glass window	2 Carp	3.40	4.706		645	232		877	1,050
1800	Prehung, incl. jambs, pulls & hardware	"	12	1.333		705	65.50		770.50	875
2500	Heaters only (incl. above), wall mounted, to 200 C.F.					800			800	880
2750	To 300 C.F.					1,075			1,075	1,175
3000	Floor standing, to 720 C.F., 10,000 watts, w/controls	1 Elec	3	2.667		2,625	151		2,776	3,125
3250	To 1,000 C.F., 16,000 watts	"	3	2.667	▼	3,900	151		4,051	4,500

13 24 26 – Steam Baths

13 24 26.50 Steam Baths and Components		Crew	Daily Output	Labor-Hours	Unit	Material	2017 Bare Costs Labor	Equipment	Total	Total Incl O&P
0010	**STEAM BATHS AND COMPONENTS**									
0020	Heater, timer & head, single, to 140 C.F.	1 Plum	1.20	6.667	Ea.	2,325	410		2,735	3,175
0500	To 300 C.F.		1.10	7.273		2,600	450		3,050	3,525
1000	Commercial size, with blow-down assembly, to 800 C.F.		.90	8.889		5,725	550		6,275	7,125
1500	To 2500 C.F.	▼	.80	10		8,000	620		8,620	9,725
2000	Multiple, motels, apts., 2 baths, w/blow-down assm., 500 C.F.	Q-1	1.30	12.308		6,675	685		7,360	8,375
2500	4 baths	"	.70	22.857	▼	10,700	1,275		11,975	13,600
2700	Conversion unit for residential tub, including door					3,750			3,750	4,100

13 28 Athletic and Recreational Special Construction

13 28 33 – Athletic and Recreational Court Walls

13 28 33.50 Sport Court		Crew	Daily Output	Labor-Hours	Unit	Material	2017 Bare Costs Labor	Equipment	Total	Total Incl O&P
0010	**SPORT COURT**									
0020	Floors, No. 2 & better maple, 25/32" thick				SF Flr.				6.05	6.65
0100	Walls, laminated plastic bonded to galv. steel studs				SF Wall				7.70	8.45
0300	Squash, regulation court in existing building, minimum				Court	38,400			38,400	42,200
0400	Maximum				"	42,800			42,800	47,100
0450	Rule of thumb for components:									
0470	Walls	3 Carp	.15	160	Court	11,500	7,875		19,375	24,800
0500	Floor	"	.25	96		9,100	4,725		13,825	17,300
0550	Lighting	2 Elec	.60	26.667		2,200	1,500		3,700	4,675
0600	Handball, racquetball court in existing building, minimum	C-1	.20	160		41,600	7,475		49,075	57,000
0800	Maximum	"	.10	320		45,000	15,000		60,000	72,500
0900	Rule of thumb for components: walls	3 Carp	.12	200		13,200	9,850		23,050	29,600
1000	Floor	▼	.25	96		9,100	4,725		13,825	17,300

13 28 Athletic and Recreational Special Construction

13 28 33 – Athletic and Recreational Court Walls

13 28 33.50 Sport Court	Crew	Daily Output	Labor-Hours	Unit	Material	2017 Bare Costs Labor	Equipment	Total	Total Incl O&P
1100 Ceiling	3 Carp	.33	72.727	Court	4,400	3,575		7,975	10,300
1200 Lighting	2 Elec	.60	26.667	↓	2,300	1,500		3,800	4,775

13 31 Fabric Structures

13 31 13 – Air-Supported Fabric Structures

13 31 13.09 Air Supported Tank Covers

		Crew	Daily Output	Labor-Hours	Unit	Material	2017 Bare Costs Labor	Equipment	Total	Total Incl O&P
0010	**AIR SUPPORTED TANK COVERS**, vinyl polyester									
0100	Scrim, double layer, with hardware, blower, standby & controls									
0200	Round, 75' diameter	B-2	4500	.009	S.F.	11.95	.35		12.30	13.65
0300	100' diameter		5000	.008		10.85	.32		11.17	12.45
0400	150' diameter		5000	.008		8.60	.32		8.92	9.95
0500	Rectangular, 20' x 20'		4500	.009		23.50	.35		23.85	26.50
0600	30' x 40'		4500	.009		23.50	.35		23.85	26.50
0700	50' x 60'	↓	4500	.009		23.50	.35		23.85	26.50
0800	For single wall construction, deduct, minimum					.80			.80	.88
0900	Maximum					2.37			2.37	2.61
1000	For maximum resistance to atmosphere or cold, add				↓	1.15			1.15	1.27
1100	For average shipping charges, add				Total	1,975			1,975	2,175

13 31 13.13 Single-Walled Air-Supported Structures

		Crew	Daily Output	Labor-Hours	Unit	Material	2017 Bare Costs Labor	Equipment	Total	Total Incl O&P
0010	**SINGLE-WALLED AIR-SUPPORTED STRUCTURES** R133113-10									
0020	Site preparation, incl. anchor placement and utilities	B-11B	1000	.016	SF Flr.	1.16	.72	.27	2.15	2.68
0030	For concrete, see Section 03 30 53.40									
0050	Warehouse, polyester/vinyl fabric, 28 oz., over 10 yr. life, welded									
0060	Seams, tension cables, primary & auxiliary inflation system,									
0070	airlock, personnel doors and liner									
0100	5000 S.F.	4 Clab	5000	.006	SF Flr.	26.50	.25		26.75	30
0250	12,000 S.F.	"	6000	.005		19.15	.21		19.36	21.50
0400	24,000 S.F.	8 Clab	12000	.005		13.50	.21		13.71	15.15
0500	50,000 S.F.	"	12500	.005	↓	12.50	.20		12.70	14.05
0700	12 oz. reinforced vinyl fabric, 5 yr. life, sewn seams,									
0710	accordion door, including liner									
0750	3000 S.F.	4 Clab	3000	.011	SF Flr.	13.20	.42		13.62	15.20
0800	12,000 S.F.	"	6000	.005		11.25	.21		11.46	12.65
0850	24,000 S.F.	8 Clab	12000	.005		9.50	.21		9.71	10.75
0950	Deduct for single layer					1.04			1.04	1.14
1000	Add for welded seams					1.50			1.50	1.65
1050	Add for double layer, welded seams included				↓	3			3	3.30
1250	Tedlar/vinyl fabric, 28 oz., with liner, over 10 yr. life,									
1260	incl. overhead and personnel doors									
1300	3000 S.F.	4 Clab	3000	.011	SF Flr.	25	.42		25.42	27.50
1450	12,000 S.F.	"	6000	.005		17.45	.21		17.66	19.50
1550	24,000 S.F.	8 Clab	12000	.005		13.50	.21		13.71	15.15
1700	Deduct for single layer				↓	2			2	2.20
2250	Greenhouse/shelter, woven polyethylene with liner, 2 yr. life,									
2260	sewn seams, including doors									
2300	3000 S.F.	4 Clab	3000	.011	SF Flr.	16	.42		16.42	18.25
2350	12,000 S.F.	"	6000	.005		14	.21		14.21	15.70
2450	24,000 S.F.	8 Clab	12000	.005		12	.21		12.21	13.50
2550	Deduct for single layer				↓	.98			.98	1.08
2600	Tennis/gymnasium, polyester/vinyl fabric, 28 oz., over 10 yr. life,									
2610	including thermal liner, heat and lights									

13 31 Fabric Structures

13 31 13 – Air-Supported Fabric Structures

13 31 13.13 Single-Walled Air-Supported Structures	Crew	Daily Output	Labor-Hours	Unit	Material	2017 Bare Costs Labor	Equipment	Total	Total Incl O&P
2650	7200 S.F.	4 Clab	6000	.005	SF Flr.	24	.21		24.21	26.50
2750	13,000 S.F.	"	6500	.005		18.25	.19		18.44	20.50
2850	Over 24,000 S.F.	8 Clab	12000	.005		16.70	.21		16.91	18.65
2860	For low temperature conditions, add					1.15			1.15	1.27
2870	For average shipping charges, add				Total	5,600			5,600	6,150
2900	Thermal liner, translucent reinforced vinyl				SF Flr.	1.15			1.15	1.27
2950	Metalized mylar fabric and mesh, double liner				"	2.37			2.37	2.61
3050	Stadium/convention center, teflon coated fiberglass, heavy weight,									
3060	over 20 yr. life, incl. thermal liner and heating system									
3100	Minimum	9 Clab	26000	.003	SF Flr.	58	.11		58.11	64
3110	Maximum	"	19000	.004	"	69	.15		69.15	76
3400	Doors, air lock, 15' long, 10' x 10'	2 Carp	.80	20	Ea.	20,700	985		21,685	24,300
3600	15' x 15'	"	.50	32		30,800	1,575		32,375	36,300
3700	For each added 5' length, add					5,500			5,500	6,050
3900	Revolving personnel door, 6' diameter, 6'-6" high	2 Carp	.80	20	↓	15,400	985		16,385	18,400

13 31 23 – Tensioned Fabric Structures

13 31 23.50 Tension Structures

		Crew	Daily Output	Labor-Hours	Unit	Material	2017 Bare Costs Labor	Equipment	Total	Total Incl O&P
0010	**TENSION STRUCTURES** Rigid steel/alum. frame, vinyl coated poly									
0100	Fabric shell, 60' clear span, not incl. foundations or floors									
0200	6,000 S.F.	B-41	1000	.044	SF Flr.	14.10	1.79	.33	16.22	18.60
0300	12,000 S.F.		1100	.040		13.45	1.63	.30	15.38	17.60
0400	80' to 99' clear span, 20,800 S.F.	↓	1220	.036		13.30	1.47	.27	15.04	17.20
0410	100' to 119' clear span, 10,000 S.F.	L-5	2175	.026		14.25	1.41	.31	15.97	18.30
0430	26,000 S.F.		2300	.024		13.25	1.33	.29	14.87	17.10
0450	36,000 S.F.		2500	.022		13.10	1.23	.27	14.60	16.70
0460	120' to 149' clear span, 24,000 S.F.		3000	.019		14.50	1.02	.22	15.74	17.90
0470	150' to 199' clear span, 30,000 S.F.	↓	6000	.009		15.05	.51	.11	15.67	17.55
0480	200' clear span, 40,000 S.F.	E-6	8000	.016	↓	18.50	.87	.25	19.62	22
0500	For roll-up door, 12' x 14', add	L-2	1	16	Ea.	5,600	690		6,290	7,225

13 34 Fabricated Engineered Structures

13 34 13 – Glazed Structures

13 34 13.13 Greenhouses

		Crew	Daily Output	Labor-Hours	Unit	Material	2017 Bare Costs Labor	Equipment	Total	Total Incl O&P
0010	**GREENHOUSES**, Shell only, stock units, not incl. 2' stub walls,									
0020	foundation, floors, heat or compartments									
0300	Residential type, free standing, 8'-6" long x 7'-6" wide	2 Carp	59	.271	SF Flr.	25	13.35		38.35	48
0400	10'-6" wide		85	.188		39.50	9.25		48.75	57.50
0600	13'-6" wide		108	.148		46	7.30		53.30	61.50
0700	17'-0" wide		160	.100		46	4.93		50.93	58
0900	Lean-to type, 3'-10" wide		34	.471		44	23		67	83.50
1000	6'-10" wide	↓	58	.276	↓	29	13.60		42.60	53
1500	Commercial, custom, truss frame, incl. equip., plumbing, elec.,									
1550	benches and controls, under 2,000 S.F.				SF Flr.	13.30			13.30	14.65
1700	Over 5,000 S.F.				"	12.20			12.20	13.40
2000	Institutional, custom, rigid frame, including compartments and									
2050	multi-controls, under 500 S.F.				SF Flr.	30			30	33
2150	Over 2,000 S.F.				"	11.10			11.10	12.25
3700	For 1/4" tempered glass, add				SF Surf	1.57			1.57	1.73
3900	Cooling, 1200 CFM exhaust fan, add				Ea.	330			330	360
4000	7850 CFM				↓	1,050			1,050	1,150

13 34 Fabricated Engineered Structures

13 34 13 – Glazed Structures

13 34 13.13 Greenhouses

		Crew	Daily Output	Labor-Hours	Unit	Material	2017 Bare Costs Labor	Equipment	Total	Total Incl O&P
4200	For heaters, 10 MBH, add				Ea.	235			235	259
4300	60 MBH, add					780			780	855
4500	For benches, 2' x 8', add					170			170	187
4600	4' x 10', add				↓	207			207	228
4800	For ventilation & humidity control w/4 integrated outlets, add				Total	255			255	280
4900	For environmental controls and automation, 8 outputs, 9 stages, add				"	1,050			1,050	1,150
5100	For humidification equipment, add				Ea.	315			315	345
5200	For vinyl shading, add				S.F.	.45			.45	.50
6000	Geodesic hemisphere, 1/8" plexiglass glazing									
6050	8' diameter	2 Carp	2	8	Ea.	6,450	395		6,845	7,700
6150	24' diameter		.35	45.714		15,500	2,250		17,750	20,500
6250	48' diameter	↓	.20	80	↓	35,000	3,950		38,950	44,400

13 34 13.19 Swimming Pool Enclosures

		Crew	Daily Output	Labor-Hours	Unit	Material	2017 Bare Costs Labor	Equipment	Total	Total Incl O&P
0010	**SWIMMING POOL ENCLOSURES** Translucent, free standing									
0020	not including foundations, heat or light									
0200	Economy	2 Carp	200	.080	SF Hor.	33.50	3.94		37.44	43
0600	Deluxe	"	70	.229		93	11.25		104.25	119
0700	For motorized roof, 40% opening, solid roof, add					21			21	23
0800	Skylight type roof, add				↓	13.50			13.50	14.85

13 34 16 – Grandstands and Bleachers

13 34 16.13 Grandstands

		Crew	Daily Output	Labor-Hours	Unit	Material	2017 Bare Costs Labor	Equipment	Total	Total Incl O&P
0010	**GRANDSTANDS** Permanent, municipal, including foundation									
0300	Steel, economy				Seat	22.50			22.50	25
0400	Steel, deluxe					25.50			25.50	28
0900	Composite, steel, wood and plastic, stock design, economy					42			42	46
1000	Deluxe				↓	76			76	83.50

13 34 16.53 Bleachers

		Crew	Daily Output	Labor-Hours	Unit	Material	2017 Bare Costs Labor	Equipment	Total	Total Incl O&P
0010	**BLEACHERS**									
0020	Bleachers, outdoor, portable, 5 tiers, 42 seats	2 Sswk	120	.133	Seat	73.50	7.25		80.75	93
0100	5 tiers, 54 seats		80	.200		71.50	10.85		82.35	96.50
0200	10 tiers, 104 seats		120	.133		81.50	7.25		88.75	102
0300	10 tiers, 144 seats	↓	80	.200	↓	71.50	10.85		82.35	96.50
0500	Permanent bleachers, aluminum seat, steel frame, 24" row									
0600	8 tiers, 80 seats	2 Sswk	60	.267	Seat	67.50	14.50		82	98.50
0700	8 tiers, 160 seats		48	.333		65.50	18.10		83.60	102
0925	15 tiers, 154 to 165 seats		60	.267		77.50	14.50		92	109
0975	15 tiers, 214 to 225 seats		60	.267		67.50	14.50		82	98.50
1050	15 tiers, 274 to 285 seats		60	.267		62.50	14.50		77	92.50
1200	Seat backs only, 30" row, fiberglass		160	.100		23	5.45		28.45	34.50
1300	Steel and wood	↓	160	.100	↓	23	5.45		28.45	34.50
1400	NOTE: average seating is 1.5' in width									

13 34 19 – Metal Building Systems

13 34 19.50 Pre-Engineered Steel Buildings

		Crew	Daily Output	Labor-Hours	Unit	Material	2017 Bare Costs Labor	Equipment	Total	Total Incl O&P
0010	**PRE-ENGINEERED STEEL BUILDINGS** R133419-10									
0100	Clear span rigid frame, 26 ga. colored roofing and siding									
0150	20' to 29' wide, 10' eave height	E-2	425	.132	SF Flr.	9.25	7.10	3.93	20.28	26
0160	14' eave height		350	.160		10	8.60	4.77	23.37	30.50
0170	16' eave height		320	.175		10.75	9.45	5.20	25.40	33
0180	20' eave height		275	.204		11.70	10.95	6.10	28.75	37.50
0190	24' eave height		240	.233		12.95	12.55	6.95	32.45	42.50
0200	30' to 49' wide, 10' eave height	↓	535	.105	↓	7.05	5.65	3.12	15.82	20.50

13 34 19.50 Pre-Engineered Steel Buildings		Crew	Daily Output	Labor-Hours	Unit	Material	2017 Bare Costs Labor	Equipment	Total	Total Incl O&P
0300	14' eave height	E-2	450	.124	SF Flr.	7.65	6.70	3.71	18.06	23.50
0400	16' eave height		415	.135		8.15	7.25	4.03	19.43	25
0500	20' eave height		360	.156		8.85	8.40	4.64	21.89	28.50
0600	24' eave height		320	.175		9.70	9.45	5.20	24.35	32
0700	50' to 100' wide, 10' eave height		770	.073		6	3.92	2.17	12.09	15.35
0900	16' eave height		600	.093		6.90	5.05	2.79	14.74	18.80
1000	20' eave height		490	.114		7.50	6.15	3.41	17.06	22
1100	24' eave height		435	.129		8.25	6.95	3.84	19.04	24.50
1200	Clear span tapered beam frame, 26 ga. colored roofing/siding									
1300	30' to 39' wide, 10' eave height	E-2	535	.105	SF Flr.	8	5.65	3.12	16.77	21.50
1400	14' eave height		450	.124		8.85	6.70	3.71	19.26	24.50
1500	16' eave height		415	.135		9.30	7.25	4.03	20.58	26.50
1600	20' eave height		360	.156		10.25	8.40	4.64	23.29	30
1700	40' wide, 10' eave height		600	.093		7.15	5.05	2.79	14.99	19.05
1800	14' eave height		510	.110		7.90	5.90	3.28	17.08	22
1900	16' eave height		475	.118		8.25	6.35	3.52	18.12	23
2000	20' eave height		415	.135		9.05	7.25	4.03	20.33	26
2100	50' to 79' wide, 10' eave height		770	.073		6.65	3.92	2.17	12.74	16.05
2200	14' eave height		675	.083		7.20	4.47	2.48	14.15	17.90
2300	16' eave height		635	.088		7.50	4.75	2.63	14.88	18.85
2400	20' eave height		490	.114		8.75	6.15	3.41	18.31	23.50
2410	80' to 100' wide, 10' eave height		935	.060		5.90	3.23	1.79	10.92	13.70
2420	14' eave height		750	.075		6.50	4.02	2.23	12.75	16.15
2430	16' eave height		685	.082		6.80	4.41	2.44	13.65	17.35
2440	20' eave height		560	.100		7.25	5.40	2.98	15.63	20
2460	101' to 120' wide, 10' eave height		950	.059		5.40	3.18	1.76	10.34	13.05
2470	14' eave height		770	.073		6	3.92	2.17	12.09	15.35
2480	16' eave height		675	.083		6.40	4.47	2.48	13.35	17
2490	20' eave height		560	.100		6.85	5.40	2.98	15.23	19.60
2500	Single post 2-span frame, 26 ga. colored roofing and siding									
2600	80' wide, 14' eave height	E-2	740	.076	SF Flr.	6	4.08	2.26	12.34	15.70
2700	16' eave height		695	.081		6.40	4.34	2.40	13.14	16.75
2800	20' eave height		625	.090		6.90	4.83	2.67	14.40	18.40
2900	24' eave height		570	.098		7.55	5.30	2.93	15.78	20
3000	100' wide, 14' eave height		835	.067		5.85	3.61	2	11.46	14.45
3100	16' eave height		795	.070		5.40	3.80	2.10	11.30	14.40
3200	20' eave height		730	.077		6.60	4.13	2.29	13.02	16.45
3300	24' eave height		670	.084		7.30	4.50	2.49	14.29	18.10
3400	120' wide, 14' eave height		870	.064		6.80	3.47	1.92	12.19	15.25
3500	16' eave height		830	.067		6.05	3.64	2.01	11.70	14.75
3600	20' eave height		765	.073		6.60	3.94	2.18	12.72	16.05
3700	24' eave height		705	.079		7.20	4.28	2.37	13.85	17.50
3800	Double post 3-span frame, 26 ga. colored roofing and siding									
3900	150' wide, 14' eave height	E-2	925	.061	SF Flr.	4.77	3.26	1.81	9.84	12.55
4000	16' eave height		890	.063		4.97	3.39	1.88	10.24	13
4100	20' eave height		820	.068		5.40	3.68	2.04	11.12	14.20
4200	24' eave height		765	.073		6	3.94	2.18	12.12	15.40
4300	Triple post 4-span frame, 26 ga. colored roofing and siding									
4400	160' wide, 14' eave height	E-2	970	.058	SF Flr.	4.64	3.11	1.72	9.47	12.05
4500	16' eave height		930	.060		4.88	3.25	1.80	9.93	12.60
4600	20' eave height		870	.064		4.80	3.47	1.92	10.19	13.05
4700	24' eave height		815	.069		5.45	3.70	2.05	11.20	14.25
4800	200' wide, 14' eave height		1030	.054		4.29	2.93	1.62	8.84	11.25

13 34 19.50 Pre-Engineered Steel Buildings	Crew	Daily Output	Labor-Hours	Unit	Material	2017 Bare Costs Labor	Equipment	Total	Total Incl O&P
4900	16' eave height	E-2	995	.056	SF Flr.	4.44	3.03	1.68	9.15	11.65
5000	20' eave height		935	.060		4.89	3.23	1.79	9.91	12.60
5100	24' eave height		885	.063		5.50	3.41	1.89	10.80	13.70
5200	Accessory items: add to the basic building cost above									
5250	Eave overhang, 2' wide, 26 ga., with soffit	E-2	360	.156	L.F.	32.50	8.40	4.64	45.54	54.50
5300	4' wide, without soffit		300	.187		28.50	10.05	5.55	44.10	54
5350	With soffit		250	.224		41.50	12.05	6.70	60.25	73
5400	6' wide, without soffit		250	.224		37	12.05	6.70	55.75	68
5450	With soffit		200	.280		49.50	15.10	8.35	72.95	88
5500	Entrance canopy, incl. frame, 4' x 4'		25	2.240	Ea.	510	121	67	698	830
5550	4' x 8'		19	2.947	"	595	159	88	842	1,000
5600	End wall roof overhang, 4' wide, without soffit		850	.066	L.F.	18.20	3.55	1.97	23.72	28
5650	With soffit		500	.112	"	29.50	6.05	3.34	38.89	46
5700	Doors, HM self-framing, incl. butts, lockset and trim									
5750	Single leaf, 3070 (3' x 7'), economy	2 Sswk	5	3.200	Opng.	615	174		789	970
5800	Deluxe		4	4		660	217		877	1,075
5825	Glazed		4	4		750	217		967	1,175
5850	3670 (3'-6" x 7')		4	4		820	217		1,037	1,250
5900	4070 (4' x 7')		3	5.333		885	290		1,175	1,450
5950	Double leaf, 6070 (6' x 7')		2	8		1,125	435		1,560	1,975
6000	Glazed		2	8		1,400	435		1,835	2,275
6050	Framing only, for openings, 3' x 7'		4	4		183	217		400	560
6100	10' x 10'		3	5.333		600	290		890	1,150
6150	For windows below, 2020 (2' x 2')		6	2.667		193	145		338	455
6200	4030 (4' x 3')		5	3.200		236	174		410	550
6250	Flashings, 26 ga., corner or eave, painted		240	.067	L.F.	4.69	3.62		8.31	11.20
6300	Galvanized		240	.067		4.43	3.62		8.05	10.90
6350	Rake flashing, painted		240	.067		5.05	3.62		8.67	11.65
6400	Galvanized		240	.067		4.74	3.62		8.36	11.25
6450	Ridge flashing, 18" wide, painted		240	.067		6.75	3.62		10.37	13.50
6500	Galvanized		240	.067		7.65	3.62		11.27	14.45
6550	Gutter, eave type, 26 ga., painted		320	.050		7.30	2.72		10.02	12.60
6650	Valley type, between buildings, painted		120	.133		13.85	7.25		21.10	27.50
6710	Insulation, rated .6 lb. density, unfaced 4" thick, R13	2 Carp	2300	.007	S.F.	.46	.34		.80	1.04
6720	6" thick, R19		2300	.007		.62	.34		.96	1.21
6730	10" thick, R30		2300	.007		1.18	.34		1.52	1.83
6750	Insulation, rated .6 lb. density, poly/scrim/foil (PSF) faced									
6760	4" thick, R13	2 Carp	2300	.007	S.F.	.69	.34		1.03	1.29
6770	6" thick, R19		2300	.007		.97	.34		1.31	1.60
6780	9-1/2" thick, R30		2300	.007		1.07	.34		1.41	1.71
6800	Insulation, rated .6 lb. density, vinyl faced 1-1/2" thick, R5		2300	.007		.37	.34		.71	.94
6850	3" thick, R10		2300	.007		.36	.34		.70	.93
6900	4" thick, R13		2300	.007		.47	.34		.81	1.05
6920	6" thick, R19		2300	.007		.61	.34		.95	1.20
6930	10" thick, R30		2300	.007		1.57	.34		1.91	2.26
6950	Foil/scrim/kraft (FSK) faced, 1-1/2" thick, R5		2300	.007		.40	.34		.74	.97
7000	2" thick, R6		2300	.007		.52	.34		.86	1.10
7050	3" thick, R10		2300	.007		.48	.34		.82	1.06
7100	4" thick, R13		2300	.007		.50	.34		.84	1.08
7110	6" thick, R19		2300	.007		.72	.34		1.06	1.32
7120	10" thick, R30		2300	.007		1	.34		1.34	1.63
7150	Metalized polyester/scrim/kraft (PSK) facing, 1-1/2" thk, R5		2300	.007		.65	.34		.99	1.25
7200	2" thick, R6		2300	.007		.77	.34		1.11	1.38

13 34 19.50 Pre-Engineered Steel Buildings	Crew	Daily Output	Labor-Hours	Unit	Material	2017 Bare Costs Labor	Equipment	Total	Total Incl O&P
7250	3" thick, R11	2 Carp	2300	.007	S.F.	.87	.34		1.21	1.49
7300	4" thick, R13		2300	.007		.99	.34		1.33	1.62
7310	6" thick, R19		2300	.007		1.16	.34		1.50	1.81
7320	10" thick, R30		2300	.007		1.29	.34		1.63	1.95
7350	Vinyl/scrim/foil (VSF), 1-1/2" thick, R5		2300	.007		.57	.34		.91	1.16
7400	2" thick, R6		2300	.007		.73	.34		1.07	1.33
7450	3" thick, R10		2300	.007		.79	.34		1.13	1.40
7500	4" thick, R13		2300	.007		.97	.34		1.31	1.60
7510	Vinyl/scrim/vinyl (VSV), 4" thick, R13		2300	.007		.57	.34		.91	1.16
7520	6" thick, R19		2300	.007		.72	.34		1.06	1.32
7530	9-1/2" thick, R30		2300	.007		.98	.34		1.32	1.61
7540	Polyprop/scrim/polyester (PSP), 4" thick, R13		2300	.007		.64	.34		.98	1.23
7550	6" thick, R19		2300	.007		.79	.34		1.13	1.40
7555	10" thick, R30		2300	.007		1.40	.34		1.74	2.07
7560	Polyprop/scrim/kraft/polyester (PSKP), 4" thick, R13		2300	.007		.59	.34		.93	1.18
7570	6" thick, R19		2300	.007		.81	.34		1.15	1.42
7580	10" thick, R19		2300	.007		1.55	.34		1.89	2.24
7585	Vinyl/scrim/polyester (VSP), 4" thick, R13		2300	.007		.60	.34		.94	1.19
7590	6" thick, R19		2300	.007		.74	.34		1.08	1.34
7600	10" thick, R30		2300	.007		1.16	.34		1.50	1.81
7635	Insulation installation, over the purlin, second layer, up to 4" thick, add						90%			
7640	Insulation installation, between the purlins, up to 4" thick, add						100%			
7650	Sash, single slide, glazed, with screens, 2020 (2' x 2')	E-1	22	1.091	Opng.	139	59	5.90	203.90	255
7700	3030 (3' x 3')		14	1.714		310	92.50	9.25	411.75	505
7750	4030 (4' x 3')		13	1.846		415	99.50	10	524.50	630
7800	6040 (6' x 4')		12	2		830	108	10.80	948.80	1,100
7850	Double slide sash, 3030 (3' x 3')		14	1.714		227	92.50	9.25	328.75	410
7900	6040 (6' x 4')		12	2		605	108	10.80	723.80	850
7950	Fixed glass, no screens, 3030 (3' x 3')		14	1.714		223	92.50	9.25	324.75	405
8000	6040 (6' x 4')		12	2		595	108	10.80	713.80	840
8050	Prefinished storm sash, 3030 (3' x 3')		70	.343		85	18.45	1.85	105.30	126
8100	Siding and roofing, see Sections 07 41 13 & 07 42 13									
8200	Skylight, fiberglass panels, to 30 S.F.	E-1	10	2.400	Ea.	130	129	13	272	365
8250	Larger sizes, add for excess over 30 S.F.	"	300	.080	S.F.	4.32	4.31	.43	9.06	12.20
8300	Roof vents, turbine ventilator, wind driven									
8350	No damper, includes base, galvanized									
8400	12" diameter	Q-9	10	1.600	Ea.	67	83.50		150.50	202
8450	20" diameter		8	2		189	105		294	370
8500	24" diameter		8	2		299	105		404	490
8600	Continuous, 26 ga., 10' long, 9" wide	2 Sswk	4	4		36.50	217		253.50	400
8650	12" wide	"	4	4		36.50	217		253.50	400

13 34 23 – Fabricated Structures

13 34 23.10 Comfort Stations

		Crew	Daily Output	Labor-Hours	Unit	Material	2017 Bare Costs Labor	Equipment	Total	Total Incl O&P
0010	**COMFORT STATIONS** Prefab., stock, w/doors, windows & fixt.									
0100	Not incl. interior finish or electrical									
0300	Mobile, on steel frame, 2 unit				S.F.	182			182	201
0350	7 unit					305			305	335
0400	Permanent, including concrete slab, 2 unit	B-12J	50	.320		249	15.20	18.20	282.40	315
0500	6 unit	"	43	.372		191	17.65	21	229.65	262
0600	Alternate pricing method, mobile, 2 fixture				Fixture	6,325			6,325	6,950
0650	7 fixture					10,900			10,900	12,000
0700	Permanent, 2 unit	B-12J	.70	22.857		20,500	1,075	1,300	22,875	25,700

13 34 Fabricated Engineered Structures

13 34 23 – Fabricated Structures

13 34 23.10 Comfort Stations	Crew	Daily Output	Labor-Hours	Unit	Material	2017 Bare Costs Labor	Equipment	Total	Total Incl O&P
0750　　　　6 unit	B-12J	.50	32	Fixture	17,400	1,525	1,825	20,750	23,500

13 34 23.15 Domes

	Crew	Daily Output	Labor-Hours	Unit	Material	Labor	Equipment	Total	Total Incl O&P
0010 **DOMES**									
0020　　Domes, rev. alum., elec. drive, for astronomy obsv. shell only, stock units									
0600　　　　10'-6" diameter	2 Carp	.25	64	Ea.	50,500	3,150		53,650	61,000
0900　　　　18'-6" diameter		.17	94.118		89,500	4,625		94,125	105,000
1200　　　　24'-6" diameter	↓	.08	200	↓	118,500	9,850		128,350	145,000
1500　　Domes, bulk storage, shell only, dual radius hemisphere, arch, steel									
1600　　　　framing, corrugated steel covering, 150' diameter	E-2	550	.102	SF Flr.	34	5.50	3.04	42.54	49.50
1700　　　　　　400' diameter	"	720	.078		27.50	4.19	2.32	34.01	40
1800　　　　Wood framing, wood decking, to 400' diameter	F-4	400	.120	↓	37	6	2.69	45.69	52.50
1900　　Radial framed wood (2" x 6"), 1/2" thick									
2000　　　　plywood, asphalt shingles, 50' diameter	F-3	2000	.020	SF Flr.	68.50	1.01	.29	69.80	77.50
2100　　　　　　60' diameter		1900	.021		59	1.06	.30	60.36	66.50
2200　　　　　　72' diameter		1800	.022		49	1.12	.32	50.44	56
2300　　　　　　116' diameter		1730	.023		33.50	1.17	.33	35	38.50
2400　　　　　　150' diameter	↓	1500	.027	↓	36	1.35	.38	37.73	42

13 34 23.16 Fabricated Control Booths

	Crew	Daily Output	Labor-Hours	Unit	Material	Labor	Equipment	Total	Total Incl O&P
0010 **FABRICATED CONTROL BOOTHS**									
0100　　Guard House, prefab conc. w/bullet resistant doors & windows, roof & wiring									
0110　　　　8' x 8', Level III	L-10	1	24	Ea.	51,000	1,325	575	52,900	59,000
0120　　　　8' x 8', Level IV	"	1	24	"	58,000	1,325	575	59,900	67,000

13 34 23.25 Garage Costs

	Crew	Daily Output	Labor-Hours	Unit	Material	Labor	Equipment	Total	Total Incl O&P
0010 **GARAGE COSTS**									
0020　　Public parking, average				Car				18,400	20,200
0300　　Residential, wood, 12' x 20', one car prefab shell, stock, economy	2 Carp	1	16	Total	5,900	790		6,690	7,700
0350　　　　Custom		.67	23.881		6,825	1,175		8,000	9,325
0400　　　　Two car, 24' x 20', economy		.67	23.881		11,000	1,175		12,175	13,900
0450　　　　　　Custom	↓	.50	32	↓	13,300	1,575		14,875	17,100

13 34 23.30 Garden House

	Crew	Daily Output	Labor-Hours	Unit	Material	Labor	Equipment	Total	Total Incl O&P
0010 **GARDEN HOUSE** Prefab wood, no floors or foundations									
0100　　6' x 6'	2 Carp	200	.080	SF Flr.	52	3.94		55.94	63
0300　　8' x 12'	"	48	.333	"	39.50	16.40		55.90	68.50

13 34 23.35 Geodesic Domes

	Crew	Daily Output	Labor-Hours	Unit	Material	Labor	Equipment	Total	Total Incl O&P
0010 **GEODESIC DOMES** Shell only, interlocking plywood panels　　R133423-30									
0400　　　　30' diameter	F-5	1.60	20	Ea.	24,400	995		25,395	28,300
0500　　　　33' diameter		1.14	28.070		25,800	1,400		27,200	30,500
0600　　　　40' diameter	↓	1	32		29,900	1,600		31,500	35,400
0700　　　　45' diameter	F-3	1.13	35.556		31,400	1,800	510	33,710	37,800
0750　　　　56' diameter		1	40		55,000	2,025	575	57,600	64,000
0800　　　　60' diameter		1	40		62,500	2,025	575	65,100	72,000
0850　　　　67' diameter	↓	.80	50	↓	87,500	2,525	715	90,740	101,000
1100　　Aluminum panel, with 6" insulation									
1200　　　　100' diameter				SF Flr.	26			26	28.50
1300　　　　500' diameter				"	25.50			25.50	28
1600　　Aluminum framed, plexiglass closure panels									
1700　　　　40' diameter				SF Flr.	64			64	70.50
1800　　　　200' diameter				"	59			59	65
2100　　Aluminum framed, aluminum closure panels									
2200　　　　40' diameter				SF Flr.	21			21	23
2300　　　　100' diameter				↓	20			20	22

13 34 23 – Fabricated Structures

13 34 23.35 Geodesic Domes		Crew	Daily Output	Labor-Hours	Unit	Material	2017 Bare Costs Labor	Equipment	Total	Total Incl O&P
2400	200' diameter				SF Flr.	20			20	22
2500	For VRP faced bonded fiberglass insulation, add								10	10
2700	Aluminum framed, fiberglass sandwich panel closure									
2800	6' diameter	2 Carp	150	.107	SF Flr.	28	5.25		33.25	39
2900	28' diameter	"	350	.046	"	25.50	2.25		27.75	31.50

13 34 23.45 Kiosks

		Crew	Daily Output	Labor-Hours	Unit	Material	Labor	Equipment	Total	Total Incl O&P
0010	**KIOSKS**									
0020	Round, advertising type, 5' diameter, 7' high, aluminum wall, illuminated				Ea.	25,600			25,600	28,200
0100	Aluminum wall, non-illuminated					24,500			24,500	27,000
0500	Rectangular, 5' x 9', 7'-6" high, aluminum wall, illuminated					27,800			27,800	30,600
0600	Aluminum wall, non-illuminated				▼	26,200			26,200	28,800

13 34 23.60 Portable Booths

		Crew	Daily Output	Labor-Hours	Unit	Material	Labor	Equipment	Total	Total Incl O&P
0010	**PORTABLE BOOTHS** Prefab. aluminum with doors, windows, ext. roof									
0100	lights wiring & insulation, 15 S.F. building, O.D., painted				S.F.	284			284	310
0300	30 S.F. building					228			228	251
0400	50 S.F. building					180			180	198
0600	80 S.F. building					156			156	172
0700	100 S.F. building				▼	133			133	146
0900	Acoustical booth, 27 Db @ 1,000 Hz, 15 S.F. floor				Ea.	3,850			3,850	4,225
1000	7' x 7'-6", including light & ventilation					7,925			7,925	8,700
1200	Ticket booth, galv. steel, not incl. foundations., 4' x 4'					5,100			5,100	5,600
1300	4' x 6'				▼	7,475			7,475	8,225

13 34 23.70 Shelters

		Crew	Daily Output	Labor-Hours	Unit	Material	Labor	Equipment	Total	Total Incl O&P
0010	**SHELTERS**									
0020	Aluminum frame, acrylic glazing, 3' x 9' x 8' high	2 Sswk	1.14	14.035	Ea.	3,175	760		3,935	4,750
0100	9' x 12' x 8' high	"	.73	21.918	"	7,450	1,200		8,650	10,200

13 34 43 – Aircraft Hangars

13 34 43.50 Hangars

		Crew	Daily Output	Labor-Hours	Unit	Material	Labor	Equipment	Total	Total Incl O&P
0010	**HANGARS** Prefabricated steel T hangars, Galv. steel roof &									
0100	walls, incl. electric bi-folding doors									
0110	not including floors or foundations, 4 unit	E-2	1275	.044	SF Flr.	13.85	2.37	1.31	17.53	20.50
0130	8 unit		1063	.053		12.60	2.84	1.57	17.01	20
0900	With bottom rolling doors, 4 unit		1386	.040		12.75	2.18	1.21	16.14	18.90
1000	8 unit	▼	966	.058	▼	11.45	3.12	1.73	16.30	19.55
1200	Alternate pricing method:									
1300	Galv. roof and walls, electric bi-folding doors, 4 plane	E-2	1.06	52.830	Plane	18,400	2,850	1,575	22,825	26,600
1500	8 plane		.91	61.538		15,000	3,325	1,825	20,150	24,000
1600	With bottom rolling doors, 4 plane		1.25	44.800		16,900	2,425	1,325	20,650	24,000
1800	8 plane	▼	.97	57.732	▼	13,700	3,100	1,725	18,525	22,100
2000	Circular type, prefab., steel frame, plastic skin, electric									
2010	door, including foundations, 80' diameter									

13 34 53 – Agricultural Structures

13 34 53.50 Silos

		Crew	Daily Output	Labor-Hours	Unit	Material	Labor	Equipment	Total	Total Incl O&P
0010	**SILOS**									
0500	Steel, factory fab., 30,000 gallon cap., painted, economy	L-5	1	56	Ea.	22,500	3,075	675	26,250	30,500
0700	Deluxe		.50	112		35,800	6,125	1,350	43,275	51,000
0800	Epoxy lined, economy		1	56		36,800	3,075	675	40,550	46,300
1000	Deluxe	▼	.50	112	▼	46,700	6,125	1,350	54,175	63,000

13 34 Fabricated Engineered Structures

13 34 63 – Natural Fiber Construction

13 34 63.50 Straw Bale Construction

		Crew	Daily Output	Labor-Hours	Unit	Material	2017 Bare Costs Labor	2017 Bare Costs Equipment	Total	Total Incl O&P
0010	**STRAW BALE CONSTRUCTION**									
2020	Straw bales in walls w/modified post and beam frame	G 2 Carp	320	.050	S.F.	6	2.46		8.46	10.35

13 36 Towers

13 36 13 – Metal Towers

13 36 13.50 Control Towers

		Crew	Daily Output	Labor-Hours	Unit	Material	Labor	Equipment	Total	Total Incl O&P
0010	**CONTROL TOWERS**									
0020	Modular 12' x 10', incl. instruments				Ea.	801,500			801,500	882,000
0500	With standard 40' tower				"	1,260,000			1,260,000	1,386,000
1000	Temporary portable control towers, 8' x 12',									
1010	complete with one position communications				Ea.				266,000	293,000

13 42 Building Modules

13 42 63 – Detention Cell Modules

13 42 63.16 Steel Detention Cell Modules

		Crew	Daily Output	Labor-Hours	Unit	Material	Labor	Equipment	Total	Total Incl O&P
0010	**STEEL DETENTION CELL MODULES**									
2000	Cells, prefab., 5' to 6' wide, 7' to 8' high, 7' to 8' deep,									
2010	bar front, cot, not incl. plumbing	E-4	1.50	21.333	Ea.	9,800	1,175	86.50	11,061.50	12,800

13 47 Facility Protection

13 47 13 – Cathodic Protection

13 47 13.16 Cathodic Prot. for Underground Storage Tanks

		Crew	Daily Output	Labor-Hours	Unit	Material	Labor	Equipment	Total	Total Incl O&P
0010	**CATHODIC PROTECTION FOR UNDERGROUND STORAGE TANKS**									
1000	Anodes, magnesium type, 9 #	R-15	18.50	2.595	Ea.	35	145	15.60	195.60	273
1010	17 #		13	3.692		74.50	206	22	302.50	415
1020	32 #		10	4.800		117	268	29	414	560
1030	48 #		7.20	6.667		161	370	40	571	780
1100	Graphite type w/epoxy cap, 3" x 60" (32 #)	R-22	8.40	4.438		120	230		350	475
1110	4" x 80" (68 #)		6	6.213		232	320		552	735
1120	6" x 72" (80 #)		5.20	7.169		1,500	370		1,870	2,200
1130	6" x 36" (45 #)		9.60	3.883		760	201		961	1,125
2000	Rectifiers, silicon type, air cooled, 28 V/10 A	R-19	3.50	5.714		2,050	325		2,375	2,725
2010	20 V/20 A		3.50	5.714		2,100	325		2,425	2,800
2100	Oil immersed, 28 V/10 A		3	6.667		2,825	380		3,205	3,675
2110	20 V/20 A		3	6.667		3,000	380		3,380	3,875
3000	Anode backfill, coke breeze	R-22	3850	.010	Lb.	.28	.50		.78	1.06
4000	Cable, HMWPE, No. 8		2.40	15.533	M.L.F.	570	805		1,375	1,825
4010	No. 6		2.40	15.533		845	805		1,650	2,125
4020	No. 4		2.40	15.533		1,275	805		2,080	2,600
4030	No. 2		2.40	15.533		1,975	805		2,780	3,375
4040	No. 1		2.20	16.945		2,700	880		3,580	4,300
4050	No. 1/0		2.20	16.945		3,500	880		4,380	5,175
4060	No. 2/0		2.20	16.945		4,975	880		5,855	6,800
4070	No. 4/0		2	18.640		6,275	965		7,240	8,375
5000	Test station, 7 terminal box, flush curb type w/lockable cover	R-19	12	1.667	Ea.	76.50	94.50		171	227
5010	Reference cell, 2" dia PVC conduit, cplg., plug, set flush	"	4.80	4.167	"	151	236		387	520

13 48 Sound, Vibration, and Seismic Control

13 48 13 – Manufactured Sound and Vibration Control Components

13 48 13.50 Audio Masking

		Crew	Daily Output	Labor-Hours	Unit	Material	2017 Bare Costs Labor	Equipment	Total	Total Incl O&P
0010	**AUDIO MASKING**, acoustical enclosure, 4" thick wall and ceiling									
0020	8# per S.F., up to 12' span	3 Carp	72	.333	SF Surf	33.50	16.40		49.90	62
0300	Better quality panels, 10.5# per S.F.		64	.375		38	18.45		56.45	70.50
0400	Reverb-chamber, 4" thick, parallel walls		60	.400		47.50	19.70		67.20	82
0600	Skewed wall, parallel roof, 4" thick panels		55	.436		54	21.50		75.50	92.50
0700	Skewed walls, skewed roof, 4" layers, 4" air space		48	.500		61	24.50		85.50	105
0900	Sound-absorbing panels, pntd. mtl., 2'-6" x 8', under 1,000 S.F.		215	.112		12.55	5.50		18.05	22.50
1100	Over 1,000 S.F.		240	.100		12.10	4.93		17.03	21
1200	Fabric faced	▼	240	.100		9.80	4.93		14.73	18.35
1500	Flexible transparent curtain, clear	3 Shee	215	.112		7.60	6.50		14.10	18.25
1600	50% foam		215	.112		10.60	6.50		17.10	21.50
1700	75% foam		215	.112		10.60	6.50		17.10	21.50
1800	100% foam	▼	215	.112	▼	10.60	6.50		17.10	21.50
3100	Audio masking system, including speakers, amplification									
3110	and signal generator									
3200	Ceiling mounted, 5,000 S.F.	2 Elec	2400	.007	S.F.	1.24	.38		1.62	1.93
3300	10,000 S.F.		2800	.006		.99	.32		1.31	1.57
3400	Plenum mounted, 5,000 S.F.		3800	.004		1.10	.24		1.34	1.57
3500	10,000 S.F.	▼	4400	.004	▼	.74	.21		.95	1.12

13 49 Radiation Protection

13 49 13 – Integrated X-Ray Shielding Assemblies

13 49 13.50 Lead Sheets

		Crew	Daily Output	Labor-Hours	Unit	Material	2017 Bare Costs Labor	Equipment	Total	Total Incl O&P
0010	**LEAD SHEETS**									
0300	Lead sheets, 1/16" thick	2 Lath	135	.119	S.F.	10.40	5.70		16.10	19.85
0400	1/8" thick		120	.133		19.60	6.40		26	31
0500	Lead shielding, 1/4" thick		135	.119		38	5.70		43.70	50
0550	1/2" thick	▼	120	.133	▼	75.50	6.40		81.90	92.50
0950	Lead headed nails (average 1 lb. per sheet)				Lb.	8.05			8.05	8.85
1000	Butt joints in 1/8" lead or thicker, 2" batten strip x 7' long	2 Lath	240	.067	Ea.	28.50	3.19		31.69	35.50
1200	X-ray protection, average radiography or fluoroscopy									
1210	room, up to 300 S.F. floor, 1/16" lead, economy	2 Lath	.25	64	Total	10,100	3,075		13,175	15,700
1500	7'-0" walls, deluxe	"	.15	106	"	12,200	5,100		17,300	21,000
1600	Deep therapy X-ray room, 250 kV capacity,									
1800	up to 300 S.F. floor, 1/4" lead, economy	2 Lath	.08	200	Total	28,300	9,575		37,875	45,300
1900	7'-0" walls, deluxe	"	.06	266	"	34,900	12,800		47,700	57,500

13 49 19 – Lead-Lined Materials

13 49 19.50 Shielding Lead

		Crew	Daily Output	Labor-Hours	Unit	Material	2017 Bare Costs Labor	Equipment	Total	Total Incl O&P
0010	**SHIELDING LEAD**									
0100	Laminated lead in wood doors, 1/16" thick, no hardware				S.F.	53.50			53.50	59
0200	Lead lined door frame, not incl. hardware,									
0210	1/16" thick lead, butt prepared for hardware	1 Lath	2.40	3.333	Ea.	825	160		985	1,150
0850	Window frame with 1/16" lead and voice passage, 36" x 60"	2 Glaz	2	8		4,275	380		4,655	5,275
0870	24" x 36" frame		4	4	▼	2,225	191		2,416	2,750
0900	Lead gypsum board, 5/8" thick with 1/16" lead		160	.100	S.F.	11.55	4.78		16.33	20
0910	1/8" lead	▼	140	.114		23.50	5.45		28.95	34.50
0930	1/32" lead	2 Lath	200	.080	▼	8.10	3.83		11.93	14.55

470

13 49 Radiation Protection

13 49 21 – Lead Glazing

13 49 21.50 Lead Glazing	Crew	Daily Output	Labor-Hours	Unit	Material	2017 Bare Costs Labor	2017 Bare Costs Equipment	Total	Total Incl O&P
0010 **LEAD GLAZING**									
0600 Lead glass, 1/4" thick, 2.0 mm LE, 12" x 16"	2 Glaz	13	1.231	Ea.	370	59		429	500
0700 24" x 36"		8	2		1,300	95.50		1,395.50	1,575
0800 36" x 60"	↓	2	8	↓	3,600	380		3,980	4,550
2000 X-ray viewing panels, clear lead plastic									
2010 7 mm thick, 0.3 mm LE, 2.3 lb./S.F.	H-3	139	.115	S.F.	248	4.89		252.89	281
2020 12 mm thick, 0.5 mm LE, 3.9 lb./S.F.		82	.195		365	8.30		373.30	415
2030 18 mm thick, 0.8 mm LE, 5.9 lb./S.F.		54	.296		420	12.60		432.60	480
2040 22 mm thick, 1.0 mm LE, 7.2 lb./S.F.		44	.364		545	15.45		560.45	625
2050 35 mm thick, 1.5 mm LE, 11.5 lb./S.F.		28	.571		835	24.50		859.50	955
2060 46 mm thick, 2.0 mm LE, 15.0 lb./S.F.	↓	21	.762	↓	1,100	32.50		1,132.50	1,250
2090 For panels 12 S.F. to 48 S.F., add crating charge				Ea.				50	50

13 49 23 – Integrated RFI/EMI Shielding Assemblies

13 49 23.50 Modular Shielding Partitions	Crew	Daily Output	Labor-Hours	Unit	Material	2017 Bare Costs Labor	2017 Bare Costs Equipment	Total	Total Incl O&P
0010 **MODULAR SHIELDING PARTITIONS**									
4000 X-ray barriers, modular, panels mounted within framework for									
4002 attaching to floor, wall or ceiling, upper portion is clear lead									
4005 plastic window panels 48"H, lower portion is opaque leaded									
4008 steel panels 36"H, structural supports not incl.									
4010 1-section barrier, 36"W x 84"H overall									
4020 0.5 mm LE panels	H-3	6.40	2.500	Ea.	8,375	106		8,481	9,400
4030 0.8 mm LE panels		6.40	2.500		9,050	106		9,156	10,100
4040 1.0 mm LE panels		5.33	3.002		10,600	127		10,727	11,800
4050 1.5 mm LE panels	↓	5.33	3.002	↓	14,100	127		14,227	15,700
4060 2-section barrier, 72"W x 84"H overall									
4070 0.5 mm LE panels	H-3	4	4	Ea.	12,100	170		12,270	13,700
4080 0.8 mm LE panels		4	4		13,500	170		13,670	15,100
4090 1.0 mm LE panels		3.56	4.494		16,500	191		16,691	18,500
5000 1.5 mm LE panels	↓	3.20	5	↓	23,500	212		23,712	26,200
5010 3-section barrier, 108"W x 84"H overall									
5020 0.5 mm LE panels	H-3	3.20	5	Ea.	18,200	212		18,412	20,300
5030 0.8 mm LE panels		3.20	5		20,200	212		20,412	22,500
5040 1.0 mm LE panels		2.67	5.993		24,800	254		25,054	27,600
5050 1.5 mm LE panels	↓	2.46	6.504	↓	35,200	276		35,476	39,200
7000 X-ray barriers, mobile, mounted within framework w/casters on									
7005 bottom, clear lead plastic window panels on upper portion,									
7010 opaque on lower, 30"W x 75"H overall, incl. framework									
7020 24"H upper w/0.5 mm LE, 48"H lower w/0.8 mm LE	1 Carp	16	.500	Ea.	4,025	24.50		4,049.50	4,475
7030 48"W x 75"H overall, incl. framework									
7040 36"H upper w/0.5 mm LE, 36"H lower w/0.8 mm LE	1 Carp	16	.500	Ea.	6,350	24.50		6,374.50	7,050
7050 36"H upper w/1.0 mm LE, 36"H lower w/1.5 mm LE	"	16	.500	"	7,525	24.50		7,549.50	8,325
7060 72"W x 75"H overall, incl. framework									
7070 36"H upper w/0.5 mm LE, 36"H lower w/0.8 mm LE	1 Carp	16	.500	Ea.	7,525	24.50		7,549.50	8,325
7080 36"H upper w/1.0 mm LE, 36"H lower w/1.5 mm LE	"	16	.500	"	9,425	24.50		9,449.50	10,400

13 49 33 – Radio Frequency Shielding

13 49 33.50 Shielding, Radio Frequency	Crew	Daily Output	Labor-Hours	Unit	Material	2017 Bare Costs Labor	2017 Bare Costs Equipment	Total	Total Incl O&P
0010 **SHIELDING, RADIO FREQUENCY**									
0020 Prefabricated, galvanized steel	2 Carp	375	.043	SF Surf	5.30	2.10		7.40	9
0040 5 oz., copper floor panel		480	.033		4.48	1.64		6.12	7.45
0050 5 oz., copper wall/ceiling panel		155	.103		4.48	5.10		9.58	12.75
0100 12 oz., copper floor panel	↓	470	.034	↓	9.05	1.68		10.73	12.50

13 49 Radiation Protection

13 49 33 – Radio Frequency Shielding

13 49 33.50 Shielding, Radio Frequency	Crew	Daily Output	Labor-Hours	Unit	Material	2017 Bare Costs Labor	Equipment	Total	Total Incl O&P
0110 12 oz., copper wall/ceiling panel	2 Carp	140	.114	SF Surf	9.05	5.65		14.70	18.55
0150 Door, copper/wood laminate, 4' x 7'	↓	1.50	10.667	Ea.	8,050	525		8,575	9,650

13 53 Meteorological Instrumentation

13 53 09 – Weather Instrumentation

13 53 09.50 Weather Station

	Crew	Daily Output	Labor-Hours	Unit	Material	Labor	Equipment	Total	Total Incl O&P
0010 **WEATHER STATION**									
0020 Remote recording, solar powered, with rain gauge & display, 400 ft range				Ea.	775			775	850
0100 1 mile range				"	1,900			1,900	2,075

Estimating Tips
General

- Many products in Division 14 will require some type of support or blocking for installation not included with the item itself. Examples are supports for conveyors or tube systems, attachment points for lifts, and footings for hoists or cranes. Add these supports in the appropriate division.

14 10 00 Dumbwaiters
14 20 00 Elevators

- Dumbwaiters and elevators are estimated and purchased in a method similar to buying a car. The manufacturer has a base unit with standard features. Added to this base unit price will be whatever options the owner or specifications require. Increased load capacity, additional vertical travel, additional stops, higher speed, and cab finish options are items to be considered. When developing an estimate for dumbwaiters and elevators, remember that some items needed by the installers may have to be included as part of the general contract.

Examples are:

- ☐ shaftway
- ☐ rail support brackets
- ☐ machine room
- ☐ electrical supply
- ☐ sill angles
- ☐ electrical connections
- ☐ pits
- ☐ roof penthouses
- ☐ pit ladders

Check the job specifications and drawings before pricing.

- Installation of elevators and handicapped lifts in historic structures can require significant additional costs. The associated structural requirements may involve cutting into and repairing finishes, moldings, flooring, etc. The estimator must account for these special conditions.

14 30 00 Escalators
and Moving Walks

- Escalators and moving walks are specialty items installed by specialty contractors. There are numerous options associated with these items. For specific options, contact a manufacturer or contractor. In a method similar to estimating dumbwaiters and elevators, you should verify the extent of general contract work and add items as necessary.

14 40 00 Lifts
14 90 00 Other
Conveying Equipment

- Products such as correspondence lifts, chutes, and pneumatic tube systems, as well as other items specified in this subdivision, may require trained installers. The general contractor might not have any choice as to who will perform the installation or when it will be performed. Long lead times are often required for these products, making early decisions in scheduling necessary.

Reference Numbers

Reference numbers are shown at the beginning of some major classifications. These numbers refer to related items in the Reference Section. The reference information may be an estimating procedure, an alternate pricing method, or technical information.

Note: Not all subdivisions listed here necessarily appear. ■

Did you know?

Our online estimating solution gives you the same access to RSMeans' data with 24/7 access:

- Quickly locate costs in the searchable database.
- Build cost lists, estimates, and reports in minutes.
- Adjust costs to any location in the U.S. and Canada with the click of a button.

Start your free trial today at
www.RSMeansOnline.com

14 11 Manual Dumbwaiters

14 11 10 – Hand Operated Dumbwaiters

14 11 10.20 Manual Dumbwaiters		Crew	Daily Output	Labor-Hours	Unit	Material	2017 Bare Costs Labor	Equipment	Total	Total Incl O&P
0010	**MANUAL DUMBWAITERS**									
0020	2 stop, hand powered, up to 75 lb. capacity	2 Elev	.75	21.333	Ea.	3,100	1,750		4,850	6,000
0100	76 lb. capacity and up		.50	32	"	6,900	2,625		9,525	11,500
0300	For each additional stop, add		.75	21.333	Stop	1,125	1,750		2,875	3,850

14 12 Electric Dumbwaiters

14 12 10 – Dumbwaiters

14 12 10.10 Electric Dumbwaiters

		Crew	Daily Output	Labor-Hours	Unit	Material	2017 Bare Costs Labor	Equipment	Total	Total Incl O&P
0010	**ELECTRIC DUMBWAITERS**									
0020	2 stop, up to 75 lb. capacity	2 Elev	.13	123	Ea.	7,625	10,100		17,725	23,400
0100	76 lb. capacity and up		.11	145	"	23,000	11,900		34,900	43,100
0600	For each additional stop, add		.54	29.630	Stop	3,400	2,425		5,825	7,375

14 21 Electric Traction Elevators

14 21 13 – Electric Traction Freight Elevators

14 21 13.10 Electric Traction Freight Elevators and Options

		Crew	Daily Output	Labor-Hours	Unit	Material	2017 Bare Costs Labor	Equipment	Total	Total Incl O&P
0010	**ELECTRIC TRACTION FREIGHT ELEVATORS AND OPTIONS** R142000-10									
0425	Electric freight, base unit, 4000 lb., 200 fpm, 4 stop, std. fin.	2 Elev	.05	320	Ea.	118,500	26,200		144,700	169,500
0450	For 5000 lb. capacity, add					6,175			6,175	6,800
0500	For 6000 lb. capacity, add					14,500			14,500	16,000
0525	For 7000 lb. capacity, add					19,500			19,500	21,500
0550	For 8000 lb. capacity, add					23,600			23,600	26,000
0575	For 10000 lb. capacity, add					32,500			32,500	35,800
0600	For 12000 lb. capacity, add					39,500			39,500	43,500
0625	For 16000 lb. capacity, add					48,000			48,000	53,000
0650	For 20000 lb. capacity, add					53,500			53,500	58,500
0675	For increased speed, 250 fpm, add					14,100			14,100	15,500
0700	300 fpm, geared electric, add					17,600			17,600	19,300
0725	350 fpm, geared electric, add					21,400			21,400	23,600
0750	400 fpm, geared electric, add					25,300			25,300	27,900
0775	500 fpm, gearless electric, add					34,900			34,900	38,400
0800	600 fpm, gearless electric, add					39,000			39,000	42,900
0825	700 fpm, gearless electric, add					46,100			46,100	50,500
0850	800 fpm, gearless electric, add					51,000			51,000	56,500
0875	For class "B" loading, add					2,800			2,800	3,075
0900	For class "C-1" loading, add					7,000			7,000	7,700
0925	For class "C-2" loading, add					8,350			8,350	9,175
0950	For class "C-3" loading, add					11,600			11,600	12,800
0975	For travel over 40 V.L.F., add	2 Elev	7.25	2.207	V.L.F.	620	181		801	950
1000	For number of stops over 4, add	"	.27	59.259	Stop	4,525	4,850		9,375	12,200

14 21 23 – Electric Traction Passenger Elevators

14 21 23.10 Electric Traction Passenger Elevators and Options

		Crew	Daily Output	Labor-Hours	Unit	Material	2017 Bare Costs Labor	Equipment	Total	Total Incl O&P
0010	**ELECTRIC TRACTION PASSENGER ELEVATORS AND OPTIONS**									
1625	Electric pass., base unit, 2000 lb., 200 fpm, 4 stop, std. fin.	2 Elev	.05	320	Ea.	96,000	26,200		122,200	145,000
1650	For 2500 lb. capacity, add					4,075			4,075	4,475
1675	For 3000 lb. capacity, add					4,550			4,550	5,000
1700	For 3500 lb. capacity, add					6,100			6,100	6,700
1725	For 4000 lb. capacity, add					7,100			7,100	7,800
1750	For 4500 lb. capacity, add					9,850			9,850	10,800

14 21 Electric Traction Elevators

14 21 23 – Electric Traction Passenger Elevators

14 21 23.10 Electric Traction Passenger Elevators and Options	Crew	Daily Output	Labor-Hours	Unit	Material	2017 Bare Costs Labor	Equipment	Total	Total Incl O&P
1775	For 5000 lb. capacity, add				Ea.	11,900			11,900	13,000
1800	For increased speed, 250 fpm, geared electric, add					3,375			3,375	3,700
1825	300 fpm, geared electric, add					6,725			6,725	7,375
1850	350 fpm, geared electric, add					8,350			8,350	9,200
1875	400 fpm, geared electric, add					11,800			11,800	13,000
1900	500 fpm, gearless electric, add					30,200			30,200	33,200
1925	600 fpm, gearless electric, add					45,400			45,400	49,900
1950	700 fpm, gearless electric, add					54,000			54,000	59,500
1975	800 fpm, gearless electric, add					59,500			59,500	65,500
2000	For travel over 40 V.L.F., add	2 Elev	7.25	2.207	V.L.F.	720	181		901	1,050
2025	For number of stops over 4, add		.27	59.259	Stop	3,175	4,850		8,025	10,800
2400	Electric hospital, base unit, 4000 lb., 200 fpm, 4 stop, std fin.		.05	320	Ea.	87,500	26,200		113,700	135,000
2425	For 4500 lb. capacity, add					6,025			6,025	6,625
2450	For 5000 lb. capacity, add					7,900			7,900	8,700
2475	For increased speed, 250 fpm, geared electric, add					3,775			3,775	4,150
2500	300 fpm, geared electric, add					7,175			7,175	7,900
2525	350 fpm, geared electric, add					8,675			8,675	9,550
2550	400 fpm, geared electric, add					12,700			12,700	13,900
2575	500 fpm, gearless electric, add					35,500			35,500	39,100
2600	600 fpm, gearless electric, add					52,000			52,000	57,500
2625	700 fpm, gearless electric, add					58,000			58,000	63,500
2650	800 fpm, gearless electric, add					64,500			64,500	71,000
2675	For travel over 40 V.L.F., add	2 Elev	7.25	2.207	V.L.F.	179	181		360	465
2700	For number of stops over 4, add	"	.27	59.259	Stop	4,700	4,850		9,550	12,400

14 21 33 – Electric Traction Residential Elevators

14 21 33.20 Residential Elevators

		Crew	Daily Output	Labor-Hours	Unit	Material	2017 Bare Costs Labor	Equipment	Total	Total Incl O&P
0010	**RESIDENTIAL ELEVATORS**									
7000	Residential, cab type, 1 floor, 2 stop, economy model	2 Elev	.20	80	Ea.	11,600	6,550		18,150	22,600
7100	Custom model		.10	160		19,600	13,100		32,700	41,000
7200	2 floor, 3 stop, economy model		.12	133		17,200	10,900		28,100	35,200
7300	Custom model		.06	266		28,100	21,800		49,900	63,500

14 24 Hydraulic Elevators

14 24 13 – Hydraulic Freight Elevators

14 24 13.10 Hydraulic Freight Elevators and Options

		Crew	Daily Output	Labor-Hours	Unit	Material	2017 Bare Costs Labor	Equipment	Total	Total Incl O&P
0010	**HYDRAULIC FREIGHT ELEVATORS AND OPTIONS**									
1025	Hydraulic freight, base unit, 2000 lb., 50 fpm, 2 stop, std. fin.	2 Elev	.10	160	Ea.	82,500	13,100		95,600	110,000
1050	For 2500 lb. capacity, add					4,100			4,100	4,500
1075	For 3000 lb. capacity, add					5,425			5,425	5,975
1100	For 3500 lb. capacity, add					8,875			8,875	9,750
1125	For 4000 lb. capacity, add					10,100			10,100	11,100
1150	For 4500 lb. capacity, add					11,700			11,700	12,900
1175	For 5000 lb. capacity, add					16,000			16,000	17,600
1200	For 6000 lb. capacity, add					16,300			16,300	17,900
1225	For 7000 lb. capacity, add					25,000			25,000	27,500
1250	For 8000 lb. capacity, add					29,700			29,700	32,600
1275	For 10000 lb. capacity, add					32,000			32,000	35,200
1300	For 12000 lb. capacity, add					38,800			38,800	42,700
1325	For 16000 lb. capacity, add					51,500			51,500	56,500
1350	For 20000 lb. capacity, add					57,500			57,500	63,000

For customer support on your Building Construction Costs with RSMeans Data, call 800.448.8182.

475

14 24 13 - Hydraulic Freight Elevators

14 24 13.10 Hydraulic Freight Elevators and Options	Crew	Daily Output	Labor-Hours	Unit	Material	2017 Bare Costs Labor	Equipment	Total	Total Incl O&P
1375	For increased speed, 100 fpm, add				Ea.	1,175			1,175	1,275
1400	125 fpm, add					3,025			3,025	3,325
1425	150 fpm, add					4,825			4,825	5,300
1450	175 fpm, add					6,750			6,750	7,425
1475	For class "B" loading, add					2,700			2,700	2,975
1500	For class "C-1" loading, add					6,900			6,900	7,575
1525	For class "C-2" loading, add					8,250			8,250	9,075
1550	For class "C-3" loading, add					11,400			11,400	12,500
1575	For travel over 20 V.L.F., add	2 Elev	7.25	2.207	V.L.F.	875	181		1,056	1,225
1600	For number of stops over 2, add	"	.27	59.259	Stop	2,200	4,850		7,050	9,675

14 24 23 - Hydraulic Passenger Elevators

14 24 23.10 Hydraulic Passenger Elevators and Options

		Crew	Daily Output	Labor-Hours	Unit	Material	2017 Bare Costs Labor	Equipment	Total	Total Incl O&P
0010	**HYDRAULIC PASSENGER ELEVATORS AND OPTIONS**									
2050	Hyd. pass., base unit, 1500 lb., 100 fpm, 2 stop, std. fin.	2 Elev	.10	160	Ea.	39,300	13,100		52,400	63,000
2075	For 2000 lb. capacity, add					845			845	925
2100	For 2500 lb. capacity, add					2,925			2,925	3,225
2125	For 3000 lb. capacity, add					4,150			4,150	4,575
2150	For 3500 lb. capacity, add					7,425			7,425	8,175
2175	For 4000 lb. capacity, add					8,950			8,950	9,850
2200	For 4500 lb. capacity, add					11,800			11,800	12,900
2225	For 5000 lb. capacity, add					16,600			16,600	18,300
2250	For increased speed, 125 fpm, add					1,175			1,175	1,275
2275	150 fpm, add					2,675			2,675	2,950
2300	175 fpm, add					5,300			5,300	5,825
2325	200 fpm, add					10,000			10,000	11,000
2350	For travel over 12 V.L.F., add	2 Elev	7.25	2.207	V.L.F.	710	181		891	1,050
2375	For number of stops over 2, add		.27	59.259	Stop	1,000	4,850		5,850	8,350
2725	Hydraulic hospital, base unit, 4000 lb., 100 fpm, 2 stop, std. fin.		.10	160	Ea.	65,000	13,100		78,100	91,000
2775	For 4500 lb. capacity, add					7,300			7,300	8,025
2800	For 5000 lb. capacity, add					10,700			10,700	11,700
2825	For increased speed, 125 fpm, add					2,175			2,175	2,400
2850	150 fpm, add					3,675			3,675	4,050
2875	175 fpm, add					6,175			6,175	6,800
2900	200 fpm, add					9,050			9,050	9,975
2925	For travel over 12 V.L.F., add	2 Elev	7.25	2.207	V.L.F.	395	181		576	705
2950	For number of stops over 2, add	"	.27	59.259	Stop	4,525	4,850		9,375	12,200

14 27 Custom Elevator Cabs and Doors

14 27 13 - Custom Elevator Cab Finishes

14 27 13.10 Cab Finishes

		Crew	Daily Output	Labor-Hours	Unit	Material	2017 Bare Costs Labor	Equipment	Total	Total Incl O&P
0010	**CAB FINISHES**									
3325	Passenger elevator cab finishes (based on 3500 lb. cab size)									
3350	Acrylic panel ceiling				Ea.	775			775	855
3375	Aluminum eggcrate ceiling					700			700	770
3400	Stainless steel doors					4,050			4,050	4,475
3425	Carpet flooring					620			620	680
3450	Epoxy flooring					470			470	520
3475	Quarry tile flooring					900			900	990
3500	Slate flooring					1,650			1,650	1,800
3525	Textured rubber flooring					655			655	720

14 27 Custom Elevator Cabs and Doors

14 27 13 – Custom Elevator Cab Finishes

14 27 13.10 Cab Finishes	Crew	Daily Output	Labor-Hours	Unit	Material	2017 Bare Costs Labor	Equipment	Total	Total Incl O&P
3550	Stainless steel walls				Ea.	4,150			4,150	4,575
3575	Stainless steel returns at door					1,200			1,200	1,325
4450	Hospital elevator cab finishes (based on 3500 lb. cab size)									
4475	Aluminum eggcrate ceiling				Ea.	700			700	770
4500	Stainless steel doors					4,050			4,050	4,475
4525	Epoxy flooring					470			470	520
4550	Quarry tile flooring					900			900	990
4575	Textured rubber flooring					655			655	720
4600	Stainless steel walls					4,700			4,700	5,150
4625	Stainless steel returns at door					910			910	1,000

14 28 Elevator Equipment and Controls

14 28 10 – Elevator Equipment and Control Options

14 28 10.10 Elevator Controls and Doors

		Crew	Daily Output	Labor-Hours	Unit	Material	2017 Bare Costs Labor	Equipment	Total	Total Incl O&P
0010	**ELEVATOR CONTROLS AND DOORS**									
2975	Passenger elevator options									
3000	2 car group automatic controls	2 Elev	.66	24.242	Ea.	3,475	1,975		5,450	6,775
3025	3 car group automatic controls		.44	36.364		5,275	2,975		8,250	10,300
3050	4 car group automatic controls		.33	48.485		9,000	3,975		12,975	15,800
3075	5 car group automatic controls		.26	61.538		13,400	5,025		18,425	22,200
3100	6 car group automatic controls		.22	72.727		20,400	5,950		26,350	31,300
3125	Intercom service		3	5.333		555	435		990	1,250
3150	Duplex car selective collective		.66	24.242		3,875	1,975		5,850	7,225
3175	Center opening 1 speed doors		2	8		2,025	655		2,680	3,200
3200	Center opening 2 speed doors		2	8		2,850	655		3,505	4,100
3225	Rear opening doors (opposite front)		2	8		4,375	655		5,030	5,800
3250	Side opening 2 speed doors		2	8		7,425	655		8,080	9,150
3275	Automatic emergency power switching		.66	24.242		1,250	1,975		3,225	4,325
3300	Manual emergency power switching		8	2		525	164		689	820
3625	Hall finishes, stainless steel doors					1,450			1,450	1,600
3650	Stainless steel frames					1,500			1,500	1,650
3675	12 month maintenance contract								3,750	4,125
3700	Signal devices, hall lanterns	2 Elev	8	2		535	164		699	830
3725	Position indicators, up to 3		9.40	1.702		340	139		479	585
3750	Position indicators, per each over 3		32	.500		94.50	41		135.50	165
3775	High speed heavy duty door opener					2,675			2,675	2,950
3800	Variable voltage, O.H. gearless machine, min.	2 Elev	.16	100		34,200	8,175		42,375	49,800
3815	Maximum		.07	228		76,000	18,700		94,700	111,500
3825	Basement installed geared machine		.33	48.485		13,500	3,975		17,475	20,800
3850	Freight elevator options									
3875	Doors, bi-parting	2 Elev	.66	24.242	Ea.	6,050	1,975		8,025	9,600
3900	Power operated door and gate	"	.66	24.242		24,800	1,975		26,775	30,300
3925	Finishes, steel plate floor					1,100			1,100	1,200
3950	14 ga. 1/4" x 4' steel plate walls					2,050			2,050	2,250
3975	12 month maintenance contract								3,750	4,125
4000	Signal devices, hall lanterns	2 Elev	8	2		520	164		684	820
4025	Position indicators, up to 3		9.40	1.702		365	139		504	610
4050	Position indicators, per each over 3		32	.500		103	41		144	174
4075	Variable voltage basement installed geared machine		.66	24.242		20,500	1,975		22,475	25,600
4100	Hospital elevator options									
4125	2 car group automatic controls	2 Elev	.66	24.242	Ea.	3,500	1,975		5,475	6,800

14 28 10.10 Elevator Controls and Doors

	14 28 10.10 Elevator Controls and Doors	Crew	Daily Output	Labor-Hours	Unit	Material	2017 Bare Costs Labor	Equipment	Total	Total Incl O&P
4150	3 car group automatic controls	2 Elev	.44	36.364	Ea.	5,275	2,975		8,250	10,300
4175	4 car group automatic controls		.33	48.485		12,700	3,975		16,675	19,900
4200	5 car group automatic controls		.26	61.538		13,000	5,025		18,025	21,800
4225	6 car group automatic controls		.22	72.727		19,900	5,950		25,850	30,800
4250	Intercom service		3	5.333		530	435		965	1,225
4275	Duplex car selective collective		.66	24.242		3,825	1,975		5,800	7,150
4300	Center opening 1 speed doors		2	8		2,025	655		2,680	3,200
4325	Center opening 2 speed doors		2	8		2,675	655		3,330	3,925
4350	Rear opening doors (opposite front)		2	8		4,400	655		5,055	5,800
4375	Side opening 2 speed doors		2	8		6,600	655		7,255	8,225
4400	Automatic emergency power switching		.66	24.242		1,225	1,975		3,200	4,275
4425	Manual emergency power switching	▼	8	2		515	164		679	810
4675	Hall finishes, stainless steel doors					1,500			1,500	1,650
4700	Stainless steel frames					1,500			1,500	1,650
4725	12 month maintenance contract								3,750	4,125
4750	Signal devices, hall lanterns	2 Elev	8	2		505	164		669	800
4775	Position indicators, up to 3		9.40	1.702		340	139		479	585
4800	Position indicators, per each over 3	▼	32	.500		94	41		135	164
4825	High speed heavy duty door opener					2,675			2,675	2,950
4850	Variable voltage, O.H. gearless machine, min.	2 Elev	.16	100		34,800	8,175		42,975	50,500
4865	Maximum		.07	228		76,000	18,700		94,700	111,500
4875	Basement installed geared machine	▼	.33	48.485	▼	17,500	3,975		21,475	25,100
5000	Drilling for piston, casing included, 18" diameter	B-48	80	.700	V.L.F.	64.50	31.50	36.50	132.50	159

14 31 Escalators

14 31 10 – Glass and Steel Escalators

14 31 10.10 Escalators

		Crew	Daily Output	Labor-Hours	Unit	Material	2017 Bare Costs Labor	Equipment	Total	Total Incl O&P
0010	**ESCALATORS**									
1000	Glass, 32" wide x 10' floor to floor height	M-1	.07	457	Ea.	88,500	35,500	715	124,715	151,500
1010	48" wide x 10' floor to floor height		.07	457		95,500	35,500	715	131,715	159,000
1020	32" wide x 15' floor to floor height		.06	533		93,500	41,400	830	135,730	166,000
1030	48" wide x 15' floor to floor height		.06	533		99,000	41,400	830	141,230	172,000
1040	32" wide x 20' floor to floor height		.05	653		99,500	50,500	1,025	151,025	186,500
1050	48" wide x 20' floor to floor height		.05	653		107,500	50,500	1,025	159,025	195,500
1060	32" wide x 25' floor to floor height		.04	800		109,500	62,000	1,250	172,750	215,000
1070	48" wide x 25' floor to floor height		.04	800		126,500	62,000	1,250	189,750	233,500
1080	Enameled steel, 32" wide x 10' floor to floor height		.07	457		96,000	35,500	715	132,215	159,500
1090	48" wide x 10' floor to floor height		.07	457		104,500	35,500	715	140,715	168,500
1110	32" wide x 15' floor to floor height		.06	533		102,000	41,400	830	144,230	175,000
1120	48" wide x 15' floor to floor height		.06	533		108,000	41,400	830	150,230	182,000
1130	32" wide x 20' floor to floor height		.05	653		108,500	50,500	1,025	160,025	196,500
1140	48" wide x 20' floor to floor height		.05	653		118,000	50,500	1,025	169,525	206,500
1150	32" wide x 25' floor to floor height		.04	800		118,500	62,000	1,250	181,750	225,000
1160	48" wide x 25' floor to floor height		.04	800		136,500	62,000	1,250	199,750	245,000
1170	Stainless steel, 32" wide x 10' floor to floor height		.07	457		101,500	35,500	715	137,715	166,000
1180	48" wide x 10' floor to floor height		.07	457		109,500	35,500	715	145,715	174,500
1500	32" wide x 15' floor to floor height		.06	533		108,000	41,400	830	150,230	181,500
1700	48" wide x 15' floor to floor height		.06	533		114,000	41,400	830	156,230	188,000
1750	32" wide x 18' floor to floor height		.05	615		128,000	47,800	960	176,760	213,500
1775	48" wide x 18' floor to floor height		.05	615		139,500	47,800	960	188,260	225,500
2300	32" wide x 25' floor to floor height	▼	.04	800	▼	132,000	62,000	1,250	195,250	239,500

For customer support on your Building Construction Costs with RSMeans Data, call 800.448.8182.

14 31 Escalators

14 31 10 – Glass and Steel Escalators

14 31 10.10 Escalators	Crew	Daily Output	Labor-Hours	Unit	Material	2017 Bare Costs Labor	Equipment	Total	Total Incl O&P
2500 48" wide x 25' floor to floor height	M-1	.04	800	Ea.	143,500	62,000	1,250	206,750	252,500

14 32 Moving Walks

14 32 10 – Moving Walkways

14 32 10.10 Moving Walks

		Crew	Daily Output	Labor-Hours	Unit	Material	2017 Bare Costs Labor	Equipment	Total	Total Incl O&P
0010	**MOVING WALKS** R143210-20									
0020	Walk, 27" tread width, minimum	M-1	6.50	4.923	L.F.	925	385	7.70	1,317.70	1,600
0100	300' to 500', maximum		4.43	7.223		1,275	560	11.25	1,846.25	2,250
0300	48" tread width walk, minimum		4.43	7.223		2,100	560	11.25	2,671.25	3,150
0400	100' to 350', maximum		3.82	8.377		2,425	650	13.05	3,088.05	3,650
0600	Ramp, 12° incline, 36" tread width, minimum		5.27	6.072		1,700	470	9.45	2,179.45	2,600
0700	70' to 90' maximum		3.82	8.377		2,525	650	13.05	3,188.05	3,750
0900	48" tread width, minimum		3.57	8.964		2,500	695	14	3,209	3,825
1000	40' to 70', maximum		2.91	10.997		3,125	855	17.15	3,997.15	4,750

14 42 Wheelchair Lifts

14 42 13 – Inclined Wheelchair Lifts

14 42 13.10 Inclined Wheelchair Lifts and Stairclimbers

		Crew	Daily Output	Labor-Hours	Unit	Material	2017 Bare Costs Labor	Equipment	Total	Total Incl O&P
0010	**INCLINED WHEELCHAIR LIFTS AND STAIRCLIMBERS**									
7700	Stair climber (chair lift), single seat, minimum	2 Elev	1	16	Ea.	5,425	1,300		6,725	7,900
7800	Maximum		.20	80		7,425	6,550		13,975	17,900
8700	Stair lift, minimum		1	16		14,700	1,300		16,000	18,100
8900	Maximum		.20	80		23,200	6,550		29,750	35,400

14 42 16 – Vertical Wheelchair Lifts

14 42 16.10 Wheelchair Lifts

		Crew	Daily Output	Labor-Hours	Unit	Material	2017 Bare Costs Labor	Equipment	Total	Total Incl O&P
0010	**WHEELCHAIR LIFTS**									
8000	Wheelchair lift, minimum	2 Elev	1	16	Ea.	7,400	1,300		8,700	10,100
8500	Maximum	"	.50	32	"	17,600	2,625		20,225	23,200

14 45 Vehicle Lifts

14 45 10 – Hydraulic Vehicle Lifts

14 45 10.10 Hydraulic Lifts

		Crew	Daily Output	Labor-Hours	Unit	Material	2017 Bare Costs Labor	Equipment	Total	Total Incl O&P
0010	**HYDRAULIC LIFTS**									
2200	Single post, 8000 lb. capacity	L-4	.40	60	Ea.	5,775	2,800		8,575	10,700
2810	Double post, 6000 lb. capacity		2.67	8.989		8,150	420		8,570	9,600
2815	9000 lb. capacity		2.29	10.480		19,300	490		19,790	22,100
2820	15,000 lb. capacity		2	12		21,800	560		22,360	24,900
2822	Four post, 26,000 lb. capacity		1.80	13.333		13,800	620		14,420	16,200
2825	30,000 lb. capacity		1.60	15		48,300	700		49,000	54,000
2830	Ramp style, 4 post, 25,000 lb. capacity		2	12		18,900	560		19,460	21,700
2835	35,000 lb. capacity		1	24		88,000	1,125		89,125	98,500
2840	50,000 lb. capacity		1	24		98,500	1,125		99,625	110,000
2845	75,000 lb. capacity		1	24		114,500	1,125		115,625	127,500
2850	For drive thru tracks, add, minimum					1,100			1,100	1,225
2855	Maximum					1,900			1,900	2,075
2860	Ramp extensions, 3' (set of 2)					995			995	1,100
2865	Rolling jack platform					3,475			3,475	3,825

14 45 Vehicle Lifts

14 45 10 – Hydraulic Vehicle Lifts

14 45 10.10 Hydraulic Lifts	Crew	Daily Output	Labor-Hours	Unit	Material	2017 Bare Costs Labor	Equipment	Total	Total Incl O&P
2870	Electric/hydraulic jacking beam				Ea.	9,275			9,275	10,200
2880	Scissor lift, portable, 6000 lb. capacity					9,100			9,100	10,000

14 91 Facility Chutes

14 91 33 – Laundry and Linen Chutes

14 91 33.10 Chutes

14 91 33.10		Crew	Daily Output	Labor-Hours	Unit	Material	2017 Bare Costs Labor	Equipment	Total	Total Incl O&P
0011	**CHUTES**, linen, trash or refuse									
0050	Aluminized steel, 16 ga., 18" diameter	2 Shee	3.50	4.571	Floor	1,700	265		1,965	2,275
0100	24" diameter		3.20	5		1,850	290		2,140	2,475
0200	30" diameter		3	5.333		2,125	310		2,435	2,800
0300	36" diameter		2.80	5.714		2,625	330		2,955	3,400
0400	Galvanized steel, 16 ga., 18" diameter		3.50	4.571		1,025	265		1,290	1,525
0500	24" diameter		3.20	5		1,150	290		1,440	1,700
0600	30" diameter		3	5.333		1,275	310		1,585	1,875
0700	36" diameter		2.80	5.714		1,525	330		1,855	2,175
0800	Stainless steel, 18" diameter		3.50	4.571		3,000	265		3,265	3,700
0900	24" diameter		3.20	5		3,150	290		3,440	3,900
1000	30" diameter		3	5.333		3,750	310		4,060	4,600
1005	36" diameter		2.80	5.714		3,950	330		4,280	4,825
1200	Linen chute bottom collector, aluminized steel		4	4	Ea.	1,425	232		1,657	1,925
1300	Stainless steel		4	4		1,825	232		2,057	2,350
1500	Refuse, bottom hopper, aluminized steel, 18" diameter		3	5.333		1,050	310		1,360	1,625
1600	24" diameter		3	5.333		1,250	310		1,560	1,850
1800	36" diameter		3	5.333		2,500	310		2,810	3,225

14 91 82 – Trash Chutes

14 91 82.10 Trash Chutes and Accessories

14 91 82.10		Crew	Daily Output	Labor-Hours	Unit	Material	2017 Bare Costs Labor	Equipment	Total	Total Incl O&P
0010	**TRASH CHUTES AND ACCESSORIES**									
2900	Package chutes, spiral type, minimum	2 Shee	4.50	3.556	Floor	2,425	206		2,631	3,000
3000	Maximum	"	1.50	10.667	"	6,325	620		6,945	7,925

14 92 Pneumatic Tube Systems

14 92 10 – Conventional, Automatic and Computer Controlled Pneumatic Tube Systems

14 92 10.10 Pneumatic Tube Systems

14 92 10.10		Crew	Daily Output	Labor-Hours	Unit	Material	2017 Bare Costs Labor	Equipment	Total	Total Incl O&P
0010	**PNEUMATIC TUBE SYSTEMS**									
0020	100' long, single tube, 2 stations, stock									
0100	3" diameter	2 Stpi	.12	133	Total	3,375	8,300		11,675	16,200
0300	4" diameter	"	.09	177	"	4,275	11,100		15,375	21,400
0400	Twin tube, two stations or more, conventional system									
0600	2-1/2" round	2 Stpi	62.50	.256	L.F.	40.50	15.95		56.45	68.50
0700	3" round		46	.348		38	21.50		59.50	74.50
0900	4" round		49.60	.323		48	20		68	83.50
1000	4" x 7" oval		37.60	.426		89	26.50		115.50	138
1050	Add for blower		2	8	System	5,200	500		5,700	6,475
1110	Plus for each round station, add		7.50	2.133	Ea.	1,350	133		1,483	1,700
1150	Plus for each oval station, add		7.50	2.133	"	1,350	133		1,483	1,700
1200	Alternate pricing method: base cost, economy model		.75	21.333	Total	5,750	1,325		7,075	8,325
1300	Custom model		.25	64	"	11,500	3,975		15,475	18,700
1500	Plus total system length, add, for economy model		93.40	.171	L.F.	8.25	10.65		18.90	25
1600	For custom model		37.60	.426	"	25	26.50		51.50	67.50

14 92 Pneumatic Tube Systems

14 92 10 – Conventional, Automatic and Computer Controlled Pneumatic Tube Systems

14 92 10.10 Pneumatic Tube Systems	Crew	Daily Output	Labor-Hours	Unit	Material	2017 Bare Costs Labor	Equipment	Total	Total Incl O&P
1800	Completely automatic system, 4" round, 15 to 50 stations	2 Stpi	.29	55.172	Station	22,600	3,425		26,025	30,000
2200	51 to 144 stations		.32	50		15,600	3,125		18,725	21,900
2400	6" round or 4" x 7" oval, 15 to 50 stations		.24	66.667		27,600	4,150		31,750	36,700
2800	51 to 144 stations		.23	69.565		21,100	4,325		25,425	29,700

For customer support on your Building Construction Costs with RSMeans Data, call 800.448.8182.

481

Division Notes

	CREW	DAILY OUTPUT	LABOR-HOURS	UNIT	BARE COSTS				TOTAL INCL O&P
					MAT.	LABOR	EQUIP.	TOTAL	

Estimating Tips

Pipe for fire protection and all uses is located in Subdivisions 21 11 13 and 22 11 13.

The labor adjustment factors listed in Subdivision 22 01 02.20 also apply to Division 21.

Many, but not all, areas in the U.S. require backflow protection in the fire system. Insurance underwriters may have specific requirements for the type of materials to be installed or design requirements based on the hazard to be protected. Local jurisdictions may have requirements not covered by code. It is advisable to be aware of any special conditions.

For your reference, the following is a list of the most applicable Fire Codes and Standards which may be purchased from the NFPA, 1 Batterymarch Park, Quincy, MA 02169-7471.

- NFPA 1: Uniform Fire Code
- NFPA 10: Portable Fire Extinguishers
- NFPA 11: Low-, Medium-, and High-Expansion Foam
- NFPA 12: Carbon Dioxide Extinguishing Systems (Also companion 12A)
- NFPA 13: Installation of Sprinkler Systems (Also companion 13D, 13E, and 13R)
- NFPA 14: Installation of Standpipe and Hose Systems
- NFPA 15: Water Spray Fixed Systems for Fire Protection
- NFPA 16: Installation of Foam-Water Sprinkler and Foam-Water Spray Systems
- NFPA 17: Dry Chemical Extinguishing Systems (Also companion 17A)
- NFPA 18: Wetting Agents
- NFPA 20: Installation of Stationary Pumps for Fire Protection
- NFPA 22: Water Tanks for Private Fire Protection
- NFPA 24: Installation of Private Fire Service Mains and their Appurtenances
- NFPA 25: Inspection, Testing and Maintenance of Water-Based Fire Protection

Reference Numbers

Reference numbers are shown at the beginning of some major classifications. These numbers refer to related items in the Reference Section. The reference information may be an estimating procedure, an alternate pricing method, or technical information.

Note: Not all subdivisions listed here necessarily appear. ■

Did you know?

Our online estimating solution gives you the same access to RSMeans' data with 24/7 access:

- Quickly locate costs in the searchable database.
- Build cost lists, estimates, and reports in minutes.
- Adjust costs to any location in the U.S. and Canada with the click of a button.

Start your free trial today at
www.RSMeansOnline.com

21 05 Common Work Results for Fire Suppression

21 05 23 – General-Duty Valves for Water-Based Fire-Suppression Piping

21 05 23.50 General-Duty Valves	Crew	Daily Output	Labor-Hours	Unit	Material	2017 Bare Costs Labor	Equipment	Total	Total Incl O&P
0010 **GENERAL-DUTY VALVES**, for water-based fire suppression									
6200 Valves and components									
6500 Check, swing, C.I. body, brass fittings, auto. ball drip									
6520 4" size	Q-12	3	5.333	Ea.	380	283		663	845
6800 Check, wafer, butterfly type, C.I. body, bronze fittings									
6820 4" size	Q-12	4	4	Ea.	1,200	212		1,412	1,625

21 11 Facility Fire-Suppression Water-Service Piping

21 11 16 – Facility Fire Hydrants

21 11 16.50 Fire Hydrants for Buildings

21 11 16.50 Fire Hydrants for Buildings	Crew	Daily Output	Labor-Hours	Unit	Material	2017 Bare Costs Labor	Equipment	Total	Total Incl O&P
0010 **FIRE HYDRANTS FOR BUILDINGS**									
3750 Hydrants, wall, w/caps, single, flush, polished brass									
3800 2-1/2" x 2-1/2"	Q-12	5	3.200	Ea.	248	170		418	530
3840 2-1/2" x 3"	"	5	3.200		500	170		670	805
3900 For polished chrome, add					20%				
3950 Double, flush, polished brass									
4000 2-1/2" x 2-1/2" x 4"	Q-12	5	3.200	Ea.	765	170		935	1,100
4040 2-1/2" x 2-1/2" x 6"	"	4.60	3.478		875	185		1,060	1,250
4200 For polished chrome, add					10%				
4350 Double, projecting, polished brass									
4400 2-1/2" x 2-1/2" x 4"	Q-12	5	3.200	Ea.	296	170		466	580
4450 2-1/2" x 2-1/2" x 6"	"	4.60	3.478	"	605	185		790	945
4460 Valve control, dbl. flush/projecting hydrant, cap &									
4470 chain, extension rod & cplg., escutcheon, polished brass	Q-12	8	2	Ea.	272	106		378	460

21 11 19 – Fire-Department Connections

21 11 19.50 Connections for the Fire-Department

21 11 19.50 Connections for the Fire-Department	Crew	Daily Output	Labor-Hours	Unit	Material	2017 Bare Costs Labor	Equipment	Total	Total Incl O&P
0010 **CONNECTIONS FOR THE FIRE-DEPARTMENT**									
0020 For fire pro. cabinets, see Section 10 44 13.53									
7140 Standpipe connections, wall, w/plugs & chains									
7160 Single, flush, brass, 2-1/2" x 2-1/2"	Q-12	5	3.200	Ea.	188	170		358	465
7180 2-1/2" x 3"	"	5	3.200	"	196	170		366	475
7240 For polished chrome, add					15%				
7280 Double, flush, polished brass									
7300 2-1/2" x 2-1/2" x 4"	Q-12	5	3.200	Ea.	655	170		825	975
7330 2-1/2" x 2-1/2" x 6"	"	4.60	3.478	"	890	185		1,075	1,250
7400 For polished chrome, add					15%				
7440 For sill cock combination, add				Ea.	96			96	105
7900 Three way, flush, polished brass									
7920 2-1/2" (3) x 4"	Q-12	4.80	3.333	Ea.	1,950	177		2,127	2,425
7930 2-1/2" (3) x 6"	"	4.60	3.478		2,000	185		2,185	2,475
8000 For polished chrome, add					9%				
8020 Three way, projecting, polished brass									
8040 2-1/2" (3) x 4"	Q-12	4.80	3.333	Ea.	850	177		1,027	1,200

21 12 Fire-Suppression Standpipes

21 12 13 – Fire-Suppression Hoses and Nozzles

21 12 13.50 Fire Hoses and Nozzles	Crew	Daily Output	Labor-Hours	Unit	Material	2017 Bare Costs Labor	Equipment	Total	Total Incl O&P
0010 **FIRE HOSES AND NOZZLES**									
0200 Adapters, rough brass, straight hose threads									
0220 One piece, female to male, rocker lugs									
0240 1" x 1"				Ea.	58			58	63.50
0320 2" x 2"					42			42	46
0380 2-1/2" x 2-1/2"					19			19	21
2200 Hose, less couplings									
2260 Synthetic jacket, lined, 300 lb. test, 1-1/2" diameter	Q-12	2600	.006	L.F.	3	.33		3.33	3.79
2270 2" diameter		2200	.007		2.45	.39		2.84	3.28
2280 2-1/2" diameter		2200	.007		5.55	.39		5.94	6.70
2290 3" diameter		2200	.007		3.10	.39		3.49	3.99
2360 High strength, 500 lb. test, 1-1/2" diameter		2600	.006		2.40	.33		2.73	3.13
2380 2-1/2" diameter		2200	.007		5.35	.39		5.74	6.50
5600 Nozzles, brass									
5620 Adjustable fog, 3/4" booster line				Ea.	140			140	154
5630 1" booster line					166			166	183
5640 1-1/2" leader line					84			84	92.50
5660 2-1/2" direct connection					184			184	202
5680 2-1/2" playpipe nozzle					222			222	245
5780 For chrome plated, add					8%				
5850 Electrical fire, adjustable fog, no shock									
5900 1-1/2"				Ea.	234			234	258
5920 2-1/2"					700			700	770
5980 For polished chrome, add					6%				
6200 Heavy duty, comb. adj. fog and str. stream, with handle									
6210 1" booster line				Ea.	212			212	234

21 12 19 – Fire-Suppression Hose Racks

21 12 19.50 Fire Hose Racks

	Crew	Daily Output	Labor-Hours	Unit	Material	Labor	Equipment	Total	Total Incl O&P
0010 **FIRE HOSE RACKS**									
2600 Hose rack, swinging, for 1-1/2" diameter hose,									
2620 Enameled steel, 50' & 75' lengths of hose	Q-12	20	.800	Ea.	69	42.50		111.50	141
2640 100' and 125' lengths of hose	"	20	.800	"	35	42.50		77.50	103

21 12 23 – Fire-Suppression Hose Valves

21 12 23.70 Fire Hose Valves

	Crew	Daily Output	Labor-Hours	Unit	Material	Labor	Equipment	Total	Total Incl O&P
0010 **FIRE HOSE VALVES**									
0080 Wheel handle, 300 lb., 1-1/2"	1 Spri	12	.667	Ea.	105	39.50		144.50	176
0090 2-1/2"	"	7	1.143	"	206	67.50		273.50	330
0100 For polished brass, add					35%				
0110 For polished chrome, add					50%				

For customer support on your Building Construction Costs with RSMeans Data, call 800.448.8182.

485

21 13 Fire-Suppression Sprinkler Systems

21 13 13 – Wet-Pipe Sprinkler Systems

21 13 13.50 Wet-Pipe Sprinkler System Components

21 13 13.50 Wet-Pipe Sprinkler System Components	Crew	Daily Output	Labor-Hours	Unit	Material	2017 Bare Costs Labor	Equipment	Total	Total Incl O&P
0010 **WET-PIPE SPRINKLER SYSTEM COMPONENTS**									
2600 Sprinkler heads, not including supply piping									
3700 Standard spray, pendent or upright, brass, 135°F to 286°F									
3730 1/2" NPT, 7/16" orifice	1 Spri	16	.500	Ea.	16	29.50		45.50	62
3740 1/2" NPT, 1/2" orifice	"	16	.500		10.40	29.50		39.90	56
3860 For wax and lead coating, add					41.50			41.50	46
3880 For wax coating, add					26			26	28.50
3900 For lead coating, add					31.50			31.50	34.50
3920 For 360°F, same cost									
3930 For 400°F	1 Spri	16	.500	Ea.	99	29.50		128.50	154
3940 For 500°F	"	16	.500	"	99	29.50		128.50	154
4500 Sidewall, horizontal, brass, 135°F to 286°F									
4520 1/2" NPT, 1/2" orifice	1 Spri	16	.500	Ea.	26.50	29.50		56	74
4540 For 360°F, same cost									
4800 Recessed pendent, brass, 135°F to 286°F									
4820 1/2" NPT, 3/8" orifice	1 Spri	10	.800	Ea.	45	47		92	121
4830 1/2" NPT, 7/16" orifice		10	.800		19.55	47		66.55	93
4840 1/2" NPT, 1/2" orifice		10	.800		15.15	47		62.15	88

21 13 16 – Dry-Pipe Sprinkler Systems

21 13 16.50 Dry-Pipe Sprinkler System Components

	Crew	Daily Output	Labor-Hours	Unit	Material	Labor	Equipment	Total	Total Incl O&P
0010 **DRY-PIPE SPRINKLER SYSTEM COMPONENTS**									
0600 Accelerator	1 Spri	8	1	Ea.	820	59		879	990
2600 Sprinkler heads, not including supply piping									
2640 Dry, pendent, 1/2" orifice, 3/4" or 1" NPT									
2700 15-1/4" to 18" length	1 Spri	14	.571	Ea.	154	33.50		187.50	220
2710 18-1/4" to 21" length		13	.615		160	36.50		196.50	231
2720 21-1/4" to 24" length		13	.615		166	36.50		202.50	237
2730 24-1/4" to 27" length		13	.615		172	36.50		208.50	244

21 13 19 – Preaction Sprinkler Systems

21 13 19.50 Preaction Sprinkler System Components

	Crew	Daily Output	Labor-Hours	Unit	Material	Labor	Equipment	Total	Total Incl O&P
0010 **PREACTION SPRINKLER SYSTEM COMPONENTS**									
3000 Preaction valve cabinet									
3100 Single interlock, pneum. release, panel, 1/2 hp comp. regul. air trim									
3110 1-1/2"	Q-12	3	5.333	Ea.	45,500	283		45,783	50,500
3120 2"		3	5.333		45,500	283		45,783	50,500
3130 2-1/2"		3	5.333		45,500	283		45,783	50,500
3140 3"		3	5.333		45,600	283		45,883	50,500
3150 4"		2	8		46,800	425		47,225	52,000
3160 6"	Q-13	4	8		48,600	450		49,050	54,000
3200 Double interlock, pneum. release, panel, 1/2 hp comp. regul. air trim									
3210 1-1/2"	Q-12	3	5.333	Ea.	46,500	283		46,783	51,500
3220 2"		3	5.333		46,500	283		46,783	51,500
3230 2-1/2"		3	5.333		46,500	283		46,783	51,500
3240 3"		3	5.333		46,600	283		46,883	51,500
3250 4"		2	8		48,500	425		48,925	54,000
3260 6"	Q-13	4	8		49,800	450		50,250	55,000

21 13 26 – Deluge Fire-Suppression Sprinkler Systems

21 13 26.50 Deluge Fire-Suppression Sprinkler Sys. Comp.

	Crew	Daily Output	Labor-Hours	Unit	Material	Labor	Equipment	Total	Total Incl O&P
0010 **DELUGE FIRE-SUPPRESSION SPRINKLER SYSTEM COMPONENTS**									
1400 Deluge system, monitoring panel w/deluge valve & trim	1 Spri	18	.444	Ea.	6,950	26		6,976	7,675
6200 Valves and components									

21 13 Fire-Suppression Sprinkler Systems

21 13 26 – Deluge Fire-Suppression Sprinkler Systems

21 13 26.50 Deluge Fire-Suppression Sprinkler Sys. Comp.	Crew	Daily Output	Labor-Hours	Unit	Material	2017 Bare Costs Labor	Equipment	Total	Total Incl O&P
7000	Deluge, assembly, incl. trim, pressure									
7020	operated relief, emergency release, gauges									
7040	2" size	Q-12	2	8	Ea.	3,975	425		4,400	5,000
7060	3" size	"	1.50	10.667	"	4,450	565		5,015	5,750

21 13 39 – Foam-Water Systems

21 13 39.50 Foam-Water System Components

		Crew	Daily Output	Labor-Hours	Unit	Material	Labor	Equipment	Total	Total Incl O&P
0010	**FOAM-WATER SYSTEM COMPONENTS**									
2600	Sprinkler heads, not including supply piping									
3600	Foam-water, pendent or upright, 1/2" NPT	1 Spri	12	.667	Ea.	216	39.50		255.50	298

21 21 Carbon-Dioxide Fire-Extinguishing Systems

21 21 16 – Carbon-Dioxide Fire-Extinguishing Equipment

21 21 16.50 CO2 Fire Extinguishing System

		Crew	Daily Output	Labor-Hours	Unit	Material	Labor	Equipment	Total	Total Incl O&P
0010	**CO$_2$ FIRE EXTINGUISHING SYSTEM**									
0042	For detectors and control stations, see Section 28 31 23.50									
0100	Control panel, single zone with batteries (2 zones det., 1 suppr.)	1 Elec	1	8	Ea.	350	455		805	1,075
0150	Multizone (4) with batteries (8 zones det., 4 suppr.)	"	.50	16		760	905		1,665	2,175
1000	Dispersion nozzle, CO$_2$, 3" x 5"	1 Plum	18	.444		67	27.50		94.50	115
2000	Extinguisher, CO$_2$ system, high pressure, 75 lb. cylinder	Q-1	6	2.667		1,275	148		1,423	1,625
2100	100 lb. cylinder	"	5	3.200		1,300	178		1,478	1,725
3000	Electro/mechanical release	L-1	4	4		188	237		425	560
3400	Manual pull station	1 Plum	6	1.333		60	82.50		142.50	190
4000	Pneumatic damper release	"	8	1		258	62		320	380

21 22 Clean-Agent Fire-Extinguishing Systems

21 22 16 – Clean-Agent Fire-Extinguishing Equipment

21 22 16.50 Clean-Agent Extinguishing Systems

		Crew	Daily Output	Labor-Hours	Unit	Material	Labor	Equipment	Total	Total Incl O&P
0010	**CLEAN-AGENT EXTINGUISHING SYSTEMS**									
0020	FM200 fire extinguishing system									
1100	Dispersion nozzle FM200, 1-1/2"	1 Plum	14	.571	Ea.	92	35.50		127.50	155
2400	Extinguisher, FM200 system, filled, with mounting bracket									
2460	26 lb. container	Q-1	8	2	Ea.	2,600	111		2,711	3,025
2480	44 lb. container		7	2.286		3,000	127		3,127	3,500
2500	63 lb. container		6	2.667		4,550	148		4,698	5,250
2520	101 lb. container		5	3.200		5,325	178		5,503	6,125
2540	196 lb. container		4	4		7,850	223		8,073	8,950
6000	FM200 system, simple nozzle layout, with broad dispersion				C.F.	1.76			1.76	1.94
6010	Extinguisher, FM200 system, filled, with mounting bracket									
6020	Complex nozzle layout and/or including underfloor dispersion				C.F.	3.50			3.50	3.85
6100	20,000 cf 2 exits, 8' clng					2.75			2.75	3.03
6200	100,000 cf 4 exits, 8' clng					2.30			2.30	2.53
6300	250,000 cf 6 exits, 8' clng					1.85			1.85	2.04
7010	HFC-227ea fire extinguishing system									
7100	Cylinders with clean-agent									
7110	Does not include pallete jack/fork lift rental fees									
7120	70 lb cyl, w/ 35 lb agent, no solenoid	Q-12	14	1.143	Ea.	2,400	60.50		2,460.50	2,750
7130	70 lb cyl w/ 70 lb agent, no solenoid		10	1.600		3,075	85		3,160	3,525
7140	70 lb cyl w/ 35 lb agent, w/ solenoid		14	1.143		2,875	60.50		2,935.50	3,250
7150	70 lb cyl w/ 70 lb agent, w/ solenoid		10	1.600		3,625	85		3,710	4,125

21 22 Clean-Agent Fire-Extinguishing Systems

21 22 16 – Clean-Agent Fire-Extinguishing Equipment

21 22 16.50 Clean-Agent Extinguishing Systems	Crew	Daily Output	Labor-Hours	Unit	Material	2017 Bare Costs Labor	Equipment	Total	Total Incl O&P
7220	250 lb cyl, w/ 125 lb agent, no solenoid	Q-12	8	2	Ea.	5,425	106		5,531	6,100
7230	250 lb cyl, w/ 250 lb agent, no solenoid		5	3.200		5,425	170		5,595	6,200
7240	250 lb cyl, w/ 125 lb agent, w/ solenoid		8	2		6,225	106		6,331	7,000
7250	250 lb cyl, w/ 250 lb agent, w/ solenoid		5	3.200		8,925	170		9,095	10,100
7320	560 lb cyl, w/ 300 lb agent, no solenoid		4	4		10,200	212		10,412	11,500
7330	560 lb cyl, w/ 560 lb agent, no solenoid		2.50	6.400		15,200	340		15,540	17,300
7340	560 lb cyl, w/ 300 lb agent, w/ solenoid		4	4		11,500	212		11,712	13,000
7350	560 lb cyl, w/ 560 lb agent, w/ solenoid		2.50	6.400		17,100	340		17,440	19,400
7420	1200 lb cyl, w/ 600 lb agent, no solenoid	Q-13	4	8		19,200	450		19,650	21,800
7430	1200 lb cyl, w/ 1200 lb agent, no solenoid		3	10.667		30,900	600		31,500	34,900
7440	1200 lb cyl, w/ 600 lb agent, w/ solenoid		4	8		40,100	450		40,550	44,800
7450	1200 lb cyl, w/ 1200 lb agent, w/ solenoid		3	10.667		63,000	600		63,600	70,000
7500	Accessories								
7510	Dispersion nozzle	1 Spri	16	.500	Ea.	95	29.50		124.50	150
7520	Agent release panel	1 Elec	4	2		60	113		173	236
7530	Maintenance switch	"	6	1.333		35	75.50		110.50	152
7540	Solenoid valve, 12v dc	1 Spri	8	1		232	59		291	345
7550	12v ac		8	1		415	59		474	545
7560	12v dc, exp.proof		8	1		390	59		449	520

21 31 Centrifugal Fire Pumps

21 31 13 – Electric-Drive, Centrifugal Fire Pumps

21 31 13.50 Electric-Drive Fire Pumps

		Crew	Daily Output	Labor-Hours	Unit	Material	2017 Bare Costs Labor	Equipment	Total	Total Incl O&P
0010	**ELECTRIC-DRIVE FIRE PUMPS** Including controller, fittings and relief valve									
3100	250 GPM, 55 psi, 15 HP, 3550 RPM, 2" pump	Q-13	.70	45.714	Ea.	14,900	2,575		17,475	20,300
3200	500 GPM, 50 psi, 27 HP, 1770 RPM, 4" pump		.68	47.059		15,300	2,650		17,950	20,900
3350	750 GPM, 50 psi, 44 HP, 1770 RPM, 5" pump		.64	50		16,100	2,800		18,900	22,000
3400	750 GPM, 100 psi, 66 HP, 3550 RPM, 4" pump		.58	55.172		18,500	3,100		21,600	25,000
5000	For jockey pump 1", 3 HP, with control, add	Q-12	2	8		2,750	425		3,175	3,675

21 31 16 – Diesel-Drive, Centrifugal Fire Pumps

21 31 16.50 Diesel-Drive Fire Pumps

		Crew	Daily Output	Labor-Hours	Unit	Material	2017 Bare Costs Labor	Equipment	Total	Total Incl O&P
0010	**DIESEL-DRIVE FIRE PUMPS** Including controller, fittings and relief valve									
0050	500 GPM, 50 psi, 27 HP, 4" pump	Q-13	.64	50	Ea.	36,900	2,800		39,700	44,900
0200	750 GPM, 50 psi, 44 HP, 5" pump		.60	53.333		38,000	3,000		41,000	46,300
0400	1000 GPM, 100 psi, 89 HP, 4" pump		.56	57.143		43,300	3,200		46,500	52,500
0700	2000 GPM, 100 psi, 167 HP, 6" pump		.34	94.118		55,000	5,300		60,300	68,500
0950	3500 GPM, 100 psi, 300 HP, 10" pump		.24	133		78,000	7,500		85,500	97,000

Estimating Tips
22 10 00 Plumbing Piping and Pumps
This subdivision is primarily basic pipe and related materials. The pipe may be used by any of the mechanical disciplines, i.e., plumbing, fire protection, heating, and air conditioning.

Note: CPVC plastic piping approved for fire protection is located in 21 11 13.

- The labor adjustment factors listed in Subdivision 22 01 02.20 apply throughout Divisions 21, 22, and 23. CAUTION: the correct percentage may vary for the same items. For example, the percentage add for the basic pipe installation should be based on the maximum height that the installer must install for that particular section. If the pipe is to be located 14' above the floor but it is suspended on threaded rod from beams, the bottom flange of which is 18' high (4' rods), then the height is actually 18' and the add is 20%. The pipe cover, however, does not have to go above the 14', and so the add should be 10%.

- Most pipe is priced first as straight pipe with a joint (coupling, weld, etc.) every 10' and a hanger usually every 10'. There are exceptions with hanger spacing such as for cast iron pipe (5')

and plastic pipe (3 per 10'). Following each type of pipe there are several lines listing sizes and the amount to be subtracted to delete couplings and hangers. This is for pipe that is to be buried or supported together on trapeze hangers. The reason that the couplings are deleted is that these runs are usually long, and frequently longer lengths of pipe are used. By deleting the couplings, the estimator is expected to look up and add back the correct reduced number of couplings.

- When preparing an estimate, it may be necessary to approximate the fittings. Fittings usually run between 25% and 50% of the cost of the pipe. The lower percentage is for simpler runs, and the higher number is for complex areas, such as mechanical rooms.

- For historic restoration projects, the systems must be as invisible as possible, and pathways must be sought for pipes, conduit, and ductwork. While installations in accessible spaces (such as basements and attics) are relatively straightforward to estimate, labor costs may be more difficult to determine when delivery systems must be concealed.

22 40 00 Plumbing Fixtures
- Plumbing fixture costs usually require two lines: the fixture itself and its "rough-in, supply, and waste."

- In the Assemblies Section (Plumbing D2010) for the desired fixture, the System Components Group at the center of the page shows the fixture on the first line. The rest of the list (fittings, pipe, tubing, etc.) will total up to what we refer to in the Unit Price section as "Rough-in, supply, waste, and vent." Note that for most fixtures we allow a nominal 5' of tubing to reach from the fixture to a main or riser.

- Remember that gas- and oil-fired units need venting.

Reference Numbers
Reference numbers are shown at the beginning of some major classifications. These numbers refer to related items in the Reference Section. The reference information may be an estimating procedure, an alternate pricing method, or technical information.

Note: Not all subdivisions listed here necessarily appear. ■

Did you know?

Our online estimating solution gives you the same access to RSMeans' data with 24/7 access:

- Quickly locate costs in the searchable database.

- Build cost lists, estimates, and reports in minutes.

- Adjust costs to any location in the U.S. and Canada with the click of a button.

Start your free trial today at
www.RSMeansOnline.com

22 01 02 – Labor Adjustments

22 01 02.10 Boilers, General	Crew	Daily Output	Labor-Hours	Unit	Material	2017 Bare Costs Labor	Equipment	Total	Total Incl O&P
0010 **BOILERS, GENERAL**, Prices do not include flue piping, elec. wiring,									
0020 gas or oil piping, boiler base, pad, or tankless unless noted									
0100 Boiler H.P.: 10 KW = 34 lb./steam/hr. = 33,475 BTU/hr.									
0150 To convert SFR to BTU rating: Hot water, 150 x SFR;									
0160 Forced hot water, 180 x SFR; steam, 240 x SFR									

22 01 02.20 Labor Adjustment Factors

	Crew	Daily Output	Labor-Hours	Unit	Material	2017 Bare Costs Labor	Equipment	Total	Total Incl O&P
0010 **LABOR ADJUSTMENT FACTORS** (For Div. 21, 22 and 23) R220102-20									
0100 Labor factors, The below are reasonable suggestions, however									
0110 each project must be evaluated for its own peculiarities, and									
0120 the adjustments be increased or decreased depending on the									
0130 severity of the special conditions.									
1000 Add to labor for elevated installation (Above floor level)									
1080 10' to 14.5' high						10%			
1100 15' to 19.5' high						20%			
1120 20' to 24.5' high						25%			
1140 25' to 29.5' high						35%			
1160 30' to 34.5' high						40%			
1180 35' to 39.5' high						50%			
1200 40' and higher						55%			
2000 Add to labor for crawl space									
2100 3' high						40%			
2140 4' high						30%			
3000 Add to labor for multi-story building									
3010 For new construction (No elevator available)									
3100 Add for floors 3 thru 10						5%			
3110 Add for floors 11 thru 15						10%			
3120 Add for floors 16 thru 20						15%			
3130 Add for floors 21 thru 30						20%			
3140 Add for floors 31 and up						30%			
3170 For existing structure (Elevator available)									
3180 Add for work on floor 3 and above						2%			
4000 Add to labor for working in existing occupied buildings									
4100 Hospital						35%			
4140 Office building						25%			
4180 School						20%			
4220 Factory or warehouse						15%			
4260 Multi dwelling						15%			
5000 Add to labor, miscellaneous									
5100 Cramped shaft						35%			
5140 Congested area						15%			
5180 Excessive heat or cold						30%			
9000 Labor factors, The above are reasonable suggestions, however									
9010 each project should be evaluated for its own peculiarities.									
9100 Other factors to be considered are:									
9140 Movement of material and equipment through finished areas									
9180 Equipment room									
9220 Attic space									
9260 No service road									
9300 Poor unloading/storage area									
9340 Congested site area/heavy traffic									

22 05 Common Work Results for Plumbing

22 05 05 – Selective Demolition for Plumbing

22 05 05.10 Plumbing Demolition

	Crew	Daily Output	Labor-Hours	Unit	Material	2017 Bare Costs Labor	Equipment	Total	Total Incl O&P
0010 **PLUMBING DEMOLITION**									
1020 Fixtures, including 10' piping									
1100 Bathtubs, cast iron	1 Plum	4	2	Ea.		124		124	187
1120 Fiberglass		6	1.333			82.50		82.50	124
1140 Steel		5	1.600			99		99	149
1200 Lavatory, wall hung		10	.800			49.50		49.50	74.50
1220 Counter top		8	1			62		62	93.50
1300 Sink, single compartment		8	1			62		62	93.50
1320 Double compartment		7	1.143			70.50		70.50	107
1400 Water closet, floor mounted		8	1			62		62	93.50
1420 Wall mounted		7	1.143			70.50		70.50	107
1500 Urinal, floor mounted		4	2			124		124	187
1520 Wall mounted		7	1.143			70.50		70.50	107
1600 Water fountains, free standing		8	1			62		62	93.50
1620 Wall or deck mounted		6	1.333			82.50		82.50	124
2000 Piping, metal, up thru 1-1/2" diameter		200	.040	L.F.		2.47		2.47	3.73
2050 2" thru 3-1/2" diameter		150	.053			3.30		3.30	4.97
2100 4" thru 6" diameter	2 Plum	100	.160			9.90		9.90	14.90
2150 8" thru 14" diameter	"	60	.267			16.50		16.50	25
2153 16" thru 20" diameter	Q-18	70	.343			19.90	.83	20.73	31
2155 24" thru 26" diameter		55	.436			25.50	1.06	26.56	39.50
2156 30" thru 36" diameter		40	.600			35	1.45	36.45	54
2160 Plastic pipe with fittings, up thru 1-1/2" diameter	1 Plum	250	.032			1.98		1.98	2.98
2162 2" thru 3" diameter	"	200	.040			2.47		2.47	3.73
2164 4" thru 6" diameter	Q-1	200	.080			4.45		4.45	6.70
2166 8" thru 14" diameter		150	.107			5.95		5.95	8.95
2168 16" diameter		100	.160			8.90		8.90	13.45
2212 Deduct for salvage, aluminum scrap				Ton				700	770
2214 Brass scrap								2,675	2,675
2216 Copper scrap								3,525	3,525
2218 Lead scrap								570	570
2220 Steel scrap								200	200
2250 Water heater, 40 gal.	1 Plum	6	1.333	Ea.		82.50		82.50	124
9470 Water softener	Q-1	2	8	"		445		445	670

22 05 23 – General-Duty Valves for Plumbing Piping

22 05 23.10 Valves, Brass

	Crew	Daily Output	Labor-Hours	Unit	Material	2017 Bare Costs Labor	Equipment	Total	Total Incl O&P
0010 **VALVES, BRASS**									
0500 Gas cocks, threaded									
0530 1/2"	1 Plum	24	.333	Ea.	15.30	20.50		35.80	48
0540 3/4"		22	.364		20	22.50		42.50	56
0550 1"		19	.421		36.50	26		62.50	79.50
0560 1-1/4"		15	.533		52	33		85	107

22 05 23.20 Valves, Bronze

	Crew	Daily Output	Labor-Hours	Unit	Material	2017 Bare Costs Labor	Equipment	Total	Total Incl O&P
0010 **VALVES, BRONZE**									
1020 Angle, 150 lb., rising stem, threaded									
1030 1/8"	1 Plum	24	.333	Ea.	134	20.50		154.50	179
1040 1/4"		24	.333		134	20.50		154.50	179
1050 3/8"		24	.333		134	20.50		154.50	179
1060 1/2"		22	.364		134	22.50		156.50	182
1070 3/4"		20	.400		200	24.50		224.50	258
1080 1"		19	.421		293	26		319	365
1100 1-1/2"		13	.615		445	38		483	550

491

For customer support on your Building Construction Costs with RSMeans Data, call 800.448.8182.

22 05 23.20 Valves, Bronze		Crew	Daily Output	Labor-Hours	Unit	Material	2017 Bare Costs Labor	Equipment	Total	Total Incl O&P
1102	Soldered same price as threaded									
1110	2"	1 Plum	11	.727	Ea.	715	45		760	855
1300	Ball									
1398	Threaded, 150 psi									
1400	1/4"	1 Plum	24	.333	Ea.	12.75	20.50		33.25	45
1430	3/8"		24	.333		12.75	20.50		33.25	45
1450	1/2"		22	.364		12.75	22.50		35.25	48
1460	3/4"		20	.400		21	24.50		45.50	60.50
1470	1"		19	.421		31	26		57	74
1480	1-1/4"		15	.533		54.50	33		87.50	110
1490	1-1/2"		13	.615		71	38		109	136
1500	2"		11	.727		86	45		131	163
1522	Solder the same price as threaded									
1750	Check, swing, class 150, regrinding disc, threaded									
1800	1/8"	1 Plum	24	.333	Ea.	70	20.50		90.50	108
1830	1/4"		24	.333		63.50	20.50		84	101
1840	3/8"		24	.333		68	20.50		88.50	106
1850	1/2"		24	.333		72.50	20.50		93	111
1860	3/4"		20	.400		96	24.50		120.50	143
1870	1"		19	.421		138	26		164	192
1880	1-1/4"		15	.533		200	33		233	270
1890	1-1/2"		13	.615		255	38		293	340
1900	2"		11	.727		345	45		390	445
1910	2-1/2"	Q-1	15	1.067		770	59.50		829.50	935
2000	For 200 lb., add					5%	10%			
2040	For 300 lb., add					15%	15%			
2850	Gate, N.R.S., soldered, 125 psi									
2900	3/8"	1 Plum	24	.333	Ea.	64	20.50		84.50	102
2920	1/2"		24	.333		55	20.50		75.50	91.50
2940	3/4"		20	.400		63	24.50		87.50	107
2950	1"		19	.421		81.50	26		107.50	130
2960	1-1/4"		15	.533		135	33		168	198
2970	1-1/2"		13	.615		152	38		190	225
2980	2"		11	.727		198	45		243	285
2990	2-1/2"	Q-1	15	1.067		490	59.50		549.50	625
3000	3"	"	13	1.231		660	68.50		728.50	835
3850	Rising stem, soldered, 300 psi									
3950	1"	1 Plum	19	.421	Ea.	183	26		209	241
3980	2"	"	11	.727		500	45		545	620
4000	3"	Q-1	13	1.231		1,650	68.50		1,718.50	1,925
4250	Threaded, class 150									
4310	1/4"	1 Plum	24	.333	Ea.	72.50	20.50		93	111
4320	3/8"		24	.333		72.50	20.50		93	111
4330	1/2"		24	.333		66	20.50		86.50	104
4340	3/4"-		20	.400		77	24.50		101.50	123
4350	1"		19	.421		104	26		130	154
4360	1-1/4"		15	.533		141	33		174	205
4370	1-1/2"		13	.615		178	38		216	253
4380	2"		11	.727		239	45		284	330
4390	2-1/2"	Q-1	15	1.067		560	59.50		619.50	705
4400	3"	"	13	1.231		775	68.50		843.50	960
4500	For 300 psi, threaded, add					100%	15%			
4540	For chain operated type, add					15%				

22 05 Common Work Results for Plumbing

22 05 23 – General-Duty Valves for Plumbing Piping

22 05 23.20 Valves, Bronze		Crew	Daily Output	Labor-Hours	Unit	Material	2017 Bare Costs Labor	Equipment	Total	Total Incl O&P
4850	Globe, class 150, rising stem, threaded									
4920	1/4"	1 Plum	24	.333	Ea.	102	20.50		122.50	143
4940	3/8"		24	.333		101	20.50		121.50	142
4950	1/2"		24	.333		101	20.50		121.50	142
4960	3/4"		20	.400		134	24.50		158.50	186
4970	1"		19	.421		203	26		229	264
4980	1-1/4"		15	.533		299	33		332	380
4990	1-1/2"		13	.615		340	38		378	435
5000	2"		11	.727		545	45		590	670
5010	2-1/2"	Q-1	15	1.067		1,225	59.50		1,284.50	1,450
5020	3"	"	13	1.231		1,750	68.50		1,818.50	2,025
5120	For 300 lb. threaded, add					50%	15%			
5600	Relief, pressure & temperature, self-closing, ASME, threaded									
5640	3/4"	1 Plum	28	.286	Ea.	206	17.65		223.65	253
5650	1"		24	.333		330	20.50		350.50	390
5660	1-1/4"		20	.400		675	24.50		699.50	785
5670	1-1/2"		18	.444		1,300	27.50		1,327.50	1,475
5680	2"		16	.500		1,400	31		1,431	1,600
5950	Pressure, poppet type, threaded									
6000	1/2"	1 Plum	30	.267	Ea.	74.50	16.50		91	107
6040	3/4"	"	28	.286	"	84	17.65		101.65	119
6400	Pressure, water, ASME, threaded									
6440	3/4"	1 Plum	28	.286	Ea.	153	17.65		170.65	195
6450	1"		24	.333		287	20.50		307.50	345
6460	1-1/4"		20	.400		445	24.50		469.50	525
6470	1-1/2"		18	.444		625	27.50		652.50	730
6480	2"		16	.500		905	31		936	1,050
6490	2-1/2"		15	.533		3,450	33		3,483	3,850
6900	Reducing, water pressure									
6920	300 psi to 25-75 psi, threaded or sweat									
6940	1/2"	1 Plum	24	.333	Ea.	435	20.50		455.50	505
6950	3/4"		20	.400		445	24.50		469.50	530
6960	1"		19	.421		690	26		716	795
6970	1-1/4"		15	.533		1,200	33		1,233	1,375
6980	1-1/2"		13	.615		1,825	38		1,863	2,050
8350	Tempering, water, sweat connections									
8400	1/2"	1 Plum	24	.333	Ea.	118	20.50		138.50	161
8440	3/4"	"	20	.400	"	137	24.50		161.50	188
8650	Threaded connections									
8700	1/2"	1 Plum	24	.333	Ea.	141	20.50		161.50	186
8740	3/4"		20	.400		845	24.50		869.50	970
8750	1"		19	.421		955	26		981	1,100
8760	1-1/4"		15	.533		1,475	33		1,508	1,675
8770	1-1/2"		13	.615		1,600	38		1,638	1,825
8780	2"		11	.727		2,425	45		2,470	2,725
8800	Water heater water & gas safety shut off									
8810	Protection against a leaking water heater									
8814	Shut off valve	1 Plum	16	.500	Ea.	186	31		217	252
8818	Water heater dam		32	.250		31.50	15.45		46.95	58
8822	Gas control wiring harness		32	.250		24	15.45		39.45	50
8830	Whole house flood safety shut off									
8834	Connections									
8838	3/4" NPT	1 Plum	12	.667	Ea.	935	41		976	1,075

493

22 05 23 – General-Duty Valves for Plumbing Piping

22 05 23.20 Valves, Bronze	Crew	Daily Output	Labor-Hours	Unit	Material	2017 Bare Costs Labor	Equipment	Total	Total Incl O&P
8842	1" NPT	1 Plum	11	.727	Ea.	960	45		1,005	1,125
8846	1-1/4" NPT	↓	10	.800	↓	995	49.50		1,044.50	1,175

22 05 23.60 Valves, Plastic

		Crew	Daily Output	Labor-Hours	Unit	Material	Labor	Equipment	Total	Total Incl O&P
0010	**VALVES, PLASTIC**									
1100	Angle, PVC, threaded									
1110	1/4"	1 Plum	26	.308	Ea.	50	19		69	83.50
1120	1/2"		26	.308		71.50	19		90.50	108
1130	3/4"		25	.320		85	19.80		104.80	124
1140	1"	↓	23	.348	↓	103	21.50		124.50	147
1150	Ball, PVC, socket or threaded, true union									
1230	1/2"	1 Plum	26	.308	Ea.	47	19		66	80.50
1240	3/4"		25	.320		55.50	19.80		75.30	91.50
1250	1"		23	.348		66.50	21.50		88	106
1260	1-1/4"		21	.381		105	23.50		128.50	152
1270	1-1/2"		20	.400		105	24.50		129.50	154
1280	2"	↓	17	.471		138	29		167	196
1360	For PVC, flanged, add				↓	100%	15%			
1650	CPVC, socket or threaded, single union									
1700	1/2"	1 Plum	26	.308	Ea.	73	19		92	109
1720	3/4"		25	.320		87.50	19.80		107.30	126
1730	1"		23	.348		101	21.50		122.50	144
1750	1-1/4"		21	.381		146	23.50		169.50	196
1760	1-1/2"	↓	20	.400		176	24.50		200.50	231
1840	For CPVC, flanged, add					65%	15%			
1880	For true union, socket or threaded, add				↓	50%	5%			
2050	Polypropylene, threaded									
2100	1/4"	1 Plum	26	.308	Ea.	37	19		56	69.50
2120	3/8"		26	.308		37	19		56	69.50
2130	1/2"		26	.308		37	19		56	69
2140	3/4"		25	.320		44	19.80		63.80	78.50
2150	1"		23	.348		51.50	21.50		73	89
2160	1-1/4"		21	.381		68.50	23.50		92	111
2170	1-1/2"		20	.400		84.50	24.50		109	131
2180	2"	↓	17	.471	↓	114	29		143	169
4850	Foot valve, PVC, socket or threaded									
4900	1/2"	1 Plum	34	.235	Ea.	59	14.55		73.55	87
4930	3/4"		32	.250		66.50	15.45		81.95	97
4940	1"		28	.286		86.50	17.65		104.15	122
4950	1-1/4"		27	.296		166	18.30		184.30	211
4960	1-1/2"	↓	26	.308	↓	166	19		185	212
6350	Y sediment strainer, PVC, socket or threaded									
6400	1/2"	1 Plum	26	.308	Ea.	51	19		70	84.50
6440	3/4"		24	.333		54.50	20.50		75	91
6450	1"		23	.348		64.50	21.50		86	104
6460	1-1/4"		21	.381		109	23.50		132.50	155
6470	1-1/2"	↓	20	.400	↓	109	24.50		133.50	157

22 05 48 – Vibration and Seismic Controls for Plumbing Piping and Equipment

22 05 48.10 Seismic Bracing Supports

		Crew	Daily Output	Labor-Hours	Unit	Material	Labor	Equipment	Total	Total Incl O&P
0010	**SEISMIC BRACING SUPPORTS**									
0020	Clamps									
0030	C-clamp, for mounting on steel beam									
0040	3/8" threaded rod	1 Skwk	160	.050	Ea.	2.41	2.57		4.98	6.60

22 05 48.10 Seismic Bracing Supports	Crew	Daily Output	Labor-Hours	Unit	Material	Labor	Equipment	Total	Total Incl O&P
0050 1/2" threaded rod	1 Skwk	160	.050	Ea.	2.45	2.57		5.02	6.65
0060 5/8" threaded rod		160	.050		3.72	2.57		6.29	8.05
0070 3/4" threaded rod	↓	160	.050	↓	4.58	2.57		7.15	9
0100 Brackets									
0110 Beam side or wall malleable iron									
0120 3/8" threaded rod	1 Skwk	48	.167	Ea.	3.74	8.60		12.34	17.30
0130 1/2" threaded rod		48	.167		3.41	8.60		12.01	16.95
0140 5/8" threaded rod		48	.167		10.35	8.60		18.95	24.50
0150 3/4" threaded rod		48	.167		12.10	8.60		20.70	26.50
0160 7/8" threaded rod	↓	48	.167	↓	12.50	8.60		21.10	27
0170 For concrete installation, add						30%			
0180 Wall, welded steel									
0190 0 size 12" wide 18" deep	1 Skwk	34	.235	Ea.	171	12.10		183.10	207
0200 1 size 18" wide 24" deep		34	.235		228	12.10		240.10	270
0210 2 size 24" wide 30" deep	↓	34	.235	↓	299	12.10		311.10	350
0300 Rod, carbon steel									
0310 Continuous thread									
0320 1/4" thread	1 Skwk	144	.056	L.F.	2.18	2.86		5.04	6.80
0330 3/8" thread		144	.056		2.33	2.86		5.19	6.95
0340 1/2" thread		144	.056		3.67	2.86		6.53	8.45
0350 5/8" thread		144	.056		5.20	2.86		8.06	10.10
0360 3/4" thread		144	.056		9.15	2.86		12.01	14.50
0370 7/8" thread	↓	144	.056	↓	11.50	2.86		14.36	17.05
0380 For galvanized, add						30%			
0400 Channel, steel									
0410 3/4" x 1-1/2"	1 Skwk	80	.100	L.F.	2.47	5.15		7.62	10.60
0420 1-1/2" x 1-1/2"		70	.114		3.16	5.90		9.06	12.55
0430 1-7/8" x 1-1/2"		60	.133		23	6.85		29.85	36
0440 3" x 1-1/2"	↓	50	.160	↓	39.50	8.25		47.75	56
0450 Spring nuts									
0460 3/8"	1 Skwk	100	.080	Ea.	1.49	4.12		5.61	8
0470 1/2"	"	80	.100	"	1.69	5.15		6.84	9.75
0500 Welding, field									
0510 Cleaning and welding plates, bars, or rods									
0520 To existing beams, columns, or trusses									
0530 1" weld	1 Skwk	144	.056	Ea.	.23	2.86		3.09	4.65
0540 2" weld		72	.111		.46	5.70		6.16	9.30
0550 3" weld		54	.148		.69	7.60		8.29	12.50
0560 4" weld		36	.222		.92	11.45		12.37	18.60
0570 5" weld		30	.267		1.15	13.70		14.85	22.50
0580 6" weld	↓	24	.333	↓	1.38	17.15		18.53	28
0600 Vibration absorbers									
0610 Hangers, neoprene flex									
0620 10-120 lb. capacity	1 Skwk	8	1	Ea.	25.50	51.50		77	107
0630 75-550 lb. capacity		8	1		40.50	51.50		92	124
0640 250-1100 lb. capacity		6	1.333		82.50	68.50		151	197
0650 1000-4000 lb. capacity	↓	6	1.333	↓	147	68.50		215.50	268
0660 Spring flex									
0670 60-450 lb. capacity	1 Skwk	8	1	Ea.	75	51.50		126.50	161
0680 85-450 lb. capacity		8	1		101	51.50		152.50	190
0690 600-900 lb. capacity		8	1		111	51.50		162.50	201
0700 1100-1300 lb. capacity	↓	6	1.333	↓	123	68.50		191.50	242
0710 Mounts, neoprene									

22 05 Common Work Results for Plumbing

22 05 48 – Vibration and Seismic Controls for Plumbing Piping and Equipment

22 05 48.10 Seismic Bracing Supports		Crew	Daily Output	Labor-Hours	Unit	Material	2017 Bare Costs Labor	Equipment	Total	Total Incl O&P
0720	135-380 lb. capacity	1 Skwk	7	1.143	Ea.	17.80	59		76.80	110
0730	250-1100 lb. capacity		7	1.143		51	59		110	147
0740	1000-4000 lb. capacity	↓	5	1.600	↓	114	82.50		196.50	252

22 05 76 – Facility Drainage Piping Cleanouts

22 05 76.10 Cleanouts

		Crew	Daily Output	Labor-Hours	Unit	Material	2017 Bare Costs Labor	Equipment	Total	Total Incl O&P
0010	**CLEANOUTS**									
0060	Floor type									
0080	Round or square, scoriated nickel bronze top									
0100	2" pipe size	1 Plum	10	.800	Ea.	172	49.50		221.50	264
0120	3" pipe size		8	1		280	62		342	405
0140	4" pipe size	↓	6	1.333	↓	280	82.50		362.50	435
0980	Round top, recessed for terrazzo									
1000	2" pipe size	1 Plum	9	.889	Ea.	187	55		242	289
1080	3" pipe size		6	1.333		280	82.50		362.50	435
1100	4" pipe size	↓	4	2		280	124		404	495
1120	5" pipe size	Q-1	6	2.667	↓	355	148		503	615

22 05 76.20 Cleanout Tees

		Crew	Daily Output	Labor-Hours	Unit	Material	2017 Bare Costs Labor	Equipment	Total	Total Incl O&P
0010	**CLEANOUT TEES**									
0100	Cast iron, B&S, with countersunk plug									
0200	2" pipe size	1 Plum	4	2	Ea.	218	124		342	425
0220	3" pipe size		3.60	2.222		238	137		375	470
0240	4" pipe size	↓	3.30	2.424		296	150		446	550
0280	6" pipe size	Q-1	5	3.200	↓	800	178		978	1,150
0500	For round smooth access cover, same price									
4000	Plastic, tees and adapters. Add plugs									
4010	ABS, DWV									
4020	Cleanout tee, 1-1/2" pipe size	1 Plum	15	.533	Ea.	10.95	33		43.95	61.50
4030	2" pipe size	Q-1	27	.593		11.95	33		44.95	62.50
4040	3" pipe size		21	.762		22.50	42.50		65	88.50
4050	4" pipe size	↓	16	1		48.50	55.50		104	138
4100	Cleanout plug, 1-1/2" pipe size	1 Plum	32	.250		1.90	15.45		17.35	25.50
4110	2" pipe size	Q-1	56	.286		2.48	15.90		.18.38	26.50
4120	3" pipe size		36	.444		4.05	24.50		28.55	42
4130	4" pipe size		30	.533		7	29.50		36.50	52.50
4180	Cleanout adapter fitting, 1-1/2" pipe size	1 Plum	32	.250		3.02	15.45		18.47	27
4190	2" pipe size	Q-1	56	.286		4.60	15.90		20.50	29
4200	3" pipe size		36	.444		11.75	24.50		36.25	50.50
4210	4" pipe size	↓	30	.533	↓	22	29.50		51.50	69
5000	PVC, DWV									
5010	Cleanout tee, 1-1/2" pipe size	1 Plum	15	.533	Ea.	8.20	33		41.20	58.50
5020	2" pipe size	Q-1	27	.593		9.60	33		42.60	60
5030	3" pipe size		21	.762		18.60	42.50		61.10	84.50
5040	4" pipe size	↓	16	1		33	55.50		88.50	121
5090	Cleanout plug, 1-1/2" pipe size	1 Plum	32	.250		1.95	15.45		17.40	25.50
5100	2" pipe size	Q-1	56	.286		2.18	15.90		18.08	26.50
5110	3" pipe size		36	.444		3.90	24.50		28.40	42
5120	4" pipe size		30	.533		5.35	29.50		34.85	51
5130	6" pipe size	↓	24	.667		18.40	37		55.40	76
5170	Cleanout adapter fitting, 1-1/2" pipe size	1 Plum	32	.250		2.63	15.45		18.08	26.50
5180	2" pipe size	Q-1	56	.286		3.38	15.90		19.28	27.50
5190	3" pipe size		36	.444		9.45	24.50		33.95	48
5200	4" pipe size	↓	30	.533		15.55	29.50		45.05	62

For customer support on your Building Construction Costs with RSMeans Data, call 800.448.8182.

22 05 Common Work Results for Plumbing

22 05 76 – Facility Drainage Piping Cleanouts

22 05 76.20 Cleanout Tees		Crew	Daily Output	Labor-Hours	Unit	Material	2017 Bare Costs Labor	Equipment	Total	Total Incl O&P
5210	6" pipe size	Q-1	24	.667	Ea.	34.50	37		71.50	94

22 07 Plumbing Insulation

22 07 19 – Plumbing Piping Insulation

22 07 19.10 Piping Insulation

			Crew	Daily Output	Labor-Hours	Unit	Material	2017 Bare Costs Labor	Equipment	Total	Total Incl O&P
0010	**PIPING INSULATION**										
0100	Rule of thumb, as a percentage of total mechanical costs					Job				10%	10%
0110	Insulation req'd. is based on the surface size/area to be covered										
0600	Pipe covering (price copper tube one size less than IPS)										
6600	Fiberglass, with all service jacket										
6840	1" wall, 1/2" iron pipe size	G	Q-14	240	.067	L.F.	.85	3.30		4.15	6.10
6870	1" iron pipe size	G		220	.073		.99	3.60		4.59	6.70
6900	2" iron pipe size	G		200	.080		1.53	3.96		5.49	7.85
6920	3" iron pipe size	G		180	.089		1.86	4.40		6.26	8.90
6940	4" iron pipe size	G		150	.107		2.48	5.30		7.78	10.95
7320	2" wall, 1/2" iron pipe size	G		220	.073		3.07	3.60		6.67	9
7440	6" iron pipe size	G		100	.160		6.40	7.95		14.35	19.30
7460	8" iron pipe size	G		80	.200		7.50	9.90		17.40	23.50
7480	10" iron pipe size	G		70	.229		8.95	11.35		20.30	27.50
7490	12" iron pipe size	G		65	.246		10.05	12.20		22.25	30
7800	For fiberglass with standard canvas jacket, deduct						5%				
7802	For fittings, add 3 L.F. for each fitting										
7804	plus 4 L.F. for each flange of the fitting										
7810	Finishes										
7812	For .016" aluminum jacket, add	G	Q-14	200	.080	S.F.	1.01	3.96		4.97	7.25
7813	For .010" stainless steel, add	G	"	160	.100	"	3.15	4.96		8.11	11.15
7879	Rubber tubing, flexible closed cell foam										
7880	3/8" wall, 1/4" iron pipe size	G	1 Asbe	120	.067	L.F.	.31	3.67		3.98	6.05
7910	1/2" iron pipe size	G		115	.070		.47	3.83		4.30	6.45
7920	3/4" iron pipe size	G		115	.070		.54	3.83		4.37	6.55
7930	1" iron pipe size	G		110	.073		.57	4		4.57	6.85
7950	1-1/2" iron pipe size	G		110	.073		.75	4		4.75	7.05
8100	1/2" wall, 1/4" iron pipe size	G		90	.089		.72	4.89		5.61	8.40
8130	1/2" iron pipe size	G		89	.090		.84	4.95		5.79	8.60
8140	3/4" iron pipe size	G		89	.090		.85	4.95		5.80	8.65
8150	1" iron pipe size	G		88	.091		.82	5		5.82	8.70
8170	1-1/2" iron pipe size	G		87	.092		1.49	5.05		6.54	9.50
8180	2" iron pipe size	G		86	.093		1.91	5.10		7.01	10.05
8200	3" iron pipe size	G		85	.094		2.37	5.20		7.57	10.65
8300	3/4" wall, 1/4" iron pipe size	G		90	.089		.89	4.89		5.78	8.60
8330	1/2" iron pipe size	G		89	.090		1.09	4.95		6.04	8.90
8340	3/4" iron pipe size	G		89	.090		1.54	4.95		6.49	9.40
8350	1" iron pipe size	G		88	.091		1.77	5		6.77	9.75
8370	1-1/2" iron pipe size	G		87	.092		2.62	5.05		7.67	10.75
8380	2" iron pipe size	G		86	.093		3.04	5.10		8.14	11.30
8400	3" iron pipe size	G		85	.094		4.77	5.20		9.97	13.30
8444	1" wall, 1/2" iron pipe size	G		86	.093		2.32	5.10		7.42	10.50
8445	3/4" iron pipe size	G		84	.095		2.81	5.25		8.06	11.25
8446	1" iron pipe size	G		84	.095		3.39	5.25		8.64	11.90
8447	1-1/4" iron pipe size	G		82	.098		3.75	5.35		9.10	12.50
8448	1-1/2" iron pipe size	G		82	.098		4.88	5.35		10.23	13.70

22 07 Plumbing Insulation

22 07 19 – Plumbing Piping Insulation

22 07 19.10 Piping Insulation		Crew	Daily Output	Labor-Hours	Unit	Material	2017 Bare Costs Labor	Equipment	Total	Total Incl O&P
8449	2" iron pipe size	G	1 Asbe	80	.100	L.F.	6.10	5.50		11.60	15.25
8450	2-1/2" iron pipe size	G	↓	80	.100	↓	7.45	5.50		12.95	16.75
8456	Rubber insulation tape, 1/8" x 2" x 30'	G				Ea.	21			21	23

22 11 Facility Water Distribution

22 11 13 – Facility Water Distribution Piping

22 11 13.14 Pipe, Brass

		Crew	Daily Output	Labor-Hours	Unit	Material	2017 Bare Costs Labor	Equipment	Total	Total Incl O&P
0010	**PIPE, BRASS**, Plain end									
0900	Field threaded, coupling & clevis hanger assembly 10' O.C.									
0920	Regular weight									
1120	1/2" diameter	1 Plum	48	.167	L.F.	7.60	10.30		17.90	24
1140	3/4" diameter		46	.174		10.10	10.75		20.85	27.50
1160	1" diameter	↓	43	.186		17.40	11.50		28.90	36.50
1180	1-1/4" diameter	Q-1	72	.222		24.50	12.35		36.85	45.50
1200	1-1/2" diameter .		65	.246		29	13.70		42.70	52.50
1220	2" diameter	↓	53	.302	↓	30	16.80		46.80	58.50

22 11 13.23 Pipe/Tube, Copper

		Crew	Daily Output	Labor-Hours	Unit	Material	2017 Bare Costs Labor	Equipment	Total	Total Incl O&P
0010	**PIPE/TUBE, COPPER**, Solder joints									
1000	Type K tubing, couplings & clevis hanger assemblies 10' O.C.									
1100	1/4" diameter	1 Plum	84	.095	L.F.	3.77	5.90		9.67	13.05
1200	1" diameter		66	.121		12.60	7.50		20.10	25
1260	2" diameter	↓	40	.200	↓	25	12.35		37.35	46
2000	Type L tubing, couplings & clevis hanger assemblies 10' O.C.									
2100	1/4" diameter	1 Plum	88	.091	L.F.	2.70	5.60		8.30	11.45
2120	3/8" diameter		84	.095		3.35	5.90		9.25	12.60
2140	1/2" diameter		81	.099		3.54	6.10		9.64	13.10
2160	5/8" diameter		79	.101		4.99	6.25		11.24	14.95
2180	3/4" diameter		76	.105		4.90	6.50		11.40	15.20
2200	1" diameter		68	.118		9.80	7.25		17.05	21.50
2220	1-1/4" diameter		58	.138		12.25	8.50		20.75	26.50
2240	1-1/2" diameter		52	.154		14.45	9.50		23.95	30.50
2260	2" diameter	↓	42	.190		19.70	11.75		31.45	39.50
2280	2-1/2" diameter	Q-1	62	.258		25.50	14.35		39.85	49.50
2300	3" diameter		56	.286		34	15.90		49.90	61
2320	3-1/2" diameter		43	.372		46.50	20.50		67	82.50
2340	4" diameter		39	.410		58	23		81	98.50
2360	5" diameter	↓	34	.471		103	26		129	154
2380	6" diameter	Q-2	40	.600		136	34.50		170.50	201
2400	8" diameter	"	36	.667		227	38.50		265.50	305
2410	For other than full hard temper, add				↓	21%				
2590	For silver solder, add						15%			
4000	Type DWV tubing, couplings & clevis hanger assemblies 10' O.C.									
4100	1-1/4" diameter	1 Plum	60	.133	L.F.	11.40	8.25		19.65	25
4120	1-1/2" diameter		54	.148		13.30	9.15		22.45	28.50
4140	2" diameter	↓	44	.182		16.50	11.25		27.75	35
4160	3" diameter	Q-1	58	.276		24.50	15.35		39.85	50
4180	4" diameter		40	.400		44.50	22.50		67	82.50
4200	5" diameter	↓	36	.444		101	24.50		125.50	149
4220	6" diameter	Q-2	42	.571	↓	137	33		170	201

22 11 Facility Water Distribution

22 11 13 – Facility Water Distribution Piping

22 11 13.44 Pipe, Steel

		Crew	Daily Output	Labor-Hours	Unit	Material	2017 Bare Costs Labor	2017 Bare Costs Equipment	Total	Total Incl O&P
0010	**PIPE, STEEL**									
0012	The steel pipe in this section does not include fittings such as ells, tees									
0014	For fittings either add a % (usually 25 to 35%) or see									
0015	the Mechanical or Plumbing Costs									
0020	All pipe sizes are to Spec. A-53 unless noted otherwise R221113-50									
0050	Schedule 40, threaded, with couplings, and clevis hanger									
0060	assemblies sized for covering, 10' O.C.									
0540	Black, 1/4" diameter	1 Plum	66	.121	L.F.	7.80	7.50		15.30	19.90
0550	3/8" diameter		65	.123		8.55	7.60		16.15	21
0560	1/2" diameter		63	.127		5.95	7.85		13.80	18.40
0570	3/4" diameter		61	.131		6.25	8.10		14.35	19.10
0580	1" diameter		53	.151		4.58	9.35		13.93	19.15
0590	1-1/4" diameter	Q-1	89	.180		5.40	10		15.40	21
0600	1-1/2" diameter		80	.200		5.95	11.15		17.10	23.50
0610	2" diameter		64	.250		12.55	13.90		26.45	35
0620	2-1/2" diameter		50	.320		16.75	17.80		34.55	45.50
0630	3" diameter		43	.372		19.60	20.50		40.10	52.50
0640	3-1/2" diameter		40	.400		25.50	22.50		48	61.50
0650	4" diameter		36	.444		22.50	24.50		47	62
1280	All pipe sizes are to Spec. A-53 unless noted otherwise									
1281	Schedule 40, threaded, with couplings and clevis hanger									
1282	assemblies sized for covering, 10' O.C.									
1290	Galvanized, 1/4" diameter	1 Plum	66	.121	L.F.	9.75	7.50		17.25	22
1300	3/8" diameter		65	.123		10.25	7.60		17.85	23
1310	1/2" diameter		63	.127		5.95	7.85		13.80	18.40
1320	3/4" diameter		61	.131		6.50	8.10		14.60	19.45
1330	1" diameter		53	.151		5.10	9.35		14.45	19.70
1340	1-1/4" diameter	Q-1	89	.180		5.95	10		15.95	21.50
1350	1-1/2" diameter		80	.200		6.75	11.15		17.90	24
1360	2" diameter		64	.250		13.60	13.90		27.50	36
1370	2-1/2" diameter		50	.320		18.55	17.80		36.35	47.50
1380	3" diameter		43	.372		22	20.50		42.50	55
1390	3-1/2" diameter		40	.400		27	22.50		49.50	63
1400	4" diameter		36	.444		25	24.50		49.50	65
2000	Welded, sch. 40, on yoke & roll hanger assy's, sized for covering, 10' O.C.									
2040	Black, 1" diameter	Q-15	93	.172	L.F.	4.21	9.55	.62	14.38	19.75
2070	2" diameter		61	.262		11.70	14.60	.95	27.25	36
2090	3" diameter		43	.372		17.65	20.50	1.35	39.50	52
2110	4" diameter		37	.432		17.85	24	1.57	43.42	58
2120	5" diameter		32	.500		37.50	28	1.82	67.32	85
2130	6" diameter	Q-16	36	.667		46	38.50	1.61	86.11	110
2140	8" diameter		29	.828		75	47.50	2	124.50	157
2150	10" diameter		24	1		96	57.50	2.42	155.92	195
2160	12" diameter		19	1.263		113	73	3.06	189.06	237

22 11 13.48 Pipe, Fittings and Valves, Steel, Grooved-Joint

		Crew	Daily Output	Labor-Hours	Unit	Material	2017 Bare Costs Labor	2017 Bare Costs Equipment	Total	Total Incl O&P
0010	**PIPE, FITTINGS AND VALVES, STEEL, GROOVED-JOINT**									
0012	Fittings are ductile iron. Steel fittings noted.									
0020	Pipe includes coupling & clevis type hanger assemblies, 10' O.C.									
1000	Schedule 40, black									
1040	3/4" diameter	1 Plum	71	.113	L.F.	7.90	6.95		14.85	19.20
1050	1" diameter		63	.127		7.70	7.85		15.55	20.50
1060	1-1/4" diameter		58	.138		8.80	8.50		17.30	22.50

For customer support on your Building Construction Costs with RSMeans Data, call 800.448.8182.

499

22 11 13.48 Pipe, Fittings and Valves, Steel, Grooved-Joint		Crew	Daily Output	Labor-Hours	Unit	Material	2017 Bare Costs Labor	Equipment	Total	Total Incl O&P
1070	1-1/2" diameter	1 Plum	51	.157	L.F.	9.45	9.70		19.15	25
1080	2" diameter	▼	40	.200		10.65	12.35		23	30.50
1090	2-1/2" diameter	Q-1	57	.281		11.70	15.60		27.30	36.50
1100	3" diameter		50	.320		14.25	17.80		32.05	42.50
1110	4" diameter		45	.356		24.50	19.80		44.30	57
1120	5" diameter	▼	37	.432		45.50	24		69.50	87
1130	6" diameter	Q-2	42	.571	▼	52	33		85	107
1800	Galvanized									
1840	3/4" diameter	1 Plum	71	.113	L.F.	8.15	6.95		15.10	19.50
1850	1" diameter		63	.127		8.40	7.85		16.25	21
1860	1-1/4" diameter		58	.138		9.75	8.50		18.25	23.50
1870	1-1/2" diameter		51	.157		10.65	9.70		20.35	26.50
1880	2" diameter	▼	40	.200		12.05	12.35		24.40	32
1890	2-1/2" diameter	Q-1	57	.281		13.35	15.60		28.95	38
1900	3" diameter		50	.320		16.40	17.80		34.20	45
1910	4" diameter		45	.356		28	19.80		47.80	60.50
1920	5" diameter	▼	37	.432		32	24		56	72
1930	6" diameter	Q-2	42	.571	▼	34.50	33		67.50	87.50
3990	Fittings: coupling material required at joints not incl. in fitting price.									
3994	Add 1 selected coupling, material only, per joint for installed price.									
4000	Elbow, 90° or 45°, painted									
4030	3/4" diameter	1 Plum	50	.160	Ea.	63.50	9.90		73.40	85
4040	1" diameter		50	.160		34	9.90		43.90	52.50
4050	1-1/4" diameter		40	.200		34	12.35		46.35	56
4060	1-1/2" diameter		33	.242		34	15		49	60
4070	2" diameter	▼	25	.320		34	19.80		53.80	67.50
4080	2-1/2" diameter	Q-1	40	.400		34	22.50		56.50	71
4090	3" diameter		33	.485		60	27		87	107
4100	4" diameter		25	.640		65	35.50		100.50	125
4110	5" diameter	▼	20	.800		155	44.50		199.50	238
4120	6" diameter	Q-2	25	.960		182	55.50		237.50	285
4250	For galvanized elbows, add				▼	26%				
4690	Tee, painted									
4700	3/4" diameter	1 Plum	38	.211	Ea.	68.50	13		81.50	94.50
4740	1" diameter		33	.242		53	15		68	80.50
4750	1-1/4" diameter		27	.296		53	18.30		71.30	85.50
4760	1-1/2" diameter		22	.364		53	22.50		75.50	92
4770	2" diameter	▼	17	.471		53	29		82	102
4780	2-1/2" diameter	Q-1	27	.593		53	33		86	108
4790	3" diameter		22	.727		72	40.50		112.50	141
4800	4" diameter		17	.941		110	52.50		162.50	200
4810	5" diameter	▼	13	1.231		256	68.50		324.50	385
4820	6" diameter	Q-2	17	1.412		295	81.50		376.50	450
4900	For galvanized tees, add				▼	24%				
4906	Couplings, rigid style, painted									
4908	1" diameter	1 Plum	100	.080	Ea.	26.50	4.94		31.44	36.50
4909	1-1/4" diameter		100	.080		24.50	4.94		29.44	34.50
4910	1-1/2" diameter		67	.119		26.50	7.40		33.90	40
4912	2" diameter	▼	50	.160		33.50	9.90		43.40	51.50
4914	2-1/2" diameter	Q-1	80	.200		38	11.15		49.15	59
4916	3" diameter		67	.239		44	13.30		57.30	68.50
4918	4" diameter		50	.320		61.50	17.80		79.30	95
4920	5" diameter	▼	40	.400	▼	79.50	22.50		102	121

22 11 Facility Water Distribution

22 11 13 – Facility Water Distribution Piping

22 11 13.48 Pipe, Fittings and Valves, Steel, Grooved-Joint		Crew	Daily Output	Labor-Hours	Unit	Material	2017 Bare Costs Labor	Equipment	Total	Total Incl O&P
4922	6" diameter	Q-2	50	.480	Ea.	105	27.50		132.50	158
4940	Flexible, standard, painted									
4950	3/4" diameter	1 Plum	100	.080	Ea.	19	4.94		23.94	28.50
4960	1" diameter		100	.080		19	4.94		23.94	28.50
4970	1-1/4" diameter		80	.100		25	6.20		31.20	37
4980	1-1/2" diameter		67	.119		27	7.40		34.40	40.50
4990	2" diameter		50	.160		29	9.90		38.90	47
5000	2-1/2" diameter	Q-1	80	.200		33.50	11.15		44.65	54
5010	3" diameter		67	.239		37.50	13.30		50.80	61
5020	3-1/2" diameter		57	.281		53.50	15.60		69.10	82
5030	4" diameter		50	.320		54	17.80		71.80	86
5040	5" diameter		40	.400		81	22.50		103.50	123
5050	6" diameter	Q-2	50	.480		96	27.50		123.50	148
5200	For galvanized couplings, add					33%				
7790	Valves: coupling material required at joints not incl. in valve price.									
7794	Add 1 selected coupling, material only, per joint for installed price.									

22 11 13.64 Pipe, Stainless Steel

		Crew	Daily Output	Labor-Hours	Unit	Material	2017 Bare Costs Labor	Equipment	Total	Total Incl O&P
0010	**PIPE, STAINLESS STEEL**									
3500	Threaded, couplings and clevis hanger assemblies, 10' O.C.									
3520	Schedule 40, type 304									
3540	1/4" diameter	1 Plum	54	.148	L.F.	9.15	9.15		18.30	24
3550	3/8" diameter		53	.151		10.65	9.35		20	26
3560	1/2" diameter		52	.154		15.20	9.50		24.70	31
3580	1" diameter		45	.178		25	11		36	44
3610	2" diameter	Q-1	57	.281		57	15.60		72.60	86
3640	4" diameter	Q-2	51	.471		146	27		173	201
3740	For small quantities, add					10%				
4250	Schedule 40, type 316									
4290	1/4" diameter	1 Plum	54	.148	L.F.	13.95	9.15		23.10	29
4300	3/8" diameter		53	.151		15.25	9.35		24.60	31
4310	1/2" diameter		52	.154		21.50	9.50		31	38
4320	3/4" diameter		51	.157		24.50	9.70		34.20	41.50
4330	1" diameter		45	.178		33	11		44	53
4360	2" diameter	Q-1	57	.281		65.50	15.60		81.10	96
4390	4" diameter	Q-2	51	.471		159	27		186	216
4490	For small quantities, add					10%				

22 11 13.74 Pipe, Plastic

		Crew	Daily Output	Labor-Hours	Unit	Material	2017 Bare Costs Labor	Equipment	Total	Total Incl O&P
0010	**PIPE, PLASTIC**									
1800	PVC, couplings 10' O.C., clevis hanger assemblies, 3 per 10'									
1820	Schedule 40									
1860	1/2" diameter	1 Plum	54	.148	L.F.	5	9.15		14.15	19.30
1870	3/4" diameter		51	.157		5.30	9.70		15	20.50
1880	1" diameter		46	.174		14.40	10.75		25.15	32
1890	1-1/4" diameter		42	.190		15.10	11.75		26.85	34.50
1900	1-1/2" diameter		36	.222		15.30	13.75		29.05	37.50
1910	2" diameter	Q-1	59	.271		16.55	15.10		31.65	41
1920	2-1/2" diameter		56	.286		10	15.90		25.90	35
1930	3" diameter		53	.302		12.20	16.80		29	39
1940	4" diameter		48	.333		30	18.55		48.55	61
1950	5" diameter		43	.372		38.50	20.50		59	73.50
1960	6" diameter		39	.410		26	23		49	63
4100	DWV type, schedule 40, couplings 10' O.C., clevis hanger assy's, 3 per 10'									

22 11 13.74 Pipe, Plastic	Crew	Daily Output	Labor-Hours	Unit	Material	2017 Bare Costs Labor	Equipment	Total	Total Incl O&P
4210 ABS, schedule 40, foam core type									
4212 Plain end black									
4214 1-1/2" diameter	1 Plum	39	.205	L.F.	13.90	12.70		26.60	34.50
4216 2" diameter	Q-1	62	.258		14.20	14.35		28.55	37
4218 3" diameter		56	.286		8.05	15.90		23.95	33
4220 4" diameter		51	.314		25	17.45		42.45	54
4222 6" diameter		42	.381		18.55	21		39.55	52.50
4240 To delete coupling & hangers, subtract									
4244 1-1/2" diam. to 6" diam.					43%	48%			
4400 PVC									
4410 1-1/4" diameter	1 Plum	42	.190	L.F.	14.15	11.75		25.90	33.50
4420 1-1/2" diameter	"	36	.222		13.85	13.75		27.60	36
4460 2" diameter	Q-1	59	.271		14.25	15.10		29.35	38.50
4470 3" diameter		53	.302		8.05	16.80		24.85	34.50
4480 4" diameter		48	.333		10.05	18.55		28.60	39
4490 6" diameter		39	.410		16.80	23		39.80	53
5300 CPVC, socket joint, couplings 10' O.C., clevis hanger assemblies, 3 per 10'									
5302 Schedule 40									
5304 1/2" diameter	1 Plum	54	.148	L.F.	6.20	9.15		15.35	20.50
5305 3/4" diameter		51	.157		7	9.70		16.70	22.50
5306 1" diameter		46	.174		17.05	10.75		27.80	35
5307 1-1/4" diameter		42	.190		18.65	11.75		30.40	38.50
5308 1-1/2" diameter		36	.222		18.55	13.75		32.30	41
5309 2" diameter	Q-1	59	.271		20	15.10		35.10	45
5310 2-1/2" diameter		56	.286		17.95	15.90		33.85	44
5311 3" diameter		53	.302		24.50	16.80		41.30	52.50
5360 CPVC, threaded, couplings 10' O.C., clevis hanger assemblies, 3 per 10'									
5380 Schedule 40									
5460 1/2" diameter	1 Plum	54	.148	L.F.	7	9.15		16.15	21.50
5470 3/4" diameter		51	.157		8.40	9.70		18.10	24
5480 1" diameter		46	.174		18.50	10.75		29.25	36.50
5490 1-1/4" diameter		42	.190		19.75	11.75		31.50	39.50
5500 1-1/2" diameter		36	.222		19.50	13.75		33.25	42
5510 2" diameter	Q-1	59	.271		21.50	15.10		36.60	46.50
5520 2-1/2" diameter		56	.286		19.20	15.90		35.10	45
5530 3" diameter		53	.302		26.50	16.80		43.30	54.50
7280 PEX, flexible, no couplings or hangers									
7282 Note: For labor costs add 25% to the couplings and fittings labor total.									
7285 For fittings see section 23 83 16.10 7000									
7300 Non-barrier type, hot/cold tubing rolls									
7310 1/4" diameter x 100'				L.F.	.49			.49	.54
7350 3/8" diameter x 100'					.55			.55	.61
7360 1/2" diameter x 100'					.61			.61	.67
7370 1/2" diameter x 500'					.61			.61	.67
7380 1/2" diameter x 1000'					.61			.61	.67
7400 3/4" diameter x 100'					.92			.92	1.01
7410 3/4" diameter x 500'					1.11			1.11	1.22
7420 3/4" diameter x 1000'					1.11			1.11	1.22
7460 1" diameter x 100'					1.90			1.90	2.09
7470 1" diameter x 300'					1.90			1.90	2.09
7480 1" diameter x 500'					1.90			1.90	2.09
7500 1-1/4" diameter x 100'					3.23			3.23	3.55
7510 1-1/4" diameter x 300'					3.23			3.23	3.55

22 11 Facility Water Distribution

22 11 13 – Facility Water Distribution Piping

22 11 13.74 Pipe, Plastic	Crew	Daily Output	Labor-Hours	Unit	Material	2017 Bare Costs Labor	2017 Bare Costs Equipment	Total	Total Incl O&P
7540	1-1/2" diameter x 100'				L.F.	4.40			4.40	4.84
7550	1-1/2" diameter x 300'					4.40			4.40	4.84
7596	Most sizes available in red or blue									
7700	Non-barrier type, hot/cold tubing straight lengths									
7710	1/2" diameter x 20'				L.F.	.61			.61	.67
7750	3/4" diameter x 20'					1.11			1.11	1.22
7760	1" diameter x 20'					1.90			1.90	2.09
7770	1-1/4" diameter x 20'					3.23			3.23	3.55
7780	1-1/2" diameter x 20'					4.40			4.40	4.84
7790	2" diameter					8.60			8.60	9.45
7796	Most sizes available in red or blue									

22 11 19 – Domestic Water Piping Specialties

22 11 19.10 Flexible Connectors

		Crew	Daily Output	Labor-Hours	Unit	Material	Labor	Equipment	Total	Total Incl O&P
0010	**FLEXIBLE CONNECTORS**, Corrugated, 7/8" O.D., 1/2" I.D.									
0050	Gas, seamless brass, steel fittings									
0200	12" long	1 Plum	36	.222	Ea.	6.20	13.75		19.95	27.50
0220	18" long		36	.222		7.70	13.75		21.45	29
0240	24" long		34	.235		9.15	14.55		23.70	32
0280	36" long		32	.250		10.90	15.45		26.35	35.50
0340	60" long		30	.267		16.45	16.50		32.95	43
2000	Water, copper tubing, dielectric separators									
2100	12" long	1 Plum	36	.222	Ea.	6.20	13.75		19.95	27.50
2260	24" long	"	34	.235	"	9.20	14.55		23.75	32

22 11 19.14 Flexible Metal Hose

		Crew	Daily Output	Labor-Hours	Unit	Material	Labor	Equipment	Total	Total Incl O&P
0010	**FLEXIBLE METAL HOSE**, Connectors, standard lengths									
0100	Bronze braided, bronze ends									
0120	3/8" diameter x 12"	1 Stpi	26	.308	Ea.	27	19.15		46.15	58.50
0160	3/4" diameter x 12"		20	.400		41	25		66	82.50
0180	1" diameter x 18"		19	.421		48.50	26		74.50	93
0200	1-1/2" diameter x 18"		13	.615		66.50	38.50		105	131
0220	2" diameter x 18"		11	.727		87	45.50		132.50	164

22 11 19.26 Pressure Regulators

		Crew	Daily Output	Labor-Hours	Unit	Material	Labor	Equipment	Total	Total Incl O&P
0010	**PRESSURE REGULATORS**									
3000	Steam, high capacity, bronze body, stainless steel trim									
3020	Threaded, 1/2" diameter	1 Stpi	24	.333	Ea.	2,450	21		2,471	2,700
3030	3/4" diameter		24	.333		2,450	21		2,471	2,700
3040	1" diameter		19	.421		2,725	26		2,751	3,050
3060	1-1/4" diameter		15	.533		3,000	33		3,033	3,350
3080	1-1/2" diameter		13	.615		3,450	38.50		3,488.50	3,850
3100	2" diameter		11	.727		4,225	45.50		4,270.50	4,725
3120	2-1/2" diameter	Q-5	12	1.333		5,275	74.50		5,349.50	5,950
3140	3" diameter	"	11	1.455		6,025	81.50		6,106.50	6,750
3500	Flanged connection, iron body, 125 lb. W.S.P.									
3520	3" diameter	Q-5	11	1.455	Ea.	6,600	81.50		6,681.50	7,375
3540	4" diameter	"	5	3.200	"	8,325	179		8,504	9,425

22 11 19.38 Water Supply Meters

		Crew	Daily Output	Labor-Hours	Unit	Material	Labor	Equipment	Total	Total Incl O&P
0010	**WATER SUPPLY METERS**									
2000	Domestic/commercial, bronze									
2020	Threaded									
2060	5/8" diameter, to 20 GPM	1 Plum	16	.500	Ea.	50.50	31		81.50	102
2080	3/4" diameter, to 30 GPM		14	.571		92	35.50		127.50	155

For customer support on your Building Construction Costs with RSMeans Data, call 800.448.8182.

503

22 11 Facility Water Distribution

22 11 19 – Domestic Water Piping Specialties

22 11 19.38 Water Supply Meters

		Crew	Daily Output	Labor-Hours	Unit	Material	2017 Bare Costs Labor	2017 Bare Costs Equipment	Total	Total Incl O&P
2100	1" diameter, to 50 GPM	1 Plum	12	.667	Ea.	140	41		181	216
2300	Threaded/flanged									
2340	1-1/2" diameter, to 100 GPM	1 Plum	8	1	Ea.	340	62		402	470
2360	2" diameter, to 160 GPM	"	6	1.333	"	465	82.50		547.50	635
2600	Flanged, compound									
2640	3" diameter, 320 GPM	Q-1	3	5.333	Ea.	2,175	297		2,472	2,850
2660	4" diameter, to 500 GPM		1.50	10.667		3,475	595		4,070	4,725
2680	6" diameter, to 1,000 GPM		1	16		5,700	890		6,590	7,625
2700	8" diameter, to 1,800 GPM	↓	.80	20	↓	8,850	1,125		9,975	11,400

22 11 19.42 Backflow Preventers

		Crew	Daily Output	Labor-Hours	Unit	Material	2017 Bare Costs Labor	2017 Bare Costs Equipment	Total	Total Incl O&P
0010	**BACKFLOW PREVENTERS**, Includes valves									
0020	and four test cocks, corrosion resistant, automatic operation									
4000	Reduced pressure principle									
4100	Threaded, bronze, valves are ball									
4120	3/4" pipe size	1 Plum	16	.500	Ea.	300	31		331	375
4140	1" pipe size		14	.571		340	35.50		375.50	430
4150	1-1/4" pipe size		12	.667		560	41		601	675
4160	1-1/2" pipe size		10	.800		645	49.50		694.50	785
4180	2" pipe size	↓	7	1.143	↓	690	70.50		760.50	865
5000	Flanged, bronze, valves are OS&Y									
5060	2-1/2" pipe size	Q-1	5	3.200	Ea.	5,100	178		5,278	5,875
5080	3" pipe size		4.50	3.556		5,800	198		5,998	6,675
5100	4" pipe size	↓	3	5.333		7,250	297		7,547	8,425
5120	6" pipe size	Q-2	3	8	↓	10,500	460		10,960	12,300
5600	Flanged, iron, valves are OS&Y									
5660	2-1/2" pipe size	Q-1	5	3.200	Ea.	2,600	178		2,778	3,150
5680	3" pipe size		4.50	3.556		2,750	198		2,948	3,325
5700	4" pipe size	↓	3	5.333		3,450	297		3,747	4,250
5720	6" pipe size	Q-2	3	8		4,975	460		5,435	6,175
5740	8" pipe size		2	12		8,150	690		8,840	10,000
5760	10" pipe size	↓	1	24	↓	10,900	1,375		12,275	14,100

22 11 19.50 Vacuum Breakers

		Crew	Daily Output	Labor-Hours	Unit	Material	2017 Bare Costs Labor	2017 Bare Costs Equipment	Total	Total Incl O&P
0010	**VACUUM BREAKERS**									
0013	See also backflow preventers Section 22 11 19.42									
1000	Anti-siphon continuous pressure type									
1010	Max. 150 PSI - 210°F									
1020	Bronze body									
1030	1/2" size	1 Stpi	24	.333	Ea.	166	21		187	214
1040	3/4" size		20	.400		166	25		191	220
1050	1" size		19	.421		170	26		196	227
1060	1-1/4" size		15	.533		335	33		368	420
1070	1-1/2" size		13	.615		415	38.50		453.50	515
1080	2" size	↓	11	.727	↓	425	45.50		470.50	540
1200	Max. 125 PSI with atmospheric vent									
1210	Brass, in-line construction									
1220	1/4" size	1 Stpi	24	.333	Ea.	129	21		150	174
1230	3/8" size	"	24	.333	↓	129	21		150	174
1260	For polished chrome finish, add					13%				
2000	Anti-siphon, non-continuous pressure type									
2010	Hot or cold water 125 PSI - 210°F									
2020	Bronze body									
2030	1/4" size	1 Stpi	24	.333	Ea.	71	21		92	110

22 11 19 – Domestic Water Piping Specialties

22 11 19.50 Vacuum Breakers

		Crew	Daily Output	Labor-Hours	Unit	Material	2017 Bare Costs Labor	Equipment	Total	Total Incl O&P
2040	3/8" size	1 Stpi	24	.333	Ea.	71	21		92	110
2050	1/2" size		24	.333		80	21		101	120
2060	3/4" size		20	.400		95	25		120	143
2070	1" size		19	.421		148	26		174	202
2080	1-1/4" size		15	.533		258	33		291	335
2090	1-1/2" size		13	.615		305	38.50		343.50	395
2100	2" size		11	.727		470	45.50		515.50	590
2110	2-1/2" size		8	1		1,350	62.50		1,412.50	1,600
2120	3" size	↓	6	1.333	↓	1,800	83		1,883	2,100
2150	For polished chrome finish, add					50%				

22 11 19.54 Water Hammer Arresters/Shock Absorbers

		Crew	Daily Output	Labor-Hours	Unit	Material	2017 Bare Costs Labor	Equipment	Total	Total Incl O&P
0010	**WATER HAMMER ARRESTERS/SHOCK ABSORBERS**									
0490	Copper									
0500	3/4" male I.P.S. For 1 to 11 fixtures	1 Plum	12	.667	Ea.	28.50	41		69.50	93
0600	1" male I.P.S. For 12 to 32 fixtures		8	1		48	62		110	147
0700	1-1/4" male I.P.S. For 33 to 60 fixtures		8	1		48.50	62		110.50	147
0800	1-1/2" male I.P.S. For 61 to 113 fixtures		8	1		70	62		132	171
0900	2" male I.P.S. For 114 to 154 fixtures		8	1		102	62		164	206
1000	2-1/2" male I.P.S. For 155 to 330 fixtures	↓	4	2	↓	310	124		434	525

22 11 19.64 Hydrants

		Crew	Daily Output	Labor-Hours	Unit	Material	2017 Bare Costs Labor	Equipment	Total	Total Incl O&P
0010	**HYDRANTS**									
0050	Wall type, moderate climate, bronze, encased									
0200	3/4" IPS connection	1 Plum	16	.500	Ea.	640	31		671	750
0300	1" IPS connection		14	.571		620	35.50		655.50	740
0500	Anti-siphon type, 3/4" connection	↓	16	.500	↓	675	31		706	790
1000	Non-freeze, bronze, exposed									
1100	3/4" IPS connection, 4" to 9" thick wall	1 Plum	14	.571	Ea.	420	35.50		455.50	520
1120	10" to 14" thick wall		12	.667		455	41		496	560
1140	15" to 19" thick wall		12	.667		510	41		551	620
1160	20" to 24" thick wall	↓	10	.800	↓	630	49.50		679.50	770
1200	For 1" IPS connection, add					15%	10%			
1240	For 3/4" adapter type vacuum breaker, add				Ea.	53			53	58.50
1280	For anti-siphon type, add				"	112			112	123
2000	Non-freeze bronze, encased, anti-siphon type									
2100	3/4" IPS connection, 5" to 9" thick wall	1 Plum	14	.571	Ea.	1,175	35.50		1,210.50	1,325
2120	10" to 14" thick wall		12	.667		1,375	41		1,416	1,550
2140	15" to 19" thick wall	↓	12	.667	↓	1,500	41		1,541	1,700
3000	Ground box type, bronze frame, 3/4" IPS connection									
3080	Non-freeze, all bronze, polished face, set flush									
3100	2 feet depth of bury	1 Plum	8	1	Ea.	870	62		932	1,050
3140	4 feet depth of bury		8	1		960	62		1,022	1,150
3180	6 feet depth of bury		7	1.143		1,200	70.50		1,270.50	1,425
3220	8 feet depth of bury	↓	5	1.600		1,525	99		1,624	1,825
3400	For 1" IPS connection, add					15%	10%			
3550	For 2" connection, add					445%	24%			
3600	For tapped drain port in box, add				↓	72			72	79
5000	Moderate climate, all bronze, polished face									
5020	and scoriated cover, set flush									
5100	3/4" IPS connection	1 Plum	16	.500	Ea.	725	31		756	840
5120	1" IPS connection	"	14	.571		1,100	35.50		1,135.50	1,250
5200	For tapped drain port in box, add				↓	72			72	79

22 11 Facility Water Distribution

22 11 23 – Domestic Water Pumps

22 11 23.10 General Utility Pumps

22 11 23.10 General Utility Pumps	Crew	Daily Output	Labor-Hours	Unit	Material	2017 Bare Costs Labor	Equipment	Total	Total Incl O&P
0010 **GENERAL UTILITY PUMPS**									
2000 Single stage									
3000 Double suction,									
3190 75 HP, to 2500 GPM	Q-3	.28	114	Ea.	20,900	6,725		27,625	33,100
3220 100 HP, to 3000 GPM		.26	123		26,000	7,250		33,250	39,500
3240 150 HP, to 4000 GPM		.24	133		30,600	7,850		38,450	45,400

22 13 Facility Sanitary Sewerage

22 13 16 – Sanitary Waste and Vent Piping

22 13 16.20 Pipe, Cast Iron

22 13 16.20 Pipe, Cast Iron	Crew	Daily Output	Labor-Hours	Unit	Material	2017 Bare Costs Labor	Equipment	Total	Total Incl O&P
0010 **PIPE, CAST IRON**, Soil, on clevis hanger assemblies, 5' O.C.									
0020 Single hub, service wt., lead & oakum joints 10' O.C.									
2120 2" diameter	Q-1	63	.254	L.F.	21	14.15		35.15	44.50
2140 3" diameter		60	.267		20	14.85		34.85	44.50
2160 4" diameter		55	.291		35	16.20		51.20	63
2180 5" diameter	Q-2	76	.316		43.50	18.20		61.70	75.50
2200 6" diameter	"	73	.329		42.50	18.95		61.45	75
2220 8" diameter	Q-3	59	.542		63.50	32		95.50	118
2240 10" diameter		54	.593		101	35		136	164
2260 12" diameter		48	.667		144	39		183	217
2320 For service weight, double hub, add					10%				
2340 For extra heavy, single hub, add					48%	4%			
2360 For extra heavy, double hub, add					71%	4%			
2400 Lead for caulking (1#/diam. in.)	Q-1	160	.100	Lb.	1.04	5.55		6.59	9.55
2420 Oakum for caulking (1/8#/diam. in.)	"	40	.400	"	4.22	22.50		26.72	38
4000 No hub, couplings 10' O.C.									
4100 1-1/2" diameter	Q-1	71	.225	L.F.	19.95	12.55		32.50	41
4120 2" diameter		67	.239		21	13.30		34.30	43
4140 3" diameter		64	.250		19.80	13.90		33.70	43
4160 4" diameter		58	.276		35	15.35		50.35	61.50

22 13 16.50 Shower Drains

22 13 16.50 Shower Drains	Crew	Daily Output	Labor-Hours	Unit	Material	2017 Bare Costs Labor	Equipment	Total	Total Incl O&P
0010 **SHOWER DRAINS**									
2780 Shower, with strainer, uniform diam. trap, bronze top									
2800 2" and 3" pipe size	Q-1	8	2	Ea.	345	111		456	550
2820 4" pipe size	"	7	2.286		380	127		507	610
2840 For galvanized body, add					135			135	149

22 13 16.60 Traps

22 13 16.60 Traps	Crew	Daily Output	Labor-Hours	Unit	Material	2017 Bare Costs Labor	Equipment	Total	Total Incl O&P
0010 **TRAPS**									
0030 Cast iron, service weight									
0050 Running P trap, without vent									
1100 2"	Q-1	16	1	Ea.	148	55.50		203.50	247
1140 3"		14	1.143		148	63.50		211.50	259
1150 4"		13	1.231		148	68.50		216.50	266
1160 6"	Q-2	17	1.412		655	81.50		736.50	845
1180 Running trap, single hub, with vent									
2080 3" pipe size, 3" vent	Q-1	14	1.143	Ea.	117	63.50		180.50	225
2120 4" pipe size, 4" vent	"	13	1.231		160	68.50		228.50	278
2300 For double hub, vent, add					10%	20%			
3000 P trap, B&S, 2" pipe size	Q-1	16	1		35	55.50		90.50	123
3040 3" pipe size	"	14	1.143		52.50	63.50		116	154

22 13 16 – Sanitary Waste and Vent Piping

22 13 16.60 Traps		Crew	Daily Output	Labor-Hours	Unit	Material	2017 Bare Costs Labor	Equipment	Total	Total Incl O&P
3350	Deep seal trap, B&S									
3400	1-1/4" pipe size	Q-1	14	1.143	Ea.	54.50	63.50		118	156
3410	1-1/2" pipe size		14	1.143		54.50	63.50		118	156
3420	2" pipe size		14	1.143		50.50	63.50		114	152
3440	3" pipe size		12	1.333		64.50	74		138.50	183
4700	Copper, drainage, drum trap									
4800	3" x 5" solid, 1-1/2" pipe size	1 Plum	16	.500	Ea.	95	31		126	152
4840	3" x 6" swivel, 1-1/2" pipe size	"	16	.500	"	144	31		175	205
5100	P trap, standard pattern									
5200	1-1/4" pipe size	1 Plum	18	.444	Ea.	58.50	27.50		86	106
5240	1-1/2" pipe size		17	.471		78	29		107	130
5260	2" pipe size		15	.533		121	33		154	183
5280	3" pipe size		11	.727		350	45		395	455
5340	With cleanout, swivel joint and slip joint									
5360	1-1/4" pipe size	1 Plum	18	.444	Ea.	99.50	27.50		127	151
5400	1-1/2" pipe size	"	17	.471	"	130	29		159	187

22 13 16.80 Vent Flashing and Caps

22 13 16.80 Vent Flashing and Caps		Crew	Daily Output	Labor-Hours	Unit	Material	2017 Bare Costs Labor	Equipment	Total	Total Incl O&P
0010	**VENT FLASHING AND CAPS**									
0120	Vent caps									
0140	Cast iron									
0180	2-1/2" - 3-5/8" pipe	1 Plum	21	.381	Ea.	35.50	23.50		59	74.50
0190	4" - 4-1/8" pipe	"	19	.421	"	56.50	26		82.50	102
0900	Vent flashing									
1000	Aluminum with lead ring									
1020	1-1/4" pipe	1 Plum	20	.400	Ea.	4.68	24.50		29.18	42.50
1030	1-1/2" pipe		20	.400		4.95	24.50		29.45	43
1040	2" pipe		18	.444		5	27.50		32.50	47
1050	3" pipe		17	.471		5.55	29		34.55	50
1060	4" pipe		16	.500		6.70	31		37.70	54
1350	Copper with neoprene ring									
1400	1-1/4" pipe	1 Plum	20	.400	Ea.	68	24.50		92.50	112
1430	1-1/2" pipe		20	.400		68	24.50		92.50	112
1440	2" pipe		18	.444		68	27.50		95.50	116
1450	3" pipe		17	.471		82	29		111	134
1460	4" pipe		16	.500		82	31		113	137

22 13 19 – Sanitary Waste Piping Specialties

22 13 19.13 Sanitary Drains

22 13 19.13 Sanitary Drains		Crew	Daily Output	Labor-Hours	Unit	Material	2017 Bare Costs Labor	Equipment	Total	Total Incl O&P
0010	**SANITARY DRAINS**									
0400	Deck, auto park, C.I., 13" top									
0440	3", 4", 5", and 6" pipe size	Q-1	8	2	Ea.	1,175	111		1,286	1,475
0480	For galvanized body, add				"	555			555	615
2000	Floor, medium duty, C.I., deep flange, 7" diam. top									
2040	2" and 3" pipe size	Q-1	12	1.333	Ea.	164	74		238	292
2080	For galvanized body, add					69			69	76
2120	With polished bronze top					255			255	281
2400	Heavy duty, with sediment bucket, C.I., 12" diam. loose grate									
2420	2", 3", 4", 5", and 6" pipe size	Q-1	9	1.778	Ea.	565	99		664	770
2460	With polished bronze top				"	795			795	875
2500	Heavy duty, cleanout & trap w/bucket, C.I., 15" top									
2540	2", 3", and 4" pipe size	Q-1	6	2.667	Ea.	6,700	148		6,848	7,600
2560	For galvanized body, add					1,200			1,200	1,325
2580	With polished bronze top					7,450			7,450	8,200

22 13 23 – Sanitary Waste Interceptors

22 13 23.10 Interceptors	Crew	Daily Output	Labor-Hours	Unit	Material	2017 Bare Costs Labor	Equipment	Total	Total Incl O&P
0010	**INTERCEPTORS**								
0150	Grease, fabricated steel, 4 GPM, 8 lb. fat capacity	1 Plum	4	2	Ea.	970	124		1,094	1,250
0200	7 GPM, 14 lb. fat capacity		4	2		1,350	124		1,474	1,650
1000	10 GPM, 20 lb. fat capacity		4	2		1,575	124		1,699	1,925
1040	15 GPM, 30 lb. fat capacity		4	2		2,350	124		2,474	2,750
1060	20 GPM, 40 lb. fat capacity	↓	3	2.667		2,875	165		3,040	3,400
1120	50 GPM, 100 lb. fat capacity	Q-1	2	8		5,275	445		5,720	6,475
1160	100 GPM, 200 lb. fat capacity	"	2	8	↓	11,900	445		12,345	13,800
1580	For seepage pan, add					7%				
3000	Hair, cast iron, 1-1/4" and 1-1/2" pipe connection	1 Plum	8	1	Ea.	365	62		427	495
3100	For chrome-plated cast iron, add					223			223	245
4000	Oil, fabricated steel, 10 GPM, 2" pipe size	1 Plum	4	2		2,150	124		2,274	2,550
4100	15 GPM, 2" or 3" pipe size		4	2		2,950	124		3,074	3,425
4120	20 GPM, 2" or 3" pipe size	↓	3	2.667		3,575	165		3,740	4,200
4220	100 GPM, 3" pipe size	Q-1	2	8		11,900	445		12,345	13,800
6000	Solids, precious metals recovery, C.I., 1-1/4" to 2" pipe	1 Plum	4	2		550	124		674	790
6100	Dental Lab., large, C.I., 1-1/2" to 2" pipe	"	3	2.667	↓	1,925	165		2,090	2,350

22 13 29 – Sanitary Sewerage Pumps

22 13 29.13 Wet-Pit-Mounted, Vertical Sewerage Pumps

		Crew	Daily Output	Labor-Hours	Unit	Material	Labor	Equipment	Total	Total Incl O&P
0010	**WET-PIT-MOUNTED, VERTICAL SEWERAGE PUMPS**									
0020	Controls incl. alarm/disconnect panel w/wire. Excavation not included									
0260	Simplex, 9 GPM at 60 PSIG, 91 gal. tank				Ea.	3,375			3,375	3,725
0300	Unit with manway, 26" I.D., 18" high					3,925			3,925	4,325
0340	26" I.D., 36" high					3,975			3,975	4,375
0380	43" I.D., 4' high				↓	4,100			4,100	4,500
3000	Indoor residential type installation									
3020	Simplex, 9 GPM at 60 PSIG, 91 gal. HDPE tank				Ea.	3,450			3,450	3,800

22 13 29.14 Sewage Ejector Pumps

		Crew	Daily Output	Labor-Hours	Unit	Material	Labor	Equipment	Total	Total Incl O&P
0010	**SEWAGE EJECTOR PUMPS**, With operating and level controls									
0100	Simplex system incl. tank, cover, pump 15' head									
0500	37 gal. PE tank, 12 GPM, 1/2 HP, 2" discharge	Q-1	3.20	5	Ea.	495	278		773	965
0510	3" discharge		3.10	5.161		540	287		827	1,025
0530	87 GPM, .7 HP, 2" discharge		3.20	5		765	278		1,043	1,250
0540	3" discharge		3.10	5.161		830	287		1,117	1,350
0600	45 gal. coated stl. tank, 12 GPM, 1/2 HP, 2" discharge		3	5.333		890	297		1,187	1,425
0610	3" discharge		2.90	5.517		925	305		1,230	1,500
0630	87 GPM, .7 HP, 2" discharge		3	5.333		1,150	297		1,447	1,700
0640	3" discharge		2.90	5.517		1,200	305		1,505	1,800
0660	134 GPM, 1 HP, 2" discharge		2.80	5.714		1,225	320		1,545	1,825
0680	3" discharge		2.70	5.926		1,300	330		1,630	1,925
0700	70 gal. PE tank, 12 GPM, 1/2 HP, 2" discharge		2.60	6.154		960	340		1,300	1,575
0710	3" discharge		2.40	6.667		1,025	370		1,395	1,675
0730	87 GPM, 0.7 HP, 2" discharge		2.50	6.400		1,225	355		1,580	1,875
0740	3" discharge		2.30	6.957		1,325	385		1,710	2,025
0760	134 GPM, 1 HP, 2" discharge		2.20	7.273		1,350	405		1,755	2,075
0770	3" discharge	↓	2	8	↓	1,425	445		1,870	2,250

22 14 Facility Storm Drainage

22 14 23 – Storm Drainage Piping Specialties

22 14 23.33 Backwater Valves

		Crew	Daily Output	Labor-Hours	Unit	Material	2017 Bare Costs Labor	Equipment	Total	Total Incl O&P
0010	**BACKWATER VALVES**, C.I. Body									
6980	Bronze gate and automatic flapper valves									
7000	3" and 4" pipe size	Q-1	13	1.231	Ea.	2,250	68.50		2,318.50	2,575
7100	5" and 6" pipe size	"	13	1.231	"	3,450	68.50		3,518.50	3,875
7240	Bronze flapper valve, bolted cover									
7260	2" pipe size	Q-1	16	1	Ea.	655	55.50		710.50	810
7300	4" pipe size	"	13	1.231	↓	3,875	68.50		3,943.50	4,375
7340	6" pipe size	Q-2	17	1.412		1,350	81.50		1,431.50	1,600

22 14 26 – Facility Storm Drains

22 14 26.13 Roof Drains

		Crew	Daily Output	Labor-Hours	Unit	Material	2017 Bare Costs Labor	Equipment	Total	Total Incl O&P
0010	**ROOF DRAINS**									
0140	Cornice, C.I., 45° or 90° outlet									
0200	3" and 4" pipe size	Q-1	12	1.333	Ea.	375	74		449	520
0260	For galvanized body, add					54			54	59.50
0280	For polished bronze dome, add				↓	45			45	49.50
3860	Roof, flat metal deck, C.I. body, 12" C.I. dome									
3890	3" pipe size	Q-1	14	1.143	Ea.	298	63.50		361.50	425
3920	6" pipe size	"	10	1.600	"	725	89		814	935
4620	Main, all aluminum, 12" low profile dome									
4640	2", 3" and 4" pipe size	Q-1	14	1.143	Ea.	355	63.50		418.50	485

22 14 26.16 Facility Area Drains

		Crew	Daily Output	Labor-Hours	Unit	Material	2017 Bare Costs Labor	Equipment	Total	Total Incl O&P
0010	**FACILITY AREA DRAINS**									
4980	Scupper floor, oblique strainer, C.I.									
5000	6" x 7" top, 2", 3" and 4" pipe size	Q-1	16	1	Ea.	310	55.50		365.50	425
5100	8" x 12" top, 5" and 6" pipe size	"	14	1.143		610	63.50		673.50	765
5160	For galvanized body, add					40%				
5200	For polished bronze strainer, add				↓	85%				

22 14 26.19 Facility Trench Drains

		Crew	Daily Output	Labor-Hours	Unit	Material	2017 Bare Costs Labor	Equipment	Total	Total Incl O&P
0010	**FACILITY TRENCH DRAINS**									
5980	Trench, floor, heavy duty, modular, C.I., 12" x 12" top									
6000	2", 3", 4", 5", & 6" pipe size	Q-1	8	2	Ea.	990	111		1,101	1,275
6100	For unit with polished bronze top	"	8	2	"	1,425	111		1,536	1,750
6600	Trench, floor, for cement concrete encasement									
6610	Not including trenching or concrete									
6640	Polyester polymer concrete									
6650	4" internal width, with grate									
6660	Light duty steel grate	Q-1	120	.133	L.F.	33	7.40		40.40	47.50
6670	Medium duty steel grate	↓	115	.139	↓	41.50	7.75		49.25	57
6680	Heavy duty iron grate	↓	110	.145	↓	45	8.10		53.10	61.50
6700	12" internal width, with grate									
6770	Heavy duty galvanized grate	Q-1	80	.200	L.F.	123	11.15		134.15	152
6800	Fiberglass									
6810	8" internal width, with grate									
6820	Medium duty galvanized grate	Q-1	115	.139	L.F.	67.50	7.75		75.25	85.50
6830	Heavy duty iron grate	"	110	.145	"	99.50	8.10		107.60	122

22 14 29 – Sump Pumps

22 14 29.13 Wet-Pit-Mounted, Vertical Sump Pumps

		Crew	Daily Output	Labor-Hours	Unit	Material	2017 Bare Costs Labor	Equipment	Total	Total Incl O&P
0010	**WET-PIT-MOUNTED, VERTICAL SUMP PUMPS**									
0400	Molded PVC base, 21 GPM at 15' head, 1/3 HP	1 Plum	5	1.600	Ea.	141	99		240	305
0800	Iron base, 21 GPM at 15' head, 1/3 HP	↓	5	1.600	↓	170	99		269	335
1200	Solid brass, 21 GPM at 15' head, 1/3 HP	↓	5	1.600	↓	300	99		399	480

22 14 29 – Sump Pumps

22 14 29.13 Wet-Pit-Mounted, Vertical Sump Pumps	Crew	Daily Output	Labor-Hours	Unit	Material	2017 Bare Costs Labor	Equipment	Total	Total Incl O&P
2000	Sump pump, single stage								
2010	25 GPM, 1 HP, 1-1/2" discharge	Q-1	1.80	8.889	Ea.	3,875	495		4,370	5,000
2020	75 GPM, 1-1/2 HP, 2" discharge		1.50	10.667		4,100	595		4,695	5,400
2030	100 GPM, 2 HP, 2-1/2" discharge		1.30	12.308		4,175	685		4,860	5,600
2040	150 GPM, 3 HP, 3" discharge		1.10	14.545		4,175	810		4,985	5,800
2050	200 GPM, 3 HP, 3" discharge		1	16		4,425	890		5,315	6,225
2060	300 GPM, 10 HP, 4" discharge	Q-2	1.20	20		4,775	1,150		5,925	7,000
2070	500 GPM, 15 HP, 5" discharge		1.10	21.818		5,425	1,250		6,675	7,875
2080	800 GPM, 20 HP, 6" discharge		1	24		6,425	1,375		7,800	9,150
2090	1000 GPM, 30 HP, 6" discharge		.85	28.235		7,050	1,625		8,675	10,200
2100	1600 GPM, 50 HP, 8" discharge		.72	33.333		11,000	1,925		12,925	15,000
2110	2000 GPM, 60 HP, 8" discharge	Q-3	.85	37.647		11,200	2,225		13,425	15,800
2202	For general purpose float switch, copper coated float, add	Q-1	5	3.200		104	178		282	385

22 14 29.16 Submersible Sump Pumps

		Crew	Daily Output	Labor-Hours	Unit	Material	2017 Bare Costs Labor	Equipment	Total	Total Incl O&P
0010	**SUBMERSIBLE SUMP PUMPS**									
7000	Sump pump, automatic									
7100	Plastic, 1-1/4" discharge, 1/4 HP	1 Plum	6.40	1.250	Ea.	150	77.50		227.50	282
7140	1/3 HP		6	1.333		217	82.50		299.50	365
7160	1/2 HP		5.40	1.481		266	91.50		357.50	430
7180	1-1/2" discharge, 1/2 HP		5.20	1.538		305	95		400	480
7500	Cast iron, 1-1/4" discharge, 1/4 HP		6	1.333		211	82.50		293.50	355
7540	1/3 HP		6	1.333		248	82.50		330.50	395
7560	1/2 HP		5	1.600		300	99		399	480

22 31 Domestic Water Softeners

22 31 13 – Residential Domestic Water Softeners

22 31 13.10 Residential Water Softeners

		Crew	Daily Output	Labor-Hours	Unit	Material	2017 Bare Costs Labor	Equipment	Total	Total Incl O&P
0010	**RESIDENTIAL WATER SOFTENERS**									
7350	Water softener, automatic, to 30 grains per gallon	2 Plum	5	3.200	Ea.	380	198		578	720
7400	To 100 grains per gallon	"	4	4	"	815	247		1,062	1,275

22 31 16 – Commercial Domestic Water Softeners

22 31 16.10 Water Softeners

		Crew	Daily Output	Labor-Hours	Unit	Material	2017 Bare Costs Labor	Equipment	Total	Total Incl O&P
0010	**WATER SOFTENERS**									
5800	Softener systems, automatic, intermediate sizes									
5820	available, may be used in multiples.									
6000	Hardness capacity between regenerations and flow									
6100	150,000 grains, 37 GPM cont., 51 GPM peak	Q-1	1.20	13.333	Ea.	6,350	740		7,090	8,100
6200	300,000 grains, 81 GPM cont., 113 GPM peak		1	16		9,150	890		10,040	11,500
6300	750,000 grains, 160 GPM cont., 230 GPM peak		.80	20		13,100	1,125		14,225	16,100
6400	900,000 grains, 185 GPM cont., 270 GPM peak		.70	22.857		21,100	1,275		22,375	25,100

22 33 Electric Domestic Water Heaters

22 33 13 – Instantaneous Electric Domestic Water Heaters

22 33 13.10 Hot Water Dispensers

		Crew	Daily Output	Labor-Hours	Unit	Material	2017 Bare Costs Labor	Equipment	Total	Total Incl O&P
0010	**HOT WATER DISPENSERS**									
0160	Commercial, 100 cup, 11.3 amp	1 Plum	14	.571	Ea.	440	35.50		475.50	540
3180	Household, 60 cup	"	14	.571	"	241	35.50		276.50	320

22 33 30 – Residential, Electric Domestic Water Heaters

22 33 30.13 Residential, Small-Capacity Elec. Water Heaters

		Crew	Daily Output	Labor-Hours	Unit	Material	2017 Bare Costs Labor	Equipment	Total	Total Incl O&P
0010	**RESIDENTIAL, SMALL-CAPACITY ELECTRIC DOMESTIC WATER HEATERS**									
1000	Residential, electric, glass lined tank, 5 yr., 10 gal., single element	1 Plum	2.30	3.478	Ea.	465	215		680	835
1040	20 gallon, single element		2.20	3.636		615	225		840	1,025
1060	30 gallon, double element		2.20	3.636		1,075	225		1,300	1,550
1080	40 gallon, double element		2	4		1,150	247		1,397	1,650
1100	52 gallon, double element		2	4		1,300	247		1,547	1,825
1180	120 gallon, double element	▼	1.40	5.714	▼	2,775	355		3,130	3,575

22 33 33 – Light-Commercial Electric Domestic Water Heaters

22 33 33.10 Commercial Electric Water Heaters

		Crew	Daily Output	Labor-Hours	Unit	Material	2017 Bare Costs Labor	Equipment	Total	Total Incl O&P
0010	**COMMERCIAL ELECTRIC WATER HEATERS**									
4000	Commercial, 100° rise. NOTE: for each size tank, a range of									
4010	heaters between the ones shown are available									
4020	Electric									
4100	5 gal., 3 kW, 12 GPH, 208 volt	1 Plum	2	4	Ea.	5,175	247		5,422	6,050
4120	10 gal., 6 kW, 25 GPH, 208 volt		2	4		5,700	247		5,947	6,650
4130	30 gal., 24 kW, 98 GPH, 208 volt		1.92	4.167		9,400	258		9,658	10,800
4136	40 gal., 36 kW, 148 GPH, 208 volt		1.88	4.255		11,300	263		11,563	12,800
4140	50 gal., 9 kW, 37 GPH, 208 volt		1.80	4.444		7,875	275		8,150	9,075
4160	50 gal., 36 kW, 148 GPH, 208 volt	▼	1.80	4.444		12,100	275		12,375	13,700
4300	200 gal., 15 kW, 61 GPH, 480 volt	Q-1	1.70	9.412		37,000	525		37,525	41,500
4320	200 gal., 120 kW, 490 GPH, 480 volt		1.70	9.412		50,500	525		51,025	56,500
4460	400 gal., 30 kW, 123 GPH, 480 volt	▼	1	16		50,500	890		51,390	57,000
5400	Modulating step control for under 90 kW, 2-5 steps	1 Elec	5.30	1.509		820	85.50		905.50	1,025
5440	For above 90 kW, 1 through 5 steps beyond standard, add		3.20	2.500		223	142		365	460
5460	For above 90 kW, 6 through 10 steps beyond standard, add		2.70	2.963		460	168		628	755
5480	For above 90 kW, 11 through 18 steps beyond standard, add	▼	1.60	5	▼	685	283		968	1,175

22 34 Fuel-Fired Domestic Water Heaters

22 34 13 – Instantaneous, Tankless, Gas Domestic Water Heaters

22 34 13.10 Instantaneous, Tankless, Gas Water Heaters

			Crew	Daily Output	Labor-Hours	Unit	Material	2017 Bare Costs Labor	Equipment	Total	Total Incl O&P
0010	**INSTANTANEOUS, TANKLESS, GAS WATER HEATERS**										
9410	Natural gas/propane, 3.2 GPM	G	1 Plum	2	4	Ea.	540	247		787	970
9420	6.4 GPM	G		1.90	4.211		770	260		1,030	1,250
9430	8.4 GPM	G		1.80	4.444		925	275		1,200	1,450
9440	9.5 GPM	G	▼	1.60	5	▼	1,075	310		1,385	1,675

22 34 30 – Residential Gas Domestic Water Heaters

22 34 30.13 Residential, Atmos, Gas Domestic Wtr Heaters

		Crew	Daily Output	Labor-Hours	Unit	Material	2017 Bare Costs Labor	Equipment	Total	Total Incl O&P
0010	**RESIDENTIAL, ATMOSPHERIC, GAS DOMESTIC WATER HEATERS**									
2000	Gas fired, foam lined tank, 10 yr., vent not incl.									
2040	30 gallon	1 Plum	2	4	Ea.	1,925	247		2,172	2,500
2100	75 gallon		1.50	5.333		2,925	330		3,255	3,725
2120	100 gallon		1.30	6.154		3,075	380		3,455	3,975
2900	Water heater, safety-drain pan, 26" round	▼	20	.400	▼	34.50	24.50		59	75.50
3000	Tank leak safety, water & gas shut off see 22 05 23.20 8800									

For customer support on your Building Construction Costs with RSMeans Data, call 800.448.8182.

511

22 34 Fuel-Fired Domestic Water Heaters

22 34 36 – Commercial Gas Domestic Water Heaters

22 34 36.13 Commercial, Atmos., Gas Domestic Water Htrs.	Crew	Daily Output	Labor-Hours	Unit	Material	2017 Bare Costs Labor	2017 Bare Costs Equipment	Total	Total Incl O&P
0010 **COMMERCIAL, ATMOSPHERIC, GAS DOMESTIC WATER HEATERS**									
6000 Gas fired, flush jacket, std. controls, vent not incl.									
6040 75 MBH input, 73 GPH	1 Plum	1.40	5.714	Ea.	4,100	355		4,455	5,025
6060 98 MBH input, 95 GPH		1.40	5.714		8,100	355		8,455	9,425
6080 120 MBH input, 110 GPH		1.20	6.667		8,325	410		8,735	9,800
6180 200 MBH input, 192 GPH		.60	13.333		11,000	825		11,825	13,400
6200 250 MBH input, 245 GPH		.50	16		11,800	990		12,790	14,500
6900 For low water cutoff, add		8	1		365	62		427	495
6960 For bronze body hot water circulator, add	▼	4	2		2,075	124		2,199	2,450

22 34 46 – Oil-Fired Domestic Water Heaters

22 34 46.10 Residential Oil-Fired Water Heaters

	Crew	Daily Output	Labor-Hours	Unit	Material	Labor	Equipment	Total	Total Incl O&P
0010 **RESIDENTIAL OIL-FIRED WATER HEATERS**									
3000 Oil fired, glass lined tank, 5 yr., vent not included, 30 gallon	1 Plum	2	4	Ea.	1,400	247		1,647	1,900
3040 50 gallon		1.80	4.444		1,700	275		1,975	2,300
3060 70 gallon	▼	1.50	5.333	▼	2,400	330		2,730	3,125

22 34 46.20 Commercial Oil-Fired Water Heaters

	Crew	Daily Output	Labor-Hours	Unit	Material	Labor	Equipment	Total	Total Incl O&P
0010 **COMMERCIAL OIL-FIRED WATER HEATERS**									
8000 Oil fired, glass lined, UL listed, std. controls, vent not incl.									
8060 140 gal., 140 MBH input, 134 GPH	Q-1	2.13	7.512	Ea.	28,200	420		28,620	31,600
8080 140 gal., 199 MBH input, 191 GPH		2	8		29,100	445		29,545	32,700
8100 140 gal., 255 MBH input, 247 GPH		1.60	10		29,900	555		30,455	33,700
8160 140 gal., 540 MBH input, 519 GPH		.96	16.667		39,700	925		40,625	45,100
8180 140 gal., 720 MBH input, 691 GPH	▼	.92	17.391		40,400	965		41,365	46,000
8280 201 gal., 1250 MBH input, 1200 GPH	Q-2	1.22	19.672		62,500	1,125		63,625	70,000
8300 201 gal., 1500 MBH input, 1441 GPH	"	1.16	20.690		67,500	1,200		68,700	76,000
8900 For low water cutoff, add	1 Plum	8	1		365	62		427	495
8960 For bronze body hot water circulator, add	"	4	2		840	124		964	1,100

22 35 Domestic Water Heat Exchangers

22 35 30 – Water Heating by Steam

22 35 30.10 Water Heating Transfer Package

	Crew	Daily Output	Labor-Hours	Unit	Material	Labor	Equipment	Total	Total Incl O&P
0010 **WATER HEATING TRANSFER PACKAGE**, Complete controls,									
0020 expansion tank, converter, air separator									
1000 Hot water, 180°F enter, 200°F leaving, 15# steam									
1010 One pump system, 28 GPM	Q-6	.75	32	Ea.	21,100	1,850		22,950	26,000
1020 35 GPM		.70	34.286		22,900	2,000		24,900	28,200
1040 55 GPM		.65	36.923		27,100	2,150		29,250	33,000
1060 130 GPM		.55	43.636		34,200	2,525		36,725	41,400
1080 255 GPM		.40	60		45,000	3,475		48,475	55,000
1100 550 GPM	▼	.30	80	▼	61,000	4,650		65,650	74,000

22 41 Residential Plumbing Fixtures

22 41 06 – Plumbing Fixtures General

22 41 06.10 Plumbing Fixture Notes	Crew	Daily Output	Labor-Hours	Unit	Material	2017 Bare Costs Labor	Equipment	Total	Total Incl O&P
0010 **PLUMBING FIXTURE NOTES**, Incl. trim fittings unless otherwise noted									
0080 For rough-in, supply, waste, and vent, see add for each type									
0122 For electric water coolers, see Section 22 47 16.10									
0160 For color, unless otherwise noted, add				Ea.	20%				

22 41 13 – Residential Water Closets, Urinals, and Bidets

22 41 13.13 Water Closets

	Crew	Daily Output	Labor-Hours	Unit	Material	Labor	Equipment	Total	Total Incl O&P
0010 **WATER CLOSETS**									
0032 For automatic flush, see Line 22 42 39.10 0972									
0150 Tank type, vitreous china, incl. seat, supply pipe w/stop, 1.6 gpf or noted									
0200 Wall hung									
0400 Two piece, close coupled	Q-1	5.30	3.019	Ea.	435	168		603	735
0960 For rough-in, supply, waste, vent and carrier	"	2.73	5.861	"	1,150	325		1,475	1,775
0999 Floor mounted									
1100 Two piece, close coupled	Q-1	5.30	3.019	Ea.	194	168		362	465
1102 Economy		5.30	3.019		108	168		276	370
1110 Two piece, close coupled, dual flush		5.30	3.019		267	168		435	545
1140 Two piece, close coupled, 1.28 gpf, ADA [G]		5.30	3.019		259	168		427	535
1960 For color, add					30%				
1980 For rough-in, supply, waste and vent	Q-1	3.05	5.246	Ea.	380	292		672	860

22 41 16 – Residential Lavatories and Sinks

22 41 16.13 Lavatories

	Crew	Daily Output	Labor-Hours	Unit	Material	Labor	Equipment	Total	Total Incl O&P
0010 **LAVATORIES**, With trim, white unless noted otherwise									
0500 Vanity top, porcelain enamel on cast iron									
0600 20" x 18"	Q-1	6.40	2.500	Ea.	320	139		459	560
0640 33" x 19" oval		6.40	2.500		480	139		619	735
0720 19" round		6.40	2.500		420	139		559	675
0860 For color, add					25%				
1000 Cultured marble, 19" x 17", single bowl	Q-1	6.40	2.500	Ea.	152	139		291	375
1040 25" x 19", single bowl	"	6.40	2.500	"	176	139		315	405
1580 For color, same price									
1900 Stainless steel, self-rimming, 25" x 22", single bowl, ledge	Q-1	6.40	2.500	Ea.	345	139		484	585
1960 17" x 22", single bowl		6.40	2.500		335	139		474	575
2600 Steel, enameled, 20" x 17", single bowl		5.80	2.759		151	153		304	400
2660 19" round		5.80	2.759		165	153		318	415
2900 Vitreous china, 20" x 16", single bowl		5.40	2.963		238	165		403	510
2960 20" x 17", single bowl		5.40	2.963		139	165		304	400
3020 19" round, single bowl		5.40	2.963		140	165		305	405
3200 22" x 13", single bowl		5.40	2.963		245	165		410	520
3560 For color, add					50%				
3580 Rough-in, supply, waste and vent for all above lavatories	Q-1	2.30	6.957	Ea.	209	385		594	815
4000 Wall hung									
4040 Porcelain enamel on cast iron, 16" x 14", single bowl	Q-1	8	2	Ea.	505	111		616	725
4180 20" x 18", single bowl		8	2		267	111		378	460
4240 22" x 19", single bowl		8	2		700	111		811	940
4580 For color, add					30%				
6000 Vitreous china, 18" x 15", single bowl with backsplash	Q-1	7	2.286	Ea.	188	127		315	400
6500 For color, add					30%				
6960 Rough-in, supply, waste and vent for above lavatories	Q-1	1.66	9.639	Ea.	390	535		925	1,250
7000 Pedestal type									
7600 Vitreous china, 27" x 21", white	Q-1	6.60	2.424	Ea.	685	135		820	960
7610 27" x 21", colored		6.60	2.424		870	135		1,005	1,150
7620 27" x 21", premium color		6.60	2.424		985	135		1,120	1,275

22 41 Residential Plumbing Fixtures

22 41 16 – Residential Lavatories and Sinks

22 41 16.13 Lavatories

		Crew	Daily Output	Labor-Hours	Unit	Material	2017 Bare Costs Labor	Equipment	Total	Total Incl O&P
7660	26" x 20", white	Q-1	6.60	2.424	Ea.	675	135		810	950
7670	26" x 20", colored		6.60	2.424		855	135		990	1,150
7680	26" x 20", premium color		6.60	2.424		970	135		1,105	1,275
7700	24" x 20", white		6.60	2.424		400	135		535	645
7710	24" x 20", colored		6.60	2.424		495	135		630	750
7720	24" x 20", premium color		6.60	2.424		560	135		695	820
7760	21" x 18", white		6.60	2.424		280	135		415	515
7770	21" x 18", colored		6.60	2.424		340	135		475	580
7990	Rough-in, supply, waste and vent for pedestal lavatories		1.66	9.639		390	535		925	1,250

22 41 16.16 Sinks

		Crew	Daily Output	Labor-Hours	Unit	Material	2017 Bare Costs Labor	Equipment	Total	Total Incl O&P
0010	**SINKS**, With faucets and drain									
2000	Kitchen, counter top style, P.E. on C.I., 24" x 21" single bowl	Q-1	5.60	2.857	Ea.	300	159		459	570
2100	31" x 22" single bowl		5.60	2.857		555	159		714	850
2200	32" x 21" double bowl		4.80	3.333		375	185		560	690
3000	Stainless steel, self rimming, 19" x 18" single bowl		5.60	2.857		610	159		769	910
3100	25" x 22" single bowl		5.60	2.857		675	159		834	985
4000	Steel, enameled, with ledge, 24" x 21" single bowl		5.60	2.857		530	159		689	825
4100	32" x 21" double bowl		4.80	3.333		525	185		710	855
4960	For color sinks except stainless steel, add					10%				
4980	For rough-in, supply, waste and vent, counter top sinks	Q-1	2.14	7.477		268	415		683	925
5000	Kitchen, raised deck, P.E. on C.I.									
5100	32" x 21", dual level, double bowl	Q-1	2.60	6.154	Ea.	445	340		785	1,000
5700	For color, add					20%				
5790	For rough-in, supply, waste & vent, sinks	Q-1	1.85	8.649		268	480		748	1,025

22 41 19 – Residential Bathtubs

22 41 19.10 Baths

		Crew	Daily Output	Labor-Hours	Unit	Material	2017 Bare Costs Labor	Equipment	Total	Total Incl O&P
0010	**BATHS**									
0100	Tubs, recessed porcelain enamel on cast iron, with trim									
0180	48" x 42"	Q-1	4	4	Ea.	2,700	223		2,923	3,275
0220	72" x 36"		3	5.333		2,775	297		3,072	3,500
2000	Enameled formed steel, 4'-6" long		5.80	2.759		490	153		643	765
4000	Soaking, acrylic, w/pop-up drain 66" x 36" x 20" deep		5.50	2.909		1,650	162		1,812	2,075
4100	60" x 42" x 20" deep		5	3.200		1,225	178		1,403	1,625
9600	Rough-in, supply, waste and vent, for all above tubs, add		2.07	7.729		415	430		845	1,100

22 41 23 – Residential Showers

22 41 23.20 Showers

		Crew	Daily Output	Labor-Hours	Unit	Material	2017 Bare Costs Labor	Equipment	Total	Total Incl O&P
0010	**SHOWERS**									
1500	Stall, with drain only. Add for valve and door/curtain									
3000	Fiberglass, one piece, with 3 walls, 32" x 32" square	Q-1	5.50	2.909	Ea.	340	162		502	620
3100	36" x 36" square		5.50	2.909		390	162		552	670
3250	64" x 65-3/4" x 81-1/2" fold. seat, whlchr.		3.80	4.211		2,300	234		2,534	2,900
4000	Polypropylene, stall only, w/molded-stone floor, 30" x 30"		2	8		640	445		1,085	1,375
4200	Rough-in, supply, waste and vent for above showers		2.05	7.805		365	435		800	1,050

22 41 23.40 Shower System Components

		Crew	Daily Output	Labor-Hours	Unit	Material	2017 Bare Costs Labor	Equipment	Total	Total Incl O&P
0010	**SHOWER SYSTEM COMPONENTS**									
4500	Receptor only									
4510	For tile, 36" x 36"	1 Plum	4	2	Ea.	370	124		494	590
4520	Fiberglass receptor only, 32" x 32"		8	1		100	62		162	204
4530	34" x 34"		7.80	1.026		120	63.50		183.50	228
4540	36" x 36"		7.60	1.053		117	65		182	226
4600	Rectangular									

22 41 Residential Plumbing Fixtures

22 41 23 – Residential Showers

22 41 23.40 Shower System Components	Crew	Daily Output	Labor-Hours	Unit	Material	2017 Bare Costs Labor	Equipment	Total	Total Incl O&P
4620	32" x 48"	1 Plum	7.40	1.081	Ea.	139	67		206	253
4630	34" x 54"		7.20	1.111		166	68.50		234.50	287
4640	34" x 60"		7	1.143		176	70.50		246.50	300
5000	Built-in, head, arm, 2.5 GPM valve		4	2		82	124		206	277
5200	Head, arm, by-pass, integral stops, handles	↓	3.60	2.222	↓	260	137		397	495

22 41 36 – Residential Laundry Trays

22 41 36.10 Laundry Sinks

		Crew	Daily Output	Labor-Hours	Unit	Material	Labor	Equipment	Total	Total Incl O&P
0010	**LAUNDRY SINKS**, With trim									
0020	Porcelain enamel on cast iron, black iron frame									
0050	24" x 21", single compartment	Q-1	6	2.667	Ea.	595	148		743	880
0100	26" x 21", single compartment	"	6	2.667	"	620	148		768	910
2000	Molded stone, on wall hanger or legs									
2020	22" x 23", single compartment	Q-1	6	2.667	Ea.	181	148		329	425
2100	45" x 21", double compartment	"	5	3.200	"	345	178		523	650
3000	Plastic, on wall hanger or legs									
3020	18" x 23", single compartment	Q-1	6.50	2.462	Ea.	146	137		283	370
3300	40" x 24", double compartment		5.50	2.909		310	162		472	590
5000	Stainless steel, counter top, 22" x 17" single compartment		6	2.667		77	148		225	310
5200	33" x 22", double compartment		5	3.200		92	178		270	370
9600	Rough-in, supply, waste and vent, for all laundry sinks	↓	2.14	7.477	↓	268	415		683	925

22 41 39 – Residential Faucets, Supplies and Trim

22 41 39.10 Faucets and Fittings

		Crew	Daily Output	Labor-Hours	Unit	Material	Labor	Equipment	Total	Total Incl O&P
0010	**FAUCETS AND FITTINGS**									
0150	Bath, faucets, diverter spout combination, sweat	1 Plum	8	1	Ea.	87	62		149	189
0200	For integral stops, IPS unions, add					111			111	122
0420	Bath, press-bal mix valve w/diverter, spout, shower head, arm/flange	1 Plum	8	1		173	62		235	284
0810	Bidet									
0812	Fitting, over the rim, swivel spray/pop-up drain	1 Plum	8	1	Ea.	219	62		281	335
1000	Kitchen sink faucets, top mount, cast spout		10	.800		78	49.50		127.50	161
1100	For spray, add	↓	24	.333	↓	17.10	20.50		37.60	50
1300	Single control lever handle									
1310	With pull out spray									
1320	Polished chrome	1 Plum	10	.800	Ea.	200	49.50		249.50	294
2000	Laundry faucets, shelf type, IPS or copper unions		12	.667		62	41		103	131
2100	Lavatory faucet, centerset, without drain	↓	10	.800	↓	64	49.50		113.50	145
2210	Porcelain cross handles and pop-up drain									
2220	Polished chrome	1 Plum	6.66	1.201	Ea.	222	74		296	355
2230	Polished brass	"	6.66	1.201	"	335	74		409	480
2260	Single lever handle and pop-up drain									
2280	Satin nickel	1 Plum	6.66	1.201	Ea.	277	74		351	415
2290	Polished chrome		6.66	1.201		198	74		272	330
2810	Automatic sensor and operator, with faucet head [G]		6.15	1.301		460	80.50		540.50	625
4000	Shower by-pass valve with union		18	.444		57.50	27.50		85	105
4200	Shower thermostatic mixing valve, concealed, with shower head trim kit	↓	8	1	↓	340	62		402	470
4220	Shower pressure balancing mixing valve									
4230	With shower head, arm, flange and diverter tub spout									
4240	Chrome	1 Plum	6.14	1.303	Ea.	375	80.50		455.50	535
4250	Satin nickel		6.14	1.303		495	80.50		575.50	660
4260	Polished graphite		6.14	1.303		505	80.50		585.50	675
5000	Sillcock, compact, brass, IPS or copper to hose	↓	24	.333	↓	10.20	20.50		30.70	42.50

For customer support on your Building Construction Costs with RSMeans Data, call 800.448.8182.

515

22 42 13 – Commercial Water Closets, Urinals, and Bidets

22 42 13.13 Water Closets

		Crew	Daily Output	Labor-Hours	Unit	Material	2017 Bare Costs Labor	Equipment	Total	Total Incl O&P
0010	**WATER CLOSETS**									
3000	Bowl only, with flush valve, seat, 1.6 gpf unless noted									
3100	Wall hung	Q-1	5.80	2.759	Ea.	835	153		988	1,150
3200	For rough-in, supply, waste and vent, single WC		2.56	6.250		1,200	350		1,550	1,850
3300	Floor mounted		5.80	2.759		296	153		449	555
3350	With wall outlet		5.80	2.759		485	153		638	765
3360	With floor outlet, 1.28 gpf ▣G		5.80	2.759		535	153		688	820
3362	With floor outlet, 1.28 gpf, ADA ▣G		5.80	2.759		555	153		708	840
3370	For rough-in, supply, waste and vent, single WC	▼	2.84	5.634	▼	435	315		750	950
3390	Floor mounted children's size, 10-3/4" high									
3392	With automatic flush sensor, 1.6 gpf	Q-1	6.20	2.581	Ea.	590	144		734	865
3396	With automatic flush sensor, 1.28 gpf		6.20	2.581		590	144		734	865
3400	For rough-in, supply, waste and vent, single WC	▼	2.84	5.634	▼	435	315		750	950

22 42 13.16 Urinals

		Crew	Daily Output	Labor-Hours	Unit	Material	2017 Bare Costs Labor	Equipment	Total	Total Incl O&P
0010	**URINALS**									
3000	Wall hung, vitreous china, with self-closing valve									
3100	Siphon jet type	Q-1	3	5.333	Ea.	279	297		576	755
3120	Blowout type		3	5.333		470	297		767	965
3140	Water saving .5 gpf ▣G		3	5.333		560	297		857	1,075
3300	Rough-in, supply, waste & vent		2.83	5.654		610	315		925	1,150
5000	Stall type, vitreous china, includes valve		2.50	6.400		755	355		1,110	1,375
6980	Rough-in, supply, waste and vent	▼	1.99	8.040	▼	395	445		840	1,100
8000	Waterless (no flush) urinal									
8010	Wall hung									
8014	Fiberglass reinforced polyester									
8020	Standard unit ▣G	Q-1	21.30	.751	Ea.	480	42		522	590
8030	ADA compliant unit	"	21.30	.751		500	42		542	615
8070	For solid color, add ▣G					60			60	66
8080	For 2" brass flange (new const.), add ▣G	Q-1	96	.167	▼	18	9.25		27.25	34
8200	Vitreous china									
8220	ADA compliant unit, 14" ▣G	Q-1	21.30	.751	Ea.	264	42		306	355
8250	ADA compliant unit, 15.5"	"	21.30	.751		320	42		362	415
8270	For solid color, add ▣G					60			60	66
8290	Rough-in, supply, waste & vent ▣G	Q-1	2.92	5.479	▼	575	305		880	1,100
8400	Trap liquid									
8410	1 quart ▣G				Ea.	20			20	22
8420	1 gallon ▣G				"	72			72	79

22 42 16 – Commercial Lavatories and Sinks

22 42 16.13 Lavatories

		Crew	Daily Output	Labor-Hours	Unit	Material	2017 Bare Costs Labor	Equipment	Total	Total Incl O&P
0010	**LAVATORIES**, With trim, white unless noted otherwise									
0020	Commercial lavatories same as residential. See Section 22 41 16									

22 42 16.40 Service Sinks

		Crew	Daily Output	Labor-Hours	Unit	Material	2017 Bare Costs Labor	Equipment	Total	Total Incl O&P
0010	**SERVICE SINKS**									
6650	Service, floor, corner, P.E. on C.I., 28" x 28"	Q-1	4.40	3.636	Ea.	1,025	202		1,227	1,450
6790	For rough-in, supply, waste & vent, floor service sinks		1.64	9.756		790	545		1,335	1,700
7000	Service, wall, P.E. on C.I., roll rim, 22" x 18"		4	4		785	223		1,008	1,200
7100	24" x 20"		4	4		870	223		1,093	1,300
8600	Vitreous china, 22" x 20"	▼	4	4		615	223		838	1,000
8960	For stainless steel rim guard, front or one side, add					58			58	63.50
8980	For rough-in, supply, waste & vent, wall service sinks	Q-1	1.30	12.308	▼	1,050	685		1,735	2,175

22 42 Commercial Plumbing Fixtures

22 42 23 - Commercial Showers

22 42 23.30 Group Showers

		Crew	Daily Output	Labor-Hours	Unit	Material	2017 Bare Costs Labor	Equipment	Total	Total Incl O&P
0010	**GROUP SHOWERS**									
6000	Group, w/pressure balancing valve, rough-in and rigging not included									
6800	Column, 6 heads, no receptors, less partitions	Q-1	3	5.333	Ea.	9,400	297		9,697	10,800
6900	With stainless steel partitions		1	16		12,100	890		12,990	14,800
7600	5 heads, no receptors, less partitions		3	5.333		6,475	297		6,772	7,575
7620	4 heads (1 handicap) no receptors, less partitions		3	5.333		5,750	297		6,047	6,775
7700	With stainless steel partitions		1	16		6,450	890		7,340	8,450
8000	Wall, 2 heads, no receptors, less partitions		4	4		2,775	223		2,998	3,375
8100	With stainless steel partitions		2	8		6,075	445		6,520	7,375

22 42 33 - Wash Fountains

22 42 33.20 Commercial Wash Fountains

		Crew	Daily Output	Labor-Hours	Unit	Material	2017 Bare Costs Labor	Equipment	Total	Total Incl O&P
0010	**COMMERCIAL WASH FOUNTAINS**									
1900	Group, foot control									
2000	Precast terrazzo, circular, 36" diam., 5 or 6 persons	Q-2	3	8	Ea.	7,425	460		7,885	8,850
2100	54" diameter for 8 or 10 persons		2.50	9.600		9,250	555		9,805	11,000
2400	Semi-circular, 36" diam. for 3 persons		3	8		6,500	460		6,960	7,850
2500	54" diam. for 4 or 5 persons		2.50	9.600		8,750	555		9,305	10,500
2700	Quarter circle (corner), 54" for 3 persons		3.50	6.857		8,025	395		8,420	9,425
3000	Stainless steel, circular, 36" diameter		3.50	6.857		6,675	395		7,070	7,950
3100	54" diameter		2.80	8.571		8,200	495		8,695	9,775
3400	Semi-circular, 36" diameter		3.50	6.857		5,125	395		5,520	6,250
3500	54" diameter		2.80	8.571		7,150	495		7,645	8,625
5610	Group, infrared control, barrier free									
5614	Precast terrazzo									
5620	Semi-circular 36" diam. for 3 persons	Q-2	3	8	Ea.	7,400	460		7,860	8,825
5630	46" diam. for 4 persons		2.80	8.571		7,975	495		8,470	9,525
5640	Circular, 54" diam. for 8 persons, button control		2.50	9.600		9,675	555		10,230	11,400
5700	Rough-in, supply, waste and vent for above wash fountains	Q-1	1.82	8.791		405	490		895	1,175
6200	Duo for small washrooms, stainless steel		2	8		2,750	445		3,195	3,700
6500	Rough-in, supply, waste & vent for duo fountains		2.02	7.921		230	440		670	920

22 42 39 - Commercial Faucets, Supplies, and Trim

22 42 39.10 Faucets and Fittings

		Crew	Daily Output	Labor-Hours	Unit	Material	2017 Bare Costs Labor	Equipment	Total	Total Incl O&P
0010	**FAUCETS AND FITTINGS**									
0840	Flush valves, with vacuum breaker									
0850	Water closet									
0860	Exposed, rear spud	1 Plum	8	1	Ea.	153	62		215	263
0870	Top spud		8	1		153	62		215	262
0880	Concealed, rear spud		8	1		206	62		268	320
0890	Top spud		8	1		164	62		226	274
0900	Wall hung		8	1		183	62		245	295
0920	Urinal									
0930	Exposed, stall	1 Plum	8	1	Ea.	153	62		215	262
0940	Wall (washout)		8	1		141	62		203	249
0950	Pedestal, top spud		8	1		143	62		205	252
0960	Concealed, stall		8	1		161	62		223	271
0970	Wall (washout)		8	1		173	62		235	285
0971	Automatic flush sensor and operator for									
0972	urinals or water closets, standard	G	1 Plum	8	1	Ea.	465	62		527	605
0980	High efficiency water saving									
0984	Water closets, 1.28 gpf	G	1 Plum	8	1	Ea.	420	62		482	560
0988	Urinals, .5 gpf	G	"	8	1	"	420	62		482	560

For customer support on your Building Construction Costs with RSMeans Data, call 800.448.8182.

517

22 42 39 – Commercial Faucets, Supplies, and Trim

22 42 39.10 Faucets and Fittings	Crew	Daily Output	Labor-Hours	Unit	Material	2017 Bare Costs Labor	Equipment	Total	Total Incl O&P
2790	Faucets for lavatories									
2800	Self-closing, center set	1 Plum	10	.800	Ea.	150	49.50		199.50	240
2810	Automatic sensor and operator, with faucet head		6.15	1.301		460	80.50		540.50	625
3000	Service sink faucet, cast spout, pail hook, hose end		14	.571		76.50	35.50		112	138

22 42 39.30 Carriers and Supports	Crew	Daily Output	Labor-Hours	Unit	Material	2017 Bare Costs Labor	Equipment	Total	Total Incl O&P
0010	**CARRIERS AND SUPPORTS**, For plumbing fixtures									
0500	Drinking fountain, wall mounted									
0600	Plate type with studs, top back plate	1 Plum	7	1.143	Ea.	96	70.50		166.50	213
0700	Top front and back plate		7	1.143		118	70.50		188.50	237
0800	Top & bottom, front & back plates, w/bearing jacks		7	1.143		217	70.50		287.50	345
3000	Lavatory, concealed arm									
3050	Floor mounted, single									
3100	High back fixture	1 Plum	6	1.333	Ea.	570	82.50		652.50	750
3200	Flat slab fixture		6	1.333		490	82.50		572.50	665
3220	Paraplegic		6	1.333		455	82.50		537.50	625
3250	Floor mounted, back to back									
3300	High back fixtures	1 Plum	5	1.600	Ea.	810	99		909	1,050
3400	Flat slab fixtures		5	1.600		1,000	99		1,099	1,250
3430	Paraplegic		5	1.600		815	99		914	1,050
3500	Wall mounted, in stud or masonry									
3600	High back fixture	1 Plum	6	1.333	Ea.	335	82.50		417.50	495
3700	Flat slab fixture	"	6	1.333	"	220	82.50		302.50	365
4600	Sink, floor mounted									
4650	Exposed arm system									
4700	Single heavy fixture	1 Plum	5	1.600	Ea.	680	99		779	900
4750	Single heavy sink with slab		5	1.600		890	99		989	1,125
4800	Back to back, standard fixtures		5	1.600		500	99		599	700
4850	Back to back, heavy fixtures		5	1.600		810	99		909	1,050
4900	Back to back, heavy sink with slab		5	1.600		1,025	99		1,124	1,300
4950	Exposed offset arm system									
5000	Single heavy deep fixture	1 Plum	5	1.600	Ea.	905	99		1,004	1,150
5100	Plate type system									
5200	With bearing jacks, single fixture	1 Plum	5	1.600	Ea.	955	99		1,054	1,200
5300	With exposed arms, single heavy fixture		5	1.600		955	99		1,054	1,200
5400	Wall mounted, exposed arms, single heavy fixture		5	1.600		345	99		444	530
6000	Urinal, floor mounted, 2" or 3" coupling, blowout type		6	1.333		605	82.50		687.50	790
6100	With fixture or hanger bolts, blowout or washout		6	1.333		435	82.50		517.50	600
6200	With bearing plate		6	1.333		480	82.50		562.50	655
6300	Wall mounted, plate type system		6	1.333		340	82.50		422.50	495
6980	Water closet, siphon jet									
7000	Horizontal, adjustable, caulk									
7040	Single, 4" pipe size	1 Plum	5.33	1.501	Ea.	965	93		1,058	1,200
7050	4" pipe size, paraplegic		5.33	1.501		565	93		658	760
7060	5" pipe size		5.33	1.501		745	93		838	960
7100	Double, 4" pipe size		5	1.600		1,050	99		1,149	1,300
7110	4" pipe size, paraplegic		5	1.600		1,050	99		1,149	1,300
7120	5" pipe size		5	1.600		1,300	99		1,399	1,600
7160	Horizontal, adjustable, extended, caulk									
7180	Single, 4" pipe size	1 Plum	5.33	1.501	Ea.	760	93		853	975
7200	5" pipe size		5.33	1.501		1,000	93		1,093	1,250
7240	Double, 4" pipe size		5	1.600		1,325	99		1,424	1,600
7260	5" pipe size		5	1.600		1,675	99		1,774	2,000

22 42 Commercial Plumbing Fixtures

22 42 39 – Commercial Faucets, Supplies, and Trim

22 42 39.30 Carriers and Supports	Crew	Daily Output	Labor-Hours	Unit	Material	2017 Bare Costs Labor	2017 Bare Costs Equipment	Total	Total Incl O&P
7400	Vertical, adjustable, caulk or thread								
7440	Single, 4" pipe size	1 Plum	5.33	1.501	Ea.	665	93		758	875
7460	5" pipe size		5.33	1.501		840	93		933	1,075
7480	6" pipe size		5	1.600		985	99		1,084	1,225
7520	Double, 4" pipe size		5	1.600		1,150	99		1,249	1,425
7540	5" pipe size		5	1.600		1,325	99		1,424	1,625
7560	6" pipe size	↓	4	2	↓	1,475	124		1,599	1,800
7600	Vertical, adjustable, extended, caulk								
7620	Single, 4" pipe size	1 Plum	5.33	1.501	Ea.	955	93		1,048	1,200
7640	5" pipe size		5.33	1.501		1,300	93		1,393	1,575
7680	6" pipe size		5	1.600		1,500	99		1,599	1,800
7720	Double, 4" pipe size		5	1.600		1,250	99		1,349	1,525
7740	5" pipe size		5	1.600		1,450	99		1,549	1,725
7760	6" pipe size	↓	4	2	↓	1,575	124		1,699	1,925
7780	Water closet, blow out								
7800	Vertical offset, caulk or thread								
7820	Single, 4" pipe size	1 Plum	5.33	1.501	Ea.	1,025	93		1,118	1,275
7840	Double, 4" pipe size	"	5	1.600	"	1,725	99		1,824	2,050
7880	Vertical offset, extended, caulk								
7900	Single, 4" pipe size	1 Plum	5.33	1.501	Ea.	1,275	93		1,368	1,550
7920	Double, 4" pipe size	"	5	1.600	"	2,000	99		2,099	2,350
7960	Vertical, for floor mounted back-outlet								
7980	Single, 4" thread, 2" vent	1 Plum	5.33	1.501	Ea.	685	93		778	890
8000	Double, 4" thread, 2" vent	"	6	1.333	"	2,025	82.50		2,107.50	2,350
8040	Vertical, for floor mounted back-outlet, extended								
8060	Single, 4" caulk, 2" vent	1 Plum	6	1.333	Ea.	685	82.50		767.50	875
8080	Double, 4" caulk, 2" vent	"	6	1.333	"	2,025	82.50		2,107.50	2,350
8200	Water closet, residential								
8220	Vertical centerline, floor mount								
8240	Single, 3" caulk, 2" or 3" vent	1 Plum	6	1.333	Ea.	655	82.50		737.50	850
8260	4" caulk, 2" or 4" vent		6	1.333		845	82.50		927.50	1,050
8280	3" copper sweat, 3" vent		6	1.333		590	82.50		672.50	775
8300	4" copper sweat, 4" vent	↓	6	1.333	↓	715	82.50		797.50	910
8400	Vertical offset, floor mount								
8420	Single, 3" or 4" caulk, vent	1 Plum	4	2	Ea.	820	124		944	1,100
8440	3" or 4" copper sweat, vent		5	1.600		820	99		919	1,050
8460	Double, 3" or 4" caulk, vent		4	2		1,400	124		1,524	1,725
8480	3" or 4" copper sweat, vent	↓	5	1.600	↓	1,400	99		1,499	1,700
9000	Water cooler (electric), floor mounted								
9100	Plate type with bearing plate, single	1 Plum	6	1.333	Ea.	395	82.50		477.50	555

For customer support on your Building Construction Costs with RSMeans Data, call 800.448.8182.

519

22 45 Emergency Plumbing Fixtures

22 45 13 – Emergency Showers

22 45 13.10 Emergency Showers	Crew	Daily Output	Labor-Hours	Unit	Material	2017 Bare Costs Labor	Equipment	Total	Total Incl O&P
0010 **EMERGENCY SHOWERS**, Rough-in not included									
5000 Shower, single head, drench, ball valve, pull, freestanding	Q-1	4	4	Ea.	380	223		603	750
5200 Horizontal or vertical supply		4	4		570	223		793	960
6000 Multi-nozzle, eye/face wash combination		4	4		680	223		903	1,075
6400 Multi-nozzle, 12 spray, shower only		4	4		2,050	223		2,273	2,575
6600 For freeze-proof, add		6	2.667		480	148		628	755

22 45 16 – Eyewash Equipment

22 45 16.10 Eyewash Safety Equipment

	Crew	Daily Output	Labor-Hours	Unit	Material	Labor	Equipment	Total	Total Incl O&P
0010 **EYEWASH SAFETY EQUIPMENT**, Rough-in not included									
1000 Eye wash fountain									
1400 Plastic bowl, pedestal mounted	Q-1	4	4	Ea.	288	223		511	650
1600 Unmounted		4	4		211	223		434	565
1800 Wall mounted		4	4		505	223		728	890
2000 Stainless steel, pedestal mounted		4	4		360	223		583	735
2200 Unmounted		4	4		278	223		501	640
2400 Wall mounted		4	4		298	223		521	665

22 45 19 – Self-Contained Eyewash Equipment

22 45 19.10 Self-Contained Eyewash Safety Equipment

	Crew	Daily Output	Labor-Hours	Unit	Material	Labor	Equipment	Total	Total Incl O&P
0010 **SELF-CONTAINED EYEWASH SAFETY EQUIPMENT**									
3000 Eye wash, portable, self-contained				Ea.	1,325			1,325	1,450

22 45 26 – Eye/Face Wash Equipment

22 45 26.10 Eye/Face Wash Safety Equipment

	Crew	Daily Output	Labor-Hours	Unit	Material	Labor	Equipment	Total	Total Incl O&P
0010 **EYE/FACE WASH SAFETY EQUIPMENT**, Rough-in not included									
4000 Eye and face wash, combination fountain									
4200 Stainless steel, pedestal mounted	Q-1	4	4	Ea.	1,100	223		1,323	1,550
4400 Unmounted		4	4		278	223		501	640
4600 Wall mounted		4	4		253	223		476	615

22 47 Drinking Fountains and Water Coolers

22 47 13 – Drinking Fountains

22 47 13.10 Drinking Water Fountains

	Crew	Daily Output	Labor-Hours	Unit	Material	Labor	Equipment	Total	Total Incl O&P
0010 **DRINKING WATER FOUNTAINS**, For connection to cold water supply									
1000 Wall mounted, non-recessed									
1400 Bronze, with no back	1 Plum	4	2	Ea.	1,100	124		1,224	1,375
1800 Cast aluminum, enameled, for correctional institutions		4	2		1,400	124		1,524	1,725
2000 Fiberglass, 12" back, single bubbler unit		4	2		1,925	124		2,049	2,300
2040 Dual bubbler		3.20	2.500		2,250	155		2,405	2,700
2400 Precast stone, no back		4	2		935	124		1,059	1,200
2700 Stainless steel, single bubbler, no back		4	2		1,075	124		1,199	1,350
2740 With back		4	2		575	124		699	820
2780 Dual handle & wheelchair projection type		4	2		720	124		844	975
2820 Dual level for handicapped type		3.20	2.500		1,575	155		1,730	1,950
3300 Vitreous china									
3340 7" back	1 Plum	4	2	Ea.	575	124		699	820
3940 For vandal-resistant bottom plate, add					77			77	84.50
3960 For freeze-proof valve system, add	1 Plum	2	4		705	247		952	1,150
3980 For rough-in, supply and waste, add	"	2.21	3.620		171	224		395	530
4000 Wall mounted, semi-recessed									
4200 Poly-marble, single bubbler	1 Plum	4	2	Ea.	980	124		1,104	1,250

22 47 Drinking Fountains and Water Coolers

22 47 13 – Drinking Fountains

22 47 13.10 Drinking Water Fountains		Crew	Daily Output	Labor-Hours	Unit	Material	2017 Bare Costs Labor	Equipment	Total	Total Incl O&P
4600	Stainless steel, satin finish, single bubbler	1 Plum	4	2	Ea.	1,325	124		1,449	1,650
4900	Vitreous china, single bubbler		4	2		920	124		1,044	1,200
5980	For rough-in, supply and waste, add	↓	1.83	4.372	↓	171	270		441	600
6000	Wall mounted, fully recessed									
6400	Poly-marble, single bubbler	1 Plum	4	2	Ea.	1,625	124		1,749	1,950
6800	Stainless steel, single bubbler		4	2		1,525	124		1,649	1,850
7560	For freeze-proof valve system, add		2	4		830	247		1,077	1,300
7580	For rough-in, supply and waste, add	↓	1.83	4.372	↓	171	270		441	600
7600	Floor mounted, pedestal type									
7700	Aluminum, architectural style, C.I. base	1 Plum	2	4	Ea.	2,250	247		2,497	2,850
7780	Wheelchair handicap unit		2	4		1,550	247		1,797	2,075
8400	Stainless steel, architectural style		2	4		1,775	247		2,022	2,325
8600	Enameled iron, heavy duty service, 2 bubblers		2	4		2,650	247		2,897	3,275
8660	4 bubblers		2	4		3,950	247		4,197	4,700
8880	For freeze-proof valve system, add		2	4		705	247		952	1,150
8900	For rough-in, supply and waste, add	↓	1.83	4.372	↓	171	270		441	600
9100	Deck mounted									
9500	Stainless steel, circular receptor	1 Plum	4	2	Ea.	435	124		559	665
9760	White enameled steel, 14" x 9" receptor		4	2		375	124		499	600
9860	White enameled cast iron, 24" x 16" receptor		3	2.667		475	165		640	770
9980	For rough-in, supply and waste, add	↓	1.83	4.372	↓	171	270		441	600

22 47 16 – Pressure Water Coolers

22 47 16.10 Electric Water Coolers		Crew	Daily Output	Labor-Hours	Unit	Material	2017 Bare Costs Labor	Equipment	Total	Total Incl O&P
0010	**ELECTRIC WATER COOLERS**									
0100	Wall mounted, non-recessed									
0140	4 GPH	Q-1	4	4	Ea.	680	223		903	1,075
0160	8 GPH, barrier free, sensor operated		4	4		1,075	223		1,298	1,500
0180	8.2 GPH		4	4		795	223		1,018	1,200
0600	8 GPH hot and cold water	↓	4	4		1,050	223		1,273	1,500
0640	For stainless steel cabinet, add					90			90	99.50
1000	Dual height, 8.2 GPH	Q-1	3.80	4.211		2,100	234		2,334	2,650
1040	14.3 GPH	"	3.80	4.211		1,325	234		1,559	1,800
1240	For stainless steel cabinet, add					188			188	207
2600	Wheelchair type, 8 GPH	Q-1	4	4		945	223		1,168	1,350
3300	Semi-recessed, 8.1 GPH		4	4		785	223		1,008	1,200
3320	12 GPH	↓	4	4	↓	905	223		1,128	1,325
4600	Floor mounted, flush-to-wall									
4640	4 GPH	1 Plum	3	2.667	Ea.	775	165		940	1,100
4680	8.2 GPH		3	2.667		815	165		980	1,150
4720	14.3 GPH		3	2.667		935	165		1,100	1,275
4960	14 GPH hot and cold water	↓	3	2.667		1,075	165		1,240	1,425
4980	For stainless steel cabinet, add					138			138	152
5000	Dual height, 8.2 GPH	1 Plum	2	4		1,200	247		1,447	1,700
5040	14.3 GPH	"	2	4		1,250	247		1,497	1,750
5120	For stainless steel cabinet, add					200			200	220
9800	For supply, waste & vent, all coolers	1 Plum	2.21	3.620	↓	171	224		395	530

For customer support on your Building Construction Costs with RSMeans Data, call 800.448.8182.

521

22 51 Swimming Pool Plumbing Systems

22 51 19 – Swimming Pool Water Treatment Equipment

22 51 19.50 Swimming Pool Filtration Equipment	Crew	Daily Output	Labor-Hours	Unit	Material	2017 Bare Costs Labor	Equipment	Total	Total Incl O&P
0010 **SWIMMING POOL FILTRATION EQUIPMENT**									
0900 Filter system, sand or diatomite type, incl. pump, 6,000 gal./hr.	2 Plum	1.80	8.889	Total	2,325	550		2,875	3,375
1020 Add for chlorination system, 800 S.F. pool		3	5.333	Ea.	179	330		509	690
1040 5,000 S.F. pool		3	5.333	"	1,850	330		2,180	2,525

22 52 Fountain Plumbing Systems

22 52 16 – Fountain Pumps

22 52 16.10 Fountain Water Pumps

	Crew	Daily Output	Labor-Hours	Unit	Material	2017 Bare Costs Labor	Equipment	Total	Total Incl O&P
0010 **FOUNTAIN WATER PUMPS**									
0100 Pump w/controls									
0200 Single phase, 100' cord, 1/2 H.P. pump	2 Skwk	4.40	3.636	Ea.	1,200	187		1,387	1,625
0300 3/4 H.P. pump		4.30	3.721		1,300	191		1,491	1,725
0400 1 H.P. pump		4.20	3.810		1,900	196		2,096	2,400
0500 1-1/2 H.P. pump		4.10	3.902		2,575	201		2,776	3,125
0600 2 H.P. pump		4	4		3,950	206		4,156	4,675
0700 Three phase, 200' cord, 5 H.P. pump		3.90	4.103		5,550	211		5,761	6,425
0800 7-1/2 H.P. pump		3.80	4.211		9,525	217		9,742	10,800
0900 10 H.P. pump		3.70	4.324		14,100	222		14,322	15,900
1000 15 H.P. pump		3.60	4.444		18,000	229		18,229	20,200
2000 DESIGN NOTE: Use two horsepower per surface acre.									

22 52 33 – Fountain Ancillary

22 52 33.10 Fountain Miscellaneous

	Crew	Daily Output	Labor-Hours	Unit	Material	2017 Bare Costs Labor	Equipment	Total	Total Incl O&P
0010 **FOUNTAIN MISCELLANEOUS**									
1300 Lights w/mounting kits, 200 watt	2 Skwk	18	.889	Ea.	1,200	45.50		1,245.50	1,400
1400 300 watt		18	.889		1,350	45.50		1,395.50	1,550
1500 500 watt		18	.889		1,525	45.50		1,570.50	1,750
1600 Color blender		12	1.333		590	68.50		658.50	755

22 66 Chemical-Waste Systems for Lab. and Healthcare Facilities

22 66 53 – Laboratory Chemical-Waste and Vent Piping

22 66 53.30 Glass Pipe

	Crew	Daily Output	Labor-Hours	Unit	Material	2017 Bare Costs Labor	Equipment	Total	Total Incl O&P
0010 **GLASS PIPE**, Borosilicate, couplings & clevis hanger assemblies, 10' O.C.									
0020 Drainage									
1100 1-1/2" diameter	Q-1	52	.308	L.F.	15.90	17.10		33	43.50
1120 2" diameter		44	.364		19.85	20		39.85	52.50
1140 3" diameter		39	.410		23	23		46	60
1160 4" diameter		30	.533		44.50	29.50		74	94
1180 6" diameter		26	.615		81.50	34		115.50	142

22 66 53.60 Corrosion Resistant Pipe

	Crew	Daily Output	Labor-Hours	Unit	Material	2017 Bare Costs Labor	Equipment	Total	Total Incl O&P
0010 **CORROSION RESISTANT PIPE**, No couplings or hangers									
0020 Iron alloy, drain, mechanical joint									
1000 1-1/2" diameter	Q-1	70	.229	L.F.	66	12.70		78.70	92
1100 2" diameter		66	.242		68	13.50		81.50	95
1120 3" diameter		60	.267		87	14.85		101.85	119
1140 4" diameter		52	.308		112	17.10		129.10	149
2980 Plastic, epoxy, fiberglass filament wound, B&S joint									
3000 2" diameter	Q-1	62	.258	L.F.	12.10	14.35		26.45	35
3100 3" diameter		51	.314		14.15	17.45		31.60	42

522

For customer support on your Building Construction Costs with RSMeans Data, call 800.448.8182.

22 66 53.60 Corrosion Resistant Pipe		Crew	Daily Output	Labor-Hours	Unit	Material	2017 Bare Costs Labor	Equipment	Total	Total Incl O&P
3120	4" diameter	Q-1	45	.356	L.F.	20	19.80		39.80	52
3140	6" diameter	↓	32	.500	↓	28.50	28		56.50	73
3980	Polyester, fiberglass filament wound, B&S joint									
4000	2" diameter	Q-1	62	.258	L.F.	13.20	14.35		27.55	36
4100	3" diameter		51	.314		17.20	17.45		34.65	45.50
4120	4" diameter		45	.356		25	19.80		44.80	58
4140	6" diameter	↓	32	.500	↓	36.50	28		64.50	82
4980	Polypropylene, acid resistant, fire retardant, Schedule 40									
5000	1-1/2" diameter	Q-1	68	.235	L.F.	8.25	13.10		21.35	29
5100	2" diameter		62	.258		11.30	14.35		25.65	34
5120	3" diameter		51	.314		23	17.45		40.45	52
5140	4" diameter	↓	45	.356	↓	29.50	19.80		49.30	62
5980	Proxylene, fire retardant, Schedule 40									
6000	1-1/2" diameter	Q-1	68	.235	L.F.	12.90	13.10		26	34
6100	2" diameter		62	.258		17.70	14.35		32.05	41
6120	3" diameter		51	.314		32	17.45		49.45	61.50
6140	4" diameter	↓	45	.356	↓	45	19.80		64.80	79.50

Division Notes

	CREW	DAILY OUTPUT	LABOR-HOURS	UNIT	BARE COSTS				TOTAL INCL O&P
					MAT.	LABOR	EQUIP.	TOTAL	

Estimating Tips

The labor adjustment factors listed in Subdivision 22 01 02.20 also apply to Division 23.

23 10 00 Facility Fuel Systems

- The prices in this subdivision for above- and below-ground storage tanks do not include foundations or hold-down slabs, unless noted. The estimator should refer to Divisions 3 and 31 for foundation system pricing. In addition to the foundations, required tank accessories, such as tank gauges, leak detection devices, and additional manholes and piping, must be added to the tank prices.

23 50 00 Central Heating Equipment

- When estimating the cost of an HVAC system, check to see who is responsible for providing and installing the temperature control system. It is possible to overlook controls, assuming that they would be included in the electrical estimate.
- When looking up a boiler, be careful on specified capacity. Some

manufacturers rate their products on output while others use input.

- Include HVAC insulation for pipe, boiler, and duct (wrap and liner).
- Be careful when looking up mechanical items to get the correct pressure rating and connection type (thread, weld, flange).

23 70 00 Central HVAC Equipment

- Combination heating and cooling units are sized by the air conditioning requirements. (See Reference No. R236000-20 for the preliminary sizing guide.)
- A ton of air conditioning is nominally 400 CFM.
- Rectangular duct is taken off by the linear foot for each size, but its cost is usually estimated by the pound. Remember that SMACNA standards now base duct on internal pressure.
- Prefabricated duct is estimated and purchased like pipe: straight sections and fittings.
- Note that cranes or other lifting equipment are not included on any

lines in Division 23. For example, if a crane is required to lift a heavy piece of pipe into place high above a gym floor, or to put a rooftop unit on the roof of a four-story building, etc., it must be added. Due to the potential for extreme variation—from nothing additional required to a major crane or helicopter—we feel that including a nominal amount for "lifting contingency" would be useless and detract from the accuracy of the estimate. When using equipment rental cost data from RSMeans, do not forget to include the cost of the operator(s).

Reference Numbers

Reference numbers are shown at the beginning of some major classifications. These numbers refer to related items in the Reference Section. The reference information may be an estimating procedure, an alternate pricing method, or technical information.

Note: Not all subdivisions listed here necessarily appear. ■

Did you know?

Our online estimating solution gives you the same access to RSMeans' data with 24/7 access:

- Quickly locate costs in the searchable database.
- Build cost lists, estimates, and reports in minutes.
- Adjust costs to any location in the U.S. and Canada with the click of a button.

Start your free trial today at
www.RSMeansOnline.com

23 05 Common Work Results for HVAC

23 05 02 – HVAC General

23 05 02.10 Air Conditioning, General	Crew	Daily Output	Labor-Hours	Unit	Material	2017 Bare Costs Labor	Equipment	Total	Total Incl O&P
0010 **AIR CONDITIONING, GENERAL** Prices are for standard efficiencies (SEER 13)									
0020　for upgrade to SEER 14 add					10%				

23 05 05 – Selective Demolition for HVAC

23 05 05.10 HVAC Demolition

		Crew	Daily Output	Labor-Hours	Unit	Material	Labor	Equipment	Total	Total Incl O&P
0010	**HVAC DEMOLITION**									
0100	Air conditioner, split unit, 3 ton	Q-5	2	8	Ea.		450		450	675
0150	Package unit, 3 ton	Q-6	3	8	"		465		465	700
0298	Boilers									
0300	Electric, up thru 148 kW	Q-19	2	12	Ea.		675		675	1,025
0310	150 thru 518 kW	"	1	24			1,350		1,350	2,025
0320	550 thru 2000 kW	Q-21	.40	80			4,625		4,625	6,950
0330	2070 kW and up	"	.30	106			6,150		6,150	9,275
0340	Gas and/or oil, up thru 150 MBH	Q-7	2.20	14.545			860		860	1,300
0350	160 thru 2000 MBH		.80	40			2,375		2,375	3,575
0360	2100 thru 4500 MBH		.50	64			3,800		3,800	5,725
0370	4600 thru 7000 MBH		.30	106			6,325		6,325	9,550
0380	7100 thru 12,000 MBH		.16	200			11,900		11,900	17,900
0390	12,200 thru 25,000 MBH	↓	.12	266	↓		15,800		15,800	23,900
1000	Ductwork, 4" high, 8" wide	1 Clab	200	.040	L.F.		1.57		1.57	2.40
1100	6" high, 8" wide		165	.048			1.90		1.90	2.91
1200	10" high, 12" wide		125	.064			2.51		2.51	3.84
1300	12"-14" high, 16"-18" wide		85	.094			3.68		3.68	5.65
1400	18" high, 24" wide		67	.119			4.67		4.67	7.15
1500	30" high, 36" wide		56	.143			5.60		5.60	8.55
1540	72" wide	↓	50	.160	↓		6.25		6.25	9.60
3000	Mechanical equipment, light items. Unit is weight, not cooling.	Q-5	.90	17.778	Ton		995		995	1,500
3600	Heavy items	"	1.10	14.545	"		815		815	1,225
5090	Remove refrigerant from system	1 Stpi	40	.200	Lb.		12.45		12.45	18.80

23 05 23 – General-Duty Valves for HVAC Piping

23 05 23.30 Valves, Iron Body

		Crew	Daily Output	Labor-Hours	Unit	Material	Labor	Equipment	Total	Total Incl O&P
0010	**VALVES, IRON BODY**									
1020	Butterfly, wafer type, gear actuator, 200 lb.									
1030	2"	1 Plum	14	.571	Ea.	105	35.50		140.50	169
1040	2-1/2"	Q-1	9	1.778		106	99		205	266
1050	3"		8	2		110	111		221	289
1060	4"	↓	5	3.200		123	178		301	405
1070	5"	Q-2	5	4.800		138	277		415	570
1080	6"	"	5	4.800	↓	156	277		433	590
1650	Gate, 125 lb., N.R.S.									
2150	Flanged									
2200	2"	1 Plum	5	1.600	Ea.	690	99		789	910
2240	2-1/2"	Q-1	5	3.200		705	178		883	1,050
2260	3"		4.50	3.556		795	198		993	1,175
2280	4"	↓	3	5.333		1,125	297		1,422	1,700
2300	6"	Q-2	3	8	↓	1,925	460		2,385	2,825
3550	OS&Y, 125 lb., flanged									
3600	2"	1 Plum	5	1.600	Ea.	455	99		554	650
3660	3"	Q-1	4.50	3.556		505	198		703	855
3680	4"	"	3	5.333		735	297		1,032	1,250
3700	6"	Q-2	3	8		1,175	460		1,635	2,000
3900	For 175 lb., flanged, add					200%	10%			
5450	Swing check, 125 lb., threaded									

23 05 Common Work Results for HVAC

23 05 23 – General-Duty Valves for HVAC Piping

23 05 23.30 Valves, Iron Body

		Crew	Daily Output	Labor-Hours	Unit	Material	2017 Bare Costs Labor	2017 Bare Costs Equipment	Total	Total Incl O&P
5500	2"	1 Plum	11	.727	Ea.	415	45		460	525
5540	2-1/2"	Q-1	15	1.067		535	59.50		594.50	675
5550	3"		13	1.231		580	68.50		648.50	740
5560	4"	↓	10	1.600	↓	920	89		1,009	1,125
5950	Flanged									
6000	2"	1 Plum	5	1.600	Ea.	415	99		514	610
6040	2-1/2"	Q-1	5	3.200		380	178		558	690
6050	3"		4.50	3.556		405	198		603	745
6060	4"	↓	3	5.333		640	297		937	1,150
6070	6"	Q-2	3	8	↓	1,100	460		1,560	1,900

23 05 23.80 Valves, Steel

		Crew	Daily Output	Labor-Hours	Unit	Material	2017 Bare Costs Labor	2017 Bare Costs Equipment	Total	Total Incl O&P
0010	**VALVES, STEEL**									
0800	Cast									
1350	Check valve, swing type, 150 lb., flanged									
1370	1"	1 Plum	10	.800	Ea.	360	49.50		409.50	470
1400	2"	"	8	1		610	62		672	765
1440	2-1/2"	Q-1	5	3.200		705	178		883	1,050
1450	3"		4.50	3.556		720	198		918	1,100
1460	4"	↓	3	5.333		1,025	297		1,322	1,575
1540	For 300 lb., flanged, add					50%	15%			
1548	For 600 lb., flanged, add				↓	110%	20%			
1950	Gate valve, 150 lb., flanged									
2000	2"	1 Plum	8	1	Ea.	630	62		692	785
2040	2-1/2"	Q-1	5	3.200		890	178		1,068	1,250
2050	3"		4.50	3.556		890	198		1,088	1,275
2060	4"	↓	3	5.333		1,125	297		1,422	1,700
2070	6"	Q-2	3	8	↓	1,800	460		2,260	2,675
3650	Globe valve, 150 lb., flanged									
3700	2"	1 Plum	8	1	Ea.	790	62		852	965
3740	2-1/2"	Q-1	5	3.200		1,000	178		1,178	1,375
3750	3"		4.50	3.556		1,000	198		1,198	1,400
3760	4"	↓	3	5.333		1,475	297		1,772	2,050
3770	6"	Q-2	3	8	↓	2,300	460		2,760	3,225
5150	Forged									
5650	Check valve, class 800, horizontal									
5698	Threaded									
5700	1/4"	1 Plum	24	.333	Ea.	90.50	20.50		111	131
5720	3/8"		24	.333		90.50	20.50		111	131
5730	1/2"		24	.333		90.50	20.50		111	131
5740	3/4"		20	.400		96.50	24.50		121	144
5750	1"		19	.421		114	26		140	165
5760	1-1/4"	↓	15	.533	↓	223	33		256	295

23 05 93 – Testing, Adjusting, and Balancing for HVAC

23 05 93.10 Balancing, Air

		Crew	Daily Output	Labor-Hours	Unit	Material	2017 Bare Costs Labor	2017 Bare Costs Equipment	Total	Total Incl O&P
0010	**BALANCING, AIR** (Subcontractor's quote incl. material and labor)									
0900	Heating and ventilating equipment									
1000	Centrifugal fans, utility sets				Ea.				420	420
1100	Heating and ventilating unit								630	630
1200	In-line fan								630	630
1300	Propeller and wall fan								119	119
1400	Roof exhaust fan								280	280
2000	Air conditioning equipment, central station				↓				910	910

23 05 Common Work Results for HVAC

23 05 93 – Testing, Adjusting, and Balancing for HVAC

23 05 93.10 Balancing, Air

		Crew	Daily Output	Labor-Hours	Unit	Material	2017 Bare Costs Labor	Equipment	Total	Total Incl O&P
2100	Built-up low pressure unit				Ea.				840	840
2200	Built-up high pressure unit								980	980
2500	Multi-zone A.C. and heating unit								630	630
2600	For each zone over one, add								140	140
2700	Package A.C. unit								350	350
2800	Rooftop heating and cooling unit								490	490
3000	Supply, return, exhaust, registers & diffusers, avg. height ceiling								84	84
3100	High ceiling								126	126
3200	Floor height								70	70

23 05 93.20 Balancing, Water

		Crew	Daily Output	Labor-Hours	Unit	Material	2017 Bare Costs Labor	Equipment	Total	Total Incl O&P
0010	**BALANCING, WATER** (Subcontractor's quote incl. material and labor)									
0050	Air cooled condenser				Ea.				256	256
0080	Boiler								515	515
0100	Cabinet unit heater								88	88
0200	Chiller								620	620
0300	Convector								73	73
0500	Cooling tower								475	475
0600	Fan coil unit, unit ventilator								132	132
0700	Fin tube and radiant panels								146	146
0800	Main and duct re-heat coils								135	135
0810	Heat exchanger								135	135
1000	Pumps								320	320
1100	Unit heater								102	102

23 07 HVAC Insulation

23 07 13 – Duct Insulation

23 07 13.10 Duct Thermal Insulation

			Crew	Daily Output	Labor-Hours	Unit	Material	2017 Bare Costs Labor	Equipment	Total	Total Incl O&P
0010	**DUCT THERMAL INSULATION**										
0100	Rule of thumb, as a percentage of total mechanical costs					Job				10%	10%
0110	Insulation req'd. is based on the surface size/area to be covered										
3000	Ductwork										
3020	Blanket type, fiberglass, flexible										
3030	Fire rated for grease and hazardous exhaust ducts										
3060	1-1/2" thick		Q-14	84	.190	S.F.	4.54	9.45		13.99	19.65
3090	Fire rated for plenums										
3100	1/2" x 24" x 25'		Q-14	1.94	8.247	Roll	167	410		577	820
3110	1/2" x 24" x 25'			98	.163	S.F.	3.35	8.10		11.45	16.25
3120	1/2" x 48" x 25'			1.04	15.385	Roll	335	760		1,095	1,550
3126	1/2" x 48" x 25'			104	.154	S.F.	3.35	7.60		10.95	15.55
3140	FSK vapor barrier wrap, .75 lb. density										
3160	1" thick	G	Q-14	350	.046	S.F.	.22	2.26		2.48	3.76
3170	1-1/2" thick	G		320	.050		.27	2.48		2.75	4.15
3180	2" thick	G		300	.053		.32	2.64		2.96	4.46
3190	3" thick	G		260	.062		.45	3.05		3.50	5.25
3200	4" thick	G		242	.066		.64	3.28		3.92	5.80
3210	Vinyl jacket, same as FSK										
3280	Unfaced, 1 lb. density										
3310	1" thick	G	Q-14	360	.044	S.F.	.24	2.20		2.44	3.68
3320	1-1/2" thick	G		330	.048		.36	2.40		2.76	4.13
3330	2" thick	G		310	.052		.44	2.56		3	4.45
3400	FSK facing, 1 lb. density										

23 07 HVAC Insulation

23 07 13 – Duct Insulation

23 07 13.10 Duct Thermal Insulation

		Crew	Daily Output	Labor-Hours	Unit	Material	2017 Bare Costs Labor	Equipment	Total	Total Incl O&P
3420	1-1/2" thick	G Q-14	310	.052	S.F.	.36	2.56		2.92	4.37
3430	2" thick	G "	300	.053	"	.44	2.64		3.08	4.59
3450	FSK facing, 1.5 lb. density									
3470	1-1/2" thick	G Q-14	300	.053	S.F.	.45	2.64		3.09	4.61
3480	2" thick	G "	290	.055	"	.54	2.73		3.27	4.84
3795	Finishes									
3800	Stainless steel woven mesh	Q-14	100	.160	S.F.	.93	7.95		8.88	13.30
3810	For .010" stainless steel, add		160	.100		3.15	4.96		8.11	11.15
3820	18 oz. fiberglass cloth, pasted on		170	.094		.84	4.66		5.50	8.15
3900	8 oz. canvas, pasted on		180	.089		.26	4.40		4.66	7.15
3940	For .016" aluminum jacket, add		200	.080		1.01	3.96		4.97	7.25
7878	Contact cement, quart can				Ea.	12.50			12.50	13.75

23 07 16 – HVAC Equipment Insulation

23 07 16.10 HVAC Equipment Thermal Insulation

		Crew	Daily Output	Labor-Hours	Unit	Material	2017 Bare Costs Labor	Equipment	Total	Total Incl O&P
0010	**HVAC EQUIPMENT THERMAL INSULATION**									
0100	Rule of thumb, as a percentage of total mechanical costs				Job				10%	10%
0110	Insulation req'd. is based on the surface size/area to be covered									
1000	Boiler, 1-1/2" calcium silicate only	G Q-14	110	.145	S.F.	3.65	7.20		10.85	15.20
1020	Plus 2" fiberglass	G "	80	.200	"	5.15	9.90		15.05	21
2000	Breeching, 2" calcium silicate									
2020	Rectangular	G Q-14	42	.381	S.F.	7.15	18.90		26.05	37.50
2040	Round	G "	38.70	.413	"	7.45	20.50		27.95	40

23 09 Instrumentation and Control for HVAC

23 09 33 – Electric and Electronic Control System for HVAC

23 09 33.10 Electronic Control Systems

		Crew	Daily Output	Labor-Hours	Unit	Material	2017 Bare Costs Labor	Equipment	Total	Total Incl O&P
0010	**ELECTRONIC CONTROL SYSTEMS**									
0020	For electronic costs, add to Section 23 09 43.10				Ea.				15%	15%

23 09 43 – Pneumatic Control System for HVAC

23 09 43.10 Pneumatic Control Systems

		Crew	Daily Output	Labor-Hours	Unit	Material	2017 Bare Costs Labor	Equipment	Total	Total Incl O&P
0010	**PNEUMATIC CONTROL SYSTEMS**									
0011	Including a nominal 50 ft. of tubing. Add control panelboard if req'd.									
0100	Heating and ventilating, split system									
0200	Mixed air control, economizer cycle, panel readout, tubing									
0220	Up to 10 tons	G Q-19	.68	35.294	Ea.	4,300	1,975		6,275	7,725
0240	For 10 to 20 tons	G	.63	37.915		4,600	2,125		6,725	8,275
0260	For over 20 tons	G	.58	41.096		5,000	2,300		7,300	8,975
0300	Heating coil, hot water, 3 way valve,									
0320	Freezestat, limit control on discharge, readout	Q-5	.69	23.088	Ea.	3,200	1,300		4,500	5,475
0500	Cooling coil, chilled water, room									
0520	Thermostat, 3 way valve	Q-5	2	8	Ea.	1,425	450		1,875	2,250
0600	Cooling tower, fan cycle, damper control,									
0620	Control system including water readout in/out at panel	Q-19	.67	35.821	Ea.	5,675	2,025		7,700	9,250
1000	Unit ventilator, day/night operation,									
1100	freezestat, ASHRAE, cycle 2	Q-19	.91	26.374	Ea.	3,125	1,475		4,600	5,675
2000	Compensated hot water from boiler, valve control,									
2100	readout and reset at panel, up to 60 GPM	Q-19	.55	43.956	Ea.	5,875	2,475		8,350	10,200
2120	For 120 GPM		.51	47.059		6,275	2,650		8,925	10,900
2140	For 240 GPM		.49	49.180		6,575	2,775		9,350	11,400
3000	Boiler room combustion air, damper to 5 S.F., controls		1.37	17.582		2,825	990		3,815	4,600

23 09 43 – Pneumatic Control System for HVAC

23 09 43.10 Pneumatic Control Systems	Crew	Daily Output	Labor-Hours	Unit	Material	2017 Bare Costs Labor	Equipment	Total	Total Incl O&P
3500	Fan coil, heating and cooling valves, 4 pipe control system	Q-19	3	8	Ea.	1,275	450		1,725	2,075
3600	Heat exchanger system controls	↓	.86	27.907		2,750	1,575		4,325	5,375
4000	Pneumatic thermostat, including controlling room radiator valve	Q-5	2.43	6.593		850	370		1,220	1,500
4060	Pump control system	Q-19	3	8		1,300	450		1,750	2,125
4500	Air supply for pneumatic control system									
4600	Tank mounted duplex compressor, starter, alternator,									
4620	piping, dryer, PRV station and filter									
4630	1/2 HP	Q-19	.68	35.139	Ea.	10,500	1,975		12,475	14,600
4660	1-1/2 HP	↓	.58	41.739		12,800	2,350		15,150	17,600
4690	5 HP	↓	.42	57.143	↓	30,500	3,225		33,725	38,400

23 13 Facility Fuel-Storage Tanks

23 13 13 – Facility Underground Fuel-Oil, Storage Tanks

23 13 13.09 Single-Wall Steel Fuel-Oil Tanks

		Crew	Daily Output	Labor-Hours	Unit	Material	2017 Bare Costs Labor	Equipment	Total	Total Incl O&P
0010	**SINGLE-WALL STEEL FUEL-OIL TANKS**									
5000	Tanks, steel ugnd., sti-p3, not incl. hold-down bars									
5500	Excavation, pad, pumps and piping not included									
5510	Single wall, 500 gallon capacity, 7 ga. shell	Q-5	2.70	5.926	Ea.	1,800	330		2,130	2,500
5520	1,000 gallon capacity, 7 ga. shell	"	2.50	6.400		2,525	360		2,885	3,325
5530	2,000 gallon capacity, 1/4" thick shell	Q-7	4.60	6.957		3,275	410		3,685	4,225
5535	2,500 gallon capacity, 7 ga. shell	Q-5	3	5.333		4,900	299		5,199	5,825
5540	5,000 gallon capacity, 1/4" thick shell	Q-7	3.20	10		6,425	595		7,020	7,975
5580	15,000 gallon capacity, 5/16" thick shell		1.70	18.824		11,500	1,125		12,625	14,400
5600	20,000 gallon capacity, 5/16" thick shell		1.50	21.333		20,200	1,275		21,475	24,200
5610	25,000 gallon capacity, 3/8" thick shell		1.30	24.615		26,500	1,450		27,950	31,400
5620	30,000 gallon capacity, 3/8" thick shell		1.10	29.091		30,900	1,725		32,625	36,600
5630	40,000 gallon capacity, 3/8" thick shell		.90	35.556		42,400	2,100		44,500	49,900
5640	50,000 gallon capacity, 3/8" thick shell	↓	.80	40	↓	47,100	2,375		49,475	55,500

23 13 13.23 Glass-Fiber-Reinfcd-Plastic, Fuel-Oil, Storage

		Crew	Daily Output	Labor-Hours	Unit	Material	2017 Bare Costs Labor	Equipment	Total	Total Incl O&P
0010	**GLASS-FIBER-REINFCD-PLASTIC, UNDERGRND. FUEL-OIL, STORAGE**									
0210	Fiberglass, underground, single wall, U.L. listed, not including									
0220	manway or hold-down strap									
0240	2,000 gallon capacity	Q-7	4.57	7.002	Ea.	6,550	415		6,965	7,850
0245	3,000 gallon capacity		3.90	8.205		7,850	485		8,335	9,375
0250	4,000 gallon capacity		3.55	9.014		9,125	535		9,660	10,800
0255	5,000 gallon capacity		3.20	10		10,100	595		10,695	12,100
0260	6,000 gallon capacity		2.67	11.985		10,400	710		11,110	12,500
0280	10,000 gallon capacity		2	16		14,100	950		15,050	16,900
0284	15,000 gallon capacity		1.68	19.048		22,600	1,125		23,725	26,500
0290	20,000 gallon capacity	↓	1.45	22.069		29,000	1,300		30,300	33,900
0500	For manway, fittings and hold-downs, add				↓	20%	15%			
1020	Fiberglass, underground, double wall, U.L. listed									
1030	includes manways, not incl. hold-down straps									
1040	600 gallon capacity	Q-5	2.42	6.612	Ea.	7,925	370		8,295	9,275
1050	1,000 gallon capacity	"	2.25	7.111		10,800	400		11,200	12,500
1060	2,500 gallon capacity	Q-7	4.16	7.692		15,700	455		16,155	17,900
1070	3,000 gallon capacity		3.90	8.205		17,700	485		18,185	20,100
1080	4,000 gallon capacity		3.64	8.791		17,900	520		18,420	20,500
1090	6,000 gallon capacity		2.42	13.223		23,300	785		24,085	26,800
1100	8,000 gallon capacity		2.08	15.385		26,000	910		26,910	29,900
1110	10,000 gallon capacity	↓	1.82	17.582	↓	30,100	1,050		31,150	34,700

23 13 Facility Fuel-Storage Tanks

23 13 13 – Facility Underground Fuel-Oil, Storage Tanks

23 13 13.23 Glass-Fiber-Reinfcd-Plastic, Fuel-Oil, Storage

		Crew	Daily Output	Labor-Hours	Unit	Material	2017 Bare Costs Labor	Equipment	Total	Total Incl O&P
1120	12,000 gallon capacity	Q-7	1.70	18.824	Ea.	37,500	1,125		38,625	43,000
2210	Fiberglass, underground, single wall, U.L. listed, including									
2220	hold-down straps, no manways									
2240	2,000 gallon capacity	Q-7	3.55	9.014	Ea.	7,025	535		7,560	8,550
2250	4,000 gallon capacity		2.90	11.034		9,600	655		10,255	11,600
2260	6,000 gallon capacity		2	16		11,300	950		12,250	13,800
2280	10,000 gallon capacity		1.60	20		15,100	1,175		16,275	18,400
2284	15,000 gallon capacity		1.39	23.022		23,500	1,375		24,875	28,000
2290	20,000 gallon capacity		1.14	28.070		30,500	1,675		32,175	36,000
3020	Fiberglass, underground, double wall, U.L. listed									
3030	includes manways and hold-down straps									
3040	600 gallon capacity	Q-5	1.86	8.602	Ea.	8,400	480		8,880	9,975
3050	1,000 gallon capacity	"	1.70	9.412		11,300	525		11,825	13,200
3060	2,500 gallon capacity	Q-7	3.29	9.726		16,100	575		16,675	18,700
3070	3,000 gallon capacity		3.13	10.224		18,100	605		18,705	20,800
3080	4,000 gallon capacity		2.93	10.922		18,400	645		19,045	21,200
3090	6,000 gallon capacity		1.86	17.204		24,300	1,025		25,325	28,300
3100	8,000 gallon capacity		1.65	19.394		26,900	1,150		28,050	31,300
3110	10,000 gallon capacity		1.48	21.622		31,000	1,275		32,275	36,000
3120	12,000 gallon capacity		1.40	22.857		38,500	1,350		39,850	44,400

23 13 23 – Facility Aboveground Fuel-Oil, Storage Tanks

23 13 23.13 Vertical, Steel, Abvground Fuel-Oil, Stor. Tanks

		Crew	Daily Output	Labor-Hours	Unit	Material	2017 Bare Costs Labor	Equipment	Total	Total Incl O&P
0010	**VERTICAL, STEEL, ABOVEGROUND FUEL-OIL, STORAGE TANKS**									
4000	Fixed roof oil storage tanks, steel (1 BBL=42 gal. w/foundation 3'D x 1'W)									
4200	5,000 barrels				Ea.				194,000	213,500
4300	24,000 barrels								333,500	367,000
4500	56,000 barrels								729,000	802,000
4600	110,000 barrels								1,060,000	1,166,000
4800	143,000 barrels								1,250,000	1,375,000
4900	225,000 barrels								1,360,000	1,496,000
5100	Floating roof gasoline tanks, steel, 5,000 barrels (w/foundation 3'D x 1'W)								204,000	225,000
5200	25,000 barrels								381,000	419,000
5400	55,000 barrels								839,000	923,000
5500	100,000 barrels								1,253,000	1,379,000
5700	150,000 barrels								1,532,000	1,685,000
5800	225,000 barrels								2,300,000	2,783,000

23 13 23.16 Horizontal, Stl, Abvgrd Fuel-Oil, Storage Tanks

		Crew	Daily Output	Labor-Hours	Unit	Material	2017 Bare Costs Labor	Equipment	Total	Total Incl O&P
0010	**HORIZONTAL, STEEL, ABOVEGROUND FUEL-OIL, STORAGE TANKS**									
3000	Steel, storage, aboveground, including cradles, coating,									
3020	fittings, not including foundation, pumps or piping									
3040	Single wall, 275 gallon	Q-5	5	3.200	Ea.	515	179		694	835
3060	550 gallon	"	2.70	5.926		4,125	330		4,455	5,050
3080	1,000 gallon	Q-7	5	6.400		6,775	380		7,155	8,025
3100	1,500 gallon		4.75	6.737		9,675	400		10,075	11,300
3120	2,000 gallon		4.60	6.957		11,700	410		12,110	13,500
3140	5,000 gallon		3.20	10		21,500	595		22,095	24,600
3150	10,000 gallon		2	16		38,700	950		39,650	44,000
3160	15,000 gallon		1.70	18.824		49,700	1,125		50,825	56,000
3170	20,000 gallon		1.45	22.069		64,500	1,300		65,800	73,000
3180	25,000 gallon		1.30	24.615		75,000	1,450		76,450	84,500
3190	30,000 gallon		1.10	29.091		90,000	1,725		91,725	101,500
3320	Double wall, 500 gallon capacity	Q-5	2.40	6.667		1,750	375		2,125	2,500

For customer support on your Building Construction Costs with RSMeans Data, call 800.448.8182.

531

23 13 Facility Fuel-Storage Tanks

23 13 23 – Facility Aboveground Fuel-Oil, Storage Tanks

23 13 23.16 Horizontal, Stl, Abvgrd Fuel-Oil, Storage Tanks	Crew	Daily Output	Labor-Hours	Unit	Material	2017 Bare Costs Labor	Equipment	Total	Total Incl O&P
3330	2000 gallon capacity	Q-7	4.15	7.711	Ea.	5,700	455		6,155	6,975
3340	4000 gallon capacity		3.60	8.889		12,500	525		13,025	14,600
3350	6000 gallon capacity		2.40	13.333		14,200	790		14,990	16,800
3360	8000 gallon capacity		2	16		16,800	950		17,750	19,900
3370	10000 gallon capacity		1.80	17.778		30,000	1,050		31,050	34,600
3380	15000 gallon capacity		1.50	21.333		40,000	1,275		41,275	45,900
3390	20000 gallon capacity		1.30	24.615		43,000	1,450		44,450	49,500
3400	25000 gallon capacity		1.15	27.826		53,500	1,650		55,150	61,500
3410	30000 gallon capacity		1	32		64,500	1,900		66,400	74,000

23 13 23.26 Horizontal, Conc., Abvgrd Fuel-Oil, Stor. Tanks

		Crew	Daily Output	Labor-Hours	Unit	Material	Labor	Equipment	Total	Total Incl O&P
0010	**HORIZONTAL, CONCRETE, ABOVEGROUND FUEL-OIL, STORAGE TANKS**									
0050	Concrete, storage, aboveground, including pad & pump									
0100	500 gallon	F-3	2	20	Ea.	10,100	1,000	286	11,386	13,000
0200	1,000 gallon	"	2	20		14,100	1,000	286	15,386	17,500
0300	2,000 gallon	F-4	2	24		18,200	1,200	540	19,940	22,400
0400	4,000 gallon		2	24		23,200	1,200	540	24,940	28,000
0500	8,000 gallon		2	24		36,400	1,200	540	38,140	42,400
0600	12,000 gallon		2	24		48,500	1,200	540	50,240	56,000

23 21 Hydronic Piping and Pumps

23 21 20 – Hydronic HVAC Piping Specialties

23 21 20.10 Air Control

		Crew	Daily Output	Labor-Hours	Unit	Material	Labor	Equipment	Total	Total Incl O&P
0010	**AIR CONTROL**									
0030	Air separator, with strainer									
0040	2" diameter	Q-5	6	2.667	Ea.	1,375	149		1,524	1,725
0080	2-1/2" diameter		5	3.200		1,525	179		1,704	1,950
0100	3" diameter		4	4		2,350	224		2,574	2,950
0120	4" diameter		3	5.333		3,375	299		3,674	4,175
0130	5" diameter	Q-6	3.60	6.667		4,300	385		4,685	5,300
0140	6" diameter	"	3.40	7.059		5,150	410		5,560	6,300

23 21 20.18 Automatic Air Vent

		Crew	Daily Output	Labor-Hours	Unit	Material	Labor	Equipment	Total	Total Incl O&P
0010	**AUTOMATIC AIR VENT**									
0020	Cast iron body, stainless steel internals, float type									
0060	1/2" NPT inlet, 300 psi	1 Stpi	12	.667	Ea.	163	41.50		204.50	242
0220	3/4" NPT inlet, 250 psi	"	10	.800		370	50		420	480
0340	1-1/2" NPT inlet, 250 psi	Q-5	12	1.333		1,150	74.50		1,224.50	1,375

23 21 20.42 Expansion Joints

		Crew	Daily Output	Labor-Hours	Unit	Material	Labor	Equipment	Total	Total Incl O&P
0010	**EXPANSION JOINTS**									
0100	Bellows type, neoprene cover, flanged spool									
0140	6" face to face, 1-1/4" diameter	1 Stpi	11	.727	Ea.	255	45.50		300.50	350
0160	1-1/2" diameter	"	10.60	.755		255	47		302	350
0180	2" diameter	Q-5	13.30	1.203		258	67.50		325.50	385
0190	2-1/2" diameter		12.40	1.290		267	72.50		339.50	405
0200	3" diameter		11.40	1.404		299	78.50		377.50	450
0480	10" face to face, 2" diameter		13	1.231		370	69		439	515
0500	2-1/2" diameter		12	1.333		390	74.50		464.50	545
0520	3" diameter		11	1.455		400	81.50		481.50	565
0540	4" diameter		8	2		455	112		567	670
0560	5" diameter		7	2.286		540	128		668	790
0580	6" diameter		6	2.667		560	149		709	840

23 21 Hydronic Piping and Pumps

23 21 20 - Hydronic HVAC Piping Specialties

23 21 20.46 Expansion Tanks

		Crew	Daily Output	Labor-Hours	Unit	Material	2017 Bare Costs Labor	Equipment	Total	Total Incl O&P
0010	**EXPANSION TANKS**									
1507	Underground fuel-oil storage tanks, see Section 23 13 13									
1512	Tank leak detection systems, see Section 28 33 33.50									
2000	Steel, liquid expansion, ASME, painted, 15 gallon capacity	Q-5	17	.941	Ea.	710	52.50		762.50	860
2020	24 gallon capacity		14	1.143		765	64		829	935
2040	30 gallon capacity		12	1.333		820	74.50		894.50	1,025
2060	40 gallon capacity		10	1.600		960	89.50		1,049.50	1,175
2080	60 gallon capacity		8	2		1,150	112		1,262	1,425
2100	80 gallon capacity		7	2.286		1,225	128		1,353	1,550
2120	100 gallon capacity		6	2.667		1,675	149		1,824	2,050
3000	Steel ASME expansion, rubber diaphragm, 19 gal. cap. accept.		12	1.333		2,700	74.50		2,774.50	3,100
3020	31 gallon capacity		8	2		3,000	112		3,112	3,475
3040	61 gallon capacity		6	2.667		4,375	149		4,524	5,025
3080	119 gallon capacity		4	4		4,525	224		4,749	5,325
3100	158 gallon capacity		3.80	4.211		6,300	236		6,536	7,275
3140	317 gallon capacity		2.80	5.714		9,500	320		9,820	11,000
3180	528 gallon capacity		2.40	6.667		15,500	375		15,875	17,600

23 21 20.58 Hydronic Heating Control Valves

		Crew	Daily Output	Labor-Hours	Unit	Material	2017 Bare Costs Labor	Equipment	Total	Total Incl O&P
0010	**HYDRONIC HEATING CONTROL VALVES**									
0050	Hot water, nonelectric, thermostatic									
0100	Radiator supply, 1/2" diameter	1 Stpi	24	.333	Ea.	68.50	21		89.50	107
0120	3/4" diameter		20	.400		71.50	25		96.50	116
0140	1" diameter		19	.421		88	26		114	136
0160	1-1/4" diameter		15	.533		125	33		158	187
0500	For low pressure steam, add					25%				

23 21 20.70 Steam Traps

		Crew	Daily Output	Labor-Hours	Unit	Material	2017 Bare Costs Labor	Equipment	Total	Total Incl O&P
0010	**STEAM TRAPS**									
0030	Cast iron body, threaded									
0040	Inverted bucket									
0050	1/2" pipe size	1 Stpi	12	.667	Ea.	157	41.50		198.50	236
0070	3/4" pipe size		10	.800		274	50		324	375
0100	1" pipe size		9	.889		420	55.50		475.50	550
0120	1-1/4" pipe size		8	1		635	62.50		697.50	795
1000	Float & thermostatic, 15 psi									
1010	3/4" pipe size	1 Stpi	16	.500	Ea.	149	31		180	211
1020	1" pipe size		15	.533		180	33		213	248
1040	1-1/2" pipe size		9	.889		315	55.50		370.50	435
1060	2" pipe size		6	1.333		755	83		838	955

23 21 20.76 Strainers, Y Type, Bronze Body

		Crew	Daily Output	Labor-Hours	Unit	Material	2017 Bare Costs Labor	Equipment	Total	Total Incl O&P
0010	**STRAINERS, Y TYPE, BRONZE BODY**									
0050	Screwed, 125 lb., 1/4" pipe size	1 Stpi	24	.333	Ea.	33.50	21		54.50	68
0070	3/8" pipe size		24	.333		36.50	21		57.50	72
0100	1/2" pipe size		20	.400		36.50	25		61.50	78
0140	1" pipe size		17	.471		60.50	29.50		90	111
0160	1-1/2" pipe size		14	.571		131	35.50		166.50	198
0180	2" pipe size		13	.615		173	38.50		211.50	248
0182	3" pipe size		12	.667		1,100	41.50		1,141.50	1,275
0200	300 lb., 2-1/2" pipe size	Q-5	17	.941		720	52.50		772.50	875
0220	3" pipe size		16	1		1,275	56		1,331	1,475
0240	4" pipe size		15	1.067		2,325	60		2,385	2,650
0500	For 300 lb. rating 1/4" thru 2", add					15%				

For customer support on your Building Construction Costs with RSMeans Data, call 800.448.8182.

533

23 21 Hydronic Piping and Pumps

23 21 20 – Hydronic HVAC Piping Specialties

23 21 20.76 Strainers, Y Type, Bronze Body

		Crew	Daily Output	Labor-Hours	Unit	Material	2017 Bare Costs Labor	Equipment	Total	Total Incl O&P
1000	Flanged, 150 lb., 1-1/2" pipe size	1 Stpi	11	.727	Ea.	475	45.50		520.50	595
1020	2" pipe size	"	8	1		645	62.50		707.50	805
1030	2-1/2" pipe size	Q-5	5	3.200		945	179		1,124	1,325
1040	3" pipe size		4.50	3.556		1,175	199		1,374	1,600
1060	4" pipe size		3	5.333		1,800	299		2,099	2,425
1100	6" pipe size	Q-6	3	8		3,475	465		3,940	4,525
1106	8" pipe size	"	2.60	9.231		3,775	535		4,310	4,950
1500	For 300 lb. rating, add					40%				

23 21 20.78 Strainers, Y Type, Iron Body

		Crew	Daily Output	Labor-Hours	Unit	Material	2017 Bare Costs Labor	Equipment	Total	Total Incl O&P
0010	**STRAINERS, Y TYPE, IRON BODY**									
0050	Screwed, 250 lb., 1/4" pipe size	1 Stpi	20	.400	Ea.	13.70	25		38.70	52.50
0070	3/8" pipe size		20	.400		13.70	25		38.70	52.50
0100	1/2" pipe size		20	.400		13.70	25		38.70	52.50
0140	1" pipe size		16	.500		21	31		52	70
0160	1-1/2" pipe size		12	.667		40.50	41.50		82	107
0180	2" pipe size		8	1		60.50	62.50		123	161
0220	3" pipe size	Q-5	11	1.455		291	81.50		372.50	445
0240	4" pipe size	"	5	3.200		495	179		674	815
0500	For galvanized body, add					50%				
1000	Flanged, 125 lb., 1-1/2" pipe size	1 Stpi	11	.727	Ea.	119	45.50		164.50	200
1020	2" pipe size	"	8	1		136	62.50		198.50	243
1040	3" pipe size	Q-5	4.50	3.556		179	199		378	495
1060	4" pipe size	"	3	5.333		296	299		595	775
1080	5" pipe size	Q-6	3.40	7.059		375	410		785	1,025
1100	6" pipe size	"	3	8		595	465		1,060	1,350
1500	For 250 lb. rating, add					20%				
2000	For galvanized body, add					50%				
2500	For steel body, add					40%				

23 21 20.88 Venturi Flow

		Crew	Daily Output	Labor-Hours	Unit	Material	2017 Bare Costs Labor	Equipment	Total	Total Incl O&P
0010	**VENTURI FLOW,** Measuring device									
0050	1/2" diameter	1 Stpi	24	.333	Ea.	292	21		313	350
0120	1" diameter		19	.421		286	26		312	355
0140	1-1/4" diameter		15	.533		355	33		388	440
0160	1-1/2" diameter		13	.615		370	38.50		408.50	465
0180	2" diameter		11	.727		380	45.50		425.50	490
0220	3" diameter	Q-5	14	1.143		535	64		599	685
0240	4" diameter	"	11	1.455		800	81.50		881.50	1,000
0280	6" diameter	Q-6	3.50	6.857		1,175	400		1,575	1,900
0500	For meter, add					2,250			2,250	2,500

23 21 23 – Hydronic Pumps

23 21 23.13 In-Line Centrifugal Hydronic Pumps

		Crew	Daily Output	Labor-Hours	Unit	Material	2017 Bare Costs Labor	Equipment	Total	Total Incl O&P
0010	**IN-LINE CENTRIFUGAL HYDRONIC PUMPS**									
0600	Bronze, sweat connections, 1/40 HP, in line									
0640	3/4" size	Q-1	16	1	Ea.	244	55.50		299.50	355
1000	Flange connection, 3/4" to 1-1/2" size									
1040	1/12 HP	Q-1	6	2.667	Ea.	635	148		783	925
1060	1/8 HP		6	2.667		1,100	148		1,248	1,425
1100	1/3 HP		6	2.667		1,225	148		1,373	1,575
1140	2" size, 1/6 HP		5	3.200		1,575	178		1,753	2,000
1180	2-1/2" size, 1/4 HP		5	3.200		1,975	178		2,153	2,450
2000	Cast iron, flange connection									
2040	3/4" to 1-1/2" size, in line, 1/12 HP	Q-1	6	2.667	Ea.	425	148		573	690

23 21 Hydronic Piping and Pumps

23 21 23 – Hydronic Pumps

23 21 23.13 In-Line Centrifugal Hydronic Pumps

	Crew	Daily Output	Labor-Hours	Unit	Material	2017 Bare Costs Labor	Equipment	Total	Total Incl O&P
2100	1/3 HP	Q-1	6	2.667	Ea.	765	148		913	1,075
2101	Pumps, circulating, 3/4" to 1-1/2" size, 1/3 HP		6	2.667		915	148		1,063	1,225
2140	2" size, 1/6 HP		5	3.200		840	178		1,018	1,200
2180	2-1/2" size, 1/4 HP		5	3.200		1,050	178		1,228	1,425
2220	3" size, 1/4 HP	↓	4	4	↓	1,075	223		1,298	1,525
2600	For nonferrous impeller, add					3%				

23 21 29 – Automatic Condensate Pump Units

23 21 29.10 Condensate Removal Pump System

			Crew	Daily Output	Labor-Hours	Unit	Material	2017 Bare Costs Labor	Equipment	Total	Total Incl O&P
0010	**CONDENSATE REMOVAL PUMP SYSTEM**										
0020	Pump with 1 gal. ABS tank										
0100	115 V										
0120	1/50 HP, 200 GPH	G	1 Stpi	12	.667	Ea.	187	41.50		228.50	268
0140	1/18 HP, 270 GPH	G		10	.800		192	50		242	286
0160	1/5 HP, 450 GPH	G	↓	8	1	↓	435	62.50		497.50	575
0200	230 V										
0240	1/18 HP, 270 GPH		1 Stpi	10	.800	Ea.	206	50		256	300
0260	1/5 HP, 450 GPH	G	"	8	1	"	490	62.50		552.50	635

23 22 Steam and Condensate Piping and Pumps

23 22 13 – Steam and Condensate Heating Piping

23 22 13.23 Aboveground Steam and Condensate Piping

		Crew	Daily Output	Labor-Hours	Unit	Material	2017 Bare Costs Labor	Equipment	Total	Total Incl O&P
0010	**ABOVEGROUND STEAM AND CONDENSATE HEATING PIPING**									
0020	Condensate meter									
0100	500 lb. per hour	1 Stpi	14	.571	Ea.	3,550	35.50		3,585.50	3,950
0140	1500 lb. per hour	"	7	1.143	"	4,175	71		4,246	4,675

23 22 23 – Steam Condensate Pumps

23 22 23.10 Condensate Return System

		Crew	Daily Output	Labor-Hours	Unit	Material	2017 Bare Costs Labor	Equipment	Total	Total Incl O&P
0010	**CONDENSATE RETURN SYSTEM**									
2000	Simplex									
2010	With pump, motor, CI receiver, float switch									
2020	3/4 HP, 15 GPM	Q-1	1.80	8.889	Ea.	6,550	495		7,045	7,950
2100	Duplex									
2110	With 2 pumps and motors, CI receiver, float switch, alternator									
2120	3/4 HP, 15 GPM, 15 gal. CI receiver	Q-1	1.40	11.429	Ea.	7,150	635		7,785	8,800
2130	1 HP, 25 GPM		1.20	13.333		8,575	740		9,315	10,600
2140	1-1/2 HP, 45 GPM		1	16		9,950	890		10,840	12,400
2150	1-1/2 HP, 60 GPM	↓	1	16	↓	11,200	890		12,090	13,700

For customer support on your Building Construction Costs with RSMeans Data, call 800.448.8182.

535

23 31 13.13 Rectangular Metal Ducts	Crew	Daily Output	Labor-Hours	Unit	Material	2017 Bare Costs Labor	Equipment	Total	Total Incl O&P
0010 **RECTANGULAR METAL DUCTS**									
0020 Fabricated rectangular, includes fittings, joints, supports,									
0021 allowance for flexible connections and field sketches.									
0030 Does not include "as-built dwgs." or insulation.									
0031 NOTE: Fabrication and installation are combined									
0040 as LABOR cost. Approx. 25% fittings assumed.									
0042 Fabrication/Inst. is to commercial quality standards									
0043 (SMACNA or equiv.) for structure, sealing, leak testing, etc.									
0050 Add to labor for elevated installation									
0051 of fabricated ductwork									
0052 10' to 15' high						6%			
0053 15' to 20' high						12%			
0054 20' to 25' high						15%			
0055 25' to 30' high						21%			
0056 30' to 35' high						24%			
0057 35' to 40' high						30%			
0058 Over 40' high						33%			
0072 For duct insulation see Line 23 07 13.10 3000									
0100 Aluminum, alloy 3003-H14, under 100 lb.	Q-10	75	.320	Lb.	3.19	17.35		20.54	30
0110 100 to 500 lb.		80	.300		1.88	16.25		18.13	27
0120 500 to 1,000 lb.		95	.253		1.81	13.70		15.51	23
0140 1,000 to 2,000 lb.		120	.200		1.77	10.85		12.62	18.55
0150 2,000 to 5,000 lb.		130	.185		1.77	10		11.77	17.25
0160 Over 5,000 lb.		145	.166		1.77	8.95		10.72	15.70
0500 Galvanized steel, under 200 lb.		235	.102		.57	5.55		6.12	9.10
0520 200 to 500 lb.		245	.098		.56	5.30		5.86	8.70
0540 500 to 1,000 lb.		255	.094		.55	5.10		5.65	8.40
0560 1,000 to 2,000 lb.		265	.091		.54	4.91		5.45	8.10
0570 2,000 to 5,000 lb.		275	.087		.53	4.73		5.26	7.85
0580 Over 5,000 lb.		285	.084		.53	4.56		5.09	7.60
1000 Stainless steel, type 304, under 100 lb.		165	.145		4.24	7.90		12.14	16.70
1020 100 to 500 lb.		175	.137		3.89	7.45		11.34	15.65
1030 500 to 1,000 lb.		190	.126		2.71	6.85		9.56	13.50
1040 1,000 to 2,000 lb.		200	.120		2.41	6.50		8.91	12.60
1050 2,000 to 5,000 lb.		225	.107		2.33	5.80		8.13	11.40
1060 Over 5,000 lb.		235	.102		2.33	5.55		7.88	11
1100 For medium pressure ductwork, add						15%			
1200 For high pressure ductwork, add						40%			
1210 For welded ductwork, add						85%			
1220 For 30% fittings, add						11%			
1224 For 40% fittings, add						34%			
1228 For 50% fittings, add						56%			
1232 For 60% fittings, add						79%			
1236 For 70% fittings, add						101%			
1240 For 80% fittings, add						124%			
1244 For 90% fittings, add						147%			
1248 For 100% fittings, add						169%			
1252 Note: Fittings add includes time for detailing and installation.									

23 31 13.19 Metal Duct Fittings

	Crew	Daily Output	Labor-Hours	Unit	Material	2017 Bare Costs Labor	Equipment	Total	Total Incl O&P
0010 **METAL DUCT FITTINGS**									
2000 Fabrics for flexible connections, with metal edge	1 Shee	100	.080	L.F.	3.44	4.64		8.08	10.90
2100 Without metal edge	"	160	.050	"	2.46	2.90		5.36	7.15

23 31 HVAC Ducts and Casings

23 31 16 – Nonmetal Ducts

23 31 16.13 Fibrous-Glass Ducts	Crew	Daily Output	Labor-Hours	Unit	Material	2017 Bare Costs Labor	Equipment	Total	Total Incl O&P
0010 **FIBROUS-GLASS DUCTS**									
3490 Rigid fiberglass duct board, foil reinf. kraft facing									
3500 Rectangular, 1" thick, alum. faced, (FRK), std. weight	Q-10	350	.069	SF Surf	.80	3.72		4.52	6.60

23 33 Air Duct Accessories

23 33 13 – Dampers

23 33 13.13 Volume-Control Dampers

	Crew	Daily Output	Labor-Hours	Unit	Material	Labor	Equipment	Total	Total Incl O&P
0010 **VOLUME-CONTROL DAMPERS**									
5990 Multi-blade dampers, opposed blade, 8" x 6"	1 Shee	24	.333	Ea.	22	19.35		41.35	53.50
5994 8" x 8"		22	.364		22.50	21		43.50	57.50
5996 10" x 10"		21	.381		27	22		49	63.50
6000 12" x 12"		21	.381		30	22		52	67
6020 12" x 18"		18	.444		40	26		66	83.50
6030 14" x 10"		20	.400		29	23		52	67.50
6031 14" x 14"		17	.471		35.50	27.50		63	81
6033 16" x 12"		17	.471		35.50	27.50		63	81
6035 16" x 16"		16	.500		44.50	29		73.50	93.50
6037 18" x 16"		15	.533		48.50	31		79.50	101
6038 18" x 18"		15	.533		52.50	31		83.50	105
6070 20" x 16"		14	.571		52.50	33		85.50	109
6072 20" x 20"		13	.615		63.50	35.50		99	124
6074 22" x 18"		14	.571		63.50	33		96.50	121
6076 24" x 16"		11	.727		62	42		104	133
6078 24" x 20"		8	1		73.50	58		131.50	170
6080 24" x 24"		8	1		86	58		144	184
6110 26" x 26"		6	1.333		96	77.50		173.50	224
6133 30" x 30"	Q-9	6.60	2.424		138	127		265	345
6135 32" x 32"		6.40	2.500		156	131		287	370
6180 48" x 36"		5.60	2.857		256	149		405	510
8000 Multi-blade dampers, parallel blade									
8100 8" x 8"	1 Shee	24	.333	Ea.	83	19.35		102.35	121
8140 16" x 10"		20	.400		104	23		127	151
8200 24" x 16"		11	.727		136	42		178	214
8260 30" x 18"		7	1.143		188	66.50		254.50	310

23 33 13.16 Fire Dampers

	Crew	Daily Output	Labor-Hours	Unit	Material	Labor	Equipment	Total	Total Incl O&P
0010 **FIRE DAMPERS**									
3000 Fire damper, curtain type, 1-1/2 hr. rated, vertical, 6" x 6"	1 Shee	24	.333	Ea.	24.50	19.35		43.85	56.50
3020 8" x 6"		22	.364		24.50	21		45.50	59.50
3240 16" x 14"		18	.444		44.50	26		70.50	88.50
3400 24" x 20"		8	1		58.50	58		116.50	153

23 33 13.28 Splitter Damper Assembly

	Crew	Daily Output	Labor-Hours	Unit	Material	Labor	Equipment	Total	Total Incl O&P
0010 **SPLITTER DAMPER ASSEMBLY**									
7000 Self locking, 1' rod	1 Shee	24	.333	Ea.	21	19.35		40.35	52.50
7020 3' rod		22	.364		30	21		51	65.50
7040 4' rod		20	.400		33.50	23		56.50	72.50
7060 6' rod		18	.444		41	26		67	84.50

For customer support on your Building Construction Costs with RSMeans Data, call 800.448.8182.

537

23 33 Air Duct Accessories

23 33 19 – Duct Silencers

23 33 19.10 Duct Silencers		Crew	Daily Output	Labor-Hours	Unit	Material	2017 Bare Costs Labor	Equipment	Total	Total Incl O&P
0010	**DUCT SILENCERS**									
9000	Silencers, noise control for air flow, duct				MCFM	58			58	63.50

23 33 33 – Duct-Mounting Access Doors

23 33 33.13 Duct Access Doors

		Crew	Daily Output	Labor-Hours	Unit	Material	Labor	Equipment	Total	Total Incl O&P
0010	**DUCT ACCESS DOORS**									
1000	Duct access door, insulated, 6" x 6"	1 Shee	14	.571	Ea.	17.80	33		50.80	70.50
1020	10" x 10"		11	.727		20.50	42		62.50	87
1040	12" x 12"		10	.800		22	46.50		68.50	95.50
1050	12" x 18"		9	.889		38.50	51.50		90	122
1070	18" x 18"		8	1		35.50	58		93.50	128
1074	24" x 18"		8	1		46	58		104	140

23 33 46 – Flexible Ducts

23 33 46.10 Flexible Air Ducts

		Crew	Daily Output	Labor-Hours	Unit	Material	Labor	Equipment	Total	Total Incl O&P
0010	**FLEXIBLE AIR DUCTS**									
1280	Add to labor for elevated installation									
1282	of prefabricated (purchased) ductwork									
1283	10' to 15' high						10%			
1284	15' to 20' high						20%			
1285	20' to 25' high						25%			
1286	25' to 30' high						35%			
1287	30' to 35' high						40%			
1288	35' to 40' high						50%			
1289	Over 40' high						55%			
1300	Flexible, coated fiberglass fabric on corr. resist. metal helix									
1400	pressure to 12" (WG) UL-181									
1500	Noninsulated, 3" diameter	Q-9	400	.040	L.F.	1.20	2.09		3.29	4.52
1540	5" diameter		320	.050		1.38	2.61		3.99	5.50
1560	6" diameter		280	.057		1.60	2.99		4.59	6.35
1580	7" diameter		240	.067		1.78	3.48		5.26	7.30
1600	8" diameter		200	.080		2.02	4.18		6.20	8.60
1640	10" diameter		160	.100		2.61	5.25		7.86	10.85
1660	12" diameter		120	.133		3.10	6.95		10.05	14.05
1900	Insulated, 1" thick, PE jacket, 3" diameter	G	380	.042		2.62	2.20		4.82	6.25
1910	4" diameter	G	340	.047		2.80	2.46		5.26	6.85
1920	5" diameter	G	300	.053		2.93	2.79		5.72	7.50
1940	6" diameter	G	260	.062		3.11	3.22		6.33	8.35
1960	7" diameter	G	220	.073		3.60	3.80		7.40	9.75
1980	8" diameter	G	180	.089		3.80	4.64		8.44	11.30
2020	10" diameter	G	140	.114		4.55	5.95		10.50	14.15
2040	12" diameter	G	100	.160		5.25	8.35		13.60	18.60

23 33 53 – Duct Liners

23 33 53.10 Duct Liner Board

		Crew	Daily Output	Labor-Hours	Unit	Material	Labor	Equipment	Total	Total Incl O&P
0010	**DUCT LINER BOARD**									
3340	Board type fiberglass liner, FSK, 1-1/2 lb. density									
3344	1" thick	G	Q-14	150	.107	S.F.	.97	5.30		6.27	9.25
3345	1-1/2" thick	G		130	.123		1.06	6.10		7.16	10.65
3346	2" thick	G		120	.133		1.24	6.60		7.84	11.60
3348	3" thick	G		110	.145		1.60	7.20		8.80	12.95
3350	4" thick	G		100	.160		1.96	7.95		9.91	14.45
3356	3 lb. density, 1" thick	G		150	.107		1.24	5.30		6.54	9.55

23 33 Air Duct Accessories

23 33 53 – Duct Liners

23 33 53.10 Duct Liner Board		Crew	Daily Output	Labor-Hours	Unit	Material	2017 Bare Costs Labor	Equipment	Total	Total Incl O&P
3358	1-1/2" thick	G Q-14	130	.123	S.F.	1.56	6.10		7.66	11.20
3360	2" thick	G	120	.133		1.90	6.60		8.50	12.35
3362	2-1/2" thick	G	110	.145		2.23	7.20		9.43	13.65
3364	3" thick	G	100	.160		2.57	7.95		10.52	15.15
3366	4" thick	G	90	.178		3.22	8.80		12.02	17.25
3370	6 lb. density, 1" thick	G	140	.114		1.75	5.65		7.40	10.75
3374	1-1/2" thick	G	120	.133		2.35	6.60		8.95	12.85
3378	2" thick	G	100	.160		2.94	7.95		10.89	15.55
3490	Board type, fiberglass liner, 3 lb. density									
3680	No finish									
3700	1" thick	G Q-14	170	.094	S.F.	.69	4.66		5.35	8
3710	1-1/2" thick	G	140	.114		1.03	5.65		6.68	9.95
3720	2" thick	G	130	.123		1.38	6.10		7.48	11
3940	Board type, non-fibrous foam									
3950	Temperature, bacteria and fungi resistant									
3960	1" thick	G Q-14	150	.107	S.F.	2.57	5.30		7.87	11.05
3970	1-1/2" thick	G	130	.123		4.19	6.10		10.29	14.10
3980	2" thick	G	120	.133		4.02	6.60		10.62	14.65

23 34 HVAC Fans

23 34 13 – Axial HVAC Fans

23 34 13.10 Axial Flow HVAC Fans

		Crew	Daily Output	Labor-Hours	Unit	Material	2017 Bare Costs Labor	Equipment	Total	Total Incl O&P
0010	**AXIAL FLOW HVAC FANS**									
0020	Air conditioning and process air handling									
1500	Vaneaxial, low pressure, 2000 CFM, 1/2 HP	Q-20	3.60	5.556	Ea.	2,250	295		2,545	2,950
1520	4,000 CFM, 1 HP		3.20	6.250		2,650	330		2,980	3,400
1540	8,000 CFM, 2 HP		2.80	7.143		3,350	380		3,730	4,275

23 34 14 – Blower HVAC Fans

23 34 14.10 Blower Type HVAC Fans

		Crew	Daily Output	Labor-Hours	Unit	Material	2017 Bare Costs Labor	Equipment	Total	Total Incl O&P
0010	**BLOWER TYPE HVAC FANS**									
2500	Ceiling fan, right angle, extra quiet, 0.10" S.P.									
2520	95 CFM	Q-20	20	1	Ea.	315	53		368	430
2540	210 CFM		19	1.053		375	56		431	495
2560	385 CFM		18	1.111		475	59		534	610
2580	885 CFM		16	1.250		935	66.50		1,001.50	1,125
2600	1,650 CFM		13	1.538		1,300	81.50		1,381.50	1,550
2620	2,960 CFM		11	1.818		1,725	96.50		1,821.50	2,050
2640	For wall or roof cap, add	1 Shee	16	.500		315	29		344	395
2660	For straight thru fan, add					10%				
2680	For speed control switch, add	1 Elec	16	.500		173	28.50		201.50	233
7500	Utility set, steel construction, pedestal, 1/4" S.P.									
7520	Direct drive, 150 CFM, 1/8 HP	Q-20	6.40	3.125	Ea.	900	166		1,066	1,250
7540	485 CFM, 1/6 HP		5.80	3.448		1,125	183		1,308	1,525
7560	1950 CFM, 1/2 HP		4.80	4.167		1,325	221		1,546	1,775
7580	2410 CFM, 3/4 HP		4.40	4.545		2,450	241		2,691	3,075
7600	3328 CFM, 1-1/2 HP		3	6.667		2,725	355		3,080	3,550
7680	V-belt drive, drive cover, 3 phase									
7700	800 CFM, 1/4 HP	Q-20	6	3.333	Ea.	1,025	177		1,202	1,400
7720	1,300 CFM, 1/3 HP		5	4		1,075	212		1,287	1,500
7740	2,000 CFM, 1 HP		4.60	4.348		1,275	231		1,506	1,750

23 34 14.10 Blower Type HVAC Fans	Crew	Daily Output	Labor-Hours	Unit	Material	2017 Bare Costs Labor	2017 Bare Costs Equipment	Total	Total Incl O&P
7760	2,900 CFM, 3/4 HP	Q-20	4.20	4.762	Ea.	1,725	253		1,978	2,250

		Crew	Daily Output	Labor-Hours	Unit	Material	Labor	Equipment	Total	Total Incl O&P
0010	**CENTRIFUGAL TYPE HVAC FANS**									
0200	In-line centrifugal, supply/exhaust booster									
0220	aluminum wheel/hub, disconnect switch, 1/4" S.P.									
0240	500 CFM, 10" diameter connection	Q-20	3	6.667	Ea.	1,450	355		1,805	2,150
0260	1,380 CFM, 12" diameter connection		2	10		1,475	530		2,005	2,425
0280	1,520 CFM, 16" diameter connection		2	10		1,525	530		2,055	2,475
0300	2,560 CFM, 18" diameter connection		1	20		1,650	1,050		2,700	3,450
0320	3,480 CFM, 20" diameter connection		.80	25		1,925	1,325		3,250	4,125
0326	5,080 CFM, 20" diameter connection		.75	26.667		2,100	1,425		3,525	4,450
3500	Centrifugal, airfoil, motor and drive, complete									
3520	1000 CFM, 1/2 HP	Q-20	2.50	8	Ea.	1,925	425		2,350	2,775
3540	2,000 CFM, 1 HP		2	10		2,175	530		2,705	3,200
3560	4,000 CFM, 3 HP		1.80	11.111		2,775	590		3,365	3,950
3580	8,000 CFM, 7-1/2 HP		1.40	14.286		4,175	760		4,935	5,750
3600	12,000 CFM, 10 HP		1	20		5,575	1,050		6,625	7,750
5000	Utility set, centrifugal, V belt drive, motor									
5020	1/4" S.P., 1200 CFM, 1/4 HP	Q-20	6	3.333	Ea.	1,875	177		2,052	2,350
5040	1520 CFM, 1/3 HP		5	4		2,475	212		2,687	3,050
5060	1850 CFM, 1/2 HP		4	5		2,450	266		2,716	3,100
5080	2180 CFM, 3/4 HP		3	6.667		2,900	355		3,255	3,750
5100	1/2" S.P., 3600 CFM, 1 HP		2	10		3,000	530		3,530	4,100
5120	4250 CFM, 1-1/2 HP		1.60	12.500		3,650	665		4,315	5,025
5140	4800 CFM, 2 HP		1.40	14.286		4,425	760		5,185	6,025
7000	Roof exhauster, centrifugal, aluminum housing, 12" galvanized									
7020	curb, bird screen, back draft damper, 1/4" S.P.									
7100	Direct drive, 320 CFM, 11" sq. damper	Q-20	7	2.857	Ea.	770	152		922	1,075
7120	600 CFM, 11" sq. damper		6	3.333		985	177		1,162	1,350
7140	815 CFM, 13" sq. damper		5	4		985	212		1,197	1,400
7160	1450 CFM, 13" sq. damper		4.20	4.762		1,325	253		1,578	1,850
7180	2050 CFM, 16" sq. damper		4	5		1,700	266		1,966	2,275
7200	V-belt drive, 1650 CFM, 12" sq. damper		6	3.333		1,425	177		1,602	1,850
7220	2750 CFM, 21" sq. damper		5	4		1,700	212		1,912	2,200
7230	3500 CFM, 21" sq. damper		4.50	4.444		1,925	236		2,161	2,475
7240	4910 CFM, 23" sq. damper		4	5		2,325	266		2,591	2,975
7260	8525 CFM, 28" sq. damper		3	6.667		3,075	355		3,430	3,950
7280	13,760 CFM, 35" sq. damper		2	10		4,325	530		4,855	5,550
7300	20,558 CFM, 43" sq. damper		1	20		8,525	1,050		9,575	11,000
7320	For 2 speed winding, add					15%				
7340	For explosion proof motor, add					605			605	665
7360	For belt driven, top discharge, add					15%				
8500	Wall exhausters, centrifugal, auto damper, 1/8" S.P.									
8520	Direct drive, 610 CFM, 1/20 HP	Q-20	14	1.429	Ea.	445	76		521	605
8540	796 CFM, 1/12 HP		13	1.538		970	81.50		1,051.50	1,200
8560	822 CFM, 1/6 HP		12	1.667		1,125	88.50		1,213.50	1,375
8580	1,320 CFM, 1/4 HP		12	1.667		1,375	88.50		1,463.50	1,650
9500	V-belt drive, 3 phase									
9520	2,800 CFM, 1/4 HP	Q-20	9	2.222	Ea.	2,025	118		2,143	2,400
9540	3,740 CFM, 1/2 HP	"	8	2.500	"	2,100	133		2,233	2,500

540

For customer support on your Building Construction Costs with RSMeans Data, call 800.448.8182.

23 34 HVAC Fans

23 34 23 – HVAC Power Ventilators

23 34 23.10 HVAC Power Circulators and Ventilators

		Crew	Daily Output	Labor-Hours	Unit	Material	2017 Bare Costs Labor	Equipment	Total	Total Incl O&P
0010	**HVAC POWER CIRCULATORS AND VENTILATORS**									
3000	Paddle blade air circulator, 3 speed switch									
3020	42", 5,000 CFM high, 3000 CFM low G	1 Elec	2.40	3.333	Ea.	135	189		324	430
3040	52", 6,500 CFM high, 4000 CFM low G	"	2.20	3.636	"	180	206		386	510
3100	For antique white motor, same cost									
3200	For brass plated motor, same cost									
3300	For light adaptor kit, add G				Ea.	41			41	45.50
6000	Propeller exhaust, wall shutter									
6020	Direct drive, one speed, .075" S.P.									
6100	653 CFM, 1/30 HP	Q-20	10	2	Ea.	202	106		308	385
6120	1033 CFM, 1/20 HP		9	2.222		300	118		418	510
6140	1323 CFM, 1/15 HP		8	2.500		335	133		468	570
6300	V-belt drive, 3 phase									
6320	6175 CFM, 3/4 HP	Q-20	5	4	Ea.	1,250	212		1,462	1,725
6340	7500 CFM, 3/4 HP		5	4		1,325	212		1,537	1,800
6360	10,100 CFM, 1 HP		4.50	4.444		1,550	236		1,786	2,050
6380	14,300 CFM, 1-1/2 HP		4	5		1,825	266		2,091	2,400
6650	Residential, bath exhaust, grille, back draft damper									
6660	50 CFM	Q-20	24	.833	Ea.	56.50	44.50		101	130
6670	110 CFM		22	.909		95	48.50		143.50	179
6680	Light combination, squirrel cage, 100 watt, 70 CFM		24	.833		108	44.50		152.50	187
6700	Light/heater combination, ceiling mounted									
6710	70 CFM, 1450 watt	Q-20	24	.833	Ea.	157	44.50		201.50	240
6800	Heater combination, recessed, 70 CFM		24	.833		65.50	44.50		110	140
6820	With 2 infrared bulbs		23	.870		102	46		148	183
6900	Kitchen exhaust, grille, complete, 160 CFM		22	.909		116	48.50		164.50	201
6910	180 CFM		20	1		97	53		150	188
6920	270 CFM		18	1.111		184	59		243	292
6930	350 CFM		16	1.250		138	66.50		204.50	253
6940	Residential roof jacks and wall caps									
6944	Wall cap with back draft damper									
6946	3" & 4" diam. round duct	1 Shee	11	.727	Ea.	27	42		69	94
6948	6" diam. round duct	"	11	.727	"	64	42		106	135
6958	Roof jack with bird screen and back draft damper									
6960	3" & 4" diam. round duct	1 Shee	11	.727	Ea.	23.50	42		65.50	90.50
6962	3-1/4" x 10" rectangular duct	"	10	.800	"	47	46.50		93.50	123
6980	Transition									
6982	3-1/4" x 10" to 6" diam. round	1 Shee	20	.400	Ea.	28.50	23		51.50	66.50

23 34 33 – Air Curtains

23 34 33.10 Air Barrier Curtains

		Crew	Daily Output	Labor-Hours	Unit	Material	Labor	Equipment	Total	Total Incl O&P
0010	**AIR BARRIER CURTAINS**, Incl. motor starters, transformers,									
0050	and door switches									
2450	Conveyor openings or service windows									
3000	Service window, 5' high x 25" wide	2 Shee	5	3.200	Ea.	350	186		536	670
3100	Environmental separation									
3110	Door heights up to 8', low profile, super quiet									
3120	Unheated, variable speed									
3130	36" wide	2 Shee	4	4	Ea.	660	232		892	1,075
3134	42" wide		3.80	4.211		690	244		934	1,125
3138	48" wide		3.60	4.444		715	258		973	1,175
3142	60" wide		3.40	4.706		740	273		1,013	1,225
3146	72" wide	Q-3	4.60	6.957		915	410		1,325	1,625

For customer support on your Building Construction Costs with RSMeans Data, call 800.448.8182.

541

23 34 HVAC Fans

23 34 33 – Air Curtains

23 34 33.10 Air Barrier Curtains		Crew	Daily Output	Labor-Hours	Unit	Material	2017 Bare Costs Labor	Equipment	Total	Total Incl O&P
3150	96" wide	Q-3	4.40	7.273	Ea.	1,375	430		1,805	2,175
3154	120" wide		4.20	7.619		1,475	450		1,925	2,300
3158	144" wide		4	8		1,800	470		2,270	2,700
3200	Door heights up to 10'									
3210	Unheated									
3230	36" wide	2 Shee	3.80	4.211	Ea.	700	244		944	1,150
3234	42" wide		3.60	4.444		720	258		978	1,175
3238	48" wide		3.40	4.706		740	273		1,013	1,225
3242	60" wide		3.20	5		1,075	290		1,365	1,625
3246	72" wide	Q-3	4.40	7.273		1,150	430		1,580	1,925
3250	96" wide		4.20	7.619		1,375	450		1,825	2,175
3254	120" wide		4	8		1,875	470		2,345	2,775
3258	144" wide		3.80	8.421		2,025	495		2,520	3,000
3300	Door heights up to 12'									
3310	Unheated									
3334	42" wide	2 Shee	3.40	4.706	Ea.	1,025	273		1,298	1,550
3338	48" wide		3.20	5		1,050	290		1,340	1,600
3342	60" wide		3	5.333		1,050	310		1,360	1,650
3346	72" wide	Q-3	4.20	7.619		1,850	450		2,300	2,725
3350	96" wide		4	8		2,025	470		2,495	2,925
3354	120" wide		3.80	8.421		2,300	495		2,795	3,275
3358	144" wide		3.60	8.889		2,650	525		3,175	3,700
3400	Door heights up to 16'									
3410	Unheated									
3438	48" wide	2 Shee	3	5.333	Ea.	1,175	310		1,485	1,775
3442	60" wide	"	2.80	5.714		1,225	330		1,555	1,850
3446	72" wide	Q-3	3.80	8.421		2,100	495		2,595	3,075
3450	96" wide		3.60	8.889		2,400	525		2,925	3,450
3454	120" wide		3.40	9.412		3,250	555		3,805	4,400
3458	144" wide		3.20	10		3,375	590		3,965	4,600
3470	Heated, electric									
3474	48" wide	2 Shee	2.90	5.517	Ea.	1,950	320		2,270	2,650
3478	60" wide	"	2.70	5.926		2,000	345		2,345	2,725
3482	72" wide	Q-3	3.70	8.649		3,425	510		3,935	4,550
3486	96" wide		3.50	9.143		3,625	540		4,165	4,775
3490	120" wide		3.30	9.697		4,850	570		5,420	6,175
3494	144" wide		3.10	10.323		4,925	605		5,530	6,350

23 37 Air Outlets and Inlets

23 37 13 – Diffusers, Registers, and Grilles

23 37 13.10 Diffusers

23 37 13.10 Diffusers		Crew	Daily Output	Labor-Hours	Unit	Material	2017 Bare Costs Labor	Equipment	Total	Total Incl O&P
0010	**DIFFUSERS**, Aluminum, opposed blade damper unless noted									
0100	Ceiling, linear, also for sidewall									
0500	Perforated, 24" x 24" lay-in panel size, 6" x 6"	1 Shee	16	.500	Ea.	158	29		187	219
0520	8" x 8"		15	.533		166	31		197	231
0530	9" x 9"		14	.571		169	33		202	237
0540	10" x 10"		14	.571		169	33		202	237
0560	12" x 12"		12	.667		176	38.50		214.50	253
0590	16" x 16"		11	.727		197	42		239	282
0600	18" x 18"		10	.800		211	46.50		257.50	305
0610	20" x 20"		10	.800		228	46.50		274.50	320

23 37 Air Outlets and Inlets

23 37 13 – Diffusers, Registers, and Grilles

		Crew	Daily Output	Labor-Hours	Unit	Material	2017 Bare Costs Labor	Equipment	Total	Total Incl O&P
23 37 13.10	**Diffusers**									
0620	24" x 24"	1 Shee	9	.889	Ea.	250	51.50		301.50	355
1000	Rectangular, 1 to 4 way blow, 6" x 6"		16	.500		42	29		71	90.50
1010	8" x 8"		15	.533		58.50	31		89.50	112
1014	9" x 9"		15	.533		52.50	31		83.50	105
1016	10" x 10"		15	.533		80.50	31		111.50	136
1020	12" x 6"		15	.533		72	31		103	127
1040	12" x 9"		14	.571		76.50	33		109.50	135
1060	12" x 12"		12	.667		70.50	38.50		109	137
1070	14" x 6"		13	.615		78.50	35.50		114	141
1074	14" x 14"		12	.667		129	38.50		167.50	201
1150	18" x 18"		9	.889		114	51.50		165.50	204
1160	21" x 21"		8	1		217	58		275	325
1170	24" x 12"		10	.800		165	46.50		211.50	252
1500	Round, butterfly damper, steel, diffuser size, 6" diameter		18	.444		11	26		37	51.50
1520	8" diameter		16	.500		11.70	29		40.70	57.50
1540	10" diameter		14	.571		14.55	33		47.55	67
1560	12" diameter		12	.667		19.20	38.50		57.70	80
1580	14" diameter		10	.800		24	46.50		70.50	97.50
2000	T bar mounting, 24" x 24" lay-in frame, 6" x 6"		16	.500		72.50	29		101.50	125
2020	8" x 8"		14	.571		72.50	33		105.50	131
2040	12" x 12"		12	.667		89	38.50		127.50	157
2060	16" x 16"		11	.727		120	42		162	197
2080	18" x 18"		10	.800		134	46.50		180.50	218
6000	For steel diffusers instead of aluminum, deduct					10%				
23 37 13.30	**Grilles**									
0010	**GRILLES**									
0020	Aluminum, unless noted otherwise									
1000	Air return, steel, 6" x 6"	1 Shee	26	.308	Ea.	19.80	17.85		37.65	49.50
1020	10" x 6"		24	.333		19.80	19.35		39.15	51.50
1080	16" x 8"		22	.364		28	21		49	63.50
1100	12" x 12"		22	.364		28	21		49	63.50
1120	24" x 12"		18	.444		37	26		63	80
1220	24" x 18"		16	.500		45.50	29		74.50	94.50
1280	36" x 24"		14	.571		78.50	33		111.50	138
3000	Filter grille with filter, 12" x 12"		24	.333		55	19.35		74.35	90
3020	18" x 12"		20	.400		71.50	23		94.50	115
3040	24" x 18"		18	.444		87	26		113	135
3060	24" x 24"		16	.500		101	29		130	156
6000	For steel grilles instead of aluminum in above, deduct					10%				
23 37 13.60	**Registers**									
0010	**REGISTERS**									
0980	Air supply									
1000	Ceiling/wall, O.B. damper, anodized aluminum									
1010	One or two way deflection, adj. curved face bars									
1020	8" x 4"	1 Shee	26	.308	Ea.	17.30	17.85		35.15	46.50
1120	12" x 12"		18	.444		29.50	26		55.50	72
1240	20" x 6"		18	.444		26.50	26		52.50	68.50
1340	24" x 8"		13	.615		35.50	35.50		71	93.50
1350	24" x 18"		12	.667		63.50	38.50		102	129
2700	Above registers in steel instead of aluminum, deduct					10%				
4000	Floor, toe operated damper, enameled steel									
4020	4" x 8"	1 Shee	32	.250	Ea.	12.70	14.50		27.20	36

23 37 13 – Diffusers, Registers, and Grilles

23 37 13.60 Registers		Crew	Daily Output	Labor-Hours	Unit	Material	2017 Bare Costs Labor	Equipment	Total	Total Incl O&P
4100	8" x 10"	1 Shee	22	.364	Ea.	15.50	21		36.50	49.50
4140	10" x 10"		20	.400		18.60	23		41.60	56
4220	14" x 14"		16	.500		44.50	29		73.50	93.50
4240	14" x 20"	↓	15	.533	↓	52	31		83	105
4980	Air return									
5000	Ceiling or wall, fixed 45° face blades									
5010	Adjustable O.B. damper, anodized aluminum									
5020	4" x 8"	1 Shee	26	.308	Ea.	16.25	17.85		34.10	45.50
5060	6" x 10"		19	.421		19.45	24.50		43.95	59
5280	24" x 24"		11	.727		89	42		131	163
5300	24" x 36"	↓	8	1	↓	142	58		200	245
6000	For steel construction instead of aluminum, deduct					10%				

23 37 15 – Louvers

23 37 15.40 HVAC Louvers

		Crew	Daily Output	Labor-Hours	Unit	Material	Labor	Equipment	Total	Total Incl O&P
0010	**HVAC LOUVERS**									
0100	Aluminum, extruded, with screen, mill finish									
1002	Brick vent, see also Section 04 05 23.19									
1100	Standard, 4" deep, 8" wide, 5" high	1 Shee	24	.333	Ea.	34.50	19.35		53.85	67.50
1200	Modular, 4" deep, 7-3/4" wide, 5" high		24	.333		37	19.35		56.35	70
1300	Speed brick, 4" deep, 11-5/8" wide, 3-7/8" high		24	.333		37	19.35		56.35	70
1400	Fuel oil brick, 4" deep, 8" wide, 5" high		24	.333	↓	63.50	19.35		82.85	99.50
2000	Cooling tower and mechanical equip., screens, light weight		40	.200	S.F.	15.85	11.60		27.45	35
2020	Standard weight		35	.229		42	13.25		55.25	67
2500	Dual combination, automatic, intake or exhaust		20	.400		57.50	23		80.50	99
2520	Manual operation		20	.400		43	23		66	83
2540	Electric or pneumatic operation		20	.400	↓	43	23		66	83
2560	Motor, for electric or pneumatic	↓	14	.571	Ea.	495	33		528	595
3000	Fixed blade, continuous line									
3100	Mullion type, stormproof	1 Shee	28	.286	S.F.	43	16.60		59.60	73
3200	Stormproof		28	.286		43	16.60		59.60	73
3300	Vertical line		28	.286	↓	51	16.60		67.60	81.50
3500	For damper to use with above, add					50%	30%			
3520	Motor, for damper, electric or pneumatic	1 Shee	14	.571	Ea.	495	33		528	595
4000	Operating, 45°, manual, electric or pneumatic		24	.333	S.F.	51.50	19.35		70.85	86
4100	Motor, for electric or pneumatic		14	.571	Ea.	495	33		528	595
4200	Penthouse, roof		56	.143	S.F.	25.50	8.30		33.80	40.50
4300	Walls		40	.200		60	11.60		71.60	83.50
5000	Thinline, under 4" thick, fixed blade	↓	40	.200	↓	24.50	11.60		36.10	45
5010	Finishes, applied by mfr. at additional cost, available in colors									
5020	Prime coat only, add				S.F.	3.40			3.40	3.74
5040	Baked enamel finish coating, add					6.30			6.30	6.90
5060	Anodized finish, add					6.80			6.80	7.50
5080	Duranodic finish, add					12.35			12.35	13.60
5100	Fluoropolymer finish coating, add					19.45			19.45	21.50
9980	For small orders (under 10 pieces), add				↓	25%				

23 37 23 – HVAC Gravity Ventilators

23 37 23.10 HVAC Gravity Air Ventilators

		Crew	Daily Output	Labor-Hours	Unit	Material	Labor	Equipment	Total	Total Incl O&P
0010	**HVAC GRAVITY AIR VENTILATORS**, Includes base									
1280	Rotary ventilators, wind driven, galvanized									
1300	4" neck diameter	Q-9	20	.800	Ea.	47.50	42		89.50	117
1340	6" neck diameter		16	1		48	52.50		100.50	133
1400	12" neck diameter	↓	10	1.600	↓	67	83.50		150.50	202

23 37 Air Outlets and Inlets

23 37 23 – HVAC Gravity Ventilators

23 37 23.10 HVAC Gravity Air Ventilators	Crew	Daily Output	Labor-Hours	Unit	Material	2017 Bare Costs Labor	Equipment	Total	Total Incl O&P
1500	24" neck diameter	Q-9	8	2	Ea.	299	105		404	490
1540	36" neck diameter	↓	6	2.667	↓	575	139		714	845
2000	Stationary, gravity, syphon, galvanized									
2160	6" neck diameter, 66 CFM	Q-9	16	1	Ea.	34	52.50		86.50	117
2240	12" neck diameter, 160 CFM		10	1.600		78.50	83.50		162	214
2340	24" neck diameter, 900 CFM		8	2		266	105		371	455
2380	36" neck diameter, 2,000 CFM		6	2.667		415	139		554	670
4200	Stationary mushroom, aluminum, 16" orifice diameter		10	1.600		650	83.50		733.50	845
4220	26" orifice diameter		6.15	2.602		960	136		1,096	1,250
4230	30" orifice diameter		5.71	2.802		1,400	146		1,546	1,775
4240	38" orifice diameter		5	3.200		2,025	167		2,192	2,475
4250	42" orifice diameter		4.70	3.404		2,675	178		2,853	3,225
4260	50" orifice diameter	↓	4.44	3.604	↓	3,175	188		3,363	3,800
5000	Relief vent									
5500	Rectangular, aluminum, galvanized curb									
5510	intake/exhaust, 0.033" SP									
5580	500 CFM, 12" x 12"	Q-9	8.60	1.860	Ea.	715	97		812	935
5600	600 CFM, 12" x 16"		8	2		800	105		905	1,050
5640	1000 CFM, 12" x 24"		6.60	2.424		895	127		1,022	1,175
5680	3000 CFM, 20" x 42"	↓	4	4	↓	1,575	209		1,784	2,050
5880	Size is throat area, volume is at 500 fpm									
7000	Note: sizes based on exhaust. Intake, with 0.125" SP									
7100	loss, approximately twice listed capacity.									

23 38 Ventilation Hoods

23 38 13 – Commercial-Kitchen Hoods

23 38 13.10 Hood and Ventilation Equipment

		Crew	Daily Output	Labor-Hours	Unit	Material	Labor	Equipment	Total	Total Incl O&P
0010	**HOOD AND VENTILATION EQUIPMENT**									
2970	Exhaust hood, sst, gutter on all sides, 4' x 4' x 2'	1 Carp	1.80	4.444	Ea.	4,950	219		5,169	5,775
2980	4' x 4' x 7'	"	1.60	5	"	7,900	246		8,146	9,050

23 41 Particulate Air Filtration

23 41 13 – Panel Air Filters

23 41 13.10 Panel Type Air Filters

			Crew	Daily Output	Labor-Hours	Unit	Material	Labor	Equipment	Total	Total Incl O&P
0010	**PANEL TYPE AIR FILTERS**										
2950	Mechanical media filtration units										
3000	High efficiency type, with frame, non-supported	G				MCFM	35			35	38.50
3100	Supported type	G				"	55			55	60.50
5500	Throwaway glass or paper media type, 12" x 36" x 1"					Ea.	2.39			2.39	2.63

23 41 16 – Renewable-Media Air Filters

23 41 16.10 Disposable Media Air Filters

		Crew	Daily Output	Labor-Hours	Unit	Material	Labor	Equipment	Total	Total Incl O&P
0010	**DISPOSABLE MEDIA AIR FILTERS**									
5000	Renewable disposable roll				C.S.F.	5.25			5.25	5.75

23 41 19 – Washable Air Filters

23 41 19.10 Permanent Air Filters

			Crew	Daily Output	Labor-Hours	Unit	Material	Labor	Equipment	Total	Total Incl O&P
0010	**PERMANENT AIR FILTERS**										
4500	Permanent washable	G				MCFM	25			25	27.50

545

23 41 Particulate Air Filtration

23 41 23 – Extended Surface Filters

23 41 23.10 Expanded Surface Filters

23 41 23.10 Expanded Surface Filters	Crew	Daily Output	Labor-Hours	Unit	Material	2017 Bare Costs Labor	Equipment	Total	Total Incl O&P
0010 **EXPANDED SURFACE FILTERS**									
4000 Medium efficiency, extended surface	[G]			MCFM	6			6	6.60

23 42 Gas-Phase Air Filtration

23 42 13 – Activated-Carbon Air Filtration

23 42 13.10 Charcoal Type Air Filtration

	Crew	Daily Output	Labor-Hours	Unit	Material	2017 Bare Costs Labor	Equipment	Total	Total Incl O&P
0010 **CHARCOAL TYPE AIR FILTRATION**									
0050 Activated charcoal type, full flow				MCFM	650			650	715
0060 Full flow, impregnated media 12" deep					225			225	248
0070 HEPA filter & frame for field erection					450			450	495
0080 HEPA filter-diffuser, ceiling install.				↓	350			350	385

23 43 Electronic Air Cleaners

23 43 13 – Washable Electronic Air Cleaners

23 43 13.10 Electronic Air Cleaners

	Crew	Daily Output	Labor-Hours	Unit	Material	2017 Bare Costs Labor	Equipment	Total	Total Incl O&P
0010 **ELECTRONIC AIR CLEANERS**									
2000 Electronic air cleaner, duct mounted									
2150 1000 CFM	1 Shee	4	2	Ea.	430	116		546	655
2200 1200 CFM		3.80	2.105		505	122		627	740
2250 1400 CFM	↓	3.60	2.222	↓	530	129		659	780

23 51 Breechings, Chimneys, and Stacks

23 51 13 – Draft Control Devices

23 51 13.13 Draft-Induction Fans

	Crew	Daily Output	Labor-Hours	Unit	Material	2017 Bare Costs Labor	Equipment	Total	Total Incl O&P
0010 **DRAFT-INDUCTION FANS**									
1000 Breeching installation									
1800 Hot gas, 600°F, variable pitch pulley and motor									
1860 8" diam. inlet, 1/4 H.P., 1 phase, 1120 CFM	Q-9	4	4	Ea.	2,150	209		2,359	2,675
1900 12" diam. inlet, 3/4 H.P., 3 phase, 2960 CFM		3	5.333		2,925	279		3,204	3,625
1980 24" diam. inlet, 7-1/2 H.P., 3 phase, 17,760 CFM	↓	.80	20	↓	8,375	1,050		9,425	10,800
2300 For multi-blade damper at fan inlet, add					20%				

23 51 23 – Gas Vents

23 51 23.10 Gas Chimney Vents

	Crew	Daily Output	Labor-Hours	Unit	Material	2017 Bare Costs Labor	Equipment	Total	Total Incl O&P
0010 **GAS CHIMNEY VENTS**, Prefab metal, U.L. listed									
0020 Gas, double wall, galvanized steel									
0080 3" diameter	Q-9	72	.222	V.L.F.	6.60	11.60		18.20	25
0100 4" diameter		68	.235		8.15	12.30		20.45	28
0120 5" diameter		64	.250		9.25	13.05		22.30	30
0140 6" diameter		60	.267		11.05	13.95		25	33.50
0160 7" diameter		56	.286		21	14.95		35.95	46
0180 8" diameter		52	.308		22.50	16.10		38.60	49.50
0200 10" diameter		48	.333		43	17.40		60.40	74
0220 12" diameter		44	.364		51.50	19		70.50	85.50
0260 16" diameter		40	.400		123	21		144	167
0300 20" diameter	Q-10	36	.667	↓	179	36		215	252

23 51 Breechings, Chimneys, and Stacks

23 51 26 – All-Fuel Vent Chimneys

23 51 26.30 All-Fuel Vent Chimneys, Double Wall, St. Stl.	Crew	Daily Output	Labor-Hours	Unit	Material	2017 Bare Costs Labor	Equipment	Total	Total Incl O&P
0010 **ALL-FUEL VENT CHIMNEYS, DOUBLE WALL, STAINLESS STEEL**									
7780 All fuel, pressure tight, double wall, 4" insulation, U.L. listed, 1400°F.									
7790 304 stainless steel liner, aluminized steel outer jacket									
7800 6" diameter	Q-9	60	.267	V.L.F.	65	13.95		78.95	93
7804 8" diameter		52	.308		75	16.10		91.10	107
7806 10" diameter		48	.333		83.50	17.40		100.90	118
7808 12" diameter		44	.364		95.50	19		114.50	134
7810 14" diameter		42	.381		107	19.90		126.90	149
7880 For 316 stainless steel liner, add				L.F.	30%				

23 52 Heating Boilers

23 52 13 – Electric Boilers

23 52 13.10 Electric Boilers, ASME

	Crew	Daily Output	Labor-Hours	Unit	Material	Labor	Equipment	Total	Total Incl O&P
0010 **ELECTRIC BOILERS, ASME**, Standard controls and trim									
1000 Steam, 6 KW, 20.5 MBH	Q-19	1.20	20	Ea.	4,100	1,125		5,225	6,200
1160 60 KW, 205 MBH		1	24		6,925	1,350		8,275	9,625
1220 112 KW, 382 MBH		.75	32		9,475	1,800		11,275	13,100
1280 222 KW, 758 MBH		.55	43.636		23,800	2,450		26,250	29,900
1380 518 KW, 1768 MBH	Q-21	.36	88.889		32,600	5,125		37,725	43,600
1480 814 KW, 2778 MBH		.25	128		40,600	7,400		48,000	56,000
1600 2,340 KW, 7984 MBH		.16	200		86,500	11,500		98,000	112,500
2000 Hot water, 7.5 KW, 25.6 MBH	Q-19	1.30	18.462		4,950	1,050		6,000	7,000
2100 90 KW, 307 MBH		1.10	21.818		6,000	1,225		7,225	8,450
2220 296 KW, 1010 MBH		.55	43.636		16,100	2,450		18,550	21,400
2500 1036 KW, 3536 MBH	Q-21	.34	94.118		36,300	5,425		41,725	48,100
2680 2400 KW, 8191 MBH		.25	128		68,000	7,400		75,400	85,500
2820 3600 KW, 12,283 MBH		.16	200		94,000	11,500		105,500	121,000

23 52 23 – Cast-Iron Boilers

23 52 23.20 Gas-Fired Boilers

	Crew	Daily Output	Labor-Hours	Unit	Material	Labor	Equipment	Total	Total Incl O&P
0010 **GAS-FIRED BOILERS**, Natural or propane, standard controls, packaged									
1000 Cast iron, with insulated jacket									
2000 Steam, gross output, 81 MBH	Q-7	1.40	22.857	Ea.	2,300	1,350		3,650	4,600
2080 203 MBH		.90	35.556		3,700	2,100		5,800	7,250
2180 400 MBH		.56	56.838		5,675	3,375		9,050	11,300
2240 765 MBH		.43	74.419		11,800	4,400		16,200	19,700
2320 1,875 MBH		.30	106		24,200	6,325		30,525	36,300
2440 4,720 MBH		.15	207		65,500	12,300		77,800	90,500
2480 6,100 MBH		.13	246		85,000	14,600		99,600	115,500
2540 6,970 MBH		.10	320		95,000	19,000		114,000	133,000
3000 Hot water, gross output, 80 MBH		1.46	21.918		1,925	1,300		3,225	4,075
3140 320 MBH		.80	40		4,525	2,375		6,900	8,550
3260 1,088 MBH		.40	80		13,000	4,750		17,750	21,500
3360 2,856 MBH		.20	160		29,700	9,475		39,175	46,900
3380 3,264 MBH		.18	179		31,500	10,700		42,200	51,000
3480 6,100 MBH		.13	250		112,500	14,800		127,300	146,000
3540 6,970 MBH		.09	359		126,500	21,300		147,800	171,500
7000 For tankless water heater, add					10%				

23 52 23 – Cast-Iron Boilers

23 52 23.30 Gas/Oil Fired Boilers

		Crew	Daily Output	Labor-Hours	Unit	Material	2017 Bare Costs Labor	Equipment	Total	Total Incl O&P
0010	**GAS/OIL FIRED BOILERS**, Combination with burners and controls, packaged									
1000	Cast iron with insulated jacket									
2000	Steam, gross output, 720 MBH	Q-7	.43	74.074	Ea.	14,200	4,400		18,600	22,200
2080	1,600 MBH		.30	107		20,100	6,350		26,450	31,700
2140	2,700 MBH		.19	165		27,000	9,825		36,825	44,500
2280	5,520 MBH		.14	235		87,000	13,900		100,900	116,500
2340	6,390 MBH		.11	296		93,000	17,600		110,600	129,000
2380	6,970 MBH	↓	.09	372	↓	98,500	22,100		120,600	142,000
2900	Hot water, gross output									
2910	200 MBH	Q-6	.62	39.024	Ea.	10,200	2,275		12,475	14,700
2920	300 MBH		.49	49.080		10,200	2,850		13,050	15,600
2930	400 MBH		.41	57.971		12,000	3,375		15,375	18,300
2940	500 MBH	↓	.36	67.039		12,900	3,900		16,800	20,100
3000	584 MBH	Q-7	.44	72.072		14,300	4,275		18,575	22,200
3060	1,460 MBH		.28	113		35,400	6,700		42,100	49,000
3160	4,088 MBH		.16	195		60,000	11,600		71,600	83,500
3300	13,500 MBH, 403.3 BHP	↓	.04	727		185,000	43,100		228,100	268,500

23 52 23.40 Oil-Fired Boilers

		Crew	Daily Output	Labor-Hours	Unit	Material	2017 Bare Costs Labor	Equipment	Total	Total Incl O&P
0010	**OIL-FIRED BOILERS**, Standard controls, flame retention burner, packaged									
1000	Cast iron, with insulated flush jacket									
2000	Steam, gross output, 109 MBH	Q-7	1.20	26.667	Ea.	2,225	1,575		3,800	4,825
2060	207 MBH		.90	35.556		3,025	2,100		5,125	6,500
2180	1,084 MBH		.38	85.106		10,500	5,050		15,550	19,200
2280	3,000 MBH		.19	170		22,800	10,100		32,900	40,300
2380	5,520 MBH		.14	235		76,000	13,900		89,900	104,500
2460	6,970 MBH	↓	.09	363	↓	97,500	21,500		119,000	139,500
3000	Hot water, same price as steam									
4000	For tankless coil in smaller sizes, add				Ea.	15%				

23 52 26 – Steel Boilers

23 52 26.40 Oil-Fired Boilers

		Crew	Daily Output	Labor-Hours	Unit	Material	2017 Bare Costs Labor	Equipment	Total	Total Incl O&P
0010	**OIL-FIRED BOILERS**, Standard controls, flame retention burner									
5000	Steel, with insulated flush jacket									
7000	Hot water, gross output, 103 MBH	Q-6	1.60	15	Ea.	1,875	870		2,745	3,400
7120	420 MBH		.70	34.483		7,425	2,000		9,425	11,200
7320	3,150 MBH	↓	.13	184	↓	30,600	10,700		41,300	49,900
7340	For tankless coil in steam or hot water, add					7%				

23 52 28 – Swimming Pool Boilers

23 52 28.10 Swimming Pool Heaters

		Crew	Daily Output	Labor-Hours	Unit	Material	2017 Bare Costs Labor	Equipment	Total	Total Incl O&P
0010	**SWIMMING POOL HEATERS**, Not including wiring, external									
0020	piping, base or pad									
0160	Gas fired, input, 155 MBH	Q-6	1.50	16	Ea.	1,975	930		2,905	3,575
0200	199 MBH		1	24		2,100	1,400		3,500	4,400
0280	500 MBH		.40	60		8,725	3,475		12,200	14,900
0400	1,800 MBH	↓	.14	171		21,100	9,950		31,050	38,200
2000	Electric, 12 KW, 4,800 gallon pool	Q-19	3	8		2,100	450		2,550	3,000
2020	15 KW, 7,200 gallon pool		2.80	8.571		2,150	480		2,630	3,075
2040	24 KW, 9,600 gallon pool		2.40	10		2,475	560		3,035	3,575
2100	57 KW, 24,000 gallon pool		1.20	20	↓	3,650	1,125		4,775	5,700

23 54 Furnaces

23 54 13 – Electric-Resistance Furnaces

23 54 13.10 Electric Furnaces	Crew	Daily Output	Labor-Hours	Unit	Material	2017 Bare Costs Labor	Equipment	Total	Total Incl O&P
0010 **ELECTRIC FURNACES**, Hot air, blowers, std. controls									
0011 not including gas, oil or flue piping									
1000 Electric, UL listed									
1100 34.1 MBH	Q-20	4.40	4.545	Ea.	530	241		771	955

23 54 16 – Fuel-Fired Furnaces

23 54 16.13 Gas-Fired Furnaces

	Crew	Daily Output	Labor-Hours	Unit	Material	2017 Bare Costs Labor	Equipment	Total	Total Incl O&P
0010 **GAS-FIRED FURNACES**									
3000 Gas, AGA certified, upflow, direct drive models									
3020 45 MBH input	Q-9	4	4	Ea.	580	209		789	955
3040 60 MBH input		3.80	4.211		595	220		815	990
3060 75 MBH input		3.60	4.444		660	232		892	1,075
3100 100 MBH input		3.20	5		690	261		951	1,150

23 54 16.16 Oil-Fired Furnaces

	Crew	Daily Output	Labor-Hours	Unit	Material	2017 Bare Costs Labor	Equipment	Total	Total Incl O&P
0010 **OIL-FIRED FURNACES**									
6000 Oil, UL listed, atomizing gun type burner									
6020 56 MBH output	Q-9	3.60	4.444	Ea.	2,475	232		2,707	3,075
6030 84 MBH output		3.50	4.571		2,525	239		2,764	3,175
6040 95 MBH output		3.40	4.706		2,550	246		2,796	3,175
6060 134 MBH output		3.20	5		2,575	261		2,836	3,225
6080 151 MBH output		3	5.333		2,750	279		3,029	3,450

23 55 Fuel-Fired Heaters

23 55 13 – Fuel-Fired Duct Heaters

23 55 13.16 Gas-Fired Duct Heaters

	Crew	Daily Output	Labor-Hours	Unit	Material	2017 Bare Costs Labor	Equipment	Total	Total Incl O&P
0010 **GAS-FIRED DUCT HEATERS**, Includes burner, controls, stainless steel									
0020 heat exchanger. Gas fired, electric ignition									
0030 Indoor installation									
0100 120 MBH output	Q-5	4	4	Ea.	3,325	224		3,549	4,025
0130 200 MBH output		2.70	5.926		4,225	330		4,555	5,150
0140 240 MBH output		2.30	6.957		4,425	390		4,815	5,475
0180 320 MBH output		1.60	10		5,175	560		5,735	6,550
0300 For powered venter and adapter, add					540			540	595
0502 For required flue pipe, see Section 23 51 23.10									
1000 Outdoor installation, with power venter									
1020 75 MBH output	Q-5	4	4	Ea.	3,700	224		3,924	4,425
1060 120 MBH output		4	4		4,075	224		4,299	4,850
1100 187 MBH output		3	5.333		5,025	299		5,324	6,000
1140 300 MBH output		1.80	8.889		7,975	500		8,475	9,525
1180 450 MBH output		1.40	11.429		9,625	640		10,265	11,600

23 55 23 – Gas-Fired Radiant Heaters

23 55 23.10 Infrared Type Heating Units

	Crew	Daily Output	Labor-Hours	Unit	Material	2017 Bare Costs Labor	Equipment	Total	Total Incl O&P
0010 **INFRARED TYPE HEATING UNITS**									
0020 Gas fired, unvented, electric ignition, 100% shutoff.									
0030 Piping and wiring not included									
0120 45 MBH	Q-5	5	3.200	Ea.	1,000	179		1,179	1,375
0160 60 MBH		4	4		1,000	224		1,224	1,450
0240 120 MBH		2	8		1,650	450		2,100	2,500
1000 Gas fired, vented, electric ignition, tubular									
1020 Piping and wiring not included, 20' to 80' lengths									
1030 Single stage, input, 60 MBH	Q-6	4.50	5.333	Ea.	1,450	310		1,760	2,075

23 55 23 – Gas-Fired Radiant Heaters

23 55 23.10 Infrared Type Heating Units	Crew	Daily Output	Labor-Hours	Unit	Material	2017 Bare Costs Labor	Equipment	Total	Total Incl O&P
1040	80 MBH	Q-6	3.90	6.154	Ea.	1,450	360		1,810	2,150
1050	100 MBH		3.40	7.059		1,450	410		1,860	2,225
1060	125 MBH		2.90	8.276		1,450	480		1,930	2,325
1070	150 MBH		2.70	8.889		1,450	515		1,965	2,375
1080	170 MBH		2.50	9.600		1,450	560		2,010	2,450
1090	200 MBH		2.20	10.909		1,650	635		2,285	2,775
1100	Note: Final pricing may vary due to									
1110	tube length and configuration package selected									
1130	Two stage, input, 60 MBH high, 45 MBH low	Q-6	4.50	5.333	Ea.	1,775	310		2,085	2,425
1140	80 MBH high, 60 MBH low		3.90	6.154		1,775	360		2,135	2,500
1150	100 MBH high, 65 MBH low		3.40	7.059		1,775	410		2,185	2,575
1160	125 MBH high, 95 MBH low		2.90	8.276		1,775	480		2,255	2,675
1170	150 MBH high, 100 MBH low		2.70	8.889		1,775	515		2,290	2,725
1180	170 MBH high, 125 MBH low		2.50	9.600		2,000	560		2,560	3,050
1190	200 MBH high, 150 MBH low		2.20	10.909		2,000	635		2,635	3,150
1220	Note: Final pricing may vary due to									
1230	tube length and configuration package selected									

23 55 33 – Fuel-Fired Unit Heaters

23 55 33.13 Oil-Fired Unit Heaters

		Crew	Daily Output	Labor-Hours	Unit	Material	Labor	Equipment	Total	Total Incl O&P
0010	**OIL-FIRED UNIT HEATERS**, Cabinet, grilles, fan, ctrl., burner, no piping									
6000	Oil fired, suspension mounted, 94 MBH output	Q-5	4	4	Ea.	5,100	224		5,324	5,975
6040	140 MBH output		3	5.333		5,375	299		5,674	6,350
6060	184 MBH output		3	5.333		5,700	299		5,999	6,725

23 55 33.16 Gas-Fired Unit Heaters

		Crew	Daily Output	Labor-Hours	Unit	Material	Labor	Equipment	Total	Total Incl O&P
0010	**GAS-FIRED UNIT HEATERS**, Cabinet, grilles, fan, ctrls., burner, no piping									
0022	thermostat, no piping. For flue see Section 23 51 23.10									
1000	Gas fired, floor mounted									
1100	60 MBH output	Q-5	10	1.600	Ea.	935	89.50		1,024.50	1,150
1140	100 MBH output		8	2		1,025	112		1,137	1,300
1180	180 MBH output		6	2.667		1,475	149		1,624	1,850
2000	Suspension mounted, propeller fan, 20 MBH output		8.50	1.882		1,575	105		1,680	1,875
2040	60 MBH output		7	2.286		1,800	128		1,928	2,200
2060	80 MBH output		6	2.667		1,975	149		2,124	2,400
2100	130 MBH output		5	3.200		2,350	179		2,529	2,850
2240	320 MBH output		2	8		4,350	450		4,800	5,475
2500	For powered venter and adapter, add					400			400	440
5000	Wall furnace, 17.5 MBH output	Q-5	6	2.667		900	149		1,049	1,225
5020	24 MBH output		5	3.200		940	179		1,119	1,300
5040	35 MBH output		4	4		980	224		1,204	1,425

23 56 Solar Energy Heating Equipment

23 56 16 – Packaged Solar Heating Equipment

23 56 16.40 Solar Heating Systems

		Crew	Daily Output	Labor-Hours	Unit	Material	2017 Bare Costs Labor	Equipment	Total	Total Incl O&P
0010	**SOLAR HEATING SYSTEMS** R235616-60									
0020	System/Package prices, not including connecting									
0030	pipe, insulation, or special heating/plumbing fixtures									
0500	Hot water, standard package, low temperature									
0540	1 collector, circulator, fittings, 65 gal. tank	G	Q-1	.50	32	Ea.	3,750	1,775		5,525	6,800
0580	2 collectors, circulator, fittings, 120 gal. tank	G		.40	40		5,125	2,225		7,350	9,000
0620	3 collectors, circulator, fittings, 120 gal. tank	G		.34	47.059		7,000	2,625		9,625	11,700
0700	Medium temperature package									
0720	1 collector, circulator, fittings, 80 gal. tank	G	Q-1	.50	32	Ea.	5,100	1,775		6,875	8,300
0740	2 collectors, circulator, fittings, 120 gal. tank	G		.40	40		6,575	2,225		8,800	10,600
0780	3 collectors, circulator, fittings, 120 gal. tank	G		.30	53.333		7,450	2,975		10,425	12,700
0980	For each additional 120 gal. tank, add	G					1,775			1,775	1,950

23 56 19 – Solar Heating Components

23 56 19.50 Solar Heating Ancillary

		Crew	Daily Output	Labor-Hours	Unit	Material	2017 Bare Costs Labor	Equipment	Total	Total Incl O&P
0010	**SOLAR HEATING ANCILLARY**									
2300	Circulators, air									
2310	Blowers									
2400	Reversible fan, 20" diameter, 2 speed	G	Q-9	18	.889	Ea.	114	46.50		160.50	196
2870	1/12 HP, 30 GPM	G	Q-1	10	1.600	"	350	89		439	520
3000	Collector panels, air with aluminum absorber plate									
3010	Wall or roof mount									
3040	Flat black, plastic glazing									
3080	4' x 8'	G	Q-9	6	2.667	Ea.	670	139		809	950
3200	Flush roof mount, 10' to 16' x 22" wide	G	"	96	.167	L.F.	134	8.70		142.70	161
3300	Collector panels, liquid with copper absorber plate									
3330	Alum. frame, 4' x 8', 5/32" single glazing	G	Q-1	9.50	1.684	Ea.	1,000	93.50		1,093.50	1,250
3390	Alum. frame, 4' x 10', 5/32" single glazing	G		6	2.667		1,150	148		1,298	1,500
3450	Flat black, alum. frame, 3.5' x 7.5'	G		9	1.778		880	99		979	1,125
3500	4' x 8'	G		5.50	2.909		1,050	162		1,212	1,400
3520	4' x 10'	G		10	1.600		1,250	89		1,339	1,500
3540	4' x 12.5'	G		5	3.200		1,275	178		1,453	1,675
3600	Liquid, full wetted, plastic, alum. frame, 4' x 10'	G		5	3.200		320	178		498	625
3650	Collector panel mounting, flat roof or ground rack	G		7	2.286		247	127		374	465
3670	Roof clamps	G		70	.229	Set	2.86	12.70		15.56	22.50
3700	Roof strap, teflon	G	1 Plum	205	.039	L.F.	25	2.41		27.41	31.50
3900	Differential controller with two sensors									
3930	Thermostat, hard wired	G	1 Plum	8	1	Ea.	102	62		164	206
4100	Five station with digital read-out	G	"	3	2.667	"	266	165		431	540
4300	Heat exchanger									
4580	Fluid to fluid package includes two circulating pumps									
4590	expansion tank, check valve, relief valve									
4600	controller, high temperature cutoff and sensors	G	Q-1	2.50	6.400	Ea.	810	355		1,165	1,425
4650	Heat transfer fluid									
4700	Propylene glycol, inhibited anti-freeze	G	1 Plum	28	.286	Gal.	15.15	17.65		32.80	43
8250	Water storage tank with heat exchanger and electric element									
8300	80 gal. with 2" x 2 lb. density insulation	G	1 Plum	1.60	5	Ea.	1,600	310		1,910	2,250
8380	120 gal. with 2" x 2 lb. density insulation	G		1.40	5.714		1,825	355		2,180	2,550
8400	120 gal. with 2" x 2 lb. density insul., 40 S.F. heat coil	G		1.40	5.714		2,325	355		2,680	3,075

For customer support on your Building Construction Costs with RSMeans Data, call 800.448.8182.

551

23 57 Heat Exchangers for HVAC

23 57 16 – Steam-to-Water Heat Exchangers

23 57 16.10 Shell/Tube Type Steam-to-Water Heat Exch.	Crew	Daily Output	Labor-Hours	Unit	Material	2017 Bare Costs Labor	Equipment	Total	Total Incl O&P
0010 **SHELL AND TUBE TYPE STEAM-TO-WATER HEAT EXCHANGERS**									
0016 Shell & tube type, 2 or 4 pass, 3/4" O.D. copper tubes,									
0020 C.I. heads, C.I. tube sheet, steel shell									
0100 Hot water 40°F to 180°F, by steam at 10 PSI									
0120 8 GPM	Q-5	6	2.667	Ea.	2,300	149		2,449	2,775
0140 10 GPM		5	3.200		3,475	179		3,654	4,100
0160 40 GPM		4	4		5,375	224		5,599	6,275
0180 64 GPM		2	8		8,250	450		8,700	9,750
0200 96 GPM		1	16		11,000	895		11,895	13,500
0220 120 GPM	Q-6	1.50	16		14,500	930		15,430	17,300

23 57 19 – Liquid-to-Liquid Heat Exchangers

23 57 19.13 Plate-Type, Liquid-to-Liquid Heat Exchangers

	Crew	Daily Output	Labor-Hours	Unit	Material	2017 Bare Costs Labor	Equipment	Total	Total Incl O&P
0010 **PLATE-TYPE, LIQUID-TO-LIQUID HEAT EXCHANGERS**									
3000 Plate type,									
3100 400 GPM	Q-6	.80	30	Ea.	39,700	1,750		41,450	46,300
3120 800 GPM	"	.50	48		68,500	2,800		71,300	79,500
3140 1200 GPM	Q-7	.34	94.118		102,000	5,575		107,575	120,500
3160 1800 GPM	"	.24	133		135,000	7,900		142,900	160,500

23 57 19.16 Shell-Type, Liquid-to-Liquid Heat Exchangers

	Crew	Daily Output	Labor-Hours	Unit	Material	2017 Bare Costs Labor	Equipment	Total	Total Incl O&P
0010 **SHELL-TYPE, LIQUID-TO-LIQUID HEAT EXCHANGERS**									
1000 Hot water 40°F to 140°F, by water at 200°F									
1020 7 GPM	Q-5	6	2.667	Ea.	2,825	149		2,974	3,350
1040 16 GPM		5	3.200		4,000	179		4,179	4,675
1060 34 GPM		4	4		6,075	224		6,299	7,025
1100 74 GPM		1.50	10.667		11,000	600		11,600	13,000

23 62 Packaged Compressor and Condenser Units

23 62 13 – Packaged Air-Cooled Refrigerant Compressor and Condenser Units

23 62 13.10 Packaged Air-Cooled Refrig. Condensing Units

	Crew	Daily Output	Labor-Hours	Unit	Material	2017 Bare Costs Labor	Equipment	Total	Total Incl O&P
0010 **PACKAGED AIR-COOLED REFRIGERANT CONDENSING UNITS**									
0020 Condensing unit									
0030 Air cooled, compressor, standard controls									
0050 1.5 ton	Q-5	2.50	6.400	Ea.	1,025	360		1,385	1,675
0500 5 ton		.60	26.667		2,075	1,500		3,575	4,550
0600 10 ton		.50	32		3,700	1,800		5,500	6,775
0700 20 ton	Q-6	.40	60		8,100	3,475		11,575	14,200

23 63 Refrigerant Condensers

23 63 13 – Air-Cooled Refrigerant Condensers

23 63 13.10 Air-Cooled Refrig. Condensers

	Crew	Daily Output	Labor-Hours	Unit	Material	2017 Bare Costs Labor	Equipment	Total	Total Incl O&P
0010 **AIR-COOLED REFRIG. CONDENSERS**									
0080 Air cooled, belt drive, propeller fan									
0240 50 ton	Q-6	.69	34.985	Ea.	10,800	2,025		12,825	14,900
0280 59 ton		.58	41.308		12,900	2,400		15,300	17,800
0320 73 ton		.47	51.173		16,900	2,975		19,875	23,100
0360 86 ton		.40	60.302		19,600	3,500		23,100	26,800
0380 88 ton		.39	61.697		21,000	3,575		24,575	28,500
1550 Air cooled, direct drive, propeller fan									

552

For customer support on your Building Construction Costs with RSMeans Data, call 800.448.8182.

23 63 Refrigerant Condensers

23 63 13 – Air-Cooled Refrigerant Condensers

	23 63 13.10 Air-Cooled Refrig. Condensers	Crew	Daily Output	Labor-Hours	Unit	Material	2017 Bare Costs Labor	Equipment	Total	Total Incl O&P
1590	1 ton	Q-5	3.80	4.211	Ea.	1,650	236		1,886	2,175
1600	1-1/2 ton		3.60	4.444		1,975	249		2,224	2,550
1620	2 ton		3.20	5		2,175	280		2,455	2,825
1640	5 ton		2	8		5,025	450		5,475	6,200
1660	10 ton		1.40	11.429		6,850	640		7,490	8,525
1690	16 ton		1.10	14.545		10,200	815		11,015	12,400
1720	26 ton	↓	.84	19.002		12,500	1,075		13,575	15,400
1760	41 ton	Q-6	.77	31.008		17,800	1,800		19,600	22,300
1800	63 ton	"	.55	44.037	↓	28,000	2,550		30,550	34,700

23 64 Packaged Water Chillers

23 64 13 – Absorption Water Chillers

23 64 13.16 Indirect-Fired Absorption Water Chillers

		Crew	Daily Output	Labor-Hours	Unit	Material	2017 Bare Costs Labor	Equipment	Total	Total Incl O&P
0010	**INDIRECT-FIRED ABSORPTION WATER CHILLERS**									
0020	Steam or hot water, water cooled									
0050	100 ton	Q-7	.13	240	Ea.	126,500	14,200		140,700	160,500
0400	420 ton	"	.10	323	"	379,000	19,200		398,200	446,000

23 64 16 – Centrifugal Water Chillers

23 64 16.10 Centrifugal Type Water Chillers

		Crew	Daily Output	Labor-Hours	Unit	Material	2017 Bare Costs Labor	Equipment	Total	Total Incl O&P
0010	**CENTRIFUGAL TYPE WATER CHILLERS**, With standard controls									
0020	Centrifugal liquid chiller, water cooled									
0030	not including water tower									
0100	2000 ton (twin 1000 ton units)	Q-7	.07	477	Ea.	739,500	28,300		767,800	856,000

23 64 19 – Reciprocating Water Chillers

23 64 19.10 Reciprocating Type Water Chillers

		Crew	Daily Output	Labor-Hours	Unit	Material	2017 Bare Costs Labor	Equipment	Total	Total Incl O&P
0010	**RECIPROCATING TYPE WATER CHILLERS**, With standard controls									
0494	Water chillers, integral air cooled condenser									
0600	100 ton cooling	Q-7	.25	129	Ea.	68,000	7,650		75,650	86,000
0980	Water cooled, multiple compressor, semi-hermetic, tower not incl.									
1000	15 ton cooling	Q-6	.36	65.934	Ea.	15,400	3,825		19,225	22,700
1020	25 ton cooling	Q-7	.41	78.049		16,000	4,625		20,625	24,600
1060	35 ton cooling		.31	101		20,400	6,050		26,450	31,500
1090	45 ton cooling		.29	111		22,700	6,575		29,275	35,000
1100	50 ton cooling		.28	113		30,000	6,750		36,750	43,200
1160	100 ton cooling		.18	179		50,500	10,700		61,200	71,500
1180	125 ton cooling		.16	196		53,000	11,600		64,600	76,000
1200	145 ton cooling	↓	.16	202	↓	59,500	12,000		71,500	83,500
1451	Water cooled, dual compressors, semi-hermetic, tower not incl.									
1500	80 ton cooling	Q-7	.14	222	Ea.	29,900	13,200		43,100	53,000
1520	100 ton cooling		.14	228		40,400	13,500		53,900	65,000
1540	120 ton cooling	↓	.14	231	↓	49,000	13,700		62,700	74,500

23 64 23 – Scroll Water Chillers

23 64 23.10 Scroll Water Chillers

		Crew	Daily Output	Labor-Hours	Unit	Material	2017 Bare Costs Labor	Equipment	Total	Total Incl O&P
0010	**SCROLL WATER CHILLERS**, With standard controls									
0480	Packaged w/integral air cooled condenser									
0482	10 ton cooling	Q-7	.34	94.118	Ea.	15,000	5,575		20,575	24,800
0490	15 ton cooling		.37	86.486		18,000	5,125		23,125	27,500
0500	20 ton cooling	↓	.34	94.118		21,800	5,575		27,375	32,400
0520	40 ton cooling	↓	.30	108	↓	32,800	6,400		39,200	45,800

23 64 Packaged Water Chillers

23 64 23 – Scroll Water Chillers

23 64 23.10 Scroll Water Chillers

		Crew	Daily Output	Labor-Hours	Unit	Material	2017 Bare Costs Labor	Equipment	Total	Total Incl O&P
0680	Scroll water cooled, single compressor, hermetic, tower not incl.									
0700	2 ton cooling	Q-5	.57	28.070	Ea.	3,525	1,575		5,100	6,250
0710	5 ton cooling		.57	28.070		4,225	1,575		5,800	7,000
0740	8 ton cooling	↓	.31	52.117		5,800	2,925		8,725	10,800
0760	10 ton cooling	Q-6	.36	67.039		6,700	3,900		10,600	13,200
0800	20 ton cooling	Q-7	.38	83.990		11,900	4,975		16,875	20,600
0820	30 ton cooling	"	.33	96.096	↓	13,400	5,700		19,100	23,300

23 64 26 – Rotary-Screw Water Chillers

23 64 26.10 Rotary-Screw Type Water Chillers

		Crew	Daily Output	Labor-Hours	Unit	Material	2017 Bare Costs Labor	Equipment	Total	Total Incl O&P
0010	**ROTARY-SCREW TYPE WATER CHILLERS**, With standard controls									
0110	Screw, liquid chiller, air cooled, insulated evaporator									
0120	130 ton	Q-7	.14	228	Ea.	94,500	13,500		108,000	124,500
0124	160 ton		.13	246		116,500	14,600		131,100	150,000
0128	180 ton		.13	250		131,000	14,800		145,800	166,500
0132	210 ton		.12	258		144,000	15,300		159,300	181,500
0136	270 ton		.12	266		164,500	15,800		180,300	205,000
0140	320 ton	↓	.12	275	↓	206,500	16,300		222,800	252,000
0200	Packaged unit, water cooled, not incl. tower									
0210	80 ton	Q-7	.14	223	Ea.	47,100	13,300		60,400	72,000
0240	200 ton		.13	251		83,000	14,900		97,900	113,500
0270	350 ton	↓	.12	275	↓	141,000	16,300		157,300	179,500
1450	Water cooled, tower not included									
1580	150 ton cooling, screw compressors	Q-7	.13	240	Ea.	69,500	14,300		83,800	98,000
1620	200 ton cooling, screw compressors		.13	250		96,000	14,800		110,800	128,000
1660	291 ton cooling, screw compressors	↓	.12	260	↓	100,000	15,400		115,400	133,500

23 65 Cooling Towers

23 65 13 – Forced-Draft Cooling Towers

23 65 13.10 Forced-Draft Type Cooling Towers

		Crew	Daily Output	Labor-Hours	Unit	Material	2017 Bare Costs Labor	Equipment	Total	Total Incl O&P
0010	**FORCED-DRAFT TYPE COOLING TOWERS**, Packaged units									
0070	Galvanized steel									
0080	Induced draft, crossflow									
0100	Vertical, belt drive, 61 tons	Q-6	90	.267	TonAC	217	15.50		232.50	263
0150	100 ton		100	.240		210	13.95		223.95	252
0200	115 ton		109	.220		182	12.80		194.80	220
0250	131 ton		120	.200		219	11.60		230.60	259
0260	162 ton	↓	132	.182	↓	177	10.55		187.55	211
1000	For higher capacities, use multiples									
1500	Induced air, double flow									
1900	Vertical, gear drive, 167 ton	Q-6	126	.190	TonAC	172	11.05		183.05	206
2000	297 ton		129	.186		106	10.80		116.80	132
2100	582 ton		132	.182		59	10.55		69.55	81
2150	849 ton		142	.169		79.50	9.80		89.30	102
2200	1016 ton	↓	150	.160	↓	80	9.30		89.30	102
3000	For higher capacities, use multiples									
3500	For pumps and piping, add	Q-6	38	.632	TonAC	109	36.50		145.50	176
4000	For absorption systems, add				"	75%	75%			
4100	Cooling water chemical feeder	Q-5	3	5.333	Ea.	400	299		699	890
5000	Fiberglass tower on galvanized steel support structure									
5010	Draw thru									

23 65 Cooling Towers

23 65 13 – Forced-Draft Cooling Towers

23 65 13.10 Forced-Draft Type Cooling Towers

		Crew	Daily Output	Labor-Hours	Unit	Material	2017 Bare Costs Labor	Equipment	Total	Total Incl O&P
5100	100 ton	Q-6	1.40	17.143	Ea.	13,900	995		14,895	16,800
5120	120 ton		1.20	20		16,200	1,150		17,350	19,600
5140	140 ton		1	24		17,500	1,400		18,900	21,300
5160	160 ton		.80	30		19,400	1,750		21,150	24,000
5180	180 ton		.65	36.923		22,100	2,150		24,250	27,500
5200	200 ton	↓	.48	50	↓	25,000	2,900		27,900	31,900
5300	For stainless steel support structure, add					30%				
5360	For higher capacities, use multiples of each size									
6000	Stainless steel									
6010	Induced draft, crossflow, horizontal, belt drive									
6100	57 ton	Q-6	1.50	16	Ea.	28,200	930		29,130	32,400
6120	91 ton		.99	24.242		34,100	1,400		35,500	39,600
6140	111 ton		.43	55.814		45,700	3,250		48,950	55,000
6160	126 ton	↓	.22	109	↓	47,400	6,350		53,750	61,500

23 73 Indoor Central-Station Air-Handling Units

23 73 13 – Modular Indoor Central-Station Air-Handling Units

23 73 13.10 Air-Handling Units

		Crew	Daily Output	Labor-Hours	Unit	Material	2017 Bare Costs Labor	Equipment	Total	Total Incl O&P
0010	**AIR-HANDLING UNITS**, Built-Up									
0100	With cooling/heating coil section, filters, mixing box									
0880	Single zone, horizontal/vertical									
0890	Constant volume									
0900	1600 CFM	Q-5	1.20	13.333	Ea.	5,150	745		5,895	6,800
0920	5000 CFM	Q-6	1.40	17.143		14,400	995		15,395	17,400
0940	11,500 CFM		1	24		25,800	1,400		27,200	30,400
0970	22,000 CFM		.60	40		50,500	2,325		52,825	59,000
1000	40,000 CFM	↓	.30	80		92,500	4,650		97,150	109,000

23 73 39 – Indoor, Direct Gas-Fired Heating and Ventilating Units

23 73 39.10 Make-Up Air Unit

		Crew	Daily Output	Labor-Hours	Unit	Material	2017 Bare Costs Labor	Equipment	Total	Total Incl O&P
0010	**MAKE-UP AIR UNIT**									
0020	Indoor suspension, natural/LP gas, direct fired,									
0032	standard control. For flue see Section 23 51 23.10									
0040	70°F temperature rise, MBH is input									
0100	75 MBH input	Q-6	3.60	6.667	Ea.	5,150	385		5,535	6,250
0160	150 MBH input		3	8		6,250	465		6,715	7,575
0220	225 MBH input		2.40	10		7,000	580		7,580	8,575
0300	400 MBH input	↓	1.60	15		16,500	870		17,370	19,500
0600	For discharge louver assembly, add					5%				
0700	For filters, add					10%				
0800	For air shut-off damper section, add				↓	30%				

For customer support on your Building Construction Costs with RSMeans Data, call 800.448.8182.

555

23 74 Packaged Outdoor HVAC Equipment

23 74 33 – Dedicated Outdoor-Air Units

23 74 33.10 Rooftop Air Conditioners		Crew	Daily Output	Labor-Hours	Unit	Material	2017 Bare Costs Labor	Equipment	Total	Total Incl O&P
0010	**ROOFTOP AIR CONDITIONERS**, Standard controls, curb, economizer									
1000	Single zone, electric cool, gas heat									
1100	3 ton cooling, 60 MBH heating R236000-20	Q-5	.70	22.857	Ea.	2,800	1,275		4,075	5,000
1120	4 ton cooling, 95 MBH heating		.61	26.403		3,325	1,475		4,800	5,875
1140	5 ton cooling, 112 MBH heating		.56	28.521		4,375	1,600		5,975	7,200
1145	6 ton cooling, 140 MBH heating		.52	30.769		5,000	1,725		6,725	8,075
1150	7.5 ton cooling, 170 MBH heating		.50	32.258		5,825	1,800		7,625	9,125
1156	8.5 ton cooling, 170 MBH heating		.46	34.783		7,075	1,950		9,025	10,700
1160	10 ton cooling, 200 MBH heating	Q-6	.67	35.982		9,150	2,100		11,250	13,300
1170	12.5 ton cooling, 230 MBH heating		.63	37.975		10,400	2,200		12,600	14,700
1190	17.5 ton cooling, 330 MBH heating		.52	45.889		13,900	2,675		16,575	19,300
1200	20 ton cooling, 360 MBH heating	Q-7	.67	47.976		29,500	2,850		32,350	36,800
1210	25 ton cooling, 450 MBH heating		.56	57.554		32,000	3,400		35,400	40,400
1220	30 ton cooling, 540 MBH heating		.47	68.376		35,700	4,050		39,750	45,400
1240	40 ton cooling, 675 MBH heating		.35	91.168		43,900	5,400		49,300	56,500
2000	Multizone, electric cool, gas heat, economizer									
2100	15 ton cooling, 360 MBH heating	Q-7	.61	52.545	Ea.	65,500	3,125		68,625	76,500
2120	20 ton cooling, 360 MBH heating		.53	60.038		70,500	3,550		74,050	83,000
2200	40 ton cooling, 540 MBH heating		.28	113		124,500	6,750		131,250	147,000
2210	50 ton cooling, 540 MBH heating		.23	142		156,000	8,425		164,425	184,000
2220	70 ton cooling, 1500 MBH heating		.16	198		167,500	11,800		179,300	202,000
2240	80 ton cooling, 1500 MBH heating		.14	228		191,000	13,500		204,500	231,000
2260	90 ton cooling, 1500 MBH heating		.13	256		196,500	15,200		211,700	239,000
2280	105 ton cooling, 1500 MBH heating		.11	290		218,000	17,200		235,200	266,000
2400	For hot water heat coil, deduct					5%				
2500	For steam heat coil, deduct					2%				
2600	For electric heat, deduct					3%	5%			

23 81 Decentralized Unitary HVAC Equipment

23 81 13 – Packaged Terminal Air-Conditioners

23 81 13.10 Packaged Cabinet Type Air-Conditioners

23 81 13.10 Packaged Cabinet Type Air-Conditioners		Crew	Daily Output	Labor-Hours	Unit	Material	2017 Bare Costs Labor	Equipment	Total	Total Incl O&P
0010	**PACKAGED CABINET TYPE AIR-CONDITIONERS**, Cabinet, wall sleeve,									
0100	louver, electric heat, thermostat, manual changeover, 208 V									
0200	6,000 BTUH cooling, 8800 BTU heat	Q-5	6	2.667	Ea.	865	149		1,014	1,175
0220	9,000 BTUH cooling, 13,900 BTU heat		5	3.200		1,275	179		1,454	1,675
0240	12,000 BTUH cooling, 13,900 BTU heat		4	4		1,375	224		1,599	1,850
0260	15,000 BTUH cooling, 13,900 BTU heat		3	5.333		1,575	299		1,874	2,200
0500	For hot water coil, increase heat by 10%, add					5%	10%			
1000	For steam, increase heat output by 30%, add					8%	10%			

23 81 19 – Self-Contained Air-Conditioners

23 81 19.20 Self-Contained Single Package

23 81 19.20 Self-Contained Single Package		Crew	Daily Output	Labor-Hours	Unit	Material	2017 Bare Costs Labor	Equipment	Total	Total Incl O&P
0010	**SELF-CONTAINED SINGLE PACKAGE**									
0100	Air cooled, for free blow or duct, not incl. remote condenser									
0110	Constant volume									
0200	3 ton cooling	Q-5	1	16	Ea.	3,725	895		4,620	5,450
0220	5 ton cooling	Q-6	1.20	20		4,325	1,150		5,475	6,500
0240	10 ton cooling	Q-7	1	32		7,675	1,900		9,575	11,300
0260	20 ton cooling		.90	35.556		12,500	2,100		14,600	16,900
0280	30 ton cooling		.80	40		25,500	2,375		27,875	31,700
0340	60 ton cooling	Q-8	.40	80		56,500	4,750	146	61,396	69,500

23 81 Decentralized Unitary HVAC Equipment

23 81 19 – Self-Contained Air-Conditioners

23 81 19.20 Self-Contained Single Package	Crew	Daily Output	Labor-Hours	Unit	Material	2017 Bare Costs Labor	Equipment	Total	Total Incl O&P
0490 For duct mounting, no price change									
0500 For steam heating coils, add				Ea.	10%	10%			
1000 Water cooled for free blow or duct, not including tower									
1010 Constant volume									
1100 3 ton cooling	Q-6	1	24	Ea.	3,625	1,400		5,025	6,075
1120 5 ton cooling	"	1	24		4,525	1,400		5,925	7,075
1140 10 ton cooling	Q-7	.90	35.556		8,550	2,100		10,650	12,600
1160 20 ton cooling		.80	40		27,700	2,375		30,075	34,100
1180 30 ton cooling		.70	45.714		36,700	2,700		39,400	44,500

23 81 23 – Computer-Room Air-Conditioners

23 81 23.10 Computer Room Units

	Crew	Daily Output	Labor-Hours	Unit	Material	2017 Bare Costs Labor	Equipment	Total	Total Incl O&P
0010 **COMPUTER ROOM UNITS**									
1000 Air cooled, includes remote condenser but not									
1020 interconnecting tubing or refrigerant									
1080 3 ton	Q-5	.50	32	Ea.	19,700	1,800		21,500	24,400
1120 5 ton		.45	35.556		21,100	2,000		23,100	26,200
1160 6 ton		.30	53.333		38,900	3,000		41,900	47,300
1200 8 ton		.27	59.259		39,300	3,325		42,625	48,300
1240 10 ton		.25	64		41,100	3,575		44,675	50,500
1260 12 ton		.24	66.667		42,600	3,725		46,325	52,500
1280 15 ton		.22	72.727		45,200	4,075		49,275	56,000
1290 18 ton		.20	80		52,000	4,475		56,475	64,000
1300 20 ton	Q-6	.26	92.308		54,000	5,375		59,375	67,500
1320 22 ton		.24	100		55,000	5,800		60,800	69,500
1360 30 ton		.21	114		67,500	6,650		74,150	84,500
2200 Chilled water, for connection to									
2220 existing chiller system of adequate capacity									
2260 5 ton	Q-5	.74	21.622	Ea.	15,000	1,200		16,200	18,300

23 81 43 – Air-Source Unitary Heat Pumps

23 81 43.10 Air-Source Heat Pumps

	Crew	Daily Output	Labor-Hours	Unit	Material	2017 Bare Costs Labor	Equipment	Total	Total Incl O&P
0010 **AIR-SOURCE HEAT PUMPS**, Not including interconnecting tubing									
1000 Air to air, split system, not including curbs, pads, fan coil and ductwork									
1012 Outside condensing unit only, for fan coil see Section 23 82 19.10									
1020 2 ton cooling, 8.5 MBH heat @ 0°F	Q-5	2	8	Ea.	1,650	450		2,100	2,475
1060 5 ton cooling, 27 MBH heat @ 0°F		.50	32		2,675	1,800		4,475	5,625
1080 7.5 ton cooling, 33 MBH heat @ 0°F		.45	35.556		4,775	2,000		6,775	8,250
1100 10 ton cooling, 50 MBH heat @ 0°F	Q-6	.64	37.500		6,275	2,175		8,450	10,200
1120 15 ton cooling, 64 MBH heat @ 0°F		.50	48		8,725	2,800		11,525	13,800
1130 20 ton cooling, 85 MBH heat @ 0°F		.35	68.571		17,500	3,975		21,475	25,300
1140 25 ton cooling, 119 MBH heat @ 0°F		.25	96		20,700	5,575		26,275	31,100
1500 Single package, not including curbs, pads, or plenums									
1520 2 ton cooling, 6.5 MBH heat @ 0°F	Q-5	1.50	10.667	Ea.	3,150	600		3,750	4,350
1580 4 ton cooling, 13 MBH heat @ 0°F		.96	16.667		4,225	935		5,160	6,050
1640 7.5 ton cooling, 35 MBH heat @ 0°F		.40	40		7,075	2,250		9,325	11,200

23 81 46 – Water-Source Unitary Heat Pumps

23 81 46.10 Water Source Heat Pumps

	Crew	Daily Output	Labor-Hours	Unit	Material	2017 Bare Costs Labor	Equipment	Total	Total Incl O&P
0010 **WATER SOURCE HEAT PUMPS**, Not incl. connecting tubing or water source									
2000 Water source to air, single package									
2100 1 ton cooling, 13 MBH heat @ 75°F	Q-5	2	8	Ea.	1,800	450		2,250	2,650
2140 2 ton cooling, 19 MBH heat @ 75°F		1.70	9.412		2,225	525		2,750	3,250
2220 5 ton cooling, 29 MBH heat @ 75°F		.90	17.778		3,300	995		4,295	5,150

For customer support on your Building Construction Costs with RSMeans Data, call 800.448.8182.

557

23 81 Decentralized Unitary HVAC Equipment

23 81 46 – Water-Source Unitary Heat Pumps

23 81 46.10 Water Source Heat Pumps	Crew	Daily Output	Labor-Hours	Unit	Material	2017 Bare Costs Labor	Equipment	Total	Total Incl O&P
3960 For supplementary heat coil, add				Ea.	10%				
4000 For increase in capacity thru use									
4020 of solar collector, size boiler at 60%									

23 82 Convection Heating and Cooling Units

23 82 16 – Air Coils

23 82 16.10 Flanged Coils

		Crew	Daily Output	Labor-Hours	Unit	Material	2017 Bare Costs Labor	Equipment	Total	Total Incl O&P
0010	**FLANGED COILS**									
0500	Chilled water cooling, 6 rows, 24" x 48"	Q-5	3.20	5	Ea.	4,275	280		4,555	5,125
1000	Direct expansion cooling, 6 rows, 24" x 48"		2.80	5.714		4,625	320		4,945	5,575
1500	Hot water heating, 1 row, 24" x 48"		4	4		1,700	224		1,924	2,225
2000	Steam heating, 1 row, 24" x 48"	↓	3.06	5.229	↓	2,400	293		2,693	3,100

23 82 16.20 Duct Heaters

		Crew	Daily Output	Labor-Hours	Unit	Material	2017 Bare Costs Labor	Equipment	Total	Total Incl O&P
0010	**DUCT HEATERS**, Electric, 480 V, 3 Ph.									
0020	Finned tubular insert, 500°F									
0100	8" wide x 6" high, 4.0 kW	Q-20	16	1.250	Ea.	775	66.50		841.50	950
0120	12" high, 8.0 kW		15	1.333		1,275	71		1,346	1,525
0140	18" high, 12.0 kW		14	1.429		1,800	76		1,876	2,100
0160	24" high, 16.0 kW		13	1.538		2,325	81.50		2,406.50	2,675
0180	30" high, 20.0 kW		12	1.667		2,825	88.50		2,913.50	3,250
0300	12" wide x 6" high, 6.7 kW		15	1.333		825	71		896	1,025
0360	24" high, 26.7 kW		12	1.667		2,400	88.50		2,488.50	2,750
0700	24" wide x 6" high, 17.8 kW		13	1.538		975	81.50		1,056.50	1,200
0760	24" high, 71.1 kW	↓	10	2	↓	3,000	106		3,106	3,450
8000	To obtain BTU multiply kW by 3413									

23 82 19 – Fan Coil Units

23 82 19.10 Fan Coil Air Conditioning

		Crew	Daily Output	Labor-Hours	Unit	Material	2017 Bare Costs Labor	Equipment	Total	Total Incl O&P
0010	**FAN COIL AIR CONDITIONING**									
0030	Fan coil AC, cabinet mounted, filters and controls									
0100	Chilled water, 1/2 ton cooling	Q-5	8	2	Ea.	555	112		667	780
0120	1 ton cooling		6	2.667		830	149		979	1,150
0140	1.5 ton cooling		5.50	2.909		835	163		998	1,175
0150	2 ton cooling		5.25	3.048		1,200	171		1,371	1,575
0180	3 ton cooling	↓	4	4	↓	1,950	224		2,174	2,500
0262	For hot water coil, add					40%	10%			
0320	1 ton cooling	Q-5	6	2.667	Ea.	1,575	149		1,724	1,975
0940	Direct expansion, for use w/air cooled condensing unit, 1.5 ton cooling		5	3.200		610	179		789	945
1000	5 ton cooling		3	5.333		1,075	299		1,374	1,625
1040	10 ton cooling	Q-6	2.60	9.231		2,000	535		2,535	3,000
1060	20 ton cooling	"	.70	34.286		3,850	2,000		5,850	7,225
1500	For hot water coil, add	↓			↓	40%	10%			

23 82 19.20 Heating and Ventilating Units

		Crew	Daily Output	Labor-Hours	Unit	Material	2017 Bare Costs Labor	Equipment	Total	Total Incl O&P
0010	**HEATING AND VENTILATING UNITS**, Classroom units									
0020	Includes filter, heating/cooling coils, standard controls									
0080	750 CFM, 2 tons cooling	Q-6	2	12	Ea.	4,250	695		4,945	5,750
0120	1250 CFM, 3 tons cooling		1.40	17.143		5,200	995		6,195	7,225
0140	1500 CFM, 4 tons cooling	↓	.80	30		5,550	1,750		7,300	8,750
0500	For electric heat, add					35%				
1000	For no cooling, deduct	↓			↓	25%	10%			

23 82 Convection Heating and Cooling Units

23 82 29 – Radiators

23 82 29.10 Hydronic Heating

		Crew	Daily Output	Labor-Hours	Unit	Material	2017 Bare Costs Labor	Equipment	Total	Total Incl O&P
0010	**HYDRONIC HEATING**, Terminal units, not incl. main supply pipe									
1000	Radiation									
1100	Panel, baseboard, C.I., including supports, no covers	Q-5	46	.348	L.F.	40	19.50		59.50	74
3000	Radiators, cast iron									
3100	Free standing or wall hung, 6 tube, 25" high	Q-5	96	.167	Section	48	9.35		57.35	67
3200	4 tube, 19" high	"	96	.167	"	31	9.35		40.35	48
3250	Adj. brackets, 2 per wall radiator up to 30 sections	1 Stpi	32	.250	Ea.	55	15.55		70.55	84
9500	To convert SFR to BTU rating: Hot water, 150 x SFR									
9510	Forced hot water, 180 x SFR; steam, 240 x SFR									

23 82 33 – Convectors

23 82 33.10 Convector Units

		Crew	Daily Output	Labor-Hours	Unit	Material	2017 Bare Costs Labor	Equipment	Total	Total Incl O&P
0010	**CONVECTOR UNITS**, Terminal units, not incl. main supply pipe									
2204	Convector, multifin, 2 pipe w/cabinet									
2210	17" H x 24" L	Q-5	10	1.600	Ea.	102	89.50		191.50	247
2214	17" H x 36" L		8.60	1.860		153	104		257	325
2218	17" H x 48" L		7.40	2.162		204	121		325	405
2222	21" H x 24" L		9	1.778		102	99.50		201.50	262
2226	21" H x 36" L		8.20	1.951		153	109		262	335
2228	21" H x 48" L	↓	6.80	2.353	↓	204	132		336	425
2240	For knob operated damper, add					140%				
2241	For metal trim strips, add	Q-5	64	.250	Ea.	13.20	14		27.20	35.50
2243	For snap-on inlet grille, add					10%	10%			
2245	For hinged access door, add	Q-5	64	.250	Ea.	36.50	14		50.50	61
2246	For air chamber, auto-venting, add	"	58	.276	"	8.20	15.45		23.65	32.50

23 82 36 – Finned-Tube Radiation Heaters

23 82 36.10 Finned Tube Radiation

		Crew	Daily Output	Labor-Hours	Unit	Material	2017 Bare Costs Labor	Equipment	Total	Total Incl O&P
0010	**FINNED TUBE RADIATION**, Terminal units, not incl. main supply pipe									
1150	Fin tube, wall hung, 14" slope top cover, with damper									
1200	1-1/4" copper tube, 4-1/4" alum. fin	Q-5	38	.421	L.F.	43.50	23.50		67	83
1250	1-1/4" steel tube, 4-1/4" steel fin	"	36	.444	"	40	25		65	81.50
1500	Note: fin tube may also require corners, caps, etc.									

23 82 39 – Unit Heaters

23 82 39.16 Propeller Unit Heaters

		Crew	Daily Output	Labor-Hours	Unit	Material	2017 Bare Costs Labor	Equipment	Total	Total Incl O&P
0010	**PROPELLER UNIT HEATERS**									
3950	Unit heaters, propeller, 115 V 2 psi steam, 60°F entering air									
4000	Horizontal, 12 MBH	Q-5	12	1.333	Ea.	375	74.50		449.50	530
4060	43.9 MBH		8	2		570	112		682	795
4140	96.8 MBH		6	2.667		830	149		979	1,150
4180	157.6 MBH		4	4		1,100	224		1,324	1,575
4240	286.9 MBH		2	8		1,725	450		2,175	2,575
4260	364 MBH		1.80	8.889		2,150	500		2,650	3,125
4270	404 MBH	↓	1.60	10	↓	2,250	560		2,810	3,325
4300	Vertical diffuser same price									
4310	Vertical flow, 40 MBH	Q-5	11	1.455	Ea.	565	81.50		646.50	745
4314	58.5 MBH		8	2		585	112		697	810
4326	131.0 MBH		4	4		880	224		1,104	1,300
4346	297.0 MBH		1.80	8.889		1,675	500		2,175	2,600
4354	420 MBH (460 V)	Q-6	1.80	13.333		2,250	775		3,025	3,650
4358	500 MBH (460 V)		1.71	14.035		2,975	815		3,790	4,500
4362	570 MBH (460 V)		1.40	17.143		4,075	995		5,070	6,000

559

For customer support on your Building Construction Costs with RSMeans Data, call 800.448.8182.

23 82 Convection Heating and Cooling Units

23 82 39 – Unit Heaters

23 82 39.16 Propeller Unit Heaters	Crew	Daily Output	Labor-Hours	Unit	Material	2017 Bare Costs Labor	Equipment	Total	Total Incl O&P
4366	620 MBH (460 V)	Q-6	1.30	18.462	Ea.	4,075	1,075		5,150	6,125
4370	960 MBH (460 V)	↓	1.10	21.818	↓	7,750	1,275		9,025	10,500

23 83 Radiant Heating Units

23 83 16 – Radiant-Heating Hydronic Piping

23 83 16.10 Radiant Floor Heating

		Crew	Daily Output	Labor-Hours	Unit	Material	2017 Bare Costs Labor	Equipment	Total	Total Incl O&P
0010	**RADIANT FLOOR HEATING**									
0100	Tubing, PEX (cross-linked polyethylene)									
0110	Oxygen barrier type for systems with ferrous materials									
0120	1/2"	Q-5	800	.020	L.F.	.98	1.12		2.10	2.77
0130	3/4"		535	.030		1.39	1.68		3.07	4.06
0140	1"	↓	400	.040	↓	2.16	2.24		4.40	5.75
0200	Non barrier type for ferrous free systems									
0210	1/2"	Q-5	800	.020	L.F.	.50	1.12		1.62	2.24
0220	3/4"		535	.030		.92	1.68		2.60	3.54
0230	1"	↓	400	.040	↓	1.58	2.24		3.82	5.10
1000	Manifolds									
1110	Brass									
1120	With supply and return valves, flow meter, thermometer,									
1122	auto air vent and drain/fill valve.									
1130	1", 2 circuit	Q-5	14	1.143	Ea.	281	64		345	405
1140	1", 3 circuit		13.50	1.185		320	66.50		386.50	455
1150	1", 4 circuit		13	1.231		350	69		419	490
1154	1", 5 circuit		12.50	1.280		415	71.50		486.50	570
1158	1", 6 circuit		12	1.333		455	74.50		529.50	615
1162	1", 7 circuit		11.50	1.391		495	78		573	665
1166	1", 8 circuit		11	1.455		500	81.50		581.50	675
1172	1", 9 circuit		10.50	1.524		590	85.50		675.50	780
1174	1", 10 circuit		10	1.600		635	89.50		724.50	835
1178	1", 11 circuit		9.50	1.684		660	94.50		754.50	865
1182	1", 12 circuit	↓	9	1.778	↓	730	99.50		829.50	955
1610	Copper manifold header (cut to size)									
1620	1" header, 12 circuit 1/2" sweat outlets	Q-5	3.33	4.805	Ea.	96	269		365	510
1630	1-1/4" header, 12 circuit 1/2" sweat outlets		3.20	5		112	280		392	550
1640	1-1/4" header, 12 circuit 3/4" sweat outlets		3	5.333		120	299		419	580
1650	1-1/2" header, 12 circuit 3/4" sweat outlets		3.10	5.161		144	289		433	595
1660	2" header, 12 circuit 3/4" sweat outlets	↓	2.90	5.517	↓	212	310		522	700
3000	Valves									
3110	Thermostatic zone valve actuator with end switch	Q-5	40	.400	Ea.	42	22.50		64.50	80.50
3114	Thermostatic zone valve actuator	"	36	.444	"	85.50	25		110.50	132
3120	Motorized straight zone valve with operator complete									
3130	3/4"	Q-5	35	.457	Ea.	138	25.50		163.50	190
3140	1"		32	.500		149	28		177	207
3150	1-1/4"	↓	29.60	.541	↓	189	30.50		219.50	254
3500	4 way mixing valve, manual, brass									
3530	1"	Q-5	13.30	1.203	Ea.	191	67.50		258.50	310
3540	1-1/4"		11.40	1.404		207	78.50		285.50	345
3550	1-1/2"		11	1.455		264	81.50		345.50	415
3560	2"		10.60	1.509		375	84.50		459.50	540
3800	Mixing valve motor, 4 way for valves, 1" and 1-1/4"		34	.471		355	26.50		381.50	430
3810	Mixing valve motor, 4 way for valves, 1-1/2" and 2"	↓	30	.533	↓	380	30		410	460

23 83 Radiant Heating Units

23 83 16 – Radiant-Heating Hydronic Piping

23 83 16.10 Radiant Floor Heating	Crew	Daily Output	Labor-Hours	Unit	Material	2017 Bare Costs Labor	Equipment	Total	Total Incl O&P
5000	Radiant floor heating, zone control panel									
5120	4 zone actuator valve control, expandable	Q-5	20	.800	Ea.	149	45		194	232
5130	6 zone actuator valve control, expandable		18	.889		238	50		288	335
6070	Thermal track, straight panel for long continuous runs, 5.333 S.F.		40	.400		29	22.50		51.50	66
6080	Thermal track, utility panel, for direction reverse at run end, 5.333 S.F.		40	.400		29	22.50		51.50	66
6090	Combination panel, for direction reverse plus straight run, 5.333 S.F.		40	.400		29	22.50		51.50	66
7000	PEX tubing fittings									
7100	Compression type									
7116	Coupling									
7120	1/2" x 1/2"	1 Stpi	27	.296	Ea.	6.30	18.45		24.75	35
7124	3/4" x 3/4"	"	23	.348	"	12.40	21.50		33.90	46
7130	Adapter									
7132	1/2" x female sweat 1/2"	1 Stpi	27	.296	Ea.	4.09	18.45		22.54	32.50
7134	1/2" x female sweat 3/4"		26	.308		4.58	19.15		23.73	34
7136	5/8" x female sweat 3/4"		24	.333		6.55	21		27.55	38.50
7140	Elbow									
7142	1/2" x female sweat 1/2"	1 Stpi	27	.296	Ea.	6.25	18.45		24.70	35
7144	1/2" x female sweat 3/4"		26	.308		7.30	19.15		26.45	37
7146	5/8" x female sweat 3/4"		24	.333		8.20	21		29.20	40.50
7200	Insert type									
7206	PEX x male NPT									
7210	1/2" x 1/2"	1 Stpi	29	.276	Ea.	2.56	17.15		19.71	29
7220	3/4" x 3/4"		27	.296		3.77	18.45		22.22	32
7230	1" x 1"		26	.308		6.35	19.15		25.50	36
7300	PEX coupling									
7310	1/2" x 1/2"	1 Stpi	30	.267	Ea.	.51	16.60		17.11	25.50
7320	3/4" x 3/4"		29	.276		.72	17.15		17.87	27
7330	1" x 1"		28	.286		1.26	17.80		19.06	28.50
7400	PEX stainless crimp ring									
7410	1/2" x 1/2"	1 Stpi	86	.093	Ea.	.44	5.80		6.24	9.25
7420	3/4" x 3/4"		84	.095		.60	5.95		6.55	9.60
7430	1" x 1"		82	.098		.85	6.05		6.90	10.10

23 83 33 – Electric Radiant Heaters

23 83 33.10 Electric Heating

		Crew	Daily Output	Labor-Hours	Unit	Material	2017 Bare Costs Labor	Equipment	Total	Total Incl O&P
0010	**ELECTRIC HEATING**, not incl. conduit or feed wiring									
1100	Rule of thumb: Baseboard units, including control	1 Elec	4.40	1.818	kW	106	103		209	271
1300	Baseboard heaters, 2' long, 350 watt		8	1	Ea.	26	56.50		82.50	114
1400	3' long, 750 watt		8	1		30	56.50		86.50	118
1600	4' long, 1000 watt		6.70	1.194		35	67.50		102.50	140
1800	5' long, 935 watt		5.70	1.404		43	79.50		122.50	166
2000	6' long, 1500 watt		5	1.600		49	90.50		139.50	190
2400	8' long, 2000 watt		4	2		59.50	113		172.50	236
2950	Wall heaters with fan, 120 to 277 volt									
3170	1000 watt	1 Elec	6	1.333	Ea.	84.50	75.50		160	206
3180	1250 watt		5	1.600		84.50	90.50		175	229
3190	1500 watt		4	2		84.50	113		197.50	263
3600	Thermostats, integral		16	.500		27	28.50		55.50	72
3800	Line voltage, 1 pole		8	1		15.65	56.50		72.15	102

For customer support on your Building Construction Costs with RSMeans Data, call 800.448.8182.

561

23 84 13.10 Humidifier Units	Crew	Daily Output	Labor-Hours	Unit	Material	2017 Bare Costs Labor	Equipment	Total	Total Incl O&P
0010 **HUMIDIFIER UNITS**									
0520 Steam, room or duct, filter, regulators, auto. controls, 220 V									
0540 11 lb. per hour	Q-5	6	2.667	Ea.	2,725	149		2,874	3,225
0560 22 lb. per hour		5	3.200		3,000	179		3,179	3,575
0580 33 lb. per hour		4	4		3,100	224		3,324	3,750
0600 50 lb. per hour		4	4		3,575	224		3,799	4,300
0620 100 lb. per hour		3	5.333		4,525	299		4,824	5,425

Estimating Tips
26 05 00 Common Work Results for Electrical

- Conduit should be taken off in three main categories—power distribution, branch power, and branch lighting—so the estimator can concentrate on systems and components, therefore making it easier to ensure all items have been accounted for.

- For cost modifications for elevated conduit installation, add the percentages to labor according to the height of installation and only to the quantities exceeding the different height levels, not to the total conduit quantities. Refer to 26 01 02.20 for labor adjustment factors.

- Remember that aluminum wiring of equal ampacity is larger in diameter than copper and may require larger conduit.

- If more than three wires at a time are being pulled, deduct percentages from the labor hours of that grouping of wires.

- When taking off grounding systems, identify separately the type and size of wire, and list each unique type of ground connection.

- The estimator should take the weights of materials into consideration when completing a takeoff. Topics to consider include: How will the materials be supported? What methods of support are available? How high will the support structure have to reach? Will the final support structure be able to withstand the total burden? Is the support material included or separate from the fixture, equipment, and material specified?

- Do not overlook the costs for equipment used in the installation. If scaffolding or highlifts are available in the field, contractors may use them in lieu of the proposed ladders and rolling staging.

26 20 00 Low-Voltage Electrical Transmission

- Supports and concrete pads may be shown on drawings for the larger equipment, or the support system may be only a piece of plywood for the back of a panelboard. In either case, they must be included in the costs.

26 40 00 Electrical and Cathodic Protection

- When taking off cathodic protection systems, identify the type and size of cable, and list each unique type of anode connection.

26 50 00 Lighting

- Fixtures should be taken off room by room using the fixture schedule, specifications, and the ceiling plan. For large concentrations of lighting fixtures in the same area, deduct the percentages from labor hours.

Reference Numbers

Reference numbers are shown at the beginning of some major classifications. These numbers refer to related items in the Reference Section. The reference information may be an estimating procedure, an alternate pricing method, or technical information.

Note: Not all subdivisions listed here necessarily appear. ∎

26 01 02 – Labor Adjustment

26 01 02.20 Labor Adjustment Factors	Crew	Daily Output	Labor-Hours	Unit	Material	2017 Bare Costs Labor	Equipment	Total	Total Incl O&P
0010 **LABOR ADJUSTMENT FACTORS** (For Div. 26, 27 and 28) R260519-90									
0100 Subtract from labor for Economy of Scale for Wire									
0110 4-5 wires						25%			
0120 6-10 wires						30%			
0130 11-15 wires						35%			
0140 over 15 wires						40%			
0150 Labor adjustment factors (For Div. 26, 27, 28 and 48)									
0200 Labor factors, The below are reasonable suggestions, however									
0210 each project must be evaluated for its own peculiarities, and									
0220 the adjustments be increased or decreased depending on the									
0230 severity of the special conditions.									
1000 Add to labor for elevated installation (above floor level)									
1010 10' to 14.5' high						10%			
1020 15' to 19.5' high						20%			
1030 20' to 24.5' high						25%			
1040 25' to 29.5' high						35%			
1050 30' to 34.5' high						40%			
1060 35' to 39.5' high						50%			
1070 40' and higher						55%			
2000 Add to labor for crawl space									
2010 3' high						40%			
2020 4' high						30%			
3000 Add to labor for multi-story building									
3100 For new construction (No elevator available)									
3110 Add for floors 3 thru 10						5%			
3120 Add for floors 11 thru 15						10%			
3130 Add for floors 16 thru 20						15%			
3140 Add for floors 21 thru 30						20%			
3150 Add for floors 31 and up						30%			
3200 For existing structure (Elevator available)									
3210 Add for work on floor 3 and above						2%			
4000 Add to labor for working in existing occupied buildings									
4010 Hospital						35%			
4020 Office building						25%			
4030 School						20%			
4040 Factory or warehouse						15%			
4050 Multi-dwelling						15%			
5000 Add to labor, miscellaneous									
5010 Cramped shaft						35%			
5020 Congested area						15%			
5030 Excessive heat or cold						30%			
6000 Labor factors, the above are reasonable suggestions, however									
6100 Each project should be evaluated for its own peculiarities									
6200 Other factors to be considered are:									
6210 Movement of material and equipment through finished areas						10%			
6220 Equipment room min security direct access w/authorization						15%			
6230 Attic space						25%			
6240 No service road						25%			
6250 Poor unloading/storage area, no hydraulic lifts or jacks						20%			
6260 Congested site area/heavy traffic						20%			
7000 Correctional facilities (no compounding division 1 adjustment factors)									
7010 Minimum security w/facilities escort						30%			
7020 Medium security w/ facilities and correctional officer escort						40%			

26 01 Operation and Maintenance of Electrical Systems

26 01 02 – Labor Adjustment

26 01 02.20 Labor Adjustment Factors	Crew	Daily Output	Labor-Hours	Unit	Material	2017 Bare Costs Labor	2017 Bare Costs Equipment	Total	Total Incl O&P
7030	Max security w/facilities & correctional officer escort (no inmate contact)					50%			

26 05 Common Work Results for Electrical

26 05 05 – Selective Demolition for Electrical

26 05 05.10 Electrical Demolition

		Crew	Daily Output	Labor-Hours	Unit	Material	2017 Bare Costs Labor	2017 Bare Costs Equipment	Total	Total Incl O&P
0010	**ELECTRICAL DEMOLITION**									
0020	Electrical dml, conduit to 10' high, including fittings & hangers									
0100	Rigid galvanized steel, 1/2" to 1" diameter	1 Elec	242	.033	L.F.		1.87		1.87	2.80
0120	1-1/4" to 2"	"	200	.040			2.26		2.26	3.39
0140	2-1/2" to 3-1/2"	2 Elec	302	.053			3		3	4.49
0160	4" to 6"	"	160	.100			5.65		5.65	8.50
0200	Electric metallic tubing (EMT), 1/2" to 1"	1 Elec	394	.020			1.15		1.15	1.72
0220	1-1/4" to 1-1/2"		326	.025			1.39		1.39	2.08
0240	2" to 3"		236	.034			1.92		1.92	2.87
0260	3-1/2" to 4"	2 Elec	310	.052			2.92		2.92	4.38
0270	Armored cable (BX) avg. 50' runs									
0280	#14, 2 wire	1 Elec	690	.012	L.F.		.66		.66	.98
0290	#14, 3 wire		571	.014			.79		.79	1.19
0300	#12, 2 wire		605	.013			.75		.75	1.12
0310	#12, 3 wire		514	.016			.88		.88	1.32
0320	#10, 2 wire		514	.016			.88		.88	1.32
0330	#10, 3 wire		425	.019			1.07		1.07	1.60
0340	#8, 3 wire		342	.023			1.32		1.32	1.98
0350	Non metallic sheathed cable (Romex)									
0360	#14, 2 wire	1 Elec	720	.011	L.F.		.63		.63	.94
0370	#14, 3 wire		657	.012			.69		.69	1.03
0380	#12, 2 wire		629	.013			.72		.72	1.08
0390	#10, 3 wire		450	.018			1.01		1.01	1.51
0400	Wiremold raceway, including fittings & hangers									
0420	No. 3000	1 Elec	250	.032	L.F.		1.81		1.81	2.71
0440	No. 4000		217	.037			2.09		2.09	3.13
0460	No. 6000		166	.048			2.73		2.73	4.09
0462	Plugmold with receptacle		114	.070			3.97		3.97	5.95
0465	Telephone/power pole		12	.667	Ea.		37.50		37.50	56.50
0470	Non-metallic, straight section		480	.017	L.F.		.94		.94	1.41
0500	Channels, steel, including fittings & hangers									
0520	3/4" x 1-1/2"	1 Elec	308	.026	L.F.		1.47		1.47	2.20
0540	1-1/2" x 1-1/2"		269	.030			1.68		1.68	2.52
0560	1-1/2" x 1-7/8"		229	.035			1.98		1.98	2.96
0600	Copper bus duct, indoor, 3 phase									
0610	Including hangers & supports									
0620	225 amp	2 Elec	135	.119	L.F.		6.70		6.70	10.05
0640	400 amp		106	.151			8.55		8.55	12.80
0660	600 amp		86	.186			10.55		10.55	15.80
0680	1000 amp		60	.267			15.10		15.10	22.50
0700	1600 amp		40	.400			22.50		22.50	34
0720	3000 amp		10	1.600			90.50		90.50	136
1300	Transformer, dry type, 1 phase, incl. removal of									
1320	supports, wire & conduit terminations									
1340	1 kVA	1 Elec	7.70	1.039	Ea.		59		59	88
1420	75 kVA	2 Elec	2.50	6.400	"		360		360	545

26 05 05.10 Electrical Demolition	Crew	Daily Output	Labor-Hours	Unit	Material	2017 Bare Costs Labor	2017 Bare Costs Equipment	Total	Total Incl O&P
1440	3 phase to 600V, primary									
1460	3 kVA	1 Elec	3.87	2.067	Ea.		117		117	175
1520	75 kVA	2 Elec	2.69	5.948			335		335	505
1550	300 kVA	R-3	1.80	11.111			630	76.50	706.50	1,025
1570	750 kVA	"	1.10	18.182	↓		1,025	125	1,150	1,700
1800	Wire, THW-THWN-THHN, removed from									
1810	in place conduit, to 10' high									
1830	#14	1 Elec	65	.123	C.L.F.		6.95		6.95	10.45
1840	#12		55	.145			8.25		8.25	12.35
1850	#10		45.50	.176			9.95		9.95	14.90
1860	#8		40.40	.198			11.20		11.20	16.80
1870	#6	↓	32.60	.245			13.90		13.90	21
1880	#4	2 Elec	53	.302			17.10		17.10	25.50
1890	#3		50	.320			18.10		18.10	27
1900	#2		44.60	.359			20.50		20.50	30.50
1910	1/0		33.20	.482			27.50		27.50	41
1920	2/0		29.20	.548			31		31	46.50
1930	3/0		25	.640			36		36	54.50
1940	4/0		22	.727			41		41	61.50
1950	250 kcmil		20	.800			45.50		45.50	68
1960	300 kcmil		19	.842			47.50		47.50	71.50
1970	350 kcmil		18	.889			50.50		50.50	75.50
1980	400 kcmil		17	.941			53.50		53.50	80
1990	500 kcmil	↓	16.20	.988	↓		56		56	84
2000	Interior fluorescent fixtures, incl. supports									
2010	& whips, to 10' high									
2100	Recessed drop-in 2' x 2', 2 lamp	2 Elec	35	.457	Ea.		26		26	39
2110	2' x 2', 4 lamp		30	.533			30		30	45
2120	2' x 4', 2 lamp		33	.485			27.50		27.50	41
2140	2' x 4', 4 lamp		30	.533			30		30	45
2160	4' x 4', 4 lamp	↓	20	.800	↓		45.50		45.50	68
2180	Surface mount, acrylic lens & hinged frame									
2200	1' x 4', 2 lamp	2 Elec	44	.364	Ea.		20.50		20.50	31
2210	6" x 4', 2 lamp		44	.364			20.50		20.50	31
2220	2' x 2', 2 lamp		44	.364			20.50		20.50	31
2260	2' x 4', 4 lamp		33	.485			27.50		27.50	41
2280	4' x 4', 4 lamp		23	.696			39.50		39.50	59
2281	4' x 4', 6 lamp	↓	23	.696	↓		39.50		39.50	59
2300	Strip fixtures, surface mount									
2320	4' long, 1 lamp	2 Elec	53	.302	Ea.		17.10		17.10	25.50
2340	4' long, 2 lamp		50	.320			18.10		18.10	27
2360	8' long, 1 lamp		42	.381			21.50		21.50	32.50
2380	8' long, 2 lamp	↓	40	.400	↓		22.50		22.50	34
2400	Pendant mount, industrial, incl. removal									
2410	of chain or rod hangers, to 10' high									
2420	4' long, 2 lamp	2 Elec	35	.457	Ea.		26		26	39
2421	4' long, 4 lamp		35	.457			26		26	39
2440	8' long, 2 lamp	↓	27	.593	↓		33.50		33.50	50.50

26 05 Common Work Results for Electrical

26 05 13 – Medium-Voltage Cables

26 05 13.16 Medium-Voltage, Single Cable		Crew	Daily Output	Labor-Hours	Unit	Material	2017 Bare Costs Labor	Equipment	Total	Total Incl O&P
0010	**MEDIUM-VOLTAGE, SINGLE CABLE** Splicing & terminations not included									
0040	Copper, XLP shielding, 5 kV, #6	2 Elec	4.40	3.636	C.L.F.	153	206		359	480
0050	#4		4.40	3.636		199	206		405	530
0100	#2		4	4		262	226		488	630
0200	#1		4	4		325	226		551	695
0400	1/0		3.80	4.211		350	238		588	735
0600	2/0		3.60	4.444		420	252		672	840
0800	4/0		3.20	5		570	283		853	1,050
1000	250 kcmil	3 Elec	4.50	5.333		680	300		980	1,200
1200	350 kcmil		3.90	6.154		885	350		1,235	1,500
1400	500 kcmil		3.60	6.667		1,050	375		1,425	1,725
1600	15 kV, ungrounded neutral, #1	2 Elec	4	4		330	226		556	700
1800	1/0		3.80	4.211		390	238		628	785
2000	2/0		3.60	4.444		445	252		697	865
2200	4/0		3.20	5		595	283		878	1,075
2400	250 kcmil	3 Elec	4.50	5.333		655	300		955	1,175
2600	350 kcmil		3.90	6.154		830	350		1,180	1,425
2800	500 kcmil		3.60	6.667		1,025	375		1,400	1,700

26 05 19 – Low-Voltage Electrical Power Conductors and Cables

26 05 19.20 Armored Cable		Crew	Daily Output	Labor-Hours	Unit	Material	2017 Bare Costs Labor	Equipment	Total	Total Incl O&P
0010	**ARMORED CABLE**									
0050	600 volt, copper (BX), #14, 2 conductor, solid	1 Elec	2.40	3.333	C.L.F.	45.50	189		234.50	335
0100	3 conductor, solid		2.20	3.636		73	206		279	390
0150	#12, 2 conductor, solid		2.30	3.478		46	197		243	345
0200	3 conductor, solid		2	4		74.50	226		300.50	420
0250	#10, 2 conductor, solid		2	4		87.50	226		313.50	435
0300	3 conductor, solid		1.60	5		121	283		404	560
0340	#8, 2 conductor, stranded		1.50	5.333		229	300		529	700
0350	3 conductor, stranded		1.30	6.154		229	350		579	770
0400	3 conductor with PVC jacket, in cable tray, #6		3.10	2.581		540	146		686	810
0450	#4	2 Elec	5.40	2.963		655	168		823	970
0500	#2		4.60	3.478		795	197		992	1,175
0550	#1		4	4		1,025	226		1,251	1,475
0600	1/0		3.60	4.444		1,100	252		1,352	1,575
0650	2/0		3.40	4.706		1,325	266		1,591	1,850
0700	3/0		3.20	5		1,775	283		2,058	2,375
0750	4/0		3	5.333		2,125	300		2,425	2,775
0800	250 kcmil	3 Elec	3.60	6.667		2,450	375		2,825	3,275
0850	350 kcmil		3.30	7.273		3,300	410		3,710	4,275
0900	500 kcmil		3	8		4,575	455		5,030	5,700
1050	5 kV, copper, 3 conductor with PVC jacket,									
1060	non-shielded, in cable tray, #4	2 Elec	380	.042	L.F.	6.85	2.38		9.23	11.10
1100	#2		360	.044		8.90	2.52		11.42	13.55
1200	#1		300	.053		11.35	3.02		14.37	17
1400	1/0		290	.055		13.10	3.12		16.22	19.10
1600	2/0		260	.062		15.15	3.48		18.63	22
2000	4/0		240	.067		20	3.77		23.77	27.50
2100	250 kcmil	3 Elec	330	.073		27.50	4.12		31.62	36.50
2150	350 kcmil		315	.076		34	4.31		38.31	44
2200	500 kcmil		270	.089		46.50	5.05		51.55	58.50
2400	15 kV, copper, 3 conductor with PVC jacket galv., steel armored									
2500	grounded neutral, in cable tray, #2	2 Elec	300	.053	L.F.	15	3.02		18.02	21

For customer support on your Building Construction Costs with RSMeans Data, call 800.448.8182.

567

26 05 19 – Low-Voltage Electrical Power Conductors and Cables

26 05 19.20 Armored Cable		Crew	Daily Output	Labor-Hours	Unit	Material	2017 Bare Costs Labor	Equipment	Total	Total Incl O&P
2600	#1	2 Elec	280	.057	L.F.	15.35	3.23		18.58	22
2800	1/0		260	.062		18.25	3.48		21.73	25
2900	2/0		220	.073		23.50	4.12		27.62	31.50
3000	4/0		190	.084		26.50	4.77		31.27	36
3100	250 kcmil	3 Elec	270	.089		29.50	5.05		34.55	40
3150	350 kcmil		240	.100		34.50	5.65		40.15	46.50
3200	500 kcmil		210	.114		46.50	6.45		52.95	61
3400	15 kV, copper, 3 conductor with PVC jacket,									
3450	ungrounded neutral, in cable tray, #2	2 Elec	260	.062	L.F.	15.55	3.48		19.03	22.50
3500	#1		230	.070		17.20	3.94		21.14	25
3600	1/0		200	.080		19.70	4.53		24.23	28.50
3700	2/0		190	.084		24	4.77		28.77	33.50
3800	4/0		160	.100		29	5.65		34.65	40.50
4000	250 kcmil	3 Elec	210	.114		34	6.45		40.45	47
4050	350 kcmil		195	.123		45	6.95		51.95	59.50
4100	500 kcmil		180	.133		54.50	7.55		62.05	71.50
9010	600 volt, copper (MC) steel clad, #14, 2 wire	R-1A	8.16	1.961	C.L.F.	52	100		152	207
9020	3 wire		7.21	2.219		74.50	113		187.50	251
9030	4 wire		6.69	2.392		103	122		225	297
9040	#12, 2 wire		7.21	2.219		46.50	113		159.50	220
9050	3 wire		6.69	2.392		79	122		201	270
9070	#10, 2 wire		6.04	2.649		108	135		243	320
9080	3 wire		5.65	2.832		142	144		286	370
9100	#8, 2 wire, stranded		3.94	4.061		194	207		401	525
9110	3 wire, stranded		3.40	4.706		272	240		512	660
9200	600 volt, copper (MC) aluminum clad, #14, 2 wire		8.16	1.961		45	100		145	200
9210	3 wire		7.21	2.219		73	113		186	250
9220	4 wire		6.69	2.392		112	122		234	305
9230	#12, 2 wire		7.21	2.219		45	113		158	219
9240	3 wire		6.69	2.392		81.50	122		203.50	273
9250	4 wire		6.04	2.649		108	135		243	320
9260	#10, 2 wire		6.04	2.649		116	135		251	330
9270	3 wire		5.65	2.832		141	144		285	370
9280	4 wire		5.23	3.059		254	156		410	515

26 05 19.35 Cable Terminations

		Crew	Daily Output	Labor-Hours	Unit	Material	2017 Bare Costs Labor	Equipment	Total	Total Incl O&P
0010	**CABLE TERMINATIONS**									
0015	Wire connectors, screw type, #22 to #14	1 Elec	260	.031	Ea.	.06	1.74		1.80	2.68
0020	#18 to #12		240	.033		.07	1.89		1.96	2.91
0035	#16 to #10		230	.035		.27	1.97		2.24	3.25
0040	#14 to #8		210	.038		.30	2.16		2.46	3.56
0045	#12 to #6		180	.044		.62	2.52		3.14	4.45
0050	Terminal lugs, solderless, #16 to #10		50	.160		.37	9.05		9.42	13.95
0100	#8 to #4		30	.267		.73	15.10		15.83	23.50
0150	#2 to #1		22	.364		1.03	20.50		21.53	32
0200	1/0 to 2/0		16	.500		1.64	28.50		30.14	44.50
0250	3/0		12	.667		3.55	37.50		41.05	60.50
0300	4/0		11	.727		4.35	41		45.35	66.50
0350	250 kcmil		9	.889		3.41	50.50		53.91	79.50
0400	350 kcmil		7	1.143		4.44	64.50		68.94	102
0450	500 kcmil		6	1.333		8.20	75.50		83.70	122
1600	Crimp 1 hole lugs, copper or aluminum, 600 volt									
1620	#14	1 Elec	60	.133	Ea.	.56	7.55		8.11	11.90

26 05 Common Work Results for Electrical

26 05 19 – Low-Voltage Electrical Power Conductors and Cables

26 05 19.35 Cable Terminations

		Crew	Daily Output	Labor-Hours	Unit	Material	2017 Bare Costs Labor	Equipment	Total	Total Incl O&P
1630	#12	1 Elec	50	.160	Ea.	.90	9.05		9.95	14.55
1640	#10		45	.178		.90	10.05		10.95	16.10
1780	#8		36	.222		1.80	12.60		14.40	21
1800	#6		30	.267		2.02	15.10		17.12	24.50
2000	#4		27	.296		2.56	16.75		19.31	28
2200	#2		24	.333		3.13	18.85		21.98	32
2400	#1		20	.400		4.61	22.50		27.11	39
2500	1/0		17.50	.457		4.98	26		30.98	44.50
2600	2/0		15	.533		6.15	30		36.15	52
2800	3/0		12	.667		6	37.50		43.50	63
3000	4/0		11	.727		6.65	41		47.65	69
3200	250 kcmil		9	.889		6.85	50.50		57.35	83
3400	300 kcmil		8	1		8.05	56.50		64.55	94
3500	350 kcmil		7	1.143		8.75	64.50		73.25	107
3600	400 kcmil		6.50	1.231		10.85	69.50		80.35	116
3800	500 kcmil		6	1.333		12.55	75.50		88.05	127

26 05 19.50 Mineral Insulated Cable

		Crew	Daily Output	Labor-Hours	Unit	Material	2017 Bare Costs Labor	Equipment	Total	Total Incl O&P
0010	**MINERAL INSULATED CABLE** 600 volt									
0100	1 conductor, #12	1 Elec	1.60	5	C.L.F.	350	283		633	810
0200	#10		1.60	5		450	283		733	920
0400	#8		1.50	5.333		500	300		800	1,000
0500	#6		1.40	5.714		595	325		920	1,150
0600	#4	2 Elec	2.40	6.667		800	375		1,175	1,450
0800	#2		2.20	7.273		1,125	410		1,535	1,875
0900	#1		2.10	7.619		1,300	430		1,730	2,075
1000	1/0		2	8		1,525	455		1,980	2,350
1100	2/0		1.90	8.421		1,825	475		2,300	2,725
1200	3/0		1.80	8.889		2,175	505		2,680	3,150
1400	4/0		1.60	10		2,525	565		3,090	3,625
1410	250 kcmil	3 Elec	2.40	10		2,850	565		3,415	3,975
1420	350 kcmil		1.95	12.308		3,250	695		3,945	4,625
1430	500 kcmil		1.95	12.308		4,175	695		4,870	5,650

26 05 19.55 Non-Metallic Sheathed Cable

		Crew	Daily Output	Labor-Hours	Unit	Material	2017 Bare Costs Labor	Equipment	Total	Total Incl O&P
0010	**NON-METALLIC SHEATHED CABLE** 600 volt									
0100	Copper with ground wire (Romex)									
0150	#14, 2 conductor	1 Elec	2.70	2.963	C.L.F.	21.50	168		189.50	275
0200	3 conductor		2.40	3.333		30.50	189		219.50	315
0220	4 conductor		2.20	3.636		44	206		250	360
0250	#12, 2 conductor		2.50	3.200		33	181		214	310
0300	3 conductor		2.20	3.636		46.50	206		252.50	360
0320	4 conductor		2	4		68.50	226		294.50	415
0350	#10, 2 conductor		2.20	3.636		50.50	206		256.50	365
0400	3 conductor		1.80	4.444		73.50	252		325.50	455
0430	#8, 2 conductor		1.60	5		79.50	283		362.50	510
0450	3 conductor		1.50	5.333		118	300		418	580
0500	#6, 3 conductor		1.40	5.714		192	325		517	695
0550	SE type SER aluminum cable, 3 RHW and									
0600	1 bare neutral, 3 #8 & 1 #8	1 Elec	1.60	5	C.L.F.	69	283		352	500
0650	3 #6 & 1 #6	"	1.40	5.714		78	325		403	570
0700	3 #4 & 1 #6	2 Elec	2.40	6.667		87.50	375		462.50	660
0750	3 #2 & 1 #4		2.20	7.273		129	410		539	755
0800	3 #1/0 & 1 #2		2	8		195	455		650	895

For customer support on your Building Construction Costs with RSMeans Data, call 800.448.8182.

26 05 19 – Low-Voltage Electrical Power Conductors and Cables

26 05 19.55 Non-Metallic Sheathed Cable		Crew	Daily Output	Labor-Hours	Unit	Material	2017 Bare Costs Labor	Equipment	Total	Total Incl O&P
0850	3 #2/0 & 1 #1	2 Elec	1.80	8.889	C.L.F.	229	505		734	1,000
0900	3 #4/0 & 1 #2/0	↓	1.60	10	↓	325	565		890	1,200

26 05 19.90 Wire

		Crew	Daily Output	Labor-Hours	Unit	Material	2017 Bare Costs Labor	Equipment	Total	Total Incl O&P
0010	**WIRE** R260519-92									
0020	600 volt, copper type THW, solid, #14	1 Elec	13	.615	C.L.F.	7	35		42	59.50
0030	#12		11	.727		10.65	41		51.65	73.50
0040	#10		10	.800		16.70	45.50		62.20	86.50
0050	Stranded, #14 R260533-22		13	.615		8.20	35		43.20	61
0100	#12		11	.727		12.55	41		53.55	75.50
0120	#10		10	.800		19.80	45.50		65.30	90
0140	#8		8	1		32.50	56.50		89	121
0160	#6	↓	6.50	1.231		55.50	69.50		125	165
0180	#4	2 Elec	10.60	1.509		85	85.50		170.50	222
0200	#3		10	1.600		106	90.50		196.50	252
0220	#2		9	1.778		133	101		234	297
0240	#1		8	2		175	113		288	365
0260	1/0		6.60	2.424		218	137		355	445
0280	2/0		5.80	2.759		274	156		430	535
0300	3/0		5	3.200		345	181		526	650
0350	4/0	↓	4.40	3.636		440	206		646	790
0400	250 kcmil	3 Elec	6	4		495	226		721	885
0420	300 kcmil		5.70	4.211		560	238		798	970
0450	350 kcmil		5.40	4.444		650	252		902	1,100
0480	400 kcmil		5.10	4.706		745	266		1,011	1,225
0490	500 kcmil	↓	4.80	5		925	283		1,208	1,450
0540	600 volt, aluminum type THHN, stranded, #6	1 Elec	8	1		36.50	56.50		93	125
0560	#4	2 Elec	13	1.231		45.50	69.50		115	154
0580	#2		10.60	1.509		61.50	85.50		147	196
0600	#1		9	1.778		90	101		191	250
0620	1/0		8	2		108	113		221	289
0640	2/0		7.20	2.222		128	126		254	330
0680	3/0		6.60	2.424		158	137		295	380
0700	4/0	↓	6.20	2.581		176	146		322	415
0720	250 kcmil	3 Elec	8.70	2.759		215	156		371	470
0740	300 kcmil		8.10	2.963		297	168		465	575
0760	350 kcmil		7.50	3.200		300	181		481	600
0780	400 kcmil		6.90	3.478		355	197		552	685
0800	500 kcmil		6	4		390	226		616	770
0850	600 kcmil		5.70	4.211		495	238		733	895
0880	700 kcmil		5.10	4.706		570	266		836	1,025
0900	750 kcmil		4.80	5		590	283		873	1,075
0910	1000 kcmil	↓	3.78	6.349		875	360		1,235	1,500
0920	600 volt, copper type THWN-THHN, solid, #14	1 Elec	13	.615		7	35		42	59.50
0940	#12		11	.727		10.65	41		51.65	73.50
0960	#10		10	.800		16.70	45.50		62.20	86.50
1000	Stranded, #14		13	.615		7.30	35		42.30	60
1200	#12		11	.727		10.45	41		51.45	73
1250	#10		10	.800		16	45.50		61.50	85.50
1300	#8		8	1		28	56.50		84.50	116
1350	#6	↓	6.50	1.231		48	69.50		117.50	157
1400	#4	2 Elec	10.60	1.509	↓	68.50	85.50		154	204

26 05 23 – Control-Voltage Electrical Power Cables

26 05 23.10 Control Cable		Crew	Daily Output	Labor-Hours	Unit	Material	2017 Bare Costs Labor	2017 Bare Costs Equipment	Total	Total Incl O&P
0010	**CONTROL CABLE**									
0020	600 volt, copper, #14 THWN wire with PVC jacket, 2 wires	1 Elec	9	.889	C.L.F.	35	50.50		85.50	114
0030	3 wires		8	1		50.50	56.50		107	141
0100	4 wires		7	1.143		60	64.50		124.50	163
0150	5 wires		6.50	1.231		75	69.50		144.50	186
0200	6 wires		6	1.333		102	75.50		177.50	225
0300	8 wires		5.30	1.509		132	85.50		217.50	273
0400	10 wires		4.80	1.667		152	94.50		246.50	310
0500	12 wires		4.30	1.860		197	105		302	375
0600	14 wires		3.80	2.105		212	119		331	415
0700	16 wires		3.50	2.286		240	129		369	460
0800	18 wires		3.30	2.424		270	137		407	505
0810	19 wires		3.10	2.581		285	146		431	535
0900	20 wires		3	2.667		299	151		450	555
1000	22 wires		2.80	2.857		330	162		492	600

26 05 26 – Grounding and Bonding for Electrical Systems

26 05 26.80 Grounding

26 05 26.80 Grounding		Crew	Daily Output	Labor-Hours	Unit	Material	2017 Bare Costs Labor	2017 Bare Costs Equipment	Total	Total Incl O&P
0010	**GROUNDING**									
0030	Rod, copper clad, 8' long, 1/2" diameter	1 Elec	5.50	1.455	Ea.	20	82.50		102.50	145
0050	3/4" diameter		5.30	1.509		34	85.50		119.50	166
0080	10' long, 1/2" diameter		4.80	1.667		22.50	94.50		117	166
0100	3/4" diameter		4.40	1.818		39	103		142	197
0130	15' long, 3/4" diameter		4	2		51	113		164	226
0390	Bare copper wire, stranded, #8		11	.727	C.L.F.	25.50	41		66.50	89.50
0400	#6		10	.800		45	45.50		90.50	118
0600	#2	2 Elec	10	1.600		93.50	90.50		184	239
0800	3/0		6.60	2.424		268	137		405	500
1000	4/0		5.70	2.807		340	159		499	615
1200	250 kcmil	3 Elec	7.20	3.333		400	189		589	725
1800	Water pipe ground clamps, heavy duty									
2000	Bronze, 1/2" to 1" diameter	1 Elec	8	1	Ea.	24	56.50		80.50	111
2100	1-1/4" to 2" diameter		8	1		33	56.50		89.50	122
2200	2-1/2" to 3" diameter		6	1.333		44.50	75.50		120	162
2800	Brazed connections, #6 wire		12	.667		16.50	37.50		54	74.50
3000	#2 wire		10	.800		22	45.50		67.50	92.50
3100	3/0 wire		8	1		33	56.50		89.50	121
3200	4/0 wire		7	1.143		37.50	64.50		102	138
3400	250 kcmil wire		5	1.600		43.50	90.50		134	184
3600	500 kcmil wire		4	2		54	113		167	230

26 05 33 – Raceway and Boxes for Electrical Systems

26 05 33.13 Conduit

26 05 33.13 Conduit		Crew	Daily Output	Labor-Hours	Unit	Material	2017 Bare Costs Labor	2017 Bare Costs Equipment	Total	Total Incl O&P
0010	**CONDUIT** To 10' high, includes 2 terminations, 2 elbows,	R260533-22								
0020	11 beam clamps, and 11 couplings per 100 L.F.									
0300	Aluminum, 1/2" diameter	1 Elec	100	.080	L.F.	1.79	4.53		6.32	8.75
0500	3/4" diameter		90	.089		2.54	5.05		7.59	10.35
0700	1" diameter		80	.100		3.48	5.65		9.13	12.35
1000	1-1/4" diameter		70	.114		4.28	6.45		10.73	14.40
1030	1-1/2" diameter		65	.123		5.45	6.95		12.40	16.45
1050	2" diameter		60	.133		7.70	7.55		15.25	19.80
1070	2-1/2" diameter		50	.160		11.50	9.05		20.55	26
1100	3" diameter	2 Elec	90	.178		15.30	10.05		25.35	32

26 05 33.13 Conduit

26 05 33.13 Conduit		Crew	Daily Output	Labor-Hours	Unit	Material	2017 Bare Costs Labor	Equipment	Total	Total Incl O&P
1130	3-1/2" diameter	2 Elec	80	.200	L.F.	21	11.30		32.30	40
1140	4" diameter		70	.229		24.50	12.95		37.45	46.50
1750	Rigid galvanized steel, 1/2" diameter	1 Elec	90	.089		2.50	5.05		7.55	10.30
1770	3/4" diameter		80	.100		4.66	5.65		10.31	13.65
1800	1" diameter		65	.123		7.10	6.95		14.05	18.30
1830	1-1/4" diameter		60	.133		5.10	7.55		12.65	16.90
1850	1-1/2" diameter		55	.145		8	8.25		16.25	21
1870	2" diameter		45	.178		10.25	10.05		20.30	26.50
1900	2-1/2" diameter		35	.229		12.95	12.95		25.90	33.50
1930	3" diameter	2 Elec	50	.320		14.95	18.10		33.05	43.50
1950	3-1/2" diameter		44	.364		18.95	20.50		39.45	52
1970	4" diameter		40	.400		22	22.50		44.50	58.50
2500	Steel, intermediate conduit (IMC), 1/2" diameter	1 Elec	100	.080		2.20	4.53		6.73	9.20
2530	3/4" diameter		90	.089		2.75	5.05		7.80	10.60
2550	1" diameter		70	.114		3.93	6.45		10.38	14
2570	1-1/4" diameter		65	.123		4.94	6.95		11.89	15.90
2600	1-1/2" diameter		60	.133		6.55	7.55		14.10	18.50
2630	2" diameter		50	.160		8.05	9.05		17.10	22.50
2650	2-1/2" diameter		40	.200		14.05	11.30		25.35	32.50
2670	3" diameter	2 Elec	60	.267		18.50	15.10		33.60	43
2700	3-1/2" diameter		54	.296		23	16.75		39.75	50.50
2730	4" diameter		50	.320		13.65	18.10		31.75	42
5000	Electric metallic tubing (EMT), 1/2" diameter	1 Elec	170	.047		.69	2.66		3.35	4.75
5020	3/4" diameter		130	.062		.98	3.48		4.46	6.30
5040	1" diameter		115	.070		1.65	3.94		5.59	7.70
5060	1-1/4" diameter		100	.080		2.66	4.53		7.19	9.70
5080	1-1/2" diameter		90	.089		3.38	5.05		8.43	11.25
5100	2" diameter		80	.100		4.24	5.65		9.89	13.15
5120	2-1/2" diameter		60	.133		8.70	7.55		16.25	21
5140	3" diameter	2 Elec	100	.160		10.20	9.05		19.25	25
5160	3-1/2" diameter		90	.178		12.85	10.05		22.90	29
5180	4" diameter		80	.200		13.85	11.30		25.15	32
9900	Add to labor for higher elevated installation									
9905	10' to 14.5' high, add						10%			
9910	15' to 20' high, add						20%			
9920	20' to 25' high, add						25%			
9930	25' to 30' high, add						35%			
9940	30' to 35' high, add						40%			
9950	35' to 40' high, add						50%			
9960	Over 40' high, add						55%			

26 05 33.16 Boxes for Electrical Systems

26 05 33.16 Boxes for Electrical Systems		Crew	Daily Output	Labor-Hours	Unit	Material	2017 Bare Costs Labor	Equipment	Total	Total Incl O&P
0010	**BOXES FOR ELECTRICAL SYSTEMS**									
0020	Pressed steel, octagon, 4"	1 Elec	20	.400	Ea.	2.22	22.50		24.72	36.50
0060	Covers, blank		64	.125		.93	7.10		8.03	11.60
0100	Extension rings		40	.200		3.68	11.30		14.98	21
0150	Square, 4"		20	.400		2.88	22.50		25.38	37
0200	Extension rings		40	.200		3.72	11.30		15.02	21
0250	Covers, blank		64	.125		1	7.10		8.10	11.70
0300	Plaster rings		64	.125		2.04	7.10		9.14	12.85
0650	Switchbox		27	.296		3.55	16.75		20.30	29
1100	Concrete, floor, 1 gang		5.30	1.509		89.50	85.50		175	227
2000	Poke-thru fitting, fire rated, for 3-3/4" floor		6.80	1.176		131	66.50		197.50	244

26 05 33 – Raceway and Boxes for Electrical Systems

	26 05 33.16 Boxes for Electrical Systems	Crew	Daily Output	Labor-Hours	Unit	Material	2017 Bare Costs Labor	Equipment	Total	Total Incl O&P
2040	For 7" floor	1 Elec	6.80	1.176	Ea.	181	66.50		247.50	299
2100	Pedestal, 15 amp, duplex receptacle & blank plate		5.25	1.524		138	86.50		224.50	280
2120	Duplex receptacle and telephone plate		5.25	1.524		138	86.50		224.50	280
2140	Pedestal, 20 amp, duplex recept. & phone plate		5	1.600		139	90.50		229.50	288
2200	Abandonment plate	↓	32	.250	↓	38.50	14.15		52.65	63.50

26 05 33.18 Pull Boxes

		Crew	Daily Output	Labor-Hours	Unit	Material	2017 Bare Costs Labor	Equipment	Total	Total Incl O&P
0010	**PULL BOXES**									
0100	Steel, pull box, NEMA 1, type SC, 6" W x 6" H x 4" D	1 Elec	8	1	Ea.	9.65	56.50		66.15	95.50
0200	8" W x 8" H x 4" D		8	1		13.20	56.50		69.70	99.50
0300	10" W x 12" H x 6" D		5.30	1.509		22.50	85.50		108	153
0400	16" W x 20" H x 8" D		4	2		84.50	113		197.50	263
0500	20" W x 24" H x 8" D		3.20	2.500		98	142		240	320
0600	24" W x 36" H x 8" D		2.70	2.963		153	168		321	420
0650	Pull box, hinged, NEMA 1, 6" W x 6" H x 4" D		8	1		11.60	56.50		68.10	98
0800	12" W x 16" H x 6" D		4.70	1.702		57	96.50		153.50	207
1000	20" W x 20" H x 6" D		3.60	2.222		99	126		225	297
1200	20" W x 20" H x 8" D		3.20	2.500		149	142		291	375
1400	24" W x 36" H x 8" D		2.70	2.963		237	168		405	510
1600	24" W x 42" H x 8" D	↓	2	4	↓	350	226		576	725
2100	Pull box, NEMA 3R, type SC, raintight & weatherproof									
2150	6" L x 6" W x 6" D	1 Elec	10	.800	Ea.	16.95	45.50		62.45	86.50
2200	8" L x 6" W x 6" D		8	1		23.50	56.50		80	111
2250	10" L x 6" W x 6" D		7	1.143		32	64.50		96.50	132
2300	12" L x 12" W x 6" D		5	1.600		46.50	90.50		137	187
2350	16" L x 16" W x 6" D		4.50	1.778		89	101		190	249
2400	20" L x 20" W x 6" D		4	2		89	113		202	268
2450	24" L x 18" W x 8" D		3	2.667		123	151		274	360
2500	24" L x 24" W x 10" D		2.50	3.200		247	181		428	545
2550	30" L x 24" W x 12" D		2	4		385	226		611	765
2600	36" L x 36" W x 12" D	↓	1.50	5.333	↓	430	300		730	925
2800	Cast iron, pull boxes for surface mounting									
3000	NEMA 4, watertight & dust tight									
3050	6" L x 6" W x 6" D	1 Elec	4	2	Ea.	267	113		380	465
3100	8" L x 6" W x 6" D		3.20	2.500		370	142		512	620
3150	10" L x 6" W x 6" D		2.50	3.200		415	181		596	730
3200	12" L x 12" W x 6" D		2.30	3.478		760	197		957	1,125
3250	16" L x 16" W x 6" D		1.30	6.154		1,100	350		1,450	1,750
3300	20" L x 20" W x 6" D		.80	10		1,525	565		2,090	2,525
3350	24" L x 18" W x 8" D		.70	11.429		2,600	645		3,245	3,825
3400	24" L x 24" W x 10" D		.50	16		4,900	905		5,805	6,750
3450	30" L x 24" W x 12" D		.40	20		6,275	1,125		7,400	8,600
3500	36" L x 36" W x 12" D	↓	.20	40	↓	5,075	2,275		7,350	9,000
6000	J.I.C. wiring boxes, NEMA 12, dust tight & drip tight									
6050	6" L x 8" W x 4" D	1 Elec	10	.800	Ea.	62.50	45.50		108	137
6100	8" L x 10" W x 4" D		8	1		77.50	56.50		134	170
6150	12" L x 14" W x 6" D		5.30	1.509		131	85.50		216.50	272
6200	14" L x 16" W x 6" D		4.70	1.702		154	96.50		250.50	315
6250	16" L x 20" W x 6" D		4.40	1.818		229	103		332	405
6300	24" L x 30" W x 6" D		3.20	2.500		335	142		477	580
6350	24" L x 30" W x 8" D		2.90	2.759		345	156		501	615
6400	24" L x 36" W x 8" D		2.70	2.963		385	168		553	675
6450	24" L x 42" W x 8" D	↓	2.30	3.478	↓	435	197		632	770

26 05 33 – Raceway and Boxes for Electrical Systems

26 05 33.18 Pull Boxes

		Crew	Daily Output	Labor-Hours	Unit	Material	2017 Bare Costs Labor	Equipment	Total	Total Incl O&P
6500	24" L x 48" W x 8" D	1 Elec	2	4	Ea.	475	226		701	860

26 05 33.23 Wireway

		Crew	Daily Output	Labor-Hours	Unit	Material	2017 Bare Costs Labor	Equipment	Total	Total Incl O&P
0010	**WIREWAY** to 10' high									
0100	NEMA 1, Screw cover w/fittings and supports, 2-1/2" x 2-1/2"	1 Elec	45	.178	L.F.	10.80	10.05		20.85	27
0200	4" x 4"	"	40	.200		11.65	11.30		22.95	30
0400	6" x 6"	2 Elec	60	.267		19.60	15.10		34.70	44
0600	8" x 8"	"	40	.400		28.50	22.50		51	65.50
4475	NEMA 3R, Screw cover w/fittings and supports, 4" x 4"	1 Elec	36	.222		14	12.60		26.60	34.50
4480	6" x 6"	2 Elec	55	.291		18.75	16.45		35.20	45
4485	8" x 8"		36	.444		29	25		54	69.50
4490	12" x 12"		18	.889		46.50	50.50		97	127

26 05 36 – Cable Trays for Electrical Systems

26 05 36.10 Cable Tray Ladder Type

		Crew	Daily Output	Labor-Hours	Unit	Material	2017 Bare Costs Labor	Equipment	Total	Total Incl O&P
0010	**CABLE TRAY LADDER TYPE** w/ftngs. & supports, 4" dp., to 15' elev.									
0160	Galvanized steel tray									
0170	4" rung spacing, 6" wide	2 Elec	98	.163	L.F.	14.80	9.25		24.05	30
0200	12" wide		86	.186		17.80	10.55		28.35	35.50
0400	18" wide		82	.195		20.50	11.05		31.55	39
0600	24" wide		78	.205		24	11.60		35.60	43.50
3200	Aluminum tray, 4" deep, 6" rung spacing, 6" wide		134	.119		14.55	6.75		21.30	26
3220	12" wide		124	.129		16.30	7.30		23.60	29
3230	18" wide		114	.140		18.15	7.95		26.10	32
3240	24" wide		106	.151		21	8.55		29.55	36

26 05 39 – Underfloor Raceways for Electrical Systems

26 05 39.30 Conduit In Concrete Slab

		Crew	Daily Output	Labor-Hours	Unit	Material	2017 Bare Costs Labor	Equipment	Total	Total Incl O&P
0010	**CONDUIT IN CONCRETE SLAB** Including terminations,									
0020	fittings and supports									
3230	PVC, schedule 40, 1/2" diameter	1 Elec	270	.030	L.F.	.49	1.68		2.17	3.05
3250	3/4" diameter		230	.035		.56	1.97		2.53	3.57
3270	1" diameter		200	.040		.76	2.26		3.02	4.23
3300	1-1/4" diameter		170	.047		1	2.66		3.66	5.10
3330	1-1/2" diameter		140	.057		1.21	3.23		4.44	6.20
3350	2" diameter		120	.067		1.53	3.77		5.30	7.35
4350	Rigid galvanized steel, 1/2" diameter		200	.040		2.37	2.26		4.63	6
4400	3/4" diameter		170	.047		4.34	2.66		7	8.75
4450	1" diameter		130	.062		6.70	3.48		10.18	12.55
4500	1-1/4" diameter		110	.073		4.78	4.12		8.90	11.40
4600	1-1/2" diameter		100	.080		7.30	4.53		11.83	14.80
4800	2" diameter		90	.089		9.05	5.05		14.10	17.50

26 05 39.40 Conduit In Trench

		Crew	Daily Output	Labor-Hours	Unit	Material	2017 Bare Costs Labor	Equipment	Total	Total Incl O&P
0010	**CONDUIT IN TRENCH** Includes terminations and fittings									
0020	Does not include excavation or backfill, see Section 31 23 16									
0200	Rigid galvanized steel, 2" diameter	1 Elec	150	.053	L.F.	8.70	3.02		11.72	14.05
0400	2-1/2" diameter	"	100	.080		12.10	4.53		16.63	20
0600	3" diameter	2 Elec	160	.100		14.10	5.65		19.75	24
0800	3-1/2" diameter		140	.114		18.35	6.45		24.80	29.50
1000	4" diameter		100	.160		21	9.05		30.05	36.50
1200	5" diameter		80	.200		42	11.30		53.30	63
1400	6" diameter		60	.267		58.50	15.10		73.60	86.50

26 05 Common Work Results for Electrical

26 05 43 – Underground Ducts and Raceways for Electrical Systems

26 05 43.10 Trench Duct

	26 05 43.10 Trench Duct	Crew	Daily Output	Labor-Hours	Unit	Material	2017 Bare Costs Labor	Equipment	Total	Total Incl O&P
0010	**TRENCH DUCT** Steel with cover									
0020	Standard adjustable, depths to 4"									
0100	Straight, single compartment, 9" wide	2 Elec	40	.400	L.F.	224	22.50		246.50	280
0200	12" wide		32	.500		270	28.50		298.50	340
0400	18" wide		26	.615		325	35		360	410
0600	24" wide		22	.727		390	41		431	490
0800	30" wide		20	.800		228	45.50		273.50	320
1000	36" wide		16	1		262	56.50		318.50	375
1200	Horizontal elbow, 9" wide		5.40	2.963	Ea.	425	168		593	720
1400	12" wide		4.60	3.478		455	197		652	795
1600	18" wide		4	4		600	226		826	1,000
1800	24" wide		3.20	5		855	283		1,138	1,375
2000	30" wide		2.60	6.154		1,125	350		1,475	1,775
2200	36" wide		2.40	6.667		1,500	375		1,875	2,225
2400	Vertical elbow, 9" wide		5.40	2.963		171	168		339	440
2600	12" wide		4.60	3.478		161	197		358	470
2800	18" wide		4	4		185	226		411	545
3000	24" wide		3.20	5		230	283		513	680
3200	30" wide		2.60	6.154		253	350		603	800
3400	36" wide		2.40	6.667		279	375		654	870
3600	Cross, 9" wide		4	4		695	226		921	1,100
3800	12" wide		3.20	5		735	283		1,018	1,225
4000	18" wide		2.60	6.154		880	350		1,230	1,500
4200	24" wide		2.20	7.273		1,100	410		1,510	1,825
4400	30" wide		2	8		1,400	455		1,855	2,225
4600	36" wide		1.80	8.889		1,750	505		2,255	2,675
4800	End closure, 9" wide		14.40	1.111		42	63		105	140
5000	12" wide		12	1.333		48	75.50		123.50	166
5200	18" wide		10	1.600		73.50	90.50		164	217
5400	24" wide		8	2		97	113		210	277
5600	30" wide		6.60	2.424		148	137		285	370
5800	36" wide		5.80	2.759		144	156		300	395
6000	Tees, 9" wide		4	4		405	226		631	790
6200	12" wide		3.60	4.444		475	252		727	895
6400	18" wide		3.20	5		605	283		888	1,100
6600	24" wide		3	5.333		870	300		1,170	1,400
6800	30" wide		2.60	6.154		1,125	350		1,475	1,775
7000	36" wide		2	8		1,475	455		1,930	2,300
7200	Riser, and cabinet connector, 9" wide		5.40	2.963		178	168		346	445
7400	12" wide		4.60	3.478		208	197		405	525
7600	18" wide		4	4		209	226		435	570
7800	24" wide		3.20	5		310	283		593	765
8000	30" wide		2.60	6.154		355	350		705	910
8200	36" wide		2	8		415	455		870	1,125
8400	Insert assembly, cell to conduit adapter, 1-1/4"	1 Elec	16	.500		71	28.50		99.50	121

26 05 43.20 Underfloor Duct

	26 05 43.20 Underfloor Duct	Crew	Daily Output	Labor-Hours	Unit	Material	2017 Bare Costs Labor	Equipment	Total	Total Incl O&P
0010	**UNDERFLOOR DUCT**									
0100	Duct, 1-3/8" x 3-1/8" blank, standard	2 Elec	160	.100	L.F.	12.70	5.65		18.35	22.50
0200	1-3/8" x 7-1/4" blank, super duct		120	.133		29.50	7.55		37.05	44
0400	7/8" or 1-3/8" insert type, 24" O.C., 1-3/8" x 3-1/8", std.		140	.114		19.90	6.45		26.35	31.50
0600	1-3/8" x 7-1/4", super duct		100	.160		34.50	9.05		43.55	51.50
0800	Junction box, single duct, 1 level, 3-1/8"	1 Elec	4	2	Ea.	445	113		558	660

For customer support on your Building Construction Costs with RSMeans Data, call 800.448.8182.

575

26 05 43 – Underground Ducts and Raceways for Electrical Systems

26 05 43.20 Underfloor Duct		Crew	Daily Output	Labor-Hours	Unit	Material	2017 Bare Costs Labor	Equipment	Total	Total Incl O&P
1000	Junction box, single duct, 1 level, 7-1/4"	1 Elec	2.70	2.963	Ea.	520	168		688	820
1200	1 level, 2 duct, 3-1/8"		3.20	2.500		585	142		727	850
1400	Junction box, 1 level, 2 duct, 7-1/4"		2.30	3.478		1,450	197		1,647	1,875
1580	Junction box, 1 level, one 3-1/8" + one 7-1/4" x same		2.30	3.478		995	197		1,192	1,400
1600	Triple duct, 3-1/8"		2.30	3.478		995	197		1,192	1,400
1800	Insert to conduit adapter, 3/4" & 1"		32	.250		37	14.15		51.15	62
2000	Support, single cell		27	.296		55	16.75		71.75	85.50
2200	Super duct		16	.500		56	28.50		84.50	104
2400	Double cell		16	.500		55	28.50		83.50	103
2600	Triple cell		11	.727		53	41		94	120
2800	Vertical elbow, standard duct		10	.800		98.50	45.50		144	176
3000	Super duct		8	1		98.50	56.50		155	193
3200	Cabinet connector, standard duct		32	.250		77.50	14.15		91.65	107
3400	Super duct		27	.296		75.50	16.75		92.25	108
3600	Conduit adapter, 1" to 1-1/4"		32	.250		74.50	14.15		88.65	103
3800	2" to 1-1/4"		27	.296		89.50	16.75		106.25	124
4000	Outlet, low tension (tele, computer, etc.)		8	1		104	56.50		160.50	199
4200	High tension, receptacle (120 volt)		8	1		108	56.50		164.50	204

26 05 80 – Wiring Connections

26 05 80.10 Motor Connections

0010	**MOTOR CONNECTIONS**									
0020	Flexible conduit and fittings, 115 volt, 1 phase, up to 1 HP motor	1 Elec	8	1	Ea.	4.92	56.50		61.42	90.50
0110	230 volt, 3 phase, 3 HP motor		6.78	1.180		6.20	67		73.20	107
0112	5 HP motor		5.47	1.463		5.05	83		88.05	130
0114	7-1/2 HP motor		4.61	1.735		8.05	98		106.05	156
0120	10 HP motor		4.20	1.905		17.20	108		125.20	181
0200	25 HP motor		2.70	2.963		23	168		191	276
0400	50 HP motor		2.20	3.636		47.50	206		253.50	365
0600	100 HP motor		1.50	5.333		107	300		407	570

26 05 90 – Residential Applications

26 05 90.10 Residential Wiring

0010	**RESIDENTIAL WIRING**									
0020	20' avg. runs and #14/2 wiring incl. unless otherwise noted									
1000	Service & panel, includes 24' SE-AL cable, service eye, meter,									
1010	Socket, panel board, main bkr., ground rod, 15 or 20 amp									
1020	1-pole circuit breakers, and misc. hardware									
1100	100 amp, with 10 branch breakers	1 Elec	1.19	6.723	Ea.	490	380		870	1,100
1110	With PVC conduit and wire		.92	8.696		530	490		1,020	1,325
1120	With RGS conduit and wire		.73	10.959		745	620		1,365	1,750
1150	150 amp, with 14 branch breakers		1.03	7.767		845	440		1,285	1,600
1170	With PVC conduit and wire		.82	9.756		930	550		1,480	1,850
1180	With RGS conduit and wire		.67	11.940		1,250	675		1,925	2,400
1200	200 amp, with 18 branch breakers	2 Elec	1.80	8.889		1,050	505		1,555	1,925
1220	With PVC conduit and wire		1.46	10.959		1,150	620		1,770	2,175
1230	With RGS conduit and wire		1.24	12.903		1,550	730		2,280	2,800
1800	Lightning surge suppressor	1 Elec	32	.250		50.50	14.15		64.65	77
2000	Switch devices									
2100	Single pole, 15 amp, Ivory, with a 1-gang box, cover plate,									
2110	Type NM (Romex) cable	1 Elec	17.10	.468	Ea.	18.70	26.50		45.20	60
2120	Type MC (BX) cable		14.30	.559		27.50	31.50		59	77.50
2130	EMT & wire		5.71	1.401		36	79.50		115.50	159
2150	3-way, #14/3, type NM cable		14.55	.550		12.40	31		43.40	60

26 05 90.10 Residential Wiring	Crew	Daily Output	Labor-Hours	Unit	Material	2017 Bare Costs Labor	Equipment	Total	Total Incl O&P
2170	Type MC cable	1 Elec	12.31	.650	Ea.	24.50	37		61.50	82
2180	EMT & wire		5	1.600		29	90.50		119.50	168
2200	4-way, #14/3, type NM cable		14.55	.550		21	31		52	69.50
2220	Type MC cable		12.31	.650		33	37		70	91.50
2230	EMT & wire		5	1.600		37.50	90.50		128	177
2250	S.P., 20 amp, #12/2, type NM cable		13.33	.600		14.75	34		48.75	67
2270	Type MC cable		11.43	.700		21	39.50		60.50	82.50
2280	EMT & wire		4.85	1.649		33	93.50		126.50	177
2290	S.P. rotary dimmer, 600W, no wiring		17	.471		32.50	26.50		59	75.50
2300	S.P. rotary dimmer, 600W, type NM cable		14.55	.550		36.50	31		67.50	87
2320	Type MC cable		12.31	.650		45.50	37		82.50	105
2330	EMT & wire		5	1.600		55	90.50		145.50	197
2350	3-way rotary dimmer, type NM cable		13.33	.600		21	34		55	74
2370	Type MC cable		11.43	.700		29.50	39.50		69	92
2380	EMT & wire		4.85	1.649		39.50	93.50		133	184
2400	Interval timer wall switch, 20 amp, 1-30 min., #12/2									
2410	Type NM cable	1 Elec	14.55	.550	Ea.	57	31		88	110
2420	Type MC cable		12.31	.650		62	37		99	123
2430	EMT & wire		5	1.600		75.50	90.50		166	219
2500	Decorator style									
2510	S.P., 15 amp, type NM cable	1 Elec	17.10	.468	Ea.	23.50	26.50		50	65.50
2520	Type MC cable		14.30	.559		32.50	31.50		64	83
2530	EMT & wire		5.71	1.401		40.50	79.50		120	164
2550	3-way, #14/3, type NM cable		14.55	.550		17.30	31		48.30	65.50
2570	Type MC cable		12.31	.650		29.50	37		66.50	87.50
2580	EMT & wire		5	1.600		34	90.50		124.50	174
2600	4-way, #14/3, type NM cable		14.55	.550		25.50	31		56.50	75
2620	Type MC cable		12.31	.650		38	37		75	97
2630	EMT & wire		5	1.600		42.50	90.50		133	183
2650	S.P., 20 amp, #12/2, type NM cable		13.33	.600		19.65	34		53.65	72.50
2670	Type MC cable		11.43	.700		26	39.50		65.50	88
2680	EMT & wire		4.85	1.649		38	93.50		131.50	182
2700	S.P., slide dimmer, type NM cable		17.10	.468		24.50	26.50		51	66.50
2720	Type MC cable		14.30	.559		33	31.50		64.50	84
2730	EMT & wire		5.71	1.401		43	79.50		122.50	166
2750	S.P., touch dimmer, type NM cable		17.10	.468		41	26.50		67.50	84.50
2770	Type MC cable		14.30	.559		50	31.50		81.50	102
2780	EMT & wire		5.71	1.401		59.50	79.50		139	185
2800	3-way touch dimmer, type NM cable		13.33	.600		51.50	34		85.50	108
2820	Type MC cable		11.43	.700		60	39.50		99.50	126
2830	EMT & wire		4.85	1.649		70	93.50		163.50	217
3000	Combination devices									
3100	S.P. switch/15 amp recpt., Ivory, 1-gang box, plate									
3110	Type NM cable	1 Elec	11.43	.700	Ea.	24	39.50		63.50	86
3120	Type MC cable		10	.800		33	45.50		78.50	104
3130	EMT & wire		4.40	1.818		42.50	103		145.50	201
3150	S.P. switch/pilot light, type NM cable		11.43	.700		24.50	39.50		64	86.50
3170	Type MC cable		10	.800		33	45.50		78.50	105
3180	EMT & wire		4.43	1.806		43	102		145	200
3190	2-S.P. switches, 2-#14/2, no wiring		14	.571		13.70	32.50		46.20	63.50
3200	2-S.P. switches, 2-#14/2, type NM cables		10	.800		26.50	45.50		72	97.50
3220	Type MC cable		8.89	.900		40.50	51		91.50	121
3230	EMT & wire		4.10	1.951		45	110		155	215

26 05 90.10 Residential Wiring	Crew	Daily Output	Labor-Hours	Unit	Material	2017 Bare Costs Labor	2017 Bare Costs Equipment	Total	Total Incl O&P
3250	3-way switch/15 amp recpt., #14/3, type NM cable	1 Elec	10	.800	Ea.	31.50	45.50		77	103
3270	Type MC cable		8.89	.900		44	51		95	125
3280	EMT & wire		4.10	1.951		48.50	110		158.50	218
3300	2-3 way switches, 2-#14/3, type NM cables		8.89	.900		40.50	51		91.50	121
3320	Type MC cable		8	1		61.50	56.50		118	153
3330	EMT & wire		4	2		55	113		168	231
3350	S.P. switch/20 amp recpt., #12/2, type NM cable		10	.800		31	45.50		76.50	103
3370	Type MC cable		8.89	.900		36	51		87	116
3380	EMT & wire		4.10	1.951		49.50	110		159.50	220
3400	Decorator style									
3410	S.P. switch/15 amp recpt., type NM cable	1 Elec	11.43	.700	Ea.	29	39.50		68.50	91.50
3420	Type MC cable		10	.800		37.50	45.50		83	110
3430	EMT & wire		4.40	1.818		47.50	103		150.50	207
3450	S.P. switch/pilot light, type NM cable		11.43	.700		29	39.50		68.50	91.50
3470	Type MC cable		10	.800		38	45.50		83.50	110
3480	EMT & wire		4.40	1.818		47.50	103		150.50	207
3500	2-S.P. switches, 2-#14/2, type NM cables		10	.800		31.50	45.50		77	103
3520	Type MC cable		8.89	.900		45	51		96	126
3530	EMT & wire		4.10	1.951		50	110		160	220
3550	3-way/15 amp recpt., #14/3, type NM cable		10	.800		36.50	45.50		82	108
3570	Type MC cable		8.89	.900		49	51		100	130
3580	EMT & wire		4.10	1.951		53	110		163	224
3650	2-3 way switches, 2-#14/3, type NM cables		8.89	.900		45.50	51		96.50	127
3670	Type MC cable		8	1		66	56.50		122.50	158
3680	EMT & wire		4	2		60	113		173	236
3700	S.P. switch/20 amp recpt., #12/2, type NM cable		10	.800		36	45.50		81.50	108
3720	Type MC cable		8.89	.900		41	51		92	122
3730	EMT & wire		4.10	1.951		54.50	110		164.50	225
4000	Receptacle devices									
4010	Duplex outlet, 15 amp recpt., Ivory, 1-gang box, plate									
4015	Type NM cable	1 Elec	14.55	.550	Ea.	10.30	31		41.30	58
4020	Type MC cable		12.31	.650		19	37		56	76
4030	EMT & wire		5.33	1.501		27.50	85		112.50	157
4050	With #12/2, type NM cable		12.31	.650		12.65	37		49.65	69
4070	Type MC cable		10.67	.750		19	42.50		61.50	84.50
4080	EMT & wire		4.71	1.699		31	96		127	178
4100	20 amp recpt., #12/2, type NM cable		12.31	.650		19.30	37		56.30	76.50
4120	Type MC cable		10.67	.750		25.50	42.50		68	91.50
4130	EMT & wire		4.71	1.699		37.50	96		133.50	186
4140	For GFI see Section 26 05 90.10 line 4300 below									
4150	Decorator style, 15 amp recpt., type NM cable	1 Elec	14.55	.550	Ea.	15.20	31		46.20	63.50
4170	Type MC cable		12.31	.650		24	37		61	81.50
4180	EMT & wire		5.33	1.501		32.50	85		117.50	163
4200	With #12/2, type NM cable		12.31	.650		17.55	37		54.55	74.50
4220	Type MC cable		10.67	.750		24	42.50		66.50	90
4230	EMT & wire		4.71	1.699		36	96		132	184
4250	20 amp recpt., #12/2, type NM cable		12.31	.650		24	37		61	81.50
4270	Type MC cable		10.67	.750		30.50	42.50		73	97
4280	EMT & wire		4.71	1.699		42.50	96		138.50	191
4300	GFI, 15 amp recpt., type NM cable		12.31	.650		23	37		60	80
4320	Type MC cable		10.67	.750		31.50	42.50		74	98
4330	EMT & wire		4.71	1.699		40	96		136	188
4350	GFI with #12/2, type NM cable		10.67	.750		25	42.50		67.50	91

26 05 90.10 Residential Wiring	Crew	Daily Output	Labor-Hours	Unit	Material	2017 Bare Costs Labor	Equipment	Total	Total Incl O&P
4370	Type MC cable	1 Elec	9.20	.870	Ea.	31.50	49		80.50	108
4380	EMT & wire		4.21	1.900		43.50	108		151.50	209
4400	20 amp recpt., #12/2, type NM cable		10.67	.750		51.50	42.50		94	120
4420	Type MC cable		9.20	.870		58	49		107	137
4430	EMT & wire		4.21	1.900		70	108		178	238
4500	Weather-proof cover for above receptacles, add	▼	32	.250	▼	2.47	14.15		16.62	23.50
4550	Air conditioner outlet, 20 amp-240 volt recpt.									
4560	30' of #12/2, 2 pole circuit breaker									
4570	Type NM cable	1 Elec	10	.800	Ea.	62	45.50		107.50	137
4580	Type MC cable		9	.889		70	50.50		120.50	153
4590	EMT & wire		4	2		80.50	113		193.50	259
4600	Decorator style, type NM cable		10	.800		67	45.50		112.50	142
4620	Type MC cable		9	.889		74.50	50.50		125	158
4630	EMT & wire	▼	4	2	▼	85	113		198	264
4650	Dryer outlet, 30 amp-240 volt recpt., 20' of #10/3									
4660	2 pole circuit breaker									
4670	Type NM cable	1 Elec	6.41	1.248	Ea.	58	70.50		128.50	170
4680	Type MC cable		5.71	1.401		64	79.50		143.50	190
4690	EMT & wire	▼	3.48	2.299	▼	74	130		204	277
4700	Range outlet, 50 amp-240 volt recpt., 30' of #8/3									
4710	Type NM cable	1 Elec	4.21	1.900	Ea.	85	108		193	255
4720	Type MC cable		4	2		126	113		239	310
4730	EMT & wire		2.96	2.703		104	153		257	345
4750	Central vacuum outlet, Type NM cable		6.40	1.250		54	71		125	165
4770	Type MC cable		5.71	1.401		67	79.50		146.50	193
4780	EMT & wire	▼	3.48	2.299	▼	80.50	130		210.50	284
4800	30 amp-110 volt locking recpt., #10/2 circ. bkr.									
4810	Type NM cable	1 Elec	6.20	1.290	Ea.	62	73		135	177
4820	Type MC cable		5.40	1.481		78.50	84		162.50	213
4830	EMT & wire	▼	3.20	2.500	▼	91	142		233	310
4900	Low voltage outlets									
4910	Telephone recpt., 20' of 4/C phone wire	1 Elec	26	.308	Ea.	8.20	17.40		25.60	35
4920	TV recpt., 20' of RG59U coax wire, F type connector	"	16	.500	"	15.75	28.50		44.25	60
4950	Door bell chime, transformer, 2 buttons, 60' of bellwire									
4970	Economy model	1 Elec	11.50	.696	Ea.	49.50	39.50		89	114
4980	Custom model		11.50	.696		99	39.50		138.50	168
4990	Luxury model, 3 buttons	▼	9.50	.842	▼	201	47.50		248.50	293
6000	Lighting outlets									
6050	Wire only (for fixture), type NM cable	1 Elec	32	.250	Ea.	6.35	14.15		20.50	28
6070	Type MC cable		24	.333		12.55	18.85		31.40	42.50
6080	EMT & wire		10	.800		20	45.50		65.50	90
6100	Box (4"), and wire (for fixture), type NM cable		25	.320		13.90	18.10		32	42.50
6120	Type MC cable		20	.400		20	22.50		42.50	56
6130	EMT & wire	▼	11	.727	▼	27.50	41		68.50	92
6200	Fixtures (use with line 6050 or 6100 above)									
6210	Canopy style, economy grade	1 Elec	40	.200	Ea.	27	11.30		38.30	46.50
6220	Custom grade		40	.200		53.50	11.30		64.80	76
6250	Dining room chandelier, economy grade		19	.421		67.50	24		91.50	110
6260	Custom grade		19	.421		315	24		339	380
6270	Luxury grade		15	.533		820	30		850	950
6310	Kitchen fixture (fluorescent), economy grade		30	.267		71.50	15.10		86.60	102
6320	Custom grade		25	.320		221	18.10		239.10	271
6350	Outdoor, wall mounted, economy grade		30	.267		25.50	15.10		40.60	50.50

26 05 90.10 Residential Wiring	Crew	Daily Output	Labor-Hours	Unit	Material	2017 Bare Costs Labor	Equipment	Total	Total Incl O&P
6360 Custom grade	1 Elec	30	.267	Ea.	120	15.10		135.10	155
6370 Luxury grade		25	.320		248	18.10		266.10	300
6410 Outdoor PAR floodlights, 1 lamp, 150 watt		20	.400		27.50	22.50		50	64
6420 2 lamp, 150 watt each		20	.400		45.50	22.50		68	84
6425 Motion sensing, 2 lamp, 150 watt each		20	.400		75.50	22.50		98	117
6430 For infrared security sensor, add		32	.250		132	14.15		146.15	166
6450 Outdoor, quartz-halogen, 300 watt flood		20	.400		33	22.50		55.50	70.50
6600 Recessed downlight, round, pre-wired, 50 or 75 watt trim		30	.267		68	15.10		83.10	97
6610 With shower light trim		30	.267		82.50	15.10		97.60	113
6620 With wall washer trim		28	.286		85	16.15		101.15	118
6630 With eye-ball trim		28	.286		84.50	16.15		100.65	117
6700 Porcelain lamp holder		40	.200		2.98	11.30		14.28	20
6710 With pull switch		40	.200		6.55	11.30		17.85	24
6750 Fluorescent strip, 2-20 watt tube, wrap around diffuser, 24"		24	.333		42	18.85		60.85	75
6760 1-34 watt tube, 48"		24	.333		100	18.85		118.85	139
6770 2-34 watt tubes, 48"		20	.400		118	22.50		140.50	164
6800 Bathroom heat lamp, 1-250 watt		28	.286		38.50	16.15		54.65	66
6810 2-250 watt lamps		28	.286		61	16.15		77.15	91.50
6820 For timer switch, see Section 26 05 90.10 line 2400									
6900 Outdoor post lamp, incl. post, fixture, 35' of #14/2									
6910 Type NM cable	1 Elec	3.50	2.286	Ea.	285	129		414	510
6920 Photo-eye, add		27	.296		27.50	16.75		44.25	55
6950 Clock dial time switch, 24 hr., w/enclosure, type NM cable		11.43	.700		73	39.50		112.50	140
6970 Type MC cable		11	.727		81.50	41		122.50	152
6980 EMT & wire		4.85	1.649		90	93.50		183.50	239
7000 Alarm systems									
7050 Smoke detectors, box, #14/3, type NM cable	1 Elec	14.55	.550	Ea.	34	31		65	84
7070 Type MC cable		12.31	.650		44.50	37		81.50	104
7080 EMT & wire		5	1.600		49	90.50		139.50	190
7090 For relay output to security system, add					10.20			10.20	11.20
8000 Residential equipment									
8050 Disposal hook-up, incl. switch, outlet box, 3' of flex									
8060 20 amp-1 pole circ. bkr., and 25' of #12/2									
8070 Type NM cable	1 Elec	10	.800	Ea.	36	45.50		81.50	108
8080 Type MC cable		8	1		43	56.50		99.50	133
8090 EMT & wire		5	1.600		57.50	90.50		148	199
8100 Trash compactor or dishwasher hook-up, incl. outlet box,									
8110 3' of flex, 15 amp-1 pole circ. bkr., and 25' of #14/2									
8120 Type NM cable	1 Elec	10	.800	Ea.	21	45.50		66.50	91
8130 Type MC cable		8	1		30.50	56.50		87	119
8140 EMT & wire		5	1.600		42.50	90.50		133	183
8150 Hot water sink dispenser hook-up, use line 8100									
8200 Vent/exhaust fan hook-up, type NM cable	1 Elec	32	.250	Ea.	6.35	14.15		20.50	28
8220 Type MC cable		24	.333		12.55	18.85		31.40	42.50
8230 EMT & wire		10	.800		20	45.50		65.50	90
8250 Bathroom vent fan, 50 CFM (use with above hook-up)									
8260 Economy model	1 Elec	15	.533	Ea.	20	30		50	67.50
8270 Low noise model		15	.533		43.50	30		73.50	92.50
8280 Custom model		12	.667		131	37.50		168.50	201
8300 Bathroom or kitchen vent fan, 110 CFM									
8310 Economy model	1 Elec	15	.533	Ea.	67.50	30		97.50	120
8320 Low noise model	"	15	.533	"	130	30		160	188
8350 Paddle fan, variable speed (w/o lights)									

26 05 Common Work Results for Electrical

26 05 90 – Residential Applications

26 05 90.10 Residential Wiring	Crew	Daily Output	Labor-Hours	Unit	Material	2017 Bare Costs Labor	Equipment	Total	Total Incl O&P
8360	Economy model (AC motor)	1 Elec	10	.800	Ea.	109	45.50		154.50	188
8362	With light kit		10	.800		150	45.50		195.50	233
8370	Custom model (AC motor)		10	.800		262	45.50		307.50	355
8372	With light kit		10	.800		305	45.50		350.50	405
8380	Luxury model (DC motor)		8	1		288	56.50		344.50	400
8382	With light kit		8	1		330	56.50		386.50	445
8390	Remote speed switch for above, add		12	.667		33.50	37.50		71	93.50
8500	Whole house exhaust fan, ceiling mount, 36", variable speed									
8510	Remote switch, incl. shutters, 20 amp-1 pole circ. bkr.									
8520	30' of #12/2, type NM cable	1 Elec	4	2	Ea.	1,350	113		1,463	1,650
8530	Type MC cable		3.50	2.286		1,350	129		1,479	1,675
8540	EMT & wire		3	2.667		1,375	151		1,526	1,725
8600	Whirlpool tub hook-up, incl. timer switch, outlet box									
8610	3' of flex, 20 amp-1 pole GFI circ. bkr.									
8620	30' of #12/2, type NM cable	1 Elec	5	1.600	Ea.	113	90.50		203.50	260
8630	Type MC cable		4.20	1.905		116	108		224	290
8640	EMT & wire		3.40	2.353		128	133		261	340
8650	Hot water heater hook-up, incl. 1-2 pole circ. bkr., box;									
8660	3' of flex, 20' of #10/2, type NM cable	1 Elec	5	1.600	Ea.	31	90.50		121.50	170
8670	Type MC cable		4.20	1.905		42	108		150	208
8680	EMT & wire		3.40	2.353		46	133		179	251
9000	Heating/air conditioning									
9050	Furnace/boiler hook-up, incl. firestat, local on-off switch									
9060	Emergency switch, and 40' of type NM cable	1 Elec	4	2	Ea.	52	113		165	227
9070	Type MC cable		3.50	2.286		65.50	129		194.50	266
9080	EMT & wire		1.50	5.333		83	300		383	540
9100	Air conditioner hook-up, incl. local 60 amp disc. switch									
9110	3' sealtite, 40 amp, 2 pole circuit breaker									
9130	40' of #8/2, type NM cable	1 Elec	3.50	2.286	Ea.	146	129		275	355
9140	Type MC cable		3	2.667		206	151		357	455
9150	EMT & wire		1.30	6.154		186	350		536	725
9200	Heat pump hook-up, 1-40 & 1-100 amp 2 pole circ. bkr.									
9210	Local disconnect switch, 3' sealtite									
9220	40' of #8/2 & 30' of #3/2									
9230	Type NM cable	1 Elec	1.30	6.154	Ea.	460	350		810	1,025
9240	Type MC cable		1.08	7.407		515	420		935	1,200
9250	EMT & wire		.94	8.511		520	480		1,000	1,300
9500	Thermostat hook-up, using low voltage wire									
9520	Heating only, 25' of #18-3	1 Elec	24	.333	Ea.	8	18.85		26.85	37.50
9530	Heating/cooling, 25' of #18-4	"	20	.400	"	10.50	22.50		33	45.50

26 09 13 – Electrical Power Monitoring

26 09 13.10 Switchboard Instruments

	26 09 13.10 Switchboard Instruments		Crew	Daily Output	Labor-Hours	Unit	Material	2017 Bare Costs Labor	Equipment	Total	Total Incl O&P
0010	**SWITCHBOARD INSTRUMENTS** 3 phase, 4 wire										
0100	AC indicating, ammeter & switch		1 Elec	8	1	Ea.	2,875	56.50		2,931.50	3,250
0200	Voltmeter & switch			8	1		2,900	56.50		2,956.50	3,250
0300	Wattmeter			8	1		4,150	56.50		4,206.50	4,650
0400	AC recording, ammeter			4	2		7,400	113		7,513	8,300
0500	Voltmeter			4	2		7,400	113		7,513	8,300
0600	Ground fault protection, zero sequence			2.70	2.963		6,525	168		6,693	7,425
0700	Ground return path			2.70	2.963		6,525	168		6,693	7,425
0800	3 current transformers, 5 to 800 amp			2	4		3,050	226		3,276	3,700
0900	1000 to 1500 amp			1.30	6.154		4,375	350		4,725	5,325
1200	2000 to 4000 amp			1	8		5,150	455		5,605	6,350
1300	Fused potential transformer, maximum 600 volt			8	1		1,150	56.50		1,206.50	1,325

26 09 13.30 Smart Metering

	26 09 13.30 Smart Metering		Crew	Daily Output	Labor-Hours	Unit	Material	2017 Bare Costs Labor	Equipment	Total	Total Incl O&P
0010	**SMART METERING**, In panel										
0100	Single phase, 120/208 volt, 100 amp	G	1 Elec	8.78	.911	Ea.	380	51.50		431.50	500
0120	200 amp	G		8.78	.911		380	51.50		431.50	500
0200	277 volt, 100 amp	G		8.78	.911		405	51.50		456.50	530
0220	200 amp	G		8.78	.911		405	51.50		456.50	530
1100	Three phase, 120/208 volt, 100 amp	G		4.69	1.706		795	96.50		891.50	1,025
1120	200 amp	G		4.69	1.706		795	96.50		891.50	1,025
1130	400 amp	G		4.69	1.706		795	96.50		891.50	1,025
1140	800 amp	G		4.69	1.706		795	96.50		891.50	1,025
1150	1600 amp	G		4.69	1.706		795	96.50		891.50	1,025
1200	277/480 volt, 100 amp	G		4.69	1.706		850	96.50		946.50	1,075
1220	200 amp	G		4.69	1.706		850	96.50		946.50	1,075
1230	400 amp	G		4.69	1.706		850	96.50		946.50	1,075
1240	800 amp	G		4.69	1.706		850	96.50		946.50	1,075
1250	1600 amp	G		4.69	1.706		850	96.50		946.50	1,075
2000	Data recorder, 8 meters	G		10.97	.729		1,600	41.50		1,641.50	1,800
2100	16 meters	G		8.53	.938		4,000	53		4,053	4,475
3000	Software package, per meter, basic	G					246			246	271
3100	Premium	G					680			680	750

26 09 23 – Lighting Control Devices

26 09 23.10 Energy Saving Lighting Devices

	26 09 23.10 Energy Saving Lighting Devices		Crew	Daily Output	Labor-Hours	Unit	Material	2017 Bare Costs Labor	Equipment	Total	Total Incl O&P
0010	**ENERGY SAVING LIGHTING DEVICES**										
0100	Occupancy sensors, passive infrared ceiling mounted	G	1 Elec	7	1.143	Ea.	79	64.50		143.50	184
0110	Ultrasonic ceiling mounted	G		7	1.143		89	64.50		153.50	195
0120	Dual technology ceiling mounted	G		6.50	1.231		165	69.50		234.50	285
0150	Automatic wall switches	G		24	.333		75	18.85		93.85	111
0160	Daylighting sensor, manual control, ceiling mounted	G		7	1.143		129	64.50		193.50	239
0170	Remote and dimming control with remote controller	G		6.50	1.231		159	69.50		228.50	279
0200	Passive infared ceiling mounted			6.50	1.231		32.50	69.50		102	140
0400	Remote power pack	G		10	.800		32.50	45.50		78	104
0450	Photoelectric control, S.P.S.T. 120 V	G		8	1		22.50	56.50		79	110
0500	S.P.S.T. 208 V/277 V	G		8	1		28.50	56.50		85	117
0550	D.P.S.T. 120 V	G		6	1.333		217	75.50		292.50	350
0600	D.P.S.T. 208 V/277 V	G		6	1.333		202	75.50		277.50	335
0650	S.P.D.T. 208 V/277 V	G		6	1.333		210	75.50		285.50	345
0660	Daylight level sensor, wall mounted, on/off or dimming	G		8	1		135	56.50		191.50	234

26 12 Medium-Voltage Transformers

26 12 19 – Pad-Mounted, Liquid-Filled, Medium-Voltage Transformers

26 12 19.10 Transformer, Oil-Filled	Crew	Daily Output	Labor-Hours	Unit	Material	2017 Bare Costs Labor	Equipment	Total	Total Incl O&P
0010 **TRANSFORMER, OIL-FILLED** primary delta or Y,									
0050 Pad mounted 5 kV or 15 kV, with taps, 277/480 V secondary, 3 phase									
0100 150 kVA	R-3	.65	30.769	Ea.	9,550	1,750	212	11,512	13,400
0200 300 kVA		.45	44.444		13,600	2,525	305	16,430	19,100
0300 500 kVA		.40	50		19,400	2,825	345	22,570	25,900
0400 750 kVA		.38	52.632		24,500	2,975	365	27,840	31,900
0500 1000 kVA		.26	76.923		29,000	4,350	530	33,880	39,100
0600 1500 kVA		.23	86.957		34,500	4,925	600	40,025	46,100
0700 2000 kVA		.20	100		43,600	5,650	690	49,940	57,000
0800 3750 kVA	▼	.16	125	▼	81,500	7,075	865	89,440	101,500

26 22 Low-Voltage Transformers

26 22 13 – Low-Voltage Distribution Transformers

26 22 13.10 Transformer, Dry-Type

26 22 13.10 Transformer, Dry-Type	Crew	Daily Output	Labor-Hours	Unit	Material	2017 Bare Costs Labor	Equipment	Total	Total Incl O&P
0010 **TRANSFORMER, DRY-TYPE**									
0050 Single phase, 240/480 volt primary, 120/240 volt secondary									
0100 1 kVA	1 Elec	2	4	Ea.	375	226		601	750
0300 2 kVA		1.60	5		580	283		863	1,050
0500 3 kVA		1.40	5.714		720	325		1,045	1,275
0700 5 kVA	▼	1.20	6.667		980	375		1,355	1,650
0900 7.5 kVA	2 Elec	2.20	7.273		1,350	410		1,760	2,100
1100 10 kVA		1.60	10		1,550	565		2,115	2,550
1300 15 kVA		1.20	13.333		2,075	755		2,830	3,400
1500 25 kVA		1	16		2,325	905		3,230	3,900
1700 37.5 kVA		.80	20		2,825	1,125		3,950	4,800
1900 50 kVA		.70	22.857		3,425	1,300		4,725	5,725
2100 75 kVA		.65	24.615		4,425	1,400		5,825	6,925
2190 480 V primary, 120/240 V secondary, nonvent., 15 kVA		1.20	13.333		1,625	755		2,380	2,925
2200 25 kVA		.90	17.778		2,425	1,000		3,425	4,150
2210 37 kVA		.75	21.333		2,875	1,200		4,075	4,950
2220 50 kVA	▼	.65	24.615	▼	3,375	1,400		4,775	5,800
2300 3 phase, 480 volt primary, 120/208 volt secondary									
2310 Ventilated, 3 kVA	1 Elec	1	8	Ea.	1,100	455		1,555	1,900
2700 6 kVA		.80	10		1,250	565		1,815	2,225
2900 9 kVA	▼	.70	11.429		1,250	645		1,895	2,350
3100 15 kVA	2 Elec	1.10	14.545		1,575	825		2,400	2,975
3300 30 kVA		.90	17.778		1,575	1,000		2,575	3,225
3500 45 kVA		.80	20		1,875	1,125		3,000	3,750
3700 75 kVA	▼	.70	22.857		2,500	1,300		3,800	4,700
3900 112.5 kVA	R-3	.90	22.222		4,150	1,250	153	5,553	6,625
4100 150 kVA		.85	23.529		5,025	1,325	162	6,512	7,700
4300 225 kVA		.65	30.769		7,325	1,750	212	9,287	10,900
4500 300 kVA		.55	36.364		9,175	2,050	251	11,476	13,500
4700 500 kVA		.45	44.444		15,300	2,525	305	18,130	20,900
4800 750 kVA	▼	.35	57.143	▼	22,200	3,225	395	25,820	29,700

26 24 Switchboards and Panelboards

26 24 13 – Switchboards

26 24 13.10 Incoming Switchboards

		Crew	Daily Output	Labor-Hours	Unit	Material	2017 Bare Costs Labor	Equipment	Total	Total Incl O&P
0010	**INCOMING SWITCHBOARDS** main service section									
0100	Aluminum bus bars, not including CT's or PT's									
0200	No main disconnect, includes CT compartment									
0300	120/208 volt, 4 wire, 600 amp	2 Elec	1	16	Ea.	4,050	905		4,955	5,800
0400	800 amp		.88	18.182		4,050	1,025		5,075	6,000
0500	1000 amp		.80	20		4,875	1,125		6,000	7,050
0600	1200 amp		.72	22.222		4,875	1,250		6,125	7,225
0700	1600 amp		.66	24.242		4,875	1,375		6,250	7,400
0800	2000 amp		.62	25.806		5,250	1,450		6,700	7,975
1000	3000 amp		.56	28.571		6,925	1,625		8,550	10,000
2000	Fused switch & CT compartment									
2100	120/208 volt, 4 wire, 400 amp	2 Elec	1.12	14.286	Ea.	2,700	810		3,510	4,175
2200	600 amp		.94	17.021		3,225	965		4,190	4,975
2300	800 amp		.84	19.048		11,000	1,075		12,075	13,600
2400	1200 amp		.68	23.529		14,300	1,325		15,625	17,700
2900	Pressure switch & CT compartment									
3000	120/208 volt, 4 wire, 800 amp	2 Elec	.80	20	Ea.	9,850	1,125		10,975	12,500
3100	1200 amp		.66	24.242		19,000	1,375		20,375	23,100
3200	1600 amp		.62	25.806		20,300	1,450		21,750	24,500
3300	2000 amp		.56	28.571		21,500	1,625		23,125	26,100
4400	Circuit breaker, molded case & CT compartment									
4600	3 pole, 4 wire, 600 amp	2 Elec	.94	17.021	Ea.	8,475	965		9,440	10,800
4800	800 amp		.84	19.048		10,100	1,075		11,175	12,800
5000	1200 amp		.68	23.529		13,800	1,325		15,125	17,200
5100	Copper bus bars, not incl. CT's or PT's, add, minimum					15%				

26 24 13.30 Distribution Switchboards Section

		Crew	Daily Output	Labor-Hours	Unit	Material	2017 Bare Costs Labor	Equipment	Total	Total Incl O&P
0010	**DISTRIBUTION SWITCHBOARDS SECTION**									
0100	Aluminum bus bars, not including breakers									
0195	120/208 or 277/480 volt, 4 wire, 400 amp	2 Elec	1.10	14.545	Ea.	1,550	825		2,375	2,925
0200	600 amp		1	16		1,450	905		2,355	2,925
0300	800 amp		.88	18.182		1,825	1,025		2,850	3,550
0400	1000 amp		.80	20		2,425	1,125		3,550	4,375
0500	1200 amp		.72	22.222		3,000	1,250		4,250	5,175
0600	1600 amp		.66	24.242		4,000	1,375		5,375	6,425
0700	2000 amp		.62	25.806		4,975	1,450		6,425	7,675
0800	2500 amp		.60	26.667		5,975	1,500		7,475	8,825
0900	3000 amp		.56	28.571		6,975	1,625		8,600	10,100
0950	4000 amp		.52	30.769		7,975	1,750		9,725	11,400

26 24 13.40 Switchboards Feeder Section

		Crew	Daily Output	Labor-Hours	Unit	Material	2017 Bare Costs Labor	Equipment	Total	Total Incl O&P
0010	**SWITCHBOARDS FEEDER SECTION** group mounted devices									
0030	Circuit breakers									
0160	FA frame, 15 to 60 amp, 240 volt, 1 pole	1 Elec	8	1	Ea.	116	56.50		172.50	212
0280	FA frame, 70 to 100 amp, 240 volt, 1 pole		7	1.143		213	64.50		277.50	330
0420	KA frame, 70 to 225 amp		3.20	2.500		1,525	142		1,667	1,900
0430	LA frame, 125 to 400 amp		2.30	3.478		4,150	197		4,347	4,875
0460	MA frame, 450 to 600 amp		1.60	5		6,175	283		6,458	7,200
0470	700 to 800 amp		1.30	6.154		8,000	350		8,350	9,325
0480	MAL frame, 1000 amp		1	8		8,300	455		8,755	9,825
0490	PA frame, 1200 amp		.80	10		16,900	565		17,465	19,500
0500	Branch circuit, fusible switch, 600 volt, double 30/30 amp		4	2		855	113		968	1,125
0550	60/60 amp		3.20	2.500		880	142		1,022	1,175
0600	100/100 amp		2.70	2.963		1,100	168		1,268	1,475

26 24 13.40 Switchboards Feeder Section

		Crew	Daily Output	Labor-Hours	Unit	Material	2017 Bare Costs Labor	Equipment	Total	Total Incl O&P
0650	Single, 30 amp	1 Elec	5.30	1.509	Ea.	695	85.50		780.50	895
0700	60 amp		4.70	1.702		825	96.50		921.50	1,050
0750	100 amp		4	2		1,050	113		1,163	1,325
0800	200 amp		2.70	2.963		1,375	168		1,543	1,750
0850	400 amp		2.30	3.478		2,525	197		2,722	3,075
0900	600 amp		1.80	4.444		3,075	252		3,327	3,750
0950	800 amp		1.30	6.154		5,150	350		5,500	6,200
1000	1200 amp	↓	.80	10	↓	5,900	565		6,465	7,350

26 24 16 – Panelboards

26 24 16.20 Panelboard and Load Center Circuit Breakers

		Crew	Daily Output	Labor-Hours	Unit	Material	2017 Bare Costs Labor	Equipment	Total	Total Incl O&P
0010	**PANELBOARD AND LOAD CENTER CIRCUIT BREAKERS**									
0050	Bolt-on, 10,000 amp I.C., 120 volt, 1 pole									
0100	15 to 50 amp	1 Elec	10	.800	Ea.	19.60	45.50		65.10	89.50
0200	60 amp		8	1		19.20	56.50		75.70	106
0300	70 amp	↓	8	1	↓	28	56.50		84.50	116
0350	240 volt, 2 pole									
0400	15 to 50 amp	1 Elec	8	1	Ea.	84	56.50		140.50	178
0500	60 amp		7.50	1.067		57.50	60.50		118	154
0600	80 to 100 amp		5	1.600		110	90.50		200.50	257
0700	3 pole, 15 to 60 amp		6.20	1.290		136	73		209	258
0800	70 amp		5	1.600		171	90.50		261.50	325
0900	80 to 100 amp		3.60	2.222		195	126		321	405
1000	22,000 amp I.C., 240 volt, 2 pole, 70 – 225 amp		2.70	2.963		610	168		778	925
1100	3 pole, 70 – 225 amp		2.30	3.478		610	197		807	965
1200	14,000 amp I.C., 277 volts, 1 pole, 15 – 30 amp		8	1		52.50	56.50		109	143
1300	22,000 amp I.C., 480 volts, 2 pole, 70 – 225 amp		2.70	2.963		550	168		718	855
1400	3 pole, 70 – 225 amp	↓	2.30	3.478	↓	770	197		967	1,150

26 24 16.30 Panelboards Commercial Applications

		Crew	Daily Output	Labor-Hours	Unit	Material	2017 Bare Costs Labor	Equipment	Total	Total Incl O&P
0010	**PANELBOARDS COMMERCIAL APPLICATIONS**									
0050	NQOD, w/20 amp 1 pole bolt-on circuit breakers									
0100	3 wire, 120/240 volts, 100 amp main lugs									
0150	10 circuits	1 Elec	1	8	Ea.	735	455		1,190	1,500
0200	14 circuits		.88	9.091		815	515		1,330	1,675
0250	18 circuits		.75	10.667		885	605		1,490	1,875
0300	20 circuits	↓	.65	12.308		995	695		1,690	2,150
0350	225 amp main lugs, 24 circuits	2 Elec	1.20	13.333		1,125	755		1,880	2,350
0400	30 circuits		.90	17.778		1,300	1,000		2,300	2,925
0450	36 circuits		.80	20		1,475	1,125		2,600	3,325
0500	38 circuits		.72	22.222		1,600	1,250		2,850	3,650
0550	42 circuits	↓	.66	24.242		1,650	1,375		3,025	3,875
0600	4 wire, 120/208 volts, 100 amp main lugs, 12 circuits	1 Elec	1	8		785	455		1,240	1,550
0650	16 circuits		.75	10.667		905	605		1,510	1,900
0700	20 circuits		.65	12.308		1,050	695		1,745	2,200
0750	24 circuits		.60	13.333		1,025	755		1,780	2,250
0800	30 circuits	↓	.53	15.094		1,300	855		2,155	2,700
0850	225 amp main lugs, 32 circuits	2 Elec	.90	17.778		1,450	1,000		2,450	3,100
0900	34 circuits		.84	19.048		1,475	1,075		2,550	3,250
0950	36 circuits		.80	20		1,525	1,125		2,650	3,375
1000	42 circuits		.68	23.529		1,700	1,325		3,025	3,875
1010	400 amp main lugs, 42 circs	↓	.68	23.529	↓	1,700	1,325		3,025	3,875
1200	NEHB, w/20 amp, 1 pole bolt-on circuit breakers									
1250	4 wire, 277/480 volts, 100 amp main lugs, 12 circuits	1 Elec	.88	9.091	Ea.	1,325	515		1,840	2,250

585

For customer support on your Building Construction Costs with RSMeans Data, call 800.448.8182.

26 24 16 – Panelboards

26 24 16.30 Panelboards Commercial Applications

		Crew	Daily Output	Labor-Hours	Unit	Material	2017 Bare Costs Labor	Equipment	Total	Total Incl O&P
1300	20 circuits	1 Elec	.60	13.333	Ea.	1,975	755		2,730	3,300
1350	225 amp main lugs, 24 circuits	2 Elec	.90	17.778		2,250	1,000		3,250	3,975
1400	30 circuits		.80	20		2,700	1,125		3,825	4,650
1448	32 circuits		4.90	3.265		3,125	185		3,310	3,725
1450	36 circuits		.72	22.222		3,125	1,250		4,375	5,325
1600	NQOD panel, w/20 amp, 1 pole, circuit breakers									
1650	3 wire, 120/240 volt with main circuit breaker									
1700	100 amp main, 12 circuits	1 Elec	.80	10	Ea.	985	565		1,550	1,925
1750	20 circuits	"	.60	13.333		1,250	755		2,005	2,500
1800	225 amp main, 30 circuits	2 Elec	.68	23.529		2,350	1,325		3,675	4,600
1801	225 amp main, 32 circuits		5	3.200		2,350	181		2,531	2,875
1850	42 circuits		.52	30.769		2,725	1,750		4,475	5,600
1900	400 amp main, 30 circuits		.54	29.630		3,250	1,675		4,925	6,100
1950	42 circuits		.50	32		3,650	1,800		5,450	6,725
2000	4 wire, 120/208 volts with main circuit breaker									
2050	100 amp main, 24 circuits	1 Elec	.47	17.021	Ea.	1,525	965		2,490	3,125
2100	30 circuits	"	.40	20		1,650	1,125		2,775	3,525
2200	225 amp main, 32 circuits	2 Elec	.72	22.222		2,750	1,250		4,000	4,900
2250	42 circuits		.56	28.571		2,825	1,625		4,450	5,550
2300	400 amp main, 42 circuits		.48	33.333		4,025	1,875		5,900	7,250
2350	600 amp main, 42 circuits		.40	40		5,950	2,275		8,225	9,950
2400	NEHB, with 20 amp, 1 pole circuit breaker									
2450	4 wire, 277/480 volts with main circuit breaker									
2500	100 amp main, 24 circuits	1 Elec	.42	19.048	Ea.	2,600	1,075		3,675	4,475
2550	30 circuits	"	.38	21.053		3,025	1,200		4,225	5,100
2600	225 amp main, 30 circuits	2 Elec	.72	22.222		3,800	1,250		5,050	6,050
2650	42 circuits	"	.56	28.571		4,675	1,625		6,300	7,575

26 24 19 – Motor-Control Centers

26 24 19.40 Motor Starters and Controls

		Crew	Daily Output	Labor-Hours	Unit	Material	2017 Bare Costs Labor	Equipment	Total	Total Incl O&P
0010	**MOTOR STARTERS AND CONTROLS**									
0050	Magnetic, FVNR, with enclosure and heaters, 480 volt									
0080	2 HP, size 00	1 Elec	3.50	2.286	Ea.	184	129		313	395
0100	5 HP, size 0		2.30	3.478		320	197		517	645
0200	10 HP, size 1		1.60	5		250	283		533	700
0300	25 HP, size 2	2 Elec	2.20	7.273		470	410		880	1,125
0400	50 HP, size 3		1.80	8.889		765	505		1,270	1,600
0500	100 HP, size 4		1.20	13.333		1,700	755		2,455	3,000
0600	200 HP, size 5		.90	17.778		3,975	1,000		4,975	5,875
0700	Combination, with motor circuit protectors, 5 HP, size 0	1 Elec	1.80	4.444		1,025	252		1,277	1,500
0800	10 HP, size 1	"	1.30	6.154		1,075	350		1,425	1,700
0900	25 HP, size 2	2 Elec	2	8		1,475	455		1,930	2,300
1000	50 HP, size 3		1.32	12.121		2,150	685		2,835	3,400
1200	100 HP, size 4		.80	20		4,675	1,125		5,800	6,825
1400	Combination, with fused switch, 5 HP, size 0	1 Elec	1.80	4.444		550	252		802	980
1600	10 HP, size 1	"	1.30	6.154		590	350		940	1,175
1800	25 HP, size 2	2 Elec	2	8		960	455		1,415	1,725
2000	50 HP, size 3		1.32	12.121		1,625	685		2,310	2,800
2200	100 HP, size 4		.80	20		2,825	1,125		3,950	4,800

26 25 13 – Bus Duct/Busway and Fittings

	26 25 13.40 Copper Bus Duct	Crew	Daily Output	Labor-Hours	Unit	Material	2017 Bare Costs Labor	Equipment	Total	Total Incl O&P
0010	**COPPER BUS DUCT** 10 ft. long									
0050	Indoor 3 pole 4 wire, plug-in, straight section, 225 amp	2 Elec	40	.400	L.F.	229	22.50		251.50	286
1000	400 amp		32	.500		385	28.50		413.50	470
1500	600 amp		26	.615		440	35		475	530
2400	800 amp		20	.800		640	45.50		685.50	770
2450	1000 amp		18	.889		300	50.50		350.50	405
2500	1350 amp		16	1		415	56.50		471.50	540
2510	1600 amp		12	1.333		470	75.50		545.50	630
2520	2000 amp		10	1.600		595	90.50		685.50	790
2550	Feeder, 600 amp		28	.571		204	32.50		236.50	273
2600	800 amp		22	.727		750	41		791	885
2700	1000 amp		20	.800		885	45.50		930.50	1,050
2800	1350 amp		18	.889		385	50.50		435.50	500
2900	1600 amp		14	1.143		1,425	64.50		1,489.50	1,675
3000	2000 amp		12	1.333		1,775	75.50		1,850.50	2,075
3100	Elbows, 225 amp		4	4	Ea.	1,575	226		1,801	2,075
3200	400 amp		3.60	4.444		1,575	252		1,827	2,100
3300	600 amp		3.20	5		1,575	283		1,858	2,150
3400	800 amp		2.80	5.714		1,700	325		2,025	2,350
3500	1000 amp		2.60	6.154		1,900	350		2,250	2,625
3600	1350 amp		2.40	6.667		1,875	375		2,250	2,650
3700	1600 amp		2.20	7.273		2,050	410		2,460	2,875
3800	2000 amp		1.80	8.889		2,525	505		3,030	3,525
4000	End box, 225 amp		34	.471		174	26.50		200.50	231
4100	400 amp		32	.500		188	28.50		216.50	250
4200	600 amp		28	.571		188	32.50		220.50	256
4300	800 amp		26	.615		188	35		223	259
4400	1000 amp		24	.667		182	37.50		219.50	258
4500	1350 amp		22	.727		179	41		220	258
4600	1600 amp		20	.800		179	45.50		224.50	264
4700	2000 amp		18	.889		219	50.50		269.50	315
4800	Cable tap box end, 225 amp		3.20	5		1,200	283		1,483	1,750
5000	400 amp		2.60	6.154		1,200	350		1,550	1,850
5100	600 amp		2.20	7.273		1,650	410		2,060	2,450
5200	800 amp		2	8		1,750	455		2,205	2,575
5300	1000 amp		1.60	10		1,750	565		2,315	2,775
5400	1350 amp		1.40	11.429		2,250	645		2,895	3,450
5500	1600 amp		1.20	13.333		2,500	755		3,255	3,875
5600	2000 amp		1	16		2,800	905		3,705	4,425
5700	Switchboard stub, 225 amp		5.40	2.963		1,250	168		1,418	1,650
5800	400 amp		4.60	3.478		1,325	197		1,522	1,750
5900	600 amp		4	4		1,375	226		1,601	1,875
6000	800 amp		3.20	5		1,675	283		1,958	2,250
6100	1000 amp		3	5.333		1,925	300		2,225	2,575
6200	1350 amp		2.60	6.154		2,400	350		2,750	3,150
6300	1600 amp		2.40	6.667		2,700	375		3,075	3,550
6400	2000 amp		2	8		3,275	455		3,730	4,275
6490	Tee fittings, 225 amp		2.40	6.667		2,175	375		2,550	2,975
6500	400 amp		2	8		2,175	455		2,630	3,075
6600	600 amp		1.80	8.889		2,175	505		2,680	3,150
6700	800 amp		1.60	10		2,250	565		2,815	3,350
6800	1350 amp		1.20	13.333		3,025	755		3,780	4,475
7000	1600 amp		1	16		3,450	905		4,355	5,150

587

26 25 13.40 Copper Bus Duct		Crew	Daily Output	Labor-Hours	Unit	Material	2017 Bare Costs Labor	Equipment	Total	Total Incl O&P
7100	2000 amp	2 Elec	.80	20	Ea.	4,100	1,125		5,225	6,200
7200	Plug-in fusible switches w/3 fuses, 600 volt, 3 pole, 30 amp	1 Elec	4	2		860	113		973	1,125
7300	60 amp		3.60	2.222		970	126		1,096	1,275
7400	100 amp		2.70	2.963		1,325	168		1,493	1,700
7500	200 amp	2 Elec	3.20	5		2,325	283		2,608	3,000
7600	400 amp		1.40	11.429		6,500	645		7,145	8,125
7700	600 amp		.90	17.778		7,850	1,000		8,850	10,100
7800	800 amp		.66	24.242		12,300	1,375		13,675	15,600
7900	1200 amp		.50	32		23,100	1,800		24,900	28,100
8000	Plug-in circuit breakers, molded case, 15 to 50 amp	1 Elec	4.40	1.818		815	103		918	1,050
8100	70 to 100 amp	"	3.10	2.581		905	146		1,051	1,225
8200	150 to 225 amp	2 Elec	3.40	4.706		2,450	266		2,716	3,100
8300	250 to 400 amp		1.40	11.429		4,300	645		4,945	5,725
8400	500 to 600 amp		1	16		5,825	905		6,730	7,750
8500	700 to 800 amp		.64	25		7,175	1,425		8,600	10,000
8600	900 to 1000 amp		.56	28.571		10,300	1,625		11,925	13,700
8700	1200 amp		.44	36.364		12,400	2,050		14,450	16,700

26 27 16.10 Cabinets		Crew	Daily Output	Labor-Hours	Unit	Material	2017 Bare Costs Labor	Equipment	Total	Total Incl O&P
0010	**CABINETS**									
7000	Cabinets, current transformer									
7050	Single door, 24" H x 24" W x 10" D	1 Elec	1.60	5	Ea.	176	283		459	620
7100	30" H x 24" W x 10" D		1.30	6.154		190	350		540	730
7150	36" H x 24" W x 10" D		1.10	7.273		252	410		662	895
7200	30" H x 30" W x 10" D		1	8		157	455		612	855
7250	36" H x 30" W x 10" D		.90	8.889		189	505		694	965
7300	36" H x 36" W x 10" D		.80	10		195	565		760	1,075
7500	Double door, 48" H x 36" W x 10" D		.60	13.333		370	755		1,125	1,525
7550	24" H x 24" W x 12" D		1	8		108	455		563	800

26 27 23.40 Surface Raceway		Crew	Daily Output	Labor-Hours	Unit	Material	2017 Bare Costs Labor	Equipment	Total	Total Incl O&P
0010	**SURFACE RACEWAY**									
0090	Metal, straight section									
0100	No. 500	1 Elec	100	.080	L.F.	1.07	4.53		5.60	8
0110	No. 700		100	.080		1.19	4.53		5.72	8.10
0400	No. 1500, small pancake		90	.089		2.20	5.05		7.25	9.95
0600	No. 2000, base & cover, blank		90	.089		2.24	5.05		7.29	10
0800	No. 3000, base & cover, blank		75	.107		4.28	6.05		10.33	13.75
1000	No. 4000, base & cover, blank		65	.123		8.15	6.95		15.10	19.45
1200	No. 6000, base & cover, blank		50	.160		10.80	9.05		19.85	25.50
2400	Fittings, elbows, No. 500		40	.200	Ea.	2.25	11.30		13.55	19.45
2800	Elbow cover, No. 2000		40	.200		3.25	11.30		14.55	20.50
2880	Tee, No. 500		42	.190		4.10	10.80		14.90	20.50
2900	No. 2000		27	.296		12.10	16.75		28.85	38.50
3000	Switch box, No. 500		16	.500		6.20	28.50		34.70	49.50
3400	Telephone outlet, No. 1500		16	.500		36	28.50		64.50	82
3600	Junction box, No. 1500		16	.500		15.75	28.50		44.25	60
3800	Plugmold wired sections, No. 2000									

26 27 Low-Voltage Distribution Equipment

26 27 23 – Indoor Service Poles

26 27 23.40 Surface Raceway	Crew	Daily Output	Labor-Hours	Unit	Material	2017 Bare Costs Labor	Equipment	Total	Total Incl O&P
4000	1 circuit, 6 outlets, 3 ft. long	1 Elec	8	1	Ea.	36.50	56.50		93	125
4100	2 circuits, 8 outlets, 6 ft. long	"	5.30	1.509	"	60	85.50		145.50	194

26 27 26 – Wiring Devices

26 27 26.10 Low Voltage Switching

		Crew	Daily Output	Labor-Hours	Unit	Material	Labor	Equipment	Total	Total Incl O&P
0010	**LOW VOLTAGE SWITCHING**									
3600	Relays, 120 V or 277 V standard	1 Elec	12	.667	Ea.	41	37.50		78.50	102
3800	Flush switch, standard		40	.200		11.50	11.30		22.80	29.50
4000	Interchangeable		40	.200		15.05	11.30		26.35	33.50
4100	Surface switch, standard		40	.200		8.20	11.30		19.50	26
4200	Transformer 115 V to 25 V		12	.667		128	37.50		165.50	198
4400	Master control, 12 circuit, manual		4	2		127	113		240	310
4500	25 circuit, motorized		4	2		141	113		254	325
4600	Rectifier, silicon		12	.667		46	37.50		83.50	107
4800	Switchplates, 1 gang, 1, 2 or 3 switch, plastic		80	.100		5	5.65		10.65	14
5000	Stainless steel		80	.100		11.35	5.65		17	21
5400	2 gang, 3 switch, stainless steel		53	.151		23	8.55		31.55	38
5500	4 switch, plastic		53	.151		10.45	8.55		19	24.50
5800	3 gang, 9 switch, stainless steel		32	.250		65.50	14.15		79.65	93

26 27 26.20 Wiring Devices Elements

		Crew	Daily Output	Labor-Hours	Unit	Material	Labor	Equipment	Total	Total Incl O&P
0010	**WIRING DEVICES ELEMENTS**									
0200	Toggle switch, quiet type, single pole, 15 amp	1 Elec	40	.200	Ea.	.54	11.30		11.84	17.55
0600	3 way, 15 amp		23	.348		1.90	19.70		21.60	31.50
0900	4 way, 15 amp		15	.533		7.75	30		37.75	53.50
1650	Dimmer switch, 120 volt, incandescent, 600 watt, 1 pole [G]		16	.500		22.50	28.50		51	67.50
2460	Receptacle, duplex, 120 volt, grounded, 15 amp		40	.200		1.26	11.30		12.56	18.35
2470	20 amp		27	.296		7.95	16.75		24.70	33.50
2490	Dryer, 30 amp		15	.533		4.46	30		34.46	50
2500	Range, 50 amp		11	.727		12.15	41		53.15	75
2600	Wall plates, stainless steel, 1 gang		80	.100		2.56	5.65		8.21	11.30
2800	2 gang		53	.151		4.33	8.55		12.88	17.55
3200	Lampholder, keyless		26	.308		16.85	17.40		34.25	44.50
3400	Pullchain with receptacle		22	.364		20.50	20.50		41	53.50

26 27 73 – Door Chimes

26 27 73.10 Doorbell System

		Crew	Daily Output	Labor-Hours	Unit	Material	Labor	Equipment	Total	Total Incl O&P
0010	**DOORBELL SYSTEM**, incl. transformer, button & signal									
0100	6" bell	1 Elec	4	2	Ea.	135	113		248	320
0200	Buzzer	"	4	2	"	109	113		222	290

26 28 Low-Voltage Circuit Protective Devices

26 28 16 – Enclosed Switches and Circuit Breakers

26 28 16.10 Circuit Breakers

		Crew	Daily Output	Labor-Hours	Unit	Material	Labor	Equipment	Total	Total Incl O&P
0010	**CIRCUIT BREAKERS** (in enclosure)									
0100	Enclosed (NEMA 1), 600 volt, 3 pole, 30 amp	1 Elec	3.20	2.500	Ea.	490	142		632	750
0200	60 amp		2.80	2.857		605	162		767	905
0400	100 amp		2.30	3.478		690	197		887	1,050
0500	200 amp		1.50	5.333		1,450	300		1,750	2,050
0600	225 amp		1.50	5.333		1,600	300		1,900	2,200
0700	400 amp	2 Elec	1.60	10		2,725	565		3,290	3,850
0800	600 amp		1.20	13.333		3,950	755		4,705	5,475
1000	800 amp		.94	17.021		5,150	965		6,115	7,100

26 28 16.20 Safety Switches		Crew	Daily Output	Labor-Hours	Unit	Material	2017 Bare Costs Labor	Equipment	Total	Total Incl O&P
0010	**SAFETY SWITCHES**									
0100	General duty 240 volt, 3 pole NEMA 1, fusible, 30 amp	1 Elec	3.20	2.500	Ea.	64	142		206	283
0200	60 amp		2.30	3.478		109	197		306	415
0300	100 amp		1.90	4.211		186	238		424	560
0400	200 amp	↓	1.30	6.154		400	350		750	960
0500	400 amp	2 Elec	1.80	8.889		980	505		1,485	1,825
0600	600 amp	"	1.20	13.333	↓	1,900	755		2,655	3,200
2900	Heavy duty, 240 volt, 3 pole NEMA 1 fusible									
2910	30 amp	1 Elec	3.20	2.500	Ea.	103	142		245	325
3000	60 amp		2.30	3.478		174	197		371	485
3300	100 amp		1.90	4.211		275	238		513	655
3500	200 amp	↓	1.30	6.154		485	350		835	1,050
3700	400 amp	2 Elec	1.80	8.889		1,225	505		1,730	2,100
3900	600 amp	"	1.20	13.333	↓	2,875	755		3,630	4,275

26 29 13.20 Control Stations		Crew	Daily Output	Labor-Hours	Unit	Material	2017 Bare Costs Labor	Equipment	Total	Total Incl O&P
0010	**CONTROL STATIONS**									
0050	NEMA 1, heavy duty, stop/start	1 Elec	8	1	Ea.	137	56.50		193.50	235
0100	Stop/start, pilot light		6.20	1.290		186	73		259	315
0200	Hand/off/automatic		6.20	1.290		101	73		174	220
0400	Stop/start/reverse	↓	5.30	1.509	↓	184	85.50		269.50	330

26 32 13.13 Diesel-Engine-Driven Generator Sets		Crew	Daily Output	Labor-Hours	Unit	Material	2017 Bare Costs Labor	Equipment	Total	Total Incl O&P
0010	**DIESEL-ENGINE-DRIVEN GENERATOR SETS**									
2000	Diesel engine, including battery, charger,									
2010	muffler, & day tank, 30 kW	R-3	.55	36.364	Ea.	11,000	2,050	251	13,301	15,500
2100	50 kW		.42	47.619		20,000	2,700	330	23,030	26,400
2200	75 kW		.35	57.143		23,100	3,225	395	26,720	30,700
2300	100 kW		.31	64.516		27,900	3,650	445	31,995	36,700
2400	125 kW		.29	68.966		29,300	3,900	475	33,675	38,700
2500	150 kW		.26	76.923		37,600	4,350	530	42,480	48,500
2501	Generator set, dsl eng in alum encl, incl btry, chgr, muf & day tank, 150 kW		.26	76.923		41,400	4,350	530	46,280	52,500
2600	175 kW		.25	80		41,900	4,525	550	46,975	53,500
2700	200 kW		.24	83.333		45,200	4,725	575	50,500	57,500
2800	250 kW		.23	86.957		48,400	4,925	600	53,925	61,500
2900	300 kW		.22	90.909		51,500	5,150	625	57,275	65,000
3000	350 kW		.20	100		59,000	5,650	690	65,340	74,000
3100	400 kW		.19	105		72,000	5,950	725	78,675	89,000
3200	500 kW	↓	.18	111	↓	91,500	6,300	765	98,565	111,000

26 32 13.16 Gas-Engine-Driven Generator Sets

		Crew	Daily Output	Labor-Hours	Unit	Material	2017 Bare Costs Labor	Equipment	Total	Total Incl O&P
0010	**GAS-ENGINE-DRIVEN GENERATOR SETS**									
0020	Gas or gasoline operated, includes battery,									
0050	charger, & muffler									
0200	7.5 kW	R-3	.83	24.096	Ea.	8,550	1,375	166	10,091	11,700

26 32 Packaged Generator Assemblies

26 32 13 – Engine Generators

26 32 13.16 Gas-Engine-Driven Generator Sets		Crew	Daily Output	Labor-Hours	Unit	Material	2017 Bare Costs Labor	Equipment	Total	Total Incl O&P
0300	11.5 kW	R-3	.71	28.169	Ea.	12,100	1,600	194	13,894	15,900
0400	20 kW		.63	31.746		14,300	1,800	219	16,319	18,600
0500	35 kW	↓	.55	36.364		17,000	2,050	251	19,301	22,100
0600	80 kW	R-13	.40	105		27,900	5,775	540	34,215	40,000
0700	100 kW		.33	127		30,600	7,000	655	38,255	44,800
0800	125 kW		.28	150		62,500	8,250	770	71,520	82,500
0900	185 kW	↓	.25	168	↓	83,000	9,250	865	93,115	106,000

26 33 Battery Equipment

26 33 43 – Battery Chargers

26 33 43.55 Electric Vehicle Charging

			Crew	Daily Output	Labor-Hours	Unit	Material	2017 Bare Costs Labor	Equipment	Total	Total Incl O&P
0010	**ELECTRIC VEHICLE CHARGING**										
0020	Level 2, wall mounted										
2200	Heavy duty	G	R-1A	15.36	1.042	Ea.	2,525	53		2,578	2,850
2210	with RFID	G		12.29	1.302		3,100	66.50		3,166.50	3,525
2300	Free standing, single connector	G		10.24	1.563		2,900	79.50		2,979.50	3,300
2310	with RFID	G		8.78	1.822		3,625	93		3,718	4,125
2320	Double connector	G		7.68	2.083		4,850	106		4,956	5,475
2330	with RFID	G	↓	6.83	2.343	↓	6,300	119		6,419	7,100

26 35 Power Filters and Conditioners

26 35 13 – Capacitors

26 35 13.10 Capacitors Indoor

		Crew	Daily Output	Labor-Hours	Unit	Material	2017 Bare Costs Labor	Equipment	Total	Total Incl O&P
0010	**CAPACITORS INDOOR**									
0020	240 volts, single & 3 phase, 0.5 kVAR	1 Elec	2.70	2.963	Ea.	470	168		638	765
0100	1.0 kVAR		2.70	2.963		565	168		733	875
0150	2.5 kVAR		2	4		640	226		866	1,050
0200	5.0 kVAR		1.80	4.444		790	252		1,042	1,250
0250	7.5 kVAR		1.60	5		885	283		1,168	1,400
0300	10 kVAR		1.50	5.333		1,025	300		1,325	1,575
0350	15 kVAR		1.30	6.154		1,325	350		1,675	1,975
0400	20 kVAR		1.10	7.273		1,600	410		2,010	2,375
0450	25 kVAR		1	8		1,825	455		2,280	2,700
1000	480 volts, single & 3 phase, 1 kVAR		2.70	2.963		425	168		593	720
1050	2 kVAR		2.70	2.963		490	168		658	790
1100	5 kVAR		2	4		620	226		846	1,025
1150	7.5 kVAR		2	4		670	226		896	1,075
1200	10 kVAR		2	4		710	226		936	1,125
1250	15 kVAR		2	4		840	226		1,066	1,250
1300	20 kVAR		1.60	5		915	283		1,198	1,425
1350	30 kVAR		1.50	5.333		1,100	300		1,400	1,675
1400	40 kVAR		1.20	6.667		1,375	375		1,750	2,075
1450	50 kVAR	↓	1.10	7.273	↓	1,600	410		2,010	2,400

For customer support on your Building Construction Costs with RSMeans Data, call 800.448.8182.

591

26 51 13 – Interior Lighting Fixtures, Lamps, and Ballasts

26 51 13.50 Interior Lighting Fixtures		Crew	Daily Output	Labor-Hours	Unit	Material	2017 Bare Costs Labor	Equipment	Total	Total Incl O&P
0010	**INTERIOR LIGHTING FIXTURES** Including lamps, mounting									
0030	hardware and connections									
0100	Fluorescent, C.W. lamps, troffer, recess mounted in grid, RS									
0130	Grid ceiling mount									
0200	Acrylic lens, 1'W x 4'L, two 40 watt	1 Elec	5.70	1.404	Ea.	51.50	79.50		131	176
0210	1'W x 4'L, three 40 watt		5.40	1.481		58.50	84		142.50	190
0300	2'W x 2'L, two U40 watt		5.70	1.404		55	79.50		134.50	180
0400	2'W x 4'L, two 40 watt		5.30	1.509		54	85.50		139.50	188
0500	2'W x 4'L, three 40 watt		5	1.600		59.50	90.50		150	202
0600	2'W x 4'L, four 40 watt		4.70	1.702		62	96.50		158.50	213
0700	4'W x 4'L, four 40 watt	2 Elec	6.40	2.500		276	142		418	515
0800	4'W x 4'L, six 40 watt		6.20	2.581		296	146		442	545
0900	4'W x 4'L, eight 40 watt		5.80	2.759		340	156		496	605
0910	Acrylic lens, 1'W x 4'L, two 32 watt T8 [G]	1 Elec	5.70	1.404		61	79.50		140.50	187
0930	2'W x 2'L, two U32 watt T8 [G]		5.70	1.404		95.50	79.50		175	224
0940	2'W x 4'L, two 32 watt T8 [G]		5.30	1.509		79	85.50		164.50	215
0950	2'W x 4'L, three 32 watt T8 [G]		5	1.600		69.50	90.50		160	213
0960	2'W x 4'L, four 32 watt T8 [G]		4.70	1.702		74	96.50		170.50	226
1000	Surface mounted, RS									
1030	Acrylic lens with hinged & latched door frame									
1100	1'W x 4'L, two 40 watt	1 Elec	7	1.143	Ea.	65.50	64.50		130	169
1110	1'W x 4'L, three 40 watt		6.70	1.194		68	67.50		135.50	176
1200	2'W x 2'L, two U40 watt		7	1.143		70.50	64.50		135	175
1300	2'W x 4'L, two 40 watt		6.20	1.290		80	73		153	197
1400	2'W x 4'L, three 40 watt		5.70	1.404		83	79.50		162.50	210
1500	2'W x 4'L, four 40 watt		5.30	1.509		85	85.50		170.50	222
1501	2'W x 4'L, 6-40W T8		5.20	1.538		85	87		172	224
1600	4'W x 4'L, four 40 watt	2 Elec	7.20	2.222		420	126		546	650
1700	4'W x 4'L, six 40 watt		6.60	2.424		455	137		592	705
1800	4'W x 4'L, eight 40 watt		6.20	2.581		475	146		621	740
1900	2'W x 8'L, four 40 watt		6.40	2.500		167	142		309	395
2000	2'W x 8'L, eight 40 watt		6.20	2.581		179	146		325	415
2100	Strip fixture									
2130	Surface mounted									
2200	4' long, one 40 watt, RS	1 Elec	8.50	.941	Ea.	31	53.50		84.50	114
2300	4' long, two 40 watt, RS		8	1		43	56.50		99.50	133
2400	4' long, one 40 watt, SL		8	1		53.50	56.50		110	144
2500	4' long, two 40 watt, SL		7	1.143		72.50	64.50		137	177
2600	8' long, one 75 watt, SL	2 Elec	13.40	1.194		55	67.50		122.50	162
2700	8' long, two 75 watt, SL	"	12.40	1.290		66.50	73		139.50	182
2800	4' long, two 60 watt, HO	1 Elec	6.70	1.194		107	67.50		174.50	219
2900	8' long, two 110 watt, HO	2 Elec	10.60	1.509		108	85.50		193.50	247
2950	High bay pendent mounted, 16" W x 4' L, four 54 watt, T5HO [G]		8.90	1.798		244	102		346	420
2952	2' W x 4' L, six 54 watt, T5HO [G]		8.50	1.882		310	107		417	500
2954	2' W x 4' L, six 32 watt, T8 [G]		8.50	1.882		179	107		286	355
3000	Strip, pendent mounted, industrial, white porcelain enamel									
3100	4' long, two 40 watt, RS	1 Elec	5.70	1.404	Ea.	53	79.50		132.50	177
3200	4' long, two 60 watt, HO	"	5	1.600		83	90.50		173.50	228
3300	8' long, two 75 watt, SL	2 Elec	8.80	1.818		98.50	103		201.50	262
3400	8' long, two 110 watt, HO	"	8	2		126	113		239	310
3470	Troffer, air handling, 2'W x 4'L with four 32 watt T8 [G]	1 Elec	4	2		115	113		228	297
3480	2'W x 2'L with two U32 watt T8 [G]		5.50	1.455		111	82.50		193.50	245
3490	Air connector insulated, 5" diameter		20	.400		68	22.50		90.50	109

26 51 13.50 Interior Lighting Fixtures

		Crew	Daily Output	Labor-Hours	Unit	Material	2017 Bare Costs Labor	Equipment	Total	Total Incl O&P
3500	6" diameter	1 Elec	20	.400	Ea.	69	22.50		91.50	110
3510	Troffer parabolic lay-in, 1'W x 4'L with one 32 W T8 [G]		5.70	1.404		117	79.50		196.50	248
3520	1'W x 4'L with two 32 W T8 [G]		5.30	1.509		140	85.50		225.50	282
3525	2'W x 2'L with two U32 W T8 [G]		5.70	1.404		118	79.50		197.50	249
3530	2'W x 4'L with three 32 W T8 [G]		5	1.600		131	90.50		221.50	280
3531	Intr fxtr, fluor, troffer prismatic lay-in, 2'w x 4'l W/three 32 W T8		5	1.600		152	90.50		242.50	305
4450	Incandescent, high hat can, round alzak reflector, prewired									
4470	100 watt	1 Elec	8	1	Ea.	64.50	56.50		121	156
4480	150 watt		8	1		105	56.50		161.50	200
4500	300 watt		6.70	1.194		242	67.50		309.50	365
4600	Square glass lens with metal trim, prewired									
4630	100 watt	1 Elec	6.70	1.194	Ea.	55.50	67.50		123	162
4700	200 watt		6.70	1.194		97.50	67.50		165	208
4800	300 watt		5.70	1.404		146	79.50		225.50	279
4900	Ceiling/wall, surface mounted, metal cylinder, 75 watt		10	.800		50.50	45.50		96	124
4920	150 watt		10	.800		81	45.50		126.50	157
5200	Ceiling, surface mounted, opal glass drum									
5300	8", one 60 watt lamp	1 Elec	10	.800	Ea.	65.50	45.50		111	140
5400	10", two 60 watt lamps		8	1		72.50	56.50		129	165
5500	12", four 60 watt lamps		6.70	1.194		103	67.50		170.50	214
6010	Vapor tight, incandescent, ceiling mounted, 200 watt		6.20	1.290		78.50	73		151.50	195
6100	Fluorescent, surface mounted, 2 lamps, 4'L, RS, 40 watt		3.20	2.500		116	142		258	340
6850	Vandalproof, surface mounted, fluorescent, two 32 watt T8 [G]		3.20	2.500		252	142		394	490
6860	Incandescent, one 150 watt		8	1		90	56.50		146.50	184
7500	Ballast replacement, by weight of ballast, to 15' high									
7520	Indoor fluorescent, less than 2 lb.	1 Elec	10	.800	Ea.	23.50	45.50		69	94
7540	Two 40W, watt reducer, 2 to 5 lb.		9.40	.851		71	48		119	151
7560	Two F96 slimline, over 5 lb.		8	1		110	56.50		166.50	206
7580	Vaportite ballast, less than 2 lb.		9.40	.851		23.50	48		71.50	98
7600	2 lb. to 5 lb.		8.90	.899		71	51		122	155
7620	Over 5 lb.		7.60	1.053		110	59.50		169.50	211
7630	Electronic ballast for two tubes		8	1		33.50	56.50		90	122
7640	Dimmable ballast one lamp [G]		8	1		94.50	56.50		151	189
7650	Dimmable ballast two-lamp [G]		7.60	1.053		100	59.50		159.50	200

26 51 13.55 Interior LED Fixtures

		Crew	Daily Output	Labor-Hours	Unit	Material	2017 Bare Costs Labor	Equipment	Total	Total Incl O&P
0010	**INTERIOR LED FIXTURES** Incl. lamps, and mounting hardware									
0100	Downlight, recess mounted, 7.5" diameter, 25 watt [G]	1 Elec	8	1	Ea.	335	56.50		391.50	450
0120	10" diameter, 36 watt [G]		8	1		360	56.50		416.50	480
0160	cylinder, 10 watts [G]		8	1		102	56.50		158.50	197
0180	20 watts [G]		8	1		166	56.50		222.50	268
1000	Troffer, recess mounted, 2' x 4', 3200 Lumens [G]		5.30	1.509		138	85.50		223.50	279
1010	4800 Lumens [G]		5	1.600		179	90.50		269.50	330
1020	6400 Lumens [G]		4.70	1.702		198	96.50		294.50	360
1100	Troffer retrofit lamp, 38 watt [G]		21	.381		238	21.50		259.50	294
1110	60 watt [G]		20	.400		340	22.50		362.50	410
1120	100 watt [G]		18	.444		510	25		535	600
1200	Troffer, volumetric recess mounted, 2' x 2' [G]		5.70	1.404		279	79.50		358.50	425
2000	Strip, surface mounted, one light bar 4' long, 3500K [G]		8.50	.941		260	53.50		313.50	365
2010	5000K [G]		8	1		260	56.50		316.50	370
2020	Two light bar 4' long, 5000K [G]		7	1.143		410	64.50		474.50	545
3000	Linear, suspended mounted, one light bar 4' long, 37 watt [G]		6.70	1.194		153	67.50		220.50	269
3010	One light bar 8' long, 74 watt [G]	2 Elec	12.20	1.311		283	74		357	420

For customer support on your Building Construction Costs with RSMeans Data, call 800.448.8182.

593

26 51 Interior Lighting

26 51 13 – Interior Lighting Fixtures, Lamps, and Ballasts

26 51 13.55 Interior LED Fixtures		Crew	Daily Output	Labor-Hours	Unit	Material	2017 Bare Costs Labor	Equipment	Total	Total Incl O&P
3020	Two light bar 4' long, 74 watt	[G] 1 Elec	5.70	1.404	Ea.	305	79.50		384.50	455
3030	Two light bar 8' long, 148 watt	[G] 2 Elec	8.80	1.818		350	103		453	540
4000	High bay, surface mounted, round, 150 watts	[G]	5.41	2.959		475	167		642	770
4010	2 bars,164 watts	[G]	5.41	2.959		445	167		612	740
4020	3 bars, 246 watts	[G]	5.01	3.197		575	181		756	900
4030	4 bars, 328 watts	[G]	4.60	3.478		700	197		897	1,075
4040	5 bars, 410 watts	[G] 3 Elec	4.20	5.716		830	325		1,155	1,400
4050	6 bars, 492 watts	[G]	3.80	6.324		935	360		1,295	1,550
4060	7 bars, 574 watts	[G]	3.39	7.075		1,025	400		1,425	1,725
4070	8 bars, 656 watts	[G]	2.99	8.029		1,125	455		1,580	1,925
5000	Track, lighthead, 6 watt	[G] 1 Elec	32	.250		54.50	14.15		68.65	81
5010	9 watt	[G] "	32	.250		61.50	14.15		75.65	88.50
6000	Garage, surface mounted, 103 watts	[G] 2 Elec	6.50	2.462		970	139		1,109	1,275
6100	pendent mounted, 80 watts	[G]	6.50	2.462		695	139		834	975
6200	95 watts	[G]	6.50	2.462		800	139		939	1,100
6300	125 watts	[G]	6.50	2.462		840	139		979	1,125

26 52 Emergency Lighting

26 52 13 – Emergency Lighting Equipments

26 52 13.10 Emergency Lighting and Battery Units

		Crew	Daily Output	Labor-Hours	Unit	Material	2017 Bare Costs Labor	Equipment	Total	Total Incl O&P
0010	**EMERGENCY LIGHTING AND BATTERY UNITS**									
0300	Emergency light units, battery operated									
0350	Twin sealed beam light, 25 watt, 6 volt each									
0500	Lead battery operated	1 Elec	4	2	Ea.	166	113		279	350
0700	Nickel cadmium battery operated		4	2		335	113		448	535
0900	Self-contained fluorescent lamp pack		10	.800		166	45.50		211.50	251

26 53 Exit Signs

26 53 13 – Exit Lighting

26 53 13.10 Exit Lighting Fixtures

		Crew	Daily Output	Labor-Hours	Unit	Material	2017 Bare Costs Labor	Equipment	Total	Total Incl O&P
0010	**EXIT LIGHTING FIXTURES**									
0080	Exit light ceiling or wall mount, incandescent, single face	1 Elec	8	1	Ea.	70	56.50		126.50	162
0100	Double face		6.70	1.194		50	67.50		117.50	156
0200	LED standard, single face	[G]	8	1		47.50	56.50		104	137
0220	Double face	[G]	6.70	1.194		51	67.50		118.50	157
0230	LED vandal-resistant, single face	[G]	7.27	1.100		212	62.50		274.50	325
0240	LED w/battery unit, single face	[G]	4.40	1.818		184	103		287	355
0260	Double face	[G]	4	2		188	113		301	375
0262	LED w/battery unit, vandal-resistant, single face	[G]	4.40	1.818		245	103		348	425
0270	Combination emergency light units and exit sign		4	2		179	113		292	365

26 54 Classified Location Lighting

26 54 13 - Classified Lighting

26 54 13.20 Explosionproof	Crew	Daily Output	Labor-Hours	Unit	Material	2017 Bare Costs Labor	Equipment	Total	Total Incl O&P
0010 **EXPLOSIONPROOF**, incl lamps, mounting hardware and connections									
6510 Incandescent, ceiling mounted, 200 watt	1 Elec	4	2	Ea.	1,250	113		1,363	1,575
6600 Fluorescent, RS, 4' long, ceiling mounted, two 40 watt	"	2.70	2.963	"	3,325	168		3,493	3,925

26 55 Special Purpose Lighting

26 55 33 - Hazard Warning Lighting

26 55 33.10 Warning Beacons

	Crew	Daily Output	Labor-Hours	Unit	Material	Labor	Equipment	Total	Total Incl O&P
0010 **WARNING BEACONS**									
0015 Surface mount with colored or clear lens									
0100 Rotating beacon									
0110 120V, 40 watt halogen	1 Elec	3.50	2.286	Ea.	101	129		230	305
0120 24V, 20 watt halogen	"	3.50	2.286	"	211	129		340	425
0200 Steady beacon									
0210 120V, 40 watt halogen	1 Elec	3.50	2.286	Ea.	101	129		230	305
0220 24V, 20W		3.50	2.286		101	129		230	305
0230 12V DC, incandescent	↓	3.50	2.286	↓	101	129		230	305
0300 Flashing beacon									
0310 120V, 40 watt halogen	1 Elec	3.50	2.286	Ea.	101	129		230	305
0320 24V, 20 watt halogen		3.50	2.286		101	129		230	305
0410 12V DC with two 6V lantern batteries	↓	7	1.143	↓	101	64.50		165.50	208

26 55 61 - Theatrical Lighting

26 55 61.10 Lights

	Crew	Daily Output	Labor-Hours	Unit	Material	Labor	Equipment	Total	Total Incl O&P
0010 **LIGHTS**									
2000 Lights, border, quartz, reflector, vented,									
2100 colored or white	1 Elec	20	.400	L.F.	178	22.50		200.50	230
2500 Spotlight, follow spot, with transformer, 2,100 watt	"	4	2	Ea.	3,300	113		3,413	3,800
2600 For no transformer, deduct					940			940	1,025
3000 Stationary spot, fresnel quartz, 6" lens	1 Elec	4	2		184	113		297	370
3100 8" lens		4	2		238	113		351	430
3500 Ellipsoidal quartz, 1,000W, 6" lens		4	2		355	113		468	560
3600 12" lens		4	2		640	113		753	870
4000 Strobe light, 1 to 15 flashes per second, quartz		3	2.667		810	151		961	1,125
4500 Color wheel, portable, five hole, motorized	↓	4	2	↓	209	113		322	400

26 56 Exterior Lighting

26 56 13 - Lighting Poles and Standards

26 56 13.10 Lighting Poles

	Crew	Daily Output	Labor-Hours	Unit	Material	Labor	Equipment	Total	Total Incl O&P
0010 **LIGHTING POLES**									
2800 Light poles, anchor base									
2820 not including concrete bases									
2840 Aluminum pole, 8' high	1 Elec	4	2	Ea.	735	113		848	975
3000 20' high	R-3	2.90	6.897		975	390	47.50	1,412.50	1,725
3200 30' high		2.60	7.692		1,850	435	53	2,338	2,750
3400 35' high		2.30	8.696		2,000	490	60	2,550	3,000
3600 40' high	↓	2	10		2,300	565	69	2,934	3,450
3800 Bracket arms, 1 arm	1 Elec	8	1		126	56.50		182.50	223
4000 2 arms		8	1		253	56.50		309.50	365
4200 3 arms		5.30	1.509		380	85.50		465.50	545
4400 4 arms	↓	4.80	1.667	↓	505	94.50		599.50	695

For customer support on your Building Construction Costs with RSMeans Data, call 800.448.8182.

595

26 56 Exterior Lighting

26 56 13 – Lighting Poles and Standards

26 56 13.10 Lighting Poles		Crew	Daily Output	Labor-Hours	Unit	Material	2017 Bare Costs Labor	Equipment	Total	Total Incl O&P
4500	Steel pole, galvanized, 8' high	1 Elec	3.80	2.105	Ea.	635	119		754	880
4600	20' high	R-3	2.60	7.692		1,025	435	53	1,513	1,850
4800	30' high		2.30	8.696		1,350	490	60	1,900	2,300
5000	35' high		2.20	9.091		1,500	515	62.50	2,077.50	2,500
5200	40' high	↓	1.70	11.765		1,850	665	81	2,596	3,125
5400	Bracket arms, 1 arm	1 Elec	8	1		188	56.50		244.50	292
5600	2 arms		8	1		289	56.50		345.50	405
5800	3 arms		5.30	1.509		225	85.50		310.50	375
6000	4 arms	↓	5.30	1.509		320	85.50		405.50	480
6462	20' high	R-3	2.90	6.897		1,025	390	47.50	1,462.50	1,775
6463	30' high		2.30	8.696		1,350	490	60	1,900	2,300
6464	35' high		2.40	8.333		1,500	470	57.50	2,027.50	2,425
6465	25' high	↓	2.70	7.407	↓	1,025	420	51	1,496	1,800

26 56 16 – Parking Lighting

26 56 16.55 Parking LED Lighting

0010	**PARKING LED LIGHTING**									
0100	Round pole mounting, 88 lamp watts	[G]	1 Elec	2	4	Ea.	1,200	226		1,426	1,650

26 56 19 – Roadway Lighting

26 56 19.20 Roadway Luminaire

0010	**ROADWAY LUMINAIRE**									
2650	Roadway area luminaire, low pressure sodium, 135 watt	1 Elec	2	4	Ea.	705	226		931	1,125
2700	180 watt	"	2	4		750	226		976	1,175
2750	Metal halide, 400 watt	2 Elec	4.40	3.636		605	206		811	975
2760	1000 watt		4	4		680	226		906	1,075
2780	High pressure sodium, 400 watt		4.40	3.636		630	206		836	1,000
2790	1000 watt	↓	4	4	↓	715	226		941	1,125

26 56 19.55 Roadway LED Luminaire

0010	**ROADWAY LED LUMINAIRE**									
0100	LED fixture, 72 LEDs, 120 V AC or 12 V DC, equal to 60 watt	[G]	1 Elec	2.70	2.963	Ea.	585	168		753	895
0110	108 LEDs, 120 V AC or 12 V DC, equal to 90 watt	[G]		2.70	2.963		690	168		858	1,000
0120	144 LEDs, 120 V AC or 12 V DC, equal to 120 watt	[G]	↓	2.70	2.963		845	168		1,013	1,175
0130	252 LEDs, 120 V AC or 12 V DC, equal to 210 watt	[G]	2 Elec	4.40	3.636		1,175	206		1,381	1,600
0140	Replaces high pressure sodium fixture, 75 watt	[G]	1 Elec	2.70	2.963		395	168		563	685
0150	125 watt	[G]		2.70	2.963		450	168		618	745
0160	150 watt	[G]		2.70	2.963		530	168		698	830
0170	175 watt	[G]		2.70	2.963		780	168		948	1,100
0180	200 watt	[G]		2.70	2.963		745	168		913	1,075
0190	250 watt	[G]	2 Elec	4.40	3.636		850	206		1,056	1,250
0200	320 watt	[G]	"	4.40	3.636	↓	930	206		1,136	1,325

26 56 23 – Area Lighting

26 56 23.10 Exterior Fixtures

0010	**EXTERIOR FIXTURES** With lamps									
0200	Wall mounted, incandescent, 100 watt	1 Elec	8	1	Ea.	37.50	56.50		94	126
0400	Quartz, 500 watt		5.30	1.509		61.50	85.50		147	196
1100	Wall pack, low pressure sodium, 35 watt		4	2		199	113		312	390
1150	55 watt	↓	4	2	↓	236	113		349	430

26 56 23.55 Exterior LED Fixtures

0010	**EXTERIOR LED FIXTURES**									
0100	Wall mounted, indoor/outdoor, 12 watt	[G]	1 Elec	10	.800	Ea.	191	45.50		236.50	278
0110	32 watt	[G]	↓	10	.800		460	45.50		505.50	575

26 56 Exterior Lighting

26 56 23 – Area Lighting

26 56 23.55 Exterior LED Fixtures

26 56 23.55 Exterior LED Fixtures		Crew	Daily Output	Labor-Hours	Unit	Material	2017 Bare Costs Labor	Equipment	Total	Total Incl O&P
0120	66 watt	G	1 Elec	10	.800	Ea.	705	45.50		750.50	845
0200	outdoor, 110 watt	G		10	.800		1,100	45.50		1,145.50	1,300
0210	220 watt	G		10	.800		1,750	45.50		1,795.50	2,000
0300	modular, type IV, 120 V, 50 lamp watts	G		9	.889		1,150	50.50		1,200.50	1,350
0310	101 lamp watts	G		9	.889		1,300	50.50		1,350.50	1,500
0320	126 lamp watts	G		9	.889		1,625	50.50		1,675.50	1,850
0330	202 lamp watts	G		9	.889		1,850	50.50		1,900.50	2,100
0340	240 V, 50 lamp watts	G		8	1		1,200	56.50		1,256.50	1,400
0350	101 lamp watts	G		8	1		1,350	56.50		1,406.50	1,550
0360	126 lamp watts	G		8	1		1,450	56.50		1,506.50	1,675
0370	202 lamp watts	G		8	1		1,900	56.50		1,956.50	2,150
0400	wall pack, glass, 13 lamp watts	G		4	2		445	113		558	660
0410	poly w/photocell, 26 lamp watts	G		4	2		297	113		410	495
0420	50 lamp watts	G		4	2		750	113		863	995
0430	replacement, 40 watts	G		4	2		365	113		478	570
0440	60 watts	G		4	2		460	113		573	675

26 56 36 – Flood Lighting

26 56 36.20 Floodlights

26 56 36.20 Floodlights		Crew	Daily Output	Labor-Hours	Unit	Material	2017 Bare Costs Labor	Equipment	Total	Total Incl O&P
0010	**FLOODLIGHTS** with ballast and lamp,									
1400	Pole mounted, pole not included									
1950	Metal halide, 175 watt	1 Elec	2.70	2.963	Ea.	231	168		399	505
2000	400 watt	2 Elec	4.40	3.636		390	206		596	740
2200	1000 watt	"	4	4		755	226		981	1,175
2340	High pressure sodium, 70 watt	1 Elec	2.70	2.963		250	168		418	525
2400	400 watt	2 Elec	4.40	3.636		315	206		521	660
2600	1000 watt	"	4	4		580	226		806	975

26 56 36.55 LED Floodlights

26 56 36.55 LED Floodlights		Crew	Daily Output	Labor-Hours	Unit	Material	2017 Bare Costs Labor	Equipment	Total	Total Incl O&P
0010	**LED FLOODLIGHTS** with ballast and lamp,									
0020	Pole mounted, pole not included									
0100	11 watt	G	1 Elec	4	2	Ea.	410	113		523	625
0110	46 watt	G		4	2		1,225	113		1,338	1,525
0120	90 watt	G		4	2		2,000	113		2,113	2,375
0130	288 watt	G		4	2		1,800	113		1,913	2,150

26 61 Lighting Systems and Accessories

26 61 23 – Lamps Applications

26 61 23.10 Lamps

26 61 23.10 Lamps		Crew	Daily Output	Labor-Hours	Unit	Material	2017 Bare Costs Labor	Equipment	Total	Total Incl O&P
0010	**LAMPS**									
0080	Fluorescent, rapid start, cool white, 2' long, 20 watt	1 Elec	1	8	C	360	455		815	1,075
0100	4' long, 40 watt		.90	8.889		275	505		780	1,050
0200	Slimline, 4' long, 40 watt		.90	8.889		1,450	505		1,955	2,350
0210	4' long, 30 watt energy saver	G	.90	8.889		1,450	505		1,955	2,350
0400	High output, 4' long, 60 watt		.90	8.889		755	505		1,260	1,600
0410	8' long, 95 watt energy saver	G	.80	10		545	565		1,110	1,450
0500	8' long, 110 watt		.80	10		545	565		1,110	1,450
0512	2' long, T5, 14 watt energy saver	G	1	8		730	455		1,185	1,475
0514	3' long, T5, 21 watt energy saver	G	.90	8.889		180	505		685	955
0516	4' long, T5, 28 watt energy saver	G	.90	8.889		180	505		685	955
0517	4' long, T5, 54 watt energy saver	G	.90	8.889		1,575	505		2,080	2,475
0560	Twin tube compact lamp	G	.90	8.889		355	505		860	1,150

For customer support on your Building Construction Costs with RSMeans Data, call 800.448.8182.

597

26 61 23.10 Lamps

26 61 23.10 Lamps	Crew	Daily Output	Labor-Hours	Unit	Material	2017 Bare Costs Labor	Equipment	Total	Total Incl O&P
0570	Double twin tube compact lamp	1 Elec	.80	10	C	725	565		1,290	1,650
0600	Mercury vapor, mogul base, deluxe white, 100 watt		.30	26.667		11,400	1,500		12,900	14,900
0700	250 watt		.30	26.667		9,300	1,500		10,800	12,500
0800	400 watt		.30	26.667		8,400	1,500		9,900	11,500
0900	1000 watt		.20	40		2,300	2,275		4,575	5,925
1000	Metal halide, mogul base, 175 watt		.30	26.667		2,450	1,500		3,950	4,950
1200	400 watt		.30	26.667		2,075	1,500		3,575	4,525
1300	1000 watt		.20	40		41,900	2,275		44,175	49,500
1350	High pressure sodium, 70 watt		.30	26.667		2,650	1,500		4,150	5,175
1380	250 watt		.30	26.667		2,225	1,500		3,725	4,700
1400	400 watt		.30	26.667		4,700	1,500		6,200	7,400
1450	1000 watt		.20	40		8,525	2,275		10,800	12,800
3000	Guards, fluorescent lamp, 4' long		1	8		1,325	455		1,780	2,125
3200	8' long		.90	8.889		2,650	505		3,155	3,650

26 61 23.55 LED Lamps

26 61 23.55 LED Lamps	Crew	Daily Output	Labor-Hours	Unit	Material	2017 Bare Costs Labor	Equipment	Total	Total Incl O&P
0010	**LED LAMPS**									
0100	LED lamp, interior, shape A60, equal to 60 watt	1 Elec	160	.050	Ea.	19	2.83		21.83	25
0110	7W LED decorative c, ca, ,f g, shape.		160	.050		34.50	2.83		37.33	41.50
0120	12V mini lamp LED		160	.050		5	2.83		7.83	9.75
0200	Globe frosted A60, equal to 60 watt		160	.050		11.40	2.83		14.23	16.80
0205	LED lamp, interior, globe		160	.050		16.15	2.83		18.98	22
0210	2.2W LED LMP		160	.050		66	2.83		68.83	77
0220	2.2W LED replacement decorative lamp		160	.050		68	2.83		70.83	78.50
0230	3.5W LED replacement decorative lamp		160	.050		68	2.83		70.83	78.50
0240	4.5W 120V LED replacement decorative lamp		160	.050		68	2.83		70.83	78.50
0250	4.5W 120V LED, 2700k replacement decorative lamp		160	.050		68	2.83		70.83	78.50
0260	4.9W 120V LED,2700k replacement decorative lamp candelabra base		160	.050		9.40	2.83		12.23	14.60
0270	4.9W 120V LED,3000k replacement decorative lamp candelabra base		160	.050		9.40	2.83		12.23	14.60
0280	5 W LED PAR 20 Parabolic Reflector Lamp FL		140	.057		60	3.23		63.23	71
0300	Globe earth, equal to 100 watt		140	.057		27.50	3.23		30.73	35.50
0305	7 W LED Reflector Lamp 3000k DIM		140	.057		58	3.23		61.23	69
0310	10 watt omni LED warm white light bulb E26 medium base 120 volt card		140	.057		17.10	3.23		20.33	23.50
0315	7 W LED Reflector Lamp WFL 2700k DIM		140	.057		58	3.23		61.23	69
0320	9 watt omni LED warm white light bulb E26 medium base 120 volt card		140	.057		9.05	3.23		12.28	14.80
1100	MR16, 3 watt, replacement of halogen lamp 25 watt		130	.062		19	3.48		22.48	26
1200	6 watt replacement of halogen lamp 45 watt		130	.062		20	3.48		23.48	27
2100	10 watt, PAR20, equal to 60 watt		130	.062		26.50	3.48		29.98	34.50
2200	15 watt, PAR30, equal to 100 watt		130	.062		73.50	3.48		76.98	86
2210	LED lamp 50 watt w/6ft pri		130	.062		610	3.48		613.48	675
2220	11 Watt reflector dimmable warm white LED light bulb with medium base		130	.062		41.50	3.48		44.98	50.50
2221	12 W A-Line LED Lamp DIM		130	.062		78	3.48		81.48	91
2225	13 Watt reflector LED warm white e26 with medium base 120 volt box		130	.062		39	3.48		42.48	48
2226	13W br30 LED lamp		130	.062		54	3.48		57.48	64.50
2227	15W 120V br30 inc LED lamp, 2700k		130	.062		52.50	3.48		55.98	63
2228	15W 120V br30 inc LED lamp, 4000k		130	.062		50	3.48		53.48	60
2230	3 Watt dimmable warm white decorative LED lamp with medium base		160	.050		12.60	2.83		15.43	18.10
2240	.43 watt night light LED daylight bulb E12 candelabra base 120 volt 2 pack		160	.050		4.72	2.83		7.55	9.45
2250	15 watt omni-directional LED warm white e26 medium base 120 volt box		160	.050		40	2.83		42.83	48
2251	11watt omni-directional LED warm white e26 medium base 120 volt box		160	.050		28	2.83		30.83	34.50
2252	10watt omni-directional LED warm white e26 medium base 120 volt box		160	.050		28	2.83		30.83	34.50
2253	7 watt omni A19 LED warm white e26 medium base 120 volt box		160	.050		20.50	2.83		23.33	26.50
2255	8 PAR 20 parabolic reflector 2700 LED lamp		160	.050		33	2.83		35.83	40.50

598

For customer support on your Building Construction Costs with RSMeans Data, call 800.448.8182.

26 61 Lighting Systems and Accessories

26 61 23 – Lamps Applications

26 61 23.55 LED Lamps

		Crew	Daily Output	Labor-Hours	Unit	Material	2017 Bare Costs Labor	Equipment	Total	Total Incl O&P
2256	8 PAR 20 Parabolic Reflector 3000 LED Lamp	1 Elec	160	.050	Ea.	33	2.83		35.83	40.50
2260	12W par38 120V LED 15 degree directional lamp		160	.050		203	2.83		205.83	227
2270	12W par38 120V LED 25 degree directional lamp		160	.050		203	2.83		205.83	227
2280	12W par38 120V LED 40 degree directional lamp		160	.050		203	2.83		205.83	227
2285	16W par38 120V 2700k LED parabolic reflector lamp		160	.050		21.50	2.83		24.33	28
2290	16W par38 120V 3000k LED parabolic reflector lamp		160	.050		21.50	2.83		24.33	28
3000	15W par30 LED daylight E26 medium base 120V box		160	.050		60	2.83		62.83	70
3100	17W LED3000k par38 100W replacement		160	.050		47	2.83		49.83	55.50
3110	17W LED par38 100W replacement		160	.050		47	2.83		49.83	55.50
3120	17W LED par38 100W replacement parabolic reflector		160	.050		47	2.83		49.83	55.50
3130	24 W LED T8 PW straight fluorescent lamp		8.89	.900		289	51		340	395
3135	Linear fluorescent LED lamp 120 V		8.89	.900		48	51		99	129
3200	30W LED2700K recessed 8055E par 38 high power		160	.050		234	2.83		236.83	261
3210	30W LED 4200K 120 degree 8055E par 38 high power		160	.050		234	2.83		236.83	261
3220	30W LED 5700K 8055E par 38 high power		160	.050		234	2.83		236.83	261
3230	50W LED2700K 8045M par38 277V high power		160	.050		340	2.83		342.83	380
3240	50W LED4200K 8045M par38 277 V high power retro fit		160	.050		340	2.83		342.83	380
3250	50W LED 5700K 8045M par38 277V high power retro fit		160	.050		340	2.83		342.83	380
8000	10 watt LED par 30 / fl 10 pk		160	.050		96	2.83		98.83	110
8010	Gen 3 Par 30 15 watt short neck power LED 120 VAC E26 80 +cri 300k dimm		160	.050		12.95	2.83		15.78	18.50
8020	12 PAR 30 2700K parabolic reflector LED lamp		160	.050		39.50	2.83		42.33	47.50
8030	12 PAR 30 3000K parabolic reflector LED lamp		160	.050		39.50	2.83		42.33	47.50
8040	3500K LED advantage T8 9 W 800LM 2ft linear 2 BD frosted		69	.116		10.45	6.55		17	21.50
8050	3500K LED litespan T8 9 W 900LM 2ft linear frosted		69	.116		15.30	6.55		21.85	26.50
8060	4000K LED advantage T8 9 W 800LM 2ft linear 2BD frosted		69	.116		10.45	6.55		17	21.50
8070	4000K LED litespan T8 9 W 900LM 2ft linear frosted		69	.116		15.30	6.55		21.85	26.50
8080	5000K LED advantage T8 9 W 800LM 2ft linear 2BD frosted		69	.116		10.45	6.55		17	21.50
8090	5000K LED litespan T8 9 W 900LM 2ft linear frosted		69	.116		15.30	6.55		21.85	26.50
8100	3500K LED advantage T8 18 W 1600LM 4ft linear 2 BD frosted		69	.116		15.30	6.55		21.85	26.50
8105	18 W LED 4ft T8 4000K frost 1600L linear lamp		69	.116		18.50	6.55		25.05	30.50
8108	18 W LED 48 inch T8 4100K 1890LM linear lamp		69	.116		45.50	6.55		52.05	60
8110	3500K LED advantage T8 18 W 1800LM 4ft linear 2 BD frosted		69	.116		19.80	6.55		26.35	32
8120	4000K LED advantage T8 18 W 1600LM 4ft linear 2 BD frosted		69	.116		15.30	6.55		21.85	26.50
8130	4000K LED litespan T8 18 W 1800LM 4ft linear frosted		69	.116		19.80	6.55		26.35	32
8140	5000K LED advantage T8 18 W 1600LM 4ft linear 2BD frosted		69	.116		15.30	6.55		21.85	26.50
8150	5000K LED litespan T8 18 W 1800LM 4ft linear 2BD frosted		69	.116		19.80	6.55		26.35	32
8200	16.5 T8 3000 IF-6U U-shape fluorescent LED lamp		65	.123		38	6.95		44.95	52.50
8210	16.5 T8 3500 IF-6U U-shape fluorescent LED lamp		65	.123		38	6.95		44.95	52.50
8220	16.5 T8 4000 IF-6U U-shape fluorescent LED lamp		65	.123		38	6.95		44.95	52.50
8230	16.5 T8 5000 IF-6U U-shape fluorescent LED lamp		65	.123		38	6.95		44.95	52.50
8240	18 W 6 inch T8 4100 K U-shape frosted fluorescent LED lamp		60	.133		34	7.55		41.55	49
8250	18 W 6 inch T8 5000 K U-shape frosted fluorescent LED lamp		60	.133		34	7.55		41.55	49
8260	Circular 12 W linear fluorescent LED 2700K MOD lamp		62.40	.128		273	7.25		280.25	310
8270	Circular 12 W linear fluorescent LED3000K MOD lamp		62.40	.128		273	7.25		280.25	310
8280	Circular 12 W linear fluorescent LED 3500K MOD lamp		62.40	.128		273	7.25		280.25	310
8290	Circular 18 W linear fluorescent LED3000K MOD lamp		62.40	.128		375	7.25		382.25	420
8300	Circular 18 W linear fluorescent LED3500K MOD lamp		62.40	.128		375	7.25		382.25	420
8310	Circular 18 W linear fluorescent LED 5000K MOD lamp		62.40	.128		375	7.25		382.25	420
8320	Circular 18 W linear fluorescent LED 4100K MOD lamp	▼	62.40	.128	▼	375	7.25		382.25	420

For customer support on your Building Construction Costs with RSMeans Data, call 800.448.8182.

599

26 71 13 – Motors Applications

26 71 13.20 Motors		Crew	Daily Output	Labor-Hours	Unit	Material	2017 Bare Costs Labor	Equipment	Total	Total Incl O&P
0010	**MOTORS** 230/460 volts, 60 HZ									
0050	Dripproof, premium efficiency, 1.15 service factor									
0060	1800 RPM, 1/4 HP	1 Elec	5.33	1.501	Ea.	252	85		337	405
0070	1/3 HP		5.33	1.501		232	85		317	380
0080	1/2 HP		5.33	1.501		194	85		279	340
0090	3/4 HP		5.33	1.501		271	85		356	425
0100	1 HP		4.50	1.778		290	101		391	470
0250	5 HP		4.50	1.778		620	101		721	830
0350	10 HP	↓	4	2		1,150	113		1,263	1,450
0450	20 HP	2 Elec	5.20	3.077	↓	1,950	174		2,124	2,400

26 71 13.40 Motors Explosion Proof

26 71 13.40 Motors Explosion Proof		Crew	Daily Output	Labor-Hours	Unit	Material	2017 Bare Costs Labor	Equipment	Total	Total Incl O&P
0010	**MOTORS EXPLOSION PROOF**, 208-230/460 volts, 60 HZ									
0020	1800 RPM, 1/4 HP	1 Elec	5	1.600	Ea.	565	90.50		655.50	760
0030	1/3 HP		5	1.600		272	90.50		362.50	435
0040	1/2 HP		5	1.600		485	90.50		575.50	665
0050	3/4 HP		5.33	1.501		430	85		515	600
0060	1 HP		4.20	1.905		555	108		663	770
0090	5 HP		4.20	1.905		890	108		998	1,125
0110	10 HP	↓	3.70	2.162		1,300	122		1,422	1,600
0130	20 HP	2 Elec	5	3.200		2,125	181		2,306	2,600
2000	3600 RPM, 1/4 HP	1 Elec	5	1.600		405	90.50		495.50	580
2010	1/3 HP		5	1.600		425	90.50		515.50	600
2020	1/2 HP		5	1.600		390	90.50		480.50	565
2030	3/4 HP		5	1.600		530	90.50		620.50	715
2040	1 HP		4.20	1.905		610	108		718	830
2070	5 HP		4.20	1.905		975	108		1,083	1,225
2090	10 HP	↓	3.70	2.162		1,275	122		1,397	1,575
2110	20 HP	2 Elec	5	3.200	↓	2,150	181		2,331	2,625

Estimating Tips
27 20 00 Data Communications
27 30 00 Voice Communications
27 40 00 Audio-Video Communications

- When estimating material costs for special systems, it is always prudent to obtain manufacturers' quotations for equipment prices and special installation requirements that may affect the total cost.

- For cost modifications for elevated tray installation, add the percentages to labor according to the height of the installation and only to the quantities exceeding the different height levels, not to the total tray quantities. Refer to 26 01 02.20 for labor adjustment factors.

- Do not overlook the costs for equipment used in the installation. If scissor lifts and boom lifts are available in the field, contractors may use them in lieu of the proposed ladders and rolling staging.

Reference Numbers
Reference numbers are shown at the beginning of some major classifications. These numbers refer to related items in the Reference Section. The reference information may be an estimating procedure, an alternate pricing method, or technical information.

Note: Not all subdivisions listed here necessarily appear. ■

Did you know?
Our online estimating solution gives you the same access to RSMeans' data with 24/7 access:

- Quickly locate costs in the searchable database.
- Build cost lists, estimates, and reports in minutes.
- Adjust costs to any location in the U.S. and Canada with the click of a button.

Start your free trial today at
www.RSMeansOnline.com

27 13 Communications Backbone Cabling

27 13 23 – Communications Optical Fiber Backbone Cabling

27 13 23.13 Communications Optical Fiber	Crew	Daily Output	Labor-Hours	Unit	Material	2017 Bare Costs Labor	Equipment	Total	Total Incl O&P
0010	**COMMUNICATIONS OPTICAL FIBER**									
0040	Specialized tools & techniques cause installation costs to vary.									
0070	Fiber optic, cable, bulk simplex, single mode	1 Elec	8	1	C.L.F.	24.50	56.50		81	112
0080	Multi mode		8	1		29.50	56.50		86	118
0090	4 strand, single mode		7.34	1.090		40.50	61.50		102	138
0095	Multi mode		7.34	1.090		50.50	61.50		112	148
0100	12 strand, single mode		6.67	1.199		78	68		146	188
0105	Multi mode		6.67	1.199		96.50	68		164.50	208
0150	Jumper				Ea.	33			33	36.50
0200	Pigtail					37			37	40.50
0300	Connector	1 Elec	24	.333		27	18.85		45.85	58
0350	Finger splice		32	.250		40	14.15		54.15	65.50
0400	Transceiver (low cost bi-directional)		8	1		480	56.50		536.50	610
0450	Rack housing, 4 rack spaces, 12 panels (144 fibers)		2	4		560	226		786	955
0500	Patch panel, 12 ports		6	1.333		300	75.50		375.50	445
1000	Cable, 62.5 microns, direct burial, 4 fiber	R-15	1200	.040	L.F.	.86	2.23	.24	3.33	4.56
1020	Indoor, 2 fiber	R-19	1000	.020		.43	1.13		1.56	2.17
1040	Outdoor, aerial/duct	"	1670	.012		.69	.68		1.37	1.78
1060	50 microns, direct burial, 8 fiber	R-22	4000	.009		1.34	.48		1.82	2.19
1080	12 fiber		4000	.009		2.18	.48		2.66	3.12
1100	Indoor, 12 fiber		759	.049		2.07	2.55		4.62	6.10
1120	Connectors, 62.5 micron cable, transmission	R-19	40	.500	Ea.	16.75	28.50		45.25	61
1140	Cable splice		40	.500		18.65	28.50		47.15	63
1160	125 micron cable, transmission		16	1.250		17.50	71		88.50	125
1180	Receiver, 1.2 mile range		20	1		247	56.50		303.50	355
1200	1.9 mile range		20	1		233	56.50		289.50	340
1220	6.2 mile range		5	4		290	227		517	660
1240	Transmitter, 1.2 mile range		20	1		271	56.50		327.50	385
1260	1.9 mile range		20	1		305	56.50		361.50	420
1280	6.2 mile range		5	4		390	227		617	765
1300	Modem, 1.2 mile range		5	4		168	227		395	525
1320	6.2 mile range		5	4		305	227		532	675
1340	1.9 mile range, 12 channel		5	4		1,925	227		2,152	2,450
1360	Repeater, 1.2 mile range		10	2		385	113		498	595
1380	1.9 mile range		10	2		480	113		593	700
1400	6.2 mile range		5	4		925	227		1,152	1,375
1420	1.2 mile range, digital		5	4		445	227		672	830

27 41 Audio-Video Systems

27 41 33 – Master Antenna Television Systems

27 41 33.10 T.V. Systems

		Crew	Daily Output	Labor-Hours	Unit	Material	2017 Bare Costs Labor	Equipment	Total	Total Incl O&P
0010	**T.V. SYSTEMS**, not including rough-in wires, cables & conduits									
0100	Master TV antenna system									
0200	VHF reception & distribution, 12 outlets	1 Elec	6	1.333	Outlet	156	75.50		231.50	285
0400	30 outlets		10	.800		155	45.50		200.50	238
0600	100 outlets		13	.615		158	35		193	225
0800	VHF & UHF reception & distribution, 12 outlets		6	1.333		233	75.50		308.50	370
1000	30 outlets		10	.800		155	45.50		200.50	238
1200	100 outlets		13	.615		158	35		193	225
1400	School and deluxe systems, 12 outlets		2.40	3.333		310	189		499	625
1600	30 outlets		4	2		270	113		383	465

27 41 Audio-Video Systems

27 41 33 – Master Antenna Television Systems

27 41 33.10 T.V. Systems		Crew	Daily Output	Labor-Hours	Unit	Material	2017 Bare Costs Labor	Equipment	Total	Total Incl O&P
1800	80 outlets	1 Elec	5.30	1.509	Outlet	259	85.50		344.50	415

27 51 Distributed Audio-Video Communications Systems

27 51 16 – Public Address Systems

27 51 16.10 Public Address System

		Crew	Daily Output	Labor-Hours	Unit	Material	Labor	Equipment	Total	Total Incl O&P
0010	**PUBLIC ADDRESS SYSTEM**									
0100	Conventional, office	1 Elec	5.33	1.501	Speaker	149	85		234	291
0200	Industrial	"	2.70	2.963	"	288	168		456	565

27 51 19 – Sound Masking Systems

27 51 19.10 Sound System

		Crew	Daily Output	Labor-Hours	Unit	Material	Labor	Equipment	Total	Total Incl O&P
0010	**SOUND SYSTEM**, not including rough-in wires, cables & conduits									
0100	Components, projector outlet	1 Elec	8	1	Ea.	46.50	56.50		103	136
0200	Microphone		4	2		115	113		228	297
0400	Speakers, ceiling or wall		8	1		131	56.50		187.50	229
0600	Trumpets		4	2		243	113		356	435
0800	Privacy switch		8	1		97	56.50		153.50	192
1000	Monitor panel		4	2		430	113		543	645
1200	Antenna, AM/FM		4	2		151	113		264	335
1400	Volume control		8	1		52	56.50		108.50	142
1600	Amplifier, 250 watts		1	8		1,425	455		1,880	2,250
1800	Cabinets		1	8		940	455		1,395	1,700
2000	Intercom, 30 station capacity, master station	2 Elec	2	8		2,300	455		2,755	3,200
2200	Remote station	1 Elec	8	1		181	56.50		237.50	285
2400	Intercom outlets		8	1		107	56.50		163.50	202
2600	Handset		4	2		355	113		468	560
2800	Emergency call system, 12 zones, annunciator		1.30	6.154		1,075	350		1,425	1,700
3000	Bell		5.30	1.509		110	85.50		195.50	249
3200	Light or relay		8	1		55	56.50		111.50	146
3400	Transformer		4	2		241	113		354	435
3600	House telephone, talking station		1.60	5		520	283		803	995
3800	Press to talk, release to listen		5.30	1.509		121	85.50		206.50	261
4000	System-on button					72.50			72.50	79.50
4200	Door release	1 Elec	4	2		129	113		242	310
4400	Combination speaker and microphone		8	1		220	56.50		276.50	325
4600	Termination box		3.20	2.500		69	142		211	288
4800	Amplifier or power supply		5.30	1.509		795	85.50		880.50	1,000
5000	Vestibule door unit		16	.500	Name	146	28.50		174.50	204
5200	Strip cabinet		27	.296	Ea.	276	16.75		292.75	330
5400	Directory		16	.500	"	130	28.50		158.50	186

For customer support on your Building Construction Costs with RSMeans Data, call 800.448.8182.

603

27 52 Healthcare Communications and Monitoring Systems

27 52 23 – Nurse Call/Code Blue Systems

27 52 23.10 Nurse Call Systems	Crew	Daily Output	Labor-Hours	Unit	Material	2017 Bare Costs Labor	Equipment	Total	Total Incl O&P
0010	**NURSE CALL SYSTEMS**								
0100	Single bedside call station	1 Elec	8	1	Ea.	239	56.50		295.50	350
0200	Ceiling speaker station		8	1		72.50	56.50		129	165
0400	Emergency call station		8	1		78	56.50		134.50	171
0600	Pillow speaker		8	1		191	56.50		247.50	295
0800	Double bedside call station		4	2		154	113		267	340
1000	Duty station		4	2		130	113		243	315
1200	Standard call button		8	1		93.50	56.50		150	188
1400	Lights, corridor, dome or zone indicator		8	1		53	56.50		109.50	144
1600	Master control station for 20 stations	2 Elec	.65	24.615	Total	3,450	1,400		4,850	5,875

27 53 Distributed Systems

27 53 13 – Clock Systems

27 53 13.50 Clock Equipments

		Crew	Daily Output	Labor-Hours	Unit	Material	2017 Bare Costs Labor	Equipment	Total	Total Incl O&P
0010	**CLOCK EQUIPMENTS**, not including wires & conduits									
0100	Time system components, master controller	1 Elec	.33	24.242	Ea.	1,675	1,375		3,050	3,875
0200	Program bell		8	1		112	56.50		168.50	208
0400	Combination clock & speaker		3.20	2.500		200	142		342	430
0600	Frequency generator		2	4		2,425	226		2,651	3,025
0800	Job time automatic stamp recorder		4	2		565	113		678	790
1600	Master time clock system, clocks & bells, 20 room	4 Elec	.20	160		6,400	9,050		15,450	20,700
1800	50 room	"	.08	400		13,500	22,600		36,100	48,800
1900	Time clock	1 Elec	3.20	2.500		450	142		592	705
2000	100 cards in & out, 1 color					10			10	11
2200	2 colors					10			10	11
2800	Metal rack for 25 cards	1 Elec	7	1.143		50	64.50		114.50	152

Estimating Tips

- When estimating material costs for electronic safety and security systems, it is always prudent to obtain manufacturers' quotations for equipment prices and special installation requirements that may affect the total cost.

- Fire alarm systems consist of control panels, annunciator panels, batteries with rack, charger, and fire alarm actuating and indicating devices. Some fire alarm systems include speakers, telephone lines, door closer controls, and other components. Be careful not to overlook the costs related to installation for these items. Also be aware of costs for integrated automation instrumentation and terminal devices, control equipment, control wiring, and programming. Insurance underwriters may have specific requirements for the type of materials to be installed or design requirements based on the hazard to be protected. Local jurisdictions may have requirements not covered by code. It is advisable to be aware of any special conditions.

- Security equipment includes items such as CCTV, access control, and other detection and identification systems to perform alert and alarm functions. Be sure to consider the costs related to installation for this security equipment, such as for integrated automation instrumentation and terminal devices, control equipment, control wiring, and programming.

Reference Numbers

Reference numbers are shown at the beginning of some major classifications. These numbers refer to related items in the Reference Section. The reference information may be an estimating procedure, an alternate pricing method, or technical information.

Note: Not all subdivisions listed here necessarily appear. ∎

Did you know?

Our online estimating solution gives you the same access to RSMeans' data with 24/7 access:

- Quickly locate costs in the searchable database.
- Build cost lists, estimates, and reports in minutes.
- Adjust costs to any location in the U.S. and Canada with the click of a button.

Start your free trial today at
www.RSMeansOnline.com

28 13 Access Control

28 13 53 – Security Access Detection

28 13 53.13 Security Access Metal Detectors	Crew	Daily Output	Labor-Hours	Unit	Material	2017 Bare Costs Labor	2017 Bare Costs Equipment	Total	Total Incl O&P
0010 **SECURITY ACCESS METAL DETECTORS**									
0240 Metal detector, hand-held, wand type, unit only				Ea.	129			129	138
0250 Metal detector, walk through portal type, single zone	1 Elec	2	4		4,000	226		4,226	4,750
0260 Multi-zone	"	2	4	↓	5,925	226		6,151	6,875

28 13 53.16 Security Access X-Ray Equipment

	Crew	Daily Output	Labor-Hours	Unit	Material	Labor	Equipment	Total	Total Incl O&P
0010 **SECURITY ACCESS X-RAY EQUIPMENT**									
0290 X-ray machine, desk top, for mail/small packages/letters	1 Elec	4	2	Ea.	3,475	113		3,588	4,000
0300 Conveyor type, incl monitor		2	4		16,000	226		16,226	17,900
0310 Includes additional features		2	4		28,000	226		28,226	31,100
0320 X-ray machine, large unit, for airports, incl monitor	2 Elec	1	16		40,000	905		40,905	45,400
0330 Full console	"	.50	32	↓	68,500	1,800		70,300	78,000

28 13 53.23 Security Access Explosive Detection Equipment

	Crew	Daily Output	Labor-Hours	Unit	Material	Labor	Equipment	Total	Total Incl O&P
0010 **SECURITY ACCESS EXPLOSIVE DETECTION EQUIPMENT**									
0270 Explosives detector, walk through portal type	1 Elec	2	4	Ea.	44,200	226		44,426	48,900
0280 Hand-held, battery operated				"				25,500	28,100

28 16 Intrusion Detection

28 16 16 – Intrusion Detection Systems Infrastructure

28 16 16.50 Intrusion Detection

	Crew	Daily Output	Labor-Hours	Unit	Material	Labor	Equipment	Total	Total Incl O&P
0010 **INTRUSION DETECTION**, not including wires & conduits									
0100 Burglar alarm, battery operated, mechanical trigger	1 Elec	4	2	Ea.	280	113		393	480
0200 Electrical trigger		4	2		335	113		448	540
0400 For outside key control, add		8	1		87	56.50		143.50	181
0600 For remote signaling circuitry, add		8	1		138	56.50		194.50	237
0800 Card reader, flush type, standard		2.70	2.963		795	168		963	1,125
1000 Multi-code		2.70	2.963		1,200	168		1,368	1,575
1200 Door switches, hinge switch		5.30	1.509		62.50	85.50		148	197
1400 Magnetic switch		5.30	1.509		77	85.50		162.50	213
1600 Exit control locks, horn alarm		4	2		213	113		326	405
1800 Flashing light alarm		4	2		244	113		357	440
2000 Indicating panels, 1 channel		2.70	2.963		271	168		439	550
2200 10 channel	2 Elec	3.20	5		1,075	283		1,358	1,600
2400 20 channel		2	8		2,500	455		2,955	3,400
2600 40 channel		1.14	14.035		4,500	795		5,295	6,150
2800 Ultrasonic motion detector, 12 volt	1 Elec	2.30	3.478		210	197		407	525
3000 Infrared photoelectric detector	"	4	2	↓	146	113		259	330

28 23 Video Surveillance

28 23 13 – Video Surveillance Control and Management Systems

28 23 13.10 Closed Circuit Television System

	Crew	Daily Output	Labor-Hours	Unit	Material	Labor	Equipment	Total	Total Incl O&P
0010 **CLOSED CIRCUIT TELEVISION SYSTEM**									
2000 Surveillance, one station (camera & monitor)	2 Elec	2.60	6.154	Total	1,500	350		1,850	2,175
2200 For additional camera stations, add	1 Elec	2.70	2.963	Ea.	835	168		1,003	1,175
2400 Industrial quality, one station (camera & monitor)	2 Elec	2.60	6.154	Total	3,100	350		3,450	3,925
2600 For additional camera stations, add	1 Elec	2.70	2.963	Ea.	1,900	168		2,068	2,325
2610 For low light, add		2.70	2.963		1,525	168		1,693	1,925
2620 For very low light, add		2.70	2.963		11,200	168		11,368	12,600
2800 For weatherproof camera station, add		1.30	6.154		1,175	350		1,525	1,800
3000 For pan and tilt, add		1.30	6.154	↓	3,025	350		3,375	3,850

28 23 Video Surveillance

28 23 13 – Video Surveillance Control and Management Systems

28 23 13.10 Closed Circuit Television System

		Crew	Daily Output	Labor-Hours	Unit	Material	2017 Bare Costs Labor	Equipment	Total	Total Incl O&P
3200	For zoom lens - remote control, add	1 Elec	2	4	Ea.	2,775	226		3,001	3,400
3400	Extended zoom lens		2	4		10,200	226		10,426	11,500
3410	For automatic iris for low light, add		2	4		2,425	226		2,651	3,025
3600	Educational T.V. studio, basic 3 camera system, black & white,									
3800	electrical & electronic equip. only	4 Elec	.80	40	Total	14,400	2,275		16,675	19,300
4000	Full console		.28	114		61,500	6,475		67,975	77,000
4100	As above, but color system		.28	114		81,500	6,475		87,975	99,000
4120	Full console		.12	266		354,000	15,100		369,100	412,000
4200	For film chain, black & white, add	1 Elec	1	8	Ea.	16,500	455		16,955	18,900
4250	Color, add		.25	32		20,100	1,800		21,900	24,800
4400	For video recorders, add		1	8		3,475	455		3,930	4,500
4600	Premium	4 Elec	.40	80		28,900	4,525		33,425	38,600

28 23 23 – Video Surveillance Systems Infrastructure

28 23 23.50 Video Surveillance Equipments

		Crew	Daily Output	Labor-Hours	Unit	Material	2017 Bare Costs Labor	Equipment	Total	Total Incl O&P
0010	**VIDEO SURVEILLANCE EQUIPMENTS**									
0200	Video cameras, wireless, hidden in exit signs, clocks, etc., incl. receiver	1 Elec	3	2.667	Ea.	200	151		351	445
0210	Accessories for video recorder, single camera		3	2.667		183	151		334	425
0220	For multiple cameras		3	2.667		1,750	151		1,901	2,150
0230	Video cameras, wireless, for under vehicle searching, complete		2	4		16,000	226		16,226	17,900
0234	Master monitor station, 3 doors x 5 color monitor with tilt feature		2	4		810	226		1,036	1,225

28 31 Fire Detection and Alarm

28 31 23 – Fire Detection and Alarm Annunciation Panels and Fire Stations

28 31 23.50 Alarm Panels and Devices

		Crew	Daily Output	Labor-Hours	Unit	Material	2017 Bare Costs Labor	Equipment	Total	Total Incl O&P
0010	**ALARM PANELS AND DEVICES**, not including wires & conduits									
2200	Intercom remote station	1 Elec	8	1	Ea.	169	56.50		225.50	271
2400	Intercom outlet		8	1		99.50	56.50		156	194
2600	Sound system, intercom handset		8	1		330	56.50		386.50	445
3600	4 zone	2 Elec	2	8		575	455		1,030	1,300
3800	8 zone		1	16		860	905		1,765	2,300
3810	5 zone		1.50	10.667		795	605		1,400	1,775
3900	10 zone		1.25	12.800		1,050	725		1,775	2,225
4000	12 zone		.67	23.988		2,150	1,350		3,500	4,400
4020	Alarm device, tamper, flow	1 Elec	8	1		240	56.50		296.50	350
4025	Fire alarm, loop expander card		16	.500		740	28.50		768.50	855
4050	Actuating device		8	1		340	56.50		396.50	460
4200	Battery and rack		4	2		415	113		528	625
4400	Automatic charger		8	1		590	56.50		646.50	735
4600	Signal bell		8	1		130	56.50		186.50	228
4610	Fire alarm signal bell 10 Inch red 20-24 V P		8	1		136	56.50		192.50	234
4800	Trouble buzzer or manual station		8	1		84	56.50		140.50	178
5425	Duct smoke and heat detector 2 wire		8	1		131	56.50		187.50	229
5430	Fire alarm duct detector controller		3	2.667		197	151		348	445
5435	Fire alarm duct detector sensor kit		8	1		75	56.50		131.50	168
5440	Remote test station for smoke detector duct type		5.30	1.509		60.50	85.50		146	195
5460	Remote fire alarm indicator light		5.30	1.509		18.55	85.50		104.05	149
5600	Strobe and horn		5.30	1.509		154	85.50		239.50	297
5610	Strobe and horn (ADA type)		5.30	1.509		154	85.50		239.50	297
5620	Visual alarm (ADA type)		6.70	1.194		104	67.50		171.50	215
5800	Fire alarm horn		6.70	1.194		61.50	67.50		129	169

For customer support on your Building Construction Costs with RSMeans Data, call 800.448.8182.

607

28 31 Fire Detection and Alarm

28 31 23 – Fire Detection and Alarm Annunciation Panels and Fire Stations

28 31 23.50 Alarm Panels and Devices	Crew	Daily Output	Labor-Hours	Unit	Material	2017 Bare Costs Labor	Equipment	Total	Total Incl O&P
6000	Door holder, electro-magnetic	1 Elec	4	2	Ea.	104	113		217	284
6200	Combination holder and closer		3.20	2.500		124	142		266	350
6600	Drill switch		8	1		375	56.50		431.50	495
6800	Master box		2.70	2.963		6,475	168		6,643	7,375
7000	Break glass station		8	1		61.50	56.50		118	153
7800	Remote annunciator, 8 zone lamp		1.80	4.444		213	252		465	610
8000	12 zone lamp	2 Elec	2.60	6.154		420	350		770	985
8200	16 zone lamp	"	2.20	7.273		380	410		790	1,025

28 31 43 – Fire Detection Sensors

28 31 43.50 Fire and Heat Detectors

0010	FIRE & HEAT DETECTORS									
5000	Detector, W/bell at rate of rise	1 Elec	8	1	Ea.	51.50	56.50		108	142

28 31 46 – Smoke Detection Sensors

28 31 46.50 Smoke Detectors

0010	SMOKE DETECTORS									
5200	Smoke detector, ceiling type	1 Elec	6.20	1.290	Ea.	130	73		203	252
5240	Smoke detector addressable type		6	1.333		224	75.50		299.50	360
5400	Duct type		3.20	2.500		330	142		472	575
5420	Duct addressable type		3.20	2.500		515	142		657	775
8300	Smoke alarm with w/integrated strobe light 120 V 16DB 60 FPM flash rate		16	.500		92	28.50		120.50	144
8310	Photoelectric smoke detector with strobe 120 V 90 DB ceiling mount		12	.667		160	37.50		197.50	233
8320	120 V, 90 DB wall mount		12	.667		159	37.50		196.50	232
8440	Fire alarm beam detector, motorised reflective, infrared optical beam	2 Elec	1.60	10		800	565		1,365	1,725

28 31 49 – Carbon-Monoxide Detection Sensors

28 31 49.50 Carbon-Monoxide Detectors

0010	CARBON-MONOXIDE DETECTORS									
8400	Smoke and carbon monoxide alarm battery operated photoelectric low profile	1 Elec	24	.333	Ea.	46	18.85		64.85	79
8410	low profile photoelectric battery powered		24	.333		28.50	18.85		47.35	60
8420	photoelectric low profile sealed lithium		24	.333		48.50	18.85		67.35	81.50
8430	Photoelectric low profile sealed lithium smoke and CO with voice combo		24	.333		38	18.85		56.85	70.50
8500	Carbon monoxide sensor, wall mount 1Mod 1 relay output smoke & heat		24	.333		530	18.85		548.85	610
8700	Carbon monoxide detector, battery operated, wall mounted		16	.500		52.50	28.50		81	100
8710	Hardwired, wall and ceiling mounted		8	1		101	56.50		157.50	196
8720	Duct mounted		8	1		345	56.50		401.50	460

28 33 Gas Detection and Alarm

28 33 33 – Gas Detection Sensors

28 33 33.50 Tank Leak Detection Systems

0010	TANK LEAK DETECTION SYSTEMS Liquid and vapor									
0100	For hydrocarbons and hazardous liquids/vapors									
0120	Controller, data acquisition, incl. printer, modem, RS232 port									
0140	24 channel, for use with all probes				Ea.	2,975			2,975	3,275
0160	9 channel, for external monitoring				"	1,200			1,200	1,325
0200	Probes									
0210	Well monitoring									
0220	Liquid phase detection				Ea.	675			675	740
0230	Hydrocarbon vapor, fixed position					645			645	705
0240	Hydrocarbon vapor, float mounted					555			555	610
0250	Both liquid and vapor hydrocarbon					550			550	605

28 33 Gas Detection and Alarm

28 33 33 – Gas Detection Sensors

28 33 33.50 Tank Leak Detection Systems	Crew	Daily Output	Labor-Hours	Unit	Material	2017 Bare Costs Labor	Equipment	Total	Total Incl O&P
0300 Secondary containment, liquid phase									
0310 Pipe trench/manway sump				Ea.	605			605	665
0320 Double wall pipe and manual sump					510			510	565
0330 Double wall fiberglass annular space					395			395	435
0340 Double wall steel tank annular space					325			325	355
0500 Accessories									
0510 Modem, non-dedicated phone line				Ea.	279			279	305
0600 Monitoring, internal									
0610 Automatic tank gauge, incl. overfill				Ea.	1,250			1,250	1,375
0620 Product line				"	1,300			1,300	1,425
0700 Monitoring, special									
0710 Cathodic protection				Ea.	700			700	770
0720 Annular space chemical monitor				"	955			955	1,050

28 39 Mass Notification Systems

28 39 10 – Notification Systems

28 39 10.10 Mass Notification System

	Crew	Daily Output	Labor-Hours	Unit	Material	2017 Bare Costs Labor	Equipment	Total	Total Incl O&P
0010 **MASS NOTIFICATION SYSTEM**									
0100 Wireless command center, 10,000 devices	2 Elec	1.33	12.030	Ea.	2,275	680		2,955	3,525
0200 Option, email notification					1,700			1,700	1,850
0210 Remote device supervision & monitor					2,425			2,425	2,650
0300 Antenna VHF or UHF, for medium range	1 Elec	4	2		87	113		200	266
0310 For high-power transmitter		2	4		990	226		1,216	1,450
0400 Transmitter, 25 watt		4	2		1,650	113		1,763	1,975
0410 40 watt		2.66	3.008		1,900	170		2,070	2,350
0420 100 watt		1.33	6.015		6,175	340		6,515	7,300
0500 Wireless receiver/control module for speaker		8	1		279	56.50		335.50	390
0600 Desktop paging controller, stand alone		4	2		760	113		873	1,000

For customer support on your Building Construction Costs with RSMeans Data, call 800.448.8182.

609

Division Notes

	CREW	DAILY OUTPUT	LABOR-HOURS	UNIT	BARE COSTS				TOTAL INCL O&P
					MAT.	LABOR	EQUIP.	TOTAL	

Estimating Tips
31 05 00 Common Work Results for Earthwork

- Estimating the actual cost of performing earthwork requires careful consideration of the variables involved. This includes items such as type of soil, whether water will be encountered, dewatering, whether banks need bracing, disposal of excavated earth, and length of haul to fill or spoil sites, etc. If the project has large quantities of cut or fill, consider raising or lowering the site to reduce costs, while paying close attention to the effect on site drainage and utilities.

- If the project has large quantities of fill, creating a borrow pit on the site can significantly lower the costs.

- It is very important to consider what time of year the project is scheduled for completion. Bad weather can create large cost overruns from dewatering, site repair, and lost productivity from cold weather.

Reference Numbers

Reference numbers are shown at the beginning of some major classifications. These numbers refer to related items in the Reference Section. The reference information may be an estimating procedure, an alternate pricing method, or technical information.

Note: Not all subdivisions listed here necessarily appear. ■

Did you know?

Our online estimating solution gives you the same access to RSMeans' data with 24/7 access:

- Quickly locate costs in the searchable database.
- Build cost lists, estimates, and reports in minutes.
- Adjust costs to any location in the U.S. and Canada with the click of a button.

Start your free trial today at **www.RSMeansOnline.com**

31 05 Common Work Results for Earthwork

31 05 13 – Soils for Earthwork

31 05 13.10 Borrow

	31 05 13.10 Borrow		Crew	Daily Output	Labor-Hours	Unit	Material	2017 Bare Costs Labor	Equipment	Total	Total Incl O&P
0010	**BORROW**	R312316-40									
0020	Spread, 200 H.P. dozer, no compaction, 2 mi. RT haul										
0200	Common borrow		B-15	600	.047	C.Y.	12.80	2.19	4.45	19.44	22.50
0700	Screened loam			600	.047		28	2.19	4.45	34.64	38.50
0800	Topsoil, weed free		↓	600	.047		25	2.19	4.45	31.64	35.50
0900	For 5 mile haul, add		B-34B	200	.040	↓		1.82	3.21	5.03	6.30

31 05 16 – Aggregates for Earthwork

31 05 16.10 Borrow

	31 05 16.10 Borrow		Crew	Daily Output	Labor-Hours	Unit	Material	2017 Bare Costs Labor	Equipment	Total	Total Incl O&P
0010	**BORROW**	R312316-40									
0020	Spread, with 200 H.P. dozer, no compaction, 2 mi. RT haul										
0100	Bank run gravel		B-15	600	.047	L.C.Y.	18.90	2.19	4.45	25.54	29
0300	Crushed stone (1.40 tons per CY), 1-1/2"			600	.047		26	2.19	4.45	32.64	36.50
0320	3/4"			600	.047		26	2.19	4.45	32.64	36.50
0340	1/2"			600	.047		29	2.19	4.45	35.64	39.50
0360	3/8"			600	.047		30.50	2.19	4.45	37.14	41.50
0400	Sand, washed, concrete			600	.047		38	2.19	4.45	44.64	49.50
0500	Dead or bank sand			600	.047		18.60	2.19	4.45	25.24	28.50
0600	Select structural fill		↓	600	.047		20	2.19	4.45	26.64	30
0900	For 5 mile haul, add		B-34B	200	.040	↓		1.82	3.21	5.03	6.30

31 05 23 – Cement and Concrete for Earthwork

31 05 23.30 Plant Mixed Bituminous Concrete

	31 05 23.30 Plant Mixed Bituminous Concrete	Crew	Daily Output	Labor-Hours	Unit	Material	2017 Bare Costs Labor	Equipment	Total	Total Incl O&P
0010	**PLANT MIXED BITUMINOUS CONCRETE**									
0020	Asphaltic concrete plant mix (145 lb. per C.F.)				Ton	66			66	72.50
0040	Asphaltic concrete less than 300 tons add trucking costs									
0050	See Section 31 23 23.20 for hauling costs									
0200	All weather patching mix, hot				Ton	66.50			66.50	73
0250	Cold patch					75			75	82.50
0300	Berm mix				↓	65			65	71.50

31 06 Schedules for Earthwork

31 06 60 – Schedules for Special Foundations and Load Bearing Elements

31 06 60.14 Piling Special Costs

	31 06 60.14 Piling Special Costs	Crew	Daily Output	Labor-Hours	Unit	Material	2017 Bare Costs Labor	Equipment	Total	Total Incl O&P
0010	**PILING SPECIAL COSTS**									
0011	Piling special costs, pile caps, see Section 03 30 53.40									
0500	Cutoffs, concrete piles, plain	1 Pile	5.50	1.455	Ea.		72		72	114
0600	With steel thin shell, add		38	.211			10.40		10.40	16.55
0700	Steel pile or "H" piles		19	.421			21		21	33
0800	Wood piles		38	.211	↓		10.40		10.40	16.55
0900	Pre-augering up to 30' deep, average soil, 24" diameter	B-43	180	.267	L.F.		11.65	14.30	25.95	33.50
0920	36" diameter		115	.417			18.25	22.50	40.75	52.50
0960	48" diameter		70	.686			30	37	67	86
0980	60" diameter	↓	50	.960	↓		42	51.50	93.50	121
1000	Testing, any type piles, test load is twice the design load									
1050	50 ton design load, 100 ton test				Ea.				14,000	15,500
1100	100 ton design load, 200 ton test								20,000	22,000
1150	150 ton design load, 300 ton test								26,000	28,500
1200	200 ton design load, 400 ton test								28,000	31,000
1250	400 ton design load, 800 ton test				↓				32,000	35,000
1500	Wet conditions, soft damp ground									
1600	Requiring mats for crane, add								40%	40%

31 06 Schedules for Earthwork

31 06 60 – Schedules for Special Foundations and Load Bearing Elements

31 06 60.14 Piling Special Costs

		Crew	Daily Output	Labor-Hours	Unit	Material	2017 Bare Costs Labor	Equipment	Total	Total Incl O&P
1700	Barge mounted driving rig, add								30%	30%

31 06 60.15 Mobilization

		Crew	Daily Output	Labor-Hours	Unit	Material	2017 Bare Costs Labor	Equipment	Total	Total Incl O&P
0010	**MOBILIZATION**									
0020	Set up & remove, air compressor, 600 CFM	A-5	3.30	5.455	Ea.		217	16.45	233.45	350
0100	1200 CFM	"	2.20	8.182			325	24.50	349.50	520
0200	Crane, with pile leads and pile hammer, 75 ton	B-19	.60	106			5,450	3,100	8,550	11,900
0300	150 ton	"	.36	177			9,075	5,175	14,250	19,900
0500	Drill rig, for caissons, to 36", minimum	B-43	2	24			1,050	1,275	2,325	3,025
0520	Maximum		.50	96			4,200	5,150	9,350	12,100
0600	Up to 84"	↓	1	48			2,100	2,575	4,675	6,025
0800	Auxiliary boiler, for steam small	A-5	1.66	10.843			430	33	463	695
0900	Large	"	.83	21.687			860	65.50	925.50	1,400
1100	Rule of thumb: complete pile driving set up, small	B-19	.45	142			7,275	4,125	11,400	15,900
1200	Large	"	.27	237	↓		12,100	6,900	19,000	26,500
1500	Mobilization, barge, by tug boat	B-83	25	.640	Mile		29.50	32.50	62	80.50

31 11 Clearing and Grubbing

31 11 10 – Clearing and Grubbing Land

31 11 10.10 Clear and Grub Site

		Crew	Daily Output	Labor-Hours	Unit	Material	2017 Bare Costs Labor	Equipment	Total	Total Incl O&P
0010	**CLEAR AND GRUB SITE**									
0020	Cut & chip light trees to 6" diam.	B-7	1	48	Acre		2,000	1,775	3,775	5,025
0150	Grub stumps and remove	B-30	2	12			580	1,125	1,705	2,125
0200	Cut & chip medium, trees to 12" diam.	B-7	.70	68.571			2,875	2,525	5,400	7,200
0250	Grub stumps and remove	B-30	1	24			1,150	2,250	3,400	4,225
0300	Cut & chip heavy, trees to 24" diam.	B-7	.30	160			6,700	5,925	12,625	16,700
0350	Grub stumps and remove	B-30	.50	48			2,325	4,500	6,825	8,450
0400	If burning is allowed, deduct cut & chip				↓				40%	40%
3000	Chipping stumps, to 18" deep, 12" diam.	B-86	20	.400	Ea.		21.50	9.20	30.70	42.50
3040	18" diameter		16	.500			27	11.45	38.45	53
3080	24" diameter		14	.571			30.50	13.10	43.60	61
3100	30" diameter		12	.667			35.50	15.30	50.80	71
3120	36" diameter		10	.800			43	18.35	61.35	85
3160	48" diameter	↓	8	1	↓		53.50	23	76.50	106
5000	Tree thinning, feller buncher, conifer									
5080	Up to 8" diameter	B-93	240	.033	Ea.		1.78	3.47	5.25	6.50
5120	12" diameter		160	.050			2.68	5.20	7.88	9.80
5240	Hardwood, up to 4" diameter		240	.033			1.78	3.47	5.25	6.50
5280	8" diameter		180	.044			2.38	4.63	7.01	8.70
5320	12" diameter	↓	120	.067	↓		3.57	6.95	10.52	13.05
7000	Tree removal, congested area, aerial lift truck									
7040	8" diameter	B-85	7	5.714	Ea.		247	140	387	530
7080	12" diameter		6	6.667			289	163	452	620
7120	18" diameter		5	8			345	195	540	745
7160	24" diameter		4	10			435	244	679	930
7240	36" diameter		3	13.333			575	325	900	1,250
7280	48" diameter	↓	2	20	↓		865	490	1,355	1,850

31 13 Selective Tree and Shrub Removal and Trimming

31 13 13 – Selective Tree and Shrub Removal

31 13 13.10 Selective Clearing	Crew	Daily Output	Labor-Hours	Unit	Material	2017 Bare Costs Labor	Equipment	Total	Total Incl O&P
0010	**SELECTIVE CLEARING**								
0020	Clearing brush with brush saw	A-1C	.25	32	Acre		1,250	118	1,368	2,050
0100	By hand	1 Clab	.12	66.667			2,600		2,600	4,000
0300	With dozer, ball and chain, light clearing	B-11A	2	8			370	690	1,060	1,325
0400	Medium clearing		1.50	10.667			495	925	1,420	1,775
0500	With dozer and brush rake, light		10	1.600			74	138	212	265
0550	Medium brush to 4" diameter		8	2			92.50	173	265.50	330
0600	Heavy brush to 4" diameter		6.40	2.500			116	216	332	415
1000	Brush mowing, tractor w/rotary mower, no removal									
1020	Light density	B-84	2	4	Acre		214	183	397	525
1040	Medium density		1.50	5.333			286	244	530	700
1080	Heavy density		1	8			430	365	795	1,050

31 13 13.20 Selective Tree Removal

31 13 13.20 Selective Tree Removal	Crew	Daily Output	Labor-Hours	Unit	Material	2017 Bare Costs Labor	Equipment	Total	Total Incl O&P
0010	**SELECTIVE TREE REMOVAL**								
0011	With tractor, large tract, firm									
0020	level terrain, no boulders, less than 12" diam. trees									
0300	300 HP dozer, up to 400 trees/acre, 0 to 25% hardwoods	B-10M	.75	16	Acre		780	2,525	3,305	3,950
0340	25% to 50% hardwoods		.60	20			975	3,175	4,150	4,950
0370	75% to 100% hardwoods		.45	26.667			1,300	4,225	5,525	6,625
0400	500 trees/acre, 0% to 25% hardwoods		.60	20			975	3,175	4,150	4,950
0440	25% to 50% hardwoods		.48	25			1,225	3,950	5,175	6,200
0470	75% to 100% hardwoods		.36	33.333			1,625	5,275	6,900	8,275
0500	More than 600 trees/acre, 0 to 25% hardwoods		.52	23.077			1,125	3,650	4,775	5,725
0540	25% to 50% hardwoods		.42	28.571			1,400	4,525	5,925	7,100
0570	75% to 100% hardwoods		.31	38.710			1,875	6,125	8,000	9,625
0900	Large tract clearing per tree									
1500	300 HP dozer, to 12" diameter, softwood	B-10M	320	.038	Ea.		1.83	5.95	7.78	9.35
1550	Hardwood		100	.120			5.85	19	24.85	30
1600	12" to 24" diameter, softwood		200	.060			2.93	9.50	12.43	14.90
1650	Hardwood		80	.150			7.30	24	31.30	37
1700	24" to 36" diameter, softwood		100	.120			5.85	19	24.85	30
1750	Hardwood		50	.240			11.70	38	49.70	60
1800	36" to 48" diameter, softwood		70	.171			8.35	27	35.35	42.50
1850	Hardwood		35	.343			16.70	54.50	71.20	85
2000	Stump removal on site by hydraulic backhoe, 1-1/2 C.Y.									
2040	4" to 6" diameter	B-17	60	.533	Ea.		23.50	12.60	36.10	49.50
2050	8" to 12" diameter	B-30	33	.727			35	68	103	128
2100	14" to 24" diameter		25	.960			46.50	90	136.50	169
2150	26" to 36" diameter		16	1.500			72.50	141	213.50	264
3000	Remove selective trees, on site using chain saws and chipper,									
3050	not incl. stumps, up to 6" diameter	B-7	18	2.667	Ea.		112	98.50	210.50	280
3100	8" to 12" diameter		12	4			168	148	316	420
3150	14" to 24" diameter		10	4.800			201	178	379	500
3200	26" to 36" diameter		8	6			251	222	473	630
3300	Machine load, 2 mile haul to dump, 12" diam. tree	A-3B	8	2			99	150	249	315

31 14 Earth Stripping and Stockpiling

31 14 13 – Soil Stripping and Stockpiling

31 14 13.23 Topsoil Stripping and Stockpiling

	Crew	Daily Output	Labor-Hours	Unit	Material	2017 Bare Costs Labor	Equipment	Total	Total Incl O&P
0010 **TOPSOIL STRIPPING AND STOCKPILING**									
0020 200 H.P. dozer, ideal conditions	B-10B	2300	.005	C.Y.		.25	.60	.85	1.05
0100 Adverse conditions	"	1150	.010			.51	1.20	1.71	2.09
0200 300 H.P. dozer, ideal conditions	B-10M	3000	.004			.20	.63	.83	1
0300 Adverse conditions	"	1650	.007			.35	1.15	1.50	1.81
0400 400 H.P. dozer, ideal conditions	B-10X	3900	.003			.15	.61	.76	.90
0500 Adverse conditions	"	2000	.006			.29	1.18	1.47	1.74
0600 Clay, dry and soft, 200 H.P. dozer, ideal conditions	B-10B	1600	.008			.37	.87	1.24	1.51
0700 Adverse conditions	"	800	.015			.73	1.73	2.46	3.01
1000 Medium hard, 300 H.P. dozer, ideal conditions	B-10M	2000	.006			.29	.95	1.24	1.49
1100 Adverse conditions	"	1100	.011			.53	1.73	2.26	2.71
1200 Very hard, 400 H.P. dozer, ideal conditions	B-10X	2600	.005			.23	.91	1.14	1.34
1300 Adverse conditions	"	1340	.009			.44	1.77	2.21	2.60
1400 Loam or topsoil, remove and stockpile on site									
1420 6" deep, 200' haul	B-10B	865	.014	C.Y.		.68	1.60	2.28	2.79
1430 300' haul		520	.023			1.13	2.66	3.79	4.64
1440 500' haul		225	.053			2.60	6.15	8.75	10.70
1450 Alternate method: 6" deep, 200' haul		5090	.002	S.Y.		.12	.27	.39	.47
1460 500' haul		1325	.009	"		.44	1.04	1.48	1.82
1500 Loam or topsoil, remove/stockpile on site									
1510 By hand, 6" deep, 50' haul, less than 100 S.Y.	B-1	100	.240	S.Y.		9.55		9.55	14.65
1520 By skid steer, 6" deep, 100' haul, 101-500 S.Y.	B-62	500	.048			2.07	.34	2.41	3.53
1530 100' haul, 501-900 S.Y.	"	900	.027			1.15	.19	1.34	1.96
1540 200' haul, 901-1100 S.Y.	B-63	1000	.040			1.66	.17	1.83	2.73
1550 By dozer, 200' haul, 1101-4000 S.Y.	B-10B	4000	.003			.15	.35	.50	.60

31 22 Grading

31 22 13 – Rough Grading

31 22 13.20 Rough Grading Sites

	Crew	Daily Output	Labor-Hours	Unit	Material	2017 Bare Costs Labor	Equipment	Total	Total Incl O&P
0010 **ROUGH GRADING SITES**									
0100 Rough grade sites 400 S.F. or less, hand labor	B-1	2	12	Ea.		480		480	730
0120 410-1000 S.F.	"	1	24			955		955	1,475
0130 1100-3000 S.F., skid steer & labor	B-62	1.50	16			690	114	804	1,175
0140 3100-5000 S.F.	"	1	24			1,025	171	1,196	1,775
0150 5100-8000 S.F.	B-63	1	40			1,650	171	1,821	2,725
0160 8100-10000 S.F.	"	.75	53.333			2,225	228	2,453	3,625
0170 8100-10000 S.F., dozer	B-10L	1	12			585	470	1,055	1,400
0200 Rough grade open sites 10000-20000 S.F., grader	B-11L	1.80	8.889			410	375	785	1,050
0210 20100-25000 S.F.		1.40	11.429			530	485	1,015	1,350
0220 25100-30000 S.F.		1.20	13.333			620	565	1,185	1,550
0230 30100-35000 S.F.		1	16			740	680	1,420	1,875
0240 35100-40000 S.F.		.90	17.778			825	755	1,580	2,075
0250 40100-45000 S.F.		.80	20			925	850	1,775	2,325
0260 45100-50000 S.F.		.72	22.222			1,025	940	1,965	2,600
0270 50100-75000 S.F.		.50	32			1,475	1,350	2,825	3,750
0280 75100-100000 S.F.		.36	44.444			2,050	1,875	3,925	5,200

31 22 Grading

31 22 16 – Fine Grading

31 22 16.10 Finish Grading		Crew	Daily Output	Labor-Hours	Unit	Material	2017 Bare Costs Labor	2017 Bare Costs Equipment	Total	Total Incl O&P
0010	**FINISH GRADING**									
0012	Finish grading area to be paved with grader, small area	B-11L	400	.040	S.Y.		1.85	1.70	3.55	4.69
0100	Large area		2000	.008			.37	.34	.71	.93
1100	Fine grade for slab on grade, machine		1040	.015			.71	.65	1.36	1.80
1150	Hand grading	B-18	700	.034			1.37	.06	1.43	2.16
3500	Finish grading lagoon bottoms	B-11L	4	4	M.S.F.		185	170	355	470

31 23 Excavation and Fill

31 23 16 – Excavation

31 23 16.13 Excavating, Trench

	EXCAVATING, TRENCH	Crew	Daily Output	Labor-Hours	Unit	Material	2017 Bare Costs Labor	2017 Bare Costs Equipment	Total	Total Incl O&P
0010	**EXCAVATING, TRENCH**									
0011	Or continuous footing									
0020	Common earth with no sheeting or dewatering included									
0050	1' to 4' deep, 3/8 C.Y. excavator	B-11C	150	.107	B.C.Y.		4.94	2.44	7.38	10.20
0060	1/2 C.Y. excavator	B-11M	200	.080			3.71	1.97	5.68	7.80
0090	4' to 6' deep, 1/2 C.Y. excavator	"	200	.080			3.71	1.97	5.68	7.80
0100	5/8 C.Y. excavator	B-12Q	250	.064			3.04	2.46	5.50	7.30
0110	3/4 C.Y. excavator	B-12F	300	.053			2.53	2.25	4.78	6.30
0300	1/2 C.Y. excavator, truck mounted	B-12J	200	.080			3.79	4.55	8.34	10.75
0500	6' to 10' deep, 3/4 C.Y. excavator	B-12F	225	.071			3.37	3	6.37	8.45
0510	1 C.Y. excavator	B-12A	400	.040			1.90	2	3.90	5.10
0600	1 C.Y. excavator, truck mounted	B-12K	400	.040			1.90	2.86	4.76	6.05
0610	1-1/2 C.Y. excavator	B-12B	600	.027			1.27	1.61	2.88	3.69
0900	10' to 14' deep, 3/4 C.Y. excavator	B-12F	200	.080			3.79	3.37	7.16	9.45
0910	1 C.Y. excavator	B-12A	360	.044			2.11	2.22	4.33	5.65
1000	1-1/2 C.Y. excavator	B-12B	540	.030			1.41	1.79	3.20	4.11
1300	14' to 20' deep, 1 C.Y. excavator	B-12A	320	.050			2.37	2.49	4.86	6.35
1310	1-1/2 C.Y. excavator	B-12B	480	.033			1.58	2.01	3.59	4.61
1320	2-1/2 C.Y. excavator	B-12S	765	.021			.99	2.01	3	3.72
1340	20' to 24' deep, 1 C.Y. excavator	B-12A	288	.056			2.64	2.77	5.41	7.05
1342	1-1/2 C.Y. excavator	B-12B	432	.037			1.76	2.23	3.99	5.15
1344	2-1/2 C.Y. excavator	B-12S	685	.023			1.11	2.24	3.35	4.16
1352	4' to 6' deep, 1/2 C.Y. excavator w/trench box	B-13H	188	.085			4.04	5.25	9.29	11.95
1354	5/8 C.Y. excavator	"	235	.068			3.23	4.22	7.45	9.55
1356	3/4 C.Y. excavator	B-13G	282	.057			2.69	2.68	5.37	7.05
1362	6' to 10' deep, 3/4 C.Y. excavator w/trench box	"	212	.075			3.58	3.56	7.14	9.35
1370	1 C.Y. excavator	B-13D	376	.043			2.02	2.34	4.36	5.65
1371	1-1/2 C.Y. excavator	B-13E	564	.028			1.35	1.85	3.20	4.09
1374	10' to 14' deep, 3/4 C.Y. excavator w/trench box	B-13G	188	.085			4.04	4.02	8.06	10.55
1375	1 C.Y. excavator	B-13D	338	.047			2.25	2.60	4.85	6.25
1376	1-1/2 C.Y. excavator	B-13E	508	.032			1.49	2.06	3.55	4.54
1381	14' to 20' deep, 1 C.Y. excavator w/trench box	B-13D	301	.053			2.52	2.92	5.44	7.05
1382	1-1/2 C.Y. excavator	B-13E	451	.035			1.68	2.32	4	5.10
1383	2-1/2 C.Y. excavator	B-13J	720	.022			1.05	2.25	3.30	4.07
1386	20' to 24' deep, 1 C.Y. excavator w/trench box	B-13D	271	.059			2.80	3.24	6.04	7.85
1387	1-1/2 C.Y. excavator	B-13E	406	.039			1.87	2.58	4.45	5.65
1388	2-1/2 C.Y. excavator	B-13J	645	.025			1.18	2.51	3.69	4.55
1391	Shoring by S.F./day trench wall protected loose mat., 4' W	B-6	3200	.008	SF Wall	.58	.32	.11	1.01	1.26
1392	Rent shoring per week per S.F. wall protected, loose mat., 4' W					1.61			1.61	1.77
1395	Hydraulic shoring, S.F. trench wall protected stable mat., 4' W	2 Clab	2700	.006		.19	.23		.42	.57

31 23 16.13 Excavating, Trench

		Crew	Daily Output	Labor-Hours	Unit	Material	2017 Bare Costs Labor	2017 Bare Costs Equipment	Total	Total Incl O&P
1397	semi-stable material, 4' W	2 Clab	2400	.007	SF Wall	.27	.26		.53	.69
1398	Rent hydraulic shoring per day/S.F. wall, stable mat., 4' W					.32			.32	.35
1399	semi-stable material					.39			.39	.43
1400	By hand with pick and shovel 2' to 6' deep, light soil	1 Clab	8	1	B.C.Y.		39		39	60
1500	Heavy soil	"	4	2	"		78.50		78.50	120
1700	For tamping backfilled trenches, air tamp, add	A-1G	100	.080	E.C.Y.		3.13	.57	3.70	5.40
1900	Vibrating plate, add	B-18	180	.133	"		5.30	.24	5.54	8.40
2100	Trim sides and bottom for concrete pours, common earth		1500	.016	S.F.		.64	.03	.67	1.01
2300	Hardpan		600	.040	"		1.59	.07	1.66	2.52
2400	Pier and spread footing excavation, add to above				B.C.Y.				30%	30%
3000	Backfill trench, F.E. loader, wheel mtd., 1 C.Y. bucket									
3020	Minimal haul	B-10R	400	.030	L.C.Y.		1.46	.78	2.24	3.08
3040	100' haul	"	200	.060			2.93	1.56	4.49	6.15
3080	2-1/4 C.Y. bucket, minimum haul	B-10T	600	.020			.98	.93	1.91	2.50
3090	100' haul	"	300	.040			1.95	1.85	3.80	5
5020	Loam & sandy clay with no sheeting or dewatering included									
5050	1' to 4' deep, 3/8 C.Y. tractor loader/backhoe	B-11C	162	.099	B.C.Y.		4.58	2.26	6.84	9.45
5060	1/2 C.Y. excavator	B-11M	216	.074			3.43	1.83	5.26	7.20
5080	4' to 6' deep, 1/2 C.Y. excavator	"	216	.074			3.43	1.83	5.26	7.20
5090	5/8 C.Y. excavator	B-12Q	276	.058			2.75	2.22	4.97	6.65
5100	3/4 C.Y. excavator	B-12F	324	.049			2.34	2.08	4.42	5.85
5130	1/2 C.Y. excavator, truck mounted	B-12J	216	.074			3.51	4.21	7.72	10
5140	6' to 10' deep, 3/4 C.Y. excavator	B-12F	243	.066			3.12	2.77	5.89	7.80
5150	1 C.Y. excavator	B-12A	432	.037			1.76	1.85	3.61	4.70
5160	1 C.Y. excavator, truck mounted	B-12K	432	.037			1.76	2.65	4.41	5.60
5170	1-1/2 C.Y. excavator	B-12B	648	.025			1.17	1.49	2.66	3.42
5190	10' to 14' deep, 3/4 C.Y. excavator	B-12F	216	.074			3.51	3.12	6.63	8.80
5200	1 C.Y. excavator	B-12A	389	.041			1.95	2.05	4	5.25
5210	1-1/2 C.Y. excavator	B-12B	583	.027			1.30	1.65	2.95	3.80
5250	14' to 20' deep, 1 C.Y. excavator	B-12A	346	.046			2.19	2.31	4.50	5.90
5260	1-1/2 C.Y. excavator	B-12B	518	.031			1.47	1.86	3.33	4.28
5270	2-1/2 C.Y. excavator	B-12S	826	.019			.92	1.86	2.78	3.45
5300	20' to 24' deep, 1 C.Y. excavator	B-12A	311	.051			2.44	2.57	5.01	6.55
5310	1-1/2 C.Y. excavator	B-12B	467	.034			1.63	2.07	3.70	4.74
5320	2-1/2 C.Y. excavator	B-12S	740	.022			1.03	2.08	3.11	3.84
5352	4' to 6' deep, 1/2 C.Y. excavator w/trench box	B-13H	205	.078			3.70	4.83	8.53	10.95
5354	5/8 C.Y. excavator	"	257	.062			2.95	3.86	6.81	8.75
5356	3/4 C.Y. excavator	B-13G	308	.052			2.46	2.45	4.91	6.45
5362	6' to 10' deep, 3/4 C.Y. excavator w/trench box	"	231	.069			3.29	3.27	6.56	8.60
5364	1 C.Y. excavator	B-13D	410	.039			1.85	2.14	3.99	5.15
5366	1-1/2 C.Y. excavator	B-13E	616	.026			1.23	1.70	2.93	3.74
5370	10' to 14' deep, 3/4 C.Y. excavator w/trench box	B-13G	205	.078			3.70	3.68	7.38	9.70
5372	1 C.Y. excavator	B-13D	370	.043			2.05	2.38	4.43	5.75
5374	1-1/2 C.Y. excavator	B-13E	554	.029			1.37	1.89	3.26	4.16
5382	14' to 20' deep, 1 C.Y. excavator w/trench box	B-13D	329	.049			2.31	2.67	4.98	6.45
5384	1-1/2 C.Y. excavator	B-13E	492	.033			1.54	2.13	3.67	4.69
5386	2-1/2 C.Y. excavator	B-13J	780	.021			.97	2.07	3.04	3.76
5392	20' to 24' deep, 1 C.Y. excavator w/trench box	B-13D	295	.054			2.57	2.98	5.55	7.20
5394	1-1/2 C.Y. excavator	B-13E	444	.036			1.71	2.36	4.07	5.20
5396	2-1/2 C.Y. excavator	B-13J	695	.023			1.09	2.33	3.42	4.22
6020	Sand & gravel with no sheeting or dewatering included									
6050	1' to 4' deep, 3/8 C.Y. excavator	B-11C	165	.097	B.C.Y.		4.49	2.22	6.71	9.30
6060	1/2 C.Y. excavator	B-11M	220	.073			3.37	1.79	5.16	7.10

31 23 Excavation and Fill

31 23 16 – Excavation

31 23 16.13 Excavating, Trench

		Crew	Daily Output	Labor-Hours	Unit	Material	2017 Bare Costs Labor	Equipment	Total	Total Incl O&P
6080	4' to 6' deep, 1/2 C.Y. excavator	B-11M	220	.073	B.C.Y.		3.37	1.79	5.16	7.10
6090	5/8 C.Y. excavator	B-12Q	275	.058			2.76	2.23	4.99	6.65
6100	3/4 C.Y. excavator	B-12F	330	.048			2.30	2.04	4.34	5.75
6130	1/2 C.Y. excavator, truck mounted	B-12J	220	.073			3.45	4.14	7.59	9.80
6140	6' to 10' deep, 3/4 C.Y. excavator	B-12F	248	.065			3.06	2.72	5.78	7.65
6150	1 C.Y. excavator	B-12A	440	.036			1.72	1.81	3.53	4.62
6160	1 C.Y. excavator, truck mounted	B-12K	440	.036			1.72	2.60	4.32	5.50
6170	1-1/2 C.Y. excavator	B-12B	660	.024			1.15	1.46	2.61	3.36
6190	10' to 14' deep, 3/4 C.Y. excavator	B-12F	220	.073			3.45	3.06	6.51	8.60
6200	1 C.Y. excavator	B-12A	396	.040			1.92	2.02	3.94	5.15
6210	1-1/2 C.Y. excavator	B-12B	594	.027			1.28	1.62	2.90	3.73
6250	14' to 20' deep, 1 C.Y. excavator	B-12A	352	.045			2.16	2.27	4.43	5.75
6260	1-1/2 C.Y. excavator	B-12B	528	.030			1.44	1.83	3.27	4.20
6270	2-1/2 C.Y. excavator	B-12S	840	.019			.90	1.83	2.73	3.38
6300	20' to 24' deep, 1 C.Y. excavator	B-12A	317	.050			2.39	2.52	4.91	6.40
6310	1-1/2 C.Y. excavator	B-12B	475	.034			1.60	2.03	3.63	4.66
6320	2-1/2 C.Y. excavator	B-12S	755	.021			1.01	2.03	3.04	3.77
6352	4' to 6' deep, 1/2 C.Y. excavator w/trench box	B-13H	209	.077			3.63	4.74	8.37	10.70
6354	5/8 C.Y. excavator	"	261	.061			2.91	3.80	6.71	8.60
6356	3/4 C.Y. excavator	B-13G	314	.051			2.42	2.41	4.83	6.35
6362	6' to 10' deep, 3/4 C.Y. excavator w/trench box	"	236	.068			3.22	3.20	6.42	8.40
6364	1 C.Y. excavator	B-13D	418	.038			1.82	2.10	3.92	5.05
6366	1-1/2 C.Y. excavator	B-13E	627	.026			1.21	1.67	2.88	3.68
6370	10' to 14' deep, 3/4 C.Y. excavator w/trench box	B-13G	209	.077			3.63	3.61	7.24	9.50
6372	1 C.Y. excavator	B-13D	376	.043			2.02	2.34	4.36	5.65
6374	1-1/2 C.Y. excavator	B-13E	564	.028			1.35	1.85	3.20	4.09
6382	14' to 20' deep, 1 C.Y. excavator w/trench box	B-13D	334	.048			2.27	2.63	4.90	6.35
6384	1-1/2 C.Y. excavator	B-13E	502	.032			1.51	2.08	3.59	4.59
6386	2-1/2 C.Y. excavator	B-13J	790	.020			.96	2.05	3.01	3.71
6392	20' to 24' deep, 1 C.Y. excavator w/trench box	B-13D	301	.053			2.52	2.92	5.44	7.05
6394	1-1/2 C.Y. excavator	B-13E	452	.035			1.68	2.31	3.99	5.10
6396	2-1/2 C.Y. excavator	B-13J	710	.023	▼		1.07	2.28	3.35	4.14
7020	Dense hard clay with no sheeting or dewatering included									
7050	1' to 4' deep, 3/8 C.Y. excavator	B-11C	132	.121	B.C.Y.		5.60	2.77	8.37	11.60
7060	1/2 C.Y. excavator	B-11M	176	.091			4.21	2.24	6.45	8.85
7080	4' to 6' deep, 1/2 C.Y. excavator	"	176	.091			4.21	2.24	6.45	8.85
7090	5/8 C.Y. excavator	B-12Q	220	.073			3.45	2.79	6.24	8.30
7100	3/4 C.Y. excavator	B-12F	264	.061			2.87	2.55	5.42	7.20
7130	1/2 C.Y. excavator, truck mounted	B-12J	176	.091			4.31	5.15	9.46	12.25
7140	6' to 10' deep, 3/4 C.Y. excavator	B-12F	198	.081			3.83	3.41	7.24	9.60
7150	1 C.Y. excavator	B-12A	352	.045			2.16	2.27	4.43	5.75
7160	1 C.Y. excavator, truck mounted	B-12K	352	.045			2.16	3.25	5.41	6.85
7170	1-1/2 C.Y. excavator	B-12B	528	.030			1.44	1.83	3.27	4.20
7190	10' to 14' deep, 3/4 C.Y. excavator	B-12F	176	.091			4.31	3.83	8.14	10.75
7200	1 C.Y. excavator	B-12A	317	.050			2.39	2.52	4.91	6.40
7210	1-1/2 C.Y. excavator	B-12B	475	.034			1.60	2.03	3.63	4.66
7250	14' to 20' deep, 1 C.Y. excavator	B-12A	282	.057			2.69	2.83	5.52	7.20
7260	1-1/2 C.Y. excavator	B-12B	422	.038			1.80	2.29	4.09	5.25
7270	2-1/2 C.Y. excavator	B-12S	675	.024			1.12	2.28	3.40	4.21
7300	20' to 24' deep, 1 C.Y. excavator	B-12A	254	.063			2.99	3.14	6.13	8
7310	1-1/2 C.Y. excavator	B-12B	380	.042			2	2.54	4.54	5.85
7320	2-1/2 C.Y. excavator	B-12S	605	.026	▼		1.25	2.54	3.79	4.70

31 23 Excavation and Fill

31 23 16 – Excavation

31 23 16.14 Excavating, Utility Trench	Crew	Daily Output	Labor-Hours	Unit	Material	2017 Bare Costs Labor	Equipment	Total	Total Incl O&P
0010 EXCAVATING, UTILITY TRENCH									
0011 Common earth									
0050 Trenching with chain trencher, 12 H.P., operator walking									
0100 4" wide trench, 12" deep	B-53	800	.010	L.F.		.51	.08	.59	.86
0150 18" deep		750	.011			.54	.08	.62	.91
0200 24" deep		700	.011			.58	.09	.67	.98
0300 6" wide trench, 12" deep		650	.012			.63	.10	.73	1.06
0350 18" deep		600	.013			.68	.11	.79	1.15
0400 24" deep		550	.015			.74	.12	.86	1.25
0450 36" deep		450	.018			.91	.14	1.05	1.53
0600 8" wide trench, 12" deep		475	.017			.86	.13	.99	1.45
0650 18" deep		400	.020			1.02	.16	1.18	1.72
0700 24" deep		350	.023			1.17	.18	1.35	1.96
0750 36" deep	↓	300	.027	↓		1.36	.21	1.57	2.29
1000 Backfill by hand including compaction, add									
1050 4" wide trench, 12" deep	A-1G	800	.010	L.F.		.39	.07	.46	.68
1100 18" deep		530	.015			.59	.11	.70	1.03
1150 24" deep		400	.020			.78	.14	.92	1.36
1300 6" wide trench, 12" deep		540	.015			.58	.11	.69	1.01
1350 18" deep		405	.020			.77	.14	.91	1.34
1400 24" deep		270	.030			1.16	.21	1.37	2.01
1450 36" deep		180	.044			1.74	.32	2.06	3.02
1600 8" wide trench, 12" deep		400	.020			.78	.14	.92	1.36
1650 18" deep		265	.030			1.18	.21	1.39	2.05
1700 24" deep		200	.040			1.57	.28	1.85	2.71
1750 36" deep	↓	135	.059	↓		2.32	.42	2.74	4.02
2000 Chain trencher, 40 H.P. operator riding									
2050 6" wide trench and backfill, 12" deep	B-54	1200	.007	L.F.		.34	.27	.61	.81
2100 18" deep		1000	.008			.41	.33	.74	.98
2150 24" deep		975	.008			.42	.33	.75	1
2200 36" deep		900	.009			.45	.36	.81	1.09
2250 48" deep		750	.011			.54	.43	.97	1.30
2300 60" deep		650	.012			.63	.50	1.13	1.50
2400 8" wide trench and backfill, 12" deep		1000	.008			.41	.33	.74	.98
2450 18" deep		950	.008			.43	.34	.77	1.03
2500 24" deep		900	.009			.45	.36	.81	1.09
2550 36" deep		800	.010			.51	.41	.92	1.22
2600 48" deep		650	.012			.63	.50	1.13	1.50
2700 12" wide trench and backfill, 12" deep		975	.008			.42	.33	.75	1
2750 18" deep		860	.009			.47	.38	.85	1.14
2800 24" deep		800	.010			.51	.41	.92	1.22
2850 36" deep		725	.011			.56	.45	1.01	1.34
3000 16" wide trench and backfill, 12" deep		835	.010			.49	.39	.88	1.17
3050 18" deep		750	.011			.54	.43	.97	1.30
3100 24" deep	↓	700	.011	↓		.58	.46	1.04	1.39
3200 Compaction with vibratory plate, add								35%	35%
5100 Hand excavate and trim for pipe bells after trench excavation									
5200 8" pipe	1 Clab	155	.052	L.F.		2.02		2.02	3.10
5300 18" pipe	"	130	.062	"		2.41		2.41	3.69

31 23 Excavation and Fill

31 23 16 – Excavation

31 23 16.16 Structural Excavation for Minor Structures

		Crew	Daily Output	Labor-Hours	Unit	Material	2017 Bare Costs Labor	Equipment	Total	Total Incl O&P
0010	**STRUCTURAL EXCAVATION FOR MINOR STRUCTURES** R312316-40									
0015	Hand, pits to 6' deep, sandy soil	1 Clab	8	1	B.C.Y.		39		39	60
0100	Heavy soil or clay		4	2			78.50		78.50	120
0300	Pits 6' to 12' deep, sandy soil		5	1.600			62.50		62.50	96
0500	Heavy soil or clay		3	2.667			104		104	160
0700	Pits 12' to 18' deep, sandy soil		4	2			78.50		78.50	120
0900	Heavy soil or clay		2	4			157		157	240
1100	Hand loading trucks from stock pile, sandy soil		12	.667			26		26	40
1300	Heavy soil or clay	▼	8	1	▼		39		39	60
1500	For wet or muck hand excavation, add to above						50%			50%
6000	Machine excavation, for spread and mat footings, elevator pits,									
6001	and small building foundations									
6030	Common earth, hydraulic backhoe, 1/2 C.Y. bucket	B-12E	55	.291	B.C.Y.		13.80	8.05	21.85	30
6035	3/4 C.Y. bucket	B-12F	90	.178			8.45	7.50	15.95	21
6040	1 C.Y. bucket	B-12A	108	.148			7.05	7.40	14.45	18.85
6050	1-1/2 C.Y. bucket	B-12B	144	.111			5.25	6.70	11.95	15.35
6060	2 C.Y. bucket	B-12C	200	.080			3.79	5.60	9.39	11.95
6070	Sand and gravel, 3/4 C.Y. bucket	B-12F	100	.160			7.60	6.75	14.35	18.95
6080	1 C.Y. bucket	B-12A	120	.133			6.30	6.65	12.95	16.90
6090	1-1/2 C.Y. bucket	B-12B	160	.100			4.74	6.05	10.79	13.85
6100	2 C.Y. bucket	B-12C	220	.073			3.45	5.10	8.55	10.85
6110	Clay, till, or blasted rock, 3/4 C.Y. bucket	B-12F	80	.200			9.50	8.45	17.95	23.50
6120	1 C.Y. bucket	B-12A	95	.168			8	8.40	16.40	21.50
6130	1-1/2 C.Y. bucket	B-12B	130	.123			5.85	7.40	13.25	17.05
6140	2 C.Y. bucket	B-12C	175	.091			4.34	6.40	10.74	13.65
6230	Sandy clay & loam, hydraulic backhoe, 1/2 C.Y. bucket	B-12E	60	.267			12.65	7.40	20.05	27.50
6235	3/4 C.Y. bucket	B-12F	98	.163			7.75	6.90	14.65	19.35
6240	1 C.Y. bucket	B-12A	116	.138			6.55	6.90	13.45	17.50
6250	1-1/2 C.Y. bucket	B-12B	156	.103	▼		4.86	6.20	11.06	14.20
9010	For mobilization or demobilization, see Section 01 54 36.50									
9020	For dewatering, see Section 31 23 19.20									
9022	For larger structures, see Bulk Excavation, Section 31 23 16.42									
9024	For loading onto trucks, add								15%	15%
9026	For hauling, see Section 31 23 23.20									
9030	For sheeting or soldier bms/lagging, see Section 31 52 16.10									
9040	For trench excavation of strip ftgs, see Section 31 23 16.13									

31 23 16.26 Rock Removal

		Crew	Daily Output	Labor-Hours	Unit	Material	2017 Bare Costs Labor	Equipment	Total	Total Incl O&P
0010	**ROCK REMOVAL**									
0015	Drilling only rock, 2" hole for rock bolts	B-47	316	.076	L.F.		3.32	4.84	8.16	10.35
0800	2-1/2" hole for pre-splitting		250	.096			4.20	6.10	10.30	13.10
4600	Quarry operations, 2-1/2" to 3-1/2" diameter	▼	240	.100	▼		4.38	6.35	10.73	13.65

31 23 16.30 Drilling and Blasting Rock

		Crew	Daily Output	Labor-Hours	Unit	Material	2017 Bare Costs Labor	Equipment	Total	Total Incl O&P
0010	**DRILLING AND BLASTING ROCK**									
0020	Rock, open face, under 1500 C.Y.	B-47	225	.107	B.C.Y.	4.01	4.67	6.80	15.48	18.95
0100	Over 1500 C.Y.		300	.080		4.01	3.50	5.10	12.61	15.35
0200	Areas where blasting mats are required, under 1500 C.Y.		175	.137		4.01	6	8.75	18.76	23
0250	Over 1500 C.Y.	▼	250	.096		4.01	4.20	6.10	14.31	17.50
0300	Bulk drilling and blasting, can vary greatly, average								9.65	12.20
0500	Pits, average								25.50	31.50
1300	Deep hole method, up to 1500 C.Y.	B-47	50	.480		4.01	21	30.50	55.51	70
1400	Over 1500 C.Y.		66	.364		4.01	15.90	23	42.91	54.50
1900	Restricted areas, up to 1500 C.Y.	▼	13	1.846		4.01	81	118	203.01	256

31 23 Excavation and Fill

31 23 16 – Excavation

31 23 16.30 Drilling and Blasting Rock

		Crew	Daily Output	Labor-Hours	Unit	Material	2017 Bare Costs Labor	Equipment	Total	Total Incl O&P
2000	Over 1500 C.Y.	B-47	20	1.200	B.C.Y.	4.01	52.50	76.50	133.01	168
2200	Trenches, up to 1500 C.Y.		22	1.091		11.65	48	69.50	129.15	162
2300	Over 1500 C.Y.		26	.923		11.65	40.50	59	111.15	139
2500	Pier holes, up to 1500 C.Y.		22	1.091		4.01	48	69.50	121.51	154
2600	Over 1500 C.Y.		31	.774		4.01	34	49.50	87.51	110
2800	Boulders under 1/2 C.Y., loaded on truck, no hauling	B-100	80	.150			7.30	12.20	19.50	24.50
2900	Boulders, drilled, blasted	B-47	100	.240		4.01	10.50	15.30	29.81	37
3100	Jackhammer operators with foreman compressor, air tools	B-9	1	40	Day		1,575	213	1,788	2,650
3300	Track drill, compressor, operator and foreman	B-47	1	24	"		1,050	1,525	2,575	3,275
3500	Blasting caps				Ea.	6.50			6.50	7.15
3700	Explosives					.50			.50	.55
3800	Blasting mats, for purchase, no mobilization, 10' x 15' x 12"					1,275			1,275	1,400
3900	Blasting mats, rent, for first day					214			214	235
4000	Per added day					64			64	70.50
4200	Preblast survey for 6 room house, individual lot, minimum	A-6	2.40	6.667			330	17.35	347.35	525
4300	Maximum	"	1.35	11.852			590	31	621	935
4500	City block within zone of influence, minimum	A-8	25200	.001	S.F.		.07		.07	.10
4600	Maximum	"	15100	.002	"		.11		.11	.17

31 23 16.42 Excavating, Bulk Bank Measure

		Crew	Daily Output	Labor-Hours	Unit	Material	2017 Bare Costs Labor	Equipment	Total	Total Incl O&P
0010	**EXCAVATING, BULK BANK MEASURE** R312316-40									
0011	Common earth piled									
0020	For loading onto trucks, add								15%	15%
0050	For mobilization and demobilization, see Section 01 54 36.50 R312316-45									
0100	For hauling, see Section 31 23 23.20									
0200	Excavator, hydraulic, crawler mtd., 1 C.Y. cap. = 100 C.Y./hr.	B-12A	800	.020	B.C.Y.		.95	1	1.95	2.54
0250	1-1/2 C.Y. cap. = 125 C.Y./hr.	B-12B	1000	.016			.76	.97	1.73	2.21
0260	2 C.Y. cap. = 165 C.Y./hr.	B-12C	1320	.012			.57	.85	1.42	1.81
0300	3 C.Y. cap. = 260 C.Y./hr.	B-12D	2080	.008			.36	1.15	1.51	1.82
0305	3-1/2 C.Y. cap. = 300 C.Y./hr.	"	2400	.007			.32	1	1.32	1.58
0310	Wheel mounted, 1/2 C.Y. cap. = 40 C.Y./hr.	B-12E	320	.050			2.37	1.39	3.76	5.15
0360	3/4 C.Y. cap. = 60 C.Y./hr.	B-12F	480	.033			1.58	1.40	2.98	3.94
0500	Clamshell, 1/2 C.Y. cap. = 20 C.Y./hr.	B-12G	160	.100			4.74	4.98	9.72	12.65
0550	1 C.Y. cap. = 35 C.Y./hr.	B-12H	280	.057			2.71	4.80	7.51	9.40
0950	Dragline, 1/2 C.Y. cap. = 30 C.Y./hr.	B-12I	240	.067			3.16	4.14	7.30	9.35
1000	3/4 C.Y. cap. = 35 C.Y./hr.	"	280	.057			2.71	3.55	6.26	8
1050	1-1/2 C.Y. cap. = 65 C.Y./hr.	B-12P	520	.031			1.46	2.59	4.05	5.05
1200	Front end loader, track mtd., 1-1/2 C.Y. cap. = 70 C.Y./hr.	B-10N	560	.021			1.04	1.01	2.05	2.71
1250	2-1/2 C.Y. cap. = 95 C.Y./hr.	B-100	760	.016			.77	1.29	2.06	2.58
1300	3 C.Y. cap. = 130 C.Y./hr.	B-10P	1040	.012			.56	1.23	1.79	2.21
1350	5 C.Y. cap. = 160 C.Y./hr.	B-10Q	1280	.009			.46	1.19	1.65	2
1500	Wheel mounted, 3/4 C.Y. cap. = 45 C.Y./hr.	B-10R	360	.033			1.62	.87	2.49	3.43
1550	1-1/2 C.Y. cap. = 80 C.Y./hr.	B-10S	640	.019			.91	.57	1.48	2.01
1600	2-1/4 C.Y. cap. = 100 C.Y./hr.	B-10T	800	.015			.73	.70	1.43	1.87
1650	5 C.Y. cap. = 185 C.Y./hr.	B-10U	1480	.008			.40	.73	1.13	1.40
1800	Hydraulic excavator, truck mtd. 1/2 C.Y. = 30 C.Y./hr.	B-12J	240	.067			3.16	3.79	6.95	9
1850	48 inch bucket, 1 C.Y. = 45 C.Y./hr.	B-12K	360	.044			2.11	3.18	5.29	6.70
3700	Shovel, 1/2 C.Y. capacity = 55 C.Y./hr.	B-12L	440	.036			1.72	1.86	3.58	4.67
3750	3/4 C.Y. capacity = 85 C.Y./hr.	B-12M	680	.024			1.12	1.53	2.65	3.38
3800	1 C.Y. capacity = 120 C.Y./hr.	B-12N	960	.017			.79	1.43	2.22	2.77
3850	1-1/2 C.Y. capacity = 160 C.Y./hr.	B-120	1280	.013			.59	1.09	1.68	2.10
3900	3 C.Y. cap. = 250 C.Y./hr.	B-12T	2000	.008			.38	.87	1.25	1.54
4000	For soft soil or sand, deduct								15%	15%

31 23 16 – Excavation

31 23 16.42 Excavating, Bulk Bank Measure	Crew	Daily Output	Labor-Hours	Unit	Material	2017 Bare Costs Labor	2017 Bare Costs Equipment	Total	Total Incl O&P
4100	For heavy soil or stiff clay, add				B.C.Y.				60%	60%
4200	For wet excavation with clamshell or dragline, add								100%	100%
4250	All other equipment, add								50%	50%
4400	Clamshell in sheeting or cofferdam, minimum	B-12H	160	.100			4.74	8.40	13.14	16.45
4450	Maximum	"	60	.267	↓		12.65	22.50	35.15	44
5000	Excavating, bulk bank measure, sandy clay & loam piled									
5020	For loading onto trucks, add								15%	15%
5100	Excavator, hydraulic, crawler mtd., 1 C.Y. cap. = 120 C.Y./hr.	B-12A	960	.017	B.C.Y.		.79	.83	1.62	2.11
5150	1-1/2 C.Y. cap. = 150 C.Y./hr.	B-12B	1200	.013			.63	.80	1.43	1.84
5300	2 C.Y. cap. = 195 C.Y./hr.	B-12C	1560	.010			.49	.72	1.21	1.53
5400	3 C.Y. cap. = 300 C.Y./hr.	B-12D	2400	.007			.32	1	1.32	1.58
5500	3.5 C.Y. cap. = 350 C.Y./hr.	"	2800	.006			.27	.86	1.13	1.35
5610	Wheel mounted, 1/2 C.Y. cap. = 44 C.Y./hr.	B-12E	352	.045			2.16	1.26	3.42	4.67
5660	3/4 C.Y. cap. = 66 C.Y./hr.	B-12F	528	.030	↓		1.44	1.28	2.72	3.59
8000	For hauling excavated material, see Section 31 23 23.20									

31 23 16.46 Excavating, Bulk, Dozer

		Crew	Daily Output	Labor-Hours	Unit	Material	Labor	Equipment	Total	Total Incl O&P
0010	**EXCAVATING, BULK, DOZER**									
0011	Open site									
2000	80 H.P., 50' haul, sand & gravel	B-10L	460	.026	B.C.Y.		1.27	1.02	2.29	3.05
2010	Sandy clay & loam		440	.027			1.33	1.06	2.39	3.19
2020	Common earth		400	.030			1.46	1.17	2.63	3.51
2040	Clay		250	.048			2.34	1.87	4.21	5.60
2200	150' haul, sand & gravel		230	.052			2.54	2.04	4.58	6.10
2210	Sandy clay & loam		220	.055			2.66	2.13	4.79	6.40
2220	Common earth		200	.060			2.93	2.34	5.27	7
2240	Clay		125	.096			4.68	3.75	8.43	11.20
2400	300' haul, sand & gravel		120	.100			4.88	3.91	8.79	11.70
2410	Sandy clay & loam		115	.104			5.10	4.07	9.17	12.20
2420	Common earth		100	.120			5.85	4.69	10.54	14.05
2440	Clay	↓	65	.185			9	7.20	16.20	21.50
3000	105 H.P., 50' haul, sand & gravel	B-10W	700	.017			.84	.87	1.71	2.23
3010	Sandy clay & loam		680	.018			.86	.89	1.75	2.29
3020	Common earth		610	.020			.96	1	1.96	2.56
3040	Clay		385	.031			1.52	1.58	3.10	4.05
3200	150' haul, sand & gravel		310	.039			1.89	1.96	3.85	5
3210	Sandy clay & loam		300	.040			1.95	2.03	3.98	5.20
3220	Common earth		270	.044			2.17	2.25	4.42	5.75
3240	Clay		170	.071			3.44	3.58	7.02	9.15
3300	300' haul, sand & gravel		140	.086			4.18	4.34	8.52	11.15
3310	Sandy clay & loam		135	.089			4.33	4.50	8.83	11.55
3320	Common earth		120	.100			4.88	5.05	9.93	12.95
3340	Clay	↓	100	.120			5.85	6.10	11.95	15.60
4000	200 H.P., 50' haul, sand & gravel	B-10B	1400	.009			.42	.99	1.41	1.72
4010	Sandy clay & loam		1360	.009			.43	1.02	1.45	1.77
4020	Common earth		1230	.010			.48	1.13	1.61	1.96
4040	Clay		770	.016			.76	1.80	2.56	3.13
4200	150' haul, sand & gravel		595	.020			.98	2.33	3.31	4.05
4210	Sandy clay & loam		580	.021			1.01	2.39	3.40	4.15
4220	Common earth		516	.023			1.13	2.68	3.81	4.67
4240	Clay		325	.037			1.80	4.26	6.06	7.40
4400	300' haul, sand & gravel		310	.039			1.89	4.46	6.35	7.75
4410	Sandy clay & loam	↓	300	.040			1.95	4.61	6.56	8

31 23 Excavation and Fill

31 23 16 – Excavation

31 23 16.46 Excavating, Bulk, Dozer

		Crew	Daily Output	Labor-Hours	Unit	Material	2017 Bare Costs Labor	Equipment	Total	Total Incl O&P
4420	Common earth	B-10B	270	.044	B.C.Y.		2.17	5.15	7.32	8.95
4440	Clay	↓	170	.071			3.44	8.15	11.59	14.15
5000	300 H.P., 50' haul, sand & gravel	B-10M	1900	.006			.31	1	1.31	1.57
5010	Sandy clay & loam		1850	.006			.32	1.03	1.35	1.61
5020	Common earth		1650	.007			.35	1.15	1.50	1.81
5040	Clay		1025	.012			.57	1.85	2.42	2.91
5200	150' haul, sand & gravel		920	.013			.64	2.06	2.70	3.24
5210	Sandy clay & loam		895	.013			.65	2.12	2.77	3.33
5220	Common earth		800	.015			.73	2.38	3.11	3.72
5240	Clay		500	.024			1.17	3.80	4.97	5.95
5400	300' haul, sand & gravel		470	.026			1.24	4.04	5.28	6.35
5410	Sandy clay & loam		455	.026			1.29	4.18	5.47	6.55
5420	Common earth		410	.029			1.43	4.63	6.06	7.25
5440	Clay	↓	250	.048			2.34	7.60	9.94	11.90
5500	460 H.P., 50' haul, sand & gravel	B-10X	1930	.006			.30	1.23	1.53	1.81
5506	Sandy clay & loam		1880	.006			.31	1.26	1.57	1.85
5510	Common earth		1680	.007			.35	1.41	1.76	2.08
5520	Clay		1050	.011			.56	2.26	2.82	3.33
5530	150' haul, sand & gravel		1290	.009			.45	1.84	2.29	2.71
5535	Sandy clay & loam		1250	.010			.47	1.89	2.36	2.79
5540	Common earth		1120	.011			.52	2.11	2.63	3.11
5550	Clay		700	.017			.84	3.38	4.22	4.99
5560	300' haul, sand & gravel		660	.018			.89	3.59	4.48	5.30
5565	Sandy clay & loam		640	.019			.91	3.70	4.61	5.45
5570	Common earth		575	.021			1.02	4.12	5.14	6.05
5580	Clay	↓	350	.034			1.67	6.75	8.42	10
6000	700 H.P., 50' haul, sand & gravel	B-10V	3500	.003			.17	1.49	1.66	1.89
6006	Sandy clay & loam		3400	.004			.17	1.53	1.70	1.95
6010	Common earth		3035	.004			.19	1.72	1.91	2.18
6020	Clay		1925	.006			.30	2.71	3.01	3.44
6030	150' haul, sand & gravel		2025	.006			.29	2.58	2.87	3.28
6035	Sandy clay & loam		1960	.006			.30	2.66	2.96	3.38
6040	Common earth		1750	.007			.33	2.98	3.31	3.79
6050	Clay		1100	.011			.53	4.74	5.27	6
6060	300' haul, sand & gravel		1030	.012			.57	5.05	5.62	6.40
6065	Sandy clay & loam		1005	.012			.58	5.20	5.78	6.60
6070	Common earth		900	.013			.65	5.80	6.45	7.40
6080	Clay	↓	550	.022	↓		1.06	9.50	10.56	12.05

31 23 16.50 Excavation, Bulk, Scrapers

		Crew	Daily Output	Labor-Hours	Unit	Material	2017 Bare Costs Labor	Equipment	Total	Total Incl O&P
0010	**EXCAVATION, BULK, SCRAPERS** R312316-40									
0100	Elev. scraper 11 C.Y., sand & gravel 1500' haul, 1/4 dozer	B-33F	690	.020	B.C.Y.		1	2.42	3.42	4.18
0150	3000' haul		610	.023			1.13	2.73	3.86	4.73
0200	5000' haul		505	.028			1.37	3.30	4.67	5.70
0300	Common earth, 1500' haul		600	.023			1.15	2.78	3.93	4.81
0350	3000' haul		530	.026			1.31	3.15	4.46	5.45
0400	5000' haul		440	.032			1.57	3.79	5.36	6.55
0410	Sandy clay & loam, 1500' haul		648	.022			1.07	2.57	3.64	4.45
0420	3000' haul		572	.024			1.21	2.91	4.12	5.05
0430	5000' haul		475	.029			1.46	3.51	4.97	6.05
0500	Clay, 1500' haul		375	.037			1.85	4.44	6.29	7.70
0550	3000' haul		330	.042			2.10	5.05	7.15	8.75
0600	5000' haul	↓	275	.051	↓		2.52	6.05	8.57	10.45

31 23 16 – Excavation

31 23 16.50 Excavation, Bulk, Scrapers	Crew	Daily Output	Labor-Hours	Unit	Material	2017 Bare Costs Labor	Equipment	Total	Total Incl O&P
1000	Self propelled scraper, 14 C.Y. 1/4 push dozer									
1050	Sand and gravel, 1500' haul	B-33D	920	.015	B.C.Y.		.75	3.16	3.91	4.61
1100	3000' haul		805	.017			.86	3.61	4.47	5.25
1200	5000' haul		645	.022			1.07	4.50	5.57	6.60
1300	Common earth, 1500' haul		800	.018			.87	3.63	4.50	5.30
1350	3000' haul		700	.020			.99	4.15	5.14	6.05
1400	5000' haul		560	.025			1.24	5.20	6.44	7.60
1420	Sandy clay & loam, 1500' haul		864	.016			.80	3.36	4.16	4.92
1430	3000' haul		786	.018			.88	3.69	4.57	5.40
1440	5000' haul		605	.023			1.14	4.80	5.94	7.05
1500	Clay, 1500' haul		500	.028			1.38	5.80	7.18	8.50
1550	3000' haul		440	.032			1.57	6.60	8.17	9.65
1600	5000' haul		350	.040			1.98	8.30	10.28	12.15
2000	21 C.Y., 1/4 push dozer, sand & gravel, 1500' haul	B-33E	1180	.012			.59	2.76	3.35	3.93
2100	3000' haul		910	.015			.76	3.58	4.34	5.10
2200	5000' haul		750	.019			.92	4.35	5.27	6.20
2300	Common earth, 1500' haul		1030	.014			.67	3.16	3.83	4.50
2350	3000' haul		790	.018			.88	4.13	5.01	5.85
2400	5000' haul		650	.022			1.06	5	6.06	7.10
2420	Sandy clay & loam, 1500' haul		1112	.013			.62	2.93	3.55	4.16
2430	3000' haul		854	.016			.81	3.82	4.63	5.45
2440	5000' haul		702	.020			.99	4.64	5.63	6.60
2500	Clay, 1500' haul		645	.022			1.07	5.05	6.12	7.20
2550	3000' haul		495	.028			1.40	6.60	8	9.35
2600	5000' haul		405	.035			1.71	8.05	9.76	11.45
2700	Towed, 10 C.Y., 1/4 push dozer, sand & gravel, 1500' haul	B-33B	560	.025			1.24	4.51	5.75	6.85
2720	3000' haul		450	.031			1.54	5.60	7.14	8.50
2730	5000' haul		365	.038			1.90	6.90	8.80	10.50
2750	Common earth, 1500' haul		420	.033			1.65	6	7.65	9.10
2770	3000' haul		400	.035			1.73	6.30	8.03	9.60
2780	5000' haul		310	.045			2.23	8.15	10.38	12.35
2785	Sandy clay & loam, 1500' haul		454	.031			1.52	5.55	7.07	8.40
2790	3000' haul		432	.032			1.60	5.85	7.45	8.90
2795	5000' haul		340	.041			2.04	7.40	9.44	11.25
2800	Clay, 1500' haul		315	.044			2.20	8	10.20	12.15
2820	3000' haul		300	.047			2.31	8.40	10.71	12.75
2840	5000' haul		225	.062			3.08	11.20	14.28	17
2900	15 C.Y., 1/4 push dozer, sand & gravel, 1500' haul	B-33C	800	.018			.87	3.18	4.05	4.81
2920	3000' haul		640	.022			1.08	3.97	5.05	6
2940	5000' haul		520	.027			1.33	4.89	6.22	7.40
2960	Common earth, 1500' haul		600	.023			1.15	4.24	5.39	6.40
2980	3000' haul		560	.025			1.24	4.54	5.78	6.85
3000	5000' haul		440	.032			1.57	5.80	7.37	8.75
3005	Sandy clay & loam, 1500' haul		648	.022			1.07	3.92	4.99	5.95
3010	3000' haul		605	.023			1.14	4.20	5.34	6.35
3015	5000' haul		475	.029			1.46	5.35	6.81	8.10
3020	Clay, 1500' haul		450	.031			1.54	5.65	7.19	8.55
3040	3000' haul		420	.033			1.65	6.05	7.70	9.15
3060	5000' haul		320	.044			2.16	7.95	10.11	12.05

31 23 19 – Dewatering

31 23 19.20 Dewatering Systems	Crew	Daily Output	Labor-Hours	Unit	Material	2017 Bare Costs Labor	Equipment	Total	Total Incl O&P
0010 **DEWATERING SYSTEMS**									
0020 Excavate drainage trench, 2' wide, 2' deep	B-11C	90	.178	C.Y.		8.25	4.07	12.32	17.05
0100 2' wide, 3' deep, with backhoe loader	"	135	.119			5.50	2.71	8.21	11.35
0200 Excavate sump pits by hand, light soil	1 Clab	7.10	1.127			44		44	67.50
0300 Heavy soil	"	3.50	2.286	↓		89.50		89.50	137
0500 Pumping 8 hr., attended 2 hrs. per day, including 20 L.F.									
0550 of suction hose & 100 L.F. discharge hose									
0600 2" diaphragm pump used for 8 hours	B-10H	4	3	Day		146	18.70	164.70	243
0650 4" diaphragm pump used for 8 hours	B-10I	4	3			146	30.50	176.50	256
0800 8 hrs. attended, 2" diaphragm pump	B-10H	1	12			585	75	660	970
0900 3" centrifugal pump	B-10J	1	12			585	85	670	985
1000 4" diaphragm pump	B-10I	1	12			585	122	707	1,025
1100 6" centrifugal pump	B-10K	1	12	↓		585	350	935	1,275
1300 CMP, incl. excavation 3' deep, 12" diameter	B-6	115	.209	L.F.	11.10	9	3.18	23.28	29.50
1400 18" diameter		100	.240	"	20	10.35	3.66	34.01	42
1600 Sump hole construction, incl. excavation and gravel, pit		1250	.019	C.F.	1.09	.83	.29	2.21	2.78
1700 With 12" gravel collar, 12" pipe, corrugated, 16 ga.		70	.343	L.F.	21	14.80	5.25	41.05	52
1800 15" pipe, corrugated, 16 ga.		55	.436		27.50	18.80	6.65	52.95	66.50
1900 18" pipe, corrugated, 16 ga.		50	.480		32	20.50	7.30	59.80	75
2000 24" pipe, corrugated, 14 ga.		40	.600	↓	38.50	26	9.15	73.65	92
2200 Wood lining, up to 4' x 4', add	↓	300	.080	SFCA	16.15	3.45	1.22	20.82	24.50
9950 See Section 31 23 19.40 for wellpoints									
9960 See Section 31 23 19.30 for deep well systems									

31 23 19.30 Wells

	Crew	Daily Output	Labor-Hours	Unit	Material	Labor	Equipment	Total	Total Incl O&P
0010 **WELLS**									
0011 For dewatering 10' to 20' deep, 2' diameter									
0020 with steel casing, minimum	B-6	165	.145	V.L.F.	38	6.25	2.22	46.47	54
0050 Average		98	.245		43	10.55	3.74	57.29	67.50
0100 Maximum	↓	49	.490	↓	48	21	7.45	76.45	93
0300 For dewatering pumps see 01 54 33 in Reference Section									
0500 For domestic water wells, see Section 33 21 13.10									

31 23 19.40 Wellpoints

	Crew	Daily Output	Labor-Hours	Unit	Material	Labor	Equipment	Total	Total Incl O&P
0010 **WELLPOINTS** R312319-90									
0011 For equipment rental, see 01 54 33 in Reference Section									
0100 Installation and removal of single stage system									
0110 Labor only, .75 labor-hours per L.F.	1 Clab	10.70	.748	LF Hdr		29.50		29.50	45
0200 2.0 labor-hours per L.F.	"	4	2	"		78.50		78.50	120
0400 Pump operation, 4 @ 6 hr. shifts									
0410 Per 24 hour day	4 Eqlt	1.27	25.197	Day		1,275		1,275	1,950
0500 Per 168 hour week, 160 hr. straight, 8 hr. double time		.18	177	Week		9,075		9,075	13,700
0550 Per 4.3 week month	↓	.04	800	Month		40,800		40,800	61,500
0600 Complete installation, operation, equipment rental, fuel &									
0610 removal of system with 2" wellpoints 5' O.C.									
0700 100' long header, 6" diameter, first month	4 Eqlt	3.23	9.907	LF Hdr	159	505		664	940
0800 Thereafter, per month		4.13	7.748		127	395		522	740
1000 200' long header, 8" diameter, first month		6	5.333		145	272		417	570
1100 Thereafter, per month		8.39	3.814		71.50	195		266.50	375
1300 500' long header, 8" diameter, first month		10.63	3.010		55.50	154		209.50	294
1400 Thereafter, per month		20.91	1.530		40	78		118	162
1600 1,000' long header, 10" diameter, first month		11.62	2.754		47.50	140		187.50	265
1700 Thereafter, per month	↓	41.81	.765		24	39		63	85.50
1900 Note: above figures include pumping 168 hrs. per week									

For customer support on your Building Construction Costs with RSMeans Data, call 800.448.8182.

625

31 23 Excavation and Fill

31 23 19 – Dewatering

31 23 19.40 Wellpoints		Crew	Daily Output	Labor-Hours	Unit	Material	2017 Bare Costs Labor	2017 Bare Costs Equipment	Total	Total Incl O&P
1910	and include the pump operator and one stand-by pump.									

31 23 23 – Fill

31 23 23.13 Backfill

		Crew	Daily Output	Labor-Hours	Unit	Material	Labor	Equipment	Total	Total Incl O&P
0010	**BACKFILL** R312323-30									
0015	By hand, no compaction, light soil	1 Clab	14	.571	L.C.Y.		22.50		22.50	34.50
0100	Heavy soil		11	.727	"		28.50		28.50	43.50
0300	Compaction in 6" layers, hand tamp, add to above		20.60	.388	E.C.Y.		15.20		15.20	23.50
0400	Roller compaction operator walking, add	B-10A	100	.120			5.85	1.83	7.68	10.90
0500	Air tamp, add	B-9D	190	.211			8.35	1.29	9.64	14.15
0600	Vibrating plate, add	A-1D	60	.133			5.20	.56	5.76	8.60
0800	Compaction in 12" layers, hand tamp, add to above	1 Clab	34	.235			9.20		9.20	14.10
0900	Roller compaction operator walking, add	B-10A	150	.080			3.90	1.22	5.12	7.25
1000	Air tamp, add	B-9	285	.140			5.55	.75	6.30	9.30
1100	Vibrating plate, add	A-1E	90	.089			3.48	.48	3.96	5.85
1300	Dozer backfilling, bulk, up to 300' haul, no compaction	B-10B	1200	.010	L.C.Y.		.49	1.15	1.64	2.01
1400	Air tamped, add	B-11B	80	.200	E.C.Y.		9	3.42	12.42	17.45
1600	Compacting backfill, 6" to 12" lifts, vibrating roller	B-10C	800	.015			.73	2.26	2.99	3.60
1700	Sheepsfoot roller	B-10D	750	.016			.78	2.45	3.23	3.87
1900	Dozer backfilling, trench, up to 300' haul, no compaction	B-10B	900	.013	L.C.Y.		.65	1.54	2.19	2.68
2000	Air tamped, add	B-11B	80	.200	E.C.Y.		9	3.42	12.42	17.45
2200	Compacting backfill, 6" to 12" lifts, vibrating roller	B-10C	700	.017			.84	2.59	3.43	4.12
2300	Sheepsfoot roller	B-10D	650	.018			.90	2.82	3.72	4.48
2350	Spreading in 8" layers, small dozer	B-10B	1060	.011	L.C.Y.		.55	1.31	1.86	2.28

31 23 23.14 Backfill, Structural

		Crew	Daily Output	Labor-Hours	Unit	Material	Labor	Equipment	Total	Total Incl O&P
0010	**BACKFILL, STRUCTURAL**									
0011	Dozer or F.E. loader									
0020	From existing stockpile, no compaction									
1000	55 H.P. wheeled loader, 50' haul, common earth	B-11C	200	.080	L.C.Y.		3.71	1.83	5.54	7.65
2000	80 H.P., 50' haul, sand & gravel	B-10L	1100	.011			.53	.43	.96	1.28
2010	Sandy clay & loam		1070	.011			.55	.44	.99	1.31
2020	Common earth		975	.012			.60	.48	1.08	1.44
2040	Clay		850	.014			.69	.55	1.24	1.65
2400	300' haul, sand & gravel		370	.032			1.58	1.27	2.85	3.79
2410	Sandy clay & loam		360	.033			1.62	1.30	2.92	3.90
2420	Common earth		330	.036			1.77	1.42	3.19	4.25
2440	Clay		290	.041			2.02	1.62	3.64	4.84
3000	105 H.P., 50' haul, sand & gravel	B-10W	1350	.009			.43	.45	.88	1.16
3010	Sandy clay & loam		1325	.009			.44	.46	.90	1.17
3020	Common earth		1225	.010			.48	.50	.98	1.28
3040	Clay		1100	.011			.53	.55	1.08	1.42
3300	300' haul, sand & gravel		465	.026			1.26	1.31	2.57	3.35
3310	Sandy clay & loam		455	.026			1.29	1.34	2.63	3.42
3320	Common earth		415	.029			1.41	1.47	2.88	3.75
3340	Clay		370	.032			1.58	1.64	3.22	4.21
4000	200 H.P., 50' haul, sand & gravel	B-10B	2500	.005			.23	.55	.78	.97
4010	Sandy clay & loam		2435	.005			.24	.57	.81	.99
4020	Common earth		2200	.005			.27	.63	.90	1.09
4040	Clay		1950	.006			.30	.71	1.01	1.24
4400	300' haul, sand & gravel		805	.015			.73	1.72	2.45	2.99
4410	Sandy clay & loam		790	.015			.74	1.75	2.49	3.05
4420	Common earth		735	.016			.80	1.88	2.68	3.28
4440	Clay		660	.018			.89	2.10	2.99	3.66

31 23 Excavation and Fill

31 23 23 – Fill

31 23 23.14 Backfill, Structural

		Crew	Daily Output	Labor-Hours	Unit	Material	2017 Bare Costs Labor	Equipment	Total	Total Incl O&P
5000	300 H.P., 50' haul, sand & gravel	B-10M	3170	.004	L.C.Y.		.18	.60	.78	.94
5010	Sandy clay & loam		3110	.004			.19	.61	.80	.96
5020	Common earth		2900	.004			.20	.66	.86	1.03
5040	Clay		2700	.004			.22	.70	.92	1.10
5400	300' haul, sand & gravel		1500	.008			.39	1.27	1.66	1.98
5410	Sandy clay & loam		1470	.008			.40	1.29	1.69	2.02
5420	Common earth		1350	.009			.43	1.41	1.84	2.21
5440	Clay		1225	.010			.48	1.55	2.03	2.44
6010	For trench backfill, see Sections 31 23 16.13 and 31 23 16.14									
6100	For compaction, see Section 31 23 23.24									

31 23 23.16 Fill By Borrow and Utility Bedding

		Crew	Daily Output	Labor-Hours	Unit	Material	2017 Bare Costs Labor	Equipment	Total	Total Incl O&P
0010	**FILL BY BORROW AND UTILITY BEDDING**									
0015	Fill by borrow, load, 1 mile haul, spread with dozer									
0020	for embankments	B-15	1200	.023	L.C.Y.	12.80	1.09	2.22	16.11	18.20
0035	Select fill for shoulders & embankments	"	1200	.023	"	20	1.09	2.22	23.31	26
0040	Fill, for hauling over 1 mile, add to above per C.Y., see Section 31 23 23.20				Mile				1.41	1.73
0049	Utility bedding, for pipe & conduit, not incl. compaction									
0050	Crushed or screened bank run gravel	B-6	150	.160	L.C.Y.	21.50	6.90	2.44	30.84	36.50
0100	Crushed stone 3/4" to 1/2"		150	.160		26	6.90	2.44	35.34	41.50
0200	Sand, dead or bank		150	.160		18.60	6.90	2.44	27.94	33.50
0500	Compacting bedding in trench	A-1D	90	.089	E.C.Y.		3.48	.37	3.85	5.75
0600	If material source exceeds 2 miles, add for extra mileage.									
0610	See Section 31 23 23.20 for hauling mileage add.									

31 23 23.17 General Fill

		Crew	Daily Output	Labor-Hours	Unit	Material	2017 Bare Costs Labor	Equipment	Total	Total Incl O&P
0010	**GENERAL FILL**									
0011	Spread dumped material, no compaction									
0020	By dozer	B-10B	1000	.012	L.C.Y.		.59	1.38	1.97	2.41
0100	By hand	1 Clab	12	.667	"		26		26	40
0500	Gravel fill, compacted, under floor slabs, 4" deep	B-37	10000	.005	S.F.	.40	.20	.02	.62	.76
0600	6" deep		8600	.006		.60	.23	.02	.85	1.03
0700	9" deep		7200	.007		1	.28	.02	1.30	1.54
0800	12" deep		6000	.008		1.40	.33	.03	1.76	2.08
1000	Alternate pricing method, 4" deep		120	.400	E.C.Y.	30	16.60	1.27	47.87	60
1100	6" deep		160	.300		30	12.45	.95	43.40	53
1200	9" deep		200	.240		30	9.95	.76	40.71	49
1300	12" deep		220	.218		30	9.05	.69	39.74	47.50
1500	For fill under exterior paving, see Section 32 11 23.23									

31 23 23.20 Hauling

		Crew	Daily Output	Labor-Hours	Unit	Material	2017 Bare Costs Labor	Equipment	Total	Total Incl O&P
0010	**HAULING**									
0011	Excavated or borrow, loose cubic yards									
0012	no loading equipment, including hauling, waiting, loading/dumping									
0013	time per cycle (wait, load, travel, unload or dump & return)									
0014	8 C.Y. truck, 15 MPH ave, cycle 0.5 miles, 10 min. wait/Ld./Uld.	B-34A	320	.025	L.C.Y.		1.14	1.22	2.36	3.06
0016	cycle 1 mile		272	.029			1.34	1.44	2.78	3.60
0018	cycle 2 miles		208	.038			1.75	1.88	3.63	4.71
0020	cycle 4 miles		144	.056			2.53	2.72	5.25	6.80
0022	cycle 6 miles		112	.071			3.25	3.49	6.74	8.75
0024	cycle 8 miles		88	.091			4.14	4.44	8.58	11.15
0026	20 MPH ave, cycle 0.5 mile		336	.024			1.08	1.16	2.24	2.92
0028	cycle 1 mile		296	.027			1.23	1.32	2.55	3.31
0030	cycle 2 miles		240	.033			1.52	1.63	3.15	4.08
0032	cycle 4 miles		176	.045			2.07	2.22	4.29	5.55

31 23 23.20 Hauling		Crew	Daily Output	Labor-Hours	Unit	Material	2017 Bare Costs Labor	Equipment	Total	Total Incl O&P
0034	cycle 6 miles	B-34A	136	.059	L.C.Y.		2.68	2.88	5.56	7.20
0036	cycle 8 miles		112	.071			3.25	3.49	6.74	8.75
0044	25 MPH ave, cycle 4 miles		192	.042			1.90	2.04	3.94	5.10
0046	cycle 6 miles		160	.050			2.28	2.44	4.72	6.15
0048	cycle 8 miles		128	.063			2.85	3.06	5.91	7.65
0050	30 MPH ave, cycle 4 miles		216	.037			1.69	1.81	3.50	4.53
0052	cycle 6 miles		176	.045			2.07	2.22	4.29	5.55
0054	cycle 8 miles		144	.056			2.53	2.72	5.25	6.80
0114	15 MPH ave, cycle 0.5 mile, 15 min. wait/Ld./Uld.		224	.036			1.63	1.75	3.38	4.37
0116	cycle 1 mile		200	.040			1.82	1.96	3.78	4.90
0118	cycle 2 miles		168	.048			2.17	2.33	4.50	5.85
0120	cycle 4 miles		120	.067			3.04	3.26	6.30	8.15
0122	cycle 6 miles		96	.083			3.80	4.07	7.87	10.20
0124	cycle 8 miles		80	.100			4.56	4.89	9.45	12.25
0126	20 MPH ave, cycle 0.5 mile		232	.034			1.57	1.69	3.26	4.22
0128	cycle 1 mile		208	.038			1.75	1.88	3.63	4.71
0130	cycle 2 miles		184	.043			1.98	2.13	4.11	5.35
0132	cycle 4 miles		144	.056			2.53	2.72	5.25	6.80
0134	cycle 6 miles		112	.071			3.25	3.49	6.74	8.75
0136	cycle 8 miles		96	.083			3.80	4.07	7.87	10.20
0144	25 MPH ave, cycle 4 miles		152	.053			2.40	2.57	4.97	6.45
0146	cycle 6 miles		128	.063			2.85	3.06	5.91	7.65
0148	cycle 8 miles		112	.071			3.25	3.49	6.74	8.75
0150	30 MPH ave, cycle 4 miles		168	.048			2.17	2.33	4.50	5.85
0152	cycle 6 miles		144	.056			2.53	2.72	5.25	6.80
0154	cycle 8 miles		120	.067			3.04	3.26	6.30	8.15
0214	15 MPH ave, cycle 0.5 mile, 20 min wait/Ld./Uld.		176	.045			2.07	2.22	4.29	5.55
0216	cycle 1 mile		160	.050			2.28	2.44	4.72	6.15
0218	cycle 2 miles		136	.059			2.68	2.88	5.56	7.20
0220	cycle 4 miles		104	.077			3.50	3.76	7.26	9.45
0222	cycle 6 miles		88	.091			4.14	4.44	8.58	11.15
0224	cycle 8 miles		72	.111			5.05	5.45	10.50	13.60
0226	20 MPH ave, cycle 0.5 mile		176	.045			2.07	2.22	4.29	5.55
0228	cycle 1 mile		168	.048			2.17	2.33	4.50	5.85
0230	cycle 2 miles		144	.056			2.53	2.72	5.25	6.80
0232	cycle 4 miles		120	.067			3.04	3.26	6.30	8.15
0234	cycle 6 miles		96	.083			3.80	4.07	7.87	10.20
0236	cycle 8 miles		88	.091			4.14	4.44	8.58	11.15
0244	25 MPH ave, cycle 4 miles		128	.063			2.85	3.06	5.91	7.65
0246	cycle 6 miles		112	.071			3.25	3.49	6.74	8.75
0248	cycle 8 miles		96	.083			3.80	4.07	7.87	10.20
0250	30 MPH ave, cycle 4 miles		136	.059			2.68	2.88	5.56	7.20
0252	cycle 6 miles		120	.067			3.04	3.26	6.30	8.15
0254	cycle 8 miles		104	.077			3.50	3.76	7.26	9.45
0314	15 MPH ave, cycle 0.5 mile, 25 min wait/Ld./Uld.		144	.056			2.53	2.72	5.25	6.80
0316	cycle 1 mile		128	.063			2.85	3.06	5.91	7.65
0318	cycle 2 miles		112	.071			3.25	3.49	6.74	8.75
0320	cycle 4 miles		96	.083			3.80	4.07	7.87	10.20
0322	cycle 6 miles		80	.100			4.56	4.89	9.45	12.25
0324	cycle 8 miles		64	.125			5.70	6.10	11.80	15.30
0326	20 MPH ave, cycle 0.5 mile		144	.056			2.53	2.72	5.25	6.80
0328	cycle 1 mile		136	.059			2.68	2.88	5.56	7.20
0330	cycle 2 miles		120	.067			3.04	3.26	6.30	8.15

31 23 23 – Fill

31 23 23.20 Hauling		Crew	Daily Output	Labor-Hours	Unit	Material	Labor	Equipment	Total	Total Incl O&P
							2017 Bare Costs			
0332	cycle 4 miles	B-34A	104	.077	L.C.Y.		3.50	3.76	7.26	9.45
0334	cycle 6 miles		88	.091			4.14	4.44	8.58	11.15
0336	cycle 8 miles		80	.100			4.56	4.89	9.45	12.25
0344	25 MPH ave, cycle 4 miles		112	.071			3.25	3.49	6.74	8.75
0346	cycle 6 miles		96	.083			3.80	4.07	7.87	10.20
0348	cycle 8 miles		88	.091			4.14	4.44	8.58	11.15
0350	30 MPH ave, cycle 4 miles		112	.071			3.25	3.49	6.74	8.75
0352	cycle 6 miles		104	.077			3.50	3.76	7.26	9.45
0354	cycle 8 miles		96	.083			3.80	4.07	7.87	10.20
0414	15 MPH ave, cycle 0.5 mile, 30 min wait/Ld./Uld.		120	.067			3.04	3.26	6.30	8.15
0416	cycle 1 mile		112	.071			3.25	3.49	6.74	8.75
0418	cycle 2 miles		96	.083			3.80	4.07	7.87	10.20
0420	cycle 4 miles		80	.100			4.56	4.89	9.45	12.25
0422	cycle 6 miles		72	.111			5.05	5.45	10.50	13.60
0424	cycle 8 miles		64	.125			5.70	6.10	11.80	15.30
0426	20 MPH ave, cycle 0.5 mile		120	.067			3.04	3.26	6.30	8.15
0428	cycle 1 mile		112	.071			3.25	3.49	6.74	8.75
0430	cycle 2 miles		104	.077			3.50	3.76	7.26	9.45
0432	cycle 4 miles		88	.091			4.14	4.44	8.58	11.15
0434	cycle 6 miles		80	.100			4.56	4.89	9.45	12.25
0436	cycle 8 miles		72	.111			5.05	5.45	10.50	13.60
0444	25 MPH ave, cycle 4 miles		96	.083			3.80	4.07	7.87	10.20
0446	cycle 6 miles		88	.091			4.14	4.44	8.58	11.15
0448	cycle 8 miles		80	.100			4.56	4.89	9.45	12.25
0450	30 MPH ave, cycle 4 miles		96	.083			3.80	4.07	7.87	10.20
0452	cycle 6 miles		88	.091			4.14	4.44	8.58	11.15
0454	cycle 8 miles		80	.100			4.56	4.89	9.45	12.25
0514	15 MPH ave, cycle 0.5 mile, 35 min wait/Ld./Uld.		104	.077			3.50	3.76	7.26	9.45
0516	cycle 1 mile		96	.083			3.80	4.07	7.87	10.20
0518	cycle 2 miles		88	.091			4.14	4.44	8.58	11.15
0520	cycle 4 miles		72	.111			5.05	5.45	10.50	13.60
0522	cycle 6 miles		64	.125			5.70	6.10	11.80	15.30
0524	cycle 8 miles		56	.143			6.50	7	13.50	17.50
0526	20 MPH ave, cycle 0.5 mile		104	.077			3.50	3.76	7.26	9.45
0528	cycle 1 mile		96	.083			3.80	4.07	7.87	10.20
0530	cycle 2 miles		96	.083			3.80	4.07	7.87	10.20
0532	cycle 4 miles		80	.100			4.56	4.89	9.45	12.25
0534	cycle 6 miles		72	.111			5.05	5.45	10.50	13.60
0536	cycle 8 miles		64	.125			5.70	6.10	11.80	15.30
0544	25 MPH ave, cycle 4 miles		88	.091			4.14	4.44	8.58	11.15
0546	cycle 6 miles		80	.100			4.56	4.89	9.45	12.25
0548	cycle 8 miles		72	.111			5.05	5.45	10.50	13.60
0550	30 MPH ave, cycle 4 miles		88	.091			4.14	4.44	8.58	11.15
0552	cycle 6 miles		80	.100			4.56	4.89	9.45	12.25
0554	cycle 8 miles		72	.111			5.05	5.45	10.50	13.60
1014	12 C.Y. truck, cycle 0.5 mile, 15 MPH ave, 15 min. wait/Ld./Uld.	B-34B	336	.024			1.08	1.91	2.99	3.74
1016	cycle 1 mile		300	.027			1.21	2.14	3.35	4.19
1018	cycle 2 miles		252	.032			1.45	2.55	4	4.99
1020	cycle 4 miles		180	.044			2.02	3.57	5.59	7
1022	cycle 6 miles		144	.056			2.53	4.46	6.99	8.75
1024	cycle 8 miles		120	.067			3.04	5.35	8.39	10.50
1025	cycle 10 miles		96	.083			3.80	6.70	10.50	13.05
1026	20 MPH ave, cycle 0.5 mile		348	.023			1.05	1.85	2.90	3.61

31 23 23.20 Hauling		Crew	Daily Output	Labor-Hours	Unit	Material	2017 Bare Costs Labor	Equipment	Total	Total Incl O&P
1028	cycle 1 mile	B-34B	312	.026	L.C.Y.		1.17	2.06	3.23	4.03
1030	cycle 2 miles		276	.029			1.32	2.33	3.65	4.55
1032	cycle 4 miles		216	.037			1.69	2.98	4.67	5.80
1034	cycle 6 miles		168	.048			2.17	3.83	6	7.50
1036	cycle 8 miles		144	.056			2.53	4.46	6.99	8.75
1038	cycle 10 miles		120	.067			3.04	5.35	8.39	10.50
1040	25 MPH ave, cycle 4 miles		228	.035			1.60	2.82	4.42	5.50
1042	cycle 6 miles		192	.042			1.90	3.35	5.25	6.55
1044	cycle 8 miles		168	.048			2.17	3.83	6	7.50
1046	cycle 10 miles		144	.056			2.53	4.46	6.99	8.75
1050	30 MPH ave, cycle 4 miles		252	.032			1.45	2.55	4	4.99
1052	cycle 6 miles		216	.037			1.69	2.98	4.67	5.80
1054	cycle 8 miles		180	.044			2.02	3.57	5.59	7
1056	cycle 10 miles		156	.051			2.34	4.12	6.46	8.05
1060	35 MPH ave, cycle 4 miles		264	.030			1.38	2.43	3.81	4.76
1062	cycle 6 miles		228	.035			1.60	2.82	4.42	5.50
1064	cycle 8 miles		204	.039			1.79	3.15	4.94	6.15
1066	cycle 10 miles		180	.044			2.02	3.57	5.59	7
1068	cycle 20 miles		120	.067			3.04	5.35	8.39	10.50
1069	cycle 30 miles		84	.095			4.34	7.65	11.99	14.95
1070	cycle 40 miles		72	.111			5.05	8.95	14	17.45
1072	40 MPH ave, cycle 6 miles		240	.033			1.52	2.68	4.20	5.25
1074	cycle 8 miles		216	.037			1.69	2.98	4.67	5.80
1076	cycle 10 miles		192	.042			1.90	3.35	5.25	6.55
1078	cycle 20 miles		120	.067			3.04	5.35	8.39	10.50
1080	cycle 30 miles		96	.083			3.80	6.70	10.50	13.05
1082	cycle 40 miles		72	.111			5.05	8.95	14	17.45
1084	cycle 50 miles		60	.133			6.05	10.70	16.75	21
1094	45 MPH ave, cycle 8 miles		216	.037			1.69	2.98	4.67	5.80
1096	cycle 10 miles		204	.039			1.79	3.15	4.94	6.15
1098	cycle 20 miles		132	.061			2.76	4.87	7.63	9.50
1100	cycle 30 miles		108	.074			3.37	5.95	9.32	11.65
1102	cycle 40 miles		84	.095			4.34	7.65	11.99	14.95
1104	cycle 50 miles		72	.111			5.05	8.95	14	17.45
1106	50 MPH ave, cycle 10 miles		216	.037			1.69	2.98	4.67	5.80
1108	cycle 20 miles		144	.056			2.53	4.46	6.99	8.75
1110	cycle 30 miles		108	.074			3.37	5.95	9.32	11.65
1112	cycle 40 miles		84	.095			4.34	7.65	11.99	14.95
1114	cycle 50 miles		72	.111			5.05	8.95	14	17.45
1214	15 MPH ave, cycle 0.5 mile, 20 min. wait/Ld./Uld.		264	.030			1.38	2.43	3.81	4.76
1216	cycle 1 mile		240	.033			1.52	2.68	4.20	5.25
1218	cycle 2 miles		204	.039			1.79	3.15	4.94	6.15
1220	cycle 4 miles		156	.051			2.34	4.12	6.46	8.05
1222	cycle 6 miles		132	.061			2.76	4.87	7.63	9.50
1224	cycle 8 miles		108	.074			3.37	5.95	9.32	11.65
1225	cycle 10 miles		96	.083			3.80	6.70	10.50	13.05
1226	20 MPH ave, cycle 0.5 mile		264	.030			1.38	2.43	3.81	4.76
1228	cycle 1 mile		252	.032			1.45	2.55	4	4.99
1230	cycle 2 miles		216	.037			1.69	2.98	4.67	5.80
1232	cycle 4 miles		180	.044			2.02	3.57	5.59	7
1234	cycle 6 miles		144	.056			2.53	4.46	6.99	8.75
1236	cycle 8 miles		132	.061			2.76	4.87	7.63	9.50
1238	cycle 10 miles		108	.074			3.37	5.95	9.32	11.65

31 23 23 – Fill

31 23 23.20 Hauling	Crew	Daily Output	Labor-Hours	Unit	Material	2017 Bare Costs Labor	Equipment	Total	Total Incl O&P
1240	25 MPH ave, cycle 4 miles	B-34B	192	.042	L.C.Y.		1.90	3.35	5.25	6.55
1242	cycle 6 miles		168	.048			2.17	3.83	6	7.50
1244	cycle 8 miles		144	.056			2.53	4.46	6.99	8.75
1246	cycle 10 miles		132	.061			2.76	4.87	7.63	9.50
1250	30 MPH ave, cycle 4 miles		204	.039			1.79	3.15	4.94	6.15
1252	cycle 6 miles		180	.044			2.02	3.57	5.59	7
1254	cycle 8 miles		156	.051			2.34	4.12	6.46	8.05
1256	cycle 10 miles		144	.056			2.53	4.46	6.99	8.75
1260	35 MPH ave, cycle 4 miles		216	.037			1.69	2.98	4.67	5.80
1262	cycle 6 miles		192	.042			1.90	3.35	5.25	6.55
1264	cycle 8 miles		168	.048			2.17	3.83	6	7.50
1266	cycle 10 miles		156	.051			2.34	4.12	6.46	8.05
1268	cycle 20 miles		108	.074			3.37	5.95	9.32	11.65
1269	cycle 30 miles		72	.111			5.05	8.95	14	17.45
1270	cycle 40 miles		60	.133			6.05	10.70	16.75	21
1272	40 MPH ave, cycle 6 miles		192	.042			1.90	3.35	5.25	6.55
1274	cycle 8 miles		180	.044			2.02	3.57	5.59	7
1276	cycle 10 miles		156	.051			2.34	4.12	6.46	8.05
1278	cycle 20 miles		108	.074			3.37	5.95	9.32	11.65
1280	cycle 30 miles		84	.095			4.34	7.65	11.99	14.95
1282	cycle 40 miles		72	.111			5.05	8.95	14	17.45
1284	cycle 50 miles		60	.133			6.05	10.70	16.75	21
1294	45 MPH ave, cycle 8 miles		180	.044			2.02	3.57	5.59	7
1296	cycle 10 miles		168	.048			2.17	3.83	6	7.50
1298	cycle 20 miles		120	.067			3.04	5.35	8.39	10.50
1300	cycle 30 miles		96	.083			3.80	6.70	10.50	13.05
1302	cycle 40 miles		72	.111			5.05	8.95	14	17.45
1304	cycle 50 miles		60	.133			6.05	10.70	16.75	21
1306	50 MPH ave, cycle 10 miles		180	.044			2.02	3.57	5.59	7
1308	cycle 20 miles		132	.061			2.76	4.87	7.63	9.50
1310	cycle 30 miles		96	.083			3.80	6.70	10.50	13.05
1312	cycle 40 miles		84	.095			4.34	7.65	11.99	14.95
1314	cycle 50 miles		72	.111			5.05	8.95	14	17.45
1414	15 MPH ave, cycle 0.5 mile, 25 min. wait/Ld./Uld.		204	.039			1.79	3.15	4.94	6.15
1416	cycle 1 mile		192	.042			1.90	3.35	5.25	6.55
1418	cycle 2 miles		168	.048			2.17	3.83	6	7.50
1420	cycle 4 miles		132	.061			2.76	4.87	7.63	9.50
1422	cycle 6 miles		120	.067			3.04	5.35	8.39	10.50
1424	cycle 8 miles		96	.083			3.80	6.70	10.50	13.05
1425	cycle 10 miles		84	.095			4.34	7.65	11.99	14.95
1426	20 MPH ave, cycle 0.5 mile		216	.037			1.69	2.98	4.67	5.80
1428	cycle 1 mile		204	.039			1.79	3.15	4.94	6.15
1430	cycle 2 miles		180	.044			2.02	3.57	5.59	7
1432	cycle 4 miles		156	.051			2.34	4.12	6.46	8.05
1434	cycle 6 miles		132	.061			2.76	4.87	7.63	9.50
1436	cycle 8 miles		120	.067			3.04	5.35	8.39	10.50
1438	cycle 10 miles		96	.083			3.80	6.70	10.50	13.05
1440	25 MPH ave, cycle 4 miles		168	.048			2.17	3.83	6	7.50
1442	cycle 6 miles		144	.056			2.53	4.46	6.99	8.75
1444	cycle 8 miles		132	.061			2.76	4.87	7.63	9.50
1446	cycle 10 miles		108	.074			3.37	5.95	9.32	11.65
1450	30 MPH ave, cycle 4 miles		168	.048			2.17	3.83	6	7.50
1452	cycle 6 miles		156	.051			2.34	4.12	6.46	8.05

631

31 23 23.20 Hauling	Crew	Daily Output	Labor-Hours	Unit	Material	2017 Bare Costs Labor	2017 Bare Costs Equipment	Total	Total Incl O&P
1454	cycle 8 miles	B-34B	132	.061	L.C.Y.		2.76	4.87	7.63	9.50
1456	cycle 10 miles		120	.067			3.04	5.35	8.39	10.50
1460	35 MPH ave, cycle 4 miles		180	.044			2.02	3.57	5.59	7
1462	cycle 6 miles		156	.051			2.34	4.12	6.46	8.05
1464	cycle 8 miles		144	.056			2.53	4.46	6.99	8.75
1466	cycle 10 miles		132	.061			2.76	4.87	7.63	9.50
1468	cycle 20 miles		96	.083			3.80	6.70	10.50	13.05
1469	cycle 30 miles		72	.111			5.05	8.95	14	17.45
1470	cycle 40 miles		60	.133			6.05	10.70	16.75	21
1472	40 MPH ave, cycle 6 miles		168	.048			2.17	3.83	6	7.50
1474	cycle 8 miles		156	.051			2.34	4.12	6.46	8.05
1476	cycle 10 miles		144	.056			2.53	4.46	6.99	8.75
1478	cycle 20 miles		96	.083			3.80	6.70	10.50	13.05
1480	cycle 30 miles		84	.095			4.34	7.65	11.99	14.95
1482	cycle 40 miles		60	.133			6.05	10.70	16.75	21
1484	cycle 50 miles		60	.133			6.05	10.70	16.75	21
1494	45 MPH ave, cycle 8 miles		156	.051			2.34	4.12	6.46	8.05
1496	cycle 10 miles		144	.056			2.53	4.46	6.99	8.75
1498	cycle 20 miles		108	.074			3.37	5.95	9.32	11.65
1500	cycle 30 miles		84	.095			4.34	7.65	11.99	14.95
1502	cycle 40 miles		72	.111			5.05	8.95	14	17.45
1504	cycle 50 miles		60	.133			6.05	10.70	16.75	21
1506	50 MPH ave, cycle 10 miles		156	.051			2.34	4.12	6.46	8.05
1508	cycle 20 miles		120	.067			3.04	5.35	8.39	10.50
1510	cycle 30 miles		96	.083			3.80	6.70	10.50	13.05
1512	cycle 40 miles		72	.111			5.05	8.95	14	17.45
1514	cycle 50 miles		60	.133			6.05	10.70	16.75	21
1614	15 MPH ave, cycle 0.5 mile, 30 min. wait/Ld./Uld.		180	.044			2.02	3.57	5.59	7
1616	cycle 1 mile		168	.048			2.17	3.83	6	7.50
1618	cycle 2 miles		144	.056			2.53	4.46	6.99	8.75
1620	cycle 4 miles		120	.067			3.04	5.35	8.39	10.50
1622	cycle 6 miles		108	.074			3.37	5.95	9.32	11.65
1624	cycle 8 miles		84	.095			4.34	7.65	11.99	14.95
1625	cycle 10 miles		84	.095			4.34	7.65	11.99	14.95
1626	20 MPH ave, cycle 0.5 mile		180	.044			2.02	3.57	5.59	7
1628	cycle 1 mile		168	.048			2.17	3.83	6	7.50
1630	cycle 2 miles		156	.051			2.34	4.12	6.46	8.05
1632	cycle 4 miles		132	.061			2.76	4.87	7.63	9.50
1634	cycle 6 miles		120	.067			3.04	5.35	8.39	10.50
1636	cycle 8 miles		108	.074			3.37	5.95	9.32	11.65
1638	cycle 10 miles		96	.083			3.80	6.70	10.50	13.05
1640	25 MPH ave, cycle 4 miles		144	.056			2.53	4.46	6.99	8.75
1642	cycle 6 miles		132	.061			2.76	4.87	7.63	9.50
1644	cycle 8 miles		108	.074			3.37	5.95	9.32	11.65
1646	cycle 10 miles		108	.074			3.37	5.95	9.32	11.65
1650	30 MPH ave, cycle 4 miles		144	.056			2.53	4.46	6.99	8.75
1652	cycle 6 miles		132	.061			2.76	4.87	7.63	9.50
1654	cycle 8 miles		120	.067			3.04	5.35	8.39	10.50
1656	cycle 10 miles		108	.074			3.37	5.95	9.32	11.65
1660	35 MPH ave, cycle 4 miles		156	.051			2.34	4.12	6.46	8.05
1662	cycle 6 miles		144	.056			2.53	4.46	6.99	8.75
1664	cycle 8 miles		132	.061			2.76	4.87	7.63	9.50
1666	cycle 10 miles		120	.067			3.04	5.35	8.39	10.50

31 23 23.20 Hauling		Crew	Daily Output	Labor-Hours	Unit	Material	2017 Bare Costs Labor	2017 Bare Costs Equipment	Total	Total Incl O&P
1668	cycle 20 miles	B-34B	84	.095	L.C.Y.		4.34	7.65	11.99	14.95
1669	cycle 30 miles		72	.111			5.05	8.95	14	17.45
1670	cycle 40 miles		60	.133			6.05	10.70	16.75	21
1672	40 MPH ave, cycle 6 miles		144	.056			2.53	4.46	6.99	8.75
1674	cycle 8 miles		132	.061			2.76	4.87	7.63	9.50
1676	cycle 10 miles		120	.067			3.04	5.35	8.39	10.50
1678	cycle 20 miles		96	.083			3.80	6.70	10.50	13.05
1680	cycle 30 miles		72	.111			5.05	8.95	14	17.45
1682	cycle 40 miles		60	.133			6.05	10.70	16.75	21
1684	cycle 50 miles		48	.167			7.60	13.40	21	26
1694	45 MPH ave, cycle 8 miles		144	.056			2.53	4.46	6.99	8.75
1696	cycle 10 miles		132	.061			2.76	4.87	7.63	9.50
1698	cycle 20 miles		96	.083			3.80	6.70	10.50	13.05
1700	cycle 30 miles		84	.095			4.34	7.65	11.99	14.95
1702	cycle 40 miles		60	.133			6.05	10.70	16.75	21
1704	cycle 50 miles		60	.133			6.05	10.70	16.75	21
1706	50 MPH ave, cycle 10 miles		132	.061			2.76	4.87	7.63	9.50
1708	cycle 20 miles		108	.074			3.37	5.95	9.32	11.65
1710	cycle 30 miles		84	.095			4.34	7.65	11.99	14.95
1712	cycle 40 miles		72	.111			5.05	8.95	14	17.45
1714	cycle 50 miles		60	.133			6.05	10.70	16.75	21
2000	Hauling, 8 C.Y. truck, small project cost per hour	B-34A	8	1	Hr.		45.50	49	94.50	123
2100	12 C.Y. Truck	B-34B	8	1			45.50	80.50	126	157
2150	16.5 C.Y. Truck	B-34C	8	1			45.50	86	131.50	163
2175	18 C.Y. 8 wheel Truck	B-34I	8	1			45.50	102	147.50	181
2200	20 C.Y. Truck	B-34D	8	1			45.50	88	133.50	165
2300	Grading at dump, or embankment if required, by dozer	B-10B	1000	.012	L.C.Y.		.59	1.38	1.97	2.41
2310	Spotter at fill or cut, if required	1 Clab	8	1	Hr.		39		39	60
9014	18 C.Y. truck, 8 wheels,15 min. wait/Ld./Uld.,15 MPH, cycle 0.5 mi.	B-34I	504	.016	L.C.Y.		.72	1.62	2.34	2.87
9016	cycle 1 mile		450	.018			.81	1.81	2.62	3.21
9018	cycle 2 miles		378	.021			.96	2.16	3.12	3.82
9020	cycle 4 miles		270	.030			1.35	3.02	4.37	5.35
9022	cycle 6 miles		216	.037			1.69	3.78	5.47	6.70
9024	cycle 8 miles		180	.044			2.02	4.53	6.55	8.05
9025	cycle 10 miles		144	.056			2.53	5.65	8.18	10.05
9026	20 MPH ave, cycle 0.5 mile		522	.015			.70	1.56	2.26	2.77
9028	cycle 1 mile		468	.017			.78	1.74	2.52	3.09
9030	cycle 2 miles		414	.019			.88	1.97	2.85	3.50
9032	cycle 4 miles		324	.025			1.12	2.52	3.64	4.47
9034	cycle 6 miles		252	.032			1.45	3.24	4.69	5.75
9036	cycle 8 miles		216	.037			1.69	3.78	5.47	6.70
9038	cycle 10 miles		180	.044			2.02	4.53	6.55	8.05
9040	25 MPH ave, cycle 4 miles		342	.023			1.07	2.38	3.45	4.23
9042	cycle 6 miles		288	.028			1.27	2.83	4.10	5.05
9044	cycle 8 miles		252	.032			1.45	3.24	4.69	5.75
9046	cycle 10 miles		216	.037			1.69	3.78	5.47	6.70
9050	30 MPH ave, cycle 4 miles		378	.021			.96	2.16	3.12	3.82
9052	cycle 6 miles		324	.025			1.12	2.52	3.64	4.47
9054	cycle 8 miles		270	.030			1.35	3.02	4.37	5.35
9056	cycle 10 miles		234	.034			1.56	3.49	5.05	6.20
9060	35 MPH ave, cycle 4 miles		396	.020			.92	2.06	2.98	3.66
9062	cycle 6 miles		342	.023			1.07	2.38	3.45	4.23
9064	cycle 8 miles		288	.028			1.27	2.83	4.10	5.05

31 23 23.20 Hauling		Crew	Daily Output	Labor-Hours	Unit	Material	2017 Bare Costs Labor	2017 Bare Costs Equipment	Total	Total Incl O&P
9066	cycle 10 miles	B-34I	270	.030	L.C.Y.		1.35	3.02	4.37	5.35
9068	cycle 20 miles		162	.049			2.25	5.05	7.30	8.95
9070	cycle 30 miles		126	.063			2.89	6.45	9.34	11.45
9072	cycle 40 miles		90	.089			4.05	9.05	13.10	16.05
9074	40 MPH ave, cycle 6 miles		360	.022			1.01	2.27	3.28	4.02
9076	cycle 8 miles		324	.025			1.12	2.52	3.64	4.47
9078	cycle 10 miles		288	.028			1.27	2.83	4.10	5.05
9080	cycle 20 miles		180	.044			2.02	4.53	6.55	8.05
9082	cycle 30 miles		144	.056			2.53	5.65	8.18	10.05
9084	cycle 40 miles		108	.074			3.37	7.55	10.92	13.40
9086	cycle 50 miles		90	.089			4.05	9.05	13.10	16.05
9094	45 MPH ave, cycle 8 miles		324	.025			1.12	2.52	3.64	4.47
9096	cycle 10 miles		306	.026			1.19	2.67	3.86	4.73
9098	cycle 20 miles		198	.040			1.84	4.12	5.96	7.30
9100	cycle 30 miles		144	.056			2.53	5.65	8.18	10.05
9102	cycle 40 miles		126	.063			2.89	6.45	9.34	11.45
9104	cycle 50 miles		108	.074			3.37	7.55	10.92	13.40
9106	50 MPH ave, cycle 10 miles		324	.025			1.12	2.52	3.64	4.47
9108	cycle 20 miles		216	.037			1.69	3.78	5.47	6.70
9110	cycle 30 miles		162	.049			2.25	5.05	7.30	8.95
9112	cycle 40 miles		126	.063			2.89	6.45	9.34	11.45
9114	cycle 50 miles		108	.074			3.37	7.55	10.92	13.40
9214	20 min. wait/Ld./Uld.,15 MPH, cycle 0.5 mi.		396	.020			.92	2.06	2.98	3.66
9216	cycle 1 mile		360	.022			1.01	2.27	3.28	4.02
9218	cycle 2 miles		306	.026			1.19	2.67	3.86	4.73
9220	cycle 4 miles		234	.034			1.56	3.49	5.05	6.20
9222	cycle 6 miles		198	.040			1.84	4.12	5.96	7.30
9224	cycle 8 miles		162	.049			2.25	5.05	7.30	8.95
9225	cycle 10 miles		144	.056			2.53	5.65	8.18	10.05
9226	20 MPH ave, cycle 0.5 mile		396	.020			.92	2.06	2.98	3.66
9228	cycle 1 mile		378	.021			.96	2.16	3.12	3.82
9230	cycle 2 miles		324	.025			1.12	2.52	3.64	4.47
9232	cycle 4 miles		270	.030			1.35	3.02	4.37	5.35
9234	cycle 6 miles		216	.037			1.69	3.78	5.47	6.70
9236	cycle 8 miles		198	.040			1.84	4.12	5.96	7.30
9238	cycle 10 miles		162	.049			2.25	5.05	7.30	8.95
9240	25 MPH ave, cycle 4 miles		288	.028			1.27	2.83	4.10	5.05
9242	cycle 6 miles		252	.032			1.45	3.24	4.69	5.75
9244	cycle 8 miles		216	.037			1.69	3.78	5.47	6.70
9246	cycle 10 miles		198	.040			1.84	4.12	5.96	7.30
9250	30 MPH ave, cycle 4 miles		306	.026			1.19	2.67	3.86	4.73
9252	cycle 6 miles		270	.030			1.35	3.02	4.37	5.35
9254	cycle 8 miles		234	.034			1.56	3.49	5.05	6.20
9256	cycle 10 miles		216	.037			1.69	3.78	5.47	6.70
9260	35 MPH ave, cycle 4 miles		324	.025			1.12	2.52	3.64	4.47
9262	cycle 6 miles		288	.028			1.27	2.83	4.10	5.05
9264	cycle 8 miles		252	.032			1.45	3.24	4.69	5.75
9266	cycle 10 miles		234	.034			1.56	3.49	5.05	6.20
9268	cycle 20 miles		162	.049			2.25	5.05	7.30	8.95
9270	cycle 30 miles		108	.074			3.37	7.55	10.92	13.40
9272	cycle 40 miles		90	.089			4.05	9.05	13.10	16.05
9274	40 MPH ave, cycle 6 miles		288	.028			1.27	2.83	4.10	5.05
9276	cycle 8 miles		270	.030			1.35	3.02	4.37	5.35

31 23 23.20 Hauling		Crew	Daily Output	Labor-Hours	Unit	Material	Labor	Equipment	Total	Total Incl O&P
9278	cycle 10 miles	B-34I	234	.034	L.C.Y.		1.56	3.49	5.05	6.20
9280	cycle 20 miles		162	.049			2.25	5.05	7.30	8.95
9282	cycle 30 miles		126	.063			2.89	6.45	9.34	11.45
9284	cycle 40 miles		108	.074			3.37	7.55	10.92	13.40
9286	cycle 50 miles		90	.089			4.05	9.05	13.10	16.05
9294	45 MPH ave, cycle 8 miles		270	.030			1.35	3.02	4.37	5.35
9296	cycle 10 miles		252	.032			1.45	3.24	4.69	5.75
9298	cycle 20 miles		180	.044			2.02	4.53	6.55	8.05
9300	cycle 30 miles		144	.056			2.53	5.65	8.18	10.05
9302	cycle 40 miles		108	.074			3.37	7.55	10.92	13.40
9304	cycle 50 miles		90	.089			4.05	9.05	13.10	16.05
9306	50 MPH ave, cycle 10 miles		270	.030			1.35	3.02	4.37	5.35
9308	cycle 20 miles		198	.040			1.84	4.12	5.96	7.30
9310	cycle 30 miles		144	.056			2.53	5.65	8.18	10.05
9312	cycle 40 miles		126	.063			2.89	6.45	9.34	11.45
9314	cycle 50 miles		108	.074			3.37	7.55	10.92	13.40
9414	25 min. wait/Ld./Uld.,15 MPH, cycle 0.5 mi.		306	.026			1.19	2.67	3.86	4.73
9416	cycle 1 mile		288	.028			1.27	2.83	4.10	5.05
9418	cycle 2 miles		252	.032			1.45	3.24	4.69	5.75
9420	cycle 4 miles		198	.040			1.84	4.12	5.96	7.30
9422	cycle 6 miles		180	.044			2.02	4.53	6.55	8.05
9424	cycle 8 miles		144	.056			2.53	5.65	8.18	10.05
9425	cycle 10 miles		126	.063			2.89	6.45	9.34	11.45
9426	20 MPH ave, cycle 0.5 mile		324	.025			1.12	2.52	3.64	4.47
9428	cycle 1 mile		306	.026			1.19	2.67	3.86	4.73
9430	cycle 2 miles		270	.030			1.35	3.02	4.37	5.35
9432	cycle 4 miles		234	.034			1.56	3.49	5.05	6.20
9434	cycle 6 miles		198	.040			1.84	4.12	5.96	7.30
9436	cycle 8 miles		180	.044			2.02	4.53	6.55	8.05
9438	cycle 10 miles		144	.056			2.53	5.65	8.18	10.05
9440	25 MPH ave, cycle 4 miles		252	.032			1.45	3.24	4.69	5.75
9442	cycle 6 miles		216	.037			1.69	3.78	5.47	6.70
9444	cycle 8 miles		198	.040			1.84	4.12	5.96	7.30
9446	cycle 10 miles		180	.044			2.02	4.53	6.55	8.05
9450	30 MPH ave, cycle 4 miles		252	.032			1.45	3.24	4.69	5.75
9452	cycle 6 miles		234	.034			1.56	3.49	5.05	6.20
9454	cycle 8 miles		198	.040			1.84	4.12	5.96	7.30
9456	cycle 10 miles		180	.044			2.02	4.53	6.55	8.05
9460	35 MPH ave, cycle 4 miles		270	.030			1.35	3.02	4.37	5.35
9462	cycle 6 miles		234	.034			1.56	3.49	5.05	6.20
9464	cycle 8 miles		216	.037			1.69	3.78	5.47	6.70
9466	cycle 10 miles		198	.040			1.84	4.12	5.96	7.30
9468	cycle 20 miles		144	.056			2.53	5.65	8.18	10.05
9470	cycle 30 miles		108	.074			3.37	7.55	10.92	13.40
9472	cycle 40 miles		90	.089			4.05	9.05	13.10	16.05
9474	40 MPH ave, cycle 6 miles		252	.032			1.45	3.24	4.69	5.75
9476	cycle 8 miles		234	.034			1.56	3.49	5.05	6.20
9478	cycle 10 miles		216	.037			1.69	3.78	5.47	6.70
9480	cycle 20 miles		144	.056			2.53	5.65	8.18	10.05
9482	cycle 30 miles		126	.063			2.89	6.45	9.34	11.45
9484	cycle 40 miles		90	.089			4.05	9.05	13.10	16.05
9486	cycle 50 miles		90	.089			4.05	9.05	13.10	16.05
9494	45 MPH ave, cycle 8 miles		234	.034			1.56	3.49	5.05	6.20

31 23 23 – Fill

31 23 23.20 Hauling		Crew	Daily Output	Labor-Hours	Unit	Material	2017 Bare Costs Labor	Equipment	Total	Total Incl O&P
9496	cycle 10 miles	B-34I	216	.037	L.C.Y.		1.69	3.78	5.47	6.70
9498	cycle 20 miles		162	.049			2.25	5.05	7.30	8.95
9500	cycle 30 miles		126	.063			2.89	6.45	9.34	11.45
9502	cycle 40 miles		108	.074			3.37	7.55	10.92	13.40
9504	cycle 50 miles		90	.089			4.05	9.05	13.10	16.05
9506	50 MPH ave, cycle 10 miles		234	.034			1.56	3.49	5.05	6.20
9508	cycle 20 miles		180	.044			2.02	4.53	6.55	8.05
9510	cycle 30 miles		144	.056			2.53	5.65	8.18	10.05
9512	cycle 40 miles		108	.074			3.37	7.55	10.92	13.40
9514	cycle 50 miles		90	.089			4.05	9.05	13.10	16.05
9614	30 min. wait/Ld./Uld.,15 MPH, cycle 0.5 mi.		270	.030			1.35	3.02	4.37	5.35
9616	cycle 1 mile		252	.032			1.45	3.24	4.69	5.75
9618	cycle 2 miles		216	.037			1.69	3.78	5.47	6.70
9620	cycle 4 miles		180	.044			2.02	4.53	6.55	8.05
9622	cycle 6 miles		162	.049			2.25	5.05	7.30	8.95
9624	cycle 8 miles		126	.063			2.89	6.45	9.34	11.45
9625	cycle 10 miles		126	.063			2.89	6.45	9.34	11.45
9626	20 MPH ave, cycle 0.5 mile		270	.030			1.35	3.02	4.37	5.35
9628	cycle 1 mile		252	.032			1.45	3.24	4.69	5.75
9630	cycle 2 miles		234	.034			1.56	3.49	5.05	6.20
9632	cycle 4 miles		198	.040			1.84	4.12	5.96	7.30
9634	cycle 6 miles		180	.044			2.02	4.53	6.55	8.05
9636	cycle 8 miles		162	.049			2.25	5.05	7.30	8.95
9638	cycle 10 miles		144	.056			2.53	5.65	8.18	10.05
9640	25 MPH ave, cycle 4 miles		216	.037			1.69	3.78	5.47	6.70
9642	cycle 6 miles		198	.040			1.84	4.12	5.96	7.30
9644	cycle 8 miles		180	.044			2.02	4.53	6.55	8.05
9646	cycle 10 miles		162	.049			2.25	5.05	7.30	8.95
9650	30 MPH ave, cycle 4 miles		216	.037			1.69	3.78	5.47	6.70
9652	cycle 6 miles		198	.040			1.84	4.12	5.96	7.30
9654	cycle 8 miles		180	.044			2.02	4.53	6.55	8.05
9656	cycle 10 miles		162	.049			2.25	5.05	7.30	8.95
9660	35 MPH ave, cycle 4 miles		234	.034			1.56	3.49	5.05	6.20
9662	cycle 6 miles		216	.037			1.69	3.78	5.47	6.70
9664	cycle 8 miles		198	.040			1.84	4.12	5.96	7.30
9666	cycle 10 miles		180	.044			2.02	4.53	6.55	8.05
9668	cycle 20 miles		126	.063			2.89	6.45	9.34	11.45
9670	cycle 30 miles		108	.074			3.37	7.55	10.92	13.40
9672	cycle 40 miles		90	.089			4.05	9.05	13.10	16.05
9674	40 MPH ave, cycle 6 miles		216	.037			1.69	3.78	5.47	6.70
9676	cycle 8 miles		198	.040			1.84	4.12	5.96	7.30
9678	cycle 10 miles		180	.044			2.02	4.53	6.55	8.05
9680	cycle 20 miles		144	.056			2.53	5.65	8.18	10.05
9682	cycle 30 miles		108	.074			3.37	7.55	10.92	13.40
9684	cycle 40 miles		90	.089			4.05	9.05	13.10	16.05
9686	cycle 50 miles		72	.111			5.05	11.35	16.40	20
9694	45 MPH ave, cycle 8 miles		216	.037			1.69	3.78	5.47	6.70
9696	cycle 10 miles		198	.040			1.84	4.12	5.96	7.30
9698	cycle 20 miles		144	.056			2.53	5.65	8.18	10.05
9700	cycle 30 miles		126	.063			2.89	6.45	9.34	11.45
9702	cycle 40 miles		108	.074			3.37	7.55	10.92	13.40
9704	cycle 50 miles		90	.089			4.05	9.05	13.10	16.05
9706	50 MPH ave, cycle 10 miles		198	.040			1.84	4.12	5.96	7.30

31 23 Excavation and Fill

31 23 23 – Fill

31 23 23.20 Hauling

		Crew	Daily Output	Labor-Hours	Unit	Material	2017 Bare Costs Labor	Equipment	Total	Total Incl O&P
9708	cycle 20 miles	B-34I	162	.049	L.C.Y.		2.25	5.05	7.30	8.95
9710	cycle 30 miles		126	.063			2.89	6.45	9.34	11.45
9712	cycle 40 miles		108	.074			3.37	7.55	10.92	13.40
9714	cycle 50 miles	↓	90	.089	↓		4.05	9.05	13.10	16.05

31 23 23.24 Compaction, Structural

			Crew	Daily Output	Labor-Hours	Unit	Material	Labor	Equipment	Total	Total Incl O&P
0010	**COMPACTION, STRUCTURAL**	R312323-30									
0020	Steel wheel tandem roller, 5 tons		B-10E	8	1.500	Hr.		73	19.05	92.05	132
0100	10 tons		B-10F	8	1.500	"		73	29.50	102.50	144
0300	Sheepsfoot or wobbly wheel roller, 8" lifts, common fill		B-10G	1300	.009	E.C.Y.		.45	1	1.45	1.78
0400	Select fill		"	1500	.008			.39	.86	1.25	1.54
0600	Vibratory plate, 8" lifts, common fill		A-1D	200	.040			1.57	.17	1.74	2.58
0700	Select fill		"	216	.037	↓		1.45	.16	1.61	2.39

31 25 Erosion and Sedimentation Controls

31 25 14 – Stabilization Measures for Erosion and Sedimentation Control

31 25 14.16 Rolled Erosion Control Mats and Blankets

			Crew	Daily Output	Labor-Hours	Unit	Material	Labor	Equipment	Total	Total Incl O&P
0010	**ROLLED EROSION CONTROL MATS AND BLANKETS**										
0020	Jute mesh, 100 S.Y. per roll, 4' wide, stapled	G	B-80A	2400	.010	S.Y.	.77	.39	.11	1.27	1.57
0070	Paper biodegradable mesh	G	B-1	2500	.010		.19	.38		.57	.80
0080	Paper mulch	G	B-64	20000	.001		.14	.03	.02	.19	.22
0100	Plastic netting, stapled, 2" x 1" mesh, 20 mil	G	B-1	2500	.010		.29	.38		.67	.91
0200	Polypropylene mesh, stapled, 6.5 oz./S.Y.	G		2500	.010		1.61	.38		1.99	2.36
0300	Tobacco netting, or jute mesh #2, stapled	G	↓	2500	.010	↓	.25	.38		.63	.87
1000	Silt fence, install and maintain, remove	G	B-62	1300	.018	L.F.	.55	.80	.13	1.48	1.96
1100	Allow 10% per month for maintenance; 6-month max life										
1200	Place and remove hay bales, staked	G	A-2	3	8	Ton	276	325	72.50	673.50	885
1250	Place and remove hay bales, staked (alt. pricing)	G	"	2500	.010	L.F.	4.61	.39	.09	5.09	5.75

31 31 Soil Treatment

31 31 16 – Termite Control

31 31 16.13 Chemical Termite Control

		Crew	Daily Output	Labor-Hours	Unit	Material	Labor	Equipment	Total	Total Incl O&P
0010	**CHEMICAL TERMITE CONTROL**									
0020	Slab and walls, residential	1 Skwk	1200	.007	SF Flr.	.32	.34		.66	.88
0100	Commercial, minimum		2496	.003		.33	.17		.50	.61
0200	Maximum		1645	.005	↓	.50	.25		.75	.92
0400	Insecticides for termite control, minimum		14.20	.563	Gal.	69	29		98	120
0500	Maximum	↓	11	.727	"	118	37.50		155.50	188

31 32 Soil Stabilization

31 32 13 – Soil Mixing Stabilization

31 32 13.30 Calcium Chloride

	Crew	Daily Output	Labor-Hours	Unit	Material	2017 Bare Costs Labor	Equipment	Total	Total Incl O&P
0010 **CALCIUM CHLORIDE**									
0020 Calcium chloride, delivered, 100 lb. bags, truckload lots				Ton	560			560	615
0030 Solution, 4 lb. flake per gallon, tank truck delivery				Gal.	1.73			1.73	1.90

31 33 Rock Stabilization

31 33 13 – Rock Bolting and Grouting

31 33 13.10 Rock Bolting

	Crew	Daily Output	Labor-Hours	Unit	Material	2017 Bare Costs Labor	Equipment	Total	Total Incl O&P
0010 **ROCK BOLTING**									
2020 Hollow core, prestressable anchor, 1" diameter, 5' long	2 Skwk	32	.500	Ea.	180	25.50		205.50	238
2025 10' long		24	.667		305	34.50		339.50	390
2060 2" diameter, 5' long		32	.500		595	25.50		620.50	695
2065 10' long		24	.667		1,225	34.50		1,259.50	1,400
2100 Super high-tensile, 3/4" diameter, 5' long		32	.500		48.50	25.50		74	93
2105 10' long		24	.667		130	34.50		164.50	195
2160 2" diameter, 5' long		32	.500		405	25.50		430.50	485
2165 10' long		24	.667		690	34.50		724.50	815
4400 Drill hole for rock bolt, 1-3/4" diam., 5' long (for 3/4" bolt)	B-56	17	.941			42.50	89	131.50	163
4405 10' long		9	1.778			80	168	248	305
4420 2" diameter, 5' long (for 1" bolt)		13	1.231			55.50	116	171.50	213
4425 10' long		7	2.286			103	216	319	395
4460 3-1/2" diameter, 5' long (for 2" bolt)		10	1.600			72	151	223	276
4465 10' long		5	3.200			144	305	449	555

31 36 Gabions

31 36 13 – Gabion Boxes

31 36 13.10 Gabion Box Systems

	Crew	Daily Output	Labor-Hours	Unit	Material	2017 Bare Costs Labor	Equipment	Total	Total Incl O&P
0010 **GABION BOX SYSTEMS**									
0400 Gabions, galvanized steel mesh mats or boxes, stone filled, 6" deep	B-13	200	.280	S.Y.	18.55	12.05	3.37	33.97	42.50
0500 9" deep		163	.344		23	14.80	4.13	41.93	52.50
0600 12" deep		153	.366		31	15.75	4.40	51.15	63
0700 18" deep		102	.549		43.50	23.50	6.60	73.60	91.50
0800 36" deep		60	.933		74	40	11.25	125.25	155

31 37 Riprap

31 37 13 – Machined Riprap

31 37 13.10 Riprap and Rock Lining

	Crew	Daily Output	Labor-Hours	Unit	Material	2017 Bare Costs Labor	Equipment	Total	Total Incl O&P
0010 **RIPRAP AND ROCK LINING**									
0011 Random, broken stone									
0100 Machine placed for slope protection	B-12G	62	.258	L.C.Y.	29.50	12.25	12.85	54.60	65
0110 3/8 to 1/4 C.Y. pieces, grouted	B-13	80	.700	S.Y.	63	30	8.40	101.40	124
0200 18" minimum thickness, not grouted	"	53	1.057	"	18.60	45.50	12.70	76.80	104
0300 Dumped, 50 lb. average	B-11A	800	.020	Ton	26	.93	1.73	28.66	32.50
0350 100 lb. average		700	.023		26	1.06	1.98	29.04	33
0370 300 lb. average		600	.027		26	1.24	2.31	29.55	33.50

31 41 Shoring

31 41 13 – Timber Shoring

31 41 13.10 Building Shoring

		Crew	Daily Output	Labor-Hours	Unit	Material	2017 Bare Costs Labor	Equipment	Total	Total Incl O&P
0010	**BUILDING SHORING**									
0020	Shoring, existing building, with timber, no salvage allowance	B-51	2.20	21.818	M.B.F.	865	880	99	1,844	2,400
1000	On cribbing with 35 ton screw jacks, per box and jack	"	3.60	13.333	Jack	65	540	60.50	665.50	960
1100	Masonry openings in walls, see Section 02 41 19.16									

31 41 16 – Sheet Piling

31 41 16.10 Sheet Piling Systems

		Crew	Daily Output	Labor-Hours	Unit	Material	2017 Bare Costs Labor	Equipment	Total	Total Incl O&P
0010	**SHEET PILING SYSTEMS**									
0020	Sheet piling, 50,000 psi steel, not incl. wales, 22 psf, left in place	B-40	10.81	5.920	Ton	1,450	305	350	2,105	2,450
0100	Drive, extract & salvage R314116-40		6	10.667		475	545	635	1,655	2,075
0300	20' deep excavation, 27 psf, left in place		12.95	4.942		1,450	253	294	1,997	2,325
0400	Drive, extract & salvage R314116-45		6.55	9.771		475	500	580	1,555	1,950
0600	25' deep excavation, 38 psf, left in place		19	3.368		1,450	172	200	1,822	2,100
0700	Drive, extract & salvage		10.50	6.095		475	310	365	1,150	1,400
0900	40' deep excavation, 38 psf, left in place		21.20	3.019		1,450	154	180	1,784	2,050
1000	Drive, extract & salvage		12.25	5.224		475	267	310	1,052	1,275
1200	15' deep excavation, 22 psf, left in place		983	.065	S.F.	16.85	3.33	3.87	24.05	28
1300	Drive, extract & salvage		545	.117		5.30	6	7	18.30	23
1500	20' deep excavation, 27 psf, left in place		960	.067		21	3.41	3.97	28.38	33
1600	Drive, extract & salvage		485	.132		6.90	6.75	7.85	21.50	27
1800	25' deep excavation, 38 psf, left in place		1000	.064		31	3.27	3.81	38.08	44
1900	Drive, extract & salvage		553	.116		9.45	5.90	6.90	22.25	27
2100	Rent steel sheet piling and wales, first month				Ton	284			284	310
2200	Per added month					28.50			28.50	31
2300	Rental piling left in place, add to rental					1,075			1,075	1,175
2500	Wales, connections & struts, 2/3 salvage					440			440	485
2700	High strength piling, 60,000 psi, add					145			145	159
2800	65,000 psi, add					217			217	239
3000	Tie rod, not upset, 1-1/2" to 4" diameter with turnbuckle					1,900			1,900	2,075
3100	No turnbuckle					1,500			1,500	1,650
3300	Upset, 1-3/4" to 4" diameter with turnbuckle					2,175			2,175	2,400
3400	No turnbuckle					1,925			1,925	2,125
3600	Lightweight, 18" to 28" wide, 7 ga., 9.22 psf, and									
3610	9 ga., 8.6 psf, minimum				Lb.	.75			.75	.83
3700	Average					.81			.81	.89
3750	Maximum					.97			.97	1.07
3900	Wood, solid sheeting, incl. wales, braces and spacers, R314116-40									
3910	drive, extract & salvage, 8' deep excavation	B-31	330	.121	S.F.	1.79	5.05	.60	7.44	10.35
4000	10' deep, 50 S.F./hr. in & 150 S.F./hr. out		300	.133		1.84	5.55	.66	8.05	11.25
4100	12' deep, 45 S.F./hr. in & 135 S.F./hr. out		270	.148		1.90	6.15	.73	8.78	12.35
4200	14' deep, 42 S.F./hr. in & 126 S.F./hr. out		250	.160		1.95	6.65	.79	9.39	13.20
4300	16' deep, 40 S.F./hr. in & 120 S.F./hr. out		240	.167		2.01	6.95	.83	9.79	13.75
4400	18' deep, 38 S.F./hr. in & 114 S.F./hr. out		230	.174		2.08	7.25	.86	10.19	14.35
4500	20' deep, 35 S.F./hr. in & 105 S.F./hr. out		210	.190		2.15	7.90	.94	10.99	15.55
4520	Left in place, 8' deep, 55 S.F./hr.		440	.091		3.22	3.78	.45	7.45	9.85
4540	10' deep, 50 S.F./hr.		400	.100		3.39	4.16	.50	8.05	10.65
4560	12' deep, 45 S.F./hr.		360	.111		3.58	4.62	.55	8.75	11.65
4565	14' deep, 42 S.F./hr.		335	.119		3.79	4.96	.59	9.34	12.40
4570	16' deep, 40 S.F./hr.		320	.125		4.03	5.20	.62	9.85	13.05
4580	18' deep, 38 S.F./hr.		305	.131		4.30	5.45	.65	10.40	13.80
4590	20' deep, 35 S.F./hr.		280	.143		4.60	5.95	.71	11.26	14.95
4700	Alternate pricing, left in place, 8' deep		1.76	22.727	M.B.F.	725	945	113	1,783	2,375
4800	Drive, extract and salvage, 8' deep		1.32	30.303	"	645	1,250	150	2,045	2,800

For customer support on your Building Construction Costs with RSMeans Data, call 800.448.8182.

639

31 41 Shoring

31 41 16 – Sheet Piling

31 41 16.10 Sheet Piling Systems	Crew	Daily Output	Labor-Hours	Unit	Material	2017 Bare Costs Labor	Equipment	Total	Total Incl O&P
5000	For treated lumber add cost of treatment to lumber								

31 43 Concrete Raising

31 43 13 – Pressure Grouting

31 43 13.13 Concrete Pressure Grouting

		Crew	Daily Output	Labor-Hours	Unit	Material	2017 Bare Costs Labor	Equipment	Total	Total Incl O&P
0010	**CONCRETE PRESSURE GROUTING**									
0020	Grouting, pressure, cement & sand, 1:1 mix, minimum	B-61	124	.323	Bag	13.70	13.50	2.75	29.95	38.50
0100	Maximum		51	.784	"	13.70	33	6.70	53.40	72.50
0200	Cement and sand, 1:1 mix, minimum		250	.160	C.F.	27.50	6.70	1.36	35.56	42
0300	Maximum		100	.400		41	16.75	3.41	61.16	74.50
0400	Epoxy cement grout, minimum		137	.292		625	12.25	2.49	639.74	705
0500	Maximum		57	.702		625	29.50	6	660.50	735
0700	Alternate pricing method: (Add for materials)									
0710	5 person crew and equipment	B-61	1	40	Day		1,675	340	2,015	2,925

31 45 Vibroflotation and Densification

31 45 13 – Vibroflotation

31 45 13.10 Vibroflotation Densification

		Crew	Daily Output	Labor-Hours	Unit	Material	2017 Bare Costs Labor	Equipment	Total	Total Incl O&P
0010	**VIBROFLOTATION DENSIFICATION**	R314513-90								
0900	Vibroflotation compacted sand cylinder, minimum	B-60	750	.075	V.L.F.		3.47	3.15	6.62	8.70
0950	Maximum		325	.172			8	7.25	15.25	20
1100	Vibro replacement compacted stone cylinder, minimum		500	.112			5.20	4.72	9.92	13.10
1150	Maximum		250	.224			10.40	9.45	19.85	26
1300	Mobilization and demobilization, minimum		.47	119	Total		5,525	5,025	10,550	13,900
1400	Maximum		.14	400	"		18,600	16,900	35,500	46,800

31 46 Needle Beams

31 46 13 – Cantilever Needle Beams

31 46 13.10 Needle Beams

		Crew	Daily Output	Labor-Hours	Unit	Material	2017 Bare Costs Labor	Equipment	Total	Total Incl O&P
0010	**NEEDLE BEAMS**									
0011	Incl. wood shoring 10' x 10' opening									
0400	Block, concrete, 8" thick	B-9	7.10	5.634	Ea.	51.50	223	30	304.50	430
0420	12" thick		6.70	5.970		62.50	236	32	330.50	465
0800	Brick, 4" thick with 8" backup block		5.70	7.018		62.50	278	37.50	378	535
1000	Brick, solid, 8" thick		6.20	6.452		51.50	255	34.50	341	485
1040	12" thick		4.90	8.163		62.50	325	43.50	431	610
1080	16" thick		4.50	8.889		84.50	350	47.50	482	685
2000	Add for additional floors of shoring	B-1	6	4		51.50	159		210.50	300

31 48 Underpinning

31 48 13 – Underpinning Piers

31 48 13.10 Underpinning Foundations	Crew	Daily Output	Labor-Hours	Unit	Material	2017 Bare Costs Labor	Equipment	Total	Total Incl O&P
0010 **UNDERPINNING FOUNDATIONS**									
0011 Including excavation,									
0020 forming, reinforcing, concrete and equipment									
0100 5' to 16' below grade, 100 to 500 C.Y.	B-52	2.30	24.348	C.Y.	330	1,100	279	1,709	2,375
0200 Over 500 C.Y.		2.50	22.400		295	1,025	256	1,576	2,150
0400 16' to 25' below grade, 100 to 500 C.Y.		2	28		360	1,275	320	1,955	2,700
0500 Over 500 C.Y.		2.10	26.667		340	1,225	305	1,870	2,550
0700 26' to 40' below grade, 100 to 500 C.Y.		1.60	35		395	1,600	400	2,395	3,300
0800 Over 500 C.Y.	▼	1.80	31.111		360	1,425	355	2,140	2,925
0900 For under 50 C.Y., add					10%	40%			
1000 For 50 C.Y. to 100 C.Y., add			▼		5%	20%			

31 52 Cofferdams

31 52 16 – Timber Cofferdams

31 52 16.10 Cofferdams

	Crew	Daily Output	Labor-Hours	Unit	Material	2017 Bare Costs Labor	Equipment	Total	Total Incl O&P
0010 **COFFERDAMS**									
0011 Incl. mobilization and temporary sheeting									
0080 Soldier beams & lagging H piles with 3" wood sheeting									
0090 horizontal between piles, including removal of wales & braces									
0100 No hydrostatic head, 15' deep, 1 line of braces, minimum	B-50	545	.206	S.F.	8.65	9.95	4.68	23.28	30
0200 Maximum		495	.226		9.60	10.95	5.15	25.70	33.50
0400 15' to 22' deep with 2 lines of braces, 10" H, minimum		360	.311		10.20	15.05	7.10	32.35	42.50
0500 Maximum		330	.339		11.55	16.40	7.75	35.70	46.50
0700 23' to 35' deep with 3 lines of braces, 12" H, minimum		325	.345		13.30	16.65	7.85	37.80	49.50
0800 Maximum		295	.380		14.45	18.35	8.65	41.45	54
1000 36' to 45' deep with 4 lines of braces, 14" H, minimum		290	.386		14.95	18.65	8.80	42.40	55
1100 Maximum		265	.423		15.75	20.50	9.65	45.90	60
1300 No hydrostatic head, left in place, 15' dp., 1 line of braces, min.		635	.176		11.55	8.55	4.02	24.12	30.50
1400 Maximum		575	.195		12.35	9.40	4.44	26.19	33
1600 15' to 22' deep with 2 lines of braces, minimum		455	.246		17.30	11.90	5.60	34.80	44
1700 Maximum		415	.270		19.25	13.05	6.15	38.45	48.50
1900 23' to 35' deep with 3 lines of braces, minimum		420	.267		20.50	12.90	6.05	39.45	49
2000 Maximum		380	.295		23	14.25	6.70	43.95	54.50
2200 36' to 45' deep with 4 lines of braces, minimum		385	.291		24.50	14.05	6.65	45.20	56.50
2300 Maximum	▼	350	.320		29	15.45	7.30	51.75	64
2350 Lagging only, 3" thick wood between piles 8' O.C., minimum	B-46	400	.120		1.92	5.35	.11	7.38	10.65
2370 Maximum		250	.192		2.89	8.55	.18	11.62	16.75
2400 Open sheeting no bracing, for trenches to 10' deep, min.		1736	.028		.87	1.23	.03	2.13	2.91
2450 Maximum	▼	1510	.032	▼	.96	1.42	.03	2.41	3.31
2500 Tie-back method, add to open sheeting, add, minimum								20%	20%
2550 Maximum								60%	60%
2700 Tie-backs only, based on tie-backs total length, minimum	B-46	86.80	.553	L.F.	15.10	24.50	.52	40.12	55.50
2750 Maximum		38.50	1.247	"	26.50	55.50	1.17	83.17	118
3500 Tie-backs only, typical average, 25' long		2	24	Ea.	665	1,075	22.50	1,762.50	2,425
3600 35' long	▼	1.58	30.380	"	885	1,350	28.50	2,263.50	3,125
6000 See also Section 31 41 16.10									

For customer support on your Building Construction Costs with RSMeans Data, call 800.448.8182.

641

31 56 23 – Lean Concrete Slurry Walls

31 56 23.20 Slurry Trench	Crew	Daily Output	Labor-Hours	Unit	Material	2017 Bare Costs Labor	Equipment	Total	Total Incl O&P
0010 **SLURRY TRENCH**									
0011 Excavated slurry trench in wet soils									
0020 backfilled with 3000 PSI concrete, no reinforcing steel									
0050 Minimum	C-7	333	.216	C.F.	8.75	9.25	3.47	21.47	27.50
0100 Maximum		200	.360	"	14.65	15.40	5.80	35.85	46
0200 Alternate pricing method, minimum		150	.480	S.F.	17.50	20.50	7.70	45.70	59
0300 Maximum		120	.600		26	25.50	9.65	61.15	78.50
0500 Reinforced slurry trench, minimum	B-48	177	.316		13.10	14.15	16.50	43.75	54
0600 Maximum	"	69	.812		43.50	36.50	42.50	122.50	150
0800 Haul for disposal, 2 mile haul, excavated material, add	B-34B	99	.081	C.Y.		3.68	6.50	10.18	12.70
0900 Haul bentonite castings for disposal, add	"	40	.200	"		9.10	16.05	25.15	31.50

31 62 13 – Concrete Piles

31 62 13.23 Prestressed Concrete Piles

	Crew	Daily Output	Labor-Hours	Unit	Material	2017 Bare Costs Labor	Equipment	Total	Total Incl O&P
0010 **PRESTRESSED CONCRETE PILES**, 200 piles									
0020 Unless specified otherwise, not incl. pile caps or mobilization									
2200 Precast, prestressed, 50' long, cylinder, 12" diam., 2-3/8" wall	B-19	720	.089	V.L.F.	25.50	4.54	2.58	32.62	38
2300 14" diameter, 2-1/2" wall		680	.094		30.50	4.81	2.74	38.05	44
2500 16" diameter, 3" wall		640	.100		39	5.10	2.91	47.01	54
2600 18" diameter, 3-1/2" wall	B-19A	600	.107		49.50	5.45	3.91	58.86	67
2800 20" diameter, 4" wall		560	.114		54.50	5.85	4.19	64.54	73.50
2900 24" diameter, 5" wall		520	.123		70	6.30	4.51	80.81	92
3100 Precast, prestressed, 40' long, 10" thick, square	B-19	700	.091		13.55	4.67	2.66	20.88	25
3200 12" thick, square		680	.094		20	4.81	2.74	27.55	32.50
3400 14" thick, square		600	.107		22	5.45	3.10	30.55	36.50
3500 Octagonal		640	.100		28	5.10	2.91	36.01	42
3700 16" thick, square		560	.114		31.50	5.85	3.32	40.67	47.50
3800 Octagonal		600	.107		33.50	5.45	3.10	42.05	49
4000 18" thick, square	B-19A	520	.123		39	6.30	4.51	49.81	57.50
4100 Octagonal	B-19	560	.114		40	5.85	3.32	49.17	57
4300 20" thick, square	B-19A	480	.133		44	6.80	4.89	55.69	64
4400 Octagonal	B-19	520	.123		44	6.30	3.58	53.88	62
4600 24" thick, square	B-19A	440	.145		57	7.45	5.35	69.80	80
4700 Octagonal	B-19	480	.133		63.50	6.80	3.87	74.17	84.50
4730 Precast, prestressed, 60' long, 10" thick, square		700	.091		14.25	4.67	2.66	21.58	26
4740 12" thick, square (60' long)		680	.094		21	4.81	2.74	28.55	33.50
4750 Mobilization for 10,000 L.F. pile job, add		3300	.019			.99	.56	1.55	2.16
4800 25,000 L.F. pile job, add		8500	.008			.38	.22	.60	.84

31 62 16 – Steel Piles

31 62 16.13 Steel Piles

	Crew	Daily Output	Labor-Hours	Unit	Material	2017 Bare Costs Labor	Equipment	Total	Total Incl O&P
0010 **STEEL PILES**									
0100 Step tapered, round, concrete filled									
0110 8" tip, 12" butt, 60 ton capacity, 30' depth	B-19	760	.084	V.L.F.	17.55	4.30	2.45	24.30	28.50
0120 60' depth with extension		740	.086		36	4.42	2.51	42.93	49
0130 80' depth with extensions		700	.091		55.50	4.67	2.66	62.83	71
0250 "H" Sections, 50' long, HP8 x 36		640	.100		16.55	5.10	2.91	24.56	29.50
0400 HP10 X 42		610	.105		19.35	5.35	3.05	27.75	33
0500 HP10 X 57		610	.105		26	5.35	3.05	34.40	40
0700 HP12 X 53		590	.108		24.50	5.55	3.15	33.20	39

31 62 Driven Piles

31 62 16 – Steel Piles

31 62 16.13 Steel Piles

		Crew	Daily Output	Labor-Hours	Unit	Material	2017 Bare Costs Labor	2017 Bare Costs Equipment	Total	Total Incl O&P
0800	HP12 X 74	B-19A	590	.108	V.L.F.	34	5.55	3.97	43.52	50.50
1000	HP14 X 73		540	.119		34.50	6.05	4.34	44.89	51.50
1100	HP14 X 89		540	.119		41	6.05	4.34	51.39	59
1300	HP14 X 102		510	.125		47	6.40	4.60	58	66.50
1400	HP14 X 117		510	.125		54	6.40	4.60	65	74.50
1600	Splice on standard points, not in leads, 8" or 10"	1 Sswl	5	1.600	Ea.	97.50	87		184.50	252
1700	12" or 14"		4	2		135	109		244	330
1900	Heavy duty points, not in leads, 10" wide		4	2		180	109		289	380
2100	14" wide		3.50	2.286		220	124		344	450

31 62 19 – Timber Piles

31 62 19.10 Wood Piles

		Crew	Daily Output	Labor-Hours	Unit	Material	2017 Bare Costs Labor	2017 Bare Costs Equipment	Total	Total Incl O&P
0010	**WOOD PILES**									
0011	Friction or end bearing, not including									
0050	mobilization or demobilization									
0100	Untreated piles, up to 30' long, 12" butts, 8" points	B-19	625	.102	V.L.F.	12.30	5.25	2.98	20.53	25
0200	30' to 39' long, 12" butts, 8" points		700	.091		12.30	4.67	2.66	19.63	23.50
0300	40' to 49' long, 12" butts, 7" points		720	.089		12.30	4.54	2.58	19.42	23.50
0400	50' to 59' long, 13" butts, 7" points		800	.080		12	4.09	2.32	18.41	22
0500	60' to 69' long, 13" butts, 7" points		840	.076		13.50	3.89	2.21	19.60	23.50
0600	70' to 80' long, 13" butts, 6" points		840	.076		15	3.89	2.21	21.10	25
0800	Treated piles, 12 lb. per C.F.,									
0810	friction or end bearing, ASTM class B									
1000	Up to 30' long, 12" butts, 8" points	B-19	625	.102	V.L.F.	11.90	5.25	2.98	20.13	24.50
1100	30' to 39' long, 12" butts, 8" points		700	.091		13.15	4.67	2.66	20.48	24.50
1200	40' to 49' long, 12" butts, 7" points		720	.089		14.05	4.54	2.58	21.17	25.50
1300	50' to 59' long, 13" butts, 7" points		800	.080		15.75	4.09	2.32	22.16	26
1400	60' to 69' long, 13" butts, 6" points	B-19A	840	.076		19.95	3.89	2.79	26.63	31
1500	70' to 80' long, 13" butts, 6" points	"	840	.076		21	3.89	2.79	27.68	32
1600	Treated piles, C.C.A., 2.5# per C.F.									
1610	8" butts, 10' long	B-19	400	.160	V.L.F.	12.50	8.20	4.65	25.35	31.50
1620	11' to 16' long		500	.128		12.50	6.55	3.72	22.77	28
1630	17' to 20' long		575	.111		12.50	5.70	3.23	21.43	26
1640	10" butts, 10' to 16' long		500	.128		12.80	6.55	3.72	23.07	28.50
1650	17' to 20' long		575	.111		12.80	5.70	3.23	21.73	26.50
1660	21' to 40' long		700	.091		12.80	4.67	2.66	20.13	24.50
1670	12" butts, 10' to 20' long		575	.111		13.20	5.70	3.23	22.13	27
1680	21' to 35' long		650	.098		13.20	5.05	2.86	21.11	25.50
1690	36' to 40' long		700	.091		13.20	4.67	2.66	20.53	24.50
1695	14" butts, to 40' long		700	.091		13.75	4.67	2.66	21.08	25.50
1700	Boot for pile tip, minimum	1 Pile	27	.296	Ea.	30	14.65		44.65	56.50
1800	Maximum		21	.381		90	18.85		108.85	129
2000	Point for pile tip, minimum		20	.400		30	19.80		49.80	64.50
2100	Maximum		15	.533		108	26.50		134.50	161
2300	Splice for piles over 50' long, minimum	B-46	35	1.371		58	61.50	1.29	120.79	161
2400	Maximum		20	2.400		73.50	107	2.26	182.76	251
2600	Concrete encasement with wire mesh and tube		331	.145	V.L.F.	73.50	6.50	.14	80.14	91.50
2700	Mobilization for 10,000 L.F. pile job, add	B-19	3300	.019			.99	.56	1.55	2.16
2800	25,000 L.F. pile job, add	"	8500	.008			.38	.22	.60	.84

31 62 23 – Composite Piles

31 62 23.13 Concrete-Filled Steel Piles	Crew	Daily Output	Labor-Hours	Unit	Material	2017 Bare Costs Labor	Equipment	Total	Total Incl O&P
0010 **CONCRETE-FILLED STEEL PILES** no mobilization or demobilization									
2600 Pipe piles, 50' lg. 8" diam., 29 lb. per L.F., no concrete	B-19	500	.128	V.L.F.	21	6.55	3.72	31.27	37.50
2700 Concrete filled		460	.139		24	7.10	4.04	35.14	42
2900 10" diameter, 34 lb. per L.F., no concrete		500	.128		23.50	6.55	3.72	33.77	40.50
3000 Concrete filled		450	.142		28.50	7.25	4.13	39.88	47.50
3200 12" diameter, 44 lb. per L.F., no concrete		475	.135		29	6.90	3.92	39.82	46.50
3300 Concrete filled		415	.154		35	7.90	4.48	47.38	55.50
3500 14" diameter, 46 lb. per L.F., no concrete		430	.149		31	7.60	4.33	42.93	50.50
3600 Concrete filled		355	.180		40	9.20	5.25	54.45	64
3800 16" diameter, 52 lb. per L.F., no concrete		385	.166		37	8.50	4.83	50.33	59.50
3900 Concrete filled		335	.191		48.50	9.75	5.55	63.80	75
4100 18" diameter, 59 lb. per L.F., no concrete		355	.180		42	9.20	5.25	56.45	66
4200 Concrete filled		310	.206		58.50	10.55	6	75.05	87
4400 Splices for pipe piles, stl., not in leads, 8" diameter	1 Sswl	5	1.600	Ea.	63.50	87		150.50	215
4410 10" diameter		4.75	1.684		78	91.50		169.50	238
4430 12" diameter		4.50	1.778		108	96.50		204.50	280
4500 14" diameter		4.25	1.882		136	102		238	320
4600 16" diameter		4	2		169	109		278	365
4650 18" diameter		3.75	2.133		280	116		396	505
4710 Steel pipe pile backing rings, w/spacer, 8" diameter		12	.667		8.60	36		44.60	70
4720 10" diameter		12	.667		11.30	36		47.30	73
4730 12" diameter		10	.800		13.60	43.50		57.10	87.50
4740 14" diameter		9	.889		15.15	48.50		63.65	97
4750 16" diameter		8	1		19.05	54.50		73.55	112
4760 18" diameter		6	1.333		20.50	72.50		93	144
4800 Points, standard, 8" diameter		4.61	1.735		101	94		195	269
4840 10" diameter		4.45	1.798		134	97.50		231.50	310
4880 12" diameter		4.25	1.882		188	102		290	375
4900 14" diameter		4.05	1.975		217	107		324	420
5000 16" diameter		3.37	2.374		305	129		434	550
5050 18" diameter		3.50	2.286		405	124		529	650
5200 Points, heavy duty, 10" diameter		2.90	2.759		222	150		372	495
5240 12" diameter		2.95	2.712		265	147		412	535
5260 14" diameter		2.95	2.712		300	147		447	580
5280 16" diameter		2.95	2.712		360	147		507	640
5290 18" diameter		2.80	2.857		490	155		645	800
5500 For reinforcing steel, add		1150	.007	Lb.	.69	.38		1.07	1.39
5700 For thick wall sections, add				"	.71			.71	.78

31 63 26 – Drilled Caissons

31 63 26.13 Fixed End Caisson Piles	Crew	Daily Output	Labor-Hours	Unit	Material	2017 Bare Costs Labor	Equipment	Total	Total Incl O&P
0010 **FIXED END CAISSON PILES** R316326-60									
0015 Including excavation, concrete, 50 lb. reinforcing									
0020 per C.Y., not incl. mobilization, boulder removal, disposal									
0100 Open style, machine drilled, to 50' deep, in stable ground, no									
0110 casings or ground water, 18" diam., 0.065 C.Y./L.F.	B-43	200	.240	V.L.F.	9.15	10.50	12.85	32.50	40.50
0200 24" diameter, 0.116 C.Y./L.F.		190	.253		16.40	11.05	13.55	41	50
0300 30" diameter, 0.182 C.Y./L.F.		150	.320		25.50	14	17.15	56.65	69
0400 36" diameter, 0.262 C.Y./L.F.		125	.384		37	16.80	20.50	74.30	88.50

31 63 Bored Piles

31 63 26 – Drilled Caissons

31 63 26.13 Fixed End Caisson Piles	Crew	Daily Output	Labor-Hours	Unit	Material	2017 Bare Costs Labor	Equipment	Total	Total Incl O&P
0500	48" diameter, 0.465 C.Y./L.F.	B-43	100	.480	V.L.F.	66	21	25.50	112.50	133
0600	60" diameter, 0.727 C.Y./L.F.		90	.533		103	23.50	28.50	155	180
0700	72" diameter, 1.05 C.Y./L.F.		80	.600		149	26	32	207	239
0800	84" diameter, 1.43 C.Y./L.F.	▼	75	.640	▼	202	28	34.50	264.50	305
1000	For bell excavation and concrete, add									
1020	4' bell diameter, 24" shaft, 0.444 C.Y.	B-43	20	2.400	Ea.	52.50	105	129	286.50	360
1040	6' bell diameter, 30" shaft, 1.57 C.Y.		5.70	8.421		185	370	450	1,005	1,250
1060	8' bell diameter, 36" shaft, 3.72 C.Y.		2.40	20		440	875	1,075	2,390	2,975
1080	9' bell diameter, 48" shaft, 4.48 C.Y.		2	24		530	1,050	1,275	2,855	3,600
1100	10' bell diameter, 60" shaft, 5.24 C.Y.		1.70	28.235		620	1,225	1,525	3,370	4,225
1120	12' bell diameter, 72" shaft, 8.74 C.Y.		1	48		1,025	2,100	2,575	5,700	7,150
1140	14' bell diameter, 84" shaft, 13.6 C.Y.	▼	.70	68.571	▼	1,600	3,000	3,675	8,275	10,400
1200	Open style, machine drilled, to 50' deep, in wet ground, pulled									
1300	casing and pumping, 18" diameter, 0.065 C.Y./L.F.	B-48	160	.350	V.L.F.	9.15	15.65	18.25	43.05	54
1400	24" diameter, 0.116 C.Y./L.F.		125	.448		16.40	20	23.50	59.90	74
1500	30" diameter, 0.182 C.Y./L.F.		85	.659		25.50	29.50	34.50	89.50	111
1600	36" diameter, 0.262 C.Y./L.F.	▼	60	.933		37	42	48.50	127.50	158
1700	48" diameter, 0.465 C.Y./L.F.	B-49	55	1.600		66	75	65.50	206.50	260
1800	60" diameter, 0.727 C.Y./L.F.		35	2.514		103	118	103	324	405
1900	72" diameter, 1.05 C.Y./L.F.		30	2.933		149	138	120	407	505
2000	84" diameter, 1.43 C.Y./L.F.	▼	25	3.520	▼	202	165	144	511	635
2100	For bell excavation and concrete, add									
2120	4' bell diameter, 24" shaft, 0.444 C.Y.	B-48	19.80	2.828	Ea.	52.50	127	147	326.50	415
2140	6' bell diameter, 30" shaft, 1.57 C.Y.		5.70	9.825		185	440	510	1,135	1,450
2160	8' bell diameter, 36" shaft, 3.72 C.Y.	▼	2.40	23.333		440	1,050	1,225	2,715	3,400
2180	9' bell diameter, 48" shaft, 4.48 C.Y.	B-49	3.30	26.667		530	1,250	1,100	2,880	3,700
2200	10' bell diameter, 60" shaft, 5.24 C.Y.		2.80	31.429		620	1,475	1,275	3,370	4,325
2220	12' bell diameter, 72" shaft, 8.74 C.Y.		1.60	55		1,025	2,575	2,250	5,850	7,550
2240	14' bell diameter, 84" shaft, 13.6 C.Y.	▼	1	88	▼	1,600	4,125	3,600	9,325	12,100
2300	Open style, machine drilled, to 50' deep, in soft rocks and									
2400	medium hard shales, 18" diameter, 0.065 C.Y./L.F.	B-49	50	1.760	V.L.F.	9.15	82.50	72	163.65	216
2500	24" diameter, 0.116 C.Y./L.F.		30	2.933		16.40	138	120	274.40	360
2600	30" diameter, 0.182 C.Y./L.F.		20	4.400		25.50	206	180	411.50	540
2700	36" diameter, 0.262 C.Y./L.F.		15	5.867		37	275	239	551	725
2800	48" diameter, 0.465 C.Y./L.F.		10	8.800		66	415	360	841	1,100
2900	60" diameter, 0.727 C.Y./L.F.		7	12.571		103	590	515	1,208	1,575
3000	72" diameter, 1.05 C.Y./L.F.		6	14.667		149	690	600	1,439	1,875
3100	84" diameter, 1.43 C.Y./L.F.	▼	5	17.600	▼	202	825	720	1,747	2,300
3200	For bell excavation and concrete, add									
3220	4' bell diameter, 24" shaft, 0.444 C.Y.	B-49	10.90	8.073	Ea.	52.50	380	330	762.50	1,000
3240	6' bell diameter, 30" shaft, 1.57 C.Y.		3.10	28.387		185	1,325	1,150	2,660	3,525
3260	8' bell diameter, 36" shaft, 3.72 C.Y.		1.30	67.692		440	3,175	2,750	6,365	8,400
3280	9' bell diameter, 48" shaft, 4.48 C.Y.		1.10	80		530	3,750	3,275	7,555	9,925
3300	10' bell diameter, 60" shaft, 5.24 C.Y.		.90	97.778		620	4,575	4,000	9,195	12,100
3320	12' bell diameter, 72" shaft, 8.74 C.Y.		.60	146		1,025	6,875	5,975	13,875	18,200
3340	14' bell diameter, 84" shaft, 13.6 C.Y.		.40	220	▼	1,600	10,300	8,975	20,875	27,500
3600	For rock excavation, sockets, add, minimum		120	.733	C.F.		34.50	30	64.50	85.50
3650	Average		95	.926			43.50	38	81.50	108
3700	Maximum	▼	48	1.833	▼		86	75	161	215
3900	For 50' to 100' deep, add				V.L.F.				7%	7%
4000	For 100' to 150' deep, add								25%	25%
4100	For 150' to 200' deep, add								30%	30%
4200	For casings left in place, add				Lb.	1.36			1.36	1.50

645

31 63 Bored Piles

31 63 26 – Drilled Caissons

31 63 26.13 Fixed End Caisson Piles

		Crew	Daily Output	Labor-Hours	Unit	Material	2017 Bare Costs Labor	Equipment	Total	Total Incl O&P
4300	For other than 50 lb. reinf. per C.Y., add or deduct				Lb.	1.26			1.26	1.39
4400	For steel "I" beam cores, add	B-49	8.30	10.602	Ton	2,100	495	435	3,030	3,550
4500	Load and haul excess excavation, 2 miles	B-34B	178	.045	L.C.Y.		2.05	3.61	5.66	7.05
4600	For mobilization, 50 mile radius, rig to 36"	B-43	2	24	Ea.		1,050	1,275	2,325	3,025
4650	Rig to 84"	B-48	1.75	32			1,425	1,675	3,100	4,000
4700	For low headroom, add								50%	50%
4750	For difficult access, add								25%	25%
5000	Bottom inspection	1 Skwk	1.20	6.667	▼		345		345	530

31 63 26.16 Concrete Caissons for Marine Construction

		Crew	Daily Output	Labor-Hours	Unit	Material	2017 Bare Costs Labor	Equipment	Total	Total Incl O&P
0010	**CONCRETE CAISSONS FOR MARINE CONSTRUCTION**									
0100	Caissons, incl. mobilization and demobilization, up to 50 miles									
0200	Uncased shafts, 30 to 80 tons cap., 17" diam., 10' depth	B-44	88	.727	V.L.F.	23.50	36.50	22.50	82.50	108
0300	25' depth		165	.388		16.85	19.40	12.10	48.35	62.50
0400	80 to 150 ton capacity, 22" diameter, 10' depth		80	.800		29.50	40	25	94.50	123
0500	20' depth		130	.492		23.50	24.50	15.35	63.35	81.50
0700	Cased shafts, 10 to 30 ton capacity, 10-5/8" diam., 20' depth		175	.366		16.85	18.30	11.40	46.55	59.50
0800	30' depth		240	.267		15.75	13.35	8.30	37.40	47.50
0850	30 to 60 ton capacity, 12" diameter, 20' depth		160	.400		23.50	20	12.45	55.95	70.50
0900	40' depth		230	.278		18.15	13.90	8.65	40.70	51
1000	80 to 100 ton capacity, 16" diameter, 20' depth		160	.400		33.50	20	12.45	65.95	81.50
1100	40' depth		230	.278		31.50	13.90	8.65	54.05	65.50
1200	110 to 140 ton capacity, 17-5/8" diameter, 20' depth		160	.400		36.50	20	12.45	68.95	84.50
1300	40' depth		230	.278		33.50	13.90	8.65	56.05	68
1400	140 to 175 ton capacity, 19" diameter, 20' depth		130	.492		39.50	24.50	15.35	79.35	99
1500	40' depth	▼	210	.305	▼	36.50	15.25	9.50	61.25	74.50
1700	Over 30' long, L.F. cost tends to be lower									
1900	Maximum depth is about 90'									

31 63 29 – Drilled Concrete Piers and Shafts

31 63 29.13 Uncased Drilled Concrete Piers

		Crew	Daily Output	Labor-Hours	Unit	Material	2017 Bare Costs Labor	Equipment	Total	Total Incl O&P
0010	**UNCASED DRILLED CONCRETE PIERS**									
0020	Unless specified otherwise, not incl. pile caps or mobilization									
0050	Cast in place augered piles, no casing or reinforcing									
0060	8" diameter	B-43	540	.089	V.L.F.	4.20	3.89	4.76	12.85	15.75
0065	10" diameter		480	.100		6.70	4.37	5.35	16.42	19.90
0070	12" diameter		420	.114		9.40	5	6.15	20.55	24.50
0075	14" diameter		360	.133		12.70	5.85	7.15	25.70	30.50
0080	16" diameter		300	.160		17.10	7	8.60	32.70	39
0085	18" diameter	▼	240	.200	▼	21	8.75	10.70	40.45	48.50
0100	Cast in place, thin wall shell pile, straight sided,									
0110	not incl. reinforcing, 8" diam., 16 ga., 5.8 lb./L.F.	B-19	700	.091	V.L.F.	9.50	4.67	2.66	16.83	20.50
0200	10" diameter, 16 ga. corrugated, 7.3 lb./L.F.		650	.098		12.45	5.05	2.86	20.36	24.50
0300	12" diameter, 16 ga. corrugated, 8.7 lb./L.F.		600	.107		16.15	5.45	3.10	24.70	29.50
0400	14" diameter, 16 ga. corrugated, 10.0 lb./L.F.		550	.116		19	5.95	3.38	28.33	34
0500	16" diameter, 16 ga. corrugated, 11.6 lb./L.F.	▼	500	.128	▼	23.50	6.55	3.72	33.77	40
0800	Cast in place friction pile, 50' long, fluted,									
0810	tapered steel, 4000 psi concrete, no reinforcing									
0900	12" diameter, 7 ga.	B-19	600	.107	V.L.F.	29.50	5.45	3.10	38.05	44.50
1000	14" diameter, 7 ga.		560	.114		32	5.85	3.32	41.17	48.50
1100	16" diameter, 7 ga.		520	.123		38	6.30	3.58	47.88	55
1200	18" diameter, 7 ga.	▼	480	.133	▼	44.50	6.80	3.87	55.17	63.50
1300	End bearing, fluted, constant diameter,									
1320	4000 psi concrete, no reinforcing									

31 63 Bored Piles

31 63 29 – Drilled Concrete Piers and Shafts

31 63 29.13 Uncased Drilled Concrete Piers	Crew	Daily Output	Labor-Hours	Unit	Material	2017 Bare Costs Labor	Equipment	Total	Total Incl O&P
1340	12" diameter, 7 ga.	B-19	600	.107	V.L.F.	31	5.45	3.10	39.55	46
1360	14" diameter, 7 ga.		560	.114		39	5.85	3.32	48.17	55.50
1380	16" diameter, 7 ga.		520	.123		45	6.30	3.58	54.88	63
1400	18" diameter, 7 ga.		480	.133		49.50	6.80	3.87	60.17	69.50

31 63 29.20 Cast In Place Piles, Adds

		Crew	Daily Output	Labor-Hours	Unit	Material	Labor	Equipment	Total	Total Incl O&P
0010	**CAST IN PLACE PILES, ADDS**									
1500	For reinforcing steel, add				Lb.	.94			.94	1.03
1700	For ball or pedestal end, add	B-19	11	5.818	C.Y.	150	297	169	616	815
1900	For lengths above 60', concrete, add	"	11	5.818	"	157	297	169	623	825
2000	For steel thin shell, pipe only				Lb.	1.50			1.50	1.65

Division Notes

	CREW	DAILY OUTPUT	LABOR-HOURS	UNIT	BARE COSTS				TOTAL INCL O&P
					MAT.	LABOR	EQUIP.	TOTAL	
	CREW	DAILY OUTPUT	LABOR-HOURS	UNIT	MAT.	LABOR	EQUIP.	TOTAL	TOTAL INCL O&P

Estimating Tips
32 01 00 Operations and Maintenance of Exterior Improvements
- Recycling of asphalt pavement is becoming very popular and is an alternative to removal and replacement. It can be a good value engineering proposal if removed pavement can be recycled, either at the project site or at another site that is reasonably close to the project site. Sections on repair of flexible and rigid pavement are included.

32 10 00 Bases, Ballasts, and Paving
- When estimating paving, keep in mind the project schedule. Also note that prices for asphalt and concrete are generally higher in the cold seasons. Lines for pavement markings, including tactile warning systems and fence lines, are included.

32 90 00 Planting
- The timing of planting and guarantee specifications often dictate the costs for establishing tree and shrub growth and a stand of grass or ground cover. Establish the work performance schedule to coincide with the local planting season. Maintenance and growth guarantees can add from 20%–100% to the total landscaping cost and can be contractually cumbersome. The cost to replace trees and shrubs can be as high as 5% of the total cost, depending on the planting zone, soil conditions, and time of year.

Reference Numbers
Reference numbers are shown at the beginning of some major classifications. These numbers refer to related items in the Reference Section. The reference information may be an estimating procedure, an alternate pricing method, or technical information.

Note: Not all subdivisions listed here necessarily appear. ■

Did you know?
Our online estimating solution gives you the same access to RSMeans' data with 24/7 access:
- Quickly locate costs in the searchable database.
- Build cost lists, estimates, and reports in minutes.
- Adjust costs to any location in the U.S. and Canada with the click of a button.

Start your free trial today at
www.RSMeansOnline.com

32 01 13 – Flexible Paving Surface Treatment

32 01 13.61 Slurry Seal (Latex Modified)

		Crew	Daily Output	Labor-Hours	Unit	Material	2017 Bare Costs Labor	Equipment	Total	Total Incl O&P
0010	**SLURRY SEAL (LATEX MODIFIED)**									
3780	Rubberized asphalt (latex) seal	B-45	5000	.003	S.Y.	2.14	.16	.18	2.48	2.79

32 01 13.64 Sand Seal

		Crew	Daily Output	Labor-Hours	Unit	Material	Labor	Equipment	Total	Total Incl O&P
0010	**SAND SEAL**									
2080	Sand sealing, sharp sand, asphalt emulsion, small area	B-91	10000	.006	S.Y.	1.52	.30	.23	2.05	2.38
2120	Roadway or large area	"	18000	.004	"	1.31	.17	.13	1.61	1.84

32 01 13.66 Fog Seal

		Crew	Daily Output	Labor-Hours	Unit	Material	Labor	Equipment	Total	Total Incl O&P
0010	**FOG SEAL**									
0012	Sealcoating, 2 coat coal tar pitch emulsion over 10,000 S.Y.	B-45	5000	.003	S.Y.	.87	.16	.18	1.21	1.40
0030	1000 to 10,000 S.Y.	"	3000	.005		.87	.26	.30	1.43	1.69
0100	Under 1000 S.Y.	B-1	1050	.023		.87	.91		1.78	2.35
0300	Petroleum resistant, over 10,000 S.Y.	B-45	5000	.003		1.34	.16	.18	1.68	1.91
0320	1000 to 10,000 S.Y.	"	3000	.005		1.34	.26	.30	1.90	2.20
0400	Under 1000 S.Y.	B-1	1050	.023		1.34	.91		2.25	2.86
0600	Non-skid pavement renewal, over 10,000 S.Y.	B-45	5000	.003		1.41	.16	.18	1.75	1.99
0620	1000 to 10,000 S.Y.	"	3000	.005		1.41	.26	.30	1.97	2.28
0700	Under 1000 S.Y.	B-1	1050	.023		1.41	.91		2.32	2.94
0800	Prepare and clean surface for above	A-2	8545	.003			.11	.03	.14	.20
1000	Hand seal asphalt curbing	B-1	4420	.005	L.F.	.64	.22		.86	1.03
1900	Asphalt surface treatment, single course, small area									
1901	0.30 gal/S.Y. asphalt material, 20#/S.Y. aggregate	B-91	5000	.013	S.Y.	1.34	.61	.46	2.41	2.90
1910	Roadway or large area		10000	.006		1.23	.30	.23	1.76	2.07
1950	Asphalt surface treatment, dbl. course for small area		3000	.021		3.03	1.01	.77	4.81	5.70
1960	Roadway or large area		6000	.011		2.73	.51	.38	3.62	4.19
1980	Asphalt surface treatment, single course, for shoulders		7500	.009		1.50	.40	.31	2.21	2.60

32 06 Schedules for Exterior Improvements

32 06 10 – Schedules for Bases, Ballasts, and Paving

32 06 10.10 Sidewalks, Driveways and Patios

		Crew	Daily Output	Labor-Hours	Unit	Material	Labor	Equipment	Total	Total Incl O&P
0010	**SIDEWALKS, DRIVEWAYS AND PATIOS** No base									
0020	Asphaltic concrete, 2" thick	B-37	720	.067	S.Y.	7	2.76	.21	9.97	12.15
0100	2-1/2" thick	"	660	.073	"	8.90	3.02	.23	12.15	14.60
0300	Concrete, 3000 psi, CIP, 6 x 6 - W1.4 x W1.4 mesh,									
0310	broomed finish, no base, 4" thick	B-24	600	.040	S.F.	1.97	1.80		3.77	4.89
0350	5" thick		545	.044		2.62	1.98		4.60	5.90
0400	6" thick		510	.047		3.06	2.12		5.18	6.60
0450	For bank run gravel base, 4" thick, add	B-18	2500	.010		.44	.38	.02	.84	1.09
0520	8" thick, add	"	1600	.015		.89	.60	.03	1.52	1.93
0550	Exposed aggregate finish, add to above, minimum	B-24	1875	.013		.12	.58		.70	1
0600	Maximum	"	455	.053		.40	2.38		2.78	4.05
1000	Crushed stone, 1" thick, white marble	2 Clab	1700	.009		.42	.37		.79	1.02
1050	Bluestone	"	1700	.009		.12	.37		.49	.69
1700	Redwood, prefabricated, 4' x 4' sections	2 Carp	316	.051		4.77	2.49		7.26	9.05
1750	Redwood planks, 1" thick, on sleepers	"	240	.067		4.77	3.28		8.05	10.30
2250	Stone dust, 4" thick	B-62	900	.027	S.Y.	3.95	1.15	.19	5.29	6.30

32 06 10.20 Steps

		Crew	Daily Output	Labor-Hours	Unit	Material	Labor	Equipment	Total	Total Incl O&P
0010	**STEPS**									
0011	Incl. excav., borrow & concrete base as required									
0100	Brick steps	B-24	35	.686	LF Riser	15.95	31		46.95	64.50
0200	Railroad ties	2 Clab	25	.640		3.54	25		28.54	42.50

650

For customer support on your Building Construction Costs with RSMeans Data, call 800.448.8182.

32 06 Schedules for Exterior Improvements

32 06 10 - Schedules for Bases, Ballasts, and Paving

32 06 10.20 Steps	Crew	Daily Output	Labor-Hours	Unit	Material	2017 Bare Costs Labor	Equipment	Total	Total Incl O&P
0300 Bluestone treads, 12" x 2" or 12" x 1-1/2"	B-24	30	.800	LF Riser	37.50	36		73.50	95.50
0500 Concrete, cast in place, see Section 03 30 53.40									
0600 Precast concrete, see Section 03 41 23.50									
4025 Steel edge strips, incl. stakes, 1/4" x 5"	B-1	390	.062	L.F.	4.26	2.45		6.71	8.45
4050 Edging, landscape timber or railroad ties, 6" x 8"	2 Carp	170	.094	"	2.24	4.64		6.88	9.55

32 11 Base Courses

32 11 23 - Aggregate Base Courses

32 11 23.23 Base Course Drainage Layers

		Crew	Daily Output	Labor-Hours	Unit	Material	2017 Bare Costs Labor	Equipment	Total	Total Incl O&P
0010	**BASE COURSE DRAINAGE LAYERS**									
0011	For roadways and large areas									
0050	Crushed 3/4" stone base, compacted, 3" deep	B-36C	5200	.008	S.Y.	2.53	.38	.79	3.70	4.22
0100	6" deep		5000	.008		5.05	.40	.82	6.27	7.05
0200	9" deep		4600	.009		7.60	.43	.89	8.92	10
0300	12" deep		4200	.010		10.10	.47	.97	11.54	12.95
0301	Crushed 1-1/2" stone base, compacted to 4" deep	B-36B	6000	.011		4.21	.51	.79	5.51	6.25
0302	6" deep		5400	.012		6.30	.56	.88	7.74	8.75
0303	8" deep		4500	.014		8.40	.67	1.06	10.13	11.45
0304	12" deep		3800	.017		12.65	.80	1.25	14.70	16.50
0350	Bank run gravel, spread and compacted									
0370	6" deep	B-32	6000	.005	S.Y.	3.69	.27	.38	4.34	4.87
0390	9" deep		4900	.007		5.55	.33	.47	6.35	7.10
0400	12" deep		4200	.008		7.35	.38	.55	8.28	9.30
6000	Stabilization fabric, polypropylene, 6 oz./S.Y.	B-6	10000	.002		.72	.10	.04	.86	.99
6900	For small and irregular areas, add						50%	50%		
7000	Prepare and roll sub-base, small areas to 2500 S.Y.	B-32A	1500	.016	S.Y.		.78	.91	1.69	2.18
8000	Large areas over 2500 S.Y.	"	3500	.007			.33	.39	.72	.94
8050	For roadways	B-32	4000	.008			.40	.58	.98	1.24

32 11 26 - Asphaltic Base Courses

32 11 26.19 Bituminous-Stabilized Base Courses

		Crew	Daily Output	Labor-Hours	Unit	Material	2017 Bare Costs Labor	Equipment	Total	Total Incl O&P
0010	**BITUMINOUS-STABILIZED BASE COURSES**									
0020	And large paved areas									
0700	Liquid application to gravel base, asphalt emulsion	B-45	6000	.003	Gal.	4.46	.13	.15	4.74	5.25
0800	Prime and seal, cut back asphalt		6000	.003	"	5.25	.13	.15	5.53	6.15
1000	Macadam penetration crushed stone, 2 gal. per S.Y., 4" thick		6000	.003	S.Y.	8.90	.13	.15	9.18	10.15
1100	6" thick, 3 gal. per S.Y.		4000	.004		13.40	.20	.22	13.82	15.25
1200	8" thick, 4 gal. per S.Y.		3000	.005		17.85	.26	.30	18.41	20.50
8900	For small and irregular areas, add						50%	50%		

32 12 Flexible Paving

32 12 16 – Asphalt Paving

32 12 16.13 Plant-Mix Asphalt Paving

32 12 16.13 Plant-Mix Asphalt Paving	Crew	Daily Output	Labor-Hours	Unit	Material	2017 Bare Costs Labor	Equipment	Total	Total Incl O&P
0010 **PLANT-MIX ASPHALT PAVING**									
0020 And large paved areas with no hauling included									
0025 See Section 31 23 23.20 for hauling costs									
0080 Binder course, 1-1/2" thick	B-25	7725	.011	S.Y.	5.40	.49	.37	6.26	7.10
0120 2" thick		6345	.014		7.20	.60	.45	8.25	9.30
0130 2-1/2" thick		5620	.016		9	.68	.50	10.18	11.50
0160 3" thick		4905	.018		10.80	.78	.58	12.16	13.65
0170 3-1/2" thick		4520	.019		12.60	.84	.63	14.07	15.85
0200 4" thick	▼	4140	.021		14.40	.92	.68	16	17.95
0300 Wearing course, 1" thick	B-25B	10575	.009		3.57	.40	.29	4.26	4.85
0340 1-1/2" thick		7725	.012		6	.55	.40	6.95	7.90
0380 2" thick		6345	.015		8.05	.67	.48	9.20	10.40
0420 2-1/2" thick		5480	.018		9.95	.77	.56	11.28	12.70
0460 3" thick		4900	.020		11.85	.86	.63	13.34	15
0470 3-1/2" thick		4520	.021		13.90	.94	.68	15.52	17.50
0480 4" thick	▼	4140	.023		15.85	1.02	.74	17.61	19.80
0500 Open graded friction course	B-25C	5000	.010	▼	2.34	.43	.49	3.26	3.77
0800 Alternate method of figuring paving costs									
0810 Binder course, 1-1/2" thick	B-25	630	.140	Ton	66	6.05	4.49	76.54	86.50
0811 2" thick		690	.128		66	5.50	4.10	75.60	85.50
0812 3" thick		800	.110		66	4.76	3.54	74.30	83.50
0813 4" thick	▼	900	.098		66	4.23	3.14	73.37	82.50
0850 Wearing course, 1" thick	B-25B	575	.167		72.50	7.35	5.35	85.20	96.50
0851 1-1/2" thick		630	.152		72.50	6.70	4.87	84.07	95
0852 2" thick		690	.139		72.50	6.15	4.44	83.09	93.50
0853 2-1/2" thick		765	.125		72.50	5.55	4.01	82.06	92.50
0854 3" thick	▼	800	.120	▼	72.50	5.30	3.83	81.63	92
1000 Pavement replacement over trench, 2" thick	B-37	90	.533	S.Y.	7.40	22	1.70	31.10	44
1050 4" thick		70	.686		14.70	28.50	2.18	45.38	62
1080 6" thick	▼	55	.873	▼	23.50	36	2.78	62.28	84

32 12 16.14 Asphaltic Concrete Paving

32 12 16.14 Asphaltic Concrete Paving	Crew	Daily Output	Labor-Hours	Unit	Material	2017 Bare Costs Labor	Equipment	Total	Total Incl O&P
0011 **ASPHALTIC CONCRETE PAVING**, parking lots & driveways									
0015 No asphalt hauling included									
0018 Use 6.05 C.Y. per inch per M.S.F. for hauling									
0020 6" stone base, 2" binder course, 1" topping	B-25C	9000	.005	S.F.	1.80	.24	.27	2.31	2.64
0025 2" binder course, 2" topping		9000	.005		2.24	.24	.27	2.75	3.12
0030 3" binder course, 2" topping		9000	.005		2.64	.24	.27	3.15	3.57
0035 4" binder course, 2" topping		9000	.005		3.04	.24	.27	3.55	4
0040 1.5" binder course, 1" topping		9000	.005		1.60	.24	.27	2.11	2.42
0042 3" binder course, 1" topping		9000	.005		2.21	.24	.27	2.72	3.09
0045 3" binder course, 3" topping		9000	.005		3.09	.24	.27	3.60	4.05
0050 4" binder course, 3" topping		9000	.005		3.48	.24	.27	3.99	4.49
0055 4" binder course, 4" topping		9000	.005		3.92	.24	.27	4.43	4.97
0300 Binder course, 1-1/2" thick		35000	.001		.60	.06	.07	.73	.83
0400 2" thick		25000	.002		.78	.09	.10	.97	1.10
0500 3" thick		15000	.003		1.20	.14	.16	1.50	1.72
0600 4" thick		10800	.004		1.57	.20	.23	2	2.28
0800 Sand finish course, 3/4" thick		41000	.001		.31	.05	.06	.42	.50
0900 1" thick	▼	34000	.001		.39	.06	.07	.52	.61
1000 Fill pot holes, hot mix, 2" thick	B-16	4200	.008		.82	.31	.15	1.28	1.55
1100 4" thick		3500	.009		1.20	.38	.18	1.76	2.10
1120 6" thick	▼	3100	.010	▼	1.62	.43	.21	2.26	2.66

652

32 12 Flexible Paving

32 12 16 – Asphalt Paving

32 12 16.14 Asphaltic Concrete Paving		Crew	Daily Output	Labor-Hours	Unit	Material	2017 Bare Costs Labor	Equipment	Total	Total Incl O&P
1140	Cold patch, 2" thick	B-51	3000	.016	S.F.	.93	.65	.07	1.65	2.10
1160	4" thick		2700	.018		1.78	.72	.08	2.58	3.15
1180	6" thick		1900	.025		2.77	1.02	.11	3.90	4.74

32 13 Rigid Paving

32 13 13 – Concrete Paving

32 13 13.25 Concrete Pavement, Highways

		Crew	Daily Output	Labor-Hours	Unit	Material	2017 Bare Costs Labor	Equipment	Total	Total Incl O&P
0010	**CONCRETE PAVEMENT, HIGHWAYS**									
0015	Including joints, finishing and curing									
0020	Fixed form, 12' pass, unreinforced, 6" thick	B-26	3000	.029	S.Y.	25.50	1.29	1.04	27.83	31
0100	8" thick		2750	.032		34.50	1.41	1.14	37.05	41.50
0110	8" thick, small area		1375	.064		34.50	2.82	2.28	39.60	45
0200	9" thick		2500	.035		39.50	1.55	1.25	42.30	47
0300	10" thick		2100	.042		43	1.84	1.49	46.33	52
0310	10" thick, small area		1050	.084		43	3.69	2.98	49.67	56.50
0400	12" thick		1800	.049		49.50	2.15	1.74	53.39	59.50
0410	Conc. pavement, w/jt., fnsh.&curing, fix form, 24' pass, unreinforced, 6"T		6000	.015		24	.65	.52	25.17	28
0430	8" thick		5500	.016		33	.70	.57	34.27	37.50
0440	9" thick		5000	.018		37.50	.77	.63	38.90	43.50
0450	10" thick		4200	.021		41.50	.92	.75	43.17	47.50
0460	12" thick		3600	.024		48	1.08	.87	49.95	55
0470	15" thick		3000	.029		62	1.29	1.04	64.33	71.50
0500	Fixed form 12' pass 15" thick		1500	.059		63	2.58	2.09	67.67	75.50
0510	For small irregular areas, add				%	10%	100%	100%		
0520	Welded wire fabric, sheets for rigid paving 2.33 lb./S.Y.	2 Rodm	389	.041	S.Y.	1.30	2.23		3.53	4.87
0530	Reinforcing steel for rigid paving 12 lb./S.Y.		666	.024		5.95	1.30		7.25	8.50
0540	Reinforcing steel for rigid paving 18 lb./S.Y.		444	.036		8.90	1.96		10.86	12.80
0620	Slip form, 12' pass, unreinforced, 6" thick	B-26A	5600	.016		24.50	.69	.59	25.78	28.50
0624	8" thick		5300	.017		33.50	.73	.62	34.85	39
0626	9" thick		4820	.018		38	.80	.68	39.48	44
0628	10" thick		4050	.022		42	.96	.81	43.77	48.50
0630	12" thick		3470	.025		48.50	1.12	.95	50.57	55.50
0632	15" thick		2890	.030		61	1.34	1.14	63.48	71
0640	Slip form, 24' pass, unreinforced, 6" thick		11200	.008		24	.35	.29	24.64	27.50
0644	8" thick		10600	.008		32	.37	.31	32.68	36.50
0646	9" thick		9640	.009		37	.40	.34	37.74	41.50
0648	10" thick		8100	.011		41	.48	.41	41.89	46
0650	12" thick		6940	.013		47	.56	.47	48.03	53.50
0652	15" thick		5780	.015		59.50	.67	.57	60.74	66.50
0700	Finishing, broom finish small areas	2 Cefi	120	.133			6.25		6.25	9.30
1000	Curing, with sprayed membrane by hand	2 Clab	1500	.011		1.14	.42		1.56	1.90
1650	For integral coloring, see Section 03 05 13.20									

For customer support on your Building Construction Costs with RSMeans Data, call 800.448.8182.

653

32 14 Unit Paving

32 14 13 – Precast Concrete Unit Paving

32 14 13.13 Interlocking Precast Concrete Unit Paving

		Crew	Daily Output	Labor-Hours	Unit	Material	2017 Bare Costs Labor	2017 Bare Costs Equipment	Total	Total Incl O&P
0010	**INTERLOCKING PRECAST CONCRETE UNIT PAVING**									
0020	"V" blocks for retaining soil	D-1	205	.078	S.F.	10.30	3.38		13.68	16.50

32 14 13.16 Precast Concrete Unit Paving Slabs

		Crew	Daily Output	Labor-Hours	Unit	Material	Labor	Equipment	Total	Total Incl O&P
0010	**PRECAST CONCRETE UNIT PAVING SLABS**									
0710	Precast concrete patio blocks, 2-3/8" thick, colors, 8" x 16"	D-1	265	.060	S.F.	8.85	2.61		11.46	13.75
0750	Exposed local aggregate, natural	2 Bric	250	.064		8.85	3.06		11.91	14.40
0800	Colors		250	.064		8.85	3.06		11.91	14.40
0850	Exposed granite or limestone aggregate		250	.064		8.85	3.06		11.91	14.40
0900	Exposed white tumblestone aggregate		250	.064		8.85	3.06		11.91	14.45

32 14 13.18 Precast Concrete Plantable Pavers

		Crew	Daily Output	Labor-Hours	Unit	Material	Labor	Equipment	Total	Total Incl O&P
0010	**PRECAST CONCRETE PLANTABLE PAVERS** (50% grass)									
0015	Subgrade preparation and grass planting not included									
0100	Precast concrete plantable pavers with topsoil, 24" x 16"	B-63	800	.050	S.F.	4.39	2.08	.21	6.68	8.25
0200	Less than 600 square feet or irregular area	"	500	.080	"	4.39	3.32	.34	8.05	10.25
0300	3/4" crushed stone base for plantable pavers, 6 inch depth	B-62	1000	.024	S.Y.	3.94	1.03	.17	5.14	6.10
0400	8 inch depth		900	.027		5.25	1.15	.19	6.59	7.70
0500	10 inch depth		800	.030		6.55	1.29	.21	8.05	9.40
0600	12 inch depth		700	.034		7.85	1.48	.24	9.57	11.15
0700	Hydro seeding plantable pavers	B-81A	20	.800	M.S.F.	11.40	33.50	19.65	64.55	84.50
0800	Apply fertilizer and seed to plantable pavers	1 Clab	8	1	"	40	39		79	104

32 14 16 – Brick Unit Paving

32 14 16.10 Brick Paving

		Crew	Daily Output	Labor-Hours	Unit	Material	Labor	Equipment	Total	Total Incl O&P
0010	**BRICK PAVING**									
0012	4" x 8" x 1-1/2", without joints (4.5 bricks/S.F.)	D-1	110	.145	S.F.	2.25	6.30		8.55	12.15
0100	Grouted, 3/8" joint (3.9 bricks/S.F.)		90	.178		2.02	7.70		9.72	14.05
0200	4" x 8" x 2-1/4", without joints (4.5 bricks/S.F.)		110	.145		2.33	6.30		8.63	12.25
0300	Grouted, 3/8" joint (3.9 bricks/S.F.)		90	.178		2.02	7.70		9.72	14.05
0455	Pervious brick paving, 4" x 8" x 3-1/4", without joints (4.5 bricks/S.F.)		110	.145		3.49	6.30		9.79	13.55
0500	Bedding, asphalt, 3/4" thick	B-25	5130	.017		.66	.74	.55	1.95	2.47
0540	Course washed sand bed, 1" thick	B-18	5000	.005		.36	.19	.01	.56	.70
0580	Mortar, 1" thick	D-1	300	.053		.80	2.31		3.11	4.43
0620	2" thick		200	.080		1.59	3.46		5.05	7.05
1500	Brick on 1" thick sand bed laid flat, 4.5 per S.F.		100	.160		2.82	6.90		9.72	13.75
2000	Brick pavers, laid on edge, 7.2 per S.F.		70	.229		3.95	9.90		13.85	19.55
2500	For 4" thick concrete bed and joints, add		595	.027		1.38	1.16		2.54	3.31
2800	For steam cleaning, add	A-1H	950	.008		.10	.33	.08	.51	.71

32 14 23 – Asphalt Unit Paving

32 14 23.10 Asphalt Blocks

		Crew	Daily Output	Labor-Hours	Unit	Material	Labor	Equipment	Total	Total Incl O&P
0010	**ASPHALT BLOCKS**									
0020	Rectangular, 6" x 12" x 1-1/4", w/bed & neopr. adhesive	D-1	135	.119	S.F.	8.05	5.15		13.20	16.75
0100	3" thick		130	.123		11.25	5.35		16.60	20.50
0300	Hexagonal tile, 8" wide, 1-1/4" thick		135	.119		8.05	5.15		13.20	16.75
0400	2" thick		130	.123		11.25	5.35		16.60	20.50
0500	Square, 8" x 8", 1-1/4" thick		135	.119		8.05	5.15		13.20	16.75
0600	2" thick		130	.123		11.25	5.35		16.60	20.50
0900	For exposed aggregate (ground finish), add					.49			.49	.54
0910	For colors, add					.49			.49	.54

32 14 Unit Paving

32 14 40 – Stone Paving

32 14 40.10 Stone Pavers

		Crew	Daily Output	Labor-Hours	Unit	Material	2017 Bare Costs Labor	2017 Bare Costs Equipment	Total	Total Incl O&P
0010	**STONE PAVERS**									
1100	Flagging, bluestone, irregular, 1" thick,	D-1	81	.198	S.F.	9.15	8.55		17.70	23
1150	Snapped random rectangular, 1" thick		92	.174		13.85	7.55		21.40	27
1200	1-1/2" thick		85	.188		16.65	8.15		24.80	31
1250	2" thick		83	.193		19.40	8.35		27.75	34.50
1300	Slate, natural cleft, irregular, 3/4" thick		92	.174		9	7.55		16.55	21.50
1350	Random rectangular, gauged, 1/2" thick		105	.152		19.50	6.60		26.10	31.50
1400	Random rectangular, butt joint, gauged, 1/4" thick	↓	150	.107	↓	21	4.62		25.62	30
1500	For interior setting, add								25%	25%
1550	Granite blocks, 3-1/2" x 3-1/2" x 3-1/2"	D-1	92	.174	S.F.	16.95	7.55		24.50	30
1600	4" to 12" long, 3" to 5" wide, 3" to 5" thick		98	.163		14.15	7.05		21.20	26.50
1650	6" to 15" long, 3" to 6" wide, 3" to 5" thick	↓	105	.152	↓	7.55	6.60		14.15	18.45

32 16 Curbs, Gutters, Sidewalks, and Driveways

32 16 13 – Curbs and Gutters

32 16 13.13 Cast-in-Place Concrete Curbs and Gutters

		Crew	Daily Output	Labor-Hours	Unit	Material	2017 Bare Costs Labor	2017 Bare Costs Equipment	Total	Total Incl O&P
0010	**CAST-IN-PLACE CONCRETE CURBS AND GUTTERS**									
0290	Forms only, no concrete									
0300	Concrete, wood forms, 6" x 18", straight	C-2	500	.096	L.F.	3.12	4.60		7.72	10.50
0400	6" x 18", radius	"	200	.240	"	3.24	11.50		14.74	21
0402	Forms and concrete complete									
0404	Concrete, wood forms, 6" x 18", straight & concrete	C-2A	500	.096	L.F.	6.45	4.56		11.01	14
0406	6" x 18", radius		200	.240		6.55	11.40		17.95	24.50
0410	Steel forms, 6" x 18", straight		700	.069		4.81	3.26		8.07	10.25
0411	6" x 18", radius	↓	400	.120		4.04	5.70		9.74	13.15
0415	Machine formed, 6" x 18", straight	B-69A	2000	.024		3.95	1.04	.60	5.59	6.55
0416	6" x 18", radius	"	900	.053	↓	3.97	2.30	1.33	7.60	9.35
0421	Curb and gutter, straight									
0422	with 6" high curb and 6" thick gutter, wood forms									
0430	24" wide, .055 C.Y. per L.F.	C-2A	375	.128	L.F.	16.60	6.10		22.70	27.50
0435	30" wide, .066 C.Y. per L.F.		340	.141		18.25	6.70		24.95	30
0440	Steel forms, 24" wide, straight		700	.069		7.65	3.26		10.91	13.35
0441	Radius		500	.096		7.25	4.56		11.81	14.95
0442	30" wide, straight		700	.069		8.95	3.26		12.21	14.75
0443	Radius	↓	500	.096		8.30	4.56		12.86	16.10
0445	Machine formed, 24" wide, straight	B-69A	2000	.024		6.60	1.04	.60	8.24	9.50
0446	Radius		900	.053		6.60	2.30	1.33	10.23	12.20
0447	30" wide, straight		2000	.024		7.65	1.04	.60	9.29	10.70
0448	Radius	↓	900	.053		7.65	2.30	1.33	11.28	13.40

32 16 13.23 Precast Concrete Curbs and Gutters

		Crew	Daily Output	Labor-Hours	Unit	Material	2017 Bare Costs Labor	2017 Bare Costs Equipment	Total	Total Incl O&P
0010	**PRECAST CONCRETE CURBS AND GUTTERS**									
0550	Precast, 6" x 18", straight	B-29	700	.080	L.F.	9.15	3.44	1.30	13.89	16.75
0600	6" x 18", radius	"	325	.172	"	10.15	7.40	2.80	20.35	25.50

32 16 13.33 Asphalt Curbs

		Crew	Daily Output	Labor-Hours	Unit	Material	2017 Bare Costs Labor	2017 Bare Costs Equipment	Total	Total Incl O&P
0010	**ASPHALT CURBS**									
0012	Curbs, asphaltic, machine formed, 8" wide, 6" high, 40 L.F./ton	B-27	1000	.032	L.F.	1.66	1.27	.31	3.24	4.11
0100	8" wide, 8" high, 30 L.F./ton		900	.036		2.22	1.41	.35	3.98	4.99
0150	Asphaltic berm, 12" W, 3" to 6" H, 35 L.F./ton, before pavement		700	.046		.04	1.81	.45	2.30	3.31
0200	12" W, 1-1/2" to 4" H, 60 L.F./ton, laid with pavement	B-2	1050	.038	↓	.02	1.51		1.53	2.34

32 16 13 - Curbs and Gutters

32 16 13.43 Stone Curbs	Crew	Daily Output	Labor-Hours	Unit	Material	2017 Bare Costs Labor	Equipment	Total	Total Incl O&P
0010 **STONE CURBS**									
1000 Granite, split face, straight, 5" x 16"	D-13	275	.175	L.F.	14.10	8.15	1.58	23.83	29.50
1100 6" x 18"	"	250	.192		18.55	8.95	1.74	29.24	36
1300 Radius curbing, 6" x 18", over 10' radius	B-29	260	.215		22.50	9.25	3.50	35.25	43
1400 Corners, 2' radius	"	80	.700	Ea.	76	30	11.40	117.40	143
1600 Edging, 4-1/2" x 12", straight	D-13	300	.160	L.F.	7.05	7.45	1.45	15.95	21
1800 Curb inlets (guttermouth) straight	B-29	41	1.366	Ea.	169	59	22	250	300
2000 Indian granite (Belgian block)									
2100 Jumbo, 10-1/2" x 7-1/2" x 4", grey	D-1	150	.107	L.F.	7.65	4.62		12.27	15.50
2150 Pink		150	.107		8.35	4.62		12.97	16.30
2200 Regular, 9" x 4-1/2" x 4-1/2", grey		160	.100		4.53	4.33		8.86	11.65
2250 Pink		160	.100		5.95	4.33		10.28	13.20
2300 Cubes, 4" x 4" x 4", grey		175	.091		3.52	3.96		7.48	9.95
2350 Pink		175	.091		3.55	3.96		7.51	10
2400 6" x 6" x 6", pink		155	.103		11.75	4.47		16.22	19.75
2500 Alternate pricing method for Indian granite									
2550 Jumbo, 10-1/2" x 7-1/2" x 4" (30 lb.), grey				Ton	435			435	480
2600 Pink					485			485	535
2650 Regular, 9" x 4-1/2" x 4-1/2" (20 lb.), grey					320			320	350
2700 Pink					415			415	460
2750 Cubes, 4" x 4" x 4" (5 lb.), grey					425			425	470
2800 Pink					455			455	500
2850 6" x 6" x 6" (25 lb.), pink					455			455	500
2900 For pallets, add					22			22	24

32 17 Paving Specialties

32 17 13 - Parking Bumpers

32 17 13.13 Metal Parking Bumpers

	Crew	Daily Output	Labor-Hours	Unit	Material	Labor	Equipment	Total	Total Incl O&P
0010 **METAL PARKING BUMPERS**									
0015 Bumper rails for garages, 12 ga. rail, 6" wide, with steel									
0020 posts 12'-6" O.C., minimum	E-4	190	.168	L.F.	19.20	9.25	.68	29.13	37
0030 Average		165	.194		24	10.65	.79	35.44	45
0100 Maximum		140	.229		29	12.55	.93	42.48	53.50
0300 12" channel rail, minimum		160	.200		24	10.95	.81	35.76	45.50
0400 Maximum		120	.267		36	14.60	1.08	51.68	65
1300 Pipe bollards, conc. filled/paint, 8' L x 4' D hole, 6" diam.	B-6	20	1.200	Ea.	610	51.50	18.30	679.80	770
1400 8" diam.		15	1.600		690	69	24.50	783.50	890
1500 12" diam.		12	2		975	86	30.50	1,091.50	1,250
2030 Folding with individual padlocks	B-2	50	.800		189	31.50		220.50	256
8000 Parking lot control, see Section 11 12 13.10									
8900 Security bollards, SS, lighted, hyd., incl. controls, group of 3	L-7	.06	509	Ea.	48,500	24,100		72,600	90,500
8910 Group of 5	"	.04	682	"	70,000	32,400		102,400	126,500

32 17 13.16 Plastic Parking Bumpers

	Crew	Daily Output	Labor-Hours	Unit	Material	Labor	Equipment	Total	Total Incl O&P
0010 **PLASTIC PARKING BUMPERS**									
1200 Thermoplastic, 6" x 10" x 6'-0"	B-2	120	.333	Ea.	62	13.20		75.20	88

32 17 13.19 Precast Concrete Parking Bumpers

	Crew	Daily Output	Labor-Hours	Unit	Material	Labor	Equipment	Total	Total Incl O&P
0010 **PRECAST CONCRETE PARKING BUMPERS**									
1000 Wheel stops, precast concrete incl. dowels, 6" x 10" x 6'-0"	B-2	120	.333	Ea.	48.50	13.20		61.70	73
1100 8" x 13" x 6'-0"	"	120	.333	"	53	13.20		66.20	78.50

32 17 Paving Specialties

32 17 13 – Parking Bumpers

32 17 13.26 Wood Parking Bumpers

		Crew	Daily Output	Labor-Hours	Unit	Material	2017 Bare Costs Labor	Equipment	Total	Total Incl O&P
0010	**WOOD PARKING BUMPERS**									
0020	Parking barriers, timber w/saddles, treated type									
0100	4" x 4" for cars	B-2	520	.077	L.F.	3	3.04		6.04	7.95
0200	6" x 6" for trucks		520	.077	"	6.30	3.04		9.34	11.60
0600	Flexible fixed stanchion, 2' high, 3" diameter	▼	100	.400	Ea.	39	15.80		54.80	67

32 17 23 – Pavement Markings

32 17 23.13 Painted Pavement Markings

		Crew	Daily Output	Labor-Hours	Unit	Material	2017 Bare Costs Labor	Equipment	Total	Total Incl O&P
0010	**PAINTED PAVEMENT MARKINGS**									
0020	Acrylic waterborne, white or yellow, 4" wide, less than 3000 L.F.	B-78	20000	.002	L.F.	.11	.10	.03	.24	.30
0200	6" wide, less than 3000 L.F.		11000	.004		.17	.18	.05	.40	.50
0500	8" wide, less than 3000 L.F.		10000	.005		.22	.19	.05	.46	.60
0600	12" wide, less than 3000 L.F.		4000	.012	▼	.33	.48	.14	.95	1.25
0620	Arrows or gore lines		2300	.021	S.F.	.20	.84	.24	1.28	1.77
0640	Temporary paint, white or yellow, less than 3000 L.F.	▼	15000	.003	L.F.	.08	.13	.04	.25	.32
0660	Removal	1 Clab	300	.027			1.04		1.04	1.60
0680	Temporary tape	2 Clab	1500	.011		.48	.42		.90	1.17
0710	Thermoplastic, white or yellow, 4" wide, less than 6000 L.F.	B-79	15000	.003		.34	.11	.09	.54	.64
0730	6" wide, less than 6000 L.F.		14000	.003		.51	.12	.10	.73	.85
0740	8" wide, less than 6000 L.F.		12000	.003		.68	.14	.11	.93	1.08
0750	12" wide, less than 6000 L.F.		6000	.007	▼	1	.27	.22	1.49	1.76
0760	Arrows		660	.061	S.F.	.57	2.46	2.03	5.06	6.60
0770	Gore lines		2500	.016		.57	.65	.53	1.75	2.21
0780	Letters	▼	660	.061	▼	.57	2.46	2.03	5.06	6.60
1000	Airport painted markings									
1050	Traffic safety flashing truck for airport painting	A-2B	1	8	Day		355	217	572	775
1100	Painting, white or yellow, taxiway markings	B-78	4000	.012	S.F.	.28	.48	.14	.90	1.20
1110	with 12 lb. beads per 100 S.F.		4000	.012		.58	.48	.14	1.20	1.52
1200	Runway markings		3500	.014		.28	.55	.16	.99	1.32
1210	with 12 lb. beads per 100 S.F.		3500	.014		.58	.55	.16	1.29	1.64
1300	Pavement location or direction signs		2500	.019		.28	.77	.22	1.27	1.73
1310	with 12 lb. beads per 100 S.F.		2500	.019	▼	.58	.77	.22	1.57	2.05
1350	Mobilization airport pavement painting	▼	4	12	Ea.		485	137	622	890
1400	Paint markings or pavement signs removal daytime	B-78B	400	.045	S.F.		1.82	.92	2.74	3.80
1500	Removal nighttime		335	.054	"		2.17	1.09	3.26	4.53
1600	Mobilization pavement paint removal	▼	4	4.500	Ea.		182	91.50	273.50	380

32 17 23.14 Pavement Parking Markings

		Crew	Daily Output	Labor-Hours	Unit	Material	2017 Bare Costs Labor	Equipment	Total	Total Incl O&P
0010	**PAVEMENT PARKING MARKINGS**									
0790	Layout of pavement marking	A-2	25000	.001	L.F.		.04	.01	.05	.07
0800	Lines on pvmt., parking stall, paint, white, 4" wide	B-78B	400	.045	Stall	5.20	1.82	.92	7.94	9.50
0825	Parking stall, small quantities	2 Pord	80	.200		10.35	8.35		18.70	24
0830	Lines on pvmt., parking stall, thermoplastic, white, 4" wide	B-79	300	.133	▼	14	5.40	4.46	23.86	28.50
1000	Street letters and numbers	B-78B	1600	.011	S.F.	.72	.46	.23	1.41	1.74

32 17 26 – Tactile Warning Surfacing

32 17 26.10 Tactile Warning Surfacing

		Crew	Daily Output	Labor-Hours	Unit	Material	2017 Bare Costs Labor	Equipment	Total	Total Incl O&P
0010	**TACTILE WARNING SURFACING**									
0100	Tactile warning tiles S.F.	2 Clab	400	.040	S.F.	17.25	1.57		18.82	21.50

For customer support on your Building Construction Costs with RSMeans Data, call 800.448.8182.

657

32 18 Athletic and Recreational Surfacing

32 18 13 – Synthetic Grass Surfacing

32 18 13.10 Artificial Grass Surfacing

	Crew	Daily Output	Labor-Hours	Unit	Material	2017 Bare Costs Labor	Equipment	Total	Total Incl O&P
0010 **ARTIFICIAL GRASS SURFACING**									
0015 Not including asphalt base or drainage,									
0020 but including cushion pad, over 50,000 S.F.									
0200 1/2" pile and 5/16" cushion pad, standard	C-17	3200	.025	S.F.	5.25	1.30		6.55	7.75
0300 Deluxe		2560	.031		5.15	1.62		6.77	8.15
0500 1/2" pile and 5/8" cushion pad, standard		2844	.028		5.15	1.46		6.61	7.90
0600 Deluxe		2327	.034		5.15	1.78		6.93	8.40
0800 For asphaltic concrete base, 2-1/2" thick,									
0900 with 6" crushed stone sub-base, add	B-25	12000	.007	S.F.	1.63	.32	.24	2.19	2.53

32 18 16 – Synthetic Resilient Surfacing

32 18 16.13 Playground Protective Surfacing

	Crew	Daily Output	Labor-Hours	Unit	Material	2017 Bare Costs Labor	Equipment	Total	Total Incl O&P
0010 **PLAYGROUND PROTECTIVE SURFACING**									
0100 Resilient rubber surface, poured in place, 4" thick, black	2 Skwk	300	.053	S.F.	11.45	2.74		14.19	16.80
0150 2" thick topping, colors	"	2800	.006		6.65	.29		6.94	7.75
0200 Wood chip mulch, 6" deep	1 Clab	300	.027		.79	1.04		1.83	2.47

32 18 23 – Athletic Surfacing

32 18 23.33 Running Track Surfacing

	Crew	Daily Output	Labor-Hours	Unit	Material	2017 Bare Costs Labor	Equipment	Total	Total Incl O&P
0010 **RUNNING TRACK SURFACING**									
0020 Running track, asphalt concrete pavement, 2-1/2"	B-37	300	.160	S.Y.	14.45	6.65	.51	21.61	26.50
0102 Surface, latex rubber system, 1/2" thick, black	B-20	115	.209		47.50	9.15		56.65	66
0152 Colors		115	.209		58	9.15		67.15	77.50
0302 Urethane rubber system, 1/2" thick, black		110	.218		35.50	9.60		45.10	53.50
0402 Color coating		110	.218		43.50	9.60		53.10	62.50

32 18 23.53 Tennis Court Surfacing

	Crew	Daily Output	Labor-Hours	Unit	Material	2017 Bare Costs Labor	Equipment	Total	Total Incl O&P
0010 **TENNIS COURT SURFACING**									
0020 Tennis court, asphalt, incl. base, 2-1/2" thick, one court	B-37	450	.107	S.Y.	30.50	4.42	.34	35.26	41
0200 Two courts		675	.071		18.20	2.95	.23	21.38	25
0300 Clay courts		360	.133		41	5.55	.42	46.97	54.50
0400 Pulverized natural greenstone with 4" base, fast dry		250	.192		39	7.95	.61	47.56	55.50
0800 Rubber-acrylic base resilient pavement		600	.080		62	3.32	.25	65.57	74
1000 Colored sealer, acrylic emulsion, 3 coats	2 Clab	800	.020		6.85	.78		7.63	8.75
1100 3 coats, 2 colors	"	900	.018		9.55	.70		10.25	11.55
1200 For preparing old courts, add	1 Clab	825	.010			.38		.38	.58
1400 Posts for nets, 3-1/2" diameter with eye bolts	B-1	3.40	7.059	Pr.	282	281		563	740
1500 With pulley & reel		3.40	7.059	"	790	281		1,071	1,300
1700 Net, 42' long, nylon thread with binder		50	.480	Ea.	255	19.10		274.10	310
1800 All metal		6.50	3.692	"	530	147		677	805
2000 Paint markings on asphalt, 2 coats	1 Pord	1.78	4.494	Court	210	187		397	515
2200 Complete court with fence, etc., asphaltic conc., minimum	B-37	.20	240		31,000	9,950	765	41,715	50,000
2300 Maximum		.16	300		61,000	12,400	955	74,355	87,000
2800 Clay courts, minimum		.20	240		34,800	9,950	765	45,515	54,000
2900 Maximum		.16	300		64,500	12,400	955	77,855	91,000

32 31 Fences and Gates

32 31 13 – Chain Link Fences and Gates

32 31 13.20 Fence, Chain Link Industrial

		Crew	Daily Output	Labor-Hours	Unit	Material	2017 Bare Costs Labor	2017 Bare Costs Equipment	Total	Total Incl O&P
0010	**FENCE, CHAIN LINK INDUSTRIAL**									
0011	Schedule 40, including concrete									
0020	3 strands barb wire, 2" post @ 10' O.C., set in concrete, 6' H									
0200	9 ga. wire, galv. steel, in concrete	B-80C	240	.100	L.F.	19.75	4.08	.94	24.77	28.50
0248	Fence, add for vinyl coated fabric				S.F.	.68			.68	.75
0300	Aluminized steel	B-80C	240	.100	L.F.	22	4.08	.94	27.02	31
0301	Fence, Wrought Iron		240	.100		30	4.08	.94	35.02	40
0500	6 ga. wire, galv. steel		240	.100		25	4.08	.94	30.02	34.50
0600	Aluminized steel		240	.100		30.50	4.08	.94	35.52	40.50
0800	6 ga. wire, 6' high but omit barbed wire, galv. steel		250	.096		20	3.92	.90	24.82	29
0900	Aluminized steel, in concrete		250	.096		24	3.92	.90	28.82	33.50
0920	8' H, 6 ga. wire, 2-1/2" line post, galv. steel, in concrete		180	.133		32	5.45	1.25	38.70	44.50
0940	Aluminized steel, in concrete		180	.133	↓	39	5.45	1.25	45.70	52.50
1100	Add for corner posts, 3" diam., galv. steel, in concrete		40	.600	Ea.	88.50	24.50	5.65	118.65	141
1200	Aluminized steel, in concrete		40	.600		88.50	24.50	5.65	118.65	141
1300	Add for braces, galv. steel		80	.300		39	12.25	2.82	54.07	64.50
1350	Aluminized steel		80	.300		50	12.25	2.82	65.07	77
1400	Gate for 6' high fence, 1-5/8" frame, 3' wide, galv. steel		10	2.400		207	98	22.50	327.50	400
1500	Aluminized steel, in concrete	↓	10	2.400	↓	209	98	22.50	329.50	405
2000	5'-0" high fence, 9 ga., no barbed wire, 2" line post, in concrete									
2010	10' O.C., 1-5/8" top rail, in concrete									
2100	Galvanized steel, in concrete	B-80C	300	.080	L.F.	21	3.27	.75	25.02	29
2200	Aluminized steel, in concrete		300	.080	"	19	3.27	.75	23.02	27
2400	Gate, 4' wide, 5' high, 2" frame, galv. steel, in concrete		10	2.400	Ea.	219	98	22.50	339.50	415
2500	Aluminized steel, in concrete		10	2.400	"	197	98	22.50	317.50	390
3100	Overhead slide gate, chain link, 6' high, to 18' wide, in concrete	↓	38	.632	L.F.	98.50	26	5.95	130.45	154
3110	Cantilever type, in concrete	B-80	48	.667		141	29	14.05	184.05	215
3120	8' high, in concrete		24	1.333		168	58.50	28	254.50	305
3130	10' high, in concrete	↓	18	1.778	↓	207	78	37.50	322.50	390
5000	Double swing gates, incl. posts & hardware, in concrete									
5010	5' high, 12' opening, in concrete	B-80C	3.40	7.059	Opng.	540	288	66.50	894.50	1,100
5020	20' opening, in concrete		2.80	8.571		620	350	80.50	1,050.50	1,300
5060	6' high, 12' opening, in concrete		3.20	7.500		470	305	70.50	845.50	1,075
5070	20' opening, in concrete		2.60	9.231		660	375	86.50	1,121.50	1,400
5080	8' high, 12' opening, in concrete	B-80	2.13	15.002		500	660	315	1,475	1,900
5090	20' opening, in concrete		1.45	22.069		710	970	465	2,145	2,775
5100	10' high, 12' opening, in concrete		1.31	24.427		840	1,075	515	2,430	3,125
5110	20' opening, in concrete		1.03	31.068		885	1,375	655	2,915	3,775
5120	12' high, 12' opening, in concrete		1.05	30.476		1,200	1,325	640	3,165	4,050
5130	20' opening, in concrete	↓	.85	37.647	↓	1,250	1,650	790	3,690	4,750
5190	For aluminized steel, add					20%				

32 31 13.25 Fence, Chain Link Residential

		Crew	Daily Output	Labor-Hours	Unit	Material	2017 Bare Costs Labor	2017 Bare Costs Equipment	Total	Total Incl O&P
0010	**FENCE, CHAIN LINK RESIDENTIAL**									
0011	Schedule 20, 11 ga. wire, 1-5/8" post									
0020	10' O.C., 1-3/8" top rail, 2" corner post, galv. stl. 3' high	B-80C	500	.048	L.F.	2.18	1.96	.45	4.59	5.90
0050	4' high		400	.060		7.70	2.45	.56	10.71	12.80
0100	6' high		200	.120	↓	9.75	4.90	1.13	15.78	19.45
0150	Add for gate 3' wide, 1-3/8" frame, 3' high		12	2	Ea.	84	81.50	18.80	184.30	237
0170	4' high		10	2.400		90	98	22.50	210.50	273
0190	6' high		10	2.400		112	98	22.50	232.50	298
0200	Add for gate 4' wide, 1-3/8" frame, 3' high		9	2.667		94	109	25	228	298
0220	4' high	↓	9	2.667	↓	101	109	25	235	305

32 31 Fences and Gates

32 31 13 – Chain Link Fences and Gates

32 31 13.25 Fence, Chain Link Residential

32 31 13.25 Fence, Chain Link Residential	Crew	Daily Output	Labor-Hours	Unit	Material	2017 Bare Costs Labor	Equipment	Total	Total Incl O&P
0240	6' high	B-80C	8	3	Ea.	127	122	28	277	355
0350	Aluminized steel, 11 ga. wire, 3' high		500	.048	L.F.	9.05	1.96	.45	11.46	13.50
0380	4' high		400	.060		8.50	2.45	.56	11.51	13.70
0400	6' high		200	.120	▼	11.60	4.90	1.13	17.63	21.50
0450	Add for gate 3' wide, 1-3/8" frame, 3' high		12	2	Ea.	104	81.50	18.80	204.30	259
0470	4' high		10	2.400		104	98	22.50	224.50	289
0490	6' high		10	2.400		129	98	22.50	249.50	315
0500	Add for gate 4' wide, 1-3/8" frame, 3' high		10	2.400		108	98	22.50	228.50	293
0520	4' high		9	2.667		125	109	25	259	330
0540	6' high		8	3	▼	134	122	28	284	365
0620	Vinyl covered, 9 ga. wire, 3' high		500	.048	L.F.	8.05	1.96	.45	10.46	12.35
0640	4' high		400	.060		8.40	2.45	.56	11.41	13.60
0660	6' high		200	.120	▼	10.45	4.90	1.13	16.48	20
0720	Add for gate 3' wide, 1-3/8" frame, 3' high		12	2	Ea.	97.50	81.50	18.80	197.80	252
0740	4' high		10	2.400		104	98	22.50	224.50	288
0760	6' high		10	2.400		124	98	22.50	244.50	310
0780	Add for gate 4' wide, 1-3/8" frame, 3' high		10	2.400		102	98	22.50	222.50	286
0800	4' high		9	2.667		106	109	25	240	310
0820	6' high	▼	8	3	▼	132	122	28	282	365
7076	Fence, for small jobs 100 L.F. fence or less w/or wo gate, add				L.F.	20%				

32 31 13.26 Tennis Court Fences and Gates

32 31 13.26 Tennis Court Fences and Gates	Crew	Daily Output	Labor-Hours	Unit	Material	2017 Bare Costs Labor	Equipment	Total	Total Incl O&P
0010	**TENNIS COURT FENCES AND GATES**									
0860	Tennis courts, 11 ga. wire, 2-1/2" post set									
0870	in concrete, 10' O.C., 1-5/8" top rail									
0900	10' high	B-80	190	.168	L.F.	23.50	7.40	3.55	34.45	40.50
0920	12' high		170	.188	"	23	8.25	3.96	35.21	42.50
1000	Add for gate 4' wide, 1-5/8" frame 7' high		10	3.200	Ea.	250	140	67.50	457.50	560
1040	Aluminized steel, 11 ga. wire 10' high		190	.168	L.F.	21.50	7.40	3.55	32.45	39
1100	12' high		170	.188	"	24.50	8.25	3.96	36.71	43.50
1140	Add for gate 4' wide, 1-5/8" frame, 7' high		10	3.200	Ea.	261	140	67.50	468.50	575
1250	Vinyl covered, 9 ga. wire, 10' high		190	.168	L.F.	22.50	7.40	3.55	33.45	39.50
1300	12' high	▼	170	.188	"	26	8.25	3.96	38.21	45.50
1310	Fence, CL, tennis court, transom gate, single, galv., 4' x 7'	B-80A	8.72	2.752	Ea.	320	108	31	459	550
1400	Add for gate 4' wide, 1-5/8" frame, 7' high	B-80	10	3.200	"	325	140	67.50	532.50	640

32 31 13.33 Chain Link Backstops

32 31 13.33 Chain Link Backstops	Crew	Daily Output	Labor-Hours	Unit	Material	2017 Bare Costs Labor	Equipment	Total	Total Incl O&P
0010	**CHAIN LINK BACKSTOPS**									
0015	Backstops, baseball, prefabricated, 30' wide, 12' high & 1 overhang	B-1	1	24	Ea.	2,500	955		3,455	4,225
0100	40' wide, 12' high & 2 overhangs	"	.75	32		6,925	1,275		8,200	9,550
0300	Basketball, steel, single goal	B-13	3.04	18.421		1,600	795	222	2,617	3,225
0400	Double goal	"	1.92	29.167	▼	1,975	1,250	350	3,575	4,475
0600	Tennis, wire mesh with pair of ends	B-1	2.48	9.677	Set	2,700	385		3,085	3,550
0700	Enclosed court	"	1.30	18.462	Ea.	9,150	735		9,885	11,200

32 31 13.53 High-Security Chain Link Fences, Gates and Sys.

32 31 13.53 High-Security Chain Link Fences, Gates and Sys.	Crew	Daily Output	Labor-Hours	Unit	Material	2017 Bare Costs Labor	Equipment	Total	Total Incl O&P
0010	**HIGH-SECURITY CHAIN LINK FENCES, GATES AND SYSTEMS**									
0100	Fence, chain link, security, 7' H, standard FE-7, incl excavation & posts	B-80C	480	.050	L.F.	45.50	2.04	.47	48.01	53.50
0200	Fence, barbed wire, security, 7' H, with 3 wire barbed wire arm	"	400	.060	"	7.85	2.45	.56	10.86	13
0300	Complete systems, including material and installation									
0310	Taunt wire fence detection system				M.L.F.				25,100	27,600
0410	Microwave fence detection system								41,300	45,400
0510	Passive magnetic fence detection system								19,500	21,400
0610	Infrared fence detection system								12,900	14,400
0710	Strain relief fence detection system				▼				25,100	27,600

32 31 Fences and Gates

32 31 13 – Chain Link Fences and Gates

32 31 13.53 High-Security Chain Link Fences, Gates and Sys.	Crew	Daily Output	Labor-Hours	Unit	Material	2017 Bare Costs Labor	Equipment	Total	Total Incl O&P
0810	Electro-shock fence detection system				M.L.F.				35,900	39,500
0910	Photo-electric fence detection system								16,300	18,000

32 31 19 – Decorative Metal Fences and Gates

32 31 19.10 Decorative Fence

		Crew	Daily Output	Labor-Hours	Unit	Material	Labor	Equipment	Total	Total Incl O&P
0010	**DECORATIVE FENCE**									
5300	Tubular picket, steel, 6' sections, 1-9/16" posts, 4' high	B-80C	300	.080	L.F.	35.50	3.27	.75	39.52	45.50
5400	2" posts, 5' high		240	.100		40.50	4.08	.94	45.52	51.50
5600	2" posts, 6' high		200	.120		48.50	4.90	1.13	54.53	62
5700	Staggered picket, 1-9/16" posts, 4' high		300	.080		35.50	3.27	.75	39.52	45.50
5800	2" posts, 5' high		240	.100		40.50	4.08	.94	45.52	51.50
5900	2" posts, 6' high		200	.120		48.50	4.90	1.13	54.53	62
6200	Gates, 4' high, 3' wide	B-1	10	2.400	Ea.	325	95.50		420.50	500
6300	5' high, 3' wide		10	2.400		400	95.50		495.50	585
6400	6' high, 3' wide		10	2.400		475	95.50		570.50	665
6500	4' wide		10	2.400		485	95.50		580.50	680

32 31 26 – Wire Fences and Gates

32 31 26.10 Fences, Misc. Metal

		Crew	Daily Output	Labor-Hours	Unit	Material	Labor	Equipment	Total	Total Incl O&P
0010	**FENCES, MISC. METAL**									
0012	Chicken wire, posts @ 4', 1" mesh, 4' high	B-80C	410	.059	L.F.	3.40	2.39	.55	6.34	8
0100	2" mesh, 6' high		350	.069		3.82	2.80	.64	7.26	9.20
0200	Galv. steel, 12 ga., 2" x 4" mesh, posts 5' O.C., 3' high		300	.080		2.78	3.27	.75	6.80	8.85
0300	5' high		300	.080		3.38	3.27	.75	7.40	9.55
0400	14 ga., 1" x 2" mesh, 3' high		300	.080		3.42	3.27	.75	7.44	9.55
0500	5' high		300	.080		4.57	3.27	.75	8.59	10.85
1000	Kennel fencing, 1-1/2" mesh, 6' long, 3'-6" wide, 6'-2" high	2 Clab	4	4	Ea.	510	157		667	800
1050	12' long		4	4		720	157		877	1,025
1200	Top covers, 1-1/2" mesh, 6' long		15	1.067		136	42		178	214
1250	12' long		12	1.333		191	52		243	290
1300	For kennel doors, see Section 08 31 13.40									
4500	Security fence, prison grade, set in concrete, 12' high	B-80	25	1.280	L.F.	61.50	56	27	144.50	183
4600	16' high	"	20	1.600	"	79	70	33.50	182.50	231

32 31 26.20 Wire Fencing, General

		Crew	Daily Output	Labor-Hours	Unit	Material	Labor	Equipment	Total	Total Incl O&P
0010	**WIRE FENCING, GENERAL**									
0015	Barbed wire, galvanized, domestic steel, hi-tensile 15-1/2 ga.				M.L.F.	101			101	112
0020	Standard, 12-3/4 ga.					114			114	125
0210	Barbless wire, 2-strand galvanized, 12-1/2 ga.					114			114	125
0500	Helical razor ribbon, stainless steel, 18" dia x 18" spacing				C.L.F.	169			169	186
0600	Hardware cloth galv., 1/4" mesh, 23 ga., 2' wide				C.S.F.	62			62	68
0700	3' wide					44			44	48
0900	1/2" mesh, 19 ga., 2' wide					31			31	34
1000	4' wide					24			24	26.50
1200	Chain link fabric, steel, 2" mesh, 6 ga., galvanized					157			157	172
1300	9 ga., galvanized					88.50			88.50	97.50
1350	Vinyl coated					84.50			84.50	93
1360	Aluminized					82			82	90
1400	2-1/4" mesh, 11.5 ga., galvanized					56.50			56.50	62
1600	1-3/4" mesh (tennis courts), 11.5 ga. (core), vinyl coated					62.50			62.50	69
1700	9 ga., galvanized					85			85	93.50
2100	Welded wire fabric, galvanized, 1" x 2", 14 ga.	2 Carp	1600	.010	S.F.	.61	.49		1.10	1.42
2200	2" x 4", 12-1/2 ga.				C.S.F.	58.50			58.50	64.50

661

For customer support on your Building Construction Costs with RSMeans Data, call 800.448.8182.

32 31 29 – Wood Fences and Gates

32 31 29.20 Fence, Wood Rail	Crew	Daily Output	Labor-Hours	Unit	Material	2017 Bare Costs Labor	Equipment	Total	Total Incl O&P
0010 **FENCE, WOOD RAIL**									
0012 Picket, No. 2 cedar, Gothic, 2 rail, 3' high	B-1	160	.150	L.F.	8.10	5.95		14.05	18.05
0050 Gate, 3'-6" wide	B-80C	9	2.667	Ea.	78	109	25	212	279
0400 3 rail, 4' high		150	.160	L.F.	9.10	6.55	1.50	17.15	21.50
0500 Gate, 3'-6" wide		9	2.667	Ea.	95	109	25	229	298
0600 Open rail, rustic, No. 1 cedar, 2 rail, 3' high		160	.150	L.F.	6.05	6.10	1.41	13.56	17.55
0650 Gate, 3' wide		9	2.667	Ea.	83	109	25	217	285
0700 3 rail, 4' high		150	.160	L.F.	7.20	6.55	1.50	15.25	19.55
0900 Gate, 3' wide		9	2.667	Ea.	103	109	25	237	305
1200 Stockade, No. 2 cedar, treated wood rails, 6' high		160	.150	L.F.	9.10	6.10	1.41	16.61	21
1250 Gate, 3' wide		9	2.667	Ea.	97	109	25	231	300
1300 No. 1 cedar, 3-1/4" cedar rails, 6' high		160	.150	L.F.	20.50	6.10	1.41	28.01	33.50
1500 Gate, 3' wide		9	2.667	Ea.	222	109	25	356	440
1520 Open rail, split, No. 1 cedar, 2 rail, 3' high		160	.150	L.F.	6.05	6.10	1.41	13.56	17.55
1540 3 rail, 4' high		150	.160		8.10	6.55	1.50	16.15	20.50
3300 Board, shadow box, 1" x 6", treated pine, 6' high		160	.150		12.65	6.10	1.41	20.16	25
3400 No. 1 cedar, 6' high		150	.160		25	6.55	1.50	33.05	39
3900 Basket weave, No. 1 cedar, 6' high	▼	160	.150	▼	34	6.10	1.41	41.51	48.50
3950 Gate, 3'-6" wide	B-1	8	3	Ea.	234	119		353	440
4000 Treated pine, 6' high		150	.160	L.F.	16.20	6.35		22.55	27.50
4200 Gate, 3'-6" wide		9	2.667	Ea.	178	106		284	360
5000 Fence rail, redwood, 2" x 4", merch. grade, 8'		2400	.010	L.F.	2.53	.40		2.93	3.39
5050 Select grade, 8'		2400	.010	"	5.55	.40		5.95	6.70
6000 Fence post, select redwood, earthpacked & treated, 4" x 4" x 6'		96	.250	Ea.	14	9.95		23.95	30.50
6010 4" x 4" x 8'		96	.250		19.25	9.95		29.20	36.50
6020 Set in concrete, 4" x 4" x 6'		50	.480		22	19.10		41.10	53.50
6030 4" x 4" x 8'		50	.480		23	19.10		42.10	55
6040 Wood post, 4' high, set in concrete, incl. concrete		50	.480		14.20	19.10		33.30	45
6050 Earth packed		96	.250		17.20	9.95		27.15	34
6060 6' high, set in concrete, incl. concrete		50	.480		17.70	19.10		36.80	49
6070 Earth packed	▼	96	.250	▼	12.10	9.95		22.05	28.50

32 32 13 – Cast-in-Place Concrete Retaining Walls

32 32 13.10 Retaining Walls, Cast Concrete

	Crew	Daily Output	Labor-Hours	Unit	Material	2017 Bare Costs Labor	Equipment	Total	Total Incl O&P
0010 **RETAINING WALLS, CAST CONCRETE**									
1800 Concrete gravity wall with vertical face including excavation & backfill									
1850 No reinforcing									
1900 6' high, level embankment	C-17C	36	2.306	L.F.	87	120	16.85	223.85	298
2000 33° slope embankment		32	2.594		100	135	18.95	253.95	340
2200 8' high, no surcharge		27	3.074		107	160	22.50	289.50	390
2300 33° slope embankment		24	3.458		130	180	25.50	335.50	450
2500 10' high, level embankment		19	4.368		153	227	32	412	555
2600 33° slope embankment	▼	18	4.611	▼	212	240	33.50	485.50	640
2800 Reinforced concrete cantilever, incl. excavation, backfill & reinf.									
2900 6' high, 33° slope embankment	C-17C	35	2.371	L.F.	78.50	123	17.30	218.80	296
3000 8' high, 33° slope embankment		29	2.862		90.50	149	21	260.50	350
3100 10' high, 33° slope embankment		20	4.150		118	216	30.50	364.50	495
3200 20' high, 500 lb. per L.F. surcharge	▼	7.50	11.067	▼	355	575	81	1,011	1,375
3500 Concrete cribbing, incl. excavation and backfill									

32 32 Retaining Walls

32 32 13 − Cast-in-Place Concrete Retaining Walls

32 32 13.10 Retaining Walls, Cast Concrete	Crew	Daily Output	Labor-Hours	Unit	Material	2017 Bare Costs Labor	Equipment	Total	Total Incl O&P
3700	12' high, open face	B-13	210	.267	S.F.	39.50	11.50	3.21	54.21	64.50
3900	Closed face	"	210	.267	"	37	11.50	3.21	51.71	61.50
4100	Concrete filled slurry trench, see Section 31 56 23.20									

32 32 23 − Segmental Retaining Walls

32 32 23.13 Segmental Conc. Unit Masonry Retaining Walls

		Crew	Daily Output	Labor-Hours	Unit	Material	Labor	Equipment	Total	Total Incl O&P
0010	**SEGMENTAL CONC. UNIT MASONRY RETAINING WALLS**									
7100	Segmental Retaining Wall system, incl. pins, and void fill									
7120	base and backfill not included									
7140	Large unit, 8" high x 18" wide x 20" deep, 3 plane split	B-62	300	.080	S.F.	12.40	3.45	.57	16.42	19.55
7150	Straight split		300	.080		12.50	3.45	.57	16.52	19.65
7160	Medium, lt. wt., 8" high x 18" wide x 12" deep, 3 plane split		400	.060		7.10	2.59	.43	10.12	12.20
7170	Straight split		400	.060		9.45	2.59	.43	12.47	14.80
7180	Small unit, 4" x 18" x 10" deep, 3 plane split		400	.060		14.20	2.59	.43	17.22	20
7190	Straight split		400	.060		10.70	2.59	.43	13.72	16.15
7200	Cap unit, 3 plane split		300	.080		13.25	3.45	.57	17.27	20.50
7210	Cap unit, straight split		300	.080		13.25	3.45	.57	17.27	20.50
7250	Geo-grid soil reinforcement 4' x 50'	2 Clab	22500	.001		.68	.03		.71	.79
7255	Geo-grid soil reinforcement 6' x 150'	"	22500	.001		.54	.03		.57	.63

32 32 26 − Metal Crib Retaining Walls

32 32 26.10 Metal Bin Retaining Walls

		Crew	Daily Output	Labor-Hours	Unit	Material	Labor	Equipment	Total	Total Incl O&P
0010	**METAL BIN RETAINING WALLS**									
0011	Aluminized steel bin, excavation									
0020	and backfill not included, 10' wide									
0100	4' high, 5.5' deep	B-13	650	.086	S.F.	27	3.71	1.04	31.75	36.50
0200	8' high, 5.5' deep		615	.091		31	3.92	1.10	36.02	41
0300	10' high, 7.7' deep		580	.097		34.50	4.16	1.16	39.82	45.50
0400	12' high, 7.7' deep		530	.106		37	4.55	1.27	42.82	49.50
0500	16' high, 7.7' deep		515	.109		39	4.68	1.31	44.99	51.50
0600	16' high, 9.9' deep		500	.112		41.50	4.82	1.35	47.67	54.50
0700	20' high, 9.9' deep		470	.119		46.50	5.15	1.43	53.08	60.50
0800	20' high, 12.1' deep		460	.122		42	5.25	1.46	48.71	55.50
0900	24' high, 12.1' deep		455	.123		44.50	5.30	1.48	51.28	58.50
1000	24' high, 14.3' deep		450	.124		52.50	5.35	1.50	59.35	67.50
1100	28' high, 14.3' deep		440	.127		54.50	5.50	1.53	61.53	70
1300	For plain galvanized bin type walls, deduct					10%				

32 32 29 − Timber Retaining Walls

32 32 29.10 Landscape Timber Retaining Walls

		Crew	Daily Output	Labor-Hours	Unit	Material	Labor	Equipment	Total	Total Incl O&P
0010	**LANDSCAPE TIMBER RETAINING WALLS**									
0100	Treated timbers, 6" x 6"	1 Clab	265	.030	L.F.	2.47	1.18		3.65	4.53
0110	6" x 8"	"	200	.040	"	5.20	1.57		6.77	8.15
0120	Drilling holes in timbers for fastening, 1/2"	1 Carp	450	.018	Inch		.88		.88	1.34
0130	5/8"	"	450	.018	"		.88		.88	1.34
0140	Reinforcing rods for fastening, 1/2"	1 Clab	312	.026	L.F.	.35	1		1.35	1.92
0150	5/8"	"	312	.026	"	.54	1		1.54	2.13
0160	Reinforcing fabric	2 Clab	2500	.006	S.Y.	2.13	.25		2.38	2.72
0170	Gravel backfill		28	.571	C.Y.	17.20	22.50		39.70	53.50
0180	Perforated pipe, 4" diameter with silt sock		1200	.013	L.F.	1.07	.52		1.59	1.98
0190	Galvanized 60d common nails	1 Clab	625	.013	Ea.	.17	.50		.67	.96
0200	20d common nails	"	3800	.002	"	.04	.08		.12	.17

32 32 Retaining Walls

32 32 36 – Gabion Retaining Walls

32 32 36.10 Stone Gabion Retaining Walls

		Crew	Daily Output	Labor-Hours	Unit	Material	2017 Bare Costs Labor	Equipment	Total	Total Incl O&P
0010	**STONE GABION RETAINING WALLS**									
4300	Stone filled gabions, not incl. excavation,									
4310	Stone, delivered, 3' wide									
4350	Galvanized, 6' high, 33° slope embankment	B-13	49	1.143	L.F.	53	49	13.75	115.75	149
4500	Highway surcharge		27	2.074		72	89.50	25	186.50	243
4600	9' high, up to 33° slope embankment		24	2.333		119	100	28	247	315
4700	Highway surcharge		16	3.500		126	151	42	319	415
4900	12' high, up to 33° slope embankment		14	4		186	172	48	406	520
5000	Highway surcharge		11	5.091		179	219	61	459	600
5950	For PVC coating, add					12%				

32 32 53 – Stone Retaining Walls

32 32 53.10 Retaining Walls, Stone

		Crew	Daily Output	Labor-Hours	Unit	Material	2017 Bare Costs Labor	Equipment	Total	Total Incl O&P
0010	**RETAINING WALLS, STONE**									
0015	Including excavation, concrete footing and									
0020	stone 3' below grade. Price is exposed face area.									
0200	Decorative random stone, to 6' high, 1'-6" thick, dry set	D-1	35	.457	S.F.	61.50	19.80		81.30	98.50
0300	Mortar set		40	.400		63.50	17.30		80.80	96.50
0500	Cut stone, to 6' high, 1'-6" thick, dry set		35	.457		64	19.80		83.80	101
0600	Mortar set		40	.400		64.50	17.30		81.80	97.50
0800	Random stone, 6' to 10' high, 2' thick, dry set		45	.356		69.50	15.40		84.90	100
0900	Mortar set		50	.320		72.50	13.85		86.35	101
1100	Cut stone, 6' to 10' high, 2' thick, dry set		45	.356		70.50	15.40		85.90	101
1200	Mortar set		50	.320		72.50	13.85		86.35	102

32 33 Site Furnishings

32 33 33 – Site Manufactured Planters

32 33 33.10 Planters

		Crew	Daily Output	Labor-Hours	Unit	Material	2017 Bare Costs Labor	Equipment	Total	Total Incl O&P
0010	**PLANTERS**									
0012	Concrete, sandblasted, precast, 48" diameter, 24" high	2 Clab	15	1.067	Ea.	655	42		697	785
0100	Fluted, precast, 7' diameter, 36" high		10	1.600		1,675	62.50		1,737.50	1,950
0300	Fiberglass, circular, 36" diameter, 24" high		15	1.067		675	42		717	810
0320	36" diameter, 27" high		12	1.333		715	52		767	865
0330	33" high		15	1.067		720	42		762	860
0335	24" diameter, 36" high		15	1.067		430	42		472	535
0340	60" diameter, 39" high		8	2		1,275	78.50		1,353.50	1,525
0400	60" diameter, 24" high		10	1.600		1,300	62.50		1,362.50	1,525
0600	Square, 24" side, 36" high		15	1.067		680	42		722	815
0610	24" side, 27" high		12	1.333		740	52		792	895
0620	24" side, 16" high		20	.800		355	31.50		386.50	440
0700	48" side, 36" high		15	1.067		1,125	42		1,167	1,300
0900	Planter/bench, 72" square, 36" high		5	3.200		1,950	125		2,075	2,350
1000	96" square, 27" high		5	3.200		2,425	125		2,550	2,850
1200	Wood, square, 48" side, 24" high		15	1.067		1,600	42		1,642	1,825
1300	Circular, 48" diameter, 30" high		10	1.600		1,075	62.50		1,137.50	1,275
1500	72" diameter, 30" high		10	1.600		1,875	62.50		1,937.50	2,175
1600	Planter/bench, 72"		5	3.200		3,625	125		3,750	4,175

32 33 Site Furnishings

32 33 43 – Site Seating and Tables

32 33 43.13 Site Seating	Crew	Daily Output	Labor-Hours	Unit	Material	2017 Bare Costs Labor	Equipment	Total	Total Incl O&P
0010 **SITE SEATING**									
0012 Seating, benches, park, precast conc., w/backs, wood rails, 4' long	2 Clab	5	3.200	Ea.	625	125		750	880
0100 8' long		4	4		1,050	157		1,207	1,400
0300 Fiberglass, without back, one piece, 4' long		10	1.600		610	62.50		672.50	770
0400 8' long		7	2.286		820	89.50		909.50	1,050
0500 Steel barstock pedestals w/backs, 2" x 3" wood rails, 4' long		10	1.600		1,325	62.50		1,387.50	1,575
0510 8' long		7	2.286		1,700	89.50		1,789.50	2,000
0515 Powder coated steel, 4" x 4" plastic slats, 6' L		8	2		535	78.50		613.50	705
0520 3" x 8" wood plank, 4' long		10	1.600		1,450	62.50		1,512.50	1,700
0530 8' long		7	2.286		1,475	89.50		1,564.50	1,750
0540 Backless, 4" x 4" wood plank, 4' square		10	1.600		1,200	62.50		1,262.50	1,425
0550 8' long		7	2.286		1,100	89.50		1,189.50	1,350
0560 Powder coated steel, with back and 2 anti-vagrant dividers, 6' long		8	2		1,125	78.50		1,203.50	1,375
0600 Aluminum pedestals, with backs, aluminum slats, 8' long		8	2		525	78.50		603.50	700
0610 15' long		5	3.200		955	125		1,080	1,250
0620 Portable, aluminum slats, 8' long		8	2		485	78.50		563.50	655
0630 15' long		5	3.200		550	125		675	795
0800 Cast iron pedestals, back & arms, wood slats, 4' long		8	2		435	78.50		513.50	600
0820 8' long		5	3.200		1,100	125		1,225	1,400
0840 Backless, wood slats, 4' long		8	2		585	78.50		663.50	765
0860 8' long		5	3.200		1,075	125		1,200	1,400
1700 Steel frame, fir seat, 10' long		10	1.600		395	62.50		457.50	530
2000 Benches, park, with back, galv. stl. frame, 4" x 4" plastic slats, 6' L	↓	7	2.286	↓	525	89.50		614.50	715

32 34 Fabricated Bridges

32 34 20 – Fabricated Pedestrian Bridges

32 34 20.10 Bridges, Pedestrian	Crew	Daily Output	Labor-Hours	Unit	Material	2017 Bare Costs Labor	Equipment	Total	Total Incl O&P
0010 **BRIDGES, PEDESTRIAN**									
0011 Spans over streams, roadways, etc.									
0020 including erection, not including foundations									
0050 Precast concrete, complete in place, 8' wide, 60' span	E-2	215	.260	S.F.	139	14.05	7.75	160.80	185
0100 100' span		185	.303		152	16.30	9.05	177.35	203
0150 120' span		160	.350		166	18.85	10.45	195.30	224
0200 150' span		145	.386		172	21	11.50	204.50	236
0300 Steel, trussed or arch spans, compl. in place, 8' wide, 40' span		320	.175		117	9.45	5.20	131.65	150
0400 50' span		395	.142		105	7.65	4.23	116.88	133
0500 60' span		465	.120		105	6.50	3.59	115.09	131
0600 80' span		570	.098		126	5.30	2.93	134.23	150
0700 100' span		465	.120		176	6.50	3.59	186.09	209
0800 120' span		365	.153		223	8.25	4.58	235.83	264
0900 150' span		310	.181		237	9.75	5.40	252.15	283
1000 160' span		255	.220		237	11.85	6.55	255.40	287
1100 10' wide, 80' span		640	.088		125	4.72	2.61	132.33	149
1200 120' span		415	.135		162	7.25	4.03	173.28	194
1300 150' span		445	.126		182	6.80	3.76	192.56	215
1400 200' span	↓	205	.273		194	14.70	8.15	216.85	247
1600 Wood, laminated type, complete in place, 80' span	C-12	203	.236		83	11.60	2.82	97.42	112
1700 130' span	"	153	.314	↓	86.50	15.35	3.74	105.59	123

32 35 Screening Devices

32 35 16 – Sound Barriers

32 35 16.10 Traffic Barriers, Highway Sound Barriers

		Crew	Daily Output	Labor-Hours	Unit	Material	2017 Bare Costs Labor	Equipment	Total	Total Incl O&P
0010	**TRAFFIC BARRIERS, HIGHWAY SOUND BARRIERS**									
0020	Highway sound barriers, not including footing									
0100	Precast concrete, concrete columns @ 30' OC, 8" T, 8' H	C-12	400	.120	L.F.	160	5.90	1.43	167.33	187
0110	12' H		265	.181		240	8.85	2.16	251.01	280
0120	16' H		200	.240		320	11.75	2.86	334.61	370
0130	20' H		160	.300		400	14.70	3.58	418.28	465
0400	Lt. Wt. composite panel, cementitious face, St. posts @ 12' OC, 8' H	B-80B	190	.168		167	7.10	1.30	175.40	196
0410	12' H		125	.256		251	10.80	1.98	263.78	295
0420	16' H		95	.337		335	14.20	2.61	351.81	395
0430	20' H		75	.427		420	17.95	3.30	441.25	490

32 84 Planting Irrigation

32 84 23 – Underground Sprinklers

32 84 23.10 Sprinkler Irrigation System

		Crew	Daily Output	Labor-Hours	Unit	Material	2017 Bare Costs Labor	Equipment	Total	Total Incl O&P
0010	**SPRINKLER IRRIGATION SYSTEM**									
0011	For lawns									
0800	Residential system, custom, 1" supply	B-20	2000	.012	S.F.	.25	.53		.78	1.09
0900	1-1/2" supply	"	1800	.013	"	.46	.59		1.05	1.41
1020	Pop up spray head w/risers, hi-pop, full circle pattern, 4"	2 Skwk	76	.211	Ea.	3.98	10.85		14.83	21
1030	1/2 circle pattern, 4"		76	.211		5.55	10.85		16.40	23
1040	6", full circle pattern		76	.211		10.25	10.85		21.10	28
1050	1/2 circle pattern, 6"		76	.211		10.25	10.85		21.10	28
1060	12", full circle pattern		76	.211		12.05	10.85		22.90	30
1070	1/2 circle pattern, 12"		76	.211		14.85	10.85		25.70	33
1080	Pop up bubbler head w/risers, hi-pop bubbler head, 4"		76	.211		3.72	10.85		14.57	20.50
1110	Impact full/part circle sprinklers, 28'-54' 25-60 PSI		37	.432		19.05	22.50		41.55	55.50
1120	Spaced 37'-49' @ 25-50 PSI		37	.432		23.50	22.50		46	60.50
1130	Spaced 43'-61' @ 30-60 PSI		37	.432		66.50	22.50		89	108
1140	Spaced 54'-78' @ 40-80 PSI		37	.432		115	22.50		137.50	161
1145	Impact rotor pop-up full/part commercial circle sprinklers									
1150	Spaced 42'-65' 35-80 PSI	2 Skwk	25	.640	Ea.	16.55	33		49.55	68.50
1160	Spaced 48'-76' 45-85 PSI	"	25	.640	"	20	33		53	72.50
1165	Impact rotor pop-up part. circle comm., 53'-75', 55-100 PSI, w/accessories									
1180	Sprinkler, premium, pop-up rotator, 50'-100'	2 Skwk	25	.640	Ea.	93.50	33		126.50	154
1250	Plastic case, 2 nozzle, metal cover		25	.640		98	33		131	159
1260	Rubber cover		25	.640		101	33		134	162
1270	Iron case, 2 nozzle, metal cover		22	.727		143	37.50		180.50	215
1280	Rubber cover		22	.727		143	37.50		180.50	215
1282	Impact rotor pop-up full circle commercial, 39'-99', 30-100 PSI									
1284	Plastic case, metal cover	2 Skwk	25	.640	Ea.	77	33		110	135
1286	Rubber cover		25	.640		102	33		135	163
1288	Iron case, metal cover		22	.727		132	37.50		169.50	203
1290	Rubber cover		22	.727		173	37.50		210.50	248
1292	Plastic case, 2 nozzle, metal cover		22	.727		99	37.50		136.50	167
1294	Rubber cover		22	.727		99	37.50		136.50	167
1296	Iron case, 2 nozzle, metal cover		20	.800		131	41		172	208
1298	Rubber cover		20	.800		136	41		177	213
1305	Electric remote control valve, plastic, 3/4"		18	.889		18.30	45.50		63.80	90.50
1310	1"		18	.889		25.50	45.50		71	98.50
1320	1-1/2"		18	.889		98.50	45.50		144	179
1330	2"		18	.889		118	45.50		163.50	201

32 84 Planting Irrigation

32 84 23 – Underground Sprinklers

32 84 23.10 Sprinkler Irrigation System	Crew	Daily Output	Labor-Hours	Unit	Material	2017 Bare Costs Labor	Equipment	Total	Total Incl O&P
1335	Quick coupling valves, brass, locking cover								
1340	Inlet coupling valve, 3/4"	2 Skwk	18.75	.853	Ea.	24.50	44		68.50	94.50
1350	1"		18.75	.853		32.50	44		76.50	103
1360	Controller valve boxes, 6" round boxes		18.75	.853		7.70	44		51.70	76
1370	10" round boxes		14.25	1.123		17.95	58		75.95	109
1380	12" square box		9.75	1.641		17.60	84.50		102.10	149
1388	Electromech. control, 14 day 3-60 min., auto start to 23/day								
1390	4 station	2 Skwk	1.04	15.385	Ea.	81	790		871	1,325
1400	7 station		.64	25		151	1,275		1,426	2,150
1410	12 station		.40	40		184	2,050		2,234	3,375
1420	Dual programs, 18 station		.24	66.667		216	3,425		3,641	5,525
1430	23 station		.16	100		238	5,150		5,388	8,175
1435	Backflow preventer, bronze, 0-175 PSI, w/valves, test cocks								
1440	3/4"	2 Skwk	6	2.667	Ea.	88.50	137		225.50	310
1450	1"		6	2.667		102	137		239	325
1460	1-1/2"		6	2.667		270	137		407	510
1470	2"		6	2.667		380	137		517	625
1475	Pressure vacuum breaker, brass, 15-150 PSI								
1480	3/4"	2 Skwk	6	2.667	Ea.	27	137		164	241
1490	1"		6	2.667		32.50	137		169.50	247
1500	1-1/2"		6	2.667		80.50	137		217.50	300
1510	2"		6	2.667		138	137		275	365

32 91 Planting Preparation

32 91 13 – Soil Preparation

32 91 13.16 Mulching

	32 91 13.16 Mulching	Crew	Daily Output	Labor-Hours	Unit	Material	2017 Bare Costs Labor	Equipment	Total	Total Incl O&P
0010	**MULCHING**									
0100	Aged barks, 3" deep, hand spread	1 Clab	100	.080	S.Y.	3.57	3.13		6.70	8.75
0150	Skid steer loader	B-63	13.50	2.963	M.S.F.	395	123	12.70	530.70	635
0200	Hay, 1" deep, hand spread	1 Clab	475	.017	S.Y.	.43	.66		1.09	1.48
0250	Power mulcher, small	B-64	180	.089	M.S.F.	48	3.70	2	53.70	60.50
0350	Large	B-65	530	.030	"	48	1.26	.98	50.24	55.50
0400	Humus peat, 1" deep, hand spread	1 Clab	700	.011	S.Y.	2.63	.45		3.08	3.58
0450	Push spreader	"	2500	.003	"	2.63	.13		2.76	3.08
0550	Tractor spreader	B-66	700	.011	M.S.F.	292	.58	.38	292.96	320
0600	Oat straw, 1" deep, hand spread	1 Clab	475	.017	S.Y.	.56	.66		1.22	1.63
0650	Power mulcher, small	B-64	180	.089	M.S.F.	62	3.70	2	67.70	76.50
0700	Large	B-65	530	.030	"	62	1.26	.98	64.24	71.50
0750	Add for asphaltic emulsion	B-45	1770	.009	Gal.	5.80	.45	.50	6.75	7.65
0800	Peat moss, 1" deep, hand spread	1 Clab	900	.009	S.Y.	2.60	.35		2.95	3.39
0850	Push spreader	"	2500	.003	"	2.60	.13		2.73	3.05
0950	Tractor spreader	B-66	700	.011	M.S.F.	289	.58	.38	289.96	320
1000	Polyethylene film, 6 mil	2 Clab	2000	.008	S.Y.	.50	.31		.81	1.03
1100	Redwood nuggets, 3" deep, hand spread	1 Clab	150	.053	"	2.78	2.09		4.87	6.25
1150	Skid steer loader	B-63	13.50	2.963	M.S.F.	310	123	12.70	445.70	540
1200	Stone mulch, hand spread, ceramic chips, economy	1 Clab	125	.064	S.Y.	6.90	2.51		9.41	11.45
1250	Deluxe	"	95	.084	"	10.60	3.30		13.90	16.70
1300	Granite chips	B-1	10	2.400	C.Y.	60	95.50		155.50	212
1400	Marble chips		10	2.400		189	95.50		284.50	355
1600	Pea gravel		28	.857		109	34		143	172
1700	Quartz		10	2.400		190	95.50		285.50	355

For customer support on your Building Construction Costs with RSMeans Data, call 800.448.8182.

667

32 91 Planting Preparation

32 91 13 – Soil Preparation

32 91 13.16 Mulching

		Crew	Daily Output	Labor-Hours	Unit	Material	2017 Bare Costs Labor	2017 Bare Costs Equipment	Total	Total Incl O&P
1800	Tar paper, 15 lb. felt	1 Clab	800	.010	S.Y.	.48	.39		.87	1.12
1900	Wood chips, 2" deep, hand spread	"	220	.036	"	1.80	1.42		3.22	4.16
1950	Skid steer loader	B-63	20.30	1.970	M.S.F.	200	82	8.45	290.45	355

32 91 13.26 Planting Beds

		Crew	Daily Output	Labor-Hours	Unit	Material	2017 Bare Costs Labor	2017 Bare Costs Equipment	Total	Total Incl O&P
0010	**PLANTING BEDS**									
0100	Backfill planting pit, by hand, on site topsoil	2 Clab	18	.889	C.Y.		35		35	53.50
0200	Prepared planting mix, by hand	"	24	.667			26		26	40
0300	Skid steer loader, on site topsoil	B-62	340	.071			3.04	.50	3.54	5.20
0400	Prepared planting mix	"	410	.059			2.52	.42	2.94	4.31
1000	Excavate planting pit, by hand, sandy soil	2 Clab	16	1			39		39	60
1100	Heavy soil or clay	"	8	2			78.50		78.50	120
1200	1/2 C.Y. backhoe, sandy soil	B-11C	150	.107			4.94	2.44	7.38	10.20
1300	Heavy soil or clay	"	115	.139			6.45	3.18	9.63	13.30
2000	Mix planting soil, incl. loam, manure, peat, by hand	2 Clab	60	.267		44	10.45		54.45	64.50
2100	Skid steer loader	B-62	150	.160		44	6.90	1.14	52.04	60.50
3000	Pile sod, skid steer loader	"	2800	.009	S.Y.		.37	.06	.43	.63
3100	By hand	2 Clab	400	.040			1.57		1.57	2.40
4000	Remove sod, F.E. loader	B-10S	2000	.006			.29	.18	.47	.64
4100	Sod cutter	B-12K	3200	.005			.24	.36	.60	.75
4200	By hand	2 Clab	240	.067			2.61		2.61	4

32 91 19 – Landscape Grading

32 91 19.13 Topsoil Placement and Grading

		Crew	Daily Output	Labor-Hours	Unit	Material	2017 Bare Costs Labor	2017 Bare Costs Equipment	Total	Total Incl O&P
0010	**TOPSOIL PLACEMENT AND GRADING**									
0400	Spread from pile to rough finish grade, F.E. loader, 1.5 C.Y.	B-10S	200	.060	C.Y.		2.93	1.81	4.74	6.45
0500	Up to 200' radius, by hand	1 Clab	14	.571			22.50		22.50	34.50
0600	Top dress by hand, 1 C.Y. for 600 S.F.	"	11.50	.696		28	27		55	72
0700	Furnish and place, truck dumped, screened, 4" deep	B-10S	1300	.009	S.Y.	3.48	.45	.28	4.21	4.82
0800	6" deep	"	820	.015	"	4.46	.71	.44	5.61	6.45

32 92 Turf and Grasses

32 92 19 – Seeding

32 92 19.13 Mechanical Seeding

		Crew	Daily Output	Labor-Hours	Unit	Material	2017 Bare Costs Labor	2017 Bare Costs Equipment	Total	Total Incl O&P
0010	**MECHANICAL SEEDING**									
0020	Mechanical seeding, 215 lb./acre	B-66	1.50	5.333	Acre	560	272	179	1,011	1,225
0100	44 lb./M.S.Y.	"	2500	.003	S.Y.	.20	.16	.11	.47	.59
0101	44 lb./M.S.Y.	1 Clab	13950	.001	S.F.	.02	.02		.04	.05
0300	Fine grading and seeding incl. lime, fertilizer & seed,									
0310	with equipment	B-14	1000	.048	S.Y.	.45	1.99	.37	2.81	3.94
0400	Fertilizer hand push spreader, 35 lb. per M.S.F.	1 Clab	200	.040	M.S.F.	9.85	1.57		11.42	13.25
0600	Limestone hand push spreader, 50 lb. per M.S.F.		180	.044		5.30	1.74		7.04	8.45
0800	Grass seed hand push spreader, 4.5 lb. per M.S.F.		180	.044		22	1.74		23.74	27
1000	Hydro or air seeding for large areas, incl. seed and fertilizer	B-81	8900	.003	S.Y.	.51	.12	.07	.70	.83
1100	With wood fiber mulch added	"	8900	.003	"	2.02	.12	.07	2.21	2.49
1300	Seed only, over 100 lb., field seed, minimum				Lb.	1.80			1.80	1.98
1400	Maximum					1.70			1.70	1.87
1500	Lawn seed, minimum					1.40			1.40	1.54
1600	Maximum					2.50			2.50	2.75
1800	Aerial operations, seeding only, field seed	B-58	50	.480	Acre	585	20.50	63	668.50	745
1900	Lawn seed		50	.480		455	20.50	63	538.50	600
2100	Seed and liquid fertilizer, field seed		50	.480		745	20.50	63	828.50	920

32 92 Turf and Grasses

32 92 19 – Seeding

32 92 19.13 Mechanical Seeding	Crew	Daily Output	Labor-Hours	Unit	Material	2017 Bare Costs Labor	Equipment	Total	Total Incl O&P
2200　　Lawn seed	B-58	50	.480	Acre	615	20.50	63	698.50	775

32 92 23 – Sodding

32 92 23.10 Sodding Systems

	Crew	Daily Output	Labor-Hours	Unit	Material	Labor	Equipment	Total	Total Incl O&P
0010 **SODDING SYSTEMS**									
0020　Sodding, 1" deep, bluegrass sod, on level ground, over 8 M.S.F.	B-63	22	1.818	M.S.F.	243	75.50	7.80	326.30	390
0200　　　4 M.S.F.		17	2.353		240	97.50	10.05	347.55	425
0300　　　1000 S.F.		13.50	2.963		290	123	12.70	425.70	520
0500　　Sloped ground, over 8 M.S.F.		6	6.667		243	277	28.50	548.50	725
0600　　　4 M.S.F.		5	8		240	330	34	604	805
0700　　　1000 S.F.		4	10		290	415	43	748	1,000
1000　　Bent grass sod, on level ground, over 6 M.S.F.		20	2		255	83	8.55	346.55	415
1100　　　3 M.S.F.		18	2.222		269	92.50	9.50	371	445
1200　　　Sodding 1000 S.F. or less		14	2.857		293	119	12.25	424.25	515
1500　　Sloped ground, over 6 M.S.F.		15	2.667		255	111	11.40	377.40	465
1600　　　3 M.S.F.		13.50	2.963		269	123	12.70	404.70	500
1700　　　1000 S.F.	▼	12	3.333	▼	293	138	14.25	445.25	545

32 93 Plants

32 93 10 – General Planting Costs

32 93 10.12 Travel

	Crew	Daily Output	Labor-Hours	Unit	Material	Labor	Equipment	Total	Total Incl O&P
0010 **TRAVEL** add to all nursery items									
0015　　10 to 20 miles one way, add				All				5%	5%
0100　　30 to 50 miles one way, add				"				10%	10%

32 93 13 – Ground Covers

32 93 13.10 Ground Cover Plants

	Crew	Daily Output	Labor-Hours	Unit	Material	Labor	Equipment	Total	Total Incl O&P
0010 **GROUND COVER PLANTS**									
0012　Plants, pachysandra, in prepared beds	B-1	15	1.600	C	84.50	63.50		148	190
0200　　Vinca minor, 1 yr., bare root, in prepared beds		12	2	"	72	79.50		151.50	201
0600　Stone chips, in 50 lb. bags, Georgia marble		520	.046	Bag	4.75	1.84		6.59	8.05
0700　　Onyx gemstone		260	.092		16	3.68		19.68	23.50
0800　　Quartz		260	.092	▼	16	3.68		19.68	23.50
0900　　Pea gravel, truckload lots	▼	28	.857	Ton	28.50	34		62.50	83.50

32 93 33 – Shrubs

32 93 33.10 Shrubs and Trees

	Crew	Daily Output	Labor-Hours	Unit	Material	Labor	Equipment	Total	Total Incl O&P
0010 **SHRUBS AND TREES**									
0011　Evergreen, in prepared beds, B & B									
0100　　Arborvitae pyramidal, 4'-5'	B-17	30	1.067	Ea.	102	46.50	25	173.50	212
0150　　　Globe, 12"-15"	B-1	96	.250		24.50	9.95		34.45	42.50
0300　　Cedar, blue, 8'-10'	B-17	18	1.778		238	77.50	42	357.50	425
0500　　Hemlock, Canadian, 2-1/2'-3'	B-1	36	.667		32	26.50		58.50	76
0550　　Holly, Savannah, 8'-10' H		9.68	2.479		264	98.50		362.50	440
0600　　Juniper, andorra, 18"-24"		80	.300		37.50	11.95		49.45	60
0620　　　Wiltoni, 15"-18"	▼	80	.300		26.50	11.95		38.45	47.50
0640　　　Skyrocket, 4-1/2'-5'	B-17	55	.582		109	25.50	13.75	148.25	174
0660　　　Blue pfitzer, 2'-2-1/2'	B-1	44	.545		39	21.50		60.50	76.50
0680　　　Ketleerie, 2-1/2'-3'		50	.480		53.50	19.10		72.60	88
0700　　Pine, black, 2-1/2'-3'		50	.480		61	19.10		80.10	96.50
0720　　　Mugo, 18"-24"	▼	60	.400		61	15.95		76.95	91.50
0740　　　White, 4'-5'	B-17	75	.427	▼	52.50	18.65	10.10	81.25	97.50

For customer support on your Building Construction Costs with RSMeans Data, call 800.448.8182.

669

32 93 33 – Shrubs

32 93 33.10 Shrubs and Trees

		Crew	Daily Output	Labor-Hours	Unit	Material	2017 Bare Costs Labor	Equipment	Total	Total Incl O&P
0800	Spruce, blue, 18"-24"	B-1	60	.400	Ea.	68.50	15.95		84.45	100
0840	Norway, 4'-5'	B-17	75	.427		85.50	18.65	10.10	114.25	134
0900	Yew, denisforma, 12"-15"	B-1	60	.400		36	15.95		51.95	64
1000	Capitata, 18"-24"		30	.800		33.50	32		65.50	86
1100	Hicksi, 2'-2-1/2'		30	.800		80	32		112	137

32 93 33.20 Shrubs

		Crew	Daily Output	Labor-Hours	Unit	Material	2017 Bare Costs Labor	Equipment	Total	Total Incl O&P
0010	**SHRUBS**									
0011	Broadleaf Evergreen, planted in prepared beds									
0100	Andromeda, 15"-18", container	B-1	96	.250	Ea.	33.50	9.95		43.45	52.50
0200	Azalea, 15"-18", container		96	.250		30.50	9.95		40.45	49
0300	Barberry, 9"-12", container		130	.185		18.15	7.35		25.50	31
0400	Boxwood, 15"-18", B & B		96	.250		44	9.95		53.95	64
0500	Euonymus, emerald gaiety, 12"-15", container		115	.209		24.50	8.30		32.80	40
0600	Holly, 15"-18", B & B		96	.250		37.50	9.95		47.45	56.50
0900	Mount laurel, 18"-24", B & B		80	.300		72.50	11.95		84.45	98.50
1000	Paxistema, 9"-12" high		130	.185		22	7.35		29.35	35.50
1100	Rhododendron, 18"-24", container		48	.500		38.50	19.90		58.40	73
1200	Rosemary, 1 gal. container		600	.040		18.25	1.59		19.84	22.50
2000	Deciduous, planted in prepared beds, amelanchier, 2'-3', B & B		57	.421		123	16.75		139.75	161
2100	Azalea, 15"-18", B & B		96	.250		30	9.95		39.95	48.50
2300	Bayberry, 2'-3', B & B		57	.421		24.50	16.75		41.25	52.50
2600	Cotoneaster, 15"-18", B & B		80	.300		27	11.95		38.95	48.50
2800	Dogwood, 3'-4', B & B	B-17	40	.800		33	35	18.95	86.95	111
2900	Euonymus, alatus compacta, 15"-18", container	B-1	80	.300		27	11.95		38.95	48
3200	Forsythia, 2'-3', container	"	60	.400		19	15.95		34.95	45.50
3300	Hibiscus, 3'-4', B & B	B-17	75	.427		49.50	18.65	10.10	78.25	94
3400	Honeysuckle, 3'-4', B & B	B-1	60	.400		27	15.95		42.95	54.50
3500	Hydrangea, 2'-3', B & B	"	57	.421		30.50	16.75		47.25	59.50
3600	Lilac, 3'-4', B & B	B-17	40	.800		29	35	18.95	82.95	106
3900	Privet, bare root, 18"-24"	B-1	80	.300		14.95	11.95		26.90	35
4100	Quince, 2'-3', B & B	"	57	.421		29	16.75		45.75	57.50
4200	Russian olive, 3'-4', B & B	B-17	75	.427		29.50	18.65	10.10	58.25	72
4400	Spirea, 3'-4', B & B	B-1	70	.343		21	13.65		34.65	44
4500	Viburnum, 3'-4', B & B	B-17	40	.800		26	35	18.95	79.95	103

32 93 43 – Trees

32 93 43.20 Trees

			Crew	Daily Output	Labor-Hours	Unit	Material	2017 Bare Costs Labor	Equipment	Total	Total Incl O&P
0010	**TREES**										
0011	Deciduous, in prep. beds, balled & burlapped (B&B)										
0100	Ash, 2" caliper	G	B-17	8	4	Ea.	193	175	94.50	462.50	585
0200	Beech, 5'-6'	G		50	.640		189	28	15.15	232.15	267
0300	Birch, 6'-8', 3 stems	G		20	1.600		168	70	38	276	335
0500	Crabapple, 6'-8'	G		20	1.600		140	70	38	248	300
0600	Dogwood, 4'-5'	G		40	.800		134	35	18.95	187.95	221
0700	Eastern redbud, 4'-5'	G		40	.800		149	35	18.95	202.95	238
0800	Elm, 8'-10'	G		20	1.600		253	70	38	361	425
0900	Ginkgo, 6'-7'	G		24	1.333		145	58.50	31.50	235	283
1000	Hawthorn, 8'-10', 1" caliper	G		20	1.600		159	70	38	267	325
1100	Honeylocust, 10'-12', 1-1/2" caliper	G		10	3.200		202	140	75.50	417.50	520
1300	Larch, 8'	G		32	1		124	43.50	23.50	191	230
1400	Linden, 8'-10', 1" caliper	G		20	1.600		141	70	38	249	305
1500	Magnolia, 4'-5'	G		20	1.600		99	70	38	207	257
1600	Maple, red, 8'-10', 1-1/2" caliper	G		10	3.200		198	140	75.50	413.50	515

For customer support on your Building Construction Costs with RSMeans Data, call 800.448.8182.

32 93 Plants

32 93 43 – Trees

32 93 43.20 Trees		Crew	Daily Output	Labor-Hours	Unit	Material	2017 Bare Costs Labor	2017 Bare Costs Equipment	Total	Total Incl O&P
1700	Mountain ash, 8'-10', 1" caliper	G	B-17	16	2	Ea.	172	87.50	47.50	307	375
1800	Oak, 2-1/2"-3" caliper	G		6	5.333		320	233	126	679	845
2100	Planetree, 9'-11', 1-1/4" caliper	G		10	3.200		231	140	75.50	446.50	550
2200	Plum, 6'-8', 1" caliper	G		20	1.600		80.50	70	38	188.50	236
2300	Poplar, 9'-11', 1-1/4" caliper	G		10	3.200		141	140	75.50	356.50	450
2500	Sumac, 2'-3'	G		75	.427		43	18.65	10.10	71.75	87
2700	Tulip, 5'-6'	G		40	.800		45.50	35	18.95	99.45	124
2800	Willow, 6'-8', 1" caliper	G		20	1.600		94	70	38	202	251

32 94 Planting Accessories

32 94 13 – Landscape Edging

32 94 13.20 Edging

		Crew	Daily Output	Labor-Hours	Unit	Material	Labor	Equipment	Total	Total Incl O&P
0010	**EDGING**									
0050	Aluminum alloy, including stakes, 1/8" x 4", mill finish	B-1	390	.062	L.F.	4	2.45		6.45	8.15
0051	Black paint		390	.062		4.64	2.45		7.09	8.85
0052	Black anodized		390	.062		5.35	2.45		7.80	9.65
0100	Brick, set horizontally, 1-1/2 bricks per L.F.	D-1	370	.043		1.37	1.87		3.24	4.39
0150	Set vertically, 3 bricks per L.F.	"	135	.119		3.54	5.15		8.69	11.80
0200	Corrugated aluminum, roll, 4" wide	1 Carp	650	.012		1.87	.61		2.48	2.99
0250	6" wide	"	550	.015		2.34	.72		3.06	3.67
0600	Railroad ties, 6" x 8"	2 Carp	170	.094		2.24	4.64		6.88	9.55
0650	7" x 9"		136	.118		2.49	5.80		8.29	11.65
0750	Redwood 2" x 4"		330	.048		2.26	2.39		4.65	6.15
0800	Steel edge strips, incl. stakes, 1/4" x 5"	B-1	390	.062		4.26	2.45		6.71	8.45
0850	3/16" x 4"	"	390	.062		3.37	2.45		5.82	7.45

32 94 50 – Tree Guying

32 94 50.10 Tree Guying Systems

		Crew	Daily Output	Labor-Hours	Unit	Material	Labor	Equipment	Total	Total Incl O&P
0010	**TREE GUYING SYSTEMS**									
0015	Tree guying including stakes, guy wire and wrap									
0100	Less than 3" caliper, 2 stakes	2 Clab	35	.457	Ea.	13.20	17.90		31.10	42
0200	3" to 4" caliper, 3 stakes	"	21	.762	"	19.75	30		49.75	67
1000	Including arrowhead anchor, cable, turnbuckles and wrap									
1100	Less than 3" caliper, 3" anchors	2 Clab	20	.800	Ea.	21.50	31.50		53	72
1200	3" to 6" caliper, 4" anchors		15	1.067		31.50	42		73.50	98.50
1300	6" caliper, 6" anchors		12	1.333		24	52		76	107
1400	8" caliper, 8" anchors		9	1.778		115	69.50		184.50	234

32 96 Transplanting

32 96 23 – Plant and Bulb Transplanting

32 96 23.23 Planting

		Crew	Daily Output	Labor-Hours	Unit	Material	Labor	Equipment	Total	Total Incl O&P
0010	**PLANTING**									
0012	Moving shrubs on site, 12" ball	B-62	28	.857	Ea.		37	6.10	43.10	63
0100	24" ball	"	22	1.091	"		47	7.80	54.80	80

32 96 23.43 Moving Trees

		Crew	Daily Output	Labor-Hours	Unit	Material	Labor	Equipment	Total	Total Incl O&P
0010	**MOVING TREES**, On site									
0300	Moving trees on site, 36" ball	B-6	3.75	6.400	Ea.		276	97.50	373.50	525
0400	60" ball	"	1	24	"		1,025	365	1,390	1,975

For customer support on your Building Construction Costs with RSMeans Data, call 800.448.8182.

671

Division Notes

	CREW	DAILY OUTPUT	LABOR-HOURS	UNIT	BARE COSTS				TOTAL INCL O&P
					MAT.	LABOR	EQUIP.	TOTAL	

Estimating Tips
33 10 00 Water Utilities
33 30 00 Sanitary Sewerage Utilities
33 40 00 Storm Drainage Utilities

- Never assume that the water, sewer, and drainage lines will go in at the early stages of the project. Consider the site access needs before dividing the site in half with open trenches, loose pipe, and machinery obstructions. Always inspect the site to establish that the site drawings are complete. Check off all existing utilities on your drawings as you locate them. Be especially careful with underground utilities because appurtenances are sometimes buried during regrading or repaving operations. If you find any discrepancies, mark up the site plan for further research. Differing site conditions can be very costly if discovered later in the project.

- See also Section 33 01 00 for restoration of pipe where removal/replacement may be undesirable. Use of new types of piping materials can reduce the overall project cost. Owners/design engineers should consider the installing contractor as a valuable source of current information on utility products and local conditions that could lead to significant cost savings.

Reference Numbers
Reference numbers are shown at the beginning of some major classifications. These numbers refer to related items in the Reference Section. The reference information may be an estimating procedure, an alternate pricing method, or technical information.

Note: Not all subdivisions listed here necessarily appear. ■

33 01 Operation and Maintenance of Utilities

33 01 10 – Operation and Maintenance of Water Utilities

33 01 10.10 Corrosion Resistance

		Crew	Daily Output	Labor-Hours	Unit	Material	2017 Bare Costs Labor	Equipment	Total	Total Incl O&P
0010	**CORROSION RESISTANCE**									
0012	Wrap & coat, add to pipe, 4" diameter				L.F.	2.18			2.18	2.40
0040	6" diameter					3.22			3.22	3.54
0060	8" diameter					3.99			3.99	4.39
0100	12" diameter					6.10			6.10	6.70
0200	24" diameter					12.70			12.70	14
0500	Coating, bituminous, per diameter inch, 1 coat, add					.23			.23	.25
0540	3 coat					.69			.69	.76
0560	Coal tar epoxy, per diameter inch, 1 coat, add					.25			.25	.28
0600	3 coat					.70			.70	.77

33 01 30 – Operation and Maintenance of Sewer Utilities

33 01 30.72 Relining Sewers

		Crew	Daily Output	Labor-Hours	Unit	Material	2017 Bare Costs Labor	Equipment	Total	Total Incl O&P
0010	**RELINING SEWERS**									
0011	With cement incl. bypass & cleaning									
0020	Less than 10,000 L.F., urban, 6" to 10"	C-17E	130	.615	L.F.	9.55	32	.77	42.32	60.50
0200	24" to 36"		90	.889		15.25	46	1.11	62.36	89
0300	48" to 72"		80	1		24.50	52	1.25	77.75	108

33 05 Common Work Results for Utilities

33 05 16 – Utility Structures

33 05 16.13 Precast Concrete Utility Structures

		Crew	Daily Output	Labor-Hours	Unit	Material	2017 Bare Costs Labor	Equipment	Total	Total Incl O&P
0010	**PRECAST CONCRETE UTILITY STRUCTURES**, 6" thick									
0050	5' x 10' x 6' high, I.D.	B-13	2	28	Ea.	1,800	1,200	335	3,335	4,200
0100	6' x 10' x 6' high, I.D.		2	28		1,875	1,200	335	3,410	4,300
0150	5' x 12' x 6' high, I.D.		2	28		1,975	1,200	335	3,510	4,400
0200	6' x 12' x 6' high, I.D.		1.80	31.111		2,225	1,350	375	3,950	4,900
0250	6' x 13' x 6' high, I.D.		1.50	37.333		2,925	1,600	450	4,975	6,150
0300	8' x 14' x 7' high, I.D.		1	56		3,150	2,400	675	6,225	7,900
0350	Hand hole, precast concrete, 1-1/2" thick									
0400	1'-0" x 2'-0" x 1'-9", I.D., light duty	B-1	4	6	Ea.	430	239		669	835
0450	4'-6" x 3'-2" x 2'-0", O.D., heavy duty	B-6	3	8	"	1,525	345	122	1,992	2,325

33 05 23 – Trenchless Utility Installation

33 05 23.19 Microtunneling

		Crew	Daily Output	Labor-Hours	Unit	Material	2017 Bare Costs Labor	Equipment	Total	Total Incl O&P
0010	**MICROTUNNELING**									
0011	Not including excavation, backfill, shoring,									
0020	or dewatering, average 50'/day, slurry method									
0100	24" to 48" outside diameter, minimum				L.F.				875	965
0110	Adverse conditions, add				%				50%	50%
1000	Rent microtunneling machine, average monthly lease				Month				97,500	107,000
1010	Operating technician				Day				630	705
1100	Mobilization and demobilization, minimum				Job				41,200	45,900
1110	Maximum				"				445,500	490,500

33 05 23.20 Horizontal Boring

		Crew	Daily Output	Labor-Hours	Unit	Material	2017 Bare Costs Labor	Equipment	Total	Total Incl O&P
0010	**HORIZONTAL BORING**									
0011	Casing only, 100' minimum,									
0020	not incl. jacking pits or dewatering									
0100	Roadwork, 1/2" thick wall, 24" diameter casing	B-42	20	3.200	L.F.	117	142	63.50	322.50	420
0200	36" diameter		16	4		215	178	79.50	472.50	600
0300	48" diameter		15	4.267		297	190	84.50	571.50	710
0500	Railroad work, 24" diameter		15	4.267		117	190	84.50	391.50	515

33 05 Common Work Results for Utilities

33 05 23 – Trenchless Utility Installation

33 05 23.20 Horizontal Boring	Crew	Daily Output	Labor-Hours	Unit	Material	2017 Bare Costs Labor	Equipment	Total	Total Incl O&P
0600	36" diameter	B-42	14	4.571	L.F.	215	203	90.50	508.50	650
0700	48" diameter	▼	12	5.333		297	237	106	640	805
0900	For ledge, add				▼				20%	20%

33 05 26 – Utility Identification

33 05 26.05 Utility Connection

		Crew	Daily Output	Labor-Hours	Unit	Material	Labor	Equipment	Total	Total Incl O&P
0010	**UTILITY CONNECTION**									
0020	Water, sanitary, stormwater, gas, single connection	B-14	1	48	Ea.	2,975	2,000	365	5,340	6,725
0030	Telecommunication	"	3	16	"	385	665	122	1,172	1,575

33 05 26.10 Utility Accessories

		Crew	Daily Output	Labor-Hours	Unit	Material	Labor	Equipment	Total	Total Incl O&P
0010	**UTILITY ACCESSORIES**									
0400	Underground tape, detectable, reinforced, alum. foil core, 2"	1 Clab	150	.053	C.L.F.	6	2.09		8.09	9.80
0500	6"	"	140	.057	"	35	2.24		37.24	42

33 11 Water Utility Distribution Piping

33 11 13 – Public Water Utility Distribution Piping

33 11 13.15 Water Supply, Ductile Iron Pipe

		Crew	Daily Output	Labor-Hours	Unit	Material	Labor	Equipment	Total	Total Incl O&P
0010	**WATER SUPPLY, DUCTILE IRON PIPE** R331113-80									
0020	Not including excavation or backfill									
2000	Pipe, class 50 water piping, 18' lengths									
2020	Mechanical joint, 4" diameter	B-21A	200	.200	L.F.	30.50	9.90	2.18	42.58	51
2040	6" diameter		160	.250		26.50	12.35	2.72	41.57	51
2060	8" diameter		133.33	.300		44.50	14.85	3.27	62.62	75
2080	10" diameter		114.29	.350		58.50	17.30	3.81	79.61	94.50
2100	12" diameter		105.26	.380		79	18.80	4.14	101.94	120
2120	14" diameter		100	.400		93	19.80	4.36	117.16	137
2140	16" diameter		72.73	.550		94.50	27	6	127.50	152
2160	18" diameter		68.97	.580		126	28.50	6.30	160.80	189
2170	20" diameter		57.14	.700		127	34.50	7.60	169.10	201
2180	24" diameter		47.06	.850		141	42	9.25	192.25	229
3000	Push-on joint, 4" diameter		400	.100		15.50	4.95	1.09	21.54	26
3020	6" diameter		333.33	.120		18.25	5.95	1.31	25.51	30.50
3040	8" diameter		200	.200		27.50	9.90	2.18	39.58	48
3060	10" diameter		181.82	.220		39	10.90	2.40	52.30	61.50
3080	12" diameter		160	.250		41	12.35	2.72	56.07	67
3100	14" diameter		133.33	.300		41	14.85	3.27	59.12	71
3120	16" diameter		114.29	.350		42	17.30	3.81	63.11	77
3140	18" diameter		100	.400		46.50	19.80	4.36	70.66	86.50
3160	20" diameter		88.89	.450		48.50	22.50	4.90	75.90	93
3180	24" diameter	▼	76.92	.520	▼	60.50	25.50	5.65	91.65	112
8000	Piping, fittings, mechanical joint, AWWA C110									
8006	90° bend, 4" diameter	B-20A	16	2	Ea.	155	96		251	315
8020	6" diameter		12.80	2.500		229	120		349	435
8040	8" diameter	▼	10.67	2.999		450	144		594	715
8060	10" diameter	B-21A	11.43	3.500		620	173	38	831	985
8080	12" diameter		10.53	3.799		880	188	41.50	1,109.50	1,300
8100	14" diameter		10	4		1,350	198	43.50	1,591.50	1,825
8120	16" diameter		7.27	5.502		1,650	272	60	1,982	2,300
8140	18" diameter		6.90	5.797		2,375	287	63	2,725	3,100
8160	20" diameter		5.71	7.005		2,800	345	76.50	3,221.50	3,675
8180	24" diameter	▼	4.70	8.511	▼	4,200	420	92.50	4,712.50	5,350

For customer support on your Building Construction Costs with RSMeans Data, call 800.448.8182.

675

33 11 13 – Public Water Utility Distribution Piping

33 11 13.15 Water Supply, Ductile Iron Pipe

		Crew	Daily Output	Labor-Hours	Unit	Material	2017 Bare Costs Labor	Equipment	Total	Total Incl O&P
8200	Wye or tee, 4" diameter	B-20A	10.67	2.999	Ea.	375	144		519	635
8220	6" diameter		8.53	3.751		565	180		745	895
8240	8" diameter	↓	7.11	4.501		895	216		1,111	1,300
8260	10" diameter	B-21A	7.62	5.249		1,175	260	57	1,492	1,750
8280	12" diameter		7.02	5.698		1,700	282	62	2,044	2,375
8300	14" diameter		6.67	5.997		2,750	297	65.50	3,112.50	3,550
8320	16" diameter		4.85	8.247		3,050	410	90	3,550	4,075
8340	18" diameter		4.60	8.696		3,825	430	94.50	4,349.50	4,950
8360	20" diameter		3.81	10.499		5,750	520	114	6,384	7,275
8380	24" diameter		3.14	12.739		8,000	630	139	8,769	9,900
8450	Decreaser, 6" x 4" diameter	B-20A	14.22	2.250		207	108		315	390
8460	8" x 6" diameter	"	11.64	2.749		310	132		442	540
8470	10" x 6" diameter	B-21A	13.33	3.001		390	148	32.50	570.50	690
8480	12" x 6" diameter		12.70	3.150		550	156	34.50	740.50	880
8490	16" x 6" diameter		10	4		890	198	43.50	1,131.50	1,325
8500	20" x 6" diameter	↓	8.42	4.751	↓	1,625	235	51.50	1,911.50	2,175
8550	Piping, butterfly valves, cast iron									
8560	4" diameter	B-20	6	4	Ea.	440	176		616	750
8570	6" diameter	"	5	4.800		600	211		811	985
8580	8" diameter	B-21	4	7		780	320	34.50	1,134.50	1,375
8590	10" diameter		3.50	8		1,100	365	39.50	1,504.50	1,800
8600	12" diameter		3	9.333		1,500	425	46	1,971	2,350
8610	14" diameter		2	14		2,900	640	69	3,609	4,250
8620	16" diameter	↓	2	14	↓	4,225	640	69	4,934	5,700

33 11 13.25 Water Supply, Polyvinyl Chloride Pipe

		Crew	Daily Output	Labor-Hours	Unit	Material	2017 Bare Costs Labor	Equipment	Total	Total Incl O&P
0010	**WATER SUPPLY, POLYVINYL CHLORIDE PIPE**									
0020	Not including excavation or backfill, unless specified									
2100	PVC pipe, Class 150, 1-1/2" diameter	Q-1A	750	.013	L.F.	.47	.83		1.30	1.77
2120	2" diameter		686	.015		.69	.91		1.60	2.13
2140	2-1/2" diameter	↓	500	.020		1.18	1.24		2.42	3.18
2160	3" diameter	B-20	430	.056	↓	1.33	2.45		3.78	5.20
3010	AWWA C905, PR 100, DR 25									
3030	14" diameter	B-20A	213	.150	L.F.	12.85	7.20		20.05	25
3040	16" diameter		200	.160		18.10	7.65		25.75	31.50
3050	18" diameter		160	.200		23	9.60		32.60	39.50
3060	20" diameter		133	.241		28.50	11.50		40	48.50
3070	24" diameter		107	.299		42	14.30		56.30	68
3080	30" diameter		80	.400		78	19.15		97.15	115
3090	36" diameter		80	.400		122	19.15		141.15	163
3100	42" diameter		60	.533		147	25.50		172.50	201
3200	48" diameter		60	.533		192	25.50		217.50	250
4520	Pressure pipe Class 150, SDR 18, AWWA C900, 4" diameter		380	.084		2.67	4.03		6.70	9.05
4530	6" diameter		316	.101		5.15	4.85		10	13
4540	8" diameter		264	.121		7.85	5.80		13.65	17.40
4550	10" diameter		220	.145		12.25	6.95		19.20	24
4560	12" diameter	↓	186	.172	↓	15.95	8.25		24.20	30
8000	Fittings with rubber gasket									
8003	Class 150, DR 18									
8006	90° Bend , 4" diameter	B-20	100	.240	Ea.	46.50	10.55		57.05	67.50
8020	6" diameter		90	.267		83	11.70		94.70	110
8040	8" diameter		80	.300		160	13.20		173.20	196
8060	10" diameter	↓	50	.480	↓	365	21		386	440

33 11 Water Utility Distribution Piping

33 11 13 – Public Water Utility Distribution Piping

33 11 13.25 Water Supply, Polyvinyl Chloride Pipe		Crew	Daily Output	Labor-Hours	Unit	Material	2017 Bare Costs Labor	Equipment	Total	Total Incl O&P
8080	12" diameter	B-20	30	.800	Ea.	470	35		505	570
8100	Tee, 4" diameter		90	.267		64	11.70		75.70	88.50
8120	6" diameter		80	.300		144	13.20		157.20	178
8140	8" diameter		70	.343		204	15.05		219.05	248
8160	10" diameter		40	.600		655	26.50		681.50	760
8180	12" diameter		20	1.200		850	52.50		902.50	1,025
8200	45° Bend, 4" diameter		100	.240		47	10.55		57.55	67.50
8220	6" diameter		90	.267		81	11.70		92.70	107
8240	8" diameter		50	.480		154	21		175	202
8260	10" diameter		50	.480		310	21		331	375
8280	12" diameter		30	.800		400	35		435	495
8300	Reducing tee 6" x 4"		100	.240		115	10.55		125.55	143
8320	8" x 6"		90	.267		165	11.70		176.70	200
8330	10" x 6"		90	.267		335	11.70		346.70	390
8340	10" x 8"		90	.267		350	11.70		361.70	405
8350	12" x 6"		90	.267		400	11.70		411.70	460
8360	12" x 8"		90	.267		415	11.70		426.70	480
8400	Tapped service tee (threaded type) 6" x 6" x 3/4"		100	.240		93	10.55		103.55	118
8430	6" x 6" x 1"		90	.267		93	11.70		104.70	120
8440	6" x 6" x 1-1/2"		90	.267		94.50	11.70		106.20	122
8450	6" x 6" x 2"		90	.267		93	11.70		104.70	120
8460	8" x 8" x 3/4"		90	.267		140	11.70		151.70	172
8470	8" x 8" x 1"		90	.267		140	11.70		151.70	172
8480	8" x 8" x 1-1/2"		90	.267		140	11.70		151.70	172
8490	8" x 8" x 2"		90	.267		140	11.70		151.70	172
8500	Repair coupling 4"		100	.240		27	10.55		37.55	45.50
8520	6" diameter		90	.267		40	11.70		51.70	62
8540	8" diameter		50	.480		97	21		118	140
8560	10" diameter		50	.480		225	21		246	280
8580	12" diameter		50	.480		330	21		351	395
8600	Plug end 4"		100	.240		23	10.55		33.55	41.50
8620	6" diameter		90	.267		40	11.70		51.70	62
8640	8" diameter		50	.480		69.50	21		90.50	109
8660	10" diameter		50	.480		107	21		128	151
8680	12" diameter		50	.480		131	21		152	178

33 12 Water Utility Distribution Equipment

33 12 19 – Water Utility Distribution Fire Hydrants

33 12 19.10 Fire Hydrants

		Crew	Daily Output	Labor-Hours	Unit	Material	2017 Bare Costs Labor	Equipment	Total	Total Incl O&P
0010	**FIRE HYDRANTS**									
0020	Mechanical joints unless otherwise noted									
1000	Fire hydrants, two way; excavation and backfill not incl.									
1100	4-1/2" valve size, depth 2'-0"	B-21	10	2.800	Ea.	1,925	128	13.80	2,066.80	2,325
1120	2'-6"		10	2.800		2,050	128	13.80	2,191.80	2,450
1140	3'-0"		10	2.800		2,200	128	13.80	2,341.80	2,600
1300	7'-0"		6	4.667		2,500	213	23	2,736	3,100
2400	Lower barrel extensions with stems, 1'-0"	B-20	14	1.714		320	75.50		395.50	465
2480	3'-0"	"	12	2		750	88		838	960

For customer support on your Building Construction Costs with RSMeans Data, call 800.448.8182.

677

33 16 Water Utility Storage Tanks

33 16 13 – Aboveground Water Utility Storage Tanks

33 16 13.13 Steel Water Storage Tanks

	33 16 13.13 Steel Water Storage Tanks	Crew	Daily Output	Labor-Hours	Unit	Material	2017 Bare Costs Labor	Equipment	Total	Total Incl O&P
0010	**STEEL WATER STORAGE TANKS**									
0910	Steel, ground level, ht./diam. less than 1, not incl. fdn., 100,000 gallons				Ea.				202,000	244,500
1000	250,000 gallons								295,500	324,000
1200	500,000 gallons								417,000	458,500
1250	750,000 gallons								538,000	591,500
1300	1,000,000 gallons								558,000	725,500
1500	2,000,000 gallons								1,043,000	1,148,000
1600	4,000,000 gallons								2,121,000	2,333,000
1800	6,000,000 gallons								3,095,000	3,405,000
1850	8,000,000 gallons								4,068,000	4,475,000
1910	10,000,000 gallons								5,050,000	5,554,500
2100	Steel standpipes, ht./diam. more than 1,100' to overflow, no fdn.									
2200	500,000 gallons				Ea.				546,500	600,500
2400	750,000 gallons								722,500	794,500
2500	1,000,000 gallons								1,060,500	1,167,000
2700	1,500,000 gallons								1,749,000	1,923,000
2800	2,000,000 gallons								2,327,000	2,559,000

33 16 13.16 Prestressed Conc. Water Storage Tanks

		Crew	Daily Output	Labor-Hours	Unit	Material	2017 Bare Costs Labor	Equipment	Total	Total Incl O&P
0010	**PRESTRESSED CONC. WATER STORAGE TANKS**									
0020	Not including fdn., pipe or pumps, 250,000 gallons				Ea.				299,000	329,500
0100	500,000 gallons								487,000	536,000
0300	1,000,000 gallons								707,000	807,500
0400	2,000,000 gallons								1,072,000	1,179,000
0600	4,000,000 gallons								1,706,000	1,877,000
0700	6,000,000 gallons								2,266,000	2,493,000
0750	8,000,000 gallons								2,924,000	3,216,000
0800	10,000,000 gallons								3,533,000	3,886,000

33 16 13.23 Plastic-Coated Fabric Pillow Water Tanks

		Crew	Daily Output	Labor-Hours	Unit	Material	2017 Bare Costs Labor	Equipment	Total	Total Incl O&P
0010	**PLASTIC-COATED FABRIC PILLOW WATER TANKS**									
7000	Water tanks, vinyl coated fabric pillow tanks, freestanding, 5,000 gallons	4 Clab	4	8	Ea.	3,700	315		4,015	4,525
7100	Supporting embankment not included, 25,000 gallons	6 Clab	2	24		13,300	940		14,240	16,100
7200	50,000 gallons	8 Clab	1.50	42.667		18,600	1,675		20,275	23,000
7300	100,000 gallons	9 Clab	.90	80		42,500	3,125		45,625	51,500
7400	150,000 gallons		.50	144		61,000	5,650		66,650	75,500
7500	200,000 gallons		.40	180		75,500	7,050		82,550	94,000
7600	250,000 gallons		.30	240		106,500	9,400		115,900	132,000

33 16 19 – Elevated Water Utility Storage Tanks

33 16 19.50 Elevated Water Storage Tanks

		Crew	Daily Output	Labor-Hours	Unit	Material	2017 Bare Costs Labor	Equipment	Total	Total Incl O&P
0010	**ELEVATED WATER STORAGE TANKS**									
0011	Not incl. pipe, pumps or foundation									
3000	Elevated water tanks, 100' to bottom capacity line, incl. painting									
3010	50,000 gallons				Ea.				185,000	204,000
3300	100,000 gallons								280,000	307,500
3400	250,000 gallons								751,500	826,500
3600	500,000 gallons								1,336,000	1,470,000
3700	750,000 gallons								1,622,000	1,783,500
3900	1,000,000 gallons								2,322,000	2,556,000

33 21 Water Supply Wells

33 21 13 – Public Water Supply Wells

33 21 13.10 Wells and Accessories

33 21 13.10 Wells and Accessories	Crew	Daily Output	Labor-Hours	Unit	Material	2017 Bare Costs Labor	Equipment	Total	Total Incl O&P
0010 **WELLS & ACCESSORIES**									
0011 Domestic									
0100 Drilled, 4" to 6" diameter	B-23	120	.333	L.F.		13.20	23.50	36.70	46
0200 8" diameter	"	95.20	.420	"		16.60	30	46.60	58.50
0400 Gravel pack well, 40' deep, incl. gravel & casing, complete									
0500 24" diameter casing x 18" diameter screen	B-23	.13	307	Total	38,400	12,200	21,900	72,500	85,000
0600 36" diameter casing x 18" diameter screen	↓	.12	333	"	39,600	13,200	23,700	76,500	90,000
0800 Observation wells, 1-1/4" riser pipe	↓	163	.245	V.L.F.	20.50	9.70	17.45	47.65	56.50
0900 For flush Buffalo roadway box, add	1 Skwk	16.60	.482	Ea.	50.50	25		75.50	93.50
1200 Test well, 2-1/2" diameter, up to 50' deep (15 to 50 GPM)	B-23	1.51	26.490	"	800	1,050	1,875	3,725	4,550
1300 Over 50' deep, add	"	121.80	.328	L.F.	21.50	13	23.50	58	69
1500 Pumps, installed in wells to 100' deep, 4" submersible									
1510 1/2 H.P.	Q-1	3.22	4.969	Ea.	730	276		1,006	1,225
1520 3/4 H.P.		2.66	6.015		855	335		1,190	1,450
1600 1 H.P.	↓	2.29	6.987		905	390		1,295	1,575
1700 1-1/2 H.P.	Q-22	1.60	10		1,800	555	360	2,715	3,200
1800 2 H.P.		1.33	12.030		1,650	670	430	2,750	3,275
1900 3 H.P.		1.14	14.035		2,075	780	500	3,355	4,000
2000 5 H.P.		1.14	14.035		2,800	780	500	4,080	4,825
3000 Pump, 6" submersible, 25' to 150' deep, 25 H.P., 249 to 297 GPM		.89	17.978		3,250	1,000	645	4,895	5,775
3100 25' to 500' deep, 30 H.P., 100 to 300 GPM	↓	.73	21.918	↓	4,125	1,225	785	6,135	7,250
8000 Steel well casing	B-23A	3020	.008	Lb.	1.29	.35	.90	2.54	2.94
9950 See Section 31 23 19.40 for wellpoints									
9960 See Section 31 23 19.30 for drainage wells									

33 21 13.20 Water Supply Wells, Pumps

33 21 13.20 Water Supply Wells, Pumps	Crew	Daily Output	Labor-Hours	Unit	Material	2017 Bare Costs Labor	Equipment	Total	Total Incl O&P
0010 **WATER SUPPLY WELLS, PUMPS**									
0011 With pressure control									
1000 Deep well, jet, 42 gal. galvanized tank									
1040 3/4 HP	1 Plum	.80	10	Ea.	1,100	620		1,720	2,150
3000 Shallow well, jet, 30 gal. galvanized tank									
3040 1/2 HP	1 Plum	2	4	Ea.	895	247		1,142	1,350

33 31 Sanitary Utility Sewerage Piping

33 31 13 – Public Sanitary Utility Sewerage Piping

33 31 13.15 Sewage Collection, Concrete Pipe

33 31 13.15 Sewage Collection, Concrete Pipe	Crew	Daily Output	Labor-Hours	Unit	Material	2017 Bare Costs Labor	Equipment	Total	Total Incl O&P
0010 **SEWAGE COLLECTION, CONCRETE PIPE**									
0020 See Section 33 41 13.60 for sewage/drainage collection, concrete pipe									

33 31 13.25 Sewage Collection, Polyvinyl Chloride Pipe

33 31 13.25 Sewage Collection, Polyvinyl Chloride Pipe	Crew	Daily Output	Labor-Hours	Unit	Material	2017 Bare Costs Labor	Equipment	Total	Total Incl O&P
0010 **SEWAGE COLLECTION, POLYVINYL CHLORIDE PIPE**									
0020 Not including excavation or backfill									
2000 20' lengths, SDR 35, B&S, 4" diameter	B-20	375	.064	L.F.	1.55	2.81		4.36	6
2040 6" diameter	↓	350	.069		3.28	3.01		6.29	8.25
2080 13' lengths, SDR 35, B&S, 8" diameter	↓	335	.072		5.90	3.15		9.05	11.35
2120 10" diameter	B-21	330	.085		10.65	3.87	.42	14.94	18.05
2160 12" diameter		320	.088		12.20	3.99	.43	16.62	19.95
2200 15" diameter	↓	240	.117		14.40	5.30	.58	20.28	24.50
4000 Piping, DWV PVC, no exc./bkfill., 10' L, Sch 40, 4" diameter	B-20	375	.064		3.18	2.81		5.99	7.80
4010 6" diameter		350	.069		6.75	3.01		9.76	12
4020 8" diameter	↓	335	.072	↓	10.55	3.15		13.70	16.50

For customer support on your Building Construction Costs with RSMeans Data, call 800.448.8182.

679

33 36 Utility Septic Tanks

33 36 13 – Utility Septic Tank and Effluent Wet Wells

33 36 13.13 Concrete Utility Septic Tank

		Crew	Daily Output	Labor-Hours	Unit	Material	2017 Bare Costs Labor	Equipment	Total	Total Incl O&P
0010	**CONCRETE UTILITY SEPTIC TANK**									
0011	Not including excavation or piping									
0015	Septic tanks, precast, 1,000 gallon	B-21	8	3.500	Ea.	1,025	160	17.25	1,202.25	1,425
0060	1,500 gallon		7	4		1,625	182	19.70	1,826.70	2,100
0100	2,000 gallon		5	5.600		2,325	255	27.50	2,607.50	2,975
0200	5,000 gallon	B-13	3.50	16		11,200	690	192	12,082	13,700
0300	15,000 gallon, 4 piece	B-13B	1.70	32.941		23,600	1,425	635	25,660	28,900
0400	25,000 gallon, 4 piece		1.10	50.909		42,400	2,200	980	45,580	51,000
0500	40,000 gallon, 4 piece		.80	70		52,500	3,025	1,350	56,875	63,500
0520	50,000 gallon, 5 piece	B-13C	.60	93.333		60,500	4,025	3,000	67,525	76,000
0640	75,000 gallon, cast in place	C-14C	.25	448		73,500	21,100	112	94,712	113,000
0660	100,000 gallon	"	.15	746		91,000	35,100	187	126,287	154,000
1150	Leaching field chambers, 13' x 3'-7" x 1'-4", standard	B-13	16	3.500		500	151	42	693	825
1200	Heavy duty, 8' x 4' x 1'-6"		14	4		284	172	48	504	625
1300	13' x 3'-9" x 1'-6"		12	4.667		1,250	201	56	1,507	1,750
1350	20' x 4' x 1'-6"		5	11.200		1,200	480	135	1,815	2,175
1400	Leaching pit, precast concrete, 3' diameter, 3' deep	B-21	8	3.500		725	160	17.25	902.25	1,050
1500	6' diameter, 3' section		4.70	5.957		815	272	29.50	1,116.50	1,350
2000	Velocity reducing pit, precast conc., 6' diameter, 3' deep		4.70	5.957		1,625	272	29.50	1,926.50	2,250

33 36 13.19 Polyethylene Utility Septic Tank

		Crew	Daily Output	Labor-Hours	Unit	Material	2017 Bare Costs Labor	Equipment	Total	Total Incl O&P
0010	**POLYETHYLENE UTILITY SEPTIC TANK**									
0015	High density polyethylene, 1,000 gallon	B-21	8	3.500	Ea.	1,325	160	17.25	1,502.25	1,725
0020	1,250 gallon		8	3.500		1,300	160	17.25	1,477.25	1,700
0025	1,500 gallon		7	4		1,525	182	19.70	1,726.70	1,975

33 36 19 – Utility Septic Tank Effluent Filter

33 36 19.13 Utility Septic Tank Effluent Tube Filter

		Crew	Daily Output	Labor-Hours	Unit	Material	2017 Bare Costs Labor	Equipment	Total	Total Incl O&P
0010	**UTILITY SEPTIC TANK EFFLUENT TUBE FILTER**									
3000	Effluent filter, 4" diameter	1 Skwk	8	1	Ea.	43	51.50		94.50	126
3020	6" diameter		7	1.143		248	59		307	365
3030	8" diameter		7	1.143		221	59		280	335
3040	8" diameter, very fine		7	1.143		460	59		519	600
3050	10" diameter, very fine		6	1.333		234	68.50		302.50	365
3060	10" diameter		6	1.333		280	68.50		348.50	415
3080	12" diameter		6	1.333		645	68.50		713.50	815
3090	15" diameter		5	1.600		1,075	82.50		1,157.50	1,300

33 36 33 – Utility Septic Tank Drainage Field

33 36 33.13 Utility Septic Tank Tile Drainage Field

		Crew	Daily Output	Labor-Hours	Unit	Material	2017 Bare Costs Labor	Equipment	Total	Total Incl O&P
0010	**UTILITY SEPTIC TANK TILE DRAINAGE FIELD**									
0015	Distribution box, concrete, 5 outlets	2 Clab	20	.800	Ea.	92.50	31.50		124	150
0020	7 outlets		16	1		92.50	39		131.50	162
0025	9 outlets		8	2		550	78.50		628.50	725
0115	Distribution boxes, HDPE, 5 outlets		20	.800		64.50	31.50		96	119
0117	6 outlets		15	1.067		65	42		107	136
0118	7 outlets		15	1.067		65	42		107	136
0120	8 outlets		10	1.600		68.50	62.50		131	172
0240	Distribution boxes, Outlet Flow Leveler	1 Clab	50	.160		2.30	6.25		8.55	12.15
0300	Precast concrete, galley, 4' x 4' x 4'	B-21	16	1.750		236	80	8.65	324.65	390
0350	HDPE infiltration chamber 12" H x 15" W	2 Clab	300	.053	L.F.	6.95	2.09		9.04	10.85
0351	12" H x 15" W End Cap	1 Clab	32	.250	Ea.	18.65	9.80		28.45	35.50
0355	chamber 12" H x 22" W	2 Clab	300	.053	L.F.	6.65	2.09		8.74	10.50
0356	12" H x 22" W End Cap	1 Clab	32	.250	Ea.	16.75	9.80		26.55	33.50

33 36 Utility Septic Tanks

33 36 33 – Utility Septic Tank Drainage Field

33 36 33.13 Utility Septic Tank Tile Drainage Field

		Crew	Daily Output	Labor-Hours	Unit	Material	2017 Bare Costs Labor	Equipment	Total	Total Incl O&P
0360	chamber 13" H x 34" W	2 Clab	300	.053	L.F.	14.15	2.09		16.24	18.80
0361	13" H x 34" W End Cap	1 Clab	32	.250	Ea.	50.50	9.80		60.30	70.50
0365	chamber 16" H x 34" W	2 Clab	300	.053	L.F.	12.70	2.09		14.79	17.15
0366	16" H x 34" W End Cap	1 Clab	32	.250	Ea.	8.15	9.80		17.95	24
0370	chamber 8" H x 16" W	2 Clab	300	.053	L.F.	9.85	2.09		11.94	14
0371	8" H x 16" W End Cap	1 Clab	32	.250	Ea.	11	9.80		20.80	27

33 36 50 – Drainage Field Systems

33 36 50.10 Drainage Field Excavation and Fill

		Crew	Daily Output	Labor-Hours	Unit	Material	2017 Bare Costs Labor	Equipment	Total	Total Incl O&P
0010	**DRAINAGE FIELD EXCAVATION AND FILL**									
2200	Septic tank & drainage field excavation with 3/4 C.Y. backhoe	B-12F	145	.110	C.Y.		5.25	4.65	9.90	13.05
2400	4' trench for disposal field, 3/4 C.Y. backhoe	"	335	.048	L.F.		2.27	2.01	4.28	5.65
2600	Gravel fill, run of bank	B-6	150	.160	C.Y.	17.20	6.90	2.44	26.54	32
2800	Crushed stone, 3/4"	"	150	.160	"	40	6.90	2.44	49.34	57

33 41 Storm Utility Drainage Piping

33 41 13 – Public Storm Utility Drainage Piping

33 41 13.40 Piping, Storm Drainage, Corrugated Metal

		Crew	Daily Output	Labor-Hours	Unit	Material	2017 Bare Costs Labor	Equipment	Total	Total Incl O&P
0010	**PIPING, STORM DRAINAGE, CORRUGATED METAL**									
0020	Not including excavation or backfill									
2000	Corrugated metal pipe, galvanized									
2020	Bituminous coated with paved invert, 20' lengths									
2040	8" diameter, 16 ga.	B-14	330	.145	L.F.	8.75	6.05	1.11	15.91	20
2060	10" diameter, 16 ga.		260	.185		9.10	7.65	1.41	18.16	23.50
2080	12" diameter, 16 ga.		210	.229		11.10	9.50	1.74	22.34	28.50
2100	15" diameter, 16 ga.		200	.240		15.30	9.95	1.83	27.08	34
2120	18" diameter, 16 ga.		190	.253		20	10.45	1.93	32.38	40
2140	24" diameter, 14 ga.		160	.300		24.50	12.45	2.29	39.24	48.50
2160	30" diameter, 14 ga.	B-13	120	.467		29.50	20	5.60	55.10	69
2180	36" diameter, 12 ga.		120	.467		34.50	20	5.60	60.10	74.50
2200	48" diameter, 12 ga.		100	.560		52	24	6.75	82.75	101
2220	60" diameter, 10 ga.	B-13B	75	.747		79.50	32	14.35	125.85	152
2240	72" diameter, 8 ga.	"	45	1.244		95	53.50	24	172.50	214
2500	Galvanized, uncoated, 20' lengths									
2520	8" diameter, 16 ga.	B-14	355	.135	L.F.	7.85	5.60	1.03	14.48	18.35
2540	10" diameter, 16 ga.		280	.171		9	7.10	1.31	17.41	22
2560	12" diameter, 16 ga.		220	.218		10	9.05	1.66	20.71	26.50
2580	15" diameter, 16 ga.		220	.218		12.50	9.05	1.66	23.21	29.50
2600	18" diameter, 16 ga.		205	.234		15.10	9.70	1.79	26.59	33.50
2620	24" diameter, 14 ga.		175	.274		22	11.35	2.09	35.44	43.50
2640	30" diameter, 14 ga.	B-13	130	.431		27	18.55	5.20	50.75	63.50
2660	36" diameter, 12 ga.		130	.431		31	18.55	5.20	54.75	68
2680	48" diameter, 12 ga.		110	.509		46.50	22	6.10	74.60	91.50
2690	60" diameter, 10 ga.	B-13B	78	.718		71.50	31	13.80	116.30	141
2780	End sections, 8" diameter	B-14	35	1.371	Ea.	73	57	10.45	140.45	179
2785	10" diameter		35	1.371		77	57	10.45	144.45	183
2790	12" diameter		35	1.371		114	57	10.45	181.45	224
2800	18" diameter		30	1.600		116	66.50	12.20	194.70	241
2810	24" diameter	B-13	25	2.240		216	96.50	27	339.50	415
2820	30" diameter		25	2.240		330	96.50	27	453.50	540
2825	36" diameter		20	2.800		465	121	33.50	619.50	735

681

33 41 13.40 Piping, Storm Drainage, Corrugated Metal	Crew	Daily Output	Labor-Hours	Unit	Material	2017 Bare Costs Labor	Equipment	Total	Total Incl O&P
2830	48" diameter	B-13	10	5.600	Ea.	935	241	67.50	1,243.50	1,475
2835	60" diameter	B-13B	5	11.200		1,625	480	215	2,320	2,775
2840	72" diameter	"	4	14	↓	1,950	605	269	2,824	3,375

33 41 13.60 Sewage/Drainage Collection, Concrete Pipe

		Crew	Daily Output	Labor-Hours	Unit	Material	2017 Bare Costs Labor	Equipment	Total	Total Incl O&P
0010	**SEWAGE/DRAINAGE COLLECTION, CONCRETE PIPE**									
0020	Not including excavation or backfill									
1000	Non-reinforced pipe, extra strength, B&S or T&G joints									
1010	6" diameter	B-14	265.04	.181	L.F.	7.55	7.50	1.38	16.43	21.50
1020	8" diameter		224	.214		8.30	8.90	1.64	18.84	24.50
1030	10" diameter		216	.222		9.20	9.20	1.70	20.10	26
1040	12" diameter		200	.240		9.95	9.95	1.83	21.73	28
1050	15" diameter		180	.267		14	11.05	2.03	27.08	34.50
1060	18" diameter		144	.333		16.95	13.80	2.54	33.29	42.50
1070	21" diameter		112	.429		18.05	17.75	3.27	39.07	50.50
1080	24" diameter	↓	100	.480	↓	19.15	19.90	3.66	42.71	55.50
2000	Reinforced culvert, class 3, no gaskets									
2010	12" diameter	B-14	150	.320	L.F.	12.65	13.25	2.44	28.34	37
2020	15" diameter		150	.320		16.35	13.25	2.44	32.04	41
2030	18" diameter		132	.364		21	15.10	2.77	38.87	49
2035	21" diameter		120	.400		22	16.60	3.05	41.65	53
2040	24" diameter	↓	100	.480		23	19.90	3.66	46.56	59.50
2045	27" diameter	B-13	92	.609		33.50	26	7.30	66.80	85
2050	30" diameter		88	.636		39.50	27.50	7.65	74.65	94
2060	36" diameter	↓	72	.778		62	33.50	9.35	104.85	129
2070	42" diameter	B-13B	72	.778		84	33.50	14.95	132.45	160
2080	48" diameter		64	.875		102	37.50	16.80	156.30	189
2090	60" diameter		48	1.167		154	50	22.50	226.50	270
2100	72" diameter		40	1.400		230	60.50	27	317.50	375
2120	84" diameter		32	1.750		315	75.50	33.50	424	500
2140	96" diameter	↓	24	2.333		380	100	45	525	625
2200	With gaskets, class 3, 12" diameter	B-21	168	.167		13.95	7.60	.82	22.37	28
2220	15" diameter		160	.175		18	8	.86	26.86	33
2230	18" diameter		152	.184		23	8.40	.91	32.31	39.50
2240	24" diameter	↓	136	.206		30	9.40	1.02	40.42	48.50
2260	30" diameter	B-13	88	.636		47.50	27.50	7.65	82.65	103
2270	36" diameter	"	72	.778		71.50	33.50	9.35	114.35	140
2290	48" diameter	B-13B	64	.875		115	37.50	16.80	169.30	203
2310	72" diameter	"	40	1.400	↓	249	60.50	27	336.50	395
2330	Flared ends, 12" diameter	B-21	31	.903	Ea.	235	41	4.45	280.45	325
2340	15" diameter		25	1.120		279	51	5.50	335.50	390
2400	18" diameter		20	1.400		325	64	6.90	395.90	460
2420	24" diameter	↓	14	2		385	91	9.85	485.85	575
2440	36" diameter	B-13	10	5.600	↓	820	241	67.50	1,128.50	1,350
3080	Radius pipe, add to pipe prices, 12" to 60" diameter				L.F.	50%				
3090	Over 60" diameter, add				"	20%				
3500	Reinforced elliptical, 8' lengths, C507 class 3									
3520	14" x 23" inside, round equivalent 18" diameter	B-21	82	.341	L.F.	43	15.55	1.68	60.23	73
3530	24" x 38" inside, round equivalent 30" diameter	B-13	58	.966		64.50	41.50	11.60	117.60	147
3540	29" x 45" inside, round equivalent 36" diameter		52	1.077		81.50	46.50	12.95	140.95	175
3550	38" x 60" inside, round equivalent 48" diameter		38	1.474		141	63.50	17.75	222.25	272
3560	48" x 76" inside, round equivalent 60" diameter		26	2.154		193	92.50	26	311.50	385
3570	58" x 91" inside, round equivalent 72" diameter	↓	22	2.545	↓	281	110	30.50	421.50	510

682

For customer support on your Building Construction Costs with RSMeans Data, call 800.448.8182.

33 41 Storm Utility Drainage Piping

33 41 13 – Public Storm Utility Drainage Piping

33 41 13.60 Sewage/Drainage Collection, Concrete Pipe	Crew	Daily Output	Labor-Hours	Unit	Material	2017 Bare Costs Labor	Equipment	Total	Total Incl O&P
3780	Concrete slotted pipe, class 4 mortar joint									
3800	12" diameter	B-21	168	.167	L.F.	28	7.60	.82	36.42	43.50
3840	18" diameter	"	152	.184	"	32	8.40	.91	41.31	49
3900	Concrete slotted pipe, Class 4 O-ring joint									
3940	12" diameter	B-21	168	.167	L.F.	28	7.60	.82	36.42	43
3960	18" diameter	"	152	.184	"	32	8.40	.91	41.31	49

33 42 Culverts

33 42 16 – Concrete Culverts

33 42 16.15 Oval Arch Culverts

		Crew	Daily Output	Labor-Hours	Unit	Material	2017 Bare Costs Labor	Equipment	Total	Total Incl O&P
0010	**OVAL ARCH CULVERTS**									
3000	Corrugated galvanized or aluminum, coated & paved									
3020	17" x 13", 16 ga., 15" equivalent	B-14	200	.240	L.F.	14.50	9.95	1.83	26.28	33
3040	21" x 15", 16 ga., 18" equivalent		150	.320		19.25	13.25	2.44	34.94	44
3060	28" x 20", 14 ga., 24" equivalent		125	.384		24.50	15.90	2.93	43.33	54.50
3080	35" x 24", 14 ga., 30" equivalent		100	.480		30	19.90	3.66	53.56	67
3100	42" x 29", 12 ga., 36" equivalent	B-13	100	.560		35.50	24	6.75	66.25	83.50
3120	49" x 33", 12 ga., 42" equivalent		90	.622		43	27	7.50	77.50	97
3140	57" x 38", 12 ga., 48" equivalent		75	.747		56	32	9	97	120
3160	Steel, plain oval arch culverts, plain									
3180	17" x 13", 16 ga., 15" equivalent	B-14	225	.213	L.F.	13.20	8.85	1.63	23.68	30
3200	21" x 15", 16 ga., 18" equivalent		175	.274		16.05	11.35	2.09	29.49	37.50
3220	28" x 20", 14 ga., 24" equivalent		150	.320		22.50	13.25	2.44	38.19	47.50
3240	35" x 24", 14 ga., 30" equivalent	B-13	108	.519		27	22.50	6.25	55.75	70.50
3260	42" x 29", 12 ga., 36" equivalent		108	.519		32	22.50	6.25	60.75	76.50
3280	49" x 33", 12 ga., 42" equivalent		92	.609		38	26	7.30	71.30	90
3300	57" x 38", 12 ga., 48" equivalent		75	.747		53.50	32	9	94.50	118
3320	End sections, 17" x 13"		22	2.545	Ea.	144	110	30.50	284.50	360
3340	42" x 29"		17	3.294	"	395	142	39.50	576.50	695
3360	Multi-plate arch, steel	B-20	1690	.014	Lb.	1.29	.62		1.91	2.38

33 44 Storm Utility Water Drains

33 44 13 – Utility Area Drains

33 44 13.13 Catch Basins

		Crew	Daily Output	Labor-Hours	Unit	Material	2017 Bare Costs Labor	Equipment	Total	Total Incl O&P
0010	**CATCH BASINS**									
0011	Not including footing & excavation									
1600	Frames & grates, C.I., 24" square, 500 lb.	B-6	7.80	3.077	Ea.	350	133	47	530	640
1700	26" D shape, 600 lb.		7	3.429		605	148	52.50	805.50	950
1800	Light traffic, 18" diameter, 100 lb.		10	2.400		200	103	36.50	339.50	420
1900	24" diameter, 300 lb.		8.70	2.759		195	119	42	356	440
2000	36" diameter, 900 lb.		5.80	4.138		630	178	63	871	1,025
2100	Heavy traffic, 24" diameter, 400 lb.		7.80	3.077		260	133	47	440	540
2200	36" diameter, 1150 lb.		3	8		860	345	122	1,327	1,600
2300	Mass. State standard, 26" diameter, 475 lb.		7	3.429		635	148	52.50	835.50	980
2400	30" diameter, 620 lb.		7	3.429		340	148	52.50	540.50	660
2500	Watertight, 24" diameter, 350 lb.		7.80	3.077		360	133	47	540	650
2600	26" diameter, 500 lb.		7	3.429		415	148	52.50	615.50	740
2700	32" diameter, 575 lb.		6	4		900	172	61	1,133	1,325
2800	3 piece cover & frame, 10" deep,									

For customer support on your Building Construction Costs with RSMeans Data, call 800.448.8182.

683

33 44 Storm Utility Water Drains

33 44 13 – Utility Area Drains

33 44 13.13 Catch Basins	Crew	Daily Output	Labor-Hours	Unit	Material	2017 Bare Costs Labor	Equipment	Total	Total Incl O&P
2900	1200 lb., for heavy equipment	B-6	3	8	Ea.	1,050	345	122	1,517	1,800
3000	Raised for paving 1-1/4" to 2" high									
3100	4 piece expansion ring									
3200	20" to 26" diameter	1 Clab	3	2.667	Ea.	174	104		278	350
3300	30" to 36" diameter	"	3	2.667	"	241	104		345	425
3320	Frames and covers, existing, raised for paving, 2", including									
3340	row of brick, concrete collar, up to 12" wide frame	B-6	18	1.333	Ea.	50.50	57.50	20.50	128.50	166
3360	20" to 26" wide frame		11	2.182		75	94	33.50	202.50	262
3380	30" to 36" wide frame	↓	9	2.667		93	115	40.50	248.50	320
3400	Inverts, single channel brick	D-1	3	5.333		104	231		335	470
3500	Concrete		5	3.200		121	139		260	345
3600	Triple channel, brick		2	8		163	345		508	710
3700	Concrete	↓	3	5.333	↓	157	231		388	525

33 46 Subdrainage

33 46 16 – Subdrainage Piping

33 46 16.25 Piping, Subdrainage, Corrugated Metal

		Crew	Daily Output	Labor-Hours	Unit	Material	2017 Bare Costs Labor	Equipment	Total	Total Incl O&P
0010	**PIPING, SUBDRAINAGE, CORRUGATED METAL**									
0021	Not including excavation and backfill									
2010	Aluminum, perforated									
2020	6" diameter, 18 ga.	B-20	380	.063	L.F.	6.50	2.77		9.27	11.40
2200	8" diameter, 16 ga.	"	370	.065		8.55	2.85		11.40	13.75
2220	10" diameter, 16 ga.	B-21	360	.078		10.70	3.55	.38	14.63	17.60
2240	12" diameter, 16 ga.		285	.098		11.95	4.48	.48	16.91	20.50
2260	18" diameter, 16 ga.	↓	205	.137	↓	17.95	6.25	.67	24.87	30
3000	Uncoated galvanized, perforated									
3020	6" diameter, 18 ga.	B-20	380	.063	L.F.	6.05	2.77		8.82	10.90
3200	8" diameter, 16 ga.	"	370	.065		8.30	2.85		11.15	13.50
3220	10" diameter, 16 ga.	B-21	360	.078		8.80	3.55	.38	12.73	15.55
3240	12" diameter, 16 ga.		285	.098		9.80	4.48	.48	14.76	18.15
3260	18" diameter, 16 ga.	↓	205	.137	↓	15	6.25	.67	21.92	27
4000	Steel, perforated, asphalt coated									
4020	6" diameter, 18 ga.	B-20	380	.063	L.F.	6.65	2.77		9.42	11.55
4030	8" diameter, 18 ga.	"	370	.065		8.75	2.85		11.60	13.95
4040	10" diameter, 16 ga.	B-21	360	.078		10.40	3.55	.38	14.33	17.30
4050	12" diameter, 16 ga.		285	.098		11.50	4.48	.48	16.46	20
4060	18" diameter, 16 ga.	↓	205	.137	↓	17.65	6.25	.67	24.57	29.50

33 46 16.40 Piping, Subdrainage, Polyvinyl Chloride

0010	**PIPING, SUBDRAINAGE, POLYVINYL CHLORIDE**									
0020	Perforated, price as solid pipe, Section 33 31 13.25									

33 49 Storm Drainage Structures

33 49 13 – Storm Drainage Manholes, Frames, and Covers

33 49 13.10 Storm Drainage Manholes, Frames and Covers

		Crew	Daily Output	Labor-Hours	Unit	Material	2017 Bare Costs Labor	2017 Bare Costs Equipment	Total	Total Incl O&P
0010	**STORM DRAINAGE MANHOLES, FRAMES & COVERS**									
0020	Excludes footing, excavation, backfill (See line items for frame & cover)									
0050	Brick, 4' inside diameter, 4' deep	D-1	1	16	Ea.	545	690		1,235	1,675
0100	6' deep		.70	22.857		775	990		1,765	2,375
0150	8' deep		.50	32		995	1,375		2,370	3,225
0200	For depths over 8', add		4	4	V.L.F.	88	173		261	365
0400	Concrete blocks (radial), 4' I.D., 4' deep		1.50	10.667	Ea.	415	460		875	1,175
0500	6' deep		1	16		555	690		1,245	1,675
0600	8' deep		.70	22.857		690	990		1,680	2,275
0700	For depths over 8', add		5.50	2.909	V.L.F.	71.50	126		197.50	273
0800	Concrete, cast in place, 4' x 4', 8" thick, 4' deep	C-14H	2	24	Ea.	560	1,150	14.15	1,724.15	2,400
0900	6' deep		1.50	32		810	1,550	18.90	2,378.90	3,250
1000	8' deep		1	48		1,175	2,325	28.50	3,528.50	4,850
1100	For depths over 8', add		8	6	V.L.F.	129	290	3.54	422.54	590
1110	Precast, 4' I.D., 4' deep	B-22	4.10	7.317	Ea.	840	340	50.50	1,230.50	1,500
1120	6' deep		3	10		1,500	465	69	2,034	2,425
1130	8' deep		2	15		1,950	695	104	2,749	3,325
1140	For depths over 8', add		16	1.875	V.L.F.	175	87	12.95	274.95	340
1150	5' I.D., 4' deep	B-6	3	8	Ea.	1,775	345	122	2,242	2,600
1160	6' deep		2	12		2,150	515	183	2,848	3,375
1170	8' deep		1.50	16		3,000	690	244	3,934	4,625
1180	For depths over 8', add		12	2	V.L.F.	380	86	30.50	496.50	580
1190	6' I.D., 4' deep		2	12	Ea.	2,475	515	183	3,173	3,725
1200	6' deep		1.50	16		3,200	690	244	4,134	4,850
1210	8' deep		1	24		3,775	1,025	365	5,165	6,150
1220	For depths over 8', add		8	3	V.L.F.	500	129	46	675	800
1250	Slab tops, precast, 8" thick									
1300	4' diameter manhole	B-6	8	3	Ea.	255	129	46	430	530
1400	5' diameter manhole		7.50	3.200		420	138	49	607	725
1500	6' diameter manhole		7	3.429		645	148	52.50	845.50	995
3800	Steps, heavyweight cast iron, 7" x 9"	1 Bric	40	.200		18.05	9.55		27.60	34.50
3900	8" x 9"		40	.200		21.50	9.55		31.05	38.50
3928	12" x 10-1/2"		40	.200		26	9.55		35.55	43.50
4000	Standard sizes, galvanized steel		40	.200		24.50	9.55		34.05	41.50
4100	Aluminum		40	.200		28	9.55		37.55	45.50
4150	Polyethylene		40	.200		31.50	9.55		41.05	49

33 51 Natural-Gas Distribution

33 51 13 – Natural-Gas Piping

33 51 13.10 Piping, Gas Service and Distribution, P.E.

		Crew	Daily Output	Labor-Hours	Unit	Material	2017 Bare Costs Labor	2017 Bare Costs Equipment	Total	Total Incl O&P
0010	**PIPING, GAS SERVICE AND DISTRIBUTION, POLYETHYLENE**									
0020	Not including excavation or backfill									
1000	60 psi coils, compression coupling @ 100', 1/2" diameter, SDR 11	B-20A	608	.053	L.F.	.47	2.52		2.99	4.35
1010	1" diameter, SDR 11		544	.059		1.07	2.82		3.89	5.45
1040	1-1/4" diameter, SDR 11		544	.059		1.53	2.82		4.35	5.95
1100	2" diameter, SDR 11		488	.066		2.13	3.14		5.27	7.10
1160	3" diameter, SDR 11		408	.078		4.43	3.76		8.19	10.55
1500	60 psi 40' joints with coupling, 3" diameter, SDR 11	B-21A	408	.098		8.85	4.85	1.07	14.77	18.25
1540	4" diameter, SDR 11		352	.114		12.75	5.60	1.24	19.59	24
1600	6" diameter, SDR 11		328	.122		31.50	6.05	1.33	38.88	45.50
1640	8" diameter, SDR 11		272	.147		43.50	7.25	1.60	52.35	61

685

33 52 16.13 Gasoline Piping		Crew	Daily Output	Labor-Hours	Unit	Material	2017 Bare Costs Labor	Equipment	Total	Total Incl O&P
0010	**GASOLINE PIPING**									
0020	Primary containment pipe, fiberglass-reinforced									
0030	Plastic pipe 15' & 30' lengths									
0040	2" diameter	Q-6	425	.056	L.F.	6.45	3.28		9.73	12.05
0050	3" diameter		400	.060		10	3.49		13.49	16.25
0060	4" diameter		375	.064		12.95	3.72		16.67	19.85
0100	Fittings									
0110	Elbows, 90° & 45°, bell-ends, 2"	Q-6	24	1	Ea.	41.50	58		99.50	133
0120	3" diameter		22	1.091		51.50	63.50		115	153
0130	4" diameter		20	1.200		66	69.50		135.50	178
0200	Tees, bell ends, 2"		21	1.143		55.50	66.50		122	161
0210	3" diameter		18	1.333		60.50	77.50		138	184
0220	4" diameter		15	1.600		79.50	93		172.50	228
0230	Flanges bell ends, 2"		24	1		31.50	58		89.50	123
0240	3" diameter		22	1.091		37	63.50		100.50	136
0250	4" diameter		20	1.200		42.50	69.50		112	152
0260	Sleeve couplings, 2"		21	1.143		11.80	66.50		78.30	113
0270	3" diameter		18	1.333		16.80	77.50		94.30	135
0280	4" diameter		15	1.600		21.50	93		114.50	164
0290	Threaded adapters, 2"		21	1.143		16.95	66.50		83.45	119
0300	3" diameter		18	1.333		32	77.50		109.50	153
0310	4" diameter		15	1.600		35.50	93		128.50	180
0320	Reducers, 2"		27	.889		25.50	51.50		77	106
0330	3" diameter		22	1.091		26	63.50		89.50	124
0340	4" diameter		20	1.200		34.50	69.50		104	143
1010	Gas station product line for secondary containment (double wall)									
1100	Fiberglass reinforced plastic pipe 25' lengths									
1120	Pipe, plain end, 3" diameter	Q-6	375	.064	L.F.	25.50	3.72		29.22	33.50
1130	4" diameter		350	.069		30.50	3.98		34.48	39.50
1140	5" diameter		325	.074		34	4.29		38.29	44
1150	6" diameter		300	.080		37	4.65		41.65	47.50
1200	Fittings									
1230	Elbows, 90° & 45°, 3" diameter	Q-6	18	1.333	Ea.	122	77.50		199.50	252
1240	4" diameter		16	1.500		158	87		245	305
1250	5" diameter		14	1.714		173	99.50		272.50	340
1260	6" diameter		12	2		192	116		308	385
1270	Tees, 3" diameter		15	1.600		160	93		253	315
1280	4" diameter		12	2		191	116		307	385
1290	5" diameter		9	2.667		299	155		454	565
1300	6" diameter		6	4		355	232		587	740
1310	Couplings, 3" diameter		18	1.333		51.50	77.50		129	174
1320	4" diameter		16	1.500		113	87		200	256
1330	5" diameter		14	1.714		202	99.50		301.50	375
1340	6" diameter		12	2		299	116		415	505
1350	Cross-over nipples, 3" diameter		18	1.333		9.90	77.50		87.40	128
1360	4" diameter		16	1.500		12.05	87		99.05	145
1370	5" diameter		14	1.714		15	99.50		114.50	167
1380	6" diameter		12	2		18.05	116		134.05	195
1400	Telescoping, reducers, concentric 4" x 3"		18	1.333		43.50	77.50		121	165
1410	5" x 4"		17	1.412		90	82		172	223
1420	6" x 5"		16	1.500		221	87		308	375

33 71 Electrical Utility Transmission and Distribution

33 71 16 – Electrical Utility Poles

33 71 16.33 Wood Electrical Utility Poles	Crew	Daily Output	Labor-Hours	Unit	Material	2017 Bare Costs Labor	Equipment	Total	Total Incl O&P
0010 **WOOD ELECTRICAL UTILITY POLES**									
0011 Excludes excavation, backfill and cast-in-place concrete									
6200 Electric & tel sitework, 20' high, treated wd., see Section 26 56 13.10	R-3	3.10	6.452	Ea.	213	365	44.50	622.50	835
6400 25' high		2.90	6.897		282	390	47.50	719.50	950
6600 30' high		2.60	7.692		375	435	53	863	1,125
6800 35' high		2.40	8.333		500	470	57.50	1,027.50	1,325
7000 40' high		2.30	8.696		670	490	60	1,220	1,550
7200 45' high		1.70	11.765		900	665	81	1,646	2,075
7400 Cross arms with hardware & insulators									
7600 4' long	1 Elec	2.50	3.200	Ea.	149	181		330	435
7800 5' long		2.40	3.333		165	189		354	465
8000 6' long		2.20	3.636		171	206		377	500

33 71 19 – Electrical Underground Ducts and Manholes

33 71 19.17 Electric and Telephone Underground

	Crew	Daily Output	Labor-Hours	Unit	Material	2017 Bare Costs Labor	Equipment	Total	Total Incl O&P
0010 **ELECTRIC AND TELEPHONE UNDERGROUND**									
0011 Not including excavation									
0200 backfill and cast in place concrete									
0400 Hand holes, precast concrete, with concrete cover									
0600 2' x 2' x 3' deep	R-3	2.40	8.333	Ea.	430	470	57.50	957.50	1,250
0800 3' x 3' x 3' deep		1.90	10.526		560	595	72.50	1,227.50	1,600
1000 4' x 4' x 4' deep		1.40	14.286		1,425	810	98.50	2,333.50	2,875
1200 Manholes, precast with iron racks & pulling irons, C.I. frame									
1400 and cover, 4' x 6' x 7' deep	B-13	2	28	Ea.	2,925	1,200	335	4,460	5,425
1600 6' x 8' x 7' deep		1.90	29.474		3,275	1,275	355	4,905	5,925
1800 6' x 10' x 7' deep		1.80	31.111		3,675	1,350	375	5,400	6,500
4200 Underground duct, banks ready for concrete fill, min. of 7.5"									
4400 between conduits, center to center									
4580 PVC, type EB, 1 @ 2" diameter	2 Elec	480	.033	L.F.	.61	1.89		2.50	3.50
4600 2 @ 2" diameter		240	.067		1.22	3.77		4.99	7
4800 4 @ 2" diameter		120	.133		2.44	7.55		9.99	14
5000 2 @ 3" diameter		200	.080		1.85	4.53		6.38	8.85
5200 4 @ 3" diameter		100	.160		3.70	9.05		12.75	17.60
5400 2 @ 4" diameter		160	.100		2.66	5.65		8.31	11.45
5600 4 @ 4" diameter		80	.200		5.30	11.30		16.60	23
5800 6 @ 4" diameter		54	.296		8	16.75		24.75	34
6200 Rigid galvanized steel, 2 @ 2" diameter		180	.089		17.45	5.05		22.50	27
6400 4 @ 2" diameter		90	.178		35	10.05		45.05	53.50
6800 2 @ 3" diameter		100	.160		28	9.05		37.05	44.50
7000 4 @ 3" diameter		50	.320		56	18.10		74.10	88.50
7200 2 @ 4" diameter		70	.229		41	12.95		53.95	65
7400 4 @ 4" diameter		34	.471		82.50	26.50		109	131
7600 6 @ 4" diameter		22	.727		124	41		165	198

33 81 Communications Structures

33 81 13 – Communications Transmission Towers

33 81 13.10 Radio Towers	Crew	Daily Output	Labor-Hours	Unit	Material	2017 Bare Costs Labor	Equipment	Total	Total Incl O&P
0010 **RADIO TOWERS**									
0020 Guyed, 50' H, 40 lb. sect., 70 MPH basic wind speed	2 Sswk	1	16	Ea.	2,775	870		3,645	4,500
0100 Wind load 90 MPH basic wind speed	"	1	16		3,675	870		4,545	5,475
0300 190' high, 40 lb. section, wind load 70 MPH basic wind speed	K-2	.33	72.727		9,525	3,750	820	14,095	17,500
0400 200' high, 70 lb. section, wind load 90 MPH basic wind speed		.33	72.727		15,800	3,750	820	20,370	24,300
0600 300' high, 70 lb. section, wind load 70 MPH basic wind speed		.20	120		25,300	6,200	1,350	32,850	39,300
0700 270' high, 90 lb. section, wind load 90 MPH basic wind speed		.20	120		27,400	6,200	1,350	34,950	41,600
0800 400' high, 100 lb. section, wind load 70 MPH basic wind speed		.14	171		38,200	8,850	1,925	48,975	58,500
0900 Self-supporting, 60' high, wind load 70 MPH basic wind speed		.80	30		4,500	1,550	340	6,390	7,825
0910 60' high, wind load 90 MPH basic wind speed		.45	53.333		5,250	2,750	600	8,600	10,900
1000 120' high, wind load 70 MPH basic wind speed		.40	60		9,925	3,100	675	13,700	16,700
1200 190' high, wind load 90 MPH basic wind speed		.20	120		28,300	6,200	1,350	35,850	42,600
2000 For states west of Rocky Mountains, add for shipping					10%				

Estimating Tips
34 11 00 Rail Tracks
This subdivision includes items that may involve either repair of existing or construction of new railroad tracks. Additional preparation work, such as the roadbed earthwork, would be found in Division 31. Additional new construction siding and turnouts are found in Subdivision 34 72. Maintenance of railroads is found under 34 01 23 Operation and Maintenance of Railways.

34 40 00 Traffic Signals
This subdivision includes traffic signal systems. Other traffic control devices such as traffic signs are found in Subdivision 10 14 53 Traffic Signage.

34 70 00 Vehicle Barriers
This subdivision includes security vehicle barriers, guide and guard rails, crash barriers, and delineators. The actual maintenance and construction of concrete and asphalt pavement are found in Division 32.

Reference Numbers
Reference numbers are shown at the beginning of some major classifications. These numbers refer to related items in the Reference Section. The reference information may be an estimating procedure, an alternate pricing method, or technical information.

Note: Not all subdivisions listed here necessarily appear. ■

Did you know?
Our online estimating solution gives you the same access to RSMeans' data with 24/7 access:
- Quickly locate costs in the searchable database.
- Build cost lists, estimates, and reports in minutes.
- Adjust costs to any location in the U.S. and Canada with the click of a button.

Start your free trial today at
www.RSMeansOnline.com

34 01 23.51 Maintenance of Railroads	Crew	Daily Output	Labor-Hours	Unit	Material	2017 Bare Costs Labor	Equipment	Total	Total Incl O&P
0010 **MAINTENANCE OF RAILROADS**									
0400 Resurface and realign existing track	B-14	200	.240	L.F.		9.95	1.83	11.78	17.20
0600 For crushed stone ballast, add	"	500	.096	"	12.65	3.98	.73	17.36	21

34 11 Rail Tracks

34 11 13 – Track Rails

34 11 13.23 Heavy Rail Track

	Crew	Daily Output	Labor-Hours	Unit	Material	2017 Bare Costs Labor	Equipment	Total	Total Incl O&P
0010 **HEAVY RAIL TRACK** R347216-10									
1000 Rail, 100 lb. prime grade				L.F.	31.50			31.50	34.50
1500 Relay rail				"	15.70			15.70	17.30

34 11 33 – Track Cross Ties

34 11 33.13 Concrete Track Cross Ties

	Crew	Daily Output	Labor-Hours	Unit	Material	2017 Bare Costs Labor	Equipment	Total	Total Incl O&P
0010 **CONCRETE TRACK CROSS TIES**									
1400 Ties, concrete, 8'-6" long, 30" O.C.	B-14	80	.600	Ea.	126	25	4.58	155.58	181

34 11 33.16 Timber Track Cross Ties

	Crew	Daily Output	Labor-Hours	Unit	Material	2017 Bare Costs Labor	Equipment	Total	Total Incl O&P
0010 **TIMBER TRACK CROSS TIES**									
1600 Wood, pressure treated, 6" x 8" x 8'-6", C.L. lots	B-14	90	.533	Ea.	54.50	22	4.07	80.57	98.50
1700 L.C.L. lots		90	.533		57.50	22	4.07	83.57	101
1900 Heavy duty, 7" x 9" x 8'-6", C.L. lots		70	.686		60	28.50	5.25	93.75	115
2000 L.C.L. lots		70	.686		60	28.50	5.25	93.75	115

34 11 33.17 Timber Switch Ties

	Crew	Daily Output	Labor-Hours	Unit	Material	2017 Bare Costs Labor	Equipment	Total	Total Incl O&P
0010 **TIMBER SWITCH TIES**									
1200 Switch timber, for a #8 switch, pressure treated	B-14	3.70	12.973	M.B.F.	3,275	540	99	3,914	4,525
1300 Complete set of timbers, 3.7 MBF for #8 switch	"	1	48	Total	12,600	2,000	365	14,965	17,400

34 11 93 – Track Appurtenances and Accessories

34 11 93.50 Track Accessories

	Crew	Daily Output	Labor-Hours	Unit	Material	2017 Bare Costs Labor	Equipment	Total	Total Incl O&P
0010 **TRACK ACCESSORIES**									
0020 Car bumpers, test	B-14	2	24	Ea.	3,425	995	183	4,603	5,500
0100 Heavy duty		2	24		6,500	995	183	7,678	8,875
0200 Derails hand throw (sliding)		10	4.800		1,425	199	36.50	1,660.50	1,900
0300 Hand throw with standard timbers, open stand & target		8	6		1,525	249	46	1,820	2,100
2400 Wheel stops, fixed		18	2.667	Pr.	1,200	111	20.50	1,331.50	1,500
2450 Hinged		14	3.429	"	1,600	142	26	1,768	2,025

34 11 93.60 Track Material

	Crew	Daily Output	Labor-Hours	Unit	Material	2017 Bare Costs Labor	Equipment	Total	Total Incl O&P
0010 **TRACK MATERIAL**									
0020 Track bolts				Ea.	3.99			3.99	4.39
0100 Joint bars				Pr.	83.50			83.50	92
0200 Spikes				Ea.	2.03			2.03	2.23
0300 Tie plates				"	13.95			13.95	15.35

34 41 Roadway Signaling and Control Equipment

34 41 13 – Traffic Signals

34 41 13.10 Traffic Signals Systems

34 41 13.10 Traffic Signals Systems	Crew	Daily Output	Labor-Hours	Unit	Material	2017 Bare Costs Labor	2017 Bare Costs Equipment	Total	Total Incl O&P
0010 **TRAFFIC SIGNALS SYSTEMS**									
0020 Component costs									
0600 Crew employs crane/directional driller as required									
1000 Vertical mast with foundation									
1010 Mast sized for single arm to 40'; no lighting or power function	R-11	.50	112	Signal	10,300	6,050	1,525	17,875	22,100
1100 Horizontal arm									
1110 Per linear foot of arm	R-11	50	1.120	Signal	206	60.50	15.15	281.65	335
1200 Traffic signal									
1210 Includes signal, bracket, sensor, and wiring	R-11	2.50	22.400	Signal	1,025	1,200	305	2,530	3,275
1300 Pedestrian signals and callers									
1310 Includes four signals with brackets and two call buttons	R-11	2.50	22.400	Signal	3,100	1,200	305	4,605	5,550
1400 Controller, design, and underground conduit									
1410 Includes miscellaneous signage and adjacent surface work	R-11	.25	224	Signal	20,600	12,100	3,025	35,725	44,300

34 71 Roadway Construction

34 71 13 – Vehicle Barriers

34 71 13.17 Security Vehicle Barriers

34 71 13.17 Security Vehicle Barriers	Crew	Daily Output	Labor-Hours	Unit	Material	2017 Bare Costs Labor	2017 Bare Costs Equipment	Total	Total Incl O&P
0010 **SECURITY VEHICLE BARRIERS**									
0020 Security planters excludes filling material									
0100 Concrete security planter, exposed aggregate 36" diam. x 30" high	B-11M	8	2	Ea.	615	92.50	49.50	757	870
0200 48" diam. x 36" high		8	2		980	92.50	49.50	1,122	1,275
0300 53" diam. x 18" high		8	2		865	92.50	49.50	1,007	1,150
0400 72" diam. x 18" high with seats		8	2		2,250	92.50	49.50	2,392	2,675
0450 84" diam. x 36" high		8	2		2,025	92.50	49.50	2,167	2,450
0500 36" x 36" x 24" high square		8	2		515	92.50	49.50	657	760
0600 36" x 36" x 30" high square		8	2		715	92.50	49.50	857	980
0700 48" L x 24" W x 30" H rectangle		8	2		585	92.50	49.50	727	840
0800 72" L x 24" W x 30" H rectangle		8	2		615	92.50	49.50	757	870
0900 96" L x 24" W x 30" H rectangle		8	2		820	92.50	49.50	962	1,100
0950 Decorative geometric concrete barrier, 96" L x 24" W x 36" H		8	2		850	92.50	49.50	992	1,125
1000 Concrete security planter, filling with washed sand or gravel <1 C.Y.		8	2		36	92.50	49.50	178	235
1050 2 C.Y. or less		6	2.667		72	124	65.50	261.50	340
1200 Jersey concrete barrier, 10' L x 2' by 0.5' W x 30" H	B-21B	16	2.500		400	107	36	543	645
1300 10 or more same site		24	1.667		400	71.50	24	495.50	575
1400 10' L x 2' by 0.5' W x 32" H		16	2.500		630	107	36	773	900
1500 10 or more same site		24	1.667		630	71.50	24	725.50	830
1600 20' L x 2' by 0.5' W x 30" H		12	3.333		690	143	47.50	880.50	1,025
1700 10 or more same site		18	2.222		690	95	32	817	940
1800 20' L x 2' by 0.5' W x 32" H		12	3.333		650	143	47.50	840.50	985
1900 10 or more same site		18	2.222		650	95	32	777	895
2000 GFRC decorative security barrier per 10 feet section including concrete		4	10		3,475	430	143	4,048	4,625
2100 Per 12 feet section including concrete		4	10		4,150	430	143	4,723	5,400
2210 GFRC decorative security barrier will stop 30 MPH, 4000 lb. vehicle									
2300 High security barrier base prep per 12 feet section on bare ground	B-11C	4	4	Ea.	23.50	185	91.50	300	410
2310 GFRC barrier base prep does not include haul away of excavated matl.									
2400 GFRC decorative high security barrier per 12 feet section w/concrete	B-21B	3	13.333	Ea.	5,425	570	191	6,186	7,050
2410 GFRC decorative high security barrier will stop 50 MPH, 15000 lb. vehicle									
2500 GFRC decorative impaler security barrier per 10 feet section w/prep	B-6	4	6	Ea.	2,125	259	91.50	2,475.50	2,825
2600 Per 12 feet section w/prep	"	4	6	"	2,550	259	91.50	2,900.50	3,300
2610 Impaler barrier should stop 50 MPH, 15000 lb. vehicle w/some penetr.									
2700 Pipe bollards, steel, concrete filled/painted, 8' L x 4' D hole, 8" diam.	B-6	10	2.400	Ea.	840	103	36.50	979.50	1,125

34 71 13 – Vehicle Barriers

34 71 13.17 Security Vehicle Barriers	Crew	Daily Output	Labor-Hours	Unit	Material	2017 Bare Costs Labor	Equipment	Total	Total Incl O&P
2710	Schedule 80 concrete bollards will stop 4000 lb. vehicle @ 30 MPH									
2800	GFRC decorative jersey barrier cover per 10 feet section excludes soil	B-6	8	3	Ea.	1,000	129	46	1,175	1,350
2900	Per 12 feet section excludes soil		8	3		1,200	129	46	1,375	1,575
3000	GFRC decorative 8" diameter bollard cover		12	2		375	86	30.50	491.50	580
3100	GFRC decorative barrier face 10 foot section excludes earth backing		10	2.400		950	103	36.50	1,089.50	1,250
3200	Drop arm crash barrier, 15000 lb. vehicle @ 50 MPH									
3205	Includes all material, labor for complete installation									
3210	12' width				Ea.				64,500	71,000
3310	24' width								86,000	95,000
3410	Wedge crash barrier, 10' width								98,000	108,000
3510	12.5' width								108,000	119,000
3520	15' width								118,500	130,000
3610	Sliding crash barrier, 12' width								178,500	196,000
3710	Sliding roller crash barrier, 20' width								198,000	217,500
3810	Sliding cantilever crash barrier, 20' width								198,000	217,500
3890	Note: Raised bollard crash barriers should be used w/tire shredders									
3910	Raised bollard crash barrier, 10' width				Ea.				38,800	42,900
4010	12' width								44,900	49,000
4110	Raised bollard crash barrier, 10' width, solar powered								46,900	52,000
4210	12' width								53,000	58,000
4310	In ground tire shredder, 16' width								43,900	47,900

34 71 13.26 Vehicle Guide Rails

		Crew	Daily Output	Labor-Hours	Unit	Material	2017 Bare Costs Labor	Equipment	Total	Total Incl O&P
0010	**VEHICLE GUIDE RAILS**									
0012	Corrugated stl., galv. stl. posts, 6'-3" O.C.	B-80	850	.038	L.F.	24	1.65	.79	26.44	30
0200	End sections, galvanized, flared		50	.640	Ea.	100	28	13.45	141.45	167
0300	Wrap around end		50	.640	"	143	28	13.45	184.45	214
0400	Timber guide rail, 4" x 8" with 6" x 8" wood posts, treated		960	.033	L.F.	12.10	1.46	.70	14.26	16.30
0600	Cable guide rail, 3 at 3/4" cables, steel posts, single face		900	.036		11.25	1.56	.75	13.56	15.55
0700	Wood posts		950	.034		13.15	1.48	.71	15.34	17.50
0900	Guide rail, steel box beam, 6" x 6"		120	.267		35	11.70	5.60	52.30	63
1100	Median barrier, steel box beam, 6" x 8"		215	.149		47.50	6.55	3.13	57.18	65.50
1400	Resilient guide fence and light shield, 6' high	B-2	130	.308		20.50	12.15		32.65	41
1500	Concrete posts, individual, 6'-5", triangular	B-80	110	.291	Ea.	64	12.75	6.10	82.85	96.50
1550	Square	"	110	.291	"	69	12.75	6.10	87.85	102
2000	Median, precast concrete, 3'-6" high, 2' wide, single face	B-29	380	.147	L.F.	57.50	6.35	2.39	66.24	75.50
2200	Double face	"	340	.165	"	66	7.10	2.68	75.78	86.50
2400	Speed bumps, thermoplastic, 10-1/2" x 2-1/4" x 48" long	B-2	120	.333	Ea.	106	13.20		119.20	136
3030	Impact barrier, UTMCD, barrel type	B-16	30	1.067	"	395	44	21.50	460.50	520

34 71 19 – Vehicle Delineators

34 71 19.13 Fixed Vehicle Delineators

		Crew	Daily Output	Labor-Hours	Unit	Material	2017 Bare Costs Labor	Equipment	Total	Total Incl O&P
0010	**FIXED VEHICLE DELINEATORS**									
0020	Crash barriers									
0100	Traffic channelizing pavement markers, layout only	A-7	2000	.012	Ea.		.65	.02	.67	1.01
0110	13" x 7-1/2" x 2-1/2" high, non-plowable, install	2 Clab	96	.167		26	6.55		32.55	38.50
0200	8" x 8" x 3-1/4" high, non-plowable, install		96	.167		25	6.55		31.55	37.50
0230	4" x 4" x 3/4" high, non-plowable, install		120	.133		2.50	5.20		7.70	10.75
0240	9-1/4" x 5-7/8" x 1/4" high, plowable, concrete pavmt.	A-2A	70	.343		5	14	5.30	24.30	33
0250	9-1/4" x 5-7/8" x 1/4" high, plowable, asphalt pavmt.	"	120	.200		4	8.15	3.09	15.24	20.50
0300	Barrier and curb delineators, reflectorized, 2" x 4"	2 Clab	150	.107		2.05	4.18		6.23	8.65
0310	3" x 5"	"	150	.107		4	4.18		8.18	10.80
0500	Rumble strip, polycarbonate									
0510	24" x 3-1/2" x 1/2" high	2 Clab	50	.320	Ea.	8.60	12.55		21.15	28.50

34 72 Railway Construction

34 72 16 – Railway Siding

34 72 16.50 Railroad Sidings	Crew	Daily Output	Labor-Hours	Unit	Material	2017 Bare Costs Labor	Equipment	Total	Total Incl O&P
0010 **RAILROAD SIDINGS** R347216-10									
0800 Siding, yard spur, level grade									
0820 100 lb. new rail	B-14	57	.842	L.F.	124	35	6.45	165.45	198
1002 Steel ties in concrete, incl. fasteners & plates									
1020 100 lb. new rail	B-14	22	2.182	L.F.	189	90.50	16.65	296.15	365

34 72 16.60 Railroad Turnouts

34 72 16.60 Railroad Turnouts	Crew	Daily Output	Labor-Hours	Unit	Material	2017 Bare Costs Labor	Equipment	Total	Total Incl O&P
0010 **RAILROAD TURNOUTS**									
2200 Turnout, #8 complete, w/rails, plates, bars, frog, switch point,									
2250 timbers, and ballast to 6" below bottom of ties									
2280 90 lb. rails	B-13	.25	224	Ea.	37,000	9,650	2,700	49,350	58,500
2290 90 lb. relay rails		.25	224		24,900	9,650	2,700	37,250	45,100
2300 100 lb. rails		.25	224		41,200	9,650	2,700	53,550	63,000
2310 100 lb. relay rails		.25	224		27,700	9,650	2,700	40,050	48,200
2320 110 lb. rails		.25	224		45,400	9,650	2,700	57,750	67,500
2330 110 lb. relay rails		.25	224		30,600	9,650	2,700	42,950	51,500
2340 115 lb. rails		.25	224		49,800	9,650	2,700	62,150	72,500
2350 115 lb. relay rails		.25	224		33,200	9,650	2,700	45,550	54,000
2360 132 lb. rails		.25	224		57,000	9,650	2,700	69,350	80,000
2370 132 lb. relay rails		.25	224		37,700	9,650	2,700	50,050	59,000

Division Notes

	CREW	DAILY OUTPUT	LABOR-HOURS	UNIT	BARE COSTS				TOTAL INCL O&P
					MAT.	LABOR	EQUIP.	TOTAL	

Estimating Tips

35 01 50 Operation and Maintenance of Marine Construction
Includes unit price lines for pile cleaning and pile wrapping for protection.

35 20 16 Hydraulic Gates
This subdivision includes various types of gates that are commonly used in waterway and canal construction. Various earthwork items and structural support are found in Division 31, and concrete work is found in Division 3.

35 20 23 Dredging
This subdivision includes barge and shore dredging systems for rivers, canals, and channels.

35 31 00 Shoreline Protection
This subdivision includes breakwaters, bulkheads, and revetments for ocean and river inlets. Additional earthwork may be required from Division 31 and concrete work from Division 3.

35 41 00 Levees
Contains information on levee construction, including the estimated cost of clay cone material.

35 49 00 Waterway Structures
This subdivision includes breakwaters and bulkheads for canals.

35 51 00 Floating Construction
This section includes floating piers, docks, and dock accessories. Fixed Pier Timber Construction is found in Section 06 13 33. Driven piles are found in Division 31, as well as sheet piling, cofferdams, and riprap.

Reference Numbers
Reference numbers are shown at the beginning of some major classifications. These numbers refer to related items in the Reference Section. The reference information may be an estimating procedure, an alternate pricing method, or technical information.

Note: Not all subdivisions listed here necessarily appear. ■

Did you know?
Our online estimating solution gives you the same access to RSMeans' data with 24/7 access:

■ Quickly locate costs in the searchable database.

■ Build cost lists, estimates, and reports in minutes.

■ Adjust costs to any location in the U.S. and Canada with the click of a button.

Start your free trial today at
www.RSMeansOnline.com

35 20 Waterway and Marine Construction and Equipment

35 20 23 – Dredging

35 20 23.13 Mechanical Dredging	Crew	Daily Output	Labor-Hours	Unit	Material	2017 Bare Costs Labor	Equipment	Total	Total Incl O&P
0010 **MECHANICAL DREDGING**									
0020 Dredging mobilization and demobilization, add to below, minimum	B-8	.53	120	Total		5,525	6,125	11,650	15,100
0100 Maximum	"	.10	640	"		29,300	32,400	61,700	80,000
0300 Barge mounted clamshell excavation into scows									
0310 Dumped 20 miles at sea, minimum	B-57	310	.155	B.C.Y.		7.05	5.85	12.90	17.15
0400 Maximum	"	213	.225	"		10.30	8.50	18.80	25
0500 Barge mounted dragline or clamshell, hopper dumped,									
0510 pumped 1000' to shore dump, minimum	B-57	340	.141	B.C.Y.		6.45	5.30	11.75	15.65
0525 All pumping uses 2000 gallons of water per cubic yard									
0600 Maximum	B-57	243	.198	B.C.Y.		9.05	7.45	16.50	22

35 20 23.23 Hydraulic Dredging

	Crew	Daily Output	Labor-Hours	Unit	Material	Labor	Equipment	Total	Total Incl O&P
0010 **HYDRAULIC DREDGING**									
1000 Hydraulic method, pumped 1000' to shore dump, minimum	B-57	460	.104	B.C.Y.		4.77	3.93	8.70	11.55
1100 Maximum		310	.155			7.05	5.85	12.90	17.15
1400 Into scows dumped 20 miles, minimum		425	.113			5.15	4.25	9.40	12.55
1500 Maximum	↓	243	.198			9.05	7.45	16.50	22
1600 For inland rivers and canals in South, deduct				↓				30%	30%

35 51 Floating Construction

35 51 13 – Floating Piers

35 51 13.23 Floating Wood Piers

	Crew	Daily Output	Labor-Hours	Unit	Material	Labor	Equipment	Total	Total Incl O&P
0010 **FLOATING WOOD PIERS**									
0020 Polyethylene encased polystyrene, no pilings included	F-3	330	.121	S.F.	30	6.15	1.74	37.89	44.50
0200 Pile supported, shore constructed, bare, 3" decking		130	.308		29	15.55	4.41	48.96	60.50
0250 4" decking		120	.333		30	16.85	4.77	51.62	64
0400 Floating, small boat, prefab, no shore facilities, minimum		250	.160		25.50	8.10	2.29	35.89	43
0500 Maximum		150	.267	↓	54	13.50	3.82	71.32	83.50
0700 Per slip, minimum (180 S.F. each)		1.59	25.157	Ea.	5,375	1,275	360	7,010	8,275
0800 Maximum	↓	1.40	28.571	"	8,675	1,450	410	10,535	12,200

Estimating Tips

Products such as conveyors, material handling cranes and hoists, as well as other items specified in this division, require trained installers. The general contractor may not have any choice as to who will perform the installation or when it will be performed. Long lead times are often required for these products, making early decisions in purchasing and scheduling necessary. The installation of this type of equipment may require the embedment of mounting hardware during construction of floors, structural walls, or interior walls/partitions. Electrical connections will require coordination with the electrical contractor.

Reference Numbers

Reference numbers are shown at the beginning of some major classifications. These numbers refer to related items in the Reference Section. The reference information may be an estimating procedure, an alternate pricing method, or technical information.

Note: Not all subdivisions listed here necessarily appear. ■

41 21 23.16 Container Piece Material Conveyors	Crew	Daily Output	Labor-Hours	Unit	Material	2017 Bare Costs Labor	Equipment	Total	Total Incl O&P
0010 **CONTAINER PIECE MATERIAL CONVEYORS**									
0020 Gravity fed, 2" rollers, 3" O.C.									
0050 10' sections with 2 supports, 600 lb. capacity, 18" wide				Ea.	480			480	530
0100 24" wide					545			545	595
0150 1400 lb. capacity, 18" wide					640			640	705
0200 24" wide					545			545	595
0350 Horizontal belt, center drive and takeup, 60 fpm									
0400 16" belt, 26.5' length	2 Mill	.50	32	Ea.	3,325	1,650		4,975	6,075
0450 24" belt, 41.5' length		.40	40		5,100	2,050		7,150	8,650
0500 61.5' length		.30	53.333		7,200	2,750		9,950	12,000
0600 Inclined belt, 10' rise with horizontal loader and									
0620 End idler assembly, 27.5' length, 18" belt	2 Mill	.30	53.333	Ea.	7,200	2,750		9,950	12,000
0700 24" belt	"	.15	106	"	8,925	5,500		14,425	17,900
3600 Monorail, overhead, manual, channel type									
3700 125 lb. per L.F.	1 Mill	26	.308	L.F.	19.40	15.85		35.25	44.50
3900 500 lb. per L.F.	"	21	.381	"	20.50	19.65		40.15	51.50
4000 Trolleys for above, 2 wheel, 125 lb. capacity				Ea.	84.50			84.50	93
4200 4 wheel, 250 lb. capacity					335			335	365
4300 8 wheel, 500 lb. capacity					775			775	850

41 22 13.10 Crane Rail

		Crew	Daily Output	Labor-Hours	Unit	Material	2017 Bare Costs Labor	Equipment	Total	Total Incl O&P
0010	**CRANE RAIL**									
0020	Box beam bridge, no equipment included	E-4	3400	.009	Lb.	1.31	.52	.04	1.87	2.34
0200	Running track only, 104 lb. per yard	"	5600	.006	"	.66	.31	.02	.99	1.27

41 22 13.13 Bridge Cranes

		Crew	Daily Output	Labor-Hours	Unit	Material	2017 Bare Costs Labor	Equipment	Total	Total Incl O&P
0010	**BRIDGE CRANES**									
0100	1 girder, 20' span, 3 ton	M-3	1	34	Ea.	26,700	2,075	138	28,913	32,600
0125	5 ton		1	34		29,300	2,075	138	31,513	35,500
0150	7.5 ton		1	34		34,700	2,075	138	36,913	41,400
0175	10 ton		.80	42.500		46,200	2,575	173	48,948	55,000
0200	15 ton		.80	42.500		59,500	2,575	173	62,248	69,500
0225	30' span, 3 ton		1	34		27,700	2,075	138	29,913	33,800
0250	5 ton		1	34		30,600	2,075	138	32,813	36,900
0275	7.5 ton		1	34		36,500	2,075	138	38,713	43,400
0300	10 ton		.80	42.500		47,900	2,575	173	50,648	56,500
0325	15 ton		.80	42.500		62,500	2,575	173	65,248	72,500
0350	2 girder, 40' span, 3 ton	M-4	.50	72		45,300	4,325	390	50,015	57,000
0375	5 ton		.50	72		47,600	4,325	390	52,315	59,500
0400	7.5 ton		.50	72		52,000	4,325	390	56,715	64,500
0425	10 ton		.40	90		61,500	5,425	485	67,410	76,000
0450	15 ton		.40	90		83,500	5,425	485	89,410	100,500
0475	25 ton		.30	120		98,500	7,225	645	106,370	119,500
0500	50' span, 3 ton		.50	72		52,000	4,325	390	56,715	64,000
0525	5 ton		.50	72		54,000	4,325	390	58,715	66,500
0550	7.5 ton		.50	72		58,000	4,325	390	62,715	70,500
0575	10 ton		.40	90		67,000	5,425	485	72,910	82,500
0600	15 ton		.40	90		87,500	5,425	485	93,410	105,000
0625	25 ton		.30	120		103,000	7,225	645	110,870	125,000

41 22 Cranes and Hoists

41 22 13 - Cranes

41 22 13.19 Jib Cranes	Crew	Daily Output	Labor-Hours	Unit	Material	2017 Bare Costs Labor	2017 Bare Costs Equipment	Total	Total Incl O&P
0010 **JIB CRANES**									
0020 Jib crane, wall cantilever, 500 lb. capacity, 8' span	2 Mill	1	16	Ea.	1,425	825		2,250	2,775
0040 12' span		1	16		1,550	825		2,375	2,925
0060 16' span		1	16		1,725	825		2,550	3,100
0080 20' span		1	16		2,200	825		3,025	3,600
0100 1000 lb. capacity, 8' span		1	16		1,525	825		2,350	2,875
0120 12' span		1	16		1,650	825		2,475	3,025
0130 16' span		1	16		2,200	825		3,025	3,625
0150 20' span		1	16		2,575	825		3,400	4,025

41 22 23 - Hoists

41 22 23.10 Material Handling

41 22 23.10 Material Handling	Crew	Daily Output	Labor-Hours	Unit	Material	2017 Bare Costs Labor	2017 Bare Costs Equipment	Total	Total Incl O&P
0010 **MATERIAL HANDLING**, cranes, hoists and lifts									
1500 Cranes, portable hydraulic, floor type, 2,000 lb. capacity				Ea.	3,925			3,925	4,325
1600 4,000 lb. capacity					4,500			4,500	4,950
1800 Movable gantry type, 12' to 15' range, 2,000 lb. capacity					3,950			3,950	4,325
1900 6,000 lb. capacity					5,975			5,975	6,575
2100 Hoists, electric overhead, chain, hook hung, 15' lift, 1 ton cap.					2,575			2,575	2,850
2200 3 ton capacity					3,575			3,575	3,925
2500 5 ton capacity					7,150			7,150	7,875
2600 For hand-pushed trolley, add					15%				
2700 For geared trolley, add					30%				
2800 For motor trolley, add					75%				
3000 For lifts over 15', 1 ton, add				L.F.	23			23	25
3100 5 ton, add				"	67.50			67.50	74
3300 Lifts, scissor type, portable, electric, 36" high, 2,000 lb.				Ea.	3,575			3,575	3,950
3400 48" high, 4,000 lb.				"	4,250			4,250	4,675

For customer support on your Building Construction Costs with RSMeans Data, call 800.448.8182.

699

Division Notes

	CREW	DAILY OUTPUT	LABOR-HOURS	UNIT	BARE COSTS				TOTAL INCL O&P
					MAT.	LABOR	EQUIP.	TOTAL	

Estimating Tips

This section involves equipment and construction costs for air noise and odor pollution control systems. These systems may be interrelated and care must be taken that the complete systems are estimated. For example, air pollution equipment may include dust and air-entrained particles that have to be collected. The vacuum systems could be noisy, requiring silencers to reduce noise pollution, and the collected solids have to be disposed of to prevent solid pollution.

Reference Numbers

Reference numbers are shown at the beginning of some major classifications. These numbers refer to related items in the Reference Section. The reference information may be an estimating procedure, an alternate pricing method, or technical information.

Note: Not all subdivisions listed here necessarily appear. ■

Did you know?

Our online estimating solution gives you the same access to RSMeans' data with 24/7 access:

- Quickly locate costs in the searchable database.
- Build cost lists, estimates, and reports in minutes.
- Adjust costs to any location in the U.S. and Canada with the click of a button.

Start your free trial today at
www.RSMeansOnline.com

44 11 16.10 Dust Collection Systems	Crew	Daily Output	Labor-Hours	Unit	Material	2017 Bare Costs Labor	Equipment	Total	Total Incl O&P
0010 **DUST COLLECTION SYSTEMS** Commercial/Industrial									
0120 Central vacuum units									
0130 Includes stand, filters and motorized shaker									
0200 500 CFM, 10" inlet, 2 HP	Q-20	2.40	8.333	Ea.	4,950	445		5,395	6,100
0220 1000 CFM, 10" inlet, 3 HP		2.20	9.091		5,200	485		5,685	6,450
0240 1500 CFM, 10" inlet, 5 HP		2	10		5,475	530		6,005	6,825
0260 3000 CFM, 13" inlet, 10 HP		1.50	13.333		15,400	710		16,110	18,000
0280 5000 CFM, 16" inlet, 2 @ 10 HP		1	20		16,400	1,050		17,450	19,600
1000 Vacuum tubing, galvanized									
1100 2-1/8" OD, 16 ga.	Q-9	440	.036	L.F.	3.37	1.90		5.27	6.60
1110 2-1/2" OD, 16 ga.		420	.038		3.70	1.99		5.69	7.10
1120 3" OD, 16 ga.		400	.040		4.72	2.09		6.81	8.40
1130 3-1/2" OD, 16 ga.		380	.042		6.70	2.20		8.90	10.70
1140 4" OD, 16 ga.		360	.044		6.85	2.32		9.17	11.10
1150 5" OD, 14 ga.		320	.050		12.35	2.61		14.96	17.55
1160 6" OD, 14 ga.		280	.057		13.80	2.99		16.79	19.75
1170 8" OD, 14 ga.		200	.080		21	4.18		25.18	29.50
1180 10" OD, 12 ga.		160	.100		40.50	5.25		45.75	52.50
1190 12" OD, 12 ga.		120	.133		53.50	6.95		60.45	69.50
1200 14" OD, 12 ga.		80	.200		61	10.45		71.45	83
1940 Hose, flexible wire reinforced rubber									
1956 3" diam.	Q-9	400	.040	L.F.	9.60	2.09		11.69	13.75
1960 4" diam.		360	.044		11.10	2.32		13.42	15.80
1970 5" diam.		320	.050		14.45	2.61		17.06	19.90
1980 6" diam.		280	.057		15.10	2.99		18.09	21
2000 90° Elbow, slip fit									
2110 2-1/8" diam.	Q-9	70	.229	Ea.	10.85	11.95		22.80	30.50
2120 2-1/2" diam.		65	.246		15.20	12.85		28.05	36.50
2130 3" diam.		60	.267		20.50	13.95		34.45	44
2140 3-1/2" diam.		55	.291		25.50	15.20		40.70	51.50
2150 4" diam.		50	.320		31.50	16.70		48.20	60
2160 5" diam.		45	.356		60	18.60		78.60	94.50
2170 6" diam.		40	.400		82.50	21		103.50	123
2180 8" diam.		30	.533		157	28		185	215
2400 45° Elbow, slip fit									
2410 2-1/8" diam.	Q-9	70	.229	Ea.	9.50	11.95		21.45	29
2420 2-1/2" diam.		65	.246		14	12.85		26.85	35
2430 3" diam.		60	.267		16.90	13.95		30.85	40
2440 3-1/2" diam.		55	.291		21.50	15.20		36.70	47
2450 4" diam.		50	.320		27.50	16.70		44.20	56
2460 5" diam.		45	.356		46.50	18.60		65.10	79.50
2470 6" diam.		40	.400		63	21		84	101
2480 8" diam.		35	.457		123	24		147	173
2800 90° TY, slip fit thru 6" diam.									
2810 2-1/8" diam.	Q-9	42	.381	Ea.	17.75	19.90		37.65	50
2820 2-1/2" diam.		39	.410		26.50	21.50		48	62
2830 3" diam.		36	.444		37	23		60	76
2840 3-1/2" diam.		33	.485		39	25.50		64.50	81.50
2850 4" diam.		30	.533		41.50	28		69.50	88.50
2860 5" diam.		27	.593		70.50	31		101.50	125
2870 6" diam.		24	.667		101	35		136	165
2880 8" diam., butt end					155			155	170
2890 10" diam., butt end					475			475	525

44 11 16.10 Dust Collection Systems	Crew	Daily Output	Labor-Hours	Unit	Material	Labor	Equipment	Total	Total Incl O&P
2900	12" diam., butt end				Ea.	475			475	525
2910	14" diam., butt end					855			855	940
2920	6" x 4" diam., butt end					106			106	116
2930	8" x 4" diam., butt end					200			200	220
2940	10" x 4" diam., butt end					263			263	289
2950	12" x 4" diam., butt end					300			300	330
3100	90° Elbow, butt end segmented									
3110	8" diam., butt end, segmented				Ea.	224			224	246
3120	10" diam., butt end, segmented					390			390	430
3130	12" diam., butt end, segmented					495			495	545
3140	14" diam., butt end, segmented					620			620	680
3200	45° Elbow, butt end segmented									
3210	8" diam., butt end, segmented				Ea.	123			123	136
3220	10" diam., butt end, segmented					282			282	310
3230	12" diam., butt end, segmented					288			288	315
3240	14" diam., butt end, segmented					585			585	645
3400	All butt end fittings require one coupling per joint.									
3410	Labor for fitting included with couplings.									
3460	Compression coupling, galvanized, neoprene gasket									
3470	2-1/8" diam.	Q-9	44	.364	Ea.	11.40	19		30.40	41.50
3480	2-1/2" diam.		44	.364		11.40	19		30.40	41.50
3490	3" diam.		38	.421		16.40	22		38.40	51.50
3500	3-1/2" diam.		35	.457		18.35	24		42.35	56.50
3510	4" diam.		33	.485		19.90	25.50		45.40	61
3520	5" diam.		29	.552		22.50	29		51.50	69
3530	6" diam.		26	.615		26.50	32		58.50	78.50
3540	8" diam.		22	.727		48.50	38		86.50	111
3550	10" diam.		20	.800		71	42		113	142
3560	12" diam.		18	.889		88.50	46.50		135	169
3570	14" diam.		16	1		125	52.50		177.50	218
3800	Air gate valves, galvanized									
3810	2-1/8" diam.	Q-9	30	.533	Ea.	117	28		145	172
3820	2-1/2" diam.		28	.571		127	30		157	186
3830	3" diam.		26	.615		135	32		167	198
3840	4" diam.		23	.696		156	36.50		192.50	228
3850	6" diam.		18	.889		217	46.50		263.50	310

Division Notes

	CREW	DAILY OUTPUT	LABOR-HOURS	UNIT	BARE COSTS				TOTAL INCL O&P
					MAT.	LABOR	EQUIP.	TOTAL	

Estimating Tips

This division contains information about water and wastewater equipment and systems, which was formerly located in Division 44. The main areas of focus are total wastewater treatment plants and components of wastewater treatment plants. Also included in this section are oil/water separators for wastewater treatment.

Reference Numbers

Reference numbers are shown at the beginning of some major classifications. These numbers refer to related items in the Reference Section. The reference information may be an estimating procedure, an alternate pricing method, or technical information.

Note: Not all subdivisions listed here necessarily appear. ■

46 07 53 – Packaged Wastewater Treatment Equipment

46 07 53.10 Biological Pkg. Wastewater Treatment Plants	Crew	Daily Output	Labor-Hours	Unit	Material	2017 Bare Costs Labor	Equipment	Total	Total Incl O&P
0010 **BIOLOGICAL PACKAGED WASTEWATER TREATMENT PLANTS**									
0011　Not including fencing or external piping									
0020　Steel packaged, blown air aeration plants									
0100　　1,000 GPD				Gal.				55	60.50
0200　　5,000 GPD								22	24
0300　　15,000 GPD								22	24
0400　　30,000 GPD								15.40	16.95
0500　　50,000 GPD								11	12.10
0600　　100,000 GPD								9.90	10.90
0700　　200,000 GPD								8.80	9.70
0800　　500,000 GPD				▽				7.70	8.45
1000　Concrete, extended aeration, primary and secondary treatment									
1010　　10,000 GPD				Gal.				22	24
1100　　30,000 GPD								15.40	16.95
1200　　50,000 GPD								11	12.10
1400　　100,000 GPD								9.90	10.90
1500　　500,000 GPD				▽				7.70	8.45
1700　Municipal wastewater treatment facility									
1720　　1.0 MGD				Gal.				11	12.10
1740　　1.5 MGD								10.60	11.65
1760　　2.0 MGD								10	11
1780　　3.0 MGD								7.80	8.60
1800　　5.0 MGD				▽				5.80	6.70
2000　Holding tank system, not incl. excavation or backfill									
2010　　Recirculating chemical water closet	2 Plum	4	4	Ea.	545	247		792	975
2100　　　For voltage converter, add	"	16	1		310	62		372	435
2200　　　For high level alarm, add	1 Plum	7.80	1.026	▽	120	63.50		183.50	228

46 07 53.20 Wastewater Treatment System

	Crew	Daily Output	Labor-Hours	Unit	Material	Labor	Equipment	Total	Total Incl O&P
0010 **WASTEWATER TREATMENT SYSTEM**									
0020　Fiberglass, 1,000 gallon	B-21	1.29	21.705	Ea.	4,325	990	107	5,422	6,400
0100　　1,500 gallon	"	1.03	27.184	"	8,650	1,250	134	10,034	11,600

Estimating Tips

- When estimating costs for the installation of electrical power generation equipment, factors to review include access to the job site, access and setting up at the installation site, required connections, uncrating pads, anchors, leveling, final assembly of the components, and temporary protection from physical damage, such as environmental exposure.

- Be aware of the costs of equipment supports, concrete pads, and vibration isolators. Cross-reference them against other trades' specifications. Also, review site and structural drawings for items that must be included in the estimates.

- It is important to include items that are not documented in the plans and specifications but must be priced. These items include, but are not limited to, testing, dust protection, roof penetration, core drilling concrete floors and walls, patching, cleanup, and final adjustments. Add a contingency or allowance for utility company fees for power hookups, if needed.

- The project size and scope of electrical power generation equipment will have a significant impact on cost. The intent of RSMeans cost data is to provide a benchmark cost so that owners, engineers, and electrical contractors will have a comfortable number with which to start a project. Additionally, there are many websites available to use for research and to obtain a vendor's quote to finalize costs.

Reference Numbers

Reference numbers are shown at the beginning of some major classifications. These numbers refer to related items in the Reference Section. The reference information may be an estimating procedure, an alternate pricing method, or technical information.

Note: Not all subdivisions listed here necessarily appear. ■

Did you know?

Our online estimating solution gives you the same access to RSMeans' data with 24/7 access:

- Quickly locate costs in the searchable database.
- Build cost lists, estimates, and reports in minutes.
- Adjust costs to any location in the U.S. and Canada with the click of a button.

Start your free trial today at
www.RSMeansOnline.com

48 15 13.50 Wind Turbines and Components		Crew	Daily Output	Labor-Hours	Unit	Material	2017 Bare Costs Labor	Equipment	Total	Total Incl O&P
0010	**WIND TURBINES & COMPONENTS**									
0500	Complete system, grid connected									
1000	20 kW, 31' dia, incl. labor & material	G			System				49,900	49,900
1010	Enhanced	G							92,000	92,000
1500	10 kW, 23' dia, incl. labor & material	G							74,000	74,000
2000	2.4 kW, 12' dia, incl. labor & material	G			↓				18,000	18,000
2900	Component system									
3000	Turbine, 400 watt, 3' dia	G	1 Elec	3.41	2.346	Ea.	550	133		683	805
3100	600 watt, 3' dia	G		2.56	3.125		710	177		887	1,050
3200	1000 watt, 9' dia	G	↓	2.05	3.902	↓	1,500	221		1,721	1,975
3400	Mounting hardware									
3500	30' guyed tower kit	G	2 Clab	5.12	3.125	Ea.	410	122		532	640
3505	3' galvanized helical earth screw	G	1 Clab	8	1		49.50	39		88.50	115
3510	Attic mount kit	G	1 Rofc	2.56	3.125		194	135		329	440
3520	Roof mount kit	G	1 Clab	3.41	2.346	↓	320	92		412	490
8900	Equipment									
9100	DC to AC inverter for, 48 V, 4,000 watt	G	1 Elec	2	4	Ea.	2,275	226		2,501	2,850

Reference Section

All the reference information is in one section, making it easy to find what you need to know . . . and easy to use the data set on a daily basis. This section is visually identified by a vertical black bar on the page edges.

In this Reference Section, we've included Equipment Rental Costs, a listing of rental and operating costs; Crew Listings, a full listing of all crews and equipment, and their costs; Historical Cost Indexes for cost comparisons over time; City Cost Indexes and Location Factors for adjusting costs to the region you are in; Reference Tables, where you will find explanations, estimating information and procedures, or technical data; Change Orders, information on pricing changes to contract documents; and an explanation of all the Abbreviations in the data set.

Table of Contents

Construction Equipment Rental Costs 711

Crew Listings 723

Historical Cost Indexes 760

City Cost Indexes 761

Location Factors 804

Reference Tables 810

R01 General Requirements 810
R02 Existing Conditions 823
R03 Concrete 826
R04 Masonry 842
R05 Metals 844
R06 Wood, Plastics & Composites 851
R07 Thermal & Moisture Protection 851
R08 Openings 853
R09 Finishes 856
R13 Special Construction 859
R14 Conveying Equipment 861

Reference Tables (cont.)

R22 Plumbing 862
R23 Heating, Ventilating & Air Conditioning 863
R26 Electrical 864
R31 Earthwork 867
R32 Exterior Improvements 871
R33 Utilities 872
R34 Transportation 872

Change Orders 873

Square Foot Costs 876

Abbreviations 877

Estimating Tips

- This section contains the average costs to rent and operate hundreds of pieces of construction equipment. This is useful information when estimating the time and material requirements of any particular operation in order to establish a unit or total cost. Bare equipment costs shown on a unit cost line include not only rental, but also operating costs for equipment under normal use.

Rental Costs

- Equipment rental rates are obtained from the following industry sources throughout North America: contractors, suppliers, dealers, manufacturers, and distributors.

- Rental rates vary throughout the country, with larger cities generally having lower rates. Lease plans for new equipment are available for periods in excess of six months, with a percentage of payments applying toward purchase.

- Monthly rental rates vary from 2% to 5% of the purchase price of the equipment depending on the anticipated life of the equipment and its wearing parts.

- Weekly rental rates are about 1/3 of the monthly rates, and daily rental rates are about 1/3 of the weekly rate.

- Rental rates can also be treated as reimbursement costs for contractor-owned equipment. Owned equipment costs include depreciation, loan payments, interest, taxes, insurance, storage, and major repairs.

Operating Costs

- The operating costs include parts and labor for routine servicing, such as the repair and replacement of pumps, filters, and worn lines. Normal operating expendables, such as fuel, lubricants, tires, and electricity (where applicable), are also included.

- Extraordinary operating expendables with highly variable wear patterns, such as diamond bits and blades, are excluded. These costs can be found as material costs in the Unit Price section.

- The hourly operating costs listed do not include the operator's wages.

Equipment Cost/Day

- Any power equipment required by a crew is shown in the Crew Listings with a daily cost.

- This daily cost of equipment needed by a crew includes both the rental cost and the operating cost and is based on dividing the weekly rental rate by 5 (number of working days in the week), and then adding the hourly operating cost times 8 (the number of hours in a day). This "Equipment Cost/Day" is shown in the far right column of the Equipment Rental section.

- If equipment is needed for only one or two days, it is best to develop your own cost by including components for daily rent and hourly operating costs. This is important when the listed Crew for a task does not contain the equipment needed, such as a crane for lifting mechanical heating/cooling equipment up onto a roof.

- If the quantity of work is less than the crew's Daily Output shown for a Unit Price line item that includes a bare unit equipment cost, the recommendation is to estimate one day's rental cost and operating cost for equipment shown in the Crew Listing for that line item.

Mobilization, Demobilization Costs

- The cost to move construction equipment from an equipment yard or rental company to the job site and back again is not included in equipment rental costs listed in the Reference Section, nor in the bare equipment cost of any unit price line item, nor in any equipment costs shown in the Crew Listings.

- Mobilization (to the site) and demobilization (from the site) costs can be found in the Unit Price section.

- If a piece of equipment is already at the job site, it is not appropriate to utilize mobilization, demobilization costs again in an estimate. ∎

01 54 33 | Equipment Rental

		UNIT	HOURLY OPER. COST	RENT PER DAY	RENT PER WEEK	RENT PER MONTH	EQUIPMENT COST/DAY	
10	**0010** **CONCRETE EQUIPMENT RENTAL** without operators R015433 -10							**10**
	0200 Bucket, concrete lightweight, 1/2 C.Y.	Ea.	.80	23.50	70	210	20.40	
	0300 1 C.Y.		.90	28	84	252	24	
	0400 1-1/2 C.Y.		1.15	38.50	115	345	32.20	
	0500 2 C.Y.		1.25	45	135	405	37	
	0580 8 C.Y.		6.15	258	775	2,325	204.20	
	0600 Cart, concrete, self-propelled, operator walking, 10 C.F.		3.00	56.50	170	510	58	
	0700 Operator riding, 18 C.F.		4.95	95	285	855	96.60	
	0800 Conveyer for concrete, portable, gas, 16" wide, 26' long		11.35	125	375	1,125	165.80	
	0900 46' long		11.70	150	450	1,350	183.60	
	1000 56' long		11.85	158	475	1,425	189.80	
	1100 Core drill, electric, 2-1/2 H.P., 1" to 8" bit diameter		1.83	67.50	203	610	55.25	
	1150 11 H.P., 8" to 18" cores		5.45	118	355	1,075	114.60	
	1200 Finisher, concrete floor, gas, riding trowel, 96" wide		10.99	147	440	1,325	175.90	
	1300 Gas, walk-behind, 3 blade, 36" trowel		1.85	20.50	62	186	27.20	
	1400 4 blade, 48" trowel		3.62	27.50	83	249	45.55	
	1500 Float, hand-operated (Bull float), 48" wide		.08	13.35	40	120	8.65	
	1570 Curb builder, 14 H.P., gas, single screw		14.35	263	790	2,375	272.80	
	1590 Double screw		15.30	320	955	2,875	313.40	
	1600 Floor grinder, concrete and terrazzo, electric, 22" path		2.69	167	500	1,500	121.50	
	1700 Edger, concrete, electric, 7" path		1.03	51.50	155	465	39.25	
	1750 Vacuum pick-up system for floor grinders, wet/dry		1.48	81.50	245	735	60.85	
	1800 Mixer, powered, mortar and concrete, gas, 6 C.F., 18 H.P.		7.85	120	360	1,075	134.80	
	1900 10 C.F., 25 H.P.		9.70	145	435	1,300	164.60	
	2000 16 C.F.		10.00	167	500	1,500	180	
	2100 Concrete, stationary, tilt drum, 2 C.Y.		7.00	235	705	2,125	197	
	2120 Pump, concrete, truck mounted, 4" line, 80' boom		28.30	965	2,900	8,700	806.40	
	2140 5" line, 110' boom		35.55	1,325	4,010	12,000	1,086	
	2160 Mud jack, 50 C.F. per hr.		6.60	133	400	1,200	132.80	
	2180 225 C.F. per hr.		8.80	152	455	1,375	161.40	
	2190 Shotcrete pump rig, 12 C.Y./hr.		13.90	232	695	2,075	250.20	
	2200 35 C.Y./hr.		15.90	248	745	2,225	276.20	
	2600 Saw, concrete, manual, gas, 18 H.P.		5.90	45	135	405	74.20	
	2650 Self-propelled, gas, 30 H.P.		11.45	103	310	930	153.60	
	2675 V-groove crack chaser, manual, gas, 6 H.P.		2.00	17.65	53	159	26.60	
	2700 Vibrators, concrete, electric, 60 cycle, 2 H.P.		.46	8.65	26	78	8.90	
	2800 3 H.P.		.59	11.65	35	105	11.70	
	2900 Gas engine, 5 H.P.		1.75	16.35	49	147	23.80	
	3000 8 H.P.		2.38	15.35	46	138	28.25	
	3050 Vibrating screed, gas engine, 8 H.P.		2.47	74.50	223	670	64.35	
	3120 Concrete transit mixer, 6 x 4, 250 H.P., 8 C.Y., rear discharge		57.25	580	1,745	5,225	807	
	3200 Front discharge		67.15	715	2,145	6,425	966.20	
	3300 6 x 6, 285 H.P., 12 C.Y., rear discharge		66.20	675	2,030	6,100	935.60	
	3400 Front discharge		69.15	715	2,145	6,425	982.20	
20	**0010** **EARTHWORK EQUIPMENT RENTAL** without operators R015433 -10							**20**
	0040 Aggregate spreader, push type, 8' to 12' wide	Ea.	2.75	25.50	77	231	37.40	
	0045 Tailgate type, 8' wide		2.65	32.50	98	294	40.80	
	0055 Earth auger, truck mounted, for fence & sign posts, utility poles		17.05	445	1,335	4,000	403.40	
	0060 For borings and monitoring wells		43.80	720	2,155	6,475	781.40	
	0070 Portable, trailer mounted		2.77	33.50	100	300	42.15	
	0075 Truck mounted, for caissons, water wells		89.45	3,100	9,285	27,900	2,573	
	0080 Horizontal boring machine, 12" to 36" diameter, 45 H.P.		22.60	207	620	1,850	304.80	
	0090 12" to 48" diameter, 65 H.P.		31.15	360	1,080	3,250	465.20	
	0095 Auger, for fence posts, gas engine, hand held		.51	6.35	19	57	7.90	
	0100 Excavator, diesel hydraulic, crawler mounted, 1/2 C.Y. cap.		23.45	425	1,280	3,850	443.60	
	0120 5/8 C.Y. capacity		31.75	600	1,800	5,400	614	
	0140 3/4 C.Y. capacity		34.90	660	1,975	5,925	674.20	
	0150 1 C.Y. capacity		46.40	710	2,135	6,400	798.20	

01 54 33 | Equipment Rental

		UNIT	HOURLY OPER. COST	RENT PER DAY	RENT PER WEEK	RENT PER MONTH	EQUIPMENT COST/DAY	
0200	1-1/2 C.Y. capacity	Ea.	54.75	880	2,635	7,900	965	20
0300	2 C.Y. capacity		61.90	1,050	3,140	9,425	1,123	
0320	2-1/2 C.Y. capacity		94.70	1,300	3,890	11,700	1,536	
0325	3-1/2 C.Y. capacity		134.25	2,200	6,635	19,900	2,401	
0330	4-1/2 C.Y. capacity		166.65	2,925	8,755	26,300	3,084	
0335	6 C.Y. capacity		214.30	3,125	9,410	28,200	3,596	
0340	7 C.Y. capacity		216.45	3,275	9,830	29,500	3,698	
0342	Excavator attachments, bucket thumbs		3.25	247	740	2,225	174	
0345	Grapples		2.80	195	585	1,750	139.40	
0346	Hydraulic hammer for boom mounting, 4000 ft lb.		12.55	355	1,060	3,175	312.40	
0347	5000 ft lb.		14.65	430	1,295	3,875	376.20	
0348	8000 ft lb.		21.60	635	1,900	5,700	552.80	
0349	12,000 ft lb.		23.65	755	2,270	6,800	643.20	
0350	Gradall type, truck mounted, 3 ton @ 15' radius, 5/8 C.Y.		52.50	815	2,450	7,350	910	
0370	1 C.Y. capacity		61.90	1,075	3,245	9,725	1,144	
0400	Backhoe-loader, 40 to 45 H.P., 5/8 C.Y. capacity		14.25	257	770	2,300	268	
0450	45 H.P. to 60 H.P., 3/4 C.Y. capacity		22.15	315	945	2,825	366.20	
0460	80 H.P., 1-1/4 C.Y. capacity		23.55	345	1,030	3,100	394.40	
0470	112 H.P., 1-1/2 C.Y. capacity		37.45	650	1,945	5,825	688.60	
0482	Backhoe-loader attachment, compactor, 20,000 lb.		5.90	143	430	1,300	133.20	
0485	Hydraulic hammer, 750 ft lb.		3.35	98.50	295	885	85.80	
0486	Hydraulic hammer, 1200 ft lb.		6.40	218	655	1,975	182.20	
0500	Brush chipper, gas engine, 6" cutter head, 35 H.P.		9.90	107	320	960	143.20	
0550	Diesel engine, 12" cutter head, 130 H.P.		26.30	325	970	2,900	404.40	
0600	15" cutter head, 165 H.P.		32.05	390	1,170	3,500	490.40	
0750	Bucket, clamshell, general purpose, 3/8 C.Y.		1.30	38.50	115	345	33.40	
0800	1/2 C.Y.		1.40	45	135	405	38.20	
0850	3/4 C.Y.		1.55	55	165	495	45.40	
0900	1 C.Y.		1.60	60	180	540	48.80	
0950	1-1/2 C.Y.		2.60	81.50	245	735	69.80	
1000	2 C.Y.		2.75	90	270	810	76	
1010	Bucket, dragline, medium duty, 1/2 C.Y.		.75	23.50	70	210	20	
1020	3/4 C.Y.		.75	24.50	74	222	20.80	
1030	1 C.Y.		.80	26.50	80	240	22.40	
1040	1-1/2 C.Y.		1.25	40	120	360	34	
1050	2 C.Y.		1.30	45	135	405	37.40	
1070	3 C.Y.		2.00	63.50	190	570	54	
1200	Compactor, manually guided 2-drum vibratory smooth roller, 7.5 H.P.		7.10	210	630	1,900	182.80	
1250	Rammer/tamper, gas, 8"		2.75	46.50	140	420	50	
1260	15"		3.10	53.50	160	480	56.80	
1300	Vibratory plate, gas, 18" plate, 3000 lb. blow		2.30	25.50	76	228	33.60	
1350	21" plate, 5000 lb. blow		2.85	33.50	100	300	42.80	
1370	Curb builder/extruder, 14 H.P., gas, single screw		14.35	263	790	2,375	272.80	
1390	Double screw		15.30	320	955	2,875	313.40	
1500	Disc harrow attachment, for tractor		.46	76.50	230	690	49.70	
1810	Feller buncher, shearing & accumulating trees, 100 H.P.		45.65	780	2,340	7,025	833.20	
1860	Grader, self-propelled, 25,000 lb.		38.80	750	2,245	6,725	759.40	
1910	30,000 lb.		39.30	605	1,820	5,450	678.40	
1920	40,000 lb.		64.30	1,225	3,670	11,000	1,248	
1930	55,000 lb.		78.85	1,825	5,455	16,400	1,722	
1950	Hammer, pavement breaker, self-propelled, diesel, 1000 to 1250 lb.		30.65	400	1,200	3,600	485.20	
2000	1300 to 1500 lb.		45.98	800	2,400	7,200	847.85	
2050	Pile driving hammer, steam or air, 4150 ft lb. @ 225 bpm		11.25	550	1,645	4,925	419	
2100	8750 ft lb. @ 145 bpm		13.75	765	2,300	6,900	570	
2150	15,000 ft lb. @ 60 bpm		14.10	820	2,465	7,400	605.80	
2200	24,450 ft lb. @ 111 bpm		15.15	910	2,730	8,200	667.20	
2250	Leads, 60' high for pile driving hammers up to 20,000 ft lb.		3.40	81.50	244	730	76	
2300	90' high for hammers over 20,000 ft lb.		5.10	142	426	1,275	126	

01 54 33 | Equipment Rental

		UNIT	HOURLY OPER. COST	RENT PER DAY	RENT PER WEEK	RENT PER MONTH	EQUIPMENT COST/DAY	
20	2350	Diesel type hammer, 22,400 ft lb.	Ea.	19.00	450	1,345	4,025	421	20
	2400	41,300 ft lb.		28.25	580	1,740	5,225	574	
	2450	141,000 ft lb.		45.65	915	2,750	8,250	915.20	
	2500	Vib. elec. hammer/extractor, 200 kW diesel generator, 34 H.P.		50.30	675	2,025	6,075	807.40	
	2550	80 H.P.		90.70	970	2,915	8,750	1,309	
	2600	150 H.P.		170.95	1,875	5,630	16,900	2,494	
	2800	Log chipper, up to 22" diameter, 600 H.P.		57.85	650	1,950	5,850	852.80	
	2850	Logger, for skidding & stacking logs, 150 H.P.		48.90	800	2,400	7,200	871.20	
	2860	Mulcher, diesel powered, trailer mounted		21.75	215	645	1,925	303	
	2900	Rake, spring tooth, with tractor		16.89	350	1,045	3,125	344.10	
	3000	Roller, vibratory, tandem, smooth drum, 20 H.P.		7.80	150	450	1,350	152.40	
	3050	35 H.P.		10.20	260	780	2,350	237.60	
	3100	Towed type vibratory compactor, smooth drum, 50 H.P.		25.30	375	1,125	3,375	427.40	
	3150	Sheepsfoot, 50 H.P.		26.25	400	1,205	3,625	451	
	3170	Landfill compactor, 220 H.P.		76.55	1,625	4,860	14,600	1,584	
	3200	Pneumatic tire roller, 80 H.P.		14.10	405	1,215	3,650	355.80	
	3250	120 H.P.		21.30	685	2,050	6,150	580.40	
	3300	Sheepsfoot vibratory roller, 240 H.P.		67.40	1,250	3,785	11,400	1,296	
	3320	340 H.P.		90.85	1,875	5,590	16,800	1,845	
	3350	Smooth drum vibratory roller, 75 H.P.		23.80	690	2,075	6,225	605.40	
	3400	125 H.P.		28.75	755	2,270	6,800	684	
	3410	Rotary mower, brush, 60", with tractor		21.35	325	975	2,925	365.80	
	3420	Rototiller, walk-behind, gas, 5 H.P.		1.50	44	132	395	38.40	
	3422	8 H.P.		2.59	80	240	720	68.70	
	3440	Scrapers, towed type, 7 C.Y. capacity		5.90	117	350	1,050	117.20	
	3450	10 C.Y. capacity		6.70	158	475	1,425	148.60	
	3500	15 C.Y. capacity		7.20	183	550	1,650	167.60	
	3525	Self-propelled, single engine, 14 C.Y. capacity		141.45	2,150	6,485	19,500	2,429	
	3550	Dual engine, 21 C.Y. capacity		164.55	2,450	7,340	22,000	2,784	
	3600	31 C.Y. capacity		217.70	3,400	10,170	30,500	3,776	
	3640	44 C.Y. capacity		267.35	4,300	12,865	38,600	4,712	
	3650	Elevating type, single engine, 11 C.Y. capacity		67.70	1,075	3,250	9,750	1,192	
	3700	22 C.Y. capacity		129.45	2,425	7,270	21,800	2,490	
	3710	Screening plant, 110 H.P. w/5' x 10' screen		19.53	385	1,155	3,475	387.25	
	3720	5' x 16' screen		24.83	500	1,495	4,475	497.65	
	3850	Shovel, crawler-mounted, front-loading, 7 C.Y. capacity		240.65	3,625	10,910	32,700	4,107	
	3855	12 C.Y. capacity		391.30	5,025	15,110	45,300	6,152	
	3860	Shovel/backhoe bucket, 1/2 C.Y.		2.50	68.50	205	615	61	
	3870	3/4 C.Y.		2.60	75	225	675	65.80	
	3880	1 C.Y.		2.70	85	255	765	72.60	
	3890	1-1/2 C.Y.		2.80	98.50	295	885	81.40	
	3910	3 C.Y.		3.15	132	395	1,175	104.20	
	3950	Stump chipper, 18" deep, 30 H.P.		6.79	215	646	1,950	183.50	
	4110	Dozer, crawler, torque converter, diesel 80 H.P.		27.95	410	1,225	3,675	468.60	
	4150	105 H.P.		34.25	555	1,670	5,000	608	
	4200	140 H.P.		48.40	765	2,300	6,900	847.20	
	4260	200 H.P.		72.10	1,350	4,035	12,100	1,384	
	4310	300 H.P.		95.10	1,900	5,695	17,100	1,900	
	4360	410 H.P.		125.20	2,275	6,830	20,500	2,368	
	4370	500 H.P.		160.05	3,425	10,310	30,900	3,342	
	4380	700 H.P.		264.25	5,175	15,520	46,600	5,218	
	4400	Loader, crawler, torque conv., diesel, 1-1/2 C.Y., 80 H.P.		32.65	510	1,535	4,600	568.20	
	4450	1-1/2 to 1-3/4 C.Y., 95 H.P.		33.60	655	1,970	5,900	662.80	
	4510	1-3/4 to 2-1/4 C.Y., 130 H.P.		53.20	920	2,760	8,275	977.60	
	4530	2-1/2 to 3-1/4 C.Y., 190 H.P.		65.10	1,275	3,800	11,400	1,281	
	4560	3-1/2 to 5 C.Y., 275 H.P.		82.45	1,450	4,340	13,000	1,528	
	4610	Front end loader, 4WD, articulated frame, diesel, 1 to 1-1/4 C.Y., 70 H.P.		18.85	270	810	2,425	312.80	
	4620	1-1/2 to 1-3/4 C.Y., 95 H.P.		23.25	293	880	2,650	362	

01 54 33 | Equipment Rental

		UNIT	HOURLY OPER. COST	RENT PER DAY	RENT PER WEEK	RENT PER MONTH	EQUIPMENT COST/DAY	
20	4650	1-3/4 to 2 C.Y., 130 H.P.	Ea.	24.90	370	1,105	3,325	420.20	**20**
	4710	2-1/2 to 3-1/2 C.Y., 145 H.P.		35.40	455	1,365	4,100	556.20	
	4730	3 to 4-1/2 C.Y., 185 H.P.		40.70	555	1,660	4,975	657.60	
	4760	5-1/4 to 5-3/4 C.Y., 270 H.P.		63.35	955	2,870	8,600	1,081	
	4810	7 to 9 C.Y., 475 H.P.		108.10	1,825	5,510	16,500	1,967	
	4870	9 to 11 C.Y., 620 H.P.		154.85	3,300	9,925	29,800	3,224	
	4880	Skid-steer loader, wheeled, 10 C.F., 30 H.P. gas		9.75	155	465	1,400	171	
	4890	1 C.Y., 78 H.P., diesel		18.80	265	795	2,375	309.40	
	4892	Skid-steer attachment, auger		.40	66.50	200	600	43.20	
	4893	Backhoe		.65	108	325	975	70.20	
	4894	Broom		.79	131	393	1,175	84.90	
	4895	Forks		.19	31.50	95	285	20.50	
	4896	Grapple		.55	92.50	277	830	59.80	
	4897	Concrete hammer		1.07	178	535	1,600	115.55	
	4898	Tree spade		1.03	172	515	1,550	111.25	
	4899	Trencher		.51	85.50	256	770	55.30	
	4900	Trencher, chain, boom type, gas, operator walking, 12 H.P.		4.45	46.50	140	420	63.60	
	4910	Operator riding, 40 H.P.		18.65	293	880	2,650	325.20	
	5000	Wheel type, diesel, 4' deep, 12" wide		80.25	860	2,575	7,725	1,157	
	5100	6' deep, 20" wide		90.50	2,000	6,030	18,100	1,930	
	5150	Chain type, diesel, 5' deep, 8" wide		27.70	400	1,195	3,575	460.60	
	5200	Diesel, 8' deep, 16" wide		116.95	2,550	7,620	22,900	2,460	
	5202	Rock trencher, wheel type, 6" wide x 18" deep		38.00	740	2,225	6,675	749	
	5206	Chain type, 18" wide x 7' deep		109.44	2,950	8,845	26,500	2,645	
	5210	Tree spade, self-propelled		11.28	267	800	2,400	250.25	
	5250	Truck, dump, 2-axle, 12 ton, 8 C.Y. payload, 220 H.P.		31.00	238	715	2,150	391	
	5300	Three axle dump, 16 ton, 12 C.Y. payload, 400 H.P.		54.70	340	1,025	3,075	642.60	
	5310	Four axle dump, 25 ton, 18 C.Y. payload, 450 H.P.		64.70	495	1,490	4,475	815.60	
	5350	Dump trailer only, rear dump, 16-1/2 C.Y.		5.50	140	420	1,250	128	
	5400	20 C.Y.		5.95	158	475	1,425	142.60	
	5450	Flatbed, single axle, 1-1/2 ton rating		22.05	68.50	205	615	217.40	
	5500	3 ton rating		26.40	98.50	295	885	270.20	
	5550	Off highway rear dump, 25 ton capacity		72.20	1,500	4,490	13,500	1,476	
	5600	35 ton capacity		78.15	1,500	4,535	13,600	1,532	
	5610	50 ton capacity		101.00	2,000	5,990	18,000	2,006	
	5620	65 ton capacity		104.00	2,050	6,160	18,500	2,064	
	5630	100 ton capacity		141.00	3,050	9,120	27,400	2,952	
	6000	Vibratory plow, 25 H.P., walking		7.45	63.50	190	570	97.60	
40	0010	**GENERAL EQUIPMENT RENTAL** without operators [R015433-10]							**40**
	0020	Aerial lift, scissor type, to 20' high, 1200 lb. capacity, electric	Ea.	3.15	55	165	495	58.20	
	0030	To 30' high, 1200 lb. capacity		3.60	71.50	215	645	71.80	
	0040	Over 30' high, 1500 lb. capacity		4.90	130	390	1,175	117.20	
	0070	Articulating boom, to 45' high, 500 lb. capacity, diesel [R015433-15]		10.05	277	830	2,500	246.40	
	0075	To 60' high, 500 lb. capacity		14.05	485	1,460	4,375	404.40	
	0080	To 80' high, 500 lb. capacity		16.90	585	1,760	5,275	487.20	
	0085	To 125' high, 500 lb. capacity		20.40	690	2,075	6,225	578.20	
	0100	Telescoping boom to 40' high, 500 lb. capacity, diesel		11.85	325	970	2,900	288.80	
	0105	To 45' high, 500 lb. capacity		12.60	330	995	2,975	299.80	
	0110	To 60' high, 500 lb. capacity		16.25	545	1,630	4,900	456	
	0115	To 80' high, 500 lb. capacity		22.45	645	1,930	5,800	565.60	
	0120	To 100' high, 500 lb. capacity		26.35	795	2,385	7,150	687.80	
	0125	To 120' high, 500 lb. capacity		29.80	940	2,815	8,450	801.40	
	0195	Air compressor, portable, 6.5 CFM, electric		.88	13.65	41	123	15.25	
	0196	Gasoline		.65	20.50	62	186	17.60	
	0200	Towed type, gas engine, 60 CFM		12.00	51.50	155	465	127	
	0300	160 CFM		13.95	53.50	160	480	143.60	
	0400	Diesel engine, rotary screw, 250 CFM		13.90	117	350	1,050	181.20	
	0500	365 CFM		18.75	143	430	1,300	236	

For customer support on your Building Construction Costs with RSMeans Data, call 800.448.8182.

715

40

		UNIT	HOURLY OPER. COST	RENT PER DAY	RENT PER WEEK	RENT PER MONTH	EQUIPMENT COST/DAY	
0550	450 CFM	Ea.	23.80	178	535	1,600	297.40	40
0600	600 CFM		41.75	247	740	2,225	482	
0700	750 CFM	▼	41.90	255	765	2,300	488.20	
0800	For silenced models, small sizes, add to rent		3%	5%	5%	5%		
0900	Large sizes, add to rent		5%	7%	7%	7%		
0930	Air tools, breaker, pavement, 60 lb.	Ea.	.50	10.35	31	93	10.20	
0940	80 lb.		.55	10.65	32	96	10.80	
0950	Drills, hand (jackhammer), 65 lb.		.60	18.35	55	165	15.80	
0960	Track or wagon, swing boom, 4" drifter		56.05	945	2,840	8,525	1,016	
0970	5" drifter		65.70	1,150	3,425	10,300	1,211	
0975	Track mounted quarry drill, 6" diameter drill		109.15	1,725	5,145	15,400	1,902	
0980	Dust control per drill		1.03	25.50	77	231	23.65	
0990	Hammer, chipping, 12 lb.		.60	26	78	234	20.40	
1000	Hose, air with couplings, 50' long, 3/4" diameter		.03	5	15	45	3.25	
1100	1" diameter		.04	6.35	19	57	4.10	
1200	1-1/2" diameter		.05	9	27	81	5.80	
1300	2" diameter		.07	12	36	108	7.75	
1400	2-1/2" diameter		.11	19	57	171	12.30	
1410	3" diameter		.14	23	69	207	14.90	
1450	Drill, steel, 7/8" x 2'		.05	8.65	26	78	5.60	
1460	7/8" x 6'		.05	9	27	81	5.80	
1520	Moil points		.02	3.67	11	33	2.35	
1525	Pneumatic nailer w/accessories		.49	32.50	97	291	23.30	
1530	Sheeting driver for 60 lb. breaker		.04	6	18	54	3.90	
1540	For 90 lb. breaker		.12	8	24	72	5.75	
1550	Spade, 25 lb.		.50	7.35	22	66	8.40	
1560	Tamper, single, 35 lb.		.55	36.50	109	325	26.20	
1570	Triple, 140 lb.		.82	54.50	164	490	39.35	
1580	Wrenches, impact, air powered, up to 3/4" bolt		.40	13	39	117	11	
1590	Up to 1-1/4" bolt		.50	24.50	73	219	18.60	
1600	Barricades, barrels, reflectorized, 1 to 99 barrels		.03	4.60	13.80	41.50	3	
1610	100 to 200 barrels		.02	3.53	10.60	32	2.30	
1620	Barrels with flashers, 1 to 99 barrels		.03	5.25	15.80	47.50	3.40	
1630	100 to 200 barrels		.03	4.20	12.60	38	2.75	
1640	Barrels with steady burn type C lights		.04	7	21	63	4.50	
1650	Illuminated board, trailer mounted, with generator		3.45	130	390	1,175	105.60	
1670	Portable barricade, stock, with flashers, 1 to 6 units		.03	5.25	15.80	47.50	3.40	
1680	25 to 50 units		.03	4.90	14.70	44	3.20	
1685	Butt fusion machine, wheeled, 1.5 HP electric, 2" - 8" diameter pipe		2.63	167	500	1,500	121.05	
1690	Tracked, 20 HP diesel, 4"-12" diameter pipe		10.89	545	1,640	4,925	415.10	
1695	83 HP diesel, 8" - 24" diameter pipe		47.24	2,400	7,215	21,600	1,821	
1700	Carts, brick, hand powered, 1000 lb. capacity		.47	78.50	235	705	50.75	
1800	Gas engine, 1500 lb., 7-1/2' lift		3.55	115	345	1,025	97.40	
1822	Dehumidifier, medium, 6 lb./hr., 150 CFM		1.11	69.50	209	625	50.70	
1824	Large, 18 lb./hr., 600 CFM		2.12	133	398	1,200	96.55	
1830	Distributor, asphalt, trailer mounted, 2000 gal., 38 H.P. diesel		10.25	360	1,075	3,225	297	
1840	3000 gal., 38 H.P. diesel		11.80	390	1,165	3,500	327.40	
1850	Drill, rotary hammer, electric		1.06	27	81	243	24.70	
1860	Carbide bit, 1-1/2" diameter, add to electric rotary hammer		.03	4.67	14	42	3.05	
1865	Rotary, crawler, 250 H.P.		142.30	2,300	6,925	20,800	2,523	
1870	Emulsion sprayer, 65 gal., 5 H.P. gas engine		2.75	104	312	935	84.40	
1880	200 gal., 5 H.P. engine		7.35	173	520	1,550	162.80	
1900	Floor auto-scrubbing machine, walk-behind, 28" path		5.03	325	980	2,950	236.25	
1930	Floodlight, mercury vapor, or quartz, on tripod, 1000 watt		.45	21.50	64	192	16.40	
1940	2000 watt		.85	43	129	385	32.60	
1950	Floodlights, trailer mounted with generator, 1 - 300 watt light		3.60	73.50	220	660	72.80	
1960	2 - 1000 watt lights		4.70	98.50	295	885	96.60	
2000	4 - 300 watt lights	▼	4.45	93.50	280	840	91.60	

01 54 33 | Equipment Rental

		UNIT	HOURLY OPER. COST	RENT PER DAY	RENT PER WEEK	RENT PER MONTH	EQUIPMENT COST/DAY	
40	2005	Foam spray rig, incl. box trailer, compressor, generator, proportioner	Ea.	28.70	520	1,565	4,700	542.60	**40**
	2015	Forklift, pneumatic tire, rough terr, straight mast, 5000 lb, 12' lift, gas		22.35	212	635	1,900	305.80	
	2025	8000 lb, 12' lift		26.00	285	855	2,575	379	
	2030	5000 lb, 12' lift, diesel		17.40	238	715	2,150	282.20	
	2035	8000 lb, 12' lift, diesel		18.45	270	810	2,425	309.60	
	2045	All terrain, telescoping boom, diesel, 5000 lb, 10' reach, 19' lift		18.25	330	985	2,950	343	
	2055	6600 lb, 29' reach, 42' lift		22.25	385	1,155	3,475	409	
	2065	10,000 lb, 31' reach, 45' lift		26.20	520	1,555	4,675	520.60	
	2070	Cushion tire, smooth floor, gas, 5000 lb capacity		9.55	83.50	250	750	126.40	
	2075	8000 lb capacity		13.65	103	310	930	171.20	
	2085	Diesel, 5000 lb capacity		8.75	90	270	810	124	
	2090	12,000 lb capacity		13.55	142	425	1,275	193.40	
	2095	20,000 lb capacity		19.75	178	535	1,600	265	
	2100	Generator, electric, gas engine, 1.5 kW to 3 kW		3.15	12.65	38	114	32.80	
	2200	5 kW		4.10	16	48	144	42.40	
	2300	10 kW		7.70	36.50	110	330	83.60	
	2400	25 kW		8.75	90	270	810	124	
	2500	Diesel engine, 20 kW		10.05	78.50	235	705	127.40	
	2600	50 kW		19.15	105	315	945	216.20	
	2700	100 kW		35.30	142	425	1,275	367.40	
	2800	250 kW		69.45	268	805	2,425	716.60	
	2850	Hammer, hydraulic, for mounting on boom, to 500 ft lb.		2.60	76.50	230	690	66.80	
	2860	1000 ft lb.		4.54	128	385	1,150	113.30	
	2900	Heaters, space, oil or electric, 50 MBH		1.39	8	24	72	15.90	
	3000	100 MBH		2.52	10.65	32	96	26.55	
	3100	300 MBH		7.36	38.50	115	345	81.90	
	3150	500 MBH		11.95	45	135	405	122.60	
	3200	Hose, water, suction with coupling, 20' long, 2" diameter		.02	3	9	27	1.95	
	3210	3" diameter		.03	4.67	14	42	3.05	
	3220	4" diameter		.03	5.35	16	48	3.45	
	3230	6" diameter		.11	18.35	55	165	11.90	
	3240	8" diameter		.34	56	168	505	36.30	
	3250	Discharge hose with coupling, 50' long, 2" diameter		.01	1.33	4	12	.90	
	3260	3" diameter		.02	2.67	8	24	1.75	
	3270	4" diameter		.02	3.67	11	33	2.35	
	3280	6" diameter		.06	9.65	29	87	6.30	
	3290	8" diameter		.18	30	90	270	19.45	
	3295	Insulation blower		.78	6.35	19	57	10.05	
	3300	Ladders, extension type, 16' to 36' long		.14	22.50	68	204	14.70	
	3400	40' to 60' long		.19	31	93	279	20.10	
	3405	Lance for cutting concrete		2.13	55.50	167	500	50.45	
	3407	Lawn mower, rotary, 22", 5 H.P.		1.35	33.50	101	305	31	
	3408	48" self-propelled		2.90	95	285	855	80.20	
	3410	Level, electronic, automatic, with tripod and leveling rod		.87	57.50	173	520	41.55	
	3430	Laser type, for pipe and sewer line and grade		.73	48.50	145	435	34.85	
	3440	Rotating beam for interior control		.80	53	159	475	38.20	
	3460	Builder's optical transit, with tripod and rod		.09	14.65	44	132	9.50	
	3500	Light towers, towable, with diesel generator, 2000 watt		4.45	93.50	280	840	91.60	
	3600	4000 watt		4.70	98.50	295	885	96.60	
	3700	Mixer, powered, plaster and mortar, 6 C.F., 7 H.P.		2.30	20	60	180	30.40	
	3800	10 C.F., 9 H.P.		2.45	32.50	98	294	39.20	
	3850	Nailer, pneumatic		.49	32.50	97	291	23.30	
	3900	Paint sprayers complete, 8 CFM		1.06	70.50	211	635	50.70	
	4000	17 CFM		1.89	126	378	1,125	90.70	
	4020	Pavers, bituminous, rubber tires, 8' wide, 50 H.P., diesel		30.50	540	1,615	4,850	567	
	4030	10' wide, 150 H.P.		97.10	1,950	5,875	17,600	1,952	
	4050	Crawler, 8' wide, 100 H.P., diesel		85.85	2,050	6,165	18,500	1,920	
	4060	10' wide, 150 H.P.		104.85	2,425	7,290	21,900	2,297	

		UNIT	HOURLY OPER. COST	RENT PER DAY	RENT PER WEEK	RENT PER MONTH	EQUIPMENT COST/DAY	
40	4070	Concrete paver, 12' to 24' wide, 250 H.P.	Ea.	96.45	1,575	4,730	14,200	1,718	**40**
	4080	Placer-spreader-trimmer, 24' wide, 300 H.P.		133.20	2,300	6,934	20,800	2,452	
	4100	Pump, centrifugal gas pump, 1-1/2" diam., 65 GPM		3.75	50	150	450	60	
	4200	2" diameter, 130 GPM		5.00	60	180	540	76	
	4300	3" diameter, 250 GPM		5.20	61.50	185	555	78.60	
	4400	6" diameter, 1500 GPM		25.80	198	595	1,775	325.40	
	4500	Submersible electric pump, 1-1/4" diameter, 55 GPM		.41	17.65	53	159	13.90	
	4600	1-1/2" diameter, 83 GPM		.45	20.50	61	183	15.80	
	4700	2" diameter, 120 GPM		1.54	25.50	76	228	27.50	
	4800	3" diameter, 300 GPM		2.78	45	135	405	49.25	
	4900	4" diameter, 560 GPM		14.43	172	515	1,550	218.45	
	5000	6" diameter, 1590 GPM		21.57	227	680	2,050	308.55	
	5100	Diaphragm pump, gas, single, 1-1/2" diameter		1.06	52	156	470	39.70	
	5200	2" diameter		4.00	65	195	585	71	
	5300	3" diameter		4.00	65	195	585	71	
	5400	Double, 4" diameter		5.90	112	335	1,000	114.20	
	5450	Pressure washer 5 GPM, 3000 psi		4.25	53.50	160	480	66	
	5460	7 GPM, 3000 psi		5.50	63.50	190	570	82	
	5500	Trash pump, self-priming, gas, 2" diameter		4.00	22.50	67	201	45.40	
	5600	Diesel, 4" diameter		7.65	96.50	290	870	119.20	
	5650	Diesel, 6" diameter		20.65	172	515	1,550	268.20	
	5655	Grout Pump		24.05	275	825	2,475	357.40	
	5700	Salamanders, L.P. gas fired, 100,000 Btu		2.72	13.65	41	123	29.95	
	5705	50,000 Btu		1.59	10.65	32	96	19.10	
	5720	Sandblaster, portable, open top, 3 C.F. capacity		.60	27.50	83	249	21.40	
	5730	6 C.F. capacity		.95	41.50	125	375	32.60	
	5740	Accessories for above		.13	22.50	67	201	14.45	
	5750	Sander, floor		.89	23.50	71	213	21.30	
	5760	Edger		.53	16	48	144	13.85	
	5800	Saw, chain, gas engine, 18" long		2.00	22.50	67	201	29.40	
	5900	Hydraulic powered, 36" long		.75	65	195	585	45	
	5950	60" long		.75	66.50	200	600	46	
	6000	Masonry, table mounted, 14" diameter, 5 H.P.		1.32	56.50	170	510	44.55	
	6050	Portable cut-off, 8 H.P.		2.15	35	105	315	38.20	
	6100	Circular, hand held, electric, 7-1/4" diameter		.23	4.67	14	42	4.65	
	6200	12" diameter		.23	8.35	25	75	6.85	
	6250	Wall saw, w/hydraulic power, 10 H.P.		8.30	61.50	185	555	103.40	
	6275	Shot blaster, walk-behind, 20" wide		4.80	305	910	2,725	220.40	
	6280	Sidewalk broom, walk-behind		2.21	74.50	224	670	62.50	
	6300	Steam cleaner, 100 gallons per hour		3.40	80	240	720	75.20	
	6310	200 gallons per hour		4.60	98.50	295	885	95.80	
	6340	Tar Kettle/Pot, 400 gallons		14.21	76.50	230	690	159.70	
	6350	Torch, cutting, acetylene-oxygen, 150' hose, excludes gases		.30	16	48	144	12	
	6360	Hourly operating cost includes tips and gas		21.00				168	
	6410	Toilet, portable chemical		.13	21.50	65	195	14.05	
	6420	Recycle flush type		.16	26.50	79	237	17.10	
	6430	Toilet, fresh water flush, garden hose,		.19	31.50	95	285	20.50	
	6440	Hoisted, non-flush, for high rise		.15	25.50	77	231	16.60	
	6465	Tractor, farm with attachment		20.65	335	1,010	3,025	367.20	
	6480	Trailers, platform, flush deck, 2 axle, 3 ton capacity		1.50	20	60	180	24	
	6500	25 ton capacity		6.05	133	400	1,200	128.40	
	6600	40 ton capacity		7.75	188	565	1,700	175	
	6700	3 axle, 50 ton capacity		8.40	208	625	1,875	192.20	
	6800	75 ton capacity		10.65	275	825	2,475	250.20	
	6810	Trailer mounted cable reel for high voltage line work		5.68	270	811	2,425	207.65	
	6820	Trailer mounted cable tensioning rig		11.27	535	1,610	4,825	412.15	
	6830	Cable pulling rig		71.16	3,000	9,030	27,100	2,375	
	6850	Portable cable/wire puller, 8000 lb max pulling capacity		3.71	167	502	1,500	130.10	

01 54 33 | Equipment Rental

		UNIT	HOURLY OPER. COST	RENT PER DAY	RENT PER WEEK	RENT PER MONTH	EQUIPMENT COST/DAY	
40								**40**
6900	Water tank trailer, engine driven discharge, 5000 gallons	Ea.	7.00	145	435	1,300	143	
6925	10,000 gallons		9.55	200	600	1,800	196.40	
6950	Water truck, off highway, 6000 gallons		82.10	785	2,350	7,050	1,127	
7010	Tram car for high voltage line work, powered, 2 conductor		6.64	147	440	1,325	141.10	
7020	Transit (builder's level) with tripod		.09	14.65	44	132	9.50	
7030	Trench box, 3000 lb., 6' x 8'		.56	93.50	280	840	60.50	
7040	7200 lb., 6' x 20'		.75	125	375	1,125	81	
7050	8000 lb., 8' x 16'		1.08	180	540	1,625	116.65	
7060	9500 lb., 8' x 20'		1.21	201	603	1,800	130.30	
7065	11,000 lb., 8' x 24'		1.27	211	633	1,900	136.75	
7070	12,000 lb., 10' x 20'		1.50	251	752	2,250	162.40	
7100	Truck, pickup, 3/4 ton, 2 wheel drive		11.80	58.50	175	525	129.40	
7200	4 wheel drive		12.00	73.50	220	660	140	
7250	Crew carrier, 9 passenger		16.70	86.50	260	780	185.60	
7290	Flat bed truck, 20,000 lb. GVW		19.50	127	380	1,150	232	
7300	Tractor, 4 x 2, 220 H.P.		27.20	198	595	1,775	336.60	
7410	330 H.P.		40.20	273	820	2,450	485.60	
7500	6 x 4, 380 H.P.		46.10	320	955	2,875	559.80	
7600	450 H.P.		56.10	385	1,155	3,475	679.80	
7610	Tractor, with A frame, boom and winch, 225 H.P.		30.10	278	835	2,500	407.80	
7620	Vacuum truck, hazardous material, 2500 gallons		12.20	293	880	2,650	273.60	
7625	5,000 gallons		12.65	410	1,230	3,700	347.20	
7650	Vacuum, HEPA, 16 gallon, wet/dry		.84	19	57	171	18.10	
7655	55 gallon, wet/dry		.82	28.50	86	258	23.75	
7660	Water tank, portable		.16	26.50	80	240	17.30	
7690	Sewer/catch basin vacuum, 14 C.Y., 1500 gallons		16.99	615	1,850	5,550	505.90	
7700	Welder, electric, 200 amp		3.81	17.35	52	156	40.90	
7800	300 amp		5.67	21.50	64	192	58.15	
7900	Gas engine, 200 amp		12.30	26.50	80	240	114.40	
8000	300 amp		14.10	28.50	85	255	129.80	
8100	Wheelbarrow, any size		.08	12.65	38	114	8.25	
8200	Wrecking ball, 4000 lb.		2.35	70	210	630	60.80	
50	**HIGHWAY EQUIPMENT RENTAL** without operators	R015433 -10						**50**
0010								
0050	Asphalt batch plant, portable drum mixer, 100 ton/hr.	Ea.	80.67	1,500	4,535	13,600	1,552	
0060	200 ton/hr.		92.19	1,600	4,810	14,400	1,700	
0070	300 ton/hr.		109.71	1,875	5,660	17,000	2,010	
0100	Backhoe attachment, long stick, up to 185 H.P., 10.5' long		.36	24	72	216	17.30	
0140	Up to 250 H.P., 12' long		.40	26.50	79	237	19	
0180	Over 250 H.P., 15' long		.55	36.50	109	325	26.20	
0200	Special dipper arm, up to 100 H.P., 32' long		1.12	74.50	223	670	53.55	
0240	Over 100 H.P., 33' long		1.39	92.50	278	835	66.70	
0280	Catch basin/sewer cleaning truck, 3 ton, 9 C.Y., 1000 gal.		38.00	390	1,175	3,525	539	
0300	Concrete batch plant, portable, electric, 200 C.Y./hr.		23.11	525	1,570	4,700	498.90	
0520	Grader/dozer attachment, ripper/scarifier, rear mounted, up to 135 H.P.		3.00	60	180	540	60	
0540	Up to 180 H.P.		3.95	90	270	810	85.60	
0580	Up to 250 H.P.		5.60	142	425	1,275	129.80	
0700	Pvmt. removal bucket, for hyd. excavator, up to 90 H.P.		1.95	53.50	160	480	47.60	
0740	Up to 200 H.P.		2.10	68.50	205	615	57.80	
0780	Over 200 H.P.		2.30	86.50	260	780	70.40	
0900	Aggregate spreader, self-propelled, 187 H.P.		50.25	690	2,075	6,225	817	
1000	Chemical spreader, 3 C.Y.		3.15	43.50	130	390	51.20	
1900	Hammermill, traveling, 250 H.P.		66.08	2,175	6,490	19,500	1,827	
2000	Horizontal borer, 3" diameter, 13 H.P. gas driven		5.65	58.50	175	525	80.20	
2150	Horizontal directional drill, 20,000 lb. thrust, 78 H.P. diesel		28.20	710	2,130	6,400	651.60	
2160	30,000 lb. thrust, 115 H.P.		34.75	1,075	3,255	9,775	929	
2170	50,000 lb. thrust, 170 H.P.		49.65	1,375	4,160	12,500	1,229	
2190	Mud trailer for HDD, 1500 gallons, 175 H.P., gas		27.45	153	460	1,375	311.60	
2200	Hydromulcher, diesel, 3000 gallon, for truck mounting		21.10	248	745	2,225	317.80	

01 54 33 | Equipment Rental

		UNIT	HOURLY OPER. COST	RENT PER DAY	RENT PER WEEK	RENT PER MONTH	EQUIPMENT COST/DAY	
50	2300	Gas, 600 gallon	Ea.	7.80	100	300	900	122.40	**50**
	2400	Joint & crack cleaner, walk behind, 25 H.P.		3.45	50	150	450	57.60	
	2500	Filler, trailer mounted, 400 gallons, 20 H.P.		9.10	212	635	1,900	199.80	
	3000	Paint striper, self-propelled, 40 gallon, 22 H.P.		6.90	157	470	1,400	149.20	
	3100	120 gallon, 120 H.P.		20.55	400	1,195	3,575	403.40	
	3200	Post drivers, 6" I-Beam frame, for truck mounting		16.90	385	1,150	3,450	365.20	
	3400	Road sweeper, self-propelled, 8' wide, 90 H.P.		39.10	645	1,940	5,825	700.80	
	3450	Road sweeper, vacuum assisted, 4 C.Y., 220 gallons		68.85	630	1,885	5,650	927.80	
	4000	Road mixer, self-propelled, 130 H.P.		44.90	840	2,520	7,550	863.20	
	4100	310 H.P.		76.55	2,225	6,650	20,000	1,942	
	4220	Cold mix paver, incl. pug mill and bitumen tank, 165 H.P.		92.95	2,500	7,535	22,600	2,251	
	4240	Pavement brush, towed		3.25	93.50	280	840	82	
	4250	Paver, asphalt, wheel or crawler, 130 H.P., diesel		92.60	2,500	7,470	22,400	2,235	
	4300	Paver, road widener, gas, 1' to 6', 67 H.P.		46.00	945	2,830	8,500	934	
	4400	Diesel, 2' to 14', 88 H.P.		57.45	1,125	3,385	10,200	1,137	
	4600	Slipform pavers, curb and gutter, 2 track, 75 H.P.		62.00	1,175	3,490	10,500	1,194	
	4700	4 track, 165 H.P.		40.30	800	2,400	7,200	802.40	
	4800	Median barrier, 215 H.P.		63.25	1,225	3,695	11,100	1,245	
	4901	Trailer, low bed, 75 ton capacity		10.80	260	780	2,350	242.40	
	5000	Road planer, walk behind, 10" cutting width, 10 H.P.		3.30	33.50	100	300	46.40	
	5100	Self-propelled, 12" cutting width, 64 H.P.		8.95	113	340	1,025	139.60	
	5120	Traffic line remover, metal ball blaster, truck mounted, 115 H.P.		46.35	840	2,515	7,550	873.80	
	5140	Grinder, truck mounted, 115 H.P.		51.30	900	2,705	8,125	951.40	
	5160	Walk-behind, 11 H.P.		3.75	53.50	160	480	62	
	5200	Pavement profiler, 4' to 6' wide, 450 H.P.		225.65	3,500	10,525	31,600	3,910	
	5300	8' to 10' wide, 750 H.P.		349.00	4,600	13,805	41,400	5,553	
	5400	Roadway plate, steel, 1" x 8' x 20'		.08	14	42	126	9.05	
	5600	Stabilizer, self-propelled, 150 H.P.		43.90	715	2,150	6,450	781.20	
	5700	310 H.P.		83.60	1,975	5,960	17,900	1,861	
	5800	Striper, truck mounted, 120 gallon paint, 460 H.P.		55.20	490	1,465	4,400	734.60	
	5900	Thermal paint heating kettle, 115 gallons		7.26	25.50	77	231	73.50	
	6000	Tar kettle, 330 gallon, trailer mounted		10.97	58.50	175	525	122.75	
	7000	Tunnel locomotive, diesel, 8 to 12 ton		29.80	610	1,835	5,500	605.40	
	7005	Electric, 10 ton		26.90	700	2,105	6,325	636.20	
	7010	Muck cars, 1/2 C.Y. capacity		2.10	25.50	76	228	32	
	7020	1 C.Y. capacity		2.30	33.50	100	300	38.40	
	7030	2 C.Y. capacity		2.45	40	120	360	43.60	
	7040	Side dump, 2 C.Y. capacity		2.65	46.50	140	420	49.20	
	7050	3 C.Y. capacity		3.60	53.50	160	480	60.80	
	7060	5 C.Y. capacity		5.15	68.50	205	615	82.20	
	7100	Ventilating blower for tunnel, 7-1/2 H.P.		2.08	53.50	160	480	48.65	
	7110	10 H.P.		2.26	55	165	495	51.10	
	7120	20 H.P.		3.58	71.50	215	645	71.65	
	7140	40 H.P.		5.75	102	305	915	107	
	7160	60 H.P.		8.83	158	475	1,425	165.65	
	7175	75 H.P.		11.60	218	655	1,975	223.80	
	7180	200 H.P.		23.76	315	950	2,850	380.10	
	7800	Windrow loader, elevating		53.10	1,375	4,145	12,400	1,254	
60	0010	**LIFTING AND HOISTING EQUIPMENT RENTAL** without operators R015433 -10							**60**
	0150	Crane, flatbed mounted, 3 ton capacity	Ea.	14.60	218	655	1,975	247.80	
	0200	Crane, climbing, 106' jib, 6000 lb. capacity, 410 fpm R312316 -45		40.06	1,725	5,160	15,500	1,352	
	0300	101' jib, 10,250 lb. capacity, 270 fpm		46.96	2,175	6,540	19,600	1,684	
	0500	Tower, static, 130' high, 106' jib, 6200 lb. capacity at 400 fpm		44.06	1,975	5,960	17,900	1,544	
	0600	Crawler mounted, lattice boom, 1/2 C.Y., 15 tons at 12' radius		37.97	755	2,270	6,800	757.75	
	0700	3/4 C.Y., 20 tons at 12' radius		50.63	945	2,835	8,500	972.05	
	0800	1 C.Y., 25 tons at 12' radius		67.50	1,250	3,780	11,300	1,296	
	0900	1-1/2 C.Y., 40 tons at 12' radius		67.95	1,275	3,845	11,500	1,313	
	1000	2 C.Y., 50 tons at 12' radius		80.20	1,725	5,168	15,500	1,675	

01 54 33 | Equipment Rental

		UNIT	HOURLY OPER. COST	RENT PER DAY	RENT PER WEEK	RENT PER MONTH	EQUIPMENT COST/DAY	
60	1100	3 C.Y., 75 tons at 12' radius	Ea.	77.65	1,700	5,120	15,400	1,645	60
	1200	100 ton capacity, 60' boom		88.00	1,825	5,500	16,500	1,804	
	1300	165 ton capacity, 60' boom		108.25	2,275	6,825	20,500	2,231	
	1400	200 ton capacity, 70' boom		143.85	2,900	8,730	26,200	2,897	
	1500	350 ton capacity, 80' boom		187.45	4,200	12,600	37,800	4,020	
	1600	Truck mounted, lattice boom, 6 x 4, 20 tons at 10' radius		39.35	1,275	3,800	11,400	1,075	
	1700	25 tons at 10' radius		42.61	1,375	4,135	12,400	1,168	
	1800	8 x 4, 30 tons at 10' radius		45.99	1,475	4,400	13,200	1,248	
	1900	40 tons at 12' radius		48.87	1,525	4,600	13,800	1,311	
	2000	60 tons at 15' radius		54.26	1,625	4,870	14,600	1,408	
	2050	82 tons at 15' radius		60.22	1,725	5,200	15,600	1,522	
	2100	90 tons at 15' radius		67.15	1,900	5,670	17,000	1,671	
	2200	115 tons at 15' radius		75.69	2,100	6,335	19,000	1,873	
	2300	150 tons at 18' radius		83.25	2,225	6,670	20,000	2,000	
	2350	165 tons at 18' radius		88.09	2,350	7,070	21,200	2,119	
	2400	Truck mounted, hydraulic, 12 ton capacity		35.35	485	1,450	4,350	572.80	
	2500	25 ton capacity		40.60	580	1,745	5,225	673.80	
	2550	33 ton capacity		49.70	775	2,330	7,000	863.60	
	2560	40 ton capacity		55.05	915	2,745	8,225	989.40	
	2600	55 ton capacity		63.75	945	2,830	8,500	1,076	
	2700	80 ton capacity		83.75	1,575	4,730	14,200	1,616	
	2720	100 ton capacity		93.00	1,700	5,085	15,300	1,761	
	2740	120 ton capacity		107.35	1,825	5,495	16,500	1,958	
	2760	150 ton capacity		113.30	2,225	6,700	20,100	2,246	
	2800	Self-propelled, 4 x 4, with telescoping boom, 5 ton		15.50	253	760	2,275	276	
	2900	12-1/2 ton capacity		26.95	365	1,100	3,300	435.60	
	3000	15 ton capacity		36.50	455	1,365	4,100	565	
	3050	20 ton capacity		29.85	525	1,575	4,725	553.80	
	3100	25 ton capacity		31.20	595	1,780	5,350	605.60	
	3150	40 ton capacity		47.40	660	1,985	5,950	776.20	
	3200	Derricks, guy, 20 ton capacity, 60' boom, 75' mast		22.29	420	1,263	3,800	430.90	
	3300	100' boom, 115' mast		35.37	725	2,170	6,500	716.95	
	3400	Stiffleg, 20 ton capacity, 70' boom, 37' mast		24.93	545	1,640	4,925	527.45	
	3500	100' boom, 47' mast		38.59	875	2,630	7,900	834.70	
	3550	Helicopter, small, lift to 1250 lb. maximum, w/pilot		94.10	3,400	10,200	30,600	2,793	
	3600	Hoists, chain type, overhead, manual, 3/4 ton		.15	.33	1	3	1.40	
	3900	10 ton		.75	6.65	20	60	10	
	4000	Hoist and tower, 5000 lb. cap., portable electric, 40' high		5.06	242	727	2,175	185.90	
	4100	For each added 10' section, add		.11	19	57	171	12.30	
	4200	Hoist and single tubular tower, 5000 lb. electric, 100' high		6.86	340	1,016	3,050	258.10	
	4300	For each added 6'-6" section, add		.20	33	99	297	21.40	
	4400	Hoist and double tubular tower, 5000 lb., 100' high		7.38	375	1,119	3,350	282.85	
	4500	For each added 6'-6" section, add		.22	36.50	109	325	23.55	
	4550	Hoist and tower, mast type, 6000 lb., 100' high		7.94	385	1,160	3,475	295.50	
	4570	For each added 10' section, add		.13	22.50	67	201	14.45	
	4600	Hoist and tower, personnel, electric, 2000 lb., 100' @ 125 fpm		16.88	1,025	3,090	9,275	753.05	
	4700	3000 lb., 100' @ 200 fpm		19.28	1,175	3,500	10,500	854.25	
	4800	3000 lb., 150' @ 300 fpm		21.38	1,300	3,920	11,800	955.05	
	4900	4000 lb., 100' @ 300 fpm		22.14	1,325	4,000	12,000	977.10	
	5000	6000 lb., 100' @ 275 fpm		23.80	1,400	4,190	12,600	1,028	
	5100	For added heights up to 500', add	L.F.	.01	1.67	5	15	1.10	
	5200	Jacks, hydraulic, 20 ton	Ea.	.05	2	6	18	1.60	
	5500	100 ton		.40	12	36	108	10.40	
	6100	Jacks, hydraulic, climbing w/50' jackrods, control console, 30 ton cap.		2.09	139	418	1,250	100.30	
	6150	For each added 10' jackrod section, add		.05	3.33	10	30	2.40	
	6300	50 ton capacity		3.36	224	672	2,025	161.30	
	6350	For each added 10' jackrod section, add		.06	4	12	36	2.90	
	6500	125 ton capacity		8.75	585	1,750	5,250	420	

01 54 33 | Equipment Rental

		UNIT	HOURLY OPER. COST	RENT PER DAY	RENT PER WEEK	RENT PER MONTH	EQUIPMENT COST/DAY	
60	6550	For each added 10' jackrod section, add	Ea.	.60	39.50	119	355	28.60	60
	6600	Cable jack, 10 ton capacity with 200' cable		1.75	117	350	1,050	84	
	6650	For each added 50' of cable, add	↓	.21	14	42	126	10.10	
70	0010	**WELLPOINT EQUIPMENT RENTAL** without operators	R015433 -10						70
	0020	Based on 2 months rental							
	0100	Combination jetting & wellpoint pump, 60 H.P. diesel	Ea.	15.32	345	1,036	3,100	329.75	
	0200	High pressure gas jet pump, 200 H.P., 300 psi	"	33.10	295	885	2,650	441.80	
	0300	Discharge pipe, 8" diameter	L.F.	.01	.56	1.68	5.05	.40	
	0350	12" diameter		.01	.83	2.48	7.45	.60	
	0400	Header pipe, flows up to 150 GPM, 4" diameter		.01	.51	1.53	4.59	.40	
	0500	400 GPM, 6" diameter		.01	.60	1.79	5.35	.45	
	0600	800 GPM, 8" diameter		.01	.83	2.48	7.45	.60	
	0700	1500 GPM, 10" diameter		.01	.87	2.61	7.85	.60	
	0800	2500 GPM, 12" diameter		.02	1.64	4.93	14.80	1.15	
	0900	4500 GPM, 16" diameter		.03	2.10	6.31	18.95	1.50	
	0950	For quick coupling aluminum and plastic pipe, add	↓	.03	2.18	6.54	19.60	1.55	
	1100	Wellpoint, 25' long, with fittings & riser pipe, 1-1/2" or 2" diameter	Ea.	.07	4.35	13.05	39	3.15	
	1200	Wellpoint pump, diesel powered, 4" suction, 20 H.P.		6.87	199	597	1,800	174.35	
	1300	6" suction, 30 H.P.		9.22	247	741	2,225	221.95	
	1400	8" suction, 40 H.P.		12.49	340	1,016	3,050	303.10	
	1500	10" suction, 75 H.P.		18.40	395	1,187	3,550	384.60	
	1600	12" suction, 100 H.P.		26.68	630	1,890	5,675	591.45	
	1700	12" suction, 175 H.P.	↓	38.17	695	2,090	6,275	723.35	
80	0010	**MARINE EQUIPMENT RENTAL** without operators	R015433 -10						80
	0200	Barge, 400 Ton, 30' wide x 90' long	Ea.	17.65	1,100	3,265	9,800	794.20	
	0240	800 Ton, 45' wide x 90' long		21.40	1,325	3,980	11,900	967.20	
	2000	Tugboat, diesel, 100 H.P.		35.00	220	660	1,975	412	
	2040	250 H.P.		71.15	395	1,190	3,575	807.20	
	2080	380 H.P.		145.20	1,200	3,575	10,700	1,877	
	3000	Small work boat, gas, 16-foot, 50 H.P.		14.75	61.50	185	555	155	
	4000	Large, diesel, 48-foot, 200 H.P.	↓	84.45	1,275	3,815	11,400	1,439	

Crew No.	Bare Costs Hr.	Bare Costs Daily	Incl. Subs O&P Hr.	Incl. Subs O&P Daily	Cost Per Labor-Hour Bare Costs	Cost Per Labor-Hour Incl. O&P
Crew A-1						
1 Building Laborer	$39.15	$313.20	$60.00	$480.00	$39.15	$60.00
1 Concrete Saw, Gas Manual		74.20		81.62	9.28	10.20
8 L.H., Daily Totals		$387.40		$561.62	$48.42	$70.20
Crew A-1A						
1 Skilled Worker	$51.45	$411.60	$79.20	$633.60	$51.45	$79.20
1 Shot Blaster, 20"		220.40		242.44	27.55	30.31
8 L.H., Daily Totals		$632.00		$876.04	$79.00	$109.51
Crew A-1B						
1 Building Laborer	$39.15	$313.20	$60.00	$480.00	$39.15	$60.00
1 Concrete Saw		153.60		168.96	19.20	21.12
8 L.H., Daily Totals		$466.80		$648.96	$58.35	$81.12
Crew A-1C						
1 Building Laborer	$39.15	$313.20	$60.00	$480.00	$39.15	$60.00
1 Chain Saw, Gas, 18"		29.40		32.34	3.67	4.04
8 L.H., Daily Totals		$342.60		$512.34	$42.83	$64.04
Crew A-1D						
1 Building Laborer	$39.15	$313.20	$60.00	$480.00	$39.15	$60.00
1 Vibrating Plate, Gas, 18"		33.60		36.96	4.20	4.62
8 L.H., Daily Totals		$346.80		$516.96	$43.35	$64.62
Crew A-1E						
1 Building Laborer	$39.15	$313.20	$60.00	$480.00	$39.15	$60.00
1 Vibrating Plate, Gas, 21"		42.80		47.08	5.35	5.88
8 L.H., Daily Totals		$356.00		$527.08	$44.50	$65.89
Crew A-1F						
1 Building Laborer	$39.15	$313.20	$60.00	$480.00	$39.15	$60.00
1 Rammer/Tamper, Gas, 8"		50.00		55.00	6.25	6.88
8 L.H., Daily Totals		$363.20		$535.00	$45.40	$66.88
Crew A-1G						
1 Building Laborer	$39.15	$313.20	$60.00	$480.00	$39.15	$60.00
1 Rammer/Tamper, Gas, 15"		56.80		62.48	7.10	7.81
8 L.H., Daily Totals		$370.00		$542.48	$46.25	$67.81
Crew A-1H						
1 Building Laborer	$39.15	$313.20	$60.00	$480.00	$39.15	$60.00
1 Exterior Steam Cleaner		75.20		82.72	9.40	10.34
8 L.H., Daily Totals		$388.40		$562.72	$48.55	$70.34
Crew A-1J						
1 Building Laborer	$39.15	$313.20	$60.00	$480.00	$39.15	$60.00
1 Cultivator, Walk-Behind, 5 H.P.		38.40		42.24	4.80	5.28
8 L.H., Daily Totals		$351.60		$522.24	$43.95	$65.28
Crew A-1K						
1 Building Laborer	$39.15	$313.20	$60.00	$480.00	$39.15	$60.00
1 Cultivator, Walk-Behind, 8 H.P.		68.70		75.57	8.59	9.45
8 L.H., Daily Totals		$381.90		$555.57	$47.74	$69.45
Crew A-1M						
1 Building Laborer	$39.15	$313.20	$60.00	$480.00	$39.15	$60.00
1 Snow Blower, Walk-Behind		62.50		68.75	7.81	8.59
8 L.H., Daily Totals		$375.70		$548.75	$46.96	$68.59

Crew No.	Bare Costs Hr.	Bare Costs Daily	Incl. Subs O&P Hr.	Incl. Subs O&P Daily	Cost Per Labor-Hour Bare Costs	Cost Per Labor-Hour Incl. O&P
Crew A-2						
2 Laborers	$39.15	$626.40	$60.00	$960.00	$40.82	$62.20
1 Truck Driver (light)	44.15	353.20	66.60	532.80		
1 Flatbed Truck, Gas, 1.5 Ton		217.40		239.14	9.06	9.96
24 L.H., Daily Totals		$1197.00		$1731.94	$49.88	$72.16
Crew A-2A						
2 Laborers	$39.15	$626.40	$60.00	$960.00	$40.82	$62.20
1 Truck Driver (light)	44.15	353.20	66.60	532.80		
1 Flatbed Truck, Gas, 1.5 Ton		217.40		239.14		
1 Concrete Saw		153.60		168.96	15.46	17.00
24 L.H., Daily Totals		$1350.60		$1900.90	$56.27	$79.20
Crew A-2B						
1 Truck Driver (light)	$44.15	$353.20	$66.60	$532.80	$44.15	$66.60
1 Flatbed Truck, Gas, 1.5 Ton		217.40		239.14	27.18	29.89
8 L.H., Daily Totals		$570.60		$771.94	$71.33	$96.49
Crew A-3A						
1 Equip. Oper. (light)	$51.00	$408.00	$77.15	$617.20	$51.00	$77.15
1 Pickup Truck, 4x4, 3/4 Ton		140.00		154.00	17.50	19.25
8 L.H., Daily Totals		$548.00		$771.20	$68.50	$96.40
Crew A-3B						
1 Equip. Oper. (medium)	$53.55	$428.40	$81.00	$648.00	$49.55	$74.85
1 Truck Driver (heavy)	45.55	364.40	68.70	549.60		
1 Dump Truck, 12 C.Y., 400 H.P.		642.60		706.86		
1 F.E. Loader, W.M., 2.5 C.Y.		556.20		611.82	74.92	82.42
16 L.H., Daily Totals		$1991.60		$2516.28	$124.47	$157.27
Crew A-3C						
1 Equip. Oper. (light)	$51.00	$408.00	$77.15	$617.20	$51.00	$77.15
1 Loader, Skid Steer, 78 H.P.		309.40		340.34	38.67	42.54
8 L.H., Daily Totals		$717.40		$957.54	$89.67	$119.69
Crew A-3D						
1 Truck Driver (light)	$44.15	$353.20	$66.60	$532.80	$44.15	$66.60
1 Pickup Truck, 4x4, 3/4 Ton		140.00		154.00		
1 Flatbed Trailer, 25 Ton		128.40		141.24	33.55	36.91
8 L.H., Daily Totals		$621.60		$828.04	$77.70	$103.51
Crew A-3E						
1 Equip. Oper. (crane)	$55.70	$445.60	$84.25	$674.00	$50.63	$76.47
1 Truck Driver (heavy)	45.55	364.40	68.70	549.60		
1 Pickup Truck, 4x4, 3/4 Ton		140.00		154.00	8.75	9.63
16 L.H., Daily Totals		$950.00		$1377.60	$59.38	$86.10
Crew A-3F						
1 Equip. Oper. (crane)	$55.70	$445.60	$84.25	$674.00	$50.63	$76.47
1 Truck Driver (heavy)	45.55	364.40	68.70	549.60		
1 Pickup Truck, 4x4, 3/4 Ton		140.00		154.00		
1 Truck Tractor, 6x4, 380 H.P.		559.80		615.78		
1 Lowbed Trailer, 75 Ton		242.40		266.64	58.89	64.78
16 L.H., Daily Totals		$1752.20		$2260.02	$109.51	$141.25

Crews - Standard

Crew A-3G

Crew A-3G	Bare Costs Hr.	Daily	Incl. Subs O&P Hr.	Daily	Cost Per Labor-Hour Bare Costs	Incl. O&P
1 Equip. Oper. (crane)	$55.70	$445.60	$84.25	$674.00	$50.63	$76.47
1 Truck Driver (heavy)	45.55	364.40	68.70	549.60		
1 Pickup Truck, 4x4, 3/4 Ton		140.00		154.00		
1 Truck Tractor, 6x4, 450 H.P.		679.80		747.78		
1 Lowbed Trailer, 75 Ton		242.40		266.64	66.39	73.03
16 L.H., Daily Totals		$1872.20		$2392.02	$117.01	$149.50

Crew A-3H

Crew A-3H	Bare Costs Hr.	Daily	Incl. Subs O&P Hr.	Daily	Cost Per Labor-Hour Bare Costs	Incl. O&P
1 Equip. Oper. (crane)	$55.70	$445.60	$84.25	$674.00	$55.70	$84.25
1 Hyd. Crane, 12 Ton (Daily)		767.80		844.58	95.97	105.57
8 L.H., Daily Totals		$1213.40		$1518.58	$151.68	$189.82

Crew A-3I

Crew A-3I	Bare Costs Hr.	Daily	Incl. Subs O&P Hr.	Daily	Cost Per Labor-Hour Bare Costs	Incl. O&P
1 Equip. Oper. (crane)	$55.70	$445.60	$84.25	$674.00	$55.70	$84.25
1 Hyd. Crane, 25 Ton (Daily)		904.80		995.28	113.10	124.41
8 L.H., Daily Totals		$1350.40		$1669.28	$168.80	$208.66

Crew A-3J

Crew A-3J	Bare Costs Hr.	Daily	Incl. Subs O&P Hr.	Daily	Cost Per Labor-Hour Bare Costs	Incl. O&P
1 Equip. Oper. (crane)	$55.70	$445.60	$84.25	$674.00	$55.70	$84.25
1 Hyd. Crane, 40 Ton (Daily)		1354.00		1489.40	169.25	186.18
8 L.H., Daily Totals		$1799.60		$2163.40	$224.95	$270.43

Crew A-3K

Crew A-3K	Bare Costs Hr.	Daily	Incl. Subs O&P Hr.	Daily	Cost Per Labor-Hour Bare Costs	Incl. O&P
1 Equip. Oper. (crane)	$55.70	$445.60	$84.25	$674.00	$51.85	$78.42
1 Equip. Oper. (oiler)	48.00	384.00	72.60	580.80		
1 Hyd. Crane, 55 Ton (Daily)		1455.00		1600.50		
1 P/U Truck, 3/4 Ton (Daily)		154.40		169.84	100.59	110.65
16 L.H., Daily Totals		$2439.00		$3025.14	$152.44	$189.07

Crew A-3L

Crew A-3L	Bare Costs Hr.	Daily	Incl. Subs O&P Hr.	Daily	Cost Per Labor-Hour Bare Costs	Incl. O&P
1 Equip. Oper. (crane)	$55.70	$445.60	$84.25	$674.00	$51.85	$78.42
1 Equip. Oper. (oiler)	48.00	384.00	72.60	580.80		
1 Hyd. Crane, 80 Ton (Daily)		2245.00		2469.50		
1 P/U Truck, 3/4 Ton (Daily)		154.40		169.84	149.96	164.96
16 L.H., Daily Totals		$3229.00		$3894.14	$201.81	$243.38

Crew A-3M

Crew A-3M	Bare Costs Hr.	Daily	Incl. Subs O&P Hr.	Daily	Cost Per Labor-Hour Bare Costs	Incl. O&P
1 Equip. Oper. (crane)	$55.70	$445.60	$84.25	$674.00	$51.85	$78.42
1 Equip. Oper. (oiler)	48.00	384.00	72.60	580.80		
1 Hyd. Crane, 100 Ton (Daily)		2439.00		2682.90		
1 P/U Truck, 3/4 Ton (Daily)		154.40		169.84	162.09	178.30
16 L.H., Daily Totals		$3423.00		$4107.54	$213.94	$256.72

Crew A-3N

Crew A-3N	Bare Costs Hr.	Daily	Incl. Subs O&P Hr.	Daily	Cost Per Labor-Hour Bare Costs	Incl. O&P
1 Equip. Oper. (crane)	$55.70	$445.60	$84.25	$674.00	$55.70	$84.25
1 Tower Crane (monthly)		1166.00		1282.60	145.75	160.32
8 L.H., Daily Totals		$1611.60		$1956.60	$201.45	$244.57

Crew A-3P

Crew A-3P	Bare Costs Hr.	Daily	Incl. Subs O&P Hr.	Daily	Cost Per Labor-Hour Bare Costs	Incl. O&P
1 Equip. Oper. (light)	$51.00	$408.00	$77.15	$617.20	$51.00	$77.15
1 A.T. Forklift, 31' reach, 45' lift		520.60		572.66	65.08	71.58
8 L.H., Daily Totals		$928.60		$1189.86	$116.08	$148.73

Crew A-3Q

Crew A-3Q	Bare Costs Hr.	Daily	Incl. Subs O&P Hr.	Daily	Cost Per Labor-Hour Bare Costs	Incl. O&P
1 Equip. Oper. (light)	$51.00	$408.00	$77.15	$617.20	$51.00	$77.15
1 Pickup Truck, 4x4, 3/4 Ton		140.00		154.00		
1 Flatbed Trailer, 3 Ton		24.00		26.40	20.50	22.55
8 L.H., Daily Totals		$572.00		$797.60	$71.50	$99.70

Crew A-3R

Crew A-3R	Bare Costs Hr.	Daily	Incl. Subs O&P Hr.	Daily	Cost Per Labor-Hour Bare Costs	Incl. O&P
1 Equip. Oper. (light)	$51.00	$408.00	$77.15	$617.20	$51.00	$77.15
1 Forklift, Smooth Floor, 8,000 Lb.		171.20		188.32	21.40	23.54
8 L.H., Daily Totals		$579.20		$805.52	$72.40	$100.69

Crew A-4

Crew A-4	Bare Costs Hr.	Daily	Incl. Subs O&P Hr.	Daily	Cost Per Labor-Hour Bare Costs	Incl. O&P
2 Carpenters	$49.25	$788.00	$75.45	$1207.20	$46.73	$71.28
1 Painter, Ordinary	41.70	333.60	62.95	503.60		
24 L.H., Daily Totals		$1121.60		$1710.80	$46.73	$71.28

Crew A-5

Crew A-5	Bare Costs Hr.	Daily	Incl. Subs O&P Hr.	Daily	Cost Per Labor-Hour Bare Costs	Incl. O&P
2 Laborers	$39.15	$626.40	$60.00	$960.00	$39.71	$60.73
.25 Truck Driver (light)	44.15	88.30	66.60	133.20		
.25 Flatbed Truck, Gas, 1.5 Ton		54.35		59.78	3.02	3.32
18 L.H., Daily Totals		$769.05		$1152.98	$42.73	$64.05

Crew A-6

Crew A-6	Bare Costs Hr.	Daily	Incl. Subs O&P Hr.	Daily	Cost Per Labor-Hour Bare Costs	Incl. O&P
1 Instrument Man	$51.45	$411.60	$79.20	$633.60	$49.63	$75.97
1 Rodman/Chainman	47.80	382.40	72.75	582.00		
1 Level, Electronic		41.55		45.70	2.60	2.86
16 L.H., Daily Totals		$835.55		$1261.31	$52.22	$78.83

Crew A-7

Crew A-7	Bare Costs Hr.	Daily	Incl. Subs O&P Hr.	Daily	Cost Per Labor-Hour Bare Costs	Incl. O&P
1 Chief of Party	$62.70	$501.60	$95.25	$762.00	$53.98	$82.40
1 Instrument Man	51.45	411.60	79.20	633.60		
1 Rodman/Chainman	47.80	382.40	72.75	582.00		
1 Level, Electronic		41.55		45.70	1.73	1.90
24 L.H., Daily Totals		$1337.15		$2023.31	$55.71	$84.30

Crew A-8

Crew A-8	Bare Costs Hr.	Daily	Incl. Subs O&P Hr.	Daily	Cost Per Labor-Hour Bare Costs	Incl. O&P
1 Chief of Party	$62.70	$501.60	$95.25	$762.00	$52.44	$79.99
1 Instrument Man	51.45	411.60	79.20	633.60		
2 Rodmen/Chainmen	47.80	764.80	72.75	1164.00		
1 Level, Electronic		41.55		45.70	1.30	1.43
32 L.H., Daily Totals		$1719.55		$2605.30	$53.74	$81.42

Crew A-9

Crew A-9	Bare Costs Hr.	Daily	Incl. Subs O&P Hr.	Daily	Cost Per Labor-Hour Bare Costs	Incl. O&P
1 Asbestos Foreman	$55.55	$444.40	$86.30	$690.40	$55.11	$85.64
7 Asbestos Workers	55.05	3082.80	85.55	4790.80		
64 L.H., Daily Totals		$3527.20		$5481.20	$55.11	$85.64

Crew A-10A

Crew A-10A	Bare Costs Hr.	Daily	Incl. Subs O&P Hr.	Daily	Cost Per Labor-Hour Bare Costs	Incl. O&P
1 Asbestos Foreman	$55.55	$444.40	$86.30	$690.40	$55.22	$85.80
2 Asbestos Workers	55.05	880.80	85.55	1368.80		
24 L.H., Daily Totals		$1325.20		$2059.20	$55.22	$85.80

Crew A-10B

Crew A-10B	Bare Costs Hr.	Daily	Incl. Subs O&P Hr.	Daily	Cost Per Labor-Hour Bare Costs	Incl. O&P
1 Asbestos Foreman	$55.55	$444.40	$86.30	$690.40	$55.17	$85.74
3 Asbestos Workers	55.05	1321.20	85.55	2053.20		
32 L.H., Daily Totals		$1765.60		$2743.60	$55.17	$85.74

Crew A-10C

Crew A-10C	Bare Costs Hr.	Daily	Incl. Subs O&P Hr.	Daily	Cost Per Labor-Hour Bare Costs	Incl. O&P
3 Asbestos Workers	$55.05	$1321.20	$85.55	$2053.20	$55.05	$85.55
1 Flatbed Truck, Gas, 1.5 Ton		217.40		239.14	9.06	9.96
24 L.H., Daily Totals		$1538.60		$2292.34	$64.11	$95.51

Crew A-10D

Crew No.	Bare Costs Hr.	Daily	Incl. Subs O&P Hr.	Daily	Cost Per Labor-Hour Bare Costs	Incl. O&P
2 Asbestos Workers	$55.05	$880.80	$85.55	$1368.80	$53.45	$81.99
1 Equip. Oper. (crane)	55.70	445.60	84.25	674.00		
1 Equip. Oper. (oiler)	48.00	384.00	72.60	580.80		
1 Hydraulic Crane, 33 Ton		863.60		949.96	26.99	29.69
32 L.H., Daily Totals		$2574.00		$3573.56	$80.44	$111.67

Crew A-11

Crew No.	Bare Costs Hr.	Daily	Incl. Subs O&P Hr.	Daily	Cost Per Labor-Hour Bare Costs	Incl. O&P
1 Asbestos Foreman	$55.55	$444.40	$86.30	$690.40	$55.11	$85.64
7 Asbestos Workers	55.05	3082.80	85.55	4790.80		
2 Chip. Hammers, 12 Lb., Elec.		40.80		44.88	.64	.70
64 L.H., Daily Totals		$3568.00		$5526.08	$55.75	$86.35

Crew A-12

Crew No.	Bare Costs Hr.	Daily	Incl. Subs O&P Hr.	Daily	Cost Per Labor-Hour Bare Costs	Incl. O&P
1 Asbestos Foreman	$55.55	$444.40	$86.30	$690.40	$55.11	$85.64
7 Asbestos Workers	55.05	3082.80	85.55	4790.80		
1 Trk-Mtd Vac, 14 CY, 1500 Gal.		505.90		556.49		
1 Flatbed Truck, 20,000 GVW		232.00		255.20	11.53	12.68
64 L.H., Daily Totals		$4265.10		$6292.89	$66.64	$98.33

Crew A-13

Crew No.	Bare Costs Hr.	Daily	Incl. Subs O&P Hr.	Daily	Cost Per Labor-Hour Bare Costs	Incl. O&P
1 Equip. Oper. (light)	$51.00	$408.00	$77.15	$617.20	$51.00	$77.15
1 Trk-Mtd Vac, 14 CY, 1500 Gal.		505.90		556.49		
1 Flatbed Truck, 20,000 GVW		232.00		255.20	92.24	101.46
8 L.H., Daily Totals		$1145.90		$1428.89	$143.24	$178.61

Crew B-1

Crew No.	Bare Costs Hr.	Daily	Incl. Subs O&P Hr.	Daily	Cost Per Labor-Hour Bare Costs	Incl. O&P
1 Labor Foreman (outside)	$41.15	$329.20	$63.05	$504.40	$39.82	$61.02
2 Laborers	39.15	626.40	60.00	960.00		
24 L.H., Daily Totals		$955.60		$1464.40	$39.82	$61.02

Crew B-1A

Crew No.	Bare Costs Hr.	Daily	Incl. Subs O&P Hr.	Daily	Cost Per Labor-Hour Bare Costs	Incl. O&P
1 Labor Foreman (outside)	$41.15	$329.20	$63.05	$504.40	$39.82	$61.02
2 Laborers	39.15	626.40	60.00	960.00		
2 Cutting Torches		24.00		26.40		
2 Sets of Gases		336.00		369.60	15.00	16.50
24 L.H., Daily Totals		$1315.60		$1860.40	$54.82	$77.52

Crew B-1B

Crew No.	Bare Costs Hr.	Daily	Incl. Subs O&P Hr.	Daily	Cost Per Labor-Hour Bare Costs	Incl. O&P
1 Labor Foreman (outside)	$41.15	$329.20	$63.05	$504.40	$43.79	$66.83
2 Laborers	39.15	626.40	60.00	960.00		
1 Equip. Oper. (crane)	55.70	445.60	84.25	674.00		
2 Cutting Torches		24.00		26.40		
2 Sets of Gases		336.00		369.60		
1 Hyd. Crane, 12 Ton		572.80		630.08	29.15	32.06
32 L.H., Daily Totals		$2334.00		$3164.48	$72.94	$98.89

Crew B-1C

Crew No.	Bare Costs Hr.	Daily	Incl. Subs O&P Hr.	Daily	Cost Per Labor-Hour Bare Costs	Incl. O&P
1 Labor Foreman (outside)	$41.15	$329.20	$63.05	$504.40	$39.82	$61.02
2 Laborers	39.15	626.40	60.00	960.00		
1 Telescoping Boom Lift, to 60'		456.00		501.60	19.00	20.90
24 L.H., Daily Totals		$1411.60		$1966.00	$58.82	$81.92

Crew B-1D

Crew No.	Bare Costs Hr.	Daily	Incl. Subs O&P Hr.	Daily	Cost Per Labor-Hour Bare Costs	Incl. O&P
2 Laborers	$39.15	$626.40	$60.00	$960.00	$39.15	$60.00
1 Small Work Boat, Gas, 50 H.P.		155.00		170.50		
1 Pressure Washer, 7 GPM		82.00		90.20	14.81	16.29
16 L.H., Daily Totals		$863.40		$1220.70	$53.96	$76.29

Crew B-1E

Crew No.	Bare Costs Hr.	Daily	Incl. Subs O&P Hr.	Daily	Cost Per Labor-Hour Bare Costs	Incl. O&P
1 Labor Foreman (outside)	$41.15	$329.20	$63.05	$504.40	$39.65	$60.76
3 Laborers	39.15	939.60	60.00	1440.00		
1 Work Boat, Diesel, 200 H.P.		1439.00		1582.90		
2 Pressure Washers, 7 GPM		164.00		180.40	50.09	55.10
32 L.H., Daily Totals		$2871.80		$3707.70	$89.74	$115.87

Crew B-1F

Crew No.	Bare Costs Hr.	Daily	Incl. Subs O&P Hr.	Daily	Cost Per Labor-Hour Bare Costs	Incl. O&P
2 Skilled Workers	$51.45	$823.20	$79.20	$1267.20	$47.35	$72.80
1 Laborer	39.15	313.20	60.00	480.00		
1 Small Work Boat, Gas, 50 H.P.		155.00		170.50		
1 Pressure Washer, 7 GPM		82.00		90.20	9.88	10.86
24 L.H., Daily Totals		$1373.40		$2007.90	$57.23	$83.66

Crew B-1G

Crew No.	Bare Costs Hr.	Daily	Incl. Subs O&P Hr.	Daily	Cost Per Labor-Hour Bare Costs	Incl. O&P
2 Laborers	$39.15	$626.40	$60.00	$960.00	$39.15	$60.00
1 Small Work Boat, Gas, 50 H.P.		155.00		170.50	9.69	10.66
16 L.H., Daily Totals		$781.40		$1130.50	$48.84	$70.66

Crew B-1H

Crew No.	Bare Costs Hr.	Daily	Incl. Subs O&P Hr.	Daily	Cost Per Labor-Hour Bare Costs	Incl. O&P
2 Skilled Workers	$51.45	$823.20	$79.20	$1267.20	$47.35	$72.80
1 Laborer	39.15	313.20	60.00	480.00		
1 Small Work Boat, Gas, 50 H.P.		155.00		170.50	6.46	7.10
24 L.H., Daily Totals		$1291.40		$1917.70	$53.81	$79.90

Crew B-1J

Crew No.	Bare Costs Hr.	Daily	Incl. Subs O&P Hr.	Daily	Cost Per Labor-Hour Bare Costs	Incl. O&P
1 Labor Foreman (inside)	$39.65	$317.20	$60.75	$486.00	$39.40	$60.38
1 Laborer	39.15	313.20	60.00	480.00		
16 L.H., Daily Totals		$630.40		$966.00	$39.40	$60.38

Crew B-1K

Crew No.	Bare Costs Hr.	Daily	Incl. Subs O&P Hr.	Daily	Cost Per Labor-Hour Bare Costs	Incl. O&P
1 Carpenter Foreman (inside)	$49.75	$398.00	$76.20	$609.60	$49.50	$75.83
1 Carpenter	49.25	394.00	75.45	603.60		
16 L.H., Daily Totals		$792.00		$1213.20	$49.50	$75.83

Crew B-2

Crew No.	Bare Costs Hr.	Daily	Incl. Subs O&P Hr.	Daily	Cost Per Labor-Hour Bare Costs	Incl. O&P
1 Labor Foreman (outside)	$41.15	$329.20	$63.05	$504.40	$39.55	$60.61
4 Laborers	39.15	1252.80	60.00	1920.00		
40 L.H., Daily Totals		$1582.00		$2424.40	$39.55	$60.61

Crew B-2A

Crew No.	Bare Costs Hr.	Daily	Incl. Subs O&P Hr.	Daily	Cost Per Labor-Hour Bare Costs	Incl. O&P
1 Labor Foreman (outside)	$41.15	$329.20	$63.05	$504.40	$39.82	$61.02
2 Laborers	39.15	626.40	60.00	960.00		
1 Telescoping Boom Lift, to 60'		456.00		501.60	19.00	20.90
24 L.H., Daily Totals		$1411.60		$1966.00	$58.82	$81.92

Crew B-3

Crew No.	Bare Costs Hr.	Daily	Incl. Subs O&P Hr.	Daily	Cost Per Labor-Hour Bare Costs	Incl. O&P
1 Labor Foreman (outside)	$41.15	$329.20	$63.05	$504.40	$44.02	$66.91
2 Laborers	39.15	626.40	60.00	960.00		
1 Equip. Oper. (medium)	53.55	428.40	81.00	648.00		
2 Truck Drivers (heavy)	45.55	728.80	68.70	1099.20		
1 Crawler Loader, 3 C.Y.		1281.00		1409.10		
2 Dump Trucks, 12 C.Y., 400 H.P.		1285.20		1413.72	53.46	58.81
48 L.H., Daily Totals		$4679.00		$6034.42	$97.48	$125.72

Crew B-3A

Crew No.	Bare Costs Hr.	Daily	Incl. Subs O&P Hr.	Daily	Cost Per Labor-Hour Bare Costs	Incl. O&P
4 Laborers	$39.15	$1252.80	$60.00	$1920.00	$42.03	$64.20
1 Equip. Oper. (medium)	53.55	428.40	81.00	648.00		
1 Hyd. Excavator, 1.5 C.Y.		965.00		1061.50	24.13	26.54
40 L.H., Daily Totals		$2646.20		$3629.50	$66.16	$90.74

Crew No.	Bare Costs		Incl. Subs O&P		Cost Per Labor-Hour	
Crew B-3B	Hr.	Daily	Hr.	Daily	Bare Costs	Incl. O&P
2 Laborers	$39.15	$626.40	$60.00	$960.00	$44.35	$67.42
1 Equip. Oper. (medium)	53.55	428.40	81.00	648.00		
1 Truck Driver (heavy)	45.55	364.40	68.70	549.60		
1 Backhoe Loader, 80 H.P.		394.40		433.84		
1 Dump Truck, 12 C.Y., 400 H.P.		642.60		706.86	32.41	35.65
32 L.H., Daily Totals		$2456.20		$3298.30	$76.76	$103.07
Crew B-3C	Hr.	Daily	Hr.	Daily	Bare Costs	Incl. O&P
3 Laborers	$39.15	$939.60	$60.00	$1440.00	$42.75	$65.25
1 Equip. Oper. (medium)	53.55	428.40	81.00	648.00		
1 Crawler Loader, 4 C.Y.		1528.00		1680.80	47.75	52.52
32 L.H., Daily Totals		$2896.00		$3768.80	$90.50	$117.78
Crew B-4	Hr.	Daily	Hr.	Daily	Bare Costs	Incl. O&P
1 Labor Foreman (outside)	$41.15	$329.20	$63.05	$504.40	$40.55	$61.96
4 Laborers	39.15	1252.80	60.00	1920.00		
1 Truck Driver (heavy)	45.55	364.40	68.70	549.60		
1 Truck Tractor, 220 H.P.		336.60		370.26		
1 Flatbed Trailer, 40 Ton		175.00		192.50	10.66	11.72
48 L.H., Daily Totals		$2458.00		$3536.76	$51.21	$73.68
Crew B-5	Hr.	Daily	Hr.	Daily	Bare Costs	Incl. O&P
1 Labor Foreman (outside)	$41.15	$329.20	$63.05	$504.40	$43.55	$66.44
4 Laborers	39.15	1252.80	60.00	1920.00		
2 Equip. Oper. (medium)	53.55	856.80	81.00	1296.00		
1 Air Compressor, 250 cfm		181.20		199.32		
2 Breakers, Pavement, 60 lb.		20.40		22.44		
2 -50' Air Hoses, 1.5"		11.60		12.76		
1 Crawler Loader, 3 C.Y.		1281.00		1409.10	26.68	29.35
56 L.H., Daily Totals		$3933.00		$5364.02	$70.23	$95.79
Crew B-5A	Hr.	Daily	Hr.	Daily	Bare Costs	Incl. O&P
1 Labor Foreman (outside)	$41.15	$329.20	$63.05	$504.40	$43.77	$66.63
6 Laborers	39.15	1879.20	60.00	2880.00		
2 Equip. Oper. (medium)	53.55	856.80	81.00	1296.00		
1 Equip. Oper. (light)	51.00	408.00	77.15	617.20		
2 Truck Drivers (heavy)	45.55	728.80	68.70	1099.20		
1 Air Compressor, 365 cfm		236.00		259.60		
2 Breakers, Pavement, 60 lb.		20.40		22.44		
8 -50' Air Hoses, 1"		32.80		36.08		
2 Dump Trucks, 8 C.Y., 220 H.P.		782.00		860.20	11.16	12.27
96 L.H., Daily Totals		$5273.20		$7575.12	$54.93	$78.91
Crew B-5B	Hr.	Daily	Hr.	Daily	Bare Costs	Incl. O&P
1 Powderman	$51.45	$411.60	$79.20	$633.60	$49.20	$74.55
2 Equip. Oper. (medium)	53.55	856.80	81.00	1296.00		
3 Truck Drivers (heavy)	45.55	1093.20	68.70	1648.80		
1 F.E. Loader, W.M., 2.5 C.Y.		556.20		611.82		
3 Dump Trucks, 12 C.Y., 400 H.P.		1927.80		2120.58		
1 Air Compressor, 365 CFM		236.00		259.60	56.67	62.33
48 L.H., Daily Totals		$5081.60		$6570.40	$105.87	$136.88

Crew No.	Bare Costs		Incl. Subs O&P		Cost Per Labor-Hour	
Crew B-5C	Hr.	Daily	Hr.	Daily	Bare Costs	Incl. O&P
3 Laborers	$39.15	$939.60	$60.00	$1440.00	$45.73	$69.41
1 Equip. Oper. (medium)	53.55	428.40	81.00	648.00		
2 Truck Drivers (heavy)	45.55	728.80	68.70	1099.20		
1 Equip. Oper. (crane)	55.70	445.60	84.25	674.00		
1 Equip. Oper. (oiler)	48.00	384.00	72.60	580.80		
2 Dump Trucks, 12 C.Y., 400 H.P.		1285.20		1413.72		
1 Crawler Loader, 4 C.Y.		1528.00		1680.80		
1 S.P. Crane, 4x4, 25 Ton		605.60		666.16	53.42	58.76
64 L.H., Daily Totals		$6345.20		$8202.68	$99.14	$128.17
Crew B-5D	Hr.	Daily	Hr.	Daily	Bare Costs	Incl. O&P
1 Labor Foreman (outside)	$41.15	$329.20	$63.05	$504.40	$43.80	$66.72
4 Laborers	39.15	1252.80	60.00	1920.00		
2 Equip. Oper. (medium)	53.55	856.80	81.00	1296.00		
1 Truck Driver (heavy)	45.55	364.40	68.70	549.60		
1 Air Compressor, 250 cfm		181.20		199.32		
2 Breakers, Pavement, 60 lb.		20.40		22.44		
2 -50' Air Hoses, 1.5"		11.60		12.76		
1 Crawler Loader, 3 C.Y.		1281.00		1409.10		
1 Dump Truck, 12 C.Y., 400 H.P.		642.60		706.86	33.39	36.73
64 L.H., Daily Totals		$4940.00		$6620.48	$77.19	$103.45
Crew B-6	Hr.	Daily	Hr.	Daily	Bare Costs	Incl. O&P
2 Laborers	$39.15	$626.40	$60.00	$960.00	$43.10	$65.72
1 Equip. Oper. (light)	51.00	408.00	77.15	617.20		
1 Backhoe Loader, 48 H.P.		366.20		402.82	15.26	16.78
24 L.H., Daily Totals		$1400.60		$1980.02	$58.36	$82.50
Crew B-6A	Hr.	Daily	Hr.	Daily	Bare Costs	Incl. O&P
.5 Labor Foreman (outside)	$41.15	$164.60	$63.05	$252.20	$45.31	$69.01
1 Laborer	39.15	313.20	60.00	480.00		
1 Equip. Oper. (medium)	53.55	428.40	81.00	648.00		
1 Vacuum Truck, 5000 Gal.		347.20		381.92	17.36	19.10
20 L.H., Daily Totals		$1253.40		$1762.12	$62.67	$88.11
Crew B-6B	Hr.	Daily	Hr.	Daily	Bare Costs	Incl. O&P
2 Labor Foremen (outside)	$41.15	$658.40	$63.05	$1008.80	$39.82	$61.02
4 Laborers	39.15	1252.80	60.00	1920.00		
1 S.P. Crane, 4x4, 5 Ton		276.00		303.60		
1 Flatbed Truck, Gas, 1.5 Ton		217.40		239.14		
1 Butt Fusion Mach., 4"-12" diam.		415.10		456.61	18.93	20.82
48 L.H., Daily Totals		$2819.70		$3928.15	$58.74	$81.84
Crew B-6C	Hr.	Daily	Hr.	Daily	Bare Costs	Incl. O&P
2 Labor Foremen (outside)	$41.15	$658.40	$63.05	$1008.80	$39.82	$61.02
4 Laborers	39.15	1252.80	60.00	1920.00		
1 S.P. Crane, 4x4, 12 Ton		435.60		479.16		
1 Flatbed Truck, Gas, 3 Ton		270.20		297.22		
1 Butt Fusion Mach., 8"-24" diam.		1821.00		2003.10	52.64	57.91
48 L.H., Daily Totals		$4438.00		$5708.28	$92.46	$118.92
Crew B-7	Hr.	Daily	Hr.	Daily	Bare Costs	Incl. O&P
1 Labor Foreman (outside)	$41.15	$329.20	$63.05	$504.40	$41.88	$64.01
4 Laborers	39.15	1252.80	60.00	1920.00		
1 Equip. Oper. (medium)	53.55	428.40	81.00	648.00		
1 Brush Chipper, 12", 130 H.P.		404.40		444.84		
1 Crawler Loader, 3 C.Y.		1281.00		1409.10		
2 Chain Saws, Gas, 36" Long		90.00		99.00	36.99	40.69
48 L.H., Daily Totals		$3785.80		$5025.34	$78.87	$104.69

Crew B-7A

Crew No.	Bare Hr.	Bare Daily	Incl. Subs O&P Hr.	Incl. Subs O&P Daily	Cost Per Labor-Hour Bare Costs	Cost Per Labor-Hour Incl. O&P
2 Laborers	$39.15	$626.40	$60.00	$960.00	$43.10	$65.72
1 Equip. Oper. (light)	51.00	408.00	77.15	617.20		
1 Rake w/Tractor		344.10		378.51		
2 Chain Saws, Gas, 18"		58.80		64.68	16.79	18.47
24 L.H., Daily Totals		$1437.30		$2020.39	$59.89	$84.18

Crew B-7B

Crew No.	Bare Hr.	Bare Daily	Incl. Subs O&P Hr.	Incl. Subs O&P Daily	Bare Costs	Incl. O&P
1 Labor Foreman (outside)	$41.15	$329.20	$63.05	$504.40	$42.41	$64.68
4 Laborers	39.15	1252.80	60.00	1920.00		
1 Equip. Oper. (medium)	53.55	428.40	81.00	648.00		
1 Truck Driver (heavy)	45.55	364.40	68.70	549.60		
1 Brush Chipper, 12", 130 H.P.		404.40		444.84		
1 Crawler Loader, 3 C.Y.		1281.00		1409.10		
2 Chain Saws, Gas, 36" Long		90.00		99.00		
1 Dump Truck, 8 C.Y., 220 H.P.		391.00		430.10	38.69	42.55
56 L.H., Daily Totals		$4541.20		$6005.04	$81.09	$107.23

Crew B-7C

Crew No.	Bare Hr.	Bare Daily	Incl. Subs O&P Hr.	Incl. Subs O&P Daily	Bare Costs	Incl. O&P
1 Labor Foreman (outside)	$41.15	$329.20	$63.05	$504.40	$42.41	$64.68
4 Laborers	39.15	1252.80	60.00	1920.00		
1 Equip. Oper. (medium)	53.55	428.40	81.00	648.00		
1 Truck Driver (heavy)	45.55	364.40	68.70	549.60		
1 Brush Chipper, 12", 130 H.P.		404.40		444.84		
1 Crawler Loader, 3 C.Y.		1281.00		1409.10		
2 Chain Saws, Gas, 36" Long		90.00		99.00		
1 Dump Truck, 12 C.Y., 400 H.P.		642.60		706.86	43.18	47.50
56 L.H., Daily Totals		$4792.80		$6281.80	$85.59	$112.18

Crew B-8

Crew No.	Bare Hr.	Bare Daily	Incl. Subs O&P Hr.	Incl. Subs O&P Daily	Bare Costs	Incl. O&P
1 Labor Foreman (outside)	$41.15	$329.20	$63.05	$504.40	$45.71	$69.38
2 Laborers	39.15	626.40	60.00	960.00		
2 Equip. Oper. (medium)	53.55	856.80	81.00	1296.00		
1 Equip. Oper. (oiler)	48.00	384.00	72.60	580.80		
2 Truck Drivers (heavy)	45.55	728.80	68.70	1099.20		
1 Hyd. Crane, 25 Ton		673.80		741.18		
1 Crawler Loader, 3 C.Y.		1281.00		1409.10		
2 Dump Trucks, 12 C.Y., 400 H.P.		1285.20		1413.72	50.63	55.69
64 L.H., Daily Totals		$6165.20		$8004.40	$96.33	$125.07

Crew B-9

Crew No.	Bare Hr.	Bare Daily	Incl. Subs O&P Hr.	Incl. Subs O&P Daily	Bare Costs	Incl. O&P
1 Labor Foreman (outside)	$41.15	$329.20	$63.05	$504.40	$39.55	$60.61
4 Laborers	39.15	1252.80	60.00	1920.00		
1 Air Compressor, 250 cfm		181.20		199.32		
2 Breakers, Pavement, 60 lb.		20.40		22.44		
2 -50' Air Hoses, 1.5"		11.60		12.76	5.33	5.86
40 L.H., Daily Totals		$1795.20		$2658.92	$44.88	$66.47

Crew B-9A

Crew No.	Bare Hr.	Bare Daily	Incl. Subs O&P Hr.	Incl. Subs O&P Daily	Bare Costs	Incl. O&P
2 Laborers	$39.15	$626.40	$60.00	$960.00	$41.28	$62.90
1 Truck Driver (heavy)	45.55	364.40	68.70	549.60		
1 Water Tank Trailer, 5000 Gal.		143.00		157.30		
1 Truck Tractor, 220 H.P.		336.60		370.26		
2 -50' Discharge Hoses, 3"		3.50		3.85	20.13	22.14
24 L.H., Daily Totals		$1473.90		$2041.01	$61.41	$85.04

Crew B-9B

Crew No.	Bare Hr.	Bare Daily	Incl. Subs O&P Hr.	Incl. Subs O&P Daily	Bare Costs	Incl. O&P
2 Laborers	$39.15	$626.40	$60.00	$960.00	$41.28	$62.90
1 Truck Driver (heavy)	45.55	364.40	68.70	549.60		
2 -50' Discharge Hoses, 3"		3.50		3.85		
1 Water Tank Trailer, 5000 Gal.		143.00		157.30		
1 Truck Tractor, 220 H.P.		336.60		370.26		
1 Pressure Washer		66.00		72.60	22.88	25.17
24 L.H., Daily Totals		$1539.90		$2113.61	$64.16	$88.07

Crew B-9D

Crew No.	Bare Hr.	Bare Daily	Incl. Subs O&P Hr.	Incl. Subs O&P Daily	Bare Costs	Incl. O&P
1 Labor Foreman (outside)	$41.15	$329.20	$63.05	$504.40	$39.55	$60.61
4 Common Laborers	39.15	1252.80	60.00	1920.00		
1 Air Compressor, 250 cfm		181.20		199.32		
2 -50' Air Hoses, 1.5"		11.60		12.76		
2 Air Powered Tampers		52.40		57.64	6.13	6.74
40 L.H., Daily Totals		$1827.20		$2694.12	$45.68	$67.35

Crew B-10

Crew No.	Bare Hr.	Bare Daily	Incl. Subs O&P Hr.	Incl. Subs O&P Daily	Bare Costs	Incl. O&P
1 Equip. Oper. (medium)	$53.55	$428.40	$81.00	$648.00	$48.75	$74.00
.5 Laborer	39.15	156.60	60.00	240.00		
12 L.H., Daily Totals		$585.00		$888.00	$48.75	$74.00

Crew B-10A

Crew No.	Bare Hr.	Bare Daily	Incl. Subs O&P Hr.	Incl. Subs O&P Daily	Bare Costs	Incl. O&P
1 Equip. Oper. (medium)	$53.55	$428.40	$81.00	$648.00	$48.75	$74.00
.5 Laborer	39.15	156.60	60.00	240.00		
1 Roller, 2-Drum, W.B., 7.5 H.P.		182.80		201.08	15.23	16.76
12 L.H., Daily Totals		$767.80		$1089.08	$63.98	$90.76

Crew B-10B

Crew No.	Bare Hr.	Bare Daily	Incl. Subs O&P Hr.	Incl. Subs O&P Daily	Bare Costs	Incl. O&P
1 Equip. Oper. (medium)	$53.55	$428.40	$81.00	$648.00	$48.75	$74.00
.5 Laborer	39.15	156.60	60.00	240.00		
1 Dozer, 200 H.P.		1384.00		1522.40	115.33	126.87
12 L.H., Daily Totals		$1969.00		$2410.40	$164.08	$200.87

Crew B-10C

Crew No.	Bare Hr.	Bare Daily	Incl. Subs O&P Hr.	Incl. Subs O&P Daily	Bare Costs	Incl. O&P
1 Equip. Oper. (medium)	$53.55	$428.40	$81.00	$648.00	$48.75	$74.00
.5 Laborer	39.15	156.60	60.00	240.00		
1 Dozer, 200 H.P.		1384.00		1522.40		
1 Vibratory Roller, Towed, 23 Ton		427.40		470.14	150.95	166.04
12 L.H., Daily Totals		$2396.40		$2880.54	$199.70	$240.04

Crew B-10D

Crew No.	Bare Hr.	Bare Daily	Incl. Subs O&P Hr.	Incl. Subs O&P Daily	Bare Costs	Incl. O&P
1 Equip. Oper. (medium)	$53.55	$428.40	$81.00	$648.00	$48.75	$74.00
.5 Laborer	39.15	156.60	60.00	240.00		
1 Dozer, 200 H.P.		1384.00		1522.40		
1 Sheepsft. Roller, Towed		451.00		496.10	152.92	168.21
12 L.H., Daily Totals		$2420.00		$2906.50	$201.67	$242.21

Crew B-10E

Crew No.	Bare Hr.	Bare Daily	Incl. Subs O&P Hr.	Incl. Subs O&P Daily	Bare Costs	Incl. O&P
1 Equip. Oper. (medium)	$53.55	$428.40	$81.00	$648.00	$48.75	$74.00
.5 Laborer	39.15	156.60	60.00	240.00		
1 Tandem Roller, 5 Ton		152.40		167.64	12.70	13.97
12 L.H., Daily Totals		$737.40		$1055.64	$61.45	$87.97

Crew B-10F

Crew No.	Bare Hr.	Bare Daily	Incl. Subs O&P Hr.	Incl. Subs O&P Daily	Bare Costs	Incl. O&P
1 Equip. Oper. (medium)	$53.55	$428.40	$81.00	$648.00	$48.75	$74.00
.5 Laborer	39.15	156.60	60.00	240.00		
1 Tandem Roller, 10 Ton		237.60		261.36	19.80	21.78
12 L.H., Daily Totals		$822.60		$1149.36	$68.55	$95.78

Crew No.	Bare Costs Hr.	Daily	Incl. Subs O&P Hr.	Daily	Cost Per Labor-Hour Bare Costs	Incl. O&P
Crew B-10G	Hr.	Daily	Hr.	Daily	Bare Costs	Incl. O&P
1 Equip. Oper. (medium)	$53.55	$428.40	$81.00	$648.00	$48.75	$74.00
.5 Laborer	39.15	156.60	60.00	240.00		
1 Sheepsfoot Roller, 240 H.P.		1296.00		1425.60	108.00	118.80
12 L.H., Daily Totals		$1881.00		$2313.60	$156.75	$192.80
Crew B-10H	Hr.	Daily	Hr.	Daily	Bare Costs	Incl. O&P
1 Equip. Oper. (medium)	$53.55	$428.40	$81.00	$648.00	$48.75	$74.00
.5 Laborer	39.15	156.60	60.00	240.00		
1 Diaphragm Water Pump, 2"		71.00		78.10		
1 -20' Suction Hose, 2"		1.95		2.15		
2 -50' Discharge Hoses, 2"		1.80		1.98	6.23	6.85
12 L.H., Daily Totals		$659.75		$970.23	$54.98	$80.85
Crew B-10I	Hr.	Daily	Hr.	Daily	Bare Costs	Incl. O&P
1 Equip. Oper. (medium)	$53.55	$428.40	$81.00	$648.00	$48.75	$74.00
.5 Laborer	39.15	156.60	60.00	240.00		
1 Diaphragm Water Pump, 4"		114.20		125.62		
1 -20' Suction Hose, 4"		3.45		3.79		
2 -50' Discharge Hoses, 4"		4.70		5.17	10.20	11.22
12 L.H., Daily Totals		$707.35		$1022.59	$58.95	$85.22
Crew B-10J	Hr.	Daily	Hr.	Daily	Bare Costs	Incl. O&P
1 Equip. Oper. (medium)	$53.55	$428.40	$81.00	$648.00	$48.75	$74.00
.5 Laborer	39.15	156.60	60.00	240.00		
1 Centrifugal Water Pump, 3"		78.60		86.46		
1 -20' Suction Hose, 3"		3.05		3.36		
2 -50' Discharge Hoses, 3"		3.50		3.85	7.10	7.81
12 L.H., Daily Totals		$670.15		$981.66	$55.85	$81.81
Crew B-10K	Hr.	Daily	Hr.	Daily	Bare Costs	Incl. O&P
1 Equip. Oper. (medium)	$53.55	$428.40	$81.00	$648.00	$48.75	$74.00
.5 Laborer	39.15	156.60	60.00	240.00		
1 Centr. Water Pump, 6"		325.40		357.94		
1 -20' Suction Hose, 6"		11.90		13.09		
2 -50' Discharge Hoses, 6"		12.60		13.86	29.16	32.07
12 L.H., Daily Totals		$934.90		$1272.89	$77.91	$106.07
Crew B-10L	Hr.	Daily	Hr.	Daily	Bare Costs	Incl. O&P
1 Equip. Oper. (medium)	$53.55	$428.40	$81.00	$648.00	$48.75	$74.00
.5 Laborer	39.15	156.60	60.00	240.00		
1 Dozer, 80 H.P.		468.60		515.46	39.05	42.95
12 L.H., Daily Totals		$1053.60		$1403.46	$87.80	$116.96
Crew B-10M	Hr.	Daily	Hr.	Daily	Bare Costs	Incl. O&P
1 Equip. Oper. (medium)	$53.55	$428.40	$81.00	$648.00	$48.75	$74.00
.5 Laborer	39.15	156.60	60.00	240.00		
1 Dozer, 300 H.P.		1900.00		2090.00	158.33	174.17
12 L.H., Daily Totals		$2485.00		$2978.00	$207.08	$248.17
Crew B-10N	Hr.	Daily	Hr.	Daily	Bare Costs	Incl. O&P
1 Equip. Oper. (medium)	$53.55	$428.40	$81.00	$648.00	$48.75	$74.00
.5 Laborer	39.15	156.60	60.00	240.00		
1 F.E. Loader, T.M., 1.5 C.Y.		568.20		625.02	47.35	52.09
12 L.H., Daily Totals		$1153.20		$1513.02	$96.10	$126.08
Crew B-10O	Hr.	Daily	Hr.	Daily	Bare Costs	Incl. O&P
1 Equip. Oper. (medium)	$53.55	$428.40	$81.00	$648.00	$48.75	$74.00
.5 Laborer	39.15	156.60	60.00	240.00		
1 F.E. Loader, T.M., 2.25 C.Y.		977.60		1075.36	81.47	89.61
12 L.H., Daily Totals		$1562.60		$1963.36	$130.22	$163.61

Crew No.	Bare Costs Hr.	Daily	Incl. Subs O&P Hr.	Daily	Cost Per Labor-Hour Bare Costs	Incl. O&P
Crew B-10P	Hr.	Daily	Hr.	Daily	Bare Costs	Incl. O&P
1 Equip. Oper. (medium)	$53.55	$428.40	$81.00	$648.00	$48.75	$74.00
.5 Laborer	39.15	156.60	60.00	240.00		
1 Crawler Loader, 3 C.Y.		1281.00		1409.10	106.75	117.43
12 L.H., Daily Totals		$1866.00		$2297.10	$155.50	$191.43
Crew B-10Q	Hr.	Daily	Hr.	Daily	Bare Costs	Incl. O&P
1 Equip. Oper. (medium)	$53.55	$428.40	$81.00	$648.00	$48.75	$74.00
.5 Laborer	39.15	156.60	60.00	240.00		
1 Crawler Loader, 4 C.Y.		1528.00		1680.80	127.33	140.07
12 L.H., Daily Totals		$2113.00		$2568.80	$176.08	$214.07
Crew B-10R	Hr.	Daily	Hr.	Daily	Bare Costs	Incl. O&P
1 Equip. Oper. (medium)	$53.55	$428.40	$81.00	$648.00	$48.75	$74.00
.5 Laborer	39.15	156.60	60.00	240.00		
1 F.E. Loader, W.M., 1 C.Y.		312.80		344.08	26.07	28.67
12 L.H., Daily Totals		$897.80		$1232.08	$74.82	$102.67
Crew B-10S	Hr.	Daily	Hr.	Daily	Bare Costs	Incl. O&P
1 Equip. Oper. (medium)	$53.55	$428.40	$81.00	$648.00	$48.75	$74.00
.5 Laborer	39.15	156.60	60.00	240.00		
1 F.E. Loader, W.M., 1.5 C.Y.		362.00		398.20	30.17	33.18
12 L.H., Daily Totals		$947.00		$1286.20	$78.92	$107.18
Crew B-10T	Hr.	Daily	Hr.	Daily	Bare Costs	Incl. O&P
1 Equip. Oper. (medium)	$53.55	$428.40	$81.00	$648.00	$48.75	$74.00
.5 Laborer	39.15	156.60	60.00	240.00		
1 F.E. Loader, W.M., 2.5 C.Y.		556.20		611.82	46.35	50.98
12 L.H., Daily Totals		$1141.20		$1499.82	$95.10	$124.99
Crew B-10U	Hr.	Daily	Hr.	Daily	Bare Costs	Incl. O&P
1 Equip. Oper. (medium)	$53.55	$428.40	$81.00	$648.00	$48.75	$74.00
.5 Laborer	39.15	156.60	60.00	240.00		
1 F.E. Loader, W.M., 5.5 C.Y.		1081.00		1189.10	90.08	99.09
12 L.H., Daily Totals		$1666.00		$2077.10	$138.83	$173.09
Crew B-10V	Hr.	Daily	Hr.	Daily	Bare Costs	Incl. O&P
1 Equip. Oper. (medium)	$53.55	$428.40	$81.00	$648.00	$48.75	$74.00
.5 Laborer	39.15	156.60	60.00	240.00		
1 Dozer, 700 H.P.		5218.00		5739.80	434.83	478.32
12 L.H., Daily Totals		$5803.00		$6627.80	$483.58	$552.32
Crew B-10W	Hr.	Daily	Hr.	Daily	Bare Costs	Incl. O&P
1 Equip. Oper. (medium)	$53.55	$428.40	$81.00	$648.00	$48.75	$74.00
.5 Laborer	39.15	156.60	60.00	240.00		
1 Dozer, 105 H.P.		608.00		668.80	50.67	55.73
12 L.H., Daily Totals		$1193.00		$1556.80	$99.42	$129.73
Crew B-10X	Hr.	Daily	Hr.	Daily	Bare Costs	Incl. O&P
1 Equip. Oper. (medium)	$53.55	$428.40	$81.00	$648.00	$48.75	$74.00
.5 Laborer	39.15	156.60	60.00	240.00		
1 Dozer, 410 H.P.		2368.00		2604.80	197.33	217.07
12 L.H., Daily Totals		$2953.00		$3492.80	$246.08	$291.07
Crew B-10Y	Hr.	Daily	Hr.	Daily	Bare Costs	Incl. O&P
1 Equip. Oper. (medium)	$53.55	$428.40	$81.00	$648.00	$48.75	$74.00
.5 Laborer	39.15	156.60	60.00	240.00		
1 Vibr. Roller, Towed, 12 Ton		605.40		665.94	50.45	55.49
12 L.H., Daily Totals		$1190.40		$1553.94	$99.20	$129.50

Crew No.	Bare Costs Hr.	Daily	Incl. Subs O&P Hr.	Daily	Cost Per Labor-Hour Bare Costs	Incl. O&P
Crew B-11A	Hr.	Daily	Hr.	Daily	Bare Costs	Incl. O&P
1 Equipment Oper. (med.)	$53.55	$428.40	$81.00	$648.00	$46.35	$70.50
1 Laborer	39.15	313.20	60.00	480.00		
1 Dozer, 200 H.P.		1384.00		1522.40	86.50	95.15
16 L.H., Daily Totals		$2125.60		$2650.40	$132.85	$165.65

Crew No.	Bare Costs Hr.	Daily	Incl. Subs O&P Hr.	Daily	Cost Per Labor-Hour Bare Costs	Incl. O&P
Crew B-11B	Hr.	Daily	Hr.	Daily	Bare Costs	Incl. O&P
1 Equipment Oper. (light)	$51.00	$408.00	$77.15	$617.20	$45.08	$68.58
1 Laborer	39.15	313.20	60.00	480.00		
1 Air Powered Tamper		26.20		28.82		
1 Air Compressor, 365 cfm		236.00		259.60		
2 -50' Air Hoses, 1.5"		11.60		12.76	17.11	18.82
16 L.H., Daily Totals		$995.00		$1398.38	$62.19	$87.40

Crew No.	Bare Costs Hr.	Daily	Incl. Subs O&P Hr.	Daily	Cost Per Labor-Hour Bare Costs	Incl. O&P
Crew B-11C	Hr.	Daily	Hr.	Daily	Bare Costs	Incl. O&P
1 Equipment Oper. (med.)	$53.55	$428.40	$81.00	$648.00	$46.35	$70.50
1 Laborer	39.15	313.20	60.00	480.00		
1 Backhoe Loader, 48 H.P.		366.20		402.82	22.89	25.18
16 L.H., Daily Totals		$1107.80		$1530.82	$69.24	$95.68

Crew No.	Bare Costs Hr.	Daily	Incl. Subs O&P Hr.	Daily	Cost Per Labor-Hour Bare Costs	Incl. O&P
Crew B-11J	Hr.	Daily	Hr.	Daily	Bare Costs	Incl. O&P
1 Equipment Oper. (med.)	$53.55	$428.40	$81.00	$648.00	$46.35	$70.50
1 Laborer	39.15	313.20	60.00	480.00		
1 Grader, 30,000 Lbs.		678.40		746.24		
1 Ripper, Beam & 1 Shank		85.60		94.16	47.75	52.52
16 L.H., Daily Totals		$1505.60		$1968.40	$94.10	$123.03

Crew No.	Bare Costs Hr.	Daily	Incl. Subs O&P Hr.	Daily	Cost Per Labor-Hour Bare Costs	Incl. O&P
Crew B-11K	Hr.	Daily	Hr.	Daily	Bare Costs	Incl. O&P
1 Equipment Oper. (med.)	$53.55	$428.40	$81.00	$648.00	$46.35	$70.50
1 Laborer	39.15	313.20	60.00	480.00		
1 Trencher, Chain Type, 8' D		2460.00		2706.00	153.75	169.13
16 L.H., Daily Totals		$3201.60		$3834.00	$200.10	$239.63

Crew No.	Bare Costs Hr.	Daily	Incl. Subs O&P Hr.	Daily	Cost Per Labor-Hour Bare Costs	Incl. O&P
Crew B-11L	Hr.	Daily	Hr.	Daily	Bare Costs	Incl. O&P
1 Equipment Oper. (med.)	$53.55	$428.40	$81.00	$648.00	$46.35	$70.50
1 Laborer	39.15	313.20	60.00	480.00		
1 Grader, 30,000 Lbs.		678.40		746.24	42.40	46.64
16 L.H., Daily Totals		$1420.00		$1874.24	$88.75	$117.14

Crew No.	Bare Costs Hr.	Daily	Incl. Subs O&P Hr.	Daily	Cost Per Labor-Hour Bare Costs	Incl. O&P
Crew B-11M	Hr.	Daily	Hr.	Daily	Bare Costs	Incl. O&P
1 Equipment Oper. (med.)	$53.55	$428.40	$81.00	$648.00	$46.35	$70.50
1 Laborer	39.15	313.20	60.00	480.00		
1 Backhoe Loader, 80 H.P.		394.40		433.84	24.65	27.11
16 L.H., Daily Totals		$1136.00		$1561.84	$71.00	$97.61

Crew No.	Bare Costs Hr.	Daily	Incl. Subs O&P Hr.	Daily	Cost Per Labor-Hour Bare Costs	Incl. O&P
Crew B-11N	Hr.	Daily	Hr.	Daily	Bare Costs	Incl. O&P
1 Labor Foreman (outside)	$41.15	$329.20	$63.05	$504.40	$46.84	$70.81
2 Equipment Operators (med.)	53.55	856.80	81.00	1296.00		
6 Truck Drivers (heavy)	45.55	2186.40	68.70	3297.60		
1 F.E. Loader, W.M., 5.5 C.Y.		1081.00		1189.10		
1 Dozer, 410 H.P.		2368.00		2604.80		
6 Dump Trucks, Off Hwy., 50 Ton		12036.00		13239.60	215.07	236.58
72 L.H., Daily Totals		$18857.40		$22131.50	$261.91	$307.38

Crew No.	Bare Costs Hr.	Daily	Incl. Subs O&P Hr.	Daily	Cost Per Labor-Hour Bare Costs	Incl. O&P
Crew B-11Q	Hr.	Daily	Hr.	Daily	Bare Costs	Incl. O&P
1 Equipment Operator (med.)	$53.55	$428.40	$81.00	$648.00	$48.75	$74.00
.5 Laborer	39.15	156.60	60.00	240.00		
1 Dozer, 140 H.P.		847.20		931.92	70.60	77.66
12 L.H., Daily Totals		$1432.20		$1819.92	$119.35	$151.66

Crew No.	Bare Costs Hr.	Daily	Incl. Subs O&P Hr.	Daily	Cost Per Labor-Hour Bare Costs	Incl. O&P
Crew B-11R	Hr.	Daily	Hr.	Daily	Bare Costs	Incl. O&P
1 Equipment Operator (med.)	$53.55	$428.40	$81.00	$648.00	$48.75	$74.00
.5 Laborer	39.15	156.60	60.00	240.00		
1 Dozer, 200 H.P.		1384.00		1522.40	115.33	126.87
12 L.H., Daily Totals		$1969.00		$2410.40	$164.08	$200.87

Crew No.	Bare Costs Hr.	Daily	Incl. Subs O&P Hr.	Daily	Cost Per Labor-Hour Bare Costs	Incl. O&P
Crew B-11S	Hr.	Daily	Hr.	Daily	Bare Costs	Incl. O&P
1 Equipment Operator (med.)	$53.55	$428.40	$81.00	$648.00	$48.75	$74.00
.5 Laborer	39.15	156.60	60.00	240.00		
1 Dozer, 300 H.P.		1900.00		2090.00		
1 Ripper, Beam & 1 Shank		85.60		94.16	165.47	182.01
12 L.H., Daily Totals		$2570.60		$3072.16	$214.22	$256.01

Crew No.	Bare Costs Hr.	Daily	Incl. Subs O&P Hr.	Daily	Cost Per Labor-Hour Bare Costs	Incl. O&P
Crew B-11T	Hr.	Daily	Hr.	Daily	Bare Costs	Incl. O&P
1 Equipment Operator (med.)	$53.55	$428.40	$81.00	$648.00	$48.75	$74.00
.5 Laborer	39.15	156.60	60.00	240.00		
1 Dozer, 410 H.P.		2368.00		2604.80		
1 Ripper, Beam & 2 Shanks		129.80		142.78	208.15	228.97
12 L.H., Daily Totals		$3082.80		$3635.58	$256.90	$302.96

Crew No.	Bare Costs Hr.	Daily	Incl. Subs O&P Hr.	Daily	Cost Per Labor-Hour Bare Costs	Incl. O&P
Crew B-11U	Hr.	Daily	Hr.	Daily	Bare Costs	Incl. O&P
1 Equipment Operator (med.)	$53.55	$428.40	$81.00	$648.00	$48.75	$74.00
.5 Laborer	39.15	156.60	60.00	240.00		
1 Dozer, 520 H.P.		3342.00		3676.20	278.50	306.35
12 L.H., Daily Totals		$3927.00		$4564.20	$327.25	$380.35

Crew No.	Bare Costs Hr.	Daily	Incl. Subs O&P Hr.	Daily	Cost Per Labor-Hour Bare Costs	Incl. O&P
Crew B-11V	Hr.	Daily	Hr.	Daily	Bare Costs	Incl. O&P
3 Laborers	$39.15	$939.60	$60.00	$1440.00	$39.15	$60.00
1 Roller, 2-Drum, W.B., 7.5 H.P.		182.80		201.08	7.62	8.38
24 L.H., Daily Totals		$1122.40		$1641.08	$46.77	$68.38

Crew No.	Bare Costs Hr.	Daily	Incl. Subs O&P Hr.	Daily	Cost Per Labor-Hour Bare Costs	Incl. O&P
Crew B-11W	Hr.	Daily	Hr.	Daily	Bare Costs	Incl. O&P
1 Equipment Operator (med.)	$53.55	$428.40	$81.00	$648.00	$45.68	$69.00
1 Common Laborer	39.15	313.20	60.00	480.00		
10 Truck Drivers (heavy)	45.55	3644.00	68.70	5496.00		
1 Dozer, 200 H.P.		1384.00		1522.40		
1 Vibratory Roller, Towed, 23 Ton		427.40		470.14		
10 Dump Trucks, 8 C.Y., 220 H.P.		3910.00		4301.00	59.60	65.56
96 L.H., Daily Totals		$10107.00		$12917.54	$105.28	$134.56

Crew No.	Bare Costs Hr.	Daily	Incl. Subs O&P Hr.	Daily	Cost Per Labor-Hour Bare Costs	Incl. O&P
Crew B-11Y	Hr.	Daily	Hr.	Daily	Bare Costs	Incl. O&P
1 Labor Foreman (outside)	$41.15	$329.20	$63.05	$504.40	$44.17	$67.34
5 Common Laborers	39.15	1566.00	60.00	2400.00		
3 Equipment Operators (med.)	53.55	1285.20	81.00	1944.00		
1 Dozer, 80 H.P.		468.60		515.46		
2 Rollers, 2-Drum, W.B., 7.5 H.P.		365.60		402.16		
4 Vibrating Plates, Gas, 21"		171.20		188.32	13.96	15.36
72 L.H., Daily Totals		$4185.80		$5954.34	$58.14	$82.70

Crew No.	Bare Costs Hr.	Daily	Incl. Subs O&P Hr.	Daily	Cost Per Labor-Hour Bare Costs	Incl. O&P
Crew B-12A	Hr.	Daily	Hr.	Daily	Bare Costs	Incl. O&P
1 Equip. Oper. (crane)	$55.70	$445.60	$84.25	$674.00	$47.42	$72.13
1 Laborer	39.15	313.20	60.00	480.00		
1 Hyd. Excavator, 1 C.Y.		798.20		878.02	49.89	54.88
16 L.H., Daily Totals		$1557.00		$2032.02	$97.31	$127.00

Crew No.	Bare Costs Hr.	Daily	Incl. Subs O&P Hr.	Daily	Cost Per Labor-Hour Bare Costs	Incl. O&P
Crew B-12B	Hr.	Daily	Hr.	Daily	Bare Costs	Incl. O&P
1 Equip. Oper. (crane)	$55.70	$445.60	$84.25	$674.00	$47.42	$72.13
1 Laborer	39.15	313.20	60.00	480.00		
1 Hyd. Excavator, 1.5 C.Y.		965.00		1061.50	60.31	66.34
16 L.H., Daily Totals		$1723.80		$2215.50	$107.74	$138.47

729

Crew No.	Bare Costs		Incl. Subs O&P		Cost Per Labor-Hour	
Crew B-12C	**Hr.**	**Daily**	**Hr.**	**Daily**	**Bare Costs**	**Incl. O&P**
1 Equip. Oper. (crane)	$55.70	$445.60	$84.25	$674.00	$47.42	$72.13
1 Laborer	39.15	313.20	60.00	480.00		
1 Hyd. Excavator, 2 C.Y.		1123.00		1235.30	70.19	77.21
16 L.H., Daily Totals		$1881.80		$2389.30	$117.61	$149.33

Crew B-12D	Hr.	Daily	Hr.	Daily	Bare Costs	Incl. O&P
1 Equip. Oper. (crane)	$55.70	$445.60	$84.25	$674.00	$47.42	$72.13
1 Laborer	39.15	313.20	60.00	480.00		
1 Hyd. Excavator, 3.5 C.Y.		2401.00		2641.10	150.06	165.07
16 L.H., Daily Totals		$3159.80		$3795.10	$197.49	$237.19

Crew B-12E	Hr.	Daily	Hr.	Daily	Bare Costs	Incl. O&P
1 Equip. Oper. (crane)	$55.70	$445.60	$84.25	$674.00	$47.42	$72.13
1 Laborer	39.15	313.20	60.00	480.00		
1 Hyd. Excavator, .5 C.Y.		443.60		487.96	27.73	30.50
16 L.H., Daily Totals		$1202.40		$1641.96	$75.15	$102.62

Crew B-12F	Hr.	Daily	Hr.	Daily	Bare Costs	Incl. O&P
1 Equip. Oper. (crane)	$55.70	$445.60	$84.25	$674.00	$47.42	$72.13
1 Laborer	39.15	313.20	60.00	480.00		
1 Hyd. Excavator, .75 C.Y.		674.20		741.62	42.14	46.35
16 L.H., Daily Totals		$1433.00		$1895.62	$89.56	$118.48

Crew B-12G	Hr.	Daily	Hr.	Daily	Bare Costs	Incl. O&P
1 Equip. Oper. (crane)	$55.70	$445.60	$84.25	$674.00	$47.42	$72.13
1 Laborer	39.15	313.20	60.00	480.00		
1 Crawler Crane, 15 Ton		757.75		833.52		
1 Clamshell Bucket, .5 C.Y.		38.20		42.02	49.75	54.72
16 L.H., Daily Totals		$1554.75		$2029.55	$97.17	$126.85

Crew B-12H	Hr.	Daily	Hr.	Daily	Bare Costs	Incl. O&P
1 Equip. Oper. (crane)	$55.70	$445.60	$84.25	$674.00	$47.42	$72.13
1 Laborer	39.15	313.20	60.00	480.00		
1 Crawler Crane, 25 Ton		1296.00		1425.60		
1 Clamshell Bucket, 1 C.Y.		48.80		53.68	84.05	92.45
16 L.H., Daily Totals		$2103.60		$2633.28	$131.47	$164.58

Crew B-12I	Hr.	Daily	Hr.	Daily	Bare Costs	Incl. O&P
1 Equip. Oper. (crane)	$55.70	$445.60	$84.25	$674.00	$47.42	$72.13
1 Laborer	39.15	313.20	60.00	480.00		
1 Crawler Crane, 20 Ton		972.05		1069.26		
1 Dragline Bucket, .75 C.Y.		20.80		22.88	62.05	68.26
16 L.H., Daily Totals		$1751.65		$2246.14	$109.48	$140.38

Crew B-12J	Hr.	Daily	Hr.	Daily	Bare Costs	Incl. O&P
1 Equip. Oper. (crane)	$55.70	$445.60	$84.25	$674.00	$47.42	$72.13
1 Laborer	39.15	313.20	60.00	480.00		
1 Gradall, 5/8 C.Y.		910.00		1001.00	56.88	62.56
16 L.H., Daily Totals		$1668.80		$2155.00	$104.30	$134.69

Crew B-12K	Hr.	Daily	Hr.	Daily	Bare Costs	Incl. O&P
1 Equip. Oper. (crane)	$55.70	$445.60	$84.25	$674.00	$47.42	$72.13
1 Laborer	39.15	313.20	60.00	480.00		
1 Gradall, 3 Ton, 1 C.Y.		1144.00		1258.40	71.50	78.65
16 L.H., Daily Totals		$1902.80		$2412.40	$118.93	$150.78

Crew B-12L	Hr.	Daily	Hr.	Daily	Bare Costs	Incl. O&P
1 Equip. Oper. (crane)	$55.70	$445.60	$84.25	$674.00	$47.42	$72.13
1 Laborer	39.15	313.20	60.00	480.00		
1 Crawler Crane, 15 Ton		757.75		833.52		
1 F.E. Attachment, .5 C.Y.		61.00		67.10	51.17	56.29
16 L.H., Daily Totals		$1577.55		$2054.63	$98.60	$128.41

Crew B-12M	Hr.	Daily	Hr.	Daily	Bare Costs	Incl. O&P
1 Equip. Oper. (crane)	$55.70	$445.60	$84.25	$674.00	$47.42	$72.13
1 Laborer	39.15	313.20	60.00	480.00		
1 Crawler Crane, 20 Ton		972.05		1069.26		
1 F.E. Attachment, .75 C.Y.		65.80		72.38	64.87	71.35
16 L.H., Daily Totals		$1796.65		$2295.64	$112.29	$143.48

Crew B-12N	Hr.	Daily	Hr.	Daily	Bare Costs	Incl. O&P
1 Equip. Oper. (crane)	$55.70	$445.60	$84.25	$674.00	$47.42	$72.13
1 Laborer	39.15	313.20	60.00	480.00		
1 Crawler Crane, 25 Ton		1296.00		1425.60		
1 F.E. Attachment, 1 C.Y.		72.60		79.86	85.54	94.09
16 L.H., Daily Totals		$2127.40		$2659.46	$132.96	$166.22

Crew B-12O	Hr.	Daily	Hr.	Daily	Bare Costs	Incl. O&P
1 Equip. Oper. (crane)	$55.70	$445.60	$84.25	$674.00	$47.42	$72.13
1 Laborer	39.15	313.20	60.00	480.00		
1 Crawler Crane, 40 Ton		1313.00		1444.30		
1 F.E. Attachment, 1.5 C.Y.		81.40		89.54	87.15	95.86
16 L.H., Daily Totals		$2153.20		$2687.84	$134.57	$167.99

Crew B-12P	Hr.	Daily	Hr.	Daily	Bare Costs	Incl. O&P
1 Equip. Oper. (crane)	$55.70	$445.60	$84.25	$674.00	$47.42	$72.13
1 Laborer	39.15	313.20	60.00	480.00		
1 Crawler Crane, 40 Ton		1313.00		1444.30		
1 Dragline Bucket, 1.5 C.Y.		34.00		37.40	84.19	92.61
16 L.H., Daily Totals		$2105.80		$2635.70	$131.61	$164.73

Crew B-12Q	Hr.	Daily	Hr.	Daily	Bare Costs	Incl. O&P
1 Equip. Oper. (crane)	$55.70	$445.60	$84.25	$674.00	$47.42	$72.13
1 Laborer	39.15	313.20	60.00	480.00		
1 Hyd. Excavator, 5/8 C.Y.		614.00		675.40	38.38	42.21
16 L.H., Daily Totals		$1372.80		$1829.40	$85.80	$114.34

Crew B-12S	Hr.	Daily	Hr.	Daily	Bare Costs	Incl. O&P
1 Equip. Oper. (crane)	$55.70	$445.60	$84.25	$674.00	$47.42	$72.13
1 Laborer	39.15	313.20	60.00	480.00		
1 Hyd. Excavator, 2.5 C.Y.		1536.00		1689.60	96.00	105.60
16 L.H., Daily Totals		$2294.80		$2843.60	$143.43	$177.72

Crew B-12T	Hr.	Daily	Hr.	Daily	Bare Costs	Incl. O&P
1 Equip. Oper. (crane)	$55.70	$445.60	$84.25	$674.00	$47.42	$72.13
1 Laborer	39.15	313.20	60.00	480.00		
1 Crawler Crane, 75 Ton		1645.00		1809.50		
1 F.E. Attachment, 3 C.Y.		104.20		114.62	109.33	120.26
16 L.H., Daily Totals		$2508.00		$3078.12	$156.75	$192.38

Crew B-12V	Hr.	Daily	Hr.	Daily	Bare Costs	Incl. O&P
1 Equip. Oper. (crane)	$55.70	$445.60	$84.25	$674.00	$47.42	$72.13
1 Laborer	39.15	313.20	60.00	480.00		
1 Crawler Crane, 75 Ton		1645.00		1809.50		
1 Dragline Bucket, 3 C.Y.		54.00		59.40	106.19	116.81
16 L.H., Daily Totals		$2457.80		$3022.90	$153.61	$188.93

Crew No.	Bare Costs		Incl. Subs O&P		Cost Per Labor-Hour	

Left Column

Crew B-12Y	Hr.	Daily	Hr.	Daily	Bare Costs	Incl. O&P
1 Equip. Oper. (crane)	$55.70	$445.60	$84.25	$674.00	$44.67	$68.08
2 Laborers	39.15	626.40	60.00	960.00		
1 Hyd. Excavator, 3.5 C.Y.		2401.00		2641.10	100.04	110.05
24 L.H., Daily Totals		$3473.00		$4275.10	$144.71	$178.13

Crew B-12Z	Hr.	Daily	Hr.	Daily	Bare Costs	Incl. O&P
1 Equip. Oper. (crane)	$55.70	$445.60	$84.25	$674.00	$44.67	$68.08
2 Laborers	39.15	626.40	60.00	960.00		
1 Hyd. Excavator, 2.5 C.Y.		1536.00		1689.60	64.00	70.40
24 L.H., Daily Totals		$2608.00		$3323.60	$108.67	$138.48

Crew B-13	Hr.	Daily	Hr.	Daily	Bare Costs	Incl. O&P
1 Labor Foreman (outside)	$41.15	$329.20	$63.05	$504.40	$43.06	$65.70
4 Laborers	39.15	1252.80	60.00	1920.00		
1 Equip. Oper. (crane)	55.70	445.60	84.25	674.00		
1 Equip. Oper. (oiler)	48.00	384.00	72.60	580.80		
1 Hyd. Crane, 25 Ton		673.80		741.18	12.03	13.24
56 L.H., Daily Totals		$3085.40		$4420.38	$55.10	$78.94

Crew B-13A	Hr.	Daily	Hr.	Daily	Bare Costs	Incl. O&P
1 Labor Foreman (outside)	$41.15	$329.20	$63.05	$504.40	$45.38	$68.92
2 Laborers	39.15	626.40	60.00	960.00		
2 Equipment Operators (med.)	53.55	856.80	81.00	1296.00		
2 Truck Drivers (heavy)	45.55	728.80	68.70	1099.20		
1 Crawler Crane, 75 Ton		1645.00		1809.50		
1 Crawler Loader, 4 C.Y.		1528.00		1680.80		
2 Dump Trucks, 8 C.Y., 220 H.P.		782.00		860.20	70.63	77.69
56 L.H., Daily Totals		$6496.20		$8210.10	$116.00	$146.61

Crew B-13B	Hr.	Daily	Hr.	Daily	Bare Costs	Incl. O&P
1 Labor Foreman (outside)	$41.15	$329.20	$63.05	$504.40	$43.06	$65.70
4 Laborers	39.15	1252.80	60.00	1920.00		
1 Equip. Oper. (crane)	55.70	445.60	84.25	674.00		
1 Equip. Oper. (oiler)	48.00	384.00	72.60	580.80		
1 Hyd. Crane, 55 Ton		1076.00		1183.60	19.21	21.14
56 L.H., Daily Totals		$3487.60		$4862.80	$62.28	$86.84

Crew B-13C	Hr.	Daily	Hr.	Daily	Bare Costs	Incl. O&P
1 Labor Foreman (outside)	$41.15	$329.20	$63.05	$504.40	$43.06	$65.70
4 Laborers	39.15	1252.80	60.00	1920.00		
1 Equip. Oper. (crane)	55.70	445.60	84.25	674.00		
1 Equip. Oper. (oiler)	48.00	384.00	72.60	580.80		
1 Crawler Crane, 100 Ton		1804.00		1984.40	32.21	35.44
56 L.H., Daily Totals		$4215.60		$5663.60	$75.28	$101.14

Crew B-13D	Hr.	Daily	Hr.	Daily	Bare Costs	Incl. O&P
1 Laborer	$39.15	$313.20	$60.00	$480.00	$47.42	$72.13
1 Equip. Oper. (crane)	55.70	445.60	84.25	674.00		
1 Hyd. Excavator, 1 C.Y.		798.20		878.02		
1 Trench Box		81.00		89.10	54.95	60.45
16 L.H., Daily Totals		$1638.00		$2121.12	$102.38	$132.57

Crew B-13E	Hr.	Daily	Hr.	Daily	Bare Costs	Incl. O&P
1 Laborer	$39.15	$313.20	$60.00	$480.00	$47.42	$72.13
1 Equip. Oper. (crane)	55.70	445.60	84.25	674.00		
1 Hyd. Excavator, 1.5 C.Y.		965.00		1061.50		
1 Trench Box		81.00		89.10	65.38	71.91
16 L.H., Daily Totals		$1804.80		$2304.60	$112.80	$144.04

Right Column

Crew B-13F	Hr.	Daily	Hr.	Daily	Bare Costs	Incl. O&P
1 Laborer	$39.15	$313.20	$60.00	$480.00	$47.42	$72.13
1 Equip. Oper. (crane)	55.70	445.60	84.25	674.00		
1 Hyd. Excavator, 3.5 C.Y.		2401.00		2641.10		
1 Trench Box		81.00		89.10	155.13	170.64
16 L.H., Daily Totals		$3240.80		$3884.20	$202.55	$242.76

Crew B-13G	Hr.	Daily	Hr.	Daily	Bare Costs	Incl. O&P
1 Laborer	$39.15	$313.20	$60.00	$480.00	$47.42	$72.13
1 Equip. Oper. (crane)	55.70	445.60	84.25	674.00		
1 Hyd. Excavator, .75 C.Y.		674.20		741.62		
1 Trench Box		81.00		89.10	47.20	51.92
16 L.H., Daily Totals		$1514.00		$1984.72	$94.63	$124.05

Crew B-13H	Hr.	Daily	Hr.	Daily	Bare Costs	Incl. O&P
1 Laborer	$39.15	$313.20	$60.00	$480.00	$47.42	$72.13
1 Equip. Oper. (crane)	55.70	445.60	84.25	674.00		
1 Gradall, 5/8 C.Y.		910.00		1001.00		
1 Trench Box		81.00		89.10	61.94	68.13
16 L.H., Daily Totals		$1749.80		$2244.10	$109.36	$140.26

Crew B-13I	Hr.	Daily	Hr.	Daily	Bare Costs	Incl. O&P
1 Laborer	$39.15	$313.20	$60.00	$480.00	$47.42	$72.13
1 Equip. Oper. (crane)	55.70	445.60	84.25	674.00		
1 Gradall, 3 Ton, 1 C.Y.		1144.00		1258.40		
1 Trench Box		81.00		89.10	76.56	84.22
16 L.H., Daily Totals		$1983.80		$2501.50	$123.99	$156.34

Crew B-13J	Hr.	Daily	Hr.	Daily	Bare Costs	Incl. O&P
1 Laborer	$39.15	$313.20	$60.00	$480.00	$47.42	$72.13
1 Equip. Oper. (crane)	55.70	445.60	84.25	674.00		
1 Hyd. Excavator, 2.5 C.Y.		1536.00		1689.60		
1 Trench Box		81.00		89.10	101.06	111.17
16 L.H., Daily Totals		$2375.80		$2932.70	$148.49	$183.29

Crew B-13K	Hr.	Daily	Hr.	Daily	Bare Costs	Incl. O&P
2 Equip. Opers. (crane)	$55.70	$891.20	$84.25	$1348.00	$55.70	$84.25
1 Hyd. Excavator, .75 C.Y.		674.20		741.62		
1 Hyd. Hammer, 4000 ft-lb		312.40		343.64		
1 Hyd. Excavator, .75 C.Y.		674.20		741.62	103.80	114.18
16 L.H., Daily Totals		$2552.00		$3174.88	$159.50	$198.43

Crew B-13L	Hr.	Daily	Hr.	Daily	Bare Costs	Incl. O&P
2 Equip. Opers. (crane)	$55.70	$891.20	$84.25	$1348.00	$55.70	$84.25
1 Hyd. Excavator, 1.5 C.Y.		965.00		1061.50		
1 Hyd. Hammer, 5000 ft-lb		376.20		413.82		
1 Hyd. Excavator, .75 C.Y.		674.20		741.62	125.96	138.56
16 L.H., Daily Totals		$2906.60		$3564.94	$181.66	$222.81

Crew B-13M	Hr.	Daily	Hr.	Daily	Bare Costs	Incl. O&P
2 Equip. Opers. (crane)	$55.70	$891.20	$84.25	$1348.00	$55.70	$84.25
1 Hyd. Excavator, 2.5 C.Y.		1536.00		1689.60		
1 Hyd. Hammer, 8000 ft-lb		552.80		608.08		
1 Hyd. Excavator, 1.5 C.Y.		965.00		1061.50	190.86	209.95
16 L.H., Daily Totals		$3945.00		$4707.18	$246.56	$294.20

731

Crew B-13N	Hr.	Daily	Hr.	Daily	Bare Costs	Incl. O&P
2 Equip. Opers. (crane)	$55.70	$891.20	$84.25	$1348.00	$55.70	$84.25
1 Hyd. Excavator, 3.5 C.Y.		2401.00		2641.10		
1 Hyd. Hammer, 12,000 ft-lb		643.20		707.52		
1 Hyd. Excavator, 1.5 C.Y.		965.00		1061.50	250.57	275.63
16 L.H., Daily Totals		$4900.40		$5758.12	$306.27	$359.88

Crew B-14	Hr.	Daily	Hr.	Daily	Bare Costs	Incl. O&P
1 Labor Foreman (outside)	$41.15	$329.20	$63.05	$504.40	$41.46	$63.37
4 Laborers	39.15	1252.80	60.00	1920.00		
1 Equip. Oper. (light)	51.00	408.00	77.15	617.20		
1 Backhoe Loader, 48 H.P.		366.20		402.82	7.63	8.39
48 L.H., Daily Totals		$2356.20		$3444.42	$49.09	$71.76

Crew B-14A	Hr.	Daily	Hr.	Daily	Bare Costs	Incl. O&P
1 Equip. Oper. (crane)	$55.70	$445.60	$84.25	$674.00	$50.18	$76.17
.5 Laborer	39.15	156.60	60.00	240.00		
1 Hyd. Excavator, 4.5 C.Y.		3084.00		3392.40	257.00	282.70
12 L.H., Daily Totals		$3686.20		$4306.40	$307.18	$358.87

Crew B-14B	Hr.	Daily	Hr.	Daily	Bare Costs	Incl. O&P
1 Equip. Oper. (crane)	$55.70	$445.60	$84.25	$674.00	$50.18	$76.17
.5 Laborer	39.15	156.60	60.00	240.00		
1 Hyd. Excavator, 6 C.Y.		3596.00		3955.60	299.67	329.63
12 L.H., Daily Totals		$4198.20		$4869.60	$349.85	$405.80

Crew B-14C	Hr.	Daily	Hr.	Daily	Bare Costs	Incl. O&P
1 Equip. Oper. (crane)	$55.70	$445.60	$84.25	$674.00	$50.18	$76.17
.5 Laborer	39.15	156.60	60.00	240.00		
1 Hyd. Excavator, 7 C.Y.		3698.00		4067.80	308.17	338.98
12 L.H., Daily Totals		$4300.20		$4981.80	$358.35	$415.15

Crew B-14F	Hr.	Daily	Hr.	Daily	Bare Costs	Incl. O&P
1 Equip. Oper. (crane)	$55.70	$445.60	$84.25	$674.00	$50.18	$76.17
.5 Laborer	39.15	156.60	60.00	240.00		
1 Hyd. Shovel, 7 C.Y.		4107.00		4517.70	342.25	376.48
12 L.H., Daily Totals		$4709.20		$5431.70	$392.43	$452.64

Crew B-14G	Hr.	Daily	Hr.	Daily	Bare Costs	Incl. O&P
1 Equip. Oper. (crane)	$55.70	$445.60	$84.25	$674.00	$50.18	$76.17
.5 Laborer	39.15	156.60	60.00	240.00		
1 Hyd. Shovel, 12 C.Y.		6152.00		6767.20	512.67	563.93
12 L.H., Daily Totals		$6754.20		$7681.20	$562.85	$640.10

Crew B-14J	Hr.	Daily	Hr.	Daily	Bare Costs	Incl. O&P
1 Equip. Oper. (medium)	$53.55	$428.40	$81.00	$648.00	$48.75	$74.00
.5 Laborer	39.15	156.60	60.00	240.00		
1 F.E. Loader, 8 C.Y.		1967.00		2163.70	163.92	180.31
12 L.H., Daily Totals		$2552.00		$3051.70	$212.67	$254.31

Crew B-14K	Hr.	Daily	Hr.	Daily	Bare Costs	Incl. O&P
1 Equip. Oper. (medium)	$53.55	$428.40	$81.00	$648.00	$48.75	$74.00
.5 Laborer	39.15	156.60	60.00	240.00		
1 F.E. Loader, 10 C.Y.		3224.00		3546.40	268.67	295.53
12 L.H., Daily Totals		$3809.00		$4434.40	$317.42	$369.53

Crew B-15	Hr.	Daily	Hr.	Daily	Bare Costs	Incl. O&P
1 Equipment Oper. (med.)	$53.55	$428.40	$81.00	$648.00	$46.92	$70.97
.5 Laborer	39.15	156.60	60.00	240.00		
2 Truck Drivers (heavy)	45.55	728.80	68.70	1099.20		
2 Dump Trucks, 12 C.Y., 400 H.P.		1285.20		1413.72		
1 Dozer, 200 H.P.		1384.00		1522.40	95.33	104.86
28 L.H., Daily Totals		$3983.00		$4923.32	$142.25	$175.83

Crew B-16	Hr.	Daily	Hr.	Daily	Bare Costs	Incl. O&P
1 Labor Foreman (outside)	$41.15	$329.20	$63.05	$504.40	$41.25	$62.94
2 Laborers	39.15	626.40	60.00	960.00		
1 Truck Driver (heavy)	45.55	364.40	68.70	549.60		
1 Dump Truck, 12 C.Y., 400 H.P.		642.60		706.86	20.08	22.09
32 L.H., Daily Totals		$1962.60		$2720.86	$61.33	$85.03

Crew B-17	Hr.	Daily	Hr.	Daily	Bare Costs	Incl. O&P
2 Laborers	$39.15	$626.40	$60.00	$960.00	$43.71	$66.46
1 Equip. Oper. (light)	51.00	408.00	77.15	617.20		
1 Truck Driver (heavy)	45.55	364.40	68.70	549.60		
1 Backhoe Loader, 48 H.P.		366.20		402.82		
1 Dump Truck, 8 C.Y., 220 H.P.		391.00		430.10	23.66	26.03
32 L.H., Daily Totals		$2156.00		$2959.72	$67.38	$92.49

Crew B-17A	Hr.	Daily	Hr.	Daily	Bare Costs	Incl. O&P
2 Labor Foremen (outside)	$41.15	$658.40	$63.05	$1008.80	$42.21	$64.75
6 Laborers	39.15	1879.20	60.00	2880.00		
1 Skilled Worker Foreman (out)	53.45	427.60	82.25	658.00		
1 Skilled Worker	51.45	411.60	79.20	633.60		
80 L.H., Daily Totals		$3376.80		$5180.40	$42.21	$64.75

Crew B-17B	Hr.	Daily	Hr.	Daily	Bare Costs	Incl. O&P
2 Laborers	$39.15	$626.40	$60.00	$960.00	$43.71	$66.46
1 Equip. Oper. (light)	51.00	408.00	77.15	617.20		
1 Truck Driver (heavy)	45.55	364.40	68.70	549.60		
1 Backhoe Loader, 48 H.P.		366.20		402.82		
1 Dump Truck, 12 C.Y., 400 H.P.		642.60		706.86	31.52	34.68
32 L.H., Daily Totals		$2407.60		$3236.48	$75.24	$101.14

Crew B-18	Hr.	Daily	Hr.	Daily	Bare Costs	Incl. O&P
1 Labor Foreman (outside)	$41.15	$329.20	$63.05	$504.40	$39.82	$61.02
2 Laborers	39.15	626.40	60.00	960.00		
1 Vibrating Plate, Gas, 21"		42.80		47.08	1.78	1.96
24 L.H., Daily Totals		$998.40		$1511.48	$41.60	$62.98

Crew B-19	Hr.	Daily	Hr.	Daily	Bare Costs	Incl. O&P
1 Pile Driver Foreman (outside)	$51.50	$412.00	$81.80	$654.40	$51.11	$79.66
4 Pile Drivers	49.50	1584.00	78.60	2515.20		
2 Equip. Oper. (crane)	55.70	891.20	84.25	1348.00		
1 Equip. Oper. (oiler)	48.00	384.00	72.60	580.80		
1 Crawler Crane, 40 Ton		1313.00		1444.30		
1 Lead, 90' High		126.00		138.60		
1 Hammer, Diesel, 22k ft-lb		421.00		463.10	29.06	31.97
64 L.H., Daily Totals		$5131.20		$7144.40	$80.17	$111.63

Crew No.	Bare Costs		Incl. Subs O&P		Cost Per Labor-Hour	

Crew B-19A	Hr.	Daily	Hr.	Daily	Bare Costs	Incl. O&P
1 Pile Driver Foreman (outside)	$51.50	$412.00	$81.80	$654.40	$51.11	$79.66
4 Pile Drivers	49.50	1584.00	78.60	2515.20		
2 Equip. Oper. (crane)	55.70	891.20	84.25	1348.00		
1 Equip. Oper. (oiler)	48.00	384.00	72.60	580.80		
1 Crawler Crane, 75 Ton		1645.00		1809.50		
1 Lead, 90' high		126.00		138.60		
1 Hammer, Diesel, 41k ft-lb		574.00		631.40	36.64	40.30
64 L.H., Daily Totals		$5616.20		$7677.90	$87.75	$119.97

Crew B-19B	Hr.	Daily	Hr.	Daily	Bare Costs	Incl. O&P
1 Pile Driver Foreman (outside)	$51.50	$412.00	$81.80	$654.40	$51.11	$79.66
4 Pile Drivers	49.50	1584.00	78.60	2515.20		
2 Equip. Oper. (crane)	55.70	891.20	84.25	1348.00		
1 Equip. Oper. (oiler)	48.00	384.00	72.60	580.80		
1 Crawler Crane, 40 Ton		1313.00		1444.30		
1 Lead, 90' High		126.00		138.60		
1 Hammer, Diesel, 22k ft-lb		421.00		463.10		
1 Barge, 400 Ton		794.20		873.62	41.47	45.62
64 L.H., Daily Totals		$5925.40		$8018.02	$92.58	$125.28

Crew B-19C	Hr.	Daily	Hr.	Daily	Bare Costs	Incl. O&P
1 Pile Driver Foreman (outside)	$51.50	$412.00	$81.80	$654.40	$51.11	$79.66
4 Pile Drivers	49.50	1584.00	78.60	2515.20		
2 Equip. Oper. (crane)	55.70	891.20	84.25	1348.00		
1 Equip. Oper. (oiler)	48.00	384.00	72.60	580.80		
1 Crawler Crane, 75 Ton		1645.00		1809.50		
1 Lead, 90' High		126.00		138.60		
1 Hammer, Diesel, 41k ft-lb		574.00		631.40		
1 Barge, 400 Ton		794.20		873.62	49.05	53.95
64 L.H., Daily Totals		$6410.40		$8551.52	$100.16	$133.62

Crew B-20	Hr.	Daily	Hr.	Daily	Bare Costs	Incl. O&P
1 Labor Foreman (outside)	$41.15	$329.20	$63.05	$504.40	$43.92	$67.42
1 Skilled Worker	51.45	411.60	79.20	633.60		
1 Laborer	39.15	313.20	60.00	480.00		
24 L.H., Daily Totals		$1054.00		$1618.00	$43.92	$67.42

Crew B-20A	Hr.	Daily	Hr.	Daily	Bare Costs	Incl. O&P
1 Labor Foreman (outside)	$41.15	$329.20	$63.05	$504.40	$47.89	$72.72
1 Laborer	39.15	313.20	60.00	480.00		
1 Plumber	61.80	494.40	93.25	746.00		
1 Plumber Apprentice	49.45	395.60	74.60	596.80		
32 L.H., Daily Totals		$1532.40		$2327.20	$47.89	$72.72

Crew B-21	Hr.	Daily	Hr.	Daily	Bare Costs	Incl. O&P
1 Labor Foreman (outside)	$41.15	$329.20	$63.05	$504.40	$45.60	$69.82
1 Skilled Worker	51.45	411.60	79.20	633.60		
1 Laborer	39.15	313.20	60.00	480.00		
.5 Equip. Oper. (crane)	55.70	222.80	84.25	337.00		
.5 S.P. Crane, 4x4, 5 Ton		138.00		151.80	4.93	5.42
28 L.H., Daily Totals		$1414.80		$2106.80	$50.53	$75.24

Crew B-21A	Hr.	Daily	Hr.	Daily	Bare Costs	Incl. O&P
1 Labor Foreman (outside)	$41.15	$329.20	$63.05	$504.40	$49.45	$75.03
1 Laborer	39.15	313.20	60.00	480.00		
1 Plumber	61.80	494.40	93.25	746.00		
1 Plumber Apprentice	49.45	395.60	74.60	596.80		
1 Equip. Oper. (crane)	55.70	445.60	84.25	674.00		
1 S.P. Crane, 4x4, 12 Ton		435.60		479.16	10.89	11.98
40 L.H., Daily Totals		$2413.60		$3480.36	$60.34	$87.01

Crew B-21B	Hr.	Daily	Hr.	Daily	Bare Costs	Incl. O&P
1 Labor Foreman (outside)	$41.15	$329.20	$63.05	$504.40	$42.86	$65.46
3 Laborers	39.15	939.60	60.00	1440.00		
1 Equip. Oper. (crane)	55.70	445.60	84.25	674.00		
1 Hyd. Crane, 12 Ton		572.80		630.08	14.32	15.75
40 L.H., Daily Totals		$2287.20		$3248.48	$57.18	$81.21

Crew B-21C	Hr.	Daily	Hr.	Daily	Bare Costs	Incl. O&P
1 Labor Foreman (outside)	$41.15	$329.20	$63.05	$504.40	$43.06	$65.70
4 Laborers	39.15	1252.80	60.00	1920.00		
1 Equip. Oper. (crane)	55.70	445.60	84.25	674.00		
1 Equip. Oper. (oiler)	48.00	384.00	72.60	580.80		
2 Cutting Torches		24.00		26.40		
2 Sets of Gases		336.00		369.60		
1 Lattice Boom Crane, 90 Ton		1671.00		1838.10	36.27	39.89
56 L.H., Daily Totals		$4442.60		$5913.30	$79.33	$105.59

Crew B-22	Hr.	Daily	Hr.	Daily	Bare Costs	Incl. O&P
1 Labor Foreman (outside)	$41.15	$329.20	$63.05	$504.40	$46.27	$70.78
1 Skilled Worker	51.45	411.60	79.20	633.60		
1 Laborer	39.15	313.20	60.00	480.00		
.75 Equip. Oper. (crane)	55.70	334.20	84.25	505.50		
.75 S.P. Crane, 4x4, 5 Ton		207.00		227.70	6.90	7.59
30 L.H., Daily Totals		$1595.20		$2351.20	$53.17	$78.37

Crew B-22A	Hr.	Daily	Hr.	Daily	Bare Costs	Incl. O&P
1 Labor Foreman (outside)	$41.15	$329.20	$63.05	$504.40	$45.32	$69.30
1 Skilled Worker	51.45	411.60	79.20	633.60		
2 Laborers	39.15	626.40	60.00	960.00		
1 Equipment Operator, Crane	55.70	445.60	84.25	674.00		
1 S.P. Crane, 4x4, 5 Ton		276.00		303.60		
1 Butt Fusion Mach., 4"-12" diam.		415.10		456.61	17.28	19.01
40 L.H., Daily Totals		$2503.90		$3532.21	$62.60	$88.31

Crew B-22B	Hr.	Daily	Hr.	Daily	Bare Costs	Incl. O&P
1 Labor Foreman (outside)	$41.15	$329.20	$63.05	$504.40	$45.32	$69.30
1 Skilled Worker	51.45	411.60	79.20	633.60		
2 Laborers	39.15	626.40	60.00	960.00		
1 Equip. Oper. (crane)	55.70	445.60	84.25	674.00		
1 S.P. Crane, 4x4, 5 Ton		276.00		303.60		
1 Butt Fusion Mach., 8"-24" diam.		1821.00		2003.10	52.42	57.67
40 L.H., Daily Totals		$3909.80		$5078.70	$97.75	$126.97

Crew B-22C	Hr.	Daily	Hr.	Daily	Bare Costs	Incl. O&P
1 Skilled Worker	$51.45	$411.60	$79.20	$633.60	$45.30	$69.60
1 Laborer	39.15	313.20	60.00	480.00		
1 Butt Fusion Mach., 2"-8" diam.		121.05		133.16	7.57	8.32
16 L.H., Daily Totals		$845.85		$1246.76	$52.87	$77.92

Crew B-23	Hr.	Daily	Hr.	Daily	Bare Costs	Incl. O&P
1 Labor Foreman (outside)	$41.15	$329.20	$63.05	$504.40	$39.55	$60.61
4 Laborers	39.15	1252.80	60.00	1920.00		
1 Drill Rig, Truck-Mounted		2573.00		2830.30		
1 Flatbed Truck, Gas, 3 Ton		270.20		297.22	71.08	78.19
40 L.H., Daily Totals		$4425.20		$5551.92	$110.63	$138.80

Crew No.	Bare Costs		Incl. Subs O&P		Cost Per Labor-Hour	

Crew B-23A

	Hr.	Daily	Hr.	Daily	Bare Costs	Incl. O&P
1 Labor Foreman (outside)	$41.15	$329.20	$63.05	$504.40	$44.62	$68.02
1 Laborer	39.15	313.20	60.00	480.00		
1 Equip. Oper. (medium)	53.55	428.40	81.00	648.00		
1 Drill Rig, Truck-Mounted		2573.00		2830.30		
1 Pickup Truck, 3/4 Ton		129.40		142.34	112.60	123.86
24 L.H., Daily Totals		$3773.20		$4605.04	$157.22	$191.88

Crew B-23B

	Hr.	Daily	Hr.	Daily	Bare Costs	Incl. O&P
1 Labor Foreman (outside)	$41.15	$329.20	$63.05	$504.40	$44.62	$68.02
1 Laborer	39.15	313.20	60.00	480.00		
1 Equip. Oper. (medium)	53.55	428.40	81.00	648.00		
1 Drill Rig, Truck-Mounted		2573.00		2830.30		
1 Pickup Truck, 3/4 Ton		129.40		142.34		
1 Centr. Water Pump, 6"		325.40		357.94	126.16	138.77
24 L.H., Daily Totals		$4098.60		$4962.98	$170.78	$206.79

Crew B-24

	Hr.	Daily	Hr.	Daily	Bare Costs	Incl. O&P
1 Cement Finisher	$46.80	$374.40	$69.60	$556.80	$45.07	$68.35
1 Laborer	39.15	313.20	60.00	480.00		
1 Carpenter	49.25	394.00	75.45	603.60		
24 L.H., Daily Totals		$1081.60		$1640.40	$45.07	$68.35

Crew B-25

	Hr.	Daily	Hr.	Daily	Bare Costs	Incl. O&P
1 Labor Foreman (outside)	$41.15	$329.20	$63.05	$504.40	$43.26	$66.00
7 Laborers	39.15	2192.40	60.00	3360.00		
3 Equip. Oper. (medium)	53.55	1285.20	81.00	1944.00		
1 Asphalt Paver, 130 H.P.		2235.00		2458.50		
1 Tandem Roller, 10 Ton		237.60		261.36		
1 Roller, Pneum. Whl., 12 Ton		355.80		391.38	32.14	35.35
88 L.H., Daily Totals		$6635.20		$8919.64	$75.40	$101.36

Crew B-25B

	Hr.	Daily	Hr.	Daily	Bare Costs	Incl. O&P
1 Labor Foreman (outside)	$41.15	$329.20	$63.05	$504.40	$44.12	$67.25
7 Laborers	39.15	2192.40	60.00	3360.00		
4 Equip. Oper. (medium)	53.55	1713.60	81.00	2592.00		
1 Asphalt Paver, 130 H.P.		2235.00		2458.50		
2 Tandem Rollers, 10 Ton		475.20		522.72		
1 Roller, Pneum. Whl., 12 Ton		355.80		391.38	31.94	35.13
96 L.H., Daily Totals		$7301.20		$9829.00	$76.05	$102.39

Crew B-25C

	Hr.	Daily	Hr.	Daily	Bare Costs	Incl. O&P
1 Labor Foreman (outside)	$41.15	$329.20	$63.05	$504.40	$44.28	$67.51
3 Laborers	39.15	939.60	60.00	1440.00		
2 Equip. Oper. (medium)	53.55	856.80	81.00	1296.00		
1 Asphalt Paver, 130 H.P.		2235.00		2458.50		
1 Tandem Roller, 10 Ton		237.60		261.36	51.51	56.66
48 L.H., Daily Totals		$4598.20		$5960.26	$95.80	$124.17

Crew B-25D

	Hr.	Daily	Hr.	Daily	Bare Costs	Incl. O&P
1 Labor Foreman (outside)	$41.15	$329.20	$63.05	$504.40	$44.49	$67.80
3 Laborers	39.15	939.60	60.00	1440.00		
2.125 Equip. Oper. (medium)	53.55	910.35	81.00	1377.00		
.125 Truck Driver (heavy)	45.55	45.55	68.70	68.70		
.125 Truck Tractor, 6x4, 380 H.P.		69.97		76.97		
.125 Dist. Tanker, 3000 Gallon		40.92		45.02		
1 Asphalt Paver, 130 H.P.		2235.00		2458.50		
1 Tandem Roller, 10 Ton		237.60		261.36	51.67	56.84
50 L.H., Daily Totals		$4808.20		$6231.95	$96.16	$124.64

Crew B-25E

	Hr.	Daily	Hr.	Daily	Bare Costs	Incl. O&P
1 Labor Foreman (outside)	$41.15	$329.20	$63.05	$504.40	$44.69	$68.07
3 Laborers	39.15	939.60	60.00	1440.00		
2.250 Equip. Oper. (medium)	53.55	963.90	81.00	1458.00		
.25 Truck Driver (heavy)	45.55	91.10	68.70	137.40		
.25 Truck Tractor, 6x4, 380 H.P.		139.95		153.94		
.25 Dist. Tanker, 3000 Gallon		81.85		90.03		
1 Asphalt Paver, 130 H.P.		2235.00		2458.50		
1 Tandem Roller, 10 Ton		237.60		261.36	51.82	57.00
52 L.H., Daily Totals		$5018.20		$6503.64	$96.50	$125.07

Crew B-26

	Hr.	Daily	Hr.	Daily	Bare Costs	Incl. O&P
1 Labor Foreman (outside)	$41.15	$329.20	$63.05	$504.40	$44.02	$67.12
6 Laborers	39.15	1879.20	60.00	2880.00		
2 Equip. Oper. (medium)	53.55	856.80	81.00	1296.00		
1 Rodman (reinf.)	54.30	434.40	83.70	669.60		
1 Cement Finisher	46.80	374.40	69.60	556.80		
1 Grader, 30,000 Lbs.		678.40		746.24		
1 Paving Mach. & Equip.		2452.00		2697.20	35.57	39.13
88 L.H., Daily Totals		$7004.40		$9350.24	$79.60	$106.25

Crew B-26A

	Hr.	Daily	Hr.	Daily	Bare Costs	Incl. O&P
1 Labor Foreman (outside)	$41.15	$329.20	$63.05	$504.40	$44.02	$67.12
6 Laborers	39.15	1879.20	60.00	2880.00		
2 Equip. Oper. (medium)	53.55	856.80	81.00	1296.00		
1 Rodman (reinf.)	54.30	434.40	83.70	669.60		
1 Cement Finisher	46.80	374.40	69.60	556.80		
1 Grader, 30,000 Lbs.		678.40		746.24		
1 Paving Mach. & Equip.		2452.00		2697.20		
1 Concrete Saw		153.60		168.96	37.32	41.05
88 L.H., Daily Totals		$7158.00		$9519.20	$81.34	$108.17

Crew B-26B

	Hr.	Daily	Hr.	Daily	Bare Costs	Incl. O&P
1 Labor Foreman (outside)	$41.15	$329.20	$63.05	$504.40	$44.82	$68.28
6 Laborers	39.15	1879.20	60.00	2880.00		
3 Equip. Oper. (medium)	53.55	1285.20	81.00	1944.00		
1 Rodman (reinf.)	54.30	434.40	83.70	669.60		
1 Cement Finisher	46.80	374.40	69.60	556.80		
1 Grader, 30,000 Lbs.		678.40		746.24		
1 Paving Mach. & Equip.		2452.00		2697.20		
1 Concrete Pump, 110' Boom		1086.00		1194.60	43.92	48.31
96 L.H., Daily Totals		$8518.80		$11192.84	$88.74	$116.59

Crew B-26C

	Hr.	Daily	Hr.	Daily	Bare Costs	Incl. O&P
1 Labor Foreman (outside)	$41.15	$329.20	$63.05	$504.40	$43.07	$65.73
6 Laborers	39.15	1879.20	60.00	2880.00		
1 Equip. Oper. (medium)	53.55	428.40	81.00	648.00		
1 Rodman (reinf.)	54.30	434.40	83.70	669.60		
1 Cement Finisher	46.80	374.40	69.60	556.80		
1 Paving Mach. & Equip.		2452.00		2697.20		
1 Concrete Saw		153.60		168.96	32.57	35.83
80 L.H., Daily Totals		$6051.20		$8124.96	$75.64	$101.56

Crew B-27

	Hr.	Daily	Hr.	Daily	Bare Costs	Incl. O&P
1 Labor Foreman (outside)	$41.15	$329.20	$63.05	$504.40	$39.65	$60.76
3 Laborers	39.15	939.60	60.00	1440.00		
1 Berm Machine		313.40		344.74	9.79	10.77
32 L.H., Daily Totals		$1582.20		$2289.14	$49.44	$71.54

Crews - Standard

Crew No.	Bare Costs		Incl. Subs O&P		Cost Per Labor-Hour	

Crew B-28	Hr.	Daily	Hr.	Daily	Bare Costs	Incl. O&P
2 Carpenters	$49.25	$788.00	$75.45	$1207.20	$45.88	$70.30
1 Laborer	39.15	313.20	60.00	480.00		
24 L.H., Daily Totals		$1101.20		$1687.20	$45.88	$70.30

Crew B-29	Hr.	Daily	Hr.	Daily	Bare Costs	Incl. O&P
1 Labor Foreman (outside)	$41.15	$329.20	$63.05	$504.40	$43.06	$65.70
4 Laborers	39.15	1252.80	60.00	1920.00		
1 Equip. Oper. (crane)	55.70	445.60	84.25	674.00		
1 Equip. Oper. (oiler)	48.00	384.00	72.60	580.80		
1 Gradall, 5/8 C.Y.		910.00		1001.00	16.25	17.88
56 L.H., Daily Totals		$3321.60		$4680.20	$59.31	$83.58

Crew B-30	Hr.	Daily	Hr.	Daily	Bare Costs	Incl. O&P
1 Equip. Oper. (medium)	$53.55	$428.40	$81.00	$648.00	$48.22	$72.80
2 Truck Drivers (heavy)	45.55	728.80	68.70	1099.20		
1 Hyd. Excavator, 1.5 C.Y.		965.00		1061.50		
2 Dump Trucks, 12 C.Y., 400 H.P.		1285.20		1413.72	93.76	103.13
24 L.H., Daily Totals		$3407.40		$4222.42	$141.97	$175.93

Crew B-31	Hr.	Daily	Hr.	Daily	Bare Costs	Incl. O&P
1 Labor Foreman (outside)	$41.15	$329.20	$63.05	$504.40	$41.57	$63.70
3 Laborers	39.15	939.60	60.00	1440.00		
1 Carpenter	49.25	394.00	75.45	603.60		
1 Air Compressor, 250 cfm		181.20		199.32		
1 Sheeting Driver		5.75		6.33		
2 -50' Air Hoses, 1.5"		11.60		12.76	4.96	5.46
40 L.H., Daily Totals		$1861.35		$2766.41	$46.53	$69.16

Crew B-32	Hr.	Daily	Hr.	Daily	Bare Costs	Incl. O&P
1 Laborer	$39.15	$313.20	$60.00	$480.00	$49.95	$75.75
3 Equip. Oper. (medium)	53.55	1285.20	81.00	1944.00		
1 Grader, 30,000 Lbs.		678.40		746.24		
1 Tandem Roller, 10 Ton		237.60		261.36		
1 Dozer, 200 H.P.		1384.00		1522.40	71.88	79.06
32 L.H., Daily Totals		$3898.40		$4954.00	$121.83	$154.81

Crew B-32A	Hr.	Daily	Hr.	Daily	Bare Costs	Incl. O&P
1 Laborer	$39.15	$313.20	$60.00	$480.00	$48.75	$74.00
2 Equip. Oper. (medium)	53.55	856.80	81.00	1296.00		
1 Grader, 30,000 Lbs.		678.40		746.24		
1 Roller, Vibratory, 25 Ton		684.00		752.40	56.77	62.44
24 L.H., Daily Totals		$2532.40		$3274.64	$105.52	$136.44

Crew B-32B	Hr.	Daily	Hr.	Daily	Bare Costs	Incl. O&P
1 Laborer	$39.15	$313.20	$60.00	$480.00	$48.75	$74.00
2 Equip. Oper. (medium)	53.55	856.80	81.00	1296.00		
1 Dozer, 200 H.P.		1384.00		1522.40		
1 Roller, Vibratory, 25 Ton		684.00		752.40	86.17	94.78
24 L.H., Daily Totals		$3238.00		$4050.80	$134.92	$168.78

Crew B-32C	Hr.	Daily	Hr.	Daily	Bare Costs	Incl. O&P
1 Labor Foreman (outside)	$41.15	$329.20	$63.05	$504.40	$46.68	$71.01
2 Laborers	39.15	626.40	60.00	960.00		
3 Equip. Oper. (medium)	53.55	1285.20	81.00	1944.00		
1 Grader, 30,000 Lbs.		678.40		746.24		
1 Tandem Roller, 10 Ton		237.60		261.36		
1 Dozer, 200 H.P.		1384.00		1522.40	47.92	52.71
48 L.H., Daily Totals		$4540.80		$5938.40	$94.60	$123.72

Crew B-33A	Hr.	Daily	Hr.	Daily	Bare Costs	Incl. O&P
1 Equip. Oper. (medium)	$53.55	$428.40	$81.00	$648.00	$49.44	$75.00
.5 Laborer	39.15	156.60	60.00	240.00		
.25 Equip. Oper. (medium)	53.55	107.10	81.00	162.00		
1 Scraper, Towed, 7 C.Y.		117.20		128.92		
1.25 Dozers, 300 H.P.		2375.00		2612.50	178.01	195.82
14 L.H., Daily Totals		$3184.30		$3791.42	$227.45	$270.82

Crew B-33B	Hr.	Daily	Hr.	Daily	Bare Costs	Incl. O&P
1 Equip. Oper. (medium)	$53.55	$428.40	$81.00	$648.00	$49.44	$75.00
.5 Laborer	39.15	156.60	60.00	240.00		
.25 Equip. Oper. (medium)	53.55	107.10	81.00	162.00		
1 Scraper, Towed, 10 C.Y.		148.60		163.46		
1.25 Dozers, 300 H.P.		2375.00		2612.50	180.26	198.28
14 L.H., Daily Totals		$3215.70		$3825.96	$229.69	$273.28

Crew B-33C	Hr.	Daily	Hr.	Daily	Bare Costs	Incl. O&P
1 Equip. Oper. (medium)	$53.55	$428.40	$81.00	$648.00	$49.44	$75.00
.5 Laborer	39.15	156.60	60.00	240.00		
.25 Equip. Oper. (medium)	53.55	107.10	81.00	162.00		
1 Scraper, Towed, 15 C.Y.		167.60		184.36		
1.25 Dozers, 300 H.P.		2375.00		2612.50	181.61	199.78
14 L.H., Daily Totals		$3234.70		$3846.86	$231.05	$274.78

Crew B-33D	Hr.	Daily	Hr.	Daily	Bare Costs	Incl. O&P
1 Equip. Oper. (medium)	$53.55	$428.40	$81.00	$648.00	$49.44	$75.00
.5 Laborer	39.15	156.60	60.00	240.00		
.25 Equip. Oper. (medium)	53.55	107.10	81.00	162.00		
1 S.P. Scraper, 14 C.Y.		2429.00		2671.90		
.25 Dozer, 300 H.P.		475.00		522.50	207.43	228.17
14 L.H., Daily Totals		$3596.10		$4244.40	$256.86	$303.17

Crew B-33E	Hr.	Daily	Hr.	Daily	Bare Costs	Incl. O&P
1 Equip. Oper. (medium)	$53.55	$428.40	$81.00	$648.00	$49.44	$75.00
.5 Laborer	39.15	156.60	60.00	240.00		
.25 Equip. Oper. (medium)	53.55	107.10	81.00	162.00		
1 S.P. Scraper, 21 C.Y.		2784.00		3062.40		
.25 Dozer, 300 H.P.		475.00		522.50	232.79	256.06
14 L.H., Daily Totals		$3951.10		$4634.90	$282.22	$331.06

Crew B-33F	Hr.	Daily	Hr.	Daily	Bare Costs	Incl. O&P
1 Equip. Oper. (medium)	$53.55	$428.40	$81.00	$648.00	$49.44	$75.00
.5 Laborer	39.15	156.60	60.00	240.00		
.25 Equip. Oper. (medium)	53.55	107.10	81.00	162.00		
1 Elev. Scraper, 11 C.Y.		1192.00		1311.20		
.25 Dozer, 300 H.P.		475.00		522.50	119.07	130.98
14 L.H., Daily Totals		$2359.10		$2883.70	$168.51	$205.98

Crew B-33G	Hr.	Daily	Hr.	Daily	Bare Costs	Incl. O&P
1 Equip. Oper. (medium)	$53.55	$428.40	$81.00	$648.00	$49.44	$75.00
.5 Laborer	39.15	156.60	60.00	240.00		
.25 Equip. Oper. (medium)	53.55	107.10	81.00	162.00		
1 Elev. Scraper, 22 C.Y.		2490.00		2739.00		
.25 Dozer, 300 H.P.		475.00		522.50	211.79	232.96
14 L.H., Daily Totals		$3657.10		$4311.50	$261.22	$307.96

735

Crew B-33H

Crew No.	Bare Costs Hr.	Bare Costs Daily	Incl. Subs O&P Hr.	Incl. Subs O&P Daily	Cost Per Labor-Hour Bare Costs	Cost Per Labor-Hour Incl. O&P
.5 Laborer	$39.15	$156.60	$60.00	$240.00	$49.44	$75.00
1 Equipment Operator (med.)	53.55	428.40	81.00	648.00		
.25 Equipment Operator (med.)	53.55	107.10	81.00	162.00		
1 S.P. Scraper, 44 C.Y.		4712.00		5183.20		
.25 Dozer, 410 H.P.		592.00		651.20	378.86	416.74
14 L.H., Daily Totals		$5996.10		$6884.40	$428.29	$491.74

Crew B-33J

Crew No.	Bare Costs Hr.	Bare Costs Daily	Incl. Subs O&P Hr.	Incl. Subs O&P Daily	Cost Per Labor-Hour Bare Costs	Cost Per Labor-Hour Incl. O&P
1 Equipment Operator (med.)	$53.55	$428.40	$81.00	$648.00	$53.55	$81.00
1 S.P. Scraper, 14 C.Y.		2429.00		2671.90	303.63	333.99
8 L.H., Daily Totals		$2857.40		$3319.90	$357.18	$414.99

Crew B-33K

Crew No.	Bare Costs Hr.	Bare Costs Daily	Incl. Subs O&P Hr.	Incl. Subs O&P Daily	Cost Per Labor-Hour Bare Costs	Cost Per Labor-Hour Incl. O&P
1 Equipment Operator (med.)	$53.55	$428.40	$81.00	$648.00	$49.44	$75.00
.25 Equipment Operator (med.)	53.55	107.10	81.00	162.00		
.5 Laborer	39.15	156.60	60.00	240.00		
1 S.P. Scraper, 31 C.Y.		3776.00		4153.60		
.25 Dozer, 410 H.P.		592.00		651.20	312.00	343.20
14 L.H., Daily Totals		$5060.10		$5854.80	$361.44	$418.20

Crew B-34A

Crew No.	Bare Costs Hr.	Bare Costs Daily	Incl. Subs O&P Hr.	Incl. Subs O&P Daily	Cost Per Labor-Hour Bare Costs	Cost Per Labor-Hour Incl. O&P
1 Truck Driver (heavy)	$45.55	$364.40	$68.70	$549.60	$45.55	$68.70
1 Dump Truck, 8 C.Y., 220 H.P.		391.00		430.10	48.88	53.76
8 L.H., Daily Totals		$755.40		$979.70	$94.42	$122.46

Crew B-34B

Crew No.	Bare Costs Hr.	Bare Costs Daily	Incl. Subs O&P Hr.	Incl. Subs O&P Daily	Cost Per Labor-Hour Bare Costs	Cost Per Labor-Hour Incl. O&P
1 Truck Driver (heavy)	$45.55	$364.40	$68.70	$549.60	$45.55	$68.70
1 Dump Truck, 12 C.Y., 400 H.P.		642.60		706.86	80.33	88.36
8 L.H., Daily Totals		$1007.00		$1256.46	$125.88	$157.06

Crew B-34C

Crew No.	Bare Costs Hr.	Bare Costs Daily	Incl. Subs O&P Hr.	Incl. Subs O&P Daily	Cost Per Labor-Hour Bare Costs	Cost Per Labor-Hour Incl. O&P
1 Truck Driver (heavy)	$45.55	$364.40	$68.70	$549.60	$45.55	$68.70
1 Truck Tractor, 6x4, 380 H.P.		559.80		615.78		
1 Dump Trailer, 16.5 C.Y.		128.00		140.80	85.97	94.57
8 L.H., Daily Totals		$1052.20		$1306.18	$131.53	$163.27

Crew B-34D

Crew No.	Bare Costs Hr.	Bare Costs Daily	Incl. Subs O&P Hr.	Incl. Subs O&P Daily	Cost Per Labor-Hour Bare Costs	Cost Per Labor-Hour Incl. O&P
1 Truck Driver (heavy)	$45.55	$364.40	$68.70	$549.60	$45.55	$68.70
1 Truck Tractor, 6x4, 380 H.P.		559.80		615.78		
1 Dump Trailer, 20 C.Y.		142.60		156.86	87.80	96.58
8 L.H., Daily Totals		$1066.80		$1322.24	$133.35	$165.28

Crew B-34E

Crew No.	Bare Costs Hr.	Bare Costs Daily	Incl. Subs O&P Hr.	Incl. Subs O&P Daily	Cost Per Labor-Hour Bare Costs	Cost Per Labor-Hour Incl. O&P
1 Truck Driver (heavy)	$45.55	$364.40	$68.70	$549.60	$45.55	$68.70
1 Dump Truck, Off Hwy., 25 Ton		1476.00		1623.60	184.50	202.95
8 L.H., Daily Totals		$1840.40		$2173.20	$230.05	$271.65

Crew B-34F

Crew No.	Bare Costs Hr.	Bare Costs Daily	Incl. Subs O&P Hr.	Incl. Subs O&P Daily	Cost Per Labor-Hour Bare Costs	Cost Per Labor-Hour Incl. O&P
1 Truck Driver (heavy)	$45.55	$364.40	$68.70	$549.60	$45.55	$68.70
1 Dump Truck, Off Hwy., 35 Ton		1532.00		1685.20	191.50	210.65
8 L.H., Daily Totals		$1896.40		$2234.80	$237.05	$279.35

Crew B-34G

Crew No.	Bare Costs Hr.	Bare Costs Daily	Incl. Subs O&P Hr.	Incl. Subs O&P Daily	Cost Per Labor-Hour Bare Costs	Cost Per Labor-Hour Incl. O&P
1 Truck Driver (heavy)	$45.55	$364.40	$68.70	$549.60	$45.55	$68.70
1 Dump Truck, Off Hwy., 50 Ton		2006.00		2206.60	250.75	275.82
8 L.H., Daily Totals		$2370.40		$2756.20	$296.30	$344.52

Crew B-34H

Crew No.	Bare Costs Hr.	Bare Costs Daily	Incl. Subs O&P Hr.	Incl. Subs O&P Daily	Cost Per Labor-Hour Bare Costs	Cost Per Labor-Hour Incl. O&P
1 Truck Driver (heavy)	$45.55	$364.40	$68.70	$549.60	$45.55	$68.70
1 Dump Truck, Off Hwy., 65 Ton		2064.00		2270.40	258.00	283.80
8 L.H., Daily Totals		$2428.40		$2820.00	$303.55	$352.50

Crew B-34I

Crew No.	Bare Costs Hr.	Bare Costs Daily	Incl. Subs O&P Hr.	Incl. Subs O&P Daily	Cost Per Labor-Hour Bare Costs	Cost Per Labor-Hour Incl. O&P
1 Truck Driver (heavy)	$45.55	$364.40	$68.70	$549.60	$45.55	$68.70
1 Dump Truck, 18 C.Y., 450 H.P.		815.60		897.16	101.95	112.15
8 L.H., Daily Totals		$1180.00		$1446.76	$147.50	$180.85

Crew B-34J

Crew No.	Bare Costs Hr.	Bare Costs Daily	Incl. Subs O&P Hr.	Incl. Subs O&P Daily	Cost Per Labor-Hour Bare Costs	Cost Per Labor-Hour Incl. O&P
1 Truck Driver (heavy)	$45.55	$364.40	$68.70	$549.60	$45.55	$68.70
1 Dump Truck, Off Hwy., 100 Ton		2952.00		3247.20	369.00	405.90
8 L.H., Daily Totals		$3316.40		$3796.80	$414.55	$474.60

Crew B-34K

Crew No.	Bare Costs Hr.	Bare Costs Daily	Incl. Subs O&P Hr.	Incl. Subs O&P Daily	Cost Per Labor-Hour Bare Costs	Cost Per Labor-Hour Incl. O&P
1 Truck Driver (heavy)	$45.55	$364.40	$68.70	$549.60	$45.55	$68.70
1 Truck Tractor, 6x4, 450 H.P.		679.80		747.78		
1 Lowbed Trailer, 75 Ton		242.40		266.64	115.28	126.80
8 L.H., Daily Totals		$1286.60		$1564.02	$160.82	$195.50

Crew B-34L

Crew No.	Bare Costs Hr.	Bare Costs Daily	Incl. Subs O&P Hr.	Incl. Subs O&P Daily	Cost Per Labor-Hour Bare Costs	Cost Per Labor-Hour Incl. O&P
1 Equip. Oper. (light)	$51.00	$408.00	$77.15	$617.20	$51.00	$77.15
1 Flatbed Truck, Gas, 1.5 Ton		217.40		239.14	27.18	29.89
8 L.H., Daily Totals		$625.40		$856.34	$78.17	$107.04

Crew B-34M

Crew No.	Bare Costs Hr.	Bare Costs Daily	Incl. Subs O&P Hr.	Incl. Subs O&P Daily	Cost Per Labor-Hour Bare Costs	Cost Per Labor-Hour Incl. O&P
1 Equip. Oper. (light)	$51.00	$408.00	$77.15	$617.20	$51.00	$77.15
1 Flatbed Truck, Gas, 3 Ton		270.20		297.22	33.77	37.15
8 L.H., Daily Totals		$678.20		$914.42	$84.78	$114.30

Crew B-34N

Crew No.	Bare Costs Hr.	Bare Costs Daily	Incl. Subs O&P Hr.	Incl. Subs O&P Daily	Cost Per Labor-Hour Bare Costs	Cost Per Labor-Hour Incl. O&P
1 Truck Driver (heavy)	$45.55	$364.40	$68.70	$549.60	$49.55	$74.85
1 Equip. Oper. (medium)	53.55	428.40	81.00	648.00		
1 Truck Tractor, 6x4, 380 H.P.		559.80		615.78		
1 Flatbed Trailer, 40 Ton		175.00		192.50	45.92	50.52
16 L.H., Daily Totals		$1527.60		$2005.88	$95.47	$125.37

Crew B-34P

Crew No.	Bare Costs Hr.	Bare Costs Daily	Incl. Subs O&P Hr.	Incl. Subs O&P Daily	Cost Per Labor-Hour Bare Costs	Cost Per Labor-Hour Incl. O&P
1 Pipe Fitter	$62.25	$498.00	$93.95	$751.60	$53.32	$80.52
1 Truck Driver (light)	44.15	353.20	66.60	532.80		
1 Equip. Oper. (medium)	53.55	428.40	81.00	648.00		
1 Flatbed Truck, Gas, 3 Ton		270.20		297.22		
1 Backhoe Loader, 48 H.P.		366.20		402.82	26.52	29.17
24 L.H., Daily Totals		$1916.00		$2632.44	$79.83	$109.69

Crew B-34Q

Crew No.	Bare Costs Hr.	Bare Costs Daily	Incl. Subs O&P Hr.	Incl. Subs O&P Daily	Cost Per Labor-Hour Bare Costs	Cost Per Labor-Hour Incl. O&P
1 Pipe Fitter	$62.25	$498.00	$93.95	$751.60	$54.03	$81.60
1 Truck Driver (light)	44.15	353.20	66.60	532.80		
1 Equip. Oper. (crane)	55.70	445.60	84.25	674.00		
1 Flatbed Trailer, 25 Ton		128.40		141.24		
1 Dump Truck, 8 C.Y., 220 H.P.		391.00		430.10		
1 Hyd. Crane, 25 Ton		673.80		741.18	49.72	54.69
24 L.H., Daily Totals		$2490.00		$3270.92	$103.75	$136.29

Crew No.	Bare Costs Hr.	Daily	Incl. Subs O&P Hr.	Daily	Cost Per Labor-Hour Bare Costs	Incl. O&P
Crew B-34R	Hr.	Daily	Hr.	Daily	Bare Costs	Incl. O&P
1 Pipe Fitter	$62.25	$498.00	$93.95	$751.60	$54.03	$81.60
1 Truck Driver (light)	44.15	353.20	66.60	532.80		
1 Equip. Oper. (crane)	55.70	445.60	84.25	674.00		
1 Flatbed Trailer, 25 Ton		128.40		141.24		
1 Dump Truck, 8 C.Y., 220 H.P.		391.00		430.10		
1 Hyd. Crane, 25 Ton		673.80		741.18		
1 Hyd. Excavator, 1 C.Y.		798.20		878.02	82.97	91.27
24 L.H., Daily Totals		$3288.20		$4148.94	$137.01	$172.87
Crew B-34S	Hr.	Daily	Hr.	Daily	Bare Costs	Incl. O&P
2 Pipe Fitters	$62.25	$996.00	$93.95	$1503.20	$56.44	$85.21
1 Truck Driver (heavy)	45.55	364.40	68.70	549.60		
1 Equip. Oper. (crane)	55.70	445.60	84.25	674.00		
1 Flatbed Trailer, 40 Ton		175.00		192.50		
1 Truck Tractor, 6x4, 380 H.P.		559.80		615.78		
1 Hyd. Crane, 80 Ton		1616.00		1777.60		
1 Hyd. Excavator, 2 C.Y.		1123.00		1235.30	108.56	119.41
32 L.H., Daily Totals		$5279.80		$6547.98	$164.99	$204.62
Crew B-34T	Hr.	Daily	Hr.	Daily	Bare Costs	Incl. O&P
2 Pipe Fitters	$62.25	$996.00	$93.95	$1503.20	$56.44	$85.21
1 Truck Driver (heavy)	45.55	364.40	68.70	549.60		
1 Equip. Oper. (crane)	55.70	445.60	84.25	674.00		
1 Flatbed Trailer, 40 Ton		175.00		192.50		
1 Truck Tractor, 6x4, 380 H.P.		559.80		615.78		
1 Hyd. Crane, 80 Ton		1616.00		1777.60	73.46	80.81
32 L.H., Daily Totals		$4156.80		$5312.68	$129.90	$166.02
Crew B-34U	Hr.	Daily	Hr.	Daily	Bare Costs	Incl. O&P
1 Truck Driver (heavy)	$45.55	$364.40	$68.70	$549.60	$48.27	$72.92
1 Equip. Oper. (light)	51.00	408.00	77.15	617.20		
1 Truck Tractor, 220 H.P.		336.60		370.26		
1 Flatbed Trailer, 25 Ton		128.40		141.24	29.06	31.97
16 L.H., Daily Totals		$1237.40		$1678.30	$77.34	$104.89
Crew B-34V	Hr.	Daily	Hr.	Daily	Bare Costs	Incl. O&P
1 Truck Driver (heavy)	$45.55	$364.40	$68.70	$549.60	$50.75	$76.70
1 Equip. Oper. (crane)	55.70	445.60	84.25	674.00		
1 Equip. Oper. (light)	51.00	408.00	77.15	617.20		
1 Truck Tractor, 6x4, 450 H.P.		679.80		747.78		
1 Equipment Trailer, 50 Ton		192.20		211.42		
1 Pickup Truck, 4x4, 3/4 Ton		140.00		154.00	42.17	46.38
24 L.H., Daily Totals		$2230.00		$2954.00	$92.92	$123.08
Crew B-34W	Hr.	Daily	Hr.	Daily	Bare Costs	Incl. O&P
5 Truck Drivers (heavy)	$45.55	$1822.00	$68.70	$2748.00	$48.24	$72.94
2 Equip. Opers. (crane)	55.70	891.20	84.25	1348.00		
1 Equip. Oper. (mechanic)	55.85	446.80	84.50	676.00		
1 Laborer	39.15	313.20	60.00	480.00		
4 Truck Tractors, 6x4, 380 H.P.		2239.20		2463.12		
2 Equipment Trailers, 50 Ton		384.40		422.84		
2 Flatbed Trailers, 40 Ton		350.00		385.00		
1 Pickup Truck, 4x4, 3/4 Ton		140.00		154.00		
1 S.P. Crane, 4x4, 20 Ton		553.80		609.18	50.94	56.03
72 L.H., Daily Totals		$7140.60		$9286.14	$99.17	$128.97

Crew No.	Bare Costs Hr.	Daily	Incl. Subs O&P Hr.	Daily	Cost Per Labor-Hour Bare Costs	Incl. O&P
Crew B-35	Hr.	Daily	Hr.	Daily	Bare Costs	Incl. O&P
1 Labor Foreman (outside)	$41.15	$329.20	$63.05	$504.40	$49.54	$75.39
1 Skilled Worker	51.45	411.60	79.20	633.60		
1 Welder (plumber)	61.80	494.40	93.25	746.00		
1 Laborer	39.15	313.20	60.00	480.00		
1 Equip. Oper. (crane)	55.70	445.60	84.25	674.00		
1 Equip. Oper. (oiler)	48.00	384.00	72.60	580.80		
1 Welder, Electric, 300 amp		58.15		63.97		
1 Hyd. Excavator, .75 C.Y.		674.20		741.62	15.26	16.78
48 L.H., Daily Totals		$3110.35		$4424.39	$64.80	$92.17
Crew B-35A	Hr.	Daily	Hr.	Daily	Bare Costs	Incl. O&P
1 Labor Foreman (outside)	$41.15	$329.20	$63.05	$504.40	$48.06	$73.19
2 Laborers	39.15	626.40	60.00	960.00		
1 Skilled Worker	51.45	411.60	79.20	633.60		
1 Welder (plumber)	61.80	494.40	93.25	746.00		
1 Equip. Oper. (crane)	55.70	445.60	84.25	674.00		
1 Equip. Oper. (oiler)	48.00	384.00	72.60	580.80		
1 Welder, Gas Engine, 300 amp		129.80		142.78		
1 Crawler Crane, 75 Ton		1645.00		1809.50	31.69	34.86
56 L.H., Daily Totals		$4466.00		$6051.08	$79.75	$108.06
Crew B-36	Hr.	Daily	Hr.	Daily	Bare Costs	Incl. O&P
1 Labor Foreman (outside)	$41.15	$329.20	$63.05	$504.40	$45.31	$69.01
2 Laborers	39.15	626.40	60.00	960.00		
2 Equip. Oper. (medium)	53.55	856.80	81.00	1296.00		
1 Dozer, 200 H.P.		1384.00		1522.40		
1 Aggregate Spreader		37.40		41.14		
1 Tandem Roller, 10 Ton		237.60		261.36	41.48	45.62
40 L.H., Daily Totals		$3471.40		$4585.30	$86.78	$114.63
Crew B-36A	Hr.	Daily	Hr.	Daily	Bare Costs	Incl. O&P
1 Labor Foreman (outside)	$41.15	$329.20	$63.05	$504.40	$47.66	$72.44
2 Laborers	39.15	626.40	60.00	960.00		
4 Equip. Oper. (medium)	53.55	1713.60	81.00	2592.00		
1 Dozer, 200 H.P.		1384.00		1522.40		
1 Aggregate Spreader		37.40		41.14		
1 Tandem Roller, 10 Ton		237.60		261.36		
1 Roller, Pneum. Whl., 12 Ton		355.80		391.38	35.98	39.58
56 L.H., Daily Totals		$4684.00		$6272.68	$83.64	$112.01
Crew B-36B	Hr.	Daily	Hr.	Daily	Bare Costs	Incl. O&P
1 Labor Foreman (outside)	$41.15	$329.20	$63.05	$504.40	$47.40	$71.97
2 Laborers	39.15	626.40	60.00	960.00		
4 Equip. Oper. (medium)	53.55	1713.60	81.00	2592.00		
1 Truck Driver (heavy)	45.55	364.40	68.70	549.60		
1 Grader, 30,000 Lbs.		678.40		746.24		
1 F.E. Loader, Crl. 1.5 C.Y.		662.80		729.08		
1 Dozer, 300 H.P.		1900.00		2090.00		
1 Roller, Vibratory, 25 Ton		684.00		752.40		
1 Truck Tractor, 6x4, 450 H.P.		679.80		747.78		
1 Water Tank Trailer, 5000 Gal.		143.00		157.30	74.19	81.61
64 L.H., Daily Totals		$7781.60		$9828.80	$121.59	$153.58

737

Crew B-36C	Bare Costs Hr.	Bare Costs Daily	Incl. Subs O&P Hr.	Incl. Subs O&P Daily	Cost Per Labor-Hour Bare Costs	Cost Per Labor-Hour Incl. O&P
1 Labor Foreman (outside)	$41.15	$329.20	$63.05	$504.40	$49.47	$74.95
3 Equip. Oper. (medium)	53.55	1285.20	81.00	1944.00		
1 Truck Driver (heavy)	45.55	364.40	68.70	549.60		
1 Grader, 30,000 Lbs.		678.40		746.24		
1 Dozer, 300 H.P.		1900.00		2090.00		
1 Roller, Vibratory, 25 Ton		684.00		752.40		
1 Truck Tractor, 6x4, 450 H.P.		679.80		747.78		
1 Water Tank Trailer, 5000 Gal.		143.00		157.30	102.13	112.34
40 L.H., Daily Totals		$6064.00		$7491.72	$151.60	$187.29

Crew B-36D	Hr.	Daily	Hr.	Daily	Bare Costs	Incl. O&P
1 Labor Foreman (outside)	$41.15	$329.20	$63.05	$504.40	$50.45	$76.51
3 Equip. Oper. (medium)	53.55	1285.20	81.00	1944.00		
1 Grader, 30,000 Lbs.		678.40		746.24		
1 Dozer, 300 H.P.		1900.00		2090.00		
1 Roller, Vibratory, 25 Ton		684.00		752.40	101.95	112.15
32 L.H., Daily Totals		$4876.80		$6037.04	$152.40	$188.66

Crew B-37	Hr.	Daily	Hr.	Daily	Bare Costs	Incl. O&P
1 Labor Foreman (outside)	$41.15	$329.20	$63.05	$504.40	$41.46	$63.37
4 Laborers	39.15	1252.80	60.00	1920.00		
1 Equip. Oper. (light)	51.00	408.00	77.15	617.20		
1 Tandem Roller, 5 Ton		152.40		167.64	3.17	3.49
48 L.H., Daily Totals		$2142.40		$3209.24	$44.63	$66.86

Crew B-37A	Hr.	Daily	Hr.	Daily	Bare Costs	Incl. O&P
2 Laborers	$39.15	$626.40	$60.00	$960.00	$40.82	$62.20
1 Truck Driver (light)	44.15	353.20	66.60	532.80		
1 Flatbed Truck, Gas, 1.5 Ton		217.40		239.14		
1 Tar Kettle, T.M.		122.75		135.03	14.17	15.59
24 L.H., Daily Totals		$1319.75		$1866.96	$54.99	$77.79

Crew B-37B	Hr.	Daily	Hr.	Daily	Bare Costs	Incl. O&P
3 Laborers	$39.15	$939.60	$60.00	$1440.00	$40.40	$61.65
1 Truck Driver (light)	44.15	353.20	66.60	532.80		
1 Flatbed Truck, Gas, 1.5 Ton		217.40		239.14		
1 Tar Kettle, T.M.		122.75		135.03	10.63	11.69
32 L.H., Daily Totals		$1632.95		$2346.97	$51.03	$73.34

Crew B-37C	Hr.	Daily	Hr.	Daily	Bare Costs	Incl. O&P
2 Laborers	$39.15	$626.40	$60.00	$960.00	$41.65	$63.30
2 Truck Drivers (light)	44.15	706.40	66.60	1065.60		
2 Flatbed Trucks, Gas, 1.5 Ton		434.80		478.28		
1 Tar Kettle, T.M.		122.75		135.03	17.42	19.17
32 L.H., Daily Totals		$1890.35		$2638.91	$59.07	$82.47

Crew B-37D	Hr.	Daily	Hr.	Daily	Bare Costs	Incl. O&P
1 Laborer	$39.15	$313.20	$60.00	$480.00	$41.65	$63.30
1 Truck Driver (light)	44.15	353.20	66.60	532.80		
1 Pickup Truck, 3/4 Ton		129.40		142.34	8.09	8.90
16 L.H., Daily Totals		$795.80		$1155.14	$49.74	$72.20

Crew B-37E	Bare Costs Hr.	Bare Costs Daily	Incl. Subs O&P Hr.	Incl. Subs O&P Daily	Cost Per Labor-Hour Bare Costs	Cost Per Labor-Hour Incl. O&P
3 Laborers	$39.15	$939.60	$60.00	$1440.00	$44.33	$67.34
1 Equip. Oper. (light)	51.00	408.00	77.15	617.20		
1 Equip. Oper. (medium)	53.55	428.40	81.00	648.00		
2 Truck Drivers (light)	44.15	706.40	66.60	1065.60		
4 Barrels w/ Flasher		13.60		14.96		
1 Concrete Saw		153.60		168.96		
1 Rotary Hammer Drill		24.70		27.17		
1 Hammer Drill Bit		3.05		3.36		
1 Loader, Skid Steer, 30 H.P.		171.00		188.10		
1 Conc. Hammer Attach.		115.55		127.11		
1 Vibrating Plate, Gas, 18"		33.60		36.96		
2 Flatbed Trucks, Gas, 1.5 Ton		434.80		478.28	16.96	18.66
56 L.H., Daily Totals		$3432.30		$4815.69	$61.29	$85.99

Crew B-37F	Hr.	Daily	Hr.	Daily	Bare Costs	Incl. O&P
3 Laborers	$39.15	$939.60	$60.00	$1440.00	$40.40	$61.65
1 Truck Driver (light)	44.15	353.20	66.60	532.80		
4 Barrels w/ Flasher		13.60		14.96		
1 Concrete Mixer, 10 C.F.		164.60		181.06		
1 Air Compressor, 60 cfm		127.00		139.70		
1 -50' Air Hose, 3/4"		3.25		3.58		
1 Spade (Chipper)		8.40		9.24		
1 Flatbed Truck, Gas, 1.5 Ton		217.40		239.14	16.70	18.36
32 L.H., Daily Totals		$1827.05		$2560.47	$57.10	$80.01

Crew B-37G	Hr.	Daily	Hr.	Daily	Bare Costs	Incl. O&P
1 Labor Foreman (outside)	$41.15	$329.20	$63.05	$504.40	$41.46	$63.37
4 Laborers	39.15	1252.80	60.00	1920.00		
1 Equip. Oper. (light)	51.00	408.00	77.15	617.20		
1 Berm Machine		313.40		344.74		
1 Tandem Roller, 5 Ton		152.40		167.64	9.70	10.67
48 L.H., Daily Totals		$2455.80		$3553.98	$51.16	$74.04

Crew B-37H	Hr.	Daily	Hr.	Daily	Bare Costs	Incl. O&P
1 Labor Foreman (outside)	$41.15	$329.20	$63.05	$504.40	$41.46	$63.37
4 Laborers	39.15	1252.80	60.00	1920.00		
1 Equip. Oper. (light)	51.00	408.00	77.15	617.20		
1 Tandem Roller, 5 Ton		152.40		167.64		
1 Flatbed Truck, Gas, 1.5 Ton		217.40		239.14		
1 Tar Kettle, T.M.		122.75		135.03	10.26	11.29
48 L.H., Daily Totals		$2482.55		$3583.41	$51.72	$74.65

Crew B-37I	Hr.	Daily	Hr.	Daily	Bare Costs	Incl. O&P
3 Laborers	$39.15	$939.60	$60.00	$1440.00	$44.33	$67.34
1 Equip. Oper. (light)	51.00	408.00	77.15	617.20		
1 Equip. Oper. (medium)	53.55	428.40	81.00	648.00		
2 Truck Drivers (light)	44.15	706.40	66.60	1065.60		
4 Barrels w/ Flasher		13.60		14.96		
1 Concrete Saw		153.60		168.96		
1 Rotary Hammer Drill		24.70		27.17		
1 Hammer Drill Bit		3.05		3.36		
1 Air Compressor, 60 cfm		127.00		139.70		
1 -50' Air Hose, 3/4"		3.25		3.58		
1 Spade (Chipper)		8.40		9.24		
1 Loader, Skid Steer, 30 H.P.		171.00		188.10		
1 Conc. Hammer Attach.		115.55		127.11		
1 Concrete Mixer, 10 C.F.		164.60		181.06		
1 Vibrating Plate, Gas, 18"		33.60		36.96		
2 Flatbed Trucks, Gas, 1.5 Ton		434.80		478.28	22.38	24.62
56 L.H., Daily Totals		$3735.55		$5149.27	$66.71	$91.95

For customer support on your Building Construction Costs with RSMeans Data, call 800.448.8182.

Crew No.		Bare Costs	Incl. Subs O&P		Cost Per Labor-Hour	

Crew B-37J

	Hr.	Daily	Hr.	Daily	Bare Costs	Incl. O&P
1 Labor Foreman (outside)	$41.15	$329.20	$63.05	$504.40	$41.46	$63.37
4 Laborers	39.15	1252.80	60.00	1920.00		
1 Equip. Oper. (light)	51.00	408.00	77.15	617.20		
1 Air Compressor, 60 cfm		127.00		139.70		
1 -50' Air Hose, 3/4"		3.25		3.58		
2 Concrete Mixers, 10 C.F.		329.20		362.12		
2 Flatbed Trucks, Gas, 1.5 Ton		434.80		478.28		
1 Shot Blaster, 20"		220.40		242.44	23.22	25.54
48 L.H., Daily Totals		$3104.65		$4267.72	$64.68	$88.91

Crew B-37K

	Hr.	Daily	Hr.	Daily	Bare Costs	Incl. O&P
1 Labor Foreman (outside)	$41.15	$329.20	$63.05	$504.40	$41.46	$63.37
4 Laborers	39.15	1252.80	60.00	1920.00		
1 Equip. Oper. (light)	51.00	408.00	77.15	617.20		
1 Air Compressor, 60 cfm		127.00		139.70		
1 -50' Air Hose, 3/4"		3.25		3.58		
2 Flatbed Trucks, Gas, 1.5 Ton		434.80		478.28		
1 Shot Blaster, 20"		220.40		242.44	16.36	18.00
48 L.H., Daily Totals		$2775.45		$3905.59	$57.82	$81.37

Crew B-38

	Hr.	Daily	Hr.	Daily	Bare Costs	Incl. O&P
1 Labor Foreman (outside)	$41.15	$329.20	$63.05	$504.40	$44.80	$68.24
2 Laborers	39.15	626.40	60.00	960.00		
1 Equip. Oper. (light)	51.00	408.00	77.15	617.20		
1 Equip. Oper. (medium)	53.55	428.40	81.00	648.00		
1 Backhoe Loader, 48 H.P.		366.20		402.82		
1 Hyd. Hammer (1200 lb.)		182.20		200.42		
1 F.E. Loader, W.M., 4 C.Y.		657.60		723.36		
1 Pvmt. Rem. Bucket		57.80		63.58	31.59	34.75
40 L.H., Daily Totals		$3055.80		$4119.78	$76.39	$102.99

Crew B-39

	Hr.	Daily	Hr.	Daily	Bare Costs	Incl. O&P
1 Labor Foreman (outside)	$41.15	$329.20	$63.05	$504.40	$41.46	$63.37
4 Laborers	39.15	1252.80	60.00	1920.00		
1 Equip. Oper. (light)	51.00	408.00	77.15	617.20		
1 Air Compressor, 250 cfm		181.20		199.32		
2 Breakers, Pavement, 60 lb.		20.40		22.44		
2 -50' Air Hoses, 1.5"		11.60		12.76	4.44	4.89
48 L.H., Daily Totals		$2203.20		$3276.12	$45.90	$68.25

Crew B-40

	Hr.	Daily	Hr.	Daily	Bare Costs	Incl. O&P
1 Pile Driver Foreman (outside)	$51.50	$412.00	$81.80	$654.40	$51.11	$79.66
4 Pile Drivers	49.50	1584.00	78.60	2515.20		
2 Equip. Oper. (crane)	55.70	891.20	84.25	1348.00		
1 Equip. Oper. (oiler)	48.00	384.00	72.60	580.80		
1 Crawler Crane, 40 Ton		1313.00		1444.30		
1 Vibratory Hammer & Gen.		2494.00		2743.40	59.48	65.43
64 L.H., Daily Totals		$7078.20		$9286.10	$110.60	$145.10

Crew B-40B

	Hr.	Daily	Hr.	Daily	Bare Costs	Incl. O&P
1 Labor Foreman (outside)	$41.15	$329.20	$63.05	$504.40	$43.72	$66.65
3 Laborers	39.15	939.60	60.00	1440.00		
1 Equip. Oper. (crane)	55.70	445.60	84.25	674.00		
1 Equip. Oper. (oiler)	48.00	384.00	72.60	580.80		
1 Lattice Boom Crane, 40 Ton		1311.00		1442.10	27.31	30.04
48 L.H., Daily Totals		$3409.40		$4641.30	$71.03	$96.69

Crew B-41

	Hr.	Daily	Hr.	Daily	Bare Costs	Incl. O&P
1 Labor Foreman (outside)	$41.15	$329.20	$63.05	$504.40	$40.67	$62.23
4 Laborers	39.15	1252.80	60.00	1920.00		
.25 Equip. Oper. (crane)	55.70	111.40	84.25	168.50		
.25 Equip. Oper. (oiler)	48.00	96.00	72.60	145.20		
.25 Crawler Crane, 40 Ton		328.25		361.07	7.46	8.21
44 L.H., Daily Totals		$2117.65		$3099.18	$48.13	$70.44

Crew B-42

	Hr.	Daily	Hr.	Daily	Bare Costs	Incl. O&P
1 Labor Foreman (outside)	$41.15	$329.20	$63.05	$504.40	$44.47	$68.80
4 Laborers	39.15	1252.80	60.00	1920.00		
1 Equip. Oper. (crane)	55.70	445.60	84.25	674.00		
1 Equip. Oper. (oiler)	48.00	384.00	72.60	580.80		
1 Welder	54.30	434.40	90.50	724.00		
1 Hyd. Crane, 25 Ton		673.80		741.18		
1 Welder, Gas Engine, 300 amp		129.80		142.78		
1 Horz. Boring Csg. Mch.		465.20		511.72	19.82	21.81
64 L.H., Daily Totals		$4114.80		$5798.88	$64.29	$90.61

Crew B-43

	Hr.	Daily	Hr.	Daily	Bare Costs	Incl. O&P
1 Labor Foreman (outside)	$41.15	$329.20	$63.05	$504.40	$43.72	$66.65
3 Laborers	39.15	939.60	60.00	1440.00		
1 Equip. Oper. (crane)	55.70	445.60	84.25	674.00		
1 Equip. Oper. (oiler)	48.00	384.00	72.60	580.80		
1 Drill Rig, Truck-Mounted		2573.00		2830.30	53.60	58.96
48 L.H., Daily Totals		$4671.40		$6029.50	$97.32	$125.61

Crew B-44

	Hr.	Daily	Hr.	Daily	Bare Costs	Incl. O&P
1 Pile Driver Foreman (outside)	$51.50	$412.00	$81.80	$654.40	$50.01	$78.09
4 Pile Drivers	49.50	1584.00	78.60	2515.20		
2 Equip. Oper. (crane)	55.70	891.20	84.25	1348.00		
1 Laborer	39.15	313.20	60.00	480.00		
1 Crawler Crane, 40 Ton		1313.00		1444.30		
1 Lead, 60' High		76.00		83.60		
1 Hammer, Diesel, 15K ft.-lbs.		605.80		666.38	31.17	34.29
64 L.H., Daily Totals		$5195.20		$7191.88	$81.18	$112.37

Crew B-45

	Hr.	Daily	Hr.	Daily	Bare Costs	Incl. O&P
1 Equip. Oper. (medium)	$53.55	$428.40	$81.00	$648.00	$49.55	$74.85
1 Truck Driver (heavy)	45.55	364.40	68.70	549.60		
1 Dist. Tanker, 3000 Gallon		327.40		360.14		
1 Truck Tractor, 6x4, 380 H.P.		559.80		615.78	55.45	60.99
16 L.H., Daily Totals		$1680.00		$2173.52	$105.00	$135.85

Crew B-46

	Hr.	Daily	Hr.	Daily	Bare Costs	Incl. O&P
1 Pile Driver Foreman (outside)	$51.50	$412.00	$81.80	$654.40	$44.66	$69.83
2 Pile Drivers	49.50	792.00	78.60	1257.60		
3 Laborers	39.15	939.60	60.00	1440.00		
1 Chain Saw, Gas, 36" Long		45.00		49.50	.94	1.03
48 L.H., Daily Totals		$2188.60		$3401.50	$45.60	$70.86

Crew B-47

	Hr.	Daily	Hr.	Daily	Bare Costs	Incl. O&P
1 Blast Foreman (outside)	$41.15	$329.20	$63.05	$504.40	$43.77	$66.73
1 Driller	39.15	313.20	60.00	480.00		
1 Equip. Oper. (light)	51.00	408.00	77.15	617.20		
1 Air Track Drill, 4"		1016.00		1117.60		
1 Air Compressor, 600 cfm		482.00		530.20		
2 -50' Air Hoses, 3"		29.80		32.78	63.66	70.02
24 L.H., Daily Totals		$2578.20		$3282.18	$107.43	$136.76

Crew B-47A	Hr.	Daily	Hr.	Daily	Bare Costs	Incl. O&P
1 Drilling Foreman (outside)	$41.15	$329.20	$63.05	$504.40	$48.28	$73.30
1 Equip. Oper. (heavy)	55.70	445.60	84.25	674.00		
1 Equip. Oper. (oiler)	48.00	384.00	72.60	580.80		
1 Air Track Drill, 5"		1211.00		1332.10	50.46	55.50
24 L.H., Daily Totals		$2369.80		$3091.30	$98.74	$128.80

Crew B-47C	Hr.	Daily	Hr.	Daily	Bare Costs	Incl. O&P
1 Laborer	$39.15	$313.20	$60.00	$480.00	$45.08	$68.58
1 Equip. Oper. (light)	51.00	408.00	77.15	617.20		
1 Air Compressor, 750 cfm		488.20		537.02		
2 -50' Air Hoses, 3"		29.80		32.78		
1 Air Track Drill, 4"		1016.00		1117.60	95.88	105.46
16 L.H., Daily Totals		$2255.20		$2784.60	$140.95	$174.04

Crew B-47E	Hr.	Daily	Hr.	Daily	Bare Costs	Incl. O&P
1 Labor Foreman (outside)	$41.15	$329.20	$63.05	$504.40	$39.65	$60.76
3 Laborers	39.15	939.60	60.00	1440.00		
1 Flatbed Truck, Gas, 3 Ton		270.20		297.22	8.44	9.29
32 L.H., Daily Totals		$1539.00		$2241.62	$48.09	$70.05

Crew B-47G	Hr.	Daily	Hr.	Daily	Bare Costs	Incl. O&P
1 Labor Foreman (outside)	$41.15	$329.20	$63.05	$504.40	$42.61	$65.05
2 Laborers	39.15	626.40	60.00	960.00		
1 Equip. Oper. (light)	51.00	408.00	77.15	617.20		
1 Air Track Drill, 4"		1016.00		1117.60		
1 Air Compressor, 600 cfm		482.00		530.20		
2 -50' Air Hoses, 3"		29.80		32.78		
1 Gunite Pump Rig		357.40		393.14	58.91	64.80
32 L.H., Daily Totals		$3248.80		$4155.32	$101.53	$129.85

Crew B-47H	Hr.	Daily	Hr.	Daily	Bare Costs	Incl. O&P
1 Skilled Worker Foreman (out)	$53.45	$427.60	$82.25	$658.00	$51.95	$79.96
3 Skilled Workers	51.45	1234.80	79.20	1900.80		
1 Flatbed Truck, Gas, 3 Ton		270.20		297.22	8.44	9.29
32 L.H., Daily Totals		$1932.60		$2856.02	$60.39	$89.25

Crew B-48	Hr.	Daily	Hr.	Daily	Bare Costs	Incl. O&P
1 Labor Foreman (outside)	$41.15	$329.20	$63.05	$504.40	$44.76	$68.15
3 Laborers	39.15	939.60	60.00	1440.00		
1 Equip. Oper. (crane)	55.70	445.60	84.25	674.00		
1 Equip. Oper. (oiler)	48.00	384.00	72.60	580.80		
1 Equip. Oper. (light)	51.00	408.00	77.15	617.20		
1 Centr. Water Pump, 6"		325.40		357.94		
1 -20' Suction Hose, 6"		11.90		13.09		
1 -50' Discharge Hose, 6"		6.30		6.93		
1 Drill Rig, Truck-Mounted		2573.00		2830.30	52.08	57.29
56 L.H., Daily Totals		$5423.00		$7024.66	$96.84	$125.44

Crew B-49	Hr.	Daily	Hr.	Daily	Bare Costs	Incl. O&P
1 Labor Foreman (outside)	$41.15	$329.20	$63.05	$504.40	$46.91	$71.92
3 Laborers	39.15	939.60	60.00	1440.00		
2 Equip. Oper. (crane)	55.70	891.20	84.25	1348.00		
2 Equip. Oper. (oilers)	48.00	768.00	72.60	1161.60		
1 Equip. Oper. (light)	51.00	408.00	77.15	617.20		
2 Pile Drivers	49.50	792.00	78.60	1257.60		
1 Hyd. Crane, 25 Ton		673.80		741.18		
1 Centr. Water Pump, 6"		325.40		357.94		
1 -20' Suction Hose, 6"		11.90		13.09		
1 -50' Discharge Hose, 6"		6.30		6.93		
1 Drill Rig, Truck-Mounted		2573.00		2830.30	40.80	44.88
88 L.H., Daily Totals		$7718.40		$10278.24	$87.71	$116.80

Crew B-50	Hr.	Daily	Hr.	Daily	Bare Costs	Incl. O&P
2 Pile Driver Foremen (outside)	$51.50	$824.00	$81.80	$1308.80	$48.35	$75.45
6 Pile Drivers	49.50	2376.00	78.60	3772.80		
2 Equip. Oper. (crane)	55.70	891.20	84.25	1348.00		
1 Equip. Oper. (oiler)	48.00	384.00	72.60	580.80		
3 Laborers	39.15	939.60	60.00	1440.00		
1 Crawler Crane, 40 Ton		1313.00		1444.30		
1 Lead, 60' High		76.00		83.60		
1 Hammer, Diesel, 15K ft.-lbs.		605.80		666.38		
1 Air Compressor, 600 cfm		482.00		530.20		
2 -50' Air Hoses, 3"		29.80		32.78		
1 Chain Saw, Gas, 36" Long		45.00		49.50	22.78	25.06
112 L.H., Daily Totals		$7966.40		$11257.16	$71.13	$100.51

Crew B-51	Hr.	Daily	Hr.	Daily	Bare Costs	Incl. O&P
1 Labor Foreman (outside)	$41.15	$329.20	$63.05	$504.40	$40.32	$61.61
4 Laborers	39.15	1252.80	60.00	1920.00		
1 Truck Driver (light)	44.15	353.20	66.60	532.80		
1 Flatbed Truck, Gas, 1.5 Ton		217.40		239.14	4.53	4.98
48 L.H., Daily Totals		$2152.60		$3196.34	$44.85	$66.59

Crew B-52	Hr.	Daily	Hr.	Daily	Bare Costs	Incl. O&P
1 Carpenter Foreman (outside)	$51.25	$410.00	$78.50	$628.00	$45.52	$69.41
1 Carpenter	49.25	394.00	75.45	603.60		
3 Laborers	39.15	939.60	60.00	1440.00		
1 Cement Finisher	46.80	374.40	69.60	556.80		
.5 Rodman (reinf.)	54.30	217.20	83.70	334.80		
.5 Equip. Oper. (medium)	53.55	214.20	81.00	324.00		
.5 Crawler Loader, 3 C.Y.		640.50		704.55	11.44	12.58
56 L.H., Daily Totals		$3189.90		$4591.75	$56.96	$82.00

Crew B-53	Hr.	Daily	Hr.	Daily	Bare Costs	Incl. O&P
1 Equip. Oper. (light)	$51.00	$408.00	$77.15	$617.20	$51.00	$77.15
1 Trencher, Chain, 12 H.P.		63.60		69.96	7.95	8.74
8 L.H., Daily Totals		$471.60		$687.16	$58.95	$85.89

Crew B-54	Hr.	Daily	Hr.	Daily	Bare Costs	Incl. O&P
1 Equip. Oper. (light)	$51.00	$408.00	$77.15	$617.20	$51.00	$77.15
1 Trencher, Chain, 40 H.P.		325.20		357.72	40.65	44.72
8 L.H., Daily Totals		$733.20		$974.92	$91.65	$121.86

Crew B-54A	Hr.	Daily	Hr.	Daily	Bare Costs	Incl. O&P
.17 Labor Foreman (outside)	$41.15	$55.96	$63.05	$85.75	$51.75	$78.39
1 Equipment Operator (med.)	53.55	428.40	81.00	648.00		
1 Wheel Trencher, 67 H.P.		1157.00		1272.70	123.61	135.97
9.36 L.H., Daily Totals		$1641.36		$2006.45	$175.36	$214.36

Crew B-54B	Hr.	Daily	Hr.	Daily	Bare Costs	Incl. O&P
.25 Labor Foreman (outside)	$41.15	$82.30	$63.05	$126.10	$51.07	$77.41
1 Equipment Operator (med.)	53.55	428.40	81.00	648.00		
1 Wheel Trencher, 150 H.P.		1930.00		2123.00	193.00	212.30
10 L.H., Daily Totals		$2440.70		$2897.10	$244.07	$289.71

Crew B-54C	Hr.	Daily	Hr.	Daily	Bare Costs	Incl. O&P
1 Laborer	$39.15	$313.20	$60.00	$480.00	$46.35	$70.50
1 Equipment Operator (med.)	53.55	428.40	81.00	648.00		
1 Wheel Trencher, 67 H.P.		1157.00		1272.70	72.31	79.54
16 L.H., Daily Totals		$1898.60		$2400.70	$118.66	$150.04

For customer support on your Building Construction Costs with RSMeans Data, call 800.448.8182.

Crew No.	Bare Costs		Incl. Subs O&P		Cost Per Labor-Hour	
	Hr.	Daily	Hr.	Daily	Bare Costs	Incl. O&P
Crew B-54D	Hr.	Daily	Hr.	Daily	Bare Costs	Incl. O&P
1 Laborer	$39.15	$313.20	$60.00	$480.00	$46.35	$70.50
1 Equipment Operator (med.)	53.55	428.40	81.00	648.00		
1 Rock Trencher, 6" Width		749.00		823.90	46.81	51.49
16 L.H., Daily Totals		$1490.60		$1951.90	$93.16	$121.99
Crew B-54E	Hr.	Daily	Hr.	Daily	Bare Costs	Incl. O&P
1 Laborer	$39.15	$313.20	$60.00	$480.00	$46.35	$70.50
1 Equipment Operator (med.)	53.55	428.40	81.00	648.00		
1 Rock Trencher, 18" Width		2645.00		2909.50	165.31	181.84
16 L.H., Daily Totals		$3386.60		$4037.50	$211.66	$252.34
Crew B-55	Hr.	Daily	Hr.	Daily	Bare Costs	Incl. O&P
2 Laborers	$39.15	$626.40	$60.00	$960.00	$40.82	$62.20
1 Truck Driver (light)	44.15	353.20	66.60	532.80		
1 Truck-Mounted Earth Auger		781.40		859.54		
1 Flatbed Truck, Gas, 3 Ton		270.20		297.22	43.82	48.20
24 L.H., Daily Totals		$2031.20		$2649.56	$84.63	$110.40
Crew B-56	Hr.	Daily	Hr.	Daily	Bare Costs	Incl. O&P
1 Laborer	$39.15	$313.20	$60.00	$480.00	$45.08	$68.58
1 Equip. Oper. (light)	51.00	408.00	77.15	617.20		
1 Air Track Drill, 4"		1016.00		1117.60		
1 Air Compressor, 600 cfm		482.00		530.20		
1 -50' Air Hose, 3"		14.90		16.39	94.56	104.01
16 L.H., Daily Totals		$2234.10		$2761.39	$139.63	$172.59
Crew B-57	Hr.	Daily	Hr.	Daily	Bare Costs	Incl. O&P
1 Labor Foreman (outside)	$41.15	$329.20	$63.05	$504.40	$45.69	$69.51
2 Laborers	39.15	626.40	60.00	960.00		
1 Equip. Oper. (crane)	55.70	445.60	84.25	674.00		
1 Equip. Oper. (light)	51.00	408.00	77.15	617.20		
1 Equip. Oper. (oiler)	48.00	384.00	72.60	580.80		
1 Crawler Crane, 25 Ton		1296.00		1425.60		
1 Clamshell Bucket, 1 C.Y.		48.80		53.68		
1 Centr. Water Pump, 6"		325.40		357.94		
1 -20' Suction Hose, 6"		11.90		13.09		
20 -50' Discharge Hoses, 6"		126.00		138.60	37.67	41.44
48 L.H., Daily Totals		$4001.30		$5325.31	$83.36	$110.94
Crew B-58	Hr.	Daily	Hr.	Daily	Bare Costs	Incl. O&P
2 Laborers	$39.15	$626.40	$60.00	$960.00	$43.10	$65.72
1 Equip. Oper. (light)	51.00	408.00	77.15	617.20		
1 Backhoe Loader, 48 H.P.		366.20		402.82		
1 Small Helicopter, w/ Pilot		2793.00		3072.30	131.63	144.80
24 L.H., Daily Totals		$4193.60		$5052.32	$174.73	$210.51
Crew B-59	Hr.	Daily	Hr.	Daily	Bare Costs	Incl. O&P
1 Truck Driver (heavy)	$45.55	$364.40	$68.70	$549.60	$45.55	$68.70
1 Truck Tractor, 220 H.P.		336.60		370.26		
1 Water Tank Trailer, 5000 Gal.		143.00		157.30	59.95	65.94
8 L.H., Daily Totals		$844.00		$1077.16	$105.50	$134.65
Crew B-59A	Hr.	Daily	Hr.	Daily	Bare Costs	Incl. O&P
2 Laborers	$39.15	$626.40	$60.00	$960.00	$41.28	$62.90
1 Truck Driver (heavy)	45.55	364.40	68.70	549.60		
1 Water Tank Trailer, 5000 Gal.		143.00		157.30		
1 Truck Tractor, 220 H.P.		336.60		370.26	19.98	21.98
24 L.H., Daily Totals		$1470.40		$2037.16	$61.27	$84.88

Crew No.	Bare Costs		Incl. Subs O&P		Cost Per Labor-Hour	
Crew B-60	Hr.	Daily	Hr.	Daily	Bare Costs	Incl. O&P
1 Labor Foreman (outside)	$41.15	$329.20	$63.05	$504.40	$46.45	$70.60
2 Laborers	39.15	626.40	60.00	960.00		
1 Equip. Oper. (crane)	55.70	445.60	84.25	674.00		
2 Equip. Oper. (light)	51.00	816.00	77.15	1234.40		
1 Equip. Oper. (oiler)	48.00	384.00	72.60	580.80		
1 Crawler Crane, 40 Ton		1313.00		1444.30		
1 Lead, 60' High		76.00		83.60		
1 Hammer, Diesel, 15K ft.-lbs.		605.80		666.38		
1 Backhoe Loader, 48 H.P.		366.20		402.82	42.16	46.38
56 L.H., Daily Totals		$4962.20		$6550.70	$88.61	$116.98
Crew B-61	Hr.	Daily	Hr.	Daily	Bare Costs	Incl. O&P
1 Labor Foreman (outside)	$41.15	$329.20	$63.05	$504.40	$41.92	$64.04
3 Laborers	39.15	939.60	60.00	1440.00		
1 Equip. Oper. (light)	51.00	408.00	77.15	617.20		
1 Cement Mixer, 2 C.Y.		197.00		216.70		
1 Air Compressor, 160 cfm		143.60		157.96	8.52	9.37
40 L.H., Daily Totals		$2017.40		$2936.26	$50.44	$73.41
Crew B-62	Hr.	Daily	Hr.	Daily	Bare Costs	Incl. O&P
2 Laborers	$39.15	$626.40	$60.00	$960.00	$43.10	$65.72
1 Equip. Oper. (light)	51.00	408.00	77.15	617.20		
1 Loader, Skid Steer, 30 H.P.		171.00		188.10	7.13	7.84
24 L.H., Daily Totals		$1205.40		$1765.30	$50.23	$73.55
Crew B-62A	Hr.	Daily	Hr.	Daily	Bare Costs	Incl. O&P
2 Laborers	$39.15	$626.40	$60.00	$960.00	$43.10	$65.72
1 Equip. Oper. (light)	51.00	408.00	77.15	617.20		
1 Loader, Skid Steer, 30 H.P.		171.00		188.10		
1 Trencher Attachment		55.30		60.83	9.43	10.37
24 L.H., Daily Totals		$1260.70		$1826.13	$52.53	$76.09
Crew B-63	Hr.	Daily	Hr.	Daily	Bare Costs	Incl. O&P
4 Laborers	$39.15	$1252.80	$60.00	$1920.00	$41.52	$63.43
1 Equip. Oper. (light)	51.00	408.00	77.15	617.20		
1 Loader, Skid Steer, 30 H.P.		171.00		188.10	4.28	4.70
40 L.H., Daily Totals		$1831.80		$2725.30	$45.80	$68.13
Crew B-63B	Hr.	Daily	Hr.	Daily	Bare Costs	Incl. O&P
1 Labor Foreman (inside)	$39.65	$317.20	$60.75	$486.00	$42.24	$64.47
2 Laborers	39.15	626.40	60.00	960.00		
1 Equip. Oper. (light)	51.00	408.00	77.15	617.20		
1 Loader, Skid Steer, 78 H.P.		309.40		340.34	9.67	10.64
32 L.H., Daily Totals		$1661.00		$2403.54	$51.91	$75.11
Crew B-64	Hr.	Daily	Hr.	Daily	Bare Costs	Incl. O&P
1 Laborer	$39.15	$313.20	$60.00	$480.00	$41.65	$63.30
1 Truck Driver (light)	44.15	353.20	66.60	532.80		
1 Power Mulcher (small)		143.20		157.52		
1 Flatbed Truck, Gas, 1.5 Ton		217.40		239.14	22.54	24.79
16 L.H., Daily Totals		$1027.00		$1409.46	$64.19	$88.09
Crew B-65	Hr.	Daily	Hr.	Daily	Bare Costs	Incl. O&P
1 Laborer	$39.15	$313.20	$60.00	$480.00	$41.65	$63.30
1 Truck Driver (light)	44.15	353.20	66.60	532.80		
1 Power Mulcher (Large)		303.00		333.30		
1 Flatbed Truck, Gas, 1.5 Ton		217.40		239.14	32.52	35.78
16 L.H., Daily Totals		$1186.80		$1585.24	$74.17	$99.08

Crew No.	Bare Costs Hr.	Daily	Incl. Subs O&P Hr.	Daily	Cost Per Labor-Hour Bare Costs	Incl. O&P
Crew B-66	Hr.	Daily	Hr.	Daily	Bare Costs	Incl. O&P
1 Equip. Oper. (light)	$51.00	$408.00	$77.15	$617.20	$51.00	$77.15
1 Loader-Backhoe, 40 H.P.		268.00		294.80	33.50	36.85
8 L.H., Daily Totals		$676.00		$912.00	$84.50	$114.00
Crew B-67	Hr.	Daily	Hr.	Daily	Bare Costs	Incl. O&P
1 Millwright	$51.55	$412.40	$75.50	$604.00	$51.27	$76.33
1 Equip. Oper. (light)	51.00	408.00	77.15	617.20		
1 R.T. Forklift, 5,000 Lb., diesel		282.20		310.42	17.64	19.40
16 L.H., Daily Totals		$1102.60		$1531.62	$68.91	$95.73
Crew B-67B	Hr.	Daily	Hr.	Daily	Bare Costs	Incl. O&P
1 Millwright Foreman (inside)	$52.05	$416.40	$76.25	$610.00	$51.80	$75.88
1 Millwright	51.55	412.40	75.50	604.00		
16 L.H., Daily Totals		$828.80		$1214.00	$51.80	$75.88
Crew B-68	Hr.	Daily	Hr.	Daily	Bare Costs	Incl. O&P
2 Millwrights	$51.55	$824.80	$75.50	$1208.00	$51.37	$76.05
1 Equip. Oper. (light)	51.00	408.00	77.15	617.20		
1 R.T. Forklift, 5,000 Lb., diesel		282.20		310.42	11.76	12.93
24 L.H., Daily Totals		$1515.00		$2135.62	$63.13	$88.98
Crew B-68A	Hr.	Daily	Hr.	Daily	Bare Costs	Incl. O&P
1 Millwright Foreman (inside)	$52.05	$416.40	$76.25	$610.00	$51.72	$75.75
2 Millwrights	51.55	824.80	75.50	1208.00		
1 Forklift, Smooth Floor, 8,000 Lb.		171.20		188.32	7.13	7.85
24 L.H., Daily Totals		$1412.40		$2006.32	$58.85	$83.60
Crew B-68B	Hr.	Daily	Hr.	Daily	Bare Costs	Incl. O&P
1 Millwright Foreman (inside)	$52.05	$416.40	$76.25	$610.00	$55.99	$83.34
2 Millwrights	51.55	824.80	75.50	1208.00		
2 Electricians	56.60	905.60	84.80	1356.80		
2 Plumbers	61.80	988.80	93.25	1492.00		
1 R.T. Forklift, 5,000 Lb., gas		305.80		336.38	5.46	6.01
56 L.H., Daily Totals		$3441.40		$5003.18	$61.45	$89.34
Crew B-68C	Hr.	Daily	Hr.	Daily	Bare Costs	Incl. O&P
1 Millwright Foreman (inside)	$52.05	$416.40	$76.25	$610.00	$55.50	$82.45
1 Millwright	51.55	412.40	75.50	604.00		
1 Electrician	56.60	452.80	84.80	678.40		
1 Plumber	61.80	494.40	93.25	746.00		
1 R.T. Forklift, 5,000 Lb., gas		305.80		336.38	9.56	10.51
32 L.H., Daily Totals		$2081.80		$2974.78	$65.06	$92.96
Crew B-68D	Hr.	Daily	Hr.	Daily	Bare Costs	Incl. O&P
1 Labor Foreman (inside)	$39.65	$317.20	$60.75	$486.00	$43.27	$65.97
1 Laborer	39.15	313.20	60.00	480.00		
1 Equip. Oper. (light)	51.00	408.00	77.15	617.20		
1 R.T. Forklift, 5,000 Lb., gas		305.80		336.38	12.74	14.02
24 L.H., Daily Totals		$1344.20		$1919.58	$56.01	$79.98
Crew B-68E	Hr.	Daily	Hr.	Daily	Bare Costs	Incl. O&P
1 Struc. Steel Foreman (inside)	$54.80	$438.40	$91.35	$730.80	$54.40	$90.67
3 Struc. Steel Workers	54.30	1303.20	90.50	2172.00		
1 Welder	54.30	434.40	90.50	724.00		
1 Forklift, Smooth Floor, 8,000 Lb.		171.20		188.32	4.28	4.71
40 L.H., Daily Totals		$2347.20		$3815.12	$58.68	$95.38

Crew No.	Bare Costs Hr.	Daily	Incl. Subs O&P Hr.	Daily	Cost Per Labor-Hour Bare Costs	Incl. O&P
Crew B-68F	Hr.	Daily	Hr.	Daily	Bare Costs	Incl. O&P
1 Skilled Worker Foreman (out)	$53.45	$427.60	$82.25	$658.00	$52.12	$80.22
2 Skilled Workers	51.45	823.20	79.20	1267.20		
1 R.T. Forklift, 5,000 Lb., gas		305.80		336.38	12.74	14.02
24 L.H., Daily Totals		$1556.60		$2261.58	$64.86	$94.23
Crew B-68G	Hr.	Daily	Hr.	Daily	Bare Costs	Incl. O&P
2 Structural Steel Workers	$54.30	$868.80	$90.50	$1448.00	$54.30	$90.50
1 R.T. Forklift, 5,000 Lb., gas		305.80		336.38	19.11	21.02
16 L.H., Daily Totals		$1174.60		$1784.38	$73.41	$111.52
Crew B-69	Hr.	Daily	Hr.	Daily	Bare Costs	Incl. O&P
1 Labor Foreman (outside)	$41.15	$329.20	$63.05	$504.40	$43.72	$66.65
3 Laborers	39.15	939.60	60.00	1440.00		
1 Equip. Oper. (crane)	55.70	445.60	84.25	674.00		
1 Equip. Oper. (oiler)	48.00	384.00	72.60	580.80		
1 Hyd. Crane, 80 Ton		1616.00		1777.60	33.67	37.03
48 L.H., Daily Totals		$3714.40		$4976.80	$77.38	$103.68
Crew B-69A	Hr.	Daily	Hr.	Daily	Bare Costs	Incl. O&P
1 Labor Foreman (outside)	$41.15	$329.20	$63.05	$504.40	$43.16	$65.61
3 Laborers	39.15	939.60	60.00	1440.00		
1 Equip. Oper. (medium)	53.55	428.40	81.00	648.00		
1 Concrete Finisher	46.80	374.40	69.60	556.80		
1 Curb/Gutter Paver, 2-Track		1194.00		1313.40	24.88	27.36
48 L.H., Daily Totals		$3265.60		$4462.60	$68.03	$92.97
Crew B-69B	Hr.	Daily	Hr.	Daily	Bare Costs	Incl. O&P
1 Labor Foreman (outside)	$41.15	$329.20	$63.05	$504.40	$43.16	$65.61
3 Laborers	39.15	939.60	60.00	1440.00		
1 Equip. Oper. (medium)	53.55	428.40	81.00	648.00		
1 Cement Finisher	46.80	374.40	69.60	556.80		
1 Curb/Gutter Paver, 4-Track		802.40		882.64	16.72	18.39
48 L.H., Daily Totals		$2874.00		$4031.84	$59.88	$84.00
Crew B-70	Hr.	Daily	Hr.	Daily	Bare Costs	Incl. O&P
1 Labor Foreman (outside)	$41.15	$329.20	$63.05	$504.40	$45.61	$69.44
3 Laborers	39.15	939.60	60.00	1440.00		
3 Equip. Oper. (medium)	53.55	1285.20	81.00	1944.00		
1 Grader, 30,000 Lbs.		678.40		746.24		
1 Ripper, Beam & 1 Shank		85.60		94.16		
1 Road Sweeper, S.P., 8' wide		700.80		770.88		
1 F.E. Loader, W.M., 1.5 C.Y.		362.00		398.20	32.62	35.88
56 L.H., Daily Totals		$4380.80		$5897.88	$78.23	$105.32
Crew B-70A	Hr.	Daily	Hr.	Daily	Bare Costs	Incl. O&P
1 Laborer	$39.15	$313.20	$60.00	$480.00	$50.67	$76.80
4 Equip. Oper. (medium)	53.55	1713.60	81.00	2592.00		
1 Grader, 40,000 Lbs.		1248.00		1372.80		
1 F.E. Loader, W.M., 2.5 C.Y.		556.20		611.82		
1 Dozer, 80 H.P.		468.60		515.46		
1 Roller, Pneum. Whl., 12 Ton		355.80		391.38	65.72	72.29
40 L.H., Daily Totals		$4655.40		$5963.46	$116.39	$149.09
Crew B-71	Hr.	Daily	Hr.	Daily	Bare Costs	Incl. O&P
1 Labor Foreman (outside)	$41.15	$329.20	$63.05	$504.40	$45.61	$69.44
3 Laborers	39.15	939.60	60.00	1440.00		
3 Equip. Oper. (medium)	53.55	1285.20	81.00	1944.00		
1 Pvmt. Profiler, 750 H.P.		5553.00		6108.30		
1 Road Sweeper, S.P., 8' wide		700.80		770.88		
1 F.E. Loader, W.M., 1.5 C.Y.		362.00		398.20	118.14	129.95
56 L.H., Daily Totals		$9169.80		$11165.78	$163.75	$199.39

For customer support on your Building Construction Costs with RSMeans Data, call 800.448.8182.

Crew No.	Bare Costs		Incl. Subs O&P		Cost Per Labor-Hour	
Crew B-72	Hr.	Daily	Hr.	Daily	Bare Costs	Incl. O&P
1 Labor Foreman (outside)	$41.15	$329.20	$63.05	$504.40	$46.60	$70.88
3 Laborers	39.15	939.60	60.00	1440.00		
4 Equip. Oper. (medium)	53.55	1713.60	81.00	2592.00		
1 Pvmt. Profiler, 750 H.P.		5553.00		6108.30		
1 Hammermill, 250 H.P.		1827.00		2009.70		
1 Windrow Loader		1254.00		1379.40		
1 Mix Paver, 165 H.P.		2251.00		2476.10		
1 Roller, Pneum. Whl., 12 Ton		355.80		391.38	175.64	193.20
64 L.H., Daily Totals		$14223.20		$16901.28	$222.24	$264.08
Crew B-73	Hr.	Daily	Hr.	Daily	Bare Costs	Incl. O&P
1 Labor Foreman (outside)	$41.15	$329.20	$63.05	$504.40	$48.40	$73.51
2 Laborers	39.15	626.40	60.00	960.00		
5 Equip. Oper. (medium)	53.55	2142.00	81.00	3240.00		
1 Road Mixer, 310 H.P.		1942.00		2136.20		
1 Tandem Roller, 10 Ton		237.60		261.36		
1 Hammermill, 250 H.P.		1827.00		2009.70		
1 Grader, 30,000 Lbs.		678.40		746.24		
.5 F.E. Loader, W.M. 1.5 C.Y.		181.00		199.10		
.5 Truck Tractor, 220 H.P.		168.30		185.13		
.5 Water Tank Trailer, 5000 Gal.		71.50		78.65	79.78	87.76
64 L.H., Daily Totals		$8203.40		$10320.78	$128.18	$161.26
Crew B-74	Hr.	Daily	Hr.	Daily	Bare Costs	Incl. O&P
1 Labor Foreman (outside)	$41.15	$329.20	$63.05	$504.40	$48.20	$73.06
1 Laborer	39.15	313.20	60.00	480.00		
4 Equip. Oper. (medium)	53.55	1713.60	81.00	2592.00		
2 Truck Drivers (heavy)	45.55	728.80	68.70	1099.20		
1 Grader, 30,000 Lbs.		678.40		746.24		
1 Ripper, Beam & 1 Shank		85.60		94.16		
2 Stabilizers, 310 H.P.		3722.00		4094.20		
1 Flatbed Truck, Gas, 3 Ton		270.20		297.22		
1 Chem. Spreader, Towed		51.20		56.32		
1 Roller, Vibratory, 25 Ton		684.00		752.40		
1 Water Tank Trailer, 5000 Gal.		143.00		157.30		
1 Truck Tractor, 220 H.P.		336.60		370.26	93.30	102.63
64 L.H., Daily Totals		$9055.80		$11243.70	$141.50	$175.68
Crew B-75	Hr.	Daily	Hr.	Daily	Bare Costs	Incl. O&P
1 Labor Foreman (outside)	$41.15	$329.20	$63.05	$504.40	$48.58	$73.68
1 Laborer	39.15	313.20	60.00	480.00		
4 Equip. Oper. (medium)	53.55	1713.60	81.00	2592.00		
1 Truck Driver (heavy)	45.55	364.40	68.70	549.60		
1 Grader, 30,000 Lbs.		678.40		746.24		
1 Ripper, Beam & 1 Shank		85.60		94.16		
2 Stabilizers, 310 H.P.		3722.00		4094.20		
1 Dist. Tanker, 3000 Gallon		327.40		360.14		
1 Truck Tractor, 6x4, 380 H.P.		559.80		615.78		
1 Roller, Vibratory, 25 Ton		684.00		752.40	108.16	118.98
56 L.H., Daily Totals		$8777.60		$10788.92	$156.74	$192.66

Crew No.	Bare Costs		Incl. Subs O&P		Cost Per Labor-Hour	
Crew B-76	Hr.	Daily	Hr.	Daily	Bare Costs	Incl. O&P
1 Dock Builder Foreman (outside)	$51.50	$412.00	$81.80	$654.40	$50.93	$79.54
5 Dock Builders	49.50	1980.00	78.60	3144.00		
2 Equip. Oper. (crane)	55.70	891.20	84.25	1348.00		
1 Equip. Oper. (oiler)	48.00	384.00	72.60	580.80		
1 Crawler Crane, 50 Ton		1675.00		1842.50		
1 Barge, 400 Ton		794.20		873.62		
1 Hammer, Diesel, 15K ft.-lbs.		605.80		666.38		
1 Lead, 60' High		76.00		83.60		
1 Air Compressor, 600 cfm		482.00		530.20		
2 -50' Air Hoses, 3"		29.80		32.78	50.87	55.96
72 L.H., Daily Totals		$7330.00		$9756.28	$101.81	$135.50
Crew B-76A	Hr.	Daily	Hr.	Daily	Bare Costs	Incl. O&P
1 Labor Foreman (outside)	$41.15	$329.20	$63.05	$504.40	$42.58	$64.99
5 Laborers	39.15	1566.00	60.00	2400.00		
1 Equip. Oper. (crane)	55.70	445.60	84.25	674.00		
1 Equip. Oper. (oiler)	48.00	384.00	72.60	580.80		
1 Crawler Crane, 50 Ton		1675.00		1842.50		
1 Barge, 400 Ton		794.20		873.62	38.58	42.44
64 L.H., Daily Totals		$5194.00		$6875.32	$81.16	$107.43
Crew B-77	Hr.	Daily	Hr.	Daily	Bare Costs	Incl. O&P
1 Labor Foreman (outside)	$41.15	$329.20	$63.05	$504.40	$40.55	$61.93
3 Laborers	39.15	939.60	60.00	1440.00		
1 Truck Driver (light)	44.15	353.20	66.60	532.80		
1 Crack Cleaner, 25 H.P.		57.60		63.36		
1 Crack Filler, Trailer Mtd.		199.80		219.78		
1 Flatbed Truck, Gas, 3 Ton		270.20		297.22	13.19	14.51
40 L.H., Daily Totals		$2149.60		$3057.56	$53.74	$76.44
Crew B-78	Hr.	Daily	Hr.	Daily	Bare Costs	Incl. O&P
1 Labor Foreman (outside)	$41.15	$329.20	$63.05	$504.40	$40.32	$61.61
4 Laborers	39.15	1252.80	60.00	1920.00		
1 Truck Driver (light)	44.15	353.20	66.60	532.80		
1 Paint Striper, S.P., 40 Gallon		149.20		164.12		
1 Flatbed Truck, Gas, 3 Ton		270.20		297.22		
1 Pickup Truck, 3/4 Ton		129.40		142.34	11.43	12.58
48 L.H., Daily Totals		$2484.00		$3560.88	$51.75	$74.19
Crew B-78A	Hr.	Daily	Hr.	Daily	Bare Costs	Incl. O&P
1 Equip. Oper. (light)	$51.00	$408.00	$77.15	$617.20	$51.00	$77.15
1 Line Rem. (Metal Balls) 115 H.P.		873.80		961.18	109.22	120.15
8 L.H., Daily Totals		$1281.80		$1578.38	$160.22	$197.30
Crew B-78B	Hr.	Daily	Hr.	Daily	Bare Costs	Incl. O&P
2 Laborers	$39.15	$626.40	$60.00	$960.00	$40.47	$61.91
.25 Equip. Oper. (light)	51.00	102.00	77.15	154.30		
1 Pickup Truck, 3/4 Ton		129.40		142.34		
1 Line Rem.,11 H.P.,Walk Behind		62.00		68.20		
.25 Road Sweeper, S.P., 8' wide		175.20		192.72	20.37	22.40
18 L.H., Daily Totals		$1095.00		$1517.56	$60.83	$84.31
Crew B-78C	Hr.	Daily	Hr.	Daily	Bare Costs	Incl. O&P
1 Labor Foreman (outside)	$41.15	$329.20	$63.05	$504.40	$40.32	$61.61
4 Laborers	39.15	1252.80	60.00	1920.00		
1 Truck Driver (light)	44.15	353.20	66.60	532.80		
1 Paint Striper, T.M., 120 Gal.		734.60		808.06		
1 Flatbed Truck, Gas, 3 Ton		270.20		297.22		
1 Pickup Truck, 3/4 Ton		129.40		142.34	23.63	25.99
48 L.H., Daily Totals		$3069.40		$4204.82	$63.95	$87.60

743

Crew B-78D

Crew No.	Hr.	Daily	Hr.	Daily	Bare Costs	Incl. O&P
2 Labor Foremen (outside)	$41.15	$658.40	$63.05	$1008.80	$40.05	$61.27
7 Laborers	39.15	2192.40	60.00	3360.00		
1 Truck Driver (light)	44.15	353.20	66.60	532.80		
1 Paint Striper, T.M., 120 Gal.		734.60		808.06		
1 Flatbed Truck, Gas, 3 Ton		270.20		297.22		
3 Pickup Trucks, 3/4 Ton		388.20		427.02		
1 Air Compressor, 60 cfm		127.00		139.70		
1 -50' Air Hose, 3/4"		3.25		3.58		
1 Breaker, Pavement, 60 lb.		10.20		11.22	19.17	21.08
80 L.H., Daily Totals		$4737.45		$6588.40	$59.22	$82.35

Crew B-78E

Crew No.	Hr.	Daily	Hr.	Daily	Bare Costs	Incl. O&P
2 Labor Foremen (outside)	$41.15	$658.40	$63.05	$1008.80	$39.90	$61.06
9 Laborers	39.15	2818.80	60.00	4320.00		
1 Truck Driver (light)	44.15	353.20	66.60	532.80		
1 Paint Striper, T.M., 120 Gal.		734.60		808.06		
1 Flatbed Truck, Gas, 3 Ton		270.20		297.22		
4 Pickup Trucks, 3/4 Ton		517.60		569.36		
2 Air Compressors, 60 cfm		254.00		279.40		
2 -50' Air Hoses, 3/4"		6.50		7.15		
2 Breakers, Pavement, 60 lb.		20.40		22.44	18.78	20.66
96 L.H., Daily Totals		$5633.70		$7845.23	$58.68	$81.72

Crew B-78F

Crew No.	Hr.	Daily	Hr.	Daily	Bare Costs	Incl. O&P
2 Labor Foremen (outside)	$41.15	$658.40	$63.05	$1008.80	$39.79	$60.91
11 Laborers	39.15	3445.20	60.00	5280.00		
1 Truck Driver (light)	44.15	353.20	66.60	532.80		
1 Paint Striper, T.M., 120 Gal.		734.60		808.06		
1 Flatbed Truck, Gas, 3 Ton		270.20		297.22		
7 Pickup Trucks, 3/4 Ton		905.80		996.38		
3 Air Compressors, 60 cfm		381.00		419.10		
3 -50' Air Hoses, 3/4"		9.75		10.73		
3 Breakers, Pavement, 60 lb.		30.60		33.66	20.82	22.90
112 L.H., Daily Totals		$6788.75		$9386.75	$60.61	$83.81

Crew B-79

Crew No.	Hr.	Daily	Hr.	Daily	Bare Costs	Incl. O&P
1 Labor Foreman (outside)	$41.15	$329.20	$63.05	$504.40	$40.55	$61.93
3 Laborers	39.15	939.60	60.00	1440.00		
1 Truck Driver (light)	44.15	353.20	66.60	532.80		
1 Paint Striper, T.M., 120 Gal.		734.60		808.06		
1 Heating Kettle, 115 Gallon		73.50		80.85		
1 Flatbed Truck, Gas, 3 Ton		270.20		297.22		
2 Pickup Trucks, 3/4 Ton		258.80		284.68	33.43	36.77
40 L.H., Daily Totals		$2959.10		$3948.01	$73.98	$98.70

Crew B-79A

Crew No.	Hr.	Daily	Hr.	Daily	Bare Costs	Incl. O&P
1.5 Equip. Oper. (light)	$51.00	$612.00	$77.15	$925.80	$51.00	$77.15
.5 Line Remov. (Grinder) 115 H.P.		475.70		523.27		
1 Line Rem. (Metal Balls) 115 H.P.		873.80		961.18	112.46	123.70
12 L.H., Daily Totals		$1961.50		$2410.25	$163.46	$200.85

Crew B-79B

Crew No.	Hr.	Daily	Hr.	Daily	Bare Costs	Incl. O&P
1 Laborer	$39.15	$313.20	$60.00	$480.00	$39.15	$60.00
1 Set of Gases		168.00		184.80	21.00	23.10
8 L.H., Daily Totals		$481.20		$664.80	$60.15	$83.10

Crew B-79C

Crew No.	Hr.	Daily	Hr.	Daily	Bare Costs	Incl. O&P
1 Labor Foreman (outside)	$41.15	$329.20	$63.05	$504.40	$40.15	$61.38
5 Laborers	39.15	1566.00	60.00	2400.00		
1 Truck Driver (light)	44.15	353.20	66.60	532.80		
1 Paint Striper, T.M., 120 Gal.		734.60		808.06		
1 Heating Kettle, 115 Gallon		73.50		80.85		
1 Flatbed Truck, Gas, 3 Ton		270.20		297.22		
3 Pickup Trucks, 3/4 Ton		388.20		427.02		
1 Air Compressor, 60 cfm		127.00		139.70		
1 -50' Air Hose, 3/4"		3.25		3.58		
1 Breaker, Pavement, 60 lb.		10.20		11.22	28.70	31.57
56 L.H., Daily Totals		$3855.35		$5204.85	$68.85	$92.94

Crew B-79D

Crew No.	Hr.	Daily	Hr.	Daily	Bare Costs	Incl. O&P
2 Labor Foremen (outside)	$41.15	$658.40	$63.05	$1008.80	$40.27	$61.59
5 Laborers	39.15	1566.00	60.00	2400.00		
1 Truck Driver (light)	44.15	353.20	66.60	532.80		
1 Paint Striper, T.M., 120 Gal.		734.60		808.06		
1 Heating Kettle, 115 Gallon		73.50		80.85		
1 Flatbed Truck, Gas, 3 Ton		270.20		297.22		
4 Pickup Trucks, 3/4 Ton		517.60		569.36		
1 Air Compressor, 60 cfm		127.00		139.70		
1 -50' Air Hose, 3/4"		3.25		3.58		
1 Breaker, Pavement, 60 lb.		10.20		11.22	27.13	29.84
64 L.H., Daily Totals		$4313.95		$5851.59	$67.41	$91.43

Crew B-79E

Crew No.	Hr.	Daily	Hr.	Daily	Bare Costs	Incl. O&P
2 Labor Foremen (outside)	$41.15	$658.40	$63.05	$1008.80	$40.05	$61.27
7 Laborers	39.15	2192.40	60.00	3360.00		
1 Truck Driver (light)	44.15	353.20	66.60	532.80		
1 Paint Striper, T.M., 120 Gal.		734.60		808.06		
1 Heating Kettle, 115 Gallon		73.50		80.85		
1 Flatbed Truck, Gas, 3 Ton		270.20		297.22		
5 Pickup Trucks, 3/4 Ton		647.00		711.70		
2 Air Compressors, 60 cfm		254.00		279.40		
2 -50' Air Hoses, 3/4"		6.50		7.15		
2 Breakers, Pavement, 60 lb.		20.40		22.44	25.08	27.59
80 L.H., Daily Totals		$5210.20		$7108.42	$65.13	$88.86

Crew B-80

Crew No.	Hr.	Daily	Hr.	Daily	Bare Costs	Incl. O&P
1 Labor Foreman (outside)	$41.15	$329.20	$63.05	$504.40	$43.86	$66.70
1 Laborer	39.15	313.20	60.00	480.00		
1 Truck Driver (light)	44.15	353.20	66.60	532.80		
1 Equip. Oper. (light)	51.00	408.00	77.15	617.20		
1 Flatbed Truck, Gas, 3 Ton		270.20		297.22		
1 Earth Auger, Truck-Mtd.		403.40		443.74	21.05	23.16
32 L.H., Daily Totals		$2077.20		$2875.36	$64.91	$89.86

Crew B-80A

Crew No.	Hr.	Daily	Hr.	Daily	Bare Costs	Incl. O&P
3 Laborers	$39.15	$939.60	$60.00	$1440.00	$39.15	$60.00
1 Flatbed Truck, Gas, 3 Ton		270.20		297.22	11.26	12.38
24 L.H., Daily Totals		$1209.80		$1737.22	$50.41	$72.38

Crew B-80B

Crew No.	Hr.	Daily	Hr.	Daily	Bare Costs	Incl. O&P
3 Laborers	$39.15	$939.60	$60.00	$1440.00	$42.11	$64.29
1 Equip. Oper. (light)	51.00	408.00	77.15	617.20		
1 Crane, Flatbed Mounted, 3 Ton		247.80		272.58	7.74	8.52
32 L.H., Daily Totals		$1595.40		$2329.78	$49.86	$72.81

Crew No.	Bare Costs		Incl. Subs O&P		Cost Per Labor-Hour	
Crew B-80C	Hr.	Daily	Hr.	Daily	Bare Costs	Incl. O&P
2 Laborers	$39.15	$626.40	$60.00	$960.00	$40.82	$62.20
1 Truck Driver (light)	44.15	353.20	66.60	532.80		
1 Flatbed Truck, Gas, 1.5 Ton		217.40		239.14		
1 Manual Fence Post Auger, Gas		7.90		8.69	9.39	10.33
24 L.H., Daily Totals		$1204.90		$1740.63	$50.20	$72.53

Crew No.	Bare Costs		Incl. Subs O&P		Cost Per Labor-Hour	
Crew B-81	Hr.	Daily	Hr.	Daily	Bare Costs	Incl. O&P
1 Laborer	$39.15	$313.20	$60.00	$480.00	$46.08	$69.90
1 Equip. Oper. (medium)	53.55	428.40	81.00	648.00		
1 Truck Driver (heavy)	45.55	364.40	68.70	549.60		
1 Hydromulcher, T.M., 3000 Gal.		317.80		349.58		
1 Truck Tractor, 220 H.P.		336.60		370.26	27.27	29.99
24 L.H., Daily Totals		$1760.40		$2397.44	$73.35	$99.89

Crew No.	Bare Costs		Incl. Subs O&P		Cost Per Labor-Hour	
Crew B-81A	Hr.	Daily	Hr.	Daily	Bare Costs	Incl. O&P
1 Laborer	$39.15	$313.20	$60.00	$480.00	$41.65	$63.30
1 Truck Driver (light)	44.15	353.20	66.60	532.80		
1 Hydromulcher, T.M., 600 Gal.		122.40		134.64		
1 Flatbed Truck, Gas, 3 Ton		270.20		297.22	24.54	26.99
16 L.H., Daily Totals		$1059.00		$1444.66	$66.19	$90.29

Crew No.	Bare Costs		Incl. Subs O&P		Cost Per Labor-Hour	
Crew B-82	Hr.	Daily	Hr.	Daily	Bare Costs	Incl. O&P
1 Laborer	$39.15	$313.20	$60.00	$480.00	$45.08	$68.58
1 Equip. Oper. (light)	51.00	408.00	77.15	617.20		
1 Horiz. Borer, 6 H.P.		80.20		88.22	5.01	5.51
16 L.H., Daily Totals		$801.40		$1185.42	$50.09	$74.09

Crew No.	Bare Costs		Incl. Subs O&P		Cost Per Labor-Hour	
Crew B-82A	Hr.	Daily	Hr.	Daily	Bare Costs	Incl. O&P
2 Laborers	$39.15	$626.40	$60.00	$960.00	$45.08	$68.58
2 Equip. Opers. (light)	51.00	816.00	77.15	1234.40		
2 Dump Truck, 8 C.Y., 220 H.P.		782.00		860.20		
1 Flatbed Trailer, 25 Ton		128.40		141.24		
1 Horiz. Dir. Drill, 20k lb. Thrust		651.60		716.76		
1 Mud Trailer for HDD, 1500 Gal.		311.60		342.76		
1 Pickup Truck, 4x4, 3/4 Ton		140.00		154.00		
1 Flatbed Trailer, 3 Ton		24.00		26.40		
1 Loader, Skid Steer, 78 H.P.		309.40		340.34	73.34	80.68
32 L.H., Daily Totals		$3789.40		$4776.10	$118.42	$149.25

Crew No.	Bare Costs		Incl. Subs O&P		Cost Per Labor-Hour	
Crew B-82B	Hr.	Daily	Hr.	Daily	Bare Costs	Incl. O&P
2 Laborers	$39.15	$626.40	$60.00	$960.00	$45.08	$68.58
2 Equip. Opers. (light)	51.00	816.00	77.15	1234.40		
2 Dump Truck, 8 C.Y., 220 H.P.		782.00		860.20		
1 Flatbed Trailer, 25 Ton		128.40		141.24		
1 Horiz. Dir. Drill, 30k lb. Thrust		929.00		1021.90		
1 Mud Trailer for HDD, 1500 Gal.		311.60		342.76		
1 Pickup Truck, 4x4, 3/4 Ton		140.00		154.00		
1 Flatbed Trailer, 3 Ton		24.00		26.40		
1 Loader, Skid Steer, 78 H.P.		309.40		340.34	82.01	90.21
32 L.H., Daily Totals		$4066.80		$5081.24	$127.09	$158.79

Crew No.	Bare Costs		Incl. Subs O&P		Cost Per Labor-Hour	
Crew B-82C	Hr.	Daily	Hr.	Daily	Bare Costs	Incl. O&P
2 Laborers	$39.15	$626.40	$60.00	$960.00	$45.08	$68.58
2 Equip. Opers. (light)	51.00	816.00	77.15	1234.40		
2 Dump Truck, 8 C.Y., 220 H.P.		782.00		860.20		
1 Flatbed Trailer, 25 Ton		128.40		141.24		
1 Horiz. Dir. Drill, 50k lb. Thrust		1229.00		1351.90		
1 Mud Trailer for HDD, 1500 Gal.		311.60		342.76		
1 Pickup Truck, 4x4, 3/4 Ton		140.00		154.00		
1 Flatbed Trailer, 3 Ton		24.00		26.40		
1 Loader, Skid Steer, 78 H.P.		309.40		340.34	91.39	100.53
32 L.H., Daily Totals		$4366.80		$5411.24	$136.46	$169.10

Crew No.	Bare Costs		Incl. Subs O&P		Cost Per Labor-Hour	
Crew B-82D	Hr.	Daily	Hr.	Daily	Bare Costs	Incl. O&P
1 Equip. Oper. (light)	$51.00	$408.00	$77.15	$617.20	$51.00	$77.15
1 Mud Trailer for HDD, 1500 Gal.		311.60		342.76	38.95	42.84
8 L.H., Daily Totals		$719.60		$959.96	$89.95	$120.00

Crew No.	Bare Costs		Incl. Subs O&P		Cost Per Labor-Hour	
Crew B-83	Hr.	Daily	Hr.	Daily	Bare Costs	Incl. O&P
1 Tugboat Captain	$53.55	$428.40	$81.00	$648.00	$46.35	$70.50
1 Tugboat Hand	39.15	313.20	60.00	480.00		
1 Tugboat, 250 H.P.		807.20		887.92	50.45	55.49
16 L.H., Daily Totals		$1548.80		$2015.92	$96.80	$126.00

Crew No.	Bare Costs		Incl. Subs O&P		Cost Per Labor-Hour	
Crew B-84	Hr.	Daily	Hr.	Daily	Bare Costs	Incl. O&P
1 Equip. Oper. (medium)	$53.55	$428.40	$81.00	$648.00	$53.55	$81.00
1 Rotary Mower/Tractor		365.80		402.38	45.73	50.30
8 L.H., Daily Totals		$794.20		$1050.38	$99.28	$131.30

Crew No.	Bare Costs		Incl. Subs O&P		Cost Per Labor-Hour	
Crew B-85	Hr.	Daily	Hr.	Daily	Bare Costs	Incl. O&P
3 Laborers	$39.15	$939.60	$60.00	$1440.00	$43.31	$65.94
1 Equip. Oper. (medium)	53.55	428.40	81.00	648.00		
1 Truck Driver (heavy)	45.55	364.40	68.70	549.60		
1 Telescoping Boom Lift, to 80'		565.60		622.16		
1 Brush Chipper, 12", 130 H.P.		404.40		444.84		
1 Pruning Saw, Rotary		6.85		7.54	24.42	26.86
40 L.H., Daily Totals		$2709.25		$3712.14	$67.73	$92.80

Crew No.	Bare Costs		Incl. Subs O&P		Cost Per Labor-Hour	
Crew B-86	Hr.	Daily	Hr.	Daily	Bare Costs	Incl. O&P
1 Equip. Oper. (medium)	$53.55	$428.40	$81.00	$648.00	$53.55	$81.00
1 Stump Chipper, S.P.		183.50		201.85	22.94	25.23
8 L.H., Daily Totals		$611.90		$849.85	$76.49	$106.23

Crew No.	Bare Costs		Incl. Subs O&P		Cost Per Labor-Hour	
Crew B-86A	Hr.	Daily	Hr.	Daily	Bare Costs	Incl. O&P
1 Equip. Oper. (medium)	$53.55	$428.40	$81.00	$648.00	$53.55	$81.00
1 Grader, 30,000 Lbs.		678.40		746.24	84.80	93.28
8 L.H., Daily Totals		$1106.80		$1394.24	$138.35	$174.28

Crew No.	Bare Costs		Incl. Subs O&P		Cost Per Labor-Hour	
Crew B-86B	Hr.	Daily	Hr.	Daily	Bare Costs	Incl. O&P
1 Equip. Oper. (medium)	$53.55	$428.40	$81.00	$648.00	$53.55	$81.00
1 Dozer, 200 H.P.		1384.00		1522.40	173.00	190.30
8 L.H., Daily Totals		$1812.40		$2170.40	$226.55	$271.30

Crew No.	Bare Costs		Incl. Subs O&P		Cost Per Labor-Hour	
Crew B-87	Hr.	Daily	Hr.	Daily	Bare Costs	Incl. O&P
1 Laborer	$39.15	$313.20	$60.00	$480.00	$50.67	$76.80
4 Equip. Oper. (medium)	53.55	1713.60	81.00	2592.00		
2 Feller Bunchers, 100 H.P.		1666.40		1833.04		
1 Log Chipper, 22" Tree		852.80		938.08		
1 Dozer, 105 H.P.		608.00		668.80		
1 Chain Saw, Gas, 36" Long		45.00		49.50	79.31	87.24
40 L.H., Daily Totals		$5199.00		$6561.42	$129.97	$164.04

745

Crew B-88

Crew B-88	Hr.	Daily	Hr.	Daily	Bare Costs	Incl. O&P
1 Laborer	$39.15	$313.20	$60.00	$480.00	$51.49	$78.00
6 Equip. Oper. (medium)	53.55	2570.40	81.00	3888.00		
2 Feller Bunchers, 100 H.P.		1666.40		1833.04		
1 Log Chipper, 22" Tree		852.80		938.08		
2 Log Skidders, 50 H.P.		1742.40		1916.64		
1 Dozer, 105 H.P.		608.00		668.80		
1 Chain Saw, Gas, 36" Long		45.00		49.50	87.76	96.54
56 L.H., Daily Totals		$7798.20		$9774.06	$139.25	$174.54

Crew B-89

Crew B-89	Hr.	Daily	Hr.	Daily	Bare Costs	Incl. O&P
1 Equip. Oper. (light)	$51.00	$408.00	$77.15	$617.20	$47.58	$71.88
1 Truck Driver (light)	44.15	353.20	66.60	532.80		
1 Flatbed Truck, Gas, 3 Ton		270.20		297.22		
1 Concrete Saw		153.60		168.96		
1 Water Tank, 65 Gal.		17.30		19.03	27.57	30.33
16 L.H., Daily Totals		$1202.30		$1635.21	$75.14	$102.20

Crew B-89A

Crew B-89A	Hr.	Daily	Hr.	Daily	Bare Costs	Incl. O&P
1 Skilled Worker	$51.45	$411.60	$79.20	$633.60	$45.30	$69.60
1 Laborer	39.15	313.20	60.00	480.00		
1 Core Drill (Large)		114.60		126.06	7.16	7.88
16 L.H., Daily Totals		$839.40		$1239.66	$52.46	$77.48

Crew B-89B

Crew B-89B	Hr.	Daily	Hr.	Daily	Bare Costs	Incl. O&P
1 Equip. Oper. (light)	$51.00	$408.00	$77.15	$617.20	$47.58	$71.88
1 Truck Driver (light)	44.15	353.20	66.60	532.80		
1 Wall Saw, Hydraulic, 10 H.P.		103.40		113.74		
1 Generator, Diesel, 100 kW		367.40		404.14		
1 Water Tank, 65 Gal.		17.30		19.03		
1 Flatbed Truck, Gas, 3 Ton		270.20		297.22	47.39	52.13
16 L.H., Daily Totals		$1519.50		$1984.13	$94.97	$124.01

Crew B-90

Crew B-90	Hr.	Daily	Hr.	Daily	Bare Costs	Incl. O&P
1 Labor Foreman (outside)	$41.15	$329.20	$63.05	$504.40	$43.96	$66.84
3 Laborers	39.15	939.60	60.00	1440.00		
2 Equip. Oper. (light)	51.00	816.00	77.15	1234.40		
2 Truck Drivers (heavy)	45.55	728.80	68.70	1099.20		
1 Road Mixer, 310 H.P.		1942.00		2136.20		
1 Dist. Truck, 2000 Gal.		297.00		326.70	34.98	38.48
64 L.H., Daily Totals		$5052.60		$6740.90	$78.95	$105.33

Crew B-90A

Crew B-90A	Hr.	Daily	Hr.	Daily	Bare Costs	Incl. O&P
1 Labor Foreman (outside)	$41.15	$329.20	$63.05	$504.40	$47.66	$72.44
2 Laborers	39.15	626.40	60.00	960.00		
4 Equip. Oper. (medium)	53.55	1713.60	81.00	2592.00		
2 Graders, 30,000 Lbs.		1356.80		1492.48		
1 Tandem Roller, 10 Ton		237.60		261.36		
1 Roller, Pneum. Whl., 12 Ton		355.80		391.38	34.83	38.31
56 L.H., Daily Totals		$4619.40		$6201.62	$82.49	$110.74

Crew B-90B

Crew B-90B	Hr.	Daily	Hr.	Daily	Bare Costs	Incl. O&P
1 Labor Foreman (outside)	$41.15	$329.20	$63.05	$504.40	$46.68	$71.01
2 Laborers	39.15	626.40	60.00	960.00		
3 Equip. Oper. (medium)	53.55	1285.20	81.00	1944.00		
1 Roller, Pneum. Whl., 12 Ton		355.80		391.38		
1 Road Mixer, 310 H.P.		1942.00		2136.20	47.87	52.66
48 L.H., Daily Totals		$4538.60		$5935.98	$94.55	$123.67

Crew B-90C

Crew B-90C	Hr.	Daily	Hr.	Daily	Bare Costs	Incl. O&P
1 Labor Foreman (outside)	$41.15	$329.20	$63.05	$504.40	$45.00	$68.38
4 Laborers	39.15	1252.80	60.00	1920.00		
3 Equip. Oper. (medium)	53.55	1285.20	81.00	1944.00		
3 Truck Drivers (heavy)	45.55	1093.20	68.70	1648.80		
3 Road Mixers, 310 H.P.		5826.00		6408.60	66.20	72.83
88 L.H., Daily Totals		$9786.40		$12425.80	$111.21	$141.20

Crew B-90D

Crew B-90D	Hr.	Daily	Hr.	Daily	Bare Costs	Incl. O&P
1 Labor Foreman (outside)	$41.15	$329.20	$63.05	$504.40	$44.10	$67.09
6 Laborers	39.15	1879.20	60.00	2880.00		
3 Equip. Oper. (medium)	53.55	1285.20	81.00	1944.00		
3 Truck Drivers (heavy)	45.55	1093.20	68.70	1648.80		
3 Road Mixers, 310 H.P.		5826.00		6408.60	56.02	61.62
104 L.H., Daily Totals		$10412.80		$13385.80	$100.12	$128.71

Crew B-90E

Crew B-90E	Hr.	Daily	Hr.	Daily	Bare Costs	Incl. O&P
1 Labor Foreman (outside)	$41.15	$329.20	$63.05	$504.40	$44.88	$68.31
4 Laborers	39.15	1252.80	60.00	1920.00		
3 Equip. Oper. (medium)	53.55	1285.20	81.00	1944.00		
1 Truck Driver (heavy)	45.55	364.40	68.70	549.60		
1 Road Mixer, 310 H.P.		1942.00		2136.20	26.97	29.67
72 L.H., Daily Totals		$5173.60		$7054.20	$71.86	$97.97

Crew B-91

Crew B-91	Hr.	Daily	Hr.	Daily	Bare Costs	Incl. O&P
1 Labor Foreman (outside)	$41.15	$329.20	$63.05	$504.40	$47.40	$71.97
2 Laborers	39.15	626.40	60.00	960.00		
4 Equip. Oper. (medium)	53.55	1713.60	81.00	2592.00		
1 Truck Driver (heavy)	45.55	364.40	68.70	549.60		
1 Dist. Tanker, 3000 Gallon		327.40		360.14		
1 Truck Tractor, 6x4, 380 H.P.		559.80		615.78		
1 Aggreg. Spreader, S.P.		817.00		898.70		
1 Roller, Pneum. Whl., 12 Ton		355.80		391.38		
1 Tandem Roller, 10 Ton		237.60		261.36	35.90	39.49
64 L.H., Daily Totals		$5331.20		$7133.36	$83.30	$111.46

Crew B-91B

Crew B-91B	Hr.	Daily	Hr.	Daily	Bare Costs	Incl. O&P
1 Laborer	$39.15	$313.20	$60.00	$480.00	$46.35	$70.50
1 Equipment Oper. (med.)	53.55	428.40	81.00	648.00		
1 Road Sweeper, Vac. Assist.		927.80		1020.58	57.99	63.79
16 L.H., Daily Totals		$1669.40		$2148.58	$104.34	$134.29

Crew B-91C

Crew B-91C	Hr.	Daily	Hr.	Daily	Bare Costs	Incl. O&P
1 Laborer	$39.15	$313.20	$60.00	$480.00	$41.65	$63.30
1 Truck Driver (light)	44.15	353.20	66.60	532.80		
1 Catch Basin Cleaning Truck		539.00		592.90	33.69	37.06
16 L.H., Daily Totals		$1205.40		$1605.70	$75.34	$100.36

Crew B-91D

Crew B-91D	Hr.	Daily	Hr.	Daily	Bare Costs	Incl. O&P
1 Labor Foreman (outside)	$41.15	$329.20	$63.05	$504.40	$45.83	$69.65
5 Laborers	39.15	1566.00	60.00	2400.00		
5 Equip. Oper. (medium)	53.55	2142.00	81.00	3240.00		
2 Truck Drivers (heavy)	45.55	728.80	68.70	1099.20		
1 Aggreg. Spreader, S.P.		817.00		898.70		
2 Truck Tractors, 6x4, 380 H.P.		1119.60		1231.56		
2 Dist. Tankers, 3000 Gallon		654.80		720.28		
2 Pavement Brushes, Towed		164.00		180.40		
2 Rollers Pneum. Whl., 12 Ton		711.60		782.76	33.34	36.67
104 L.H., Daily Totals		$8233.00		$11057.30	$79.16	$106.32

Crews - Standard

Crew B-92

Crew B-92	Bare Costs Hr.	Daily	Incl. Subs O&P Hr.	Daily	Cost Per Labor-Hour Bare Costs	Incl. O&P
1 Labor Foreman (outside)	$41.15	$329.20	$63.05	$504.40	$39.65	$60.76
3 Laborers	39.15	939.60	60.00	1440.00		
1 Crack Cleaner, 25 H.P.		57.60		63.36		
1 Air Compressor, 60 cfm		127.00		139.70		
1 Tar Kettle, T.M.		122.75		135.03		
1 Flatbed Truck, Gas, 3 Ton		270.20		297.22	18.05	19.85
32 L.H., Daily Totals		$1846.35		$2579.70	$57.70	$80.62

Crew B-93

Crew B-93	Bare Costs Hr.	Daily	Incl. Subs O&P Hr.	Daily	Cost Per Labor-Hour Bare Costs	Incl. O&P
1 Equip. Oper. (medium)	$53.55	$428.40	$81.00	$648.00	$53.55	$81.00
1 Feller Buncher, 100 H.P.		833.20		916.52	104.15	114.57
8 L.H., Daily Totals		$1261.60		$1564.52	$157.70	$195.57

Crew B-94A

Crew B-94A	Bare Costs Hr.	Daily	Incl. Subs O&P Hr.	Daily	Cost Per Labor-Hour Bare Costs	Incl. O&P
1 Laborer	$39.15	$313.20	$60.00	$480.00	$39.15	$60.00
1 Diaphragm Water Pump, 2"		71.00		78.10		
1 -20' Suction Hose, 2"		1.95		2.15		
2 -50' Discharge Hoses, 2"		1.80		1.98	9.34	10.28
8 L.H., Daily Totals		$387.95		$562.23	$48.49	$70.28

Crew B-94B

Crew B-94B	Bare Costs Hr.	Daily	Incl. Subs O&P Hr.	Daily	Cost Per Labor-Hour Bare Costs	Incl. O&P
1 Laborer	$39.15	$313.20	$60.00	$480.00	$39.15	$60.00
1 Diaphragm Water Pump, 4"		114.20		125.62		
1 -20' Suction Hose, 4"		3.45		3.79		
2 -50' Discharge Hoses, 4"		4.70		5.17	15.29	16.82
8 L.H., Daily Totals		$435.55		$614.59	$54.44	$76.82

Crew B-94C

Crew B-94C	Bare Costs Hr.	Daily	Incl. Subs O&P Hr.	Daily	Cost Per Labor-Hour Bare Costs	Incl. O&P
1 Laborer	$39.15	$313.20	$60.00	$480.00	$39.15	$60.00
1 Centrifugal Water Pump, 3"		78.60		86.46		
1 -20' Suction Hose, 3"		3.05		3.36		
2 -50' Discharge Hoses, 3"		3.50		3.85	10.64	11.71
8 L.H., Daily Totals		$398.35		$573.66	$49.79	$71.71

Crew B-94D

Crew B-94D	Bare Costs Hr.	Daily	Incl. Subs O&P Hr.	Daily	Cost Per Labor-Hour Bare Costs	Incl. O&P
1 Laborer	$39.15	$313.20	$60.00	$480.00	$39.15	$60.00
1 Centr. Water Pump, 6"		325.40		357.94		
1 -20' Suction Hose, 6"		11.90		13.09		
2 -50' Discharge Hoses, 6"		12.60		13.86	43.74	48.11
8 L.H., Daily Totals		$663.10		$864.89	$82.89	$108.11

Crew C-1

Crew C-1	Bare Costs Hr.	Daily	Incl. Subs O&P Hr.	Daily	Cost Per Labor-Hour Bare Costs	Incl. O&P
3 Carpenters	$49.25	$1182.00	$75.45	$1810.80	$46.73	$71.59
1 Laborer	39.15	313.20	60.00	480.00		
32 L.H., Daily Totals		$1495.20		$2290.80	$46.73	$71.59

Crew C-2

Crew C-2	Bare Costs Hr.	Daily	Incl. Subs O&P Hr.	Daily	Cost Per Labor-Hour Bare Costs	Incl. O&P
1 Carpenter Foreman (outside)	$51.25	$410.00	$78.50	$628.00	$47.90	$73.38
4 Carpenters	49.25	1576.00	75.45	2414.40		
1 Laborer	39.15	313.20	60.00	480.00		
48 L.H., Daily Totals		$2299.20		$3522.40	$47.90	$73.38

Crew C-2A

Crew C-2A	Bare Costs Hr.	Daily	Incl. Subs O&P Hr.	Daily	Cost Per Labor-Hour Bare Costs	Incl. O&P
1 Carpenter Foreman (outside)	$51.25	$410.00	$78.50	$628.00	$47.49	$72.41
3 Carpenters	49.25	1182.00	75.45	1810.80		
1 Cement Finisher	46.80	374.40	69.60	556.80		
1 Laborer	39.15	313.20	60.00	480.00		
48 L.H., Daily Totals		$2279.60		$3475.60	$47.49	$72.41

Crew C-3

Crew C-3	Bare Costs Hr.	Daily	Incl. Subs O&P Hr.	Daily	Cost Per Labor-Hour Bare Costs	Incl. O&P
1 Rodman Foreman (outside)	$56.30	$450.40	$86.75	$694.00	$50.35	$77.34
4 Rodmen (reinf.)	54.30	1737.60	83.70	2678.40		
1 Equip. Oper. (light)	51.00	408.00	77.15	617.20		
2 Laborers	39.15	626.40	60.00	960.00		
3 Stressing Equipment		31.20		34.32		
.5 Grouting Equipment		80.70		88.77	1.75	1.92
64 L.H., Daily Totals		$3334.30		$5072.69	$52.10	$79.26

Crew C-4

Crew C-4	Bare Costs Hr.	Daily	Incl. Subs O&P Hr.	Daily	Cost Per Labor-Hour Bare Costs	Incl. O&P
1 Rodman Foreman (outside)	$56.30	$450.40	$86.75	$694.00	$54.80	$84.46
3 Rodmen (reinf.)	54.30	1303.20	83.70	2008.80		
3 Stressing Equipment		31.20		34.32	.97	1.07
32 L.H., Daily Totals		$1784.80		$2737.12	$55.77	$85.53

Crew C-4A

Crew C-4A	Bare Costs Hr.	Daily	Incl. Subs O&P Hr.	Daily	Cost Per Labor-Hour Bare Costs	Incl. O&P
2 Rodmen (reinf.)	$54.30	$868.80	$83.70	$1339.20	$54.30	$83.70
4 Stressing Equipment		41.60		45.76	2.60	2.86
16 L.H., Daily Totals		$910.40		$1384.96	$56.90	$86.56

Crew C-5

Crew C-5	Bare Costs Hr.	Daily	Incl. Subs O&P Hr.	Daily	Cost Per Labor-Hour Bare Costs	Incl. O&P
1 Rodman Foreman (outside)	$56.30	$450.40	$86.75	$694.00	$53.89	$82.63
4 Rodmen (reinf.)	54.30	1737.60	83.70	2678.40		
1 Equip. Oper. (crane)	55.70	445.60	84.25	674.00		
1 Equip. Oper. (oiler)	48.00	384.00	72.60	580.80		
1 Hyd. Crane, 25 Ton		673.80		741.18	12.03	13.24
56 L.H., Daily Totals		$3691.40		$5368.38	$65.92	$95.86

Crew C-6

Crew C-6	Bare Costs Hr.	Daily	Incl. Subs O&P Hr.	Daily	Cost Per Labor-Hour Bare Costs	Incl. O&P
1 Labor Foreman (outside)	$41.15	$329.20	$63.05	$504.40	$40.76	$62.11
4 Laborers	39.15	1252.80	60.00	1920.00		
1 Cement Finisher	46.80	374.40	69.60	556.80		
2 Gas Engine Vibrators		56.50		62.15	1.18	1.29
48 L.H., Daily Totals		$2012.90		$3043.35	$41.94	$63.40

Crew C-7

Crew C-7	Bare Costs Hr.	Daily	Incl. Subs O&P Hr.	Daily	Cost Per Labor-Hour Bare Costs	Incl. O&P
1 Labor Foreman (outside)	$41.15	$329.20	$63.05	$504.40	$42.81	$65.14
5 Laborers	39.15	1566.00	60.00	2400.00		
1 Cement Finisher	46.80	374.40	69.60	556.80		
1 Equip. Oper. (medium)	53.55	428.40	81.00	648.00		
1 Equip. Oper. (oiler)	48.00	384.00	72.60	580.80		
2 Gas Engine Vibrators		56.50		62.15		
1 Concrete Bucket, 1 C.Y.		24.00		26.40		
1 Hyd. Crane, 55 Ton		1076.00		1183.60	16.06	17.67
72 L.H., Daily Totals		$4238.50		$5962.15	$58.87	$82.81

Crew C-7A

Crew C-7A	Bare Costs Hr.	Daily	Incl. Subs O&P Hr.	Daily	Cost Per Labor-Hour Bare Costs	Incl. O&P
1 Labor Foreman (outside)	$41.15	$329.20	$63.05	$504.40	$41.00	$62.56
5 Laborers	39.15	1566.00	60.00	2400.00		
2 Truck Drivers (heavy)	45.55	728.80	68.70	1099.20		
2 Conc. Transit Mixers		1964.40		2160.84	30.69	33.76
64 L.H., Daily Totals		$4588.40		$6164.44	$71.69	$96.32

Crew C-7B

Crew C-7B	Bare Costs Hr.	Daily	Incl. Subs O&P Hr.	Daily	Cost Per Labor-Hour Bare Costs	Incl. O&P
1 Labor Foreman (outside)	$41.15	$329.20	$63.05	$504.40	$42.58	$64.99
5 Laborers	39.15	1566.00	60.00	2400.00		
1 Equipment Operator, Crane	55.70	445.60	84.25	674.00		
1 Equipment Oiler	48.00	384.00	72.60	580.80		
1 Conc. Bucket, 2 C.Y.		37.00		40.70		
1 Lattice Boom Crane, 165 Ton		2119.00		2330.90	33.69	37.06
64 L.H., Daily Totals		$4880.80		$6530.80	$76.26	$102.04

For customer support on your Building Construction Costs with RSMeans Data, call 800.448.8182.

Crew No.	Bare Costs Hr.	Daily	Incl. Subs O&P Hr.	Daily	Cost Per Labor-Hour Bare Costs	Incl. O&P
Crew C-7C						
1 Labor Foreman (outside)	$41.15	$329.20	$63.05	$504.40	$43.00	$65.63
5 Laborers	39.15	1566.00	60.00	2400.00		
2 Equipment Operators (med.)	53.55	856.80	81.00	1296.00		
2 F.E. Loaders, W.M., 4 C.Y.		1315.20		1446.72	20.55	22.61
64 L.H., Daily Totals		$4067.20		$5647.12	$63.55	$88.24
Crew C-7D						
1 Labor Foreman (outside)	$41.15	$329.20	$63.05	$504.40	$41.49	$63.44
5 Laborers	39.15	1566.00	60.00	2400.00		
1 Equip. Oper. (medium)	53.55	428.40	81.00	648.00		
1 Concrete Conveyer		189.80		208.78	3.39	3.73
56 L.H., Daily Totals		$2513.40		$3761.18	$44.88	$67.16
Crew C-8						
1 Labor Foreman (outside)	$41.15	$329.20	$63.05	$504.40	$43.68	$66.18
3 Laborers	39.15	939.60	60.00	1440.00		
2 Cement Finishers	46.80	748.80	69.60	1113.60		
1 Equip. Oper. (medium)	53.55	428.40	81.00	648.00		
1 Concrete Pump (Small)		806.40		887.04	14.40	15.84
56 L.H., Daily Totals		$3252.40		$4593.04	$58.08	$82.02
Crew C-8A						
1 Labor Foreman (outside)	$41.15	$329.20	$63.05	$504.40	$42.03	$63.71
3 Laborers	39.15	939.60	60.00	1440.00		
2 Cement Finishers	46.80	748.80	69.60	1113.60		
48 L.H., Daily Totals		$2017.60		$3058.00	$42.03	$63.71
Crew C-8B						
1 Labor Foreman (outside)	$41.15	$329.20	$63.05	$504.40	$42.43	$64.81
3 Laborers	39.15	939.60	60.00	1440.00		
1 Equip. Oper. (medium)	53.55	428.40	81.00	648.00		
1 Vibrating Power Screed		64.35		70.78		
1 Roller, Vibratory, 25 Ton		684.00		752.40		
1 Dozer, 200 H.P.		1384.00		1522.40	53.31	58.64
40 L.H., Daily Totals		$3829.55		$4937.98	$95.74	$123.45
Crew C-8C						
1 Labor Foreman (outside)	$41.15	$329.20	$63.05	$504.40	$43.16	$65.61
3 Laborers	39.15	939.60	60.00	1440.00		
1 Cement Finisher	46.80	374.40	69.60	556.80		
1 Equip. Oper. (medium)	53.55	428.40	81.00	648.00		
1 Shotcrete Rig, 12 C.Y./hr		250.20		275.22		
1 Air Compressor, 160 cfm		143.60		157.96		
4 -50' Air Hoses, 1"		16.40		18.04		
4 -50' Air Hoses, 2"		31.00		34.10	9.19	10.11
48 L.H., Daily Totals		$2512.80		$3634.52	$52.35	$75.72
Crew C-8D						
1 Labor Foreman (outside)	$41.15	$329.20	$63.05	$504.40	$44.52	$67.45
1 Laborer	39.15	313.20	60.00	480.00		
1 Cement Finisher	46.80	374.40	69.60	556.80		
1 Equipment Oper. (light)	51.00	408.00	77.15	617.20		
1 Air Compressor, 250 cfm		181.20		199.32		
2 -50' Air Hoses, 1"		8.20		9.02	5.92	6.51
32 L.H., Daily Totals		$1614.20		$2366.74	$50.44	$73.96

Crew No.	Bare Costs Hr.	Daily	Incl. Subs O&P Hr.	Daily	Cost Per Labor-Hour Bare Costs	Incl. O&P
Crew C-8E						
1 Labor Foreman (outside)	$41.15	$329.20	$63.05	$504.40	$42.73	$64.97
3 Laborers	39.15	939.60	60.00	1440.00		
1 Cement Finisher	46.80	374.40	69.60	556.80		
1 Equipment Oper. (light)	51.00	408.00	77.15	617.20		
1 Shotcrete Rig, 35 C.Y./hr		276.20		303.82		
1 Air Compressor, 250 cfm		181.20		199.32		
4 -50' Air Hoses, 1"		16.40		18.04		
4 -50' Air Hoses, 2"		31.00		34.10	10.52	11.57
48 L.H., Daily Totals		$2556.00		$3673.68	$53.25	$76.53
Crew C-10						
1 Laborer	$39.15	$313.20	$60.00	$480.00	$44.25	$66.40
2 Cement Finishers	46.80	748.80	69.60	1113.60		
24 L.H., Daily Totals		$1062.00		$1593.60	$44.25	$66.40
Crew C-10B						
3 Laborers	$39.15	$939.60	$60.00	$1440.00	$42.21	$63.84
2 Cement Finishers	46.80	748.80	69.60	1113.60		
1 Concrete Mixer, 10 C.F.		164.60		181.06		
2 Trowels, 48" Walk-Behind		91.10		100.21	6.39	7.03
40 L.H., Daily Totals		$1944.10		$2834.87	$48.60	$70.87
Crew C-10C						
1 Laborer	$39.15	$313.20	$60.00	$480.00	$44.25	$66.40
2 Cement Finishers	46.80	748.80	69.60	1113.60		
1 Trowel, 48" Walk-Behind		45.55		50.10	1.90	2.09
24 L.H., Daily Totals		$1107.55		$1643.70	$46.15	$68.49
Crew C-10D						
1 Laborer	$39.15	$313.20	$60.00	$480.00	$44.25	$66.40
2 Cement Finishers	46.80	748.80	69.60	1113.60		
1 Vibrating Power Screed		64.35		70.78		
1 Trowel, 48" Walk-Behind		45.55		50.10	4.58	5.04
24 L.H., Daily Totals		$1171.90		$1714.49	$48.83	$71.44
Crew C-10E						
1 Laborer	$39.15	$313.20	$60.00	$480.00	$44.25	$66.40
2 Cement Finishers	46.80	748.80	69.60	1113.60		
1 Vibrating Power Screed		64.35		70.78		
1 Cement Trowel, 96" Ride-On		175.90		193.49	10.01	11.01
24 L.H., Daily Totals		$1302.25		$1857.88	$54.26	$77.41
Crew C-10F						
1 Laborer	$39.15	$313.20	$60.00	$480.00	$44.25	$66.40
2 Cement Finishers	46.80	748.80	69.60	1113.60		
1 Telescoping Boom Lift, to 60'		456.00		501.60	19.00	20.90
24 L.H., Daily Totals		$1518.00		$2095.20	$63.25	$87.30
Crew C-11						
1 Struc. Steel Foreman (outside)	$56.30	$450.40	$93.85	$750.80	$53.98	$88.19
6 Struc. Steel Workers	54.30	2606.40	90.50	4344.00		
1 Equip. Oper. (crane)	55.70	445.60	84.25	674.00		
1 Equip. Oper. (oiler)	48.00	384.00	72.60	580.80		
1 Lattice Boom Crane, 150 Ton		2000.00		2200.00	27.78	30.56
72 L.H., Daily Totals		$5886.40		$8549.60	$81.76	$118.74

Crew No.	Bare Costs		Incl. Subs O&P		Cost Per Labor-Hour	
Crew C-12	Hr.	Daily	Hr.	Daily	Bare Costs	Incl. O&P
1 Carpenter Foreman (outside)	$51.25	$410.00	$78.50	$628.00	$48.98	$74.85
3 Carpenters	49.25	1182.00	75.45	1810.80		
1 Laborer	39.15	313.20	60.00	480.00		
1 Equip. Oper. (crane)	55.70	445.60	84.25	674.00		
1 Hyd. Crane, 12 Ton		572.80		630.08	11.93	13.13
48 L.H., Daily Totals		$2923.60		$4222.88	$60.91	$87.98
Crew C-13	Hr.	Daily	Hr.	Daily	Bare Costs	Incl. O&P
1 Struc. Steel Worker	$54.30	$434.40	$90.50	$724.00	$52.62	$85.48
1 Welder	54.30	434.40	90.50	724.00		
1 Carpenter	49.25	394.00	75.45	603.60		
1 Welder, Gas Engine, 300 amp		129.80		142.78	5.41	5.95
24 L.H., Daily Totals		$1392.60		$2194.38	$58.02	$91.43
Crew C-14	Hr.	Daily	Hr.	Daily	Bare Costs	Incl. O&P
1 Carpenter Foreman (outside)	$51.25	$410.00	$78.50	$628.00	$48.26	$73.70
5 Carpenters	49.25	1970.00	75.45	3018.00		
4 Laborers	39.15	1252.80	60.00	1920.00		
4 Rodmen (reinf.)	54.30	1737.60	83.70	2678.40		
2 Cement Finishers	46.80	748.80	69.60	1113.60		
1 Equip. Oper. (crane)	55.70	445.60	84.25	674.00		
1 Equip. Oper. (oiler)	48.00	384.00	72.60	580.80		
1 Hyd. Crane, 80 Ton		1616.00		1777.60	11.22	12.34
144 L.H., Daily Totals		$8564.80		$12390.40	$59.48	$86.04
Crew C-14A	Hr.	Daily	Hr.	Daily	Bare Costs	Incl. O&P
1 Carpenter Foreman (outside)	$51.25	$410.00	$78.50	$628.00	$49.40	$75.64
16 Carpenters	49.25	6304.00	75.45	9657.60		
4 Rodmen (reinf.)	54.30	1737.60	83.70	2678.40		
2 Laborers	39.15	626.40	60.00	960.00		
1 Cement Finisher	46.80	374.40	69.60	556.80		
1 Equip. Oper. (medium)	53.55	428.40	81.00	648.00		
1 Gas Engine Vibrator		28.25		31.07		
1 Concrete Pump (Small)		806.40		887.04	4.17	4.59
200 L.H., Daily Totals		$10715.45		$16046.92	$53.58	$80.23
Crew C-14B	Hr.	Daily	Hr.	Daily	Bare Costs	Incl. O&P
1 Carpenter Foreman (outside)	$51.25	$410.00	$78.50	$628.00	$49.30	$75.41
16 Carpenters	49.25	6304.00	75.45	9657.60		
4 Rodmen (reinf.)	54.30	1737.60	83.70	2678.40		
2 Laborers	39.15	626.40	60.00	960.00		
2 Cement Finishers	46.80	748.80	69.60	1113.60		
1 Equip. Oper. (medium)	53.55	428.40	81.00	648.00		
1 Gas Engine Vibrator		28.25		31.07		
1 Concrete Pump (Small)		806.40		887.04	4.01	4.41
208 L.H., Daily Totals		$11089.85		$16603.72	$53.32	$79.83
Crew C-14C	Hr.	Daily	Hr.	Daily	Bare Costs	Incl. O&P
1 Carpenter Foreman (outside)	$51.25	$410.00	$78.50	$628.00	$47.05	$72.01
6 Carpenters	49.25	2364.00	75.45	3621.60		
2 Rodmen (reinf.)	54.30	868.80	83.70	1339.20		
4 Laborers	39.15	1252.80	60.00	1920.00		
1 Cement Finisher	46.80	374.40	69.60	556.80		
1 Gas Engine Vibrator		28.25		31.07	.25	.28
112 L.H., Daily Totals		$5298.25		$8096.68	$47.31	$72.29

Crew No.	Bare Costs		Incl. Subs O&P		Cost Per Labor-Hour	
Crew C-14D	Hr.	Daily	Hr.	Daily	Bare Costs	Incl. O&P
1 Carpenter Foreman (outside)	$51.25	$410.00	$78.50	$628.00	$49.00	$74.98
18 Carpenters	49.25	7092.00	75.45	10864.80		
2 Rodmen (reinf.)	54.30	868.80	83.70	1339.20		
2 Laborers	39.15	626.40	60.00	960.00		
1 Cement Finisher	46.80	374.40	69.60	556.80		
1 Equip. Oper. (medium)	53.55	428.40	81.00	648.00		
1 Gas Engine Vibrator		28.25		31.07		
1 Concrete Pump (Small)		806.40		887.04	4.17	4.59
200 L.H., Daily Totals		$10634.65		$15914.92	$53.17	$79.57
Crew C-14E	Hr.	Daily	Hr.	Daily	Bare Costs	Incl. O&P
1 Carpenter Foreman (outside)	$51.25	$410.00	$78.50	$628.00	$48.29	$73.98
2 Carpenters	49.25	788.00	75.45	1207.20		
4 Rodmen (reinf.)	54.30	1737.60	83.70	2678.40		
3 Laborers	39.15	939.60	60.00	1440.00		
1 Cement Finisher	46.80	374.40	69.60	556.80		
1 Gas Engine Vibrator		28.25		31.07	.32	.35
88 L.H., Daily Totals		$4277.85		$6541.48	$48.61	$74.33
Crew C-14F	Hr.	Daily	Hr.	Daily	Bare Costs	Incl. O&P
1 Labor Foreman (outside)	$41.15	$329.20	$63.05	$504.40	$44.47	$66.74
2 Laborers	39.15	626.40	60.00	960.00		
6 Cement Finishers	46.80	2246.40	69.60	3340.80		
1 Gas Engine Vibrator		28.25		31.07	.39	.43
72 L.H., Daily Totals		$3230.25		$4836.27	$44.86	$67.17
Crew C-14G	Hr.	Daily	Hr.	Daily	Bare Costs	Incl. O&P
1 Labor Foreman (outside)	$41.15	$329.20	$63.05	$504.40	$43.81	$65.92
2 Laborers	39.15	626.40	60.00	960.00		
4 Cement Finishers	46.80	1497.60	69.60	2227.20		
1 Gas Engine Vibrator		28.25		31.07	.50	.55
56 L.H., Daily Totals		$2481.45		$3722.68	$44.31	$66.48
Crew C-14H	Hr.	Daily	Hr.	Daily	Bare Costs	Incl. O&P
1 Carpenter Foreman (outside)	$51.25	$410.00	$78.50	$628.00	$48.33	$73.78
2 Carpenters	49.25	788.00	75.45	1207.20		
1 Rodman (reinf.)	54.30	434.40	83.70	669.60		
1 Laborer	39.15	313.20	60.00	480.00		
1 Cement Finisher	46.80	374.40	69.60	556.80		
1 Gas Engine Vibrator		28.25		31.07	.59	.65
48 L.H., Daily Totals		$2348.25		$3572.68	$48.92	$74.43
Crew C-14L	Hr.	Daily	Hr.	Daily	Bare Costs	Incl. O&P
1 Carpenter Foreman (outside)	$51.25	$410.00	$78.50	$628.00	$45.85	$70.07
6 Carpenters	49.25	2364.00	75.45	3621.60		
4 Laborers	39.15	1252.80	60.00	1920.00		
1 Cement Finisher	46.80	374.40	69.60	556.80		
1 Gas Engine Vibrator		28.25		31.07	.29	.32
96 L.H., Daily Totals		$4429.45		$6757.48	$46.14	$70.39
Crew C-14M	Hr.	Daily	Hr.	Daily	Bare Costs	Incl. O&P
1 Carpenter Foreman (outside)	$51.25	$410.00	$78.50	$628.00	$47.84	$72.96
2 Carpenters	49.25	788.00	75.45	1207.20		
1 Rodman (reinf.)	54.30	434.40	83.70	669.60		
2 Laborers	39.15	626.40	60.00	960.00		
1 Cement Finisher	46.80	374.40	69.60	556.80		
1 Equip. Oper. (medium)	53.55	428.40	81.00	648.00		
1 Gas Engine Vibrator		28.25		31.07		
1 Concrete Pump (Small)		806.40		887.04	13.04	14.35
64 L.H., Daily Totals		$3896.25		$5587.72	$60.88	$87.31

Crew C-15

Crew No.	Bare Costs Hr.	Bare Costs Daily	Incl. Subs O&P Hr.	Incl. Subs O&P Daily	Cost Per Labor-Hour Bare Costs	Cost Per Labor-Hour Incl. O&P
1 Carpenter Foreman (outside)	$51.25	$410.00	$78.50	$628.00	$46.12	$70.26
2 Carpenters	49.25	788.00	75.45	1207.20		
3 Laborers	39.15	939.60	60.00	1440.00		
2 Cement Finishers	46.80	748.80	69.60	1113.60		
1 Rodman (reinf.)	54.30	434.40	83.70	669.60		
72 L.H., Daily Totals		$3320.80		$5058.40	$46.12	$70.26

Crew C-16

Crew No.	Bare Costs Hr.	Bare Costs Daily	Incl. Subs O&P Hr.	Incl. Subs O&P Daily	Cost Per Labor-Hour Bare Costs	Cost Per Labor-Hour Incl. O&P
1 Labor Foreman (outside)	$41.15	$329.20	$63.05	$504.40	$43.68	$66.18
3 Laborers	39.15	939.60	60.00	1440.00		
2 Cement Finishers	46.80	748.80	69.60	1113.60		
1 Equip. Oper. (medium)	53.55	428.40	81.00	648.00		
1 Gunite Pump Rig		357.40		393.14		
2 -50' Air Hoses, 3/4"		6.50		7.15		
2 -50' Air Hoses, 2"		15.50		17.05	6.78	7.45
56 L.H., Daily Totals		$2825.40		$4123.34	$50.45	$73.63

Crew C-16A

Crew No.	Bare Costs Hr.	Bare Costs Daily	Incl. Subs O&P Hr.	Incl. Subs O&P Daily	Cost Per Labor-Hour Bare Costs	Cost Per Labor-Hour Incl. O&P
1 Laborer	$39.15	$313.20	$60.00	$480.00	$46.58	$70.05
2 Cement Finishers	46.80	748.80	69.60	1113.60		
1 Equip. Oper. (medium)	53.55	428.40	81.00	648.00		
1 Gunite Pump Rig		357.40		393.14		
2 -50' Air Hoses, 3/4"		6.50		7.15		
2 -50' Air Hoses, 2"		15.50		17.05		
1 Telescoping Boom Lift, to 60'		456.00		501.60	26.11	28.72
32 L.H., Daily Totals		$2325.80		$3160.54	$72.68	$98.77

Crew C-17

Crew No.	Bare Costs Hr.	Bare Costs Daily	Incl. Subs O&P Hr.	Incl. Subs O&P Daily	Cost Per Labor-Hour Bare Costs	Cost Per Labor-Hour Incl. O&P
2 Skilled Worker Foremen (out)	$53.45	$855.20	$82.25	$1316.00	$51.85	$79.81
8 Skilled Workers	51.45	3292.80	79.20	5068.80		
80 L.H., Daily Totals		$4148.00		$6384.80	$51.85	$79.81

Crew C-17A

Crew No.	Bare Costs Hr.	Bare Costs Daily	Incl. Subs O&P Hr.	Incl. Subs O&P Daily	Cost Per Labor-Hour Bare Costs	Cost Per Labor-Hour Incl. O&P
2 Skilled Worker Foremen (out)	$53.45	$855.20	$82.25	$1316.00	$51.90	$79.86
8 Skilled Workers	51.45	3292.80	79.20	5068.80		
.125 Equip. Oper. (crane)	55.70	55.70	84.25	84.25		
.125 Hyd. Crane, 80 Ton		202.00		222.20	2.49	2.74
81 L.H., Daily Totals		$4405.70		$6691.25	$54.39	$82.61

Crew C-17B

Crew No.	Bare Costs Hr.	Bare Costs Daily	Incl. Subs O&P Hr.	Incl. Subs O&P Daily	Cost Per Labor-Hour Bare Costs	Cost Per Labor-Hour Incl. O&P
2 Skilled Worker Foremen (out)	$53.45	$855.20	$82.25	$1316.00	$51.94	$79.92
8 Skilled Workers	51.45	3292.80	79.20	5068.80		
.25 Equip. Oper. (crane)	55.70	111.40	84.25	168.50		
.25 Hyd. Crane, 80 Ton		404.00		444.40		
.25 Trowel, 48" Walk-Behind		11.39		12.53	5.07	5.57
82 L.H., Daily Totals		$4674.79		$7010.23	$57.01	$85.49

Crew C-17C

Crew No.	Bare Costs Hr.	Bare Costs Daily	Incl. Subs O&P Hr.	Incl. Subs O&P Daily	Cost Per Labor-Hour Bare Costs	Cost Per Labor-Hour Incl. O&P
2 Skilled Worker Foremen (out)	$53.45	$855.20	$82.25	$1316.00	$51.99	$79.97
8 Skilled Workers	51.45	3292.80	79.20	5068.80		
.375 Equip. Oper. (crane)	55.70	167.10	84.25	252.75		
.375 Hyd. Crane, 80 Ton		606.00		666.60	7.30	8.03
83 L.H., Daily Totals		$4921.10		$7304.15	$59.29	$88.00

Crew C-17D

Crew No.	Bare Costs Hr.	Bare Costs Daily	Incl. Subs O&P Hr.	Incl. Subs O&P Daily	Cost Per Labor-Hour Bare Costs	Cost Per Labor-Hour Incl. O&P
2 Skilled Worker Foremen (out)	$53.45	$855.20	$82.25	$1316.00	$52.03	$80.02
8 Skilled Workers	51.45	3292.80	79.20	5068.80		
.5 Equip. Oper. (crane)	55.70	222.80	84.25	337.00		
.5 Hyd. Crane, 80 Ton		808.00		888.80	9.62	10.58
84 L.H., Daily Totals		$5178.80		$7610.60	$61.65	$90.60

Crew C-17E

Crew No.	Bare Costs Hr.	Bare Costs Daily	Incl. Subs O&P Hr.	Incl. Subs O&P Daily	Cost Per Labor-Hour Bare Costs	Cost Per Labor-Hour Incl. O&P
2 Skilled Worker Foremen (out)	$53.45	$855.20	$82.25	$1316.00	$51.85	$79.81
8 Skilled Workers	51.45	3292.80	79.20	5068.80		
1 Hyd. Jack with Rods		100.30		110.33	1.25	1.38
80 L.H., Daily Totals		$4248.30		$6495.13	$53.10	$81.19

Crew C-18

Crew No.	Bare Costs Hr.	Bare Costs Daily	Incl. Subs O&P Hr.	Incl. Subs O&P Daily	Cost Per Labor-Hour Bare Costs	Cost Per Labor-Hour Incl. O&P
.125 Labor Foreman (outside)	$41.15	$41.15	$63.05	$63.05	$39.37	$60.34
1 Laborer	39.15	313.20	60.00	480.00		
1 Concrete Cart, 10 C.F.		58.00		63.80	6.44	7.09
9 L.H., Daily Totals		$412.35		$606.85	$45.82	$67.43

Crew C-19

Crew No.	Bare Costs Hr.	Bare Costs Daily	Incl. Subs O&P Hr.	Incl. Subs O&P Daily	Cost Per Labor-Hour Bare Costs	Cost Per Labor-Hour Incl. O&P
.125 Labor Foreman (outside)	$41.15	$41.15	$63.05	$63.05	$39.37	$60.34
1 Laborer	39.15	313.20	60.00	480.00		
1 Concrete Cart, 18 C.F.		96.60		106.26	10.73	11.81
9 L.H., Daily Totals		$450.95		$649.31	$50.11	$72.15

Crew C-20

Crew No.	Bare Costs Hr.	Bare Costs Daily	Incl. Subs O&P Hr.	Incl. Subs O&P Daily	Cost Per Labor-Hour Bare Costs	Cost Per Labor-Hour Incl. O&P
1 Labor Foreman (outside)	$41.15	$329.20	$63.05	$504.40	$42.16	$64.21
5 Laborers	39.15	1566.00	60.00	2400.00		
1 Cement Finisher	46.80	374.40	69.60	556.80		
1 Equip. Oper. (medium)	53.55	428.40	81.00	648.00		
2 Gas Engine Vibrators		56.50		62.15		
1 Concrete Pump (Small)		806.40		887.04	13.48	14.83
64 L.H., Daily Totals		$3560.90		$5058.39	$55.64	$79.04

Crew C-21

Crew No.	Bare Costs Hr.	Bare Costs Daily	Incl. Subs O&P Hr.	Incl. Subs O&P Daily	Cost Per Labor-Hour Bare Costs	Cost Per Labor-Hour Incl. O&P
1 Labor Foreman (outside)	$41.15	$329.20	$63.05	$504.40	$42.16	$64.21
5 Laborers	39.15	1566.00	60.00	2400.00		
1 Cement Finisher	46.80	374.40	69.60	556.80		
1 Equip. Oper. (medium)	53.55	428.40	81.00	648.00		
2 Gas Engine Vibrators		56.50		62.15		
1 Concrete Conveyer		189.80		208.78	3.85	4.23
64 L.H., Daily Totals		$2944.30		$4380.13	$46.00	$68.44

Crew C-22

Crew No.	Bare Costs Hr.	Bare Costs Daily	Incl. Subs O&P Hr.	Incl. Subs O&P Daily	Cost Per Labor-Hour Bare Costs	Cost Per Labor-Hour Incl. O&P
1 Rodman Foreman (outside)	$56.30	$450.40	$86.75	$694.00	$54.56	$84.03
4 Rodmen (reinf.)	54.30	1737.60	83.70	2678.40		
.125 Equip. Oper. (crane)	55.70	55.70	84.25	84.25		
.125 Equip. Oper. (oiler)	48.00	48.00	72.60	72.60		
.125 Hyd. Crane, 25 Ton		84.22		92.65	2.01	2.21
42 L.H., Daily Totals		$2375.93		$3621.90	$56.57	$86.24

Crew C-23

Crew No.	Bare Costs Hr.	Bare Costs Daily	Incl. Subs O&P Hr.	Incl. Subs O&P Daily	Cost Per Labor-Hour Bare Costs	Cost Per Labor-Hour Incl. O&P
2 Skilled Worker Foremen (out)	$53.45	$855.20	$82.25	$1316.00	$51.93	$79.66
6 Skilled Workers	51.45	2469.60	79.20	3801.60		
1 Equip. Oper. (crane)	55.70	445.60	84.25	674.00		
1 Equip. Oper. (oiler)	48.00	384.00	72.60	580.80		
1 Lattice Boom Crane, 90 Ton		1671.00		1838.10	20.89	22.98
80 L.H., Daily Totals		$5825.40		$8210.50	$72.82	$102.63

Crew C-23A

Crew No.	Bare Costs Hr.	Bare Costs Daily	Incl. Subs O&P Hr.	Incl. Subs O&P Daily	Cost Per Labor-Hour Bare Costs	Cost Per Labor-Hour Incl. O&P
1 Labor Foreman (outside)	$41.15	$329.20	$63.05	$504.40	$44.63	$67.98
2 Laborers	39.15	626.40	60.00	960.00		
1 Equip. Oper. (crane)	55.70	445.60	84.25	674.00		
1 Equip. Oper. (oiler)	48.00	384.00	72.60	580.80		
1 Crawler Crane, 100 Ton		1804.00		1984.40		
3 Conc. Buckets, 8 C.Y.		612.60		673.86	60.41	66.46
40 L.H., Daily Totals		$4201.80		$5377.46	$105.05	$134.44

Crews - Standard

Crew C-24

Crew C-24	Hr.	Daily	Hr.	Daily	Bare Costs	Incl. O&P
2 Skilled Worker Foremen (out)	$53.45	$855.20	$82.25	$1316.00	$51.93	$79.66
6 Skilled Workers	51.45	2469.60	79.20	3801.60		
1 Equip. Oper. (crane)	55.70	445.60	84.25	674.00		
1 Equip. Oper. (oiler)	48.00	384.00	72.60	580.80		
1 Lattice Boom Crane, 150 Ton		2000.00		2200.00	25.00	27.50
80 L.H., Daily Totals		$6154.40		$8572.40	$76.93	$107.16

Crew C-25

Crew C-25	Hr.	Daily	Hr.	Daily	Bare Costs	Incl. O&P
2 Rodmen (reinf.)	$54.30	$868.80	$83.70	$1339.20	$43.20	$69.13
2 Rodmen Helpers	32.10	513.60	54.55	872.80		
32 L.H., Daily Totals		$1382.40		$2212.00	$43.20	$69.13

Crew C-27

Crew C-27	Hr.	Daily	Hr.	Daily	Bare Costs	Incl. O&P
2 Cement Finishers	$46.80	$748.80	$69.60	$1113.60	$46.80	$69.60
1 Concrete Saw		153.60		168.96	9.60	10.56
16 L.H., Daily Totals		$902.40		$1282.56	$56.40	$80.16

Crew C-28

Crew C-28	Hr.	Daily	Hr.	Daily	Bare Costs	Incl. O&P
1 Cement Finisher	$46.80	$374.40	$69.60	$556.80	$46.80	$69.60
1 Portable Air Compressor, Gas		17.60		19.36	2.20	2.42
8 L.H., Daily Totals		$392.00		$576.16	$49.00	$72.02

Crew C-29

Crew C-29	Hr.	Daily	Hr.	Daily	Bare Costs	Incl. O&P
1 Laborer	$39.15	$313.20	$60.00	$480.00	$39.15	$60.00
1 Pressure Washer		66.00		72.60	8.25	9.07
8 L.H., Daily Totals		$379.20		$552.60	$47.40	$69.08

Crew C-30

Crew C-30	Hr.	Daily	Hr.	Daily	Bare Costs	Incl. O&P
1 Laborer	$39.15	$313.20	$60.00	$480.00	$39.15	$60.00
1 Concrete Mixer, 10 C.F.		164.60		181.06	20.57	22.63
8 L.H., Daily Totals		$477.80		$661.06	$59.73	$82.63

Crew C-31

Crew C-31	Hr.	Daily	Hr.	Daily	Bare Costs	Incl. O&P
1 Cement Finisher	$46.80	$374.40	$69.60	$556.80	$46.80	$69.60
1 Grout Pump		357.40		393.14	44.67	49.14
8 L.H., Daily Totals		$731.80		$949.94	$91.47	$118.74

Crew C-32

Crew C-32	Hr.	Daily	Hr.	Daily	Bare Costs	Incl. O&P
1 Cement Finisher	$46.80	$374.40	$69.60	$556.80	$42.98	$64.80
1 Laborer	39.15	313.20	60.00	480.00		
1 Crack Chaser Saw, Gas, 6 H.P.		26.60		29.26		
1 Vacuum Pick-Up System		60.85		66.94	5.47	6.01
16 L.H., Daily Totals		$775.05		$1132.99	$48.44	$70.81

Crew D-1

Crew D-1	Hr.	Daily	Hr.	Daily	Bare Costs	Incl. O&P
1 Bricklayer	$47.75	$382.00	$73.45	$587.60	$43.27	$66.55
1 Bricklayer Helper	38.80	310.40	59.65	477.20		
16 L.H., Daily Totals		$692.40		$1064.80	$43.27	$66.55

Crew D-2

Crew D-2	Hr.	Daily	Hr.	Daily	Bare Costs	Incl. O&P
3 Bricklayers	$47.75	$1146.00	$73.45	$1762.80	$44.63	$68.61
2 Bricklayer Helpers	38.80	620.80	59.65	954.40		
.5 Carpenter	49.25	197.00	75.45	301.80		
44 L.H., Daily Totals		$1963.80		$3019.00	$44.63	$68.61

Crew D-3

Crew D-3	Hr.	Daily	Hr.	Daily	Bare Costs	Incl. O&P
3 Bricklayers	$47.75	$1146.00	$73.45	$1762.80	$44.41	$68.29
2 Bricklayer Helpers	38.80	620.80	59.65	954.40		
.25 Carpenter	49.25	98.50	75.45	150.90		
42 L.H., Daily Totals		$1865.30		$2868.10	$44.41	$68.29

Crew D-4

Crew D-4	Hr.	Daily	Hr.	Daily	Bare Costs	Incl. O&P
1 Bricklayer	$47.75	$382.00	$73.45	$587.60	$44.09	$67.47
2 Bricklayer Helpers	38.80	620.80	59.65	954.40		
1 Equip. Oper. (light)	51.00	408.00	77.15	617.20		
1 Grout Pump, 50 C.F./hr.		132.80		146.08	4.15	4.57
32 L.H., Daily Totals		$1543.60		$2305.28	$48.24	$72.04

Crew D-5

Crew D-5	Hr.	Daily	Hr.	Daily	Bare Costs	Incl. O&P
1 Bricklayer	47.75	382.00	73.45	587.60	47.75	73.45
8 L.H., Daily Totals		$382.00		$587.60	$47.75	$73.45

Crew D-6

Crew D-6	Hr.	Daily	Hr.	Daily	Bare Costs	Incl. O&P
3 Bricklayers	$47.75	$1146.00	$73.45	$1762.80	$43.51	$66.91
3 Bricklayer Helpers	38.80	931.20	59.65	1431.60		
.25 Carpenter	49.25	98.50	75.45	150.90		
50 L.H., Daily Totals		$2175.70		$3345.30	$43.51	$66.91

Crew D-7

Crew D-7	Hr.	Daily	Hr.	Daily	Bare Costs	Incl. O&P
1 Tile Layer	$45.85	$366.80	$68.05	$544.40	$41.08	$60.95
1 Tile Layer Helper	36.30	290.40	53.85	430.80		
16 L.H., Daily Totals		$657.20		$975.20	$41.08	$60.95

Crew D-8

Crew D-8	Hr.	Daily	Hr.	Daily	Bare Costs	Incl. O&P
3 Bricklayers	$47.75	$1146.00	$73.45	$1762.80	$44.17	$67.93
2 Bricklayer Helpers	38.80	620.80	59.65	954.40		
40 L.H., Daily Totals		$1766.80		$2717.20	$44.17	$67.93

Crew D-9

Crew D-9	Hr.	Daily	Hr.	Daily	Bare Costs	Incl. O&P
3 Bricklayers	$47.75	$1146.00	$73.45	$1762.80	$43.27	$66.55
3 Bricklayer Helpers	38.80	931.20	59.65	1431.60		
48 L.H., Daily Totals		$2077.20		$3194.40	$43.27	$66.55

Crew D-10

Crew D-10	Hr.	Daily	Hr.	Daily	Bare Costs	Incl. O&P
1 Bricklayer Foreman (outside)	$49.75	$398.00	$76.50	$612.00	$48.00	$73.46
1 Bricklayer	47.75	382.00	73.45	587.60		
1 Bricklayer Helper	38.80	310.40	59.65	477.20		
1 Equip. Oper. (crane)	55.70	445.60	84.25	674.00		
1 S.P. Crane, 4x4, 12 Ton		435.60		479.16	13.61	14.97
32 L.H., Daily Totals		$1971.60		$2829.96	$61.61	$88.44

Crew D-11

Crew D-11	Hr.	Daily	Hr.	Daily	Bare Costs	Incl. O&P
1 Bricklayer Foreman (outside)	$49.75	$398.00	$76.50	$612.00	$45.43	$69.87
1 Bricklayer	47.75	382.00	73.45	587.60		
1 Bricklayer Helper	38.80	310.40	59.65	477.20		
24 L.H., Daily Totals		$1090.40		$1676.80	$45.43	$69.87

Crew D-12

Crew D-12	Hr.	Daily	Hr.	Daily	Bare Costs	Incl. O&P
1 Bricklayer Foreman (outside)	$49.75	$398.00	$76.50	$612.00	$43.77	$67.31
1 Bricklayer	47.75	382.00	73.45	587.60		
2 Bricklayer Helpers	38.80	620.80	59.65	954.40		
32 L.H., Daily Totals		$1400.80		$2154.00	$43.77	$67.31

Crews - Standard

Crew D-13

Crew No.	Bare Costs Hr.	Daily	Incl. Subs O&P Hr.	Daily	Cost Per Labor-Hour Bare Costs	Incl. O&P
1 Bricklayer Foreman (outside)	$49.75	$398.00	$76.50	$612.00	$46.67	$71.49
1 Bricklayer	47.75	382.00	73.45	587.60		
2 Bricklayer Helpers	38.80	620.80	59.65	954.40		
1 Carpenter	49.25	394.00	75.45	603.60		
1 Equip. Oper. (crane)	55.70	445.60	84.25	674.00		
1 S.P. Crane, 4x4, 12 Ton		435.60		479.16	9.07	9.98
48 L.H., Daily Totals		$2676.00		$3910.76	$55.75	$81.47

Crew D-14

Crew No.	Bare Costs Hr.	Daily	Incl. Subs O&P Hr.	Daily	Cost Per Labor-Hour Bare Costs	Incl. O&P
3 Bricklayers	$47.75	$1146.00	$73.45	$1762.80	$45.51	$70.00
1 Bricklayer Helper	38.80	310.40	59.65	477.20		
32 L.H., Daily Totals		$1456.40		$2240.00	$45.51	$70.00

Crew E-1

Crew No.	Bare Costs Hr.	Daily	Incl. Subs O&P Hr.	Daily	Cost Per Labor-Hour Bare Costs	Incl. O&P
1 Welder Foreman (outside)	$56.30	$450.40	$93.85	$750.80	$53.87	$87.17
1 Welder	54.30	434.40	90.50	724.00		
1 Equip. Oper. (light)	51.00	408.00	77.15	617.20		
1 Welder, Gas Engine, 300 amp		129.80		142.78	5.41	5.95
24 L.H., Daily Totals		$1422.60		$2234.78	$59.27	$93.12

Crew E-2

Crew No.	Bare Costs Hr.	Daily	Incl. Subs O&P Hr.	Daily	Cost Per Labor-Hour Bare Costs	Incl. O&P
1 Struc. Steel Foreman (outside)	$56.30	$450.40	$93.85	$750.80	$53.89	$87.53
4 Struc. Steel Workers	54.30	1737.60	90.50	2896.00		
1 Equip. Oper. (crane)	55.70	445.60	84.25	674.00		
1 Equip. Oper. (oiler)	48.00	384.00	72.60	580.80		
1 Lattice Boom Crane, 90 Ton		1671.00		1838.10	29.84	32.82
56 L.H., Daily Totals		$4688.60		$6739.70	$83.72	$120.35

Crew E-3

Crew No.	Bare Costs Hr.	Daily	Incl. Subs O&P Hr.	Daily	Cost Per Labor-Hour Bare Costs	Incl. O&P
1 Struc. Steel Foreman (outside)	$56.30	$450.40	$93.85	$750.80	$54.97	$91.62
1 Struc. Steel Worker	54.30	434.40	90.50	724.00		
1 Welder	54.30	434.40	90.50	724.00		
1 Welder, Gas Engine, 300 amp		129.80		142.78	5.41	5.95
24 L.H., Daily Totals		$1449.00		$2341.58	$60.38	$97.57

Crew E-3A

Crew No.	Bare Costs Hr.	Daily	Incl. Subs O&P Hr.	Daily	Cost Per Labor-Hour Bare Costs	Incl. O&P
1 Struc. Steel Foreman (outside)	$56.30	$450.40	$93.85	$750.80	$54.97	$91.62
1 Struc. Steel Worker	54.30	434.40	90.50	724.00		
1 Welder	54.30	434.40	90.50	724.00		
1 Welder, Gas Engine, 300 amp		129.80		142.78		
1 Telescoping Boom Lift, to 40'		288.80		317.68	17.44	19.19
24 L.H., Daily Totals		$1737.80		$2659.26	$72.41	$110.80

Crew E-4

Crew No.	Bare Costs Hr.	Daily	Incl. Subs O&P Hr.	Daily	Cost Per Labor-Hour Bare Costs	Incl. O&P
1 Struc. Steel Foreman (outside)	$56.30	$450.40	$93.85	$750.80	$54.80	$91.34
3 Struc. Steel Workers	54.30	1303.20	90.50	2172.00		
1 Welder, Gas Engine, 300 amp		129.80		142.78	4.06	4.46
32 L.H., Daily Totals		$1883.40		$3065.58	$58.86	$95.80

Crew E-5

Crew No.	Bare Costs Hr.	Daily	Incl. Subs O&P Hr.	Daily	Cost Per Labor-Hour Bare Costs	Incl. O&P
2 Struc. Steel Foremen (outside)	$56.30	$900.80	$93.85	$1501.60	$54.21	$88.75
5 Struc. Steel Workers	54.30	2172.00	90.50	3620.00		
1 Equip. Oper. (crane)	55.70	445.60	84.25	674.00		
1 Welder	54.30	434.40	90.50	724.00		
1 Equip. Oper. (oiler)	48.00	384.00	72.60	580.80		
1 Lattice Boom Crane, 90 Ton		1671.00		1838.10		
1 Welder, Gas Engine, 300 amp		129.80		142.78	22.51	24.76
80 L.H., Daily Totals		$6137.60		$9081.28	$76.72	$113.52

Crew E-6

Crew No.	Bare Costs Hr.	Daily	Incl. Subs O&P Hr.	Daily	Cost Per Labor-Hour Bare Costs	Incl. O&P
3 Struc. Steel Foremen (outside)	$56.30	$1351.20	$93.85	$2252.40	$54.16	$88.78
9 Struc. Steel Workers	54.30	3909.60	90.50	6516.00		
1 Equip. Oper. (crane)	55.70	445.60	84.25	674.00		
1 Welder	54.30	434.40	90.50	724.00		
1 Equip. Oper. (oiler)	48.00	384.00	72.60	580.80		
1 Equip. Oper. (light)	51.00	408.00	77.15	617.20		
1 Lattice Boom Crane, 90 Ton		1671.00		1838.10		
1 Welder, Gas Engine, 300 amp		129.80		142.78		
1 Air Compressor, 160 cfm		143.60		157.96		
2 Impact Wrenches		37.20		40.92	15.48	17.03
128 L.H., Daily Totals		$8914.40		$13544.16	$69.64	$105.81

Crew E-7

Crew No.	Bare Costs Hr.	Daily	Incl. Subs O&P Hr.	Daily	Cost Per Labor-Hour Bare Costs	Incl. O&P
1 Struc. Steel Foreman (outside)	$56.30	$450.40	$93.85	$750.80	$54.21	$88.75
4 Struc. Steel Workers	54.30	1737.60	90.50	2896.00		
1 Equip. Oper. (crane)	55.70	445.60	84.25	674.00		
1 Equip. Oper. (oiler)	48.00	384.00	72.60	580.80		
1 Welder Foreman (outside)	56.30	450.40	93.85	750.80		
2 Welders	54.30	868.80	90.50	1448.00		
1 Lattice Boom Crane, 90 Ton		1671.00		1838.10		
2 Welder, Gas Engine, 300 amp		259.60		285.56	24.13	26.55
80 L.H., Daily Totals		$6267.40		$9224.06	$78.34	$115.30

Crew E-8

Crew No.	Bare Costs Hr.	Daily	Incl. Subs O&P Hr.	Daily	Cost Per Labor-Hour Bare Costs	Incl. O&P
1 Struc. Steel Foreman (outside)	$56.30	$450.40	$93.85	$750.80	$53.98	$88.13
4 Struc. Steel Workers	54.30	1737.60	90.50	2896.00		
1 Welder Foreman (outside)	56.30	450.40	93.85	750.80		
4 Welders	54.30	1737.60	90.50	2896.00		
1 Equip. Oper. (crane)	55.70	445.60	84.25	674.00		
1 Equip. Oper. (oiler)	48.00	384.00	72.60	580.80		
1 Equip. Oper. (light)	51.00	408.00	77.15	617.20		
1 Lattice Boom Crane, 90 Ton		1671.00		1838.10		
4 Welder, Gas Engine, 300 amp		519.20		571.12	21.06	23.17
104 L.H., Daily Totals		$7803.80		$11574.82	$75.04	$111.30

Crew E-9

Crew No.	Bare Costs Hr.	Daily	Incl. Subs O&P Hr.	Daily	Cost Per Labor-Hour Bare Costs	Incl. O&P
2 Struc. Steel Foremen (outside)	$56.30	$900.80	$93.85	$1501.60	$54.16	$88.78
5 Struc. Steel Workers	54.30	2172.00	90.50	3620.00		
1 Welder Foreman (outside)	56.30	450.40	93.85	750.80		
5 Welders	54.30	2172.00	90.50	3620.00		
1 Equip. Oper. (crane)	55.70	445.60	84.25	674.00		
1 Equip. Oper. (oiler)	48.00	384.00	72.60	580.80		
1 Equip. Oper. (light)	51.00	408.00	77.15	617.20		
1 Lattice Boom Crane, 90 Ton		1671.00		1838.10		
5 Welder, Gas Engine, 300 amp		649.00		713.90	18.13	19.94
128 L.H., Daily Totals		$9252.80		$13916.40	$72.29	$108.72

Crew E-10

Crew No.	Bare Costs Hr.	Daily	Incl. Subs O&P Hr.	Daily	Cost Per Labor-Hour Bare Costs	Incl. O&P
1 Welder Foreman (outside)	$56.30	$450.40	$93.85	$750.80	$55.30	$92.17
1 Welder	54.30	434.40	90.50	724.00		
1 Welder, Gas Engine, 300 amp		129.80		142.78		
1 Flatbed Truck, Gas, 3 Ton		270.20		297.22	25.00	27.50
16 L.H., Daily Totals		$1284.80		$1914.80	$80.30	$119.68

Crew E-11

Crew No.	Bare Costs Hr.	Daily	Incl. Subs O&P Hr.	Daily	Cost Per Labor-Hour Bare Costs	Incl. O&P
2 Painters, Struc. Steel	$43.00	$688.00	$72.65	$1162.40	$44.04	$70.61
1 Building Laborer	39.15	313.20	60.00	480.00		
1 Equip. Oper. (light)	51.00	408.00	77.15	617.20		
1 Air Compressor, 250 cfm		181.20		199.32		
1 Sandblaster, Portable, 3 C.F.		21.40		23.54		
1 Set Sand Blasting Accessories		14.45		15.90	6.78	7.46
32 L.H., Daily Totals		$1626.25		$2498.36	$50.82	$78.07

Crew No.	Bare Costs		Incl. Subs O&P		Cost Per Labor-Hour	

Crew E-11A	Hr.	Daily	Hr.	Daily	Bare Costs	Incl. O&P
2 Painters, Struc. Steel	$43.00	$688.00	$72.65	$1162.40	$44.04	$70.61
1 Building Laborer	39.15	313.20	60.00	480.00		
1 Equip. Oper. (light)	51.00	408.00	77.15	617.20		
1 Air Compressor, 250 cfm		181.20		199.32		
1 Sandblaster, Portable, 3 C.F.		21.40		23.54		
1 Set Sand Blasting Accessories		14.45		15.90		
1 Telescoping Boom Lift, to 60'		456.00		501.60	21.03	23.14
32 L.H., Daily Totals		$2082.25		$2999.95	$65.07	$93.75

Crew E-11B	Hr.	Daily	Hr.	Daily	Bare Costs	Incl. O&P
2 Painters, Struc. Steel	$43.00	$688.00	$72.65	$1162.40	$41.72	$68.43
1 Building Laborer	39.15	313.20	60.00	480.00		
2 Paint Sprayer, 8 C.F.M.		101.40		111.54		
1 Telescoping Boom Lift, to 60'		456.00		501.60	23.23	25.55
24 L.H., Daily Totals		$1558.60		$2255.54	$64.94	$93.98

Crew E-12	Hr.	Daily	Hr.	Daily	Bare Costs	Incl. O&P
1 Welder Foreman (outside)	$56.30	$450.40	$93.85	$750.80	$53.65	$85.50
1 Equip. Oper. (light)	51.00	408.00	77.15	617.20		
1 Welder, Gas Engine, 300 amp		129.80		142.78	8.11	8.92
16 L.H., Daily Totals		$988.20		$1510.78	$61.76	$94.42

Crew E-13	Hr.	Daily	Hr.	Daily	Bare Costs	Incl. O&P
1 Welder Foreman (outside)	$56.30	$450.40	$93.85	$750.80	$54.53	$88.28
.5 Equip. Oper. (light)	51.00	204.00	77.15	308.60		
1 Welder, Gas Engine, 300 amp		129.80		142.78	10.82	11.90
12 L.H., Daily Totals		$784.20		$1202.18	$65.35	$100.18

Crew E-14	Hr.	Daily	Hr.	Daily	Bare Costs	Incl. O&P
1 Welder Foreman (outside)	$56.30	$450.40	$93.85	$750.80	$56.30	$93.85
1 Welder, Gas Engine, 300 amp		129.80		142.78	16.23	17.85
8 L.H., Daily Totals		$580.20		$893.58	$72.53	$111.70

Crew E-16	Hr.	Daily	Hr.	Daily	Bare Costs	Incl. O&P
1 Welder Foreman (outside)	$56.30	$450.40	$93.85	$750.80	$55.30	$92.17
1 Welder	54.30	434.40	90.50	724.00		
1 Welder, Gas Engine, 300 amp		129.80		142.78	8.11	8.92
16 L.H., Daily Totals		$1014.60		$1617.58	$63.41	$101.10

Crew E-17	Hr.	Daily	Hr.	Daily	Bare Costs	Incl. O&P
1 Struc. Steel Foreman (outside)	$56.30	$450.40	$93.85	$750.80	$55.30	$92.17
1 Structural Steel Worker	54.30	434.40	90.50	724.00		
16 L.H., Daily Totals		$884.80		$1474.80	$55.30	$92.17

Crew E-18	Hr.	Daily	Hr.	Daily	Bare Costs	Incl. O&P
1 Struc. Steel Foreman (outside)	$56.30	$450.40	$93.85	$750.80	$54.55	$89.27
3 Structural Steel Workers	54.30	1303.20	90.50	2172.00		
1 Equipment Operator (med.)	53.55	428.40	81.00	648.00		
1 Lattice Boom Crane, 20 Ton		1075.00		1182.50	26.88	29.56
40 L.H., Daily Totals		$3257.00		$4753.30	$81.42	$118.83

Crew E-19	Hr.	Daily	Hr.	Daily	Bare Costs	Incl. O&P
1 Struc. Steel Foreman (outside)	$56.30	$450.40	$93.85	$750.80	$53.87	$87.17
1 Structural Steel Worker	54.30	434.40	90.50	724.00		
1 Equip. Oper. (light)	51.00	408.00	77.15	617.20		
1 Lattice Boom Crane, 20 Ton		1075.00		1182.50	44.79	49.27
24 L.H., Daily Totals		$2367.80		$3274.50	$98.66	$136.44

Crew E-20	Hr.	Daily	Hr.	Daily	Bare Costs	Incl. O&P
1 Struc. Steel Foreman (outside)	$56.30	$450.40	$93.85	$750.80	$53.94	$87.90
5 Structural Steel Workers	54.30	2172.00	90.50	3620.00		
1 Equip. Oper. (crane)	55.70	445.60	84.25	674.00		
1 Equip. Oper. (oiler)	48.00	384.00	72.60	580.80		
1 Lattice Boom Crane, 40 Ton		1311.00		1442.10	20.48	22.53
64 L.H., Daily Totals		$4763.00		$7067.70	$74.42	$110.43

Crew E-22	Hr.	Daily	Hr.	Daily	Bare Costs	Incl. O&P
1 Skilled Worker Foreman (out)	$53.45	$427.60	$82.25	$658.00	$52.12	$80.22
2 Skilled Workers	51.45	823.20	79.20	1267.20		
24 L.H., Daily Totals		$1250.80		$1925.20	$52.12	$80.22

Crew E-24	Hr.	Daily	Hr.	Daily	Bare Costs	Incl. O&P
3 Structural Steel Workers	$54.30	$1303.20	$90.50	$2172.00	$54.11	$88.13
1 Equipment Operator (med.)	53.55	428.40	81.00	648.00		
1 Hyd. Crane, 25 Ton		673.80		741.18	21.06	23.16
32 L.H., Daily Totals		$2405.40		$3561.18	$75.17	$111.29

Crew E-25	Hr.	Daily	Hr.	Daily	Bare Costs	Incl. O&P
1 Welder Foreman (outside)	$56.30	$450.40	$93.85	$750.80	$56.30	$93.85
1 Cutting Torch		12.00		13.20	1.50	1.65
8 L.H., Daily Totals		$462.40		$764.00	$57.80	$95.50

Crew E-26	Hr.	Daily	Hr.	Daily	Bare Costs	Incl. O&P
1 Struc. Steel Foreman (outside)	$56.30	$450.40	$93.85	$750.80	$55.57	$91.25
1 Struc. Steel Worker	54.30	434.40	90.50	724.00		
1 Welder	54.30	434.40	90.50	724.00		
.25 Electrician	56.60	113.20	84.80	169.60		
.25 Plumber	61.80	123.60	93.25	186.50		
1 Welder, Gas Engine, 300 amp		129.80		142.78	4.64	5.10
28 L.H., Daily Totals		$1685.80		$2697.68	$60.21	$96.35

Crew E-27	Hr.	Daily	Hr.	Daily	Bare Costs	Incl. O&P
1 Struc. Steel Foreman (outside)	$56.30	$450.40	$93.85	$750.80	$53.94	$87.90
5 Struc. Steel Workers	54.30	2172.00	90.50	3620.00		
1 Equip. Oper. (crane)	55.70	445.60	84.25	674.00		
1 Equip. Oper. (oiler)	48.00	384.00	72.60	580.80		
1 Hyd. Crane, 12 Ton		572.80		630.08		
1 Hyd. Crane, 80 Ton		1616.00		1777.60	34.20	37.62
64 L.H., Daily Totals		$5640.80		$8033.28	$88.14	$125.52

Crew F-3	Hr.	Daily	Hr.	Daily	Bare Costs	Incl. O&P
4 Carpenters	$49.25	$1576.00	$75.45	$2414.40	$50.54	$77.21
1 Equip. Oper. (crane)	55.70	445.60	84.25	674.00		
1 Hyd. Crane, 12 Ton		572.80		630.08	14.32	15.75
40 L.H., Daily Totals		$2594.40		$3718.48	$64.86	$92.96

Crew F-4	Hr.	Daily	Hr.	Daily	Bare Costs	Incl. O&P
4 Carpenters	$49.25	$1576.00	$75.45	$2414.40	$50.12	$76.44
1 Equip. Oper. (crane)	55.70	445.60	84.25	674.00		
1 Equip. Oper. (oiler)	48.00	384.00	72.60	580.80		
1 Hyd. Crane, 55 Ton		1076.00		1183.60	22.42	24.66
48 L.H., Daily Totals		$3481.60		$4852.80	$72.53	$101.10

Crew F-5	Hr.	Daily	Hr.	Daily	Bare Costs	Incl. O&P
1 Carpenter Foreman (outside)	$51.25	$410.00	$78.50	$628.00	$49.75	$76.21
3 Carpenters	49.25	1182.00	75.45	1810.80		
32 L.H., Daily Totals		$1592.00		$2438.80	$49.75	$76.21

For customer support on your Building Construction Costs with RSMeans Data, call 800.448.8182.

Crew No.	Bare Costs Hr.	Daily	Incl. Subs O&P Hr.	Daily	Cost Per Labor-Hour Bare Costs	Incl. O&P
Crew F-6						
2 Carpenters	$49.25	$788.00	$75.45	$1207.20	$46.50	$71.03
2 Building Laborers	39.15	626.40	60.00	960.00		
1 Equip. Oper. (crane)	55.70	445.60	84.25	674.00		
1 Hyd. Crane, 12 Ton		572.80		630.08	14.32	15.75
40 L.H., Daily Totals		$2432.80		$3471.28	$60.82	$86.78

Crew No.	Bare Costs Hr.	Daily	Incl. Subs O&P Hr.	Daily	Cost Per Labor-Hour Bare Costs	Incl. O&P
Crew F-7						
2 Carpenters	$49.25	$788.00	$75.45	$1207.20	$44.20	$67.72
2 Building Laborers	39.15	626.40	60.00	960.00		
32 L.H., Daily Totals		$1414.40		$2167.20	$44.20	$67.72

Crew No.	Bare Costs Hr.	Daily	Incl. Subs O&P Hr.	Daily	Cost Per Labor-Hour Bare Costs	Incl. O&P
Crew G-1						
1 Roofer Foreman (outside)	$45.15	$361.20	$76.75	$614.00	$40.28	$68.46
4 Roofers Composition	43.15	1380.80	73.35	2347.20		
2 Roofer Helpers	32.10	513.60	54.55	872.80		
1 Application Equipment		183.60		201.96		
1 Tar Kettle/Pot		159.70		175.67		
1 Crew Truck		185.60		204.16	9.44	10.39
56 L.H., Daily Totals		$2784.50		$4415.79	$49.72	$78.85

Crew No.	Bare Costs Hr.	Daily	Incl. Subs O&P Hr.	Daily	Cost Per Labor-Hour Bare Costs	Incl. O&P
Crew G-2						
1 Plasterer	$45.90	$367.20	$68.85	$550.80	$41.37	$62.48
1 Plasterer Helper	39.05	312.40	58.60	468.80		
1 Building Laborer	39.15	313.20	60.00	480.00		
1 Grout Pump, 50 C.F./hr.		132.80		146.08	5.53	6.09
24 L.H., Daily Totals		$1125.60		$1645.68	$46.90	$68.57

Crew No.	Bare Costs Hr.	Daily	Incl. Subs O&P Hr.	Daily	Cost Per Labor-Hour Bare Costs	Incl. O&P
Crew G-2A						
1 Roofer Composition	$43.15	$345.20	$73.35	$586.80	$38.13	$62.63
1 Roofer Helper	32.10	256.80	54.55	436.40		
1 Building Laborer	39.15	313.20	60.00	480.00		
1 Foam Spray Rig, Trailer-Mtd.		542.60		596.86		
1 Pickup Truck, 3/4 Ton		129.40		142.34	28.00	30.80
24 L.H., Daily Totals		$1587.20		$2242.40	$66.13	$93.43

Crew No.	Bare Costs Hr.	Daily	Incl. Subs O&P Hr.	Daily	Cost Per Labor-Hour Bare Costs	Incl. O&P
Crew G-3						
2 Sheet Metal Workers	$58.05	$928.80	$88.85	$1421.60	$48.60	$74.42
2 Building Laborers	39.15	626.40	60.00	960.00		
32 L.H., Daily Totals		$1555.20		$2381.60	$48.60	$74.42

Crew No.	Bare Costs Hr.	Daily	Incl. Subs O&P Hr.	Daily	Cost Per Labor-Hour Bare Costs	Incl. O&P
Crew G-4						
1 Labor Foreman (outside)	$41.15	$329.20	$63.05	$504.40	$39.82	$61.02
2 Building Laborers	39.15	626.40	60.00	960.00		
1 Flatbed Truck, Gas, 1.5 Ton		217.40		239.14		
1 Air Compressor, 160 cfm		143.60		157.96	15.04	16.55
24 L.H., Daily Totals		$1316.60		$1861.50	$54.86	$77.56

Crew No.	Bare Costs Hr.	Daily	Incl. Subs O&P Hr.	Daily	Cost Per Labor-Hour Bare Costs	Incl. O&P
Crew G-5						
1 Roofer Foreman (outside)	$45.15	$361.20	$76.75	$614.00	$39.13	$66.51
2 Roofers Composition	43.15	690.40	73.35	1173.60		
2 Roofer Helpers	32.10	513.60	54.55	872.80		
1 Application Equipment		183.60		201.96	4.59	5.05
40 L.H., Daily Totals		$1748.80		$2862.36	$43.72	$71.56

Crew No.	Bare Costs Hr.	Daily	Incl. Subs O&P Hr.	Daily	Cost Per Labor-Hour Bare Costs	Incl. O&P
Crew G-6A						
2 Roofers Composition	$43.15	$690.40	$73.35	$1173.60	$43.15	$73.35
1 Small Compressor, Electric		15.25		16.77		
2 Pneumatic Nailers		46.60		51.26	3.87	4.25
16 L.H., Daily Totals		$752.25		$1241.64	$47.02	$77.60

Crew No.	Bare Costs Hr.	Daily	Incl. Subs O&P Hr.	Daily	Cost Per Labor-Hour Bare Costs	Incl. O&P
Crew G-7						
1 Carpenter	$49.25	$394.00	$75.45	$603.60	$49.25	$75.45
1 Small Compressor, Electric		15.25		16.77		
1 Pneumatic Nailer		23.30		25.63	4.82	5.30
8 L.H., Daily Totals		$432.55		$646.01	$54.07	$80.75

Crew No.	Bare Costs Hr.	Daily	Incl. Subs O&P Hr.	Daily	Cost Per Labor-Hour Bare Costs	Incl. O&P
Crew H-1						
2 Glaziers	$47.80	$764.80	$72.75	$1164.00	$51.05	$81.63
2 Struc. Steel Workers	54.30	868.80	90.50	1448.00		
32 L.H., Daily Totals		$1633.60		$2612.00	$51.05	$81.63

Crew No.	Bare Costs Hr.	Daily	Incl. Subs O&P Hr.	Daily	Cost Per Labor-Hour Bare Costs	Incl. O&P
Crew H-2						
2 Glaziers	$47.80	$764.80	$72.75	$1164.00	$44.92	$68.50
1 Building Laborer	39.15	313.20	60.00	480.00		
24 L.H., Daily Totals		$1078.00		$1644.00	$44.92	$68.50

Crew No.	Bare Costs Hr.	Daily	Incl. Subs O&P Hr.	Daily	Cost Per Labor-Hour Bare Costs	Incl. O&P
Crew H-3						
1 Glazier	$47.80	$382.40	$72.75	$582.00	$42.45	$65.15
1 Helper	37.10	296.80	57.55	460.40		
16 L.H., Daily Totals		$679.20		$1042.40	$42.45	$65.15

Crew No.	Bare Costs Hr.	Daily	Incl. Subs O&P Hr.	Daily	Cost Per Labor-Hour Bare Costs	Incl. O&P
Crew H-4						
1 Carpenter	$49.25	$394.00	$75.45	$603.60	$45.86	$70.16
1 Carpenter Helper	37.10	296.80	57.55	460.40		
.5 Electrician	56.60	226.40	84.80	339.20		
20 L.H., Daily Totals		$917.20		$1403.20	$45.86	$70.16

Crew No.	Bare Costs Hr.	Daily	Incl. Subs O&P Hr.	Daily	Cost Per Labor-Hour Bare Costs	Incl. O&P
Crew J-1						
3 Plasterers	$45.90	$1101.60	$68.85	$1652.40	$43.16	$64.75
2 Plasterer Helpers	39.05	624.80	58.60	937.60		
1 Mixing Machine, 6 C.F.		134.80		148.28	3.37	3.71
40 L.H., Daily Totals		$1861.20		$2738.28	$46.53	$68.46

Crew No.	Bare Costs Hr.	Daily	Incl. Subs O&P Hr.	Daily	Cost Per Labor-Hour Bare Costs	Incl. O&P
Crew J-2						
3 Plasterers	$45.90	$1101.60	$68.85	$1652.40	$43.95	$65.78
2 Plasterer Helpers	39.05	624.80	58.60	937.60		
1 Lather	47.90	383.20	70.90	567.20		
1 Mixing Machine, 6 C.F.		134.80		148.28	2.81	3.09
48 L.H., Daily Totals		$2244.40		$3305.48	$46.76	$68.86

Crew No.	Bare Costs Hr.	Daily	Incl. Subs O&P Hr.	Daily	Cost Per Labor-Hour Bare Costs	Incl. O&P
Crew J-3						
1 Terrazzo Worker	$45.95	$367.60	$68.20	$545.60	$42.38	$62.90
1 Terrazzo Helper	38.80	310.40	57.60	460.80		
1 Floor Grinder, 22" Path		121.50		133.65		
1 Terrazzo Mixer		180.00		198.00	18.84	20.73
16 L.H., Daily Totals		$979.50		$1338.05	$61.22	$83.63

Crew No.	Bare Costs Hr.	Daily	Incl. Subs O&P Hr.	Daily	Cost Per Labor-Hour Bare Costs	Incl. O&P
Crew J-4						
2 Cement Finishers	$46.80	$748.80	$69.60	$1113.60	$44.25	$66.40
1 Laborer	39.15	313.20	60.00	480.00		
1 Floor Grinder, 22" Path		121.50		133.65		
1 Floor Edger, 7" Path		39.25		43.17		
1 Vacuum Pick-Up System		60.85		66.94	9.23	10.16
24 L.H., Daily Totals		$1283.60		$1837.36	$53.48	$76.56

Crews - Standard

Crew J-4A

	Bare Costs Hr.	Daily	Incl. Subs O&P Hr.	Daily	Cost Per L.H. Bare Costs	Incl. O&P
2 Cement Finishers	$46.80	$748.80	$69.60	$1113.60	$42.98	$64.80
2 Laborers	39.15	626.40	60.00	960.00		
1 Floor Grinder, 22" Path		121.50		133.65		
1 Floor Edger, 7" Path		39.25		43.17		
1 Vacuum Pick-Up System		60.85		66.94		
1 Floor Auto Scrubber		236.25		259.88	14.31	15.74
32 L.H., Daily Totals		$1833.05		$2577.24	$57.28	$80.54

Crew J-4B

	Bare Costs Hr.	Daily	Incl. Subs O&P Hr.	Daily	Cost Per L.H. Bare Costs	Incl. O&P
1 Laborer	$39.15	$313.20	$60.00	$480.00	$39.15	$60.00
1 Floor Auto Scrubber		236.25		259.88	29.53	32.48
8 L.H., Daily Totals		$549.45		$739.88	$68.68	$92.48

Crew J-6

	Bare Costs Hr.	Daily	Incl. Subs O&P Hr.	Daily	Cost Per L.H. Bare Costs	Incl. O&P
2 Painters	$41.70	$667.20	$62.95	$1007.20	$43.39	$65.76
1 Building Laborer	39.15	313.20	60.00	480.00		
1 Equip. Oper. (light)	51.00	408.00	77.15	617.20		
1 Air Compressor, 250 cfm		181.20		199.32		
1 Sandblaster, Portable, 3 C.F.		21.40		23.54		
1 Set Sand Blasting Accessories		14.45		15.90	6.78	7.46
32 L.H., Daily Totals		$1605.45		$2343.16	$50.17	$73.22

Crew J-7

	Bare Costs Hr.	Daily	Incl. Subs O&P Hr.	Daily	Cost Per L.H. Bare Costs	Incl. O&P
2 Painters	$41.70	$667.20	$62.95	$1007.20	$41.70	$62.95
1 Floor Belt Sander		21.30		23.43		
1 Floor Sanding Edger		13.85		15.23	2.20	2.42
16 L.H., Daily Totals		$702.35		$1045.87	$43.90	$65.37

Crew K-1

	Bare Costs Hr.	Daily	Incl. Subs O&P Hr.	Daily	Cost Per L.H. Bare Costs	Incl. O&P
1 Carpenter	$49.25	$394.00	$75.45	$603.60	$46.70	$71.03
1 Truck Driver (light)	44.15	353.20	66.60	532.80		
1 Flatbed Truck, Gas, 3 Ton		270.20		297.22	16.89	18.58
16 L.H., Daily Totals		$1017.40		$1433.62	$63.59	$89.60

Crew K-2

	Bare Costs Hr.	Daily	Incl. Subs O&P Hr.	Daily	Cost Per L.H. Bare Costs	Incl. O&P
1 Struc. Steel Foreman (outside)	$56.30	$450.40	$93.85	$750.80	$51.58	$83.65
1 Struc. Steel Worker	54.30	434.40	90.50	724.00		
1 Truck Driver (light)	44.15	353.20	66.60	532.80		
1 Flatbed Truck, Gas, 3 Ton		270.20		297.22	11.26	12.38
24 L.H., Daily Totals		$1508.20		$2304.82	$62.84	$96.03

Crew L-1

	Bare Costs Hr.	Daily	Incl. Subs O&P Hr.	Daily	Cost Per L.H. Bare Costs	Incl. O&P
1 Electrician	$56.60	$452.80	$84.80	$678.40	$59.20	$89.03
1 Plumber	61.80	494.40	93.25	746.00		
16 L.H., Daily Totals		$947.20		$1424.40	$59.20	$89.03

Crew L-2

	Bare Costs Hr.	Daily	Incl. Subs O&P Hr.	Daily	Cost Per L.H. Bare Costs	Incl. O&P
1 Carpenter	$49.25	$394.00	$75.45	$603.60	$43.17	$66.50
1 Carpenter Helper	37.10	296.80	57.55	460.40		
16 L.H., Daily Totals		$690.80		$1064.00	$43.17	$66.50

Crew L-3

	Bare Costs Hr.	Daily	Incl. Subs O&P Hr.	Daily	Cost Per L.H. Bare Costs	Incl. O&P
1 Carpenter	$49.25	$394.00	$75.45	$603.60	$53.29	$81.14
.5 Electrician	56.60	226.40	84.80	339.20		
.5 Sheet Metal Worker	58.05	232.20	88.85	355.40		
16 L.H., Daily Totals		$852.60		$1298.20	$53.29	$81.14

Crew L-3A

	Bare Costs Hr.	Daily	Incl. Subs O&P Hr.	Daily	Cost Per L.H. Bare Costs	Incl. O&P
1 Carpenter Foreman (outside)	$51.25	$410.00	$78.50	$628.00	$53.52	$81.95
.5 Sheet Metal Worker	58.05	232.20	88.85	355.40		
12 L.H., Daily Totals		$642.20		$983.40	$53.52	$81.95

Crew L-4

	Bare Costs Hr.	Daily	Incl. Subs O&P Hr.	Daily	Cost Per L.H. Bare Costs	Incl. O&P
2 Skilled Workers	$51.45	$823.20	$79.20	$1267.20	$46.67	$71.98
1 Helper	37.10	296.80	57.55	460.40		
24 L.H., Daily Totals		$1120.00		$1727.60	$46.67	$71.98

Crew L-5

	Bare Costs Hr.	Daily	Incl. Subs O&P Hr.	Daily	Cost Per L.H. Bare Costs	Incl. O&P
1 Struc. Steel Foreman (outside)	$56.30	$450.40	$93.85	$750.80	$54.79	$90.09
5 Struc. Steel Workers	54.30	2172.00	90.50	3620.00		
1 Equip. Oper. (crane)	55.70	445.60	84.25	674.00		
1 Hyd. Crane, 25 Ton		673.80		741.18	12.03	13.24
56 L.H., Daily Totals		$3741.80		$5785.98	$66.82	$103.32

Crew L-5A

	Bare Costs Hr.	Daily	Incl. Subs O&P Hr.	Daily	Cost Per L.H. Bare Costs	Incl. O&P
1 Struc. Steel Foreman (outside)	$56.30	$450.40	$93.85	$750.80	$55.15	$89.78
2 Structural Steel Workers	54.30	868.80	90.50	1448.00		
1 Equip. Oper. (crane)	55.70	445.60	84.25	674.00		
1 S.P. Crane, 4x4, 25 Ton		605.60		666.16	18.93	20.82
32 L.H., Daily Totals		$2370.40		$3538.96	$74.08	$110.59

Crew L-5B

	Bare Costs Hr.	Daily	Incl. Subs O&P Hr.	Daily	Cost Per L.H. Bare Costs	Incl. O&P
1 Struc. Steel Foreman (outside)	$56.30	$450.40	$93.85	$750.80	$56.26	$87.69
2 Structural Steel Workers	54.30	868.80	90.50	1448.00		
2 Electricians	56.60	905.60	84.80	1356.80		
2 Steamfitters/Pipefitters	62.25	996.00	93.95	1503.20		
1 Equip. Oper. (crane)	55.70	445.60	84.25	674.00		
1 Equip. Oper. (oiler)	48.00	384.00	72.60	580.80		
1 Hyd. Crane, 80 Ton		1616.00		1777.60	22.44	24.69
72 L.H., Daily Totals		$5666.40		$8091.20	$78.70	$112.38

Crew L-6

	Bare Costs Hr.	Daily	Incl. Subs O&P Hr.	Daily	Cost Per L.H. Bare Costs	Incl. O&P
1 Plumber	$61.80	$494.40	$93.25	$746.00	$60.07	$90.43
.5 Electrician	56.60	226.40	84.80	339.20		
12 L.H., Daily Totals		$720.80		$1085.20	$60.07	$90.43

Crew L-7

	Bare Costs Hr.	Daily	Incl. Subs O&P Hr.	Daily	Cost Per L.H. Bare Costs	Incl. O&P
2 Carpenters	$49.25	$788.00	$75.45	$1207.20	$47.41	$72.37
1 Building Laborer	39.15	313.20	60.00	480.00		
.5 Electrician	56.60	226.40	84.80	339.20		
28 L.H., Daily Totals		$1327.60		$2026.40	$47.41	$72.37

Crew L-8

	Bare Costs Hr.	Daily	Incl. Subs O&P Hr.	Daily	Cost Per L.H. Bare Costs	Incl. O&P
2 Carpenters	$49.25	$788.00	$75.45	$1207.20	$51.76	$79.01
.5 Plumber	61.80	247.20	93.25	373.00		
20 L.H., Daily Totals		$1035.20		$1580.20	$51.76	$79.01

Crew L-9

	Bare Costs Hr.	Daily	Incl. Subs O&P Hr.	Daily	Cost Per L.H. Bare Costs	Incl. O&P
1 Labor Foreman (inside)	$39.65	$317.20	$60.75	$486.00	$44.57	$69.70
2 Building Laborers	39.15	626.40	60.00	960.00		
1 Struc. Steel Worker	54.30	434.40	90.50	724.00		
.5 Electrician	56.60	226.40	84.80	339.20		
36 L.H., Daily Totals		$1604.40		$2509.20	$44.57	$69.70

Crew No.	Bare Costs		Incl. Subs O&P		Cost Per Labor-Hour	
	Hr.	Daily	Hr.	Daily	Bare Costs	Incl. O&P
Crew L-10					**$55.43**	**$89.53**
1 Struc. Steel Foreman (outside)	$56.30	$450.40	$93.85	$750.80		
1 Structural Steel Worker	54.30	434.40	90.50	724.00		
1 Equip. Oper. (crane)	55.70	445.60	84.25	674.00		
1 Hyd. Crane, 12 Ton		572.80		630.08	23.87	26.25
24 L.H., Daily Totals		$1903.20		$2778.88	$79.30	$115.79
Crew L-11	Hr.	Daily	Hr.	Daily	Bare Costs	Incl. O&P
					$46.25	**$71.70**
2 Wreckers	$39.15	$626.40	$62.70	$1003.20		
1 Equip. Oper. (crane)	55.70	445.60	84.25	674.00		
1 Equip. Oper. (light)	51.00	408.00	77.15	617.20		
1 Hyd. Excavator, 2.5 C.Y.		1536.00		1689.60		
1 Loader, Skid Steer, 78 H.P.		309.40		340.34	57.67	63.44
32 L.H., Daily Totals		$3325.40		$4324.34	$103.92	$135.14
Crew M-1	Hr.	Daily	Hr.	Daily	Bare Costs	Incl. O&P
					$77.71	**$116.04**
3 Elevator Constructors	$81.80	$1963.20	$122.15	$2931.60		
1 Elevator Apprentice	65.45	523.60	97.70	781.60		
5 Hand Tools		50.00		55.00	1.56	1.72
32 L.H., Daily Totals		$2536.80		$3768.20	$79.28	$117.76
Crew M-3	Hr.	Daily	Hr.	Daily	Bare Costs	Incl. O&P
					$60.80	**$91.27**
1 Electrician Foreman (outside)	$58.60	$468.80	$87.80	$702.40		
1 Common Laborer	39.15	313.20	60.00	480.00		
.25 Equipment Operator (med.)	53.55	107.10	81.00	162.00		
1 Elevator Constructor	81.80	654.40	122.15	977.20		
1 Elevator Apprentice	65.45	523.60	97.70	781.60		
.25 S.P. Crane, 4x4, 20 Ton		138.45		152.29	4.07	4.48
34 L.H., Daily Totals		$2205.55		$3255.49	$64.87	$95.75
Crew M-4	Hr.	Daily	Hr.	Daily	Bare Costs	Incl. O&P
					$60.21	**$90.41**
1 Electrician Foreman (outside)	$58.60	$468.80	$87.80	$702.40		
1 Common Laborer	39.15	313.20	60.00	480.00		
.25 Equipment Operator, Crane	55.70	111.40	84.25	168.50		
.25 Equip. Oper. (oiler)	48.00	96.00	72.60	145.20		
1 Elevator Constructor	81.80	654.40	122.15	977.20		
1 Elevator Apprentice	65.45	523.60	97.70	781.60		
.25 S.P. Crane, 4x4, 40 Ton		194.05		213.46	5.39	5.93
36 L.H., Daily Totals		$2361.45		$3468.36	$65.60	$96.34
Crew Q-1	Hr.	Daily	Hr.	Daily	Bare Costs	Incl. O&P
					$55.63	**$83.92**
1 Plumber	$61.80	$494.40	$93.25	$746.00		
1 Plumber Apprentice	49.45	395.60	74.60	596.80		
16 L.H., Daily Totals		$890.00		$1342.80	$55.63	$83.92
Crew Q-1A	Hr.	Daily	Hr.	Daily	Bare Costs	Incl. O&P
					$62.20	**$93.85**
.25 Plumber Foreman (outside)	$63.80	$127.60	$96.25	$192.50		
1 Plumber	61.80	494.40	93.25	746.00		
10 L.H., Daily Totals		$622.00		$938.50	$62.20	$93.85
Crew Q-1C	Hr.	Daily	Hr.	Daily	Bare Costs	Incl. O&P
					$54.93	**$82.95**
1 Plumber	$61.80	$494.40	$93.25	$746.00		
1 Plumber Apprentice	49.45	395.60	74.60	596.80		
1 Equip. Oper. (medium)	53.55	428.40	81.00	648.00		
1 Trencher, Chain Type, 8' D		2460.00		2706.00	102.50	112.75
24 L.H., Daily Totals		$3778.40		$4696.80	$157.43	$195.70
Crew Q-2	Hr.	Daily	Hr.	Daily	Bare Costs	Incl. O&P
					$57.68	**$87.03**
2 Plumbers	$61.80	$988.80	$93.25	$1492.00		
1 Plumber Apprentice	49.45	395.60	74.60	596.80		
24 L.H., Daily Totals		$1384.40		$2088.80	$57.68	$87.03

Crew No.	Bare Costs		Incl. Subs O&P		Cost Per Labor-Hour	
	Hr.	Daily	Hr.	Daily	Bare Costs	Incl. O&P
Crew Q-3					**$58.84**	**$88.78**
1 Plumber Foreman (inside)	$62.30	$498.40	$94.00	$752.00		
2 Plumbers	61.80	988.80	93.25	1492.00		
1 Plumber Apprentice	49.45	395.60	74.60	596.80		
32 L.H., Daily Totals		$1882.80		$2840.80	$58.84	$88.78
Crew Q-4	Hr.	Daily	Hr.	Daily	Bare Costs	Incl. O&P
					$58.84	**$88.78**
1 Plumber Foreman (inside)	$62.30	$498.40	$94.00	$752.00		
1 Plumber	61.80	494.40	93.25	746.00		
1 Welder (plumber)	61.80	494.40	93.25	746.00		
1 Plumber Apprentice	49.45	395.60	74.60	596.80		
1 Welder, Electric, 300 amp		58.15		63.97	1.82	2.00
32 L.H., Daily Totals		$1940.95		$2904.76	$60.65	$90.77
Crew Q-5	Hr.	Daily	Hr.	Daily	Bare Costs	Incl. O&P
					$56.02	**$84.55**
1 Steamfitter	$62.25	$498.00	$93.95	$751.60		
1 Steamfitter Apprentice	49.80	398.40	75.15	601.20		
16 L.H., Daily Totals		$896.40		$1352.80	$56.02	$84.55
Crew Q-6	Hr.	Daily	Hr.	Daily	Bare Costs	Incl. O&P
					$58.10	**$87.68**
2 Steamfitters	$62.25	$996.00	$93.95	$1503.20		
1 Steamfitter Apprentice	49.80	398.40	75.15	601.20		
24 L.H., Daily Totals		$1394.40		$2104.40	$58.10	$87.68
Crew Q-7	Hr.	Daily	Hr.	Daily	Bare Costs	Incl. O&P
					$59.26	**$89.44**
1 Steamfitter Foreman (inside)	$62.75	$502.00	$94.70	$757.60		
2 Steamfitters	62.25	996.00	93.95	1503.20		
1 Steamfitter Apprentice	49.80	398.40	75.15	601.20		
32 L.H., Daily Totals		$1896.40		$2862.00	$59.26	$89.44
Crew Q-8	Hr.	Daily	Hr.	Daily	Bare Costs	Incl. O&P
					$59.26	**$89.44**
1 Steamfitter Foreman (inside)	$62.75	$502.00	$94.70	$757.60		
1 Steamfitter	62.25	498.00	93.95	751.60		
1 Welder (steamfitter)	62.25	498.00	93.95	751.60		
1 Steamfitter Apprentice	49.80	398.40	75.15	601.20		
1 Welder, Electric, 300 amp		58.15		63.97	1.82	2.00
32 L.H., Daily Totals		$1954.55		$2925.97	$61.08	$91.44
Crew Q-9	Hr.	Daily	Hr.	Daily	Bare Costs	Incl. O&P
					$52.25	**$79.97**
1 Sheet Metal Worker	$58.05	$464.40	$88.85	$710.80		
1 Sheet Metal Apprentice	46.45	371.60	71.10	568.80		
16 L.H., Daily Totals		$836.00		$1279.60	$52.25	$79.97
Crew Q-10	Hr.	Daily	Hr.	Daily	Bare Costs	Incl. O&P
					$54.18	**$82.93**
2 Sheet Metal Workers	$58.05	$928.80	$88.85	$1421.60		
1 Sheet Metal Apprentice	46.45	371.60	71.10	568.80		
24 L.H., Daily Totals		$1300.40		$1990.40	$54.18	$82.93
Crew Q-11	Hr.	Daily	Hr.	Daily	Bare Costs	Incl. O&P
					$55.27	**$84.61**
1 Sheet Metal Foreman (inside)	$58.55	$468.40	$89.65	$717.20		
2 Sheet Metal Workers	58.05	928.80	88.85	1421.60		
1 Sheet Metal Apprentice	46.45	371.60	71.10	568.80		
32 L.H., Daily Totals		$1768.80		$2707.60	$55.27	$84.61
Crew Q-12	Hr.	Daily	Hr.	Daily	Bare Costs	Incl. O&P
					$53.10	**$80.40**
1 Sprinkler Installer	$59.00	$472.00	$89.35	$714.80		
1 Sprinkler Apprentice	47.20	377.60	71.45	571.60		
16 L.H., Daily Totals		$849.60		$1286.40	$53.10	$80.40

For customer support on your Building Construction Costs with RSMeans Data, call 800.448.8182.

Crews - Standard

Crew Q-13

Crew Q-13	Hr.	Daily	Hr.	Daily	Bare Costs	Incl. O&P
1 Sprinkler Foreman (inside)	$59.50	$476.00	$90.10	$720.80	$56.17	$85.06
2 Sprinkler Installers	59.00	944.00	89.35	1429.60		
1 Sprinkler Apprentice	47.20	377.60	71.45	571.60		
32 L.H., Daily Totals		$1797.60		$2722.00	$56.17	$85.06

Crew Q-14

Crew Q-14	Hr.	Daily	Hr.	Daily	Bare Costs	Incl. O&P
1 Asbestos Worker	$55.05	$440.40	$85.55	$684.40	$49.55	$77.00
1 Asbestos Apprentice	44.05	352.40	68.45	547.60		
16 L.H., Daily Totals		$792.80		$1232.00	$49.55	$77.00

Crew Q-15

Crew Q-15	Hr.	Daily	Hr.	Daily	Bare Costs	Incl. O&P
1 Plumber	$61.80	$494.40	$93.25	$746.00	$55.63	$83.92
1 Plumber Apprentice	49.45	395.60	74.60	596.80		
1 Welder, Electric, 300 amp		58.15		63.97	3.63	4.00
16 L.H., Daily Totals		$948.15		$1406.77	$59.26	$87.92

Crew Q-16

Crew Q-16	Hr.	Daily	Hr.	Daily	Bare Costs	Incl. O&P
2 Plumbers	$61.80	$988.80	$93.25	$1492.00	$57.68	$87.03
1 Plumber Apprentice	49.45	395.60	74.60	596.80		
1 Welder, Electric, 300 amp		58.15		63.97	2.42	2.67
24 L.H., Daily Totals		$1442.55		$2152.76	$60.11	$89.70

Crew Q-17

Crew Q-17	Hr.	Daily	Hr.	Daily	Bare Costs	Incl. O&P
1 Steamfitter	$62.25	$498.00	$93.95	$751.60	$56.02	$84.55
1 Steamfitter Apprentice	49.80	398.40	75.15	601.20		
1 Welder, Electric, 300 amp		58.15		63.97	3.63	4.00
16 L.H., Daily Totals		$954.55		$1416.77	$59.66	$88.55

Crew Q-17A

Crew Q-17A	Hr.	Daily	Hr.	Daily	Bare Costs	Incl. O&P
1 Steamfitter	$62.25	$498.00	$93.95	$751.60	$55.92	$84.45
1 Steamfitter Apprentice	49.80	398.40	75.15	601.20		
1 Equip. Oper. (crane)	55.70	445.60	84.25	674.00		
1 Hyd. Crane, 12 Ton		572.80		630.08		
1 Welder, Electric, 300 amp		58.15		63.97	26.29	28.92
24 L.H., Daily Totals		$1972.95		$2720.84	$82.21	$113.37

Crew Q-18

Crew Q-18	Hr.	Daily	Hr.	Daily	Bare Costs	Incl. O&P
2 Steamfitters	$62.25	$996.00	$93.95	$1503.20	$58.10	$87.68
1 Steamfitter Apprentice	49.80	398.40	75.15	601.20		
1 Welder, Electric, 300 amp		58.15		63.97	2.42	2.67
24 L.H., Daily Totals		$1452.55		$2168.36	$60.52	$90.35

Crew Q-19

Crew Q-19	Hr.	Daily	Hr.	Daily	Bare Costs	Incl. O&P
1 Steamfitter	$62.25	$498.00	$93.95	$751.60	$56.22	$84.63
1 Steamfitter Apprentice	49.80	398.40	75.15	601.20		
1 Electrician	56.60	452.80	84.80	678.40		
24 L.H., Daily Totals		$1349.20		$2031.20	$56.22	$84.63

Crew Q-20

Crew Q-20	Hr.	Daily	Hr.	Daily	Bare Costs	Incl. O&P
1 Sheet Metal Worker	$58.05	$464.40	$88.85	$710.80	$53.12	$80.94
1 Sheet Metal Apprentice	46.45	371.60	71.10	568.80		
.5 Electrician	56.60	226.40	84.80	339.20		
20 L.H., Daily Totals		$1062.40		$1618.80	$53.12	$80.94

Crew Q-21

Crew Q-21	Hr.	Daily	Hr.	Daily	Bare Costs	Incl. O&P
2 Steamfitters	$62.25	$996.00	$93.95	$1503.20	$57.73	$86.96
1 Steamfitter Apprentice	49.80	398.40	75.15	601.20		
1 Electrician	56.60	452.80	84.80	678.40		
32 L.H., Daily Totals		$1847.20		$2782.80	$57.73	$86.96

Crew Q-22

Crew Q-22	Hr.	Daily	Hr.	Daily	Bare Costs	Incl. O&P
1 Plumber	$61.80	$494.40	$93.25	$746.00	$55.63	$83.92
1 Plumber Apprentice	49.45	395.60	74.60	596.80		
1 Hyd. Crane, 12 Ton		572.80		630.08	35.80	39.38
16 L.H., Daily Totals		$1462.80		$1972.88	$91.42	$123.31

Crew Q-22A

Crew Q-22A	Hr.	Daily	Hr.	Daily	Bare Costs	Incl. O&P
1 Plumber	$61.80	$494.40	$93.25	$746.00	$51.52	$78.03
1 Plumber Apprentice	49.45	395.60	74.60	596.80		
1 Laborer	39.15	313.20	60.00	480.00		
1 Equip. Oper. (crane)	55.70	445.60	84.25	674.00		
1 Hyd. Crane, 12 Ton		572.80		630.08	17.90	19.69
32 L.H., Daily Totals		$2221.60		$3126.88	$69.42	$97.72

Crew Q-23

Crew Q-23	Hr.	Daily	Hr.	Daily	Bare Costs	Incl. O&P
1 Plumber Foreman (outside)	$63.80	$510.40	$96.25	$770.00	$59.72	$90.17
1 Plumber	61.80	494.40	93.25	746.00		
1 Equip. Oper. (medium)	53.55	428.40	81.00	648.00		
1 Lattice Boom Crane, 20 Ton		1075.00		1182.50	44.79	49.27
24 L.H., Daily Totals		$2508.20		$3346.50	$104.51	$139.44

Crew R-1

Crew R-1	Hr.	Daily	Hr.	Daily	Bare Costs	Incl. O&P
1 Electrician Foreman	$57.10	$456.80	$85.55	$684.40	$52.92	$79.28
3 Electricians	56.60	1358.40	84.80	2035.20		
2 Electrician Apprentices	45.30	724.80	67.85	1085.60		
48 L.H., Daily Totals		$2540.00		$3805.20	$52.92	$79.28

Crew R-1A

Crew R-1A	Hr.	Daily	Hr.	Daily	Bare Costs	Incl. O&P
1 Electrician	$56.60	$452.80	$84.80	$678.40	$50.95	$76.33
1 Electrician Apprentice	45.30	362.40	67.85	542.80		
16 L.H., Daily Totals		$815.20		$1221.20	$50.95	$76.33

Crew R-1B

Crew R-1B	Hr.	Daily	Hr.	Daily	Bare Costs	Incl. O&P
1 Electrician	$56.60	$452.80	$84.80	$678.40	$49.07	$73.50
2 Electrician Apprentices	45.30	724.80	67.85	1085.60		
24 L.H., Daily Totals		$1177.60		$1764.00	$49.07	$73.50

Crew R-1C

Crew R-1C	Hr.	Daily	Hr.	Daily	Bare Costs	Incl. O&P
2 Electricians	$56.60	$905.60	$84.80	$1356.80	$50.95	$76.33
2 Electrician Apprentices	45.30	724.80	67.85	1085.60		
1 Portable cable puller, 8000 lb.		130.10		143.11	4.07	4.47
32 L.H., Daily Totals		$1760.50		$2585.51	$55.02	$80.80

Crew R-2

Crew R-2	Hr.	Daily	Hr.	Daily	Bare Costs	Incl. O&P
1 Electrician Foreman	$57.10	$456.80	$85.55	$684.40	$53.31	$79.99
3 Electricians	56.60	1358.40	84.80	2035.20		
2 Electrician Apprentices	45.30	724.80	67.85	1085.60		
1 Equip. Oper. (crane)	55.70	445.60	84.25	674.00		
1 S.P. Crane, 4x4, 5 Ton		276.00		303.60	4.93	5.42
56 L.H., Daily Totals		$3261.60		$4782.80	$58.24	$85.41

Crew R-3

Crew R-3	Hr.	Daily	Hr.	Daily	Bare Costs	Incl. O&P
1 Electrician Foreman	$57.10	$456.80	$85.55	$684.40	$56.62	$84.99
1 Electrician	56.60	452.80	84.80	678.40		
.5 Equip. Oper. (crane)	55.70	222.80	84.25	337.00		
.5 S.P. Crane, 4x4, 5 Ton		138.00		151.80	6.90	7.59
20 L.H., Daily Totals		$1270.40		$1851.60	$63.52	$92.58

For customer support on your Building Construction Costs with RSMeans Data, call 800.448.8182.

Crew No.	Bare Costs		Incl. Subs O&P		Cost Per Labor-Hour	
Crew R-4	**Hr.**	**Daily**	**Hr.**	**Daily**	**Bare Costs**	**Incl. O&P**
1 Struc. Steel Foreman (outside)	$56.30	$450.40	$93.85	$750.80	$55.16	$90.03
3 Struc. Steel Workers	54.30	1303.20	90.50	2172.00		
1 Electrician	56.60	452.80	84.80	678.40		
1 Welder, Gas Engine, 300 amp		129.80		142.78	3.25	3.57
40 L.H., Daily Totals		$2336.20		$3743.98	$58.41	$93.60
Crew R-5	**Hr.**	**Daily**	**Hr.**	**Daily**	**Bare Costs**	**Incl. O&P**
1 Electrician Foreman	$57.10	$456.80	$85.55	$684.40	$49.55	$74.96
4 Electrician Linemen	56.60	1811.20	84.80	2713.60		
2 Electrician Operators	56.60	905.60	84.80	1356.80		
4 Electrician Groundmen	37.10	1187.20	57.55	1841.60		
1 Crew Truck		185.60		204.16		
1 Flatbed Truck, 20,000 GVW		232.00		255.20		
1 Pickup Truck, 3/4 Ton		129.40		142.34		
.2 Hyd. Crane, 55 Ton		215.20		236.72		
.2 Hyd. Crane, 12 Ton		114.56		126.02		
.2 Earth Auger, Truck-Mtd.		80.68		88.75		
1 Tractor w/Winch		407.80		448.58	15.51	17.07
88 L.H., Daily Totals		$5726.04		$8098.16	$65.07	$92.02
Crew R-6	**Hr.**	**Daily**	**Hr.**	**Daily**	**Bare Costs**	**Incl. O&P**
1 Electrician Foreman	$57.10	$456.80	$85.55	$684.40	$49.55	$74.96
4 Electrician Linemen	56.60	1811.20	84.80	2713.60		
2 Electrician Operators	56.60	905.60	84.80	1356.80		
4 Electrician Groundmen	37.10	1187.20	57.55	1841.60		
1 Crew Truck		185.60		204.16		
1 Flatbed Truck, 20,000 GVW		232.00		255.20		
1 Pickup Truck, 3/4 Ton		129.40		142.34		
.2 Hyd. Crane, 55 Ton		215.20		236.72		
.2 Hyd. Crane, 12 Ton		114.56		126.02		
.2 Earth Auger, Truck-Mtd.		80.68		88.75		
1 Tractor w/Winch		407.80		448.58		
3 Cable Trailers		622.95		685.25		
.5 Tensioning Rig		206.07		226.68		
.5 Cable Pulling Rig		1187.50		1306.25	38.43	42.27
88 L.H., Daily Totals		$7742.56		$10316.34	$87.98	$117.23
Crew R-7	**Hr.**	**Daily**	**Hr.**	**Daily**	**Bare Costs**	**Incl. O&P**
1 Electrician Foreman	$57.10	$456.80	$85.55	$684.40	$40.43	$62.22
5 Electrician Groundmen	37.10	1484.00	57.55	2302.00		
1 Crew Truck		185.60		204.16	3.87	4.25
48 L.H., Daily Totals		$2126.40		$3190.56	$44.30	$66.47
Crew R-8	**Hr.**	**Daily**	**Hr.**	**Daily**	**Bare Costs**	**Incl. O&P**
1 Electrician Foreman	$57.10	$456.80	$85.55	$684.40	$50.18	$75.84
3 Electrician Linemen	56.60	1358.40	84.80	2035.20		
2 Electrician Groundmen	37.10	593.60	57.55	920.80		
1 Pickup Truck, 3/4 Ton		129.40		142.34		
1 Crew Truck		185.60		204.16	6.56	7.22
48 L.H., Daily Totals		$2723.80		$3986.90	$56.75	$83.06
Crew R-9	**Hr.**	**Daily**	**Hr.**	**Daily**	**Bare Costs**	**Incl. O&P**
1 Electrician Foreman	$57.10	$456.80	$85.55	$684.40	$46.91	$71.27
1 Electrician Lineman	56.60	452.80	84.80	678.40		
2 Electrician Operators	56.60	905.60	84.80	1356.80		
4 Electrician Groundmen	37.10	1187.20	57.55	1841.60		
1 Pickup Truck, 3/4 Ton		129.40		142.34		
1 Crew Truck		185.60		204.16	4.92	5.41
64 L.H., Daily Totals		$3317.40		$4907.70	$51.83	$76.68

Crew No.	Bare Costs		Incl. Subs O&P		Cost Per Labor-Hour	
Crew R-10	**Hr.**	**Daily**	**Hr.**	**Daily**	**Bare Costs**	**Incl. O&P**
1 Electrician Foreman	$57.10	$456.80	$85.55	$684.40	$53.43	$80.38
4 Electrician Linemen	56.60	1811.20	84.80	2713.60		
1 Electrician Groundman	37.10	296.80	57.55	460.40		
1 Crew Truck		185.60		204.16		
3 Tram Cars		423.30		465.63	12.69	13.95
48 L.H., Daily Totals		$3173.70		$4528.19	$66.12	$94.34
Crew R-11	**Hr.**	**Daily**	**Hr.**	**Daily**	**Bare Costs**	**Incl. O&P**
1 Electrician Foreman	$57.10	$456.80	$85.55	$684.40	$54.05	$81.29
4 Electricians	56.60	1811.20	84.80	2713.60		
1 Equip. Oper. (crane)	55.70	445.60	84.25	674.00		
1 Common Laborer	39.15	313.20	60.00	480.00		
1 Crew Truck		185.60		204.16		
1 Hyd. Crane, 12 Ton		572.80		630.08	13.54	14.90
56 L.H., Daily Totals		$3785.20		$5386.24	$67.59	$96.18
Crew R-12	**Hr.**	**Daily**	**Hr.**	**Daily**	**Bare Costs**	**Incl. O&P**
1 Carpenter Foreman (inside)	$49.75	$398.00	$76.20	$609.60	$46.47	$71.77
4 Carpenters	49.25	1576.00	75.45	2414.40		
4 Common Laborers	39.15	1252.80	60.00	1920.00		
1 Equip. Oper. (medium)	53.55	428.40	81.00	648.00		
1 Steel Worker	54.30	434.40	90.50	724.00		
1 Dozer, 200 H.P.		1384.00		1522.40		
1 Pickup Truck, 3/4 Ton		129.40		142.34	17.20	18.92
88 L.H., Daily Totals		$5603.00		$7980.74	$63.67	$90.69
Crew R-13	**Hr.**	**Daily**	**Hr.**	**Daily**	**Bare Costs**	**Incl. O&P**
1 Electrician Foreman	$57.10	$456.80	$85.55	$684.40	$55.01	$82.59
3 Electricians	56.60	1358.40	84.80	2035.20		
.25 Equip. Oper. (crane)	55.70	111.40	84.25	168.50		
1 Equipment Oiler	48.00	384.00	72.60	580.80		
.25 Hydraulic Crane, 33 Ton		215.90		237.49	5.14	5.65
42 L.H., Daily Totals		$2526.50		$3706.39	$60.15	$88.25
Crew R-15	**Hr.**	**Daily**	**Hr.**	**Daily**	**Bare Costs**	**Incl. O&P**
1 Electrician Foreman	$57.10	$456.80	$85.55	$684.40	$55.75	$83.65
4 Electricians	56.60	1811.20	84.80	2713.60		
1 Equipment Oper. (light)	51.00	408.00	77.15	617.20		
1 Telescoping Boom Lift, to 40'		288.80		317.68	6.02	6.62
48 L.H., Daily Totals		$2964.80		$4332.88	$61.77	$90.27
Crew R-15A	**Hr.**	**Daily**	**Hr.**	**Daily**	**Bare Costs**	**Incl. O&P**
1 Electrician Foreman	$57.10	$456.80	$85.55	$684.40	$49.93	$75.38
2 Electricians	56.60	905.60	84.80	1356.80		
2 Common Laborers	39.15	626.40	60.00	960.00		
1 Equip. Oper. (light)	51.00	408.00	77.15	617.20		
1 Telescoping Boom Lift, to 40'		288.80		317.68	6.02	6.62
48 L.H., Daily Totals		$2685.60		$3936.08	$55.95	$82.00
Crew R-18	**Hr.**	**Daily**	**Hr.**	**Daily**	**Bare Costs**	**Incl. O&P**
.25 Electrician Foreman	$57.10	$114.20	$85.55	$171.10	$49.68	$74.43
1 Electrician	56.60	452.80	84.80	678.40		
2 Electrician Apprentices	45.30	724.80	67.85	1085.60		
26 L.H., Daily Totals		$1291.80		$1935.10	$49.68	$74.43
Crew R-19	**Hr.**	**Daily**	**Hr.**	**Daily**	**Bare Costs**	**Incl. O&P**
.5 Electrician Foreman	$57.10	$228.40	$85.55	$342.20	$56.70	$84.95
2 Electricians	56.60	905.60	84.80	1356.80		
20 L.H., Daily Totals		$1134.00		$1699.00	$56.70	$84.95

Crews - Standard

Crew No.	Bare Costs		Incl. Subs O & P		Cost Per Labor-Hour	

Crew R-21	Hr.	Daily	Hr.	Daily	Bare Costs	Incl. O&P
1 Electrician Foreman	$57.10	$456.80	$85.55	$684.40	$56.65	$84.89
3 Electricians	56.60	1358.40	84.80	2035.20		
.1 Equip. Oper. (medium)	53.55	42.84	81.00	64.80		
.1 S.P. Crane, 4x4, 25 Ton		60.56		66.62	1.85	2.03
32.8 L.H., Daily Totals		$1918.60		$2851.02	$58.49	$86.92

Crew R-22	Hr.	Daily	Hr.	Daily	Bare Costs	Incl. O&P
.66 Electrician Foreman	$57.10	$301.49	$85.55	$451.70	$51.82	$77.63
2 Electricians	56.60	905.60	84.80	1356.80		
2 Electrician Apprentices	45.30	724.80	67.85	1085.60		
37.28 L.H., Daily Totals		$1931.89		$2894.10	$51.82	$77.63

Crew R-30	Hr.	Daily	Hr.	Daily	Bare Costs	Incl. O&P
.25 Electrician Foreman (outside)	$58.60	$117.20	$87.80	$175.60	$46.02	$69.77
1 Electrician	56.60	452.80	84.80	678.40		
2 Laborers (Semi-Skilled)	39.15	626.40	60.00	960.00		
26 L.H., Daily Totals		$1196.40		$1814.00	$46.02	$69.77

Crew R-31	Hr.	Daily	Hr.	Daily	Bare Costs	Incl. O&P
1 Electrician	$56.60	$452.80	$84.80	$678.40	$56.60	$84.80
1 Core Drill, Electric, 2.5 H.P.		55.25		60.77	6.91	7.60
8 L.H., Daily Totals		$508.05		$739.17	$63.51	$92.40

Crew W-41E	Hr.	Daily	Hr.	Daily	Bare Costs	Incl. O&P
.5 Plumber Foreman (outside)	$63.80	$255.20	$96.25	$385.00	$53.14	$80.55
1 Plumber	61.80	494.40	93.25	746.00		
1 Laborer	39.15	313.20	60.00	480.00		
20 L.H., Daily Totals		$1062.80		$1611.00	$53.14	$80.55

Historical Cost Indexes

The table below lists both the RSMeans® historical cost index based on Jan. 1, 1993 = 100 as well as the computed value of an index based on Jan. 1, 2017 costs. Since the Jan. 1, 2017 figure is estimated, space is left to write in the actual index figures as they become available through the quarterly *RSMeans Construction Cost Indexes*.

To compute the actual index based on Jan. 1, 2017 = 100, divide the historical cost index for a particular year by the actual Jan. 1, 2017 construction cost index. Space has been left to advance the index figures as the year progresses.

Year	Historical Cost Index Jan. 1, 1993 = 100		Current Index Based on Jan. 1, 2017 = 100		Year	Historical Cost Index Jan. 1, 1993 = 100	Current Index Based on Jan. 1, 2017 = 100		Year	Historical Cost Index Jan. 1, 1993 = 100	Current Index Based on Jan. 1, 2017 = 100	
	Est.	Actual	Est.	Actual		Actual	Est.	Actual		Actual	Est.	Actual
Oct 2017*					July 2002	128.7	61.7		July 1984	82.0	39.3	
July 2017*					2001	125.1	60.0		1983	80.2	38.4	
April 2017*					2000	120.9	58.0		1982	76.1	36.5	
Jan 2017*	208.5		100.0	100.0	1999	117.6	56.4		1981	70.0	33.6	
July 2016		207.3	99.4		1998	115.1	55.2		1980	62.9	30.2	
2015		206.2	98.9		1997	112.8	54.1		1979	57.8	27.7	
2014		204.9	98.3		1996	110.2	52.9		1978	53.5	25.7	
2013		201.2	96.5		1995	107.6	51.6		1977	49.5	23.7	
2012		194.6	93.3		1994	104.4	50.1		1976	46.9	22.5	
2011		191.2	91.7		1993	101.7	48.8		1975	44.8	21.5	
2010		183.5	88.0		1992	99.4	47.7		1974	41.4	19.9	
2009		180.1	86.4		1991	96.8	46.4		1973	37.7	18.1	
2008		180.4	86.5		1990	94.3	45.2		1972	34.8	16.7	
2007		169.4	81.2		1989	92.1	44.2		1971	32.1	15.4	
2006		162.0	77.7		1988	89.9	43.1		1970	28.7	13.8	
2005		151.6	72.7		1987	87.7	42.1		1969	26.9	12.9	
2004		143.7	68.9		1986	84.2	40.4		1968	24.9	11.9	
2003		132.0	63.3		1985	82.6	39.6		1967	23.5	11.3	

Adjustments to Costs

The "Historical Cost Index" can be used to convert national average building costs at a particular time to the approximate building costs for some other time.

Example:

Estimate and compare construction costs for different years in the same city.

To estimate the national average construction cost of a building in 1970, knowing that it cost $900,000 in 2017:

INDEX in 1970 = 28.7

INDEX in 2017 = 208.5

Note: The city cost indexes for Canada can be used to convert U.S. national averages to local costs in Canadian dollars.

Example:

To estimate and compare the cost of a building in Toronto, ON in 2017 with the known cost of $600,000 (US$) in New York, NY in 2017:

INDEX Toronto = 110.6

INDEX New York = 134.6

$$\frac{\text{INDEX Toronto}}{\text{INDEX New York}} \times \text{Cost New York} = \text{Cost Toronto}$$

$$\frac{110.6}{134.6} \times \$600,000 = .822 \times \$600,000 = \$493,200$$

The construction cost of the building in Toronto is $493,200 (CN$).

Time Adjustment Using the Historical Cost Indexes:

$$\frac{\text{Index for Year A}}{\text{Index for Year B}} \times \text{Cost in Year B} = \text{Cost in Year A}$$

$$\frac{\text{INDEX 1970}}{\text{INDEX 2017}} \times \text{Cost 2017} = \text{Cost 1970}$$

$$\frac{28.7}{208.5} \times \$900,000 = .138 \times \$900,000 = \$124,200$$

The construction cost of the building in 1970 was $124,200.

*Historical Cost Index updates and other resources are provided on the following website: **http://info.thegordiangroup.com/RSMeans.html**

How to Use the City Cost Indexes

What you should know before you begin

RSMeans City Cost Indexes (CCI) are an extremely useful tool for when you want to compare costs from city to city and region to region.

This publication contains average construction cost indexes for 731 U.S. and Canadian cities covering over 930 three-digit zip code locations, as listed directly under each city.

Keep in mind that a City Cost Index number is a percentage ratio of a specific city's cost to the national average cost of the same item at a stated time period.

In other words, these index figures represent relative construction factors (or, if you prefer, multipliers) for material and installation costs, as well as the weighted average for Total In Place costs for each CSI MasterFormat division. Installation costs include both labor and equipment rental costs. When estimating equipment rental rates only for a specific location, use 01 54 33 EQUIPMENT RENTAL COSTS in the Reference Section.

The 30 City Average Index is the average of 30 major U.S. cities and serves as a national average.

Index figures for both material and installation are based on the 30 major city average of 100 and represent the cost relationship as of July 1, 2016. The index for each division is computed from representative material and labor quantities for that division. The weighted average for each city is a weighted total of the components listed above it. It does not include relative productivity between trades or cities.

As changes occur in local material prices, labor rates, and equipment rental rates (including fuel costs), the impact of these changes should be accurately measured by the change in the City Cost Index for each particular city (as compared to the 30 city average).

Therefore, if you know (or have estimated) building costs in one city today, you can easily convert those costs to expected building costs in another city.

In addition, by using the Historical Cost Index, you can easily convert national average building costs at a particular time to the approximate building costs for some other time. The City Cost Indexes can then be applied to calculate the costs for a particular city.

Quick calculations

Location Adjustment Using the City Cost Indexes:

$$\frac{\text{Index for City A}}{\text{Index for City B}} \times \text{Cost in City B} = \text{Cost in City A}$$

Time Adjustment for the National Average Using the Historical Cost Index:

$$\frac{\text{Index for Year A}}{\text{Index for Year B}} \times \text{Cost in Year B} = \text{Cost in Year A}$$

Adjustment from the National Average:

$$\frac{\text{Index for City A}}{100} \times \text{National Average Cost} = \text{Cost in City A}$$

Since each of the other RSMeans data sets contains many different items, any *one* item multiplied by the particular city index may give incorrect results. However, the larger the number of items compiled, the closer the results should be to actual costs for that particular city.

The City Cost Indexes for Canadian cities are calculated using Canadian material and equipment prices and labor rates in Canadian dollars. Therefore, indexes for Canadian cities can be used to convert U.S. national average prices to local costs in Canadian dollars.

How to use this section

1. Compare costs from city to city.

In using the RSMeans Indexes, remember that an index number is not a fixed number but a ratio: It's a percentage ratio of a building component's cost at any stated time to the national average cost of that same component at the same time period. Put in the form of an equation:

$$\frac{\text{Specific City Cost}}{\text{National Average Cost}} \times 100 = \text{City Index Number}$$

Therefore, when making cost comparisons between cities, do not subtract one city's index number from the index number of another city and read the result as a percentage difference. Instead, divide one city's index number by that of the other city. The resulting number may then be used as a multiplier to calculate cost differences from city to city.

The formula used to find cost differences between cities for the purpose of comparison is as follows:

$$\frac{\text{City A Index}}{\text{City B Index}} \times \text{City B Cost (Known)} = \text{City A Cost (Unknown)}$$

In addition, you can use RSMeans CCI to calculate and compare costs division by division between cities using the same basic formula. (Just be sure that you're comparing similar divisions.)

2. Compare a specific city's construction costs with the national average.

When you're studying construction location feasibility, it's advisable to compare a prospective project's cost index with an index of the national average cost.

For example, divide the weighted average index of construction costs of a specific city by that of the 30 City Average, which = 100.

$$\frac{\text{City Index}}{100} = \% \text{ of National Average}$$

As a result, you get a ratio that indicates the relative cost of construction in that city in comparison with the national average.

3. Convert U.S. national average to actual costs in Canadian City.

$$\frac{\text{Index for Canadian City}}{100} \times \text{National Average Cost} = \text{Cost in Canadian City in \$ CAN}$$

For customer support on your Building Construction Costs with RSMeans Data, call 800.448.8182.

761

4. Adjust construction cost data based on a national average.
When you use a source of construction cost data which is based on a national average (such as RSMeans cost data), it is necessary to adjust those costs to a specific location.

$$\frac{\text{City Index}}{100} \times \frac{\text{Cost Based on}}{\text{National Average Costs}} = \frac{\text{City Cost}}{\text{(Unknown)}}$$

5. When applying the City Cost Indexes to demolition projects, use the appropriate division installation index. For example, for removal of existing doors and windows, use the Division 8 (Openings) index.

What you might like to know about how we developed the Indexes

The information presented in the CCI is organized according to the Construction Specifications Institute (CSI) MasterFormat 2014 classification system.

To create a reliable index, RSMeans researched the building type most often constructed in the United States and Canada. Because it was concluded that no one type of building completely represented the building construction industry, nine different types of buildings were combined to create a composite model.

The exact material, labor, and equipment quantities are based on detailed analyses of these nine building types, and then each quantity is weighted in proportion to expected usage. These various material items, labor hours, and equipment rental rates are thus combined to form a composite building representing as closely as possible the actual usage of materials, labor, and equipment in the North American building construction industry.

The following structures were chosen to make up that composite model:

1. Factory, 1 story
2. Office, 2–4 stories
3. Store, Retail
4. Town Hall, 2–3 stories
5. High School, 2–3 stories
6. Hospital, 4–8 stories
7. Garage, Parking
8. Apartment, 1–3 stories
9. Hotel/Motel, 2–3 stories

For the purposes of ensuring the timeliness of the data, the components of the index for the composite model have been streamlined. They currently consist of:

- specific quantities of 66 commonly used construction materials;
- specific labor-hours for 21 building construction trades; and
- specific days of equipment rental for 6 types of construction equipment (normally used to install the 66 material items by the 21 trades.) Fuel costs and routine maintenance costs are included in the equipment cost.

Material and equipment price quotations are gathered quarterly from cities in the United States and Canada. These prices and the latest negotiated labor wage rates for 21 different building trades are used to compile the quarterly update of the City Cost Index.

The 30 major U.S. cities used to calculate the national average are:

Atlanta, GA
Baltimore, MD
Boston, MA
Buffalo, NY
Chicago, IL
Cincinnati, OH
Cleveland, OH
Columbus, OH
Dallas, TX
Denver, CO
Detroit, MI
Houston, TX
Indianapolis, IN
Kansas City, MO
Los Angeles, CA

Memphis, TN
Milwaukee, WI
Minneapolis, MN
Nashville, TN
New Orleans, LA
New York, NY
Philadelphia, PA
Phoenix, AZ
Pittsburgh, PA
St. Louis, MO
San Antonio, TX
San Diego, CA
San Francisco, CA
Seattle, WA
Washington, DC

What the CCI does not indicate

The weighted average for each city is a total of the divisional components weighted to reflect typical usage. It does not include the productivity variations between trades or cities.

In addition, the CCI does not take into consideration factors such as the following:

- managerial efficiency
- competitive conditions
- automation
- restrictive union practices
- unique local requirements
- regional variations due to specific building codes

DIVISION		UNITED STATES 30 CITY AVERAGE			ALABAMA ANNISTON 362			BIRMINGHAM 350 - 352			BUTLER 369			DECATUR 356			DOTHAN 363		
		MAT.	INST.	TOTAL	MAT.	INST.	TOTAL	MAT.	INST.	TOTAL	MAT.	INST.	TOTAL	MAT.	INST.	TOTAL	MAT.	INST.	TOTAL
015433	CONTRACTOR EQUIPMENT		100.0	100.0		106.4	106.4		106.6	106.6		103.9	103.9		106.4	106.4		103.9	103.9
0241, 31 - 34	SITE & INFRASTRUCTURE, DEMOLITION	100.0	100.0	100.0	86.9	94.1	91.9	94.4	94.3	94.4	100.6	89.4	92.8	87.6	93.5	91.7	98.2	89.8	92.4
0310	Concrete Forming & Accessories	100.0	100.0	100.0	90.2	68.6	71.5	94.4	69.1	72.5	86.6	69.1	71.5	97.0	64.3	68.7	95.6	70.5	73.9
0320	Concrete Reinforcing	100.0	100.0	100.0	86.2	72.4	79.2	95.7	72.2	83.7	91.2	71.9	81.4	89.5	67.9	78.5	91.2	71.9	81.4
0330	Cast-in-Place Concrete	100.0	100.0	100.0	88.4	69.0	81.0	97.9	69.3	87.0	86.2	69.3	79.7	96.1	68.5	85.6	86.2	69.4	79.8
03	CONCRETE	100.0	100.0	100.0	93.5	71.1	83.2	93.3	71.4	83.2	94.5	71.4	83.8	93.2	68.2	81.6	93.8	72.1	83.7
04	MASONRY	100.0	100.0	100.0	94.0	73.4	81.3	91.5	68.0	77.0	98.9	73.5	83.2	89.4	72.4	78.9	100.3	73.5	83.8
05	METALS	100.0	100.0	100.0	99.2	96.6	98.4	98.9	96.6	98.2	98.1	97.3	97.9	101.3	94.6	99.3	98.2	97.5	97.9
06	WOOD, PLASTICS & COMPOSITES	100.0	100.0	100.0	89.2	69.2	78.2	94.7	69.2	80.7	84.0	69.2	75.9	102.4	63.4	81.0	95.6	70.8	82.0
07	THERMAL & MOISTURE PROTECTION	100.0	100.0	100.0	95.8	67.0	83.8	99.0	68.3	86.3	95.8	69.8	85.0	97.7	68.7	85.7	95.8	70.4	85.2
08	OPENINGS	100.0	100.0	100.0	95.9	68.4	89.7	103.1	68.4	95.2	96.0	68.3	89.7	108.0	63.8	98.0	96.0	69.1	89.9
0920	Plaster & Gypsum Board	100.0	100.0	100.0	88.0	68.8	75.3	91.0	68.8	76.3	85.1	68.8	74.3	93.9	62.9	73.3	95.3	70.5	78.8
0950, 0980	Ceilings & Acoustic Treatment	100.0	100.0	100.0	75.0	68.8	70.9	83.8	68.8	73.8	75.0	68.8	70.9	80.3	62.9	68.7	75.0	70.5	72.0
0960	Flooring	100.0	100.0	100.0	86.3	82.7	85.3	97.8	74.4	91.0	91.1	82.7	88.6	92.1	82.7	89.3	96.4	82.7	92.4
0970, 0990	Wall Finishes & Painting/Coating	100.0	100.0	100.0	89.3	61.1	72.7	91.0	61.1	73.4	89.3	49.9	66.1	86.5	63.8	73.1	89.3	80.3	84.0
09	FINISHES	100.0	100.0	100.0	81.7	70.9	75.8	89.1	69.3	78.2	84.6	69.7	76.4	85.7	67.6	75.8	87.4	73.9	80.0
COVERS	DIVS. 10 - 14, 25, 28, 41, 43, 44, 46	100.0	100.0	100.0	100.0	74.0	94.1	100.0	85.2	96.7	100.0	81.1	95.7	100.0	73.0	93.9	100.0	74.0	94.1
21, 22, 23	FIRE SUPPRESSION, PLUMBING & HVAC	100.0	100.0	100.0	100.8	52.1	79.2	100.0	62.5	83.4	97.9	67.8	84.6	100.0	66.6	85.2	97.9	66.4	83.9
26, 27, 3370	ELECTRICAL, COMMUNICATIONS & UTIL.	100.0	100.0	100.0	92.6	61.8	76.3	96.7	60.3	77.4	94.5	69.2	81.1	92.4	66.0	78.4	93.4	78.1	85.3
MF2014	WEIGHTED AVERAGE	100.0	100.0	100.0	95.7	70.0	84.3	97.5	71.7	86.1	96.0	74.3	86.4	97.5	72.3	86.3	96.1	75.7	87.1

DIVISION		ALABAMA EVERGREEN 364			GADSDEN 359			HUNTSVILLE 357 - 358			JASPER 355			MOBILE 365 - 366			MONTGOMERY 360 - 361		
		MAT.	INST.	TOTAL	MAT.	INST.	TOTAL	MAT.	INST.	TOTAL	MAT.	INST.	TOTAL	MAT.	INST.	TOTAL	MAT.	INST.	TOTAL
015433	CONTRACTOR EQUIPMENT		103.9	103.9		106.4	106.4		106.4	106.4		106.4	106.4		103.9	103.9		103.9	103.9
0241, 31 - 34	SITE & INFRASTRUCTURE, DEMOLITION	101.0	89.4	92.9	93.4	94.1	93.9	87.3	93.9	91.9	93.2	94.1	93.8	93.3	90.5	91.4	91.7	89.6	90.3
0310	Concrete Forming & Accessories	83.2	70.2	72.0	88.8	69.0	71.7	97.0	66.0	70.2	94.2	68.9	72.3	94.7	70.1	73.4	93.7	68.5	71.9
0320	Concrete Reinforcing	91.3	71.9	81.4	95.0	72.2	83.4	89.5	71.7	80.5	89.5	72.5	80.9	88.9	71.4	80.0	96.9	72.0	84.3
0330	Cast-in-Place Concrete	86.2	69.3	79.7	96.1	69.2	85.9	93.6	67.8	83.8	106.5	69.3	92.3	90.3	69.2	82.3	87.4	68.4	80.2
03	CONCRETE	94.9	71.9	84.2	97.7	71.4	85.5	92.0	69.5	81.5	101.2	71.4	87.4	90.2	71.8	81.6	88.3	70.9	80.2
04	MASONRY	98.9	73.5	83.2	87.8	67.7	75.5	90.7	65.4	75.1	85.4	73.7	78.2	97.1	65.4	77.6	94.3	64.0	75.6
05	METALS	98.1	97.2	97.9	99.3	96.6	98.5	101.3	96.2	99.8	99.2	96.8	98.5	100.1	96.5	99.0	98.9	96.5	98.2
06	WOOD, PLASTICS & COMPOSITES	80.5	70.8	75.2	93.2	69.2	80.0	102.4	66.4	82.6	99.6	69.2	82.9	94.2	70.8	81.4	91.5	69.2	79.3
07	THERMAL & MOISTURE PROTECTION	95.8	70.7	85.4	97.9	68.2	85.6	97.7	66.8	84.8	97.9	65.0	84.3	95.4	67.7	83.9	94.1	66.9	82.8
08	OPENINGS	96.0	69.1	89.9	104.4	68.2	96.2	107.7	67.2	98.5	104.3	68.4	96.2	98.8	68.9	92.1	97.0	68.4	90.5
0920	Plaster & Gypsum Board	84.4	70.5	75.1	86.8	68.8	74.8	93.9	65.9	75.3	90.7	68.8	76.2	92.1	70.5	77.7	89.6	68.8	75.3
0950, 0980	Ceilings & Acoustic Treatment	75.0	70.5	72.0	77.8	68.8	71.8	82.2	65.9	71.3	77.8	68.8	71.8	80.3	70.5	73.8	83.6	68.8	73.8
0960	Flooring	88.9	82.7	87.1	88.1	74.4	84.1	92.1	74.4	86.9	90.3	82.7	88.1	96.0	65.3	87.1	94.1	65.3	85.7
0970, 0990	Wall Finishes & Painting/Coating	89.3	49.8	66.1	86.5	61.1	71.6	86.5	63.9	73.2	86.5	61.1	71.6	92.6	48.9	66.9	89.7	61.1	72.9
09	FINISHES	83.9	70.6	76.6	83.5	69.2	75.7	86.1	67.2	75.8	84.6	71.0	77.2	87.5	66.9	76.2	87.0	67.0	76.1
COVERS	DIVS. 10 - 14, 25, 28, 41, 43, 44, 46	100.0	74.0	94.1	100.0	85.2	96.6	100.0	84.1	96.4	100.0	74.9	94.3	100.0	86.2	96.9	100.0	84.8	96.6
21, 22, 23	FIRE SUPPRESSION, PLUMBING & HVAC	97.9	67.8	84.6	102.2	64.1	85.3	100.0	62.8	83.5	102.2	62.5	84.6	99.8	65.1	84.5	99.9	65.7	84.8
26, 27, 3370	ELECTRICAL, COMMUNICATIONS & UTIL.	91.9	69.2	79.9	92.4	60.3	75.4	93.3	65.5	78.6	92.1	60.3	75.2	95.2	58.1	75.5	95.6	78.1	86.3
MF2014	WEIGHTED AVERAGE	95.7	74.3	86.2	97.5	71.9	86.2	97.4	71.5	86.0	97.9	72.1	86.5	96.6	71.2	85.4	95.8	73.6	86.0

DIVISION		ALABAMA PHENIX CITY 368			SELMA 367			TUSCALOOSA 354			ALASKA ANCHORAGE 995 - 996			FAIRBANKS 997			JUNEAU 998		
		MAT.	INST.	TOTAL	MAT.	INST.	TOTAL	MAT.	INST.	TOTAL	MAT.	INST.	TOTAL	MAT.	INST.	TOTAL	MAT.	INST.	TOTAL
015433	CONTRACTOR EQUIPMENT		103.9	103.9		103.9	103.9		106.4	106.4		117.0	117.0		117.0	117.0		117.0	117.0
0241, 31 - 34	SITE & INFRASTRUCTURE, DEMOLITION	104.8	89.8	94.3	98.1	89.7	92.3	87.9	94.1	92.2	127.7	132.1	130.8	124.9	132.1	129.9	143.2	132.1	135.4
0310	Concrete Forming & Accessories	90.2	69.2	72.0	87.7	69.2	71.7	97.0	69.1	72.8	126.6	120.4	121.2	133.9	120.5	122.3	132.8	120.4	122.1
0320	Concrete Reinforcing	91.2	66.5	78.7	91.2	72.5	81.7	89.5	72.2	80.7	148.5	117.0	132.5	150.8	117.0	133.6	145.2	117.0	130.9
0330	Cast-in-Place Concrete	86.2	69.3	79.7	86.2	69.3	79.7	97.5	69.2	86.7	114.0	119.0	115.9	119.4	119.3	119.3	126.6	119.0	123.7
03	CONCRETE	97.7	70.5	85.1	93.3	71.5	83.2	93.8	71.4	83.4	122.9	118.5	120.8	114.5	118.7	116.4	127.6	118.5	123.4
04	MASONRY	98.8	73.4	83.2	102.2	73.4	84.5	89.6	67.7	76.1	178.8	122.3	144.0	188.5	122.3	147.8	168.5	122.3	140.1
05	METALS	98.1	95.0	97.2	98.1	97.2	97.8	100.5	96.7	99.4	120.2	104.7	115.5	124.3	104.9	118.4	122.9	104.7	117.3
06	WOOD, PLASTICS & COMPOSITES	88.9	69.1	78.0	86.0	69.2	76.8	102.4	69.2	84.2	114.6	119.4	117.3	129.6	119.4	124.0	124.4	119.4	121.6
07	THERMAL & MOISTURE PROTECTION	96.2	69.7	85.2	95.7	69.8	84.9	97.7	68.2	85.5	166.3	117.1	145.8	181.7	118.2	155.4	187.0	117.1	158.0
08	OPENINGS	95.9	66.8	89.3	95.9	68.4	89.7	107.7	68.4	98.8	126.8	117.4	124.6	135.0	117.7	131.1	127.8	117.4	125.4
0920	Plaster & Gypsum Board	89.3	68.7	75.6	87.4	68.8	75.0	93.9	68.8	77.2	140.1	119.8	126.6	166.9	119.8	135.6	153.0	119.8	130.9
0950, 0980	Ceilings & Acoustic Treatment	75.0	68.7	70.8	75.0	68.8	70.9	82.2	68.8	73.3	123.9	119.8	121.1	127.0	119.8	122.2	136.2	119.8	125.3
0960	Flooring	93.0	82.7	90.0	91.5	82.7	89.0	92.1	74.4	86.9	125.9	123.2	125.1	124.3	123.2	124.0	133.2	123.2	130.3
0970, 0990	Wall Finishes & Painting/Coating	89.3	80.3	84.0	89.3	61.1	72.7	86.5	61.1	71.6	107.1	117.0	112.9	110.3	121.7	117.0	112.8	117.0	115.3
09	FINISHES	86.2	73.0	79.0	84.8	70.9	77.2	86.1	69.2	76.9	126.3	121.1	123.5	129.6	121.6	125.2	132.7	121.1	126.4
COVERS	DIVS. 10 - 14, 25, 28, 41, 43, 44, 46	100.0	85.2	96.6	100.0	74.5	94.2	100.0	85.2	96.6	100.0	110.4	102.4	100.0	110.4	102.4	100.0	110.4	102.4
21, 22, 23	FIRE SUPPRESSION, PLUMBING & HVAC	97.9	66.4	83.9	97.9	66.5	84.0	100.0	66.6	85.2	100.6	105.4	102.7	100.6	109.0	104.3	100.6	105.4	102.8
26, 27, 3370	ELECTRICAL, COMMUNICATIONS & UTIL.	94.0	68.7	80.6	93.1	78.1	85.2	92.9	60.3	75.6	112.9	112.6	112.7	120.2	112.6	116.2	108.1	112.6	110.5
MF2014	WEIGHTED AVERAGE	96.6	74.0	86.6	95.8	75.2	86.7	97.5	72.5	86.4	119.7	115.0	117.6	122.3	115.9	119.4	121.5	115.0	118.6

For customer support on your Building Construction Costs with RSMeans Data, call 800.448.8182.

763

DIVISION		ALASKA			ARIZONA														
		KETCHIKAN			CHAMBERS			FLAGSTAFF			GLOBE			KINGMAN			MESA/TEMPE		
		999			865			860			855			864			852		
		MAT.	INST.	TOTAL	MAT.	INST.	TOTAL	MAT.	INST.	TOTAL	MAT.	INST.	TOTAL	MAT.	INST.	TOTAL	MAT.	INST.	TOTAL
015433	CONTRACTOR EQUIPMENT		117.0	117.0		91.9	91.9		91.9	91.9		92.9	92.9		91.9	91.9		92.9	92.9
0241, 31 - 34	SITE & INFRASTRUCTURE, DEMOLITION	183.3	132.1	147.6	71.7	97.0	89.3	91.2	96.9	95.2	102.9	98.5	99.8	71.6	96.9	89.3	92.7	97.8	96.2
0310	Concrete Forming & Accessories	124.7	120.4	121.0	99.5	70.4	74.3	104.9	67.3	72.4	95.9	70.7	74.1	97.7	70.3	74.0	99.1	70.5	74.4
0320	Concrete Reinforcing	112.9	117.0	115.0	102.8	82.6	92.6	102.7	82.6	92.5	106.7	82.7	94.5	103.0	82.6	92.6	107.4	82.6	94.8
0330	Cast-in-Place Concrete	241.6	119.0	194.9	87.5	70.4	81.0	87.6	70.4	81.0	86.0	70.7	80.2	87.3	70.3	80.8	86.7	70.0	80.3
03	CONCRETE	195.2	118.5	159.6	94.6	72.5	84.3	115.4	71.1	94.8	107.5	72.9	91.5	94.3	72.4	84.1	98.2	72.4	86.3
04	MASONRY	196.3	122.3	150.8	97.6	59.4	74.1	97.7	59.4	74.1	110.2	59.5	79.0	97.6	59.4	74.1	110.4	59.2	78.9
05	METALS	124.5	104.7	118.4	100.7	74.3	92.6	101.2	74.1	92.9	96.1	76.9	90.2	101.4	74.0	93.0	96.4	74.9	89.8
06	WOOD, PLASTICS & COMPOSITES	120.3	119.4	119.8	100.5	72.7	85.3	106.3	68.6	85.6	95.5	72.9	83.1	98.7	72.7	83.1	98.7	72.9	84.5
07	THERMAL & MOISTURE PROTECTION	186.5	117.1	157.7	95.0	73.0	85.9	96.9	71.8	86.5	100.8	73.4	89.4	95.0	73.0	85.9	100.7	71.5	88.6
08	OPENINGS	130.7	117.4	127.6	106.3	70.0	98.0	106.4	72.3	98.7	95.5	70.3	89.8	106.5	72.4	98.7	95.6	72.4	90.4
0920	Plaster & Gypsum Board	155.6	119.8	131.8	88.1	72.1	77.5	91.3	67.9	75.7	88.4	72.1	77.6	81.1	72.1	75.1	90.4	72.1	78.3
0950, 0980	Ceilings & Acoustic Treatment	120.3	119.8	120.0	101.9	72.1	82.1	102.8	67.9	79.5	94.0	72.1	79.5	102.8	72.1	82.4	94.0	72.1	79.5
0960	Flooring	124.3	123.2	124.0	93.7	60.6	84.1	96.1	60.6	85.8	101.2	60.6	89.5	92.4	60.6	83.2	102.9	60.6	90.7
0970, 0990	Wall Finishes & Painting/Coating	110.3	117.0	114.2	91.1	65.2	75.9	91.1	65.2	75.9	96.3	65.2	78.0	91.1	65.2	75.9	96.3	65.2	78.0
09	FINISHES	130.7	121.1	125.4	92.3	67.7	78.8	95.5	65.3	79.0	98.8	67.8	81.9	91.1	67.7	78.3	98.5	67.8	81.7
COVERS	DIVS. 10 - 14, 25, 28, 41, 43, 44, 46	100.0	110.4	102.4	100.0	83.9	96.4	100.0	83.4	96.2	100.0	84.2	96.4	100.0	83.8	96.3	100.0	84.2	96.4
21, 22, 23	FIRE SUPPRESSION, PLUMBING & HVAC	98.6	105.4	101.6	97.6	77.7	88.8	100.2	77.9	90.3	96.5	77.8	88.2	97.6	77.7	88.8	100.1	77.9	90.3
26, 27, 3370	ELECTRICAL, COMMUNICATIONS & UTIL.	120.2	112.6	116.2	103.4	88.5	95.5	102.3	63.9	82.0	98.2	64.0	80.0	103.5	63.9	82.5	95.0	63.9	78.5
MF2014	WEIGHTED AVERAGE	133.4	115.0	125.3	98.3	76.2	88.6	102.3	72.3	89.1	99.3	73.3	87.8	98.3	72.9	87.1	98.4	73.0	87.2

DIVISION		ARIZONA												ARKANSAS					
		PHOENIX			PRESCOTT			SHOW LOW			TUCSON			BATESVILLE			CAMDEN		
		850,853			863			859			856 - 857			725			717		
		MAT.	INST.	TOTAL	MAT.	INST.	TOTAL	MAT.	INST.	TOTAL	MAT.	INST.	TOTAL	MAT.	INST.	TOTAL	MAT.	INST.	TOTAL
015433	CONTRACTOR EQUIPMENT		93.8	93.8		91.9	91.9		92.9	92.9		92.9	92.9		93.3	93.3		93.3	93.3
0241, 31 - 34	SITE & INFRASTRUCTURE, DEMOLITION	93.2	98.0	96.5	78.5	96.9	91.3	105.2	97.9	100.1	88.2	97.8	94.9	73.7	91.0	85.8	80.7	91.0	87.9
0310	Concrete Forming & Accessories	100.0	68.0	72.3	101.0	70.4	74.5	103.3	70.5	75.0	99.5	67.7	72.0	84.7	58.5	62.1	82.2	58.5	61.7
0320	Concrete Reinforcing	105.5	82.7	93.9	102.7	82.6	92.5	107.4	82.6	94.8	88.9	82.6	85.7	92.8	66.2	79.3	101.5	65.9	83.4
0330	Cast-in-Place Concrete	86.8	70.2	80.5	87.6	70.3	81.0	86.0	70.0	79.9	89.3	69.9	81.9	70.6	76.3	72.8	80.9	76.3	79.1
03	CONCRETE	97.8	71.4	85.6	100.3	72.4	87.3	110.0	72.4	92.6	96.8	71.1	84.9	76.7	66.9	72.1	83.1	66.8	75.5
04	MASONRY	97.6	59.2	74.0	97.7	59.3	74.1	110.3	59.3	78.9	95.6	59.2	73.2	94.4	43.4	63.0	111.8	37.7	66.1
05	METALS	97.8	75.7	91.0	101.2	73.9	92.9	95.9	75.1	89.5	97.1	74.7	90.3	100.5	74.3	92.5	104.6	73.9	95.2
06	WOOD, PLASTICS & COMPOSITES	99.6	69.4	83.0	101.8	72.9	85.9	103.0	72.9	86.5	99.0	69.3	82.7	93.4	60.0	75.1	93.3	60.0	75.0
07	THERMAL & MOISTURE PROTECTION	99.7	70.1	87.4	95.6	73.2	86.3	101.0	72.9	89.4	101.8	65.1	86.5	103.9	59.7	85.6	96.8	58.1	80.7
08	OPENINGS	97.4	72.8	91.8	106.4	70.3	98.2	94.8	72.4	89.7	92.1	72.7	87.7	99.9	55.4	89.8	102.8	57.3	92.5
0920	Plaster & Gypsum Board	92.5	68.4	76.5	88.3	72.3	77.7	92.4	72.1	78.9	95.3	68.4	77.4	82.0	59.3	66.9	85.1	59.3	68.0
0950, 0980	Ceilings & Acoustic Treatment	101.6	68.4	79.5	101.1	72.3	82.0	94.0	72.1	79.5	94.9	68.4	77.3	84.1	59.3	67.6	79.6	59.3	66.1
0960	Flooring	103.2	60.6	90.8	94.6	60.6	84.8	104.8	65.3	93.3	94.2	60.6	84.4	89.5	54.3	79.3	98.4	36.4	80.5
0970, 0990	Wall Finishes & Painting/Coating	96.3	67.4	79.3	91.1	65.2	75.9	96.3	65.2	78.0	97.4	65.2	78.5	93.7	36.7	60.2	96.4	47.2	67.5
09	FINISHES	100.5	66.0	81.7	92.9	67.8	79.2	100.6	68.6	83.1	96.7	65.7	79.8	80.5	54.9	66.5	83.3	52.7	66.6
COVERS	DIVS. 10 - 14, 25, 28, 41, 43, 44, 46	100.0	84.0	96.4	100.0	83.8	96.3	100.0	84.2	96.4	100.0	83.8	96.3	100.0	68.8	92.9	100.0	68.8	92.9
21, 22, 23	FIRE SUPPRESSION, PLUMBING & HVAC	100.0	77.9	90.2	100.2	77.7	90.3	96.5	77.7	88.2	100.0	71.6	87.5	96.7	51.5	76.7	96.5	57.9	79.4
26, 27, 3370	ELECTRICAL, COMMUNICATIONS & UTIL.	100.4	63.9	81.1	102.0	63.9	81.8	95.3	63.9	78.7	97.2	63.9	79.6	95.9	62.5	78.2	95.2	58.8	75.9
MF2014	WEIGHTED AVERAGE	98.8	72.7	87.3	99.8	72.8	87.9	99.5	73.2	87.9	97.2	71.0	85.6	93.6	60.9	79.1	96.3	60.9	80.7

DIVISION		ARKANSAS																	
		FAYETTEVILLE			FORT SMITH			HARRISON			HOT SPRINGS			JONESBORO			LITTLE ROCK		
		727			729			726			719			724			720 - 722		
		MAT.	INST.	TOTAL	MAT.	INST.	TOTAL	MAT.	INST.	TOTAL	MAT.	INST.	TOTAL	MAT.	INST.	TOTAL	MAT.	INST.	TOTAL
015433	CONTRACTOR EQUIPMENT		93.3	93.3		93.3	93.3		93.3	93.3		93.3	93.3		118.2	118.2		93.3	93.3
0241, 31 - 34	SITE & INFRASTRUCTURE, DEMOLITION	73.1	91.1	85.6	78.2	91.0	87.2	78.6	91.0	87.3	84.0	91.1	88.9	98.3	109.7	106.3	86.9	91.1	89.8
0310	Concrete Forming & Accessories	80.2	58.7	61.6	99.7	58.4	64.0	89.2	58.4	62.5	79.7	58.6	61.5	88.3	58.7	62.7	93.7	59.3	63.9
0320	Concrete Reinforcing	92.8	60.9	76.6	94.0	63.7	78.6	92.5	66.2	79.1	99.5	66.2	82.6	89.7	66.4	77.8	99.5	67.7	83.4
0330	Cast-in-Place Concrete	70.6	76.4	72.8	80.7	76.1	79.0	78.3	76.2	77.5	82.8	76.4	80.3	76.9	77.7	77.2	78.2	76.6	77.6
03	CONCRETE	76.4	66.1	71.6	83.8	66.3	75.7	83.4	66.8	75.7	86.5	66.9	77.4	80.9	68.5	75.1	84.7	67.6	76.8
04	MASONRY	85.6	42.9	59.3	92.6	48.8	65.6	94.8	44.5	63.8	83.8	43.1	58.7	86.9	40.9	58.5	88.4	68.2	75.9
05	METALS	100.5	73.5	92.3	102.8	73.8	94.0	101.7	74.2	93.3	104.5	74.4	95.3	96.9	89.9	94.8	103.1	76.8	95.1
06	WOOD, PLASTICS & COMPOSITES	89.9	60.0	73.5	109.7	60.0	82.4	99.3	60.0	77.7	90.7	60.0	73.8	97.6	60.4	77.2	96.6	60.0	76.5
07	THERMAL & MOISTURE PROTECTION	104.7	59.5	86.0	105.1	61.1	86.8	104.3	59.9	85.9	97.0	59.6	81.5	109.7	58.9	88.6	99.3	66.5	85.7
08	OPENINGS	99.9	57.3	90.2	101.9	55.9	91.5	100.7	54.2	90.2	102.8	57.2	92.5	105.6	58.5	94.9	98.1	57.4	88.9
0920	Plaster & Gypsum Board	81.4	59.3	66.7	88.3	59.3	69.0	87.3	59.3	68.7	83.8	59.3	67.5	96.1	59.3	71.6	92.7	59.3	70.5
0950, 0980	Ceilings & Acoustic Treatment	84.1	59.3	67.6	85.8	59.3	68.2	85.8	59.3	68.2	79.6	59.3	66.1	88.1	59.3	68.9	90.8	59.3	69.8
0960	Flooring	86.5	53.2	76.8	96.4	53.8	84.0	91.8	54.3	80.9	97.3	53.2	84.5	64.7	48.2	59.9	96.2	71.9	89.2
0970, 0990	Wall Finishes & Painting/Coating	93.7	49.2	67.6	93.7	50.0	68.1	93.7	31.7	57.3	96.4	47.2	67.5	83.0	45.0	60.7	92.0	47.2	65.7
09	FINISHES	79.6	55.8	66.6	84.1	56.5	69.0	82.7	54.2	67.1	83.2	55.8	68.2	78.4	54.6	65.4	87.0	60.0	72.2
COVERS	DIVS. 10 - 14, 25, 28, 41, 43, 44, 46	100.0	68.8	92.9	100.0	79.1	95.3	100.0	69.9	93.2	100.0	68.8	93.0	100.0	67.8	92.7	100.0	79.2	95.3
21, 22, 23	FIRE SUPPRESSION, PLUMBING & HVAC	96.7	62.4	81.5	100.3	48.1	77.2	96.7	49.3	75.7	96.5	52.4	77.0	100.6	50.5	78.5	100.0	52.5	79.0
26, 27, 3370	ELECTRICAL, COMMUNICATIONS & UTIL.	89.8	56.1	71.9	93.2	55.1	73.0	94.5	56.1	74.1	97.4	66.1	80.8	99.0	62.6	79.7	99.9	66.1	82.0
MF2014	WEIGHTED AVERAGE	92.4	62.3	79.1	96.1	60.2	80.2	94.9	59.6	79.3	95.6	61.8	80.7	95.6	63.7	81.5	96.4	65.7	82.8

ARKANSAS / CALIFORNIA

DIVISION		PINE BLUFF 716 MAT.	INST.	TOTAL	RUSSELLVILLE 728 MAT.	INST.	TOTAL	TEXARKANA 718 MAT.	INST.	TOTAL	WEST MEMPHIS 723 MAT.	INST.	TOTAL	ALHAMBRA 917-918 MAT.	INST.	TOTAL	ANAHEIM 928 MAT.	INST.	TOTAL
015433	CONTRACTOR EQUIPMENT		93.3	93.3		93.3	93.3		94.6	94.6		118.2	118.2		97.3	97.3		102.4	102.4
0241, 31 - 34	SITE & INFRASTRUCTURE, DEMOLITION	86.0	91.0	89.5	74.9	91.0	86.2	97.1	92.9	94.2	105.6	109.4	108.2	99.3	110.2	106.9	100.9	110.8	107.8
0310	Concrete Forming & Accessories	79.4	58.8	61.5	85.4	58.3	62.0	85.5	58.5	62.2	94.0	59.0	63.7	120.0	137.3	135.0	106.7	137.6	133.4
0320	Concrete Reinforcing	101.4	66.9	83.9	93.4	66.1	79.5	101.0	66.1	83.3	89.7	62.9	76.1	103.6	126.1	115.0	96.0	125.9	111.2
0330	Cast-in-Place Concrete	82.8	76.3	80.3	74.0	76.2	74.8	90.2	76.3	84.9	80.6	77.7	79.5	87.4	127.8	102.8	88.6	131.5	104.9
03	CONCRETE	87.2	67.1	77.9	79.6	66.7	73.6	85.5	66.9	76.9	87.7	68.0	78.6	99.4	130.7	113.9	99.3	132.2	114.6
04	MASONRY	118.6	54.2	79.0	91.9	41.9	61.1	98.3	42.3	63.8	75.4	40.7	54.1	112.7	143.6	131.7	79.5	138.4	115.8
05	METALS	105.3	75.5	96.2	100.5	74.0	92.4	96.9	74.2	90.0	96.0	89.0	93.9	92.0	109.3	97.3	107.5	111.1	108.6
06	WOOD, PLASTICS & COMPOSITES	90.2	60.0	73.6	94.7	60.0	75.6	97.7	60.0	77.0	103.3	60.4	79.7	98.9	134.4	118.4	99.4	134.8	118.8
07	THERMAL & MOISTURE PROTECTION	97.1	62.2	82.6	105.0	59.2	86.0	97.7	59.3	81.8	110.2	58.8	88.9	100.1	131.4	113.1	106.2	134.2	117.8
08	OPENINGS	104.0	55.7	93.1	99.9	56.5	90.0	109.1	56.5	97.2	103.0	56.8	92.6	91.3	133.3	100.8	107.6	133.5	113.5
0920	Plaster & Gypsum Board	83.5	59.3	67.4	82.0	59.3	66.9	86.5	59.3	68.4	98.4	59.3	72.4	95.5	135.6	122.2	105.7	135.6	125.6
0950, 0980	Ceilings & Acoustic Treatment	79.6	59.3	66.1	84.1	59.3	67.6	83.0	59.3	67.2	86.2	59.3	68.3	108.3	135.6	126.5	109.2	135.6	126.8
0960	Flooring	97.0	53.8	84.4	89.0	54.3	78.9	99.3	45.9	83.8	67.1	48.2	61.6	103.2	116.3	107.0	101.6	116.3	105.8
0970, 0990	Wall Finishes & Painting/Coating	96.4	46.8	67.3	93.7	33.9	58.6	96.4	42.5	64.7	83.0	47.2	62.0	107.4	120.8	115.3	91.8	120.8	108.8
09	FINISHES	83.1	56.4	68.5	80.7	54.5	66.4	85.5	54.0	68.3	79.7	55.1	66.3	103.7	131.5	118.9	99.7	131.8	117.2
COVERS	DIVS. 10 - 14, 25, 28, 41, 43, 44, 46	100.0	79.1	95.3	100.0	68.8	92.9	100.0	68.8	92.9	100.0	69.7	93.1	100.0	117.0	103.8	100.0	117.6	104.0
21, 22, 23	FIRE SUPPRESSION, PLUMBING & HVAC	100.1	50.7	78.3	96.7	51.3	76.6	100.1	56.6	80.9	97.1	65.4	83.1	96.4	125.6	109.3	100.0	130.1	113.3
26, 27, 3370	ELECTRICAL, COMMUNICATIONS & UTIL.	95.3	64.9	79.2	93.2	56.1	73.5	97.3	57.9	76.4	100.6	63.1	80.7	120.7	130.2	125.7	90.7	111.5	101.7
MF2014	WEIGHTED AVERAGE	98.4	62.9	82.7	93.6	59.8	78.7	97.0	61.4	81.3	95.2	66.8	82.6	99.8	127.1	111.8	100.3	125.5	111.4

CALIFORNIA

DIVISION		BAKERSFIELD 932-933 MAT.	INST.	TOTAL	BERKELEY 947 MAT.	INST.	TOTAL	EUREKA 955 MAT.	INST.	TOTAL	FRESNO 936-938 MAT.	INST.	TOTAL	INGLEWOOD 903-905 MAT.	INST.	TOTAL	LONG BEACH 906-908 MAT.	INST.	TOTAL
015433	CONTRACTOR EQUIPMENT		100.3	100.3		104.1	104.1		100.1	100.1		100.3	100.3		99.0	99.0		99.0	99.0
0241, 31 - 34	SITE & INFRASTRUCTURE, DEMOLITION	97.8	108.5	105.3	106.4	110.5	109.2	112.4	107.5	109.0	102.1	107.9	106.2	92.0	106.7	102.3	99.4	106.7	104.5
0310	Concrete Forming & Accessories	104.5	137.3	132.8	116.0	164.4	157.8	115.8	149.5	144.9	104.0	148.7	142.7	108.9	137.6	133.8	103.7	137.7	133.1
0320	Concrete Reinforcing	96.5	125.8	111.4	89.7	127.7	109.0	104.4	127.3	116.0	81.0	126.7	104.2	97.8	126.2	112.2	97.0	126.2	111.8
0330	Cast-in-Place Concrete	87.9	130.5	104.1	114.5	131.9	121.1	96.1	128.1	108.3	95.4	127.6	107.6	84.3	130.5	101.9	95.9	130.5	109.1
03	CONCRETE	90.9	131.6	109.8	105.8	144.5	123.8	110.1	136.5	122.4	95.0	135.8	113.9	93.1	131.8	111.1	103.0	131.9	116.4
04	MASONRY	93.7	137.8	120.9	115.3	157.0	141.0	105.0	153.7	135.0	98.5	146.4	128.0	80.3	143.7	119.3	89.9	143.7	123.0
05	METALS	107.9	110.3	108.6	110.2	114.2	111.4	107.3	113.0	109.0	108.2	111.0	109.1	102.8	111.1	105.4	102.7	111.2	105.3
06	WOOD, PLASTICS & COMPOSITES	91.8	134.8	115.4	106.8	169.8	141.4	113.4	152.3	134.8	103.4	152.3	130.3	99.2	134.9	118.8	93.5	134.9	116.2
07	THERMAL & MOISTURE PROTECTION	100.9	126.6	111.6	103.4	154.5	124.6	109.9	147.0	125.3	92.4	132.2	109.0	101.5	131.5	114.0	101.8	133.2	114.8
08	OPENINGS	95.0	131.1	103.2	95.8	157.1	109.7	106.7	139.3	114.1	96.9	140.6	106.8	93.4	133.5	102.5	93.4	133.5	102.5
0920	Plaster & Gypsum Board	94.0	135.6	121.6	110.3	171.1	150.7	110.9	153.6	139.3	92.8	153.6	133.2	100.9	135.6	124.0	97.3	135.6	122.8
0950, 0980	Ceilings & Acoustic Treatment	96.1	135.6	122.4	99.6	171.1	147.2	113.5	153.6	140.1	92.7	153.6	133.2	104.6	135.6	125.2	104.6	135.6	125.2
0960	Flooring	107.4	116.3	110.0	120.6	130.9	123.6	105.5	130.9	112.9	110.2	109.5	110.0	106.0	116.3	109.0	103.0	116.3	106.9
0970, 0990	Wall Finishes & Painting/Coating	95.1	110.5	104.2	112.8	164.1	142.9	93.2	129.7	114.6	107.3	116.3	112.6	106.7	120.8	115.0	106.7	120.8	115.0
09	FINISHES	95.2	130.7	114.6	109.4	159.9	137.0	104.9	145.6	127.1	97.0	140.3	120.7	104.1	131.7	119.2	103.2	131.7	118.8
COVERS	DIVS. 10 - 14, 25, 28, 41, 43, 44, 46	100.0	114.8	103.3	100.0	130.9	107.0	100.0	127.7	106.3	100.0	127.6	106.2	100.0	117.9	104.1	100.0	117.9	104.1
21, 22, 23	FIRE SUPPRESSION, PLUMBING & HVAC	100.0	123.1	110.2	96.6	156.6	123.1	96.4	123.1	108.2	100.1	124.4	110.9	96.2	123.0	108.0	96.2	125.7	109.2
26, 27, 3370	ELECTRICAL, COMMUNICATIONS & UTIL.	105.3	107.9	106.7	102.4	160.4	133.2	97.2	123.9	111.4	95.6	103.3	99.7	100.5	130.2	116.3	100.2	130.2	116.1
MF2014	WEIGHTED AVERAGE	99.4	122.6	109.6	103.3	147.7	123.0	103.6	130.6	115.5	99.5	126.1	111.3	97.3	126.7	110.3	99.0	127.3	111.5

CALIFORNIA

DIVISION		LOS ANGELES 900-902 MAT.	INST.	TOTAL	MARYSVILLE 959 MAT.	INST.	TOTAL	MODESTO 953 MAT.	INST.	TOTAL	MOJAVE 935 MAT.	INST.	TOTAL	OAKLAND 946 MAT.	INST.	TOTAL	OXNARD 930 MAT.	INST.	TOTAL
015433	CONTRACTOR EQUIPMENT		101.3	101.3		100.1	100.1		100.1	100.1		100.3	100.3		104.1	104.1		99.2	99.2
0241, 31 - 34	SITE & INFRASTRUCTURE, DEMOLITION	99.6	108.8	106.0	108.8	107.5	107.9	103.4	107.6	106.3	95.3	108.5	104.5	111.4	110.5	110.7	102.4	106.5	105.3
0310	Concrete Forming & Accessories	105.7	138.0	133.7	105.5	149.2	143.3	101.9	148.9	142.6	116.2	137.1	134.3	104.8	164.4	156.3	107.3	137.7	133.6
0320	Concrete Reinforcing	98.6	126.1	112.6	104.4	126.6	115.6	108.1	126.5	117.5	96.9	125.8	111.6	91.8	127.7	110.0	95.2	125.8	110.7
0330	Cast-in-Place Concrete	91.0	131.2	106.3	107.4	127.9	115.2	96.2	127.8	108.2	82.5	130.4	100.8	108.6	131.9	117.5	95.1	131.9	109.3
03	CONCRETE	98.6	132.3	114.2	110.9	136.2	122.6	101.6	136.0	117.6	86.9	131.5	107.6	106.2	144.5	124.0	98.0	132.2	112.3
04	MASONRY	95.3	141.6	123.8	105.9	142.4	128.4	103.1	142.4	127.3	96.6	137.8	122.0	122.4	157.0	143.8	99.6	136.4	122.3
05	METALS	109.5	111.7	110.1	106.8	111.9	108.3	103.5	111.6	106.0	105.4	110.2	106.8	105.8	114.2	108.4	103.2	110.7	105.5
06	WOOD, PLASTICS & COMPOSITES	99.3	135.0	118.9	101.0	152.3	129.2	96.7	152.3	127.3	102.5	134.8	120.3	95.6	169.8	136.4	97.7	134.8	118.1
07	THERMAL & MOISTURE PROTECTION	100.9	133.1	114.3	109.4	136.8	120.8	108.9	134.0	119.4	98.2	124.2	109.0	102.1	154.5	123.8	101.5	132.4	114.3
08	OPENINGS	100.5	133.6	108.0	106.0	141.8	114.1	104.7	140.7	112.9	91.6	131.1	100.5	95.9	157.1	109.7	94.3	133.5	103.2
0920	Plaster & Gypsum Board	100.7	135.6	123.9	103.5	153.6	136.8	105.7	153.6	137.5	102.6	135.6	124.5	104.9	171.1	148.9	96.7	135.6	122.6
0950, 0980	Ceilings & Acoustic Treatment	114.8	135.6	128.7	112.6	153.6	139.9	109.2	153.6	138.7	94.1	135.6	121.7	102.3	171.1	148.1	96.5	135.6	122.5
0960	Flooring	103.5	116.3	107.2	101.2	121.5	107.1	101.5	121.3	107.3	111.2	116.3	112.7	114.0	130.9	118.9	104.0	116.3	106.8
0970, 0990	Wall Finishes & Painting/Coating	105.8	120.8	114.6	93.2	134.4	117.4	93.2	127.0	113.0	91.7	109.1	101.9	112.8	164.1	142.9	91.7	115.3	105.6
09	FINISHES	105.5	132.0	120.0	101.9	144.4	125.1	101.1	143.6	124.3	97.3	130.5	115.4	107.6	159.9	136.2	94.7	131.2	114.7
COVERS	DIVS. 10 - 14, 25, 28, 41, 43, 44, 46	100.0	118.3	104.1	100.0	127.7	106.3	100.0	127.6	106.2	100.0	114.8	103.4	100.0	130.9	107.0	100.0	117.8	104.0
21, 22, 23	FIRE SUPPRESSION, PLUMBING & HVAC	100.0	130.2	113.3	96.4	125.7	109.4	100.0	124.3	110.7	96.4	122.5	108.0	100.1	156.6	125.1	100.0	130.1	113.3
26, 27, 3370	ELECTRICAL, COMMUNICATIONS & UTIL.	98.8	129.3	115.0	94.0	121.4	108.5	96.3	106.7	101.9	94.1	110.8	102.9	101.6	160.4	132.8	99.4	113.7	107.0
MF2014	WEIGHTED AVERAGE	101.6	128.2	113.4	102.8	129.2	114.5	101.6	126.6	112.6	96.5	122.7	108.1	103.5	147.7	123.1	98.9	125.1	110.4

For customer support on your Building Construction Costs with RSMeans Data, call 800.448.8182.

765

City Cost Indexes

CALIFORNIA

DIVISION		PALM SPRINGS 922			PALO ALTO 943			PASADENA 910 - 912			REDDING 960			RICHMOND 948			RIVERSIDE 925		
		MAT.	INST.	TOTAL	MAT.	INST.	TOTAL	MAT.	INST.	TOTAL	MAT.	INST.	TOTAL	MAT.	INST.	TOTAL	MAT.	INST.	TOTAL
015433	CONTRACTOR EQUIPMENT		101.2	101.2		104.1	104.1		97.3	97.3		100.1	100.1		104.1	104.1		101.2	101.2
0241, 31 - 34	SITE & INFRASTRUCTURE, DEMOLITION	92.3	108.8	103.8	102.7	110.5	108.1	96.0	110.2	105.9	133.8	107.5	115.5	110.7	110.4	110.5	99.4	108.7	105.9
0310	Concrete Forming & Accessories	103.3	137.6	133.0	103.0	164.3	156.1	108.5	137.3	133.4	108.4	149.2	143.7	118.8	164.1	158.0	107.3	137.6	133.5
0320	Concrete Reinforcing	109.8	126.0	118.0	89.7	127.4	108.9	104.5	126.1	115.5	139.4	126.6	132.9	89.7	127.3	108.8	106.6	125.8	116.4
0330	Cast-in-Place Concrete	84.4	131.5	102.4	96.9	131.9	110.2	82.9	127.8	100.0	115.3	127.9	120.1	111.4	131.7	119.1	91.9	131.4	106.9
03	CONCRETE	93.6	132.2	111.5	95.4	144.5	118.2	94.8	130.7	111.4	122.4	136.2	128.8	108.1	144.3	124.9	99.7	132.1	114.7
04	MASONRY	77.6	140.9	116.0	98.1	151.5	131.0	97.4	143.6	125.8	134.8	142.4	139.5	115.1	151.5	137.5	78.5	138.1	115.2
05	METALS	108.2	111.3	109.2	103.4	113.8	106.6	92.1	109.3	97.4	102.6	111.9	105.4	103.5	113.5	106.5	107.6	111.0	108.7
06	WOOD, PLASTICS & COMPOSITES	94.4	134.8	116.6	93.2	169.8	135.3	85.8	134.4	112.6	111.6	152.3	134.0	110.1	169.8	142.9	99.4	134.8	118.8
07	THERMAL & MOISTURE PROTECTION	105.5	134.9	117.7	101.2	154.6	123.4	99.7	131.4	112.9	135.2	136.8	135.9	101.9	150.6	122.1	106.4	133.2	117.5
08	OPENINGS	103.2	133.5	110.1	95.9	157.1	109.7	91.3	133.3	100.8	119.3	141.8	124.4	95.9	157.1	109.7	106.2	133.5	112.4
0920	Plaster & Gypsum Board	101.0	135.6	124.0	103.3	171.1	148.4	89.2	135.6	120.1	104.9	153.6	137.2	111.8	171.1	151.2	104.9	135.6	125.3
0950, 0980	Ceilings & Acoustic Treatment	105.9	135.6	125.7	100.4	171.1	147.5	108.3	135.6	126.5	137.6	153.6	148.2	100.4	171.1	147.5	113.4	135.6	128.2
0960	Flooring	103.8	122.6	109.2	112.6	130.9	117.9	97.2	116.3	102.7	98.1	121.8	105.0	123.0	130.9	125.3	105.4	116.3	108.5
0970, 0990	Wall Finishes & Painting/Coating	90.2	120.8	108.2	112.8	164.1	142.9	107.4	120.8	115.3	104.0	134.4	121.9	112.8	164.1	142.9	90.2	120.8	108.2
09	FINISHES	98.2	132.8	117.1	106.0	159.9	135.5	100.9	131.5	117.6	107.8	144.5	127.9	110.9	159.9	137.7	101.2	131.8	117.9
COVERS	DIVS. 10 - 14, 25, 28, 41, 43, 44, 46	100.0	117.6	104.0	100.0	131.0	107.0	100.0	117.0	103.8	100.0	127.7	106.3	100.0	130.9	107.0	100.0	117.6	104.0
21, 22, 23	FIRE SUPPRESSION, PLUMBING & HVAC	96.4	125.5	109.3	96.6	156.3	123.0	96.4	122.9	108.1	100.5	125.7	111.6	96.6	154.7	122.3	100.0	130.1	113.3
26, 27, 3370	ELECTRICAL, COMMUNICATIONS & UTIL.	93.7	113.8	104.4	101.5	169.1	137.4	117.4	130.2	124.2	99.0	121.4	110.9	102.1	139.5	121.9	90.5	110.9	101.3
MF2014	WEIGHTED AVERAGE	98.2	125.1	110.1	99.4	148.3	121.0	97.7	126.5	110.4	109.8	129.2	118.4	102.7	143.7	120.8	100.2	125.2	111.2

CALIFORNIA

DIVISION		SACRAMENTO 942,956 - 958			SALINAS 939			SAN BERNARDINO 923 - 924			SAN DIEGO 919 - 921			SAN FRANCISCO 940 - 941			SAN JOSE 951		
		MAT.	INST.	TOTAL	MAT.	INST.	TOTAL	MAT.	INST.	TOTAL	MAT.	INST.	TOTAL	MAT.	INST.	TOTAL	MAT.	INST.	TOTAL
015433	CONTRACTOR EQUIPMENT		102.8	102.8		100.3	100.3		101.2	101.2		98.9	98.9		108.5	108.5		101.9	101.9
0241, 31 - 34	SITE & INFRASTRUCTURE, DEMOLITION	92.2	116.6	109.2	116.0	108.2	110.5	78.3	108.7	99.5	103.1	104.0	103.8	113.5	111.2	111.9	134.9	102.4	112.2
0310	Concrete Forming & Accessories	103.0	151.7	145.1	110.6	152.1	146.5	111.2	137.5	133.9	108.7	124.4	122.3	104.6	165.3	157.1	108.0	164.2	156.6
0320	Concrete Reinforcing	85.3	126.8	106.4	95.6	127.0	111.6	106.6	125.9	116.4	101.2	125.7	113.6	104.6	127.9	116.4	95.5	127.4	111.7
0330	Cast-in-Place Concrete	91.9	128.8	105.9	94.9	128.2	107.6	63.5	131.4	89.4	90.9	113.7	99.6	111.3	132.8	119.5	111.7	131.2	119.1
03	CONCRETE	97.1	137.4	115.8	104.5	137.7	119.9	73.7	132.1	100.8	100.0	120.1	109.3	109.2	145.7	126.1	107.9	144.5	124.9
04	MASONRY	99.0	145.8	127.8	96.7	148.3	128.4	84.9	139.0	118.2	89.3	130.0	114.4	123.0	158.2	144.7	133.1	151.6	144.5
05	METALS	101.3	107.9	103.3	108.2	112.8	109.6	107.6	111.0	108.6	109.5	110.1	109.7	111.1	120.3	113.9	101.8	118.7	106.9
06	WOOD, PLASTICS & COMPOSITES	89.4	155.7	125.8	102.1	155.5	131.5	102.9	134.8	120.4	99.6	121.5	111.6	95.6	169.9	136.4	109.4	169.5	142.5
07	THERMAL & MOISTURE PROTECTION	111.5	137.8	122.4	98.9	143.6	117.5	104.6	134.4	116.9	104.7	117.8	110.1	103.2	155.0	124.7	105.4	152.1	124.8
08	OPENINGS	109.1	142.5	116.7	95.5	149.2	107.6	103.3	132.8	110.0	98.8	124.6	104.6	100.3	157.1	113.1	96.1	156.9	109.9
0920	Plaster & Gypsum Board	100.6	156.8	138.0	97.6	156.8	137.0	106.3	135.6	125.8	95.9	121.9	113.2	106.9	171.1	149.6	101.8	171.1	147.9
0950, 0980	Ceilings & Acoustic Treatment	100.4	156.8	138.0	94.1	156.8	135.8	109.2	135.6	126.8	115.9	121.9	119.9	108.9	171.1	150.3	108.4	171.1	150.1
0960	Flooring	112.6	121.5	115.2	105.0	130.9	112.5	107.4	116.3	110.0	105.9	116.3	108.9	114.0	130.9	118.9	94.8	130.9	105.3
0970, 0990	Wall Finishes & Painting/Coating	109.4	140.1	127.4	92.6	164.1	134.6	90.2	120.8	108.2	104.0	120.8	113.9	112.8	173.9	148.7	93.5	164.1	134.9
09	FINISHES	104.8	147.1	127.9	96.7	151.2	126.5	99.6	131.8	117.2	106.6	122.1	115.1	109.3	161.1	137.6	100.0	159.7	132.6
COVERS	DIVS. 10 - 14, 25, 28, 41, 43, 44, 46	100.0	128.5	106.4	100.0	128.0	106.3	100.0	112.4	102.8	100.0	113.9	103.1	100.0	131.2	107.1	100.0	130.4	106.9
21, 22, 23	FIRE SUPPRESSION, PLUMBING & HVAC	100.0	125.6	111.3	96.4	131.8	112.1	96.4	127.3	110.1	100.0	126.3	111.6	100.1	190.7	140.1	100.0	159.5	126.3
26, 27, 3370	ELECTRICAL, COMMUNICATIONS & UTIL.	97.2	111.0	104.5	95.1	123.9	110.4	93.7	111.1	103.0	103.5	102.9	103.2	101.7	172.1	139.0	99.2	168.1	135.8
MF2014	WEIGHTED AVERAGE	101.0	129.1	113.4	100.2	133.2	114.8	95.8	124.5	108.5	102.0	118.1	109.1	105.6	157.7	128.6	103.6	148.6	123.5

CALIFORNIA

DIVISION		SAN LUIS OBISPO 934			SAN MATEO 944			SAN RAFAEL 949			SANTA ANA 926 - 927			SANTA BARBARA 931			SANTA CRUZ 950		
		MAT.	INST.	TOTAL	MAT.	INST.	TOTAL	MAT.	INST.	TOTAL	MAT.	INST.	TOTAL	MAT.	INST.	TOTAL	MAT.	INST.	TOTAL
015433	CONTRACTOR EQUIPMENT		100.3	100.3		104.1	104.1		103.3	103.3		101.2	101.2		100.3	100.3		101.9	101.9
0241, 31 - 34	SITE & INFRASTRUCTURE, DEMOLITION	108.0	108.6	108.4	108.5	110.4	109.9	104.5	116.4	112.8	90.7	108.7	103.3	102.4	108.6	106.7	134.5	102.2	112.0
0310	Concrete Forming & Accessories	118.0	137.6	135.0	109.0	164.3	156.8	113.8	164.3	157.5	111.4	137.5	133.9	108.1	137.7	133.7	108.0	152.4	146.4
0320	Concrete Reinforcing	96.9	125.8	111.6	89.7	127.6	108.9	90.5	127.6	109.3	110.4	126.0	118.3	95.2	125.8	110.7	117.9	127.1	122.6
0330	Cast-in-Place Concrete	102.0	130.6	112.9	107.8	131.8	117.0	125.5	131.0	127.6	81.0	131.4	100.2	95.7	130.7	109.0	111.0	130.1	118.3
03	CONCRETE	102.7	131.8	116.2	104.7	144.5	123.1	127.1	140.0	134.9	91.1	132.1	110.1	95.0	131.9	112.1	110.5	138.8	123.6
04	MASONRY	98.2	136.9	122.1	114.8	154.9	139.5	94.3	154.9	131.6	74.5	141.3	115.6	96.9	136.9	121.6	137.1	148.4	144.1
05	METALS	106.1	110.6	107.5	103.3	114.0	106.6	104.4	110.7	106.3	107.7	111.2	108.7	103.7	110.7	105.8	109.2	117.4	111.8
06	WOOD, PLASTICS & COMPOSITES	104.6	134.8	121.2	100.4	169.8	138.5	96.6	169.5	136.7	104.9	134.8	121.3	97.7	134.8	118.1	109.4	155.6	134.8
07	THERMAL & MOISTURE PROTECTION	99.0	132.1	112.7	101.7	153.7	123.2	106.0	152.9	125.5	105.8	135.0	117.9	98.5	131.5	112.2	104.9	146.0	122.0
08	OPENINGS	93.5	131.1	102.0	95.9	157.1	109.7	106.6	156.9	118.0	102.5	133.5	109.5	95.3	133.7	104.0	97.4	149.3	109.1
0920	Plaster & Gypsum Board	103.6	135.6	124.9	108.2	171.1	150.0	109.9	171.1	150.6	108.0	135.6	126.4	96.7	135.6	122.6	110.4	156.8	141.3
0950, 0980	Ceilings & Acoustic Treatment	94.1	135.6	121.7	100.4	171.1	147.5	107.2	171.1	149.7	109.2	135.6	126.8	96.5	135.6	122.5	110.1	156.8	141.2
0960	Flooring	112.2	116.3	113.4	116.3	130.9	120.5	128.3	130.9	129.1	107.9	116.3	110.3	104.1	116.3	107.6	98.9	130.9	108.2
0970, 0990	Wall Finishes & Painting/Coating	91.7	113.0	104.3	112.8	164.1	142.9	108.8	158.6	138.1	90.2	120.8	108.2	91.7	115.3	105.6	93.7	164.1	135.0
09	FINISHES	98.7	130.8	116.3	108.2	159.9	136.5	111.4	159.1	137.5	101.1	131.7	117.8	95.2	131.2	114.9	102.8	151.3	129.3
COVERS	DIVS. 10 - 14, 25, 28, 41, 43, 44, 46	100.0	124.8	105.6	100.0	130.9	107.0	100.0	130.2	106.8	100.0	117.6	104.0	100.0	117.8	104.0	100.0	128.4	106.4
21, 22, 23	FIRE SUPPRESSION, PLUMBING & HVAC	96.4	127.4	110.1	96.6	155.7	122.7	96.5	183.8	135.1	96.4	122.1	107.8	100.0	130.1	113.3	100.0	132.2	114.2
26, 27, 3370	ELECTRICAL, COMMUNICATIONS & UTIL.	94.1	113.4	104.3	101.5	160.8	132.9	98.5	123.9	111.9	93.7	112.9	103.9	93.0	112.9	103.6	98.3	123.9	111.9
MF2014	WEIGHTED AVERAGE	99.3	124.7	110.5	101.8	147.3	121.9	104.9	148.2	124.0	97.9	124.1	109.5	98.2	125.2	110.1	105.6	133.6	118.0

For customer support on your Building Construction Costs with RSMeans Data, call 800.448.8182.

City Cost Indexes

		CALIFORNIA															COLORADO		
DIVISION		SANTA ROSA			STOCKTON			SUSANVILLE			VALLEJO			VAN NUYS			ALAMOSA		
		954			952			961			945			913 - 916			811		
		MAT.	INST.	TOTAL	MAT.	INST.	TOTAL	MAT.	INST.	TOTAL	MAT.	INST.	TOTAL	MAT.	INST.	TOTAL	MAT.	INST.	TOTAL
015433	CONTRACTOR EQUIPMENT		100.7	100.7		100.1	100.1		100.1	100.1		103.3	103.3		97.3	97.3		94.0	94.0
0241, 31 - 34	SITE & INFRASTRUCTURE, DEMOLITION	105.1	107.5	106.8	103.1	107.6	106.2	142.3	107.5	118.0	93.2	116.4	109.3	113.9	110.2	111.3	141.3	89.6	105.2
0310	Concrete Forming & Accessories	104.3	163.6	155.6	105.9	151.3	145.2	109.6	145.0	140.3	104.2	163.2	155.2	115.5	137.3	134.3	105.3	54.9	61.7
0320	Concrete Reinforcing	105.3	127.7	116.7	108.1	126.6	117.5	139.4	126.6	132.9	91.6	127.5	109.8	104.5	126.1	115.5	111.1	63.3	86.8
0330	Cast-in-Place Concrete	105.5	129.5	114.7	93.7	127.9	106.7	104.9	127.9	113.6	100.0	130.1	111.5	87.4	127.8	102.8	97.6	72.7	88.1
03	CONCRETE	110.8	143.5	125.9	100.7	137.2	117.6	125.0	134.3	129.3	101.9	143.1	121.0	109.8	130.7	119.5	114.1	63.5	90.6
04	MASONRY	103.7	157.1	136.6	103.0	142.4	127.2	133.2	142.4	138.9	71.7	153.8	122.3	112.7	143.6	131.7	132.9	61.7	89.1
05	METALS	108.0	115.0	110.1	103.8	111.7	106.2	101.4	111.8	104.5	104.4	110.0	106.1	91.2	109.3	96.8	103.0	77.0	95.1
06	WOOD, PLASTICS & COMPOSITES	95.8	169.3	136.2	102.1	155.5	131.4	113.3	146.8	131.8	86.7	169.5	132.2	94.0	134.4	116.2	98.4	51.9	72.8
07	THERMAL & MOISTURE PROTECTION	106.3	153.4	125.8	109.4	135.2	120.1	137.0	134.1	135.8	104.1	152.4	124.1	100.8	131.4	113.5	105.9	67.8	90.1
08	OPENINGS	104.2	156.8	116.1	104.7	142.4	113.3	120.2	138.8	124.4	108.5	155.8	119.2	91.2	133.3	100.7	96.7	59.0	88.2
0920	Plaster & Gypsum Board	102.7	171.1	148.2	105.7	156.8	139.7	105.7	148.0	133.8	104.6	171.1	148.8	93.6	135.6	121.5	77.8	50.5	59.6
0950, 0980	Ceilings & Acoustic Treatment	109.2	171.1	150.4	116.8	156.8	143.5	130.1	148.0	142.0	109.1	171.1	150.4	105.7	135.6	125.6	98.5	50.5	66.6
0960	Flooring	104.2	116.0	107.6	101.5	121.3	107.3	98.5	121.5	105.2	122.1	130.9	124.7	100.3	116.3	105.0	112.8	73.1	101.3
0970, 0990	Wall Finishes & Painting/Coating	90.2	158.6	130.4	93.2	130.7	115.2	104.0	134.4	121.9	109.8	158.6	138.5	107.4	120.8	115.3	105.9	62.1	80.2
09	FINISHES	100.2	156.1	130.7	102.7	145.8	126.3	107.6	141.2	126.0	108.0	158.8	135.8	103.2	131.4	118.6	101.3	57.9	77.6
COVERS	DIVS. 10 - 14, 25, 28, 41, 43, 44, 46	100.0	129.3	106.6	100.0	128.0	106.3	100.0	127.1	106.1	100.0	129.6	106.8	100.0	117.0	103.8	100.0	83.7	96.3
21, 22, 23	FIRE SUPPRESSION, PLUMBING & HVAC	96.4	183.1	134.8	100.0	124.3	110.8	96.9	125.7	109.6	100.1	143.6	119.3	96.4	122.9	108.1	96.4	66.1	83.0
26, 27, 3370	ELECTRICAL, COMMUNICATIONS & UTIL.	94.0	121.4	108.6	96.3	113.7	105.5	99.4	121.4	111.1	94.7	129.4	113.1	117.4	130.2	124.2	98.2	66.8	81.5
MF2014	WEIGHTED AVERAGE	102.3	147.1	122.1	101.7	128.2	113.4	109.5	128.3	117.8	100.5	139.9	117.9	100.9	126.5	112.2	103.9	67.3	87.8

		COLORADO																	
DIVISION		BOULDER			COLORADO SPRINGS			DENVER			DURANGO			FORT COLLINS			FORT MORGAN		
		803			808 - 809			800 - 802			813			805			807		
		MAT.	INST.	TOTAL	MAT.	INST.	TOTAL	MAT.	INST.	TOTAL	MAT.	INST.	TOTAL	MAT.	INST.	TOTAL	MAT.	INST.	TOTAL
015433	CONTRACTOR EQUIPMENT		96.4	96.4		94.2	94.2		98.3	98.3		94.0	94.0		96.4	96.4		96.4	96.4
0241, 31 - 34	SITE & INFRASTRUCTURE, DEMOLITION	97.6	97.2	97.3	99.4	92.1	94.3	98.3	101.3	100.4	134.4	89.6	103.1	109.9	97.1	101.0	99.7	96.9	97.7
0310	Concrete Forming & Accessories	106.4	74.5	78.8	96.6	59.0	64.1	103.0	62.7	68.1	111.8	54.9	62.6	103.8	71.7	76.0	106.9	54.6	61.7
0320	Concrete Reinforcing	101.2	64.3	82.5	100.4	65.3	82.6	100.4	65.2	82.6	111.1	63.2	86.8	101.3	64.3	82.5	101.4	63.4	82.1
0330	Cast-in-Place Concrete	109.6	73.4	95.8	112.6	72.3	97.2	106.5	71.9	93.3	112.3	72.7	97.2	123.7	72.0	104.0	107.5	72.0	94.0
03	CONCRETE	111.0	72.7	93.3	115.0	65.6	92.1	108.4	67.0	89.2	115.9	63.6	91.6	122.9	71.0	98.8	109.3	63.1	87.9
04	MASONRY	98.1	66.8	78.9	99.4	64.4	77.9	101.3	64.6	78.7	119.3	61.1	83.5	116.3	64.3	84.3	113.5	64.3	83.2
05	METALS	97.1	76.5	90.8	100.1	77.0	93.0	102.5	76.5	94.5	103.0	77.0	95.1	98.3	76.5	91.6	96.8	76.1	90.5
06	WOOD, PLASTICS & COMPOSITES	109.2	77.0	91.5	99.9	57.6	76.7	107.0	62.2	82.4	107.2	51.9	76.8	106.8	74.5	89.0	109.2	51.4	77.5
07	THERMAL & MOISTURE PROTECTION	100.2	73.5	89.1	100.9	69.9	88.0	99.4	69.7	87.1	105.9	67.6	90.0	100.4	73.9	89.4	100.1	70.0	87.6
08	OPENINGS	100.1	73.0	94.0	104.5	62.6	95.0	105.0	65.1	96.0	103.7	59.0	93.6	100.1	71.6	93.7	100.1	58.7	90.7
0920	Plaster & Gypsum Board	116.0	76.8	89.9	100.0	56.7	71.2	111.9	61.6	78.5	90.3	50.5	63.8	110.4	74.1	86.3	116.0	50.5	72.4
0950, 0980	Ceilings & Acoustic Treatment	85.5	76.8	79.7	93.1	56.7	68.9	96.1	61.6	73.1	98.5	50.5	66.6	85.5	74.1	77.9	85.5	50.5	62.2
0960	Flooring	108.6	73.1	98.3	100.0	73.1	92.2	104.9	73.1	95.7	118.3	73.1	105.2	104.7	73.1	95.6	109.0	73.1	98.6
0970, 0990	Wall Finishes & Painting/Coating	101.0	60.9	77.4	100.7	48.2	69.9	101.0	60.9	77.4	105.9	47.7	71.8	101.0	60.9	77.4	101.0	75.7	86.1
09	FINISHES	102.6	73.0	86.4	99.4	59.5	77.6	102.9	63.7	81.5	103.9	56.3	77.9	101.2	70.9	84.6	102.6	59.0	78.8
COVERS	DIVS. 10 - 14, 25, 28, 41, 43, 44, 46	100.0	86.3	96.9	100.0	83.5	96.3	100.0	83.8	96.3	100.0	83.7	96.3	100.0	85.2	96.7	100.0	82.7	96.1
21, 22, 23	FIRE SUPPRESSION, PLUMBING & HVAC	96.4	73.2	86.2	100.1	74.8	88.9	99.9	75.6	89.2	96.4	73.4	86.2	100.0	71.9	87.6	96.4	71.9	85.6
26, 27, 3370	ELECTRICAL, COMMUNICATIONS & UTIL.	94.1	79.5	86.3	97.3	76.6	86.3	98.9	81.5	89.7	97.6	63.5	79.5	94.1	79.5	86.3	94.5	77.3	85.4
MF2014	WEIGHTED AVERAGE	99.7	76.0	89.2	102.1	71.8	88.7	102.2	74.2	89.8	104.3	68.2	88.3	103.2	74.8	90.7	100.3	70.9	87.3

		COLORADO																	
DIVISION		GLENWOOD SPRINGS			GOLDEN			GRAND JUNCTION			GREELEY			MONTROSE			PUEBLO		
		816			804			815			806			814			810		
		MAT.	INST.	TOTAL	MAT.	INST.	TOTAL	MAT.	INST.	TOTAL	MAT.	INST.	TOTAL	MAT.	INST.	TOTAL	MAT.	INST.	TOTAL
015433	CONTRACTOR EQUIPMENT		97.3	97.3		96.4	96.4		97.3	97.3		96.4	96.4		95.7	95.7		94.0	94.0
0241, 31 - 34	SITE & INFRASTRUCTURE, DEMOLITION	150.8	97.7	113.8	111.3	97.2	101.5	133.8	97.9	108.8	96.1	97.1	96.8	144.0	93.6	108.9	125.2	89.5	100.3
0310	Concrete Forming & Accessories	102.1	54.6	61.0	99.0	61.1	66.2	110.5	76.3	80.9	101.6	75.9	79.4	101.6	54.8	61.1	107.6	59.1	65.6
0320	Concrete Reinforcing	109.9	63.4	86.3	101.4	64.3	82.6	110.3	64.4	87.0	101.2	64.3	82.5	109.8	63.2	86.2	106.2	65.0	85.3
0330	Cast-in-Place Concrete	97.6	71.5	87.7	107.6	73.4	94.6	108.0	73.1	94.7	103.4	72.0	91.4	97.6	72.1	87.9	96.9	72.5	87.6
03	CONCRETE	119.4	63.0	93.3	121.6	66.7	96.1	112.4	73.5	94.3	105.7	72.9	90.5	110.0	63.3	88.3	102.3	65.6	85.3
04	MASONRY	103.7	64.2	79.4	116.4	66.8	85.9	140.2	67.6	95.5	109.9	64.3	81.8	112.4	61.1	80.8	99.8	60.7	75.7
05	METALS	102.7	76.8	94.8	97.0	76.4	90.7	104.3	77.4	96.0	98.2	76.6	91.6	101.8	76.8	94.1	105.9	77.5	97.2
06	WOOD, PLASTICS & COMPOSITES	93.7	51.6	70.6	101.8	58.9	78.2	104.9	79.5	90.9	104.1	80.1	90.9	94.9	51.8	71.2	100.9	57.9	77.3
07	THERMAL & MOISTURE PROTECTION	105.8	68.5	90.3	101.2	70.0	88.3	104.9	75.4	92.6	99.8	75.8	89.8	106.0	67.6	90.0	104.3	67.4	89.0
08	OPENINGS	102.7	58.8	92.8	100.1	63.1	91.7	103.4	74.3	96.8	100.1	74.5	94.3	103.9	58.9	93.7	98.6	62.8	90.5
0920	Plaster & Gypsum Board	120.3	50.5	73.9	108.1	58.2	74.9	132.4	79.1	97.0	108.7	79.9	89.6	77.2	50.5	59.4	81.9	56.7	65.1
0950, 0980	Ceilings & Acoustic Treatment	97.7	50.5	66.3	85.5	58.2	67.3	97.7	79.1	85.3	85.5	79.9	81.8	98.5	50.5	66.6	106.2	56.7	73.2
0960	Flooring	111.9	73.1	100.7	102.4	73.1	93.9	117.5	73.1	104.7	103.6	73.1	94.8	115.5	73.1	103.2	114.1	73.1	102.2
0970, 0990	Wall Finishes & Painting/Coating	105.9	57.4	77.4	101.0	60.9	77.4	105.9	60.9	79.5	101.0	60.9	77.4	105.9	47.7	71.8	105.9	44.7	70.0
09	FINISHES	107.3	57.1	79.9	101.0	62.4	79.9	108.8	74.5	90.0	99.8	74.2	85.8	102.1	56.2	77.0	102.0	59.4	78.7
COVERS	DIVS. 10 - 14, 25, 28, 41, 43, 44, 46	100.0	83.0	96.1	100.0	84.3	96.5	100.0	86.8	97.0	100.0	85.8	96.8	100.0	83.3	96.2	100.0	84.2	96.4
21, 22, 23	FIRE SUPPRESSION, PLUMBING & HVAC	96.4	67.3	83.5	96.4	73.2	86.2	99.9	73.6	88.3	100.0	71.9	87.6	96.4	77.8	88.2	99.9	63.3	83.7
26, 27, 3370	ELECTRICAL, COMMUNICATIONS & UTIL.	94.9	63.5	78.3	94.5	79.5	86.5	97.3	54.5	74.6	94.1	79.5	86.3	97.3	51.6	73.1	98.2	66.8	81.5
MF2014	WEIGHTED AVERAGE	104.2	67.8	88.1	102.1	73.0	89.2	106.2	73.2	91.6	100.2	75.7	89.4	103.0	67.7	87.4	101.9	67.4	86.7

DIVISION		COLORADO			CONNECTICUT														
		SALIDA			BRIDGEPORT			BRISTOL			HARTFORD			MERIDEN			NEW BRITAIN		
		812			066			060			061			064			060		
		MAT.	INST.	TOTAL	MAT.	INST.	TOTAL	MAT.	INST.	TOTAL	MAT.	INST.	TOTAL	MAT.	INST.	TOTAL	MAT.	INST.	TOTAL
015433	CONTRACTOR EQUIPMENT		95.7	95.7		99.8	99.8		99.8	99.8		99.8	99.8		100.2	100.2		99.8	99.8
0241, 31 - 34	SITE & INFRASTRUCTURE, DEMOLITION	134.1	93.6	105.9	102.4	104.0	103.5	101.6	104.0	103.3	97.1	104.0	101.9	99.6	104.8	103.2	101.8	104.0	103.3
0310	Concrete Forming & Accessories	110.4	54.6	62.1	105.3	119.0	117.1	105.3	118.9	117.0	104.2	118.9	116.9	105.0	118.9	117.0	105.8	118.8	117.1
0320	Concrete Reinforcing	109.5	63.2	86.0	115.4	140.1	127.9	115.4	140.0	127.9	110.7	140.0	125.6	115.4	140.0	127.9	115.4	140.0	127.9
0330	Cast-in-Place Concrete	111.8	72.1	96.7	93.1	128.7	106.7	87.2	128.7	103.0	91.4	128.7	105.6	83.9	128.7	101.0	88.6	128.7	103.9
03	CONCRETE	110.7	63.2	88.7	96.3	125.2	109.7	93.4	125.2	108.1	93.5	125.2	108.2	91.8	125.2	107.3	94.1	125.1	108.5
04	MASONRY	141.5	61.7	92.4	115.4	131.2	125.1	106.5	131.2	121.7	105.8	131.2	121.4	106.2	131.2	121.6	108.9	131.2	122.6
05	METALS	101.5	76.6	93.9	93.0	115.9	100.0	93.0	115.8	99.9	97.6	115.8	103.2	90.3	115.8	98.1	89.4	115.7	97.5
06	WOOD, PLASTICS & COMPOSITES	101.9	51.8	74.4	110.6	117.4	114.4	110.6	117.4	114.4	95.3	117.4	107.4	110.6	117.4	114.4	110.6	117.4	114.4
07	THERMAL & MOISTURE PROTECTION	104.8	67.8	89.4	101.1	123.5	110.4	101.2	122.0	109.8	106.4	122.0	112.8	101.2	122.0	109.8	101.2	119.7	108.9
08	OPENINGS	96.8	58.9	88.2	95.7	122.3	101.7	95.7	122.3	101.7	94.0	122.3	100.4	98.0	122.3	103.5	95.7	122.3	101.7
0920	Plaster & Gypsum Board	77.5	50.5	59.5	115.6	117.5	116.8	115.6	117.5	116.8	108.1	117.5	114.3	116.8	117.5	117.2	115.6	117.5	116.8
0950, 0980	Ceilings & Acoustic Treatment	98.5	50.5	66.6	92.7	117.5	109.2	92.7	117.5	109.2	94.9	117.5	109.9	95.8	117.5	110.2	92.7	117.5	109.2
0960	Flooring	121.1	73.1	107.2	92.9	129.4	103.5	92.9	124.7	102.1	96.0	129.4	105.7	92.9	124.7	102.1	92.9	124.7	102.1
0970, 0990	Wall Finishes & Painting/Coating	105.9	47.7	71.8	95.6	128.9	115.2	95.6	127.7	114.4	97.8	127.7	115.4	95.6	128.9	115.2	95.6	127.7	114.4
09	FINISHES	102.7	56.2	77.3	93.8	121.4	108.9	93.9	120.5	108.4	93.8	121.3	108.8	94.7	120.6	108.9	93.9	120.5	108.4
COVERS	DIVS. 10 - 14, 25, 28, 41, 43, 44, 46	100.0	83.3	96.2	100.0	112.6	102.9	100.0	112.6	102.9	100.0	112.6	102.9	100.0	112.6	102.9	100.0	112.6	102.9
21, 22, 23	FIRE SUPPRESSION, PLUMBING & HVAC	96.4	63.5	81.8	100.0	116.0	107.1	100.0	116.0	107.1	100.0	116.0	107.1	96.5	116.0	105.1	100.0	116.0	107.1
26, 27, 3370	ELECTRICAL, COMMUNICATIONS & UTIL.	97.5	62.4	78.9	100.1	109.1	104.9	100.1	109.1	104.9	99.5	109.1	104.8	100.0	108.5	104.5	100.2	109.1	104.9
MF2014	WEIGHTED AVERAGE	103.6	66.2	87.0	98.3	118.0	107.0	97.5	117.8	106.5	97.9	117.9	106.8	96.3	117.8	105.8	97.1	117.7	106.2

DIVISION		CONNECTICUT																	
		NEW HAVEN			NEW LONDON			NORWALK			STAMFORD			WATERBURY			WILLIMANTIC		
		065			063			068			069			067			062		
		MAT.	INST.	TOTAL	MAT.	INST.	TOTAL	MAT.	INST.	TOTAL	MAT.	INST.	TOTAL	MAT.	INST.	TOTAL	MAT.	INST.	TOTAL
015433	CONTRACTOR EQUIPMENT		100.2	100.2		100.2	100.2		99.8	99.8		99.8	99.8		99.8	99.8		99.8	99.8
0241, 31 - 34	SITE & INFRASTRUCTURE, DEMOLITION	101.7	104.6	103.7	94.3	104.7	101.5	102.2	104.0	103.5	102.8	104.0	103.7	102.2	104.0	103.5	102.2	103.9	103.4
0310	Concrete Forming & Accessories	105.0	118.9	117.0	105.0	118.9	117.0	105.3	119.3	117.4	105.3	119.3	117.4	105.3	118.9	117.1	105.3	118.9	117.0
0320	Concrete Reinforcing	115.4	140.0	127.9	90.5	140.0	115.6	115.4	140.2	128.0	115.4	140.2	128.0	115.4	140.0	127.9	115.4	140.0	127.9
0330	Cast-in-Place Concrete	90.2	128.7	104.8	76.9	128.7	96.6	91.6	130.1	106.3	93.1	130.1	107.2	93.1	128.7	106.7	87.0	128.7	102.9
03	CONCRETE	107.3	125.2	115.6	82.5	125.2	102.3	95.5	125.9	109.6	96.3	125.9	110.0	96.3	125.2	109.7	93.3	125.2	108.1
04	MASONRY	106.8	131.2	121.8	105.2	131.2	121.2	106.1	132.6	122.4	107.0	132.6	122.7	107.0	131.2	121.9	106.4	131.2	121.7
05	METALS	89.7	115.8	97.6	89.4	115.8	97.4	93.0	116.3	100.1	93.0	116.3	100.1	93.0	115.8	100.0	92.7	115.8	99.8
06	WOOD, PLASTICS & COMPOSITES	110.6	117.4	114.4	110.6	117.4	114.4	110.6	117.4	114.4	110.6	117.4	114.4	110.6	117.4	114.4	110.6	117.4	114.4
07	THERMAL & MOISTURE PROTECTION	101.3	121.3	109.6	101.1	122.0	109.8	101.2	124.1	110.7	101.2	124.1	110.7	101.2	121.3	109.6	101.4	121.9	109.9
08	OPENINGS	95.7	122.3	101.7	98.0	122.3	103.5	95.7	122.3	101.7	95.7	122.3	101.7	95.7	122.3	101.7	98.1	122.3	103.5
0920	Plaster & Gypsum Board	115.6	117.5	116.8	115.6	117.5	116.8	115.6	117.5	116.8	115.6	117.5	116.8	115.6	117.5	116.8	115.6	117.5	116.8
0950, 0980	Ceilings & Acoustic Treatment	92.7	117.5	109.2	90.8	117.5	108.5	92.7	117.5	109.2	92.7	117.5	109.2	92.7	117.5	109.2	90.8	117.5	108.5
0960	Flooring	92.9	129.4	103.5	92.9	129.4	103.5	92.9	124.7	102.1	92.9	124.7	102.1	92.9	129.4	103.5	92.9	128.3	103.2
0970, 0990	Wall Finishes & Painting/Coating	95.6	128.9	115.2	95.6	128.9	115.2	95.6	128.9	115.2	95.6	128.9	115.2	95.6	128.9	115.2	95.6	128.9	115.2
09	FINISHES	93.9	121.4	108.9	93.0	121.4	108.6	93.9	120.6	108.5	93.9	120.6	108.5	93.8	121.4	108.9	93.6	121.2	108.7
COVERS	DIVS. 10 - 14, 25, 28, 41, 43, 44, 46	100.0	112.6	102.9	100.0	112.6	102.9	100.0	112.8	102.9	100.0	112.8	102.9	100.0	112.6	102.9	100.0	112.6	102.9
21, 22, 23	FIRE SUPPRESSION, PLUMBING & HVAC	100.0	116.0	107.1	96.5	116.0	105.1	100.0	116.1	107.1	100.0	116.1	107.1	100.0	116.0	107.1	100.0	116.0	107.1
26, 27, 3370	ELECTRICAL, COMMUNICATIONS & UTIL.	100.0	108.5	104.5	96.8	110.5	104.1	100.1	159.5	131.6	100.1	162.7	133.3	99.6	114.4	107.4	100.1	111.4	106.1
MF2014	WEIGHTED AVERAGE	98.7	117.8	107.1	94.4	118.1	104.9	97.8	125.2	109.9	97.9	125.6	110.2	97.8	118.6	107.0	97.7	118.2	106.8

DIVISION		D.C.			DELAWARE									FLORIDA					
		WASHINGTON			DOVER			NEWARK			WILMINGTON			DAYTONA BEACH			FORT LAUDERDALE		
		200 - 205			199			197			198			321			333		
		MAT.	INST.	TOTAL	MAT.	INST.	TOTAL	MAT.	INST.	TOTAL	MAT.	INST.	TOTAL	MAT.	INST.	TOTAL	MAT.	INST.	TOTAL
015433	CONTRACTOR EQUIPMENT		107.2	107.2		124.1	124.1		124.1	124.1		124.3	124.3		103.9	103.9		96.6	96.6
0241, 31 - 34	SITE & INFRASTRUCTURE, DEMOLITION	105.0	96.0	98.7	104.6	115.6	112.3	104.3	114.4	111.4	103.5	114.8	111.4	100.2	90.2	93.2	92.4	77.4	81.9
0310	Concrete Forming & Accessories	98.5	72.2	75.8	99.3	102.5	102.0	101.0	102.3	102.1	100.0	102.3	102.0	99.9	62.6	67.6	99.3	58.0	63.6
0320	Concrete Reinforcing	99.8	84.4	92.0	100.1	115.3	107.8	98.1	115.3	106.8	100.1	115.3	107.8	89.6	61.7	75.4	83.9	56.5	70.0
0330	Cast-in-Place Concrete	112.2	80.2	100.0	101.7	105.2	103.0	88.6	105.1	94.9	96.3	105.1	99.7	86.0	66.7	78.6	90.7	63.7	80.4
03	CONCRETE	106.9	78.4	93.6	99.0	106.8	102.6	93.4	106.7	99.6	96.4	106.7	101.2	87.4	65.7	77.3	90.2	61.7	77.0
04	MASONRY	104.7	74.9	86.3	102.9	98.4	100.1	104.5	98.4	100.7	103.1	98.4	100.2	91.2	70.5	78.4	96.6	65.7	77.5
05	METALS	98.4	96.9	97.9	98.3	124.9	106.5	99.7	124.8	107.3	98.3	124.8	106.4	101.9	91.1	98.6	99.4	88.5	96.1
06	WOOD, PLASTICS & COMPOSITES	93.1	70.5	80.7	91.1	101.6	96.8	96.6	101.6	99.3	87.5	101.6	95.2	100.8	61.2	79.1	90.6	58.1	72.8
07	THERMAL & MOISTURE PROTECTION	101.4	83.9	94.1	102.2	112.5	106.5	106.2	112.5	108.8	102.6	112.5	106.7	97.8	69.1	85.9	100.8	63.7	85.4
08	OPENINGS	96.9	73.9	91.7	88.0	111.9	93.5	92.1	111.9	96.6	85.7	111.9	91.6	95.5	60.1	87.5	96.3	57.2	87.5
0920	Plaster & Gypsum Board	105.6	69.6	81.7	97.3	101.5	100.1	99.2	101.5	100.8	96.7	101.5	99.9	88.7	60.6	70.0	97.7	57.4	70.9
0950, 0980	Ceilings & Acoustic Treatment	106.3	69.6	81.9	95.9	101.5	99.7	92.5	101.5	98.5	93.4	101.5	98.8	80.6	60.6	67.3	84.6	57.4	66.5
0960	Flooring	95.3	77.1	90.0	93.4	106.0	97.1	91.8	106.0	95.9	93.7	106.0	97.3	103.5	78.7	96.3	100.8	78.7	94.4
0970, 0990	Wall Finishes & Painting/Coating	106.8	76.3	88.9	91.1	112.7	103.8	89.2	112.7	103.0	86.9	112.7	102.1	98.1	61.8	76.8	90.9	64.3	75.3
09	FINISHES	97.2	72.7	83.8	91.1	103.4	97.8	91.0	103.4	97.8	90.5	103.4	97.5	90.9	65.0	76.7	91.4	62.3	75.5
COVERS	DIVS. 10 - 14, 25, 28, 41, 43, 44, 46	100.0	95.6	99.0	100.0	105.9	101.3	100.0	105.9	101.3	100.0	105.9	101.3	100.0	82.5	96.0	100.0	82.0	95.9
21, 22, 23	FIRE SUPPRESSION, PLUMBING & HVAC	100.0	88.4	94.9	100.0	120.5	109.1	100.1	120.5	109.2	100.0	120.5	109.1	99.9	72.5	87.7	99.9	57.6	81.2
26, 27, 3370	ELECTRICAL, COMMUNICATIONS & UTIL.	98.7	101.2	100.0	98.4	113.4	106.3	98.7	113.4	106.4	98.5	113.4	106.4	94.4	63.3	77.9	94.3	68.3	80.5
MF2014	WEIGHTED AVERAGE	100.2	86.1	94.0	97.6	112.1	104.0	97.9	111.9	104.1	97.0	112.0	103.6	96.4	71.8	85.6	96.6	66.4	83.2

FLORIDA

DIVISION		FORT MYERS 339,341 MAT.	INST.	TOTAL	GAINESVILLE 326,344 MAT.	INST.	TOTAL	JACKSONVILLE 320,322 MAT.	INST.	TOTAL	LAKELAND 338 MAT.	INST.	TOTAL	MELBOURNE 329 MAT.	INST.	TOTAL	MIAMI 330-332,340 MAT.	INST.	TOTAL
015433	CONTRACTOR EQUIPMENT		103.9	103.9		103.9	103.9		103.9	103.9		103.9	103.9		103.9	103.9		96.6	96.6
0241, 31 - 34	SITE & INFRASTRUCTURE, DEMOLITION	103.9	89.8	94.0	107.8	90.0	95.4	100.3	90.0	93.1	105.8	90.0	94.8	106.9	90.1	95.2	93.8	77.4	82.4
0310	Concrete Forming & Accessories	94.9	58.0	63.0	94.8	56.3	61.4	99.6	56.3	62.1	91.0	60.5	64.7	96.0	63.0	67.5	104.1	58.1	64.3
0320	Concrete Reinforcing	84.8	56.5	70.5	95.1	56.4	75.5	89.6	56.4	72.8	86.9	75.7	81.3	90.6	65.1	77.7	90.0	56.5	73.0
0330	Cast-in-Place Concrete	94.6	63.7	82.8	98.8	64.9	85.9	86.8	64.9	78.5	96.7	64.8	84.6	103.7	66.2	89.4	87.6	63.8	78.5
03	CONCRETE	90.8	61.7	77.3	98.1	61.3	81.1	87.8	61.3	75.5	92.5	66.6	80.5	98.6	66.3	83.6	88.4	61.7	76.0
04	MASONRY	90.1	65.7	75.1	104.5	67.6	81.8	90.9	67.6	76.6	106.5	67.3	82.4	88.8	69.6	77.0	96.8	65.7	77.6
05	METALS	101.6	88.5	97.6	100.9	88.9	97.2	100.4	88.9	96.9	101.6	95.7	99.8	110.8	92.3	105.1	99.7	88.6	96.3
06	WOOD, PLASTICS & COMPOSITES	87.5	58.1	71.4	94.8	54.7	72.7	100.8	54.7	75.5	82.7	60.4	70.4	96.4	62.2	77.6	95.9	58.1	75.2
07	THERMAL & MOISTURE PROTECTION	100.6	64.0	85.4	98.1	64.6	84.2	98.0	64.5	84.1	100.6	65.0	85.8	98.3	66.6	85.1	100.5	63.7	85.2
08	OPENINGS	97.6	57.2	88.4	95.0	55.3	86.1	95.5	55.3	86.4	97.5	62.9	89.7	94.7	61.4	87.2	98.3	57.2	89.0
0920	Plaster & Gypsum Board	93.8	57.4	69.6	85.5	53.9	64.5	88.7	53.9	65.6	90.5	59.7	70.1	85.5	61.6	69.6	93.7	57.4	69.6
0950, 0980	Ceilings & Acoustic Treatment	80.1	57.4	65.0	75.0	53.9	61.0	80.6	53.9	62.8	80.1	59.7	66.5	79.8	61.6	67.7	89.7	57.4	68.2
0960	Flooring	97.6	76.9	91.6	100.8	78.3	94.3	103.5	78.7	96.3	95.3	77.0	90.0	101.0	78.7	94.5	99.8	78.7	93.7
0970, 0990	Wall Finishes & Painting/Coating	95.0	64.2	76.9	98.1	61.8	76.8	98.1	64.2	78.2	95.0	64.2	76.9	98.1	81.6	88.4	88.9	64.3	74.4
09	FINISHES	91.0	61.9	75.1	89.2	60.4	73.5	90.9	60.8	74.4	90.0	63.6	75.6	90.0	67.5	77.7	91.7	62.3	75.6
COVERS	DIVS. 10 - 14, 25, 28, 41, 43, 44, 46	100.0	77.0	94.8	100.0	80.8	95.6	100.0	78.9	95.2	100.0	77.8	95.0	100.0	82.3	96.0	100.0	82.0	95.9
21, 22, 23	FIRE SUPPRESSION, PLUMBING & HVAC	97.9	57.6	80.1	98.6	60.9	81.9	99.9	60.9	82.6	97.9	57.8	80.2	99.9	74.9	88.8	99.9	57.9	81.3
26, 27, 3370	ELECTRICAL, COMMUNICATIONS & UTIL.	96.3	59.5	76.8	94.7	59.6	76.1	94.1	64.0	78.1	94.5	65.0	78.9	95.5	62.6	78.0	98.0	71.4	83.9
MF2014	WEIGHTED AVERAGE	96.9	65.9	83.2	98.0	66.7	84.2	96.2	67.3	83.4	97.6	68.8	84.9	99.3	72.7	87.5	97.1	66.9	83.8

FLORIDA

DIVISION		ORLANDO 327-328,347 MAT.	INST.	TOTAL	PANAMA CITY 324 MAT.	INST.	TOTAL	PENSACOLA 325 MAT.	INST.	TOTAL	SARASOTA 342 MAT.	INST.	TOTAL	ST. PETERSBURG 337 MAT.	INST.	TOTAL	TALLAHASSEE 323 MAT.	INST.	TOTAL
015433	CONTRACTOR EQUIPMENT		103.9	103.9		103.9	103.9		103.9	103.9		103.9	103.9		103.9	103.9		103.9	103.9
0241, 31 - 34	SITE & INFRASTRUCTURE, DEMOLITION	99.0	90.2	92.8	111.1	89.5	96.0	111.5	89.9	96.4	112.5	90.0	96.8	107.4	89.7	95.1	98.3	90.0	92.5
0310	Concrete Forming & Accessories	102.9	63.1	68.4	98.9	43.0	50.5	96.8	66.2	70.4	97.5	60.5	65.5	98.6	59.9	65.1	101.3	56.4	62.4
0320	Concrete Reinforcing	92.8	62.9	77.6	93.6	68.8	81.0	96.0	68.8	82.2	89.8	73.3	81.4	86.9	73.3	80.0	96.8	56.5	76.3
0330	Cast-in-Place Concrete	100.9	66.6	87.8	91.3	60.8	79.7	112.8	65.9	94.9	102.4	64.8	88.1	97.8	62.9	84.5	90.3	64.9	80.6
03	CONCRETE	94.2	66.1	81.1	96.3	56.1	77.7	106.0	68.3	88.5	96.8	66.1	82.6	94.0	65.2	80.7	90.2	61.4	76.8
04	MASONRY	96.5	70.5	80.5	95.5	60.5	74.0	115.9	69.4	87.3	94.3	67.3	77.7	146.7	64.1	95.8	93.3	67.6	77.5
05	METALS	99.8	91.3	97.2	101.7	92.9	99.0	102.7	93.1	99.8	103.5	94.8	100.8	102.4	94.7	100.1	98.9	89.0	95.9
06	WOOD, PLASTICS & COMPOSITES	92.7	62.2	75.9	99.6	40.6	67.2	97.6	67.4	81.1	98.0	60.4	77.3	92.0	61.3	75.1	100.1	54.7	75.1
07	THERMAL & MOISTURE PROTECTION	98.6	69.2	86.4	98.3	58.5	81.8	98.3	67.3	85.4	99.0	65.0	84.9	100.8	63.2	85.2	99.5	64.6	85.0
08	OPENINGS	99.2	60.9	90.6	93.5	50.5	83.7	93.5	65.2	87.1	99.1	62.4	90.8	96.3	62.9	88.7	98.2	55.3	88.5
0920	Plaster & Gypsum Board	96.7	61.6	73.3	87.8	39.4	55.6	95.5	67.0	76.6	95.0*	59.7	71.6	96.1	60.7	72.6	96.8	53.9	68.3
0950, 0980	Ceilings & Acoustic Treatment	88.8	61.6	70.7	79.8	39.4	52.9	79.8	67.0	71.3	84.7	59.7	68.1	82.0	60.7	67.8	88.1	53.9	65.3
0960	Flooring	97.7	78.7	92.2	103.0	76.9	95.4	98.6	78.7	92.8	106.0	76.9	97.6	99.6	76.9	93.0	99.2	78.7	93.2
0970, 0990	Wall Finishes & Painting/Coating	96.8	64.2	77.7	98.1	61.8	76.8	98.1	61.8	76.8	98.6	64.2	78.4	95.0	64.3	76.9	95.2	64.2	77.0
09	FINISHES	92.3	65.9	77.8	91.5	50.1	68.9	91.0	68.4	78.7	96.3	63.6	78.5	92.6	63.4	76.6	91.5	60.8	74.7
COVERS	DIVS. 10 - 14, 25, 28, 41, 43, 44, 46	100.0	82.6	96.1	100.0	69.3	93.1	100.0	75.2	94.4	100.0	77.8	95.0	100.0	75.0	94.4	100.0	77.1	94.8
21, 22, 23	FIRE SUPPRESSION, PLUMBING & HVAC	100.0	58.1	81.5	99.9	48.4	77.1	99.9	52.9	79.1	99.8	57.8	81.3	99.9	56.1	80.5	99.9	68.3	85.9
26, 27, 3370	ELECTRICAL, COMMUNICATIONS & UTIL.	98.4	64.3	80.3	93.2	59.6	75.4	96.8	53.6	73.9	96.6	64.0	79.3	94.5	63.8	78.2	99.0	59.6	78.1
MF2014	WEIGHTED AVERAGE	98.1	69.1	85.3	97.7	60.7	81.4	100.4	67.1	85.7	99.4	68.5	85.8	100.6	67.5	86.0	97.2	68.2	84.4

FLORIDA / GEORGIA

DIVISION		TAMPA 335-336,346 MAT.	INST.	TOTAL	WEST PALM BEACH 334,349 MAT.	INST.	TOTAL	ALBANY 317,398 MAT.	INST.	TOTAL	ATHENS 306 MAT.	INST.	TOTAL	ATLANTA 300-303,399 MAT.	INST.	TOTAL	AUGUSTA 308-309 MAT.	INST.	TOTAL
015433	CONTRACTOR EQUIPMENT		103.9	103.9		96.6	96.6		97.6	97.6		96.8	96.8		95.8	95.8		96.8	96.8
0241, 31 - 34	SITE & INFRASTRUCTURE, DEMOLITION	107.8	90.2	95.5	89.3	77.4	81.0	100.6	80.3	86.4	101.6	96.9	98.3	98.1	94.8	95.8	94.7	97.2	96.5
0310	Concrete Forming & Accessories	101.8	61.6	67.1	103.0	58.0	64.1	95.3	62.4	66.8	92.4	45.9	52.2	96.3	73.3	76.4	93.5	73.8	76.4
0320	Concrete Reinforcing	83.9	75.7	79.7	86.3	54.2	70.0	90.6	71.7	81.0	94.0	66.2	79.9	93.3	71.0	82.0	94.4	68.2	81.1
0330	Cast-in-Place Concrete	95.6	66.5	84.5	86.2	63.7	77.7	86.8	69.6	80.3	106.7	70.6	93.0	106.7	72.0	93.5	100.8	71.2	89.5
03	CONCRETE	92.7	67.6	81.1	87.0	61.3	75.1	88.7	68.5	79.4	102.5	59.4	82.5	99.9	73.1	87.5	95.1	72.5	84.6
04	MASONRY	97.6	70.2	80.7	96.1	65.7	77.4	93.8	80.8	85.8	80.5	81.1	80.8	94.4	67.8	78.0	94.6	81.5	86.5
05	METALS	101.5	95.7	99.7	98.5	87.7	95.2	100.2	99.6	100.0	96.7	81.4	92.0	97.6	84.6	93.6	96.4	83.1	92.3
06	WOOD, PLASTICS & COMPOSITES	96.0	60.4	76.4	95.6	58.1	75.0	86.1	60.6	72.1	93.3	38.6	63.3	97.2	74.8	84.9	94.8	76.2	84.5
07	THERMAL & MOISTURE PROTECTION	101.0	66.3	86.6	100.6	64.3	85.5	99.8	71.7	88.1	95.8	71.5	85.7	95.6	72.4	86.0	95.5	75.8	87.3
08	OPENINGS	97.5	62.9	89.7	95.9	56.7	87.0	87.9	66.2	83.0	92.5	52.7	83.5	97.9	76.7	93.1	92.5	74.1	88.3
0920	Plaster & Gypsum Board	98.4	59.7	72.7	102.1	57.4	72.4	103.6	60.0	74.6	94.7	37.3	56.5	97.0	74.3	81.9	96.0	75.8	82.6
0950, 0980	Ceilings & Acoustic Treatment	84.6	59.7	68.0	80.1	57.4	65.0	81.3	60.0	67.1	95.0	37.3	56.6	95.0	74.3	81.2	95.9	75.8	82.6
0960	Flooring	100.8	77.0	93.9	102.8	78.7	95.8	103.4	88.5	99.1	98.3	86.8	94.9	99.9	86.8	96.1	98.5	86.8	95.1
0970, 0990	Wall Finishes & Painting/Coating	95.0	64.2	76.9	90.9	61.8	73.8	92.3	95.6	94.2	94.9	95.6	95.3	94.9	95.6	95.3	94.9	76.1	83.9
09	FINISHES	93.8	64.4	77.7	91.4	62.0	75.4	94.8	70.0	81.3	96.2	56.8	74.7	96.6	78.2	86.5	96.0	76.8	85.5
COVERS	DIVS. 10 - 14, 25, 28, 41, 43, 44, 46	100.0	80.9	95.7	100.0	82.0	95.9	100.0	83.8	96.3	100.0	79.1	95.3	100.0	86.1	96.9	100.0	85.4	96.7
21, 22, 23	FIRE SUPPRESSION, PLUMBING & HVAC	99.9	59.3	81.9	97.9	57.4	80.0	99.9	70.2	86.8	96.5	68.7	84.2	100.0	70.5	87.0	100.1	69.6	86.6
26, 27, 3370	ELECTRICAL, COMMUNICATIONS & UTIL.	94.3	63.8	78.1	95.4	68.3	81.0	96.7	63.4	79.1	100.4	69.2	83.9	99.5	72.5	85.2	101.1	67.1	83.1
MF2014	WEIGHTED AVERAGE	98.1	69.7	85.5	95.6	66.2	82.6	96.1	73.8	86.3	96.8	70.0	84.9	98.5	75.9	88.5	97.1	76.1	87.8

For customer support on your Building Construction Costs with RSMeans Data, call 800.448.8182.

769

City Cost Indexes

GEORGIA

DIVISION		COLUMBUS 318-319			DALTON 307			GAINESVILLE 305			MACON 310-312			SAVANNAH 313-314			STATESBORO 304		
		MAT.	INST.	TOTAL	MAT.	INST.	TOTAL	MAT.	INST.	TOTAL	MAT.	INST.	TOTAL	MAT.	INST.	TOTAL	MAT.	INST.	TOTAL
015433	CONTRACTOR EQUIPMENT		97.6	97.6		113.0	113.0		96.8	96.8		107.8	107.8		98.5	98.5		100.0	100.0
0241, 31 - 34	SITE & INFRASTRUCTURE, DEMOLITION	100.5	80.5	86.5	101.2	102.6	102.2	101.4	96.9	98.3	102.1	96.1	97.9	100.5	81.9	87.5	102.2	82.8	88.7
0310	Concrete Forming & Accessories	95.3	70.8	74.1	85.2	49.5	54.3	96.0	43.2	50.4	94.9	48.2	54.5	96.5	72.7	75.9	80.0	54.2	57.6
0320	Concrete Reinforcing	90.4	71.8	81.0	93.5	63.3	78.2	93.8	66.1	79.7	91.6	71.8	81.5	97.3	68.3	82.6	93.0	37.0	64.6
0330	Cast-in-Place Concrete	86.5	69.8	80.2	103.6	69.6	90.7	112.2	70.4	96.2	85.3	71.1	79.9	90.2	69.8	82.4	106.5	69.3	92.3
03	CONCRETE	88.5	72.5	81.1	101.7	61.3	83.0	104.3	58.1	82.9	88.1	62.7	76.3	89.9	72.6	81.9	102.0	58.7	81.9
04	MASONRY	93.8	81.1	86.0	81.8	81.0	81.3	89.1	81.1	84.1	106.4	81.2	90.9	88.7	81.4	84.2	83.6	81.0	82.0
05	METALS	99.9	100.5	100.1	97.9	96.3	97.4	96.0	80.8	91.3	95.3	100.3	96.8	101.5	86.3	96.9	101.5	86.3	96.9
06	WOOD, PLASTICS & COMPOSITES	86.1	71.8	78.3	76.7	44.9	59.2	97.0	35.8	63.4	92.9	41.4	64.6	91.8	74.5	82.3	70.6	50.9	59.8
07	THERMAL & MOISTURE PROTECTION	99.7	74.5	89.2	97.8	71.3	86.8	95.7	71.2	85.5	98.1	72.2	87.3	97.3	74.1	87.7	96.4	70.8	85.8
08	OPENINGS	87.9	72.5	84.4	93.3	55.4	84.7	92.4	51.1	83.1	87.6	55.7	80.4	93.4	73.2	88.8	94.3	51.2	84.5
0920	Plaster & Gypsum Board	103.6	71.5	82.2	83.4	43.7	57.0	97.0	34.4	55.4	109.5	40.2	63.5	100.1	74.3	82.9	83.0	50.0	61.1
0950, 0980	Ceilings & Acoustic Treatment	81.3	71.5	74.8	107.3	43.7	65.0	95.0	34.4	54.7	76.4	40.2	52.3	90.6	74.3	79.7	103.6	50.0	67.9
0960	Flooring	103.4	86.8	98.6	99.4	86.8	95.7	99.9	86.8	96.1	81.6	88.9	83.7	98.8	88.9	95.9	117.2	86.8	108.4
0970, 0990	Wall Finishes & Painting/Coating	92.3	95.6	94.2	85.8	74.6	79.3	94.9	95.6	95.3	94.4	74.6	82.8	91.5	80.6	85.1	92.6	74.6	82.1
09	FINISHES	94.7	76.3	84.6	104.9	58.2	79.4	96.8	55.2	74.1	84.1	56.6	69.1	93.4	76.7	84.3	108.4	61.7	82.9
COVERS	DIVS. 10 - 14, 25, 28, 41, 43, 44, 46	100.0	85.0	96.6	100.0	26.8	83.4	100.0	37.6	85.9	100.0	81.8	95.9	100.0	85.0	96.6	100.0	44.0	87.3
21, 22, 23	FIRE SUPPRESSION, PLUMBING & HVAC	100.0	67.6	85.6	96.6	63.7	82.0	96.5	67.4	83.6	100.0	69.9	86.6	100.0	67.6	85.7	97.1	67.1	83.8
26, 27, 3370	ELECTRICAL, COMMUNICATIONS & UTIL.	96.9	72.5	84.0	108.7	67.4	86.8	100.4	72.4	85.6	95.8	63.2	78.5	101.1	74.3	86.9	100.4	57.1	77.4
MF2014	WEIGHTED AVERAGE	96.0	76.5	87.4	98.5	69.5	85.7	97.4	68.3	84.5	94.9	71.8	84.7	96.2	76.8	87.6	98.9	66.9	84.7

DIVISION		GEORGIA VALDOSTA 316			WAYCROSS 315			HAWAII HILO 967			HONOLULU 968			STATES & POSS., GUAM 969			IDAHO BOISE 836 - 837		
		MAT.	INST.	TOTAL	MAT.	INST.	TOTAL	MAT.	INST.	TOTAL	MAT.	INST.	TOTAL	MAT.	INST.	TOTAL	MAT.	INST.	TOTAL
015433	CONTRACTOR EQUIPMENT		97.6	97.6		97.6	97.6		100.7	100.7		100.7	100.7		168.4	168.4		98.8	98.8
0241, 31 - 34	SITE & INFRASTRUCTURE, DEMOLITION	110.0	80.3	89.3	106.8	81.5	89.2	156.2	106.9	121.8	164.8	106.9	124.4	202.8	104.2	134.0	87.6	97.7	94.6
0310	Concrete Forming & Accessories	85.2	43.6	49.2	87.3	67.3	70.0	111.8	121.2	119.9	124.5	121.2	121.6	114.4	55.4	63.4	101.2	81.6	84.2
0320	Concrete Reinforcing	92.5	62.0	77.0	92.5	62.2	77.1	145.1	112.9	128.7	168.8	112.9	140.4	262.3	28.5	143.6	104.8	77.5	90.9
0330	Cast-in-Place Concrete	85.0	69.6	79.1	96.0	69.3	85.8	193.9	121.6	166.4	155.9	121.6	142.9	168.1	102.2	143.0	88.4	85.3	87.2
03	CONCRETE	93.7	58.4	77.3	96.5	68.8	83.7	157.7	118.9	139.7	148.9	118.9	135.0	160.7	68.0	117.7	99.0	82.3	91.3
04	MASONRY	99.3	81.4	88.3	100.2	81.0	88.4	155.9	113.7	129.9	142.4	113.7	124.7	221.4	37.4	108.1	128.3	87.9	103.5
05	METALS	99.4	95.4	98.2	98.5	91.0	96.2	108.7	101.9	106.6	121.3	101.9	115.3	139.9	78.0	121.0	107.4	81.2	99.4
06	WOOD, PLASTICS & COMPOSITES	74.3	35.5	53.0	75.9	68.5	71.8	115.3	122.6	119.3	129.7	122.6	125.8	127.1	58.1	89.2	93.0	80.5	86.1
07	THERMAL & MOISTURE PROTECTION	99.9	69.8	87.4	99.7	72.4	88.4	135.5	116.8	127.7	154.1	116.8	138.6	158.8	62.4	118.8	95.7	84.1	90.9
08	OPENINGS	84.4	49.9	76.6	84.7	65.2	80.3	117.3	118.4	117.5	132.0	118.4	128.9	121.9	46.9	104.9	97.4	74.8	92.3
0920	Plaster & Gypsum Board	95.7	34.2	54.9	95.7	68.1	77.4	108.1	123.0	118.0	152.4	123.0	132.8	216.4	46.0	103.1	92.2	80.0	84.1
0950, 0980	Ceilings & Acoustic Treatment	78.7	34.2	49.1	76.8	68.1	71.0	127.6	123.0	124.5	137.2	123.0	127.7	242.0	46.0	111.6	99.5	80.0	86.5
0960	Flooring	97.1	88.5	94.6	98.3	86.8	95.0	114.8	126.3	118.1	134.4	126.3	132.0	135.3	41.9	108.2	96.6	90.3	94.7
0970, 0990	Wall Finishes & Painting/Coating	92.3	95.6	94.2	92.3	74.6	81.9	99.1	139.8	123.0	107.1	139.8	126.3	104.0	33.6	62.7	96.0	42.1	64.3
09	FINISHES	92.2	55.4	72.1	91.8	72.0	81.0	112.1	124.7	118.9	127.2	124.7	125.8	184.9	51.5	112.0	94.6	79.5	86.3
COVERS	DIVS. 10 - 14, 25, 28, 41, 43, 44, 46	100.0	80.7	95.6	100.0	51.9	89.1	100.0	111.1	102.5	100.0	111.1	102.5	100.0	70.1	93.2	100.0	86.4	96.9
21, 22, 23	FIRE SUPPRESSION, PLUMBING & HVAC	100.0	69.9	86.6	97.7	64.6	83.1	100.5	106.5	103.1	100.5	106.5	103.1	103.7	35.3	73.4	100.0	73.0	88.1
26, 27, 3370	ELECTRICAL, COMMUNICATIONS & UTIL.	95.0	56.0	74.3	99.3	57.1	76.9	105.6	121.2	113.9	107.1	121.2	114.6	152.4	38.9	92.2	97.7	69.5	82.8
MF2014	WEIGHTED AVERAGE	96.2	68.0	83.8	96.3	70.4	84.9	117.7	113.9	116.0	122.0	113.9	118.5	140.4	54.4	102.4	101.1	79.7	91.6

DIVISION		IDAHO COEUR D'ALENE 838			IDAHO FALLS 834			LEWISTON 835			POCATELLO 832			TWIN FALLS 833			ILLINOIS BLOOMINGTON 617		
		MAT.	INST.	TOTAL	MAT.	INST.	TOTAL	MAT.	INST.	TOTAL	MAT.	INST.	TOTAL	MAT.	INST.	TOTAL	MAT.	INST.	TOTAL
015433	CONTRACTOR EQUIPMENT		93.4	93.4		98.8	98.8		93.4	93.4		98.8	98.8		98.8	98.8		106.9	106.9
0241, 31 - 34	SITE & INFRASTRUCTURE, DEMOLITION	86.4	91.6	90.0	85.1	97.6	93.8	93.2	92.4	92.7	88.4	97.6	94.8	95.5	97.1	96.6	95.5	101.9	100.0
0310	Concrete Forming & Accessories	112.2	79.1	83.6	95.0	80.0	82.0	117.5	79.8	84.9	101.3	81.3	84.0	102.6	54.3	60.8	85.0	119.6	114.9
0320	Concrete Reinforcing	112.9	92.3	102.5	106.8	77.5	91.9	112.9	92.6	102.6	105.2	77.4	91.1	107.2	77.0	91.9	96.5	103.8	100.2
0330	Cast-in-Place Concrete	95.5	83.4	90.9	84.1	84.1	84.1	99.3	83.9	93.4	90.8	85.2	88.7	93.2	83.9	89.7	95.7	118.5	104.4
03	CONCRETE	105.5	83.1	95.1	90.8	81.1	86.3	109.1	83.6	97.3	98.0	82.1	90.6	106.0	69.4	89.1	95.2	117.1	105.4
04	MASONRY	130.0	81.8	100.3	123.2	85.8	100.2	130.4	85.1	102.5	125.7	87.9	102.5	128.6	85.8	102.2	117.3	124.6	121.8
05	METALS	100.6	86.2	96.2	115.7	81.1	105.1	100.0	86.8	96.0	115.8	80.9	105.1	115.8	80.6	105.1	99.2	125.6	107.2
06	WOOD, PLASTICS & COMPOSITES	98.1	77.7	86.9	87.2	79.6	83.0	103.4	77.7	89.3	93.0	80.5	86.2	94.0	45.3	67.2	86.8	117.6	103.7
07	THERMAL & MOISTURE PROTECTION	149.9	83.6	122.4	94.9	76.6	87.3	150.3	86.7	123.9	95.5	76.7	87.7	96.3	78.9	89.1	98.6	116.5	106.0
08	OPENINGS	114.7	75.2	105.8	100.4	74.3	94.5	108.4	76.4	101.1	98.1	68.3	91.4	101.2	45.8	88.7	94.7	122.2	100.9
0920	Plaster & Gypsum Board	166.2	77.2	107.1	77.4	79.0	78.5	167.9	77.2	107.6	79.2	80.0	79.7	80.9	43.7	56.2	92.8	118.1	109.6
0950, 0980	Ceilings & Acoustic Treatment	130.7	77.2	95.1	99.4	79.0	85.9	130.7	77.2	95.1	106.2	80.0	88.7	101.9	43.7	63.2	86.5	118.1	107.5
0960	Flooring	136.8	79.1	120.1	96.2	90.3	94.5	140.3	89.8	125.6	99.9	90.3	97.1	101.2	90.3	98.0	88.6	122.7	98.5
0970, 0990	Wall Finishes & Painting/Coating	115.1	73.1	90.4	96.0	42.1	64.3	115.1	73.1	90.4	95.9	42.1	64.3	96.0	42.1	64.3	87.5	141.7	119.3
09	FINISHES	159.1	78.2	114.9	91.8	78.4	84.5	160.5	80.6	116.8	95.1	79.5	86.6	95.5	58.3	75.2	88.9	122.8	107.4
COVERS	DIVS. 10 - 14, 25, 28, 41, 43, 44, 46	100.0	97.5	99.4	100.0	87.8	97.2	100.0	99.7	99.5	100.0	86.4	96.9	100.0	84.0	96.4	100.0	107.9	101.8
21, 22, 23	FIRE SUPPRESSION, PLUMBING & HVAC	99.6	81.2	91.4	101.0	70.7	87.6	101.0	84.3	93.6	99.9	73.0	88.0	99.9	69.6	86.5	96.4	106.7	101.0
26, 27, 3370	ELECTRICAL, COMMUNICATIONS & UTIL.	90.6	88.4	89.4	90.3	68.7	78.8	88.6	79.8	83.9	95.3	68.7	81.2	91.7	58.2	73.9	95.1	96.7	95.9
MF2014	WEIGHTED AVERAGE	109.0	83.7	97.8	100.7	78.3	90.8	109.1	84.1	98.1	102.1	79.0	91.9	103.5	70.8	89.1	97.1	113.1	104.2

ILLINOIS

DIVISION		CARBONDALE 629			CENTRALIA 628			CHAMPAIGN 618 - 619			CHICAGO 606 - 608			DECATUR 625			EAST ST. LOUIS 620 - 622		
		MAT.	INST.	TOTAL	MAT.	INST.	TOTAL	MAT.	INST.	TOTAL	MAT.	INST.	TOTAL	MAT.	INST.	TOTAL	MAT.	INST.	TOTAL
015433	CONTRACTOR EQUIPMENT		115.9	115.9		115.9	115.9		107.8	107.8		100.7	100.7		107.8	107.8		115.9	115.9
0241, 31 - 34	SITE & INFRASTRUCTURE, DEMOLITION	96.1	103.0	100.9	96.4	103.7	101.5	104.6	103.0	103.5	98.6	106.3	104.0	92.9	103.0	99.9	98.4	103.8	102.2
0310	Concrete Forming & Accessories	93.3	109.2	107.0	95.0	113.9	111.3	91.6	115.8	112.6	97.8	158.8	150.6	95.5	119.5	116.3	90.7	115.1	111.8
0320	Concrete Reinforcing	95.9	103.7	99.8	95.9	103.9	99.9	96.5	98.6	97.6	100.4	147.7	124.4	94.0	99.3	96.7	95.8	104.1	100.0
0330	Cast-in-Place Concrete	85.4	105.5	93.1	85.9	119.2	98.6	110.9	113.0	111.7	109.7	155.9	127.3	93.3	116.7	102.2	87.3	120.1	99.8
03	CONCRETE	82.7	108.5	94.7	83.2	115.5	98.2	107.8	112.6	110.0	103.3	155.1	127.3	93.3	115.6	103.7	84.2	116.4	99.1
04	MASONRY	79.1	113.6	100.3	79.2	116.2	101.9	142.3	121.3	129.3	103.2	165.8	141.8	74.7	116.6	100.5	79.4	121.1	105.1
05	METALS	98.8	132.5	109.1	98.9	134.8	109.9	99.2	121.4	106.0	95.6	144.3	110.5	102.5	121.0	108.2	99.9	135.3	110.8
06	WOOD, PLASTICS & COMPOSITES	91.2	106.0	99.3	93.6	111.7	103.6	93.2	114.0	104.6	100.0	157.4	131.6	91.5	119.1	106.7	88.7	112.4	101.7
07	THERMAL & MOISTURE PROTECTION	96.0	103.9	99.3	96.0	110.7	102.1	99.3	115.4	106.0	102.4	150.2	122.2	101.8	112.2	106.1	96.0	110.2	101.9
08	OPENINGS	89.5	118.3	96.0	89.5	121.4	96.7	95.3	116.7	100.2	104.2	167.5	118.5	100.7	119.8	105.0	89.6	121.8	96.9
0920	Plaster & Gypsum Board	97.3	106.2	103.2	98.3	112.1	107.5	94.7	114.4	107.8	98.0	159.0	138.5	99.7	119.7	113.0	96.3	112.7	107.2
0950, 0980	Ceilings & Acoustic Treatment	84.9	106.2	99.0	84.9	112.1	103.0	86.5	114.4	105.1	97.0	159.0	138.3	90.8	119.7	110.0	84.9	112.7	103.4
0960	Flooring	114.3	116.7	115.0	115.2	112.4	114.4	92.0	111.1	97.5	91.3	158.7	110.9	102.5	111.7	105.2	113.2	112.4	113.0
0970, 0990	Wall Finishes & Painting/Coating	97.8	106.8	103.1	97.8	93.0	95.0	87.5	101.7	95.9	89.8	162.6	132.5	89.6	113.7	103.8	97.8	111.4	105.8
09	FINISHES	94.1	109.1	102.3	94.6	111.0	103.6	90.9	113.7	103.4	93.6	159.9	129.9	94.0	118.1	107.2	93.8	114.3	105.0
COVERS	DIVS. 10 - 14, 25, 28, 41, 43, 44, 46	100.0	106.6	101.5	100.0	107.4	101.7	100.0	106.7	101.5	100.0	127.3	106.2	100.0	107.3	101.6	100.0	107.9	101.8
21, 22, 23	FIRE SUPPRESSION, PLUMBING & HVAC	96.4	106.9	101.1	96.4	95.1	95.8	96.4	106.1	100.7	100.0	139.7	117.5	100.0	98.4	99.3	100.0	98.7	99.4
26, 27, 3370	ELECTRICAL, COMMUNICATIONS & UTIL.	97.4	110.2	104.2	98.9	108.7	104.1	98.3	96.8	97.5	96.6	139.1	119.1	101.6	92.8	97.0	98.5	103.1	100.9
MF2014	WEIGHTED AVERAGE	93.7	111.0	101.3	93.9	110.4	101.2	100.8	110.2	104.9	99.5	146.0	120.0	97.9	108.6	102.6	94.9	111.5	102.3

ILLINOIS

DIVISION		EFFINGHAM 624			GALESBURG 614			JOLIET 604			KANKAKEE 609			LA SALLE 613			NORTH SUBURBAN 600 - 603		
		MAT.	INST.	TOTAL	MAT.	INST.	TOTAL	MAT.	INST.	TOTAL	MAT.	INST.	TOTAL	MAT.	INST.	TOTAL	MAT.	INST.	TOTAL
015433	CONTRACTOR EQUIPMENT		107.8	107.8		106.9	106.9		98.1	98.1		98.1	98.1		106.9	106.9		98.1	98.1
0241, 31 - 34	SITE & INFRASTRUCTURE, DEMOLITION	96.6	102.0	100.4	98.0	101.5	100.5	98.7	104.2	102.5	92.6	103.5	100.2	97.4	102.5	100.9	97.9	104.2	102.3
0310	Concrete Forming & Accessories	100.2	114.5	112.6	91.5	117.1	113.7	99.6	162.4	154.0	93.1	145.7	138.6	105.7	125.9	123.2	99.0	158.1	150.1
0320	Concrete Reinforcing	97.0	92.9	95.0	96.0	103.7	99.9	100.4	139.8	120.4	101.2	136.3	119.0	96.1	135.2	115.9	100.4	147.6	124.4
0330	Cast-in-Place Concrete	93.0	109.6	99.3	98.5	108.6	102.4	109.6	154.3	126.6	102.2	141.6	117.2	98.4	128.4	109.8	109.7	152.8	126.1
03	CONCRETE	94.1	109.6	101.3	98.2	112.6	104.9	103.4	154.6	127.2	97.2	142.1	118.0	99.0	129.2	113.0	103.4	153.6	126.7
04	MASONRY	83.0	117.0	103.9	117.5	124.6	121.9	106.4	166.2	143.2	102.7	152.7	133.5	117.5	150.6	137.9	103.2	164.5	140.9
05	METALS	99.7	115.5	104.6	99.2	124.7	107.0	93.8	138.5	107.5	93.8	135.9	106.6	99.2	145.2	113.3	94.8	142.2	109.3
06	WOOD, PLASTICS & COMPOSITES	93.8	114.0	104.9	93.1	115.5	105.4	101.6	162.3	135.0	94.8	144.3	122.0	107.7	123.3	116.3	100.0	157.4	131.6
07	THERMAL & MOISTURE PROTECTION	101.3	111.3	105.4	98.8	112.0	104.3	102.3	150.0	122.1	101.4	141.7	118.1	98.9	127.1	110.6	102.6	147.9	123.4
08	OPENINGS	95.4	114.9	99.8	94.7	116.1	99.5	101.7	167.6	116.7	94.8	156.6	108.8	94.7	135.0	103.8	101.8	167.4	116.7
0920	Plaster & Gypsum Board	99.6	114.4	109.4	94.7	116.0	108.9	95.0	164.0	140.9	92.7	145.5	127.8	101.7	124.0	116.5	98.0	159.0	138.5
0950, 0980	Ceilings & Acoustic Treatment	84.9	114.4	104.5	86.5	116.0	106.1	97.0	164.0	141.6	97.0	145.5	129.3	86.5	124.0	111.5	97.0	159.0	138.3
0960	Flooring	103.6	111.1	105.8	91.8	122.7	100.7	91.0	158.7	110.6	87.7	159.7	108.6	98.7	126.0	106.6	91.3	158.7	110.9
0970, 0990	Wall Finishes & Painting/Coating	89.6	111.4	102.4	87.5	101.7	95.9	88.1	181.6	143.0	88.1	141.7	119.6	87.5	141.7	119.3	89.8	162.6	132.5
09	FINISHES	93.1	114.4	104.8	90.3	117.2	105.0	93.1	165.0	132.4	91.4	148.3	122.5	93.3	126.4	111.4	93.6	159.6	129.7
COVERS	DIVS. 10 - 14, 25, 28, 41, 43, 44, 46	100.0	79.7	95.4	100.0	104.8	101.1	100.0	127.9	106.3	100.0	120.5	104.6	100.0	106.3	101.4	100.0	124.3	105.5
21, 22, 23	FIRE SUPPRESSION, PLUMBING & HVAC	96.5	105.3	100.4	96.4	104.1	99.8	100.0	135.5	115.7	96.4	131.1	111.8	96.4	125.2	109.1	99.9	134.6	115.3
26, 27, 3370	ELECTRICAL, COMMUNICATIONS & UTIL.	99.3	110.2	105.1	96.0	88.7	92.1	96.0	130.5	114.3	90.9	129.8	111.5	93.1	129.8	112.6	95.7	135.5	116.8
MF2014	WEIGHTED AVERAGE	96.4	109.5	102.2	97.8	109.4	102.9	99.0	143.9	118.8	95.6	136.3	113.6	98.0	129.1	111.7	99.0	143.5	118.7

ILLINOIS

DIVISION		PEORIA 615 - 616			QUINCY 623			ROCK ISLAND 612			ROCKFORD 610 - 611			SOUTH SUBURBAN 605			SPRINGFIELD 626 - 627		
		MAT.	INST.	TOTAL	MAT.	INST.	TOTAL	MAT.	INST.	TOTAL	MAT.	INST.	TOTAL	MAT.	INST.	TOTAL	MAT.	INST.	TOTAL
015433	CONTRACTOR EQUIPMENT		106.9	106.9		107.8	107.8		106.9	106.9		106.9	106.9		98.1	98.1		107.8	107.8
0241, 31 - 34	SITE & INFRASTRUCTURE, DEMOLITION	98.3	102.0	100.8	95.5	102.5	100.4	96.1	101.0	99.5	97.8	103.0	101.5	97.9	103.9	102.1	96.6	103.0	101.1
0310	Concrete Forming & Accessories	94.4	123.1	119.2	98.0	116.4	113.9	93.0	103.1	101.7	98.8	132.8	128.2	99.0	158.1	150.1	95.9	119.8	116.5
0320	Concrete Reinforcing	93.5	103.8	98.7	96.7	85.2	90.8	96.0	97.0	96.5	88.5	129.9	109.5	100.4	147.6	124.4	97.1	99.8	98.5
0330	Cast-in-Place Concrete	95.5	118.0	104.1	93.1	103.3	97.0	96.4	102.1	98.6	97.8	133.0	111.2	109.7	152.7	126.1	88.5	113.7	98.1
03	CONCRETE	95.4	118.6	106.1	93.8	107.0	99.9	96.1	102.8	99.2	96.1	132.9	113.1	103.4	153.5	126.6	91.5	113.8	101.9
04	MASONRY	116.7	125.4	122.1	106.8	111.8	109.9	117.3	102.3	108.1	90.6	143.6	123.2	103.2	164.5	140.9	85.8	124.3	109.5
05	METALS	101.8	125.6	109.1	99.8	113.7	104.0	99.2	120.0	105.6	101.8	141.5	114.0	94.8	142.2	109.3	100.2	121.7	106.8
06	WOOD, PLASTICS & COMPOSITES	100.6	122.1	112.4	91.4	119.1	106.6	94.6	102.3	98.9	100.5	129.8	116.6	100.0	157.4	131.6	90.8	119.1	106.4
07	THERMAL & MOISTURE PROTECTION	99.5	115.4	106.1	101.3	107.1	103.7	98.7	101.4	99.8	101.9	134.2	115.3	102.6	147.9	121.4	103.8	116.6	109.1
08	OPENINGS	100.7	125.8	106.4	96.2	115.2	100.5	94.7	106.9	97.5	100.7	140.4	109.7	101.8	167.4	116.7	100.3	119.8	104.7
0920	Plaster & Gypsum Board	98.5	122.7	114.6	98.3	119.7	112.5	94.7	102.4	99.9	98.5	130.7	119.9	98.0	159.0	138.5	99.3	119.7	112.8
0950, 0980	Ceilings & Acoustic Treatment	91.6	122.7	112.3	84.9	119.7	108.0	86.5	102.4	97.1	91.6	130.7	117.6	97.0	159.0	138.3	94.6	119.7	111.3
0960	Flooring	95.5	122.7	103.4	102.5	114.7	106.0	93.0	94.6	93.5	95.5	126.0	104.4	91.3	158.7	110.9	105.9	111.7	107.6
0970, 0990	Wall Finishes & Painting/Coating	87.5	141.7	119.3	89.6	113.7	103.8	87.5	101.7	95.9	87.5	149.6	124.0	89.8	162.6	132.5	87.8	113.7	103.0
09	FINISHES	93.0	125.6	110.8	92.6	117.1	106.0	90.6	101.1	96.3	93.0	133.2	115.0	93.6	159.6	129.7	96.6	118.2	108.4
COVERS	DIVS. 10 - 14, 25, 28, 41, 43, 44, 46	100.0	108.7	102.0	100.0	81.5	95.8	100.0	101.0	100.2	100.0	116.5	103.7	100.0	124.3	105.5	100.0	107.3	101.7
21, 22, 23	FIRE SUPPRESSION, PLUMBING & HVAC	100.0	103.1	101.3	96.5	100.3	98.2	96.4	98.9	97.5	100.0	117.3	107.7	99.9	134.6	115.3	100.0	104.0	101.7
26, 27, 3370	ELECTRICAL, COMMUNICATIONS & UTIL.	97.1	99.1	98.2	96.6	82.6	89.2	88.2	96.3	92.5	97.3	132.5	116.0	95.7	135.5	116.8	104.2	91.7	97.6
MF2014	WEIGHTED AVERAGE	99.7	113.4	105.8	97.3	103.9	100.2	96.8	102.4	99.2	98.6	128.9	112.0	99.0	143.5	118.6	98.4	110.4	103.7

For customer support on your Building Construction Costs with RSMeans Data, call 800.448.8182.

771

INDIANA

DIVISION		ANDERSON 460 MAT.	INST.	TOTAL	BLOOMINGTON 474 MAT.	INST.	TOTAL	COLUMBUS 472 MAT.	INST.	TOTAL	EVANSVILLE 476-477 MAT.	INST.	TOTAL	FORT WAYNE 467-468 MAT.	INST.	TOTAL	GARY 463-464 MAT.	INST.	TOTAL
015433	CONTRACTOR EQUIPMENT		98.3	98.3		84.6	84.6		84.6	84.6		115.7	115.7		98.3	98.3		98.3	98.3
0241, 31 - 34	SITE & INFRASTRUCTURE, DEMOLITION	98.6	94.6	95.8	87.3	93.3	91.5	83.9	93.1	90.3	92.9	123.7	114.4	99.6	94.3	95.9	99.2	98.2	98.5
0310	Concrete Forming & Accessories	96.1	81.4	83.3	100.6	81.4	84.0	94.6	79.4	81.5	94.0	85.3	86.4	94.2	75.5	78.0	96.1	117.9	114.9
0320	Concrete Reinforcing	96.3	85.8	91.0	91.8	85.4	88.6	92.2	85.4	88.8	100.5	79.3	89.7	96.3	77.8	86.9	96.3	114.2	105.4
0330	Cast-in-Place Concrete	102.1	79.4	93.4	99.0	78.7	91.2	98.5	76.5	90.2	94.6	88.2	92.1	108.5	75.7	96.0	106.7	118.2	111.1
03	CONCRETE	94.4	82.0	88.7	101.1	80.8	91.7	100.4	79.1	90.5	101.3	85.3	93.9	97.4	76.7	87.8	96.7	116.9	106.1
04	MASONRY	91.3	78.3	83.3	93.0	74.2	81.4	92.9	74.1	81.3	88.4	82.6	84.8	94.6	75.4	82.7	92.7	115.4	106.7
05	METALS	96.6	90.6	94.7	100.0	75.4	92.5	100.0	74.7	92.3	92.8	84.2	90.2	96.6	87.4	93.8	96.6	107.4	99.9
06	WOOD, PLASTICS & COMPOSITES	96.5	81.7	88.4	111.0	81.8	94.9	106.1	79.1	91.3	91.7	85.0	88.0	96.3	75.5	84.9	93.9	116.6	106.4
07	THERMAL & MOISTURE PROTECTION	110.6	77.8	97.0	97.0	79.7	89.8	96.4	82.5	90.6	101.3	86.5	95.2	110.4	74.8	95.6	109.1	110.4	109.6
08	OPENINGS	95.2	79.8	91.7	100.9	79.8	96.1	97.0	78.4	92.8	94.8	80.2	91.5	95.2	72.4	90.0	95.2	118.4	100.4
0920	Plaster & Gypsum Board	105.9	81.4	89.6	98.7	82.0	87.6	96.1	79.3	84.9	94.4	84.1	87.6	105.3	75.0	85.2	99.0	117.3	111.2
0950, 0980	Ceilings & Acoustic Treatment	87.2	81.4	83.4	77.3	82.0	80.4	77.3	79.3	78.6	81.1	84.1	83.1	87.2	75.0	79.1	87.2	117.3	107.3
0960	Flooring	98.0	79.5	92.6	102.4	81.5	96.4	97.1	79.5	92.0	96.7	74.5	90.3	98.0	75.7	91.5	98.0	114.1	102.7
0970, 0990	Wall Finishes & Painting/Coating	94.0	70.1	80.0	87.1	80.9	83.5	87.1	81.1	83.6	92.5	88.8	90.3	94.0	71.3	80.7	94.0	121.6	110.2
09	FINISHES	92.7	80.1	85.8	91.7	81.4	86.1	89.7	79.5	84.1	90.3	83.7	86.7	92.4	75.1	82.9	91.6	117.8	105.9
COVERS	DIVS. 10 - 14, 25, 28, 41, 43, 44, 46	100.0	90.8	97.9	100.0	89.7	97.7	100.0	89.4	97.6	100.0	95.8	99.0	100.0	88.3	97.4	100.0	105.6	101.3
21, 22, 23	FIRE SUPPRESSION, PLUMBING & HVAC	100.0	78.1	90.3	99.7	78.3	90.2	96.1	78.1	88.1	100.0	80.2	91.2	100.0	73.7	88.3	100.0	106.9	103.1
26, 27, 3370	ELECTRICAL, COMMUNICATIONS & UTIL.	87.1	88.8	88.0	99.8	89.7	94.5	99.0	88.7	93.6	96.0	88.3	91.9	87.7	78.5	82.8	98.6	115.9	107.8
MF2014	WEIGHTED AVERAGE	96.2	83.3	90.5	98.8	81.6	91.2	97.1	80.8	89.9	96.4	87.1	92.3	96.7	78.4	88.7	97.5	111.8	103.8

INDIANA

DIVISION		INDIANAPOLIS 461-462 MAT.	INST.	TOTAL	KOKOMO 469 MAT.	INST.	TOTAL	LAFAYETTE 479 MAT.	INST.	TOTAL	LAWRENCEBURG 470 MAT.	INST.	TOTAL	MUNCIE 473 MAT.	INST.	TOTAL	NEW ALBANY 471 MAT.	INST.	TOTAL
015433	CONTRACTOR EQUIPMENT		85.6	85.6		98.3	98.3		84.6	84.6		105.2	105.2		96.8	96.8		94.5	94.5
0241, 31 - 34	SITE & INFRASTRUCTURE, DEMOLITION	99.1	93.4	95.1	95.0	94.5	94.7	84.8	93.1	90.6	82.8	109.1	101.2	87.2	93.7	91.7	79.8	95.6	90.8
0310	Concrete Forming & Accessories	97.2	84.5	86.2	99.4	78.3	81.1	92.2	77.2	79.2	91.1	78.2	79.9	92.1	80.8	82.3	89.4	76.6	78.3
0320	Concrete Reinforcing	95.9	85.8	90.7	87.1	85.7	86.4	91.8	85.7	88.7	91.0	77.9	84.4	101.5	85.7	93.5	92.4	80.3	86.3
0330	Cast-in-Place Concrete	96.9	85.2	92.4	101.1	83.5	94.4	99.1	82.5	92.8	92.6	75.8	86.2	104.0	78.5	94.3	95.7	75.5	88.0
03	CONCRETE	96.1	84.3	90.6	91.4	82.0	87.1	100.6	80.2	91.2	93.6	78.0	86.4	99.5	81.5	91.1	99.1	77.2	88.9
04	MASONRY	94.1	79.2	84.9	91.0	77.1	82.4	98.5	78.4	86.1	77.2	74.0	75.2	94.9	78.4	84.7	84.0	68.8	74.7
05	METALS	95.0	73.8	88.5	93.2	90.4	92.3	98.5	75.4	91.4	94.9	86.2	92.2	101.8	90.5	98.3	97.0	82.0	92.4
06	WOOD, PLASTICS & COMPOSITES	95.3	85.0	89.7	99.6	77.2	87.3	103.4	76.0	88.3	90.1	78.0	83.5	105.2	81.2	92.0	91.9	77.2	83.9
07	THERMAL & MOISTURE PROTECTION	102.8	82.0	94.1	109.6	78.6	96.7	96.4	81.0	90.0	102.1	78.6	92.4	99.5	79.2	91.1	88.3	73.1	82.0
08	OPENINGS	100.1	81.6	95.9	90.6	77.3	87.6	95.5	76.7	91.2	96.7	75.1	91.8	94.3	79.5	90.9	94.3	76.7	90.3
0920	Plaster & Gypsum Board	97.5	84.7	89.0	110.5	76.8	88.1	93.7	76.1	82.0	73.3	78.0	76.4	94.4	81.4	85.8	92.1	77.0	82.1
0950, 0980	Ceilings & Acoustic Treatment	92.6	84.7	87.4	87.2	76.8	80.3	73.9	76.1	75.3	84.4	78.0	80.2	77.3	81.4	80.0	81.1	77.0	78.4
0960	Flooring	98.8	79.5	93.2	102.2	87.6	98.0	96.1	79.5	91.3	71.3	79.5	73.6	96.1	79.5	91.3	94.2	57.2	83.5
0970, 0990	Wall Finishes & Painting/Coating	92.9	82.7	86.9	94.0	72.3	81.3	87.1	82.8	84.6	88.0	73.6	79.5	87.1	70.1	77.2	92.5	67.1	77.6
09	FINISHES	93.9	83.5	88.2	94.4	79.2	86.1	88.4	77.9	82.6	80.2	78.0	79.0	88.9	79.5	83.7	89.5	71.8	79.8
COVERS	DIVS. 10 - 14, 25, 28, 41, 43, 44, 46	100.0	91.9	98.2	100.0	90.2	97.8	100.0	89.3	97.6	100.0	87.8	97.3	100.0	89.7	97.7	100.0	86.9	97.0
21, 22, 23	FIRE SUPPRESSION, PLUMBING & HVAC	99.9	80.1	91.1	96.4	79.6	89.0	96.1	78.6	88.4	97.2	73.1	86.5	99.7	77.9	90.1	96.4	77.3	88.0
26, 27, 3370	ELECTRICAL, COMMUNICATIONS & UTIL.	102.9	89.7	95.9	91.6	81.2	86.1	98.5	82.2	89.9	93.9	74.7	83.8	91.8	79.1	85.0	94.5	77.4	85.4
MF2014	WEIGHTED AVERAGE	98.2	83.3	91.6	94.4	82.2	89.0	96.8	80.3	89.5	93.5	79.5	87.3	97.2	81.7	90.4	94.8	77.8	87.3

INDIANA / IOWA

DIVISION		SOUTH BEND 465-466 MAT.	INST.	TOTAL	TERRE HAUTE 478 MAT.	INST.	TOTAL	WASHINGTON 475 MAT.	INST.	TOTAL	BURLINGTON 526 MAT.	INST.	TOTAL	CARROLL 514 MAT.	INST.	TOTAL	CEDAR RAPIDS 522-524 MAT.	INST.	TOTAL
015433	CONTRACTOR EQUIPMENT		110.3	110.3		115.7	115.7		115.7	115.7		102.9	102.9		102.9	102.9		99.6	99.6
0241, 31 - 34	SITE & INFRASTRUCTURE, DEMOLITION	97.4	95.2	95.9	94.7	123.9	115.1	94.6	124.1	115.2	99.3	98.0	98.4	88.1	98.6	95.4	100.5	97.1	98.1
0310	Concrete Forming & Accessories	98.4	81.6	83.8	94.9	82.6	84.3	95.7	83.0	84.7	95.2	84.7	86.1	82.9	82.7	82.7	101.2	86.9	88.8
0320	Concrete Reinforcing	97.1	84.9	90.9	100.5	85.9	93.1	92.9	85.8	89.3	94.6	85.2	89.8	95.3	77.2	86.1	95.3	81.0	88.0
0330	Cast-in-Place Concrete	96.8	82.2	91.2	91.6	83.3	88.4	99.6	88.4	95.4	107.0	58.2	88.4	107.0	59.7	89.0	107.2	84.6	98.6
03	CONCRETE	91.0	84.0	87.7	104.4	83.6	94.7	110.2	85.5	98.8	97.4	76.4	87.7	96.3	74.6	86.2	97.5	85.8	92.1
04	MASONRY	95.1	77.9	84.5	96.3	78.3	85.2	88.5	80.1	83.3	101.4	77.5	86.7	103.1	76.6	86.8	107.2	82.8	92.2
05	METALS	98.9	106.1	101.1	93.6	86.9	91.5	88.1	86.9	87.7	88.7	95.5	90.7	88.7	92.5	89.9	91.0	95.0	92.2
06	WOOD, PLASTICS & COMPOSITES	94.5	81.2	87.2	93.7	82.9	87.8	94.1	81.7	87.3	89.1	83.9	86.3	76.5	87.1	82.3	95.3	87.8	91.2
07	THERMAL & MOISTURE PROTECTION	103.4	82.8	94.9	101.4	81.8	93.3	101.3	85.2	94.6	105.1	81.4	95.3	105.5	74.2	92.5	106.2	85.0	97.4
08	OPENINGS	94.9	79.0	91.3	95.2	80.5	91.9	92.0	80.3	89.4	93.8	86.3	92.1	98.2	82.6	94.7	98.7	83.0	95.1
0920	Plaster & Gypsum Board	93.6	80.9	85.2	94.4	82.0	86.2	94.8	80.8	85.5	109.4	83.6	92.2	104.8	86.8	92.8	113.6	87.8	96.5
0950, 0980	Ceilings & Acoustic Treatment	90.6	80.9	84.1	81.1	82.0	81.7	76.8	80.8	79.4	94.4	83.6	87.2	94.4	86.8	89.3	96.9	87.8	90.8
0960	Flooring	97.7	87.6	94.7	96.7	80.2	91.9	97.6	69.6	89.5	98.7	73.2	91.3	92.6	84.3	90.2	113.0	89.3	106.1
0970, 0990	Wall Finishes & Painting/Coating	90.1	85.9	87.6	92.5	80.2	85.3	92.5	88.8	90.3	98.0	90.8	93.8	98.0	77.6	86.0	99.3	72.4	83.5
09	FINISHES	92.1	83.1	87.2	90.3	81.7	85.6	90.0	80.8	85.0	96.0	83.5	89.2	92.0	83.5	87.4	101.3	86.1	93.0
COVERS	DIVS. 10 - 14, 25, 28, 41, 43, 44, 46	100.0	92.6	98.3	100.0	92.6	98.3	100.0	95.4	99.0	100.0	91.1	98.0	100.0	66.8	92.5	100.0	92.8	98.4
21, 22, 23	FIRE SUPPRESSION, PLUMBING & HVAC	99.9	74.9	88.9	100.0	79.0	90.7	96.4	81.2	89.7	96.7	80.1	89.3	96.7	74.5	86.9	100.2	80.1	91.3
26, 27, 3370	ELECTRICAL, COMMUNICATIONS & UTIL.	97.0	86.5	91.4	94.2	90.1	92.0	94.7	88.3	91.3	101.1	76.1	87.8	101.9	81.4	91.0	98.6	81.7	89.6
MF2014	WEIGHTED AVERAGE	96.9	84.6	91.5	97.2	86.2	92.4	95.5	87.0	91.7	96.2	82.7	90.3	96.0	80.6	89.2	98.6	85.6	92.8

City Cost Indexes

IOWA

DIVISION		COUNCIL BLUFFS 515 MAT.	INST.	TOTAL	CRESTON 508 MAT.	INST.	TOTAL	DAVENPORT 527-528 MAT.	INST.	TOTAL	DECORAH 521 MAT.	INST.	TOTAL	DES MOINES 500-503,509 MAT.	INST.	TOTAL	DUBUQUE 520 MAT.	INST.	TOTAL
015433	CONTRACTOR EQUIPMENT		98.9	98.9		102.9	102.9		102.9	102.9		102.9	102.9		104.9	104.9		98.4	98.4
0241, 31 - 34	SITE & INFRASTRUCTURE, DEMOLITION	103.5	92.9	96.1	94.5	99.7	98.1	99.5	99.6	99.6	97.9	97.6	97.7	100.4	101.3	101.0	98.3	94.3	95.5
0310	Concrete Forming & Accessories	82.3	77.1	77.8	79.7	86.0	85.1	100.7	96.2	96.8	92.8	76.8	79.0	98.0	87.6	89.0	83.7	79.6	80.1
0320	Concrete Reinforcing	97.2	77.6	87.3	95.1	76.7	85.8	95.3	98.4	96.9	94.6	71.7	83.0	99.8	81.7	90.6	94.0	80.4	87.1
0330	Cast-in-Place Concrete	111.5	80.7	99.8	112.3	64.3	94.0	103.3	92.0	99.0	104.1	81.0	95.3	95.5	86.1	91.9	105.0	99.1	102.8
03	CONCRETE	99.9	79.4	90.4	98.9	77.6	89.0	95.6	95.6	95.6	95.4	78.3	87.5	92.1	86.8	89.6	94.4	87.4	91.2
04	MASONRY	108.4	75.7	88.2	104.1	74.9	86.1	103.9	87.2	93.7	123.4	72.0	91.8	91.0	88.3	89.4	108.2	72.2	86.0
05	METALS	96.2	92.5	95.1	96.4	92.1	92.5	91.0	102.8	94.6	88.8	89.8	89.1	98.6	97.0	98.1	89.6	94.1	91.0
06	WOOD, PLASTICS & COMPOSITES	75.5	78.0	76.8	72.3	87.1	80.4	95.4	97.1	96.3	86.3	76.6	81.0	90.1	87.1	88.4	77.0	78.8	78.0
07	THERMAL & MOISTURE PROTECTION	105.5	72.4	91.8	105.7	81.8	95.8	105.6	92.3	100.1	105.3	74.4	92.5	97.4	86.6	92.9	105.8	81.4	95.7
08	OPENINGS	97.7	78.9	93.5	105.2	84.4	100.5	98.7	96.9	98.3	96.8	79.1	92.8	96.1	87.4	94.1	97.8	81.8	94.1
0920	Plaster & Gypsum Board	104.8	77.7	86.8	102.5	86.8	92.1	113.6	97.1	102.6	108.1	76.1	86.8	95.2	86.8	89.6	104.8	78.5	87.3
0950, 0980	Ceilings & Acoustic Treatment	94.4	77.7	83.3	87.2	86.8	87.0	96.9	97.1	97.0	94.4	76.1	82.2	91.5	86.8	88.4	94.4	78.5	83.8
0960	Flooring	91.0	85.1	89.3	83.1	73.2	80.3	101.3	94.3	99.3	98.1	73.2	90.9	92.6	94.3	93.1	103.5	73.2	94.8
0970, 0990	Wall Finishes & Painting/Coating	93.9	64.9	76.9	87.4	79.1	82.5	98.0	96.0	96.8	98.0	86.6	91.3	88.8	86.6	87.5	98.5	88.4	92.6
09	FINISHES	92.7	77.4	84.4	85.6	83.4	84.4	97.8	96.2	97.0	95.6	77.4	85.6	90.3	89.0	89.6	96.6	78.9	86.9
COVERS	DIVS. 10 - 14, 25, 28, 41, 43, 44, 46	100.0	90.3	97.8	100.0	69.5	93.1	100.0	95.7	99.0	100.0	83.5	96.3	100.0	94.1	98.7	100.0	91.3	98.0
21, 22, 23	FIRE SUPPRESSION, PLUMBING & HVAC	100.2	74.4	88.8	96.4	82.4	90.2	100.2	92.4	96.8	96.7	73.9	86.6	99.8	83.4	92.6	100.2	75.0	89.0
26, 27, 3370	ELECTRICAL, COMMUNICATIONS & UTIL.	104.1	81.9	92.4	92.6	81.4	86.6	96.7	92.7	94.6	98.6	51.2	73.5	103.9	83.4	93.0	102.6	77.7	89.4
MF2014	WEIGHTED AVERAGE	99.4	80.4	91.0	96.5	82.9	90.5	97.6	94.7	96.3	97.1	75.5	87.6	97.3	88.4	93.3	97.7	81.5	90.5

IOWA

DIVISION		FORT DODGE 505 MAT.	INST.	TOTAL	MASON CITY 504 MAT.	INST.	TOTAL	OTTUMWA 525 MAT.	INST.	TOTAL	SHENANDOAH 516 MAT.	INST.	TOTAL	SIBLEY 512 MAT.	INST.	TOTAL	SIOUX CITY 510-511 MAT.	INST.	TOTAL
015433	CONTRACTOR EQUIPMENT		102.9	102.9		102.9	102.9		98.4	98.4		98.9	98.9		102.9	102.9		102.9	102.9
0241, 31 - 34	SITE & INFRASTRUCTURE, DEMOLITION	103.2	96.4	98.4	103.3	97.5	99.3	98.8	92.1	94.1	102.1	94.2	96.6	109.5	97.7	101.2	111.1	98.6	102.4
0310	Concrete Forming & Accessories	80.3	77.2	77.6	84.4	76.5	77.6	90.7	76.9	78.8	84.0	59.5	62.8	84.4	39.9	45.9	101.2	74.4	78.0
0320	Concrete Reinforcing	95.1	67.6	81.2	95.0	76.6	85.7	94.6	85.2	89.8	97.2	68.0	82.4	97.2	62.5	79.6	95.3	91.2	93.2
0330	Cast-in-Place Concrete	105.4	44.1	82.0	105.4	75.9	94.1	107.7	67.7	92.5	107.8	80.6	97.4	105.6	58.8	87.8	106.3	55.5	87.0
03	CONCRETE	94.3	65.0	80.7	94.5	77.3	86.5	97.0	76.2	87.3	97.3	69.7	84.5	96.3	52.5	76.0	96.9	71.9	85.3
04	MASONRY	102.9	53.4	72.4	116.6	70.7	88.3	104.7	56.9	75.3	108.0	75.7	88.1	127.3	53.7	82.0	101.1	66.9	80.0
05	METALS	92.7	86.2	90.7	92.8	91.9	92.5	88.6	95.0	90.6	95.2	88.4	93.1	88.9	85.3	87.8	91.0	97.7	93.0
06	WOOD, PLASTICS & COMPOSITES	72.7	87.1	80.6	76.5	76.6	76.6	83.7	83.7	83.7	77.0	54.4	64.6	77.7	36.8	55.3	95.3	75.6	84.5
07	THERMAL & MOISTURE PROTECTION	105.0	68.3	89.8	104.5	74.8	92.1	106.0	71.2	91.6	104.8	68.3	89.7	105.1	55.2	84.4	105.6	69.8	90.7
08	OPENINGS	99.2	70.2	92.6	91.5	80.5	89.0	98.2	82.8	94.7	89.3	59.8	82.7	94.9	42.3	83.0	98.7	77.7	93.9
0920	Plaster & Gypsum Board	102.5	86.8	92.1	87.2	76.1	79.8	105.8	83.6	91.0	104.8	53.5	70.7	104.8	35.2	58.5	113.6	75.1	88.0
0950, 0980	Ceilings & Acoustic Treatment	87.2	86.8	87.0	87.2	76.1	79.8	94.4	83.6	87.2	94.4	53.5	67.2	94.4	35.2	55.0	96.9	75.1	82.4
0960	Flooring	84.4	73.2	81.2	86.4	73.2	82.6	106.8	73.2	97.1	91.8	78.3	87.9	93.5	73.2	87.7	101.4	73.2	93.2
0970, 0990	Wall Finishes & Painting/Coating	87.4	86.6	86.9	87.4	86.6	86.9	98.5	86.6	91.5	93.9	64.9	76.9	98.0	60.9	76.2	98.0	65.7	79.0
09	FINISHES	87.4	78.8	82.7	88.0	77.0	82.0	97.8	77.8	86.8	92.8	62.3	76.1	95.3	46.7	68.7	99.4	73.2	85.1
COVERS	DIVS. 10 - 14, 25, 28, 41, 43, 44, 46	100.0	83.3	96.2	100.0	87.9	97.3	100.0	83.7	96.3	100.0	63.5	91.7	100.0	76.9	94.8	100.0	89.6	97.6
21, 22, 23	FIRE SUPPRESSION, PLUMBING & HVAC	96.4	71.3	85.3	96.4	80.3	89.3	96.7	73.4	86.4	96.7	87.9	92.8	96.7	67.8	83.9	100.2	78.5	90.6
26, 27, 3370	ELECTRICAL, COMMUNICATIONS & UTIL.	99.1	75.2	86.4	98.2	51.2	73.2	100.9	74.7	87.0	98.6	81.3	89.4	98.6	51.2	73.5	98.6	74.5	85.8
MF2014	WEIGHTED AVERAGE	96.2	73.9	86.4	96.1	76.9	87.6	96.9	77.1	88.2	96.6	77.7	88.3	97.4	61.8	81.7	98.3	78.6	89.6

IOWA / KANSAS

DIVISION		SPENCER 513 MAT.	INST.	TOTAL	WATERLOO 506-507 MAT.	INST.	TOTAL	BELLEVILLE 669 MAT.	INST.	TOTAL	COLBY 677 MAT.	INST.	TOTAL	DODGE CITY 678 MAT.	INST.	TOTAL	EMPORIA 668 MAT.	INST.	TOTAL
015433	CONTRACTOR EQUIPMENT		102.9	102.9		102.9	102.9		108.8	108.8		108.8	108.8		108.8	108.8		106.9	106.9
0241, 31 - 34	SITE & INFRASTRUCTURE, DEMOLITION	109.5	96.3	100.3	108.5	97.3	100.7	108.7	98.5	101.6	108.3	99.2	101.9	110.7	98.1	101.9	100.8	95.9	97.4
0310	Concrete Forming & Accessories	90.6	39.7	46.6	96.3	57.4	62.7	97.1	57.6	63.0	103.3	60.5	66.3	96.3	61.9	66.6	87.9	70.0	72.4
0320	Concrete Reinforcing	97.2	63.6	80.2	95.8	80.7	88.1	97.4	57.1	77.0	96.9	57.1	76.7	94.5	57.0	75.4	96.1	57.7	76.6
0330	Cast-in-Place Concrete	105.6	69.7	92.0	112.9	77.7	99.5	114.0	89.2	104.5	107.6	93.1	102.1	109.4	92.7	103.0	110.2	93.3	103.7
03	CONCRETE	96.6	56.3	78.0	100.3	70.1	86.3	111.8	70.2	92.5	107.6	72.8	91.5	108.9	73.2	92.4	104.2	77.2	91.7
04	MASONRY	127.3	53.7	82.0	103.7	78.1	87.9	93.0	57.0	70.8	103.2	63.0	78.5	113.7	61.8	81.8	99.0	71.2	81.8
05	METALS	88.8	85.6	87.9	95.0	94.1	94.7	101.2	85.1	96.3	101.6	85.9	96.8	103.0	84.4	97.3	100.9	86.7	96.6
06	WOOD, PLASTICS & COMPOSITES	83.5	36.8	57.9	89.6	49.5	67.6	98.5	55.3	74.7	107.4	55.8	79.1	99.4	58.2	76.8	89.8	68.0	77.9
07	THERMAL & MOISTURE PROTECTION	106.1	55.9	85.2	104.7	76.5	93.0	97.3	65.0	83.9	98.4	67.6	85.6	98.4	67.2	85.4	95.5	80.3	89.2
08	OPENINGS	106.2	42.6	91.8	91.9	62.8	85.3	98.4	59.5	89.6	103.6	59.8	93.7	103.6	57.0	93.0	96.3	65.0	89.2
0920	Plaster & Gypsum Board	105.8	35.2	58.9	111.0	48.2	69.3	95.4	54.1	67.9	100.6	54.6	70.0	95.0	57.1	69.8	92.7	67.2	75.8
0950, 0980	Ceilings & Acoustic Treatment	94.4	35.2	55.0	89.7	48.2	62.1	81.5	54.1	63.2	78.9	54.6	62.8	78.9	57.1	64.4	81.5	67.2	72.0
0960	Flooring	96.5	73.2	89.7	91.8	83.6	89.4	94.5	69.8	87.4	92.5	69.8	85.9	88.5	69.8	83.1	89.5	69.8	83.8
0970, 0990	Wall Finishes & Painting/Coating	98.0	60.9	76.2	87.4	86.6	86.9	87.5	82.6	84.6	95.0	82.6	87.7	95.0	82.6	87.7	87.5	82.6	84.6
09	FINISHES	96.3	45.4	68.5	91.6	63.7	76.3	89.1	61.6	74.0	88.7	63.5	74.9	86.9	64.9	74.9	86.2	70.7	77.7
COVERS	DIVS. 10 - 14, 25, 28, 41, 43, 44, 46	100.0	76.9	94.8	100.0	88.5	97.4	100.0	87.9	97.3	100.0	90.2	97.8	100.0	90.5	97.8	100.0	91.5	98.1
21, 22, 23	FIRE SUPPRESSION, PLUMBING & HVAC	96.7	71.4	85.5	100.0	79.4	90.9	96.3	71.3	85.3	96.4	69.6	84.6	100.0	70.5	86.9	96.3	76.3	87.5
26, 27, 3370	ELECTRICAL, COMMUNICATIONS & UTIL.	100.4	51.2	74.3	94.6	61.8	77.2	99.3	68.0	82.7	99.1	75.7	86.7	96.0	71.1	82.8	96.6	71.9	83.5
MF2014	WEIGHTED AVERAGE	99.1	62.9	83.1	97.5	75.5	87.8	99.5	71.1	86.9	100.2	73.3	88.3	101.4	72.6	88.7	97.7	77.1	88.6

For customer support on your Building Construction Costs with RSMeans Data, call 800.448.8182.

773

		KANSAS																
	DIVISION	FORT SCOTT			HAYS			HUTCHINSON			INDEPENDENCE			KANSAS CITY			LIBERAL		
		667			676			675			673			660 - 662			679		
		MAT.	INST.	TOTAL	MAT.	INST.	TOTAL	MAT.	INST.	TOTAL	MAT.	INST.	TOTAL	MAT.	INST.	TOTAL	MAT.	INST.	TOTAL
015433	CONTRACTOR EQUIPMENT		107.8	107.8		108.8	108.8		108.8	108.8		108.8	108.8		105.3	105.3		108.8	108.8
0241, 31 - 34	SITE & INFRASTRUCTURE, DEMOLITION	97.6	95.9	96.5	113.7	98.7	103.3	92.7	99.2	97.2	112.0	99.2	103.1	92.0	96.0	94.8	113.2	98.7	103.1
0310	Concrete Forming & Accessories	105.5	84.0	86.9	100.7	59.7	65.2	90.7	57.2	61.8	112.4	72.2	77.7	102.0	99.2	99.6	96.8	49.6	56.0
0320	Concrete Reinforcing	95.4	93.5	94.5	94.5	57.1	75.5	94.5	57.0	75.5	93.9	65.6	79.5	92.6	102.2	97.4	95.8	57.0	76.1
0330	Cast-in-Place Concrete	102.2	88.1	96.8	84.9	89.4	86.6	78.6	92.7	84.0	109.9	93.0	103.5	87.5	100.3	92.4	84.9	89.0	86.4
03	CONCRETE	99.6	87.9	94.2	99.8	71.2	86.5	82.6	71.1	77.3	110.1	79.4	95.9	91.9	100.6	95.9	101.7	66.4	85.4
04	MASONRY	100.4	57.8	74.2	112.9	57.1	78.5	103.3	61.8	77.8	100.3	66.8	79.6	101.1	98.7	99.6	111.6	55.4	77.0
05	METALS	100.9	98.3	100.1	101.2	85.6	96.4	101.0	84.5	95.9	100.9	87.4	96.8	109.1	106.1	108.2	101.5	84.4	96.3
06	WOOD, PLASTICS & COMPOSITES	108.5	91.0	98.9	104.2	58.2	78.9	94.4	51.7	70.9	117.7	71.2	92.1	104.4	99.4	101.7	100.0	44.7	69.6
07	THERMAL & MOISTURE PROTECTION	96.4	75.6	87.8	98.7	65.3	84.9	97.2	66.5	84.5	98.4	79.8	90.7	96.1	98.7	97.2	98.9	62.8	83.9
08	OPENINGS	96.3	86.0	93.9	103.5	57.0	93.0	103.5	53.5	92.2	101.3	65.4	93.2	97.6	97.6	97.6	103.6	49.6	91.4
0920	Plaster & Gypsum Board	97.7	90.7	93.1	97.9	57.1	70.8	94.0	50.4	65.0	107.5	70.4	82.8	91.4	99.4	96.7	95.6	43.2	60.8
0950, 0980	Ceilings & Acoustic Treatment	81.5	90.7	87.6	78.9	57.1	64.4	78.9	50.4	60.0	78.9	70.4	73.3	81.5	99.4	93.4	78.9	43.2	55.2
0960	Flooring	104.5	69.6	94.4	91.2	69.8	85.0	85.5	69.8	80.9	97.0	69.6	89.0	84.1	98.2	88.2	88.8	69.8	83.3
0970, 0990	Wall Finishes & Painting/Coating	89.2	82.6	85.3	95.0	82.6	87.7	95.0	82.6	87.7	95.0	82.6	87.7	94.7	99.2	97.4	95.0	82.6	87.7
09	FINISHES	91.6	82.4	86.5	88.5	63.3	74.7	84.3	61.1	71.6	91.2	73.0	81.3	86.6	99.3	93.5	87.7	55.4	70.1
COVERS	DIVS. 10 - 14, 25, 28, 41, 43, 44, 46	100.0	91.0	98.0	100.0	88.3	97.3	100.0	89.7	97.7	100.0	91.9	98.2	100.0	98.4	99.6	100.0	86.8	97.0
21, 22, 23	FIRE SUPPRESSION, PLUMBING & HVAC	96.3	69.4	84.4	96.4	67.2	83.5	96.4	70.5	85.0	96.4	73.1	86.1	99.8	97.4	98.8	96.4	68.6	84.1
26, 27, 3370	ELECTRICAL, COMMUNICATIONS & UTIL.	95.8	71.9	83.1	98.0	66.1	81.1	93.1	66.1	78.8	95.3	71.9	82.9	101.5	95.6	98.4	96.0	71.1	82.8
MF2014	WEIGHTED AVERAGE	97.7	79.3	89.6	99.7	70.3	86.7	95.5	71.0	84.7	100.0	76.9	89.8	99.0	98.8	98.9	99.6	68.8	86.0

		KANSAS									KENTUCKY								
	DIVISION	SALINA			TOPEKA			WICHITA			ASHLAND			BOWLING GREEN			CAMPTON		
		674			664 - 666			670 - 672			411 - 412			421 - 422			413 - 414		
		MAT.	INST.	TOTAL	MAT.	INST.	TOTAL	MAT.	INST.	TOTAL	MAT.	INST.	TOTAL	MAT.	INST.	TOTAL	MAT.	INST.	TOTAL
015433	CONTRACTOR EQUIPMENT		108.8	108.8		106.9	106.9		108.8	108.8		101.8	101.8		94.5	94.5		101.1	101.1
0241, 31 - 34	SITE & INFRASTRUCTURE, DEMOLITION	101.2	98.7	99.5	95.4	95.2	95.3	97.2	101.1	99.9	115.2	84.0	93.4	79.9	95.5	90.8	88.9	96.8	94.4
0310	Concrete Forming & Accessories	92.6	57.8	62.5	98.9	68.3	72.4	98.5	61.5	66.5	87.1	94.5	93.5	85.7	79.6	80.4	89.0	75.3	77.1
0320	Concrete Reinforcing	93.9	57.2	75.3	92.1	104.1	98.2	92.1	80.5	86.2	93.9	95.5	94.7	91.1	78.1	84.5	92.0	94.3	93.2
0330	Cast-in-Place Concrete	95.0	89.5	92.9	91.9	86.7	89.9	88.9	98.4	92.5	87.0	97.2	90.9	86.4	73.1	81.3	96.3	71.1	86.7
03	CONCRETE	96.2	70.4	84.2	93.5	82.3	88.3	92.0	79.1	86.0	94.8	96.8	95.7	93.3	77.5	86.0	96.9	77.7	88.0
04	MASONRY	129.3	57.1	84.8	94.7	60.8	73.8	100.4	71.5	82.6	95.4	93.5	94.2	97.6	72.1	81.9	94.1	59.0	72.5
05	METALS	102.8	85.8	97.6	105.2	103.4	104.6	105.2	93.0	101.4	96.0	110.7	100.5	97.7	84.6	93.7	97.0	90.9	95.1
06	WOOD, PLASTICS & COMPOSITES	95.9	55.3	73.6	99.9	71.0	84.0	101.3	51.7	74.0	74.5	92.9	84.6	87.0	79.9	83.1	85.6	80.0	82.5
07	THERMAL & MOISTURE PROTECTION	97.9	65.0	84.2	100.2	75.8	90.1	97.3	69.9	85.9	92.4	92.0	92.3	88.3	80.7	85.1	101.4	71.0	88.8
08	OPENINGS	103.5	55.4	92.6	103.8	77.4	97.8	107.0	58.7	96.0	93.6	93.3	93.5	94.3	81.2	91.3	95.6	82.4	92.7
0920	Plaster & Gypsum Board	94.0	54.1	67.5	95.7	70.2	78.8	93.4	50.4	64.8	61.6	92.9	82.4	88.2	79.7	82.6	88.2	79.0	82.1
0950, 0980	Ceilings & Acoustic Treatment	78.9	54.1	62.4	87.0	70.2	75.8	82.8	50.4	61.2	76.3	92.9	87.4	81.1	79.7	80.2	81.1	79.0	79.7
0960	Flooring	87.0	69.8	82.0	95.7	69.8	88.2	98.1	69.8	89.9	76.7	98.6	83.1	92.0	78.7	88.2	94.1	37.5	77.7
0970, 0990	Wall Finishes & Painting/Coating	95.0	82.6	87.7	92.0	75.3	82.2	95.3	71.0	81.1	94.7	94.0	94.3	92.5	89.9	91.0	92.5	57.4	71.9
09	FINISHES	85.5	61.6	72.4	91.1	68.5	78.8	90.6	62.3	75.1	77.5	95.4	87.3	88.1	80.5	84.0	88.9	66.2	76.5
COVERS	DIVS. 10 - 14, 25, 28, 41, 43, 44, 46	100.0	86.4	96.9	100.0	83.0	96.2	100.0	91.5	98.1	100.0	83.5	96.3	100.0	89.8	97.7	100.0	49.5	88.6
21, 22, 23	FIRE SUPPRESSION, PLUMBING & HVAC	100.0	69.2	86.4	100.0	70.9	87.1	99.8	77.7	90.0	96.2	87.5	92.3	100.0	79.2	90.8	96.4	75.9	87.4
26, 27, 3370	ELECTRICAL, COMMUNICATIONS & UTIL.	95.7	74.3	84.4	100.2	71.9	85.2	99.9	74.3	86.3	92.2	95.7	94.0	94.8	79.2	86.5	92.1	95.6	94.0
MF2014	WEIGHTED AVERAGE	100.2	71.4	87.5	99.4	77.0	89.5	99.7	77.2	89.8	94.3	93.7	94.0	95.5	80.7	88.9	95.5	78.3	87.9

		KENTUCKY																	
	DIVISION	CORBIN			COVINGTON			ELIZABETHTOWN			FRANKFORT			HAZARD			HENDERSON		
		407 - 409			410			427			406			417 - 418			424		
		MAT.	INST.	TOTAL	MAT.	INST.	TOTAL	MAT.	INST.	TOTAL	MAT.	INST.	TOTAL	MAT.	INST.	TOTAL	MAT.	INST.	TOTAL
015433	CONTRACTOR EQUIPMENT		101.1	101.1		105.2	105.2		94.5	94.5		101.1	101.1		101.1	101.1		115.7	115.7
0241, 31 - 34	SITE & INFRASTRUCTURE, DEMOLITION	91.1	97.3	95.4	84.3	109.3	101.7	74.3	95.2	88.9	88.9	97.6	94.9	86.6	98.0	94.5	82.7	123.1	110.9
0310	Concrete Forming & Accessories	84.5	73.9	75.3	84.4	73.3	74.8	80.4	77.4	77.8	94.1	75.4	77.9	85.5	75.7	77.0	92.1	80.9	82.4
0320	Concrete Reinforcing	102.2	62.6	82.1	90.6	80.2	85.3	91.6	81.1	86.3	107.6	81.0	94.1	92.4	62.8	77.4	91.2	78.2	84.6
0330	Cast-in-Place Concrete	88.5	75.5	83.5	92.2	76.8	86.3	78.1	70.4	75.2	87.6	77.4	83.7	92.6	73.9	85.5	76.3	89.3	81.3
03	CONCRETE	89.6	73.1	82.0	95.2	76.8	86.7	85.2	76.2	81.0	91.9	77.6	85.3	93.7	73.4	84.2	90.3	83.6	87.2
04	MASONRY	91.5	64.1	74.6	109.2	76.3	88.9	81.0	66.5	72.1	87.0	70.5	76.9	92.8	61.2	73.3	102.1	78.7	87.4
05	METALS	96.7	78.9	91.3	94.8	93.2	94.3	96.8	86.0	93.5	99.6	86.4	95.6	97.0	79.2	91.5	87.9	84.9	87.0
06	WOOD, PLASTICS & COMPOSITES	72.4	74.6	73.6	83.8	70.0	76.2	82.6	79.9	81.1	86.5	74.6	79.9	82.8	80.0	81.3	89.5	79.6	84.0
07	THERMAL & MOISTURE PROTECTION	106.1	73.3	92.5	102.4	79.8	93.0	87.7	72.7	81.5	104.0	75.5	92.2	101.3	72.9	89.5	100.7	80.5	92.3
08	OPENINGS	89.5	62.4	83.4	97.6	76.0	92.7	94.3	79.0	90.8	93.0	77.4	89.5	96.0	75.1	91.2	92.4	79.8	89.6
0920	Plaster & Gypsum Board	93.1	73.4	80.0	70.8	69.7	70.1	87.5	79.7	82.4	93.6	73.4	80.2	87.5	79.0	81.9	91.2	78.5	82.8
0950, 0980	Ceilings & Acoustic Treatment	77.6	73.4	74.8	83.6	69.7	74.4	81.1	79.7	80.2	86.1	73.4	77.7	81.1	79.0	79.7	76.8	78.5	78.0
0960	Flooring	90.8	37.5	75.3	68.8	82.9	72.9	89.3	78.2	86.1	96.7	49.5	83.0	92.3	38.2	76.6	95.7	79.0	90.9
0970, 0990	Wall Finishes & Painting/Coating	94.2	66.0	77.6	88.0	73.4	79.4	92.5	69.0	78.7	96.6	72.6	82.5	92.5	57.4	71.9	92.5	91.3	91.8
09	FINISHES	85.5	66.0	74.8	79.1	74.4	76.6	86.9	76.5	81.2	89.9	70.0	79.1	88.1	66.9	76.5	88.2	81.8	84.7
COVERS	DIVS. 10 - 14, 25, 28, 41, 43, 44, 46	100.0	88.6	97.4	100.0	89.5	97.6	100.0	75.1	94.4	100.0	60.1	91.0	100.0	50.3	88.8	100.0	59.5	90.8
21, 22, 23	FIRE SUPPRESSION, PLUMBING & HVAC	96.5	74.9	86.9	97.2	78.4	88.9	96.6	75.8	87.4	100.1	77.6	90.1	96.4	76.4	87.6	96.6	79.4	89.0
26, 27, 3370	ELECTRICAL, COMMUNICATIONS & UTIL.	91.2	95.6	93.5	96.0	76.5	85.6	91.9	79.2	85.2	100.1	77.5	88.1	92.1	95.6	94.0	94.2	77.8	85.5
MF2014	WEIGHTED AVERAGE	93.5	77.3	86.3	95.5	81.1	89.2	92.2	78.0	85.9	96.4	77.7	88.1	94.9	76.9	87.0	93.2	83.3	88.8

City Cost Indexes

KENTUCKY

DIVISION		LEXINGTON 403-405 MAT.	INST.	TOTAL	LOUISVILLE 400-402 MAT.	INST.	TOTAL	OWENSBORO 423 MAT.	INST.	TOTAL	PADUCAH 420 MAT.	INST.	TOTAL	PIKEVILLE 415-416 MAT.	INST.	TOTAL	SOMERSET 425-426 MAT.	INST.	TOTAL
015433	CONTRACTOR EQUIPMENT		101.1	101.1		94.5	94.5		115.7	115.7		115.7	115.7		101.8	101.8		101.1	101.1
0241, 31 - 34	SITE & INFRASTRUCTURE, DEMOLITION	93.1	99.2	97.4	87.8	95.4	93.1	92.9	123.7	114.4	85.5	122.9	111.6	126.7	83.0	96.2	79.3	97.4	91.9
0310	Concrete Forming & Accessories	96.0	76.3	78.9	94.7	78.7	80.8	90.5	80.6	81.9	88.5	83.6	84.3	96.1	80.5	82.6	86.5	74.3	76.0
0320	Concrete Reinforcing	112.2	83.7	97.8	107.6	83.7	95.5	91.2	80.4	85.7	91.8	81.7	86.7	94.5	95.3	94.9	91.6	62.9	77.0
0330	Cast-in-Place Concrete	90.5	80.7	86.8	87.7	72.0	81.7	89.0	88.2	88.7	81.3	85.4	82.9	95.7	92.4	94.4	76.3	93.2	82.8
03	CONCRETE	92.7	79.6	86.6	92.0	77.6	85.3	102.5	83.5	93.7	94.9	84.1	89.9	108.6	88.8	99.4	80.2	79.4	79.8
04	MASONRY	89.7	73.4	79.6	87.9	70.2	77.0	93.4	76.1	82.8	96.1	80.9	86.8	92.6	85.8	88.4	87.7	67.1	75.0
05	METALS	99.3	87.8	95.8	100.4	86.4	96.1	89.3	87.0	88.6	86.5	87.0	86.6	96.0	109.5	100.1	96.9	79.5	91.6
06	WOOD, PLASTICS & COMPOSITES	86.3	74.6	79.9	86.6	79.9	82.9	87.5	79.6	83.1	85.3	84.8	85.0	83.1	79.4	81.1	83.3	74.6	78.5
07	THERMAL & MOISTURE PROTECTION	106.4	77.3	94.3	103.4	74.3	91.3	101.3	79.9	92.5	100.7	84.7	94.1	93.2	83.2	89.0	100.7	74.3	89.7
08	OPENINGS	89.8	77.0	86.9	83.8	79.0	82.7	92.4	80.4	89.7	91.7	83.9	90.0	94.2	78.3	90.6	95.0	68.8	89.1
0920	Plaster & Gypsum Board	102.5	73.4	83.2	94.0	79.7	84.5	89.9	78.5	82.3	88.9	83.9	85.6	64.2	79.0	74.0	87.5	73.4	78.2
0950, 0980	Ceilings & Acoustic Treatment	81.0	73.4	76.0	89.0	79.7	82.8	76.8	78.5	78.0	76.8	83.9	81.5	76.3	79.0	78.1	81.1	73.4	76.0
0960	Flooring	96.0	60.9	85.8	95.9	78.2	90.8	95.1	78.7	90.3	94.0	54.0	82.4	81.1	98.6	86.2	92.6	37.5	76.6
0970, 0990	Wall Finishes & Painting/Coating	94.2	71.1	80.6	95.9	69.0	80.1	92.5	91.3	91.8	92.5	77.9	83.9	94.7	71.2	80.9	92.5	69.0	78.7
09	FINISHES	89.1	72.2	79.9	90.3	77.7	83.4	88.4	81.1	84.4	87.6	77.6	82.1	80.1	83.1	81.7	87.5	66.3	75.9
COVERS	DIVS. 10 - 14, 25, 28, 41, 43, 44, 46	100.0	91.8	98.1	100.0	89.3	97.6	100.0	96.8	99.3	100.0	56.8	90.2	100.0	49.8	88.6	100.0	88.6	97.4
21, 22, 23	FIRE SUPPRESSION, PLUMBING & HVAC	100.1	77.8	90.2	100.1	78.5	90.6	100.0	80.4	91.3	96.6	80.3	89.4	96.2	83.2	90.4	96.6	75.2	87.1
26, 27, 3370	ELECTRICAL, COMMUNICATIONS & UTIL.	93.9	79.2	86.1	99.8	79.2	88.9	94.3	77.8	85.5	96.6	78.6	87.1	95.2	68.5	81.0	92.6	95.6	94.2
MF2014	WEIGHTED AVERAGE	95.7	80.0	88.8	95.5	79.9	88.6	95.5	84.6	90.7	93.4	83.8	89.1	96.8	83.3	90.8	92.7	78.9	86.6

LOUISIANA

DIVISION		ALEXANDRIA 713-714 MAT.	INST.	TOTAL	BATON ROUGE 707-708 MAT.	INST.	TOTAL	HAMMOND 704 MAT.	INST.	TOTAL	LAFAYETTE 705 MAT.	INST.	TOTAL	LAKE CHARLES 706 MAT.	INST.	TOTAL	MONROE 712 MAT.	INST.	TOTAL
015433	CONTRACTOR EQUIPMENT		94.6	94.6		92.6	92.6		93.1	93.1		93.1	93.1		92.6	92.6		94.6	94.6
0241, 31 - 34	SITE & INFRASTRUCTURE, DEMOLITION	103.5	93.3	96.4	102.3	92.2	95.2	99.7	91.6	94.0	100.7	93.6	95.7	101.4	92.9	95.5	103.5	93.3	96.4
0310	Concrete Forming & Accessories	81.3	63.2	65.7	96.7	75.3	78.2	79.0	58.7	61.4	96.7	69.2	72.9	97.6	68.5	72.5	80.9	62.5	65.0
0320	Concrete Reinforcing	102.9	64.5	83.4	91.8	54.4	72.8	90.9	54.2	72.3	92.2	54.5	73.1	92.2	54.6	73.1	101.7	64.0	82.6
0330	Cast-in-Place Concrete	94.3	67.2	84.0	94.2	76.9	87.6	92.6	66.7	82.7	92.1	67.7	82.8	97.0	69.7	86.6	94.3	68.0	84.3
03	CONCRETE	90.9	65.6	79.1	91.9	72.5	82.9	90.0	61.4	76.8	91.0	66.5	79.7	93.4	67.0	81.2	90.7	65.4	79.0
04	MASONRY	116.2	63.6	83.8	88.8	63.6	73.3	94.4	56.5	71.1	94.4	62.2	74.6	93.8	65.6	76.4	110.7	55.9	76.9
05	METALS	95.7	75.4	89.5	99.9	70.3	90.8	89.6	69.3	83.4	88.9	70.2	83.2	88.9	70.6	83.3	95.6	74.4	89.1
06	WOOD, PLASTICS & COMPOSITES	93.2	62.3	76.3	103.2	77.4	89.1	87.1	58.6	71.5	106.9	71.4	87.4	105.2	68.4	85.0	92.6	62.3	76.0
07	THERMAL & MOISTURE PROTECTION	98.2	68.0	85.7	97.8	70.0	86.3	97.2	64.9	83.8	97.7	68.3	85.5	97.4	69.3	85.7	98.2	65.5	84.6
08	OPENINGS	110.7	65.3	100.4	102.2	70.7	95.1	97.3	57.3	88.2	101.0	58.0	91.2	101.0	65.2	92.9	110.7	58.5	98.9
0920	Plaster & Gypsum Board	84.0	61.7	69.1	99.6	77.2	84.7	99.4	57.8	71.8	107.2	71.0	83.1	107.2	67.9	81.1	83.7	61.7	69.0
0950, 0980	Ceilings & Acoustic Treatment	80.4	61.7	67.9	95.6	77.2	83.4	97.9	57.8	71.3	96.3	71.0	79.4	97.2	67.9	77.7	80.4	61.7	67.9
0960	Flooring	97.3	64.5	87.8	93.2	60.1	83.6	91.7	58.6	82.1	100.7	64.5	90.2	100.7	66.9	90.9	96.8	52.4	83.9
0970, 0990	Wall Finishes & Painting/Coating	96.4	61.0	75.6	92.6	74.7	82.1	100.1	42.9	66.5	100.1	51.4	71.5	100.1	79.2	87.8	96.4	66.3	78.7
09	FINISHES	84.5	62.9	72.7	89.3	73.0	80.4	92.8	56.6	73.0	96.2	66.8	80.1	96.3	69.4	81.6	84.3	60.8	71.5
COVERS	DIVS. 10 - 14, 25, 28, 41, 43, 44, 46	100.0	80.7	95.6	100.0	87.4	97.1	100.0	80.1	95.5	100.0	85.4	96.7	100.0	86.2	96.9	100.0	82.2	96.0
21, 22, 23	FIRE SUPPRESSION, PLUMBING & HVAC	100.1	63.7	84.0	100.0	65.0	84.5	96.6	61.7	81.2	100.2	63.1	83.8	100.2	64.9	84.6	100.1	55.9	80.6
26, 27, 3370	ELECTRICAL, COMMUNICATIONS & UTIL.	94.2	66.6	79.6	98.8	60.2	78.3	93.3	72.4	82.2	94.5	64.0	78.3	94.0	66.4	79.4	96.1	60.1	77.0
MF2014	WEIGHTED AVERAGE	98.3	68.3	85.1	97.7	70.0	85.5	94.2	65.4	81.5	96.0	67.8	83.6	96.3	69.6	84.5	98.2	64.3	83.2

LOUISIANA / MAINE

DIVISION		NEW ORLEANS 700-701 MAT.	INST.	TOTAL	SHREVEPORT 710-711 MAT.	INST.	TOTAL	THIBODAUX 703 MAT.	INST.	TOTAL	AUGUSTA 043 MAT.	INST.	TOTAL	BANGOR 044 MAT.	INST.	TOTAL	BATH 045 MAT.	INST.	TOTAL
015433	CONTRACTOR EQUIPMENT		87.6	87.6		94.6	94.6		93.1	93.1		99.8	99.8		99.8	99.8		99.8	99.8
0241, 31 - 34	SITE & INFRASTRUCTURE, DEMOLITION	99.3	94.2	95.7	105.8	92.7	96.6	102.0	93.6	96.2	91.5	99.3	96.9	94.6	99.6	98.1	92.5	99.3	97.2
0310	Concrete Forming & Accessories	97.3	70.5	74.2	95.1	45.6	52.3	91.1	68.7	71.7	100.9	72.7	76.5	94.4	78.3	80.5	90.1	72.9	75.3
0320	Concrete Reinforcing	91.8	54.5	72.9	103.1	54.1	78.3	90.9	52.8	71.6	101.0	79.8	90.2	90.9	81.1	85.9	89.9	81.0	85.4
0330	Cast-in-Place Concrete	95.4	69.8	85.6	97.3	62.9	84.2	99.4	54.6	82.3	92.1	102.7	96.1	74.4	110.3	88.1	74.4	102.8	85.2
03	CONCRETE	96.0	67.3	82.7	94.4	54.4	75.8	95.1	61.5	79.5	93.2	85.1	89.4	86.3	90.4	88.2	86.3	85.4	85.9
04	MASONRY	99.2	61.8	76.2	99.9	49.0	68.6	118.1	55.7	79.7	104.9	54.6	73.9	120.7	89.5	101.5	128.0	75.3	95.5
05	METALS	98.8	62.4	87.6	99.9	71.0	91.1	89.6	68.9	83.3	97.7	90.2	95.4	87.9	92.3	89.3	86.4	91.9	88.1
06	WOOD, PLASTICS & COMPOSITES	100.8	73.0	85.5	100.1	44.1	69.3	94.4	70.8	81.4	95.3	76.8	85.1	98.2	76.8	86.4	92.8	76.8	84.0
07	THERMAL & MOISTURE PROTECTION	93.6	68.9	83.4	95.8	60.0	81.0	97.2	66.3	84.4	103.7	64.4	87.4	100.7	82.9	93.3	100.7	70.8	88.3
08	OPENINGS	103.1	65.9	94.7	108.0	43.4	93.4	101.9	60.4	92.5	102.9	80.6	97.8	98.2	80.2	94.1	98.2	80.6	94.2
0920	Plaster & Gypsum Board	97.4	72.5	80.9	93.4	42.9	59.8	100.7	70.4	80.6	111.4	75.7	87.7	112.7	75.7	88.1	108.5	75.7	86.7
0950, 0980	Ceilings & Acoustic Treatment	102.5	72.5	82.5	85.4	42.9	57.1	97.9	70.4	79.6	96.9	75.7	82.8	84.0	75.7	78.5	83.1	75.7	78.2
0960	Flooring	103.5	64.5	92.2	101.6	56.2	88.4	98.0	40.4	81.3	94.2	48.5	81.0	90.4	102.2	93.9	88.6	48.5	77.0
0970, 0990	Wall Finishes & Painting/Coating	102.0	63.1	79.1	88.6	39.6	59.8	101.3	44.0	67.7	93.7	92.7	93.1	94.0	43.6	64.4	94.0	92.7	93.3
09	FINISHES	100.4	68.9	83.2	88.8	45.7	65.3	95.1	60.9	76.4	93.0	69.2	80.0	90.1	78.6	83.8	88.7	69.3	78.1
COVERS	DIVS. 10 - 14, 25, 28, 41, 43, 44, 46	100.0	85.7	96.8	100.0	80.5	95.6	100.0	85.6	96.8	100.0	98.8	99.7	100.0	103.2	100.7	100.0	98.8	99.7
21, 22, 23	FIRE SUPPRESSION, PLUMBING & HVAC	100.1	64.6	84.4	99.9	58.7	81.7	96.6	62.8	81.7	99.9	62.2	83.2	100.1	68.9	86.3	96.6	62.2	81.4
26, 27, 3370	ELECTRICAL, COMMUNICATIONS & UTIL.	96.7	72.3	83.8	101.5	66.2	82.8	92.1	70.6	80.7	100.7	78.8	89.1	99.2	74.7	86.2	97.2	78.8	87.4
MF2014	WEIGHTED AVERAGE	99.1	69.3	86.0	99.4	60.2	82.1	96.8	66.5	83.4	98.6	75.3	88.3	96.1	82.5	90.1	95.0	77.8	87.4

775

City Cost Indexes

MAINE

DIVISION		HOULTON 047 MAT.	INST.	TOTAL	KITTERY 039 MAT.	INST.	TOTAL	LEWISTON 042 MAT.	INST.	TOTAL	MACHIAS 046 MAT.	INST.	TOTAL	PORTLAND 040-041 MAT.	INST.	TOTAL	ROCKLAND 048 MAT.	INST.	TOTAL
015433	CONTRACTOR EQUIPMENT		99.8	99.8		99.8	99.8		99.8	99.8		99.8	99.8		99.8	99.8		99.8	99.8
0241, 31-34	SITE & INFRASTRUCTURE, DEMOLITION	94.4	100.1	98.4	83.7	99.4	94.6	92.1	99.6	97.3	93.8	100.1	98.2	92.1	99.6	97.3	90.2	100.1	97.1
0310	Concrete Forming & Accessories	98.4	77.6	80.4	90.3	73.5	75.8	100.3	78.3	81.3	95.3	77.6	80.0	101.6	78.3	81.4	96.4	77.6	80.1
0320	Concrete Reinforcing	90.9	79.5	85.1	86.6	81.1	83.8	111.7	81.1	96.2	90.9	79.5	85.1	101.0	81.1	90.9	90.9	79.5	85.1
0330	Cast-in-Place Concrete	74.4	110.1	88.0	73.7	103.3	85.0	75.9	110.3	89.0	74.4	110.0	88.0	92.1	110.3	99.1	75.9	110.0	88.9
03	CONCRETE	87.2	89.8	88.4	81.8	85.9	83.7	86.6	90.5	88.4	86.7	89.8	88.1	92.5	90.5	91.5	84.3	89.8	86.8
04	MASONRY	102.4	63.0	78.1	117.7	76.0	92.0	103.0	89.5	94.7	102.4	63.0	78.1	108.5	89.5	96.8	95.8	63.0	75.6
05	METALS	86.6	89.2	87.6	83.0	92.2	85.8	91.2	92.3	91.5	86.7	89.8	87.6	97.8	92.3	96.1	86.5	89.8	87.5
06	WOOD, PLASTICS & COMPOSITES	102.1	76.8	88.2	96.1	76.8	85.5	104.0	76.8	89.0	99.1	76.8	86.8	95.4	76.8	85.2	99.9	76.8	87.2
07	THERMAL & MOISTURE PROTECTION	100.8	70.3	88.1	101.7	70.3	88.7	100.5	82.9	93.2	100.7	69.5	87.8	103.9	82.9	95.2	100.4	69.5	87.6
08	OPENINGS	98.3	80.6	94.2	97.5	84.4	94.5	101.5	80.2	96.7	98.3	80.6	94.3	102.1	80.2	97.2	98.2	80.6	94.2
0920	Plaster & Gypsum Board	115.1	75.7	88.9	104.8	75.7	85.4	118.0	75.7	89.9	113.4	75.7	88.3	111.9	75.7	87.8	113.4	75.7	88.3
0950, 0980	Ceilings & Acoustic Treatment	83.1	75.7	78.2	95.8	75.7	82.4	92.5	75.7	81.3	83.1	75.7	78.2	95.1	75.7	82.2	83.1	75.7	78.2
0960	Flooring	91.8	45.4	78.3	90.4	50.6	78.9	93.4	102.2	96.0	90.9	45.3	77.7	95.0	102.2	97.1	91.3	45.4	78.0
0970, 0990	Wall Finishes & Painting/Coating	94.0	92.7	93.3	87.9	106.8	99.0	94.0	43.6	64.4	94.0	92.7	93.3	96.6	43.6	65.5	94.0	92.7	93.3
09	FINISHES	90.6	72.0	80.4	90.1	71.3	79.9	93.0	78.6	85.1	90.1	72.0	80.2	93.8	78.6	85.5	89.8	72.0	80.1
COVERS	DIVS. 10-14, 25, 28, 41, 43, 44, 46	100.0	103.2	100.7	100.0	99.1	99.8	100.0	103.2	100.7	100.0	103.2	100.7	100.0	103.2	100.7	100.0	103.2	100.7
21, 22, 23	FIRE SUPPRESSION, PLUMBING & HVAC	96.6	68.9	84.3	96.5	72.5	85.9	100.1	69.0	86.3	96.6	68.9	84.3	99.9	69.0	86.2	96.6	68.9	84.3
26, 27, 3370	ELECTRICAL, COMMUNICATIONS & UTIL.	101.2	78.8	89.3	94.6	78.8	86.2	101.3	78.8	89.3	101.2	78.8	89.3	103.3	78.8	90.3	101.2	78.8	89.3
MF2014	WEIGHTED AVERAGE	94.6	78.9	87.7	93.0	80.6	87.5	96.6	83.1	90.6	94.4	78.9	87.6	99.0	83.1	92.0	93.7	78.9	87.1

MAINE / MARYLAND

DIVISION		WATERVILLE 049 MAT.	INST.	TOTAL	ANNAPOLIS 214 MAT.	INST.	TOTAL	BALTIMORE 210-212 MAT.	INST.	TOTAL	COLLEGE PARK 207-208 MAT.	INST.	TOTAL	CUMBERLAND 215 MAT.	INST.	TOTAL	EASTON 216 MAT.	INST.	TOTAL
015433	CONTRACTOR EQUIPMENT		99.8	99.8		107.0	107.0		106.1	106.1		112.7	112.7		107.0	107.0		107.0	107.0
0241, 31-34	SITE & INFRASTRUCTURE, DEMOLITION	94.2	99.3	97.7	105.2	95.2	98.2	103.7	98.5	100.0	102.0	99.6	100.4	96.5	96.0	96.2	103.8	92.3	95.8
0310	Concrete Forming & Accessories	89.6	72.7	75.0	101.7	77.8	81.0	98.7	78.1	80.9	85.4	72.9	74.5	92.5	87.1	87.8	90.4	75.4	77.4
0320	Concrete Reinforcing	90.9	79.8	85.2	98.4	90.9	94.6	101.0	90.9	95.9	99.8	87.6	93.6	89.0	85.7	87.3	88.3	87.2	87.7
0330	Cast-in-Place Concrete	74.4	102.7	85.2	114.9	75.7	99.9	112.0	77.2	98.8	113.3	75.1	98.7	95.6	82.2	90.5	106.1	68.8	91.9
03	CONCRETE	87.7	85.1	86.5	102.6	80.9	92.5	107.4	81.1	95.2	106.7	78.2	93.5	91.7	86.4	89.2	100.0	76.6	89.1
04	MASONRY	113.1	54.6	77.1	101.8	72.0	83.5	108.2	72.1	86.0	119.7	69.9	89.0	104.9	90.8	96.2	119.4	41.7	71.5
05	METALS	86.6	90.2	87.7	98.3	106.2	100.7	97.4	101.0	98.5	85.4	110.5	93.1	94.3	105.0	97.5	94.5	99.9	96.2
06	WOOD, PLASTICS & COMPOSITES	92.3	76.8	83.8	95.6	79.8	86.9	98.1	80.0	88.2	79.1	73.9	76.3	88.5	86.6	87.4	86.6	83.4	84.8
07	THERMAL & MOISTURE PROTECTION	100.8	64.4	85.7	103.8	81.1	94.4	101.4	81.8	93.2	101.6	80.9	93.0	100.3	82.1	92.8	100.5	71.6	88.5
08	OPENINGS	98.3	80.6	94.2	99.3	88.1	96.8	98.2	88.2	95.9	91.5	81.1	89.1	98.3	89.3	96.2	96.5	78.1	92.4
0920	Plaster & Gypsum Board	108.5	75.7	86.7	102.3	79.5	87.1	102.8	79.5	87.3	97.4	73.3	81.4	103.7	86.4	92.2	103.7	83.2	90.1
0950, 0980	Ceilings & Acoustic Treatment	83.1	75.7	78.2	90.7	79.5	83.2	107.1	79.5	88.7	97.8	73.3	81.5	99.2	86.4	90.7	99.2	83.2	88.5
0960	Flooring	88.3	48.5	76.8	91.1	79.8	87.8	95.4	79.8	90.9	88.5	78.0	85.5	88.6	96.5	90.9	87.7	47.9	76.2
0970, 0990	Wall Finishes & Painting/Coating	94.0	92.7	93.3	90.5	75.7	81.8	104.0	75.7	87.4	106.8	75.7	88.5	99.2	92.6	95.3	99.2	75.7	85.4
09	FINISHES	88.8	69.2	78.1	89.0	77.6	82.8	100.6	77.8	88.1	92.2	73.4	81.9	94.3	89.8	91.8	94.5	71.5	81.9
COVERS	DIVS. 10-14, 25, 28, 41, 43, 44, 46	100.0	98.6	99.7	100.0	88.3	97.4	100.0	88.9	97.5	100.0	83.8	96.3	100.0	92.9	98.4	100.0	73.0	93.9
21, 22, 23	FIRE SUPPRESSION, PLUMBING & HVAC	96.6	62.2	81.4	100.0	82.8	92.4	100.0	82.9	92.4	100.0	82.9	92.4	96.3	73.4	86.2	96.3	76.4	87.5
26, 27, 3370	ELECTRICAL, COMMUNICATIONS & UTIL.	101.2	78.8	89.3	99.0	92.4	95.5	98.8	92.4	95.4	98.3	101.2	99.8	96.5	82.1	88.9	95.9	65.8	79.9
MF2014	WEIGHTED AVERAGE	94.9	75.3	86.3	99.2	85.6	93.2	100.7	85.5	94.0	98.6	86.0	92.0	96.3	86.6	92.0	98.0	74.2	87.5

MARYLAND / MASSACHUSETTS

DIVISION		ELKTON 219 MAT.	INST.	TOTAL	HAGERSTOWN 217 MAT.	INST.	TOTAL	SALISBURY 218 MAT.	INST.	TOTAL	SILVER SPRING 209 MAT.	INST.	TOTAL	WALDORF 206 MAT.	INST.	TOTAL	BOSTON 020-022, 024 MAT.	INST.	TOTAL
015433	CONTRACTOR EQUIPMENT		107.0	107.0		107.0	107.0		107.0	107.0		104.0	104.0		104.0	104.0		103.5	103.5
0241, 31-34	SITE & INFRASTRUCTURE, DEMOLITION	90.4	94.9	93.5	94.6	96.1	95.6	103.8	92.4	95.8	89.9	91.2	90.8	96.2	90.9	92.5	100.8	104.0	103.1
0310	Concrete Forming & Accessories	96.5	92.9	93.4	91.5	82.2	83.4	104.8	53.5	60.4	94.1	71.1	74.2	101.5	69.4	73.8	102.6	140.3	135.2
0320	Concrete Reinforcing	88.3	113.5	101.1	89.0	85.7	87.3	88.3	67.1	77.5	98.6	87.5	92.9	99.2	87.5	93.3	103.3	146.7	125.4
0330	Cast-in-Place Concrete	85.9	75.0	81.8	91.0	82.3	87.8	106.1	66.9	91.2	116.0	78.1	101.6	129.9	75.4	109.1	97.1	141.0	113.8
03	CONCRETE	84.3	91.4	87.6	88.0	84.2	86.2	100.8	62.6	83.1	104.7	78.2	92.4	115.6	76.5	97.4	99.6	140.9	118.8
04	MASONRY	103.3	58.8	75.9	110.5	90.8	98.4	119.1	45.9	74.0	119.2	75.0	92.0	101.9	70.1	82.3	105.5	149.1	132.4
05	METALS	94.6	113.8	100.4	94.4	105.0	97.6	94.6	92.6	93.9	89.6	106.1	94.7	89.6	106.3	94.7	96.6	131.7	107.3
06	WOOD, PLASTICS & COMPOSITES	93.5	101.2	97.7	87.8	79.8	83.4	103.4	55.5	77.0	85.4	69.7	76.7	92.1	69.7	79.8	104.5	141.2	124.6
07	THERMAL & MOISTURE PROTECTION	100.0	79.8	91.6	100.7	87.3	95.1	100.7	69.0	87.6	105.2	87.9	98.1	105.8	85.7	97.4	99.7	137.8	115.5
08	OPENINGS	96.5	101.7	97.7	96.5	82.4	93.3	96.8	64.2	89.4	84.2	78.2	82.8	84.8	78.8	83.4	99.9	145.1	110.1
0920	Plaster & Gypsum Board	107.3	101.5	103.5	103.7	79.5	87.6	114.3	54.5	74.5	103.0	69.7	80.9	106.0	69.7	81.9	103.5	142.0	129.1
0950, 0980	Ceilings & Acoustic Treatment	99.2	101.5	100.7	100.1	79.5	86.4	99.2	54.5	69.5	106.3	69.7	81.9	106.3	69.7	81.9	102.0	142.0	128.6
0960	Flooring	90.2	54.3	79.8	88.2	96.5	90.6	94.1	60.0	84.2	94.4	78.0	89.6	97.9	78.0	92.1	97.8	161.0	116.2
0970, 0990	Wall Finishes & Painting/Coating	99.2	75.7	85.4	99.2	75.7	85.4	99.2	75.7	85.4	114.7	75.7	91.8	114.7	75.7	91.8	101.8	156.9	134.2
09	FINISHES	94.9	85.5	89.8	94.2	83.9	88.6	97.9	56.6	75.3	93.0	71.2	81.1	94.8	70.7	81.6	100.4	146.8	125.8
COVERS	DIVS. 10-14, 25, 28, 41, 43, 44, 46	100.0	59.8	90.9	100.0	92.2	98.2	100.0	82.1	96.0	100.0	83.3	96.2	100.0	83.8	96.3	100.0	117.2	103.9
21, 22, 23	FIRE SUPPRESSION, PLUMBING & HVAC	96.3	81.3	89.7	99.9	87.3	94.3	96.3	74.0	86.5	96.6	85.1	91.5	96.6	82.4	90.3	100.1	126.4	111.7
26, 27, 3370	ELECTRICAL, COMMUNICATIONS & UTIL.	97.7	91.5	94.4	96.3	82.1	88.8	94.7	65.1	79.0	95.7	101.2	98.6	93.3	101.2	97.5	101.3	129.4	116.2
MF2014	WEIGHTED AVERAGE	95.2	86.8	91.5	96.6	88.3	92.9	98.4	68.8	85.4	96.0	85.6	91.4	96.7	84.2	91.2	99.9	133.4	114.7

MASSACHUSETTS

DIVISION		BROCKTON 023			BUZZARDS BAY 025			FALL RIVER 027			FITCHBURG 014			FRAMINGHAM 017			GREENFIELD 013		
		MAT.	INST.	TOTAL	MAT.	INST.	TOTAL	MAT.	INST.	TOTAL	MAT.	INST.	TOTAL	MAT.	INST.	TOTAL	MAT.	INST.	TOTAL
015433	CONTRACTOR EQUIPMENT		102.4	102.4		102.4	102.4		103.3	103.3		99.8	99.8		101.7	101.7		99.8	99.8
0241, 31 - 34	SITE & INFRASTRUCTURE, DEMOLITION	94.5	104.5	101.5	84.8	104.5	98.5	93.5	104.6	101.2	85.2	104.4	98.6	81.9	104.2	97.5	88.8	103.4	98.9
0310	Concrete Forming & Accessories	100.7	126.4	123.0	98.2	126.0	122.3	100.7	126.3	122.8	94.8	126.2	122.0	102.6	126.5	123.3	93.1	112.1	109.5
0320	Concrete Reinforcing	102.6	146.0	124.6	82.3	123.1	103.0	102.6	123.2	113.0	86.3	145.6	116.4	86.3	146.3	116.8	90.0	119.7	105.1
0330	Cast-in-Place Concrete	91.5	137.5	109.0	76.1	137.3	99.4	88.5	137.8	107.3	79.5	137.4	101.6	79.5	137.6	101.6	81.7	126.2	98.7
03	CONCRETE	95.0	133.2	112.7	80.5	128.9	102.9	93.6	129.2	110.1	79.5	132.8	104.2	82.1	133.3	105.8	82.9	117.9	99.1
04	MASONRY	102.6	133.6	121.7	95.5	141.3	123.7	103.3	141.3	126.7	103.3	137.8	124.6	110.1	138.4	127.5	108.2	122.4	117.0
05	METALS	93.1	128.1	103.8	88.1	118.8	97.5	93.1	119.2	101.1	90.6	124.8	101.1	90.6	128.4	102.2	92.9	110.7	98.4
06	WOOD, PLASTICS & COMPOSITES	103.8	126.0	116.0	100.9	126.0	114.7	103.8	126.3	116.2	100.5	126.0	114.6	106.7	125.8	117.2	98.6	110.5	105.1
07	THERMAL & MOISTURE PROTECTION	100.5	129.6	112.6	99.4	129.2	111.7	100.3	128.5	112.0	99.7	125.7	110.5	99.8	130.8	112.7	99.8	112.0	104.8
08	OPENINGS	99.6	134.0	107.4	95.4	124.0	101.9	99.6	124.2	105.1	100.4	133.8	108.0	91.2	133.9	100.8	100.8	114.9	104.0
0920	Plaster & Gypsum Board	91.7	126.3	114.7	87.8	126.3	113.4	91.7	126.3	114.7	108.3	126.3	120.3	111.0	126.3	121.2	109.3	110.3	110.0
0950, 0980	Ceilings & Acoustic Treatment	104.9	126.3	119.1	91.5	126.3	114.7	104.9	126.3	119.1	90.4	126.3	114.3	90.4	126.3	114.3	98.9	110.3	106.5
0960	Flooring	95.1	161.0	114.2	92.9	161.0	112.6	94.1	161.0	113.5	91.5	161.0	111.7	93.4	161.0	113.0	90.7	133.3	103.1
0970, 0990	Wall Finishes & Painting/Coating	93.9	140.0	120.9	93.9	140.0	120.9	93.9	140.0	120.9	90.7	140.0	119.6	91.6	140.0	120.0	90.7	116.4	105.8
09	FINISHES	94.5	134.9	116.6	90.0	134.9	114.5	94.3	135.1	116.5	89.8	134.9	114.4	90.5	134.7	114.7	91.7	116.7	105.4
COVERS	DIVS. 10 - 14, 25, 28, 41, 43, 44, 46	100.0	114.0	103.2	100.0	114.0	103.2	100.0	114.6	103.3	100.0	108.6	102.0	100.0	113.6	103.1	100.0	106.0	101.4
21, 22, 23	FIRE SUPPRESSION, PLUMBING & HVAC	100.3	107.6	103.6	96.8	107.4	101.5	100.3	107.4	103.5	96.9	108.6	102.1	96.9	121.6	107.9	96.9	100.5	98.5
26, 27, 3370	ELECTRICAL, COMMUNICATIONS & UTIL.	99.8	103.0	101.5	97.0	103.0	100.2	99.7	103.0	101.4	100.0	100.7	100.4	96.6	122.2	110.2	100.0	97.7	98.8
MF2014	WEIGHTED AVERAGE	97.8	120.3	107.7	92.6	119.2	104.3	97.6	119.3	107.2	94.2	120.0	105.6	93.5	126.5	108.1	95.5	109.1	101.5

MASSACHUSETTS

DIVISION		HYANNIS 026			LAWRENCE 019			LOWELL 018			NEW BEDFORD 027			PITTSFIELD 012			SPRINGFIELD 010 - 011		
		MAT.	INST.	TOTAL	MAT.	INST.	TOTAL	MAT.	INST.	TOTAL	MAT.	INST.	TOTAL	MAT.	INST.	TOTAL	MAT.	INST.	TOTAL
015433	CONTRACTOR EQUIPMENT		102.4	102.4		102.4	102.4		99.8	99.8		103.3	103.3		99.8	99.8		99.8	99.8
0241, 31 - 34	SITE & INFRASTRUCTURE, DEMOLITION	90.9	104.5	100.4	94.0	104.5	101.3	92.9	104.4	100.9	92.0	104.6	100.8	93.9	103.0	100.2	93.3	103.4	100.3
0310	Concrete Forming & Accessories	92.4	126.0	121.5	103.9	126.9	123.8	100.2	126.8	123.2	100.7	126.3	122.8	100.2	109.7	108.5	100.4	112.1	110.5
0320	Concrete Reinforcing	82.3	123.1	103.0	106.8	138.0	122.6	107.7	137.8	123.0	102.6	123.2	113.0	89.4	121.9	105.9	107.7	119.7	113.8
0330	Cast-in-Place Concrete	83.5	137.3	104.0	92.0	137.6	109.4	83.7	137.6	104.2	78.0	137.8	100.8	91.3	118.3	101.6	87.1	126.2	102.0
03	CONCRETE	86.5	128.9	106.2	95.9	132.0	112.7	87.9	131.7	108.2	88.5	129.2	107.4	89.1	114.5	100.9	89.5	117.9	102.7
04	MASONRY	101.5	141.3	126.0	116.1	141.3	131.6	102.2	137.8	124.1	101.5	141.3	126.0	102.8	116.5	111.2	102.4	122.4	114.7
05	METALS	89.6	118.8	98.5	93.3	125.6	103.2	93.3	122.1	102.1	93.1	119.2	101.1	93.1	111.5	98.7	95.9	110.7	100.4
06	WOOD, PLASTICS & COMPOSITES	94.5	126.0	111.8	106.9	126.0	117.4	106.4	126.0	117.2	103.8	126.3	116.2	106.4	110.5	108.6	106.4	110.5	108.6
07	THERMAL & MOISTURE PROTECTION	99.9	129.2	112.0	100.4	131.8	113.4	100.2	130.8	112.9	100.3	128.5	112.0	100.3	108.6	103.7	100.2	112.0	105.1
08	OPENINGS	95.9	124.0	102.3	95.0	131.6	103.3	101.9	131.6	108.6	99.6	124.2	105.1	101.9	115.5	105.0	101.9	114.9	104.8
0920	Plaster & Gypsum Board	83.5	126.3	112.0	113.6	126.3	122.0	113.6	126.3	122.0	91.7	126.3	114.7	113.6	110.3	111.4	113.6	110.3	111.4
0950, 0980	Ceilings & Acoustic Treatment	98.1	126.3	116.9	100.8	126.3	117.8	100.8	126.3	117.8	104.9	126.3	119.1	100.8	110.3	107.1	100.8	110.3	107.1
0960	Flooring	90.1	161.0	110.7	93.8	161.0	113.3	93.8	161.0	113.3	94.1	161.0	113.5	94.3	133.3	105.6	93.3	133.3	104.9
0970, 0990	Wall Finishes & Painting/Coating	93.9	140.0	120.9	90.8	140.0	119.7	90.7	140.0	119.6	93.9	140.0	120.9	90.7	116.4	105.8	91.9	116.4	106.3
09	FINISHES	90.3	134.9	114.6	93.8	134.9	116.3	93.8	134.9	116.2	94.1	135.1	116.5	94.0	115.2	105.6	93.7	116.7	106.3
COVERS	DIVS. 10 - 14, 25, 28, 41, 43, 44, 46	100.0	114.0	103.2	100.0	114.1	103.2	100.0	114.1	103.2	100.0	114.6	103.3	100.0	104.0	100.9	100.0	106.0	101.4
21, 22, 23	FIRE SUPPRESSION, PLUMBING & HVAC	100.3	107.4	103.4	100.1	124.1	110.7	100.1	124.1	110.7	100.3	107.4	103.5	100.1	97.5	98.9	100.1	100.5	100.3
26, 27, 3370	ELECTRICAL, COMMUNICATIONS & UTIL.	97.4	103.0	100.4	99.1	127.6	114.2	99.6	122.1	111.5	100.6	103.0	101.8	99.6	97.7	98.6	99.6	97.7	98.6
MF2014	WEIGHTED AVERAGE	94.9	119.2	105.6	97.9	127.6	111.1	97.0	126.1	109.9	96.9	119.3	106.8	97.2	107.1	101.6	97.7	109.1	102.7

MASSACHUSETTS / MICHIGAN

DIVISION		WORCESTER 015 - 016			ANN ARBOR 481			BATTLE CREEK 490			BAY CITY 487			DEARBORN 481			DETROIT 482		
		MAT.	INST.	TOTAL	MAT.	INST.	TOTAL	MAT.	INST.	TOTAL	MAT.	INST.	TOTAL	MAT.	INST.	TOTAL	MAT.	INST.	TOTAL
015433	CONTRACTOR EQUIPMENT		99.8	99.8		114.1	114.1		100.4	100.4		114.1	114.1		114.1	114.1		95.9	95.9
0241, 31 - 34	SITE & INFRASTRUCTURE, DEMOLITION	93.3	104.4	101.0	82.3	97.9	93.2	93.4	86.0	88.3	73.7	96.9	89.9	82.0	98.0	93.2	97.6	99.8	99.2
0310	Concrete Forming & Accessories	100.8	126.2	122.8	98.2	108.5	107.1	98.4	80.6	83.0	98.3	84.4	86.3	98.1	109.4	107.9	100.8	109.3	108.2
0320	Concrete Reinforcing	107.7	145.6	127.0	97.6	105.4	101.6	94.8	79.2	86.9	97.6	104.6	101.2	97.6	105.5	101.6	99.4	105.4	102.5
0330	Cast-in-Place Concrete	86.6	137.4	105.9	87.7	102.7	93.4	84.9	94.1	88.4	83.9	87.2	85.2	85.8	104.0	92.7	104.4	103.6	104.1
03	CONCRETE	89.3	132.7	109.4	90.7	106.8	98.1	88.2	84.7	86.6	88.9	90.4	89.6	89.7	107.6	98.0	102.4	105.8	104.0
04	MASONRY	102.0	137.8	124.1	99.4	101.4	100.6	99.2	80.9	87.9	99.0	82.4	88.8	99.2	103.7	102.0	95.0	103.7	100.4
05	METALS	96.0	124.7	104.8	99.3	118.0	105.1	103.5	81.8	96.9	99.9	114.8	104.5	99.4	118.1	105.1	98.7	94.3	97.4
06	WOOD, PLASTICS & COMPOSITES	106.9	126.0	117.4	91.4	110.6	101.9	89.4	79.2	83.8	91.4	84.7	87.7	91.4	110.6	101.9	97.9	110.5	104.9
07	THERMAL & MOISTURE PROTECTION	100.2	125.7	110.8	103.7	103.3	103.5	100.2	79.2	91.5	101.1	86.8	95.2	102.0	107.3	104.2	101.3	107.1	103.7
08	OPENINGS	101.9	133.8	109.1	96.8	104.5	98.5	91.9	77.1	88.6	96.8	88.8	95.0	96.8	104.5	98.5	98.8	104.5	100.1
0920	Plaster & Gypsum Board	113.6	126.3	122.0	105.1	110.5	108.7	98.0	75.1	82.8	105.1	83.9	91.0	105.1	110.5	108.7	98.7	110.5	106.5
0950, 0980	Ceilings & Acoustic Treatment	100.8	126.3	117.8	87.2	110.5	102.7	79.8	75.1	76.7	88.1	83.9	85.3	87.2	110.5	102.7	96.8	110.5	105.9
0960	Flooring	93.8	153.3	111.1	93.0	110.0	97.9	93.9	70.4	87.1	93.0	74.3	87.6	92.5	104.4	96.0	97.9	104.4	99.8
0970, 0990	Wall Finishes & Painting/Coating	90.7	140.0	119.6	83.9	101.8	94.4	89.8	79.5	83.8	83.9	81.3	82.4	83.9	99.9	93.3	89.1	99.9	95.4
09	FINISHES	93.8	133.3	115.4	91.5	108.6	100.8	88.3	78.6	83.0	91.3	81.7	86.0	91.3	108.0	100.4	97.2	107.9	103.1
COVERS	DIVS. 10 - 14, 25, 28, 41, 43, 44, 46	100.0	108.7	102.0	100.0	103.0	100.7	100.0	95.3	98.9	100.0	89.1	97.5	100.0	103.8	100.9	100.0	103.6	100.8
21, 22, 23	FIRE SUPPRESSION, PLUMBING & HVAC	100.1	108.6	103.8	100.1	95.2	97.9	100.1	85.8	93.8	100.1	80.9	91.6	100.1	104.0	101.8	100.0	104.5	102.0
26, 27, 3370	ELECTRICAL, COMMUNICATIONS & UTIL.	99.6	105.2	102.5	97.5	103.5	100.7	94.9	82.2	88.1	96.6	92.4	94.4	97.5	97.7	97.6	98.8	97.6	98.2
MF2014	WEIGHTED AVERAGE	97.7	120.4	107.7	97.0	103.6	99.9	96.5	83.1	90.6	96.4	89.3	93.3	96.8	105.1	100.5	99.3	102.9	100.9

For customer support on your Building Construction Costs with RSMeans Data, call 800.448.8182.

777

DIVISION		MICHIGAN																
		FLINT			GAYLORD			GRAND RAPIDS			IRON MOUNTAIN			JACKSON			KALAMAZOO		
		484 - 485			497			493,495			498 - 499			492			491		
		MAT.	INST.	TOTAL	MAT.	INST.	TOTAL	MAT.	INST.	TOTAL	MAT.	INST.	TOTAL	MAT.	INST.	TOTAL	MAT.	INST.	TOTAL
015433	CONTRACTOR EQUIPMENT		114.1	114.1		108.5	108.5		100.4	100.4		95.0	95.0		108.5	108.5		100.4	100.4
0241, 31 - 34	SITE & INFRASTRUCTURE, DEMOLITION	71.6	97.0	89.3	87.9	83.0	84.5	92.2	86.1	87.9	96.4	92.3	93.5	110.7	85.4	93.0	93.7	86.0	88.3
0310	Concrete Forming & Accessories	101.1	85.9	88.0	96.9	74.6	77.6	97.6	81.0	83.3	88.1	81.1	82.1	93.4	84.0	85.3	98.4	80.3	82.7
0320	Concrete Reinforcing	97.6	105.0	101.3	88.3	89.8	89.1	99.4	79.3	89.2	88.2	86.4	87.3	85.8	104.6	95.4	94.8	79.2	86.9
0330	Cast-in-Place Concrete	88.3	90.9	89.3	84.7	82.5	83.9	89.1	95.7	91.6	100.3	71.2	89.2	84.6	88.3	86.0	86.6	96.1	90.2
03	CONCRETE	91.1	92.4	91.7	84.9	81.9	83.5	92.2	85.4	89.1	93.5	79.2	86.9	80.0	90.4	84.8	91.4	85.2	88.6
04	MASONRY	99.4	91.1	94.3	110.1	77.4	89.9	92.0	83.1	86.5	95.0	82.0	87.0	89.2	84.1	86.0	97.7	81.6	87.8
05	METALS	99.4	115.5	104.3	105.1	112.3	107.3	100.5	81.6	94.7	104.5	91.1	100.4	105.3	112.9	107.6	103.5	81.6	96.8
06	WOOD, PLASTICS & COMPOSITES	94.8	84.8	89.3	82.9	74.1	78.1	92.4	79.9	85.5	78.5	81.7	80.3	81.7	82.6	82.2	89.4	79.2	83.8
07	THERMAL & MOISTURE PROTECTION	101.4	90.9	97.0	98.7	72.3	87.7	101.9	73.4	90.1	102.6	77.8	92.3	98.0	86.6	93.3	100.2	80.0	91.8
08	OPENINGS	96.8	88.2	94.8	92.2	81.8	89.9	99.1	78.9	94.5	99.2	72.6	93.2	91.3	86.5	90.2	91.9	74.2	87.9
0920	Plaster & Gypsum Board	107.1	84.0	91.7	97.8	72.3	80.9	97.9	75.7	83.2	52.7	81.8	72.0	96.2	81.0	86.1	98.0	75.1	82.8
0950, 0980	Ceilings & Acoustic Treatment	87.2	84.0	85.0	79.0	72.3	74.5	91.5	75.7	81.0	77.9	81.8	80.5	79.0	81.0	80.3	79.8	75.1	76.7
0960	Flooring	93.0	85.6	90.9	87.0	82.2	85.6	96.1	76.6	90.5	106.5	90.2	101.8	85.7	74.0	82.3	93.9	74.0	88.2
0970, 0990	Wall Finishes & Painting/Coating	83.9	80.0	81.6	86.1	82.5	84.0	91.1	79.2	84.1	104.1	69.2	83.6	86.1	99.9	94.2	89.8	79.5	83.8
09	FINISHES	91.0	84.7	87.5	87.7	76.6	81.7	92.3	79.8	85.5	88.1	81.7	84.6	88.7	83.2	85.7	88.3	78.6	83.0
COVERS	DIVS. 10 - 14, 25, 28, 41, 43, 44, 46	100.0	90.2	97.8	100.0	82.7	96.1	100.0	90.1	97.7	100.0	81.2	95.7	100.0	99.5	99.9	100.0	89.6	97.7
21, 22, 23	FIRE SUPPRESSION, PLUMBING & HVAC	100.1	86.7	94.2	96.8	73.9	86.7	100.1	82.4	92.3	96.7	80.5	89.6	96.8	86.7	92.3	100.1	77.7	90.2
26, 27, 3370	ELECTRICAL, COMMUNICATIONS & UTIL.	97.5	94.1	95.7	92.6	75.1	83.3	100.0	74.6	86.5	99.2	83.4	90.8	96.7	102.6	99.9	94.7	80.3	87.1
MF2014	WEIGHTED AVERAGE	96.7	92.4	94.8	95.7	80.7	89.1	97.8	81.5	90.6	97.7	82.6	91.0	95.0	91.4	93.4	96.8	80.9	89.8

DIVISION		MICHIGAN															MINNESOTA		
		LANSING			MUSKEGON			ROYAL OAK			SAGINAW			TRAVERSE CITY			BEMIDJI		
		488 - 489			494			480,483			486			496			566		
		MAT.	INST.	TOTAL	MAT.	INST.	TOTAL	MAT.	INST.	TOTAL	MAT.	INST.	TOTAL	MAT.	INST.	TOTAL	MAT.	INST.	TOTAL
015433	CONTRACTOR EQUIPMENT		114.1	114.1		100.4	100.4		93.2	93.2		114.1	114.1		95.0	95.0		101.8	101.8
0241, 31 - 34	SITE & INFRASTRUCTURE, DEMOLITION	92.5	97.1	95.7	91.3	86.0	87.6	86.4	96.9	93.7	74.7	96.8	90.1	82.0	91.7	88.8	95.9	100.6	99.2
0310	Concrete Forming & Accessories	96.3	85.3	86.7	98.7	78.8	81.4	93.5	108.8	106.7	98.3	84.2	86.1	88.1	74.8	76.6	86.7	84.8	85.1
0320	Concrete Reinforcing	99.4	104.6	102.1	95.4	79.0	87.1	88.4	90.7	89.6	97.6	104.6	101.1	89.5	78.5	84.0	97.4	95.8	96.6
0330	Cast-in-Place Concrete	93.6	90.1	92.3	84.6	90.6	86.9	76.8	103.4	86.9	86.7	87.1	86.8	78.2	78.9	78.5	101.5	101.4	101.5
03	CONCRETE	92.0	91.8	91.9	86.6	82.6	84.7	78.0	103.0	89.6	90.2	90.3	90.2	76.9	77.6	77.2	92.6	93.9	93.2
04	MASONRY	98.3	87.2	91.5	96.3	77.3	84.6	92.9	101.4	98.1	100.8	82.4	89.5	93.2	78.8	84.3	101.2	100.4	100.7
05	METALS	100.5	114.6	104.8	101.1	81.1	95.0	102.8	91.6	99.4	99.4	114.6	104.1	104.5	87.3	99.2	92.0	115.4	99.2
06	WOOD, PLASTICS & COMPOSITES	90.9	84.0	87.1	86.2	77.6	81.5	87.2	110.6	100.0	87.8	84.7	86.1	78.5	74.9	76.5	66.9	80.4	74.4
07	THERMAL & MOISTURE PROTECTION	98.9	84.0	92.7	99.1	73.5	88.5	99.7	104.3	101.6	102.3	87.1	96.0	101.7	70.4	88.7	107.0	95.0	102.0
08	OPENINGS	99.6	87.0	96.7	91.2	77.4	88.1	96.6	104.4	98.4	94.8	88.8	93.4	99.2	67.6	92.1	99.1	101.2	99.6
0920	Plaster & Gypsum Board	97.3	83.1	87.9	77.1	73.4	74.7	102.8	110.5	107.9	105.1	83.9	91.0	52.7	74.8	67.4	108.9	80.2	89.9
0950, 0980	Ceilings & Acoustic Treatment	89.0	83.1	85.1	80.7	73.4	75.8	86.4	110.5	102.4	87.2	83.9	85.0	77.9	74.8	75.9	119.6	80.2	93.4
0960	Flooring	97.5	76.6	91.4	92.6	76.6	88.0	89.9	101.8	93.4	93.0	74.3	87.6	106.5	88.3	101.2	95.1	110.6	99.6
0970, 0990	Wall Finishes & Painting/Coating	92.8	79.2	84.8	88.0	78.0	82.1	85.1	99.9	93.8	83.9	81.3	82.4	104.1	42.0	67.6	93.9	110.5	103.7
09	FINISHES	92.3	82.5	87.0	84.4	77.4	80.6	90.3	107.2	99.5	91.2	81.7	86.0	87.1	74.0	79.9	99.5	91.5	95.1
COVERS	DIVS. 10 - 14, 25, 28, 41, 43, 44, 46	100.0	98.7	99.7	100.0	89.4	97.6	100.0	100.1	100.0	100.0	89.1	97.5	100.0	79.7	95.4	100.0	97.2	99.4
21, 22, 23	FIRE SUPPRESSION, PLUMBING & HVAC	100.0	86.2	93.8	99.9	81.8	91.9	96.8	100.9	98.6	100.1	80.4	91.4	96.7	79.2	89.0	96.9	81.1	89.9
26, 27, 3370	ELECTRICAL, COMMUNICATIONS & UTIL.	100.3	88.0	93.8	95.1	76.9	85.5	99.6	100.0	99.8	95.6	89.8	92.5	94.3	75.1	84.1	104.8	104.7	104.8
MF2014	WEIGHTED AVERAGE	98.0	90.7	94.8	95.2	80.3	88.6	95.0	101.0	97.7	96.3	88.8	93.0	94.5	78.7	87.5	97.3	95.8	96.6

DIVISION		MINNESOTA																	
		BRAINERD			DETROIT LAKES			DULUTH			MANKATO			MINNEAPOLIS			ROCHESTER		
		564			565			556 - 558			560			553 - 555			559		
		MAT.	INST.	TOTAL	MAT.	INST.	TOTAL	MAT.	INST.	TOTAL	MAT.	INST.	TOTAL	MAT.	INST.	TOTAL	MAT.	INST.	TOTAL
015433	CONTRACTOR EQUIPMENT		104.5	104.5		101.8	101.8		104.8	104.8		104.5	104.5		107.9	107.9		104.8	104.8
0241, 31 - 34	SITE & INFRASTRUCTURE, DEMOLITION	98.2	105.7	103.4	94.1	100.9	98.8	97.7	103.9	102.0	94.9	105.4	102.2	97.4	107.8	104.7	97.8	102.8	101.3
0310	Concrete Forming & Accessories	87.8	87.1	87.2	83.7	86.4	86.0	99.3	98.5	98.6	95.9	91.0	91.7	99.5	113.1	111.2	100.6	94.9	95.7
0320	Concrete Reinforcing	96.1	95.9	96.0	97.4	95.8	96.6	96.3	96.0	96.1	96.0	103.1	99.6	95.3	103.7	99.5	93.8	103.2	98.6
0330	Cast-in-Place Concrete	110.4	105.6	108.6	98.6	104.1	100.7	97.3	102.2	99.2	101.7	106.2	103.4	101.5	112.8	105.8	99.1	96.5	98.1
03	CONCRETE	96.6	96.4	96.5	90.3	95.5	92.7	95.7	100.4	97.9	92.1	99.6	95.6	98.9	112.0	105.0	94.1	98.2	96.0
04	MASONRY	126.1	108.0	115.0	125.7	105.3	113.1	100.7	108.1	105.2	114.0	101.6	106.4	106.4	117.5	113.2	100.6	104.1	102.7
05	METALS	93.2	115.2	99.9	92.0	115.1	99.0	99.7	117.3	105.1	93.0	118.8	100.9	99.1	122.4	106.2	99.3	120.7	105.9
06	WOOD, PLASTICS & COMPOSITES	83.8	80.2	81.8	64.2	80.4	73.1	88.8	96.5	93.0	92.6	87.8	90.0	93.6	110.6	103.0	94.6	93.0	93.7
07	THERMAL & MOISTURE PROTECTION	105.4	102.4	104.2	106.9	101.3	104.6	106.1	104.0	105.2	105.8	91.5	99.9	104.8	116.7	109.7	109.8	94.6	103.5
08	OPENINGS	86.0	101.1	89.4	99.1	101.2	99.6	107.3	104.2	106.6	90.5	107.8	94.4	100.7	120.3	105.1	101.2	110.6	103.4
0920	Plaster & Gypsum Board	94.8	80.2	85.1	107.9	80.2	89.5	96.2	96.9	96.6	99.4	88.0	91.9	99.0	111.2	107.1	103.3	93.2	96.6
0950, 0980	Ceilings & Acoustic Treatment	56.4	80.2	72.3	119.6	80.2	93.4	98.2	96.9	97.3	56.4	88.0	77.5	98.5	111.2	107.0	92.7	93.2	93.0
0960	Flooring	93.9	110.6	98.7	93.7	110.6	98.6	99.4	118.5	104.9	95.9	115.5	101.6	98.5	110.6	102.0	97.1	81.0	92.4
0970, 0990	Wall Finishes & Painting/Coating	88.2	110.5	101.3	93.9	88.8	90.9	88.6	104.5	98.0	99.5	102.0	101.0	104.8	125.5	117.0	89.5	103.1	97.5
09	FINISHES	83.7	93.0	88.7	98.8	90.3	94.2	94.0	102.7	98.8	85.5	96.2	91.4	98.4	114.0	106.9	92.7	93.4	93.1
COVERS	DIVS. 10 - 14, 25, 28, 41, 43, 44, 46	100.0	98.9	99.8	100.0	98.7	99.7	100.0	90.4	97.8	100.0	98.2	99.6	100.0	105.4	101.2	100.0	99.6	99.9
21, 22, 23	FIRE SUPPRESSION, PLUMBING & HVAC	95.9	84.4	90.8	96.9	83.5	90.9	99.9	96.4	98.3	95.9	85.4	91.3	100.0	108.0	103.5	100.0	92.9	96.8
26, 27, 3370	ELECTRICAL, COMMUNICATIONS & UTIL.	102.3	102.1	102.2	104.6	68.3	85.3	100.4	102.2	101.3	109.2	91.1	99.6	101.1	111.7	106.7	100.1	91.1	95.3
MF2014	WEIGHTED AVERAGE	96.1	98.1	97.0	98.0	92.1	95.4	99.8	102.6	101.1	96.3	97.3	96.7	100.1	112.8	105.7	98.9	98.9	98.9

MINNESOTA / MISSISSIPPI

DIVISION		SAINT PAUL 550-551			ST. CLOUD 563			THIEF RIVER FALLS 567			WILLMAR 562			WINDOM 561			BILOXI 395		
		MAT.	INST.	TOTAL	MAT.	INST.	TOTAL	MAT.	INST.	TOTAL	MAT.	INST.	TOTAL	MAT.	INST.	TOTAL	MAT.	INST.	TOTAL
015433	CONTRACTOR EQUIPMENT		104.8	104.8		104.5	104.5		101.8	101.8		104.5	104.5		104.5	104.5		104.5	104.5
0241, 31 - 34	SITE & INFRASTRUCTURE, DEMOLITION	95.1	104.6	101.7	93.4	106.3	102.4	94.9	100.5	98.8	92.8	105.7	101.8	86.7	104.9	99.4	103.5	90.5	94.4
0310	Concrete Forming & Accessories	99.3	113.6	111.7	85.1	102.4	100.0	87.4	83.9	84.4	84.9	81.1	81.6	89.3	75.3	77.2	96.3	64.7	69.0
0320	Concrete Reinforcing	97.1	103.6	100.4	96.1	103.1	99.7	97.8	95.7	96.7	95.7	103.0	99.4	95.7	102.4	99.1	83.2	51.2	67.0
0330	Cast-in-Place Concrete	103.6	111.5	106.6	97.4	107.8	101.4	100.6	84.1	94.3	98.9	82.3	92.6	85.6	87.6	86.3	115.1	61.8	94.8
03	CONCRETE	99.0	111.8	104.9	88.1	105.3	96.1	91.3	87.4	89.5	88.1	86.9	87.5	79.1	86.1	82.3	101.5	63.3	83.7
04	MASONRY	99.7	117.4	110.6	110.1	104.3	106.5	101.1	100.4	100.7	114.2	104.4	108.2	125.3	86.4	101.3	99.4	62.2	76.5
05	METALS	99.7	122.3	106.6	93.8	119.5	101.7	92.1	114.4	98.9	93.0	118.5	100.8	92.9	117.2	100.3	93.1	87.1	91.3
06	WOOD, PLASTICS & COMPOSITES	94.7	111.4	103.9	81.3	99.9	91.5	67.9	80.4	74.8	81.0	72.2	76.2	85.0	69.9	76.7	102.9	69.7	84.7
07	THERMAL & MOISTURE PROTECTION	104.9	115.9	109.4	105.6	103.3	104.6	107.9	88.7	99.9	105.4	97.9	102.3	105.3	83.7	96.4	97.8	64.7	84.0
08	OPENINGS	99.8	120.7	104.5	91.0	114.5	96.3	99.1	101.2	99.6	88.0	99.2	90.6	91.6	98.0	93.1	101.0	59.1	91.5
0920	Plaster & Gypsum Board	97.1	112.2	107.1	94.8	100.5	98.6	108.6	80.2	89.7	94.8	72.0	79.6	94.8	69.6	78.1	97.0	69.3	78.6
0950, 0980	Ceilings & Acoustic Treatment	97.0	112.2	107.1	56.4	100.5	85.7	119.6	80.2	93.4	56.4	72.0	66.8	56.4	69.6	65.2	87.6	69.3	75.4
0960	Flooring	92.9	110.6	98.0	90.6	110.6	96.4	94.7	110.6	99.3	91.9	106.8	96.3	94.6	106.8	98.1	94.6	77.7	89.7
0970, 0990	Wall Finishes & Painting/Coating	91.3	125.5	111.4	99.5	125.5	114.8	93.9	88.8	90.9	93.9	88.8	90.9	93.9	102.0	96.8	84.9	49.9	64.4
09	FINISHES	91.7	114.4	104.1	83.1	106.1	95.7	99.3	89.1	93.7	83.1	85.2	84.3	83.4	82.9	83.1	88.4	66.3	76.3
COVERS	DIVS. 10 - 14, 25, 28, 41, 43, 44, 46	100.0	105.0	101.1	100.0	94.6	98.8	100.0	97.1	99.3	100.0	98.0	99.5	100.0	94.5	98.8	100.0	71.1	93.5
21, 22, 23	FIRE SUPPRESSION, PLUMBING & HVAC	99.9	109.8	104.3	99.5	104.8	101.8	96.9	80.9	89.8	95.9	102.0	98.6	95.9	81.9	89.7	99.9	51.3	78.4
26, 27, 3370	ELECTRICAL, COMMUNICATIONS & UTIL.	99.7	113.0	106.8	102.3	113.0	108.0	101.7	68.3	84.0	102.3	81.8	91.4	109.2	91.1	99.6	101.0	57.6	78.0
MF2014	WEIGHTED AVERAGE	99.0	113.2	105.3	95.5	107.6	100.9	96.8	89.3	93.5	94.5	96.4	95.3	94.9	90.4	92.9	98.2	64.7	83.4

MISSISSIPPI

DIVISION		CLARKSDALE 386			COLUMBUS 397			GREENVILLE 387			GREENWOOD 389			JACKSON 390 - 392			LAUREL 394		
		MAT.	INST.	TOTAL	MAT.	INST.	TOTAL	MAT.	INST.	TOTAL	MAT.	INST.	TOTAL	MAT.	INST.	TOTAL	MAT.	INST.	TOTAL
015433	CONTRACTOR EQUIPMENT		104.5	104.5		104.5	104.5		104.5	104.5		104.5	104.5		104.5	104.5		104.5	104.5
0241, 31 - 34	SITE & INFRASTRUCTURE, DEMOLITION	96.6	89.9	91.9	102.2	90.7	94.1	102.1	91.0	94.4	99.3	89.6	92.5	98.4	91.1	93.3	107.4	89.9	95.2
0310	Concrete Forming & Accessories	86.5	47.0	52.3	84.7	48.9	53.7	83.1	66.2	68.5	95.9	46.9	53.5	94.6	66.7	70.4	84.8	63.5	66.3
0320	Concrete Reinforcing	96.9	65.4	80.9	89.1	36.3	62.3	97.4	52.1	74.4	96.9	44.5	70.3	92.4	51.3	71.5	89.7	33.2	61.1
0330	Cast-in-Place Concrete	101.4	51.7	82.5	117.4	57.0	94.4	104.5	58.0	86.8	109.1	51.3	87.1	100.3	66.7	87.5	114.8	53.0	91.3
03	CONCRETE	95.2	54.1	76.2	103.0	51.8	79.2	100.8	62.6	83.1	101.6	50.1	77.8	95.7	65.7	81.8	105.6	56.3	82.8
04	MASONRY	94.8	42.4	62.5	126.4	48.7	78.6	141.0	71.0	97.9	95.4	42.3	62.7	105.2	71.0	84.1	122.3	45.3	74.9
05	METALS	93.4	88.6	92.0	90.3	80.2	87.2	94.4	86.9	92.1	93.4	78.7	88.9	100.0	87.1	96.0	90.3	77.2	86.3
06	WOOD, PLASTICS & COMPOSITES	85.0	47.9	64.6	88.9	48.8	66.9	81.8	67.9	74.2	97.5	47.9	70.3	97.7	67.9	81.3	90.0	69.7	78.9
07	THERMAL & MOISTURE PROTECTION	96.6	52.1	78.1	97.7	52.1	78.8	97.1	66.6	84.4	97.0	54.4	79.3	96.1	67.9	84.4	97.9	54.2	79.8
08	OPENINGS	96.3	51.0	86.0	100.6	44.3	87.8	96.0	58.3	87.5	96.3	44.6	84.6	102.2	58.1	92.2	97.4	55.6	88.0
0920	Plaster & Gypsum Board	87.5	46.9	60.5	88.0	47.8	61.3	87.2	67.4	74.1	98.4	46.9	64.2	88.8	67.4	74.6	88.0	69.3	75.6
0950, 0980	Ceilings & Acoustic Treatment	82.1	46.9	58.7	82.3	47.8	59.4	85.0	67.4	73.3	82.1	46.9	58.7	90.9	67.4	75.3	82.3	69.3	73.7
0960	Flooring	100.1	47.7	84.9	88.5	53.5	78.4	98.3	47.7	83.7	106.0	47.7	89.1	91.8	77.7	87.7	87.1	47.7	75.7
0970, 0990	Wall Finishes & Painting/Coating	94.3	49.9	68.3	84.9	49.9	64.4	94.3	49.9	68.3	94.3	49.9	68.3	85.6	49.9	64.7	84.9	49.9	64.4
09	FINISHES	90.5	47.1	66.8	84.3	49.5	65.3	91.1	61.2	74.8	94.1	47.1	68.4	87.4	67.5	76.5	84.4	60.6	71.4
COVERS	DIVS. 10 - 14, 25, 28, 41, 43, 44, 46	100.0	51.3	89.0	100.0	52.4	89.2	100.0	73.8	94.1	100.0	51.3	89.0	100.0	73.8	94.1	100.0	36.5	85.6
21, 22, 23	FIRE SUPPRESSION, PLUMBING & HVAC	98.3	53.9	78.7	97.9	55.5	79.2	100.0	59.3	82.0	98.3	54.1	78.8	99.9	59.3	82.0	97.9	43.8	74.0
26, 27, 3370	ELECTRICAL, COMMUNICATIONS & UTIL.	96.4	45.1	69.2	98.2	47.8	71.5	96.4	57.3	75.7	96.4	42.3	67.7	102.4	57.3	78.5	99.9	57.3	77.3
MF2014	WEIGHTED AVERAGE	95.8	56.4	78.4	98.1	56.8	79.8	99.5	66.7	85.0	97.1	54.3	78.2	99.0	67.9	85.3	98.2	57.2	80.1

MISSISSIPPI / MISSOURI

DIVISION		MCCOMB 396			MERIDIAN 393			TUPELO 388			BOWLING GREEN 633			CAPE GIRARDEAU 637			CHILLICOTHE 646		
		MAT.	INST.	TOTAL	MAT.	INST.	TOTAL	MAT.	INST.	TOTAL	MAT.	INST.	TOTAL	MAT.	INST.	TOTAL	MAT.	INST.	TOTAL
015433	CONTRACTOR EQUIPMENT		104.5	104.5		104.5	104.5		104.5	104.5		112.3	112.3		112.3	112.3		106.4	106.4
0241, 31 - 34	SITE & INFRASTRUCTURE, DEMOLITION	95.1	89.7	91.4	99.2	90.8	93.4	94.0	89.8	91.1	88.4	95.3	93.2	90.0	95.2	93.6	108.5	94.2	98.5
0310	Concrete Forming & Accessories	84.7	48.4	53.3	81.9	65.5	67.7	83.6	48.7	53.5	95.1	91.8	92.3	87.6	79.7	80.8	84.7	98.5	96.7
0320	Concrete Reinforcing	90.2	34.9	62.2	89.1	51.3	69.9	94.8	44.5	69.3	106.5	93.0	99.7	107.8	79.0	93.2	104.0	94.9	99.4
0330	Cast-in-Place Concrete	101.9	51.2	82.6	109.1	65.0	92.3	101.4	72.3	90.3	91.8	99.6	94.8	90.9	88.9	90.1	98.5	95.9	97.5
03	CONCRETE	92.1	49.2	72.2	96.7	64.6	81.8	94.8	58.2	77.8	96.1	96.4	96.3	95.3	84.8	90.4	105.9	95.9	101.3
04	MASONRY	127.7	41.8	74.8	99.0	67.9	79.8	129.6	45.0	77.5	116.3	95.9	103.7	113.2	78.9	92.1	103.7	94.6	98.1
05	METALS	90.5	76.5	86.2	91.3	87.1	90.0	93.3	78.9	88.9	94.6	118.7	102.0	95.6	111.3	100.4	91.8	109.3	97.2
06	WOOD, PLASTICS & COMPOSITES	88.9	50.5	67.8	86.4	67.9	76.2	82.4	48.8	63.9	97.8	92.2	94.7	90.9	77.7	83.7	94.9	100.6	98.0
07	THERMAL & MOISTURE PROTECTION	97.2	54.5	79.5	97.4	66.4	84.6	96.6	55.8	79.6	100.2	99.4	100.2	100.2	84.3	93.6	93.1	96.6	94.5
08	OPENINGS	100.6	44.3	87.9	100.3	58.1	90.7	96.2	48.1	85.3	99.9	98.2	99.5	99.9	76.0	94.5	92.9	97.8	94.0
0920	Plaster & Gypsum Board	88.0	49.6	62.5	88.0	67.4	74.3	87.2	47.8	61.0	99.3	92.2	94.6	98.6	77.4	84.5	99.3	100.4	100.0
0950, 0980	Ceilings & Acoustic Treatment	82.3	49.6	60.5	84.2	67.4	73.0	82.1	47.8	59.3	91.6	92.2	92.0	91.6	77.4	82.2	90.2	100.4	97.0
0960	Flooring	88.5	47.7	76.7	87.1	77.7	84.4	98.6	47.7	83.8	99.4	97.8	98.9	95.9	82.9	92.1	95.9	104.9	98.5
0970, 0990	Wall Finishes & Painting/Coating	84.9	49.9	64.4	84.9	49.9	64.4	94.3	48.2	67.2	100.2	102.8	101.8	100.2	71.4	83.3	95.4	86.6	90.2
09	FINISHES	83.7	48.5	64.5	83.9	66.7	74.5	90.0	48.1	67.1	100.5	94.2	97.1	99.4	78.5	87.9	99.8	99.1	99.4
COVERS	DIVS. 10 - 14, 25, 28, 41, 43, 44, 46	100.0	54.4	89.7	100.0	72.8	93.8	100.0	52.4	89.2	100.0	84.7	96.5	100.0	91.8	98.1	100.0	86.0	96.8
21, 22, 23	FIRE SUPPRESSION, PLUMBING & HVAC	97.9	53.3	78.2	99.9	57.7	81.2	98.5	55.1	79.3	96.4	96.3	96.4	100.0	95.1	97.8	96.6	97.6	97.0
26, 27, 3370	ELECTRICAL, COMMUNICATIONS & UTIL.	96.7	48.1	70.9	99.9	58.0	77.7	96.1	47.9	70.6	103.2	81.7	91.8	103.2	91.0	96.7	95.9	79.2	87.0
MF2014	WEIGHTED AVERAGE	96.4	54.9	78.1	96.5	67.0	83.5	97.3	57.1	79.6	98.7	95.8	97.4	99.2	89.5	95.0	97.5	95.2	96.5

For customer support on your Building Construction Costs with RSMeans Data, call 800.448.8182.

779

		MISSOURI																
	DIVISION	COLUMBIA 652			FLAT RIVER 636			HANNIBAL 634			HARRISONVILLE 647			JEFFERSON CITY 650 - 651			JOPLIN 648		
		MAT.	INST.	TOTAL	MAT.	INST.	TOTAL	MAT.	INST.	TOTAL	MAT.	INST.	TOTAL	MAT.	INST.	TOTAL	MAT.	INST.	TOTAL
015433	CONTRACTOR EQUIPMENT		115.9	115.9		112.3	112.3		112.3	112.3		106.4	106.4		115.9	115.9		110.0	110.0
0241, 31 - 34	SITE & INFRASTRUCTURE, DEMOLITION	96.7	100.2	99.1	91.0	95.2	94.0	86.3	95.2	92.5	99.6	95.3	96.6	96.1	100.8	99.4	108.8	99.1	102.1
0310	Concrete Forming & Accessories	84.2	83.8	83.8	101.5	88.1	89.9	93.3	84.6	85.8	82.1	103.9	100.9	94.6	82.1	83.8	95.8	76.2	78.8
0320	Concrete Reinforcing	91.2	109.7	100.6	107.8	101.8	104.8	105.9	92.9	99.3	103.6	107.7	105.7	98.5	101.5	100.0	107.4	87.7	97.4
0330	Cast-in-Place Concrete	86.0	88.1	86.8	94.8	93.0	94.1	86.9	98.9	91.5	100.9	104.0	102.1	91.1	88.1	89.9	106.7	78.0	95.8
03	CONCRETE	84.8	91.8	88.1	99.2	94.0	96.8	92.4	92.9	92.6	101.6	105.2	103.3	91.8	89.6	90.8	104.5	80.1	93.2
04	MASONRY	147.1	75.9	103.2	113.5	78.6	92.0	107.7	94.9	99.8	98.1	102.5	100.8	104.1	75.9	86.7	96.9	82.4	88.0
05	METALS	102.4	124.2	109.0	94.5	121.1	102.6	94.6	118.3	101.9	92.3	115.1	99.2	102.0	121.0	107.8	94.8	102.1	97.0
06	WOOD, PLASTICS & COMPOSITES	85.9	81.6	83.6	105.8	87.9	96.0	96.1	83.3	89.1	91.7	104.8	98.9	94.6	79.4	86.2	105.5	75.3	88.9
07	THERMAL & MOISTURE PROTECTION	94.9	84.8	90.7	101.0	94.2	98.2	100.7	95.7	98.6	92.3	103.5	96.9	102.3	84.6	94.9	92.2	78.9	86.7
08	OPENINGS	99.7	89.2	97.4	99.9	98.4	99.6	99.9	87.1	97.0	92.9	106.3	96.0	96.5	85.8	94.0	94.0	79.7	90.8
0920	Plaster & Gypsum Board	93.1	81.2	85.2	105.2	87.8	93.7	98.9	83.1	88.4	94.9	104.7	101.5	97.4	78.8	85.1	105.7	74.4	84.9
0950, 0980	Ceilings & Acoustic Treatment	90.3	81.2	84.2	91.6	87.8	89.1	91.6	83.1	86.0	90.2	104.7	99.9	93.3	78.8	83.7	91.1	74.4	80.0
0960	Flooring	94.7	74.4	88.8	102.7	82.9	97.0	98.7	97.8	98.4	91.4	74.4	86.5	100.5	74.4	92.9	121.3	74.4	107.7
0970, 0990	Wall Finishes & Painting/Coating	95.7	80.1	86.5	100.2	76.4	86.3	100.2	100.2	100.2	99.8	105.9	103.4	89.1	80.1	83.8	94.9	80.8	86.6
09	FINISHES	88.6	81.2	84.6	102.5	85.0	93.0	100.1	87.8	93.4	97.4	98.5	98.0	92.0	79.8	85.4	106.5	76.7	90.2
COVERS	DIVS. 10 - 14, 25, 28, 41, 43, 44, 46	100.0	97.2	99.4	100.0	92.8	98.4	100.0	83.4	96.3	100.0	88.0	97.3	100.0	96.9	99.3	100.0	84.5	96.5
21, 22, 23	FIRE SUPPRESSION, PLUMBING & HVAC	99.9	96.9	98.6	96.4	95.8	96.1	96.4	95.8	96.1	96.5	101.1	98.5	99.9	97.5	98.8	100.1	70.8	87.2
26, 27, 3370	ELECTRICAL, COMMUNICATIONS & UTIL.	94.3	85.1	89.4	108.0	98.4	102.9	101.9	81.7	91.2	102.5	98.7	100.5	99.5	85.1	91.8	93.8	75.1	83.9
MF2014	WEIGHTED AVERAGE	99.0	92.5	96.1	99.7	95.0	97.6	97.6	93.5	95.8	96.9	101.9	99.1	98.3	91.7	95.4	98.8	80.8	90.9

		MISSOURI																
	DIVISION	KANSAS CITY 640 - 641			KIRKSVILLE 635			POPLAR BLUFF 639			ROLLA 654 - 655			SEDALIA 653			SIKESTON 638		
		MAT.	INST.	TOTAL	MAT.	INST.	TOTAL	MAT.	INST.	TOTAL	MAT.	INST.	TOTAL	MAT.	INST.	TOTAL	MAT.	INST.	TOTAL
015433	CONTRACTOR EQUIPMENT		106.2	106.2		101.8	101.8		104.2	104.2		115.9	115.9		105.3	105.3		104.2	104.2
0241, 31 - 34	SITE & INFRASTRUCTURE, DEMOLITION	101.5	96.0	97.7	90.5	90.4	90.4	77.6	94.5	89.4	95.6	100.0	98.7	95.0	95.3	95.2	81.0	94.6	90.5
0310	Concrete Forming & Accessories	94.8	104.8	103.5	85.6	78.2	79.2	85.9	81.8	82.4	91.6	96.8	96.1	89.4	81.2	82.3	87.1	78.7	79.9
0320	Concrete Reinforcing	102.0	111.2	106.7	106.9	84.1	95.3	110.1	78.9	94.3	91.6	84.5	88.0	90.4	107.0	98.8	109.4	79.0	93.9
0330	Cast-in-Place Concrete	99.2	105.8	101.7	94.7	86.7	91.7	72.9	87.8	78.6	88.0	96.2	91.1	91.9	86.7	89.9	77.9	87.9	81.7
03	CONCRETE	100.7	106.9	103.6	111.9	83.5	98.8	86.0	84.6	85.3	86.6	96.0	90.9	100.9	88.9	95.3	89.8	83.3	86.8
04	MASONRY	100.2	104.6	102.9	120.2	85.5	98.8	111.5	76.1	89.7	119.7	87.9	100.1	126.2	83.9	100.1	111.1	76.1	89.5
05	METALS	102.3	117.0	106.8	94.3	103.0	97.0	94.9	100.1	96.5	101.8	114.4	105.7	100.7	113.1	104.5	95.3	100.8	97.0
06	WOOD, PLASTICS & COMPOSITES	104.7	104.7	104.7	83.7	75.8	79.4	82.9	82.4	82.6	93.2	99.1	96.4	86.2	79.4	82.5	84.4	77.7	80.7
07	THERMAL & MOISTURE PROTECTION	92.4	106.1	98.1	107.6	91.4	100.9	105.8	84.4	96.9	95.2	95.0	95.1	101.1	91.4	97.1	105.9	85.0	97.2
08	OPENINGS	100.0	107.1	101.6	105.4	80.3	99.7	106.4	78.6	100.1	99.7	89.0	97.3	104.7	89.5	101.3	106.4	76.1	99.5
0920	Plaster & Gypsum Board	103.3	104.7	104.3	94.3	75.4	81.7	94.6	82.2	86.4	95.1	99.1	97.7	88.8	78.8	82.2	96.3	77.4	83.7
0950, 0980	Ceilings & Acoustic Treatment	99.5	104.7	103.0	89.9	75.4	80.2	91.6	82.2	85.4	90.3	99.1	96.2	90.3	78.8	82.7	91.6	77.4	82.2
0960	Flooring	97.9	104.9	100.0	77.0	74.4	76.2	90.7	82.9	88.4	98.5	74.4	91.5	76.9	74.4	76.2	91.2	82.9	88.8
0970, 0990	Wall Finishes & Painting/Coating	99.8	109.5	105.5	95.9	80.1	86.6	95.0	71.4	81.2	95.7	98.2	97.2	95.7	104.2	100.7	95.0	71.4	81.2
09	FINISHES	102.4	105.3	104.0	99.5	77.2	87.3	99.1	80.7	89.0	90.2	93.6	92.0	87.4	81.6	84.3	99.8	78.3	88.0
COVERS	DIVS. 10 - 14, 25, 28, 41, 43, 44, 46	100.0	101.0	100.2	100.0	82.7	96.1	100.0	91.4	98.1	100.0	86.5	96.9	100.0	90.6	97.9	100.0	88.1	97.3
21, 22, 23	FIRE SUPPRESSION, PLUMBING & HVAC	100.0	101.7	100.7	96.4	94.6	95.6	96.4	93.6	95.2	96.3	97.1	96.7	96.3	93.6	95.1	96.4	93.7	95.2
26, 27, 3370	ELECTRICAL, COMMUNICATIONS & UTIL.	104.2	101.0	102.5	102.2	80.0	90.4	102.4	91.0	96.3	92.7	78.2	85.0	93.9	98.7	96.4	101.4	133.3	118.3
MF2014	WEIGHTED AVERAGE	100.9	104.4	102.5	101.4	87.1	95.1	97.5	88.2	93.4	97.0	94.1	95.7	99.4	92.7	96.4	98.1	93.5	96.1

		MISSOURI									MONTANA								
	DIVISION	SPRINGFIELD 656 - 658			ST. JOSEPH 644 - 645			ST. LOUIS 630 - 631			BILLINGS 590 - 591			BUTTE 597			GREAT FALLS 594		
		MAT.	INST.	TOTAL	MAT.	INST.	TOTAL	MAT.	INST.	TOTAL	MAT.	INST.	TOTAL	MAT.	INST.	TOTAL	MAT.	INST.	TOTAL
015433	CONTRACTOR EQUIPMENT		107.8	107.8		106.4	106.4		112.1	112.1		102.0	102.0		101.8	101.8		101.8	101.8
0241, 31 - 34	SITE & INFRASTRUCTURE, DEMOLITION	97.7	97.4	97.5	103.2	93.2	96.2	90.3	98.0	95.6	94.2	98.7	97.3	99.9	97.4	98.2	103.6	98.4	100.0
0310	Concrete Forming & Accessories	97.9	78.5	81.1	94.6	92.4	92.7	101.6	104.2	103.9	96.6	67.4	71.4	84.1	65.7	68.2	96.5	65.7	69.9
0320	Concrete Reinforcing	87.9	110.0	99.1	100.8	107.4	104.1	98.6	103.0	100.8	102.3	81.7	91.8	110.9	81.6	96.1	102.3	81.6	91.8
0330	Cast-in-Place Concrete	93.4	77.4	87.3	99.1	101.3	100.0	90.9	101.7	95.0	118.5	70.9	100.3	130.4	69.4	107.2	137.8	69.4	111.7
03	CONCRETE	96.9	84.9	91.3	100.5	99.1	99.8	94.9	104.4	99.3	100.6	72.1	87.4	104.7	70.8	89.0	109.8	70.8	91.7
04	MASONRY	96.0	82.8	87.9	100.1	94.2	96.5	93.7	107.0	101.9	130.7	75.1	96.5	126.6	72.8	93.5	131.0	72.8	95.2
05	METALS	107.1	110.3	108.1	98.5	114.5	103.4	100.3	123.0	107.3	105.5	89.2	100.5	99.6	89.0	96.4	103.1	89.0	98.8
06	WOOD, PLASTICS & COMPOSITES	93.7	77.7	84.9	105.5	91.9	98.0	105.6	103.7	104.6	86.8	65.1	74.9	74.9	64.2	69.0	87.9	64.2	74.9
07	THERMAL & MOISTURE PROTECTION	99.3	79.4	91.0	92.7	93.6	93.1	100.8	101.5	101.1	110.3	71.3	94.1	110.1	69.6	93.3	110.7	65.0	91.8
08	OPENINGS	107.2	92.3	103.8	98.2	99.2	98.4	100.3	107.3	101.9	97.0	67.1	90.2	95.2	66.6	88.8	98.2	66.6	91.0
0920	Plaster & Gypsum Board	96.1	77.1	83.5	107.5	91.5	96.8	106.2	104.1	104.8	113.3	64.5	80.8	112.3	63.5	79.9	121.9	63.5	83.1
0950, 0980	Ceilings & Acoustic Treatment	90.3	77.1	81.6	98.7	91.5	93.9	97.5	104.1	101.9	89.2	64.5	72.7	95.1	63.5	74.1	96.8	63.5	74.6
0960	Flooring	97.5	74.4	90.8	101.0	104.9	102.2	102.4	97.8	101.1	99.4	81.1	94.1	96.6	81.8	92.3	103.9	81.1	97.3
0970, 0990	Wall Finishes & Painting/Coating	90.1	105.9	99.4	95.4	81.8	87.4	100.2	108.5	105.1	98.6	72.2	83.1	97.5	72.2	82.6	97.5	72.2	82.6
09	FINISHES	92.4	81.0	86.2	103.4	93.4	98.0	103.7	103.5	103.6	94.3	70.0	81.0	94.3	69.0	80.5	98.4	68.8	82.2
COVERS	DIVS. 10 - 14, 25, 28, 41, 43, 44, 46	100.0	94.2	98.7	100.0	97.9	99.5	100.0	101.8	100.4	100.0	93.9	98.6	100.0	93.0	98.4	100.0	93.0	98.4
21, 22, 23	FIRE SUPPRESSION, PLUMBING & HVAC	100.0	72.8	87.9	100.1	87.9	94.7	100.0	104.8	102.1	100.1	72.0	87.7	100.2	68.9	86.4	100.2	68.9	86.4
26, 27, 3370	ELECTRICAL, COMMUNICATIONS & UTIL.	98.2	72.4	84.5	102.5	98.5	100.4	105.8	91.1	98.0	99.4	73.9	85.9	107.0	70.1	87.4	98.8	71.3	84.2
MF2014	WEIGHTED AVERAGE	100.5	83.6	93.0	100.1	96.1	98.3	99.8	104.0	101.7	101.8	76.4	90.5	101.7	74.5	89.6	103.2	74.6	90.5

MONTANA

DIVISION		HAVRE 595		HELENA 596		KALISPELL 599		MILES CITY 593		MISSOULA 598		WOLF POINT 592	
		MAT.	INST.	TOTAL	MAT.	INST.	TOTAL	MAT.	INST.	TOTAL	MAT.	INST.	TOTAL	MAT.	INST.	TOTAL	MAT.	INST.	TOTAL
015433	CONTRACTOR EQUIPMENT		101.8	101.8		101.8	101.8		101.8	101.8		101.8	101.8		101.8	101.8		101.8	101.8
0241, 31 - 34	SITE & INFRASTRUCTURE, DEMOLITION	107.1	97.2	100.2	93.6	97.4	96.3	90.1	97.4	95.2	96.3	96.8	96.7	82.8	97.1	92.8	113.4	97.0	101.9
0310	Concrete Forming & Accessories	77.3	63.8	65.6	100.1	59.6	65.1	87.1	65.8	68.7	95.1	62.3	66.7	87.1	65.5	68.4	87.9	62.5	65.9
0320	Concrete Reinforcing	111.8	81.6	96.5	118.0	80.4	98.9	113.9	80.4	96.9	111.4	81.6	96.3	112.8	80.3	96.3	113.0	80.4	96.5
0330	Cast-in-Place Concrete	140.5	66.6	112.3	101.9	69.5	89.6	113.2	69.6	96.6	124.0	64.3	101.2	96.1	69.3	85.9	138.9	64.7	110.6
03	CONCRETE	112.8	69.0	92.5	96.9	67.9	83.5	94.3	70.7	83.4	101.5	67.6	85.8	82.9	70.5	77.1	116.0	67.6	93.5
04	MASONRY	127.7	67.9	90.9	118.2	73.2	90.5	125.4	73.6	93.5	132.9	63.9	90.4	152.5	73.3	103.7	134.2	64.5	91.3
05	METALS	96.0	89.0	93.8	101.8	88.5	97.7	95.8	88.6	93.6	95.2	89.0	93.3	96.3	88.4	93.9	95.3	88.5	93.2
06	WOOD, PLASTICS & COMPOSITES	67.4	64.2	65.6	90.3	55.9	71.4	77.8	64.2	70.3	85.3	64.2	73.7	77.8	64.2	70.3	77.6	64.2	70.2
07	THERMAL & MOISTURE PROTECTION	110.5	63.1	90.8	104.5	65.6	88.3	109.7	72.6	94.3	110.1	61.3	89.8	109.3	69.7	92.9	111.0	61.6	90.5
08	OPENINGS	95.8	66.6	89.2	99.0	61.8	90.6	95.8	66.3	89.1	95.3	66.6	88.8	95.3	66.3	88.7	95.3	66.3	88.7
0920	Plaster & Gypsum Board	108.0	63.5	78.4	108.2	55.0	72.8	112.3	63.5	79.9	121.1	63.5	82.8	112.3	63.5	79.9	115.9	63.5	81.1
0950, 0980	Ceilings & Acoustic Treatment	95.1	63.5	74.1	99.4	55.0	69.9	95.1	63.5	74.1	93.4	63.5	73.5	95.1	63.5	74.1	93.4	63.5	73.5
0960	Flooring	94.0	81.1	90.2	105.1	81.1	98.1	98.5	81.8	93.6	103.7	81.1	97.2	98.5	81.8	93.6	100.0	81.1	94.5
0970, 0990	Wall Finishes & Painting/Coating	97.5	72.2	82.6	97.8	72.2	82.7	97.5	72.2	82.6	97.5	72.2	82.6	97.5	57.1	73.8	97.5	72.2	82.6
09	FINISHES	93.5	67.6	79.4	98.3	64.1	79.6	94.3	69.1	80.5	97.1	66.6	80.4	93.7	67.3	79.3	96.6	66.8	80.3
COVERS	DIVS. 10 - 14, 25, 28, 41, 43, 44, 46	100.0	91.3	98.0	100.0	92.2	98.2	100.0	93.1	98.4	100.0	90.0	97.7	100.0	93.0	98.4	100.0	90.2	97.8
21, 22, 23	FIRE SUPPRESSION, PLUMBING & HVAC	96.6	66.4	83.3	100.0	69.1	86.3	96.6	64.5	82.4	96.6	64.9	82.6	100.2	64.2	84.3	96.6	65.2	82.7
26, 27, 3370	ELECTRICAL, COMMUNICATIONS & UTIL.	98.8	70.2	83.6	105.5	70.1	86.8	103.4	69.5	85.4	98.8	75.2	86.3	104.6	69.5	86.0	98.8	75.2	86.3
MF2014	WEIGHTED AVERAGE	100.7	72.7	88.3	101.0	73.0	88.7	98.4	73.5	87.4	99.5	72.3	87.5	99.0	73.1	87.5	101.7	72.4	88.8

NEBRASKA

DIVISION		ALLIANCE 693		COLUMBUS 686		GRAND ISLAND 688		HASTINGS 689		LINCOLN 683 - 685		MCCOOK 690	
		MAT.	INST.	TOTAL	MAT.	INST.	TOTAL	MAT.	INST.	TOTAL	MAT.	INST.	TOTAL	MAT.	INST.	TOTAL	MAT.	INST.	TOTAL
015433	CONTRACTOR EQUIPMENT		99.7	99.7		106.9	106.9		106.9	106.9		106.9	106.9		106.9	106.9		106.9	106.9
0241, 31 - 34	SITE & INFRASTRUCTURE, DEMOLITION	99.1	101.5	100.8	101.5	94.8	96.8	106.3	95.6	98.8	105.1	94.9	98.0	91.7	95.6	94.4	102.5	94.7	97.1
0310	Concrete Forming & Accessories	88.1	55.9	60.3	97.3	75.3	78.3	96.9	72.0	75.4	100.2	74.9	78.4	97.1	74.0	77.1	93.7	55.8	60.9
0320	Concrete Reinforcing	111.5	81.6	96.3	100.9	87.5	94.1	100.4	73.9	86.9	100.4	74.6	87.3	99.8	74.0	86.7	103.8	70.1	86.7
0330	Cast-in-Place Concrete	105.6	76.7	94.6	107.6	80.0	97.1	114.0	64.9	95.3	114.0	61.1	93.8	89.5	71.5	82.6	114.1	57.3	92.4
03	CONCRETE	118.6	68.5	95.4	102.5	80.2	92.2	107.4	71.2	90.6	107.6	71.3	90.7	90.7	74.4	83.2	107.4	60.6	85.7
04	MASONRY	112.3	71.1	86.9	118.3	74.5	91.3	110.8	72.1	86.9	120.4	83.1	97.4	98.5	72.0	82.2	107.5	70.9	84.9
05	METALS	103.2	82.1	96.8	94.9	97.7	95.7	96.5	93.7	95.7	97.5	93.3	96.2	98.5	93.9	97.1	97.8	89.9	95.4
06	WOOD, PLASTICS & COMPOSITES	86.6	51.7	67.4	99.5	76.4	86.8	98.8	71.6	83.9	102.3	76.4	88.1	99.2	74.2	85.5	95.2	51.7	71.3
07	THERMAL & MOISTURE PROTECTION	104.3	65.1	88.1	102.7	80.2	93.4	102.9	79.0	93.0	102.9	81.7	94.1	99.8	78.8	91.1	99.1	75.6	89.3
08	OPENINGS	94.3	59.2	86.4	94.0	75.9	89.9	94.0	70.5	88.7	94.0	72.5	89.1	103.1	68.2	95.2	94.8	56.4	86.1
0920	Plaster & Gypsum Board	82.7	50.4	61.2	94.6	75.8	82.1	93.9	70.9	78.6	95.6	75.8	82.4	103.6	73.5	83.6	94.7	50.4	65.3
0950, 0980	Ceilings & Acoustic Treatment	90.5	50.4	63.8	83.3	75.8	78.3	83.3	70.9	75.0	83.3	75.8	78.3	94.1	73.5	80.4	86.5	50.4	62.5
0960	Flooring	94.0	76.2	88.8	88.1	93.1	89.5	87.8	99.5	91.2	89.2	93.1	90.3	100.4	88.4	96.9	92.8	76.2	88.0
0970, 0990	Wall Finishes & Painting/Coating	154.6	53.9	95.5	75.9	62.7	68.2	75.9	66.6	70.4	75.9	62.7	68.2	94.7	81.3	86.8	87.5	47.0	63.7
09	FINISHES	93.5	58.4	74.3	86.2	77.1	81.2	86.3	75.4	80.4	87.0	76.4	81.2	94.7	77.6	85.4	91.3	57.2	72.6
COVERS	DIVS. 10 - 14, 25, 28, 41, 43, 44, 46	100.0	61.7	91.3	100.0	85.6	96.7	100.0	88.8	97.5	100.0	85.6	96.7	100.0	89.1	97.5	100.0	61.0	91.2
21, 22, 23	FIRE SUPPRESSION, PLUMBING & HVAC	96.6	72.5	85.9	96.5	75.6	87.3	100.1	78.7	90.6	96.5	75.2	87.1	99.9	78.7	90.5	96.4	75.3	87.1
26, 27, 3370	ELECTRICAL, COMMUNICATIONS & UTIL.	93.1	67.3	79.4	91.5	82.3	86.7	90.3	67.4	78.1	89.7	83.6	86.4	100.4	67.4	82.9	95.2	67.3	80.4
MF2014	WEIGHTED AVERAGE	100.9	71.1	87.7	97.1	81.3	90.1	98.4	77.6	89.2	98.3	80.4	90.4	98.2	78.3	89.4	98.4	70.8	86.2

NEBRASKA / NEVADA

DIVISION		NORFOLK 687		NORTH PLATTE 691		OMAHA 680 - 681		VALENTINE 692		CARSON CITY 897		ELKO 898	
		MAT.	INST.	TOTAL	MAT.	INST.	TOTAL	MAT.	INST.	TOTAL	MAT.	INST.	TOTAL	MAT.	INST.	TOTAL	MAT.	INST.	TOTAL
015433	CONTRACTOR EQUIPMENT		95.8	95.8		106.9	106.9		95.8	95.8		99.4	99.4		98.8	98.8		98.8	98.8
0241, 31 - 34	SITE & INFRASTRUCTURE, DEMOLITION	84.0	94.0	90.9	103.8	94.7	97.4	91.3	94.7	93.7	87.2	100.1	96.2	87.7	98.5	95.2	70.1	97.2	89.0
0310	Concrete Forming & Accessories	83.3	74.3	75.5	96.3	70.7	74.1	95.9	75.4	78.2	84.2	55.2	59.2	108.1	80.2	84.0	113.8	77.0	81.9
0320	Concrete Reinforcing	101.0	64.9	82.7	103.2	73.8	88.3	99.5	74.1	86.6	103.8	64.6	83.9	105.9	109.7	107.8	113.7	90.9	102.1
0330	Cast-in-Place Concrete	108.3	60.1	89.9	114.1	62.0	94.3	89.8	79.8	86.0	100.7	55.1	83.4	95.3	82.0	90.2	92.3	75.0	85.7
03	CONCRETE	101.1	68.3	85.9	107.5	69.5	89.9	90.8	77.3	84.5	105.0	58.0	83.2	97.2	86.1	92.1	95.0	78.9	87.5
04	MASONRY	125.1	74.5	93.9	95.3	70.9	80.3	99.5	75.6	84.8	107.4	70.9	84.9	117.1	70.7	88.5	123.8	69.5	90.4
05	METALS	98.4	78.6	92.3	96.9	91.4	95.2	98.5	83.8	94.0	109.5	78.3	99.9	108.2	92.7	103.5	111.8	83.1	103.0
06	WOOD, PLASTICS & COMPOSITES	82.7	75.9	79.0	97.1	71.6	83.1	94.5	75.9	84.3	81.0	51.2	64.7	91.2	80.3	85.2	102.0	77.1	88.3
07	THERMAL & MOISTURE PROTECTION	102.5	78.6	92.6	99.0	77.8	90.2	98.6	80.0	90.9	99.9	74.2	89.2	106.6	79.4	95.3	103.3	73.2	90.8
08	OPENINGS	95.6	69.9	89.8	94.1	69.9	88.6	100.9	75.4	95.1	96.5	56.4	87.5	98.1	80.8	94.2	101.7	72.7	95.1
0920	Plaster & Gypsum Board	94.7	75.8	82.1	94.7	70.9	78.9	101.4	75.8	84.4	96.1	50.4	65.7	98.3	79.7	86.0	101.3	76.4	84.8
0950, 0980	Ceilings & Acoustic Treatment	96.1	75.8	82.1	86.5	70.9	76.1	95.3	75.8	82.3	101.9	50.4	67.7	96.9	79.7	85.5	97.4	76.4	83.4
0960	Flooring	109.0	93.1	104.4	93.9	93.1	93.7	100.4	85.1	96.0	119.8	74.0	106.5	102.7	71.1	93.5	108.0	71.1	97.3
0970, 0990	Wall Finishes & Painting/Coating	132.2	62.7	91.4	87.5	62.7	73.0	105.0	66.1	82.2	154.2	65.0	101.8	97.3	78.1	86.0	96.8	75.7	84.4
09	FINISHES	102.8	76.4	88.4	91.6	73.6	81.7	97.9	76.2	86.0	111.9	58.4	82.6	95.7	78.2	86.1	95.8	75.7	84.8
COVERS	DIVS. 10 - 14, 25, 28, 41, 43, 44, 46	100.0	84.5	96.5	100.0	63.2	91.7	100.0	88.2	97.3	100.0	59.1	90.7	100.0	81.8	95.9	100.0	92.9	98.4
21, 22, 23	FIRE SUPPRESSION, PLUMBING & HVAC	96.2	75.3	86.9	100.0	74.8	88.8	99.9	76.0	89.3	96.0	74.7	86.6	100.0	70.1	86.8	98.4	75.5	88.3
26, 27, 3370	ELECTRICAL, COMMUNICATIONS & UTIL.	90.5	82.3	86.2	93.4	67.3	79.6	98.8	82.9	90.4	90.6	67.3	78.3	104.1	91.0	97.2	101.3	89.3	94.9
MF2014	WEIGHTED AVERAGE	98.6	77.3	89.2	98.2	75.1	88.0	98.0	79.8	90.0	100.9	69.7	87.1	101.6	81.8	92.8	101.5	80.1	92.0

For customer support on your Building Construction Costs with RSMeans Data, call 800.448.8182.

781

City Cost Indexes

NEVADA / NEW HAMPSHIRE

		ELY 893			LAS VEGAS 889 - 891			RENO 894 - 895			CHARLESTON 036			CLAREMONT 037			CONCORD 032 - 033		
DIVISION		MAT.	INST.	TOTAL	MAT.	INST.	TOTAL	MAT.	INST.	TOTAL	MAT.	INST.	TOTAL	MAT.	INST.	TOTAL	MAT.	INST.	TOTAL
015433	CONTRACTOR EQUIPMENT		98.8	98.8		98.8	98.8		98.8	98.8		99.8	99.8		99.8	99.8		99.8	99.8
0241, 31 - 34	SITE & INFRASTRUCTURE, DEMOLITION	76.1	98.6	91.8	79.5	101.0	94.5	75.8	98.5	91.6	85.6	97.9	94.2	79.4	97.9	92.3	91.5	102.0	98.9
0310	Concrete Forming & Accessories	106.6	102.5	103.1	107.6	105.7	106.0	102.7	80.3	83.4	87.9	81.3	82.2	93.9	81.0	82.8	99.7	93.4	94.3
0320	Concrete Reinforcing	112.4	92.9	102.5	103.5	118.3	111.0	106.2	117.2	111.8	86.6	84.0	85.3	86.6	83.9	85.3	101.2	84.0	92.5
0330	Cast-in-Place Concrete	99.1	98.6	98.9	96.0	106.5	100.0	104.8	82.0	96.1	88.2	89.5	88.7	81.0	103.1	89.4	104.0	114.3	107.9
03	CONCRETE	103.3	99.3	101.4	98.5	107.7	102.8	102.9	87.5	95.7	90.5	84.9	87.9	83.2	89.5	86.1	98.1	99.0	98.5
04	MASONRY	129.4	80.6	99.3	116.2	96.0	103.9	122.9	70.7	90.8	97.6	80.2	86.9	97.8	80.2	87.0	106.1	95.9	99.8
05	METALS	111.7	93.5	106.1	120.4	100.9	114.4	113.4	95.2	107.8	89.9	89.9	89.9	89.9	89.7	89.8	96.2	91.2	94.7
06	WOOD, PLASTICS & COMPOSITES	93.0	104.2	99.1	91.0	104.2	98.3	87.3	80.3	83.4	94.6	87.9	90.9	100.3	87.9	93.5	94.3	92.9	93.5
07	THERMAL & MOISTURE PROTECTION	103.8	97.7	101.3	117.1	97.9	109.1	103.3	79.4	93.4	101.3	75.6	90.6	101.1	75.6	90.5	105.5	94.5	100.9
08	OPENINGS	101.6	92.6	99.6	100.6	108.2	102.3	99.4	82.5	95.6	97.5	85.4	94.7	98.6	85.4	95.6	97.1	88.2	95.0
0920	Plaster & Gypsum Board	96.7	104.3	101.8	91.9	104.3	100.2	87.0	79.7	82.2	104.8	87.1	93.0	106.1	87.1	93.5	104.4	92.2	96.3
0950, 0980	Ceilings & Acoustic Treatment	97.4	104.3	102.0	105.0	104.3	104.6	102.5	79.7	87.3	95.8	87.1	90.0	95.8	87.1	90.0	102.3	92.2	95.6
0960	Flooring	105.3	71.1	95.4	96.4	99.4	97.3	101.7	71.1	92.8	88.8	100.0	92.1	91.4	100.0	93.9	97.9	102.5	99.2
0970, 0990	Wall Finishes & Painting/Coating	96.8	113.1	106.4	99.4	117.1	109.8	96.8	78.1	85.8	87.9	93.9	91.4	87.9	93.9	91.4	92.0	93.9	93.1
09	FINISHES	94.9	98.4	96.8	93.3	105.7	100.1	93.3	78.2	85.0	88.8	86.9	87.8	89.3	86.9	88.0	92.4	95.0	93.9
COVERS	DIVS. 10 - 14, 25, 28, 41, 43, 44, 46	100.0	75.0	94.3	100.0	102.1	100.5	100.0	81.8	95.9	100.0	88.3	97.3	100.0	88.3	97.3	100.0	107.5	101.7
21, 22, 23	FIRE SUPPRESSION, PLUMBING & HVAC	98.4	100.6	99.3	100.3	99.6	100.0	100.1	70.1	86.9	96.5	60.6	80.6	96.5	59.7	80.3	99.9	84.5	93.1
26, 27, 3370	ELECTRICAL, COMMUNICATIONS & UTIL.	101.5	100.8	101.1	105.9	112.4	109.3	102.0	91.0	96.2	95.9	50.4	71.8	95.9	50.4	71.8	96.8	81.1	88.5
MF2014	WEIGHTED AVERAGE	102.9	96.1	99.9	104.1	103.6	103.9	102.7	82.3	93.7	94.3	76.0	86.2	93.4	76.4	85.9	98.0	91.7	95.2

NEW HAMPSHIRE / NEW JERSEY

		KEENE 034			LITTLETON 035			MANCHESTER 031			NASHUA 030			PORTSMOUTH 038			ATLANTIC CITY 082,084		
DIVISION		MAT.	INST.	TOTAL	MAT.	INST.	TOTAL	MAT.	INST.	TOTAL	MAT.	INST.	TOTAL	MAT.	INST.	TOTAL	MAT.	INST.	TOTAL
015433	CONTRACTOR EQUIPMENT		99.8	99.8		99.8	99.8		99.8	99.8		99.8	99.8		99.8	99.8		97.2	97.2
0241, 31 - 34	SITE & INFRASTRUCTURE, DEMOLITION	92.9	98.2	96.6	79.6	98.2	92.6	92.3	102.0	99.1	94.3	102.0	99.7	87.8	102.1	97.8	91.5	105.8	101.5
0310	Concrete Forming & Accessories	92.5	83.0	84.2	105.2	83.0	86.0	99.7	93.6	94.4	100.9	93.6	94.6	89.4	93.5	93.0	112.0	145.5	140.9
0320	Concrete Reinforcing	86.6	84.0	85.3	87.4	84.0	85.7	101.2	84.1	92.5	108.3	84.1	96.0	86.6	84.1	85.3	76.8	136.5	107.1
0330	Cast-in-Place Concrete	88.6	105.3	95.0	79.5	104.6	89.1	104.0	114.4	107.9	83.9	114.4	95.5	79.5	114.4	92.8	82.0	141.0	104.5
03	CONCRETE	90.2	91.1	90.7	82.7	90.9	86.5	98.0	99.1	98.5	90.2	99.1	94.3	82.8	99.1	90.3	86.8	140.6	111.8
04	MASONRY	100.4	83.6	90.0	110.6	83.6	94.0	102.8	95.9	98.5	102.8	95.9	98.5	98.0	95.9	96.7	113.5	142.7	131.5
05	METALS	90.5	90.3	90.5	90.6	90.3	90.5	97.7	91.3	95.8	96.0	91.3	94.6	92.0	91.5	91.9	92.9	114.7	99.6
06	WOOD, PLASTICS & COMPOSITES	98.7	87.9	92.8	109.5	87.9	97.6	94.5	92.9	93.6	107.9	92.9	99.7	95.7	92.9	94.1	118.4	146.3	133.4
07	THERMAL & MOISTURE PROTECTION	101.8	77.2	91.6	101.2	77.2	91.2	104.9	94.5	100.6	102.2	94.5	99.0	101.8	104.0	102.7	100.6	136.8	115.6
08	OPENINGS	96.1	89.2	94.6	99.5	85.4	96.3	98.4	91.9	96.9	100.8	91.6	98.7	101.0	81.5	96.6	97.0	141.0	106.9
0920	Plaster & Gypsum Board	105.1	87.1	93.1	118.9	87.1	97.8	104.4	92.2	96.3	114.6	92.2	99.7	104.8	92.2	96.4	116.0	147.2	136.7
0950, 0980	Ceilings & Acoustic Treatment	95.8	87.1	90.0	95.8	87.1	90.0	102.3	92.2	95.6	107.8	92.2	97.4	96.8	92.2	93.7	88.7	147.2	127.6
0960	Flooring	91.0	100.0	93.6	102.4	100.0	101.7	95.1	102.5	97.2	95.0	102.5	97.2	89.0	102.5	92.9	98.6	163.0	117.3
0970, 0990	Wall Finishes & Painting/Coating	87.9	108.1	99.8	87.9	93.9	91.4	94.0	108.1	102.3	87.9	108.9	100.2	87.9	108.9	100.2	86.0	144.6	120.4
09	FINISHES	90.7	89.3	89.9	94.3	87.8	90.7	93.4	96.6	95.2	95.6	96.7	96.2	89.7	96.7	93.5	92.4	150.2	124.0
COVERS	DIVS. 10 - 14, 25, 28, 41, 43, 44, 46	100.0	89.5	97.6	100.0	96.1	99.1	100.0	107.5	101.7	100.0	107.5	101.7	100.0	107.5	101.7	100.0	119.8	104.5
21, 22, 23	FIRE SUPPRESSION, PLUMBING & HVAC	96.5	63.9	82.1	96.5	70.4	85.0	99.9	84.5	93.1	100.1	84.5	93.2	100.1	84.5	93.2	99.7	130.6	113.4
26, 27, 3370	ELECTRICAL, COMMUNICATIONS & UTIL.	95.9	60.2	77.0	97.0	53.3	73.8	96.9	81.1	88.5	98.3	81.1	89.2	96.4	81.1	88.3	90.1	144.7	119.0
MF2014	WEIGHTED AVERAGE	94.7	79.9	88.2	94.8	80.2	88.3	98.4	92.1	95.6	97.8	92.1	95.3	95.1	91.9	93.7	95.8	134.7	113.0

NEW JERSEY

		CAMDEN 081			DOVER 078			ELIZABETH 072			HACKENSACK 076			JERSEY CITY 073			LONG BRANCH 077		
DIVISION		MAT.	INST.	TOTAL	MAT.	INST.	TOTAL	MAT.	INST.	TOTAL	MAT.	INST.	TOTAL	MAT.	INST.	TOTAL	MAT.	INST.	TOTAL
015433	CONTRACTOR EQUIPMENT		97.2	97.2		99.8	99.8		99.8	99.8		99.8	99.8		97.2	97.2		96.8	96.8
0241, 31 - 34	SITE & INFRASTRUCTURE, DEMOLITION	92.7	105.8	101.8	96.5	106.3	103.3	100.4	106.3	104.5	97.4	107.1	104.1	88.7	107.0	101.4	92.7	106.9	102.6
0310	Concrete Forming & Accessories	102.7	145.4	139.7	101.4	146.3	140.2	114.3	146.3	142.0	101.4	146.3	140.3	105.9	146.4	141.0	106.4	146.0	140.7
0320	Concrete Reinforcing	100.8	133.8	117.5	84.6	151.9	118.8	84.6	151.9	118.8	84.6	151.9	118.8	109.8	151.9	131.2	84.6	151.8	118.7
0330	Cast-in-Place Concrete	79.5	141.3	103.1	87.8	136.3	106.3	75.4	136.3	98.6	85.8	136.3	105.0	68.5	136.4	94.3	76.1	141.2	100.9
03	CONCRETE	87.3	140.2	111.8	89.0	142.4	113.8	85.4	142.4	111.8	87.4	142.4	112.9	83.3	142.2	110.6	87.0	143.7	113.3
04	MASONRY	103.5	142.7	127.6	95.4	143.3	124.9	113.2	143.3	131.7	99.5	143.3	126.4	91.1	143.3	123.2	105.2	143.3	128.6
05	METALS	98.6	113.1	103.0	90.8	124.8	101.2	92.3	124.8	102.2	90.9	124.8	101.3	96.6	121.7	104.3	90.9	121.3	100.2
06	WOOD, PLASTICS & COMPOSITES	107.1	146.3	128.6	102.7	146.3	126.7	117.7	146.3	133.4	102.7	146.3	126.7	103.5	146.3	127.0	104.5	146.3	127.5
07	THERMAL & MOISTURE PROTECTION	100.5	137.9	116.0	102.9	139.9	118.2	103.1	139.9	118.4	102.6	135.9	116.5	102.4	139.9	117.9	102.5	132.8	115.1
08	OPENINGS	99.3	139.9	108.5	104.4	144.1	113.4	102.8	144.1	112.1	102.2	144.1	111.7	100.9	144.1	110.7	96.8	144.1	107.5
0920	Plaster & Gypsum Board	111.5	147.2	135.2	107.2	147.2	133.7	114.4	147.2	136.2	107.2	147.2	133.7	110.2	147.2	134.8	109.2	147.2	134.4
0950, 0980	Ceilings & Acoustic Treatment	98.2	147.2	130.8	85.2	147.2	126.4	87.0	147.2	127.0	85.2	147.2	126.4	94.7	147.2	129.6	85.2	147.2	126.4
0960	Flooring	94.5	163.0	114.3	86.7	189.9	116.7	92.1	189.9	120.5	86.7	189.9	116.7	87.9	189.9	117.5	88.1	187.5	116.9
0970, 0990	Wall Finishes & Painting/Coating	86.0	144.6	120.4	87.0	144.6	120.8	87.0	144.6	120.8	87.0	144.6	120.8	87.1	144.6	120.8	87.1	144.6	120.9
09	FINISHES	92.4	150.2	124.0	87.9	155.0	124.6	91.4	155.0	126.1	87.8	155.0	124.5	90.4	155.0	125.7	88.8	154.6	124.8
COVERS	DIVS. 10 - 14, 25, 28, 41, 43, 44, 46	100.0	119.8	104.5	100.0	131.6	107.1	100.0	131.6	107.1	100.0	131.6	107.1	100.0	131.6	107.1	100.0	119.9	104.5
21, 22, 23	FIRE SUPPRESSION, PLUMBING & HVAC	100.0	130.6	113.5	99.8	137.9	116.7	100.1	137.9	116.8	99.8	137.9	116.7	100.1	137.9	116.8	99.8	137.9	116.7
26, 27, 3370	ELECTRICAL, COMMUNICATIONS & UTIL.	94.6	139.8	118.6	90.5	143.6	118.7	91.1	143.6	119.0	90.5	144.6	119.2	95.0	144.6	121.3	90.2	137.5	115.3
MF2014	WEIGHTED AVERAGE	97.0	133.7	113.2	95.4	138.5	114.5	96.5	138.5	115.1	95.2	138.6	114.4	95.5	138.4	114.5	94.8	137.0	113.4

NEW JERSEY

DIVISION		NEW BRUNSWICK 088 - 089 MAT.	INST.	TOTAL	NEWARK 070 - 071 MAT.	INST.	TOTAL	PATERSON 074 - 075 MAT.	INST.	TOTAL	POINT PLEASANT 087 MAT.	INST.	TOTAL	SUMMIT 079 MAT.	INST.	TOTAL	TRENTON 085 - 086 MAT.	INST.	TOTAL
015433	CONTRACTOR EQUIPMENT		96.8	96.8		99.8	99.8		99.8	99.8		96.8	96.8		99.8	99.8		96.8	96.8
0241, 31 - 34	SITE & INFRASTRUCTURE, DEMOLITION	104.1	106.9	106.1	100.6	106.3	104.6	99.2	107.1	104.7	105.7	106.9	106.5	98.1	106.3	103.8	90.8	106.9	102.0
0310	Concrete Forming & Accessories	105.9	146.4	140.9	103.4	146.3	140.5	103.5	146.3	140.5	100.1	146.0	139.8	104.3	146.3	140.6	103.1	145.7	140.0
0320	Concrete Reinforcing	77.7	151.9	115.3	108.0	151.9	130.3	109.8	151.9	131.2	77.7	151.8	115.3	84.6	151.9	118.8	102.0	119.6	111.0
0330	Cast-in-Place Concrete	101.4	141.3	116.6	96.8	136.3	111.9	87.3	136.3	106.0	101.4	141.2	116.5	72.9	136.3	97.1	99.8	141.0	115.5
03	CONCRETE	103.2	143.9	122.1	96.8	142.4	117.9	92.2	142.4	115.5	102.9	143.7	121.8	82.7	142.4	110.3	97.4	137.9	116.2
04	MASONRY	112.0	143.3	131.2	98.5	143.3	126.1	95.9	143.3	125.1	100.1	143.3	126.7	98.5	143.3	126.1	104.0	143.3	128.2
05	METALS	93.0	121.6	101.8	98.3	124.8	106.4	92.1	124.8	102.1	93.0	121.3	101.7	90.8	124.8	101.2	98.3	109.4	101.7
06	WOOD, PLASTICS & COMPOSITES	111.9	146.3	130.8	96.0	146.3	123.7	105.3	146.3	127.9	104.5	146.3	127.4	106.5	146.3	128.4	98.3	146.3	124.7
07	THERMAL & MOISTURE PROTECTION	100.9	139.1	116.8	104.4	139.9	119.1	103.0	135.9	116.6	101.0	136.9	115.9	103.3	139.9	118.5	101.0	137.1	116.0
08	OPENINGS	92.1	144.1	103.9	102.7	144.1	112.1	107.5	144.1	115.8	93.9	145.2	105.5	108.5	144.1	116.6	100.5	136.2	108.6
0920	Plaster & Gypsum Board	113.4	147.2	135.8	103.7	147.2	132.6	110.2	147.2	134.8	108.4	147.2	134.2	109.2	147.2	134.4	105.3	147.2	133.1
0950, 0980	Ceilings & Acoustic Treatment	88.7	147.2	127.6	97.2	147.2	130.4	94.7	147.2	129.6	88.7	147.2	127.6	85.2	147.2	126.4	99.6	147.2	131.3
0960	Flooring	96.0	189.9	123.2	92.6	189.9	120.8	87.9	189.9	117.5	93.3	163.0	113.5	88.1	189.9	117.6	97.9	187.5	123.9
0970, 0990	Wall Finishes & Painting/Coating	86.0	144.6	120.4	88.4	144.6	121.4	87.0	144.6	120.8	86.0	144.6	120.4	87.0	144.6	120.8	89.7	144.6	121.9
09	FINISHES	92.4	155.0	126.6	90.3	155.0	125.7	90.5	155.0	125.8	90.9	150.3	123.4	88.9	155.0	125.0	93.0	154.6	126.6
COVERS	DIVS. 10 - 14, 25, 28, 41, 43, 44, 46	100.0	131.4	107.1	100.0	131.6	107.1	100.0	131.6	107.1	100.0	119.4	104.4	100.0	131.6	107.1	100.0	119.9	104.5
21, 22, 23	FIRE SUPPRESSION, PLUMBING & HVAC	99.7	137.9	116.6	100.1	137.9	116.8	100.1	137.9	116.8	99.7	137.9	116.6	99.8	137.9	116.7	100.0	137.6	116.6
26, 27, 3370	ELECTRICAL, COMMUNICATIONS & UTIL.	90.7	142.3	118.1	98.4	144.6	122.9	95.0	143.6	120.8	90.1	137.5	115.2	91.1	143.6	119.0	97.9	137.7	119.0
MF2014	WEIGHTED AVERAGE	97.6	138.3	115.6	98.7	138.7	116.4	97.2	138.5	115.5	96.9	136.6	114.4	95.5	138.5	114.5	98.6	134.8	114.6

NEW JERSEY / NEW MEXICO

DIVISION		VINELAND 080,083 MAT.	INST.	TOTAL	ALBUQUERQUE 870 - 872 MAT.	INST.	TOTAL	CARRIZOZO 883 MAT.	INST.	TOTAL	CLOVIS 881 MAT.	INST.	TOTAL	FARMINGTON 874 MAT.	INST.	TOTAL	GALLUP 873 MAT.	INST.	TOTAL
015433	CONTRACTOR EQUIPMENT		97.2	97.2		112.7	112.7		112.7	112.7		112.7	112.7		112.7	112.7		112.7	112.7
0241, 31 - 34	SITE & INFRASTRUCTURE, DEMOLITION	95.8	105.8	102.8	90.9	103.8	99.9	110.4	103.8	105.8	97.3	103.8	101.8	97.6	103.8	101.9	107.1	103.8	104.8
0310	Concrete Forming & Accessories	97.2	145.7	139.1	100.0	65.7	70.3	99.5	65.7	70.3	99.5	65.6	70.2	100.0	65.7	70.3	100.0	65.7	70.3
0320	Concrete Reinforcing	76.8	136.5	107.1	106.0	70.0	87.7	113.0	70.0	91.2	114.3	70.0	91.8	116.0	70.0	92.6	111.0	70.0	90.2
0330	Cast-in-Place Concrete	88.4	141.3	108.6	89.2	70.2	82.0	92.5	70.2	84.0	92.4	70.2	83.9	90.0	70.2	82.5	84.7	70.2	79.2
03	CONCRETE	91.3	140.8	114.3	95.2	69.4	83.3	118.1	69.4	95.5	105.8	69.4	88.9	98.9	69.4	85.2	105.5	69.4	88.8
04	MASONRY	100.7	143.3	126.9	108.9	60.8	79.2	109.3	60.8	79.4	109.3	60.8	79.4	118.6	60.8	83.0	103.0	60.8	77.0
05	METALS	92.9	114.7	99.6	106.6	91.2	101.9	104.4	91.2	100.4	104.1	91.0	100.1	104.2	91.2	100.2	103.4	91.2	99.6
06	WOOD, PLASTICS & COMPOSITES	101.3	146.3	126.0	95.0	66.6	79.4	93.0	66.6	78.5	93.0	66.6	78.5	95.1	66.6	79.5	95.1	66.6	79.5
07	THERMAL & MOISTURE PROTECTION	100.4	137.1	115.6	98.1	73.1	87.7	101.0	73.1	89.4	99.7	73.1	88.7	98.3	73.1	87.9	99.5	73.1	88.5
08	OPENINGS	93.5	141.0	104.2	98.7	66.8	91.4	96.9	66.8	90.1	97.1	66.8	90.2	101.0	66.8	93.3	101.1	66.8	93.3
0920	Plaster & Gypsum Board	106.8	147.2	133.6	100.6	65.4	77.2	77.5	65.4	69.5	77.5	65.4	69.5	92.7	65.4	74.6	92.7	65.4	74.6
0950, 0980	Ceilings & Acoustic Treatment	88.7	147.2	127.6	96.6	65.4	75.9	98.5	65.4	76.5	98.5	65.4	76.5	94.5	65.4	75.1	94.5	65.4	75.1
0960	Flooring	92.3	163.0	112.8	93.3	66.7	85.6	101.4	66.7	91.4	101.4	66.7	91.4	95.0	66.7	86.8	95.0	66.7	86.8
0970, 0990	Wall Finishes & Painting/Coating	86.0	144.6	120.4	98.7	54.3	72.6	96.0	54.3	71.5	96.0	54.3	71.5	93.2	54.3	70.3	93.2	54.3	70.3
09	FINISHES	89.6	150.4	122.8	91.9	64.5	76.9	95.5	64.5	78.6	94.2	64.5	78.0	90.8	64.5	76.4	92.1	64.5	77.0
COVERS	DIVS. 10 - 14, 25, 28, 41, 43, 44, 46	100.0	120.0	104.5	100.0	86.5	97.0	100.0	86.5	97.0	100.0	86.5	97.0	100.0	86.5	97.0	100.0	86.5	97.0
21, 22, 23	FIRE SUPPRESSION, PLUMBING & HVAC	99.7	130.9	113.5	100.1	69.0	86.4	97.8	69.0	85.1	97.8	68.7	85.0	100.0	69.0	86.3	97.8	69.0	85.1
26, 27, 3370	ELECTRICAL, COMMUNICATIONS & UTIL.	90.1	144.7	119.0	90.7	74.4	82.1	92.1	74.4	82.7	89.7	74.4	81.6	88.5	74.4	81.0	87.7	74.4	80.7
MF2014	WEIGHTED AVERAGE	95.1	134.8	112.6	98.9	73.7	87.8	101.7	73.7	89.4	99.5	73.7	88.1	99.6	73.7	88.2	99.3	73.7	88.0

NEW MEXICO

DIVISION		LAS CRUCES 880 MAT.	INST.	TOTAL	LAS VEGAS 877 MAT.	INST.	TOTAL	ROSWELL 882 MAT.	INST.	TOTAL	SANTA FE 875 MAT.	INST.	TOTAL	SOCORRO 878 MAT.	INST.	TOTAL	TRUTH/CONSEQUENCES 879 MAT.	INST.	TOTAL
015433	CONTRACTOR EQUIPMENT		86.8	86.8		112.7	112.7		112.7	112.7		112.7	112.7		112.7	112.7		86.8	86.8
0241, 31 - 34	SITE & INFRASTRUCTURE, DEMOLITION	97.4	81.9	86.6	96.9	103.8	101.7	99.5	103.8	102.5	100.6	103.8	102.8	93.2	103.8	100.6	113.1	81.9	91.3
0310	Concrete Forming & Accessories	96.0	64.6	68.8	100.0	65.7	70.3	99.5	65.7	70.3	99.6	65.7	70.3	100.0	65.7	70.3	97.7	64.6	69.0
0320	Concrete Reinforcing	110.0	69.8	89.6	112.9	70.0	91.1	114.3	70.0	91.8	105.5	70.0	87.5	115.1	70.0	92.2	108.3	69.8	88.8
0330	Cast-in-Place Concrete	87.3	62.6	77.9	87.5	70.2	81.0	92.9	70.2	84.0	92.9	70.2	84.3	85.8	70.2	79.9	94.0	62.6	82.0
03	CONCRETE	83.4	65.8	75.2	96.4	69.4	83.9	106.6	69.4	89.4	95.4	69.4	83.4	95.3	69.4	83.3	88.7	65.9	78.1
04	MASONRY	104.5	60.4	77.4	103.3	60.8	77.1	120.9	60.8	83.9	96.2	60.8	74.4	103.2	60.8	77.1	100.4	60.4	75.8
05	METALS	102.8	83.0	96.8	103.1	91.2	99.4	105.3	91.2	101.0	100.4	91.2	97.6	103.4	91.2	99.6	102.9	83.0	96.8
06	WOOD, PLASTICS & COMPOSITES	81.8	65.5	72.9	95.1	66.6	79.5	93.0	66.6	78.5	97.9	66.6	80.7	95.1	66.6	79.5	86.2	65.5	74.8
07	THERMAL & MOISTURE PROTECTION	87.0	68.1	79.2	97.9	73.1	87.6	99.9	73.1	88.8	100.0	73.1	88.8	97.9	73.1	87.6	86.6	68.1	78.9
08	OPENINGS	92.6	66.2	86.6	97.4	66.8	90.5	96.9	66.8	90.1	98.7	66.8	91.5	97.3	66.8	90.4	90.9	66.2	85.3
0920	Plaster & Gypsum Board	75.8	65.4	68.9	92.7	65.4	74.6	77.5	65.4	69.5	106.3	65.4	79.2	92.7	65.4	74.6	94.1	65.4	75.0
0950, 0980	Ceilings & Acoustic Treatment	85.3	65.4	72.1	94.5	65.4	75.1	98.5	65.4	76.5	95.7	65.4	75.6	94.5	65.4	75.1	83.3	65.4	71.4
0960	Flooring	133.1	66.7	113.9	95.0	66.7	86.8	101.4	66.7	91.4	103.3	66.7	92.7	95.0	66.7	86.8	125.6	66.7	108.5
0970, 0990	Wall Finishes & Painting/Coating	84.9	54.3	66.9	93.2	54.3	70.3	96.0	54.3	71.5	99.4	54.3	72.9	93.2	54.3	70.3	85.3	54.3	67.1
09	FINISHES	104.3	63.7	82.1	90.7	64.5	76.4	94.3	64.5	78.0	96.4	64.5	79.0	90.6	64.5	76.3	102.9	63.7	81.4
COVERS	DIVS. 10 - 14, 25, 28, 41, 43, 44, 46	100.0	83.9	96.4	100.0	86.5	97.0	100.0	86.5	97.0	100.0	86.5	97.0	100.0	86.5	97.0	100.0	83.9	96.4
21, 22, 23	FIRE SUPPRESSION, PLUMBING & HVAC	100.5	68.7	86.4	97.8	69.0	85.1	99.9	69.0	86.3	99.9	69.0	86.3	97.8	69.0	85.1	97.8	68.7	84.9
26, 27, 3370	ELECTRICAL, COMMUNICATIONS & UTIL.	91.5	74.2	82.3	90.3	74.4	81.9	91.1	74.4	82.3	102.1	74.4	87.4	88.2	74.4	80.9	92.0	74.4	82.7
MF2014	WEIGHTED AVERAGE	96.8	70.3	85.1	97.6	73.7	87.0	101.0	73.7	88.9	99.1	73.7	87.9	97.2	73.7	86.8	96.9	70.3	85.2

For customer support on your Building Construction Costs with RSMeans Data, call 800.448.8182.

783

DIVISION		NEW MEXICO TUCUMCARI 884 MAT.	INST.	TOTAL	NEW YORK ALBANY 120 - 122 MAT.	INST.	TOTAL	BINGHAMTON 137 - 139 MAT.	INST.	TOTAL	BRONX 104 MAT.	INST.	TOTAL	BROOKLYN 112 MAT.	INST.	TOTAL	BUFFALO 140 - 142 MAT.	INST.	TOTAL
015433	CONTRACTOR EQUIPMENT		112.7	112.7		117.3	117.3		122.6	122.6		108.7	108.7		113.7	113.7		97.4	97.4
0241, 31 - 34	SITE & INFRASTRUCTURE, DEMOLITION	96.9	103.8	101.7	84.7	105.8	99.4	96.3	94.1	94.8	95.4	115.1	109.2	121.1	126.8	125.1	97.9	100.1	99.5
0310	Concrete Forming & Accessories	99.5	65.6	70.2	102.2	106.4	105.8	102.9	93.0	94.4	100.4	198.3	185.1	109.4	187.4	176.9	103.1	122.3	119.7
0320	Concrete Reinforcing	112.1	70.0	90.7	111.0	120.5	115.8	100.0	105.3	102.7	99.9	173.5	137.3	101.4	224.2	163.8	106.1	113.9	110.0
0330	Cast-in-Place Concrete	92.4	70.2	83.9	76.5	115.2	91.2	106.7	109.2	107.7	89.9	176.4	122.9	108.8	174.6	133.9	110.9	122.0	115.2
03	CONCRETE	105.0	69.4	88.5	86.2	113.1	98.7	96.2	103.2	99.5	93.1	185.3	135.8	107.7	187.6	144.7	105.3	120.0	112.1
04	MASONRY	120.8	60.8	83.8	91.9	116.4	107.0	108.2	107.6	107.8	91.5	177.7	144.6	118.9	177.6	155.1	106.2	124.7	117.6
05	METALS	104.1	91.0	100.1	97.7	129.9	107.5	91.7	135.8	105.2	87.1	171.8	113.0	99.4	168.7	120.6	99.2	106.2	101.3
06	WOOD, PLASTICS & COMPOSITES	93.0	66.6	78.5	96.9	103.3	100.4	106.1	89.3	96.8	100.1	204.1	157.3	107.5	189.2	152.4	104.9	122.8	114.7
07	THERMAL & MOISTURE PROTECTION	99.7	73.1	88.6	108.0	109.7	108.7	108.1	97.3	103.6	106.3	169.3	132.4	108.4	167.2	132.8	102.2	115.0	107.5
08	OPENINGS	96.8	66.8	90.0	97.2	105.6	99.1	92.4	94.0	92.8	94.2	205.2	119.4	90.0	194.8	113.8	101.4	114.9	104.5
0920	Plaster & Gypsum Board	77.5	65.4	69.5	103.4	103.2	103.3	107.4	88.7	94.9	96.2	206.7	169.6	103.6	191.7	162.1	106.9	123.4	117.8
0950, 0980	Ceilings & Acoustic Treatment	98.5	65.4	76.5	94.3	103.2	100.2	92.3	88.7	89.9	82.5	206.7	165.1	88.1	191.7	157.0	100.7	123.4	115.8
0960	Flooring	101.4	66.7	91.4	96.8	108.5	100.2	105.0	99.5	103.4	102.9	190.4	128.3	112.1	190.4	134.4	97.5	118.8	103.7
0970, 0990	Wall Finishes & Painting/Coating	96.0	54.3	71.5	92.4	105.6	99.2	89.3	101.5	96.5	108.8	162.9	140.6	116.9	162.9	143.9	98.8	118.0	110.0
09	FINISHES	94.1	64.5	77.9	92.4	105.6	99.6	93.6	94.3	94.0	95.7	194.8	149.9	106.8	186.0	150.1	101.4	122.1	112.7
COVERS	DIVS. 10 - 14, 25, 28, 41, 43, 44, 46	100.0	86.5	97.0	100.0	101.9	100.4	100.0	99.4	99.9	100.0	135.6	108.0	100.0	133.2	107.5	100.0	107.3	101.7
21, 22, 23	FIRE SUPPRESSION, PLUMBING & HVAC	97.8	68.7	85.0	100.0	107.2	103.2	100.6	98.0	99.5	100.2	170.9	131.5	99.7	170.7	131.1	100.0	101.8	100.8
26, 27, 3370	ELECTRICAL, COMMUNICATIONS & UTIL.	92.1	74.4	82.7	95.7	110.1	103.4	99.2	103.5	101.5	90.8	186.2	141.4	98.6	186.2	145.1	101.7	105.5	103.7
MF2014	WEIGHTED AVERAGE	100.1	73.6	88.4	96.0	110.9	102.6	97.4	103.0	99.9	94.8	175.1	130.3	101.9	174.2	133.8	101.3	111.3	105.7

DIVISION		NEW YORK ELMIRA 148 - 149 MAT.	INST.	TOTAL	FAR ROCKAWAY 116 MAT.	INST.	TOTAL	FLUSHING 113 MAT.	INST.	TOTAL	GLENS FALLS 128 MAT.	INST.	TOTAL	HICKSVILLE 115,117,118 MAT.	INST.	TOTAL	JAMAICA 114 MAT.	INST.	TOTAL
015433	CONTRACTOR EQUIPMENT		125.3	125.3		113.7	113.7		113.7	113.7		117.3	117.3		113.7	113.7		113.7	113.7
0241, 31 - 34	SITE & INFRASTRUCTURE, DEMOLITION	100.8	94.3	96.3	124.4	126.8	126.1	124.5	126.8	126.1	74.7	105.4	96.1	113.9	125.9	122.3	118.6	126.8	124.3
0310	Concrete Forming & Accessories	84.5	98.7	96.8	95.2	198.2	184.3	99.3	198.2	184.9	87.6	98.9	97.3	91.6	166.6	156.5	99.3	198.2	184.9
0320	Concrete Reinforcing	106.4	107.2	106.8	101.4	224.2	163.8	103.2	224.2	164.6	106.6	112.4	109.6	101.4	224.1	163.7	101.4	224.2	163.8
0330	Cast-in-Place Concrete	99.6	108.6	103.0	117.7	174.6	139.4	117.7	174.6	139.4	73.8	112.4	88.5	100.0	171.1	127.1	108.8	174.6	133.9
03	CONCRETE	91.6	106.0	98.3	114.0	192.5	150.4	114.5	192.5	150.7	80.2	107.3	92.8	99.5	176.9	135.4	107.1	192.5	146.7
04	MASONRY	104.4	108.6	107.0	123.5	177.6	156.8	117.5	177.6	154.5	98.1	111.6	106.4	113.2	172.6	149.8	121.2	177.6	156.0
05	METALS	94.3	138.4	107.8	99.4	168.8	120.6	99.4	168.8	120.6	91.5	126.3	102.2	100.8	165.7	120.7	99.4	168.8	120.6
06	WOOD, PLASTICS & COMPOSITES	84.3	97.3	91.4	91.5	203.8	153.2	96.0	203.8	155.2	87.5	95.8	92.0	88.2	164.5	130.2	96.0	203.8	155.2
07	THERMAL & MOISTURE PROTECTION	104.3	98.0	101.7	108.3	168.8	133.4	108.4	168.8	133.5	101.4	106.6	103.6	108.0	162.3	130.5	108.2	168.8	133.3
08	OPENINGS	99.3	98.9	99.2	88.7	202.8	114.5	88.7	202.8	114.5	90.6	100.1	92.7	89.1	181.3	110.0	88.7	202.8	114.5
0920	Plaster & Gypsum Board	103.7	97.1	99.3	92.5	206.7	168.4	94.8	206.7	169.2	95.0	95.5	95.4	92.1	166.4	141.5	94.8	206.7	169.2
0950, 0980	Ceilings & Acoustic Treatment	95.8	97.1	96.7	77.1	206.7	163.4	77.1	206.7	163.4	83.1	95.5	91.4	76.2	166.4	136.2	77.1	206.7	163.4
0960	Flooring	90.9	102.6	94.3	106.7	190.4	131.0	108.4	190.4	132.1	87.9	108.5	93.9	105.6	176.8	126.3	108.4	190.4	132.1
0970, 0990	Wall Finishes & Painting/Coating	96.0	93.3	94.4	116.9	162.9	143.9	116.9	162.9	143.9	93.0	99.2	96.6	116.9	162.9	143.9	116.9	162.9	143.9
09	FINISHES	94.5	98.8	96.8	101.8	194.6	152.5	102.6	194.6	152.9	84.9	99.9	93.1	100.4	167.9	137.3	102.2	194.6	152.7
COVERS	DIVS. 10 - 14, 25, 28, 41, 43, 44, 46	100.0	100.9	100.2	100.0	134.7	107.9	100.0	134.7	107.9	100.0	98.6	99.7	100.0	128.7	106.5	100.0	134.7	107.9
21, 22, 23	FIRE SUPPRESSION, PLUMBING & HVAC	96.7	95.9	96.3	96.1	170.7	129.1	96.1	170.7	129.1	96.7	105.5	100.6	99.7	161.1	126.9	96.1	170.7	129.1
26, 27, 3370	ELECTRICAL, COMMUNICATIONS & UTIL.	98.3	106.6	102.7	105.9	186.2	148.5	105.9	186.2	148.5	90.8	110.1	101.0	98.0	146.2	123.6	97.0	186.2	144.3
MF2014	WEIGHTED AVERAGE	96.8	104.6	100.3	102.3	176.5	135.1	102.1	176.5	135.0	91.4	107.7	98.6	99.8	160.8	126.8	100.3	176.5	134.0

DIVISION		NEW YORK JAMESTOWN 147 MAT.	INST.	TOTAL	KINGSTON 124 MAT.	INST.	TOTAL	LONG ISLAND CITY 111 MAT.	INST.	TOTAL	MONTICELLO 127 MAT.	INST.	TOTAL	MOUNT VERNON 105 MAT.	INST.	TOTAL	NEW ROCHELLE 108 MAT.	INST.	TOTAL
015433	CONTRACTOR EQUIPMENT		93.6	93.6		113.7	113.7		113.7	113.7		113.7	113.7		108.7	108.7		108.7	108.7
0241, 31 - 34	SITE & INFRASTRUCTURE, DEMOLITION	102.2	94.9	97.1	147.0	122.3	129.7	122.2	126.8	125.4	141.9	122.8	128.6	100.5	111.4	108.1	100.2	111.4	108.1
0310	Concrete Forming & Accessories	84.6	91.6	90.6	89.0	131.7	126.0	103.8	198.2	185.5	96.5	131.7	127.0	90.5	143.6	136.4	105.8	143.6	138.5
0320	Concrete Reinforcing	106.6	113.3	110.0	107.1	158.3	133.1	101.4	224.2	163.8	106.2	158.3	132.6	98.9	172.4	136.2	99.0	172.4	136.3
0330	Cast-in-Place Concrete	103.3	86.8	97.0	99.7	147.7	118.0	112.3	174.6	136.1	93.4	147.7	114.0	100.4	150.6	119.5	100.4	150.6	119.5
03	CONCRETE	94.6	93.8	94.2	100.4	141.5	119.4	110.3	192.5	148.4	95.6	141.5	116.9	102.8	150.3	124.8	102.1	150.3	124.4
04	MASONRY	113.8	104.8	108.3	112.9	155.7	139.3	115.7	177.6	153.8	105.7	155.7	136.5	97.4	157.3	134.2	97.4	157.3	134.2
05	METALS	91.7	103.3	95.2	99.4	136.0	110.6	99.4	168.8	120.6	99.4	136.0	110.6	86.9	142.3	103.8	87.1	142.2	104.0
06	WOOD, PLASTICS & COMPOSITES	82.9	88.8	86.2	88.7	125.9	109.1	101.8	203.8	157.9	95.5	125.9	112.2	91.0	139.2	117.5	107.4	139.2	124.9
07	THERMAL & MOISTURE PROTECTION	103.8	96.3	100.7	125.5	147.0	134.4	108.3	168.8	133.4	125.1	147.0	134.2	107.3	148.8	124.5	107.3	148.8	124.5
08	OPENINGS	99.1	96.3	98.5	92.8	143.9	104.4	88.7	202.8	114.5	88.4	143.9	101.0	94.2	154.7	107.9	94.3	154.7	108.0
0920	Plaster & Gypsum Board	88.6	88.4	88.4	97.2	126.7	116.8	99.4	206.7	170.7	97.9	126.7	117.0	91.6	139.9	123.7	104.1	139.9	127.9
0950, 0980	Ceilings & Acoustic Treatment	92.4	88.4	89.7	73.8	126.7	109.0	77.1	206.7	163.4	73.8	126.7	109.0	80.8	139.9	120.2	80.8	139.9	120.2
0960	Flooring	93.3	102.6	96.0	105.6	163.4	122.3	110.1	190.4	133.4	108.1	163.4	124.1	94.0	173.8	117.1	102.1	173.8	122.7
0970, 0990	Wall Finishes & Painting/Coating	96.8	96.9	96.8	123.0	127.5	125.6	116.9	162.9	143.9	123.0	127.5	125.6	107.1	162.9	139.9	107.1	162.9	139.9
09	FINISHES	92.5	93.5	93.0	99.1	136.7	119.6	103.5	194.6	153.3	99.5	136.7	119.8	92.6	150.9	124.4	96.6	150.9	126.3
COVERS	DIVS. 10 - 14, 25, 28, 41, 43, 44, 46	100.0	99.6	99.9	100.0	119.4	104.4	100.0	134.7	107.9	100.0	119.4	104.4	100.0	123.5	105.3	100.0	120.4	104.6
21, 22, 23	FIRE SUPPRESSION, PLUMBING & HVAC	96.5	89.9	93.6	96.6	126.9	110.0	99.7	170.7	131.1	96.6	123.1	108.3	96.8	141.6	116.6	96.8	141.6	116.6
26, 27, 3370	ELECTRICAL, COMMUNICATIONS & UTIL.	97.3	96.3	96.8	92.1	124.2	109.1	97.5	186.2	144.6	92.1	124.2	109.1	89.1	153.6	123.3	89.1	153.6	123.3
MF2014	WEIGHTED AVERAGE	96.9	95.7	96.4	100.1	134.2	115.2	101.5	176.5	134.6	98.6	133.4	114.0	95.1	145.1	117.2	95.5	145.0	117.4

784

NEW YORK

DIVISION		NEW YORK 100 - 102			NIAGARA FALLS 143			PLATTSBURGH 129			POUGHKEEPSIE 125 - 126			QUEENS 110			RIVERHEAD 119		
		MAT.	INST.	TOTAL	MAT.	INST.	TOTAL	MAT.	INST.	TOTAL	MAT.	INST.	TOTAL	MAT.	INST.	TOTAL	MAT.	INST.	TOTAL
015433	CONTRACTOR EQUIPMENT		106.9	106.9		93.6	93.6		97.7	97.7		113.7	113.7		113.7	113.7		113.7	113.7
0241, 31 - 34	SITE & INFRASTRUCTURE, DEMOLITION	102.6	112.8	109.7	104.5	96.4	98.9	113.2	102.8	106.0	142.9	122.5	128.7	117.3	126.8	124.0	115.2	126.0	122.7
0310	Concrete Forming & Accessories	104.5	200.4	187.4	84.5	121.8	116.8	93.4	98.3	97.7	89.0	180.6	168.2	91.9	198.2	183.9	96.3	166.7	157.2
0320	Concrete Reinforcing	105.7	231.8	169.7	105.2	114.4	109.8	110.9	114.4	112.7	107.1	158.4	133.1	103.2	224.2	164.6	103.4	224.1	164.6
0330	Cast-in-Place Concrete	101.1	183.0	132.3	106.8	130.9	116.0	90.3	107.6	96.9	96.6	145.1	115.0	103.4	174.6	130.5	101.6	171.4	128.2
03	CONCRETE	103.1	197.6	146.9	96.8	122.8	108.9	92.9	104.0	98.0	97.8	162.5	127.8	102.8	192.5	144.4	100.1	177.0	135.8
04	MASONRY	100.6	182.2	150.8	120.9	134.8	129.5	92.1	104.9	100.0	105.3	150.4	133.0	109.7	177.6	151.5	119.1	173.0	152.3
05	METALS	98.1	171.8	120.6	94.3	104.9	97.3	95.5	101.4	97.3	99.4	136.7	110.8	99.4	168.8	120.6	101.3	165.9	121.1
06	WOOD, PLASTICS & COMPOSITES	104.1	204.1	159.0	82.9	117.0	101.6	93.9	95.3	94.7	88.7	194.0	146.6	88.4	203.8	151.8	93.1	164.5	132.4
07	THERMAL & MOISTURE PROTECTION	106.4	172.0	133.6	103.9	118.3	109.8	119.4	101.7	112.1	125.4	152.0	136.5	108.0	168.8	133.2	108.9	162.4	131.1
08	OPENINGS	100.8	204.7	124.3	99.1	111.9	102.0	98.3	100.5	98.8	92.8	181.2	112.8	88.7	202.8	114.5	89.1	181.3	110.0
0920	Plaster & Gypsum Board	102.2	206.7	171.7	88.6	117.4	107.7	114.8	94.6	101.4	97.2	196.7	163.3	92.1	206.7	168.3	93.3	166.4	141.9
0950, 0980	Ceilings & Acoustic Treatment	98.5	206.7	170.5	92.4	117.4	109.0	101.0	94.6	96.8	73.8	196.7	155.5	77.1	206.7	163.4	77.0	166.4	136.5
0960	Flooring	104.3	190.4	129.2	93.3	118.8	100.7	111.6	108.5	110.7	105.6	168.1	123.7	105.6	190.4	130.2	106.7	176.8	127.1
0970, 0990	Wall Finishes & Painting/Coating	108.8	159.1	138.3	96.8	118.7	109.7	118.5	97.9	106.4	123.0	127.5	125.6	116.9	162.9	143.9	116.9	162.9	143.9
09	FINISHES	100.7	195.5	152.5	92.7	121.3	108.3	97.4	99.5	98.6	98.9	176.1	141.1	100.8	194.6	152.0	101.0	167.6	137.4
COVERS	DIVS. 10 - 14, 25, 28, 41, 43, 44, 46	100.0	143.4	109.8	100.0	110.3	102.3	100.0	99.5	99.9	100.0	125.1	105.7	100.0	134.7	107.9	100.0	128.8	106.5
21, 22, 23	FIRE SUPPRESSION, PLUMBING & HVAC	100.1	173.2	132.4	96.5	107.0	101.2	96.6	105.6	100.6	96.6	126.6	109.9	99.7	170.7	131.1	99.9	161.3	127.1
26, 27, 3370	ELECTRICAL, COMMUNICATIONS & UTIL.	97.3	186.2	144.5	95.8	104.2	100.2	89.5	95.4	92.6	92.1	131.4	112.9	98.0	186.2	144.8	99.7	146.2	124.3
MF2014	WEIGHTED AVERAGE	100.3	177.9	134.6	97.9	113.1	104.6	96.7	101.9	99.0	99.3	145.2	119.6	99.9	176.5	133.7	100.6	160.9	127.3

NEW YORK

DIVISION		ROCHESTER 144 - 146			SCHENECTADY 123			STATEN ISLAND 103			SUFFERN 109			SYRACUSE 130 - 132			UTICA 133 - 135		
		MAT.	INST.	TOTAL	MAT.	INST.	TOTAL	MAT.	INST.	TOTAL	MAT.	INST.	TOTAL	MAT.	INST.	TOTAL	MAT.	INST.	TOTAL
015433	CONTRACTOR EQUIPMENT		122.0	122.0		117.3	117.3		108.7	108.7		108.7	108.7		117.3	117.3		117.3	117.3
0241, 31 - 34	SITE & INFRASTRUCTURE, DEMOLITION	92.0	110.6	104.9	85.1	105.8	99.5	104.6	115.1	111.9	97.4	110.8	106.8	94.9	105.0	101.9	73.4	104.2	94.9
0310	Concrete Forming & Accessories	102.1	104.3	104.0	104.7	106.4	106.2	89.9	187.8	174.6	99.0	151.8	144.7	102.0	97.5	98.1	103.2	93.0	94.4
0320	Concrete Reinforcing	109.8	107.1	108.4	105.5	120.5	113.1	99.9	224.3	163.0	99.0	158.5	129.2	101.1	105.2	103.2	101.1	113.0	107.1
0330	Cast-in-Place Concrete	98.9	109.6	103.0	89.1	115.2	99.0	100.4	176.5	129.4	97.4	149.1	117.1	99.1	111.8	103.9	90.7	108.0	97.3
03	CONCRETE	97.8	107.9	102.5	93.3	113.1	102.5	104.5	188.6	143.5	99.2	151.0	123.2	98.9	105.1	101.8	96.7	103.2	99.7
04	MASONRY	101.5	111.5	107.6	94.7	116.4	108.1	103.5	177.7	149.2	96.9	153.2	131.6	100.8	113.2	108.5	91.8	107.1	101.2
05	METALS	99.5	124.2	107.1	95.5	129.9	106.0	85.2	172.0	111.8	85.2	137.0	101.1	95.1	121.5	103.2	93.2	124.4	102.8
06	WOOD, PLASTICS & COMPOSITES	96.0	103.1	99.9	105.6	103.3	104.3	89.6	189.5	144.5	100.0	154.1	129.7	102.3	93.0	97.2	102.3	89.4	95.2
07	THERMAL & MOISTURE PROTECTION	103.3	105.5	104.2	102.9	109.7	105.7	106.7	167.7	132.0	107.2	148.6	124.4	102.7	102.4	102.6	91.1	100.5	95.0
08	OPENINGS	103.4	102.7	103.3	96.3	105.6	98.4	94.2	197.2	117.5	94.3	159.4	109.0	94.8	94.4	94.7	97.6	94.4	96.9
0920	Plaster & Gypsum Board	104.9	103.3	103.8	104.3	103.2	103.6	91.6	191.7	158.1	95.5	155.3	135.3	98.8	92.6	94.7	98.8	89.0	92.3
0950, 0980	Ceilings & Acoustic Treatment	99.0	103.3	101.8	89.7	103.2	98.7	82.5	191.7	155.2	80.8	155.3	130.4	92.3	92.6	92.5	92.3	89.0	90.1
0960	Flooring	95.8	111.3	100.3	95.6	108.5	99.3	97.9	190.4	124.7	97.8	181.8	122.1	93.5	97.8	94.8	91.3	97.9	93.2
0970, 0990	Wall Finishes & Painting/Coating	98.1	104.8	102.1	93.0	100.6	97.5	108.8	162.9	140.6	107.1	135.2	123.6	93.3	105.2	100.3	86.7	105.2	97.5
09	FINISHES	97.1	105.9	101.9	90.5	105.6	98.8	94.4	186.3	144.6	93.9	156.7	128.3	92.7	97.7	95.4	90.9	94.4	92.8
COVERS	DIVS. 10 - 14, 25, 28, 41, 43, 44, 46	100.0	102.8	100.6	100.0	101.9	100.4	100.0	134.0	107.7	100.0	123.2	105.2	100.0	101.1	100.3	100.0	96.3	99.2
21, 22, 23	FIRE SUPPRESSION, PLUMBING & HVAC	100.0	92.7	96.8	100.2	106.7	103.1	100.2	170.9	131.5	96.8	130.2	111.6	100.3	98.3	99.4	100.3	97.5	99.0
26, 27, 3370	ELECTRICAL, COMMUNICATIONS & UTIL.	101.3	97.4	99.2	95.2	110.1	103.1	90.8	186.2	141.4	95.5	124.2	110.7	99.2	105.9	102.7	97.4	105.9	101.9
MF2014	WEIGHTED AVERAGE	99.8	104.5	101.9	96.3	110.8	102.7	96.5	173.9	130.7	95.1	138.8	114.4	97.8	104.4	100.7	95.9	102.9	99.0

DIVISION		NEW YORK									NORTH CAROLINA								
		WATERTOWN 136			WHITE PLAINS 106			YONKERS 107			ASHEVILLE 287 - 288			CHARLOTTE 281 - 282			DURHAM 277		
		MAT.	INST.	TOTAL	MAT.	INST.	TOTAL	MAT.	INST.	TOTAL	MAT.	INST.	TOTAL	MAT.	INST.	TOTAL	MAT.	INST.	TOTAL
015433	CONTRACTOR EQUIPMENT		117.3	117.3		108.7	108.7		108.7	108.7		103.2	103.2		103.2	103.2		108.7	108.7
0241, 31 - 34	SITE & INFRASTRUCTURE, DEMOLITION	81.2	104.7	97.6	95.1	111.4	106.5	102.1	111.5	108.6	96.9	82.4	86.8	98.7	82.6	87.5	98.6	91.8	93.8
0310	Concrete Forming & Accessories	87.6	100.1	98.4	104.1	143.6	138.3	104.3	155.1	148.3	93.9	65.2	69.1	97.5	64.7	69.2	98.4	65.2	69.7
0320	Concrete Reinforcing	101.7	105.3	103.5	99.0	172.4	136.3	103.0	172.5	138.3	99.3	67.5	83.2	103.8	65.7	84.4	102.7	65.7	83.9
0330	Cast-in-Place Concrete	105.6	111.9	108.0	89.2	150.6	112.6	99.7	150.9	119.2	108.2	72.4	94.6	110.3	71.6	95.6	93.6	72.4	85.5
03	CONCRETE	109.5	106.3	108.0	92.7	150.3	119.4	102.2	155.6	126.9	102.6	69.8	87.4	102.2	69.0	86.8	95.5	69.5	83.4
04	MASONRY	93.0	112.9	105.3	96.3	157.3	133.8	100.2	157.3	135.3	91.9	74.4	81.1	95.8	73.0	81.7	88.0	74.4	79.7
05	METALS	93.3	121.5	101.9	86.7	142.3	103.7	94.7	142.8	109.4	95.5	92.7	94.6	96.2	92.1	95.0	111.6	92.1	105.6
06	WOOD, PLASTICS & COMPOSITES	84.9	97.2	91.7	105.2	139.2	123.9	105.1	154.1	132.0	91.0	63.8	76.1	91.9	63.8	76.4	95.2	63.8	77.9
07	THERMAL & MOISTURE PROTECTION	91.4	103.9	96.6	107.0	148.8	124.3	107.3	151.7	125.7	103.9	67.2	88.6	98.2	66.5	85.0	104.7	67.2	89.1
08	OPENINGS	97.6	98.9	97.9	94.3	154.7	108.0	98.1	162.4	112.7	93.9	63.4	87.0	96.2	63.0	88.7	101.0	63.0	92.4
0920	Plaster & Gypsum Board	89.9	97.0	94.6	98.5	139.9	126.0	101.9	155.3	137.4	106.6	62.6	77.4	104.9	62.6	76.8	98.7	62.6	74.7
0950, 0980	Ceilings & Acoustic Treatment	92.3	97.0	95.4	80.8	139.9	120.2	96.9	155.3	135.8	81.3	62.6	68.8	85.3	62.6	70.2	83.7	62.6	69.6
0960	Flooring	84.2	97.9	88.2	100.4	173.8	121.7	99.9	190.4	126.1	99.6	81.8	94.4	98.7	81.8	93.8	103.1	81.8	96.9
0970, 0990	Wall Finishes & Painting/Coating	86.7	99.1	93.9	107.1	162.9	139.9	107.1	162.9	139.9	108.2	60.9	80.4	100.5	60.9	77.2	102.4	60.9	78.0
09	FINISHES	88.2	99.3	94.2	94.8	150.9	125.4	98.9	162.9	133.9	92.8	67.9	79.2	91.0	67.5	78.2	92.2	67.9	78.9
COVERS	DIVS. 10 - 14, 25, 28, 41, 43, 44, 46	100.0	101.1	100.3	100.0	123.5	105.3	100.0	133.0	107.5	100.0	85.6	96.8	100.0	85.2	96.6	100.0	85.6	96.8
21, 22, 23	FIRE SUPPRESSION, PLUMBING & HVAC	100.3	90.7	96.0	100.4	141.6	118.6	100.4	141.8	118.7	100.5	62.1	83.5	99.9	62.1	83.2	100.5	62.1	83.5
26, 27, 3370	ELECTRICAL, COMMUNICATIONS & UTIL.	99.2	95.4	97.2	89.1	153.6	123.3	95.6	169.1	134.6	100.9	60.5	79.5	100.0	62.6	80.1	99.0	62.2	79.4
MF2014	WEIGHTED AVERAGE	97.5	101.9	99.5	94.8	145.1	117.0	99.0	150.4	121.7	98.0	70.3	85.8	98.1	70.2	85.8	100.3	71.1	87.4

City Cost Indexes

NORTH CAROLINA

DIVISION		ELIZABETH CITY 279			FAYETTEVILLE 283			GASTONIA 280			GREENSBORO 270,272 - 274			HICKORY 286			KINSTON 285		
		MAT.	INST.	TOTAL	MAT.	INST.	TOTAL	MAT.	INST.	TOTAL	MAT.	INST.	TOTAL	MAT.	INST.	TOTAL	MAT.	INST.	TOTAL
015433	CONTRACTOR EQUIPMENT		113.0	113.0		108.7	108.7		103.2	103.2		108.7	108.7		108.7	108.7		108.7	108.7
0241, 31 - 34	SITE & INFRASTRUCTURE, DEMOLITION	102.6	93.6	96.3	96.3	91.7	93.1	96.8	82.7	87.0	98.5	91.9	93.9	95.9	90.4	92.1	94.9	90.3	91.7
0310	Concrete Forming & Accessories	84.0	65.2	67.7	93.5	64.7	68.6	100.3	65.3	70.0	98.2	65.2	69.7	90.2	65.0	68.4	86.4	64.3	67.3
0320	Concrete Reinforcing	100.4	73.1	86.6	103.2	65.7	84.2	99.7	65.7	82.4	101.4	65.6	83.2	99.3	65.6	82.2	98.8	65.6	81.9
0330	Cast-in-Place Concrete	93.8	74.1	86.3	113.5	71.6	97.5	105.8	72.4	93.1	92.9	72.4	85.1	108.1	72.4	94.5	104.4	71.5	91.9
03	CONCRETE	95.7	71.3	84.4	104.6	69.0	88.1	101.1	69.5	86.5	95.0	69.5	83.2	102.4	69.4	87.1	99.2	68.7	85.1
04	MASONRY	100.5	73.0	83.5	95.1	73.0	81.5	96.3	74.4	82.8	84.0	74.4	78.1	80.5	74.4	76.7	87.5	73.0	78.6
05	METALS	98.1	95.4	97.3	115.3	92.1	108.2	96.0	92.1	94.8	104.3	92.1	100.5	95.5	91.9	94.4	94.3	91.7	93.5
06	WOOD, PLASTICS & COMPOSITES	80.4	64.5	71.6	90.3	63.8	75.7	98.8	63.8	79.6	94.9	63.8	77.8	86.2	63.8	73.9	83.2	63.8	72.5
07	THERMAL & MOISTURE PROTECTION	103.9	66.4	88.3	103.3	66.5	88.1	104.1	67.2	88.8	104.5	67.2	89.0	104.2	67.2	88.9	104.1	66.5	88.5
08	OPENINGS	97.8	65.1	90.4	94.0	63.0	87.0	97.6	63.0	89.8	101.0	63.0	92.4	93.9	63.0	86.9	94.0	63.0	87.0
0920	Plaster & Gypsum Board	92.3	62.6	72.6	110.8	62.6	78.8	112.7	62.6	79.4	100.2	62.6	75.2	106.6	62.6	77.4	106.4	62.6	77.3
0950, 0980	Ceilings & Acoustic Treatment	83.7	62.6	69.6	83.0	62.6	69.4	84.7	62.6	70.0	83.7	62.6	69.6	81.3	62.6	68.8	84.7	62.6	70.0
0960	Flooring	94.9	81.8	91.1	99.8	81.8	94.6	103.1	81.8	96.9	103.1	81.8	96.9	99.5	81.8	94.4	96.6	81.8	92.3
0970, 0990	Wall Finishes & Painting/Coating	102.4	60.9	78.0	108.2	60.9	80.4	108.2	60.9	80.4	102.4	60.9	78.0	108.2	60.9	80.4	108.2	60.9	80.4
09	FINISHES	89.2	68.0	77.6	93.8	67.5	79.5	95.3	67.9	80.3	92.4	67.9	79.0	93.0	67.9	79.3	92.6	67.5	78.9
COVERS	DIVS. 10 - 14, 25, 28, 41, 43, 44, 46	100.0	89.7	97.7	100.0	85.2	96.6	100.0	85.7	96.8	100.0	84.9	96.6	100.0	85.6	96.7	100.0	85.1	96.6
21, 22, 23	FIRE SUPPRESSION, PLUMBING & HVAC	96.8	61.1	81.0	100.2	61.4	83.0	100.5	62.1	83.5	100.4	62.1	83.4	96.9	62.1	81.5	96.9	60.9	81.0
26, 27, 3370	ELECTRICAL, COMMUNICATIONS & UTIL.	98.7	34.2	64.5	100.2	62.2	80.0	100.4	62.6	80.3	98.0	60.5	78.1	98.5	62.6	79.4	98.3	60.5	78.2
MF2014	WEIGHTED AVERAGE	97.2	67.8	84.2	101.7	70.7	88.0	98.8	70.5	86.3	98.7	70.9	86.4	96.4	71.1	85.2	96.0	70.2	84.6

DIVISION		NORTH CAROLINA MURPHY 289			RALEIGH 275 - 276			ROCKY MOUNT 278			WILMINGTON 284			WINSTON-SALEM 271			NORTH DAKOTA BISMARCK 585		
		MAT.	INST.	TOTAL	MAT.	INST.	TOTAL	MAT.	INST.	TOTAL	MAT.	INST.	TOTAL	MAT.	INST.	TOTAL	MAT.	INST.	TOTAL
015433	CONTRACTOR EQUIPMENT		103.2	103.2		108.7	108.7		108.7	108.7		103.2	103.2		108.7	108.7		101.8	101.8
0241, 31 - 34	SITE & INFRASTRUCTURE, DEMOLITION	98.1	80.9	86.1	98.2	91.8	93.7	100.9	91.8	94.5	98.1	82.3	87.1	98.8	91.9	94.0	99.7	100.0	99.9
0310	Concrete Forming & Accessories	101.0	64.5	69.4	97.3	64.6	69.0	90.3	64.6	68.1	95.3	64.7	68.8	100.1	65.2	69.9	107.1	75.0	79.3
0320	Concrete Reinforcing	98.8	67.5	82.9	103.6	65.6	84.3	100.4	65.6	82.8	100.1	65.7	82.6	101.4	65.6	83.2	96.4	91.5	93.9
0330	Cast-in-Place Concrete	111.9	71.6	96.5	96.3	71.6	86.9	91.7	71.6	84.1	107.8	71.6	94.0	95.2	73.3	86.9	99.1	78.1	91.1
03	CONCRETE	105.7	69.2	88.8	93.6	68.9	82.2	96.6	68.9	83.7	102.6	69.0	87.0	96.2	69.8	84.0	93.9	79.7	87.4
04	MASONRY	83.6	73.0	77.1	82.3	73.0	76.6	77.4	73.0	74.7	81.1	73.0	76.1	84.2	74.4	78.2	105.2	80.2	89.8
05	METALS	93.4	92.5	93.1	96.8	92.0	95.3	97.3	92.0	95.7	95.0	92.1	94.1	101.5	92.1	98.6	101.3	92.9	98.7
06	WOOD, PLASTICS & COMPOSITES	99.4	63.8	79.8	92.0	63.8	76.5	86.9	63.8	74.2	92.8	63.8	76.9	94.9	63.8	77.8	97.4	70.7	82.8
07	THERMAL & MOISTURE PROTECTION	104.1	66.5	88.5	99.1	66.5	85.6	104.4	66.5	88.7	103.9	66.5	88.4	104.5	67.2	89.0	108.9	83.3	98.3
08	OPENINGS	93.8	63.4	87.0	99.3	63.0	91.1	97.0	63.0	89.3	94.0	63.0	87.0	101.0	63.0	92.4	106.4	80.7	100.6
0920	Plaster & Gypsum Board	111.9	62.6	79.1	91.8	62.6	72.4	94.2	62.6	73.2	108.7	62.6	78.0	100.2	62.6	75.2	97.7	70.3	79.5
0950, 0980	Ceilings & Acoustic Treatment	81.3	62.6	68.8	85.3	62.6	70.2	82.0	62.6	69.1	83.0	62.6	69.4	83.7	62.6	69.6	110.3	70.3	83.7
0960	Flooring	103.4	81.8	97.2	96.2	81.8	92.0	98.6	81.8	93.7	100.4	81.8	95.0	103.1	81.8	96.9	91.8	48.1	79.2
0970, 0990	Wall Finishes & Painting/Coating	108.2	60.9	80.4	97.1	60.9	75.8	102.4	60.9	78.0	108.2	60.9	80.4	102.4	54.2	74.1	94.5	60.0	74.3
09	FINISHES	94.9	67.5	79.9	89.1	67.5	77.3	90.3	67.5	77.9	93.7	67.5	79.4	92.4	67.2	78.6	95.6	68.9	81.0
COVERS	DIVS. 10 - 14, 25, 28, 41, 43, 44, 46	100.0	85.1	96.6	100.0	85.1	96.6	100.0	85.2	96.6	100.0	85.2	96.6	100.0	85.6	96.8	100.0	89.4	97.6
21, 22, 23	FIRE SUPPRESSION, PLUMBING & HVAC	96.9	61.4	81.2	99.9	61.3	82.9	96.8	61.2	81.1	100.5	61.4	83.2	100.4	62.1	83.4	100.0	77.4	90.0
26, 27, 3370	ELECTRICAL, COMMUNICATIONS & UTIL.	101.9	60.5	79.9	101.0	40.0	68.6	100.9	62.2	80.3	101.2	60.5	79.6	98.0	60.5	78.1	101.6	76.0	88.0
MF2014	WEIGHTED AVERAGE	97.2	69.7	85.1	96.7	67.6	83.9	96.3	70.6	85.0	97.6	69.7	85.3	98.4	70.8	86.2	100.5	80.5	91.7

NORTH DAKOTA

DIVISION		DEVILS LAKE 583			DICKINSON 586			FARGO 580 - 581			GRAND FORKS 582			JAMESTOWN 584			MINOT 587		
		MAT.	INST.	TOTAL	MAT.	INST.	TOTAL	MAT.	INST.	TOTAL	MAT.	INST.	TOTAL	MAT.	INST.	TOTAL	MAT.	INST.	TOTAL
015433	CONTRACTOR EQUIPMENT		101.8	101.8		101.8	101.8		101.8	101.8		101.8	101.8		101.8	101.8		101.8	101.8
0241, 31 - 34	SITE & INFRASTRUCTURE, DEMOLITION	105.6	99.9	101.7	113.4	99.9	104.0	97.7	99.1	98.7	109.3	99.9	102.8	104.7	99.9	101.4	106.9	100.0	102.0
0310	Concrete Forming & Accessories	102.5	74.5	78.2	91.8	74.4	76.8	100.7	70.4	74.5	95.7	74.4	77.3	93.5	74.7	77.3	91.6	75.0	77.2
0320	Concrete Reinforcing	97.6	91.7	94.6	98.5	91.4	94.9	97.4	92.2	94.8	96.1	91.7	93.9	98.2	92.2	95.1	99.5	91.7	95.5
0330	Cast-in-Place Concrete	118.9	77.9	103.3	101.9	77.9	96.3	97.8	78.9	90.6	107.6	77.9	96.3	117.4	78.0	102.4	107.6	86.0	99.3
03	CONCRETE	103.0	79.5	92.1	101.9	79.4	91.5	95.3	78.1	87.3	101.6	79.7	91.4	101.6	79.7	91.4	98.1	82.5	90.9
04	MASONRY	122.3	85.4	99.5	124.0	80.2	97.0	109.0	79.2	90.7	115.9	85.4	97.1	135.8	92.0	108.8	115.0	83.6	95.7
05	METALS	97.3	92.8	95.9	97.3	92.4	95.8	99.8	94.0	98.0	97.3	92.8	95.9	97.3	93.5	96.1	97.6	93.2	96.2
06	WOOD, PLASTICS & COMPOSITES	92.8	70.7	80.7	80.9	70.7	75.3	92.0	70.7	80.3	85.0	70.7	77.2	82.7	70.7	76.1	80.6	70.7	75.2
07	THERMAL & MOISTURE PROTECTION	108.5	83.5	98.1	109.1	82.4	98.1	107.0	80.9	96.2	108.8	83.5	98.3	108.4	85.8	99.0	108.5	84.3	98.5
08	OPENINGS	100.3	80.7	95.9	100.3	80.7	95.9	100.2	80.7	95.8	98.9	80.7	94.8	100.3	80.7	95.9	99.1	80.7	94.9
0920	Plaster & Gypsum Board	121.8	70.3	87.5	112.6	70.3	84.5	98.9	70.3	79.9	113.9	70.3	84.9	113.6	70.3	84.8	112.6	70.3	84.5
0950, 0980	Ceilings & Acoustic Treatment	104.7	70.3	81.8	104.7	70.3	81.8	102.7	70.3	81.1	104.7	70.3	81.8	104.7	70.3	81.8	104.7	70.3	81.8
0960	Flooring	99.1	48.1	84.3	92.9	48.1	79.4	103.0	48.1	87.1	94.1	48.1	80.8	92.9	48.1	79.9	91.9	48.1	79.2
0970, 0990	Wall Finishes & Painting/Coating	92.8	60.0	73.6	92.8	60.0	73.6	94.1	70.4	80.2	92.8	70.4	79.6	92.8	60.0	73.6	92.8	60.0	73.6
09	FINISHES	98.9	68.9	82.5	96.5	68.9	81.4	97.8	66.8	80.9	96.7	70.0	82.1	95.9	68.9	81.1	95.6	68.9	81.0
COVERS	DIVS. 10 - 14, 25, 28, 41, 43, 44, 46	100.0	56.5	90.2	100.0	56.6	90.2	100.0	85.1	96.6	100.0	56.6	90.2	100.0	86.8	97.0	100.0	89.4	97.6
21, 22, 23	FIRE SUPPRESSION, PLUMBING & HVAC	96.7	82.1	90.3	96.7	74.7	87.0	100.0	75.7	89.2	100.3	73.3	88.4	96.7	73.4	86.4	100.3	73.4	88.4
26, 27, 3370	ELECTRICAL, COMMUNICATIONS & UTIL.	97.5	49.3	71.9	106.5	73.5	89.0	101.4	66.8	83.0	101.1	64.6	81.8	97.5	49.3	71.9	104.5	77.0	89.9
MF2014	WEIGHTED AVERAGE	100.4	77.3	90.2	101.1	78.5	91.1	99.9	78.1	90.3	100.4	77.7	90.4	100.5	77.2	90.2	100.4	80.6	91.6

For customer support on your Building Construction Costs with RSMeans Data, call 800.448.8182.

Block 1

DIVISION		NORTH DAKOTA WILLISTON 588			OHIO AKRON 442 - 443			OHIO ATHENS 457			OHIO CANTON 446 - 447			OHIO CHILLICOTHE 456			OHIO CINCINNATI 451 - 452		
		MAT.	INST.	TOTAL	MAT.	INST.	TOTAL	MAT.	INST.	TOTAL	MAT.	INST.	TOTAL	MAT.	INST.	TOTAL	MAT.	INST.	TOTAL
015433	CONTRACTOR EQUIPMENT		101.8	101.8		92.8	92.8		88.2	88.2		92.8	92.8		99.6	99.6		93.9	93.9
0241, 31 - 34	SITE & INFRASTRUCTURE, DEMOLITION	107.3	97.6	100.5	97.2	100.6	99.6	115.5	89.9	97.6	97.4	100.3	99.4	101.2	101.0	101.1	97.8	97.2	97.4
0310	Concrete Forming & Accessories	97.6	74.2	77.3	101.7	86.8	88.8	93.1	90.0	90.4	101.7	77.3	80.6	95.5	82.2	84.0	97.3	71.7	75.1
0320	Concrete Reinforcing	100.4	91.6	95.9	97.7	90.9	94.3	101.4	84.3	92.7	97.7	79.4	88.4	98.1	79.0	88.4	104.0	78.5	91.1
0330	Cast-in-Place Concrete	107.6	77.8	96.2	102.8	91.8	98.6	109.4	87.6	101.1	103.8	89.3	98.3	99.3	84.7	93.7	91.5	75.7	85.5
03	CONCRETE	99.4	79.3	90.1	100.1	88.7	94.8	106.0	87.6	97.5	100.6	81.6	91.8	100.3	82.9	92.2	94.6	74.7	85.3
04	MASONRY	109.1	83.6	93.4	94.8	90.5	92.1	84.2	83.9	84.0	95.5	80.7	86.4	90.9	94.2	92.9	90.4	78.8	83.3
05	METALS	97.4	92.4	95.9	98.5	80.6	93.0	103.6	78.2	95.8	98.5	75.7	91.5	95.5	86.5	92.7	97.9	81.1	92.7
06	WOOD, PLASTICS & COMPOSITES	86.4	70.7	77.8	108.0	85.7	95.8	85.2	93.5	89.7	108.4	75.8	90.5	98.3	79.4	87.9	100.7	69.7	83.7
07	THERMAL & MOISTURE PROTECTION	108.7	83.4	98.2	104.7	93.6	100.1	99.7	89.7	95.5	105.7	89.1	98.8	101.8	87.9	96.1	99.6	80.2	91.5
08	OPENINGS	100.4	80.7	95.9	109.9	86.1	104.6	98.6	86.4	95.9	103.6	73.8	96.9	89.8	75.8	86.6	97.9	71.2	91.8
0920	Plaster & Gypsum Board	113.9	70.3	84.9	103.0	85.1	91.1	95.1	93.1	93.8	104.0	75.0	84.7	97.8	79.2	85.4	99.7	69.3	79.5
0950, 0980	Ceilings & Acoustic Treatment	104.7	70.3	81.8	93.4	85.1	87.9	102.3	93.1	96.2	93.4	75.0	81.2	96.8	79.2	85.1	97.6	69.3	78.8
0960	Flooring	95.1	48.1	81.5	93.7	79.0	89.5	122.4	78.4	109.7	93.9	68.2	86.4	99.9	78.4	93.7	101.0	78.0	94.4
0970, 0990	Wall Finishes & Painting/Coating	92.8	60.0	73.6	93.4	92.4	92.8	106.8	90.5	97.2	93.4	73.6	81.8	104.0	80.0	89.9	104.0	72.3	85.4
09	FINISHES	96.9	68.9	81.6	96.7	85.6	90.6	101.6	88.4	94.4	96.9	74.4	84.6	98.6	80.9	88.9	99.1	72.3	84.5
COVERS	DIVS. 10 - 14, 25, 28, 41, 43, 44, 46	100.0	56.6	90.2	100.0	94.0	98.6	100.0	90.7	97.9	100.0	91.6	98.1	100.0	90.5	97.9	100.0	86.7	97.0
21, 22, 23	FIRE SUPPRESSION, PLUMBING & HVAC	96.7	74.7	87.0	100.0	89.9	95.5	96.5	76.8	87.8	100.0	78.3	90.4	97.1	89.5	93.7	100.0	75.6	89.2
26, 27, 3370	ELECTRICAL, COMMUNICATIONS & UTIL.	101.6	74.5	87.2	99.3	87.3	92.9	101.2	93.0	96.9	98.5	87.9	92.9	101.0	78.5	89.1	99.7	74.8	86.5
MF2014	WEIGHTED AVERAGE	99.5	78.8	90.4	100.4	88.8	95.3	100.2	85.2	93.6	99.8	81.8	91.8	97.1	86.4	92.4	98.1	77.7	89.1

Block 2

DIVISION		OHIO CLEVELAND 441			OHIO COLUMBUS 430 - 432			OHIO DAYTON 453 - 454			OHIO HAMILTON 450			OHIO LIMA 458			OHIO LORAIN 440		
		MAT.	INST.	TOTAL	MAT.	INST.	TOTAL	MAT.	INST.	TOTAL	MAT.	INST.	TOTAL	MAT.	INST.	TOTAL	MAT.	INST.	TOTAL
015433	CONTRACTOR EQUIPMENT		91.1	91.1		91.1	91.1		92.4	92.4		99.6	99.6		91.8	91.8		92.8	92.8
0241, 31 - 34	SITE & INFRASTRUCTURE, DEMOLITION	97.1	96.0	96.3	101.3	92.0	94.8	96.6	100.2	99.1	96.6	100.5	99.3	108.8	90.0	95.6	96.5	100.7	99.4
0310	Concrete Forming & Accessories	101.7	91.4	92.8	99.5	77.5	80.4	97.3	69.8	73.5	97.4	72.1	75.5	93.1	79.6	81.4	101.7	78.1	81.3
0320	Concrete Reinforcing	98.2	91.4	94.7	99.2	80.3	89.6	104.0	80.7	92.2	104.0	78.5	91.1	101.4	80.9	91.0	97.7	91.3	94.5
0330	Cast-in-Place Concrete	100.8	99.3	100.2	92.0	82.7	88.5	85.1	79.1	82.8	91.2	76.4	85.6	100.5	89.0	96.1	97.8	94.6	96.6
03	CONCRETE	99.2	93.5	96.6	94.3	79.8	87.6	91.5	75.1	83.9	94.4	75.4	85.6	98.7	83.1	91.4	97.7	85.9	92.2
04	MASONRY	99.4	101.1	100.5	93.6	87.2	89.7	89.9	77.2	82.1	90.3	79.7	83.7	114.5	80.5	93.5	91.3	95.7	94.0
05	METALS	100.0	83.7	95.0	98.0	80.2	92.5	97.1	77.9	91.2	97.1	85.9	93.7	103.6	80.9	96.7	99.1	81.9	93.8
06	WOOD, PLASTICS & COMPOSITES	107.0	89.0	97.1	98.2	76.4	86.2	102.0	67.3	83.0	100.7	69.8	83.7	85.1	79.2	81.9	108.0	74.0	89.4
07	THERMAL & MOISTURE PROTECTION	102.3	98.9	100.9	98.5	86.3	93.5	106.2	78.7	94.8	101.9	80.4	93.0	99.2	86.0	93.7	105.6	94.9	101.2
08	OPENINGS	99.8	87.9	97.1	98.5	74.4	93.0	97.3	70.4	91.2	94.9	71.2	89.6	98.7	73.5	93.0	103.6	79.7	98.2
0920	Plaster & Gypsum Board	102.3	88.4	93.1	97.9	75.7	83.1	99.7	66.8	77.8	99.7	69.3	79.5	95.1	78.5	84.0	103.0	73.2	83.2
0950, 0980	Ceilings & Acoustic Treatment	91.7	88.4	89.5	93.4	75.7	81.6	98.6	66.8	77.4	97.6	69.3	78.8	101.3	78.5	86.1	93.4	73.2	79.9
0960	Flooring	93.5	91.1	92.8	98.9	78.4	93.0	103.6	85.8	98.4	101.0	78.0	94.4	121.3	78.1	108.8	93.9	91.1	93.1
0970, 0990	Wall Finishes & Painting/Coating	93.4	94.7	94.1	100.1	80.0	88.3	104.0	72.9	85.7	104.0	72.3	85.4	106.9	73.2	87.1	93.4	94.7	94.1
09	FINISHES	96.3	91.5	93.7	96.5	77.5	86.1	100.0	72.5	84.9	99.0	72.6	84.6	100.6	78.5	88.5	96.8	81.4	88.4
COVERS	DIVS. 10 - 14, 25, 28, 41, 43, 44, 46	100.0	98.3	99.6	100.0	89.0	97.5	100.0	86.3	96.9	100.0	87.1	97.1	100.0	88.6	97.4	100.0	94.6	98.8
21, 22, 23	FIRE SUPPRESSION, PLUMBING & HVAC	100.0	92.3	96.6	100.0	86.0	93.8	100.9	75.0	89.4	100.6	75.9	89.7	96.5	83.5	90.7	100.0	86.5	94.0
26, 27, 3370	ELECTRICAL, COMMUNICATIONS & UTIL.	98.8	98.2	98.5	98.4	84.3	90.9	98.4	78.4	87.8	98.7	78.7	88.1	101.6	78.4	89.3	98.6	83.0	90.3
MF2014	WEIGHTED AVERAGE	99.5	93.7	96.9	98.0	83.4	91.5	97.8	77.8	89.0	97.7	79.2	89.5	100.6	81.8	92.3	99.3	86.9	93.8

Block 3

DIVISION		OHIO MANSFIELD 448 - 449			OHIO MARION 433			OHIO SPRINGFIELD 455			OHIO STEUBENVILLE 439			OHIO TOLEDO 434 - 436			OHIO YOUNGSTOWN 444 - 445		
		MAT.	INST.	TOTAL	MAT.	INST.	TOTAL	MAT.	INST.	TOTAL	MAT.	INST.	TOTAL	MAT.	INST.	TOTAL	MAT.	INST.	TOTAL
015433	CONTRACTOR EQUIPMENT		92.8	92.8		92.8	92.8		92.4	92.4		96.8	96.8		96.8	96.8		92.8	92.8
0241, 31 - 34	SITE & INFRASTRUCTURE, DEMOLITION	93.2	100.4	98.2	96.9	96.5	96.7	96.9	100.2	99.2	142.5	105.1	116.4	100.4	97.0	98.1	97.1	100.3	99.3
0310	Concrete Forming & Accessories	90.7	77.0	78.9	95.7	77.6	80.0	97.3	74.2	77.3	97.2	82.5	84.5	99.5	89.0	90.4	101.7	80.2	83.1
0320	Concrete Reinforcing	88.9	80.0	84.4	91.4	80.5	85.9	104.0	80.7	92.2	89.1	95.4	92.3	99.2	83.6	91.3	97.7	85.7	91.6
0330	Cast-in-Place Concrete	95.2	84.3	91.0	84.0	82.5	83.5	87.4	79.2	84.3	91.2	83.4	88.2	92.0	93.4	92.5	101.8	90.1	97.3
03	CONCRETE	92.3	79.8	86.5	86.5	79.8	83.4	92.6	77.1	85.4	90.5	84.6	87.7	94.3	89.5	92.1	99.6	84.2	92.5
04	MASONRY	93.9	93.2	93.5	95.3	90.7	92.4	90.1	77.2	82.2	82.9	91.7	88.3	101.6	93.7	96.7	95.1	86.9	90.1
05	METALS	99.3	76.6	92.4	97.0	79.7	91.7	97.0	78.1	91.2	93.5	82.1	90.0	97.7	86.5	94.3	98.5	78.5	92.4
06	WOOD, PLASTICS & COMPOSITES	96.0	74.0	83.9	94.4	76.4	84.5	103.6	73.0	86.8	90.0	80.9	85.0	98.2	88.7	93.0	108.0	78.9	92.0
07	THERMAL & MOISTURE PROTECTION	103.9	92.0	98.9	98.1	90.7	95.0	106.1	79.3	95.0	110.0	88.4	101.0	98.2	94.0	96.5	105.9	90.4	99.5
08	OPENINGS	104.2	70.7	96.6	93.1	71.7	88.3	95.4	73.5	90.4	93.4	82.0	90.8	95.8	85.5	93.5	103.6	78.7	98.0
0920	Plaster & Gypsum Board	96.6	73.2	81.0	95.9	75.7	82.5	99.7	72.6	81.7	95.0	79.9	84.9	97.9	88.4	91.6	103.0	78.1	86.5
0950, 0980	Ceilings & Acoustic Treatment	94.3	73.2	80.2	93.4	75.7	81.6	98.6	72.6	81.3	90.5	79.9	83.4	93.4	88.4	90.1	93.4	78.1	83.2
0960	Flooring	88.9	92.4	89.9	97.4	92.4	96.0	103.6	85.8	98.4	125.8	83.9	113.7	97.7	87.4	94.7	93.9	81.0	90.2
0970, 0990	Wall Finishes & Painting/Coating	93.4	77.9	84.3	100.1	77.9	87.1	104.0	72.9	85.7	112.5	88.3	98.3	100.2	88.9	93.6	93.4	81.6	86.5
09	FINISHES	94.3	79.4	86.2	95.4	80.1	87.0	100.0	75.8	86.8	112.9	82.9	96.5	96.1	88.5	92.0	96.8	79.9	87.6
COVERS	DIVS. 10 - 14, 25, 28, 41, 43, 44, 46	100.0	92.0	98.2	100.0	88.5	97.4	100.0	87.0	97.1	100.0	87.5	97.2	100.0	95.2	98.9	100.0	92.1	98.2
21, 22, 23	FIRE SUPPRESSION, PLUMBING & HVAC	96.4	85.9	91.8	96.4	86.0	91.8	100.9	81.3	92.2	96.9	89.7	93.7	100.0	94.4	97.5	100.0	82.2	92.1
26, 27, 3370	ELECTRICAL, COMMUNICATIONS & UTIL.	96.2	72.4	83.6	92.7	72.4	81.9	98.4	84.3	90.9	87.6	104.0	96.3	98.5	109.9	104.6	98.6	87.5	92.7
MF2014	WEIGHTED AVERAGE	97.4	83.0	91.0	94.8	82.7	89.4	97.7	80.9	90.3	96.7	90.3	93.8	98.0	94.1	96.3	99.6	84.8	93.1

City Cost Indexes

DIVISION		OHIO ZANESVILLE 437-438 MAT.	INST.	TOTAL	OKLAHOMA ARDMORE 734 MAT.	INST.	TOTAL	CLINTON 736 MAT.	INST.	TOTAL	DURANT 747 MAT.	INST.	TOTAL	ENID 737 MAT.	INST.	TOTAL	GUYMON 739 MAT.	INST.	TOTAL
015433	CONTRACTOR EQUIPMENT		92.8	92.8		83.6	83.6		82.7	82.7		82.7	82.7		82.7	82.7		82.7	82.7
0241, 31 - 34	SITE & INFRASTRUCTURE, DEMOLITION	100.0	96.5	97.6	99.0	98.4	98.6	100.6	96.8	97.9	95.6	94.3	94.7	102.2	96.8	98.4	104.9	96.6	99.1
0310	Concrete Forming & Accessories	92.9	77.4	79.5	94.9	58.7	63.5	93.4	58.7	63.3	85.2	58.3	61.9	97.1	58.8	64.0	100.5	58.6	64.3
0320	Concrete Reinforcing	90.9	93.9	92.4	91.0	65.9	78.3	91.6	65.9	78.5	95.5	66.3	80.7	91.0	65.9	78.2	91.6	63.3	77.2
0330	Cast-in-Place Concrete	88.5	82.3	86.2	98.4	74.1	89.1	95.1	74.1	87.1	92.7	74.0	85.6	95.1	74.2	87.1	95.1	73.9	87.0
03	CONCRETE	90.2	81.9	86.4	91.5	65.1	79.3	90.9	65.1	78.9	89.1	65.0	77.9	91.3	65.2	79.2	94.0	64.5	80.3
04	MASONRY	92.6	80.3	85.0	97.3	57.0	72.5	121.6	57.0	81.8	90.8	65.3	75.1	103.1	57.0	74.7	99.6	56.9	73.3
05	METALS	98.4	84.0	94.0	107.2	61.2	93.1	107.3	61.2	93.2	92.3	61.4	82.9	108.7	61.3	94.2	107.8	59.7	93.1
06	WOOD, PLASTICS & COMPOSITES	90.3	76.4	82.7	104.4	58.8	79.3	103.5	58.8	78.9	92.5	58.8	74.0	106.8	58.8	80.4	110.3	58.8	82.0
07	THERMAL & MOISTURE PROTECTION	98.2	84.7	92.6	106.6	65.6	89.6	106.7	65.6	89.7	100.0	66.9	86.3	106.8	65.6	89.7	107.2	65.6	89.9
08	OPENINGS	93.1	77.9	89.7	103.4	58.0	93.1	103.4	58.0	93.1	97.5	58.0	88.5	104.6	58.0	94.0	103.5	57.4	93.1
0920	Plaster & Gypsum Board	92.6	75.7	81.4	88.9	58.2	68.5	88.5	58.2	68.4	77.9	58.2	64.8	89.2	58.2	68.6	89.2	58.2	68.6
0950, 0980	Ceilings & Acoustic Treatment	93.4	75.7	81.6	83.4	58.2	66.6	83.4	58.2	66.6	78.8	58.2	65.1	83.4	58.2	66.6	83.4	58.2	66.6
0960	Flooring	95.3	78.4	90.4	94.6	60.7	84.8	93.5	60.7	84.0	95.0	54.8	83.3	95.3	75.0	89.4	96.9	60.7	86.4
0970, 0990	Wall Finishes & Painting/Coating	100.1	80.0	88.3	87.0	45.0	62.3	87.0	45.0	62.3	94.9	45.0	65.6	87.0	45.0	62.3	87.0	45.0	62.3
09	FINISHES	94.6	77.3	85.1	84.1	56.7	69.1	83.9	56.7	69.1	84.2	55.5	68.5	84.7	59.7	71.0	85.5	58.1	70.5
COVERS	DIVS. 10 - 14, 25, 28, 41, 43, 44, 46	100.0	84.4	96.5	100.0	78.8	95.2	100.0	78.8	95.2	100.0	78.8	95.2	100.0	78.8	95.2	100.0	78.8	95.2
21, 22, 23	FIRE SUPPRESSION, PLUMBING & HVAC	96.4	87.7	92.6	96.8	66.8	83.5	96.8	66.8	83.5	96.8	66.8	83.5	100.3	66.8	85.5	96.8	65.5	82.9
26, 27, 3370	ELECTRICAL, COMMUNICATIONS & UTIL.	92.8	75.9	83.9	94.4	71.3	82.1	95.4	71.3	82.6	96.7	63.7	79.2	95.4	71.3	82.6	97.1	59.0	76.9
MF2014	WEIGHTED AVERAGE	95.3	82.8	89.8	98.1	66.7	84.2	99.4	66.6	84.9	94.1	66.1	81.7	99.8	67.0	85.3	99.2	64.5	83.9

DIVISION		OKLAHOMA LAWTON 735 MAT.	INST.	TOTAL	MCALESTER 745 MAT.	INST.	TOTAL	MIAMI 743 MAT.	INST.	TOTAL	MUSKOGEE 744 MAT.	INST.	TOTAL	OKLAHOMA CITY 730-731 MAT.	INST.	TOTAL	PONCA CITY 746 MAT.	INST.	TOTAL
015433	CONTRACTOR EQUIPMENT		83.6	83.6		82.7	82.7		94.6	94.6		94.6	94.6		83.9	83.9		82.7	82.7
0241, 31 - 34	SITE & INFRASTRUCTURE, DEMOLITION	98.1	98.4	98.3	88.8	96.6	94.3	90.1	93.7	92.6	90.4	93.7	92.7	96.5	99.0	98.2	96.2	96.8	96.6
0310	Concrete Forming & Accessories	100.5	58.8	64.4	83.2	43.7	49.1	96.7	58.5	63.7	101.2	58.3	64.1	97.3	59.0	64.1	92.0	58.7	63.2
0320	Concrete Reinforcing	91.2	65.9	78.4	95.2	65.8	80.3	93.8	65.9	79.6	94.6	65.5	79.8	100.7	65.9	83.0	94.6	65.9	80.0
0330	Cast-in-Place Concrete	91.9	74.2	85.2	81.3	73.8	78.4	85.2	75.2	81.4	86.2	75.1	82.0	92.0	74.4	85.3	95.2	74.2	87.2
03	CONCRETE	87.7	65.2	77.3	79.9	58.3	69.9	84.4	66.5	76.1	86.1	66.3	76.9	90.8	65.4	79.0	91.3	65.1	79.2
04	MASONRY	99.5	57.0	73.3	109.0	56.9	76.9	93.8	57.1	71.2	111.1	51.4	74.3	101.9	57.4	74.5	86.5	57.0	68.8
05	METALS	112.8	61.3	97.1	92.3	60.4	82.5	92.2	76.9	87.6	93.7	76.1	88.3	105.1	61.3	91.7	92.2	61.2	82.7
06	WOOD, PLASTICS & COMPOSITES	109.4	58.8	81.6	90.2	39.0	62.1	104.7	58.9	79.5	108.9	58.9	81.4	97.7	58.8	76.3	100.4	58.8	77.5
07	THERMAL & MOISTURE PROTECTION	106.6	65.6	89.6	99.6	62.5	84.2	100.0	65.6	85.7	100.3	63.7	85.1	97.4	66.0	84.4	100.2	65.9	85.9
08	OPENINGS	106.4	58.0	95.4	97.4	47.1	86.1	97.4	58.1	88.5	98.7	58.1	89.5	103.1	58.0	92.9	97.4	58.0	88.5
0920	Plaster & Gypsum Board	90.8	58.2	69.1	76.9	37.9	51.0	83.8	58.2	66.8	85.8	58.2	67.4	95.4	58.2	70.7	82.8	58.2	66.4
0950, 0980	Ceilings & Acoustic Treatment	90.1	58.2	68.9	78.8	37.9	51.6	78.8	58.2	65.1	87.3	58.2	67.9	93.4	58.2	70.0	78.8	58.2	65.1
0960	Flooring	97.3	75.0	90.8	93.8	60.7	84.2	101.4	54.8	87.9	103.9	35.2	84.0	94.4	75.0	88.7	98.2	60.7	87.4
0970, 0990	Wall Finishes & Painting/Coating	87.0	45.0	62.3	94.9	45.0	65.6	94.9	45.0	65.6	94.9	45.0	65.6	90.3	45.0	63.7	94.9	45.0	65.6
09	FINISHES	86.3	59.7	71.8	83.1	45.1	62.4	86.2	55.6	69.5	89.1	52.0	68.8	88.2	59.8	72.7	85.9	57.7	70.5
COVERS	DIVS. 10 - 14, 25, 28, 41, 43, 44, 46	100.0	78.8	95.2	100.0	76.7	94.7	100.0	79.3	95.3	100.0	79.3	95.3	100.0	79.0	95.2	100.0	78.8	95.2
21, 22, 23	FIRE SUPPRESSION, PLUMBING & HVAC	100.3	66.8	85.5	96.8	62.2	81.5	96.8	62.3	81.5	100.3	62.3	83.5	100.1	67.0	85.5	96.8	62.3	81.5
26, 27, 3370	ELECTRICAL, COMMUNICATIONS & UTIL.	97.2	71.3	83.4	95.1	66.2	79.8	96.5	66.3	80.5	94.6	72.8	83.0	102.5	71.3	85.9	94.6	69.4	81.2
MF2014	WEIGHTED AVERAGE	100.2	67.1	85.6	93.4	61.6	79.3	93.8	66.2	81.6	96.1	66.0	82.8	99.3	67.3	85.1	94.2	65.5	81.5

DIVISION		OKLAHOMA POTEAU 749 MAT.	INST.	TOTAL	SHAWNEE 748 MAT.	INST.	TOTAL	TULSA 740-741 MAT.	INST.	TOTAL	WOODWARD 738 MAT.	INST.	TOTAL	OREGON BEND 977 MAT.	INST.	TOTAL	EUGENE 974 MAT.	INST.	TOTAL
015433	CONTRACTOR EQUIPMENT		93.3	93.3		82.7	82.7		94.6	94.6		82.7	82.7		100.3	100.3		100.3	100.3
0241, 31 - 34	SITE & INFRASTRUCTURE, DEMOLITION	77.1	89.0	85.4	99.3	96.8	97.6	96.7	92.7	93.9	100.9	96.8	98.0	107.9	103.4	104.7	97.8	103.4	101.7
0310	Concrete Forming & Accessories	89.9	58.2	62.5	85.1	58.6	62.2	101.3	58.4	64.2	93.5	46.0	52.4	110.8	94.9	97.0	107.1	95.0	96.6
0320	Concrete Reinforcing	95.7	65.8	80.5	94.6	63.3	78.7	94.9	65.9	80.1	91.0	65.8	78.2	94.8	103.7	99.3	99.0	103.7	101.4
0330	Cast-in-Place Concrete	85.2	75.1	81.3	98.3	74.1	89.1	93.9	75.1	86.7	95.1	74.1	87.1	105.7	99.5	103.3	102.3	99.5	101.3
03	CONCRETE	86.6	66.3	77.2	93.0	64.7	79.8	91.4	66.4	79.8	91.1	59.4	76.4	106.5	97.6	102.4	98.0	97.7	97.9
04	MASONRY	94.2	57.1	71.3	110.4	57.0	77.5	94.7	57.1	71.5	92.5	57.0	70.6	105.8	96.4	100.0	102.7	96.4	98.8
05	METALS	92.3	76.6	87.5	92.2	60.2	82.4	96.6	76.8	90.6	107.4	61.1	93.2	97.6	92.5	96.0	98.3	92.7	96.6
06	WOOD, PLASTICS & COMPOSITES	97.1	58.9	76.1	92.4	58.8	73.9	108.1	58.9	81.1	103.6	41.7	69.6	102.2	94.5	98.0	98.1	94.5	96.1
07	THERMAL & MOISTURE PROTECTION	100.1	65.6	85.8	100.2	65.4	85.7	100.3	65.7	86.0	106.8	64.0	89.0	110.5	95.6	104.3	109.6	99.3	105.3
08	OPENINGS	97.4	58.1	88.5	97.4	57.4	88.4	100.4	58.1	90.8	103.4	48.6	91.0	101.4	96.5	100.2	101.6	96.5	100.5
0920	Plaster & Gypsum Board	80.9	58.2	65.8	77.9	58.2	64.8	85.8	58.2	67.4	88.5	40.6	56.7	115.2	94.2	101.2	113.2	94.2	100.6
0950, 0980	Ceilings & Acoustic Treatment	78.8	58.2	65.1	78.8	58.2	65.1	87.3	58.2	67.9	83.4	40.6	54.9	88.1	94.2	92.2	89.1	94.2	92.5
0960	Flooring	97.5	54.8	85.1	95.0	60.7	85.1	102.8	51.2	87.8	93.5	57.9	83.2	104.2	99.3	102.8	102.4	99.3	101.5
0970, 0990	Wall Finishes & Painting/Coating	94.9	45.0	65.6	94.9	45.0	65.6	94.9	45.0	65.6	87.0	45.0	62.3	94.2	72.2	81.3	94.2	72.2	81.3
09	FINISHES	83.9	55.6	68.4	84.4	56.7	69.3	88.9	54.9	70.3	84.0	46.2	63.3	98.9	93.0	95.7	97.2	93.0	94.9
COVERS	DIVS. 10 - 14, 25, 28, 41, 43, 44, 46	100.0	79.1	95.3	100.0	78.8	95.2	100.0	79.2	95.3	100.0	77.0	94.8	100.0	98.9	99.7	100.0	98.9	99.7
21, 22, 23	FIRE SUPPRESSION, PLUMBING & HVAC	96.8	62.3	81.5	96.8	66.8	83.5	100.3	64.1	84.3	96.8	66.8	83.5	96.5	101.9	98.9	100.1	101.9	100.9
26, 27, 3370	ELECTRICAL, COMMUNICATIONS & UTIL.	94.7	66.3	79.7	96.7	71.3	83.2	96.7	66.3	80.6	97.0	71.3	83.3	100.6	95.0	97.6	99.2	95.0	97.0
MF2014	WEIGHTED AVERAGE	93.3	65.8	81.2	95.7	66.4	82.7	97.0	66.5	83.5	98.1	63.8	82.9	100.6	97.3	99.2	99.7	97.5	98.7

For customer support on your Building Construction Costs with RSMeans Data, call 800.448.8182.

City Cost Indexes

		OREGON																
	DIVISION	KLAMATH FALLS			MEDFORD			PENDLETON			PORTLAND			SALEM			VALE		
		976			975			978			970 - 972			973			979		
		MAT.	INST.	TOTAL	MAT.	INST.	TOTAL	MAT.	INST.	TOTAL	MAT.	INST.	TOTAL	MAT.	INST.	TOTAL	MAT.	INST.	TOTAL
015433	CONTRACTOR EQUIPMENT		100.3	100.3		100.3	100.3		97.7	97.7		100.3	100.3		100.3	100.3		97.7	97.7
0241, 31 - 34	SITE & INFRASTRUCTURE, DEMOLITION	112.0	103.3	106.0	105.6	103.3	104.0	104.5	96.2	98.7	100.3	103.4	102.4	92.1	103.4	99.9	91.5	96.1	94.7
0310	Concrete Forming & Accessories	103.4	94.6	95.8	102.4	94.6	95.6	103.9	95.1	96.3	108.3	94.9	96.7	107.0	94.9	96.5	110.8	93.9	96.1
0320	Concrete Reinforcing	94.8	103.6	99.3	96.4	103.6	100.1	93.9	103.7	98.9	99.7	103.7	101.7	107.2	103.7	105.4	91.6	103.5	97.7
0330	Cast-in-Place Concrete	105.7	101.9	104.2	105.7	99.4	103.3	106.5	100.3	104.1	105.2	99.5	103.0	96.6	99.5	97.7	83.8	102.9	91.1
03	CONCRETE	109.3	98.3	104.2	104.0	97.5	101.0	90.7	98.1	94.1	99.5	97.7	98.7	95.4	97.6	96.5	75.7	98.3	86.2
04	MASONRY	120.1	96.4	105.5	99.7	96.4	97.7	110.3	96.5	101.8	104.5	96.4	99.5	109.1	96.4	101.3	108.4	96.5	101.1
05	METALS	97.6	92.2	96.0	97.9	92.2	96.1	104.6	93.2	101.1	99.4	92.6	97.3	105.3	92.5	101.4	104.5	92.0	100.7
06	WOOD, PLASTICS & COMPOSITES	93.2	94.5	93.9	92.1	94.5	93.5	94.9	94.6	94.8	99.1	94.5	96.6	93.6	94.5	94.1	103.2	94.6	98.5
07	THERMAL & MOISTURE PROTECTION	110.7	91.8	102.9	110.3	91.8	102.7	103.2	95.3	99.9	109.5	95.6	103.8	106.0	95.6	101.7	102.5	92.4	98.3
08	OPENINGS	101.4	96.5	100.3	104.5	96.5	102.7	97.9	96.5	97.6	99.3	96.5	98.7	101.3	96.5	100.2	97.8	87.6	95.5
0920	Plaster & Gypsum Board	110.0	94.2	99.5	109.4	94.2	99.3	96.7	94.2	95.0	113.1	94.2	100.5	109.1	94.2	99.2	102.9	94.2	97.1
0950, 0980	Ceilings & Acoustic Treatment	95.7	94.2	94.7	102.4	94.2	96.9	63.5	94.2	83.9	91.0	94.2	93.1	96.1	94.2	94.8	63.5	94.2	83.9
0960	Flooring	100.9	99.3	100.4	100.2	99.3	100.0	70.1	99.3	78.6	100.1	99.3	99.9	103.8	99.3	102.5	72.4	99.3	80.2
0970, 0990	Wall Finishes & Painting/Coating	94.2	67.0	78.2	94.2	67.0	78.2	86.5	74.1	79.2	94.0	74.1	82.3	93.8	72.2	81.1	86.5	72.2	78.1
09	FINISHES	99.2	92.4	95.5	99.5	92.4	95.6	70.2	93.3	82.8	96.9	93.2	94.9	95.9	93.0	94.3	70.8	93.1	83.0
COVERS	DIVS. 10 - 14, 25, 28, 41, 43, 44, 46	100.0	98.8	99.7	100.0	98.8	99.7	100.0	101.4	100.3	100.0	98.9	99.8	100.0	98.9	99.7	100.0	99.1	99.8
21, 22, 23	FIRE SUPPRESSION, PLUMBING & HVAC	96.5	101.9	98.9	100.1	101.9	100.9	98.6	108.5	103.0	100.1	101.8	100.8	100.1	101.9	100.9	98.6	69.6	85.8
26, 27, 3370	ELECTRICAL, COMMUNICATIONS & UTIL.	99.2	78.0	87.9	102.8	78.0	89.6	91.5	96.2	94.0	99.3	106.2	103.0	106.5	95.0	100.4	91.5	68.7	79.4
MF2014	WEIGHTED AVERAGE	101.6	94.8	98.6	101.3	94.7	98.4	96.5	98.6	97.4	100.0	98.9	99.5	101.2	97.3	99.5	94.3	85.8	90.5

		PENNSYLVANIA																	
	DIVISION	ALLENTOWN			ALTOONA			BEDFORD			BRADFORD			BUTLER			CHAMBERSBURG		
		181			166			155			167			160			172		
		MAT.	INST.	TOTAL	MAT.	INST.	TOTAL	MAT.	INST.	TOTAL	MAT.	INST.	TOTAL	MAT.	INST.	TOTAL	MAT.	INST.	TOTAL
015433	CONTRACTOR EQUIPMENT		117.3	117.3		117.3	117.3		114.8	114.8		117.3	117.3		117.3	117.3		116.4	116.4
0241, 31 - 34	SITE & INFRASTRUCTURE, DEMOLITION	93.6	103.9	100.8	96.5	103.3	101.2	106.2	100.6	102.3	92.7	102.2	99.3	87.8	104.0	99.1	89.3	100.6	97.2
0310	Concrete Forming & Accessories	101.3	113.6	111.9	85.8	85.4	85.5	83.5	82.0	82.2	88.1	103.8	101.7	87.3	91.1	90.6	88.9	82.2	83.1
0320	Concrete Reinforcing	101.1	114.3	107.8	97.9	105.2	101.6	99.7	103.0	101.4	100.0	105.0	102.6	98.6	118.7	108.8	93.7	113.1	103.5
0330	Cast-in-Place Concrete	89.8	107.2	96.5	100.1	92.8	97.3	115.6	87.6	104.9	95.6	96.9	96.1	88.4	101.1	93.2	96.6	98.3	97.2
03	CONCRETE	93.4	112.4	102.2	89.4	92.9	91.1	106.4	89.2	98.4	95.4	102.6	98.8	81.3	100.8	90.3	99.4	94.8	97.3
04	MASONRY	95.7	101.6	99.3	99.2	89.9	93.5	110.5	86.5	95.7	96.5	84.8	89.3	101.3	98.2	99.4	102.4	85.0	91.7
05	METALS	95.4	124.0	104.1	89.6	117.0	98.0	98.3	115.0	103.4	93.3	116.9	100.5	89.3	123.5	99.8	91.7	119.3	100.1
06	WOOD, PLASTICS & COMPOSITES	101.7	115.1	109.1	81.0	83.8	82.5	86.3	81.6	83.7	87.3	109.5	99.5	82.3	88.3	85.6	88.5	81.7	84.8
07	THERMAL & MOISTURE PROTECTION	102.7	114.6	107.7	101.6	94.3	98.6	100.4	90.4	96.2	102.6	93.4	98.8	101.2	97.0	99.4	94.8	88.1	92.0
08	OPENINGS	94.8	111.4	98.6	88.5	87.1	88.2	96.5	85.8	94.1	94.8	101.1	96.2	88.5	97.1	90.4	91.6	86.6	90.5
0920	Plaster & Gypsum Board	96.8	115.4	109.2	88.7	83.2	85.1	96.9	81.1	86.4	89.8	109.6	103.0	88.7	87.9	88.2	104.3	81.1	88.8
0950, 0980	Ceilings & Acoustic Treatment	83.8	115.4	104.8	87.2	83.2	84.6	95.0	81.1	85.7	87.3	109.6	102.1	88.1	87.9	88.0	85.9	81.1	82.7
0960	Flooring	93.5	95.9	94.2	87.0	104.9	92.2	91.3	82.6	88.8	87.6	104.9	92.6	87.9	82.6	86.4	93.6	82.6	90.4
0970, 0990	Wall Finishes & Painting/Coating	93.3	108.5	102.2	88.8	110.5	101.5	96.8	113.1	106.4	93.3	107.3	101.5	88.8	113.1	103.1	91.8	107.3	100.9
09	FINISHES	90.6	110.4	101.4	88.5	91.0	89.9	97.1	85.2	90.6	88.7	105.3	97.7	88.5	91.2	90.0	89.2	84.6	86.7
COVERS	DIVS. 10 - 14, 25, 28, 41, 43, 44, 46	100.0	104.9	101.1	100.0	98.4	99.6	100.0	96.8	99.3	100.0	100.7	100.2	100.0	100.5	100.1	100.0	95.2	98.9
21, 22, 23	FIRE SUPPRESSION, PLUMBING & HVAC	100.3	114.8	106.7	99.7	85.8	93.5	96.4	82.4	90.2	96.7	95.4	96.1	96.2	98.0	97.0	96.7	89.0	93.3
26, 27, 3370	ELECTRICAL, COMMUNICATIONS & UTIL.	98.5	100.6	99.6	89.0	113.5	102.0	95.2	113.5	104.9	92.2	113.5	103.5	89.6	112.0	101.5	88.6	88.6	88.6
MF2014	WEIGHTED AVERAGE	96.7	110.2	102.6	93.3	96.6	94.8	99.2	93.7	96.8	94.9	102.1	98.1	91.4	102.2	96.2	94.5	92.5	93.6

		PENNSYLVANIA																	
	DIVISION	DOYLESTOWN			DUBOIS			ERIE			GREENSBURG			HARRISBURG			HAZLETON		
		189			158			164 - 165			156			170 - 171			182		
		MAT.	INST.	TOTAL	MAT.	INST.	TOTAL	MAT.	INST.	TOTAL	MAT.	INST.	TOTAL	MAT.	INST.	TOTAL	MAT.	INST.	TOTAL
015433	CONTRACTOR EQUIPMENT		94.5	94.5		114.8	114.8		117.3	117.3		114.8	114.8		116.4	116.4		117.3	117.3
0241, 31 - 34	SITE & INFRASTRUCTURE, DEMOLITION	107.3	89.9	95.1	111.0	100.7	103.8	93.4	103.2	100.3	102.2	102.6	102.5	89.4	101.8	98.1	87.0	102.7	98.0
0310	Concrete Forming & Accessories	84.9	129.8	123.7	83.0	82.9	82.9	100.7	89.3	90.8	89.9	95.2	94.5	99.2	89.9	91.2	82.3	89.1	88.2
0320	Concrete Reinforcing	97.8	145.9	122.2	99.1	105.1	102.1	100.0	106.3	103.2	99.1	118.5	108.9	102.7	114.2	108.5	98.1	114.1	106.2
0330	Cast-in-Place Concrete	84.8	133.5	103.4	111.5	96.1	105.6	98.4	94.5	96.9	107.3	100.6	104.7	95.1	101.8	97.6	84.8	100.8	90.9
03	CONCRETE	89.0	133.0	109.4	107.9	92.8	100.9	94.5	95.5	95.0	101.1	102.4	101.7	94.4	99.8	96.9	86.2	99.0	92.1
04	MASONRY	99.1	124.4	114.7	111.0	86.9	96.2	88.4	93.5	91.6	121.7	96.0	105.9	96.4	89.3	92.1	108.8	94.6	100.0
05	METALS	93.0	121.1	101.6	98.3	115.2	103.5	89.8	116.9	98.1	98.2	121.7	105.4	98.0	122.9	105.6	95.1	121.4	103.1
06	WOOD, PLASTICS & COMPOSITES	83.1	130.6	109.2	85.3	81.6	83.3	97.8	87.1	91.9	92.2	94.1	93.3	96.0	89.1	92.2	81.7	86.5	84.3
07	THERMAL & MOISTURE PROTECTION	100.1	129.2	112.2	100.7	92.3	97.2	102.1	94.0	98.8	100.3	97.0	98.9	97.2	106.0	100.9	102.1	103.2	102.6
08	OPENINGS	97.2	136.0	106.0	96.5	85.8	94.1	88.7	91.0	89.2	96.5	100.3	97.3	95.9	90.9	94.8	95.4	95.2	95.3
0920	Plaster & Gypsum Board	87.5	131.3	116.6	95.9	81.1	86.0	96.8	86.6	90.0	98.0	93.9	95.3	108.7	88.7	95.4	87.9	86.0	86.6
0950, 0980	Ceilings & Acoustic Treatment	83.1	131.3	115.2	95.0	81.1	85.7	83.8	86.6	85.7	94.2	93.9	94.0	95.4	88.7	90.9	84.8	86.0	85.6
0960	Flooring	77.7	136.6	94.7	91.1	104.9	95.1	93.9	104.9	97.1	95.1	82.6	91.5	96.2	90.7	94.6	84.8	94.7	87.6
0970, 0990	Wall Finishes & Painting/Coating	92.9	148.7	125.6	96.8	113.1	106.4	98.7	93.6	95.7	96.8	113.1	106.4	93.8	90.3	91.8	93.3	110.9	103.6
09	FINISHES	82.0	133.0	109.9	97.4	89.5	93.0	91.6	92.1	91.9	97.9	94.6	96.1	92.9	89.5	91.0	86.6	92.0	89.6
COVERS	DIVS. 10 - 14, 25, 28, 41, 43, 44, 46	100.0	116.4	103.7	100.0	97.3	99.4	100.0	100.1	100.0	100.0	101.0	100.2	100.0	97.4	99.4	100.0	99.7	99.9
21, 22, 23	FIRE SUPPRESSION, PLUMBING & HVAC	96.2	129.3	110.8	96.4	83.1	90.6	99.7	101.5	100.5	96.4	91.1	94.1	100.0	93.0	96.9	96.7	96.7	96.7
26, 27, 3370	ELECTRICAL, COMMUNICATIONS & UTIL.	91.7	130.3	112.2	95.8	113.5	105.2	90.7	94.8	92.9	95.8	113.5	105.2	95.5	91.1	93.2	93.1	94.1	93.6
MF2014	WEIGHTED AVERAGE	94.0	126.0	108.2	99.6	95.0	97.6	93.2	98.5	95.5	99.2	101.3	100.1	96.9	96.7	96.8	94.5	98.7	96.3

City Cost Indexes

PENNSYLVANIA

DIVISION		INDIANA 157 MAT.	INST.	TOTAL	JOHNSTOWN 159 MAT.	INST.	TOTAL	KITTANNING 162 MAT.	INST.	TOTAL	LANCASTER 175-176 MAT.	INST.	TOTAL	LEHIGH VALLEY 180 MAT.	INST.	TOTAL	MONTROSE 188 MAT.	INST.	TOTAL
015433	CONTRACTOR EQUIPMENT		114.8	114.8		114.8	114.8		117.3	117.3		116.4	116.4		117.3	117.3		117.3	117.3
0241, 31-34	SITE & INFRASTRUCTURE, DEMOLITION	100.2	101.5	101.1	106.7	101.9	103.4	90.4	103.7	99.7	81.2	101.8	95.6	90.9	104.0	100.0	89.6	103.3	99.1
0310	Concrete Forming & Accessories	84.1	92.2	91.1	83.0	85.3	85.0	87.3	96.4	95.2	91.1	90.2	90.3	94.6	131.0	126.1	83.4	93.0	91.7
0320	Concrete Reinforcing	98.3	118.5	108.6	99.7	118.3	109.2	98.6	118.6	108.8	93.3	114.2	103.9	98.1	112.5	105.4	102.8	114.2	108.6
0330	Cast-in-Place Concrete	105.2	99.0	102.8	116.7	92.5	107.5	91.9	99.6	94.8	82.2	101.8	89.7	91.9	111.0	99.1	90.1	99.3	93.6
03	CONCRETE	98.5	100.4	99.4	107.3	95.0	101.6	83.9	102.6	92.6	87.3	99.9	93.1	92.6	121.0	105.8	91.2	100.2	95.4
04	MASONRY	107.0	95.5	99.9	107.9	90.2	97.0	104.0	95.4	98.7	108.5	89.8	97.0	95.7	113.5	106.7	95.7	98.6	97.5
05	METALS	98.3	120.8	105.2	98.3	119.9	104.9	99.4	122.9	99.6	91.7	123.0	101.3	95.0	121.7	103.2	93.4	121.7	102.0
06	WOOD, PLASTICS & COMPOSITES	87.0	91.5	89.5	85.3	83.7	84.4	82.3	96.6	90.1	91.2	89.1	90.1	93.6	134.4	116.0	82.5	90.1	86.7
07	THERMAL & MOISTURE PROTECTION	100.1	96.6	98.7	100.4	92.6	97.2	101.3	97.2	99.6	94.2	106.3	99.2	102.5	116.6	108.3	102.1	92.8	98.3
08	OPENINGS	96.5	94.4	96.0	96.5	90.2	95.1	88.5	101.6	91.5	91.6	90.9	91.4	95.3	121.6	101.3	92.0	94.8	92.6
0920	Plaster & Gypsum Board	97.2	91.3	93.3	95.7	83.2	87.4	88.7	96.3	93.8	106.3	88.7	94.6	90.6	135.2	120.2	88.5	89.7	89.3
0950, 0980	Ceilings & Acoustic Treatment	95.0	91.3	92.5	94.2	83.2	86.9	88.1	96.3	93.6	85.9	88.7	87.8	84.8	135.2	118.3	87.3	89.7	88.9
0960	Flooring	92.0	104.9	95.8	91.1	82.6	88.6	87.9	104.9	92.8	94.6	101.4	96.6	90.3	99.7	93.1	85.5	104.9	91.1
0970, 0990	Wall Finishes & Painting/Coating	96.8	113.1	106.4	96.8	110.5	104.9	88.8	113.1	103.1	91.8	90.3	90.9	93.3	63.7	75.9	93.3	110.9	103.6
09	FINISHES	96.9	96.4	96.7	96.7	87.1	91.5	88.6	99.4	94.5	89.2	91.5	90.4	88.9	119.5	105.6	87.6	96.4	92.4
COVERS	DIVS. 10 - 14, 25, 28, 41, 43, 44, 46	100.0	100.0	100.0	100.0	98.3	99.6	100.0	100.7	100.2	100.0	97.6	99.5	100.0	106.1	101.4	100.0	101.4	100.3
21, 22, 23	FIRE SUPPRESSION, PLUMBING & HVAC	96.4	83.0	90.5	96.4	81.8	89.9	96.2	94.2	95.3	96.7	93.2	95.2	96.7	114.2	104.4	96.7	99.6	98.0
26, 27, 3370	ELECTRICAL, COMMUNICATIONS & UTIL.	95.8	113.5	105.2	95.8	113.5	105.2	89.0	113.5	102.0	90.0	97.2	93.8	93.1	140.2	118.1	92.2	98.7	95.7
MF2014	WEIGHTED AVERAGE	98.0	99.0	98.4	99.2	95.8	97.7	91.9	102.9	96.8	93.2	97.9	95.3	95.0	119.6	105.9	93.8	101.0	97.0

PENNSYLVANIA

DIVISION		NEW CASTLE 161 MAT.	INST.	TOTAL	NORRISTOWN 194 MAT.	INST.	TOTAL	OIL CITY 163 MAT.	INST.	TOTAL	PHILADELPHIA 190-191 MAT.	INST.	TOTAL	PITTSBURGH 150-152 MAT.	INST.	TOTAL	POTTSVILLE 179 MAT.	INST.	TOTAL
015433	CONTRACTOR EQUIPMENT		117.3	117.3		100.8	100.8		117.3	117.3		96.8	96.8		106.0	106.0		116.4	116.4
0241, 31-34	SITE & INFRASTRUCTURE, DEMOLITION	88.2	104.2	99.3	97.3	100.8	99.8	86.8	102.4	97.7	103.0	97.9	99.4	105.8	100.1	101.8	84.2	101.7	96.4
0310	Concrete Forming & Accessories	87.3	99.2	97.6	86.1	130.2	124.2	87.3	95.9	94.8	103.2	138.9	134.1	97.4	99.6	99.3	82.3	91.3	90.1
0320	Concrete Reinforcing	97.4	101.1	99.3	96.8	145.9	121.7	98.6	98.7	98.6	105.3	146.4	126.2	100.3	118.8	109.7	92.6	113.2	103.1
0330	Cast-in-Place Concrete	89.2	101.1	93.7	83.7	133.7	102.8	86.7	99.2	91.5	93.5	137.2	110.1	111.5	101.2	107.6	87.6	102.8	93.4
03	CONCRETE	81.7	101.4	90.8	87.9	133.3	109.0	80.1	98.8	88.8	99.3	138.4	117.4	104.5	103.9	104.2	90.9	100.5	95.4
04	MASONRY	100.9	98.2	99.3	111.4	124.9	119.7	100.3	91.4	94.8	99.7	135.3	121.6	101.7	100.6	101.0	101.7	94.3	97.1
05	METALS	89.4	117.4	98.0	95.5	122.4	103.7	89.4	115.5	97.4	97.9	121.4	105.1	99.7	112.8	103.7	91.9	121.2	100.9
06	WOOD, PLASTICS & COMPOSITES	82.3	99.4	91.7	81.7	130.5	108.5	82.3	96.6	90.1	98.4	139.7	121.1	101.0	99.5	100.1	81.2	88.3	85.1
07	THERMAL & MOISTURE PROTECTION	101.2	108.1	104.0	105.8	129.5	115.6	101.1	95.1	98.6	102.5	137.3	116.9	100.5	99.2	100.0	94.4	105.6	99.0
08	OPENINGS	88.5	95.7	90.1	87.4	136.0	98.4	88.5	97.5	90.6	101.1	141.0	110.1	100.3	103.2	100.9	91.6	96.1	92.7
0920	Plaster & Gypsum Board	88.7	99.3	95.7	86.5	131.3	116.3	88.7	96.3	93.8	100.1	140.6	127.0	102.3	99.3	100.3	101.0	87.9	92.3
0950, 0980	Ceilings & Acoustic Treatment	88.1	99.3	95.6	87.4	131.3	116.6	88.1	96.3	93.6	99.8	140.6	126.9	95.0	99.3	97.9	85.9	87.9	87.2
0960	Flooring	87.9	82.6	86.4	87.0	136.6	101.4	87.9	106.4	93.3	101.6	147.6	115.0	99.0	106.4	101.1	90.4	101.4	93.6
0970, 0990	Wall Finishes & Painting/Coating	88.8	113.1	103.1	90.5	148.7	124.6	88.8	113.1	103.1	96.7	148.7	127.2	96.8	113.1	106.4	91.8	110.9	103.0
09	FINISHES	88.5	97.7	93.5	86.4	132.9	111.8	88.4	99.7	94.6	100.8	141.8	123.2	100.0	101.9	101.0	87.4	94.4	91.2
COVERS	DIVS. 10 - 14, 25, 28, 41, 43, 44, 46	100.0	101.7	100.4	100.0	116.6	103.7	100.0	100.7	100.2	100.0	119.5	104.4	100.0	101.8	100.4	100.0	98.1	99.6
21, 22, 23	FIRE SUPPRESSION, PLUMBING & HVAC	96.2	101.9	98.7	96.4	128.3	110.5	96.2	97.4	96.7	100.2	134.0	115.1	100.0	102.0	100.9	96.7	98.6	97.5
26, 27, 3370	ELECTRICAL, COMMUNICATIONS & UTIL.	89.6	103.4	96.9	93.2	153.9	125.4	91.4	112.0	102.3	97.2	153.9	127.3	98.3	113.5	106.4	88.2	93.4	91.0
MF2014	WEIGHTED AVERAGE	91.5	102.6	96.4	94.3	130.1	110.1	91.4	101.4	95.8	99.6	134.5	115.0	100.6	104.5	102.3	93.0	99.5	95.9

PENNSYLVANIA

DIVISION		READING 195-196 MAT.	INST.	TOTAL	SCRANTON 184-185 MAT.	INST.	TOTAL	STATE COLLEGE 168 MAT.	INST.	TOTAL	STROUDSBURG 183 MAT.	INST.	TOTAL	SUNBURY 178 MAT.	INST.	TOTAL	UNIONTOWN 154 MAT.	INST.	TOTAL
015433	CONTRACTOR EQUIPMENT		124.1	124.1		117.3	117.3		116.4	116.4		117.3	117.3		117.3	117.3		114.8	114.8
0241, 31-34	SITE & INFRASTRUCTURE, DEMOLITION	101.7	115.2	111.1	94.1	103.3	100.5	84.5	101.8	96.6	88.6	103.9	99.3	96.6	102.7	100.9	100.8	102.6	102.1
0310	Concrete Forming & Accessories	102.9	92.6	94.0	101.5	93.1	94.2	86.3	85.5	85.6	88.9	98.0	96.7	94.6	88.4	89.2	77.3	95.1	92.7
0320	Concrete Reinforcing	98.1	144.5	121.7	101.1	111.3	106.2	99.3	105.3	102.3	101.4	114.3	107.9	95.2	114.1	104.8	99.1	118.5	108.9
0330	Cast-in-Place Concrete	75.0	102.3	85.4	93.8	99.5	95.9	90.4	92.8	91.4	88.4	106.1	95.1	95.7	100.1	97.4	105.2	100.6	103.4
03	CONCRETE	86.9	106.5	96.0	95.4	100.0	94.3	95.5	93.2	94.3	89.9	104.8	96.8	94.7	98.5	96.4	98.2	102.3	100.1
04	MASONRY	100.6	96.1	97.8	96.1	98.9	97.8	101.6	90.4	94.7	93.7	113.2	105.7	102.2	87.2	93.0	123.1	96.0	106.4
05	METALS	95.8	134.8	107.8	97.3	122.7	105.1	93.1	117.1	100.4	95.1	121.9	103.3	91.7	121.9	100.9	98.1	121.6	105.3
06	WOOD, PLASTICS & COMPOSITES	99.2	89.1	93.6	101.7	90.1	95.3	89.1	83.8	86.2	88.2	90.1	89.2	89.2	88.3	88.7	79.9	94.1	87.7
07	THERMAL & MOISTURE PROTECTION	106.0	108.9	107.2	102.6	96.8	100.2	101.8	103.9	102.7	102.3	101.5	101.9	95.4	98.5	96.7	100.0	97.0	98.8
08	OPENINGS	91.8	104.4	94.7	94.8	95.6	95.0	91.8	87.1	90.7	95.4	97.7	95.9	91.7	93.9	92.2	96.4	100.2	97.3
0920	Plaster & Gypsum Board	96.7	88.7	91.4	98.8	89.7	92.7	90.4	83.2	85.6	89.2	89.7	89.5	100.2	87.9	92.0	94.1	93.9	93.9
0950, 0980	Ceilings & Acoustic Treatment	80.5	88.7	86.0	92.3	89.7	90.5	83.9	83.2	83.5	83.1	89.7	87.5	82.5	87.9	86.1	94.2	93.9	94.0
0960	Flooring	91.8	82.6	89.1	93.5	94.7	93.8	90.9	82.6	88.5	88.2	99.7	91.5	91.3	104.9	95.2	88.1	82.6	86.5
0970, 0990	Wall Finishes & Painting/Coating	89.2	108.5	100.5	93.3	110.9	103.6	93.3	110.5	103.4	93.3	110.9	103.4	91.8	110.9	103.0	96.8	111.0	105.1
09	FINISHES	88.1	91.3	89.8	92.6	94.7	93.7	88.0	87.1	87.5	87.6	98.7	93.6	88.3	93.3	91.0	95.2	94.3	94.7
COVERS	DIVS. 10 - 14, 25, 28, 41, 43, 44, 46	100.0	100.3	100.1	100.0	101.5	100.3	100.0	96.7	99.3	100.0	105.7	101.3	100.0	98.1	99.6	100.0	101.0	100.2
21, 22, 23	FIRE SUPPRESSION, PLUMBING & HVAC	100.1	111.0	105.0	100.3	99.8	100.1	96.7	92.6	94.9	96.7	105.8	100.8	96.7	90.5	93.9	96.4	85.7	91.7
26, 27, 3370	ELECTRICAL, COMMUNICATIONS & UTIL.	99.6	97.2	98.3	98.5	98.7	98.6	91.4	113.5	103.1	93.1	146.7	121.5	88.5	95.6	92.3	92.7	113.5	103.7
MF2014	WEIGHTED AVERAGE	96.0	106.1	100.5	97.4	101.0	99.0	94.5	97.8	95.9	94.4	111.9	102.1	94.0	96.8	95.2	98.2	100.1	99.0

PENNSYLVANIA

DIVISION		WASHINGTON 153 MAT.	INST.	TOTAL	WELLSBORO 169 MAT.	INST.	TOTAL	WESTCHESTER 193 MAT.	INST.	TOTAL	WILKES-BARRE 186-187 MAT.	INST.	TOTAL	WILLIAMSPORT 177 MAT.	INST.	TOTAL	YORK 173-174 MAT.	INST.	TOTAL
015433	CONTRACTOR EQUIPMENT		114.8	114.8		117.3	117.3		100.8	100.8		117.3	117.3		117.3	117.3		116.4	116.4
0241, 31 - 34	SITE & INFRASTRUCTURE, DEMOLITION	100.9	102.8	102.2	96.2	102.7	100.7	103.4	101.7	102.2	86.6	103.2	98.2	87.4	103.2	98.4	85.0	101.8	96.7
0310	Concrete Forming & Accessories	84.2	97.1	95.4	87.5	88.2	88.1	93.2	130.6	125.6	91.7	91.7	91.7	91.1	88.7	89.0	85.8	88.1	87.8
0320	Concrete Reinforcing	99.1	118.8	109.1	99.3	113.2	106.3	95.8	145.0	120.8	100.0	114.1	107.2	94.5	114.2	104.5	95.2	113.2	104.3
0330	Cast-in-Place Concrete	105.2	100.7	103.5	94.8	94.0	94.5	92.7	134.8	108.7	84.8	99.3	90.3	81.0	95.4	86.5	88.0	101.7	93.2
03	CONCRETE	98.6	103.3	100.8	97.8	96.1	97.0	95.7	133.7	113.3	87.0	99.7	92.9	82.4	97.0	89.2	92.0	98.8	95.2
04	MASONRY	107.4	100.6	103.3	102.9	87.7	93.5	106.1	127.1	119.0	109.1	98.6	102.6	93.6	91.0	92.0	103.3	89.8	95.0
05	METALS	98.0	122.2	105.4	93.2	121.6	101.9	95.5	121.7	103.5	93.3	122.7	102.3	91.7	122.6	101.1	93.3	122.9	102.3
06	WOOD, PLASTICS & COMPOSITES	87.0	96.5	92.2	86.8	88.3	87.6	88.4	130.5	111.6	90.5	88.3	89.3	85.8	88.3	87.2	84.6	86.5	85.6
07	THERMAL & MOISTURE PROTECTION	100.1	98.9	99.6	102.8	91.0	97.9	106.2	126.0	114.4	102.1	96.5	99.7	94.8	93.0	94.1	94.4	106.0	99.2
08	OPENINGS	96.4	101.6	97.6	94.8	93.9	94.6	87.4	135.8	98.3	92.0	94.7	92.6	91.7	93.9	92.2	91.6	89.4	91.1
0920	Plaster & Gypsum Board	97.0	96.3	96.6	89.0	87.9	88.3	87.5	131.3	116.6	90.1	87.9	88.6	101.0	87.9	92.3	101.3	86.0	91.1
0950, 0980	Ceilings & Acoustic Treatment	94.2	96.3	95.6	83.9	87.9	86.5	87.4	131.3	116.6	87.3	87.9	87.7	85.9	87.9	87.2	85.0	86.0	85.6
0960	Flooring	92.1	82.6	89.3	87.2	101.4	91.3	90.0	136.6	103.5	89.1	94.7	90.7	90.1	92.0	90.6	91.8	101.4	94.6
0970, 0990	Wall Finishes & Painting/Coating	96.8	111.0	105.1	93.3	110.9	103.6	90.5	144.8	122.4	93.3	108.5	102.2	91.8	108.5	101.6	91.8	90.3	90.9
09	FINISHES	96.8	95.7	96.2	88.1	93.1	90.8	87.9	133.4	112.7	88.7	93.3	91.2	88.0	90.5	89.3	87.7	89.9	88.9
COVERS	DIVS. 10 - 14, 25, 28, 41, 43, 44, 46	100.0	101.2	100.3	100.0	95.7	99.0	100.0	114.5	103.3	100.0	101.2	100.3	100.0	97.1	99.4	100.0	97.2	99.4
21, 22, 23	FIRE SUPPRESSION, PLUMBING & HVAC	96.4	98.0	97.1	96.7	89.8	93.7	96.4	129.4	111.0	96.7	99.6	98.0	96.7	92.4	94.8	100.2	93.2	97.1
26, 27, 3370	ELECTRICAL, COMMUNICATIONS & UTIL.	95.2	113.5	104.9	92.2	96.0	94.2	93.1	112.3	103.3	93.1	94.1	93.6	89.0	77.9	83.1	90.0	83.4	86.5
MF2014	WEIGHTED AVERAGE	97.9	103.8	100.5	95.6	96.1	95.8	95.3	124.7	108.3	94.1	100.0	96.7	91.8	94.5	93.0	94.5	95.6	94.9

DIVISION		PUERTO RICO SAN JUAN 009 MAT.	INST.	TOTAL	RHODE ISLAND NEWPORT 028 MAT.	INST.	TOTAL	PROVIDENCE 029 MAT.	INST.	TOTAL	SOUTH CAROLINA AIKEN 298 MAT.	INST.	TOTAL	BEAUFORT 299 MAT.	INST.	TOTAL	CHARLESTON 294 MAT.	INST.	TOTAL
015433	CONTRACTOR EQUIPMENT		87.9	87.9		102.1	102.1		102.1	102.1		108.3	108.3		108.3	108.3		108.3	108.3
0241, 31 - 34	SITE & INFRASTRUCTURE, DEMOLITION	132.0	90.3	102.9	89.3	103.6	99.2	89.3	103.6	99.3	122.8	90.3	100.1	118.1	89.7	98.3	103.3	90.0	94.0
0310	Concrete Forming & Accessories	93.1	17.1	27.4	100.6	122.3	119.4	100.6	122.3	119.3	98.6	64.5	69.1	97.5	39.5	47.3	96.6	64.4	68.7
0320	Concrete Reinforcing	193.7	13.3	102.2	102.6	112.7	107.7	103.8	112.7	108.3	99.3	63.1	81.0	98.3	26.8	62.0	98.1	64.3	81.0
0330	Cast-in-Place Concrete	96.0	28.3	70.2	74.5	119.1	91.5	92.4	119.1	102.6	84.6	66.7	77.8	84.6	66.3	77.6	99.3	66.7	86.8
03	CONCRETE	99.7	21.5	63.4	86.8	119.0	101.7	92.8	119.0	105.0	105.8	66.8	87.7	102.9	49.2	78.0	97.6	66.9	83.3
04	MASONRY	89.9	15.9	44.4	97.0	117.9	109.9	101.0	117.9	111.4	79.2	65.8	70.9	93.6	65.8	76.4	94.9	65.8	76.9
05	METALS	112.9	35.9	89.4	93.1	110.9	98.5	97.6	110.9	101.7	97.3	91.2	95.4	97.3	81.3	92.4	99.1	91.5	96.8
06	WOOD, PLASTICS & COMPOSITES	103.3	16.8	55.7	103.7	122.8	114.2	98.5	122.8	111.9	95.3	67.6	80.1	94.0	35.0	61.5	92.9	67.6	79.0
07	THERMAL & MOISTURE PROTECTION	129.1	21.3	84.4	100.0	117.4	107.2	103.0	117.4	109.0	99.7	65.9	85.7	99.4	56.4	81.6	98.5	64.0	84.2
08	OPENINGS	154.4	14.8	122.8	99.6	120.8	104.4	100.8	120.8	105.4	95.5	63.9	88.4	95.5	40.0	83.0	99.4	64.5	91.5
0920	Plaster & Gypsum Board	168.8	14.5	66.3	90.9	123.0	112.2	96.6	123.0	114.1	99.3	66.5	77.5	102.6	33.0	56.3	104.1	66.5	79.1
0950, 0980	Ceilings & Acoustic Treatment	240.9	14.5	90.3	94.9	123.0	113.6	99.9	123.0	115.3	83.8	66.5	72.3	86.3	33.0	50.8	86.3	66.5	73.1
0960	Flooring	213.3	15.1	155.9	94.1	127.9	103.9	92.7	127.9	102.9	102.0	95.1	100.0	103.6	80.3	96.9	103.2	81.8	97.0
0970, 0990	Wall Finishes & Painting/Coating	194.9	16.4	90.1	93.9	118.5	108.3	90.2	118.5	106.8	98.3	63.6	77.9	98.3	35.2	61.2	98.3	63.6	77.9
09	FINISHES	201.0	16.8	100.3	92.1	123.7	109.4	90.3	123.7	108.6	94.0	70.4	81.1	95.0	45.6	68.0	93.2	68.1	79.5
COVERS	DIVS. 10 - 14, 25, 28, 41, 43, 44, 46	100.0	16.2	81.1	100.0	107.5	101.7	100.0	107.5	101.7	100.0	72.0	93.7	100.0	78.2	95.1	100.0	71.9	93.7
21, 22, 23	FIRE SUPPRESSION, PLUMBING & HVAC	104.2	13.5	64.1	100.3	108.6	104.0	100.1	108.6	103.9	97.0	56.1	78.9	97.0	55.9	78.8	100.5	56.8	81.2
26, 27, 3370	ELECTRICAL, COMMUNICATIONS & UTIL.	124.0	17.7	67.6	100.6	99.1	99.8	100.4	99.1	99.7	100.8	63.9	81.2	104.8	33.5	67.0	103.0	59.3	79.8
MF2014	WEIGHTED AVERAGE	121.1	24.3	78.3	96.2	112.3	103.3	97.9	112.3	104.2	98.2	68.6	85.1	98.9	56.4	80.1	99.1	67.8	85.2

DIVISION		SOUTH CAROLINA COLUMBIA 290-292 MAT.	INST.	TOTAL	FLORENCE 295 MAT.	INST.	TOTAL	GREENVILLE 296 MAT.	INST.	TOTAL	ROCK HILL 297 MAT.	INST.	TOTAL	SPARTANBURG 293 MAT.	INST.	TOTAL	SOUTH DAKOTA ABERDEEN 574 MAT.	INST.	TOTAL
015433	CONTRACTOR EQUIPMENT		108.3	108.3		108.3	108.3		108.3	108.3		108.3	108.3		108.3	108.3		101.8	101.8
0241, 31 - 34	SITE & INFRASTRUCTURE, DEMOLITION	102.1	90.0	93.6	112.4	89.8	96.7	107.6	89.4	94.9	105.3	89.2	94.1	107.4	89.4	94.8	102.7	95.1	97.4
0310	Concrete Forming & Accessories	95.6	64.4	68.6	84.2	64.4	67.0	96.1	64.3	68.6	94.2	64.0	68.0	99.0	64.3	69.0	96.9	37.2	45.2
0320	Concrete Reinforcing	103.6	64.3	83.6	97.7	64.3	80.8	97.7	64.3	80.7	98.5	64.2	81.1	97.7	63.9	80.5	101.9	41.2	71.1
0330	Cast-in-Place Concrete	100.4	66.7	87.5	84.6	66.6	77.7	84.5	66.6	77.7	84.5	66.5	77.7	84.5	66.6	77.7	108.7	65.6	92.3
03	CONCRETE	95.5	66.9	82.2	97.1	66.8	83.1	95.6	66.8	82.3	93.5	66.6	81.0	95.8	66.8	82.3	101.2	49.4	77.2
04	MASONRY	88.5	65.8	74.5	79.3	65.7	70.9	76.8	65.8	70.0	101.1	65.8	79.3	79.3	65.8	70.9	115.3	53.1	77.0
05	METALS	96.2	91.5	94.8	98.0	91.4	96.0	98.0	91.5	96.0	97.3	90.3	95.1	98.0	91.3	96.0	97.8	68.0	88.7
06	WOOD, PLASTICS & COMPOSITES	93.0	67.6	79.1	80.2	67.6	73.3	92.5	67.6	78.8	91.2	67.6	78.2	96.4	67.6	80.6	92.7	35.8	61.4
07	THERMAL & MOISTURE PROTECTION	94.6	64.0	81.9	98.8	64.0	84.4	98.7	64.0	84.3	98.6	59.6	82.4	98.8	64.0	84.4	102.7	51.1	81.2
08	OPENINGS	97.7	64.5	90.2	95.6	64.5	88.5	95.5	64.5	88.5	95.5	64.5	88.5	95.5	64.4	88.5	95.8	35.1	82.1
0920	Plaster & Gypsum Board	96.5	66.5	76.6	93.6	66.5	75.6	98.4	66.5	77.2	98.0	66.5	77.1	101.0	66.5	78.1	112.0	34.3	60.4
0950, 0980	Ceilings & Acoustic Treatment	88.8	66.5	74.0	84.7	66.5	72.6	83.8	66.5	72.3	83.8	66.5	72.3	83.8	66.5	72.3	91.5	34.3	53.5
0960	Flooring	92.7	80.4	89.2	94.1	80.3	90.1	100.8	80.3	94.8	99.6	80.3	94.0	102.2	80.3	95.9	98.3	46.4	83.3
0970, 0990	Wall Finishes & Painting/Coating	92.2	63.6	75.4	89.3	63.6	73.3	98.3	63.6	77.9	98.3	63.6	77.9	98.3	63.6	77.9	93.6	35.6	59.5
09	FINISHES	88.8	67.8	77.4	89.9	67.9	77.8	91.9	67.9	78.8	91.3	67.8	78.5	92.7	67.9	79.1	94.0	38.4	63.6
COVERS	DIVS. 10 - 14, 25, 28, 41, 43, 44, 46	100.0	72.0	93.7	100.0	72.0	93.7	100.0	71.9	93.7	100.0	71.9	93.7	100.0	71.9	93.7	100.0	38.7	86.1
21, 22, 23	FIRE SUPPRESSION, PLUMBING & HVAC	100.0	56.8	80.9	100.5	56.8	81.2	100.5	56.8	81.2	97.0	56.0	78.9	100.5	56.8	81.2	100.2	43.5	75.1
26, 27, 3370	ELECTRICAL, COMMUNICATIONS & UTIL.	102.8	64.7	82.6	100.7	64.7	81.6	103.1	62.0	81.3	103.1	62.0	81.3	103.1	62.0	81.3	104.3	73.8	88.1
MF2014	WEIGHTED AVERAGE	97.1	68.5	84.5	97.3	68.5	84.5	97.3	68.1	84.4	97.3	67.6	84.2	97.6	68.1	84.5	100.1	54.7	80.1

For customer support on your Building Construction Costs with RSMeans Data, call 800.448.8182.

791

City Cost Indexes

SOUTH DAKOTA

DIVISION		MITCHELL 573			MOBRIDGE 576			PIERRE 575			RAPID CITY 577			SIOUX FALLS 570-571			WATERTOWN 572		
		MAT.	INST.	TOTAL	MAT.	INST.	TOTAL	MAT.	INST.	TOTAL	MAT.	INST.	TOTAL	MAT.	INST.	TOTAL	MAT.	INST.	TOTAL
015433	CONTRACTOR EQUIPMENT		101.8	101.8		101.8	101.8		101.8	101.8		101.8	101.8		102.8	102.8		101.8	101.8
0241, 31 - 34	SITE & INFRASTRUCTURE, DEMOLITION	99.4	95.1	96.4	99.4	95.1	96.4	102.6	95.4	97.6	101.0	95.3	97.0	92.5	97.1	95.7	99.3	95.1	96.3
0310	Concrete Forming & Accessories	96.1	36.8	44.8	86.7	37.3	43.9	99.2	39.9	47.9	104.7	50.2	57.5	102.9	56.1	62.4	83.4	36.4	42.8
0320	Concrete Reinforcing	101.3	50.0	75.3	103.9	41.2	72.1	102.9	95.7	99.3	95.3	95.9	95.6	102.9	96.0	99.4	98.5	41.2	69.5
0330	Cast-in-Place Concrete	105.6	40.8	80.9	105.6	43.4	81.9	103.5	48.5	82.5	104.8	47.7	83.0	86.6	49.4	72.4	105.6	44.4	82.3
03	CONCRETE	98.9	42.1	72.5	98.6	41.8	72.3	96.4	54.2	76.8	98.2	58.6	79.8	88.5	61.8	76.1	97.7	41.8	71.8
04	MASONRY	103.0	52.5	71.9	111.8	54.6	76.6	114.1	55.3	77.9	111.9	55.7	77.3	93.8	57.8	71.6	139.6	56.6	88.5
05	METALS	96.8	67.5	87.9	96.9	68.0	88.1	101.3	86.8	96.8	99.5	87.4	95.8	98.5	87.5	95.1	96.8	67.8	88.0
06	WOOD, PLASTICS & COMPOSITES	91.7	36.3	61.2	81.3	36.0	56.4	98.0	36.5	64.2	96.4	50.8	71.3	96.1	57.9	75.1	77.7	35.8	54.7
07	THERMAL & MOISTURE PROTECTION	102.4	49.4	80.4	102.5	51.5	81.3	103.6	51.1	81.8	103.1	53.0	82.3	102.0	55.3	82.6	102.3	51.9	81.4
08	OPENINGS	94.9	35.6	81.5	97.5	34.7	83.3	101.3	47.7	89.2	99.6	55.5	89.6	101.9	59.3	92.3	94.9	34.8	81.3
0920	Plaster & Gypsum Board	110.4	34.9	60.2	105.1	34.6	58.2	101.4	35.1	57.4	111.4	49.8	70.4	99.9	57.0	71.4	102.8	34.3	57.3
0950, 0980	Ceilings & Acoustic Treatment	87.3	34.9	52.4	91.5	34.6	53.6	96.1	35.1	55.5	93.2	49.8	64.3	87.2	57.0	67.1	87.3	34.3	52.0
0960	Flooring	97.9	46.4	82.9	93.1	49.2	80.4	100.0	33.1	80.6	97.6	72.8	90.4	101.5	73.3	93.3	91.7	46.4	78.6
0970, 0990	Wall Finishes & Painting/Coating	93.6	38.9	61.5	93.6	39.9	62.1	98.5	97.9	98.1	93.6	97.9	96.1	100.9	97.9	99.1	93.6	35.6	59.5
09	FINISHES	92.6	38.5	63.0	91.3	39.0	62.8	95.9	43.1	67.1	94.0	59.2	75.0	93.8	63.8	77.4	89.7	37.8	61.4
COVERS	DIVS. 10 - 14, 25, 28, 41, 43, 44, 46	100.0	35.8	85.5	100.0	43.1	87.1	100.0	75.3	94.4	100.0	76.3	94.6	100.0	77.8	95.0	100.0	38.6	86.1
21, 22, 23	FIRE SUPPRESSION, PLUMBING & HVAC	96.6	39.3	71.3	96.6	62.6	81.6	100.0	70.2	86.8	100.2	69.4	86.6	100.0	58.5	81.7	96.6	42.3	72.6
26, 27, 3370	ELECTRICAL, COMMUNICATIONS & UTIL.	102.5	42.2	70.5	104.3	43.0	71.8	108.4	49.7	77.3	100.4	49.7	73.5	103.0	72.6	86.9	101.5	47.3	72.8
MF2014	WEIGHTED AVERAGE	97.8	48.3	75.9	98.5	53.9	78.8	101.3	62.0	83.9	99.9	65.2	84.6	97.8	67.8	84.6	99.0	50.0	77.4

TENNESSEE

DIVISION		CHATTANOOGA 373 - 374			COLUMBIA 384			COOKEVILLE 385			JACKSON 383			JOHNSON CITY 376			KNOXVILLE 377 - 379		
		MAT.	INST.	TOTAL	MAT.	INST.	TOTAL	MAT.	INST.	TOTAL	MAT.	INST.	TOTAL	MAT.	INST.	TOTAL	MAT.	INST.	TOTAL
015433	CONTRACTOR EQUIPMENT		109.2	109.2		103.9	103.9		103.9	103.9		110.4	110.4		102.8	102.8		102.8	102.8
0241, 31 - 34	SITE & INFRASTRUCTURE, DEMOLITION	103.0	99.5	100.5	87.4	89.8	89.0	92.9	86.7	88.6	95.9	100.1	98.9	109.7	86.0	93.1	89.5	88.9	89.1
0310	Concrete Forming & Accessories	97.2	57.2	62.6	83.3	63.6	66.2	83.5	33.8	40.5	90.5	43.6	50.0	83.4	61.5	64.5	95.3	66.2	70.1
0320	Concrete Reinforcing	89.4	67.4	78.2	85.1	63.9	74.3	85.1	63.4	74.1	85.1	66.6	75.7	90.0	63.9	76.8	89.4	63.9	76.5
0330	Cast-in-Place Concrete	99.4	61.7	85.1	90.5	63.2	80.1	102.5	41.6	79.3	100.1	70.1	88.7	80.0	51.4	69.1	93.4	65.0	82.6
03	CONCRETE	96.7	62.4	80.8	93.8	65.2	80.6	104.1	44.3	76.4	94.7	59.1	78.2	106.6	60.2	85.1	94.1	67.0	81.6
04	MASONRY	104.1	49.3	70.4	119.2	58.2	81.6	113.6	40.6	68.7	119.4	44.1	73.0	118.4	42.0	71.3	81.6	52.8	63.9
05	METALS	98.9	88.9	95.8	94.5	89.6	93.0	94.6	88.4	92.7	96.8	89.7	94.7	96.0	87.4	93.4	99.3	87.9	95.8
06	WOOD, PLASTICS & COMPOSITES	104.2	58.9	79.3	72.2	64.8	68.1	72.4	31.7	50.0	88.3	42.8	63.3	77.5	67.7	72.1	90.1	67.7	77.8
07	THERMAL & MOISTURE PROTECTION	99.7	60.4	83.4	94.5	63.3	81.5	95.0	51.5	76.9	97.0	57.3	80.5	95.1	55.4	78.7	93.0	63.9	80.9
08	OPENINGS	98.6	53.7	88.5	92.0	54.4	83.5	92.0	37.2	79.6	99.1	46.8	87.3	94.9	63.9	87.9	92.5	58.5	84.8
0920	Plaster & Gypsum Board	79.7	58.3	65.5	85.4	64.2	71.3	85.4	30.3	48.8	87.4	41.7	57.0	101.8	67.2	78.8	108.6	67.2	81.1
0950, 0980	Ceilings & Acoustic Treatment	103.5	58.3	73.4	77.0	64.2	68.5	77.0	30.3	45.9	85.5	41.7	56.3	99.8	67.2	78.1	100.6	67.2	78.4
0960	Flooring	97.7	54.0	85.0	83.7	57.2	76.0	83.8	53.0	74.8	83.1	57.5	75.7	92.7	46.5	79.3	97.9	49.6	83.9
0970, 0990	Wall Finishes & Painting/Coating	98.2	60.1	75.8	84.9	60.1	70.4	84.9	60.1	70.4	86.6	60.1	71.1	95.3	62.6	76.1	95.3	60.1	74.7
09	FINISHES	96.1	56.8	74.6	86.3	62.0	73.0	86.8	38.8	60.6	86.5	46.6	64.7	100.3	59.4	77.9	93.6	62.5	76.6
COVERS	DIVS. 10 - 14, 25, 28, 41, 43, 44, 46	100.0	39.9	86.4	100.0	71.3	93.5	100.0	38.3	86.1	100.0	65.6	92.2	100.0	79.1	95.3	100.0	83.0	96.2
21, 22, 23	FIRE SUPPRESSION, PLUMBING & HVAC	100.2	57.1	81.1	97.7	75.1	87.7	97.7	66.7	84.0	100.1	60.1	82.4	99.9	57.5	81.1	99.9	62.0	83.1
26, 27, 3370	ELECTRICAL, COMMUNICATIONS & UTIL.	103.6	88.9	95.8	93.5	54.8	73.0	95.2	63.4	78.3	99.7	55.8	76.4	94.0	44.2	67.6	100.1	57.9	77.7
MF2014	WEIGHTED AVERAGE	99.6	67.1	85.3	95.4	68.7	83.6	96.8	57.7	79.5	98.3	61.3	81.9	99.8	60.7	82.5	96.3	66.4	83.1

TENNESSEE / TEXAS

DIVISION		MCKENZIE 382			MEMPHIS 375, 380 - 381			NASHVILLE 370 - 372			ABILENE 795 - 796			AMARILLO 790 - 791			AUSTIN 786 - 787		
		MAT.	INST.	TOTAL	MAT.	INST.	TOTAL	MAT.	INST.	TOTAL	MAT.	INST.	TOTAL	MAT.	INST.	TOTAL	MAT.	INST.	TOTAL
015433	CONTRACTOR EQUIPMENT		103.9	103.9		104.1	104.1		103.4	103.4		94.6	94.6		94.6	94.6		93.7	93.7
0241, 31 - 34	SITE & INFRASTRUCTURE, DEMOLITION	92.5	86.8	88.5	91.3	97.5	95.6	98.9	95.9	96.8	98.1	93.5	94.9	95.7	92.8	93.7	97.9	92.5	94.1
0310	Concrete Forming & Accessories	91.5	36.2	43.6	96.1	66.5	70.5	96.9	66.5	70.6	99.3	63.8	68.6	96.1	54.8	60.4	95.3	54.9	60.4
0320	Concrete Reinforcing	85.2	66.7	75.8	96.0	68.6	82.1	95.6	64.6	79.9	98.3	51.6	74.6	106.7	51.6	78.7	96.4	49.0	72.4
0330	Cast-in-Place Concrete	100.3	54.8	83.0	90.0	78.3	85.5	91.0	66.6	81.7	87.7	69.9	80.9	79.0	69.8	75.5	92.7	69.5	83.8
03	CONCRETE	102.6	50.5	78.4	93.8	72.2	83.8	94.0	67.4	81.6	86.4	64.6	76.3	85.1	60.5	73.7	90.2	60.0	76.2
04	MASONRY	118.1	48.5	75.3	93.5	56.7	70.8	90.4	68.7	77.0	96.9	63.5	76.3	96.2	63.5	76.0	99.1	62.2	76.4
05	METALS	94.6	89.3	93.0	100.3	85.9	95.9	100.4	85.1	95.7	104.4	71.7	94.4	99.6	71.5	91.0	96.4	71.3	88.7
06	WOOD, PLASTICS & COMPOSITES	80.7	34.2	55.2	97.8	67.1	80.9	98.7	66.9	81.2	105.0	65.7	83.4	100.7	53.6	74.8	94.9	54.1	72.5
07	THERMAL & MOISTURE PROTECTION	95.0	48.1	75.5	97.0	67.1	84.6	94.1	68.3	83.4	99.6	67.1	86.1	95.7	65.4	83.1	97.4	66.4	84.5
08	OPENINGS	92.0	39.5	80.2	98.1	63.1	90.2	98.4	63.4	90.5	102.4	60.4	92.9	104.4	53.8	92.9	105.4	51.8	93.2
0920	Plaster & Gypsum Board	88.3	32.8	51.4	93.7	66.4	75.6	94.0	66.1	75.5	84.9	65.1	71.7	93.8	52.7	66.5	89.6	53.2	65.4
0950, 0980	Ceilings & Acoustic Treatment	77.0	32.8	47.6	97.8	66.4	76.9	98.7	66.1	77.0	89.1	65.1	73.1	95.9	52.7	67.2	92.4	53.2	66.3
0960	Flooring	86.7	52.7	78.1	98.5	78.0	92.6	93.8	76.5	88.8	99.3	71.7	91.3	100.7	67.6	91.1	94.0	67.6	86.4
0970, 0990	Wall Finishes & Painting/Coating	84.9	47.0	62.7	90.8	60.7	73.2	97.7	71.7	82.4	94.2	55.0	71.2	89.7	55.0	69.3	95.7	47.1	67.2
09	FINISHES	88.0	40.2	61.9	95.2	67.9	80.3	98.5	68.7	82.2	86.2	64.7	74.4	90.2	56.8	71.9	88.5	55.8	70.6
COVERS	DIVS. 10 - 14, 25, 28, 41, 43, 44, 46	100.0	25.5	83.2	100.0	83.0	96.2	100.0	83.3	96.2	100.0	81.2	95.7	100.0	77.7	95.0	100.0	79.1	95.3
21, 22, 23	FIRE SUPPRESSION, PLUMBING & HVAC	97.7	60.3	81.2	100.0	72.0	87.6	99.9	78.0	90.2	100.3	50.1	78.1	100.0	54.1	79.7	100.0	59.4	82.0
26, 27, 3370	ELECTRICAL, COMMUNICATIONS & UTIL.	95.0	60.0	76.4	101.6	65.7	82.6	100.6	63.4	80.9	99.7	57.6	77.4	101.9	63.8	81.7	99.0	61.0	78.8
MF2014	WEIGHTED AVERAGE	96.9	57.4	79.5	98.2	72.2	86.7	98.4	73.6	87.4	98.1	63.7	82.9	97.5	63.2	82.4	97.5	63.6	82.5

TEXAS

DIVISION		BEAUMONT 776 - 777			BROWNWOOD 768			BRYAN 778			CHILDRESS 792			CORPUS CHRISTI 783 - 784			DALLAS 752 - 753		
		MAT.	INST.	TOTAL	MAT.	INST.	TOTAL	MAT.	INST.	TOTAL	MAT.	INST.	TOTAL	MAT.	INST.	TOTAL	MAT.	INST.	TOTAL
015433	CONTRACTOR EQUIPMENT		99.0	99.0		94.6	94.6		99.0	99.0		94.6	94.6		105.2	105.2		105.1	105.1
0241, 31 - 34	SITE & INFRASTRUCTURE, DEMOLITION	87.9	97.4	94.5	106.5	93.4	97.4	79.1	98.1	92.4	109.3	91.8	97.1	144.2	88.7	105.5	107.3	95.3	98.9
0310	Concrete Forming & Accessories	104.5	57.4	63.8	101.0	63.4	68.4	82.6	61.9	64.7	97.6	63.3	67.9	100.2	54.7	60.9	95.6	64.5	68.7
0320	Concrete Reinforcing	92.3	62.9	77.4	93.8	51.0	72.0	94.9	52.0	73.1	98.4	51.0	74.3	85.3	47.6	66.2	94.2	51.7	72.6
0330	Cast-in-Place Concrete	92.8	62.8	81.4	108.3	63.0	91.1	74.3	62.8	69.9	89.9	63.0	79.7	113.6	70.0	97.0	118.8	72.0	100.9
03	CONCRETE	96.2	61.3	80.0	99.4	61.9	82.0	80.2	61.5	71.5	94.5	61.8	79.3	98.9	60.9	81.3	106.5	66.2	87.8
04	MASONRY	103.5	59.4	76.4	130.7	54.2	83.6	145.0	54.9	89.5	100.9	54.2	72.2	87.3	51.8	65.5	103.9	63.6	79.1
05	METALS	95.1	76.4	89.3	94.8	70.5	87.4	95.0	73.7	88.5	102.0	70.5	92.4	93.9	86.2	91.5	99.3	82.3	94.1
06	WOOD, PLASTICS & COMPOSITES	115.3	56.8	83.1	105.0	65.7	83.4	80.2	62.9	70.7	104.4	65.7	83.1	124.3	54.6	86.0	96.7	65.9	79.8
07	THERMAL & MOISTURE PROTECTION	103.5	65.4	87.7	96.9	63.2	82.9	95.3	64.7	82.6	100.2	61.8	84.3	103.5	64.3	87.3	93.2	69.3	83.3
08	OPENINGS	95.7	57.9	87.2	98.2	60.4	89.6	96.3	58.7	87.8	97.6	60.4	89.2	104.2	51.8	92.3	100.3	60.5	91.3
0920	Plaster & Gypsum Board	97.4	56.0	69.9	85.3	65.1	71.8	85.4	62.3	70.0	84.5	65.1	71.6	98.6	53.6	68.7	91.1	65.1	73.8
0950, 0980	Ceilings & Acoustic Treatment	96.6	56.0	69.6	76.1	65.1	68.8	88.6	62.3	71.1	87.4	65.1	72.5	91.1	53.6	66.1	93.2	65.1	74.5
0960	Flooring	119.3	65.9	103.8	83.3	57.9	76.0	89.2	59.8	80.7	97.6	57.9	86.1	109.2	57.9	94.3	96.0	67.6	87.8
0970, 0990	Wall Finishes & Painting/Coating	91.6	56.3	70.9	92.9	55.0	70.7	89.6	59.4	71.9	94.2	55.0	71.2	109.4	47.1	72.8	102.2	55.0	74.5
09	FINISHES	94.8	58.4	74.9	78.9	61.8	69.6	82.2	61.0	70.7	86.6	61.8	73.0	100.6	53.9	75.0	97.1	64.0	79.0
COVERS	DIVS. 10 - 14, 25, 28, 41, 43, 44, 46	100.0	83.5	96.3	100.0	79.0	95.2	100.0	80.6	95.6	100.0	79.0	95.2	100.0	82.7	96.1	100.0	81.9	95.9
21, 22, 23	FIRE SUPPRESSION, PLUMBING & HVAC	100.1	65.3	84.7	96.6	48.1	75.2	96.6	64.8	82.5	96.7	54.1	77.9	100.1	56.9	81.0	100.0	61.5	83.0
26, 27, 3370	ELECTRICAL, COMMUNICATIONS & UTIL.	96.9	68.0	81.6	94.7	50.0	71.0	94.9	68.8	81.1	99.7	60.4	78.9	95.1	57.1	74.9	96.5	63.6	79.0
MF2014	WEIGHTED AVERAGE	97.6	67.3	84.2	97.5	60.3	81.1	95.0	67.0	82.7	97.9	62.8	82.4	99.7	62.5	83.3	100.3	68.4	86.2

TEXAS

DIVISION		DEL RIO 788			DENTON 762			EASTLAND 764			EL PASO 798 - 799,885			FORT WORTH 760 - 761			GALVESTON 775		
		MAT.	INST.	TOTAL	MAT.	INST.	TOTAL	MAT.	INST.	TOTAL	MAT.	INST.	TOTAL	MAT.	INST.	TOTAL	MAT.	INST.	TOTAL
015433	CONTRACTOR EQUIPMENT		93.7	93.7		104.7	104.7		94.6	94.6		94.6	94.6		94.6	94.6		112.6	112.6
0241, 31 - 34	SITE & INFRASTRUCTURE, DEMOLITION	124.1	91.3	101.2	105.7	85.0	91.2	109.5	91.4	96.9	98.0	91.2	93.3	101.9	94.3	96.6	103.8	96.7	98.8
0310	Concrete Forming & Accessories	96.8	55.1	60.7	108.5	62.5	68.7	101.7	63.2	68.4	97.9	59.0	64.2	98.7	63.8	68.5	92.4	58.8	63.4
0320	Concrete Reinforcing	86.0	47.5	66.5	95.2	51.0	72.8	94.0	51.4	72.4	103.7	51.0	76.9	100.9	51.0	75.6	94.3	61.6	77.7
0330	Cast-in-Place Concrete	122.4	62.6	99.6	83.8	62.7	75.7	114.5	63.0	94.9	84.0	62.0	75.6	99.7	63.2	85.8	98.7	63.9	85.4
03	CONCRETE	119.7	57.3	90.8	78.3	62.5	71.0	104.2	61.9	84.6	87.2	59.4	74.3	94.5	62.2	79.5	97.6	63.3	81.7
04	MASONRY	102.8	54.0	72.8	140.1	51.7	85.6	98.8	62.3	76.3	87.6	53.9	66.8	98.0	54.2	71.0	100.0	55.0	72.3
05	METALS	93.8	67.7	85.8	94.5	86.6	92.1	94.7	71.0	87.4	99.6	69.2	90.3	96.4	71.0	88.6	96.7	94.3	95.9
06	WOOD, PLASTICS & COMPOSITES	102.3	54.4	76.0	116.5	65.8	88.6	112.2	65.7	86.6	90.3	60.3	73.8	99.1	65.7	80.7	97.8	58.5	76.2
07	THERMAL & MOISTURE PROTECTION	100.0	63.5	84.9	94.8	62.9	81.6	97.3	65.4	84.1	98.1	61.8	83.0	93.2	64.5	81.3	94.4	65.0	82.2
08	OPENINGS	99.4	51.0	88.5	115.8	60.0	103.2	70.6	60.4	68.3	97.9	53.5	87.8	101.4	60.4	92.1	100.5	58.6	91.0
0920	Plaster & Gypsum Board	94.9	53.6	67.4	89.9	65.1	73.4	85.3	65.1	71.8	93.8	59.6	71.1	88.8	65.1	73.0	92.0	57.6	69.2
0950, 0980	Ceilings & Acoustic Treatment	87.5	53.6	64.9	82.0	65.1	70.7	76.1	65.1	68.8	90.7	59.6	70.0	87.9	65.1	72.7	92.9	57.6	69.4
0960	Flooring	93.7	57.9	83.3	78.9	57.9	72.8	105.5	57.9	91.7	96.8	57.9	85.5	101.5	57.9	88.8	104.6	59.8	91.6
0970, 0990	Wall Finishes & Painting/Coating	96.6	45.3	66.5	103.3	52.1	73.2	94.3	55.0	71.3	88.9	49.8	70.1	95.3	55.2	71.7	100.4	57.2	75.0
09	FINISHES	92.8	54.2	71.7	78.0	60.9	68.7	85.9	61.8	72.7	88.7	57.9	71.9	88.4	61.8	73.9	91.9	58.3	73.5
COVERS	DIVS. 10 - 14, 25, 28, 41, 43, 44, 46	100.0	79.2	95.3	100.0	77.6	94.9	100.0	77.2	94.8	100.0	78.1	95.0	100.0	81.2	95.7	100.0	84.0	96.4
21, 22, 23	FIRE SUPPRESSION, PLUMBING & HVAC	96.5	62.2	81.3	96.6	57.7	79.4	96.6	48.1	75.1	99.9	64.7	84.3	99.9	58.6	81.6	96.5	64.9	82.5
26, 27, 3370	ELECTRICAL, COMMUNICATIONS & UTIL.	97.4	64.8	80.1	96.9	63.6	79.2	94.6	63.5	78.1	99.3	54.1	75.3	96.2	63.5	78.9	96.5	64.6	79.6
MF2014	WEIGHTED AVERAGE	100.5	62.8	83.9	97.5	64.8	83.0	94.1	62.9	80.3	96.3	62.8	81.5	97.3	64.7	82.9	97.3	68.3	84.5

TEXAS

DIVISION		GIDDINGS 789			GREENVILLE 754			HOUSTON 770 - 772			HUNTSVILLE 773			LAREDO 780			LONGVIEW 756		
		MAT.	INST.	TOTAL	MAT.	INST.	TOTAL	MAT.	INST.	TOTAL	MAT.	INST.	TOTAL	MAT.	INST.	TOTAL	MAT.	INST.	TOTAL
015433	CONTRACTOR EQUIPMENT		93.7	93.7		105.1	105.1		102.3	102.3		99.0	99.0		93.7	93.7		95.7	95.7
0241, 31 - 34	SITE & INFRASTRUCTURE, DEMOLITION	108.9	91.9	97.1	100.3	88.1	91.8	102.5	96.9	98.6	94.6	98.1	97.1	102.3	92.6	95.5	99.4	94.9	96.2
0310	Concrete Forming & Accessories	94.1	54.4	59.7	87.1	61.6	65.0	94.5	59.0	63.8	90.0	58.5	62.8	96.9	55.7	61.3	83.5	63.4	66.1
0320	Concrete Reinforcing	86.6	48.5	67.3	94.7	51.0	72.6	94.1	52.3	72.9	95.1	52.0	73.3	86.0	48.2	66.8	93.6	51.0	72.0
0330	Cast-in-Place Concrete	103.6	62.0	87.7	108.1	64.1	91.4	95.7	64.9	84.0	102.3	63.1	87.3	87.5	70.2	80.9	125.8	63.0	101.8
03	CONCRETE	96.0	57.1	78.0	98.0	62.4	81.5	95.2	61.2	79.4	104.3	60.1	83.8	90.4	60.5	76.5	115.9	61.8	90.8
04	MASONRY	110.6	54.2	74.8	161.3	54.2	95.4	99.7	58.9	74.6	142.9	55.0	88.8	96.0	63.6	76.0	157.2	54.8	94.2
05	METALS	93.3	70.3	86.3	96.9	84.9	93.2	99.3	78.1	92.8	95.0	74.1	88.6	96.2	71.0	88.5	89.7	69.4	83.5
06	WOOD, PLASTICS & COMPOSITES	101.3	54.4	75.6	87.9	63.1	74.3	100.1	58.6	77.3	88.8	58.4	72.1	102.3	54.4	76.0	82.7	65.7	73.4
07	THERMAL & MOISTURE PROTECTION	100.6	62.6	84.8	93.2	63.1	80.7	93.9	66.5	82.6	96.4	64.2	83.0	99.3	67.1	85.9	94.8	62.8	81.5
08	OPENINGS	98.5	51.9	88.0	98.5	59.0	89.5	103.1	56.3	92.5	96.3	56.2	87.2	99.5	51.7	88.7	88.9	60.0	82.4
0920	Plaster & Gypsum Board	93.9	53.6	67.1	85.2	62.3	70.0	94.1	57.6	69.9	89.4	57.6	68.3	95.7	53.6	67.7	84.2	65.1	71.5
0950, 0980	Ceilings & Acoustic Treatment	87.5	53.6	64.9	89.0	62.3	71.2	97.9	57.6	71.1	88.6	57.6	68.0	90.9	53.6	66.0	84.8	65.1	71.7
0960	Flooring	93.3	57.9	83.0	91.0	57.9	81.4	106.1	59.8	92.7	93.8	57.9	83.4	93.5	67.6	86.0	96.4	57.9	85.2
0970, 0990	Wall Finishes & Painting/Coating	96.6	47.1	67.5	102.2	55.0	74.5	100.4	59.4	76.3	89.6	59.4	71.9	96.6	47.1	67.5	92.4	52.1	68.7
09	FINISHES	91.4	53.7	70.8	93.4	60.4	75.4	99.6	58.6	77.2	85.0	58.0	70.3	91.9	56.3	72.5	99.0	61.5	78.5
COVERS	DIVS. 10 - 14, 25, 28, 41, 43, 44, 46	100.0	74.8	94.3	100.0	79.2	95.3	100.0	84.2	96.4	100.0	75.1	94.4	100.0	79.3	95.3	100.0	79.1	95.3
21, 22, 23	FIRE SUPPRESSION, PLUMBING & HVAC	96.6	62.3	81.4	96.5	59.9	80.3	100.0	64.8	84.4	96.6	64.7	82.5	100.1	62.3	83.4	96.4	58.4	79.6
26, 27, 3370	ELECTRICAL, COMMUNICATIONS & UTIL.	94.2	60.3	76.2	93.3	63.6	77.5	98.5	68.8	82.8	94.9	64.1	78.5	97.5	61.9	78.6	93.8	54.2	72.8
MF2014	WEIGHTED AVERAGE	97.0	62.1	81.6	99.9	65.5	84.7	99.3	67.5	85.2	98.6	65.5	84.0	97.1	64.7	82.8	100.1	63.1	83.7

For customer support on your Building Construction Costs with RSMeans Data, call 800.448.8182.

TEXAS

DIVISION		LUBBOCK 793-794 MAT.	INST.	TOTAL	LUFKIN 759 MAT.	INST.	TOTAL	MCALLEN 785 MAT.	INST.	TOTAL	MCKINNEY 750 MAT.	INST.	TOTAL	MIDLAND 797 MAT.	INST.	TOTAL	ODESSA 797 MAT.	INST.	TOTAL
015433	CONTRACTOR EQUIPMENT		107.6	107.6		95.7	95.7		105.3	105.3		105.1	105.1		107.6	107.6		94.6	94.6
0241, 31 - 34	SITE & INFRASTRUCTURE, DEMOLITION	123.2	92.5	101.8	94.1	96.2	95.6	149.1	88.9	107.1	96.6	88.1	90.7	126.4	90.5	101.4	98.3	94.3	95.5
0310	Concrete Forming & Accessories	98.2	55.0	60.9	86.4	58.4	62.2	101.3	55.3	61.5	86.2	61.6	64.9	102.3	63.7	68.9	99.3	64.0	68.8
0320	Concrete Reinforcing	99.5	51.6	75.2	95.3	61.4	78.1	85.5	47.6	66.3	94.7	51.0	72.6	100.6	51.6	75.7	98.3	51.6	74.6
0330	Cast-in-Place Concrete	87.8	71.0	81.4	112.5	62.6	93.5	123.5	64.5	101.0	101.4	64.1	87.2	93.4	70.9	84.8	87.7	70.0	80.9
03	CONCRETE	85.2	62.1	74.5	106.9	61.2	85.7	106.8	59.3	84.8	92.8	62.4	78.7	89.4	65.9	78.5	86.4	64.7	76.3
04	MASONRY	96.2	63.6	76.1	120.7	54.7	80.1	102.5	54.1	72.7	173.3	54.2	100.0	113.1	63.6	82.6	96.9	63.5	76.3
05	METALS	108.3	87.6	102.0	97.0	72.0	89.4	93.7	85.8	91.3	96.8	84.9	93.2	106.7	87.5	100.8	103.8	71.8	94.0
06	WOOD, PLASTICS & COMPOSITES	104.2	53.8	76.5	90.3	58.5	72.8	123.1	54.6	85.4	87.0	63.1	73.9	108.9	65.8	85.2	105.0	65.7	83.4
07	THERMAL & MOISTURE PROTECTION	88.9	66.4	79.6	94.6	61.5	80.9	103.3	64.4	87.2	93.0	63.1	80.6	89.2	67.7	80.3	99.6	66.7	86.0
08	OPENINGS	110.0	53.9	97.3	67.6	59.0	65.6	103.4	51.0	91.5	98.4	59.0	89.5	111.5	60.5	100.0	102.4	60.4	92.9
0920	Plaster & Gypsum Board	85.1	52.7	63.6	83.3	57.6	66.2	99.3	53.6	68.9	85.2	62.3	70.0	87.0	65.1	72.4	84.9	65.1	71.7
0950, 0980	Ceilings & Acoustic Treatment	89.9	52.7	65.2	80.5	57.6	65.3	90.9	53.6	66.0	89.0	62.3	71.2	88.2	65.1	72.8	89.1	65.1	73.1
0960	Flooring	92.6	67.6	85.4	128.5	57.9	108.0	108.6	56.3	93.4	90.5	57.9	81.1	93.9	67.6	86.3	99.3	67.6	90.1
0970, 0990	Wall Finishes & Painting/Coating	105.1	55.0	75.7	92.4	55.0	70.5	109.4	45.3	71.8	102.2	55.0	74.5	105.1	55.0	75.7	94.2	55.0	71.2
09	FINISHES	88.5	56.9	71.2	107.3	58.0	80.4	100.9	53.9	75.2	93.1	60.4	75.2	89.3	63.6	75.3	86.2	63.8	74.0
COVERS	DIVS. 10 - 14, 25, 28, 41, 43, 44, 46	100.0	80.2	95.5	100.0	75.4	94.4	100.0	79.5	95.4	100.0	79.2	95.3	100.0	79.3	95.3	100.0	79.0	95.3
21, 22, 23	FIRE SUPPRESSION, PLUMBING & HVAC	99.7	50.0	77.7	96.4	61.4	80.9	96.6	58.1	79.5	96.5	59.9	80.3	96.1	50.2	75.8	100.3	56.2	80.8
26, 27, 3370	ELECTRICAL, COMMUNICATIONS & UTIL.	98.4	63.2	79.8	95.1	66.5	79.9	94.9	34.3	62.7	93.3	63.6	77.5	98.4	63.2	79.8	99.8	63.2	80.4
MF2014	WEIGHTED AVERAGE	99.7	64.1	84.0	96.7	65.0	82.7	100.7	59.5	82.5	99.7	65.5	84.6	100.4	65.8	85.1	98.0	65.7	83.7

TEXAS

DIVISION		PALESTINE 758 MAT.	INST.	TOTAL	SAN ANGELO 769 MAT.	INST.	TOTAL	SAN ANTONIO 781-782 MAT.	INST.	TOTAL	TEMPLE 765 MAT.	INST.	TOTAL	TEXARKANA 755 MAT.	INST.	TOTAL	TYLER 757 MAT.	INST.	TOTAL
015433	CONTRACTOR EQUIPMENT		95.7	95.7		94.6	94.6		98.1	98.1		94.6	94.6		95.7	95.7		95.7	95.7
0241, 31 - 34	SITE & INFRASTRUCTURE, DEMOLITION	99.6	95.2	96.5	102.6	93.9	96.5	102.8	96.0	98.1	90.5	94.0	92.9	88.4	97.7	94.9	98.7	95.0	96.1
0310	Concrete Forming & Accessories	78.0	63.5	65.4	101.3	55.4	61.6	96.6	56.0	61.5	104.9	54.8	61.6	93.4	64.2	68.1	88.2	63.6	66.9
0320	Concrete Reinforcing	92.9	51.0	71.6	93.6	51.0	72.0	92.2	48.2	69.9	93.8	48.5	70.8	92.7	51.6	71.9	93.6	51.0	72.0
0330	Cast-in-Place Concrete	102.8	63.0	87.6	102.2	69.8	89.9	88.0	71.6	81.7	83.7	62.5	75.6	103.6	70.0	90.8	123.5	63.1	100.5
03	CONCRETE	108.1	61.8	86.6	94.8	60.6	79.0	91.3	61.2	77.3	80.9	57.5	70.0	98.9	64.7	83.0	115.2	61.9	90.5
04	MASONRY	114.8	54.8	77.9	127.1	54.9	82.6	91.6	63.7	74.4	139.2	53.6	86.5	177.7	63.5	107.4	167.5	54.9	98.2
05	METALS	96.8	69.5	88.4	95.0	70.8	87.6	99.3	72.8	91.2	94.7	70.5	87.3	89.6	70.8	83.9	96.7	69.6	88.4
06	WOOD, PLASTICS & COMPOSITES	82.2	65.7	73.1	105.3	54.5	77.3	100.1	54.6	75.1	114.9	54.5	81.7	92.9	65.7	78.0	91.7	65.7	77.4
07	THERMAL & MOISTURE PROTECTION	95.1	63.9	82.2	96.7	64.3	83.3	97.3	68.1	85.2	96.4	62.7	82.4	94.4	67.6	83.3	94.9	63.9	82.1
08	OPENINGS	67.5	60.4	65.9	98.2	54.2	88.3	98.4	51.8	87.9	67.3	53.6	64.2	88.9	60.4	82.4	67.4	60.4	65.9
0920	Plaster & Gypsum Board	81.0	65.1	70.4	85.3	53.6	64.2	95.0	53.6	67.5	85.3	53.6	64.2	87.9	65.1	72.7	83.3	65.1	71.2
0950, 0980	Ceilings & Acoustic Treatment	80.5	65.1	70.2	76.1	53.6	61.1	96.9	53.6	68.1	76.1	53.6	61.1	84.8	65.1	71.7	80.5	65.1	70.2
0960	Flooring	118.8	57.9	101.1	83.5	57.9	76.0	102.5	67.6	92.4	107.2	57.9	92.9	105.3	67.6	94.4	130.9	57.9	109.7
0970, 0990	Wall Finishes & Painting/Coating	92.4	55.0	70.5	92.9	55.0	70.7	99.4	47.1	68.7	94.3	47.1	66.6	92.4	55.0	70.5	92.4	55.0	70.5
09	FINISHES	104.6	61.9	81.3	78.7	55.3	65.9	100.4	56.5	76.4	85.1	54.0	68.1	101.6	63.9	81.0	108.3	61.9	82.9
COVERS	DIVS. 10 - 14, 25, 28, 41, 43, 44, 46	100.0	79.1	95.3	100.0	79.7	95.4	100.0	79.7	95.4	100.0	77.3	94.9	100.0	81.4	95.8	100.0	81.3	95.8
21, 22, 23	FIRE SUPPRESSION, PLUMBING & HVAC	96.4	58.6	79.7	96.6	50.0	76.0	100.0	65.0	84.5	96.6	53.9	77.7	96.4	57.6	79.3	96.4	61.5	81.0
26, 27, 3370	ELECTRICAL, COMMUNICATIONS & UTIL.	91.3	52.4	70.7	98.8	59.2	77.8	99.0	61.9	79.3	95.8	56.4	74.9	95.0	59.1	76.0	93.8	58.2	74.9
MF2014	WEIGHTED AVERAGE	96.0	63.0	81.4	97.1	60.8	81.0	98.1	65.8	83.9	92.4	60.3	78.2	99.1	65.7	84.4	100.1	64.5	84.4

TEXAS / **UTAH**

DIVISION		VICTORIA 779 MAT.	INST.	TOTAL	WACO 766-767 MAT.	INST.	TOTAL	WAXAHACHIE 751 MAT.	INST.	TOTAL	WHARTON 774 MAT.	INST.	TOTAL	WICHITA FALLS 763 MAT.	INST.	TOTAL	LOGAN 843 MAT.	INST.	TOTAL
015433	CONTRACTOR EQUIPMENT		111.2	111.2		94.6	94.6		105.1	105.1		112.6	112.6		94.6	94.6		98.0	98.0
0241, 31 - 34	SITE & INFRASTRUCTURE, DEMOLITION	108.7	93.9	98.4	99.3	94.5	95.9	98.4	88.1	91.2	113.8	95.9	101.3	100.1	94.3	96.0	99.8	95.2	96.6
0310	Concrete Forming & Accessories	91.9	55.5	60.4	103.3	64.2	69.4	86.2	63.5	66.5	86.9	57.2	61.2	103.3	63.8	69.1	104.9	59.1	65.3
0320	Concrete Reinforcing	90.5	47.2	68.6	93.5	49.1	70.9	94.7	51.0	72.5	94.2	52.0	72.8	93.5	51.4	72.1	105.2	86.8	95.9
0330	Cast-in-Place Concrete	110.9	63.9	93.0	90.6	70.2	82.8	107.0	64.1	90.7	113.9	63.7	94.8	96.8	63.2	84.0	84.3	70.2	78.9
03	CONCRETE	105.5	59.3	84.1	87.5	64.4	76.8	96.9	63.2	81.3	109.8	60.8	87.0	90.5	62.2	77.4	106.7	68.6	89.0
04	MASONRY	118.3	56.0	79.3	96.9	63.6	76.4	161.9	54.2	95.6	101.2	54.8	72.6	97.4	60.5	74.7	110.9	58.4	78.5
05	METALS	95.1	89.2	93.3	96.9	71.7	89.2	96.9	84.8	93.2	96.7	89.1	94.3	96.9	71.4	89.1	107.1	84.3	100.1
06	WOOD, PLASTICS & COMPOSITES	101.3	54.5	75.6	113.1	65.7	87.0	87.0	65.9	75.4	91.4	56.9	72.5	113.1	65.7	87.0	84.7	56.7	69.3
07	THERMAL & MOISTURE PROTECTION	98.2	64.6	84.3	97.2	67.9	85.1	93.2	63.3	80.8	94.8	64.9	82.4	97.2	64.9	83.8	97.9	64.8	84.2
08	OPENINGS	100.3	53.0	89.6	78.8	59.8	74.5	98.5	60.5	89.9	100.5	55.4	90.3	78.8	60.4	74.7	92.7	61.4	85.6
0920	Plaster & Gypsum Board	88.9	53.6	65.4	85.9	65.1	72.0	85.6	65.1	72.0	87.7	56.0	66.6	85.9	65.1	72.0	77.7	55.5	62.9
0950, 0980	Ceilings & Acoustic Treatment	93.7	53.6	67.0	78.6	65.1	69.6	90.7	65.1	73.6	92.9	56.0	68.3	78.6	65.1	69.6	99.4	55.5	70.2
0960	Flooring	103.0	57.9	89.9	106.4	67.6	95.2	90.5	57.9	81.1	101.6	70.6	92.6	107.3	67.5	95.8	101.4	61.0	89.7
0970, 0990	Wall Finishes & Painting/Coating	100.6	59.4	76.4	94.3	55.0	71.3	102.2	55.0	74.5	100.4	55.0	76.3	96.8	55.0	72.3	96.0	57.8	73.6
09	FINISHES	89.5	55.8	71.1	86.0	63.8	73.9	93.7	62.0	76.4	91.1	59.4	73.8	86.4	63.8	74.1	95.0	58.3	74.9
COVERS	DIVS. 10 - 14, 25, 28, 41, 43, 44, 46	100.0	74.9	94.3	100.0	81.2	95.7	100.0	79.5	95.4	100.0	75.2	94.4	100.0	79.0	95.3	100.0	83.5	96.3
21, 22, 23	FIRE SUPPRESSION, PLUMBING & HVAC	96.6	64.6	82.5	100.2	60.1	82.5	96.5	53.4	77.4	96.5	62.0	81.3	100.2	53.5	79.6	99.9	66.1	84.9
26, 27, 3370	ELECTRICAL, COMMUNICATIONS & UTIL.	101.3	54.6	76.6	98.9	57.1	76.7	93.3	63.6	77.5	100.3	64.1	81.1	100.6	57.0	77.5	96.7	71.0	83.1
MF2014	WEIGHTED AVERAGE	99.5	64.7	84.1	94.2	65.7	81.6	99.7	64.5	84.2	99.4	66.4	84.8	94.8	63.6	81.0	100.8	69.5	87.0

DIVISION		UTAH											VERMONT					
		OGDEN			PRICE			PROVO			SALT LAKE CITY			BELLOWS FALLS			BENNINGTON		
		842,844			845			846 - 847			840 - 841			051			052		
		MAT.	INST.	TOTAL	MAT.	INST.	TOTAL	MAT.	INST.	TOTAL	MAT.	INST.	TOTAL	MAT.	INST.	TOTAL	MAT.	INST.	TOTAL
015433	CONTRACTOR EQUIPMENT		98.0	98.0		97.0	97.0		97.0	97.0		97.9	97.9		99.8	99.8		99.8	99.8
0241, 31 - 34	SITE & INFRASTRUCTURE, DEMOLITION	87.4	95.2	92.8	97.2	93.1	94.3	96.0	93.5	94.3	87.0	95.1	92.7	90.4	102.1	98.6	89.8	102.1	98.4
0310	Concrete Forming & Accessories	105.0	59.1	65.3	107.5	46.5	54.7	106.7	59.1	65.5	107.4	59.1	65.6	96.9	107.0	105.7	94.5	106.8	105.1
0320	Concrete Reinforcing	104.9	86.8	95.7	112.8	86.7	99.5	113.8	86.8	100.1	107.2	86.8	96.9	86.1	81.9	83.9	86.1	81.8	83.9
0330	Cast-in-Place Concrete	85.6	70.2	79.7	84.4	58.5	74.5	84.4	70.2	79.0	93.5	70.2	84.6	87.6	113.6	97.5	87.6	113.5	97.5
03	CONCRETE	95.6	68.6	83.1	107.9	58.9	85.2	106.2	68.6	88.8	115.7	68.7	93.9	87.2	104.4	95.2	87.0	104.2	95.0
04	MASONRY	105.0	58.4	76.3	116.6	56.5	79.6	116.7	58.4	80.8	119.5	58.4	81.8	106.0	87.9	94.9	117.6	87.9	99.3
05	METALS	107.6	84.3	100.5	104.3	82.9	97.8	105.2	84.3	98.8	111.0	84.3	102.9	90.1	88.4	89.6	90.1	88.1	89.5
06	WOOD, PLASTICS & COMPOSITES	84.7	56.7	69.3	87.7	41.7	62.4	86.0	56.7	69.9	86.4	56.7	70.1	108.7	113.0	111.1	105.9	113.0	109.8
07	THERMAL & MOISTURE PROTECTION	96.8	64.8	83.5	99.7	60.0	83.2	99.8	64.8	85.2	104.3	64.8	87.9	97.3	89.5	94.1	97.3	85.5	92.4
08	OPENINGS	92.7	61.4	85.6	96.6	49.4	86.0	96.7	61.4	88.7	94.6	61.4	87.1	100.6	101.3	100.7	100.6	101.3	100.7
0920	Plaster & Gypsum Board	77.7	55.5	62.9	80.3	40.1	53.6	78.0	55.5	63.1	90.7	55.5	67.3	105.6	112.9	110.4	103.9	112.9	109.9
0950, 0980	Ceilings & Acoustic Treatment	99.4	55.5	70.2	99.4	40.1	59.9	99.4	55.5	70.2	92.8	55.5	68.0	91.2	112.9	105.6	91.2	112.9	105.6
0960	Flooring	99.3	61.0	88.2	102.6	45.6	86.1	102.3	61.0	90.3	103.5	61.0	91.2	96.5	96.5	96.5	95.7	96.5	95.9
0970, 0990	Wall Finishes & Painting/Coating	96.0	57.8	73.6	96.0	57.8	73.6	96.0	57.8	73.6	99.2	66.3	79.9	89.3	106.1	99.2	89.3	106.1	99.2
09	FINISHES	93.1	58.3	74.1	96.2	45.9	68.7	95.7	58.3	75.2	95.5	59.2	75.7	90.6	106.1	99.0	90.1	106.1	98.8
COVERS	DIVS. 10 - 14, 25, 28, 41, 43, 44, 46	100.0	83.5	96.3	100.0	56.0	90.0	100.0	83.5	96.3	100.0	83.5	96.3	100.0	96.3	99.2	100.0	96.2	99.1
21, 22, 23	FIRE SUPPRESSION, PLUMBING & HVAC	99.9	66.1	84.9	98.1	65.0	83.4	99.9	66.1	84.9	100.1	66.1	85.0	96.7	91.9	94.6	96.7	91.9	94.6
26, 27, 3370	ELECTRICAL, COMMUNICATIONS & UTIL.	97.0	71.0	83.2	102.3	71.0	85.7	97.4	71.0	83.4	99.6	71.0	84.4	102.8	86.2	94.0	102.8	59.0	79.6
MF2014	WEIGHTED AVERAGE	98.7	69.5	85.8	101.5	64.2	85.0	101.3	69.4	87.2	103.4	69.6	88.5	95.6	95.4	95.5	96.1	91.4	94.0

DIVISION		VERMONT																
		BRATTLEBORO			BURLINGTON			GUILDHALL			MONTPELIER			RUTLAND			ST. JOHNSBURY		
		053			054			059			056			057			058		
		MAT.	INST.	TOTAL	MAT.	INST.	TOTAL	MAT.	INST.	TOTAL	MAT.	INST.	TOTAL	MAT.	INST.	TOTAL	MAT.	INST.	TOTAL
015433	CONTRACTOR EQUIPMENT		99.8	99.8		99.8	99.8		99.8	99.8		99.8	99.8		99.8	99.8		99.8	99.8
0241, 31 - 34	SITE & INFRASTRUCTURE, DEMOLITION	91.4	102.1	98.9	95.0	102.0	99.9	89.4	97.4	95.0	93.6	102.0	99.4	93.9	102.0	99.5	89.5	101.0	97.5
0310	Concrete Forming & Accessories	97.1	106.9	105.6	98.9	84.3	86.2	95.0	99.8	99.2	98.9	105.7	104.8	97.4	84.3	86.0	93.6	100.3	99.4
0320	Concrete Reinforcing	85.2	81.8	83.5	105.8	81.7	93.5	86.9	81.6	84.2	105.8	81.7	93.6	107.6	81.7	94.5	85.2	81.7	83.4
0330	Cast-in-Place Concrete	90.4	113.6	99.2	104.6	112.5	107.6	84.9	104.2	92.3	104.5	112.6	107.5	85.8	112.5	96.0	84.9	104.4	92.3
03	CONCRETE	89.2	104.3	96.2	98.9	93.7	96.5	84.8	97.8	90.8	98.8	103.4	100.9	89.9	93.7	91.7	84.5	98.1	90.8
04	MASONRY	117.2	87.9	99.2	115.6	86.6	97.8	117.5	72.1	89.6	111.9	86.6	96.3	96.5	86.6	90.4	147.0	72.1	100.9
05	METALS	90.1	88.3	89.5	97.6	87.4	94.5	90.1	87.1	89.2	95.7	87.6	93.2	95.8	87.4	93.2	90.1	87.5	89.3
06	WOOD, PLASTICS & COMPOSITES	109.0	113.0	111.2	101.1	84.0	91.7	105.5	113.0	109.6	95.4	113.0	105.1	109.1	84.0	95.3	100.9	113.0	107.5
07	THERMAL & MOISTURE PROTECTION	97.4	85.5	92.5	103.8	83.9	95.6	97.2	78.3	89.3	104.6	87.1	97.3	97.5	83.9	91.9	97.1	78.3	89.3
08	OPENINGS	100.6	101.3	100.7	105.4	81.6	100.0	100.6	97.5	99.9	106.0	97.5	104.1	103.8	81.6	98.8	100.6	97.5	99.9
0920	Plaster & Gypsum Board	105.6	112.9	110.4	104.9	83.1	90.4	111.5	112.9	112.4	104.5	112.9	110.1	106.0	83.1	90.8	112.8	112.9	112.9
0950, 0980	Ceilings & Acoustic Treatment	91.2	112.9	105.6	97.1	83.1	87.8	91.2	112.9	105.6	97.3	112.9	107.7	95.7	83.1	87.3	91.2	112.9	105.6
0960	Flooring	96.6	96.5	96.6	102.9	96.5	101.0	99.5	96.5	98.6	104.4	96.5	102.1	96.5	96.5	96.5	102.9	96.5	101.0
0970, 0990	Wall Finishes & Painting/Coating	89.3	106.1	99.2	99.2	92.1	95.1	89.3	92.1	91.0	95.6	92.1	93.6	89.3	92.1	91.0	89.3	92.1	91.0
09	FINISHES	90.7	106.1	99.1	94.6	87.2	90.6	92.2	100.6	96.8	94.8	104.2	99.9	91.7	87.2	89.2	93.4	100.6	97.3
COVERS	DIVS. 10 - 14, 25, 28, 41, 43, 44, 46	100.0	96.3	99.2	100.0	92.6	98.3	100.0	90.8	97.9	100.0	95.8	99.0	100.0	92.6	98.3	100.0	90.9	97.9
21, 22, 23	FIRE SUPPRESSION, PLUMBING & HVAC	96.7	91.9	94.6	100.0	68.8	86.2	96.7	61.4	81.1	96.4	68.8	84.2	100.3	68.8	86.4	96.7	61.4	81.1
26, 27, 3370	ELECTRICAL, COMMUNICATIONS & UTIL.	102.8	86.2	94.0	102.7	57.0	78.5	102.8	59.0	79.6	101.9	57.0	78.1	102.8	57.0	78.5	102.8	59.0	79.6
MF2014	WEIGHTED AVERAGE	96.5	95.2	95.9	100.7	80.8	91.9	96.0	80.7	89.3	99.3	85.5	93.2	97.8	80.8	90.3	97.5	81.1	90.3

DIVISION		VERMONT			VIRGINIA													
		WHITE RIVER JCT.			ALEXANDRIA			ARLINGTON			BRISTOL			CHARLOTTESVILLE			CULPEPER		
		050			223			222			242			229			227		
		MAT.	INST.	TOTAL	MAT.	INST.	TOTAL	MAT.	INST.	TOTAL	MAT.	INST.	TOTAL	MAT.	INST.	TOTAL	MAT.	INST.	TOTAL
015433	CONTRACTOR EQUIPMENT		99.8	99.8		110.1	110.1		108.7	108.7		108.7	108.7		113.1	113.1		108.7	108.7
0241, 31 - 34	SITE & INFRASTRUCTURE, DEMOLITION	94.0	101.1	99.0	114.7	94.0	100.2	124.8	92.0	101.9	109.0	90.6	96.1	113.3	93.1	99.2	112.4	91.5	97.8
0310	Concrete Forming & Accessories	91.7	100.8	99.6	92.9	71.7	74.5	91.9	71.9	74.6	88.0	40.2	46.6	86.4	47.2	52.5	83.4	71.2	72.9
0320	Concrete Reinforcing	86.1	81.7	83.9	85.6	83.6	84.6	96.6	83.6	90.0	96.5	64.5	80.3	95.9	69.8	82.7	96.6	83.5	90.0
0330	Cast-in-Place Concrete	90.4	105.2	96.0	105.4	79.0	95.4	102.6	81.1	94.4	102.2	46.4	80.9	106.2	76.3	94.8	105.0	80.5	95.7
03	CONCRETE	90.8	98.6	94.4	101.8	77.9	90.7	106.4	78.8	93.6	102.6	49.1	77.8	103.2	63.6	84.8	100.2	78.1	90.0
04	MASONRY	131.9	73.6	96.0	93.3	74.2	81.6	108.9	74.3	87.6	97.6	48.8	67.5	123.0	52.4	79.6	109.7	74.1	87.8
05	METALS	90.1	87.5	89.3	100.8	101.5	101.0	99.4	102.9	100.5	98.3	91.0	96.0	98.5	96.1	97.8	98.6	100.7	99.2
06	WOOD, PLASTICS & COMPOSITES	103.0	113.0	108.5	93.8	70.0	80.7	90.5	70.0	79.3	83.8	37.5	58.4	82.4	42.8	60.7	81.2	70.0	75.1
07	THERMAL & MOISTURE PROTECTION	97.5	79.0	89.8	103.2	82.2	94.5	105.3	82.0	95.6	104.6	57.6	85.1	104.2	65.3	88.1	104.4	81.6	95.0
08	OPENINGS	100.6	97.5	99.9	97.8	74.2	92.5	95.9	74.2	91.0	98.9	43.8	86.4	97.1	52.2	86.9	97.4	73.9	92.1
0920	Plaster & Gypsum Board	102.6	112.9	109.5	103.9	69.0	80.7	100.6	69.0	79.6	97.1	35.6	56.2	97.1	40.4	59.4	97.3	69.0	78.5
0950, 0980	Ceilings & Acoustic Treatment	91.2	112.9	105.6	90.9	69.0	76.3	89.2	69.0	75.8	88.3	35.6	53.3	88.3	40.4	56.4	89.2	69.0	75.8
0960	Flooring	94.5	96.5	95.1	100.6	76.7	93.7	99.0	76.7	92.5	95.3	59.0	84.8	93.9	59.0	83.8	93.9	76.7	88.9
0970, 0990	Wall Finishes & Painting/Coating	89.3	92.1	91.0	118.5	76.4	93.8	118.5	76.4	93.8	104.9	51.5	73.5	104.9	60.9	79.1	118.5	60.7	84.5
09	FINISHES	89.9	100.9	95.9	96.2	72.3	83.1	96.1	72.3	83.1	92.7	43.9	66.0	92.3	49.1	68.7	93.0	70.5	80.7
COVERS	DIVS. 10 - 14, 25, 28, 41, 43, 44, 46	100.0	91.4	98.1	100.0	89.2	97.6	100.0	88.8	97.5	100.0	64.5	92.0	100.0	83.6	96.3	100.0	88.8	97.5
21, 22, 23	FIRE SUPPRESSION, PLUMBING & HVAC	96.7	62.2	81.4	100.4	85.7	93.9	100.4	85.7	93.9	96.9	48.1	75.3	96.9	67.5	83.9	96.9	70.8	85.3
26, 27, 3370	ELECTRICAL, COMMUNICATIONS & UTIL.	102.8	57.0	78.5	98.0	99.5	98.8	95.6	99.5	97.7	97.7	39.4	66.7	97.6	72.4	84.2	100.2	99.5	99.8
MF2014	WEIGHTED AVERAGE	97.4	81.3	90.3	99.8	85.2	93.3	100.7	85.3	93.9	98.5	54.4	79.0	99.8	68.0	85.7	99.0	81.5	91.3

For customer support on your Building Construction Costs with RSMeans Data, call 800.448.8182.

795

VIRGINIA

DIVISION		FAIRFAX 220-221			FARMVILLE 239			FREDERICKSBURG 224-225			GRUNDY 246			HARRISONBURG 228			LYNCHBURG 245		
		MAT.	INST.	TOTAL	MAT.	INST.	TOTAL	MAT.	INST.	TOTAL	MAT.	INST.	TOTAL	MAT.	INST.	TOTAL	MAT.	INST.	TOTAL
015433	CONTRACTOR EQUIPMENT		108.7	108.7		113.1	113.1		108.7	108.7		108.7	108.7		108.7	108.7		108.7	108.7
0241, 31-34	SITE & INFRASTRUCTURE, DEMOLITION	123.5	91.9	101.4	108.0	92.4	97.1	111.9	91.5	97.7	106.8	89.4	94.7	120.5	90.0	99.3	107.6	90.8	95.9
0310	Concrete Forming & Accessories	86.3	71.7	73.7	98.4	45.6	52.7	86.3	70.4	72.6	91.2	34.4	42.1	82.3	40.2	45.9	88.0	57.8	61.9
0320	Concrete Reinforcing	96.6	83.6	90.0	93.0	45.4	68.8	97.3	83.6	90.3	95.2	45.8	70.1	96.6	58.8	77.5	95.9	64.6	80.0
0330	Cast-in-Place Concrete	102.6	92.3	98.6	104.6	57.8	86.8	104.1	79.3	94.7	102.1	45.4	80.5	102.6	55.7	84.7	102.1	59.3	85.8
03	CONCRETE	106.1	82.5	95.2	99.1	52.0	77.3	99.9	77.4	89.5	101.2	42.6	74.0	103.8	51.3	79.4	101.1	61.5	82.7
04	MASONRY	108.8	74.1	87.5	106.1	53.8	73.9	108.8	71.9	86.1	98.3	44.9	65.4	106.8	47.7	70.4	114.3	52.4	76.2
05	METALS	98.7	102.6	99.9	94.1	82.7	90.6	98.6	101.6	99.5	98.3	76.4	91.6	98.5	88.5	95.5	98.4	91.7	96.4
06	WOOD, PLASTICS & COMPOSITES	83.8	70.0	76.2	94.1	41.4	65.2	83.8	69.8	76.1	86.4	32.9	57.0	80.3	36.6	56.3	83.8	58.8	70.1
07	THERMAL & MOISTURE PROTECTION	105.0	77.3	93.5	105.1	52.8	83.4	104.4	80.6	94.5	104.6	43.7	79.3	104.8	61.7	86.9	104.4	64.2	87.7
08	OPENINGS	95.9	73.9	90.9	97.8	40.2	84.7	97.1	73.6	91.8	98.9	32.3	83.8	97.4	46.3	85.8	97.4	55.5	87.9
0920	Plaster & Gypsum Board	97.3	69.0	78.5	106.4	38.9	61.6	97.3	68.8	78.4	97.1	30.9	53.1	97.1	34.7	55.6	97.1	57.5	70.8
0950, 0980	Ceilings & Acoustic Treatment	89.2	69.0	75.8	85.0	38.9	54.3	89.2	68.8	75.6	88.3	30.9	50.1	88.3	34.7	52.6	88.3	57.5	67.8
0960	Flooring	95.7	76.7	90.2	98.3	53.5	85.3	95.7	76.7	90.2	96.7	29.9	77.3	93.6	65.7	85.5	95.3	59.0	84.8
0970, 0990	Wall Finishes & Painting/Coating	118.5	76.4	93.8	103.3	60.9	78.4	118.5	60.7	84.5	104.9	33.3	62.8	118.5	51.5	79.1	104.9	51.5	73.5
09	FINISHES	94.6	72.2	82.4	93.2	48.3	68.7	93.5	69.8	80.6	92.9	32.7	60.0	93.4	44.5	66.7	92.5	56.5	72.8
COVERS	DIVS. 10 - 14, 25, 28, 41, 43, 44, 46	100.0	88.8	97.5	100.0	47.2	88.1	100.0	84.0	96.4	100.0	41.4	86.7	100.0	64.3	91.9	100.0	68.1	92.8
21, 22, 23	FIRE SUPPRESSION, PLUMBING & HVAC	96.9	85.6	91.9	97.0	47.0	74.9	96.9	84.1	91.2	96.9	61.1	81.0	96.9	68.8	84.4	96.9	67.2	83.7
26, 27, 3370	ELECTRICAL, COMMUNICATIONS & UTIL.	98.8	99.5	99.2	93.9	46.3	68.7	95.8	99.5	97.7	97.7	42.8	68.6	97.8	93.6	95.6	98.7	50.5	73.2
MF2014	WEIGHTED AVERAGE	99.9	85.5	93.6	97.5	55.1	78.8	98.6	83.8	92.0	98.4	51.8	77.8	99.4	66.5	84.8	99.1	64.8	83.9

VIRGINIA

DIVISION		NEWPORT NEWS 236			NORFOLK 233-235			PETERSBURG 238			PORTSMOUTH 237			PULASKI 243			RICHMOND 230-232		
		MAT.	INST.	TOTAL	MAT.	INST.	TOTAL	MAT.	INST.	TOTAL	MAT.	INST.	TOTAL	MAT.	INST.	TOTAL	MAT.	INST.	TOTAL
015433	CONTRACTOR EQUIPMENT		113.1	113.1		113.7	113.7		113.1	113.1		113.0	113.0		108.7	108.7		113.1	113.1
0241, 31-34	SITE & INFRASTRUCTURE, DEMOLITION	106.8	93.5	97.5	105.8	94.5	98.0	110.5	93.4	98.6	105.4	92.8	96.6	106.1	89.7	94.7	101.4	93.4	95.8
0310	Concrete Forming & Accessories	97.7	63.0	67.7	100.2	63.0	68.0	91.3	65.0	68.6	87.3	50.3	55.3	91.2	36.5	43.9	96.2	65.0	69.2
0320	Concrete Reinforcing	92.8	66.2	79.3	101.7	66.2	83.7	92.4	70.0	81.0	92.4	66.2	79.1	95.2	58.6	76.6	101.7	70.0	85.6
0330	Cast-in-Place Concrete	101.7	79.0	93.0	102.3	79.0	93.4	108.0	79.2	97.0	100.7	63.5	86.5	102.1	46.2	80.8	100.0	79.2	92.0
03	CONCRETE	96.3	71.0	84.6	96.6	71.0	84.7	101.6	72.6	88.2	95.2	59.9	78.8	101.2	46.3	75.7	95.3	72.6	84.8
04	MASONRY	100.3	55.9	73.0	98.1	55.9	72.1	114.8	60.3	81.2	106.4	51.1	72.4	93.1	45.8	64.0	99.4	60.3	75.3
05	METALS	96.4	94.5	95.8	98.0	94.5	96.9	94.2	96.5	94.9	95.4	94.2	95.0	98.3	87.2	94.9	98.0	96.5	97.6
06	WOOD, PLASTICS & COMPOSITES	93.1	62.1	76.1	91.7	62.1	75.5	85.4	64.3	73.8	82.3	48.4	63.6	86.4	34.5	57.9	92.7	64.3	77.1
07	THERMAL & MOISTURE PROTECTION	105.0	68.0	89.7	102.9	68.0	88.4	105.0	70.8	90.8	105.1	63.6	87.9	104.6	49.4	81.7	102.1	70.7	89.1
08	OPENINGS	98.1	58.2	89.1	96.1	58.2	87.6	97.5	64.0	89.9	98.2	50.7	87.5	98.9	39.1	85.4	98.5	64.0	90.7
0920	Plaster & Gypsum Board	107.4	60.2	76.0	99.9	60.2	73.5	100.7	62.4	75.3	101.2	46.0	64.5	97.1	32.5	54.2	102.7	62.4	75.9
0950, 0980	Ceilings & Acoustic Treatment	89.2	60.2	69.9	93.5	60.2	71.3	85.5	62.4	70.3	89.2	46.0	60.5	88.3	32.5	51.2	88.8	62.4	71.3
0960	Flooring	98.3	59.0	86.9	93.3	59.0	83.4	94.0	63.2	85.1	90.8	59.0	81.6	96.7	59.0	85.8	93.4	63.2	84.6
0970, 0990	Wall Finishes & Painting/Coating	103.3	60.9	78.4	100.9	60.9	77.4	103.3	60.9	78.4	103.3	60.9	78.4	104.9	51.5	73.5	96.0	60.9	75.4
09	FINISHES	94.0	61.6	76.3	91.2	61.6	75.0	91.5	63.7	76.3	91.0	51.6	69.5	92.9	40.7	64.4	91.3	63.7	76.2
COVERS	DIVS. 10 - 14, 25, 28, 41, 43, 44, 46	100.0	87.4	97.1	100.0	87.3	97.1	100.0	87.6	97.2	100.0	68.5	92.9	100.0	41.6	86.8	100.0	87.6	97.2
21, 22, 23	FIRE SUPPRESSION, PLUMBING & HVAC	100.5	64.6	84.6	100.1	65.7	84.9	97.0	70.1	85.1	100.5	62.8	83.9	96.9	60.0	80.5	100.0	70.1	86.7
26, 27, 3370	ELECTRICAL, COMMUNICATIONS & UTIL.	96.8	65.4	80.1	99.2	61.6	79.3	96.9	72.4	83.9	95.0	61.6	77.3	97.7	58.1	76.7	101.0	72.4	85.8
MF2014	WEIGHTED AVERAGE	98.3	69.9	85.8	98.1	69.7	85.5	98.3	73.5	87.4	97.8	64.5	83.1	98.1	56.8	79.9	98.3	73.5	87.3

DIVISION		VIRGINIA									WASHINGTON								
		ROANOKE 240-241			STAUNTON 244			WINCHESTER 226			CLARKSTON 994			EVERETT 982			OLYMPIA 985		
		MAT.	INST.	TOTAL	MAT.	INST.	TOTAL	MAT.	INST.	TOTAL	MAT.	INST.	TOTAL	MAT.	INST.	TOTAL	MAT.	INST.	TOTAL
015433	CONTRACTOR EQUIPMENT		108.7	108.7		113.1	113.1		108.7	108.7		92.7	92.7		103.5	103.5		103.5	103.5
0241, 31-34	SITE & INFRASTRUCTURE, DEMOLITION	106.7	90.8	95.6	109.9	91.8	97.3	119.1	91.6	99.9	105.4	90.4	94.9	93.1	111.8	106.1	91.3	111.9	105.7
0310	Concrete Forming & Accessories	97.6	58.1	63.4	90.9	47.2	53.1	84.7	71.4	73.2	114.2	68.6	74.7	117.1	102.1	104.1	106.6	102.7	103.2
0320	Concrete Reinforcing	96.3	64.8	80.3	95.9	44.5	69.8	96.0	83.6	89.7	110.9	91.7	101.1	110.8	109.3	110.0	115.9	109.4	112.6
0330	Cast-in-Place Concrete	116.0	59.5	94.5	106.2	49.3	84.5	102.6	68.2	89.5	89.5	89.4	89.5	98.2	107.3	101.7	84.5	108.1	93.5
03	CONCRETE	105.5	61.7	85.2	102.5	49.8	78.1	103.3	74.0	89.7	99.2	80.2	90.4	93.4	104.6	98.6	86.6	105.2	95.2
04	MASONRY	99.6	55.0	72.1	109.4	48.7	72.0	104.1	74.5	85.9	102.3	84.0	91.0	114.0	99.5	105.1	108.3	102.3	104.6
05	METALS	100.5	92.1	98.0	98.5	85.6	94.5	98.6	101.3	99.4	93.9	85.0	91.2	104.7	95.5	101.9	104.7	95.7	102.0
06	WOOD, PLASTICS & COMPOSITES	94.6	58.8	75.0	86.4	46.9	64.7	82.4	70.0	75.6	107.8	63.2	83.3	111.9	102.1	106.5	96.5	102.1	99.6
07	THERMAL & MOISTURE PROTECTION	104.4	68.0	89.3	104.2	50.2	81.8	104.9	80.0	94.6	159.1	83.7	127.8	111.9	104.0	108.7	110.6	104.3	108.0
08	OPENINGS	97.8	55.5	88.2	97.4	44.9	85.5	99.0	73.4	93.2	118.4	68.2	107.0	106.7	103.5	106.0	110.4	104.1	109.0
0920	Plaster & Gypsum Board	103.9	57.5	73.1	97.1	44.5	62.2	97.3	69.0	78.5	149.2	62.0	91.2	112.3	102.2	105.6	101.6	102.2	102.0
0950, 0980	Ceilings & Acoustic Treatment	90.9	57.5	68.7	88.3	44.5	59.2	89.2	69.0	75.8	105.1	62.0	76.4	104.8	102.2	103.1	104.6	102.2	103.0
0960	Flooring	100.6	59.0	88.6	96.2	35.0	78.5	95.1	76.7	89.8	91.4	73.4	86.2	109.4	104.4	107.9	98.9	109.2	101.9
0970, 0990	Wall Finishes & Painting/Coating	104.9	51.5	73.5	104.9	32.7	62.5	118.5	82.6	97.4	81.4	76.9	78.7	92.9	95.2	94.2	90.2	96.8	94.1
09	FINISHES	95.1	57.3	74.4	92.7	43.3	65.7	94.0	72.8	82.4	108.6	68.8	86.8	105.2	101.7	103.3	96.5	103.2	100.2
COVERS	DIVS. 10 - 14, 25, 28, 41, 43, 44, 46	100.0	68.1	92.8	100.0	67.5	92.6	100.0	88.8	97.5	100.0	95.7	99.0	100.0	101.9	100.4	100.0	102.5	100.6
21, 22, 23	FIRE SUPPRESSION, PLUMBING & HVAC	100.4	64.8	84.7	96.9	56.0	78.8	96.9	85.6	91.9	97.2	83.0	90.9	100.2	97.0	98.8	99.9	97.8	99.0
26, 27, 3370	ELECTRICAL, COMMUNICATIONS & UTIL.	97.7	58.2	76.7	96.5	74.1	84.6	96.2	99.5	97.9	87.5	95.0	91.5	104.2	95.9	99.8	102.2	98.4	100.2
MF2014	WEIGHTED AVERAGE	100.2	65.8	85.0	98.9	60.5	81.9	99.2	84.4	92.7	102.0	82.9	93.5	102.6	100.4	101.6	100.7	101.5	101.1

WASHINGTON

DIVISION		RICHLAND 993 MAT.	INST.	TOTAL	SEATTLE 980 - 981,987 MAT.	INST.	TOTAL	SPOKANE 990 - 992 MAT.	INST.	TOTAL	TACOMA 983 - 984 MAT.	INST.	TOTAL	VANCOUVER 986 MAT.	INST.	TOTAL	WENATCHEE 988 MAT.	INST.	TOTAL
015433	CONTRACTOR EQUIPMENT		92.7	92.7		104.2	104.2		92.7	92.7		103.5	103.5		99.7	99.7		103.5	103.5
0241, 31 - 34	SITE & INFRASTRUCTURE, DEMOLITION	107.4	91.2	96.1	98.4	111.6	107.6	106.8	91.2	95.9	96.3	111.9	107.2	106.4	98.7	101.0	105.8	110.4	109.0
0310	Concrete Forming & Accessories	114.3	79.9	84.5	107.8	103.0	103.6	119.9	79.3	84.8	107.2	102.7	103.3	107.8	93.5	95.5	109.1	77.4	81.7
0320	Concrete Reinforcing	106.3	91.8	99.0	115.9	109.5	112.7	107.0	91.7	99.3	109.4	109.4	109.4	110.3	109.0	109.7	110.3	92.1	101.1
0330	Cast-in-Place Concrete	89.7	85.0	87.9	102.8	107.9	104.7	93.3	84.8	90.0	100.9	108.1	103.7	112.7	98.9	107.5	103.0	76.8	93.0
03	CONCRETE	98.8	83.9	91.9	99.8	105.4	102.4	101.0	83.5	92.9	95.1	105.2	99.8	104.9	98.0	101.7	102.6	80.0	92.1
04	MASONRY	103.4	84.5	91.7	117.6	103.4	108.9	103.9	84.5	92.0	113.8	102.3	106.7	113.0	94.6	101.7	117.2	90.3	100.6
05	METALS	94.2	87.8	92.3	105.0	97.6	102.7	96.6	87.3	93.7	106.5	95.7	103.2	104.1	96.5	101.8	104.0	88.0	99.1
06	WOOD, PLASTICS & COMPOSITES	107.9	77.3	91.1	101.9	102.0	102.0	116.5	77.3	95.0	101.5	102.1	101.8	92.9	93.4	93.1	103.1	75.0	87.7
07	THERMAL & MOISTURE PROTECTION	160.2	83.8	128.5	109.2	105.1	107.5	156.4	85.2	126.9	111.7	104.3	108.6	111.8	96.5	105.4	111.1	90.1	102.4
08	OPENINGS	116.3	74.7	106.9	102.9	104.1	103.2	117.0	72.6	106.9	107.5	104.1	106.8	103.7	96.4	102.0	107.0	74.4	99.6
0920	Plaster & Gypsum Board	149.2	76.5	100.9	106.7	102.2	103.7	139.6	76.5	97.7	110.5	102.2	105.0	108.6	93.5	98.6	113.4	74.3	87.4
0950, 0980	Ceilings & Acoustic Treatment	111.7	76.5	88.3	114.6	102.2	106.3	107.0	76.5	86.7	108.2	102.2	104.2	105.2	93.5	97.4	100.3	74.3	83.0
0960	Flooring	91.7	80.7	88.5	102.9	99.8	102.0	90.7	90.3	90.6	101.6	106.9	103.2	107.7	97.3	104.7	104.8	57.4	91.1
0970, 0990	Wall Finishes & Painting/Coating	81.4	76.9	78.8	102.0	96.8	99.0	81.5	73.7	76.9	92.9	96.8	95.2	95.7	74.6	83.3	92.9	75.5	82.7
09	FINISHES	110.2	78.6	93.0	106.4	101.2	103.6	107.2	80.3	92.7	103.4	102.7	103.1	101.7	92.1	96.4	104.2	72.1	86.6
COVERS	DIVS. 10 - 14, 25, 28, 41, 43, 44, 46	100.0	97.3	99.4	100.0	97.3	99.4	100.0	97.3	99.4	100.0	102.5	100.6	100.0	95.9	99.1	100.0	96.4	99.2
21, 22, 23	FIRE SUPPRESSION, PLUMBING & HVAC	100.9	110.3	105.0	100.1	115.4	106.8	100.8	83.0	92.9	100.2	97.8	99.1	100.3	91.2	96.3	96.7	84.9	91.5
26, 27, 3370	ELECTRICAL, COMMUNICATIONS & UTIL.	85.4	95.0	90.5	103.9	110.6	107.5	83.6	78.7	81.0	104.0	98.4	101.0	109.5	98.4	103.6	104.9	95.9	100.1
MF2014	WEIGHTED AVERAGE	102.6	91.4	97.6	103.2	107.1	104.9	102.9	83.3	94.3	103.0	101.5	102.4	104.0	95.2	100.1	103.3	86.9	96.0

		WASHINGTON			WEST VIRGINIA														
DIVISION		YAKIMA 989 MAT.	INST.	TOTAL	BECKLEY 258 - 259 MAT.	INST.	TOTAL	BLUEFIELD 247 - 248 MAT.	INST.	TOTAL	BUCKHANNON 262 MAT.	INST.	TOTAL	CHARLESTON 250 - 253 MAT.	INST.	TOTAL	CLARKSBURG 263 - 264 MAT.	INST.	TOTAL
015433	CONTRACTOR EQUIPMENT		103.5	103.5		108.7	108.7		108.7	108.7		108.7	108.7		108.7	108.7		108.7	108.7
0241, 31 - 34	SITE & INFRASTRUCTURE, DEMOLITION	98.8	111.2	107.4	100.8	92.9	95.3	100.7	92.9	95.2	106.9	92.8	97.1	100.2	93.9	95.8	107.6	92.8	97.3
0310	Concrete Forming & Accessories	107.7	98.0	99.3	89.7	88.1	88.3	87.9	88.2	88.1	87.3	88.6	88.4	103.2	89.3	91.2	84.7	88.6	88.0
0320	Concrete Reinforcing	109.9	91.9	100.7	93.4	85.8	89.6	94.9	95.2	95.1	95.5	95.3	95.4	101.1	85.8	93.3	95.5	96.1	95.8
0330	Cast-in-Place Concrete	107.8	84.7	99.7	102.3	95.4	99.7	99.9	95.4	98.2	99.6	94.9	97.8	99.0	114.8	105.0	109.2	93.1	103.0
03	CONCRETE	99.7	92.0	96.1	98.1	91.3	95.0	96.9	93.0	95.1	100.0	93.0	96.8	96.9	98.6	97.7	104.1	92.5	98.7
04	MASONRY	106.6	83.2	92.2	94.7	91.7	92.9	94.9	91.7	92.9	107.1	90.5	96.9	90.3	93.0	92.0	111.1	90.5	98.4
05	METALS	104.6	88.5	99.7	103.3	103.9	103.5	98.5	107.2	101.2	98.7	107.5	101.4	101.8	104.2	102.6	98.7	107.7	101.4
06	WOOD, PLASTICS & COMPOSITES	101.9	102.1	102.0	85.4	87.4	86.5	85.6	87.4	86.6	84.9	88.3	86.8	97.1	87.4	91.8	81.6	88.3	85.2
07	THERMAL & MOISTURE PROTECTION	111.8	86.5	101.3	108.4	89.0	100.4	104.3	89.0	98.0	104.6	90.0	98.6	105.5	89.5	98.9	104.5	89.7	98.4
08	OPENINGS	107.0	100.4	105.5	96.5	83.1	93.4	99.3	85.3	96.2	99.4	85.8	96.3	96.6	83.1	93.6	99.4	86.0	96.3
0920	Plaster & Gypsum Board	110.1	102.2	104.8	98.9	86.8	90.9	96.4	86.8	90.0	96.8	87.7	90.8	101.0	86.8	91.6	94.1	87.7	89.9
0950, 0980	Ceilings & Acoustic Treatment	102.2	102.2	102.2	81.6	86.8	85.1	86.6	86.8	86.8	88.3	87.7	87.9	90.8	86.8	88.2	88.3	87.7	87.9
0960	Flooring	102.6	59.0	90.0	98.2	96.9	97.8	93.0	96.9	94.1	92.7	98.7	94.5	100.5	96.9	99.5	91.7	98.7	93.7
0970, 0990	Wall Finishes & Painting/Coating	92.9	73.7	81.6	95.9	86.3	90.3	104.9	86.3	93.9	104.9	89.2	95.7	91.4	91.9	91.7	104.9	89.2	95.7
09	FINISHES	102.7	87.9	94.6	90.9	89.6	90.2	90.9	89.6	90.2	91.8	90.5	91.1	93.5	90.8	92.0	91.0	90.5	90.7
COVERS	DIVS. 10 - 14, 25, 28, 41, 43, 44, 46	100.0	99.3	99.8	100.0	93.7	98.6	100.0	93.7	98.6	100.0	92.9	98.4	100.0	94.0	98.7	100.0	92.9	98.4
21, 22, 23	FIRE SUPPRESSION, PLUMBING & HVAC	100.2	109.9	104.5	97.0	89.3	93.6	96.9	88.1	93.0	96.9	92.1	94.8	100.2	90.8	96.1	96.9	92.1	94.7
26, 27, 3370	ELECTRICAL, COMMUNICATIONS & UTIL.	107.0	95.0	100.7	94.0	89.3	91.5	96.6	89.3	92.7	98.0	94.3	96.0	99.1	89.3	93.9	98.0	94.3	96.0
MF2014	WEIGHTED AVERAGE	103.2	96.6	100.2	97.8	91.4	95.0	97.3	91.7	94.8	98.7	93.3	96.3	98.6	93.1	96.2	99.3	93.3	96.7

WEST VIRGINIA

DIVISION		GASSAWAY 266 MAT.	INST.	TOTAL	HUNTINGTON 255 - 257 MAT.	INST.	TOTAL	LEWISBURG 249 MAT.	INST.	TOTAL	MARTINSBURG 254 MAT.	INST.	TOTAL	MORGANTOWN 265 MAT.	INST.	TOTAL	PARKERSBURG 261 MAT.	INST.	TOTAL
015433	CONTRACTOR EQUIPMENT		108.7	108.7		108.7	108.7		108.7	108.7		108.7	108.7		108.7	108.7		108.7	108.7
0241, 31 - 34	SITE & INFRASTRUCTURE, DEMOLITION	104.3	92.8	96.3	105.5	93.9	97.4	116.4	92.8	100.0	104.7	93.6	97.0	101.7	93.7	96.1	109.9	93.8	98.7
0310	Concrete Forming & Accessories	86.7	94.1	93.1	102.0	89.3	91.0	85.0	87.8	87.4	89.7	79.9	81.2	85.0	88.7	88.2	89.6	85.5	86.1
0320	Concrete Reinforcing	95.5	95.1	95.3	94.8	94.1	94.5	95.5	95.0	95.3	93.4	95.1	94.3	95.5	96.1	95.8	94.9	95.0	94.9
0330	Cast-in-Place Concrete	104.3	112.8	107.5	111.6	90.4	103.5	100.0	95.3	98.2	107.2	82.3	97.7	99.6	94.9	97.8	101.8	91.2	97.8
03	CONCRETE	100.4	101.7	101.0	103.5	91.6	98.0	106.9	92.7	100.3	101.9	84.4	93.8	96.7	93.2	95.1	102.2	90.3	96.7
04	MASONRY	112.1	87.6	97.0	93.5	94.5	94.1	97.7	88.8	92.2	96.2	90.5	92.7	129.7	90.5	105.5	84.0	87.4	86.1
05	METALS	98.6	107.3	101.3	105.7	107.1	106.1	98.6	106.6	101.0	103.7	100.7	102.8	98.7	107.9	101.5	99.3	107.0	101.7
06	WOOD, PLASTICS & COMPOSITES	84.1	95.7	90.5	96.2	87.6	91.5	81.9	87.4	84.9	85.4	77.6	81.1	81.9	88.3	85.4	85.0	83.3	84.0
07	THERMAL & MOISTURE PROTECTION	104.3	89.7	98.3	108.5	89.8	100.8	105.2	88.2	98.2	108.7	86.0	99.3	104.4	90.0	98.4	104.5	89.0	98.0
08	OPENINGS	97.5	89.9	95.7	95.8	85.2	93.4	99.3	85.3	96.2	98.3	70.4	92.0	100.6	86.0	97.3	98.3	83.1	94.8
0920	Plaster & Gypsum Board	96.1	95.4	95.6	106.2	87.1	93.5	94.1	86.8	89.3	99.3	76.8	84.3	94.1	87.7	89.9	97.1	82.6	87.5
0950, 0980	Ceilings & Acoustic Treatment	88.3	95.4	93.0	83.3	87.1	85.8	88.3	86.8	87.3	83.3	76.8	79.0	88.3	87.7	87.9	88.3	82.6	84.5
0960	Flooring	92.5	96.9	93.8	105.9	100.0	104.2	91.8	96.9	93.3	98.2	96.9	97.8	91.8	98.7	93.8	95.8	96.9	96.1
0970, 0990	Wall Finishes & Painting/Coating	104.9	91.9	97.2	95.9	86.3	90.3	104.9	68.7	83.6	95.9	89.2	92.0	104.9	89.2	95.7	104.9	86.3	93.9
09	FINISHES	91.3	94.8	93.2	94.5	90.9	92.5	92.1	87.7	89.7	91.5	83.5	87.1	90.6	90.9	90.7	92.8	87.4	89.9
COVERS	DIVS. 10 - 14, 25, 28, 41, 43, 44, 46	100.0	93.7	98.6	100.0	94.0	98.6	100.0	69.1	93.0	100.0	86.5	96.9	100.0	89.2	97.6	100.0	93.0	98.4
21, 22, 23	FIRE SUPPRESSION, PLUMBING & HVAC	96.9	87.7	92.8	100.6	90.7	96.2	96.9	89.3	93.5	97.0	84.9	91.7	96.9	92.1	94.8	100.3	90.1	95.8
26, 27, 3370	ELECTRICAL, COMMUNICATIONS & UTIL.	98.0	89.3	93.4	97.8	93.4	95.5	93.9	89.3	91.5	100.0	78.0	88.3	98.2	94.3	96.1	98.1	80.4	88.7
MF2014	WEIGHTED AVERAGE	98.7	93.4	96.3	100.4	93.2	97.2	98.9	90.6	95.3	99.4	85.8	93.4	99.3	93.4	96.7	98.7	89.7	94.7

DIVISION		WEST VIRGINIA									WISCONSIN								
		PETERSBURG 268			ROMNEY 267			WHEELING 260			BELOIT 535			EAU CLAIRE 547			GREEN BAY 541 - 543		
		MAT.	INST.	TOTAL	MAT.	INST.	TOTAL	MAT.	INST.	TOTAL	MAT.	INST.	TOTAL	MAT.	INST.	TOTAL	MAT.	INST.	TOTAL
015433	CONTRACTOR EQUIPMENT		108.7	108.7		108.7	108.7		108.7	108.7		103.4	103.4		104.1	104.1		101.8	101.8
0241, 31 - 34	SITE & INFRASTRUCTURE, DEMOLITION	100.9	93.6	95.8	103.8	93.7	96.7	110.6	93.7	98.8	96.7	108.2	104.7	96.8	105.5	102.9	100.1	101.2	100.8
0310	Concrete Forming & Accessories	88.3	80.0	81.2	84.3	80.4	80.9	91.2	88.6	88.9	98.1	94.8	95.3	97.4	108.7	107.2	106.2	108.2	108.0
0320	Concrete Reinforcing	94.9	95.2	95.0	95.5	95.2	95.4	94.3	95.8	95.0	97.3	133.0	115.4	94.2	111.7	103.1	92.2	107.2	99.8
0330	Cast-in-Place Concrete	99.6	82.4	93.0	104.3	82.5	96.0	101.8	84.0	95.0	105.3	100.7	103.6	99.6	101.3	100.3	103.0	105.5	103.9
03	CONCRETE	96.8	84.8	91.3	100.3	85.1	93.2	102.2	89.3	96.2	98.5	103.7	101.0	94.6	106.6	100.2	97.6	107.1	102.0
04	MASONRY	101.9	90.5	94.9	98.0	90.5	93.4	110.2	85.8	95.2	98.7	105.7	103.0	93.1	101.9	98.5	124.6	101.3	110.3
05	METALS	98.8	106.7	101.2	98.8	107.1	101.3	99.5	107.2	101.8	95.9	112.2	100.9	94.8	105.8	98.2	97.3	104.9	99.6
06	WOOD, PLASTICS & COMPOSITES	85.8	77.6	81.3	81.1	77.6	79.2	86.4	88.3	87.4	94.5	91.6	92.9	98.4	111.2	105.5	103.0	111.2	107.5
07	THERMAL & MOISTURE PROTECTION	104.4	86.2	96.8	104.5	86.9	97.2	104.8	88.0	97.8	100.1	96.3	98.5	104.5	100.1	102.6	106.7	88.6	99.2
08	OPENINGS	100.6	80.0	96.0	100.6	80.0	95.9	99.1	86.0	96.1	96.6	107.4	99.1	102.1	109.6	103.8	98.5	109.4	101.0
0920	Plaster & Gypsum Board	96.8	76.8	83.5	93.8	76.8	82.5	97.1	87.7	90.9	96.7	91.8	93.4	110.1	111.9	111.3	106.5	111.9	110.1
0950, 0980	Ceilings & Acoustic Treatment	88.3	76.8	80.7	88.3	76.8	80.7	88.3	87.7	87.9	83.7	91.8	89.1	86.7	111.9	103.5	81.1	111.9	101.6
0960	Flooring	93.6	98.7	95.1	91.6	98.7	93.6	96.7	98.7	97.3	93.1	125.5	102.5	84.8	112.9	92.9	101.2	126.0	108.4
0970, 0990	Wall Finishes & Painting/Coating	104.9	89.2	95.7	104.9	89.2	95.7	104.9	89.2	95.7	98.0	106.5	103.0	87.7	81.4	84.0	97.2	85.2	90.2
09	FINISHES	91.5	84.3	87.6	90.7	84.3	87.2	93.1	90.7	91.8	92.7	101.3	97.4	89.2	107.4	99.2	93.7	110.2	102.7
COVERS	DIVS. 10 - 14, 25, 28, 41, 43, 44, 46	100.0	56.8	90.2	100.0	91.7	98.1	100.0	84.4	96.5	100.0	96.9	99.3	100.0	99.2	99.8	100.0	99.2	99.8
21, 22, 23	FIRE SUPPRESSION, PLUMBING & HVAC	96.9	87.6	92.8	96.9	87.4	92.7	100.4	92.3	96.8	100.1	96.7	98.6	100.1	88.2	94.8	100.4	86.5	94.2
26, 27, 3370	ELECTRICAL, COMMUNICATIONS & UTIL.	101.4	78.0	89.0	100.6	78.0	88.6	95.3	94.3	94.8	99.8	87.4	93.2	104.6	87.1	95.3	98.7	83.4	90.6
MF2014	WEIGHTED AVERAGE	98.4	86.6	93.2	98.5	87.7	93.7	100.0	92.2	96.5	97.9	100.0	99.1	98.0	99.2	98.5	100.0	97.9	99.1

DIVISION		WISCONSIN																	
		KENOSHA 531			LA CROSSE 546			LANCASTER 538			MADISON 537			MILWAUKEE 530,532			NEW RICHMOND 540		
		MAT.	INST.	TOTAL	MAT.	INST.	TOTAL	MAT.	INST.	TOTAL	MAT.	INST.	TOTAL	MAT.	INST.	TOTAL	MAT.	INST.	TOTAL
015433	CONTRACTOR EQUIPMENT		101.3	101.3		104.1	104.1		103.4	103.4		103.4	103.4		90.4	90.4		104.5	104.5
0241, 31 - 34	SITE & INFRASTRUCTURE, DEMOLITION	102.3	105.5	104.5	90.6	105.6	101.1	95.8	108.0	104.3	93.9	108.9	104.3	94.1	98.5	97.2	95.6	105.9	102.8
0310	Concrete Forming & Accessories	106.2	111.4	110.7	84.3	108.5	105.3	97.4	99.0	98.8	103.3	98.2	98.9	98.7	115.3	113.1	92.7	95.8	95.3
0320	Concrete Reinforcing	97.1	111.9	104.6	93.9	104.1	99.1	98.8	103.8	101.3	100.3	104.2	102.3	102.1	112.0	107.2	91.2	111.3	101.4
0330	Cast-in-Place Concrete	114.6	109.3	112.6	89.6	101.5	94.1	104.7	100.3	103.0	95.3	100.7	97.3	102.2	112.7	104.9	103.7	81.1	95.1
03	CONCRETE	103.5	110.5	106.7	86.0	105.3	94.9	98.2	100.4	99.2	93.8	100.2	96.8	99.0	112.7	105.4	92.5	93.8	93.1
04	MASONRY	96.2	113.2	106.7	92.2	101.9	98.2	98.8	99.0	98.9	96.7	102.9	100.5	100.8	118.1	111.5	119.5	98.9	106.8
05	METALS	96.8	105.4	99.4	94.7	103.8	97.5	93.5	100.7	95.7	99.0	102.0	99.9	95.7	95.8	95.8	95.1	104.9	98.1
06	WOOD, PLASTICS & COMPOSITES	98.1	111.2	105.3	84.5	111.2	99.2	93.8	98.7	96.5	94.7	96.3	95.6	98.5	114.4	107.2	88.0	94.6	91.6
07	THERMAL & MOISTURE PROTECTION	100.3	108.9	103.9	103.9	99.8	102.2	99.9	95.5	98.1	98.5	102.5	100.1	99.4	112.8	104.9	105.6	90.5	99.3
08	OPENINGS	91.3	110.4	95.6	102.1	102.4	102.2	92.6	95.6	93.2	102.6	100.5	102.1	96.8	112.1	100.2	88.2	93.8	89.4
0920	Plaster & Gypsum Board	86.6	111.9	103.4	105.0	111.9	109.6	95.9	99.1	98.0	106.1	96.6	99.8	97.0	114.8	108.8	95.6	95.0	95.2
0950, 0980	Ceilings & Acoustic Treatment	83.7	111.9	102.5	85.9	111.9	103.2	80.3	99.1	92.8	91.3	96.6	94.8	91.8	114.8	107.1	55.6	95.0	81.8
0960	Flooring	110.2	118.3	112.6	78.5	118.3	90.0	92.6	105.5	96.4	95.8	114.0	101.1	94.2	117.6	101.0	94.0	110.0	98.7
0970, 0990	Wall Finishes & Painting/Coating	108.7	121.1	116.0	87.7	82.8	84.8	98.0	104.1	101.6	93.4	106.5	101.1	102.7	123.3	114.8	99.5	87.0	92.2
09	FINISHES	97.3	114.2	106.5	86.0	108.7	98.4	91.7	101.3	96.9	92.7	101.9	97.7	98.6	117.0	108.6	84.3	98.0	91.8
COVERS	DIVS. 10 - 14, 25, 28, 41, 43, 44, 46	100.0	101.4	100.3	100.0	99.2	99.8	100.0	89.3	97.6	100.0	100.2	100.1	100.0	103.6	100.8	100.0	91.6	98.1
21, 22, 23	FIRE SUPPRESSION, PLUMBING & HVAC	100.3	97.4	99.0	100.1	88.1	94.8	96.5	88.0	92.8	100.0	97.7	98.9	99.9	106.2	102.7	95.9	88.3	92.6
26, 27, 3370	ELECTRICAL, COMMUNICATIONS & UTIL.	100.2	99.9	100.0	104.9	87.1	95.4	99.5	87.1	92.9	99.6	96.6	98.0	100.7	101.9	101.3	102.5	87.1	94.3
MF2014	WEIGHTED AVERAGE	98.6	105.8	101.8	96.3	98.7	97.4	96.1	95.8	96.0	98.3	100.5	99.3	98.6	107.9	102.7	95.9	94.6	95.3

DIVISION		WISCONSIN																	
		OSHKOSH 549			PORTAGE 539			RACINE 534			RHINELANDER 545			SUPERIOR 548			WAUSAU 544		
		MAT.	INST.	TOTAL	MAT.	INST.	TOTAL	MAT.	INST.	TOTAL	MAT.	INST.	TOTAL	MAT.	INST.	TOTAL	MAT.	INST.	TOTAL
015433	CONTRACTOR EQUIPMENT		101.8	101.8		103.4	103.4		103.4	103.4		101.8	101.8		104.5	104.5		101.8	101.8
0241, 31 - 34	SITE & INFRASTRUCTURE, DEMOLITION	91.9	100.8	98.1	86.7	107.8	101.4	96.4	109.5	105.5	104.0	100.8	101.8	92.3	105.6	101.6	87.7	102.1	97.7
0310	Concrete Forming & Accessories	88.9	92.1	91.6	89.8	94.1	93.5	98.4	111.5	109.7	86.6	92.8	92.0	90.7	91.9	91.7	88.4	107.9	105.3
0320	Concrete Reinforcing	92.4	99.4	95.9	98.9	104.0	101.4	97.3	111.9	104.7	92.5	97.8	95.2	91.2	103.9	97.7	92.5	103.8	98.3
0330	Cast-in-Place Concrete	95.5	95.7	95.6	89.9	101.0	94.1	103.3	109.1	105.5	108.3	98.5	104.6	97.7	103.5	99.9	89.0	97.4	92.2
03	CONCRETE	87.6	95.0	91.1	85.9	98.4	91.7	97.6	110.4	103.5	99.5	96.1	97.9	87.2	98.5	92.4	82.8	103.5	92.4
04	MASONRY	106.6	100.8	103.1	88.7	113.2	107.6	98.7	113.2	107.6	125.0	97.9	108.3	118.7	104.2	109.8	106.1	98.0	101.1
05	METALS	95.3	100.6	96.9	94.1	101.1	96.2	97.5	105.5	99.9	95.2	100.5	96.8	96.0	103.6	98.3	95.0	103.2	97.5
06	WOOD, PLASTICS & COMPOSITES	84.0	91.7	88.2	84.8	91.6	88.5	94.8	111.2	103.8	81.7	91.7	87.2	86.4	90.2	88.5	83.5	111.2	98.8
07	THERMAL & MOISTURE PROTECTION	105.7	83.7	96.6	99.3	91.1	95.9	100.2	108.5	103.6	106.6	84.8	97.5	105.2	91.4	99.5	105.6	86.3	97.6
08	OPENINGS	94.9	91.4	94.1	92.7	97.5	93.8	96.6	110.4	99.7	94.9	90.2	93.8	87.7	91.7	88.6	95.1	102.8	96.9
0920	Plaster & Gypsum Board	94.4	91.8	92.6	89.2	91.8	90.9	96.7	111.9	106.8	94.4	91.8	92.6	95.4	90.5	92.1	94.4	111.9	106.0
0950, 0980	Ceilings & Acoustic Treatment	81.1	91.8	88.2	82.9	91.8	88.8	83.7	111.9	102.5	81.1	91.8	88.2	56.4	90.5	79.1	81.1	111.9	101.6
0960	Flooring	91.9	118.3	99.6	88.8	115.1	96.5	93.1	118.3	100.4	91.3	111.9	97.3	95.1	127.5	104.5	91.8	111.9	97.7
0970, 0990	Wall Finishes & Painting/Coating	93.9	85.2	88.8	88.0	104.1	101.6	98.0	121.1	111.5	93.9	85.2	88.8	88.2	104.8	98.0	93.9	85.2	88.8
09	FINISHES	88.4	96.7	92.9	89.6	98.9	94.7	92.7	114.2	104.4	89.2	94.8	92.3	83.7	100.0	92.6	88.1	107.5	98.7
COVERS	DIVS. 10 - 14, 25, 28, 41, 43, 44, 46	100.0	89.1	97.5	100.0	90.0	97.7	100.0	101.4	100.3	100.0	90.2	97.8	100.0	89.8	97.7	100.0	99.2	99.8
21, 22, 23	FIRE SUPPRESSION, PLUMBING & HVAC	96.9	81.8	90.2	96.5	97.5	97.0	100.1	97.5	98.9	96.9	87.3	92.6	95.9	92.3	94.3	96.9	87.7	92.8
26, 27, 3370	ELECTRICAL, COMMUNICATIONS & UTIL.	103.1	78.3	90.0	103.4	96.6	99.8	99.6	103.5	101.7	102.4	79.7	90.4	107.7	101.3	104.3	104.3	82.4	92.7
MF2014	WEIGHTED AVERAGE	95.9	90.7	93.6	94.6	98.6	96.3	98.1	106.6	101.8	98.6	91.9	95.7	95.7	98.5	96.9	95.3	96.4	95.8

City Cost Indexes

WYOMING

DIVISION		CASPER 826 MAT.	INST.	TOTAL	CHEYENNE 820 MAT.	INST.	TOTAL	NEWCASTLE 827 MAT.	INST.	TOTAL	RAWLINS 823 MAT.	INST.	TOTAL	RIVERTON 825 MAT.	INST.	TOTAL	ROCK SPRINGS 829-831 MAT.	INST.	TOTAL
015433	CONTRACTOR EQUIPMENT		98.8	98.8		98.8	98.8		98.8	98.8		98.8	98.8		98.8	98.8		98.8	98.8
0241, 31 - 34	SITE & INFRASTRUCTURE, DEMOLITION	97.3	95.7	96.2	93.2	95.7	95.0	85.4	95.4	92.4	99.6	95.4	96.7	92.9	95.3	94.6	89.6	95.4	93.6
0310	Concrete Forming & Accessories	102.8	53.4	60.1	103.9	68.5	73.3	94.6	75.8	78.4	98.8	75.9	79.0	93.6	64.4	68.4	101.0	65.4	70.2
0320	Concrete Reinforcing	111.9	83.8	97.6	107.4	83.8	95.4	115.3	84.3	99.6	115.0	84.3	99.4	116.0	84.2	99.9	116.0	84.0	99.8
0330	Cast-in-Place Concrete	97.1	75.8	89.0	91.7	75.8	85.7	92.6	75.8	86.2	92.7	70.4	84.2	92.7	65.0	82.1	92.6	65.0	82.1
03	CONCRETE	100.2	67.4	85.0	97.8	74.2	86.9	98.1	77.5	88.6	111.7	75.7	95.0	106.3	68.7	88.9	98.6	69.1	84.9
04	MASONRY	105.2	55.1	74.4	108.7	55.1	75.7	105.2	63.8	79.7	105.2	63.8	79.7	105.2	63.8	79.8	170.3	59.4	102.0
05	METALS	103.0	80.9	96.2	105.5	81.0	98.0	101.8	81.6	95.6	101.8	81.6	95.6	101.9	81.3	95.6	102.6	81.0	96.0
06	WOOD, PLASTICS & COMPOSITES	93.5	48.1	68.5	91.3	68.5	78.8	82.5	78.6	80.3	86.1	78.6	81.9	81.5	63.4	71.6	90.8	64.8	76.5
07	THERMAL & MOISTURE PROTECTION	106.0	64.2	88.6	106.0	66.4	88.4	102.3	67.7	87.9	103.8	77.9	93.1	103.2	65.3	87.5	102.4	68.5	88.3
08	OPENINGS	103.3	60.1	93.5	104.2	71.3	96.7	108.3	76.8	101.1	107.9	76.8	100.9	108.1	68.5	99.2	108.7	69.2	99.7
0920	Plaster & Gypsum Board	102.5	46.7	65.4	91.7	67.6	75.7	88.5	78.0	81.5	88.9	78.0	81.6	88.5	62.4	71.2	100.1	63.8	76.0
0950, 0980	Ceilings & Acoustic Treatment	103.2	46.7	65.6	95.1	67.6	76.8	97.0	78.0	84.3	97.0	78.0	84.3	97.0	62.4	74.0	97.0	63.8	74.9
0960	Flooring	112.6	73.8	101.4	111.7	73.8	100.8	105.1	47.2	88.3	108.1	47.2	90.5	104.6	63.9	92.8	110.6	47.2	92.2
0970, 0990	Wall Finishes & Painting/Coating	94.8	58.7	73.6	100.9	58.7	76.1	97.5	58.7	74.7	97.5	58.7	74.7	97.5	58.7	74.7	97.5	58.7	74.7
09	FINISHES	101.2	56.4	76.7	98.2	68.3	81.9	93.4	68.7	79.9	95.7	68.7	81.0	94.1	63.3	77.3	96.6	60.7	77.0
COVERS	DIVS. 10 - 14, 25, 28, 41, 43, 44, 46	100.0	86.2	96.9	100.0	88.4	97.4	100.0	90.6	97.9	100.0	90.6	97.9	100.0	83.0	96.1	100.0	81.2	95.8
21, 22, 23	FIRE SUPPRESSION, PLUMBING & HVAC	100.0	71.5	87.4	100.0	71.5	87.4	98.1	71.5	86.3	98.1	71.5	86.3	98.1	71.5	86.3	99.9	71.5	87.3
26, 27, 3370	ELECTRICAL, COMMUNICATIONS & UTIL.	101.2	62.0	80.4	103.5	65.0	83.0	102.2	83.1	92.1	102.2	69.3	84.8	102.2	64.1	82.0	100.3	83.1	91.2
MF2014	WEIGHTED AVERAGE	101.4	68.3	86.8	101.5	72.0	88.5	100.1	76.4	89.6	102.3	74.5	90.0	101.3	71.0	87.9	104.2	73.0	90.4

WYOMING / CANADA

DIVISION		SHERIDAN 828 MAT.	INST.	TOTAL	WHEATLAND 822 MAT.	INST.	TOTAL	WORLAND 824 MAT.	INST.	TOTAL	YELLOWSTONE NAT'L PA 821 MAT.	INST.	TOTAL	BARRIE, ONTARIO MAT.	INST.	TOTAL	BATHURST, NEW BRUNSWICK MAT.	INST.	TOTAL
015433	CONTRACTOR EQUIPMENT		98.8	98.8		98.8	98.8		98.8	98.8		98.8	98.8		104.3	104.3		103.8	103.8
0241, 31 - 34	SITE & INFRASTRUCTURE, DEMOLITION	93.3	95.7	95.0	90.3	95.4	93.9	87.5	95.4	93.0	87.7	95.4	93.0	118.8	101.1	106.5	105.3	96.6	99.2
0310	Concrete Forming & Accessories	101.7	64.8	69.8	96.5	51.6	57.6	96.6	65.4	69.6	96.6	65.4	69.6	126.3	88.5	93.6	104.8	62.2	67.9
0320	Concrete Reinforcing	116.0	84.3	99.9	115.3	83.7	99.3	116.0	84.2	99.9	118.0	84.2	100.9	177.8	90.4	133.5	147.5	59.9	103.0
0330	Cast-in-Place Concrete	95.8	75.8	88.2	96.7	62.7	83.8	92.6	65.0	82.1	92.6	65.0	82.1	159.5	87.7	132.2	118.0	59.8	95.8
03	CONCRETE	106.3	72.6	90.7	103.0	62.0	84.0	98.3	69.1	84.8	98.6	69.1	84.9	144.2	89.0	118.6	116.8	61.9	91.4
04	MASONRY	105.5	63.8	79.9	105.6	55.1	74.5	105.2	63.8	79.8	105.3	63.8	79.8	169.9	96.0	124.4	161.6	61.8	100.1
05	METALS	105.3	81.6	98.0	101.7	80.3	95.2	101.9	81.3	95.6	102.5	80.7	95.8	108.3	94.6	104.1	103.7	75.6	95.1
06	WOOD, PLASTICS & COMPOSITES	92.5	63.4	76.6	84.1	46.4	63.4	84.2	64.8	73.5	84.2	64.8	73.5	118.3	87.2	101.2	98.7	62.4	78.7
07	THERMAL & MOISTURE PROTECTION	103.5	68.3	88.9	102.5	59.5	84.7	102.3	66.9	87.6	101.7	65.6	86.8	114.2	92.4	105.2	109.9	62.6	90.3
08	OPENINGS	108.9	68.5	99.7	106.8	59.2	96.0	108.5	69.2	99.6	101.5	68.9	94.1	92.6	86.8	91.3	86.4	54.8	79.3
0920	Plaster & Gypsum Board	112.5	62.4	79.2	88.5	44.9	59.5	88.5	63.8	72.1	88.7	63.8	72.2	151.5	86.8	108.5	126.6	61.2	83.2
0950, 0980	Ceilings & Acoustic Treatment	99.8	62.4	74.9	97.0	44.9	62.3	97.0	63.8	74.9	97.8	63.8	75.2	92.7	86.8	88.7	108.8	61.2	77.1
0960	Flooring	109.1	63.9	96.0	106.7	46.4	89.3	106.7	47.2	89.5	106.7	47.2	89.5	123.9	92.3	114.8	105.2	43.9	87.4
0970, 0990	Wall Finishes & Painting/Coating	99.7	58.7	75.7	97.5	58.7	74.7	97.5	58.7	74.7	97.5	58.7	74.7	107.7	89.1	96.8	113.9	51.0	76.9
09	FINISHES	101.1	63.6	80.6	94.2	49.8	69.9	93.9	60.7	75.7	94.1	60.8	75.9	112.3	89.5	99.9	107.6	57.6	80.3
COVERS	DIVS. 10 - 14, 25, 28, 41, 43, 44, 46	100.0	87.8	97.3	100.0	80.0	95.5	100.0	81.7	95.9	100.0	81.8	95.9	139.2	69.5	123.5	131.1	61.8	115.5
21, 22, 23	FIRE SUPPRESSION, PLUMBING & HVAC	98.1	71.5	86.3	98.1	71.5	86.3	98.1	71.5	86.3	98.1	71.5	86.3	103.1	98.3	101.0	103.2	68.0	87.6
26, 27, 3370	ELECTRICAL, COMMUNICATIONS & UTIL.	104.8	64.1	83.2	102.2	83.1	92.1	102.2	83.1	92.1	101.1	93.3	97.0	115.9	89.7	102.0	112.4	60.4	84.9
MF2014	WEIGHTED AVERAGE	103.0	71.9	89.2	100.7	69.2	86.8	100.2	73.4	88.4	99.5	74.8	88.6	117.1	92.7	106.3	109.8	66.2	90.5

CANADA

DIVISION		BRANDON, MANITOBA MAT.	INST.	TOTAL	BRANTFORD, ONTARIO MAT.	INST.	TOTAL	BRIDGEWATER, NOVA SCOTIA MAT.	INST.	TOTAL	CALGARY, ALBERTA MAT.	INST.	TOTAL	CAP-DE-LA-MADELEINE, QUEBEC MAT.	INST.	TOTAL	CHARLESBOURG, QUEBEC MAT.	INST.	TOTAL
015433	CONTRACTOR EQUIPMENT		106.6	106.6		104.3	104.3		103.6	103.6		123.7	123.7		104.4	104.4		104.4	104.4
0241, 31 - 34	SITE & INFRASTRUCTURE, DEMOLITION	135.8	99.7	110.6	118.8	101.4	106.7	103.1	98.5	99.9	123.5	117.3	119.2	98.6	100.0	99.6	98.6	100.0	99.6
0310	Concrete Forming & Accessories	148.3	71.1	81.5	125.6	95.9	100.0	97.4	73.0	76.3	123.6	100.6	103.7	131.6	85.2	91.5	131.6	85.2	91.5
0320	Concrete Reinforcing	195.9	56.7	125.3	171.4	89.1	129.6	145.8	49.6	97.0	138.1	79.0	108.1	145.8	75.4	110.1	145.8	75.4	110.1
0330	Cast-in-Place Concrete	124.8	75.1	105.9	139.8	109.1	128.1	143.6	71.3	116.1	155.3	107.9	137.2	112.8	93.6	105.5	112.8	93.6	105.5
03	CONCRETE	136.0	70.9	105.8	131.7	99.4	116.8	130.0	69.2	101.8	137.5	99.9	120.0	115.7	86.8	102.3	115.7	86.8	102.3
04	MASONRY	227.8	65.3	127.7	168.3	100.4	126.5	163.8	70.5	106.4	197.8	91.5	132.3	164.4	82.1	113.7	164.4	82.1	113.7
05	METALS	117.3	81.0	106.2	102.5	95.3	100.3	101.8	78.3	94.6	123.6	100.0	116.3	100.9	87.7	96.9	100.9	87.7	96.9
06	WOOD, PLASTICS & COMPOSITES	151.4	72.1	107.8	120.2	94.7	106.2	91.0	72.7	80.9	96.7	100.6	98.9	130.9	85.2	105.8	130.9	85.2	105.8
07	THERMAL & MOISTURE PROTECTION	136.9	73.7	110.7	118.3	98.1	109.9	112.9	72.5	96.1	122.4	98.3	112.4	111.6	90.3	102.8	111.6	90.3	102.8
08	OPENINGS	102.5	64.2	93.8	90.4	92.9	90.9	84.5	65.7	80.3	84.3	86.9	84.9	91.7	77.5	88.5	91.7	77.5	88.5
0920	Plaster & Gypsum Board	113.1	70.9	85.0	116.6	94.5	101.9	122.4	71.9	88.8	115.2	99.8	105.0	144.8	84.7	104.8	144.8	84.7	104.8
0950, 0980	Ceilings & Acoustic Treatment	114.4	70.9	85.4	99.5	94.5	96.1	99.5	71.9	81.1	142.4	99.8	114.0	99.5	84.7	89.6	99.5	84.7	89.6
0960	Flooring	138.8	65.5	117.5	119.2	92.3	111.4	100.7	62.3	89.5	124.7	88.7	114.3	119.2	90.9	111.0	119.2	90.9	111.0
0970, 0990	Wall Finishes & Painting/Coating	121.2	57.5	83.8	112.9	97.8	104.0	112.9	63.1	83.6	121.3	112.0	115.8	112.9	88.9	98.8	112.9	88.9	98.8
09	FINISHES	121.7	69.0	92.9	109.1	95.8	101.9	103.8	70.3	85.5	124.6	100.2	111.3	112.4	87.0	98.5	112.4	87.0	98.5
COVERS	DIVS. 10 - 14, 25, 28, 41, 43, 44, 46	131.1	64.4	116.0	131.1	71.6	117.7	131.1	65.0	116.2	131.1	97.5	123.5	131.1	81.0	119.8	131.1	81.0	119.8
21, 22, 23	FIRE SUPPRESSION, PLUMBING & HVAC	103.3	82.9	94.2	103.2	101.2	102.3	103.2	83.6	94.5	105.2	93.1	99.9	103.5	88.5	96.9	103.5	88.5	96.9
26, 27, 3370	ELECTRICAL, COMMUNICATIONS & UTIL.	117.7	68.3	91.5	111.1	89.2	99.5	116.2	63.5	88.2	109.7	101.8	105.5	110.5	70.6	89.3	110.5	70.6	89.3
MF2014	WEIGHTED AVERAGE	123.4	75.1	102.0	113.0	96.6	105.8	111.0	74.7	95.0	119.8	98.6	110.4	110.4	85.1	99.2	110.4	85.1	99.2

CANADA

DIVISION		CHARLOTTETOWN, PRINCE EDWARD ISLAND			CHICOUTIMI, QUEBEC			CORNER BROOK, NEWFOUNDLAND			CORNWALL, ONTARIO			DALHOUSIE, NEW BRUNSWICK			DARTMOUTH, NOVA SCOTIA		
		MAT.	INST.	TOTAL	MAT.	INST.	TOTAL	MAT.	INST.	TOTAL	MAT.	INST.	TOTAL	MAT.	INST.	TOTAL	MAT.	INST.	TOTAL
015433	CONTRACTOR EQUIPMENT		117.9	117.9		104.4	104.4		104.6	104.6		104.3	104.3		103.8	103.8		103.6	103.6
0241, 31 - 34	SITE & INFRASTRUCTURE, DEMOLITION	128.4	106.2	112.9	102.2	99.5	100.3	140.2	96.7	109.9	116.8	100.9	105.7	101.6	96.6	98.1	126.4	98.5	106.9
0310	Concrete Forming & Accessories	119.5	56.7	65.2	134.6	92.0	97.7	124.4	58.8	67.6	123.3	88.7	93.4	103.3	62.4	67.9	113.7	73.0	78.5
0320	Concrete Reinforcing	156.7	48.4	101.7	111.3	93.4	102.2	179.7	49.5	113.6	171.4	88.8	129.4	149.2	59.9	103.9	187.8	49.6	117.6
0330	Cast-in-Place Concrete	149.5	59.0	115.0	111.1	97.2	105.8	145.3	67.6	115.7	125.7	99.4	115.7	121.0	59.9	97.7	140.2	71.3	114.0
03	CONCRETE	137.6	57.8	100.6	107.4	94.1	101.3	170.7	61.5	120.1	124.8	92.7	109.9	123.0	62.0	94.8	151.8	69.2	113.5
04	MASONRY	182.9	59.0	106.6	163.9	91.7	119.4	223.6	60.2	123.0	167.2	92.1	120.9	163.1	61.8	100.7	238.4	70.5	135.0
05	METALS	125.8	80.8	112.0	103.5	92.5	100.1	117.7	75.6	104.9	102.4	93.9	99.8	97.1	75.8	90.6	118.1	78.3	105.9
06	WOOD, PLASTICS & COMPOSITES	98.0	56.2	75.0	131.6	92.6	110.2	129.7	57.8	90.2	118.5	88.0	101.7	97.9	62.4	78.4	117.9	72.7	93.0
07	THERMAL & MOISTURE PROTECTION	130.7	59.9	101.3	110.3	98.8	105.5	141.8	62.0	108.7	118.1	92.5	107.5	114.4	62.6	92.9	138.8	72.5	111.3
08	OPENINGS	82.8	48.9	75.1	90.5	78.1	87.7	109.0	54.9	96.8	91.7	86.4	90.5	87.3	54.8	79.9	92.8	65.7	86.7
0920	Plaster & Gypsum Board	116.2	54.6	75.2	144.8	92.2	109.8	147.0	56.5	86.9	173.3	87.6	116.4	130.5	61.2	84.5	142.6	71.9	95.6
0950, 0980	Ceilings & Acoustic Treatment	126.5	54.6	78.7	107.9	92.2	97.5	115.3	56.5	76.2	102.0	87.6	92.5	102.8	61.2	75.1	122.1	71.9	88.7
0960	Flooring	113.7	58.5	97.7	121.9	90.9	112.9	119.6	52.4	100.1	119.2	91.0	111.0	106.7	67.1	95.2	114.2	62.3	99.2
0970, 0990	Wall Finishes & Painting/Coating	121.7	42.4	75.1	113.9	105.2	108.8	121.1	60.0	85.2	112.9	91.1	100.1	116.6	51.0	78.0	121.1	63.1	87.0
09	FINISHES	118.7	55.8	84.4	114.6	94.1	103.4	121.4	57.7	86.5	117.5	89.6	102.3	108.7	62.4	83.4	119.1	70.3	92.4
COVERS	DIVS. 10 - 14, 25, 28, 41, 43, 44, 46	131.1	61.7	115.4	131.1	84.5	120.6	131.1	61.5	115.4	131.1	69.1	117.1	131.1	61.8	115.5	131.1	65.0	116.2
21, 22, 23	FIRE SUPPRESSION, PLUMBING & HVAC	103.6	61.9	85.2	103.2	80.8	93.3	103.3	68.8	88.0	103.5	99.1	101.5	103.2	68.1	87.7	103.3	83.6	94.5
26, 27, 3370	ELECTRICAL, COMMUNICATIONS & UTIL.	111.5	51.4	79.6	109.3	86.2	97.0	115.3	55.7	83.7	111.7	90.2	100.3	113.1	57.1	83.4	120.0	63.5	90.0
MF2014	WEIGHTED AVERAGE	119.0	63.4	94.4	109.7	89.3	100.7	128.1	65.4	100.4	113.0	93.0	104.1	109.8	66.4	90.6	124.5	74.7	102.5

CANADA

DIVISION		EDMONTON, ALBERTA			FORT MCMURRAY, ALBERTA			FREDERICTON, NEW BRUNSWICK			GATINEAU, QUEBEC			GRANBY, QUEBEC			HALIFAX, NOVA SCOTIA		
		MAT.	INST.	TOTAL	MAT.	INST.	TOTAL	MAT.	INST.	TOTAL	MAT.	INST.	TOTAL	MAT.	INST.	TOTAL	MAT.	INST.	TOTAL
015433	CONTRACTOR EQUIPMENT		123.8	123.8		106.8	106.8		118.7	118.7		104.4	104.4		104.4	104.4		117.3	117.3
0241, 31 - 34	SITE & INFRASTRUCTURE, DEMOLITION	134.4	117.4	122.5	123.9	102.2	108.7	110.0	107.5	108.3	98.4	100.0	99.5	98.9	100.0	99.7	106.8	109.0	108.3
0310	Concrete Forming & Accessories	122.2	100.6	103.5	123.4	93.9	97.9	124.9	63.0	71.4	131.6	85.1	91.4	131.6	85.0	91.3	118.4	83.6	88.3
0320	Concrete Reinforcing	139.1	79.0	108.6	158.6	78.9	118.1	146.9	60.1	102.8	154.2	75.4	114.2	154.2	75.4	114.2	156.7	66.7	111.0
0330	Cast-in-Place Concrete	160.6	107.9	140.5	185.6	104.4	154.6	124.6	60.7	100.3	111.3	93.5	104.5	115.0	93.5	106.8	112.5	83.9	101.6
03	CONCRETE	140.1	99.9	121.4	152.1	95.1	125.7	124.6	63.3	96.2	116.0	86.7	102.4	117.8	86.7	103.4	119.0	81.8	101.7
04	MASONRY	197.6	91.5	132.3	208.7	88.5	134.7	178.1	62.2	107.3	164.3	82.1	113.7	164.6	82.1	113.8	181.8	85.9	122.7
05	METALS	123.6	100.0	116.4	127.3	90.6	116.1	122.6	86.0	111.4	100.9	87.6	96.9	101.1	87.5	97.0	124.1	92.5	114.4
06	WOOD, PLASTICS & COMPOSITES	100.1	100.6	100.4	114.6	93.6	103.0	111.8	62.7	84.8	130.9	85.2	105.8	130.9	85.2	105.8	99.8	82.9	90.5
07	THERMAL & MOISTURE PROTECTION	126.1	98.3	114.5	126.8	95.4	113.8	125.2	62.9	99.3	111.6	90.3	102.8	111.6	88.7	102.1	125.2	84.4	108.3
08	OPENINGS	82.3	87.0	83.4	91.7	83.1	89.7	85.9	53.9	78.7	91.7	72.9	87.4	91.7	72.9	87.4	90.4	73.4	86.5
0920	Plaster & Gypsum Board	112.9	99.8	104.2	117.4	93.0	101.2	121.0	61.2	81.3	114.5	84.7	94.7	114.5	84.7	94.7	113.7	82.1	92.7
0950, 0980	Ceilings & Acoustic Treatment	137.5	99.8	112.4	107.1	93.0	97.7	128.2	61.2	83.6	99.5	84.7	89.6	99.5	84.7	89.6	120.4	82.1	94.9
0960	Flooring	120.7	88.7	111.4	119.2	88.7	110.3	115.9	70.0	102.6	119.2	90.9	111.0	119.2	90.9	111.0	106.7	83.9	100.1
0970, 0990	Wall Finishes & Painting/Coating	112.2	112.0	112.1	113.0	93.9	101.8	118.9	64.9	87.2	112.9	88.9	98.8	112.9	88.9	98.8	116.9	89.4	100.8
09	FINISHES	118.6	100.2	108.5	112.1	93.3	101.8	119.3	64.7	89.5	108.1	87.0	96.6	108.1	87.0	96.6	112.5	84.8	97.3
COVERS	DIVS. 10 - 14, 25, 28, 41, 43, 44, 46	131.1	97.5	123.5	131.1	94.9	122.9	131.1	62.6	115.6	131.1	81.0	119.8	131.1	81.0	119.8	131.1	69.7	117.2
21, 22, 23	FIRE SUPPRESSION, PLUMBING & HVAC	105.2	93.1	99.9	103.7	97.8	101.1	103.4	77.4	91.9	103.5	88.5	96.8	103.2	88.5	96.7	104.8	79.7	93.7
26, 27, 3370	ELECTRICAL, COMMUNICATIONS & UTIL.	109.8	101.8	105.6	105.5	83.8	94.0	111.7	75.4	92.4	110.5	70.5	89.3	111.1	70.5	89.6	110.3	85.0	96.9
MF2014	WEIGHTED AVERAGE	119.8	98.6	110.4	122.1	92.8	109.2	116.4	73.3	97.4	110.1	84.9	98.9	110.3	84.8	99.0	116.1	85.0	102.4

CANADA

DIVISION		HAMILTON, ONTARIO			HULL, QUEBEC			JOLIETTE, QUEBEC			KAMLOOPS, BRITISH COLUMBIA			KINGSTON, ONTARIO			KITCHENER, ONTARIO		
		MAT.	INST.	TOTAL	MAT.	INST.	TOTAL	MAT.	INST.	TOTAL	MAT.	INST.	TOTAL	MAT.	INST.	TOTAL	MAT.	INST.	TOTAL
015433	CONTRACTOR EQUIPMENT		120.6	120.6		104.4	104.4		104.4	104.4		107.8	107.8		106.6	106.6		106.5	106.5
0241, 31 - 34	SITE & INFRASTRUCTURE, DEMOLITION	110.0	115.8	114.1	98.4	100.0	99.5	99.0	100.0	99.7	121.3	103.8	109.1	116.8	104.9	108.5	98.0	106.3	103.6
0310	Concrete Forming & Accessories	122.0	94.3	98.0	131.6	85.1	91.4	131.6	85.2	91.5	123.3	88.5	93.2	123.4	88.8	93.4	114.2	91.1	94.2
0320	Concrete Reinforcing	137.6	99.4	118.2	154.2	75.4	114.2	145.8	75.4	110.1	114.3	78.9	96.4	171.4	88.8	129.4	97.8	99.6	98.7
0330	Cast-in-Place Concrete	119.2	101.0	112.3	111.3	93.5	104.5	115.9	93.6	107.4	100.1	98.2	99.5	125.7	99.4	115.7	116.2	101.1	110.4
03	CONCRETE	116.8	98.3	108.2	116.0	86.7	102.4	117.2	86.8	103.1	124.9	90.5	108.9	126.7	92.8	110.9	102.7	96.3	99.7
04	MASONRY	171.5	99.1	126.9	164.3	82.1	113.7	164.7	82.1	113.8	171.6	90.3	121.5	174.1	92.1	123.7	145.3	102.3	118.8
05	METALS	120.2	107.7	116.3	101.1	87.6	97.0	101.1	87.7	97.0	103.1	90.1	99.1	103.8	93.9	100.8	111.7	98.6	107.7
06	WOOD, PLASTICS & COMPOSITES	99.5	93.2	96.0	130.9	85.2	105.8	130.9	85.2	105.8	102.9	87.0	94.1	118.5	88.1	101.8	107.9	89.1	97.6
07	THERMAL & MOISTURE PROTECTION	124.0	101.7	114.8	111.6	90.3	102.8	111.6	90.3	102.8	127.9	87.7	111.2	118.1	93.7	108.0	115.2	101.4	109.4
08	OPENINGS	86.7	93.3	88.2	91.7	72.9	87.4	91.7	77.5	88.5	88.4	84.2	87.5	91.7	86.1	90.4	83.2	89.3	84.5
0920	Plaster & Gypsum Board	123.7	92.6	103.1	114.5	84.7	94.7	144.8	84.7	104.8	102.2	86.2	91.6	176.2	87.8	117.4	105.6	88.7	94.4
0950, 0980	Ceilings & Acoustic Treatment	126.1	92.6	103.8	99.5	84.7	89.6	99.5	84.7	89.6	99.5	86.2	90.6	114.7	87.8	96.8	101.4	88.7	93.0
0960	Flooring	116.1	97.8	110.8	119.2	90.9	111.0	119.2	90.9	111.0	117.8	52.0	98.8	119.2	91.0	111.0	106.4	97.8	103.9
0970, 0990	Wall Finishes & Painting/Coating	106.6	101.8	103.7	112.9	88.9	98.8	112.9	88.9	98.8	112.9	81.3	94.3	112.9	84.2	96.0	106.8	96.9	101.0
09	FINISHES	116.2	95.6	105.0	108.1	87.0	96.6	112.4	87.0	98.5	108.6	81.2	93.6	120.6	88.9	103.3	101.7	92.5	96.7
COVERS	DIVS. 10 - 14, 25, 28, 41, 43, 44, 46	131.1	94.6	122.9	131.1	81.0	119.8	131.1	81.0	119.8	131.1	90.4	121.9	131.1	69.1	117.1	131.1	93.3	122.6
21, 22, 23	FIRE SUPPRESSION, PLUMBING & HVAC	105.4	88.4	97.9	103.2	88.5	96.7	103.2	88.5	96.7	103.2	92.0	98.2	103.5	99.2	101.6	104.4	91.9	98.9
26, 27, 3370	ELECTRICAL, COMMUNICATIONS & UTIL.	105.8	99.7	102.5	112.3	70.5	90.2	111.1	70.6	89.6	114.4	80.2	96.3	111.8	88.9	99.6	109.6	102.7	106.0
MF2014	WEIGHTED AVERAGE	114.3	98.1	107.1	110.2	84.9	99.0	110.6	85.1	99.3	112.7	88.8	102.2	114.1	93.1	104.8	107.8	97.0	103.0

City Cost Indexes

CANADA

DIVISION		LAVAL, QUEBEC			LETHBRIDGE, ALBERTA			LLOYDMINSTER, ALBERTA			LONDON, ONTARIO			MEDICINE HAT, ALBERTA			MONCTON, NEW BRUNSWICK		
		MAT.	INST.	TOTAL	MAT.	INST.	TOTAL	MAT.	INST.	TOTAL	MAT.	INST.	TOTAL	MAT.	INST.	TOTAL	MAT.	INST.	TOTAL
015433	CONTRACTOR EQUIPMENT		104.4	104.4		106.8	106.8		106.8	106.8		118.6	118.6		106.8	106.8		103.8	103.8
0241, 31 - 34	SITE & INFRASTRUCTURE, DEMOLITION	98.9	100.0	99.6	116.2	102.8	106.8	116.1	102.1	106.4	107.6	112.7	111.1	114.8	102.2	106.0	104.7	97.0	99.3
0310	Concrete Forming & Accessories	131.9	85.1	91.4	124.8	94.0	98.1	122.9	83.9	89.2	121.8	92.3	96.3	124.8	83.9	89.4	104.8	64.8	70.2
0320	Concrete Reinforcing	154.2	75.4	114.2	158.6	78.9	118.1	158.6	78.9	118.1	138.0	99.6	118.5	158.6	78.9	118.1	147.5	65.3	105.8
0330	Cast-in-Place Concrete	115.0	93.5	106.8	139.2	104.4	125.9	129.1	100.6	118.3	124.3	103.4	116.3	129.1	100.6	118.3	113.6	77.8	100.0
03	CONCRETE	117.9	86.7	103.4	129.7	95.1	113.7	124.7	89.3	108.3	119.3	98.1	109.5	124.9	89.3	108.4	114.7	70.7	94.3
04	MASONRY	164.6	82.1	113.8	182.3	88.5	124.6	163.7	81.9	113.3	175.7	103.3	131.1	163.7	81.9	113.3	161.3	63.4	101.0
05	METALS	101.0	87.6	96.9	121.8	90.7	112.3	103.1	90.5	99.2	120.0	105.9	115.8	103.2	90.5	99.3	103.7	86.0	98.3
06	WOOD, PLASTICS & COMPOSITES	131.0	85.2	105.9	118.0	93.6	104.6	114.6	83.6	97.5	106.2	90.3	97.5	118.0	83.6	99.1	98.7	64.4	79.8
07	THERMAL & MOISTURE PROTECTION	112.2	90.3	103.1	124.0	95.4	112.1	120.5	90.7	108.1	125.4	101.4	115.4	127.2	90.7	112.0	114.3	67.0	94.7
08	OPENINGS	91.7	72.9	87.4	91.7	83.1	89.7	91.7	77.6	88.5	81.0	89.7	83.0	91.7	77.6	88.5	86.4	62.2	80.9
0920	Plaster & Gypsum Board	114.8	84.7	94.8	106.7	93.0	97.6	102.7	82.7	89.4	131.6	89.7	103.7	105.0	82.7	90.2	126.6	63.3	84.5
0950, 0980	Ceilings & Acoustic Treatment	99.5	84.7	89.6	107.1	93.0	97.7	99.5	82.7	88.3	132.0	89.7	103.8	99.5	82.7	88.3	108.8	63.3	78.5
0960	Flooring	119.2	90.9	111.0	119.2	88.7	110.3	119.2	88.7	110.3	113.3	97.8	108.8	119.2	88.7	110.3	105.2	68.7	94.6
0970, 0990	Wall Finishes & Painting/Coating	112.9	88.9	98.8	112.8	102.5	106.8	113.0	79.9	93.5	113.2	103.9	107.8	112.8	79.9	93.5	113.9	52.2	77.7
09	FINISHES	108.2	87.0	96.6	109.7	94.2	101.2	107.7	84.2	94.8	119.0	94.3	105.5	107.8	84.2	94.9	107.6	64.2	83.9
COVERS	DIVS. 10 - 14, 25, 28, 41, 43, 44, 46	131.1	81.0	119.8	131.1	94.9	122.9	131.1	91.5	122.2	131.1	94.4	122.8	131.1	91.5	122.2	131.1	63.5	115.9
21, 22, 23	FIRE SUPPRESSION, PLUMBING & HVAC	104.5	88.5	97.4	103.5	94.5	99.5	103.5	94.4	99.5	105.4	90.3	98.7	103.2	91.1	97.8	103.2	70.4	88.7
26, 27, 3370	ELECTRICAL, COMMUNICATIONS & UTIL.	109.6	70.5	88.9	106.9	83.8	94.6	104.7	83.8	93.6	104.9	102.7	103.7	104.7	83.8	93.6	116.5	62.3	87.8
MF2014	WEIGHTED AVERAGE	110.5	84.9	99.2	116.7	92.3	105.9	111.5	88.9	101.5	114.3	98.5	107.3	111.7	88.2	101.3	110.1	70.6	92.6

CANADA

DIVISION		MONTREAL, QUEBEC			MOOSE JAW, SASKATCHEWAN			NEW GLASGOW, NOVA SCOTIA			NEWCASTLE, NEW BRUNSWICK			NORTH BAY, ONTARIO			OSHAWA, ONTARIO		
		MAT.	INST.	TOTAL	MAT.	INST.	TOTAL	MAT.	INST.	TOTAL	MAT.	INST.	TOTAL	MAT.	INST.	TOTAL	MAT.	INST.	TOTAL
015433	CONTRACTOR EQUIPMENT		120.1	120.1		102.8	102.8		103.6	103.6		103.8	103.8		104.3	104.3		106.5	106.5
0241, 31 - 34	SITE & INFRASTRUCTURE, DEMOLITION	112.3	110.7	111.2	116.3	96.2	102.2	120.4	98.5	105.1	105.3	96.6	99.2	136.3	100.5	111.3	109.1	105.1	106.3
0310	Concrete Forming & Accessories	121.4	92.4	96.3	107.5	58.7	65.3	113.6	73.0	78.5	104.8	62.4	68.1	149.2	86.1	94.7	119.8	91.0	94.9
0320	Concrete Reinforcing	141.2	93.5	117.0	111.9	63.9	87.5	179.7	49.6	113.7	147.5	59.9	103.1	212.2	88.3	149.3	155.0	94.4	124.2
0330	Cast-in-Place Concrete	131.7	98.5	119.0	125.3	67.9	103.4	140.2	71.3	114.0	118.0	59.9	95.9	135.1	85.8	116.3	134.4	88.5	116.9
03	CONCRETE	123.2	95.4	110.3	109.8	63.8	88.5	150.7	69.2	112.9	116.8	62.1	91.4	152.5	86.8	122.1	125.1	91.0	109.3
04	MASONRY	171.6	91.7	122.4	162.4	60.7	99.7	223.2	70.5	129.2	161.6	61.8	100.1	229.5	88.1	142.4	148.3	94.7	115.3
05	METALS	124.5	102.4	117.8	99.9	77.4	93.0	115.7	78.3	104.3	103.7	75.8	95.2	116.5	93.6	109.5	103.1	97.2	101.3
06	WOOD, PLASTICS & COMPOSITES	104.2	93.0	98.0	100.4	57.3	76.7	117.9	72.7	93.0	98.7	62.4	78.7	154.4	86.6	117.1	114.5	90.0	101.0
07	THERMAL & MOISTURE PROTECTION	118.4	98.5	110.1	111.0	64.4	91.7	138.8	72.5	111.3	114.3	62.6	92.9	145.6	88.7	122.0	116.1	93.2	106.6
08	OPENINGS	86.3	80.3	84.9	87.6	55.1	80.2	92.8	65.7	86.7	86.4	54.8	79.3	100.8	83.9	97.0	88.5	90.7	89.0
0920	Plaster & Gypsum Board	121.2	92.2	101.9	99.6	56.1	70.7	140.9	71.9	95.0	126.6	61.2	83.2	134.7	86.2	102.4	108.8	89.7	96.1
0950, 0980	Ceilings & Acoustic Treatment	133.8	92.2	106.1	99.5	56.1	70.6	114.4	71.9	86.1	108.8	61.2	77.1	114.4	86.2	95.6	98.0	89.7	92.5
0960	Flooring	117.1	93.1	110.1	108.9	56.7	93.8	114.2	62.3	99.2	105.2	67.1	94.1	138.8	91.0	124.9	109.4	100.4	106.8
0970, 0990	Wall Finishes & Painting/Coating	108.2	105.2	106.4	112.9	64.9	84.7	121.1	63.1	87.0	113.9	51.0	76.9	121.1	90.4	103.1	106.8	111.6	109.6
09	FINISHES	116.8	94.8	104.8	104.0	58.7	79.2	117.3	70.3	91.6	107.6	62.4	82.9	124.5	87.9	104.5	103.0	94.8	98.5
COVERS	DIVS. 10 - 14, 25, 28, 41, 43, 44, 46	131.1	85.4	120.8	131.1	61.5	115.4	131.1	65.0	116.2	131.1	61.8	115.5	131.1	67.9	116.8	131.1	93.0	122.5
21, 22, 23	FIRE SUPPRESSION, PLUMBING & HVAC	105.2	80.9	94.5	103.5	73.9	90.4	103.3	83.6	94.5	103.2	68.1	87.6	103.3	97.1	100.5	104.4	102.7	103.7
26, 27, 3370	ELECTRICAL, COMMUNICATIONS & UTIL.	106.4	86.2	95.7	113.3	60.9	85.5	115.6	63.5	87.9	111.8	60.4	84.5	115.8	90.1	102.2	110.7	92.5	101.1
MF2014	WEIGHTED AVERAGE	115.7	91.5	105.0	108.7	67.9	90.7	122.4	74.7	101.4	109.9	66.8	90.9	125.5	90.8	110.2	110.5	96.4	104.2

CANADA

DIVISION		OTTAWA, ONTARIO			OWEN SOUND, ONTARIO			PETERBOROUGH, ONTARIO			PORTAGE LA PRAIRIE, MANITOBA			PRINCE ALBERT, SASKATCHEWAN			PRINCE GEORGE, BRITISH COLUMBIA		
		MAT.	INST.	TOTAL	MAT.	INST.	TOTAL	MAT.	INST.	TOTAL	MAT.	INST.	TOTAL	MAT.	INST.	TOTAL	MAT.	INST.	TOTAL
015433	CONTRACTOR EQUIPMENT		119.1	119.1		104.3	104.3		104.3	104.3		106.6	106.6		102.8	102.8		107.8	107.8
0241, 31 - 34	SITE & INFRASTRUCTURE, DEMOLITION	107.1	113.2	111.3	118.8	100.9	106.3	118.8	100.8	106.2	117.4	99.6	105.0	111.7	96.3	101.0	124.8	103.8	110.2
0310	Concrete Forming & Accessories	123.1	92.4	96.5	126.3	84.7	90.3	125.6	87.2	92.4	125.0	70.7	78.0	107.5	58.5	65.1	113.0	83.4	87.4
0320	Concrete Reinforcing	130.1	99.6	114.6	177.8	90.4	133.5	171.4	88.8	129.5	158.6	56.7	106.9	116.9	63.9	90.0	114.3	78.9	96.4
0330	Cast-in-Place Concrete	132.3	105.4	122.0	159.5	81.8	129.9	139.8	87.4	119.8	129.1	74.5	108.3	113.6	67.9	96.1	125.7	98.2	115.3
03	CONCRETE	122.2	98.9	111.4	144.2	85.2	116.8	131.7	88.0	111.4	118.6	70.5	96.3	104.7	63.7	85.7	136.5	88.2	114.1
04	MASONRY	170.3	105.9	130.6	169.9	93.6	122.9	168.3	94.7	123.0	166.9	64.2	103.7	161.5	60.7	99.4	173.7	90.3	122.3
05	METALS	122.9	106.6	117.9	108.3	94.4	104.0	102.5	94.1	100.0	103.2	80.9	96.4	100.0	77.2	93.0	103.1	90.1	99.1
06	WOOD, PLASTICS & COMPOSITES	105.7	89.1	96.6	118.3	83.3	99.1	120.2	85.3	101.0	117.9	72.1	92.7	100.4	57.3	76.7	102.9	79.9	90.3
07	THERMAL & MOISTURE PROTECTION	128.0	102.6	117.5	114.2	89.6	104.0	118.3	94.5	108.4	111.5	73.2	95.6	110.9	63.3	91.1	121.9	86.9	107.4
08	OPENINGS	90.9	89.0	90.4	92.6	83.3	90.5	90.4	85.7	89.3	91.7	64.2	85.4	86.5	55.1	79.4	88.4	80.3	86.6
0920	Plaster & Gypsum Board	136.2	88.5	104.5	151.5	82.8	105.9	116.6	84.8	95.5	104.6	70.9	82.2	99.6	56.1	70.7	102.2	78.9	86.7
0950, 0980	Ceilings & Acoustic Treatment	131.1	88.5	102.7	92.7	82.8	86.1	99.5	84.8	89.7	99.5	70.9	80.4	99.5	56.1	70.6	99.5	78.9	85.8
0960	Flooring	107.9	93.2	103.7	123.9	92.3	114.8	112.9	91.0	111.0	119.2	65.5	103.6	108.9	56.7	93.8	114.5	71.3	102.0
0970, 0990	Wall Finishes & Painting/Coating	114.0	98.5	104.9	107.7	89.1	96.8	112.9	92.7	101.0	113.0	57.5	80.4	112.9	55.3	79.1	112.9	81.3	94.3
09	FINISHES	119.1	92.8	104.7	112.3	86.7	98.3	109.1	88.5	97.9	107.8	68.7	86.4	104.0	57.6	78.7	107.6	80.4	92.7
COVERS	DIVS. 10 - 14, 25, 28, 41, 43, 44, 46	131.1	93.0	122.5	139.2	68.3	123.2	131.1	69.3	117.2	131.1	64.0	116.0	131.1	61.5	115.4	131.1	89.6	121.8
21, 22, 23	FIRE SUPPRESSION, PLUMBING & HVAC	105.5	92.0	99.5	103.1	97.0	100.4	103.2	100.6	102.1	103.2	82.4	94.0	103.5	66.6	87.2	103.2	92.0	98.2
26, 27, 3370	ELECTRICAL, COMMUNICATIONS & UTIL.	107.8	103.5	105.5	117.3	88.9	102.2	111.1	89.7	99.8	113.0	58.9	84.3	113.3	60.9	85.5	111.3	80.2	94.8
MF2014	WEIGHTED AVERAGE	116.3	99.2	108.8	117.3	90.9	105.6	113.0	92.7	104.1	111.4	73.4	94.6	107.8	66.1	89.4	113.8	88.1	102.4

		CANADA																
DIVISION		QUEBEC CITY, QUEBEC			RED DEER, ALBERTA			REGINA, SASKATCHEWAN			RIMOUSKI, QUEBEC			ROUYN-NORANDA, QUEBEC			SAINT HYACINTHE, QUEBEC		
		MAT.	INST.	TOTAL	MAT.	INST.	TOTAL	MAT.	INST.	TOTAL	MAT.	INST.	TOTAL	MAT.	INST.	TOTAL	MAT.	INST.	TOTAL
015433	CONTRACTOR EQUIPMENT		120.7	120.7		106.8	106.8		121.5	121.5		104.4	104.4		104.4	104.4		104.4	104.4
0241, 31 - 34	SITE & INFRASTRUCTURE, DEMOLITION	113.9	110.1	111.2	114.8	102.2	106.0	120.9	112.3	114.9	98.8	99.5	99.3	98.4	100.0	99.5	98.9	100.0	99.6
0310	Concrete Forming & Accessories	123.7	92.6	96.8	139.6	83.9	91.4	120.9	89.1	93.4	131.6	92.0	97.3	131.6	85.1	91.4	131.6	85.1	91.4
0320	Concrete Reinforcing	145.8	93.5	119.3	158.6	78.9	118.1	144.5	87.8	115.7	110.0	93.4	101.6	154.2	75.4	114.2	154.2	75.4	114.2
0330	Cast-in-Place Concrete	127.6	98.7	116.6	129.1	100.6	118.3	142.6	93.8	124.0	117.0	97.2	109.5	111.3	93.5	104.5	115.0	93.5	106.8
03	CONCRETE	122.0	95.7	109.8	125.7	89.3	108.8	130.3	91.3	112.2	113.0	94.1	104.3	116.0	86.7	102.4	117.8	86.7	103.4
04	MASONRY	166.1	91.7	120.3	163.7	81.9	113.3	188.8	91.2	128.7	164.1	91.7	119.5	164.3	82.1	113.7	164.6	82.1	113.8
05	METALS	124.5	103.8	118.2	103.2	90.5	99.3	126.5	100.6	118.6	100.7	92.5	98.2	101.1	87.6	97.0	101.1	87.6	97.0
06	WOOD, PLASTICS & COMPOSITES	112.2	93.0	101.6	118.0	83.6	99.1	105.6	89.7	96.8	130.9	92.6	109.8	130.9	85.2	105.8	130.9	85.2	105.8
07	THERMAL & MOISTURE PROTECTION	118.7	98.6	110.4	137.2	90.7	117.9	127.1	87.3	110.6	111.6	98.8	106.3	111.6	90.3	102.8	111.9	90.3	102.9
08	OPENINGS	89.2	88.0	89.0	91.7	77.6	88.5	87.8	78.9	85.8	91.2	78.1	88.3	91.7	72.9	87.4	91.7	72.9	87.4
0920	Plaster & Gypsum Board	123.6	92.2	102.7	105.0	82.7	90.2	125.8	88.9	101.3	144.6	92.2	109.8	114.3	84.7	94.6	114.3	84.7	94.6
0950, 0980	Ceilings & Acoustic Treatment	132.9	92.2	105.8	99.5	82.7	88.3	139.1	88.9	105.7	98.6	92.2	94.4	98.6	84.7	89.3	98.6	84.7	89.3
0960	Flooring	117.5	93.1	110.4	121.5	88.7	112.0	120.1	97.8	113.7	120.6	90.9	112.0	119.2	90.9	111.0	119.2	90.9	111.0
0970, 0990	Wall Finishes & Painting/Coating	126.4	105.2	114.0	112.8	79.9	93.5	119.2	72.2	91.6	116.2	105.2	109.7	112.9	88.9	98.8	112.9	88.9	98.8
09	FINISHES	118.7	94.8	105.6	108.5	84.2	95.2	126.0	89.8	106.2	112.9	94.1	102.6	107.9	87.0	96.5	107.9	87.0	96.5
COVERS	DIVS. 10 - 14, 25, 28, 41, 43, 44, 46	131.1	85.4	120.8	131.1	91.5	122.2	131.1	71.2	117.6	131.1	84.5	120.6	131.1	81.0	119.8	131.1	81.0	119.8
21, 22, 23	FIRE SUPPRESSION, PLUMBING & HVAC	105.2	81.0	94.5	103.2	91.1	97.8	104.9	82.5	95.0	103.2	80.8	93.3	103.2	88.5	96.7	100.3	88.5	95.1
26, 27, 3370	ELECTRICAL, COMMUNICATIONS & UTIL.	109.7	86.2	97.2	104.7	83.8	93.6	112.9	89.7	100.6	111.1	86.2	97.9	111.1	70.5	89.6	111.8	70.5	89.9
MF2014	WEIGHTED AVERAGE	116.2	92.0	105.5	112.1	88.2	101.6	119.8	90.2	106.7	110.0	89.3	100.9	110.1	84.9	98.9	109.8	84.9	98.8

		CANADA																	
DIVISION		SAINT JOHN, NEW BRUNSWICK			SARNIA, ONTARIO			SASKATOON, SASKATCHEWAN			SAULT STE MARIE, ONTARIO			SHERBROOKE, QUEBEC			SOREL, QUEBEC		
		MAT.	INST.	TOTAL	MAT.	INST.	TOTAL	MAT.	INST.	TOTAL	MAT.	INST.	TOTAL	MAT.	INST.	TOTAL	MAT.	INST.	TOTAL
015433	CONTRACTOR EQUIPMENT		103.8	103.8		104.3	104.3		102.8	102.8		104.3	104.3		104.4	104.4		104.4	104.4
0241, 31 - 34	SITE & INFRASTRUCTURE, DEMOLITION	105.4	98.4	100.5	117.3	100.9	105.9	112.5	98.4	102.7	107.1	100.5	102.5	98.9	100.0	99.7	99.0	100.0	99.7
0310	Concrete Forming & Accessories	125.7	60.1	75.9	124.3	94.3	90.4	107.1	80.7	91.2	113.3	90.3	93.4	131.6	85.1	91.4	131.6	85.2	91.5
0320	Concrete Reinforcing	147.5	65.8	106.0	122.0	90.2	105.9	119.6	87.7	103.4	110.0	88.9	99.3	154.2	75.4	114.2	145.8	75.4	110.1
0330	Cast-in-Place Concrete	116.3	79.6	102.3	129.0	100.9	118.2	124.3	91.9	111.9	115.9	86.0	104.5	115.0	93.5	106.8	115.9	93.6	107.4
03	CONCRETE	117.4	72.8	96.7	119.9	96.0	108.8	111.1	89.8	101.2	105.3	88.9	97.7	117.8	86.7	103.4	117.2	86.8	103.1
04	MASONRY	183.0	72.8	115.1	179.9	97.4	129.1	172.1	91.1	122.3	165.1	96.9	123.1	164.6	82.1	113.8	164.7	82.1	113.8
05	METALS	103.6	87.4	98.6	102.5	94.5	100.0	97.8	89.6	95.3	101.7	94.8	99.6	100.9	87.6	96.9	101.1	87.7	97.0
06	WOOD, PLASTICS & COMPOSITES	120.8	66.3	90.9	119.1	93.6	105.1	97.9	89.3	93.2	107.0	91.5	98.5	130.9	85.2	105.8	130.9	85.2	105.8
07	THERMAL & MOISTURE PROTECTION	114.6	73.7	97.6	118.4	98.2	110.0	110.6	86.7	100.7	117.1	93.4	107.3	111.6	90.3	102.8	111.6	90.3	102.8
08	OPENINGS	86.3	62.2	80.9	92.9	89.6	92.1	87.1	78.7	85.2	84.7	87.9	85.4	91.7	72.9	87.4	91.7	77.5	88.5
0920	Plaster & Gypsum Board	139.3	65.3	90.1	139.9	93.4	109.0	115.6	88.9	97.9	107.8	91.2	96.8	114.3	84.7	94.6	144.6	84.7	104.8
0950, 0980	Ceilings & Acoustic Treatment	113.8	65.3	81.5	103.7	93.4	96.8	119.7	88.9	99.2	99.5	91.2	94.0	98.6	84.7	89.3	98.6	84.7	89.3
0960	Flooring	117.4	68.7	103.3	119.2	99.5	113.5	111.4	97.8	107.5	111.8	96.4	107.4	119.2	90.9	111.0	119.2	90.9	111.0
0970, 0990	Wall Finishes & Painting/Coating	113.9	83.0	95.7	112.9	104.8	108.1	116.6	90.2	101.1	112.9	97.0	103.6	112.9	88.9	98.8	112.9	88.9	98.8
09	FINISHES	114.1	69.4	89.6	113.3	96.8	104.3	112.2	91.5	100.9	105.0	92.5	98.2	107.9	87.0	96.5	112.2	87.0	98.4
COVERS	DIVS. 10 - 14, 25, 28, 41, 43, 44, 46	131.1	64.8	116.1	131.1	70.8	117.5	131.1	70.7	117.5	131.1	92.0	122.3	131.1	81.0	119.8	131.1	81.0	119.8
21, 22, 23	FIRE SUPPRESSION, PLUMBING & HVAC	103.2	80.5	93.1	103.2	106.7	104.7	104.6	82.4	94.8	103.2	94.8	99.5	103.5	88.5	96.8	103.2	88.5	96.7
26, 27, 3370	ELECTRICAL, COMMUNICATIONS & UTIL.	119.2	88.9	103.1	113.9	91.9	102.2	112.0	89.7	100.2	112.7	90.1	100.7	111.1	70.5	89.6	111.1	70.6	89.6
MF2014	WEIGHTED AVERAGE	112.5	78.8	97.6	113.0	97.2	106.0	109.6	88.0	100.1	108.2	93.3	101.6	110.4	84.9	99.1	110.6	85.1	99.3

		CANADA																	
DIVISION		ST CATHARINES, ONTARIO			ST JEROME, QUEBEC			ST JOHNS, NEWFOUNDLAND			SUDBURY, ONTARIO			SUMMERSIDE, PRINCE EDWARD ISLAND			SYDNEY, NOVA SCOTIA		
		MAT.	INST.	TOTAL	MAT.	INST.	TOTAL	MAT.	INST.	TOTAL	MAT.	INST.	TOTAL	MAT.	INST.	TOTAL	MAT.	INST.	TOTAL
015433	CONTRACTOR EQUIPMENT		104.3	104.3		104.4	104.4		117.3	117.3		104.3	104.3		103.6	103.6		103.6	103.6
0241, 31 - 34	SITE & INFRASTRUCTURE, DEMOLITION	98.2	102.6	101.3	98.4	100.0	99.5	121.2	108.1	112.0	98.4	102.2	101.0	130.9	95.8	106.4	116.0	98.5	103.8
0310	Concrete Forming & Accessories	112.1	97.6	99.5	131.6	85.1	91.4	124.9	90.2	94.9	107.9	92.4	94.5	113.9	56.4	64.2	113.6	73.0	78.5
0320	Concrete Reinforcing	98.7	99.6	99.2	154.2	75.4	114.2	173.2	77.5	124.6	99.5	98.9	99.2	177.3	48.4	111.8	179.7	49.6	113.7
0330	Cast-in-Place Concrete	111.0	103.0	108.0	111.3	93.5	104.5	140.8	101.8	125.9	111.9	99.5	107.2	133.1	58.3	104.6	108.2	71.3	94.1
03	CONCRETE	100.1	99.9	100.0	116.0	86.7	102.4	136.5	92.9	116.3	100.4	96.1	98.4	159.3	56.8	111.8	135.2	69.2	104.6
04	MASONRY	144.9	104.7	120.1	164.3	82.1	113.7	182.3	96.8	129.6	144.9	100.9	117.8	222.4	59.0	121.8	220.5	70.5	128.2
05	METALS	103.0	98.6	101.7	101.1	87.6	97.0	129.9	98.7	120.4	102.5	96.5	100.7	115.7	71.4	102.2	115.7	78.3	104.3
06	WOOD, PLASTICS & COMPOSITES	105.7	97.1	101.0	130.9	85.2	105.8	105.0	88.3	95.9	101.9	91.5	96.2	118.3	55.9	84.0	117.9	72.7	93.0
07	THERMAL & MOISTURE PROTECTION	115.1	106.7	111.6	111.6	90.3	102.8	140.3	99.9	123.5	114.5	100.0	108.5	138.0	61.4	106.2	138.8	72.5	111.3
08	OPENINGS	82.7	94.1	85.3	91.7	72.9	87.4	87.3	78.1	85.2	83.3	88.4	84.5	104.9	48.7	92.1	92.8	65.7	86.7
0920	Plaster & Gypsum Board	97.0	97.0	97.0	114.3	84.7	94.6	136.6	87.5	104.1	98.8	91.2	93.8	141.5	54.6	83.8	140.9	71.9	95.0
0950, 0980	Ceilings & Acoustic Treatment	98.0	97.0	97.4	98.6	84.7	89.3	124.7	87.5	100.0	93.0	91.2	91.8	114.4	54.6	74.6	114.4	71.9	86.1
0960	Flooring	105.1	94.6	102.0	119.2	90.9	111.0	118.0	54.2	99.5	103.2	96.4	101.2	114.3	58.5	98.1	114.2	62.3	99.2
0970, 0990	Wall Finishes & Painting/Coating	106.8	115.0	111.6	112.9	88.9	98.8	116.6	105.0	109.8	106.8	97.9	101.6	121.1	42.4	74.9	121.1	63.1	87.0
09	FINISHES	99.5	99.0	99.2	107.9	87.0	96.5	121.6	85.0	101.6	98.1	93.6	95.6	118.4	55.6	84.1	117.3	70.3	91.6
COVERS	DIVS. 10 - 14, 25, 28, 41, 43, 44, 46	131.1	72.5	117.9	131.1	81.0	119.8	131.1	71.3	117.6	131.1	93.4	122.6	131.1	61.0	115.3	131.1	65.0	116.2
21, 22, 23	FIRE SUPPRESSION, PLUMBING & HVAC	104.4	91.7	98.8	103.2	88.5	96.7	105.0	87.2	97.1	103.1	91.1	97.8	103.3	61.8	84.9	103.3	83.6	94.5
26, 27, 3370	ELECTRICAL, COMMUNICATIONS & UTIL.	111.1	103.5	107.1	111.7	70.5	89.9	115.5	88.2	101.0	111.1	104.0	107.3	114.7	51.3	81.1	115.6	63.5	87.9
MF2014	WEIGHTED AVERAGE	105.9	98.2	102.5	110.1	84.9	99.0	121.1	91.0	107.8	105.5	96.4	101.5	125.1	61.5	97.0	120.3	74.7	100.2

City Cost Indexes

CANADA

DIVISION		THUNDER BAY, ONTARIO			TIMMINS, ONTARIO			TORONTO, ONTARIO			TROIS RIVIERES, QUEBEC			TRURO, NOVA SCOTIA			VANCOUVER, BRITISH COLUMBIA		
		MAT.	INST.	TOTAL	MAT.	INST.	TOTAL	MAT.	INST.	TOTAL	MAT.	INST.	TOTAL	MAT.	INST.	TOTAL	MAT.	INST.	TOTAL
015433	CONTRACTOR EQUIPMENT		104.3	104.3		104.3	104.3		118.6	118.6		104.4	104.4		103.6	103.6		134.3	134.3
0241, 31 - 34	SITE & INFRASTRUCTURE, DEMOLITION	103.1	102.4	102.6	118.8	100.5	106.0	123.7	113.4	116.5	116.3	100.0	104.9	103.3	98.5	100.0	113.0	122.7	119.8
0310	Concrete Forming & Accessories	119.8	95.8	99.0	125.6	86.1	91.5	120.9	103.7	106.0	157.2	85.2	94.9	97.4	73.1	76.3	123.8	93.9	98.0
0320	Concrete Reinforcing	88.2	98.4	93.4	171.4	88.3	129.2	137.8	102.8	120.1	179.7	75.4	126.8	145.8	49.6	97.0	128.1	84.4	105.9
0330	Cast-in-Place Concrete	122.2	102.2	114.6	139.8	85.8	119.2	115.7	114.8	115.3	112.0	93.6	105.0	145.1	71.3	117.0	126.4	101.9	117.0
03	CONCRETE	107.7	98.5	103.4	131.7	86.8	110.9	115.0	107.8	111.7	137.6	86.8	114.0	130.7	69.2	102.2	121.6	96.2	109.8
04	MASONRY	145.5	104.9	120.5	168.3	88.1	118.9	171.3	111.0	134.2	224.8	82.1	136.9	163.9	70.5	106.4	166.2	96.1	123.0
05	METALS	103.0	96.7	101.1	102.5	93.6	99.8	122.1	108.2	117.8	114.9	87.7	106.6	101.8	78.3	94.6	124.5	107.1	119.2
06	WOOD, PLASTICS & COMPOSITES	114.5	94.4	103.5	120.2	86.6	101.7	103.6	101.5	102.4	168.7	85.2	122.8	91.0	72.7	80.9	105.0	91.5	97.6
07	THERMAL & MOISTURE PROTECTION	115.4	102.6	110.1	118.3	88.7	106.0	133.5	110.4	123.9	137.1	90.3	117.7	112.9	72.5	96.1	131.8	92.1	115.3
08	OPENINGS	81.9	91.4	84.1	90.4	83.9	88.9	85.2	99.9	88.6	102.5	77.5	96.8	84.5	65.7	80.3	84.2	87.7	85.0
0920	Plaster & Gypsum Board	124.9	94.2	104.5	116.6	86.2	96.4	119.1	101.2	107.2	169.3	84.7	113.1	122.4	71.9	88.8	108.9	90.3	96.5
0950, 0980	Ceilings & Acoustic Treatment	93.0	94.2	93.8	99.5	86.2	90.6	120.8	101.2	107.8	113.6	84.7	94.3	99.5	71.9	81.1	129.9	90.3	103.5
0960	Flooring	109.4	101.3	107.0	119.2	91.0	111.0	111.6	103.7	109.3	138.8	90.9	124.9	100.7	62.3	89.5	127.5	91.9	117.2
0970, 0990	Wall Finishes & Painting/Coating	106.8	99.0	102.2	112.9	90.4	99.7	104.5	111.6	108.7	121.1	88.9	102.2	112.9	63.1	83.6	114.8	94.9	103.1
09	FINISHES	103.9	97.2	100.2	109.1	87.9	97.5	112.7	104.5	108.2	128.3	87.0	105.7	103.8	70.3	85.5	120.1	93.7	105.7
COVERS	DIVS. 10 - 14, 25, 28, 41, 43, 44, 46	131.1	72.7	117.9	131.1	67.9	116.8	131.1	97.9	123.6	131.1	81.0	119.8	131.1	65.0	116.2	131.1	95.6	123.1
21, 22, 23	FIRE SUPPRESSION, PLUMBING & HVAC	104.4	92.0	98.9	103.2	97.1	100.5	105.1	100.0	102.8	103.3	88.5	96.7	103.2	83.6	94.5	105.3	84.6	96.1
26, 27, 3370	ELECTRICAL, COMMUNICATIONS & UTIL.	109.6	102.1	105.7	112.7	90.1	100.7	105.6	105.8	105.7	115.9	70.6	91.8	110.8	63.5	85.7	104.5	81.6	92.3
MF2014	WEIGHTED AVERAGE	107.2	97.2	102.8	113.2	90.8	103.3	114.5	105.6	110.6	123.0	85.1	106.3	110.6	74.7	94.8	115.6	93.9	106.0

CANADA

DIVISION		VICTORIA, BRITISH COLUMBIA			WHITEHORSE, YUKON			WINDSOR, ONTARIO			WINNIPEG, MANITOBA			YARMOUTH, NOVA SCOTIA			YELLOWKNIFE, NWT		
		MAT.	INST.	TOTAL	MAT.	INST.	TOTAL	MAT.	INST.	TOTAL	MAT.	INST.	TOTAL	MAT.	INST.	TOTAL	MAT.	INST.	TOTAL
015433	CONTRACTOR EQUIPMENT		111.1	111.1		126.7	126.7		104.3	104.3		121.4	121.4		103.6	103.6		124.9	124.9
0241, 31 - 34	SITE & INFRASTRUCTURE, DEMOLITION	123.5	108.5	113.0	144.9	113.4	122.9	94.6	102.4	100.0	113.4	109.4	110.6	120.2	98.5	105.1	157.1	117.7	129.7
0310	Concrete Forming & Accessories	112.2	92.8	95.4	135.0	60.6	70.7	119.8	93.6	97.2	127.9	68.3	76.4	113.6	73.0	78.5	135.5	81.2	88.5
0320	Concrete Reinforcing	117.0	84.2	100.4	163.4	65.8	113.9	96.7	99.5	98.1	151.1	61.8	105.8	179.7	49.6	113.7	159.7	68.4	113.4
0330	Cast-in-Place Concrete	128.8	99.8	117.7	187.5	73.3	144.0	113.7	103.2	109.7	148.6	74.0	120.2	138.8	71.3	113.1	186.9	91.0	150.4
03	CONCRETE	140.0	93.8	118.6	159.5	67.7	120.4	101.6	98.1	100.0	134.5	70.6	104.9	150.0	69.2	112.6	165.2	83.5	127.3
04	MASONRY	175.1	96.1	126.5	255.8	62.3	136.6	145.0	102.9	119.1	185.5	68.3	113.3	223.1	70.5	129.1	251.4	74.0	142.1
05	METALS	99.2	89.9	96.3	131.7	91.1	119.3	103.0	96.6	101.1	128.5	88.6	116.3	115.7	78.3	104.3	129.1	94.1	118.4
06	WOOD, PLASTICS & COMPOSITES	101.7	91.0	95.8	124.1	59.3	88.5	114.5	92.6	102.5	113.0	69.1	88.9	117.9	72.7	93.0	126.2	82.8	102.3
07	THERMAL & MOISTURE PROTECTION	120.4	92.0	108.6	145.5	65.9	112.5	115.2	101.8	109.6	125.6	72.8	103.7	138.8	72.5	111.3	149.5	82.6	121.7
08	OPENINGS	88.6	83.1	87.4	106.2	56.6	95.0	81.7	90.5	83.7	82.1	62.1	77.6	92.8	65.7	86.7	99.0	70.1	92.4
0920	Plaster & Gypsum Board	108.9	90.3	96.5	154.1	57.3	89.8	109.8	92.4	98.2	117.1	67.4	84.1	140.9	71.9	95.0	158.8	81.5	107.4
0950, 0980	Ceilings & Acoustic Treatment	101.9	90.3	94.2	155.4	57.3	90.1	93.0	92.4	92.6	140.6	67.4	91.9	114.4	71.9	86.1	149.2	81.5	104.2
0960	Flooring	117.2	71.3	103.9	133.4	58.3	111.6	109.4	98.6	106.2	115.5	70.4	102.4	114.2	62.3	99.2	131.4	87.6	118.7
0970, 0990	Wall Finishes & Painting/Coating	116.6	92.5	102.4	124.0	56.5	84.4	106.8	99.9	102.8	111.6	55.3	78.5	121.1	63.1	87.0	123.2	80.7	98.2
09	FINISHES	110.8	89.4	99.1	142.4	59.5	97.1	101.5	95.2	98.1	121.1	67.7	91.9	117.3	70.3	91.6	139.0	81.8	107.8
COVERS	DIVS. 10 - 14, 25, 28, 41, 43, 44, 46	131.1	70.5	117.4	131.1	64.0	116.0	131.1	71.7	117.7	131.1	66.2	116.4	131.1	65.0	116.2	131.1	67.5	116.7
21, 22, 23	FIRE SUPPRESSION, PLUMBING & HVAC	103.2	82.6	94.1	104.0	75.2	91.2	104.4	91.3	98.6	105.2	66.0	87.9	103.3	83.6	94.5	103.9	92.9	99.0
26, 27, 3370	ELECTRICAL, COMMUNICATIONS & UTIL.	111.9	81.0	95.5	127.0	61.8	92.4	114.2	105.5	109.6	111.8	66.7	87.9	115.6	63.5	87.9	129.1	83.8	105.0
MF2014	WEIGHTED AVERAGE	113.9	88.8	102.8	134.4	71.9	106.8	106.4	96.9	102.2	119.2	72.7	98.7	122.3	74.7	101.3	133.2	86.9	112.8

For customer support on your Building Construction Costs with RSMeans Data, call 800.448.8182.

Costs shown in RSMeans cost data publications are based on national averages for materials and installation. To adjust these costs to a specific location, simply multiply the base cost by the factor and divide by 100 for that city. The data is arranged alphabetically by state and postal zip code numbers. For a city not listed, use the factor for a nearby city with similar economic characteristics.

STATE/ZIP	CITY	MAT.	INST.	TOTAL
ALABAMA				
350-352	Birmingham	97.5	71.7	86.1
354	Tuscaloosa	97.5	72.5	86.4
355	Jasper	97.9	72.1	86.5
356	Decatur	97.5	72.3	86.3
357-358	Huntsville	97.4	71.5	86.0
359	Gadsden	97.5	71.9	86.2
360-361	Montgomery	95.8	73.6	86.0
362	Anniston	95.7	70.0	84.3
363	Dothan	96.1	75.7	87.1
364	Evergreen	95.7	74.3	86.2
365-366	Mobile	96.6	71.2	85.4
367	Selma	95.8	75.2	86.7
368	Phenix City	96.6	74.0	86.6
369	Butler	96.0	74.3	86.4
ALASKA				
995-996	Anchorage	119.7	115.0	117.6
997	Fairbanks	122.3	115.9	119.4
998	Juneau	121.5	115.0	118.6
999	Ketchikan	133.4	115.0	125.3
ARIZONA				
850,853	Phoenix	98.8	72.7	87.3
851,852	Mesa/Tempe	98.4	73.0	87.2
855	Globe	99.3	73.3	87.8
856-857	Tucson	97.2	71.0	85.6
859	Show Low	99.5	73.2	87.9
860	Flagstaff	102.3	72.3	89.1
863	Prescott	99.8	72.8	87.9
864	Kingman	98.3	72.9	87.1
865	Chambers	98.3	76.2	88.6
ARKANSAS				
716	Pine Bluff	98.4	62.9	82.7
717	Camden	96.3	60.9	80.7
718	Texarkana	97.0	61.4	81.3
719	Hot Springs	95.6	61.8	80.7
720-722	Little Rock	96.4	65.7	82.8
723	West Memphis	95.2	66.8	82.6
724	Jonesboro	95.6	63.7	81.5
725	Batesville	93.6	60.9	79.1
726	Harrison	94.9	59.6	79.3
727	Fayetteville	92.4	62.3	79.1
728	Russellville	93.6	59.8	78.7
729	Fort Smith	96.1	60.2	80.2
CALIFORNIA				
900-902	Los Angeles	101.6	128.2	113.4
903-905	Inglewood	97.3	126.7	110.3
906-908	Long Beach	99.0	127.3	111.5
910-912	Pasadena	97.7	126.5	110.4
913-916	Van Nuys	100.9	126.5	112.2
917-918	Alhambra	99.8	127.1	111.8
919-921	San Diego	102.0	118.1	109.1
922	Palm Springs	98.2	125.1	110.1
923-924	San Bernardino	95.8	124.5	108.5
925	Riverside	100.2	125.2	111.2
926-927	Santa Ana	97.9	124.1	109.5
928	Anaheim	100.3	125.5	111.4
930	Oxnard	98.9	125.1	110.4
931	Santa Barbara	98.2	125.2	110.1
932-933	Bakersfield	99.4	122.6	109.6
934	San Luis Obispo	99.3	124.7	110.5
935	Mojave	96.5	122.7	108.1
936-938	Fresno	99.5	126.1	111.3
939	Salinas	100.2	133.2	114.8
940-941	San Francisco	105.6	157.7	128.6
942,956-958	Sacramento	101.0	129.1	113.4
943	Palo Alto	99.4	148.3	121.0
944	San Mateo	101.8	147.3	121.9
945	Vallejo	100.5	139.9	117.9
946	Oakland	103.5	147.7	123.1
947	Berkeley	103.3	147.7	123.0
948	Richmond	102.7	143.7	120.8
949	San Rafael	104.9	148.2	124.0
950	Santa Cruz	105.6	133.6	118.0
CALIFORNIA (CONT'D)				
951	San Jose	103.6	148.6	123.5
952	Stockton	101.7	128.2	113.4
953	Modesto	101.6	126.6	112.6
954	Santa Rosa	102.3	147.1	122.1
955	Eureka	103.6	130.6	115.5
959	Marysville	102.8	129.2	114.5
960	Redding	109.8	129.2	118.4
961	Susanville	109.5	128.3	117.8
COLORADO				
800-802	Denver	102.2	74.2	89.8
803	Boulder	99.7	76.0	89.2
804	Golden	102.1	73.0	89.2
805	Fort Collins	103.2	74.8	90.7
806	Greeley	100.2	75.7	89.4
807	Fort Morgan	100.3	70.9	87.3
808-809	Colorado Springs	102.1	71.8	88.7
810	Pueblo	101.9	67.4	86.7
811	Alamosa	103.9	67.3	87.8
812	Salida	103.6	66.2	87.0
813	Durango	104.3	68.2	88.3
814	Montrose	103.0	67.7	87.4
815	Grand Junction	106.2	73.2	91.6
816	Glenwood Springs	104.2	67.8	88.1
CONNECTICUT				
060	New Britain	97.1	117.7	106.2
061	Hartford	97.9	117.9	106.8
062	Willimantic	97.7	118.2	106.8
063	New London	94.4	118.1	104.9
064	Meriden	96.3	117.8	105.8
065	New Haven	98.7	117.8	107.1
066	Bridgeport	98.3	118.0	107.0
067	Waterbury	97.8	118.6	107.0
068	Norwalk	97.8	125.2	109.9
069	Stamford	97.9	125.6	110.2
D.C.				
200-205	Washington	100.2	86.1	94.0
DELAWARE				
197	Newark	97.9	111.9	104.1
198	Wilmington	97.0	112.0	103.6
199	Dover	97.6	112.1	104.0
FLORIDA				
320,322	Jacksonville	96.2	67.3	83.4
321	Daytona Beach	96.4	71.8	85.6
323	Tallahassee	97.2	68.2	84.4
324	Panama City	97.7	60.7	81.4
325	Pensacola	100.4	67.1	85.7
326,344	Gainesville	98.0	66.7	84.2
327-328,347	Orlando	98.1	69.1	85.3
329	Melbourne	99.3	72.7	87.5
330-332,340	Miami	97.1	66.9	83.8
333	Fort Lauderdale	96.6	66.4	83.2
334,349	West Palm Beach	95.6	66.2	82.6
335-336,346	Tampa	98.1	69.7	85.5
337	St. Petersburg	100.6	67.5	86.0
338	Lakeland	97.6	68.8	84.9
339,341	Fort Myers	96.9	65.9	83.2
342	Sarasota	99.4	68.5	85.8
GEORGIA				
300-303,399	Atlanta	98.5	75.9	88.5
304	Statesboro	98.9	66.9	84.7
305	Gainesville	97.4	68.3	84.5
306	Athens	96.8	70.0	84.9
307	Dalton	98.5	69.5	85.7
308-309	Augusta	97.1	76.1	87.8
310-312	Macon	94.9	71.8	84.7
313-314	Savannah	96.2	76.8	87.6
315	Waycross	96.3	70.4	84.9
316	Valdosta	96.2	68.0	83.8
317,398	Albany	96.1	73.8	86.3
318-319	Columbus	96.0	76.5	87.4

STATE/ZIP	CITY	MAT.	INST.	TOTAL
HAWAII				
967	Hilo	117.7	113.9	116.0
968	Honolulu	122.0	113.9	118.5
STATES & POSS.				
969	Guam	140.4	54.4	102.4
IDAHO				
832	Pocatello	102.1	79.0	91.9
833	Twin Falls	103.5	70.8	89.1
834	Idaho Falls	100.7	78.3	90.8
835	Lewiston	109.1	84.1	98.1
836-837	Boise	101.1	79.7	91.6
838	Coeur d'Alene	109.0	83.7	97.8
ILLINOIS				
600-603	North Suburban	99.0	143.5	118.7
604	Joliet	99.0	143.9	118.8
605	South Suburban	99.0	143.5	118.6
606-608	Chicago	99.5	146.0	120.0
609	Kankakee	95.6	136.3	113.6
610-611	Rockford	98.6	128.9	112.0
612	Rock Island	96.8	102.4	99.2
613	La Salle	98.0	129.1	111.7
614	Galesburg	97.8	109.4	102.9
615-616	Peoria	99.7	113.4	105.8
617	Bloomington	97.1	113.1	104.2
618-619	Champaign	100.8	110.2	104.9
620-622	East St. Louis	94.9	111.5	102.3
623	Quincy	97.3	103.9	100.2
624	Effingham	96.4	109.5	102.2
625	Decatur	97.9	108.6	102.6
626-627	Springfield	98.4	110.4	103.7
628	Centralia	93.9	110.4	101.2
629	Carbondale	93.7	111.0	101.3
INDIANA				
460	Anderson	96.2	83.3	90.5
461-462	Indianapolis	98.2	83.3	91.6
463-464	Gary	97.5	111.8	103.8
465-466	South Bend	96.9	84.6	91.5
467-468	Fort Wayne	96.7	78.4	88.7
469	Kokomo	94.4	82.2	89.0
470	Lawrenceburg	93.5	79.5	87.3
471	New Albany	94.8	77.8	87.3
472	Columbus	97.1	80.8	89.9
473	Muncie	97.2	81.7	90.4
474	Bloomington	98.8	81.6	91.2
475	Washington	95.5	87.0	91.7
476-477	Evansville	96.4	87.1	92.3
478	Terre Haute	97.2	86.2	92.4
479	Lafayette	96.8	80.3	89.5
IOWA				
500-503,509	Des Moines	97.3	88.4	93.3
504	Mason City	96.1	76.9	87.6
505	Fort Dodge	96.2	73.9	86.4
506-507	Waterloo	97.5	75.5	87.8
508	Creston	96.5	82.9	90.5
510-511	Sioux City	98.3	78.6	89.6
512	Sibley	97.4	61.8	81.7
513	Spencer	99.1	62.9	83.1
514	Carroll	96.0	80.6	89.2
515	Council Bluffs	99.4	80.4	91.0
516	Shenandoah	96.6	77.7	88.3
520	Dubuque	97.7	81.5	90.5
521	Decorah	97.1	75.5	87.6
522-524	Cedar Rapids	98.6	85.6	92.8
525	Ottumwa	96.9	77.1	88.2
526	Burlington	96.2	82.7	90.3
527-528	Davenport	97.6	94.7	96.3
KANSAS				
660-662	Kansas City	99.0	98.8	98.9
664-666	Topeka	99.4	77.0	89.5
667	Fort Scott	97.7	79.3	89.6
668	Emporia	97.7	77.1	88.6
669	Belleville	99.5	71.1	86.9
670-672	Wichita	99.7	77.2	89.8
673	Independence	100.0	76.9	89.8
674	Salina	100.2	71.4	87.5
675	Hutchinson	95.5	71.0	84.7
676	Hays	99.7	70.3	86.7
677	Colby	100.2	73.3	88.3

STATE/ZIP	CITY	MAT.	INST.	TOTAL
KANSAS (CONT'D)				
678	Dodge City	101.4	72.6	88.7
679	Liberal	99.6	68.8	86.0
KENTUCKY				
400-402	Louisville	95.5	79.9	88.6
403-405	Lexington	95.7	80.0	88.8
406	Frankfort	96.4	77.7	88.1
407-409	Corbin	93.5	77.3	86.3
410	Covington	95.5	81.1	89.2
411-412	Ashland	94.3	93.7	94.0
413-414	Campton	95.5	78.3	87.9
415-416	Pikeville	96.8	83.3	90.8
417-418	Hazard	94.9	76.9	87.0
420	Paducah	93.4	83.8	89.1
421-422	Bowling Green	95.5	80.7	88.9
423	Owensboro	95.5	84.6	90.7
424	Henderson	93.2	83.3	88.8
425-426	Somerset	92.7	78.9	86.6
427	Elizabethtown	92.2	78.0	85.9
LOUISIANA				
700-701	New Orleans	99.1	69.3	86.0
703	Thibodaux	96.8	66.5	83.4
704	Hammond	94.2	65.4	81.5
705	Lafayette	96.0	67.8	83.6
706	Lake Charles	96.3	69.6	84.5
707-708	Baton Rouge	97.7	70.0	85.5
710-711	Shreveport	99.4	60.2	82.1
712	Monroe	98.2	64.3	83.2
713-714	Alexandria	98.3	68.3	85.1
MAINE				
039	Kittery	93.0	80.6	87.5
040-041	Portland	99.0	83.1	92.0
042	Lewiston	96.6	83.1	90.6
043	Augusta	98.6	75.3	88.3
044	Bangor	96.1	82.5	90.1
045	Bath	95.0	77.8	87.4
046	Machias	94.4	78.9	87.6
047	Houlton	94.6	78.9	87.7
048	Rockland	93.7	78.9	87.1
049	Waterville	94.9	75.3	86.3
MARYLAND				
206	Waldorf	96.7	84.2	91.2
207-208	College Park	96.8	86.0	92.0
209	Silver Spring	96.0	85.6	91.4
210-212	Baltimore	100.7	85.5	94.0
214	Annapolis	99.2	85.6	93.2
215	Cumberland	96.3	86.6	92.0
216	Easton	98.0	74.2	87.5
217	Hagerstown	96.6	88.3	92.9
218	Salisbury	98.4	68.8	85.4
219	Elkton	95.2	86.8	91.5
MASSACHUSETTS				
010-011	Springfield	97.7	109.1	102.7
012	Pittsfield	97.2	107.1	101.6
013	Greenfield	95.5	109.1	101.5
014	Fitchburg	94.2	120.0	105.6
015-016	Worcester	97.7	120.4	107.7
017	Framingham	93.5	126.5	108.1
018	Lowell	97.0	126.1	109.9
019	Lawrence	97.9	127.6	111.1
020-022, 024	Boston	99.9	133.4	114.7
023	Brockton	97.8	120.3	107.7
025	Buzzards Bay	92.6	119.2	104.3
026	Hyannis	94.9	119.2	105.6
027	New Bedford	96.9	119.3	106.8
MICHIGAN				
480,483	Royal Oak	95.0	101.0	97.7
481	Ann Arbor	97.0	103.6	99.9
482	Detroit	99.3	102.9	100.9
484-485	Flint	96.7	92.4	94.8
486	Saginaw	96.3	88.8	93.0
487	Bay City	96.4	89.3	93.3
488-489	Lansing	98.0	90.7	94.8
490	Battle Creek	96.5	83.1	90.6
491	Kalamazoo	96.8	80.9	89.8
492	Jackson	95.0	91.4	93.4
493,495	Grand Rapids	97.8	81.5	90.6
494	Muskegon	95.2	80.3	88.6

STATE/ZIP	CITY	MAT.	INST.	TOTAL	STATE/ZIP	CITY	MAT.	INST.	TOTAL
MICHIGAN (CONT'D)					**NEW HAMPSHIRE (CONT'D)**				
496	Traverse City	94.5	78.7	87.5	032-033	Concord	98.0	91.7	95.2
497	Gaylord	95.7	80.7	89.1	034	Keene	94.7	79.9	88.2
498-499	Iron Mountain	97.7	82.6	91.0	035	Littleton	94.8	80.2	88.3
					036	Charleston	94.3	76.0	86.2
MINNESOTA					037	Claremont	93.4	76.4	85.9
550-551	Saint Paul	99.0	113.2	105.3	038	Portsmouth	95.1	91.9	93.7
553-555	Minneapolis	100.1	112.8	105.7					
556-558	Duluth	99.8	102.6	101.1	**NEW JERSEY**				
559	Rochester	98.9	98.9	98.9	070-071	Newark	98.7	138.7	116.4
560	Mankato	96.3	97.3	96.7	072	Elizabeth	96.5	138.5	115.1
561	Windom	94.9	90.4	92.9	073	Jersey City	95.5	138.4	114.5
562	Willmar	94.5	96.4	95.3	074-075	Paterson	97.2	138.5	115.5
563	St. Cloud	95.5	107.6	100.9	076	Hackensack	95.2	138.6	114.4
564	Brainerd	96.1	98.1	97.0	077	Long Branch	94.8	137.0	113.4
565	Detroit Lakes	98.0	92.1	95.4	078	Dover	95.4	138.5	114.5
566	Bemidji	97.3	95.8	96.6	079	Summit	95.5	138.5	114.5
567	Thief River Falls	96.8	89.3	93.5	080,083	Vineland	95.1	134.8	112.6
					081	Camden	97.0	133.7	113.2
MISSISSIPPI					082,084	Atlantic City	95.8	134.7	113.0
386	Clarksdale	95.8	56.4	78.4	085-086	Trenton	98.6	134.8	114.6
387	Greenville	99.5	66.7	85.0	087	Point Pleasant	96.9	136.6	114.4
388	Tupelo	97.3	57.1	79.6	088-089	New Brunswick	97.6	138.3	115.6
389	Greenwood	97.1	54.3	78.2					
390-392	Jackson	99.0	67.9	85.3	**NEW MEXICO**				
393	Meridian	96.5	67.0	83.5	870-872	Albuquerque	98.9	73.7	87.8
394	Laurel	98.2	57.2	80.1	873	Gallup	99.3	73.7	88.0
395	Biloxi	98.2	64.7	83.4	874	Farmington	99.6	73.7	88.2
396	McComb	96.4	54.9	78.1	875	Santa Fe	99.1	73.7	87.9
397	Columbus	98.1	56.8	79.8	877	Las Vegas	97.6	73.7	87.0
					878	Socorro	97.2	73.7	86.8
MISSOURI					879	Truth/Consequences	96.9	70.3	85.2
630-631	St. Louis	99.8	104.0	101.7	880	Las Cruces	96.8	70.3	85.1
633	Bowling Green	98.7	95.8	97.4	881	Clovis	99.5	73.6	88.1
634	Hannibal	97.6	93.5	95.8	882	Roswell	101.0	73.7	88.9
635	Kirksville	101.4	87.1	95.1	883	Carrizozo	101.7	73.7	89.4
636	Flat River	99.7	95.0	97.6	884	Tucumcari	100.1	73.6	88.4
637	Cape Girardeau	99.2	89.5	95.0					
638	Sikeston	98.1	93.5	96.1	**NEW YORK**				
639	Poplar Bluff	97.5	88.2	93.4	100-102	New York	100.3	177.9	134.6
640-641	Kansas City	100.9	104.4	102.5	103	Staten Island	96.5	173.9	130.7
644-645	St. Joseph	100.1	96.1	98.3	104	Bronx	94.8	175.1	130.3
646	Chillicothe	97.5	95.2	96.5	105	Mount Vernon	95.1	145.1	117.2
647	Harrisonville	96.9	101.9	99.1	106	White Plains	94.8	145.1	117.0
648	Joplin	98.8	80.8	90.9	107	Yonkers	99.0	150.4	121.7
650-651	Jefferson City	98.3	91.7	95.4	108	New Rochelle	95.5	145.0	117.4
652	Columbia	99.0	92.5	96.1	109	Suffern	95.1	138.8	114.4
653	Sedalia	99.4	92.7	96.4	110	Queens	99.9	176.5	133.7
654-655	Rolla	97.0	94.1	95.7	111	Long Island City	101.5	176.5	134.6
656-658	Springfield	100.5	83.6	93.0	112	Brooklyn	101.9	174.2	133.8
					113	Flushing	102.1	176.5	135.0
MONTANA					114	Jamaica	100.3	176.5	134.0
590-591	Billings	101.8	76.4	90.5	115,117,118	Hicksville	99.8	160.8	126.8
592	Wolf Point	101.7	72.4	88.8	116	Far Rockaway	102.3	176.5	135.1
593	Miles City	99.5	72.3	87.5	119	Riverhead	100.6	160.9	127.3
594	Great Falls	103.2	74.6	90.5	120-122	Albany	96.0	110.9	102.6
595	Havre	100.7	72.7	88.3	123	Schenectady	96.3	110.8	102.7
596	Helena	101.0	73.0	88.7	124	Kingston	100.1	134.2	115.2
597	Butte	101.7	74.5	89.6	125-126	Poughkeepsie	99.3	145.2	119.6
598	Missoula	99.0	73.1	87.5	127	Monticello	98.6	133.4	114.0
599	Kalispell	98.4	73.5	87.4	128	Glens Falls	91.4	107.7	98.6
					129	Plattsburgh	96.7	101.9	99.0
NEBRASKA					130-132	Syracuse	97.8	104.4	100.7
680-681	Omaha	98.0	79.8	90.0	133-135	Utica	95.9	102.9	99.0
683-685	Lincoln	98.2	78.3	89.4	136	Watertown	97.5	101.9	99.5
686	Columbus	97.1	81.3	90.1	137-139	Binghamton	97.4	103.0	99.9
687	Norfolk	98.6	77.3	89.2	140-142	Buffalo	101.3	111.3	105.7
688	Grand Island	98.4	77.6	89.2	143	Niagara Falls	97.9	113.1	104.6
689	Hastings	98.3	80.4	90.4	144-146	Rochester	99.8	104.5	101.9
690	McCook	98.4	70.8	86.2	147	Jamestown	96.9	95.7	96.4
691	North Platte	98.2	75.1	88.0	148-149	Elmira	96.8	104.6	100.3
692	Valentine	100.9	69.7	87.1					
693	Alliance	100.9	71.1	87.7	**NORTH CAROLINA**				
					270,272-274	Greensboro	98.7	70.9	86.4
NEVADA					271	Winston-Salem	98.4	70.8	86.2
889-891	Las Vegas	104.1	103.6	103.9	275-276	Raleigh	96.7	67.6	83.9
893	Ely	102.9	96.1	99.9	277	Durham	100.3	71.1	87.4
894-895	Reno	102.7	82.3	93.7	278	Rocky Mount	96.3	70.6	85.0
897	Carson City	101.6	81.8	92.8	279	Elizabeth City	97.2	67.8	84.2
898	Elko	101.5	80.1	92.0	280	Gastonia	98.8	70.5	86.3
					281-282	Charlotte	98.1	70.2	85.8
NEW HAMPSHIRE					283	Fayetteville	101.7	70.7	88.0
030	Nashua	97.8	92.1	95.3	284	Wilmington	97.6	69.7	85.3
031	Manchester	98.4	92.1	95.6	285	Kinston	96.0	70.2	84.6

Location Factors - Commercial

STATE/ZIP	CITY	MAT.	INST.	TOTAL	STATE/ZIP	CITY	MAT.	INST.	TOTAL
NORTH CAROLINA (CONT'D)					**PENNSYLVANIA (CONT'D)**				
286	Hickory	96.4	71.1	85.2	177	Williamsport	91.8	94.5	93.0
287-288	Asheville	98.0	70.3	85.8	178	Sunbury	94.0	96.8	95.2
289	Murphy	97.2	69.7	85.1	179	Pottsville	93.0	99.5	95.9
					180	Lehigh Valley	95.0	119.6	105.9
NORTH DAKOTA					181	Allentown	96.7	110.2	102.6
580-581	Fargo	99.9	78.1	90.3	182	Hazleton	94.5	98.7	96.3
582	Grand Forks	100.4	77.7	90.4	183	Stroudsburg	94.4	111.9	102.1
583	Devils Lake	100.4	77.3	90.2	184-185	Scranton	97.4	101.0	99.0
584	Jamestown	100.5	77.2	90.2	186-187	Wilkes-Barre	94.1	100.0	96.7
585	Bismarck	100.5	80.5	91.7	188	Montrose	93.8	101.0	97.0
586	Dickinson	101.1	78.5	91.1	189	Doylestown	94.0	126.0	108.2
587	Minot	100.4	80.6	91.6	190-191	Philadelphia	99.6	134.5	115.0
588	Williston	99.5	78.8	90.4	193	Westchester	95.3	124.7	108.3
					194	Norristown	94.3	130.1	110.1
OHIO					195-196	Reading	96.0	106.1	100.5
430-432	Columbus	98.0	83.4	91.5					
433	Marion	94.8	82.7	89.4	**PUERTO RICO**				
434-436	Toledo	98.0	94.1	96.3	009	San Juan	121.1	24.3	78.3
437-438	Zanesville	95.3	82.8	89.8					
439	Steubenville	96.7	90.3	93.8	**RHODE ISLAND**				
440	Lorain	99.3	86.9	93.8	028	Newport	96.2	112.3	103.3
441	Cleveland	99.5	93.7	96.9	029	Providence	97.9	112.3	104.2
442-443	Akron	100.4	88.8	95.3					
444-445	Youngstown	99.6	84.8	93.1	**SOUTH CAROLINA**				
446-447	Canton	99.8	81.8	91.8	290-292	Columbia	97.1	68.5	84.5
448-449	Mansfield	97.4	83.0	91.0	293	Spartanburg	97.6	68.1	84.5
450	Hamilton	97.7	79.2	89.5	294	Charleston	99.1	67.8	85.2
451-452	Cincinnati	98.1	77.7	89.1	295	Florence	97.3	68.5	84.5
453-454	Dayton	97.8	77.8	89.0	296	Greenville	97.3	68.1	84.4
455	Springfield	97.7	80.9	90.3	297	Rock Hill	97.3	67.6	84.2
456	Chillicothe	97.1	86.4	92.4	298	Aiken	98.2	68.6	85.1
457	Athens	100.2	85.2	93.6	299	Beaufort	98.9	56.4	80.1
458	Lima	100.6	81.8	92.3					
					SOUTH DAKOTA				
OKLAHOMA					570-571	Sioux Falls	97.8	67.8	84.6
730-731	Oklahoma City	99.3	67.3	85.1	572	Watertown	99.0	50.0	77.4
734	Ardmore	98.1	66.7	84.2	573	Mitchell	97.8	48.3	75.9
735	Lawton	100.2	67.1	85.6	574	Aberdeen	100.1	54.7	80.1
736	Clinton	99.4	66.6	84.9	575	Pierre	101.3	62.0	83.9
737	Enid	99.8	67.0	85.3	576	Mobridge	98.5	53.9	78.8
738	Woodward	98.1	63.8	82.9	577	Rapid City	99.9	65.2	84.6
739	Guymon	99.2	64.5	83.9					
740-741	Tulsa	97.0	66.5	83.5	**TENNESSEE**				
743	Miami	93.8	66.2	81.6	370-372	Nashville	98.4	73.6	87.4
744	Muskogee	96.1	66.0	82.8	373-374	Chattanooga	99.6	67.1	85.3
745	McAlester	93.4	61.6	79.3	375,380-381	Memphis	98.2	72.2	86.7
746	Ponca City	94.2	65.5	81.5	376	Johnson City	99.8	60.7	82.5
747	Durant	94.1	66.1	81.7	377-379	Knoxville	96.3	66.4	83.1
748	Shawnee	95.7	66.4	82.7	382	McKenzie	96.9	57.4	79.5
749	Poteau	93.3	65.8	81.2	383	Jackson	98.3	61.3	81.9
					384	Columbia	95.4	68.7	83.6
OREGON					385	Cookeville	96.8	57.7	79.5
970-972	Portland	100.0	98.9	99.5					
973	Salem	101.2	97.3	99.5	**TEXAS**				
974	Eugene	99.7	97.5	98.7	750	McKinney	99.7	65.5	84.6
975	Medford	101.3	94.7	98.4	751	Waxahackie	99.7	64.5	84.2
976	Klamath Falls	101.6	94.8	98.6	752-753	Dallas	100.3	68.4	86.2
977	Bend	100.6	97.3	99.2	754	Greenville	99.9	65.5	84.7
978	Pendleton	96.5	98.6	97.4	755	Texarkana	99.1	65.7	84.4
979	Vale	94.3	85.8	90.5	756	Longview	100.1	63.1	83.7
					757	Tyler	100.1	64.5	84.4
PENNSYLVANIA					758	Palestine	96.0	63.0	81.4
150-152	Pittsburgh	100.6	104.5	102.3	759	Lufkin	96.7	65.0	82.7
153	Washington	97.9	103.8	100.5	760-761	Fort Worth	97.3	64.7	82.9
154	Uniontown	98.2	100.1	99.0	762	Denton	97.5	64.8	83.0
155	Bedford	99.2	93.7	96.8	763	Wichita Falls	94.8	63.6	81.0
156	Greensburg	99.2	101.3	100.1	764	Eastland	94.1	62.9	80.3
157	Indiana	98.0	99.0	98.4	765	Temple	92.4	60.3	78.2
158	Dubois	99.6	95.0	97.6	766-767	Waco	94.2	65.7	81.6
159	Johnstown	99.2	95.8	97.7	768	Brownwood	97.5	60.3	81.1
160	Butler	91.4	102.2	96.2	769	San Angelo	97.1	60.8	81.0
161	New Castle	91.5	102.6	96.4	770-772	Houston	99.3	67.5	85.2
162	Kittanning	91.9	102.9	96.8	773	Huntsville	98.6	65.5	84.0
163	Oil City	91.4	101.4	95.8	774	Wharton	99.4	66.4	84.8
164-165	Erie	93.2	98.5	95.5	775	Galveston	97.3	68.3	84.5
166	Altoona	93.3	96.6	94.8	776-777	Beaumont	97.6	67.3	84.2
167	Bradford	94.9	102.1	98.1	778	Bryan	95.0	67.0	82.7
168	State College	94.5	97.8	95.9	779	Victoria	99.5	64.7	84.1
169	Wellsboro	95.6	96.1	95.8	780	Laredo	97.1	64.7	82.8
170-171	Harrisburg	96.9	96.7	96.8	781-782	San Antonio	98.1	65.8	83.9
172	Chambersburg	94.5	92.5	93.6	783-784	Corpus Christi	99.7	62.5	83.3
173-174	York	94.5	95.6	94.9	785	McAllen	100.7	59.5	82.5
175-176	Lancaster	93.2	97.9	95.3	786-787	Austin	97.5	63.6	82.5

807

Location Factors - Commercial

STATE/ZIP	CITY	MAT.	INST.	TOTAL
TEXAS (CONT'D)				
788	Del Rio	100.5	62.8	83.9
789	Giddings	97.0	62.1	81.6
790-791	Amarillo	97.5	63.2	82.4
792	Childress	97.9	62.8	82.4
793-794	Lubbock	99.7	64.1	84.0
795-796	Abilene	98.1	63.7	82.9
797	Midland	100.4	65.8	85.1
798-799,885	El Paso	96.3	62.8	81.5
UTAH				
840-841	Salt Lake City	103.4	69.6	88.5
842,844	Ogden	98.7	69.5	85.8
843	Logan	100.8	69.5	87.0
845	Price	101.5	64.2	85.0
846-847	Provo	101.3	69.4	87.2
VERMONT				
050	White River Jct.	97.4	81.3	90.3
051	Bellows Falls	95.6	95.4	95.5
052	Bennington	96.1	91.4	94.0
053	Brattleboro	96.5	95.2	95.9
054	Burlington	100.7	80.8	91.9
056	Montpelier	99.3	85.5	93.2
057	Rutland	97.8	80.8	90.3
058	St. Johnsbury	97.5	81.1	90.3
059	Guildhall	96.0	80.7	89.3
VIRGINIA				
220-221	Fairfax	99.9	85.5	93.6
222	Arlington	100.7	85.3	93.9
223	Alexandria	99.8	85.2	93.3
224-225	Fredericksburg	98.6	83.8	92.0
226	Winchester	99.2	84.4	92.7
227	Culpeper	99.0	81.5	91.3
228	Harrisonburg	99.4	66.5	84.8
229	Charlottesville	99.8	68.0	85.7
230-232	Richmond	98.3	73.5	87.3
233-235	Norfolk	98.1	69.7	85.5
236	Newport News	98.3	69.9	85.8
237	Portsmouth	97.8	64.5	83.1
238	Petersburg	98.3	73.5	87.4
239	Farmville	97.5	55.1	78.8
240-241	Roanoke	100.2	65.8	85.0
242	Bristol	98.5	54.4	79.0
243	Pulaski	98.1	56.8	79.9
244	Staunton	98.9	60.5	81.9
245	Lynchburg	99.1	64.8	83.9
246	Grundy	98.4	51.8	77.8
WASHINGTON				
980-981,987	Seattle	103.2	107.1	104.9
982	Everett	102.6	100.4	101.6
983-984	Tacoma	103.0	101.5	102.4
985	Olympia	100.7	101.5	101.1
986	Vancouver	104.0	95.2	100.1
988	Wenatchee	103.3	86.9	96.0
989	Yakima	103.2	96.6	100.2
990-992	Spokane	102.9	83.3	94.3
993	Richland	102.6	91.4	97.6
994	Clarkston	102.0	82.9	93.5
WEST VIRGINIA				
247-248	Bluefield	97.3	91.7	94.8
249	Lewisburg	98.9	90.6	95.3
250-253	Charleston	98.6	93.1	96.2
254	Martinsburg	99.4	85.8	93.4
255-257	Huntington	100.4	93.2	97.2
258-259	Beckley	97.8	91.4	95.0
260	Wheeling	100.0	92.2	96.5
261	Parkersburg	98.7	89.7	94.7
262	Buckhannon	98.7	93.3	96.3
263-264	Clarksburg	99.3	93.3	96.7
265	Morgantown	99.3	93.4	96.7
266	Gassaway	98.7	93.4	96.3
267	Romney	98.5	87.7	93.7
268	Petersburg	98.4	86.6	93.2
WISCONSIN				
530,532	Milwaukee	98.6	107.9	102.7
531	Kenosha	98.6	105.8	101.8
534	Racine	98.1	106.6	101.8
535	Beloit	97.9	100.6	99.1
537	Madison	98.3	100.5	99.3

STATE/ZIP	CITY	MAT.	INST.	TOTAL
WISCONSIN (CONT'D)				
538	Lancaster	96.1	95.8	96.0
539	Portage	94.6	98.6	96.3
540	New Richmond	95.9	94.6	95.3
541-543	Green Bay	100.0	97.9	99.1
544	Wausau	95.3	96.4	95.8
545	Rhinelander	98.6	91.9	95.7
546	La Crosse	96.3	98.7	97.4
547	Eau Claire	98.0	99.2	98.5
548	Superior	95.7	98.5	96.9
549	Oshkosh	95.9	90.7	93.6
WYOMING				
820	Cheyenne	101.5	72.0	88.5
821	Yellowstone Nat'l Park	99.5	74.8	88.6
822	Wheatland	100.7	69.2	86.8
823	Rawlins	102.3	74.5	90.0
824	Worland	100.2	73.4	88.4
825	Riverton	101.3	71.0	87.9
826	Casper	101.4	68.3	86.8
827	Newcastle	100.1	76.4	89.6
828	Sheridan	103.0	71.9	89.2
829-831	Rock Springs	104.2	73.0	90.4
CANADIAN FACTORS (reflect Canadian currency)				
ALBERTA				
	Calgary	119.8	98.6	110.4
	Edmonton	119.8	98.6	110.4
	Fort McMurray	122.1	92.8	109.2
	Lethbridge	116.7	92.3	105.9
	Lloydminster	111.5	88.9	101.5
	Medicine Hat	111.7	88.2	101.3
	Red Deer	112.1	88.2	101.6
BRITISH COLUMBIA				
	Kamloops	112.7	88.8	102.2
	Prince George	113.8	88.1	102.4
	Vancouver	115.6	93.9	106.0
	Victoria	113.9	88.8	102.8
MANITOBA				
	Brandon	123.4	75.1	102.0
	Portage la Prairie	111.4	73.4	94.6
	Winnipeg	119.2	72.7	98.7
NEW BRUNSWICK				
	Bathurst	109.8	66.2	90.5
	Dalhousie	109.8	66.4	90.6
	Fredericton	116.4	73.3	97.4
	Moncton	110.1	70.6	92.6
	Newcastle	109.9	66.8	90.9
	St. John	112.5	78.8	97.6
NEWFOUNDLAND				
	Corner Brook	128.1	65.4	100.4
	St. Johns	121.1	91.0	107.8
NORTHWEST TERRITORIES				
	Yellowknife	133.2	86.9	112.8
NOVA SCOTIA				
	Bridgewater	111.0	74.7	95.0
	Dartmouth	124.5	74.7	102.5
	Halifax	116.1	85.0	102.4
	New Glasgow	122.4	74.7	101.4
	Sydney	120.3	74.7	100.2
	Truro	110.6	74.7	94.8
	Yarmouth	122.3	74.7	101.3
ONTARIO				
	Barrie	117.1	92.7	106.3
	Brantford	113.0	96.6	105.8
	Cornwall	113.0	93.0	104.1
	Hamilton	114.3	98.1	107.1
	Kingston	114.1	93.1	104.8
	Kitchener	107.8	97.0	103.0
	London	114.3	98.5	107.3
	North Bay	125.5	90.8	110.2
	Oshawa	110.5	96.4	104.2
	Ottawa	116.3	99.2	108.8
	Owen Sound	117.3	90.9	105.6
	Peterborough	113.0	92.7	104.1
	Sarnia	113.0	97.2	106.0
	Sault Ste. Marie	108.2	93.3	101.6

Location Factors - Commercial

STATE/ZIP	CITY	MAT.	INST.	TOTAL
	St. Catharines	105.9	98.2	102.5
	Sudbury	105.5	96.4	101.5
	Thunder Bay	107.2	97.2	102.8
	Timmins	113.2	90.8	103.3
	Toronto	114.5	105.6	110.6
	Windsor	106.4	96.9	102.2
PRINCE EDWARD ISLAND				
	Charlottetown	119.0	63.4	94.4
	Summerside	125.1	61.5	97.0
QUEBEC				
	Cap-de-la-Madeleine	110.4	85.1	99.2
	Charlesbourg	110.4	85.1	99.2
	Chicoutimi	109.7	89.3	100.7
	Gatineau	110.1	84.9	98.9
	Granby	110.3	84.8	99.0
	Hull	110.2	84.9	99.0
	Joliette	110.6	85.1	99.3
	Laval	110.5	84.9	99.2
	Montreal	115.7	91.5	105.0
	Quebec City	116.2	92.0	105.5
	Rimouski	110.0	89.3	100.9
	Rouyn-Noranda	110.1	84.9	98.9
	Saint-Hyacinthe	109.8	84.9	98.8
	Sherbrooke	110.4	84.9	99.1
	Sorel	110.6	85.1	99.3
	Saint-Jerome	110.1	84.9	99.0
	Trois-Rivieres	123.0	85.1	106.3
SASKATCHEWAN				
	Moose Jaw	108.7	67.9	90.7
	Prince Albert	107.8	66.1	89.4
	Regina	119.8	90.2	106.7
	Saskatoon	109.6	88.0	100.1
YUKON				
	Whitehorse	134.4	71.9	106.8

For customer support on your Building Construction Costs with RSMeans Data, call 800.448.8182.

R011105-05   Tips for Accurate Estimating

1. Use pre-printed or columnar forms for orderly sequence of dimensions and locations and for recording telephone quotations.

2. Use only the front side of each paper or form except for certain pre-printed summary forms.

3. Be consistent in listing dimensions: For example, length x width x height. This helps in rechecking to ensure that, the total length of partitions is appropriate for the building area.

4. Use printed (rather than measured) dimensions where given.

5. Add up multiple printed dimensions for a single entry where possible.

6. Measure all other dimensions carefully.

7. Use each set of dimensions to calculate multiple related quantities.

8. Convert foot and inch measurements to decimal feet when listing. Memorize decimal equivalents to .01 parts of a foot (1/8″ equals approximately .01′).

9. Do not "round off" quantities until the final summary.

10. Mark drawings with different colors as items are taken off.

11. Keep similar items together, different items separate.

12. Identify location and drawing numbers to aid in future checking for completeness.

13. Measure or list everything on the drawings or mentioned in the specifications.

14. It may be necessary to list items not called for to make the job complete.

15. Be alert for: Notes on plans such as N.T.S. (not to scale); changes in scale throughout the drawings; reduced size drawings; discrepancies between the specifications and the drawings.

16. Develop a consistent pattern of performing an estimate. For example:
    a. Start the quantity takeoff at the lower floor and move to the next higher floor.
    b. Proceed from the main section of the building to the wings.
    c. Proceed from south to north or vice versa, clockwise or counterclockwise.
    d. Take off floor plan quantities first, elevations next, then detail drawings.

17. List all gross dimensions that can be either used again for different quantities, or used as a rough check of other quantities for verification (exterior perimeter, gross floor area, individual floor areas, etc.).

18. Utilize design symmetry or repetition (repetitive floors, repetitive wings, symmetrical design around a center line, similar room layouts, etc.). Note: Extreme caution is needed here so as not to omit or duplicate an area.

19. Do not convert units until the final total is obtained. For instance, when estimating concrete work, keep all units to the nearest cubic foot, then summarize and convert to cubic yards.

20. When figuring alternatives, it is best to total all items involved in the basic system, then total all items involved in the alternates. Therefore you work with positive numbers in all cases. When adds and deducts are used, it is often confusing whether to add or subtract a portion of an item; especially on a complicated or involved alternate.

R011105-50   Metric Conversion Factors

Description: This table is primarily for converting customary U.S. units in the left hand column to SI metric units in the right hand column. In addition, conversion factors for some commonly encountered Canadian and non-SI metric units are included.

	If You Know		Multiply By		To Find
Length	Inches	x	25.4[a]	=	Millimeters
	Feet	x	0.3048[a]	=	Meters
	Yards	x	0.9144[a]	=	Meters
	Miles (statute)	x	1.609	=	Kilometers
Area	Square inches	x	645.2	=	Square millimeters
	Square feet	x	0.0929	=	Square meters
	Square yards	x	0.8361	=	Square meters
Volume (Capacity)	Cubic inches	x	16,387	=	Cubic millimeters
	Cubic feet	x	0.02832	=	Cubic meters
	Cubic yards	x	0.7646	=	Cubic meters
	Gallons (U.S. liquids)[b]	x	0.003785	=	Cubic meters[c]
	Gallons (Canadian liquid)[b]	x	0.004546	=	Cubic meters[c]
	Ounces (U.S. liquid)[b]	x	29.57	=	Milliliters[c, d]
	Quarts (U.S. liquid)[b]	x	0.9464	=	Liters[c, d]
	Gallons (U.S. liquid)[b]	x	3.785	=	Liters[c, d]
Force	Kilograms force[d]	x	9.807	=	Newtons
	Pounds force	x	4.448	=	Newtons
	Pounds force	x	0.4536	=	Kilograms force[d]
	Kips	x	4448	=	Newtons
	Kips	x	453.6	=	Kilograms force[d]
Pressure, Stress, Strength (Force per unit area)	Kilograms force per square centimeter[d]	x	0.09807	=	Megapascals
	Pounds force per square inch (psi)	x	0.006895	=	Megapascals
	Kips per square inch	x	6.895	=	Megapascals
	Pounds force per square inch (psi)	x	0.07031	=	Kilograms force per square centimeter[d]
	Pounds force per square foot	x	47.88	=	Pascals
	Pounds force per square foot	x	4.882	=	Kilograms force per square meter[d]
Flow	Cubic feet per minute	x	0.4719	=	Liters per second
	Gallons per minute	x	0.0631	=	Liters per second
	Gallons per hour	x	1.05	=	Milliliters per second
Bending Moment Or Torque	Inch-pounds force	x	0.01152	=	Meter-kilograms force[d]
	Inch-pounds force	x	0.1130	=	Newton-meters
	Foot-pounds force	x	0.1383	=	Meter-kilograms force[d]
	Foot-pounds force	x	1.356	=	Newton-meters
	Meter-kilograms force[d]	x	9.807	=	Newton-meters
Mass	Ounces (avoirdupois)	x	28.35	=	Grams
	Pounds (avoirdupois)	x	0.4536	=	Kilograms
	Tons (metric)	x	1000	=	Kilograms
	Tons, short (2000 pounds)	x	907.2	=	Kilograms
	Tons, short (2000 pounds)	x	0.9072	=	Megagrams[e]
Mass per Unit Volume	Pounds mass per cubic foot	x	16.02	=	Kilograms per cubic meter
	Pounds mass per cubic yard	x	0.5933	=	Kilograms per cubic meter
	Pounds mass per gallon (U.S. liquid)[b]	x	119.8	=	Kilograms per cubic meter
	Pounds mass per gallon (Canadian liquid)[b]	x	99.78	=	Kilograms per cubic meter
Temperature	Degrees Fahrenheit	(F-32)/1.8		=	Degrees Celsius
	Degrees Fahrenheit	(F+459.67)/1.8		=	Degrees Kelvin
	Degrees Celsius	C+273.15		=	Degrees Kelvin

[a]The factor given is exact
[b]One U.S. gallon = 0.8327 Canadian gallon
[c]1 liter = 1000 milliliters = 1000 cubic centimeters
1 cubic decimeter = 0.001 cubic meter

[d]Metric but not SI unit
[e]Called "tonne" in England and
"metric ton" in other metric countries

R011105-60   Weights and Measures

Measures of Length
1 Mile = 1760 Yards = 5280 Feet
1 Yard = 3 Feet = 36 inches
1 Foot = 12 Inches
1 Mil = 0.001 Inch
1 Fathom = 2 Yards = 6 Feet
1 Rod = 5.5 Yards = 16.5 Feet
1 Hand = 4 Inches
1 Span = 9 Inches
1 Micro-inch = One Millionth Inch or 0.000001 Inch
1 Micron = One Millionth Meter + 0.00003937 Inch

Surveyor's Measure
1 Mile = 8 Furlongs = 80 Chains
1 Furlong = 10 Chains = 220 Yards
1 Chain = 4 Rods = 22 Yards = 66 Feet = 100 Links
1 Link = 7.92 Inches

Square Measure
1 Square Mile = 640 Acres = 6400 Square Chains
1 Acre = 10 Square Chains = 4840 Square Yards =
   43,560 Sq. Ft.
1 Square Chain = 16 Square Rods = 484 Square Yards =
   4356 Sq. Ft.
1 Square Rod = 30.25 Square Yards = 272.25 Square Feet = 625 Square
Lines
1 Square Yard = 9 Square Feet
1 Square Foot = 144 Square Inches
An Acre equals a Square 208.7 Feet per Side

Cubic Measure
1 Cubic Yard = 27 Cubic Feet
1 Cubic Foot = 1728 Cubic Inches
1 Cord of Wood = 4 x 4 x 8 Feet = 128 Cubic Feet
1 Perch of Masonry = 16½ x 1½ x 1 Foot = 24.75 Cubic Feet

Avoirdupois or Commercial Weight
1 Gross or Long Ton = 2240 Pounds
1 Net or Short Ton = 2000 Pounds
1 Pound = 16 Ounces = 7000 Grains
1 Ounce = 16 Drachms = 437.5 Grains
1 Stone = 14 Pounds

Power
1 British Thermal Unit per Hour = 0.2931 Watts
1 Ton (Refrigeration) = 3.517 Kilowatts
1 Horsepower (Boiler) = 9.81 Kilowatts
1 Horsepower (550 ft-lb/s) = 0.746 Kilowatts

Shipping Measure
For Measuring Internal Capacity of a Vessel:
   1 Register Ton = 100 Cubic Feet

For Measurement of Cargo:
   Approximately 40 Cubic Feet of Merchandise is considered a Shipping
   Ton, unless that bulk would weigh more than 2000 Pounds, in which case
   Freight Charge may be based upon weight.

40 Cubic Feet = 32.143 U.S. Bushels = 31.16 Imp. Bushels

Liquid Measure
1 Imperial Gallon = 1.2009 U.S. Gallon = 277.42 Cu. In.
1 Cubic Foot = 7.48 U.S. Gallons

R011110-10   Architectural Fees

Tabulated below are typical percentage fees by project size, for good professional architectural service. Fees may vary from those listed depending upon degree of design difficulty and economic conditions in any particular area.

Rates can be interpolated horizontally and vertically. Various portions of the same project requiring different rates should be adjusted proportionately. For alterations, add 50% to the fee for the first $500,000 of project cost and add 25% to the fee for project cost over $500,000.

Architectural fees tabulated below include Structural, Mechanical and Electrical Engineering Fees. They do not include the fees for special consultants such as kitchen planning, security, acoustical, interior design, etc.

Civil Engineering fees are included in the Architectural fee for project sites requiring minimal design such as city sites. However, separate Civil Engineering fees must be added when utility connections require design, drainage calculations are needed, stepped foundations are required, or provisions are required to protect adjacent wetlands.

Building Types	Total Project Size in Thousands of Dollars						
	100	250	500	1,000	5,000	10,000	50,000
Factories, garages, warehouses, repetitive housing	9.0%	8.0%	7.0%	6.2%	5.3%	4.9%	4.5%
Apartments, banks, schools, libraries, offices, municipal buildings	12.2	12.3	9.2	8.0	7.0	6.6	6.2
Churches, hospitals, homes, laboratories, museums, research	15.0	13.6	12.7	11.9	9.5	8.8	8.0
Memorials, monumental work, decorative furnishings	—	16.0	14.5	13.1	10.0	9.0	8.3

R011110-30   Engineering Fees

Typical **Structural Engineering Fees** based on type of construction and total project size. These fees are included in Architectural Fees.

Type of Construction	Total Project Size (in thousands of dollars)			
	$500	$500-$1,000	$1,000-$5,000	Over $5000
Industrial buildings, factories & warehouses	Technical payroll times 2.0 to 2.5	1.60%	1.25%	1.00%
Hotels, apartments, offices, dormitories, hospitals, public buildings, food stores		2.00%	1.70%	1.20%
Museums, banks, churches and cathedrals		2.00%	1.75%	1.25%
Thin shells, prestressed concrete, earthquake resistive		2.00%	1.75%	1.50%
Parking ramps, auditoriums, stadiums, convention halls, hangars & boiler houses		2.50%	2.00%	1.75%
Special buildings, major alterations, underpinning & future expansion	↓	Add to above 0.5%	Add to above 0.5%	Add to above 0.5%

For complex reinforced concrete or unusually complicated structures, add 20% to 50%.

Typical **Mechanical and Electrical Engineering Fees** are based on the size of the subcontract. The fee structure for both is shown below. These fees are included in Architectural Fees.

Type of Construction	Subcontract Size							
	$25,000	$50,000	$100,000	$225,000	$350,000	$500,000	$750,000	$1,000,000
Simple structures	6.4%	5.7%	4.8%	4.5%	4.4%	4.3%	4.2%	4.1%
Intermediate structures	8.0	7.3	6.5	5.6	5.1	5.0	4.9	4.8
Complex structures	10.1	9.0	9.0	8.0	7.5	7.5	7.0	7.0

For renovations, add 15% to 25% to applicable fee.

General Requirements   **R0121 Allowances**

R012153-60   Security Factors

Contractors entering, working in, and exiting secure facilities often lose productive time during a normal workday. The recommended allowances in this section are intended to provide for the loss of productivity by increasing labor costs. Note that different costs are associated with searches upon entry only and searches upon entry and exit. Time spent in a queue is unpredictable and not part of these allowances. Contractors should plan ahead for this situation.

Security checkpoints are designed to reflect the level of security required to gain access or egress. An extreme example is when contractors, along with any materials, tools, equipment, and vehicles, must be physically searched and have all materials, tools, equipment, and vehicles inventoried and documented prior to both entry and exit.

Physical searches without going through the documentation process represent the next level and take up less time.

Electronic searches—passing through a detector or x-ray machine with no documentation of materials, tools, equipment, and vehicles—take less time than physical searches.

Visual searches of materials, tools, equipment, and vehicles represent the next level of security.

Finally, access by means of an ID card or displayed sticker takes the least amount of time.

Another consideration is if the searches described above are performed each and every day, or if they are performed only on the first day with access granted by ID card or displayed sticker for the remainder of the project. The figures for this situation have been calculated to represent the initial check-in as described and subsequent entry by ID card or displayed sticker for up to 20 days on site. For the situation described above, where the time period is beyond 20 days, the impact on labor cost is negligible.

There are situations where tradespeople must be accompanied by an escort and observed during the work day. The loss of freedom of movement will slow down productivity for the tradesperson. Costs for the observer have not been included. Those costs are normally born by the owner.

General Requirements   R0121 Allowances

R012157-20   Construction Time Requirements

The table below lists the construction durations for various building types along with their respective project sizes and project values. Design time runs 25% to 40% of construction time.

Building Type	Size S.F.	Project Value	Construction Duration
Industrial/Warehouse	100,000	$8,000,000	14 months
	500,000	$32,000,000	19 months
	1,000,000	$75,000,000	21 months
Offices/Retail	50,000	$7,000,000	15 months
	250,000	$28,000,000	23 months
	500,000	$58,000,000	34 months
Institutional/Hospitals/Laboratory	200,000	$45,000,000	31 months
	500,000	$110,000,000	52 months
	750,000	$160,000,000	55 months
	1,000,000	$210,000,000	60 months

General Requirements   R0129 Payment Procedures

R012909-80   Sales Tax by State

State sales tax on materials is tabulated below (5 states have no sales tax). Many states allow local jurisdictions, such as a county or city, to levy additional sales tax.

Some projects may be sales tax exempt, particularly those constructed with public funds.

State	Tax (%)	State	Tax (%)	State	Tax (%)	State	Tax (%)
Alabama	4	Illinois	6.25	Montana	0	Rhode Island	7
Alaska	0	Indiana	7	Nebraska	5.5	South Carolina	6
Arizona	5.6	Iowa	6	Nevada	6.85	South Dakota	4
Arkansas	6.5	Kansas	6.5	New Hampshire	0	Tennessee	7
California	7.5	Kentucky	6	New Jersey	7	Texas	6.25
Colorado	2.9	Louisiana	4	New Mexico	5.125	Utah	5.95
Connecticut	6.35	Maine	5.5	New York	4	Vermont	6
Delaware	0	Maryland	6	North Carolina	4.75	Virginia	5.3
District of Columbia	5.75	Massachusetts	6.25	North Dakota	5	Washington	6.5
Florida	6	Michigan	6	Ohio	5.75	West Virginia	6
Georgia	4	Minnesota	6.875	Oklahoma	4.5	Wisconsin	5
Hawaii	4	Mississippi	7	Oregon	0	Wyoming	4
Idaho	6	Missouri	4.225	Pennsylvania	6	Average	5.09 %

Sales Tax by Province (Canada)

GST - a value-added tax, which the government imposes on most goods and services provided in or imported into Canada. PST - a retail sales tax, which five of the provinces impose on the prices of most goods and some

services. QST - a value-added tax, similar to the federal GST, which Quebec imposes. HST - Three provinces have combined their retail sales taxes with the federal GST into one harmonized tax.

Province	PST (%)	QST (%)	GST(%)	HST(%)
Alberta	0	0	5	0
British Columbia	7	0	5	0
Manitoba	8	0	5	0
New Brunswick	0	0	0	13
Newfoundland	0	0	0	13
Northwest Territories	0	0	5	0
Nova Scotia	0	0	0	15
Ontario	0	0	0	13
Prince Edward Island	0	0	0	14
Quebec	0	9.975	5	0
Saskatchewan	5	0	5	0
Yukon	0	0	5	0

R012909-85   Unemployment Taxes and Social Security Taxes

State unemployment tax rates vary not only from state to state, but also with the experience rating of the contractor. The federal unemployment tax rate is 6.0% of the first $7,000 of wages. This is reduced by a credit of up to 5.4% for timely payment to the state. The minimum federal unemployment tax is 0.6% after all credits.

Social security (FICA) for 2017 is estimated at time of publication to be 7.65% of wages up to $118,500.

R012909-86   Unemployment Tax by State

Information is from the U.S. Department of Labor, state unemployment tax rates.

State	Tax (%)	State	Tax (%)	State	Tax (%)	State	Tax (%)
Alabama	7.1	Illinois	8.15	Montana	6.12	Rhode Island	9.79
Alaska	5.4	Indiana	7.5	Nebraska	5.4	South Carolina	6.03
Arizona	7.79	Iowa	7.5	Nevada	5.4	South Dakota	9.5
Arkansas	6.8	Kansas	7.4	New Hampshire	6	Tennessee	10
California	6.2	Kentucky	10	New Jersey	7	Texas	7.49
Colorado	8.2	Louisiana	6.2	New Mexico	6.4	Utah	7.3
Connecticut	6.8	Maine	6.86	New York	8.9	Vermont	8.4
Delaware	8.2	Maryland	9	North Carolina	5.76	Virginia	6.54
District of Columbia	7.2	Massachusetts	11.13	North Dakota	9.7	Washington	5.84
Florida	5.4	Michigan	10.3	Ohio	8.6	West Virginia	7.5
Georgia	5.4	Minnesota	9.1	Oklahoma	5.5	Wisconsin	12
Hawaii	5.8	Mississippi	5.6	Oregon	5.4	Wyoming	10
Idaho	5.4	Missouri	9.75	Pennsylvania	10.89	Median	7.3 %

R012909-90   Overtime

One way to improve the completion date of a project or eliminate negative float from a schedule is to compress activity duration times. This can be achieved by increasing the crew size or working overtime with the proposed crew.

To determine the costs of working overtime to compress activity duration times, consider the following examples. Below is an overtime efficiency and cost chart based on a five, six, or seven day week with an eight through twelve hour day. Payroll percentage increases for time and one half and double times are shown for the various working days.

Days per Week	Hours per Day	Production Efficiency					Payroll Cost Factors	
		1st Week	2nd Week	3rd Week	4th Week	Average 4 Weeks	@ 1-1/2 Times	@ 2 Times
5	8	100%	100%	100%	100%	100%	1.000	1.000
	9	100	100	95	90	96	1.056	1.111
	10	100	95	90	85	93	1.100	1.200
	11	95	90	75	65	81	1.136	1.273
	12	90	85	70	60	76	1.167	1.333
6	8	100	100	95	90	96	1.083	1.167
	9	100	95	90	85	93	1.130	1.259
	10	95	90	85	80	88	1.167	1.333
	11	95	85	70	65	79	1.197	1.394
	12	90	80	65	60	74	1.222	1.444
7	8	100	95	85	75	89	1.143	1.286
	9	95	90	80	70	84	1.183	1.365
	10	90	85	75	65	79	1.214	1.429
	11	85	80	65	60	73	1.240	1.481
	12	85	75	60	55	69	1.262	1.524

General Requirements

R0131 Project Management & Coordination

R013113-40   Builder's Risk Insurance

Builder's risk insurance is insurance on a building during construction. Premiums are paid by the owner or the contractor. Blasting, collapse and underground insurance would raise total insurance costs.

R013113-50 General Contractor's Overhead

There are two distinct types of overhead on a construction project: Project overhead and main office overhead. Project overhead includes those costs at a construction site not directly associated with the installation of construction materials. Examples of project overhead costs include the following:

1. Superintendent
2. Construction office and storage trailers
3. Temporary sanitary facilities
4. Temporary utilities
5. Security fencing
6. Photographs
7. Cleanup
8. Performance and payment bonds

The above project overhead items are also referred to as general requirements and therefore are estimated in Division 1. Division 1 is the first division listed in the CSI MasterFormat but it is usually the last division estimated. The sum of the costs in Divisions 1 through 49 is referred to as the sum of the direct costs.

All construction projects also include indirect costs. The primary components of indirect costs are the contractor's main office overhead and profit. The amount of the main office overhead expense varies depending on the following:

1. Owner's compensation
2. Project managers' and estimators' wages
3. Clerical support wages
4. Office rent and utilities
5. Corporate legal and accounting costs
6. Advertising
7. Automobile expenses
8. Association dues
9. Travel and entertainment expenses

These costs are usually calculated as a percentage of annual sales volume. This percentage can range from 35% for a small contractor doing less than $500,000 to 5% for a large contractor with sales in excess of $100 million.

R013113-55 Installing Contractor's Overhead

Installing contractors (subcontractors) also incur costs for general requirements and main office overhead.

Included within the total incl. overhead and profit costs is a percent mark-up for overhead that includes:

1. Compensation and benefits for office staff and project managers
2. Office rent, utilities, business equipment, and maintenance
3. Corporate legal and accounting costs

4. Advertising
5. Vehicle expenses (for office staff and project managers)
6. Association dues
7. Travel, entertainment
8. Insurance
9. Small tools and equipment

817

R013113-60   Workers' Compensation Insurance Rates by Trade

The table below tabulates the national averages for workers' compensation insurance rates by trade and type of building. The average "Insurance Rate" is multiplied by the "% of Building Cost" for each trade. This produces the "Workers' Compensation" cost by % of total labor cost, to be added for each trade by building type to determine the weighted average workers' compensation rate for the building types analyzed.

Trade	Insurance Rate (% Labor Cost) Range			Insurance Rate (% Labor Cost) Average	% of Building Cost Office Bldgs.	% of Building Cost Schools & Apts.	% of Building Cost Mfg.	Workers' Compensation Office Bldgs.	Workers' Compensation Schools & Apts.	Workers' Compensation Mfg.
Excavation, Grading, etc.	3.3 %	to	20.7%	9.0%	4.8%	4.9%	4.5%	0.43%	0.44%	0.41%
Piles & Foundations	6.3	to	34.3	14.5	7.1	5.2	8.7	1.03	0.75	1.26
Concrete	4.5	to	32.5	12.4	5.0	14.8	3.7	0.62	1.84	0.46
Masonry	4.4	to	43.4	14.5	6.9	7.5	1.9	1.00	1.09	0.28
Structural Steel	6.3	to	71.0	24.4	10.7	3.9	17.6	2.61	0.95	4.29
Miscellaneous & Ornamental Metals	4.4	to	28.9	11.3	2.8	4.0	3.6	0.32	0.45	0.41
Carpentry & Millwork	4.7	to	39.7	13.9	3.7	4.0	0.5	0.51	0.56	0.07
Metal or Composition Siding	6.3	to	92.8	19.1	2.3	0.3	4.3	0.44	0.06	0.82
Roofing	6.3	to	108.8	30.7	2.3	2.6	3.1	0.71	0.80	0.95
Doors & Hardware	3.9	to	39.7	11.6	0.9	1.4	0.4	0.10	0.16	0.05
Sash & Glazing	4.5	to	29.4	12.9	3.5	4.0	1.0	0.45	0.52	0.13
Lath & Plaster	3.2	to	42.3	10.7	3.3	6.9	0.8	0.35	0.74	0.09
Tile, Marble & Floors	2.9	to	22.4	9.1	2.6	3.0	0.5	0.24	0.27	0.05
Acoustical Ceilings	2.6	to	43.8	8.7	2.4	0.2	0.3	0.21	0.02	0.03
Painting	4.6	to	35.9	11.7	1.5	1.6	1.6	0.18	0.19	0.19
Interior Partitions	4.7	to	39.7	13.9	3.9	4.3	4.4	0.54	0.60	0.61
Miscellaneous Items	2.1	to	97.7	11.9	5.2	3.7	9.7	0.62	0.44	1.16
Elevators	1.4	to	14.4	5.0	2.1	1.1	2.2	0.11	0.06	0.11
Sprinklers	2.3	to	23.4	7.1	0.5	—	2.0	0.04	—	0.14
Plumbing	1.8	to	17.8	6.6	4.9	7.2	5.2	0.32	0.48	0.34
Heat., Vent., Air Conditioning	3.3	to	19.2	8.8	13.5	11.0	12.9	1.19	0.97	1.14
Electrical	2.2	to	14.4	5.5	10.1	8.4	11.1	0.56	0.46	0.61
Total	1.4 %	to	108.8%	—	100.0%	100.0%	100.0%	12.58%	11.85%	13.60%
					Overall Weighted Average	12.68%				

Workers' Compensation Insurance Rates by States

The table below lists the weighted average Workers' Compensation base rate for each state with a factor comparing this with the national average of 12.6%.

State	Weighted Average	Factor	State	Weighted Average	Factor	State	Weighted Average	Factor
Alabama	17.8%	141	Kentucky	10.5%	83	North Dakota	7.2%	57
Alaska	10.6	84	Louisiana	19.2	152	Ohio	7.7	61
Arizona	11.2	89	Maine	11.3	90	Oklahoma	11.1	88
Arkansas	7.0	56	Maryland	12.3	98	Oregon	10.0	79
California	29.4	233	Massachusetts	11.4	90	Pennsylvania	19.5	155
Colorado	7.8	62	Michigan	9.2	73	Rhode Island	13.8	110
Connecticut	20.9	166	Minnesota	19.7	156	South Carolina	15.6	124
Delaware	14.1	112	Mississippi	11.9	94	South Dakota	12.5	99
District of Columbia	10.2	81	Missouri	13.5	107	Tennessee	10.2	81
Florida	9.6	76	Montana	8.2	65	Texas	7.3	58
Georgia	31.3	248	Nebraska	14.1	112	Utah	8.3	66
Hawaii	8.6	68	Nevada	8.4	67	Vermont	12.0	95
Idaho	9.7	77	New Hampshire	13.8	110	Virginia	7.9	63
Illinois	23.7	188	New Jersey	15.3	121	Washington	9.8	78
Indiana	4.8	38	New Mexico	14.7	117	West Virginia	5.7	45
Iowa	15.2	121	New York	18.4	146	Wisconsin	14.1	112
Kansas	7.2	57	North Carolina	16.5	131	Wyoming	6.4	51
			Weighted Average for U.S. is	12.7% of payroll = 100%				

The weighted average skilled worker rate for 35 trades is 12.6%. For bidding purposes, apply the full value of Workers' Compensation directly to total labor costs, or if labor is 38%, materials 42% and overhead and profit 20% of total cost, carry 38/80 x 12.6% = 6.0% of cost (before overhead and profit)

into overhead. Rates vary not only from state to state but also with the experience rating of the contractor.

Rates are the most current available at the time of publication.

R013113-80 Performance Bond

This table shows the cost of a performance bond for a construction job scheduled to be completed in 12 months. Add 1% of the premium cost per month for jobs requiring more than 12 months to complete. The rates are "standard" rates offered to contractors that the bonding company considers financially sound and capable of doing the work. Preferred rates are offered by some bonding companies based upon financial strength of the contractor. Actual rates vary from contractor to contractor and from bonding company to bonding company. Contractors should prequalify through a bonding agency before submitting a bid on a contract that requires a bond.

Contract Amount	Building Construction Class B Projects			Highways & Bridges					
				Class A New Construction			Class A-1 Highway Resurfacing		
First $ 100,000 bid	$25.00 per M			$15.00 per M			$9.40 per M		
Next 400,000 bid	$ 2,500	plus $15.00	per M	$ 1,500	plus $10.00	per M	$ 940	plus $7.20	per M
Next 2,000,000 bid	8,500	plus 10.00	per M	5,500	plus 7.00	per M	3,820	plus 5.00	per M
Next 2,500,000 bid	28,500	plus 7.50	per M	19,500	plus 5.50	per M	15,820	plus 4.50	per M
Next 2,500,000 bid	47,250	plus 7.00	per M	33,250	plus 5.00	per M	28,320	plus 4.50	per M
Over 7,500,000 bid	64,750	plus 6.00	per M	45,750	plus 4.50	per M	39,570	plus 4.00	per M

General Requirements | R0151 Temporary Utilities

R015113-65 Temporary Power Equipment

Cost data for the temporary equipment was developed utilizing the following information.

1) Re-usable material-services, transformers, equipment and cords are based on new purchase and prorated to three projects.
2) PVC feeder includes trench and backfill.
3) Connections include disconnects and fuses.
4) Labor units include an allowance for removal.
5) No utility company charges or fees are included.
6) Concrete pads or vaults are not included.
7) Utility company conduits not included.

819

R015423-10   Steel Tubular Scaffolding

On new construction, tubular scaffolding is efficient up to 60' high or five stories. Above this it is usually better to use a hung scaffolding if construction permits. Swing scaffolding operations may interfere with tenants. In this case, the tubular is more practical at all heights.

In repairing or cleaning the front of an existing building the cost of tubular scaffolding per S.F. of building front increases as the height increases above the first tier. The first tier cost is relatively high due to leveling and alignment.

The minimum efficient crew for erecting and dismantling is three workers. They can set up and remove 18 frame sections per day up to 5 stories high. For 6 to 12 stories high, a crew of four is most efficient. Use two or more on top and two on the bottom for handing up or hoisting. They can

also set up and remove 18 frame sections per day. At 7' horizontal spacing, this will run about 800 S.F. per day of erecting and dismantling. Time for placing and removing planks must be added to the above. A crew of three can place and remove 72 planks per day up to 5 stories. For over 5 stories, a crew of four can place and remove 80 planks per day.

The table below shows the number of pieces required to erect tubular steel scaffolding for 1000 S.F. of building frontage. This area is made up of a scaffolding system that is 12 frames (11 bays) long by 2 frames high.

For jobs under twenty-five frames, add 50% to rental cost. Rental rates will be lower for jobs over three months duration. Large quantities for long periods can reduce rental rates by 20%.

Description of Component	Number of Pieces for 1000 S.F. of Building Front	Unit
5' Wide Standard Frame, 6'-4" High	24	Ea.
Leveling Jack & Plate	24	
Cross Brace	44	
Side Arm Bracket, 21"	12	
Guardrail Post	12	
Guardrail, 7' section	22	
Stairway Section	2	
Stairway Starter Bar	1	
Stairway Inside Handrail	2	
Stairway Outside Handrail	2	
Walk-Thru Frame Guardrail	2	

Scaffolding is often used as falsework over 15' high during construction of cast-in-place concrete beams and slabs. Two foot wide scaffolding is generally used for heavy beam construction. The span between frames depends upon the load to be carried with a maximum span of 5'.

Heavy duty shoring frames with a capacity of 10,000#/leg can be spaced up to 10' O.C. depending upon form support design and loading.

Scaffolding used as horizontal shoring requires less than half the material required with conventional shoring.

On new construction, erection is done by carpenters.

Rolling towers supporting horizontal shores can reduce labor and speed the job. For maintenance work, catwalks with spans up to 70' can be supported by the rolling towers.

R015423-20   Pump Staging

Pump staging is generally not available for rent. The table below shows the number of pieces required to erect pump staging for 2400 S.F. of building

frontage. This area is made up of a pump jack system that is 3 poles (2 bays) wide by 2 poles high.

Item	Number of Pieces for 2400 S.F. of Building Front	Unit
Aluminum pole section, 24' long	6	Ea.
Aluminum splice joint, 6' long	3	
Aluminum foldable brace	3	
Aluminum pump jack	3	
Aluminum support for workbench/back safety rail	3	
Aluminum scaffold plank/workbench, 14" wide x 24' long	4	
Safety net, 22' long	2	
Aluminum plank end safety rail	2	

The cost in place for this 2400 S.F. will depend on how many uses are realized during the life of the equipment.

R015433-10   Contractor Equipment

Rental Rates shown elsewhere in the book pertain to late model high quality machines in excellent working condition, rented from equipment dealers. Rental rates from contractors may be substantially lower than the rental rates from equipment dealers depending upon economic conditions; for older, less productive machines, reduce rates by a maximum of 15%. Any overtime must be added to the base rates. For shift work, rates are lower. Usual rule of thumb is 150% of one shift rate for two shifts; 200% for three shifts.

For periods of less than one week, operated equipment is usually more economical to rent than renting bare equipment and hiring an operator.

Costs to move equipment to a job site (mobilization) or from a job site (demobilization) are not included in rental rates, nor in any Equipment costs on any Unit Price line items or crew listings. These costs can be found elsewhere. If a piece of equipment is already at a job site, it is not appropriate to utilize mob/demob costs in an estimate again.

Rental rates vary throughout the country with larger cities generally having lower rates. Lease plans for new equipment are available for periods in excess of six months with a percentage of payments applying toward purchase.

Rental rates can also be treated as reimbursement costs for contractor-owned equipment. Owned equipment costs include depreciation, loan payments, interest, taxes, insurance, storage, and major repairs.

Monthly rental rates vary from 2% to 5% of the cost of the equipment depending on the anticipated life of the equipment and its wearing parts. Weekly rates are about 1/3 the monthly rates and daily rental rates are about 1/3 the weekly rates.

The hourly operating costs for each piece of equipment include costs to the user such as fuel, oil, lubrication, normal expendables for the equipment, and a percentage of the mechanic's wages chargeable to maintenance. The hourly operating costs listed do not include the operator's wages.

The daily cost for equipment used in the standard crews is figured by dividing the weekly rate by five, then adding eight times the hourly operating cost to give the total daily equipment cost, not including the operator. This figure is in the right hand column of the Equipment listings under Equipment Cost/Day.

Pile Driving rates shown for the pile hammer and extractor do not include leads, cranes, boilers or compressors. Vibratory pile driving requires an added field specialist during set-up and pile driving operation for the electric model. The hydraulic model requires a field specialist for set-up only. Up to 125 reuses of sheet piling are possible using vibratory drivers. For normal conditions, crane capacity for hammer type and size is as follows.

Crane Capacity	Hammer Type and Size		
	Air or Steam	Diesel	Vibratory
25 ton	to 8,750 ft.-lb.		70 H.P.
40 ton	15,000 ft.-lb.	to 32,000 ft.-lb.	170 H.P.
60 ton	25,000 ft.-lb.		300 H.P.
100 ton		112,000 ft.-lb.	

Cranes should be specified for the job by size, building and site characteristics, availability, performance characteristics, and duration of time required.

Backhoes & Shovels rent for about the same as equivalent size cranes but maintenance and operating expenses are higher. The crane operator's rate must be adjusted for high boom heights. Average adjustments: for 150' boom add 2% per hour; over 185', add 4% per hour; over 210', add 6% per hour; over 250', add 8% per hour and over 295', add 12% per hour.

Tower Cranes of the climbing or static type have jibs from 50' to 200' and capacities at maximum reach range from 4,000 to 14,000 pounds. Lifting capacities increase up to maximum load as the hook radius decreases.

Typical rental rates, based on purchase price, are about 2% to 3% per month.

Erection and dismantling run between 500 and 2000 labor hours. Climbing operation takes 10 labor hours per 20' climb. Crane dead time is about 5 hours per 40' climb. If crane is bolted to side of the building add cost of ties and extra mast sections. Climbing cranes have from 80' to 180' of mast while static cranes have 80' to 800' of mast.

Truck Cranes can be converted to tower cranes by using tower attachments. Mast heights over 400' have been used.

A single 100' high material **Hoist and Tower** can be erected and dismantled in about 400 labor hours; a double 100' high hoist and tower in about 600 labor hours. Erection times for additional heights are 3 and 4 labor hours per vertical foot respectively up to 150', and 4 to 5 labor hours per vertical foot over 150' high. A 40' high portable Buck hoist takes about 160 labor hours to erect and dismantle. Additional heights take 2 labor hours per vertical foot to 80' and 3 labor hours per vertical foot for the next 100'. Most material hoists do not meet local code requirements for carrying personnel.

A 150' high **Personnel Hoist** requires about 500 to 800 labor hours to erect and dismantle. Budget erection time at 5 labor hours per vertical foot for all trades. Local code requirements or labor scarcity requiring overtime can add up to 50% to any of the above erection costs.

Earthmoving Equipment: The selection of earthmoving equipment depends upon the type and quantity of material, moisture content, haul distance, haul road, time available, and equipment available. Short haul cut and fill operations may require dozers only, while another operation may require excavators, a fleet of trucks, and spreading and compaction equipment. Stockpiled material and granular material are easily excavated with front end loaders. Scrapers are most economically used with hauls between 300' and 1-1/2 miles if adequate haul roads can be maintained. Shovels are often used for blasted rock and any material where a vertical face of 8' or more can be excavated. Special conditions may dictate the use of draglines, clamshells, or backhoes. Spreading and compaction equipment must be matched to the soil characteristics, the compaction required and the rate the fill is being supplied.

R015433-15   Heavy Lifting

Hydraulic Climbing Jacks

The use of hydraulic heavy lift systems is an alternative to conventional type crane equipment. The lifting, lowering, pushing, or pulling mechanism is a hydraulic climbing jack moving on a square steel jackrod from 1-5/8" to 4" square, or a steel cable. The jackrod or cable can be vertical or horizontal, stationary or movable, depending on the individual application. When the jackrod is stationary, the climbing jack will climb the rod and push or pull the load along with itself. When the climbing jack is stationary, the jackrod is movable with the load attached to the end and the climbing jack will lift or lower the jackrod with the attached load. The heavy lift system is normally operated by a single control lever located at the hydraulic pump.

The system is flexible in that one or more climbing jacks can be applied wherever a load support point is required, and the rate of lift synchronized.

Economic benefits have been demonstrated on projects such as: erection of ground assembled roofs and floors, complete bridge spans, girders and trusses, towers, chimney liners and steel vessels, storage tanks, and heavy machinery. Other uses are raising and lowering offshore work platforms, caissons, tunnel sections and pipelines.

R015436-50   Mobilization

Costs to move rented construction equipment to a job site from an equipment dealer's or contractor's yard (mobilization) or off the job site (demobilization), are not included in the rental or operating rates, nor in the equipment cost on a unit price line or in a crew listing. These costs can be found consolidated in the Mobilization section of the data and elsewhere in particular site work sections. If a piece of

equipment is already on the job site, it is not appropriate to include mob/demob costs in a new estimate that requires use of that equipment. The following table identifies approximate sizes of rented construction equipment that would be hauled on a towed trailer. Because this listing is not all-encompassing, the user can infer as to what size trailer might be required for a piece of equipment not listed.

3-ton Trailer	20-ton Trailer	40-ton Trailer	50-ton Trailer
20 H.P. Excavator	110 H.P. Excavator	200 H.P. Excavator	270 H.P. Excavator
50 H.P. Skid Steer	165 H.P. Dozer	300 H.P. Dozer	Small Crawler Crane
35 H.P. Roller	150 H.P. Roller	400 H.P. Scraper	500 H.P. Scraper
40 H.P. Trencher	Backhoe	450 H.P. Art. Dump Truck	500 H.P. Art. Dump Truck

R024119-10   Demolition Defined

Whole Building Demolition - Demolition of the whole building with no concern for any particular building element, component, or material type being demolished. This type of demolition is accomplished with large pieces of construction equipment that break up the structure, load it into trucks and haul it to a disposal site, but disposal or dump fees are not included. Demolition of below-grade foundation elements, such as footings, foundation walls, grade beams, slabs on grade, etc., is not included. Certain mechanical equipment containing flammable liquids or ozone-depleting refrigerants, electric lighting elements, communication equipment components, and other building elements may contain hazardous waste, and must be removed, either selectively or carefully, as hazardous waste before the building can be demolished.

Foundation Demolition - Demolition of below-grade foundation footings, foundation walls, grade beams, and slabs on grade. This type of demolition is accomplished by hand or pneumatic hand tools, and does not include saw cutting, or handling, loading, hauling, or disposal of the debris.

Gutting - Removal of building interior finishes and electrical/mechanical systems down to the load-bearing and sub-floor elements of the rough building frame, with no concern for any particular building element, component, or material type being demolished. This type of demolition is accomplished by hand or pneumatic hand tools, and includes loading into trucks, but not hauling, disposal or dump fees, scaffolding, or shoring. Certain mechanical equipment containing flammable liquids or ozone-depleting refrigerants, electric lighting elements, communication equipment components, and other building elements may contain hazardous waste, and must be removed, either selectively or carefully, as hazardous waste, before the building is gutted.

Selective Demolition - Demolition of a selected building element, component, or finish, with some concern for surrounding or adjacent elements, components, or finishes (see the first Subdivision (s) at the beginning of appropriate Divisions). This type of demolition is accomplished by hand or pneumatic hand tools, and does not include handling, loading,

storing, hauling, or disposal of the debris, scaffolding, or shoring. "Gutting" methods may be used in order to save time, but damage that is caused to surrounding or adjacent elements, components, or finishes may have to be repaired at a later time.

Careful Removal - Removal of a piece of service equipment, building element or component, or material type, with great concern for both the removed item and surrounding or adjacent elements, components or finishes. The purpose of careful removal may be to protect the removed item for later re-use, preserve a higher salvage value of the removed item, or replace an item while taking care to protect surrounding or adjacent elements, components, connections, or finishes from cosmetic and/or structural damage. An approximation of the time required to perform this type of removal is 1/3 to 1/2 the time it would take to install a new item of like kind (see Reference Number R220105-10). This type of removal is accomplished by hand or pneumatic hand tools, and does not include loading, hauling, or storing the removed item, scaffolding, shoring, or lifting equipment.

Cutout Demolition - Demolition of a small quantity of floor, wall, roof, or other assembly, with concern for the appearance and structural integrity of the surrounding materials. This type of demolition is accomplished by hand or pneumatic hand tools, and does not include saw cutting, handling, loading, hauling, or disposal of debris, scaffolding, or shoring.

Rubbish Handling - Work activities that involve handling, loading or hauling of debris. Generally, the cost of rubbish handling must be added to the cost of all types of demolition, with the exception of whole building demolition.

Minor Site Demolition - Demolition of site elements outside the footprint of a building. This type of demolition is accomplished by hand or pneumatic hand tools, or with larger pieces of construction equipment, and may include loading a removed item onto a truck (check the Crew for equipment used). It does not include saw cutting, hauling or disposal of debris, and, sometimes, handling or loading.

R024119-20   Dumpsters

Dumpster rental costs on construction sites are presented in two ways:

The cost per week rental includes the delivery of the dumpster; its pulling or emptying once per week, and its final removal. The assumption is made that the dumpster contractor could choose to empty a dumpster by simply bringing in an empty unit and removing the full one. These costs also include the disposal of the materials in the dumpster.

The alternate pricing can be used when actual planned conditions are not approximated by the weekly numbers. For example, these lines can be used when a dumpster is needed for 4 weeks and will need to be emptied 2 or 3 times per week. Conversely the alternate pricing lines can be used when a dumpster will be rented for several weeks or months but needs to be emptied only a few times over this period.

R024119-30   Rubbish Handling Chutes

To correctly estimate the cost of rubbish handling chute systems, the individual components must be priced separately. First choose the size of the system; a 30-inch diameter chute is quite common, but the sizes range from 18 to 36 inches in diameter. The 30-inch chute comes in a standard weight and two thinner weights. The thinner weight chutes are sometimes chosen for cost savings, but they are more easily damaged.

There are several types of major chute pieces that make up the chute system. The first component to consider is the top chute section (top intake hopper) where the material is dropped into the chute at the highest point. After determining the top chute, the intermediate chute pieces called the regular chute sections are priced. Next, the number of chute control door sections (intermediate intake hoppers) must be determined. In the more complex systems, a chute control door section is provided at each floor level. The last major component to consider is bolt down frames; these are usually provided at every other floor level.

There are a number of accessories to consider for safe operation and control. There are covers for the top chute and the chute door sections. The top

chute can have a trough that allows for better loading of the chute. For the safest operation, a chute warning light system can be added that will warn the other chute intake locations not to load while another is being used. There are dust control devices that spray a water mist to keep down the dust as the debris is loaded into a dumpster. There are special breakaway cords that are used to prevent damage to the chute if the dumpster is removed without disconnecting from the chute. There are chute liners that can be installed to protect the chute structure from physical damage from rough abrasive materials. Warning signs can be posted at each floor level that is provided with a chute control door section.

In summary, a complete rubbish handling chute system will include one top section, several intermediate regular sections, several intermediate control door (intake hopper) sections and bolt down frames at every other floor level starting with the top floor. If so desired, the system can also include covers and a light warning system for a safer operation. The bottom of the chute should always be above the Dumpster and should be tied off with a breakaway cord to the Dumpster.

R026510-20   Underground Storage Tank Removal

Underground storage tank removal can be divided into two categories: non-leaking and leaking. Prior to removing an underground storage tank, tests should be made, with the proper authorities present, to determine whether a tank has been leaking or the surrounding soil has been contaminated.

To safely remove liquid underground storage tanks:

1. Excavate to the top of the tank.
2. Disconnect all piping.
3. Open all tank vents and access ports.
4. Remove all liquids and/or sludge.
5. Purge the tank with an inert gas.
6. Provide access to the inside of the tank and clean out the interior using proper personal protective equipment (PPE).
7. Excavate soil surrounding the tank using proper PPE for on-site personnel.
8. Pull and properly dispose of the tank.
9. Clean up the site of all contaminated material.
10. Install new tanks or close the excavation.

R028213-20   Asbestos Removal Process

Asbestos removal is accomplished by a specialty contractor who understands the federal and state regulations regarding the handling and disposal of the material. The process of asbestos removal is divided into many individual steps. An accurate estimate can be calculated only after all the steps have been priced.

The steps are generally as follows:

1. Obtain an asbestos abatement plan from an industrial hygienist.
2. Monitor the air quality in and around the removal area and along the path of travel between the removal area and transport area. This establishes the background contamination.
3. Construct a two part decontamination chamber at entrance to removal area.
4. Install a HEPA filter to create a negative pressure in the removal area.
5. Install wall, floor and ceiling protection as required by the plan, usually 2 layers of fireproof 6 mil polyethylene.
6. Industrial hygienist visually inspects work area to verify compliance with plan.
7. Provide temporary supports for conduit and piping affected by the removal process.
8. Proceed with asbestos removal and bagging process. Monitor air quality as described in Step #2. Discontinue operations when contaminate levels exceed applicable standards.
9. Document the legal disposal of materials in accordance with EPA standards.
10. Thoroughly clean removal area including all ledges, crevices and surfaces.
11. Post abatement inspection by industrial hygienist to verify plan compliance.
12. Provide a certificate from a licensed industrial hygienist attesting that contaminate levels are within acceptable standards before returning area to regular use.

R028319-60   Lead Paint Remediation Methods

Lead paint remediation can be accomplished by the following methods.

1. Abrasive blast
2. Chemical stripping
3. Power tool cleaning with vacuum collection system
4. Encapsulation
5. Remove and replace
6. Enclosure

Each of these methods has strengths and weaknesses depending on the specific circumstances of the project. The following is an overview of each method.

1. **Abrasive blasting** is usually accomplished with sand or recyclable metallic blast. Before work can begin, the area must be contained to ensure the blast material with lead does not escape to the atmosphere. The use of vacuum blast greatly reduces the containment requirements. Lead abatement equipment that may be associated with this work includes a negative air machine. In addition, it is necessary to have an industrial hygienist monitor the project on a continual basis. When the work is complete, the spent blast sand with lead must be disposed of as a hazardous material. If metallic shot was used, the lead is separated from the shot and disposed of as hazardous material. Worker protection includes disposable clothing and respiratory protection.

2. **Chemical stripping** requires strong chemicals to be applied to the surface to remove the lead paint. Before the work can begin, the area under/adjacent to the work area must be covered to catch the chemical and removed lead. After the chemical is applied to the painted surface it is usually covered with paper. The chemical is left in place for the specified period, then the paper with lead paint is pulled or scraped off. The process may require several chemical applications. The paper with chemicals and lead paint adhered to it, plus the containment and loose scrapings collected by a HEPA (high efficiency particulate air filter) vac, must be disposed of as hazardous material. The chemical stripping process usually requires a neutralizing agent and several wash downs after the paint is removed. Worker protection includes a neoprene or other compatible protective clothing and respiratory

protection with face shield. An industrial hygienist is required intermittently during the process.

3. **Power tool cleaning** is accomplished using shrouded needle blasting guns. The shrouding with different end configurations is held up against the surface to be cleaned. The area is blasted with hardened needles and the shroud captures the lead with a HEPA vac and deposits it in a holding tank. An industrial hygienist monitors the project. Protective clothing and a respirator are required until air samples prove otherwise. When the work is complete the lead must be disposed of as hazardous material.

4. **Encapsulation** is a method that leaves the well bonded lead paint in place after the peeling paint has been removed. Before the work can begin, the area under/adjacent to the work must be covered to catch the scrapings. The scraped surface is then washed with a detergent and rinsed. The prepared surface is covered with approximately 10 mils of paint. A reinforcing fabric can also be embedded in the paint covering. The scraped paint and containment must be disposed of as hazardous material. Workers must wear protective clothing and respirators.

5. **Removing and replacing** are effective ways to remove lead paint from windows, gypsum walls and concrete masonry surfaces. The painted materials are removed and new materials are installed. Workers should wear a respirator and tyvek suit. The demolished materials must be disposed of as hazardous waste if it fails the TCLP (toxicity characteristic leachate process) test.

6. **Enclosure** is the process that permanently seals lead painted materials in place. This process has many applications such as covering lead painted drywall with new drywall, covering exterior construction with tyvek paper then residing, or covering lead painted structural members with aluminum or plastic. The seams on all enclosing materials must be securely sealed. An industrial hygienist monitors the project, and protective clothing and a respirator are required until air samples prove otherwise.

All the processes require clearance monitoring and wipe testing as required by the hygienist.

825

R031113-10   Wall Form Materials

Aluminum Forms

Approximate weight is 3 lbs. per S.F.C.A. Standard widths are available from 4″ to 36″ with 36″ most common. Standard lengths of 2′, 4′, 6′ to 8′ are available. Forms are lightweight and fewer ties are needed with the wider widths. The form face is either smooth or textured.

Metal Framed Plywood Forms

Manufacturers claim over 75 reuses of plywood and over 300 reuses of steel frames. Many specials such as corners, fillers, pilasters, etc. are available. Monthly rental is generally about 15% of purchase price for first month and 9% per month thereafter with 90% of rental applied to purchase for the first month and decreasing percentages thereafter. Aluminum framed forms cost 25% to 30% more than steel framed.

After the first month, extra days may be prorated from the monthly charge. Rental rates do not include ties, accessories, cleaning, loss of hardware or freight in and out. Approximate weight is 5 lbs. per S.F. for steel; 3 lbs. per S.F. for aluminum.

Forms can be rented with option to buy.

Plywood Forms, Job Fabricated

There are two types of plywood used for concrete forms.
1. Exterior plyform is completely waterproof. This is face oiled to facilitate stripping. Ten reuses can be expected with this type with 25 reuses possible.
2. An overlaid type consists of a resin fiber fused to exterior plyform. No oiling is required except to facilitate cleaning. This is available in both high density (HDO) and medium density overlaid (MDO). Using HDO, 50 reuses can be expected with 200 possible.

Plyform is available in 5/8″ and 3/4″ thickness. High density overlaid is available in 3/8″, 1/2″, 5/8″ and 3/4″ thickness.

5/8″ thick is sufficient for most building forms, while 3/4″ is best on heavy construction.

Plywood Forms, Modular, Prefabricated

There are many plywood forming systems without frames. Most of these are manufactured from 1-1/8″ (HDO) plywood and have some hardware attached. These are used principally for foundation walls 8′ or less high. With care and maintenance, 100 reuses can be attained with decreasing quality of surface finish.

Steel Forms

Approximate weight is 6-1/2 lbs. per S.F.C.A. including accessories. Standard widths are available from 2″ to 24″, with 24″ the most common. Standard lengths are from 2′ to 8′, with 4′ the most common. Forms are easily ganged into modular units.

Forms are usually leased for 15% of the purchase price per month prorated daily over 30 days.

Rental may be applied to sale price, and usually rental forms are bought. With careful handling and cleaning 200 to 400 reuses are possible.

Straight wall gang forms up to 12′ x 20′ or 8′ x 30′ can be fabricated. These crane handled forms usually lease for approx. 9% per month.

Individual job analysis is available from the manufacturer at no charge.

R031113-30   Slipforms

The slipform method of forming may be used for forming circular silo and multi-celled storage bin type structures over 30′ high, and building core shear walls over eight stories high. The shear walls usually enclose elevator shafts, stairwells, mechanical spaces, and toilet rooms. Reuse of the form on duplicate structures will reduce the height necessary and spread the cost of building the form. Slipform systems can be used to cast chimneys, towers, piers, dams, underground shafts or other structures capable of being extruded.

Slipforms are usually 4′ high and are raised semi-continuously by jacks climbing on rods which are embedded in the concrete. The jacks are powered by a hydraulic, pneumatic, or electric source and are available in 3, 6, and 22 ton capacities. Interior work decks and exterior scaffolds must be provided for placing inserts, embedded items, reinforcing steel, and

concrete. Scaffolds below the form for finishers may be required. The interior work decks are often used as roof slab forms on silos and bin work. Form raising rates will range from 6″ to 20″ per hour for silos; 6″ to 30″ per hour for buildings; and 6″ to 48″ per hour for shaft work.

Reinforcing bars and stressing strands are usually hoisted by a crane or gin pole, and the concrete material can be hoisted by a crane, winch-powered skip, or pumps. The slipform system is operated on a continuous 24-hour day when a monolithic structure is desired. For least cost, the system is operated only during normal working hours.

Placing concrete will range from 0.5 to 1.5 labor-hours per C.Y. Bucks, blockouts, keyways, weldplates, etc. are extra.

R031113-40 Forms for Reinforced Concrete

Design Economy

Avoid many sizes in proportioning beams and columns.

From story to story avoid changing column dimensions. Gain strength by adding steel or using a richer mix. If a change in size of column is necessary, vary one dimension only to minimize form alterations. Keep beams and columns the same width.

From floor to floor in a multi-story building vary beam depth, not width, as that will leave the slab panel form unchanged. It is cheaper to vary the strength of a beam from floor to floor by means of a steel area than by 2″ changes in either width or depth.

Cost Factors

Material includes the cost of lumber, cost of rent for metal pans or forms if used, nails, form ties, form oil, bolts and accessories.

Labor includes the cost of carpenters to make up, erect, remove and repair, plus common labor to clean and move. Having carpenters remove forms minimizes repairs.

Improper alignment and condition of forms will increase finishing cost. When forms are heavily oiled, concrete surfaces must be neutralized before finishing. Special curing compounds will cause spillages to spall off in first frost. Gang forming methods will reduce costs on large projects.

Materials Used

Boards are seldom used unless their architectural finish is required. Generally, steel, fiberglass and plywood are used for contact surfaces. Labor on plywood is 10% less than with boards. The plywood is backed up with

2 × 4′s at 12″ to 32″ O.C. Walers are generally 2 - 2 × 4′s. Column forms are held together with steel yokes or bands. Shoring is with adjustable shoring or scaffolding for high ceilings.

Reuse

Floor and column forms can be reused four or possibly five times without excessive repair. Remember to allow for 10% waste on each reuse.

When modular sized wall forms are made, up to twenty uses can be expected with exterior plyform.

When forms are reused, the cost to erect, strip, clean and move will not be affected. 10% replacement of lumber should be included and about one hour of carpenter time for repairs on each reuse per 100 S.F.

The reuse cost for certain accessory items normally rented on a monthly basis will be lower than the cost for the first use.

After the fifth use, new material required plus time needed for repair prevent the form cost from dropping further; it may go up. Much depends on care in stripping, the number of special bays, changes in beam or column sizes and other factors.

Costs for multiple use of formwork may be developed as follows:

2 Uses	3 Uses	4 Uses
$\dfrac{\text{(1st Use + Reuse)}}{2} = \text{avg. cost/2 uses}$	$\dfrac{\text{(1st Use + 2 Reuses)}}{3} = \text{avg. cost/3 uses}$	$\dfrac{\text{(1st use + 3 Reuses)}}{4} = \text{avg. cost/4 uses}$

R031113-60 Formwork Labor-Hours

Item	Unit	Hours Required Fabricate	Hours Required Erect & Strip	Hours Required Clean & Move	Total Hours 1 Use	Multiple Use 2 Use	Multiple Use 3 Use	Multiple Use 4 Use
Beam and Girder, interior beams, 12″ wide	100 S.F.	6.4	8.3	1.3	16.0	13.3	12.4	12.0
Hung from steel beams		5.8	7.7	1.3	14.8	12.4	11.6	11.2
Beam sides only, 36″ high		5.8	7.2	1.3	14.3	11.9	11.1	10.7
Beam bottoms only, 24″ wide		6.6	13.0	1.3	20.9	18.1	17.2	16.7
Box out for openings		9.9	10.0	1.1	21.0	16.6	15.1	14.3
Buttress forms, to 8′ high		6.0	6.5	1.2	13.7	11.2	10.4	10.0
Centering, steel, 3/4″ rib lath			1.0		1.0			
3/8″ rib lath or slab form	▼		0.9		0.9			
Chamfer strip or keyway	100 L.F.		1.5		1.5	1.5	1.5	1.5
Columns, fiber tube 8″ diameter			20.6		20.6			
12″			21.3		21.3			
16″			22.9		22.9			
20″			23.7		23.7			
24″			24.6		24.6			
30″	▼		25.6		25.6			
Columns, round steel, 12″ diameter			22.0		22.0	22.0	22.0	22.0
16″			25.6		25.6	25.6	25.6	25.6
20″			30.5		30.5	30.5	30.5	30.5
24″	▼	▼	37.7		37.7	37.7	37.7	37.7
Columns, plywood 8″ x 8″	100 S.F.	7.0	11.0	1.2	19.2	16.2	15.2	14.7
12″ x 12″		6.0	10.5	1.2	17.7	15.2	14.4	14.0
16″ x 16″		5.9	10.0	1.2	17.1	14.7	13.8	13.4
24″ x 24″		5.8	9.8	1.2	16.8	14.4	13.6	13.2
Columns, steel framed plywood 8″ x 8″			10.0	1.0	11.0	11.0	11.0	11.0
12″ x 12″			9.3	1.0	10.3	10.3	10.3	10.3
16″ x 16″			8.5	1.0	9.5	9.5	9.5	9.5
24″ x 24″			7.8	1.0	8.8	8.8	8.8	8.8
Drop head forms, plywood		9.0	12.5	1.5	23.0	19.0	17.7	17.0
Coping forms		8.5	15.0	1.5	25.0	21.3	20.0	19.4
Culvert, box			14.5	4.3	18.8	18.8	18.8	18.8
Curb forms, 6″ to 12″ high, on grade		5.0	8.5	1.2	14.7	12.7	12.1	11.7
On elevated slabs	▼	6.0	10.8	1.2	18.0	15.5	14.7	14.3
Edge forms to 6″ high, on grade	100 L.F.	2.0	3.5	0.6	6.1	5.6	5.4	5.3
7″ to 12″ high	100 S.F.	2.5	5.0	1.0	8.5	7.8	7.5	7.4
Equipment foundations		10.0	18.0	2.0	30.0	25.5	24.0	23.3
Flat slabs, including drops		3.5	6.0	1.2	10.7	9.5	9.0	8.8
Hung from steel		3.0	5.5	1.2	9.7	8.7	8.4	8.2
Closed deck for domes		3.0	5.8	1.2	10.0	9.0	8.7	8.5
Open deck for pans		2.2	5.3	1.0	8.5	7.9	7.7	7.6
Footings, continuous, 12″ high		3.5	3.5	1.5	8.5	7.3	6.8	6.6
Spread, 12″ high		4.7	4.2	1.6	10.5	8.7	8.0	7.7
Pile caps, square or rectangular		4.5	5.0	1.5	11.0	9.3	8.7	8.4
Grade beams, 24″ deep		2.5	5.3	1.2	9.0	8.3	8.0	7.9
Lintel or Sill forms		8.0	17.0	2.0	27.0	23.5	22.3	21.8
Spandrel beams, 12″ wide		9.0	11.2	1.3	21.5	17.5	16.2	15.5
Stairs			25.0	4.0	29.0	29.0	29.0	29.0
Trench forms in floor		4.5	14.0	1.5	20.0	18.3	17.7	17.4
Walls, Plywood, at grade, to 8′ high		5.0	6.5	1.5	13.0	11.0	9.7	9.5
8′ to 16′		7.5	8.0	1.5	17.0	13.8	12.7	12.1
16′ to 20′		9.0	10.0	1.5	20.5	16.5	15.2	14.5
Foundation walls, to 8′ high		4.5	6.5	1.0	12.0	10.3	9.7	9.4
8′ to 16′ high		5.5	7.5	1.0	14.0	11.8	11.0	10.6
Retaining wall to 12′ high, battered		6.0	8.5	1.5	16.0	13.5	12.7	12.3
Radial walls to 12′ high, smooth		8.0	9.5	2.0	19.5	16.0	14.8	14.3
2′ chords		7.0	8.0	1.5	16.5	13.5	12.5	12.0
Prefabricated modular, to 8′ high		—	4.3	1.0	5.3	5.3	5.3	5.3
Steel, to 8′ high		—	6.8	1.2	8.0	8.0	8.0	8.0
8′ to 16′ high		—	9.1	1.5	10.6	10.3	10.2	10.2
Steel framed plywood to 8′ high		—	6.8	1.2	8.0	7.5	7.3	7.2
8′ to 16′ high	▼	—	9.3	1.2	10.5	9.5	9.2	9.0

For customer support on your Building Construction Costs with RSMeans Data, call 800.448.8182.

R032110-10 Reinforcing Steel Weights and Measures

Bar Designation No.**	Nominal Weight Lb./Ft.	U.S. Customary Units			SI Units			
		Nominal Dimensions*			Nominal Dimensions*			
		Diameter in.	Cross Sectional Area, in.^{2}	Perimeter in.	Nominal Weight kg/m	Diameter mm	Cross Sectional Area, cm^2	Perimeter mm
3	.376	.375	.11	1.178	.560	9.52	.71	29.9
4	.668	.500	.20	1.571	.994	12.70	1.29	39.9
5	1.043	.625	.31	1.963	1.552	15.88	2.00	49.9
6	1.502	.750	.44	2.356	2.235	19.05	2.84	59.8
7	2.044	.875	.60	2.749	3.042	22.22	3.87	69.8
8	2.670	1.000	.79	3.142	3.973	25.40	5.10	79.8
9	3.400	1.128	1.00	3.544	5.059	28.65	6.45	90.0
10	4.303	1.270	1.27	3.990	6.403	32.26	8.19	101.4
11	5.313	1.410	1.56	4.430	7.906	35.81	10.06	112.5
14	7.650	1.693	2.25	5.320	11.384	43.00	14.52	135.1
18	13.600	2.257	4.00	7.090	20.238	57.33	25.81	180.1

* The nominal dimensions of a deformed bar are equivalent to those of a plain round bar having the same weight per foot as the deformed bar.
** Bar numbers are based on the number of eighths of an inch included in the nominal diameter of the bars.

R032110-20 Metric Rebar Specification - ASTM A615-81

Grade 300 (300 MPa* = 43,560 psi; +8.7% vs. Grade 40)				
Grade 400 (400 MPa* = 58,000 psi; –3.4% vs. Grade 60)				
Bar No.	Diameter mm	Area mm^2	Equivalent in.^{2}	Comparison with U.S. Customary Bars
10M	11.3	100	.16	Between #3 & #4
15M	16.0	200	.31	#5 (.31 in.^{2})
20M	19.5	300	.47	#6 (.44 in.^{2})
25M	25.2	500	.78	#8 (.79 in.^{2})
30M	29.9	700	1.09	#9 (1.00 in.^{2})
35M	35.7	1000	1.55	#11 (1.56 in.^{2})
45M	43.7	1500	2.33	#14 (2.25 in.^{2})
55M	56.4	2500	3.88	#18 (4.00 in.^{2})

* MPa = megapascals

R032110-40   Weight of Steel Reinforcing Per Square Foot of Wall (PSF)

Reinforced weights: The table below suggests the weights per square foot for reinforcing steel in walls. Weights are approximate and will be the same for all grades of steel bars. For bars in two directions, add weights for each size and spacing.

C/C Spacing in Inches	#3 Wt. (PSF)	#4 Wt. (PSF)	#5 Wt. (PSF)	#6 Wt. (PSF)	#7 Wt. (PSF)	#8 Wt. (PSF)	#9 Wt. (PSF)	#10 Wt. (PSF)	#11 Wt. (PSF)
2"	2.26	4.01	6.26	9.01	12.27				
3"	1.50	2.67	4.17	6.01	8.18	10.68	13.60	17.21	21.25
4"	1.13	2.01	3.13	4.51	6.13	8.10	10.20	12.91	15.94
5"	.90	1.60	2.50	3.60	4.91	6.41	8.16	10.33	12.75
6"	.752	1.34	2.09	3.00	4.09	5.34	6.80	8.61	10.63
8"	.564	1.00	1.57	2.25	3.07	4.01	5.10	6.46	7.97
10"	.451	.802	1.25	1.80	2.45	3.20	4.08	5.16	6.38
12"	.376	.668	1.04	1.50	2.04	2.67	3.40	4.30	5.31
18"	.251	.445	.695	1.00	1.32	1.78	2.27	2.86	3.54
24"	.188	.334	.522	.751	1.02	1.34	1.70	2.15	2.66
30"	.150	.267	.417	.600	.817	1.07	1.36	1.72	2.13
36"	.125	.223	.348	.501	.681	.890	1.13	1.43	1.77
42"	.107	.191	.298	.429	.584	.753	.97	1.17	1.52
48"	.094	.167	.261	.376	.511	.668	.85	1.08	1.33

R032110-50   Minimum Wall Reinforcement Weight (PSF)

This table lists the approximate minimum wall reinforcement weights per S.F. according to the specification of .12% of gross area for vertical bars and .20% of gross area for horizontal bars.

Location	Wall Thickness	Bar Size	Horizontal Steel Spacing C/C	Sq. In. Req'd per S.F.	Total Wt. per S.F.	Bar Size	Vertical Steel Spacing C/C	Sq. In. Req'd per S.F.	Total Wt. per S.F.	Horizontal & Vertical Steel Total Weight per S.F.
Both Faces	10"	#4	18"	.24	.89#	#3	18"	.14	.50#	1.39#
	12"	#4	16"	.29	1.00	#3	16"	.17	.60	1.60
	14"	#4	14"	.34	1.14	#3	13"	.20	.69	1.84
	16"	#4	12"	.38	1.34	#3	11"	.23	.82	2.16
	18"	#5	17"	.43	1.47	#4	18"	.26	.89	2.36
One Face	6"	#3	9"	.15	.50	#3	18"	.09	.25	.75
	8"	#4	12"	.19	.67	#3	11"	.12	.41	1.08
	10"	#5	15"	.24	.83	#4	16"	.14	.50	1.34

R032110-70   Bend, Place, and Tie Reinforcing

Placing and tying by rodmen for footings and slabs run from nine hrs. per ton for heavy bars to fifteen hrs. per ton for light bars. For beams, columns, and walls, production runs from eight hrs. per ton for heavy bars to twenty hrs. per ton for light bars. The overall average for typical reinforced concrete buildings is about fourteen hrs. per ton. These production figures include the time for placing accessories and usual inserts, but not their material cost (allow 15% of the cost of delivered bent rods). Equipment handling is necessary for the larger-sized bars so that installation costs for the very heavy bars will not decrease proportionately.

Installation costs for splicing reinforcing bars include allowance for equipment to hold the bars in place while splicing as well as necessary scaffolding for iron workers.

R032110-80   Shop-Fabricated Reinforcing Steel

The material prices for reinforcing, shown in the unit cost sections of the data set, are for 50 tons or more of shop-fabricated reinforcing steel and include:

1. Mill base price of reinforcing steel
2. Mill grade/size/length extras
3. Mill delivery to the fabrication shop
4. Shop storage and handling
5. Shop drafting/detailing
6. Shop shearing and bending
7. Shop listing
8. Shop delivery to the job site

Both material and installation costs can be considerably higher for small jobs consisting primarily of smaller bars, while material costs may be slightly lower for larger jobs.

Concrete — R0322 Welded Wire Fabric Reinforcing

R032205-30   Common Stock Styles of Welded Wire Fabric

This table provides some of the basic specifications, sizes, and weights of welded wire fabric used for reinforcing concrete.

New Designation		Old Designation		Steel Area per Foot				Approximate Weight per 100 S.F.	
Spacing — Cross Sectional Area (in.) — (Sq. in. 100)		Spacing — Wire Gauge (in.) — (AS & W)		Longitudinal		Transverse			
				in.	cm	in.	cm	lbs	kg
Rolls	6 x 6 — W1.4 x W1.4	6 x 6 — 10 x 10		.028	.071	.028	.071	21	9.53
	6 x 6 — W2.0 x W2.0	6 x 6 — 8 x 8	1	.040	.102	.040	.102	29	13.15
	6 x 6 — W2.9 x W2.9	6 x 6 — 6 x 6		.058	.147	.058	.147	42	19.05
	6 x 6 — W4.0 x W4.0	6 x 6 — 4 x 4		.080	.203	.080	.203	58	26.91
	4 x 4 — W1.4 x W1.4	4 x 4 — 10 x 10		.042	.107	.042	.107	31	14.06
	4 x 4 — W2.0 x W2.0	4 x 4 — 8 x 8	1	.060	.152	.060	.152	43	19.50
	4 x 4 — W2.9 x W2.9	4 x 4 — 6 x 6		.087	.227	.087	.227	62	28.12
	4 x 4 — W4.0 x W4.0	4 x 4 — 4 x 4		.120	.305	.120	.305	85	38.56
Sheets	6 x 6 — W2.9 x W2.9	6 x 6 — 6 x 6		.058	.147	.058	.147	42	19.05
	6 x 6 — W4.0 x W4.0	6 x 6 — 4 x 4		.080	.203	.080	.203	58	26.31
	6 x 6 — W5.5 x W5.5	6 x 6 — 2 x 2	2	.110	.279	.110	.279	80	36.29
	4 x 4 — W1.4 x W1.4	4 x 4 — 4 x 4		.120	.305	.120	.305	85	38.56

NOTES: 1. Exact W—number size for 8 gauge is W2.1
2. Exact W—number size for 2 gauge is W5.4

The above table was compiled with the following excerpts from the WRI Manual of Standard Practices, 7th Edition, Copyright 2006. Reproduced with permission of the Wire Reinforcement Institute, Inc.:

1. Chapter 3, page 7, Table 1 Common Styles of Metric Wire Reinforcement (WWR) With Equivalent US Customary Units
2. Chapter 6, page 19, Table 5 Customary Units
3. Chapter 6, Page 23, Table 7 Customary Units (in.) Welded Plain Wire Reinforcement
4. Chapter 6, Page 25 Table 8 Wire Size Comparison
5. Chapter 9, Page 30, Table 9 Weight of Longitudinal Wires Weight (Mass) Estimating Tables
6. Chapter 9, Page 31, Table 9M Weight of Longitudinal Wires Weight (Mass) Estimating Tables
7. Chapter 9, Page 32, Table 10 Weight of Transverse Wires Based on 62″ lengths of transverse wire (60″ width plus 1″ overhand each side)
8. Chapter 9, Page 33, Table 10M Weight of Transverse Wires

Concrete — R0330 Cast-In-Place Concrete

R033053-50   Industrial Chimneys

Foundation requirements in C.Y. of concrete for various sized chimneys.

Size Chimney	2 Ton Soil	3 Ton Soil	Size Chimney	2 Ton Soil	3 Ton Soil	Size Chimney	2 Ton Soil	3 Ton Soil
75' x 3'-0″	13 C.Y.	11 C.Y.	160' x 6'-6″	86 C.Y.	76 C.Y.	300' x 10'-0″	325 C.Y.	245 C.Y.
85' x 5'-6″	19	16	175' x 7'-0″	108	95	350' x 12'-0″	422	320
100' x 5'-0″	24	20	200' x 6'-0″	125	105	400' x 14'-0″	520	400
125' x 5'-6″	43	36	250' x 8'-0″	230	175	500' x 18'-0″	725	575

R033105-10 Proportionate Quantities

The tables below show both quantities per S.F. of floor areas as well as form and reinforcing quantities per C.Y. Unusual structural requirements would increase the ratios below. High strength reinforcing would reduce the steel weights. Figures are for 3000 psi concrete and 60,000 psi reinforcing unless specified otherwise.

Type of Construction	Live Load	Span	Per S.F. of Floor Area				Per C.Y. of Concrete		
			Concrete	Forms	Reinf.	Pans	Forms	Reinf.	Pans
Flat Plate	50 psf	15 Ft.	.46 C.F.	1.06 S.F.	1.71 lb.		62 S.F.	101 lb.	
		20	.63	1.02	2.40		44	104	
		25	.79	1.02	3.03		35	104	
	100	15	.46	1.04	2.14		61	126	
		20	.71	1.02	2.72		39	104	
		25	.83	1.01	3.47		33	113	
Flat Plate (waffle construction) 20" domes	50	20	.43	1.00	2.10	.84 S.F.	63	135	53 S.F.
		25	.52	1.00	2.90	.89	52	150	46
		30	.64	1.00	3.70	.87	42	155	37
	100	20	.51	1.00	2.30	.84	53	125	45
		25	.64	1.00	3.20	.83	42	135	35
		30	.76	1.00	4.40	.81	36	160	29
Waffle Construction 30" domes	50	25	.69	1.06	1.83	.68	42	72	40
		30	.74	1.06	2.39	.69	39	87	39
		35	.86	1.05	2.71	.69	33	85	39
		40	.78	1.00	4.80	.68	35	165	40
Flat Slab (two way with drop panels)	50	20	.62	1.03	2.34		45	102	
		25	.77	1.03	2.99		36	105	
		30	.95	1.03	4.09		29	116	
	100	20	.64	1.03	2.83		43	119	
		25	.79	1.03	3.88		35	133	
		30	.96	1.03	4.66		29	131	
	200	20	.73	1.03	3.03		38	112	
		25	.86	1.03	4.23		32	133	
		30	1.06	1.03	5.30		26	135	
One Way Joists 20" Pans	50	15	.36	1.04	1.40	.93	78	105	70
		20	.42	1.05	1.80	.94	67	120	60
		25	.47	1.05	2.60	.94	60	150	54
	100	15	.38	1.07	1.90	.93	77	140	66
		20	.44	1.08	2.40	.94	67	150	58
		25	.52	1.07	3.50	.94	55	185	49
One Way Joists 8" x 16" filler blocks	50	15	.34	1.06	1.80	.81 Ea.	84	145	64 Ea.
		20	.40	1.08	2.20	.82	73	145	55
		25	.46	1.07	3.20	.83	63	190	49
	100	15	.39	1.07	1.90	.81	74	130	56
		20	.46	1.09	2.80	.82	64	160	48
		25	.53	1.10	3.60	.83	56	190	42
One Way Beam & Slab	50	15	.42	1.30	1.73		84	111	
		20	.51	1.28	2.61		68	138	
		25	.64	1.25	2.78		53	117	
	100	15	.42	1.30	1.90		84	122	
		20	.54	1.35	2.69		68	154	
		25	.69	1.37	3.93		54	145	
	200	15	.44	1.31	2.24		80	137	
		20	.58	1.40	3.30		65	163	
		25	.69	1.42	4.89		53	183	
Two Way Beam & Slab	100	15	.47	1.20	2.26		69	130	
		20	.63	1.29	3.06		55	131	
		25	.83	1.33	3.79		43	123	
	200	15	.49	1.25	2.70		41	149	
		20	.66	1.32	4.04		54	165	
		25	.88	1.32	6.08		41	187	

R033105-10 Proportionate Quantities (cont.)

4000 psi Concrete and 60,000 psi Reinforcing—Form and Reinforcing Quantities per C.Y.					
Item	Size	Forms	Reinforcing	Minimum	Maximum
Columns (square tied)	10" x 10"	130 S.F.C.A.	#5 to #11	220 lbs.	875 lbs.
	12" x 12"	108	#6 to #14	200	955
	14" x 14"	92	#7 to #14	190	900
	16" x 16"	81	#6 to #14	187	1082
	18" x 18"	72	#6 to #14	170	906
	20" x 20"	65	#7 to #18	150	1080
	22" x 22"	59	#8 to #18	153	902
	24" x 24"	54	#8 to #18	164	884
	26" x 26"	50	#9 to #18	169	994
	28" x 28"	46	#9 to #18	147	864
	30" x 30"	43	#10 to #18	146	983
	32" x 32"	40	#10 to #18	175	866
	34" x 34"	38	#10 to #18	157	772
	36" x 36"	36	#10 to #18	175	852
	38" x 38"	34	#10 to #18	158	765
	40" x 40"	32	#10 to #18	143	692

Item	Size	Form	Spiral	Reinforcing	Minimum	Maximum
Columns (spirally reinforced)	12" diameter	34.5 L.F.	190 lbs.	#4 to #11	165 lbs.	1505 lb.
		34.5	190	#14 & #18	—	1100
	14"	25	170	#4 to #11	150	970
		25	170	#14 & #18	800	1000
	16"	19	160	#4 to #11	160	950
		19	160	#14 & #18	605	1080
	18"	15	150	#4 to #11	160	915
		15	150	#14 & #18	480	1075
	20"	12	130	#4 to #11	155	865
		12	130	#14 & #18	385	1020
	22"	10	125	#4 to #11	165	775
		10	125	#14 & #18	320	995
	24"	9	120	#4 to #11	195	800
		9	120	#14 & #18	290	1150
	26"	7.3	100	#4 to #11	200	729
		7.3	100	#14 & #18	235	1035
	28"	6.3	95	#4 to #11	175	700
		6.3	95	#14 & #18	200	1075
	30"	5.5	90	#4 to #11	180	670
		5.5	90	#14 & #18	175	1015
	32"	4.8	85	#4 to #11	185	615
		4.8	85	#14 & #18	155	955
	34"	4.3	80	#4 to #11	180	600
		4.3	80	#14 & #18	170	855
	36"	3.8	75	#4 to #11	165	570
		3.8	75	#14 & #18	155	865
	40"	3.0	70	#4 to #11	165	500
		3.0	70	#14 & #18	145	765

R033105-10   Proportionate Quantities (cont.)

3000 psi Concrete and 60,000 psi Reinforcing—Form and Reinforcing Quantities per C.Y.

Item	Type	Loading	Height	C.Y./L.F.	Forms/C.Y.	Reinf./C.Y.
Retaining Walls	Cantilever	Level Backfill	4 Ft.	0.2 C.Y.	49 S.F.	35 lbs.
			8	0.5	42	45
			12	0.8	35	70
			16	1.1	32	85
			20	1.6	28	105
		Highway Surcharge	4	0.3	41	35
			8	0.5	36	55
			12	0.8	33	90
			16	1.2	30	120
			20	1.7	27	155
		Railroad Surcharge	4	0.4	28	45
			8	0.8	25	65
			12	1.3	22	90
			16	1.9	20	100
			20	2.6	18	120
	Gravity, with Vertical Face	Level Backfill	4	0.4	37	None
			7	0.6	27	
			10	1.2	20	
		Sloping Backfill	4	0.3	31	
			7	0.8	21	
			10	1.6	15	↓

	Span	Live Load in Kips per Linear Foot							
		Under 1 Kip		2 to 3 Kips		4 to 5 Kips		6 to 7 Kips	
		Forms	Reinf.	Forms	Reinf.	Forms	Reinf.	Forms	Reinf.
Beams	10 Ft.	—	—	90 S.F.	170 #	85 S.F.	175 #	75 S.F.	185 #
	16	130 S.F.	165 #	85	180	75	180	65	225
	20	110	170	75	185	62	200	51	200
	26	90	170	65	215	62	215	—	—
	30	85	175	60	200	—	—	—	—

Item	Size	Type	Forms per C.Y.	Reinforcing per C.Y.
Spread Footings	Under 1 C.Y.	1,000 psf soil	24 S.F.	44 lbs.
		5,000	24	42
		10,000	24	52
	1 C.Y. to 5 C.Y.	1,000	14	49
		5,000	14	50
		10,000	14	50
	Over 5 C.Y.	1,000	9	54
		5,000	9	52
		10,000	9	56
Pile Caps (30 Ton Concrete Piles)	Under 5 C.Y.	shallow caps	20	65
		medium	20	50
		deep	20	40
	5 C.Y. to 10 C.Y.	shallow	14	55
		medium	15	45
		deep	15	40
	10 C.Y. to 20 C.Y.	shallow	11	60
		medium	11	45
		deep	12	35
	Over 20 C.Y.	shallow	9	60
		medium	9	45
		deep	10	40

R033105-10 Proportionate Quantities (cont.)

		3000 psi Concrete and 60,000 psi Reinforcing — Form and Reinforcing Quantities per C.Y.				
Item	Size	Pile Spacing	50 T Pile	100 T Pile	50 T Pile	100 T Pile
Pile Caps (Steel H Piles)	Under 5 C.Y.	24" O.C.	24 S.F.	24 S.F.	75 lbs.	90 lbs.
		30"	25	25	80	100
		36"	24	24	80	110
	5 C.Y. to 10 C.Y.	24"	15	15	80	110
		30"	15	15	85	110
		36"	15	15	75	90
	Over 10 C.Y.	24"	13	13	85	90
		30"	11	11	85	95
		36"	10	10	85	90

		8" Thick		10" Thick		12" Thick		15" Thick	
	Height	Forms	Reinf.	Forms	Reinf.	Forms	Reinf.	Forms	Reinf.
Basement Walls	7 Ft.	81 S.F.	44 lbs.	65 S.F.	45 lbs.	54 S.F.	44 lbs.	41 S.F.	43 lbs.
	8		44		45		44		43
	9		46		45		44		43
	10		57		45		44		43
	12		83		50		52		43
	14		116		65		64		51
	16				86		90		65
	18	↓		↓		↓	106	↓	70

R033105-20 Materials for One C.Y. of Concrete

This is an approximate method of figuring quantities of cement, sand and coarse aggregate for a field mix with waste allowance included.

With crushed gravel as coarse aggregate, to determine barrels of cement required, divide 10 by total mix; that is, for 1:2:4 mix, 10 divided by 7 = 1-3/7 barrels. If the coarse aggregate is crushed stone, use 10-1/2 instead of 10 as given for gravel.

To determine tons of sand required, multiply barrels of cement by parts of sand and then by 0.2; that is, for the 1:2:4 mix, as above, 1-3/7 x 2 x .2 = .57 tons.

Tons of crushed gravel are in the same ratio to tons of sand as parts in the mix, or 4/2 x .57 = 1.14 tons.

1 bag cement = 94#	1 C.Y. sand or crushed gravel = 2700#	1 C.Y. crushed stone = 2575#
4 bags = 1 barrel	1 ton sand or crushed gravel = 20 C.F.	1 ton crushed stone = 21 C.F.

Average carload of cement is 692 bags; of sand or gravel is 56 tons.

Do not stack stored cement over 10 bags high.

R033105-30 Metric Equivalents of Cement Content for Concrete Mixes

94 Pound Bags per Cubic Yard	Kilograms per Cubic Meter	94 Pound Bags per Cubic Yard	Kilograms per Cubic Meter
1.0	55.77	7.0	390.4
1.5	83.65	7.5	418.3
2.0	111.5	8.0	446.2
2.5	139.4	8.5	474.0
3.0	167.3	9.0	501.9
3.5	195.2	9.5	529.8
4.0	223.1	10.0	557.7
4.5	251.0	10.5	585.6
5.0	278.8	11.0	613.5
5.5	306.7	11.5	641.3
6.0	334.6	12.0	669.2
6.5	362.5	12.5	697.1

a. If you know the cement content in pounds per cubic yard,
   multiply by .5933 to obtain kilograms per cubic meter.

b. If you know the cement content in 94 pound bags per cubic yard,
   multiply by 55.77 to obtain kilograms per cubic meter.

835

R033105-40   **Metric Equivalents of Common Concrete Strengths**
(to convert other psi values to megapascals, multiply by 0.006895)

U.S. Values psi	SI Values Megapascals	Non-SI Metric Values kgf/cm²*
2000	14	140
2500	17	175
3000	21	210
3500	24	245
4000	28	280
4500	31	315
5000	34	350
6000	41	420
7000	48	490
8000	55	560
9000	62	630
10,000	69	705

* kilograms force per square centimeter

R033105-50   **Quantities of Cement, Sand and Stone for One C.Y. of Concrete per Various Mixes**

This table can be used to determine the quantities of the ingredients for smaller quantities of site mixed concrete.

Concrete (C.Y.)	Mix = 1:1:1-3/4			Mix = 1:2:2.25			Mix = 1:2.25:3			Mix = 1:3:4		
	Cement (sacks)	Sand (C.Y.)	Stone (C.Y.)	Cement (sacks)	Sand (C.Y.)	Stone (C.Y.)	Cement (sacks)	Sand (C.Y.)	Stone (C.Y.)	Cement (sacks)	Sand (C.Y.)	Stone (C.Y.)
1	10	.37	.63	7.75	.56	.65	6.25	.52	.70	5	.56	.74
2	20	.74	1.26	15.50	1.12	1.30	12.50	1.04	1.40	10	1.12	1.48
3	30	1.11	1.89	23.25	1.68	1.95	18.75	1.56	2.10	15	1.68	2.22
4	40	1.48	2.52	31.00	2.24	2.60	25.00	2.08	2.80	20	2.24	2.96
5	50	1.85	3.15	38.75	2.80	3.25	31.25	2.60	3.50	25	2.80	3.70
6	60	2.22	3.78	46.50	3.36	3.90	37.50	3.12	4.20	30	3.36	4.44
7	70	2.59	4.41	54.25	3.92	4.55	43.75	3.64	4.90	35	3.92	5.18
8	80	2.96	5.04	62.00	4.48	5.20	50.00	4.16	5.60	40	4.48	5.92
9	90	3.33	5.67	69.75	5.04	5.85	56.25	4.68	6.30	45	5.04	6.66
10	100	3.70	6.30	77.50	5.60	6.50	62.50	5.20	7.00	50	5.60	7.40
11	110	4.07	6.93	85.25	6.16	7.15	68.75	5.72	7.70	55	6.16	8.14
12	120	4.44	7.56	93.00	6.72	7.80	75.00	6.24	8.40	60	6.72	8.88
13	130	4.82	8.20	100.76	7.28	8.46	81.26	6.76	9.10	65	7.28	9.62
14	140	5.18	8.82	108.50	7.84	9.10	87.50	7.28	9.80	70	7.84	10.36
15	150	5.56	9.46	116.26	8.40	9.76	93.76	7.80	10.50	75	8.40	11.10
16	160	5.92	10.08	124.00	8.96	10.40	100.00	8.32	11.20	80	8.96	11.84
17	170	6.30	10.72	131.76	9.52	11.06	106.26	8.84	11.90	85	9.52	12.58
18	180	6.66	11.34	139.50	10.08	11.70	112.50	9.36	12.60	90	10.08	13.32
19	190	7.04	11.98	147.26	10.64	12.36	118.76	9.84	13.30	95	10.64	14.06
20	200	7.40	12.60	155.00	11.20	13.00	125.00	10.40	14.00	100	11.20	14.80
21	210	7.77	13.23	162.75	11.76	13.65	131.25	10.92	14.70	105	11.76	15.54
22	220	8.14	13.86	170.05	12.32	14.30	137.50	11.44	15.40	110	12.32	16.28
23	230	8.51	14.49	178.25	12.88	14.95	143.75	11.96	16.10	115	12.88	17.02
24	240	8.88	15.12	186.00	13.44	15.60	150.00	12.48	16.80	120	13.44	17.76
25	250	9.25	15.75	193.75	14.00	16.25	156.25	13.00	17.50	125	14.00	18.50
26	260	9.64	16.40	201.52	14.56	16.92	162.52	13.52	18.20	130	14.56	19.24
27	270	10.00	17.00	209.26	15.12	17.56	168.76	14.04	18.90	135	15.02	20.00
28	280	10.36	17.64	217.00	15.68	18.20	175.00	14.56	19.60	140	15.68	20.72
29	290	10.74	18.28	224.76	16.24	18.86	181.26	15.08	20.30	145	16.24	21.46

R033105-65 Field-Mix Concrete

Presently most building jobs are built with ready-mixed concrete except at isolated locations and some larger jobs requiring over 10,000 C.Y. where land is readily available for setting up a temporary batch plant.

The most economical mix is a controlled mix using local aggregate proportioned by trial to give the required strength with the least cost of material.

R033105-70 Placing Ready-Mixed Concrete

For ground pours allow for 5% waste when figuring quantities.

Prices in the front of the data set assume normal deliveries. If deliveries are made before 8 A.M. or after 5 P.M. or on Saturday afternoons add 30%. Negotiated discounts for large volumes are not included in prices in front of the data set.

For the lower floors without truck access, concrete may be wheeled in rubber-tired buggies, conveyer handled, crane handled or pumped. Pumping is economical if there is top steel. Conveyers are more efficient for thick slabs.

At higher floors the rubber-tired buggies may be hoisted by a hoisting tower and wheeled to the location. Placement by a conveyer is limited to three

floors and is best for high-volume pours. Pumped concrete is best when the building has no crane access. Concrete may be pumped directly as high as thirty-six stories using special pumping techniques. Normal maximum height is about fifteen stories.

The best pumping aggregate is screened and graded bank gravel rather than crushed stone.

Pumping downward is more difficult than pumping upward. The horizontal distance from pump to pour may increase preparation time prior to pour. Placing by cranes, either mobile, climbing or tower types, continues as the most efficient method for high-rise concrete buildings.

R033105-85 Lift Slabs

The cost advantage of the lift slab method is due to placing all concrete, reinforcing steel, inserts and electrical conduit at ground level and in reduction of formwork. Minimum economical project size is about 30,000 S.F. Slabs may be tilted for parking garage ramps.

It is now used in all types of buildings and has gone up to 22 stories high in apartment buildings. The current trend is to use post-tensioned flat plate slabs with spans from 22' to 35'. Cylindrical void forms are used when deep slabs are required. One pound of prestressing steel is about equal to seven pounds of conventional reinforcing.

To be considered cured for stressing and lifting, a slab must have attained 75% of design strength. Seven days are usually sufficient with four to five days possible if high early strength cement is used. Slabs can be stacked using two coats of a non-bonding agent to insure that slabs do not stick to each other. Lifting is done by companies specializing in this work. Lift rate is 5' to 15' per hour with an average of 10' per hour. Total areas up to 33,000 S.F. have been lifted at one time. 24 to 36 jacking columns are common. Most economical bay sizes are 24' to 28' with four to fourteen stories most efficient. Continuous design reduces reinforcing steel cost. Use of post-tensioned slabs allows larger bay sizes.

R033543-10   Polished Concrete Floors

A polished concrete floor has a glossy mirror-like appearance and is created by grinding the concrete floor with finer and finer diamond grits, similar to sanding wood, until the desired level of reflective clarity and sheen are achieved. The technical term for this type of polished concrete is bonded abrasive polished concrete. The basic piece of equipment used in the polishing process is a walk-behind planetary grinder for working large floor areas. This grinder drives diamond-impregnated abrasive discs, which progress from coarse- to fine-grit discs.

The process begins with the use of very coarse diamond segments or discs bonded in a metallic matrix. These segments are coarse enough to allow the removal of pits, blemishes, stains, and light coatings from the floor surface in preparation for final smoothing. The condition of the original concrete surface will dictate the grit coarseness of the initial grinding step which will generally end up being a three- to four-step process using ever finer grits. The purpose of this initial grinding step is to remove surface coatings and blemishes and to cut down into the cream for very fine aggregate exposure, or deeper into the fine aggregate layer just below the cream layer, or even deeper into the coarse aggregate layer. These initial grinding steps will progress up to the 100/120 grit. If wet grinding is done, a waste slurry is produced that must be removed between grit changes and disposed of properly. If dry grinding is done, a high performance vacuum will pick up the dust during grinding and collect it in bags which must be disposed of properly.

The process continues with honing the floor in a series of steps that progresses from 100-grit to 400-grit diamond abrasive discs embedded in a plastic or resin matrix. At some point during, or just prior to, the honing step, one or two coats of stain or dye can be sprayed onto the surface to give color to the concrete, and two coats of densifier/hardener must be applied to the floor surface and allowed to dry. This sprayed-on densifier/hardener will penetrate about 1/8″ into the concrete to make the surface harder, denser and more abrasion-resistant.

The process ends with polishing the floor surface in a series of steps that progresses from resin-impregnated 800-grit (medium polish) to 1500-grit (high polish) to 3000-grit (very high polish), depending on the desired level of reflective clarity and sheen.

The Concrete Polishing Association of America (CPAA) has defined the flooring options available when processing concrete to a desired finish. The first category is aggregate exposure, the grinding of a concrete surface with bonded abrasives, in as many abrasive grits necessary, to achieve one of the following classes:

A. Cream – very little surface cut depth; little aggregate exposure

B. Fine aggregate (salt and pepper) – surface cut depth of 1/16″; fine aggregate exposure with little or no medium aggregate exposure at random locations

C. Medium aggregate – surface cut depth of 1/8″; medium aggregate exposure with little or no large aggregate exposure at random locations

D. Large aggregate – surface cut depth of 1/4″; large aggregate exposure with little or no fine aggregate exposure at random locations

The second CPAA defined category is reflective clarity and sheen, the polishing of a concrete surface with the minimum number of bonded abrasives as indicated to achieve one of the following levels:

1. Ground – flat appearance with none to very slight diffused reflection; none to very low reflective sheen; using a minimum total of 4 grit levels up 100-grit

2. Honed – matte appearance with or without slight diffused reflection; low to medium reflective sheen; using a minimum total of 5 grit levels up to 400-grit

3. Semi-polished – objects being reflected are not quite sharp and crisp but can be easily identified; medium to high reflective sheen; using a minimum total of 6 grit levels up to 800-grit

4. Highly-polished – objects being reflected are sharp and crisp as would be seen in a mirror-like reflection; high to highest reflective sheen; using a minimum total of up to 8 grit levels up to 1500-grit or 3000-grit

The CPAA defines reflective clarity as the degree of sharpness and crispness of the reflection of overhead objects when viewed 5′ above and perpendicular to the floor surface. Reflective sheen is the degree of gloss reflected from a surface when viewed at least 20′ from and at an angle to the floor surface. These terms are relatively subjective. The final outcome depends on the internal makeup and surface condition of the original concrete floor, the experience of the floor polishing crew, and the expectations of the owner. Before the grinding, honing, and polishing work commences on the main floor area, it might be beneficial to do a mock-up panel in the same floor but in an out of the way place to demonstrate the sequence of steps with increasingly fine abrasive grits and to demonstrate the final reflective clarity and reflective sheen. This mock-up panel will be within the area of, and part of, the final work.

©Concrete Polishing Association of America "Glossary."

R034105-30   Prestressed Precast Concrete Structural Units

Type	Location	Depth	Span in Ft.		Live Load Lb. per S.F.
Double Tee	Floor	28" to 34"	60 to 80		50 to 80
8' to 10'	Roof	12" to 24"	30 to 50		40
	Wall	Width 8'	Up to 55' high		Wind
Multiple Tee	Roof	8" to 12"	15 to 40		40
8'	Floor	8" to 12"	15 to 30		100
Plank	Roof		Roof	Floor	
or	or	4"	13	12	40 for Roof
		6"	22	18	
		8"	26	25	
		10"	33	29	100 for Floor
	Floor	12"	42	32	
Single Tee	Roof	28"	40		
8' to 10'		32"	80		40
		36"	100		
		48"	120		
AASHTO Girder	Bridges	Type 4	100		Highway
		5	110		
		6	125		
Box Beam	Bridges	15"	40		Highway
4'		27"	to		
		33"	100		

The majority of precast projects today utilize double tees rather than single tees because of speed and ease of installation. As a result casting beds at manufacturing plants are normally formed for double tees. Single tee projects will therefore require an initial set up charge to be spread over the individual single tee costs.

For floors, a 2" to 3" topping is field cast over the shapes. For roofs, insulating concrete or rigid insulation is placed over the shapes.

Member lengths up to 40' are standard haul, 40' to 60' require special permits and lengths over 60' must be escorted. Excessive width and/or length can add up to 100% on hauling costs.

Large heavy members may require two cranes for lifting which would increase erection costs by about 45%. An eight man crew can install 12 to 20 double tees, or 45 to 70 quad tees or planks per day.

Grouting of connections must also be included.

Several system buildings utilizing precast members are available. Heights can go up to 22 stories for apartment buildings. The optimum design ratio is 3 S.F. of surface to 1 S.F. of floor area.

R034136-90   Prestressed Concrete, Post-Tensioned

In post-tensioned concrete the steel tendons are tensioned after the concrete has reached about 3/4 of its ultimate strength. The cableways are grouted after tensioning to provide bond between the steel and concrete. If bond is to be prevented, the tendons are coated with a corrosion-preventative grease and wrapped with waterproofed paper or plastic. Bonded tendons are usually used when ultimate strength (beams & girders) are controlling factors.

High strength concrete is used to fully utilize the steel, thereby reducing the size and weight of the member. A plasticizing agent may be added to reduce water content. Maximum size aggregate ranges from 1/2" to 1-1/2" depending on the spacing of the tendons.

The types of steel commonly used are bars and strands. Job conditions determine which is best suited. Bars are best for vertical prestresses since they are easy to support. The trend is for steel manufacturers to supply a finished package, cut to length, which reduces field preparation to a minimum.

Bars vary from 3/4" to 1-3/8" diameter. The table below gives time in labor-hours per tendon for placing, tensioning and grouting (if required) a 75' beam. Tendons used in buildings are not usually grouted; tendons for bridges usually are grouted. For strands the table indicates the labor-hours per pound for typical prestressed units 100' long. Simple span beams usually require one- end stressing regardless of lengths. Continuous beams are usually stressed from two ends. Long slabs are poured from the center outward and stressed in 75' increments after the initial 150' center pour.

Length	100' Beam		75' Beam		100' Slab	
Type Steel	Strand		Bars		Strand	
Diameter	0.5"		3/4"	1-3/8"	0.5"	0.6"
Number	4	12	1	1	1	1
Force in Kips	100	300	42	143	25	35
Preparation & Placing Cables	3.6	7.4	0.9	2.9	0.9	1.1
Stressing Cables	2.0	2.4	0.8	1.6	0.5	0.5
Grouting, if required	2.5	3.0	0.6	1.3		
Total Labor Hours	8.1	12.8	2.3	5.8	1.4	1.6
Prestressing Steel Weights (Lbs.)	215	640	115	380	53	74
Labor-hours per Lb. Bonded	0.038	0.020	0.020	0.015		
Non-bonded					0.026	0.022

Labor Hours per Tendon and per Pound of Prestressed Steel

Flat slab construction — 4000 psi concrete with span-to-depth ratio between 36 and 44. Two way post-tensioned steel averages 1.0 lb. per S.F. for 24' to 28' bays (usually strand) and additional reinforcing steel averages .5 lb. per S.F.

Pan and joist construction — 4000 psi concrete with span-to-depth ratio between 28 to 30. Post-tensioned steel averages .8 lb. per S.F. and reinforcing steel about 1.0 lb. per S.F. Placing and stressing average 40 hours per ton of total material.

Beam construction — 4000 to 5000 psi concrete. Steel weights vary greatly.

Labor cost per pound goes down as the size and length of the tendon increase. The primary economic consideration is the cost per kip for the member.

Post-tensioning becomes feasible for beams and girders over 30' long; for continuous two-way slabs over 20' clear; and for transferring upper building loads over longer spans at lower levels. Post-tension suppliers will provide engineering services at no cost to the user. Substantial economies are possible by using post-tensioned lift slabs.

Concrete | R0345 Precast Architectural Concrete

R034513-10   Precast Concrete Wall Panels

Panels are either solid or insulated with plain, colored or textured finishes. Transportation is an important cost factor. Prices shown in the unit cost section of the data set are based on delivery within 50 miles of a plant including fabricators' overhead and profit. Engineering data is available from fabricators to assist with construction details. Usual minimum job size for economical use of panels is about 5000 S.F. Small jobs can double the prices shown. For large, highly repetitive jobs, deduct up to 15% from the prices shown.

2″ thick panels cost about the same as 3″ thick panels, and maximum panel size is less. For building panels faced with granite, marble or stone, add the material prices from those unit cost sections to the plain panel price shown. There is a growing trend toward aggregate facings and broken rib finishes rather than plain gray concrete panels.

No allowance has been made in the unit cost section for supporting steel framework. On one story buildings, panels may rest on grade beams and require only wind bracing and fasteners. On multi-story buildings panels can span from column to column and floor to floor. Plastic-designed steel-framed structures may have large deflections which slow down erection and raise costs.

Large panels are more economical than small panels on a S.F. basis. When figuring areas include all protrusions, returns, etc. Overhangs can triple erection costs. Panels over 45′ have been produced. Larger flat units should be prestressed. Vacuum lifting of smooth finish panels eliminates inserts and can speed erection.

Concrete | R0347 Site-Cast Concrete

R034713-20   Tilt Up Concrete Panels

The advantage of tilt up construction is in the low cost of forms and the placing of concrete and reinforcing. Panels up to 75′ high and 5-1/2″ thick have been tilted using strongbacks. Tilt up has been used for one to five story buildings and is well-suited for warehouses, stores, offices, schools and residences.

The panels are cast in forms on the floor slab. Most jobs use 5-1/2″ thick solid reinforced concrete panels. Sandwich panels with a layer of insulating materials are also used. Where dampness is a factor, lightweight aggregate is used. Optimum panel size is 300 to 500 S.F.

Slabs are usually poured with 3000 psi concrete which permits tilting seven days after pouring. Slabs may be stacked on top of each other and are separated from each other by either two coats of bond breaker or a film of polyethylene. Use of high early-strength cement allows tilting two days after a pour. Tilting up is done with a roller outrigger crane with a capacity of at least 1-1/2 times the weight of the panel at the required reach. Exterior precast columns can be set at the same time as the panels; interior precast columns can be set first and the panels clipped directly to them. The use of cast-in-place concrete columns is diminishing due to shrinkage problems. Structural steel columns are sometimes used if crane rails are planned. Panels can be clipped to the columns or lowered between the flanges. Steel channels with anchors may be used as edge forms for the slab. When the panels are lifted the channels form an integral steel column to take structural loads. Roof loads can be carried directly by the panels for wall heights to 14′.

Requirements of local building codes may be a limiting factor and should be checked. Building floor slabs should be poured first and should be a minimum of 5″ thick with 100% compaction of soil or 6″ thick with less than 100% compaction.

Setting times as fast as nine minutes per panel have been observed, but a safer expectation would be four panels per hour with a crane and a four-man setting crew. If a crane erects from inside a building, some provision must be made to get the crane out after walls are erected. Good yarding procedure is important to minimize delays. Equalizing three-point lifting beams and self-releasing pick-up hooks speed erection. If panels must be carried to their final location, setting time per panel will be increased and erection costs may approach the erection cost range of architectural precast wall panels. Placing panels into slots formed in continuous footers will speed erection.

Reinforcing should be with #5 bars with vertical bars on the bottom. If surface is to be sandblasted, stainless steel chairs should be used to prevent rust staining.

Use of a broom finish is popular since the unavoidable surface blemishes are concealed.

Precast columns run from three to five times the C.Y. price of the panels only.

Concrete | R0352 Lightweight Concrete Roof Insulation

R035216-10   Lightweight Concrete

Lightweight aggregate concrete is usually purchased ready mixed, but it can also be field mixed.

Vermiculite or Perlite comes in bags of 4 C.F. under various trade names. The weight is about 8 lbs. per C.F. For insulating roof fill use 1:6 mix. For a structural deck use 1:4 mix over gypsum boards, steeltex, steel centering, etc., supported by closely spaced joists or bulb trees. For structural slabs use 1:3:2 vermiculite sand concrete over steeltex, metal lath, steel centering, etc., on joists spaced 2′-0″ O.C. for maximum L.L. of 80 P.S.F. Use same mix

for slab base fill over steel flooring or regular reinforced concrete slab when tile, terrazzo or other finish is to be laid over.

For slabs on grade use 1:3:2 mix when tile, etc., finish is to be laid over. If radiant heating units are installed use a 1:6 mix for a base. After coils are in place, cover with a regular granolithic finish (mix 1:3:2) to a minimum depth of 1-1/2″ over top of units.

Reinforce all slabs with 6 × 6 or 10 × 10 welded wire mesh.

Masonry | R0401 Maintenance of Masonry

R040130-10   Cleaning Face Brick

On smooth brick a person can clean 70 S.F. an hour; on rough brick 50 S.F. per hour. Use one gallon muriatic acid to 20 gallons of water for 1000

S.F. Do not use acid solution until wall is at least seven days old, but a mild soap solution may be used after two days.

Time has been allowed for cleanup in brick prices.

Masonry | R0405 Common Work Results for Masonry

R040513-10   Cement Mortar (material only)

Type N - 1:1:6 mix by volume. Use everywhere above grade except as noted below. - 1:3 mix using conventional masonry cement which saves handling two separate bagged materials.

Type M - 1:1/4:3 mix by volume, or 1 part cement, 1/4 (10% by wt.) lime, 3 parts sand. Use for heavy loads and where earthquakes or hurricanes may occur. Also for reinforced brick, sewers, manholes and everywhere below grade.

Mix Proportions by Volume and Compressive Strength of Mortar

Where Used	Mortar Type	Allowable Proportions by Volume				Compressive Strength @ 28 days
		Portland Cement	Masonry Cement	Hydrated Lime	Masonry Sand	
Plain Masonry	M	1	1	—	6	
		1	—	1/4	3	2500 psi
	S	1/2	1	—	4	
		1	—	1/4 to 1/2	4	1800 psi
	N	—	1	—	3	
		1	—	1/2 to 1-1/4	6	750 psi
	O	—	1	—	3	
		1	—	1-1/4 to 2-1/2	9	350 psi
	K	1	—	2-1/2 to 4	12	75 psi
Reinforced Masonry	PM	1	1	—	6	2500 psi
	PL	1	—	1/4 to 1/2	4	2500 psi

Note: The total aggregate should be between 2.25 to 3 times the sum of the cement and lime used.

The labor cost to mix the mortar is included in the productivity and labor cost of unit price lines in unit cost sections for brickwork, blockwork and stonework.

The material cost of mixed mortar is included in the material cost of those same unit price lines and includes the cost of renting and operating a 10 C.F. mixer at the rate of 200 C.F. per day.

There are two types of mortar color used. One type is the inert additive type with about 100 lbs. per M brick as the typical quantity required. These colors are also available in smaller-batch-sized bags (1 lb. to 15 lb.) which can be placed directly into the mixer without measuring. The other type is premixed and replaces the masonry cement. Dark green color has the highest cost.

R040519-50   Masonry Reinforcing

Horizontal joint reinforcing helps prevent wall cracks where wall movement may occur and in many locations is required by code. Horizontal joint reinforcing is generally not considered to be structural reinforcing and an unreinforced wall may still contain joint reinforcing.

Reinforcing strips come in 10' and 12' lengths and in truss and ladder shapes, with and without drips. Field labor runs between 2.7 to 5.3 hours per 1000 L.F. for wall thicknesses up to 12".

The wire meets ASTM A82 for cold drawn steel wire and the typical size is 9 ga. sides and ties with 3/16" diameter also available. Typical finish is mill galvanized with zinc coating at .10 oz. per S.F. Class I (.40 oz. per S.F.) and Class III (.80 oz. per S.F.) are also available, as is hot dipped galvanizing at 1.50 oz. per S.F.

R042110-10   Economy in Bricklaying

Have adequate supervision. Be sure bricklayers are always supplied with materials so there is no waiting. Place experienced bricklayers at corners and openings.

Use only screened sand for mortar. Otherwise, labor time will be wasted picking out pebbles. Use seamless metal tubs for mortar as they do not leak or catch the trowel. Locate stack and mortar for easy wheeling.

Have brick delivered for stacking. This makes for faster handling, reduces chipping and breakage, and requires less storage space. Many dealers will deliver select common in 2' × 3' × 4' pallets or face brick packaged. This affords quick handling with a crane or forklift and easy tonging in units of ten, which reduces waste.

Use wider bricks for one wythe wall construction. Keep scaffolding away from the wall to allow mortar to fall clear and not stain the wall.

On large jobs develop specialized crews for each type of masonry unit.

Consider designing for prefabricated panel construction on high rise projects.

Avoid excessive corners or openings. Each opening adds about 50% to the labor cost for area of opening.

Bolting stone panels and using window frames as stops reduce labor costs and speed up erection.

R042110-20   Common and Face Brick

Common building brick manufactured according to ASTM C62 and facing brick manufactured according to ASTM C216 are the two standard bricks available for general building use.

Building brick is made in three grades: SW, where high resistance to damage caused by cyclic freezing is required; MW, where moderate resistance to cyclic freezing is needed; and NW, where little resistance to cyclic freezing is needed. Facing brick is made in only the two grades SW and MW. Additionally, facing brick is available in three types: FBS, for general use; FBX, for general use where a higher degree of precision and lower permissible variation in size than FBS are needed; and FBA, for general use to produce characteristic architectural effects resulting from non-uniformity in size and texture of the units.

In figuring the material cost of brickwork, an allowance of 25% mortar waste and 3% brick breakage was included. If bricks are delivered palletized

with 280 to 300 per pallet, or packaged, allow only 1-1/2% for breakage. Packaged or palletized delivery is practical when a job is big enough to have a crane or other equipment available to handle a package of brick. This is so on all industrial work but not always true on small commercial buildings.

The use of buff and gray face is increasing, and there is a continuing trend to the Norman, Roman, Jumbo and SCR brick.

Common red clay brick for backup is not used that often. Concrete block is the most usual backup material with occasional use of sand lime or cement brick. Building brick is commonly used in solid walls for strength and as a fire stop.

Brick panels built on the ground and then crane erected to the upper floors have proven to be economical. This allows the work to be done under cover and without scaffolding.

R042110-50   Brick, Block & Mortar Quantities

Running Bond						For Other Bonds Standard Size Add to S.F. Quantities in Table to Left		
Number of Brick per S.F. of Wall - Single Wythe with 3/8" Joints				C.F. of Mortar per M Bricks, Waste Included				
Type Brick	Nominal Size (incl. mortar) L H W	Modular Coursing	Number of Brick per S.F.	3/8" Joint	1/2" Joint	Bond Type	Description	Factor
Standard	8 x 2-2/3 x 4	3C=8"	6.75	8.1	10.3	Common	full header every fifth course	+20%
Economy	8 x 4 x 4	1C=4"	4.50	9.1	11.6		full header every sixth course	+16.7%
Engineer	8 x 3-1/5 x 4	5C=16"	5.63	8.5	10.8	English	full header every second course	+50%
Fire	9 x 2-1/2 x 4-1/2	2C=5"	6.40	550 # Fireclay	—	Flemish	alternate headers every course	+33.3%
Jumbo	12 x 4 x 6 or 8	1C=4"	3.00	22.5	29.2		every sixth course	+5.6%
Norman	12 x 2-2/3 x 4	3C=8"	4.50	11.2	14.3	Header = W x H exposed		+100%
Norwegian	12 x 3-1/5 x 4	5C=16"	3.75	11.7	14.9	Rowlock = H x W exposed		+100%
Roman	12 x 2 x 4	2C=4"	6.00	10.7	13.7	Rowlock stretcher = L x W exposed		+33.3%
SCR	12 x 2-2/3 x 6	3C=8"	4.50	21.8	28.0	Soldier = H x L exposed		—
Utility	12 x 4 x 4	1C=4"	3.00	12.3	15.7	Sailor = W x L exposed		-33.3%

Concrete Blocks Nominal Size		Approximate Weight per S.F.		Blocks per 100 S.F.	Mortar per M block, waste included	
		Standard	Lightweight		Partitions	Back up
2"	x 8" x 16"	20 PSF	15 PSF	113	27 C.F.	36 C.F.
4"		30	20		41	51
6"		42	30		56	66
8"		55	38		72	82
10"		70	47		87	97
12"		85	55		102	112

Brick & Mortar Quantities
©Brick Industry Association. 2009 Feb. Technical Notes on Brick Construction 10:
    Dimensioning and Estimating Brick Masonry. Reston (VA): BIA. Table 1 Modular Brick Sizes and Table 4 Quantity Estimates for Brick Masonry.

R042210-20   Concrete Block

The material cost of special block such as corner, jamb and head block can be figured at the same price as ordinary block of equal size. Labor on specials is about the same as equal-sized regular block.

Bond beams and 16″ high lintel blocks are more expensive than regular units of equal size. Lintel blocks are 8″ long and either 8″ or 16″ high.

Use of a motorized mortar spreader box will speed construction of continuous walls.

Hollow non-load-bearing units are made according to ASTM C129 and hollow load-bearing units according to ASTM C90.

R050516-30   Coating Structural Steel

On field-welded jobs, the shop-applied primer coat is necessarily omitted. All painting must be done in the field and usually consists of red oxide rust inhibitive paint or an aluminum paint. The table below shows paint coverage and daily production for field painting.

See Division 05 05 13.50 for hot-dipped galvanizing and Division 09 97 13.23 for field-applied cold galvanizing and other paints and protective coatings.

See Division 05 01 10.51 for steel surface preparation treatments such as wire brushing, pressure washing and sand blasting.

Type Construction	Surface Area per Ton	Coat	One Gallon Covers		In 8 Hrs. Person Covers		Average per Ton Spray	
			Brush	Spray	Brush	Spray	Gallons	Labor-hours
Light Structural	300 S.F. to 500 S.F.	1st	500 S.F.	455 S.F.	640 S.F.	2000 S.F.	0.9 gals.	1.6 L.H.
		2nd	450	410	800	2400	1.0	1.3
		3rd	450	410	960	3200	1.0	1.0
Medium	150 S.F. to 300 S.F.	All	400	365	1600	3200	0.6	0.6
Heavy Structural	50 S.F. to 150 S.F.	1st	400	365	1920	4000	0.2	0.2
		2nd	400	365	2000	4000	0.2	0.2
		3rd	400	365	2000	4000	0.2	0.2
Weighted Average	225 S.F.	All	400	365	1350	3000	0.6	0.6

R050521-20   Welded Structural Steel

Usual weight reductions with welded design run 10% to 20% compared with bolted or riveted connections. This amounts to about the same total cost compared with bolted structures since field welding is more expensive than bolts. For normal spans of 18′ to 24′ figure 6 to 7 connections per ton.

Trusses — For welded trusses add 4% to weight of main members for connections. Up to 15% less steel can be expected in a welded truss compared to one that is shop bolted. Cost of erection is the same whether shop bolted or welded.

General — Typical electrodes for structural steel welding are E6010, E6011, E60T and E70T. Typical buildings vary between 2# to 8# of weld rod per

ton of steel. Buildings utilizing continuous design require about three times as much welding as conventional welded structures. In estimating field erection by welding, it is best to use the average linear feet of weld per ton to arrive at the welding cost per ton. The type, size and position of the weld will have a direct bearing on the cost per linear foot. A typical field welder will deposit 1.8# to 2# of weld rod per hour manually. Using semiautomatic methods can increase production by as much as 50% to 75%.

R050523-10   High Strength Bolts

Common bolts (A307) are usually used in secondary connections (see Division 05 05 23.10).

High strength bolts (A325 and A490) are usually specified for primary connections such as column splices, beam and girder connections to columns, column bracing, connections for supports of operating equipment or of other live loads which produce impact or reversal of stress, and in structures carrying cranes of over 5-ton capacity.

Allow 20 field bolts per ton of steel for a 6 story office building, apartment house or light industrial building. For 6 to 12 stories allow 25 bolts per ton, and above 12 stories, 30 bolts per ton. On power stations, 20 to 25 bolts per ton are needed.

R051223-10   Structural Steel

The bare material prices for structural steel, shown in the unit cost sections of the data set, are for 100 tons of shop-fabricated structural steel and include:

1. Mill base price of structural steel
2. Mill scrap/grade/size/length extras
3. Mill delivery to a metals service center (warehouse)
4. Service center storage and handling
5. Service center delivery to a fabrication shop
6. Shop storage and handling
7. Shop drafting/detailing
8. Shop fabrication

9. Shop coat of primer paint
10. Shop listing
11. Shop delivery to the job site

In unit cost sections of the data set that contain items for field fabrication of steel components, the bare material cost of steel includes:

1. Mill base price of structural steel
2. Mill scrap/grade/size/length extras
3. Mill delivery to a metals service center (warehouse)
4. Service center storage and handling
5. Service center delivery to the job site

R051223-20   Steel Estimating Quantities

One estimate on erection is that a crane can handle 35 to 60 pieces per day. Say the average is 45. With usual sizes of beams, girders, and columns, this would amount to about 20 tons per day. The type of connection greatly affects the speed of erection. Moment connections for continuous design slow down production and increase erection costs.

Short open web bar joists can be set at the rate of 75 to 80 per day, with 50 per day being the average for setting long span joists.

After main members are calculated, add the following for usual allowances: base plates 2% to 3%; column splices 4% to 5%; and miscellaneous details 4% to 5%, for a total of 10% to 13% in addition to main members.

The ratio of column to beam tonnage varies depending on type of steels used, typical spans, story heights and live loads.

It is more economical to keep the column size constant and to vary the strength of the column by using high strength steels. This also saves floor space. Buildings have recently gone as high as ten stories with 8″ high strength columns. For light columns under W8X31 lb. sections, concrete filled steel columns are economical.

High strength steels may be used in columns and beams to save floor space and to meet head room requirements. High strength steels in some sizes sometimes require long lead times.

Round, square and rectangular columns, both plain and concrete filled, are readily available and save floor area, but are higher in cost per pound than rolled columns. For high unbraced columns, tube columns may be less expensive.

Below are average minimum figures for the weights of the structural steel frame for different types of buildings using A36 steel, rolled shapes and simple joints. For economy in domes, rise to span ratio = .13. Open web joist framing systems will reduce weights by 10% to 40%. Composite design can reduce steel weight by up to 25% but additional concrete floor slab thickness may be required. Continuous design can reduce the weights up to 20%. There are many building codes with different live load requirements and different structural requirements, such as hurricane and earthquake loadings, which can alter the figures.

Structural Steel Weights per S.F. of Floor Area									
Type of Building	No. of Stories	Avg. Spans	L.L. #/S.F.	Lbs. Per S.F.	Type of Building	No. of Stories	Avg. Spans	L.L. #/S.F.	Lbs. Per S.F.
Steel Frame Mfg.	1	20′x20′	40	8	Apartments	2-8	20′x20′	40	8
		30′x30′		13		9-25			14
		40′x40′		18	Office	to 10	Various	80	10
Parking garage	4	Various	80	8.5		20			18
Domes (Schwedler)*	1	200′	30	10		30			26
		300′		15		over 50			35

845

R051223-25   Common Structural Steel Specifications

ASTM A992 (formerly A36, then A572 Grade 50) is the all-purpose carbon grade steel widely used in building and bridge construction.

The other high-strength steels listed below may each have certain advantages over ASTM A992 structural carbon steel, depending on the application. They have proven to be economical choices where, due to lighter members, the reduction of dead load and the associated savings in shipping cost can be significant.

ASTM A588 atmospheric weathering, high-strength, low-alloy steels can be used in the bare (uncoated) condition, where exposure to normal atmosphere causes a tightly adherant oxide to form on the surface, protecting the steel from further oxidation. ASTM A242 corrosion-resistant, high-strength, low-alloy steels have enhanced atmospheric corrosion resistance of at least two times that of carbon structural steels with copper, or four times that of carbon structural steels without copper. The reduction or elimination of maintenance resulting from the use of these steels often offsets their higher initial cost.

Steel Type	ASTM Designation	Minimum Yield Stress in KSI	Shapes Available
Carbon	A36	36	All structural shape groups, and plates & bars up through 8″ thick
	A529	50	Structural shape group 1, and plates & bars up through 2″ thick
High-Strength Low-Alloy Quenched & Self-Tempered	A913	50	All structural shape groups
		60	
		65	
		70	
High-Strength Low-Alloy Columbium-Vanadium	A572	42	All structural shape groups, and plates & bars up through 6″ thick
		50	All structural shape groups, and plates & bars up through 4″ thick
		55	Structural shape groups 1 & 2, and plates & bars up through 2″ thick
		60	Structural shape groups 1 & 2, and plates & bars up through 1-1/4″ thick
		65	Structural shape group 1, and plates & bars up through 1-1/4″ thick
High-Strength Low-Alloy Columbium-Vanadium	A992	50	All structural shape groups
Weathering High-Strength Low-Alloy	A242	42	Structural shape groups 4 & 5, and plates & bars over 1-1/2″ up through 4″ thick
		46	Structural shape group 3, and plates & bars over 3/4″ up through 1-1/2″ thick
		50	Structural shape groups 1 & 2, and plates & bars up through 3/4″ thick
Weathering High-Strength Low-Alloy	A588	42	Plates & bars over 5″ up through 8″ thick
		46	Plates & bars over 4″ up through 5″ thick
		50	All structural shape groups, and plates & bars up through 4″ thick
Quenched and Tempered Low-Alloy	A852	70	Plates & bars up through 4″ thick
Quenched and Tempered Alloy	A514	90	Plates & bars over 2-1/2″ up through 6″ thick
		100	Plates & bars up through 2-1/2″ thick

R051223-30   High Strength Steels

The mill price of high strength steels may be higher than A992 carbon steel but their proper use can achieve overall savings through total reduced weights. For columns with L/r over 100, A992 steel is best; under 100, high strength steels are economical. For heavy columns, high strength steels are economical when cover plates are eliminated. There is no economy using high strength steels for clip angles or supports or for beams where deflection governs. Thinner members are more economical than thick.

The per ton erection and fabricating costs of the high strength steels will be higher than for A992 since the same number of pieces, but less weight, will be installed.

R051223-35 Common Steel Sections

The upper portion of this table shows the name, shape, common designation and basic characteristics of commonly used steel sections. The lower portion explains how to read the designations used for the above illustrated common sections.

Shape & Designation	Name & Characteristics	Shape & Designation	Name & Characteristics
W	W Shape Parallel flange surfaces	MC	Miscellaneous Channel Infrequently rolled by some producers
S	American Standard Beam (I Beam) Sloped inner flange	L	Angle Equal or unequal legs, constant thickness
M	Miscellaneous Beams Cannot be classified as W, HP or S; infrequently rolled by some producers	T	Structural Tee Cut from W, M or S on center of web
C	American Standard Channel Sloped inner flange	HP	Bearing Pile Parallel flanges and equal flange and web thickness

Common drawing designations follow:

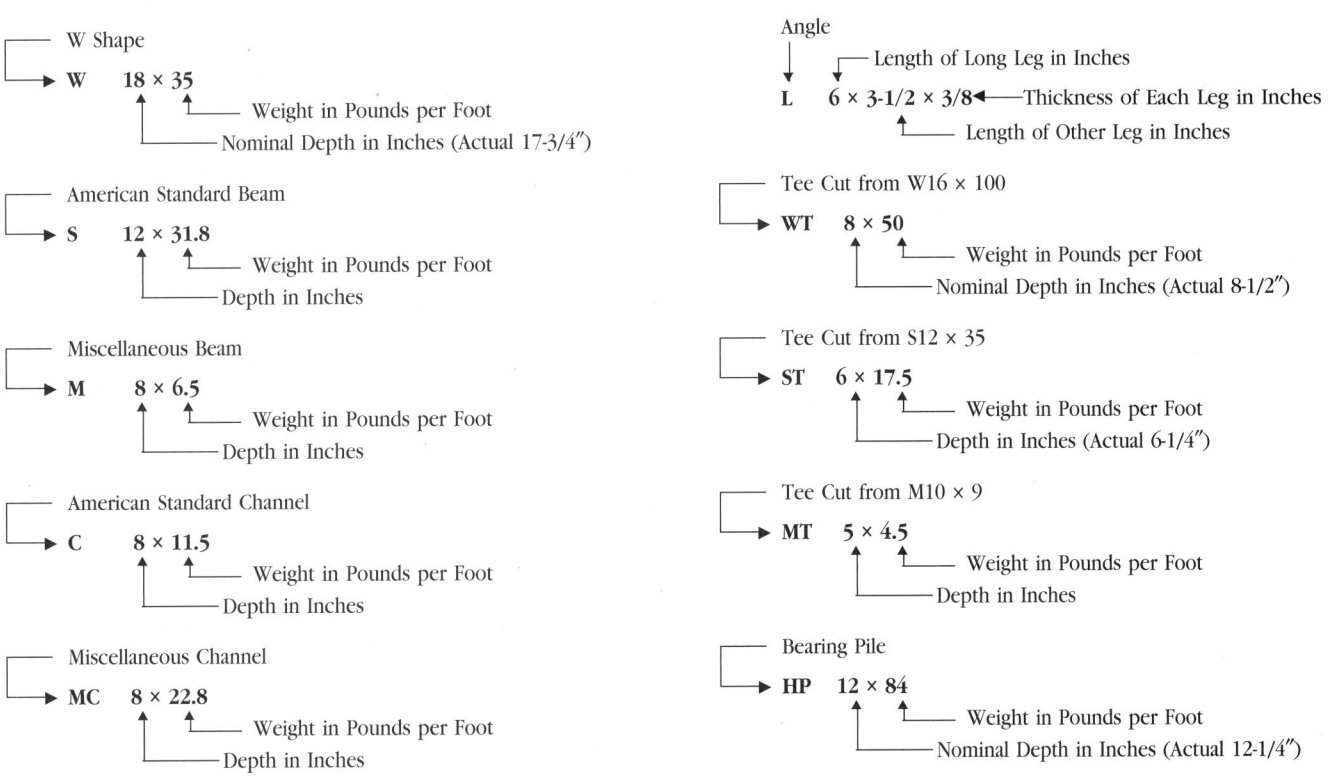

W Shape
W 18 × 35
— Weight in Pounds per Foot
— Nominal Depth in Inches (Actual 17-3/4″)

American Standard Beam
S 12 × 31.8
— Weight in Pounds per Foot
— Depth in Inches

Miscellaneous Beam
M 8 × 6.5
— Weight in Pounds per Foot
— Depth in Inches

American Standard Channel
C 8 × 11.5
— Weight in Pounds per Foot
— Depth in Inches

Miscellaneous Channel
MC 8 × 22.8
— Weight in Pounds per Foot
— Depth in Inches

Angle
L 6 × 3-1/2 × 3/8 ← Length of Long Leg in Inches
— Thickness of Each Leg in Inches
— Length of Other Leg in Inches

Tee Cut from W16 × 100
WT 8 × 50
— Weight in Pounds per Foot
— Nominal Depth in Inches (Actual 8-1/2″)

Tee Cut from S12 × 35
ST 6 × 17.5
— Weight in Pounds per Foot
— Depth in Inches (Actual 6-1/4″)

Tee Cut from M10 × 9
MT 5 × 4.5
— Weight in Pounds per Foot
— Depth in Inches

Bearing Pile
HP 12 × 84
— Weight in Pounds per Foot
— Nominal Depth in Inches (Actual 12-1/4″)

R051223-45 Installation Time for Structural Steel Building Components

The following tables show the expected average installation times for various structural steel shapes. Table A presents installation times for columns, Table B for beams, Table C for light framing and bolts, and Table D for structural steel for various project types.

Table A		
Description	Labor-Hours	Unit
Columns		
Steel, Concrete Filled		
3-1/2″ Diameter	.933	Ea.
6-5/8″ Diameter	1.120	Ea.
Steel Pipe		
3″ Diameter	.933	Ea.
8″ Diameter	1.120	Ea.
12″ Diameter	1.244	Ea.
Structural Tubing		
4″ x 4″	.966	Ea.
8″ x 8″	1.120	Ea.
12″ x 8″	1.167	Ea.
W Shape 2 Tier		
W8 x 31	.052	L.F.
W8 x 67	.057	L.F.
W10 x 45	.054	L.F.
W10 x 112	.058	L.F.
W12 x 50	.054	L.F.
W12 x 190	.061	L.F.
W14 x 74	.057	L.F.
W14 x 176	.061	L.F.

Table B				
Description	Labor-Hours	Unit	Labor-Hours	Unit
Beams, W Shape				
W6 x 9	.949	Ea.	.093	L.F.
W10 x 22	1.037	Ea.	.085	L.F.
W12 x 26	1.037	Ea.	.064	L.F.
W14 x 34	1.333	Ea.	.069	L.F.
W16 x 31	1.333	Ea.	.062	L.F.
W18 x 50	2.162	Ea.	.088	L.F.
W21 x 62	2.222	Ea.	.077	L.F.
W24 x 76	2.353	Ea.	.072	L.F.
W27 x 94	2.581	Ea.	.067	L.F.
W30 x 108	2.857	Ea.	.067	L.F.
W33 x 130	3.200	Ea.	.071	L.F.
W36 x 300	3.810	Ea.	.077	L.F.

Table C		
Description	Labor-Hours	Unit
Light Framing		
Angles 4″ and Larger	.055	lbs.
Less than 4″	.091	lbs.
Channels 8″ and Larger	.048	lbs.
Less than 8″	.072	lbs.
Cross Bracing Angles	.055	lbs.
Rods	.034	lbs.
Hanging Lintels	.069	lbs.
High Strength Bolts in Place		
3/4″ Bolts	.070	Ea.
7/8″ Bolts	.076	Ea.

Table D				
Description	Labor-Hours	Unit	Labor-Hours	Unit
Apartments, Nursing Homes, etc.				
1-2 Stories	4.211	Piece	7.767	Ton
3-6 Stories	4.444	Piece	7.921	Ton
7-15 Stories	4.923	Piece	9.014	Ton
Over 15 Stories	5.333	Piece	9.209	Ton
Offices, Hospitals, etc.				
1-2 Stories	4.211	Piece	7.767	Ton
3-6 Stories	4.741	Piece	8.889	Ton
7-15 Stories	4.923	Piece	9.014	Ton
Over 15 Stories	5.120	Piece	9.209	Ton
Industrial Buildings				
1 Story	3.478	Piece	6.202	Ton

R051223-50 Subpurlins

Bulb tee subpurlins are structural members designed to support and reinforce a variety of roof deck systems such as precast cement fiber roof deck tiles, monolithic roof deck systems, and gypsum or lightweight concrete over formboard. Other uses include interstitial service ceiling systems, wall panel systems, and joist anchoring in bond beams. See the Unit Price section for pricing on a square foot basis at 32-5/8″ O.C. Maximum span is based on a 3-span condition with a total allowable vertical load of 40 psf.

R051223-80 Dimensions and Weights of Sheet Steel

| Gauge No. | Approximate Thickness | | | | Weight | | |
	Inches (in fractions) Wrought Iron	Inches (in decimal parts) Wrought Iron	Steel	Millimeters Steel	per S.F. in Ounces	per S.F. in Lbs.	per Square Meter in Kg.
0000000	1/2"	.5	.4782	12.146	320	20.000	97.650
000000	15/32"	.46875	.4484	11.389	300	18.750	91.550
00000	7/16"	.4375	.4185	10.630	280	17.500	85.440
0000	13/32"	.40625	.3886	9.870	260	16.250	79.330
000	3/8"	.375	.3587	9.111	240	15.000	73.240
00	11/32"	.34375	.3288	8.352	220	13.750	67.130
0	5/16"	.3125	.2989	7.592	200	12.500	61.030
1	9/32"	.28125	.2690	6.833	180	11.250	54.930
2	17/64"	.265625	.2541	6.454	170	10.625	51.880
3	1/4"	.25	.2391	6.073	160	10.000	48.820
4	15/64"	.234375	.2242	5.695	150	9.375	45.770
5	7/32"	.21875	.2092	5.314	140	8.750	42.720
6	13/64"	.203125	.1943	4.935	130	8.125	39.670
7	3/16"	.1875	.1793	4.554	120	7.500	36.320
8	11/64"	.171875	.1644	4.176	110	6.875	33.570
9	5/32"	.15625	.1495	3.797	100	6.250	30.520
10	9/64"	.140625	.1345	3.416	90	5.625	27.460
11	1/8"	.125	.1196	3.038	80	5.000	24.410
12	7/64"	.109375	.1046	2.657	70	4.375	21.360
13	3/32"	.09375	.0897	2.278	60	3.750	18.310
14	5/64"	.078125	.0747	1.897	50	3.125	15.260
15	9/128"	.0713125	.0673	1.709	45	2.813	13.730
16	1/16"	.0625	.0598	1.519	40	2.500	12.210
17	9/160"	.05625	.0538	1.367	36	2.250	10.990
18	1/20"	.05	.0478	1.214	32	2.000	9.765
19	7/160"	.04375	.0418	1.062	28	1.750	8.544
20	3/80"	.0375	.0359	.912	24	1.500	7.324
21	11/320"	.034375	.0329	.836	22	1.375	6.713
22	1/32"	.03125	.0299	.759	20	1.250	6.103
23	9/320"	.028125	.0269	.683	18	1.125	5.490
24	1/40"	.025	.0239	.607	16	1.000	4.882
25	7/320"	.021875	.0209	.531	14	.875	4.272
26	3/160"	.01875	.0179	.455	12	.750	3.662
27	11/640"	.0171875	.0164	.417	11	.688	3.357
28	1/64"	.015625	.0149	.378	10	.625	3.052

R053100-10   Decking Descriptions

General - All Deck Products

A steel deck is made by cold forming structural grade sheet steel into a repeating pattern of parallel ribs. The strength and stiffness of the panels are the result of the ribs and the material properties of the steel. Deck lengths can be varied to suit job conditions, but because of shipping considerations, are usually less than 40 feet. Standard deck width varies with the product used but full sheets are usually 12″, 18″, 24″, 30″, or 36″. The deck is typically furnished in a standard width with the ends cut square. Any cutting for width, such as at openings or for angular fit, is done at the job site.

The deck is typically attached to the building frame with arc puddle welds, self-drilling screws, or powder or pneumatically driven pins. Sheet to sheet fastening is done with screws, button punching (crimping), or welds.

Composite Floor Deck

After installation and adequate fastening, a floor deck serves several purposes. It (a) acts as a working platform, (b) stabilizes the frame, (c) serves as a concrete form for the slab, and (d) reinforces the slab to carry the design loads applied during the life of the building. Composite decks are distinguished by the presence of shear connector devices as part of the deck. These devices are designed to mechanically lock the concrete and deck together so that the concrete and the deck work together to carry subsequent floor loads. These shear connector devices can be rolled-in embossments, lugs, holes, or wires welded to the panels. The deck profile can also be used to interlock concrete and steel.

Composite deck finishes are either galvanized (zinc coated) or phosphatized/painted. Galvanized deck has a zinc coating on both the top and bottom surfaces. The phosphatized/painted deck has a bare (phosphatized) top surface that will come into contact with the concrete. This bare top surface can be expected to develop rust before the concrete is placed. The bottom side of the deck has a primer coat of paint.

A composite floor deck is normally installed so the panel ends do not overlap on the supporting beams. Shear lugs or panel profile shapes often prevent a tight metal to metal fit if the panel ends overlap; the air gap caused by overlapping will prevent proper fusion with the structural steel supports when the panel end laps are shear stud welded.

Adequate end bearing of the deck must be obtained as shown on the drawings. If bearing is actually less in the field than shown on the drawings, further investigation is required.

Roof Deck

A roof deck is not designed to act compositely with other materials. A roof deck acts alone in transferring horizontal and vertical loads into the building frame. Roof deck rib openings are usually narrower than floor deck rib openings. This provides adequate support of the rigid thermal insulation board.

A roof deck is typically installed to endlap approximately 2″ over supports. However, it can be butted (or lapped more than 2″) to solve field fit problems. Since designers frequently use the installed deck system as part of the horizontal bracing system (the deck as a diaphragm), any fastening substitution or change should be approved by the designer. Continuous perimeter support of the deck is necessary to limit edge deflection in the finished roof and may be required for diaphragm shear transfer.

Standard roof deck finishes are galvanized or primer painted. The standard factory applied paint for roof decks is a primer paint and is not intended to weather for extended periods of time. Field painting or touching up of abrasions and deterioration of the primer coat or other protective finishes is the responsibility of the contractor.

Cellular Deck

A cellular deck is made by attaching a bottom steel sheet to a roof deck or composite floor deck panel. A cellular deck can be used in the same manner as a floor deck. Electrical, telephone, and data wires are easily run through the chase created between the deck panel and the bottom sheet.

When used as part of the electrical distribution system, the cellular deck must be installed so that the ribs line up and create a smooth cell transition at abutting ends. The joint that occurs at butting cell ends must be taped or otherwise sealed to prevent wet concrete from seeping into the cell. Cell interiors must be free of welding burrs, or other sharp intrusions, to prevent damage to wires.

When used as a roof deck, the bottom flat plate is usually left exposed to view. Care must be maintained during erection to keep good alignment and prevent damage.

A cellular deck is sometimes used with the flat plate on the top side to provide a flat working surface. Installation of the deck for this purpose requires special methods for attachment to the frame because the flat plate, now on the top, can prevent direct access to the deck material that is bearing on the structural steel. It may be advisable to treat the flat top surface to prevent slipping.

A cellular deck is always furnished galvanized or painted over galvanized.

Form Deck

A form deck can be any floor or roof deck product used as a concrete form. Connections to the frame are by the same methods used to anchor floor and roof decks. Welding washers are recommended when welding a deck that is less than 20 gauge thickness.

A form deck is furnished galvanized, prime painted, or uncoated. A galvanized deck must be used for those roof deck systems where a form deck is used to carry a lightweight insulating concrete fill.

Wood, Plastics & Comp. | R0611 Wood Framing

R061110-30 Lumber Product Material Prices

The price of forest products fluctuates widely from location to location and from season to season depending upon economic conditions. The bare material prices in the unit cost sections of the data set show the National Average material prices in effect Jan. 1 of this data year. It must be noted that lumber prices in general may change significantly during the year.

Availability of certain items depends upon geographic location and must be checked prior to firm-price bidding.

Wood, Plastics & Comp. | R0616 Sheathing

R061636-20 Plywood

There are two types of plywood used in construction: interior, which is moisture-resistant but not waterproofed, and exterior, which is waterproofed.

The grade of the exterior surface of the plywood sheets is designated by the first letter: A, for smooth surface with patches allowed; B, for solid surface with patches and plugs allowed; C, which may be surface plugged or may have knot holes up to 1″ wide; and D, which is used only for interior type plywood and may have knot holes up to 2-1/2″ wide. "Structural Grade" is specifically designed for engineered applications such as box beams. All CC & DD grades have roof and floor spans marked on them.

Underlayment-grade plywood runs from 1/4″ to 1-1/4″ thick. Thicknesses 5/8″ and over have optional tongue and groove joints which eliminate the need for blocking the edges. Underlayment 19/32″ and over may be referred to as Sturd-i-Floor.

The price of plywood can fluctuate widely due to geographic and economic conditions.

Typical uses for various plywood grades are as follows:

AA-AD Interior — cupboards, shelving, paneling, furniture

BB Plyform — concrete form plywood

CDX — wall and roof sheathing

Structural — box beams, girders, stressed skin panels

AA-AC Exterior — fences, signs, siding, soffits, etc.

Underlayment — base for resilient floor coverings

Overlaid HDO — high density for concrete forms & highway signs

Overlaid MDO — medium density for painting, siding, soffits & signs

303 Siding — exterior siding, textured, striated, embossed, etc.

Thermal & Moist. Protec. | R0731 Shingles & Shakes

R073126-20 Roof Slate

16″, 18″ and 20″ are standard lengths, and slate usually comes in random widths. For standard 3/16″ thickness use 1-1/2″ copper nails. Allow for 3% breakage.

Thermal & Moist. Protec. | R0751 Built-Up Bituminous Roofing

R075113-20 Built-Up Roofing

Asphalt is available in kegs of 100 lbs. each; coal tar pitch in 560 lb. kegs. Prepared roofing felts are available in a wide range of sizes, weights and characteristics. However, the most commonly used are #15 (432 S.F. per roll, 13 lbs. per square) and #30 (216 S.F. per roll, 27 lbs. per square).

Inter-ply bitumen varies from 24 lbs. per sq. (asphalt) to 30 lbs. per sq. (coal tar) per ply, MF4@ 25%. Flood coat bitumen also varies from 60 lbs. per sq. (asphalt) to 75 lbs. per sq. (coal tar), MF4@ 25%. Expendable equipment (mops, brooms, screeds, etc.) runs about 16% of the bitumen cost. For new, inexperienced crews this factor may be much higher.

A rigid insulation board is typically applied in two layers. The first is mechanically attached to nailable decks or spot or solid mopped to non-nailable decks; the second layer is then spot or solid mopped to the first layer. Membrane application follows the insulation, except in protected membrane roofs, where the membrane goes down first and the insulation on top, followed with ballast (stone or concrete pavers). Insulation and related labor costs are NOT included in prices for built-up roofing.

Reference Tables

Thermal & Moist. Protec. | **R0752 Modified Bituminous Membrane Roofing**

R075213-30 Modified Bitumen Roofing

The cost of modified bitumen roofing is highly dependent on the type of installation that is planned. Installation is based on the type of modifier used in the bitumen. The two most popular modifiers are atactic polypropylene (APP) and styrene butadiene styrene (SBS). The modifiers are added to heated bitumen during the manufacturing process to change its characteristics. A polyethylene, polyester or fiberglass reinforcing sheet is then sandwiched between layers of this bitumen. When completed, the result is a pre-assembled, built-up roof that has increased elasticity and weatherability. Some manufacturers include a surfacing material such as ceramic or mineral granules, metal particles or sand.

The preferred method of adhering SBS-modified bitumen roofing to the substrate is with hot-mopped asphalt (much the same as built-up roofing). This installation method requires a tar kettle/pot to heat the asphalt, as well as the labor, tools and equipment necessary to distribute and spread the hot asphalt.

The alternative method for applying APP and SBS modified bitumen is as follows. A skilled installer uses a torch to melt a small pool of bitumen off the membrane. This pool must form across the entire roll for proper adhesion. The installer must unroll the roofing at a pace slow enough to melt the bitumen, but fast enough to prevent damage to the rest of the membrane.

Modified bitumen roofing provides the advantages of both built-up and single-ply roofing. Labor costs are reduced over those of built-up roofing because only a single ply is necessary. The elasticity of single-ply roofing is attained with the reinforcing sheet and polymer modifiers. Modifieds have some self-healing characteristics and because of their multi-layer construction, they offer the reliability and safety of built-up roofing.

R078413-30 Firestopping

Firestopping is the sealing of structural, mechanical, electrical and other penetrations through fire-rated assemblies. The basic components of firestop systems are safing insulation and firestop sealant on both sides of wall penetrations and the top side of floor penetrations.

Pipe penetrations are assumed to be through concrete, grout, or joint compound and can be sleeved or unsleeved. Costs for the penetrations and sleeves are not included. An annular space of 1″ is assumed. Escutcheons are not included.

A metallic pipe is assumed to be copper, aluminum, cast iron or similar metallic material. An insulated metallic pipe is assumed to be covered with a thermal insulating jacket of varying thickness and materials.

A non-metallic pipe is assumed to be PVC, CPVC, FR Polypropylene or similar plastic piping material. Intumescent firestop sealants or wrap strips are included. Collars on both sides of wall penetrations and a sheet metal plate on the underside of floor penetrations are included.

Ductwork is assumed to be sheet metal, stainless steel or similar metallic material. Duct penetrations are assumed to be through concrete, grout or joint compound. Costs for penetrations and sleeves are not included. An annular space of 1/2″ is assumed.

Multi-trade openings include costs for sheet metal forms, firestop mortar, wrap strips, collars and sealants as necessary.

Structural penetrations joints are assumed to be 1/2″ or less. CMU walls are assumed to be within 1-1/2″ of the metal deck. Drywall walls are assumed to be tight to the underside of metal decking.

Metal panel, glass or curtain wall systems include a spandrel area of 5′ filled with mineral wool foil-faced insulation. Fasteners and stiffeners are included.

Openings · R0813 Metal Doors

R081313-20 Steel Door Selection Guide

Standard steel doors are classified into four levels, as recommended by the Steel Door Institute in the chart below. Each of the four levels offers a range of construction models and designs to meet architectural requirements for preference and appearance, including full flush, seamless, and stile & rail. Recommended minimum gauge requirements are also included.

For complete standard steel door construction specifications and available sizes, refer to the Steel Door Institute Technical Data Series, ANSI A250.8-98 (SDI-100), and ANSI A250.4-94 Test Procedure and Acceptance Criteria for Physical Endurance of Steel Door and Hardware Reinforcements.

Level		Model	Construction	For Full Flush or Seamless		
				Min. Gauge	Thickness (in)	Thickness (mm)
I	Standard Duty	1	Full Flush	20	0.032	0.8
		2	Seamless			
II	Heavy Duty	1	Full Flush	18	0.042	1.0
		2	Seamless			
III	Extra Heavy Duty	1	Full Flush	16	0.053	1.3
		2	Seamless			
		3	*Stile & Rail			
IV	Maximum Duty	1	Full Flush	14	0.067	1.6
		2	Seamless			

*Stiles & rails are 16 gauge; flush panels, when specified, are 18 gauge

Openings · R0851 Metal Windows

R085123-10 Steel Sash

An ironworker crew will erect 25 S.F. or 1.3 sash unit per hour, whichever is less.

A mechanic will point 30 L.F. per hour.

A painter will paint 90 S.F. per coat per hour.

A glazier production depends on light size.

Allow 1 lb. special steel sash putty per 16" x 20" light.

Openings · R0852 Wood Windows

R085216-10 Window Estimates

To ensure a complete window estimate, be sure to include the material and labor costs for each window, as well as the material and labor costs for an interior wood trim set.

853

R087110-10 Hardware Finishes

This table describes hardware finishes used throughout the industry. It also shows the base metal and the respective symbols in the three predominate systems of identification. Many of these are used in pricing descriptions in Division Eight.

US″	BMHA*	CDN^	Base	Description
US P	600	CP	Steel	Primed for Painting
US 1B	601	C1B	Steel	Bright Black Japanned
US 2C	602	C2C	Steel	Zinc Plated
US 2G	603	C2G	Steel	Zinc Plated
US 3	605	C3	Brass	Bright Brass, Clear Coated
US 4	606	C4	Brass	Satin Brass, Clear Coated
US 5	609	C5	Brass	Satin Brass, Blackened, Satin Relieved, Clear Coated
US 7	610	C7	Brass	Satin Brass, Blackened, Bright Relieved, Clear Coated
US 9	611	C9	Bronze	Bright Bronze, Clear Coated
US 10	612	C10	Bronze	Satin Bronze, Clear Coated
US 10A	641	C10A	Steel	Antiqued Bronze, Oiled and Lacquered
US 10B	613	C10B	Bronze	Antiqued Bronze, Oiled
US 11	616	C11	Bronze	Satin Bronze, Blackened, Satin Relieved, Clear Coated
US 14	618	C14	Brass/Bronze	Bright Nickel Plated, Clear Coated
US 15	619	C15	Brass/Bronze	Satin Nickel, Clear Coated
US 15A	620	C15A	Brass/Bronze	Satin Nickel Plated, Blackened, Satin Relieved, Clear Coated
US 17A	621	C17A	Brass/Bronze	Nickel Plated, Blackened, Relieved, Clear Coated
US 19	622	C19	Brass/Bronze	Flat Black Coated
US 20	623	C20	Brass/Bronze	Statuary Bronze, Light
US 20A	624	C20A	Brass/Bronze	Statuary Bronze, Dark
US 26	625	C26	Brass/Bronze	Bright Chromium
US 26D	626	C26D	Brass/Bronze	Satin Chromium
US 20	627	C27	Aluminum	Satin Aluminum Clear
US 28	628	C28	Aluminum	Anodized Dull Aluminum
US 32	629	C32	Stainless Steel	Bright Stainless Steel
US 32D	630	C32D	Stainless Steel	Stainless Steel
US 3	632	C3	Steel	Bright Brass Plated, Clear Coated
US 4	633	C4	Steel	Satin Brass, Clear Coated
US 7	636	C7	Steel	Satin Brass Plated, Blackened, Bright Relieved, Clear Coated
US 9	637	C9	Steel	Bright Bronze Plated, Clear Coated
US 5	638	C5	Steel	Satin Brass Plated, Blackened, Bright Relieved, Clear Coated
US 10	639	C10	Steel	Satin Bronze Plated, Clear Coated
US 10B	640	C10B	Steel	Antique Bronze, Oiled
US 10A	641	C10A	Steel	Antiqued Bronze, Oiled and Lacquered
US 11	643	C11	Steel	Satin Bronze Plated, Blackened, Bright Relieved, Clear Coated
US 14	645	C14	Steel	Bright Nickel Plated, Clear Coated
US 15	646	C15	Steel	Satin Nickel Plated, Clear Coated
US 15A	647	C15A	Steel	Nickel Plated, Blackened, Bright Relieved, Clear Coated
US 17A	648	C17A	Steel	Nickel Plated, Blackened, Relieved, Clear Coated
US 20	649	C20	Steel	Statuary Bronze, Light
US 20A	650	C20A	Steel	Statuary Bronze, Dark
US 26	651	C26	Steel	Bright Chromium Plated
US 26D	652	C26D	Steel	Satin Chromium Plated

* - BMHA Builders Hardware Manufacturing Association
″ - US Equivalent
^ - Canadian Equivalent
Japanning is imitating Asian lacquer work

For customer support on your Building Construction Costs with RSMeans Data, call 800.448.8182.

R088110-10   Glazing Productivity

Some glass sizes are estimated by the "united inch" (height + width). The table below shows the number of lights glazed in an eight-hour period by the crew size indicated, for glass up to 1/4″ thick. Square or nearly square lights are more economical on a S.F. basis. Long slender lights will have a high S.F. installation cost. For insulated glass reduce production by 33%. For 1/2″ float glass reduce production by 50%. Production time for glazing with two glaziers per day averages: 1/4″ float glass 120 S.F.; 1/2″ float glass 55 S.F.; 1/2″ insulated glass 95 S.F.; 3/4″ insulated glass 75 S.F.

Glazing Method	United Inches per Light							
	40″	60″	80″	100″	135″	165″	200″	240″
Number of Men in Crew	1	1	1	1	2	3	3	4
Industrial sash, putty	60	45	24	15	18	—	—	—
With stops, putty bed	50	36	21	12	16	8	4	3
Wood stops, rubber	40	27	15	9	11	6	3	2
Metal stops, rubber	30	24	14	9	9	6	3	2
Structural glass	10	7	4	3	—	—	—	—
Corrugated glass	12	9	7	4	4	4	3	—
Storefronts	16	15	13	11	7	6	4	4
Skylights, putty glass	60	36	21	12	16	—	—	—
Thiokol set	15	15	11	9	9	6	3	2
Vinyl set, snap on	18	18	13	12	12	7	5	4
Maximum area per light	2.8 S.F.	6.3 S.F.	11.1 S.F.	17.4 S.F.	31.6 S.F.	47 S.F.	69 S.F.	100 S.F.

R092000-50 Lath, Plaster and Gypsum Board

Gypsum board lath is available in 3/8″ thick × 16″ wide × 4′ long sheets as a base material for multi-layer plaster applications. It is also available as a base for either multi-layer or veneer plaster applications in 1/2″ and 5/8″ thick–4′ wide × 8′, 10′ or 12′ long sheets. Fasteners are screws or blued ring shank nails for wood framing and screws for metal framing.

Metal lath is available in diamond mesh patterns with flat or self-furring profiles. Paper backing is available for applications where excessive plaster waste needs to be avoided. A slotted mesh ribbed lath should be used in areas where the span between structural supports is greater than normal. Most metal lath comes in 27″ × 96″ sheets. Diamond mesh weighs 1.75, 2.5 or 3.4 pounds per square yard, slotted mesh lath weighs 2.75 or 3.4 pounds per square yard. Metal lath can be nailed, screwed or tied in place.

Many **accessories** are available. Corner beads, flat reinforcing strips, casing beads, control and expansion joints, furring brackets and channels are some examples. Note that accessories are not included in plaster or stucco line items.

Plaster is defined as a material or combination of materials that when mixed with a suitable amount of water, forms a plastic mass or paste. When applied to a surface, the paste adheres to it and subsequently hardens, preserving in a rigid state the form or texture imposed during the period of elasticity.

Gypsum plaster is made from ground calcined gypsum. It is mixed with aggregates and water for use as a base coat plaster.

Vermiculite plaster is a fire-retardant plaster covering used on steel beams, concrete slabs and other heavy construction materials. Vermiculite is a group name for certain clay minerals, hydrous silicates or aluminum, magnesium and iron that have been expanded by heat.

Perlite plaster is a plaster using perlite as an aggregate instead of sand. Perlite is a volcanic glass that has been expanded by heat.

Gauging plaster is a mix of gypsum plaster and lime putty that when applied produces a quick drying finish coat.

Veneer plaster is a one or two component gypsum plaster used as a thin finish coat over special gypsum board.

Keenes cement is a white cementitious material manufactured from gypsum that has been burned at a high temperature and ground to a fine powder. Alum is added to accelerate the set. The resulting plaster is hard and strong and accepts and maintains a high polish, hence it is used as a finishing plaster.

Stucco is a Portland cement based plaster used primarily as an exterior finish.

Plaster is used on both interior and exterior surfaces. Generally it is applied in multiple-coat systems. A three-coat system uses the terms scratch, brown and finish to identify each coat. A two-coat system uses base and finish to describe each coat. Each type of plaster and application system has attributes that are chosen by the designer to best fit the intended use.

Gypsum Plaster Quantities for 100 S.Y.	2 Coat, 5/8″ Thick		3 Coat, 3/4″ Thick		
	Base	Finish	Scratch	Brown	Finish
	1:3 Mix	2:1 Mix	1:2 Mix	1:3 Mix	2:1 Mix
Gypsum plaster	1,300 lb.		1,350 lb.	650 lb.	
Sand	1.75 C.Y.		1.85 C.Y.	1.35 C.Y.	
Finish hydrated lime		340 lb.			340 lb.
Gauging plaster		170 lb.			170 lb.

Vermiculite or Perlite Plaster Quantities for 100 S.Y.	2 Coat, 5/8″ Thick		3 Coat, 3/4″ Thick		
	Base	Finish	Scratch	Brown	Finish
Gypsum plaster	1,250 lb.		1,450 lb.	800 lb.	
Vermiculite or perlite	7.8 bags		8.0 bags	3.3 bags	
Finish hydrated lime		340 lb.			340 lb.
Gauging plaster		170 lb.			170 lb.

Stucco–Three-Coat System Quantities for 100 S.Y.	On Wood Frame	On Masonry
Portland cement	29 bags	21 bags
Sand	2.6 C.Y.	2.0 C.Y.
Hydrated lime	180 lb.	120 lb.

R092910-10   Levels of Gypsum Drywall Finish

In the past, contract documents often used phrases such as "industry standard" and "workmanlike finish" to specify the expected quality of gypsum board wall and ceiling installations. The vagueness of these descriptions led to unacceptable work and disputes.

In order to resolve this problem, four major trade associations concerned with the manufacture, erection, finish, and decoration of gypsum board wall and ceiling systems developed an industry-wide *Recommended Levels of Gypsum Board Finish.*

The finish of gypsum board walls and ceilings for specific final decoration is dependent on a number of factors. A primary consideration is the location of the surface and the degree of decorative treatment desired. Painted and unpainted surfaces in warehouses and other areas where appearance is normally not critical may simply require the taping of wallboard joints and 'spotting' of fastener heads. Blemish-free, smooth, monolithic surfaces often intended for painted and decorated walls and ceilings in habitated structures, ranging from single-family dwellings through monumental buildings, require additional finishing prior to the application of the final decoration.

Other factors to be considered in determining the level of finish of the gypsum board surface are (1) the type of angle of surface illumination (both natural and artificial lighting), and (2) the paint and method of application or the type and finish of wallcovering specified as the final decoration. Critical lighting conditions, gloss paints, and thin wall coverings require a higher level of gypsum board finish than heavily textured surfaces which are subsequently painted or surfaces which are to be decorated with heavy grade wall coverings.

The following descriptions were developed by the Association of the Wall and Ceiling Industries-International (AWCI), Ceiling & Interior Systems Construction Association (CISCA), Gypsum Association (GA), and Painting and Decorating Contractors of America (PDCA) as a guide.

Level 0: Used in temporary construction or wherever the final decoration has not been determined. Unfinished. No taping, finishing or corner beads are required. Also could be used where non-predecorated panels will be used in demountable-type partitions that are to be painted as a final finish.

Level 1: Frequently used in plenum areas above ceilings, in attics, in areas where the assembly would generally be concealed, or in building service corridors and other areas not normally open to public view. Some degree of sound and smoke control is provided; in some geographic areas, this level is referred to as "fire-taping," although this level of finish does not typically meet fire-resistant assembly requirements. Where a fire resistance rating is required for the gypsum board assembly, details of construction should be in accordance with reports of fire tests of assemblies that have met the requirements of the fire rating acceptable.

All joints and interior angles shall have tape embedded in joint compound. Accessories are optional at specifier discretion in corridors and other areas with pedestrian traffic. Tape and fastener heads need not be covered with joint compound. Surface shall be free of excess joint compound. Tool marks and ridges are acceptable.

Level 2: It may be specified for standard gypsum board surfaces in garages, warehouse storage, or other similar areas where surface appearance is not of primary importance.

All joints and interior angles shall have tape embedded in joint compound and shall be immediately wiped with a joint knife or trowel, leaving a thin coating of joint compound over all joints and interior angles. Fastener heads and accessories shall be covered with a coat of joint compound. Surface shall be free of excess joint compound. Tool marks and ridges are acceptable.

Level 3: Typically used in areas receiving heavy texture (spray or hand applied) finishes before final painting, or where commercial-grade (heavy duty) wall coverings are to be applied as the final decoration. This level of finish should not be used where smooth painted surfaces or where lighter weight wall coverings are specified. The prepared surface shall be coated with a drywall primer prior to the application of final finishes.

All joints and interior angles shall have tape embedded in joint compound and shall be immediately wiped with a joint knife or trowel, leaving a thin coating of joint compound over all joints and interior angles. One additional coat of joint compound shall be applied over all joints and interior angles. Fastener heads and accessories shall be covered with two separate coats of joint compound. All joint compounds shall be smooth and free of tool marks and ridges. The prepared surface shall be covered with a drywall primer prior to the application of the final decoration.

Level 4: This level should be used where residential grade (light duty) wall coverings, flat paints, or light textures are to be applied. The prepared surface shall be coated with a drywall primer prior to the application of final finishes. Release agents for wall coverings are specifically formulated to minimize damage if coverings are subsequently removed.

The weight, texture, and sheen level of the wall covering material selected should be taken into consideration when specifying wall coverings over this level of drywall treatment. Joints and fasteners must be sufficiently concealed if the wall covering material is lightweight, contains limited pattern, has a glossy finish, or has any combination of these features. In critical lighting areas, flat paints applied over light textures tend to reduce joint photographing. Gloss, semi-gloss, and enamel paints are not recommended over this level of finish.

All joints and interior angles shall have tape embedded in joint compound and shall be immediately wiped with a joint knife or trowel, leaving a thin coating of joint compound over all joints and interior angles. In addition, two separate coats of joint compound shall be applied over all flat joints and one separate coat of joint compound applied over interior angles. Fastener heads and accessories shall be covered with three separate coats of joint compound. All joint compounds shall be smooth and free of tool marks and ridges. The prepared surface shall be covered with a drywall primer like Sheetrock® First Coat prior to the application of the final decoration.

Level 5: The highest quality finish is the most effective method to provide a uniform surface and minimize the possibility of joint photographing and of fasteners showing through the final decoration. This level of finish is required where gloss, semi-gloss, or enamel is specified; when flat joints are specified over an untextured surface; or where critical lighting conditions occur. The prepared surface shall be coated with a drywall primer prior to the application of the final decoration.

All joints and interior angles shall have tape embedded in joint compound and be immediately wiped with a joint knife or trowel, leaving a thin coating of joint compound over all joints and interior angles. Two separate coats of joint compound shall be applied over all flat joints and one separate coat of joint compound applied over interior angles. Fastener heads and accessories shall be covered with three separate coats of joint compound.

A thin skim coat of joint compound shall be trowel applied to the entire surface. Excess compound is immediately troweled off, leaving a film or skim coating of compound completely covering the paper. As an alternative to a skim coat, a material manufactured especially for this purpose may be applied such as Sheetrock® Tuff-Hide primer surfacer. The surface must be smooth and free of tool marks and ridges. The prepared surface shall be covered with a drywall primer prior to the application of the final decoration.

R096613-10 Terrazzo Floor

The table below lists quantities required for 100 S.F. of 5/8" terrazzo topping, either bonded or not bonded.

Description	Bonded to Concrete 1-1/8" Bed, 1:4 Mix	Not Bonded 2-1/8" Bed and 1/4" Sand
Portland cement, 94 lb. Bag	6 bags	8 bags
Sand	10 C.F.	20 C.F.
Divider strips, 4' squares	50 L.F.	50 L.F.
Terrazzo fill, 50 lb. Bag	12 bags	12 bags
15 Lb. tarred felt		1 C.S.F.
Mesh 2 x 2 #14 galvanized		1 C.S.F.
Crew J-3	0.77 days	0.87 days

2' × 2' panels require 1.00 L.F. divider strip per S.F.
3' × 3' panels require 0.67 L.F. divider strip per S.F.
4' × 4' panels require 0.50 L.F. divider strip per S.F.
5' × 5' panels require 0.40 L.F. divider strip per S.F.
6' × 6' panels require 0.33 L.F. divider strip per S.F.

R097223-10 Wall Covering

The table below lists the quantities required for 100 S.F. of wall covering.

Description	Medium-Priced Paper	Expensive Paper
Paper	1.6 dbl. rolls	1.6 dbl. rolls
Wall sizing	0.25 gallon	0.25 gallon
Vinyl wall paste	0.6 gallon	0.6 gallon
Apply sizing	0.3 hour	0.3 hour
Apply paper	1.2 hours	1.5 hours

Most wallpapers now come in double rolls only.
To remove old paper, allow 1.3 hours per 100 S.F.

R099100-10 Painting Estimating Techniques

Proper estimating methodology is needed to obtain an accurate painting estimate. There is no known reliable shortcut or square foot method. The following steps should be followed:

- List all surfaces to be painted, with an accurate quantity (area) of each. Items having similar surface condition, finish, application method and accessibility may be grouped together.
- List all the tasks required for each surface to be painted, including surface preparation, masking, and protection of adjacent surfaces. Surface preparation may include minor repairs, washing, sanding and puttying.
- Select the proper Means line for each task. Review and consider all adjustments to labor and materials for type of paint and location of work. Apply the height adjustment carefully. For instance, when applying the adjustment for work over 8' high to a wall that is 12' high, apply the adjustment only to the area between 8' and 12' high, and not to the entire wall.

When applying more than one percent (%) adjustment, apply each to the base cost of the data, rather than applying one percentage adjustment on top of the other.

When estimating the cost of painting walls and ceilings remember to add the brushwork for all cut-ins at inside corners and around windows and doors as a LF measure. One linear foot of cut-in with a brush equals one square foot of painting.

All items for spray painting include the labor for roll-back.

Deduct for openings greater than 100 SF or openings that extend from floor to ceiling and are greater than 5' wide. Do not deduct small openings.

The cost of brushes, rollers, ladders and spray equipment are considered part of a painting contractor's overhead, and should not be added to the estimate. The cost of rented equipment such as scaffolding and swing staging should be added to the estimate.

R099100-20 Painting

Item	Coat	One Gallon Covers			In 8 Hours a Laborer Covers			Labor-Hours per 100 S.F.		
		Brush	Roller	Spray	Brush	Roller	Spray	Brush	Roller	Spray
Paint wood siding	prime	250 S.F.	225 S.F.	290 S.F.	1150 S.F.	1300 S.F.	2275 S.F.	.695	.615	.351
	others	270	250	290	1300	1625	2600	.615	.492	.307
Paint exterior trim	prime	400	—	—	650	—	—	1.230	—	—
	1st	475	—	—	800	—	—	1.000	—	—
	2nd	520	—	—	975	—	—	.820	—	—
Paint shingle siding	prime	270	255	300	650	975	1950	1.230	.820	.410
	others	360	340	380	800	1150	2275	1.000	.695	.351
Stain shingle siding	1st	180	170	200	750	1125	2250	1.068	.711	.355
	2nd	270	250	290	900	1325	2600	.888	.603	.307
Paint brick masonry	prime	180	135	160	750	800	1800	1.066	1.000	.444
	1st	270	225	290	815	975	2275	.981	.820	.351
	2nd	340	305	360	815	1150	2925	.981	.695	.273
Paint interior plaster or drywall	prime	400	380	495	1150	2000	3250	.695	.400	.246
	others	450	425	495	1300	2300	4000	.615	.347	.200
Paint interior doors and windows	prime	400	—	—	650	—	—	1.230	—	—
	1st	425	—	—	800	—	—	1.000	—	—
	2nd	450	—	—	975	—	—	.820	—	—

Special Construction | R1311 Swimming Pools

R131113-20 Swimming Pools

Pool prices given per square foot of surface area include pool structure, filter and chlorination equipment, pumps, related piping, ladders/steps, maintenance kit, skimmer and vacuum system. Decks and electrical service to equipment are not included.

Residential in-ground pool construction can be divided into two categories: vinyl lined and gunite. Vinyl lined pool walls are constructed of different materials including wood, concrete, plastic or metal. The bottom is often graded with sand over which the vinyl liner is installed. Vermiculite or soil cement bottoms may be substituted for an added cost.

Gunite pool construction is used both in residential and municipal installations. These structures are steel reinforced for strength and finished with a white cement limestone plaster.

Municipal pools will have a higher cost because plumbing codes require more expensive materials, chlorination equipment and higher filtration rates.

Municipal pools greater than 1,800 S.F. require gutter systems to control waves. This gutter may be formed into the concrete wall. Often a vinyl/stainless steel gutter or gutter/wall system is specified, which will raise the pool cost.

Competition pools usually require tile bottoms and sides with contrasting lane striping, which will also raise the pool cost.

Special Construction | R1331 Fabric Structures

R133113-10   Air Supported Structures

Air supported structures are made from fabrics that can be classified into two groups: temporary and permanent. Temporary fabrics include nylon, woven polyethylene, vinyl film, and vinyl coated dacron. These have lifespans that range from five to fifteen plus years. The cost per square foot includes a fabric shell, tension cables, primary and back-up inflation systems and doors. The lower cost structures are used for construction shelters, bulk storage and pond covers. The more expensive are used for recreational structures and warehouses.

Permanent fabrics are teflon coated fiberglass. The life of this structure is twenty plus years. The high cost limits its application to architectural designed structures which call for a clear span covered area, such as stadiums and convention centers. Both temporary and permanent structures are available in translucent fabrics which eliminate the need for daytime lighting.

Areas to be covered vary from 10,000 S.F. to any area up to 1000 feet wide by any length. Height restrictions range from a maximum of 1/2 of the

width to a minimum of 1/6 of the width. Erection of even the largest of the temporary structures requires no more than a week.

Centrifugal fans provide the inflation necessary to support the structure during application of live loads. Airlocks are usually used at large entrances to prevent loss of static pressure. Some manufacturers employ propeller fans which generate sufficient airflow (30,000 CFM) to eliminate the need for airlocks. These fans may also be automatically controlled to resist high wind conditions, regulate humidity (air changes), and provide cooling and heat.

Insulation can be provided with the addition of a second or even third interior liner, creating a dead air space with an "R" value of four to nine. Some structures allow for the liner to be collapsed into the outer shell to enable the internal heat to melt accumulated snow. For cooling or air conditioning, the exterior face of the liner can be aluminized to reflect the sun's heat.

R133113-90   Seismic Bracing

Sometimes referred to as anti-sway bracing, this support system is required in earthquake areas. The individual components must be assembled to

make a required system.

Example			Additionally, height factors must be taken into account. Add the following percentages to labor for elevated installations:	
	C-Clamp 3/8" rod	2 ea.	15' to 20' high	10%
	Rod, continuous thread 3/8"	10 L.F.	21' to 25' high	20%
	Field, weld 1"	2 ea.	26' to 30' high	30%
			31' to 35' high	40%
			36' to 40' high	50%
			41' to 50' high	60%

Special Construction | R1334 Fabricated Engineered Structures

R133419-10   Pre-Engineered Steel Buildings

These buildings are manufactured by many companies and normally erected by franchised dealers throughout the U.S. The four basic types are: Rigid Frames, Truss type, Post and Beam and the Sloped Beam type. The most popular roof slope is a low pitch of 1" in 12". The minimum economical area of these buildings is about 3000 S.F. of floor area. Bay sizes are usually 20' to 24' but can go as high as 30' with heavier girts and purlins. Eave heights are usually 12' to 24' with 18' to 20' most typical.

Material prices shown in the Unit Price section are bare costs for the building shell only and do not include floors, foundations, anchor bolts, interior finishes or utilities. Costs assume at least three bays of 24' each, a 1" in 12" roof slope, and they are based on a 30 psf roof load and a 20 psf wind

load (wind load is a function of wind speed, building height, and terrain characteristics; this should be determined by a Registered Structural Engineer) and no unusual requirements. Costs include the structural frame, 26 ga. non-insulated colored corrugated or ribbed roofing and siding panels, fasteners, closures, trim and flashing but no allowance for insulation, doors, windows, skylights, gutters or downspouts. Very large projects would generally cost less for materials than the prices shown. For roof panel substitutions and wall panel substitutions, see appropriate Unit Price sections.

Conditions at the site, weather, shape and size of the building, and labor availability will affect the erection cost of the building.

R133423-30   Dome Structures

Steel — The four types are Lamella, Schwedler, Arch and Geodesic. For maximum economy, the rise should be about 15 to 20% of the diameter. Most common diameters are in the 200' to 300' range. Lamella domes weigh about 5 P.S.F. of floor area less than Schwedler domes. The Schwedler dome weight in lbs. per S.F. approaches .046 times the diameter. Domes below 125' diameter weigh .07 times the diameter and the cost per ton of steel is higher. See R051223-20 for estimating weight.

Wood — Small domes are of sawn lumber. Larger ones are laminated. In larger sizes, triaxial and triangular cost about the same; radial domes cost more. Radial domes are economical in the 60' to 70' diameter range. The most economical range of all types is 80' to 200' diameters. Diameters can

run over 400'. All costs are quoted above the foundation. Prices include 2" decking and a tension tie ring in place.

Plywood — Stock prefab geodesic domes are available with diameters from 24' to 60'.

Fiberglass — Aluminum framed translucent sandwich panels with spans from 5' to 45' are commercially available.

Aluminum — Stressed skin aluminum panels form geodesic domes with spans ranging from 82' to 232'. An aluminum space truss, triangulated or nontriangulated, with aluminum or clear acrylic closure panels can be used for clear spans of 40' to 415'.

R142000-10 Freight Elevators

Capacities run from 2,000 lbs. to over 100,000 lbs. with 3,000 lbs. to 10,000 lbs. most common. Travel speeds are generally lower and the control is less intricate than on passenger elevators. Costs in the Unit Price sections are for hydraulic and geared elevators.

R142000-20 Elevator Selective Costs See R142000-40 for cost development.

A. Base Unit	Passenger		Freight		Hospital	
	Hydraulic	Electric	Hydraulic	Electric	Hydraulic	Electric
Capacity	1,500 lb.	2,000 lb.	2,000 lb.	4,000 lb.	4,000 lb.	4,000 lb.
Speed	100 F.P.M.	200 F.P.M.	100 F.P.M.	200 F.P.M.	100 F.P.M.	200 F.P.M.
#Stops/Travel Ft.	2/12	4/40	2/20	4/40	2/20	4/40
Push Button Oper.	Yes	Yes	Yes	Yes	Yes	Yes
Telephone Box & Wire	"	"	"	"	"	"
Emergency Lighting	"	"	No	No	"	"
Cab	Plastic Lam. Walls	Plastic Lam. Walls	Painted Steel	Painted Steel	Plastic Lam. Walls	Plastic Lam. Walls
Cove Lighting	Yes	Yes	No	No	Yes	Yes
Floor	V.C.T.	V.C.T.	Wood w/Safety Treads	Wood w/Safety Treads	V.C.T.	V.C.T.
Doors, & Speedside Slide	Yes	Yes	Yes	Yes	Yes	Yes
Gates, Manual	No	No	No	No	No	No
Signals, Lighted Buttons	Car and Hall	Car and Hall	Car and Hall	Car and Hall	Car and Hall	Car and Hall
O.H. Geared Machine	N.A.	Yes	N.A.	Yes	N.A.	Yes
Variable Voltage Contr.	"	"	N.A.	"	"	"
Emergency Alarm	Yes	"	Yes	"	Yes	"
Class "A" Loading	N.A.	N.A.	"	"	N.A.	N.A.

R142000-30 Passenger Elevators

Electric elevators are used generally but hydraulic elevators can be used for lifts up to 70′ and where large capacities are required. Hydraulic speeds are limited to 200 F.P.M. but cars are self leveling at the stops. On low rises, hydraulic installation runs about 15% less than standard electric types but on higher rises this installation cost advantage is reduced. Maintenance of hydraulic elevators is about the same as the electric type but the underground portion is not included in the maintenance contract.

In electric elevators there are several control systems available, the choice of which will be based upon elevator use, size, speed and cost criteria. The two types of drives are geared for low speeds and gearless for 450 F.P.M. and over.

The tables on the preceding pages illustrate typical installed costs of the various types of elevators available.

R142000-40 Elevator Cost Development

To price a new car or truck from the factory, you must start with the manufacturer's basic model, then add or exchange optional equipment and features. The same is true for pricing elevators.

Requirement: One-passenger elevator, five-story hydraulic, 2,500 lb. capacity, 12′ floor to floor, speed 150 F.P.M., emergency power switching and maintenance contract.

Example:

Description	Adjustment
A. Base Elevator: Hydraulic Passenger, 1500 lb. Capacity, 100 fpm, 2 Stops, Standard Finish	1 Ea.
B. Capacity Adjustment (2,500 lb.)	1 Ea.
C. Excess Travel Adjustment: 48′ Total Travel (4 x 12′) minus 12′ Base Unit Travel =	36 V.L.F.
D. Stops Adjustment: 5 Total Stops minus 2 Stops (Base Unit) =	3 Stops
E. Speed Adjustment (150 F.P.M.)	1 Ea.
F. Options:	
1. Intercom Service	1 Ea.
2. Emergency Power Switching, Automatic	1 Ea.
3. Stainless Steel Entrance Doors	5 Ea.
4. Maintenance Contract (12 Months)	1 Ea.
5. Position Indicator for main floor level (none indicated in Base Unit)	1 Ea.

Conveying Equipment | R1432 Moving Walks

R143210-20 Moving Ramps and Walks

These are a specialized form of conveyor 3' to 6' wide with capacities of 3,600 to 18,000 persons per hour. Maximum speed is 140 F.P.M. and normal incline is 0° to 15°.

Local codes will determine the maximum angle. Outdoor units would require additional weather protection.

Plumbing | R2201 Operation & Maintenance of Plumbing

R220102-20 Labor Adjustment Factors

Labor Adjustment Factors are provided for Divisions 21, 22, and 23 to assist the mechanical estimator account for the various complexities and special conditions of any particular project. While a single percentage has been entered on each line of Division 22 01 02.20, it should be understood that these are just suggested midpoints of ranges of values commonly used by mechanical estimators. They may be increased or decreased depending on the severity of the special conditions.

The group for "existing occupied buildings" has been the subject of requests for explanation. Actually there are two stages to this group: buildings that are existing and "finished" but unoccupied, and those that also are occupied. Buildings that are "finished" may result in higher labor costs due to the

workers having to be more careful not to damage finished walls, ceilings, floors, etc. and may necessitate special protective coverings and barriers. Also corridor bends and doorways may not accommodate long pieces of pipe or larger pieces of equipment. Work above an already hung ceiling can be very time consuming. The addition of occupants may force the work to be done on premium time (nights and/or weekends), eliminate the possible use of some preferred tools such as pneumatic drivers, powder charged drivers, etc. The estimator should evaluate the access to the work area and just how the work is going to be accomplished to arrive at an increase in labor costs over "normal" new construction productivity.

R220105-10 Demolition (Selective vs. Removal for Replacement)

Demolition can be divided into two basic categories.

One type of demolition involves the removal of material with no concern for its replacement. The labor-hours to estimate this work are found under "Selective Demolition" in the Fire Protection, Plumbing and HVAC Divisions. It is selective in that individual items or all the material installed as a system or trade grouping such as plumbing or heating systems are removed. This may be accomplished by the easiest way possible, such as sawing, torch cutting, or sledge hammering as well as simple unbolting.

The second type of demolition is the removal of some items for repair or replacement. This removal may involve careful draining, opening of unions,

disconnecting and tagging electrical connections, capping pipes/ducts to prevent entry of debris or leakage of the material contained as well as transporting the item away from its in-place location to a truck/dumpster. An approximation of the time required to accomplish this type of demolition is to use half of the time indicated as necessary to install a new unit. For example: installation of a new pump might be listed as requiring 6 labor-hours so if we had to estimate the removal of the old pump we would allow an additional 3 hours for a total of 9 hours. That is, the complete replacement of a defective pump with a new pump would be estimated to take 9 labor-hours.

Plumbing | R2211 Facility Water Distribution

R221113-50 Pipe Material Considerations

1. Malleable fittings should be used for gas service.
2. Malleable fittings are used where there are stresses/strains due to expansion and vibration.
3. Cast fittings may be broken as an aid to disassembling heating lines frozen by long use, temperature and minerals.
4. A cast iron pipe is extensively used for underground and submerged service.
5. Type M (light wall) copper tubing is available in hard temper only and is used for nonpressure and less severe applications than K and L.

Domestic/Imported Pipe and Fittings Costs

The prices shown in this publication for steel/cast iron pipe and steel, cast iron, and malleable iron fittings are based on domestic production sold at the normal trade discounts. The above listed items of foreign manufacture may be available at prices 1/3 to 1/2 of those shown. Some imported items after minor machining or finishing operations are being sold as domestic to further complicate the system.

6. Type L (medium wall) copper tubing, available hard or soft for interior service.
7. Type K (heavy wall) copper tubing, available in hard or soft temper for use where conditions are severe. For underground and interior service.
8. Hard drawn tubing requires fewer hangers or supports but should not be bent. Silver brazed fittings are recommended, but soft solder is normally used.
9. Type DMV (very light wall) copper tubing designed for drainage, waste and vent plus other non-critical pressure services.

Caution: Most pipe prices in this data set also include a coupling and pipe hangers which for the larger sizes can add significantly to the per foot cost and should be taken into account when comparing "book cost" with the quoted supplier's cost.

Heating, Ventilating & A.C. R2356 Solar Energy Heating Equipment

R235616-60 Solar Heating (Space and Hot Water)

Collectors should face as close to due South as possible, but variations of up to 20 degrees on either side of true South are acceptable. Local climate and collector type may influence the choice between east or west deviations. Obviously they should be located so they are not shaded from the sun's rays. Incline collectors at a slope of latitude minus 5 degrees for domestic hot water and latitude plus 15 degrees for space heating.

Flat plate collectors consist of a number of components as follows: Insulation to reduce heat loss through the bottom and sides of the collector. The enclosure which contains all the components in this assembly is usually weatherproof and prevents dust, wind and water from coming in contact with the absorber plate. The cover plate usually consists of one or more layers of a variety of glass or plastic and reduces the reradiation by creating an air space which traps the heat between the cover and the absorber plates.

The absorber plate must have a good thermal bond with the fluid passages. The absorber plate is usually metallic and treated with a surface coating which improves absorptivity. Black or dark paints or selective coatings are used for this purpose, and the design of this passage and plate combination helps determine a solar system's effectiveness.

Heat transfer fluid passage tubes are attached above and below or integral with an absorber plate for the purpose of transferring thermal energy from the absorber plate to a heat transfer medium. The heat exchanger is a device for transferring thermal energy from one fluid to another.

Piping and storage tanks should be well insulated to minimize heat losses.

Size domestic water heating storage tanks to hold 20 gallons of water per user, minimum, plus 10 gallons per dishwasher or washing machine. For domestic water heating an optimum collector size is approximately 3/4 square foot of area per gallon of water storage. For space heating of residences and small commercial applications the collector is commonly sized between 30% and 50% of the internal floor area. For space heating of large commercial applications, collector areas less than 30% of the internal floor area can still provide significant heat reductions.

A supplementary heat source is recommended for Northern states for December through February.

The solar energy transmission per square foot of collector surface varies greatly with the material used. Initial cost, heat transmittance and useful life are obviously interrelated.

Heating, Ventilating & A.C. R2360 Central Cooling Equipment

R236000-20 Air Conditioning Requirements

BTUs per hour per S.F. of floor area and S.F. per ton of air conditioning.

Type of Building	BTU/Hr per S.F.	S.F. per Ton	Type of Building	BTU/Hr per S.F.	S.F. per Ton	Type of Building	BTU/Hr per S.F.	S.F. per Ton
Apartments, Individual	26	450	Dormitory, Rooms	40	300	Libraries	50	240
Corridors	22	550	Corridors	30	400	Low Rise Office, Exterior	38	320
Auditoriums & Theaters	40	300/18*	Dress Shops	43	280	Interior	33	360
Banks	50	240	Drug Stores	80	150	Medical Centers	28	425
Barber Shops	48	250	Factories	40	300	Motels	28	425
Bars & Taverns	133	90	High Rise Office—Ext. Rms.	46	263	Office (small suite)	43	280
Beauty Parlors	66	180	Interior Rooms	37	325	Post Office, Individual Office	42	285
Bowling Alleys	68	175	Hospitals, Core	43	280	Central Area	46	260
Churches	36	330/20*	Perimeter	46	260	Residences	20	600
Cocktail Lounges	68	175	Hotel, Guest Rooms	44	275	Restaurants	60	200
Computer Rooms	141	85	Corridors	30	400	Schools & Colleges	46	260
Dental Offices	52	230	Public Spaces	55	220	Shoe Stores	55	220
Dept. Stores, Basement	34	350	Industrial Plants, Offices	38	320	Shop'g. Ctrs., Supermarkets	34	350
Main Floor	40	300	General Offices	34	350	Retail Stores	48	250
Upper Floor	30	400	Plant Areas	40	300	Specialty	60	200

*Persons per ton

12,000 BTU = 1 ton of air conditioning

R260519-90 Wire

Wire quantities are taken off by either measuring each cable run or by extending the conduit and raceway quantities times the number of conductors in the raceway. Ten percent should be added for waste and tie-ins. Keep in mind that the unit of measure of wire is C.L.F. not L.F. as in raceways so the formula would read:

$$\frac{\text{(L.F. Raceway x No. of Conductors) x 1.10}}{100} = \text{C.L.F.}$$

Price per C.L.F. of wire includes:
1. Setting up wire coils or spools on racks
2. Attaching wire to pull in means
3. Measuring and cutting wire
4. Pulling wire into a raceway
5. Identifying and tagging

Price does not include:
1. Connections to breakers, panelboards, or equipment
2. Splices

Job Conditions: Productivity is based on new construction to a height of 15′ using rolling staging in an unobstructed area. Material staging is assumed to be within 100′ of work being performed.

Economy of Scale: If more than three wires at a time are being pulled, deduct the following percentages from the labor of that grouping:

4-5 wires	25%
6-10 wires	30%
11-15 wires	35%
over 15	40%

If a wire pull is less than 100′ in length and is interrupted several times by boxes, lighting outlets, etc., it may be necessary to add the following lengths to each wire being pulled:

Junction box to junction box	2 L.F.
Lighting panel to junction box	6 L.F.
Distribution panel to sub panel	8 L.F.
Switchboard to distribution panel	12 L.F.
Switchboard to motor control center	20 L.F.
Switchboard to cable tray	40 L.F.

Measure of Drops and Riser: It is important when taking off wire quantities to include the wire for drops to electrical equipment. If heights of electrical equipment are not clearly stated, use the following guide:

	Bottom A.F.F.	Top A.F.F.	Inside Cabinet
Safety switch to 100A	5′	6′	2′
Safety switch 400 to 600A	4′	6′	3′
100A panel 12 to 30 circuit	4′	6′	3′
42 circuit panel	3′	6′	4′
Switch box	3′	3′6″	1′
Switchgear	0′	8′	8′
Motor control centers	0′	8′	8′
Transformers - wall mount	4′	8′	2′
Transformers - floor mount	0′	12′	4′

R260519-92 Minimum Copper and Aluminum Wire Size Allowed for Various Types of Insulation

	Minimum Wire Sizes								
	Copper		Aluminum			Copper		Aluminum	
Amperes	THW THWN or XHHW	THHN XHHW *	THW XHHW	THHN XHHW *	Amperes	THW THWN or XHHW	THHN XHHW *	THW XHHW	THHN XHHW *
15A	#14	#14	#12	#12	195	3/0	2/0	250kcmil	4/0
20	#12	#12	#10	#10	200	3/0	3/0	250kcmil	4/0
25	#10	#10	#10	#10	205	4/0	3/0	250kcmil	4/0
30	#10	#10	#8	#8	225	4/0	3/0	300kcmil	250kcmil
40	#8	#8	#8	#8	230	4/0	4/0	300kcmil	250kcmil
45	#8	#8	#6	#8	250	250kcmil	4/0	350kcmil	300kcmil
50	#8	#8	#6	#6	255	250kcmil	4/0	400kcmil	300kcmil
55	#6	#8	#4	#6	260	300kcmil	4/0	400kcmil	350kcmil
60	#6	#6	#4	#6	270	300kcmil	250kcmil	400kcmil	350kcmil
65	#6	#6	#4	#4	280	300kcmil	250kcmil	500kcmil	350kcmil
75	#4	#6	#3	#4	285	300kcmil	250kcmil	500kcmil	400kcmil
85	#4	#4	#2	#3	290	350kcmil	250kcmil	500kcmil	400kcmil
90	#3	#4	#2	#2	305	350kcmil	300kcmil	500kcmil	400kcmil
95	#3	#4	#1	#2	310	350kcmil	300kcmil	500kcmil	500kcmil
100	#3	#3	#1	#2	320	400kcmil	300kcmil	600kcmil	500kcmil
110	#2	#3	1/0	#1	335	400kcmil	350kcmil	600kcmil	500kcmil
115	#2	#2	1/0	#1	340	500kcmil	350kcmil	600kcmil	500kcmil
120	#1	#2	1/0	1/0	350	500kcmil	350kcmil	700kcmil	500kcmil
130	#1	#2	2/0	1/0	375	500kcmil	400kcmil	700kcmil	600kcmil
135	1/0	#1	2/0	1/0	380	500kcmil	400kcmil	750kcmil	600kcmil
150	1/0	#1	3/0	2/0	385	600kcmil	500kcmil	750kcmil	600kcmil
155	2/0	1/0	3/0	3/0	420	600kcmil	500kcmil		700kcmil
170	2/0	1/0	4/0	3/0	430		500kcmil		750kcmil
175	2/0	2/0	4/0	3/0	435		600kcmil		750kcmil
180	3/0	2/0	4/0	4/0	475		600kcmil		

*Dry Locations Only

Notes:
1. Size #14 to 4/0 is in AWG units (American Wire Gauge).
2. Size 250 to 750 is in kcmil units (Thousand Circular Mils).
3. Use next higher ampere value if exact value is not listed in table.
4. For loads that operate continuously increase ampere value by 25% to obtain proper wire size.
5. Table R260519-92 has been written for estimating only. It is based on an ambient temperature of 30°C (86° F). For ambient temperature other than 30°C (86° F), ampacity correction factors will be applied.

Reprinted with permission from NFPA 70-2014, *National Electrical Code®*, Copyright © 2013, National Fire Protection Association, Quincy, MA. This reprinted material is not the complete and official position of the NFPA on the referenced subject, which is represented solely by the standard in its entirety. NFPA 70®, *National Electrical Code* and *NEC®* are registered trademarks of the National Fire Protection Association, Quincy, MA.

Reference Tables

R260533-22   Conductors in Conduit

The table below lists the maximum number of conductors for various sized conduit using THW, TW or THWN insulations.

Copper Wire Size	1/2" TW	1/2" THW	1/2" THWN	3/4" TW	3/4" THW	3/4" THWN	1" TW	1" THW	1" THWN	1-1/4" TW	1-1/4" THW	1-1/4" THWN	1-1/2" TW	1-1/2" THW	1-1/2" THWN	2" TW	2" THW	2" THWN	2-1/2" TW	2-1/2" THW	2-1/2" THWN	3" THW	3" THWN	3-1/2" THW	3-1/2" THWN	4" THW	4" THWN
#14	9	6	13	15	10	24	25	16	39	44	29	69	60	40	94	99	65	154	142	93		143		192			
#12	7	4	10	12	8	18	19	13	29	35	24	51	47	32	70	78	53	114	111	76	164	117		157			
#10	5	4	6	9	6	11	15	11	18	26	19	32	36	26	44	60	43	73	85	61	104	95	160	127		163	
#8	2	1	3	4	3	5	7	5	9	12	10	16	17	13	22	28	22	36	40	32	51	49	79	66	106	85	136
#6		1	1		2	4		4	6		7	11		10	15		16	26		23	37	36	57	48	76	62	98
#4		1	1		1	2		3	4		5	7		7	9		12	16		17	22	27	35	36	47	47	60
#3		1	1		1	1		2	3		4	6		6	8		10	13		15	19	23	29	31	39	40	51
#2		1	1		1	1		2	3		4	5		5	7		9	11		13	16	20	25	27	33	34	43
#1					1	1		1	1		3	3		4	5		6	8		9	12	14	18	19	25	25	32
1/0					1	1		1	1		2	3		3	4		5	7		8	10	12	15	16	21	21	27
2/0					1	1		1	1		1	2		3	3		5	6		7	8	10	13	14	17	18	22
3/0					1	1		1	1		1	1		2	3		4	5		6	7	9	11	12	14	15	18
4/0						1		1	1		1	1		1	2		3	4		5	6	7	9	10	12	13	15
250 kcmil								1	1		1	1		1	1		2	3		4	4	6	7	8	10	10	12
300								1	1		1	1		1	1		2	3		3	4	5	6	7	8	9	11
350									1		1	1		1	1		1	2		3	3	4	5	6	7	8	9
400											1	1		1	1		1	1		2	3	4	5	5	6	7	8
500											1	1		1	1		1	1		1	2	3	4	4	5	6	7
600												1		1	1		1	1		1	1	3	3	4	4	5	5
700														1	1		1	1		1	1	2	3	3	4	4	5
750														1	1		1	1		1	1	2	2	3	3	4	4

R312316-40　Excavating

The selection of equipment used for structural excavation and bulk excavation or for grading is determined by the following factors.
1. Quantity of material
2. Type of material
3. Depth or height of cut
4. Length of haul
5. Condition of haul road
6. Accessibility of site
7. Moisture content and dewatering requirements
8. Availability of excavating and hauling equipment

Some additional costs must be allowed for hand trimming the sides and bottom of concrete pours and other excavation below the general excavation.

Number of B.C.Y. per truck = 1.5 C.Y. bucket × 8 passes = 12 loose C.Y.

$$= 12 \times \frac{100}{118} = 10.2 \text{ B.C.Y. per truck}$$

Truck Haul Cycle:

Load truck, 8 passes	=	4 minutes
Haul distance, 1 mile	=	9 minutes
Dump time	=	2 minutes
Return, 1 mile	=	7 minutes
Spot under machine	=	1 minute
		23 minute cycle

Add the mobilization and demobilization costs to the total excavation costs. When equipment is rented for more than three days, there is often no mobilization charge by the equipment dealer. On larger jobs outside of urban areas, scrapers can move earth economically provided a dump site or fill area and adequate haul roads are available. Excavation within sheeting bracing or cofferdam bracing is usually done with a clamshell and production

When planning excavation and fill, the following should also be considered.
1. Swell factor
2. Compaction factor
3. Moisture content
4. Density requirements

A typical example for scheduling and estimating the cost of excavation of a 15′ deep basement on a dry site when the material must be hauled off the site is outlined below.

Assumptions:
1. Swell factor, 18%
2. No mobilization or demobilization
3. Allowance included for idle time and moving on job
4. No dewatering, sheeting, or bracing
5. No truck spotter or hand trimming

Fleet Haul Production per day in B.C.Y.

$$4 \text{ trucks} \times \frac{50 \text{ min. hour}}{23 \text{ min. haul cycle}} \times 8 \text{ hrs.} \times 10.2 \text{ B.C.Y.}$$

$$= 4 \times 2.2 \times 8 \times 10.2 = 718 \text{ B.C.Y./day}$$

is low, since the clamshell may have to be guided by hand between the bracing. When excavating or filling an area enclosed with a wellpoint system, add 10% to 15% to the cost to allow for restricted access. When estimating earth excavation quantities for structures, allow work space outside the building footprint for construction of the foundation and a slope of 1:1 unless sheeting is used.

867

R312316-45   Excavating Equipment

The table below lists theoretical hourly production in C.Y./hr. bank measure for some typical excavation equipment. Figures assume 50 minute hours, 83% job efficiency, 100% operator efficiency, 90° swing and properly sized hauling units, which must be modified for adverse digging and loading conditions. Actual production costs in the front of the data set average about 50% of the theoretical values listed here.

Equipment	Soil Type	B.C.Y. Weight	% Swell	1 C.Y.	1-1/2 C.Y.	2 C.Y.	2-1/2 C.Y.	3 C.Y.	3-1/2 C.Y.	4 C.Y.
Hydraulic Excavator	Moist loam, sandy clay	3400 lb.	40%	165	195	200	275	330	385	440
"Backhoe"	Sand and gravel	3100	18	140	170	225	240	285	330	380
15' Deep Cut	Common earth	2800	30	150	180	230	250	300	350	400
	Clay, hard, dense	3000	33	120	140	190	200	240	260	320
Power Shovel Optimum Cut (Ft.)	Moist loam, sandy clay	3400	40	170 (6.0)	245 (7.0)	295 (7.8)	335 (8.4)	385 (8.8)	435 (9.1)	475 (9.4)
	Sand and gravel	3100	18	165 (6.0)	225 (7.0)	275 (7.8)	325 (8.4)	375 (8.8)	420 (9.1)	460 (9.4)
	Common earth	2800	30	145 (7.8)	200 (9.2)	250 (10.2)	295 (11.2)	335 (12.1)	375 (13.0)	425 (13.8)
	Clay, hard, dense	3000	33	120 (9.0)	175 (10.7)	220 (12.2)	255 (13.3)	300 (14.2)	335 (15.1)	375 (16.0)
Drag Line Optimum Cut (Ft.)	Moist loam, sandy clay	3400	40	130 (6.6)	180 (7.4)	220 (8.0)	250 (8.5)	290 (9.0)	325 (9.5)	385 (10.0)
	Sand and gravel	3100	18	130 (6.6)	175 (7.4)	210 (8.0)	245 (8.5)	280 (9.0)	315 (9.5)	375 (10.0)
	Common earth	2800	30	110 (8.0)	160 (9.0)	190 (9.9)	220 (10.5)	250 (11.0)	280 (11.5)	310 (12.0)
	Clay, hard, dense	3000	33	90 (9.3)	130 (10.7)	160 (11.8)	190 (12.3)	225 (12.8)	250 (13.3)	280 (12.0)

				Wheel Loaders				Track Loaders		
				3 C.Y.	4 C.Y.	6 C.Y.	8 C.Y.	2-1/4 C.Y.	3 C.Y.	4 C.Y.
Loading Tractors	Moist loam, sandy clay	3400	40	260	340	510	690	135	180	250
	Sand and gravel	3100	18	245	320	480	650	130	170	235
	Common earth	2800	30	230	300	460	620	120	155	220
	Clay, hard, dense	3000	33	200	270	415	560	110	145	200
	Rock, well-blasted	4000	50	180	245	380	520	100	130	180

R312319-90   Wellpoints

A single stage wellpoint system is usually limited to dewatering an average 15' depth below normal ground water level. Multi-stage systems are employed for greater depth with the pumping equipment installed only at the lowest header level. Ejectors with unlimited lift capacity can be economical when two or more stages of wellpoints can be replaced or when horizontal clearance is restricted, such as in deep trenches or tunneling projects, and where low water flows are expected. Wellpoints are usually spaced on 2-1/2' to 10' centers along a header pipe. Wellpoint spacing, header size, and pump size are all determined by the expected flow as dictated by soil conditions.

In almost all soils encountered in wellpoint dewatering, the wellpoints may be jetted into place. Cemented soils and stiff clays may require sand wicks about 12" in diameter around each wellpoint to increase efficiency and eliminate weeping into the excavation. These sand wicks require 1/2 to 3 C.Y. of washed filter sand and are installed by using a 12" diameter steel casing and hole puncher jetted into the ground 2' deeper than the wellpoint. Rock may require predrilled holes.

Labor required for the complete installation and removal of a single stage wellpoint system is in the range of 3/4 to 2 labor-hours per linear foot of header, depending upon jetting conditions, wellpoint spacing, etc.

Continuous pumping is necessary except in some free draining soil where temporary flooding is permissible (as in trenches which are backfilled after each day's work). Good practice requires provision of a stand-by pump during the continuous pumping operation.

Systems for continuous trenching below the water table should be installed three to four times the length of expected daily progress to ensure uninterrupted digging, and header pipe size should not be changed during the job.

For pervious free draining soils, deep wells in place of wellpoints may be economical because of lower installation and maintenance costs. Daily production ranges between two to three wells per day, for 25' to 40' depths, to one well per day for depths over 50'.

Detailed analysis and estimating for any dewatering problem is available at no cost from wellpoint manufacturers. Major firms will quote "sufficient equipment" quotes or their affiliates will offer lump sum proposals to cover complete dewatering responsibility.

Description for 200' System with 8" Header		Quantities
Equipment & Material	Wellpoints 25' long, 2" diameter @ 5' O.C.	40 Each
	Header pipe, 8" diameter	200 L.F.
	Discharge pipe, 8" diameter	100 L.F.
	8" valves	3 Each
	Combination jetting & wellpoint pump (standby)	1 Each
	Wellpoint pump, 8" diameter	1 Each
	Transportation to and from site	1 Day
	Fuel for 30 days x 60 gal./day	1800 Gallons
	Lubricants for 30 days x 16 lbs./day	480 Lbs.
	Sand for points	40 C.Y.
Labor	Technician to supervise installation	1 Week
	Labor for installation and removal of system	300 Labor-hours
	4 Operators straight time 40 hrs./wk. for 4.33 wks.	693 Hrs.
	4 Operators overtime 2 hrs./wk. for 4.33 wks.	35 Hrs.

R312323-30   Compacting Backfill

Compaction of fill in embankments, around structures, in trenches, and under slabs is important to control settlement. Factors affecting compaction are:
1. Soil gradation
2. Moisture content
3. Equipment used
4. Depth of fill per lift
5. Density required

Production Rate:

$$\frac{1.75' \text{ plate width} \times 50 \text{ F.P.M.} \times 50 \text{ min./hr.} \times .67' \text{ lift}}{27 \text{ C.F. per C.Y.}} = 108.5 \text{ C.Y./hr.}$$

Production Rate for 4 Passes:

$$\frac{108.5 \text{ C.Y.}}{4 \text{ passes}} = 27.125 \text{ C.Y./hr.} \times 8 \text{ hrs.} = 217 \text{ C.Y./day}$$

Example:

Compact granular fill around a building foundation using a 21" wide x 24" vibratory plate in 8" lifts. Operator moves at 50 F.P.M. working a 50 minute hour to develop 95% Modified Proctor Density with 4 passes.

869

Earthwork   R3141 Shoring

R314116-40   Wood Sheet Piling

Wood sheet piling may be used for depths to 20' where there is no ground water. If moderate ground water is encountered Tongue & Groove sheeting will help to keep it out. When considerable ground water is present, steel sheeting must be used.

For estimating purposes on trench excavation, sizes are as follows:

Depth	Sheeting	Wales	Braces	B.F. per S.F.
To 8'	3 x 12's	6 x 8's, 2 line	6 x 8's, @ 10'	4.0 @ 8'
8' x 12'	3 x 12's	10 x 10's, 2 line	10 x 10's, @ 9'	5.0 average
12' to 20'	3 x 12's	12 x 12's, 3 line	12 x 12's, @ 8'	7.0 average

Sheeting to be toed in at least 2' depending upon soil conditions. A five person crew with an air compressor and sheeting driver can drive and brace 440 SF/day at 8' deep, 360 SF/day at 12' deep, and 320 SF/day at 16' deep.

For normal soils, piling can be pulled in 1/3 the time to install. Pulling difficulty increases with the time in the ground. Production can be increased by high pressure jetting.

R314116-45   Steel Sheet Piling

Limiting weights are 22 to 38#/S.F. of wall surface with 27#/S.F. average for usual types and sizes. (Weights of piles themselves are from 30.7#/L.F. to 57#/L.F. but they are 15" to 21" wide.) Lightweight sections 12" to 28" wide from 3 ga. to 12 ga. thick are also available for shallow excavations. Piles may be driven two at a time with an impact or vibratory hammer (use vibratory to pull) hung from a crane without leads. A reasonable estimate of the life of steel sheet piling is 10 uses with up to 125 uses possible if a vibratory hammer is used. Used piling costs from 50% to 80% of new piling depending on location and market conditions. Sheet piling and H piles can be rented for about 30% of the delivered mill price for the first month and 5% per month thereafter. Allow 1 labor-hour per pile for cleaning and trimming after driving. These costs increase with depth and hydrostatic head. Vibratory drivers are faster in wet granular soils and are excellent for pile extraction. Pulling difficulty increases with the time in the ground and may cost more than driving. It is often economical to abandon the sheet piling, especially if it can be used as the outer wall form. Allow about 1/3 additional length or more for toeing into ground. Add bracing, waler and strut costs. Waler costs can equal the cost per ton of sheeting.

Earthwork   R3145 Vibroflotation & Densification

R314513-90   Vibroflotation and Vibro Replacement Soil Compaction

Vibroflotation is a proprietary system of compacting sandy soils in place to increase relative density to about 70%. Typical bearing capacities attained will be 6000 psf for saturated sand and 12,000 psf for dry sand. Usual range is 4000 to 8000 psf capacity. Costs in the front of the data set are for a vertical foot of compacted cylinder 6' to 10' in diameter.

Vibro replacement is a proprietary system of improving cohesive soils in place to increase bearing capacity. Most silts and clays above or below the water table can be strengthened by installation of stone columns.

The process consists of radial displacement of the soil by vibration. The created hole is then backfilled in stages with coarse granular fill which is thoroughly compacted and displaced into the surrounding soil in the form of a column.

The total project cost would depend on the number and depth of the compacted cylinders. The installing company guarantees relative soil density of the sand cylinders after compaction and the bearing capacity of the soil after the replacement process. Detailed estimating information is available from the installer at no cost.

R316326-60 Caissons

The three principal types of caissons are:

(1) Belled Caissons, which except for shallow depths and poor soil conditions, are generally recommended. They provide more bearing than shaft area. Because of its conical shape, no horizontal reinforcement of the bell is required.

(2) Straight Shaft Caissons are used where relatively light loads are to be supported by caissons that rest on high value bearing strata. While the shaft is larger in diameter than for belled types this is more than offset by the savings in time and labor.

(3) Keyed Caissons are used when extremely heavy loads are to be carried. A keyed or socketed caisson transfers its load into rock by a combination of end-bearing and shear reinforcing of the shaft. The most economical shaft often consists of a steel casing, a steel wide flange core and concrete. Allowable compressive stresses of .225 f'c for concrete, 16,000 psi for the wide flange core, and 9,000 psi for the steel casing are commonly used. The usual range of shaft diameter is 18″ to 84″. The number of sizes specified for any one project should be limited due to the problems of casing and auger storage. When hand work is to be performed, shaft diameters should not be less than 32″. When inspection of borings is required a minimum shaft diameter of 30″ is recommended. Concrete caissons are intended to be poured against earth excavation so permanent forms, which add to cost, should not be used if the excavation is clean and the earth is sufficiently impervious to prevent excessive loss of concrete.

Soil Conditions for Belling		
Good	Requires Handwork	Not Recommended
Clay	Hard Shale	Silt
Sandy Clay	Limestone	Sand
Silty Clay	Sandstone	Gravel
Clayey Silt	Weathered Mica	Igneous Rock
Hard-pan		
Soft Shale		
Decomposed Rock		

Exterior Improvements R3292 Turf & Grasses

R329219-50 Seeding

The type of grass is determined by light, shade and moisture content of soil plus intended use. Fertilizer should be disked 4″ before seeding. For steep slopes disk five tons of mulch and lay two tons of hay or straw on surface per acre after seeding. Surface mulch can be staked, lightly disked or tar emulsion sprayed. Material for mulch can be wood chips, peat moss, partially rotted hay or straw, wood fibers and sprayed emulsions. Hemp seed blankets with fertilizer are also available. For spring seeding, watering is necessary. Late fall seeding may have to be reseeded in the spring. Hydraulic seeding, power mulching, and aerial seeding can be used on large areas.

R331113-80 Piping Designations

There are several systems currently in use to describe pipe and fittings. The following paragraphs will help to identify and clarify classifications of piping systems used for water distribution.

Piping may be classified by schedule. Piping schedules include 5S, 10S, 10, 20, 30, Standard, 40, 60, Extra Strong, 80, 100, 120, 140, 160 and Double Extra Strong. These schedules are dependent upon the pipe wall thickness. The wall thickness of a particular schedule may vary with pipe size.

Ductile iron pipe for water distribution is classified by Pressure Classes such as Class 150, 200, 250, 300 and 350. These classes are actually the rated water working pressure of the pipe in pounds per square inch (psi). The pipe in these pressure classes is designed to withstand the rated water working pressure plus a surge allowance of 100 psi.

The American Water Works Association (AWWA) provides standards for various types of **plastic pipe.** C-900 is the specification for polyvinyl chloride (PVC) piping used for water distribution in sizes ranging from 4″ through 12″. C-901 is the specification for polyethylene (PE) pressure pipe, tubing and fittings used for water distribution in sizes ranging from 1/2″ through 3″. C-905 is the specification for PVC piping sizes 14″ and greater.

PVC pressure-rated pipe is identified using the standard dimensional ratio (SDR) method. This method is defined by the American Society for Testing and Materials (ASTM) Standard D 2241. This pipe is available in SDR numbers 64, 41, 32.5, 26, 21, 17, and 13.5. A pipe with an SDR of 64 will have the thinnest wall while a pipe with an SDR of 13.5 will have the thickest wall. When the pressure rating (PR) of a pipe is given in psi, it is based on a line supplying water at 73 degrees F.

The National Sanitation Foundation (NSF) seal of approval is applied to products that can be used with potable water. These products have been tested to ANSI/NSF Standard 14.

Valves and strainers are classified by American National Standards Institute (ANSI) Classes. These Classes are 125, 150, 200, 250, 300, 400, 600, 900, 1500 and 2500. Within each class there is an operating pressure range dependent upon temperature. Design parameters should be compared to the appropriate material dependent, pressure-temperature rating chart for accurate valve selection.

R347216-10 Single Track R.R. Siding

The costs for a single track RR siding in the Unit Price Section include the components shown in the table below.

Description of Component	Qty. per L.F. of Track	Unit
Ballast, 1-1/2″ crushed stone	.667	C.Y.
6″ x 8″ x 8′-6″ Treated timber ties, 22″ O.C.	.545	Ea.
Tie plates, 2 per tie	1.091	Ea.
Track rail	2.000	L.F.
Spikes, 6″, 4 per tie	2.182	Ea.
Splice bars w/ bolts, lock washers & nuts, @ 33′ O.C.	.061	Pair
Crew B-14 @ 57 L.F./Day	.018	Day

R347216-20 Single Track, Steel Ties, Concrete Bed

The costs for a R.R. siding with steel ties and a concrete bed in the Unit Price section include the components shown in the table below.

Description of Component	Qty. per L.F. of Track	Unit
Concrete bed, 9′ wide, 10″ thick	.278	C.Y.
Ties, W6x16 x 6′-6″ long, @ 30″ O.C.	.400	Ea.
Tie plates, 4 per tie	1.600	Ea.
Track rail	2.000	L.F.
Tie plate bolts, 1″, 8 per tie	3.200	Ea.
Splice bars w/bolts, lock washers & nuts, @ 33′ O.C.	.061	Pair
Crew B-14 @ 22 L.F./Day	.045	Day

Change Orders

Change Order Considerations

A change order is a written document usually prepared by the design professional and signed by the owner, the architect/engineer, and the contractor. A change order states the agreement of the parties to: an addition, deletion, or revision in the work; an adjustment in the contract sum, if any; or an adjustment in the contract time, if any. Change orders, or "extras", in the construction process occur after execution of the construction contract and impact architects/engineers, contractors, and owners.

Change orders that are properly recognized and managed can ensure orderly, professional, and profitable progress for everyone involved in the project. There are many causes for change orders and change order requests. In all cases, change orders or change order requests should be addressed promptly and in a precise and prescribed manner. The following paragraphs include information regarding change order pricing and procedures.

The Causes of Change Orders

Reasons for issuing change orders include:

- Unforeseen field conditions that require a change in the work
- Correction of design discrepancies, errors, or omissions in the contract documents
- Owner-requested changes, either by design criteria, scope of work, or project objectives
- Completion date changes for reasons unrelated to the construction process
- Changes in building code interpretations, or other public authority requirements that require a change in the work
- Changes in availability of existing or new materials and products

Procedures

Properly written contract documents must include the correct change order procedures for all parties—owners, design professionals, and contractors—to follow in order to avoid costly delays and litigation.

Being "in the right" is not always a sufficient or acceptable defense. The contract provisions requiring notification and documentation must be adhered to within a defined or reasonable time frame.

The appropriate method of handling change orders is by a written proposal and acceptance by all parties involved. Prior to starting work on a project, all parties should identify their

authorized agents who may sign and accept change orders, as well as any limits placed on their authority.

Time may be a critical factor when the need for a change arises. For such cases, the contractor might be directed to proceed on a "time and materials" basis, rather than wait for all paperwork to be processed—a delay that could impede progress. In this situation, the contractor must still follow the prescribed change order procedures including, but not limited to, notification and documentation.

Lack of documentation can be very costly, especially if legal judgments are to be made, and if certain field personnel are no longer available. For time and material change orders, the contractor should keep accurate daily records of all labor and material allocated to the change.

Owners or awarding authorities who do considerable and continual building construction (such as the federal government) realize the inevitability of change orders for numerous reasons, both predictable and unpredictable. As a result, the federal government, the American Institute of Architects (AIA), the Engineers Joint Contract Documents Committee (EJCDC), and other contractor, legal, and technical organizations have developed standards and procedures to be followed by all parties to achieve contract continuance and timely completion, while being financially fair to all concerned.

Pricing Change Orders

When pricing change orders, regardless of their cause, the most significant factor is when the change occurs. The need for a change may be perceived in the field or requested by the architect/engineer *before* any of the actual installation has begun, or may evolve or appear *during* construction when the item of work in question is partially installed. In the latter cases, the original sequence of construction is disrupted, along with all contiguous and supporting systems. Change orders cause the greatest impact when they occur *after* the installation has been completed and must be uncovered, or even replaced. Post-completion changes may be caused by necessary design changes, product failure, or changes in the owner's requirements that are not discovered until the building or the systems begin to function.

Specified procedures of notification and record keeping must be adhered to and enforced regardless of the stage of construction: *before, during,* or *after* installation. Some bidding documents anticipate change orders by requiring that unit prices including overhead and profit percentages—for additional as well as deductible changes—be listed. Generally these unit prices do not fully take into account the ripple effect, or impact on other trades, and should be used for general guidance only.

When pricing change orders, it is important to classify the time frame in which the change occurs. There are two basic time frames for change orders: *pre-installation change orders*, which occur before the start of construction, and *post-installation change orders*, which involve reworking after the original installation. Change orders that occur between these stages may be priced according to the extent of work completed using a combination of techniques developed for pricing *pre-* and *post-installation* changes.

Factors To Consider When Pricing Change Orders

As an estimator begins to prepare a change order, the following questions should be reviewed to determine their impact on the final price.

General

- *Is the change order work* pre-installation *or* post-installation?

    Change order work costs vary according to how much of the installation has been completed. Once workers have the project scoped in their minds, even though they have not started, it can be difficult to refocus. Consequently they may spend more than the normal amount of time understanding the change. Also, modifications to work in place, such as trimming or refitting, usually take more time than was initially estimated. The greater the amount of work in place, the more reluctant workers are to change it. Psychologically they may resent the change and as a result the rework takes longer than normal. Post-installation change order estimates must include demolition of existing work as required to accomplish the change. If the work is performed at a later time, additional obstacles, such as building finishes, may be present which must be protected. Regardless of whether the change occurs

pre-installation or post-installation, attempt to isolate the identifiable factors and price them separately. For example, add shipping costs that may be required pre-installation or any demolition required post-installation. Then analyze the potential impact on productivity of psychological and/or learning curve factors and adjust the output rates accordingly. One approach is to break down the typical workday into segments and quantify the impact on each segment.

Change Order Installation Efficiency

The labor-hours expressed (for new construction) are based on average installation time, using an efficiency level. For change order situations, adjustments to this efficiency level should reflect the daily labor-hour allocation for that particular occurrence.

- *Will the change substantially delay the original completion date?*

A significant change in the project may cause the original completion date to be extended. The extended schedule may subject the contractor to new wage rates dictated by relevant labor contracts. Project supervision and other project overhead must also be extended beyond the original completion date. The schedule extension may also put installation into a new weather season. For example, underground piping scheduled for October installation was delayed until January. As a result, frost penetrated the trench area, thereby changing the degree of difficulty of the task. Changes and delays may have a ripple effect throughout the project. This effect must be analyzed and negotiated with the owner.

- *What is the net effect of a deduct change order?*

In most cases, change orders resulting in a deduction or credit reflect only bare costs. The contractor may retain the overhead and profit based on the original bid.

Materials

- *Will you have to pay more or less for the new material, required by the change order, than you paid for the original purchase?*

The same material prices or discounts will usually apply to materials purchased for change orders as new construction. In some

instances, however, the contractor may forfeit the advantages of competitive pricing for change orders. Consider the following example:

A contractor purchased over $20,000 worth of fan coil units for an installation and obtained the maximum discount. Some time later it was determined the project required an additional matching unit. The contractor has to purchase this unit from the original supplier to ensure a match. The supplier at this time may not discount the unit because of the small quantity, and he is no longer in a competitive situation. The impact of quantity on purchase can add between 0% and 25% to material prices and/or subcontractor quotes.

- *If materials have been ordered or delivered to the job site, will they be subject to a cancellation charge or restocking fee?*

Check with the supplier to determine if ordered materials are subject to a cancellation charge. Delivered materials not used as a result of a change order may be subject to a restocking fee if returned to the supplier. Common restocking charges run between 20% and 40%. Also, delivery charges to return the goods to the supplier must be added.

Labor

- *How efficient is the existing crew at the actual installation?*

Is the same crew that performed the initial work going to do the change order? Possibly the change consists of the installation of a unit identical to one already installed; therefore, the change should take less time. Be sure to consider this potential productivity increase and modify the productivity rates accordingly.

- *If the crew size is increased, what impact will that have on supervision requirements?*

Under most bargaining agreements or management practices, there is a point at which a working foreman is replaced by a nonworking foreman. This replacement increases project overhead by adding a nonproductive worker. If additional workers are added to accelerate the project or to perform changes while maintaining the schedule, be sure to add additional supervision time if warranted. Calculate the

hours involved and the additional cost directly if possible.

- *What are the other impacts of increased crew size?*

The larger the crew, the greater the potential for productivity to decrease. Some of the factors that cause this productivity loss are: overcrowding (producing restrictive conditions in the working space) and possibly a shortage of any special tools and equipment required. Such factors affect not only the crew working on the elements directly involved in the change order, but other crews whose movements may also be hampered. As the crew increases, check its basic composition for changes by the addition or deletion of apprentices or nonworking foreman, and quantify the potential effects of equipment shortages or other logistical factors.

- *As new crews, unfamiliar with the project, are brought onto the site, how long will it take them to become oriented to the project requirements?*

The orientation time for a new crew to become 100% effective varies with the site and type of project. Orientation is easiest at a new construction site and most difficult at existing, very restrictive renovation sites. The type of work also affects orientation time. When all elements of the work are exposed, such as concrete or masonry work, orientation is decreased. When the work is concealed or less visible, such as existing electrical systems, orientation takes longer. Usually orientation can be accomplished in one day or less. Costs for added orientation should be itemized and added to the total estimated cost.

- *How much actual production can be gained by working overtime?*

Short term overtime can be used effectively to accomplish more work in a day. However, as overtime is scheduled to run beyond several weeks, studies have shown marked decreases in output. The following chart shows the effect of long term overtime on worker efficiency. If the anticipated change requires extended overtime to keep the job on schedule, these factors can be used as a guide to predict the impact on time and cost. Add project overhead, particularly supervision, that may also be incurred.

Days per Week	Hours per Day	Production Efficiency					Payroll Cost Factors	
		1st Week	2nd Week	3rd Week	4th Week	Average 4 Weeks	@ 1-1/2 Times	@ 2 Times
5	8	100%	100%	100%	100%	100%	100%	100%
	9	100	100	95	90	96.25	105.6	111.1
	10	100	95	90	85	92.50	110.0	120.0
	11	95	90	75	65	81.25	113.6	127.3
	12	90	85	70	60	76.25	116.7	133.3
6	8	100	100	95	90	96.25	108.3	116.7
	9	100	95	90	85	92.50	113.0	125.9
	10	95	90	85	80	87.50	116.7	133.3
	11	95	85	70	65	78.75	119.7	139.4
	12	90	80	65	60	73.75	122.2	144.4
7	8	100	95	85	75	88.75	114.3	128.6
	9	95	90	80	70	83.75	118.3	136.5
	10	90	85	75	65	78.75	121.4	142.9
	11	85	80	65	60	72.50	124.0	148.1
	12	85	75	60	55	68.75	126.2	152.4

Effects of Overtime

Caution: Under many labor agreements, Sundays and holidays are paid at a higher premium than the normal overtime rate.

The use of long-term overtime is counterproductive on almost any construction job; that is, the longer the period of overtime, the lower the actual production rate. Numerous studies have been conducted, and while they have resulted in slightly different numbers, all reach the same conclusion. The figure above tabulates the effects of overtime work on efficiency.

As illustrated, there can be a difference between the *actual* payroll cost per hour and the *effective* cost per hour for overtime work. This is due to the reduced production efficiency with the increase in weekly hours beyond 40. This difference between actual and effective cost results from overtime work over a prolonged period. Short-term overtime work does not result in as great a reduction in efficiency and, in such cases, effective cost may not vary significantly from the actual payroll cost. As the total hours per week are increased on a regular basis, more time is lost due to fatigue, lowered morale, and an increased accident rate.

As an example, assume a project where workers are working 6 days a week, 10 hours per day. From the figure above (based on productivity studies), the average effective productive hours over a 4-week period are:

$$0.875 \times 60 = 52.5$$

Depending upon the locale and day of week, overtime hours may be paid at time and a half or double time. For time and a half, the overall (average) *actual* payroll cost (including regular and overtime hours) is determined as follows:

$$\frac{40 \text{ reg. hrs.} + (20 \text{ overtime hrs.} \times 1.5)}{60 \text{ hrs.}} = 1.167$$

Based on 60 hours, the payroll cost per hour will be 116.7% of the normal rate at 40 hours per week. However, because the effective production (efficiency) for 60 hours is reduced to the equivalent of 52.5 hours, the effective cost of overtime is calculated as follows:

For time and a half:

$$\frac{40 \text{ reg. hrs.} + (20 \text{ overtime hrs.} \times 1.5)}{52.5 \text{ hrs.}} = 1.33$$

The installed cost will be 133% of the normal rate (for labor).

Thus, when figuring overtime, the actual cost per unit of work will be higher than the apparent overtime payroll dollar increase, due to the reduced productivity of the longer work week. These efficiency calculations are true only for those cost factors determined by hours worked. Costs that are applied weekly or monthly, such as equipment rentals, will not be similarly affected.

Equipment

- *What equipment is required to complete the change order?*

Change orders may require extending the rental period of equipment already on the job site, or the addition of special equipment brought in to accomplish the change work. In either case, the additional rental charges and operator labor charges must be added.

Summary

The preceding considerations and others you deem appropriate should be analyzed and applied to a change order estimate. The impact of each should be quantified and listed on the estimate to form an audit trail.

Change orders that are properly identified, documented, and managed help to ensure the orderly, professional, and profitable progress of the work. They also minimize potential claims or disputes at the end of the project.

Square Foot Costs

Back by customer demand!

You asked and we listened. For customer convenience and estimating ease, we have made the 2017 Square Foot Costs available for download at **http://info.thegordiangroup.com/2017-Cost-Data-Companion.html**. You will also find sample estimates, an RSMeans data overview video and book registration form to receive quarterly data updates throughout 2017.

Estimating Tips

- The cost figures available in the download were derived from hundreds of projects contained in the RSMeans database of completed construction projects. They include the contractor's overhead and profit. The figures have been adjusted to January of the current year.

- These projects were located throughout the U.S. and reflect a tremendous variation in square foot (S.F.) costs. This is due to differences, not only in labor and material costs, but also in individual owners' requirements. For instance, a bank in a large city would have different features than one in a rural area. This is true of all the different types of buildings analyzed. Therefore, caution should be exercised when using these square foot costs. For example, for courthouses, costs in the database are local courthouse costs and will not apply to the larger, more elaborate federal courthouses.

- None of the figures "go with" any others. All individual cost items were computed and tabulated separately. Thus, the sum of the median figures for plumbing, HVAC, and electrical will not normally total up to the total mechanical and electrical costs arrived at by separate analysis and tabulation of the projects.

- Each building was analyzed as to total and component costs and percentages. The figures were arranged in ascending order with the results tabulated as shown. The 1/4 column shows that 25% of the projects had lower costs and 75% had higher. The 3/4 column shows that 75% of the projects had lower costs and 25% had higher. The median column shows that 50% of the projects had lower costs and 50% had higher.

- Square foot costs are useful in the conceptual stage when no details are available. As soon as details become available in the project design, the square foot approach should be discontinued and the project priced as to its particular components. When more precision is required, or for estimating the replacement cost of specific buildings, the current edition of *RSMeans Square Foot Costs* should be used.

- In using the figures in this section, it is recommended that the median column be used for preliminary figures if no additional information is available. The median figures, when multiplied by the total city construction cost index figures (see City Cost Indexes) and then multiplied by the project size modifier at the end of this section, should present a fairly accurate base figure, which would then have to be adjusted in view of the estimator's experience, local economic conditions, code requirements, and the owner's particular requirements. There is no need to factor the percentage figures, as these should remain constant from city to city.

- The editors of this data would greatly appreciate receiving cost figures on one or more of your recent projects, which would then be included in the averages for next year. All cost figures received will be kept confidential, except that they will be averaged with other similar projects to arrive at square foot cost figures for next year.

See the website above for details and the discount available for submitting one or more of your projects.

A	Area Square Feet; Ampere	Brk., brk	Brick
AAFES	Army and Air Force Exchange Service	brkt	Bracket
		Brs.	Brass
ABS	Acrylonitrile Butadiene Stryrene; Asbestos Bonded Steel	Brz.	Bronze
		Bsn.	Basin
A.C., AC	Alternating Current; Air-Conditioning; Asbestos Cement; Plywood Grade A & C	Btr.	Better
		BTU	British Thermal Unit
		BTUH	BTU per Hour
		Bu.	Bushels
ACI	American Concrete Institute	BUR	Built-up Roofing
ACR	Air Conditioning Refrigeration	BX	Interlocked Armored Cable
ADA	Americans with Disabilities Act	°C	Degree Centigrade
AD	Plywood, Grade A & D	c	Conductivity, Copper Sweat
Addit.	Additional	C	Hundred; Centigrade
Adj.	Adjustable	C/C	Center to Center, Cedar on Cedar
af	Audio-frequency	C-C	Center to Center
AFFF	Aqueous Film Forming Foam	Cab	Cabinet
AFUE	Annual Fuel Utilization Efficiency	Cair.	Air Tool Laborer
AGA	American Gas Association	Cal.	Caliper
Agg.	Aggregate	Calc	Calculated
A.H., Ah	Ampere Hours	Cap.	Capacity
A hr.	Ampere-hour	Carp.	Carpenter
A.H.U., AHU	Air Handling Unit	C.B.	Circuit Breaker
A.I.A.	American Institute of Architects	C.C.A.	Chromate Copper Arsenate
AIC	Ampere Interrupting Capacity	C.C.F.	Hundred Cubic Feet
Allow.	Allowance	cd	Candela
alt., alt	Alternate	cd/sf	Candela per Square Foot
Alum.	Aluminum	CD	Grade of Plywood Face & Back
a.m.	Ante Meridiem	CDX	Plywood, Grade C & D, exterior glue
Amp.	Ampere		
Anod.	Anodized	Cefi.	Cement Finisher
ANSI	American National Standards Institute	Cem.	Cement
		CF	Hundred Feet
APA	American Plywood Association	C.F.	Cubic Feet
Approx.	Approximate	CFM	Cubic Feet per Minute
Apt.	Apartment	CFRP	Carbon Fiber Reinforced Plastic
Asb.	Asbestos	c.g.	Center of Gravity
A.S.B.C.	American Standard Building Code	CHW	Chilled Water; Commercial Hot Water
Asbe.	Asbestos Worker		
ASCE	American Society of Civil Engineers	C.I., CI	Cast Iron
A.S.H.R.A.E.	American Society of Heating, Refrig. & AC Engineers	C.I.P., CIP	Cast in Place
		Circ.	Circuit
ASME	American Society of Mechanical Engineers	C.L.	Carload Lot
		CL	Chain Link
ASTM	American Society for Testing and Materials	Clab.	Common Laborer
		Clam	Common Maintenance Laborer
Attchmt.	Attachment	C.L.F.	Hundred Linear Feet
Avg., Ave.	Average	CLF	Current Limiting Fuse
AWG	American Wire Gauge	CLP	Cross Linked Polyethylene
AWWA	American Water Works Assoc.	cm	Centimeter
Bbl.	Barrel	CMP	Corr. Metal Pipe
B&B, BB	Grade B and Better; Balled & Burlapped	CMU	Concrete Masonry Unit
		CN	Change Notice
B&S	Bell and Spigot	Col.	Column
B.&W.	Black and White	CO_2	Carbon Dioxide
b.c.c.	Body-centered Cubic	Comb.	Combination
B.C.Y.	Bank Cubic Yards	comm.	Commercial, Communication
BE	Bevel End	Compr.	Compressor
B.F.	Board Feet	Conc.	Concrete
Bg. cem.	Bag of Cement	Cont., cont	Continuous; Continued, Container
BHP	Boiler Horsepower; Brake Horsepower	Corr.	Corrugated
		Cos	Cosine
B.I.	Black Iron	Cot	Cotangent
bidir.	bidirectional	Cov.	Cover
Bit., Bitum.	Bituminous	C/P	Cedar on Paneling
Bit., Conc.	Bituminous Concrete	CPA	Control Point Adjustment
Bk.	Backed	Cplg.	Coupling
Bkrs.	Breakers	CPM	Critical Path Method
Bldg., bldg	Building	CPVC	Chlorinated Polyvinyl Chloride
Blk.	Block	C.Pr.	Hundred Pair
Bm.	Beam	CRC	Cold Rolled Channel
Boil.	Boilermaker	Creos.	Creosote
bpm	Blows per Minute	Crpt.	Carpet & Linoleum Layer
BR	Bedroom	CRT	Cathode-ray Tube
Brg., brng.	Bearing	CS	Carbon Steel, Constant Shear Bar Joist
Brhe.	Bricklayer Helper		
Bric.	Bricklayer		

Csc	Cosecant
C.S.F.	Hundred Square Feet
CSI	Construction Specifications Institute
CT	Current Transformer
CTS	Copper Tube Size
Cu	Copper, Cubic
Cu. Ft.	Cubic Foot
cw	Continuous Wave
C.W.	Cool White; Cold Water
Cwt.	100 Pounds
C.W.X.	Cool White Deluxe
C.Y.	Cubic Yard (27 cubic feet)
C.Y./Hr.	Cubic Yard per Hour
Cyl.	Cylinder
d	Penny (nail size)
D	Deep; Depth; Discharge
Dis., Disch.	Discharge
Db	Decibel
Dbl.	Double
DC	Direct Current
DDC	Direct Digital Control
Demob.	Demobilization
d.f.t.	Dry Film Thickness
d.f.u.	Drainage Fixture Units
D.H.	Double Hung
DHW	Domestic Hot Water
DI	Ductile Iron
Diag.	Diagonal
Diam., Dia	Diameter
Distrib.	Distribution
Div.	Division
Dk.	Deck
D.L.	Dead Load; Diesel
DLH	Deep Long Span Bar Joist
dlx	Deluxe
Do.	Ditto
DOP	Dioctyl Phthalate Penetration Test (Air Filters)
Dp., dp	Depth
D.P.S.T.	Double Pole, Single Throw
Dr.	Drive
DR	Dimension Ratio
Drink.	Drinking
D.S.	Double Strength
D.S.A.	Double Strength A Grade
D.S.B.	Double Strength B Grade
Dty.	Duty
DWV	Drain Waste Vent
DX	Deluxe White, Direct Expansion
dyn	Dyne
e	Eccentricity
E	Equipment Only; East; Emissivity
Ea.	Each
EB	Encased Burial
Econ.	Economy
E.C.Y	Embankment Cubic Yards
EDP	Electronic Data Processing
EIFS	Exterior Insulation Finish System
E.D.R.	Equiv. Direct Radiation
Eq.	Equation
EL	Elevation
Elec.	Electrician; Electrical
Elev.	Elevator; Elevating
EMT	Electrical Metallic Conduit; Thin Wall Conduit
Eng.	Engine, Engineered
EPDM	Ethylene Propylene Diene Monomer
EPS	Expanded Polystyrene
Eqhv.	Equip. Oper., Heavy
Eqlt.	Equip. Oper., Light
Eqmd.	Equip. Oper., Medium
Eqmm.	Equip. Oper., Master Mechanic
Eqol.	Equip. Oper., Oilers
Equip.	Equipment
ERW	Electric Resistance Welded

877

E.S.	Energy Saver	H	High Henry	Lath.	Lather
Est.	Estimated	HC	High Capacity	Lav.	Lavatory
esu	Electrostatic Units	H.D., HD	Heavy Duty; High Density	lb.; #	Pound
E.W.	Each Way	H.D.O.	High Density Overlaid	L.B., LB	Load Bearing; L Conduit Body
EWT	Entering Water Temperature	HDPE	High Density Polyethylene Plastic	L. & E.	Labor & Equipment
Excav.	Excavation	Hdr.	Header	lb./hr.	Pounds per Hour
excl	Excluding	Hdwe.	Hardware	lb./L.F.	Pounds per Linear Foot
Exp., exp	Expansion, Exposure	H.I.D., HID	High Intensity Discharge	lbf/sq.in.	Pound-force per Square Inch
Ext., ext	Exterior; Extension	Help.	Helper Average	L.C.L.	Less than Carload Lot
Extru.	Extrusion	HEPA	High Efficiency Particulate Air	L.C.Y.	Loose Cubic Yard
f.	Fiber Stress		Filter	Ld.	Load
F	Fahrenheit; Female; Fill	Hg	Mercury	LE	Lead Equivalent
Fab., fab	Fabricated; Fabric	HIC	High Interrupting Capacity	LED	Light Emitting Diode
FBGS	Fiberglass	HM	Hollow Metal	L.F.	Linear Foot
F.C.	Footcandles	HMWPE	High Molecular Weight	L.F. Nose	Linear Foot of Stair Nosing
f.c.c.	Face-centered Cubic		Polyethylene	L.F. Rsr	Linear Foot of Stair Riser
f'c.	Compressive Stress in Concrete;	HO	High Output	Lg.	Long; Length; Large
	Extreme Compressive Stress	Horiz.	Horizontal	L & H	Light and Heat
F.E.	Front End	H.P., HP	Horsepower; High Pressure	LH	Long Span Bar Joist
FEP	Fluorinated Ethylene Propylene	H.P.F.	High Power Factor	L.H.	Labor Hours
	(Teflon)	Hr.	Hour	L.L., LL	Live Load
F.G.	Flat Grain	Hrs./Day	Hours per Day	L.L.D.	Lamp Lumen Depreciation
F.H.A.	Federal Housing Administration	HSC	High Short Circuit	lm	Lumen
Fig.	Figure	Ht.	Height	lm/sf	Lumen per Square Foot
Fin.	Finished	Htg.	Heating	lm/W	Lumen per Watt
FIPS	Female Iron Pipe Size	Htrs.	Heaters	LOA	Length Over All
Fixt.	Fixture	HVAC	Heating, Ventilation & Air-	log	Logarithm
FJP	Finger jointed and primed		Conditioning	L-O-L	Lateralolet
Fl. Oz.	Fluid Ounces	Hvy.	Heavy	long.	Longitude
Flr.	Floor	HW	Hot Water	L.P., LP	Liquefied Petroleum; Low Pressure
FM	Frequency Modulation;	Hyd.; Hydr.	Hydraulic	L.P.F.	Low Power Factor
	Factory Mutual	Hz	Hertz (cycles)	LR	Long Radius
Fmg.	Framing	I.	Moment of Inertia	L.S.	Lump Sum
FM/UL	Factory Mutual/Underwriters Labs	IBC	International Building Code	Lt.	Light
Fdn.	Foundation	I.C.	Interrupting Capacity	Lt. Ga.	Light Gauge
FNPT	Female National Pipe Thread	ID	Inside Diameter	L.T.L.	Less than Truckload Lot
Fori.	Foreman, Inside	I.D.	Inside Dimension; Identification	Lt. Wt.	Lightweight
Foro.	Foreman, Outside	I.F.	Inside Frosted	L.V.	Low Voltage
Fount.	Fountain	I.M.C.	Intermediate Metal Conduit	M	Thousand; Material; Male;
fpm	Feet per Minute	In.	Inch		Light Wall Copper Tubing
FPT	Female Pipe Thread	Incan.	Incandescent	M²CA	Meters Squared Contact Area
Fr	Frame	Incl.	Included; Including	m/hr.; M.H.	Man-hour
F.R.	Fire Rating	Int.	Interior	mA	Milliampere
FRK	Foil Reinforced Kraft	Inst.	Installation	Mach.	Machine
FSK	Foil/Scrim/Kraft	Insul., insul	Insulation/Insulated	Mag. Str.	Magnetic Starter
FRP	Fiberglass Reinforced Plastic	I.P.	Iron Pipe	Maint.	Maintenance
FS	Forged Steel	I.P.S., IPS	Iron Pipe Size	Marb.	Marble Setter
FSC	Cast Body; Cast Switch Box	IPT	Iron Pipe Threaded	Mat; Mat'l.	Material
Ft., ft	Foot; Feet	I.W.	Indirect Waste	Max.	Maximum
Ftng.	Fitting	J	Joule	MBF	Thousand Board Feet
Ftg.	Footing	J.I.C.	Joint Industrial Council	MBH	Thousand BTU's per hr.
Ft lb.	Foot Pound	K	Thousand; Thousand Pounds;	MC	Metal Clad Cable
Furn.	Furniture		Heavy Wall Copper Tubing, Kelvin	MCC	Motor Control Center
FVNR	Full Voltage Non-Reversing	K.A.H.	Thousand Amp. Hours	M.C.F.	Thousand Cubic Feet
FVR	Full Voltage Reversing	kcmil	Thousand Circular Mils	MCFM	Thousand Cubic Feet per Minute
FXM	Female by Male	KD	Knock Down	M.C.M.	Thousand Circular Mils
Fy.	Minimum Yield Stress of Steel	K.D.A.T.	Kiln Dried After Treatment	MCP	Motor Circuit Protector
g	Gram	kg	Kilogram	MD	Medium Duty
G	Gauss	kG	Kilogauss	MDF	Medium-density fibreboard
Ga.	Gauge	kgf	Kilogram Force	M.D.O.	Medium Density Overlaid
Gal., gal.	Gallon	kHz	Kilohertz	Med.	Medium
Galv., galv	Galvanized	Kip	1000 Pounds	MF	Thousand Feet
GC/MS	Gas Chromatograph/Mass	KJ	Kilojoule	M.F.B.M.	Thousand Feet Board Measure
	Spectrometer	K.L.	Effective Length Factor	Mfg.	Manufacturing
Gen.	General	K.L.F.	Kips per Linear Foot	Mfrs.	Manufacturers
GFI	Ground Fault Interrupter	Km	Kilometer	mg	Milligram
GFRC	Glass Fiber Reinforced Concrete	KO	Knock Out	MGD	Million Gallons per Day
Glaz.	Glazier	K.S.F.	Kips per Square Foot	MGPH	Million Gallons per Hour
GPD	Gallons per Day	K.S.I.	Kips per Square Inch	MH, M.H.	Manhole; Metal Halide; Man-Hour
gpf	Gallon per Flush	kV	Kilovolt	MHz	Megahertz
GPH	Gallons per Hour	kVA	Kilovolt Ampere	Mi.	Mile
gpm, GPM	Gallons per Minute	kVAR	Kilovar (Reactance)	MI	Malleable Iron; Mineral Insulated
GR	Grade	KW	Kilowatt	MIPS	Male Iron Pipe Size
Gran.	Granular	KWh	Kilowatt-hour	mj	Mechanical Joint
Grnd.	Ground	L	Labor Only; Length; Long;	m	Meter
GVW	Gross Vehicle Weight		Medium Wall Copper Tubing	mm	Millimeter
GWB	Gypsum Wall Board	Lab.	Labor	Mill.	Millwright
		lat	Latitude	Min., min.	Minimum, Minute

Misc.	Miscellaneous	PCM	Phase Contrast Microscopy	SBS	Styrene Butadiere Styrene
ml	Milliliter, Mainline	PDCA	Painting and Decorating	SC	Screw Cover
M.L.F.	Thousand Linear Feet		Contractors of America	SCFM	Standard Cubic Feet per Minute
Mo.	Month	P.E., PE	Professional Engineer;	Scaf.	Scaffold
Mobil.	Mobilization		Porcelain Enamel;	Sch., Sched.	Schedule
Mog.	Mogul Base		Polyethylene; Plain End	S.C.R.	Modular Brick
MPH	Miles per Hour	P.E.C.I.	Porcelain Enamel on Cast Iron	S.D.	Sound Deadening
MPT	Male Pipe Thread	Perf.	Perforated	SDR	Standard Dimension Ratio
MRGWB	Moisture Resistant Gypsum	PEX	Cross Linked Polyethylene	S.E.	Surfaced Edge
	Wallboard	Ph.	Phase	Sel.	Select
MRT	Mile Round Trip	P.I.	Pressure Injected	SER, SEU	Service Entrance Cable
ms	Millisecond	Pile.	Pile Driver	S.F.	Square Foot
M.S.F.	Thousand Square Feet	Pkg.	Package	S.F.C.A.	Square Foot Contact Area
Mstz.	Mosaic & Terrazzo Worker	Pl.	Plate	S.F. Flr.	Square Foot of Floor
M.S.Y.	Thousand Square Yards	Plah.	Plasterer Helper	S.F.G.	Square Foot of Ground
Mtd., mtd., mtd	Mounted	Plas.	Plasterer	S.F. Hor.	Square Foot Horizontal
Mthe.	Mosaic & Terrazzo Helper	plf	Pounds Per Linear Foot	SFR	Square Feet of Radiation
Mtng.	Mounting	Pluh.	Plumber Helper	S.F. Shlf.	Square Foot of Shelf
Mult.	Multi; Multiply	Plum.	Plumber	S4S	Surface 4 Sides
MUTCD	Manual on Uniform Traffic Control	Ply.	Plywood	Shee.	Sheet Metal Worker
	Devices	p.m.	Post Meridiem	Sin.	Sine
M.V.A.	Million Volt Amperes	Pntd.	Painted	Skwk.	Skilled Worker
M.V.A.R.	Million Volt Amperes Reactance	Pord.	Painter, Ordinary	SL	Saran Lined
MV	Megavolt	pp	Pages	S.L.	Slimline
MW	Megawatt	PP, PPL	Polypropylene	Sldr.	Solder
MXM	Male by Male	P.P.M.	Parts per Million	SLH	Super Long Span Bar Joist
MYD	Thousand Yards	Pr.	Pair	S.N.	Solid Neutral
N	Natural; North	P.E.S.B.	Pre-engineered Steel Building	SO	Stranded with oil resistant inside
nA	Nanoampere	Prefab.	Prefabricated		insulation
NA	Not Available; Not Applicable	Prefin.	Prefinished	S-O-L	Socketolet
N.B.C.	National Building Code	Prop.	Propelled	sp	Standpipe
NC	Normally Closed	PSF, psf	Pounds per Square Foot	S.P.	Static Pressure; Single Pole; Self-
NEMA	National Electrical Manufacturers	PSI, psi	Pounds per Square Inch		Propelled
	Assoc.	PSIG	Pounds per Square Inch Gauge	Spri.	Sprinkler Installer
NEHB	Bolted Circuit Breaker to 600V.	PSP	Plastic Sewer Pipe	spwg	Static Pressure Water Gauge
NFPA	National Fire Protection Association	Pspr.	Painter, Spray	S.P.D.T.	Single Pole, Double Throw
NLB	Non-Load-Bearing	Psst.	Painter, Structural Steel	SPF	Spruce Pine Fir; Sprayed
NM	Non-Metallic Cable	P.T.	Potential Transformer		Polyurethane Foam
nm	Nanometer	P. & T.	Pressure & Temperature	S.P.S.T.	Single Pole, Single Throw
No.	Number	Ptd.	Painted	SPT	Standard Pipe Thread
NO	Normally Open	Ptns.	Partitions	Sq.	Square; 100 Square Feet
N.O.C.	Not Otherwise Classified	Pu	Ultimate Load	Sq. Hd.	Square Head
Nose.	Nosing	PVC	Polyvinyl Chloride	Sq. In.	Square Inch
NPT	National Pipe Thread	Pvmt.	Pavement	S.S.	Single Strength; Stainless Steel
NQOD	Combination Plug-on/Bolt on	PRV	Pressure Relief Valve	S.S.B.	Single Strength B Grade
	Circuit Breaker to 240V.	Pwr.	Power	sst, ss	Stainless Steel
N.R.C., NRC	Noise Reduction Coefficient/	Q	Quantity Heat Flow	Sswk.	Structural Steel Worker
	Nuclear Regulator Commission	Qt.	Quart	Sswl.	Structural Steel Welder
N.R.S.	Non Rising Stem	Quan., Qty.	Quantity	St.; Stl.	Steel
ns	Nanosecond	Q.C.	Quick Coupling	STC	Sound Transmission Coefficient
NTP	Notice to Proceed	r	Radius of Gyration	Std.	Standard
nW	Nanowatt	R	Resistance	Stg.	Staging
OB	Opposing Blade	R.C.P.	Reinforced Concrete Pipe	STK	Select Tight Knot
OC	On Center	Rect.	Rectangle	STP	Standard Temperature & Pressure
OD	Outside Diameter	recpt.	Receptacle	Stpi.	Steamfitter, Pipefitter
O.D.	Outside Dimension	Reg.	Regular	Str.	Strength; Starter; Straight
ODS	Overhead Distribution System	Reinf.	Reinforced	Strd.	Stranded
O.G.	Ogee	Req'd.	Required	Struct.	Structural
O.H.	Overhead	Res.	Resistant	Sty.	Story
O&P	Overhead and Profit	Resi.	Residential	Subj.	Subject
Oper.	Operator	RF	Radio Frequency	Subs.	Subcontractors
Opng.	Opening	RFID	Radio-frequency Identification	Surf.	Surface
Orna.	Ornamental	Rgh.	Rough	Sw.	Switch
OSB	Oriented Strand Board	RGS	Rigid Galvanized Steel	Swbd.	Switchboard
OS&Y	Outside Screw and Yoke	RHW	Rubber, Heat & Water Resistant;	S.Y.	Square Yard
OSHA	Occupational Safety and Health		Residential Hot Water	Syn.	Synthetic
	Act	rms	Root Mean Square	S.Y.P.	Southern Yellow Pine
Ovhd.	Overhead	Rnd.	Round	Sys.	System
OWG	Oil, Water or Gas	Rodm.	Rodman	t.	Thickness
Oz.	Ounce	Rofc.	Roofer, Composition	T	Temperature; Ton
P.	Pole; Applied Load; Projection	Rofp.	Roofer, Precast	Tan	Tangent
p.	Page	Rohe.	Roofer Helpers (Composition)	T.C.	Terra Cotta
Pape.	Paperhanger	Rots.	Roofer, Tile & Slate	T & C	Threaded and Coupled
P.A.P.R.	Powered Air Purifying Respirator	R.O.W.	Right of Way	T.D.	Temperature Difference
PAR	Parabolic Reflector	RPM	Revolutions per Minute	TDD	Telecommunications Device for
P.B., PB	Push Button	R.S.	Rapid Start		the Deaf
Pc., Pcs.	Piece, Pieces	Rsr	Riser	T.E.M.	Transmission Electron Microscopy
P.C.	Portland Cement; Power Connector	RT	Round Trip	temp	Temperature, Tempered, Temporary
P.C.F.	Pounds per Cubic Foot	S.	Suction; Single Entrance; South	TFFN	Nylon Jacketed Wire

TFE	Tetrafluoroethylene (Teflon)	U.L., UL	Underwriters Laboratory	w/	With
T. & G.	Tongue & Groove;	Uld.	Unloading	W.C., WC	Water Column; Water Closet
	Tar & Gravel	Unfin.	Unfinished	W.F.	Wide Flange
Th., Thk.	Thick	UPS	Uninterruptible Power Supply	W.G.	Water Gauge
Thn.	Thin	URD	Underground Residential	Wldg.	Welding
Thrded	Threaded		Distribution	W. Mile	Wire Mile
Tilf.	Tile Layer, Floor	US	United States	W-O-L	Weldolet
Tilh.	Tile Layer, Helper	USGBC	U.S. Green Building Council	W.R.	Water Resistant
THHN	Nylon Jacketed Wire	USP	United States Primed	Wrck.	Wrecker
THW.	Insulated Strand Wire	UTMCD	Uniform Traffic Manual For Control	WSFU	Water Supply Fixture Unit
THWN	Nylon Jacketed Wire		Devices	W.S.P.	Water, Steam, Petroleum
T.L., TL	Truckload	UTP	Unshielded Twisted Pair	WT., Wt.	Weight
T.M.	Track Mounted	V	Volt	WWF	Welded Wire Fabric
Tot.	Total	VA	Volt Amperes	XFER	Transfer
T-O-L	Threadolet	VAT	Vinyl Asbestos Tile	XFMR	Transformer
tmpd	Tempered	V.C.T.	Vinyl Composition Tile	XHD	Extra Heavy Duty
TPO	Thermoplastic Polyolefin	VAV	Variable Air Volume	XHHW	Cross-Linked Polyethylene Wire
T.S.	Trigger Start	VC	Veneer Core	XLPE	Insulation
Tr.	Trade	VDC	Volts Direct Current	XLP	Cross-linked Polyethylene
Transf.	Transformer	Vent.	Ventilation	Xport	Transport
Trhv.	Truck Driver, Heavy	Vert.	Vertical	Y	Wye
Trlr	Trailer	V.F.	Vinyl Faced	yd	Yard
Trlt.	Truck Driver, Light	V.G.	Vertical Grain	yr	Year
TTY	Teletypewriter	VHF	Very High Frequency	Δ	Delta
TV	Television	VHO	Very High Output	%	Percent
T.W.	Thermoplastic Water Resistant	Vib.	Vibrating	~	Approximately
	Wire	VLF	Vertical Linear Foot	Ø	Phase; diameter
UCI	Uniform Construction Index	VOC	Volatile Organic Compound	@	At
UF	Underground Feeder	Vol.	Volume	#	Pound; Number
UGND	Underground Feeder	VRP	Vinyl Reinforced Polyester	<	Less Than
UHF	Ultra High Frequency	W	Wire; Watt; Wide; West	>	Greater Than
U.I.	United Inch			Z	Zone

Index

A

Abandon catch basin 29
Abatement asbestos 41
  equipment asbestos 40
ABC extinguisher portable 392
  extinguisher wheeled 392
Aboveground storage tank
concrete 532
Abrasive aluminum oxide 50
  floor 80
  floor tile 335
  silicon carbide 50, 80
  terrazzo 345
  tile 333
  tread 335
ABS DWV pipe 501
Absorber shock 505
Absorption testing 15
  water chiller 553
A/C packaged terminal 556
Accelerator sprinkler system 486
Access card key 309
  control card type 309
  control fingerprint scanner ... 309
  control video camera 607
  door and panel 278
  door basement 84
  door duct 538
  door fire rated 278
  door floor 278
  door metal 278
  door roof 256
  door stainless steel 278
  floor 349
  flooring 349
  road and parking area 22
Accessory anchor bolt 64
  bathroom 333, 386, 387
  door 310
  drainage 252
  drywall 332
  duct 537
  fireplace 389
  gypsum board 332
  masonry 99
  plaster 325
  rebar 68
  reinforcing 68
  roof 256
  steel wire rope 136
  toilet 386
Accordion door 275
Acid etch finish concrete 81
  proof floor 340, 347
  resistant pipe 523
Acoustic ceiling board 336
Acoustical barrier 353
  batt 353
  block 108
  booth 468
  ceiling 336, 337
  door 283
  enclosure 468, 470
  folding door 384
  folding partition 384
  glass 315
  metal deck 143
  panel 384
  panel ceiling 336
  partition 383, 384
  room 459
  sealant 262, 330
  treatment 353
  underlayment 320
  wall tile 337

Acrylic floor 347
  plexiglass 313
  rubber roofing 247
  sign 376
  wall coating 368
  wallcovering 350
  wood block 341
Adhesive cement 344
  EPDM 225
  neoprene 225
  PVC 225
  removal floor 82
  roof 243
  wallpaper 350
Adjustable astragal 308
  jack post 128
Adjustment factor labor 490, 564
Admixture cement 81
  concrete 50
Adobe brick 111
Aeration sewage 706
Aerator 522
Aerial photography 14
  seeding 668
  survey 28
Agent form release 51
Aggregate exposed 81, 650
  lightweight 51
  masonry 97
  spreader 712, 719
  stone 669
  testing 15
Agricultural equipment 431
Aid staging 21
Air balancing 527
  barrier curtain 541
  cleaner electronic 546
  compressor 715
  compressor control system ... 530
  compressor dental 426
  compressor portable 715
  conditioner cooling & heating . 556
  conditioner direct expansion .. 558
  conditioner fan coil 558
  conditioner gas heat 556
  conditioner general 526
  conditioner packaged term. ... 556
  conditioner receptacle 579
  conditioner removal 526
  conditioner rooftop 556
  conditioner self-contained 556
  conditioner thru-wall 556
  conditioner ventilating 536
  conditioner wiring 581
  conditioning 863
  conditioning computer 557
  conditioning fan 539
  conditioning ventilating .. 529, 536,
    537, 539, 540, 545, 553, 555, 557
  control 532
  cooled belt drive condenser .. 552
  cooled condensing unit 552
  cooled direct drive condenser . 553
  curtain 541
  entraining agent 50
  filter 545, 546
  filter mechanical media 546
  filter permanent washable 545
  filter roll type 545
  filter washable 545
  filtration 40
  filtration charcoal type 546
  handling troffer 592
  hose 716
  lock 462
  make-up unit 555

quality 411
register 543
return grille 543
sampling 40
separator 532
spade 716
supply pneumatic control ... 530
supply register 543
supported building 461
supported storage tank cover . 461
supported structure demo 452
supported structures 860
tube system 480
unit make-up 555
vent automatic 532
vent convector or baseboard .. 559
vent roof 256
wall 384
Air-compressor mobilization 613
Aircraft cable 138
Airfoil fan centrifugal 540
Air-handling unit 555, 558
Airless sprayer 40
Airplane hangar 468
Airport painted marking 657
Air-source heat pump 557
Alarm device 607
  exit control 606
  panel and device 607
  residential 580
  valve sprinkler 484
All fuel chimney 547
Alley bowling 423
Alloy steel chain 162
Altar 430
Alteration fee 10
Alternating tread ladder 155
Alternative fastening 231
Aluminum astragal 308
  bench 665
  blind 434
  cable tray 574
  ceiling tile 337
  column 164, 212
  column base 161, 164
  column cap 164
  column cover 161
  commercial door 270
  conduit 571
  coping 117
  cross 430
  curtain wall glazed 288
  diffuser perforated 542
  directory board 375
  dome 467
  door 265, 284
  door double acting swing ... 285
  downspout 251
  drip edge 255
  ductwork 536
  edging 671
  entrance 287
  expansion joint 255
  extrusion 135
  fence 659, 660
  flagpole 398
  flashing 239, 249, 507
  foil 229, 232, 353
  framing 135
  grating 157
  gravel stop 251
  grille 543
  gutter 253
  handrail 350
  joist shoring 61
  louver 315, 544

louver door 316
mansard 238
mesh grating 157
nail 171
operating louver 544
oxide abrasive 50
pipe 681, 684
plank grating 157
plank scaffold 21
register 543
reglet 254
rivet 125
roof 237
roof panel 237
salvage 491
sash 289
screen/storm door 265
screw 239
security driveway gate 282
service entrance cable 569
shape structural 135
sheet metal 251
shelter 468
shingle 233
siding 238
siding paint 357
siding panel 238
sign 376
sliding door 279
steeple 397
storefront 288
storm door 265
structural 135
threshold 305
tile 236, 332
transom 287
trench cover 159
tube frame 285
weatherstrip 308
weld rod 127
window 289
window demolition 264
window impact resistant 289
Aluminum-framed door 287
Ammeter 582
Analysis petrographic 15
  sieve 15
Anchor bolt 64, 98
  bolt accessory 64
  bolt screw 69
  bolt sleeve 64
  brick 98
  buck 98
  bumper 405
  chemical 89, 121
  dovetail 67
  epoxy 89, 121
  expansion 121
  framing 172
  hollow wall 121
  lead screw 122
  machinery 67
  masonry 98
  metal nailing 121
  nailing 121
  nylon nailing 121
  partition 98
  plastic screw 122
  rafter 172
  rigid 98
  roof safety 21
  screw 69, 121
  sill 172
  steel 98
  stone 98
  super high-tensile 638

Index

toggle-bolt 121
wedge 122
Anechoic chamber 459
Angle corner 129
curb edging 129
fiberglass 216
valve bronze 491
valve plastic 494
Anti-siphon device 504
Anti-slip floor treatment 363
Apartment call system 603
hardware 300
Appliance 410
plumbing 510, 511
residential 410, 580
Approach ramp 350
Arch culvert oval 683
laminated 193
radial 193
Architectural fee 10, 812
Area clean-up 43
window well 397
Ark 430
Armored cable 567
Arrow 657
Articulating boom lift 715
Artificial grass surfacing 658
turf 658
Asbestos abatement 41
abatement equipment 40
area decontamination 43
demolition 42
disposal 43
encapsulation 44
remediation plan/method 40
removal 42
removal process 824
Ash conveyor 428
receiver 387, 443
urn 443
Ashlar stone 114
veneer 112
Asphalt base sheet 243
block 654
block floor 654
coating 224
concrete 612
curb 655
cutting 34
distributor 716
felt 243, 244
flashing 249
flood coat 243
mix design 15
pavement 658
paver 717
paving 652
plant portable 719
primer 222, 344
roof cold-applied built-up 243
roof shingle 233
sealcoat rubberized 650
sheathing 192
shingle 233
sidewalk 650
testing 15
Asphaltic binder 652
concrete 652
concrete paving 652
cover board 230
emulsion 667
pavement 652
wearing course 652
Aspirator dental 426
Astragal adjustable 308
aluminum 308
exterior moulding 308
magnetic 308
molding 206
one piece 308
overlapping 308
rubber 308
split 308
steel 308
Astronomy observation dome 467
Athletic backstop 660
carpet indoor 349
equipment 422
paving 658
pole 423
post 425
Atomizer water 41
Attachment ripper 719
Attachments skidsteer 715
Attic stair 412
Audio masking 470
Audiometric room 459
Auditorium chair 448
Auger boring 29
earth 712
hole 28
Augering 612
Auto park drain 507
Autoclave 425
aerated concrete block 104
Automatic fire-suppression 487
flush 517
opener 299
opener commercial 299
opener industrial 299
operator 299
teller 409
wall switch 582
washing machine 411
Automotive equipment 404
lift 479
spray painting booth 404
Autopsy equipment 428
Autoscrub machine 716
Awning canvas 396
fabric 396
window 290, 396
window aluminum 289
window metal-clad 292
window vinyl clad 293

B

Backer rod 260
rod polyethylene 64
Backerboard 224
cementitious 328
Backfill 626, 627
compaction 869
dozer 626
planting pit 668
structural 626
trench 617, 619
Backflow preventer 504
Backhoe 621, 713
bucket 714
extension 719
Backsplash 439
countertop 440
Backstop athletic 660
baseball 660
basketball 422, 660
chain link 660
electric 422
handball court 424
squash court 424
tennis court 660
Backup block 104
Backwater valve 509
Baffle roof 257
Bag disposable 41
glove 42
Baked enamel door 268
enamel frame 267
Balanced entrance door 287
Balancing air 527
water 528
Balcony fire escape 154
Bale hay 637
Baler 429
Ball valve bronze 492
valve plastic 494
valve polypropylene 494
wrecking 719
Ballast fixture 593
railroad 690
Ballistic door wood 282
Baluster wood stair 209
Balustrade painting 364
Bamboo flooring 340
Band joist framing 148
riser 420
Bank counter 409
equipment 409
run gravel 612
window 409
Banquet booth 445
Baptistry 430
Bar bell 421
chair reinforcing 69
front 469
grab 386
grating tread 160
grouted 73
joist painting 141
masonry reinforcing 98
restaurant 445
tie reinforcing 68
towel 387
tub 386
ungrouted 73
Zee 332
Barbed wire fence 659
Barber chair 407
equipment 407
Barge construction 722
Bariatric equipment 426
lift 426
Bark mulch 667
mulch redwood 667
Barn 431
Barrel bolt 299
Barrels flasher 716
reflectorized 716
Barricade 23, 692
flasher 716
tape 24
Barrier acoustical 353
and enclosure 23
crash 692
delineator 692
dust 23
highway sound traffic 666
impact 692
median 692
moisture 225
parking 657
plenum 353
security 406
separation 41, 46
slipform paver 720
weather 232
X-ray 471
Base cabinet 437
carpet 349
ceramic tile 333
column 172
course 651
course drainage layer 651
gravel 651
masonry 110
molding 196
plate column 134
quarry tile 335
resilient 343
road 651
screed 69
sheet 243, 244
sheet asphalt 243
sink 437
stabilizer 720
stone 114, 651
terrazzo 345
vanity 438
wall 343
wood 196
Baseball backstop 660
scoreboard 422
Baseboard demolition 170
heat 559
heat electric 561
heater electric 561
Basement bulkhead stair 84
Basement/utility window 290
Baseplate scaffold 19
shoring 19
Basic finish material 318
Basketball backstop 422, 660
equipment 422
scoreboard 422
Batch trial 15
Bath paraffin 426
steam 460
whirlpool 426, 457
Bathroom 514
accessory 333, 386, 387
exhaust fan 541
faucet 515
fixture 513, 514
heater & fan 541
soaking 514
Bathtub bar 386
enclosure 389
removal 491
residential 514
Batt acoustical 353
insulation 227
thermal 353
Bay wood wind. double hung 293
Beacon warning 595
Bead blast demo 319
board insulation 458
casing 332
corner 325, 332
parting 207
Beam & girder framing 181
and girder formwork 52
bond 97, 105
bondcrete 327
bottom formwork 52
castellated 134
ceiling 212
concrete 75
concrete placement 78
concrete placement grade 79
decorative 212
drywall 328
fiberglass wide flange 216

fireproofing 257
formwork side 53
hanger 172
laminated 193
load-bearing stud wall box 144
mantel 212
plaster 326
precast 85
precast concrete 84
precast concrete tee 85
reinforcing 71
reinforcing bolster 68
side formwork 53
soldier 641
test 15
wood 181, 189
Bearing pad 126
Bed molding 199
Bedding brick 654
pipe 627
Belgian block 656
Bell 421
& spigot pipe 506
bar 421
caisson 645
church 421
signal 607
system 589
Bellow expansion joint 532
Belt material handling 698
Bench 665
aluminum 665
fiberglass 665
folding 448
greenhouse 463
park 665
planter 664
plastic 393
player 665
wood 665
work 420
Bentonite 225, 642
panel waterproofing 225
Berm pavement 655
road 655
Bevel glass 311
siding 240
Bicycle rack 423, 449
trainer 421
Bi-fold door 269, 274
Bin part 394
retaining wall metal 663
Binder asphaltic 652
Biological safety cabinet 420
Biometric identity access 309
Bi-passing closet door 274
Birch exterior door 277
hollow core door 275
molding 206
paneling 208
solid core door 274
wood frame 213
Bird control film 314
control netting 399
Bit core drill 90
Bituminous block 654
coating 224, 674
concrete curb 655
paver 717
Bituminous-stabilized base course 651
Blackboard 384
Blanket curing 83
insulation 528
insulation wall 227
sound attenuation 353
Blast curtain 435
demo bead 319
floor shot 319
Blaster shot 718
Blasting 620
cap 621
mat 621
water 95
Bleacher 463
outdoor 463
stadium 463
telescoping 448
Blind exterior 214
structural bolt 126
vertical 435
vinyl 214
window 434
wood 436
Block acoustical 108
asphalt 654
autoclave concrete 104
backup 104
belgian 656
bituminous 654
Block, brick and mortar 843
Block cap 109
chimney 104
column 105
concrete 104-108, 844
concrete bond beam 105
concrete exterior 107
corner 109
decorative concrete 105
filler 366
floor 342
floor wood 341
glass 110
glazed 109
glazed concrete 109
grooved 105
ground face 106
hexagonal face 106
high strength 107
insulation 105
insulation insert concrete 105
lightweight 108
lintel 108
manhole 685
partition 108
patio 114, 244
profile 106
reflective 110
scored 106
scored split face 106
sill 109
slotted 108
slump 106
solar screen 109
split rib 106
wall removal 31
Blocking carpentry 180
steel 180
wood 180, 184
Blood pressure unit 425
Bloodbank refrigeration 419
Blower insulation 717
pneumatic tube 480
Blown-in cellulose 229
fiberglass 229
insulation 229
Blueboard 327
partition 321
Bluegrass sod 669
Bluestone 112
sidewalk 650
sill 116
step 651
Board & batten siding 240
batten fence 662
bulletin 374
ceiling 336
control 375, 421
directory 375, 376
dock 405
drain 225
gypsum 328
high abuse gypsum 331
insulation 226
insulation foam 226
paneling 209
partition gypsum 321
partition NLB gypsum 321
sheathing 191
siding wood 240
valance 438
verge 205
Boat work 722
Body harness 21
Boiler 547
control system 529
demolition 526
electric 547
electric steam 547
gas fired 547
gas/oil combination 548
general 490
hot water 547, 548
insulation 529
mobilization 613
oil fired 548
removal 526
steam 547, 548
Bollard 128
pipe 656
security 691
Bolster beam reinforcing 68
slab reinforcing 68
Bolt & hex nut steel 123
anchor 64, 98
barrel 299
blind structural 126
flush 299
high strength 124
remove 120
steel 172
Bolt-on circuit-breaker 585
Bond beam 97, 105
performance 14, 819
Bondcrete 327
Bonding agent 50
surface 97
Bookcase 395
Bookshelf 447
Boom lift articulating 715
lift telescoping 715
Booster fan 540
Boot pile 643
Booth acoustical 468
banquet 445
fixed 445
mounted 446
painting 404
portable 468
ticket 468
Border light 595
Bored pile 644
Borer horizontal 719
Boring and exploratory drilling 28
auger 29
cased 28
horizontal 674
machine horizontal 712
Borosilicate pipe 522
Borrow 612, 627
Bosun chair 22
Bottle storage 413
vial 47
Boulder excavation 621
Boundary and survey marker 28
Bow & bay metal-clad window 293
window 293
Bowl toilet 513
Bowling alley 423
Bowstring truss 194
Box & wiring device 573
beam load-bearing stud wall 144
buffalo 679
culvert formwork 54
distribution 680
electrical 573
electrical pull 573
flow leveler distribution 680
locker 392
mail 393
pull 573
safety deposit 409
stair 209
steel mesh 638
storage 17
termination 603
trench 719
utility 674
vent 99
Boxed header/beam 148
header/beam framing 148
Boxing ring 422
Brace cross 130
Bracing 180
framing 151
let-in 180
load-bearing stud wall 143
metal 180
metal joist 147
roof rafter 151
shoring 19
support seismic 494
wood 180
Bracket roof 21
scaffold 19
sidewall 21
Braided bronze hose 503
Branch circuit fusible switch 584
Brass hinge 305
pipe 498
salvage 491
screw 171
valve 491
Brazed connection 571
Break glass station 608
tie concrete 81
Breaker circuit 589
pavement 716
vacuum 504
Breeching draft inducer 546
insulation 529
Brick 102
adobe 111
anchor 98
bedding 654
Brick, block and mortar 843
Brick cart 716
catch basin 685
chimney simulated 389
cleaning 95
common 102
common building 102
concrete 109
demolition 96, 318

economy 102
edging 671
engineer 102
face 102, 843
floor 340, 654
flooring miscellaneous 340
manhole 685
molding 203, 206
oversized 101
paving 654
removal 30
sand 51
saw 94
sidewalk 654
sill 116
simulated 117
stair 112
step 650
structural 102
testing 15
veneer 99
veneer demolition 96
veneer thin 101
vent 544
vent box 99
wall 111
wall panel 111
wash 95
Bricklaying 843
Bridge 665
  concrete 665
  crane 698
  pedestrian 665
  sidewalk 20
Bridging 181
  framing 151
  joist 140
  load-bearing stud wall 144
  metal joist 147
  roof rafter 151
Broiler 415
Bronze angle valve 491
  ball valve 492
  body strainer 533
  cross 430
  globe valve 493
  letter 376
  plaque 376
  push-pull plate 304
  swing check valve 492
  valve 491
Broom cabinet 437
  finish concrete 80
  sidewalk 718
Brown coat 345
Brownstone 114
Brush chipper 713
  clearing 614
  cutter 713, 714
  mowing 614
Bubbler 520
  drinking 521
Buck anchor 98
  rough 183
Bucket backhoe 714
  clamshell 713
  concrete 712
  dragline 713
  excavator 719
  pavement 719
Buffalo box 679
Buggy concrete 80, 712
Builder's risk insurance 816
Building air supported 461
  component deconstruction 36
  deconstruction 35

demo footing & foundation 31
demolition 30
demolition selective 31
directory 375, 603
greenhouse 462
hangar 468
hardware 300
insulation 229
model 10
moving 38
paper 232
permit 14
portable 17
prefabricated 462
relocation 38
shoring 639
sprinkler 486
temporary 17
tension 462
Built-in range 410
  shower 515
Built-up roof 243
  roofing 851
  roofing component 243
  roofing system 243
Bulb tee subpurlin 87
Bulbtee subpurlin 135
Bulk bank measure excavating . 621
  dozer excavating 622
  drilling and blasting 620
  scraper excavation 623
  storage dome 467
Bulkhead door 278
  moveable 456
Bulkhead/cellar door 278
Bulldozer 626, 714
Bullet resistant window 409
Bulletin board 374
  board cork 374
Bulletproof glass 315
Bullfloat concrete 712
Bullnose block 107
Bumper car 656, 657
  dock 405
  door 300, 304
  parking 657
  railroad 690
  wall 304, 385
Buncher feller 713
Bunk house trailer 17
Burglar alarm indicating panel .. 606
Burial cell 38
Burlap curing 83
  rubbing concrete 81
Bus-duct cable tap box 587
  copper 587
  fitting 587
  plug in 587
  switch 588
  tap box 587
Bush hammer 81
  hammer finish concrete 81
Butt fusion machine 716
Butterfly valve 526, 676
Butyl caulking 261
  expansion joint 254
  flashing 250
  rubber 260
  waterproofing 225

C

Cab finish elevator 476
Cabinet & casework
paint/coating 361
  base 437
  broom 437
  corner base 437
  corner wall 437
  current transformer 588
  defibrillator 391
  demolition 170
  electric 588
  electrical hinged 573
  filler 438
  fire equipment 391
  hardboard 437
  hardware 438
  hinge 438
  hose rack 391
  hospital 440
  hotel 388
  key 310
  kitchen 437
  kitchen metal 440
  laboratory 439
  medicine 388
  metal 440
  nurse station 440
  oven 437
  pharmacy 420
  portable 413
  school 439
  shower 388
  stain 361
  storage 439
  strip 603
  transformer 588
  valve 391
  varnish 361
  wall 437, 439, 440
  wardrobe 395
Cable aircraft 138
  armored 567
  control 571
  copper 567, 571
  copper shielded 567
  crimp termination 568
  electric 567, 568
  electrical 570, 571
  guide rail 692
  jack 722
  MC 568
  medium-voltage 567
  mineral insulated 569
  pulling 718
  PVC jacket 571
  railing 164
  safety railing 138
  sheathed nonmetallic 569
  sheathed Romex 569
  tap box bus-duct 587
  tensioning 718
  termination 568
  trailer 718
  tray 574
  tray aluminum 574
  tray ladder type 574
  wire THWN 571
Cable/wire puller 718
Cafe door 276
Caisson bell 645
  concrete 644
  displacement 646
  foundation 644
Caissons 871

Calcium chloride 50, 638
Call system apartment 603
  system emergency 603
  system nurse 604
Camera TV closed circuit 606
Canopy 396
  entrance 396, 465
  entry 396
  framing 127
  framing wood 186
  metal 396
  patio/deck 396
Cant roof 186, 244
Cantilever needle beam 640
  retaining wall 662
Canvas awning 396
Cap blasting 621
  block 109
  concrete placement pile 79
  dowel 72
  pile 77
  post 172
Capacitor indoor 591
Car bumper 656, 657
  tram 719
Carbon dioxide extinguisher ... 487
  monoxide detector 608
Carborundum rub finish 81
Card catalog file 447
  key access 309
Carillon 421
Carousel compactor 429
Carpentry finish 437
Carpet base 349
  computer room 350
  felt pad 348
  floor 348
  maintenance 318
  nylon 348
  pad 348
  pad commercial grade 348
  padding 348
  removal 318
  sheet 348
  stair 349
  tile 348
  tile removal 318
  urethane pad 348
  wool 349
Carport 397
Carrel 447
Carrier ceiling 324
  channel 338
  fixture support 518
Carrier/support fixture 518
Cart brick 716
  concrete 80, 712
  disposal 429
  food delivery 415
  mounted extinguisher 392
Carving stone 114
Case display 439
  exhibit 439
  good metal 444
  good wood 444
  refrigerated 412
Cased boring 28
Casement window ... 289-291, 293
  window metal-clad 292
  window vinyl 296
Casework custom 437
  demo 434
  demolition 170
  educational 439
  ground 188
  hospital 439

metal 310, 439
  painting 361
  varnish 361
Cash register 407
Casing bead 332
  molding 197
Cast in place concrete 75, 78
  in place concrete curb 655
  in place pile 646
  in place pile add 647
  in place terrazzo 345
  iron bench 665
  iron damper 390
  iron grate cover 159
  iron manhole cover 683
  iron pipe 506
  iron pull box 573
  iron radiator 559
  iron trap 506
  trim lock 303
Castellated beam 134
Caster scaffold 19
Casting construction 161
  fiberglass 216
Cast-iron column base 161
  weld rod 127
Catch basin 683
  basin brick 685
  basin cleaner 719
  basin precast 685
  basin removal 29
  basin vacuum 719
  door 438
Cathodic protection 469
Catwalk scaffold 20
Caulking 261
  lead 506
  oakum 506
  polyurethane 261
  sealant 261
Cavity wall 111
  wall grout 98
  wall insulation 228, 229
  wall reinforcing 99
Cedar closet 208
  fence 662
  paneling 209
  post 213
  roof deck 190
  roof plank 190
  shingle 234
  siding 240
Ceiling 336
  acoustical 336, 337
  beam 212
  board 336
  board acoustic 336
  board fiberglass 336
  bondcrete 327
  bracing seismic 338
  carrier 324
  demolition 318
  diffuser 542
  drywall 329
  fan 539
  framing 182
  furring 188, 322, 324
  heater 411
  insulation 226, 229
  lath 324, 325
  linear metal 338
  luminous 337, 597
  molding 199
  painting 365, 366
  panel 336
  panel metal 337, 339

plaster 326
polystyrene 458
register 543
stair 412
support 127
suspended 324, 337
suspension system 337
tile 337
translucent 338
wood 338
woven wire 382
Cell equipment 431
  prison 469
Cellar door 278
  wine 413
Cellular concrete 76
  decking 143
Cellulose blown-in 229
  insulation 228
Cement adhesive 344
  admixture 81
  color 97
  content 835
  flashing 222
  grout 98, 640
  keene 326
  liner 674
  masonry 96
  masonry unit 104, 106-108
  mortar 334, 335, 842
  parging 224
  plaster 458
  Portland 51
  terrazzo portland 345
  testing 15
  underlayment self-leveling 88
  vermiculite 258
Cementitious backerboard 328
  waterproofing 225
  wood fiber plank 87
Center bulb waterstop 63
Central vacuum 428
  vacuum unit 702
Centrifugal airfoil fan 540
  liquid chiller 553
  pump 718
  type HVAC fan 540
Ceramic coating 368
  mulch 667
  tile 333, 334
  tile accessory 332
  tile countertop 441
  tile demolition 319
  tile floor 334
  tile repair 332
  tile waterproofing membrane . . 336
  tiling 332-336
Certification welding 16
Chain core rope 406
  fall hoist 721
  hoist 699
  hoist door 283
  link backstop 660
  link fence 24, 659, 661
  link fence high-security 660
  link fence paint 356
  link gate 660
  link industrial fence 659
  saw 718
  steel 162
  trencher 619, 715
Chair 446
  auditorium 448
  barber 407
  bosun 22
  dental 426

hydraulic 426
life guard 456
lift 479
locker bench 393
molding 206
movie 418
office 445
portable 420
rail demolition 170
seating 448
Chalkboard 372
  electric 373
  fixed 372
  freestanding 374
  liquid chalk 372
  metal 373, 384
  portable 374
  sliding 373
  swing leaf 373
  wall hung 372
Chamber anechoic 459
  decontamination 41, 46
  echo 470
  infiltration 680
Chamfer strip 60
Channel carrier 338
  curb edging 129
  frame 267
  furring 322, 332
  grating tread 160
  metal frame 267
  siding 240
  steel 325
Channelizing traffic 692
Charcoal type air filter 546
Charge disposal 43
  powder 125
Charging desk 447
Check floor 305
  rack clothes 407
  swing valve 492
  valve 484
  valve steel 527
Checkered plate 159
  plate cover 159
  plate platform 159
Checking bag garment 407
Checkout counter 407
  scanner 407
  supermarket 407
Checkroom equipment 407
Chemical anchor 89, 121
  cleaning masonry 95
  dry extinguisher 391
  removal graffiti 368
  spreader 719
  termite control 637
  toilet 718
  water closet 706
Chest 446
Chilled water coil 558
Chiller centrifugal type water 553
  water 553
Chimney 101
  accessory 389
  all fuel 547
  all fuel vent 547
  block 104
  brick 101
  demolition 95
  fireplace 389
  flue 116
  foundation 75
  metal 389
  positive pressure 547
  screen 389

simulated brick 389
vent 546
Chip quartz 51
  silica 51
  white marble 51
  wood 668
Chipper brush 713
  log 714
  stump 714
Chipping hammer 716
  stump 613
Chloride accelerator concrete 78
Chlorination system 522
Chlorosulfunated polyethylene . . . 38
Church bell 421
  equipment 430
  equipment demolition 402
  pew 448
Chute 480
  linen 480
  package 480
  refuse 480
  system 33
Cinema equipment 418
Circuit-breaker 589
  bolt-on 585
  feeder section 584
Circular rotating entrance door . . 287
  saw 718
  slipforms 826
Circulating pump 534
Clamp water pipe ground 571
Clamshell 621, 696
  bucket 713
Clapboard painting 357
Classroom heating & cooling 558
  seating 448
Clay fire 116
  tennis court 658
  tile 235
  tile coping 117
Clean control joint 63
  room 458
  tank 39
Cleaner catch basin 719
  crack 720
  steam 718
Cleaning & disposal equip. 417
  brick 95
  face brick 842
  masonry 94
  metal 120
  up 25
Cleanout door 390
  floor 496
  pipe 496
  PVC 496
  tee 496
Clean-up area 43
Clear and grub site 613
  grub 613
Clearing brush 614
  selective 614
  site 613
Clevis hook 162
Climber mast 21
Climbing crane 720
  hydraulic jack 721
Clip plywood 172
  wire rope 136
Clock equipment 604
  timer 580
  wall 443
Closed circuit television sys. 606
  circuit TV 409
  deck grandstand 463

Index

Closer concealed 300
  door 300
  door electric automatic 301
  door electro magnetic 301
  electronic 301
  floor 300
  holder 301
Closet cedar 208
  door 269, 274
  pole 206
  rod 395
  water 513, 516
Closing coordinator door 305
Closure trash 449
Cloth hardware 661
Clothes check rack 407
  dryer commercial 408
CMU 104
  pointing 94
Coal tar pitch 222, 243
Coat brown 345
  glaze 365
  hook 387
  rack 395
  rack & wardrobe 395
  scratch 345
Coated roofing foamed 247
Coating bituminous 224, 674
  ceramic 368
  concrete wall 368
  elastomeric 368
  epoxy 81, 368, 674
  exterior steel 369
  flood 244
  intumescent 368
  polyethylene 674
  reinforcing 72
  roof 222
  rubber 226
  silicone 226
  spray 224
  structural steel 844
  stucco wall 368
  trowel 224
  wall 368
  water repellent 226
  waterproofing 224
Cock gas 491
Coffee urn 414
Cofferdam 641
  excavation 622
Cohesion testing 15
Coil cooling 558
  flanged 558
Coiling door and grille 280
  grill 280
Coiling-type counter door 280
Coin operated gate 406
CO₂ extinguisher 392
  fire extinguishing system 487
Coin-operated washer & dryer . . . 408
Cold mix paver 720
  roofing 243
  storage door 281
  storage room 458
Collection box lobby 394
  concrete pipe sewage 679
  sewage/drainage 682
  system dust 702
Collector solar energy system . . . 551
Colonial door 274, 277
  wood frame 213
Color floor 81
  wheel 595
Coloring concrete 51
Column 212

accessory formwork 60
aluminum 164, 212
base 172
base aluminum 161, 164
base cast-iron 161
base plate 134
block 105
bondcrete 327
brick 101
cap aluminum 164
concrete 75
concrete block 105
concrete placement 78
cover 161
cover aluminum 161
cover stainless 161
demolition 96
drywall 328
fireproof 257
formwork 53
framing 182
lally 128
laminated wood 194
lath 324
lightweight 128
ornamental 164
plaster 326
plywood formwork 53
precast 85
precast concrete 85
reinforcing 71
removal 96
round fiber tube formwork 53
round fiberglass formwork 53
round steel formwork 53
steel framed formwork 54
stone 114
structural 128
structural shape 129
wood 182, 189, 212
Combination device 577
  storm door 274, 277
Comfort station 466
Command dog 24
Commercial automatic opener . . . 299
  dishwasher 417
  floor door 278
  folding partition 384
  gas water heater 512
  greenhouse 462
  half glass door 268
  lavatory 516
  mail box 393
  metal door 269
  oil-fired water heater 512
  refrigeration 412, 413, 458
  safe 408
  toilet accessory 386
  water heater 511, 512
  water heater electric 511
Commissioning 26
Common brick 102
  building brick 102
  nail 171
Communicating lockset . . . 302, 303
Communication optical fiber 602
  system 603
Compact fill 627
Compaction 637
  backfill 869
  soil 626
  structural 637
  test Proctor 16
  vibroflotation 870
Compactor 429
  earth 713

landfill 714
  residential 410
  waste 429
Compartment toilet 378
Compensation workers 12
Composite decking 185, 218
  decking woodgrained 218
  door 268, 276
  fabrication 218
  insulation 231
  metal deck 141
  rafter 185
  railing 219
  railing encased 219
Composition floor 347
  flooring 341, 347
  flooring removal 318
Compound dustproofing 50
Compressed air equipment
automotive 404
Compression seal 260
Compressive strength 15
Compressor air 715
  tank mounted 530
Computer air conditioning 557
  floor 349
  room carpet 350
Concealed closer 300
Concrete 832-836
  acid etch finish 81
  admixture 50
  asphalt 612
  asphaltic 652
  beam 75
  block 104-108, 844
  block autoclave 104
  block back-up 104
  block bond beam 105
  block column 105
  block decorative 105
  block demolition 32
  block exterior 107
  block foundation 107
  block glazed 109
  block grout 98
  block insulation 229
  block insulation insert 105
  block interlocking 107
  block lintel 108
  block painting 365
  block partition 108
  block planter 654
  break tie 81
  brick 109
  bridge 665
  broom finish 80
  bucket 712
  buggy 80, 712
  bullfloat 712
  burlap rubbing 81
  bush hammer finish 81
  caisson 644
  cart 80, 712
  cast in place 75, 78
  catch basin 685
  cellular 76
  chloride accelerator 78
  coloring 51
  column 75
  conversion 836
  conveyer 712
  coping 117
  core drilling 89
  crane and bucket 79
  cribbing 662
  curb 655

curing 83, 653
curing compound 83
cutout 32
cylinder 15
dampproofing 52
demo 29
demolition 32, 50
direct chute 79
distribution box 680
drilling 91
equipment rental 712
fiber reinforcing 78
field mix 837
finish 653
float finish 80, 81
floor edger 712
floor grinder 712
floor patching 50
floor staining 339
floors 838
form liner 58
forms 826, 827
furring 187
granolithic finish 88
grinding 82
grout 98
hand hole 687
hand mix 77
hand trowel finish 80
hardener 81
hydrodemolition 29
impact drilling 91
integral color 50
integral topping 88
integral waterproofing 51
joist 75
lance 717
lift slab 837
lightweight 76, 78, 841
lightweight insulating 87
machine trowel finish 80
manhole 685, 687
materials 835
membrane curing 83
mix design 15
mixer 712
mixes 835, 836
non-chloride accelerator 78
pan tread 80
paper curing 83
patio block 654
pavement highway 653
paver 718
pier and shaft drilled 646
pile prestressed 642
pipe 682
pipe removal 29
placement beam 78
placement column 78
placement footing 79
placement slab 79
placement wall 79
placing 78, 837
plantable paver precast 654
planter 664
polishing 82
post 692
pressure grouting 640
prestressed 840
prestressed precast 839
processing 82
pump 712
pumped 78
ready mix 78
reinforcing 71
removal 29

Index

retaining wall 662
retaining wall segmental 663
retarder 78
sand 51
sandblast finish 81
saw 712
saw blade 89
scarify 319
septic tank 680
shingle 235
short load 78
sidewalk 650
slab 75
slab conduit 574
slab cutting 89
slab X-ray 16
small quantity 77
spreader 718
stair 77
stair finish 81
stengths 836
tank 678
tank aboveground storage 532
testing 15
tie railroad 690
tile 235
tilt-up 841
topping 88
track cross tie 690
trowel 712
truck 712
truck holding 78
unit paving slab precast 654
utility vault 674
vertical shoring 61
vibrator 712
volumetric mixed 78
wall 76
wall coating 368
wall demolition 31
wall panel tilt-up 86
wall patching 50
water reducer 78
water storage tank 678
wheeling 80
winter 78
Concrete-filled steel pile 644
Condensate pump 535
removal pump 535
return pump 535
steam meter 535
Condenser 552
air cooled belt drive 552
air cooled direct drive 553
Condensing unit air cooled 552
Conditioner general air 526
Conductive floor 344, 346
rubber & vinyl flooring 344
terrazzo 346
Conductor 567
& grounding 567-570
Conductors 866
Conduit & fitting flexible 576
aluminum 571
concrete slab 574
electrical 571
high installation 572
in slab PVC 574
in trench electrical 574
in trench steel 574
intermediate steel 572
raceway 571
rigid in slab 574
rigid steel 572
Cone traffic 23
Confessional 430
Connection brazed 571
fire-department 484
motor 576
standpipe 484
utility 675
Connector flexible 503
gas 503
joist 172
timber 172
water copper tubing 503
wire 568
Consolidation test 16
Construction aid 18
barge 722
casting 161
management fee 10
photography 14
sump hole 625
temporary 23
time 10, 813, 814
Containment hazardous waste 38
Contaminated soil 39
Contingency 11
Continuous hinge 306
Contractor overhead 25
scale 400
Control & motor starter 586
air 532
board 375, 421
cable 571
damper volume 537
hand-off-automatic 590
joint 63, 99
joint clean 63
joint PVC 99
joint rubber 99
joint sawn 63
mirror traffic 399
radiator supply 533
station 590
station stop-start 590
system air compressor 530
system boiler 529
system cooling 529
system electronic 529
system pneumatic 529
system split system 529
system ventilator 529
tower 469
valve heating 533
Controller time system 604
Convection oven 415
Convector hydronic heating 559
Conveyor 698
ash 428
door 541
material handling 698
Cooking equipment 410, 415
range 410
Cooler 459
beverage 413
water 521
Cooling coil 558
control system 529
tower 554
tower fiberglass 554
tower forced-draft 554
tower louver 544
tower stainless 555
Coping 113, 117
aluminum 117
clay tile 117
concrete 117
removal 96
terra cotta 103, 117
Copper bus-duct 587
cable 567, 571
downspout 252
drum trap 507
DWV tubing 498
flashing 249, 250, 507
gravel stop 251
gutter 253
pipe 498
reglet 254
rivet 125
roof 248
salvage 491
shielded cable 567
wall covering 351
wire 570
Core drill 15, 712
drill bit 90
drilling concrete 89
testing 15
Coreboard 330
Cork floor 340
insulation 458
tile 340
tile flooring 340
wall covering 350
wall tile 350
wallpaper 350
Corner base cabinet 437
bead 325, 332
block 109
guard 129, 385
protection 385
wall cabinet 437
Cornerboard PVC 217
Cornice drain 509
molding 199, 202
painting 364
stone 114
Corridor sign 376
Corrosion resistance 674
resistant pipe 522
Corrugated metal pipe 681, 684
pipe 684
roof tile 235
siding 237-239
steel pipe 684
Cost control 14
mark-up 13
Cot prison 446
Counter bank 409
checkout 407
door 280
door coiling-type 280
flashing 254
hospital 439
top 440
top demolition 170
Countertop backsplash 440
ceramic tile 441
engineered stone 442
granite 441
laboratory 441
laminated 440
maple 440
marble 441
plastic-laminate-clad 440
postformed 441
sink 514
solid surface 441
stainless steel 434, 440
Course drainage layer base 651
wearing 652
Court air supported 461
handball 460
paddle tennis 424
racquetball 460
squash 460
Cove base ceramic tile 334
base terrazzo 345
molding 199
Cover aluminum trench 159
board asphaltic 230
board gypsum 230
cast iron grate 159
checkered plate 159
column 161
manhole 683
pool 456
stadium 462
stair tread 24
tank 461
trench 159
walkway 397
Covering wall 858
CPVC pipe 502
valve 494
Crack cleaner 720
filler 720
Crane 699
and bucket concrete 79
bridge 698
climbing 720
crawler 720
crew daily 18
hydraulic 699, 721
jib 699
material handling 699
mobilization 613
overhead bridge 698
rail 698
scale 399
tower 18, 720
truck mounted 721
Crash barrier 692
Crawler crane 720
drill rotary 716
shovel 714
Crematory 428
Crew daily crane 18
forklift 18
survey 25
Cribbing concrete 662
Crimp termination cable 568
Critical path schedule 14
Cross brace 130
wall 430
Crown molding 199, 206
Crushed stone 612
stone sidewalk 650
CSPE roof 245
Cubicle 382
curtain 382
detention 298
shower 514
toilet 381
track & hardware 382
Cultured stone 117
Culvert end 682
reinforced 682
Cupola 397
Cupric oxychloride floor 347
Curb 655
and gutter 655
asphalt 655
builder 712
concrete 655
edging 129, 656
edging angle 129
edging channel 129
extruder 713
granite 113, 656
inlet 656

precast 655
removal 29
roof 186
seal 650
slipform paver 720
terrazzo 345, 346
Curing blanket 83
compound concrete 83
concrete 83, 653
concrete membrane 83
concrete paper 83
concrete water 83
paper 232
Current transformer 582
transformer cabinet 588
Curtain air 541
air barrier 541
blast 435
cubicle 382
damper fire 537
divider 423
gymnasium divider 423
rod 386
sound 470
stage 421
track 421
vinyl 386
wall 288
wall glass 288
wall glazed aluminum 288
Custom architectural door 273
casework 437
Cut stone curb 656
stone trim 114
Cutoff pile 612
Cutout counter 441
demolition 31
slab 32
Cutter brush 713, 714
Cutting asphalt 34
block 440
concrete floor saw 89
concrete slab 89
concrete wall saw 89
masonry 34
steel 122
torch 122, 718
wood 34
Cylinder concrete 15
lockset 302
recore 302, 304

D

Daily crane crew 18
Dairy case 412
Damper 537
assembly splitter 537
fire curtain 537
fireplace 390
foundation vent 99
multi-blade 537
volume-control 537
Dampproofing 225
Darkroom 459
door 282
equipment 408
lead lined 459
revolving 459
Dasher hockey 457
ice rink 457
Day gate 283
Daylighting sensor 582
Dead lock 303
Deadbolt 303, 305
Deadlocking latch 303
Deaerator 532
Deciduous shrub 670
Deck drain 507
edge form 143
framing 184, 215
metal floor 143
roof 190, 192
slab form 142
wood 190
Decking 850
cellular 143
composite 185, 218
floor 141-143
form 142
roof 142
woodgrained composite 218
Deconstruction building 35
building component 36
material handling 37
Decontamination asbestos area . . . 43
chamber 41, 46
enclosure 41, 43, 46
equipment 41
Decorative beam 212
block 105
fence 661
Decorator device 577
switch 577
Deep freeze 410
seal trap 507
therapy room 470
Deep-longspan joist 138
Defibrillator cabinet 391
Dehumidifier 716
Delicatessen case 412
Delineator barrier 692
Delivery charge 22
Deluge sprinkler monitoring panel 486
valve assembly sprinkler 487
Demo air supported structure . . . 452
casework 434
concrete 29
footing & foundation building . . 31
garden house 452
geodesic dome 452
greenhouse 452
hangar 452
lightning protection 452
pre-engineered steel building . 453
roof window 265
silo 453
skylight 265
sound control 453
special purpose room 453
specialty 372
storage tank 454
swimming pool 454
tension structure 455
thermal & moist. protection . . . 222
X-ray/radio freq protection . . . 455
Demolish decking 167
remove pavement and curb 29
Demolition 823
air compressor 402
asbestos 42
bank equipment 402
barber equipment 402
baseboard 170
boiler 526
bowling alley 403
brick 96, 318
brick veneer 96
cabinet 170
casework 170
ceiling 318
central vacuum 402
ceramic tile 319
checkout counter 402
chimney 95
church equipment 402
column 96
commercial dishwasher 403
concrete 32, 50
concrete block 32
concrete wall 31
cooking equipment 403
crematory 402
darkroom equipment 403
dental equipment 403
detection equipment 402
disappearing stairway 403
dishwasher hood 403
disposal 32
door 264
drywall 318, 319
ductwork 526
electrical 565
equipment 402
explosive/implosive 31
fencing 30
fire protection system 403
fireplace 96
flashing 42
flooring 318
food case 402
food delivery cart 403
food dispensing equip 403
food preparation equip 403
food storage equipment 403
food storage shelving 403
framing 167
fume hood 403
garbage disposal 403
glass 264
granite 96
gutter 222
gutting 34
hammer 713
health club equipment 403
hood & ventilation equip 403
HVAC 526
hydraulic gate 402
ice machine 403
implosive 31
joist 167
lath 318
laundry equipment 402
library equipment 402
loading dock equipment 402
lubrication equip auto 402
masonry 32, 95, 319
medical equipment 403
metal 120
metal stud 319
millwork 170
millwork & trim 170
mold contaminated area 46
movable wall 319
movie equipment 402
paneling 170
parking equipment 402
partition 319
pavement 29
physical therapy 403
plaster 318, 319
plenum 319
plumbing 491
plywood 318, 319
post 168
rafter 168
railing 170
residential water heater 403
roofing 42, 223
saw cutting 34
selective building 31
sewage pumping system 402
sewage treatment 402
shooting range 403
site 29
sound system 402
spray paint booth 402
stage equipment 402
steel 120
steel window 264
stucco 319
subfloor 319
surgical light 403
terra cotta 320
terrazzo 319
tile 318
torch cutting 35
trim 170
truss 170
vault day gate 402
vault door and frame 402
vocational shop equip 403
wall 319
wall and partition 319
waste handling equipment 402
wastewater treatment 403
water softener 403
whirlpool bath 403
window 264
window & door 264
wood 318
wood framing 166
Demountable partition 382, 383
Dental chair 426
equipment 426
metal interceptor 508
office equipment 426
Dentil crown molding 200
Depository night 409
Derail railroad 690
Derrick crane guyed 721
crane stiffleg 721
Desk 446
Detection intrusion 606
leak 608
probe leak 608
system 606
tape 675
Detector carbon monoxide 608
explosive 606
infrared 606
metal 606
motion 606
smoke 608
temperature rise 608
ultrasonic motion 606
Detention cubicle 298
equipment 431
Detour sign 24
Developing tank 408
Device & box wiring 573
alarm 607
anti-siphon 504
combination 577
decorator 577
GFI 578
receptacle 578
residential 576
wiring 573, 589
Dewater 624, 625
pumping 625
Dewatering 625, 869

Index

equipment	722	truck	405	folding	384	stainless steel (access)	278
Diamond lath	324	Dog command	24	folding accordion	275	steel .. 267, 268, 283, 284, 465, 853
plate tread	160	Dome	399, 467	frame	266, 267	steel louver	315
Diaper station	386	drain	509	frame exterior wood	213	steel louver fire-rated	316
Diaphragm pump	718	geodesic	467	frame grout	98	stop	286, 304
Diesel generator	590	security	399	frame guard	386	storm	265
generator set	590	structures	860	frame interior	213	swinging glass	287
hammer	714	Domed metal-framed skylight	298	frame lead lined	470	switch burglar alarm	606
tugboat	722	Door & window interior		french exterior	276	threshold	214
Diesel-engine-driven generator	590	paint	362, 363	garage	283, 284	torrified	275
Diffuser ceiling	542	accessory	310	glass	265, 279, 286, 287	traffic	285
opposed blade damper	542	accordion	275	hand carved	270	varnish	361
perforated aluminum	542	acoustical	283	handle	438	vault	283
rectangular	543	acoustical folding	384	hangar	468	vertical lift	284
steel	543	air lock	462	hardware	300	weatherstrip	309
T-bar mount	543	aluminum	265, 284	hardware accessory	305	weatherstrip garage	309
Digital movie equipment	418	aluminum commercial	270	hollow metal	267	wire partition	382
Dimensions and weights of		aluminum double acting		hollow metal exterior	269	wood	270, 271, 273
sheet steel	849	swing	285	industrial	280	wood ballistic	282
Dimmer switch	577, 589	aluminum louver	316	interior pre-hung	277	wood double acting swing	285
Direct chute concrete	79	aluminum-framed	287	jamb molding	207	wood fire	272
expansion A/C	558	aluminum-framed entrance	287	kennel	279	wood panel	273
Directional drill horizontal	719	and grille coiling	280	kick plate	307	wood storm	277
sign	377	and panel access	278	knob	303	Doorbell system	589
Directory	375	and window material	264	labeled	268, 272	Dormer gable	187
board	375, 376	balanced entrance	287	louver	315	Dormitory furniture	446
building	375, 603	bell residential	579	louvered	275, 277	Double acting swing door	285
Disappearing stair	412	bi-fold	269, 274	mahogany	270	hung aluminum sash window	289
Disc harrow	713	bi-passing closet	274	metal	265, 269, 284, 465	hung bay wood window	293
Discharge hose	717	birch exterior	277	metal access	278	hung steel sash window	290
Dishwasher commercial	417	birch hollow core	275	metal fire	268	hung window	294
residential	410	birch solid core	274	metal toilet component	379	hung wood window	291
Diskette safe	408	blind	214	mirror	313	precast concrete tee beam	85
Dispenser hot water	511	bulkhead	278	molding	207	wall pipe	686
napkin	387	bulkhead/cellar	278	moulded	273	weight hinge	306
soap	387	bullet resistant	409	opener	264, 281, 284, 299	Dovetail anchor	67
ticket	404	bumper	300, 304	operator	264, 299, 300	Dowel cap	72
toilet tissue	387	cafe	276	overhead	283, 284	reinforcing	72
towel	386	casing PVC	217	overhead commercial	283	sleeve	72
Dispensing equipment food	416	catch	438	panel	273, 274	Downspout	251, 252, 397
Dispersion nozzle	487	chain hoist	283	paneled	269	aluminum	251
Displacement caisson	646	cleanout	390	partition	383	copper	252
Display case	439	closer	300	passage	275, 277	demolition	222
Disposable bag	41	closet	269, 274	patio	280	elbow	252
Disposal	31	closing coordinator	305	plastic laminate	271	lead coated copper	252
asbestos	43	cold storage	281	plastic-laminate toilet	380	steel	252
cart	429	colonial	274, 277	prefinished	271	strainer	251
charge	43	combination storm	274, 277	pre-hung	276, 277	Dozer	714
field	680	commercial floor	278	protector	300	backfill	626
garbage	410, 417	commercial half glass	268	refrigerator	281	excavation	622
or salvage value	35	composite	268, 276	release	603	Draft inducer breeching	546
waste	45, 429	conveyor	541	removal	264	Draft-induction fan	546
Distiller medical	425	counter	280	residential	213, 269, 274	Dragline	621, 696
water	425	custom architectural	273	residential garage	284	bucket	713
Distribution box	680	darkroom	282	residential steel	269	Drain	506, 509
box concrete	680	darkroom revolving	282	residential storm	265	board	225
box flow leveler	680	demolition	264	revolving	282, 287, 462	deck	507
box HDPE	680	double	269	revolving entrance	287	dome	509
pipe water	675, 676	double acting swing	285	rolling	280, 288	facility trench	509
switchboard	584	duct access	538	roof	256	floor	507
Distributor asphalt	716	dutch	274	sauna	460	main	509
Ditching	625, 642	dutch oven	390	seal	405	pipe	625
Divider curtain	423	electric automatic closer	301	sectional	283	roof	509
strip terrazzo	345	electro magnetic closer	301	sectional overhead	284	sanitary	507
Diving board	456	entrance	265, 274, 277, 286	security vault	283	scupper	509
stand	456	entrance strip	285	shower	388	sediment bucket	507
Dock board	405	exterior	276	sidelight	269, 276	shower	506
board magnesium	405	exterior pre-hung	277	sill	207, 213, 305	trench	509
bumper	405	fiberglass	276, 284	sliding aluminum	279	Drainage accessory	252
equipment	405	fire	270, 273, 281, 608	sliding glass	279, 280	field	680
floating	696	fire rated access	278	sliding glass vinyl clad	279	pipe	522, 679, 681, 682
leveler	405	fireplace	390	sliding panel	288	trap	506, 509
light	406	flexible	285	sliding vinyl clad	279	Drapery hardware	435
loading	405	floor	279	special	278, 283	Drawer kitchen	437
seal	405	flush	271, 274	stain	361	pass-thru	431
shelter	405	flush wood	271	stainless steel	277	security	431

track 438
type cabinet 439
Dredge 696
hydraulic 696
Dressing unit 446
Drill auger 29
console dental 426
core 15, 712
earth 28, 644
hole 638
quarry 716
rig 28
rig mobilization 613
rock 28, 620
shop 420
steel 716
track 716
wood 172, 663
Drilled concrete pier and shaft . . 646
concrete pier uncased 646
pier 645
Drilling and blasting bulk 620
and blasting rock 620
concrete 91
concrete impact 91
horizontal 674
plaster 322
quarry 620
rig 712
rock bolt 620
steel 122
Drinking bubbler 521
fountain 520, 521
fountain deck rough-in 521
fountain floor rough-in 521
fountain handicap 520
fountain support 518
fountain wall rough-in 520
Drip edge 255
edge aluminum 255
Dripproof motor 600
Driven pile 642
Driver post 720
sheeting 716
Drive-up window 409
Driveway 650
gate security 282
removal 29
security gate 282
security gate opener 283
security gate solar panel 283
Drum trap copper 507
Dry fall painting 366
pipe sprinkler head 486
wall leaded 470
Drycleaner 407
Dryer commercial clothes 408
hand 386
industrial 407
receptacle 579
residential 411
vent 411
Dry-pipe sprinkler system 486
Dry-type transformer 583
Drywall 328
accessory 332
column 328
cutout 32
demolition 318, 319
finish 857
frame 266
gypsum 328, 331
gypsum high abuse 331
high abuse 331
nail 171
painting 365, 366

partition 321
partition NLB 321
prefinished 329
removal 319
screw 332
Duck tarpaulin 23
Duct accessory 537
connection fabric flexible 536
electric 575, 587
electrical underfloor 575
fire rated blanket 528
fitting trench 575
fitting underfloor 575
flexible 538
flexible insulated 538
flexible noninsulated 538
furnace 549
grille 350
heater electric 558
humidifier 562
HVAC 536
insulation 528
liner 539
liner non-fibrous 539
mechanical 536
silencer 538
steel trench 575
thermal insulation 528
trench 575
underfloor 575
underground 687
utility 687
Ductile iron fitting 675
iron fitting mech joint 676
iron grooved joint 500, 501
iron pipe 675
Ductwork 536
aluminum 536
demolition 526
fabric coated flexible 538
fabricated 536
fiberglass 537
galvanized 536
metal 536
rectangular 536
rigid 536
Dumbbell 421
waterstop 62
Dumbwaiter electric 474
manual 474
Dump charge 34
truck 715
Dumpster 33
Dumpsters 823
Dumptruck offhighway 715
Duplex receptacle 589
Dust barrier 23
collection system 702
collector shop 420
Dustproofing 81
compound 50
Dutch door 274
oven door 390
DWV pipe ABS 501
PVC pipe 502, 679
tubing copper 498

E

Earth auger 712
compactor 713
drill 28, 644
rolling 637
scraper 714
vibrator 617, 626, 637

Earthwork 615
equipment 626
equipment rental 712
haul 627
Eave overhang 465
Ecclesiastical equipment 430
Echo chamber 470
Economy brick 102
Edge drip 255
Edger concrete floor 712
Edging 671
aluminum 671
curb 129, 656
Educational casework 439
T.V. studio 607
Efficiency flush valve high 517
Efflorescence testing 15
Effluent-filter septic system 680
EIFS 231
Ejector pump 508
Elastomeric coating 368
liquid flooring 347
membrane 333
roof 247
sheet waterproofing 225
waterproofing 225
Elbow downspout 252
pipe 500
Electric appliance 410
backstop 422
ballast 593
baseboard heater 561
boiler 547
cabinet 588
cable 567, 568
capacitor 591
chalkboard 373
commercial water heater 511
duct 575, 587
duct heater 558
dumbwaiter 474
feeder 587
fixture 592
fixture exterior 596
furnace 549
generator 717
heater 411
heating 547, 561
hinge 306
hoist 699
lamp 597
log 390
metallic tubing 572
meter 582
motor 600
panelboard 585
pool heater 548
service 589
stair 412
switch 589, 590
traction freight elevator 474
utility 687
vehicle charging 591
water cooler 521
water heater residential 511
Electrical & telephone
underground 687
box 573
cable 570, 571
conduit 571
demolition 565
fee 10
laboratory 419
metering smart 582
pole 687
tubing 571

wire 570
Electrically tinted window film . . 314
Electricity temporary 17
Electronic air cleaner 546
closer 301
control system 529
directory board 375
markerboard 374
Electrostatic painting 365
Elevated conduit 572
floor 75
installation add 490
pipe add 490
slab 75
slab formwork 54
slab reinforcing 71
water storage tank 678
water tank 678
Elevating scraper 623
Elevator 474, 861
cab finish 476
control 477
electric traction freight 474
fee 10
freight 474
hydraulic freight 475
option 477
passenger 474
residential 475
shaft wall 320
Embossed print door 273
Emergency call system 603
equipment laboratory 419
eyewash 520
lighting 594
lighting unit 594
shower 520
Employer liability 12
EMT 572
Emulsion adhesive 344
asphaltic 667
pavement 650
penetration 651
sprayer 716
Encapsulation asbestos 44
pipe 44
Encased railing 219
Encasement pile 643
Encasing steel beam formwork . . . 52
Enclosed bus assembly 587
Enclosure acoustical 468, 470
bathtub 389
commercial telephone 378
decontamination 41, 43, 46
shower 388
swimming pool 463
telephone 378
End culvert 682
grain block floor 341
Endwall overhang 465
Energy circulator air solar 551
saving lighting device 582
system controller solar 551
system heat exchanger solar . . 551
Engineer brick 102
Engineered stone countertop . . . 442
Engineering fee 10, 813
Engraved panel signage 376
Entrance aluminum 287
and storefront 287
canopy 396, 465
door 265, 274, 277, 286
door aluminum-framed 287
door fiberous glass 276
floor mat 443
frame 213

890

Index

lock ... 304
screen ... 378
strip ... 285
Entry canopy ... 396
EPDM adhesive ... 225
flashing ... 250
roofing ... 245
Epoxy anchor ... 89, 121
coated reinforcing ... 72
coating ... 81, 368, 674
fiberglass wound pipe ... 522
floor ... 346, 347
grating ... 217
grout ... 333, 335, 640
lined silo ... 468
terrazzo ... 346
wall coating ... 368
welded wire ... 73
Equipment ... 407, 712
agricultural ... 431
athletic ... 422
automotive ... 404
bank ... 409
barber ... 407
basketball ... 422
cell ... 431
checkroom ... 407
cinema ... 418
cleaning & disposal ... 417
clock ... 604
darkroom ... 408
demolition ... 402
dental ... 426
dental office ... 426
detention ... 431
earthwork ... 626
ecclesiastical ... 430
exercise ... 422
fire hose ... 485
fire-extinguishing ... 487
food preparation ... 414
foundation formwork ... 55
gymnasium ... 421, 422
health club ... 421
hospital ... 425
ice skating ... 457
installation ... 403
insulation ... 529
insurance ... 12
laboratory ... 419, 439
laboratory emergency ... 419
laundry ... 407, 408
loading dock ... 405
lube ... 404
lubrication ... 404
medical ... 425-428
medical sterilizing ... 425
mobilization ... 646
pad ... 76
parking ... 404
parking collection ... 404
parking control ... 404
parking gate ... 404
parking ticket ... 404
playfield ... 423
playground ... 423, 424, 660
refrigerated storage ... 413
refrigeration ... 457
rental ... 712, 821, 822
rental concrete ... 712
rental earthwork ... 712
rental general ... 715
rental highway ... 719
rental lifting ... 720
rental marine ... 722
rental wellpoint ... 722
safety eye/face ... 520
security ... 431
security and vault ... 408
shop ... 420
stage ... 420
swimming pool ... 456
theater and stage ... 420
waste handling ... 428, 429
Erosion control synthetic ... 637
Escalator ... 478
Escape fire ... 154
Estimate electrical heating ... 561
Estimates window ... 853
Estimating ... 810
Evergreen shrub ... 669
tree ... 669
Exam equipment medical ... 425
light ... 425
Excavating bulk bank measure ... 621
bulk dozer ... 622
equipment ... 868
trench ... 616
utility trench ... 619
Excavation ... 616, 621, 867
boulder ... 621
bulk scraper ... 623
cofferdam ... 622
dozer ... 622
hand ... 617, 619, 620, 626
machine ... 626
planting pit ... 668
scraper ... 623
septic tank ... 681
structural ... 620
tractor ... 617
trench ... 616, 619
Excavator bucket ... 719
hydraulic ... 712
Exchanger heat ... 552
Exercise equipment ... 422
ladder ... 422
rope ... 422
weight ... 421
Exhaust hood ... 411, 419
vent ... 544
Exhauster roof ... 540
Exhibit case ... 439
Exit control alarm ... 606
device panic ... 301
light ... 594
lighting ... 594
Expansion anchor ... 121
joint ... 63, 254, 325, 532
joint assembly ... 262
joint bellow ... 532
joint butyl ... 254
joint floor ... 262
joint neoprene ... 254
joint roof ... 254
shield ... 121
tank ... 533
tank steel ... 533
Expense office ... 13
Explosionproof fixture ... 595
lighting ... 595
motor ... 600
Explosive ... 621
detection equipment ... 606
detector ... 606
Explosive/implosive demolition ... 31
Exposed aggregate ... 81, 650
aggregate coating ... 368
Extension backhoe ... 719
ladder ... 717
Exterior blind ... 214
concrete block ... 107
door ... 276
door frame ... 213
door hollow metal ... 269
fixture lighting ... 596
floodlamp ... 597
insulation ... 231
insulation finish system ... 231
LED fixture ... 596
lighting fixture ... 596
molding ... 200
moulding astragal ... 308
paint door & window ... 358
plaster ... 327
pre-hung door ... 277
PVC molding ... 217
residential door ... 274
roll-up shutter ... 395
shutter ... 214
siding painting ... 357
signage ... 376
steel coating ... 369
surface preparation ... 354
tile ... 334
trim ... 200
trim paint ... 359
wood door frame ... 213
wood frame ... 213
Extinguisher cart mounted ... 392
chemical dry ... 391
CO_2 ... 392
fire ... 391, 487
FM200 fire ... 487
installation ... 392
portable ABC ... 392
pressurized fire ... 392
standard ... 391
wheeled ABC ... 392
Extra work ... 13
Extruder curb ... 713
Extrusion aluminum ... 135
Eye bolt screw ... 69
wash fountain ... 520
wash portable ... 520
Eye/face wash safety equipment ... 520
Eyewash emergency ... 520
safety equipment ... 520

F

Fabric awning ... 396
flashing ... 249
pillow water tank ... 678
stabilization ... 651
stile ... 308
structure ... 461
waterproofing ... 224
welded wire ... 73
wire ... 661
Fabricated ductwork ... 536
Fabrication composite ... 218
Fabric-backed flashing ... 250
Face brick ... 102, 843
brick cleaning ... 842
wash fountain ... 520
Facial scanner unit ... 309
Facility trench drain ... 509
Facing panel ... 115
stone ... 114
tile structural ... 103
Factor security ... 11
Fall painting dry ... 366
Fan air conditioning ... 539
bathroom exhaust ... 541
booster ... 540
ceiling ... 539
centrifugal type HVAC ... 540
coil air conditioning ... 558
draft-induction ... 546
HVAC axial flow ... 539
induced draft ... 546
in-line ... 540
kitchen exhaust ... 541
paddle ... 580
propeller exhaust ... 541
residential ... 580
roof ... 540
utility set ... 539, 540
vaneaxial ... 539
ventilation ... 580
wall ... 541
wall exhaust ... 540
wiring ... 580
Farm type siding ... 238
Fascia board demolition ... 167
metal ... 251
PVC ... 217
wood ... 186, 203
Fast food equipment ... 414
Fastener timber ... 171
wood ... 171
Fastening alternative ... 231
Faucet & fitting ... 515, 517
bathroom ... 515
laundry ... 515
lavatory ... 515
Fee architectural ... 10
engineering ... 10
Feeder electric ... 587
section branch circuit ... 584
section circuit-breaker ... 584
section frame ... 584
section switchboard ... 584
stock ... 431
Feller buncher ... 713
Felt ... 244
asphalt ... 243, 244
carpet pad ... 348
tarred ... 244
waterproofing ... 224
Fence ... 356
aluminum ... 659, 660
and gate ... 659
board batten ... 662
cedar ... 662
chain link ... 24, 661
chain link industrial ... 659
chain link residential ... 659
decorative ... 661
helical topping ... 661
mesh ... 659
metal ... 659
misc metal ... 661
picket paint ... 356
plywood ... 24
security ... 661
snow ... 39
steel ... 659, 661
temporary ... 24
tennis court ... 660
treated pine ... 662
tubular ... 661
wire ... 24, 661
wood rail ... 662
wrought iron ... 659
Fencing demolition ... 30
wire ... 661
Fertilizer ... 668
Fiber cement siding ... 242
optic cable ... 602
reinforcing concrete ... 78
steel ... 74

synthetic 74
Fiberboard insulation 230
Fiberglass angle 216
    area wall cap 397
    bench 665
    blown-in 229
    casting 216
    ceiling board 336
    cooling tower 554
    cross 430
    door 276, 284
    ductwork 537
    flagpole 398
    flat sheet 216
    floor grating 217
    formboard 59
    grating 216
    handrail 216
    insulation .. 226, 227, 229, 465, 528
    panel 23, 238, 353
    panel skylight 466
    planter 664
    reinforced ceiling 458
    reinforced plastic panel 351
    round bar 216
    round tube 216
    shade 436
    single hung window 298
    square bar 216
    square tube 216
    stair tread 216
    steeple 397
    tank 530
    threaded rod 216
    trash receptacle 449
    trench drain 509
    wall lamination 368
    waterproofing 224
    wide flange beam 216
    window 298
    wool 228
Field disposal 680
    drainage 680
    mix concrete 837
    office 17
    office expense 17
    personnel 12
    sample 47
    seeding 668
Fieldstone 112
Fill 626, 627
    by borrow 627
    by borrow & utility bedding .. 627
    floor 76
    gravel 612, 627
Filler block 366
    cabinet 438
    crack 720
    joint 260
    strip 308
Fillet welding 123
Film bird control 314
    equipment 408, 418
    polyethylene 667
    safety 314
    security 314
Filter air 546
    grille 543
    mechanical media 545
    stone 638
    swimming pool 522
Filtration air 40
    equipment 456
Fin tube radiation 559
Fine grade 668
Finish carpentry 437

concrete 653
floor 342, 367
grading 616
keene cement 326
lime 97
nail 171
refrigeration 458
wall 351
Finishing floor concrete 80
    wall concrete 81
Fir column 212
    floor 342
    molding 206
    roof deck 190
    roof plank 190
Fire brick 116
    call pullbox 608
    clay 116
    damper curtain type 537
    door 270, 273, 281, 608
    door frame 266
    door metal 268
    door wood 272
    equipment cabinet 391
    escape 154
    escape balcony 154
    escape stair 155
    extinguisher 391, 487
    extinguisher portable 391
    extinguishing system 487
    horn 607
    hose 485
    hose adapter 485
    hose equipment 485
    hose nozzle 485
    hose rack 485
    hose storage cabinet 391
    hose valve 485
    hydrant 677
    hydrant building 484
    hydrant remove 29
    protection 391
    pump 488
    rated blanket duct 528
    rated tile 103
    resistant drywall 329, 330
    resistant glass 311
    resistant wall 321
    retardant pipe 523
Firebrick 116
Fire-department connection 484
Fire-extinguishing equipment .. 487
Fireplace accessory 389
    box 117
    built-in 390
    chimney 389
    damper 390
    demolition 96
    door 390
    form 389
    free standing 389
    mantel 212
    mantel beam 212
    masonry 117
    prefabricated 389
Fireproofing 257
    plaster 257
    plastic 257
    sprayed cementitious 257
Fire-rated door steel louver 316
Firestop wood 183
Firestopping 258, 852
Fire-suppression automatic 487
Fitting bus-duct 587
    ductile iron 675
    grooved joint 499

grooved joint pipe 500
poke-thru 572
PVC 676
underfloor duct 576
waterstop 63
Fixed blade louver 544
    booth 445
    chalkboard 372
    end caisson pile 644
    roof tank 531
    tackboard 374
    vehicle delineator 692
Fixture ballast 593
    bathroom 513, 514
    carrier/support 518
    electric 592
    explosionproof 595
    fluorescent 592
    incandescent 593
    incandescent vaportight 593
    interior light 592
    plumbing .. 508, 513, 520, 534, 679
    removal 491
    residential 579
    sodium high pressure 597
    support handicap 518
    vandalproof 593
Flagging 340, 655
    slate 655
Flagpole 398
    aluminum 398
    fiberglass 398
    foundation 398
    steel internal halyard 398
    structure mounted 398
    wall 398
    wood 398
Flange beam fiberglass wide 216
Flanged coil 558
Flasher barrels 716
    barricade 716
Flashing 249-251
    aluminum 239, 249, 507
    asphalt 249
    butyl 250
    cement 222
    copper 249, 250, 507
    counter 254
    demolition 42
    EPDM 250
    fabric 249
    fabric-backed 250
    laminated sheet 250
    masonry 249
    mastic-backed 250
    membrane 244
    metal 465
    neoprene 250
    paperbacked 250
    plastic sheet 250
    PVC 250
    self-adhering 251
    sheet metal 249
    stainless 249
    valley 233
    vent 507
Flat seam sheet metal roofing ... 248
    sheet fiberglass 216
Flatbed truck 715, 719
    truck crane 720
Flexible conduit & fitting 576
    connector 503
    door 285
    duct 538
    duct connection fabric 536
    ductwork fabric coated 538

insulated duct 538
metal hose 503
noninsulated duct 538
plastic netting 399
sign 376
Float finish concrete 80, 81
    glass 310
Floater equipment 12
Floating dock 696
    floor 339
    pin 305
    roof tank 531
    wood pier 696
Flood coating 244
Floodlamp exterior 597
Floodlight LED 597
    pole mounted 597
    trailer 716
    tripod 716
Floor 340
    abrasive 80
    access 349
    acid proof 340
    acrylic 347
    adhesive removal 82
    asphalt block 654
    athletic resilient 344
    brick 340, 654
    carpet 348
    ceramic tile 334
    check 305
    cleaning 25
    cleanout 496
    closer 300
    color 81
    composition 347
    concrete finishing 80
    conductive 344, 346
    cork 340
    decking 141-143
    door 279
    door commercial 278
    drain 507
    elevated 75
    end grain block 341
    epoxy 346, 347
    expansion joint 262
    fill 76
    fill lightweight 76
    finish 367
    flagging 655
    framing removal 32
    grating 217
    grating aluminum 157
    grating fiberglass 217
    grating steel 157
    grinding 82
    hardener 50
    hatch 279
    heating radiant 560
    honing 82
    insulation 226
    marble 114
    mastic 347
    mat 443
    mat entrance 443
    nail 171
    neoprene 347
    oak 342
    paint 364
    paint & coating interior 363
    paint removal 82
    parquet 341
    pedestal 349
    plank 189
    plate stair 154

For customer support on your Building Construction Costs with RSMeans Data, call 800.448.8182.

plywood 190
polyacrylate 346
polyester 347
polyethylene 349
polyurethane 347
portable 420
quarry tile 335
refrigeration 458
register 543
removal 31
rubber 343, 658
rubber and vinyl sheet 343
sander 718
scupper 509
sealer 51
slate 115
sleeper 186
stain 364
subfloor 190
terrazzo 345, 858
tile 345
tile or terrazzo base 345
tile terrazzo 345
topping 81, 88
transition strip 348
treatment anti-slip 363
underlayment 191
varnish 364, 367
vinyl 344
vinyl sheet 343
wood 341, 342
wood athletic 342
wood composition 341
Flooring bamboo 340
  composition 341, 347
  conductive rubber & vinyl 344
  demolition 318
  elastomeric liquid 347
  masonry 340
  miscellaneous brick 340
  quartz 347
  treatment 339
  wood strip 341
Flow meter 534
Flue chimney 116
  chimney metal 546
  liner 101, 116
  prefab metal 546
  screen 389
  tile 116
Fluid applied membrane
  air barrier 233
  heat transfer 551
Fluorescent fixture 592
Fluoroscopy room 470
Flush automatic 517
  bolt 299
  door 271, 274
  high efficiency toilet (het) 517
  metal door 269
  tube framing 285
  valve 517
  wood door 271
Flying truss shoring 61
FM200 fire extinguisher 487
Foam board insulation 226
  core DWV ABS pipe 502
  insulation 229
  pipe covering 497
  roofing 247
  spray rig 717
Foamed coated roofing 247
  in place insulation 228
Foam-water system component .. 487
Fog seal 650
Foil aluminum 229, 232, 353

metallic 229
Folder laundry 407
Folding accordion door 275
  accordion partition 384
  bench 448
  door 384
  door shower 388
  gate 382
  table 445
Food delivery cart 415
  dispensing equipment 416
  mixer 414
  preparation equipment 414
  service equipment 415
  storage metal shelving ... 414
  warmer 416
Foot valve 494
Football goalpost 425
  scoreboard 422
Footing concrete placement ... 79
  formwork continuous 55
  formwork spread 56
  keyway 55
  keyway formwork 55
  reinforcing 71
  removal 31
  spread 76
Forced-draft cooling tower ... 554
Forest stewardship council ... 180
Forged valve 527
Forklift 717
  crew 18
Form deck edge 143
  decking 142
  fireplace 389
  liner concrete 58
  material 826
  release agent 51
Formblock 105
Formboard fiberglass 59
  roof deck 59
  wood fiber 59
Forms 827
  concrete 827
Formwork beam and girder 52
  beam bottom 52
  beam side 53
  box culvert 54
  column 53
  column accessory 60
  column plywood 53
  column round fiber tube ... 53
  column round fiberglass ... 53
  column round steel 53
  column steel framed 54
  continuous footing 55
  elevated slab 54
  encasing steel beam 52
  equipment foundation 55
  footing keyway 55
  gang wall 58
  gas station 56
  gas station island 56
  girder 52
  grade beam 56
  hanger accessory 60
  insert concrete 67
  insulating concrete 59
  interior beam 52
  labor hours 828
  light base 56
  mat foundation 56
  oil 62
  patch 62
  pile cap 55
  plywood 52, 54

radial wall 57
retaining wall 58
shoring concrete 61
side beam 53
sign base 56
slab blockout 57
slab box out opening 55
slab bulkhead 55, 56
slab curb 55, 56
slab edge 55, 56
slab flat plate 54
slab haunch 76
slab joist dome 54
slab joist pan 54
slab on grade 56
slab screed 57
slab thickened edge 76
slab trench 57
slab turndown 76
slab void 55, 57
slab with drop panel 54
sleeve 61
snap-tie 61
spandrel beam 52
spread footing 56
stair 59
stake 62
steel framed plywood 58
upstanding beam 53
wall 57
wall accessory 62
wall boxout 57
wall brick shelf 57
wall bulkhead 57
wall buttress 57
wall corbel 57
wall lintel 58
wall pilaster 58
wall plywood 57
wall prefab plywood 58
wall sill 58
wall steel framed 58
Foundation caisson 644
  chimney 75
  concrete block 107
  flagpole 398
  mat 76
  mat concrete placement 79
  pile 646
  scale pit 399
  underpin 641
  vent 99
  wall 107
Fountain 457
  drinking 520, 521
  eye wash 520
  face wash 520
  indoor 457
  lighting 522
  outdoor 457
  pump 522
  wash 517
  water pump 522
  yard 457
Frame baked enamel 267
  door 266, 267
  drywall 266
  entrance 213
  exterior wood door 213
  fire door 266
  grating 159
  labeled 266
  metal 266
  metal butt 267
  scaffold 19
  shoring 61

steel 266
trench grating 159
welded 266
window 292
Framing aluminum 135
  anchor 172
  band joist 148
  beam & girder 181
  boxed header/beam 148
  bracing 151
  bridging 151
  canopy 127
  ceiling 182
  column 182
  deck 184, 215
  demolition 167
  heavy 189
  joist 182
  laminated 193
  ledger 186
  lightweight 129
  lightweight angle 129
  lightweight channel 130
  lightweight junior beam 130
  lightweight tee 130
  lightweight zee 130
  load-bearing stud partition 145
  load-bearing stud wall ... 145
  metal joist 149
  metal roof parapet 151
  miscellaneous wood 183
  NLB partition 323
  open web joist wood 182
  opening 465
  pipe support 131
  porch 184
  removal 166
  roof 186
  roof metal rafter 151
  roof metal truss 153
  roof rafter 185
  roof soffit 152
  roof truss 193
  sill & ledger 186
  sleeper 186
  slotted channel 130
  soffit & canopy 186
  stair stringer 183
  steel 133
  steel tee 130
  steel zee 130
  suspended ceiling 182
  timber 189
  treated lumber 187
  tube 285
  wall 187
  web stiffener 149
  window 465
  window wall 285
  wood 180
  wood beam & girder 181
  wood joist 182
  wood plate 187
  wood sleeper 186
  wood soffit & canopy 187
  wood stair stringer 183
  wood stub wall 187
Freestanding chalkboard 374
Freeze deep 410
Freezer 412, 413, 459
Freight elevator 474
  elevator hydraulic 475
  elevators 861
French exterior door 276
Friction pile 643, 646
Frieze PVC 218

Front end loader 621
  vault 283
FRP panel 351
Fryer 415
Fuel storage tank 531
  tank 530
Full vision glass 311
Fume hood 419
Furnace duct 549
  electric 549
  gas 549
  gas fired 549
  hot air 549
  oil fired 549
  wall 550
Furnishing library 447
  site 449
Furniture 444
  dormitory 446
  hospital 447
  hotel 445
  library 447
  office 444, 445
  restaurant 445
  school 446
Furring and lathing 322
  ceiling 188, 322, 324
  channel 322, 332
  concrete 187
  masonry 187
  metal 322
  steel 322
  wall 187, 322
  wood 187
Fusible link closer 300
  switch branch circuit 584
Fusion machine butt 716

G

Gabion box 638
  retaining wall stone 664
Gable dormer 187
Galley septic 680
Galvanized ductwork 536
  plank grating 158
  reinforcing 72
  roof 237
  steel reglet 254
  welded wire 73
Galvanizing 369
  lintel 131
  metal in field 369
  metal in shop 120
Gang wall formwork 58
Gantry crane 699
Garage 467
  door 283, 284
  door residential 284
  door weatherstrip 309
  public parking 467
  residential 467
Garbage disposal 410, 417
Garden house 467
  house demo 452
Garment checking bag 407
Gas cock 491
  connector 503
  fired boiler 547
  fired furnace 549
  fired infrared heater 549
  fired space heater 550
  furnace 549
  generator set 590
  heat air conditioner 556

incinerator 428
  log 389
  pipe 685
  station formwork 56
  station island formwork 56
  station tank 530
  stop valve 491
  vent 546
  water heater commercial 512
  water heater instantaneous 511
  water heater residential 511
  water heater tankless 511
Gas-fired duct heater 549
  heating 549, 550
Gasket joint 260
  neoprene 260
Gas/oil combination boiler 548
Gasoline generator 591
  piping 686
  tank 531
Gate chain link 660
  day 283
  fence 659
  folding 382
  opener driveway security 283
  parking 404
  security 382
  security driveway 282
  slide 659
  swing 659
  valve 526, 527
  valve soldered 492
Gauging plaster 326
General air conditioner 526
  boiler 490
  contractor's overhead 817
  equipment rental 715
  fill 627
Generator diesel 590
  diesel-engine-driven 590
  electric 717
  emergency 590
  gasoline 591
  natural gas 590
  set 590
  steam 425
Geodesic dome 467
  dome demo 452
  hemisphere greenhouse 463
Geo-grid 663
Geogrid 663
GFI receptacle 578
Girder formwork 52
  reinforcing 71
  wood 181, 189
Girt steel 134
Glass 310, 312, 313
  acoustical 315
  and glazing 315
  bead 285
  bevel 311
  block 110
  bulletin board 374
  bulletproof 315
  curtain wall 288
  demolition 264
  door 265, 279, 286, 287
  door astragal 308
  door shower 388
  door sliding 280
  door swinging 287
  fiber rod reinforcing 72
  fire resistant 311
  float 310
  full vision 311
  heat reflective 312

insulating 311
laminated 315
lead 471
lined water heater 411
low emissivity 311
mirror 313, 387
mosaic 335
mosaic sheet 335
obscure 312
patterned 312
pipe 522
reduce heat transfer 311
reflective 313
safety 311
sandblast 311
sheet 312
shower stall 388
solar film 314
spandrel 312
tempered 310
tile 313
tinted 310
window 312
window wall 286
wire 312
Glassware sterilizer 419
  washer 419
Glaze coat 365
Glazed aluminum curtain wall ... 288
  block 109
  brick 102
  ceramic tile 333
  concrete block 109
  wall coating 368
Glazing 310, 311
  application 311
  plastic 313
  polycarbonate 313
  productivity 855
Globe valve 527
  valve bronze 493
Glove bag 42
  box 419
Glued laminated 193
Goalpost 425
  football 425
  soccer 425
Golf shelter 424
  tee surface 349
Gore line 657
Grab bar 386
Gradall 621, 713
Grade beam conc. placement 79
  beam formwork 56
  fine 668
Grader motorized 713
Grading 615, 633
  rough 615
  site 615
Graffiti chemical removal 368
  resistant treatment 367
Grandstand 463
Granite 112
  building 112
  chip 667
  conductive floor 347
  countertop 441
  curb 113, 656
  demolition 96
  indian 656
  paver 113
  paving block 655
  reclaimed or antique 113
  sidewalk 655
Granolithic finish concrete 88
Grass cloth wallpaper 351

lawn 668
  seed 668
  surfacing artificial 658
Grating aluminum 157
  aluminum floor 157
  aluminum mesh 157
  aluminum plank 157
  area wall 397
  area way 397
  fiberglass 216
  fiberglass floor 217
  floor 217
  frame 159
  galvanized plank 158
  plank 158
  stainless bar 158
  stainless plank 159
  stair 154
  steel 157
  steel floor 157
  steel mesh 158
Gravel base 651
  fill 612, 627
  pack well 679
  pea 667
  roof 51
  stop 251
Gravity retaining wall 662
Grease interceptor 508
  trap 508
Green roof membrane 236
  roof soil mixture 236
  roof system 236
Greenhouse 462
  air supported 461
  commercial 462
  cooling 462
  demo 452
  geodesic hemisphere 463
  residential 462
Grid spike 172
Griddle 415
Grill coiling 280
Grille air return 543
  aluminum 543
  decorative wood 212
  duct 350
  filter 543
  painting 364
  plastic 543
  side coiling 280
  top coiling 280
  window 294
Grinder concrete floor 712
  shop 420
  system pump 508
Grinding concrete 82
  floor 82
Grooved block 105
  joint ductile iron 500, 501
  joint fitting 499
  joint pipe 499
Ground 188
  box hydrant 505
  clamp water pipe 571
  cover plant 669
  face block 105
  fault protection 582
  rod 571
  socket 423
  water monitoring 16
  wire 571
Grounding 571
  & conductor 567-570
  wire brazed 571
Group shower 517

Index

wash fountain 517
Grout 97
  cavity wall 98
  cement 98, 640
  concrete 98
  concrete block 98
  door frame 98
  epoxy 333, 335, 640
  metallic non-shrink 88
  non-metallic non-shrink ... 88
  pump 718
  tile 333
  topping 88
  wall 98
Grouted bar 73
  strand 73
Guard corner 129, 385
  gutter 254
  house 467
  lamp 598
  service 24
  snow 256
  wall 129, 385
  wall & corner 385
  window 162
Guardrail scaffold 19
  temporary 24
Guide rail 692
  rail cable 692
  rail removal 30
  rail timber 692
  rail vehicle 692
Guide/guard rail 692
Gunite 83
  drymix 83
  mesh 83
Gutter 253, 465
  aluminum 253
  copper 253
  demolition 222
  guard 254
  lead coated copper 253
  monolithic 655
  stainless 253
  steel 253
  strainer 254
  valley 465
  vinyl 253
  wood 253
Gutting 34
  demolition 34
Guyed derrick crane 721
  tower 688
Gym floor underlayment 342
  mat 422
Gymnasium divider curtain .. 423
  equipment 421, 422
  floor 342, 349
Gypsum block demolition 32
  board 328
  board accessory 332
  board high abuse 331
  board leaded 470
  board partition 321
  board partition NLB 321
  board removal 319
  board system 320
  cover board 230
  drywall 328, 331
  fabric wallcovering 351
  high abuse drywall 331
  lath 324
  lath nail 171
  partition 383
  partition NLB 321
  plaster 326

restoration 318
roof deck 87
shaft wall 320
sheathing 193
sound dampening panel 330
underlayment poured 88
wallboard repair 318
weatherproof 193

H

H pile 642
Hair interceptor 508
Half round molding 206
Halotron 392
Hammer bush 81
  chipping 716
  demolition 713
  diesel 714
  drill rotary 716
  hydraulic 713, 717
  pile 713
  pile mobilization 613
  vibratory 714
Hammermill 719
Hand carved door 270
  clearing 614
  dryer 386
  excavation 617, 619, 620, 626
  hole 674
  hole concrete 687
  scanner unit 309
  split shake 235
  trowel finish concrete ... 80
Handball court 460
  court backstop 424
Handball/squash court 424
Handicap drinking fountain .. 520
  fixture support 518
  lever 305
  opener 299
  ramp 76
  water cooler 521
Handle door 438
Handling material 25, 699
  waste 428, 429
Hand-off-automatic control .. 590
Handrail 385
  aluminum 350
  fiberglass 216
  wood 206, 211
Hangar 468
  demo 452
Hanger accessory formwork ... 60
  beam 172
  joist 172
Hanging lintel 130
  wire 338
Hardboard cabinet 437
  molding 208
  overhead door 284
  paneling 208
  tempered 208
  underlayment 191
Hardener concrete 81
  floor 50
Hardware 299
  apartment 300
  cabinet 438
  cloth 661
  door 300
  drapery 435
  finishes 854
  motel/hotel 300
  window 300

Hardwood carrel 447
  floor stage 420
  grille 212
Harness body 21
Harrow disc 713
Hasp 305
Hat and coat strip 387
  rack 395
Hatch floor 279
  roof 256
  smoke 256
Haul earthwork 627
Hauling 627
  cycle 627
Haunch slab 76
Hay 667
  bale 637
Hazardous waste cleanup 40
  waste containment 38
  waste disposal 40
  waste handling 429
HDPE 38
  distribution box 680
  infiltration chamber 680
Head sprinkler 486
Header load-bearing stud wall .. 144
  pipe wellpoint 722
  wood 187
Header/beam boxed 148
Headrail metal toilet component . 379
  plastic-laminate toilet .. 380
Health club equipment 421
Hearth 117
Heat baseboard 559
  electric baseboard 561
  exchanger 552
  exchanger plate-type 552
  exchanger shell-type 552
  greenhouse 463
  pump 557
  pump air-source 557
  pump residential 581
  recovery 428
  reflective glass 312
  temporary 52
  therapy 426
  transfer fluid 551
  transfer package 512
Heated pad 52
Heater & fan bathroom 541
  electric 411
  floor mounted space 550
  gas fired infrared 549
  gas residential water ... 511
  gas water 512
  gas-fired duct 549
  infrared 549
  oil fired space 550
  sauna 460
  space 717
  swimming pool 548
  terminal 549
  tubular infrared 549
  unit 550, 558
  warm air 549
  water 411, 581
  water residential electric .. 511
Heating & cooling classroom 558
  control valve 533
  electric 547, 561
  estimate electrical 561
  gas-fired 549, 550
  hot air 549, 558
  hydronic 548, 559
  insulation 497, 529
  kettle 720

solar 551
  steam-to-water 552
  subsoil 457
Heavy construction 665
  duty shoring 19
  framing 189
  lifting 821
  rail track 690
  timber 189
Helicopter 721
Hemlock column 212
Hex bolt steel 123
Hexagonal face block 106
High abuse drywall 331
  build coating 368
  chair reinforcing 68
  efficiency flush valve .. 517
  efficiency toilet (het) flush 517
  efficiency urinal 517
  installation conduit 572
  rib lath 324
  strength block 107
  strength bolt 124, 844
  strength decorative block . 105
  strength steel 134, 846
High-security chain link fence ... 660
Highway equipment rental 719
  paver 655
  sign 378
Hinge 305
  brass 305
  cabinet 438
  continuous 306
  electric 306
  hospital 306
  paumelle 306
  prison 306
  residential 306
  school 306
  security 306
  special 306
  stainless steel 305
  steel 305
  wide throw 306
Hip rafter 186
Hockey dasher 457
  scoreboard 422
Hoist 698
  chain 699
  chain fall 721
  electric 699
  overhead 699
  personnel 721
  tower 721
Holder closer 301
  screed 69
Holding tank 706
Holdown 172
Hole drill 638
Hollow core door 271, 275
  metal door 267
  metal frame 266
  metal stud partition 325
  precast concrete plank ... 84
  wall anchor 121
Honed block 107
Honing floor 82
Hood and ventilation equip. .. 545
  exhaust 411, 419
  fume 419
  range 411
Hook clevis 162
  coat 387
  robe 387
Hopper refuse 480
Horizontal borer 719

895

Index

boring 674
boring machine 712
directional drill 719
drilling 674
Horn fire 607
Hose adapter fire 485
air 716
bibb sillcock 515
braided bronze 503
discharge 717
equipment 485
fire 485
metal flexible 503
nozzle 485
rack 485
rack cabinet 391
suction 717
valve cabinet 391
water 717
Hospital cabinet 439, 440
casework 439
door hardware 300
equipment 425
furniture 447
hinge 306
kitchen equipment 416
partition 382
tip pin 305
Hot air furnace 549
air heating 549, 558
water boiler 547, 548
water dispenser 511
water heating 547, 559
water-steam exchange 512, 552
water-water exchange 552
Hotel cabinet 388
furniture 445
lockset 302, 303
House garden 467
guard 467
safety flood shut-off 493
telephone 603
Housewrap 232
Hubbard tank 426
Humidification equipment 463
Humidifier 412, 562
duct 562
room 562
Humus peat 667
HVAC axial flow fan 539
demolition 526
duct 536
equipment insulation 529
louver 544
piping specialty 532, 534
power circulator & ventilator . . 541
Hydrant building fire 484
fire 677
ground box 505
removal 29
remove fire 29
wall 505
water 505
Hydrated lime 97
Hydraulic chair 426
crane 699, 721
dredge 696
excavator 712
hammer 713, 717
jack 721
jack climbing 721
jacking 821
lift 479
seeding 668
Hydrodemolition 29
concrete 29

Hydromulcher 719
Hydronic heating 548, 559
heating control valve 533
heating convector 559
Hypalon neoprene roofing 247

I

Ice machine 417
rink dasher 457
skating equipment 457
Icemaker 410, 417
I-joist wood & composite 193
Impact barrier 692
wrench 716
Impact-resistant aluminum
window 289
Implosive demolition 31
Incandescent fixture 593
Incinerator gas 428
municipal 428
waste 428
Inclined ramp 479
Incubator 419
Indian granite 656
Indicating panel burglar alarm . . 606
Indirect-fired water chiller 553
Indoor athletic carpet 349
fountain 457
Induced draft fan 546
Industrial address system 603
chimneys 831
door 278, 280
dryer 407
equipment installation 403
folding partition 384
lighting 592
railing 156
safety fixture 520
window 290
Inert gas 39
Infiltration chamber 680
chamber HDPE 680
Infrared broiler 415
detector 606
heater 549
heater gas fired 549
heater tubular 549
Inlet curb 656
In-line fan 540
Insecticide 637
Insert concrete formwork 67
slab lifting 69
Inspection technician 16
Installation add elevated 490
extinguisher 392
Instantaneous gas water heater . . 511
Instrument switchboard 582
Insulated glass spandrel 312
panel 231
precast wall panel 86
Insulating concrete formwork . . . 59
glass 311
sheathing 190
Insulation 226, 229
batt 227
blanket 528
blower 717
blown-in 229
board 226
boiler 529
breeching 529
building 229
cavity wall 228
ceiling 226, 229

cellulose 228
composite 231
duct 528
duct thermal 528
equipment 529
exterior 231
fiberglass . . 226, 227, 229, 465, 528
finish system exterior 231
floor 226
foam 229
foamed in place 228
heating 497, 529
HVAC equipment 529
insert concrete block 105
isocyanurate 226
loose fill 228
masonry 229
mineral fiber 228
pipe 497
piping 497
polystyrene 226, 228, 458
reflective 229
refrigeration 458
removal 42, 223
rigid 226
roof 230, 465
roof deck 230
shingle 233
spray 229
sprayed-on 229
subsoil 457
vapor barrier 232
vermiculite 228
wall 226, 465
wall blanket 227
Insurance 12, 816
builder risk 12
equipment 12
public liability 12
Intake-exhaust louver 544
vent 545
Integral color concrete 50
topping concrete 88
waterproofing 52
waterproofing concrete 51
Interceptor 508
grease 508
hair 508
metal recovery 508
oil 508
Intercom 603
Interior beam formwork 52
door frame 213
LED fixture 593
light fixture 592
lighting 592, 593
paint door & window 361
paint wall & ceiling 366
planter 448
pre-hung door 277
residential door 274
shutter 436
wood door frame 213
Interlocking concrete block 107
precast concrete unit 654
Intermediate metallic conduit . . 572
Interval timer 577
Intrusion detection 606
system 606
Intumescent coating 368
Invert manhole 684
Inverted bucket steam trap 533
Iron alloy mechanical joint pipe . 522
body valve 526
Ironer laundry 408
Ironing center 389

Ironspot brick 340
Irrigation system sprinkler 666
Isocyanurate insulation 226
IV track system 382

J

Jack cable 722
hydraulic 721
ladder 21
post adjustable 128
pump 19
roof 186
screw 639
Jackhammer 621, 716
Jacking 674
Jet water system 679
Jeweler safe 409
Jib crane 699
Job condition 11
Jockey pump fire 488
Joint assembly expansion 262
control 63, 99
expansion 63, 254, 325, 532
filler 260
gasket 260
push-on 675
reinforcing 98
roof 255
sealant replacement 222
sealer 260, 261
Jointer shop 420
Joist bridging 140
concrete 75
connector 172
deep-longspan 138
demolition 167
framing 182
hanger 172
longspan 139
metal framing 149
open web bar 140
precast concrete 85
removal 167
truss 141
web stiffener 149
wood 182, 193
Jumbo brick 102
Junction box 572
Jute mesh 637

K

Keene cement 326
Kennel door 279
fence 661
Ketone ethylene ester roofing . . . 246
Kettle 415
heating 720
tar 718, 720
Key cabinet 310
keeper 394
Keyless lock 303
Keyway footing 55
Kick plate 306
plate door 307
Kiln vocational 420
King brick 102
Kiosk 468
Kitchen appliance 410
cabinet 437
equipment 415
equipment fee 10
exhaust fan 541

metal cabinet 440
sink 514
sink faucet 515
sink residential 514
unit commercial 412
K-lath 324
Knob door 303
Kraft paper 232

L

Labeled door 268, 272
frame 266
Labor adjustment factor 490, 564
formwork 828
modifier 490
Laboratory analytical service 47
cabinet 439
casework metal 439
countertop 441
equipment 419, 439
safety equipment 419
sink 419
table 419
test equipment 419
Ladder alternating tread 155
exercise 422
extension 717
inclined metal 155
jack 21
monkey 423
rolling 20
ship 155
swimming pool 456
towel 387
type cable tray 574
vertical metal 155
Lag screw 125
screw shield 121
Lagging 641
Lally column 128
Laminated beam 193
countertop 440
epoxy & fiberglass 368
framing 193
glass 315
glued 193
lead 470
roof deck 190
sheet flashing 250
veneer member 194
wood 193
Lamp electric 597
guard 598
LED 598
metal halide 598
post 163, 595
Lampholder 589
Lamphouse 418
Lance concrete 717
Landfill compactor 714
Landing metal pan 154
stair 346
Landscape fee 10
surface 349
Laser level 717
Latch deadlocking 303
set 302, 303
Lateral arm awning retractable 396
Latex caulking 261
underlayment 343
Lath demolition 318
gypsum 324
metal 320, 324
Lath, plaster and gypsum board 856
Lath rib 324
Lathe shop 420
Lattice molding 206
Lauan door 275
Laundry equipment 407, 408
faucet 515
folder 407
ironer 408
presser 408
sink 515
spreader 407
Lava stone 113
Lavatory commercial 516
faucet 515
pedestal type 513
removal 491
residential 513
sink 513
support 518
vanity top 513
wall hung 513
Lawn grass 668
mower 717
seed 668
Lazy susan 437
Leaching field chamber 680
pit 680
Lead barrier 353
caulking 506
coated copper downspout 252
coated copper gutter 253
coated downspout 252
flashing 249
glass 471
gypsum board 470
lined darkroom 459
lined door frame 470
paint encapsulation 44
paint remediation 44
paint remediation methods 825
paint removal 45
plastic 471
roof 248
salvage 491
screw anchor 122
sheet 470
shielding 470
testing 44
Leads pile 713
Leak detection 608
detection probe 608
detection tank 608
Lean-to type greenhouse 462
Lectern 430
Lecture hall seating 448
LED fixture exterior 596
fixture interior 593
floodlight 597
lamp 598
lighting parking 596
luminaire roadway 596
Ledger framing 186
wood 186
Lens movie 418
Let-in bracing 180
Letter sign 376
slot 394
Level laser 717
Leveler dock 405
truck 405
Leveling jack shoring 19
Lever handicap 305
Lexan 313
Liability employer 12
insurance 817
Library equipment demolition 402
furnishing 447
furniture 447
shelf 447
Life guard chair 456
Lift 479
automotive 479
bariatric 426
hydraulic 479
scissor 699, 715
slab 77, 837
wheelchair 479
Lifter platform 405
Lifting equipment rental 720
Light base formwork 56
border 595
dental 426
dock 406
exam 425
exit 594
fixture interior 592, 595
fixture troffer 593
loading dock 406
nurse call 604
pole 687
pole aluminum 595
pole steel 596
shield 692
stand 40
strobe 595
support 128
temporary 17
tower 717
underwater 456
Lighting darkroom 408
emergency 594
exit 594
explosionproof 595
exterior fixture 596
fixture exterior 596
fountain 522
incandescent 593
industrial 592
interior 592, 593
outlet 579
pole 595
residential 579
stage 595
strip 592
surgical 425
temporary 17
theatrical 595
unit emergency 594
Lightning protection demo 452
suppressor 576
Lightweight aggregate 51
angle framing 129
block 108
channel framing 130
column 128
concrete 76, 78, 841
floor fill 76
framing 129
insulating concrete 87
junior beam framing 130
natural stone 113
tee framing 130
Zee framing 130
Lime 97
finish 97
hydrated 97
Limestone 113, 668
coping 117
Line gore 657
remover traffic 720
Linen chute 480
collector 480
wallcovering 351
Liner cement 674
duct 539
flue 101, 116
non-fibrous duct 539
pipe 674
Lint collector 407
Lintel 113
block 108
concrete block 108
galvanizing 131
hanging 130
precast concrete 87
steel 130
Liquid chiller centrifugal 553
chiller screw 554
Load test pile 612
Load-bearing stud partition
framing 145
stud wall bracing 143
stud wall bridging 144
stud wall framing 145
stud wall header 144
Loader front end 621
skidsteer 715
tractor 714
vacuum 41
wheeled 714
windrow 720
Loading dock 405
dock equipment 405
dock light 406
Loam 612, 615
Lobby collection box 394
Lock electric release 303
entrance 304
keyless 303
time 283
tubular 303
Locker metal 392
plastic 393
steel 392
wall mount 392
wire mesh 392
Locking receptacle 579
Lockset communicating 302, 303
cylinder 302
hotel 302, 303
mortise 303
Locomotive tunnel 720
Log chipper 714
electric 390
gas 389
skidder 714
Longspan joist 139
Loose fill insulation 228
Louver 315
aluminum 315, 544
aluminum operating 544
coating 544
cooling tower 544
door 315
fixed blade 544
HVAC 544
intake-exhaust 544
midget 315
mullion type 544
redwood 316
ventilation 316
wall 316
wood 212
Louvered door 275, 277
Lowbed trailer 720
Low-voltage silicon rectifier 589
switching 589
switchplate 589

transformer 589
Lube equipment 404
Lubrication equipment 404
Lug terminal 568
Lumber 181
  core paneling 208
  plastic 215
  product prices 851
  recycled plastic 214
  structural plastic 215
  treated 186
Luminaire roadway 596
Luminous ceiling 337, 597

M

Macadam 651
  penetration 651
Machine autoscrub 716
  excavation 626
  screw 125
  trowel finish concrete 80
  welding 719
  X-ray 606
Machinery anchor 67
Magazine shelving 447
Magnesium dock board 405
  oxychloride 258
Magnetic astragal 308
  motor starter 586
  particle test 16
Mahogany door 270
Mail box 393
  box call system 603
  box commercial 393
  slot 394
Main drain 509
  office expense 13
Maintenance carpet 318
  railroad 690
  railroad track 690
Make-up air unit 555
Mall front 288
Management fee construction 10
Manhole 685
  brick 685
  concrete 685, 687
  cover 683
  electric service 687
  frame and cover 683
  invert 684
  raise 684
  removal 29
  step 685
Man-made soil mix 236
Mansard aluminum 238
Mantel beam 212
  fireplace 212
Manual dumbwaiter 474
Map rail 374
Maple countertop 440
  floor 342
Marble 114
  chip 667
  chip white 51
  coping 117
  countertop 441
  floor 114
  screen 381
  shower stall 388
  sill 116
  soffit 114
  stair 114
  synthetic 341
  tile 341

Marina small boat 696
Marine equipment rental 722
Marker boundary and survey 28
Markerboard electronic 374
Mark-up cost 13
Masking 95
Mason scaffold 19
Masonry accessory 99
  aggregate 97
  anchor 98
  base 110
  brick 101
  cement 96
  cleaning 94
  color 97
  cutting 34
  demolition 32, 95, 319
  fireplace 117
  flashing 249
  flooring 340
  furring 187
  insulation 229
  manhole 685
  nail 171
  painting 365
  panel 111
  panel pre-fabricated 111
  pointing 94
  reinforcing 98, 842
  removal 30, 96
  restoration 97
  saw 718
  sawing 94
  selective demolition 95
  sill 116
  stabilization 94
  step 650
  testing 15
  toothing 32
  ventilator 99
  wall 104, 664
  wall tie 98
  waterproofing 97
Mass notification system 609
Massage table 422
Mast climber 21
Mastic floor 347
Mastic-backed flashing 250
Mat blasting 621
  concrete placement foundation . 79
  floor 443
  foundation 76
  foundation formwork 56
  gym 422
  wall 422
Material door and window 264
  handling 25, 699
  handling belt 698
  handling conveyor 698
  handling deconstruction 37
  handling system 698
  removal/salvage 35
Materials concrete 835
MC cable 568
Meat case 412
Mechanical 862
  dredging 696
  duct 536
  equipment demolition 526
  fee 10
  media filter 545
  seeding 668
Median barrier 692
  precast 692
Medical distiller 425
  equipment 425-428

exam equipment 425
  sterilizer 425
  sterilizing equipment 425
  waste cart 429
  waste disposal 429
  waste sanitizer 429
  X-ray 427
Medicine cabinet 388
Medium-voltage cable 567
Membrane elastomeric 333
  flashing 244
  protected 247
  roofing 243, 247
  waterproofing 224
Mercury vapor lamp 598
Mesh fence 659
  gunite 83
  partition 382
  security 325
  stucco 324
Metal bin retaining wall 663
  bookshelf 447
  bracing 180
  butt frame 267
  cabinet 440
  cabinet school 439
  canopy 396
  case good 444
  casework 310, 439
  ceiling linear 338
  ceiling panel 337, 339
  chalkboard 373, 384
  chimney 389
  cleaning 120
  deck 142
  deck acoustical 143
  deck composite 141
  deck ventilated 143
  demolition 120
  detector 606
  door 265, 269, 284, 465
  door frame 465
  door residential 269
  ductwork 536
  facing panel 239
  fascia 251
  fence 659
  fire door 270
  flashing 465
  flexible hose 503
  floor deck 143
  flue chimney 546
  frame 266
  frame channel 267
  framing parapet 151
  furring 322
  halide lamp 598
  in field galvanizing 369
  in field painting 369
  in shop galvanizing 120
  interceptor dental 508
  joist bracing 147
  joist bridging 147
  joist framing 149
  laboratory casework 439
  ladder inclined 155
  lath 320, 324
  locker 392
  molding 325
  nailing anchor 121
  overhead door 284
  paint & protective coating 120
  pan ceiling 337
  pan landing 154
  pan stair 154
  parking bumper 656

pipe 681
pipe removal 491
plate stair 154
pressure washing 120
rafter framing 151
recovery interceptor 508
roof 237
roof parapet framing 151
roof truss 134
sandblasting 120
sash 290
screen 290
sheet 248
shelf 394
shelving food storage 414
shingle 234
siding 237, 238
sign 376
soffit 242
steam cleaning 120
stud 320
stud demolition 319
stud NLB 323
support assembly 322
threshold 305
tile 332
toilet component door 379
toilet component headrail 379
toilet component panel 379
toilet component pilaster 379
toilet partition 378
trash receptacle 443
truss framing 153
water blasting 120
window 289, 290, 466
wire brushing 120
Metal-clad double hung wind. . . . 292
  window bow & bay 293
  window picture & sliding 292
Metal-framed skylight 298
Metallic conduit intermediate . . . 572
  foil 229
  hardener 81
  non-shrink grout 88
Meter electric 582
  flow 534
  steam condensate 535
  venturi flow 534
  water supply 503
  water supply domestic 503
Metric conversion 836
  conversion factors 811
  rebar specs 829
Microphone 603
Microtunneling 674
Microwave oven 410
Mill construction 189
  extra steel 134
Millwork 196
  & trim demolition 170
  demolition 170
Mineral fiber ceiling 336, 337
  fiber insulation 228
  fiberboard panel 353
  insulated cable 569
  roof 248
Minor site demolition 29
Mirror 313, 387
  door 313
  glass 313, 387
  plexiglass 313
  wall 313
Miscellaneous painting 356, 361
Mix design asphalt 15
  planting pit 668
Mixed bituminous conc. plant . . . 612

Mixer concrete ... 712
food ... 414
mortar ... 712, 717
plaster ... 717
road ... 720
Mixes concrete ... 835, 836
Mixing valve ... 515
Mobile shelving ... 447
X-ray ... 427
Mobilization ... 613, 646
air-compressor ... 613
equipment ... 640, 646
or demobilization ... 22
Model building ... 10
Modification to cost ... 11
Modified bitumen roof ... 244
bitumen roofing ... 852
bituminous barrier sheet ... 233
bituminous membrane SBS ... 245
Modifier labor ... 490
Modular office system ... 383
playground ... 424
Modulus of elasticity ... 15
Moil point ... 716
Moisture barrier ... 225
content test ... 16
Mold abatement ... 46
abatement work area ... 46
contaminated area demolition ... 46
Molding base ... 196
bed ... 199
birch ... 206
brick ... 206
casing ... 197
ceiling ... 199
chair ... 206
cornice ... 199, 202
cove ... 199
crown ... 199
dentil crown ... 200
exterior ... 200
hardboard ... 208
metal ... 325
pine ... 202
soffit ... 207
trim ... 206
window and door ... 207
wood transition ... 342
Money safe ... 409
Monitor support ... 128
Monitoring sampling ... 47
Monkey ladder ... 423
Monolithic gutter ... 655
terrazzo ... 345
Monorail ... 698
Monument survey ... 28
Mop holder strip ... 387
roof ... 244
sink ... 516
Mortar ... 97
admixture ... 97
Mortar, brick and block ... 843
Mortar cement ... 334, 335, 842
masonry cement ... 97
mixer ... 712, 717
pigment ... 97
Portland cement ... 97
restoration ... 97
sand ... 97
testing ... 15
thinset ... 335
Mortise lockset ... 303
Mortuary equipment ... 428
refrigeration ... 428
Mosaic glass ... 335
Moss peat ... 667
Motel/hotel hardware ... 300
Motion detector ... 606
Motor ... 600
connection ... 576
dripproof ... 600
electric ... 600
explosionproof ... 600
starter ... 586
starter & control ... 586
starter enclosed & heated ... 586
starter magnetic ... 586
starter w/circuit protector ... 586
starter w/fused switch ... 586
support ... 128
Motorized grader ... 713
roof ... 463
Moulded door ... 273
Mounted booth ... 446
Mounting board plywood ... 192
Movable louver blind ... 436
office partition ... 383
wall demolition ... 319
Moveable bulkhead ... 456
Movie equipment ... 418
equipment digital ... 418
lens ... 418
projector ... 418
screen ... 417
Moving building ... 38
ramp ... 479
ramps and walks ... 862
shrub ... 671
stair & walk ... 478
structure ... 38
tree ... 671
walk ... 479
Mower lawn ... 717
Mowing brush ... 614
Muck car tunnel ... 720
Mud pump ... 712
sill ... 186
trailer ... 719
Mulch bark ... 667
ceramic ... 667
stone ... 667
Mulcher power ... 714
Mulching ... 667
Mullion type louver ... 544
vertical ... 286
Multi-blade damper ... 537
Multi-channel rack enclosure ... 602
Multizone air conditioner
rooftop ... 556
Municipal incinerator ... 428
Muntin window ... 294
Mushroom ventilator stationary ... 545
Music room ... 459
Mylar tarpaulin ... 23

N

Nail ... 171
common ... 171
lead head ... 470
stake ... 62
Nailer pneumatic ... 716, 717
wood ... 183
Nailing anchor ... 121
Napkin dispenser ... 387
Natural fiber wall-covering ... 351
gas generator ... 590
Needle beam cantilever ... 640
Neoprene adhesive ... 225
expansion joint ... 254
flashing ... 250
floor ... 347
gasket ... 260
roof ... 247
waterproofing ... 225
Net safety ... 18
tennis court ... 658
Netting bird control ... 399
flexible plastic ... 399
Newel wood stair ... 210
Newspaper rack ... 447
Night depository ... 409
No hub pipe ... 506
Non-chloride accelerator
concrete ... 78
Non-destructive testing ... 16
Non-metallic non-shrink grout ... 88
Non-removable pin ... 305
Norwegian brick ... 102
Nosing rubber ... 343
safety ... 343
stair ... 343, 345
Nozzle dispersion ... 487
fire hose ... 485
fog ... 485
playpipe ... 485
Nurse call light ... 604
call system ... 604
speaker station ... 604
station cabinet ... 439, 440
Nursery item travel ... 669
Nursing home bed ... 447
Nut remove ... 120
Nylon carpet ... 348
nailing anchor ... 121

O

Oak door frame ... 213
floor ... 342
molding ... 206
paneling ... 208
threshold ... 214
Oakum caulking ... 506
Obscure glass ... 312
Observation well ... 679
Occupancy sensor ... 582
Offhighway dumptruck ... 715
Office & storage space ... 17
chair ... 445
expense ... 13
field ... 17
floor ... 349
furniture ... 444, 445
overhead ... 13
partition ... 383
partition movable ... 383
safe ... 408
system modular ... 383
trailer ... 17
Oil fired boiler ... 548
fired furnace ... 549
fired space heater ... 550
formwork ... 62
interceptor ... 508
storage tank ... 531
Oil-filled transformer ... 583
Oil-fired water heater ... 512
water heater commercial ... 512
water heater residential ... 512
Olive knuckle hinge ... 306
Omitted work ... 13
One piece astragal ... 308
One-way vent ... 256
Onyx ... 345
Open web bar joist ... 140
Opener automatic ... 299
door ... 264, 281, 284, 299
driveway security gate ... 283
handicap ... 299
industrial automatic ... 299
Opening framing ... 465
roof frame ... 130
Operable partition ... 384
Operating room equipment ... 427
Operator automatic ... 299
Option elevator ... 477
Ornamental aluminum rail ... 164
column ... 164
glass rail ... 164
railing ... 164
steel rail ... 164
wrought iron rail ... 164
OSB faced panel ... 188
OSHA testing ... 43
Outdoor bleacher ... 463
fountain ... 457
Outlet box steel ... 572
lighting ... 579
Outrigger wall pole ... 398
Oval arch culvert ... 683
Oven ... 410, 415
cabinet ... 437
convection ... 415
microwave ... 410
Overbed table ... 447
Overhang eave ... 465
endwall ... 465
Overhaul ... 34
Overhead & profit ... 13
bridge crane ... 698
commercial door ... 283
contractor ... 25, 817
door ... 283, 284
hoist ... 699
office ... 13
Overlapping astragal ... 308
Overlay face door ... 271, 272
Overpass ... 665
Oversized brick ... 101
Overtime ... 12, 13
Oxygen lance cutting ... 35

P

P trap ... 507
trap running ... 506
Package chute ... 480
receiver ... 409
Packaged terminal air
conditioner ... 556
Packaging waste ... 43
Pad bearing ... 126
carpet ... 348
commercial grade carpet ... 348
equipment ... 76
heated ... 52
vibration ... 126
Padding carpet ... 348
Paddle blade air circulator ... 541
fan ... 580
tennis court ... 424
Paint & coating ... 354
& coating interior floor ... 363
& protective coating metal ... 120
aluminum siding ... 357
and protective coating ... 120
chain link fence ... 356
door & window exterior ... 358
door & window interior ... 361-363
encapsulation lead ... 44

exterior miscellaneous 356
fence picket 356
floor 364
floor concrete 363, 364
floor wood 364
interior miscellaneous 364
remediation lead 44
removal 45
removal floor 82
removal lead 45
siding 357
sprayer 717
striper 720
trim exterior 359
wall & ceiling interior 365, 366
wall masonry exterior 360
Paint/coating cabinet &
casework 361
Painted marking airport 657
pavement marking 657
Painting 859
balustrade 364
bar joist 141
booth 404
casework 361
ceiling 365, 366
clapboard 357
concrete block 365
cornice 364
decking 357
drywall 365, 366
electrostatic 365
exterior siding 357
grille 364
masonry 365
metal in field 369
miscellaneous 356, 361
parking stall 657
pavement 657
pipe 364
plaster 365, 366
railing 356
reflective 657
shutter 356
siding 357
stair stringer 357
steel 369
steel siding 357
stucco 357
swimming pool 456
temporary road 657
tennis court 658
thermoplastic 657
trellis/lattice 357
trim 364
truss 364
wall 357, 366
window 361
Palladian window 294
Pallet rack 394
Pan shower 249
slab 75
stair metal 154
tread concrete 80
Panel acoustical 384
and device alarm 607
board residential 576
brick wall 111
ceiling 336
ceiling acoustical 336
door 273, 274
facing 115
fiberglass 23, 238, 353
fiberglass refrigeration ... 458
FRP 351
insulated 231

masonry 111
metal facing 239
metal toilet component ... 379
mineral fiberboard 353
OSB faced 188
plastic-laminate toilet .. 380
portable 384
precast concrete double wall .. 85
prefabricated 231
sandwich 238
shearwall 188
sound 470
sound absorbing 353
sound dampening gypsum ... 330
spandrel 312
steel roofing 237
structural 188
structural insulated 188
system 209
vision 298
wall 117
wood folding 436
woven wire 382
Panelboard 585
electric 585
w/circuit-breaker 585
Paneled door 269
pine door 275
Paneling 208
birch 208
board 209
cedar 209
cutout 32
demolition 170
hardboard 208
plywood 208
redwood 209
wood 208
Panelized shingle 234
Panic exit device 301
Paper building 232
sheathing 232
Paperbacked flashing 250
Paperhanging 350
Paperholder 387
Paraffin bath 426
Parallel bar 422, 426
Parapet metal framing 151
Parging cement 224
Park bench 665
Parking barrier 657
barrier precast 656
bumper 657
bumper metal 656
bumper plastic 656
bumper wood 657
collection equipment 404
control equipment 404
equipment 404
garage public 467
gate 404
gate equipment 404
LED lighting 596
lot paving 652
marking pavement 657
stall painting 657
ticket equipment 404
Parquet floor 341
wood 341
Part bin 394
Particle board underlayment 191
core door 271
Parting bead 207
Partition 382, 383
acoustical 383, 384
anchor 98

block 108
blueboard 321
bulletproof 409
concrete block 108
demolition 319
demountable 382, 383
door 383
drywall 321
folding accordion 384
folding leaf 384
framing load bearing stud 145
framing NLB 323
gypsum 383
hospital 382
mesh 382
movable office 383
NLB drywall 321
NLB gypsum 321
office 383
operable 384
plaster 320
portable 384
refrigeration 458
shower 114, 388
sliding 384
steel 383, 384
support 127
thin plaster 321
tile 103
toilet 114, 378-380
toilet stone 381
wall 321
wall NLB 321
wire 382
wire mesh 382
wood frame 183
woven wire 382
Passage door 275, 277
Passenger elevator 474
elevators 861
Pass-thru drawer 431
Patch core hole 15
formwork 62
roof 222
Patching asphalt 612
concrete floor 50
concrete wall 50
Patient care equipment 426
nurse call 604
Patio 651
block 114, 244
block concrete 654
door 280
Patio/deck canopy 396
Patterned glass 312
Paumelle hinge 306
Pavement 654
asphalt 658
asphaltic 652
berm 655
breaker 716
bucket 719
demolition 29
emulsion 650
highway concrete 653
marking 656
marking painted 657
painting 657
parking marking 657
planer 720
profiler 720
replacement 652
sealer 658
slate 115
widener 720
Paver asphalt 717

bituminous 717
cold mix 720
concrete 718
floor 340
highway 655
roof 257
shoulder 720
tile exposed aggregate ... 654
Paving asphalt 652
asphaltic concrete 652
athletic 658
block granite 655
brick 654
parking lot 652
Pea gravel 667
stone 51
Peastone 669
Peat humus 667
moss 667
Pedestal floor 349
type lavatory 513
type seating 448
Pedestrian bridge 665
Peephole 304
Pegboard 208
Penetration macadam 651
test 15
Penthouse roof louver 544
Perforated aluminum pipe ... 684
pipe 684
PVC pipe 684
Performance bond 14, 819
Perlite insulation 226
plaster 326
sprayed 368
Permit building 14
Personal respirator 40
Personnel field 12
hoist 721
protection 18
Petrographic analysis 15
Pew church 448
sanctuary 448
Pex pipe 502
tubing 502, 560
tubing fitting 561
Pharmacy cabinet 420
Phone booth 378
Photoelectric control 582
Photography 14
aerial 14
construction 14
time lapse 14
Physician's scale 425
PIB roof 246
Picket railing 156
Pickup truck 719
Pick-up vacuum 712
Picture window 291
window aluminum sash 289
window steel sash 290
window vinyl 297
Pier brick 101
Pigment mortar 97
Pilaster metal toilet component .. 379
plastic-laminate toilet ... 380
toilet partition 381
wood column 212
Pile 646
boot 643
bored 644
cap 77
cap concrete placement ... 79
cap formwork 55
cutoff 612
driven 642

900

Index

driving 821, 822
driving mobilization 613
encasement 643
foundation 646
friction 643, 646
H 641, 642
hammer 713
high strength 639
leads 713
lightweight 639
load test 612
mobilization hammer 613
pipe 646
point 643, 644
point heavy duty 644
precast 642
prestressed 642, 646
sod 668
splice 643, 644
steel 642
steel sheet 639
step tapered 642
testing 612
timber 643
treated 643
wood 643
wood sheet 639
Piling sheet 639, 870
special cost 612
Pillow tank 678
Pin powder 125
Pine door 274
door frame 213
fireplace mantel 212
floor 342
molding 202
roof deck 190
shelving 395
siding 240
stair tread 209
Pipe & fitting 518
& fitting backflow preventer 504
& fitting backwater valve 509
& fitting bronze 493
& fitting copper 498
& fitting grooved joint 500
& fitting hydrant 505
& fitting iron body 527
& fitting polypropylene 494, 523
& fitting PVC 501
& fitting steel 499
& fitting trap 507
acid resistant 523
add elevated 490
aluminum 681, 684
and fittings 862
bedding 627
bedding trench 619
bollard 656
brass 498
cast iron 506
cleanout 496
concrete 682
copper 498
corrosion resistant 522
corrugated 684
corrugated metal 681, 684
covering 497
covering fiberglass 497
CPVC 502
double wall 686
drain 625
drainage 522, 679, 681, 682
ductile iron 675
DWV PVC 502, 679
elbow 500
encapsulation 44
epoxy fiberglass wound 522
fire retardant 523
fitting grooved joint 500
foam core DWV ABS 502
gas 685
glass 522
grooved joint 499
insulation 497
insulation removal 42
iron alloy mechanical joint 522
liner 674
metal 681
no hub 506
painting 364
perforated aluminum 684
PEX 502
pile 646
plastic 501, 502
polyethylene 685
polypropylene 523
proxylene 523
PVC 501, 676, 679
rail aluminum 156
rail galvanized 156
rail stainless 156
rail steel 156
rail wall 156
railing 156
reinforced concrete 682
relay 625
removal 29
removal metal 491
removal plastic 491
sewage 679, 681, 682
sewage collection PVC 679
shock absorber 505
single hub 506
sleeve plastic 61
soil 506
stainless steel 501
steel 499, 681
subdrainage 684
support framing 131
tee 500
water 675
weld joint 499
wrapping 674
Piping designations 872
Piping gas service polyethylene 685
gasoline 686
insulation 497
specialty HVAC 532, 534
storm drainage 681
Pit excavation 620
leaching 680
scale 399
sump 625
test 28
Pitch coal tar 243
emulsion tar 650
pocket 256
Pivoted window 290
Placing concrete 78, 837
reinforcment 830
Plain tube framing 285
Plan remediation 40, 44
Planer pavement 720
shop 420
Plank floor 189
grating 158
hollow precast concrete 84
precast concrete nailable 84
precast concrete roof 84
precast concrete slab 84
roof 190
scaffolding 20
Plant and bulb transplanting 671
and planter 448
bed preparation 668
ground cover 669
mixed bituminous concrete 612
screening 714
Plantation shutter 436
Planter 448, 664
bench 664
concrete 664
fiberglass 664
interior 448
Planting 671
Plant-mix asphalt paving 652
Plaque bronze 376
Plaster 320
accessory 325
beam 326
ceiling 326
cement 458
column 326
cutout 32
demolition 318, 319
drilling 322
gauging 326
ground 188
gypsum 326
mixer 717
painting 365, 366
partition 320
partition thin 321
perlite 326
soffit 326
thincoat 328
venetian 327
vermiculite 326
wall 326
Plasterboard 328
Plastic angle valve 494
ball valve 494
bench 393
faced hardboard 208
fireproofing 257
glazing 313
grille 543
laminate door 271
lead 471
locker 393
lumber 215
lumber structural 215
matrix terrazzo 345
parking bumper 656
pipe 501, 502
pipe removal 491
railing 217
screw anchor 122
sheet flashing 250
sign 376
skylight 298
toilet compartment 381
toilet partition 380
trench drain 509
valve 494
window 295
Plastic-laminate toilet compartment 379
toilet component 380
toilet door 380
toilet headrail 380
toilet panel 380
toilet pilaster 380
Plastic-laminate-clad countertop 440
Plate checkered 159
roadway 720
shear 172
steel 131
stiffener 131
wall switch 589
wood 187
Plate-type heat exchanger 552
Platform checkered plate 159
lifter 405
telescoping 420
tennis 424
trailer 718
Plating zinc 171
Player bench 665
Playfield equipment 423
Playground equip. 423, 424, 660
modular 424
protective surfacing 658
slide 424
surface 349
whirler 424
Plenum barrier 353
demolition 319
Plexiglass 313
acrylic 313
mirror 313
Plow vibrator 715
Plug in bus-duct 587
in circuit-breaker 588
in switch 588
wall 99
Plugmold raceway 589
Plumbing 504
appliance 510, 511
demolition 491
fixture 508, 513, 520, 534, 679
fixture removal 491
laboratory 419
Plywood 851
clip 172
demolition 318, 319
fence 24
floor 190
formwork 52, 54
formwork steel framed 58
joist 193
mounting board 192
paneling 208
sheathing roof & wall 192
shelving 395
sidewalk 24
siding 240
sign 376
soffit 207
subfloor 190
underlayment 191
Pneumatic control system 529
nailer 716, 717
tube 409
tube system 480
Pocket door 300
door frame 214
pitch 256
Point heavy duty pile 644
moil 716
pile 643, 644
Pointing CMU 94
masonry 94
Poisoning soil 637
Poke-thru fitting 572
Pole aluminum light 595
athletic 423
closet 206
cross arm 687
electrical 687
light 687
lighting 595
portable decorative 406

steel light . . . 596
telephone . . . 687
utility . . . 595, 687
wood . . . 687
wood electrical utility . . . 687
Polished concrete . . . 838
Polishing concrete . . . 82
Polyacrylate floor . . . 346
terrazzo . . . 346
Polycarbonate glazing . . . 313
Polyester floor . . . 347
room darkening shade . . . 437
Polyethylene backer rod . . . 64
coating . . . 674
film . . . 667
floor . . . 349
pipe . . . 685
pool cover . . . 456
septic tank . . . 680
tarpaulin . . . 23
waterproofing . . . 224, 225
Polymer trench drain . . . 509
Polyolefin roofing thermoplastic . 246
Polypropylene pipe . . . 523
shower . . . 514
siding . . . 242
valve . . . 494
Polystyrene blind . . . 214
ceiling . . . 458
insulation . . . 226, 228, 458
Polysulfide caulking . . . 261
Polyurethane caulking . . . 261
floor . . . 347
varnish . . . 368
Polyvinyl chloride (PVC) . . . 38
soffit . . . 207, 242
tarpaulin . . . 23
Polyvinyl-chloride roof . . . 246
Pool accessory . . . 456
cover . . . 456
cover polyethylene . . . 456
filtration swimming . . . 522
heater electric . . . 548
swimming . . . 455, 456
Porcelain tile . . . 333
Porch framing . . . 184
Portable air compressor . . . 715
asphalt plant . . . 719
booth . . . 468
building . . . 17
cabinet . . . 413
chalkboard . . . 374
eye wash . . . 520
fire extinguisher . . . 391
floor . . . 342
panel . . . 384
partition . . . 384
scale . . . 400
stage . . . 420
Portland cement . . . 51
cement terrazzo . . . 345
Positive pressure chimney . . . 547
Post and panel signage . . . 377
athletic . . . 425
cap . . . 172
cedar . . . 213
concrete . . . 692
demolition . . . 168
driver . . . 720
fence . . . 659
lamp . . . 163, 595
pedestrian traffic control . . . 406
recreational . . . 423
shore . . . 61
sign . . . 378
tennis court . . . 658
Postal specialty . . . 393
Postformed countertop . . . 441
Post-tensioned concrete . . . 840
slab on grade . . . 74
Potable water softener . . . 510
water treatment . . . 510
Potters wheel . . . 420
Poured gypsum underlayment . . . 88
Powder actuated tool . . . 125
charge . . . 125
pin . . . 125
Power equipment . . . 819
mulcher . . . 714
temporary . . . 17
trowel . . . 712
wiring . . . 589
Preaction valve cabinet . . . 486
Preblast survey . . . 621
Precast beam . . . 85
bridge . . . 665
catch basin . . . 685
column . . . 85
concrete beam . . . 84
concrete channel slab . . . 84
concrete column . . . 85
concrete joist . . . 85
concrete lintel . . . 87
concrete nailable plank . . . 84
concrete plantable paver . . . 654
concrete roof plank . . . 84
concrete slab plank . . . 84
concrete stair . . . 84
concrete tee beam . . . 85
concrete tee beam double . . . 85
concrete tee beam quad . . . 85
concrete tee beam single . . . 85
concrete unit paving slab . . . 654
concrete wall . . . 86
concrete wall panel tilt-up . . . 86
concrete window sill . . . 87
conrete . . . 841
coping . . . 117
curb . . . 655
median . . . 692
members . . . 839
parking barrier . . . 656
pile . . . 642
receptor . . . 388
septic tank . . . 680
tee . . . 85
terrazzo . . . 345
wall . . . 841
wall panel . . . 86
wall panel insulated . . . 86
Pre-engineered steel building . . . 463
steel building demo . . . 453
steel buildings . . . 860
Prefab metal flue . . . 546
Prefabricated building . . . 462
comfort station . . . 466
fireplace . . . 389
Pre-fabricated masonry panel . . . 111
Prefabricated panel . . . 231
wood stair . . . 209
Prefinished door . . . 271
drywall . . . 329
floor . . . 342
hardboard paneling . . . 208
shelving . . . 395
Preformed roof panel . . . 237
roofing & siding . . . 237
Pre-hung door . . . 276, 277
Preparation exterior surface . . . 354
interior surface . . . 355
plant bed . . . 668
Presser laundry . . . 408
Pressure grouting cement . . . 640
reducing valve water . . . 493
regulator . . . 503
regulator steam . . . 503
relief valve . . . 493
switch switchboard . . . 584
valve relief . . . 493
wash . . . 355
washer . . . 718
washing metal . . . 120
Pressurized fire extinguisher . . . 392
Prestressed concrete . . . 840
concrete pile . . . 642
pile . . . 642, 646
precast concrete . . . 839
Prestressing steel . . . 73
Preventer backflow . . . 504
Prices lumber products . . . 851
Prime coat . . . 651
Primer asphalt . . . 222, 344
steel . . . 369
Prison cell . . . 469
cot . . . 446
fence . . . 661
hinge . . . 306
toilet . . . 431
Processing concrete . . . 82
Proctor compaction test . . . 16
Produce case . . . 412
Product piping . . . 686
Productivity glazing . . . 855
Profile block . . . 106
Profiler pavement . . . 720
Progress schedule . . . 14
Project overhead . . . 13
sign . . . 25
Projected window aluminum . . . 289
window steel . . . 290
window steel sash . . . 290
Projection screen . . . 373, 417, 418
Projector movie . . . 418
Propeller exhaust fan . . . 541
unit heater . . . 559
Property line survey . . . 28
Protected membrane . . . 247
Protection corner . . . 385
fire . . . 391
slope . . . 637
stile . . . 308
temporary . . . 25
termite . . . 637
winter . . . 23, 52
worker . . . 41
Protector door . . . 300
Proxylene pipe . . . 523
P&T relief valve . . . 493
PTAC unit . . . 556
Public address system . . . 603
Pull box . . . 573
box electrical . . . 573
door . . . 438
plate . . . 304
Pulpit church . . . 430
Pump . . . 535, 679
centrifugal . . . 718
circulating . . . 534
concrete . . . 712
condensate . . . 535
condensate removal . . . 535
condensate return . . . 535
contractor . . . 625
diaphragm . . . 718
fire . . . 488
fire jockey . . . 488
fountain . . . 522
general utility . . . 506
grinder system . . . 508
grout . . . 718
heat . . . 557
in-line centrifugal . . . 534
jack . . . 19
mud . . . 712
operator . . . 625
sewage ejector . . . 508
shallow well . . . 679
shotcrete . . . 712
staging . . . 18, 820
submersible . . . 510, 679, 718
sump . . . 411, 509
trash . . . 718
water . . . 534, 679, 718
water supply well . . . 679
wellpoint . . . 722
Pumped concrete . . . 78
Pumping . . . 625
dewater . . . 625
Purlin roof . . . 189
steel . . . 134
Push plate . . . 304
Push-on joint . . . 675
Push-pull plate . . . 304
Putlog scaffold . . . 20
Putting surface . . . 349
Puttying . . . 354
PVC adhesive . . . 225
blind . . . 435
cleanout . . . 496
conduit in slab . . . 574
control joint . . . 99
cornerboard . . . 217
door casing . . . 217
fascia . . . 217
fitting . . . 676
flashing . . . 250
frieze . . . 218
gravel stop . . . 251
molding exterior . . . 217
pipe . . . 501, 676, 679
pipe perforated . . . 684
rake . . . 218
roof . . . 246
sheet . . . 225, 349
siding . . . 241
soffit . . . 218
trim . . . 217
underground duct . . . 687
valve . . . 494
waterstop . . . 62

Q

Quad precast concrete tee beam . 85
Quarry drill . . . 716
drilling . . . 620
tile . . . 335
Quarter round molding . . . 207
Quartz . . . 667
chip . . . 51
flooring . . . 347
Quoin . . . 113

R

Raceway . . . 572, 574
conduit . . . 571
plugmold . . . 589
surface . . . 588
trench duct . . . 575
wiremold . . . 588
Rack bicycle . . . 423, 449

coat 395
hat 395
hose 485
pallet 394
Racquetball court 460
Radial arch 193
  wall formwork 57
Radiant floor heating 560
Radiation fin tube 559
Radiator cast iron 559
  supply control 533
  thermostat control system . 530
Radio frequency security
shielding 339
  frequency shielding 471
  tower 688
Radiography test 16
Radiology equipment 427
Rafter anchor 172
  composite 185
  demolition 168
  framing metal 151
  metal bracing 151
  metal bridging 151
  tie 186
  wood 185, 186
Rail aluminum pipe 156
  crane 698
  crash 385
  dock shelter 405
  galvanized pipe 156
  guide 692
  guide/guard 692
  map 374
  ornamental aluminum 164
  ornamental glass 164
  ornamental steel 164
  ornamental wrought iron ... 164
  stainless pipe 156
  steel pipe 156
  trolley 385
  wall pipe 156
Railing cable 164
  church 430
  composite 219
  demolition 170
  encased 219
  encased composite 219
  industrial 156
  ornamental 164
  picket 156
  pipe 156
  plastic 217
  wood 206, 209, 211
  wood stair 210
Railroad ballast 690
  bumper 690
  concrete tie 690
  derail 690
  maintenance 690
  siding 693, 872
  tie 651, 671
  tie step 650
  timber switch tie 690
  timber tie 690
  track accessory 690
  track heavy rail 690
  track maintenance 690
  track material 690
  track removal 30
  turnout 693
  wheel stop 690
Raise manhole 684
  manhole frame 684
Raised floor 350
Rake PVC 218

tractor 714
Rammer/tamper 713
Ramp approach 350
  handicap 76
  moving 479
  temporary 22
Ranch plank floor 342
Range cooking 410
  hood 411
  receptacle 579, 589
  restaurant 415
  shooting 423
Ratio water cement 15
Razor wire 661
Reach-in refrigeration 413
Reading table 447
Ready mix concrete 78, 837
Rebar accessory 68
Receiver ash 387, 443
  trash 443
Receptacle air conditioner .. 579
  device 578
  dryer 579
  duplex 589
  GFI 578
  locking 579
  range 579, 589
  telephone 579
  television 579
  trash 449
  waste 387
  weatherproof 579
Receptor precast 388
  shower 388, 514, 517
  terrazzo 388
Recessed mat 443
Reciprocating water chiller . 553
Recirculating chemical toilet 706
Reclaimed or antique granite . 113
Recorder videotape 607
Recore cylinder 302, 304
Recovery heat 428
Recreational post 423
Rectangular diffuser 543
  ductwork 536
Rectifier low-voltage silicon 589
Recycled plastic lumber 214
  rubber tire tile 443
Red bag 429
Reduce heat transfer glass .. 311
Redwood bark mulch 667
  cupola 397
  louver 316
  paneling 209
  siding 240
  wine cellar 413
Refinish floor 342
Reflective block 110
  glass 313
  insulation 229
  painting 657
  sign 377
Reflectorized barrels 716
Refrigerant removal 526
Refrigerated case 412
  storage equipment 413
  wine cellar 413
Refrigeration 458
  bloodbank 419
  commercial 412, 413
  equipment 457
  floor 458
  insulation 458
  mortuary 428
  panel fiberglass 458
  partition 458

reach-in 413
residential 410
walk-in 458
Refrigerator door 281
Refuse chute 480
  hopper 480
Register air supply 543
  cash 407
  return 544
  steel 543
  wall 544
Reglet aluminum 254
  galvanized steel 254
Regulator pressure 503
  steam pressure 503
Reinforced concrete pipe 682
  culvert 682
  plastic panel fiberglass .. 351
  PVC roof 246
Reinforcement 830
  welded wire 831
Reinforcing 830
  accessory 68
  bar chair 69
  bar splicing 70
  bar tie 68
  beam 71
  chair subgrade 69
  coating 72
  column 71
  concrete 71
  dowel 72
  elevated slab 71
  epoxy coated 72
  footing 71
  galvanized 72
  girder 71
  glass fiber rod 72
  high chair 68
  joint 98
  masonry 98
  metric 829
  slab 71
  sorting 72
  spiral 71
  steel 829, 830
  steel fiber 74
  synthetic fiber 74
  testing 15
  tie wire 70
  wall 71
Relay pipe 625
Release door 603
Relief valve P&T 493
  valve self-closing 493
  valve temperature 493
  vent ventilator 545
Relining sewer 674
Remediation plan 40, 44
  plan/method asbestos 40
Remote power pack 582
Removal air conditioner 526
  asbestos 42
  bathtub 491
  block wall 31
  boiler 526
  catch basin 29
  concrete 29
  concrete pipe 29
  curb 29
  driveway 29
  fixture 491
  floor 31
  guide rail 30
  hydrant 29
  insulation 42, 223

lavatory 491
masonry 30, 96
paint 45
pipe 29
pipe insulation 42
plumbing fixture 491
railroad track 30
refrigerant 526
shingle 223
sidewalk 30
sink 491
sod 668
steel pipe 29
stone 30
stump 614
tank 39
tree 613, 614
urinal 491
utility line 29
vat 42
water closet 491
water fountain 491
water heater 491
water softener 491
window 264
Removal/salvage material 35
Remove bolt 120
  nut 120
Rendering 10
Renovation tread cover 161
Rental equipment 712
Repellent water 366
Replacement joint sealant ... 222
  pavement 652
Resaturant roof 222
Residential alarm 580
  appliance 410, 580
  application 576
  bathtub 514
  closet door 269
  device 576
  dishwasher 410
  door 213, 269, 274
  door bell 579
  dryer 411
  elevator 475
  fan 580
  fixture 579
  folding partition 384
  garage 467
  gas water heater 511
  greenhouse 462
  gutting 34
  heat pump 581
  hinge 306
  kitchen sink 514
  lavatory 513
  lighting 579
  oil-fired water heater 512
  overhead door 284
  panel board 576
  refrigeration 410
  roof jack 541
  service 576
  sink 514
  smoke detector 580
  stair 209
  storm door 265
  switch 576
  transition 541
  wall cap 541
  wash bowl 513
  washer 411
  water heater 411, 512, 581
  water heater electric 511
  wiring 576, 580

903

Resilient base 343
  floor athletic 344
  pavement 658
Resistance corrosion 674
Respirator 41
  personal 40
Resquared shingle 234
Restaurant furniture 445
  range 415
Restoration gypsum 318
  masonry 97
  mortar 97
  window 45
Retaining wall 77, 664
  wall cast concrete 662
  wall concrete segmental 663
  wall formwork 58
  wall segmental 663
  wall stone 664
  wall stone gabion 664
  wall timber 663
Retarder concrete 78
  vapor 232
Retractable lateral arm awning . . . 396
Return register 544
Revolving darkroom 459
  dome 467
  door 282, 287, 462
  door darkroom 282
  entrance door 287
Rewind table 418
Rib lath 324
Ribbed siding 238
  waterstop 62
Ridge board 186
  cap 233, 237
  flashing 465
  roll 238
  shingle slate 234
  vent 255
Rig drill 28
Rigid anchor 98
  conduit in trench 574
  in slab conduit 574
  insulation 226
  joint sealant 261
  metal-framed skylight 298
Ring boxing 422
  split 172
  toothed 173
Ripper attachment 719
Riprap and rock lining 638
Riser pipe wellpoint 722
  rubber 343
  stair 346
  terrazzo 346
  wood stair 210
River stone 51
Rivet 125
  aluminum 125
  copper 125
  stainless 125
  steel 125
  tool 125
Road base 651
  berm 655
  mixer 720
  sign 378
  sweeper 720
  temporary 22
Roadway LED luminaire 596
  luminaire 596
  plate 720
Robe hook 387
Rock bolt drilling 620
  bolting 638

drill 28, 620
  removal 620
  trencher 715
Rod backer 260
  closet 395
  curtain 386
  ground 571
  shower 386
  tie 130, 639
  weld 127
Roll ridge 238
  roof 243
  roofing 248
  type air filter 545
Roller compaction 626
  sheepsfoot 626, 637, 714
  tandem 714
  vibratory 714
Rolling door 280, 288
  earth 637
  ladder 20
  service door 281
  tower scaffold 21
Roll-up exterior shutter 395
  shutter 395
Romex copper 569
Roof accessory 256
  adhesive 243
  aluminum 237
  baffle 257
  beam 193
  bracket 21
  built-up 243
  cant 186, 244
  clay tile 235
  coating 222
  cold-applied built-up asphalt . . . 243
  copper 248
  CSPE 245
  curb 186
  deck 190
  deck formboard 59
  deck gypsum 87
  deck insulation 230
  deck laminated 190
  deck wood 190
  decking 142
  drain 509
  elastomeric 247
  exhauster 540
  expansion joint 254
  fan 540
  fiberglass 238
  fill 841
  frame opening 130
  framing 186
  framing removal 32
  gravel 51
  hatch 256
  hatch removal 223
  insulation 230, 465
  jack 186
  jack residential 541
  joint 255
  lead 248
  membrane green 236
  metal 237
  metal rafter framing 151
  metal tile 236
  metal truss framing 153
  mineral 248
  modified bitumen 244
  mop 244
  nail 171
  panel aluminum 237
  panel preformed 237

patch 222
paver 257
paver and support 257
PIB 246
polyvinyl-chloride 246
purlin 189
PVC 246
rafter bracing 151
rafter bridging 151
rafter framing 185
reinforced PVC 246
resaturant 222
roll 243
safety anchor 21
sheathing 192
sheet metal 248
shingle 233
shingle asphalt 233
slate 234, 851
soffit framing 152
soil mixture green 236
specialty prefab 251
stainless steel 248
steel 237
stressed skin 131
system green 236
thermoplastic polyolefin 246
tile 235
TPO 246
truss 193, 194
vent 255, 466
ventilator 256
walkway 244
window demo 265
zinc 248
Roofing & siding preformed 237
  built-up 851
  cold 243
  demolition 42, 223
  EPDM 245
  flat seam sheet metal 248
  ketone ethylene ester 246
  membrane 243, 247
  modified bitumen 244
  roll 248
  single-ply 245
  SPF 247
  system built-up 243
Rooftop air conditioner 556
  multizone air conditioner 556
Room acoustical 459
  audiometric 459
  clean 458
  humidifier 562
Rope decorative 406
  exercise 422
  safety line 21
  steel wire 137
Rotary crawler drill 716
  hammer drill 716
Rototiller 714
Rough buck 183
  grading 615
  grading site 615
  stone wall 112
Rough-in drinking fount. deck . . . 521
  drinking fountain floor 521
  drinking fountain wall 520
  sink countertop 514
  sink raised deck 514
  sink service floor 516
  sink service wall 516
  tub 514
Round bar fiberglass 216
  diffuser 543
  rail fence 662

table 445
tube fiberglass 216
Rowing machine 421
Rub finish carborundum 81
Rubber and vinyl sheet floor 343
  astragal 308
  base 343
  coating 226
  control joint 99
  dock bumper 405
  floor 343, 658
  floor tile 344
  nosing 343
  pavement 658
  pipe insulation 497
  riser 343
  sheet 343
  stair 343
  threshold 305
  tile 344
  waterproofing 225
  waterstop 63
Rubberized asphalt sealcoat 650
Rubbing wall 81
Rubbish handling 33
  handling chutes 823
Rumble strip 692
Run gravel bank 612
Running track 342
  track surfacing 658
  trap 506
Rupture testing 15

S

Safe commercial 408
  diskette 408
  office 408
Safety cabinet biological 420
  deposit box 409
  equipment laboratory 419
  eye/face equipment 520
  film 314
  fixture industrial 520
  flood shut-off house 493
  glass 311
  line rope 21
  net 18
  nosing 343
  railing cable 138
  shower 520
  shut-off flood whole house . . . 493
  switch 590
  water and gas shut-off 493
Salamander 718
Sales tax 12, 814, 815
Salvage or disposal value 35
Sample field 47
Sampling air 40
Sanctuary pew 448
Sand 97
  brick 51
  concrete 51
  fill 612
  screened 97
  seal 650
Sandblast finish concrete 81
  glass 311
  masonry 95
Sandblasting equipment 718
  metal 120
Sander floor 718
Sanding 354
  floor 342
Sandstone 114

flagging 655
Sandwich panel 238
  wall panel 239
Sanitary base cove 334
  drain 507
Sash 466
  aluminum 289
  metal 290
  prefinished storm 466
  security 290
  steel 290
  wood 291
Sauna 460
  door 460
Saw blade concrete 89
  brick 94
  chain 718
  circular 718
  concrete 712
  cutting concrete floor 89
  cutting concrete wall 89
  cutting demolition 34
  cutting slab 89
  masonry 718
  shop 420
  table 420
Sawing masonry 94
Sawn control joint 63
SBS modified bituminous
  membrane 245
Scaffold aluminum plank 21
  baseplate 19
  bracket 19
  caster 19
  catwalk 20
  frame 19
  guardrail 19
  mason 19
  putlog 20
  rolling tower 21
  specialty 20
  stairway 19
  wood plank 19, 21
Scaffolding 820
  plank 20
  tubular 19
Scale 399
  commercial 399
  contractor 400
  crane 399
  physician's 425
  pit 399
  portable 400
  truck 399, 400
  warehouse 399
Scanner checkout 407
  unit facial 309
Scarify concrete 319
Schedule 14
  board 375
  critical path 14
  progress 14
School cabinet 439
  door hardware 300
  equipment 422
  furniture 446
  hinge 306
  panic device door hardware . . 300
  T.V. 607
Scissor gate 382
  lift 699, 715
Scoreboard baseball 422
  basketball 422
  football 422
  hockey 422
Scored block 106

split face block 106
Scraper earth 714
  elevating 623
  excavation 623
  self propelled 624
Scratch coat 345
Screed base 69
Screed, gas engine, 8HP
  vibrating 712
Screed holder 69
Screen chimney 389
  entrance 378
  fence 662
  metal 290
  molding 206
  projection 373, 417, 418
  security 163, 290
  sight 383
  squirrel and bird 389
  urinal 381
  window 290, 295
  wood 295
Screened sand 97
Screening plant 714
Screen/storm door aluminum . . . 265
Screw aluminum 239
  anchor 69, 121
  anchor bolt 69
  brass 171
  drywall 332
  eye bolt 69
  jack 639
  lag 125
  liquid chiller 554
  machine 125
  sheet metal 171
  steel 171
  water chiller 554
  wood 171
Scroll water chiller 553
Scrub station 427
Scupper drain 509
  floor 509
Seal compression 260
  curb 650
  dock 405
  door 405
  fog 650
  grout 332
  pavement 650
  security 308
Sealant 224
  acoustical 262, 330
  caulking 261
  replacement joint 222
  rigid joint 261
  tape 261
Sealcoat 650
Sealer floor 51
  joint 260, 261
  pavement 658
Seamless floor 347
Seating church 430
  movie 418
Secondary treatment plant 706
Sectional door 283
  overhead door 284
Security and vault equipment . . . 408
  barrier 406
  bollard 691
  dome 399
  door and frame 282
  drawer 431
  driveway gate 282
  driveway gate aluminum 282
  driveway gate steel 282

driveway gate wood 282
  equipment 431
  factor 11
  fence 661
  film 314
  gate 382
  gate driveway 282
  gate opener driveway 283
  hinge 306
  mesh 325
  planter 691
  sash 290
  screen 163, 290
  seal 308
  turnstile 406
  vault door 283
  vehicle barrier 691
  X-ray equipment 606
Sediment bucket drain 507
  strainer Y valve 494
Seeding 668, 871
See-saw 424
Segmental concrete retaining
  wall 663
  retaining wall 663
Seismic bracing 860
  bracing support 494
  ceiling bracing 338
Selective clearing 614
  demolition equipment 402
  demolition fencing 30
  demolition masonry 95
  tree removal 614
Self-adhering flashing 251
Self-closing relief valve 493
Self-contained air conditioner . . . 556
Self-propelled scraper 624
Self-supporting tower 688
Sentry dog 24
Separation barrier 41, 46
Separator air 532
Septic galley 680
  system 680
  system chamber 681
  system effluent-filter 680
  tank 680
  tank concrete 680
  tank polyethylene 680
  tank precast 680
Service door 280
  electric 589
  entrance cable aluminum 569
  laboratory analytical 47
  residential 576
  sink 516
  sink faucet 518
  station equipment 404
Sewage aeration 706
  collection concrete pipe 679
  collection PVC pipe 679
  ejector pump 508
  municipal waste water 706
  pipe 679, 681, 682
  treatment plant 706
Sewage/drainage collection 682
Sewer relining 674
Shade 436
  polyester room darkening 437
Shaft wall 320
Shake wood 235
Shape structural aluminum 135
Shear connector welded 126
  plate 172
  test 16
  wall 192
Shearwall panel 188

Sheathed nonmetallic cable 569
  Romex cable 569
Sheathing 191, 192
  asphalt 192
  board 191
  gypsum 193
  insulating 190
  paper 232
  roof 192
  roof & wall plywood 192
  wall 192
Sheepsfoot roller 626, 637, 714
Sheet carpet 348
  glass 312
  glass mosaic 335
  lead 470
  metal 248
  metal aluminum 251
  metal flashing 249
  metal roof 248
  metal screw 171, 239
  modified bituminous barrier . . 233
  piling 639, 870
  steel pile 642
Sheeting 639
  driver 716
  open 641
  tie back 641
  wale 639
  wood 625, 639, 870
Shelf bathroom 387
  bin 394
  library 447
  metal 394
Shellac door 361
Shell-type heat exchanger 552
Shelter aluminum 468
  dock 405
  golf 424
  rail dock 405
  temporary 52
Shelving 395
  food storage metal 414
  mobile 447
  pine 395
  plywood 395
  prefinished 395
  refrigeration 459
  steel 394
  storage 394
  wood 395
Shield expansion 121
  lag screw 121
  light 692
Shielded copper cable 567
Shielding lead 470
  radio frequency 471
  radio frequency security 339
Shift work 12
Shingle 233
  aluminum 233
  asphalt 233
  cedar 234
  concrete 235
  metal 234
  panelized 234
  removal 223
  roof 233
  slate 234
  steel 234
  strip 233
  wood 234
Ship ladder 155
Shock absorber 505
  absorber pipe 505
  absorbing door 264, 285

Shooting range 423
Shop drill 420
   equipment 420
Shore post 61
Shoring 639, 640
   aluminum joist 61
   baseplate 19
   bracing 19
   concrete formwork 61
   concrete vertical 61
   flying truss 61
   frame 61
   heavy duty 19
   leveling jack 19
   slab 20
   steel beam 61
   temporary 639
Short load concrete 78
Shot blast floor 319
   blaster 718
Shotcrete 83
   pump 712
Shoulder paver 720
Shovel 621
   crawler 714
Shower arm 515
   built-in 515
   by-pass valve 515
   cabinet 388
   cubicle 514
   door 388
   drain 506
   emergency 520
   enclosure 388
   glass door 388
   group 517
   pan 249
   partition 114, 388
   polypropylene 514
   receptor 388, 514, 517
   rod 386
   safety 520
   stall 514
   surround 388
Shower/tub control set 515
Shower-tub valve spout set ... 515
Shredder 429
   compactor 429
Shrinkage test 15
Shrub and tree 669
   broadleaf evergreen 670
   deciduous 670
   evergreen 669
   moving 671
Shut-off flood whole house
safety 493
   safety water and gas 493
   water heater safety 493
Shutter 436
   exterior 214
   exterior roll-up 395
   interior 436
   plantation 436
   wood 214
   wood interior 436
Side coiling grille 280
Sidelight 213
   door 269, 276
Sidewalk 340, 654
   asphalt 650
   brick 654
   bridge 20
   broom 718
   concrete 650
   driveway and patio 650
   removal 30

   temporary 24
Sidewall bracket 21
Siding aluminum 238
   bevel 240
   cedar 240
   fiber cement 242
   fiberglass 238
   metal 237, 238
   nail 171
   paint 357
   painting 357
   panel aluminum 238
   plywood 240
   polypropylene 242
   railroad 693, 872
   redwood 240
   removal 223
   ribbed 238
   stain 240
   steel 239
   vinyl 241
   wood 240
   wood board 240
Sieve analysis 15
Sign 25, 376, 377
   acrylic 376
   aluminum 376
   base formwork 56
   corridor 376
   detour 24
   directional 377
   flexible 376
   highway 378
   letter 376
   metal 376
   plastic 376
   plywood 376
   post 378
   project 25
   reflective 377
   road 378
   stainless 376
   street 376
   traffic 377
Signage custom wayfinding 377
   engraved panel 376
   exterior 376
   post and panel 377
Signal bell 607
   system traffic 691
Silencer duct 538
Silica chip 51
Silicon carbide abrasive 50, 80
Silicone 260
   caulking 261
   coating 226
   water repellent 226
Sill 113
   & ledger framing 186
   anchor 172
   block 109
   door 207, 213, 305
   masonry 116
   mud 186
   precast concrete window ... 87
   quarry tile 335
   stone 112, 115
   window 292
   wood 186
Sillcock hose bibb 515
Silo 468
   demo 453
Silt fence 637
Simulated brick 117
   stone 117, 118
Single hub pipe 506

hung aluminum sash window . 289
hung aluminum window 289
precast concrete tee beam 85
zone rooftop unit 556
Single-ply roofing 245
Sink barber 407
   base 437
   countertop 514
   countertop rough-in 514
   faucet kitchen 515
   kitchen 514
   laboratory 419
   laundry 515
   lavatory 513
   mop 516
   raised deck rough-in 514
   removal 491
   residential 514
   service 516
   service floor rough-in ... 516
   service wall rough-in 516
   support 518
Site clear and grub 613
   clearing 613
   demolition 29
   furnishing 449
   grading 615
   improvement 665, 666
   preparation 28
   rough grading 615
Skidder log 714
Skidsteer attachments 715
   loader 715
Skylight demo 265
   domed metal-framed 298
   fiberglass panel 466
   metal-framed 298
   plastic 298
   removal 223
   rigid metal-framed 298
Skyroof 288
Slab blockout formwork 57
   box out opening formwork . 55
   bulkhead formwork 55, 56
   concrete 75
   concrete placement 79
   curb formwork 55, 56
   cutout 32
   edge form 143
   edge formwork 55, 56
   elevated 75
   flat plate formwork 54
   haunch 76
   haunch formwork 76
   joist dome formwork 54
   joist pan formwork 54
   lift 77, 837
   lifting insert 69
   on grade 76
   on grade formwork 56
   on grade post-tensioned ... 74
   on grade removal 30
   pan 75
   precast concrete channel 84
   reinforcing 71
   reinforcing bolster 68
   saw cutting 89
   screed formwork 57
   shoring 20
   stamping 82
   textured 77
   thickened edge 76
   thickened edge formwork 76
   trench formwork 57
   turndown 76
   turndown formwork 76

void formwork 55, 57
waffle 75
with drop panel formwork 54
Slate 115
   flagging 655
   pavement 115
   roof 234, 851
   shingle 234
   sidewalk 655
   sill 116
   stair 115
   tile 341
Slatwall 209, 352
Sleeper floor 186
   framing 186
   wood 186
Sleeve anchor bolt 64
   dowel 72
   formwork 61
   plastic pipe 61
Slide gate 659
   playground 424
   swimming pool 456
Sliding aluminum door 279
   chalkboard 373
   door shower 388
   glass door 279
   glass vinyl clad door ... 279
   mirror 388
   panel door 288
   partition 384
   vinyl clad door 279
   window 292
   window aluminum 289
Slipform paver barrier 720
   paver curb 720
Slipforms 826
Slope protection 637
   protection rip-rap 638
Slot letter 394
Slotted block 108
   channel framing 130
   pipe 683
Slump block 106
Slurry seal (latex modified) 650
   trench 642
Smart electrical metering 582
Smoke detector 608
   hatch 256
   vent 256
   vent chimney 546
Snap-tie formwork 61
Snow fence 39
   guard 256
Soaking bathroom 514
   tub 514
Soap dispenser 387
   holder 387
Soccer goalpost 425
Socket ground 423
   wire rope 136
Sod 669
   tennis court 658
Sodding 669
Sodium high pressure fixture 597
   low pressure fixture 596
Soffit & canopy framing 186
   drywall 328
   framing wood 186
   marble 114
   metal 242
   molding 207
   plaster 326
   plywood 207
   polyvinyl 207, 242
   PVC 218

steel . . . 465
stucco . . . 327
wood . . . 207
Softener water . . . 510
Soil compaction . . . 626
decontamination . . . 39
mix man-made . . . 236
pipe . . . 506
poisoning . . . 637
sample . . . 29
stabilization . . . 638
tamping . . . 626
test . . . 16
treatment . . . 637
Solar energy . . . 551
energy circulator air . . . 551
energy system collector . . . 551
energy system controller . . . 551
energy sys. heat exchanger . . . 551
film glass . . . 314
heating . . . 551, 863
heating system . . . 551
panel driveway security gate . . . 283
screen block . . . 109
Soldier beam . . . 641
Solid surface countertop . . . 441
wood door . . . 273
Sorting reinforcing . . . 72
Sound absorbing panel . . . 353
attenuation blanket . . . 353
control demo . . . 453
curtain . . . 470
dampening panel gypsum . . . 330
movie . . . 418
panel . . . 470
proof enclosure . . . 459
system . . . 603
system speaker . . . 603
Source heat pump water . . . 557
Space heater . . . 717
heater floor mounted . . . 550
office & storage . . . 17
Spade air . . . 716
tree . . . 715
Spandrel beam formwork . . . 52
glass . . . 312
Spanish roof tile . . . 235
Speaker movie . . . 418
sound system . . . 603
station nurse . . . 604
Special construction . . . 470
door . . . 278, 283
hinge . . . 306
nurse call system . . . 604
purpose room demo . . . 453
Specialty . . . 386
demo . . . 372
piping HVAC . . . 532, 534
scaffold . . . 20
telephone . . . 378
Specific gravity . . . 16
gravity testing . . . 15
Speed bump . . . 692
SPF roofing . . . 247
Spike grid . . . 172
unit traffic . . . 405
Spinner ventilator . . . 544
Spiral reinforcing . . . 71
stair . . . 163, 209
Spire church . . . 397
Splice pile . . . 643, 644
Splicer movie . . . 418
Splicing reinforcing bar . . . 70
Split astragal . . . 308
rail fence . . . 662
rib block . . . 106
ring . . . 172
system control system . . . 529
Splitter damper assembly . . . 537
Spotlight . . . 595
Spotter . . . 633
Spray coating . . . 224
insulation . . . 229
painting booth automotive . . . 404
rig foam . . . 717
substrate . . . 43
Sprayed cementitious fireproofing . . . 257
Sprayed-on insulation . . . 229
Sprayer airless . . . 40
emulsion . . . 716
paint . . . 717
Spread footing . . . 76
footing formwork . . . 56
Spreader aggregate . . . 712, 719
chemical . . . 719
concrete . . . 718
laundry . . . 407
Spring bolt astragal . . . 308
bronze weatherstrip . . . 308
hinge . . . 306
Sprinkler head . . . 486
irrigation system . . . 666
monitoring panel deluge . . . 486
system . . . 486
system accelerator . . . 486
system dry-pipe . . . 486
underground . . . 666
Square bar fiberglass . . . 216
tube fiberglass . . . 216
Squash court . . . 460
court backstop . . . 424
Stabilization fabric . . . 651
masonry . . . 94
soil . . . 638
Stabilizer base . . . 720
Stacked bond block . . . 106
Stadium . . . 463
bleacher . . . 463
cover . . . 462
Stage curtain . . . 421
equipment . . . 420
equipment demolition . . . 402
lighting . . . 595
portable . . . 420
Staging aid . . . 21
pump . . . 18
swing . . . 22
Stain cabinet . . . 361
door . . . 361
floor . . . 364
lumber . . . 194
siding . . . 240
truss . . . 364
Staining concrete floor . . . 339
Stainless bar grating . . . 158
column cover . . . 161
cooling tower . . . 555
duct . . . 536
flashing . . . 249
gutter . . . 253
plank grating . . . 159
reglet . . . 254
rivet . . . 125
screen . . . 378
sign . . . 376
steel corner guard . . . 385
steel cot . . . 446
steel countertop . . . 434, 440
steel cross . . . 430
steel door . . . 277
steel door and frame . . . 277
steel downspout . . . 252
steel gravel stop . . . 251
steel hinge . . . 305
steel pipe . . . 501
steel roof . . . 248
steel shelf . . . 387
steel storefront . . . 288
weld rod . . . 127
Stair & walk moving . . . 478
basement . . . 209
basement bulkhead . . . 84
brick . . . 112
carpet . . . 349
ceiling . . . 412
climber . . . 479
concrete . . . 77
disappearing . . . 412
electric . . . 412
finish concrete . . . 81
fire escape . . . 155
floor plate . . . 154
formwork . . . 59
grating . . . 154
landing . . . 346
marble . . . 114
metal pan . . . 154
metal plate . . . 154
nosing . . . 343, 345
pan tread . . . 80
part wood . . . 209
precast concrete . . . 84
prefabricated wood . . . 209
railroad tie . . . 651
removal . . . 169
residential . . . 209
riser . . . 346
riser vinyl . . . 343
rubber . . . 343
slate . . . 115
spiral . . . 163, 209
stringer . . . 346
stringer framing . . . 183
stringer wood . . . 183
temporary protection . . . 24
terrazzo . . . 345
tread . . . 112, 160, 343
tread and riser . . . 343
tread fiberglass . . . 216
tread insert . . . 62
tread terrazzo . . . 346
tread tile . . . 335
wood . . . 209
Stairlift wheelchair . . . 479
Stairway door hardware . . . 300
scaffold . . . 19
Stairwork handrail . . . 211
Stake formwork . . . 62
nail . . . 62
subgrade . . . 70
Stall shower . . . 514
toilet . . . 378, 380, 381
type urinal . . . 516
urinal . . . 516
Stamping slab . . . 82
texture . . . 82
Standard extinguisher . . . 391
Standing seam . . . 248
Standpipe connection . . . 484
steel . . . 678
Starter board & switch . . . 584, 585
motor . . . 586
Station control . . . 590
diaper . . . 386
hospital . . . 426
transfer . . . 429
weather . . . 472
Stationary ventilator . . . 545
ventilator mushroom . . . 545
Steam bath . . . 460
bath residential . . . 460
boiler . . . 547, 548
boiler electric . . . 547
clean masonry . . . 95
cleaner . . . 718
cleaning metal . . . 120
condensate meter . . . 535
humidifier . . . 562
jacketed kettle . . . 415
pressure valve . . . 533
regulator pressure . . . 503
trap . . . 533
Steamer . . . 415
Steam-to-water heating . . . 552
Steel anchor . . . 98
astragal . . . 308
ballistic door . . . 282
beam shoring . . . 61
beam W-shape . . . 131
bin wall . . . 663
blocking . . . 180
bolt . . . 172
bolt & hex nut . . . 123
bridge . . . 665
bridging . . . 181
building components . . . 848
building pre-engineered . . . 463
chain . . . 162
channel . . . 325
conduit in slab . . . 574
conduit in trench . . . 574
conduit intermediate . . . 572
conduit rigid . . . 572
corner guard . . . 129
cutting . . . 122
demolition . . . 120
diffuser . . . 543
dome . . . 467
door . . 267, 268, 283, 284, 465, 853
door and frame stainless . . . 277
door residential . . . 269
downspout . . . 252
drill . . . 716
drilling . . . 122
edging . . . 651, 671
estimating . . . 845
expansion tank . . . 533
fence . . . 659, 661
fiber . . . 74
fiber reinforcing . . . 74
flashing . . . 249
frame . . . 266
framing . . . 133
furring . . . 322
girt . . . 134
grating . . . 157
gravel stop . . . 251
gutter . . . 253
hex bolt . . . 123
high strength . . . 134
hinge . . . 305
lath . . . 324
lintel . . . 130
locker . . . 392
louver door . . . 315
member structural . . . 131
mesh box . . . 638
mesh grating . . . 158
mill extra . . . 134
outlet box . . . 572
painting . . . 369
partition . . . 383, 384
pile . . . 642

907

For customer support on your Building Construction Costs with RSMeans Data, call 800.448.8182.

pile concrete-filled 644
pile sheet 642
pipe 499, 681
pipe corrugated 684
pipe removal 29
plate 131
prestressing 73
primer 369
project structural 133
purlin 134
register 543
reinforcement 830
reinforcing 829, 830
rivet 125
roof 237
roofing panel 237
salvage 491
sash 290, 853
screw 171
sections 847
security driveway gate 282
sheet pile 639
shelving 394
shingle 234
siding 239
siding demolition 224
silo 468
standpipe 678
structural 845
stud NLB 323
tank 533
tank above ground 531
testing 15
tower 688
underground duct 687
underground storage tank 530
valve 527
water storage tank 678
water tank ground level 678
weld rod 127
well casing 679
window 289, 290, 466
window demolition 264
wire rope 137
wire rope accessory 136
Steeple 397
aluminum 397
tip pin 305
Step 378
bluestone 651
brick 650
manhole 685
masonry 650
railroad tie 650
stone 113
tapered pile 642
Sterilizer barber 407
dental 426
glassware 419
medical 425
Stiffener joist web 149
plate 131
Stiffleg derrick crane 721
Stile fabric 308
protection 308
Stock feeder 431
Stockpiling of soil 615
Stone aggregate 669
anchor 98
ashlar 114
base 114, 651
cast 117
column 114
crushed 612
cultured 117
curb cut 656
fill 612
filter 638
floor 114
gabion retaining wall 664
ground cover 669
mulch 667
paver 113, 655
pea 51
removal 30
retaining wall 664
river 51
sill 112, 115
simulated 117, 118
step 113
stool 115
toilet compartment 381
tread 112
trim cut 114
wall 112, 664
Stool cap 207
doctor 426
stone 115
window 114, 116
Stop door 207
gravel 251
Stop-start control station 590
Storage bottle 413
box 17
cabinet 439
dome 467
dome bulk 467
metal shelving food 414
room cold 458
shelving 394
tank 530, 678
tank concrete aboveground 532
tank cover 461
tank demo 454
tank steel underground 530
tank underground 530
Storefront aluminum 288
Storm door 265
door residential 265
drainage manhole frame 685
drainage piping 681
sash prefinished 466
window 295
window aluminum residential 295
Stove 391, 415
woodburning 391
Strainer bronze body 533
downspout 251
gutter 254
roof 252
wire 252
Y type bronze body 533
Y type iron body 534
Strand grouted 73
ungrouted 73
Stranded wire 570
Strap tie 173
Straw 667
bale construction 469
Street sign 376
Streetlight 596
Strength compressive 15
Stressed skin roof 131
Stringer stair 346
stair terrazzo 346
Strip cabinet 603
chamfer 60
entrance 285
filler 308
floor 342
footing 76
lighting 592
rumble 692
shingle 233
soil 615
Striper paint 720
Stripping topsoil 615
Strobe light 595
Structural aluminum 135
backfill 626
brick 102
column 128
compaction 637
excavation 620
fabrication cost 134
face tile 103
facing tile 103
fee 10
framing 846
insulated panel 188
panel 188
shape column 129
steel 845
steel member 131
steel project 133
tile 103
welding 123, 134
Structure fabric 461
moving 38
tension 462
Stub switchboard 587
Stucco 327
demolition 319
mesh 324
painting 357
wall coating 368
Stud demolition 169
metal 320
NLB metal 323
NLB steel 323
NLB wall 323
partition 183, 325
partition framing load bearing 145
wall 183, 187, 321
wall blocking load-bearing 144
wall box beam load-bearing 144
wall bracing load-bearing 143
wall framing load bearing 145
wall header load-bearing 144
wall wood 321
welded 126
wood 187, 320
Stump chipper 714
chipping 613
removal 614
Subcontractor O&P 13
Subdrainage pipe 684
system 684
Subfloor 190
adhesive 191
demolition 319
plywood 190
wood 191
Subgrade reinforcing chair 69
stake 70
Submersible pump 510, 679, 718
sump pump 510
Subpurlin bulb tee 87
bulbtee 135
Subpurlins 848
Subsoil heating 457
Subsurface exploration 28
Suction hose 717
Sump hole construction 625
pit 625
pump 411, 509
pump submersible 510
Super high-tensile anchor 638
Supermarket checkout 407
scanner 407
Supply ductile iron pipe water 675
Support carrier fixture 518
ceiling 127
drinking fountain 518
framing pipe 131
lavatory 518
light 128
monitor 128
motor 128
partition 127
sink 518
urinal 518
water closet 518
water cooler 519
X-ray 128
Suppressor lightning 576
Surface bonding 97
landscape 349
playground 349
preparation exterior 354
preparation interior 355
raceway 588
Surfacing 650
Surfactant 43
Surgery equipment 427
table 425
Surgical lighting 425
Surround shower 388
tub 389
Surveillance system TV 606
Survey aerial 28
crew 25
monument 28
preblast 621
property line 28
stake 25
topographic 28
Suspended ceiling 324, 337
ceiling framing 182
tile 337
Suspension mounted heater 550
system ceiling 337
Sweeper road 720
Swell testing 15
Swimming pool 455, 456
pool blanket 456
pool demo 454
pool enclosure 463
pool equipment 456
pool filter 522
pool filtration 522
pool hydraulic lift 457
pool ladder 456
pool painting 456
pool ramp 457
pools 859
Swing 424
check steel valve 527
check valve 492, 526
check valve bronze 492
clear hinge 306
gate 659
staging 22
Switch box 572
bus-duct 588
decorator 577
dimmer 577, 589
electric 589, 590
general duty 590
residential 576
safety 590
toggle 589
Switchboard distribution 584
electric 590

feeder section 584
incoming 584
instrument 582
pressure switch 584
stub 587
w/CT compartment 584
Switching low-voltage 589
Switchplate low-voltage 589
Synthetic erosion control 637
fiber 74
fiber reinforcing 74
marble 341
turf 349
System chamber septic 681
component foam-water 487
control 529
fire extinguishing 487
grinder pump 508
septic 680
sprinkler 486, 487
subdrainage 684
tube 480
T.V. 602
T.V. surveillance 606
UHF 602
VHF 602

T

Table folding 445
laboratory 419
massage 422
overbed 447
physical therapy 426
reading 447
rewind 418
round 445
saw 420
surgery 425
Tackboard fixed 374
Tactile warning surfacing 657
Tamper 617, 716
Tamping soil 626
Tandem roller 714
Tank above ground steel 531
aboveground storage conc. .. 532
clean 39
cover 461
darkroom 408
developing 408
disposal 39
expansion 533
fiberglass 530
fixed roof 531
gasoline 531
holding 706
horizontal above ground 531
Hubbard 426
leak detection 608
pillow 678
removal 39
septic 680
steel 533
steel underground storage ... 530
storage 530, 678
testing 16
water 678, 719
water storage solar 551
Tankless gas water heater 511
Tap box bus-duct 587
Tape barricade 24
detection 675
sealant 261
temporary 657
underground 675

Tar kettle 718, 720
paper 668
pitch emulsion 650
roof 222
Target range 423
Tarpaulin 23
duck 23
Mylar 23
polyethylene 23
polyvinyl 23
Tarred felt 244
Tax 12
sales 12
social security 12
unemployment 12
Taxes 815, 816
T-bar mount diffuser 543
Teak floor 341
molding 207
paneling 208
Technician inspection 16
Tee cleanout 496
framing steel 130
pipe 500
precast 85
precast concrete beam 85
Telephone enclosure 378
enclosure commercial 378
house 603
manhole 687
pole 687
receptacle 579
specialty 378
Telescoping bleacher 448
boom lift 715
platform 420
Television equipment 603
receptacle 579
system 602
Teller automatic 409
window 409
Temperature relief valve 493
rise detector 608
Tempered glass 310
glass greenhouse 462
hardboard 208
Tempering valve 493
valve water 493
Temporary barricade 23
building 17
construction 17, 23
electricity 17
facility 23
fence 24
guardrail 24
heat 52
light 17
lighting 17
power 17
protection 25
ramp 22
road 22
road painting 657
shelter 52
shoring 639
tape 657
toilet 718
utility 17
Tennis court air supported 461
court backstop 660
court clay 658
court fence 660, 661
court fence and gate 660
court net 658
court painting 658
court post 658

court sod 658
court surface 348
court surfacing 658
Tensile test 15
Tension structure 462
structure demo 455
Terminal A/C packaged 556
air conditioner packaged ... 556
heater 549
lug 568
Termination box 603
cable 568
Termite control chemical 637
protection 637
Terne coated flashing 249
Terra cotta 103, 104
cotta coping 103, 117
cotta demolition 32, 320
cotta tile 104
Terrazzo 345
abrasive 345
base 345
conductive 346
cove base 345
curb 345, 346
demolition 319
epoxy 346
floor 345, 858
floor tile 345
monolithic 345
plastic matrix 345
polyacrylate 346
precast 345
receptor 388
riser 346
stair 345
venetian 345
wainscot 345, 346
Test beam 15
equipment laboratory 419
load pile 612
moisture content 16
pile load 612
pit 28
soil 16
ultrasonic 16
well 679
Testing 15
lead 44
OSHA 43
pile 612
sulfate soundness 15
tank 16
Textile wall covering 350
Texture stamping 82
Textured slab 77
Theater and stage equipment .. 420
Theatrical lighting 595
Thermal & moisture protection
demo 222
batt 353
Thermometer 419
Thermoplastic 313
painting 657
polyolefin roof 246
polyolefin roofing 246
Thermostat control
system radiator 530
integral 561
wire 581
Thickened edge slab 76
Thimble wire rope 136
Thin brick veneer 101
plaster partition 321
Thincoat plaster 328
Thinning tree 613

Thinset ceramic tile 333
mortar 335
Thin-set tile 333, 334
Threaded rod fiberglass 216
Threshold 305
aluminum 305
door 214
stone 114
wood 207
Thru-wall air conditioner 556
Ticket booth 468
dispenser 404
Tie back sheeting 641
rafter 186
railroad 651, 671
rod 130, 639
strap 173
wall 98
wire reinforcing 70
Tier locker 392
Tile 332, 336, 343
abrasive 333
aluminum 236, 332
carpet 348
ceiling 337
ceramic 333, 334
clay 235
concrete 235
cork 340
cork wall 350
demolition 318
exposed aggregate paver 654
exterior 334
fire rated 103
floor 345
flooring cork 340
flue 116
glass 313, 335
grout 332, 333
marble 341
metal 332
or terrazzo base floor 345
partition 103
porcelain 333
quarry 335
recycled rubber tire 443
roof 235
roof metal 236
rubber 344
slate 341
stainless steel 332
stair tread 335
structural 103
terra cotta 104
vinyl composition 344
wall 333
waterproofing membrane
ceramic 336
window sill 335
Tiling ceramic 332-336
Tilt-up concrete 841
concrete wall panel 86
precast concrete wall panel ... 86
Timber connector 172
fastener 171
framing 189
guide rail 692
heavy 189
laminated 193
pile 643
retaining wall 663
switch tie 690
switch tie railroad 690
tie railroad 690
Time lapse photography 14
lock 283

909

system controller 604
Timer clock 580
  interval 577
Tin ceiling 339
Tinted glass 310
  window film electrically 314
Toaster 415
Toggle switch 589
Toggle-bolt anchor 121
Toilet 706
  accessory 386
  accessory commercial 386
  bowl 513
  chemical 718
  compartment 378
  compartment plastic 381
  compartment plastic-laminate . 379
  compartment stone 381
  partition 114, 378-380
  partition removal 320
  prison 431
  stall 378, 380, 381
  stone partition 381
  temporary 718
  tissue dispenser 387
Tool powder actuated 125
  rivet 125
Toothed ring 173
Toothing masonry 32
Top coiling grille 280
  demolition counter 170
  dressing 668
  vanity 442
Topographic survey 28
Topographical survey 28
Topping concrete 88
  epoxy 347
  floor 81, 88
  grout 88
Topsoil 612, 615
  placement and grading 668
  stripping 615
Torch cutting 122, 718
  cutting demolition 35
Torrified door 275
Towel bar 387
  dispenser 386
Tower control 469
  cooling 554
  crane 18, 720, 821, 822
  hoist 721
  light 717
  radio 688
TPO roof 246
Track accessory 690
  accessory railroad 690
  curtain 421
  drawer 438
  drill 716
  heavy rail railroad 690
  material railroad 690
  running 342
  traverse 435
Tractor 626
  loader 714
  rake 714
  truck 719
Traffic barrier highway sound . . . 666
  channelizing 692
  cone 23
  control mirror 399
  door 285
  line 657
  line remover 720
  sign 377
  signal system 691

spike unit 405
Trailer bunk house 17
  floodlight 716
  lowbed 720
  mud 719
  office 17
  platform 718
  truck 718
  water 719
Trainer bicycle 421
Tram car 719
Transceiver 602
Transfer station 429
Transformer 583
  & bus-duct 583, 588
  cabinet 588
  current 582
  dry-type 583
  fused potential 582
  low-voltage 589
  oil-filled 583
Transition molding wood 342
  residential 541
  strip floor 348
Translucent ceiling 338
Transom aluminum 287
  lite frame 267
  window 294
Transplanting plant and bulb . . . 671
Trap cast iron 506
  deep seal 507
  drainage 506, 507
  grease 508
  inverted bucket steam 533
  rock surface 81
  steam 533
Trash closure 449
  compactor 417
  pump 718
  receiver 443
  receptacle 449
  receptacle fiberglass 449
  receptacle metal 443
Travel nursery item 669
Traverse 435
  track 435
Travertine 114
Tray cable 574
Tread abrasive 335
  and riser stair 343
  bar grating 160
  channel grating 160
  cover renovation 161
  diamond plate 160
  insert stair 62
  rubber 343
  stair 112, 160, 343
  stair pan 80
  stone 114, 115
  vinyl 343
  wood stair 211
Treadmill 422
Treated lumber 186
  lumber framing 187
  pile 643
  pine fence 662
Treatment acoustical 353
  equipment 426
  flooring 339
  plant secondary 706
  plant sewage 706
  potable water 510
Tree deciduous 670
  evergreen 669
  guying system 671
  moving 671

removal 613, 614
removal selective 614
spade 715
thinning 613
Trench backfill 617, 619
  box 719
  cover 159
  disposal field 681
  drain 509
  drain fiberglass 509
  drain plastic 509
  drain polymer 509
  duct 575
  duct fitting 575
  duct raceway 575
  duct steel 575
  excavating 616
  excavation 616, 619
  grating frame 159
  slurry 642
  utility 619
Trencher chain 619, 715
  rock 715
  wheel 715
Trenching 620, 625
Trial batch 15
Trim demolition 170
  exterior 200, 201
  exterior paint 359
  molding 206
  painting 364
  PVC 217
  tile 334
  wood 203
Triple weight hinge 306
Tripod floodlight 716
Troffer air handling 592
  light fixture 593
Trolley rail 385
Trowel coating 224
  concrete 712
  power 712
Truck concrete 712
  crane flatbed 720
  dock 405
  dump 715
  flatbed 715, 719
  holding concrete 78
  leveler 405
  loading 620
  mounted crane 721
  pickup 719
  scale 399, 400
  tractor 719
  trailer 718
  vacuum 719
  winch 719
Truss bowstring 194
  demolition 170
  framing metal 153
  joist 141
  metal roof 134
  painting 364
  plate 173
  roof 193, 194
  stain 364
  varnish 364
Tub bar 386
  rough-in 514
  soaking 514
  surround 389
Tube framing 285
  pneumatic 409
  system 480
  system air 480
  system pneumatic 480

Tubing copper 498
  electric metallic 572
  electrical 571
  fitting PEX 561
  PEX 502, 560
Tubular fence 661
  lock 303
  scaffolding 19
Tugboat diesel 722
Tumbler holder 387
Tunnel locomotive 720
  muck car 720
  ventilator 720
Turbine wind 708
Turf artificial 658
  synthetic 349
Turnbuckle wire rope 137
Turndown slab 76
Turned column 212
Turnout railroad 693
Turnstile 406
  security 406
TV closed circuit camera 606
  system 602
Two piece astragal 308
Tying wire 61

U

UHF system 602
Ultrasonic cleaner 427
  motion detector 606
  test 16
Uncased drilled concrete pier . . . 646
Underdrain 685
Undereave vent 316
Underfloor duct 575
  duct electrical 575
  duct fitting 575
Underground duct 687
  electrical & telephone 687
  sprinkler 666
  storage tank 530
  storage tank removal 824
  storage tank steel 530
  tape 675
Underlayment acoustical 320
  gym floor 342
  hardboard 191
  latex 343
  self-leveling cement 88
Underpin foundation 641
Underwater light 456
Undisturbed soil 16
Unemployment tax 12
Ungrouted bar 73
  strand 73
Unit air-handling 555, 558
  commercial kitchen 412
  heater 550, 558
  heater propeller 559
Upstanding beam formwork 53
Urethane foam 353
  insulation 458
  wall coating 368
Urinal 516
  high efficiency 517
  removal 491
  screen 381
  stall 516
  stall type 516
  support 518
  wall hung 516
  waterless 516
  watersaving 516

Urn ash . . . 443
Utensil washer medical . . . 425
washer-sanitizer . . . 419
Utility accessory . . . 675
box . . . 674
connection . . . 675
duct . . . 687
electric . . . 687
line removal . . . 29
pole . . . 595, 687
pole wood electrical . . . 687
set fan . . . 539, 540
sitework . . . 687
temporary . . . 17
trench . . . 619
trench excavating . . . 619

V

Vacuum breaker . . . 504
catch basin . . . 719
central . . . 428
cleaning . . . 428
loader . . . 41
pick-up . . . 712
truck . . . 719
unit central . . . 702
wet/dry . . . 719
Valance board . . . 438
Valley flashing . . . 233
gutter . . . 465
rafter . . . 186
Valve . . . 526
assembly sprinkler deluge . . . 487
backwater . . . 509
brass . . . 491
bronze . . . 491
bronze angle . . . 491
butterfly . . . 526, 676
cabinet . . . 391
check . . . 484
CPVC . . . 494
fire hose . . . 485
flush . . . 517
foot . . . 494
forged . . . 527
gas stop . . . 491
gate . . . 526, 527
globe . . . 527
heating control . . . 533
hot water radiator . . . 533
iron body . . . 526
mixing . . . 515
plastic . . . 494
plastic angle . . . 494
polypropylene . . . 494
polypropylene ball . . . 494
PVC . . . 494
relief pressure . . . 493
shower by-pass . . . 515
soldered gate . . . 492
spout set shower-tub . . . 515
sprinkler alarm . . . 484
steam pressure . . . 533
steel . . . 527
swing check . . . 492, 526
swing check steel . . . 527
tempering . . . 493
water pressure . . . 493
Y sediment strainer . . . 494
Vandalproof fixture . . . 593
Vaneaxial fan . . . 539
Vanity base . . . 438
top . . . 442
top lavatory . . . 513
Vapor barrier . . . 232, 459
barrier sheathing . . . 192
retarder . . . 232
Vaportight fixture incandescent . . . 593
Varnish cabinet . . . 361
casework . . . 361
door . . . 361
floor . . . 364, 367
polyurethane . . . 368
truss . . . 364
Vat removal . . . 42
Vault day gate demolition . . . 402
door . . . 283
front . . . 283
Vaulting side horse . . . 422
VCT removal . . . 319
Vehicle barrier security . . . 691
charging electric . . . 591
guide rail . . . 692
Veneer ashlar . . . 112
brick . . . 99
core paneling . . . 208
granite . . . 112
member laminated . . . 194
removal . . . 96
Venetian plaster . . . 327
terrazzo . . . 345
Vent automatic air . . . 532
box . . . 99
brick . . . 544
chimney . . . 546
chimney all fuel . . . 547
convector or baseboard air . . . 559
dryer . . . 411
exhaust . . . 544
flashing . . . 507
foundation . . . 99
gas . . . 546
intake-exhaust . . . 545
metal chimney . . . 546
one-way . . . 256
ridge . . . 255
ridge strip . . . 316
roof . . . 255, 466
smoke . . . 256
Ventilated metal deck . . . 143
Ventilating air conditioning . 529, 536, 537, 539, 540, 545, 553, 555, 557
Ventilation equip. and hood . . . 545
fan . . . 580
louver . . . 316
Ventilator control system . . . 529
masonry . . . 99
mushroom stationary . . . 545
relief vent . . . 545
roof . . . 256
spinner . . . 544
stationary mushroom . . . 545
tunnel . . . 720
Venturi flow meter . . . 534
Verge board . . . 205
Vermiculite cement . . . 258
insulation . . . 228
plaster . . . 326
Vertical blind . . . 435
lift door . . . 284
metal ladder . . . 155
Very low density polyethylene . . . 38
VHF system . . . 602
Vial bottle . . . 47
Vibrating screed, gas engine, 8HP . . . 712
Vibration pad . . . 126
Vibrator concrete . . . 712
earth . . . 617, 626, 637
plate . . . 626
plow . . . 715
Vibratory hammer . . . 714
roller . . . 714
Vibroflotation . . . 640
compaction . . . 870
Video surveillance . . . 607
Videotape recorder . . . 607
Vinyl blind . . . 214
casement window . . . 296
clad premium window . . . 294
coated fence . . . 660
composition floor . . . 344
composition tile . . . 344
curtain . . . 386
downspout . . . 252
faced wallboard . . . 328
floor . . . 344
gutter . . . 253
lead barrier . . . 353
sheet floor . . . 343
siding . . . 241
stair riser . . . 343
stair tread . . . 343
tread . . . 343
wall coating . . . 368
wallpaper . . . 350
window . . . 296
window double hung . . . 296
window single hung . . . 295
Vision panel . . . 298
Vocational kiln . . . 420
shop equipment . . . 420
Voltmeter . . . 582
Volume control damper . . . 537
Volume-control damper . . . 537

W

Waffle slab . . . 75
Wainscot ceramic tile . . . 333
molding . . . 207
quarry tile . . . 335
terrazzo . . . 345, 346
Wale . . . 639
sheeting . . . 639
Walk . . . 650
moving . . . 479
Walk-in refrigeration . . . 458
Walkway cover . . . 397
roof . . . 244
Wall & corner guard . . . 385
accessory formwork . . . 62
aluminum . . . 238
and partition demolition . . . 319
base . . . 343
blocking load-bearing stud . . . 144
box beam load-bearing stud . . . 144
boxout formwork . . . 57
bracing load-bearing stud . . . 143
brick . . . 111
brick shelf formwork . . . 57
bulkhead formwork . . . 57
bumper . . . 304, 385
buttress formwork . . . 57
cabinet . . . 437, 439, 440
canopy . . . 396
cap residential . . . 541
cast concrete retaining . . . 662
cavity . . . 111
ceramic tile . . . 333
clock . . . 443
coating . . . 368
concrete . . . 76
concrete finishing . . . 81
concrete placement . . . 79
concrete segmental retaining . . . 663
corbel formwork . . . 57
covering . . . 858
covering cork . . . 350
covering textile . . . 350
cross . . . 430
curtain . . . 288
cutout . . . 32
demolition . . . 319
drywall . . . 328
exhaust fan . . . 540
fan . . . 541
finish . . . 351
flagpole . . . 398
forms . . . 826
formwork . . . 57
foundation . . . 107
framing . . . 187
framing load bearing stud . . . 145
framing removal . . . 32
furnace . . . 550
furring . . . 187, 322
grout . . . 98
guard . . . 129, 385
header load-bearing stud . . . 144
heater . . . 411
hung lavatory . . . 513
hung urinal . . . 516
hydrant . . . 505
insulation . . . 226, 465
lath . . . 324
lintel formwork . . . 58
louver . . . 316
masonry . . . 104, 664
mat . . . 422
mirror . . . 313
NLB partition . . . 321
painting . . . 357, 366
panel . . . 117
panel brick . . . 111
panel precast . . . 86
panel precast concrete double . . . 85
panel woven wire . . . 382
paneling . . . 208
partition . . . 321
pilaster formwork . . . 58
plaster . . . 326
plug . . . 99
plywood formwork . . . 57
precast concrete . . . 86, 841
prefab plywood formwork . . . 58
register . . . 544
reinforcing . . . 71, 830
retaining . . . 77, 664
rubbing . . . 81
screen . . . 383
shaft . . . 320
shear . . . 192
sheathing . . . 192
sill formwork . . . 58
steel bin . . . 663
steel framed formwork . . . 58
stone . . . 112, 664
stone gabion retaining . . . 664
stucco . . . 327
stud . . . 183, 187, 320, 321
stud NLB . . . 323
switch plate . . . 589
tie . . . 98
tie masonry . . . 98
tile . . . 333
tile cork . . . 350
urn ash receiver . . . 387
window . . . 286, 288
wood stud . . . 321
Wallboard repair gypsum . . . 318

Wallcovering 351
acrylic 350
gypsum fabric 351
Wall-covering natural fiber 351
Wallguard 385
Wallpaper 858
cork 350
grass cloth 351
vinyl 350
Walnut floor 341
Wardrobe 395
cabinet 395
Warehouse 461
scale 399
Warm air heater 549
Warning beacon 595
surfacing tactile 657
Wash bowl residential 513
brick 95
fountain 517
fountain group 517
pressure 355
safety equipment eye/face 520
Washable air filter 545
air filter permanent 545
Washer 173
& dryer coin-operated 408
and extractor 408
commercial 408
pressure 718
residential 411
Washer-sanitizer utensil 419
Washing machine automatic 411
Waste cart medical 429
cleanup hazardous 40
compactor 429
disposal 45, 429
disposal hazardous 40
handling 428, 429
handling equipment 428, 429
incinerator 428
packaging 43
receptacle 387
Wastewater treatment system 706
Watchdog 24
Watchman service 24
Water atomizer 41
balancing 528
blasting 95
blasting metal 120
cement ratio 15
chiller 553
chiller absorption 553
chiller centrifugal type 553
chiller indirect-fired 553
chiller reciprocating 553
chiller screw 554
chiller scroll 553
closet 513, 516
closet chemical 706
closet removal 491
closet support 518, 519
coil chilled 558
cooler 521
cooler electric 521
cooler handicap 521
cooler support 519
copper tubing connector 503
curing concrete 83
dispenser hot 511
distiller 425
distribution pipe 675, 676
effect testing 15
fountain removal 491
hammer arrester 505
heater 411, 581
heater commercial 511, 512
heater gas 512
heater gas residential 511
heater oil-fired 512
heater removal 491
heater residential 411, 512, 581
heater residential electric 511
heater safety shut-off 493
heating hot 547, 559
hose 717
hydrant 505
pipe 675
pipe ground clamp 571
pressure reducing valve 493
pressure relief valve 493
pressure valve 493
pump 534, 679, 718
pump fire 488
pump fountain 522
pumping 625
reducer concrete 78
repellent 366
repellent coating 226
repellent silicone 226
softener 510
softener potable 510
softener removal 491
source heat pump 557
storage solar tank 551
supply domestic meter 503
supply ductile iron pipe 675
supply meter 503
supply PVC pipe 676
supply well pump 679
tank 678, 719
tank elevated 678
tank ground level steel 678
tempering valve 493
trailer 719
treatment 706
well 679
Waterless urinal 516
Waterproofing 225
butyl 225
cementitious 225
coating 224
demolition 224
elastomeric 225
elastomeric sheet 225
integral 52
masonry 97
membrane 224
neoprene 225
rubber 225
Watersaving urinal 516
Waterstop center bulb 63
dumbbell 62
fitting 63
PVC 62
ribbed 62
rubber 63
Water-water exchange hot 552
Watt meter 582
Wayfinding signage custom 377
Wearing course 652
Weather barrier 232
barrier or wrap 232
station 472
Weatherproof receptacle 579
Weatherstrip 281
aluminum 308
door 309
spring bronze 308
window 310
zinc 309
Weatherstripping 309
Web stiffener framing 149
Wedge anchor 122
Weight exercise 421
lifting multi station 422
Weights and measures 812
Weld joint pipe 499
rod 127
rod aluminum 127
rod cast-iron 127
rod stainless 127
rod steel 127
Welded frame 266
shear connector 126
structural steel 844
stud 126
wire epoxy 73
wire fabric 73, 831
wire galvanized 73
Welder arc 420
Welding certification 16
fillet 123
machine 719
structural 123, 134
Well 625
& accessory 679
area window 397
casing steel 679
gravel pack 679
pump 679
pump shallow 679
water 679
Wellpoint 625
discharge pipe 722
equipment rental 722
header pipe 722
pump 722
riser pipe 722
Wellpoints 869
Wet/dry vacuum 719
Wheel color 595
guard 386
potters 420
stop railroad 690
trencher 715
Wheelbarrow 719
Wheelchair lift 479
stairlift 479
Wheeled loader 714
Whirler playground 424
Whirlpool bath 426, 457
Wide flange beam fiberglass 216
throw hinge 306
Widener pavement 720
Winch truck 719
Wind turbine 708
Window 289
& door demolition 264
aluminum 289
aluminum awning 289
aluminum projected 289
aluminum residential storm 295
aluminum sliding 289
and door molding 207
awning 290, 396
bank 409
basement/utility 290
blind 434
bow & bay metal-clad 293
bullet resistant 409
casement 289-291, 293
casement wood 291
demolition 264
double hung 294
double hung aluminum sash 289
double hung bay wood 293
double hung steel sash 290
double hung vinyl 296
double hung wood 291
drive-up 409
estimates 853
fiberglass 298
fiberglass single hung 298
frame 292
framing 465
glass 312
grille 294
guard 162
hardware 300
impact resistant aluminum 289
industrial 290
metal 289, 290, 466
metal-clad awning 292
metal-clad casement 292
metal-clad double hung 292
muntin 294
painting 361
palladian 294
picture 291
picture & sliding metal-clad 292
pivoted 290
plastic 295
precast concrete sill 87
removal 264
restoration 45
sash wood 292
screen 290, 295
sill 116
sill marble 114
sill tile 335
single hung aluminum 289
single hung aluminum sash 289
single hung vinyl 295
sliding 292
sliding wood 292
steel 289, 290, 466
steel projected 290
steel sash picture 290
steel sash projected 290
stool 114, 116
storm 295
teller 409
transom 294
trim set 207
vinyl 296
vinyl casement 296
vinyl clad awning 293
vinyl clad premium 294
wall 286
wall aluminum 286
wall framing 285
weatherstrip 310
well area 397
wood 290
wood awning 290
wood bow 293
wood double hung 291
Windrow loader 720
Wine cellar 413
Winter concrete 78
protection 23, 52
Wire 864, 865
brushing metal 120
connector 568
copper 570
electrical 570
fabric welded 73
fence 24, 661
fencing 661
glass 312
ground 571
hanging 338
mesh 327, 661

mesh locker 392
mesh partition 382
partition 382
razor 661
reinforcement 831
rope clip 136
rope socket 136
rope thimble 136
rope turnbuckle 137
strainer 252
stranded 570
thermostat 581
THW 570
THWN-THHN 570
tying 61
Wiremold raceway 588
Wireway 574
Wiring air conditioner 581
device 573, 589
device & box 573
fan 580
method 583
power 589
residential 576, 580
Wood & composite I-joist 193
awning window 290
ballistic door 282
base 196
beam 181, 189
beam & girder framing 181
bench 665
blind 436
block floor 341
block floor demolition 319
blocking 180, 184
bow window 293
bracing 180
bridge 665
cabinet school 439
canopy framing 186
case good 444
ceiling 338
chip 668
column 182, 189, 212
composition floor 341
cupola 397
cutting 34
deck 190
demolition 318
dome 467

door 270, 271, 273
door frame exterior 213
door frame interior 213
door residential 274
double acting swing door 285
double hung window 291
electrical utility pole 687
fascia 186, 203
fastener 171
fiber formboard 59
fiber plank cementitious 87
fiber sheathing 192
fiber underlayment 191
firestop 183
flagpole 398
floor 341, 342
floor demolition 319
folding panel 436
folding partition 384
framing 180
framing demolition 166
framing miscellaneous 183
framing open web joist 182
framing plate 187
furring 187
girder 181
gutter 253
handrail 206, 211
header 187
interior shutter 436
joist 182, 193
joist framing 182
laminated 193
ledger 186
louver 212
nailer 183
overhead door 284
panel door 273
paneling 208
parking bumper 657
parquet 341
partition 183
pier floating 696
pile 643
plank scaffold 19, 21
planter 664
plate 187
pole 687
rafter 185, 186
rail fence 662

railing 206, 209, 211
roof deck 190
roof truss 193
sash 291
screen 295
screw 171
security driveway gate 282
shade 436
shake 235
sheet piling 870
sheeting 625, 639, 870
shelving 395
shingle 234
shutter 214
sidewalk 650
siding 240
siding demolition 224
sill 186
sleeper 186
sleeper framing 186
soffit 207
soffit & canopy framing 187
soffit framing 186
stair 209
stair baluster 209
stair newel 210
stair part 209
stair railing 210
stair riser 210
stair stringer 183
stair stringer framing 183
stair tread 211
storm door 277
strip flooring 341
stub wall framing 187
stud 187, 320
subfloor 191
threshold 207
trim 203
veneer wall covering 351
window 290
window casement 291
window demolition 264
window double hung bay 293
window sash 292
window sliding 292
Woodburning stove 391
Wool carpet 349
fiberglass 228
Work boat 722

extra 13
Worker protection 41
Workers compensation 12, 818
Woven wire partition 382
Wrapping pipe 674
Wrecking ball 719
Wrench impact 716
Wrestling mat 422
Wrought iron fence 659

X

X-ray barrier 471
concrete slab 16
dental 427
equipment 427, 606
equipment security 606
machine 606
medical 427
mobile 427
protection 470
support 128
X-ray/radio freq prot. demo 455

Y

Y sediment strainer valve 494
type bronze body strainer 533
type iron body strainer 534
Yard fountain 457
Yellow pine floor 342

Z

Z bar suspension 338
Zee bar 332
framing steel 130
Zinc divider strip 345
plating 171
roof 248
terrazzo strip 345
weatherstrip 309

Division Notes

	CREW	DAILY OUTPUT	LABOR-HOURS	UNIT	BARE COSTS				TOTAL INCL O&P
					MAT.	LABOR	EQUIP.	TOTAL	

Division Notes

	CREW	DAILY OUTPUT	LABOR-HOURS	UNIT	BARE COSTS				TOTAL INCL O&P
					MAT.	LABOR	EQUIP.	TOTAL	

Division Notes

	CREW	DAILY OUTPUT	LABOR-HOURS	UNIT	BARE COSTS				TOTAL INCL O&P
					MAT.	LABOR	EQUIP.	TOTAL	

Division Notes

		CREW	DAILY OUTPUT	LABOR-HOURS	UNIT	BARE COSTS				TOTAL INCL O&P
						MAT.	LABOR	EQUIP.	TOTAL	
					CREW	DAILY OUTPUT	LABOR-HOURS	UNIT	MAT.	LABOR	EQUIP.	TOTAL	TOTAL INCL O&P

Division Notes

	CREW	DAILY OUTPUT	LABOR-HOURS	UNIT	BARE COSTS				TOTAL INCL O&P
					MAT.	LABOR	EQUIP.	TOTAL	

Division Notes

	CREW	DAILY OUTPUT	LABOR-HOURS	UNIT	BARE COSTS				TOTAL INCL O&P
					MAT.	LABOR	EQUIP.	TOTAL	

Other Products & Services

tradition of excellence in construction cost information
nd services since 1942

Table of Contents
Annual Cost Guides
Online Estimating Solution
Seminars and Professional Development

r more information visit our website at www.RSMeans.com | Unit prices according to the latest MasterFormat

Cost Data Selection Guide

The following table provides definitive information on the content of each cost data publication. The number of lines of data provided in each unit price or assemblies division, as well as the number of crews, is listed for each data set. The presence of other elements such as reference tables, square foot models, equipment rental costs, historical cost indexes, and city cost indexes, is also indicated. You can use the table to help select the RSMeans data set that has the quantity and type of information you most need in your work.

Unit Cost Divisions	Building Construction	Mechanical	Electrical	Commercial Renovation	Square Foot	Site Work Landsc.	Green Building	Interior	Concrete Masonry	Open Shop	Heavy Construction	Light Commercial	Facilities Construction	Plumbing	Residential
1	595	414	431	535	0	527	207	334	476	594	527	278	1071	424	180
2	777	280	86	733	0	992	207	399	214	776	734	481	1221	287	274
3	1691	340	230	1084	0	1483	986	354	2037	1691	1693	482	1791	316	389
4	960	21	0	920	0	725	180	615	1158	928	615	533	1175	0	447
5	1901	158	155	1093	0	852	1799	1106	729	1901	1037	979	1918	204	746
6	2453	18	18	2111	0	110	589	1528	281	2449	123	2141	2125	22	2662
7	1596	215	128	1634	0	580	763	532	523	1593	26	1329	1697	227	1049
8	2131	80	44	2711	0	255	1140	1793	105	2119	0	2301	2954	0	1540
9	2006	86	45	1830	0	309	452	2092	394	1948	15	1671	2255	54	1440
10	1063	17	10	657	0	217	27	873	138	1063	30	554	1154	234	224
11	1096	201	166	540	0	135	56	924	29	1063	0	231	1116	164	110
12	548	0	2	299	0	219	147	1551	14	515	0	273	1574	23	217
13	744	149	158	253	0	369	122	254	78	720	267	109	760	115	104
14	273	36	0	223	0	0	0	257	0	273	0	12	293	16	6
21	130	0	41	37	0	0	0	296	0	130	0	121	669	689	259
22	1168	7559	160	1227	0	1572	1066	849	20	1157	1681	875	7507	9411	719
23	1179	6972	579	930	0	157	900	779	38	1173	110	880	5211	1920	469
26	1505	490	10389	1160	0	812	639	1158	55	1437	596	1360	10301	399	636
27	94	0	419	83	0	0	0	71	0	94	39	67	378	0	56
28	125	79	193	107	0	0	27	90	0	110	0	56	182	57	41
31	1510	733	610	806	0	3262	288	7	1217	1455	3276	604	1569	660	613
32	814	49	8	882	0	4410	353	405	292	785	1827	421	1688	142	468
33	529	1076	538	252	0	2173	41	0	236	522	2157	128	1697	1283	154
34	107	0	47	4	0	190	0	0	31	62	212	0	136	0	0
35	18	0	0	0	0	327	0	0	0	18	442	0	84	0	0
41	60	0	0	33	0	8	0	22	0	61	31	0	68	14	0
44	75	79	0	0	0	0	0	0	0	0	0	0	75	75	0
46	23	16	0	0	0	274	261	0	0	23	264	0	33	33	0
48	12	0	25	0	0	0	25	0	0	12	17	12	12	0	12
Totals	25183	19068	14482	20144	0	19958	10275	16289	8065	24672	15719	15898	50714	16769	12815

Assem Div	Building Construction	Mechanical	Electrical	Commercial Renovation	Square Foot	Site Work Landscape	Assemblies	Green Building	Interior	Concrete Masonry	Heavy Construction	Light Commercial	Facilities Construction	Plumbing	Asm Div	Residential
A		15	0	188	150	577	598	0	0	536	571	154	24	0	1	378
B		0	0	848	2503	0	5662	56	329	1976	368	2095	174	0	2	211
C		0	0	647	929	0	1310	0	1632	146	0	818	251	0	3	588
D		1069	941	712	1861	72	2540	330	827	0	0	1348	1107	1088	4	850
E		0	0	86	261	0	301	0	5	0	0	258	5	0	5	391
F		0	0	0	114	0	143	0	0	0	0	114	0	0	6	357
G		527	447	318	311	3363	791	0	0	534	1349	205	293	677	7	307
															8	760
															9	80
															10	0
															11	0
															12	0
Totals		1611	1388	2799	6129	4012	11345	386	2793	3192	2288	4992	1854	1765		3922

Reference Section	Building Construction Costs	Mechanical	Electrical	Commercial Renovation	Square Foot	Site Work Landscape	Assem.	Green Building	Interior	Concrete Masonry	Open Shop	Heavy Construction	Light Commercial	Facilities Construction	Plumbing	Resi.
Reference Tables	yes	yes	yes	yes	no	yes	yes	yes	yes	yes	yes	yes	yes	yes	yes	yes
Models					111			25					50			28
Crews	578	578	578	556		578		578	578	578	555	578	555	556	578	555
Equipment Rental Costs	yes	yes	yes	yes		yes		yes	yes	yes	yes	yes	yes	yes	yes	yes
Historical Cost Indexes	yes	yes	yes	yes	yes	yes	yes	yes	yes	yes	yes	yes	yes	yes	yes	no
City Cost Indexes	yes	yes	yes	yes	yes	yes	yes	yes	yes	yes	yes	yes	yes	yes	yes	yes

Online   Book   CD   eBook   Online add-on available
for additional fee

Annual Cost Guides

Unit prices according to the latest MasterFormat

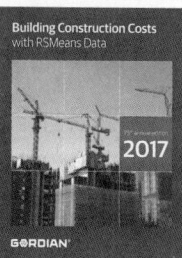

Building Construction Costs with RSMeans Data 2017

Offers you unchallenged unit price reliability in an easy-to-use format. Whether used for verifying complete, finished estimates or for periodic checks, it supplies more cost facts better and faster than any comparable source. The City Cost Indexes and Location Factors cover more than 930 areas for indexing to any project location in North America. Order and get RSMeans Quarterly Update Service FREE.

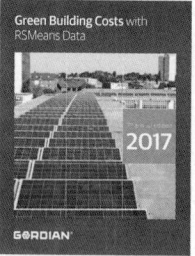

Green Building Costs with RSMeans Data 2017

Estimate, plan, and budget the costs of green building for both new commercial construction and renovation work with this seventh edition of *Green Building Costs with RSMeans Data*. More than 10,000 unit costs for a wide array of green building products plus assemblies costs. Easily identified cross references to LEED and Green Globes building rating systems criteria.

Mechanical Costs with RSMeans Data 2017

Total unit and systems price guidance for mechanical construction—materials, parts, fittings, and complete labor cost information. Includes prices for piping, heating, air conditioning, ventilation, and all related construction.

Plus new 2017 unit costs for:

- Thousands of installed HVAC/controls, sub-assemblies and assemblies
- "On-site" Location Factors for more than 930 cities and towns in the U.S. and Canada
- Crews, labor, and equipment

Facilities Construction Costs with RSMeans Data 2017

For the maintenance and construction of commercial, industrial, municipal, and institutional properties. Costs are shown for new and remodeling construction and are broken down into materials, labor, equipment, and overhead and profit. Special emphasis is given to sections on mechanical, electrical, furnishings, site work, building maintenance, finish work, and demolition.

More than 49,800 unit costs, plus assemblies costs and comprehensive Reference Section are included.

Square Foot Costs with RSMeans Data 2017

Accurate and Easy-to-Use

- Updated price information based on nationwide figures from suppliers, estimators, labor experts, and contractors.
- Green building models
- Realistic graphics, offering true-to-life illustrations of building projects
- Extensive information on using square foot cost data, including sample estimates and alternate pricing methods

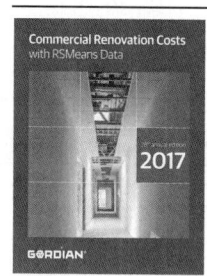

Commercial Renovation Costs with RSMeans Data 2017

Commercial/Multi-family Residential

Use this valuable tool to estimate commercial and multi-family residential renovation and remodeling.

Includes: Updated costs for hundreds of unique methods, materials, and conditions that only come up in repair and remodeling, PLUS:

- Unit costs for more than 19,600 construction components
- Installed costs for more than 2,700 assemblies
- More than 930 "on-site" localization factors for the U.S. and Canada

Electrical Costs with RSMeans Data 2017

Pricing information for every part of electrical cost planning. More than 13,800 unit and systems costs with design tables; clear specifications and drawings; engineering guides; illustrated estimating procedures; complete labor-hour and materials costs for better scheduling and procurement; and the latest electrical products and construction methods.

- A variety of special electrical systems, including cathodic protection
- Costs for maintenance, demolition, HVAC/mechanical, specialties, equipment, and more

Electrical Change Order Costs with RSMeans Data 2017

Electrical Change Order Costs with RSMeans Data provides you with electrical unit prices exclusively for pricing change orders. Analyze and check your own change order estimates against those prepared when using RSMeans cost data. It also covers productivity analysis and change order cost justifications. With useful information for calculating the effects of change orders and dealing with their administration.

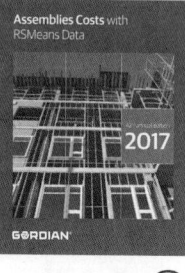

Assemblies Costs with RSMeans Data 2017

Assemblies Costs with RSMeans Data takes the guesswork out of preliminary or conceptual estimates. Now you don't have to try to calculate the assembled cost by working up individual component costs. We've done all the work for you.

Presents detailed illustrations, descriptions, specifications, and costs for every conceivable building assembly—over 350 types in all—arranged in the easy-to-use UNIFORMAT II system. Each illustrated "assembled" cost includes a complete grouping of materials and associated installation costs, including the installing contractor's overhead and profit.

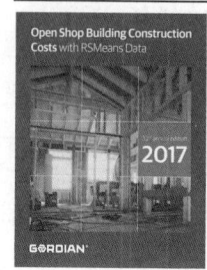

Open Shop Building Construction Costs with RSMeans Data 2017

The latest costs for accurate budgeting and estimating of new commercial and residential construction, renovation work, change orders, and cost engineering.

Open Shop "BCCD" with RSMeans Data will assist you to:

- Develop benchmark prices for change orders.
- Plug gaps in preliminary estimates and budgets.
- Estimate complex projects.
- Substantiate invoices on contracts.
- Price ADA-related renovations.

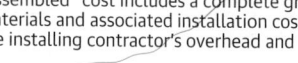

For more information visit our website at www.RSMeans.com

Residential Costs with RSMeans Data 2017

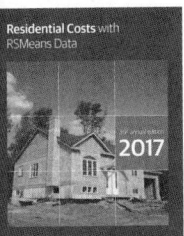

Contains square foot costs for 28 basic home models with the look of today, plus hundreds of custom additions and modifications you can quote right off the page. Includes more than 3,900 costs for 89 residential systems. Complete with blank estimating forms, sample estimates, and step-by-step instructions.

Contains line items for cultured stone and brick, PVC trim, lumber, and TPO roofing.

Site Work & Landscape Costs with RSMeans Data 2017

Includes unit and assemblies costs for earthwork, sewerage, piped utilities, site improvements, drainage, paving, trees and shrubs, street openings/repairs, underground tanks, and more. Contains more than 60 types of assemblies costs for accurate conceptual estimates.

Includes:

- Estimating for infrastructure improvements
- Environmentally-oriented construction
- ADA-mandated handicapped access
- Hazardous waste line items

Facilities Maintenance & Repair Costs with RSMeans Data 2017

Facilities Maintenance & Repair Costs with RSMeans Data gives you a complete system to manage and plan your facility repair and maintenance costs and budget efficiently. Guidelines for auditing a facility and developing an annual maintenance plan. Budgeting is included, along with reference tables on cost and management, and information on frequency and productivity of maintenance operations.

The only nationally recognized source of maintenance and repair costs. Developed in cooperation with the Civil Engineering Research Laboratory (CERL) of the Army Corps of Engineers.

Light Commercial Costs with RSMeans Data 2017

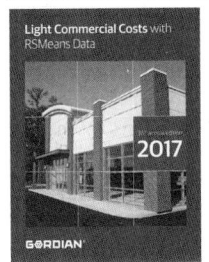

Specifically addresses the light commercial market, which is a specialized niche in the construction industry. Aids you, the owner/designer/contractor, in preparing all types of estimates—from budgets to detailed bids. Includes new advances in methods and materials.

The Assemblies section allows you to evaluate alternatives in the early stages of design/planning.

More than 15,500 unit costs ensure that you have the prices you need, when you need them.

Concrete & Masonry Costs with RSMeans Data 2017

Provides you with cost facts for virtually all concrete/masonry estimating needs, from complicated formwork to various sizes and face finishes of brick and block—all in great detail. The comprehensive Unit Price Section contains more than 8,000 selected entries. Also contains an Assemblies [Cost] Section, and a detailed Reference Section that supplements the cost data.

Labor Rates for the Construction Industry with RSMeans Data 2017

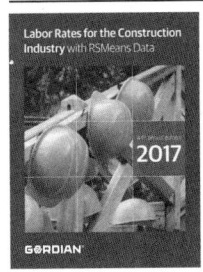

Complete information for estimating labor costs, making comparisons, and negotiating wage rates by trade for more than 300 U.S. and Canadian cities. With 46 construction trades in each city and historical wage rates included for comparison. Each city chart lists the county and is alphabetically arranged with handy visual flip tabs for quick reference.

Square Foot Renovation Costs with RSMeans Data 2017

For conceptual estimating of renovation projects, *Square Foot Renovation Costs with RSMeans Data* provides whole building and assembly level renovation costs at varying levels for 44 building models. Renovation costs are unique to each model and adjust with the intensity of renovation work being represented. Levels of renovation are:

Level 1 – Cosmetics
Level 2 – Plus Non-Masonry Partitions
Level 3 – Plus Masonry Partitions
Level 4 – Plus Bathrooms, Kitchens
Level 5 – Plus Equip, Ext Doors & Windows

Interior Costs with RSMeans Data 2017

Provides you with prices and guidance needed to make accurate interior work estimates. Contains costs on materials, equipment, hardware, custom installations, furnishings, and labor for new and remodel commercial and industrial interior construction, including updated information on office furnishings, as well as reference information.

Heavy Construction Costs with RSMeans Data 2017

A comprehensive guide to heavy construction costs. Includes costs for highly specialized projects such as tunnels, dams, highways, airports, and waterways. Information on labor rates, equipment, and materials costs is included. Features unit price costs, systems costs, and numerous reference tables for costs and design.

Plumbing Costs with RSMeans Data 2017

Comprehensive unit prices and assemblies for plumbing, irrigation systems, commercial and residential fire protection, point-of-use water heaters, and the latest approved materials. This publication and its companion, *Mechanical Costs with RSMeans Data*, provide full-range cost estimating coverage for all the mechanical trades.

Contains updated costs for potable water, radiant heat systems, and high efficiency fixtures.

Unit prices according to the latest MasterFormat

Annual Cost Guides

For more information visit our website at www.RSMeans.com

Annual Cost Guides (Continued)
Construction Cost Indexes with RSMeans Data 2017
What materials and labor costs will change unexpectedly this year? By how much?
- Breakdowns for 318 major cities
- National averages for 30 key cities
- Expanded five major city indexes

 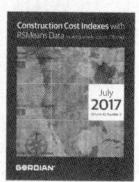  

Our Online Estimating Solution
Competitive Cost Estimates Made Easy

Our online estimating solution is a web-based service that provides accurate and up-to-date cost information to help you build competitive estimates or budgets in less time.

Quick, intuitive, easy to use and automatically updated, you'll gain instant access to hundreds of thousands of material, labor, and equipment costs from RSMeans' comprehensive database, delivering the information you need to build budgets and competitive estimates every time.

With our online estimating solutions, you can perform quick searches to locate specific costs and adjust costs to reflect prices in your geographic area. Tag and store your favorites for fast access to your frequently used line items and assemblies and clone estimates to save time. System notifications will alert you as updated data becomes available. This data is automatically updated throughout the year.

Our visual, interactive estimating features help you create, manage, save and share estimates with ease! You'll enjoy increased flexibility with customizable advanced reports. Easily edit custom report templates and import your company logo onto your estimates.

	Core	Advanced	Complete
Unit Prices	✓	✓	✓
Assemblies	X	✓	✓
Sq Ft Models	X	X	✓
Editable Sq Foot Models	X	X	✓
Editable Assembly Components	X	✓	✓
Custom Cost Data	X	✓	✓
User Defined Components	X	✓	✓
Advanced Reporting & Customization	X	✓	✓
Union Labor Type	✓	✓	✓
Open Shop*	Available for add-on at $249.99	Available for add-on at $249.99	✓
History (3-year lookback)**	Available for add-on at $249.99	Available for add-on at $249.99	Available for add-on at $249.99
Full History (2007)	Available for add-on at $499.99	Available for add-on at $499.99	Available for add-on at $499.99
Life Cycle Costing***	X	Available for add-on at $999.99	Available for add-on at $999.99
Alerts	Available for add-on at $299.99	Available for add-on at $299.99	Available for add-on at $299.99

*Open shop included in Complete Library & Bundles   **3-year history look back included with complete library
***Not available as add-on with Residential data set

Continue to check our website at www.RSMeans.com for more product offerings.

Online Estimating Solution   Unit prices according to the latest MasterFormat

Estimate with Precision

Find everything you need to develop complete, accurate estimates.

- Verified costs for construction materials
- Equipment rental costs
- Crew sizing, labor hours and labor rates
- Localized costs for U.S. and Canada

Save Time & Increase Efficiency

Make cost estimating and calculating faster and easier than ever with secure, online estimating tools.

- Quickly locate costs in the searchable database
- Create estimates in minutes with RSMeans cost lines
- Tag and store favorites for fast access to frequently used items

Improve Planning & Decision-Making

Back your estimates with complete, accurate and up-to-date cost data for informed business decisions.

- Verify construction costs from third parties
- Check validity of subcontractor proposals
- Evaluate material and assembly alternatives

Increase Profits

Use our online estimating solution to estimate projects quickly and accurately, so you can gain an edge over your competition.

- Create accurate and competitive bids
- Minimize the risk of cost overruns
- Reduce variability

30 Day Free Trial

Register for a free trial at www.RSMeans.com/Trial

- Access to unit prices, building assemblies and facilities repair and remodeling items covering every category of construction
- Powerful tools to customize, save and share cost lists, estimates and reports with ease
- Access to square foot models for quick conceptual estimates
- Allows up to 10 users to evaluate the full range of collaborative functionality.

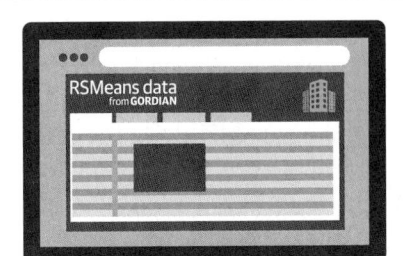

2017 Seminar Schedule

Note: call for exact dates and details.

Location	Dates	Location	Dates
Seattle, WA	January and August	San Francisco, CA	June
Dallas/Ft. Worth, TX	January	Bethesda, MD	June
Austin, TX	February	El Segundo, CA	August
Anchorage, AK	March and September	Dallas, TX	September
Las Vegas, NV	March	Raleigh, NC	October
New Orleans, LA	March	Salt Lake City, UT	October
Washington, DC	April and September	Baltimore, MD	November
Phoenix, AZ	April	Orlando, FL	November
Toronto	May	San Diego, CA	December
Denver, CO	May	San Antonio, TX	December

📞 877-620-6245

Gordian also offers a suite of online RSMeans data self-paced offerings. Check our website www.RSMeans.com for more information.

Professional Development

eLearning Training Sessions

Learn how to use our RSMeans online estimating solution from the convenience of your home or office. Our eLearning training sessions let you join a training conference call and share the instructors' desktops, so you can view the presentation and step-by-step instructions on your own computer screen. The live webinars are held from 4 p.m. to 5 p.m. or 6 p.m. to 7 p.m. eastern standard time, with a one-hour break for lunch.

For these sessions, you must have a computer with high speed Internet access and a compatible Web browser. Learn more at www.RSMeansOnline.com or call for a schedule: 781-422-5115. Webinars are generally held on selected Wednesdays each month.

Visit RSMeans.com/Online for more details on data titles in online format.

For more information visit our website at www.RSMeans.com

Training for our Online Estimating Solution

Construction estimating is vital to the decision-making process at each state of every project. Our online solution works the way you do. It's systematic, flexible and intuitive. In this one-day class you will see how you can estimate any phase of any project faster and better.

Some of what you'll learn:
- Customizing our online estimating solution
- Making the most of RSMeans "Circle Reference" numbers
- How to integrate your cost data
- Generating reports, exporting estimates to MS Excel, sharing, collaborating and more

Also available: Online estimating solution training webinar or a self-paced training program.

Facilities Construction Estimating

In this two-day course, professionals working in facilities management can get help with their daily challenges to establish budgets for all phases of a project.

Some of what you'll learn:
- Determining the full scope of a project
- Identifying the scope of risks and opportunities
- Creative solutions to estimating issues
- Organizing estimates for presentation and discussion
- Special techniques for repair/remodel and maintenance projects
- Negotiating project change orders

Who should attend: facility managers, engineers, contractors, facility tradespeople, planners, and project managers.

Construction Cost Estimating: Concepts and Practice

This one-day introductory course to improve estimating skills and effectiveness starts with the details of interpreting bid documents and ends with the summary of the estimate and bid submission.

Some of what you'll learn:
- Using the plans and specifications to create estimates
- The takeoff process—deriving all tasks with correct quantities
- Developing pricing using various sources; how subcontractor pricing fits in
- Summarizing the estimate to arrive at the final number
- Formulas for area and cubic measure, adding waste and adjusting productivity to specific projects
- Evaluating subcontractors' proposals and prices
- Adding insurance and bonds
- Understanding how labor costs are calculated
- Submitting bids and proposals

Who should attend: project managers, architects, engineers, owners' representatives, contractors, and anyone who's responsible for budgeting or estimating construction projects.

Maintenance & Repair Estimating for Facilities

This two-day course teaches attendees how to plan, budget, and estimate the cost of ongoing and preventive maintenance and repair for existing buildings and grounds.

Some of what you'll learn:
- The most financially favorable maintenance, repair, and replacement scheduling and estimating
- Auditing and value engineering facilities
- Preventive planning and facilities upgrading
- Determining both in-house and contract-out service costs
- Annual, asset-protecting M&R plan

Who should attend: facility managers, maintenance supervisors, buildings and grounds superintendents, plant managers, planners, estimators, and others involved in facilities planning and budgeting.

Practical Project Management for Construction Professionals

In this two-day course, acquire the essential knowledge and develop the skills to effectively and efficiently execute the day-to-day responsibilities of the construction project manager.

Some of what you'll learn:
- General conditions of the construction contract
- Contract modifications: change orders and construction change directives
- Negotiations with subcontractors and vendors
- Effective writing: notification and communications
- Dispute resolution: claims and liens

Who should attend: architects, engineers, owners' representatives, and project managers.

Mechanical & Electrical Estimating

This two-day course teaches attendees how to prepare more accurate and complete mechanical/electrical estimates, avoid the pitfalls of omission and double-counting, and understand the composition and rationale within the RSMeans mechanical/electrical database.

Some of what you'll learn:
- The unique way mechanical and electrical systems are interrelated
- M&E estimates—conceptual, planning, budgeting, and bidding stages
- Order of magnitude, square foot, assemblies, and unit price estimating
- Comparative cost analysis of equipment and design alternatives

Who should attend: architects, engineers, facilities managers, mechanical and electrical contractors, and others who need a highly reliable method for developing, understanding, and evaluating mechanical and electrical contracts.

Professional Development

Unit Price Estimating

This interactive two-day seminar teaches attendees how to interpret project information and process it into final, detailed estimates with the greatest accuracy level.

The most important credential an estimator can take to the job is the ability to visualize construction and estimate accurately.

Some of what you'll learn:
- Interpreting the design in terms of cost
- The most detailed, time-tested methodology for accurate pricing
- Key cost drivers—material, labor, equipment, staging, and subcontracts
- Understanding direct and indirect costs for accurate job cost accounting and change order management

Who should attend: corporate and government estimators and purchasers, architects, engineers, and others who need to produce accurate project estimates.

Conceptual Estimating Using the CD

This two-day class uses the leading industry data and a powerful software package to develop highly accurate conceptual estimates for your construction projects. All attendees must bring a laptop computer loaded with the current year Square Foot Costs with RSMeans Data and Assemblies Costs with RSMeans Data.

Some of what you'll learn:
- Introduction to conceptual estimating
- Types of conceptual estimates
- Helpful hints
- Order of magnitude estimating
- Square foot estimating
- Assemblies estimating

Who should attend: architects, engineers, contractors, construction estimators, owners' representatives, and anyone looking for an electronic method for performing square foot estimating.

Training for our CD Estimating Solution

This one-day course helps users become more familiar with the functionality of the CD. Each menu, icon, screen, and function found in the program is explained in depth. Time is devoted to hands-on estimating exercises.

Some of what you'll learn:
- Searching the database using all navigation methods
- Exporting RSMeans data to your preferred spreadsheet format
- Viewing crews, assembly components, and much more
- Automatically regionalizing the database

This training session requires you to bring a laptop computer to class.

When you register for this course you will receive an outline for your laptop requirements.

Also offered as a self-paced training program!

Assessing Scope of Work for Facility Construction Estimating

This two-day practical training program addresses the vital importance of understanding the scope of projects in order to produce accurate cost estimates for facility repair and remodeling.

Some of what you'll learn:
- Discussions of site visits, plans/specs, record drawings of facilities, and site-specific lists
- Review of CSI divisions, including means, methods, materials, and the challenges of scoping each topic
- Exercises in scope identification and scope writing for accurate estimating of projects
- Hands-on exercises that require scope, take-off, and pricing

Who should attend: corporate and government estimators, planners, facility managers, and others who need to produce accurate project estimates.

Facilities Estimating Using the CD

Combines hands-on skill building with best estimating practices and real-life problems. Brings you up-to-date with key concepts and provides tips, pointers, and guidelines to save time and avoid cost oversights and errors in this two-day class.

Some of what you'll learn:
- Estimating process concepts
- Customizing and adapting RSMeans cost data
- Establishing scope of work to account for all known variables
- Budget estimating: when, why, and how
- Site visits: what to look for—what you can't afford to overlook
- How to estimate repair and remodeling variables

This training session requires you to bring a laptop computer to class.

Who should attend: facility managers, architects, engineers, contractors, facility tradespeople, planners, project managers, and anyone involved with JOC, SABRE, or IDIQ.

Building Systems and the Construction Process

This one-day course was written to assist novices and those outside the industry in obtaining a solid understanding of the construction process - from both a building systems and construction administration approach.

Some of what you'll learn:
- Various systems used and how components come together to create a building
- Start with foundation and end with the physical systems of the structure such as HVAC and Electrical
- Focus on the process from start of design through project closeout

This training session requires you to bring a laptop computer to class.

Who should attend: building professionals or novices to help make the crossover to the construction industry; suited for anyone responsible for providing high level oversight on construction projects.

Professional Development

Visit RSMeans.com/Online for more details on data titles in online format.

For more information visit our website at www.RSMeans.com

Self-Paced Professional Development Courses

Training and professional development courses on RSMeans data are now offered in a convenient self-paced format allowing more flexibility to learn around your busy schedule.

These self-paced training courses can be completed over the course of 45 days and most are comprised of 10 lessons complete with documentation, video presentation, assessment quizzes and certificate of completion.

To view and register for the currently offered self-paced course, visit https://rsmeans.myabsorb.com.

Registration Information

Register early and save up to $100!
Register 30 days before the start date of a seminar and save $100 off your total fee. Note: This discount can be applied only once per order. It cannot be applied to team discount registrations or any other special offer.

How to register
By Phone
Register by phone at 877-620-6245

Online
Register online at
www.RSMeans.com/products/seminars.aspx

Note: Purchase Orders or Credits Cards are required to register.

Two-day seminar registration fee - $995. One-day CD or online training registration fee - $395.

One-Day Construction Cost Estimating or Building Systems and the Construction Process - $600.

Government pricing
All federal government employees save off the regular seminar price. Other promotional discounts cannot be combined with the government discount.

Team discount program
For over five attendee registrations. Call for pricing: 781-422-5115

Refund policy
Cancellations will be accepted up to ten business days prior to the seminar start. There are no refunds for cancellations received later than ten working days prior to the first day of the seminar. A $150 processing fee will be applied for all cancellations. Written notice of the cancellation is required. Substitutions can be made at any time before the session starts. No-shows are subject to the full seminar fee.

Note: Pricing subject to change.

AACE approved courses
Many seminars described and offered here have been approved for 14 hours (1.4 recertification credits) of credit by the AACE International Certification Board toward meeting the continuing education requirements for recertification as a Certified Cost Engineer/Certified Cost Consultant.

AIA Continuing Education
We are registered with the AIA Continuing Education System (AIA/CES) and are committed to developing quality learning activities in accordance with the CES criteria. Many seminars meet the AIA/CES criteria for Quality Level 2. AIA members may receive 14 learning units (LUs) for each two-day RSMeans course.

Daily course schedule
The first day of each seminar session begins at 8:30 a.m. and ends at 4:30 p.m. The second day begins at 8:00 a.m. and ends at 4:00 p.m. Participants are urged to bring a hand-held calculator since many actual problems will be worked out in each session.

Continental breakfast
Your registration includes the cost of a continental breakfast and a morning and afternoon refreshment break. These informal segments allow you to discuss topics of mutual interest with other seminar attendees. (You are free to make your own lunch and dinner arrangements.)

Hotel/transportation arrangements
We arrange to hold a block of rooms at most host hotels. To take advantage of special group rates when making your reservation, be sure to mention that you are attending the RSMeans Institute data seminar. You are, of course, free to stay at the lodging place of your choice. (Hotel reservations and transportation arrangements should be made directly by seminar attendees.)

Important
Class sizes are limited, so please register as soon as possible.

Professional Development